ISOTOPE GEOCHEMISTRY

FROM THE TREATISE ON GEOCHEMISTRY

Editors

H. D. Holland

University of Pennsylvania, Philadelphia, PA, USA

K. K. Turekian

Yale University, New Haven, CT, USA

ELSEVIER

Amsterdam • Boston • Heidelberg • London • New York • Oxford
Paris • San Diego • San Francisco • Singapore • Sydney • Tokyo
Academic Press is an imprint of Elsevier

ACADEMIC PRESS

Academic Press is an imprint of Elsevier
32 Jamestown Road, London NW1 7BY, UK
Radarweg 29, PO Box 211, 1000 AE Amsterdam, The Netherlands
30 Corporate Drive, Suite 400, Burlington, MA 01803, USA
525 B Street, Suite 1900, San Diego, CA 92101-4495, USA

First edition 2011

British Library Cataloguing in Publication Data
A catalogue record for this book is available from the British Library

Library of Congress Cataloging-in-Publication Data
A catalog record for this book is available from the Library of Congress

ISBN: 978-0-08-096710-3

For information on all Academic Press publications
visit our website at elsevierdirect.com

CONTENTS

Introduction

The recognition that there were multiple isotopes of an element was established experimentally by J. J. Thomson at Cambridge University in 1914. This discovery was soon followed by the development by F. W. Aston of the 'positive ray spectrograph,' the precursor of the mass spectrometer developed later. Among the first to develop a mass spectrometer in which the positive ion beams were measured directly using electronic means rather than photographic means was A. O. C. Nier.

Starting in 1938, Nier measured lead isotopic compositions of lead ores. The origins of the geochemical utilization of mass spectrometry, it can be argued, started with the engagement of Nier in measuring isotopes of geologic importance as his Pb isotope results became the basis of the so-called Holmes–Houtermans attempt at estimating the age of Earth from Pb isotopic variations among the lead ores. A review of Nier's contributions in developing the mass spectrometer, which made isotope geochemistry possible, has recently been published by DeLaeter and Kurz (2006). A memoir of the National Academy of Sciences also contains information on the contributions of Nier to mass spectrometry (Reynolds, 1994).

Although H. C. Urey received the Nobel Prize for discovering heavy water (based on deuterium enriched water), the initial measurements were made by physical methods. Urey's subsequent statistical mechanical calculations (Urey, 1947) resulted in his proposal that, for the light isotopes, fractionation of the isotopes between phases would have a temperature dependence and this could be recorded in solid phases such as calcareous marine shells.

This post-World War II paper of Urey's corresponded in time with the creation of Nier's latest and most geochemically applicable isotope ratio instrument allowing for the measurement of small variations in isotope ratios. Urey, together with his associates, consequently developed the oxygen isotope paleothermometer for fossil carbonate deposits.

At the University of Chicago, where Urey was a professor, a number of other investigators developed the initial approaches to measuring a variety of the light isotopes including those that were later used in geochronometry (Nier had discovered the isotope ^{40}K earlier and L. T. Aldrich, in his laboratory in Minnesota, had shown that it was radioactive which led to the ^{40}K–^{40}Ar dating tool). The idea caught on and mass spectrometry was harnessed around the world in studying the natural variations of oxygen, carbon, hydrogen and sulfur isotopes as well as measurements of the radiogenic isotopes.

It is the development of these areas of exploration, extended and amplified, that this collection of articles from the Treatise on Geochemistry represents.

The articles are grouped into the following general categories: Light stable isotopes, radiogenic tracers, noble gases, and radioactive tracers. The first three groups depend on mass spectrometric measurements. The section on radioactive tracers employs both radioactive counting techniques and the newly developed accelerator mass spectrometric techniques.

Light stable isotopes: There are two areas in which the light stable isotopes have had extensive studies: planetary (including meteorites) systems and systems involving the atmosphere and hydrosphere. There are three articles that deal with the gross planetary aspects of the isotopes of oxygen, carbon, and sulfur. Two articles address oxygen isotope variations in meteorites and carbon isotope variations in carbonaceous chondrites. These are relatable to Earth's composition. The third article deals with the peculiar property of Earth in recycling elements in subduction zones in the plate tectonics process.

The major section of the light stable isotope section deals with the hydrosphere and the atmosphere. It starts with a discussion of nonmass dependent isotopic fractionation of oxygen and sulfur isotopes. This phenomenon was first identified in meteorites as seen in the article cited earlier in the discussion of meteoritic oxygen isotopes, but was extended to processes especially in the atmosphere on Earth (and other planets). Next is the variation of carbon in atmospheric carbon dioxide followed by the variation of oxygen and hydrogen isotope variations in ice cores from the polar glacial accumulations recording climate variations over the past 800 000 years. Isotopic variations in the hydrosphere aside from the study of ice cores are discussed next. The oxygen isotopic record in calcareous deposits in deep sea sediment cores discussed in the next article echoes the initial work on isotopic fractionation initiated in Urey's laboratory. The study of sulfur isotope variations in sedimentary rocks as recorders of varying environments brings in another useful isotope system. A summary article on the use of light stable isotopic variations in sedimentary systems in general addresses not only the work done using the isotope systems discussed above but also the isotope systems of nitrogen, boron, and calcium. The application of these isotopic systems to large-scale processes such as the global oxygen cycle and the more local issues of tropospheric boundary layer ozone formation round out this section.

Radioactive tracers: A range of radioactive tracers has been used to understand Earth surface processes. The first and most powerful has been radiocarbon, discussed in the first article in this section. Other cosmogenic nuclides and products of the radioactive system associated with radon have been used in the study of atmospheric problems and this is the second article in this section.

Noble gases: The isotopes of the noble gases provide information on planetary formation and planetary degassing. These are discussed for meteorites. The distribution of the noble gases in planetary formation then follows. This approach provides the basis for the discussion of the noble gases as tracers of the processes involving Earth's mantle.

Radiogenic isotopes: The use of radiogenic isotopes in radioactive geochronometry has been discussed in a separate volume. These isotopic systems can also act as tracers of sources and processes independent of their chronological utility. It is this aspect that is included in this collection. First, sampling mantle heterogeneity using the radiogenic isotopes of a number of long-lived radioactive parents is explored. Proceeding to the Earth's surface the next article deals with the use of radiogenic isotopes in weathering and hydrology. The variations of various isotopic signatures in the ocean and in marine deposits can be used to reconstruct climate history of the marine realm. Finally the long-term history of Earth surface processes using an array of radiogenic isotopes is covered and linked to other isotopic tracers.

The diversity of isotope systems and range of problems addressable with these varied systems is a rewarding consequence of the development of the mass analysis of a variety of isotopes in natural materials started in 1938.

K.K. Turekian

REFERENCES

DeLaeter J. and Kurz M. D. (2006) Alfred Nier and the sector field mass spectrometer. *Journal of Mass Spectrometry* **41**, 847–854.
Reynolds J. H. (1994) Alfred Otto Carl Nier. Memoir of the National Academy of Sciences.
Urey H. C. (1947) The thermodynamic properties of isotopic substances. *Journal of the Chemical Society (London)* **69**, 562–581.

CONTRIBUTORS

T. A. Abrajano Jr.
Rensselaer Polytechnic Institute, Troy, NY, USA

G. E. Bebout
Lehigh University, Bethlehem, PA, USA

J. D. Blum
The University of Michigan, Ann Arbor, MI, USA

R. Bopp
Rensselaer Polytechnic Institute, Troy, NY, USA

W. S. Broecker
Lamont-Doherty Earth Observatory, Palisades, NY, USA

N. Clauer
Centre de Géochimie de la Surface, CNRS—Université Louis Pasteur, Strasbourg, France

R. N. Clayton
University of Chicago, IL, USA

D. H. Doctor
United States Geological Survey, Menlo Park, CA, USA

Y. Erel
The Hebrew University, Jerusalem, Israel

I. Gilmour
The Open University, Milton Keynes, UK

M. B. Goldhaber
US Geological Survey, Denver, CO, USA

S. L. Goldstein
Columbia University, Palisades, NY, USA

W. C. Graustein
Yale University, New Haven, CT, USA

S. R. Hemming
Columbia University, Palisades, NY, USA

D. R. Hilton
University of California San Diego, La Jolla, CA, USA

A. W. Hofmann
Max-Planck-Institut für Chemie, Mainz, Germany
Lamont-Doherty Earth Observatory, Palisades, NY, USA

J. Jouzel
Institut Pierre Simon Laplace, Saclay, France

C. Kendall
United States Geological Survey, Menlo Park, CA, USA

D. W. Lea
University of California, Santa Barbara, CA, USA

A. Lerman
Northwestern University, Evanston, IL, USA

V. O'Malley
Enterprise Ireland, Glasnevin, Republic of Ireland

S. T. Petsch
University of Massachusetts, Amherst, MA, USA

F. A. Podosek
Washington University, St. Louis, MO, USA

D. Porcelli
University of Oxford, UK

G. E. Ravizza
University of Hawaii, Manoa, HI, USA

J. Song
Rensselaer Polytechnic Institute, Troy, NY, USA

M. H. Thiemens
University of California, San Diego, La Jolla, CA, USA

K. K. Turekian
Yale University, New Haven, CT, USA

D. Yakir
Weizmann Institute of Science, Rehovot, Israel

B. Yan
Rensselaer Polytechnic Institute, Troy, NY, USA

J. C. Zachos
University of California, Santa Cruz, CA, USA

A. LIGHT STABLE ISOTOPES

a'. Planetary

1

Oxygen Isotopes in Meteorites

R. N. Clayton

University of Chicago, IL, USA

1.1 INTRODUCTION

Oxygen isotope abundance variations in meteorites are very useful in elucidating chemical and physical processes that occurred during the formation of the solar system (Clayton, 1993). On Earth, the mean abundances of the three stable isotopes are ^{16}O: 99.76%, ^{17}O: 0.039%, and ^{18}O: 0.202%. It is conventional to express variations in abundances of the isotopes in terms of isotopic ratios, relative to an arbitrary standard, called SMOW (for standard mean ocean water), as follows:

$$\delta^{18}O = \left[\frac{(^{18}O/^{16}O)_{sample}}{(^{18}O/^{16}O)_{SMOW}} - 1\right] \times 1{,}000$$

$$\delta^{17}O = \left[\frac{(^{17}O/^{16}O)_{sample}}{(^{17}O/^{16}O)_{SMOW}} - 1\right] \times 1{,}000$$

The isotopic composition of any sample can then be represented by one point on a "three-isotope plot," a graph of $\delta^{17}O$ versus $\delta^{18}O$. It will be seen that such plots are invaluable in interpreting meteoritic data. Figure 1 shows schematically the effect of various processes on an initial composition at the center of the diagram. Almost all terrestrial materials lie along a "fractionation" trend; most meteoritic materials lie near a line of "^{16}O addition" (or subtraction).

The three isotopes of oxygen are produced by nucleosynthesis in stars, but by different nuclear processes in different stellar environments. The principal isotope, ^{16}O, is a primary isotope (capable of being produced from hydrogen and helium alone), formed in massive stars (>10 solar masses), and ejected by supernova explosions. The two rare isotopes are secondary nuclei (produced in stars from nuclei formed in an earlier generation of stars), with ^{17}O coming primarily from low- and intermediate-mass stars (<8 solar masses), and ^{18}O coming primarily from high-mass stars (Prantzos *et al.*, 1996). These differences in type of stellar source result in large

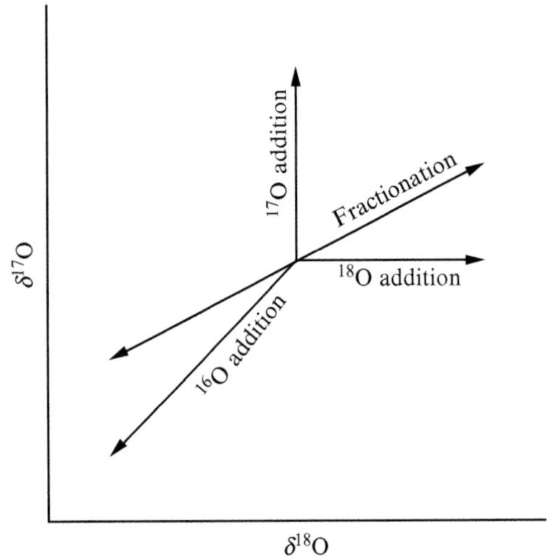

Figure 1 Schematic representation of various isotopic processes shown on an oxygen three-isotope plot. Almost all terrestrial materials plot along a line of "fractionation"; most primitive meteoritic materials plot near a line of "^{16}O addition."

observable variations in stellar isotopic abundances as functions of age, size, metallicity, and galactic location (Prantzos *et al.*, 1996). In their paper reporting the discovery of ^{18}O in the Earth's atmosphere, Giauque and Johnston (1929) refer to nonuniform distribution of oxygen isotopes as a "remote possibility," whereas Manian *et al.* (1934) sought to find variations in oxygen isotope abundances in meteorites as evidence for an origin outside the solar system.

In addition to the abundance variations due to nuclear processes, there are important isotopic variations produced within molecular clouds, the precursors to later star-formation. The most important process is isotopic self-shielding in the UV photodissociation of CO (van Dishoeck and Black, 1988). This process results from the large differences in abundance between $C^{16}O$, on the one hand, and $C^{17}O$ and $C^{18}O$ on the other. Photolysis of CO occurs by absorption of stellar UV radiation in the wavelength range 90–100 nm. The reaction proceeds by a predissociation mechanism, in which the excited electronic state lives long enough to have well-defined vibrational and rotational energy levels. As a consequence, the three isotopic species—$C^{16}O$, $C^{17}O$, and $C^{18}O$—absorb at different wavelengths, corresponding to the isotope shift in vibrational frequencies. Because of their different number densities, the abundant $C^{16}O$ becomes optically thick in the outermost part of the cloud (nearest to the external source of UV radiation), while the rare $C^{17}O$ and $C^{18}O$ remain optically thin, and hence dissociate at a greater rate in the cloud interior. The differences in chemical reactivity between $C^{16}O$ *molecules* and ^{17}O and ^{18}O *atoms* may lead to isotopically selective reaction products. This scenario has been suggested to explain meteoritic isotope patterns, as discussed below (Yurimoto and Kuramoto, 2002).

Stable isotope abundances in meteoritic material provide an opportunity to evaluate the thoroughness of mixing of isotopes of diverse stellar sources. Molybdenum presents a good test case: it has seven stable isotopes, derived from at least three types of stellar sources, corresponding to the r-process, s-process, and p-process. Presolar silicon carbide grains, extracted from primitive meteorites, contain molybdenum that has been subject to s-process neutron capture in red-giant stars, resulting in large enrichments of isotopes at masses 95, 96, 97, 98, and severe depletions (up to 100%) of isotopes at masses 92 and 94 (p-process) and 100 (r-process) (Nicolussi *et al.*, 1998). Complementary patterns have been found in whole-rock samples of several meteorites, with $>$1,000-fold smaller amplitude, suggesting the preservation of a small fraction of the initial isotopic heterogeneity (Yin *et al.*, 2002; Dauphas *et al.*, 2002). Oxygen is another element for which primordial isotopic heterogeneity might be preserved. This is discussed further below.

It would be highly desirable to have samples of oxygen-rich mineral grains that have formed in stellar atmospheres and have recorded the nucleosynthetic processes in individual stars. Similar samples are already available for carbon-rich grains, in the form of SiC and graphite, primarily from asymptotic giant branch (AGB) stars and supernovae (Anders and Zinner, 1993). These presolar grains have provided a wealth of detailed information concerning nucleosynthesis of carbon, nitrogen, silicon, calcium, titanium, and heavier elements. It is thought that such carbon-rich minerals should form only in environments with $C/O > 1$, as in the late stages of AGB evolution, or in carbon-rich layers of supernovae. By analogy, one would expect to form oxide and silicate minerals in environments with $C/O < 1$, as is common for most stars. Indeed there is evidence in infrared spectra for the formation of Al_2O_3 (corundum) and silicates, such as olivine (Speck *et al.*, 2000) around evolved oxygen-rich stars. However, searches for such grains in meteorites have yielded only a very small population of corundum grains, a few grains of spinel and hibonite, and no silicates (Nittler *et al.*, 1997). The observed oxygen isotopic compositions of presolar corundum grains show clear evidence of nuclear processes in red-giant stars, and have had significant impact on the theory of these stars (Boothroyd and Sackmann, 1999).

There are several possible reasons for the failure to recognize and analyze large populations of oxygen-rich presolar grains:

(i) they may not exist: oxygen ejected in supernova explosions may not condense into mineral grains on the short timescale available;
(ii) they may be smaller in size than can be detected by applicable techniques ($\sim$0.1 μm); and
(iii) they may be destroyed in the laboratory procedures used to isolate other types of presolar grains.

1.2 ISOTOPIC ABUNDANCES IN THE SUN AND SOLAR NEBULA

1.2.1 Isotopes of Light Elements in the Sun

For most of the chemical elements, the relative abundances of their stable isotopes in the Sun and solar nebula are well known, so that any departures from those values that may be found in meteorites and planetary materials can then be interpreted in terms of planet-forming processes. This is best illustrated for the noble gases: neon, argon, krypton, and xenon. The solar isotopic abundances are known through laboratory mass-spectrometric analysis of solar wind extracted from lunar soils (Eberhardt *et al.*, 1970) and gas-rich meteorites. Noble gases in other meteorites and in the atmospheres of Earth and Mars show many substantial differences from the solar composition, due to a variety of nonsolar processes, e.g., excesses of ^{40}Ar and ^{129}Xe due to radioactive decay of ^{40}K and ^{129}I, excesses of ^{134,136}Xe due to fission of ^{244}Pu, excesses of ^{21,22}Ne and ^{38}Ar due to cosmic-ray spallation reactions, and mass-dependent isotopic fractionation due to gas loss in a gravitational field (Pepin, 1991). For many of the nonvolatile elements, such as magnesium and iron, it is thought that terrestrial and solar isotopic abundances are almost identical, so that small differences in meteoritic materials can serve as tracers of local solar nebular processes, such as evaporation of molten silicate droplets in space (Davis *et al.*, 1990).

For a few very important elements, notably hydrogen, helium, carbon, nitrogen, and oxygen, neither of these procedures can be used to infer accurate values for the isotopic abundances in the Sun and solar nebula. Hydrogen is a special case, in that the Sun's initial complement of deuterium (^{2}H) has been destroyed early in solar history by nuclear reaction, so that today's solar-wind hydrogen is deuterium free (Epstein and Taylor, 1970). Attempts to infer the D/H ratio in the solar nebula, based on measurements in comets, meteorites, and planets, have left an uncertainty of a factor of 2 in the primordial ratio. Helium isotope ratios (^{4}He/^{3}He) are variable in the solar wind (Gloeckler and Geiss, 1998), and are strongly modified in solid solar system bodies by radioactive production of ^{4}He from uranium and thorium, and by cosmic-ray production of both isotopes.

Carbon isotope abundances (^{12}C/^{13}C ratios) are measurable in solar energetic particles (by spacecraft) (Leske *et al.*, 2001), and in molecules in the Sun and comets (Wyckoff *et al.*, 2000). All values are consistent with a mean terrestrial ratio of 89, but with analytical uncertainties of $\sim$10%. The ^{12}C/^{13}C ratio in the present near-solar interstellar medium has been found to be 70 ± 10, interpreted as reflecting "galactic chemical evolution" over the age of the solar system (Prantzos *et al.*, 1996). Isotopic measurement of solar-wind carbon in lunar soils has not produced definitive results due to lunar, meteoritic, and terrestrial background contamination.

Nitrogen isotope abundances (^{14}N/^{15}N ratios) in the Sun and solar nebula have presented problems of measurement and interpretation for decades (Kerridge, 1993). The terrestrial atmospheric ratio is 272, but meteoritic and lunar ratios vary by about a factor of 3 (Owen *et al.*, 2001). Estimates of the nitrogen isotope ratio of the Sun and solar system range from $\sim$30% lower than terrestrial, measured in the solar wind (Kallenbach *et al.*, 1998) to $\sim$30% higher than terrestrial, measured in ammonia in Jupiter (Owen *et al.*, 2001). The lunar regolith should be an excellent collector of solar wind, since the indigenous nitrogen content of lunar rocks is very low. A range in ^{14}N/^{15}N of over 35% has been found for the implanted nitrogen, which correlates in concentration with solar-wind-derived noble gases. However, there is reason to believe that $\sim$80–90% of this nitrogen has come from unidentified nonsolar sources (Hashizume *et al.*, 2000). The net result is that the isotopic composition of nitrogen in the Sun and in the solar nebula is very poorly known.

The observational constraints on the solar isotopic abundances of oxygen are also poor. The only solar-wind measurement, by the Advanced Composition Explorer, yielded a ratio of ^{16}O/^{18}O consistent with the terrestrial value, with 20% uncertainty (Wimmer-Schweingruber *et al.*, 2001). An earlier spectroscopic measurement of the solar photosphere gave a similar result (Harris *et al.*, 1987). No information is available on the solar ^{17}O abundance. The very limited state of knowledge of the solar isotope abundances of carbon, nitrogen, and oxygen illustrates the importance of the NASA Genesis mission to collect a pure solar-wind sample and return it to Earth for laboratory measurement.

1.2.2 Oxygen Isotopic Composition of the Solar Nebula

In the absence of an accurate independent measurement of oxygen isotope abundances in the Sun, the original nebular composition must be inferred from measurements on meteorites and planets. Spectroscopic measurements of planetary atmospheres have revealed no departures from terrestrial values, within rather large analytical uncertainties. Meteorites have significant advantages over other solar system bodies for tracing early nebular processes: (i) laboratory isotopic analyses are ~1,000 times more precise than remote sensing methods, and (ii) meteorites contain components that have been unchanged since their formation in the nebula 4.5 billion years ago.

For isotopic studies, oxygen has two major advantages over other light elements, such as hydrogen, carbon, and nitrogen: (i) it has three stable isotopes, rather than two, and (ii) it is a major constituent of both the solid and gaseous components of the solar nebula. The latter point is illustrated by Table 1, showing the solar system abundances of the major rock-forming elements. Over a broad range of temperatures, below the temperature of condensation of silicates, and above the temperature of condensation of ices, ~25% of the oxygen in the nebula is in the form of solid anhydrous minerals, and 75% is in gaseous molecules, primarily CO and H_2O. A major question is whether or not the solid and gaseous reservoirs had the same oxygen isotopic composition. Evidence was presented earlier for thorough isotopic homogenization of heavy elements such as molybdenum, but it is not clear that any initial isotopic differences between gas and solid would also be erased. Thus, it is not yet established whether the oxygen isotopic heterogeneity discussed below was inherited by the nebula or was generated *within* the nebula by local processes.

1.2.2.1 Primitive nebular materials

Chondritic meteorites are sedimentary rocks composed primarily of chondrules, typically submillimeter-sized spherules believed to have been molten droplets in the solar nebula, formed by melting of dust in a brief, local heating event. During the high-temperature stage, with a duration of some hours, the droplets could interact chemically and undergo isotopic exchange with the surrounding gas. Thus, isotopic analyses of individual chondrules can provide information about both the dust and gas components of the nebula. Figure 2 shows oxygen isotopic compositions of chondrules from the three major groups—ordinary (O), carbonaceous (C), and enstatite (E) chondrites—and one minor group—Rumuruti-type (R)-chondrites. The three-isotope graph has two useful properties: (i) samples related to one another by ordinary mass-dependent isotopic fractionation lie on a line of slope = 0.52, like the line labeled terrestrial fractionation (TF), and (ii) samples that are two-component mixtures lie on a straight line connecting the compositions of the end-members.

Figure 2 shows that chondrule compositions do not lie on a mass-dependent fractionation line, thus indicating isotopic heterogeneity in the nebula. Chondrules from different chondrite classes occupy different regions of the diagram, and for each class, they form near-linear arrays that are considerably steeper than a mass-dependent fractionation line. For comparison, it can be noted that analogous three-isotope graphs for iron in various meteorite types are strictly mass dependent, and show no evidence for nebular heterogeneity (Zhu *et al.*, 2001).

Another group of primitive objects with a direct link to the solar nebula are the calcium–aluminum-rich inclusions (CAIs), that range in size from a few μm to >1 cm. They are found in all types of primitive chondrites but are rare in all but the CV carbonaceous chondrites. Their

Table 1 Atomic abundances of major elements (relative to silicon).[a]

H	28,840
He	2,455
C	7.08
N	2.45
O	14.13
Mg	1.02
Si	1.00
Fe	0.85

[a] Source: Palme and Jones (Treatise on Geochemistry).

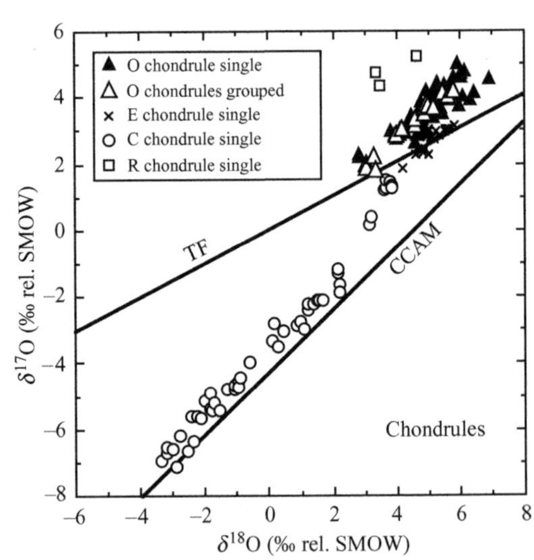

Figure 2 Oxygen isotopic compositions of chondrules from all classes of chondritic meteorites: ordinary (O), enstatite (E), carbonaceous (C), and Rumuruti-type (R). The TF line and carbonaceous chondrite anhydrous mineral (CCAM) line are shown for reference in this and many subsequent figures. Equations for these lines are: TF—$\delta^{17} = 0.52\delta^{18}$ and CCAM—$\delta^{17} = 0.941\delta^{18}-4.00$ (sources Clayton *et al.*, 1983, 1984, 1991; Weisberg *et al.*, 1991).

bulk chemical compositions correspond to the most refractory 5% of condensable solar matter (Grossman, 1973). They may represent direct condensates from the nebular gas, followed, in many cases, by further chemical and isotopic interaction with the gas. Their radiometric ages have been measured with high precision (Allègre *et al.*, 1995), and indicate solidification earlier than any other solar system rocks (excluding the presolar dust grains). Thus, the oxygen isotope abundances in CAIs may provide the best guide to the composition of the nebular gas.

CAIs exhibit a specific, characteristic pattern of oxygen isotope abundances. *Within* an individual CAI, different minerals have different isotopic compositions, with all data points falling on a straight line in the three-isotope graph. This behavior is illustrated in Figure 3, based on analyses of physically separated minerals from the Allende (CV3) carbonaceous chondrite (Clayton *et al.*, 1977). Each analysis represents a large number of grains. Figure 4 shows data obtained by ion microprobe analysis, where each point represents only one grain (Aléon *et al.*, 2002; Fagan *et al.*, 2002; Itoh *et al.*, 2002; Jones *et al.*, 2002; Krot *et al.*, 2002). The line labeled "CCAM" is the same in Figures 3 and 4. Although the ion microprobe data have larger analytical uncertainties, leading to greater scatter in the data, it is clear that the same pattern exists at both the microscopic and macroscopic level. Within individual CAIs, the sequence of isotopic composition, in terms of ^{16}O-enrichment, is spinel ≧ pyroxene > olivine > melilite = anorthite. A straight line on the three-isotope graph is indicative of some sort of two-component mixture. The fact that the range of variation in the individual-grain studies is almost the same as the range in the bulk-sample studies shows that the end-members do not lie much beyond the observed range of variation. All studies, as of early 2003, reveal an ^{16}O-rich end-member near −45‰ for both $\delta^{17}O$ and $\delta^{18}O$, frequently represented by spinel, the most refractory of the CAI phases. The most obvious interpretation is that this end-member represents the composition of the primary nebular gas, from which the CAIs originally condensed. Subsequent reaction and isotopic exchange with an isotopically modified gas could then yield the observed heterogeneities on a micrometer to millimeter scale (Clayton *et al.*, 1977; Clayton, 2002).

Another argument that the ^{16}O-rich end-member was a ubiquitous component of primitive solids is that it is found in many different chemical forms (different minerals) in many classes of meteorites: CAIs and amoeboid olivine aggregates (AOAs) from Efremovka (CV3) (Aléon *et al.*, 2002; Fagan *et al.*, 2002), AOA from a CO chondrite, Y 81020 (Itoh *et al.*, 2002), and CAI and AOA from CM and CR chondrites (Krot *et al.*, 2002). This isotopic composition can also serve as an end-member for the chondrule mixing line in Figure 2.

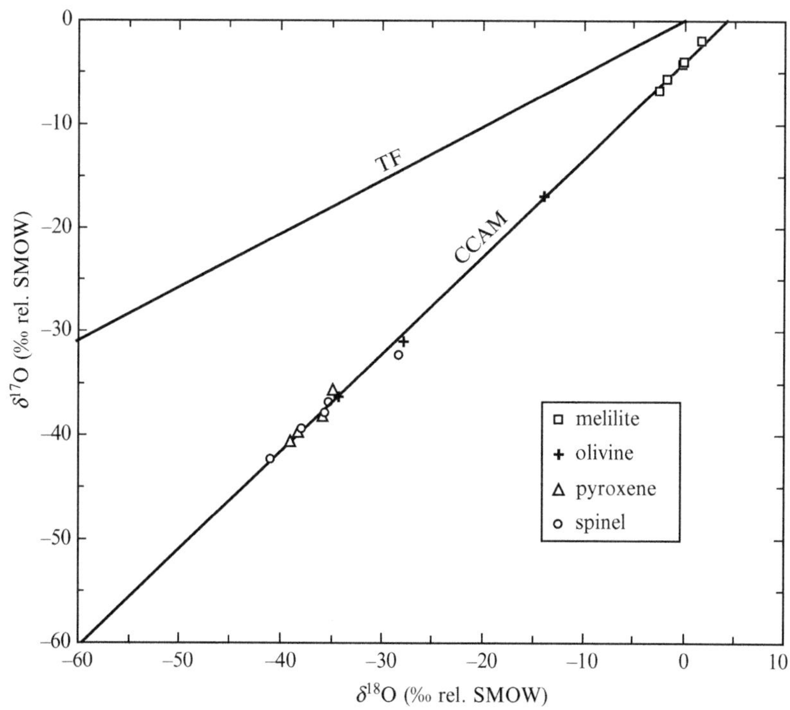

Figure 3 Oxygen isotopic compositions of physically separated minerals from several Allende CAIs. These points were used to define the CCAM line (source Clayton *et al.*, 1977).

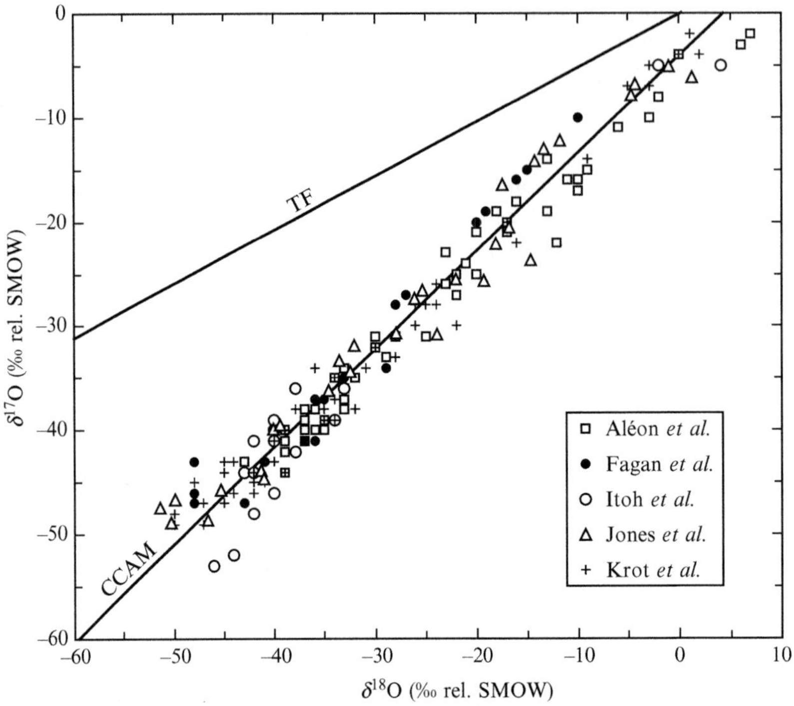

Figure 4 Ion microprobe oxygen isotope analyses of single grains in several carbonaceous chondrites. Analytical uncertainties are typically about ±2‰. The CCAM line is shown for reference sources: are noted in the figure.

If the ^{16}O-rich composition is indeed the isotopic composition of the primordial solar nebula, the consequences for solar system formation are profound. As noted above, materials with the ^{16}O-rich composition are ubiquitous, but they are also rather rare, never amounting to more than a few percent of the host meteorite. The implication is that all the other material in the inner solar system has undergone some process that changed its ^{16}O-abundance by 4–5%. This must have been a major chemical or physical process that must leave evidence in forms other than the isotopic composition of oxygen.

1.2.2.2 Sources of oxygen isotopic heterogeneity in the solar nebula

The oxygen isotopic patterns in primitive materials, shown in Figures 2–4, imply some sort of two-component mixing involving one component enriched in ^{16}O relative to the other. Several models have been proposed for the origin of these components. They can be subdivided into two categories: (i) reservoirs that were inherited from the molecular cloud from which the nebula was presumed to have come, and (ii) reservoirs that were generated within a nebula that was initially isotopically homogeneous. Two versions of the inheritance model have been proposed: (i) the dust and gas components had different proportions of supernova-produced ^{16}O, with a higher ^{16}O-abundance in the dust (Clayton *et al.*, 1977), and (ii) the gas component was depleted in ^{16}O by photochemical processes in the molecular cloud, whereas the dust was not (Yurimoto and Kuramoto, 2002). Two versions of the local generation model have been proposed: (i) a gas-phase mass-independent fractionation reaction, as has been observed in the laboratory for synthesis of O_3 from O_2 (Heidenreich and Thiemens, 1983), and (ii) isotopic self-shielding in the photolysis of CO during the accretion of the Sun (Clayton, 2002).

On the basis of analyses of bulk CAIs or separated minerals, the inheritance of ^{16}O-rich condensates from supernovae appears plausible. However, the magnitude of the isotopic difference between the putative dust and gas reservoirs is a free parameter, so that the model has no predictive power. Furthermore, the observed oxygen isotope anomalies in presolar oxide grains are best understood as resulting from processing in red giant stars, and do not show ^{16}O-excesses.

Inheritance of ^{16}O-poor gas from the presolar molecular cloud is based on the well-known phenomenon of photochemical self-shielding, in which photolysis of CO in the cloud interior affects preferentially the less-abundant isotopic species: $C^{17}O$ and $C^{18}O$. Yurimoto and Kuramoto (2002) proposed that the ^{17}O and ^{18}O atoms thus formed can react with hydrogen to form water ice. When a portion of the cloud collapses to form the solar nebula, the ice evaporates to form a

^{16}O-depleted gas reservoir. This is a plausible scenario if the dust and ice components can maintain their isotopic distinction during nebular collapse and heating.

The mass-independent isotope fractionation in the gas-phase synthesis of O_3 from O_2 produces a slope-1 line on a three-isotope plot, with ozone being depleted in ^{16}O, and residual oxygen being enriched (Heidenreich and Thiemens, 1983). This process occurs naturally in the Earth's stratosphere, and the heavy-isotope excess can be chemically transferred to other atmospheric molecules (Thiemens, 1999). It is not clear how the isotopic anomalies could have been transmitted to meteoritic materials.

The process of photochemical self-shielding, which selects isotopes on the basis of their abundances, rather than their mass, might also have occurred within an initially homogeneous solar nebula, with the nascent Sun as the UV light source (Clayton, 2002). Because the density of the solar nebula was many orders of magnitude greater than the density of its precursor molecular cloud, the depth of penetration of solar UV was only a small fraction of 1 AU. The X-wind model of growth of the Sun from an accretion disk (Shu *et al.*, 1996) provides a plausible setting for the self-shielding, in which accretionary matter is irradiated by the hot, early Sun at a distance of <0.1 AU, and a sizable fraction of the irradiated, chemically processed material is carried outward by the magnetically driven X-wind, to form the rocky parts of the inner solar system. The self-shielding mechanism can only act to increase the ^{17}O and ^{18}O abundances in the processed material, so that the primary solar system isotopic composition must have been at the ^{16}O-rich end of the mixing lines of Figures 3 and 4. In this picture, the ^{16}O-rich compositions at the lower end of the mixing line in Figure 4 represent condensates from the unaltered nebular gas, while the less-^{16}O-rich compositions are produced by chemical interaction with atomic oxygen, enriched in ^{17}O and ^{18}O by the photolysis of CO. The direct measurement of the oxygen isotope abundances in the solar wind, as sampled by the Genesis spacecraft, should allow a definitive choice from the proposed models.

1.3 OXYGEN ISOTOPES IN CHONDRITES

Chondritic meteorites, characterized by their relatively unfractionated chemical compositions, and usually consisting of chondrules and some matrix, can be subdivided into classes, as follows: carbonaceous chondrites—Vigarano-type (CV), Ornans-type (CO), Mighei-type (CM), Renazzo-type (CR), Karoonda-type (CK), Bencubbin-type (CB), and ALH 85085-type (CH); ordinary chondrites—high-iron (H), low-iron (L), and low-iron, low-metal (LL); and enstatite chondrites—high-iron (EH), low-iron (EL); R-chondrites, characterized by olivine with very high ferrous iron content.

Figure 5 shows schematically the locations and ranges of whole-rock oxygen isotopic compositions of the various chondrite classes. The distribution is intrinsically two dimensional, so that there is no single parameter, such as distance

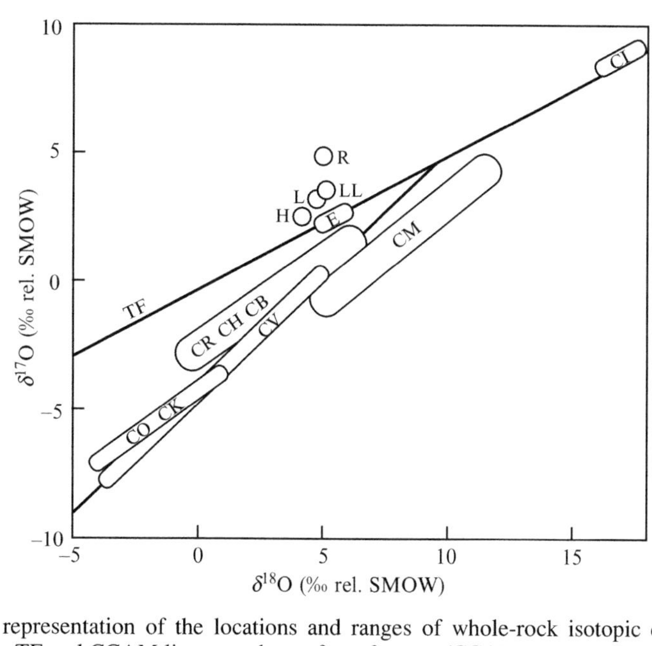

Figure 5 Schematic representation of the locations and ranges of whole-rock isotopic compositions of major chondrite classes. The TF and CCAM lines are shown for reference (CCAM mostly hidden by the CV balloon).

from the Sun, by which to order the groups. It is noteworthy that the ranges of compositions within each carbonaceous chondrite group are much larger than the ranges for the ordinary (O) and enstatite (E) groups. The large ranges for CM and CR–CH–CB groups are the result of varying degrees of low-temperature aqueous alteration, which produces phyllosilicates enriched in the heavy isotopes. It will be shown below that the CO and CM groups form a continuous trend on the three-isotope graph, and that the variations within the CO group may also be due to small degrees of aqueous alteration. The range of ~10‰ in the CV chondrites is not associated with phyllosilicate formation, but may represent internal isotopic heterogeneity due to the presence of ^{16}O-rich refractory phases. The whole-rock isotopic compositions of CV meteorites fall along the same CCAM line shown in Figure 3 for mineral analyses of refractory inclusions.

The isotopic composition of enstatite chondrites is indistinguishable from that of the Earth, which has led to suggestions that such material was a major "building block" for our planet (Javoy *et al.*, 1986). It should be noted that the Earth does not occupy a special place on the diagram: it cannot be considered as "normal" or as an end-member of a mixing line.

The most abundant stony meteorites are the ordinary chondrites (H, L, and LL). Whole-rock oxygen isotopic compositions are similar, but resolvable, for the three iron groups (Clayton *et al.*, 1991). Remarkably, analyses of individual chondrules from these meteorites show no grouping corresponding to the classification of the host meteorites (Figure 6). All ordinary chondrite chondrules appear to be derived from a single population. It is possible that the differences in whole-rock composition result from a size sorting of chondrules, as it is known that mean chondrule sizes increase in the order: H < L < LL (Rubin, 1989).

The isotopic compositions of the various carbonaceous chondrite groups imply some genetic relations among them. The CK data fall within the CO range, suggesting that CK are metamorphosed from CO-like precursors. The CR, CH, and CB meteorites have many properties in common, although the three groups differ in texture and metal content (Weisberg *et al.*, 2001).

The characteristic feature of CM chondrules is the coexistence of roughly equal amounts of high-temperature anhydrous silicates (olivine and pyroxene) and low-temperature hydrous clay minerals. It is generally believed that the clay minerals were formed by aqueous alteration of the high-temperature phases, either in space or in the parent body. Figure 7 shows that the phyllosilicate matrix is systematically enriched in the heavy isotopes of oxygen, relative to the whole rock. The tie-lines between whole-rock compositions and matrix compositions have slopes of ~0.7, implying that the water reservoir had a composition with $\Delta^{17}O$ more positive than the silicate reservoir. Clayton and Mayeda (1999) showed that the observed patterns can be accounted for with a simple closed-system hydration reaction at temperatures near 0 °C, with water/rock ratios in the range 0.4–0.6 (in terms of oxygen atoms).

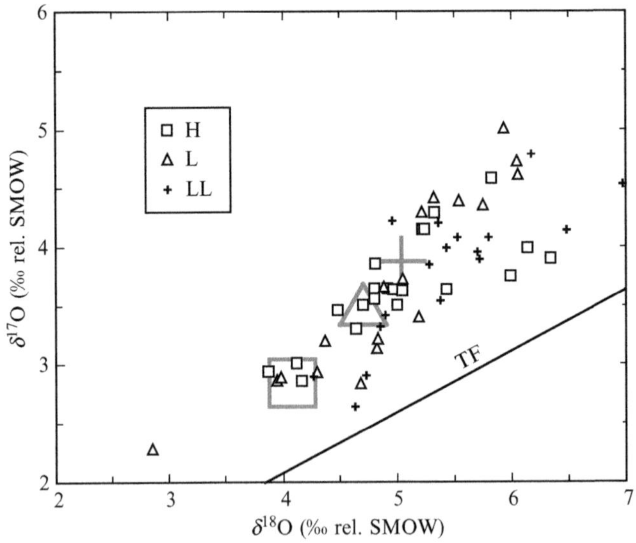

Figure 6 Oxygen isotopic compositions of individual chondrules from ordinary chondrites, with symbols showing the H, L, or LL group of the parent meteorite. The large gray symbols show the mean compositions of H, L, and LL whole rocks. There is no correlation between chondrule composition and parent composition, indicating that all ordinary chondrite chondrules are drawn from the same population. The TF line is shown for reference. The least-squares line fit to the chondrule data has a slope of 0.69 (source Clayton *et al.*, 1991).

Since the phyllosilicates in CM chondrites are a few per mil enriched in heavy isotopes relative to the whole rock, material balance requires that the unanalyzed residual anhydrous silicates must be depleted in heavy isotopes by a comparable amount. This puts their composition into the range of CO chondrites. Figure 8 shows the relationship between CO and CM chondrites, which apparently represents different water/rock ratios in the aqueous alteration of a common CO-like precursor. The genetic association of CO and CM chondrites was discussed from a chemical and textural viewpoint by Rubin and Wasson (1986).

1.3.1 Thermal Metamorphism of Chondrites

1.3.1.1 Ordinary chondrites

Most ordinary chondrites have been thermally metamorphosed within their parent bodies, as evidenced by mineral recrystallization and chemical homogenization, from which a petrographic scale, from 3 (least metamorphosed) to 6 (most metamorphosed), has been constructed (Van Schmus and Wood, 1967). This metamorphism occurred in an almost closed system with respect to oxygen, resulting in less than 0.5‰ variation in $\delta^{18}O$ for

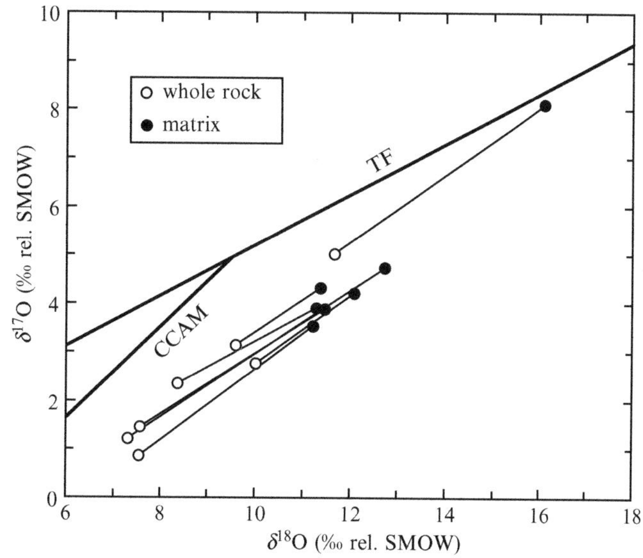

Figure 7 Whole-rock and separated matrix isotopic compositions for several CM chondrites. The matrix is always enriched in the heavier isotopes as a consequence of low-temperature aqueous alteration (source Clayton and Mayeda, 1999).

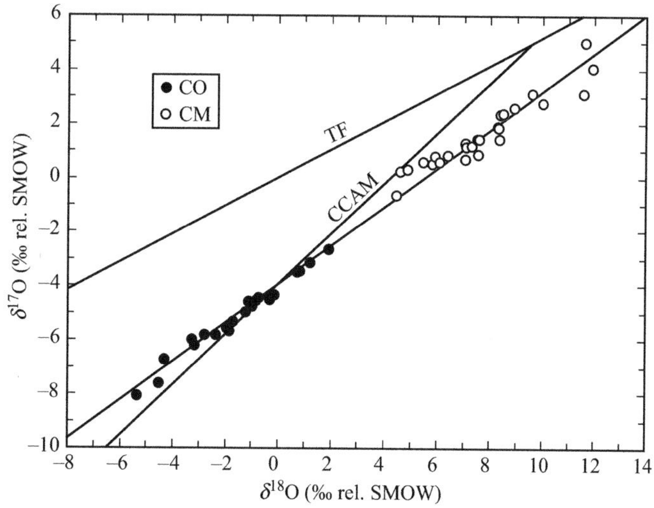

Figure 8 Whole-rock isotopic compositions of CO and CM chondrites. The line is a least-squares fit to all the data. The large range within each group reflects variations in water/rock ratio, which increases from lower left to upper right. The anhydrous precursors to both CO and CM groups should lie at the lower end of the CO group (source Clayton and Mayeda, 1999).

different metamorphic grades within each iron group (H, L, and LL) (Clayton *et al.*, 1991). This observation is in accord with the inferred anhydrous state of the O-chondrite parent bodies.

During metamorphism and recrystallization, oxygen isotopes are redistributed among mineral phases, according to the mass-dependent equilibrium fractionations corresponding to the peak metamorphic temperature. The measured mineral-pair fractionations (usually for major minerals: olivine, pyroxene, and feldspar) can then be used for metamorphic thermometry, yielding temperatures of 600 °C for an L4 chondrite, and 850 ± 50 °C for several type-5 and type-6 chondrites (Clayton *et al.*, 1991). Isotopic equilibration, even in type-6 chondrites, involves oxygen atom transport only over distances of a few millimeters (Olsen *et al.*, 1981).

1.3.1.2 CO chondrites

The CO3 meteorites have been subdivided into metamorphic grades from 3.0 to 3.7, with peak temperatures in the range 450–600 °C (Rubin, 1998). In contrast to the metamorphism of the ordinary chondrites, the CO metamorphism probably occurred in the presence of water, and the system was not closed with respect to oxygen. The least metamorphosed CO meteorites, ALH 77307 and Y 81020, are the most ^{16}O-rich; Loongana 001 and HH 073, classified as 3.8 or 4, are the most ^{16}O-poor. There is not, however, a simple one-to-one correspondence between metamorphic grade and isotopic composition. Since most of the CO chondrites are finds, some may have been altered by terrestrial weathering.

1.3.1.3 CM and CI chondrites

Some CM and CI chondrites have been heated to temperatures sufficient to cause dehydration of phyllosilicates (Akai, 1990). Reflectance spectra of many asteroids suggest that metamorphosed carbonaceous chondrite material is common (Hiroi *et al.*, 1993). Oxygen isotopic compositions of dehydrated CM and CI chondrites occupy two separate regions in the three-isotope graph (Figure 9), one falling near the terrestrial line with δ^{18}O and δ^{17}O values more positive than the unmetamorphosed CI chondrites, and the other falling near the CM chondrite region, but lower in δ^{17}O by ~1‰. Thus, neither group has isotopic compositions corresponding to those of any unmetamorphosed meteorites. The process of thermal dehydration itself probably introduces some isotopic fractionation. However, in rocks as chemically complex as CM and CI chondrites, it is not possible to predict either the magnitude or direction of such fractionation effects (Clayton *et al.*, 1997).

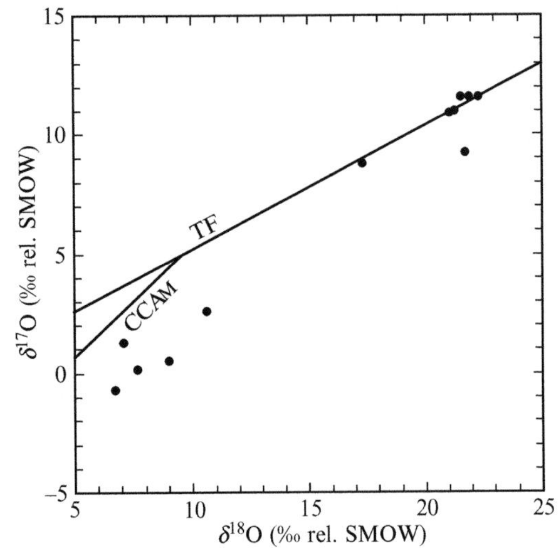

Figure 9 Oxygen isotopic compositions of metamorphosed CM- and CI-like materials, identified by their reflectance spectra, which show dehydration of phyllosilicates (Akai, 1990). The upper group may have had CI-like precursors, and the lower group may have had CM-like precursors, but the isotopic effects of metamorphism are uncertain (source Clayton and Mayeda, 1999).

Another class of objects that have apparently gone through a cycle of hydration and dehydration are the dark inclusions commonly found in CV chondrites (Clayton and Mayeda, 1999). These have a wide range of isotopic compositions, enriched in the heavy oxygen isotopes, reflecting primarily the hydration step.

1.4 OXYGEN ISOTOPES IN ACHONDRITES

The achondritic meteorites can be subdivided into the differentiated achondrites: igneous rocks from parent bodies that were extensively melted, and the undifferentiated, or primitive, achondrites from parent bodies that underwent little melting.

1.4.1 Differentiated (Evolved) Achondrites

Oxygen isotope variations in differentiated bodies—such as Earth, Moon, Mars—and the parent body of the HED group (for howardites, eucrites, and diogenites) show characteristic mass-dependent fractionation lines, with a slope of 0.52 on the three-isotope plot (Figure 10). In general, different bodies define separate, parallel trends with different Δ^{17}O ($\Delta^{17}O = \delta^{17}O - 0.52\ \delta^{18}O$), reflecting the different bulk isotopic composition of the parent body. These separate fractionation lines are useful in demonstrating that diverse lithologies are derived from a common source reservoir. For example, the SNC meteorites (for Shergotty, Nakhla, and Chassigny) range

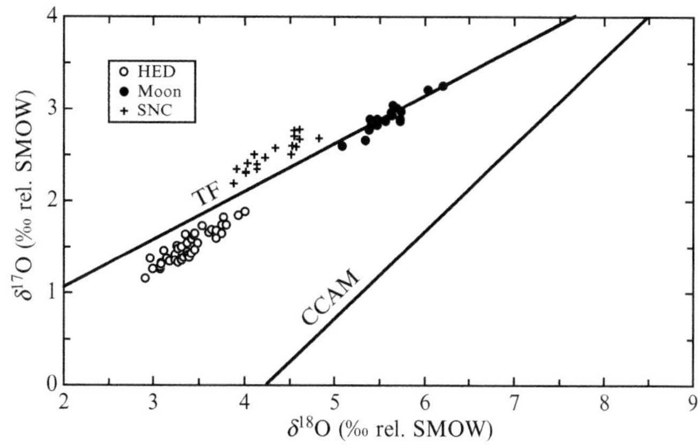

Figure 10 Oxygen isotopic compositions of whole-rock meteorites from differentiated bodies: HED, possibly asteroid Vesta; SNC, possibly Mars; and Moon. Each body produces a slope-1/2 mass-dependent fractionation line, with values of $\Delta^{17}O$ characteristic of the whole source planet. The isotopic compositions of lunar meteorites are indistinguishable from those of terrestrial mantle rocks. Note that HED and SNC lie on opposite sides of TF. This and Figure 11 are drawn to the same scale for comparison (source Clayton and Mayeda, 1996).

from basalts to peridotite, but all have the same value of $\Delta^{17}O = 0.30$‰, supporting other evidence for a common parent body. This same property identified ALH 84001 as a member of this group, even though its petrography and age are unique (Clayton and Mayeda, 1996). Similarly, the HED meteorites have a uniform value of $\Delta^{17}O = -0.25$‰, again implying a common parent body, perhaps the asteroid Vesta. Caution must be applied, however, in that accidental coincidences may occur. For example, Earth, Moon, and the aubrite (enstatite achondrite) parent body have indistinguishable values of $\Delta^{17}O = 0$‰ (Wiechert *et al.*, 2001), but are obviously separate bodies.

Besides the HED group, other achondrites that also have $\Delta^{17}O$ close to -0.25‰ are the mesosiderites, main-group pallasites, and IIIAB iron meteorites. It remains a subject of controversy whether this similarity is an accident, whether it implies a single parent body, or whether it implies several related parent bodies derived from some common reservoir.

Differentiated, or evolved, achondrite types are rare. Other than the HED group, there are only two other known asteroidal sources, one for the angrites (six known meteorites) (Mittlefehldt *et al.*, 2002) and one for the unique basalt NWA 011 (Yamaguchi *et al.*, 2002). Oxygen isotopic compositions of angrites are barely resolvable from those of the HED group (Clayton and Mayeda, 1996), whereas that of NWA 011 is distinctly different, and falls near the CR chondrite group.

The pallasites, coarse-grained stony-iron meteorites, also form three oxygen isotope groups: the main group, the Eagle Station pallasites (three members), and the pyroxene pallasites (two members) (Clayton and Mayeda, 1996).

In contrast to the small number of differentiated parent bodies represented by evolved achondritic meteorites, the number of parent bodies inferred from the chemical compositions of iron meteorites may be as large as 50 (Wasson, 1990). Of the 13 major iron meteorite groups, 10 appear to be from cores of differentiated meteorites. Many additional cores are inferred from the "ungrouped" irons, which make up ~15% of iron meteorites. It is a puzzle why we appear to sample many more cores than mantles of these asteroids.

Some iron meteorites contain oxygen-bearing phases: chromite, phosphates, and silicates, which can be isotopically analyzed to search for genetic associations with stony meteorite groups. There is a clear isotopic and chemical connection between the group IAB irons and winonaites (one class of primitive achondrite, discussed below). An equally strong association is found for IIIAB irons and main-group pallasites (Clayton and Mayeda, 1996). Oxygen-isotope links are also found between IIE irons and H-group ordinary chondrites, and between group IVA irons and L- or LL-group ordinary chondrites. Several ungrouped irons have oxygen isotopes in the range of CV carbonaceous chondrites (Clayton and Mayeda, 1996). The close association of some iron meteorites with chondrites suggests an origin as impact melts, rather than asteroidal cores.

1.4.2 Undifferentiated (Primitive) Achondrites

A combination of chemical, textural, and oxygen isotopic information permits recognition of several groups of achondrites that may have undergone a small degree of partial melting, but were never part of a differentiation process on an asteroidal scale (Goodrich and Delaney, 2000).

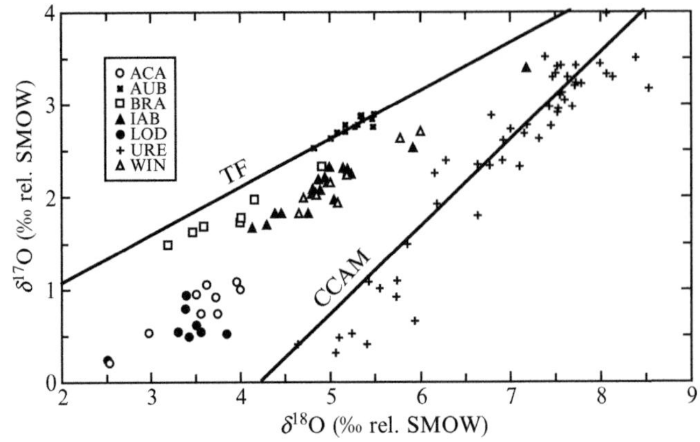

Figure 11 Oxygen isotopic compositions of whole-rock "primitive" achondrites. Data for a given class are much more scattered than data for differentiated achondrites shown in Figure 10, as a result of incomplete melting and homogenization. Several genetic associations are implied by the data: (i) aubrites and enstatite chondrites; (ii) acapulcoites and lodranites; (iii) IAB irons and winonaites; and (iv) ureilites and dark inclusions in carbonaceous chondrites. This and Figure 10 are drawn to the same scale for comparison (source Clayton and Mayeda, 1996).

They argue that the distinction between "evolved" achondrites and "primitive" achondrites is that the former are melts, while the latter are residues of partial melting. These residues may retain chemical and isotopic heterogeneities inherited from their unmelted precursors. By this criterion, the primitive achondrites include the aubrites, lodranites/ acapulcoites, brachinites, winonaites, and ureilites. Their oxygen isotopic compositions are shown in Figure 11.

The aubrites (enstatite achondrites) are clearly closely related to the enstatite chondrites, discussed earlier. They also share the property of being highly reduced, and have identical oxygen isotopic compositions (Clayton *et al.*, 1984). Experimental studies by McCoy *et al.* (1999) show that partial melting of an E-chondrite can yield aubritic material by removal of a basaltic melt and a metal-sulfide melt. This is the best known instance of a genetic connection between chondrites and achondrites.

The brachinites are a small group of olivine-rich residues, with oxygen isotopic compositions overlapping those of the HED achondrites, although no genetic association has been proposed (Clayton and Mayeda, 1996).

The lodranites and acapulcoites are almost certainly from the same parent body (McCoy *et al.*, 1997). Both groups are residues of partial melting, with the lodranites having undergone more extensive melting and recrystallization. Their oxygen isotopic compositions overlap and form a blob on the three-isotope graph, rather than a fractionation line that would indicate complete melting. The mean isotopic composition of this group falls within the range of CR and CH chondrites, but no genetic association has been proposed.

The winonaites and silicates in IAB irons share the same range of oxygen isotopic composition, and probably have a common parent body. A model for generation of winonaites and IAB–IIICD irons in the same body (completely independent of oxygen isotope evidence) was presented by Kracher (1985).

The ureilites are the most remarkable of the primitive achondrites. They are coarse-grained, carbon-bearing ultramafic rocks, and are residues of extensive partial melting that has removed a basaltic component. Their oxygen isotopic compositions are highly variable, and extend along an extrapolation of the CCAM line. The isotopic variability is probably inherited from heterogeneous carbonaceous chondrite precursors (Clayton and Mayeda, 1988).

1.5 SUMMARY AND CONCLUSIONS

Oxygen isotope abundances display a remarkable variability on all scales studied, from micrometers to planetary dimensions. Two types of processes have been identified and associated with some of this variability: (i) a nebular interaction between condensates and the ambient gas, probably at a temperature sufficiently high that mass-dependent fractionation effects were relatively small, and (ii) low-temperature aqueous alteration, probably occurring within the parent bodies. The first process begins with ^{16}O-rich condensates, as are seen in CAIs, and enriches them in ^{17}O and ^{18}O along the CCAM line. The second process modifies these materials, especially in the carbonaceous chondrites, with further enrichment of the two heavier isotopes, but on lines of shallower slope as a consequence of large low-temperature fractionation effects. Although these two processes may be adequate to account for the observed variability in the chondritic meteorites, they do not seem to provide an explanation for the range of compositions of the achondrites. Some *ad hoc* explanations have been proposed, such as the existence of additional oxygen reservoirs or time-variable oxygen sources (Wasson, 2000), but these are basically untestable.

Many questions remain unanswered. What was the anhydrous precursor for the CR–CH–CB group? What were the precursors of the metamorphosed CM and CI meteorites? What was the relationship between ureilites and carbonaceous chondrites? Why are so few differentiated parent bodies represented by achondrites? Why is the isotopic composition of the Earth identical to that of the Moon, but different from that of Mars? What is the relationship (if any) between the Earth and the enstatite meteorites? Some of these questions may be successfully addressed once we have accurate fine-scaled chronology so as to put solar nebular events into the correct time sequence.

REFERENCES

Akai J. (1990) Mineralogical evidence of heating events in Antarctic carbonaceous chondrites. *Proc. NIPR Symp. Antarct. Meteorit.* **3**, 55–68.

Aléon J., Krot A. N., McKeegan K. D., MacPherson G. J., and Ulyanov A. A. (2002) Oxygen isotopic composition of fine-grained Ca–Al-rich inclusions in the reduced CV3 chondrite Efremovka. *Lunar Planet. Sci.* **XXXIII**, #1426. The Lunar and Planetary Institute, Houston (CD-ROM).

Allègre C. J., Manhès G., and Göpel C. (1995) The age of the Earth. *Geochim. Cosmochim. Acta* **59**, 1445–1446.

Anders E. and Zinner E. (1993) Interstellar grains in primitive meteorites: diamond, silicon carbide and graphite. *Meteoritics* **28**, 490–514.

Boothroyd A. I. and Sackmann I. J. (1999) The CNO isotopes: deep circulation in red giants and first and second dredge-up. *Astrophys. J.* **510**, 232–250.

Clayton R. N. (1993) Oxygen isotopes in meteorites. *Ann. Rev. Earth Planet. Sci.* **21**, 115–149.

Clayton R. N. (2002) Self-shielding in the solar nebula. *Nature* **415**, 860–861.

Clayton R. N. and Mayeda T. K. (1988) Formation of ureilites by nebular processes. *Geochim. Cosmochim. Acta* **52**, 1313–1318.

Clayton R. N. and Mayeda T. K. (1996) Oxygen isotope studies of achondrites. *Geochim. Cosmochim. Acta* **60**, 1999–2017.

Clayton R. N. and Mayeda T. K. (1999) Oxygen isotope studies of carbonaceous chondrites. *Geochim. Cosmochim. Acta* **63**, 2089–2104.

Clayton R. N., Onuma N., Grossman L., and Mayeda T. K. (1977) Distribution of the presolar component in Allende and other carbonaceous chondrites. *Earth Planet. Sci. Lett.* **34**, 209–224.

Clayton R. N., Onuma N., Ikeda Y., Mayeda T. K., Hutcheon I. D., Olsen I. D., and Molini-Velsko C. (1983) Oxygen isotopic compositions of chondrules in Allende and ordinary chondrites. In *Chondrules and their Origins* (ed. E. A. King). Lunar and Planetary Institute, Houston, pp. 37–43.

Clayton R. N., Mayeda T. K., and Rubin A. E. (1984) Oxygen isotopic compositions of enstatite chondrites and aubrites. *J. Geophys. Res.* **89**, C245–C249.

Clayton R. N., Mayeda T. K., Goswami J. N., and Olsen E. J. (1991) Oxygen isotope studies of ordinary chondrites. *Geochim. Cosmochim. Acta* **35**, 2317–2338.

Clayton R. N., Mayeda T. K., Hiroi T., Zolensky M. E., and Lipschutz M. E. (1997) Oxygen isotopes in laboratory-heated CI and CM chondrites. *Meteorit. Planet. Sci.* **32**, A30.

Dauphas N., Marty B., and Reisberg L. (2002) Molybdenum nucleosynthetic dichotomy revealed in primitive meteorites. *Astrophys. J.* **569**, L139–L142.

Davis A. M., Hashimoto A., Clayton R. N., and Mayeda T. K. (1990) Isotope mass fractionation during evaporation of Mg_2SiO_4. *Nature* **347**, 655–658.

Eberhardt P., Geiss J., Graf H., Grögler N., Krähenbühl U., Schwaller H., Schwarzmüller H., and Stettler A. (1970) Trapped solar wind noble gases, exposure age, and K/Ar-age in Apollo 11 lunar fine material. *Proc. Apollo 11 Lunar Sci. Conf., Pergamon, New York*, pp. 1037–1070.

Epstein S. and Taylor H. P. (1970) The concentration and isotopic composition of hydrogen, carbon, and silicon in Apollo 11 lunar rocks and minerals. *Proc. Apollo 11 Lunar Sci. Conf., Pergamon, New York*, pp. 1085–1096.

Fagan T. J., Yurimoto H., Krot A. N., and Keil K. (2002) Constraints on oxygen isotopic evolution from an amoeboid olivine aggregate and Ca, Al-rich inclusions from the CV3 Efremovka. *Lunar Planet. Sci.* **XXXIII**, #1507. The Lunar and Planetary Institute, Houston (CD-ROM).

Giauque W. F. and Johnston H. L. (1929) An isotope of oxygen, mass 18. Interpretation of the atmospheric absorption bands. *J. Am. Chem. Soc.* **51**, 1436–1441.

Gloeckler G. and Geiss J. (1998) Measurement of the abundance of helium-3 in the Sun and in the local interstellar cloud with SWICS on Ulysses. *Space Sci. Rev.* **84**, 275–284.

Goodrich C. A. and Delaney J. S. (2000) Fe/Mg–Fe/Mn relations of meteorites and primary heterogeneity of primitive achondrite parent bodies. *Geochim. Cosmochim. Acta* **64**, 149–160.

Grossman L. (1973) Refractory trace elements in Ca–Al-rich inclusions in the Allende meteorite. *Geochim. Cosmochim. Acta* **37**, 1119–1140.

Harris M. J., Lambert D. L., and Goldman A. (1987) The $^{12}C/^{13}C$ and $^{16}O/^{18}O$ ratios in the solar photosphere. *Mon. Not. Roy. Astron. Soc.* **224**, 237–255.

Hashizume K., Chaussidon M., Marty B., and Robert F. (2000) Solar wind record on the Moon: deciphering presolar from planetary nitrogen. *Science* **290**, 1142–1145.

Heidenreich J. E. and Thiemens M. H. (1983) A non-mass-dependent isotope effect in the production of ozone from molecular oxygen. *J. Chem. Phys.* **78**, 892–895.

Hiroi T., Pieters C. M., Zolensky M. E., and Lipschutz M. E. (1993) Evidence of thermal metamorphism in the C, G, B, and F asteroids. *Science* **261**, 1016–1018.

Itoh S., Rubin A. E., Kojima H., Wasson J. T., and Yurimoto H. (2002) Amoeboid olivine aggregates and AOA-bearing chondrule from Y 81020 CO 3.0 chondrite: distribution of oxygen and magnesium isotopes. *Lunar Planet. Sci.* **XXXIII**, #1490. The Lunar and Planetary Institute, Houston (CD-ROM).

Javoy M., Pineau F., and Delorme H. (1986) Carbon and nitrogen isotopes in the mantle. *Chem. Geol.* **57**, 41–62.

Jones R. H., Leshin L. A., and Guan Y. (2002) Heterogeneity and ^{16}O-enrichments in oxygen isotope ratios of olivine from chondrules in the Mokoia CV3 chondrite. *Lunar Planet. Sci.* **XXXIII**, #1571. The Lunar and Planetary Institute, Houston (CD-ROM).

Kallenbach R., Geiss J., Ipavich F. M., Gloeckler G., Bochsler P., Gliem P., Hefti S., Hilchenbach M., and Hovestadt D. (1998) *Astrophys. Isotopic composition of solar-wind nitrogen: first* in situ *determination with the CELIAS/MTOF spectrometer on board SOHO. J.* **507**, L185–L188.

Kerridge J. F. (1993) Long-term compositional variation in solar corpuscular radiation: evidence from nitrogen isotopes in the lunar regolith. *Rev. Geophys.* **31**, 423–437.

Kracher A. (1985) The evolution of partially differentiated planetesimals: evidence from iron meteorite groups IAB and IIICD. *J. Geophys. Res.* **90**, C689–C698.

Krot A. N., Aléon J., and McKeegan K. D. (2002) Mineralogy, petrography and oxygen-isotopic compositions of Ca–Al-rich inclusions and amoeboid olivine aggregates in the CR carbonaceous chondrites. *Lunar Planet. Sci.* **XXXIII**, #1412. The Lunar and Planetary Institute, Houston (CD-ROM).

Leske R. A., Mewaldt R. A., Cohen C. M. S., Christian E. R., Cummings A. C., Slocum A. C., Stone E. C., von Rosenvinge T. T., and Wiedenbeck M. E. (2001) Isotopic abundances in the solar corona as inferred from ACE measurements of solar energetic particles. In *Solar and Galactic Composition* (ed. R. F. Wimmer-Schweingruber). AIP Conf. Proc. 598, Melville, NY, pp. 127–132.

Manian S. H., Urey H. C., and Bleakney W. (1934) An investigation of the relative abundance of the oxygen isotopes O^{16} : O^{18} in stone meteorites. *J. Am. Chem. Soc.* **56**, 2601–2609.

McCoy T. J., Keil K., Clayton R. N., Mayeda T. K., Bogard D. D., Garrison D. D., and Wieler R. (1997) A petrologic and isotopic study of lodranites: evidence for early formation as partial melt residues from heterogeneous precursors. *Geochim. Cosmochim. Acta* **61**, 623–637.

McCoy T. J., Dickinson T. L., and Lofgren G. E. (1999) Partial melting of the Indarch (EH4) meteorite: a textural, chemical, and phase relations view of melting and melt migration. *Meteorit. Planet. Sci.* **34**, 735–746.

Mittlefehldt D. W., Kilgore M., and Lee M. T. (2002) Petrology and geochemistry of d'Orbigny, geochemistry of Sahara 99555, and the origin of angrites. *Meteorit. Planet. Sci.* **37**, 345–369.

Nicolussi G. K., Pellin M. J., Lewis R. S., Davis A. M., Amari S., and Clayton R. N. (1998) Molybdenum isotopic composition of individual presolar silicon carbide grains from the Murchison meteorite. *Geochim. Cosmochim. Acta* **62**, 1093–1104.

Nittler L. R., Alexander C. M. O'D., Gao X., Walker R. M., and Zinner E. (1997) Stellar sapphires: the properties and origins of presolar Al_2O_3 in meteorites. *Astrophys. J.* **483**, 475–495.

Olsen E. J., Mayeda T. K., and Clayton R. N. (1981) Cristobalite-pyroxene in an L6 chondrite: implications for metamorphism. *Earth Planet. Sci. Lett.* **56**, 82–88.

Owen T., Mahaffy P. R., Niemann H. B., Atreya S., and Wong M. (2001) Protosolar nitrogen. *Astrophys. J.* **553**, L77–L79.

Pepin R. O. (1991) On the origin and early evolution of terrestrial planet atmospheres and meteoritic volatiles. *Icarus* **92**, 2–79.

Prantzos N., Aubert O., and Audouze J. (1996) Evolution of carbon and oxygen isotopes in the Galaxy. *Astron. Astrophys.* **309**, 760–774.

Rubin A. E. (1989) Size-frequency distributions of chondrules in CO3 chondrites. *Meteoritics* **24**, 179–189.

Rubin A. E. (1998) Correlated petrologic and geochemical characteristics of CO3 chondrites. *Meteorit. Planet. Sci.* **33**, 385–391.

Rubin A. E. and Wasson J. T. (1986) Chondrules in the Murray CM2 meteorite and compositional differences between CM–CO and ordinary chondrite chondrules. *Geochim. Cosmochim. Acta* **50**, 307–315.

Shu F. H., Shang H., and Lee T. (1996) Toward an astrophysical theory of chondrites. *Science* **271**, 1545–1552.

Speck A. K., Barlow M. J., Sylvester R. J., and Hofmeister A. M. (2000) Dust features in the 10 μm infrared spectra of oxygen-rich evolved stars. *Astron. Astrophys.* **146**, 437–464.

Thiemens M. H. (1999) Atmospheric science—mass-independent isotope effects in planetary atmospheres and the early solar system. *Science* **283**, 341–345.

van Dishoeck E. F. and Black J. H. (1988) The photodissociation and chemistry of interstellar CO. *Astrophys. J.* **334**, 771–802.

Van Schmus W. R. and Wood J. A. (1967) A chemical-petrologic classification for the chondritic meteorites. *Geochim. Cosmochim. Acta* **31**, 747–765.

Wasson J. T. (1990) Ungrouped iron meteorites in Antarctica: origin of anomalously high abundance. *Science* **249**, 900–902.

Wasson J. T. (2000) Oxygen isotopic evolution of the solar nebula. *Rev. Geophys.* **38**, 491–512.

Weisberg M. K., Prinz M., Kojima H., Yanai K., Clayton R. N., and Mayeda T. K. (1991) The Carlisle Lakes-type chondrites: a new grouplet with high $\Delta^{17}O$ and evidence for nebular oxidation. *Geochim. Cosmochim. Acta* **55**, 2657–2669.

Weisberg M. K., Prinz M., Clayton R. N., Mayeda T. K., Sugiura N., Zashu N., and Ebihara M. (2001) A new metal-rich chondrite grouplet. *Meteorit. Planet. Sci.* **36**, 401–418.

Wiechert U., Halliday A. N., Lee D.-C., Snyder G. A., Taylor L. A., and Rumble D. (2001) Oxygen isotopes and the Moon-forming giant impact. *Science* **294**, 345–348.

Wimmer-Schweingruber R. F., Bochsler P., and Gloeckler G. (2001) The isotopic composition of oxygen in the fast solar wind. *Geophys. Res. Lett.* **28**, 2763–2766.

Wyckoff S., Kleine M., Peterson B. A., Wehinger P. A., and Ziurys L. M. (2000) Carbon isotope abundances in comets. *Astrophys. J.* **535**, 991–999.

Yamaguchi A., Clayton R. N., Mayeda T. K., Ebihara M., Oura Y., Miura Y., Haramura H., Misawa K., Kojima H., and Nagao K. (2002) A new source of basaltic meteorites inferred from Northwest Africa 011. *Science* **196**, 334–336.

Yin Q., Jacobsen S. B., and Yamashita K. (2002) Diverse supernova sources of pre-solar material inferred from molybdenum isotopes in meteorites. *Nature* **415**, 881–885.

Yurimoto H. and Kuramoto K. (2002) A possible scenario introducing heterogeneous oxygen isotopic distribution in protoplanetary disks. *Meteorit. Planet. Sci. Suppl.* **37**, A153.

Zhu X. K., Guo Y., O'Nions R. K., Young E. D., and Ash R. D. (2001) Isotopic homogeneity of iron in the early solar nebula. *Nature* **412**, 311–313.

Isotope Geochemistry
ISBN: 978-0-08-096710-3

pp. 1–16

2

Structural and Isotopic Analysis of Organic Matter in Carbonaceous Chondrites

I. Gilmour

The Open University, Milton Keynes, UK

2.1 INTRODUCTION

The most ancient organic molecules available for study in the laboratory are those carried to Earth by infalling carbonaceous chondrite meteorites. All the classes of compounds normally considered to be of biological origin are represented in carbonaceous meteorites and, aside from some terrestrial contamination; it is safe to assume that these organic species were produced by nonbiological methods of synthesis. In effect, carbonaceous chondrites are a natural laboratory containing organic molecules that are the product of ancient chemical evolution. Understanding the sources of organic molecules in meteorites and the chemical processes that led to their formation has been the primary research goal. Circumstellar space, the solar nebulae, and asteroidal meteorite parent bodies have all been suggested as environments where organic matter may have been formed. Determination of the provenance of meteoritic organic matter requires detailed structural and isotopic information, and the fall of the Murchison CM2 chondrite in 1969 enabled the first systematic organic analyses to be performed on comparatively pristine samples of extraterrestrial organic material. Prior to that, extensive work had been undertaken on the organic matter in a range of meteorite samples galvanized, in part, by the controversial debate in the early 1960s on possible evidence for former life in the Orgueil carbonaceous chondrite (Fitch *et al.*, 1962; Meinschein *et al.*, 1963). It was eventually demonstrated that the suggested biogenic material was terrestrial contamination (Fitch and Anders, 1963; Anders *et al.*, 1964); however, the difficulties created by contamination have posed a continuing problem in the analysis and interpretation of organic material in meteorites (e.g., Watson *et al.*, 2003); this has significant implications for the return of extraterrestrial samples by space missions. Hayes (1967) extensively reviewed data acquired prior to the availability of Murchison samples.

Developments in the analysis of meteoritic organic matter have largely been driven by progress in analytical capabilities. The limited availability of samples, often restricted to a few grams at most, has presented a series of analytical challenges and significant advances were made in the late 1960s and early 1970s when the coupling of gas chromatography with electron impact mass spectrometry (GCMS) enabled detailed structural information to be obtained on individual compounds (e.g., Hayes and Biemann, 1968). Light-element stable-isotope measurements of meteoritic organic matter can provide important information on its origins and the potential of such measurements has long been recognized (Boato, 1954). Indeed, meteoritic research has led to significant improvements in stable-isotope-ratio mass spectrometry (see Pillinger (1984), for a review). The 1970s saw the start of extensive quantitative analysis of solvent extractable compounds from Murchison together with the first attempts to resolve isotopic heterogeneities in the stable isotopes of carbon, hydrogen and nitrogen and this work has been the subject of regular detailed reviews (Anders *et al.*, 1973; Hayatsu and Anders, 1981; Mullie and Reisse, 1987; Cronin and Chang, 1993; Sephton and Gilmour, 2001b).

The principal focus of this review is on the analysis of the organic matter in the Murchison CM2 chondrite, together with data from other meteorites, where it can be shown that they have not been compromised by terrestrial contamination. It primarily covers work undertaken since 1980, a period that has seen the increasing use of stable-isotopic techniques to elucidate the sources of meteoritic organic matter, improved methods to study the structure of organic matter such as NMR, and the first *in situ* examinations of organic matter in meteorites; these approaches have provided significant advances in our understanding of the processes involved in the synthesis of extraterrestrial organic matter.

2.2 ORGANIC MATERIAL IN CARBONACEOUS CHONDRITES

The carbonaceous chondrites contain up to 5% carbon in a variety of forms, including organic matter, carbonates, and minor amounts of "exotic" presolar grain material such as diamond, graphite, and silicon carbide. For a review of the classification of presolar grains and carbonaceous meteorites. Less than 25% of the organic matter in carbonaceous chondrites is present as relatively low-molecular-weight "free" compounds which can be extracted with common organic solvents. The remaining 75% or so is present as high-molecular-weight macromolecular materials that persist after prolonged treatment of crushed samples of meteorites with organic solvents and acids such as HF–HCl that remove, or partially remove, silicates and other minerals (Hayatsu and Anders, 1981). These two components require different analytical strategies and are reviewed separately.

2.3 EXTRACTABLE ORGANIC MATTER

2.3.1 Abundance and Distribution of Extractable Compounds

2.3.1.1 Carboxylic acids

Early identifications of long-chain monocarboxylic acids (fatty acids) are generally believed to have been terrestrial contaminants (e.g., Smith and Kaplan, 1970). However, short-chain monocarboxylic acids ($<C_{10}$) constitute the most abundant class of extractable compounds in Murchison (Figure 1 and Table 1; Yuen and Kvenvolden, 1973; Lawless and Yuen, 1979; Yuen *et al.*, 1984). They show complete structural diversity with all C_2–C_5 isomers present and with equal concentrations of both branched- and straight-chain isomers; there is a general decrease in the abundance of short-chain monocarboxylic acids with increasing carbon number.

The distribution of monocarboxylic acids has been investigated in other CM2s, most notably Antarctic samples (Shimoyama *et al.*, 1986, 1989; Naraoka *et al.*, 1987; Naraoka *et al.*, 1999). Where present, the abundances of monocarboxylic acids ranged from 9.5 ppm (Yamato-791198) to 191 ppm (Asuka-881458). As with Murchison structural diversity is apparent, however, Yamato-74662 and

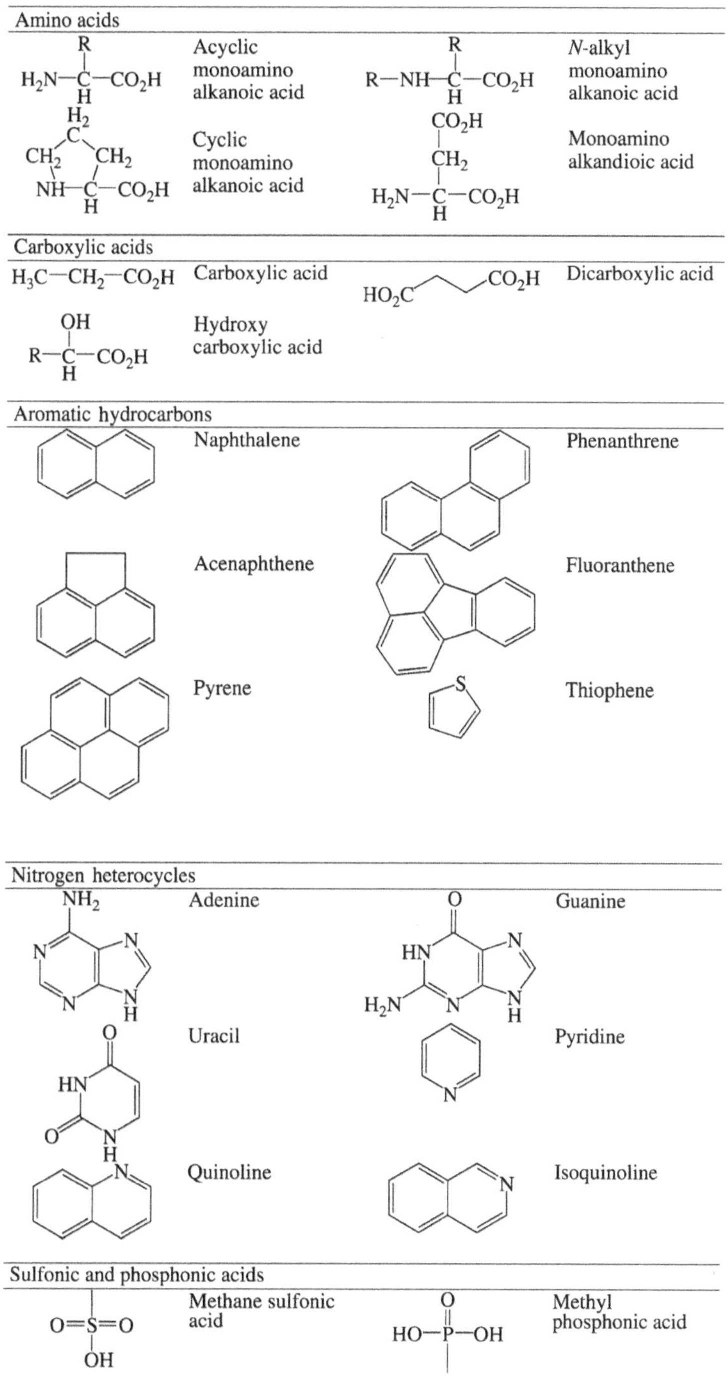

Figure 1 Representative structures of classes of organic compounds identified in carbonaceous chondrites.

Asuka-881458 showed an increasing predominance of straight chain over branched isomers (Shimoyama *et al.*, 1989; Naraoka *et al.*, 1999). Some Antarctic samples were devoid of monocarboxylic acids, which may indicate that they have lost carboxylic acids due to leaching by Antarctic ice.

A survey undertaken by Lawless *et al.* (1974) identified 17 dicarboxylic acids (Figure 1) in Murchison including 15 saturated and two unsaturated aliphatic compounds. As with monocarboxylic acids, complete structural diversity was observed. Cronin *et al.* (1993) examined the distribution of hydroxymonocarboxylic acids, dicarboxylic acids, and hydroxydicarboxylic acids in detail and identified at least 40 dicarboxylic acids with chain lengths up to C_9 and with most chain and substitution position isomers represented at each carbon number. Cronin and Chang (1993) suggested that the dicarboxylic acids are present in Murchison as carboxylate dianions, an interpretation supported by the

Table 1 Abundances of major classes of organic compounds in the Murchison (CM2) carbonaceous chondrite.

Compounds	*Abundance* (ppm)	*References*
Carbon dioxide	106	1
Carbon monoxide	0.06	1
Methane	0.14	1
Hydrocarbons		
Aliphatic	12–35	2
Aromatic	15–29	3
Carboxylic acids		
Monocarboxylic	332	4,1
Dicarboxylic	25.7	5
α-Hydroxycarboxylic	14.6	6
Amino acids	60	7
Alcohols	11	8
Aldehydes	11	8
Ketones	16	8
Sugars and related compounds	~60	9
Ammonia	19	10
Amines	8	11
Urea	25	12
Basic *N*-heterocycles	0.05–0.5	13
Pyridinecarboxylic acids	>7	14
Dicarboximides	>50	14
Pyrimidines	0.06	15
Purines	1.2	16
Benzothiophenes	0.3	17
Sulfonic acids	67	18
Phosphonic acids	1.5	19

References: 1. Yuen *et al.* (1984), 2. Kvenvolden *et al.* (1970), 3. Pering and Ponnamperuma (1971), 4. Lawless and Yuen (1979), 5. Lawless *et al.* (1974), 6. Peltzer *et al.* (1984), 7. Cronin *et al.* (1988), 8. Jungclaus *et al.* (1976b), 9. Cooper *et al.* (2001), 10. Pizzarello *et al.* (1994), 11. Jungclaus *et al.* (1976a), 12. Hayatsu *et al.* (1975), 13. Stoks and Schwartz (1982), 14. Pizzarello *et al.* (2001), 15. Stoks and Schwartz (1979), 16. Stoks and Schwartz (1981), 17. Shimoyama and Katsumata (2001), 18. Cooper *et al.* (1997), and 19. Cooper *et al.* (1992).

presence of the calcium salt of oxalic acid (Lawless *et al.*, 1974). Analyses of dicarboxylic acids in the Tagish Lake CI chondrite revealed a homologous series of saturated and unsaturated C_3–C_{10} acids (Pizzarello and Huang, 2002). Linear saturated acids were predominant and showed decreasing abundance with increasing chain length. In all, Pizzarello and Huang (2002) found 44 dicarboxylic acids, with succinic acid the most abundant. The distribution of Tagish Lake dicarboxylic acids showed good compound-to-compound correspondence with those observed in Murchison.

Hydroxy acids (α-hydroxy-carboxylic acids, Figure 1) were first reported in Murchison in the late 1970s (Peltzer and Bada, 1978). These compounds are racemic and correspond structurally to the more abundant amino acids. The presence of hydroxy acids is significant in two respects: (i) They are useful thermometers and indicate that Murchison has not experienced temperatures that would cause the pyrolytic breakdown of hydroxyacids, many of which will decompose at temperatures of less than 120 °C. (ii) The presence of α-hydroxy acids along with the structurally similar α-amino acids suggests the formation of both classes of compounds by a Strecker synthesis, the aqueous-phase component of amino acid synthesis by a Miller–Urey process (Peltzer and Bada, 1978). The α-hydroxydicarboxylic acids reported by Cronin *et al.* (1993) had not been found in previous studies of hydroxy acids in meteorites (Peltzer and Bada, 1978). Each class of hydroxy acids is numerous with carbon chains up to C_8 or C_9 and many, if not all, chain and substitution position isomers represented at each carbon number. The α-hydroxycarboxylic acids and α-hydroxydicarboxylic acids correspond structurally to many of the known meteoritic α-aminocarboxylic acids and α-aminodicarboxylic acids, respectively, a fact that supports the proposal that a Strecker synthesis was involved in the formation of both classes of compounds.

2.3.1.2 Sulfonic and phosphonic acids

Cooper *et al.* (1992, 1997) identified a homologous series of C_1–C_4 alkyl phosphonic acids and alkanesulfonic acids in water extracts from Murchison. Isomeric diversity was apparent with five of the eight possible C_1–C_4 alkyl phosphonic and seven of the eight C_1–C_4 alkanesulfonic acids identified. The relative abundances of both classes of compound decreased exponentially with increasing carbon number.

2.3.1.3 Amino acids

In comparison to living organisms and terrestrial sedimentary organic matter, the distribution of amino acids (Figure 1) reported in meteorites is unusual. Their abundance and distribution has been extensively investigated in the Murchison CM2 meteorite and to date more than 70 amino acids have been identified in hot-water extracts and many others have been partially characterized (e.g., Kvenvolden *et al.*, 1970; Nagy, 1975; Cronin *et al.*, 1981; Engel and Nagy, 1982; Cronin *et al.*, 1985, 1988; Cronin and Pizzarello, 1986). Murchison amino acids include eight of the protein amino acids and 11 others that are biologically common. The total amino acid concentration reported for Murchison is around 60 ppm (Table 1, Cronin *et al.*, 1988). However, differences in the concentrations of amino acids (Table 2) have been observed for separate stones of the Murchison CM2 meteorite that have been attributed to varying levels of terrestrial contamination and to different extraction procedures. Shock and Schulte (1990) have suggested that most of the differences

Table 2 Carbon, nitrogen, and hydrogen stable isotope compositions and abundances of individual amino acids in the Murchison meteorite.

Compound(s)	*Formula*	*Abundance* (nmol g^{-1})	$\delta^{13}C$ (‰)	*Refs.*	$\delta^{15}N$ (‰)	*Refs.*	δD (‰)	*Refs.*
Glycine	$C_2H_5NO_2$	28.1–31.0	+22	1	+37	2		
D-alanine	$C_3H_7NO_2$	12.9–17.1	+30	1	+60	2		
L-alanine	$C_3H_7NO_2$		+27	1	+57	2		
β-alanine	$C_3H_7NO_2$	5.7–8.1			+61	2		
Glycine + alanine			+41	3			+1,072	3
Sarcosine	$C_3H_7NO_2$				+129	2		
α-aminoisobutyric acid	$C_4H_9NO_2$	15.0–19.0	+5	1	+184	2	+67	3
D,L-aspartic acid	$C_4H_7NO_4$	1.0–3.9	+4	3	+61	2	+214	3
L-glutamic acid	$C_5H_9NO_4$		+6	1	+58	2	+523	3
D-glutamic	$C_5H_9NO_4$	1.9–4.6			+60	2		
D,L-proline	$C_5H_9NO_2$				+50	2		
Isovaline	$C_5H_{11}NO_2$		+17	1				
Isovaline + valine		4.6–7.5	+30	3			+713	
L-leucine	$C_6H_{13}NO_2$	0.8–1.6			+60	2		

References: 1. Engel *et al.* (1990) and Silfer (1991), 2. Engel and Macko (1997), and 3. Pizzarello *et al.* (1991).

in amino acid distributions and abundances reported for Murchison can, based on solubility data, be explained by differences in extractions procedures (Bada *et al.*, 1983; Cronin *et al.*, 1988). This conclusion is supported by the observation that the use of acid hydrolysis and repeated extractions of a sample of Murchison enabled the recovery of additional amino acids (Cronin, 1976a,b). It is also not possible to exclude minor heterogeneities in amino acid distribution between stones or the possible decomposition of amino acids with prolonged residence time on Earth (Cronin *et al.*, 1988). Nonetheless, there is reasonable agreement between studies of different stones and Table 2 reports the ranges in abundance for amino acids.

Amino acid abundances in several other CM chondrites have also been analyzed, though in less detail compared to Murchison (e.g., Lawless *et al.*, 1972, 1973; Cronin and Moore, 1976; Cronin *et al.*, 1981, 1995). Amino acid distribution in other CM chondrites is apparently similar to that of Murchison, although analysis of several samples is compromised by terrestrial contamination. Antarctic CM2s revealed an amino acid distribution and abundance similar to Murchison and are generally believed to have suffered less terrestrial biologically derived addition to the amino acid abundances, although low abundances in some samples may indicate that they have lost amino acids due to water leaching in Antarctic ice (Shimoyama *et al.*, 1979, 1985; Shimoyama and Harada, 1984). There have been limited analyses of the amino acids distribution of CI chondrites. Orgueil (CI1) apparently contains 11 amino acids, six of which were nonprotein (Lawless *et al.*, 1972). Analyses of both Orgueil and Ivuna (CI1) using high-performance liquid chromatography suggest that the amino acid distribution in the CIs is distinct from CM2s, notably β-alanine is relatively more abundant in Orgueil and Ivuna compared with Murchison while α-aminoisobutyric acid is only present in trace amounts (Ehrenfreund *et al.*, 2001). Analysis of water-soluble extracts of Tagish Lake showed that amino acids were less abundant in this possible CI2 chondrite by almost three orders of magnitude (<0.1 ppm) compared with Murchison (Kminek *et al.*, 2002; Pizzarello *et al.*, 2001). Kminek *et al.* (2002) concluded that their extremely low abundance and similarities between amino acids in Tagish Lake and in its ice-melt water indicated that the amino acids detected in the lake were terrestrial contaminants.

Cronin *et al.* (1995) summarized the characteristics of the abundance, distribution, and structure of Murchison amino acids as follows:

(i) Two simple structural types of amino acids have been identified: monoamino monocarboxylic acids and monoamino dicarboxylic acids. Two variations on these structural types occur: n-alkyl secondary amino acids (e.g., sarcosine) and cyclic secondary amino acids.

(ii) There is complete structural diversity. All of the possible α-amino monocarboxylic acids through C_7 have been identified, as have all of the chain and amino position isomers through C_5.

(iii) Homologous series of amino acids show exponential declines in concentration with increasing carbon number, each addition of a carbon atom to the series corresponding to a decrease in concentration of around 70%.

(iv) Branched-chain isomers predominate. At each carbon number, branched isomers are more abundant than straight-chain isomers.

(v) Enantiomers occur in approximately equal amounts. The majority of protein and nonprotein amino acids occur as racemic mixtures. However, there is evidence that there may be slight L-excesses in some protein and nonprotein amino acids, although these results remain controversial.

(vi) The amino acids coexist with a closely matched set of hydroxy acids.

(vii) The amino acids have unusually high $\delta^{13}C$, $\delta^{15}N$, and δD values (see below).

(viii) Acid hydrolysis of aqueous extracts of Murchison indicates that a substantial fraction of the amino acids exist as acid-labile precursors.

Engel and Nagy (1982) detected L-excesses in five protein amino acids from Murchison (alanine, glutamic acid, proline, aspartic acid, and leucine) that they argued were indigenous to the meteorite. This conclusion was criticized, largely on the grounds that the sampling procedure had not completely excluded terrestrial contaminants (Bada *et al.*, 1983). Earlier controversy over the possibility of enantiomeric excesses in meteorites had largely centered on the possible presence of enantiomeric excesses in Orgueil (Nagy *et al.*, 1964), although these results had proved to be difficult to confirm (Hayatsu, 1965) and no consensus had been reached (Hayes, 1967). Subsequent stable-isotopic evidence (see below) strengthened the argument that the L-excesses in protein amino acids in Murchison were indigenous. However, the possibility of terrestrial contamination, which would preferentially lead to an apparent L-excess in protein amino acids, led to an examination of the enantiomeric distribution of nonprotein amino acids (Cronin and Pizzarello, 1997; Pizzarello and Cronin, 1998, 2000). Small excesses (2–9%) in the α-methyl-α-amino acids in Murchison were observed, together with smaller excesses (1–6%) in the same compounds in Murray (Cronin and Pizzarello, 1997; Pizzarello and Cronin, 2000). It was argued that these L-excesses were indigenous to Murchison on the grounds that the excesses were observed in four amino acids not known in nature, there was no apparent correlation between L-excess and potential terrestrial contaminants, and that the analytical procedures adopted reduced the risk of co-elution of other amino acids during the chromatographic analysis.

2.3.1.4 Aromatic hydrocarbons

Early investigations of the distribution of aromatic hydrocarbons in meteorites employed several analytical approaches including gas chromatography, spectroscopic techniques (IR, UV, and fluorescence) and mass spectrometry, and reported a wide range of aromatic and polycyclic aromatic molecules with higher molecular weights predominating (Hayes, 1967). However, the development of GCMS in the early 1970s led to a more systematic series of investigations employing both solvent and thermal extraction, particularly of Murchison (Oró *et al.*, 1971; Pering and Ponnamperuma, 1971; Studier *et al.*, 1972; Levy *et al.*, 1973) although the relative abundances obtained for different aromatic compounds varied considerably. This almost certainly reflects differences in the analytical approach, in particular problems associated with the loss of more volatile aromatic species when solvent extracts are dried prior to analysis. More recent investigations of aromatic hydrocarbons identified the three-ring polycyclic aromatic hydrocarbons (PAHs) fluoranthene and pyrene as the most abundant compounds in solvent extracts of Murchison and a number of Antarctic CM2s (Basile *et al.*, 1984; Naraoka *et al.*, 1988; Shimoyama *et al.*, 1989; Krishnamurthy *et al.*, 1992; Gilmour and Pillinger, 1994). However, when Sephton *et al.* (1998) used supercritical CO_2 as an extraction solvent to minimize the loss of volatile material, they obtained a positive correlation between the abundance of aromatic compounds and their volatility. Higher-molecular-weight PAHs are apparently relatively minor components concentrated in previous studies as a result of the loss of more volatile species.

2.3.1.5 Heterocyclic compounds

A number of early studies reported the presence of urine and pyrimidine bases in water extracts of carbonaceous chondrites, although contamination was often suspected (Hayes, 1967). Investigations of nitrogen heterocycles in the late 1970s and early 1980s (Stoks and Schwartz, 1979, 1981, 1982) confirmed the presence of several nitrogen bases and extended earlier studies, at the same time identifying several potential terrestrial contaminants (Hayatsu, 1964; Hayatsu *et al.*, 1968, 1975; van der Velden and Schwartz, 1977). Several classes of basic and neutral nitrogen heterocycles have since been identified in Murchison including:

(i) Purines: xanthine, hypoxanthine, guanine, and adenine (Stoks and Schwartz, 1981, 1982).

(ii) Pyrimidines: uracil (Stoks and Schwartz, 1979).

(iii) Quinolines and isoquinolines with 0–4 carbon atoms as methyl or ethyl side chains (Stoks and Schwartz, 1982).

(iv) Pyridines: C_3-alkyl pyridines and C_5-alkyl pyridines, carboxylated pyridines (Hayatsu *et al.*, 1975; Stoks and Schwartz, 1982; Pizzarello *et al.*, 2001).

The quinolines, isoquinolines, and pyridines are structurally diverse with a large number of alkyl-substituted isomers. Direct isotopic measurements have not been made on nitrogen heterocycles.

Modern GCMS analysis and sample preparation techniques have overcome many the analytical problems associated with the investigation of organic sulfur compounds associated with potential reactions between elemental sulfur and organic matter during analysis (Hayes, 1967). Shimoyama and Katsumata (2001) examined the distribution of aromatic thiophenes in Murchison detecting benzothiophene, dibenzothiophene, alky-substituted dibenzothiophenes, and benzonaphthothiophenes at concentrations levels of 0.3 ppm (Table 1) confirming earlier studies (Basile *et al.*, 1984).

2.3.1.6 *Aliphatic hydrocarbons*

Considerable importance was attached to the presence of long-chain *n*-alkanes in carbonaceous chondrites in studies during the 1960s and 1970s. Analyses of the Orgueil and other meteorites in the early 1960s had identified the presence of saturated hydrocarbons (e.g., Nagy *et al.*, 1961; Meinschein *et al.*, 1963) and it was suggested, somewhat controversially, that these compounds were evidence for biogenic activity on the meteorites' parent bodies. However, when Oró *et al.* (1966) analyzed both interior and exterior portions of Orgueil, they observed a decrease in the abundance of alkanes away from the surface of the meteorite together with the presence of the biologically derived isoprenoid hydrocarbons pristane and phytane in a total of 19 meteorites. Such results were strongly suggestive that alkanes were the result of terrestrial contamination. The problem of contamination was demonstrated in 1969 when the Allende (CV3) chondrite was subjected to organic analysis within seven days of its fall. Both *n*-alkanes and isoprenoids were found concentrated at its surface (Han *et al.*, 1969).

Although there was considerable evidence of a terrestrial source for *n*-alkanes, analysis of the Murchison meteorite revived arguments in favor of an extraterrestrial origin for these compounds. Analysis of Murchison revealed a similar distribution of hydrocarbons to those found in Orgueil and Murray (Studier *et al.*, 1972). Studier *et al.* (1968) had identified *n*-alkanes as the principal hydrocarbon species above C_{10} in Orgueil and Murray, while below C_{10} aliphatic hydrocarbons were markedly deficient and benzene and alkylbenzenes were predominant. They obtained a similar hydrocarbon distribution using a Fischer–Tropsch type (FTT) synthesis from CO and H_2 in the presence of iron meteorite powder and suggested that catalytic reactions of this type may have occurred on a large scale in the solar nebula, converting CO to less volatile carbon compounds.

However, a key feature of catalyzed reactions such as Fischer–Tropsch is their structural selectivity. More recent analyses of hydrocarbons of interior and exterior portions of the Murchison, Murray, and Allende meteorites together with the highly metamorphosed New Concord chondrite have again indicated that *n*-alkanes are preferentially concentrated toward the surface of these meteorites strongly suggesting that they are terrestrial contaminants (Cronin and Pizzarello, 1990). However, indigenous branched and cyclic aliphatic hydrocarbons in Murchison appear to show complete structural diversity within homologous series arguing strongly against a Fischer–Tropsch origin (Cronin and Pizzarello, 1990). Furthermore, isotopic fractionations accompanying Fischer–Tropsch reactions in the laboratory (Yuen *et al.*, 1990) do not explain the isotopic distributions observed in *n*-alkanes in meteoritic material (Sephton *et al.*, 2001).

2.3.1.7 *Amines and amides*

Pizzarello *et al.* (1994) undertook a detailed investigation of nitrogenous compounds in Murchison isolating volatile bases and identifying a series of aliphatic amines, confirming and extending earlier work (Hayatsu *et al.*, 1975; Jungclaus *et al.*, 1976a). Significant ^{13}C, ^{15}N, and D-enrichments were observed in the isolated volatile bases, confirming their extraterrestrial origin and suggestive of an interstellar origin (Table 3). As with other groups of compounds in Murchison, the aliphatic amines showed decreasing abundance with increasing carbon number and almost complete structural diversity through C_5. Branched-chain isomers were more abundant than straight-chain isomers. Pizzarello *et al.* (1994) suggested that aliphatic amines could have originated by two possible routes: direct incorporation from molecular clouds, supported by the detection of methylamine in interstellar environments, or via the decarboxylation of α-amino acids, supported by the observation that most of the aliphatic amines could be produced by the decarboxylation of known α-amino acids in Murchison.

Cooper and Cronin (1995) detected a wide range of linear and cyclic amides in water extracts of Murchison, extending the previous positive detection of guanylurea (Hayatsu *et al.*, 1968) These included many mono- and dicarboxylic acid amides, hydroxyacid amides, and other amides with no known terrestrial source. These compounds were characterized by a structural diversity of isomers up to C_8 and a decline in abundance with increasing carbon number.

2.3.1.8 *Alcohols, aldehydes, ketones, and sugar-related compounds*

Alcohols, aldehydes, and ketones have not been the subject of extensive investigation for 20–25 years since initial studies of their distribution and abundance in Murchison, Murray, and Orgueil (Meinschein *et al.*, 1963; Studier *et al.*, 1965; Hayes and Biemann, 1968; Jungclaus *et al.*, 1976b; Basile *et al.*, 1984). The abundances of these compounds in Murchison are reported in Table 1. Junglaus *et al.* (1976b) positively

Table 3 Carbon, nitrogen, and hydrogen stable isotope compositions of organic fractions from the Murchison meteorite.

Fraction	$\delta^{13}C$ (‰)	$\delta^{15}N$ (‰)	δD (‰)	*References*
Benzene-methanol extract	+5			1
Methanol extract	+7	+88	+406	2
	+4		+957	3
Volatile hydrocarbons				
Freeze thaw disaggregation	0		−92	3
Hot water extract	+6		+217	3
H_2SO_4 treament	+17		+410	3
Isolated fractions				
Aliphatic fraction[a]	−11.5/−5		+264/+103	3
Aromatic fraction[a]	−5.5/−5		+407/+244	3
Polar fraction[a]	+6/+5		+946/+751	3
Amino acids		+102		4
	+23	+90	+1,370	5
	+26		+1,137	6
Monocarboxylic acids	+7	−1	+377	5
	−1		+652	3
Hydroxy acids	+4		+573	7
Dicarboxylic acids	+6		+357	7
Volatile bases (ammonia, amines)	+22	+93	+1,221	4
Neutral polyhydroxylated compounds	−6		+119	8

References: 1. Chang *et al.* (1978), 2. Becker and Epstein (1982), 3. Krishnamurthy *et al.* (1992), 4. Pizzarello *et al.* (1994), 5. Epstein *et al.* (1987), 6. Pizzarello *et al.* (1991), 7. Cronin *et al.* (1993), and 8. Cooper *et al.* (2001).

[a] Analyses performed on two samples held, respectively, at Chicago and Arizona State University.

identified C_1–C_4 alcohols, and C_2, C_4, and C_5 carbonyl compounds. The aromatic ketones fluoren-9-one, anthracene-dione, phenanthrenedione, benzanthracen-7-one, and anthracen-9(10*H*)-one were identified in Murchison as part of an investigation of aromatic hydrocarbons (Basile *et al.*, 1984).

Sugar-related compounds were first reported in ethanol extracts of Murray and Orgueil (Degens and Bajor, 1962; Kaplan *et al.*, 1963); however, it has been only since early 2000s that these compounds have received further investigation. Cooper *et al.* (2001) identified polyhydroxylated (polyols) sugars, sugar alcohols, and sugar acids in Murchison at abundances similar to those for amino acids (Table 1). Their indigeneity was confirmed through stable-isotope measurements of an isolated polyol fraction, which gave a $\delta^{13}C$ value of −6‰ and a δD value of +119‰ (Table 3). Cooper *et al.* (2001) found marked decreases in the abundances of higher-molecular-weight polyols and almost complete isomeric diversity.

2.3.2 Stable-isotopic Investigations of Classes of Organic Compounds

Stable-isotope studies from the 1960s through the early 1980s focused on making isotopic measurements on broad groupings of meteoritic organic compounds, e.g., HF/HCl acid residues, which concentrated the macromolecular component of the organic matter, or on solvent extracts (Krouse and Modzeleski, 1970; Kvenvolden *et al.*, 1970; Smith and Kaplan, 1970; Becker and Epstein, 1982; Robert and Epstein, 1982; Yang and Epstein, 1983). Such studies were successful in establishing that meteoritic organic matter contained material of probable interstellar origin (Yang and Epstein, 1985). Subsequent work began the process of attempting to isolate chemically identifiable components; using chromatographic techniques it was possible to obtain fractions with broadly similar chemical structures. This led to isotopic measurements being obtained on assemblages of compounds, predominantly from the Murchison meteorite, including aliphatic and aromatic hydrocarbons (Krishnamurthy *et al.*, 1992), carboxylic acids (Epstein *et al.*, 1987; Pizzarello *et al.*, 1991; Krishnamurthy *et al.*, 1992; Cronin *et al.*, 1993), amino acids (Epstein *et al.*, 1987; Pizzarello *et al.*, 1991), and polar hydrocarbons (Pizzarello *et al.*, 1994); the data for Murchison are summarized in Table 3. These investigations confirmed that a significant proportion of the solvent extractable organic matter was indigenous to meteorites since significant enrichments in ^{13}C, ^{15}N, and D were detected. The deuterium enrichments also provided strong evidence for the role of interstellar processes in the origin of meteoritic organic matter.

2.3.3 Compound-specific Isotopic Studies

The ability to measure the isotopic compositions of individual compounds in a complex mixture, termed compound-specific isotope analysis (CSIA), has been a long-held goal of stable-isotope mass spectrometry, and its potential to identify indigenous and contaminant material in meteorites was recognized well before analytical developments permitted such measurements on a routine basis (Pillinger, 1982). The determination of the stable-isotope compositions of individual compounds provides a powerful means of elucidating the reaction mechanisms and possible source environments from which the organic constituents have formed. Using preparative chromatography to isolate fractions containing simple mixtures or individual compounds, which are subsequently converted to a form suitable for isotope-ratio mass spectrometry, it has been possible to obtain $\delta^{13}C$ and δD measurements on C_2–C_5 amino acids (Table 2; Pizzarello *et al.*, 1991) and $\delta^{13}C$, δD, $\delta^{34}S$, and $\delta^{33}S$ of C_1–C_3 sulfonic acids (Table 4; Cooper *et al.*, 1997). However, it is the advent of combined gas chromatography isotope-ratio mass spectrometry (GC-IRMS) (Hayes *et al.*, 1990) that has enabled the comparatively rapid determination of the isotopic compositions of individual compounds to high levels of precision. Since the approach relies on the ability of analytes to be separated chromatographically, exploiting gas–stationary phase partitioning, volatile meteoritic compounds with hydrocarbon skeletons and relatively few functional groups have been the main focus of study. As of early 2000s, work on Murchison and a limited number of other meteorites has provided carbon stable-isotopic compositions for CO, CO_2, C_2–C_5 aliphatic hydrocarbons (Table 5; Yuen *et al.*, 1984), C_2–C_5 carboxylic acids (Table 5; Yuen *et al.*, 1984), dicarboxylic acids (Table 5, Pizzarello *et al.*, 2001; Pizzarello and Huang, 2002), C_6–C_{20} aromatic compounds (Table 6; Yuen *et al.*, 1984; Naraoka *et al.*, 2000; Gilmour and Pillinger, 1994; Sephton *et al.*, 1998), and C_{12}–C_{18} *n*-alkanes (Sephton *et al.*, 2001). GC-IRMS analyses of the less volatile polar compounds present analytical difficulties since derivatives of these compounds have to be chemically modified to increase their volatility and make them amenable to gas chromatographic analysis. It is then necessary to correct the $\delta^{13}C$ value obtained for the derivatized compound in order to determine the true $\delta^{13}C$ value of the molecule being studied. Such procedures have enabled carbon and nitrogen isotopic compositions to be determined on C_2–C_6 amino acids from the Murchison meteorite as trifluoroacetic acid-isopropyl ester derivatives (Table 2; Engel and Macko, 1997; Engel *et al.*, 1990).

2.3.3.1 Carbon

Figure 2 shows that the C_1–C_5 aliphatic hydrocarbons, amino acids, carboxylic acids, and sulfonic acids from the Murchison meteorite appear to follow a common trend when their $\delta^{13}C$ values are plotted against carbon number. $\delta^{13}C$ values generally decrease as the amount of carbon in the molecules increases. This trend has been interpreted as the result of a kinetic isotope effect during the sequential formation of higher-molecular-weight compounds from simpler precursors (Yuen *et al.*, 1984). The more reactive ^{12}C is preferentially added during the synthesis of the carbon skeleton of these compounds.

Extractable C_6–C_{22} aromatic compounds from the Murchison (CM2) and Asuka-881458 (CM2) meteorites also appear to follow a systematic trend when their $\delta^{13}C$ values are plotted against carbon number (Figure 3), although benzene and toluene from Murchison and naphthalene and biphenyl from Asuka-881458 are clear outliers. The $\delta^{13}C$ values of many of the aromatic compounds extracted (e.g., Asuka-881458 fluoranthene: −8.3‰) are markedly more ^{13}C-enriched than typical terrestrial PAHs confirming the indigenous nature of these compounds. The predominant aromatic trend is consistent that observed for C_1–C_5 compounds from Murchison and indicates an origin by a synthetic process progressively adding ^{12}C to the carbon skeleton with kinetic isotopic fractionation determining the distribution of carbon isotopes between compounds rather than thermodynamic equilibrium (Gilmour and Pillinger, 1994; Naraoka *et al.*, 2000). The $\delta^{13}C$ values obtained for extractable aromatic hydrocarbons in CM2 meteorites display a significant amount of isotopic heterogeneity with a range in $\delta^{13}C$ values of over 20‰. Compounds, with relatively high molecular weights, but which differ by only one or two carbon atoms also display significant differences in their $\delta^{13}C$ values. This has led to the

Table 4 Carbon, hydrogen, and sulfur isotopic compositions of individual sulfonic acids from the Murchison (CM2) meteorite.

Compound	$\delta^{13}C$ (‰)	δD (‰)	$\delta^{33}S$ (‰)	$\delta^{34}S$ (‰)	$\delta^{36}S$ (‰)	$\Delta^{33}S$
Methanesulfonic acid	+29.8	+483	+7.63	+11.27	+22.5	+2
Ethanesulfonic acid	+9.1	+787	+0.33	+1.13	+0.8	−0.24
Propanesulfonic acid	−0.4	+536	+0.20	+1.20	+2.1	−0.40
1-Methylethanesulfonic acid	−0.9	+852	+0.32	+0.68	+2.9	−0.02

Source: Cooper *et al.* (1997).

Table 5 Carbon isotope compositions of CO, CO_2, carboxylic acids, dicarboxylic acids, and volatile hydrocarbons in the Murchison (CM2), Tagish Lake (CI), and Orgueil (CI1) meteorites together with hydrogen isotope compositions for dicarboxylic acids.

Compound	*Murchison* $\delta^{13}C$ (‰)	*Tagish Lake* $\delta^{13}C$ (‰)	*Orgueil* $\delta^{13}C$ (‰)	*Murchison* δD (‰)	*Tagish Lake* δD (‰)	*Orgueil* δD (‰)
CO[a]	−32.0 ± 2.0					
CO_2[a]	+29.1 ± 0.2					
Monocarboxylic acids[a]						
Acetic acid	2					
Propionic acid	+17.4 ± 0.2					
Isobutyric acid	+16.9 ± 0.2					
Butyric acid	+11.0 ± 0.3					
Isovaleric acid	+8.0 ± 4.5					
Valeric acid	+4.5 ± 0.2					
Dicarboxylic acids[b]						
Succinic acid	+22.5 ± 0.6	+22.5 ± 0.6	−23.8	+1,124	+1,116	+389
Methylsuccinic acid	+15.4 ± 3.0	+15.4 ± 3.0	−19.5	+1,106	+1,112	+1,225
Glutaric acid	+22.9 ± 1.5	+22.9 ± 1.5	−20.4	+1,387	+1,322	+795
2-Methyl glutaric	+27.9 ± 1.0	+18.6 ± 0.7	−10.3	+1,463	+1,263	+1,551
3-Methyl glutaric	+19.1	+12.6				
Adipic	+21.4	+5.5 ± 0.9				
Volatile hydrocarbons[a]						
Methane	+9.2 ± 1.0					
Ethane	+3.7 ± 0.1					
Ethene	+0.1 ± 0.4					
Propane	+1.2 ± 0.1					
Isobutane	+4.4 ± 0.1					
Butane	+2.4 ± 0.1					

[a] Yuen *et al.* (1984).
[b] Pizzarello and Huang (2002).

suggestion that during the synthetic processes that led to bond formation, isotopic fractionation was at its most extreme, implying that synthesis took place in a low-temperature environment such as interstellar space (Sephton and Gilmour, 2000). The isotopic heterogeneity displayed by aromatic compounds in Murchison and Asuka-881458 may also contain evidence for different synthetic pathways. There is a 7.5‰ difference in $\delta^{13}C$ values between PAHs isomers containing a five-carbon ring (e.g., fluoranthene) and those without (e.g., pyrene), which has been interpreted as evidence of two possible pathways for the formation of PAHs (Gilmour and Pillinger, 1994; Naraoka *et al.*, 2000).

The $\delta^{13}C$ values for the C_{12}–C_{26} *n*-alkanes from six chondrites are shown in Figure 4 (Sephton *et al.*, 2001). None of the *n*-alkanes exhibit either the ^{13}C-enrichments or systematic isotopic trends that apparently characterize indigenous organic matter in meteorites. Most of the $\delta^{13}C$ values are similar both in value and in the trends shown within homologous series to $\delta^{13}C$ variations observed for terrestrial petroleum products or other terrestrial fossil hydrocarbons. These features confirm the long-held suspicion that these molecules are contaminants from the terrestrial environment added to the meteorite following its fall to Earth (Cronin and Pizzarello, 1990; Sephton *et al.*, 2001). It is interesting to note, however, that *n*-alkanes with negative $\delta^{13}C$ values were recorded in the Tagish Lake meteorite, despite samples being held at temperatures below freezing and, once collected, in clean conditions (Pizzarello *et al.*, 2001).

Engel *et al.* (1990) determined the $\delta^{13}C$ values for individual amino acid enantiomers to attempt to ascertain whether the reported L-excess in Murchison amino acids (Engel and Nagy, 1982) was a consequence of terrestrial contamination, one of the original applications envisaged for CSIA of meteoritic organics (Pillinger, 1982). Similar, nonterrestrial $\delta^{13}C$ values were obtained for L- and D-alanine of +27‰ and +30‰, respectively (Table 2), apparently confirming the indigenous nature of these acids. They argued that the more negative $\delta^{13}C$ value for the L-enantiomer could not be explained by terrestrial contamination and that this excess was indigenous to the meteorite.

2.3.3.2 Nitrogen

The $\delta^{15}N$ values of individual amino acid enantiomers in Murchison have also been used to try and ascertain whether the reported L-excess is indigenous (Engel and Macko, 1997). Engel and Macko (1997) found that the L- and D-enantiomers

Table 6 Carbon stable isotope compositions of solvent extractable aromatic compounds in carbonaceous chondrites.

Compound	*Orgueil*[a] $\delta^{13}C$ (‰)	*Tagish Lake*[b] $\delta^{13}C$ (‰)	*Cold Bokkeveld*[a] $\delta^{13}C$ (‰)	*Murchison* $\delta^{13}C$ (‰)	*Asuka-991458*[f] $\delta^{13}C$ (‰)
Benzene				-28.7 ± 0.2[c]	
Toluene			-24	-28.8 ± 1.1[d]	
Naphthalene				-12.6 ± 2.3[d]	-26.2 ± 1.2
1-Methylnaphthalene				-11.1[d]	
2-Methylnaphthalene				-5.8[d]	
Biphenyl					-25.9 ± 0.9
Phenanthrene	-23	-25	-27.0 ± 1.2	-7.5 ± 1.2[e]	-12.9 ± 0.9
Methyl-phenanthenes					-13.4 ± 0.8
Fluoranthene	-19			-5.9 ± 1.2[e]	-8.3 ± 0.5
Pyrene	-17			-13.1 ± 1.3[e]	-15.8 ± 0.8
Chrysene	-23			-14.5 ± 2.2[e]	
Benzo[ghi]fluoranthene	-15			-14.2 ± 2.2[e]	-15.8 ± 0.8
Benzanthracene	-15				
Benz[a]anthracene, Chrysene and triphenylene					-11.7 ± 2.4
Benzo[e]pyrene				-22.3 ± 4.1[e]	
Benzopyrenes and perylene					-19.1 ± 1.5
Benzo[f]fluoranthene				-15.4 ± 3.3[e]	
Benzofluoranthenes					-14.0 ± 0.5
Dibenzanthracenes	-17				
Benzo[ghi]perylene					-25.2 ± 0.5

[a] Sephton and Gilmour (2000).
[b] Pizzarello *et al.* (2001).
[c] Yuen *et al.* (1984).
[d] Sephton *et al.* (1998).
[e] Gilmour and Pillinger (1994).
[f] Naraoka *et al.* (2000).

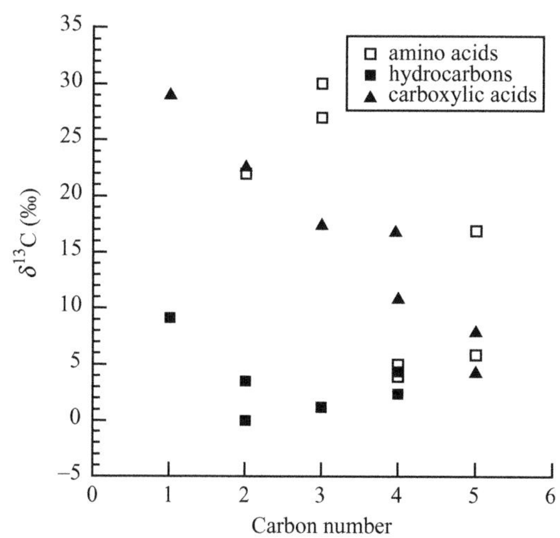

Figure 2 Carbon stable-isotope compositions of low-molecular-weight hydrocarbons, amino acids, and monocarboxylic acids from the Murchison meteorite plotted against carbon number. Carbon number 1 denotes methane and CO_2, 2 denotes ethane, ethanoic acid, glycine, etc. (source Yuen *et al.*, 1984).

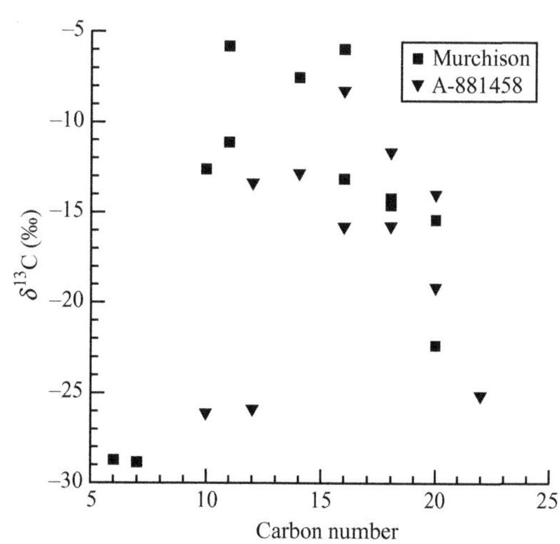

Figure 3 Carbon stable-isotope compositions of solvent extractable aromatic and PAHs plotted against carbon number from the Murchison and Asuka-881458 CM2 carbonaceous chondrites (sources Yuen *et al.*, 1984; Gilmour and Pillinger, 1994; Sephton *et al.*, 1998; Naraoka *et al.*, 2000).

of alanine and glutamic acid have significantly higher $\delta^{15}N$ values (ca. +60‰, Table 2) than their terrestrial counterparts (ca. −10‰ to +20‰) and argued that the excess of L-enantiomers over D-enantiomers is extraterrestrial in origin and not the result of terrestrial contamination. The observed ^{15}N-enrichments in the C_2–C_6 amino acids in Murchison (e.g., $\delta^{15}N = +184$‰ for

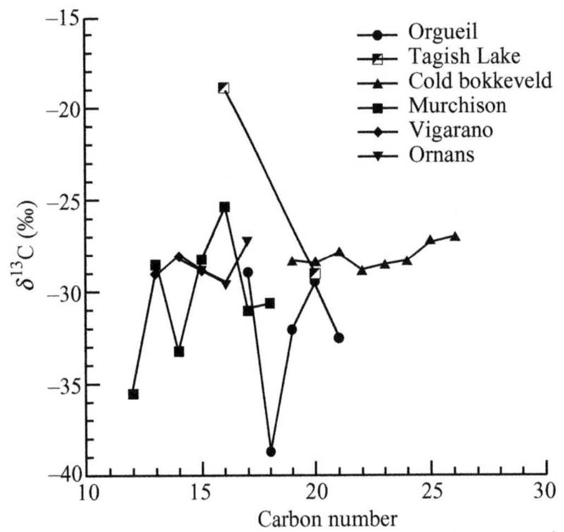

Figure 4 Carbon stable-isotope compositions of solvent extractable *n*-alkanes from the Orgueil (CI), Cold Bokkeveld (CM2), Murchison (CM2), Vigarano (CV3), Ornans (CO), and Tagish Lake carbonaceous chondrites plotted against carbon number (sources Sephton *et al.*, 2001; Pizzarello *et al.*, 2001).

α-aminoisobutyric acid, Table 2) suggest an interstellar source for these compounds or their precursors (Engel and Macko, 1997). However, Pizzarello and Cronin (1998) considered that other meteoritic amino acids may be co-eluting with L-alanine, thereby contributing to both the L-excess and $\delta^{15}N$ determinations. In parallel work, these authors examined the enantiomers of nonprotein amino acids and observed an L-excesses in the α-methyl-α-amino acids in Murchison and Murray reported above (Cronin and Pizzarello, 1997, 1999; Pizzarello and Cronin, 1998, 2000).

2.3.3.3 *Hydrogen*

Compound-specific hydrogen isotopic measurements of amino and sulfonic acids display significant deuterium enrichments suggesting an interstellar origin for these compounds or their precursors (Pizzarello *et al.*, 1991; Cooper *et al.*, 1997). Cooper *et al.* (1997) proposed that the relatively constant δD values for different sulfonic acids (Table 4) implies that the hydrogenation of their unsaturated precursors occurred within a pool of nearly uniform deuterium enrichment.

2.3.3.4 *Sulfur*

Cooper *et al.* (1997) determined the sulfur isotope compositions for a homologous series of sulfonic acids from Murchison (Table 4). They observed a nonmass-dependant enrichment in ^{33}S ($\delta^{33}S = +2$) in methanesulfonic acid that was attributed to the gas-phase UV-irradiation of CS_2 in an interstellar environment, prior to the production of the methanesulfonic acid precursor.

2.4 MACROMOLECULAR MATERIAL

Anders and Kerridge (1988) considered establishing the origins of the macromolecular organic matter in meteorites a daunting task, and studies of the macromolecular organic matter in carbonaceous chondrites have presented a series of analytical challenges. The elemental composition of the Murchison macromolecule has been determined as $C_{100}H_{71}N_3O_{12}S_2$ based on elemental analysis (Hayatsu *et al.*, 1977) and revised to $C_{100}H_{48}N_{1.8}O_{12}S_2$ based on pyrolytic release studies (Zinner, 1988). Such large macromolecules amenable to direct study using NMR, while other analytical approaches attempt to break the structure down, using techniques such as pyrolysis or chemical degradation, into fragments that are easier to study.

2.4.1 Structural Studies Using Pyrolysis and Chemical Degradation

The majority of the carbon in the Murchison macromolecular material is present within aromatic ring systems. This aromatic nature has been revealed by a series of pyrolysis studies of meteorites such as Orgueil (CI1), Murchison (CM2), Murray (CM2), and Allende (CV3) in which the macromolecular material was thermally fragmented to produce benzene, toluene, alkylbenzenes, naphthalene, alkylnaphthalenes, and PAHs with molecular weights up to around 200–300 amu (Simmonds *et al.*, 1969; Studier *et al.*, 1972; Levy *et al.*, 1973; Bandurski and Nagy, 1976; Holtzer and Oró, 1977; Murae, 1995; Kitajima *et al.*, 2002). Further identification of the aromatic units in the Murchison macromolecular material was achieved by Hayatsu *et al.* (1977), who used sodium dichromate oxidation to selectively remove aliphatic side chains in the macromolecular material and thereby isolate and release the aromatic cores present. These studies have revealed a dominance of single-ring aromatic entities but also a significant amount of two- to four-ring aromatic cores bound to the macromolecular material by a number of aliphatic linkages.

Sephton *et al.* (1998, 1999, 2000) used hydrous pyrolysis followed by supercritical extraction to examine insoluble organic matter in Orgueil (CI1), Murchison (CM2), and Cold Bokkeveld (CM2). The hydrous pyrolysates obtained for the three meteorites displayed a remarkable degree of qualitative similarity suggesting that the macromolecular materials in different carbonaceous chondrites are apparently composed of essentially the same aromatic structural units, predominantly one to three ring alkyl-substituted aromatic structures. Significant quantitative differences were observed, however, and these were interpreted as indications of the different parent body histories of the three meteorites (Sephton *et al.*, 2000).

Aliphatic hydrocarbon moieties are present in significant amounts within the Murchison macromolecular material and several pyrolysis studies have indicated that these entities exist within or around the aromatic network as hydroaromatic rings and short alkyl substituents or bridging groups (e.g., Hayatsu *et al.*, 1977; Holtzer and Oró, 1977; Levy *et al.*, 1973).

Several pyrolysis experiments have released oxygen-containing moieties such as phenols (Studier *et al.*, 1972; Hayatsu *et al.*, 1977; Sephton *et al.*, 1998), benzene carboxylic acids (Studier *et al.*, 1972), propanone (Levy *et al.*, 1973; Biemann, 1974), and methylfuran (Holtzer and Oró, 1977) from the Murchison macromolecular material. Sodium dichromate oxidation of the macromolecular material liberated the cyclic diaryl ether dibenzofuran and aromatic ketones such as fluorenone, benzophenone, and anthraquinone. Each of these organic units appears to be bound into the macromolecular network by two to four aliphatic linkages (Hayatsu *et al.*, 1977). Hayatsu *et al.* (1980) used alkaline cupric oxide to selectively cleave organic moieties incorporated into the Murchison macromolecular material by ether groups. In this way, these authors established the presence of significant amounts of phenolic species bound to the macromolecular material by ether linkages.

Thiophenes are common pyrolysis products of the Murchison macromolecular material and thiophene, methylthiophene, dimethylthiophene, and benzothiophene have been detected in this way (Biemann, 1974; Holtzer and Oró, 1977; Sephton *et al.*, 1998). Sodium dichromate oxidation has revealed the presence of substituted benzothiophene and dibenzothiophene moieties within the macromolecular material (Hayatsu *et al.*, 1980).

Pyrolysis has also led to the tentative identifications of the nitrogen heterocyclics, cyanuric acid (Studier *et al.*, 1972), and alkylpyridines (Hayatsu *et al.*, 1977) in the Murchison macromolecular material. Similar types of analyses have revealed acetonitrile and benzonitrile (Levy *et al.*, 1973; Holtzer and Oró, 1977). Substituted pyridine, quinoline, and carbazole were observed in sodium dichromate oxidation of the macromolecular material (Hayatsu *et al.*, 1977).

2.4.2 NMR and Electron Spin Resonance Studies

Cronin *et al.* (1987) first employed ^{13}C NMR spectroscopy to detect aromatic carbon within the Orgueil, Murchison, and Allende macromolecular materials. These studies indicated degrees of aromaticity of 47% and 40% for macromolecular material in Orgueil and Murchison, respectively; however, Cronin *et al.* concluded that they were underestimating the contribution of nonprotonated aromatic carbon and invoked the presence of extensive polycyclic aromatic sheets significantly larger than the one to four ring aromatic entities isolated by Hayatsu *et al.* (1977). Gardinier *et al.* (2000) revised the levels of aromatic carbon in Orgueil to between 69% and 78% and in Murchison to between 61% and 68% and estimated the function group abundance for the Murchison macromolecular material.

Cody *et al.* (2002) undertook an extensive NMR investigation using double- and single-resonance solid-state (^1H and ^{13}C) NMR to study Murchison macromolecular material. This indicated that it was a complex organic solid composed of a wide range of organic (aromatic and aliphatic) functional groups, including numerous oxygen-containing functional groups (Figure 5 and Table 7). They refined the estimates of aromatic carbon within the Murchison organic residue to 61–66%. Aside from the presence of interstellar diamond, they found no evidence for significant amounts of large polycyclic aromatic sheets, concluding that such structures comprised a maximum of 10%, and probably much less, of the macromolecular material. Using both single-pulse and cross-polarized spectra, Cody *et al.* (2002) concluded that the fraction of aromatic carbon directly bonded to hydrogen is low (~30%), indicating that the aromatic molecules in the Murchison organic residue are highly substituted and estimated the H/C ratio at 0.53–0.63, compared with and H/C ratio of 0.53 determined using elemental analysis (Zinner, 1988).

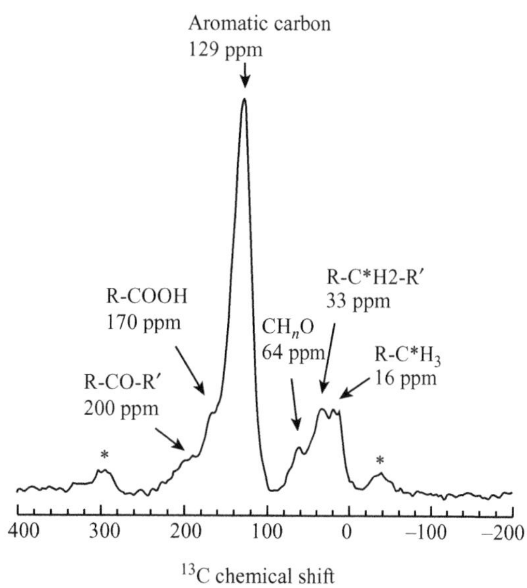

Figure 5 Cross polarization NMR spectrum of organic macromolecular material from the Murchison (CM2) chondrite. Prominent peaks and shoulders are assigned to probable functional groups identified by their respective chemical shifts in ppm (source Cody *et al.*, 2002).

Table 7 Estimates of functional group abundance by cross-polarization NMR.

Carbon type	*% of total*	*Hydrogen content*[a] ($C_{100}H_n$)	*Oxygen content*[b] ($C_{100}O_m$)
CH_3	8.0	24	0
CH, CH_2	8.7	8.7–17.4	0
$CH_{n=1-2}$, $OH_{i=0.1}$	5.5	0–16.5	5.5 (max)
Aromatic (C-H, C-R)	53.7	0–53.7	0
Aromatic (C-O)	7.6	0–7.6	7.6 (max)
R-COOH,R′	7.4	0–7.4	14.8 (max)
R-CO-R′	9.1	0	9.1
	Aromatic fraction (F_a)=0.61	$n = 32.7–126.6$	$m = 26.8–37.0$

Source: Cody *et al.* (2002).
[a] Uncertainties in identifying specific functional groups results in a range of *n*. The lower limit considers all aromatic carbon to be nonprotonated, the oxygen-substituted carbon to be a tertiary ether, the carboxylate to be an ester, and the carbonyl to be a ketone. The upper limit is derived by assuming that all the aromatic carbon is protonated, the oxygen-substituted carbon is a primary alcohol, and the carboxylate is an acid.
[b] The maximum value of *m* is derived by assuming that all oxygen-substituted alkyl carbon is in the form of alcohol, all the oxygen-substituted aromatic carbon is hydroxyl, and the carboxylate is in the form of free acid. The minimum value of *m* is derived by assuming that all carboxylate is linked to aromatic carbon via aromatic esters, the remainder of aromatic oxygen is linked to aliphatic carbon via alkyl aryl ethers, and that the remainder of the aliphatic oxygen linked as aliphatic ethers.

NMR data suggest that the range of oxygen-containing organic functionality in Murchison is substantial with a wide range in O/C ratio possible, depending on whether various oxygen-containing organic functional groups exist as free acids and hydroxyls or are linked as esters (Gardinier *et al.*, 2000; Cody *et al.*, 2002). The lower O/C values are consistent with elemental analyses, requiring that oxygen-containing functional groups in the Murchison macromolecule are highly linked. NMR data also indicated a significant proportion of tertiary (methyne, CH) carbon, suggesting that aliphatic carbon chains within the Murchison organic macromolecule are highly branched (Cody *et al.*, 2002). In contrast to Murchison and Orgueil macromolecular materials, NMR of insoluble organic matter in Tagish Lake (a possible CI2) is extremely aromatic (Pizzarello *et al.*, 2001).

Binet *et al.* (2002) have undertaken an initial electron paramagnetic resonance study to examine the distribution of free radicals in Murchison and Orgueil macromolecular material. They suggest that there are radical-rich regions, which could represent regions of pristine interstellar organic matter preserved within the macromolecular material.

2.4.3 Stable-isotopic Studies

Measurements of bulk carbon, nitrogen, and hydrogen isotopic composition indicate that chondritic macromolecular material contains significant enrichments in some of the heavier isotopes. Kerridge (1985) reviewed the contents of carbon, hydrogen, and nitrogen and $\delta^{13}C$, $\delta^{15}N$, and δD for 25 whole-rock samples of carbonaceous chondrites, updating and reviewing many earlier studies (e.g., Boato, 1954; Smith and Kaplan, 1970; Kung and Clayton, 1978; Kolodny *et al.*, 1980; Robert and Epstein, 1982; Yang and Epstein, 1983).

Two analytical approaches have been adopted in attempts to obtain stable-isotopic information on insoluble organic matter in carbonaceous chondrites: stepped-combustion analysis (e.g., Kerridge, 1983; Swart *et al.*, 1983) and CSIA of pyrolysis products (e.g., Sephton *et al.*, 1998). Stepped-combustion analysis has proved to be more successful in providing information on the major-elemental constituents of chondritic organic matter, i.e., carbon, hydrogen, nitrogen, and oxygen, whereas CSIA has started to yield detailed carbon isotopic and structural information.

Kerridge *et al.* (1987) undertook a high-resolution stepped-combustion experiment of Murchison insoluble organic matter that indicated significant variations in $\delta^{13}C$, $\delta^{15}N$, and δD and attempted to relate stable-isotopic composition to structural moieties within the insoluble organic matter (summarized in Figure 6), based on the observation that aliphatic-rich organic matter combusts at lower temperatures than aromatic-rich. High-sensitivity stable-isotope measurements combined with stepped combustion (Pillinger, 1984, 1987) have been extensively applied to the study of presolar grains in meteorites, and Alexander *et al.* (1998) applied this technique to the study of nitrogen and carbon abundances and isotopic compositions of acid-insoluble carbonaceous material in 13 chondritic meteorites. They found a range in $\delta^{15}N$ values for organic material from −40‰ to + 260‰, with the most anomalous nitrogen being associated with the petrologically most primitive meteorites. This suggested that two basic nitrogen-containing components were present within the organic material: one with $\delta^{15}N$ values of between ~0‰ and − 40‰ and a second ^{15}N-enriched component with $\delta^{15}N$ values >+260‰. The ^{15}N-enriched component was interpreted as being presolar and comprising 40–70% of the total nitrogen released in the experiment.

The combined use of pyrolysis techniques followed by CSIA measurements (both offline hydrous pyrolysis and online pyrolysis-GC-IRMS)

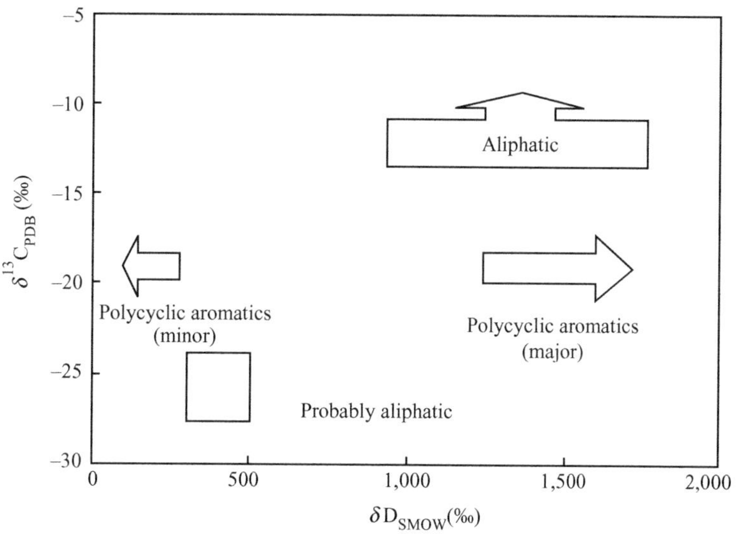

Figure 6 Carbon and hydrogen stable-isotope compositions of discrete moieties identified using stepped combustion of macromolecular material from the Murchison (CM2) chondrite (source Kerridge *et al.*, 1987).

enables the isotopic compositions of fragments of the macromolecular material to be determined. This approach has been applied to the study of the stable carbon isotope distribution of insoluble organic matter in three carbonaceous chondrites: Orgueil (CI1), Murchison (CM2), and Cold Bokkeveld (CM2) (Sephton, 1998, 2000; Sephton and Gilmour, 2001a), and both isotopic and structural information has been obtained for the macromolecular material in these meteorites (Table 8 and Figure 7). Relatively large and systematic differences in $\delta^{13}C$ values are observed for molecules that differ by only one or two carbon atoms, with compounds from Cold Bokkeveld displaying the widest range in $\delta^{13}C$ values (~27‰). Systematic differences in $\delta^{13}C$ values are apparent with increasing carbon number: $\delta^{13}C$ values in Murchison become more positive with increasing carbon number, while $\delta^{13}C$ values in Cold Bokkeveld become more negative with increasing carbon number above C_8, a similar trend is observed in Orgueil. There are strong similarities in $\delta^{13}C$ values between free and macromolecular aromatic moieties in Murchison, and, to a lesser extent, Cold Bokkeveld, suggesting a genetic relationship between the two. Based on these data, Sephton *et al.* (2000) advocate that the free aromatic compounds in these meteorites were derived from macromolecular material via parent body processes. The systematic shift in $\delta^{13}C$ values with carbon number have been interpreted as resulting from kinetic isotope effects during bond formation and destruction in the aromatic carbon skeletons of these compounds with the relatively large differences in $\delta^{13}C$ values, indicating that significant isotopic fractionations have occurred such as might be expected in low-temperature interstellar environments (Sephton and Gilmour, 2000).

Comparing data for Murchison in Figures 3 and 7, it is apparent that two trends are evident: $\delta^{13}C$ values for low-molecular-weight compounds become more positive with increasing carbon number, while $\delta^{13}C$ values for high-molecular-weight compounds become more negative, i.e., signatures of both bond formation (synthesis of higher homologs from lower ones) and bond destruction (cracking of higher homologs to form lower ones). Sephton and Gilmour (2000) suggest that these trends imply that the carbon skeletons were produced in the restrictive environment of the icy organic-rich mantles of interstellar grains where radiation-induced reactions simultaneously create and destroy organic matter to produce material with an intermediate level of complexity. Once formed, these complex organic residues would be available to participate in the formation of the carbonaceous chondrites following the collapse of the interstellar cloud.

2.5 *IN SITU* EXAMINATION OF METEORITIC ORGANIC MATTER

Early attempts at the examination of organic matter *in situ* in meteorites employed fluorescence and suggested that organic matter was coated on the surfaces of mineral grains (Alpern and Benkeiri, 1973). More recently, Pearson *et al.* (2002) used an organic labeling technique to map the distribution of organic matter in the Murchison (CM2), Ivuna (CI1), Orgueil (CI1), and Tagish Lake chondrites. A strong association was observed between the distribution of organic matter and hydrous clay minerals, suggesting that the production of clays by aqueous processes influenced the distribution of organic matter in meteorites.

Table 8 Carbon stable isotope compositions of individual aromatic molecules released by the pyrolysis of macromolecular materials in carbonaceous chondrites.

Compound	*Orgueil (CI1)*[a] $\delta^{13}C_{PDB}$ (‰)	*Murchison (CM2)* $\delta^{13}C_{PDB}$ (‰)	*Cold Bokkeveld (CM2)*[b] $\delta^{13}C_{PDB}$ (‰)
Toluene	−25.0 ± 1.1	−24.6 ± 0.2[b]	−22.9 ± 2.3
		−1.3 ± 2.0[c]	
		−5.4[c]	
Ethylbenzene	−10.0 ± 0.6	−21.9 ± 1.4[b]	+2.5 ± 2.3
m-Xylene	−17.0 ± 0.5	−19.6[b]	+4.0 ± 3.8
		−21.7 ± 0.4[c]	
		−20.3[c]	
p-Xylene	−14.0 ± 1.5	−17.8[b]	+2.4 ± 0.3
C_3-alkylbenzene	−22.5 ± 0.7	−18.1[b]	−1.0 ± 2.0
Benzaldehyde		−23.7 ± 0.6[c]	
		−24.0 ± 0.3[c]	
		−24.1[b]	
Phenol			−14.0 ± 1.5
C_3-alkylbenzene	−14.5 ± 1.8		−8.0 ± 0.4
C_4-alkylbenzene	−23.6 ± 1.1		−12.2 ± 1.1
2-Methylphenol	−12.8 ± 0.9	−10.3[b]	−3.9 ± 2.2
3-Methylphenol		−10.4[b]	−3.4
		−13.9 ± 0.9[c]	
		−18.5 ± 1.9[c]	
Naphthalene	−14.3 ± 1.4	−6.5 ± 2.5[b]	−6.1 ± 1.3
		−5.5 ± 0.5[c]	
		−6.4 ± 0.8[c]	
		−15.8[b]	
Benzothiophene			
2-Methylnaphthalene	−17.5 ± 1.5	−5.6 ± 2.1[b]	−12.1
		−6.4 ± 0.8[c]	
1-Methylnaphthalene	−18.7 ± 0.4	−7.2 ± 2.0[b]	−15.1
		−7.1[c]	
Acenaphthene		−5.9 ± 1.7[b]	

[a] Sephton *et al.* (2000).
[b] Sephton *et al.* (1998).
[c] Sephton and Gilmour (2001b).

The development of two-step laser mass spectrometry in which organic molecules are first desorbed and then ionized using a laser before detection in a mass spectrometer led to the first *in situ* identifications of PAHs in chondrites (Hahn *et al.*, 1988; Kovalenko *et al.*, 1991, 1992). The high spatial resolution of this approach has enabled its application to the study of organic matter in interplanetary dust particles collected by aircraft and in micrometeoritic material recovered from Antarctica. Clemett *et al.* (1993) examined eight stratospherically collected interplanetary dust particles and identified PAHs and their alkylated derivatives including high-mass PAHs not observed in similar studies of meteorites. The same approach has been used to recognize apparently indigenous PAHs in Antarctic micrometeorites (Clemett *et al.*, 1998). The apparent lack of a spatial relationship between organic compounds and carbonate minerals in the martian meteorite ALH 84001 has been used to infer that PAHs present in that meteorite are terrestrial in origin and not a remnant of martian biogenic activity (Stephan *et al.*, 2003).

Messenger *et al.* (1998) used two-step laser mass spectrometry to identify PAHs in individual circumstellar graphite grains extracted from the Murchison (CM2) and Acfer 094 meteorites. Some 70% of the grains studied had appreciable concentrations of PAHs (500–5,000 ppm), and in several cases correlated isotopic anomalies were observed between PAHs (phenanthrene, C_1 and C_2 alkyl phenanthrene, chrysene, and C_3 alkyl chrysene) and their parent grains. These correlations were most evident for ^{13}C-depleted grains. This isotopic linkage between specific molecules and the circumstellar grains they are associated with, suggests a genetic relationship between the two, and the authors suggested that PAH-like material was produced in the gas phase prior to the formation of graphite in a circumstellar environment.

2.6 ENVIRONMENTS OF FORMATION

Hypotheses for the origin of meteoritic organic matter must account for its molecular and isotopic composition and be consistent with models of meteorite petrogenesis; consequently, a number of potential environments have been considered (Table 9). Until the early 1990s, the favored hypothesis involved the catalytic hydrogenation

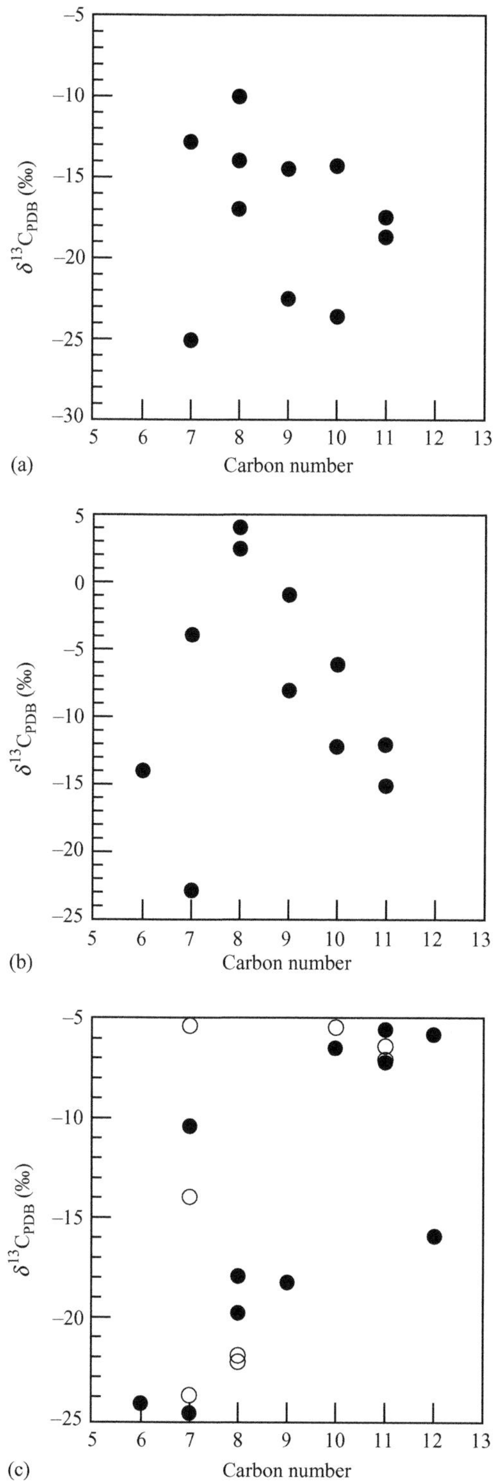

Figure 7 Carbon stable-isotope compositions of individual aromatic and polycyclic aromatic compounds, plotted against carbon number. Compounds were obtained by hydrous pyrolysis of macromolecular material from: (a) Orgueil (CI), (b) Cold Bokkeveld (CM2), and (c) by hydrous pyrolysis (closed symbols) and online pyrolysis GC-IRMS (open symbols) of Murchison (CM2) (sources (a) Sephton *et al.*, 2000; (b) Sephton *et al.*, 1998; (c) Sephton and Gilmour, 2001a).

of CO in the solar nebula. However, a characteristic of such catalytic reactions is their structural selectivity. FTT synthesis, in particular, produces a structurally selective suite of hydrocarbons and other compounds that, initially, were believed to closely resemble those observed in meteorites (Studier *et al.*, 1968; Hayatsu *et al.*, 1971, 1972; Yoshino *et al.*, 1971). However, the recognition of structural diversity in indigenous hydrocarbons in Murchison (Cronin and Pizzarello, 1990) together with inability of the isotopic fractionations associated with FTT reactions (Yuen *et al.*, 1990) to explain the isotopic distributions observed in meteorites indicates that catalytic synthesis is not the primary process involved in the synthesis of meteoritic organics. Catalytic reactions such as FTT also have problems in accounting for the preservation of deuterium-rich interstellar material in meteorites, and petrographic evidence indicates that the catalysts necessary to trigger the reaction in the primitive solar nebula were formed much later on the meteorite parent body (Kerridge *et al.*, 1979; Bunch and Chang, 1980). This all suggests that events in the solar nebula and on the meteorite parent body are more likely to amount to the secondary processing of pre-existing interstellar organic matter than primary synthesis. It is worth noting, however, that FTT reactions primarily fail to account for the properties of solvent-extractable organic matter in the Murchison meteorite, which does not exclude this process having operated elsewhere in the early solar system. Fischer–Tropsch is the only feasible thermally driven pathway available in the solar nebula to convert CO into other forms of carbon, and models suggest that the process may have been most efficient in the vicinity of the nebula equivalent to the present position of the asteroid belt (Kress and Tielens, 2001). Further studies of catalytic synthesis reactions using catalysts appropriate to specific nebular and asteroidal environments may indicate that such processes

Table 9 Sources and processes potentially involved in the production of meteoritic organic matter.

Ion–molecule reactions in interstellar clouds
Radiation chemistry in interstellar grain mantles
Condensation in stellar outflows
Equilibrium reactions in the solar nebula
Surface catalysis (Fischer–Tropsch) in the solar nebula
Kinetically controlled reactions in the solar nebula
Radiation chemistry (Miller–Urey) in the nebula
Photochemistry in nebular surface regions
Liquid-phase reactions on parent asteroid
Surface catalysis (Fischer–Tropsch) on asteroid
Radiation chemistry (Miller–Urey) in asteroid atmosphere
Thermal processing during asteroid metamorphism

Source: Kerridge (1999).

remain a significant mechanism for the production of extraterrestrial organic matter (Llorca and Casanova, 2000).

It has become increasingly apparent that the production of meteoritic organic matter must have involved a combination of different processes taking place in a wide range of environments. The detailed structural and isotopic analysis of meteoritic organic matter reviewed in this chapter has led to the development of new models for the origin of organic compounds in meteorites that envisage a distinctly different set of processes from the synthesis models proposed by earlier workers (e.g., Yoshino *et al.*, 1971; Miller *et al.*, 1976). Foremost has been the development of the so-called interstellar parent-body hypothesis. Petrographic evidence indicates that carbonaceous chondrites experienced a period of aqueous activity which, when combined with evidence for interstellar organics, led to the development of a model in which meteoritic organic matter is envisaged as being the product of parent body aqueous and thermal processing of reactive, volatile precursors such as water, HCN, NH_3, and ^{13}C-, ^{15}N-, and D-rich interstellar organics (Cronin and Chang, 1993). The model has superceded earlier ones that envisaged a primarily nebular origin for meteoritic organic matter on the strength of its ability to account for a number of the features common to meteoritic organics that have been reviewed in this chapter. These include:

(i) The proportion of amino to hydroxy acids observed in carbonaceous chondrites broadly matches that predicted by the Strecker cyanohydrin synthesis, a set of aqueous reactions in which cyanide, ammonia, and aldehydes and/or ketones are converted to amino acids and hydroxy acids (Peltzer *et al.*, 1984). There is also a rough correlation between abundance and the extent of aqueous alteration (Cronin, 1989), although the abundances of amino acids in meteorites are much lower than those predicted by a Strecker model (Cronin and Chang, 1993).

(ii) The structural diversity observed in virtually all classes of organic compounds, a predominance of branched-chain isomers and an exponential decline in abundance with increasing molecular weight. These observations are not consistent with structurally selective syntheses, but rather suggest the random production of precursor compounds.

(iii) Systematic decreases in $\delta^{13}C$ values with increasing molecular weight indicative of the synthesis of higher-molecular-weight homologs from lower ones (i.e., carbon chains are constructed by the addition of C atoms). Large differences in $\delta^{13}C$ values between deuterium-enriched substantial molecules such as PAHs indicate that such species were synthesized under conditions, such as the low temperatures of dense interstellar cloud environments, where isotopic fractionations were maximized. (Sephton and Gilmour, 2000). Some PAHs also appear to have been derived from circumstellar environments (Messenger *et al.*, 1998).

(iv) Deuterium enrichments in a number of the organic constituents, e.g., amino acids, volatile bases, and components of the macromolecular material. This deuterium enrichment is believed to a signature of ion–molecule reactions in the interstellar medium (Kolodny *et al.*, 1980; Robert and Epstein, 1982; Kerridge, 1983). Ionization of simple gaseous compounds in interstellar clouds (e.g., CH_4, CH_2O, H_2O, N_2, and NH_3) by cosmic radiation leads to fragmentation and the production of ions that react with neutral molecules to produce deuterium-enriched products. Continued ion–molecule interactions are thought to produce increasingly deuterium-enriched and complex organic matter that condenses on to dust grains (Robert and Epstein, 1982). Within interstellar clouds simple gas-phase molecules condense as icy mantles around silicate dust grains (Greenburg, 1984). When these mantles are subjected to increased temperatures or periods of UV radiation, thermally or photolytically driven polymerization reactions occur, which results in complex organic products that may also be deuterium-enriched (Sandford, 1996). Once formed, these deuterium-enriched complex organic residues would be available to participate in the formation of the carbonaceous chondrites following the collapse of the interstellar cloud.

(v) The widespread distribution, in both our own and other galaxies, of a 3.4 μm absorption feature attributed to saturated aliphatic hydrocarbons, suggesting that the organic component of interstellar dust is available for incorporation in newly forming planetary systems (e.g., Pendleton *et al.*, 1994; Pendleton, 1995).

(vi) A physical, not biological, enantiomeric excess. The cause of the chiral excess is not known, although the selective destruction of one enantiomer by UV circularly polarized light (UVCPL) from neutron stars in a presolar cloud has been invoked (Bonner and Rubenstein, 1987). The infall of icy/organic interstellar grains into the protosolar nebula would have resulted in their eventual incorporation into progressively larger planetesimals, culminating in the formation of the asteroidal parent bodies of the carbonaceous chondrites. The internal heating of the Murchison parent body by the decay of short-lived nuclides is thought to have initiated periods of aqueous alteration as liquid water was present at temperatures of less than 20 °C (Clayton and Mayeda, 1984). This would have resulted in the reaction of interstellar molecules with water and with each other to form more complex organics. Further organic synthesis could have occurred as hot fluids reacted on the surfaces of mineral catalysts during transport from the interior of the parent body. Continued heating of simpler organic molecules may then have resulted in the production of the macromolecular organic matter. Alternatively, it has also been suggested that the macromolecular material could have been produced in the solar nebula via the gas-phase pyrolysis of simple interstellar cloud-derived hydrocarbons (e.g., C_2H_2, CH_4) at temperatures of 900–1,100 K (Morgan *et al.*, 1991) followed by polymerization to form of high-molecular-weight aromatic structures.

REFERENCES

Alexander C. M. O'D., Russell S. S., Arden J. W., Ash R. D., Grady M. M., and Pillinger C. T. (1998) The origin of chondritic macromolecular organic matter: a carbon and nitrogen isotope study. *Meteorit. Planet. Sci.* **33**, 603–622.

Alpern B. and Benkeiri Y. (1973) Distribution de la matière organique de la météorite d'Orgueil par microscopie en fluorescence. *Earth. Planet. Sci. Lett.* **19**, 422–428.

Anders E. and Kerridge J. F. (1988) Future directions in meteorite research. In *Meteorites and the Early Solar System* (eds. J. F. Kerridge and M. S. Matthews). University of Arizona Press, Tucson, pp. 1149–1154.

Anders E., DuFresne E. R., Hayatsu R., Cavaille A., DuFresne A., and Fitch F. W. (1964) Contaminated meteorite. *Science* **146**, 1157–1161.

Anders E., Hayatsu R., and Studier M. H. (1973) Organic compounds in meteorites. *Science* **182**, 781–789.

Bada J. L., Cronin J. R., Ho M. S., Kvenvolden K. A., Lawless J. G., Miller S. L., Oro J., and Steinberg S. (1983) On the reported optical activity of amino acids in the Murchison meteorite. *Nature* **301**, 494–496.

Bandurski E. L. and Nagy B. (1976) The polymer-like material in the Orgueil meteorite. *Geochim. Cosmochim. Acta* **40**, 1397–1406.

Basile B. P., Middleditch B. S., and Oró J. (1984) Polycyclic aromatic hydrocarbons in the Murchison meteorite. *Org. Geochem.* **5**, 211–216.

Becker R. H. and Epstein S. (1982) Carbon, hydrogen, and nitrogen isotopes in solvent-extractable organic matter from carbonaceous chondrites. *Geochim. Cosmochim. Acta* **46**, 97–103.

Biemann K. (1974) Test result on the Viking gas chromatograph-mass spectrometer experiment. *Origins Life* **5**, 417–430.

Binet L., Gourier D., Derenne S., and Robert F. (2002) Heterogeneous distribution of paramagnetic radicals in insoluble organic matter from the Orgueil and Murchison meteorites. *Geochim. Cosmochim. Acta* **66**, 4177–4186.

Boato G. (1954) The isotopic composition of hydrogen and carbon in the carbonaceous chondrites. *Geochim. Cosmochim. Acta* **6**, 209–220.

Bonner W. A. and Rubenstein E. (1987) Supernovae, neutron stars and biomolecular chirality. *Biosystems* **20**, 99–111.

Bunch T. E. and Chang S. (1980) Carbonaceous chondrites: II. Carbonaceous chondrite phyllosilicates and light element geochemistry as indicators of parent body processes and surface conditions. *Geochim. Cosmochim. Acta* **44**, 1543–1577.

Chang S., Mack R., and Lennon K. (1978) Carbon chemistry of separated phases of Murchison and Allende. *Lunar Planet. Sci.* **IX**. The Lunar and Planetary Institute, Houston, pp. 157–159.

Clayton R. N. and Mayeda T. K. (1984) The oxygen isotope record in Murchison and other carbonaceous chondrites. *Earth Planet. Sci. Lett.* **67**, 151–161.

Clemett S. J., Maechling C. R., Zare R. N., Swan P. D., and Walker R. M. (1993) Identification of complex aromatic molecules in individual interplanetary dust particles. *Science* **262**, 721–725.

Clemett S. J., Chillier X. D. F., Gillette S., Zare R. N., Maurette M., Engrand C., and Kurat G. (1998) Observation of indigenous polycyclic aromatic hydrocarbons in 'giant' carbonaceous Antarctic micrometeorites. *Origins Life Evol. Biosphere* **28**, 425–448.

Cody G. D., Alexander C. M. O'D., and Tera F. (2002) Solid-state (^1H and ^{13}C) nuclear magnetic resonance spectroscopy of insoluble organic residue in the Murchison meteorite: a self-consistent quantitative analysis. *Geochim. Cosmochim. Acta* **66**, 1851–1865.

Cooper G. and Cronin J. R. (1995) Linear and cyclic aliphatic carboxamides of the Murchison meteorite; hydrolyzable derivatives of amino acids and other carboxylic acids. *Geochim. Cosmochim. Acta* **59**, 1003–1015.

Cooper G., Kimmich N., Belisle W., Sarinana J., Brabham K., and Garrel L. (2001) Carbonaceous meteorites as a source of sugar-related organic compounds for the early Earth. *Nature* **414**, 879–882.

Cooper G. W., Onwo W. M., and Cronin J. R. (1992) Alkyl phosphonic acids and sulfonic acids in the Murchison meteorite. *Geochim. Cosmochim. Acta* **56**, 4109–4115.

Cooper G. W., Thiemens M. H., Jackson T. L., and Chang S. (1997) Sulfur and hydrogen isotope anomalies in meteorite sulfonic acids. *Science* **277**, 1072–1074.

Cronin J. R. (1976a) Acid-labile amino acid precursors in the Murchison meteorite: I. Chromatographic fractionation. *Origins Life* **7**, 337–342.

Cronin J. R. (1976b) Acid-labile amino acid precursors in the Murchison meteorite: II. A search for peptides and amino acyl amindes. *Origins Life* **7**, 343–348.

Cronin J. R. (1989) Origin of organic compounds in carbonaceous chondrites. *Adv. Space. Res.* **9**, 59–64.

Cronin J. R. and Chang S. (1993) Organic matter in meteorites: molecular and isotopic analysis of the Murchison meteorite. In *The Chemistry of Life's Origins* (ed. J. M. Greenberg). Kluwer, Boston, pp. 209–258.

Cronin J. R. and Moore C. B. (1976) Amino acids of the Nogoya and Mokoia carbonaceous chondrites. *Geochim. Cosmochim. Acta* **40**, 853–857.

Cronin J. R. and Pizzarello S. (1986) Amino acids of the Murchison meteorite: 3. 7 carbon acyclic primary alpha-amino alkanoic acids. *Geochim. Cosmochim. Acta* **50**, 2419–2427.

Cronin J. R. and Pizzarello S. (1990) Aliphatic hydrocarbons of the Murchison meteorite. *Geochim. Cosmochim. Acta* **54**, 2859–2868.

Cronin J. R. and Pizzarello S. (1997) Enantiomeric excesses in meteoritic amino acids. *Science* **275**, 951–955.

Cronin J. R. and Pizzarello S. (1999) Amino acid enantiomer excesses in meteorites: origin and significance. *Adv. Space Res.* **23**, 293–299.

Cronin J. R., Gandy W. E., and Pizzarello S. (1981) Amino acids of the Murchison meteorite: 1. 6 carbon acyclic primary alpha-amino alkanoic acids. *J. Molec. Evol.* **17**, 265–272.

Cronin J. R., Pizzarello S., and Yuen G. U. (1985) Amino acids of the Murchison meteorite: 2. 5 carbon acyclic primary beta-amino, gamma-amino and delta-amino alkanoic acids. *Geochim. Cosmochim. Acta* **49**, 2259–2265.

Cronin J. R., Pizzarello S., and Fyre J. S. (1987) ^{13}C NMR spectroscopy of the insoluble carbon of carbonaceous chondrites. *Geochim. Cosmochim. Acta* **51**, 229–303.

Cronin J. R., Pizzarello S., and Cruikshank D. P. (1988) Organic matter in carbonaceous chondrites, planetary satellites, asteroids and comets. In *Meteorites and the Early Solar System* (eds. J. F. Kerridge and M. S. Matthews). University of Arizona Press, Tucson, pp. 819–857.

Cronin J. R., Pizzarello S., Epstein S., and Krishnamurthy R. V. (1993) Molecular and isotopic analyses of the hydroxy-acids, dicarboxylic-acids, and hydroxydicarboxylic acids of the Murchison meteorite. *Geochim. Cosmochim. Acta* **57**, 4745–4752.

Cronin J. R., Cooper G., and Pizzarello S. (1995) Characteristics and formation of amino acids and hydroxy acids of the Murchison meteorite. *Adv. Space Res.* **3**, 91–97.

Degens E. T. and Bajor M. (1962) Amino acids and sugars in the Brudeheim and Murray meteorites. *Naturwiss.* **49**, 605–606.

Ehrenfreund P., Glavin D. P., Botta O., Cooper G., and Bada J. L. (2001) Extraterrestrial amino acids in Orgueil and Ivuna: tracing the parent body of CI type carbonaceous chondrites. *Proc. Natl. Acad. Sci.* **98**, 2138–2141.

Engel M. H. and Macko S. A. (1997) Isotopic evidence for extraterrestrial non-racemic amino acids in the Murchison meteorite. *Nature* **389**, 265–267.

Engel M. H. and Nagy B. (1982) Distribution and enantiomeric composition of amino acids in the Murchison meteorite. *Nature* **296**, 837–840.

Engel M. H., Macko S. A., and Silfer J. A. (1990) Carbon isotope composition of individual amino acids in the Murchison meteorite. *Nature* **348**, 47–49.

Epstein S., Krishnamurthy R. V., Cronin J. R., Pizzarello S., and Yuen G. U. (1987) Unusual stable isotope ratios in amino acid and carboxylic acid extracts from the Murchison meteorite. *Nature* **326**, 477–479.

Fitch F. W. and Anders E. (1963) Observations on the nature of the organized elements in carbonaceous chondrites. *Ann. NY Acad. Sci.* **108**, 495–513.

Fitch F. W., Schwarcz H. P., and Anders E. (1962) 'Organized elements' in carbonaceous chondrites. *Nature* **193**, 1123–1125.

Gardinier A., Derenne S., Robert F., Behar F., Largeau C., and Maquet J. (2000) Solid state CP/MAS ^{13}C NMR of the insoluble organic matter of the Orgueil and Murchison meteorites: quantitative study. *Earth. Planet. Sci. Lett.* **184**, 9–21.

Gilmour I. and Pillinger C. T. (1994) Isotopic compositions of individual polycyclic aromatic hydrocarbons from the Murchison meteorite. *Mon. Not. Roy. Astron. Soc.* **269**, 235–240.

Greenburg J. M. (1984) Chemical evolution in space. *Origins Life* **14**, 25–36.

Hahn J. H., Zenobi R., Bada J. L., and Zare R. N. (1988) Application of two-step laser mass spectrometry to cosmogeochemistry: direct analysis of meteorites. *Science* **239**, 1523–1525.

Han J., Simoneit B. R., Burlingame A. L., and Calvin M. (1969) Organic analysis on the Pueblito de Allende meteorite. *Nature* **222**, 364–365.

Hayatsu R. (1964) Orgeuil meteorite: organic nitrogen contents. *Science* **146**, 1291–1293.

Hayatsu R. (1965) Optical activity in the Orgueil meteorite. *Science* **149**, 443–447.

Hayatsu R. and Anders E. (1981) Organic compounds in meteorites and their origins. *Top. Curr. Chem.* **99**, 1–37.

Hayatsu R., Studier M. H., Oda A., Fuse K., and Anders E. (1968) Origin of organic matter in early solar system: II. Nitrogen compounds. *Geochim. Cosmochim. Acta* **32**, 175–190.

Hayatsu R., Studier M. H., and Anders E. (1971) Origin of organic matter in early solar system: IV. Amino acids: confirmation of catalytic synthesis by mass spectrometry. *Geochim. Cosmochim. Acta* **35**, 939–951.

Hayatsu R., Studier M. H., Matsuoka S., and Anders E. (1972) Origin of organic matter in early solar system: VI. Catalytic synthesis of nitriles, nitrogen bases and porphyrin-like pigments. *Geochim. Cosmochim. Acta* **36**, 555–571.

Hayatsu R., Studier M. H., Moore L. P., and Anders E. (1975) Purines and triazines in the Murchison meteorite. *Geochim. Cosmochim. Acta* **39**, 471–488.

Hayatsu R., Matsuoka S., Scott R. G., Studier M. H., and Anders E. (1977) Origin of organic matter in the early solar system: VII. The organic polymer in carbonaceous chondrites. *Geochim. Cosmochim. Acta* **41**, 1325–1339.

Hayatsu R., Scott R. G., Studier M. H., Lewis R. S., and Anders E. (1980) Carbynes in meteorites: detection, low temperature origin and implications for interstellar molecules. *Science* **209**, 1515–1518.

Hayes J. M. (1967) Organic constituents of meteorites—a review. *Geochim. Cosmochim. Acta* **31**, 1395–1440.

Hayes J. M. and Biemann K. (1968) High resolution mass spectrometric investigations of the organic constituents of the Murray and Holbrook chondrites. *Geochim. Cosmochim. Acta* **32**, 239–269.

Hayes J. M., Freeman K. H., Popp B. N., and Hoham C. H. (1990) Compound-specific isotopic analyses—a novel tool for reconstruction of ancient biogeochemical processes. *Org. Geochem.* **16**, 1115–1128.

Holtzer G. and Oró J. (1977) Pyrolysis of organic compounds in the presence of ammonia: the Viking Mars Lander site alteration experiment. *Org. Geochem.* **1**, 37–52.

Jungclaus G., Cronin J. R., Moore C. B., and Yuen G. U. (1976a) Aliphatic amines in the Murchison meteorite. *Nature* **261**, 126–128.

Jungclaus G. A., Yuen G. U., Moore C. B., and Lawless J. G. (1976b) Evidence for the presence of low molecular weight alcohols and carbonyl compounds in the Murchison meteorite. *Meteoritics* **11**, 231–237.

Kaplan I. R., Degens E. T., and Reuter J. H. (1963) Organic compounds in stony meteorites. *Geochim. Cosmochim. Acta* **27**, 805–834.

Kerridge J. F. (1983) Isotopic composition of carbonaceous-chondrite kerogen: evidence for an interstellar origin of organic matter in meteorites. *Earth. Planet. Sci. Lett.* **64**, 186–200.

Kerridge J. F. (1985) Carbon, hydrogen, and nitrogen in carbonaceous chondrites—abundances and isotopic compositions in bulk samples. *Geochim. Cosmochim. Acta* **49**, 1707–1714.

Kerridge J. F. (1999) Formation and processing of organics in the early solar system. *Space Sci. Rev.* **90**, 275–288.

Kerridge J. F., Mackay A. L., and Boynton W. V. (1979) Magnetite in CI carbonaceous chondrites: origin by aqueous activity on a planetesimal surface. *Science* **205**, 395–397.

Kerridge J. F., Chang S., and Shipp R. (1987) Isotopic characterization of kerogen-like material in the Murchison carbonaceous chondrite. *Geochim. Cosmochim. Acta* **51**, 2527–2540.

Kitajima F., Nakamura T., Takaoka N., and Murae T. (2002) Evaluating the thermal metamorphism of CM chondrites by using the pyrolytic behavior of carbonaceous macromolecular matter. *Geochim. Cosmochim. Acta* **66**, 163–172.

Kminek G., Botta O., Glavin D. P., and Bada J. L. (2002) Amino acids in the Tagish Lake meteorite. *Meteorit. Planet. Sci.* **37**, 697–701.

Kolodny Y., Kerridge J. F., and Kaplan I. R. (1980) Deuterium in carbonaceous chondrites. *Earth Planet. Sci. Lett.* **46**, 149–158.

Kovalenko L. J., Philippoz J. M., Bucenell J. R., Zenobi R., and Zare R. N. (1991) Organic chemical analysis on a microscopic scale using two-step laser desorption/laser ionization mass spectrometry. *Space Sci. Rev.* **56**, 191–195.

Kovalenko L. J., Maechling C. R., Clemett S. J., Philippoz J. M., Zare R. N., and Alexaner C. M. O. D. (1992) Microscopic organic analysis using two step laser mass spectrometry: application to meteoritic acid residues. *Anal. Chem.* **64**, 682–690.

Kress M. E. and Tielens A. G. G. M. (2001) The role of Fischer-Tropsch catalysis in solar nebular chemistry. *Meteorit. Planet. Sci.* **36**, 75–91.

Krishnamurthy R. V., Epstein S., Cronin J. R., Pizzarello S., and Yuen G. U. (1992) Isotopic and molecular analyses of hydrocarbons and monocarboxylic acids of the Murchison meteorite. *Geochim. Cosmochim. Acta* **56**, 4045–4058.

Krouse H. R. and Modzeleski V. E. (1970) $^{13}C/^{12}C$ abundances in components of carbonaceous chondrites and terrestrial samples. *Geochim. Cosmochim. Acta* **34**, 459–474.

Kung C. C. and Clayton R. N. (1978) Nitrogen abundances and isotopic composition in stony meteorites. *Earth Planet. Sci. Lett.* **38**, 421–435.

Kvenvolden K., Lawless J., Peterson E., Flors J., Ponnamperuma C., Kaplan I. R., and Moore C. (1970) Evidence for extraterrestrial amino acids and hydrocarbons in the Murchison meteorite. *Nature* **228**, 923–936.

Lawless J. G. and Yuen G. U. (1979) Quantification of monocarboxylic acids in the Murchison carbonaceous meteorite. *Nature* **282**, 396–398.

Lawless J. G., Kvenvolden K. A., Peterson E., Ponnamperuma C., and Jarosewich E. (1972) Evidence for amino-acids of extraterrestrial origin in the Orgueil meteorite. *Nature* **236**, 66–67.

Lawless J. G., Peterson E., and Kvenvolden K. A. (1973) Amino acids in meteorites. *Astrophys. Space Sci. Lib.* **40**, 167–168.

Lawless J. G., Zeitman B., Pereira W. E., Summons R. E., and Duffield A. M. (1974) Dicarboxylic acids in the Murchison meteorite. *Nature* **251**, 40–42.

Levy R. L., Grayson M. A., and Wolf C. J. (1973) The organic analysis of the Murchison meteorite. *Geochim. Cosmochim. Acta* **37**, 467–483.

Llorca J. and Casanova I. (2000) Reaction between H_2, CO, and H_2S over Fe, Ni metal in the solar nebula: experimental evidence for the formation of sulfur bearing organic molecules and sulfides. *Meteorit. Planet. Sci.* **35**, 841–848.

Meinschein W. G., Nagy B., and Hennessy D. J. (1963) Evidence in meteorites of former life. *Ann. NY Acad. Sci.* **108**, 553–579.

Messenger S., Amari S., Gao X., Walker R. M., Clemett S. J., Chillier X. D. F., Zare R. N., and Lewis R. S. (1998) Indigenous polycyclic aromatic hydrocarbons in circumstellar graphite grains from primitive meteorites. *Astrophys. J.* **501**, 284–295.

Miller S. L., Urey H. C., and Oró J. (1976) Origin of organic compounds on the primitive Earth and in meteorites. *J. Molec. Evol.* **9**, 59–72.

Morgan W. A., Feigelson E. D., Wang H., and Frenklach M. (1991) A new mechanism for the formation of meteoritic kerogen-like material. *Science* **252**, 109–112.

Mullie F. and Reisse J. (1987) Organic matter in carbonaceous chondrites. *Top. Curr. Chem.* **137**, 83–117.

Murae T. (1995) Characterization of extraterrestrial high-molecular-weight organic-matter by pyrolysis-gas chromatography mass-spectrometry. *J. Anal. Appl. Pyrol.* **32**, 65–73.

Nagy B., Meinschein W. G., and Hennessy D. J. (1961) Mass spectrometric analysis of the Orgueil meteorite: evidence for biogenic hydrocarbons. *Ann. NY Acad. Sci.* **93**, 534–552.

Nagy B., Claus G., Colombo U., Gazzarrini F., Modzeleski V. E., Murphy M. T. J., and Rouser G. (1964) Optical activity in saponified organic matter isolated from the interior of the Orgueil meteorite. *Nature* **202**, 228–233.

Nagy B. J. (1975) *Carbonaceous Meteorites*. Elsevier, Amsterdam.

Naraoka H., Shimoyama A., Komiya M., Yamamoto H., and Harada K. (1987) Carboxylic acids and hydrocarbons in Antarctic carbonaceous chondrites. In *Twelfth Symposium on Antarctic Meteorites*. National Institute of Polar Research, Tokyo, pp. 9–11.

Naraoka H., Shimoyama A., Komiya M., Yamamoto H., and Harada K. (1988) Hydrocarbons in the Yamato-791198 carbonaceous chondrite from Antarctica. *Chem. Lett.* **1988**, 831–934.

Naraoka H., Shimoyama A., and Harada K. (1999) Molecular distribution of monocarboxylic acids in Asuka chondrites from Antarctica. *Origins Life Evol. Biosphere* **29**, 187–201.

Naraoka H., Shimoyama A., and Harada K. (2000) Isotopic evidence from an Antarctic carbonaceous chondrite for two reaction pathways of extraterrestrial PAH formation. *Earth Planet. Sci. Lett.* **184**, 1–7.

Oró J., Nooner D. W., Zlatkis A., and Wisktrom S. A. (1966) Paraffinic hydrocarbons in Orgueil, Murray, Mokoia, and other meteorites. *Life Sci. Space Res.* **4**, 63–100.

Oró J., Gibert J., Lichtenstein H., Wikstrom S., and Flory D. A. (1971) Amino-acids, aliphatic and aromatic hydrocarbons in the Murchison meteorite. *Nature* **230**, 105–106.

Pearson V. K., Sephton M., Kearsley A. T., Bland P. A., Franchi I. A., and Gilmour I. (2002) Clay mineral-organic matter relationships in the early solar system. *Meteorit. Planet. Sci.* **37**, 1829–1833.

Peltzer E. T. and Bada J. L. (1978) α-Hydroxycarboxylic acids in the Murchison meteorite. *Nature* **272**, 443–444.

Peltzer E. T., Bada J. L., Schlesinger G., and Miller S. L. (1984) The chemical conditions on the parent body of the Murchison meteorite: some conclusions based on amino, hydroxy and dicarboxylic acids. *Adv. Space Res.* **4**, 69–74.

Pendleton Y. J. (1995) Laboratory comparisons of organic materials to interstellar dust and the Murchison meteorite. *Planet. Space Sci.* **43**, 1359–1364.

Pendleton Y. J., Sandford S., Allamandola L., Tielens A. G. G. M., and Sellgren K. (1994) Near infrared absorption spectroscopy of interstellar hydrocarbon grains. *Astrophys. J.* **437**, 683–696.

Pering K. L. and Ponnamperuma C. (1971) Aromatic hydrocarbons in the Murchison meteorite. *Science* **173**, 237–239.

Pillinger C. (1982) Not quite full circle? Non-racemic amino acids in the Murchison meteorite. *Nature* **296**, 802.

Pillinger C. T. (1984) Light element stable isotopes in meteorites; from grams to picograms. *Geochim. Cosmochim. Acta* **48**, 2739–2766.

Pillinger C. T. (1987) Stable isotope measurements of meteorites and cosmic dust particles. *Phil. Trans. Roy. Soc. London A* **323**, 313–322.

Pizzarello S. and Cronin J. R. (1998) Alanine enantiomers in the Murchison Meteorite. *Nature* **394**, 236.

Pizzarello S. and Cronin J. R. (2000) Non-racemic amino acids in the Murray and Murchison meteorites. *Geochim. Cosmochim. Acta* **64**, 329–338.

Pizzarello S. and Huang Y. (2002) Molecular and isotopic analyses of Tagish Lake alkyl dicarboxylic acids. *Meteorit. Planet. Sci.* **37**, 687–696.

Pizzarello S., Krishnamurthy R. V., Epstein S., and Cronin J. R. (1991) Isotopic analyses of amino acids from the Murchison meteorite. *Geochim. Cosmochim. Acta* **55**, 905–910.

Pizzarello S., Feng X., Epstein S., and Cronin J. R. (1994) Isotopic analyses of nitrogenous compounds from the Murchison meteorite; ammonia, amines, amino acids, and polar hydrocarbons. *Geochim. Cosmochim. Acta* **58**, 5579–5587.

Pizzarello S., Huang Y. S., Becker L., Poreda R. J., Nieman R. A., Cooper G., and Williams M. (2001) The organic content of the Tagish Lake meteorite. *Science* **293**, 2236–2239.

Robert F. and Epstein S. (1982) The concentration and isotopic composition of hydrogen, carbon, and nitrogen in carbonaceous meteorites. *Geochim. Cosmochim. Acta* **46**, 81–95.

Sandford S. A. (1996) The inventory of interstellar materials available for the formation of the solar-system. *Meteorit. Planet. Sci.* **31**, 449–476.

Sephton M. A. and Gilmour I. (2000) Aromatic moeities in meteorites: relics of interstellar grain processes? *Astrophys. J.* **540**, 588–591.

Sephton M. and Gilmour I. (2001a) Pyrolysis-gas chromatography-isotope ratio mass spectrometry of macromolecular material in meteorites. *Planet. Space Sci.* **49**, 465–471.

Sephton M. A. and Gilmour I. (2001b) Compound specific isotope analysis of the organic constituents in carbonaceous chondrites. *Mass Spec. Rev.* **20**, 111–120.

Sephton M. A., Pillinger C. T., and Gilmour I. (1998) $\delta^{13}C$ of free and macromolecular polyaromatic structures in the

Murchison meteorite. *Geochim. Cosmochim. Acta* **62**, 1821–1828.

Sephton M. A., Pillinger C. T., and Gilmour I. (1999) Small-scale hydrous pyrolysis of macromolecular material in meteorites. *Planet. Space Sci* **47**, 181–187.

Sephton M. A., Pillinger C. T., and Gilmour I. (2000) Aromatic moeities in meteoritic macromolecular materials: analysis by hydrous pyrolysis and $\delta^{13}C$ of individual compounds. *Geochim. Cosmochim. Acta* **64**, 321–328.

Sephton M. A., Pillinger C. T., and Gilmour I. (2001) Normal alkanes in meteorites: molecular $\delta^{13}C$ values indicate an origin by terrestrial contamination. *Precamb. Res.* **106**, 45.

Shimoyama A. and Harada K. (1984) Amino acid depleted carbonaceous chondrites (C2) from Antarctica. *Geochem. J.* **18**, 281–286.

Shimoyama A. and Katsumata H. (2001) Polynuclear aromatic thiophenes in the murchison carbonaceous chondrite. *Chem. Lett.* **2001**, 202–203.

Shimoyama A., Ponnamperuma C., and Yanai K. (1979) Amino acids in the Yamato carbonaceous chondrite from Antarctica. *Nature* **282**, 394–396.

Shimoyama A., Harada K., and Yanai K. (1985) Amino acids from the Yamato-791198 carbonaceous chondrite from Antarctica. *Chem. Lett.* **1985**, 1183–1186.

Shimoyama A., Naraoka H., Yamamoto H., and Harada K. (1986) Carboxylic acids in the Yamato-791198 carbonaceous chondrite from Antarctica. *Chem. Lett.* **1986**, 1561–1564.

Shimoyama A., Naraoka H., Komiya M., and Harada K. (1989) Analyses of carboxylic acids and hydrocarbons in Antarctic carbonaceous chondrites, Yamato-74662 and Yamato-793321. *Geochem. J.* **23**, 181–193.

Shock E. L. and Schulte M. D. (1990) Amino acid synthesis in carbonaceous meteorites by aqueous alteration of polycyclic aromatic hydrocarbons. *Nature* **343**, 728–731.

Silfer J. A. (1991) Stable carbon and nitrogen isotope signatures of amino acids as molecular probes in geologic systems. PhD Thesis, University of Oklahoma.

Simmonds P. G., Shulman G. P., and Stembridge C. H. (1969) Organic analysis by pyrolysis gas chromatography-mass spectrometry. A candidate experiment for the biological exploration of Mars. *J. Chromatog. Sci.* **7**, 36–41.

Smith J. W. and Kaplan I. R. (1970) Endogenous carbon in carbonaceous meteorites. *Science* **167**, 1367–1370.

Stephan T., Jessberger E. K., Heiss C. H., and Rost R. (2003) TOF-SIMS analysis of polycyclic aromatic hydrocarbons in Alan Hills 84001. *Meteorit. Planet. Sci.* **38**, 109–116.

Stoks P. G. and Schwartz A. W. (1979) Uracil in carbonaceous meteorites. *Nature* **282**, 709–710.

Stoks P. G. and Schwartz A. W. (1981) Nitrogen-heterocyclic compounds in meteorites: significance and mechanisms of formation. *Geochim. Cosmochim. Acta* **45**, 563–569.

Stoks P. G. and Schwartz A. W. (1982) Basic nitrogen-heterocyclic compounds in the Murchison Meteorite. *Geochim. Cosmochim. Acta* **46**, 309–315.

Studier M. H., Hayatsu R., and Anders E. (1965) Organic compounds in carbonaceous chondrites. *Science* **149**, 1455–1459.

Studier M. H., Hayatsu R., and Anders E. (1968) Origin of organic matter in early solar system—I. Hydrocarbons. *Geochim. Cosmochim. Acta* **32**, 151–173.

Studier M. H., Hayatsu R., and Anders E. (1972) Origin of organic matter in Early Solar System—V: Further studies of meteoritic hydrocarbons and discussion of their origin. *Geochim. Cosmochim. Acta* **36**, 189–215.

Swart P. K., Grady M. M., and Pillinger C. T. (1983) A method for the identification and elimination of contamination during carbon isotopic analyses of extraterrestrial samples. *Meteoritics* **18**, 137–154.

van der Velden W. and Schwartz A. W. (1977) Search for purines and pyrimidines in the Murchison meteorite. *Geochim. Cosmochim. Acta* **41**, 961–968.

Watson J. S., Pearson V. K., Gilmour I., and Sephton M. (2003) Contamination by sesquiterpenoid derivatives in the Orgueil carbonaceous chondrite. *Org. Geochem.* **34**, 37–47.

Yang J. and Epstein S. (1983) Interstellar organic matter in meteorites. *Geochim. Cosmochim. Acta* **47**, 2199–2216.

Yang J. and Epstein S. (1985) A search for presolar organic matter in meteorite. *Geophys. Res. Lett.* **12**, 73–76.

Yoshino D., Hayatsu R., and Anders E. (1971) Origin of organic matter in early solar system: III. Amino acids: catalytic synthesis. *Geochim. Cosmochim. Acta* **35**, 927–938.

Yuen G. U. and Kvenvolden K. A. (1973) Monocarboxylic acids in Murray and Murchison carbonaceous meteorites. *Nature* **246**, 301–303.

Yuen G., Blair N., Des Marais D. J., and Chang S. (1984) Carbon isotope composition of low molecular weight hydrocarbons and mono carboxylic acids from Murchison meteorite. *Nature* **307**, 252–254.

Yuen G. U., Pecore J. A., Kerridge J. F., Pinnavaia T. J., Rightor E. G., Flores J., K. M. W., Mariner R., DesMarais D. J., and Chang S. (1990) Carbon isotopic fractionation in Fischer-Tropsch type reactions. *Lunar. Planet. Sci.* **21**, 1367–1368.

Zinner E. (1988) Interstellar cloud material in meteorites. In *Meteorites and the Early Solar System* (eds. J. F. Kerridge and M. S. Mathews). University of Arizona Press, Tucson, pp. 956–983.

Isotope Geochemistry
ISBN: 978-0-08-096710-3

pp. 17–38

3

Trace Element and Isotopic Fluxes/Subducted Slab

G. E. Bebout

Lehigh University, Bethlehem, PA, USA

3.1 INTRODUCTION

Subduction zones figure prominently in virtually all chemical geodynamic models of recent and long-term crust–mantle evolution (Allègre, 1989; Zindler and Hart, 1986; Javoy, 1998; Berner, 1999; also see Bebout, 1995). A thorough understanding of the subduction pathway (see Figure 1) is required to address many aspects of Earth's large-scale material cycling, including the origin of arc magmas, continental crust formation, and the geochemical evolution of the mantle. Over the last decades, considerable effort has been expended to trace, and even mass balance, elements and isotopes across individual arc-trench systems using seafloor inputs and mostly volcanic arc outputs (e.g., Plank and Langmuir, 1993; Hilton *et al.*, 2002; George *et al.*, 2005; Li and Bebout, 2005.

Subduction of oceanic lithosphere initiates a complex continuum of diagenetic and metamorphic reactions and related geochemical effects in the down-going slab. For some chemical components, this history profoundly influences the elemental inventory and, in some cases, the isotopic composition of the slab to depths beneath volcanic arcs and beyond. These seafloor lithologies are also subjected to deformation processes that significantly alter their physical state and, in the case of sedimentary accretion and underplating, halt their delivery to depths beyond forearcs. In addition, subduction erosion from the forearc hanging walls can impact geochemical evolution at greater depths (see Scholl and von Huene, in press). All of these physical, petrologic, and geochemical processes are thought to vary considerably among Earth's modern subduction zones (see Stern, 2002).

Whereas case studies and syntheses of the geochemistry of arc magmas abound (see Pearce and Peate, 1995; Elliott, 2003; Figure 2a), the study of the geochemistry of slab metamorphic rocks (certain high-pressure (HP) and ultrahigh-pressure (UHP) metamorphic suites) has lagged behind, likely due, in part, to the perceived complexity of the metamorphic field record. In some studies of arc geochemistry, seafloor compositions ("inputs") are simply compared with arc magmas (the "outputs"), with little or no discussion of the possible effects of subduction-zone chemical processing, whereas in others, specific chemical changes are attributed to metamorphism, providing some exciting and perhaps testable hypotheses. Many studies of HP and UHP metamorphic rocks present cursory geochemical datasets, most commonly major element compositions and concentrations of selected trace elements. However, the number of studies that directly test the influence of metamorphism on the geochemistry of the subducting slab is still surprisingly small.

This chapter provides a synthesis of current knowledge regarding the changes in the physical, petrologic, and geochemical state of subducting

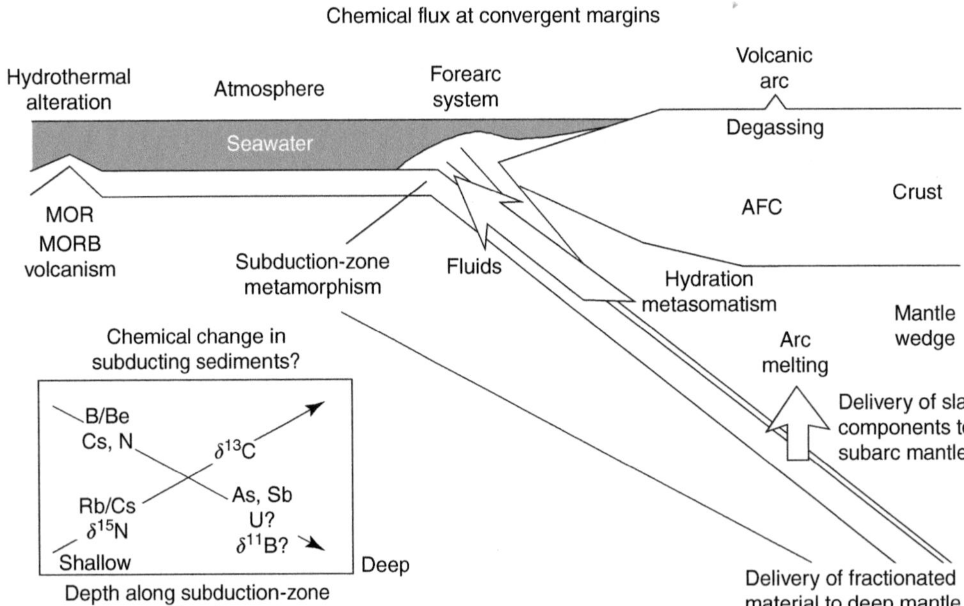

Figure 1 Schematic illustration of a continental margin-type subduction zone, showing possible sites of trace-element and isotopic fractionation. The large arrow along the slab–mantle interface indicates transfer, toward the surface, of volatile components released by subduction-zone metamorphism. Subduction could deliver, to various parts of the mantle (e.g., the forearc mantle wedge, the subarc mantle wedge, and deeper parts of the mantle), C–O–H–S–N fluids, melts, and residual mineral reservoirs strongly fractionated isotopically and chemically relative to initial compositions of the subducted rock reservoir. The inset plot demonstrates the possible evolution of chemical parameters discussed in this chapter with progressive metamorphism of sedimentary rocks during subduction. MOR, mid-ocean ridge; MORB, mid-ocean ridge basalt; AFC, assimilation-fractional crystallization. Reproduced by permission of Elsevier from Bebout (1995).

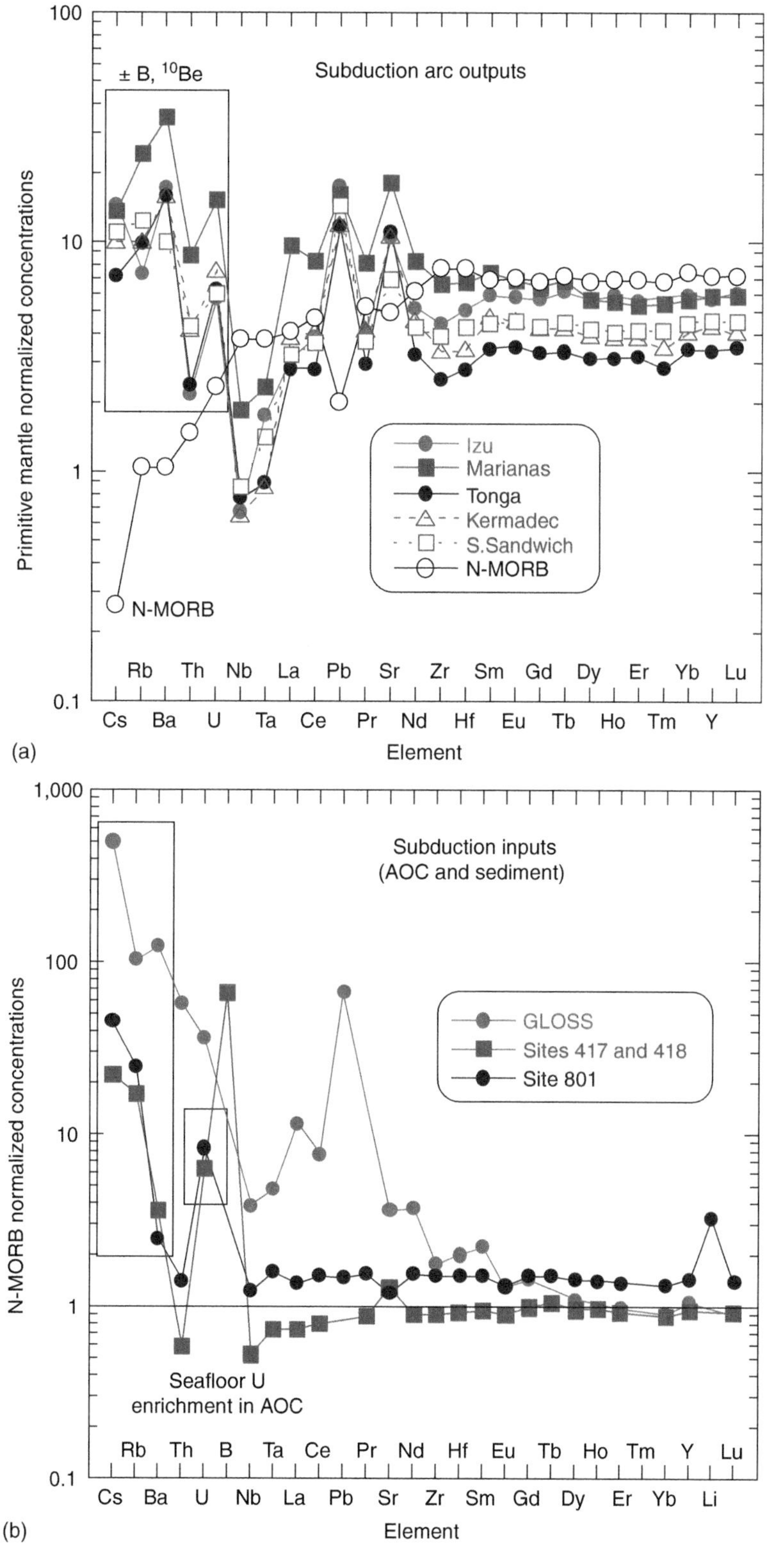

Figure 2 Chemical characteristics of subduction-zone arc outputs, based on erupted basalts (a) and inputs, based on the geochemistry of AOC and sediment (b). Data in (a) and (b) are normalized to primitive mantle and N-MORB, respectively (both from Hofmann, 1988). Data for arc basalt outputs are from Elliott (2003; electronic supplement), the GLOSS (GLOSS composite) is from Plank and Langmuir (1998), and the data for altered oceanic crust at ODP Sites 417/418 and 801 are from Staudigel *et al.* (1996) and Kelley *et al.* (2003), respectively. The boxes in (a) and (b) highlight the similarities in the seafloor enrichments in basalts and the enrichments in arc basalts relative to N-MORB.

oceanic crust that result from diagenesis and metamorphism along the subduction pathway. The focus in this contribution is largely on the chemical and isotopic tracers commonly employed to elucidate chemical cycling in subduction zones based on study of arc lavas, and on discussion of metamorphic suites for which it is possible to evaluate chemical changes in subducting oceanic crust and overlying sediments, for discussion of UHP metamorphic geochemistry of continental lithologies in collisional orogens such as Dabieshan-Su-Lu Belt and Kokchetav Massif). Emphasis is placed on elucidating connections between the metamorphic geochemical record and other subduction-related products such as décollement fluids, forearc serpentinite seamounts and mud volcanoes, arc lavas, and deep-mantle geochemical signatures that may result from incorporation of subducted oceanic crust and sediment. For a number of important chemical tracers (e.g., boron, lithium, nitrogen, organic carbon, and nitrogen, perhaps large-ion lithophile elements (LILE), particularly cesium, rubidium, barium, and strontium), significant changes occur during the earliest stages of subduction due to pore water expulsion and diagenesis (and for mafic rocks, possibly hydration). Discussion of the metamorphic record is focused primarily on HP and UHP suites in the circum-Pacific, Cyclades, and European Alps, which have seen the most geochemical study focused on issues of ocean-to-mantle cycling. This review is meant to complement a number of chapters in the *Treatise on Geochemistry* namely chapters 17,18.

3.2 THE SEAFLOOR, AS IT ENTERS THE TRENCHES

At first glance, one might consider the subducting oceanic lithosphere to be rather uniform from place to place. On average, it is composed of, a 5–7 km thick section of oceanic crust, dominantly basaltic and gabbroic in composition, with an ultramafic lithospheric mantle section and a sedimentary veneer of varying thickness. However, geochemical data for oceanic crust, either drilled on the seafloor or accessed in on-land ophiolitic sequences, demonstrates a huge diversity of chemical compositions that relate to the nature and duration of fluid–rock interaction on the seafloor (see Alt and Teagle, 2003; Kelley *et al.*, 2003; Staudigel *et al.*, 1996; Figure 2b).

The mid- to lower parts of the oceanic crust are generally thought to have little or no impact on the slab flux (i.e., the material ascending from the slab to the overlying mantle wedge), as they contain fewer volatile phases and thus do not experience devolatilization reactions and the related fluid mobility necessary for element transport. However, the extent of hydration in the uppermost mantle of subducting oceanic lithosphere is unclear (see Peacock, 2001; Kerrick, 2002). If such hydration occurs, it could profoundly increase the amount of H_2O being subducted into the deep Earth, possibly causing flux melting of the slab (see Ranero *et al.*, 2005; Kessel *et al.*, 2005), and have great consequences for the deeper geochemistry of subduction zones.

Sedimentary sequences just outboard of trenches have thicknesses ranging from <100 m to >4 km (but mostly < 1 km). Factors that influence their chemical compositions (e.g., residence time on the seafloor, water depth, which controls carbonate solubility and deposition, proximity to continental/arc sediment sources, and even ocean currents, which affect organic nutrient budgets) are highly variable. Our knowledge of the lithology and compositions of subducting sediments, largely based on analysis of deep sea drilling program/ocean drilling program (DSDP/ODP) sediment cores, is summarized by Rea and Ruff (1996), Plank and Langmuir (1998), and Jarrard (2003). Plank and Langmuir (1998) discuss the sedimentological, diagenetic, and organic processes dictating the concentrations of major and trace elements of particular interest in chemical-cycling studies (e.g., U–Th–Pb; LILE). Approximately 76 wt.% of the globally subducting sediment (GLOSS) composite of Plank and Langmuir (1998) is composed of "terrigenous" sediment with its largest accumulation near continental margins, and the remainder of this composite consists of calcium carbonate (7 wt.%), opal (10 wt.%), and mineral-bound H_2O (7 wt.%). The GLOSS composite provides an extremely useful guide to the composition of average subducting sediment for use in global chemical mass-balance studies, and Plank and Langmuir (1998) also provide estimates of sediment input into individual modern trenches.

A quick examination of a seafloor bathymetric map reminds us of the large amount of relief on the seafloor, related largely to the presence of ridges and associated transform fault segments, and oceanic plateaus produced by magmatic activity attributed to hotspots. This roughness at the top of the subducting oceanic slab could promote off-scraping/underplating of the areas of higher relief (e.g., ocean–island basalt; Stern, 2002) and representation of these parts of the slab in paleosubduction suites (Maruyama and Liou, 1989; Cloos, 1993). In many margins, convergence has a significant obliquity, potentially complicating efforts to assess chemical cycling by simple comparison of arc volcanic rocks and offshore subduction sections and, in some margins, the trench sediment section varies significantly along strike of the trench. Transient interactions, such as ridge–trench encounters (e.g., Shervais *et al.*, 2005) might disproportionately affect the tectonometamorphic record of ancient subduction episodes.

3.3 RECENT THERMAL MODELS OF SUBDUCTION

A brief discussion of the thermal evolution of subduction zones is appropriate here, as many of the geochemical changes discussed in this chapter occur at the relatively high-pressure/temperature (*P–T*) conditions in subduction zones. Increasingly high-resolution thermal models, employing variable upper-mantle viscosity and associated subduction-induced "corner flow," indicate that the slab–mantle interface and shallowest parts of the subducting oceanic crust may be warmer than predicted in earlier models, perhaps at or near temperatures required for partial melting of both sedimentary and mafic rocks (see van Keken *et al.*, 2002; Rupke *et al.*, 2004; Peacock, 2003; Kelemen *et al.*, 2003; Peacock *et al.*, 2005). As discussed below, this revelation has profound implications for the nature of slab-derived "fluids" and element mobilities. Figure 3 (from Peacock, 2003) presents representative calculated *P–T* paths for subducting lithologies, demonstrating that these models are highly dependent on the ways in which mantle wedge viscosity is handled (see the discussions by van Keken *et al.*, 2002; Peacock *et al.*, 2005). On a finer scale, Dumitru (1991) matched estimated geothermal gradients for the Franciscan paleoaccretionary margin to thermal models of forearc regions, concluding that the most important variables in these models are the subduction rate, the age of the subducting slab, and frictional heating (see also Peacock, 1992).

3.4 EARLY-STAGE PROCESSING OF SEDIMENTS AND PORE WATERS IN TRENCH AND SHALLOW FOREARC SETTINGS (<15 km)

At shallow levels in subduction zones, complex physical partitioning of sediments and sedimentary rocks into off-scraped, underplated, and more deeply subducted fractions, impacts the sedimentary input contributing to deeper subduction-zone geochemistry. It appears that, in many modern subduction zones, sediments are not accreted in the forearc, but rather, subduction erosion may convey large volumes of forearc material of varying lithology to great depths (see Scholl and von Huene, in press). The deep subduction of eroded materials has not yet been adequately considered in geochemical models balancing subduction-zone inputs and outputs. During these earliest stages of subduction, sediments experience large amounts of compaction and deformation, resulting in mechanical expulsion of pore waters (see Moore and Vrolijk, 1992). Sediment porosities of >70 vol.% have been measured outboard of trenches, and for some convergent margins, initial

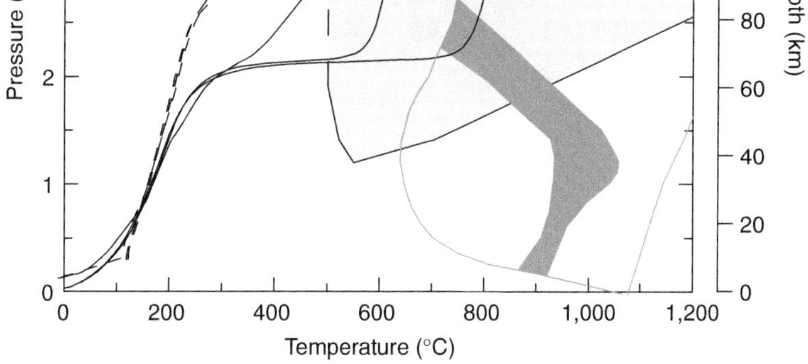

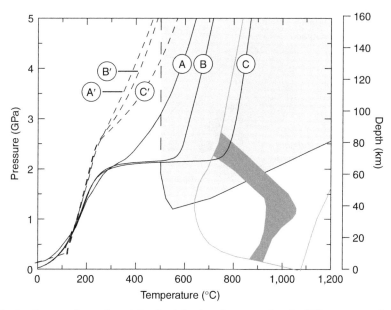

Figure 3 Synthesis of recent thermal models of subduction, incorporating, in different ways, the viscosity in the mantle wedge (figure 6 fromPeacock, 2003). The warmest path for the top of the slab (labeled, "C") is for a mantle wedge having an olivine rheology. Paths "A" and "B," both also for the top of the slab, correspond to use of isoviscous rheology, at differing model resolution (5 km for "A" and 400 km for "B," also see van Keken *et al.*, 2002; Peacock *et al.*, 2005). Paths A′, B′, and C′ (broken lines) are for the base of the slab in each corresponding model. The lighter-shaded region is the eclogite-facies stability field, the lighter-shaded, grayish lines are the water-saturated and dry partial melting reactions, and the darker-shaded region is for hornblende dehydration melting (see Peacock, 2003).

compaction and dewatering produce significant thinning beneath the forearc within a few kilometers of the trench (e.g., the seafloor sedimentary section drilled offshore of the Central American trench (Site 1039) and beneath the forearc wedge (Site 1040) by ODP Leg 170 offshore of Costa Rica; Kimura *et al.*, 1997). Apparently, this thinning of the sediment section can be accomplished without significant change in many of its trace-element concentrations (see Morris *et al.*, 2002, who presented ^{10}Be data for the young sediment beneath the décollement in the Costa Rica forearc).

3.4.1 Mineralogical and Compositional Changes During Early Subduction of Sediment

Reactions among clay minerals (smectite–illite; Srodon, 1999; Brown *et al.*, 2001; Moore *et al.*, 2001; cf. Vrolijk, 1990), particularly those producing low-grade metamorphic micas (see Frey, 1987), result in the release of significant amounts of mineral-bound H_2O (Figure 4) and produce relatively low-chloride fluids in accretionary wedges (see discussions of the influence of reactions among smectite and illite by Kastner *et al.*, 1991, Brown *et al.*, 2001, and Moore *et al.*, 2001). At shallow levels of accretionary prisms, this metamorphic fluid is likely to be dwarfed by the large amounts of expelled pore waters (see Moore and Vrolijk, 1992); however, the loss of structurally bound H_2O during clay mineral transitions to low-grade micas has important implications for the more deeply subducted H_2O budget dominated by the hydroxyl contents of minerals such as the micas (see Rupke *et al.*, 2002, 2004). Clay minerals also host many of the trace elements of most interest to those studying arc lava "slab additions" (boron, lithium, nitrogen, rubidium, cesium, and barium; see discussion below of sediment versus metasedimentary Rb/Cs). Illite, in particular, is an initial primary host for LILE such as cesium, barium, and rubidium, and for nitrogen and boron (for discussion of the incorporation of

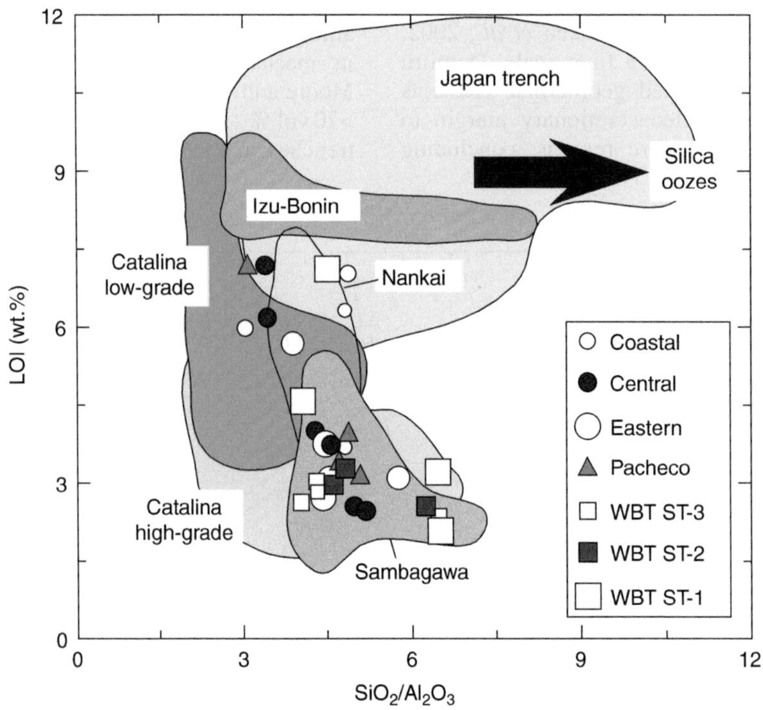

Figure 4 LOI and SiO_2/Al_2O_3 for HP paleoaccretionary metasedimentary rocks and their likely seafloor sedimentary equivalents (figure from Sadofsky and Bebout, 2003). Note that LOI correlates with SiO_2/Al_2O_3 for the Catalina Schist (data from Bebout *et al.*, 1999), Franciscan Complex Coast Ranges, California (Coastal, Central, and Eastern Belts; data from Sadofsky and Bebout, 2003), Diablo Range, California (Pacheco Pass; data from Sadofsky and Bebout, 2003), Western Baja Terrane (Baja, Mexico; data from Sadofsky and Bebout, 2003), and Sambagawa Belt (Japan; data from Nakano and Nakamura, 2001). Catalina Schist samples are broken down into low-grade facies (lawsonite-albite and lawsonite-blueschist) and high-grade facies (epidote-blueschist, epidote-amphibolite, and amphibolite). Representative marine sediments from the Japan Trench (data from Murdmaa *et al.*, 1980), Nankai trench (data from Pickering *et al.*, 1993), and Izu-Bonin forearc (data from Plank and Langmuir, 1998) are higher in LOI than the metamorphosed sediments, reflecting their higher but varying clay mineral contents. The extremely high-LOI, high-SiO_2/Al_2O_3 samples are probably due to dilution of the greywacke signal by pelagic siliceous oozes.

organic nitrogen and boron into diagenetic illite, see Compton *et al.*, 1992 and Williams *et al.*, 2001, respectively). Other dehydration reactions in shallow parts of accretionary prisms include the transformation of opal-A (which contains ~11 wt.% H_2O) to opal-CT (opal which shows the beginning of formation of domains of short-range ordering) to quartz (Moore and Vrolijk, 1992) as siliceous microfossil skeletons recrystallize. The smectite–illite and opal–quartz transformations occur over a temperature range of ~60–160 °C, depending on heating rates (Moore and Vrolijk, 1992).

Shallow reactions involving organic matter appear to be responsible for the release of large amounts of methane (biogenic methane and methane produce abiologically by heating) in many forearcs (see Moore and Vrolijk, 1992). However, the similarity in the reduced (organic) C concentrations and isotopic compositions (and C/N ratios) between sediments outboard of trenches and those in forearc metasedimentary rocks (representing subduction to 5 km or more) indicates small amounts of loss of reduced C over large volumes of sediment to produce the methane that is observed to seep out along structures. Diagenetic conversion of organic nitrogen into ammonium in clay minerals, such as illite, begins on the seafloor (see Sadofsky and Bebout, 2004a; Li and Bebout, 2005) and probably continues during very shallow subduction. This transfer of organic nitrogen into clays, and then into low-grade micas, appears to be remarkably efficient, as the C/N and nitrogen concentrations and isotopic compositions in forearc metasedimentary rocks subducted to depths of up to ~40 km are similar to those in their seafloor sediment protoliths (see Bebout, 1995; Sadofsky and Bebout, 2003).

Small amounts (< 5 wt.%) of finely disseminated calcite occur in metasedimentary rocks in the Coastal Belt of the Franciscan Complex, California (representing sediments underthrust to ≤ 5 km depths; see Blake *et al.*, 1987), but are absent in somewhat higher-grade units representing subduction to up to 30 km (Sadofsky and Bebout, 2003). This difference could have resulted from minor decarbonation or dissolution of calcite in aqueous fluids, producing not only a decrease in CO_2 content, but also decreases in whole-rock CaO and strontium. Moreover, loosely adsorbed boron, with relatively high-$\delta^{11}B$, is believed to be lost during early-stage mineral reactions and pore water expulsion, resulting in a decrease in whole-rock $\delta^{11}B$ of subducting sediments (see discussion by Bebout and Nakamura, 2003). Sadofsky and Bebout (2003) argued for a similar loss of small amounts of nitrogen with relatively high $\delta^{15}N$ to mechanically expelled pore waters at low grades in the Franciscan Complex and suggested that this was more loosely adsorbed nitrogen, perhaps occurring as nitrate.

3.4.2 Records of Fluid and Element Mobility in Modern and Ancient Accretionary Prisms

There is a wealth of geochemical data for pore fluids in accretionary prism sediments and more deformed zones, including décollement. These provide a record of element mobilization, in some cases related to reactions of the types discussed above that produce low-salinity waters, which locally dilute the more saline pore waters (Elderfield *et al.*, 1990; Kastner *et al.*, 1991; Moore and Vrolijk, 1992; also see discussions of forearc mud volcanoes in You *et al.*, 2004; Godon *et al.*, 2004). In addition to the dehydration or breakdown of clay minerals and opal-A (amorphous opal), influences on pore water chlorinity include the dissociation of methane gas hydrates (clathrates; see Kvenvolden, 1993), clay membrane ion filtration, biogenic or thermogenic combustion of organic matter, dissolution of minerals, mixing of fluids from different sources, and decarbonation reactions and related reduction in CO_2 content. Fluids sampled in the décollement on ODP drilling legs commonly have decreased chlorinity, variable $^{87}Sr/^{86}Sr$, $\delta^{18}O$, and Sr concentration, increased methane, silica content, and Mg^{2+}/Ca^{2+}, and decreased Ca^{2+} content relative to fluids away from such structures (Kastner *et al.*, 1991; cf. Elderfield *et al.*, 1990; for discussion of trace elements in pore waters in and away from décollement see Zheng and Kastner, 1997). Solomon *et al.* (2006) documented liberation of barium from barite at shallow levels beneath the décollement in the Costa Rica forearc (ODP Site 1040), and discussed the possible implications of this barium release for deeper subduction barium cycling. During the earliest stages of subduction, a large fraction of the sediment barium inventory can be tied up in the sulfate mineral barite, and sulfate depletion releases this fraction leaving only the barium component contained in the silicate and carbonate fractions of the sediment.

Vein networks in exposures of paleoaccretionary rocks formed at <15 km depths record large-scale fluid-mediated mass (and heat) transfer, and more specifically, document the mobility of methane-bearing, low-chlorinity, H_2O-rich fluids (see Moore and Vrolijk, 1992; Fisher, 1996). The methane is considered to be thermo- or biogenic in origin, and these sources can be distinguished on the basis of C isotope signatures (Ritger *et al.*, 1987). The specifics of vein formation, and the origin of the vein minerals (generally quartz- and/or calcite-rich), for example, introduced via advection of mineral components or diffusion into fractures from nearby host rocks, are debated for individual cases (see Fisher, 1996). Nevertheless, extensive vein networks are generally considered to reflect upward migration of fluids generated by a combination of compaction and dehydration of more deeply buried sediments (see Sadofsky and Bebout, 2004b). This is supported by the relatively lower salinities and elevated methane concentrations in fluid inclusions within vein minerals (e.g., Vrolijk *et al.*, 1988), evidence consistent with the "freshening" observed in faults and décollement at shallower levels (Brown *et al.*, 2001; Moore *et al.*, 2001). Based partly on the strontium-isotope compositions of veins, Sample and Reid (1998) argued that some of the fluids originate in the subducting slab (cf. Breeding and Ague, 2002).

Serpentinite seamounts in the Marianas margin are produced by the release of fluids from subducting sediments and altered oceanic crust (AOC) in shallow parts of the forearc and the reaction of these fluids with the forearc ultramafic mantle wedge. In addition, mafic blueschist-facies clasts in these settings provide a confirmation of the *P–T* estimates from thermal models of the modern Marianas slab interface (see Maekawa *et al.*, 1993), and a glimpse at the subducting oceanic crust from these depths in the forearc that complements the field record of mafic rocks in paleoaccretionary complexes such as the Franciscan Complex. Studies examining the major and trace-element and isotopic compositions of these serpentinite muds (and ultramafic clasts contained therein) in these seamounts have attempted to estimate the compositions of the fluids that generated the serpentinite at depth and have invoked additions of trace elements such as boron, lithium, arsenic, and cesium to these fluids from subducting oceanic crust and sediment (see Ryan *et al.*, 1996; Benton *et al.*, 2001, 2004; Savov *et al.*, 2005). One series of seamounts in the Marianas forearc allows an estimate of progressive fluid and element loss as a function of distance from the trench (depth to the slab; Mottl *et al.*, 2004) and provides a link with studies of devolatilization and fluid flow in paleoaccretionary complexes derived from these same depths. Mottl *et al.* (2004) interpreted trends in pore water Na/Cl, B, Cs, and especially K and Rb, across this suite of seamounts (Figure 5) as reflecting their derivation from the top of the subducting plate, driven by increasing temperature, and suggested that the

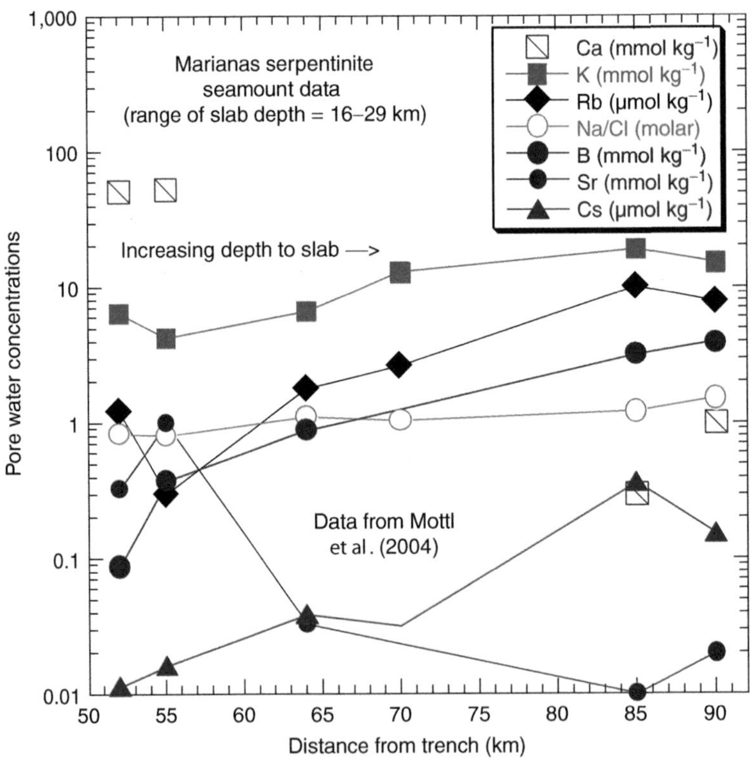

Figure 5 Variations in pore water chemistry in Marianas forearc serpentinites, as a function of distance from the trench, thought to represent progressive element loss from the subducting slab at 16–29 km depths. A number of compatibilities exist between these data and the data for paleoaccretionary rocks in the Catalina Schist and Franciscan Complex (both in California), Western Baja Terrane (Baja, Mexico), and Sambagawa (Japan), including the evidence for Sr loss at shallow levels, and increasing metamorphic losses of B, Rb, and Cs with increasing depth in the forearc. Vein geochemistry and, for B and Sr, whole-rock trends demonstrate the mobility of these elements in the forearc metasedimentary suites representing 5–30 km (see Bebout *et al.*, 1999; Nakano and Nakamura, 2001; Sadofsky and Bebout, 2003). Data from Mottl *et al.* (2004).

trends reflect dominantly upward flow with minimal lateral mixing. The range of depth inferred for the top of the slab beneath these seamounts (16–29 km) falls within the depth ranges estimate for peak metamorphism of the Franciscan Complex, Western Baja Terrane, and Catalina Schist, for which the chemical effects of metasedimentary devolatilization have been determined (Bebout *et al.*, 1993, 1999; Sadofsky and Bebout, 2003). The mobility of the elements deduced from these studies is compatible with the results of sediment-fluid experiments of You *et al.* (1996), indicating enhanced mobility of boron, NH_4, arsenic, beryllium, cesium, lithium, lead, and rubidium in forearc fluids. However, Bebout *et al.* (1999) and Sadofsky and Bebout (2003) demonstrated that, in cool subduction zones represented by most of the forearc metasedimentary suites, these fluid-mobile elements (boron, cesium, arsenic, and antimony) are largely retained in metasedimentary rocks subducted to depths of up to 40 km at concentration levels similar to those in seafloor sediments. Thus, it appears that, as for organic carbon and nitrogen, the mobility of these elements at shallower levels represents a relatively small proportion of the inventory found in sediment in some forearcs.

3.5 FOREARC TO SUBARC CHEMICAL CHANGES IN HIGH-*P/T* METAMORPHIC SUITES (15–100 km)

3.5.1 Lithologies

Subduction zone HP to UHP metamorphic suites contain a potpourri of variably metamorphosed mafic rocks crudely corresponding in chemical composition to altered seafloor basalt and gabbro and, in many cases, even preserving pillow and other fabrics indicative of submarine extrusion (Barnicoat, 1988; Kullerud *et al.*, 1990). The structural setting of these rocks varies widely from exposure in relatively intact ophiolitic sequences, such as in the Tethyan ophiolites in the Western Alps, to occurrences as blocks of meter to kilometer sizes occurring within mélange with large sedimentary and/or ultramafic components (such as in the Franciscan, Japan, the Cyclades, northern Turkey, Sulawesi, and New Caledonia; see Moore, 1984; Cloos, 1986; Dixon and Ridley, 1987; Sorensen and Barton, 1987; Bebout and Barton, 1993; Okamura, 1991; Parkinson, 1996; Breeding *et al.*, 2004; Altherr *et al.*, 2004; Spandler *et al.*, 2004). Whereas some of the mafic rocks appear to have derived, at least in part, from gabbroic protoliths developed in "Atlantic-type" seafloor systems such as those exposed in the Alps (see Tricart and Lemoine, 1991; Chalot-Prat, 2005), the more dismembered tectonic blocks in circum-Pacific complexes are generally metabasaltic, with mid-ocean ridge basalt (MORB)-like or, to a lesser extent, island-arc basalt (IAB)-, or ocean-island basalt (OIB)- like chemical characteristics. These tectonic blocks are commonly rimmed by what are interpreted as being metasomatic rinds, representing chemical interactions between the blocks, their metasedimentary or ultramafic surroundings, and metamorphic fluids (e.g., Moore, 1984; Sorensen and Grossman, 1989, 1993; Sorensen *et al.*, 1997; Breeding *et al.*, 2004; Altherr *et al.*, 2004). Some low-grade metasedimentary mélanges contain mafic blocks showing a range of peak metamorphic grade (see Cloos, 1986). Thus, the blocks and mélange matrix cannot be assumed to have achieved peak metamorphism at the same pressure and temperature conditions. Metabasaltic blocks can also occur within quartzofeldspathic gneiss or marble (e.g., Trescolmen, Adula Nappe, Zack *et al.*, 2002; for discussion of suites representing subduction of continental crust, such as the Dabieshan-Su-Lu Belt, see Li *et al.*, 2000). Given the wide range of textural settings and presumed origins of the mafic rocks in HP and UHP metamorphic suites, and the huge range in chemical compositions of these lithologies and their presumed seafloor protoliths, it can be difficult to identify chemical change that is unequivocally related to prograde subduction history. Furthermore, for the tectonic blocks floating in a chemically disparate mélange matrix with which they have metasomatically exchanged, it can be difficult to infer geochemical evolution in coherent oceanic crustal slabs experiencing prograde metamorphism in subduction zones. Less attention has been paid to the study of subduction-related chemical alteration in the UHP mafic eclogites in the Dabieshan-Su-Lu Belt and Kokchetav Massif representing deep subduction of more continental lithological packages. However, Yamamoto *et al.* (2002) provide a preliminary analysis of element loss in Kokchetav rocks that were metamorphosed at $\leq$6.0 GPa and 950 °C. Generally, the eclogites in these terranes are regarded as being more "continental," not obviously having had MORB protoliths (see Jahn, 1999). It is possible that detailed study of the prograde chemical evolution (evolution during the increases in pressure and temperature related to underthrusting) of these eclogites will provide insights into the evolution of similar lithologies subducted in ocean–ocean settings.

Largely metasedimentary paleoaccretionary complexes, such as those in the circum-Pacific, probably do not sample the lower, more pelagic part of the sediment sections (deposited in the deep sea far from the trench, at relatively low sedimentation rates), thus biasing the field record relative to what is/was more deeply subducted. Instead, these complexes are dominantly terrigenous sediment derived from arcs and continental crust that were deposited into the trench (see Sadofsky and Bebout, 2004b). In some forearcs, sediment sections below the décollement have been drilled (e.g., ODP Site 1040; Kimura *et al.*, 1997; Morris *et al.*, 2002), providing information on the earliest diagenetic processes that occur as pelagic sediments enter subduction zones. In more coherent exposures (not obviously representing tectonic blocks in melange) of HP and UHP metaophiolite exhumed during initial collision (e.g., the Jurassic ophiolites of the Western Alps; see Jolivet *et al.*, 2003), it is possible to identify related seafloor sediment. One example is the siliceous and Mn- and carbonate-rich metasediment in contact with the Jurassic ophiolite at the UHP Lago di Cignana locality (see Reinecke, 1998).

3.5.2 Prograde Metamorphic Reactions and Overprinting During Exhumation

Both continuous and discontinuous metamorphic reactions occur along *P–T* paths traversing the lawsonite–blueschist, epidote–blueschist and eclogite facies. It is not possible to do justice to this topic in this chapter; however, provides an excellent synthesis of the field, experimental, and theoretical constraints on prograde metamorphic reaction history in subducting mafic and sedimentary lithologies. Several recent studies calculated phase equilibria in KMASH and NKMASH for HP and UHP metapelitic rocks (see Proyer, 2003; Wei and Powell, 2004; also see Liou *et al.*, 1987; El-Shazly and Liou, 1991), and Proyer (2003), in particular, considered the extent of exhumation-related reaction expected for various metapelitic bulk compositions.

Interpreting each HP and UHP suite in terms of subduction history carries with it its own challenges associated with tectonic and lithologic complexity. One key factor when selecting suites best suited for geochemical work is the thermal history that the rocks experienced during exhumation (see the *P–T* diagram in Figure 6). Extents of exhumation-related overprinting are, in general, related to the degrees of heating during decompression, the duration of any heating (or cooling) during this decompression, the volatile content of the rocks prior to exhumation, and the bulk composition of the rocks. The latter dictates whether or not certain decompression-related reactions occur (see Proyer, 2003). The greatest prospects for survival of prograde reactions and geochemistry are found in mafic rocks that experienced relatively rapid, down-T exhumation paths. Such paths are commonly attributed to exhumation during continued, active underthrusting to explain the sustained high-*P–T* conditions. For a number of HP and UHP suites, metasedimentary rocks experience far greater overprinting during exhumation than adjacent metabasaltic rocks (Reinecke, 1998; van der Klauw *et al.*, 1997; Fitzherbert *et al.*, 2005). This is because the metasedimentary rocks show greater pervasive deformation and open-system behavior than the metabasalts and experience different decompression-related mineral reaction histories. Miller *et al.* (2002) (also see van der Klauw *et al.*, 1997) provide a summary of fluid processes associated with the exhumation of HP and UHP metamorphic rocks.

3.5.3 Geochemical Studies of HP and UHP Mafic Rocks

A number of subduction-zone processes could influence the compositions of subduction zone HP and UHP mafic lithologies relative to those of AOC drilled some distance from the trench (e.g., Sites 801, 1149, 504B). These include hydration and metasomatism related to plate bending in the trench (see the discussion by Ranero *et al.*, 2005), volume strain (involving selective element removal; see Bebout and Barton, 2002), devolatilization and other fluid–rock interactions along the slab–mantle interface (including interaction with sediment-derived fluids, development of metasomatic rinds on tectonic blocks), and for the higher grade lithologies, extraction of partial melts. Recent work that has attempted to identify chemical signatures of subduction-zone metamorphism in HP and UHP metabasaltic and metagabbroic rocks, and the chemical contributions of subducted oceanic crust to arc source regions and the deeper mantle, is briefly mentioned here. Slab additions to the mantle wedge beneath arcs are generally presumed to occur via a "fluid" phase, variably described as being hydrous fluid, silicate melt, or supercritical liquid (see Kessel *et al.*, 2005, and references therein). Study of the geochemistry of subduction zone HP and UHP mafic rocks thus provides insight into slab fluid production and mobility, and possible element loss in these fluids.

3.5.3.1 Whole-rock geochemistry of HP and UHP eclogites

One of the largest obstacles to demonstrating chemical fluxes in subduction-related metabasalts, relative to their protoliths, is the need to identify variations caused by either magmatism or seafloor alteration (see chapter 18). Generally, the first step in geochemical studies of mafic HP and UHP metamorphic rocks is to compare the major and trace-element compositions of these rocks to those of seafloor basalts. This is done with varying rigor, involving comparisons of the compositions of the HP/UHP rocks with compositions of both fresh normal MORB (N-MORB) and enriched MORB (E-MORB), various AOC composites, and other

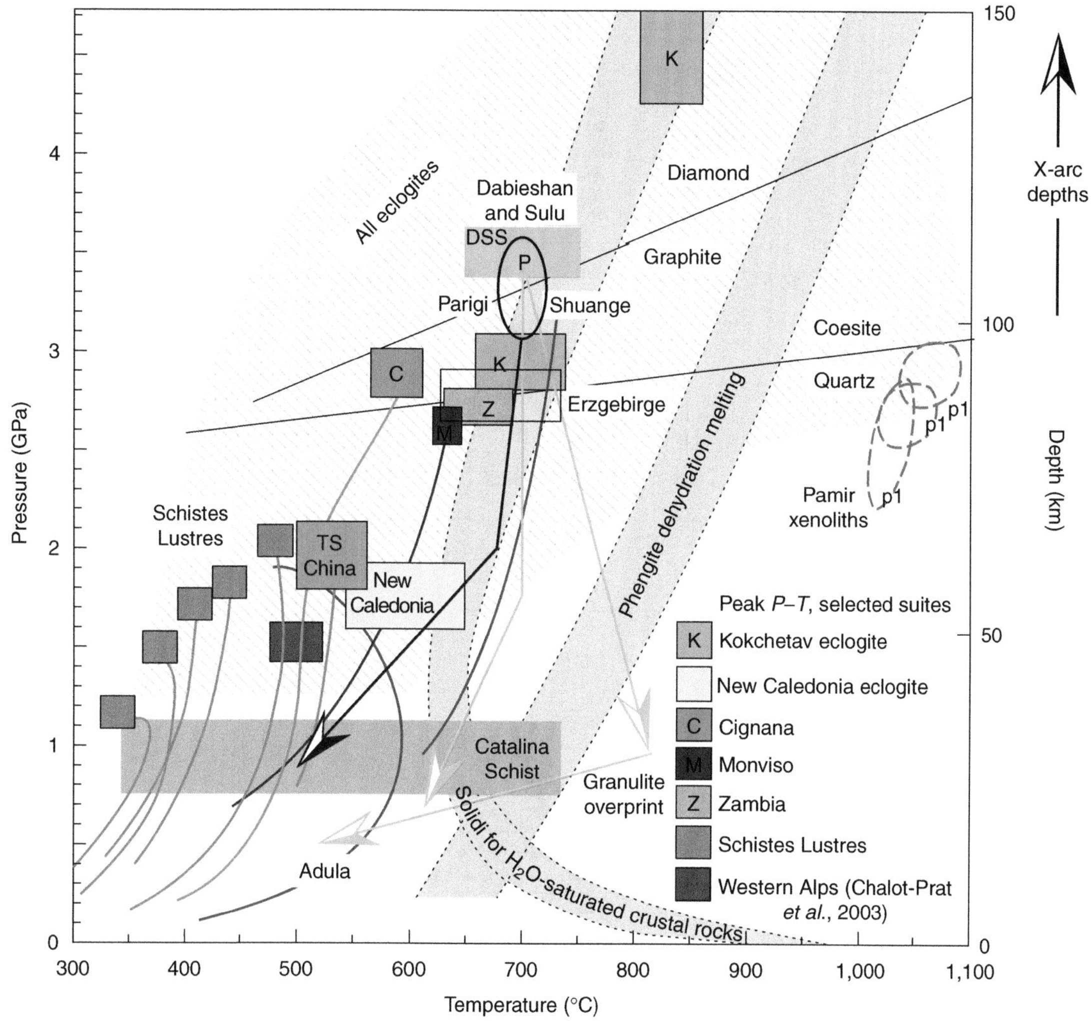

Figure 6 The peak *P–T* conditions for metamorphic suites discussed in this chapter, some representative exhumation *P–T* paths, the positions of the wet and dehydration solidi for partial melting of mafic and sedimentary lithologies, and peak metamorphic conditions of some other selected UHP suites (figure modified after Hacker, in press; also note the patterned field for "All eclogites," and the graphite–diamond and quartz–coesite equilibria). Colored rectangles are peak *P–T* conditions of the suites discussed in this chapter (mostly from Hacker, in press; references for peak *P–T* estimates provided therein; p, Pamir in dashed-oval regions; see Hacker *et al.*, 2005, for discussion of these partially melted metasedimentary rocks). For the Schistes Lustres, the multiple fields are for individual units showing a wide range in grade (from Agard *et al.*, 2002; see data for these units in Figure 12), and an estimated exhumation path is shown for each. Schematic exhumation paths for the Dabie-Sulu UHP rocks (light blue, straight-line segments) are from Zheng *et al.* (2003); also see *P–T* path for the Shuange UHP locality in Dabie. Figure 9 provides greater detail regarding the peak *P–T* and varying prograde *P–T* paths of the units of the Catalina Schist. The *P–T* for the Western Alps is for the internal units (see Chalot-Prat *et al.*, 2003), and peak metamorphic *P–T* estimates for Tianshan HP rocks are from Gao and Klemd (2001). The black, unfilled oval region is for the Parigi UHP locality of the Dora Maira Massif (see discussion in Hermann *et al.*, in press; Sharp *et al.*, 1993), and the black, unfilled rectangular region is for the Erzgebirge UHP rocks studied by Massone and Kopp (2005). The Adula rocks (exhumation path only) were investigated by Zack *et al.* (2001, 2002, 2003).

estimates. In most studies, the metamorphic rocks are deemed to be derived from seafloor basaltic or gabbroic rocks, but in some cases either OIB or IAB protoliths are considered (Becker *et al.*, 2000; John *et al.*, 2004; Saha *et al.*, 2005; Usui *et al.*, 2006). The more sophisticated approaches involve comparison of the variation among samples in a suite with the variation expected for magmatic differentiation beneath spreading centers (e.g., Chalot-Prat *et al.*, 2003; Spandler *et al.*, 2004; John *et al.*, 2004). In the Western Alps, it is possible to examine relatively coherent tracts of Jurassic ophiolites that experienced a wide range of metamorphic grade, from very low-grade external nappes that "escaped subduction" (see Chalot-Prat, 2005) to the internal nappes that experienced up to

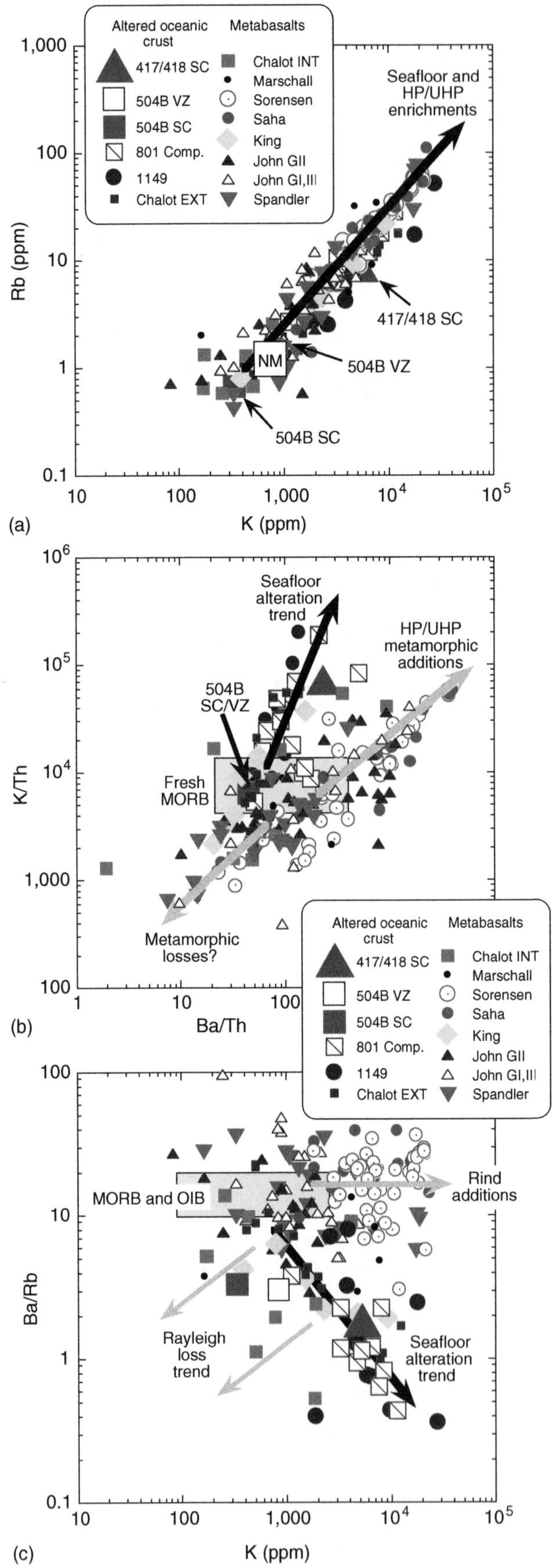
Altered oceanic crust
Metabasalts
417/418 SC
504B VZ
504B SC
801 Comp.
1149
Chalot EXT
Chalot INT
Marschall
Sorensen
Saha
King
John GII
John GI,III
Spandler
Seafloor and HP/UHP enrichments
417/418 SC
504B VZ
NM
504B SC
Rb (ppm)
K (ppm)
(a)
Seafloor alteration trend
HP/UHP metamorphic additions
504B SC/VZ
Fresh MORB
Metamorphic losses?
K/Th
Ba/Th
(b)
MORB and OIB
Rind additions
Rayleigh loss trend
Seafloor alteration trend
Ba/Rb
K (ppm)
(c)

eclogite-facies metamorphic conditions (Chalot-Prat *et al.*, 2003; van der Klauw *et al.*, 1997).

The suites that have been studied in detail, and for which data are compiled in this section, occur in very different structural–textural settings (e.g., as tectonic blocks, closely juxtaposed with mélange matrix lithologies with which they have interacted, or as more coherent and larger tracts of identifiably oceanic crustal lithologies). Fluid–rock interaction, and associated metasomatism during subduction-zone metamorphism, is expected to differ dramatically for rocks from these different settings. For the mafic blocks in mélange, it is difficult to establish the relationship between the blocks in time, space, and composition, and comparisons with protoliths are necessarily less direct (Sorensen *et al.*, 1997). In such cases, one is often concentrating on characterizing metasomatic additions (e.g., LILE) resulting from block–matrix–fluid exchange (see Sorensen and Grossman, 1989, 1993; Nelson, 1995; Bebout and Barton, 2002; Saha *et al.*, 2005) and, to a lesser extent, the possibility of element depletions related to devolatilization. Nonetheless, blocks investigated in these studies generally show geochemical affinities with MOR basalts (see figure 3 in Sorensen *et al.*, 1997), and occasionally OIB or IAB (see MacPherson *et al.*, 1990; Saha *et al.*, 2005). The evidence for element enrichments in these blocks and their metasomatic rinds provides an extremely valuable record of the mobilities of volatiles (e.g., H_2O) and major and trace elements in the HP and UHP metamorphic fluids (see Sorensen *et al.*, 1997).

In general, the HP and UHP mafic rocks scatter about average N-MORB concentrations for many moderately to highly compatible trace elements (one exception is lithium), but show large enrichments in many of the trace elements that are highly incompatible during mantle melting (cesium, rubidium, barium, uranium, thorium, lead, strontium, and potassium; the "highly nonconservative" elements of Pearce and Peate, 1995). In this way, the mafic HP and UHP rocks resemble various AOC composites (Staudigel *et al.*, 1996; Kelley *et al.*, 2003; see figure 7 in Bebout, 1995). Enrichments of these same highly incompatible trace elements in arc lavas are attributed to additions from slab-derived fluids (see Pearce and Peate, 1995; Elliott, 2003), and are the focus of the compilations and discussion in this section.

Ratios of some LILE have the potential for distinguishing between the effects of seafloor hydrothermal alteration and fluid–rock interactions during subduction-zone metamorphism (see discussion of LILE mobility by Zack *et al.*, 2001; Figure 7). Zack *et al.* (2001) suggested that it might be necessary for eclogites to be infiltrated by fluids from external sources in order to produce LILE depletions dramatic enough to be demonstrable given the protolith variability issue (see Bebout *et al.*, in press, and discussion below of how this argument pertains to metasedimentary rocks in the Catalina Schist). However, Figure 7a shows that, for the HP and UHP suites that have been studied in detail, the enrichments of potassium and rubidium due to seafloor alteration are difficult to distinguish from those due to subduction enrichments or depletions (a similar relationship exists for cesium enrichments). Any addition or subtraction of potassium, rubidium, and cesium due to the passage of fluids during HP/UHP metamorphism is obscured by this overlap with the trends for seafloor alteration. In contrast, barium is generally not as enriched as potassium, rubidium, and cesium during seafloor hydrothermal alteration (see Staudigel *et al.*, 1996; Kelley *et al.*, 2003; see Figure 2a), yet many metabasaltic rocks in HP suites have high barium concentrations. In some cases these enrichments have been attributed to HP metasomatic additions by fluids having previously interacted with sedimentary lithologies (see Sorensen *et al.*, 1997; compilation of Catalina Schist metabasaltic data in figure 7 of Bebout, 1995). Figure 7b shows relationships

Figure 7 Relationships among the LILE and Th in HP and UHP metamorphosed basalt and gabbro, compared to altered oceanic basalts. This figure illustrates the difficulties in deconvoluting seafloor and subduction-zone metamorphism effects (e.g., for K and Rb data; see (a)), and the possible use of Ba–K enrichments to distinguish between these two histories ((b) and (c)). The data plotted in these figures are from Spandler *et al.* (2004; for New Caledonia), John *et al.* (2004; for Zambia; data broken into groups, Group II and Groups I and III); King *et al.* (2004) and King (2006; for Lago di Cignana), Sorensen *et al.* (1997; tectonic blocks), Saha *et al.* (2005; tectonic blocks), Marschall *et al.* (2006; Syros), Chalot-Prat *et al.* (2003; Alps internal units), Chalot-Prat (2005, external units), Staudigel *et al.* (1996; ODP Sites 417 and 418 supercomposite), Peucker-Ehrenbrink *et al.* (2003; ODP Site 504B supercomposite and composite for volcanic zone), and Kelley *et al.* (2003; ODP Sites 801 and 1149). NM, N-MORB (from Hofmann, 1988). In (b) and (c), note the distinct trends for samples of seafloor altered basalt ("Seafloor alteration trend") and for HP and UHP additions ("Rind additions" and "HP/UHP metamorphic additions"). Note that this figure and Figure 8 do not for the most part distinguish between textural varieties and structural settings for these samples, and are intended to show gross trends only for seafloor alteration and HP/UHP metamorphic processes. On these plots, the rinds, some of which have undergone significant amounts of deformation and volume strain (see Sorensen and Grossman, 1989; Sorensen *et al.*, 1997; Bebout and Barton, 2002), are regarded as endmember cases of fluid–rock interaction (i.e., rocks known to have experienced metasomatic alteration during HP metamorphism). On (c), the trend for Rayleigh loss (labeled arrows pointing to the lower left) was calculated by Zack *et al.* (2001).

between potassium and barium (here normalized to thorium, which shows enrichment in some samples; see Figure 8d) indicating that enrichments due to hydrothermal alteration on the seafloor can be distinguished from those that occur during subduction-zone metamorphism. In these plots, some HP and UHP metamorphic samples show ratios of potassium and barium concentrations to thorium concentrations lower than those of fresh and altered MORB (and even OIB, see figure 4b in Becker *et al.*, 2000), perhaps indicating loss of potassium and barium from some samples, relative to thorium, during subduction-zone metamorphism (see Becker *et al.*, 2000; Spandler *et al.*, 2004). Becker *et al.* (2000) also noted lower potassium contents in subduction-zone metabasalts relative to seafloor basalts with similar elevation in $^{87}Sr/^{86}Sr$ attributed to seafloor alteration, and concluded that this relationship represents potassium loss. Figure 7c provides another view of the barium–potassium enrichments (and possible losses in a smaller number of samples), incorporating rubidium concentrations, and demonstrates trends of metasomatic addition (most prominently displayed by the rind data), seafloor alteration, and Rayleigh loss to dehydration fluids (the latter trends calculated by Zack *et al.*, 2001, using recently published fluid-mica partition coefficients by Melzer and Wunder, 2001). Based on a more complete consideration of the heterogeneity of oceanic crust protoliths, a consensus has recently developed that the HP and UHP metamorphic loss of elements such as barium, potassium, and uranium from such rocks is generally relatively restricted, but that significant localized losses can occur in zones of concentrated fluid–rock interaction (see Bebout and Barton, 2002; Chalot-Prat *et al.*, 2003; Spandler *et al.*, 2004, John *et al.*, 2004).

Extensive enrichment of uranium during seafloor hydrothermal alteration of oceanic crust (see Staudigel *et al.*, 1996; Kelley *et al.*, 2003) causes large fractionations of uranium from other trace elements of similar incompatibility during mantle melting (thorium, niobium, cerium, and lead). Fluxes of these elements in subduction zones are of particularinterest in studies of arc lavas and deep-mantle geochemical heterogeneity (see Section 3.8). In particular, large fluxes of uranium and lead, off the slab, are invoked to explain the elevated concentrations of these elements in arc lavas. Figures 8a and 8b illustrate the relationships among cerium and lead concentrations for MORB and OIB, variably altered seafloor basalt, and selected HP and UHP metabasaltic and metagabbroic suites. Several metasomatic rinds developed on Franciscan mafic blocks (see Saha *et al.*, 2005), and one group of eclogites for which John *et al.* (2004) invoked enhanced fluid–rock interaction (their Group IID eclogites), show some decrease in Ce concentration and associated reduction in Ce/Pb (Figure 8a). John *et al.* (2004) proposed that some combination of seafloor alteration, subduction-zone fluid–rock interactions, and possibly also exhumation effects, produced significant lead enrichment (see the trends toward the origin at low 1/Pb, one for the Group IID eclogites), and related decreased Ce/Pb in the UHP eclogites from Zambia (Figure 8b). This correlated Pb enrichment and decrease in Ce/Pb is noted for only a small number of samples of seafloor-altered basalt (possibly some samples from ODP Site 1149), and for low-grade metamorphosed seafloor basaltic rocks from the external units in the Alps (studied by Chalot-Prat, 2005, and not directly representative of seafloor altered basalt), perhaps indicating that the Pb enrichment documented by John *et al.* (2004) resulted from subduction-zone metamorphism (or postpeak metamorphic fluid–rock interactions).

The trend of uranium enrichment of AOC can be seen in Figure 8c (also see the data for AOC in this figure). A decrease in Nb/U is also observed in nearly all of the HP and UHP eclogites. Very few of the metamafic rocks contain lower uranium concentrations or higher Nb/U than that found in AOC, and it is likely that extreme loss due to metamorphism would be necessary to discern a metamorphic effect. In addition, the Th/U of AOC will be lowered due to hydrothermal alteration (see Staudigel *et al.*, 1996). Figure 8d compares Th/U of fresh MORB, AOC and seafloor sediment, and some representative metamorphic data. A number of the metamafic samples have higher Th/U than that of AOC and MORB, perhaps due to uranium loss (see the data for the Group IID eclogires of John *et al.*, 2004). Interestingly, high-grade Catalina Schist metasedimentary rocks (amphibolite facies) have Th/U higher than that of their lower-grade equivalents (in the Catalina Schist and in the Franciscan Complex and Western Baja), likely related to uranium loss with or without some thorium enrichment during partial melting reactions (see Figure 8d; Bebout *et al.*, 1999). Studies of HP mélanges have demonstrated decoupling of thorium and uranium by combinations of metasomatic and mechanical mixing (see Breeding *et al.*, 2004; King *et al.*, 2006; see below), and it is possible that these processes cause significant fractionation of these elements at depth in subduction zones.

In summary, the potassium, barium, rubidium, thorium, uranium, lead, and cerium inventories in subducting altered oceanic crust appear to be largely retained to depths of at least 90–100 km, with some potassium, barium, rubidium, lead, and thorium enrichment occurring during subduction-zone metamorphism (see Spandler *et al.*, 2004; John *et al.*, 2004; Figures 7, 8b, and 8d). These elements could thus, at greater depths, be available

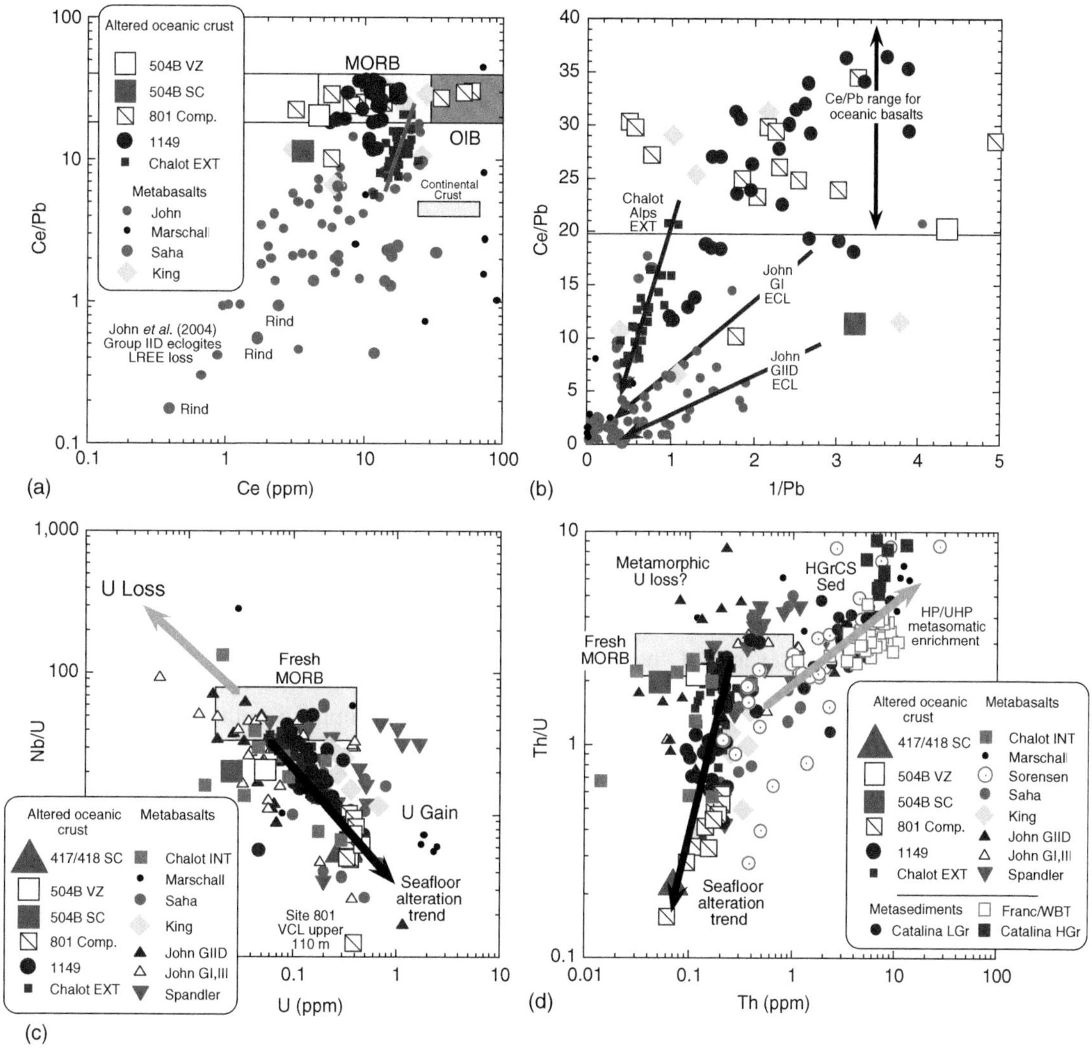

Figure 8 Relationships among Th, U, Pb, Ce, and Nb in HP and UHP metamorphosed basalt and gabbro, compared to altered seafloor basalts. Varying overlap in the two data sets produces difficulties in deconvoluting seafloor and subduction-zone metamorphism effects. In (a) samples with low Ce concentrations and correspondingly lower Ce/Pb are either metasomatic rinds on tectonic blocks (data from Saha *et al.*, 2005) or more extensively metasomatized Group II eclogites of John *et al.* (2004). In (b), many of the HP and UHP rocks show enrichments in Pb (some also with decreased Ce concentrations; see (a)) and resulting lowered Ce/Pb, and the data for some eclogites (Groups I and IID of John *et al.*, 2004; "John GI ECL" and "John GIID ECL") produce enrichment trends toward higher Pb concentrations and lower Ce/Pb, as do the very low-grade metabasalts from the Alps external units investigated by Chalot-Prat *et al.* (2003; "Chalot Alps EXT"; ophiolitic and not directly representative of seafloor basalts; trend indicated by light blue line). Note the nearly complete overlap in Nb–U data for the seafloor altered basalts and HP and UHP metamorphosed equivalents (c). In the plot of Th concentrations versus Th/U in (d), seafloor altered basalts show a distinct trend toward lower Th resulting from U enrichment relative to Th, whereas metasomatic rinds (some of the data from "Sorensen" and "Saha") are enriched in Th relative to seafloor altered basalts. Also shown in (d) are Th–U data for low- and high-grade metasedimentary rocks of the Catalina Schist (data from Bebout *et al.*, 1999). High-grade Catalina Schist metasedimentary rocks have higher Th/U than lower-grade equivalents, likely related to U loss with or without some Th enrichment during partial melting. The data plotted in these figures are from Spandler *et al.* (2004; for New Caledonia), John *et al.* (2004; for Zambia; data broken into groups, Group II and Groups I and III); King *et al.* (2004) and King (2006; for Lago di Cignana), Sorensen *et al.* (1997; tectonic blocks and rinds), Saha *et al.* (2005, tectonic blocks and rinds), Marschall *et al.* (2006; for Syros), Bebout *et al.* (1999; Catalina Schist metasedimentary rocks, broken into low-grade (LGr) and high-grade (HGr)), Sadofsky and Bebout (2003; for Franciscan Complex and Western Baja Terrane, metasedimentary rocks), Chalot-Prat *et al.* (2003; for Alps internal units), Chalot-Prat (2005; for Alps external units), Staudigel *et al.* (1996; ODP Sites 417 and 418 supercomposite), Peucker-Ehrenbrink *et al.* (2003; ODP Site 504B supercomposite and composite for volcanic zone), and Kelley *et al.* (2003; ODP Sites 801 and 1149). NM, N-MORB (from Hofmann, 1988). In (d) note the distinct trends for samples of seafloor altered basalt ("Seafloor alteration trend") and for HP and UHP additions ("HP/UHP metamorphic additions").

in the generation of arc lavas and perhaps partly subducted into the mantle beyond subarc depths. There is some indication that uranium, barium, potassium, and LREE are lost from some rocks during dehydration in the deep forearc, perhaps particularly in zones of higher fluid flux (see Figures 7b, 8a, and 8d; cf. Becker *et al.*, 2000; Spandler *et al.*, 2004; John *et al.*, 2004). Small degrees of loss (e.g., those proposed by Zack *et al.*, 2001) could be obscured by the large compositional ranges that are observed in the protoliths. The relationships in Figures 7 and 8 highlight the need to assess degrees of element mobility through careful field sampling and petrologic study, such as in the recent study of John *et al.* (2004), who demonstrated fractionation of certain trace elements in vein envelopes representing zones of higher fluid flux.

3.5.3.2 Summary and remaining important issues

This section highlights several of the more significant conclusions reached through geochemical study of subduction-zone chemical cycling based on certain HP and UHP eclogites, and includes brief discussion of a couple of the most limiting remaining uncertainties in reconstructing the chemical effects of prograde subduction on oceanic crust based on study of such rocks. Perhaps the, when the range of possible protoliths is fully considered important conclusion to be drawn from study of subduction-zone eclogites is that, when the range of possible protoliths is fully considered, most show little evidence for loss of trace elements for which slab additions to arc source regions are invoked (e.g., boron, lithium, cesium, barium, uranium, and thorium; see Chalot-Prat *et al.*, 2003; Spandler *et al.*, 2004; Philippot and Selverstone, 1991; John *et al.*, 2004; Marschall *et al.*, 2006). Similarly, such rocks commonly show little or no stable isotope or fluid inclusion evidence for open-system behavior (Nadeau *et al.*, 1993; Barnicoat and Cartwright, 1995; Cartwright and Barnicoat, 1999) and instead show the preservation of isotopic signatures obtained on the seafloor. John *et al.* (2004) suggested some localized element mobility, perhaps related to high-fluid flux zones, which resulted in fractionation of light rare earth element (LREE) from heavy rare earth element (HREE) and high-field strength element (HFSE) in some eclogites. Moreover, for lower-P paleoaccretionary systems (e.g., Catalina Schist), fluid and element mobility is heterogeneous, and enhanced in zones of higher permeability related to deformation (see Bebout, 1991b, 1997), with the more coherent nonmélange zones appearing to have behaved as closed systems.

Many HP and UHP eclogites in fact show enrichments in the elements thought to be removed from subducting oceanic crust beneath arcs. Some mafic blocks contained in metasedimentary or metaultramafic mélange experienced hydration and alteration of some of their major and trace-element and isotope signatures due to transfer from the surrounding sedimentary and ultramafic lithologies (Bebout and Barton, 1989, 1993; Bebout, 1997; Sorensen and Grossman, 1989, 1993; Nelson, 1995; Sorensen *et al.*, 1997; Zack *et al.*, 2001; Saha *et al.*, 2005). The element enrichments in metasomatic rinds, and veins and their metasomatic envelopes, serve as excellent records of the relative mobility of major and trace elements in the HP and UHP fluids.

When loss-on-ignition (LOI) data (presumably representing H_2O contents) are compared for blueschist- and eclogite-facies rocks, as expected, eclogite-facies rocks do generally have lower H_2O contents consistent with significant H_2O loss over the blueschist–eclogite transition. Perhaps more interesting than this systematic difference, explained simply by the less hydrous mineral assemblages, is the question of whether or not HP mafic rocks are hydrated during early stages of metamorphism i.e., whether their H_2O contents are, at least at shallow levels of subduction zones, higher than those of their seafloor protoliths. As with the comparisons of major and trace-element, and isotope compositions, the problem in testing this hypothesis relates to the large variability in composition of the presumed seafloor protolith sections (see Staudigel *et al.*, 1996).

These several conclusions point to what is certainly one of the key remaining questions pertaining to the use of the geochemistry of HP and UHP eclogites to reconstruct deep subduction-zone chemical evolution, that is, the extent to which the compositions of these eclogites are indeed representative of processes operating "normally" along the slab–mantle interface. Whereas a number of the ophiolitic sequences in the Italian and French Alps (and elsewhere) are relatively coherent and obviously representative of seafloor basalt, mafic blocks in mélanges do not obviously represent samples of the intact subducted slab and may show enhanced metasomatism due to their occurrence in zones of high fluid flux. Unvertainty remains regarding the volumetric significance of mélange-like zones at depth in modern subduction zones.

Yamamoto *et al.* (2002) compared the major and trace-element compositions of UHP Kokchetav Massif metabasites, ranging in grade from amphibolite and quartz–amphibole eclogite, to coesite–zoisite eclogite and diamond eclogite (P–T for the latter two groups of samples are 2.8–3.1 GPa, 660–740 °C, 6.0 GPa, 950 °C), and found loss of LILE with increasing metamorphic grade. A more thorough examination of these and other eclogites representing such extreme metamorphic conditions is warranted. As noted above (see Figure 6), a number of UHP suites record peak *P–T* conditions that straddle or are above the H_2O-saturated solidi, and could provide insight into the chemical effects of melt release that may occur beneath some modern arcs (see the thermal models in Figure 3; data presented by Becker *et al.*, 2000 for rocks with peak *P–T* above the wet solidi).

3.5.4 Geochemical Studies of HP and UHP Metasedimentary Rocks

HP and UHP metasedimentary rocks are the subject of fewer geochemical studies than corresponding mafic rocks, as most occur in paleoaccretionary complexes, representing relatively shallow levels of accretionary systems. Moreover, metasedimentary suites derived from deeper regions of subduction zones show more extensive retrograde overprints related to exhumation than mafic rocks in the same suite (see Proyer, 2003; Fitzherbert *et al.*, 2005), thus complicating efforts to identify changes due to prograde, subduction-zone metamorphism.

The Catalina Schist, and similar forearc metasedimentary rocks in the Franciscan Complex and Western Baja Terrane, have been extensively studied in order to define the degree of devolatilization and related chemical change that occur during subduction to depths of ~40 km. With the exception of the higher-grade units in the Catalina Schist (epidote-amphibolite and amphibolite facies), these rocks experienced subduction along a range of relatively cool *P–T* paths like those experienced in thermally mature modern subduction zones (Sadofsky and Bebout, 2003; Figure 9). Metasedimentary rocks in these suites are compositionally similar to the terrigenous component in GLOSS (Figure 10), which constitutes ~76 wt.% of this composite (Plank and Langmuir, 1998). Although these HP metasediments have much lower LOI (mostly representing H_2O contents;

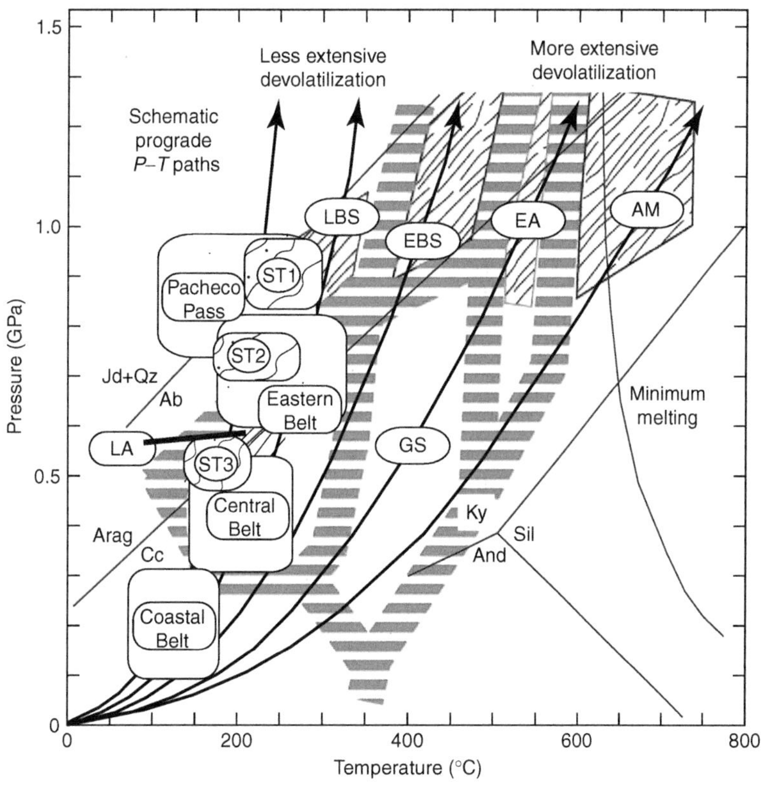

Figure 9 *P–T* diagram showing the peak metamorphic conditions and prograde paths of the units of the Catalina Schist (from Grove and Bebout, 1995; see sources of metamorphic facies stability fields and some key reactions therein). Also shown are the estimates of peak metamorphic conditions of the units of the Coast and Diablo Ranges Franciscan Complex, California, and the Western Baja Terrane (Mexico; see Sadofsky and Bebout (2003) for sources of *P–T* data). Note that the higher-grade units of the Catalina Schist (epidote-amphibolite (EA) and amphibolite (AM) were peak-metamorphosed at anomalously high temperatures relative to those in most modern subduction zones (see review of recent thermal models by Peacock, 2003), whereas the low-grade units of the Catalina Schist (lawsonite-albite (LA), lawsonite-blueschist (LBS), and even epidote-blueschist (EBS)), and the various units of the Franciscan Complex in the Coast and Diablo Ranges, California ("Coastal Belt," Central Belt," "Eastern Belt," and "Pacheco Pass"), and in the Western Baja Terrane (Baja, Mexico; "ST1," "ST2," and "ST3"), are more compatible in their peak *P–T* with conditions in modern margins.

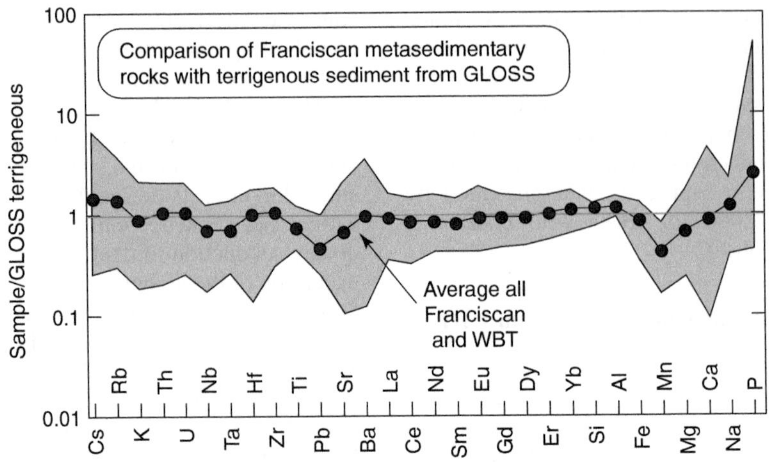

Figure 10 Plot of trace elements, in order of increasing compatibility during partial melting in the mantle (to the right; normalized to the GLOSS composite of Plank and Langmuir, 1998), for metasedimentary rocks of the Franciscan Complex and Western Baja Terrane (from Sadofsky and Bebout, 2003). The data are normalized to the mean composition of "terrigenous sediment" within GLOSS.

see Figure 4) than those of seafloor sediments, they retain much of their original inventories of even relatively fluid-mobile elements (cesium, boron, arsenic, and antimony) except in tectonometamorphic units in the Catalina Schist representing prograde metamorphism along prograde *P–T* paths warmer than those experienced in most modern subduction zones (Figure 11, Sadofsky and Bebout, 2003).

The protoliths of the Schistes Lustres, exposed in NW Italy, are believed to represent deep-sea sediment somewhat similar to the section at ODP Site 765, containing a significant carbonate component (Banda forearc; see Deville *et al.*, 1992; Agard *et al.*, 2002; and references therein). Together with the UHP metasedimentary rocks at Lago di Cignana (Reinecke, 1998), they have been studied to reconstruct their devolatilization histories to depths of ~90 km in relatively cool subduction zones (see Busigny *et al.*, 2003; Bebout *et al.*, 2004). Busigny *et al.* (2003), who measured their LILE concentrations, nitrogen concentrations, and isotope compositions, determined that the Schistes Lustres, and even the Lago di Cignana rocks (representing somewhat greater depths at higher temperatures) resemble their presumed sedimentary protoliths. These authors concluded that, in the relatively cool subduction zone that produced these rocks, the original element concentrations and the nitrogen-isotope compositions were preserved to depths of ~90 km. Mahlen *et al.* (2005) concluded that prograde metamorphism did not noticeably change the rare earth element (REE) concentrations or disrupt the Rb–Sr systematics in similar rocks. One important issue, however, in the work on this suite, and most other UHP metamorphic suites, concerns the effects of exhumation history on the prograde records of devolatilization (see Bebout *et al.*, 2004 and discussion of exhumation history by Agard *et al.*, 2002, in which argon data indicate some neoblastic mica growth during retrogression). For the traverse of the Schistes Lustres studied by Busigny *et al.* (2003), and in Lago di Cignana metasedimentary rocks, ion microprobe (secondary ion mass spectrometry (SIMS)) analyses of micas showing textures and mineral chemistry consistent with prograde growth show only subtle trends in relatively fluid-sensitive element concentrations and ratios (e.g., B/Li) as a function of increasing metamorphic grade (Figure 12, from Bebout *et al.*, 2004), an observation crudely consistent with the conclusions of Busigny *et al.* (2003). For the Cignana metasedimentary rocks, Reinecke (1998) concluded that little or no record is preserved of the prograde history, and that what exists is in the form of mineral inclusions and mineral chemistry in a small number of particularly "robust" phases (garnet, tourmaline, zircon, apatite, and dolomite). The lack of an obvious modification in whole-rock trace-element concentrations, despite significant overprinting during exhumation, would imply that the rocks behaved largely as closed systems during this later history, experiencing only redistribution of trace elements among the minerals in the evolving mineral assemblages. Boron-isotope analyses of Cignana metasedimentary tourmalines containing coesite inclusions identified zones near tourmaline rims that are clearly exhumation-related and possibly represent open-system behavior (Bebout and Nakamura, 2003).

The similarity in Cs/Rb of seafloor sediment, including Sites 765 and 1149 (sampling seafloor sediments over the full range of terrigenous, pelagic, and carbonate compositions), the Schistes

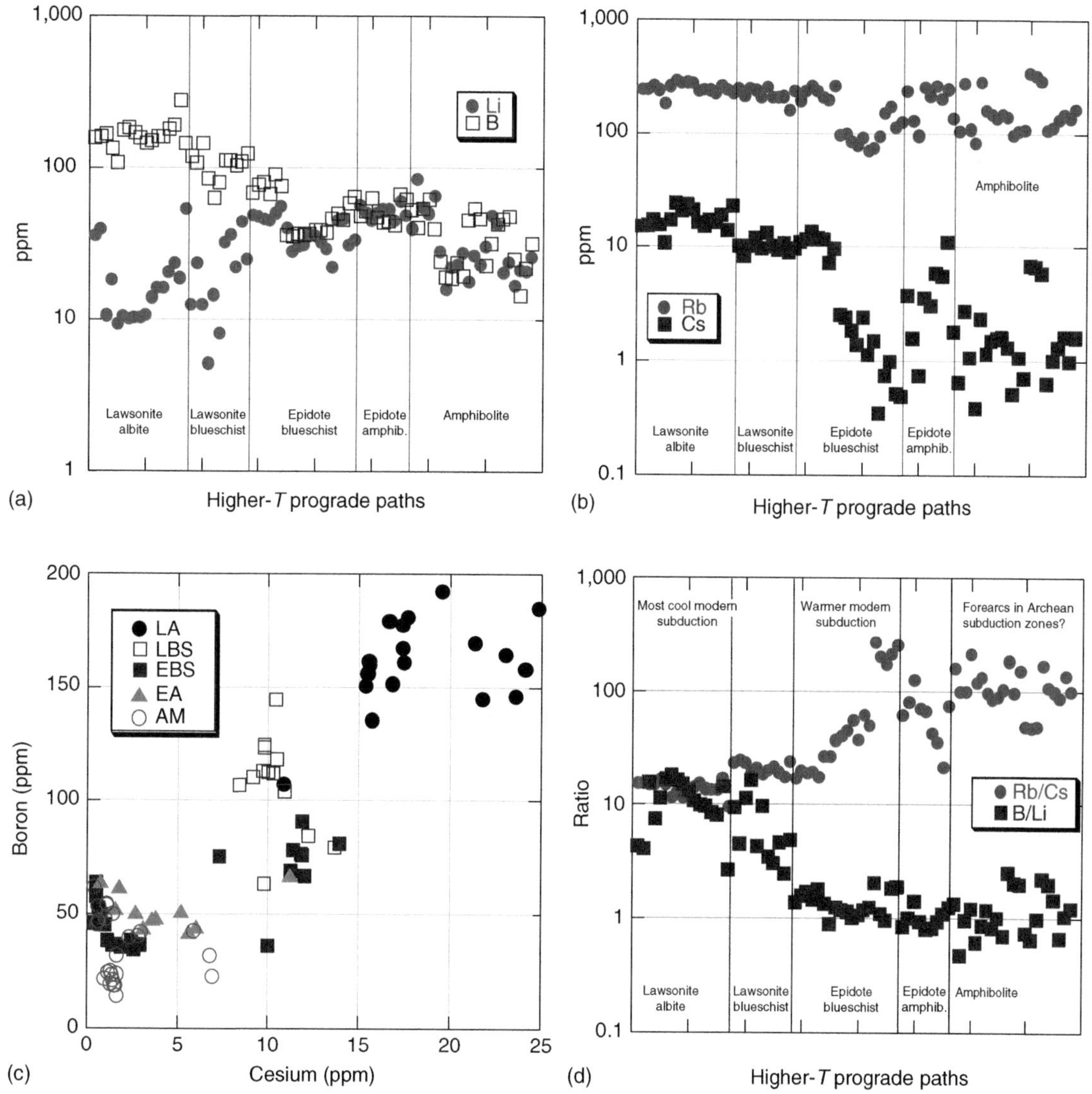

Figure 11 Plots illustrating variable concentrations of B, Li, Cs, and Rb in white micas in Catalina Schist metasedimentary rocks with increasing metamorphic grade (to the right in (a), (b), and (d)). (a) Plot of B and Li white-mica concentrations versus grade, demonstrating decreasing white-mica B concentrations, and increasing white-mica Li concentrations, in rocks experiencing paths into metamorphic grades of epidote-blueschist and higher. (b) Plot of Rb and Cs white-mica concentrations versus grade, demonstrating relatively uniform Rb concentrations across grades, and decreasing Cs concentrations in white micas in rocks experiencing paths into metamorphic grades of epidote-blueschist and higher. (c) Plot of Cs versus B in white micas in the Catalina Schist metasedimentary rocks, demonstrating correlated decreases in concentration in higher-grade units. LA, lawsonite-albite; LBS, lawsonite–blueschist amphibolite; EA, epidote-amphibolite; EBS, epidote-blueschist (d) Changes in Rb/Cs and B/Li with increasing metamorphic grade. The metamorphic temperatures, for these depths (see Figure 9), are compatible with those expected for modern cooler and warmer subduction, and the high temperatures represented by the epidote-amphibolite and amphibolite units could be similar to those experienced in Archean subduction zones. These data thus indicate efficient retention of B and Cs to 40 km depths in cooler modern subduction zones, but significant B and Cs loss in the warmer modern margins and in Archean subduction zones. In these rocks, B, Cs, and Rb are nearly entirely sited in the white micas, whereas Li residency is more complex, with significant concentrations also in chlorite in the lower-grade units (see Bebout *et al.*, 1999, in press).

Lustres, and the lower-grade metasedimentary rocks of the Catalina Schist and other circum-Pacific paleoaccretionary suites, suggests that the Rb–Cs signature of seafloor sediments remains largely intact to depths of up to ~90 km in relatively cool subduction zones (see Figure 13). Higher-grade metasedimentary rocks from the Catalina Schist (labeled, "LBS," "EBS," "EA," and "AM" in Figure 13) have lower Cs/Rb and experienced more extensive devolatilization along

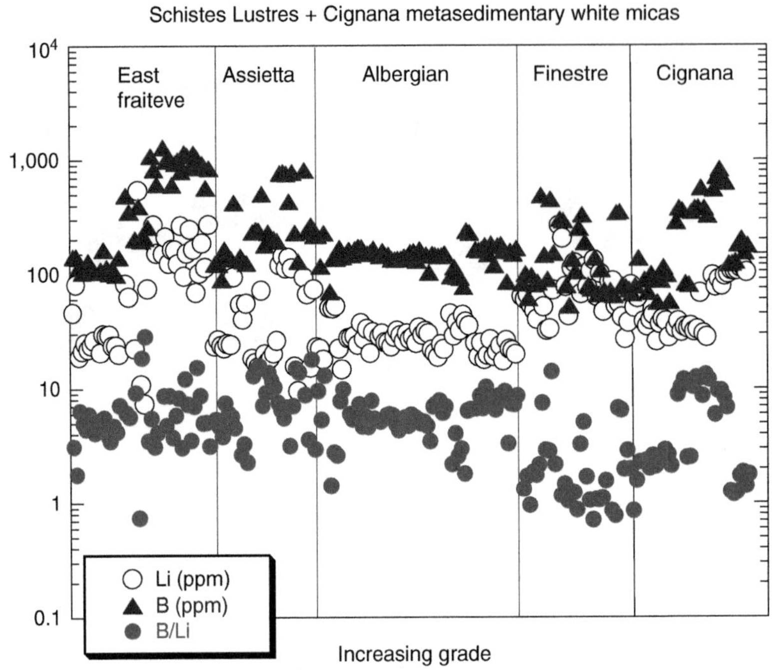

Figure 12 Concentrations of B and Li, and B/Li ratios, in phengite (data for paragonite excluded, where present in these samples, as much of it appears late-stage and exhumation-related based on textures) as a function of increasing metamorphic grade (to the right) in exposures of the Schistes Lustres in the Cottian Alps (see Agard *et al.*, 2002). Not included in this plot are data for phengite occurring as inclusions in garnet in rocks from Lago di Cignana; these phengites have higher B and related higher B/Li. Data from Bebout *et al.* (2004).

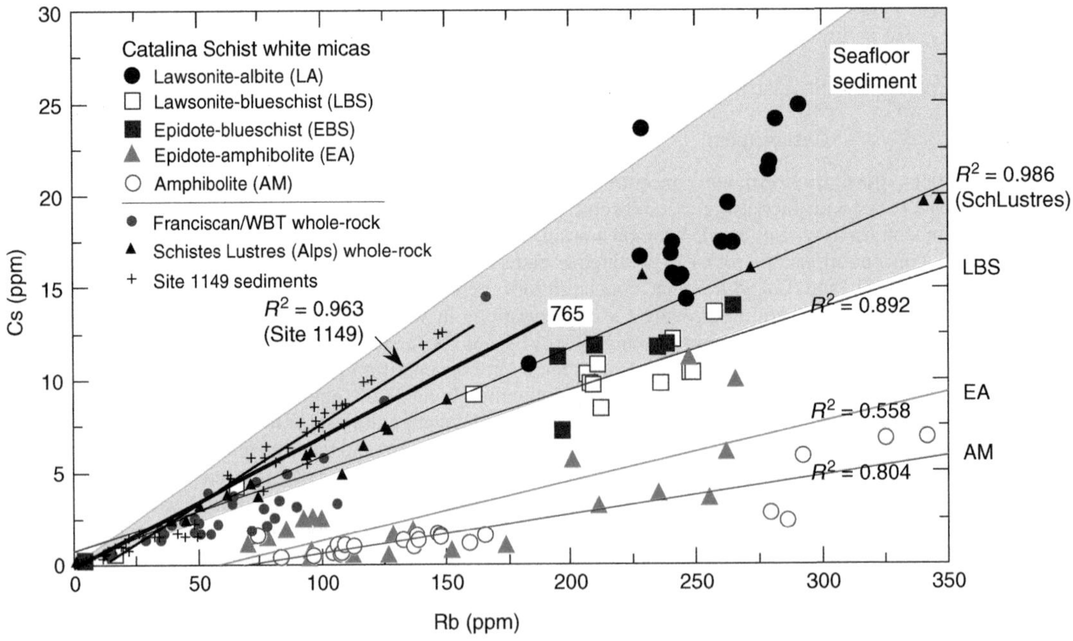

Figure 13 Comparison of Catalina Schist Rb–Cs white-mica compositions with the range of Rb/Cs of seafloor sediments (shaded region), the latter from Ben Othman *et al.* (1989) and Plank and Langmuir (1998) with data for seafloor sediment from ODP Site 765 (Banda forearc), from Plank and Ludden (1992) and data for ODP Site 1149 (Izu-Bonin) from T. Plank and K. Kelley (personal communication). Also shown, for comparison, are data for Franciscan and Western Baja Terrane ("WBT") metasedimentary rocks (from Sadofsky and Bebout, 2003) and for the Schistes Lustres (from Busigny *et al.*, 2003).

warmer prograde *P–T* paths (see Figures 9 and 11; discussion of the tectonometamorphic history of these units by Grove and Bebout, 1995).

Deeply subducted metasediments also afford an analysis of the cycling of organic carbon (and nitrogen) into the mantle, and some considerable attention has been paid to this "organic connection" (see Bebout and Fogel, 1992; Bebout, 1995; Sadofsky and Bebout, 2003; Busigny *et al.*, 2003). As discussed above in sections dealing with diagenesis, early-stage heating of organic matter in forearcs leads to the production of relatively devolatilized, amorphous carbonaceous matter that largely retains the organic/reduced carbon-isotope signatures of the seafloor sediment. Progressive recrystallization toward graphite occurs with heating and time, as documented by Beyssac *et al.* (2002), and this graphitization promotes carbon-isotope exchange with other organic-reduced carbon reservoirs in the same rocks, primarily carbonate. Interestingly, nitrogen initially bound in clays and organic matter in sediments is redistributed rather efficiently into low-grade white micas (see Bebout, 1995), and whole-rock C/N ratios of seafloor sediment appear to be retained, despite this mineralogical separation of the carbon and nitrogen organic components. Sadofsky and Bebout (2003) (also see Bebout, 1995) found little modification of the organic C–N signature in metasedimentary rocks subducted to depths of up to $\sim$40 km in a paleoaccretionary prism. As discussed in later sections, any disruption of the organic C–N reservoir conveyed into subduction zones could affect convergent margin C–N mass balance and potentially the carbon- and nitrogen-isotope budgets in the deep mantle (see discussions in Sadofsky and Bebout, 2003; Li and Bebout, 2005; Bebout and Kump, 2005).

3.5.5 HP and UHP Ultramafic Rocks and Mélange

There has been growing interest in the compositions of hydrated ultramafic rocks and the degree to which these compositions can be used to infer slab fluxes of trace elements at greater depths in subduction zones, including arc source regions (see Hattori and Guillot, 2003; Tenthorey and Hermann, 2004; Scambelluri *et al.*, 2004a, b; Chalot-Prat *et al.*, 2003). Straub and Layne (2003) suggested that "wedge serpentinite" dragged beneath the Izu volcanic front is the source of a number of trace elements (fluorine, chlorine, H_2O, lithium, and isotopically heavy boron) via dehydration and release to the subarc mantle (also see Hattori and Guillot, 2003; Scambelluri *et al.*, 2004b). Deeply subducted, variably serpentinized oceanic lithospheric mantle has increasingly been called upon to produce large amounts of H_2O that could convey slab and sediment chemical signatures into arc source regions (e.g., Rupke *et al.*, 2002, 2004; Ranero *et al.*, 2005; see Sharp and Barnes, 2004 for discussion of possible Cl subduction in seafloor serpentinites; Fruh-Green *et al.*, 2001 for discussion of O–H isotope compositions of HP ultramafic rocks in the Western Alps).

Mélange formation processes involving coeval metasomatic and mechanical mixing, are yet another means by which rocks can be "hybridized" at depth in subduction zones (see Bebout and Barton, 1989, 1993, 2002; Breeding *et al.*, 2004; King *et al.*, 2006). Figures 14 and 15 present whole-rock trace-element data for hybrid mélange rocks from the Catalina Schist (lawsonite–albite, lawsonite-blueschist, and amphibolite facies), demonstrating the extremely wide range of rock compositions that can result from this process. These mélange rocks bear little resemblance to any of their sedimentary, mafic, and ultramafic protolith mixing compositions, and any fluids interacting with them could have inherited chemical compositions reflecting hybridized sources (see discussion by King *et al.*, 2006). The production of such lithologies at the slab–mantle interface could not only strongly impact the geochemical evolution in this region, but also stabilize large amounts of volatiles to great depths in high-variance assemblages containing minerals such as chlorite, talc, and amphibole (Bebout and Barton, 2002). Relative to a simple (mechanical) mixture of known mélange-mixing components (mafic, ultramafic, and sedimentary), some of these rocks are depleted in certain trace elements (e.g., cesium, rubidium, and barium), suggesting the removal of these elements in metasomatizing fluids (see Bebout and Barton, 2002; King *et al.*, 2006).

3.5.6 Metasomatic Rinds on Blocks in Tectonic Mélange

Metasomatic rinds on mafic, ultramafic, and sedimentary blocks floating in HP mélange of varying compositions allow one to characterize metasomatism that can be directly related to subduction-zone processes (see Sorensen and Grossman, 1989, 1993; Sorensen *et al.*, 1997; Bebout and Barton, 1993, 2002; Breeding *et al.*, 2004; Saha *et al.*, 2005; see data in Figures 7 and 8). Bebout and Barton (2002) demonstrated that the matrix in the high-grade mélange in the Catalina Schist was in part produced by incorporation of rinds developed on these three lithologies (Figure 14), thus, that rind formation represents initial stages of the mixing leading to mélange matrix formation. Breeding *et al.* (2004) demonstrated that metasomatic exchange between sedimentary rock and adjacent ultramafic mélange matrix resulted in dramatic depletions, in the altered sedimentary rock, in many trace elements of interest in studies of arc lavas (Figure 16). These depletions were related to the destabilization of

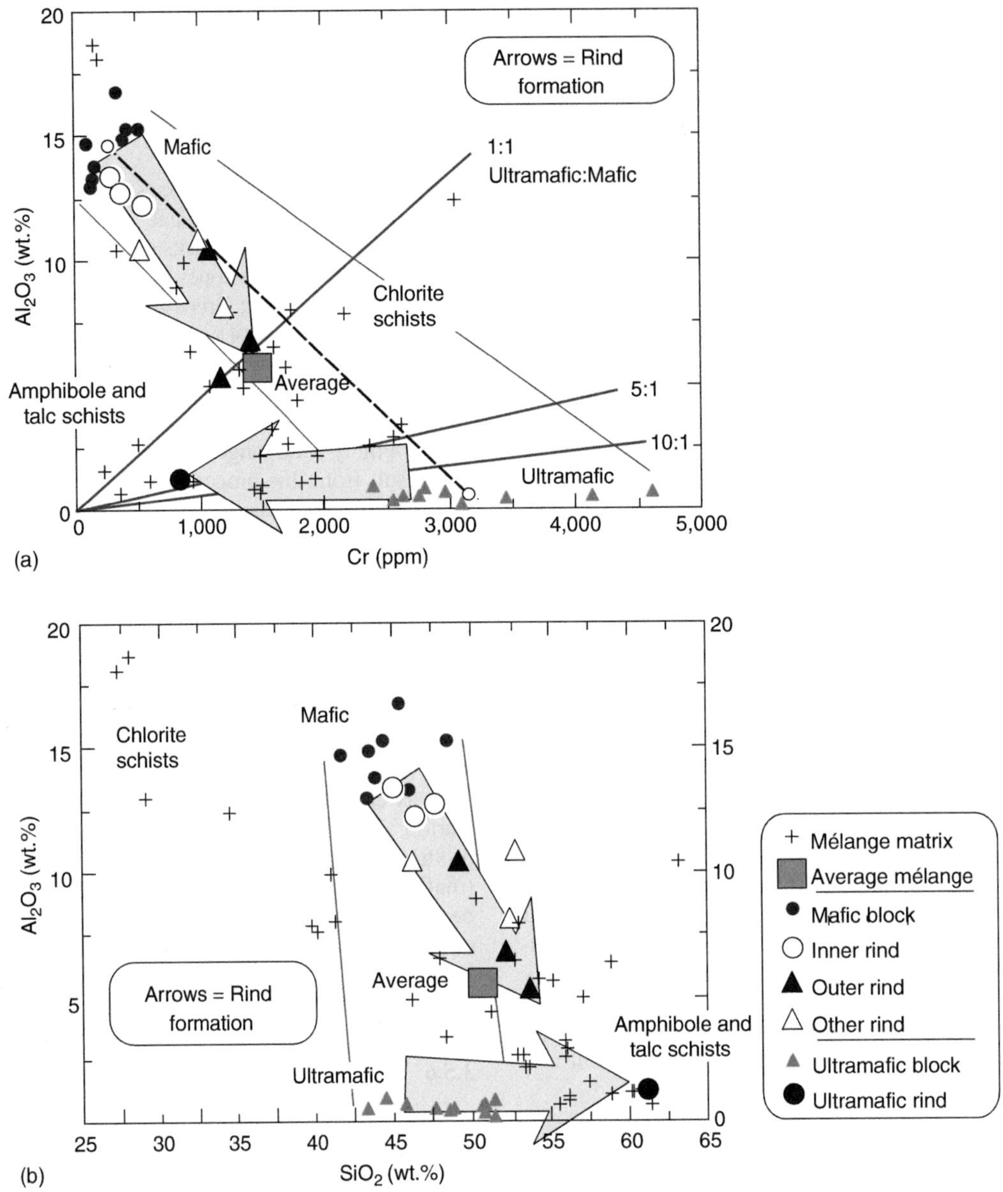

Figure 14 Cr–Al–Si compositions for mélange matrix and likely mélange mixing components in the amphibolite unit of the Catalina Schist. In both (a) and (b), the large arrows indicate the trends from tectonic block compositions observed in metasomatic rinds on the mafic and ultramafic blocks. Reproduced with permission from Bebout and Barton (2002).

key host phases such as mica and by interaction of the metasedimentary phases with mélange-equilibrated fluids that were strongly depleted in elements such as the LILE, uranium, and lead.

3.5.7 Records of Fluid/Element Mobility in Veins, Isotopic Compositions, and Fluid Inclusions

Metamorphic veins provide an additional, albeit complex, record of fluid and element mobility during various stages of metamorphism of HP and UHP rocks. Veins leave behind high-variance mineral assemblages that are, in general, regarded as fractionated residues of the fluids that produced them (see Bebout and Barton, 1993; Becker *et al.*, 1999). However, vein mineralogy, trace element, and stable isotope compositions can be used to estimate fluid compositions (given the availability of constraints on the conditions of their formation), and combined with their Sr–Nd–Pb–Os-isotope compositions, can be used to trace fluid sources. There has been great interest in whether or not vein networks in HP and UHP eclogites and metasedimentary rocks reflect large-scale fluid transport or local-scale segregations. Two common observations are that

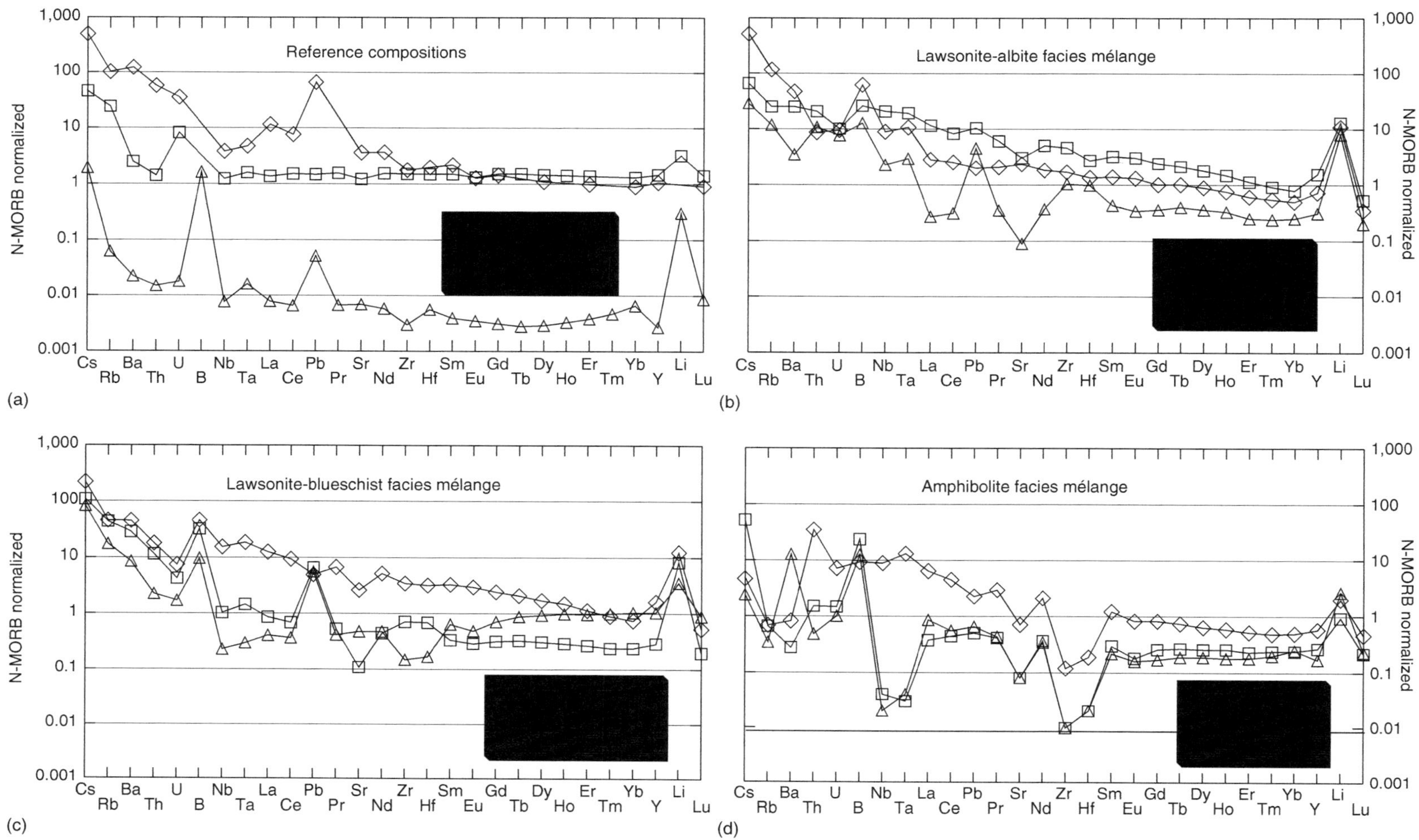

Figure 15 Concentrations of trace elements in Catalina Schist mélange matrix (lawsonite–albite (b), lawsonite-blueschist (c), and amphibolite (d) grades) (see Grove and Bebout, 1995 for discussion of tectonometamorphic evolution), relative to the N-MORB composition of Hofmann (1988), the GLOSS sediment composite from Plank and Langmuir (1998), the altered oceanic crust composite for Site 417 from Staudigel *et al*. (1996), and data for the Horoman peridotite from Tanimoto (2003), with elements in order of increasing compatibility during mantle melting to the right (from King *et al*., 2006; King, 2006). Note that the compositions of the mélange matrix samples bear little resemblance to compositions for the presumed mixing components involved in their production (see compositions of altered oceanic crust, seafloor sediment, and peridotite in (a)).

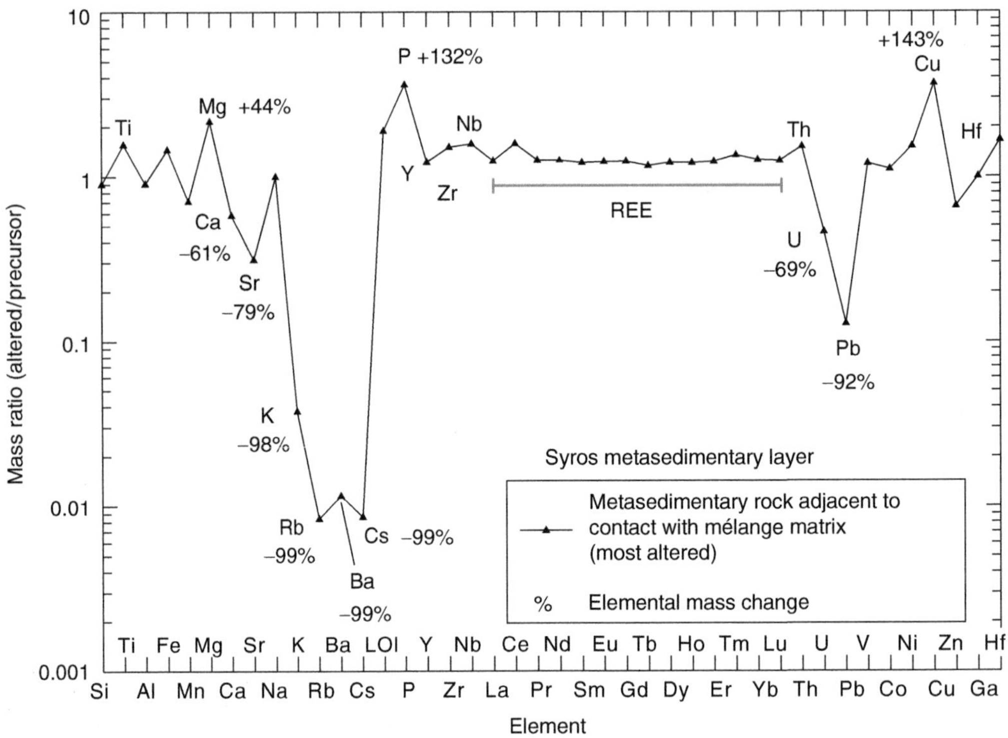

Figure 16 Plot of trace-element concentrations in metasedimentary rock having undergone metasomatic exchange with ultramafic mélange matrix, relative to the concentrations in the same layer but away from the interface with this mélange matrix (the "precursor"). Reproduced by permission of Geological Society of America from Breeding *et al.* (2004).

(1) trace elements enriched in veins are not obviously depleted in nearby host-rocks, and that the veins are in trace-element (and isotopic) equilibrium with host-rocks, and (2) some detailed studies of vein–vein haloes–host rocks clearly indicate relatively local-scale diffusion of trace elements into veins (Philippot and Selverstone, 1991; Widmer and Thompson, 2001). These observations suggest that elements primarily diffused into the fractures on a local scale, rather than being introduced from external sources. However, they do not preclude movement of fluids along the fractures (after all, once in the fractures, the fluid is gone from the system). Both Molina *et al.* (2004) and Brunsmann *et al.* (2000, 2001) argued for local-scale element transport of unknown scale. Interestingly, vein mineralogy commonly includes phases rich in elements thought to be relatively immobile (titanium, aluminum, and zirconium), demonstrating at least local-scale diffusive transfer of these elements, likely via an intergranular fluid.

Spandler and Hermann (2006) review the utility of vein mineralogy and major and trace-element compositions for evaluating mobility in and from subducting oceanic crust. They suggest that fluids released during the blueschist-to-eclogite transition at 50–80 km could be trapped in impermeable eclogitic rocks and dragged beneath arcs to contribute to their source regions. Widmer and Thompson (2001) discussed the expectations for local versus open-system redistribution and concluded that the veins they investigated (in the Zermatt-Saas Zone, Switzerland) are local-scale segregation phenomena, not channel ways for larger-scale fluid transport (also see the stable isotope study of Alpine veins by Cartwright and Barnicoat, 1999). John and Schenk (2003) argued that the blueschist-eclogite and gabbro-eclogite transitions are kinetically controlled, and that channelized infiltrating fluids, producing veins with large grains of omphacite, kyanite, and garnet, enhanced these transitions in rocks from Zambia (also see evidence for LREE fractionation in these infiltrated rocks in John *et al.*, 2004). John *et al.* (2004) suggested that the fluids were produced by dehydration in underlying hydrated mantle rocks in the subducting slab. Because fluid pathways are zones where mineral replacements are the most kinetically favored (see John *et al.*, 2004), veins and their haloes in host-rocks can provide the best records of exhumation history (also see van der Klauw *et al.*, 1997).

Isotopes can, in some cases, provide a clear record of open- and closed-system behavior and fluid sources. To utilize stable isotope compositions for this purpose, estimates of vein formation temperatures are required. In some cases, veins and shear zone rocks (mélange) are clearly out of isotopic equilibrium with the more coherent (less deformed) host rocks and an external fluid composition and source can be obtained, given estimates of temperatures during the metasomatism (e.g., amphibolite unit in the Catalina Schist; Bebout, 1991a). Stable isotopes can also be homogenized over significant volumes of rock, consistent with fluid communication (mixing) over such scales and perhaps large-scale fluid mobility (see evidence for oxygen- and nitrogen-isotope homogenization in Bebout, 1991b, 1997). In contrast with these observations, Barnicoat and Cartwright (1995) and Cartwright and Barnicoat (1999) used oxygen-isotope data for metabasaltic rocks, including pillow basalts, and suggested that fluid flow was extremely restricted, likely channelized, resulting in the preservation of oxygen-isotope signatures of seafloor hydrothermal alteration. These observations are consistent with those of other studies in which closed-system behavior was proposed in either metasedimentary or mafic compositions (e.g., Getty and Selverstone, 1994; Nadeau *et al.*, 1993; Henry *et al.*, 1996; Busigny *et al.*, 2003). The preservation of low-$\delta^{18}O$ signatures related to near-surface processes in UHP rocks in China attests to the ability of deeply subducted rocks to retain their presubduction isotopic compositions despite extensive metamorphic recrystallization (see Rumble and Yui, 1998; Zheng *et al.*, 1998), indicating limited or no infiltration of external fluids or isotopic exchange with the mantle during subduction. Stable isotope behavior similar to that observed for a Western Alps transect (including the Schistes Lustres, Cottian Alps; see Henry *et al.*, 1996), showing similarity in vein and host-rock isotopic compositions for the same minerals, was interpreted by Sadofsky and Bebout (2004b) for data from the Fransican Complex to be the result of transport of fluid through vast expanses of similar rock, resulting in apparent internal/local control.

Hermann *et al.* (in press) emphasized that the trace-element concentrations of veins in HP and UHP metamorphic rocks cannot be used as indicators of element mobility, other than in an extremely qualitative way. They stressed that one must use mineral-fluid partitioning data to estimate fluid compositions from vein trace-element concentrations. Figure 17 presents one intriguing example of vein–host-rock relations in a HP

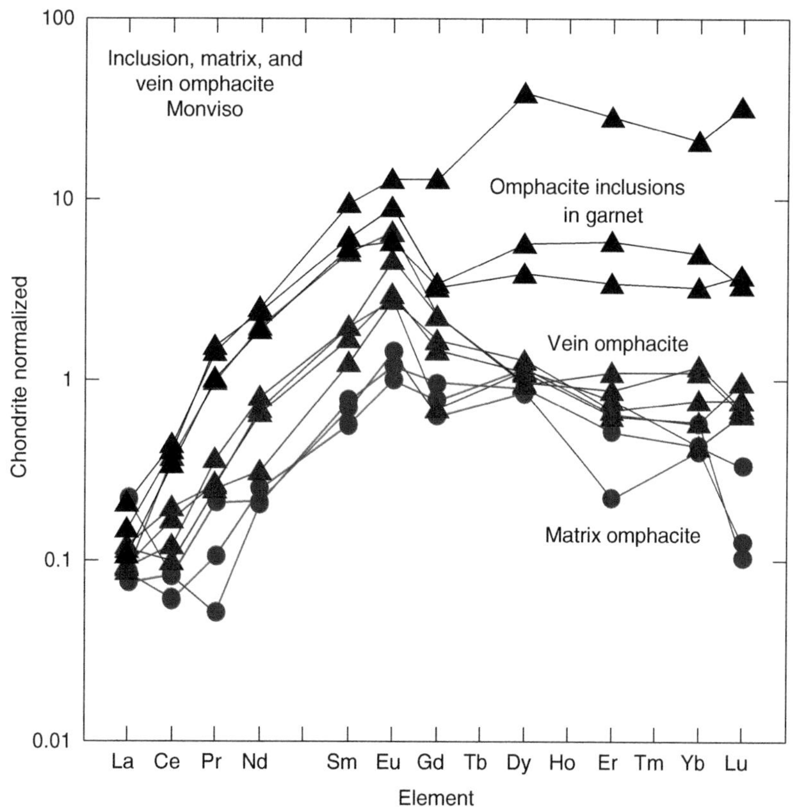

Figure 17 REE plots (normalized to chrondritic values of McDonough and Sun, 1995) for omphacite in Monviso garnet-clinopyroxene metagabbros and omphacite in a vein in the same sample. Note that the REE pattern for the vein omphacite more closely resembles the pattern for omphacite in the matrix outside garnets and reflecting HREE depletion by garnet growth (data from King *et al.*, 2004; King, 2006), presumably related to relatively late-stage production of this vein.

eclogite from Monviso, Italy (see studies of the same locality by Philippot and Selverstone, 1991; Nadeau *et al.*, 1993; Rubatto and Hermann, 2003). In this sample, garnet growth depleted the rock in HREE. Thus, omphacite included in the garnet has higher HREE, reflecting its growth before the majority of the garnet grew, whereas omphacite in the matrix is strongly depleted in HREE, due to its crystallization following garnet growth. Omphacite occurring in a vein in this same sample also has depleted HREE and may reflect control of vein chemistry by the host-rocks after garnet growth. Rubatto and Hermann (2003) also presented trace-element data for veins in mafic rocks from this locality, and Hermann *et al.* (in press) calculated fluid compositions using the trace-element concentrations of vein omphacite and previously published partition coefficients (from Green and Adam, 2003). The calculated fluid had relatively elevated concentrations of Cs, Rb, Ba, Sr, and Pb, but the concentrations of most other elements were below 1 ppm.

Growing, but still modest, numbers of strontium-, neodymium-, lead- (and Re–Os, Lu–Hf) isotopic studies of HP and UHP mafic rocks provide further constraints on the sources of trace-element enrichments observed in these rocks relative to their protoliths. They also can provide insights into the degrees of open- versus closed-system behavior, both on the seafloor and during subduction-zone metamorphism (e.g., Nelson, 1995; Becker, 2000; Lapen *et al.*, 2003; John *et al.*, 2004). Relatively few studies have evaluated isotope records of fluid–rock interactions at scales larger than the hand-specimen or outcrop scale. Strontium-, neodymium-, and lead-isotope data for mafic blocks (and in some cases, their metasomatic rinds) in California tectonic mélange are interpreted as reflecting varying combinations of seafloor alteration, metasomatic interactions with mélange matrix, and possible IAB protoliths (see Nelson, 1995; Saha *et al.*, 2005).

Variations in fluid inclusion compositions at small scales are consistent with local-scale exchange between the fluids and rocks during or after their formation and possible decrepitation (see Philippot and Selverstone, 1991; Gao and Klemd, 2001; Hermann *et al.*, in press). Touret and Frezzotti (2003) summarize the record of fluid–rock interactions in subduction zones using fluid inclusions. Recently, trace-element analyses of individual inclusions have provided constraints on trace-element mobility, at least on the local scale (see Scambelluri *et al.*, 2004a), and fluid-inclusion salinities are directly relevant to discussions of chlorine cycling (see Philippot *et al.*, 1998). In a number of studies, at least parts of the fluid-inclusion record are directly associated with exhumation history (e.g., El-Shazly and Sisson, 1999; Franz *et al.*, 2001; Fu *et al.*, 2002).

3.5.8 Prograde Chemical Changes and Open- versus Closed-System Behavior

Because HP and UHP metamorphic suites experience protracted prograde and exhumation paths, with exhumation commonly strongly overprinting the prograde assemblages, it is often necessary to identify ways to distinguish the mineral assemblages and textures and geochemistry associated with these two histories. One approach to identifying chemical change that can be directly associated with prograde subduction-zone history involves the exploitation of porphyroblast mineral inclusions and chemical and isotopic zoning. Garnet porphyroblasts are thought to experience slow volume diffusion and thus preserve prograde records through even protracted HT exhumation histories (these phases can also contain useful mineral inclusions; see King *et al.*, 2004; cf. Kurahashi *et al.*, 2001). Figure 18 demonstrates one way of doing this, by examining core-to-rim major and trace-element (and mineral inclusion) variations in garnets in UHP and metabasaltic rocks at the Lago di Cignana locality (from King *et al.*, 2004; King, 2006). In metasedimentary samples from this locality, highly overprinted by exhumation-related mineral assemblages, these records in phases such as garnet and tourmaline are the only remaining geochemical signal of the prograde history (see discussion in Bebout and Nakamura, 2003; Bebout *et al.*, 2004). However, the overprinting in mineral assemblages does not necessarily imply that the rocks were open systems during exhumation (see Bebout *et al.*, 2004).

Obviously, one of the keys to understanding any whole-rock geochemical change is the establishment of open- or closed-system behavior during various stages of metamorphism, including the exhumation stage. This can be complicated greatly by the heterogeneities in likely protoliths (see above) and by recrystallization and overprinting related to exhumation (see discussion for the Schistes Lustres and Lago di Cignana metasedimentary rocks by Bebout *et al.*, 2004). As noted above, metasomatic rinds on blocks in ultramafic or sedimentary mélange provide one more clear record of element mobility during HP and UHP metamorphism. Another involves the use of ratios among elements of varying fluid mobility that are strongly sequestered into the same mineral phase (e.g., K/Cs and Rb/Cs in micas) to effectively normalize whole-rock heterogeneity (see Bebout and Fogel, 1992; Bebout *et al.*, 1999; see Figures 11 and 13).

Others have tested the degree to which the mineral phases in HP and UHP rocks fall on mineral isochrons (Glodny *et al.*, 2003; John *et al.*, 2004; Usui *et al.*, 2006; discussion in Lapen *et al.*, 2003). It is not clear how sensitive a measure of

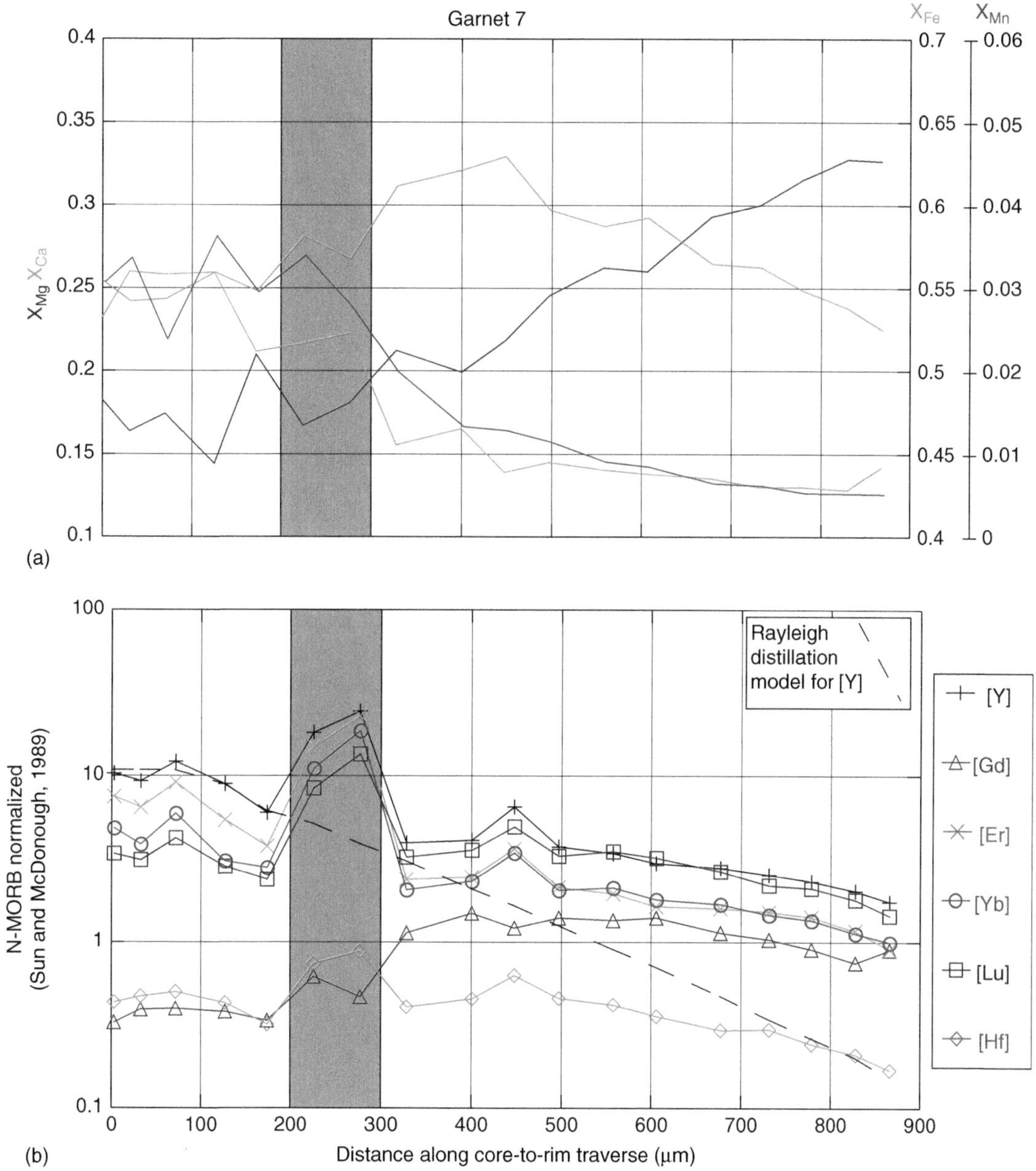

Figure 18 Major and trace-element zoning in a core-to-rim traverse (rim to the right) in a garnet in UHP metabasalt from Lago di Cignana (from King *et al.*, 2004; King, 2006; see similar approach of Usui *et al.*, in press). Note the anomaly in both major element (a) and trace-element (b) compositions that King *et al.* (2004) attribute to a prograde metamorphic reaction involving breakdown of clinozoisite and titanite to stabilize grossularite and rutile (see discussion by Skora *et al.*, in press). Trace-element concentrations are normalized to N-MORB values of Sun and McDonough (1989).

open-system behavior this type of test can be; Glodny *et al.* (2003) suggested that eclogitic rocks preserve their Rb/Sr isotopic signatures as long as they are devoid of free fluids and that fluid–rock interactions cause Sr redistribution. Usui *et al.* (2006) suggested that xenolithic eclogites, in which lawsonite breaks down to zoisite, released fluids that led to strontium–neodymium–lead isotopic reequilibration. In contrast, jadeite–clinopyroxenite xenoliths that initially lacked lawsonite or significant amounts of zoisite, did not experience this release of fluid and thus preserve older ages.

3.5.9 Mineral Chemistries of HP and UHP Suites

Recently published microanalytical work highlights the value of obtaining chemical and isotopic data at the single-mineral (and intracrystalline) scales using microanalytical methods (e.g., SIMS or laser ablation-inductively coupled plama mass spectrometry (LA-ICPMS)). Such work allows recovery of mineral–mineral partitioning data for individual trace elements and isotopes of interest. It also allows exploitation of disequilibrium features such as porphyroblast chemical/isotopic zoning, and the chemical compositions of mineral inclusions in porphyroblasts, and evaluation of whole-rock effects of exhumation by analyses of prograde and exhumation-related parageneses identified based on textural evidence (see Zack *et al.*, 2001; Spandler *et al.*, 2003, 2004; Bebout and Nakamura, 2003; King *et al.*, 2004; Usui *et al.*, 2006; earlier work by Messiga *et al.*, 1995; Tribuzio *et al.*, 1996; Sassi *et al.*, 2000; Wiesli *et al.*, 2001).

Micas, particularly phengitic white micas, are the primary hosts of a number of trace elements that are of interest in subduction tracer studies (LILE, boron, nitrogen, and lithium, the latter for which clinopyroxene and chlorite are also major mineral hosts). White micas carry these elements to great depths in subduction zones (see Domanik and Holloway, 1996) in mafic, sedimentary, and ultramafic lithologies (for ultramafic mélange, see Bebout, 1997; King *et al.*, 2006; see Figures 11–13). Paragonite (sodium-rich white mica) is also an extremely important mineral host, particularly during exhumation (see Giorgetti *et al.*, 2000; Poli and Schmidt, 1997; Zack *et al.*, 2001), incorporating elements such as boron and strontium (Bebout *et al.*, 2004). Studies by Moran *et al.* (1992), Bebout and Fogel (1992), Bebout *et al.* (1993, 1999, in press), Domanik *et al.* (1993), Zack *et al.* (2001), and Busigny *et al.* (2003), among many others, demonstrate that the whole-rock budgets of LILE, boron, and nitrogen, are nearly completely contained within white micas.

A number of other phases can exert strong control over whole-rock trace-element budgets. First, lawsonite and zoisite/clinozoisite are extremely important hosts for LREE and strontium (Okamoto and Maruyama, 1999; Pawley, 1994, see Figure 19); thus, these phases are of extreme interest to those attempting to link mineral reactions in subducting lithologies with arc outputs (see King *et al.*, 2004; Moriguti *et al.*, 2004; Feineman *et al.*, in press). Lawsonite in eclogite is preserved in only a very small number of suites, as it requires relatively cool (and/or rapid) exhumation trajectories (see recent reports of such rocks by Zack *et al.*, 2004; Altherr *et al.*, 2004; Usui *et al.*, 2006; Whitney and Davis, 2006; Tsujimori *et al.*, in press). At the Lago di

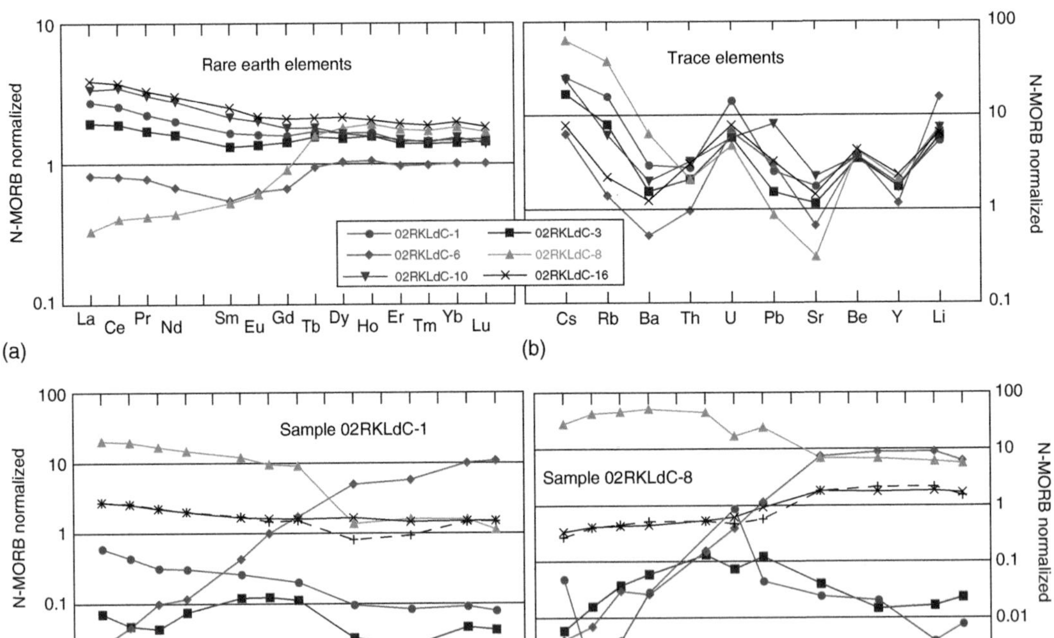

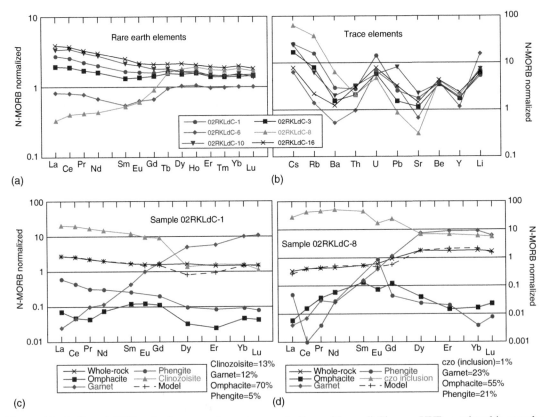

Figure 19 Bulk rock REE and other trace-element concentrations of Lago di Cignana UHP metabasaltic samples ((a) and (b)), both normalized to N-MORB of Hofmann (1988), and ((c) and (d)) mass balance of REE in two of these samples using mineral compositions and modes to reconstruct the whole-rock REE pattern. Data are from King *et al.* (2004) (also see King, 2006).

Cignana locality, which experienced a relatively low-temperature exhumation *P–T* path (see path in Figure 6), lawsonite that is inferred to have been stable in HP/UHP mineral assemblages is pseudomorphed by intergrowths of orthozoisite, paragonite, and quartz (Reinecke, 1998). Usui *et al.* (2006) reported lawsonite, both as inclusions in garnet and as matrix phases, in xenolithic eclogites from the Colorado Plateau, which they interpret as being fragments of the subducted Farallon plate (see Usui *et al.*, 2003). Zoisite and clinozoisite are comparatively common in HP/UHP mafic (and sedimentary) lithologies (e.g., Brunsmann *et al.*, 2000, 2001). Figure 19 demonstrates the strong relationship between whole-rock strontium concentration and the modal abundances and strontium concentrations of clinozoisite for a suite of UHP eclogites from Lago di Cignana (data from King *et al.*, 2004; King, 2006; see similar data for other suites in Spandler *et al.*, 2003; Usui *et al.*, 2006).

Garnet in HP and UHP metamorphic rocks exhibits characteristic enrichments in HREE, in many cases dominating the whole-rock budget for these elements (see Figure 19). Rutile generally contains most of the niobium and tantalum present in the rock (Sorensen and Grossman, 1989; Zack *et al.*, 2002; Barth *et al.*, 2001; Spandler *et al.*, 2003). Omphacite accounts for much of the lithium in eclogites (Zack *et al.*, 2002; King *et al.*, 2004), and up to 100 ppm lithium in cores of garnet has been reported for some Cignana metasedimentary rocks in which the other primary lithium hosts are phengite and paragonite (Bebout *et al.*, 2004). Sorensen and Grossman (1989) (also see Sorensen, 1991) demonstrated the strong impact that small modal amounts of allanite can have on LREE budgets in garnet-amphibolites (also see Spandler *et al.*, 2003).

3.5.10 Forearc to Subarc: Summary and Outstanding Questions

The previous sections provide a look at the geochemical consequences of important mineral reactions during subduction across the forearc geochemical "filter," and this allows insight regarding what the inventory of various trace-element and isotopic tracers is beneath the volcanic front, i.e., the geochemical menu available for addition by various "fluids" to the subarc mantle, which is addressed in the next section. To sum up:

1. Consensus is developing that, in the relatively cool subduction-zone conditions predicted by thermal models for most modern subduction zones, subducting sediments (and probably also oceanic crustal rocks) can largely retain their inventories of even the more fluid-mobile elements, to at least 90 km (Lago di Cignana; Busigny *et al.*, 2003; Bebout *et al.*, 2004). In cases where young, warm oceanic lithosphere is being subducted (e.g., Cascadia; Kirby *et al.*, 1996), related to ridge–trench encounters, and in the Archean (see Staudigel and King, 1992; Foley *et al.*, 2003; Martin, 1986), higher geothermal gradients may generate greater forearc devolatilization leading to greater loss of fluid and fluid-mobile elements to the mantle wedge. The data for HP and UHP eclogites in Figures 7 and 8 indicate that a number of the elements of greatest interest to those studying arcs are actually enriched rather than lost, certainly arguing against any strong depletions related to prograde metamorphism to depths of up to ~90 km (cf. Chalot-Prat *et al.*, 2003; Spandler *et al.*, 2004; John *et al.*, 2004; Marschall *et al.*, 2006; see the contrasting conclusion by Becker *et al.*, 2000).
2. Exhumation paths can dictate the extent to which prograde chemical history is preserved in HP and UHP suites, and thus the usefulness of these suites for study of subduction-zone chemical cycling (see the great variability in exhumation paths in Figure 6). Warmer exhumation paths, in some cases producing peak temperatures after significant decompression (e.g., path for Adula nappe; from Jolivet *et al.*, 2003; exhumation paths for some rocks in the Dabieshan) increase the likelihood of significant overprinting during exhumation. Whether prograde parageneses and chemical records can be extracted must be considered on a suite-by-suite basis (e.g., for Trescolmen, see Zack *et al.*, 2001, 2002). Lawsonite–eclogites in which lawsonite has escaped decompression-related replacements (such as those experienced by the UHP rocks at Lago di Cignana; see Reinecke, 1998; van der Klauw *et al.*, 1997) are certainly prime candidates for detailed study of prograde reaction and chemical histories (Zack *et al.*, 2004).
3. Eclogites may experience closed-system metamorphism, but protolith heterogeneity prohibits identification of any minor element loss during devolatilization, and some eclogites could have had relatively "dry" protoliths that did not experience alteration on the seafloor). It is as yet unclear whether it is possible to lose very small amounts of fluids and trace elements, over large volumes of subducting rocks, to produce the fluid and trace-element budget in the mantle wedge partly delivered toward the surface in arcs. Such loss could easily be obscured in the large protolith heterogeneity incorporated into any study of trace-element loss—in studies comparing higher-grade rocks with lower-grade or unmetamorphosed equivalents, only the extremely fluid-mobile elements (boron, cesium, arsenic, and antimony,

for the Catalina Schist metasedimentary suite) show a clear record of whole-rock loss. A mass balance of this type, involving calculation of element loss using published partition coefficients, and evaluating the extents of loss that could be achieved without observing them, has not yet been conducted (see the brief consideration of this problem for boron and beryllium and B/Be of released fluids in Bebout *et al.*, 1993; the analysis of the Catalina Schist LILE systematics by Zack *et al.*, 2001). Another question relates to whether more pronounced trace-element loss from such rocks, mostly concentrated in zones of higher fluid flux (see John *et al.*, 2004; Spandler *et al.*, 2004; see the discussion of the slab–mantle interface and devolatilization models below), could produce sufficient trace-element mobility to satisfy the arc record of slab additions.

4. Veins in many cases appear to reflect only local-scale element redistribution and it is unknown whether larger-scale removal of fluid-mobile elements can occur in subducting slabs. Although there is, in some cases, clear demonstration that fluid–rock interactions were relatively local-scale in relatively closed systems (e.g., Philippot and Selverstone, 1991; Getty and Selverstone, 1994; see Scambelluri and Philippot, 2001), some open-system behavior may occur at/near peak *P–T* conditions or along the prograde *P–T* path. Relatively, small fluid fluxes within large masses of chemically similar rock can be difficult to document using trace elements and stable isotopes and tracts of rock that contain vein systems in local equilibrium with host-rocks (Gray *et al.*, 1991; Henry *et al.*, 1996; Sadofsky and Bebout, 2004b) do not preclude infiltration by externally derived fluids. Integrated over large volumes of subducting sedimentary and mafic rocks, relatively small fluid fluxes could result in sufficient element flux to explain volcanic arc compositions. Larger-scale fluid and element transfer appears to be concentrated along zones of structural weakness (e.g., fractures, mélange), likely resulting in the disproportionate influence of such zones on the chemical characteristics of any fluids traversing the slab–mantle interface beneath arcs.
5. The geochemical significance of mélange at depth in subduction zones is not clear. The scarcity of appropriate exposures has made it difficult to evaluate the volumetric significance of mélange zones at depth along the slab–mantle interface. Zones of low seismic velocity at or near the top of the subducting oceanic lithosphere may consist of extensively hydrated assemblages (see Abers, 2000); alternatively, they may represent hydrated hanging wall lithologies. The generation of layers rich in hydrous minerals, perhaps in part a mechanical mixing zone, could also promote aseismic behavior at great depths in subduction zones (Peacock and Hyndman, 1999). Further attention should be paid to mélange zones, in field, geophysical, and theoretical studies, to evaluate their potential for producing deep subduction-zone "fluids" with hybridized mafic–sedimentary–ultramafic compositions.

3.6 THE DEEP FOREARC AND SUBARC SLAB–MANTLE INTERFACE

Relatively few detailed geochemical studies have thus far been conducted on UHP rocks directly representing the slab–mantle interface at depths corresponding to those beneath and across arcs (see compilation in Figure 6), largely leaving us to speculate regarding the structural evolution and reaction histories, and related production of "fluids," in slab and sediments subducted to these depths. Figure 20 presents a conceptual model for forearc to subarc processes at/near the slab–mantle interface from Breeding *et al.* (2004) (see similar sketches by Bebout and Barton, 1989, 2002; Hermann *et al.*, in press). It is likely that the slab–mantle interface is a region of high, but heterogeneous, fluid flux. There are abundant devolatilization (for most margins, particularly dehydration) reactions in mafic, sedimentary, and even ultramafic rocks capable of producing large amounts of fluid), and arc lava geochemistry bears testament to the presence of large amounts of H_2O-rich "fluid" beneath arcs. Active deformation should lead to heterogeneous fluid transport along zones of structural weakness. Relatively coherent tracts of mafic HP and UHP rocks for which limited, or no, fluid mobility is proposed (e.g., metamorphosed Jurassic ophiolite in the Alps), likelyrepresent the stronger, less permeable lithologies within a lithologic package characterized by strong contrasts in rheology and heterogeneous permeability structure (see discussions by Philippot, 1993; Philippot and Selverstone, 1991; Getty and Selverstone, 1994; Spandler *et al.*, 2003, 2004; Chalot-Prat *et al.*, 2003; for isotopic evidence for limited mobility in coherent rocks in the Catalina Schist, see Bebout, 1991b; Bebout and Fogel, 1992). Zones of mixing among sedimentary and hydrated mafic and ultramafic rocks likely exist along the slab–mantle interface. Roughness in the subducting slab, produced by seafloor and trench deformation and in the form of seamounts, would promote the off-scraping of mafic lithologies and incorporation into sedimentary–ultramafic mixing zones. The enhanced fluid flux in such zones would produce talc, chlorite, and amphibole, which are weak and easily

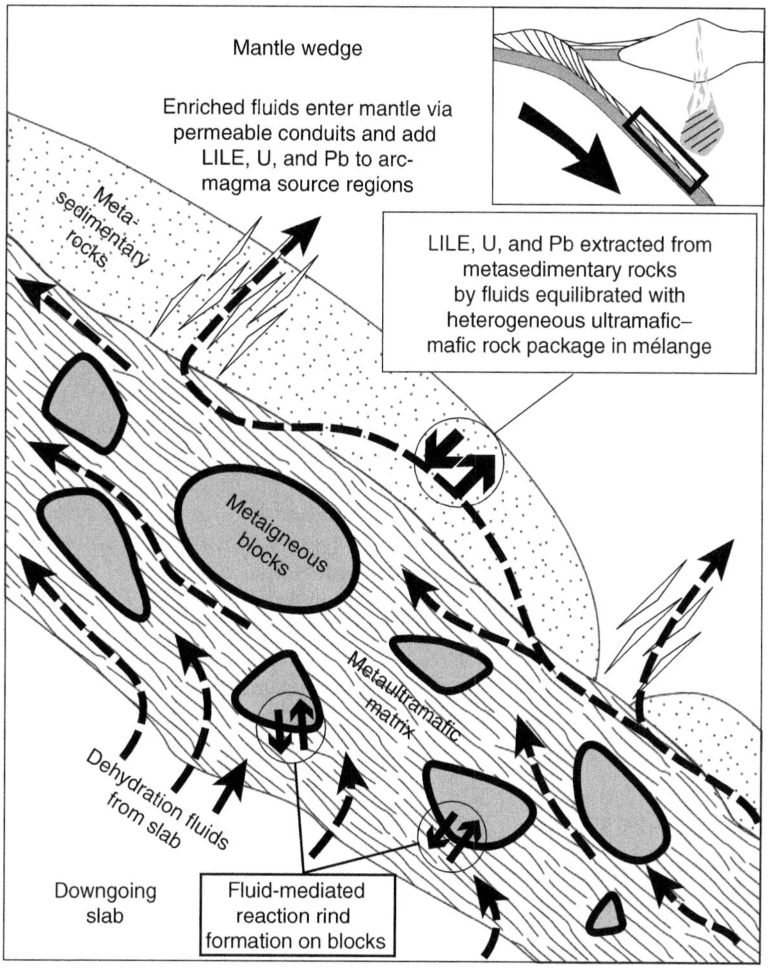

Figure 20 A view of the slab–mantle interface region (from Breeding *et al.*, 2004; also see sketches by Bebout and Barton, 1989, 2002; Hermann *et al.*, in press). See text for discussion. Reproduced by permission of Geological Society of America from Breeding *et al.* (2004).

deformed thus further enhancing fluid transport (see Peacock and Hyndman, 1999; Ague, in press). These zones of preferential fluid flow would also be more intensely metasomatized, leading to their disproportionate representation in any "fluid" (hydrous fluid, silicate melt, or supercritical liquid) emanating from them into the overlying mantle wedge or upward along the subduction-zone thrust (cf. John *et al.*, 2004; Spandler *et al.*, 2004). Hanging-wall mantle wedge rocks nearest the slab–mantle interface would become hydrated and metasomatically altered, weakening them and leading to their incorporation into the mixing zones (Peacock and Hyndman, 1999; see Bebout and Barton, 2002).

If the upper-mantle section in subducting slabs is hydrated on the seafloor or in forearc regions (see Ranero *et al.*, 2005), far greater amounts of fluid could ascend into mixing zones at the slab–mantle interface (see Hermann *et al.*, in press). The H_2O-rich fluid produced by dehydration of these ultramafic lithologies will cause metasomatic exchange with disparate lithologies and drive decarbonation reactions in more carbonate-rich systems (Gorman *et al.*, 2006), such as in the modern Central America and Banda convergent margins. In convergent margins where HP and UHP rocks are exhumed by later collision (Alps, Himalaya), mélange zones are likely to be reactivated and strongly overprinted by fluid–rock interactions during exhumation, making them more difficult to identify and investigate for their prograde/peak metamorphic geochemistry. Thus, the more structurally intact, less-permeable lithologies tend be the best preserved in these orogens, and consequently show less evidence for large-scale fluid flow. Blocks that are tectonically incorporated into mélange zones provide a record of fluid and metasomatic processes at the slab–mantle interface, but unfortunately do not record processes operating in the subducting slab.

Zack *et al.* (2001) used published mica-fluid partitioning data (Melzer and Wunder, 2001) along with LILE concentrations of the Catalina Schist metasedimentary rocks to suggest that an external source of H_2O-rich fluid is required to produce the observed fractionations of element ratios in these rocks (particularly Cs/K). As illustrated in Figure 20, from Breeding *et al.* (2004), such fluid fluxing through metasedimentary rocks, which are either underplated to the hanging-wall or incorporated as tabular bodies in mélange zones, is expected if large amounts of fluid ascend from dehydrating mafic and ultramafic rocks at depth in the subducting slab (see Ranero *et al.*, 2005).

The complexity of metasomatism along upward (and lateral) fluid flow paths into the mantle wedge immediately overlying subducting slabs is widely acknowledged and we have much to learn regarding the mineral–fluid reactions and resulting geochemistry that occur in this region (see discussion by King *et al.*, 2003). Studies of intact tracts (see Paquin *et al.*, 2004; Scambelluri *et al.*, 2006, and references therein), ultramafic xenoliths found (rarely) in arc volcanic rocks (Maury *et al.*, 1992; Schiano *et al.*, 1995; Laurora *et al.*, 2001; Widom *et al.*, 2003) and hydrated forearc mantle wedge in serpentinite seamounts (Ryan *et al.*, 1996; Benton *et al.*, 2001, 2004; Savov *et al.*, 2005) have yielded insights into the reactions and processes occurring in the wedge.

3.7 SLAB-ARC CONNECTIONS

3.7.1 Brief Discussion of the Geochemistry of Arc Basalts and Inferred Slab Additions

There is general consensus, based on the geochemistry of arc basalts and their differences relative to various MORB compositions, regarding the most likely additions to arc source regions via "fluids" from the slab (see recent discussions by Pearce and Peate, 1995; Elliott, 2003; Tatsumi, 2005; Hermann *et al.*, in press). Cesium, rubidium, radium, barium, thorium, uranium, potassium, the light REEs, lead, and strontium are enriched in arc lavas relative to MORB (Figure 2a; in some cases, also boron, lithium, ^{10}Be; Pearce and Peate, 1995; Elliott, 2003, and references therein). In general, subducting oceanic crust is enriched in the elements that are also enriched in arc lavas relative to MORB, and various combinations of hydrous fluids, silicate melts, and supercritical liquids are called upon to transfer these elements into arc source regions. A smaller number of these elements (cesium, boron, arsenic, antimony, lithium, and lead) are regarded as being particularly "fluid-mobile" and likely to be either lost to hydrous fluids during forearc devolatilization or transported in hydrous fluids into arc source regions (see Moran *et al.*, 1992; Noll *et al.*, 1996; Leeman, 1996; Ryan *et al.*, 1995, 1996). Enrichments of other elements, in arc lavas, that appear to be less mobile in hydrous fluids (e.g., thorium and beryllium) are thought to require transport in a silicate melt (Johnson and Plank, 1999; for discussion of thorium immobility in hydrous fluids see Brenan *et al.*, 1995a). This dichotomy has led to the "sediments melt, oceanic crust dehydrates" conclusion reached in a number of studies of arc geochemistry.

The evidence for mobility of LILE elements such as Ba, Rb, Cs, and Sr, and possibly also U and Pb, demonstrated in studies of metasomatic rind formation, mélange evolution, and evident in the data for metabasalts and metagabbros summarized in Figures 7 and 8, is certainly in general compatible with the "slab additions" of these elements invoked in studies of arc magmagenesis. The data of Becker *et al.* (2000), although complicated by the lack of control on protoliths, indicate LILE (particularly K, Rb, and Ba) loss during eclogite-facies metamorphism, which may for some samples reflect melt removal, as a number of the samples these authors analyzed have peak *P–T* near or above the wet solidus. Silicate melts (and supercritical liquids; see Kessel *et al.*, 2005; Hermann *et al.*, in press) are presumably more effective than hydrous fluids in removing elements such as the LILE (particularly rubidium, barium, and strontium; see discussion on granulite potassium, rubidium, and cesium by Hart and Reid, 1991).

From an arc geochemist's perspective, the forearc can be viewed as a metamorphic "filter" that can either prevent or allow the retention of relatively fluid-mobile components to greater depths in the subduction zone (for noble gases, forearc metamorphism may represent a "barrier" to deep recycling, as proposed by Staudacher and Allègre, 1988). Studies of metasedimentary rocks appear to demonstrate remarkable retention of elements such as boron and cesium to depths of up to 90 km in relatively cool subduction zones (Bebout *et al.*, 1999, 2004; Busigny *et al.*, 2003; and discussion above). With the exception of the studies by Becker *et al.* (2000) and Arculus *et al.* (1999), which were limited somewhat by insufficient consideration of protolith heterogeneity, the previous work on HP/UHP metabasaltic and metagabbroic rocks similarly points to the retention of many of the elements of interest as "slab tracers" to depths beneath the volcanic front ($\sim$100 km or more; Gill, 1981; Figures 7 and 8). The correlations between concentrations of certain trace elements (particularly potassium, barium, strontium, and thorium) in subducting sediments with average concentrations of these elements in the overlying arc lavas (Plank and Langmuir, 1993; Plank, 2005) are also in general compatible with the retention of the bulk of these elements (at least in the sediments) to depths beneath arcs.

For metabasaltic rocks, the depth to the blueschist–eclogite transition (which is strongly dependent on the thermal structure of a given subduction zone), and the breakdown of lawsonite within the eclogite facies, likely play key roles in the geochemistry of slab fluids (Peacock, 2003; Hacker *et al.*, 2003a, b; see discussion of fluids at this transition by Manning, 1998). However, the rather large number of recently published slab *P–T* paths, with various assumptions made regarding frictional heating and, particularly, mantle wedge viscosity, leaves one with the impression that it is not possible to pinpoint the temperature of the slab–mantle interface in any given arc (see Peacock *et al.*, 2005). Some recent models suggest that the slab–mantle interface is, for a number of margins, warm enough for sediment melting and nearly warm enough to approach the wet solidus of altered basalt (Peacock *et al.*, 2005). Peacock (2003) demonstrates that the blueschist (or for the hotter Nankai subduction zone, the greenschist)–eclogite transition, for rocks at or near the slab–mantle interface, could occur at depths ranging from ~50 to 110 km (see Figure 3). The majority of the eclogites that have been investigated for their geochemistry have estimated peak metamorphic pressures within, but at the lower end of this range (peak pressures corresponding to ~50–90 km, i.e., mostly not in the coesite stability field; see Figure 6). Models of sediment devolatilization also indicate significant dehydration and decarbonation in the forearc (Kerrick and Connolly, 2001; Gorman *et al.*, 2006), with an apparent locus of fluid loss in the 60–80 km depth range.

Arc lavas overlying cooler subducting slabs (e.g., Marianas and Kuriles) have higher concentrations of Cs, B, As, and Sb than the arc lavas from "warmer" subduction zones (e.g., for Cascadia, Kirby *et al.*, 1996; discussion of arc data by Noll *et al.*, 1996). This is manifest in the arc lavas from cool subduction zones showing higher ratios of these elements to less fluid-mobile elements having similar incompatibility during partial melting in the mantle wedge (e.g., B/Be, Cs/Th, As/Ce, and Sb/Ce; see Bebout *et al.*, 1999). The lower ratios in margins such as Cascadia are presumed to reflect the removal of these elements via metamorphic devolatilization in the relatively warm forearc (see Bebout *et al.*, 1999; Figure 11d), making these elements unavailable for delivery into the corresponding arc.

3.7.2 Cross-Arc Variations

3.7.2.1 Devolatilization

The cross-arc variations observed in some isotope and trace-element ratios (see review by Figure 21) provide tantalizing evidence that

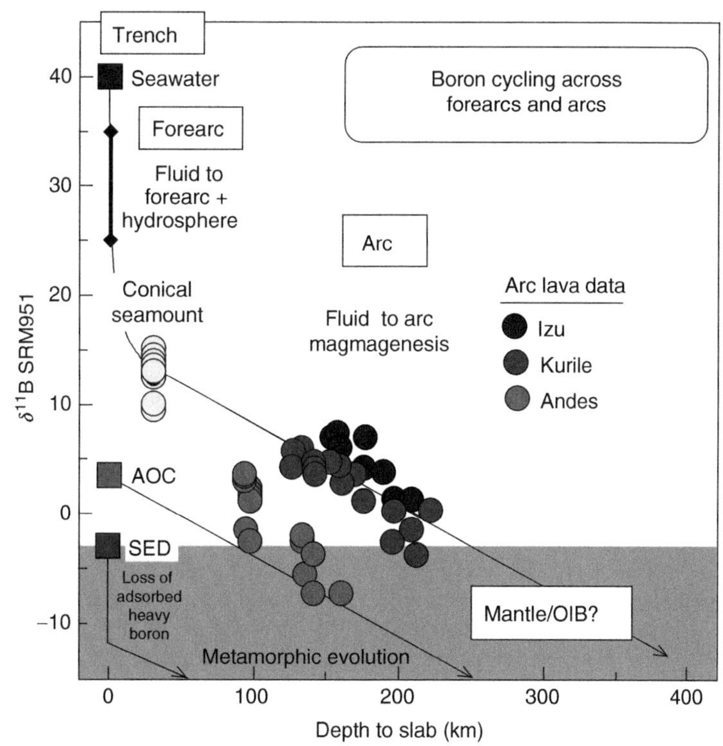

Figure 21 Cross-arc boron-isotope evolution, taking into account compositions of near-trench fluids, seawater, altered oceanic crust, and seafloor sediment (before and after loss of high-δ^{11}B adsorbed B) (Ishikawa and Nakamura, 1994; Ishikawa and Tera, 1997; Rosner *et al.*, 2003), and approximate trends in δ^{11}B (arrows) with prograde metamorphism of AOC and sediment (from the model of Bebout and Nakamura, 2003). Loss of relatively isotopically "heavy" B from metamorphosing sediment and altered oceanic crust, in varying proportions, could produce the across-arc trends. Modified with permission from Benton *et al.* (2001).

slab metamorphism results in the "distillation" of some components (resembling the losses in warmer units of the Catalina Schist), as it traverses the subarc region. Such observations lead to hypotheses that can be tested by further work on UHP metamorphic suites. These across-arc variations have been interpreted to diminishing fluid additions, changing fluid compositions, and/or changes in the proportions of hydrous fluids and silicate melts toward the back-arc (see discussion in Bebout *et al.*, 1999). The recently proposed transitions in the physical and chemical transport properties of "fluids" from subducting slabs (see Kessel *et al.*, 2005; Hermann *et al.*, in press) will need to be considered in future geochemical field studies of UHP suites that potentially represent the residues of fluid and element removal at >100 km depths in subduction zones. This could be achieved by investigating UHP suites straddling and above the wet solidi in pressure and temperature, and representing a range of depths consistent with those of the slab–mantle interface across arcs, and attempting to identify any differing styles of element loss (or gain) related to interactions with these "fluids" of differing types.

3.7.2.2 Boron and lithium isotopes

Among the tracers showing variation across arcs, boron and lithium and their isotopes are particularly noteworthy. Because both are fluid-mobile, these elements are thought to be great "fluid tracers," but with different fluid-mineral partition coefficients (See Marschall *et al.*, 2006). Boron is extremely incompatible in the mantle wedge and, in arcs, it provides a record of very recent boron flux off the slab (Figure 21), whereas lithium is believed to be more compatible and to reside in the mantle wedge for longer periods of time (see Tomascak *et al.*, 2002; Benton *et al.*, 2004). This difference in compatibility in the mantle wedge explains why lithium does not show cross-arc variations as pronounced as those for boron (Moriguti *et al.*, 2004). Moreover, a boron-rich phase, tourmaline, occurs in HP and UHP metasedimentary suites (Nakano and Nakamura, 2001; Bebout and Nakamura, 2003), whereas lithium has no analogous phase in which it is a primary constituent; this likely produces a differing overall behavior in subducting slabs in response to complex mineral reactions. Wunder *et al.* (2005, 2006) published experimentally determined boron-isotope fluid-mineral fractionation factors that appear wholly consistent with the variations observed across arcs, and lithium-isotope fractionations that appear to be consistent with the shifts to lower δ^7Li observed for eclogites as a result of devolatilization (Zack *et al.*, 2003).

Figure 21 presents some current thinking regarding trench-to-subarc (and beyond) cycling of boron, and associated boron-isotope shifts (from Benton *et al.*, 2001). This model sets forth some testable hypotheses regarding the evolution of boron concentrations and isotopic compositions in seafloor rocks subducted beneath arcs, beginning with very early loss of isotopically heavy adsorbed boron in forearc sediments (see You *et al.*, 1995). According to this model, isotopically heavy boron is lost from subducting rocks (presumably both AOC and sediment; see Bebout and Nakamura, 2003), more significantly in some margins than others (depending on thermal evolution), and this loss is likely to be primarily from micas (Bebout *et al.*, 1993). Recently published fractionation factors for boron isotopes between mica and fluid (Wunder *et al.*, 2005) confirm the previous expectations regarding the direction and magnitude of this fractionation (see Peacock and Hervig, 1999). Calculations of boron-isotope evolution for forearc sedimentary and mafic rocks, with boron housed in mica, were presented by Bebout and Nakamura (2003; also see Nakano and Nakamura, 2001 and Marschall *et al.*, in press), along with evidence from isotopically zoned tourmaline for this prograde evolution. At greater depths beneath arcs, boron-isotopic evolution (and boron cycling) should continue to be strongly dictated by reactions involving mica (and tourmaline), whether they are in equilibrium with hydrous fluids, silicate melts, or supercritical liquids; this processing beyond the forearc should result in a further decrease in δ^{11}B (observed for a number of across-arc suites; see Ishikawa and Nakamura, 1994; Ishikawa and Tera, 1997; Rosner *et al.*, 2003; Figure 21). Based on aqueous fluid-mineral partitioning experiments, Brenan *et al.* (1998) calculated that B/Be and B/Nb should both be reduced considerably by the time the slab is subucted to 200 km, providing a possible explanation for observed cross-arc trends in these ratios.

The lithium cycle in convergent margins, starting on the seafloor with early diagenetic processes, has recently been discussed by Moriguti and Nakamura (1998), Tomascak *et al.* (2002), Benton *et al.* (2004), Elliott *et al.* (2004), and Magna *et al.* (in press). Benton *et al.* (2004; also see Tomascak *et al.*, 2002) suggested that, whereas B/Be and boron isotopes in arc lavas record the latest fluid additions to the subarc mantle, lithium isotopes reflect the aggregate longer-term modification and heterogeneity in the mantle source—this factor smoothes or "erases" the subduction input signal for lithium isotopes (see Wunder *et al.*, 2006). Zack *et al.* (2003) presented data for eclogites from the Alps that suggest shifts to lower δ^7Li in these deeply subducted mafic rocks, which they related to metamorphic devolatilization. Lithium is generally thought to be more compatible in phases such as chlorite, the micas, and in eclogites, clinopyroxene (see Bebout *et al.*, 1999, in press). Bebout *et al.* (in press; also see Bebout and Fogel, 1992) documented little or no loss of lithium from metasedimentary rocks that experienced significant loss of boron, cesium, and nitrogen (data for white-mica boron and cesium in Figure 11). The model for fractionation set forth by Zack *et al.* (2003) for eclogites requires testing in studies of other suites, particularly suites for which there are better constraints on the fraction of lithium lost.

3.7.3 Carbon and Nitrogen Input and Arc Output

An example of the exciting synergy possible among studies of oceanic crustal inputs, subduction zone metamorphic rocks, and volcanic gas outputs is provided by the work on C–N cycling at the Costa Rica subduction zone. Recent work by Fischer *et al.* (2002), Hilton *et al.* (2002), Snyder *et al.* (2003), Shaw *et al.* (2003), and Zimmer *et al.* (2004) provide an assessment of the volcanic gas flux of carbon and nitrogen, including estimates of the proportions of endmember sources contributing to the gas compositions. Studies of AOC and sedimentary inputs to subduction zones and studies of metamorphic devolatilization (Bebout *et al.*, 1999; Sadofsky and Bebout, 2004a; Kerrick and Connolly, 2001; Li and Bebout, 2005) can be merged with the volcanic gas record to ascertain efficiencies of carbon and nitrogen return to the atmosphere through subduction zones. Studies of HP and UHP suites indicate that a considerable fraction of subducted carbonate (particularly when pure) is very deeply subducted (examples include Lago di Cignana, Schistes Lustres, and Dabieshan (Reinecke, 1998; Agard *et al.*, 2002; Mahlen *et al.*, 2005). It appears that graphite is also quite stable; in the Catalina Schist, even partially melted rocks retain most of their sedimentary organic carbon as graphite (Bebout, 1995). Sadofsky and Bebout (2004a) and Li and Bebout (2005) discuss the large effect on calculated arc carbon and nitrogen return, relative to subduction inputs, when the estimates of C–N input are based on analyses of seafloor sediment just outboard of the Izu-Bonin and Central America trenches. For the Central America margin, Fischer *et al.* (2002) suggested that arc volcanic nitrogen output is approximately equal to the sedimentary subduction input of nitrogen, using estimates of sedimentary nitrogen concentration from Bebout (1995). Use of estimates of sedimentary nitrogen subduction input derived from study of the sedimentary section outboard of the Central American trench (at ODP Site 1039) indicates that arc return is nearer 50% of the sedimentary input (see Li and Bebout, 2005).

There are some indications that in relatively cool subduction zones, carbon- and nitrogen-isotope compositions of seafloor lithologies are preserved to at least 90 km depth. As noted above, the deeply subducted Schistes Lustres appear to preserve $\delta^{15}N$ like that of their seafloor protoliths (Busigny *et al.*, 2003), and in the Catalina Schist, only rocks subducted along warmer *P–T* paths show significant change in nitrogen-isotope composition (Bebout and Fogel, 1992). Carbonate-rich lithologies appear to preserve seafloor-like carbon- and oxygen-isotope compositions, except in zones infiltrated by fluids from adjacent disparate lithologies, or where metasedimentary carbon reservoirs isotopically exchange during heating, resulting in evolving $\delta^{13}C$ (see discussion below; Busigny *et al.*, 2003). Studies of arc volcanic gases employing noble gases have estimated the proportions of carbon- and nitrogen-isotope sources for the gases using endmember isotopic compositions based on seafloor compositions shifted isotopically by metamorphism (see Hilton *et al.*, 2002). At present it appears that this approach is justified, although the shifts in $\delta^{15}N$ have thus far been documented only for HP metasedimentary rocks thought to have been metamorphosed along *P–T* paths warmer than those experienced in most modern convergent margins (the high-grade Catalina Schist metasedimentary rocks; Bebout and Fogel, 1992; Bebout *et al.*, 1999).

3.7.4 Other Stable Isotope Tracers (Oxygen, Hydrogen, and Chlorine)

It is generally accepted that the oxygen-isotope signatures of ocean-floor basalts are preserved through subduction-zone metamorphism, and conveyed into the deep mantle; however, the oxygen-isotopic compositions of mafic rocks in HP and UHP metamorphic suites can locally be modified by subduction-zone fluid–rock interactions (e.g., addition of sedimentary signatures, see above). Devolatilization is not expected to appreciably shift the oxygen-isotope composition of the rocks, as the majority of the oxygen reservoir in subducting rocks is retained in the residue. However, melting and dehydration of various lithologies (sedimentary, mafic, and ultramafic) could produce "fluids" of varying $\delta^{18}O$ that could be identifiable in studies of volcanic arcs. Based on oxygen-isotope compositions of olivine and plagioclase phenocrysts in arc basalts and basaltic andesites, Eiler *et al.* (2005) argued for varying contributions of low-$\delta^{18}O$, solute-rich aqueous fluids derived from dehydration of hydrothermally altered basaltic crust and partial melts of high-$\delta^{18}O$ sediments to the lavas of the Central American arc (see study of along-strike chemical variations by Carr *et al.*, 2003). Bindeman *et al.* (2005) obtained oxygen-isotope compositions for a suite of arc lavas, from various localities, previously hypothesized as reflect slab melting, and suggested that the $\delta^{18}O$ values of these lavas probably reflect varying degrees of slab and sediment melting, additions of slab-derived aqueous fluids, or crustal contamination in the over-riding plate.

Bebout (1991a) suggested that the hydrogen-isotope compositions of hydrous minerals in the Catalina Schist metasomatic rocks could reflect the entrainment of seawater-like signatures to great depth via a complex hydration–dehydration cycle beginning on the seafloor, during fluid–rock interactions at mid-ocean ridges and during diagenesis (calculated δD ranges from −30‰ to 0‰; also see Magaritz and Taylor, 1976, who considered the possibility of deep seawater entrainment into subduction zones). Oceanic igneous rocks are hydrated and undergo hydrogen-isotope exchange on the seafloor over approximately the same temperature range at which hydrogen is released as H_2O in subduction zones. This could result in release of H_2O-rich fluids in subduction zones that have hydrogen-isotope compositions near those of seawater (Bebout, 1991a). Earlier work by Poreda (1985), and later work by O'Leary *et al.* (2002) invoke additions of "fluids" with δD heavier than the mantle values near −85‰ that appear to be attributable in some way to a multistage process of seafloor hydration (mafic or ultramafic) and subduction-zone dehydration or melting.

Chlorine-isotope studies, aimed at defining the ocean-mantle Cl subduction cycle, are still mostly on the horizon, with early studies mostly identifying the dominant reservoirs for Cl subduction, the magnitude of Cl flux at convergent margins and mechanisms of Cl mobility and release into arc lavas (see discussion by Sharp and Barnes, 2004). Insufficient work, other than that on fluid inclusions and shallower fluids in accretionary prisms (for the latter, see Spivack *et al.*, 2002), has been directed at the subduction-zone metamorphic pathway to greater depths in the forearc and beneath arcs (in response to metamorphic reactions in subducting sediment and oceanic crust) and its possible effects on the size and isotopic changes of Cl being added to arc source regions or the deeper mantle (see discussion by Hermann *et al.*, in press). In their study of seafloor serpentinites, Sharp and Barnes (2004) identified water-soluble and water-insoluble fractions of chlorine with somewhat different chlorine-isotope compositions, and suggested that the chlorine subduction flux in serpentinite in oceanic crust could be as much as twice that estimated by Ito *et al.* (1983), who took into account sediment and oceanic crust containing no serpentinite (also see Jarrard, 2003). A number of authors have argued for the large impact of serpentinite dehydration in the overall slab release of chlorine (Philippot *et al.*, 1998; Scambelluri and Philippot, 2001; Sharp and Barnes, 2004).

3.7.5 The Nature of the Slab-Derived "Fluid" in Arc Magmas

As noted briefly above, it has been common for those researching slab additions to arc magmas to call upon varying combinations of hydrous fluid and silicate melts (with significant amounts of dissolved H_2O) as transport agents. Melts (generally of sediment) have been invoked to explain enrichments in Th relative to LREE and enrichments in light/middle REE and ^{10}Be (subducted in only young seafloor sediment; see discussion in). Recent experimentally determined datasets for element partitioning among aqueous fluids (of varying chlorinity), hydrous melts, and "supercritical liquids" demonstrate the capability of these "fluids" to convey these trace elements from subducting slab-crust lithologies into the mantle wedge (e.g., Ayers and Watson, 1991; Brenan *et al.*, 1995b; Keppler, 1996; Tatsumi and Kogiso, 1997; Johnson and Plank, 1999; Kessel *et al.*, 2005). These experimental studies have produced compelling arguments that these "fluids" are capable of producing many of the distinctive signatures of arc magmatism (e.g., the low Ce/Pb and low-HFSE concentrations of arc lavas; see Brenan *et al.*, 1994, 1995a; Ayers and Watson, 1991).

Discussion of the physicochemical characteristics of the slab "fluid(s)" (see Manning, 2004) ties in with the recent development of thermal models of subduction, which indicate that the slab–mantle interface, containing sediments and uppermost AOC, could be at or near the wet solidus for both lithologies. A comparison of calculated peak metamorphic *P–T* for eclogites (see Figure 6) and recently published thermal models for subduction (see models from Peacock, 2003, in Figure 3) demonstrates that the peak *P–T* conditions for a number of UHP suites fall near or above the wet solidi, and geochemical work on such rocks can potentially yield insight regarding the nature of "fluids" produced from such rocks at these conditions (see Hacker *et al.*, 2005). Hermann *et al.* (in press) provide a discussion of the current debate regarding whether the slab "fluid" is an alkali-chloride aqueous fluid (Keppler, 1996), a silicate melt, or a transitional supercritical liquid (Kessel *et al.*, 2005). More dilute aqueous solutions (e.g., present at the blueschist to eclogite transition at ~60 km depth) are regarded as being relatively ineffective in removing trace elements, based on experimentally derived partitioning data (see Hermann *et al.*, in press), whereas silicate melts and supercritical liquids appear more viable as agents for transfer of slab components into arc source regions and appear to be able to explain many of the specific enrichments invoked in studies of arc magmatism.

The Catalina Schist provides an unusual opportunity to examine evidence for partial melting of mafic and sedimentary rocks at amphibolite-facies conditions in a subduction zone (650–750 °C; 0.8–1.2 GPa; see Figures 6 and 9; Sorensen and Barton, 1987; Bebout, 1989; Bebout and Barton, 1993). Mafic blocks surrounded by largely metaultramafic mélange show migmatitic textures (Sorensen and Barton, 1987; Sorensen and Grossman, 1989), as do more coherent tracts of metagabbroic and metasedimentary rocks. Moreover, the mélange matrix contains plagioclase-rich (trondhjemitic) pegmatites. Because O–H isotopic data indicate infiltration of the mélange and the more permeable parts of the more coherent rocks by H_2O-rich fluids (Bebout and Barton, 1989; Bebout, 1991a), the amphibolite-facies unit appears to preserve evidence for mass transfer via both hydrous fluids and silicate melts, the latter represented by the pegmatites. This provides a possible analog to the models in which sedimentary rocks at or near the slab–mantle interface experience infiltration by fluids generated at greater depths, resulting in element stripping by hydrous fluids or silicate melts (see Breeding *et al.*, 2004; Hermann *et al.*, in press).

3.8 BEYOND ARCS

As discussed in chapter 18, it has become relatively routine to associate certain geochemical signatures (high-μ ($\mu = (^{238}U/^{204}Pb)_{t=0}$) (HIMU), enriched mantle (EM)-1 EM-2, and others) of the deep-mantle sampled by OIBs to varying combinations of subducted oceanic sediment, oceanic lithosphere, and oceanic island and plateau materials that may have experienced considerable ageing during long-term residence in the deep mantle (see Hart and Staudigel, 1989). These studies generally conclude that some subduction-zone chemical change is also required (Hart and Staudigel, 1989; Kelley *et al.*, 2005; Elliott *et al.*, 1999; Bach *et al.*, 2003; Chauvel *et al.*, 1995). A large amount of chemical change can be imagined as the result of complex dehydration and melting processes across deeper parts of the forearc and beneath arcs. The challenge in future studies is to identify these changes in exhumed HP and UHP rocks, and how these changes relate to the proposed transitions in subduction "fluids" from more dilute aqueous solutions, to "transitional fluids," then to hydrous silicate melts (see discussion by Hermann *et al.*, in press). The work thus far on metamorphic geochemistry in subduction zones has mostly been conducted on suites representing subduction to $\leq$100 km in *P–T* regimes dominantly generating hydrous fluids (Figure 6). It remains unclear whether certain element losses from subducting rocks over the full forearc to subarc depth range can produce appropriate chemically fractionated residues that match the source characteristics of OIBs (Figure 1). In particular, the role of supercritical liquids or silicate melts needs to be more thoroughly evaluated if these are shown to be important agents of slab flux beneath arcs (i.e., at >100 km depths). The greater LILE depletions in HP and UHP suites that have experienced partial melting (Becker *et al.*, 2000; Yamamoto *et al.*, 2002; LILE data for granulites in Hart and Reid, 1991; also see Hacker *et al.*, 2005, and a comparison of hydrous-fluid–rock and melt-rock partition coefficients by Kessel *et al.*, 2005) are a reminder that we must conduct further geochemical studies on UHP suites on or above wet and dehydration solidi (Figure 6).

It has long been recognized that low-temperature hydrothermal alteration on the seafloor shifts the $\delta^{18}O$ of shallow oceanic crust to higher values, and at greater depths and higher temperatures to lower values (Gregory and Taylor, 1981). This record of isotopic change can be preserved in HP and UHP metamorphic suites, suggesting that it can be conveyed to great depths in subduction zones without significant alteration (see Cartwright and Barnicoat, 1999). Eiler *et al.* (2000b) demonstrated correlations between $\delta^{18}O$ and other indicators of heterogeneous sources in MORB and proposed that the variations in oxygen-isotope composition reflect varying influence of recycled materials.

Bebout and Kump (2005) considered how the bulk isotopic composition of subducting carbon (sedimentary oxidized and reduced carbon, carbonate in AOC), the carbon inputs and outputs for arcs and whole-mantle, and the effects of varying extents of deep carbon subduction (i.e., retention during metamorphism) influence the mantle carbon udget (Figure 22). Metamorphic modifications of endmembers could have significant consequences for this mass balance (e.g., estimating proportions of inputs from sediments, AOC, mantle wedge, deeper mantle). The common occurrence of marble in UHP suites (Becker and Altherr, 1992; Zhang and Liou, 1996; Reinecke, 1998; Ogasawara *et al.*, 2000; Zhang *et al.*, 2003; Ohta *et al.*, 2003), with oxygen- and carbon- isotope compositions resembling those of seafloor carbonate, is evidence for the preservation of relatively pure carbonate lithologies to well beyond 100 km depths. Such preservation is predicted by models of decarbonation, depending on the extents of infiltration by H_2O-rich fluids capable of driving decarbonation reactions (Kerrick and Connolly, 2001; Gorman *et al.*, 2006). Data for higher-grade lithologies containing both oxidized (carbonate) and reduced (graphite) carbon indicate significant carbon-isotope exchange of the two reservoirs during heating. The related evolution in carbon-isotope compositions of the two reservoirs, together with loss of carbon during devolatilization, could affect the carbon-isotope mass balance for reduced–oxidized carbon entering the deep mantle (see Agrinier *et al.*, 1985). Diamonds provide a possible, but highly convoluted, record of long-term cycling of carbon (and nitrogen) into the deep mantle (see Cartigny, 2005) and there has been considerable debate regarding the extent to which diamond carbon- (and nitrogen-) isotope compositions in some cases reflect the subduction of organic signatures (see Deines, 2002).

The U/Th/Pb system (Hart and Staudigel, 1989; Elliott *et al.*, 1999), and its possible

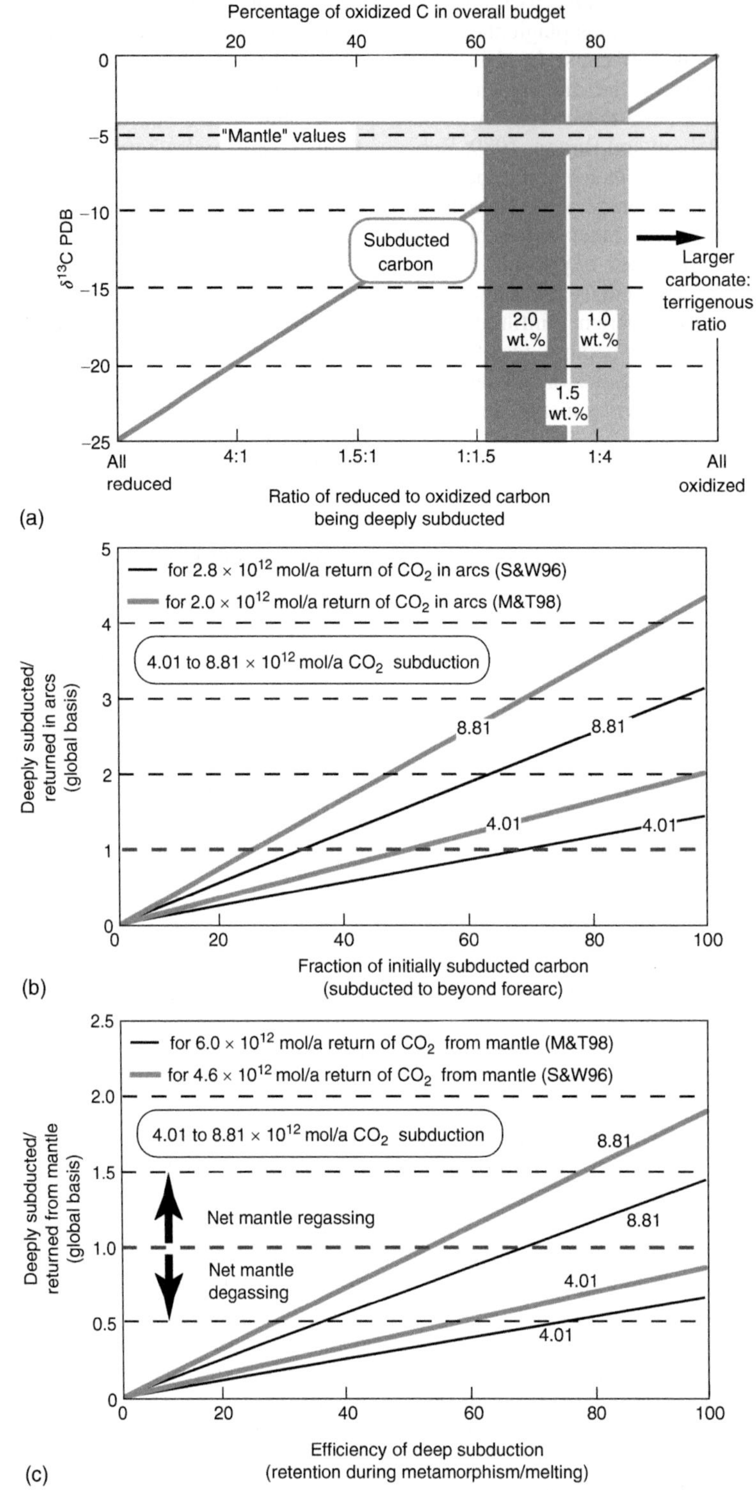

Figure 22 Carbon cycling into and out of the mantle. (a) Demonstration that, using the best estimates of modern C subduction in AOC and sediment, the canonical 1:4 ratio of reduced/oxidized C, estimated based on the mantle $\delta^{13}C$, is roughly achieved (depending on what reduced/organic C estimate is used for the "terrigenous" component in GLOSS). In (b), (c), an estimate of global C subduction in AOC and sediment is compared with two recently published estimates of C outgassing in MORB, OIB, and subduction zones (Marty and Tolstikhin, 1998; Sano and Williams, 1996). From the comparison in (a), it appears that 40±20% of subducted C (here as CO_2) is returned via arc volcanic degassing. From (c), it appears that within error, the mantle could be near steady state with respect to its C budget (i.e., subduction input nearly matches output in MORB, OIB, and volcanic arcs, but with some large uncertainties). From Bebout and Kump (2005); also see Sadofsky and Bebout (2003).

disruption by subduction-zone metamorphism, has been of great interest in chemical geodynamic models (see Miller *et al.*, 1994; Brenan *et al.*, 1995a; Bach *et al.*, 2003; Kelley *et al.*, 2005; Keppler (1996) demonstrated that Pb partitions more strongly into alkali-chloride aqueous fluid than thorium and uranium. Bach *et al.* (2003) concluded that precursors for HIMU (mantle sources characterized by high μ values, with $\mu = (^{238}U/^{204}Pb)_{t\ =\ 0}$) can be produced from AOC after 1–2 Ga, if ~80–90% lead, 40–55% rubidium, 40% strontium, and 35–40% uranium are removed during dehydration of these lithologies. Kelley *et al.* (2005) concluded that uranium and lead have different "depth-loss distribution" through subduction zones, with lead loss from AOC at shallower levels and more uranium loss at greater depths. Chauvel *et al.* (1995) suggested that subducted sulfides house the bulk of the lead and that desulfidization reactions account for the large amounts of lead addition to arcs (Figure 2a). Experiments by Brenan *et al.* (1995a) indicated that dehydration of basaltic rocks (at 900 °C, 2.0 GPa) can produce aqueous fluids with excess uranium relative to thorium (cf. data for hydrous fluids in Kessel *et al.*, 2005). Only a few studies of HP and UHP rocks have addressed these issues (see Becker *et al.*, 2000; John *et al.*, 2004; Spandler *et al.*, 2004; Figure 8) and the results are somewhat inconclusive, pointing to a number of experiments that are needed on deeply subducted mafic and sedimentary rocks. The possible change in Th/U in HP and UHP mafic rocks, relative to either fresh or altered MORB (Figure 8d), warrants further investigation. The study of sedimentary–ultramafic metasomatic exchange by Breeding *et al.* (2004; Figure 16) seemingly demonstrates another way in which thorium, uranium, and lead could be fractionated along the slab–mantle interface.

Metamorphic disruption of the rubidium–strontium isotope system is expected in subduction zones, given that rubidium and strontium partitioning differs significantly in relevant fluid–solid systems (see Kessel *et al.*, 2005). Mahlen *et al.* (2005) suggested that Rb–Sr systematics were not disrupted by HP and UHP metamorphism of metasedimentary rocks in the Zermatt–Saas complex of the Western Alps. The Sm–Nd system should be more robust than the Rb–Sr system during metamorphism, a hypothesis that is testable by focused study of HP and UHP metamorphic suites (see Bebout, in press). Work on mélange formation documents robust Sm–Nd behavior in even highly metasomatized rocks in these zones of enhanced fluid flux (King *et al.*, 2006).

Finally, the changes to the Re/Os system due to subduction zone metamorphism have been considered by Becker (2000), who analyzed samples from various Alpine suites and suggested possible Re loss relative to Os during dehydration. Like the systems discussed above, "subduction processing" is invoked to varying degrees to explain aspects of crust–mantle Re–Os mixing (see discussion by Peucker-Ehrenbrink *et al.*, 2003). Some studies present evidence for extraction of Os from the subducting slab (e.g., Brandon *et al.*, 1996; McInnes *et al.*, 1999), and work by Becker (2000) on metamorphic rocks indicated extraction of Re. However, other work indicated no significant loss of Os from the subducting slab (Becker, 2000) and that Os is not added significantly to arc rocks (Borg *et al.*, 2000; Woodland *et al.*, 2002; Chesley *et al.*, 2002). Further work directed at the HP and UHP suites is required to evaluate possible metamorphic Re and Os mobility in fluids and the possible modification of deeply subducted reservoirs.

3.9 SOME FINAL COMMENTS

A large number of insights regarding long-term crust–mantle mixing, continental crust formation, and even ocean chemistry, can be gained by a better understanding of trench to deep-mantle chemical cycling. Studies of B, Li, C, and N cycling into subduction zones stand out as more broadly incorporating the full set of trench-to-subarc processes (beginning with sedimentation, diagenesis, and hydrothermal alteration on the seafloor). The reader will be struck by the relatively small number of studies presenting more comprehensive chemical datasets for HP and UHP metamorphic rocks—this field is poised to advance rapidly (see Bebout, in press). Further work should be partly aimed at the study of fluid and element mobility at slab–mantle interface depths corresponding to those across arcs (~100–250 km), and identification of any differences in element loss as functions of changing nature of any "fluid" emanating from these deeply subducted rocks and potentially contributing to arc magmatism. Microanalytical methods should be further applied in attempts to extract information regarding element redistribution and isotope fractionation at scales of hand specimens, particularly in cases where exhumation-related overprinting is more extreme limiting whole-rock geochemical work. Some combination of geophysical and field geochemical approaches should be aimed at evaluating the volumetric significance of mélange-like mixing zones at the slab–mantle interface at subarc depths and the geochemical impact of fluid interactions along these zones. This field will, in general benefit from a better merging of field geochemical approaches with knowledge of mineral chemistry and reactions, kinetics and disequilibrium, prograde and exhumation-related *P–T* paths, fluid flow and fluid–rock interactions, and experiments bearing on the physical and chemical properties of fluids at HP and UHP metamorphic conditions.

Recent geochemical work on subduction-zone chemical cycling has increasingly been integrated with experimental, theoretical, and geophysical observations (e.g., Kirby *et al.*, 1996; Hacker *et al.*, 2003a, b; Peacock *et al.*, 2005), and this approach is likely to be the most productive. Geophysical constraints are necessary to provide context and a sense of scale, as geochemical research on the complex metamorphic record can "miss the forest for the trees" if it is focused only on the small scales without appreciation of how observations at that scale fit into the larger picture. The recent MARGINS Subduction Factory and SEIZE endeavors, which focus on the Central America arc-trench system, have led to successful merging of diverse geophysical, theoretical, and geochemical observations aimed at the problem of subduction-zone energy and mass transfer. The geochemical research conducted on the Central America margin demonstrates the need to consider mass balance on an individual convergent margin basis, and the hazards of using global estimates of the geochemistry of subducting lithologies known to show great variation on the seafloor and among individual margins. In this type of approach, insights gained through study of paleosubduction suites (e.g., degrees to which elements are retained or lost during HP to UHP metamorphism) can be used to help reconcile estimated subduction inputs and outputs.

ACKNOWLEDGMENTS

I would like to acknowledge recent collaborations with, and support from E. Nakamura, T. Moriguti, and K. Kobayashi (and other colleagues at the Institute for Study of the Earth's Interior, Misasa, Mark Barton, Marty Grove, Japan). Thanks also go to my other recent collaborators, Robbie King, Long Li, Seth Sadofsky, Mark Barton, Marty Grove, Jeff Ryan, Bill Leeman, Marilyn Fogel, Ann Bebout, Philippe Agard, Bas van der Klauw, Colin Graham, and Lee Kump. Finally, I would like to thank Roberta Rudnick for her assistance (and patience) as the editor for this manuscript and John Ayers and Roberta Rudnick for their very useful and constructive reviews.

REFERENCES

Abers G. A. (2000) Hydrated subducted crust at 100–250 km depth. *Earth Planet. Sci. Lett.* **176**, 323–330.

Agard P., Monie P., Jolivet L., and Goffe B. (2002) Exhumation of the Schistes Lustres complex: *in situ* laser probe $^{40}Ar/^{39}Ar$ constraints and implications for the Western Alps. *J. Meta. Geol.* **20**, 599–618.

Agrinier P., Javoy M., Smith D. C., and Pineau F. (1985) Carbon and oxygen isotopes in eclogites, amphibolites, veins and marbles from the Western Gneiss Region, Norway. *Chem. Geol.* **52**, 145–162.

Ague J. J. Models of permeability contrasts in subduction zone melange: implications for gradients in fluid fluxes, Syros and Tinos Islands, Greece. *Chem. Geol.* (in press).

Allègre C. J. (1989) Mantle cycling: process and time scales. In *Crust/Mantle Recycling at Convergence Zones* (eds. S. R. Hart and L. Gulen). NATO (N. Atlantic Treaty Org.), ASI (Adv. Stud. Inst.) Ser. Kluwer, Dordrecht, pp. 1–14.

Alt J. C. and Teagle D. A. H. (2003) Hydrothermal alteration of upper oceanic crust formed at a fast-spreading ridge: mineral, chemical, and isotopic evidence from ODP Site 801. *Chem. Geol.* **201**, 191–211.

Altherr R., Topuz G., Marschall H., Zack T., and Ludwig T. (2004) Evolution of a tourmaline-bearing lawsonite eclogite from the Elekdag area (Central Pontides, N Turkey): evidence for infiltration of slab-derived B-rich fluids during exhumation. *Contrib. Mineral. Petrol.* **148**, 409–425.

Arculus R. J., Lapierre H., and Jaillard E. (1999) Geochemical window into subduction and accretion processes: Raspas metamorphic complex, Ecuador. *Geology* **27**, 547–550.

Ayers J. C. and Watson E. B. (1991) Solubility of apatite, monazite, zircon, and rutile in supercritical aqueous fluids with implications for subduction zone geochemistry. *Phil. Trans. Roy. Soc. Lond.* **335**, 365–375.

Bach B., Peucker-Ehrenbrink B., Hart S. R., and Blusztajn J. S. (2003) Geochemistry of hydrothermally altered oceanic crust: DSDP/ODP Hole 504B—implications for seawater-crust exchange budgets and Sr- and Pb-isotopic evolution of the mantle. *Geochem. Geophys. Geosyst.* **4**, 8904, doi:10.1029/2002GC000419.

Barnicoat A. C. (1988) Zoned high-pressure assemblages in pillow lavas of the Zermatt-Saas ophiolite zone, Switzerland. *Lithos* **21**, 227–236.

Barnicoat A. C. and Cartwright I. (1995) Focused fluid flow during subduction: oxygen isotope data from high-pressure ophiolites of the Western Alps. *Earth Planet. Sci. Lett.* **132**, 53–61.

Barth M. G., Rudnick R. L., Horn I., McDonough W. F., Spicuzza M. J., Valley J. W., and Haggerty S. E. (2001) Geochemistry of xenolithic eclogites from West Africa. Part I: A link between low MgO eclogites and Archean crust formation. *Geochim. Cosmochim. Acta* **65**, 1499–1527.

Bebout G. E. (1989) Geological and geochemical investigations of fluid flow and mass transfer during subduction-zone metamorphism. PhD Dissertation, University of California, Los Angeles, CA, 370pp.

Bebout G. E. (1991a) Field-based evidence for devolatilization in subduction zones: implications for arc magmatism. *Science* **251**, 413–416.

Bebout G. E. (1991b) Geometry and mechanisms of fluid flow at 15 to 45 kilometer depths of an early Cretaceous accretionary complex. *Geophy. Res. Lett.* **18**, 923–926.

Bebout G. E. (1995) The impact of subduction-zone metamorphism on mantle-ocean chemical cycling. *Chem. Geol.* **126**, 191–218.

Bebout G. E. (1997) Nitrogen isotope tracers of high-temperature fluid–rock interactions: case study of the Catalina Schist, California. *Earth Planet. Sci. Lett.* **151**, 77–91.

Bebout G. E. Metamorphic chemical geodynamics of subduction zones. *Earth Planet. Sci. Lett.* (in press).

Bebout G. E., Agard P., King R., and Nakamura E. (2004) *Record of forearc devolatilization in the Schistes Lustres HP–UHP metasedimentary unit Western Alps (Extended Abstract)*. Proceedings of the Eleventh Water-Rock Symposium, Saratoga Springs, June/July.

Bebout G. E. and Barton M. D. (1989) Fluid flow and metasomatism in a subduction zone hydrothermal system: Catalina Schist Terrane, California. *Geology* **17**, 976–980.

Bebout G. E. and Barton M. D. (1993) Metasomatism during subduction: products and possible paths in the Catalina Schist, California. *Chem. Geol.* **108**, 61–92.

Bebout G. E. and Barton M. D. (2002) Tectonic and metasomatic mixing in a high-T, subduction-zone mélange—insights into the geochemical evolution of the slab–mantle interface. *Chem. Geol.* **187**, 79–106.

Bebout G. E., Bebout A. E., and Graham C. M. Cycling of B, Li, and LILE (K, Cs, Rb, Ba, Sr) into subduction zones: SIMS evidence from micas in high-P/T metasedimentary rocks. *Chem. Geol.* (in press).

Bebout G. E. and Fogel M. L. (1992) Nitrogen-isotope compositions of metasedimentary rocks in the Catalina Schist, California: implications for metamorphic devolatilization history. *Geochim. Cosmochim. Acta* **56**, 2139–2149.

Bebout G. E. and Kump L. R. (2005) *Sensitivity of global carbon cycling models to changing subduction fluxes*. Proceedings of the Goldschmidt Conference, Moscow, Idaho.

Bebout G. E. and Nakamura E. (2003) Record in metamorphic tourmalines of subduction-zone devolatilization and boron cycling. *Geology* **31**, 407–410.

Bebout G. E., Ryan J. G., and Leeman W. P. (1993) B–Be systematics in subduction-related metamorphic rocks: characterization of the subducted component. *Geochim. Cosmochim. Acta* **57**, 2227–2237.

Bebout G. E., Ryan J. G., Leeman W. P., and Bebout A. E. (1999) Fractionation of trace elements during subduction-zone metamorphism: impact of convergent margin thermal evolution. *Earth Planet. Sci. Lett.* **171**, 63–81.

Becker H. (2000) Re–Os fractionation in eclogites and blueschists and the implications for recycling of oceanic crust into the mantle. *Earth Planet. Sci. Lett.* **177**, 2300–2887.

Becker H. and Altherr R. (1992) Evidence from ultra-high pressure marbles for recycling of sediments into the mantle. *Nature* **358**, 745–748.

Becker H., Carlson R. W., and Jochum K. P. (2000) Trace element fractionation during dehydration of eclogites from high-pressure terranes and the implications for element fluxes in subduction zones. *Chem. Geol.* **163**, 65–99.

Becker H., Jochum K. P., and Carlson R. W. (1999) Constraints from high-pressure veins in eclogites on the composition of hydrous fluids in subduction zones. *Chem. Geol.* **160**, 291–308.

Ben Othman D., White W. M., and Patchett J. (1989) The geochemistry of marine sediments, island arc magma genesis, and crust-mantle recycling. *Earth Planet. Sci. Lett.* **94**, 1–21.

Benton L. D., Ryan J. G., and Savov I. P. (2004) Lithium abundance and isotope systematics of forearc serpentinites, Conical Seamount, Mariana forearc: insights into the mechanics of slab–mantle exchange during subduction. *Geochem. Geophys. Geosyst.* **5**(8), Q08J12, doi:10.1029/2004GC000708.

Benton L. D., Ryan J. G., and Tera F. (2001) Boron isotope systematics of slab fluids as inferred from a serpentinite seamount, Mariana forearc. *Earth Planet. Sci. Lett.* **187**, 273–282.

Berner R. A. (1999) A new look at the long-term carbon cycle. *GSA Today* **9**, 1–6.

Beyssac O., Rouzaud J.-N., Goffe B., Brunet F., and Chopin C. (2002) Graphitization in a high-pressure, low-temperature metamorphic gradient: a Raman microspectroscopy and HRTEM study. *Contrib. Mineral. Petrol.* **143**, 19–31, doi:10.1007/s00410-001-0324-7.

Bindeman I. N., Eiler J. M., Yogodzinski G. M., Tatsumi Y., Stern C. R., Grove T. L., Portnyagin M., Hoernle K., and Danyushevsky L. V. (2005) Oxygen isotope evidence for slab melting in modern and ancient subduction zones. *Earth Planet. Sci. Lett.* **235**, 480–496.

Blake M. C., Jayko A. S., McLaughlin R. J., and Underwood M. B. (1987) Metamorphic and tectonic evolution of the Franciscan Complex, Northern California. In *Metamorphism and Crustal Evolution of the Western United States, Rubey Metamorphism and Crustal Evolution of the Western United States, Rubey* (ed. W. G. Ernst). Prentice-Hall, Englewood Cliffs, NJ, vol. 7, pp. 1035–1060.

Borg L. E., Brandon A. D., Clynne M. A., and Walker R. J. (2000) Re–Os isotopic systematics of primitive lavas from the Lassen region of the Cascade arc, California. *Earth Planet. Sci. Lett.* **177**, 301–317.

Brandon A. D., Creaser R. A., Shirey S. B., and Carlson R. W. (1996) Osmium recycling in subduction zones. *Science* **272**, 861–864.

Breeding C. M. and Ague J. J. (2002) Slab-derived fluids and quartz-vein formation in an accretionary prism, Otago Schist, New Zealand. *Geology* **30**, 499–502.

Breeding C. M., Ague J. J., and Brocker M. (2004) Fluid–metasedimentary interactions in subduction zone mélange: implications for the chemical composition of arc magmas. *Geology* **32**, 1041–1044.

Brenan J. M., Ryerson F. J., and Shaw H. F. (1998) The role of aqueous fluids in the slab-to-mantle transfer of boron, beryllium, and lithium during subduction; experiments and models. *Geochim. Cosmochim. Acta* **62**, 3337–3347.

Brenan J. M., Shaw H. F., Phinney D. L., and Ryerson F. J. (1994) Rutile-aqueous fluid partitioning of Nb, Ta, Hf, Zr, U, and Th: implications for high field strength element depletions in island-arc basalts. *Earth Planet. Sci. Lett.* **128**, 327–339.

Brenan J. M., Shaw H. F., and Ryerson F. J. (1995a) Experimental evidence for the origin of lead enrichment in convergent margin magmas. *Nature* **378**, 54–56.

Brenan J. M., Shaw H. F., Ryerson F. J., and Phinney D. L. (1995b) Mineral-aqueous fluid partitioning of trace elements at 900 °C and 2.0 GPa: constraints on the trace element geochemistry of mantle and deep crustal fluids. *Geochim. Cosmochim. Acta* **59**, 3331–3350.

Brown K. M., Saffer D. M., and Bekins B. A. (2001) Smectite diagenesis, pore-water freshening, and fluid flow at the toe of the Nankai wedge. *Earth Planet. Sci. Lett.* **194**, 97–109.

Brunsmann A., Franz G., and Erzinger J. (2001) REE mobilization during small-scale high-pressure fluid-rock interaction and zoisite/fluid partitioning of La to Eu. *Geochim. Cosmochim. Acta* **65**, 559–570.

Brunnsmann A., Franz G., Erzinger J., and Landwehr D. (2000) Zoisite- and clinozoisite-segregations in metabasites (Tauern Window, Austria) as evidence for high-pressure fluid-rock interaction. *J. Meta. Geol.* **18**, 1–21.

Busigny V., Cartigny P., Philippot P., Ader M., and Javoy M. (2003) Massive recycling of nitrogen and other fluid-mobile elements (K, Rb, Cs, H) in a cold slab environment: evidence from HP to UHP oceanic metasediments of the Schistes Lustres nappe (Western Alps, Europe). *Earth Planet. Sci. Lett.* **215**, 27–42.

Carr M. J., Feigenson M. D., Patino L. C., and Walker J. A. (2003) Volcanism and geochemistry in Central America: progress and problems. In *Inside the Subduction Factory* (ed. J. Eiler). Am. Geophys. Un. Geophys. Monogr. AGU, Washington, DC, vol. 138, pp. 153–174.

Cartigny P. (2005) Stable isotopes and the origin of diamond. *Elements* **1**, 79–84.

Cartwright I. and Barnicoat A. C. (1999) Stable isotope geochemistry of Alpine ophiolites: a window to ocean-floor hydrothermal alteration and constraints on fluid-rock interaction during high-pressure metamorphism. *Int. J. Earth Sci.* **88**, 219–235.

Chalot-Prat F. (2005) An undeformed ophiolite in the Alps: field and geochemical evidence for a link between volcanism and shallow plate processes. In *Plates, Plumes, and Paradigms* (eds. G. R. Natland, D. C. Presnall, and D. L. Anderson). *Geol. Soc. Am. Spec. Pap.* **388**, 751–780.

Chalot-Prat F., Ganne J., and Lombard A. (2003) No significant element transfer from the oceanic plate to the mantle wedge during subduction and exhumation of the Tethys lithosphere (Western Alps). *Lithos* **69**, 69–103.

Chauvel C., Goldstein S. L., and Hofmann A. W. (1995) Hydration and dehydration of oceanic crust controls Pb evolution in the mantle. *Chem. Geol.* **126**, 65–75.

Chesley J., Ruiz J., Righter K., Ferrari L., and Gomez-Tuena A. (2002) Source contamination versus assimilation: an example from the Trans-Mexican Volcanic Arc. *Earth Planet. Sci. Lett.* **195**, 211–221.

Cloos M. (1986) Blueschists in the Franciscan Complex of California: petrotectonic constraints on uplift mechanisms. In *Blueschists and Eclogites* (eds. B. W. Evans and E. H. Brown). Geol. Soc. Am. Mem, vol. 164, pp. 77–93.

Cloos M. (1993) Lithospheric buoyancy and collisional orogenesis; subduction of oceanic plateaus, continental margins, island arcs, spreading ridges, and seamounts. *Geol. Soc. Am. Bull.* **105**, 715–737.

Compton J. S., Williams L. B., and Ferrell R. E., Jr. (1992) Mineralization of organogenic ammonium in the Monterey Formation, Santa Maria and San Joaquin basins, California, USA. *Geochim. Cosmochim. Acta* **56**, 1979–1991.

Deines P. (2002) The carbon isotope geochemistry of mantle xenoliths. *Earth-Sci. Rev.* **58**, 247–278.

Deville E., Fudral S., Lagabrielle Y., Marthaler M., and Sartori M. (1992) From oceanic closure to continental collision; a synthesis of the "Schistes Lustres" metamorphic complex of the Western Alps. *Geol. Soc. Am. Bull.* **104**, 127–139.

Dixon J. E. and Ridley J. R. (1987) Syros. In *Chemical Transport in Metasomatic Processes* (ed. H. C. Helgeson). Reidel, Boston, pp. 489–501.

Domanik K. J., Hervig R. L., and Peacock S. M. (1993) Beryllium and boron in subduction zone minerals: an ion microprobe study. *Geochim. Cosmochim. Acta* **57**, 4997–5010.

Domanik K. J. and Holloway J. R. (1996) The stability and composition of phengitic muscovite and associated phases from 5.5–11 GPa; implications for deeply subducted sediments. *Geochim. Cosmochim. Acta* **60**, 4133–4150.

Dumitru T. A. (1991) Effects of subduction parameters on geothermal gradients in forearcs, with an application to Franciscan subduction in California. *J. Geophys. Res.* **96** (B2), 621–641.

Eiler J. M., Carr M. J., Reagan M., and Stolper E. (2005) Oxygen isotope constraints on the sources of Central American arc lavas. *Geochem. Geophys. Geosyst.* **6**, Q07007, doi:10.1029/2004GC000804.

Eiler J., Crawford A., Elliott T., Farley K. A., Valley J. V., and Stolper E. M. (2000a) Oxygen isotope geochemistry of oceanic arc lavas. *J. Petrol.* **41**, 229–256.

Eiler J., Schiano P., Kitchen N., and Stolper E. M. (2000b) Oxygen-isotope evidence for recycled crust in the sources of mid-ocean-ridge basalts. *Nature* **403**, 530–534.

Elderfield H., Kastner M., and Martin J. B. (1990) Compositions and sources of fluids in sediments of the Peru subduction zone. *J. Geophys. Res.* **95**, 8819–8827.

Elliott T. (2003) Tracers of the slab. In *Inside the Subduction Factory* (ed. J. Eiler). Am. Geophys. Un. Geophys. Monogr. AGU, Washington, DC, vol. 138, pp. 23–45.

Elliott T., Jeffcoate A., and Bouman C. (2004) The terrestrial Li isotope cycle: light-weight constraints on mantle convection. *Earth Planet. Sci. Lett.* **220**, 231–245.

Elliott T., Zindler A., and Bourdon B. (1999) Exploring the kappa conundrum: the role of recycling in the lead isotope evolution of the mantle. *Earth Planet. Sci. Lett.* **169**, 129–145.

El-Shazly A. K. and Liou J. G. (1991) Glaucophane chloritoid-bearing assemblages from NE Oman: petrologic significance and a petrogenetic grid for high P metapelites. *Contrib. Mineral. Petrol.* **107**, 180–201.

El-Shazly A. K. and Sisson V. B. (1999) Retrograde evolution of eclogite facies rocks from NE Oman: evidence from fluid inclusions and petrological data. *Chem. Geol.* **154**, 193–223.

Feineman M. D., Ryerson F. J., DePaolo D. J., and Plank, T. Zoisite-aqueous fluid trace element partitioning and the affect of temperature on slab-derived fluid composition. *Chem. Geol.* (in press).

Fischer T. P., Hilton D. R., Zimmer M. M., Shaw A. M., Sharp Z. D., and Walker J. A. (2002) Subduction and recycling of nitrogen along the Central American Margin. *Science* **297**, 1154–1157.

Fisher D. M. (1996) Fabrics and veins in the forearc: a record of cyclic fluid flow at depths of <15 km. In *Subduction: Top to Bottom* (eds. G. E. Bebout, D. W. Scholl, S. H. Kirby, and J. P. Platt). Am. Geophys. Un. Geophys. Monogr. AGU, Washington, DC, vol. 96, pp. 75–89.

Fitzherbert J. A., Clarke G. L., and Powell R. (2005) Preferential retrogression of high-P metasediments and the preservation of blueschist to eclogite facies metabasite during exhumation, Diahot terrane, NE New Caledonia. *Lithos* **83**, 67–96.

Foley S. F., Buhre S., and Jacob D. E. (2003) Evolution of the Archaean crust by delamination and shallow subduction. *Nature* **421**, 249–252.

Franz L., Romer R. L., Klemd R., Schmid R., Oberhansli R., Wagner T., and Shuwen D. (2001) Eclogite-facies quartz veins within metabasites of the Dabie Shan (eastern China): pressure–temperature–time-deformation path, composition of the fluid phase and fluid flow during exhumation of high-pressure rocks. *Contrib. Mineral. Petrol.* **141**, 322–346.

Frey M. (1987) Very low-grade metamorphism of clastic sedimentary rocks. In *Low Temperature Metamorphism*. Blackie Acad, New York, pp. 9–58.

Fruh-Green G. L., Scambelluri M., and Vallis F. (2001) O–H isotope ratios of high pressure ultramafic rocks: implications for fluid sources and mobility in the subducted hydrous mantle. *Contrib. Mineral. Petrol.* **141**, 145–159.

Fu B., Zheng Y.-F., and Touret J. L. R. (2002) Petrological, isotopic and fluid inclusion studies of eclogites from Sujiahe, NW Dabie Shan (China). *Chem. Geol.* **187**, 107–128.

Gao J. and Klemd R. (2001) Primary fluids entrapped at blueschist to eclogite transition: evidence from the Tianshan meta-subduction complex in northwestern China. *Contrib. Mineral. Petrol.* **142**, 1–14.

George R., Turner S., Morris J., Plank T., Hawkesworth C., and Ryan J. (2005) Pressure–temperature–time paths of sediment recycling beneath the Tonga-Kermadec arc. *Earth Planet. Sci. Lett.* **233**, 195–211.

Getty S. R. and Selverstone J. (1994) Stable isotopic and trace element evidence for restricted fluid migration in 2 GPa eclogites. *J. Meta. Geol.* **12**, 747–760.

Gieskes J., Blanc G., Vrolijk P., Moore J. C., Mascle A., Taylor E., Andrieff P., Alvarez F., Barnes R., Beck C., Behrmann J., Brown K., Clark M., Dolan J., Fisher A., Hounslow M. W., McLellan P., Moran K., Ogawa Y., Sakai T., Schoonmaker J., Wilkins R., and Williams C. (1989) Hydrogeochemistry in the Barbados accretionary complex: Leg 110 ODP. *Palaeogeogr. Palaeoclimatol. Palaeoecol.* **71**, 83–96.

Gill J. B. (1981) *Orogenic Andesites and Plate Tectonics*. Springer, Berlin, 390pp.

Giorgetti G., Tropper P., Essene E. J., and Peacor D. R. (2000) Characterization of non-equilibrium and equilibrium occurrences of paragonite/muscovite intergrowths in an eclogite from the Sesia-Lanzo Zone (Western Alps, Italy). *Contrib. Mineral. Petrol.* **138**, 326–336.

Glodny J., Austrheim H., Molina J. F., Rusin A. I., and Seward D. (2003) Rb/Sr record of fluid–rock interaction in eclogites: the Marun-Keu complex, Polar Urals, Russia. *Geochim. Cosmochim. Acta* **67**, 4353–4371.

Godon A., Jendrzejewski N., Castrec-Rouelle D. A., Pineau F., Boulegue J., and Javoy M. (2004) Origin and evolution of fluids from mud volcanoes in the Barbados accretionary complex. *Geochim. Cosmochim. Acta* **68**, 2153–2165.

Gorman P. J., Kerrick D. M., and Connolly J. A. D. (2006) Modeling open system metamorphic decarbonation of subducting slabs. *Geochem. Geophys. Geosyst.* **7**, Q04007, doi:10.1029/2005GC001125.

Gray D. R., Gregory R. T., and Durney D. W. (1991) Rock-buffered fluid–rock interaction in deformed quartz-rich turbidite sequences eastern Australia. *J. Geophys. Res.* **93**, 4625–4656.

Green T. H. and Adam J. (2003) Experimentally determined trace element characteristics of aqueous fluid from dehydrated mafic oceanic crust at 3.0 GPa, 650–750 °C. *Eur. J. Mineral.* **15**, 815–830.

Gregory R. T. and Taylor H. P. (1981) An oxygen isotope profile in a section of Cretaceous oceanic-crust, Samail Ophiolite, Oman—evidence for delta-18-O buffering of the oceans by deep (less-than 5 km) seawater-hydrothermal circulation at mid-ocean ridges. *J. Geophys. Res.* **86**(B4), 2737–2755.

Grove M. and Bebout G. E. (1995) Cretaceous tectonic evolution of coastal southern California: insights from the Catalina Schist. *Tectonics* **14**, 1290–1308.

Hacker B. R. Pressures and temperatures of ultrahigh-pressure metamorphism: implications for UHP tectonics and H_2O in subducting slabs. *Int. Geol. Rev.* (in press).

Hacker B. R., Abers G. A., and Peacock S. M. (2003a) Subduction factory. 1: Theoretical mineralogy, densities, seismic wave speeds, and H_2O contents. *J. Geophys. Res.* **108** (B1): 2029, doi:10.1029/2001JB001127.

Hacker B. R., Luffi P., Lutkov V., Minaev V., Ratschbacher L., Plank T., Ducea M., Patino-Douce A., McWilliams M., and Metcalf J. (2005) Near-ultrahigh pressure processing of continental crust: Miocene crustal xenoliths from the Pamir. *J. Petrol.* **46**, 1661–1687.

Hacker B. R., Peacock S. M., Abers G. A., and Holloway S. D. (2003b) Subduction factory. 2: Are intermediate-depth earthquakes in subducting slabs linked to metamorphic dehydration reactions. *J. Geophys. Res.* **108**(B1), 2030, doi:10.1029/2001JB001129.

Hart S. R. and Reid M. R. (1991) Rb/Cs fractionation: a link between granulite metamorphism and the S-process. *Geochim. Cosmochim. Acta* **55**, 2379–2382.

Hart S. R. and Staudigel H. (1989) Isotopic characterization and identification of recycled components. In *Crust/Mantle Recycling at Convergence Zones* (eds. S. R. Hart and L. Gulen). NATO (N. Atlantic Treaty Org.), ASI (Adv. Stud. Inst.) Ser. Kluwer, Dordrecht, pp. 15–28.

Hattori K. H. and Guillot S. (2003) Volcanic fronts form as a consequence of serpentinite dehydration in the forearc mantle wedge. *Geology* **31**, 525–528.

Henry C., Burkhard M., and Goffe B. (1996) Evolution of synmetamorphic veins and their wallrocks through a Western Alps transect: no evidence for large-scale fluid flow. Stable isotope, major- and trace-element systematics. *Chem. Geol.* **127**, 81–109.

Hermann J., Spandler C., Hack A., and Korsakov A. V. Aqueous fluids and hydrous melts in high-pressure and ultra-high pressure rocks: implications for element transfer in subduction zones. *Lithos.* (in press).

Hilton D. R., Fischer T. P., and Marty, B. (2002) Noble gases and volatile recycling at subduction zones. In *Noble Gases in Geochemistry and Cosmochemistry* (eds. D. Porcelli, C. J. Ballentine, and R. Wieler). *Rev. Mineral. Geochem.*, Mineralogical Society of America and Geochemical Society, Washington, DC vol. 47, pp. 319–370.

Hofmann A. W. (1988) Chemical differentiation of the Earth: the relationship between mantle, continental crust, and oceanic crust. *Earth Planet. Sci. Lett.* **90**, 297–314.

Ishikawa T. and Nakamura E. (1994) Origin of the slab component in arc lavas from across-arc variation of B and Pb isotopes. *Nature* **370**, 205–208.

Ishikawa T. and Tera F. (1997) Source, composition and distribution of the fluid in the Kurile mantle wedge: constraints from across-arc variations of B/Nb and B isotopes. *Earth Planet. Sci. Lett.* **152**, 123–138.

Ito E., Harris D. M., and Anderson A. T., Jr. (1983) Alteration of oceanic crust and geologic cycling of chlorite and water. *Geochim. Cosmochim. Acta* **47**, 1613–1624.

Jahn B. M. (1999) Sm–Nd isotope tracer study of UHP metamorphic rocks: implications for continental subduction and collisional tectonics. *Int. Geol. Rev.* **41**, 859–885.

Jarrard R. D. (2003) Subduction fluxes of water, carbon dioxide, chlorine, and potassium. *Geochem. Geophys. Geosyst.* **5**, doi:10.1029/2002GC000392.

Javoy M. (1998) The birth of the Earth's atmosphere: the behaviour and fate of its major elements. *Chem. Geol.* **147**, 11–25.

John T. and Schenk V. (2003) Partial eclogitisation of gabbroic rocks in a late Precambrian subduction zone (Zambia): progade metamorphism triggered by fluid infiltration. *Contrib. Mineral. Petrol.* **146**, 174–191, doi:10.1007/s00410-003-0492-8.

John T., Scherer E. E., Haase K., and Schenk V. (2004) Trace element fractionation during fluid-induced eclogitization in a subducting slab: trace element and Lu–Hf–Sm–Nd isotope systematics. *Earth Planet. Sci. Lett.* **227**, 441–456.

Johnson M. C. and Plank T. (1999) Dehydration and melting experiments constrain the fate of subducted sediments. *Geochem. Geophys. Geosyst.* **1**, paper number 1999GC000014.

Jolivet L., Faccena C., Goffe B., Burov E., and Agard P. (2003) Subduction tectonics and exhumation of high-pressure metamorphic rocks in the Mediterranean orogens. *Am. J. Sci.* **303**, 353–409.

Kastner M., Elderfield H., and Martin J. B. (1991) Fluids in convergent margins: what do we know about their composition, origin, role in diagenesis and importance for oceanic chemical fluxes? *Philos. Trans. Roy. Soc. Lond.* **335**, 243–259.

Kelemen P. B., Rilling J. L., Parmentier E. M., Mehl L., and Hacker B. R. (2003) Thermal structure due to solid-state flow in the mantle wedge. In *Inside the Subduction Factory* (ed. J. Eiler). Am. Geophys. Un. Geophys. Monogr. AGU, Washington, DC, vol. 138, pp. 293–311.

Kelley K. A., Plank T., Farr L., Ludden J., and Staudigel H. (2005) Subduction cycling of U, Th, and Pb. *Earth Planet. Sci. Lett.* **234**, 369–383.

Kelley K. A., Plank T., Ludden J., and Staudigel H. (2003) The composition of altered oceanic crust and ODP Sites 801 and 1149. *Geochem. Geophys. Geosyst.* **4**, 8910, doi:10.1029/2002GC000435.

Keppler H. (1996) Constraints from partitioning experiments on the composition of subduction-zone fluids. *Nature* **380**, 237–240.

Kerrick D. M. (2002) Serpentinite seduction. *Science* **298**, 1344–1345.

Kerrick D. M. and Connolly J. A. D. (2001) Metamorphic devolatilization of subducted marine sediments and the transport of volatiles into the Earth's mantle. *Nature* **411**, 293–296.

Kessel R., Schmidt M. W., Ulmer P., and Pettke T. (2005) Trace element signature of subduction-zone fluids, melts and supercritical fluids at 120–180 km depths. *Nature* **437**/**29**, 724–727.

Kimura G., Silver E. A., Blum, P., *et al.* (1997) Proceedings of the ODP, Initial Reports, 170, College Station, TX (Ocean Drilling Program).

King R. L. (2006) Mélange a trois: metamorphic controls on recycling and mass transfer within subduction zones. PhD Dissertation, Lehigh University.

King R. L., Bebout G. E., Kobayashi K., Nakamura E., and van der Klauw S. N. G. C. (2004) Ultrahigh-pressure metabasaltic garnets as probes into deep subduction-zone chemical cycling. *Geochem. Geophys. Geosyst.* **5**(12): Q12J14, doi:10.1029/2004GC000746.

King R. L., Bebout G. E., Moriguti T., and Nakamura E. (2006) Elemental mixing systematics and Sr–Nd isotope geochemistry of mélange formation: obstacles to identification of fluid sources to arc volcanics. *Earth Planet. Sci. Lett.* **246**, 288–304.

King R. L., Kohn M. J., and Eiler J. M. (2003) Constraints on the petrologic structure of the subduction zone slab–mantle interface from the Franciscan Complex exotic ultramafic blocks. *Geol. Soc. Am. Bull.* **115**, 1097–1109.

Kirby S., Engdahl E. R., and Denlinger R. (1996) Intraslab earthquakes and arc volcanism: dual expressions of crustal and upper mantle metamorphism in subducting slabs. In *Subduction: Top to Bottom* (eds. G. E. Bebout, D. W. Scholl,

S. H. Kirby, and J. P. Platt), Am. Geophys. Un. Geophys. Monogr. AGU, Washington, DC, vol. 96, pp. 195–214.

Kullerud K., Stephens M. B., and Zachrisson E. (1990) Pillow lavas as protoliths for eclogites: evidence from a late Precambrian–Cambrian continental margin, Seve Nappes, Scandinavian Caledonides. *Contri. Mineral. Petrol.* **105**, 1–10.

Kurahashi E., Nakajima Y., and Ogasawara Y. (2001) Coesite inclusions and prograde compositional zonation of garnet in eclogite from Zekou in the Su-Lu ultrahigh-pressure terrane, eastern China. *J. Mineral. Petrol. Sci.* **96**, 100–108.

Kvenvolden K. A. (1993) Gas hydrates—geological perspective and global change. *Rev. Geophys.* **31/2**, 173–187.

Lapen T. J., Johnson C. M., Baumgartner L. P., Mahlen N. J., Beard B. L., and Amato J. M. (2003) Burial rates during prograde metamorphism of an ultra-high-pressure terrane: an example from Lago di Cignana, Western Alps, Italy. *Earth Planet. Sci. Lett.* **215**, 57–72.

Laurora A., Mazzucchelli M., Rivalenti G., Vannucci R., Zanetti A., Barbieri M. A., and Cingolani C. A. (2001) Metasomatism and melting in carbonated peridotite xenoliths from the mantle wedge: the Gobernador Gregores case (Southern Patagonia). *J. Petrol.* **42**, 69–87.

Leeman W. P. (1996) Boron and other fluid-mobile element systematics in volcanic arc lavas: implications for subduction processes. In *Subduction: Top to Bottom* (eds. G. E. Bebout, D. W. Scholl, S. H. Kirby, and J. P. Platt), Am. Geophys. Un. Geophys. Monogr. AGU, Washington, DC, vol. 96, pp. 269–276.

Li L. and Bebout G. E. (2005) Carbon and nitrogen geochemistry of sediments in the Central American convergent margin: insights regarding paleoproductivity and carbon and nitrogen subduction fluxes. *J. Geophys. Res.* **110**, B11202, doi:10.1029/2004JB003276.

Li S., Jagoutz E., Chen Y., and Li Q. (2000) Sm–Nd and Rb–Sr isotopic chronology and cooling history of ultrahigh pressure rocks and their country rocks at Shuanghe in the Dabie Mountains, Central China. *Geochim. Cosmochim. Acta* **64**, 1077–1093.

Liou J. G., Maruyama S., and Cho M. (1987) Very low-grade metamorphism of volcanic and volcaniclastic rocks—mineral assemblages and mineral facies. In *Low Temperature Metamorphism* (ed. M. Frey). Blackie Academy, New York, pp. 59–113.

MacPherson G. J., Phipps S. P., and Grossman J. N. (1990) Diverse sources for igneous blocks in Franciscan mélanges, California Coast Ranges. *J. Geol.* **98**, 845–862.

Maekawa H., Shozui M., Ishii T., Fryer P., and Pearce J. A. (1993) Blueschist metamorphism in an active subduction zone. *Nature* **364**, 520–523.

Magaritz M. and Taylor H. (1976) Oxygen, hydrogen and carbon isotope studies of the Franciscan formation, Coast Ranges, California. *Geochim. Cosmochim. Acta.* **40**, 215–234.

Magna T., Wiechert U., Grove T. L., and Halliday A. N. Lithium isotope fractionation in the southern Cascadia subduction zone. *Earth Planet. Sci. Lett.* (in press).

Mahlen N. J., Johnson C. M., Baumgartner L. P., and Beard B. L. (2005) Provenance of Jurassic Tethyan sediments in the HP/UHP Zermatt-Saas ophiolite, Western Alps. *Geol. Soc. Am.* **117**, 530–544.

Manning C. E. (1998) Fluid composition at the blueschist–eclogite transition in the model system Na_2O–MgO–Al_2O_3–SiO_2–H_2O–HCl, Schweiz. *Mineral. Petrogr. Mitt.* **78**, 225–242.

Manning C. E. (2004) The chemistry of subduction-zone fluids. *Earth Planet. Sci. Lett.* **223**, 1–16.

Marschall H. R., Altherr R., Ludwig T., Kalt A., Gmeling K., and Kasztovszky Z. (2006) Partitioning and budget of Li, Be, and B in high-pressure metamorphic rocks. *Geochim. Cosmochim. Acta* **70**, 4750–4769.

Marschall H. R., Altherr R., and Rupke L. Squeezing out the slab–modeling the release of Li, Be, and B during progressive high-pressure metamorphism. *Chem. Geol.* (in press).

Martin H. (1986) Effect of steeper Archean geothermal gradient on geochemistry of subduction-zone magmas. *Geology* **14**, 753–756.

Marty B. and Tolstikhin I. N. (1998) CO_2 fluxes from mid-ocean ridges, arcs and plumes. *Chem. Geol.* **145**, 233–248.

Maruyama S. and Liou J. G. (1989) Possible depth limit for underplating by a seamount. *Tectonophysics* **160**, 327–337.

Massone H.-J. and Kopp J. (2005) A low-variance mineral assemblage with talc and phengite in an eclogite from the Saxonian Erzgebirge, Central Europe, and its P–T evolution. *J. Petrol.* **46**, 355–375.

Maury R. C., Defant M. J., and Joron J. L. (1992) Metasomatism of the sub-arc mantle inferred from trace elements in Philippine xenoliths. *Nature* **360**, 661–663.

McDonough W. F. and Sun S.-S. (1995) The composition of the Earth. *Chem. Geol.* **120**, 223–253.

McInnes G. I. A., McBride J. S., Evans N. J., Lambert D. D., and Andrew A. S. (1999) Osmium isotope constraints on ore metal recycling in subduction zones. *Science* **286**, 512–516.

Melzer S. and Wunder B. (2001) K–Rb–Cs partitioning between phlogopite and fluid: experiments and consequences for the LILE signatures of island arc basalts. *Lithos* **59**, 69–90.

Messiga B., Tribuzio R., Bottazzi P., and Ottolini L. (1995) An ion microprobe study on trace element composition of clinopyroxenes from blueschist and eclogitized Fe-Ti-gabbros, Ligurian Alps, northwestern Italy: some petrologic considerations. *Geochim. Cosmochim. Acta* **59**, 59–75.

Miller D. M., Goldstein S. L., and Langmuir C. H. (1994) Cerium/lead and lead isotope ratios in arc magmas and the enrichment of lead in the continents. *Nature* **368**, 514–520.

Miller J. A., Buick I. S., Cartwright I., and Barnicoat A. (2002) Fluid processes during the exhumation of high-P metamorphic belts. *Mineral. Mag.* **66**, 93–119.

Molina J. F., Poli S., Austrheim H., Glodny J., and Rusin A. (2004) Eclogite-facies vein systems in the Marun–Keu complex (Polar Urals, Russia): textural, chemical and thermal constraints for patterns of fluid flow in the lower crust. *Contrib. Mineral. Petrol.* **147**, 484–504.

Moore D. E. (1984) Metamorphic history of a high-grade blueschist exotic block from the Franciscan Complex, California. *J. Petrol.* **25**, 126–150.

Moore G. F., Taira A., Klaus A., *et al.* (2001) New insights into deformation and fluid flow processes in the Nankai Trough accretionary prism: results of Ocean Drilling Program Let 190. *Geochem. Geophys. Geosyst.* **2**(Oct), 2001GC0000166.

Moore J. C. and Volijk P. (1992) Fluids in accretionary prisms. *Rev. Geophys.* **30**, 113–135.

Moran A. E., Sisson V. B., and Leeman W. P. (1992) Boron depletion during progressive metamorphism: implications for subduction processes. *Earth Planet. Sci. Lett.* **111**, 331–349.

Moriguti T. and Nakamura E. (1998) Across-arc variation of Li isotopes in lavas and implications for crust/mantle recycling at subduction zones. *Earth Planet. Sci. Lett.* **163**, 167–174.

Moriguti T., Shibata T., and Nakamura E. (2004) Lithium, boron and lead isotope and trace element systematics of Quaternary basaltic volcanic rocks in northeastern Japan: mineralogical controls on slab-derived fluid composition. *Chem. Geol.* **212**, 81–100.

Morris J. D., Valentine R., and Harrison T. (2002) ^{10}Be imaging of sediment accretion and subduction along the northeast Japan and Costa Rica convergent margins. *Geology* **30**, 59–62.

Mottl M. J., Wheat C. G., Fryer P., Gharib J., and Martin J. B. (2004) Chemistry of springs across the Mariana forearc shows progressive devolatilization of the subducting slab. *Geochim. Cosmochim. Acta* **68**, 4915–4933.

Murdmaa I., Gordeev V., Kuzima T., Turanskaya N., and Mikhailov M. (1980) Geochemistry of the Japan trench sediments recovered on deep sea drilling project legs 56 and 57. *Init. Rep. DSDP*, **56–57**, 1213–1232.

Nadeau S., Philippot P., and Pineau F. (1993) Fluid inclusion and mineral isotopic compositions (H–C–O) in eclogitic rocks as tracers of local fluid migration during high-pressure metamorphism. *Earth Planet. Sci. Lett.* **114**, 431–448.

Nakano T. and Nakamura E. (2001) Boron isotope geochemistry of metasedimentary rocks and tourmalines in a subduction-zone metamorphic suite. *Phys. Earth Planet. Int.* **127**, 233–252.

Nelson B. K. (1995) Fluid flow in subduction zones: evidence from Nd- and Sr-isotope variations in metabasalts of the Franciscan Complex, California. *Contrib. Mineral. Petrol.* **119**, 247–262.

Noll P. D., Newsom H. E., Leeman W. P., and Ryan J. (1996) The role of hydrothermal fluids in the production of subduction zone magmas: evidence from siderophile and chalcophile trace elements and boron. *Geochim. Cosmochim. Acta* **60**, 587–611.

Ogasawara Y., Ohta M., Fukasawa K., Katayama I., and Maruyama S. (2000) Diamond-bearing and diamond-free metacarbonate rocks from Kumdy-Kol and Kokchetav Massif, northern Kazakhstan. *Isl. Arc* **9**, 400–416.

Ohta M., Mock T., Ogasawara Y., and Rumble D. (2003) Oxygen, carbon, and strontium isotope geochemistry of diamond-bearing carbonate rocks from Kumdy-Kol Kokchetav Massif, Kazakhstan. *Lithos* **70**, 77–90.

Okamoto K. and Maruyama S. (1999) The high-pressure synthesis of lawsonite in the MORB+H_2O system. *Am. Mineral.* **84**, 362–373.

Okamura Y. (1991) Large-scale mélange formation due to seamount subduction: an example from the Mesozoic accretionary complex in central Japan. *J. Geol.* **99**, 661–674.

O'Leary J., Kitchen N., and Eiler J. (2002) *Hydrogen isotope geochemistry of Mariana Trough lavas (abstract).* Fall Am. Geophys. Un., San Francisco, December, 2002.

Paquin J., Altherr R., and Ludwig T. (2004) Li–Be–B systematics in the ultrahigh-pressure garnet peridotite from Alpe Arami (Central Swiss Alps): implications for slab-to-mantle wedge transfer. *Earth Planet. Sci. Lett.* **218**, 507–519.

Parkinson C. D. (1996) The origin and significance of metamorphosed tectonic blocks in mélange: evidence from Sulawesi Indonesia. *Terra Nova* **8**, 312–323.

Pawley A. R. (1994) The pressure and temperature stability limits of lawsonite: implications for H_2O recycling in subduction zones. *Contrib. Mineral. Petrol.* **118**, 99–108.

Peacock S. M. (1992) Blueschist-facies metamorphism, shear heating, and *P–T–t* paths in subduction shear zones. *J. Geophys. Res.* **97**, 17693–17707.

Peacock S. M. (2001) Are the lower planes of double seismic zones caused by serpentine dehydration in subducting oceanic mantle? *Geology* **29**, 299–302.

Peacock S. M. (2003) Thermal structure and metamorphic evolution of subducting slabs. In *Inside the Subduction Factory* (ed. J. Eiler). Am. Geophys. Un. Geophys. Monogr. AGU, Washington, DC, vol. 138, pp. 7–22.

Peacock S. M. and Hervig R. L. (1999) Boron isotopic composition of subduction-zone metamorphic rocks. *Chem. Geol.* **160**, 281–290.

Peacock S. M. and Hyndman R. D. (1999) Hydrous minerals in the mantle wedge and the maximum depth of subduction zone earthquakes. *Geophys. Res. Lett.* **26**, 2517–2520.

Peacock S. M., van Keken P. E., Holloway S. D., Hacker B. R., Abers G. A., and Fergason R. L. (2005) Thermal structure of the Costa Rica–Nicaragua subduction zone. *Phys. Earth Planet. Int.* **149**, 187–200.

Pearce J. A. and Peate D. W. (1995) Tectonic implications of the composition of volcanic arc magmas. *Ann. Rev. Earth Planet. Sci.* **23**, 251–285.

Peucker-Ehrenbrink B., Bach W., Hart S. R., Blusztajn J. S., and Abbruzzese T. (2003) Rhenium–osmium isotope systematics and platinum group element concentrations in oceanic crust from DSDP/ODP Sites 504 and 417/418. *Geochem. Geophys. Geosyst.* **4**, 8911, doi: 10.1029/2002GC000414.

Philippot P. (1993) Fluid–melt–rock interaction in mafic eclogites and coesite-bearing metasediments: constraints on volatile recycling during subduction. *Chem. Geol.* **108**, 93–112.

Philippot P., Agrinier P., and Scambelluri M. (1998) Chlorine cycling during subduction of altered oceanic crust. *Earth Planet. Sci. Lett.* **161**, 33–44.

Philippot P. and Selverstone J. (1991) Trace-element-rich brines in eclogitic veins: implications for fluid compositions and transport during subduction. *Contrib. Mineral. Petrol.* **106**, 417–430.

Pickering K. T., Marsh N. G., and Dickie B. (1993) Data report: inorganic major, trace, and rare earth element analyses of the muds and mudstones from Site 808. *Init Rep. DSDP* **131**, 427–450.

Plank T. (2005) Constraints from thorium/lanthanum on sediment recycling at subduction zones and the evolution of the continents. *J. Petrol.* **46**, 921–944.

Plank T. and Langmuir C. (1998) The chemical composition of subducting sediment and its consequences for the crust and mantle. *Chem. Geol.* **145**, 325–394.

Plank T. and Langmuir C. H. (1993) Tracing trace elements from sediment input to volcanic output at subduction zones. *Nature* **362**, 739–743.

Plank T. and Ludden J. (1992) Geochemistry of sediments in the Argo Abyssal Plain at ODP Site 765: a continental margin reference section for sediment recycling in subduction zones. *Proc. ODP Sci. Res.* **123**, 167–189.

Poli S. and Schmidt M. W. (1997) The high pressure stability of hydrous phases in orogenic belts: an experimental approach on eclogite forming processes. *Tectonophysics* **273**, 169–184.

Poreda R. (1985) Helium-3 and deuterium in back-arc basalts: Lau Basin and the Mariana Trough. *Earth Planet Sci. Lett.* **73**, 244–254.

Proyer A. (2003) Metamorphism of pelites in NKFMASH—a new petrogenetic grid with implications for the preservation of high-pressure mineral assemblages during exhumation. *J. Meta. Geol.* **21**, 493–509.

Ranero C. R., Villasenor A., Phipps Morgan J., and Weinrebe W. (2005) Relationship between bend-faulting at trenches and intermediate-depth seismicity. *Geochem. Geophys. Geosyst.* **6**(12): Q12002, doi:10.1029/2005GC000997.

Rea D. K. and Ruff L. J. (1996) Composition and mass flux of sedimentary materials entering the World's subduction zones: implications for global sediment budgets, great earthquakes, and volcanism. *Earth Planet. Sci. Lett.* **140**, 1–12.

Reinecke T. (1998) Prograde high- to ultrahigh-pressure metamorphism and exhumation of oceanic sediments at Lago di Cignana, Zermatt-Saas Zone, Western Alps. *Lithos* **42**, 147–189.

Ritger S., Carson B., and Suess E. (1987) Methane-derived authigenic carbonates formed by subduction-induced pore-water expulsion along the Oregon/Washington margin. *Geol. Soc. Am. Bull.* **98**, 147–156.

Rosner M., Erzinger J., Franz G., and Trumbull R. B. (2003) Slab-derived boron isotope signatures in arc volcanic rocks from the Central Andew and evidence for boron isotope fractionation during progressive slab dehydration. *Geochem. Geophys. Geosyst.* **4**, doi:10.1029/2002GC000438.

Rubatto D. and Hermann J. (2003) Zircon formation during fluid circulation in eclogites (Monviso, Western Alps): implications for Zr and Hf budget in subduction zones. *Geochim. Cosmochim. Acta* **67**, 2173–2187.

Rumble D. and Yui T.-F. (1998) The Qinglongshan oxygen and hydrogen isotope anomaly near Donghai in Jiangsu Province, China. *Geochim. Cosmochim. Acta* **62**, 3307–3321.

Rupke L. H., Morgan J. P., Hort M., and Connolly J. A. D. (2002) Are the regional differences in Central American arc lavas due to differing basaltic versus peridotitic slab sources of fluids? *Geology* **30**, 1035–1038.

Rupke L. H., Morgan J. P., Hort M., and Connolly J. A. D. (2004) Serpentine and the subduction zone water cycle. *Earth Planet. Sci. Lett.* **223**, 17–34.

Ryan J., Morris J., Bebout G., Leeman B., Tera F. (1996) Describing chemical fluxes in subduction zones: insights from "depth-profiling" studies of arc and forearc rocks. In *Subduction: Top to Bottom* (eds. G. E. Bebout, D. W. Scholl, S. H. Kirby, and J. P. Platt), Am. Geophys. Un. Geophys. Monogr., vol. 96, pp. 263–268.

Ryan J. G., Morris J. D., Tera F., Leeman W. P., and Tsvetkov A. (1995) Cross-arc geochemical variations in the Kurile island arc as a function of slab depth. *Science* **270**, 625–628.

Sadofsky S. J. and Bebout G. E. (2003) Record of forearc devolatilization in low-T, high-P/T metasedimentary suites: significance for models of convergent margin chemical cycling. *Geochem. Geophys. Geosyst.* **4**(4).

Sadofsky S. J. and Bebout G. E. (2004a) Nitrogen geochemistry of subducting sediments: new results from the Izu–Bonin–Mariana Margin and insights regarding global nitrogen subduction. *Geochem., Geophys. Geosyst.* **5**, Q03I15, doi:10.1029/2003GC000543.

Sadofsky S. J. and Bebout G. E. (2004b) Field and isotopic evidence for fluid mobility in the Franciscan Complex: forearc paleohydrogeology to depths of 30 km. *Int. Geol. Rev.* **46**, 1053–1088.

Saha A., Basu A. R., Wakabayashi J., and Wortman G. L. (2005) Geochemical evidence for a subducted infant arc in Franciscan high-grade-metamorphic tectonic blocks. *Geol. Soc. Am. Bull.* **117**, 1318–1335.

Sample J. C. and Reid M. R. (1998) Contrasting hydrogeologic regimes along strike-slip and thrust faults in the Oregon convergent margin: evidence from the chemistry of syntectonic carbonate cements and veins. *Geol. Soc. Am. Bull.* **110**, 48–59.

Sano Y. and Williams S. N. (1996) Fluxes of mantle subducted carbon along convergent margins. *Geophys. Res. Lett.* **23**, 2749–2752.

Sassi R., Harte B., Carswell D. A., and Yujing H. (2000) Trace element distribution in Central Dabie eclogites. *Contrib. Mineral. Petrol.* **139**, 298–315.

Savov I. P., Ryan J. G., D'Antonio M., Kelley K., and Mattie P. (2005) Geochemistry of serpentinized peridotites from the Mariana forearc Conical Seamount, ODP Leg 125: implications for the elemental recycling at subduction zones. *Geochem. Geophys. Geosyst.* **6**(4): Q04J15, doi:10.1029/2004GC000777.

Scambelluri M., Fiebig J., Malaspina N., Muntener O., and Pettke T. (2004a) Serpentinite subduction: implications for fluid processes and trace-element recycling. *Int. Geol. Rev.* **46**, 595–613.

Scambelluri M., Hermann J., Morten L., and Rampone E. (2006) Melt- versus fluid-induced metasomatism in spinel to garnet wedge peridotites (Ulten Zone, Eastern Italian Alps): clues from trace element and Li abundances. *Contrib. Mineral. Petrol.* **151**, 372–394, doi:10.1007/s00410-006-0064-9.

Scambelluri M., Muntener O., Ottolini L., Pettke T., and Vannucci R. (2004b) The fate of B, Cl, and Li in the subducted oceanic mantle and in the antigorite breakdown fluids. *Earth Planet. Sci. Lett.* **222**, 217–234.

Scambelluri M. and Philippot P. (2001) Deep fluids in subduction zones. *Lithos* **55**, 213–227.

Schiano P., Clocchiatti R., Shimizu N., Maury R. C., Jochum K. P., and Hofmann A. W. (1995) Hydrous, silica-rich melts in the subarc mantle and their relationships with erupted arc lavas. *Nature* **377**, 595–600.

Scholl D. W., and von Huene R. Crustal recycling at modern subduction zones applied to the past–issues of growth and preservation of continental basement, mantle geochemistry, and supercontinent reconstruction. In *The 4D Framework of Continental Crust* (eds. R. D. Hatcher, Jr., M. P. Carlson, J. H. McBride, and J. R. Martinez Catalan), Geol. Soc. Amer. Spec. Paper. (in press).

Sharp Z. D. and Barnes J. D. (2004) Water-soluble chlorides in mass seafloor serpentinites: a source of chloride in subduction zones. *Earth Planet. Sci. Lett.* **226**, 243–254.

Sharp Z. D., Essene E. J., and Hunziker J. C. (1993) Stable isotope geochemistry and phase equilibria of coesite-bearing whiteschists, Dora-Maira Massif, Western Alps. *Contrib. Mineral. Petrol.* **114**, 1–12.

Shaw A. M., Hilton D. R., Fischer T. P., Walker J. A., and Alvarado G. E. (2003) Contrasting He-C relationships in Nicaragua and Costa Rica: insights into C cycling through subduction zones. *Earth Planet. Sci. Lett.* **214**, 499–513.

Shervais J. W., Zoglmann Schuman M. M., and Hanan B. B. (2005) The Stonyford volcanic complex: a forearc seamount in the northern California Coast Ranges. *J. Petrol.* **46**, 2091–2128.

Skora S., Baumgartner L. P., Mahlen N. J., Johnson C. M., Pilet S., and Hellebrand E. Diffusion-limited REE uptake by eclogite garnets and its consequences for Lu–Hf and Sm–Nd geochronology. *Contrib. Mineral. Petrol.* (in press).

Snyder G., Poreda R., Fehn U., and Hunt A. (2003) Sources of nitrogen and methane in Central American geothermal settings: noble gas and ^{129}I evidence for crustal and magmatic volatile components. *Geochem. Geophys., Geosyst.* **4**.

Solomon E. A., Kastner M., and Robertson G. (2006) Barium cycling at the Costa Rica convergent margin. In: *Proceedings of the ODP*, Sci. Results 205 (eds. J. M. Morris, H. W. Villinger, and A. Klaus). pp. 1–21.

Sorensen S. S. (1991) Petrogenetic significance of zoned allanite in garnet amphibolites from a paleo-subduction zone: Catalina Schist, southern California. *Am. Mineral.* **76**, 589–601.

Sorensen S. S. and Barton M. D. (1987) Metasomatism and partial melting in a subduction complex, Catalina Schist, southern California. *Geology* **15**, 115–118.

Sorensen S. S. and Grossman J. N. (1989) Enrichment of trace elements in garnet amphibolites from a paleo-subduction zone: Catalina Schist, southern California. *Geochim. Cosmochim. Acta* **53**, 3155–3178.

Sorensen S. S. and Grossman J. N. (1993) Accessory minerals and subduction zone metasomatism: a geochemical comparison of two melanges (Washington and California, USA). *Chem. Geol.* **110**, 269–297.

Sorensen S. S., Grossman J. N., and Perfit M. R. (1997) Phengite-hosted LILE-enrichment in eclogite and related rocks: implications for fluid-mediated mass transfer in subduction zones and arc magma genesis. *J. Petrol.* **38**, 3–34.

Spandler C. and Hermann J. (2006) High-pressure veins in eclogite from New Caledonia and their significance for fluid migration in subduction zones. *Lithos* **89**, 135–153.

Spandler C., Hermann J., Arculus R., and Mavrogenes J. (2003) Redistribution of trace elements during prograde metamorphism from lawsonite blueschist to eclogite facies; implications for deep subduction processes. *Contrib. Mineral. Petrol.*, online article, **146**, 205–222, doi:10.1007/s00410-003-495-5.

Spandler C., Hermann J., Arculus R., and Mavrogenes J. (2004) Geochemical heterogeneity and elemental mobility in deeply subducted oceanic crust; insights from high-pressure mafic rocks from New Caledonia. *Chem. Geol.* **206**, 21–42.

Spivack A. J., Kastner M., and Ransom B. (2002) Elemental and isotopic chloride geochemistry and fluid flow in the Nankai Trough. *Geophys. Res. Lett.* **29**, 6-1–6-4.

Srodon J. (1999) Nature of mixed-layer clays and mechanisms of their formation and alteration. *Ann. Rev. Earth Planet. Sci.* **27**, 19–53.

Staudacher T. and Allègre C. J. (1988) Recycling of oceanic crust and sediments: the noble gas subduction barrier. *Earth Planet. Sci. Lett.* **89**, 173–183.

Staudigel H. and King S. D. (1992) Ultrafast subduction: the key to slab recycling efficiency and mantle differentiation? *Earth Planet. Sci. Lett.* **109**, 517–530.

Staudigel H., Plank T., White W. M., and Schminke H.-U. (1996) Geochemical fluxes during seafloor alteration of the upper oceanic crust: DSDP Sites 417 and 418. In *Subduction: Top to Bottom* (eds. G. E. Bebout, D. W. Scholl,

S. H. Kirby, and J. P. Platt). Am. Geophys. Un. Geophys. Monogr. AGU, Washington, DC, vol. 96, pp. 19–38.

Stern R. J. (2002) Subduction zones. *Rev. Geophys.* **40**(4): 1012, doi:10.1029/2001RG000108.

Straub S. M. and Layne, G. D. (2003) Decoupling of fluids and fluid-mobile elements during shallow subduction: evidence from halogen-rich andesite melt inclusions from the Izu arc volcanic front. *Geochem. Geophys. Geosyst.* **4**, doi:10.1029/2002GC000349.

Sun S.-S. and McDonough W. F. (1989) Chemical and isotope systematics of oceanic basalts: implications for mantle composition and processes, In *Magmatism in the Ocean Basins* (eds. A. D. Saunders and M. J. Norry). Geol. Soc. Lond., pp. 313–345.

Tanaka R. and Nakamura E. (2005) Boron isotopic constraints on the source of Hawaiian shield lavas. *Geochim. Cosmochim. Acta* **69**, 3385–3399.

Tanimoto M. (2003) Petrology and geochemistry of the Horoman peridotite complex, Hokkaido, northern Japan. MS Thesis, University of Okayama at Misasa, Misasa, Japan, 51pp.

Tatsumi Y. (2005) The subduction factory: how it operates in the evolving Earth. *GSA Today* **15**, 4–10.

Tatsumi Y. and Kogiso T. (1997) Trace element transport during dehydration processes in the subducted oceanic curst. 2: Origin of chemical and physical characteristics in arc magmatism. *Earth Planet. Sci. Lett.* **148**, 207–221.

Tenthorey E. and Hermann J. (2004) Composition of fluids during serpentinite breakdown in subduction zones: evidence for limited boron mobility. *Geology* **32**, 865–868.

Tomascak P. B., Widom E., Benton L. D., Goldstein S. L., and Ryan J. G. (2002) The control of lithium budgets in island arcs. *Earth Planet. Sci. Lett.* **196**, 227–238.

Touret J. L. R. and Frezzotti M.-L. (2003) Fluid inclusions in high pressure and ultrahigh pressure metamorphic rocks. In *Ultrahigh Pressure Metamorphism* (eds. D. A. Carswell and R. Compagnoni). Eotvos University Press, Budapest, pp. 467–487.

Tribuzio R., Messiga B., Vannucci R., and Bottazzi P. (1996) Rare earth element redistribution during high-pressure-low-temperature metamorphism in ophiolitic Fe-gabbros (Liguria, northwestern Italy): implications for light REE mobility in subduction zones. *Geology* **24**, 711–714.

Tricart P. and Lemoine M. (1991) The Queyras ophiolite west of Monte Viso (Western Alps): indicator of a peculiar ocean floor in the Mesozoic Tethys. *J. Geodynam.* **13**, 163–181.

Tsujimori T., Sisson V. B., Liou J. G., Harlow G. E., and Sorensen S. S. Very-low-temperature record of the subduction process: a review of worldwide lawsonite eclogites. *Lithos.* (in press).

Usui T., Kobayashi K., Nakamura E., and Helmstaedt H. Trace element fractionation in deep subduction zones inferred from lawsonite-eclogite xenolith from the Colorado Plateau. *Chem. Geol.* (in press).

Usui T., Nakamura E., and Helmstaedt H. (2006) Petrology and geochemistry of eclogite xenoliths from the Colorado Plateau: implications for the evolution of subducted oceanic crust. *J. Petrol.* **47**, 929–964.

Usui T., Nakamura E., Kobayashi K., Maruyama S., and Helmstaedt H. (2003) Fate of the subducted Farallon plate inferrerd from eclogite xenoliths in the Colorado Plateau. *Geology* **31**, 589–592.

van der Klauw S. N. G. C., Reinecke T., and Stockhert B. (1997) Exhumation of ultrahigh-pressure metamorphic oceanic crust from Lago di Cignana, Piemontese zone, Western Alps: the structural record in metabasites. *Lithos* **41**, 79–102.

van Keken P. E., Kiefer B., and Peacock S. M. (2002) High-resolution models of subduction zones: implications for mineral dehydration reactions and the transport of water into the deep mantle. *Geochem. Geophys. Geosyst.* **3**, 1056, doi:10.1029/2001GC000256.

Vrolijk P. (1990) On the mechanical role of smectite in subduction zones. *Geology* **18**, 703–707.

Vrolijk P., Myers G., and Moore J. C. (1988) Warm fluid migration along tectonic mélanges in the Kodiak accretionary complex, Alaska. *J. Geophys. Res.* **93**, 10313–10324.

Wei C. and Powell R. (2004) Calculated phase relations in high-pressure metapelites in the system NKFMASH (Na_2O–K_2O–FeO–MgO–Al_2O_3–SiO_2–H_2O). *J. Petrol.* **45**, 183–202.

Whitney D. L. and Davis P. B. (2006) Why is lawsonite eclogite so rare? Metamorphism and preservation of lawsonite eclogite, Sivrihisar, Turkey. *Geology* **34**, 473–476.

Widmer T. and Thompson A. B. (2001) Local origin of high pressure vein material in eclogite facies rocks of the Zermatt-Saas zone, Switzerland. *Am. J. Sci.* **301**, 627–656.

Widom E., Kepezhinskas P., and Defant M. (2003) The nature of metasomatism in the sub-arc mantle wedge: evidence from Re–Os isotopes in Kamchatka peridotite xenoliths. *Chem. Geol.* **196**, 283–306.

Wiesli R. A., Taylor L. A., Valley J. W., Tromsdorff V., and Kurosawa M. (2001) Geochemistry of eclogites and metapelites from Trescolmen, Central Alps, as observed from major and trace elements and oxygen isotopes. *Int. Geol. Rev.* **43**, 95–119.

Williams L. B., Hervig R. L., and Hutcheon I. (1999) Boron isotope geochemistry during diagenesis. Part II: Applications to organic-rich sediments. *Geochim. Cosmochim. Acta* **65**, 1783–1794.

Woodland S. J., Pearson G., and Thirlwall M. F. (2002) A platinum group element and Re–Os isotope investigation of siderophile element recycling in subduction zones: comparison of Grenada, Less Antilles Arc, and the Izu-Bonin Arc. *J. Petrol.* **43**, 171–198.

Wunder B., Meixner A., Romer R. L., and Heinrich W. (2006) Temperature-dependent isotopic fractionation of lithium between clinopyroxene and high-pressure hydrous fluids. *Contrib. Mineral. Petrol.* **151**, 112–120.

Wunder B., Meixner A., Romer R. L., Wirth R., and Heinrich W. (2005) The geochemical cycle of boron: constraints from boron isotope partitioning experiments between mica and fluid. *Lithos* **84**, 206–216.

Yamamoto J., Maruyama S., Parkinson C. D., and Katayama I. (2002) Geochemical characteristics of metabasites from the Kokchetav Massif: subduction zone metasomatism along an intermediate geotherm. In *The Diamond-Bearing Kokchetav Massif, Kazakhstan* (eds. C. D. Parkinson, I. Katayama, J. G. Liou, and S. Maruyama). Universal Academy Press, Tokyo, Japan, pp. 363–372.

You C.-F., Castillo P. R., Gieskes J. M., Chan L. H., and Spivack A. J. (1996) Trace element behavior in hydrothermal experiments: implications for fluid processes at shallow depths in subduction zones. *Earth Planet. Sci. Lett.* **140**, 41–52.

You C.-F., Gieskes J. M., Lee T., Yui T.-F., and Chen H.-W. (2004) Geochemistry of mud volcano fluids in the Taiwan accretionary prism. *Appl. Geochem.* **19**, 695–707.

You C.-F., Spivack A. J., Gieskes J. M., Rosenbauer R., and Bischoff J. L. (1995) Experimental study of boron geochemistry: implications for fluid processes in subduction zones. *Geochim. Cosmochim. Acta* **59**, 2435–2442.

Zack T., Foley S. F., and Rivers T. (2002) Equilibrium and disequilibrium trace element partitioning in hydrous eclogites (Trescolmen, Central Alps). *J. Petrol.* **43**, 1947–1974.

Zack T., Rivers T., Brumm R., and Kronz A. (2004) Cold subduction of oceanic crust: implications from a lawsonite eclogite from the Dominican Republic. *Eur. J. Mineral.* **16**, 909–916.

Zack T., Rivers T., and Foley S. F. (2001) Cs–Rb–Ba systematics in phengite and amphibole: an assessment of fluid mobility at 2.0 GPa in eclogites from Trescolmen, Central Alps. *Contrib. Mineral. Petrol.* **140**, 651–669.

Zack T., Tomascak P. B., Rudnick R. L., Dalpe C., and McDonough W. F. (2003) Extremely light Li in orogenic eclogites: the role of isotope fractionation during dehydration in subducted oceanic crust. *Earth Planet. Sci. Lett.* **208**, 279–290.

Zhang L., Ellis D. J., Arculus R. J., Jiang W., and Wei C. (2003) "Forbidden zone" subduction of sediments to 150 km depth—the reaction of dolomite to magnesite +aragonite in the UHPM metapelites from western Tianshan, China. *J. Meta. Geol.* **21**, 523–529.

Zhang R. Y. and Liou J. G. (1996) Significance of coesite inclusions in dolomite from eclogite in the southern Dabie mountains China. *Am. Mineral.* **80**, 181–186.

Zheng Y. and Kastner M. (1997) 12. Pore-fluid trace metal concentrations: implications for fluid–rock interactions in the Barbados accretionary prism. In *Proc. Oc. Drill. Prog., Sci. Res.* (eds. T. H. Shipley, Y. Ogawa, P. Blum and J. M. Bahr). vol. 156, pp. 163–170.

Zheng Y.-F., Fu B., Gong B., and Li L. (2003) Stable isotope geochemistry of ultrahigh pressure metamorphic rocks from the Dabie-Sulu orogen in China: implications for geodynamics and fluid regime. *Earth-Sci. Rev.* **62**, 105–161.

Zheng Y.-F., Fu B., Li Y., Xiao Y., and Shuguang Li. (1998) Oxygen and hydrogen isotope geochemistry of ultrahigh-pressure eclogites from the Dabie Mountains and the Sulu terrane. *Earth Planet. Sci. Lett.* **155**, 113–129.

Zimmer M. M., Fischer T. P., Hilton D. R., Alvarado G. E., Sharp Z. D., and Walker J. A. (2004) Nitrogen systematics and gas fluxes of subduction zones: insights from Costa Rica volatiles. *Geochem. Geophys. Geosyst.* **5**, Q05J11, doi:10.1029/2003GC000651.

Zindler A. and Hart S. (1986) Chemical geodynamics. *Ann. Rev. Earth Planet. Sci.* **14**, 493–571.

Isotope Geochemistry
ISBN: 978-0-08-096710-3

pp. 39–86

A. LIGHT STABLE ISOTOPES

a''. Atmosphere and hydrosphere

4

Nonmass-Dependent Isotopic Fractionation Processes: Mechanisms and Recent Observations in Terrestrial and Extraterrestrial Environments

M. H. Thiemens

University of California, San Diego, La Jolla, CA, USA

4.1 GENERAL INTRODUCTION

There were several key observations, which were made almost around the same time that led to studies in the fields of stable isotope geochemistry and cosmochemistry. First, it was the physicochemical formalism of isotope effects, particularly the development of the theoretical treatment that permitted calculation of the position of equilibrium in isotopic-exchange reactions. Urey (1947) and Bigeleisen and Mayer (1947) demonstrated that the position of isotope exchange between two differing isotopically substituted molecules may be calculated with high precision from knowledge of the isotopic reduced partition functions. The difference in chemical behavior for isotopically substituted molecules in this specific instance arises from quantum and statistical mechanical effects. A vibrating molecule is energetically represented as a near-classical harmonic oscillator. Thus, in the case of isotopically substituted molecules, the quantized energy levels vary, with the heavier isotopic species possessing lower vibrational frequencies. As a consequence of this specific feature, the heavy isotopically substituted molecules possess slightly stronger bond strengths, and may be calculated with high precision. The vibrational frequencies may also be measured spectroscopically. The vibrational energies are temperature-dependent and, consequently, the partitioning of isotopes between two molecules is temperature-dependent and the observed isotopic partitioning between the molecules may be used as a temperature measure of the equilibrium process of interest. The equilibrium partitioning between isotopically substituted molecules and the quantifiable isotopic partitioning is in fact a cornerstone of igneous and paleothermometry.

The basic principles of the physical chemistry of the processes were first described by Urey (1947) and Bigeleisen and Mayer (1947). With the nearly simultaneous development of the isotope-ratio mass spectrometer by Nier *et al.* (1947), the potential for application of stable isotope measurements of natural environmental samples to detail natural processes emerged. For example, isotopic fractionation processes are now observed and quantified in chemical kinetics, diffusion, evaporation–condensation, crystallization, and biologic processes (e.g., photosynthesis, respiration, nitrogen fixation, sulfate reduction, and transpiration). The isotopic compositions that arise as a superposition of these varying processes directly and sensitively reflect the processes that produce them. A detailed understanding of the physical–chemical laws is fundamental in articulating the mechanisms responsible in the relevant fractionation processes.

Although, conventional isotope effects vary widely in the physicochemical basis for their origin, they all are dependent upon relative isotopic mass differences. The first quantitative discussion of the mass dependence of isotope effects was developed by Hulston and Thode (1965). They demonstrated that conventional isotope effects alter the isotope ratios in a manner strictly dependent upon relative mass differences. Specifically, the changes in $\delta^{33}S$ and $\delta^{34}S$ are shown to be highly correlated such that the change in $\delta^{33}S$ is half that of $\delta^{34}S$. The mass range in the $\delta^{33}S$ measurement is 1 amu (33–32 amu) and for $\delta^{34}S$ is 2 amu (34–32 amu). Thus,

$$\delta^{33}S = 0.5\delta^{34}S$$

is observed and arises due to the magnitude of the relative isotopic mass differences. The actual coefficient may vary slightly from 0.5, depending upon the molecular mass. The focus of the paper of Hulston and Thode was to utilize meteoritic multi-isotope-ratio measurements as a mechanism to resolve nuclear (nucleosynthetic or spallogenic) processes from nonnuclear processes in meteorites. The fundamental assumption in this work was based upon the assumption that all conventional physicochemical processes produce correlated mass-dependent arrays, and any deviation from this relation must reflect a nuclear process. The core assumption for this assertion was that any physical or chemical process may not lead to a variance in stable isotope ratio that does not depend upon mass differences.

Clayton *et al.* (1973) observed that the high-temperature calcium–aluminum-rich inclusions (CAIs) present in the carbonaceous chondrite Allende possess an oxygen-isotope composition of $\delta^{17}O \approx \delta^{18}O$, rather than the expected mass-dependent $\delta^{17}O \approx 0.5\delta^{18}O$. Figure 1 displays the slope 1 line in a three-isotope plot of oxygen. The slope 1/2 line represents the solid terrestrial oxygen-isotopic reservoir of the Earth and Moon, with the origin being that of standard mean ocean water (SMOW).

It was suggested that this anomalous isotopic composition must derive from a nuclear, rather than chemical process. In general, an equality $\delta^{17}O = \delta^{18}O$ may arise in two ways: either by alteration of ^{17}O and ^{18}O by equal amounts or, by addition/subtraction of pure ^{16}O. Models for supernova processes had shown that for certain conditions, essentially pure ^{16}O may be produced. Clayton *et al.* also argued that it is unlikely that ^{17}O and ^{18}O would be equally altered: thus the supernova event was deemed the most plausible mechanism to account for the Allende isotopic observations.

It is now known that these basic assumptions are incorrect. Thiemens and Heidenreich (1983) demonstrated that, in the process of ozoneformation from molecular oxygen, ozone is equally enriched in ^{17}O and ^{18}O, or $\delta^{17}O = \delta^{18}O$. Figure 2 displays the results of the first experiments.

This observation, displayed in Figure 1, has several consequences. First, the fundamental assumption that only a nuclear process may produce a mass-independent isotopic composition is invalid. Clearly nucleosynthetic components are present in meteorites, but the distinction between chemical and nuclear is not as straightforward as originally thought. Most significantly, the relation $\delta^{17}O = \delta^{18}O$ is identical to that for the Allende CAIs. Second, these observations immediately raise the issue as to the source of the anomalous oxygen-isotopic composition observed in the CAI of Allende. As discussed in Thiemens and Heidenreich (1983), the meteoritic oxygen-isotopic compositions more likely result from chemical processes. In a recent review article (Thiemens, 2006), the issues associated with the source of oxygen anomalies is discussed in detail. The literature on meteoritic oxygen-isotopic observations is extensive, and most importantly, now includes high-spatial-resolution analyses, derived from the development of secondary ion-microprobes

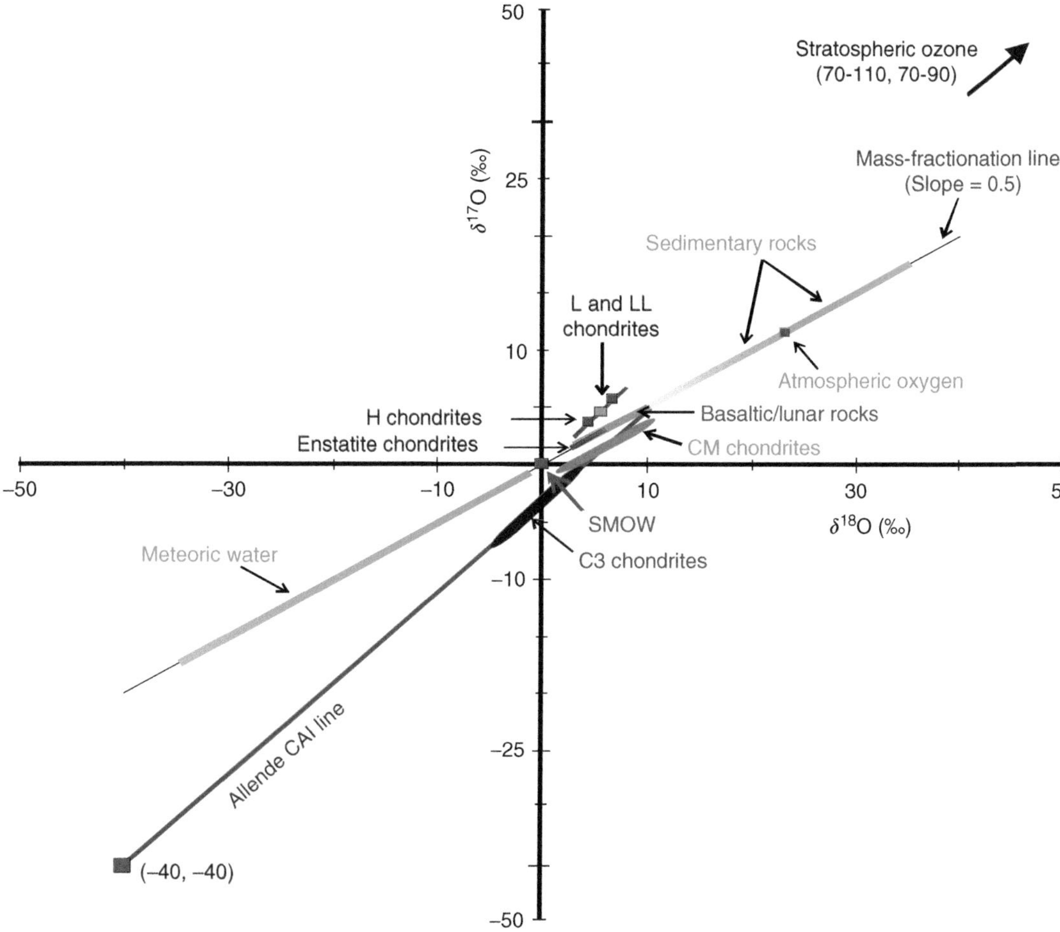

Figure 1 Three-isotope oxygen-isotopic composition of meteoritic and some terrestrial reservoirs. The slope 1 line reflects the calcium–aluminum inclusion (CAI) line first observed by Clayton *et al.* (1973).

(SIMs), nano-SIMs, and continuous-flow isotope-ratio mass spectrometry. These new techniques allow multi-oxygen-isotopic analysis at micrometer (and smaller) spatial scales. An outcome of these observations has been that there is no observational support for a nuclear source of the meteorite oxygen-isotopic compositions, which would have included identification of ^{16}O-carrier grains or, a direct correlation with; other supernova-produced isotopes. It is well documented that there are grains of highly anomalous oxygen present in meteorites (e.g., Nittler, 2003); however, they have neither the material amount needed to explain the observed bulk level-isotopic compositions, nor the correct; isotopic composition and a chemical and/or photochemical process is now generally accepted.

Contributions from photochemical isotopic self-shielding in the nebula by an early active Sun (Thiemens and Heidenreich, 1983; Clayton, 2002; Lyons and Young, 2005) or from chemical reactions (e.g., Thiemens, 1999, 2006; Marcus, 2004) are all mechanisms under consideration at present. These models all attempt to describe the overall oxygen-isotopic compositions of meteorites. Krot *et al.* (2005) have developed a model that describes the evolution of oxygen isotopes in the inner solar nebula. Figure 3 depicts the oxygen-isotopic compositions of bulk level meteorites. It is to be noted that they possess compositions that lie above and below the bulk Earth, Mars, and the Moon. A thorough and balanced review of the source of meteoritic oxygen isotopes is provided by Yurimoto *et al.* (2006). Marcus (2004) has recently developed a quantum mechanical model based upon symmetry considerations in CAI formation that accounts for the observed oxygen-isotopic compositions. The model invokes assumptions regarding the reaction parameters that require experiments.

There are other facets of chemical mass-independent processes and their applications in nature which will be a focus of this chapter. These observations include for example:

1. They are highly ubiquitous on Earth and offer a mechanism by which an extraordinary range of natural processes may be investigated.

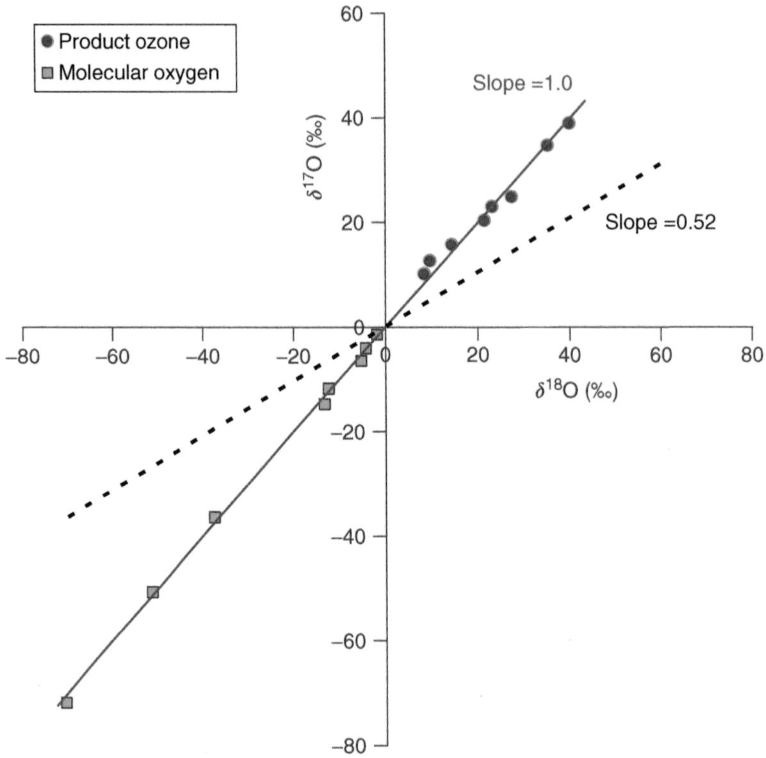

Figure 2 A three-isotope plot demonstrating the oxygen-isotopic composition of ozone produced from molecular oxygen at liquid nitrogen temperature from the experiments of Thiemens and Heidenreich (1983). It should be noted that the slope of this line is 1.0, thus reproducing the slope of the CAI line depicted in Figure 1. The starting molecular-isotopic composition is at $\delta^{18}O = \delta^{17}O = 0$.

2. The physicochemical fractionation process itself has opened a new avenue of study of physical–chemical processes, particularly of gas-phase reactions and molecular transition state theory.
3. Isotopic compositional observations of secondary minerals from martian meteorites have opened a new mechanism to understand atmospheric–regolithic processes on Mars.
4. Mass-independent isotopic compositions produced by photochemical processes have provided a mechanism to detail the evolution of oxygen in the Earth's earliest atmosphere.
5. The oxygen-isotopic composition of aerosols extracted from ice cores provides a mean to follow the changes in the Earth's oxidative capacity.
6. Present-day aerosol formational processes and their transport may be uniquely resolved from their mass-independent isotopic compositions. Such measurements are particularly important for assessing their potential role to the radiative features of the atmosphere.

There are several review papers on mass-independent chemical processes and their applications. Thiemens (1999, 2006), Weston (1999), Brenninkmeijer *et al.* (2003), Babikov *et al.* (2003a, b), Gao and Marcus, (2002a, b), and Mauersberger *et al.* (2005) reviewed the progress in understanding the physical chemistry of gas-phase mass-independent processes. Thiemens *et al.* (2001) and Thiemens (2006) reviewed the observations of mass-independent isotopic composition in various solid reservoirs of Earth and Mars, including both oxygen and sulfur isotopes. The advances in the theory and experimental basis of the mass-independent isotopic fractionation process have been significant. Although there is as yet no agreement among physical chemists upon the precise quantum level origin of the effect, there is a sufficient theoretical basis that the applications in nature are now much better understood.

4.1.1 The Physical Chemistry of Mass-Independent Isotope Effects

When the mass-independent isotopic fractionation chemical process was first discovered by Thiemens and Heidenreich (1983), there existed no physical–chemical mechanism that accounted for the ozone observations. In that paper, a mechanism based upon optical self-shielding was proposed. Although this mechanism may not account for the experimental results, there are potential cosmochemical environments where self-shielding

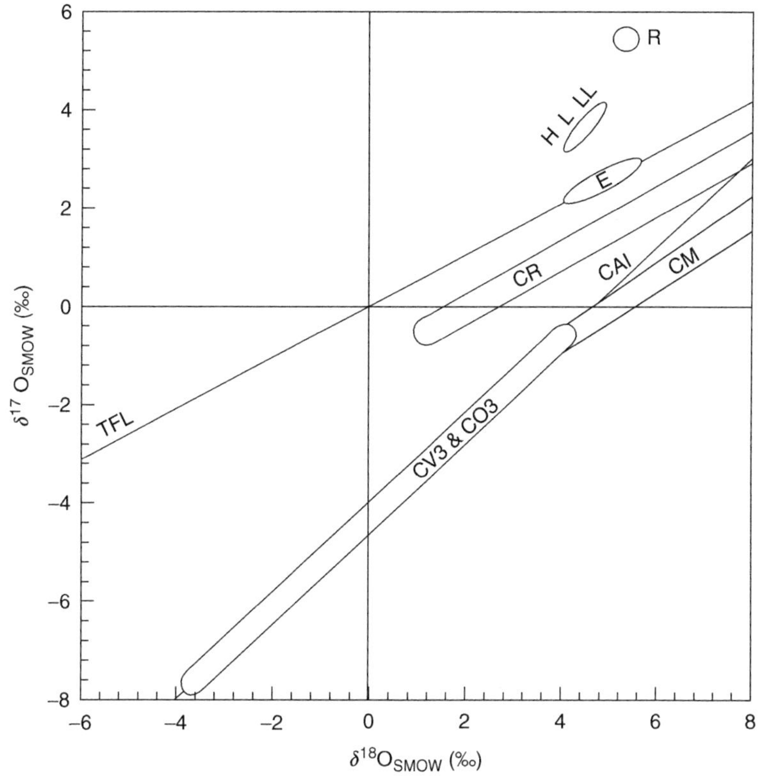

Figure 3 A three-isotope plot of the oxygen-isotopic composition of meteoritic material. The symbols denoted by terrestrial fractionation line (TFL), CV3, CO3, CM, and CR are classes of carbonaceous chondrites, enstatite chondrites (E), ordinary chondrites (H, L, LL), Rumaruti chondrites (R), and calcium–aluminum-rich inclusions (CAIs).

may be operative, as will be discussed in this chapter.

Heidenreich and Thiemens (1986) suggested a theory for the mass-independent theory based upon molecular symmetry considerations. In particular, the differential chemistry for asymmetric $^{16}O^{16}O^{17}O$, $^{16}O^{16}O^{18}O$ versus symmetric $^{16}O^{16}O^{16}O$ was proposed as the basis for the observed mass-independent effect. In this consideration, an equal $\delta^{17}O$, $\delta^{18}O$ fractionation occurs as a result of the identical chemistry for the asymmetric isotopomeric species with respect to symmetric $^{16}O^{16}O^{16}O$. It was in fact a rotational symmetry effect that originally led to the discovery of oxygen isotopes. In the observations of O_2 absorption through the atmosphere, with the Sun as the background irradiance source, it was observed that there are extra rotational absorption lines for $^{16}O^{18}O$ as compared to $^{16}O^{16}O$ demonstrating the existence of the heavy oxygen isotope. This arises because of the well-known line doubling in rotational spectra for asymmetric species. Heidenreich and Thiemens (1986) suggested that the rotational symmetry produces a longer lifetime for the asymmetric isotopomer and, in turn, a greater probability of stabilization. This hypothesis was based upon the following assumptions:

1. The probability of an excited O_3^* intermediate energetically quenching via collisions is determined by the product of its lifetime (τ) and its collisional frequency.
2. The collisional frequency is well known from classical kinetic theory to derive from mass-dependent effects, such as velocity, therefore the source of the mass independence does not arise from the collisional terms for stabilization.
3. The lifetimes of the ^{17}O and ^{18}O species ($^{16}O^{16}O^{17}O$, $^{16}O^{16}O^{18}O$) are equal and longer than symmetric $^{16}O^{16}O^{16}O$. This leads to a product equally enriched in ^{17}O and ^{18}O, that is, $\delta^{17}O = \delta^{18}O$.

The exact mechanism arises in the process of inverse predissociation, as discussed in detail by Herzberg (1966). During an atom–molecule collision, the reactants interact with one another subject to the relevant potential energy surface. The lifetime of this excited intermediate is on the order of molecular vibrational periods, or $\sim 10^{-13}$ s. The lifetime is a complex function of the chemical reaction dynamics, which in turn depends on the number of available states. In this specific instance, there is a state dependence for the isotopically substituted species. Ozone of pure ^{16}O has a C_{2v} symmetry and has half the rotational complement of the asymmetric isotopomers. As a result, it was suggested that the extended lifetime for the asymmetric species leads to a greater probability of stabilization. While these assumptions are valid for a gas-phase molecular reaction, they do not sufficiently account for the totality of the experimental ozone-isotopic observations. Reviews by Weston (1999), Thiemens (1999, 2006), and Mauersberger *et al.* (2005) have detailed the physical–chemical mechanisms.

A comprehensive quantum mechanical model for the effect has been developed by Marcus and his colleagues at the California Institute of Technology. The Gao and Marcus (2001, 2002a, b) model accounts for many of the experimental observations and utilizes classical quantum mechanical RRKM theory in its development. Statistical RRKM theory quantitatively describes the energetics of gas-phase atom–molecule encounters and the relevant parameters, which lead to either stabilization and product formation or redissociation to atomic and molecular species. This is a well-developed theory and will not be described in detail here. An important application of this theory is that it determines individual rate constants for isotopically substituted species. In a conventional RRKM approach, all parameters are mass-dependent, such as collisional frequency, bond strength, and vibrational and rotational energies. Gao and Marcus (2001, 2002a, b) provide an expression that captures an apparent symmetry dependence. During the stabilization process of the excited intermediate (e.g., O_3^*), there is a partial dependence upon the rate at which excess energy is dispersed. If this does not happen sufficiently rapidly, stabilization does not occur and the excited O_3^* will redissociate to the original O and O_2 reactants. Part of this energy-dispersal process depends upon symmetry factors. The greatest contribution is the standard mass-dependent energy process.

In the new isotopic model, there is an inclusion of an "η effect," which is a modest deviation from the statistical density of states for symmetric versus asymmetric species (Gao and Marcus, 2002a,b). There exists a partitioning of energy that derives from very slight differences in zero-point energies for the exit channels for the dissociation of the asymmetric ozone isotopomers. These exit channels do not appear in isotopically enriched experiments (Hathorn and Marcus, 1999, 2000; Gao and Marcus, 2002a, b). For the source of the mass-independent isotope effect, this then leaves only the "η effect." As stated by Gao and Marcus (2002a), this "can be regarded as symmetry driven." In a more recent paper, Gao and Marcus (2002b) have examined the effect of pressure in the context of the experimental data. There exist extensive experimental data (reviewed by Thiemens, 1999, 2002; Thiemens *et al.*, 2001, and Weston, 1999) that describe these details, in particular the effects of pressure, temperature, and oxygen source dependence. An important advancement has been the utilization of isotopic enrichment to determine the explicit rate constants of individual isotopomeric reactions. One of the first experiments was reported by Anderson *et al.* (1997), a striking feature of which was the unexpected observation that the relative rates for $^{16}O+^{16}O^{16}O$ is slower by a factor 50% (greater) from the reaction of $^{16}O+^{18}O^{18}O$. This is consistent with the models of Gao and Marcus (2001, 2002), which predict a greater rate for the asymmetric species. In addition, including the "η effect" quantitatively accounts for the difference in magnitude for the different measured rate constants. Other recent observations by Mauersberger *et al.* (1999) and Janssen *et al.* (1999, 2001, 2002) report the rate differences for a number of the isotopically substituted reactions. These measurements have been of great utility in developing a theory for the mass-independent isotopic effect in ozone. Guenther *et al.* (1999, 2000) have further resolved the effect of pressure and third-body composition on the isotopic fractionation process.

The most recent theoretical advances in resolving the quantum basis of the mass-independent effect as described by Marcus and collaborators have been of fundamental importance. This is true for both understanding the fundamental chemical physics along with the applications in nature. There remain, however, significant challenges to understanding the physical–chemical mechanism. As Janssen (2001) concluded, there are both theoretical and experimental obstacles. For example, the temperature dependence for the individual isotopomeric reactions needs to be determined. There also exist other reactions for which rate constants are presently unknown, such as those for the ^{17}O variants. The culmination of the theory of ozone-isotopic fractionations are discussed in greater detail in the reviews of Brenninkmeijer *et al.* (2003), Thiemens (2006), and Mauersberger *et al.* (2005).

This chapter focuses upon some recent observations of mass-independent isotopic processes in nature. As discussed by Thiemens *et al.* (2001) and Thiemens (2002, 2006), there exist other mass-independent isotope effects in nature that derive from nonozone reactions. For example, CO_2 photolysis produces a large mass-independent isotope effect that, in part, may account for observations in the SNC (martian) meteorites and the synthesis of their secondary minerals. UV photolysis of SO_2 produces new isotopic fractionation effect associated with its electronic state photophysical processes, which has occurred in the martian atmosphere as well as the Earth's earliest atmosphere. An accompanying mass-independent sulfur-isotopic composition determines the evolution of oxygen in the Earth's earliest atmosphere.

4.2 OBSERVATIONS IN NATURE

There now exist numerous observations of mass-independent isotopic compositions in nature. Most of these have recently been reviewed and will not be repeated here. When the first laboratory measurements of the mass-independent isotope effect were reported by Thiemens and Heidenreich (1983), their occurrence in nature was not expected, except possibly for the early solar system with respect to production of the observed meteoritic CAI data. It is significant to note that at present, all oxygen-bearing molecules in the atmosphere (except water) possess mass-independent isotopic compositions. These molecules include O_2, O_3, CO_2, CO, N_2O, H_2O_2, ClO_4, and aerosol nitrate and sulfate. Figure 4 displays the mass-independent isotopic composition of most of the molecules that have mass-independent isotopic compositions associated with them.

Mass-independent sulfur-isotopic compositions are also observed in aerosol (solid) sulfates and nitrates and sulfide and sulfate minerals from the pre-Cambrian, Miocene, and present-day volcanic sulfates, Antarctica dry valley sulfates, Namibian gypretes, and Chilean nitrates. In addition, martian (SNC meteorites) carbonates and sulfates possess both mass-independent sulfur- and oxygen-isotopic compositions. These studies have been reviewed recently (Thiemens *et al.*, 2001; Thiemens, 1999, 2006) and will also be discussed in this text.

In all of the examples cited above, insights into terrestrial atmospheric or martian cycles were obtained that could not have been obtained otherwise. In the remainder of this chapter, recent progress in individual natural processes and mass-independent chemistry are reported.

4.2.1 Atmospheric Ozone

The importance of ozone in the Earth's atmosphere is well established. Its presence in the stratosphere serves as a shield of UV light, which is vital for the sustenance of life, particularly land-based, on Earth. In the troposphere, ozone serves as a source of electronically excited atomic oxygen $O(^1D)$, which subsequently, via the reaction

$$O(^1D) + H_2O \rightarrow 2OH$$

is responsible for creation of tropospheric hydroxyl radicals. The hydroxyl free radical (OH) serves as the dominant mediator of the lifetime of most reduced molecular species in the troposphere; thus, the OH radical serves as an important controlling agent of the oxidative capacity of the atmosphere.

Mauersberger (1981), utilizing *in situ* mass spectrometric measurements, demonstrated that ozone possesses a large ^{18}O enrichment. The ^{17}O-isotopic composition was not determined and the mass-independent isotopic composition could not be detected. As reviewed by Thiemens

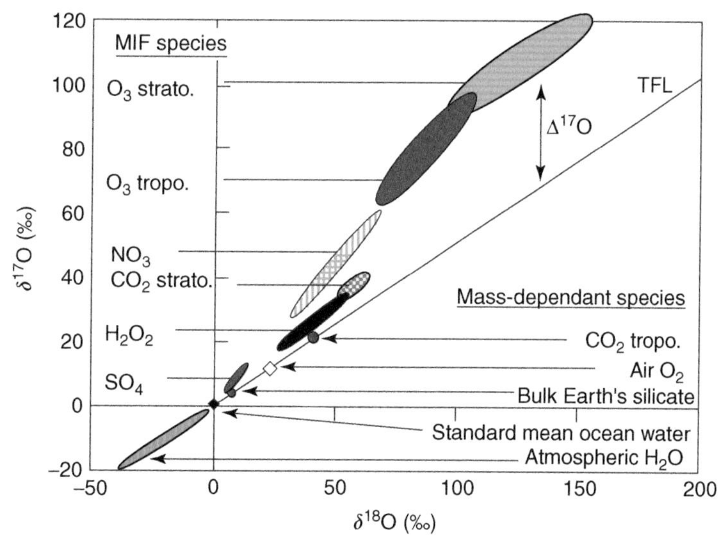

Figure 4 The mass-independent oxygen-isotopic composition of atmospheric oxygen-bearing species. It is to be particularly noted that essentially all oxygen-bearing species in the atmosphere, with the possible exception of water, are mass-independent in composition with respect to the solid Earth oxygen-isotopic reservoir. Molecular oxygen, the second most abundant gas of the Earth's atmosphere possesses a mass-independent isotopic composition arising from its steady-state interaction with ozone and carbon dioxide (see text for discussion). The data also reflect that solids (nitrate and sulfate) also possess mass-independent isotopic compositions.

(1999), Weston (1999), and Johnston and Thiemens (2003), there now exists an extensive literature on stratospheric ozone-isotopic measurements obtained by different measurement techniques. Measurements by Mauersberger (1987) confirmed that stratospheric ozone was mass-independently fractionated as first observed in the 1983 laboratory experiments of Thiemens and Heidenreich. Return ozone-isotopic analysis by Schueler *et al.* (1990) demonstrated that stratospheric ozone possessed an isotopic composition entirely consistent with laboratory observations.

Tropospheric ozone has also been studied for its $\delta^{17}O$, $\delta^{18}O$ isotopic-composition. Krankowsky *et al.* (1995) developed a collection and isotopic analytical procedure that revealed that tropospheric ozone possesses a $\delta^{17}O$, $\delta^{18}O$ enrichment of 70–90‰ (with respect to air molecular oxygen). Their observations are completely consistent with laboratory observations, including the dependence on temperature and pressure. Johnston and Thiemens (1997) reported ozone-isotopic measurements from several locations in the southwestern United States, including rural, urban, and maritime-dominated. The tropospheric measurements have shown that these measurements may provide a mechanism to develop further resolution of the rather complex tropospheric ozone cycle and particularly the complex NO_x interactions. Recent measurements of tropospheric sulfate and nitrate aerosols by the La Jolla group demonstrate that the complexities of the NO_x–SO_x–O_3 cycle (chemical transformation and transport) are significantly better understood from $\delta^{17}O$, $\delta^{18}O$ measurements. This has opened up the possibility that the isotopic composition of polar ice nitrate and sulfate could provide a unique means by which paleo-ozone and oxidative levels may be quantified.

There have been a new series of papers on stratospheric ozone. Krankowsky *et al.* (2000) and Johnson *et al.* (2001) have discussed in detail a range of stratospheric ozone-isotopes. Johnston and Thiemens (2003) have reviewed the literature on atmospheric ozone-isotopic measurements. Krankowsky *et al.* (2000) reported the isotopic composition of ozone collected from four balloon flights in the 22–33 km range. The heavy-isotope enrichment extends between 70‰ and 110‰, somewhat greater than the reported tropospheric range. The actual collection process of stratospheric ozone for return analysis is extraordinarily challenging. Details of the procedure for collection have been reported by Stehr *et al.* (1996). The collection apparatus consists of a cryogenic device, with air passed through a condenser at 80 K and a second one maintained at 63 K. The first trap removes the water (ppm level in the stratosphere) and CO_2, and the second trap collects ozone. At the relevant flow and pressure regime, only ozone is collected in the second trap, all other stratospheric gases are noncondensible. The data reported by Krankowsky *et al.* (2000) utilizing this sampling device at Aire-sur-l'Adour France (43.7 °N, 0.3 °W) and Kiruna, Sweden (67.9 °N, 21.1 °E) reveal a range in heavy-isotope enrichment. There are several significant conclusions drawn by these measurements.

- The large ^{18}O ozone enrichments reported earlier (Mauersberger, 1981) were not confirmed by the more precise measurements and appear to have been a measurement error.
- The new data are well within laboratory observations, particularly in consideration of the temperature and pressure dependence.
- The data suggest that the early *in situ* measurements may have had a problem associated with collection and/or analysis procedures.
- The data further confirm that ozone-isotopic measurements ($\delta^{17}O$, $\delta^{18}O$) provide a means by which stratospheric photochemistry and transport may be elucidated.
- Johnson *et al.* (2001) employed the Smithsonian astrophysical far-infrared spectrometer on seven balloon flights to measure stratospheric ozone-isotopic compositions. There are two important aspects to these observations. First, they confirm and are consistent with the mass spectrometric and laboratory observations. Second, the measurements provide structural information, namely, the asymmetric to symmetric ratio: $^{50}O_3$ (or ^{18}O) to $^{49}O_3$ (^{17}O).

The Smithsonian instrument (FIRS-2) is a remote-sensing Fourier transform spectrometer that detects molecular thermal emission in the atmosphere. Johnson *et al.* (2001) report data taken from seven balloon flights from 1980 to 1997. The last flight is from 68 °N; the others were between 30° and 35 °N. An important feature of the spectrometer is its ability to resolve spectra at 0.004 cm^{-1}, which allows for resolution of $^{16}O^{16}O^{16}O$, $^{17}O^{16}O^{16}O$, $^{16}O^{17}O^{16}O$, $^{18}O^{16}O^{16}O$, and $^{16}O^{18}O^{16}O$ species. It is observed, over the altitudinal range 25–35 km, that the average enhancements for symmetric, asymmetric, and total $^{50}O_3$ are 61 ± 18‰, 122 ± 10‰, and 102 ± 9‰, respectively (Johnson *et al.*, 2001). The average enhancements of ^{17}O enrichment ($^{49}O_3$) are 16 ± 76‰, 80 ± 52‰, and 73 ± 43‰, respectively. These values are in agreement with previous measurements by other techniques. Most importantly, the data provide information on isotopomers. Irion *et al.* (1996) utilized a solar occultation spectrometer to provide vertical profiles for both $^{16}O^{18}O^{16}O$ and $^{16}O^{16}O^{18}O$. Their work demonstrated that there is a significant difference between the symmetric and asymmetric species. The enhancements, on average, revealed an enhancement in the asymmetric over symmetric species by a factor of 1.7. The recent measurements of Johnson *et al.* (2001) also indicate that asymmetric ozone is significantly more enriched than the symmetric one. These enrichment factors are in agreement with laboratory measurements (Mauersberger *et al.*, 1999; Anderson *et al.*, 1989, 1997; Janssen *et al.*, 1999).

In conclusion, with respect to atmospheric ozone isotopes, there have been significant advancements in their measurement. As in, for example, nuclear astrophysics, observations in nature provide important insight into fundamental physics or chemistry, which often is not easily possible under experimental conditions. However, significant details and observations remain pending, including ozone vertical and seasonal isotopic variability. The ozone-enrichment process is also known to cascade through other atmospheric molecular species. Lyons (2001) has developed a model that quantitatively details the transferal of the ozone-isotopic anomaly through molecules, particularly nitrate and sulfate. There are two new and important applications to earth science of the stratospheric ozone phenomena. First is the interaction with carbon dioxide in the stratosphere. Second, a heavy-isotope enrichment in stratospheric ozone creates a compensating depletion in the precursor molecule oxygen. Although by material balance consideration, the effect is small, it creates a unique tracer of primary biological productivity in the ocean, freshwater, and from ice core measurements of paleoglobal productivity. Recent advances in atmospheric CO_2 and O_2 applications will be reviewed.

4.2.2 Stratospheric Carbon Dioxide

Prior to 1991, numerous isotopic measurements of stratospheric carbon dioxide were made, though only for $\delta^{18}O$. A mass-independent isotopic composition was later observed in samples from 26 to 35.5 km, with a deviation from the expected isotopic composition (Thiemens *et al.*, 1991). It has been established that the mass independence arises from isotopic exchange between electronically excited atomic oxygen, $O(^1D)$, and CO_2:

$$O(^1D) + CO_2 = CO_2 + O(^3P)$$

In this process, the ^{17}O, ^{18}O enrichment of atomic oxygen, inherited from stratospheric ozone photolysis, is passed on to CO_2 via exchange, as proposed by Yung *et al.* (1991, 1997).

A fundamentally important but unresolved issue associated with the exchange process is that it occurs via a short-lived transition state, CO_3^*. From the laboratory experiments of Wen and Thiemens (1993), it is apparent that the isotopic composition of CO_2 at steady state is not simply derived from statistical transfer. This has subsequently been addressed by Barth and Zahn (1997), Johnston *et al.* (2000), Lämmerzahl *et al.* (2002). On the basis of existing observations and models, it is not possible to totally account for all of the important aspects of the exchange process.

Utilizing a rocket-borne cryogenic whole-air collection system, Thiemens *et al.* (1995) demonstrated that stratospheric and mesospheric CO_2 possess a mass-independent isotopic composition that derives from the interaction of $O(^1D)$. This is confirmed by the observed inverse correlation of the magnitude of the $\delta^{17}O$, $\delta^{18}O$ anomaly with CH_4 and N_2O concentrations. Methane and nitrous oxide are both (in part) removed from the stratosphere by reaction with $O(^1D)$, and the enhancement in $\delta^{17}O$, $\delta^{18}O$ from interaction with the same species. This has the significance of providing a new, quantitative measure of atomic oxygen, in large part a driving factor in upper atmospheric chemistry.

Alexander *et al.* (2001) advanced the applications of stratospheric CO_2 $\delta^{17}O$, $\delta^{18}O$ measurements. The observations were reported for samples collected from within the Arctic polar vortex. The samples were obtained utilizing a balloon-borne cryogenic air sampler on February 11, 1997 (Strunk *et al.*, 2000). The samples were obtained in an altitude range extending from 12 to 21.1 km. Aside from these representing the first vortex $\delta^{17}O$, $\delta^{18}O$ measurements, concentrations of SF_6, CCl3F (CFC-11), CCl2F2 (CFC-12), and Cl_2FCClF_2 (CFC-113) were measured. An important aspect of the work of Alexander *et al.* (2001) is that a CO_2 oxygen-isotopic relation with SF_6 number density is clearly observed. Sulfur hexafluoride has a lifetime in excess of a thousand years and is only removed in the mesosphere and above by highly energetic processes such as photodecomposition and/or electron attachment. This long lifetime renders SF_6 as a valuable tool as a conservative tracer of stratospheric air mass movement, much in the same way as chloride is used to track oceanic currents and water masses. The observation of an inverse relation between the magnitude of mass independence and SF_6 concentration now combines an age factor to the chemical information obtained from the isotopic measurements and provides a measure of the age of the air mass plus the integrated odd oxygen chemical activity. Future measurements will be instrumental in providing new details of upper atmospheric chemistry and dynamics. The same is true for the combined measurement of fluorocarbons and CO_2 isotopes.

More recent measurements have been reported by Lämmerzahl *et al.* (2002). This work reports simultaneous CO_2 and O_3 isotope ratios from eight balloon flights from Kiruna, Sweden and Aire-sur-l'Adour, France. There is a correlation observed between $\delta^{17}O$ and $\delta^{18}O$ of stratospheric CO_2. The observed $\delta^{17}O/\delta^{18}O$ ratio is 1.71 ± 0.03, independent of the altitudinal range sampled. The data are reported with respect to their enrichment above tropospheric CO_2 ($\delta^{18}O = 41$, $\delta^{17}O = 21$ per mil). This assumes that tropospheric CO_2 is single-valued, which is not the case. $\delta^{18}O$ is well known to vary due to its temperature-dependent isotopic exchange with H_2O. The observations are significant in that they include both CO_2 and O_3 isotopes and aid in providing a quantitative model for establishing the physical–chemical isotopic relation between stratospheric CO_2 and O_3. Hoag *et al.* (2005) have developed a model that utilizes the triple oxygen isotope model for carbon dioxide in the troposphere to provide a new and highly unique means to trace terrestrial gross carbon fluxes.

Another important application of the CO_2- isotopic measurements is their use as a measure of stratosphere–troposphere mixing. Stratospheric CO_2 is mass-independently fractionated due to its coupling with ozone, while tropospheric is mass-dependent because of its equilibrium with water. As originally described by Urey (1947), this is a purely mass-dependent process and has been confirmed by laboratory measurements (Thiemens *et al.*, 1995). This feature provides an ideal marker for the two individual atmospheric reservoirs. A major advance in this regard was recently reported by Boering *et al.* (2004). In this work, samples were acquired by the NASA ER-2 aircraft in the lower stratosphere. In this work, both the $^{17}\Delta O$ of CO_2 and N_2O mixing ratio were determined. From the tight correlation between the 17-O-isotopic anomaly in carbon dioxide and the nitrous oxide mixing ratio, a precise value for the stratosphere–troposphere mixing flux was determined for the carbon dioxide oxygen-isotopic anomaly. This work provides a precise measure of this value, which is needed to determine the gross carbon exchange between the atmosphere and biosphere on interannual to glacial–interglacial time periods.

An important aspect of the $\delta^{17}O$, $\delta^{18}O$-isotopic observations is the actual isotopic measurement analytical technique, which is quite difficult. In the first measurements, Thiemens *et al.* (1991) reported a technique that quantitatively converts CO_2 into CF_4 and O_2. In this technique, CO_2 is reacted with BrF_5 at 800 °C for 48 h. At the end of the reaction, O_2 is separated from CF_4 by using a molecular sieve 13× powder at the melting point of pure ethyl alcohol. This step is vital as trace amounts of CF_4 may cause error in the $\delta^{17}O$ measurements, presumably due to ion–molecule reactions in the source of the mass spectrometer. Conversion into O_2 is required because measurement of CO_2 has isobaric and uncorrectable interference from the contribution at mass 45 from ^{13}C. A second technique has recently been developed that converts CO_2 into methane and water, with subsequent chemical conversion of water into HF and O_2 by reaction with fluorine (Brenninkmeijer and Röckmann, 1998). Subsequently, Assonov and Brenninkmeijer (2001) have reported another new technique. In this technique, CO_2 is measured at high precision for the $\delta^{13}C$, $\delta^{18}O$-isotopic composition. Following isotopic measurement, CO_2 is exchanged with solid CeO_2, which causes exchange of the oxygen isotopes. In the second isotopic measurement, only the 45/44 mass ratio is altered due to the exchange of oxygen isotopes. Since the $^{13}C/^{12}C$ ratio does not alter, the $\delta^{17}O$ may be determined by calculation. This technique has several advantages: (1) the mass 44, 45, 46 ratios are commonly done on isotope ratio mass spectrometers and (2) it is fast and safe, not requiring BrF_5.

4.3 AEROSOL SULFATE: PRESENT EARTH ATMOSPHERE

Thiemens *et al.* (2001) and Thiemens (2006) have reviewed the progress in measurement of the mass-independent oxygen-isotopic composition of atmospheric aerosol sulfate. The importance of sulfate in the Earth's atmosphere and environment is well established. Aerosol sulfate alters radiative processes due to its role in increasing the Earth's albedo and as a cloud condensation nucleus (CCN). While these important roles are recognized, there remain significant gaps in understanding the role of aerosols in these processes. Sulfate and nitrate aerosols are also known to be agents in increasing the incidence of cardiovascular disease. In addition, nitrate is known to be an agent in alteration of terrestrial biodiversity. As a consequence, quantification of sources, transformation mechanisms, and chemical interactions are quite important. Stable isotope ratio measurements have been of limited utility in addressing these issues in the past due to the nonspecificity of single-isotope ratio measurements such as $\delta^{34}S$ or $\delta^{18}O$.

Another environmental issue with respect to sulfur is that 70–80% of the atmospheric sulfur species present in the northern hemisphere is anthropogenic (Rasch *et al.*, 2000). A gap in understanding the atmospheric sulfur cycle arises from the inability to adequately define the oxidation process of SO_2 to sulfate (Kasibhatla *et al.*, 1997). In particular, model calculations underestimate the observed sulfate concentrations in northern latitudes, and it appears that enhanced heterogeneous chemical activity could be important in bringing together model calculations and observations. Measurement of oxygen and sulfur mass-independent isotopic compositions of aerosols has provided new insights into these cycles (Lelieveld *et al.*, 1997; Kasibhatla *et al.*, 1997).

It is now well known that mass-independent isotopic compositions are observed in atmospheric aerosol sulfate (Lee and Thiemens, 1997; Lee *et al.*, 1998, 2001a, b, 2002; Savarino *et al.*, 2000; Johnson *et al.*, 2001). The details of these papers are reviewed in Thiemens *et al.* (2001) and Thiemens (2002, 2006). The measurements of the $\delta^{17}O$, $\delta^{18}O$-isotopic composition of aerosol sulfate have revealed the following:

1. Coupled with isotopic measurements of atmospheric H_2O_2 (Savarino and Thiemens, 1993, 1999) and knowledge of reaction rate constants, pH reaction rate dependence, and isotopic fractionation factors (Savarino *et al.*, 2000), the homogenous and heterogeneous reaction pathways may be quantified. This has been a goal for decades and provides a new means to supplement modeling efforts.
2. Source processes of aerosol sulfates may be identified from the specific mass-independent isotopic signatures. For example, during the Indian Ocean Experiment (INDOEX), the La Jolla group demonstrated that the Inter-tropical Convergence Zone is a source of new aerosol particles (Alexander *et al.*, 2005). This was previously unrecognized and a definition of a new aerosol generation process is of major significance in global climate models. Alexander *et al.* (2005) have reported these measurements and developed a new isotopic model that is capable of tracing the source and transformation process of these sulfate particles in the troposphere over the Indian Ocean. The model, coupled with the isotope measurements, has revealed the strength of the combined approach to resolve the highly complex interplay of marine particulates with the precursor sulfur-bearing species and the strong alkalinity dependence in the heterogenous chemical transformational processes. Of particular importance has been the unique recognition of the importance of these new particle formation processes globally. These particulates are most important in global climate models and prior to the measurements and model, were not recognized.
3. Mass-independent sulfate laboratory measurements have further elucidated the free-radical oxidation mechanism of liquid-phase SO_2 to sulfate.
4. The mass-independent isotopic composition of sulfate may be used to quantify the relative contribution of O_3 and H_2O_2 in sulfur oxidative processes. This is a significant parameter in determining transport phenomena, particularly over long (hemispheric) ranges. Furthermore, this quantification may be used to model changes in the oxidative capacity of the Earth, both present and in the past.
5. In specific locations, the isotopic measurements provide unique process information. For example, Lee and Thiemens (1997) used combined radioactive ^{35}S (87.2 d half-life) and $\delta^{17}O$, $\delta^{18}O$-isotopic measurements to recognize that in winter months at the White Mountain Research Station (12,500 ft), there is a significant upper atmospheric source of sulfate. Johnson *et al.* (2001) have used the $\delta^{17}O$, $\delta^{18}O$ measurements to provide new insight into surface–water interactions and sources in a high alpine region of the Rocky Mountains.

The most recent studies of sulfate have furthered our understanding of the natural sulfur cycle. Lee *et al.* (2002) performed sulfur- and oxygen-isotopic measurements of sulfates produced in controlled combustion processes. These experimental observations were conducted at the Centre de Recherches Atmospheriques in Lannemezan, France in a dark combustion chamber. This system was specifically constructed to study aerosol and gas emissions derived from fuel combustion and emission. The chamber has a volume of $\sim$160 m^3, which effectively minimizes wall effects and maximizes mixing. The collections are done in the dark to minimize any potential photochemical effects. Fossil fuel and vegetation burns were performed, including Savanna grasses, Lamto grass, rice grass, hay, diesel fuel, and charcoal. All of these processes produce sulfur-and oxygen-isotopic compositions that are strictly mass-dependent. This experimentally eliminates these processes as the source of the observed mass-independent isotopic composition of aerosol sulfate.

4.3.1 Other Mass-Independent Sulfate-Isotopic Compositions

There are several other natural repositories of mass-independent sulfates, all of which have recently been reviewed (Thiemens *et al.*, 2001; Thiemens, 2002, 2006). Namibian desert gypsum ($CaSO_4$ $2H_2O$) is a novel example. Bao *et al.* (2000a, b) reported mass-independent oxygen-isotopic compositions of the massive (3×10^4 km^2) gypsum deposit of Namibia. In these measurements, isotopic trends, as a function of distance from the ocean, were observed. In a careful analysis of the various parameters, Bao *et al.* (2000a, b, 2001b) determined that the source of the oxygen-isotopic anomalies derives from the atmospheric photooxidation of dimethyl sulfide from the nearby upwelling oceanic region. These measurements therefore have provided a new mechanism to examine paleo-oceanic variations of biologic activity on tens of millions of year timescales.

Bao *et al.* (2000b) extracted sulfate from vertical profiles of Antarctic dry valley samples and utilized the oxygen-isotopic composition to resolve sources. The variance in the mass-independent oxygen-isotopic composition was defined as the source of the anomalous sulfate composition and source. In particular, the source, particle-size dependence, and relative amounts of sulfate from sea salt aerosols and biogenic sources were resolved. Observations of a mass-independent isotopic composition in sulfate from Miocene volcanogenic samples also displayed large mass-independent isotopic compositions. These results are particularly intriguing. The anomaly is quite large and is specifically associated with the Miocene event; however, other volcanic activities do not possess any mass-independent isotopic composition. Clearly, in this there remain some outstanding questions, and further measurements are needed to advance our understanding of the origin of these anomalies and how they may be exploited to further understand the natural processes associated with these events. The details of these measurements are reviewed in Thiemens *et al.* (2001) and Thiemens (2002, 2006). For applications of sulfate mass-independent isotopic compositions as a means for understanding the process of desert varnish foundation, Bao *et al.* (2001a, b, 2003, 2004, 2005) report details of the measurements and applications.

Recently, sulfur inclusions from diamonds were measured for their sulfur multi-isotopic composition by Farquhar *et al.* (2002). It was demonstrated that various populations of sulfide inclusions in diamonds from the Orpa kimberlite pipe in the Kaapvaal-Zimbabwe craton of Botswana possess mass-independent sulfur-isotopic compositions. The implication of this remarkable observation is that the anomalous isotopic fractionation pattern was produced via atmospheric photochemistry in the Archean (to be discussed) with subsequent transfer to the Earth's surface and sequestration into subducting material. The actual process is initiated with volcanic emissions of sulfur-bearing molecules in the reducing, oxygen-free atmosphere of the Archean. Photochemistry in the UV region of these molecules produces mass-independent sulfur-isotopic compositions via the process observed in present-day sulfate aerosols and laboratory experiments. This also partitions the mass-independent sulfur-isotopic anomaly between elemental and oxidized redox states and, of differing sign. The photolytic products are transferred to the surface of the Earth and enter the geological reservoir. The species undergo lithification and conversion to either sulfate (presumably oceanic) of reduced sedimentary sulfides, possibly bacterial in marginal seas and estuaries. These species are then metamorphosed and recycled into the mantle. They are eventually returned to the Earth's surface as diamonds. The remarkable aspect of the observations is that these samples are actually a record of photochemistry of the Earth's earliest atmosphere, with stable storage in diamonds.

4.4 ATMOSPHERIC MOLECULAR OXYGEN

It has been known for decades that photosynthesis and respiration predominantly establish the steady-state levels of O_2 and CO_2 in the Earth's atmosphere. From this standpoint, it might be argued that these processes are among the most important of all global biogeochemical cycles. For this reason, careful and precise establishment of the rates of photosynthetic and primary productivity in the world's oceans is essential. This rate, as reviewed by Bender (2000), is intimately linked to the primary productivity of the oceans, carbon cycling, and CO_2 levels in the ocean and atmosphere. Traditionally, in spite of this enormous importance, quantification has been difficult. In part, this stems from the magnitude of the oxygen reservoir and the hysteresis for production of a measurable change. Luz *et al.* (1999) reported an entirely new mechanism to evaluate biospheric productivity. This new technique is based upon isotopic material balance constraints on the creation of anomalous mass-independent oxygen-isotopic reservoirs in ozone, CO_2, and O_2. This anomaly is transferred from O_3 to stratospheric CO_2, which occurs in two ways: first, as a statistical isotopic exchange and, second, due to the isotopic effort associated with creation of the short-lived CO_3^* transition state.

The magnitude of the anomaly is measurable in CO_2 ($\sim$360 ppm) but not immediately for O_2 ($\sim$20%) as the steady-state effect on the $\delta^{17}O$, $\delta^{18}O$ of atmospheric O_2 is not measurable. However, there is a significantly longer lifetime for O_2 than CO_2 (factor of hundreds), which is eventually lost by photosynthesis in the upper ocean. As a result, the positive ^{17}O, ^{18}O massindependent anomaly in CO_2 produces a small but measurable isotopic anomaly in O_2, which accrues over time because of the differential lifetimes. The consequence of the field and laboratory measurements is that measurement of the mass-independent O_2-isotopic composition in oceanic vertical profiles provides a new, quantitative means by which primary productivity may be evaluated. The O_2 anomaly is only removed by respiration in the oceans. Thus, its rate of disappearance is directly linked to primary productivity. Measurement of $\delta^{17}O$, $\delta^{18}O$ of O_2 trapped in polar ice may also be used as a means of determining global biospheric primary productivity. Luz and Barkan (2000) furthered these studies by developing a new methodology for determining oceanic production rates on week timescales.

Subsequent to the review by Thiemens (2002), there have been further advances in the utilization of the molecular O_2-isotopic composition (Angert *et al.*, 2001). Though they did not measure $\delta^{17}O$, they have determined the oxygen-isotopic fractionation by respiration and diffusion in soils, measurements critical in quantifying the atmospheric O_2 cycle. The well-known ^{18}O enhancement of atmospheric O_2 is utilized to determine paleo-variations in the relative O_2 contributing proportions of marine and terrestrial sources. Quantification of the oxygen cycle is mandatory, in order to resolve the tightly coupled carbon cycle. The experiments of Angert *et al.* (2001) provide high-precision isotopic measurements requisite for modeling both the global carbon and oxygen biogeochemical cycles. More recently, Angert *et al.* (2002) have reported one of the most extensive evaluations of the $\delta^{17}O$, $\delta^{18}O$ for the individual processes associated with photosynthesis and respiration. This is a key component in extending the utilization of mass-independent O_2-isotopic compositions as a means to estimate changes in global productivity. In particular, Angert *et al.* (2002) have evaluated the values of θ for the most important processes that determine the isotopic composition of molecular oxygen:

$$\theta \equiv \frac{\ln(^{17}\alpha)}{\ln(^{18}\alpha)}$$

which follows the relation described by Mook (2000). In this instance, $^{17}\alpha$ and $^{18}\alpha$ are the isotopic fractionation factors.

The value of θ has been determined for the dark respiration factor, that is, the cytochrome pathway (0.516 ± 0.001) and its mechanistic alternative (0.514 ± 0.001). A modestly higher value for diffusion in air (0.521 ± 0.001) was determined. The precise determination of these θ factors leads to the determination of the steady-state value for diffusion and respiration in the atmosphere. The θ value of photorespiration (0.506 ± 0.003) is less than for dark respiration. These evaluations represent a major step forward in the development of $\delta^{17}O$, $\delta^{18}O$ measurements of atmospheric O_2 as a means to study global rates of photorespiration, both at present and on glacial timescales. In addition, as Angert *et al.* (2002) discuss in detail, the individual processes associated with photorespiration have been quantitatively resolved. A final important aspect is that these results will ultimately lead to a closure of the difference between the modeled and measured Dole effect. This has been an important issue since the first report of the Dole effect.

In summary, the O_2 $\delta^{17}O$, $\delta^{18}O$ measurements have significantly advanced our understanding of the oxygen and carbon global biogeochemical cycles on present and past timescales. These are enhancements in understanding made possible by the utilization of, albeit small, mass-independent isotopic composition of O_2. Finally, it is seen that the exact value of θ, or mass fractionation law, may be determined at extraordinarily high precision. The very slight variances in mass fractionation laws have now led to new applications which are reviewed in a subsequent section, this also includes the possible application of ^{17}O in the hydrologic cycle (Angert *et al.*, 2004).

4.5 THE ATMOSPHERIC AEROSOL NITRATE AND THE NITROGEN CYCLE

It is well established that atmospheric nitrate expresses its importance in a number of vital biogeochemical cycles. These include

1. its role as a major sink for nearly all NO_x species;
2. its increasingly important role as an acid agent in the environment;
3. participation as a heterogeneous reactant in polar stratosphere clouds (PSCs) and destruction of ozone in the Antarctic polar vortex;
4. as a source of fixed nitrogen to key ecosystems (Paerl, 1993); and
5. as a possible agent in radiative transfer in the Earth's atmosphere.

In spite of these highly significant roles, there remain serious limitations in the quantification of the nitrogen cycle. This includes, for example,

- depositional rates (Kendall and McDonnell, 1998; Cress *et al.*, 1995);
- chemical transformation mechanisms in the troposphere (lower and upper);
- source variability and identification; and
- long-range transport importance.

It has been shown that atmospheric aerosol nitrate possesses one of the largest mass-independent oxygen-isotopic compositions observed in nature (Michalski and Thiemens, 2000; Michalski *et al.*, 2002, 2003, 2004a, b). An important advancement in atmospheric nitrate studies has been the development of the ability to perform the $\delta^{17}O$, $\delta^{18}O$ measurements in aerosol nitrate (Michalski *et al.*, 2002). In the process of conversion of nitrate into O_2 for mass spectrometer isotope ratio analysis, high purity is essential. Ion chromatography is utilized to simultaneously isolate and concentrate nitrate. The nitrate is subsequently reacted under a proper pH regime to convert into HNO_3. Reaction with Ag_2O quantitatively converts nitric acid into $AgNO_3$, which is filtered and dried. $AgNO_3$ is thermally decomposed to O_2, NO_2, and Ag at a constant chemical and isotopic branching ratio. Following a final purification process, the O_2 is measured for the $\delta^{17}O$, $\delta^{18}O$ composition. A variety of standards have been analyzed, including USGS-35, a sample of $NaNO_3$ from the Atacama desert of Chile and IAEA-N3, an internationally distributed standard sample KNO_3, with a range of reported $\delta^{18}O$ values (Revesz *et al.*, 1997; Kornexyl *et al.*, 1999).

As Michalski and Thiemens (2003) have shown, aerosol nitrate possesses an extraordinarily large mass-independent isotopic composition, second only to ozone in magnitude. As for aerosol sulfate, this isotopic signature has been shown to provide a new means to elucidate source and chemical-transformation processes. This has proved to be an important technique by which the nitrate biogeochemical cycle may be understood further. For example, the massive mass-independent isotopic signature observed in Chilean desert nitrates uniquely reveals that these nitrates must be atmospherically derived since all other sources (by measurement) possess mass-dependent isotopic compositions. In addition, these measurements, coupled with contemporary aerosol nitrate measurements reveal that the oxygen-isotopic signatures are stable on million year timescales (Michalski *et al.*, 2004a, b). This is particularly valuable, as this permits measurement of nitrate in polar ice samples to examine paleo-variations in nitrate and in general, NO_x chemistry. As discussed in the next section, when linked to sulfate-oxygen isotopic measurements, an entirely new mechanism to study paleo-atmospheric oxidative capacity variation is available.

Finally, in the case of present-day aerosol nitrate-isotopic measurements, samples collected as a function of particle size provide another level of detection in the resolution of sources and atmospheric transformation mechanisms. Large particles (1–10 µm) typically are crustal or oceanic sea spray, depending upon where the particular samples are collected. Small particles (<0.1 µm) generally are gas-to-particle conversion process products. Using combined multi-isotope ratio measurements and size-fractionated collection procedures, it is possible to provide sophisticated details of atmospheric aerosol fates.

4.6 PALEO-OXYGEN VARIATIONS AND POLAR ICE MEASUREMENTS

As discussed earlier, the mass-independent isotopic compositions of sulfate and nitrate aerosols have enhanced our understanding of formation processes. For example, Savarino *et al.* (2000, 2003a, b) and Lee *et al.* (2001a, b) have demonstrated that the sulfate massindependent oxygen-isotopic composition quantitatively describes the proportional contributions of gas-(OH radical) and liquid-phase (O_3, H_2O_2) oxidative processes. These proportions are vital parameters in evaluating paleo-climate variations. In addition, evaluation of the mechanism of aerosol formation for both nitrate and sulfate will influence the indirect aerosol component of climate models. For these reasons, $\delta^{17}O$, $\delta^{18}O$ measurements of ice core sulfate and nitrate represent a new horizon in climate research.

Alexander *et al.* (1999, 2000, 2002a, b, c, 2003, 2004) and Savarino *et al.* (2002) have reported mass-independent isotopic compositions of sulfate in polar ice samples. In Alexander *et al.* (2002b), the oxygen-isotopic composition of sulfate in eight samples from the Vostok Antarctic ice core sample were reported. These samples spanned sufficient temporal range to include a full climate cycle. These data clearly demonstrated that during the glacial period the relative proportion of gas versus aqueous-phase oxidation was significantly greater than during interglacial periods. The work confirmed that the oxidative capacity of the atmosphere might be identified from such measurements. This work also highlights the relative importance of cloud processing, an important parameter that has been historically difficult to assess.

In addition, such measurements coupled with CH_4, H_2O_2 concentration, and MSA/SO_4 ratio measurements of polar ice species, provide a powerful tool to evaluate the variation of paleo-climates and define the quantitative coupling between atmospheric chemistry and climate. For example, Alexander *et al.* (2002a, b, c) have measured the $\delta^{17}O$ and $\delta^{18}O$ of both sulfate and nitrate from samples at Greenland ice core site A (70° 45′ N, 35° 57.5′, elevation 3,145 m). These samples were particularly interesting because, temporally, they surround the Industrial Revolution. The years spanned by these measurements are from around 1692 to 1976. The details of ice decontamination and chemistry are presented in Alexander *et al.* (2002a, b, c, 2005). These measurements have distinctly shown that biomass-burning events prior to the Industrial Revolution altered the atmospheric oxidative mechanisms in the northern hemisphere. This work is significant in that (1) it further demonstrates how $\delta^{17}O$, $\delta^{18}O$ measurements of sulfate and nitrate provide a new means by which the oxidative capacity of the atmosphere on extended timescales may be determined; and (2) the biomass-burning emissions were highly significant, even in comparison with fossil fuel combustion. The results underscore the need to include precise values for the magnitude of biomass burning in paleo-climatological models. Finally, the newest results depict the importance of mass-independent isotopic compositional measurements in ice core studies as a complement to concentration and other stable isotope ratio measurements.

4.7 MASS-INDEPENDENT ISOTOPIC COMPOSITIONS IN SOLIDS

4.7.1 Sulfur-Isotopic Fractionations in the Earth's Earliest Atmosphere and Mars

Farquhar *et al.* (2000b) reported high-precision sulfur isotope ratio measurements from martian (SNC) meteorites. Sulfur is an interesting element on the martian surface because it is abundant, exists in the atmosphere and regolith, but has a poorly described origin and evolution. Establishing the connection between the reservoirs, as well as the mechanism of transfer is of interest in the evaluation of potential for life on Mars and other planets. It was observed that the SNC meteorites possess a mass-independent sulfur-isotopic composition. Laboratory-based photolysis experiments established that the anomalous isotopic compositions may be established by UV photolysis of sulfur dioxide, which was emitted by sporadic volcanogenic activities. These observations are significant in that they provide a clear mechanistic link between the martian atmosphere and the surface.

Measurement of sulfide and sulfate samples from the Earth's pre-Cambrium led to the discovery of mass-independent isotopic compositions of both positive and negative signs. It was also revealed that between 2090 and 2450 Ma, a major change occurred in the Earth's atmosphere. As in the case of Mars, the isotopic anomalies in the pre-Cambrium result from short UV photolysis of sulfur dioxide. The requirement of short UV light clearly indicates that there were low levels of oxygen and ozone, otherwise there would be no photolysis and production of the isotopic anomaly. In the present-day atmosphere, due to the presence of the ozone–oxygen atmosphere, there is no short-wavelength photolysis of sulfur dioxide. The observation that the anomaly disappears at nearly precisely the same time as the global outbreak of banded iron formations, which are associated with development of an oxidizing atmosphere is completely consistent with this hypothesis. The observation is of the most fundamental importance as what the sulfur-isotopic anomaly clearly depicts is the origin and evolution of microbial processes in the Earth's earliest environment. This is perhaps the best record of these processes developed to date.

Farquhar *et al.* (2001) performed a careful set of laboratory experiments to detail the isotopic fractionation processes that occur at varying wavelengths of UV light. These experiments demonstrated that the sulfur-isotopic anomalies observed in pre-Cambrium samples may be reproduced by 193 nm photolysis of sulfur dioxide, including both positive and negative sulfur reservoirs. These measurements facilitated the calculation of the oxygen–ozone levels that may have existed in the pre-Cambrium. It was determined that O_2 levels could not have exceeded a few percent present levels or there would have been insufficient penetration of UV light to the surface for production of the isotopic anomalies by 193 nm light. Carbon dioxide levels could not have been higher than about 0.8 bar as well. As Farquhar *et al.* (2001) concluded, these are the first estimates for establishment of molecular oxygen and carbon dioxide levels based upon a quantifiable parameter. Subsequent to these measurements there have been numerous measurements and theoretical considerations (Kasting, 2001; Wing *et al.*, 2002; Pavlov and Kasting, 2002; Runnegar *et al.*, 2002; Mojzsis *et al.*, 2003). The measurements are reviewed in Thiemens *et al.* (2001), Farquhar and Wing (2003), and Thiemens (2006).

4.7.2 Other Sulfur-Isotopic Fractionation Processes

There are few examples of mass-independent isotopic fractionations, as yet, associated with the formation of solids. The first example was reported by Colman *et al.* (1996), who demonstrated that in the irradiation (313 nm) of gas-phase CS_2, a solid aerosol $(CS_2)_x$ was generated that is highly mass-independently fractionated, with a ^{33}S excess of 5–10‰, and a ^{36}S deficit of 61–84‰. An interesting aspect of these observations was that the optical properties of this solid were quite similar to those observed for solids produced by comet Shoemaker–Levy 9(SL9) during the collision with the outer atmosphere of Jupiter. Colman *et al.* (1996) suggested that the mechanism responsible for the observed sulfur-isotopic fractionation process involves the differing rates of nonradiative transfer from the initial absorbing rate to the final reactive state. In processes such as this, the rate is predominantly dependent upon Franck–Condon factors and, hence, is not directly dependant upon mass.

In a follow-up investigation, Zmolek *et al.* (1999) reported the results of experiments that further detailed the mechanism responsible for the production of the photopolymeric $(CS_2)_x$ solid from CS_2 photolysis. In these experiments, it was demonstrated that the sulfur-isotopic composition in products from photolysis of $^{12}CS_2$ and $^{13}CS_2$ differ significantly. This observation immediately rules out any involvement of a symmetry-dependent fractionation effect. Zmolek *et al.* (1999) suggest Franck–Condon and vibronic coupling effects as the source of the observed effect. It is specifically the nonradiative decay and intersystem crossing rates for the lowest excited states which give rise to the highly anomalous sulfur-isotopic composition observed in the solid product.

A similar photochemical fractionation process was reported by Bhattacharya *et al.* (2000). Photodissociation of CO_2 by UV light at 185 nm produces CO and O_2 enriched in ^{17}O by more than 100‰. As reported, this dissociation arises via a spin-forbidden process during the singlet to triplet transition. The mass independence derives from the reliance of the reaction rate on essentially resonant spin–orbit coupling of the low-energy vibrational states of the $^{16}O^{12}C^{17}O$ molecule of the singlet with the triplet state. This hypothesis was confirmed by demonstration that the alteration of the oxygen-isotopic fractionation results by simple substitution of ^{13}C for ^{12}C. These experiments also demonstrate that the mass-independent isotopic fractionation process is restricted neither to symmetry nor to ozone.

A completely new variety of mass-independent isotopic fractionation has been reported by Miller *et al.* (2002). It was observed that in the thermal decomposition of naturally occurring carbonates (calcium and magnesium), the product CO_2 and the residual solid oxides possess mass-independent isotopic compositions. The product CO_2 is depleted in ^{17}O by ~0.2‰ and the residual CaO or MgO enriched by a comparable factor. Numerous control experiments were performed in two different laboratories to demonstrate that the effect is associated with the thermolysis of the carbonate and not a result of secondary gas-phase reaction or experimental artifact. The effect is also observed for an extended suite of carbonate samples of widely varying $\delta^{18}O$-isotopic composition. This is the first report of a mass-independent effect associated with a solid reactant. As discussed by Miller *et al.* (2002), the physical–chemical mechanism responsible for the observed effect is unknown, though of considerable interest. That work discusses the limited likelihood of detecting this effect in nature and, therefore, at present the effect is restricted to a significant result in the fundamental physical chemistry of mass-independent isotope effects. Future experiments are needed to determine the physical–chemical basis for this effect in solids. Measurements on other solids are needed, particularly for molecules possessing oxygen bound in an equilateral triangular arrangement. Theoretical considerations of this effect are also needed if the potential occurrence in nature is to be critically explored. As Miller *et al.* discuss, detection of calcite and dolomite in dust shells surrounding evolved stars (Kemper *et al.*, 2002) may provide a cosmochemical environment where an anomalous isotopic fractionation process could occur. As discussed in the next section, it is well established that preserved interstellar grains occur in meteoritic material.

4.8 MASS-INDEPENDENT ISOTOPIC FRACTIONATION PROCESSES IN THE EARLY SOLAR SYSTEM

The suggestion by Clayton *et al.* (1973) that the $\delta^{17}O = \delta^{18}O$-isotopic composition observed in the meteorite Allende CAIs was due to a nucleosynthetic process was based upon the inability of a chemical process to produce a mass-independent isotopic composition. With the discovery by Thiemens and Heidenreich (1983) of a chemical process capable of producing the same $\delta^{17}O = \delta^{18}O$-isotopic composition, the fundamental assumption was shown to be incorrect and new potential models for the production of oxygen-isotopic reservoirs needed to be developed. As discussed in Thiemens (1999, 2006), one of the fundamental requirements for the nucleosynthetic mechanism was that it requires that the ^{16}O carrier be added as a dust grain. Clayton *et al.* (1973) hypothesized that this should produce a correlation with another element, such as magnesium and silicon.

After a quarter of century of measurements, no such correlation exists and oxygen remains the only element with meteoritic anomalies present at the bulk level, and in possession of three or more stable isotopes. Thiemens (1999, 2006), Yurimoto *et al.* (2006) have reviewed potential chemical mechanisms by which the meteoritic anomaly may be produced. Of particular significance is the position of oxygen in the periodic chart. It is the only element capable of producing a symmetry isotopic fractionation, namely, it coordinates most rock-forming elements. Any other element, except noble gases, is coordinated by oxygen, such as silicates (SiO_4) the major rock-forming minerals. Elements such as hydrogen, nitrogen, and carbon could ostensibly produce mass-independent symmetry effects; however, they only possess two stable isotopes and observation of such effects is impossible.

Recently, Marcus (2004) has developed an elegant model for the production of the meteoritic mass-independent isotopic reservoirs via chemical reactions. As discussed also by Thiemens (1988, 1999, 2006), it is the general nature of the symmetry property of the molecule that gives rise to the mass-independent isotopic fractionation. The recent model of Marcus (2004) is significant in that it takes a quantum mechanical model (the Marcus model discussed in previous sections) and applies it to a nebular chemistry network that ends with the production of solid and stable CAIs. A significant feature of the model is that the chemical reactions occur between the metal oxide and atomic oxide/hydroxide species on grain surfaces. This step results in an "entropic factor," or concentration of reactants on a grain surface that would otherwise be kinetically restrained in a purely homogenous (gas-phase) reaction network. Following the surficial reactions producing a symmetry isotopic fractionation, there is simultaneous preservation of the anomaly in the solid (CAI progenitor), and gas volatilization of oxygen, which ultimately becomes other meteoritic material of positive anomaly. This model thus is capable of producing the various oxygen reservoirs without recourse to varying oxygen-isotopic reservoirs in time and space. Experiments directly testing the significant reactions are needed, but are of extraordinary difficulty due to the requirement of gas-phase production of metal oxides.

Sulfur has the possibility of production of a symmetry effect, but its multiple valence states and inherent volatility render it difficult to maintain its original isotopic signature due to secondary-exchange reactions. The only meteoritic class that has combined low sulfate content and is isotopically anomalous at the bulk level are the Ureilites. These are achondritic meteorites, made of mostly olivines and pyroxenes. They also possess graphite, diamond, sulfides, metal, and minor silicates (Mittlefehldt *et al.*, 1998). The source of these meteorites has remained enigmatic, and based upon their bulk isotopic composition, possess highly unusual oxygen (Clayton and Mayeda, 1988). An excellent review of this highly unusual class of meteorites is provided by Goodrich *et al.* (1987). Thiemens *et al.* (1994) and Farquhar *et al.* (2000a) have reported that the sulfur-isotopic composition of ureilities is mass-independent, with a slight ^{33}S excess. As discussed by Farquhar *et al.* (2000a), it is possible that this ^{33}S enrichment may be derived from a chemical process. These particular meteorites have exceedingly low sulfur content, and perhaps these features may allow for preservation of a nebular signature. In a series of recent high-precision sulfur-isotopic measurements, it was reported that several achondritic meteorites possess mass-independent isotopic compositions (Rai *et al.*, 2005). Four achondritic meteorite groups are observed to possess mass-independent isotopic compositions. Figure 5 is a four-isotope sulfur plot of the data reported by Rai *et al.* (2005). The capital delta (Δ) is a measure of the deviation from mass dependence, as expressed in the case of oxygen isotopes. A zero value denotes purely mass-dependent fractionation. It is important to note that the precision and accuracy of measurement of sulfur isotopes is significantly better than for oxygen because of the use of sulfur hexafluoride as the analyte gas. Sulfur hexafluoride gas is extraordinarily stable, easily purified, and has essentially zero background. For this reason, the Δ33 and Δ36 for sulfur is measurable in the hundredths of per mil range. This is highly significant because this is the range where significant science exists.

As discussed by Rai *et al.* (2005), the only viable explanation at present for the source of the isotopic anomalies is photochemical production. For sulfur, there are numerous laboratory experiments investigating sulfur-isotopic fractionation mechanisms and to date, the only mass independent sulfur-isotopic fractionation observed is for short-wavelength, gas-phase photochemistry. There also exists a theoretical understanding of these processes. There are at present, no theoretical suggestions of other potential mass-independent fractionation processes.

As Figure 5 displays, the mass-independent isotopic compositions are consistent with production by a photochemical process. There are no other chemical processes known in either laboratory experiments or in nature that are capable of production of such an array. It was suggested by Rai *et al.* that the actual process was a gas-phase irradiation of sulfur species by an early active sun, such as an X-wind. Figure 6 displays a potential model. In this model, the early phase of solar evolution results in the production of copius

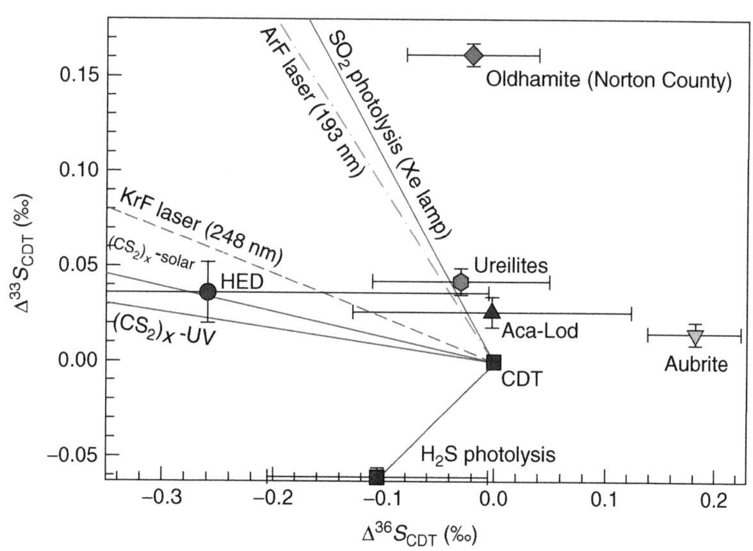

Figure 5 A plot of the deviation from mass independence for the four stable isotopes of sulfur. In this plot, mass-dependent isotopic compositions have an ΔS value of 0. The reference lines indicate isotopic fractionation trajectories for laboratory-based photochemistry experiments, as discussed in the text.

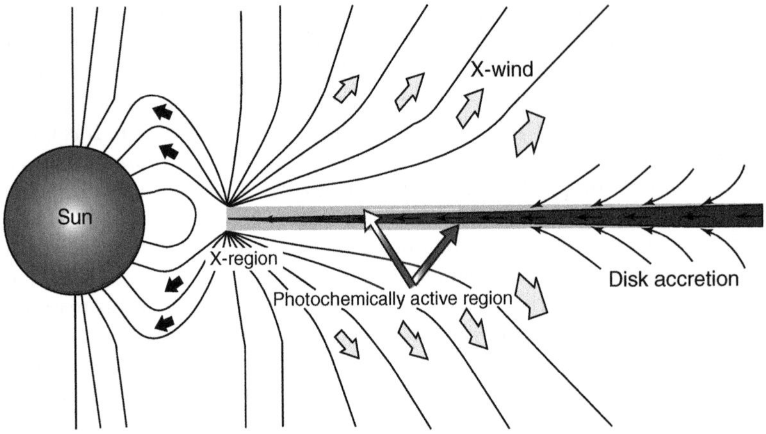

Figure 6 An illustration indicating how gas-phase photolysis by the protosun could produce mass-independent sulfur-isotopic anomalies.

amounts of short-wavelength UV light as the Sun undergoes ignition and proceeds toward the main sequence. During this time, there is convective mixing in the solar nebula that may result in the photochemistry of sulfur-bearing molecules, which produce the observed mass-independent sulfur-isotopic anomaly. The sulfur anomaly is produced in a reducing environment, consistent with its observation in the more reduced meteoritic classes. It is also possible that during this photolytic event, oxygen was also photochemically reacted.

In this scenario, the UV photolysis effect could produce the sulfur-isotopic anomalies observed in the meteorites. This is essentially the mechanism by which sulfur anomalies are produced in the present atmosphere, the Archean, and on Mars. The most significant aspect is that this is the first direct evidence of photo chemistry in the proto-nebular environment. The data also would require that planets had not as yet condensed, which may have implications for the production of the oxygen-isotopic anomalies as well as organic molecules. Cooper *et al.* (1997) have previously reported mass-independent sulfur-isotopic compositions in organic molecules extracted and isolated from the Murchison meteorite. Specifically, methyl, propyl, and ethyl sulfonic acids possess mass-independent sulfur-isotopic compositions. This also suggests that these organic species were created by photochemistry in the nebula, rather than on planetary mechanisms. This is significant as it suggests that there was a process prior to formation of the planetary bodies that produced organic species. If this is demonstrated to be the case, then the delivery of organics to the prebiotic Earth would be governed by photochemical processes rather than thermodynamic equilibrium ones.

It is now well known that there exist preserved interstellar grains in meteoritic material (see, e.g., Nittler *et al.*, 1999; Nittler, 2003; Wasserburg, 2006). There is extensive literature on this subject and will not be reviewed here. The most recent review by Wasserburg *et al.* (2006) is particularly important for establishing a most precise link between short-lived nuclei and their potential sources, such as AGB stars. These measurements are of particular importance in establishing chronologies and the recognition of intervention of varying astrophysical processes in the earliest history of the solar system. With respect to the oxygen nucleosynthetic model there are two key points. First, although there are literally thousands of isotopic measurements of the oxygen-isotopic composition of interstellar grains, there is no evidence for the presence of a ^{16}O carrier. Second, the concentration of these grains is quite low, typically parts per million or less. Therefore, there is a material balance issue as well. Given that the oxygen-isotopic anomalies exist at the bulk level in the major meteoritic element there is insufficient anomalous oxygen-isotopic atoms to produce the observed compositions. Finally, as discussed by Wasserburg *et al.* (2006), a key consideration is establishment of the bulk isotopic composition of the sun. Without the knowledge of the bulk oxygen-isotopic reservoirs, it is difficult to establish whether any given physical or chemical process is an enrichment or depletion event.

It is clear that nucleosynthetic models may not account for the observed meteoritic oxygen-isotopic anomalies, and chemical production processes are needed. Clayton (2002) abandoned the nucleosynthetic model as an explanation for the observed meteoritic oxygen-isotopic anomalies and proposed that isotopic self-shielding in the nebula accounts for the observed meteoritic-isotopic anomalies. The concept of self-shielding in the nebula as a mechanism to account for the oxygen-isotopic anomalies was first discussed in Thiemens and Heidenreich (1983). In addition, the possibility that the ^{15}N anomalies could arise from this process was also discussed by Thiemens and Heidenreich (1983). The restriction on isotopic self-shielding for oxygen is that, subsequent to production of an anomalous oxygen-isotopic reservoir in the photodissociative event, its isotopic character is removed by isotopic exchange before it may be preserved in a stable product. This issue was discussed in great detail by Navon and Wasserburg (1985).

A further issue is that the $\delta^{17}O = \delta^{18}O$ composition that would be produced by self-shielding only occurs over a restricted carbon monoxide column density. At greater capacities, the different $^{18}O/^{17}O$ abundance ratio (5.5) becomes significant and varying isotopic compositions are produced. In these regions, the $\delta^{17}O = \delta^{18}O$ fractionation is not produced. A new model of photochemical self-shielding by Lyons and Young (2003, 2004, 2005) circumvents many of the issues by allowing the photochemical self-shielding to occur at the outer edges of the presolar nebula and at low temperatures. This kinetic approach is detailed mechanistically and includes an extended reaction network and a sophisticated kinetic analysis. The occurrence of the self-shielding process at low temperatures sequesters the anomaly before it may be removed by secondary isotope exchange in ice. A difficulty of the model is that there must be an unspecified process by which the ice is converted into solid and further theoretical considerations are needed. It should be noted that the development of a kinetic model for the early solar system is an important advancement as it is becoming clear that the relevant processes are kinetically, rather than thermodynamically governed.

Given that the anomalous oxygen-isotopic effect is known to occur in the actual chemical reaction step, the issue of exchange is moot, as it is the reaction leading to a stable, solid product that produces the anomaly (Thiemens, 1999; Marcus, 2004). The recent successful models of Marcus and colleagues are of particular importance, as they also demonstrate that the source of the $\delta^{17}O = \delta^{18}O$ fractionation is a general feature of a gas-phase reaction for oxygen and not restricted to ozone formation. It is also this aspect that accounts for the issue as to why, of all the elements on the periodic chart, oxygen is unique. It is now clear that chemical production mechanisms are responsible for the production of meteoritic oxygen-isotopic anomalies. This is a major solar system process, since oxygen is the major element and the isotope effect is large, existing at the bulk level. What is presently needed are relevant experiments examining in detail the isotopic fractionations associated with the formation of solid oxides as well as further mechanistically defined models that lead to solids. Such experiments have proved to be formidably difficult due to the required reaction conditions; however, the need persists and such investigations are underway.

4.9 CONCLUDING COMMENTS

Since the discovery of a chemically produced mass-independent isotope effect by Thiemens and Heidenreich (1983), it has been demonstrated that there exists an extensive range of applications and observations in nature. Resolution of the actual physical–chemical mechanism has significantly advanced our understanding of gas-phase transition states and chemical reactions. Mass-independent oxygen-isotopic compositions have been observed in essentially all atmospheric molecules, except water. This includes O_2, O_3, CO_2, N_2O, H_2O_2, and aerosol sulfate and nitrate. In each specific instance, our understanding of its biogeochemical cycle has been enhanced as a result of the particular and specific isotopic signature. Measurements of sulfate and nitrate in ice core samples have provided new details of the past oxidative capacity of the Earth's atmosphere. Oxygen-isotopic compositions of ancient sulfates (e.g., Miocene) have provided new paleo-atmospheric information on the desert and nearby ocean in Namibia, volcanogenic processes, Antarctic dry valley sources, and desert varnish formation processes. Sulfur mass-independent isotopic anomalies in the Earth's oldest rocks (3.8×10^9 years) have provided a new means by which the evolution of oxygen–ozone, and consequently life has evolved until $\sim 2.2 \times 10^9$ years BP.

There continue to be new and novel applications of the mass-independent isotopic fractionation processes in nature. For example, Angert *et al.* (2004) have recently discussed the role and potential applicability of ^{17}O effects in the hydrologic cycle. The isotopic fractionation associated with precipitation and the temperature independency may make high-precision measurements of water an important new and unique tracer of the hydrologic cycle. This application and may also extend to measurements of the three-isotope composition of ice at high precision and may provide important new insight into the relative contributions of kinetic isotope effects, evaporation/condensation, and relative humidity on the hydrologic cycle. This work is an important advancement and is likely to advance understanding of both the hydrologic cycle and climate change.

Measurement of sulfur- and oxygen-isotopic anomalies in secondary minerals from martian meteorites have provided new insights into crust–atmosphere interactions on Mars. The observation of mass-independent sulfur-isotopic anomalies in the Earth's pre-Cambrium has opened an entirely new field for the study of the origin and evolution of oxygen. Finally, the chemical mass-independent process appears to be responsible for the production of the anomalous oxygen-isotopic compositions observed in meteorites and thus was a major process in the formation of the solar system.

There remain many new horizons in mass-independent chemistry and its applications. This includes theory, laboratory experiments, and new applications in nature, extending from Earth to the solar system, and in time, from the present to 4.55 Ga.

ACKNOWLEDGMENTS

P.I. gratefully acknowledges the generous support of the NSF (Atmospheric Chemistry and Polar Programs) and NASA (Cosmochemistry and Origins of Solar Systems). The author wishes to acknowledge his interactions and collaborations with his students and postdoctoral associates who have made most of the research possible. R.N. Clayton is thanked for comments that aided in preparation of this version of the manuscript. I wish to thank Vinai Rai and Subrata Chakraborty with their help and patience with creation of the figures for this text.

REFERENCES

Alexander B., Park R. J., Jacob D., Li Q. B., yantosca R. M., Savarino J., Lee C. C. W., and Thiemens M. H. (2005) Sulfate formation in sea-salt aerosols: constraints from oxygen isotopes. *J. Geophys. Res.* **110**, 10307.

Alexander B., Savarino J., Barkov N. I., Delmas R. J., and Thiemens M. H. (2002b) Climate driven changes in the oxidation pathways of atmospheric sulfate. *Geophys. Res. Lett.* **29**(14), 30.

Alexander B., Savarino J., Farquhar J., Thiemens M. H., Jourdain B., and Legrand M. (1999) Recent variations in the mass independent fractionation of the oxygen isotopes of sulfate: measurements from Greenland ice cores. *EOS, Trans., AGU* **80**, F134.

Alexander B., Savarino J., Kreutz K. J., and Thiemens M. H. (2004) Impact of preindustrial biomass-burning emissions on the oxidative pathways of tropospheric sulfur and nitrogen. *J. Geophys. Res.* **109**, 8030–8038.

Alexander B., Savarino J., Sneed S., Kreutz K., and Thiemens M. H. (2002a) Climate changes from measurements of atmospheric sulfate. *EOS, Trans., AGU*.

Alexander B., Savarino J., Thiemens M. H., and Delmas R. (2000) Variations in the oxygen isotopic composition of sulfate from the last interglacial period to the Holocene: implications of a mass-independent anomaly. *EOS, Trans., AGU* **81**, F38.

Alexander B., Thiemens M. H., Farquhar J., Kaufman A. J., Savarino J., and Delmas R. J. (2003) East Antarctic ice core sulfur isotope measurements over a complete glacial-interglacial cycle. *J. Geophys. Res. D* **24**, 18-1–18-7.

Alexander B., Vollmer M. K., Jackson T., Weiss R. F., and Thiemens M. H. (2001) Stratospheric CO_2 isotopic anomalies and SF_6 and CFC tracer concentrations in the Arctic polar vortex. *Geophys. Res. Lett.* **28**, 4103.

Anderson S. M., Hüsebusch D., and Mauersberger K. (1997) Surprising rate coefficients for four isotopic variants of $O = O_2$ M. *J. Chem. Phys.* **107**, 5385–5392.

Anderson S. M., Morton J., and Mauersberger K. (1989) Laboratory measurements of ozone isotopomers by tunable diode laser absorption spectroscopy. *Chem. Phys. Lett.* **156**, 175180.

Angert A., Cappa C. D., and DePaolo D. J. (2004) Kinetic ^{17}O effects in the hydrologic cycle: indirect evidence and implications. *Geochim. Cosmochim. Acta* **68**, 3487–3495.

Angert A., Luz B., and Yakir D. (2001) Fractionation of oxygen isotopes by respiration and diffusion in soils and its implications for the isotopic composition of atmospheric O_2. *Global Biogeochem. Cycles* **15**, 871–880.

Angert A., Rachmileuitch S., Barcan E., and Luz B. (2003) Effects of photorespiration, the cytochrome pathway, and the alternative pathway on the triple isotopic composition of atmospheric O_2. *Global Biogeochem. Cycles* **17**, 30-1–30-14.

Assonov S. S. and Brenninkmeijer C. A. M. (2001) A new method to determine the ^{17}O isotopic abundance in CO_2 using oxygen isotopic exchange with a solid oxide. *Rapid Commun. Mass Spectrom.* **15**, 2426–2437.

Babikov D., Kendrick B. K., Walker R. B., Pack R. T., Fleurat-Lesard P., and Schinke R. (2003a) Formation of ozone: metastable states and anomalous isotope effect. *J. Chem. Phys.* **119**, 2577–2589.

Babikov D., Kendrick B. K., Walker R. B., Schinke R., and Pack R. T. (2003b) Quantum origin of an anomalous isotope effect in ozone formation. *Chem. Phys. Lett.* **372**, 689–691.

Bao H. (2005) Sulfate in modern playa settings and in ash beds in hyperacid deserts: implication for the origin of ^{17}O-anomalous sulfate in an Oligocene ash bed. *Chem. Geol.* **127**, 34–38.

Bao H., Campbell D. A., Böckheim J. G., and Thiemens M. H. (2000b) Origins of sulphate in Antarctic dry-valley soils as deduced from anomalous ^{17}O compositions. *Nature* **407**, 499.

Bao H., Jenkins K., Khachaturyan M., and Diaz G. C. (2004) Different sulfate sources and their post-depositional migration in Atacama soils. *Earth Planet. Sci. Lett.* **224**, 577–587.

Bao H., Michalski G. M., and Thiemens M. H. (2001b) Sulfate oxygen-17 anomalies in desert varnishes. *Geochim. Cosmochim. Acta* **65**, 2029–2036.

Bao H. and Reheis M. C. (2003) Multiple oxygen and sulfur isotopic analysis on water soluble sulfate in bulk atmospheric deposition from the southwestern United States. *J. Geophys. Res.* **108**, 9-1–9-9.

Bao H., Thiemens M. H., Farquhar J., Campbell D. A., Lee C. C.-W., Heine K., and Loope D. B. (2000a) Anomalous ^{17}O compositions in massive sulphate deposits on earth. *Nature* **406**, 176–178.

Bao H., Thiemens M. H., and Heine K. (2001a) Oxygen-17 excesses of the central Namib gypretes: spatial distribution. *Earth Planet. Sci. Lett.* **192**, 125–135.

Bao H., Thiemens M. H., Loope D. B., and Yuan X.-L. (2003) Sulfate oxygen-17 anomlay in an Oligocene ash bed in mid-North America: was it dry fogs? *Geophys. Res. Lett.* **30**, 1-1–1-4.

Barth V. and Zahn A. (1997) Oxygen isotope composition of carbon dioxide in the middle atmosphere. *J. Chem. Phys.* **102**(12), 995–1001.

Bender M. L. (2000) *Science* **288**, 5473.

Bhattacharya S., Savarino J., and Thiemens M. H. (2000) A new class of oxygen isotopic fractionation in photodissociation of carbon dioxide: potential implications for atmospheres of mars and earth. *Geophys. Res. Lett.* **27**, 1459–1462.

Bigeleisen J. and Mayer M. G. (1947) Calculation of equilibrium constants for isotopic exchange reactions. *J. Chem. Phys.* **15**, 261–267.

Boering K., Jackson T., Hoag K. J., Cole A. S., Perri M. J., and Atlas E. (2004) Observations of the anomalous oxygen isotopic composition of carbon dioxide in the lower stratosphere and the flux of the anomaly to the troposphere. *Geophys. Res. Lett.* **31**, L03109.

Brenninkmeijer C. A. M., Janssen C., Kaiser K., Rockmann T., Rhee T. S., and Assonov S. S. (2003) Isotope effects in the chemistry of atmospheric trace compounds. *Chem. Rev.* **103**, 5125–5161.

Brenninkmeijer C. A. M. and Röckmann T. (1998) A rapid method for the preparation of O_2 from CO_2 for mass spectrometric measurement of $^{17}O/^{16}O$ ratio. *Rapid Commun. Mass Spectrom.* **12**, 479.

Clayton R. N. (2002) *Lunar Planet. Sci. Conf.* **XXXIII**, 1326.

Clayton R. N., Grossman L., and Mayeda T. K. (1973) A component of primitive nuclear composition in carbonaceous meteorites. *Science* **182**, 485–488.

Clayton R. N. and Mayeda T. K. (1988) *Geochim. Cosmochim. Acta* **52**.

Colman J. J., Xu J. P., Thiemens M. H., and Trogler W. C. (1996) Photopolymerization and mass-independent sulfur isotope fractionations in carbon disulfide. *Science* **273**, 774–778.

Cooper G. W., Thiemens M. H., Jackson T. L., and Chang S. (1997) Sulfur and hydrogen isotope anomalies in meteorite sulfonic acids. *Science* **277**, 1072–1074.

Cress R. G., Williams M. W., and Sievering H. (1995) *IAHS Publ.* **228**, 33.

Farquhar J., Bao H., and Thiemens M. (2000c) Atmospheric influence of earth's earliest sulfur cycle. *Science* **289**, 756–758.

Farquhar J., Jackson T. L., and Thiemens M. H. (2000a) *Geochim. Cosmochim. Acta* **64**, 1819.

Farquhar J., Savarino J., Airieau S., and Thiemens M. H. (2001) Observation of wavelength-sensitive mass-independent isotope effects during SO_2 photolysis: implications for the early atmosphere. *J. Geophys. Res.* **106**, 32829–32839.

Farquhar J., Savarino J., Jackson T. L., and Thiemens M. H. (2000b) Evidence of atmospheric sulphur in the martian regolith from sulphur isotopes in meteorites. *Nature* **404**, 50–53.

Farquhar J. and Wing B. A. (2003) Multiple sulfur isotopes and the evolution of the atmosphere. *Earth Planet. Sci. Lett.* **213**, 1–13.

Gao Y. Q. and Marcus R. A. (2001) Strange and unconventional isotope effects in ozone formation. *Science* **293**, 259.

Gao Y. Q. and Marcus R. A. (2002a) On the theory of the strange and unconventional isotopic effects in ozone formation. *J. Chem. Phys.* **116**, 137–154.

Gao Y. Q. and Marcus R. A. (2002b) A theoretical study of ozone isotopic effects using a modified ab initio potential energy surface. *J. Chem. Phys.* **117**, 1536–1543.

Goodrich C. A., Jones J. H., and Berkley S. L. (1987) *Geochim. Cosmochim. Acta* **51**, 2255.

Guenther J., Erbacher B., Krankowsky D., and Mauersberger K. (1999) Pressure dependence of two relative ozone formation rate coefficients. *Chem. Phys. Lett.* **306**, 209–213.

Guenther J., Krankowsky D., and Mauersberger K. (2000) Third-body dependence of rate coefficients for ozone formation in O_{16}–O_{18} mixtures. *Chem. Phys. Lett.* **324**, 31.

Hathorn B. C. and Marcus R. A. (2000) An intramolecular theory of the mass-independent isotope effect for ozone. II. Numerical implementation at low pressures using a loose transition state. *J. Chem. Phys.* **113**, 5586–5589.

Hathorn R. C. and Marcus R. A. (1999) An intramolecular theory of the mass-independent isotope effect for ozone. *J. Chem. Phys.* **111**, 4087–4099.

Heidenreich J. E., III and Thiemens M. H. (1986) A non-mass-dependent oxygen isotope effect in the production of ozone from molecular oxygen: the role of molecular symmetry in isotope chemistry. *J. Chem. Phys.* **84**, 2129–2136.

Herzberg G. (1966) *Molecular Spectra and Molecular Structure. III: Electronic Spectra and Electronic Structures of Polyatomic Molecules.* Van Nostrand Reinhold, New York.

Hoag K. J., Still C. J., Fung I., and Boering K. A. (2005) Triple isotope composition of tropospheric carbon dioxide as a tracer of terrestrial gross carbon fluxes. *Geophys. Res. Lett.* **32**, L02802–L02805.

Hulston J. R. and Thode H. G. (1965) Variations in the S^{33}, S^{34}, and S^{36} contents of meteorites and their relation to chemical and nuclear effects. *J. Geophys. Res.* **70**, 3475–3484.

Irion F. W., Gunson M. R., Rinsland C. P., Yung Y. L., Abrams M. C., Change A. Y., and Goldman A. (1996) Heavy ozone enrichments from ATMOS infrared solar spectra. *Geophys. Res. Lett.* **23**, 2377–2380.

Janssen C. (2001) Does symmetry drive isotopic anomalies in ozone isotopomer formation? *Science* **294**, 951.

Janssen C., Günther J., Krankowsky K., and Mauersberger K. (1999) Relative formation rates of $^{50}O_3$, and $^{52}O_3$ in ^{16}O–^{18}O mixtures. *J. Chem. Phys.* **111**, 7179–7182.

Janssen C., Guenther J., Krankowsky D., and Mauersberger K. (2002) Temperature dependence of ozone rate coefficients and isotopologue fractionation in ^{16}O–^{18}O oxygen mixtures. *Chem. Phys. Lett.* **367**, 34–38.

Janssen C., Guenther J., Mauersberger K., and Krankowsky D. (2001) Kinetic origin of the ozone isotope effect: a critical analysis of enrichments and rate coefficients. *Phys. Chem. Chem. Phys.* **3**, 4718–4721.

Johnson C. A., Mast M. A., and Kester C. L. (2001) Use of $^{17}O/^{16}O$ to trace atmospherically-deposited sulfate in surface waters: a case study in alpine watersheds in the rocky mountains. *Geophys. Res. Lett.* **28**, 4483–4486.

Johnston J. C., Röckmann T., and Brenninkmeijer C. A. M. (2000) CO_2+ $O(^1D)$ isotopic exchange: laboratory and modeling studies. *J. Geophys. Res.* **105**(15), 15213–15229.

Johnston J. C. and Thiemens M. H. (1997) The isotopic composition of tropospheric ozone in three environments. *J. Geophys. Res.* **105**, 25395–25404.

Johnston J. C. and Thiemens M. H. (2004) Mass independently fractionated ozone in the earth's atmosphere and in the laboratory. In *Handbook of Stable Isotope Techniques* (ed. P. deGroot). Elsevier, Holland.

Kasibhatla P., Chameides W. L., and St. John J. (1997) *J. Geophys. Res.* **102**(D3), 3737.

Kasting J. F. (2001) Earth history—the rise of oxygen. *Science* **293**, 819–820.

Kemper F., Jager C., Waters L., Henning T., Molster T. J., Banlow M. D., Lim T., and de Koter A. (2002) *Nature* **415**, 295.

Kendall C. and McDonnell J. J. (1998) *Isotope Tracks in Catchment Hydrology.*

Kornexyl B. E., Gehre M., Hofling R., and Werner R. A. (1999) *Rapid Commun. Mass Spectrom.* **13**, 1685.

Krankowsky P., Bartecki F., Klees G. G., Mauersberger K., and Stehr J. (1995) Measurement of heavy ozone enrichment in tropospheric ozone. *Geophys. Res. Lett.* **22**, 1713–1716.

Krankowsky P., Lämmerzahl P., and Mauersberger K. (2000) Isotopic measurements of stratospheric ozone. *Geophys. Res. Lett.* **27**, 2593–2596.

Krot A. N., Hutcheon I. D., Yurimoto H., Cuzzi J. N., and McKeegan K. D. (2005) Evolution of oxygen isotopic composition in the inner solar nebula. *Astrophys. J.* **622**, 1333–1342.

Lämmerzahl P., Rockmann T., Brenninkmeijer C. M. M., Krankowsky D., and Mauersberger K. (2002) Oxygen isotope composition of stratospheric carbon dioxide. *Geophys. Res. Lett.* **29**, 1–23.

Lee C. C.-W., Savarino J. P., and Thiemens M. H. (1998) Multiple stable isotopic studies of sulfate and hydrogen peroxide collected from rain water: a new way to investigate *in-situ* S (IV) oxidation chemistry by dissolved H_2O_2 in aqueous solution. *EOS, Trans., AGU* **79**, 91.

Lee C. C.-W. and Thiemens M. H. (1997) Multiple stabble oxygen isotopic studies of atmospheric sulfate aerosols. *EOS, Trans., AGU* **78**, F111.

Lee C.-W., Savarino J., Cachier H., and Thiemens M. H. (2002) Sulfur (^{32}S, ^{33}S, ^{34}S, ^{36}S) and oxygen (^{16}O, ^{17}O, ^{18}O) isotopic ratios of primary sulfate produced from combustion processes. *Tellus* **54B**, 193–200.

Lee C.-W., Savarino J., and Thiemens M. H. (2001a) Mass independent oxygen isotopic composition of atmospheric sulfate: origin and implications for the present and past atmospheres of earth and mars. *Geophys. Res. Lett.* **28**, 1783.

Lee C.-W., Savarino J., and Thiemens M. H. (2001b) The $\delta^{17}O$ and $\delta^{18}O$ measurements of atmospheric sulfate from a coastal and high plane region: a mass independent isotopic anomaly. *J. Geophys. Res.* **106**, 17359–17373.

Lelieveld J., Roelofs G. J., Ganzeweld L., Feichter J., and Rodhe H. (1997) *Phil. Trans. Roy. Soc. Lond., Ser. B* **352**, 149.

Luz B. and Barkan E. (2000) Assessment of oceanic productivity with the triple-isotope composition of dissolved oxygen. *Science* **288**, 2028–2031.

Luz B., Barkan E., Bender M. L., Thiemens M. H., and Boering K. A. (1999) Triple-isotope composition of atmospheric oxygen as a tracer of biosphere productivity. *Nature* **400**, 547–549.

Lyons J. R. (2001) Transfer of mass-independent fractionation in other oxygen-containing radicals in the atmosphere. *Geophys. Res. Lett.* **28**, 3231–3234.

Lyons J. R. and Young E. D. (2003) Towards an evaluation of self-shielding at the x-point as the origin of the oxygen isotope anomaly in CAIs. *Lunar Planet. Sci.* **XXXIV** (abstract).

Lyons J. R. and Young E. D. (2004) Evolution of oxygen isotopes in the solar nebula. *Lunar Planet. Sci. Conf.* **XXXV**.

Lyons J. R. and Young E. D. (2005) CO self-shielding as the origin of oxygen isotope anomalies in the early solar nebula. *Nature* **435**, 317–320.

Marcus R. A. (2004) Mass-independent isotope effect in the earliest processed solids in the the solar system: a possible chemical mechanism. *J. Chem. Phys.* **121**, 8201–8218.

Mauersberger K. (1981) Measurement of heavy ozone in the stratosphere. *Geophys. Res. Lett.* **8**, 935–938.

Mauersberger K. (1987) Ozone isotope measurements in the stratosphere. *Geophys. Res. Lett.* **14**, 80–83.

Mauersberger K., Erbacher B., Krankowsky D., Günther J., and Nickel R. (1999) Ozone isotope enrichment: isotopomer-specific rate coefficients. *Science* **283**, 370–372.

Mauersberger K., Krankowsky D., Janssen C., and Schinke R. (2005) Assessment of the ozone isotope effect. In *Advances in Atomic, Molecular, and Optical Physics* (eds. B. Benderson and H. Walther). Elsevier, Amsterdam, vol. 50.

Michalski G., Bohlke J. K., and Thiemens M. H. (2004a) Long term atmospheric deposition as the source of nitrate and other salts in the Atacama desert, Chile: new evidence from mass-independent oxygen isotopic compositions. *Geochim. Cosmochim. Acta* **68**, 4023–4038.

Michalski G., Meixner T., Fenn M., Hernandez L., Sirulnik A., Allen E., and Thiemens M. (2004b) Tracing atmospheric nitrate deposition in a complex semiarid ecosystem using $\delta^{17}O$. *Environ. Sci. Technol.* **38**, 2175–2181.

Michalski G., Savarino J., Bohlke J. K., and Thiemens M. (2002) Determination of the total oxygen isotopic composition of nitrate and the calibration of a $\Delta^{17}O$ nitrate reference material. *Anal. Chem.* **74**, 4989–4993.

Michalski G., Scott Z., Kailiing M., and Thiemens M. H. (2003) First measurements and modeling of $\delta^{17}O$ in atmospheric nitrate. *Geophys. Res. Lett.* **30**, 14-1–14-5.

Michalski G. and Thiemens M. H. (2000) *EOS, Trans., AGU* **81**, 120.

Miller M. F., Franchi I. A., Thiemens M. H., Jackson T. L., Brass A., Kurat G., and Dilliger C. T. (2002) Mass-independent fractionation of oxygen isotopes during thermal decomposition of carbonates. *Proc. Natl. Acad. Sci.* **99**, 10988–10993.

Mittlefehldt D. W., McCoy T. J., Goodrich C. A., and Kracher E. A. (1998) Non-chondritic meteorite from asteroidal bodies. *Planet. Mater.* **36**, 4-1–4-195.

Mojzsis S. J., Coath O. J., Greenwood J. P., McKeegan K. D., and Harrison T. M. (2003) Mass-independent isotope effects in the Archean (2.5–3.8 Ga) sedimentary sulfides determined by ion-microprobe multicollection. *Geochim. Cosmochim. Acta* **67**, 1635–1678.

Mook W. O. (2000) *Environmental Isotopes in the Hydrological Cycle: Principles and Applications: Vol. I. Introduction—Theory, Methods Review.* UUESCO/IAEA, Paris.

Navon O. and Wasserburg G. J. (1985) Self-shielding in O_2—a possible explanation for oxygen isotopic anomalies in meteorites. *Earth Planet. Sci. Lett.* **73**, 1–16.

Nier A. O., Ney F. P., and Ingraham M. G. (1947) A null method for the comparison of two ion currents in a mass spectrometer. *Rev. Sci. Instr.* **18**, 294–297.

Nittler L. O. D., Alexander C., Gao X., Walker R. M., and Zinner E. K. (1998) Meteoritic oxide grain from a supernova found. *Nature* **393**, 222–225.

Nittler L. R. (2003) Presolar stardust in meteorites: recent advances and scientific frontiers. *Earth Planet. Sci. Lett.* **209**, 259–273.

Paerl H. W. (1993) Interaction of nitrogen and carbon cycles in the marine environment. *Can. J. Fish. Aquatic. Sci.* **50**, 2254.

Pavlov A. A. and Kasting J. F. (2002) Mass-independent fractionation of sulfur isotopes in Archaen sediments. Strong evidence for an anoxic Archean atmosphere. *Astrobiology* **2**, 27–41.

Rai V. K., Jackson T. L., and Thiemens M. H. (2005) Photochemical mass-independent sulfur isotopes in achondritic meteorites. *Science* **309**, 1062–1064.

Rasch P. J., Barth M. C., Kiehl J. T., Schwartz S. E., and Benkovitz C. M. (2000) *J. Geophys. Res.* **105**, 1367.

Revesz K., Boehlice J. K., and Yoshinavi T. (1997) *Anal. Chem.* **69**, 4375.

Runnegar B., Coath C. D., Lyons J. R., and McKeegan K. D. (2002) Mass-independent and mass-dependent sulfur processing throughout the Archean. *Geochim. Cosmochim. Acta* **66**, A655–A670.

Savarino J., Alexander B., Sneed S., Kreutz K., and Thiemens M. H. (2002) *EOS, Trans., AGU* **83**(19), 545.

Savarino J., Bekki S., Cole-Dai J., and Thiemens M. H. (2003b) Evidence from mass independent oxygen isotopic compositions of dramatic changes in atmospheric oxidation following massive volcanic eruptions. *J. Geophys. Res.* **108**, 71-1–71-7.

Savarino J., Lee C. C.-W., and Thiemens M. H. (2000) Laboratory oxygen isotopic study of sulfur (IV) oxidation: origin of mass-independent oxygen isotopic anomaly in atmospheric sulfates and sulfate mineral deposits on earth. *J. Geophys. Res.* **105**, 29079–29090.

Savarino J., Romero A., Cole-Dai J., Bekki S., and Thiemens M. H. (2003a) UV induced mass-independent sulfur isotope fractionation in stratospheric volcanic sulfate. *Geophys. Res. Lett.* **30**, 11-1–11-4.

Savarino J. and Thiemens M. H. (1993) Analytical procedure to determine both $\delta^{18}O$ and $\delta^{17}O$ of H_2O_2 in natural water and first measurements. *Atmos. Environ.* **33**, 3683–3690.

Savarino J. and Thiemens M. H. (1999) Mass-independent oxygen isotope (^{16}O,^{17}O,^{18}O) fractionation found in Hx, Ox, reactions. *J. Phys. Chem.* **103**, 9221.

Schueler B., Morton J., and Mauersberger K. (1990) Measurement of isotopic abundances in collected stratospheric ozone samples. *Geophys. Res. Lett.* **17**, 1295–1298.

Stehr J., Krankowsky D., and Mauersberger K. (1996) Collection and analysis of atmospheric ozone samples. *J. Atmos. Chem.* **24**, 317.

Strunk M., Engel A., Schmidt U., Volk C. M., Wetter T., Levin I., and Glatzel-Mattheiz H. (2000) *Geophys. Res. Lett.* **27**, 341.

Thiemens M. H. (1999) Mass-independent isotope effects in planetary atmospheres and the early solar system. *Science* **283**, 341–345.

Thiemens M. H. (2002) Mass-independent isotope effects and their use in understanding natural processes. *Israel J. Chem.* **42**, 43–52.

Thiemens M. H. (2006) History and applications of mass independent isotope effects. *Annu. Rev. Earth Planet. Sci.* (accepted).

Thiemens M. H., Brearly A., Jackson T., and Bobias G. (1994) *Meteoritics* **29**, 540.

Thiemens M. H. and Heidenreich J. E. (1983) The mass-independent fractionation of oxygen: a novel isotope effect and its possible cosmochemical consequences. *Science* **219**, 1073–1075.

Thiemens M. H., Jackson T., Zipf E. C., Erdman P. N., and Vansgmond C. (1995) Carbon dioxide and oxygen isotope anomalies in the mesosphere and stratosphere. *Science* **270**, 969–972.

Thiemens M. H. and Jackson T. L. (1988) Pressure dependency for heavy isotope fractionation in ozone formation. *Geophys. Res. Lett.* **15**, 639–642.

Thiemens M. H., Jackson T. L., Mauersberger K., Schuler B., and Morton J. (1991) Oxygen isotope fractionation in stratospheric CO_2. *J. Geophys. Res. Lett.* **18**, 669.

Thiemens M. H., Savarino J., Farquhar J., and Bao H. (2001) Mass-independent isotopic compositions in terrestrial and extraterrestrial solids and their applications. *Acc. Chem. Res.* **34**, 645–652.

Urey H. C. (1947) The thermodynamic properties of isotopic substances. *J. Chem. Soc.* **47**, 562–581.

Wasserburg G. J., Busso M., Gallino R., and Nollett K. M. (2006) Short-lived nuclei in the early solar system: possible AGB sources. *Nucl. Phys. A* (in press).

Wen J. S. and Thiemens M. H. (1993) Multi-isotope study of the $O(^1D)+CO_2$ exchange and stratospheric consequences. *J. Geophys. Res.* **98**, 12801–12808.

Weston R. E. (1999) Anomalous or mass-independent isotope effects. *Chem. Rev.* **99**, 2115–2136.

Wing B. A., Brabson E., Farquhar J., Kaufman A. J., Rumble D., and Bekker A. (2002) $\Delta^{33}S$, $\delta^{34}S$, and $\delta^{13}C$ constraints on the paleoprotozoic atmosphere during he earliest Huronian glaciation. *Geochim. Cosmochim. Acta* **66**, 840–850.

Yung Y. L., DeMore W. B., and Pinto J. P. (1991) Isotope exchange between carbon dioxide and ozone via O(1D) in the stratosphere. *Geophys. Rev. Lett.* **18**, 13–16.

Yung Y. L., Lee A. Y. T., Irion W. B., DeMore W. B., and Wen J. (1997) Carbon dioxide in the atmosphere: isotope exchange with ozone and its use as a tracer in the middle atmosphere. *J. Geophys. Res.* **102**, 10857–10866.

Yurimoto H., Kuramoto K., Krot A. N., Scott E. R. D., Cuzzi J. N., Thiemens M. H., and Lyons J. R. (2006) Origin and evolution of oxygen isotopic compositions of the solar system. *Proto Stars Planets V* (accepted).

Zmolek P., Xu X., Jackson T., Thiemens M. H., and Trogler W. C. (1999) Large mass independent sulfur isotope fractionations during the photopolymerization of $^{13}CS_2$ and $^{12}CS_2$. *J. Phys. Chem.* **103**, 2477–2480.

Isotope Geochemistry
ISBN: 978-0-08-096710-3

pp. 87–112

5
The Stable Isotopic Composition of Atmospheric CO_2

D. Yakir

Weizmann Institute of Science, Rehovot, Israel

5.1 INTRODUCTION

When a bean leaf was sealed in a closed chamber under a lamp (Rooney, 1988), in two hours the atmospheric CO_2 in the microcosm reached an isotopic steady state with a ^{13}C abundance astonishingly similar to the global mean value of atmospheric CO_2 at that time (−7.5‰ in the $\delta^{13}C$ notation introduced below). Almost concurrently, another research group sealed a suspension of asparagus cells in a different type of microcosm in which within about two hours the atmospheric O_2 reached an isotopic steady state with ^{18}O enrichment relative to water in the microcosm that was, too, remarkably similar to the global-scale offset between atmospheric O_2 and mean ocean water (21‰ versus 23.5‰ in the $\delta^{18}O$ notation introduced below; Guy *et al.*, 1987). These classic experiments capture some of the foundations underlying the isotopic composition of atmospheric CO_2 and O_2. First, in both cases the biological system rapidly imposed a unique isotopic value on the microcosms' atmosphere via their massive photosynthetic and respiratory exchange of CO_2 and O_2. Second, in both cases the biological system acted on materials with isotopic signals previously formed by the global carbon and hydrological cycles. That is, the bean leaf introduced its previously formed organic matter (the source of the CO_2 respired into microcosm's atmosphere), and the asparagus cells were introduced complete with local tap water (from which photosynthesis released molecular oxygen). Therefore, while the isotopic composition of the biological system used was slave to long-term processes, intense metabolic processes centered on few specific enzymes (Yakir, 2002) dictated the short-term atmospheric composition.

In a similar vein, on geological timescales of millions of years, the atmosphere and its isotopic composition are integral parts of essentially a single dynamic ocean–atmosphere–biosphere system. This dynamic system exchanges material, such as carbon and oxygen, with the sediments and the lithosphere via slow processes that roughly follow the cycle of: weathering of rock and carbon uptake from the atmosphere, transport to the ocean, sedimentation, plate tectonics, metamorphism, and volcanism—leading to carbon release back to the atmosphere. But on a shorter timescale of years to millennia, the very slow geological processes retreat to the background, against which other massive fluxes control the rapid exchange of carbon and oxygen within the ocean–atmosphere–biosphere system. It is this timescale that is relevant to the well-being of our human society, and is a major focus in much of the research on the carbon cycle.

Isotopes were discovered in 1911 (Urey, 1948) and the implications of isotopic substitution in chemical reactions were realized sometime later (Bigeleisen, 1965). In practice, the use of stable isotopes in geochemistry and biogeochemistry (e.g., Craig, 1953, 1954) awaited the development of the isotope ratio mass spectrometer (McKinney *et al.*, 1950; Nier, 1947) that provided the necessary precision. Over the 50 years following this breakthrough, the application of stable isotopes has made tremendous progress in the scope of applications, as well as in the resolution and precision of the measurements. The carbon isotopic composition of rocks and sediments was measured intensively since the early 1950s (Hoefs, 1987). Isotope hydrology caught up quickly (Clark and Fritz, 1997), followed by the application of stable isotopes in biology and ecology (Rundel *et al.*, 1988; Griffiths, 1998; Ehleringer *et al.*, 1992). Today, stable isotope measurements have become an indispensable and integral part of atmospheric measurement programs (e.g., Francey *et al.*, 2001; Masarie *et al.*, 2001; Trolier *et al.*, 1996). Efforts to develop analytical and numerical models that incorporate the cycling of stable isotopes in CO_2 expanded in parallel (e.g., Bolin, 1981; Ciais *et al.*, 1997a,b; Enting *et al.*, 1995). Recently, the consideration of mass-independent isotope phenomena in nature (Thiemens, 1999; see Chapter 4, and of triple stable isotopes in geochemistry (e.g., Blunier *et al.*, 2002; Luz *et al.*, 1999; Luz and Barkan, 2000) greatly extended the potential of stable isotope applications.

The chemical and isotopic composition of the atmosphere has drawn particular attention in climate-related research both because it is the most accessible component in the tightly coupled land–ocean–atmosphere system, and because the chemical composition of the atmosphere influences climate, particularly via the concentrations of the radiatively active greenhouse gases, such as CO_2, O_3, CH_4, N_2O, and water vapor. Information obtained by measurements of the atmospheric concentration of these gases alone is limited; the additional measurements of the stable isotopic composition provide information that cannot be obtained otherwise. Isotopic fractionations during chemical, physical, and biological process in the ocean, land, and the atmosphere result in unique natural labels. Tracing these labels in time and space allows us both to identify specific fluxes of these gases, and to gain insights into the processes influencing the observed fluxes. Quantitative use of ^{18}O and ^{13}C in CO_2 must rely on precise observations, on experimentation addressing the isotope effects underlying these observations, and on modeling that tests basic assumptions and extends applications beyond our measuring capabilities. Progress is still needed on all of these fronts. But the importance of this still developing science of stable isotopes in environmental research is indisputable.

5.2 METHODOLOGY AND TERMINOLOGY

Elements in nature come in forms called isotopes that differ only in the number of their neutrons. Most isotopes are stable and can be distinguished from their counterparts simply by their masses. Remarkably, isotopes are associated with a few simple and mass-dependent traits that result in a wide range of useful isotopic signals in natural processes. Coupled with the invention of the isotope ratio mass spectrometer in 1940s (McKinney *et al.*, 1950; Nier, 1947) stable isotope signals provide the basis for application of stable isotopes to environmental sciences. Stable isotopes are denoted by their atomic mass such as ^{13}C and ^{12}C for the two stable isotopes of carbon, and ^{18}O, ^{17}O, and ^{16}O for the stable isotopes of oxygen. Because the heavy isotope is normally rare (e.g., ~1.1% for ^{13}C, 0.2% for ^{18}O, and 0.04% for ^{17}O), routine measurements of the absolute isotopic concentrations is difficult and not reliable. Alternatively, the ratio, R, of the rare to the abundant isotopes is measured, such as

$$^{13}R = {}^{13}C/{}^{12}C \qquad (1)$$

for carbon or $^{18}R = {}^{18}O/{}^{16}O$ for oxygen, etc. Note that isotope ratios are distinct from isotope concentrations and, for example, ^{13}C concentration $[^{13}C]$ can be defined as $^{13}R'$,

$$^{13}R' = {}^{13}C/[{}^{12}C + {}^{13}C + {}^{14}C] \approx {}^{13}R/(1 + {}^{13}R) \quad (2)$$

The value of $^{13}R'$ can be used to estimate, for example, ^{13}C content in the atmosphere, which is currently about 4 ppm.

Isotope ratios are normally not measured on the pure element, but rather on a gas containing the element that is convenient for the mass-spectrometric analysis. Carbon and oxygen isotope ratios are normally measured in CO_2, which means $^{13}R = {}^{13}CO_2/{}^{12}CO_2$ and $^{18}R = C^{18}O^{16}O/C^{16}O^{16}O$ based on measurements of the mass ratios 45/44 and 46/44 (the mass to charge ratio, *m*/*z*, of the ionized form of the molecules is determined in practice). This is significant because there can be contributions to these mass units other than the measured molecules. For CO_2, corrections must be made, for example, for the occurrence of $C^{17}O^{16}O$ (mass 45; the "Craig correction"; Craig (1957), Mook and Grootes (1973), and Santrock *et al.* (1985)), and contaminations such as by N_2O (mass 44; Craig (1963), Friedli and Siegenthaler (1988), Mook *et al.* (1974), and Mook and Van der Hoek (1983)). To reduce the effects of contaminants, gas samples are carefully purified, by bulk separation and/or by gas chromatography, prior to the mass spectrometric analysis. Furthermore, producing the convenient gas for isotopic analysis, such as CO_2, often involves a chemical reaction that modifies the isotopic ratio in question. In this case, the modifications must be carefully characterized and corrected for as part of the isotopic analysis. Such is the case, for example, in producing CO_2 from carbonates by acid treatment (Hoefs, 1987). For the determination of ^{18}O in water, the conventional practice is to first equilibrate a subsample with pure CO_2, determine the isotopic ratio of the CO_2 and reconstruct from it the isotopic ratio of the water (Gat, 1996; Gat and Gonfiantini, 1981; Mook, 2000).

To reduce variations due to instrumental drifts and instabilities, the common procedure in stable isotope analyses is to compare the ratio in a sample to that of a standard that is assumed to have a constant and known isotopic ratio. Isotope ratios are therefore reported in their normalized difference form as "delta values," where

$$\delta = \frac{R_{\text{sample}} - R_{\text{standard}}}{R_{\text{standard}}} = \frac{R_{\text{sample}}}{R_{\text{standard}}} - 1 \quad (3)$$

Note that δ values are practical, dimensionless parameters. Since values of δ in nature are small, they are generally given in ‰ (per mil)—which is a 10^3 factor, not a unit. Further, only the arbitrary choice of standard determines whether a δ value of a sample is positive or negative (Mook, 2000).

To allow efficient comparisons of results among different labs around the world, a series of common international standards has been established and all samples are normalized to such standards. For carbon, the common standard was originally a carbonate of the Pee Dee belemnite carbonate formation in South Carolina. Naturally, as the original standards are exhausted, appropriate substitutes have been introduced. The two main sources for international standards today are the International Atomic Energy Association (IAEA) in Vienna, and the US National Institute of Standards and Technology (NIST). Two main international standards will be used in this chapter; the first, Vienna Standard Mean Ocean Water (V-SMOW), is used primarily for water isotopes (oxygen and hydrogen). The second, the Vienna Pee-Dee Belemnite (V-PDB), is used primarily for carbon isotopes (Allison *et al.*, 1995; Coplen, 1995; Coplen *et al.*, 1983; Craig, 1957). As mentioned above, CO_2 is produced from carbonates by a chemical reaction with associated isotopic modification. Accordingly, the V-PDB-CO_2 standard is defined based on the isotopic ratio of CO_2 produced from PDB, rather than the original ratio of the carbonate. It turned out that the $^{18}O/^{16}O$ ratio of V-PDB-CO_2 is nearly identical (0.3‰ difference) to the $^{18}O/^{16}O$ ratio of CO_2 in equilibrium with seawater at 25 °C (i.e., V-SMOW-CO_2). Expressing $\delta^{18}O$ values of water on the SMOW scale and that of CO_2 in equilibrium with the same water on the V-PDB-CO_2 scale is convenient, producing nearly the same $\delta^{18}O$ values (Figure 1).

Modifications in isotopic ratios considered in this chapter occur mainly due to "kinetic" and "equilibrium" isotope fractionations. A kinetic isotope fractionation, in a one-way process, occurs because a molecule containing the lighter isotope has faster translational velocities, allowing it to move preferentially, at higher rates, through the molecules of the medium, leaving behind residue that becomes progressively enriched in the heavy isotope. A good example of this effect is molecular diffusion, in which case the isotopic discrimination can be accurately predicted based on the reduced masses of the gas species of interest and that of the media through which it diffuses (Craig, 1953). Kinetic isotope effects also occur because chemical reactions are sensitive to differences in dissociation energies of molecules. It is usually easier to form, or break, the bonds of molecules containing the lighter isotope, because the vibrational frequency of such bonds tends to be higher. This leads to a difference in the rate of reaction for the isotopic species, and to isotopic fractionation in which molecules containing lighter isotopes will be preferentially incorporated in the product of incomplete reactions, while the heavy isotopes will become enriched in the unreacted residue.

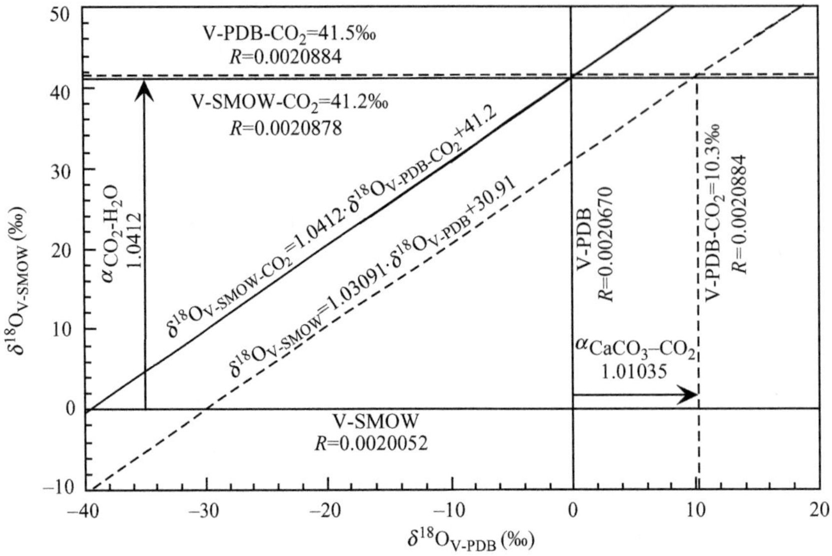

Figure 1 Relationships between stable isotope scales. Measured isotope ratios are normalized to international standards, such as V-PDB or V-SMOW. Oxygen isotope ratios may be related to the V-SMOW, V-PDB, or the V-PDB-CO_2 scale (see text for detail). It is particularly convenient to compare oxygen isotope ratios of CO_2 on the V-PDB scale with the oxygen isotope ratio of water on the V-SMOW scale. The difference between the scales almost exactly compensates for the oxygen isotopic equilibrium fractionation between CO_2 and water at 25 °C.

Of course, if reactions are complete and if all reactants are converted to product, no fractionation will be observed.

An equilibrium (thermodynamic) isotope effect is observed when a chemical or physical equilibrium between two compounds or phases is permitted. In this case, the equilibrium fractionation is the net effect of the two unidirectional isotope fractionations. Equilibrium fractionation usually results in the heavy isotope accumulating preferentially where it is bound more strongly. Vapor–liquid equilibrium fractionation reflects the fact that molecules containing the light isotopes are more volatile. As a result, the heavy isotopes are preferentially concentrated in the liquid phase. Both kinetic and equilibrium fractionations can be predicted from knowledge of the binding energies of atoms and molecules (Bigeleisen and Mayer, 1947; Bigeleisen and Wolfsberg, 1958; Urey, 1947), but often must be determined empirically (e.g., Friedman and O'Neil, 1977).

The isotope *fractionation factor,* α, is defined as the ratio of two isotopic ratios:

$$\alpha_{a-b} = \frac{R_b}{R_a} \qquad (4)$$

where a and b are often, but not always, defined as source (reactant) and product, respectively. Fractionation factors greater or smaller than 1 indicate isotopic enrichment or depletion, respectively, in going from a to b. In general, isotope fractionation factors are small and α is ~1. Therefore, *fractionation,* ε, the deviation of α from 1, is often used:

$$\varepsilon = \alpha - 1 \qquad (5)$$

As ε values are small, they are generally given as ε‰ (i.e., $\varepsilon 10^3$). Note first that fractionation factors are multiplicative ($\alpha_{total} = \alpha_1\alpha_2 \cdots \alpha_n$), while the fractionation ε is approximately additive ($\varepsilon_{total} \approx \varepsilon_1 + \varepsilon_2 + \cdots + \varepsilon_i$). Second, ε is often negative, reflecting isotopic depletion in a process, and $\varepsilon_{a-b} = -\varepsilon_{b-a}/(1+\varepsilon_{b-a}) \approx -\varepsilon_{b-a}$.

Some variations in the isotopic nomenclature should be noted. For example, the widely adopted model of Farquhar *et al.* (1982) for ^{13}C fractionation uses the definition $\alpha = R_{source}/R_{product}$ and the deviations from 1 as *discrimination*, Δ. While having the same numerical value, Δ is normally positive (i.e., $\Delta = -\varepsilon$). In the literature, Δ is often used for the overall discrimination of a system, while specific discrimination steps are assigned symbols such as "a" for steps involving diffusion, "b" for irreversible biochemical reactions and "e" for steps which show an equilibrium fractionation. Note that fractionations or discriminations are independent of reference materials. They are easily related to the measured delta values according to

$$\Delta = \frac{\delta_{source} - \delta_{product}}{1 + \delta_{product}} \qquad (6)$$

5.3 CARBON-13 IN ATMOSPHERIC CO_2

The isotopic composition of carbon has been an important research tool in the study of the carbon cycle on all timescales. In crude terms, we can observe that crustal carbon, derived from the Earth's mantle (inherited, in turn, from the parent solar nebula) has a bulk $\delta^{13}C$ value of around −5‰, resulting in carbonate sediments having $\delta^{13}C$ value around 0‰, and following the evolution of photosynthetic organisms discriminating against ^{13}C in organic matter with $\delta^{13}C$ values of around −25‰, leaving a small atmospheric CO_2 pool with a $\delta^{13}C$ of around −7‰ (Schidlowski, 1988). Global isotopic budgets require, therefore, that the partitioning of carbon between the organic and inorganic components of the dominant sedimentary carbon pools be of the order of 20% and 80%, respectively (i.e., 0.2(−25‰) + 0.8(0‰) = −5‰). The isotopic partitioning among the major carbon cycle components thus provides a powerful indicator for carbon isotopic composition of the atmosphere, as well as of changes in carbon fluxes and underlying processes, over the past 3–4 billion years (Beerling and Woodward, 1998; Hayes *et al.*, 1999; Holland, 1984; Raven, 1998; Schidlowski, 1988). Focusing on the more recent geological past, carbon isotopic compositions in marine and terrestrial sediments have also been an important paleoclimatic indicator in investigating the details of the glacial cycles (Leuenberger *et al.*, 1992) and the Holocene (Feng, 1999; Francey *et al.*, 1999; Friedli *et al.*, 1986; Hemming *et al.*, 1998; Keeling *et al.*, 1979; Marino *et al.*, 1992). The use of carbon isotopes in studying the current carbon cycle is greatly enhanced by taking advantage of direct access to atmospheric CO_2 (Keeling, 1961; Keeling *et al.*, 1980, 1979; Mook *et al.*, 1983; Figure 2).

A pressing question in any attempt to predict future variations in the atmospheric CO_2 concentrations is determining what fraction of the CO_2 input to the atmosphere remains in the atmosphere (Broecker *et al.*, 1979), is absorbed by the ocean, or is taken up the land biosphere (Ciais *et al.*, 1995a; Enting *et al.*, 1995; Francey *et al.*, 1995; Heimann and Maier-Reimer, 1996; Morimoto *et al.*, 2000; Tans *et al.*, 1990, 1993; Battle *et al.*, 2000; Still *et al.*, 2003). As is the case in the geological timescale studies, carbon isotopes offer a simple and powerful approach

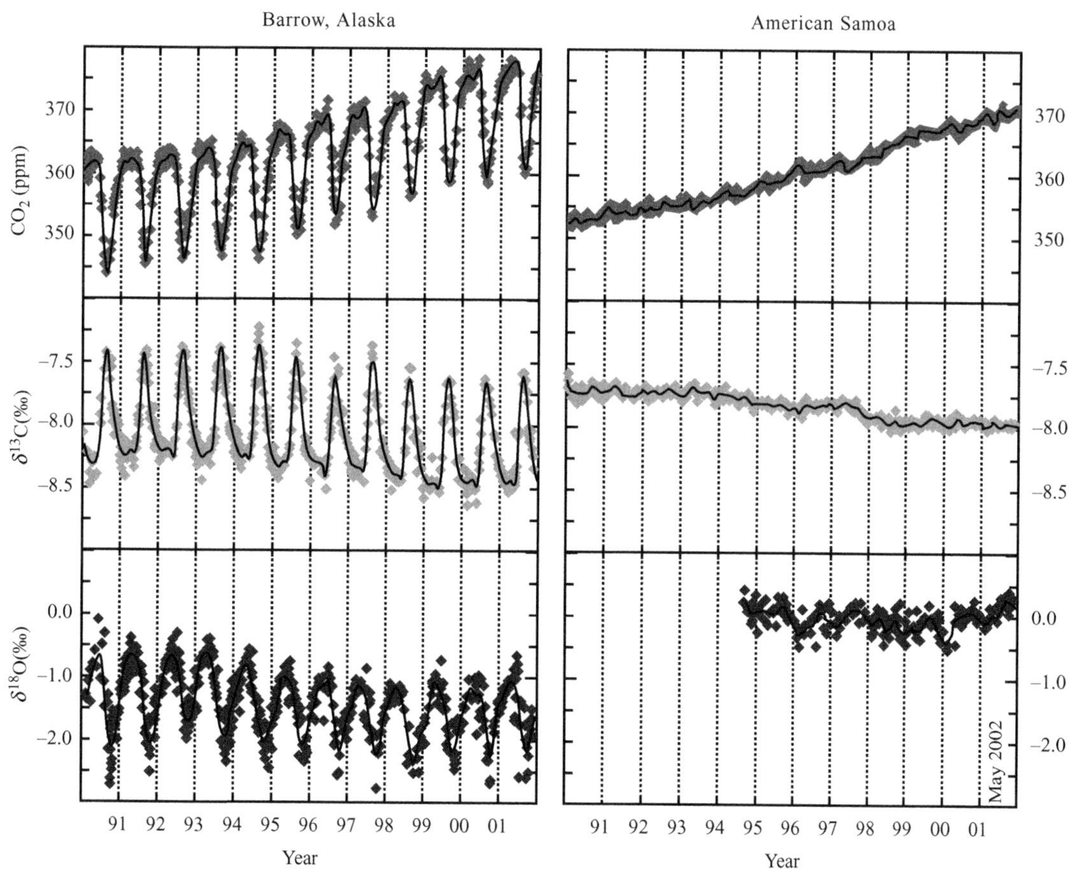

Figure 2 Seasonal and long-term trends in concentrations, $\delta^{13}C$ and $\delta^{18}O$ values of atmospheric CO_2 measured at high latitudes in the northern (Barrow, Alaska) and southern (American Samoa) hemispheres by the NOAA/CMDL-CU flask network (www.cmdl.noaa.gov/ccgg/index.html). The terrestrial biosphere, concentrated in the northern hemisphere, dominants the seasonal cycle. Fossil fuel emission dominates the long-term trends in CO_2 and in $\delta^{13}C$ and their interhemispheric gradient.

that is based, to a great extent, on the existence of the minute but easily detected biological discrimination against ^{13}C in photosynthesis. This isotopic signal is useful in two ways:

(i) There is a major difference in the expression of the photosynthetic discrimination against ^{13}C during CO_2 exchange in the atmosphere–land, or in the atmosphere–ocean systems. While exchange of CO_2 with land plants strongly reflect the photosynthetic discrimination against ^{13}C, CO_2 exchange with the ocean is based predominantly on physical equilibrium, as most of the biological processes occur with respect to the dissolved inorganic carbon in the ocean water. As a result, the two exchange fluxes are isotopically distinct and their relative contribution to changes in the atmospheric $\delta^{13}C$ can be elucidated (Ciais *et al.*, 1995b; Enting *et al.*, 1995; Francey *et al.*, 1995; Fung *et al.*, 1997; Tans *et al.*, 1993).

(ii) The $\delta^{13}C$ values of fossil fuel, and CO_2 from land use changes are imprinted with photosynthetically derived, ^{13}C-depleted carbon. The rapid addition of anthropogenic, ^{13}C-depleted CO_2 to the atmosphere, has resulted in a decreasing trend of atmospheric $\delta^{13}C$ of $\sim$1.5‰ over the past $\sim$100 years (e.g., Francey *et al.*, 1999). This trend slowly infiltrates into the other carbon pools on land and in the ocean, providing a powerful tracer for the fate of this carbon.

Tans (1980) laid out a useful approach to assess the isotopic mass balance of atmospheric CO_2 as a function of the fluxes, isotopic compositions and isotopic fractionations involved in the transfer of CO_2 between the atmosphere, the ocean and the biosphere. The simple formulation shown below involves some approximation that is justified in light of the uncertainties in the system (Tans, 1980). Accordingly, the mean temporal change in atmospheric CO_2 content (C_a) can be described by

$$\frac{d}{dt}C_a = F_{net-o} + F_{net-l} \tag{7}$$

indicating that temporal changes in C_a, are influenced by the net exchange of all fluxes, in the atmosphere–ocean, F_{net-o}, and in the atmosphere–land, F_{net-o}, systems. Temporal changes in the ^{13}C content in atmospheric CO_2 are, accordingly,

$$\frac{d}{dt}{}^{13}C = F_{ne-ot}R'_{net-o} + F_{net-l}R'_{net-l} \tag{8}$$

where R' indicates the isotopic concentrations associated with a net flux. It is immediately apparent that a significant difference in the isotopic ratios for the ocean and the land respectively, should permit the use of the two equation system (Equations (7) and (8)), together with observed variations in atmospheric CO_2 and ^{13}C, to partition the exchange of CO_2 between the atmosphere and the Earth into its oceanic and land components. It turns out that the terrestrial discrimination is $\sim$10 times larger than that in the ocean discrimination. However, unperturbed (such as by anthropogenic effects), the biosphere is normally near steady state and the net fluxes depicted above must be near zero. The associated isotopic impact on the atmosphere, averaged over time (e.g., annually) is faint (often a few 1/100s of a per mil in $\delta^{13}C$ values), requiring extremely high sampling density and analytical precision. These capabilities, continuously improving over the fourth quarter of the twentieth century, have become routine only in recent years (Francey *et al.*, 1999; Masarie *et al.*, 2001; Mook, 1986; Thoning *et al.*, 1989; Trolier *et al.*, 1996). Technological progress notwithstanding, the introduction of stable isotope measurements combined with more traditional CO_2 concentration measurement methods offered a major advance in our ability to probe the modern carbon cycle (e.g., Battle *et al.*, 2000; Ciais *et al.*, 1995b; Enting *et al.*, 1995; Ito, 2003; Still *et al.*, 2003).

In reality, we recognize that the net exchange fluxes in Equations (7) and (8) are more complex, such that Equation (7) is more commonly broken down according to

$$\frac{d}{dt}C_a = \overbrace{[F_{ER} - F_A]}^{\text{land}} + \overbrace{[F_{oa} - F_{ao}]}^{\text{ocean}} + \overbrace{[F_{ff} - F_{bb}]}^{\text{man}} + \overbrace{[F_v - F_W]}^{\text{geology}} \tag{9}$$

where C_a is the atmospheric pool of carbon in CO_2 (e.g., moles, or Pg of carbon in CO_2, where $Pg = 10^{15}$ g) and F denotes a flux (usually measured in Pg C, on annual basis; and negative fluxes, by convention, are out of the atmosphere). The atmospheric CO_2 mass balance depicted above reflects the major components of the global carbon cycle:

(i) The long-term control on atmospheric CO_2 is achieved by the uptake and dissolution of atmospheric CO_2 in surface waters, and its participation in the weathering/sedimentation of carbonate rocks (F_W) and CO_2 released to the atmosphere via tectonic processes and volcanism (F_V). Although providing communication between the atmosphere and the vast carbon pools in the Earth system, it involves less than 1% of the fluxes into or out of the atmosphere on an annual scale, and are either treated separately or neglected in this context.

(ii) The dominant components of the short-term CO_2 balance of the atmosphere are gross primary production (GPP) and respiration (R) by the land biosphere, and the air–sea gas exchange

(F_{oa} and F_{ao} for the ocean–atmosphere and atmosphere–ocean fluxes, respectively). While by far the largest components of the budget, these fluxes are also near equilibrium, with only small effects on the atmospheric budget when integrated over the annual cycle and the entire globe.

(iii) The main components of the human perturbation of the contemporary carbon cycle are fossil fuel burning (F_{ff}), and biomass burning and land-use change (pooled under F_{bb}), which release CO_2 to the atmosphere. Only part of this CO_2 stays in the atmosphere (about one half); the rest is taken up by the land (plant and soil) or by the ocean. The main anthropogenic flux of fossil fuels constitutes essentially an acceleration of the biological carbon cycle, "respiring" over a few centuries vast quantities of carbon that accumulated in the Earth sediments over million of years. The absorption of anthropogenically derived CO_2 on land and in the ocean "carry over" some of the imbalances into the natural two-way fluxes between the atmosphere and Earth.

Such processes can induce feedback effects on atmospheric CO_2 concentrations, and consequently on climate, over the years to millennia timescales (e.g., Cox *et al.*, 2000), which are motivating much of the research in this field.

It follows from the above that the mean temporal change in the ^{13}C content of atmospheric CO_2 can be described by

$$\frac{d}{dt}{}^{13}C = [F_{ER}R'_{ER} - F_{A}R'_{A}] + [F_{oa}R'_{oa} - F_{ao}R'_{ao}] + [F_{ff}R'_{ff} + F_{bb}R'_{bb}] + [F_{V}R'_{V} - F_{W}R'_{W}] \qquad (10)$$

where ^{13}C is the average atmospheric $^{13}CO_2$ concentrations (~1.1% of CO_2) and $R' = {}^{13}C/({}^{12}C+{}^{13}C)$ for each specific flux component. The terms in four square brackets on the right-hand side correspond to the land, ocean, anthropogenic, and the geologic contributions. The change in atmospheric ^{13}C can be separated into the temporal change of two components, isotopic ratio and CO_2 concentrations:

$$\frac{d}{dt}{}^{13}C = \frac{d}{dt}(C_a R'_a) = C_a \frac{d}{dt}R'_a + R'_a \frac{d}{dt}C_a \qquad (11)$$

We focus here on the change in isotopic composition, R', which is normally based on measured values, where $R = {}^{13}C/{}^{12}C$, introducing a small approximation (usually less than 0.2‰; see Section 5.2 for R to R' conversion). Further, R is conventionally reported in the delta notation (see Section 5.2). By introducing the delta notation, rearranging Equations (7)–(11), ignoring the negligible geological fluxes, and omitting terms not very different from 1, a more practical description of the isotopic mass balance of atmospheric CO_2 is obtained, which is often used for considering the contemporary ^{13}C balance of the atmosphere (Tans, 1980; Tans *et al.*, 1993; Fung *et al.*, 1997):

$$\frac{d\delta_a}{dt} = \frac{1}{C_a}[F_{ER}(\delta_{ER} - \delta_a) - F_{Ph}(\delta_{Ph} - \delta_a) + F_{oa}(\delta_{oa} - \delta_a) - F_{ao}(\delta_{ao} - \delta_a) + F_{ff}(\delta_{ff} - \delta_a) + F_{bb}(\delta_{bb} - \delta_a)] \qquad (12)$$

This isotopic mass balance approach indicates some of the potential advantages in adding the isotopic analysis to atmospheric measurements. Isotopically constrained, measurements of changes in δ_a and C_a may allow us to solve for more than one unknown flux. Alternatively, when flux estimates are sufficiently constrained, solving for isotopic signals can provide valuable information on the processes underlying observed changes. Equation (12) also indicates that temporal changes in δ_a values are influenced by the isotopic "forcing" of the various CO_2 sinks and sources, i.e., the relative size of the flux times the deviations of its isotopic signature from that of the mean atmosphere. For example (using IPCC flux and pool size values), a typical anthropogenic flux of 7.1 Pg C yr^{-1} having a mean $\delta^{13}C$ value of −28‰ into an atmospheric pool of ~730 Pg C with a mean $\delta^{13}C$ value of −8‰ will have a "forcing" of $(F_{ff}/C_a)(\delta_{ff}-\delta_a) = (7.1/730)(-28+8) = -0.2$‰ yr^{-1} (Figure 3). Similarly, if 120 Pg C yr^{-1} flux having a mean $\delta^{13}C$ value of −26‰ is taken up (negative flux) in land photosynthesis, its "forcing" would be ~+3‰ yr^{-1}. Such forcing is nearly balanced by a respiration and biomass burning of ~119 Pg C yr^{-1} with $\delta^{13}C$ value of ~−25.7‰. A net terrestrial sink of 2 Pg C yr^{-1} (as estimated) would create a net forcing of +0.05‰. Such isotopic signals are small but within the current precision of the isotopic analyses of atmospheric CO_2 (e.g., Trolier *et al.*, 1996), provided appropriate long-term calibrations are maintained (e.g., Masarie *et al.*, 2001). Comparison of such estimated forcing with observation provides critical information on the balancing act in the atmosphere CO_2 budget.

In Equation (12), the isotopic signals of the individual fluxes are required, but are not normally measured directly. Instead, the isotopic composition of the source CO_2 (often atmospheric CO_2) is measured with precision, and is combined with extensive knowledge of the isotopic fractionations in the transfer process involved. Much effort must be invested therefore to develop this process-based knowledge of isotopic fractionation for the quantitative use of the isotope approach. For example, the isotopic composition of the CO_2

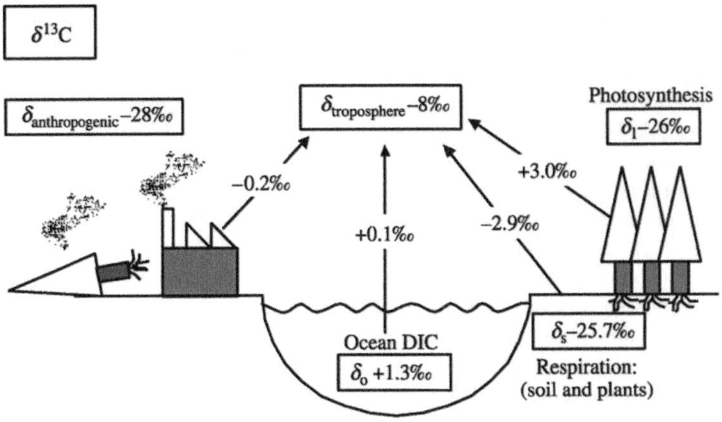

$$\frac{d(\delta_{troposphere})}{dt} = +3.0-2.9+0.1-0.2 \approx 0$$

Figure 3 Annual mean "forcings" on the $\delta^{13}C$ value of atmospheric CO_2. Each specific forcing component is the product of the flux involved, F, diluted by the atmospheric carbon pool, C_a, and the difference in $\delta^{13}C$ values between that of the flux and the mean atmosphere (i.e., forcing $=(F/C_a)(\delta_x-\delta_a)$). Values are order of magnitude estimates, as discussed in the text, and involve significant uncertainties. Typical values for the carbon reservoirs involved are given in boxes. The land biosphere has the strongest forcing, with contrasting effects of photosynthesis and respiration. Net photosynthetic carbon sink tends to enrich the atmosphere in ^{13}C. The ocean forcing is almost an order of magnitude smaller than the land effect, providing a basis for distinguishing between land and ocean sink and sources. Fossil fuel emission tends to dilute the atmospheric ^{13}C. Although the different forcing nearly balance, in reality, the fossil fuel effect dominates the atmospheric forcing resulting with a net trend of $\sim -0.02‰\ yr^{-1}$ (see Figure 4).

flux from the atmosphere to the ocean, R_{ao}, will reflect the isotopic composition of atmospheric CO_2, R_a, and the isotopic fractionation of CO_2 associated with air–sea transfer, α_{ao}:

$$R_{ao} = \alpha_{ao} R_a \tag{13}$$

In introducing the delta notation, we also adopt the term ε, (where $\varepsilon = \alpha - 1$), and the close approximation of the $\delta^{13}C$ of the CO_2 flux into the ocean:

$$\delta_{ao} = \delta_a + \varepsilon_{ao} \tag{14}$$

Similar terms can be derived for other flux components based on knowledge of the isotopic composition of the source material (e.g., atmospheric CO_2, ocean-surface inorganic carbon, soil organic carbon) and knowledge of the associated, process-based, isotopic fractionation (e.g., ε_P for photosynthesis, ε_R for respiration). Note also that terms such as "$\delta_a + \varepsilon_{ao}$" in Equation (14) are also approximations of "Δ."

5.3.1 Fossil Fuel Input

The term fossil fuel usually refers to coal and lignite, petroleum and natural gas, and usually includes also the production of cement. Cement is produced from limestone with a $\delta^{13}C$ value close to zero, because the V-PDB standard itself is derived from limestone. The other fuel components are all derived from organic matter depleted in ^{13}C. Adopted values for coal and lignite are −24‰ and −27‰, while natural gas is −41‰ with relatively large uncertainty. In particular, values of methane range from −20‰ to −75‰ with an estimated mean value of −43‰. Other less abundant gases such as ethane, propane, and butane (~16% of total carbon from natural gas) are less depleted in ^{13}C. Historical trends in the production and consumption of the various components of fossil fuel, result in a temporal change in the $\delta^{13}C$ value of the combined, global-scale, fossil fuel flux from ~−24.0‰ in 1850 to ~−28.5‰ for 1991, partly reflecting the shift in consumption pattern from coal through petroleum to natural gas (Andres *et al.*, 2000; Tans, 1981). It is generally assumed that the fossil fuel combustion does not involve additional fractionation, and mean $\delta^{13}C$ values of −28‰ to −29‰ are often assigned to Equation (12) as δ_{ff}.

5.3.2 Exchange with the Ocean

The existence and direction of a CO_2 flux between the atmosphere and the ocean can be determined by the difference in partial pressure of CO_2 (p_{CO_2}) in seawater and in the overlying atmosphere. p_{CO_2} in seawater depends on its solubility and the concentration of dissolved CO_2 (with solubility inversely related to temperature and salinity). CO_2 gas diffusing into water enters the so-called carbonate system and is partitioned among several species that, combined, are termed "dissolved inorganic carbon" (DIC). It first goes into a dissolved or aqueous form (CO_2(aq)) and is rapidly hydrated to form carbonic acid (H_2CO_3). Carbonic acid is dissociated first to bicarbonate (HCO_3^-) and further to carbonate (CO_3^{2-}):

$$\begin{aligned} &CO_2(\text{atm}) \overset{H_2O}{\longleftrightarrow} CO_2(\text{aq}) + H_2CO_3 \overset{H^+}{\longleftrightarrow} \\ &HCO_3^- \overset{H^+}{\longleftrightarrow} CO_3^{2-} \end{aligned} \quad (15)$$

An alternative pathway to bicarbonate is possible due to reaction of CO_2 and OH^- in water but is normally less important (Skirrow, 1975). The proton concentration is also influenced by total alkalinity and water dissociation, which, in turn, will influence the details of the simplified chemical process depicted above. The relative concentrations of the DIC species in Equation (15) are mainly a function of pH, temperature, and salinity. In seawater, bicarbonate is the dominant species (Skirrow, 1975), and bicarbonate and carbonate are the main components of alkalinity (Broecker and Peng, 1974). The chemical equilibrium for each of the steps in the DIC system in Equation (15) is determined from the specific temperature sensitive reaction constants (K_T). Note that in the complex system of seawater, empirical K_T values, rather than thermodynamic theoretical values, are usually adopted. Such reaction constants are of course sensitive to the effects of molecular mass, with molecules containing heavy isotopes favoring slower reactions rates, and are therefore associated with temperature-sensitive isotopic fractionations. Typical equilibrium fractionation values for the carbonate system in dilute solution at 25 °C are (Deuser and Degens, 1967; Mook, 1986; Mook *et al.*, 1974):

$$CO_2(\text{atm}) \overset{-1.1‰}{\longleftrightarrow} H_2CO_3 \overset{+9.0‰}{\longleftrightarrow} HCO_3^- \overset{-0.4‰}{\longleftrightarrow} CO_3^{2-} \quad (16)$$

Equilibrium isotopic fractions reflect the combined, unidirectional kinetic isotopic fractionations. In considering the one-way fluxes as in Equation (12), estimates of the one-way kinetic fractionations are needed. Knowledge of kinetic effects is obtained from controlled experiment with pure CO_2 and water or with salt solutions, but for the chemically complex system that is ocean-water, empirical values are adopted. For example, laboratory experiments show that the hydration of aqueous CO_2 to bicarbonate involves fractionation of ~13‰ and the dehydration reaction fractionate by ~22‰ (O'Leary *et al.*, 1992). The difference between these two kinetic fractionations, 9‰, corresponds to the equilibrium fractionation depicted above (Marlier and O'Leary, 1984; O'Leary *et al.*, 1992). In practice, it was estimated that for CO_2 taken up by the sea surface, and for CO_2 released from seawater DIC, the kinetic fractionations are smaller (Inoue and Sugimura, 1985; Mook, 1986; Siegenthaler and Münich, 1981) and are (for 20 °C, pH = 8.2 and S = 35):

$$\begin{aligned} ^{13}\varepsilon_k(\text{sea}\longrightarrow\text{air}) &= -10.3 \pm 0.3\% \\ ^{13}\varepsilon_k(\text{air}\longrightarrow\text{sea}) &= -2.0 \pm 0.2\% \end{aligned} \quad (17)$$

The sea to air fractionation, $^{13}\varepsilon_{oa}$, starting from DIC, includes both an equilibrium and a kinetic one, while the air to sea, $^{13}\varepsilon_{ao}$, is purely a kinetic fractionation. These isotopic fractionations are applied in Equation (12) to estimate the $\delta^{13}C$ value of the CO_2 flux from the atmosphere to the ocean, δ_{ao}, ($\delta_{ao} = \delta_a + {}^{13}\varepsilon_{ao}$). Isotopic fractionation due to diffusion in water may apply too, but are expected to be small (O'Leary, 1984).

Note that due to the kinetic effects discussed here, the flux into the ocean tends to increase the atmospheric $\delta^{13}C$ and the flux to the atmosphere tends to deplete the atmospheric $\delta^{13}C$ (Broecker and Maier-Reimer, 1992; Kroopnic, 1985; Kroopnick, 1980). At equilibrium, the net effect should result with a depletion of atmospheric $\delta^{13}C$ by ~8‰ relative to the ocean. As an example, ocean surface water (mean temperature 20 °C) having DIC with $\delta^{13}C$ values of, say, +1.3‰, will be in equilibrium with atmospheric CO_2 having a $\delta^{13}C$ value of −7.0‰, with $\delta^{13}C$ values for the one-way fluxes of: $\delta_{ao} \approx -7.0\text{–}2.0 = -9.0‰$, $\delta_{oa} \approx +1.3\text{–}10.3 = -9.0‰$ (Mook, 1986). In reality, the ocean–atmosphere system is not at full equilibrium (Broecker and Peng (1974) and see below). The most obvious reason is that the residence time of surface water is approximately an order of magnitude smaller than the time needed for isotopic equilibrium with the atmosphere (~1 yr versus 10 yr, respectively; Lynch-Stieglitz *et al.*, 1995; Quay *et al.*, 2003). The picture is further complicated as the $\delta^{13}C$ values of ocean surface DIC reflect a balance between several factors, including the thermodynamic effects, ocean circulation, and biological processes (Gruber *et al.*, 1999).

The mean fractionation values discussed above, and consequently the ocean–atmosphere isotopic fluxes, vary significantly both on spatial and temporal scales. Isotopic fractionations are temperature sensitive (~0.1‰ °C^{-1}; Mook, 1986) and surface ocean temperature varies in the range of 30 °C (Kalnay *et al.*, 1996), resulting with ~3‰ range in the thermodynamic effect. Further, CO_2 exchange between the ocean and the atmosphere is not uniformly distributed. Ocean circulation patterns, biological activity, and large spatial variations in atmospheric turbulence and consequently in the air–sea transfer coefficient, result in a unique pattern of ocean regions with net CO_2 degassing and others with net CO_2 absorption (Tans *et al.*, 1990).

5.3.3 Ocean Biology

Biological activities also increase the carbon content of the deep ocean at the expense of the surface ocean and atmosphere (Gildor and Follows, 2002; Sarmiento and Bender, 1994). This, in turn, is the net effect of two contrasting processes: first the production of organic matter that consumes DIC and second, production of calcium carbonate skeletal material that reduces alkalinity (e.g., consuming Ca^{2+}) and increases p_{CO_2} in surface water. By influencing the exchange of CO_2 between the ocean and the atmosphere, biology can of course affect atmospheric ^{13}C through the kinetic isotope effects discussed above. But probably the main effect of ocean biology on the ocean–atmosphere isotopic exchange is through its effect on the $\delta^{13}C$ value of surface ocean DIC. While, as noted above, atmospheric CO_2 tends to deplete surface ocean DIC in ^{13}C, biological processes have the opposite effect.

Photosynthesis in the ocean discriminates against ^{13}C in a similar way to land photosynthesis, with mean discrimination estimated at 22.0‰ (e.g., Tans *et al.*, 1993), with typical values of $\delta^{13}C$ of marine organic matter in the range of −20‰ in low- and mid-latitudes to −30‰ in the southern ocean (Goericke and Fry, 1994; Rau *et al.*, 1989). This can change, however, under conditions where CO_2 availability dramatically decreases (e.g., algal bloom). Under such conditions, some marine photosynthetic organisms can induce a CO_2 pump that, similar to C_4 plants on land, concentrate CO_2 to high levels in a specific compartment where it is almost quantitatively consumed, greatly reducing isotopic discrimination (Falkowski, 1991; Raven, 1992, 1998; Sharkey and Berry, 1985). As marine photosynthesis increases in summer and decreases in winter, it causes some seasonality in surface ocean $\delta^{13}C$ values (but not nearly as much as the seasonality on land). Consequently, the $\delta^{13}C$ values of DIC in surface water will tend to be higher in summer than in winter, as marine organisms take up ^{12}C preferentially, enriching ^{13}C in the DIC left behind. This may be important when estimates of ocean $\delta^{13}C$ values are based on samplings done during a particular period. For example, Ciais *et al.* (1995b) estimated that global mean summer versus winter ^{13}C enrichment of surface ocean DIC of 0.5‰, and particularly large effect are observed in high latitudes (Gruber *et al.*, 1999). Isotopic fractionation is also associated with carbonate shell formation in marine organisms, but this effect is very small (Emrich and Vogel, 1970; Zhang *et al.*, 1995). The relatively slow air–sea gas exchange diminishes the seasonality effect on the atmosphere.

Biologically produced, ^{13}C-depleted, organic matter can also influence the $\delta^{13}C$ of surface ocean DIC via large-scale ocean circulation. Cold water in the deep ocean accumulates large quantities of carbon derived from respiration and mineralization of organic matter. Upwelling of carbon-rich deep water in some regions will tend to reduce the $\delta^{13}C$ of surface water DIC, and consequently also of atmospheric CO_2 (Lynch-Stieglitz *et al.*, 1995). As discussed in the next section, the large-scale and slow turnover rates of organic matter in the global ocean introduces another important aspect in the isotopic interactions between the ocean and the atmosphere. The penetration of any long-term trend in the $\delta^{13}C$ values of atmospheric CO_2, such as observed today, critically depends on ocean biology that provides the major mechanism to "pump" carbon across the ocean depths.

5.3.4 Atmosphere–Ocean Disequilibrium

Isotopic methods such as based on Equation (12), interpret changes in atmospheric $\delta^{13}C$ mainly as an indicator for terrestrial CO_2 exchange because of its dominant fractionation (Figure 3). Any ^{13}C enrichment of the atmosphere is ascribed to net biotic uptake, and ^{13}C depletion to biotic CO_2 release. To implement such an approach, changes in atmospheric ^{13}C that are not due to the changes in biotic sink or source, must be accounted for. Such is the case with respect to the ~1.5‰ decrease in atmospheric $\delta^{13}C$ since the beginning of the industrial period (~−0.02‰yr^{-1} on average; Feng, 1999; Francey *et al.*, 1995) due to the emission of ^{13}C-depleted carbon from fossil fuel burning and land use changes. This atmospheric change is redistributed in the atmosphere–ocean–land system, but over a significantly slower timescale. In the ocean, the delay is due to the slow rate of ocean–atmosphere isotopic equilibrium, and because the $\delta^{13}C$ value of surface ocean DIC is influenced by the remineraliztion of the ocean total organic matter with slow turnover rates. Therefore, new carbon entering the ocean remains slightly more depleted in ^{13}C relative to the older carbon released from the ocean to the atmosphere and $\delta_{ao} > \delta_{oa}$. This signal, termed the isotopic ocean disequilibriuim, D_o (Figure 4), is difficult to measure directly on a large scale, but must be accurately estimated in atmospheric isotopic budgets (Equation (12); Gruber *et al.*, 1999; Heimann and Maier-Reimer, 1996; Quay *et al.*, 1992, 2003; Tans *et al.*, 1993; Körtzinger and Quay, 2003). A 0.1‰ error in estimating this disequilibrium could result in an error in the order of 0.5 Pg C yr^{-1} in estimating carbon sinks and sources in the biosphere.

The disequilibrium term can be defined (Fung *et al.*, 1997; Tans *et al.*, 1993) by comparing the atmospheric $\delta^{13}C$ value, δ_a, with that expected at

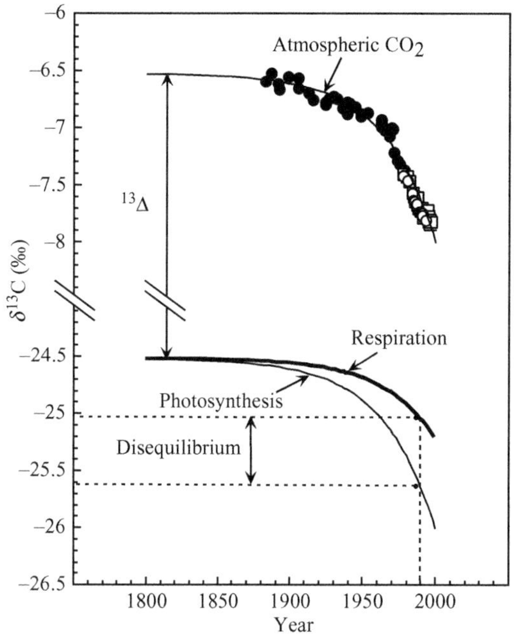

Figure 4 The global "disequilibrium" effect. $\delta^{13}C$ value of CO_2 currently fixed into plants (associated with photosynthetic discrimination, $^{13}\Delta$) is lower than that of older CO_2 respired back to the atmospheric CO_2 (no fractionation is assumed). This is due to the rapid decrease in atmospheric $\delta^{13}C$ associated with fossil fuel emissions, on the one hand, and to the slow turnover of carbon in the biosphere, on the other hand. A similar disequilibrium occurs in the ocean where the atmospheric trend influences the $\delta^{13}C$ values of newly formed DIC, while the ocean mean DIC pool lags behind this equilibrium values due to slow turnover rates (not shown). The atmospheric trend shown is based on the best fit line to the data of Francey *et al.* (1999); the land organic matter trend is obtained by applying global mean $^{13}\Delta = 18‰$, and moving it back in time by 27 yr, the first order estimate of global mean soil carbon turnover time. The resulting ~0.6‰ disequilibrium for the 1990s is within the range of current estimates for both land and ocean.

thermodynamic equilibrium with actual ocean surface DIC, δ_a^{eq}, according to

$$D_o = \left(\delta_a^{eq} - \delta_a\right) \quad (18)$$

where $\delta_a^{eq} = \delta_{DIC} + {}^{13}\varepsilon_{eq}$ (with $\varepsilon_{eq} = \varepsilon_{oa} + \varepsilon_{ao}$). The atmosphere–ocean isotopic exchange depicted in Equation (12) can be substituted by terms that distinguish changes in the atmospheric $\delta^{13}C$ value brought about by the net fluxes and those caused by the "isoflux" that reflect the isotopic disequilibrium (Ciais *et al.*, 1995b; Tans *et al.*, 1993):

$$F_{oa}(\delta_{oa} - \delta_a) - F_{ao}(\delta_{ao} - \delta_a) \approx (F_{ao} - F_{oa})\varepsilon_{ao} + F_{oa}D_o \quad (19)$$

where the disequilibrium effect on the atmosphere is estimated from the isoflux $F_{oa}D_o$ Pg C‰ yr^{-1} ("isoflux" combines flux and its isotopic composition, but since the δ notation is often employed, it is equivalent to, not the actual, isotopic flux). Note that D_o can be defined either relative to the atmosphere or relative to a surface reservoir (i.e., $D_o^* = (\delta_{DIC}^{eq} - \delta_{DIC})$; Fung *et al.*, 1997), with the choice of the definition selected for convenience and data availability. There have been a few approaches to estimating the ocean disequilibrium. Focusing on the air–sea isotopic disequilibrium yielded lower estimates (Tans *et al.*, 1993) than attempting to estimate the change over time of the $\delta^{13}C$ of total ocean column DIC (Quay *et al.*, 1992, 2003). A combination approach has also been recently proposed (Heimann and Maier-Reimer, 1996). Estimates of D_o values reported in the literature range between 0.43‰ and 0.79‰ (Gruber *et al.*, 1999; Heimann and Maier-Reimer, 1996; Quay *et al.*, 1992, 2003; Tans *et al.*, 1993; Wittenberg and Esser, 1997). Considerable uncertainties are still associated with these estimates. In all cases, it is also associated with uncertainties regarding the values and constancy of the fractionation in the sea–air exchange and F_{oa}.

5.3.5 Photosynthetic CO_2 Uptake on Land

The largest annual flux of CO_2 from the atmosphere is the supply of substrate for land plants photosynthesis. The entire land biosphere is estimated to take up $\sim 10^{17}$gC annually (the adopted value is 120 Pg C yr^{-1}; IPCC, 2001). Typically, a single illuminated green leaf takes up only $\sim 10^{-7}$ g C s^{-1}. But, notably, it is the processes that take place at the leaf scale that are ultimately responsible for the effect of land photosynthesis on the isotopic composition of atmospheric CO_2 (Park and Epstein, 1960). The key isotope effect is the preferential uptake of $^{12}CO_2$ by the primary photosynthetic enzymes, Rubisco (Ribulose bi-Phosphate Carboxylase/Oxygenase) and PEP-c (Phospho-Enol-Pyruvate-carboxylase). Because such large quantities of CO_2 are involved, this enzyme discrimination has a measurable impact on the CO_2 remaining in the atmosphere. It is clearly evident in seasonal timescale as photosynthetic removal of ^{13}C-depleted carbon peaks in the northern hemisphere summer months (Figure 2). Similar effects are also observed on geological timescales, when large quantities of ^{13}C-depleted photosynthates are deposited and stored in the soil and sediments (Hayes *et al.*, 1999; Schidlowski, 1988). But, notably, the extent of discrimination against ^{13}C in the two carboxylation enzymes is very different (29‰ versus 7‰ for Rubisco and PEP-c) respectively (Guy *et al.*, 1993; O'Leary, 1981; Roeske and O'Leary, 1984). Most plants, including all trees, use Rubisco as the primary photosynthetic enzyme, producing 3-carbon sugars as the primary product and are therefore called C_3 plants. But ~23% of global plant productivity (Still *et al.*, 2003), mostly in grassland, savannas, and agriculture, is due to PEP-c as the primary enzyme, acting on bicarbonate as a substrate and producing 4-carbon organic acids, and are called C_4 plants. The C_4 pathway is a relatively late achievement in the evolution of flowering plants (probably associated with declining atmospheric CO_2 over geological times) and is based on a specific, compartmentalized, leaf anatomy designed primarily to counteract the low affinity of Rubisco for its substrate, CO_2, and the unfavorable effects of photorespiration in C_3 plants, where part of the newly fixed carbon is lost to the atmosphere as CO_2. Succulent plants with crassulacean acid metabolism (CAM) form yet another, minor, photosynthetic group that can fix carbon by both the RuBP and PEP carboxylase reactions, but at different times. Photosynthetic bacteria utilize additional forms of photosynthesis with low fractionations, approaching that of C_4 plants. This can be due to a different type of carboxylation enzymes, such as in photosynthetic sulfur bacteria, or because of inducible CO_2 concentration mechanism (CCM, such as in cyanobacteria) that "pump" CO_2 into an internal compartment where it is consumed almost quantitatively (Kaplan and Reinhold, 1999). Note that in C_4 plants and CCM containing microorganisms, Rubisco ultimately fixes the CO_2 concentrated in the specialized compartments, but since CO_2 is almost quantitatively consumed in this case, little fractionation is expressed. The combined, net discrimination of land plants (Δ_{Ph} in Equation (12)) is sensitive to the proportional contributions of the various photosynthetic pathways, largely to the C_3/C_4 distribution (Lloyd and Farquhar, 1994; Fung *et al.*, 1997; Still *et al.*, 2003). As discussed below, within the major photosynthetic groups, the extent to which the enzyme discrimination is expressed also depends on physiological and environmental parameters.

Our current and rather extensive knowledge of ^{13}C discrimination during photosynthesis is based primarily on the pioneering works of O'Leary (1981), Vogel (1980, 1993), and Farquhar *et al.* (1982), and with more recent updates such as Farquhar and Lloyd (1993), Evans *et al.* (1986), and Lloyd and Farquhar (1994). These works provide similar analytical treatment to isotopic fractionation during photosynthesis but are expressed in different ways (often reflecting variations among chemical, geological, and physiological nomenclatures; see Section 5.2). In photosynthesis, CO_2 diffuses into active green leaves where a carboxylation reaction incorporates the carbon into an organic substance. The simplified two-step diffusion–carboxylation sequence in photosynthesis is useful to lay down the principles underlying estimates of the ^{13}C discrimination involved (Figure 5). The first step is a reversible process, followed by the irreversible carboxylation of an acceptor (cf. Berry, 1988; Farquhar *et al.*, 1982; Vogel, 1993):

$$[CO_2]_a \underset{F_3}{\overset{F_1}{\longleftrightarrow}} [CO_2]_c \overset{F_2}{\rightarrow} R-COOH \qquad (20)$$

and considering the isotope ratios:

$$R_a \underset{F_3}{\overset{F_1}{\longleftrightarrow}} R_c \overset{F_2}{\rightarrow} R_P \qquad (21)$$

where, $F_1-F_3=F_2$, F represent flux and R isotopic ratio and subscripts a, c, and p represent atmosphere, leaf chloroplasts, and photosynthetic products, respectively. The two-step *fractionations* are $\alpha_a=R_a/R_c$ for the diffusion step and $\alpha_b=R_c/R_A$ for the enzymatic assimilation (A) step, and the overall photosynthetic fractionation is then $\alpha_P=R_a/R_A$. Using these definitions in Equations (20) and (21) lead to

$$F_1\frac{R_S}{\alpha_a}-F_3\frac{R_c}{\alpha_a}=F_2\frac{R_c}{\alpha_b} \qquad (22)$$

Substituting $R_s=R_p\alpha_P$, and $R_c=R_a\alpha_b$ rearranges to

$$\alpha_P=\frac{F_2}{F_1}\alpha_a+\frac{F_3}{F_1}\alpha_b \qquad (23)$$

The unidirectional fluxes involved in the simplified sequence above are proportional to the gross CO_2 concentration gradient times the conductance, g, or the reaction rate constant, k, for each step, such that we can define $F_1=gc_a$, $F_3=gc_c$ and $F_2=k_bc_c\sim g(c_a-c_c)$, where c_a and c_c are respectively the ambient and chloroplast concentrations of CO_2. As noted above it is common to use *discriminations*, $\alpha-1$, represented by a, b, and Δ for the diffusion step, the enzymatic step, and the overall sequence, respectively. These terms can be substituted into Equation (23) that rearranges to

$$\Delta_P=\frac{c_a-c_c}{c_a}a+\frac{c_c}{c_a}b=a+(b-a)\frac{c_c}{c_a} \qquad (24)$$

This equation is now widely used to estimate isotopic discrimination in photosynthesis, and shows the two-step diffusion/reaction nature of the process and the dependence of each step on the local CO_2 concentration gradient. Typical values used in this model for C_3 plants are $a=4.4‰$ and $b=29‰$, $c_c/c_a\sim0.6$ yielding Δ_p of ~18‰. For C_4 plants, $c_c/c_a\sim0.4$ and $b=7‰$, leading to Δ_p of ~5‰.

A more rigorous treatment of Δ_p is summarized by Farquhar and Lloyd (1993) and Lloyd and Farquhar (1994). In this case, it is recognized that CO_2 diffusion into the leaf is going through a complex diffusion and dissolution pathway. The biochemical part considers the fact that while Rubisco discrimination in its isolated form is ~29‰ (Guy *et al.*, 1993; Roeske and O'Leary, 1984),

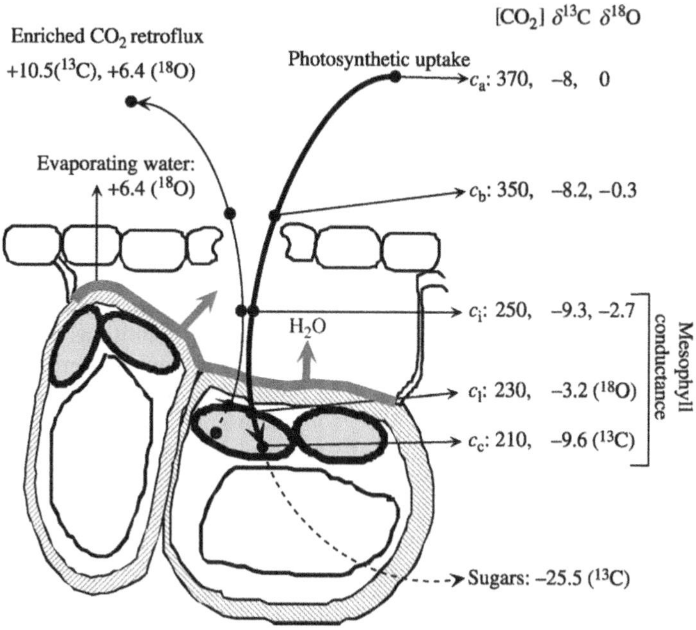

Figure 5 A simplified cross-section in a leaf indicating the CO_2 diffusion pathway between the atmosphere and the chloroplast. Changes in CO_2 concentrations and the $\delta^{13}C$ and $\delta^{18}O$ values along this pathway are indicated for the atmosphere, c_a, the leaf boundary layer, c_b, substomatal internal air spaces, c_i, the chloroplast surface, c_l and the center of the chloroplast, c_c, are indicated (pathway between c_i and c_c constitutes the internal, mesophyll, resistance to CO_2 diffusion, inadequately constrained at present). Values for $\delta^{18}O$ are based on $\delta^{18}O$ of liquid water at the evaporating surfaces (near the chloroplast membranes) of +6.4‰. The arrows indicate the one-way CO_2 fluxes into the leaf (controlled by c_a), and retro-diffusion out of the leaf (controlled by c_c and c_l in the case of ^{13}C and ^{18}O, respectively). The difference between the two one-way fluxes constitutes CO_2 assimilation into sugars (values for $[CO_2]$ in ppm, for isotopes in δ‰).

~5% of the flux goes via PEPc even in C_3 plants, potentially reducing the apparent discrimination by 1–2‰. In plants that belong to the C_4 type, all the CO_2 assimilation is first by PEPc. The primary product is then transported to a specialized, CO_2 impermeable, bundle-sheath compartment, where it is de-carboxylated and the concentrated CO_2 re-assimilated by Rubisco. If all the CO_2 in the bundle sheath is re-assimilated, Rubisco's discrimination will not be expressed. But a certain degree of "leakiness" associated with possible discrimination must be considered (Farquhar, 1983; Henderson *et al.*, 1992). The overall discrimination by leaves of C_4 plants can be estimated according to Farquhar (1983):

$$\Delta = a + [b_4 + \phi(b_3 - s) - a]c_c/c_a \qquad (25)$$

where b_4, the effective biochemical fractionation in the C_4 process is ~12‰, ϕ is leakiness factor (~0.2) and the associated fractionation, s(~1.8‰), and c_i is the internal CO_2 concentration.

Concurrently to the assimilation in leaves there is always CO_2 production in respiration and photorespiration (the first occurs all the time and involves the mitochondria, the latter occurs only in the light and involves also the oxygenase activity of Rubisco in the chloroplasts). Any discrimination in the respiration process will be reflected in the isotopic composition of the CO_2 produced and mixed into the leaf internal CO_2 pool and its effect must, therefore, be subtracted from Δ_P. The combined effect of the two respiratory components can be subtracted from Equation (24) as (Farquhar and Lloyd, 1993; Farquhar *et al.*, 1982)

$$-d = e\frac{R_d/k}{c_a} + f\frac{\Gamma^*}{c_a} \qquad (26)$$

where R_d is the rate of dark respiration in presence of light, k is the carboxylation efficiency, Γ^* is the compensation point (the CO_2 concentration at which respiration exactly matches photosynthesis) and e and f are the discriminations associated with respiration and photorespiration. Typical values used are $e=0$‰ (Lin and Ehleringer, 1997), $f=7$‰ (Rooney, 1988), and $d \approx -0.5$‰ for C_3 plants and $d \approx 0$‰ for C_4 plants. Recent studies indicate, however, that the discrimination associated with dark respiration, e, may be different from zero and as large as 6‰ (Tcherkez *et al.*, 2003). It may reflect significant $\delta^{13}C$ heterogeneity among organic compounds (Duranceau *et al.*, 1999; Ghashghaie *et al.*, 2001), as well as intra-molecular ^{13}C heterogeneity of specific substrates (Gleixner and Schmidt, 1997; Melzer and Schmidt, 1987; Rinaldi *et al.*, 1974).

The complex diffusion-reaction process responsible for the overall discrimination by C_3 plants (Figure 5) can therefore be summarized according to Farquhar and Lloyd (1993):

$$
\begin{array}{l}
\text{— — — — — — — — gas phase — — — — — / — — — —} \\
CO_2\text{-atm.} \overset{a_b}{\longleftrightarrow} \text{leaf-surface} \overset{a}{\longleftrightarrow} \text{cell-walls} \\
\quad 370\ \text{ppm} \qquad\qquad 350\ \text{ppm} \qquad\qquad 230\ \text{ppm} \\
\text{— — — — liquid phase — — — — — — — — — — — /} \\
\overset{e_s+a_1}{\longleftrightarrow} \text{chloroplasts} \overset{b}{\longrightarrow} \text{A-}CO_2 \overset{-d}{\longrightarrow} CO_2(R) \\
\qquad\qquad 210\ \text{ppm}
\end{array}
$$

and estimated according to

$$\Delta_p = \xi_b a_b + \xi_a a + \xi_1 (e_s + a_1) + \frac{c_c}{c_a} b_3 - d \quad (27)$$

where ξ is the weighting factor for each discrimination step, based on the associated CO_2 drawdown (i.e., $\xi_n = (c_n - c_{n-1})/c_a$ and c_n is the CO_2 concentration at a step n(1–4) along the diffusion pathway from the mixed atmosphere across the leaf laminar boundary layer; the leaf internal stagnant internal air spaces; and then across the liquid phase to the site of carboxylation (Figure 5). Accordingly, the diffusion discriminations are a_b(~2.9‰), a(4.4‰), and a_l[0.7‰; (O'Leary, 1984)], for the laminar boundary layer, air spaces and liquid phase respectively. In going from air into liquid there is also an equilibrium fractionation e_s [1.1‰; (Mook *et al.*, 1974)]. For C_4 plants the term c_c must be replaced with c_m, for the CO_2 concentrations in the mesophyll cytoplasm where PEP-c is active (Farquhar, 1983) (with typical values $c_m \approx 125$ ppm) and b_3 is replaced with the appropriate expression from Equation (25). This will reflect the large differences in Δ_P for C_3 and C_4 plants, with typical values, based on the above examples, of Δ_P of ~18‰ and ~5‰ for C_3 and C_4 plants respectively. Inspection of Equation (27) clearly indicates the sensitivity of Δ_P to estimates of c_c, which can vary with environmental conditions, and among plant species, but cannot be measured directly. Best estimates are obtained by leaf-scale gas exchange measurements (Evans *et al.*, 1986).

In the framework of Equation (12), we are interested in global-scale estimates of Δ_p (or $\delta_a + \varepsilon_p$). Such estimates are often obtained either by forward modeling that scale up processes discussed above (e.g., Buchmann and Kaplan, 2001; Fung *et al.*, 1997; Lloyd and Farquhar, 1994; Kaplan *et al.*, 2002; Ito, 2003; Still *et al.*, 2003), as well as by inverse modeling constrained by observations (e.g., Bousquet *et al.*, 1999; Keeling *et al.*, 1989; Quay *et al.*, 1992; Tans *et al.*, 1993; Francey *et al.*, 1995; Enting *et al.*, 1995; Ciais *et al.*, 1995a,b; Battle *et al.*, 2000; Randerson *et al.*, 2002; Still *et al.*, 2003). Global-scale estimates of plant discrimination are still associated with large uncertainties and range between ~15‰ and ~20‰ in the studies mentioned above, with the lowest values estimated by the process-based models and the highest by the inverse studies.

Considering the large differences in discrimination between C_3 and C_4 plants it is clearly important to the consideration of the relative contributions of C_3 and C_4 vegetations to global GPP and global Δ_P. This is not a trivial task and introduces significant uncertainty to the isotopic budget. The C_4 photosynthetic pathway is advantageous under low ambient CO_2 concentrations and in hot and dry conditions because it concentrates CO_2, and allows efficient CO_2 uptake when leaf stomatal conductance is low, thereby reducing water loss. It was therefore shown that the occurrence of C_4 vegetation can be predicted based on climatic conditions (Collatz *et al.*, 1998; Ehleringer *et al.*, 1997) and can continuously change (e.g., Schwartz *et al.*, 1996). Detailed vegetation maps are also often used as the basis for estimating C_4 plant distribution, but for CO_2 flux studies, estimates must also be based on the proportional productivity, and not on species distribution (Still *et al.*, 2003). Estimates indicate that C_4 vegetation contributes 20–25% to the global terrestrial productivity (Lloyd and Farquhar, 1994; Fung *et al.*, 1997; Still *et al.*, 2003; Ito, 2003). This proportion may have been larger during glacial times, and may decrease in future, high CO_2 world (Collatz *et al.*, 1998).

5.3.6 CO_2 Release in Respiration

Atmospheric CO_2 enters the land biosphere through leaves, and an equivalent, but not necessarily identical, flux leaves the biosphere as respiration (Raich and Potter, 1995; Raich and Schlesinger, 1992). Since practically all carbon is eventually respired, it can be expected that the $\delta^{13}C$ value of respired CO_2 by an ecosystem, δ_{ER}, will be near that of the photosynthesis flux, $\delta_{ER} = \delta_a - \varepsilon_p$, where δ_a is the $\delta^{13}C$ of atmospheric CO_2 and ε_p is the fractionations associated with photosynthesis. On annual or shorter timescales, however, this is not the case, because respiration acts on carbon that is undergoing chemical and isotopic modifications and is distributed among several distinct pools with widely different turnover rates, ranging from less than a year to many hundreds years (Trumbore, 2000; Trumbore *et al.*, 1995), and reality tends to be better described by

$$\delta_{ER} = \delta_a - \varepsilon_p - \varepsilon_R \quad (28)$$

where ε_R is the apparent fractionation associated with respiration. Much uncertainty is also associated with the possible isotopic fractionation involved in respiration, ε_R. Conventionally, it is assumed to be zero (Lin and Ehleringer, 1997), but recent studies in leaves indicate possible fractionations (Tcherkez *et al.*, 2003).

A basic distinction regarding respiration is made between autotrophic and heterotrophic respiration. Autotrophic respiration is carried out by the living plants and can be subdivided into aboveground (stem, branches, leaves) and belowground (roots, often considered together with their closely associated bacteria in the rhizosphere (Hanson *et al.*, 2000; Hogberg *et al.*, 2001; Rochette and Flanagan, 1997)). Autotrophic respiration consumes newly assimilated carbon with short turnover rates from few days (Ekblad and Högberg, 2001; Bowling *et al.*, 2002; Pataki *et al.*, 2003) to ~50 yr, mainly covering the range for herbaceous and woody tissues in different geographical locations (Ciais *et al.*, 1999). The $\delta^{13}C$ value of autotrophic respiration is generally assumed to be similar to that of recent photosynthates. The $\delta^{13}C$ values of plant organic matter vary significantly due to environmental and physiological parameters (e.g., Broadmeadow and Griffiths, 1993; Francey and Farquhar, 1982). In addition, there are fractionations during plant metabolism that result with changes in the $\delta^{13}C$ value of the living biomass, both on a spatial (higher values in top versus bottom of canopies, and in roots versus leaves etc. (Brugnoli *et al.*, 1998)) and on temporal (early and late season carbon; Hemming *et al.*, 2001) scales. The living biomass eventually dies, and the organic matter is translocated, transformed and decomposed in a complex chain of reactions (Boutton and Yamasaki, 1996).

Heterotrophic respiration processes dead plant material by microorganisms, which themselves are made of the carbon they consume (Gleixner *et al.*, 1993). The dead plant materials begin as litter, both above- and belowground, but are transformed into various forms of soil organic matter. Such transformation involves the loss of the more labile material as CO_2 and accumulation of residual material with longer turnover time, a process that is generally associated with increasing $\delta^{13}C$ values of soil organic matter (Balesdent and Mariotti, 1996; Ehleringer *et al.*, 2000; Ladyman and Harkness, 1980). Distinction among the specific compartments of soil organic carbon is based primarily on turnover rates (Schimel *et al.*, 1996; Trumbore, 2000), and is the basis for various schemes that are based on modeling soil processes and used to describe soil respiration and its the $\delta^{13}C$ values (e.g., Ciais *et al.*, 1999; Fung *et al.*, 1997; Potter *et al.*, 1993; Randerson *et al.*, 1996; Schimel *et al.*, 1994), with similar underlying principles. For example, soil organic matter can be broadly separated into detrital, microbial, slow, and passive carbon. Detrital and microbial pools consist of metabolic and structural carbon compounds rapidly decomposed and recycled to the atmosphere in less than 10 years. The slow carbon pool consists of more resistant compounds, with high lignin content, that turn over in 10–100 years (Bird *et al.*, 1996). The passive carbon pool turns over in more than 100 years. The turnover time of each pool can vary and is influenced by factors such as temperature, precipitation and soil moisture, soil texture and microbial community. Note that although soil organic matter shows wide variations in age, up to thousands of years, the respiration flux is strongly dominated by short-lived soil organic matter components. The mean age of soil respired CO_2, τ, can be approximated from independent estimates of the global soil carbon pool, SC, and the heterotrophic respiration flux, F_h (1,500 Pg C yr^{-1} and 55 Pg C yr^{-1} respectively; IPCC): $\tau = SC/F_h = 27$ yr, which involves significant uncertainty (e.g., Trumbore, 2000; Ito, 2003).

For the atmospheric ^{13}C budget (Equation (12)) an estimate that integrates the $\delta^{13}C$ value of ecosystem respiration is required. Autotrophic respiration $\delta^{13}C$ values may be reasonably estimated from Equation (28), although there may be short time gaps (ca. 5 d) between corresponding photosynthesis and respiration signals (Bowling *et al.*, 2002; Pataki *et al.*, 2003). The $\delta^{13}C$ value of heterotrophic respiration, however, cannot be estimated with Equation (28), and must integrate the relative contributions of the multiple carbon pools in the soil and their unique $\delta^{13}C$ value (Ciais *et al.*, 1995b, 1999; Fung *et al.*, 1997).

It is important to note that the variations in the $\delta^{13}C$ values of soil organic matter, while considered as complications above, can serve as useful tracers in partitioning the contributions of specific components to the respiration flux or in estimating the turnover rates of such components (Balesdent and Mariotti, 1996; Cerling and Wang, 1996; Ehleringer *et al.*, 2000). Such studies are often based on measurements of soil CO_2, which is enriched in ^{13}C, by 4.4‰ relative to the mean $\delta^{13}C$ value of soil organic matter or that of the soil respired CO_2, due to diffusion fractionations (Cerling *et al.*, 1991; Davidson, 1995). Note that this diffusion fractionation, while important for ^{18}O in CO_2 studies as discussed below, does not enter into the atmospheric $\delta^{13}C$ budget because soils are normally near isotopic steady state such that the ^{13}C value of soil respired CO_2 is similar to that of the source carbon (Figure 6(a)).

5.3.7 The Land Disequilibrium

Based on the discussion in the previous section it is apparent that plant photosynthesis produces organic matter from today's atmosphere, but respiration releases carbon that has been stored for different periods in plant tissues and soil organic matter. It was also noted already that over the industrial era, the $\delta^{13}C$ of the atmosphere has been declining at a mean rate of ~0.02‰ yr^{-1}. The carbon respired today is, therefore, enriched

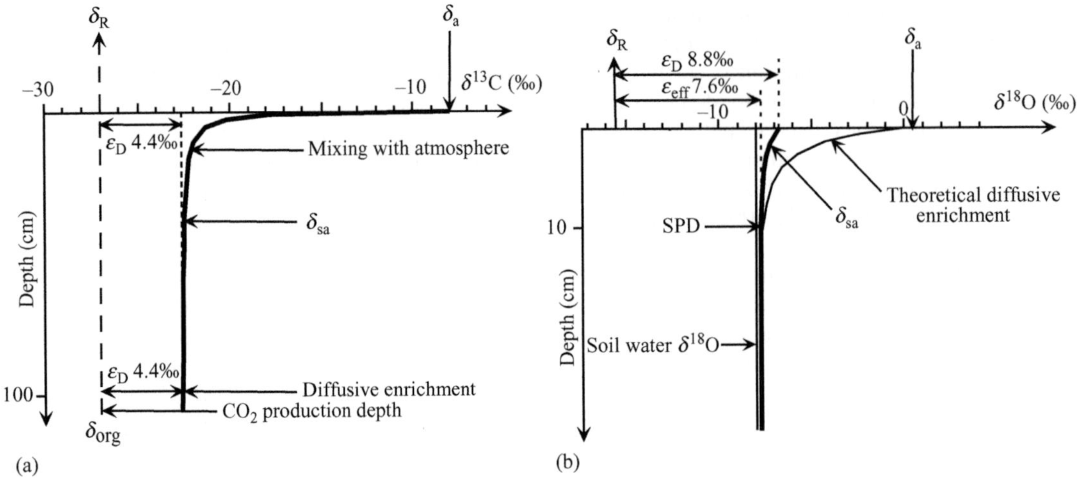

Figure 6 A schematic representation of the evolution of (a) $\delta^{13}C$ and (b) $\delta^{18}O$ values of soil respired CO_2. Decomposition of soil organic matter produces CO_2 with similar isotopic composition. Diffusion in the soil, with fractionation ε_D, results in ^{13}C-depleted respired CO_2 preferentially escaping to the atmosphere (δ_R), leaving behind ^{13}C-enriched CO_2 in the soil air (δ_{sa}). This process reaches steady state when the respiration flux has the same $\delta^{13}C$ value as the source material. Mixing with ^{13}C-depleted atmospheric CO_2 (δ_a), influence the shape of the δ_{sa} profile near the soil surface. The $\delta^{18}O$ of CO_2 reflect the $\delta^{18}O$ of soil water up to a setting point depth (SPD) that represents the minimum effective depth at which CO_2 equilibrate with soil water before escaping to the atmosphere. Near the surface, ^{18}O-depleted CO_2 escapes by diffusion to the atmosphere with a fractionation, ε_D, leaving behind ^{18}O-enriched CO_2 (with a profile that could follow the thin solid line leading to a steady-state enrichment of soil CO_2 of $\varepsilon_D = 8.8$‰). However, if the soil near the surface is not completely dry, some isotopic exchange can occur leading to a profile such as depicted by the thick solid line. If isotopic equilibrium is catalyzed and occurs up to the surface itself, the profile would follow the waterline. While a diffusion fractionation of $\varepsilon_D = 8.8$ takes place in all scenarios, a smaller "effective fractionation" is observed when comparing the $\delta^{18}O$ of soil respired CO_2 with that of measured soil water at depth, or of precipitation water. (This simplified diagram ignores additional complications due to soil water enrichment near the surface).

in ^{13}C, compared to current photosynthate, to an extent that depends on the residence time of the carbon in the biosphere (Figure 4). For example, adopting the estimated global mean τ of 27 yr mentioned above, soil respired CO_2 could be $\sim 27 \times 0.02 = 0.5$‰ more enriched in ^{13}C with respect to current atmospheric CO_2. Under such circumstances, an isotopic flux between the land biosphere and the atmosphere will exist even if there is no net CO_2 flux. Similar to the case for the ocean, this represents the land disequilibrium, D_L. Note that D_L will tend to enrich the atmosphere in ^{13}C, in the same way that photosynthetic CO_2 uptake would (due to discrimination against ^{13}C). It is obvious, therefore, that ignoring D_L would result with over-estimating ^{13}C-derived estimate of terrestrial carbon sink (or under-estimate a source). The error can be significant, in the order of 0.6 Pg C yr^{-1} globally, and proportionally larger in regions such as the tropics (Ciais *et al.*, 1999).

In a similar manner to the ocean, the land disequilibrium can be defined as

$$D_L = \left(\delta_{al}^{eq} - \delta_a\right) \tag{29}$$

where δ_{al}^{eq} is the $\delta^{13}C$ value of the atmosphere that would be in isotopic equilibrium with the biosphere (i.e., $\delta_{al}^{eq} = (\delta_b - \varepsilon_p)$, where δ_b is the mean $\delta^{13}C$ of the biosphere and ε_p is the photosynthetic fractionation) (Enting *et al.*, 1995; Ciais, 1995; Fung *et al.*, 1997; Tans *et al.*, 1993). Global-scale modeling of this effect suggests that it is actually in the range of 0.20–0.33‰ (Fung *et al.*, 1997; Quay *et al.*, 1992; Tans *et al.*, 1993), with large latitudinal variations between ~0.1‰ and ~0.6‰ (Ciais *et al.*, 1999). Interestingly, large-scale land use changes in the tropics release large quantities of relatively old (~50 yr) carbon to the atmosphere, enhancing the land disequilibrium. The significant discrepancy between the first-order estimate of $D_L = 0.5$‰ made above and the recent model estimates of 0.20–0.33‰ may reflect the difficulties in accurately estimating turnover rates of the various soil carbon pools and its global distribution, as well as in estimating the total amount of carbon in soils. Modifying the way carbon turnover is treated in models, e.g., from a "pipe" approach to a "well mixed pools" approach, reduced global D_L values, but only by ~0.1‰ (Ciais *et al.*, 1999). Further reduction in the terrestrial disequilibrium may arise from the replacement of C_3 forest subjected to de-forestation by C_4 grassland, with a rather large effect on the local scale (Ciais *et al.*, 1999). Replacing C_3 with

C_4 vegetation would result, initially, with photosynthetic uptake with ^{13}C of ~-12.4‰ and respiration ^{13}C of ~-26‰, in contrast to the fossil fuel effect in C_3 stands where uptake is at ~-26.0‰ and release at ~-25.7‰.

5.3.8 Ecosystem Discrimination

The apparent complexity in the respiratory ^{13}C signal, discussed in the previous section, provides an exciting aspect for ecosystem-scale studies. The differences in $\delta^{13}C$ values between the photosynthetic (A) and respiratory (R) CO_2 fluxes allow the partitioning of net ecosystem exchange (NEE, where NEE = A + R) into its two components, which is critical to understanding the land ecosystem response to change (Bowling *et al.*, 2001; Raupach, 2001; Yakir and Wang, 1996; Lloyd *et al.*, 2001; Styles *et al.*, 2002). Furthermore, ecosystem-scale measurements can provide experimentally based constrain on plant and soil discrimination, parameters that contribute the main uncertainties to the isotopic approach depicted in Equation (12).

To estimate isotopic discrimination at the ecosystem, local or regional scales, an estimate of the ^{13}C exchange flux, or the $\delta^{13}C$ value of NEE, at the relevant scale is required. Such estimates are obtained by sampling air in and above plant canopies, or in the and above the atmospheric boundary layer (ABL). A simple and powerful approach employs a two-member mixing model as first proposed by Keeling (1958, 1961). The equation used in the Keeling approach is derived from the basic assumption that the atmospheric concentration of a substance in an ecosystem reflects the combination of some background amount of the substance that is already present in the atmosphere and some amount of substance that is added or removed by a source or sink in the ecosystem:

$$C_E = C_a + C_s \tag{30}$$

where C_E, C_a, and C_s are the concentrations of the substance in the ecosystem, in the atmosphere and that contributed by ecosystem sources respectively. Isotope ratios of these different components can be expressed by a simple mass balance equation:

$$C_E\delta_E = C_a\delta_a + C_s\delta_s \tag{31}$$

where δ_E, δ_a, and δ_s represent the isotopic composition of the substance in the ecosystem, in the atmosphere and of the sources respectively. Combining Equations (30) and (31):

$$\delta_E = C_a\cdot(\delta_a - \delta_s)\cdot(1/C_E) + \delta_s \tag{32}$$

Equation (32) describes a linear relationship, with an intercept δ_s, which represents the $\delta^{13}C$ signature of the net sources/sinks in the ecosystem. Note that even if the ecosystem source/sink is composed of several different sub-sources/sinks the Keeling type plot can still be used as long as the relative contribution of each of these subcomponents remain fixed.

This relationship was first used by Keeling (1958, 1961) to interpret carbon isotope ratios of ambient CO_2 and to identify the sources that contribute to increases in atmospheric CO_2 concentrations at a regional basis. It is now widely applied on a wide range of studies and on a wide range of scales (Figure 7; and see Pataki *et al.*, 2003). Miller and Tans (2003) proposed some modifications to the conventional approach by re-arranging Equation (31) to allow consideration of changes in $\delta a\ Ca$ (e.g., over the seasonal cycle) and derive δs from the slope of the best fit line. Perhaps the most common application of the Keeling expression is to identify the isotopic composition of respired CO_2 in forest ecosystems (Bowling *et al.*, 2002; Buchmann *et al.*, 1997a,b; Harwood *et al.*, 1999; Pataki *et al.*, 2003; Quay *et al.*, 1989; Sternberg, 1989; 1997). The derivation of the isotopic composition of respired CO_2 has been used to determine ecosystem carbon isotope fractionation (Δ_e) by the following equation:

$$\Delta_e = (\delta_{trop} - \delta_{resp})/(1 + \delta_{resp}) \tag{33}$$

where δ_{trop}, δ_{resp} represent the $\delta^{13}C$ values of tropospheric and respired CO_2 (Buchmann *et al.*, 1998; Buchmann and Kaplan, 2001; Flanagan *et al.*, 1996). Note that in the equations, the traditional use of δ_{resp} (Keeling, 1961) is applied for both nighttime and daytime measurements. This term would be better represented by δ_s (Equation (16)), which includes effects of both respiration and photosynthesis and will be equal to δ_{resp} only for nighttime measurements. Note also that recycling of respired CO_2 within canopies can influence estimates of the δ values of respired CO_2 (Lloyd *et al.*, 1996; Sternberg, 1989). A modified equation ($\Delta_e = \delta_{atm} - \delta_{resp}$) was also used by Bakwin *et al.* (1998) to estimate regional-scale biological discrimination (Bakwin *et al.*, 1995, 1998a,b; Miller and Tans, 2003; Miller *et al.*, 2003). A more general approach to canopy-scale inverse methods, and in particular the potential in the inverse Langrangian method is discussed by Raupach (2001).

Whole ecosystem and regional-scale discrimination is also assessed by evaluating the isotopic composition of CO_2 in the atmospheric boundary layer (ABL; also called planetary boundary layer (PBL), as well as convective boundary layer (CBL), when considered during times of convective enhancement by surface heating during the day). The major advantage of the ABL budget is in its integration of surface fluxes and discrimination over large areas (10^4–10^6 km^2 compared with

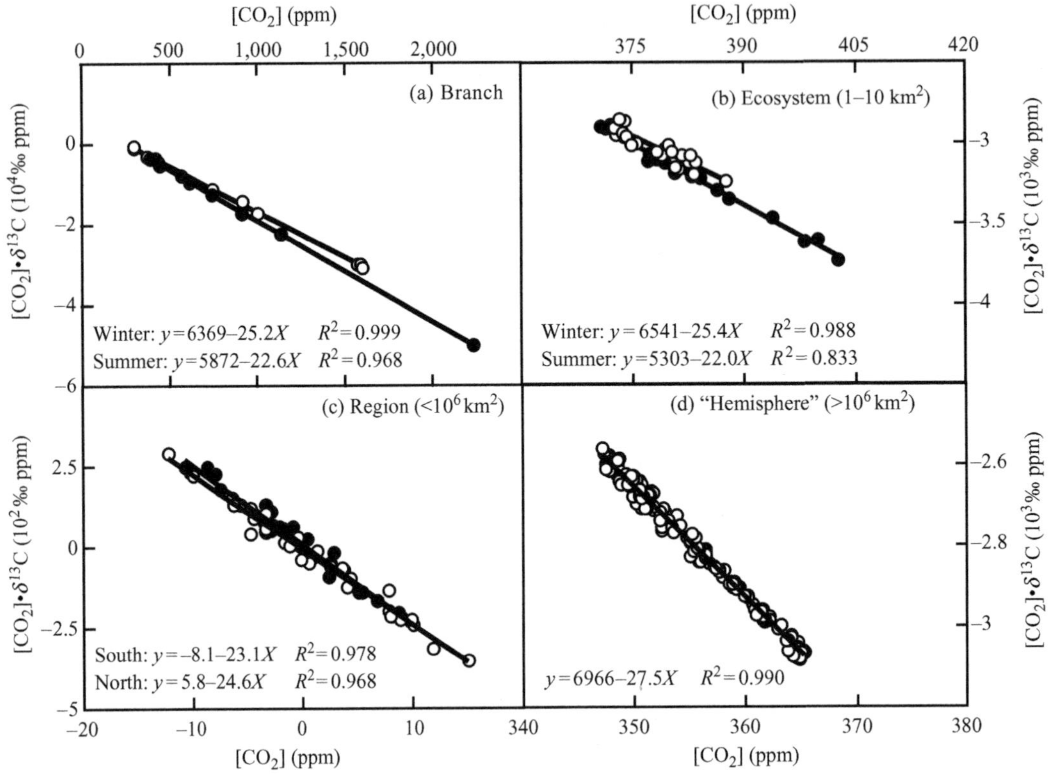

Figure 7 Modified Keeling plots identifying the ^{13}C signatures in CO_2 exchanged with the atmosphere on different spatial and timescales. A series of air samples is analyzed for changes in CO_2 concentration and its $\delta^{13}C$, and plotted in such a way that the slope of the best-fit lines provides estimate of the ^{13}C signature of the surface flux (see text for detail). Single branches are sealed for a few minutes in a chamber and allowed to respire in the dark while air samples are withdrawn. The results show seasonal change in the ^{13}C signature of respired CO_2 between the wet winter and the dry summer in the semi-arid Yatir forest, Israel. Nocturnal canopy air samples taken at about the same time in the Yatir forest show that ecosystem-scale respiration reflected the same ^{13}C signature as a single branch (D. Hemming and K. Maseyk, unpublished). Tall, ~400 m, TV towers are used to collect air samples that represent regional-scale footprints. Data from a tall tower in Wisconsin (Bakwin, unpublished results; see Bakwin *et al.*, 1998) indicate the possible influence of the extensive corn cultivation (C_4 plants with low ^{13}C discrimination) to the south of the tower on the predominantly C_3 forest area around the tower site. Background troposphere air samples are taken in permanent network stations around the world on a weekly basis by flushing air through glass flasks, which are then sent for analysis in a central laboratory. Data for the summer months (June through August) from Chemya, Alaska (NOAA/CMDL-CU network) collected during 1993–2002 show a typical mid to high latitude biospheric ^{13}C signature. The range of values shown here (−22.6‰ to −27.5‰) covers most of the observed range in the biological ^{13}C signatures. Data for the tall tower were de-trended to remove long-term trends and seasonal cycle in CO_2 and ^{13}C (see Figure 2) and only the "residual" variations in CO_2 and ^{13}C were considered, enhancing the influence of the local biology on the results; the ten years long data set from Chemya were "de-trended" only for the long-term mean trends in CO_2 and $\delta^{13}C$ enhancing large-scale effects (see Bakwin *et al.*, 1998a,b; Miller and Tans, 2003).

1–10 km^2 for measurements associated with canopy flux towers). The ABL approach relies on the turbulence and efficient mixing throughout the ABL height (~1–2 km), and on the ABL very distinctive upper limit, below the free troposphere (Figure 8). Few measurements over time within the homogeneous ABL (usually relying on small aircrafts), and knowledge of air transport through its top, provide an efficient way to report large-scale changes in the storage of constituents, such as CO_2 and $^{13}CO_2$, in the ABL that, in turn must reflect surface fluxes and isotopic discrimination (Lloyd *et al.*, 1996, 2001; Nakazawa *et al.*, 1997; Styles *et al.*, 2002a; Ramonet *et al.*, 2002). Whole ecosystem and regional-scale discrimination greatly increase the window of observation and provide greater integration of the mosaic of surface activities. This approach can help, for example, in quantifying the contribution of C_3 or C_4 vegetation to productivity such as in tropical areas (Lloyd *et al.*, 1996; Miranda *et al.*, 1997). It is also valuable, in constraining large-scale estimates of discrimination associated with photosynthesis and respiration (e.g., Fung *et al.*, 1997). But note that

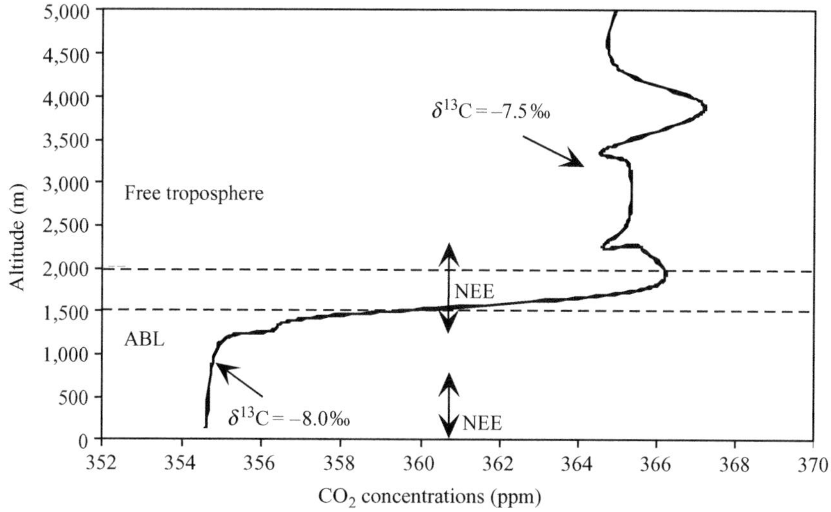

Figure 8 A representative atmospheric boundary layer (ABL) profile of CO_2 (obtained by airplane) during a summer mid-afternoon at a forest study site in Wisconsin (Helliker and Berry, unpublished). Representative $\delta^{13}C$ values for the mixed ABL and the free troposphere are indicated. The distinct ABL and its homogeneity provide a basis to estimate changes in CO_2 storage and its isotopic composition in the entire ABL column above large surface areas. Alternatively, the large concentration and isotopic gradients at the top of the ABL provide a robust basis for estimating the flux across the top of the ABL, which is assumed to be similar to that at the bottom of the ABL (i.e., the land surface NEE; see text).

canopy and CBL-scale net discrimination must still be partitioned to photosynthetic and respiration components for mass balance analysis such as based on Equation (12).

An elegant simplification of the ABL approach to estimate regional-scale fluxes and isotopic discrimination has been proposed by Helliker and Berry unpublished. Instead of following changes in storage within the ABL, this approach takes advantage of the sharp concentration gradients at top of the distinctive ABL boundary (Figure 8). Such easily measured gradients in, say, CO_2 and $^{13}CO_2$, when time averaged over weeks to months, permit precise estimates of the net flux across the top of the ABL and, based on conservation of mass, must reflect the integrated fluxes from/to the surface (e.g., NEE for CO_2). As done in small-scale studies (cf. Equations (20)–(23)), the net CO_2 exchange flux across the top of the ABL can be simply estimated from: $\mathrm{NEE} = \{[CO_2]_{ABL} - [CO_2]_{troposphere}\}/r_{CO_{2-ABL}}$, where $r_{CO_{2-ABL}}$ is the resistance to CO_2 transport within the ABL that must be independently estimated (for example from radon or meteorological re-analysis products). Similarly, the $\delta^{13}C$ of NEE, and consequently regional-scale discrimination (see above), can also be obtained from

$$\delta^{13}C_{NEE} = \{[CO_2]_{ABL}\delta^{13}C_{ABL} - [CO_2]_{troposphere}\delta^{13}C_{troposphere}\}/([CO_2]_{ABL} - [CO_2]_{troposphere}) \quad (34)$$

As discussed below, knowledge of NEE and $\delta^{13}C_{NEE}$ are prerequisites for the partitioning of NEE into its photosynthetic and respiratory components. Preliminary results of using this approach gave close agreement with surface-based, eddy covariance and isotopic measurements.

5.3.9 Incorporating Isotopes in Flux Measurements

The dynamics of canopy-scale net fluxes of water and CO_2 exchanged between vegetation and the atmosphere are routinely measured today with micrometeorological methods (e.g., with eddy covariance; www.daac.ornl.gov/FLUXNET/fluxnet.html). Combining these methods with isotopic measurements allows to partition a net flux into its gross flux components. The approach here is similar to that used on the global scale to asses, for example, ocean versus land sink/source relationships (Equation (12)), but focuses instead on the local, ecosystem, scale and considers only ecosystem respiration and photosynthesis. In simplified terms and expression, this was first employed by Yakir and Wang (1996) over crop fields, and can be developed by assuming F_N as a net flux of CO_2 (say, photosynthetic uptake and soil respiration combined) such that $F_N = F_1 + F_2$. Each flux carries a unique isotopic identity, δ_1, δ_2, and δ_N (e.g., the δ values of CO_2 taken up in photosynthesis, released in soil respiration and of the combined net flux, respectively) and isotopic mass balance takes the form

$$F_N \cdot \delta_N = F_1 \cdot \delta_1 + F_2 \cdot \delta_2 \quad (35)$$

By rearrangement we obtain estimates of the component gross fluxes:

$$F_1 = F_N \cdot \phi_1;\ F_2 = F_N \cdot \phi_2 \quad (36)$$

where $\phi_1 = (\delta_N - \delta_2)/(\delta_1 - \delta_2)$ and $\phi_2 = (\delta_1 - \delta_N)/(\delta_1 - \delta_2)$. Thus, knowledge of the net flux, F_N, and the isotopic signatures δ_N, δ_1, and δ_2 allows us to derive an estimate of the component gross fluxes. F_N and δ_N may be directly estimated by micrometeorological techniques, values of δ_N can also be estimated from Keeling type plots. But the specific isotopic signatures, δ_1 and δ_2, must be independently estimated from plant and soil samples, or from leaf-scale gas exchange and soil chamber measurements. Sampling and estimates of δ_1 and δ_2 must also consider heterogeneity in the ecosystem and be representative of the relevant scale (determined by the type of micrometeorological measurement).

Micrometeorological measurements of trace gas exchange are commonly made using the flux-gradient or eddy correlation techniques. The flux-gradient technique combines the eddy diffusivity (K_c) and concentration gradient with height (dx/dz) to calculate the flux according to

$$F_N = K_c \cdot (dx/dz) \quad (37)$$

Eddy diffusivity is estimated, for example, by wind profile measurements as described by Baldocchi *et al.* (1988). The gradient dx/dz is directly measured by the appropriate instruments at different heights above the canopy and averaged over an appropriate time interval (e.g., 30 min). The main advantages of the gradient approach are that large gradients can be observed over several meters between sampling heights, and that it does not require fast responding instruments. Measurements inside plant canopies, however, are complicated by possible counter-gradient transport and spatial heterogeneity (Baldocchi *et al.*, 1988). This flux-gradient technique can be readily adapted to include isotopic measurements. Air samples, or air moisture samples, are collected over the same time intervals and shipped to a stable isotope laboratory. The water and/or CO_2 samples are analyzed for isotopic composition to provide time integrated mean values. The concentration and isotopic data can then be used to produce a Keeling-plot, which may give an estimate of the isotopic identity of the net flux, δ_N (cf. δ_s in Equation (32)). Sampling of soil, stem and leaf water, and soil organics representative of the underlying vegetation can provide the necessary information to estimate the relevant isotopic signatures of the soil and leaf exchange flux (δ_1 and δ_2 for either water or CO_2). More directly, some of these signatures can be obtained from leaf-scale gas-exchange and/or isotopic measurements carried out concomitantly in the field (e.g., Harwood *et al.*, 1998; Wang *et al.*, 1998), or by Keeling plots for data obtained within the canopies to determine isotopic identities of soil fluxes. Yakir and Wang (1996) have successfully used the gradient approach to partition net ecosystem exchange of CO_2 into photosynthetic assimilation and respiration for several crop species. Such studies provide confidence in the feasibility of the flux-isotope approach, which is supported also by other canopy-scale isotopic studies (Buchmann and Ehleringer, 1998; Flanagan *et al.*, 1997; Moreira, 1998).

The flux-profile relationships utilized above crop fields and grasslands (cf. Cellier and Brunet, 1992) may not be valid, and are therefore not generally accepted, above forests. In forests, the eddy correlation method offers a more direct approach to flux measurements. In this case, a fast responding sonic anemometer is used to indicate the velocity and direction of the vertical air movement component (up or down eddies). A fast responding analyzer records the concentration of the trace gas of interest in conjunction with anemometer measurements. Once data are collected over an appropriate time interval (e.g., 30min), the concentration measurements can be separated to those related to eddies moving up from the canopy or down into the canopy. The net eddy flux is then calculated according to

$$F_N = \overline{\rho w' c'} \quad (38)$$

where ρ is density of dry air, w is the vertical wind velocity component (positive and negative values correspond to up/down) and c is the mixing ratio or mole fraction of the chemical constituent with respect to dry air. An overbar indicates a time-averaged quantity and a prime indicates deviation from the mean. Note that using this equation requires several assumptions about the nature of the turbulence and the nature of the landscape (Baldocchi *et al.*, 1988). Given that the exchange between the canopy and the atmosphere can be measured with this technique, then it follows that NEE, which considers also the canopy storage flux can be described as (Baldocchi, 1997; Wofsy *et al.*, 1993):

$$\mathrm{NEE} = A + R = \overline{\rho w' c'} + dc_i/dt \quad (39)$$

where A and R are the photosynthetic and ecosystem respiration fluxes and dc_i/dt is the change in canopy storage. Numerous studies have used the eddy correlation approach to provide estimates of F_N or NEE (but note that these measurements in some cases are difficult to validate with actual changes in biomass, e.g., Rochette *et al.*, 1999).

Fast responding isotope analyzers with the required precision are not available at present and the eddy correlation technique cannot be directly combined with *in situ* stable isotope measurements to estimate δ_N (Equations (34)–(36)). Alternatively, independent methods, such as the Keeling plot, with certain caveats, can be used to estimate δ_N. This requires that the data used to construct the Keeling plot represent the same time interval and footprint as the eddy correlation system. Isotopic data for Keeling plot estimates can be obtained from slow flask sampling and laboratory analysis. Bowling *et al.* (1999a) have recently argued that in the range and precision of data obtained for Keeling plots, the relationships between CO_2 concentration and $\delta^{18}O$ or $\delta^{13}C$ are practically linear such that the equation $\delta = mc + b$ (termed reciprocal Keeling equation, where δ and c are the measured isotopic and concentration values) can be used to predict the isotopic composition based on the component concentration. If the data are representative, such a relationship can be used to produce the isotopic flux of $^{13}CO_2$, F_{13}:

$$F_{13} = \overline{\rho\omega'[c(mc + b)]'} \quad (40)$$

where c is measured by the eddy correlation system and the constants m and b are derived from a reciprocal Keeling plot based on concurrent flask sampling. Assuming the ^{13}C signature of the biological system, δ_N, is a constant at the measurement scale, F_{13} correspond to $F_N\delta_N$.

There are difficulties in precise estimates of the components of Equation (40). Recent studies have addressed the differences between the Keeling relationship as determined from slow, whole-air flask sampling, and that of eddy covariance sampling (Bowling *et al.*, 1999b). Additionally, isotopic analyses are incorporated into studies using conditional sampling techniques (relaxed eddy accumulation, REA), where fast responding analytical instruments are replaced with a fast responding valve system and an accumulator that collect separate air samples from "updraft" and "downdraft" eddies over a time interval; Bowling *et al.*, 1999b, 1998; Businger and Oncley, 1990).

5.4 OXYGEN-18 IN CO_2

As stable isotope measurements of atmospheric CO_2 were added to the arsenal of environmental measurements, modern, triple-collector isotope ratio mass spectrometers have been used. Notably, these instruments routinely recorded the contributions of molecular masses 44, 45, and 46 in CO_2 (representing $^{12}C^{16}O_2$, $^{13}C^{16}O_2$, and $^{12}C^{16}O^{18}O$ and ignoring or as correcting for minor contributions from ^{17}O and doubly labeled molecules). Initially, however, attention was focused mainly on the ratio of 45/44, providing information on ^{13}C content in CO_2 (e.g., Keeling, 1958, 1961; Mook *et al.*, 1983), while the 46/44 ratios, and the ^{18}O content in CO_2, were hardly addressed (Bottinga and Craig, 1969; Mook *et al.*, 1983). This has changed noticeably in 1987 when two studies noted that the neglected ratios report a strong, large-scale and unique signal (Francey and Tans, 1987; Friedli *et al.*, 1987). These studies showed that the observed latitudinal and altitudinal gradients in atmospheric $C^{18}O^{16}O$ are quantitatively consistent with the expected effects of $C^{18}O^{16}O$ exchange between the atmosphere and the land biosphere, and that there are, in fact, no other fluxes of this magnitude that offer alternative explanations.

It is now recognized that the ^{18}O signal in CO_2 reflects a unique coupling between the global hydrological and carbon cycles, and can provide important tracer of sink and sources of CO_2 particularly in the terrestrial biosphere. The use of ^{18}O relies on the fact that CO_2 readily dissolves in water, allowing CO_2–H_2O oxygen exchange to occur. Water in leaves is highly enriched in ^{18}O, relative to soil water, due to evaporative fractionation (Dongmann, 1974; Förstel, 1978). Consequently, CO_2–H_2O exchange in leaves (associated with photosynthesis) or in soil and trunks (associated with soil, roots and plant respiration) produces contrasting ^{18}O signals in the CO_2 which is released to the atmosphere (Ciais *et al.*, 1997a,b; Farquhar *et al.*, 1993; Yakir and Wang, 1996). Because the enzyme carbonic anhydrase (CA) is present in all plant leaves and rapidly catalyzes CO_2 hydration and isotopic exchange, in spite of the short residence time of CO_2 in leaves, CO_2 involved in photosynthesis is extensively re-labeled by ^{18}O -enriched leaf water. Such differences should allow identification of CO_2 sources and sinks in the ecosystem and that may allow estimates of the individual photosynthetic and respiratory fluxes. Indeed the limited ability to partition net ecosystem productivity (NEP) into gross primary productivity (GPP; equivalent to net leaf assimilation, A) and ecosystem respiration (R; including aboveground plant respiration and belowground root and microbial respiration) has been a critical limitation in obtaining insights into the causes of the observed inter-annual variations of global NEP, as well as in explaining past variations and predict future levels of terrestrial carbon storage (Tans and White, 1998). The potential in using ^{18}O in CO_2 as a tracer of gross fluxes in the terrestrial biosphere has been experimentally demonstrated in field-scale measurements (Flanagan *et al.*, 1997; Flanagan and Varney, 1995; Yakir and Wang, 1996), and in global-scale modeling (Ciais *et al.*, 1997a,b; Farquhar *et al.*, 1993; Ishizawa *et al.*, 2002; Peylin *et al.*, 1996, 1999), constrained by the accumulating atmospheric ^{18}O data (Conway *et al.*, 1994; Masarie *et al.*, 2001; Trolier *et al.*, 1996).

The primary control on the $\delta^{18}O$ of CO_2 is the $\delta^{18}O$ of the liquid water with which it was last in contact. CO_2 isotopically equilibrates with water according to the following reaction:

$$H_2{}^{18}O_{(l)} + CO_{2(g)} \longleftrightarrow H^+ + [HCO_2{}^{18}O]^-_{(aq)} \longleftrightarrow H_2O_{(l)} + CO^{18}O_{(g)} \quad (41)$$

The temperature-dependent value for the equilibrium fractionation ε_{eq} (where $\varepsilon_{eq} = (\alpha_{eq} - 1) \times 1{,}000$) between the oxygen in the CO_2 and water is

$$\varepsilon_{eq}(T) = 17{,}604/T - 17.93 \quad (42)$$

where $d\varepsilon_{eq}/dT = -0.20‰\,°C^{-1}$, so that at 25 °C, $\varepsilon_{eq} = 41.11‰$ (Brenninkmeijer *et al.*, 1983). Water must be in the liquid phase for the hydration reaction to occur (Gemery *et al.*, 1996). Equilibration does not occur in the atmosphere because the liquid water content is generally too small, given the slow rate for the uncatalyzed forward reaction (0.012 s^{-1}, at 25 °C). The rate constant for the isotopic reaction is one-third the rate of the chemical reaction because of the three oxygen atoms present in the intermediate bicarbonate species (Mills and Urey, 1940). With the presence of carbonic anhydrase, ubiquitous in leaves, equilibrium (Equation (41)) can be reached nearly instantaneously (with turnover rate of up to $10^6\,s^{-1}$, (Silverman, 1982), and typical rates of 100–1,400 μmol $CO_2\,m^{-2}\,s^{-1}$ on leaf area basis; Gillon and Yakir, 2001). The quantity of water usually involved in CO_2–water interaction is many orders of magnitude greater than that of the CO_2 present, so isotopically equilibrated CO_2 will take on the oxygen isotopic ratio of the water in which it is dissolved (plus the temperature-dependent equilibrium fractionation), regardless of its initial $\delta^{18}O$ value.

CO_2 hydration is associated with all fluxes between the atmosphere and the Earth surface, and the ^{18}O–CO_2 budget may be expressed, as done for ^{13}C, as a mass balance with respect to the annual atmospheric trend (Ciais *et al.*, 1997a; Farquhar *et al.*, 1993; Miller *et al.*, 1999):

$$\begin{aligned}\frac{d\delta_a}{dt} = \frac{1}{c_a}[&F_L(\delta_l - \delta_a + \varepsilon_l) + F_R(\delta_a - \delta_s + \varepsilon_{eff}) \\ &+ F_I(\delta_a - \delta_s) + F_{oa}(\delta_o - \delta_a) \\ &+ \varepsilon_W(F_{oa} - F_{ao}) + F_{ff}(\delta_{ff} - \delta_a) \\ &+ F_{bb}(\delta_{bb} - \delta_a) + F_{str}(\delta_a - \delta_{str})]\end{aligned} \quad (43)$$

where

$$F_L = \frac{c_l}{c_a - c_l}A, \quad A = F_{la} - F_{al} \quad \text{and}$$

$$F_I = vc'_a$$

and c refers to concentrations, F to one-way flux to or from the atmosphere, δ to isotopic ratios, A to net photosynthetic assimilation (equivalent to GPP; and F_L the "retro-diffusion" flux from the leaf associated with GPP), and v to "piston velocity" of CO_2 diffusion into the soil and equilibrating with soil water (Tans, 1998). Subscripts used are: R for respiration (autotrophic and heterotrophic), A for assimilation, a for atmosphere (c'_a refers to concentrations near the soil surface), o for oceans, ff for fossil fuels, bb for biomass burning, l for leaves, I for atmospheric invasion into soils (allowing for oxygen exchange with soil water, but with no net flux), s for soil and str for stratosphere; ε refers to the kinetic fractionations associated with fluxes between reservoirs (ε_W for water to air, and ε_{eff} represents an effective diffusion fractionation in the soil–atmosphere interface).

Note that in Equation (43) δ_a and c_a are known accurately by measurement (Conway *et al.*, 1994; Trolier *et al.*, 1996). The fossil fuel flux is well known and its oxygen isotopic composition is assumed to be that of atmospheric O_2 used in its combustion (+23.5‰ versus SMOW; Kroopnic and Craig, 1972). The biomass burning flux is relatively small and its isotopic signature is also assumed to be that of atmospheric O_2. Although the oceanic exchange flux (F_{oa}) is large, the net oceanic exchange ($F_{oa} - F_{ao}$) is near zero, and the isotopic disequilibrium between the ocean and atmosphere ($\delta_o - \delta_a$) is relatively small, and so is the overall oceanic influence on the atmospheric ^{18}O (i.e., $F_{oa}(\delta_o - \delta_a)/c_a$). The same is true for the troposphere–stratosphere exchange, where F_{str} is large (~95 Pg C yr^{-1}, but with no net flux) and the fractionation is uncertain at present but likely small (Blunier *et al.*, 2002). It is the leaf and soil (respiration and invasion) components that dominate the uncertainty and the magnitude of the ^{18}O mass balance. Soil and leaf isotopic fluxes each contribute roughly five times more to the atmospheric signal of $\delta^{18}O$ than either the oceanic or fossil fuel components, which are the next most significant terms. Combining estimates of NEP (from ^{13}C or O_2) with accurate measurements and modeling of the ^{18}O signals of soils and leaves should allow the use of Equation (43) to better estimate F_R and F_A and consequently GPP and R.

5.4.1 The Soil Component

The $\delta^{18}O$ of CO_2 from both root respiration and decomposition of soil organic matter is strongly influenced by the oxygen isotopic composition of the water with which it is in contact. The $\delta^{18}O$ value of precipitation (Jouzel *et al.*, 1987 and this volume; Mathieu and Bariac, 1996b; Mathieu *et al.*, 2002) is the single-most important environmental control on the oxygen isotopic composition of CO_2. It transforms into soil water and feeds plants (Gat, 1996; Leroux *et al.*, 1995). The isotopic composition of precipitation is influenced by fractionation in the hydrological cycle (Craig, 1961; Dansgaard, 1964; Epstein and Mayeda,

1953; Gat, 1996), which is simulated in detail by global GCMs. Jouzel *et al.* (1987) and Yurtsever and Gat (1981), employed a simpler empirical approach using data from the global precipitation survey conducted by the IAEA to develop a multiple regression analysis that predicts the $\delta^{18}O$ of precipitation, δ_p, based on the two main influencing parameters, temperature, T (°C), and the amount of precipitation, P (mm month^{-1}) approximated by

$$\delta_p = -11.78 + (0.418 \pm 0.033)T - 0.0084 \pm 0.0048)P \quad (44)$$

Generally, δ_p decreases in going from the equator (around 0‰ on the SMOW scale) toward the poles (~−25‰), and from the coast into continental regions (Craig, 1961; Gat, 1996; Rozanski *et al.*, 1993, 1992).

In the simplest case, the $\delta^{18}O$ value of rainwater directly translates to the $\delta^{18}O$ value of soil water and, after equilibrium and kinetic fractionations are taken into account, into $\delta^{18}O$ of the CO_2 flux from the soil to the atmosphere (δ_{sa}). Even in such simple cases, seasonal variations in soil water occur also due to input from snowmelt water or floodwater, introducing seasonal, indirect precipitation signals (Hesterberg and Siegenthaler, 1991; Miller *et al.*, 1999). A quantitative treatment of the conversion from the $\delta^{18}O$ of soil water to δ_s was first proposed by Hesterberg and Siegenthaler (1991), which was further tested by Stern *et al.* (1999, 2001); (see also Amundson *et al.*, 1998). Tans (1998) expanded this treatment by considering additional factors influencing δ_{sa} and Miller *et al.* (1999) experimentally tested and quantified such parameters. The effect of soils to atmosphere flux, F_{sa}, on the $\delta^{18}O$ value of atmospheric CO_2, $F_{sa}(\delta_a - \delta_{sa})$ can be summarized as:

$$\frac{1}{c_a}F_{sa}(\delta_a - \delta_{sa}) = \frac{1}{c_a}F_{sa}\Delta_s = \frac{1}{c_a}[F_R(\delta_a - \delta_s + \varepsilon_{eff}) + F_I(\delta_a - \delta_s)] \quad (45)$$

Representing the two components of the soil–atmosphere flux, respiration and atmospheric invasion, the associated CO_2 in equilibrium with soil water, δ_s, and the effective diffusion fractionation in going from the soil into the atmosphere, ε_{eff}. These components, and what influence them is briefly discussed below.

Water in drying soils can become highly enriched in ^{18}O near the surface relative to its initial water source, producing steep isotopic gradients in soil water (enrichment of 10–15‰ are common; Allison and Barnes, 1983; Allison *et al.*, 1983; Barnes and Allison, 1983; Mathieu and Bariac, 1996a,b). Since it is difficult to estimate this gradient, it poses a potential complication in estimating δ_s. Miller *et al.* (1999), however, showed that although soil drying can have an effect on δ_s values, the ^{18}O enrichment in the top 5–10 cm of the soil cannot impart all of its isotopic signature to the CO_2 leaving the soil. This is because diffusion of CO_2 out of the soil occurs more quickly than uncatalyzed hydration. The half time for isotopic equilibration of CO_2 in water in soil pores (with typical values of air porosity and water content) is ~110 s, while the mean time expected for CO_2 molecules to diffuse through 1 cm of soil is only ~14 s. In the top 5–10 cm, where the $\delta^{18}O$ of soil water changes by at least 1‰ cm^{-1}, the diffusing CO_2 cannot fully equilibrate before escaping to the atmosphere. The lower water content of the isotopically enriched near-surface soil further inhibits the ability of the CO_2 to attain isotopic equilibrium with the soil water. In Equation (45), δ_s is defined as a weighted average of CO_2 isotopically equilibrated with soil water along a depth range. In practice, it can be regarded as the isotopic value of the CO_2 in equilibrium with soil water at a specific "setting point depth" (SPD) such that ($\delta_s + \varepsilon_D f$) is consistent with the CO_2 respired out of the soil. Miller *et al.* (1999) found that SPD ranged between 5 cm and 15 cm. Above 5 cm, CO_2 escapes from the soil faster than it can be isotopically hydrated; below 15 cm, the isotopic composition of CO_2 is reset as it diffuses toward the surface. Therefore, the steepest gradient in the $\delta^{18}O$ of soil water, often observed near the surface, does not play a large role in the isotopic composition of respired CO_2. While these results simplify estimates of δ_s, Kesselmeier and Hubert (2002) and Kesselmeier *et al.* (1999) reported some evidence for carbonic anhydrase catalytic effects in soils. In cases where catalysis is present, the value of δ_s would still be influenced by the gradients near the soil surface, and accelerate the effect of atmospheric invasion into the soil (see below). Carbonic anhydrase, can increase the hydration rate of CO_2 by several orders of magnitude (Reed and Graham, 1981) allowing CO_2 to equilibrate more fully prior to leaving the soil. In practice, the δ_s was parameterized in different models to reflect either surface soil water (Ciais *et al.*, 1997a) soil water at depth (Flanagan *et al.*, 1997; Flanagan and Varney, 1995), or mean precipitation (Farquhar *et al.*, 1993).

The type of soil in which CO_2 hydration occurs may also be important. Implicit in the hydration reaction is the assumption that all the water in soils is chemically and physically unbound, except for hydrogen bonding between water molecules. However, this assumption may be quite limited. Clays and organic matter have large surface areas to which some fraction of the soil water may be strongly adsorbed (Hsieh *et al.*, 1998; Sposito *et al.*, 1999) altering the equilibrium fractionation, ε_{eq}. Water adsorbed to a surface probably contains ^{18}O bound more strongly than when in solution, reducing the value of ε_{eq} (Miller *et al.*, 1999). Indications for such effects can be noticed in several studies using low moisture soils (below ~10%; Hsieh *et al.*, 1998; McConville *et al.*, 1999; Miller *et al.*, 1999; Scrimgeour, 1995), and was recently addressed by Isaac (2001). The hydration rate, k_h, when water is adsorbed to the soil may also be influenced. First, the large surface to volume ratio imposed on water in the soil can significantly enhance the hydration rates (Isaac, 2001). But this effect can be reversed when the additional dissociation step from the surface required prior to the hydration reaction become a significant component in the exchange process. Different components of the same soil might have widely varying adsorptive energies for water resulting in a range of k_h and ε_{eq} values.

Respired CO_2 exit the soil largely via molecular diffusion, resulting in the more rapid escape of the lighter CO_2 molecule with an expected kinetic fractionation, ε_D, of 8.8 ‰ between soil CO_2 at the soil–air interface and the flux of CO_2 into the atmosphere. This theoretical value of 8.8‰ is based on simple kinetic theory and the reduced mass of CO_2 and air (Craig, 1953) but has not been measured, and it does not consider any effect of pressure gradients between the soil and the atmosphere. ε_D can only be fully expressed if CO_2 and water are in complete isotopic equilibrium until the soil surface. Since CO_2 diffuses out of the soil faster than it can equilibrate with soil water, CO_2 remaining in the soil will be enriched in ^{18}O, relative to its equilibrium value with water, as also predicted for ^{13}C in soils (Cerling, 1984; Cerling *et al.*, 1991; Davidson, 1995). This near-surface enrichment diminishes the observed kinetic fractionation, measured relative to CO_2 in equilibrium with soil water at lower depths (Figure 6(b)). Global syntheses of $\delta^{18}O$ of CO_2 have made simple assumptions regarding the controls on $\delta^{18}O$ of respired CO_2. Both Farquhar *et al.* (1993) and Ciais *et al.* (1997a) solved versions of Equation (43) for the kinetic fractionation of CO_2 diffusing out of soil, ε_D, assuming that all other terms were well known. These studies arrived at different results: 5.0‰ (Ciais *et al.*, 1997a) and 7.6‰ (Farquhar *et al.*, 1993). Miller *et al.* (1999), using an experimental approach, estimated the effective diffusion discrimination $\varepsilon_{eff} = \varepsilon_D f$, such that f will change as a function of the rates of CO_2 hydration and diffusion which are, in turn, functions of the temperature, tortuosity, and the air and water content of the soil. Catalysis of the hydration reaction, e.g., by carbonic anhydrase activity, near the soil surface would render f close to one. As expected, estimates of ε_{eff} correlated strongly with soil moisture content in the top 5 cm, but never reach $f = 1$ in realistic soil moisture levels. A mean value of $\varepsilon_{eff} = 7.2$‰ was proposed as a robust estimate that represents a wide range of environmental conditions. While the discussion above focuses on CO_2 diffusion, CO_2 can also leave the soil by advection due to pressure gradient. This aspect is rarely considered, and its isotopic effect was estimated to be small (Stern *et al.*, 1999).

Notably, the $\delta^{18}O$ of soil water affects more than just the isotopic composition of respired CO_2. It also affects the isotopic composition of ambient CO_2, which makes contact with soil water by diffusing into and out of the soil. Depending upon its residence time in the soil, the diffusing ambient CO_2 can partially or fully equilibrate with the soil water. This diffusion–equilibration-retro-diffusion process is termed the "invasion effect" (Tans, 1998) and was demonstrated experimentally by Miller *et al.* (1999). The impact of this effect on atmospheric $\delta^{18}O$ is defined in Equation (43) as $F_I(\delta_a - \delta_s)/c_a$. This non-biological invasion flux, F_I, is a function of the CO_2 concentration near the soil surface, c'_a and the speed with which CO_2 in the air above the soil diffuses into the soil and equilibrates with soil water, v (the "piston velocity"; with values for "typical" soil of ~0.012 cm s^{-1} and effective penetration of invading CO_2 of ~3 cm):

$$F_I = c'_a v \tag{46}$$

The piston velocity was defined by Tans (1998) as $v = \sqrt{\theta_a B \theta_w \kappa D_{air} k_{h\text{-}iso}}$, where, θ_a and θ_w are the volume fractions of air and water in the soil, B is the Bunsen solubility coefficient, κ is the soil tortuosity, D_{air} is diffusivity of CO_2 in air, and $k_{h\text{-}iso}$ is the rate constant for the oxygen isotopic equilibration of CO_2 and water. Invasion does not distinguish the source of the invading air, original invading background or re-invading soil-respired CO_2, but the more invasion there is, the more the CO_2 above the soil approaches equilibrium with the $\delta^{18}O$ of the soil water. While in Equation (45) the invasion effect is estimated as a discrete variable, Stern *et al.* (2001) suggested to include it as a correction to ε_{eff} that would then vary between ~1‰ and ~10‰ among biomes and environmental conditions.

Gillon and Yakir (2001) included the invasion effect in their global ^{18}O budget of atmospheric CO_2, with relatively small overall effect (<1‰). Stern *et al.* (2001), using simulation models, estimate that while the invasion flux is usually significant, its effect on the $\delta^{18}O$ of atmospheric CO_2 is highly variable among different biomes. Generally, if the top several cm of the soil are bone-dry, or if the soil surface is covered by several cm of dry litter, dry sand, or even snow, v and F_I will be small because of their dependency on θ_w. The invasion term scales with the difference between δ_a and δ_s, so soil water with low or high near-surface $\delta^{18}O$ values may magnify the effect of CO_2 diffusion into the soil. However, in deserts or other biomes where large differences between δ_a and δ_s are accompanied by low water content, the invasion effect likely will be small. In order for a significant amount of ambient CO_2 to invade from a given parcel of air, the CO_2 residence time in the well-mixed surface layer should be large, and the height of the layer should be small. Usually, if the residence time is not greater than tens of seconds, or if $h > 1$ m, the contribution of invasion to the $\delta^{18}O$ of atmospheric CO_2 will be small. In most settings, we would expect this to be the case. One exception may be at night, near the forest floor, where temperature inversions allow respired CO_2 to accumulate. Additionally, catalysis of CO_2 hydration, such as by the presence of carbonic anhydrase near the soil surface, would greatly accelerate the invasion process. Isaac (2001) reported a new and little explored aspect of the invasion effect in dry soils. Dry soils, especially with high clay content, have 1–7% (w/w) residual water that is not removed even by oven drying at 100 °C, but that still show hydration of CO_2 in rates faster than similar quantities in the pure water form. Further, as expected for bound water effect, the observed equilibrium fractionation was smaller than expected for pure water, with the discrepancy increasing inversely to the residual soil water content.

Order of magnitude estimates of soils impact on the $\delta^{18}O$ of atmospheric CO_2 can be obtained using Equation (43) by assuming global respiration flux of ~100 Pg, kinetic fractionation of −7.2‰, mean δs values of −7.9 and a global invasion flux of, say, 7 Pg, to yield −2.5‰ yr^{-1} (Figure 9). As discussed below, on global scale, this value is similar in magnitude but opposite in sign to that of the vegetation.

5.4.2 The Leaf Component

CO_2 diffusing into leaves during photosynthesis dissolves and rapidly exchanges its oxygen with water in the chloroplast, a process that depends on catalysis by the enzyme carbonic anhydrase (Reed and Graham, 1981; Silverman, 1982), because of the short residence time of CO_2 in leaves (seconds or less). About one-third of the CO_2 diffusing into the leaves is fixed in C_3 photosynthesis. The remaining two-third diffuse back to the atmosphere after isotopic exchange with chloroplast water. The retro-diffusion CO_2 flux is enriched in ^{18}O relative to the soil water that feeds the plant because of the evaporative enrichment of leaf water (Gonfiantini *et al.*, 1965).

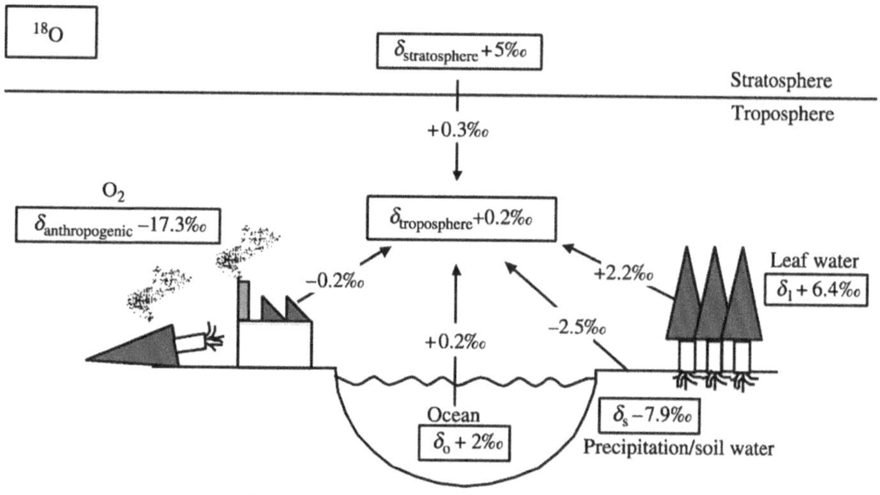

$$\frac{d(\delta_{troposphere})}{dt} = +2.2 - 2.5 + 0.2 - 0.2 + 0.3 \approx 0$$

Figure 9 Annual mean "forcings" on the $\delta^{18}O$ value of atmospheric CO_2. Each specific forcing component is the product of the flux involved, F, diluted by the atmospheric carbon pool, C_a, and the difference in $\delta^{18}O$ values between that of the flux and the mean atmosphere (i.e., forcing = $[(F/C_a)(\delta_x - \delta_a)]$. Values are order of magnitude estimates, as discussed in the text, and involve significant uncertainties. Typical values for the reservoirs involved are given in boxes and reflect the $\delta^{18}O$ value of the water in that reservoir, accept for the anthropogenic reservoir that is represented by the $\delta^{18}O$ value of atmospheric oxygen. The land biosphere has the strongest forcing, with contrasting effects of photosynthesis and respiration. Land photosynthesis tends to enrich the atmosphere in ^{18}O and soil respiration has the opposite effect, providing a basis for partitioning photosynthesis and respiration fluxes. The ocean forcing is an order of magnitude smaller. Fossil fuel emission tends to decrease ^{18}O, but the effect is "washed away" by the massive exchange of oxygen in CO_2 with water and long-term trend in atmospheric $\delta^{18}O$-CO_2 is not usually observed.

During evaporation, water containing the lighter isotope are preferentially lost and the water remaining in the leaves is enriched in the heavy isotopes (both ^{18}O and deuterium; Craig and Gordon, 1965). Because of the normally large ratio of the transpiration flux through the leaf to the leaf water volume, it is usually assumed that leaves are always near isotopic steady state with respect to environmental conditions (but see Harwood *et al.*, 1998, 1999; Wang and Yakir, 1995; Wang *et al.*, 1998). At isotopic steady state, the ^{18}O enrichment of leaf water is usually estimated with the widely used evaporative enrichment model of Craig and Gordon (1965), adapted to leaves (e.g., Dongmann *et al.*, 1974; Farquhar and Lloyd, 1993; Flanagan *et al.*, 1991, 1993; Yakir, 1992)

$$\delta_{\mathrm{lw}} = \delta_{\mathrm{i}}\varepsilon_{\mathrm{LV}} + \varepsilon_{\mathrm{K}} + h(\delta_{\mathrm{vap}} - \delta_{I} - \varepsilon_{\mathrm{K}}) \qquad (47)$$

where h is relative humidity at the leaf surface (equivalent to e_a/e_l, the vapor pressure ratio in the air and in the leaf), ε_{LV} the fractionation for liquid to vapor phase transition of water, ε_K the kinetic fractionation in the diffusion of water vapor across the stomatal cavity and the leaf boundary layer, and subscripts lw, I, and vap representing leaf water, input water from the soil, and atmospheric water vapor, respectively. Note that Equation (47) does not estimate the isotopic composition of bulk leaf water, which is a complex system that involves marked isotopic heterogeneity (Helliker and Ehleringer, 2000; Roden and Ehleringer, 1999; White, 1988; Yakir, 1998). Rather, it is assumed that δ_{lw} represents the isotopic enrichment of water at the evaporating surfaces inside the leaves (Figure 5), and that the chloroplasts are always near these surfaces to facilitate gas exchange. Carbonic anhydrase, in turn, is located predominantly in the chloroplasts, and δ_{lw} provides therefore a reasonable estimate of the specific $\delta^{18}O$ of water fraction with which CO_2 equilibrate in the leaves, and δ_l used in Equation (43) will simply be $\delta_l = \delta_{\mathrm{lw}} + \varepsilon_{\mathrm{eq}}$, where $\varepsilon_{\mathrm{eq}}$ is the water–CO_2 equilibrium fractionation. In diffusing out of the leaf, a diffusional fractionation, ε_l must also be considered, such that the final $\delta^{18}O$ value of the CO_2 reaching the atmosphere is $\delta_l + \varepsilon_l$. Note that considerable uncertainty is still associated with estimating the $\delta^{18}O$ value of leaf water on large scale and current estimates ranged between 4.4‰ and ~8.8‰ (Bender *et al.*, 1985, 1994; Dongmann, 1974; Förstel, 1978).

Consider that the net effect of the leaf flux on atmospheric ^{18}O, i.e., the term $(\delta_l - \delta_a + \varepsilon_l)$ in Equation (43), represents discrimination against ^{18}O. As CO_2 is removed from the atmosphere by the photosynthesis flux, F_A, the remaining CO_2 in the atmosphere becomes enriched in ^{18}O due to the leaf water effect. Unlike the case for ^{13}C, the apparent discrimination, $^{18}\Delta = (\delta_l - \delta_a + \varepsilon_l)$, is not reflected in the $\delta^{18}O$ of plant organic matter (rather, the access ^{16}O is washed away by water), but it can be treated mathematical in the same way (Farquhar and Lloyd, 1993) according to

$$^{18}\Delta = \frac{R_a}{R_l} - 1 = \frac{\bar{a} + \xi'\cdot(\delta_l - \delta_a)}{1 - \xi'\cdot(\delta_l - \delta_a)/1{,}000} \approx \bar{a} + \xi'\cdot(\delta_l - \delta_a) \qquad (48)$$

where R_a and R_A are the oxygen isotope ratios of CO_2 in the air and in the CO_2 flux taken up by assimilation, $\bar{a}$ is the weighted average fractionation during diffusion of CO_2 from the atmosphere to the chloroplast (similar to ε_l above with values of 8.8‰ in stagnant air and 0.8‰ in solution), $\xi' = c_l/(c_a - c_l)$ and c_a and c_l are the CO_2 concentrations in the atmosphere and in the site of oxygen exchange between CO_2 and water in the leaves, respectively (ξ' represents the retro-diffusion flux back to the atmosphere after ^{18}O exchange with leaf water), and δ_a and δ_l are the $\delta^{18}O$ values of CO_2 in the atmosphere and in equilibrium with chloroplast water, respectively.

Estimates of $^{18}\Delta$ using Equation (48) rely on the assumption of complete isotopic equilibrium between CO_2 and water in the leaf chloroplasts, due to the action of carbonic anhydrase. This assumption has recently been tested, and revised (Gillon and Yakir, 2001, 2000a,b). The extent of CO_2–H_2O equilibrium may be derived from Mills and Urey (1940) as

$$\theta_{\mathrm{eq}} = 1 - \mathrm{e}^{-k\tau/3} \qquad (49)$$

which describes the fractional approach to full equilibrium (where $\theta_{\mathrm{eq}} = 1$), as a function of the number of hydration reactions achieved per CO_2 molecule, $k\tau$ (divided by 3 for the isotopic exchange). This "coefficient of CO_2 hydration" is estimated for a leaf by calculating the rate constant k from biochemical measurements of carbonic anhydrase activity, and the residence time of CO_2 in the leaf, τ, from leaf-scale photosynthetic flux measurements of CO_2 (Gillon and Yakir, 2000a,b). A survey of θ_{eq} in a wide range of plant species showed large variations in the activity of carbonic anhydrase among major plant groups that cause variations in θ_{eq} ($\theta_{\mathrm{eq}} = 1$ reflect full equilibrium). On average, θ_{eq} was 0.93 for C3 trees, 0.70 for C3 grasses and 0.38 for C4 plants, with a global weighted mean of $\theta_{\mathrm{eq}} = 0.78$. Put simply, this indicates that, in contrast to earlier assumptions, only ~80% of the diffusional CO_2 backflux from plants to the atmosphere reflects the leaf water signal relevant to atmospheric $C^{18}OO$ budgets. Ignoring such incomplete equilibrium could result in ~20% underestimation of the gross CO_2 exchange with plants derived from atmospheric measurements of $\delta^{18}O$ values (Ciais *et al.*, 1997b; Farquhar *et al.*, 1993; Francey and Tans, 1987; Yakir and Wang, 1996).

Accordingly, the model depicted in Equation (48) was extended to allow for these variations:

$$^{18}\Delta = \frac{\bar{a} + \varepsilon\left(\theta_{eq}(\delta_l - \delta_a) - (1-\theta_{eq})\bar{a}/(\xi'+1)\right)}{1 - \varepsilon\left(\theta_{eq}(\delta_l - \delta_a) - (1-\theta_{eq})\bar{a}/(\xi'+1)\right)/1{,}000} \approx \bar{a} + \xi'[\theta_{eq}(\delta_l - \delta_a) - (1-\theta_{eq})\bar{a}/(\xi'+1)] \qquad (50)$$

where θ_{eq} represents the extent of ^{18}O equilibrium between CO_2 and water in leaves. The extent of equilibrium, θ_{eq}, can be estimated from direct measurements of carbonic anhydrase activity in leaves, or using leaf-scale measurements of $^{18}\Delta$ and solving Equation (50) for θ_{eq} (Gillon and Yakir, 2000b). Farquhar and Lloyd (1993) also proposed an indirect estimate of the extent of equilibrium related to the ratio, ρ, of Rubisco to carbonic anhydrase activities.

Note that Equations (48) and (50) can also be solved for δ_l and the $\delta^{18}O$ of leaf water. In this case, Equation (50) will yield a higher δ_l value than Equation (48). On a global scale, Gillon and Yakir (2001) estimated θ_{eq} to be 0.8, which would increase global estimates of global mean $\delta^{18}O$ value of leaf water of ~+4.4‰ (Bender *et al.*, 1994; Farquhar *et al.*, 1993) by ~2‰, providing better consistency with estimates based on studies of the $\delta^{18}O$ of atmospheric O_2 (see also Beerling, 1999).

Equation (50) indicates that $^{18}\Delta$ is regulated in leaves by three main factors; the isotopic composition of water (as reflected in δ_l), the CO_2 concentration at the CO_2–H_2O exchange sites (c_l), as well as by the extent of isotopic equilibrium between CO_2 and H_2O (θ_{eq}). The recent estimate of global mean θ_{eq} of $\theta_{eq} \approx 0.8$ (Gillon and Yakir, 2001) reflected low θ_{eq} in grasses and C4 plants ($\theta_{eq} \approx 0.4$) and high θ_{eq} in non-grasses C3 plants ($\theta_{eq} > 0.9$). An important source of uncertainty is the CO_2 concentration inside leaves (c_l), which cannot be directly measured. Recent studies of c_l also provide another demonstration of the implications of processes at cellular and sub-cellular scales to large-scale studies using stable isotopes. Leaf-scale measurements indicated that the CO_2 drawdown from the sub-stomata cavities (c_i, readily estimated in leaf-scale measurements and often shows c_i/c_a ratios of 0.6–0.7) to the chloroplast, c_c, generally yields $(c_i - c_c)/c_a \approx 0.2$ and $c_c/c_a \approx 0.55$, (Epron *et al.*, 1995; Lloyd *et al.*, 1992; Loreto *et al.*, 1992; Von Caemmerer and Evans, 1991). In C_4 plants internal CO_2 concentrations are much lower (c_i/c_a is typically ~0.3–0.4 with further drawdown to the mesophyll cells where CO_2 is fixed by the enzyme PEP-carboxylase). Such estimates of c_c rely mostly on ^{13}C discrimination by leaves and represent the CO_2 concentrations at the site of Rubisco, the primary photosynthetic enzyme, in the chloroplast. But CO_2 hydration is controlled by carbonic anhydrase, and the relevant site of CO_2–H_2O equilibration is at the limit of its activity, the chloroplast or cellular membrane. The CO_2 concentrations at this point, c_l, are distinct from those at the Rubisco site, c_c, by 0–40 ppm, (depending on plant species and rates of activities) due to large internal resistances to CO_2 diffusion. It was generally observed that the CO_2 concentration at the site of CO_2–H_2O equilibrium in leaves (c_l) was at about mid-way between c_i and c_c, $(c_i - c_l)/c_a \approx 0.1$, significantly influencing estimates of $^{18}\Delta$ using Equation (50).

Values of $^{18}\Delta$ are highly variable among plant species and environmental conditions. Modeled $^{18}\Delta$ values for different biomes can range from −20‰ to +32‰ (Farquhar *et al.*, 1993). Only few comparisons of such modeled $^{18}\Delta$ values were compared with actual, measured ecosystem-scale $^{18}\Delta$. Making such comparison in the Amazon basin (cf. Sternberg *et al.*, 1998) indicated that further comparisons with other ecosystems are necessary. Nevertheless, ^{18}O in CO_2 is potentially powerful tracer of CO_2 fluxes, distinct from ^{13}C and influenced by different process. The ^{18}O signal in CO_2 is comparable in magnitude to that of ^{13}C on a global scale (Figures 3 and 9), and often larger on a local scale. Using global mean estimates of $^{18}\Delta = 13.7$‰ (Farquhar *et al.*, 1993), GPP of 120 Pg C y^{-1} and atmospheric carbon pool of ~750 Pg (IPCC), an order of magnitude estimate of the annual scale land vegetation impact on the $\delta^{18}O$ of atmospheric CO_2 can be obtained from, $F_A {}^{18}\Delta/c_a = (120 \times 13.7)/730 = +2.2$‰ yr^{-1} (Figure 9).

5.4.3 The Minor Components

5.4.3.1 The ocean component

The Ocean has an enrichment effect on the $\delta^{18}O$ of atmospheric CO_2. This is the result of two opposing effects: first, the variations in the $\delta^{18}O$ of ocean water (δ_w), and second, temperature effects on the CO_2–H_2O equilibrium fractionation.

Mean ocean water $\delta^{18}O$, δ_w, between 60° N and 60° S, is near 0‰, but there are spatial variations due to salinity/fresh water effects. δ_w decreases where large amounts of freshwater are delivered to the ocean, because of the isotopic fractionation in the hydrological cycle (global mean precipitation is ~−7.9‰, polar ice ~−25‰). In particular, δ_w is depleted by ~1‰ at high latitudes around Antarctica and Greenland because of the massive discharge of icebergs, and near estuaries of large rivers.

In contrast, the isotopic equilibrium fractionation between CO_2 and water is inversely correlated with temperature (see above) and has the effect of increasing $\delta^{18}O$ of CO_2 in equilibrium with ocean water at high latitudes. The overall results of the opposing salinity and temperature effects is an increase in $\delta^{18}O$ of CO_2 in equilibrium with ocean water (δ_o) as a function of latitude from around 0‰ in the tropics (PDB-CO_2 scale, see Introduction), to a maximum value around 5‰ near sea ice margins (Ciais *et al.*, 1997a; Ishizawa *et al.*, 2002). Note that the fractionation associated with diffusion and hydration across the sea–atmosphere interface is small, $\varepsilon_w = 0.8$‰. Global mean $\delta^{18}O$ of CO_2 in equilibrium with ocean water is ~+2‰, the ocean isotopic forcing on δ_a, using Equation (43), and assuming ~90 Pg C yr^{-1} and 88 Pg C yr^{-1} gross ocean uptake and release fluxes (IPCC, 2001), the ocean impact on δ_a is ~+0.2‰ yr^{-1} (Figure 9).

5.4.3.2 Anthropogenic emissions

Both fossil fuels and burned biomass are combusted with atmospheric oxygen, and it is generally assumed that the resulting CO_2 acquire its $\delta^{18}O$ value of −17.3‰ PDB-CO_2 (+23.5 SMOW; Blunier *et al.*, 2002). Note that in both cases the one-way surface to atmosphere fluxes are considered for their unique ^{18}O effect on the atmosphere. Net biomass burning, and deforestation, fluxes include CO_2 uptake through re-growth that largely balance the emission. So while only the net loss of carbon contribute to changes in atmospheric $\delta^{13}C$ and c_a, the gross burning rate with its unique ^{18}O signal constitute the main effect on atmospheric $\delta^{18}O$ (for ^{18}O, re-growth can be considered as part of the land vegetation GPP; Gillon and Yakir (2001)). Assuming an anthropogenic flux (fossil fuels plus combusted biomass) of ~9.4 Pg C yr^{-1} (IPCC, 2001), the order of magnitude of the fossil fuel forcing on the $\delta^{18}O$ of atmospheric CO_2, is ~−0.2‰ yr^{-1} (Figure 9).

5.4.3.3 Troposphere–Stratosphere exchange

Stratospheric CO_2 is known to posses an unusual isotopic composition, enriched in both ^{18}O and ^{17}O (Gamo *et al.*, 1989; Thiemens, 1999; Thiemens *et al.*, 1991; Wen and Thiemens, 1993; Yung *et al.*, 1991). ^{18}O enrichment of ~2‰ is observed in CO_2 at altitudes of 20–25 km, and of more than 10‰ at altitudes of 30–35 km (Gamo *et al.*, 1989; Thiemens and Jackson, 1990). This enrichment is likely to pass on from ozone ^{18}O enrichment. It is based on the formation of a short-lived CO_3 transition state, which is created by the interaction of CO_2 with electronically exited $O(^1D)$, derived from photolysis of O_3 in the stratosphere. The CO_3 intermediate breaks apart to CO_2 randomly losing oxygen, which is processed back to O_2. This mechanism therefore effectively transfers the ^{18}O (and ^{17}O) enrichment from ozone to CO_2.

With respect to the troposphere, the stratospheric ^{18}O enrichments can be regarded as "invasion effect" (similar to that considered for soils). While there is no net flux involved, it depends on the gross exchange flux of CO_2 in and out of the stratosphere and the ^{18}O fractionation that CO_2 undergoes in the stratosphere, associated with the O_3–CO_2 oxygen exchange. The troposphere–stratosphere CO_2 exchange flux can be estimated from Appenzeller *et al.* (1996) to be ~100 Pg C yr^{-1}. The ^{18}O fractionation is not known at present but from recent estimates of model runs can be assumed to be small (Blunier *et al.*, 2002). Alternatively, Peylin *et al.* (1996) used the observed altitudinal gradient in the $\delta^{18}O$ of CO_2 from near 0‰ (PDB-CO_2 scale), to ~2‰ at 19 km (Gamo *et al.*, 1989) and ~10‰ at 35 km (Thiemens and Jackson, 1990), or stratospheric average of, say, +5‰). Fitting this gradient, the overall stratospheric effect on the annual troposphere $\delta^{18}O$ budget was ~+0.3‰.

5.4.4 Spatial and Temporal Patterns

Although some measurements were available as early as the 1970s and before (see Francey *et al.*, 1999 for a recent summary of stable isotope air sampling efforts), large-scale, high precision measurements of the $\delta^{18}O$ of atmospheric CO_2 picked up in the early 1990s, carried out by centers such as the NOAA-CU, Scripps-SIO, Groningen University-CIO, and CSRIO. These measurements show several prominent features. First, the long-term global mean $\delta^{18}O$ of atmospheric CO_2 is near steady state (i.e., $d(\delta^{18}O)/dt = 0$). Notably, however, there were periods with a significant temporal trends, such as the >0.5‰ decreasing trend during the 1990s. Second, a large, nearly 2‰, Arctic–Antarctic gradient is observed in annual mean values of the $\delta^{18}O$ of atmospheric CO_2 (Ciais *et al.*, 1997b). Third, a similarly large (ca. 1.5‰) seasonal cycle is observed at high northern latitudes, with a maximum in early summer and a minimum in early winter. Fourth, the seasonal ^{18}O cycle is clearly off phase (by about a month or more) with respect to the seasonal cycle in CO_2 concentrations, and in ^{13}C (Figure 2). Most of these features have now been modeled, initially with simple models (Farquhar *et al.*, 1993; Francey and Tans, 1987; Ishizawa *et al.*, 2002) and then with GCMs and 3D transport models coupled to biosphere models such as Sib2 (Ciais *et al.*, 1997a,b; Ishizawa *et al.*, 2002; Peylin *et al.*, 1999; 1996). Although significant uncertainties remain in many of the models parameters, the reasonable agreement of model output and observations indicates that some of the major processes underlying the observations are captured.

The strong global ^{18}O gradient, dominated by ~−1.2‰ gradient in the northern hemisphere, between Mauna Loa and Pt. Barrow, was noted by Francey and Tans (1987). They pointed out that a large northern hemisphere isotope exchange flux, in the order of 200 Pg C yr^{-1}, is required to maintain such gradient against the vigorous mixing of the atmosphere. Such flux is consistent with the retro-diffusion flux from leaves during photosynthesis (Figure 5), complemented by CO_2 exchange with soil water. Combined, these fluxes largely reflect the isotopic gradient in precipitation water on land. This early notion has remained a valid basis for the observed meridional gradient. Later refinements indicated that it is, in fact, the respiration of the biosphere, mostly above 30° N, that is the dominant control over the mean latitudinal gradient (Ciais *et al.*, 1997b; Farquhar *et al.*, 1993; Peylin *et al.*, 1999).

Seasonality in the land biosphere is a well known phenomenon and it is clearly reflected in the CO_2 and ^{13}C records, with a minimum in CO_2 concentration and maximum in ^{13}C, nearly coinciding with peak net ecosystem CO_2 uptake associated with peak photosynthetic activity (Figures 2(a) and (b); Trolier *et al.*, 1996). Early studies, such as that of Francey and Tans, were not able to clearly detect a seasonal cycle in atmospheric ^{18}O, although it would clearly be consistent with ^{18}O exchange with the terrestrial biosphere. With increasing precision and frequency of measurements, a strong seasonal cycle in ^{18}O became apparent (Figure 2(c); Cuntz *et al.*, 2002; Flanagan *et al.*, 1997; Peylin *et al.*, 1999), which also showed interhemispheric gradient in its amplitude. While in the southern hemisphere seasonality in ^{18}O was barely detected (~0.2‰), it reaches amplitude of ca. 1.5‰, and in some cases more than 2‰, in mid- and high-latitudes in the northern hemisphere, consistent with the distribution of land ecosystem activities around the globe (Peylin *et al.*, 1999; Cuntz *et al.*, 2002). Peylin *et al.* (1999), examining the factors influencing the phase and amplitude of the $\delta^{18}O$ seasonal cycle, observed that the large seasonal cycle at high latitudes is mainly due to the respiratory fluxes (with its negative $\delta^{18}O$ values) of all extra-tropical ecosystems.

In contrast with seasonality in c_a, which is clearly dominated by net ecosystem CO_2 exchange (the relative contributions of photosynthesis and respiration), seasonality in the $\delta^{18}O$ of CO_2 is more complex. It couples variations in CO_2 fluxes with variations in the $\delta^{18}O$ of precipitation, in leaf discrimination and in soil discrimination. This coupled effect can be best discussed in terms of isofluxes or isotopic "forcing" (forcing = isoflux/atm. pool) on the atmospheric ^{18}O of the gross CO_2 flux and the ^{18}O discrimination associated with it (i.e., Pg C‰; cf. Figure 7). Local δ_a is influenced by the opposing effects of the generally ^{18}O enriched photosynthesis isoflux (reflecting ^{18}O enrichment of leaf water), and the invariably depleted soil isoflux (reflecting ^{18}O depleted soil water) ($F_A{}^{18}\Delta_A + F_{sa}(\delta_{sa} - \delta_a)$). The simulation results of Peylin *et al.* (1999) indicated in simple terms (Figure 10) that the soil forcing is generally greater than that of photosynthesis, and therefore dominates the seasonal cycle. The possible reason for the relatively small photosynthetic isoflux in many locations is that while soil water is strongly depleted with increasing latitude, concurrent decrease in relative humidity and temperature result also in strong leaf water enrichments, diminishing the difference between the $\delta^{18}O$ of CO_2 equilibrated with leaf water and that of the atmosphere, reducing leaf discrimination ($^{18}\Delta_A$, Equations (48) and (50)). The simulation results demonstrate two additional interesting features associated with the seasonal cycle. First, ecosystems with similar net ecosystem exchange CO_2 flux can have very different effect on the $\delta^{18}O$ of atmospheric CO_2. For example, more continental Siberian ecosystems receive more depleted precipitation, as compared, for example, with Canadian ecosystems (3–7‰ difference). With similar net ecosystem CO_2 exchange and leaf water enrichments during the growing season, the Siberian ecosystems will have a much larger soil isoflux during the growing season. Second, due to efficient lower troposphere mixing, associated with weak vertical transport at extra-tropical latitudes, local δ_a is often dominated by distant ecosystems with large isoflux (but not necessarily large CO_2 flux). Such is the case of the disproportionate effects of the Northern Siberian taiga on regional and the global seasonal ^{18}O cycle (but not on that in CO_2). While most northern hemisphere ecosystems have seasonal amplitude in isoflux of ca. 10Pg C‰, the Northern Siberian taiga has ~20 Pg C‰. Note, therefore, that there is unique relationship between the isoflux and the CO_2 flux at any given location (Figure 9 and 10).

The phase of the ^{18}O seasonal cycle also reflects the interplay between the photosynthetic and the soil isoflux, and not the seasonality in net ecosystem CO_2 exchange. Leaf discrimination generally peaks in the spring due to high leaf water enrichment and high leaf internal CO_2 concentration (i.e., high retrodiffusion flux from the leaves to the atmosphere). During the summer, leaf discrimination decreases, often because reduced leaf water enrichment associated with increasing relative humidity, and because of lower leaf internal CO_2 concentrations. As a result, leaf isofluxes show a weak dominance early in the growing season, bringing about a seasonal ^{18}O peak in early summer. As the leaf isoflux slowly diminishes, the respiration isoflux, that has lagged photosynthesis due to the slow increase in soil temperature, takes over bringing about a seasonal minimum in late summer/early

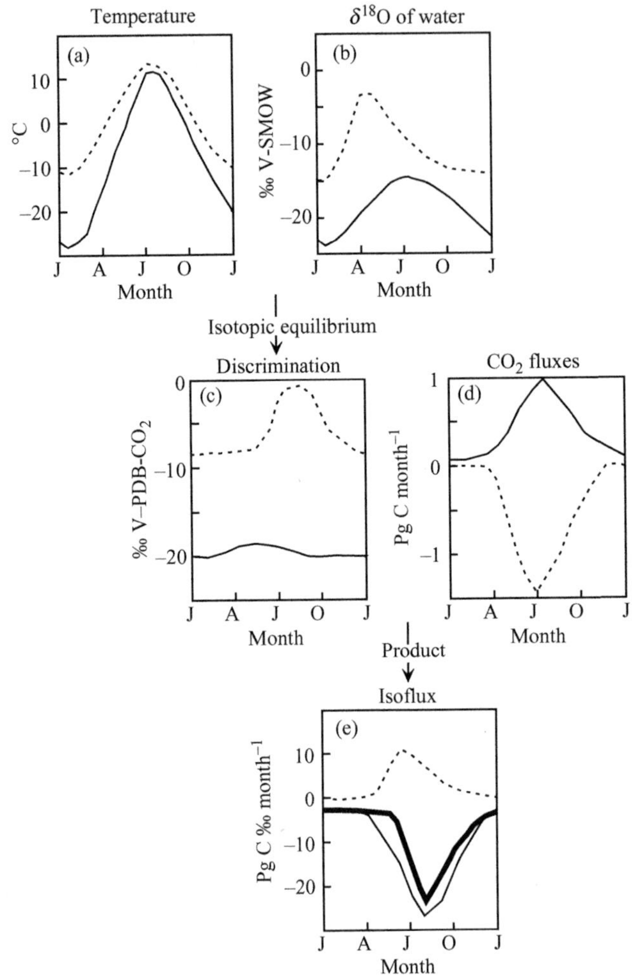

Figure 10 Development of the ^{18}O seasonal cycle in a high latitude forest biome. Water is enriched in ^{18}O due to evaporation under decreasing relative humidity in the spring and summer (top right panel; broken and solid lines refer to leaves and soil, respectively). This is transuded to CO_2 via isotopic exchange the magnitude of which is inversely correlated with temperature (top left panel). Combined, water ^{18}O content and temperature-sensitive isotopic exchange, are the major controls over the apparent ^{18}O discrimination associated with soil and leaves (middle left panel). The effect on atmospheric CO_2 is weighted by the CO_2 fluxes involved (middle right panel; Thick solid line is the combined, net, effect). The flux carrying the ^{18}O signature is termed isoflux; using isotopic ratios would yield actual ^{18}O flux, using, as often done, $\delta^{18}O$ values provide good indicators of the ^{18}O flux. Note that isoflux taken up by leaf photosynthesis (negative flux in middle right panel) actually enrich the atmosphere in ^{18}O (bottom panel). The interacting effects of the seasonal patterns in evaporating enrichment, temperature and flux result in a negative ^{18}O signature of the biome on the atmosphere that reaches its seasonal peak about one months later that the peak in the actual flux (after Peylin *et al.*, 1999).

fall. Notably, this minimum, dominating the seasonal cycle, arrives about one month after the minima in CO_2 concentrations associated with the seasonal peak photosynthesis. Here too, dominant regions, such as Siberia with their large soil isoflux, have disproportionate effect in controlling the phase of the northern hemisphere seasonal cycle and the timing of its minimum (Peylin *et al.*, 1999). Detailed understanding of the processes underlying the ^{18}O pattern discussed above, should allow using "inverse techniques," and the $\delta^{18}O$ and CO_2 data from global sampling networks to derive separately values for assimilation and respiration fluxes of large ecosystems.

ACKNOWLEDGMENTS

This work would not have been completed without the help of Hagit Affek in preparing the figures and the literature list. John Miller, Leo Sternberg, and Ralph Keeling provided helpful comments. Unpublished data were kindly provided by Peter Bakwin and NOAA/CMDL, and Joe Berry.

REFERENCES

Allison C. E., Francey R. J., and Meijer H. A. (1995) Recommendations for the reporting of stable isotope measurements of carbon and oxygen in CO_2 gas. In *References and Intercomparison Materials for Stable Isotopes of Light Elements, Proceeding of a Consultation Meeting Held in Vienna, December 1–3, 1993.* IAEA, pp. 155–162.

Allison G. B. and Barnes C. J. (1983) Estimation of Evaporation from non-vegetated surfaces using natural deuterium. *Nature* **301**(5896), 143–145.

Allison G. B., Barnes C. J., and Hughes M. W. (1983) The distribution of deuterium and ^{18}O in dry soils: 2. Experimental. *J. Hydrol.* **64**(1–4), 377–397.

Amundson R., Stern L., Baisden T., and Wang Y. (1998) The isotopic composition of soil and soil-respired CO_2. *Geoderma* **82**(1–3), 83–114.

Andres R. J., Marland G., Boden T., and Bischof S. (2000) Carbon dioxide emissions from fossil fuel consumption and cement manufacture, 1751–1991, and an estimate of their isotopic composition and latitudinal distribution. In *The Carbon Cycle* (eds. T. M. L. Wigley and D. S. Schimel). Cambridge University Press, , pp. 53–62.

Appenzeller C., Holton J. R., and Rosenlof K. H. (1996) Seasonal variation of mass transport across the tropopause. *J. Geophys. Res. Atmos.* **101**(D10), 15071–15078.

Bakwin P., Zhao C., Ussler W., III, Tans P. P., and Quesnell E. (1995) Measurements of carbon dioxide on a very tall tower. *Tellus* **47B**, 535–549.

Bakwin P. S., Tans P. P., White J. W. C., and Andres R. J. (1998a) Determination of the isotopic ($^{13}C/^{12}C$) discrimination by terrestrial biology from a global network of observations. *Global Biogeochem. Cycles* **12**(3), 555–562.

Bakwin P. S., Tans P. P., Hurst D. F., and Zhao C. (1998b) Measurements of carbon dioxide on very tall towers: results of the NOAA/CMDL program. *Tellus* **50B**, 401–415.

Baldocchi D. (1997) Measuring and modelling carbon dioxide and water vapor exchange over a temperate broad-leaved forest during the 1995 summer drought. *Plant Cell Environ.* **20**(9), 1108–1122.

Baldocchi D. D., Hicks B. B., and Meyers T. P. (1988) Measuring biosphere-atmosphere exchanges of biologically related gases with micrometeorological methods. *Ecology* **69**(5), 1331–1340.

Balesdent J. and Mariotti A. (1996) Measurement of soil organic matter turnover using ^{13}C natural abundance. In *Mass Spectrometry of Soils* (eds. T. W. Boutton and S. I. Yamasaki). Dekker.

Barnes C. J. and Allison G. B. (1983) The distribution of deuterium and ^{18}O in dry soils: 1. Theory. *J. Hydrol.* **60**(1–4), 141–156.

Battle M., Bender M. L., Tans P. P., White J. W. C., Ellis J. T., Conway T., and Francey R. J. (2000) Global carbon sinks and their variability inferred from atmospheric O_2 and $\delta^{13}C$. *Science* **287**(5462), 2467–2470.

Beerling D. J. (1999) The influence of vegetation activity on the dole effect and its implications for changes in biospheric productivity in the mid-holocene. *Proc. Roy. Soc. of London Ser. B-biol. Sci.* **266**(1419), 627–632.

Beerling D. J. and Woodward F. I. (1998) Modelling changes in the plant function over the phanerozoic. In *Stable Isotopes: Integration of Biological, Ecological and Geochemical Processes* (ed. H. Griffiths). BIOS Scientific Publishers Limited, , pp. 347–361.

Bender M., Labeyrie L. D., Raynaud D., and Lorius C. (1985) Isotopic composition of atmospheric O_2 in ice linked with deglaciation and global primary productivity. *Nature* **318** (6044), 349–352.

Bender M., Sowers T., and Labeyrie L. (1994) The dole effect and its variations during the last 130,000 years as measured in the Vostok ice core. *Global Biogeochem. Cycles* **8**(3), 363–376.

Berry J. A. (1988) Studied of mechanisms affecting the fractionation of carbon isotopes in photosynthesis. In *Stable Isotopes in Ecological Research,* vol. 68, (eds. P. W. Rundel, J. R. Ehleringer, and K. A. Nagy). Springer, pp. 82–94.

Bigeleisen J. (1965) Chemistry of isotopes. *Science* **147**(3657), 463–471.

Bigeleisen J. and Mayer M. (1947) Calculation of equilibrium constants for isotopic exchange reactions. *J. Chem. Phys.* **15**, 261–267.

Bigeleisen J. and Wolfsberg M. (1958) Theoretical and experimental aspects of isotope effects in chemical kinetics. *Adv. Chem. Phys.* **1**, 15–76.

Bird M. I., Chivas A. R., and Head J. (1996) A latitudinal gradient in carbon turnover times in forest soils. *Nature* **381**(6578), 143–146.

Blunier T., Barnett B., Bender M. L., and Hendricks M. B. (2002) Biological oxygen productivity during the last 60,000 years from triple oxygen isotope measurements. *Global Biogeochem. Cycles* **16**(3), article no. 1029.

Bolin B. (1981) *Carbon Cycle Modelling. SCOPE 16.* Wiley.

Bottinga Y. and Craig H. (1969) Oxygen isotope fractionation between CO_2 and water and isotopic composition of marine atmospheric CO_2. *Earth Planet. Sci. Lett.* **5**(5)285-and.

Bousquet P., Peylin P., Ciais P., Ramonet M., and Monfray P. (1999) Inverse modelling of annual atmospheric CO_2 sources and sinks: 2. Sensitivity study. *J. Geophys. Res. Atmos.* **104**(D21), 26179–26193.

Boutton T. W. and Yamasaki S. (eds.) (1996) *Mass Spectrometry of soils.* Dekker, New York.

Bowling D. R., Turnipseed A. A., Delany A. C., Baldocchi D. D., Greenberg J. P., and Monson R. K. (1998) The use of relaxed eddy accumulation to measure biosphere-atmosphere exchange of isoprene and of other biological trace gasses. *Oecologia* **116**(3), 306–315.

Bowling D. R., Baldocchi D. D., and Monson R. K. (1999a) Dynamics of isotopic exchange of carbon dioxide in a Tennessee deciduous forest. *Global Biogeochem. Cycles* **13**(4), 903–922.

Bowling D. R., Delany A. C., Turnipseed A. A., Baldocchi D. D., and Monson R. K. (1999b) Modification of the relaxed eddy accumulation technique to maximize measured scalar mixing ratio differences in updrafts and downdrafts. *J. Geophys. Res. Atmos.* **104**(D8), 9121–9133.

Bowling D. R., Tans P. P., and Monson R. K. (2001) Partitioning net ecosystem carbon exchange with isotopic fluxes of CO_2. *Global Change Biol.* **7**(2), 127–145.

Bowling D. R., McDowell N. G., Bond B. J., Law B. E., and Ehleringer J. R. (2002) ^{13}C content of ecosystem respiration is linked to precipitation and vapor pressure deficit. *Oecologia* **131**(1), 113–124.

Brenninkmeijer C. A. M., Kraft P., and Mook W. G. (1983) Oxygen isotope fractionation between CO_2 and H_2O. *Isotope Geosci.* **1**(2), 181–190.

Broadmeadow M. S. J. and Griffiths H. (1993) Carbon isotope discrimination and the coupling of CO_2 fluxes within canopies. In *Stable Isotopes and Plant Carbon-water Relations* (eds. J. R. Ehleringer, A. E. Hall, and G. D. Farquhar). Academic Press, , pp. 109–130.

Broecker W. S. and Maier-Reimer E. (1992) The influence of air and sea exchange on the carbon isotope distribution in the sea. *Global Biogeochem. Cycles* **6**(3), 315–320.

Broecker W. S. and Peng T. H. (1974) Tracers in the Sea.

Broecker W. S., Takahashi T., Simpson H. J., and Peng T. H. (1979) Fate of fossil-fuel carbon-dioxide and the global carbon budget. *Science* **206**(4417), 409–418.

Brugnoli E., Scartazza A., Lauteri M., Monteverdi M. C., and Máguas C. (1998) Carbon isotope discrimination in structural and non-structural carbohydrates in relation to productivity and adaptation to unfavorable conditions. In *Stable Isotopes: Integration of Biological, Ecological and Geochemical Processes* (ed. H. Griffiths). BIOS Scientific Publishers, pp. 133–146.

Buchmann N. and Ehleringer J. R. (1998) CO_2 concentration profiles, and carbon and oxygen isotopes in C_3, and C_4 crop canopies. *Agri. Forest Meteorol.* **89**(1), 45–58.

Buchmann N. and Kaplan J. O. (2001) Carbon isotope discrimination of terrestrial ecosystems—how well do observed and modeled results match? In *Global Biogeochemical Cycles in the Climate System* (eds. E. D. Schulze, M. Heimann, S. Harrison, E. Holland, J. Lloyd, I. C. Prentice, and D. Schimel). Academic Press, , pp. 253–266.

Buchmann N., Guehl J. M., Barigah T. S., and Ehleringer J. R. (1997a) Interseasonal comparison of CO_2 concentrations, isotopic composition, and carbon dynamics in an Amazonian Rainforest (French Guiana). *Oecologia* **110**(1), 120–131.

Buchmann N., Kao W. Y., and Ehleringer J. (1997b) Influence of stand structure on Ccarbon-13 of vegetation, soils, and canopy air within deciduous and Evergreen Forests in Utah, US. *Oecologia* **110**(1), 109–119.

Buchmann N., Brooks J. R., Flanagan L. B., and Ehleringer J. (1998) Carbon isotope discrimination of terrestrial ecosystems. In *Stable Isotopes: Integration of Biological, Ecological, and Geochemical Processes* (ed. H. Griffiths). BIOS Scientific Publishers, pp. 203–222.

Businger J. A. and Oncley S. P. (1990) Flux measurement with conditional sampling. *J. Atmos. Ocean. Technol.* **7**(2), 349–352.

Cellier P. and Brunet Y. (1992) Flux gradient relationships above tall plant canopies. *Agri. Forest Meteorol.* **58**(1–2), 93–117.

Cerling T. E. (1984) The stable isotopic composition of modern soil carbonate and its relationship to climate. *Earth Planet. Sci. Lett.* **71**(2), 229–240.

Cerling T. E. and Wang Y. (1996) Stable carbon and oxygen isotopes in soil CO_2 and soil carbonate: theory, practice, and application to some prairie soils of upper Midwestern North America. In *Mass Spectrometry of Soils* (eds. T. W. Boutton and S. I. Yamasaki). Dekker, pp. 113–131.

Cerling T. E., Solomon D. K., Quade J., and Bowman J. R. (1991) On the isotopic composition of carbon in soil carbon-dioxides. *Geochim. Cosmochim. Acta* **55**(11), 3403–3405.

Ciais P., Tans P. P., Trolier M., White J. W. C., and Francey R. J. (1995a) A large northern-hemisphere terrestrial CO_2 sink indicated by the $^{13}C/^{12}C$ ratio of atmospheric CO_2. *Science* **269**(5227), 1098–1102.

Ciais P., Tans P. P., White J. W. C., Trolier M., Francey R. J., Berry J. A., Randall D. R., Sellers P. J., Collatz J. G., and Schimel D. S. (1995b) Partitioning of ocean and land uptake of CO_2 as inferred by $\delta^{13}C$ measurements from the NOAA climate monitoring and diagnostics laboratory global air sampling network. *J. Geophys. Res. Atmos.* **100**(D3), 5051–5070.

Ciais P., Denning A. S., Tans P. P., Berry J. A., Randall D. A., Collatz G. J., Sellers P. J., White J. W. C., Trolier M., Meijer H. A. J., Francey R. J., Monfray P., and Heimann M. (1997a) A three-dimensional synthesis study of $\delta^{18}O$ in atmospheric CO_2: 1. Surface fluxes. *J. Geophys. Res. Atmos.* **102**(D5), 5857–5872.

Ciais P., Tans P. P., Denning A. S., Francey R. J., Trolier M., Meijer H. A. J., White J. W. C., Berry J. A., Randall D. A., Collatz G. J., Sellers P. J., Monfray P., and Heimann M. (1997b) A three-dimensional synthesis study of $\delta^{18}O$ in atmospheric CO_2: 2. Simulations with the TM2 transport model. *J. Geophys. Res. Atmos.* **102**(D5), 5873–5883.

Ciais P., Friedlingstein P., Schimel D. S., and Tans P. P. (1999) A global calculation of the $\delta^{13}C$ of soil respired carbon: implications for the biospheric uptake of anthropogenic CO_2. *Global Biogeochem. Cycles* **13**(2), 519–530.

Clark I. and Fritz P. (1997) *Environmental Isotopes in Hydrology*. Lewis Publishers.

Collatz G. J., Berry J. A., and Clark J. S. (1998) Effects of climate and atmospheric CO_2 partial pressure on the global distribution of C_4 grasses: present, past, and future. *Oecologia* **114**(4), 441–454.

Conway T. J., Tans P. P., Waterman L. S., and Thoning K. W. (1994) Evidence for interannual variability of the carbon cycle from the national oceanic and atmospheric administration climate monitoring and diagnostics laboratory global air sampling network. *J. Geophys. Res. Atmos.* **99**(D11), 22831–22855.

Coplen T. B. (1995) Discontinuance of SMOW and PDB. *Nature* **375**(6529)285–285.

Coplen T. B., Kendall C., and Hopple J. (1983) Comparison of stable isotope reference samples. *Nature* **302**, 236–238.

Cox P. M., Betts R. A., Jones C. D., Spall S. A., and Totterdel I. J. (2000) Acceleration of global warming due to carbon-cycle feedbacks in a coupled climate model. *Nature* **408**, 184–187.

Craig H. (1953) The geochemistry of the stable carbon isotopes. *Geochim. Cosmochim. Acta* **3**(53–92).

Craig H. (1954) Carbon-13 variations in sequoia rings and the atmosphere. *Science* **119**, 141–143.

Craig H. (1957) Isotopic standards for carbon and oxygen and correction factors for mass-spectrometric analysis of carbon dioxide. *Geochim. Cosmochim. Acta* **12**, 133–149.

Craig H. (1961) Isotopic variations in meteoric waters. *Science* **133**, 1702–1703.

Craig H. (1963) The effects of atmospheric NO_2 on the measured isotopic composition of atmospheric CO_2. *Geochim. Cosmochim. Acta* **27**(549–551).

Craig H. and Gordon L. I. (1965) Deuterium and oxygen-18 variations in the ocean and marine atmosphere. In *Stable Isotopes in Oceano-graphic Studies and Paleo-Temperatures*, 9–130.

Cuntz M., Ciais P., and Hoffmann G. (2002) Modelling the continental effect of oxygen isotopes over Eurasia. *Tellus Ser. B: Chem. Phys. Meteorol.* **54**(5), 895–909.

Dansgaard W. (1964) Stable isotopes in precipitation. *Tellus* **16** (4), 436–468.

Davidson G. R. (1995) The stable isotopic composition and measurement of carbon in soil CO_2. *Geochim. Cosmochim. Acta* **59**(12), 2485–2489.

Deuser W. G. and Degens E. T. (1967) Carbon isotope fractionation in system CO_2(Gas)–CO_2(Aqueous)–HCO_3^-(Aqueous). *Nature* **215**(5105), 1033.

Dongmann G. (1974) Contribution of land photosynthesis to stationary enrichment of ^{18}O in atmosphere. *Radiat. Environ. Biophys.* **11**(3), 219–225.

Dongmann G., Nurnberg H. W., Forstel H., and Wagener K. (1974) Enrichment of $H_2{}^{18}O$ in leaves of transpiring plants. *Radiat. Environ. Biophys.* **11**(1), 41–52.

Duranceau M., Ghashghaie J., Badeck F., Deleens E., and Cornic G. (1999) $\delta^{13}C$ of CO_2 Respired in the dark in relation to $\delta^{13}C$ of leaf carbohydrates in *Phaseolus vulgaris* L—under progressive drought. *Plant Cell Environ.* **22**(5), 515–523.

Ehleringer J. R., Hall A. E., and Farquhar G. D. (1992) Stable isotope and plant water relations. Academic Press, San Diego.

Ehleringer J. R., Cerling T. E., and Helliker B. R. (1997) C_4 photosynthesis, atmospheric CO_2 and climate. *Oecologia* **112**(3), 285–299.

Ehleringer J. R., Buchmann N., and Flanagan L. B. (2000) Carbon isotope ratios in belowground carbon cycle processes. *Ecol. Appl.* **10**(2), 412–422.

Ekblad A. and Högberg P. (2001) Natural abundance C-13 in CO_2 respired from forest soils reveals speed of link between tree photosynthesis and root respiration. *Oecologia* **127**, 305–308.

Emrich K. and Vogel J. C. (1970) Carbon isotope fractionation during precipitation of calcium carbonate. *Earth Planet. Sci. Lett.* **8**(5), 363–371.

Enting I. G., Trudinger C. M., and Francey R. J. (1995) A synthesis inversion of the concentration and $\delta^{13}C$ of atmospheric CO_2. *Tellus Ser. B: Chem. Phys. Meteorol.* **47**(1–2), 35–52.

Epron D., Godard D., Cornic G., and Genty B. (1995) Limitation of net CO_2 assimilation rate by internal resistances to

CO_2 transfer in the leaves of 2 tree species (*Fagus SylvaticaL* and *Castanea-sativa* Mill). *Plant Cell Environ.* **18**(1), 43–51.

Epstein S. and Mayeda T. (1953) Variations of ^{18}O content of waters from natural sources. *Geochim. Cosmochim. Acta* **4**, 213–224.

Evans J. R., Sharkey T. D., Berry J. A., and Farquhar G. D. (1986) Carbon isotope discrimination measured concurrently with gas-exchange to investigate CO_2 diffusion in leaves of higher-plants. *Austral. J. Plant Physiol.* **13**(2), 281–292.

Falkowski P. G. (1991) Species variability in the fractionation of ^{13}C and ^{12}C by marine phytoplankton. *J. Plankton Res.* **13**, 21–28.

Farquhar G. D. (1983) On the nature of carbon isotope discrimination in C_4 species. *Austral. J. Plant Physiol.* **10**(2), 205–226.

Farquhar G. D. and Lloyd J. (1993) Carbon and oxygen isotope effects in the exchange of carbon dioxide between terrestrial plants and the atmosphere. In *Stable Isotopes and Plant Carbon—Water Relations* (eds. J. R. Ehleringer, A. E. Hall, and G. D. Farquhar). Academic Press, pp. 47–70.

Farquhar G. D., Olseary M. H., and Berry J. A. (1982) On the relationship between carbon isotope discrimination and the intercellular carbon dioxide concentration in leaves. *Austral. J. Plant Physiol.* **9**(2), 121–137.

Farquhar G. D., Lloyd J., Taylor J. A., Flanagan L. B., Syvertsen J. P., Hubick K. T., Wong S. C., and Ehleringer J. R. (1993) Vegetation effects on the isotope composition of oxygen in atmospheric CO_2. *Nature* **363** (6428), 439–443.

Feng X. H. (1999) Trends in intrinsic water-use efficiency of natural trees for the past 100–200 years: a response to atmospheric CO_2 concentration. *Geochim. Cosmochim. Acta* **63**(13/14), 1891–1903.

Flanagan L. B. and Varney G. T. (1995) Influence of vegetation and soil CO_2 exchange on the concentration and stable oxygen-isotope ratio of atmospheric CO_2 within a Pinus-resinosa canopy. *Oecologia* **101**(1), 37–44.

Flanagan L. B., Bain J. F., and Ehleringer J. R. (1991) Stable oxygen and hydrogen isotope composition of leaf water in C_3 and C_4 plant-species under field conditions. *Oecologia* **88**(3), 394–400.

Flanagan L. B., Marshall J. D., and Ehleringer J. R. (1993) Photosynthetic gas-exchange and the stable-isotope composition of leaf water—comparison of a xylem-tapping mistletoe and its host. *Plant Cell Environ.* **16**(6), 623–631.

Flanagan L. B., Brooks J. R., Varney G. T., Berry S. C., and Ehleringer J. R. (1996) Carbon isotope discrimination during photosynthesis and the isotope ratio of respired CO_2 in boreal forest ecosystems. *Global Biogeochem. Cycles* **10**(4), 629–640.

Flanagan L. B., Brooks J. R., Varney G. T., and Ehleringer J. R. (1997) Discrimination against $C^{18}O^{16}O$ during photosynthesis and the oxygen isotope ratio of respired CO_2 in Bboreal forest ecosystems. *Global Biogeochem. Cycles* **11**(1), 83–98.

Förstel H. (1978) The enrichment of ^{18}O in leaf water under natural conditions. *Radiat. Environ. Biophys.* **15**, 323–344.

Francey R. J. and Farquhar G. D. (1982) An explanation of $^{13}C/^{12}C$ variations in tree rings. *Nature* **297**(5861), 28–31.

Francey R. J. and Tans P. P. (1987) Latitudinal variation in oxygen-18 of atmospheric CO_2. *Nature* **327**(6122), 495–497.

Francey R. J., Tans P. P., Allison C. E., Enting I. G., White J. W. C., and Trolier M. (1995) Changes in oceanic and terrestrial carbon uptake since 1982. *Nature* **373**(6512), 326–330.

Francey R. J., Allison C. E., Etheridge D. M., Trudinger C. M., Enting I. G., Leuenberger M., Langenfelds R. L., Michel E., and Steele L. P. (1999) A 1000-year high precision record of $\delta^{13}C$ in atmospheric CO_2. *Tellus Ser. B: Chem. Phys. Meteorol.* **51**(2), 170–193.

Francey R. J., Rayner P. J., and Allison C. E. (2001) Constraining the global carbon budget from global to regional scales—the measurement challenge. In *Global Biogeochemical Cycles in the Climate System* (eds. E. D. Schulze, M. Heimann, S. Harrison, E. Holland, J. Lloyd, I. C. Prentice, and D. Schimel). Academic Press, pp. 245–252.

Friedli H. and Siegenthaler U. (1988) Influence of N_2O on isotope analyses in CO_2 and mass-spectrometric determination of N_2O in air samples. *Tellus* **40B**, 129–133.

Friedli H., Lötscher H., Oeschger H., Siegenthaler U., and Stauffer B. (1986) Ice core record of the $^{13}C/^{12}C$ ratio of atmospheric CO_2 in the past two centuries. *Nature* **324**, 237–238.

Friedli H., Siegenthaler U., Rauber D., and Oeschger H. (1987) Measurements of concentration, $^{13}C/^{12}C$ and $^{18}O/^{16}O$ ratios of tropospheric carbon dioxide over Switzerland. *Tellus* **39B**, 80–88.

Friedman I. and O'Neil J. R. (1977) Compilation of stable isotope fractionation factors of geological interest. In *Data of Geochemistry*, 6th. edn. Geological Survey Professional Paper 44-KK.

Fung I., Field C. B., Berry J. A., Thompson M. V., Randerson J. T., Malmstrom C. M., Vitousek P. M., Collatz G. J., Sellers P. J., Randall D. A., Denning A. S., Badeck F., and John J. (1997) Carbon-13 exchanges between the atmosphere and biosphere. *Global Biogeochem. Cycles* **11**(4), 507–533.

Gamo T., Tsutsumi M., Sakai H., Nakazawa T., Tanaka M., Honda H., and Kubo H. (1989) Carbon and oxygen isotopic ratios of carbon dioxide of a stratospheric profile over Japan. *Tellus* **41B**, 127–133.

Gat J. R. (1996) Oxygen and hydrogen isotopes in the hydrologic cycle. *Ann. Rev. Earth Planet. Sci.* **24**, 225–262.

Gat J. R. and Gonfiantini R. (1981) *Stable Isotope Hydrology: Deuterium and Oxygen-18 in the Water Cycle.* Technical Reports Series No. 210, IAEA.

Gemery P. A., Trolier M., and White J. W. C. (1996) Oxygen isotope exchange between carbon dioxide and water following atmospheric sampling using glass flasks. *J. Geophys. Res. Atmos.* **101**(D9), 14415–14420.

Ghashghaie J., Duranceau M., Badeck F. W., Cornic G., Adeline M. T., and Deleens E. (2001) $\delta^{13}C$ of CO_2 Respired in the dark in relation to $\delta^{13}C$ of leaf metabolites: comparison between *Nicotiana sylvestris* and *Helianthus annuus* under drought. *Plant Cell Environ.* **24**(5), 505–515.

Gildor H. and Follows M. J. (2002) Two-way interactions between ocean biota and climate mediated by biogeochemical cycles. *Israel J. Chem.* **42**, 15–27.

Gillon J. and Yakir D. (2001) Influence of carbonic anhydrase activity in terrestrial vegetation on the ^{18}O content of atmospheric CO_2. *Science* **291**(5513), 2584–2587.

Gillon J. S. and Yakir D. (2000a) Internal conductance to CO_2 diffusion and $C^{18}OO$ discrimination in C_3 leaves. *Plant Physiol.* **123**(1), 201–213.

Gillon J. S. and Yakir D. (2000b) Naturally low carbonic anhydrase activity in C_4 and C_3 plants limits discrimination against $C^{18}OO$ during photosynthesis. *Plant Cell Environ.* **23**(9), 903–915.

Gleixner G. and Schmidt H. L. (1997) Carbon isotope effects on the fructose-1, 6-bisphosphate aldolase reaction, origin for non-statistical ^{13}C distribution in carbohydrates. *J. Biol. Chem.* **272**(9), 5382–5387.

Gleixner G., Danier H. J., Werner R. A., and Schmidt H. L. (1993) Correlations between the ^{13}C content of primary and secondary plant-products in different cell compartments and that in decomposing basidiomycetes. *Plant Physiol.* **102**(4), 1287–1290.

Goericke R. and Fry B. (1994) Variations of marine plankton $\delta^{13}C$ with latitude, temperature, and dissolved CO_2 in the world ocean. *Global Biogeochem. Cycles* **8**(1), 85–90.

Gonfiantini R., Gratziu S., and Tongiorgi E. (1965) Oxygen isotope composition of water in leaves. In *Isotopes and Radiation in Soil-plant Nutrition Studies.* IAEA pp. 405–410.

Gruber N., Keeling C. D., Bacastow R. B., Guenther P. R., Lueker T. J., Wahlen M., Meijer H. A. J., Mook W. G., and Stocker T. F. (1999) Spatiotemporal patterns of carbon-13 in the global surface oceans and the oceanic Seuss effect. *Global Biogeochem. Cycles* **13**(2), 307–335.

Guy R. D., Fogel M. F., Berry J. A., and Hoering T. C. (1987) Isotope fractionation during oxygen production and consumption by plants. In *Progress in Photosynthetic Research III* (ed. J. Biggins), pp. 597–600.

Guy R. D., Fogel M. L., and Berry J. A. (1993) Photosynthetic fractionation of the stable isotopes of oxygen and carbon. *Plant Physiol.* **101**(1), 37–47.

Griffiths H. (1998) Stable isotopes. In *Integration of Biological, Ecological and Geochemical Processes.* Bios Scientific publishers, Oxford.

Hanson P. J., Edwards N. T., Garten C. T., and Andrews J. A. (2000) Separating root and soil microbial contributions to soil respiration: a review of methods and observations. *Biogeochemistry* **48**(1), 115–146.

Harwood K. G., Gillon J. S., Griffiths H., and Broadmeadow M. S. J. (1998) Diurnal variation of $\Delta^{13}CO_2$, $\Delta C^{18}O^{16}O$ and evaporative site enrichment of $\delta H_2{}^{18}O$ in *Piper aduncum* under field conditions in Trinidad. *Plant Cell Environ.* **21**(3), 269–283.

Harwood K. G., Gillon J. S., Roberts A., and Griffiths H. (1999) Determinants of isotopic coupling of CO_2 and water vapor within a *Quercus petraea* forest canopy. *Oecologia* **119**(1), 109–119.

Hayes J. M., Strauss H., and Kaufman A. J. (1999) The abundance of ^{13}C in marine organic matter and isotopic fractionation in the global biogeochemical cycle of carbon during the past 800 Ma. *Chem. Geol.* **161**(1–3), 103–125.

Heimann M. and Maier-Reimer E. (1996) On the relations between the oceanic uptake of CO_2 and its carbon isotopes. *Global Biogeochem. Cycles* **10**, 89–110.

Helliker B. R. and Ehleringer J. R. (2000) Establishing a grassland signature in veins: ^{18}O in the leaf water of C_3 and C_4 grasses. *In Proceedings of the National Academy of Sciences of the United States of America* **97**(14), 7894–7898.

Hemming D. L., Switsur V. R., Waterhouse J. S., and Heaton T. H. E. (1998) Climate variations and the stable carbon isotope composition of tree ring cellulose: an intercomparison of *quercus robur, fagus sylvatica* and *pinus silvestris. Tellus* **50B**, 25–33.

Hemming D. L., Fritts H., Leavitt S. W., Wright W., Long A., and Shashkin A. (2001) Modelling tree-ring $\delta^{13}C$. *Dendrochronologia* **19**(1), 23–38.

Henderson S. A., Voncaemmerer S., and Farquhar G. D. (1992) Short-term measurements of carbon isotope discrimination in several C_4 species. *Austral. J. Plant Physiol.* **19**(3), 263–285.

Hesterberg R. and Siegenthaler U. (1991) Production and stable isotopic composition of CO_2 in a soil near Bern, Switzerland. *Tellus Ser. B: Chem. Phys. Meteorol.* **43**(2), 197–205.

Hoefs J. (1987) *Stable Isotope Ceochemistry.* Springer.

Hogberg P., Nordgren A., Buchmann N., Taylor A. F. S., Ekblad A., Hogberg M. N., Nyberg G., Ottosson-Lofvenius M., and Read D. J. (2001) Large-scale forest girdling shows that current photosynthesis drives soil respiration. *Nature* **411**(6839), 789–792.

Holland E. (1984) *The Chemical Evolution of the Atmosphere and Oceans.* Princeton University Press.

Hsieh J. C. C., Savin S. M., Kelly E. F., and Chadwick O. A. (1998) Measurement of soil–water $\delta^{18}O$ values by direct equilibration with CO_2. *Geoderma* **82**(1–3), 255–268.

Inoue H. and Sugimura Y. (1985) Carbon isotopic fractionation during the CO_2 exchange process between air and seawater under equilibrium and kinetic conditions. *Geochim. Cosmochim. Acta* **49**(11), 2453–2460.

IPCC (2001) Climate change 2001. The Scientific Basis (eds. J. T. Houghton, Y. Ding, D. J. Griggs, M. Noguer, P. J. van der Linden, X. Dai, K. Maskell, and C. A. Johnson). Cambridge University Press.

Isaac M. (2001) *The Oxygen Isotopic Signal of CO_2 in Low Moisture Soils.* MSc, Weizmann Institute of Science.

Ishizawa M., Nakazawa T., and Higuchi K. (2002) A multi-box model study of the role of the biospheric metabolism in the recent decline of δ^8O in atmospheric CO_2. *Tellus Ser. B: Chem. Phys. Meteorol.* **54**(4), 307–324.

Ito A. (2003) A global-scale simulation of the CO_2 exchange between the atmosphere and the terrestrial biosphere with a mechanistic model including stable carbon isotopes, 1953–1999. *Tellus* **55B**, 596–612.

Jouzel J., Russell G. L., Suozzo R. J., Koster R. D., White J. W. C., and Broecker W. S. (1987) Simulations of the HDO and $H_2^{18}O$ atmospheric cycles using the Nasa Giss general-circulation model—the seasonal cycle for present-day conditions. *J. Geophys. Res. Atmos.* **92**(D12), 14739–14760.

Kalnay E., Kanamitsu M., Kistler R., Collins W., Deaven D., Gandin L., Iredell M., Saha S., White G., Woollen J., Zhu Y., Chelliah M., Ebisuzaki W., Higgins W., Janowiak J., Mo K. C., Ropelewski C., Wang J., Leetmaa A., Reynolds R., Jenne R., and Joseph D. (1996) The NCEP/NCAR 40-year reanalysis project. *Bull. Am. Meteorol. Soc.* **77**(3), 437–471.

Kaplan A. and Reinhold L. (1999) CO_2 concentrating mechanisms in photosynthetic microorganisms. *Ann. Rev. Plant Physiol. Plant Mol. Biol.* **50**, 539–570.

Kaplan J. O., Prentice I. C., and Buchmann N. (2002) The stable carbon isotope composition of the terrestrial biosphere: modelling at scales from the leaf to the globe. *Global Biogeochem. Cycles* **16**(4), 1060.

Keeling C. D. (1958) The concentration and isotopic abundance of carbon dioxide in rural areas. *Geochim. Cosmochim. Acta* **13**, 299–313.

Keeling C. D. (1961) The concentration and isotopic abundance of atmospheric carbon dioxide in rural and marine air. *Geochim. Cosmochim. Acta* **24**, 277–298.

Keeling C. D., Mook W. G., and Tans P. P. (1979) Recent trends in the ^{13}C–^{12}C ratio of atmospheric carbon-dioxide. *Nature* **277**(5692), 121–123.

Keeling C. D., Bacastow R. B., and Tans P. P. (1980) Predicted shift in the ^{13}C–^{12}C ratio of atmospheric carbon-dioxide. *Geophys. Res. Lett.* **7**(7), 505–508.

Kesselmeier J. and Hubert A. (2002) Exchange of reduced volatile sulfur compounds between leaf litter and the atmosphere. *Atmos. Environ.* **36**(29), 4679–4686.

Kesselmeier J., Teusch N., and Kuhn U. (1999) Controlling variables for the uptake of atmospheric carbonyl sulfide by soil. *J. Geophys. Res. Atmos.* **104**(D9), 11577–11584.

Körtzinger A. and Quay P. (2003) Relationship between anthropogenic CO_2 and the ^{13}C Seuss effect in the North Atlantic Ocean. *Global Biogeochem. Cycles* **17**(1), 1005.

Kroopnic P. (1985) The distribution of ^{13}C on TCO_2 in the world ocean. *Deep Sea Res.* **32**, 57–84.

Kroopnic P. and Craig H. (1972) Atmospheric oxygen—isotopic composition and solubility fractionation. *Science* **175**(4017), 54–55.

Kroopnick P. (1980) The distribution of ^{13}C in the atlantic-ocean. *Earth Planet Sci. Lett.* **49**(2), 469–484.

Ladyman S. J. and Harkness D. D. (1980) Carbon isotope measurements as an index of soil development. *Radiocarbon* **22**, 885.

Leroux X., Bariac T., and Mariotti A. (1995) Spatial partitioning of the soil–water resource between grass and shrub components in a West-African humid Savanna. *Oecologia* **104**(2), 147–155.

Leuenberger M., Siegenthaler U., and Langway C. C. (1992) Carbon isotope composition of atmospheric CO_2 during the last ice-age from an Antarctic ice core. *Nature* **357**(6378), 488–490.

Lin G. H. and Ehleringer J. R. (1997) Carbon isotopic fractionation does not occur during dark respiration in C_3 and C_4. *Plant Physiol.* **114**(1), 391–394.

Lloyd J. and Farquhar G. D. (1994) ^{13}C Discrimination during CO_2 assimilation by the terrestrial biosphere. *Oecologia* **99** (3–4), 201–215.

Lloyd J., Syvertsen J. P., Kriedemann P. E., and Farquhar G. D. (1992) Low conductance's for CO_2 diffusion from stomata to the sites of carboxylation in leaves of woody species. *Plant Cell Environ.* **15**(8), 873–899.

Lloyd J., Kruijt B., Hollinger D. Y., Grace J., Francey R. J., Wong S. C., Kelliher F. M., Miranda A. C., Farquhar G. D., Gash J. H. C., Vygodskaya N. N., Wright I. R., Miranda H. S., and Schulze E. D. (1996) Vegetation effects on the isotopic composition of atmospheric CO_2 at local and regional scales: theoretical aspects and a comparison between rain forest in Amazonia and a Boreal forest in Siberia. *Austral. J. Plant Physiol.* **23**(3), 371–399.

Lloyd J., Francey R. J., Sogachev A., Mollicone D., Raupach M. R., Sogachev A., Arneth A., Byers J. N., Kelliher F. M., Rebmann C., Valentini R., Wong S. C., Bauer G., and Schulze E. D. (2001) Vertical profiles, boundary layer budgets, and regional flux estimates for CO_2 and its $^{13}C/^{12}C$ ratio and for water vapour above a forest/bog mosaic in central Siberia. *Global Biogeochem. Cycles* **15**, 267–284.

Loreto F., Harley P. C., Dimarco G., and Sharkey T. D. (1992) Estimation of mesophyll conductance to CO_2 flux by 3 different methods. *Plant Physiol.* **98**(4), 1437–1443.

Lynch-Stieglitz J., Stocker T. F., Broecker W. S., and Fairbanks R. G. (1995) The influence of air-sea exchange on the isotopic composition of oceanic carbon—observations and modelling. *Global Biogeochem. Cycles* **9**(4), 653–665.

Luz B. and Barkan E. (2000) Assessment of oceanic productivity with the triple-isotope composition of dissolved oxygen. *Science* **288**, 2028–2031.

Luz B., Barkan E., Bender M. L., Thiemens M. H., and Boering K. A. (1999) Triple-isotope composition of atmospheric oxygen as a tracer of biosphere productivity. *Nature* **400**, 547–550.

Marino B. D., McElroy M. B., Salawitch R. J., and Spaulding W. G. (1992) Glacial-to-interglacial variations in the carbon isotopic composition of atmospheric CO_2. *Nature* **357**(6378), 461–466.

Marlier J. F. and O'Leary M. H. (1984) Carbon kinetic isotope effects on the hydration of carbon-dioxide and the dehydration of bicarbonate ion. *J. Am. Chem. Soc.* **106**(18), 5054–5057.

Masarie K. A., Langenfelds R. L., Allison C. E., Conway T. J., Dlugokencky E. J., Francey R. J., Novelli P. C., Steele L. P., Tans P. P., Vaughn B., and White J. W. C. (2001) NOAA/CSIRO flask air intercomparison experiment: a strategy for directly assessing consistency among atmospheric measurements made by independent laboratories. *J. Geophys. Res. Atmos.* **106**(D17), 20445–20464.

Mathieu R. and Bariac T. (1996a) An isotopic study (^2H and ^{18}O) of water movements in clayey soils under a semi-arid climate. *Water Resour. Res.* **32**(4), 779–789.

Mathieu R. and Bariac T. (1996b) A numerical model for the simulation of stable isotope profiles in drying soils. *J. Geophys. Res. Atmos.* **101**(D7), 12685–12696.

Mathieu R., Pollard D., Cole J. E., White J. W. C., Webb R. S., and Thompson S. L. (2002) Simulation of stable water isotope variations by the GENESIS GCM for modern conditions. *J. Geophys. Res. Atmos.* **107**(D4)article no. 4037.

McConville C., Kalin R. M., and Flood D. (1999) Direct equilibration of soil water for $\delta^{18}O$ analysis and its application to tracer studies. *Rapid Commun. Mass Spectrom.* **13** (13), 1339–1345.

McKinney C. R., McCrea J. M., Epstein S., Allen H. A., and Urey H. C. (1950) Improvements in mass spectrometers for the measurement of small differences in isotope abundance ratios. *Rev. Sci. Instr.* **21**(8), 724–730.

Melzer E. and Schmidt H. L. (1987) Carbon isotope effects on the pyruvate dehydrogenase reaction and their importance for relative carbon-13 depletion in lipids. *J. Biol. Chem.* **262** (17), 8159–8164.

Miller J. B. and Tans P. P. (2003) Calculating isotopic discrimination from atmospheric measurements at various scales. *Tellus* **55B**, 207–214.

Miller J. B., Yakir D., White J. W. C., and Tans P. P. (1999) Measurement of $^{18}O/^{16}O$ in the soil-atmosphere CO_2 flux. *Global Biogeochem. Cycles* **13**(3), 761–774.

Miller J. B., Tans P. P., White J. W. C., Conway T. J., and Vaughn B. W. (2003) The atmospheric signal of terrestrial isotopic discrimination and its implication for carbon fluxes. *Tellus* **55B**, 197–206.

Mills G. A. and Urey H. C. (1940) The kinetics of isotopic exchange between carbon dioxide, bicarbonate ion, carbonate ion and water. *J. Am. Chem. Soc.* **62**, 1019–1028.

Miranda A. C., Miranda H. S., Lloyd J., Grace J., Francey R. J., McIntyre J. A., Meir P., Riggan P., Lockwood R., and Brass J. (1997) Fluxes of carbon, water, and energy over Brazilian Cerrado: an analysis using eddy covariance and stable isotopes. *Plant Cell Environ.* **20**(3), 315–328.

Mook W. G. (1986) ^{13}C in atmospheric CO_2. *Neth. J. Sea Res.* **20**(2–3), 211–223.

Mook W. G. (2000) *Environmental Isotopes in the Hydrological Cycle: Principles and Applications.* IAEA.

Mook W. G. and Grootes P. M. (1973) The measuring procedure and corrections for the high-precision mass-spectrometric analysis of isotopic abundance ratios, especially referring to carbon, oxygen and nitrogen. *Int. J. Mass Spectrom. Ion Phys.* **12**, 198–273.

Mook W. G. and Van der Hoek S. (1983) The N_2O correction in the carbon and oxygen isotopic analysis of atmospheric CO_2. *Isotopic Geosci.* **1**, 169–176.

Mook W. G., Bommerso Jc., and Staverma Wh. (1974) Carbon isotope fractionation between dissolved bicarbonate and gaseous carbon-dioxide. *Earth Planet. Sci. Lett.* **22**(2), 169–176.

Mook W. G., Koopmans M., Carter A. F., and Keeling C. D. (1983) Seasonal, latitudinal, and secular variations in the abundance and isotopic-ratios of atmospheric carbon-dioxide: 1. Results from land stations. *J. Geophys. Res.: Ocean. Atmos.* **88**(C15), 915–933.

Moreira M. Z. (1998) Contribution of vegetation to the water cycle of the Amazon basin: an isotopic study of plant transpiration and its water source. PhD Thesis, University of Miami.

Morimoto S., Nakazawa T., Higuchi K., and Aoki S. (2000) Latitudinal distribution of atmospheric CO_2 sources and sinks inferred by $\delta^{13}C$ measurements from 1985 to 1991. *J. Geophys. Res. Atmos.* **105**(D19), 24315–24326.

Nakazawa T., Sugawara S., Inoue G., Machida T., Makshyutov S., and Mukai H. (1997) Aircraft measurements of the concentrations of CO_2, CH_4, N_2O, and CO and the carbon and oxygen isotopic ratios of CO_2 in the troposphere over Russia. *J. Geophys. Res. Atmos.* **102**(D3), 3843–3859.

Nier A. O. (1947) A mass spectrometer for isotope and gas analysis. *Rev. Sci. Instr.* **19**(6), 398–411.

O'Leary M. H. (1981) Carbon isotope fractionation in plants. *Phytochemistry* **20**(4), 553–567.

O'Leary M. H. (1984) Measurement of the isotope fractionation associated with diffusion of carbon-dioxide in aqueous-solution. *J. Phys. Chem.* **88**(4), 823–825.

O'Leary M. H., Madhavan S., and Paneth P. (1992) Physical and chemical basis of carbon isotope fractionation in plants. *Plant Cell Environ.* **15**(9), 1099–1104.

Park R. and Epstein S. (1960) Carbon isotope fractionation during photosynthesis. *Geochim. Cosmochim. Acta* **44**, 5–15.

Pataki D. E., Ehleringer J. R., Flanagan L. B., Yakir D., Bowling D. R., Still C. J., Buchmann N., Kaplan J. O., and Berry J. A. (2003) The application and interpretation of keeling plots in terrestrial carbon cycle research. *Global Biogeochem. Cycles* **17**(1), 1022.

Peylin P., Ciais P., Tans P. P., Six K., Berry J. A., and Denning A. S. (1996) ^{18}O in atmospheric CO_2 simulated by a 3-d transport model: a sensitivity study to vegetation and soil fractionation factors. *Phys. Chem. Earth* **21**(5–6), 463–469.

Peylin P., Ciais P., Denning A. S., Tans P. P., Berry J. A., and White J. W. C. (1999) A 3-dimensional study of $\delta^{18}O$ in atmospheric CO_2: contribution of different land ecosystems. *Tellus Ser. B: Chem. Phys. Meteorol.* **51**(3), 642–667.

Potter C. S., Randerson J. T., Field C. B., Matson P. A., Vitousek P. M., Mooney H. A., and Klooster S. A. (1993) Terrestrial ecosystem production—a process model-based on global satellite and surface data. *Global Biogeochem. Cycles* **7**(4), 811–841.

Quay P., King S., Wilbur D., Wofsy S., and Richey J. (1989) $^{13}C/^{12}C$ of atmospheric CO_2 in the Amazon basin—forest and river sources. *J. Geophys. Res. Atmos.* **94**(D15), 18327–18336.

Quay P. D., Tilbrook B., and Wong C. S. (1992) Oceanic uptake of fossil-fuel CO_2—^{13}C evidence. *Science* **256** (5053), 74–79.

Quay P., Sonnerup R., Westby T., Stutsman J., and McNichol A. (2003) Changes in the $^{13}C/^{12}C$ of dissolved inorganic carbon in the ocean as a tracer of anthropogenic CO_2 uptake. *Global Biogeochem. Cycles* **17**(1), 1004.

Raich J. W. and Potter C. S. (1995) Global patterns of carbon-dioxide emissions from soils. *Global Biogeochem. Cycles* **9** (1), 23–36.

Raich J. W. and Schlesinger W. H. (1992) The global carbon-dioxide flux in soil respiration and its relationship to vegetation and climate. *Tellus Ser. B: Chem. Phys. Meteorol.* **44** (2), 81–99.

Ramonet M., Ciais P., Nepomniachii I., Sidrov K., Neubert R. E. M., Lanendöfer U., Picard D., Kazan V., Biaud S., Gust M., Kolle O., and Schulze E.-D. (2002) Three years of aircraft-based trace gas measurements over the Fyodorovskye southern taiga forest, 300 km north-west of Moscow. *Tellus* **54B**, 713–734.

Randerson J. T., Thompson M. V., Malmstrom C. M., Field C. B., and Fung I. Y. (1996) Substrate limitations for heterotrophs: implications for models that estimate the seasonal cycle of atmospheric CO_2. *Global Biogeochem. Cycles* **10**(4), 585–602.

Randerson J. T., Still C. J., Balle J. J., Fung I. Y., Doney S. C., Tans P. P., Conway T. J., White J. W. C., Vaughn B., Suits N., and Denning A. S. (2002) Carbon isotope discrimination of arctic and boreal biomes inferred from remote atmospheric measurements and a biosphere-atmosphere model. *Global Biogeochem. Cycles* **16**(3), 1028.

Rau G. H., Takahashi T., and Marais D. J. D. (1989) Latitudinal variations in plankton $\delta^{13}C$—implications for CO_2 and productivity in past oceans. *Nature* **341**(6242), 516–518.

Raupach M. (2001) Inferring biogeochemical sources and sinks from atmospheric concentrations: general consideration and applications in vegetation canopies. In *Global Biogeochemical Cycles in the Climate System* (eds. E. D. Schulze, M. Heimann, S. Harrison, E. Holland, J. Lloyd, I. C. Prentice, and D. Schimel). Academic Press, pp. 41–60.

Raven J. A. (1992) Present and potential uses of the natural abundance of stable isotopes in plant science, with illustrations from the marine environment: commissioned review. *Plant Cell Environ.* **15**(9), 1083–1090.

Raven J. A. (1998) Phylogeny, palaeoatmospheres and the evolution of phototrophy. In *Stable Isotopes: Integration of Biological, Ecological, and Geochemical Processes* (ed. H. Griffiths). BIOS Scientific Publishers, pp. 323–346.

Reed M. L. and Graham D. (1981) Carbonic anhydrase in plants: distribution and possible physiological roles. In *Progress in Phytochemistry* (eds. L. Reinhold, J. Harborn, and T. Swain). Pergamon, pp. 47–94.

Rinaldi G., Meinschein W. G., and Hayes J. M. (1974) Intramolecular carbon isotopic distribution in biologically produced acetoin. *Biomed. Mass Spectrom.* **1**, 415–417.

Rochette P. and Flanagan L. B. (1997) Quantifying rhizosphere respiration in a corn crop under field conditions. *Soil Sci. Soc. Am. J.* **61**(2), 466–474.

Rochette P., Flanagan L. B., and Gregorich E. G. (1999) Separating soil respiration into plant and soil components using analyses of the natural abundance of carbon-13. *Soil Sci. Soc. Am. J.* **63**(5), 1207–1213.

Roden J. S. and Ehleringer J. R. (1999) Observations of hydrogen and oxygen isotopes in leaf water confirm the Craig-Gordon model under wide ranging environmental conditions. *Plant Physiol.* **120**(4), 1165–1173.

Roeske C. A. and O'Leary M. H. (1984) Carbon isotope effects on the enzyme-catalyzed carboxylation of ribulose bisphosphate. *Biochemistry* **23**, 6275–6284.

Rooney M. A. (1988) Short-term carbon isotope fractionation by plants. PhD Thesis, University of Wisconsin, Madison.

Rozanski K., Araguasaraguas L., and Gonfiantini R. (1992) Relation between long-term trends of ^{18}O isotope composition of precipitation and climate. *Science* **258**(5084), 981–985.

Rozanski K., Araguás-Araguás L., and Gonfiantini R. (1993) Isotopic patterns in modern global precipitation. In *Climate Change in Continental Isotopic Records* (eds. P. K. Swart, K. C. Lohmann, J. McKenzie, and S. Savin). American Geophysical Union, pp. 1–36.

Rundel P. W., Ehleringer J. R., and Nagy K. A. (1988) *Stable Isotopes in Ecological Research. Ecological Studies 68,* Springer, NY.

Santrock J., Studley S. A., and Hayes J. M. (1985) Isotopic analyses based on the mass-spectrum of carbon-dioxide. *Analyt. Chem.* **57**(7), 1444–1448.

Sarmiento J. L. and Bender M. (1994) Carbon biogeochemistry and climate change. *Photosyn. Res.* **39**(3), 209–234.

Schidlowski M. (1988) A 3,800-million-year isotopic record of life from carbon in sedimentary rocks. *Nature* **333**(6171), 313–318.

Schimel D. S., Braswell B. H., Holland E. A., McKeown R., Ojima D. S., Painter T. H., Parton W. J., and Townsend A. R. (1994) Climatic, edaphic, and biotic controls over storage and turnover of carbon in soils. *Global Biogeochem. Cycles* **8**(3), 279–293.

Schimel D. S., Braswell B. H., McKeown R., Ojima D. S., Parton W. J., and Pulliam W. (1996) Climate and nitrogen controls on the geography and timescales of terrestrial biogeochemical cycling. *Global Biogeochem. Cycles* **10**(4), 677–692.

Schwartz D., deForesta H., Mariotti A., Balesdent J., Massimba J. P., and Girardin C. (1996) Present dynamics of the Savanna-forest boundary in the Congolese mayombe: a pedological, botanical and isotopic (^{13}C and ^{14}C) study. *Oecologia* **106**(4), 516–524.

Scrimgeour C. M. (1995) Measurement of plant and soil–water isotope composition by direct equilibration methods. *J. Hydrol.* **172**(1–4), 261–274.

Sharkey T. D. and Berry J. A. (1985) Carbon isotope fractionation of algal as influenced by an inducible CO_2 concentrating mechanism. In *Inorganic Carbon Uptake by Aquatic Photosynthetic Organisms* (eds. W. J. Lucas and J. A. Berry). American Society of Plant Physiologists, pp. 389–401.

Siegenthaler U. and Münich K. O. (1981) $^{13}C/^{12}C$ fractionation during CO_2 transfer from air to sea. In *Carbon Cycle Modelling,* SCOPE 16 (ed. B. Bolin). Wiley, pp. 249–257.

Silverman D. N. (1982) Carbonic-anhydrase—^{18}O exchange catalyzed by an enzyme with rate contributing proton transfer steps. *Meth. Enzymol.* **87**, 732–752.

Skirrow G. (1975) The dissolved gases—carbon dioxide. In *In* Chemical Oceanography (eds. J. P. Riley and G. Skirrow). Academic Press, , vol. 2, pp. 1–192.

Sposito G., Skipper N. T., Sutton R., Park S. H., Soper A. K., and Greathouse J. A. (1999) Surface geochemistry of the clay minerals. *Proceedings of the National Academy of Sciences of the United States of America* **96**(7), 3358–3364.

Stern L., Baisden W. T., and Amundson R. (1999) Processes controlling the oxygen isotope ratio of soil CO_2: analytic and numerical modelling. *Geochim. Cosmochim. Acta* **63**(6), 799–814.

Stern L. A., Amundson R., and Baisden W. T. (2001) Influence of soils on oxygen isotope ratio of atmospheric CO_2. *Global Biogeochem. Cycles* **15**(3), 753–759.

Sternberg L. (1989) A model to estimate carbon-dioxide recycling in forests using $^{13}C/^{12}C$ ratios and concentrations of ambient carbon-dioxide. *Agri. Forest Meteorol.* **48**(1–2), 163–173.

Sternberg L. D. L. (1997) Interpretation of recycling indexes—comment. *Austral. J. Plant Physiol.* **24**(3), 395–398.

Sternberg L. D., Moreira M. Z., Martinelli L. A., Victoria R. L., Barbosa E. M., Bonates L. C. M., and Nepstad D. (1998) The relationship between $^{18}O/^{16}O$ and $^{13}C/^{12}C$ ratios of ambient CO_2 in two Amazonian tropical forests. *Tellus Ser. B: Chem. Phys. Meteorol.* **50**(4), 366–376.

Still C. J., Berry J. A., Collatz G. J., and DeFries R. S. (2003) Global distribution of C_3 and C_4 vegetation: carbon cycle implications. *Global Biogeochem. Cycles* **17**(1)10.1029/2001GB001807.

Styles J. M., Lloyd J., Zolotoukhine D., Lanbon K. A., Tchebakova N., Francey R. J., Arneth A., Salamakho D., Kolle O., and Schulze E. D. (2002a) Estimates of regional surface carbon dioxide exchange and carbon and oxygen isotope discrimination during photosynthesis from concentration profiles in the atmospheric boundary layer. *Tellus* **54B**, 768–783.

Tans P. P. (1980) On calculating the transfer of ^{13}C in reservoir models of the carbon-cycle. *Tellus* **32**(5), 464–469.

Tans P. P. (1981) $^{13}C/^{12}C$ of industrial CO_2. In *Carbon Cycle Modelling,* SCOPE 16 (ed. B. Bolin). Wiley, pp. 127–129.

Tans P. P. (1998) Oxygen isotopic equilibrium between carbon dioxide and water in soils. *Tellus Ser. B: Chem. Phys. Meteorol.* **50**(2), 163–178.

Tans P. P. and White J. W. C. (1998) The global carbon cycle—in balance, with a little help from the plants. *Science* **281**(5374), 183–184.

Tans P. P., Fung I. Y., and Takahashi T. (1990) Observational constraints on the global atmospheric CO_2 budget. *Science* **247**(4949), 1431–1438.

Tans P. P., Berry J. A., and Keeling R. F. (1993) Oceanic $^{13}C/^{12}C$ observations: a new window on ocean CO_2 uptake. *Global Biogeochem. Cycles* **7**(2), 353–368.

Tcherkez G., Nogues S., Bleton J., Cornic G., Badeck F., and Ghashghaie J. (2003) Metabolic origin of carbon isotope composition of leaf dark-respired CO_2 in French bean. *Plant Physiol.* **131**(1), 237–244.

Thiemens M. H. (1999) Atmosphere science—mass-independent isotope effects in planetary atmospheres and the early solar system. *Science* **283**(5400), 341–345.

Thiemens M. H. and Jackson T. (1990) Pressure dependency for heavy isotope enhancement in ozone formation. *Geophys. Res. Lett.* **17**(6), 717–719.

Thiemens M. H., Jackson T., Mauersberger K., Schueler B., and Morton J. (1991) Oxygen isotope fractionation in stratospheric CO_2. *Geophys. Res. Lett.* **18**(4), 669–672.

Thoning K. W., Tans P. P., and Komyr W. D. (1989) Atmospheric carbon dioxide at Mauna Loa observatory: 2. Analysis of the NOAA GMCC data, 1974, 1985. *J. Geophys. Res.* **94**(8549–8565).

Trolier M., White J. W. C., Tans P. P., Masarie K. A., and Gemery P. A. (1996) Monitoring the isotopic composition of atmospheric CO_2: measurements from the NOAA global air sampling network. *J. Geophys. Res. Atmos.* **101**(D20), 25897–25916.

Trumbore S. (2000) Age of soil organic matter and soil respiration: radiocarbon constraints on belowground C dynamics. *Ecol. Appl.* **10**(2), 399–411.

Trumbore S. E., Davidson E. A., Decamargo P. B., Nepstad D. C., and Martinelli L. A. (1995) Belowground cycling of carbon in forests and pastures of eastern Amazonia. *Global Biogeochem. Cycles* **9**(4), 515–528.

Urey H. C. (1947) The Thermodynamic properties of isotopic substances. *J. Chem. Soc.* 562.

Urey H. C. (1948) Oxygen isotopes in nature and in the laboratory. *Science* **108**, 489–496.

Vogel J. C. (1980) Fractionation of the carbon isotopes during photosynthesis. In *Sitzungsberichte der Heidelberger Akademie der Wissenschaften, Mathematisch-Naturwissenschaftliche Klasse Jahrgang 1980. 3.* Abhandlung Springer, pp. 111–135.

Vogel J. C. (1993) Variability of carbon isotope fractionation during photosynthesis. In *Stable Isotopes and Plant Carbon: Water Relations* (eds. J. R. Ehleringer, A. E. Hall, and G. D. Farquhar). Academic Press, pp. 29–46.

Von Caemmerer S. and Evans J. R. (1991) Determination of the average partial–pressure of CO_2 in chloroplasts from leaves of several C_3 plants. *Austral. J. Plant Physiol.* **18**(3), 287–305.

Wang X. F. and Yakir D. (1995) Temporal and spatial variations in the oxygen-18 content of leaf water in different plant species. *Plant Cell Environ.* **18**(12), 1377–1385.

Wang X. F., Yakir D., and Avishai M. (1998) Non-climatic variations in the oxygen isotopic compositions of plants. *Global Change Biol.* **4**(8), 835–849.

Wen J. and Thiemens M. H. (1993) Multi-isotope study of the $O(^1D) + CO_2$ exchange and stratospheric consequences. *J. Geophys. Res. Atmos.* **98**(D7), 12801–12808.

White J. W. C. (1988) Stable hydrogen isotope ratio in plants: a review of current theory and some potential applications. In *Stable Isotopes in Ecological Research* (eds. P. W. Rundel, J. R. Ehleringer, and K. A. Nagi). Springer, pp. 142–162.

Wittenberg U. and Esser G. (1997) Evaluation of the isotopic disequilibrium in the terrestrial biosphere by a global carbon isotope model. *Tellus Ser. B: Chem. Phys. Meteorol.* **49**(3), 263–269.

Wofsy S. C., Goulden M. L., Munger J. W., Fan S. M., Bakwin P. S., Daube B. C., Bassow S. L., and Bazzaz F. A. (1993) Net exchange of CO_2 in a midlatitude forest. *Science* **260**(5112), 1314–1317.

Yakir D. (1992) Variations in the natural abundance of oxygen-18 and deuterium in plant carbohydrates: commissioned review. *Plant Cell Environ.* **15**(9), 1005–1020.

Yakir D. (1998) Oxygen-18 of leaf water: a crossroad for plant-associated isotopic signals. In *Stable Isotopes: Integration of Biological, Ecological and Geochemical Processes* (ed. H. Griffiths). BIOS Scientific Publishers Limited, pp. 147–183.

Yakir D. (2002) Global enzymes: sphere of influence. *Nature* **416**, 795.

Yakir D. and Wang X. F. (1996) Fluxes of CO_2 and water between terrestrial vegetation and the atmosphere estimated from isotope measurements. *Nature* **380**(6574), 515–517.

Yung Y. L., Demore W. B., and Pinto J. P. (1991) Isotopic exchange between carbon-dioxide and ozone via $O(^1D)$ in the stratosphere. *Geophys. Res. Lett.* **18**(1), 13–16.

Yurtsever Y. and Gat J. R. (1981) Atmospheric waters. In *Stable Isotope Hydrology,* Tech. Rep. Ser. 210 (eds. J. R. Gat and R. Gonfiantini). IAEA, pp. 103–42.

Zhang J., Quay P. D., and Wilbur D. O. (1995) Carbon-isotope fractionation during gas-water exchange and dissolution of CO_2. *Geochim. Cosmochim. Acta* **59**(1), 107–114.

Isotope Geochemistry
ISBN: 978-0-08-096710-3

pp. 113–150

6

Water Stable Isotopes: Atmospheric Composition and Applications in Polar Ice Core Studies

J. Jouzel

Institut Pierre Simon Laplace, Saclay, France

NOMENCLATURE

$Corr_{ocean}$	correction due to change in oceanic δ
d	deuterium excess defined as $d = \delta D - 8\delta^{18}O$
D	deuterium (heavy hydrogen, ^2H) when used in the formula of water: HDO
D	diffusion coefficient of water in air (formulas (3) and (4))
D_i	diffusion coefficient of water isotopes in air
F	ratio of the remaining to the initial amount of water vapor in an air mass
h	relative humidity
H	hydrogen
k	kinetic fractionation parameter
n_L	number of moles of liquid
n'_L	number of moles of liquid for isotopic species
n_p	number of moles of precipitation
n'_p	number of moles of precipitation for isotopic species
n_v	number of moles of water vapor
n'_v	number of moles of water vapor for isotopic species

O	oxygen
R	atomic ratios (D/H and $^{18}O/^{16}O$, respectively)
R_{slopes}	S_{spat}/S_{temp}
S_i	saturation ratio with respect to ice
S_{spat}	spatial δ/T slope
S_{temp}	temporal δ/T slope
t	time
t_a	isotopic relaxation time (drop or droplet)
T_c	condensation temperature
T_s	surface temperature site
$T_{s(inv)}$	surface temperature site determined with the inversion of simple isotopic model
$T_{s(spat)}$	surface temperature site determined with S_{spat}
T_w	moisture source temperature
V-SMOW	Vienna Standard Mean Ocean Water with D/H and $^{18}O/^{16}O$ atomic ratios of T_s
α	isotopic fractionation coefficient
α_c	isotopic fractionation coefficient when condensation occurs
α_k	kinetic fractionation coefficient
α_L	isotopic fractionation coefficient between liquid and vapor
α_s	isotopic fractionation coefficient between solid and vapor
δ	unit of isotopic ratio defined as $(R_{sample} - R_{SMOW})/R_{SMOW}$ and expressed in per mil (defined for deuterium, δD, and oxygen-18, δ^{18}O)
δD, δ^{18}O	isotopic content in per mil with respect to the Standard Mean Ocean Water
δ_L	isotopic content of a liquid cloud (stands either for δD or for δ^{18}O)
δ_p	isotopic content of a precipitation (stands either for δD or for δ^{18}O)
δ_v	isotopic content of water vapor (stands either for δD or for δ^{18}O)
Δ	difference between two time periods
Δ_{age}	age difference between the ice and entrapped air bubbles

6.1 INTRODUCTION

Natural waters formed of ~99.7% of $H_2{}^{16}O$ are also constituted of other stable isotopic molecules, mainly $H_2{}^{18}O$ (~2‰), $H_2{}^{17}O$ (~0.5‰), and $HD^{16}O$ (~0.3‰), where H and D (deuterium) correspond to ^{1}H and ^{2}H, respectively. Owing to slight differences in physical properties of these molecules, essentially their saturation vapor pressure, and their molecular diffusivity in air, fractionation processes occur at each phase change of the water except sublimation and melting of compact ice. As a result, the distribution of these water isotopes varies both spatially and temporally in the atmosphere, in the precipitation, and, in turn, in the various reservoirs of the hydrosphere and of the cryosphere. These isotopic variations have applications in such fields as climatology and cloud physics. More importantly, they are at the origin of two now well-established disciplines: isotope hydrology and isotope paleoclimatology. The various aspects dealing with isotope hydrology are reviewed by Kendall (see Chapter 7). In this chapter, we focus on this field known as "isotope paleoclimatology." As the behavior of $H_2{}^{17}O$ in the atmospheric water is very similar to that of $H_2{}^{18}O$ (more abundant and easier to precisely determine), isotope paleoclimatology is only based on the changes in concentrations of HDO and $H_2{}^{18}O$. These concentrations are given with respect to a standard as $\delta=(R_{sample} - R_{SMOW})/R_{SMOW}$ and expressed in per mil δ units (δD and δ^{18}O, respectively). In this definition, R_{sample} and R_{SMOW} are the isotopic ratios of the sample and of the Vienna Standard Mean Ocean Water (V-SMOW) with D/H and $^{18}O/^{16}O$ atomic ratios of 155.76×10^{-6} and 2005.2×10^{-6}, respectively (Hageman *et al.*, 1970; Baerstchi, 1976; Gonfiantini, 1978).

The use of water stable isotopes in paleoclimatology is based on the fact that their present-day distribution in precipitation is strongly related to climatological parameters. Of primary interest is the linear relationship between annual values of δD and δ^{18}O and mean annual temperature at the precipitation site, T_s, that is observed at middle and high latitudes (Figure 1). This relationship, which, as discussed in Section 6.3, is well explained by both simple and complex isotopic models, has given rise to the notion of "isotopic paleothermometer." In a conventional approach, the present-day spatial relationship between the isotopic concentration of the precipitation δ_p (where δ_p stands either for δD or for δ^{18}O of the precipitation, which can indifferently be used as paleothermometers) and T_s, defined over a certain region, is assumed to hold in time throughout this region. In this approach, it is assumed that the temporal slope, which applies to the isotope–temperature relationship through different climates over time at a single geographic location and should be used to interpret isotopic variations, observed at this site in terms of temperature changes, and the spatial slope ($S_{spat} = d\delta_p/dT_s$) are similar. A so-called "modern analogue method" is thus used, similar to that adopted in most other methods for reconstructing paleoclimates. Of course, the fact that present-day isotope concentrations and local temperatures are correlated is not sufficient to validate this critical assumption. Such factors as the evaporative origin and the seasonality of precipitation can also affect δD and δ^{18}O. If these factors change markedly under

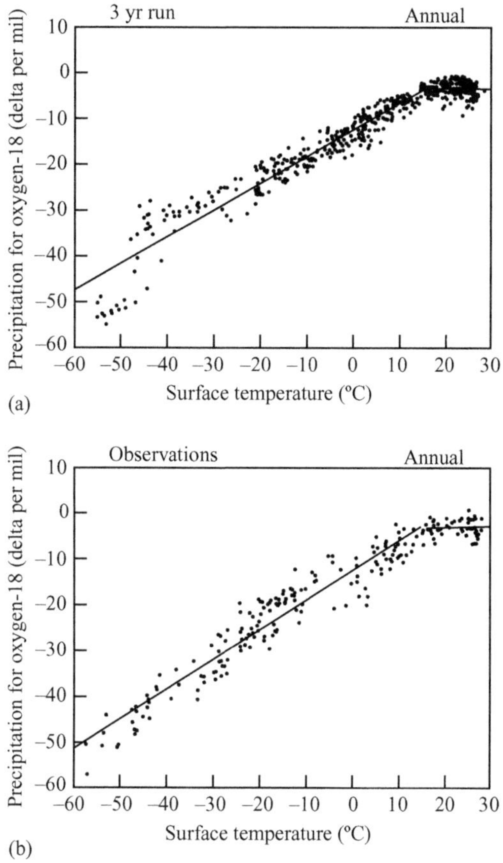

Figure 1 Annual $\delta^{18}O$ in precipitation versus annual surface temperature for: (a) 3 yr run and (b) observations as simulated by the NASA/GISS isotopic GCM (after Jouzel *et al.*, 1987a).

different climates, the spatial slope can no longer be taken as a reliable surrogate of the temporal slope for interpreting the isotopic signal. For example, there is now ample evidence that temporal slopes are considerably lower (by up to a factor of 2) than the observed present-day spatial slope, for Greenland sites.

Present-day δ_p distributions are characterized by two other interesting large-scale properties. First, there is no clear relationship between δ_p and the temperature of the site in tropical and equatorial regions. There, δ_p is more significantly influenced by precipitation amount. Second (see Figure 2), δD and $\delta^{18}O$ are linearly related to each other throughout the world with a slope of ~8 and a deuterium excess ($d = \delta D - 8\delta^{18}O$) of ~10‰ (Craig, 1961; Dansgaard, 1964). Although there has been a lot of potential interest in characterizing modifications in precipitation patterns such as those linked with changes in moonsonal activity or in the El Niño Southern Oscillation, the first property has, up to now, only been exploited in a very limited number of studies. In contrast, the fact that the deuterium excess of a precipitation is influenced by the conditions prevailing in the

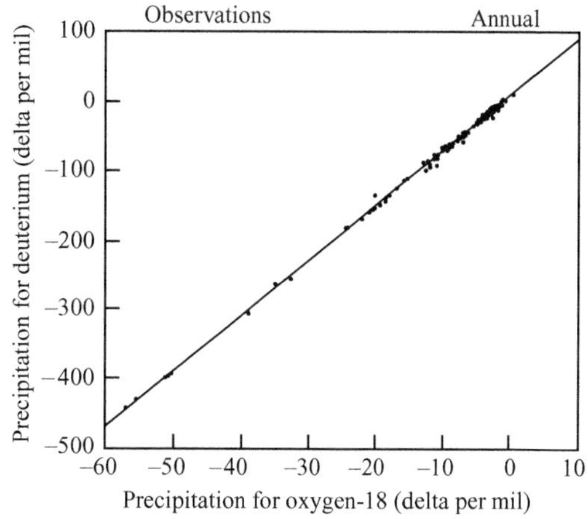

Figure 2 Annual δD in precipitation versus annual $\delta^{18}O$ in precipitation (after Jouzel *et al.*, 1987a).

oceanic moisture source region (temperature, relative humidity, and, to a lesser degree, wind speed) is now widely used to reconstruct the changes in the temperature of the evaporative source, T_w.

A large variety of isotope paleodata is available. Isotope signatures are measured directly in ice cores, groundwaters, and fluid inclusions in speleothems, and indirectly in precipitated calcite, tree ring cellulose, and other organic materials, particularly those in lake sediments. Polar ice cores are particularly suited for paleoclimate reconstructions. First, δ_p and T_s are strongly correlated in these regions, as illustrated in Figure 3 for Greenland and Antarctica. Second, they give direct access to the precipitation with little postdepositional change at least when the signal can be averaged over a certain number of years. Third, they provide continuous and potentially very detailed sequences with the longest record covering the last 420 ka (thousands of years) at the Vostok site in Antarctica (and possibly even older periods in the recently drilled EPICA core at the Dome C site also in Antarctica). Fourth, they allow measurements of both δD and $\delta^{18}O$ and thus are ideal to exploit the deuterium excess as an additional paleoindicator. In turn, although information from other archives and from tropical ice cores is now available, the focus of our chapter will be on the paleoclimatic interpretation of deep ice core isotopic profiles from Greenland and Antarctica.

This chapter is organized as follows. We first summarize what is known about present-day distribution of δD and $\delta^{18}O$ in atmospheric water vapor and in precipitation and how this distribution relates with climatic parameters (Section 6.2). The next section deals with the physics of isotopes, i.e., how differences in physical properties affect the isotopic concentration in the various water phases. In Section 6.4, we review the various models that allow us to describe isotopic distributions in water vapor and precipitation, and

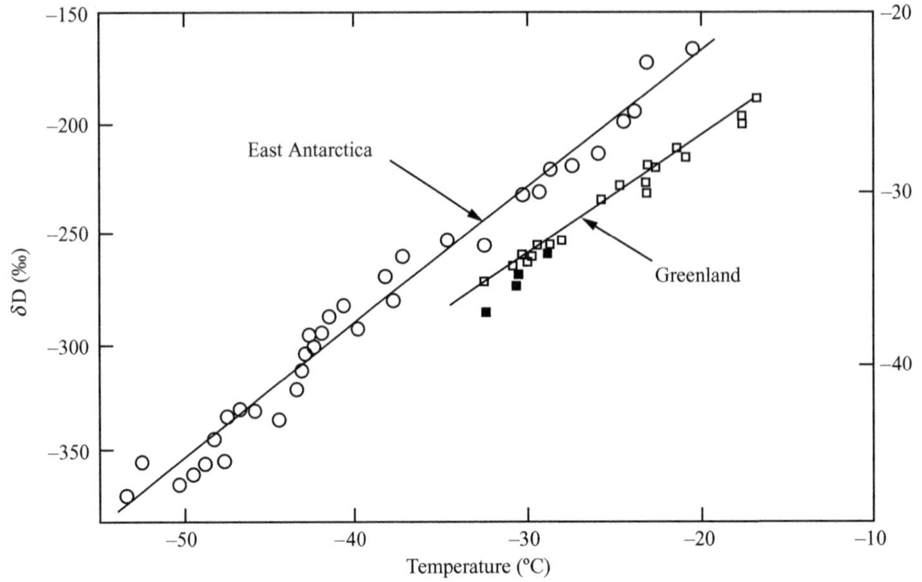

Figure 3 Isotope content of snow versus local temperature (annual average). Antarctic data (δD, left scale) are from Lorius and Merlivat (1977) and Greenland data ($\delta^{18}O$, right scale) are from Johnsen *et al.* (1989).

their ability to account for present-day observations. The next three sections are dedicated to isotopic paleodata and their paleoclimatic interpretation. First (Section 6.5), we review available isotope ice core data and then (Sections 6.6 and 6.7) the status of their interpretation in terms of paleotemperature change. In these two sections, we focus on Greenland and Antarctic deep ice cores, dedicating Section 6.6 to the conventional approach and a large part of Section 6.7 to the calibration of the "isotopic paleothermometer" through the comparison with other methods allowing one to estimate temperature changes in polar regions. The conventional approach underestimates temperature changes in Greenland, whereas available results tend to support its use for Antarctica. Our understanding of why the situation differs between both ice sheets is discussed in the light of model results and of the additional information that can be derived from the deuterium-excess parameter. Finally, we highlight some aspects of isotope modeling and the prospects of isotope climatology for the coming years.

6.2 PRESENT-DAY OBSERVATIONS

Precise measurements of the natural abundance of deuterium and ^{18}O in meteoric waters started shortly after World War II and the interest of such measurements for studying various aspects of the hydrological cycle was soon realized (Dansgaard, 1953; Friedman, 1953; Epstein and Mayeda, 1953). In 1961, a first attempt to summarize data was published by Craig (1961), who defined the Meteoric Water Line (MWL: $\delta D = 8\delta^{18}O + 10$) on which falls nonevaporated continental precipitation (Figure 2). The same year, IAEA and WMO initiated a worldwide survey of the isotope composition of monthly precipitation that has been in operation since then (Rozanski *et al.*, 1993). This survey has been the basis of several comprehensive studies in which the isotopic composition of global precipitation was discussed (Dansgaard, 1964; Friedman *et al.*, 1964; Craig and Gordon, 1965; Merlivat and Jouzel, 1979; Gat, 1980; Yurtserver and Gat, 1981; Rozanski *et al.*, 1992, 1993). The reader interested in a detailed description of available data should refer to the very comprehensive review of Rozanski *et al.* (1993). For our purpose, oriented towards a model/data comparison, we limit ourselves to describe the main characteristics of observed distributions. With this in mind, we successively examine the temporal and spatial patterns and then the relationship with meteorological parameters.

This presentation of data based on $\delta^{18}O$ concentrations would also apply for δD as these two parameters are very strongly correlated. Analysis of the long-term annual mean δD and $\delta^{18}O$ values of all IAEA/WMO network confirms that the MWL line defined by Craig (1961) is a good approximation of the locus of points representing average isotopic composition of freshwaters worldwide (Rozanski *et al.*, 1993). Beyond this general relationship, the analysis of both δD and $\delta^{18}O$ in a given precipitation, however, brings additional information about the processes leading to its formation (Section 6.8). The deuterium excess, defined by Dansgaard (1964) as $d = \delta D - 8\delta^{18}O$, is a very useful parameter to examine this information.

Studies of individual precipitation events revealed that stable isotope composition collected during single events may vary dramatically (see Rozanski *et al*., 1993, and references therein). However, examination of IAEA/WMO data shows that a clear temporal pattern (also seen in Greenland and Antarctic snow) emerges after averaging on a monthly basis. As illustrated in Figure 4 for a few

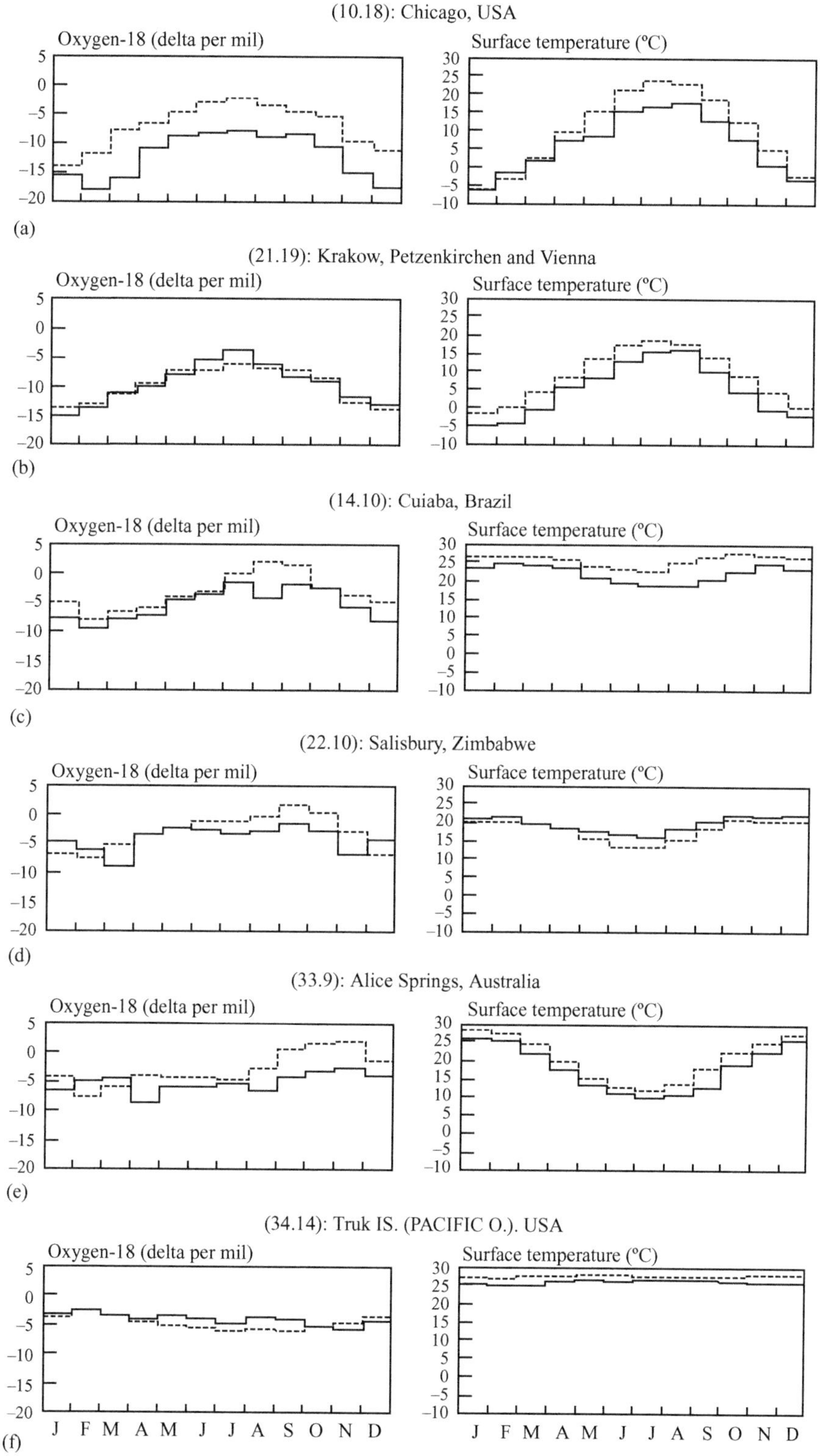

Figure 4 Seasonal cycles of $\delta^{18}O$ in precipitation for selected sites along with the surface temperature seasonal cycle. The continuous lines correspond to observations and the dotted lines to the seasonal cycles as simulated by the NASA/GISS isotopic GCM in the corresponding grid box (after Jouzel *et al*., 1987a).

selected sites, precipitation at continental mid- and high-latitude stations exhibits a seasonal cycle with that isotopically depleted in winter and enriched in summer, whereas there is a general absence of a defined seasonal cycle for island stations. These seasonal differences are due to several factors (Rozanski *et al.*, 1993): (i) seasonally changing temperature at mid- and high latitudes, with only minor fluctuations in the tropics (Figure 7 shows that a simple Rayleigh model predicts a decrease of the isotopic content of a precipitation when its temperature of formation decreases); (ii) seasonally modulated evapotranspiration flux over the continents induces seasonal differences in the atmospheric water balance; and (iii) seasonally changing source areas of the vapor and/or different storm trajectories. Over continental areas, there is a gradual enhancement of the seasonal variations with increasing distance from the coast with, for example (Rozanski *et al.*, 1993), an amplitude of the $\delta^{18}O$ signal of 2.5‰ at Valentia station in Ireland and of ~10‰ at Moskow, 3,200 km inland. This is largely due to an increase of the seasonal temperature range when going inland. Note also that the deuterium excess may vary seasonally in a given site which, in turn, indicates that the local $\delta D/\delta^{18}O$ slope differs from the value of 8 that applies to a global scale. For example, in Vienna *d* is higher in winter than in summer, which results in a slope lower than 8 (Rozanski *et al.*, 1993).

Figure 5 displays the annual average $\delta^{18}O$ of precipitation obtained from IAEA/WMO network observations and complementary data (Jouzel *et al.*, 1987a). There is a clear latitudinal pattern, with $\delta^{18}O$ decreasing as one approaches the poles. This latitudinal pattern is modulated by a continental effect, i.e., at a given latitude, $\delta^{18}O$ decreases when moving inland. Another aspect of these spatial variations, which is of practical significance for hydrological applications, is the altitude effect (in a given region $\delta^{18}O$ at higher altitudes generally will be more negative). The magnitude of the altitude effect depends on local climate and topography, with gradients in $\delta^{18}O$ of 0.15–0.50‰ per 100 m (Yurtserver and Gat, 1981).

There is a strong correlation between $\delta^{18}O$ and surface temperature fields between mid- and high latitudes (Figure 1), which does not hold true for tropical and equatorial regions. The $\delta^{18}O$ surface temperature plot of Figure 3 illustrates the strength of this correlation for Greenland and Antarctic sites, whereas Figure 1 was established using data available at IAEA/WMO stations and complementary data from polar sites (Jouzel *et al.*, 1987a). For temperatures below 15 °C, the $\delta^{18}O/T_s$ gradient is of 0.64‰/°C and the correlation coefficient is of 0.96. Another way to relate $\delta^{18}O$ to temperature is to obtain the $\delta^{18}O/T_s$ from monthly values at a single site. This seasonal slope is well defined over continental areas, but its value is generally lower from that calculated from spatial $\delta^{18}O/T_s$ gradient (Rozanski *et al.*, 1992). The relationship between mean annual values of temperature and precipitation $\delta^{18}O$ essentially vanishes above ~15 °C. In contrast,

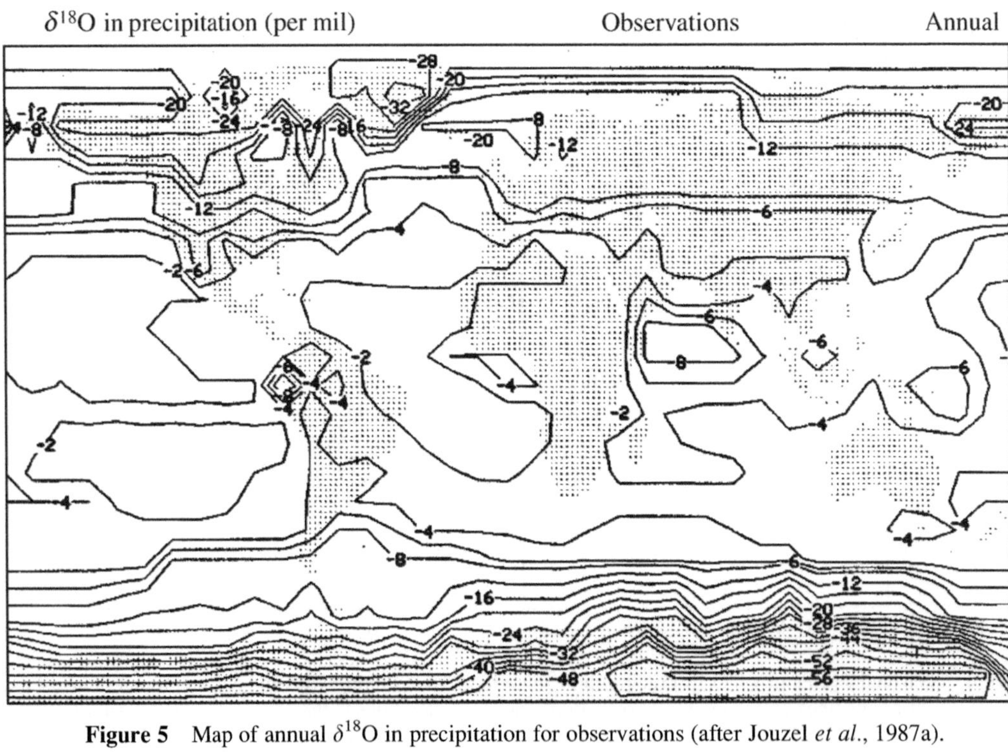

Figure 5 Map of annual $\delta^{18}O$ in precipitation for observations (after Jouzel *et al.*, 1987a).

for tropical and equatorial regions, there is, at least for oceanic areas, some relationship between the annual amount of precipitation and its $\delta^{18}O$, the so-called "amount effect" (Dansgaard, 1964), with rainy regions having low $\delta^{18}O$ and dry regions having high $\delta^{18}O$.

6.3 PHYSICS OF WATER ISOTOPES

6.3.1 Fractionation Processes

The saturation vapor pressures of HDO and $H_2{}^{18}O$ are lower than those of $H_2{}^{16}O$, both over liquid and solid phases. These differences play an important role in the course of the atmospheric water cycle as they cause fractionation effects at vapor/liquid and vapor/solid phase changes, with the condensed phase in equilibrium with vapor being enriched in heavy isotopes. The fractionation coefficient α is defined as the ratio of D/H or $^{18}O/^{16}O$ in the condensed phase to the value of the same parameter in the vapor phase. As reviewed in Jouzel (1986), many determinations of α have been made for the liquid/vapor fractionation at temperature greater than 0 °C. In contrast, there are only a few determinations for low temperatures, which are important in cloud physics, because supercooled drops and droplets commonly exist in natural clouds down to at least −15 °C (Rogers, 1979) and even below, and for the vapor/solid equilibrium. Fractionation coefficients measured by Merlivat and Nief (1967) and by Majoube (1971a,b), generally adopted in isotopic studies dealing with natural processes, are given in Table 1 for temperatures of 20 °C, 0 °C, and −20 °C. The $\alpha-1$ ratios reported in this table illustrate that the equilibrium isotopic effect is 8–10 times higher for HDO than for $H_2{}^{18}O$.

Differences in molecular diffusivities of water molecules in air, $D_{H_2{}^{16}O}$, D_{HDO}, and $D_{H_2{}^{18}O}$ give rise to what is known as the kinetic isotopic effect. Theoretical and experimental determinations have been obtained by several authors (Craig and Gordon, 1965; Ehhalt and Knott, 1965; Merlivat, 1978). Molecular diffusivity ratios D_i ($D_{HDO}/D_{H_2{}^{16}O}$ and $D_{HDO}/D_{H_2{}^{18}O}$) are practically independent of temperature. They equal 0.9755 and 0.9723, respectively (Merlivat, 1978). The isotopic kinetic effect thus has almost the same value for HDO as for $H_2{}^{18}O$ (the ratio of $1-D_i$ is equal to 0.88). The difference between the relative importance of equilibrium and kinetic effects for HDO and $H_2{}^{18}O$ (8–10 compared to 0.88) is the basic reason for the different behavior between the two isotopes, when nonequilibrium processes take place.

The fractionation effects occurring between liquid and solid phases are linked to variations of other physical properties such as heat capacities and latent heat of fusion. The values generally adopted in studies where these processes are involved are 1.0192 and 1.003 (O'Neil, 1968) for HDO and $H_2{}^{18}O$, respectively.

6.3.2 Growth of Individual Elements

We now summarize how these slightly different physical properties affect the isotopic composition of water during the growth of individual elements (droplets, drops, ice crystals, hailstones, etc.). In a first stage, the liquid phase of clouds is composed of water droplets formed by condensation of water vapor on a nucleus by heterogeneous nucleation. After this initial stage, growth of droplets and drops is mainly due to collision and coalescence. Larger elements fall faster than smaller ones and capture those lying in their paths (Rogers, 1979). The equations describing how the isotopic concentration of droplets and drops evolves are easy to establish, as they are the same for three water molecules, but for the values of the saturation vapor pressure or of the molecular diffusivity in air (Friedman *et al.*, 1962; Jouzel *et al.*, 1975; Stewart, 1975). One interesting aspect is that the ratio of molecular diffusivities no longer plays a role in the absence of supersaturation ($S=1$), when the droplet or drop is neither losing nor gaining weight. The isotopic content of the liquid phase δ_L is

$$\delta_L = \alpha(1+\delta_v) - 1 + \{(1+\delta_0) - \alpha(1+\delta_v)\} \times \exp(-t/t_a) \quad (1)$$

where δ_0 is the isotopic concentration in the liquid at time $t=0$ and δ_v the isotopic concentration in

Table 1 $HDO/H_2{}^{16}O$ and $H_2{}^{18}O/H_2{}^{16}O$ fractionation coefficients as a function of temperature and of phase change along with the α–1 ratio.

Temperature	*Liquid/vapor equilibrium*			*Solid/vapor equilibrium*		
	α_D	$\alpha^{18}O$	$(\alpha_D-1)/(\alpha^{18}O-1)$	α_D	$\alpha^{18}O$	$(\alpha_D-1)/(\alpha^{18}O-1)$
+20	1.0850	1.0098	8.7			
0	1.1123	1.0117	9.6	1.1330	1.0152	8.8
−20	1.1492	1.0141	10.6	1.1744	1.0187	9.2

After Jouzel (1986).

the vapor. In this equation, t_a is the adjustment or relaxation time, which increases as the square of the radius and decreases with decreasing temperatures. Typical values of t_a are of a few seconds for droplets and of up to half an hour for large supercooled drops. Jouzel *et al.* (1975) have demonstrated that droplets with a radius of less than 30 μm can be considered at any time in isotopic equilibrium with the vapor phase:

$$\delta_L = \alpha(1 + \delta_v) - 1 \qquad (2)$$

The complete equation applicable when growth is due to vapor transfer, and collection of droplets (coaslescence) has been derived by Jouzel *et al.* (1975). It allows one to derive an equivalent isotopic relaxation time, t'_a similar to t_a, which accounts both for condensation and coaslescence. For a drop of 1 mm radius, t'_a can be up to 15 times lower than t_a. This indicates that the collection of droplets plays an important role in maintaining a drop relatively close to isotopic equilibrium during its evolution in a cloud. Under these conditions, the deuterium content of a single drop deviates by less than 10% from isotopic equilibrium during its ascent in typical hailcloud (Jouzel, 1986). As shown in Figure 6, this deviation can be considerably higher when a population of drops is considered (Federer *et al.*, 1982), owing to the fact that in that case a drop grows from smaller drops not in isotopic equilibrium. In summary, the isotopic equilibrium between the liquid and the vapor phase is only reached for a cloud composed of small droplets. The fact that saturation is very close 1 from means that kinetic effects are not involved.

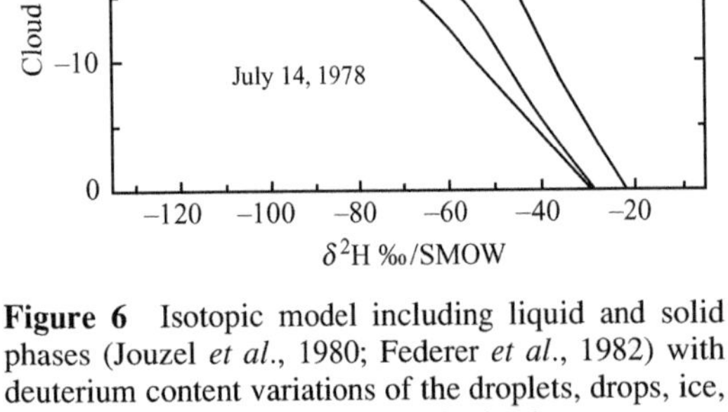

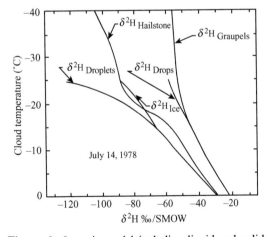

Figure 6 Isotopic model including liquid and solid phases (Jouzel *et al.*, 1980; Federer *et al.*, 1982) with deuterium content variations of the droplets, drops, ice, and graupels as a function of cloud temperature. $\delta^2H_{Hailstone}$ is the deuterium content of the water collected by hailstones.

Kinetic effects have to be taken into account when drops fall below cloud base in a subsaturated environment and they become important for precipitation falling through a dry atmosphere. This process has been the subject of both experimental and theoretical studies (Ehhalt *et al.*, 1963; Dansgaard, 1964; Ehhalt and Knott, 1965; Stewart, 1975). Evaporation results in an isotopic enrichment of the drop both for δD and $\delta^{18}O$. However, as the kinetic effect has relatively minor influence with respect to the equilibrium effect for δD, whereas the two effects have comparable sizes for $\delta^{18}O$, evaporation proceeds on a $\delta D/\delta^{18}O$ line with a slope <8, characteristic of the MWL. From isotopic growth equations (Stewart, 1975), it follows that a drop evaporating without collecting droplets evolves along on a slope close to 4. There is one interesting case where kinetic effects should be considered for a liquid phase inside a cloud. It concerns the growth of hail and hailstones in which it is possible to distinguish three types of deposits depending on thermodynamical conditions (Mason, 1971): (i) porous ice formed when the captured droplets freeze rapidly; (ii) compact ice formed when the droplets have time to spread over the surface and form a continuous film before freezing; and (iii) spongy ice produced when the heat exchange is not sufficiently rapid to allow all of the deposited water to freeze. In this latter case, only a fraction of the collected water freezes immediately and produces the skeletal framework of ice which retains the unfrozen water, whereby the whole mixture is maintained at 0 °C. This liquid phase, then much warmer than its environment, is subject to evaporation, and its δD and $\delta^{18}O$ concentrations are modified accordingly (again along a $\delta D/\delta^{18}O$ line close to 4 as first studied by Bailey *et al.* (1969) from analyses of samples of accreted ice formed in an icing tunnel). This effect can also be observed for natural hailstones (Jouzel and Merlivat, 1980; Jouzel *et al.*, 1985). Indeed, hailstones are natural samplers of isotopic conditions prevailing in large convective hailclouds. In turn, the isotopic distribution in hail gives unique insight in mechanisms of formation involved. Such hailstone studies, very interesting from the viewpoint of isotopes and isotopic models, were most active in the 1960s and 1970s (see Jouzel (1986) for a review).

The fact that kinetic effects have to be taken into account when ice crystals formed from water vapor deposition has long been overlooked. It was assumed that snow is formed at isotopic equilibrium (Dansgaard, 1964). However, considering snow is formed in a supersaturated environment with respect to the saturation vapor pressure over ice, and kinetic effects have to be considered in addition to the equilibrium fractionation with respect to ice (Jouzel and Merlivat, 1984; Fisher, 1991). Jouzel and Merlivat (1984) showed that the isotopic concentration of the solid phase, δ_s, can be simply expressed as

$$\delta_s = \alpha_s \alpha_k (1 + \delta_v) - 1 \quad (3)$$

In this equation, α_s is the equilibrium fractionation between solid and vapor, and

$$\alpha_k = \frac{S_i D_i \alpha_S}{D \alpha_S (S_i - 1) + D_i} \quad (4)$$

is the kinetic fractionation coefficient, with S_i being the supersaturation of water vapor with respect to ice. The validity of this formulation can be easily demonstrated from laboratory experiments consisting in measuring the condensate formed by water vapor deposited on a cold surface (Jouzel and Merlivat, 1984). In natural conditions α_k depends first on the supersaturation over ice (α_k is equal to 1 when the environment is just saturated with respect to ice, and decreases when supersaturation over ice increases). It depends, to a lesser degree, on pressure, temperature, and on the shape of the growing ice crystal.

Finally, some processes do not give rise to any isotopic fractionation. This is the case when supercooled drops or droplets freeze, because this freezing is instantaneous and there is no time for vapor exchange during the process. It is also the case during melting of compact ice because of the low diffusivity of water which ensures that there is no isotopic homogeneization in ice. In contrast, an isotopic fractionation is associated with the freezing of water when the process is sufficiently slow for isotopes to homogeneize, at least partially, in the liquid phase, as it is the case in many natural processes (Jouzel and Souchez, 1982; Jouzel *et al.*, 1999).

6.3.3 Isotopic Processes in Clouds

The focus in this review is on water isotopes in present-day precipitation and in paleoprecipitation. From an isotopic point of view, understanding and modeling isotopic processes as they occur in clouds, where precipitation is formed, is key to link the growth of individual elements to global-scale distributions. The complexity of cloud isotopic processes depends on the type of cloud considered.

The mechanism of cloud formation (Mason, 1971) is the cooling of moist air below its dewpoint, with the additional requirement for atmospheric clouds, that the air also contains aerosol particles which can serve as condensation nuclei. There are three types of clouds. The stratiform and cumuliform clouds formed, respectively, in stable and in convectively unstable atmospheres, whereas the cirriform clouds are the ice clouds which are, in general, higher and more tenuous, and less clearly reveal the kind of air motion which leads to their formation. As a result of the different rates of cooling and in aerosol content, liquid water contents range from less than 0.01 g m^{-3} to 10 g m^{-3}, with low values in stratiform clouds (within the narrow range 0.05–0.25 g m^{-3}) and higher values in cumulus clouds (up to 10 g m^{-3} in localized regions of large cumulus congestus). The ice content in cirriform clouds can be very low in cold regions such as central Antarctica.

Cumulus cloud development is conveniently described in terms of a parcel of air undergoing expansion while being lifted vertically. Its ultimate stage of development is the cumulonimbus, in which hail and hailstones are generated. Very schematically, the condensed phase of a cumulonimbus cloud can be divided into four types of elements: cloud droplets, large drops, ice crystals, and large ice particles (including hail and hailstones). There is mixing of the ascending parcel with the outside air, but in large cumulus clouds, there probably exists a central zone referred to as the updraft core which is unaffected by entrainment (Chilshom, 1973). It is more difficult to discuss stratiform clouds, as the point of origin and the subsequent path of an air parcel are less distinct than for cumulus. However, conceptually at least, stratus can be treated in the same way as cumulus. The same can be said for isotopic modeling: the microphysical processes which lead to fractionation effects, both in stratiform and cirriform clouds, are completely described in the more complete cumulus. Before briefly mentioning how the complexity of isotopic processes is taken into account in cumulus, we first describe the simplest case of a cloud where water vapor and water droplets coexist.

One often assumes that the condensed phase is immediately removed after its formation and that it leaves the air parcel in isotopic equilibrium with the vapor phase (Dansgaard, 1964; Friedman *et al.*, 1964; Taylor, 1972). An opposite approach is to consider the cloud as a closed system. Between these two extreme cases, a variable proportion of the condensed phase, assumed to be in isotopic equilibrium with the vapor, can be kept in the cloud (Craig and Gordon, 1965). To derive the equations for calculating the isotopic concentration in the vapor, δ_v, and in the precipitation δ_p, we use the mass conservation equation for water and water-isotopes (Equations (5) and (6)), and the assumption that the isotopic composition of the liquid kept in the cloud (Equation (7)) and of the precipitation leaving the cloud (Equation (8)) are in isotopic equilibrium with the vapor. Denoting by n_v, n_L, and n_p, the number of moles of vapor, water kept in the cloud, and precipitation, respectively, and n'_v, n'_L, and n'_p, the corresponding number of moles of water isotopes, we can write

$$dn_v + dn_L + dn_p = 0 \quad (5)$$

$$\mathrm{d}n_{\mathrm{v}}^{'} + \mathrm{d}n_{\mathrm{L}}^{'} + \mathrm{d}n_{\mathrm{p}}^{'} = 0 \quad (6)$$

$$n_{\mathrm{L}}^{'}/n_{\mathrm{L}} = \alpha_{\mathrm{L}} n_{\mathrm{v}}^{'}/n_{\mathrm{v}} \quad (7)$$

$$\mathrm{d}n_{\mathrm{p}}^{'}/\mathrm{d}n_{\mathrm{p}} = \alpha_{\mathrm{L}} n_{\mathrm{v}}^{'}/n_{\mathrm{v}} \quad (8)$$

where α_L is the fractionation coefficient between liquid and vapor. After logarithmic differentiation of Equation (7), the combination of these equations and of the one defining the δ values allows us to derive δ_v (see, e.g., Legrand *et al.* (1994)), and then δ_p applying Equation (2):

$$\frac{d\delta_{\mathrm{v}}}{1+\delta_{\mathrm{v}}} = \frac{(\alpha-1)\mathrm{d}n_{\mathrm{v}} + n_{\mathrm{L}}\mathrm{d}\alpha_{\mathrm{L}}}{n_{\mathrm{v}} + \alpha n_{\mathrm{L}}} \quad (9)$$

If no liquid is kept within the cloud (n_L=0), these equations correspond to the classical Rayleigh model, and to a closed system if all liquid is kept. The isotopic composition of the precipitation can be calculated all along the condensation process, if the amount of water leaving the cloud at each step is known. The largest range of δ-variation corresponds to the immediate removal of the condensate (Rayleigh process), and there is a decrease of this range by about a factor of 2, when all liquid formed is kept in the cloud (Jouzel, 1986).

A further complexity arises in cumulus clouds owing to the presence of drops and ice particles which are not in isotopic equilibrium due to the relative fall-speed of these elements with respect to water vapor and cloud droplets. Furthermore, at a given cloud level, the distance to isotopic equilibrium depends on the size and cloud history of these large elements (drops and ice particles). To treat this complexity, Jouzel *et al.* (1980) and Federer *et al.* (1982) used a one-dimensional steady-state cloud model that takes into account mixing with outside air, precipitation fallout, and interactions between water vapor, cloud droplets, large drops, and ice particles. It accounts both for equilibrium and kinetic effects that govern the composition of the droplets and ice crystals, respectively. It includes additional terms corresponding to the isotopic transfer between vapor and drops, calculated for a population of drops assuming a Marshall–Palmer drop distribution (Kessler, 1969), and to the mixing with outside air. This equation is very general and applicable to any cloud model provided that steady-state conditions exist. As an illustration, the variation of the deuterium content versus temperature of a hailstorm is given in Figure 6 with the curves relative to droplets, drops, cloud ice, graupels (large ice particles), and to the water collected by hailstones. As expected, large drops which are not in isotopic equilibrium with water vapor are enriched in heavy isotopes with respect to cloud droplets.

Gedzelman and Arnold (1994) built on this isotopic approach, but with a more realistic two-dimensional, non-steady-state, cloud model. The model was run for several idealized, classical stratiform and convective storm situations and the resulting isotope ratios of precipitation and water vapor estimated and compared to observations. The model reproduces many of the salient features of isotope meteorology when applied to snowstorms, stratiform rain, and convective precipitation. Also noticeable is the fact that isotope ratios are particularly low when the rain derives from a recirculation process in which air previously charged by vapor from falling rain subsequently rises. This provides a reasonable explanation for extraordinary low isotope ratios observed in some hurricanes and organized thunderstorms.

6.4 MODELING THE WATER ISOTOPE ATMOSPHERIC CYCLE

Water isotopes have been incorporated into a hierarchy of models, including dynamically simple Rayleigh-type models, and more complex models using atmospheric general circulation models (GCMs) or two-dimensional models. Through a short description of simple distillation models, we first examine the main climate parameters or processes that influence the global distribution of isotopes in precipitation. We then review the current state of development of isotopic GCMs and their performance in simulating present-day distribution of water stable isotopes in precipitation.

6.4.1 Rayleigh-type Models

A Rayleigh model (Dansgaard, 1964) considers the isotopic fractionation occurring in an isolated air parcel traveling from an oceanic source towards a polar region. The condensed phase is assumed to form in isotopic equilibrium with the surrounding vapor and to be removed immediately from the parcel (n_L=0 in Equation (9)):

$$\frac{\mathrm{d}\delta_{\mathrm{v}}}{1+\delta_{\mathrm{v}}} = (\alpha-1)\frac{\mathrm{d}n_{\mathrm{v}}}{n_{\mathrm{v}}} \quad (10)$$

This equation can be integrated under isothermal conditions (α which stands here for α_L being then constant). In this case, the isotope content of the precipitation is a unique function of the initial isotopic composition of the water, δ_0, and of the ratio, F, of the initial vapor mass within the air parcel and of the water-vapor mass remaining when the precipitation forms:

$$\delta_{\mathrm{p}} = \alpha\Big((1+\delta_0)F^{(\alpha-1)}\Big) - 1 \quad (11)$$

However, in nature, condensation processes are not isothermal and mainly occur by cooling of air masses. An exact integration of Equation (10) is no longer possible but δ_p is very well approximated by

$$\delta_p = \alpha_c\left((1+\delta_0)F^{(\alpha_m-1)}\right) - 1 \qquad (12)$$

in which α_m is the mean α_L value, and α_c its value when condensation occurs. The parcel's water vapor content is proportional to the saturation vapor pressure, a function of temperature and phase change, and is inversely proportional to the air pressure. Thus, in this simple model, the isotope content of precipitation can be expressed as a function of the initial water vapor isotopic concentration and of the initial and final condensation temperatures and air pressures.

Merlivat and Jouzel (1979) showed that the initial isotope concentration in an air parcel above the ocean, δ_{v0}, may, under certain simplifying assumptions, be expressed as a function of the fractionation coefficient α at sea surface temperature, T_w, of relative humidity, h, of a kinetic fractionation parameter k depending on wind speed (Merlivat, 1978), and of the isotopic composition of the sea surface, δ_{ocean}, as

$$\delta_{v0} = \frac{(1+\delta_{ocean})(1-k)}{\alpha(1-kh)} - 1 \qquad (13)$$

They extended the Rayleigh model by allowing some of the liquid condensate to remain in the parcel (Craig and Gordon, 1965) and by accounting for the kinetic fractionation associated with snow formation by inverse sublimation in a supersaturated environment (Jouzel and Merlivat, 1984). In the extended model, then, the isotope content of precipitation essentially depends on the evaporative source conditions, h and T_w, on the proportion of liquid kept in the cloud, and on the condensation temperature, T_c. Figure 7 illustrates the observed global-scale decrease of δ_p with T_c and thus with the surface temperature at the precipitation site, T_s, which varies with T_c over most of the globe (polar regions, particularly Antarctica, feature a strong temperature inversion and thus are an exception (not an exception to the assertion that T_c varies with T_s, just $T_c > T_s$). The straight line in Figure 7 representing Antarctic data (Dumont d'Urville–Dôme *c*-axis) is reported with respect to both T_s and the temperature at the inversion level, T_i, which is very close to T_c (Robin, 1977). The observed slope obtained in the latter case (1.12‰/°C) is very close to that predicted by the Rayleigh-type model assuming a source temperature of 20 °C (1.1–1.2‰/°C). Thus, given adequate values for its parameters, the Rayleigh-type model can reproduce the data observed over East Antarctica and, more generally, can reproduce the relationship between δ_p and local temperature for mid- and high latitudes.

Figure 7 also shows the influence of sea surface temperature, T_w, on δ_p. Merlivat and Jouzel

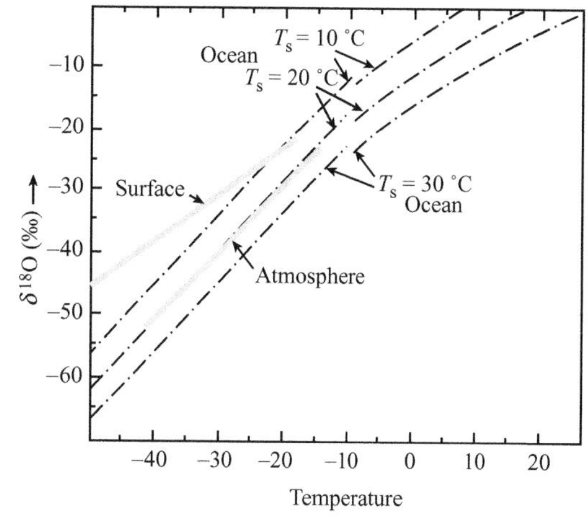

Figure 7 $\delta^{18}O$ in precipitation as calculated with a Rayleigh model. The three sets of curves are for different initial sea surface temperatures (T_w). The approach of Merlivat and Jouzel (1979) is used for the liquid phase, and that of Jouzel and Merlivat (1984) is used for snow formation. The solid lines correspond to East Antarctic data plotted with respect to either T_s (surface temperature) or T_i (temperature in the atmosphere at the inversion level).

(1979), Johnsen *et al.* (1989), and Petit *et al.* (1991) show that source conditions (temperature and humidity) also influence the relative amounts of HDO and $H_2{}^{18}O$ in the parcel and thus the deuterium excess in precipitation. The model reproduces deuterium-excess values observed in Greenland (Johnsen *et al.*, 1989) and in Antarctica, where d becomes higher than 15‰ in central regions (Petit *et al.*, 1991; Dahe *et al.*, 1994). These results as well as those concerning the isotope–temperature relationship were further confirmed by Ciais and Jouzel (1994), who introduced mixed clouds into the Rayleigh-type model, thereby allowing supercooled liquid droplets and ice crystals to coexist between ~−15 °C and −40 °C. Such coexistence could be important because the differing saturation conditions over water and ice allows liquid droplets to evaporate while water vapor condenses on ice crystals. Accounting for the associated isotopic fractionation processes in the mixed clouds, however, did not significantly modify the simulated δD and $\delta^{18}O$ of the condensed phase.

Jouzel and Koster (1996) later pointed out to a limitation of the formulation proposed by Merlivat and Jouzel (1979) to estimate the isotopic composition of water vapor above the ocean (Equation (13)). This equation is derived by assuming that the mean annual δ value of the vapor over a given oceanic region can be equated by the mean value of the water which evaporates over this region. This closure equation, which was made by Merlivat and Jouzel (1979) because they were addressing a global-scale problem, is however not valid on a regional scale, for which it has later been often used. The water vapor over a given region is indeed also influenced by distant sources of water vapor transported by the atmospheric circulation. To be correctly treated the problem thus needs to fully account for the transport of water isotopes in the atmosphere as described in the Section 6.4.2. Using such an isotopic GCM, Jouzel and Koster (1996) showed that Equation (13) is a correct approximation for global-scale problems such as the simulation of the relationship between δD and $\delta^{18}O$. They, however, recommended that the initial conditions for simple models be preferably taken from the GCM estimates of δ_{v0} over the ocean.

The same physical principles are utilized to develop isotopic models which better account for the transport of air masses at a regional scale, such as done by Fisher (1992) using a regional stable isotope model coupled to a zonally averaged global model. Other authors such as Eriksson (1965) and more recently Hendricks *et al.* (2000) considered the transport of water both by advective and eddy diffusive processes, the latter inducing less fractionation.

As discussed in Section 6.3.3, Rayleigh-type fractionation models cannot account for the complexity of large convective systems, such as those occurring in the tropics, for which δ_p depends on precipitation amount rather than temperature. Despite such limitations, they are able to reproduce the basic behavior of δD and $\delta^{18}O$ in precipitation, at least in mid- and high latitudes, where large convective systems do not dominate precipitation production.

6.4.2 Isotope Modeling with GCMs

Atmospheric GCMs simulate the time evolution of various atmospheric fields (wind speed, temperature, surface pressure, and specific humidity), discretized over the globe, through the integration of the basic physical equations: the hydrostatic equation of motion, the thermodynamic equation of state, the mass continuity equation, and a water vapor transport equation. To reproduce the observed regime of atmospheric circulation, these equations are supplemented with parametrizations for radiative transfer, surface fluxes of momentum, latent heat and sensible heat, latent heat release through condensation, and various internal processes that operate at scales not resolved by the relatively coarse mesh size of the model. These latter processes include turbulence in the boundary layer and cumulus convection, which drives convective precipitation and which redistributes momentum, heat, and water vapor over an atmospheric column. Parametrizations for nonconvective precipitation are also included, as are treatments of heat and water storage in land and ice reservoirs. A full discussion of general circulation modeling, of course, is beyond the scope of this chapter.

The incorporation of the HDO and $H_2{}^{18}O$ cycles into a GCM involves following the two isotopes through every stage of the GCMs water cycle. Simply put, the model transports the water isotopes between the atmospheric grid boxes and among the surface reservoirs with the same processes used to transport regular water. Isotopic fractionation, including both equilibrium and kinetic effects, is accounted for at every change of phase, i.e., during surface evaporation, atmospheric condensation, and re-evaporation of falling precipitation. The formulations implemented for isotopic fractionation distinguish between convective and nonconvective systems and are largely based on what is used in, or has been learnt from, the simple Rayleigh-type models described above. Although other parts of the hydrological cycle, such as surface hydrological processes and water vapor transport, do not involve fractionation, they still must be extended to include water isotopes. Indeed, a realistic transport scheme for advecting water vapor and isotopes between grid boxes is absolutely critical to the reproduction of observed isotope fields (Joussaume *et al.*, 1984; Jouzel *et al.*, 1991). In particular, the occurrence of negative water mass, which is not a serious problem for some GCMs, is catastrophic for isotope modeling.

Joussaume *et al.* (1984) pioneered GCM isotope modeling, producing global fields of δD and $\delta^{18}O$ for present-day January climate using a low-resolution version (32 points in longitude, 24 points in latitude) of the LMD GCM (Laboratoire de Météorologie Dynamique, Paris). Jouzel *et al.* (1987a) generated a full annual cycle of isotope fields with the $8° \times 10°$ (36 points in longitude, 24 points in latitude) GISS GCM (NASA Goddard Institute for Space Studies, New York) and examined the robustness of the approach through an extensive sensitivity study (Jouzel *et al.*, 1991). Simulations using finer spatial resolutions have also been performed for February and August with the LMD model (Joussaume and Jouzel, 1993) and for the full annual cycle with the GISS model (Charles *et al.*, 1994). Water isotopes have then been incorporated into a third model, the ECHAM GCM (Hoffmann and Heimann, 1993), which is the Hamburg version of the European Centre for Medium-Range Weather Forecast GCM. Both coarse grid and fine grid versions of the model have been used. Simulations for present-day climate have been produced using the ECHAM 3 version at two resolutions (T42, T21) corresponding to an LMD, ECHAM, and GISS models, i.e., LMD 4 (Andersen, 1997), ECHAM 4 (Werner *et al.*, 2000; Werner and Heimann, 2002), and the $4° \times 5°$ version of the GISS model

(Cole *et al.*, 1999; Charles *et al.*, 2001). Two other isotopic GCMs, based, respectively, on the GENESIS NCAR GCM (Mathieu *et al.*, 2002) and on the CSIRO GCM (Noone and Simmonds, 2002a,b), have been implemented, whereas new isotopic versions of the GISS and LMD GCMs are under development.

The first aim of these simulations is to determine how well isotope GCMs can reproduce observed present-day isotope distributions. The comparison with present-day data is indeed quite satisfying as illustrated from the detailed model/data comparison discussed in Hoffmann *et al.* (2000) for the GISS and ECHAM isotopic models. Figure 8 shows the annual mean $\delta^{18}O$ for the two models. The isotopic composition in the mid- and high latitudes is dominated by the "temperature effect," generating precipitation isotopically more

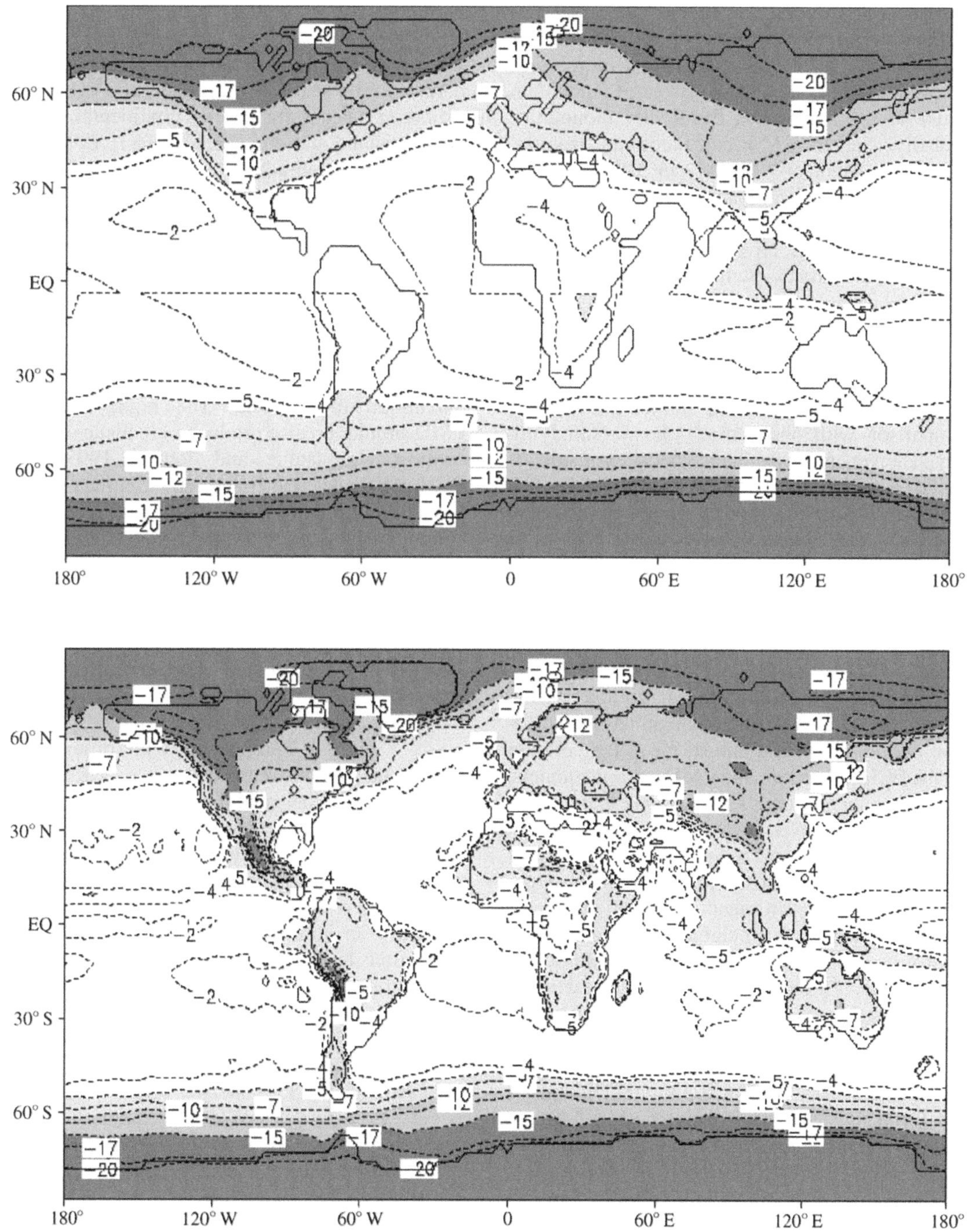

Figure 8 Plot of present-day annual average $\delta^{18}O$ in precipitation as simulated by the GISS and ECHAM isotopic models (after Hoffmann *et al.*, 2000).

and more depleted with lower temperatures. Both models calculate the most depleted precipitation over the ice sheets of Greenland (about −30‰) and Antarctica (about −55‰) which is in good agreement with observations, i.e., −55.5‰ at Vostok, East Antarctica (Lorius *et al.*, 1985) and −35‰ at Summit, central Greenland (Dansgaard *et al.*, 1993; Grootes *et al.*, 1993). The relatively low values over the interior of North America and Siberia demonstrate the strong influence that continentality has on the rainout of air masses and the corresponding isotopic depletion of rain. Owing to its finer orographic resolution the ECHAM simulates a strong uplift and, therefore, a more complete rainout of air masses at large mountain chains as the Andes and the Rocky mountains. Globally, the ECHAM calculated an isotope-laltitude gradient of −0.13‰ per 100 m, which is a little too low compared with observations. Considering the large differences between the two models in the physical parametrization of sub-grid-scale processes, the resemblance of the simulated water isotope is quite astonishing. Obviously the principal processes controlling the isotopic composition of rain, such as the transport of air masses and the rainout of vapor as a result of cooling, are simulated in a similar and, as the comparison with observations shows, satisfying way. The general resemblance between results of the GISS and the ECHAM model and their excellent correspondence with observations is also confirmed by the consideration of the zonal mean $\delta^{18}O$ values in precipitation (Hoffmann *et al.*, 2000). The difference between the two models and the corresponding zonal means of $\delta^{18}O$ of precipitation from the IAEA network hardly exceeds 1‰ over nearly all latitudes.

Jouzel *et al.* (2000) have compared the simulated relationships between annual mean isotope concentration and annual mean temperature for the ECHAM, GISS, and LMD isotopic models. The linear regressions between these parameters are performed over two temperature ranges, namely, temperatures above and temperatures below 15 °C (0 °C for ECHAM). The upper range encompasses tropical sites for which the isotope content of precipitation is controlled mostly by the amount of precipitation and not by temperature. Isotopic GCMs are successful in simulating this effect. However, predicted slopes between $\delta^{18}O$ and the amount of precipitation may be lower than the observed ones as seen for tropical islands with the ECHAM 3 model (Hoffmann and Heimann, 1997; Hoffmann *et al.*, 1998) and for tropical and equatorial regions taken as a whole with the GISS model (Jouzel *et al.*, 1987a). In the lower range, the models correctly simulate the observed linear relationship between $\delta^{18}O$ and temperature, as illustrated in Figure 1 for the GISS model. The observed and predicted gradients are within ~10%, except for the ECHAM model, which predicts a slightly lower value (Jouzel *et al.*, 2000). At the regional level, recent simulations performed by Werner *et al.* (1998) and Werner and Heimann (2002) using ECHAM 4 have clearly shown that the quality of an isotopic GCM in terms of its ability to simulate correctly the observed isotopic distribution can be excellent when using a high-resolution model.

A successful isotope GCM must also reproduce the observed seasonal cycles of $\delta^{18}O$. Seasonal cycles generated with the GISS 8°×10° model, the first to simulate a full seasonal cycle (Jouzel *et al.*, 1987a), are generally realistic, as shown in Figure 4. The seasonal amplitude of $\delta^{18}O$ in precipitation is generally larger at continental stations than at island stations (IAEA, 1981, 1992), and these larger amplitudes, though slightly underestimated in Canada and Siberia, are well simulated in central North America. Simulated amplitudes for Western Europe, Northeast America, and Southeast Asia are also quite reasonable. The GCM reproduces the general absence of a defined seasonal cycle over the island stations, the northern hemisphere character of the cycles in South America and southern Africa, and the early spring maximum in Australia. The higher-resolution LMD simulation also produces a realistic seasonal contrast (Joussaume and Jouzel, 1993). The ECHAM model (Hoffmann and Heimann, 1993) produced mixed results, with realistic cycles simulated over central North America and poorer cycles simulated over the western Pacific and central Europe. Some of the coarse grid ECHAM model's deficiencies in this regard (e.g., over the western Pacific) can easily be explained by deficiencies in the simulated climate itself, whereas others require a different explanation, such as low model resolution. As summarized in Hoffmann *et al.* (2000), a comparison with observations reveals that for all regions the seasonal gradient is lower than the spatial gradient for both the observations and the model results. This fact is usually interpreted as the influence of seasonally changing conditions in the main source regions, and model results of the seasonal gradients are generally in fair agreement with observations (deviations less than 25%).

The observed linear relationship between δD and $\delta^{18}O$ (Figure 2) is also well reproduced, both the $\delta D/\delta^{18}O$ slope and the intercept being predicted correctly; for example, the relationship obtained in the GISS model $\delta D = 8.06\delta^{18}O + 10.4$ is very close to the MWL. This model also captures some of the regional characteristics of the $\delta D/\delta^{18}O$ relationship as the lower slope observed for tropical islands (Jouzel *et al.*, 1987a). The agreement between observed and predicted deuterium-excess values is also relatively good (the difference does not exceed a few per mil) when one considers that d is a second-order parameter and that the observations are less complete than those of $\delta^{18}O$ alone. Globally, the deuterium excess simulated with the ECHAM model agrees also fairly well with observations (Hoffmann *et al.*, 1998).

Finally, the GCM approach is particularly well suited for examining the link between the evaporative origin of a precipitation mass and its isotope content. Water evaporating from a well-defined source region on the Earth's surface can be "tagged" in the GCM and followed through the atmosphere until it precipitates. Through this approach, the relative contributions of many different evaporative regions to a given region's precipitation can be quantified exactly. Joussaume *et al.* (1986) determined the evaporative contribution of 10 global divisions to local continental precipitation in the LMD model. The GISS model was used to determine the sources of local precipitation in the northern hemisphere (Koster *et al.*, 1986) and the differences in the sources of Sahelian precipitation during wet and dry years (Druyan and Koster, 1989). Using the GISS model, Koster *et al.* (1993) found that a region's $\delta^{18}O$ in precipitation is significantly related to the extent of continental water recycling in the region. Koster *et al.* (1992) showed that the deuterium content of Antarctic precipitation decreases as the temperature of the evaporative source for the water increases, by about the amount predicted by simple Rayleigh-type models. The situation appears to be more complex over Greenland. Charles *et al.* (1994) performed a similar experiment with the $4°\times5°$ version of the GISS model, focusing on Greenland. As for Antarctica, several evaporative source regions contribute to the precipitation and the isotope contents of the different contributions vary significantly. For example, moisture from the North Pacific source arrives at the Greenland coast with a $\delta^{18}O$ value ~15‰ lower than its North Atlantic counterpart, a difference attributed to the fact that North Pacific moisture is advected along a much colder path before reaching Greenland (Charles *et al.*, 1994). The tagging of moisture sources has been further exploited both with the GISS (Armengaud *et al.*, 1998; Delaygue *et al.*, 2000a,b) and the ECHAM (Werner *et al.*, 2001) models (see also Section 6.8.2).

These modeling efforts have a further common objective, the reconstruction of paleoclimatic isotope fields to help in the interpretation of paleodata. This important application of water isotope models will be examined in Section 6.8.3 after reviewing existing ice core isotopic records (Section 6.5), their conventional temperature interpretation (Section 6.6), and how this interpretation compares with other estimates of temperature changes (Section 6.7).

6.5 ICE CORE ISOTOPIC RECORDS

Since the early 1950s, ice cores have provided a wealth of information about past climatic and environmental changes. They give access to paleoclimate series that include local temperature and precipitation rate, moisture sources conditions, wind strength and aerosol fluxes of marine, volcanic, terrestrial, cosmogenic, and anthropogenic origin. Elemental composition of the air bubbles gives access to changes in greenhouse gases and in other gaseous compounds, whereas their isotopic composition also contains climate related information. One such example concerns the isotopic composition of the O_2 in air which is related to the hydrological cycle in the tropics and to sea-level change.

Ice cores from Greenland, Antarctica, and other glaciers encompass a variety of timescales. The longer timescales, in which we are interested in here, are covered by deep ice cores, the number of which is still very limited. In Greenland (Figure 9), there are only six cores reaching back the last glacial maximum (LGM) 20 kyr ago: Camp Century (Dansgaard *et al.*, 1969), Dye 3 (Dansgaard *et al.*, 1982), GRIP (Johnsen *et al.*, 1992a, 1997; Dansgaard *et al.*, 1993), GISP 2 (Grootes *et al.*, 1993), Renland (Johnsen *et al.*, 1992b), and the ongoing North GRIP project (Johnsen *et al.*, 2001).

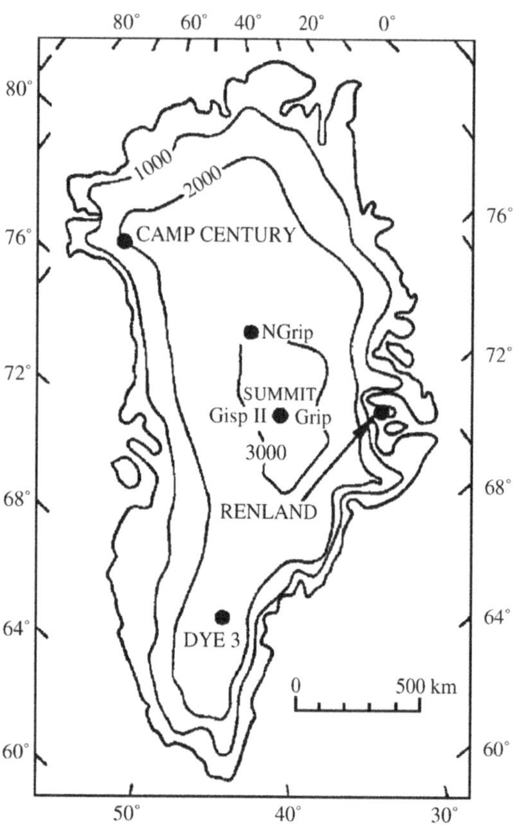

Figure 9 Drilling sites in Greenland.

Deep drilling in Antarctica (Figure 10) started with Byrd (Dansgaard *et al.*, 1969; Epstein and Sharp, 1970) and was then followed by Dome C (Lorius *et al.*, 1979) and the series of cores drilled at the Vostok station, the last one being stopped at a depth record of 3,623 m (Petit *et al.*, 1999). Three other East Antarctic inland cores reached back the LGM at Dome B (Jouzel *et al.*, 1995), Dome F (Watanabe *et al.*, 2003), and Dome C (Jouzel *et al.*, 2001), in the frame of the ongoing European Project for Ice Coring in Antarctica (EPICA). The LGM period is also covered by three other cores drilled in more coastal sites at Taylor Dome (Steig *et al.*, 1998), Law Dome (Morgan *et al.*, 2002), and Siple Dome (Severinghaus *et al.*, 2003). Recently, several tropical ice cores have been obtained that reached back this period either in the Himalayas or in the Andes (Thompson *et al.*, 1995, 1998; Ramirez *et al.*, 2003).

Either δD or $\delta^{18}O$ profiles can be indifferently used as a climatic record. Different choices have been made by various teams. The climate reconstruction is based on the interpretation of the δD profile for Vostok, Dome B, old Dome C and EPICA Dome C, and on $\delta^{18}O$ for all the other cores. Interestingly, measuring both isotopes on the same core brings additional information about the changes affecting the oceanic sources of Antarctic precipitation through the deuterium-excess parameter. Most of the ice core projects now include measurements of both isotopes. Based on the Vostok core results, we illustrate in Section 6.8 how this co-isotopic approach can be used.

Various methods are used to date ice cores each of them having advantages and drawbacks. They fall into four categories: (i) layer counting, (ii) glaciological modeling, (iii) use of time markers and correlation with other dated time series, and (iv) comparison with insolation changes (i.e., orbital tuning). Layer counting based on a multi-parametric approach is extensively used for Greenland cores and for high accumulation Antarctic sites. It is not feasible in low accumulation areas such as central Antarctica where other approaches must be employed. For example, the Vostok core has been dated combining an ice flow and an accumulation model (Petit *et al.*, 1999), but other methods (orbital tuning, comparison with other series, and inverse modeling) have been applied (see Parrenin *et al.*, 2001, and references therein). Finally, the age of the gas is younger than that of the ice due to the fact that air bubbles are trapped when firn closes off at depth. The ice age–gas age difference, Δ_{age}, is currently estimated as a function of temperature and accumulation through a firnification model (Barnola *et al.*, 1991; Schwander *et al.*, 1997).

The longest published isotopic records are those of Vostok which cover the last four climatic

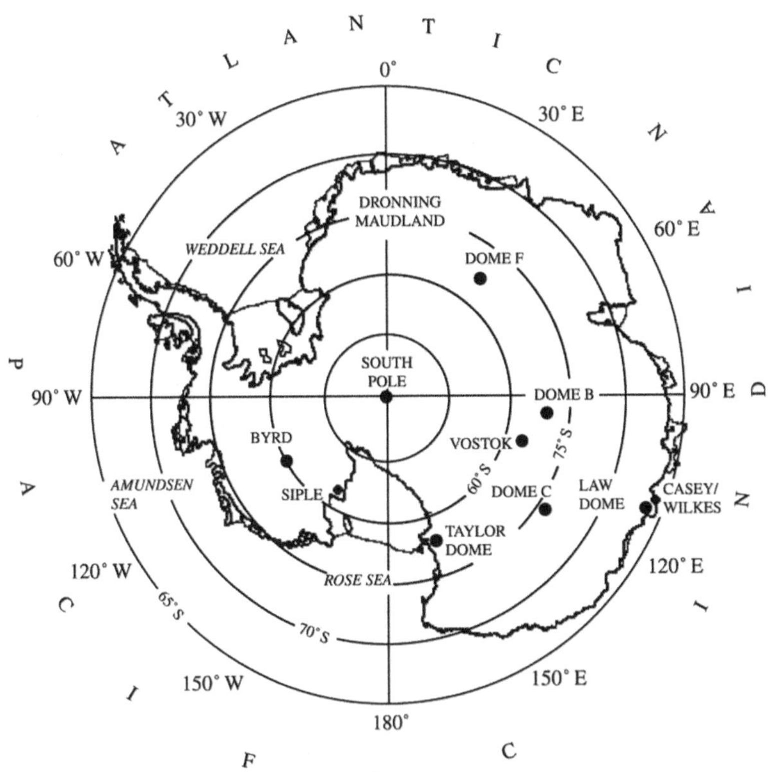

Figure 10 Drilling sites in Antactica.

cycles extending back to ~420 kyr BP (Petit *et al.*, 1999). The next longest is currently Dome F which extends back to ~320 kyr BP (Watanabe *et al.*, 2003), but the timescale covered in Antarctica should be soon extended to at least five climatic cycles, thanks to the new EPICA Dome C drilling. This core reached a depth of 2,871 m in January 2002 with an estimated age of 520 kyr BP (unpublished). Central Greenland records from GRIP (Dansgaard *et al.*, 1993) and GISP 2 (Grootes *et al.*, 1993) provide continuous and reliable isotope climate records back to ~100 kyr BP, but not beyond because of ice flow disturbances due to the proximity of the bedrock (Grootes *et al.*, 1993; Taylor *et al.*, 1993). Other cores cover the last glacial period or, at least, most of it (Byrd, Taylor Dome, Law Dome, Siple Dome, Camp Century, Dye 3, Renland, and North GRIP), and some cores only extend over the LGM or slightly beyond (old Dome C, Dome B, and tropical records). Taken together these ice core isotopic profiles have the following main characteristics.

First, the glacial–interglacial isotopic change, taken between the LGM and the more recent period (i.e., the last thousand years or so), is relatively similar for all these records. The range of variations is remarkably small, if we limit the comparison to inland sites. One exception is Camp Century, with the likely explanation that part of the signal results from a glacial–interglacial increase of the altitude of the Greenland ice sheet at this site (Raynaud and Lorius, 1973). That tropical ice cores show shifts similar to the ones observed in polar areas might be due to the fact that these cores are formed from snow that condenses at the end of the Rayleigh process, as for polar snow. Although these cores are in tropical regions where isotopic concentrations are linked with the amount of precipitation (Section 6.2), the δ_p values of high elevation precipitation are most probably controlled by temperature changes.

Second, the minimum isotopic values reached during the coldest parts of the glacial periods (glacial maxima and cold interstadials) are remarkably similar all along the longest Vostok and Dome F records, whereas highest values can be more variable from an interglacial to the next. For example, the last interglacial $\delta^{18}O$ was ~2‰ higher than the recent Holocene which was also true for the interglacial corresponding to marine stage 9 dated ~320 kyr BP (Petit *et al.*, 1999). The two long Antarctic cores Vostok and Dome F show a high degree of similarity both concerning those large changes and interglacials and smaller events, which indeed is remarkable given the distance (~1,500 km) between the two sites (Watanabe *et al.*, 2003). In the more recent period, i.e., since the LGM, similarities are remarkable for East Antarctic inland cores (Jouzel *et al.*, 2001), but there may be some differences between these records and those obtained at sites such as Taylor Dome (Steig *et al.*, 1998) and Siple Dome (Severinghaus *et al.*, 2003).

Third, the most important characteristic of the Greenland records deals with the existence of rapid climatic changes during the last glacial period and the last transition. These "Dansgaard–Oeschger" events were discovered in the Camp Century and Dye 3 Greenland cores (Dansgaard *et al.*, 1984). Rapid isotopic changes, often more than half of those corresponding to the glacial–interglacial difference and taking place in a few decades, are followed by a slower cooling and a generally rapid return to glacial conditions. The existence and characteristics of these events were fully confirmed at GRIP (Dansgaard *et al.*, 1993) and GISP 2 (Grootes *et al.*, 1993): 24 of such Dansgaard–Oeschger (D/O) events lasting between 500 yr and 2,000 yr occurred during the last glacial period and the last transition. These events (Jouzel *et al.*, 1994; Blunier *et al.*, 1998), and probably all of them (Bender *et al.*, 1994, 1999; Blunier and Brook, 2001), have smooth counterparts in the Antarctic record with, in general, the onset of isotopic changes preceding the onset in Greenland by 1,500–3,000 yr, whereas their maxima are apparently coincident. Raisbeck *et al.* (2002) have, however, pointed out that the situation can differ from at least one of these D/O events, i.e., D/O event 10 that appears to be in phase with its Antarctic counterpart.

In this review, dealing with the temperature interpretation of isotopic ice core records, we will focus mainly on the Vostok core in central East Antarctica and on the GRIP and GISP 2 cores from central Greenland. The reason is that the effort undertaken to calibrate the isotopic paleothermometer through other estimates of temperature changes has so far concentrated on these three cores. We first discuss the conventional approach (Section 6.6) and then examine it in the light of other available temperature change estimates (Section 6.7).

6.6 THE CONVENTIONAL APPROACH FOR INTERPRETING WATER ISOTOPES IN ICE CORES

Before examining the "conventional approach" of interpreting ice core δ profiles, based on the assumption that the spatial and temporal slopes are similar (see Introduction), we note here that the notion of temporal slope covers different timescales. The seasonal slope is deduced from the comparison of the isotope and temperature yearly cycles at a given site (e.g., van Ommen and Morgan, 1997). The short-term (interannual) slope is based on the comparison of mean annual isotope and temperature values at sites where temperature records are available (see, e.g., Jouzel *et al.* (1983)), whereas the long-term slope mainly refers to glacial–interglacial changes. As discussed in Jouzel *et al.* (1997), these three types of slopes generally differ. It is also important to note that some uncertainty is associated with the estimation of the spatial slope for a given region and that this spatial slope may vary from one region to another. For example, slopes of 6.04‰/°C for δD and 0.75‰/°C for $\delta^{18}O$ (defined with respect to surface temperature, T_s) are used for interpreting Dome C and Vostok isotopic profiles. Recent estimates (Delmotte, 1997) suggest that these values are not defined to be better than ±10%; they clearly may be higher in other regions of Antarctica, e.g., up to 1‰/°C for $\delta^{18}O$ in some areas of Antarctica.

In this review, we will deal mainly with the comparison between the spatial slope (defined in the region where the site is located) and the long-term temporal slope. We will thus focus on the temperature interpretation of large isotopic changes associated with glacial–interglacial changes, or occurring during glacial periods. Hereafter, the ratio of the spatial to the temporal slopes is denoted as $R_{slopes} = S_{spat}/S_{temp}$, e.g., when $R_{slopes} > 1$, the true temperature change $\Delta T_{s(spat)}$ is larger by a factor R_{slopes} than the one estimated by the conventional approach and vice versa (Δ denoting the difference between two different periods). An illustration of the conventional approach is provided by the Vostok temperature profile derived from the deuterium record (Figure 11) with

$$\Delta T_{s(spat)} = (\Delta\delta D - Corr_{ocean})/6.04 \qquad (14)$$

In this equation, $Corr_{ocean}$ is a correction applied for the change, $\Delta\delta D_{ocean}$, of the deuterium composition of the ocean resulting from the waxing and the waning of the continental ice sheets during the last four climatic cycles. It is calculated as $Corr_{ocean} = 8\Delta\delta^{18}O_{ocean}$ using the $\delta^{18}O$ oceanic record from Bassinot *et al.* (1994) after appropriate scaling. Note that this oceanic correction was incorrectly applied in Petit *et al.* (1999) and in previous estimates of the Vostok temperature record (Jouzel *et al.*, 1987b, 1993, 1996). It did not accounted for the fact that the influence of any isotopic change at the ocean surface weakens as an air mass becomes isotopically depleted. In a Rayleigh model describing the isotopic behavior of an air mass from its oceanic origin to the precipitation site, the isotopic content of a precipitation (here δD_{ice}) can be written as $1 + \delta D_{ice} = f(1 + \delta D_{ocean})$, where f is a function of climatological parameters and fractionation coefficients only. Applying this equation for present-day $\delta D_{ocean(0)}$ and for a certain period in the past, $\delta D_{ocean(t)} = \delta D_{ocean(0)} + \Delta\delta D_{ocean}$, shows that $Corr_{ocean}$ equals $\Delta\delta D_{ocean}(1 + \delta D_{ice})/(1 + \Delta\delta D_{ocean})$ and not $\Delta\delta D_{ocean}$ as previously done for interpreting Vostok records. As illustrated in Figure 12, this would have only a minor impact if the temperature record had been inferred from the $\delta^{18}O$ ice record ($1+\delta^{18}O_{ice}$ is very close to 1). This is no longer true for the deuterium correction as $1+\delta D_{ice}$ is between 0.5 and 0.6 in central East Antarctica. Consequently, the glacial–interglacial deuterium oceanic correction which in Petit *et al.* (1999) is slighly below 10‰ for the Vostok site decreases, when correctly applied changes to ~5.5‰. This reasoning based on a simple model is fully confirmed using an isotopic GCM as checked by Delaygue (2000), who performed two experiments which differ only by a change in the deuterium (9.6‰) and oxygen-18 (1.2‰) of surface oceanic waters, with the isotopic version of the NASA/GISS GCM (Jouzel *et al.*, 1987a). Values are quite similar to those calculated from the above formula (Figure 12) which can thus effectively be

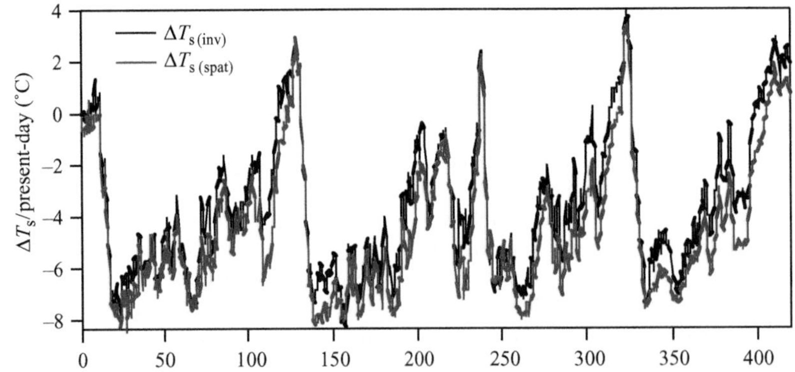

Figure 11 Vostok temperature changes from present-day values back to 420 kyr BP, estimated either $\Delta T_{s(spat)}$ (in red) by the conventional approach based on the δD profile alone (Petit *et al.*, 1999) accounting correctly for the oceanic correction (see text), or $\Delta T_{s(inv)}$ (in green) from the inverse method based on the use of deuterium excess to account for moisture source changes (source Vimeux *et al.*, 2002).

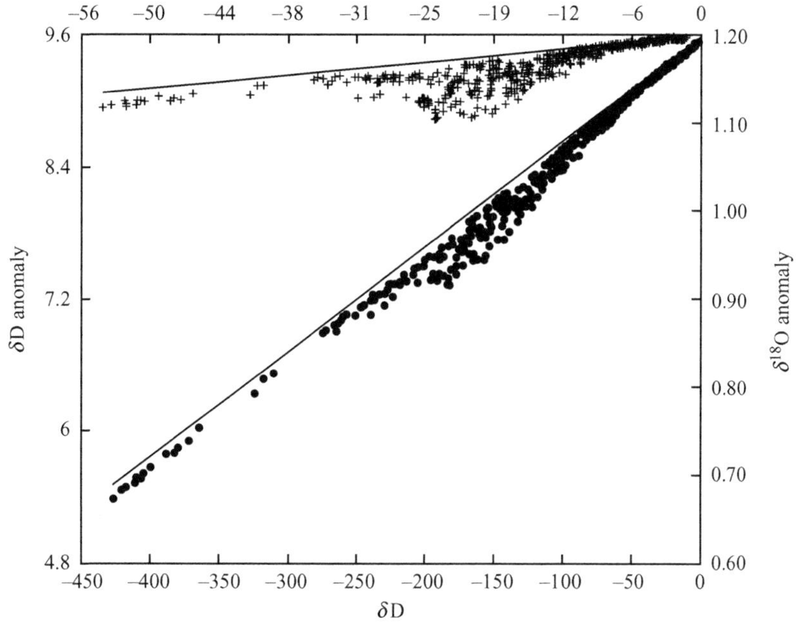

Figure 12 This plot illustrates how the oceanic correction can be estimated from a simple linear relationship directly derived from a Rayleigh-type model. The points represent the oceanic correction as calculated from the experiments (see legend of Figure 1) performed by Delaygue (2000), all model grid points being represented both for δD and $\delta^{18}O$. The two lines are directly derived from a Rayleigh-type model (see text).

used to estimate $Corr_{ocean}$. This is expected because all atmospheric fractionation processes are independent of the isotopic concentrations themselves (small departures with respect to this line are largely due to the contribution of water vapor evaporating over continents).

Of course, the fact that present-day isotope concentrations and local temperature are strongly correlated, as illustrated in Figure 3 for Greenland and Antarctica, does not validate the critical assumption that the present-day spatial slope can be used as a surrogate for the temporal slope. Glaciologists are well aware of, and have systematically pointed out to, the limitations of this conventional approach. As noted by many authors (see Jouzel *et al.*, 1997, and references therein), application of Rayleigh models first points to the combined influence of the temperature of the oceanic source (T_w) and of the temperature of condensation (T_c) on the isotopic content of a precipitation. Strictly speaking, applying the conventional approach ($R_{slopes} = 1$) requires that there is no change in the temperature of the oceanic source (Figure 13). In contrast, a parallel change in T_c and T_w will, e.g., result in an increase of R_{slopes} (Aristarain *et al.*, 1986; Boyle, 1997). To illustrate this point, we use the formulation derived by Stenni *et al.* (2001) for the EPICA Dome C site in central East Antarctica. We assume (Jouzel and Merlivat, 1984) that the snow is formed just above the inversion layer which allows us to relate ΔT_c and ΔT_s through $\Delta T_c = 0.67\Delta T_s$ (Jouzel and Merlivat, 1984). Using the mixed cloud isotopic model (Ciais and Jouzel, 1994) allows us to express the deuterium change at the site, $\Delta\delta D$, as

$$\Delta\delta D = 7.6\Delta T_s - 3.6\Delta T_w + Corr_{ocean} \quad (15)$$

A deuterium change of 60‰ (including the oceanic correction), typical of the glacial–interglacial transition in central Antarctica, will be interpreted as a $\Delta T_{s(spat)}$ of 7.9 °C if the dependence on the source temperature is neglected as done in the conventional interpretation. In contrast, if we assume that the change in ΔT_s is accompanied by a concurrent change in ΔT_w of half its size (assuming a polar amplification of the oceanic change by a factor of 2), and account for it, the resulting estimate of ΔT_s increases to 10.3 °C ($R_{slopes} = 1.31$; see Figure 13).

Many other factors can obviously influence R_{slopes}. They can be linked to other source characteristics controlling the evaporation kinetics such as relative humidity and wind speed (Merlivat and Jouzel, 1979) or to the microphysical processes prevailing in clouds such as the saturation value at snow formation (Fisher, 1991). They can result from changes in the seasonality and intermittency of precipitation fallout, which both bias the temperature sampling by this precipitation, or the wind erosion (Gallée *et al.*, 2001) which can affect the isotopic signal in a different way for an interglacial than for a glacial. Also, as the spatial slope is defined with respect to T_s, any change in the strength of the vertical inversion, between, e.g., a glacial and an interglacial, will

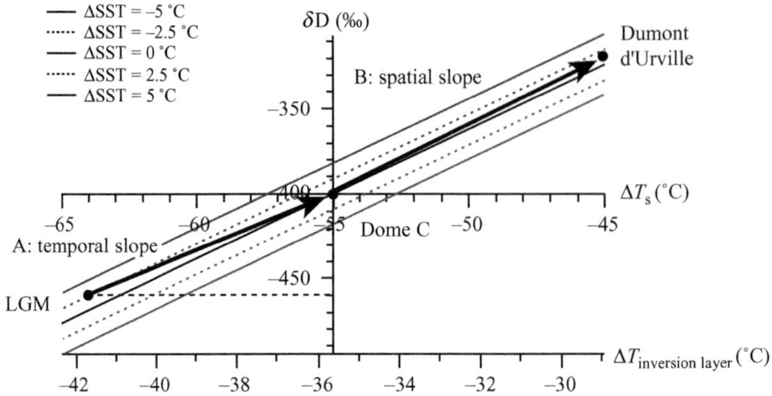

Figure 13 The influence of the source temperature on the isotopic content of the Antarctic precipitation. Line B corresponds to the observed present-day spatial slope between Dumont d'Urville and Dome C expressed with respect to the temperature of snow formation (i.e., above the inversion layer). Line A represents the temporal slope assuming that the temperature change at the oceanic source is half of that at the Dome C site.

influence R_{slopes}. Changes in cyclonic activity (Holdsworth, 2001) and in the ratio between advection and eddy transport (Hendricks *et al.*, 2000; Noone and Simmonds, 2002b) can also play a role.

6.7 ESTIMATES OF TEMPERATURE CHANGES IN GREENLAND AND ANTARCTICA

Despite the many factors potentially influencing the temporal slope, the conventional approach has been used both for Antarctic and Greenland cores until independent estimates of temperature changes became available. However, this was somewhat a surprise in our community when the interpretation of the borehole-temperature profile clearly showed that the present-day spatial slope underestimates glacial–interglacial temperature changes in central Greenland by a factor 2 or so (Cuffey *et al.*, 1995; Johnsen *et al.*, 1995). A second method, based on the detection of anomalies in the isotopic composition of nitrogen and argon in the entrapped air bubbles, has later been exploited (Severinghaus *et al.*, 1998). The use of these two approaches has also been explored for Antarctica where other methods, such as those aiming to derive temperature information from dating constraints, bring useful information. We now summarize these various estimates of temperature changes for Greenland and Antarctica.

6.7.1 Greenland

Borehole paleothermometry was the first method allowing precise estimates of glacial–interglacial changes. Because of heat diffusion, conversion of the depth–temperature record of a borehole-to a surface-temperature history involves some difficulties. The borehole-temperature record of old, high-frequency events is lost entirely. More than one possible climate history would produce a given measured borehole profile, although a range of plausible surface-temperature histories can be excluded. A very powerful technique was suggested by Paterson and Clarke (1978). This involves calibrating the isotopic paleothermometer against the borehole profile using an inverse procedure to adjust the constants in the calibration.

This exercise has been independently conducted for the GISP 2 (Cuffey *et al.*, 1995) and GRIP (Johnsen *et al.*, 1995) fluid-filled deep boreholes. The two teams assumed different forms for the calibration curves: Cuffey *et al.* (1995) assumed a linear relation between isotope and temperature but allowed some time variations, while Johnsen *et al.* (1995) kept constant this linear coefficient but allowed for an additional term proportional to $\delta 2$. They, however, ended up with similar leading results: (i) the isotopic ratios are an excellent paleothermometer over all time intervals studied and (ii) the calibration is different for recent or short times, i.e., for the last few centuries to millenia (~0.5‰/°C and 0.6‰/°C at GISP 2 and GRIP, respectively, for δ^{18}O) than for older or longer times (~0.33‰/°C and 0.23‰/°C, respectively). The isotopic GRIP and GISP 2 profiles indicate a temperature increase from glacial maximum to Holocene higher than 20 °C and up to 25 °C at Summit, taking into account the lapse-rate effect of thickening associated with accumulation increase at the end of the ice age and the change in seawater δ^{18}O. These large temperature changes were essentially confirmed by Dahl-Jensen *et al.* (1998), who have calculated an independent inverse solution for the GRIP borehole.

Borehole paleothermometry does not allow direct calibration of the isotopic changes at such rapid climatic reorganizations as the termination of the Younger Dryas cold event (Firestone, 1995) or the onset and termination of the stadial phases of the D/O oscillations. This limitation did not hold true for a new method developed by Severinghaus *et al.* (1996, 1998) and Severinghaus and Brook (1999) using a temperature change indicator in the gas phase. Whereas heat diffusion prevents borehole paleothermometry to retain information about the numerous abrupt climatic changes documented in the Greenland isotopic record, this rapidity (a few decades or less) allows the isotopic analysis of air bubbles to provide estimates of abrupt temperature changes.

Due to compaction, the density of snow increases with depth. The entrapped air is thus younger than the ice matrix with the age difference, Δ_{age}, depending both on accumulation and temperature (in central Greenland Δ_{age} varies between 200 yr and 900 yr). Air composition is very slightly modified by physical processes in them the gravitational and thermal fractionation. The latter depends on the thermal diffusion sensitivity and on the temperature difference between the surface and the close-off depth. Gases diffuse ~10 times faster than heat and in the case of a rapid change, this temperature difference is temporarily modified which causes a detectable anomaly in the isotopic composition of nitrogen and argon. Both the depth of this anomaly, which allows estimate of Δ_{age} by comparison with the ice $\delta^{18}O$ record, and its strength provide estimates of rapid changes. Focusing first on the rapid warming at the end of the Younger Dryas, 11.5 kyr BP (Preboreal transition), Severinghaus *et al.* (1996, 1998) demonstrated that thermally driven isotopic anomalies are detectable in ice core air bubbles. The inferred Δ_{age} value indicated that the Younger Dryas was 15 ± 3 °C colder than today, a value about twice larger than ΔT_{spat}. Severinghaus and Brook (1999) then studied the abrupt warming that terminated the glacial period 14.6 kyr BP and led to the Bølling. The warming, directly estimated from the size of the isotopic anomalies, amounts to ~10 °C again about twice larger than ΔT_{spat}.

The abrupt warmings that marked the start of the numerous D/O events during the last glacial period are also larger than initially thought, as shown by Lang *et al.* (1999) from a detailed study of the $^{15}N/^{14}N$ anomaly associated with D/O events 19 (~71 kyr BP). These authors elegantly combine a depth/age model which assumes that accumulation depends on temperature, and a firnification model, to simulate this $^{15}N/^{14}N$ anomaly. The best fit with observations is obtained assuming an abrupt warming of 16 °C, 60% higher than ΔT_{spat}. This model estimate agrees with preliminary estimates proposed for D/O events between 40 kyr BP and 20 kyr BP based on the assumption that Δ_{age} can be evaluated assuming that methane and temperature changes associated with D/O events are coeval (Schwander *et al.*, 1997).

All the results derived either from borehole paleothermometry or from isotopic anomalies show that the use of the spatial slope underestimate temperature changes in Central Greenland, thus challenging the conventional approach. This is illustrated in Figure 14 (Jouzel, 1999) which includes an additional result concerning the rapid cooling that occurred 8.2 kyr BP (Leuenberger *et al.*, 1999). Temperature changes are consistently higher than ΔT_{spat}, by a factor of ~2 for the LGM, for the Bølling and Preboreal transitions, and for the 8.2 kyr BP cooling, and by ~50% for the investigated D/O events. Note, however, that a recent study of D/O events 12 which combines argon and nitrogen measurements, thus allowing to separate the gravitational and thermal anomalies, indicates warmings only 20–30% larger than derived from the conventional approach (Caillon, 2001).

6.7.2 Antarctica

In Antarctica, both paleothermometry and the use of nitrogen and argon isotopes pose some problems. First, the low accumulation that prevails at East Antarctic inland sites such as Vostok erases the glacial–interglacial surface temperature signal at the depth of the LGM. Second, unlike central Greenland, central Antarctica did not experience abrupt temperature changes, and gas isotopic anomalies due to thermal diffusion are, in principle, difficult to detect. Thus, there is no perfect alternative to calibrate the isotopic paleothermometer there. Still there are useful arguments coming from the isotopic composition of the air bubbles (Caillon *et al.*, 2001) and from constraints with respect to ice core chronologies (Parrenin *et al.*, 2001). As recently reviewed by Jouzel *et al.* (2003), they converge towards the idea that the observed present-day spatial slope can be used to interpret Antarctic isotopic profiles.

A straightforward application of an inverse method clearly shows that no useful information can be directly retrieved from borehole paleothermometry at low accumulation sites like Vostok (Rommelaëre, 1997). To overcome this problem, Salamatin *et al.* (1998 and references therein) developed an inverse procedure based on the assumption that the inferable components of the surface temperature at Vostok can be expressed as a sum of harmonics of Milankovitch periods. Doing so, the conventional approach of using the present-day spatial slope is confirmed in general (Salamatin *et al.*, 1998), with, however, a significant mismatch between modeled and borehole temperatures. This mismatch decreases noticeably if surface temperature is assumed to undergo more intensive precession oscillations than temperature

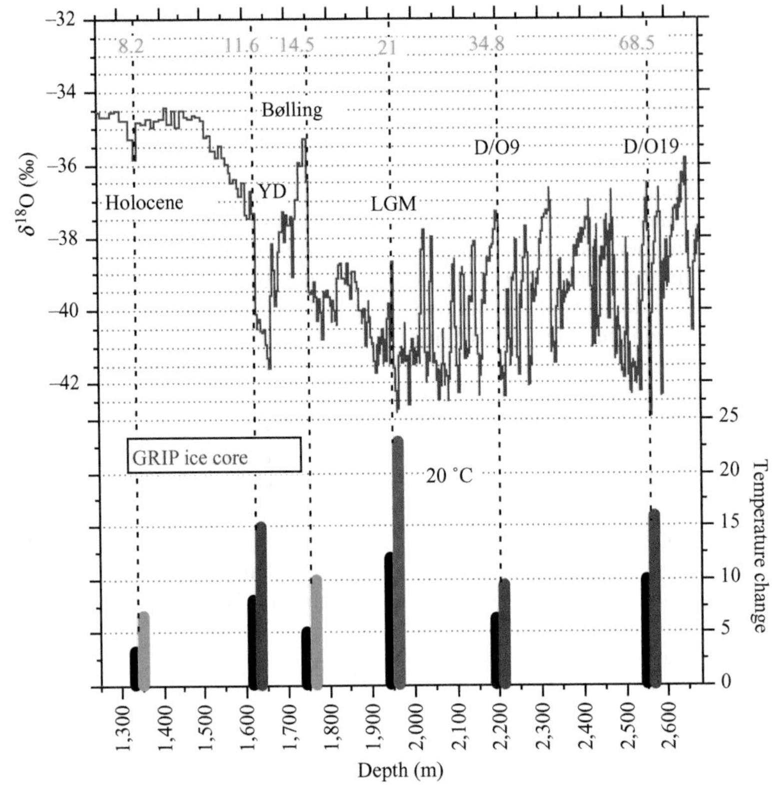

Figure 14 The bars compare, for the time periods discussed in the text, the absolute temperature changes, $\Delta T^{18}O$, derived from the Greenland $\delta^{18}O$ record using the present-day spatial slope of 0.67‰/°C (black bars) with independent estimates from paleothermometry (blue bar), from estimates of Δ_{age} (red bars), and from the thermal isotopic anomaly (green bars). Absolute values are reported for the 8,200 BP event which corresponds to a cooling. We have accounted for a 1‰ $\delta^{18}O$ oceanic change when calculating the present-day/LGM temperature difference. The upper continuous curve corresponds to the GRIP $\delta^{18}O$ record (average 100 yr values reported on a linear depth scale; the GRIP and GISP2 $\delta^{18}O$ records are similar once accounted for slight depth adjustments).

at the inversion level. With this additional assumption, it is inferred that surface temperatures changes were larger by ~30% ($R_{slopes} = 1.3$) and even by up to 50%, depending on assumptions made for the inversion procedure. The fact that Vostok paleothermometry provides such high estimates of changes in glacial–interglacial temperature is now often cited. However, there is currently no clear argument going into support of this additional assumption on which this higher estimate is based.

Although the Antarctic climate is not characterized by such abrupt changes as observed in Greenland, Caillon *et al.* (2001) have undertaken a detailed study of the most rapid isotopic warming event that occurred between 107 kyr BP and 108 kyr BP. They successfully measured a small but detectable anomaly in both nitrogen and argon isotopic compositions, resulting possibly from a gravitational signal due to a change in the firn thickness. The position of this anomaly gives a direct estimate of the close-off depth, from which it is inferred that the use of the spatial slope slightly underestimates temperature changes but by no more than 20 ± 15% (i.e., $R_{slopes} = 1.20$).

In a different study, Blunier *et al.* (in press) proposed to compare the isotopic and methane profiles between Byrd and Vostok in order to constrain the Δ_{age} at Vostok. They first assumed a large glacial–interglacial temperature (15 °C) and estimated Δ_{age} for two scenarios with different formulations of the accumulation. The experimentally deduced Δ_{age} is not compatible with either of those two scenarios. Instead, calculations made assuming the conventional temperature interpretation, and accumulation as in Petit *et al.* (1999), provide Δ_{age} estimates in agreement with experimentally deduced values (within the error bars). These authors (Blunier *et al.*, in press) conclude that the conventional interpretation is the more probable one.

One way to derive ice core chronologies (Ritz, 1992; Petit *et al.*, 1999; Schwander *et al.*, 2001) is to combine an accumulation history and an ice flow model. As a first approximation, the accumulation rate can be estimated as being proportional to the derivative of the water vapor saturation pressure above the precipitation site (Jouzel *et al.*, 1987b). This reasoning is quite certainly too simplistic. Accumulation and temperature are clearly linked over Antarctica, but the proportionality with the derivative of the saturation vapor pressure is not warranted and other parameters than temperature such as the intensity of the atmospheric circulation may influence accumulation rates. There is no guarantee either on the use of the spatial slope as a surrogate of the temporal slope for interpreting isotopic profiles. With this in mind, Parrenin *et al.* (2001) recently developed an inverse method they first applied to the Vostok ice core. Rather than assuming a linear relationship between the ice deuterium content and temperature with a prescribed slope, they simply use a second-order relationship with two free parameters, thus making no assumption on the amplitude of the glacial–interglacial temperature change. They also assumed that the present-day accumulation upstream of Vostok is a second-order function of the distance to Vostok and discussed the possible influence of atmospheric circulation changes. In turn (see Parrenin *et al.* (2001) for a detailed description), the application of the inverse method provides information both on temperature and accumulation changes. Simply assuming that the number of precessional cycles can be correctly counted in the Vostok profiles, they showed that the use of the spatial slope slightly underestimates temperature changes but probably by no more than 10–20% over the full range of observed δD changes. In this same line, we note the result derived by Schwander *et al.* (2001) from their dating of the EPICA Dome C. These authors inferred that using the spatial slope also underestimates the glacial–interglacial surface temperature change at this site by ~20% (depending on the reference taken for the S_{spat}).

6.8 DISCUSSION

Taken together, this ensemble of results clearly challenges the conventional approach of interpreting ice core isotopic profiles in Greenland with probably a larger bias for the glacial–interglacial change than for the warmings associated with D/O events. However, the results tend to support it in Antarctica, with the exception of borehole paleothermometry. We will now try to understand why the situation differs between both polar ice sheets. To do so, we separately examine the influence of two key parameters, the origin and the seasonality of the precipitation, and then discuss estimates of the temporal slope derived from experiments performed with isotopic GCMs.

6.8.1 Influence of the Seasonality of the Precipitation

There is practically no information on the seasonality of Antarctic and Greenland precipitation for a period such as the LGM and, even for present-day, only limited data for inland Antarctica (Ekaykin *et al.*, 2002). This information can only be derived from GCM experiments as closely examined by Krinner *et al.* (1997). For this purpose, these authors implemented the LMDs stretched-grid GCM adapted to have a high resolution in polar regions with various diagnoses providing the precipitation weighted temperature of the model layers where the precipitation forms. The difference between the glacial–interglacial change in the condensation temperature and ΔT_s gives thus an indication of the bias introduced by seasonality and inversion when interpreting the isotopic signal of polar precipitation. The bias due to seasonality is large for Greenland, where the model does not simulate a clear seasonality of present-day precipitation but shows a clear summer precipitation maximum for the LGM. These seasonality changes can largely explain the fact that using the present-day spatial slope for interpreting GRIP and GISP 2 isotopic profiles underestimates glacial–interglacial temperature changes by a factor of 2 (Krinner *et al.*, 1997). This conclusion is supported by experiments performed using the GENESIS/NCAR model (Fawcett *et al.*, 1997). Unlike Krinner *et al.* (1997) and Fawcett *et al.* (1997), Werner *et al.* (2000, 2001) used a GCM implemented with water isotopes (ECHAM) to examine this problem. They clearly showed that the conventional approach is biased due to a substantially increased seasonality of the precipitation during the LGM. During the glacial winter a much more zonal circulation prevents the effective transport of moisture to the Greenland ice sheet, and therefore reduces the contribution of isotopically strongly depleted winter snow to the annual mean isotope signal (Werner *et al.*, 2000). Significant changes are seen in the seasonal cycle of water from polar seas and the North Atlantic which under LGM climate is no longer transported to Greenland during autumn and winter, but during summer season (Werner *et al.*, 2001).

In contrast to Greenland, the experiments of Krinner *et al.* (1997) indicate that the condensation temperature seasonal cycle remains close to the modern level on the East Antarctic Plateau, and that there is only a weak bias due to seasonality (and little influence of other local parameters such as the intermittency of the precipitation and the strength of the inversion). Using the same diagnoses, Delaygue *et al.* (2000b) arrived at a similar conclusion for the GISS model: glacial conditions decrease the winter contribution to annual precipitation inducing a limited 15% decrease of R_{slopes}. Werner *et al.* (2001) also noted that in the ECHAM model, glacial–interglacial changes in the seasonal distribution of precipitation are much smaller for Antarctica than for Greenland, and have thus less influence on R_{slopes}. In turn, whereas seasonality changes are a plausible explanation of the underestimation of Greenland temperature changes, all available GCM experiments point to a limited impact of these processes for central Antarctica.

6.8.2 Influence of the Origin of the Precipitation

There are two complementary ways to assess the influence of the origin of a precipitation on its isotopic content. First, the combined measurement of both δD and $\delta^{18}O$ enables the calculation of a second-order isotopic parameter, the deuterium excess ($d = \delta D - 8\delta^{18}O$) which (see Section 6.4.1) depends on the temperature and relative humidity of the evaporative source (and, to a lesser degree, on the wind speed). In turn, this parameter contains information about conditions prevailing in these source regions and it has been applied, as of early 2000s, only for Antarctic sites (Cuffey and Vimeux, 2001; Stenni *et al.*, 2001; Vimeux *et al.*, 2002), to correct the conventional approach for source temperature changes and to provide an estimate of those changes. Second, it is possible to perform GCM experiments in which the origin of the precipitation, and in most experiments its isotopic composition, is tagged and then followed from its source to the precipitation site. This approach has now been applied in several experiments addressing the relationship between the origin and the isotopic content of the Antarctic or Greenland precipitation for present-day (Koster *et al.*, 1992; Armengaud *et al.*, 1998; Noone and Simmonds, 2002b) and glacial climates (Charles *et al.*, 1994; Delaygue *et al.*, 2000a,b; Werner *et al.*, 2001).

To illustrate the first approach (Figure 11), Jouzel *et al.* (2003) compared the Vostok temperature records as reconstructed using the conventional approach (see above) and applying a source correction derived from the deuterium-excess profile (Vimeux *et al.*, 2002). Basically, a linear inversion procedure is applied. It allows for the extraction of both ΔT_s, denoted hereafter $\Delta T_{s(inv)}$, and ΔT_w from these two sets of parameters (Cuffey and Vimeux, 2001; Stenni *et al.*, 2001). Unlike the conventional approach, which relies on present-day observations, this inversion is thus a purely model-based approach. The inversion procedure in which there is an attempt to account for the moisture source changes (Vimeux *et al.*, 2002) and the conventional approach in which the influence of those changes is ignored, provide results very close to each other. For example, the estimates of the successive glacial–interglacial changes are slightly larger when estimated by the conventional approach but by no more than ~10% on the average (with some differences from one cycle to another). As previously noted by Cuffey and Vimeux (2001), the most noticeable difference occurs during glacial inceptions with the consequence that applying a deuterium-excess correction improves the degree of covariation between carbon dioxide and temperature.

Experiments performed with the isotopic version of the NASA/GISS model lead to a similar conclusion (Delaygue *et al.*, 2000b). Unlike simple models that generally consider a unique source, the GCM enables the explicit tagging of the moisture providing from multiple sources (19 in each hemisphere in this particular study). This approach confirms the simple model result that moisture originating from warmer sources provides lower δ_p but shows additional impact of atmospheric circulation on δ_p which also depends on the distance between the source and the precipitation site. However, due to changes in the contributions from those various sources, the Antarctic mean source temperature does not significantly change between modern and glacial climates (Delaygue *et al.*, 2000a). This results in a relatively limited increase of R_{slopes} (10–30%).

While changes in seasonality now appear to provide an explanation for the underestimation of Greenland temperatures, the focus there was initially on the origin of the Greenland precipitation. This is based largely on Boyle's (1997) article, where a brief description popularized the idea that a parallel change in source and site temperatures explains that the temporal slope is lower than the spatial slope. The earlier article published by Charles *et al.* (1994), showing that moisture from the North Pacific source arrives at the Greenland coast with a $\delta^{18}O$ value ~15‰ lower than its North Atlantic counterpart (see Section 6.4.2), should also retain attention. Charles *et al.* (1994) attributed the lower $\delta^{18}O$ to the fact that North Pacific moisture is advected along a much colder path before reaching Greenland. As an extreme example they pointed out that a $\delta^{18}O$ anomaly of 7‰ would be generated at a Greenland site if no climate change (including temperature) other than a shift from a pure North Atlantic contribution to an even mixture of North Atlantic and Pacific moisture occurred. This example opened a debate on the relative extents to which local temperature changes and changes in the evaporative sources of precipitation define the isotope shifts recorded in Greenland ice cores. The example, however, must not be misinterpreted, as the GCM results do not suggest such extreme moisture source changes for central Greenland actually occurred. Indeed, the present-day/LGM change in the contribution of North Pacific moisture is only 2%, not 50% as in the above example (see also Charles *et al.* (1995)). A simple calculation that accounts for the simulated changes in moisture sources shows that in the GCM, even a 30‰ difference between the isotope contents of Pacific-derived and Atlantic-derived water translates into only a 1‰ or 2‰ net present-day/LGM change, which is still relatively small with respect to the 5–12‰ shifts recorded in Greenland cores.

To explain Greenland observations, Boyle (1997) pointed out the key role of cooler tropics which, as opposed to the classical CLIMAP (1981) reconstruction, has gained large support. This simple explanation is very attractive but, as noted above, Greenland results can be understood only by calling upon changes in the seasonality of the precipitation in central Greenland. In this respect, the results obtained by Delaygue *et al.* (2000b) should be considered as preliminary over Greenland, because the performances of the coarse GISS model are not satisfying (Jouzel *et al.*, 1987a). Still they reinforce the role of local parameters (i.e., seasonal cycle) at the expense of the tropical cooling proposed by Boyle (1997). This is confirmed by Werner *et al.* (2000), who showed, as predicted by Boyle (1997), that cooler sea surface temperature (SST) shifts the temperature–isotope relationship over Greenland, but the effect is small owing to a parallel change in the geographic origin of the vapor.

6.8.3 Estimating the Temporal Slope from Isotopic GCMs

Experiments and probes designed to identify separately the influence of specific factors on R_{slopes}, such as the origin and seasonality of the precipitation have been performed only recently. One initial objective of isotopic GCMs was to provide a direct comparison between spatial and temporal slopes by simulating different climatic periods and this was done as soon as the ability of isotopic GCMs to reproduce the main characteristics of δD and $\delta^{18}O$ in precipitation has been judged to be sufficient. This approach was pioneered by Joussaume and Jouzel (1993) from present-day and LGM simulations using the LMD isotopic model. However, the simulation was limited to a perpetual January and July which did not enable a reliable estimate of R_{slopes} at the yearly scale. Various present-day and LGM experiments covering several years were then performed both with the NASA/GISS (Jouzel *et al.*, 1994; Delaygue *et al.*, 2000b; Charles *et al.*, 2001) and the ECHAM Hamburg (Hoffmann *et al.*, 1997; Werner *et al.*, 2000, 2001) isotopic models.

The first comparison between present-day and LGM simulations showed indeed a low temporal slope (0.43‰/°C in $\delta^{18}O$) over Greenland (Jouzel *et al.*, 1994), but these authors did not infer from this result that glacial–interglacial change was underestimated applying the conventional approach. One reason was the fact that the two simulations then available (LMD model and coarse grid version of the GISS model) showed differences on a regional basis (Joussaume and Jouzel, 1993). Clearly not too much weight could be given to a conclusion which is model dependent. Rather, Jouzel *et al.* (1994) emphasized one result that is common to both models, i.e., that over polar ice sheets, the predicted temporal and spatial slopes are within 30%, for a given model. Indeed, such a cautious approach finds justification in the fact that simulations performed with versions of the GISS model differing only by their resolution showed that the fine grid version predicts higher temporal slopes over Greenland than the coarse grid version. The results obtained later using the ECHAM model were of better quality in general, over Greenland in particular. Indeed using this model, Werner *et al.* (2000) convincingly showed that the temporal slope is 60% smaller than the model spatial slope in the grid box enclosing Summit ($R_{slopes} = 2.5$), illustrating that the observed discrepancy between borehole and isotope temperatures is clearly reproduced in those simulations.

As far as Antarctica is concerned, the NASA/GISS model predicts spatial slopes lower than temporal slopes over most Antarctic grid points; however, whereas the difference is limited over East Antarctica ($R_{slopes} = 0.80$), it is much larger for West Antarctica ($R_{slopes} = 0.6$). However, in this latter region there are large changes in the prescribed topography of the ice cap which are probably not realistic and bias the estimate of R_{slopes} (which does not account for local changes such as the altitude of the site). Note also that the present-day spatial slope predicted with the NASA/GISS model is higher than the one observed over East Antarctica quite probably because of the very weak simulated inversion strength. In turn, the comparison with data is good when the temperature of condensation, T_c, is considered but deteriorates when T_s is taken into account (Jouzel *et al.*, 1994). Although this explanation is satisfying from an isotopic point of view, as it is T_c and not T_s that governs the isotopic content of the precipitation, this discrepancy between observed and predicted inversion strength should be kept in mind. In contrast, ECHAM results, obtained with a more realistic topography and with higher spatial resolution, show an excellent agreement between observed and predicted S_{spat}, both for East and West Antarctica (Hoffmann and Heimann, 1997). Comparison with the LGM run (Hoffmann *et al.*, 2000) provides estimates of $R_{slopes} \sim 0.9$. One can thus be confident that temporal and spatial slopes are relatively close to each other in Antarctica when the glacial–modern change is considered. These present-day and LGM ECHAM experiments have now been completed by a series of isotopic simulations with various boundary conditions corresponding to climates intermediate between present-day and the LGM. These new experiments

(Hoffmann, personal communication) generally confirm the results presented in Hoffmann *et al.* (2000).

Finally, although not based on a GCM approach, the recent modeling performed by Hendricks *et al.* (2000) is worth mentioning in this section. These authors have developed a one-dimensional model of meridional water vapor transport to evaluate the factors that control spatial and temporal variations of δD and $\delta^{18}O$ in global precipitation. They found a good agreement between S_{spat} and S_{temp} for inland sites such as Vostok and South Pole but significantly lower S_{temp} for more coastal regions. Interestingly, this feature of S_{temp} increasing inland is also seen, but much less pronounced, in GCM experiments (Hoffmann *et al.*, 2000; Jouzel *et al.*, 2000).

6.9 CONCLUSION

The main objective of this chapter was to critically review the paleoclimatic interpretation of large isotopic changes recorded in deep ice cores from Greenland and Antarctica. For this purpose, we have combined various sources of information derived from water isotope models and from independent estimates of temperature changes in polar regions. Before doing so, we have shown how simple isotopic models can be useful tools to understand the distribution of water isotopes (δD and $\delta^{18}O$) in precipitation and assessed the ability of isotopic GCMs to correctly simulate present-day isotopic fields both spatially and seasonally. This success justifies the use of these complex isotopic models for simulating isotopic distributions under different climates. These simulations provide model estimates of the temporal isotope/temperature slope to be used for interpreting isotopic paleodata. They offer the possibility of examining the influence of such factors as the seasonality and, provided the different water sources are tagged, the origin of the precipitation. Through this modeling approach we have now a reasonable explanation of why, as shown by other estimates of temperature change, the situation differs between Antarctica and Greenland with respect to the validity of the conventional approach of interpreting ice core isotopic profiles.

In Antarctica, both the origin and the seasonality of the precipitation have a low influence on the temporal slope (Sections 6.8.1 and 6.8.2). Estimates based on the comparison of LGM and present-day isotopic GCM experiments give temperature changes slightly lower than the conventional approach ($R_{slopes} = \sim 0.8$ and 0.88 with the NASA/GISS and ECHAM models, respectively). Instead, slightly higher temperature changes are suggested by the information inferred for Vostok both from ice core chronologies ($R_{slopes} = \sim 1.2$) and gas-age–ice-age differences ($R_{slopes} = \sim 1.15$–1.2). The paleothermometer calibration based on the borehole-temperature profile provides estimates outside this range ($R_{slopes} \sim 1.3$ and up to 1.5). We suggest that this discrepancy results from assumptions made to invert this temperature profile not fully satisfied, most probably the existence of more intensive precession oscillations at the surface than in the atmosphere during glacial. We, however, note that other empirical estimates (dating and gas age–ice age constraints) point to slightly stronger temperature changes. Also we keep in mind that the lowest estimate of R_{slopes} (0.80) is obtained with the low-resolution NASA/GISS GCM, which does not provide a fully satisfying picture of isotopic distribution for present day (Section 6.8.3). In turn, we propose a value of 1.1 ± 0.2 as our best current estimate of R_{slopes}. We thus conclude that the present-day spatial slope can be taken, within $10 \pm 20\%$, as a surrogate of the temporal slope to interpret isotopic profiles from the East Antarctic Plateau (Jouzel *et al.*, 2003).

In Greenland, there is no doubt that the conventional approach underestimates temperature changes (Figure 14). At first sight, a good candidate to explain this underestimation would be that source and site temperatures have varied simultaneously (Boyle, 1997). Rather, all GCM experiments point to a key role of seasonality and it is this explanation which now appears more likely. In Greenland, seasonality has indeed a strong impact due to the large asymmetrical sea-ice and atmospheric circulation changes associated with a glacial–interglacial transition, whereas East Antarctica is more circularly symmetric with no such large associated changes.

We have focused our model/data comparison on the three cores for which we have independent estimates of temperature changes, Vostok, GRIP, and GISP 2. Our conclusion about the validity of the conventional approach can be easily extended to other cores from the East Antarctic Plateau (EPICA and old Dome C, Dome B, and Dome F), which are remarkably similar for their glacial–interglacial changes (Jouzel *et al.*, 2001; Watanabe *et al.*, 2003). We will be more cautious for other cores from near coastal sites of East Antarctica (Taylor Dome, Law Dome) or from West Antarctica (Byrd and Siple Dome), as isotopic profiles may either be influenced by changes in moisture sources conditions or/and be affected by regional changes in the elevation of the ice cap. In the same way, GRIP and GISP 2 results cannot be extended to other Greenland cores. We are also not inclined to extend our conclusions to shorter timescales as the Holocene. Both model (Cole *et al.*, 1999) and empirical estimates (Jouzel *et al.*, 1997) show that such "short-term" isotopic variability differs from that associated with the cooling during the last ice age.

ACKNOWLEDGMENTS

The author would like to thank many colleagues for very stimulating discussions on the topics reviewed in this chapter, in particular, Wally Broecker, Nicolas Caillon, Gilles Delaygue, Georg Hoffmann, Sigfus Johnsen, Sylvie Joussaume, Randy Koster, Claude Lorius, Valérie Masson, Liliane Merlivat, Frédéric Parennin, Jean-Robert Petit, Gary Russell, Barbara Stenni, Bob Suozzo, Françoise Vimeux, and Jim White. This work is supported in France by the Programme National d'Études de la Dynamique du Climat (PNEDC) and by the European program Pole-Ocean-Pole (POP EVK2-2000-00089).

REFERENCES

Andersen U. (1997) Modeling the stable water isotopes in precipitation using the LMD.5.3 atmospheric general circulation model. PhD Thesis, NBIAPG, University of Copenhagen, 100pp.

Aristarain A. J., Jouzel J., and Pourchet M. (1986) Past Antarctic peninsula climate (1850–1980) deduced from an ice core isotope record. *Clim. Change* **8**, 69–89.

Armengaud A., Koster R., Jouzel J., and Ciais P. (1998) Deuterium excess in Greenland snow: analysis with simple and complex models. *J. Geophys. Res.* **103**, 8653–8947.

Baerstchi P. (1976) Absolute ^{18}O content of Standard Mean Ocean Water. *Earth Planet. Sci. Lett.* **31**, 341–344.

Bailey I. H., Hulston J. R., Macklin W. C., and Stewart J. R. (1969) On the isotopic composition of hailstones. *J. Atmos. Sci.* **26**, 689–694.

Barnola J. M., Pimienta P., Raynaud D., and Korotkevich Y. S. (1991) CO_2 climate relationship as deduced from the Vostok ice core: a re-examination based on new measurements and on a re-evaluation of the air dating. *Tellus* **43B**, 83–91.

Bassinot F. C., Labeyrie L. D., Vincent E., Quidelleur X., Shackleton N. J., and Lancelot Y. (1994) The astronomical theory of climate and the age of the Brunhes-Matuyama magnetic reversal. *Earth Planet. Sci. Lett.* **126**, 91–108.

Bender M., Sowers T., Dickson M. L., Orchado J., Grootes P., Mayewski P. A., and Meese D. A. (1994) Climate connection between Greenland and Antarctica during the last 100,000 years. *Nature* **372**, 663–666.

Bender M., Malaize B., Orchado J., Sowers T., and Jouzel J. (1999) High precision correlations of Greenland and Antarctic ice core records over the last 100 kyr. In *Geophysical Monograph, 112, Mechanisms of Global Climate Change at Millenial Timescales* (eds. P. U. Clark, R. S. Webb, and L. D. Keigwin). American Geophysical Union, pp. 149–164.

Blunier T. and Brook E. J. (2001) Timing of millenial-scale climate change in Antarctica and Greenland during the last glacial period. *Science* **291**, 109–112.

Blunier T., Chappellaz J., Schwander J., Dallenbäch A., Stauffer B., Stocker T., Raynaud D., Jouzel J., Clausen H. B., Hammer C. U., and Johnsen S. J. (1998) Asynchrony of Antarctic and Greenland climate change during the last glacial period. *Nature* **394**, 739–743.

Blunier T., Schwander J., Chappellaz J., and Parrenin F. Antarctic last glacial temperature deduced from Δage. *Earth Planet. Sci. Lett.* (in press).

Boyle E. A. (1997) Cool tropical temperatures shift the global δ^{18}O-T relationship: an explanation for the ice core δ^{18}O borehole thermometry conflict? *Geophys. Res. Lett.* **24**, 273–276.

Caillon N. (2001) Composition isotopique de l'air piégé dans les glaces polaires: outil de paléothermométrie. Thèse de Doctorat de l'Université Paris VI, University Pierre et Marie Curie, Paris, 269pp.

Caillon N., severinghaus J. P., Barnola J. M., Chappellaz J. C., Jouzel J., and Parrenin F. (2001) Estimation of temperature change and of gas age–ice age difference, 108 kyr BP, at Vostok, Antarctica. *J. Geophys. Res.* **106**, 31893–31901.

Charles C., Rind D., Jouzel J., Koster R., and Fairbanks R. (1994) Glacial interglacial changes in moisture sources for Greenland: influences on the ice core record of climate. *Science* **261**, 508–511.

Charles C., Rind D., Jouzel J., Koster R., and Fairbanks R. (1995) Seasonal precipitation timing and ice core records. *Science* **269**, 247–248.

Charles C., Rind D., Healy R., and Webb R. (2001) Tropical cooling and the isotopic content of precipitation in general circulation model simulations of the ice age climate. *Clim. Dyn.* **17**, 489–502.

Chilshom A. J. (1973) Alberta hailstorms: 1. Radar case studies models. *Meteorol. Monogr.* **14**, 37–95.

Ciais P. and Jouzel J. (1994) Deuterium and Oxygen 18 in precipitation: an isotopic model including mixed cloud processes. *J. Geophys. Res.* **99**, 16793–16803.

CLIMAP (1981) *Seasonal Reconstructions of the Earth's Surface at the Last Glacial Maximum.* Geol. Soc. Am.

Cole J., Rind D., Jouzel J., Webb R. S., and Healy R. (1999) Global controls on interannual variability of precipitation δ^{18}O: the relative roles of temperature, precipitation amount, and vapor source region. *J. Geophys. Res.* **104**, 14223–14235.

Craig H. (1961) Isotopic variations in meteoric waters. *Science* **133**, 1702–1703.

Craig H. and Gordon A. (1965) Deuterium and oxygen 18 variations in the ocean and the marine atmosphere. In *Stable Isotopes in Oceanic Studies and Paleotemperatures.* Consiglio nazionalé delle Ricerche, Laboratorio di Geologia Nucleare, Pisa, Italy, PP. 9–130.

Cuffey K. M. and Vimeux F. (2001) Covariation of carbon dioxide and temperature from the Vostok ice core after deuterium-excess correction. *Nature* **421**, 523–527.

Cuffey K. M., Clow G. D., Alley R. B., Stuiver M., Waddington E. D., and Saltus R. W. (1995) Large Arctic temperature change at the Winconsin-Holocene glacial transition. *Science* **270**, 455–458.

Dahe Q., Petit J. R., Jouzel J., and Stievenard M. (1994) Distribution of stable isotopes in surface snow along the route of the 1990 International Trans-Antarctica Expedition. *J. Glaciol.* **40**, 107–118.

Dahl-Jensen D., Mosegaard K., Gundestrup N., Clow G. D., Johnsen S. J., Hansen A. W., and Balling N. (1998) Past temperatures directly from the Greenland ice sheet. *Science* **282**, 268–271.

Dansgaard W. (1953) The abundance of ^{18}O in atmospheric water and water vapour. *Tellus* **5**, 461–469.

Dansgaard W. (1964) Stable isotopes in precipitation. *Tellus* **16**, 436–468.

Dansgaard W., Johnsen S. J., Moller J., and Langway C. C. J. (1969) One thousand centuries of climatic record from Camp Century on the Greenland ice sheet. *Science* **166**, 377–381.

Dansgaard W., Clausen H. B., Gundestrup N., Hammer C. U., Johnsen S. J., Krinstindottir P., and Reeh N. (1982) A new Greenland deep ice core. *Science* **218**, 1273–1277.

Dansgaard W., Johnsen S., Clausen H. B., Dahl-Jensen D., Gundestrup N., Hammer C. U., and Oeschger H. (1984) North Atlantic climatic oscillations revealed by deep Greenland ice cores. In *Climate Processes and Climate Sensitivity* (eds. J. E. Hansen and T. Takahashi). American Geophysical Union, Washington, DC, pp. 288–298.

Dansgaard W., Johnsen S. J., Clausen H. B., Dahl-Jensen D., Gunderstrup N. S., Hammer C. U., Steffensen J. P., Sveinbjörnsdottir A., Jouzel J., and Bond G. (1993) Evidence for general instability of past climate from a 250-kyr ice-core record. *Nature* **364**, 218–220.

Delaygue G. (2000) Relationship between the oceanic surface and the isotopic content of Antarctic precipitation: simulations for different climates. PhD Thesis, University of Aix-Marseille.

Delaygue G., Masson V., Jouzel J., Koster R. D., and Healy C. (2000a) The origin of the Antarctic precipitation: a modelling approach. *Tellus* **27**, 19–36.
Delaygue G., Jouzel J., Masson V., Koster R. D., and Bard E. (2000b) Validity of the isotopic thermometer in central Antarctica: limited impact of glacial precipitation seasonality and moisture origin. *Geophys. Res. Lett.* **27**, 2677–2680.
Delmotte M. (1997) Enregistrements climatiques à Law-Dome: variabilité pour les périodes récentes et pour la déglaciation. Thèse de doctorat, Université Joseph Fourier.
Druyan L. M. and Koster R. D. (1989) Sources of Sahel precipitation for simulated drought and rainy seasons. *J. Climate* **2**, 1348–1466.
Ehhalt D. H. and Knott K. (1965) Kinetische isotopentrennungbei der verdampfung von wasser. *Tellus* **17**, 389–397.
Ehhalt D. H., Knott K., Nagel J. F., and Vogel J. C. (1963) Deuterium and oxygen 18 in rain water. *J. Geophys. Res.* **68**, 3775–3780.
Ekaykin A. A., Lipenkov V. Y., Barkov N. I., Petit J. R., and Masson-Delmotte V. (2002) Spatial and temporal variability in isotope composition of recent snow in the vicinity of Vostok Station: implications for ice-core record interpretation. *Ann. Glaciol.* **35**, 181–186.
Epstein S. and Mayeda T. (1953) Variations of ^{18}O content of waters from natural sources. *Geochim. Cosmochim. Acta* **4**, 213–224.
Epstein S. and Sharp R. P. (1970) Antarctic ice sheet: stable isotope analyses of Byrd station cores and interhemispheric climatic implications. *Science* **16**, 1570–1572.
Eriksson E. (1965) Deuterium and oxygen-18 in deuterium and other natural waters: some theoretical considerations. *Tellus* **16**, 498–512.
Fawcett P. J., Agustdottir A. M., Alley R. B., and Shuman C. A. (1997) The Younger Dryas termination and north Atlantic deepwater formation: insight from climate model simulations and Greenland ice core data. *Paleocanography* **12**, 23–38.
Federer B., Brichet N., and Jouzel J. (1982) Stable isotopes in hailstones: Part I. The isotopic cloud model. *J. Atmos. Sci.* **39**, 1323–1335.
Firestone J. (1995) Resolving the Younger Dryas event through borehole thermometry. *J. Glaciol.* **41**, 39–50.
Fisher D. A. (1991) Remarks on the deuterium excess in precipitation in cold regions. *Tellus* **43B**, 401–407.
Fisher D. A. (1992) Stable isotope simulations using a regional stable isotope model coupled to a zonally averaged global model. *Cold Reg. Sci. Technol.* **21**, 61–77.
Friedman I. (1953) Deuterium content of natural waters and other substances. *Geochim. Cosmochim. Acta* **4**, 89–103.
Friedman I., Machta L., and Soller R. (1962) Water vapor exchange between a water droplet and its environment. *J. Geophys. Res.* **67**, 2761–2766.
Friedman I., Redfield A. C., Schoen B., and Harris J. (1964) The variation of the deuterium content of natural waters in the hydrologic cycle. *Rev. Geophys.* **2**, 177–224.
Gallée H., Guyomarc'h G., and Brun E. (2001) Impact of snow drift on the Antarctic ice sheet surface mass balance: possible sensitivity to snow-surface properties. *Boundary-Layer Meteorol.* **99**, 1–19.
Gat J. (1980) The isotopes of hydrogen and oxygen in precipitation. In *Handbook of Environmental Isotopes Geochemistry: Volume 1. "The Terrestrial Environment A"* (eds. P. Fritz and J. C. Fontes). Elsevier, , vol. 1, pp. 22–46.
Gedzelman S. D. and Arnold R. (1994) Modeling the isotopic content of precipitation. *J. Geophys. Res.* **99**, 10455–10471.
Gonfiantini R. (1978) Standards for stable isotope measurements in natural compounds. *Nature* **271**, 534–536.
Grootes P. M., Stuiver M., White J. W. C., Johnsen S. J., and Jouzel J. (1993) Comparison of the oxygen isotope records from the GISP2 and GRIP Greenland ice cores. *Nature* **366**, 552–554.
Hageman R., Nief G., and Roth E. (1970) Absolute isotopic scale for deuterium analysis of natural waters. Absolute D/H ratio for SMOW. *Tellus* **22**, 712–715.
Hendricks M. B., De Paolo D. J., and Cohen R. C. (2000) Space and time variation of δD and ^{18}O in precipitation: can paleotemperature be estimated from ice cores? *Global Biogeochem. Cycles* **14**, 851–861.
Holdsworth G. (2001) Calibration changes in the isotopic thermometer according to different climatic states. *Geophys. Res. Lett.* **28**, 2625–2628.
Hoffmann G. and Heimann M. (1993) *Water tracers in the ECHAM general circulation model.* In *Isotope Techniques in the Study of Past and Current Environmental Changes in the Hydrosphere and the Atmosphere.* IAEA, Vienna, pp. 3–14.
Hoffmann G. and Heimann M. (1997) Water isotope modeling in the Asian monsoon region. *Quat. Int.* **37**, 115–128.
Hoffmann G., Werner M., and Heimann M. (1998) Water isotope module of the ECHAM atmospheric general circulation model: a study on time scales from days to several years. *J. Geophys. Res.* **103**, 16871–16896.
Hoffmann G., Masson V., and Jouzel J. (2000) Stable water isotopes in atmospheric general circulation models. *Hydrol. Process.* **14**, 1385–1406.
IAEA. (1981) *Statistical treatment of environmental isotope data in precipitation.* IAEA, Technical Report Series no. 206, Vienna.
IAEA. (1992) *Statistical treatment of data on environmental isotopes in precipitation.* IAEA, Technical Report Series no. 331, Vienna.
Johnsen S. J., Dansgaard W., and White J. W. (1989) The origin of Arctic precipitation under present and glacial conditions. *Tellus* **41**, 452–469.
Johnsen S. J., Clausen H. B., Dansgaard W., Fuhrer K., Gunderstrup N. S., Hammer C. U., Iverssen P., Jouzel J., Stauffer B., and Steffensen J. P. (1992a) Irregular glacial interstadials recorded in a new Greenland ice core. *Nature* **359**, 311–313.
Johnsen S. J., Claussen H. B., Dansgaard W., Gundestrup N. S., Hansson N. S., Jonsson P., Steffenssen J. P., and Sveinbjörnsdottir A. E. (1992b) A "deep" ice core from East Greenland. *Medellelser om Groenland Geosci.* **29**, 1–22.
Johnsen S. J., Dahl-Jensen D., Dansgaard W., and Gundestrup N. (1995) Greenland paleotemperatures derived from GRIP bore hole temperature and ice core isotope profiles. *Tellus* **47B**, 624–629.
Johnsen S. J., Clausen H. B., Dansgard W., Gunderstrup N. S., Sveinbjörnsdottir A., Jouzel J., Hammer C. U., Anderssen U., Fisher D., and White J. (1997) The $\delta^{18}O$ record along the GRIP deep ice core. *J. Geophys. Res.* **102**, 26397–26410.
Johnsen S., Dahl-Jensen D., Gundestrup N., Steffenssen J. P., Clausen H. B., Miller H., Masson-Delmotte V., Sveinbjörnsdottir A. E., and White J. (2001) Oxygen isotope and palaeotemperature records from six Greenland ice-core stations: Camp Century, Dye-3, GRIP, GISP2, Renland and NorthGRIP. *J. Quat. Sci.* **16**, 299–307.
Joussaume S. and Jouzel J. (1993) Paleoclimatic tracers: an investigation using an atmospheric General Circulation Model under ice age conditions: 2. Water isotopes. *J. Geophys. Res.* **98**, 2807–2830.
Joussaume S., Jouzel J., and Sadourny R. (1984) A general circulation model of water isotope cycles in the atmosphere. *Nature* **311**, 24–29.
Joussaume S. J., Sadourny R., and Vignal C. (1986) Origin of precipitating water in a numerical simulation of the July climate. *Ocean–Air Interact.* **1**, 43–56.
Jouzel J. (1986) Isotopes in cloud physics: multisteps and multistages processes. In *Handbook of Environmental Isotopes Geochemistry: Volume 2: "The Terrestrial Environment B"*. (eds. P. Fritz and J. C. Fontes). Elsevier, vol. 2, pp. 61–112.
Jouzel J. (1999) Calibrating the isotopic paleothermometer. *Science* **286**, 910–911.
Jouzel J. and Koster R. (1996) A reconsideration of the initial conditions used for stable water isotopes models. *J. Geophys. Res.* **101**, 22933–22938.

Jouzel J. and Merlivat L. (1980) Growth regime of hailstones as deduced from simultaneous deuterium and oxygen 18 measurements. In *VIIIème International Conference on Cloud Physics.* Conference Proceedings AIMPA, Clermont-Ferrand, pp. 253–256.

Jouzel J. and Merlivat L. (1984) Deuterium and oxygen 18 in precipitation: modeling of the isotopic effects during snow formation. *J. Geophys. Res.* **89**, 11749–11757.

Jouzel J. and Souchez R. A. (1982) Melting-refreezing at the glacier sole and the isotopic composition of the ice. *J. Glaciol.* **28**, 35–42.

Jouzel J., Merlivat L., and Roth E. (1975) Isotopic study of hail. *J. Geophys. Res.* **80**, 5015–5030.

Jouzel J., Brichet N., Thalmann B., and Federer B. (1980) A numerical cloud model to interpret the isotope content of hailstones. In *VIIIème International Conference on Cloud Physics.* Conference AIMPA, Clermont-Ferrand, France, pp. 249–252.

Jouzel J., Merlivat L., Petit J. R., and Lorius C. (1983) Climatic information over the last century deduced from a detailed isotopic record in the South Pole snow. *J. Geophys. Res.* **88**, 2693–2703.

Jouzel J., Merlivat L., and Federer B. (1985) Isotopic study of hail. The δD-$\delta^{18}O$ relationship and growth history of large hailstones. *Quart. J. Roy. Meteorol. Soc.* **111**, 495–514.

Jouzel J., Russell G. L., Suozzo R. J., Koster R. D., White J. W. C., and Broecker W. S. (1987a) Simulations of the HDO and $H_2{}^{18}O$ atmospheric cycles using the NASA/GISS general circulation model: the seasonal cycle for present-day conditions. *J. Geophys. Res.* **92**, 14739–14760.

Jouzel J., Lorius C., Petit J. R., Genthon C., Barkov N. I., Kotlyakov V. M., and Petrov V. M. (1987b) Vostok ice core: a continuous isotope temperature record over the last climatic cycle (160,000 years). *Nature* **329**, 402–408.

Jouzel J., Koster R. D., Suozzo R. J., Russell G. L., White J. W., and Broecker W. S. (1991) Simulations of the HDO and $H_2{}^{18}O$ atmospheric cycles using the NASA GISS General Circulation Model: sensitivity experiments for present day conditions. *J. Geophys. Res.* **96**, 7495–7507.

Jouzel J., Barkov N. I., Barnola J. M., Bender M., Chappelaz J., Genthon C., Kotlyakov V. M., Lipenkov V., Lorius C., Petit J. R., Raynaud D., Raisbeck G., Ritz C., Sowers T., Stievenard M., Yiou F., and Yiou P. (1993) Extending the Vostok ice-core record of paleoclimate to the penultimate glacial period. *Nature* **364**, 407–412.

Jouzel J., Koster R. D., Suozzo R. J., and Russell G. L. (1994) Stable water isotope behaviour during the LGM: a GCM analysis. *J. Geophys. Res.* **99**, 25791–25801.

Jouzel J., Vaikmae R., Petit J. R., Martin M., Duclos Y., Stievenard M., Lorius C., Toots M., Mélières M. A., Burckle L. H., Barkov N. I., and Kotlyakov V. M. (1995) The two-step shape and timing of the last deglaciation in Antarctica. *Clim. Dyn.* **11**, 151–161.

Jouzel J., Waelbroeck C., Malaizé B., Bender M., Petit J. R., Barkov N. I., Barnola J. M., King T., Kotlyakov V. M., Lipenkov V., Lorius C., Raynaud D., Ritz C., and Sowers T. (1996) Climatic interpretation of the recently extended Vostok ice records. *Clim. Dyn.* **12**, 513–521.

Jouzel J., Alley R. B., Cuffey K. M., Dansgaard W., Grootes P., Hoffmann G., Johnsen S. J., Koster R. D., Peel D., Shuman C. A., Stievenard M., Stuiver M., and White J. (1997) Validity of the temperature reconstruction from ice cores. *J. Geophys. Res.* **102**, 26471–26487.

Jouzel J., Petit J. R., Souchez R., Barkov N. I., Lipenkov V. Y., Raynaud D., Stievenard M., Vassiliev N. I., Verbeke V., and Vimeux F. (1999) More than 200 m thick of lake ice above the subglacial Lake Vostok, Antarctica. *Science* 2138–2141.

Jouzel J., Hoffmann G., Koster R. D., and Masson V. (2000) Water isotopes in precipitation: data/model comparison for present-day and past climates. *Quat. Sci. Rev.* **19**, 363–379.

Jouzel J., Masson V., Cattani O., Falourd S., Stievenard M., Stenni B., Longinelli A., Johnsen S. J., Steffenssen J. P., Petit J. R., Schwander J., Souchez R., and Barkov N. I. (2001) A new 27 Ky high resolution East Antarctic climate record. *Geophys. Res. Lett.* **28**, 3199–3202.

Jouzel J., Vimeux F., Caillon N., Delaygue G., Hoffmann G., Masson V., and Parrenin F. (2003) Magnitude of the isotope/temperature scaling for interpretation of central Antarctic ice cores. *J. Geophys. Res.* **108**(D12), doi:10.1029/2002JD002677.

Kessler E. (1969) On the distribution and continuity of water substance in atmospheric circulation. *Meterol. Monogr.* **10**, 84.

Koster R. D., Jouzel J., Suozzo R., Russel G., Broecker W., Rind D., and Eagleson P. S. (1986) Global sources of local precipitations as determined by the NASA/GISS GCM. *Geophys. Res. Lett.* **43**, 121–124.

Koster R. D., Jouzel J., Suozzo R. J., and Russel G. L. (1992) Origin of July Antarctic precipitation and its influence on deuterium content: a GCM analysis. *Clim. Dyn.* **7**, 195–203.

Koster R. D., De Valpine P., and Jouzel J. (1993) Continental water recycling and stable water isotope concentration. *Geophys. Res. Lett.* **20**, 2215–2218.

Krinner G., Genthon C., and Jouzel J. (1997) GCM analysis of local influences on ice core δ signals. *Geophys. Res. Lett.* **24**, 2825–2828.

Lang C., Leuenberger M., Schwander J., and Johnsen S. J. (1999) 16 °C rapid temperature variation in central Greenland 70,000 years ago. *Science* **286**, 934–937.

Legrand M., Jouzel J., and Raynaud D. (1994) Past climate and trace gas content of the atmosphere inferred from polar ice cores. In *ERCA Publication* (eds.), pp. 453–477.

Leuenberger M., Lang C., and Schwander J. (1999) $\delta^{15}N$ measurements as a calibration tool for the paleothermometer and gas-ice age differences: a case study for the 8200 B.P. event on GRIP ice. *J. Geophys. Res.* **104**(D18), 22163–22170.

Lorius C. and Merlivat L. (1977) Distribution of mean surface stable isotope values in East Antarctica. Observed changes with depth in a coastal area. In *Isotopes and Impurities in Snow and Ice. Proceedings of the Grenoble Symposium Aug./Sep. 1975* (eds. IAHS), IAHS, vol. 118, pp. 125–137.

Lorius C., Merlivat L., Jouzel J., and Pourchet M. (1979) A 30,000 yr isotope climatic record from Antarctic ice. *Nature* **280**, 644–648.

Lorius C., Jouzel J., Ritz C., Merlivat L., Barkov N. I., Korotkevitch Y. S., and Kotlyakov V. M. (1985) A 150,000-year climatic record from Antarctic ice. *Nature* **316**, 591–596.

Majoube M. (1971a) Fractionement en Oxygène 18 et en deutérium entre l'eau et sa vapeur. *J. Chim. Phys.* **10**, 1423–1436.

Majoube M. (1971b) Fractionement en Oxygène 18 et en deutérium entre la glace et sa vapeur. *J. Chim. Phys.* **68**, 635–636.

Mason B. J. (1971) The Physics of Clouds (ed. Oxford Press).

Mathieu R., Pollard D., Cole J., Webb R., White J. W. C., and Thompson S. (2002) Simulation of stable water isotope variations by the GENESIS GCM for present-day conditions. *J. Geophys. Res.* **107**, D4, 10.1029.

Merlivat L. (1978) Molecular diffusivities of $H_2{}^{16}O$, $HD^{16}O$, and $H_2{}^{18}O$ in gases. *J. Chem. Phys.* **69**, 2864–2871.

Merlivat L. and Jouzel J. (1979) Global climatic interpretation of the deuterium–oxygen 18 relationship for precipitation. *J. Geophys. Res.* **84**, 5029–5033.

Merlivat L. and Nief G. (1967) Fractionnement isotopique lors des changements d'état solide-vapeur et liquide-vapeur de l'eau à des températures inférieures à 0 °C. *Tellus* **19**, 122–127.

Morgan V., Delmotte M., van Ommen T., Jouzel J., Chappellaz J., Woon S., Masson-Delmotte V., and Raynaud D. (2002) The timing of events in the last deglaciation from a coastal east Antarctic core. *Science* **297**, 1862–1864.

Noone D. and Simmonds I. (2002a) Associations between $\delta^{18}O$ of water and climate parameters in a simulation of atmospheric circulation for 1979–1995. *J. Climate.* **15**(22), 3150–3169.

Noone D. and Simmonds I. (2002b) Annular variations in moisture transport mechanisms and the abundance of $\delta^{18}O$ in Antarctic snow. *J. Geophys. Res.* **107**(D24), 4742, doi: 10.1029/2002JD002262.

O'Neil (1968) Hydrogen and oxygen isotope fractionation between ice and water. *J. Chem. Phys.* **72**, 3683–3684.

Parrenin F., Jouzel J., Waelbroeck C., Ritz C., and Barnola J. M. (2001) Dating the Vostok ice core by an inverse method. *J. Geophys. Res.* **106**, 31837–31851.

Paterson W. S. B. and Clarke G. K. C. (1978) Comparison of theoretical and observed temperature profiles in Devon Island ice cap, Cananda. *Geophys. J. Roy. Astron. Soc.* **55**, 615–632.

Petit J. R., White J. W. C., Young N. W., Jouzel J., and Korotkevich Y. S. (1991) Deuterium excess in recent Antarctic snow. *J. Geophys. Res.* **96**, 5113–5122.

Petit J. R., Jouzel J., Raynaud D., Barkov N. I., Barnola J. M., Basile I., Bender M., Chappellaz J., Davis J., Delaygue G., Delmotte M., Kotyakov V. M., Legrand M., Lipenkov V. Y., Lorius C., Pépin L., Ritz C., Saltzman E., and Stievenard M. (1999) Climate and atmospheric history of the past 420000 years from the Vostok ice core, Antarctica. *Nature* **399**, 429–436.

Raisbeck G. M., Yiou F., and Jouzel J. (2002) Cosmogenic 10Be as a high resolution correlation tool for climate records. In *Goldschmidt Conference*, Conference Davos, Switzerland.

Ramirez E., Hoffmann G., Taupin J. L., Francou B., Ribstein P., Caillon N., Landais A., Petit J. R., Pouyaud B., Schotterer U., and Stiévenard M. (2003) A new Andean deep ice core from the Illimani (6,350 m), Bolivia. *Earth Planet. Sci. Lett.* **212**, 337–350.

Raynaud D. and Lorius C. (1973) Climatic implications of total gas content in ice at Camp Century. *Nature* **243**, 283–284.

Ritz C. (1992) Un modele thermo-mecanique d'evolution pour le bassin glaciaire Antarctique Vostok-Glacier Byrd: sensibilite aux valeurs des parametres mal connus. Thèse d'Etat, Univ. de Grenoble.

Robin G. d. Q. (1977) Ice cores and climatic changes. *Phil. Trans. Roy. Soc. London, Ser. B* **280**, 143–168.

Rogers R. R. (1979) *A Short Course in Clouds Physics*. (ed. Pergamon).

Rommelaëre V. (1997) Trois problèmes inverses en glaciologie. PhD Thesis (in french), University Grenoble I, France.

Rozanski K., Araguas-Araguas L., and Gonfiantini R. (1992) Relation between long-term trends of oxygen-18 isotope composition of precipitation and climate. *Science* **258**, 981–985.

Rozanski K., Araguas-Araguas L., and Gonfiantini R. (1993) Isotopic pattern in modern global precipitation. In *Climate Change in Continental Isotopic Records*, Geophysical Monograph 78 (eds. P. K. Swart, K C L J McKenzie, and S. Savin). American Geophysical Union, Washington, DC, pp. 1–37.

Salamatin A. N., Lipenkov V. Y., Barkov N. I., Jouzel J., Petit J. R., and Raynaud D. (1998) Ice core age dating and paleothermometer calibration on the basis of isotopes and temperature profiles from deep boreholes at Vostok station (East Antarctica). *J. Geophys. Res.* **103**, 8963–8977.

Schwander J., Sowers T., Barnola J. M., Blunier T., Malaizé B., and Fuchs A. (1997) Age scale of the air in the summit ice: implication for glacial-interglacial temperature change. *J. Geophys. Res.* **D16**, 19483–19494.

Schwander J., Jouzel J., Hammer C. U., Petit J. R., Udisti R., and Wolff E. (2001) A tentative chronology of the EPICA Dome C ice core. *Geophys. Res. Lett.* **28**, 4243–4246.

Severinghaus J. P. and Brook E. (1999) Simultaneous tropical-Arctic abrupt climate change at the end of the last glacial period inferred from trapped air in polar ice. *Science* **286**, 930–934.

Severinghaus J. P., Brook E. J., Sowers T. and Alley R. B. (1996) Gaseous thermal diffusion as a gas-phase stratigraphic marker of abrupt warmings in ice core climate records. *EOS Supplement*, AGU Spring meeting, 157.

Severinghaus J. P., Sowers T., Brook E., Alley R. B., and Bender M. L. (1998) Timing of abrupt climate change at the end of the Younger Dryas interval from thermally fractionated gases in polar ice. *Nature* **391**, 141–146.

Severinghaus J. P., Grachev A., Luz B., and Caillon N. (2003) A method for precise measurement of argon 40/36 and krypton/argon ratios in trapped air in polar ice with applications to past firn thickness and abrupt climate change in Greenland and at Siple Dome, Antarctica. *Geochim. Cosmochim. Acta* **67**(3), 325–343.

Steig E., Brook E. J., White J. W. C., Sucher C. M., Bender M. L., Lehman S. J., Morse D. L., Waddigton E. D., and Clow G. D. (1998) Synchronous climate changes in Antarctica and the North Atlantic. *Science* **282**, 92–95.

Stenni B., Masson V., Johnsen S. J., Jouzel J., Longinelli A., Monnin E., Roethlisberger R., and Selmo E. (2001) An oceanic cold reversal during the last deglaciation. *Science* **293**, 2074–2077.

Stewart M. K. (1975) Stable isotope fractionation due to evaporation and isotopic exchange of falling water drops: application to atmospheric processes and evaporation of lakes. *J. Geophys. Res.* **80**, 1133–1146.

Taylor C. B. (1972) *The Vertical Distribution of the Isotopic Concentrations of Tropospheric Water Vapour over Continental Europe and their Relationship to Tropospheric Structure.* Report: INS-R, 107. N. Z. Dep. Sci. Ind.; Res., Inst. Nucl. Sci.

Taylor K. C., Hammer C. U., Alley R. B., Clausen H. B., Dahl-Jensen D., Gow A. J., Gundestrup N. S., Kipfstuhl J., Moore J. C., and Waddington E. D. (1993) Electrical conductivity measurements from the GISP2 and GRIP Greenland ice cores. *Nature* **366**, 549–552.

Thompson L. G., Mosley-Thompson E., Davis M. E., Lin P.-N., Henderson K. A., Cole-Dai J., Bolzan J. F., and Liu K.-B. (1995) Late Glacial stage and Holocene tropical- ice core records from Huascaran Peru. *Science* **269**, 46–50.

Thompson L. G., Davis M. E., Mosley-Thompson E., Sowers T. A., Henderson K. A., Zagorodnov V. S., Lin P.-N., Mikhalenko V. N., Campen R. K., Bolzan J. F., Cole-Dai J., and Francou B. (1998) A 25,000-tear tropical climate history from bolivian ice cores. *Science* **282**, 1858–1864.

van Ommen T. D. and Morgan V. (1997) Calibrating the ice core paleothermometer using seasonality. *J. Geophys. Res.* **102**, 9351–9357.

Vimeux F., Cuffey K., and Jouzel J. (2002) New insights into southern hemisphere temperature changes from Vostok ice cores using deuterium excess correction. *Earth. Planet. Sci. Lett.* **203**, 829–843.

Watanabe O., Jouzel J., Johnsen S., Parrenin F., Shoji H., and Yoshida (2003) Homogeneous climate variability accross East Antarctica over the last three glacial cycles. *Nature* **422**, 509–512.

Werner M., Heimann M. and Hoffmann G. (1998) Stable water isotopes in Greenland ice cores: ECHAM4 model simulations versus field measurements. In *International Symposium on Isotope Techniques in the Study of Past and Current Environmental Changes in the Hydrosphere and the Atmosphere.* IAEA Conference Vienna (Austria), pp. 603–612.

Werner M., Mikolajewicz U., Heimann M., and Hoffmann G. (2000) Borehole versus isotope temperatures on Greenland: seasonality does matter. *Geophys. Res. Lett.* **27**, 723–726.

Werner M., Heimann M., and Hoffmann G. (2001) Isotopic composition and origin of polar precipitation in present and glacial climate simulations. *Tellus* **53B**, 53–71.

Werner M. and Heimann M. (2002) Modeling interannual variability of water isotopes in Greenland and Antarctica. *J. Geophys. Res.* **107**, No. D1, p. ACL-1 1-13.

Yurtserver Y. and Gat J. (1981) Atmospheric waters. In *Stable Isotope Hydrology: Deuterium and Oxygen18 in the Water Cycle*. IAEA, Vienna, pp. 103–142.

Isotope Geochemistry
ISBN: 978-0-08-096710-3

pp. 151–180

7

Stable Isotope Applications in Hydrologic Studies

C. Kendall and D. H. Doctor

United States Geological Survey, Menlo Park, CA, USA

7.1 INTRODUCTION

The topic of stream flow generation has received considerable attention over the last two decades, first in response to concern about "acid rain" and more recently in response to the increasingly serious contamination of surface and shallow groundwaters by anthropogenic contaminants. Many sensitive, low-alkalinity streams in North America and Europe are already acidified. Still more streams that are not yet chronically acidic may undergo acidic episodes in response to large rainstorms and/or spring snowmelt. These acidic events can seriously damage local ecosystems. Future climate changes may exacerbate the situation by affecting biogeochemical controls on the transport of water, nutrients, and other materials from land to freshwater ecosystems.

New awareness of the potential danger to water supplies posed by the use of agricultural chemicals and urban industrial development has also focused attention on the nature of rainfall-runoff and recharge processes and the mobility of various solutes, especially nitrate and pesticides, in shallow systems. Dumping and spills of other potentially toxic materials are also of concern because these chemicals may eventually reach streams and other public water supplies. A better understanding of hydrologic flow paths and solute sources is required to determine the potential impact of contaminants on water supplies, develop management practices to preserve water quality, and devise remediation plans for sites that are already polluted.

Isotope tracers have been extremely useful in providing new insights into hydrologic processes, because they integrate small-scale variability to give an effective indication of catchment-scale processes. The main purpose of this chapter is to provide an overview of recent research into the use of naturally occurring stable isotopes to track the movement of water and solutes in hydrological systems where the waters are relatively fresh: soils, surface waters, and shallow groundwaters. For more information on shallow-system applications, the reader is referred to Kendall and McDonnell (1998). For information on groundwater systems, see Cook and Herczeg (2000).

7.1.1 Environmental Isotopes as Tracers

Environmental isotopes are naturally occurring (or, in some cases, anthropogenically produced) isotopes whose distributions in the hydrosphere can assist in the solution of hydrological and biogeochemical problems. Typical uses of environmental isotopes include the identification of sources of water and solutes, determination of water flow paths, assessment of biologic cycling of nutrients within the ecosystem, and testing flow path, water budget, and geochemical models developed using hydrologic or geochemical data.

Environmental isotopes can be used as tracers of waters and solutes in shallow low-temperature environments because:

(i) Waters that were recharged at different times, were recharged in different locations, or that followed different flow paths are often isotopically distinct; in other words, they have distinctive "fingerprints."

(ii) Unlike most chemical tracers, environmental isotopes are relatively conservative in reactions with the bedrock and soil materials. This is especially true of oxygen and hydrogen isotopes in water; the waters mentioned above retain their distinctive fingerprints until they mix with other waters.

(iii) Solutes in the water that are derived from atmospheric sources are often isotopically distinct from solutes derived from geologic and biologic sources within the catchment.

(iv) Both biological cycling of solutes and water/rock reactions often change isotopic ratios in the solutes in predictable and recognizable directions; these processes often can be reconstructed from the isotopic compositions.

The applications of environmental isotopes as hydrologic tracers in low-temperature, freshwater systems fall into two main categories: (i) tracers of the water itself (water isotope hydrology) and (ii) tracers of the solutes in the water (solute isotope biogeochemistry). These classifications are by no means universal but they are useful conceptually and often eliminate confusion when comparing results using different tracers.

Water isotope hydrology focuses on the isotopes that form water molecules: the oxygen isotopes (oxygen-16, oxygen-17, and oxygen-18) and the hydrogen isotopes (protium, deuterium, and tritium). These isotopes are ideal tracers of water sources and movement because they are constituents of water molecules, not something that is dissolved in the water like other tracers that are commonly used in hydrology (e.g., dissolved species such as chloride).

Oxygen and hydrogen isotopes are generally used to determine the source of the water (e.g., precipitation versus groundwater in streams, recharge of evaporated lake water versus snowmelt water in groundwater). In catchment research, the main use is for determining the contributions of "old" and "new" water to high flow (storm and snowmelt runoff) events in streams. "Old" water is defined as the water that existed in a catchment prior to a particular storm or snowmelt period. Old water includes groundwater, soil water, and surface water. "New" water is either rainfall or snowmelt, and is defined as the water that triggers the particular storm or snowmelt runoff event.

Solute isotope biogeochemistry focuses on isotopes of constituents that are dissolved in the water or are carried in the gas phase. The most commonly studied solute isotopes are the isotopes of carbon, nitrogen, and sulfur. Less commonly investigated stable, nonradiogenic isotopes include lithium, chloride, boron, and iron.

Although the literature contains numerous case studies involving the use of solutes (and sometimes solute isotopes) to trace water sources and flow paths, such applications include an implicit assumption that these solutes are transported conservatively with the water. Unlike the isotopes in the water molecules, the ratios of solute isotopes can be significantly altered by reaction with geological and/or biological materials as the water moves through the system. While the utility of solutes in the evaluation of rainfall-runoff processes has been repeatedly demonstrated, in a strict sense, *solute isotopes only trace solutes.* Solute isotopes also provide information on the reactions that are responsible for their presence in the water and the flow paths implied by their presence.

7.1.2 Isotope Fundamentals

7.1.2.1 Basic principles

Isotopes are atoms of the same element that have different numbers of neutrons but the same number of protons and electrons. The difference in the number of neutrons between the various isotopes of an element means that the various isotopes have different masses. The superscript number to the left of the element abbreviation indicates the number of protons plus neutrons in the isotope. For example, among the hydrogen isotopes, deuterium (denoted as ^2H or D) has one neutron and one proton. This is approximately twice the mass of protium (^1H), whereas tritium (^3H) has approximately three times the mass of protium.

The stable isotopes have nuclei that do not decay to other isotopes on geologic timescales, but may themselves be produced by the decay of radioactive isotopes. Radioactive (unstable) isotopes have nuclei that spontaneously decay over time to form other isotopes. For example, ^{14}C, a radioisotope of carbon, is produced in the atmosphere by the interaction of cosmic-ray neutrons with stable ^{14}N. With a half-life of ~5,730 yr, ^{14}C decays back to ^{14}N by emission of a beta particle. The stable ^{14}N produced by radioactive decay is called "radiogenic" nitrogen. This chapter focuses on stable, nonradiogenic isotopes. For a more thorough discussion of the fundamentals of isotope geochemistry, see Clark and Fritz (1997) and Kendall and McDonnell (1998).

7.1.2.2 Isotope fractionation

For elements of low atomic numbers, the mass differences between the isotopes of an element are large enough for many physical, chemical, and biological processes or reactions to "fractionate" or change the relative proportions of various isotopes. Two different types of processes—equilibrium isotope effects and kinetic isotope effects—cause isotope fractionation. As a consequence of fractionation processes, waters and solutes often develop unique isotopic compositions (ratios of heavy to light isotopes) that may be indicative of their source or of the processes that formed them.

Equilibrium isotope-exchange reactions involve the redistribution of isotopes of an element among various species or compounds in a closed, well-mixed system at chemical equilibrium. At isotope equilibrium, the forward and backward reaction rates of any particular isotope are identical. This does not mean that the isotopic compositions of two compounds at equilibrium are identical, but only that the ratios of the different isotopes in each compound are constant for a particular temperature. During equilibrium reactions, the heavier isotope generally preferentially accumulates in the species or compound with the higher oxidation state. For example, during sulfide oxidation, the sulfate becomes enriched in ^{34}S relative to sulfide (i.e., has a more positive $\delta^{34}S$ value); consequently, the residual sulfide becomes depleted in ^{34}S.

Chemical, physical, and biological processes can be viewed as either reversible *equilibrium* reactions or irreversible unidirectional *kinetic* reactions. In systems out of chemical and isotopic equilibrium, forward and backward reaction rates are not identical, and isotope reactions may, in fact, be unidirectional if reaction products become physically isolated from the reactants. Such reaction rates are dependent on the ratios of the masses of the isotopes and their vibrational energies, and hence are called *kinetic* isotope fractionations.

The magnitude of a kinetic isotope fractionation depends on the reaction pathway, the reaction rate, and the relative bond energies of the bonds being severed or formed by the reaction. Kinetic fractionations, especially unidirectional ones, are usually larger than the equilibrium fractionation factor for the same reaction in most low-temperature environments. As a rule, bonds between the lighter isotopes are broken more easily than equivalent bonds of heavier isotopes. Hence, the light isotopes react faster and become concentrated in the products, causing the residual reactants to become enriched in the heavy isotopes. In contrast, reversible equilibrium reactions can produce products heavier or lighter than the original reactants.

Biological processes are generally unidirectional and are excellent examples of "kinetic" isotope reactions. Organisms preferentially use the lighter isotopic species because of the lower energy "costs," resulting in significant fractionations between the substrate (which generally becomes isotopically heavier) and the biologically mediated product (which generally becomes isotopically lighter). The magnitude of the fractionation depends on the reaction pathway utilized and the relative energies of the bonds being severed and formed by the reaction. In general, slower reaction steps show greater isotopic fractionation than faster steps because the organism has time to be more selective. Kinetic reactions can result in fractionations very different from, and typically larger than, the equivalent equilibrium reaction.

Many reactions can take place either under purely equilibrium conditions or be affected by an additional kinetic isotope fractionation. For example, although evaporation can take place under purely equilibrium conditions (i.e., at 100% humidity when the air is still), more typically the products become partially isolated from the reactants (e.g., the resultant vapor is blown downwind). Under these conditions, the isotopic compositions of the water and vapor are affected by an additional kinetic isotope fractionation of variable magnitude.

The partitioning of stable isotopes between two substances A and B can be expressed by using the isotopic fractionation factor α (alpha):

$$\alpha = R_p / R_s \tag{1}$$

where R_p and R_s are the ratios of the heavy to light isotope (e.g., $^2H/^1H$ or $^{18}O/^{16}O$) in the *product* and *substrate* (reactant), respectively. Values for α tend to be very close to 1. An isotope enrichment factor, ε, can be defined as

$$\varepsilon_{p-s} = (\alpha - 1) \times 1{,}000 \tag{2}$$

7.1.2.3 Rayleigh fractionation

The Rayleigh equation is an exponential relation that describes the partitioning of isotopes between two reservoirs as one reservoir decreases in size. The equations can be used to describe an isotope fractionation process if: (i) the material is continuously removed from a mixed system containing molecules of two or more isotopic species (e.g., water with ^{18}O and ^{16}O, or sulfate with ^{34}S and ^{32}S); (ii) the fractionation accompanying the removal process at any instant is described by the fractionation factor α; and (iii) α does not change during the process. The general form of the Rayleigh equation is

$$R = R_o f^{(\alpha-1)} \tag{3}$$

where R_o is the isotope ratio (e.g., $^{18}O/^{16}O$) of the original substrate, and f is the fraction of remaining substrate. A commonly used approximate version of the equation is

$$\delta \cong \delta_o + \varepsilon \ln(f) \tag{4}$$

where δ_o is the initial isotopic composition. This approximation is valid for δ_o values near 0 and positive ϵ values less than $\sim +10‰$.

In a strict sense, the term "Rayleigh fractionation" should only be used for chemically *open* systems where the isotopic species removed at every instant are in thermodynamic and isotopic equilibrium with those remaining in the system at the moment of removal. Furthermore, such an "ideal" Rayleigh distillation is one where the reactant reservoir is finite and well mixed, and does not re-react with the product (Clark and Fritz, 1997). However, the term "Rayleigh fractionation" is commonly applied to equilibrium closed systems and kinetic fractionations as well because the situations may be computationally identical. Isotopic fractionations are strongly affected by whether a system is open or closed.

7.1.2.4 Terminology

Stable isotopic compositions are normally reported as delta (δ) values in parts per thousand (denoted as ‰ or permil, per mil, per mille, or per mill) enrichments or depletions relative to a standard of known composition. The term "δ" is spelled and pronounced delta, not del. δ values are calculated by

$$\delta(\text{in‰}) = (R_{\text{sample}}/R_{\text{standard}} - 1) \times 1{,}000 \quad (5)$$

where "R" is the ratio of the heavy to light isotope. A positive delta value means that the isotopic ratio of the sample is higher than that of the standard; a negative value means that the isotope ratio of the sample is lower than that of the standard.

Stable oxygen and hydrogen isotopic ratios are normally reported relative to the "standard mean ocean water" (SMOW) standard (Craig, 1961b) or the equivalent Vienna-SMOW (V-SMOW) standard. Carbon, nitrogen, and sulfur stable isotope ratios are reported relative to the Pee Dee Belemnite (PDB) or Vienna-PDB (VPDB), ambient air (AIR), and Canyon Diablo Troilite (CDT) standards, respectively, as defined later. The use of the "V" before SMOW or PDB indicates that the measurements were calibrated on normalized per mil scales (Coplen, 1996).

The δ values for stable isotopes are determined using isotope ratio mass spectrometry, typically either using gas-source stable isotope mass spectrometers (e.g., hydrogen, carbon, nitrogen, oxygen, sulfur, chlorine, and bromine) or solid-source mass spectrometry (e.g., lithium, boron, and iron). The analytical precisions are small relative to the ranges in δ values that occur in natural earth systems. Typical one standard deviation (1σ) analytical precisions for oxygen, carbon, nitrogen, and sulfur isotopes are in the 0.05–0.2‰ range; typical precisions for hydrogen isotopes are generally poorer, from 0.2‰ to 1.0‰, because of the low ^{2}H:^{1}H ratio of natural materials.

7.1.3 Causes of Isotopic Variation

7.1.3.1 Isotope fractionations during phase changes

The most important phase changes in hydrological systems are associated with water as it condenses, evaporates, and melts. At equilibrium, the isotope fractionation between two coexisting phases is a function of temperature. As the phase change proceeds, the δ values of both the reactant and product change in a regular fashion. For example, as water evaporates, the $\delta^{18}O$ and $\delta^{2}H$ values of the residual water increase, and the δ values of the resulting vapor also increase. The lower the humidity, the greater the kinetic fractionation associated with the phase change. The changes in their δ values can be described using the Rayleigh equations. The Rayleigh equation applies to an open system from which material is removed continuously under condition of a constant fractionation factor.

The isotope enrichment achieved can be very different in closed versus open systems. For example, Figure 1 shows the changes in the $\delta^{18}O$ of water and vapor during both open-system (solid lines) and closed-system (dashed lines) evaporation with a constant fractionation factor α_{l-v}=1.010 (i.e., the newly formed vapor is always 10‰ lower than the residual water). During open-system evaporation where the vapor is continuously removed (i.e., isolated from the water), as evaporation progresses (i.e., $f \rightarrow 0$), the $\delta^{18}O$ of the remaining water (solid line A) becomes higher. The $\delta^{18}O$ of the instantaneously formed vapor (solid line B) describes a curve parallel to that of the remaining water, but lower than it (for all values of f) by the precise amount dictated by the fractionation factor for ambient

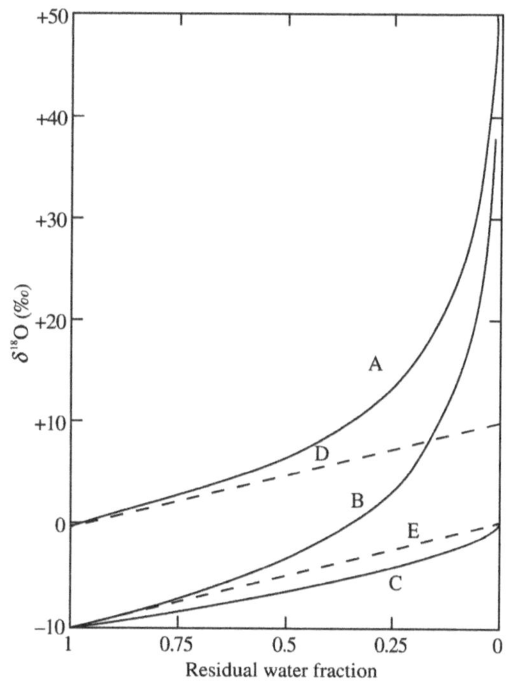

Figure 1 Isotopic change under open- and closed-system Rayleigh conditions for evaporation with a fractionation factor $\alpha = 1.01$ for an initial liquid composition of $\delta^{18}O = 0$. The $\delta^{18}O$ of the remaining water (solid line A), the instantaneous vapor being removed (solid line B), and the accumulated vapor being removed (solid line C) all increase during single-phase, open-system, evaporation under equilibrium conditions. The $\delta^{18}O$ of water (dashed line D) and vapor (dashed line E) in a two-phase closed system also increase during evaporation, but much less than in an open system; for a closed system, the δ values of the instantaneous and cumulative vapor are identical (after Gat and Gonfiantini, 1981).

temperature, in this case by 10‰. For higher temperatures, the α value would be smaller and the curves closer together. The integrated curve, giving the isotopic composition of the accumulated vapor thus removed, is shown as solid line C. Mass-balance considerations require that the isotope content of the total accumulated vapor approaches the initial water $\delta^{18}O$ value as $f \rightarrow 0$. Hence, any process that can be modeled as a Rayleigh fractionation will not exhibit fractionation between the product and source if the process proceeds to completion (with 100% yield). This is a rare occurrence in nature since hydrologic systems are neither commonly completely open nor completely closed systems.

The dashed lines in Figure 1 show the $\delta^{18}O$ of vapor (E) and water (D) during equilibrium evaporation in a closed system (i.e., where the vapor and water are in contact for the entire phase change). Note that the $\delta^{18}O$ of vapor in the open system where the vapor is continuously removed (line B) is always higher than the $\delta^{18}O$ of vapor in a closed system where the vapor (line E) and water (line D) remain in contact. In both cases, the evaporation takes place under equilibrium conditions with α=1.010, but the cumulative vapor in the closed system remains in equilibrium with the water during the entire phase change. As a rule, fractionations in a true "open-system" Rayleigh process create a much larger range in the isotopic compositions of the products and reactants than in closed systems. This is because of the lack of back-reactions in open systems. Natural processes will produce fractionations between these two "ideal" cases.

7.1.3.2 *Mixing of waters and/or solutes*

Waters mix conservatively with respect to their isotopic compositions. In other words, the isotopic compositions of mixtures are intermediate between the compositions of the end-members. Despite the terminology (the δ notation and units of ‰) and common negative values, the compositions can be treated just like any other chemical constituent (e.g., chloride content) for making mixing calculations. For example, if two streams with known discharges (Q_1, Q_2) and known $\delta^{18}O$ values ($\delta^{18}O_1$, $\delta^{18}O_2$) merge and become well mixed, the $\delta^{18}O$ of the combined flow (Q_T) can be calculated from

$$Q_T = Q_1 + Q_2 \tag{6}$$

$$\delta^{18}O_T Q_T = \delta^{18}O_1 Q_1 + \delta^{18}O_2 Q_2 \tag{7}$$

Another example: any mixing proportions of two waters with known $\delta^{18}O$ and δ^2H values will fall along a tie line between the compositions of the end-members on a $\delta^{18}O$ versus δ^2H plot.

What is *not* so obvious is that on many types of X–Y plots, mixtures of two end-members will not necessarily plot along straight lines but instead along *hyperbolic curves* (Figure 2(a)). This has been explained very elegantly by Faure (1986), using the example of $^{87}Sr/^{86}Sr$ ratios. The basic principle is that mixtures of two components that have different isotope ratios (e.g., $^{87}Sr/^{86}Sr$ or $^{15}N/^{14}N$) and different concentrations of the element in question (e.g., strontium or nitrogen) form hyperbolas when plotted on diagrams with coordinates of isotope ratios versus concentration. As the difference between the elemental concentrations of two components (end-members) approaches 0, the hyperbolas flatten to lines. The hyperbolas are concave or convex, depending on whether the component with the higher isotope ratio has a higher or lower concentration than the other component. Mixing hyperbolas can be transformed into straight lines by plotting isotope ratios versus the inverse of concentration (1/*C*), as shown in

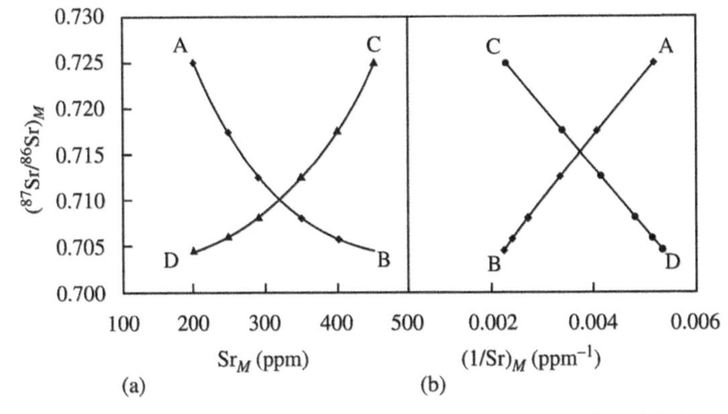

Figure 2 (a) Hyperbolas formed by the mixing of components (waters or minerals) A and B with different Sr concentrations and Sr isotope ratios ($^{87}Sr/^{86}Sr$). If the concentrations of Sr in A and B are identical, the mixing relation would be a straight line; otherwise, the mixing relations are either concave or convex curves, as shown. (b) Plotting the reciprocals of the strontium concentrations transforms the mixing hyperbolas into straight lines. If the curves in (a) were the result of some fractionation process (e.g., radioactive decay) that is an exponential relation, plotting the reciprocals of the Sr concentrations would not produce straight lines (after Faure, 1986).

Figure 2(b). One way to avoid curved lines on plots of δ values versus solute concentrations is to plot molality instead of molarity for the solute. Mixing and fractionation lines on $\delta^{18}O$ versus δ^2H plots are generally straight lines, because the concentrations of oxygen and hydrogen in water are essentially constant except for extremely saline brines.

Although the topic is rarely discussed, the isotope concentration of a sample is not necessarily equal to the isotope activity. For example, the isotope activity coefficient for water-O or water-H can be positive or negative, depending on solute type, molality, and temperature of the solution. The isotopic compositions of waters and solutes can be affected significantly by the concentration and types of salts, because the isotopic compositions of waters in the hydration spheres of salts and in regions farther from the salts are different (see Horita (1989) for a good discussion of this topic). In general, the only times when it is important to consider isotope activities are for low pH, high SO_4, and/or high magnesium brines because the activity and concentration δ values of these waters (δ_a and δ_c) are significantly different. For example, the difference ($\delta^2H_a-\delta^2H_c$) between the activity and concentration δ values for sulfuric acid solutions in mine tailings is $\sim$+16‰ for 2 m solutions. For normal saline waters (e.g., seawater), the activity coefficients for $\delta^{18}O$ and δ^2H are essentially equal to 1.

Virtually all laboratories report $\delta^{18}O$ activities (not concentrations) for water samples. The δ^2H of waters may be reported in either concentration or activity δ values, depending on the method used for preparing the samples for analysis. Methods that involve quantitative conversion of the H in H_2O to H_2 produce δ_c values. Methods that analyze H_2O by equilibrating it with H_2 (or with CO_2), and then analyzing the equilibrated gas for isotopic composition produce δ_a values. "Equilibrate" in this case means letting the liquid and gas reach isotopic equilibrium at a constant, known temperature. $\delta^{13}C$, $\delta^{15}N$, and $\delta^{34}S$ preparation methods do not involve equilibration and, hence, these are δ_c values. To avoid confusion, laboratories and research papers should always report the method used.

7.1.3.3 Geochemical and biological reactions

Reactions in shallow systems that frequently occur under equilibrium isotope conditions include (i) solute addition and precipitation (e.g., oxygen exchange between water and carbonate, sulfate, and other oxides, carbonate dissolution and precipitation) and (ii) gas dissolution and exchange (e.g., exchange of hydrogen between water and dissolved H_2S and CH_4, oxygen isotope exchange between water and CO_2). However, most of the reactions occurring in the soil zone, in shallow groundwater, and in surface waters are biologically mediated, and hence produce nonequilibrium isotope fractionations. These reactions may behave like Rayleigh fractionations in that there may be negligible back-reaction between the reactant and product, regardless of whether the system is open or closed. This arises due to the slow rates of many inorganic isotopic equilibrium exchange reactions at near-surface environmental conditions (i.e., low temperature, low pressure, pH 6–9).

Nonequilibrium isotopic fractionations typically result in larger ranges of isotopic compositions between reactants and products than those under equivalent equilibrium conditions. An example of this process is biologically mediated denitrification (reduction) of nitrate to N_2 in groundwater. The N_2 produced may be lost to the atmosphere, but even if it remains in contact with the residual nitrate, the gas does not re-equilibrate with the nitrate nor is there a biologically mediated back-reaction.

Graphical methods are commonly used for determining whether the data support an interpretation of mixing of two potential sources or fractionation of a single source. Implicit in such efforts is often the idea that mixing will produce a "straight line" connecting the compositions of the two proposed end-members whereas fractionation will produce a "curve." However, as shown in Figure 3(a), both mixing and fractionation (in this case, denitrification) can produce curves (Mariotti *et al.*, 1988), although both relations can look linear for small ranges of concentrations.

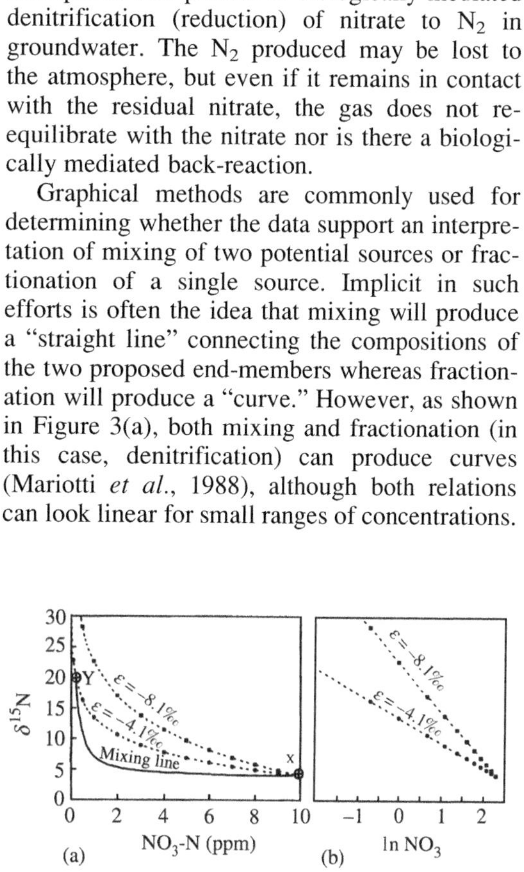

Figure 3 (a) Theoretical evolution of the $\delta^{15}N$ and the nitrate-N concentration during mixing (solid line) of two waters X and Y, and during an isotope fractionating process (e.g., denitrification of water X with an NO_3 concentration of 10 ppm). Denitrification for $\varepsilon = -4.1$‰ results in a curve (dashed line) that ends at Y. Two different enrichment factors are compared: $\varepsilon = -4.1$‰ and $\varepsilon = -8.1$‰. The data points represent successive 0.1 increments of mixing or denitrification progress. (b) Plotting the natural log of the concentrations for a fractionation process yields straight lines, different for different ε values (after Mariotti *et al.*, 1988).

However, the equations describing mixing and fractionation processes are different and, under favorable conditions, the process responsible for the curve can be identified. This is because Rayleigh fractionations are exponential relations (Equation (3)), and plotting δ values versus the natural log of concentration (C) will produce a straight line (Figure 3(b)). If an exponential relation is not observed and a straight line is produced on a δ versus $1/C$ plot (as in Figure 2(b)), this supports the contention that the data are produced by simple mixing of two end-members.

7.2 TRACING THE HYDROLOGICAL CYCLE

In most low-temperature, near-surface environments, stable hydrogen and oxygen isotopes behave conservatively. This means that as water molecules move through the subsurface, chemical exchange between the water and oxygen and hydrogen in the organic and inorganic materials through which flow occurs will have a negligible effect on the overall isotope ratios of the water (an exception is within certain geothermal systems, especially in carbonates, where large amounts of oxygen exchange between the water and rock can occur). Water isotopes are useful tracers of water flow paths, especially in confined groundwater systems dominated by Darcian flow in which a source of water with a distinctive isotopic composition forms a "plume" in the subsurface. This phenomenon forms the basis of the science of *isotope hydrology*, or the use of stable water isotopes in hydrological studies.

Although tritium also exhibits insignificant reaction with geologic materials, it does change in concentration over time because it is radioactive and decays with a half-life of ~12.4 yr. The main processes that dictate the oxygen and hydrogen isotopic compositions of waters in a catchment are: (i) phase changes that affect the water above or near the ground surface (evaporation, condensation, and melting) and (ii) simple mixing at or below the ground surface.

Oxygen and hydrogen isotopes can be used to determine the source of groundwaters and surface waters. For example, $\delta^{18}O$ and δ^2H can be used to determine the contributions of old and new water to a stream or shallow groundwater during periods of high runoff because the rain or snowmelt (new water) that triggers the runoff is often isotopically different from the water already in the catchment (old water). This section briefly explains why waters from different sources often have different isotopic compositions. For more detailed discussions of these and other environmental isotopes, the reader can consult texts such as Kendall and McDonnell (1998) and Cook and Herczeg (2000).

7.2.1 Deuterium and Oxygen-18

7.2.1.1 *Basic principles*

Craig (1961a) observed that $\delta^{18}O$ and δ^2H (or δD) values of precipitation that has not been evaporated are linearly related by

$$\delta^2H = 8\delta^{18}O + 10 \tag{8}$$

This equation, known as the "global meteoric waterline" (GMWL), is based on precipitation data from locations around the globe. The slope and intercept of the "local meteoric waterline" (LMWL) for rain from a specific catchment or basin can be different from the GMWL. The *deuterium excess* (d excess, or d) parameter has been defined to describe these different meteoric waterlines (MWLs), such that

$$d = \delta^2H - 8\delta^{18}O \tag{9}$$

On $\delta^{18}O$ versus δ^2H plots, water that has evaporated from open surfaces (e.g., ponds and lakes), or mixed with evaporated water, plots below the MWL along a trajectory typically with a slope between 2 and 5 (Figure 4). The slope of the evaporation line and the isotopic evolution of the reservoir are strongly dependent upon the humidity under which evaporation occurs, as well as temperature and wind speed (Clark and Fritz, 1997), with lower slopes found in arid regions.

7.2.1.2 *Precipitation*

The two main factors that control the isotopic character of precipitation at a given location are the temperature of condensation of the precipitation and the degree of rainout of the air mass (the ratio of water vapor that has already condensed into precipitation to the initial amount of water vapor in the air mass). The $\delta^{18}O$ and δ^2H of precipitation are also influenced by altitude, distance inland along different storm tracks, environmental conditions at the source of the vapor, latitude, and humidity. Progressive rainout as clouds move across the continent causes successive rainstorms to have increasingly lower δ values (Figure 5). At any point along the storm trajectory (i.e., for some specific fraction f of the total original vapor mass), $\delta^{18}O$ and δ^2H of the residual fraction of vapor in the air mass can be calculated using the Rayleigh equation.

At a given location, the seasonal variations in $\delta^{18}O$ and δ^2H values of precipitation and the weighted average annual $\delta^{18}O$ and δ^2H values of precipitation do not vary greatly from year to year. This happens because the annual range and sequence of climatic conditions (temperatures, vapor source, direction of air mass movement, etc.) generally vary within a predictable range. Rain in the summer months has higher δ values

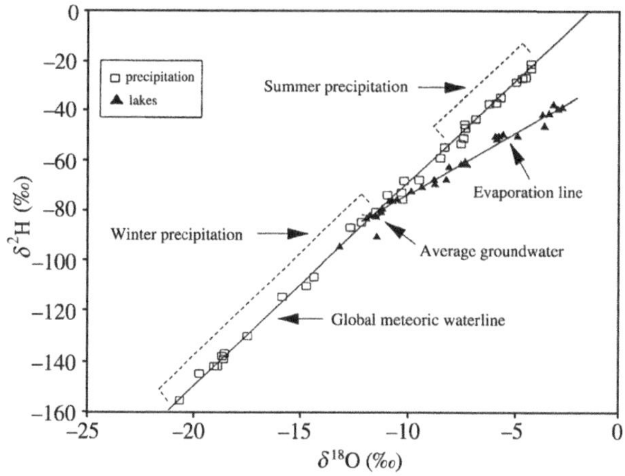

Figure 4 Isotopic compositions (δ^2H versus $\delta^{18}O$) of precipitation and lake samples from northern Wisconsin (USA). The precipitation samples define the LMWL ($\delta D = 8.03\delta^{18}O + 10.95$), which is virtually identical to the GMWL. The lake samples plot along an evaporation line with a slope of 5.2 that intersects the LMWL at the composition of average local groundwater, with additional data (after Kendall *et al.*, 1995a).

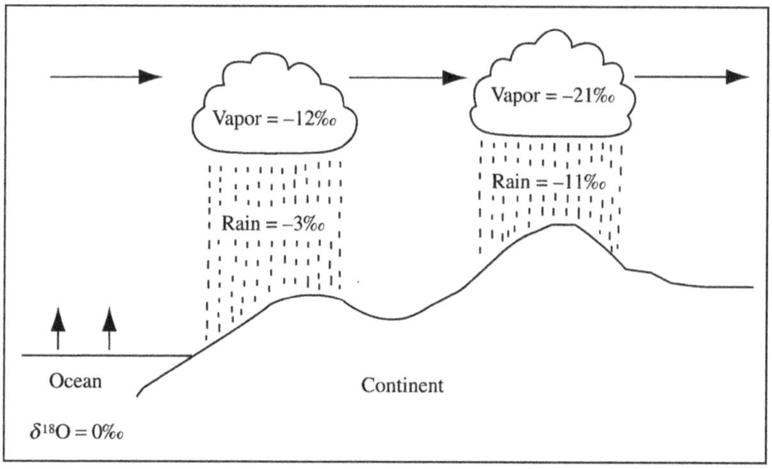

Figure 5 Origin of variations in the $\delta^{18}O$ of meteoric waters as moisture rains out of air masses moving across the continents.

than rain in the winter months due to the difference in mean seasonal air temperatures.

Superimposed on the seasonal cycles in precipitation δ values are storm-to-storm and intra-storm variations in the $\delta^{18}O$ and δ^2H values of precipitation (Figure 6). These variations may be as large as the seasonal variations. It is this potential difference in δ values between the relatively uniform old water and variable new water that permits isotope hydrologists to determine the contributions of old and new water to a stream during periods of high runoff (Sklash *et al.*, 1976). See Section 7.2.3 for more details about temporal and spatial variability in the $\delta^{18}O$ and δ^2H of precipitation.

The only long-term regional network for collection and analysis of precipitation for δ^2H and $\delta^{18}O$ in the USA (and world) was established by a collaboration between the International Atomic Energy Agency (IAEA) and the World Meteorological Organization (WMO). The discovery that changes in the $\delta^{18}O$ of precipitation over mid- and high-latitude regions during 1970s to 1990s closely followed changes in surface air temperature (Rozanski *et al.*, 1992) has increased the interest in revitalizing a global network of isotopes in precipitation (GNIP), sites where precipitation will be collected and analyzed for isotopic composition.

7.2.1.3 Shallow groundwaters

Shallow groundwater $\delta^{18}O$ and δ^2H values reflect the local average precipitation values but are modified to some extent by selective recharge and fractionation processes that may alter the $\delta^{18}O$

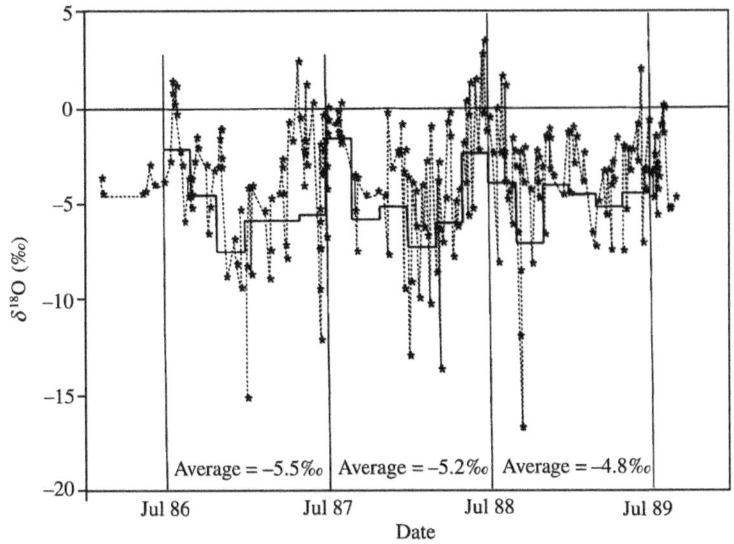

Figure 6 The $\delta^{18}O$ of storms at Panola Mountain, Georgia (USA) shows a 20‰ range in composition. Superimposed on the interstorm variability is the expected seasonal oscillation between heavy $\delta^{18}O$ values in the summer and lighter values in the winter caused by seasonal changes in temperature, storm track, and rain-out. The volume-weighted $\delta^{18}O$ values for two-month intervals are shown by heavy solid lines. The volume-weighted averages for the years 1987, 1988, and 1989 were each calculated from July 1 to June 30.

and δ^2H values of the precipitation before the water reaches the saturated zone (Gat and Tzur, 1967). These processes include (i) evaporation of rain during infiltration, (ii) selective recharge (e.g., only from major storms or during snowmelt), (iii) interception of rain water (Kendall, 1993; DeWalle and Swistock, 1994) and snow (Claassen and Downey, 1995) by the tree canopy, (iv) exchange of infiltrating water with atmospheric vapor (Kennedy *et al.*, 1986), and (v) various post-depositional processes (e.g., differential melting of snowpack or evaporation during infiltration).

The isotopic fractionation during evaporation is mainly a function of ambient humidity. Evaporation under almost 100% humidity conditions is more or less equivalent to evaporation under closed-system conditions (i.e., isotopic equilibrium is possible), and data for waters plot along a slope of 8 (i.e., along the LMWL). Evaporation at 0% humidity describes open-system evaporation. These two contrasting humidities result in the different fractionations (Figure 1). An open-surface water reservoir that is strongly evaporated will show changing $\delta^{18}O$ and δ^2H values along a trajectory with a slope of typically 5, beginning at the initial composition of the water on the LMWL, and extending out below the LMWL (Figure 4).

Once the rain or snowmelt passes into the saturated zone, the δ values of the subsurface water moving along a flow path change only by mixing with waters that have different isotopic compositions. The conventional wisdom regarding the isotopic composition of groundwater has long been that homogenizing effects of recharge and dispersive processes produce groundwater with isotopic values that approach uniformity in time and space, approximating a damped reflection of the precipitation over a period of years (Brinkmann *et al.*, 1963). Although generally the case for porous media aquifers that are dominated by Darcian flow, this may not always hold true, particularly in shallow groundwater systems where preferential flow paths are common, hydraulic conductivities are variable, and waters may have a large range of residence times (see Section 7.2.3.2). Thus, one should always test the assumption of homogeneity by collecting water over reasonable spatial and temporal scales.

7.2.1.4 Deep groundwaters and paleorecharge

Deep groundwaters can be very old, often having been recharged thousands of years in the past. Since the $\delta^{18}O$ and δ^2H values of precipitation contributing to aquifer recharge are strongly dependent on the temperature and humidity of the environment, and these δ values generally behave conservatively in the subsurface, the δ values of deep groundwaters reflect the past climatic conditions under which the recharge took place. Thus, water isotopic composition can be a powerful addition to groundwater dating tools for the identification of paleorecharge. These tracers can be especially useful for the identification of groundwater recharged during the last glacial period of the Pleistocene, when average Earth surface temperatures were dramatically lower than during the Holocene.

Confined aquifers with relatively high porosity (such as sandstones layered between clay aquitards) are the best groundwater systems for the preservation of paleorecharge signals in the isotopic composition of the groundwater. Note that water isotopes do not provide an "age" of the groundwater sampled. Instead, they provide a tool by which to distinguish between modern recharge and recharge from some other time in the past, or from some other source.

An independent means of establishing groundwater age is always needed before making conclusions about the history of recharge from stable water isotope data (Fontes, 1981). Estimation of flow velocities and recharge rates via physically based groundwater flow models may lead to theoretical "ages" of groundwater in such settings. However, stable isotopic data should not be used to estimate ages without employing a chemically based means of groundwater dating, such as is afforded through a variety of radiometric dating techniques as well as noble gas methods.

Several studies and a number of reviews have demonstrated the use of water isotopes for identifying paleorecharge groundwater. These include Fontes (1981), Dutton and Simpkins (1989), Fontes *et al.* (1991, 1993), Deák and Coplen (1996), Coplen *et al.* (2000), and Bajjali and Abu-Jaber (2001). The basic principle underlying the technique is that precipitation formed under temperature conditions that were much cooler than at present will be reflected in old groundwater that has significantly lower δ values. On a plot of δ^{18}O versus δ^{2}H, precipitation from cooler climates has more negative δ values than precipitation from warmer climates. In general, locations and times with cooler and more humid conditions have lower *d* excess values.

Climate changes in arid regions are often readily recognized in paleowaters. Strong variability in humidity causes large shifts in the deuterium excess (or *d* excess), and corresponding shifts in the vertical position of the LMWL (Clark and Fritz, 1997). The value of the *d* excess in water vapor increases as the relative humidity under which evaporation takes place decreases. Since air moisture deficit is positively correlated with temperature, *d* excess should be positively correlated with temperature. High *d* excess in groundwater indicates that the recharge was derived partly from water evaporated under conditions of low relative humidity.

Comparison of paleowater isotopic data must be made to the modern LMWL; if a long-term record of local precipitation is not available, an approximation can be obtained from the nearest station of GNIP and Isotope Hydrology Information System (ISOHIS), established by the Isotope Hydrology Section of the IAEA, Vienna, Austria.

The combined use of chemical indicators and stable isotopes is especially powerful for identifying paleowaters. Chlorine concentrations and chloride isotopes are often applied to estimate paleorecharge processes, whereas noble gas techniques have been applied to the derivation of paleotemperatures of recharge for very old groundwaters (Stute and Schlosser, 2000). Saline paleowaters related to evaporitic periods in the geologic past can be recognized by δ values that plot significantly below the MWL because of evaporative enrichment. Secondary brines resulting from the dissolution of salts by modern recharge will plot on or very near to the MWL (Fontes, 1981).

Only in very limited cases may the isotopes of water be useful as proxies for paleotemperatures (e.g., in high-latitude ice cores). The isotopic content of precipitation depends upon many factors, not just temperature. Thus, the lack of a significant difference between the δ values of modern precipitation and groundwaters dated to times corresponding with glacial maxima does not necessarily imply a poor age-dating technique. Humidity, isotopic content of water vapor sources, and precipitation distribution all play a role in the isotopic composition of meteoric recharge. In the absence of independent age-dating methods, altitude effects may be misinterpreted as paleorecharge signals, particularly in cases where rivers having high-altitude recharge areas are responsible for a large portion of groundwater recharge taking place at lower elevations downstream. One cannot expect a simple relationship between the stable isotopic content of meteoric recharge and average temperatures for past periods of different climate. Also, dispersion and molecular diffusion may cause blurring of differences in isotopic composition between Holocene and pre-Holocene groundwaters.

7.2.1.5 Surface waters

Figure 7 shows the spatial distribution of δ^{18}O and δ^{2}H values of surface waters in the USA (Kendall and Coplen, 2001). These maps were generated by the analysis of more than 4,800 depth- and width-integrated stream samples from 391 sites sampled bimonthly or quarterly for 2.5–3 yr between 1984 and 1987 by the USGS (Coplen and Kendall, 2000). Spatial distributions of δ^{18}O and δ^{2}H values are very similar to each other and closely match topographic contours. The ability of this data set to serve as a proxy for the isotopic composition of modern precipitation in the USA is supported by the excellent agreement between the river data set and the compositions of adjacent precipitation monitoring sites, the strong spatial coherence of the distributions of δ^{18}O and δ^{2}H, the good correlations of the isotopic compositions with climatic parameters, and the good agreement between the "national" MWL generated from

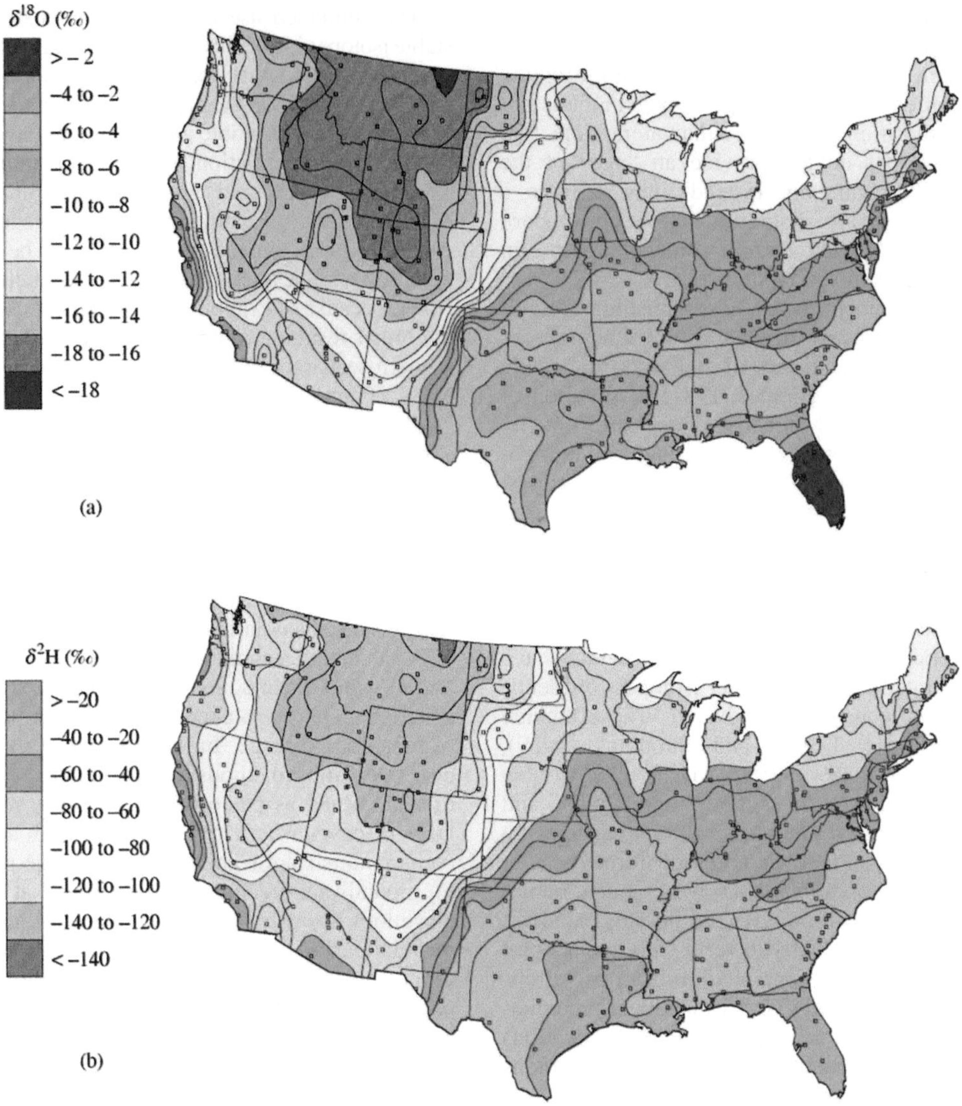

Figure 7 Spatial variation in the: (a) $\delta^{18}O$ and (b) $\delta^{2}H$ of rivers in the USA (source Kendall and Coplen, 2001).

unweighted analyses of samples from the 48 contiguous states of $\delta^2H = 8.11\ \delta delta;^{18}O + 8.99$ ($r^2 = 0.98$) with the unweighted global MWL of the IAEA-GNIP sites of $\delta^2H = 8.17\delta^{18}O + 10.35$ (Kendall and Coplen, 2001).

Although many isotope hydrologists would agree that such maps should be constructed from1 long-term annual averages of precipitation compositions, there are some advantages to stream samples as indicators of precipitation compositions. First, stream samples are relatively easy to obtain by taking advantage of existing long-term river monitoring networks. Second, stream water is a better spatial and temporal integrator of the isotopic composition of precipitation intercepted by a large drainage basin than recent precipitation collected at a single location in the basin. Third, since groundwater is probably the dominant source of stream flow in most basins, the river provides information on the composition of waters that have infiltrated the soils to recharge the groundwater system. Finally, the remains of biota that lived in the soil, lakes, and streams that are used to reconstruct paleohydrology and paleoclimates record the integrated isotopic signal of recharge water, not the isotopic compositions of the precipitation.

Water in most rivers has two main components: (i) recent precipitation that has reached the river either by surface runoff, channel precipitation, or by rapid flow through shallow subsurface flow paths; and (ii) groundwater. The relative contributions of these sources differ in each watershed or basin, and depend on the physical setting of the drainage basin (e.g., topography, soil type, depth to bedrock, vegetation, fractures, etc.), climatic parameters (e.g., precipitation amount, seasonal variations in precipitation, temperature, potential evapotranspiration, etc.), and human activities (e.g., dams, reservoirs, irrigation usages, clearing for agriculture, channel restructuring, etc.).

The $\delta^{18}O$ and $\delta^{2}H$ of rivers will reflect how the relative amounts of precipitation and groundwater vary with time, and how the isotopic compositions of the sources themselves change over time. Seasonal variations will be larger in streams where recent precipitation is the main source of flow, and smaller in streams where groundwater is the dominant source. As the basin size increases, the isotopic compositions of river waters are also increasingly affected by evaporation (Gat and Tzur, 1967). Local precipitation events are an important component of river water in the headwaters of large basins. For example, the average amount of new water in small, forested watersheds during storms is ~40%, although during storms the percentage can be higher (Genereux and Hooper, 1998). However, in the lower reaches, local additions of precipitation can be of minor importance (Friedman *et al.*, 1964; Salati *et al.*, 1979), except during floods (Criss, 1999). For example, analysis of long-term tritium records for seven large (5,000–75,000 km^2) drainage basins in the USA indicated that ~60% ± 20% of the river water was less than 1 yr old (Michel, 1992).

The dual nature of river water (partly recent precipitation, partly groundwater) can be exploited for studying regional hydrology or climatology. Under favorable circumstances, knowledge of the isotopic compositions of the major water sources can be used to quantify the time-varying contributions of these sources to river water (Sklash *et al.*, 1976; Kendall and McDonnell, 1998). Alternatively, if the isotopic composition of base flow is thought to be a good representation of mean annual precipitation (Fritz, 1981), then the $\delta^{18}O$ and $\delta^{2}H$ of rivers sampled during low flow can integrate the composition of rain over the drainage areas and be useful for assessing regional patterns in precipitation related to climate.

7.2.2 Tritium

Tritium (^3H) is a radiogenic and radioactive isotope of hydrogen with a half-life of 12.4 yr. It is an excellent tracer for determining timescales for the mixing and flow of waters because it is considered to be relatively conservative geochemically, and is ideally suited for studying processes that occur on a timescale of less than 100 yr. Tritium content is expressed in tritium units (TU), where 1 TU equals 1 ^3H atom in 10^{18} atoms of hydrogen.

Prior to the advent of atmospheric testing of thermonuclear devices in 1952, the tritium content of precipitation was probably in the range of 2–8 TU (Thatcher, 1962); this background concentration is produced by cosmic ray spallation. While elevated tritium levels have been measured in the atmosphere since 1952, tritium produced by thermonuclear testing ("bomb tritium") has been the dominant source of tritium in precipitation. A peak concentration of several thousand TU was recorded in precipitation in the northern hemisphere in 1963, the year that the atmospheric test ban treaty was signed. After 1963, the tritium levels in precipitation began to decline gradually because of radioactive decay and the cessation of atmospheric testing.

The simplest use of tritium is to check whether detectable concentrations are present in the water. Although pre-bomb atmospheric tritium concentrations are not well known, waters derived exclusively from precipitation before 1953 would have maximum tritium concentrations of ~0.1–0.4 TU by 2003. For waters with higher tritium contents, some fraction of the water must have been derived since 1953; thus, the tritium concentration can be a useful marker for recharge since the advent of nuclear testing.

Distinct "old" and "new" water tritium values are required for storm and snowmelt runoff hydrograph separation. In small (first-order) catchments where the average residence time (the average time it takes for precipitation to enter the ground and travel to the stream) of the old water is on the order of months, the old and new water tritium concentrations will not likely be distinguishable. However, in some larger catchments with longer residence times, old and new waters may be distinctive as a result of the gradual decline in precipitation tritium values since 1963 and the even more gradual decline in groundwater tritium values.

Tritium measurements are frequently used to calculate recharge rates, rates or directions of subsurface flow, and residence times. For these purposes, the seasonal, yearly, and spatial variations in the tritium content of precipitation must be accurately assessed. This is difficult to do because of the limited data available, especially before the 1960s. For a careful discussion of how to calculate the input concentration at a specific location, see Michel (1989) and Plummer *et al.* (1993). Several different approaches (e.g., piston-flow, reservoir, compartment, and advective-dispersive models) to modeling tritium concentrations in groundwater are discussed by Plummer *et al.* (1993). The narrower topic of using environmental isotopes to determine residence time is discussed briefly below.

If the initial concentration of ^3H in the atmosphere is not known, an alternative dating method is to analyze waters for both ^3H and ^3He. Because ^3H decays to ^3He, it is possible to use the tritiogenic ^3He component of ^3He in groundwater as a quantitative tracer of the age of the water since it was separated from the atmosphere.

7.2.3 Determination of Runoff Mechanisms

A major uncertainty in hydrologic and chemical modeling of catchments has been the quantification of the contributions of water and solutes from various hydrologic pathways. Isotope hydrograph separation apportions storm and snowmelt hydrographs into contributing components based on the distinctive isotopic signatures carried by the two or more water components (e.g., new precipitation, soil water, and groundwater). The basic principle behind isotope hydrograph separation is that, if the isotopic compositions of the sources of water contributing to stream flow during periods of high runoff are known and are different, then the relative amounts of each source can be determined. Since these studies began in the 1960s, the overwhelming conclusion is that old water is by far the dominant source of runoff in humid, temperate environments (Sklash *et al.*, 1976; Pearce *et al.*, 1986; Bishop, 1991). Buttle (1998) provides a succinct review of watershed hydrology.

7.2.3.1 Isotope hydrograph separation and mixing models

Until the 1970s, the term "hydrograph separation" meant a graphical technique that had been used for decades in predicting runoff volumes and timing. For example, the graphical separation technique introduced by Hewlett and Hibbert (1967) is commonly applied to storm hydrographs (i.e., stream flow versus time graphs) from forested catchments to quantify "quick flow" and "delayed flow" contributions. According to this technique, the amounts of quick flow and delayed flow in stream water can be determined simply by considering the shape and timing of the discharge hydrograph.

Isotope hydrograph separation apportions storm and snowmelt hydrographs into contributing components based on the distinctive isotopic signatures carried by the old and new water components. Hence, the method allows the calculation of the relative contributions of new precipitation and older groundwater to stream flow. A major limitation of the tracer method is that we cannot directly determine how the water reaches the stream (i.e., geographic source of the water) nor the actual runoff generation mechanism from knowledge of the temporal sources of water. In a very real sense, the isotope hydrograph separation technique is still a "black box" method that provides little direct information about what is actually going on in the subsurface. However, in combination with other types of information (e.g., estimates of saturated areas or chemical solute compositions of water along specific flow paths), runoff mechanisms can sometimes be inferred.

Hydrograph separations are typically performed with only one isotope, both to save on the cost of analyses and because of the high correlation coefficient between $\delta^{18}O$ and $\delta^{2}H$ ($r^2 > 0.95$; Gat, 1980). However, analysis of both isotopes will often prove very beneficial, and is highly recommended.

Two-component mixing models. Isotope hydrograph separation normally involves a two-component mixing model for the stream. The model assumes that water in the stream at any time during storm or snowmelt runoff is a mixture of two components: new water and old water.

During base-flow conditions (the low-flow conditions that occur between periods of storm and snowmelt runoff), the water in a stream is dominated by old water. The chemical and isotopic character of stream water at a given location during base flow represents an integration of the old-water discharged from upstream. During storm and snowmelt runoff events, however, new water is added to the stream. If the old and new water components are chemically or isotopically different, the stream water becomes changed by the addition of the new water. The extent of this change is a function of the relative contributions by the old and new water components.

The contributions of old and new water in the stream at any time can be calculated by solving the mass-balance equations for the water and tracer fluxes in the stream, provided that the stream, old water, and new water tracer concentrations are known:

$$Q_s = Q_o + Q_n \tag{10}$$

$$C_s Q_s = C_o Q_o + C_n Q_n \tag{11}$$

where Q is discharge, C refers to tracer concentration, and the subscripts "s," "o," and "n" indicate the stream, the old water components, and the new water components, respectively. If stream samples are taken at a stream gauging station, the actual volumetric contributions of old and new water can be determined. If no discharge measurements are available, old and new water contributions can be expressed as percentages of total discharge.

Although the simple two-component mixing model approach to stream hydrograph separation does not directly identify the actual runoff generation mechanisms, the model can sometimes allow the hydrologist to evaluate the importance of a given conversion process in a catchment. For example, if a rapid conversion mechanism (partial-area overland flow, saturation overland flow, or perhaps subsurface flow through macropores) is the dominant conversion process contributing to a storm runoff hydrograph, the isotopic content of the stream will generally reflect mostly new water in the stream. Conversely, if a slow conversion mechanism (Darcian subsurface flow) is dominant in producing the storm runoff, the isotopic content of the stream should indicate mostly old water in the stream. Evaluation of the flow processes can be enhanced by applying the mixing model to water collected from subsurface runoff collection pits, overland flow, and macropores.

Sklash and Farvolden (1982) listed five assumptions that must hold for reliable hydrograph separations using environmental isotopes:

(i) the old water component for an event can be characterized by a single isotopic value or variations in its isotopic content can be documented;

(ii) the new water component can be characterized by a single isotopic value or variations in its isotopic content can be documented;

(iii) the isotopic content of the old water component is significantly different from that of the new water component;

(iv) vadose water contributions to the stream are negligible during the event or they must be accounted for (use an additional tracer if isotopically different from groundwater); and

(v) surface water storage (channel storage, ponds, swamps, etc.) contributions to the stream are negligible during the runoff event.

The utility of the mixing equations for a given high runoff period is a function mainly of the magnitude of $(\delta_o - \delta_n)$ relative to the analytical error of the isotopic measurements, and the extent to which the aforementioned assumptions are indeed valid (Pearce *et al.*, 1986; Genereux, 1998). Clearly the relative amounts of new versus old water are affected by many environmental parameters such as: size of the catchment, soil thickness, ratio of rainfall rate to infiltration rate, steepness of the watershed slopes, vegetation, antecedent moisture conditions, permeability of the soil, amount of macropores, and storage capacity of the catchment.

Figure 8 shows an example of a two-component hydrograph separation for four catchments at Sleepers River Watershed, Vermont USA (Shanley *et al.*, 2001). During the snowmelt period, groundwater appeared to have an approximately constant $\delta^{18}O$ value of -11.7 ± 0.3‰, whereas snowmelt collected in snow lysimeters ranged from −20‰ to −14‰. The stream samples had $\delta^{18}O$ values intermediate between the snowmelt and groundwater. Two-component isotope separations showed that the meltwater inputs to the stream ranged from 41% to 74%, and generally

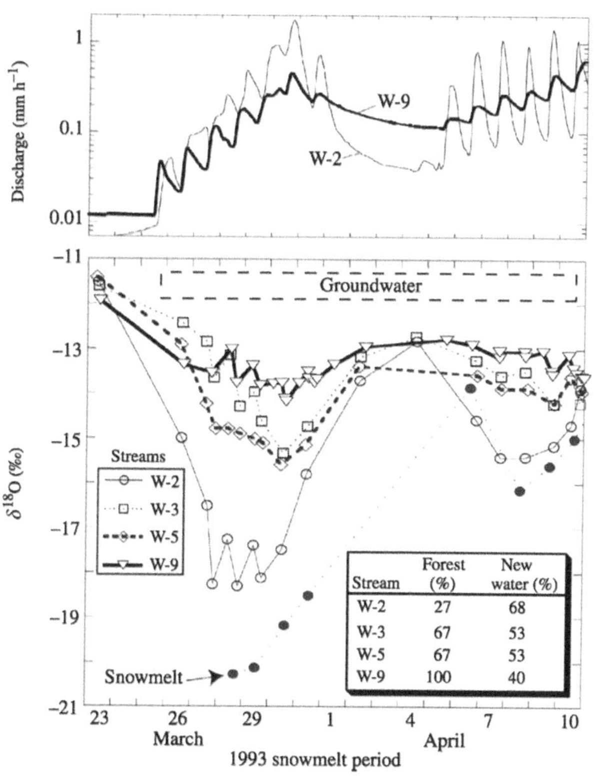

Figure 8 Early spring samples from four streams at Sleepers River Watershed, Vermont, have $\delta^{18}O$ values intermediate between the compositions of snowmelt collected in pan lysimeters and groundwater. Diurnal fluctuations in discharge correlate with diurnal changes in $\delta^{18}O$, especially at W-2. W-2, a 59 ha agricultural basin, shows much greater contributions from snowmelt than the other three catchments. The three mixed agricultural/forested nested catchments—W-9 (47-ha), W-3 (837-ha), and W-5 (11,125)—show increasing contributions from new snowmelt as scale increases (after Shanley *et al.*, 2001).

increased with catchment size (41–11,125 ha). Another multicatchment isotope hydrograph study showed no correlation of basin size and the percent contribution of subsurface flow to stream flow (Sueker *et al.*, 2000). This study found that subsurface flow positively correlated with the amount of surficial material in the basin and negatively correlated with basin slope.

Three-component mixing models. Some isotope hydrograph separation studies have shown that the stream $\delta^{18}O$ values fall outside of the mixing line defined by the two "end-members" (DeWalle *et al.*, 1988). Other studies have shown that stream δ values fall off the mixing line when both $\delta^{18}O$ and δ^2H are used for fingerprinting (Kennedy *et al.*, 1986). If the stream water is not collinear with the two suspected end-members, either the end-members are inaccurate or there must be another component (probably soil water) that plots above or below the mixing line for the other components (Kennedy *et al.*, 1986). In this case, three-component models using silica or some other chemical tracer can be used to estimate the relative contributions of the sources to stream water (DeWalle *et al.*, 1988; Wels *et al.*, 1991; Hinton *et al.*, 1994; Genereux and Hooper, 1998).

Figure 9 shows that stream water samples at the Mattole River in California plot above the mixing line defined by the two supposed sources of the stream water, namely, pre-storm base flow and new rain (Kennedy *et al.*, 1986). As often is the case, the mixing line is collinear with the LMWL. Thus, no mixture of waters from these two sources of water can produce the stream water. This means that either (i) there is an additional source of water (above the line) which contributes significantly to stream flow or (ii) some process in the catchment has caused waters to shift in isotopic composition. This third source of water is often soil water. Although no soil-water samples were collected during the January 1972 storm at the Mattole River, the composition of the last big storm to hit the area (which presumably was a major source of recharge to the soil zone) was analyzed for $\delta^{18}O$ and δ^2H. The stream samples plot along a mixing line connecting the average composition of the earlier storm and the average composition of January rain. Hence, the stream samples on Figure 9 were probably mainly a mixture of soil water recharged during the previous storm and new rainwater, with smaller contributions from base flow.

If only $\delta^{18}O$ or δ^2H had been analyzed, the stream water would have plotted intermediate between the two end-members (consider the projections of these water compositions onto the $\delta^{18}O$ and δ^2H scales) and a two-component isotope hydrograph separation would have seemed feasible. Only when both isotopes were analyzed did it become apparent that there was a mass-balance problem. Hence, we recommend analysis of both oxygen and hydrogen isotopes on a subset of stream water and end-member samples to ensure that an end-member or process is not overlooked.

Sometimes it is obvious that the two-component model is inadequate: rain combined with glacial meltwater (Behrens *et al.*, 1979), for example. For rain-on-snow events where snow cores are used to assess the new-water composition, a two-component model is also inadequate (Wallis *et al.*, 1979). However, if the catchment is snow-covered, overland flow is minimal, the surface area of the stream is small, and several snowmelt lysimeters are used to catch sequential samples of new water derived from infiltration of

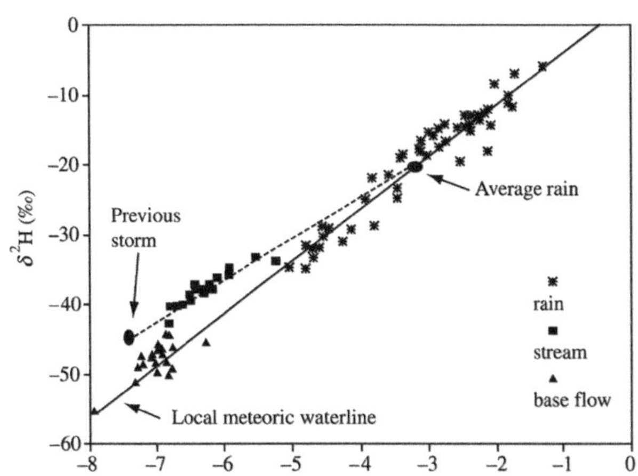

Figure 9 Storm flow samples from the Mattole River, California, have $\delta^{18}O$ and δD values that do not plot along the mixing line (solid line) between the two supposed sources: new rain and pre-storm base flow. Therefore, there must be another source of water, possibly soil water derived from a previous storm, contributing significant amounts of water to storm flow. The stream samples plot along a mixing line (dashed line) between the composition of average new rain and the composition of previous rain (after Kennedy *et al.*, 1986).

both rain and melting snow, a two-component model may be sufficient. It is difficult to judge *a priori* when the isotopic difference between soil water and groundwater is sufficient to require a three-component model (Kennedy *et al.*, 1986; DeWalle *et al.*, 1988; Swistock *et al.*, 1989; Hinton *et al.*, 1994).

DeWalle *et al.* (1988) used a three-component model for the special case where the discharge of one component was known. Wels *et al.* (1991) used a three-component model for the case where the isotopic compositions of new overland and new subsurface flow were identical, and the silica contents of new and old subsurface flow were identical. Ogunkoya and Jenkins (1993) used a model where one component was time varying. Hinton *et al.* (1994) listed five assumptions required for three-component models using two tracers.

7.2.3.2 Temporal and spatial variability in end-members

The universal applicability of the simple two-component mixing model has been frequently challenged (Kennedy *et al.*, 1986; Rodhe, 1987; Sklash, 1990; McDonnell *et al.*, 1991; Bishop, 1991; Kendall and McDonnell, 1993). These challenges address the five assumptions governing the use of the two-component mixing model given in Sklash and Farvolden (1982) and listed above.

First, the new water component is rarely constant in isotopic composition during events. Temporal variations in rain isotopic composition have been found to have an appreciable effect on hydrograph separations (McDonnell *et al.*, 1990). Second, investigators have concluded that in some catchments, soil water is an important contributor to storm runoff and is isotopically and/or chemically distinct from groundwater (Kennedy *et al.*, 1986; DeWalle *et al.*, 1988; Hooper *et al.*, 1990); in these cases, a three-component mixing model may be useful. In some catchments, the shallow soil and groundwaters are so heterogeneous in isotopic composition that isotope hydrograph separations are ill-advised (McDonnell *et al.*, 1991; Ogunkoya and Jenkins, 1991).

None of these challenges to the simple two-component mixing model approach has caused any significant change in the basic conclusion of the vast majority of the isotope and chemical hydrograph studies to date, namely, that *most storm flow in humid-temperate environments is old water* (Bishop, 1991). There is, however, a need to address the potential impact of natural isotopic variability on the use of isotopes as tracers of water sources. Most of these concerns can be addressed by applying more sophisticated two- and three-component mixing models that allow for variable isotopic signatures (e.g., Harris *et al.*, 1995).

The "new" water component. The isotopic composition of rain and snow can vary both spatially and temporally during storms. In general, rain becomes progressively more depleted in ^{18}O and ^{2}H during an event because these isotopes preferentially rain out early in the storm. Successive frontal and convective storms may have more complex isotopic variations. Intrastorm rainfall compositions have been observed to vary by as much as 90‰ in δ^2H at the Maimai watershed in New Zealand (McDonnell *et al.*, 1990), 16‰ in $\delta^{18}O$ in Pennsylvania, USA (Pionke and DeWalle, 1992) and 15‰ in $\delta^{18}O$ in Georgia, USA (Kendall, 1993).

Snowmelt also varies in isotopic composition during the melt period and is commonly different from that of melted snow cores (Hooper and Shoemaker, 1986; Stichler, 1987; Taylor *et al.*, 2001). Part of the changes in snowmelt isotopic composition can be attributed to distinct isotopic layers in the snow, rain events on snow, and isotopic fractionation during melting. In general, the snowmelt has low $\delta^{18}O$ values early in the season but the $\delta^{18}O$ values become progressively higher as the pack melts (Shanley *et al.*, 2001; Taylor *et al.*, 2001).

The variations in "new" water isotopic content are assessed by collecting sequential samples of rain or through fall during rainstorms, and samples of meltwater from snow lysimeters during snowmelt. The method of integrating this information into the separation equations depends on the magnitude and rate of temporal variations in the "new" water isotopic content and the size of the catchment. If a varying "new" water value is used, and the catchment is large enough that travel times are significant, one must account for the travel time to the stream and to the monitoring point.

In forested areas, rain intercepted by the tree canopy can have a different isotopic composition than rain in open areas (Saxena, 1986; Kendall, 1993; DeWalle and Swistock, 1994). On average, through fall during individual storms at a site in Georgia (USA) had $\delta^{18}O$ values 0.5‰ higher and δ^2H values 3.0‰ higher than rain. Site-specific differences in canopy species, density, and microclimate resulted in an average $\delta^{18}O$ range of 0.5‰ among 32 collectors over a 0.04 km^2 area for the same storm, with a maximum range of 1.2‰ (Kendall, 1993). Other sites appear to have little spatial variability (Swistock *et al.*, 1989). Because of this potential for spatial variability, putting out several large collectors and combining the through fall for isotopic analysis is advisable (DeWalle *et al.*, 1988).

Different rainfall weighting methods can substantially affect estimates of new/old waters in storm runoff in basins with large contributions of new water (McDonnell *et al.*, 1990). Use of sequential rain values is probably the best choice in very responsive catchments or in catchments with high proportions of overland flow. When rain intensities are low and soils drain slowly, current rain may not infiltrate very rapidly and thus use of the cumulative approach (i.e., running average) is probably more realistic (Kendall *et al.*, 2001a).

In large watersheds and watersheds with significant elevation differences, both the "old" water and "new" water δ values may show large spatial variability; generally, the higher elevations will have lower δ values. Such large or steep watersheds would require many sampling stations to adequately assess the variability in old and new waters. Spatial variations in the $\delta^{18}O$ of up to 5‰ were observed for rain storms in a 2,500 km^2, 35-station network in Alabama, USA (Kendall and McDonnell, 1993). Multiple collectors are especially critical for assessing temporal and spatial variability in both the timing and composition of snowmelt. Consequently, simple isotope hydrograph separation techniques are best suited for small first- and second-order catchments.

Preferential storage of different time fractions of rain in the shallow soil zone can be important because of the large amounts of rain potentially going into storage. For a storm that dropped 12 cm of rain over a 1 day period at Hydrohill, a 500 m^2 artificial catchment in China, 45% of the total rain went into storage (Kendall *et al.*, 2001a), with almost 100% of the first 2 cm of rain going into storage before runoff began.

The "old" groundwater component. Groundwater in catchments can frequently be heterogeneous in isotopic composition due to variable residence times. For example, groundwater at Hydrohill catchment showed a 5‰ range in $\delta^{18}O$ after a storm (Kendall *et al.*, 2001a). Groundwaters at the Sleepers River watershed in Vermont (USA) showed a 4‰ isotopic stratification in most piezometer nests during snowmelt (McGlynn *et al.*, 1999). Consequently, even if a large number of wells are sampled to characterize the potential variability in the groundwater composition, base flow is probably a better integrator of the water that actually discharges into the stream than any average of groundwater compositions. Hence, it may be safer to use pre-storm base flow, or interpolated intrastorm base flow (Hooper and Shoemaker, 1986), or ephemeral springs as the old-water end-member. If groundwater is used for the end-member, researchers need to check the effects of several possible choices for composition on their separations.

One major assumption in sampling base flow to represent the "old" water component is that the source of groundwater flow to the stream during storms and snowmelt is the same as the source during base flow conditions. It is possible that ephemeral springs remote from the stream or deeper groundwater flow systems may contribute differently during events than between events, and if their isotopic signatures differ from that of base flow, the assumed "old" water isotopic value may be incorrect. Although the occurrence of such situations could be tested by hydrometric monitoring and isotopic analyses of these features, their quantification could be difficult. Recent tracer studies showing the wide range of groundwater residence times in small catchments illustrate the problems with simple source assumptions (McDonnell *et al.*, 1999; Kirchner *et al.*, 2001; McGlynn *et al.*, 2003).

Soil water. The role of soil water in storm flow generation in streams has been intensively studied using the isotope tracing technique (DeWalle *et al.*, 1988; Buttle and Sami, 1990). Most isotope studies have lumped soil-water and groundwater contributions together, which may mask the importance of either component alone. However, Kennedy *et al.* (1986) observed that as much as 50% of the stream water in the Mattole catchment was probably derived from rain from a previous storm stored in the soil zone (Figure 9). If there is a statistically significant difference between the isotopic compositions of soil and groundwater, then they need to be considered separate components (McDonnell *et al.*, 1991).

The isotopic composition of soil water may be modified by a number of processes including direct evaporation, exchange with the atmosphere, and mixing with infiltrating water (Gat and Tzur, 1967). Although transpiration removes large amounts of soil water, it is believed to be non-fractionating (Zimmerman *et al.*, 1967). Wenner *et al.* (1991) found that the variable composition of the percolating rain became largely homogenized at depths as shallow as 30 cm due to mixing with water already present in the soil.

Intrastorm variability of soil water of up to 1.5‰ in $\delta^{18}O$ was observed during several storms at Panolan Mountain Georgia, and a 15‰ range in δ^2H was seen during a single storm at Maimai, New Zealand (Kendall and McDonnell, 1993). Analysis of over 1,000 water samples at Maimai (McDonnell *et al.*, 1991) showed a systematic trend in soil-water composition in both downslope and downprofile directions. Multivariate cluster analysis also revealed three distinct soil water groupings with respect to soil depth and catchment position, indicating that the soil water reservoir is poorly mixed on the timescale of storms.

If the isotopic variability of rain, through fall, meltwater, soil water, and groundwater is significant at the catchment scale (i.e., if hill-slope waters that are variable in composition actually reach the stream during the storm event) or if transit times are long and/or variable, then simple two- and three-component, constant composition, mixing models may not provide realistic interpretations of the system hydrology. One approach is to develop models with variable end-members (Harris *et al.*, 1995). Another possible solution is to include alternative, independent, chemical or isotopic methods for determining the relative amounts of water flowing along different subsurface flow paths. Recent comparisons of isotopic and chemical approaches include Rice and Hornberger (1998) and Kendall *et al.* (2001a).

Additional isotopic tracers include radon-222 (Genereux and Hemond, 1990; Genereux *et al.*, 1993); beryllium-7 and sulfur-35 (Cooper *et al.*, 1991); carbon-13 (Kendall *et al.*, 1992); strontium-87 (Bullen and Kendall, 1998); and nitrogen-15 (Burns and Kendall, 2002). In particular, carbon isotopes, possibly combined with strontium isotopes, appear to be useful for distinguishing between deep and shallow flow paths and, combined with oxygen isotopes, for distinguishing among water sources. Finally, a number of mathematical models can be employed that incorporate the variability of input tracer concentrations, and which attempt to reproduce observed output concentrations (Kirchner *et al.*, 2001; McGlynn *et al.*, 2003). Some of these techniques will be discussed in the following section.

7.2.4 Estimation of Mean Residence Time

Arguably the most important quantity that can be estimated in hydrologic studies is the age of the water that has been sampled. Water age has utility for determining hydrological processes in catchments (as discussed previously), as well as direct implications for management of larger groundwater resources via determination of aquifer sustainability and migration rates of contaminants. Thus, the estimation of mean residence time (MRT) of groundwater in a catchment or in an aquifer is an important goal. The discussion above demonstrated that the stable isotopes of the water molecule can be used to determine—qualitatively—the relative ages of water masses moving through a watershed using the mean input concentration of the tracer (i.e., mean isotopic composition of precipitation). While this method is imperfect, the signals observed in isotope data collected in time series are clear and informative. These signals can be used in a more quantitative manner to calibrate flow and transport models. While it is beyond the scope of this chapter to thoroughly review modeling techniques in hydrology, it is worth noting some of the recent advances in the combined use of geochemical data and hydrometric data for improving and calibrating predictive models.

One approach taken to model flow in catchments by using isotopic tracers is the *lumped-parameter* modeling approach. The approach is based upon the use of a mathematical convolution integral for the prediction of tracer output at sampling points that are assumed to integrate flow processes occurring over some (undefined) spatial and (defined) temporal scale. The convolution is performed between a defined tracer input function observed or estimated from tracer data in time series (such as the isotopic composition of precipitation) and a chosen model function representative of the manner in which transit times are distributed in the system according to a steady-state equation of the general form

$$C_o(t) = \int_0^\infty C_i(t-\tau)g(\tau)\mathrm{d}\tau \qquad (12)$$

where $C_o(t)$ and $C_i(t)$ are output and input tracer concentrations, τ is the residence time, and $g(\tau)$ is the system response or weighting function for the tracer (Unnikrishna *et al.*, 1995; Maloszewski and Zuber, 1996). Commonly used forms of the weighting function ($g(\tau)$) include the piston flow model (PFM), the exponential model (EM), the dispersion model (DM), and the combined exponential-piston flow model (EPM). Details of these weighting function models are provided in Maloszewski and Zuber (1996).

In the lumped-parameter approach, there is no notion *a priori* about the geometry of the flow paths in the system. This contrasts with the deterministic or distributed modeling approach, which strives to apply governing equations of flow to an estimated geometric flow field. In the latter approach, it is assumed that the properties controlling flow (such as hydraulic conductivity, transmissivity, porosity, etc.) can be adequately estimated at all points in the catchment model, regardless of scale or depth. In addition, deterministic models are founded in Darcian flow, which in fact may contribute in small degree to measured tracer responses in surface water and shallow groundwater systems. The lumped-parameter approach, however, makes no assumption about the geometry of the system. Instead, the factors that give rise to observed (measured) tracer output responses to observed (measured) tracer inputs are lumped together within a small number of model parameters.

These model parameters may or may not have any direct physical meaning, depending largely upon individual interpretation. The parameters do, however, represent a real means by which to characterize responses, predict flows, and facilitate comparisons among systems. These models are sometimes called "black box" models, reflecting the fact that nothing is assumed known about the flow system, and the entire system is treated as being homogeneous with respect to the distribution of flow lines. Some who espouse their use have softened the moniker to "gray-box" models, in order to reflect the notion that significant knowledge about the system being studied is required in order to (i) choose an appropriate model formulation, (ii) make meaningful interpretations of model output, and (iii) to apply the models wisely. In any case, it is clear that, where tracer inputs and outputs from a hydrologic system are measurable on the timescale of their major variability, lumped-parameter modeling has been the most widely applied approach. Recent reviews and applications of lumped-parameter modeling for MRT estimation in catchments have been conducted by Maloszewski and Zuber (1996), Unnikrishna *et al.* (1995), McGuire *et al.* (2002), and McGlynn *et al.* (2003).

Using the lumped-parameter modeling approach with $\delta^{18}O$ data, McGuire *et al.* (2002) calculated MRTs of 9.5 and 4.8 months for two watersheds in the central Appalachian Mountains of Pennsylvania (USA). Soil-water residence time at a depth of 100 cm was estimated to be ~2 months, indicating that the shallow groundwater was rapidly mobilized in response to precipitation events. In the steeper Maimai catchment of New Zealand, Stewart and McDonnell (1991) reported MRTs calculated using δ^2H data ranging from 12 d to greater than 100 d. Shallow suction lysimeters showed the most rapid responses (12–15 d). MRTs increased with increasing soil depth, such that lysimeters at 40 cm depth showed MRTs of 50–70 d, and lysimeters at 800 cm depth showed residence times of ~100 d or more.

7.3 CARBON, NITROGEN, AND SULFUR SOLUTE ISOTOPES

This section presents a discussion of the fundamentals of three major solute isotope systems: carbon, nitrogen, and sulfur. Recent comprehensive reviews of the geochemistry of several other solute isotope tracers in hydrologic systems include those by Faure (1986), Kendall and McDonnell (1998), and Cook and Herczeg (2000). As discussed above, water isotopes often provide useful information about residence times and relative contributions from different water sources; these data can then be used to make hypotheses about water flow paths. Solute isotopes can provide an alternative, independent isotopic method for determining the relative amounts of water flowing along various subsurface flow paths. However, the least ambiguous use of solute isotopes is for tracing the relative contributions of potential solute sources to groundwater and surface water.

7.3.1 Carbon

Over the last several decades, the decline in alkalinity in many streams in Europe and in northeastern USA as a result of acid deposition has been a subject of much concern (Likens *et al.*, 1979). The concentration of bicarbonate, the major anion buffering the water chemistry of surface waters and the main component of dissolved inorganic carbon (DIC) in most stream waters, is a measure of the "reactivity" of the watersheds and reflects the neutralization of carbonic and other acids by reactions with silicate and carbonate minerals encountered by the acidic waters during their residence in watersheds (Garrels and Mackenzie, 1971). Under favorable conditions, carbon isotopes of DIC can be valuable tools by which to understand the biogeochemical reactions controlling carbonate alkalinity in groundwater and watersheds (Mills, 1988; Kendall *et al.*, 1992).

The $\delta^{13}C$ of DIC can also be a useful tracer of the seasonal and discharge-related contributions of different hydrologic flow paths to surface stream flow (Kendall *et al.*, 1992) and to shallow groundwater discharge from springs in carbonate and fractured rock terrains (Deines *et al.*, 1974; Staniaszek and Halas, 1986; Doctor, 2002). In many carbonate-poor catchments, waters along shallow flow paths in the soil zone have characteristically low $\delta^{13}C_{DIC}$ values reflecting the dominance of soil CO_2 contributions of DIC; waters along deeper flow paths within less weathered materials have intermediate $\delta^{13}C_{DIC}$ values characteristic of carbonic-acid weathering of carbonates (Bullen and Kendall, 1998). In carbonate terrains, the extent of chemical evolution of the $\delta^{13}C$ of carbon species in groundwater can be used to distinguish between groundwaters traveling along different flow paths (Deines *et al.*, 1974).

The carbon isotopic composition of dissolved organic carbon (DOC) has also been successfully applied to studies of hydrologic and biogeochemical processes in watersheds. Humic and fulvic acids are the primary components of DOC in natural waters, and thus contribute to their overall acidity. The carbon isotopic composition of DOC reflects its source, and can provide insight into processes that control the loading of DOC to streams and groundwater. In addition, DOC is a primary nutrient and hydrologic processes that affect the fluxes of DOC are important for a better understanding of the carbon cycle.

7.3.1.1 Carbon isotope fundamentals

Carbon has two stable naturally occurring isotopes: ^{12}C and ^{13}C, and ratios of these isotopes are reported in ‰ relative to the standard Vienna Pee Dee Belemnite (V-PDB) scale. Carbonate rocks typically have $\delta^{13}C$ values of 0 ± 5‰. There is a bimodal distribution in the $\delta^{13}C$ values of terrestrial plant organic matter resulting from differences in the photosynthetic reaction utilized by the plant, with $\delta^{13}C$ values for C_3 and C_4 plants averaging ~−25‰ and −12‰, respectively (Deines, 1980). Soil organic matter has a $\delta^{13}C$ comparable to that of the source plant material, and changes in C_3 and C_4 vegetation types will result in a corresponding change in the $\delta^{13}C$ value of the soil organic matter. Therefore, recent changes in the relative proportion of C_3 and C_4 plants can be detected by measuring the carbon isotopic composition of the current plant community and soil organic matter. The $\delta^{13}C$ values of DIC in catchment waters are generally in the range of −5‰ to −25‰. Values more negative than −30‰ usually indicate the presence of oxidized methane. The $\delta^{18}O$ of DIC is not a useful tracer of alkalinity sources or processes because it exchanges rapidly with oxygen in water.

The primary reactions that produce DIC are: (i) weathering of carbonate minerals by acidic rain or other strong acids; (ii) weathering of silicate minerals by carbonic acid produced by the dissolution of biogenic soil CO_2 by infiltrating rain water; and (iii) weathering of carbonate minerals by carbonic acid. The first and second reactions produce DIC identical in $\delta^{13}C$ to the composition of either the reacting carbonate or carbonic acid, respectively, and the third reaction produces DIC with a $\delta^{13}C$ value exactly intermediate between the compositions of the carbonate and the carbonic acid. Consequently, without further information, DIC produced solely by the third reaction is identical to DIC produced in equal amounts from the first and second reactions.

If the $\delta^{13}C$ values of the reacting carbon-bearing species are known and the $\delta^{13}C$ of the stream DIC determined, in theory we can calculate the relative contributions of these two sources of carbon to the production of stream DIC and carbonate alkalinity, assuming that: (i) there are no other sources or sinks for carbon, and (ii) calcite dissolution occurs under closed-system conditions (Kendall, 1993). With additional chemical or isotopic information, the $\delta^{13}C$ values can be used to estimate proportions of DIC derived from the three reactions listed above. Other processes that may complicate the interpretation of stream water $\delta^{13}C$ values include CO_2 degassing, carbonate precipitation and dissolution, exchange with atmospheric CO_2, carbon uptake and release by aquatic organisms, oxidation of DOC, methanogenesis, and methane oxidation. Correlation of variations in $\delta^{13}C$ with major ion chemistry (especially HCO_3^-), P_{CO_2}, redox conditions, and with other isotope tracers such as $\delta^{34}S$, $^{87}Sr/^{86}Sr$, and ^{14}C may provide evidence for such processes (Clark and Fritz, 1997; Bullen and Kendall, 1998).

7.3.1.2 $\delta^{13}C$ of soil CO_2 and DIC

The biosphere, and particularly a number of soil processes, have a tremendous influence on the $\delta^{13}C$ of DIC in hydrologic systems. Soil CO_2 is comprised mainly of a mixture of atmospherically derived ($\delta^{13}C = -7‰$ to −8‰) and microbially respired CO_2. Respiration is a type of biologic oxidation of organic matter, and as such produces CO_2 with approximately the same $\delta^{13}C$ value as the organic matter (Park and Epstein, 1961); hence, areas dominated by C_3 plants should have soil CO_2 with $\delta^{13}C$ values around −25‰ to −30‰.

Several workers have demonstrated that diffusion of CO_2 gas through the soil causes a fractionation in the CO_2 that exits from the soil surface (Dörr and Münnich, 1980; Cerling *et al.*, 1991; Davidson, 1995; Amundson *et al.*, 1998). These workers draw an important distinction between soil CO_2 (present in the soil) and soil respired CO_2 (gas that has passed across the soil surface and into the atmosphere). Simultaneous measurements of shallow soil CO_2 (i.e., < 0.5 m) and soil respired CO_2 show that the $\delta^{13}C$ of soil respired CO_2 is often ~4‰ lower than soil CO_2 (Cerling *et al.*, 1991; Davidson, 1995). This magnitude of fractionation is similar to what would be expected by diffusion processes in air (4.4‰), given the theoretical diffusion coefficients of $^{12}CO_2$ and $^{13}CO_2$ (Cerling *et al.*, 1991; Davidson, 1995).

While diffusion of CO_2 across the soil-atmosphere boundary may play a dominant role in influencing the $\delta^{13}C$ of shallow soil CO_2, it is generally recognized that the $\delta^{13}C$ values of soil CO_2 depend upon several additional factors, including P_{CO_2} of the soil zone and of the atmosphere, depth of soil CO_2 sampling, and differences in $\delta^{13}C$ of soil-respired CO_2 from the decomposition of organic matter, the latter of which is occasionally observed to have progressively higher $\delta^{13}C$ values with increasing depth (Davidson, 1995; Amoitte-Suchet *et al.*, 1999). Seasonal changes in the P_{CO_2} of soils are likely to be a controlling factor on the isotopic composition of soil CO_2, and therefore ought to be measured with the appropriate frequency (Rightmire, 1978). Given that diffusion processes and mixing with atmospheric CO_2 are most likely to influence the $\delta^{13}C$ of soil CO_2 in the upper 0.5 m of soils, sampling soil CO_2 at a depth of greater than 0.5 m is recommended for estimating the $\delta^{13}C$ of DIC in soil waters when direct soil water sampling is not feasible.

The $\delta^{13}C$ of soil CO_2 can also be affected by fermentation of organic matter, followed by methanogenesis, which produces methane with $\delta^{13}C$ values ranging from −52‰ to −80‰ (Stevens and Rust, 1982). As fermentation progresses, the $\delta^{13}C$ of the residual CO_2 or DIC by-products becomes progressively higher (Carothers and Kharaka, 1980); $\delta^{13}C$ values higher than +10‰ are not uncommon. Thus, very positive $\delta^{13}C$ of DIC can be a useful indicator of methanogenesis in natural waters. Oxidation of methane, alternatively, produces CO_2 with approximately the same composition as the original methane, with $\delta^{13}C$ values generally much lower than −30‰.

The evolution of the isotopic compositions of carbon-bearing substances in uncontaminated systems where carbon is derived from carbonate minerals and soil CO_2 is bounded between two limiting cases: (i) open systems, where carbonate reacts with water in contact with a gas phase having a constant P_{CO_2}, and (ii) closed systems, where the water is isolated from the CO_2 reservoir before carbonate dissolution (Deines *et al.*, 1974; Clark and Fritz, 1997). Both of the extremes assume water residence times long enough for significant isotope exchange between the gas and the aqueous phase to take place.

If DIC in soil water or groundwater exchanges isotopically with other carbon-bearing species, then the isotopic signatures may be blurred and conservative mixing of two distinctive end-member compositions cannot be assumed. The predominant inorganic carbon species at typical soil pH values of ~5 is carbonic acid. The equilibrium isotope fractionation between CO_2 and carbonic acid at 25 °C is 1‰ (Deines *et al.*, 1974). Assuming isotopic equilibrium among all DIC-bearing species (H_2CO_3, HCO_3^-, CO_3^{2-}), the total DIC resulting from the dissolution of calcite ($\delta^{13}C = 0$‰) by carbonic acid ($\delta^{13}C = -22$‰) has a $\delta^{13}C = -11$‰. If this dissolution occurs under open-system conditions in the vadose zone, the DIC would exchange with the soil CO_2 reservoir (−21‰) and reach a $\delta^{13}C$ value of ~−22‰, thus eliminating any carbon isotopic evidence that half the DIC was derived from dissolution of calcite. The addition of strontium isotopic measurements can sometimes shed light on the proportion of carbonate-derived DIC. If the $\delta^{87}Sr$ of calcite were distinctive relative to the $\delta^{87}Sr$ of other catchment minerals, dissolution of the calcite may have left its "signature" in the $\delta^{87}Sr$ of the water (Bullen and Kendall, 1998).

The carbon in subsurface waters that flow into streams is not in chemical and isotopic equilibrium with the atmosphere. For example, CO_2 concentrations in the soil zone are often as high as 5%. Therefore, because the atmospheric concentration is ~0.03%, CO_2 is rapidly lost as soil water and shallow groundwater seeps into a stream. Laboratory experiments performed by Mook (1968) indicate that the $\delta^{13}C$ of DIC rapidly increases by ~0.5‰ during degassing. Furthermore, isotopic exchange between DIC and atmospheric CO_2 is inevitable. In streams with pH values from 5 to 6 and temperatures of 20 °C, the equilibrium $\delta^{13}C$ of stream DIC should be around −8‰. Hence, if the residence time of water in the stream is long enough, the $\delta^{13}C$ of DIC will gradually approach −8‰. However, equilibrium isotopic exchange between stream DIC and atmospheric CO_2 does not appear to be achieved in the first- and second-order streams of forested catchments due to short stream-water residence times.

Additional in-stream processes can affect the $\delta^{13}C$ of DIC (Figure 10). Assimilation of DIC by aquatic organisms through photosynthesis produces organic material with a $\delta^{13}C$ ~30‰ lower than the carbon utilized (Rau, 1979; Mook and Tan, 1991), resulting in an increase in the $\delta^{13}C$ of the remaining DIC. Dissolution of carbonate minerals in-stream will also tend to increase the $\delta^{13}C$ of DIC. In contrast, precipitation of calcite will cause a decrease in the $\delta^{13}C$ of the remaining DIC, due to the equilibrium fractionation between calcite and DIC of ~2‰.

If any of these soil-zone or in-stream processes is a significant source or sink of carbon, it may complicate the interpretation of stream $\delta^{13}C$ values (Kendall, 1993). However, correlation of

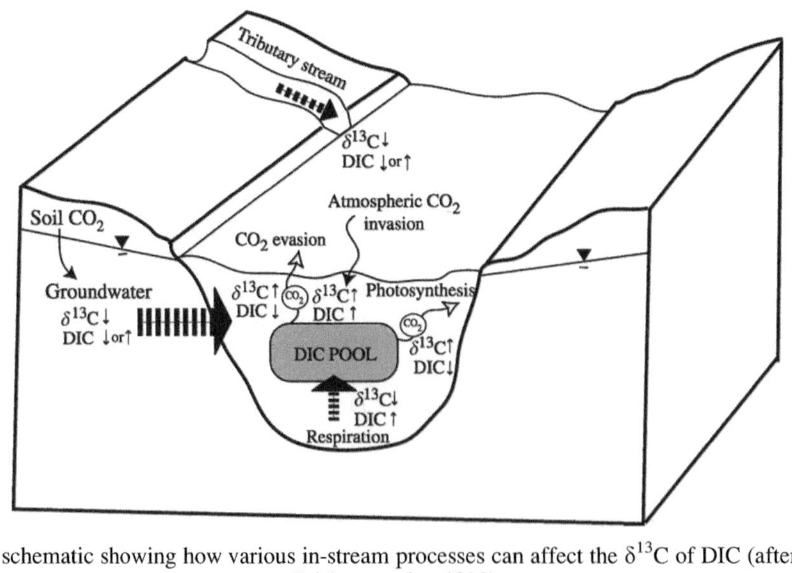

Figure 10 A schematic showing how various in-stream processes can affect the $\delta^{13}C$ of DIC (after Atekwana and Krishnamurthy, 1998).

variations in $\delta^{13}C$ with changes in hydrology, chemistry, or other isotopes such as strontium may provide evidence as to whether such processes are significant. For example, lack of any systematic increase in $\delta^{13}C$ downstream, particularly in the summer when flow is slow, would argue against significant exchange of stream DIC with atmospheric CO_2. Similarly, low pH of stream water would rule out precipitation of calcite as a means to decrease $\delta^{13}C$ of stream DIC. Finally, a strong positive correlation between $\delta^{13}C$ and DIC concentration of stream water, and a typically negative correlation between these parameters and $\delta^{87}Sr$ together, strongly suggest calcite dissolution (Bullen and Kendall, 1998).

7.3.1.3 *Tracing sources of carbonate alkalinity in rivers*

Several studies in the last decade have focused on the origin and cycling of carbon in large river systems using stable isotope techniques (e.g., Taylor and Fox, 1996; Yang *et al.*, 1996; Atekwana and Krishnamurthy, 1998; Aucour *et al.*, 1999; Karim and Veizer, 2000; Kendall *et al.*, 2001b; Hélie *et al.*, 2002). These studies have shown that carbon isotopes are useful indicators of biogeochemical reactions taking place within rivers along their courses of flow, especially concerning the relative amounts of aquatic photosynthesis and respiration.

The processes of primary concern that affect the $\delta^{13}C$ signatures of DIC in large rivers are: (i) CO_2 exchange with the atmosphere, (ii) dissolution/precipitation of carbonate minerals, and (iii) photosynthesis and respiration *in situ*. In particular, mixing between tributary waters of different DIC concentrations and $\delta^{13}C$ values has a great effect on the overall DIC composition of a large river. The studies to date generally suggest that upstream reaches and tributaries are responsible for the primary pool of DIC supplied to the main stem of a large river, with *in situ* recycling of that DIC pool controlling $\delta^{13}C$ compositions further downstream. Smaller tributaries tend to exhibit lower $\delta^{13}C$ of DIC, while further downstream in the main river stem positive shifts in the $\delta^{13}C$ of DIC occur due to photosynthetic uptake of CO_2 and equilibration with the atmosphere. The effects of these processes can vary seasonally. For example, riverine DIC can decrease while pool $\delta^{13}C_{DIC}$ increases during the summer because of photosynthesis, whereas DIC concentrations can increase and $\delta^{13}C_{DIC}$ decrease during the late fall as photosynthesis declines and in-stream decay and respiration increases (Kendall, 1993; Atekwana and Krishnamurthy, 1998; Bullen and Kendall, 1991).

Several attempts have been made to model these processes. Taylor and Fox (1996) studied the Waimakariri River in New Zealand. They modeled equilibrium of DIC with atmospheric CO_2 using a chemical mass-balance model that accounts for the kinetics of CO_2 equilibration between the aqueous and gas phase. Their results show that the measured $\delta^{13}C$ values of the river water cannot be explained solely by equilibrium with atmospheric CO_2, and that an addition of biogenic CO_2 is necessary to account for the measurements. While their model represents a step forward in kinetic considerations of atmospheric CO_2 exchange in river systems, their model does not account for DIC uptake by phytoplankton via photosynthesis, which would cause biological recycling and increases in $\delta^{13}C$ of the residual riverine DIC pool.

Aucour *et al.* (1999) applied a diffusion-based model to account for their observations of the $\delta^{13}C$ value of DIC in the Rhône River, France. Their model predicts that 15–60 min are required to establish equilibrium between dissolved and atmospheric CO_2 in the Rhône. Using $\delta^{13}C = -12.5‰$ for the starting riverine DIC and $\delta^{13}C = -8‰$ for atmospheric CO_2, they predict that it would take between 0.2 d and 2.0 d to attain the observed average $\delta^{13}C$ value of $-5‰$ in the downstream river DIC, depending upon the mean depth of the river. In addition, Aucour *et al.* (1999) observed an inverse trend between $\delta^{13}C$ of DIC and the concentration of DIC in the Rhône River. They attribute this to conservative mixing between the main stem Rhône and its tributaries.

Weiler and Nriagu (1973), Yang *et al.* (1996), and Hélie *et al.* (2002) studied the seasonal changes in $\delta^{13}C$ values of the DIC in the St. Lawrence River system in eastern Canada. They found that the Great Lakes, which are the upstream source of the St. Lawrence River, act as a strong biogeochemical buffer to the $\delta^{13}C_{DIC}$ value of the water in the river's upstream reaches. However, the main stem river downstream shows pronounced seasonal variability in the $\delta^{13}C$ of DIC, with high $\delta^{13}C$ values near isotopic equilibrium with atmospheric CO_2 in the summer during low flow, and very low $\delta^{13}C$ values during the winter and spring high flow periods (Hélie *et al.*, 2002). Hélie *et al.* (2002) report that the $\delta^{13}C$ of DIC in the Great Lakes outflow average $-0.5‰$ to $-1.0‰$, whereas further downstream two large tributaries have a strong influence on the $\delta^{13}C$ of DIC. These rivers—the Ottawa River and the Mascouche River—exhibit much lower $\delta^{13}C$ values than the main stem St. Lawrence, on average $-8.8‰$ and $-12.1‰$, respectively. Hélie *et al.* (2002) conclude that during low water levels in the summer, the Great Lakes provide nearly 80% of the river flow, and the $\delta^{13}C$ of DIC reflects that of atmospheric equilibrium with water in the Great Lakes. During higher flows in the other seasons, the $\delta^{13}C$ of the downstream St. Lawrence River is derived from the influx of the tributaries, which provide a pool of DIC with low $\delta^{13}C$ to the river. In addition, a striking increase in the P_{CO_2} (in excess of atmospheric equilibrium) of the river water was observed during the winter months, indicating a greater proportion of groundwater discharge contribution to the river. In contrast, very low P_{CO_2} values (near atmospheric equilibrium) were observed during the summer, when photosynthetic uptake of CO_2 is at its peak.

7.3.1.4 *Tracing sources of carbonate alkalinity in catchments*

As discussed previously, the headwater tributaries of large river systems in temperate climates have been shown to contain carbonate alkalinity that has a lower $\delta^{13}C$ than in the main stem portion of the river downstream. The most likely explanation for this observation is that soil water and/or groundwater with DIC with a low $\delta^{13}C$ derived from terrestrial biogenic CO_2 is being supplied to the low-order streams. Mixing between tributaries and the main stem of large rivers plays a dominant role in controlling the $\delta^{13}C$ value of DIC; biological recycling of carbon in-stream and, to a lesser extent, equilibration with atmospheric CO_2 are also important.

Under favorable conditions, carbon isotopes can be used to understand the biogeochemical reactions controlling alkalinity in watersheds (Mills, 1988; Kendall *et al.*, 1992). Figure 11 shows that Hunting Creek (in Maryland, USA) shows a strong seasonal change in alkalinity, with concentrations greater than 600 μeq L^{-1} in the summer and less than 400 μeq L^{-1} in the winter. The $\delta^{13}C$ values for weekly stream samples also show strong seasonality, with lower values in the summer (−13‰) and higher values in the winter (−5‰), inversely correlated with alkalinity. Carbonates found in fractures in the altered basalt bedrock have $\delta^{13}C$ values of ∼−5‰ and soil-derived CO_2 is ∼−21‰. Therefore, the dominant reaction controlling stream DIC in the winter appears to be strong acid weathering of carbonates, whereas the dominant reaction in the summer appears to be carbonic acid weathering of carbonates (Kendall, 1993). Of course, with just $\delta^{13}C$ values the possibility that the summer $\delta^{13}C$ values instead reflect approximately equal contributions of DIC produced by strong acid weathering of calcite (−5‰) and carbonic acid weathering of silicates (−21‰) cannot be ruled out. However, the presence of carbonate in the bedrock strongly influences weathering reactions, and the stream chemistry during the summer suggests that silicate weathering has only a minor effect.

Although variations in stream $\delta^{18}O$ during storm events can easily be interpreted in terms of

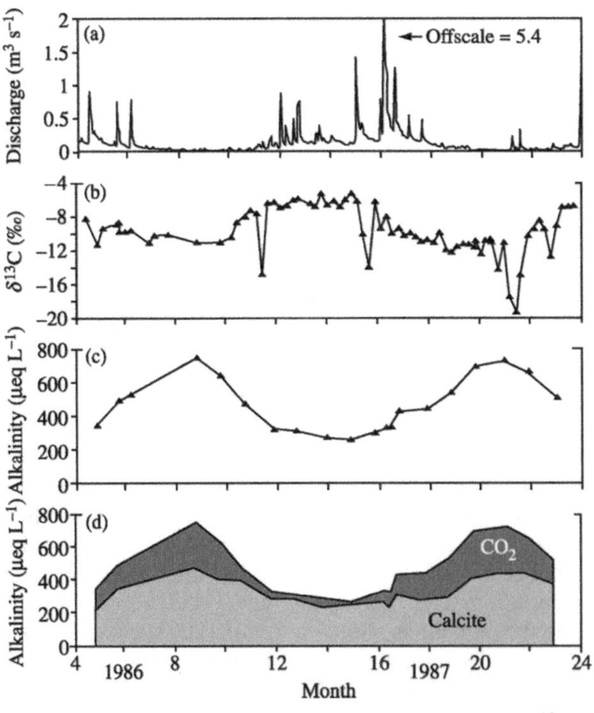

Figure 11 Determining seasonal changes in the sources of alkalinity by using the $\delta^{13}C$ of stream DIC: (a) discharge at Hunting Creek in the Catoctin Mountains, Maryland (USA), 1986–1987; (b) $\delta^{13}C$ of stream DIC collected weekly; (c) alkalinity; and (d) estimation of the relative contributions of carbon from CO_2 and calcite to stream alkalinity using $\delta^{13}C$ of calcite = −5‰ and $\delta^{13}C$ of CO_2 = −21‰; shaded areas show the relative proportions of carbon sources (after Kendall, 1993).

relative contributions of "new" and "old" waters in the two catchments, the $\delta^{18}O$ data do not reveal how the water is delivered to the stream or how these flow paths change during the event. Because $\delta^{13}C$ of DIC has been shown to be sensitive to differences in reactions occurring along shallow versus deep flow paths (Kendall *et al.*, 1992), samples were analyzed for both $\delta^{13}C$ and $\delta^{18}O$. During storm events, the $\delta^{18}O$ shows small gradual changes in water sources in response to changes in rain intensity, in contrast to rapid oscillations in $\delta^{13}C$. For example, during a single three-day storm event in March 1987, $\delta^{13}C$ in both streams oscillated rapidly over a 3‰ range, whereas $\delta^{18}O$ varied only gradually and less than 1‰ (Figure 12). The oscillations in $\delta^{13}C$ apparently reflect more complex changes in flow paths, as well as mixing among waters derived from various flow paths, than is apparent based on $\delta^{18}O$ alone (Kendall *et al.*, 1992).

7.3.1.5 Carbon-14

The radioactive isotope of carbon, ^{14}C, is continuously being produced by reaction of cosmic ray neutrons with ^{14}N in the atmosphere and decays with a half-life of 5,730 yr. ^{14}C concentrations are usually reported as specific activities (disintegrations per minute per gram of carbon relative to a standard) or as percentages of modern ^{14}C concentrations. Under favorable conditions, ^{14}C can be used to date carbon-bearing materials. However, because the "age" of a mixture of waters of different residence times is not very meaningful, the least ambiguous hydrologic use of ^{14}C is as a tracer of carbon sources. The ^{14}C of DIC can be a valuable check on conclusions derived using $\delta^{13}C$ values. For example, if the $\delta^{13}C$ values suggest that the dominant source of carbon is from carbonic acid weathering of silicates, the ^{14}C activities should be high reflecting the young age of the soil CO_2 (Schiff *et al.*, 1990). For more information.

7.3.1.6 Sources of dissolved organic carbon

The use of carbon isotopes to study DOC is becoming more prevalent due to technological advances in mass spectrometry. DOC generally occurs in natural waters in low concentrations, typically ranging between 0.5 ppm and 10 ppm carbon (Thurman, 1985). Thus, several liters to tens of liters of water were once necessary to extract enough DOC for conventional dual gas-inlet isotopic analysis. Today, automated total organic carbon analyzers (TOCs) are commercially available, and have been successfully interfaced with continuous flow isotope ratio mass spectrometers (CF-IRMS) for stable isotopic measurements of samples containing ppb concentrations of DOC (e.g., St-Jean, 2003).

^{14}C combined with $\delta^{13}C$ has been used to study the origin, transport, and fate of DOC in streams and shallow groundwater in forested catchments (Schiff *et al.*, 1990; Wassenaar *et al.*, 1991; Aravena *et al.*, 1992; Schiff *et al.*, 1997; Palmer *et al.*, 2001). Two dominant pools tend to contribute to DOC in surface waters: (i) a younger, more labile pool derived from recent organic matter in soils; and (ii) an older, more refractory pool held within groundwater (Schiff *et al.*, 1997). DOC that is mobilized into streams during storm events is generally young compared to that of groundwater feeding the stream during base flow, indicating

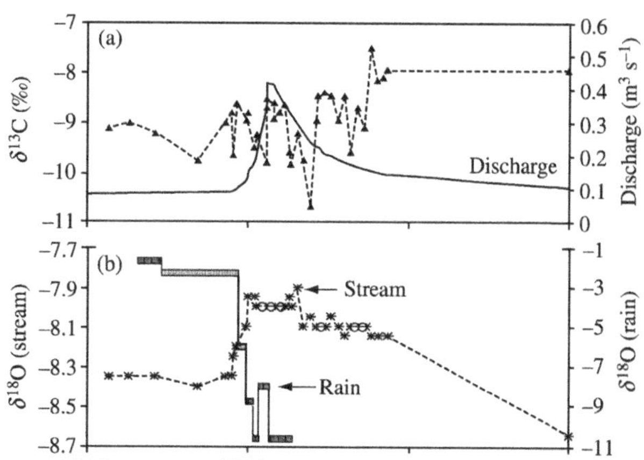

Figure 12 Rain and stream $\delta^{18}O$ and stream $\delta^{13}C$ for a storm March 16–20, 1990 at Hunting Creek, Maryland, USA: (a) $\delta^{13}C$ values and discharge and (b) $\delta^{18}O$ values of rain and stream water (note difference in the scales). Although the gradual shifts in stream $\delta^{18}O$ suggest equally gradual changes in water sources during the storm, the $\delta^{13}C$ values suggest rapidly fluctuating contributions of old waters of different $\delta^{13}C$ values in response to small changes in rainfall intensity.

that extensive cycling of the labile DOC takes place in catchment soils. Schiff *et al.* (1990) report DOC turnover times of less than 40 yr in streams, lakes, and wetlands of Canada as determined from ^{14}C measurements. Palmer *et al.* (2001) report similarly young ages of DOC (post-1955) in soil pore waters and stream waters.

In general, the DOC of streams in forested catchments is formed in soil organic horizons, while minimal amounts of stream DOC are accounted for by through fall or atmospheric deposition (Schiff *et al.*, 1990). Riparian flow paths can account for the greatest proportion of DOC exported to headwater streams (Hinton *et al.*, 1998). The $\delta^{13}C$ of stream DOC can also serve to distinguish between terrigenous-derived DOC and that derived *in situ* in streams in cases where decomposition of phytoplankton is a major DOC source (Wang *et al.*, 1998).

Aravena and Wassenaar (1993) found that while most high molecular weight DOC in shallow unconfined groundwaters has its source in recent organic carbon present in the upper soil, deeper confined aquifers may obtain a significant amount of DOC from the degradation of buried sedimentary organic sources (peats or carbonaceous shales) *in situ*. This may be accompanied by strongly methanogenic conditions in the aquifer.

Much work on characterizing changes in the isotopic composition of DOC and DIC within aquatic systems lies ahead. Carbon isotopes promise to shed much light on the primary biogeochemical processes involved in the carbon cycle.

7.3.2 Nitrogen

Recent concern about the potential danger to water supplies posed by the use of agricultural chemicals has focused attention on the mobility of various solutes, especially nitrate and pesticides, in shallow hydrologic systems. Nitrate concentrations in public water supplies have risen above acceptable levels in many areas of the world, largely as a result of overuse of fertilizers and contamination by human and animal waste. The World Health Organization and the United States Environmental Protection Agency have set a limit of $10\ mg\ L^{-1}$ nitrate (as N) for drinking water because high-nitrate water poses a health risk, especially for children, who can contract methemoglobinemia (blue-baby disease). High concentrations of nitrate in rivers and lakes can cause eutrophication, often followed by fish-kills due to oxygen depletion. Increased atmospheric loads of anthropogenic nitric and sulfuric acids have caused many sensitive, low-alkalinity streams in North America and Europe to become acidified. Still more streams that are not yet chronically acidic could undergo acidic episodes in response to large rain storms and/or spring snowmelt. These acidic "events" can seriously damage sensitive local ecosystems. Future climate changes may exacerbate the situation by affecting biogeochemical controls on the transport of water, nutrients, and other materials from land to freshwater ecosystems. Nitrogen isotope ($\delta^{15}N$) data can provide needed information about nitrogen sources and sinks. Furthermore, nitrate $\delta^{18}O$ and $\delta^{17}O$ are promising new tools for determining nitrate sources and reactions, and complement conventional uses of $\delta^{15}N$.

7.3.2.1 Nitrogen isotope fundamentals

There are two stable isotopes of N: ^{14}N and ^{15}N. Since the average abundance of ^{15}N in air is a very constant 0.366% (Junk and Svec, 1958), air (AIR) is used as the standard for reporting $\delta^{15}N$ values. Most terrestrial materials have $\delta^{15}N$ values between −20‰ and +30‰. The dominant source of nitrogen in most natural ecosystems is the atmosphere ($\delta^{15}N = 0$‰). Plants fixing N_2 from the atmosphere have $\delta^{15}N$ values of ~0‰ to +2‰, close to the $\delta^{15}N$ value of atmospheric N_2. Most plants have $\delta^{15}N$ values in the range of −5‰ to +2‰ (Fry, 1991). Other sources of nitrogen include fertilizers produced from atmospheric nitrogen with $\delta^{15}N$ values of 0 ± 3‰, and animal manure with nitrate $\delta^{15}N$ values generally in the range of +10‰ to +25‰; rock sources of nitrogen are generally negligible.

Biologically mediated reactions (e.g., assimilation, nitrification, and denitrification) strongly control nitrogen dynamics in the soil, as briefly described below. These reactions almost always result in ^{15}N enrichment of the substrate and ^{15}N depletion of the product.

7.3.2.2 Oxygen isotopes of nitrate

The evaluation of sources and cycling of nitrate in the environment can be aided by analysis of the oxygen isotopic composition of nitrate. Oxygen has three stable isotopes: ^{16}O, ^{17}O, and ^{18}O. The $\delta^{18}O$ and $\delta^{17}O$ values of nitrate are reported in per mil (‰) relative to the standard SMOW or VSMOW. There is almost an 80‰ range in $\delta^{18}O$ values, corresponding to a 30‰ "normal" range in $\delta^{15}N$ values (Figure 13). Most of the spread in $\delta^{18}O$ values is caused by precipitation samples. However, there is also considerable variability in $\delta^{18}O$ values produced by nitrification of ammonium and organic matter, and in the $\delta^{18}O$ values of nitrate present in soils, streams, and groundwaters that contain mixtures of atmospheric and microbial nitrate. Quantifying mixtures of sources of nitrate and identifying transformations of inorganic nitrogen in the subsurface can be difficult

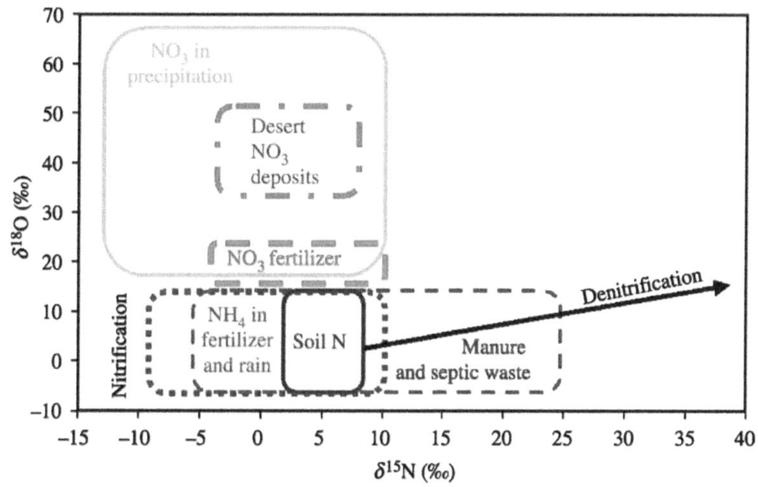

Figure 13 Typical ranges of nitrate $\delta^{18}O$ and $\delta^{15}N$ values for major nitrate reservoirs (after Kendall, 1998).

with only one isotope tracer ($\delta^{15}N$). $\delta^{18}O$ and $\delta^{17}O$ of NO_3 provide two additional tracers for deciphering the nitrogen cycle.

7.3.2.3 *Tracing atmospheric sources of nitrate*

Using $\delta^{15}N$. Complex chemical reactions in the atmosphere result in a large range of $\delta^{15}N$ values of nitrogen-bearing gases and solutes depending on the compound involved, the season, meteorological conditions, ratio of NH_4^+ to NO_3^- in the precipitation, types of anthropogenic inputs, proximity to pollution sources, distance from ocean, etc. (Hübner, 1986). Natural atmospheric sources of nitrogen-bearing gases (e.g., N_2O, HNO_3, NH_3, etc.) include volatilization of ammonia from soils and animal waste (with fractionations as large as −40‰), nitrification and denitrification in soils and surface waters, and production in thunderstorms from atmospheric N_2. Anthropogenic sources include chemical processing and combustion of fossil fuels in automobiles and power plants. The $\delta^{15}N$ values of atmospheric NO_3^- and NH_4^+ are usually in the range of −15‰ to +15‰ (Figure 13). Extremely low $\delta^{15}N$ values for NO_3^- can be expected near chemical plants because of sorption of NO_x gases (with high $\delta^{15}N$ values) in exhaust scrubbers (Hübner, 1986). In general, the NO_3^- in rain appears to have a higher $\delta^{15}N$ value than the coexisting NH_4^+. Although precipitation often contains subequal quantities of ammonium and nitrate, most of the atmospheric nitrogen that reaches the soil surface is in the form of nitrate, because ammonium is preferentially retained by the canopy relative to atmospheric nitrate (Garten and Hanson, 1990). Soil nitrate is preferentially assimilated by tree roots relative to soil ammonium (Nadelhoffer *et al.*, 1988).

Considerable attention has been given to nitrogen oxides (and sulfur oxides) in the atmosphere because of their contributions to acid rain. Several studies suggest that different isotopic techniques can be used to differentiate among different types of atmospherically derived nitrate and ammonium. For example, Heaton (1990) found that NO_x derived from vehicle exhaust had $\delta^{15}N$ values of −2‰ to −11‰, whereas power plant exhaust was +6‰ to +13‰. The differences in $\delta^{15}N$ values were not caused by differences in the $\delta^{15}N$ of the fuels combusted; instead, they reflect differences in the nature of the combustion and NO_x reaction processes.

Using $\delta^{18}O$. There are limited data on the $\delta^{18}O$ of nitrate in atmospheric deposition, with little known about possible spatial or temporal variability, or their causes. The bimodal distribution of nitrate $\delta^{18}O$ values in precipitation in North America presents moderate evidence of at least two sources and/or processes affecting the compositions. The lower mode is centered around values of +22‰ to +28‰, and the higher mode has values centering around +56‰ to +64‰ (Kendall, 1998). Several studies in the USA (e.g., Kendall, 1998; Williard, 1999; Pardo *et al.*, in press) have observed seasonal and spatial differences in the $\delta^{18}O$ of nitrate in precipitation (much higher in winter than summer). These differences may be partially related to variations in the contributions of nitrate from different sources (low δ values partially from vehicle exhaust, and high δ values partially from power plant exhaust). Or the dominant cause may be seasonal changes in atmospheric oxidative processes (Michalski *et al.*, in review). If atmospheric nitrate from power plant plumes and dispersed vehicle sources generally forms via different oxidative pathways or different extents of penetration into the atmosphere, it could account for the apparent correlation of $\delta^{18}O$ values and types of anthropogenic sources.

Further evidence of the correlation of power plant emissions and high nitrate $\delta^{18}O$ values is the observation that all reported nitrate $\delta^{18}O$ values of precipitation in Bavaria, which has high concentrations of nitrate in precipitation, many acid-rain-damaged forests, and is downwind of the highly industrialized parts of Central Europe, are >50‰; whereas samples from Muensterland, farther from the pollution sources in Central Europe, have considerably lower $\delta^{18}O$ values (Kendall, 1998). A recent study of nitrate isotopes in Greenland glaciers has verified that pre-industrial $\delta^{18}O$ values were in the range of +20‰ to +30‰ (Savarino *et al.*, 2002), suggesting that relatively recent changes in atmospheric chemistry related to anthropogenic emissions of reactive NO_x are responsible for the high ozone $\delta^{18}O$ values that apparently cause the high $\delta^{18}O$ of nitrate.

Using $\delta^{17}O$. In all terrestrial materials, there is a constant relation between the $\delta^{18}O$ and $\delta^{17}O$ values of any given substance, because isotope fractionations are mass dependent. On a three-isotope plot with $\delta^{17}O$ on the *x*-axis and $\delta^{18}O$ on the *y*-axis (Figure 14), mass *dependent* fractionations (MDF) for nitrate result in values defined by the relation: $\delta^{17}O = 0.52\delta^{18}O$, whereas mass *independent* fractionations (MIF) cause values that deviate from this relation. For nitrate, the mass-independent relation is described by $\Delta^{17}O = \delta^{17}O - 0.52\delta^{18}O$ (Michalski *et al.*, 2002). The term "Δ" is pronounced "cap delta," which is short for "capital delta." Hence, MDF results in $\Delta^{17}O = 0$, whereas an MIF results in $\Delta^{17}O \neq 0$. $\delta^{17}O$ and $\Delta^{17}O$ values reflect $^{17}O/^{16}O$ ratios, and are reported in per mil relative to SMOW.

Recent investigations show that ozone and many oxides derived from high atmospheric processes have "excess ^{17}O" (beyond the $\delta^{17}O$ expected from the $\delta^{18}O$ value) derived from a mass-independent fractionation. Further work shows that all atmospherically derived nitrate is labeled by characteristic nonterrestrial $\delta^{17}O$ signatures (Michalski *et al.*, 2002; Michalski *et al.*, in review). One especially noteworthy feature of $\Delta^{17}O$ values is that they are not affected by any terrestrial fractionating processes such as denitrification and assimilation. Seasonal variations in atmospheric $\Delta^{17}O$ (from +20‰ to +30‰) observed in southern California were explained by a shift from nitric acid production by the $OH + NO_2$ reaction, which is predominant in the spring and summer, to N_2O_5 hydrolysis reactions that dominate in the winter (Michalski *et al.*, in review). This shift is driven largely by temperature variability and NO_x concentrations (Michalski *et al.*, in review), both of which should vary in plume (power plant) and dispersed (automobile) conditions. Ice-core samples from a glacier in Greenland showed higher $\Delta^{17}O$ values during the 1880s due to large biomass burning events in North America (Savarino *et al.*, 2002). The collection of preliminary findings suggests that atmospheric source characterization may be possible when $\delta^{17}O$ is considered in combination with $\delta^{18}O$ and $\delta^{15}N$ measurements, if nitrogen sources from different combustion processes (i.e., with different oxidation chemistries and perhaps degrees of NO_x penetration and recycling in the stratosphere) have different $\delta^{17}O$ values.

7.3.2.4 Fertilizer and animal waste sources of nitrogen

Synthetic fertilizers have $\delta^{15}N$ values that are uniformly low reflecting their atmospheric source (Figure 13), generally in the range of −4‰ to +4‰; however, some fertilizer samples have shown a total range of −8‰ to +7‰ (see compilations by Hübner, 1986; Macko and Ostrom, 1994). Organic fertilizers (which include the so-called "green" fertilizers such as cover crops and plant composts, and liquid and solid animal waste) generally have higher $\delta^{15}N$ values and a much

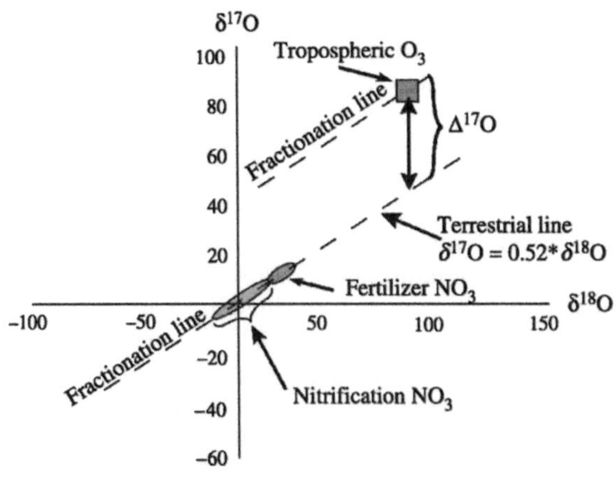

Figure 14 Schematic of relationship between $\delta^{18}O$ and $\delta^{17}O$ values (after Michalski *et al.*, 2002).

wider range of compositions (generally +2 to +30‰) than inorganic fertilizers because of their more diverse origins. Note that the $\delta^{15}N$ of nitrate in fertilized soils may not be the same as the fertilizer. Amberger and Schmidt (1987) determined that nitrate fertilizers have distinctive $\delta^{18}O$ and $\delta^{15}N$ values. All three oxygens in fertilizer nitrate are derived from atmospheric O_2 (+22 to +24‰), and hence the $\delta^{18}O$ of the nitrate is in this range.

It has often been observed that consumers (microbes to invertebrates) are 2 – 3‰ enriched in ^{15}N relative to their diet. The increases in $\delta^{15}N$ in animal tissue and solid waste relative to diet are due mainly to the excretion of low $\delta^{15}N$ organics in urine or its equivalent (Wolterink *et al.*, 1979). Animal waste products may be further enriched in ^{15}N because of volatilization of ^{15}N-depleted ammonia, and subsequent oxidation of much of the residual waste material may result in nitrate with a high $\delta^{15}N$ (Figure 13). By this process, animal waste with a typical $\delta^{15}N$ value of about +5‰ is converted to nitrate with $\delta^{15}N$ values generally in the range of +10‰ to +20‰ (Kreitler, 1975, 1979); human and other animal waste become isotopically indistinguishable under most circumstances (an exception is Fogg *et al.*, 1998).

7.3.2.5 Soil sources of nitrogen

The $\delta^{15}N$ of total soil nitrogen is affected by many factors including soil depth, vegetation, climate, particle size, cultural history, etc.; however, two factors, drainage and influence of litter, have a consistent and major influence on the $\delta^{15}N$ values (Shearer and Kohl, 1988). Soils on lower slopes and near saline seeps have higher $\delta^{15}N$ values than well-drained soils (Karamanos *et al.*, 1981), perhaps because the greater denitrification in more boggy areas results in high-$\delta^{15}N$ residual nitrate. Surface soils beneath bushes and trees often have lower $\delta^{15}N$ values than those in open areas, presumably as the result of litter deposition (Shearer and Kohl, 1988). Fractionations during litter decomposition in forests result in surface soils with lower $\delta^{15}N$ values than deeper soils (Nadelhoffer and Fry, 1988). Gormly and Spalding (1979) attributed the inverse correlation of nitrate-$\delta^{15}N$ and nitrate concentration beneath agricultural fields to increasing denitrification with depth.

7.3.2.6 Processes affecting nitrogen isotopic compositions

Irreversible (unidirectional) kinetic fractionation effects involving metabolic nitrogen transformations are generally more important than equilibrium fractionation effects in low-temperature environments. Many biological processes consist of a number of steps (e.g., nitrification: organic-N → NH_4^+ → NO_2^- → NO_3^-). Each step has the potential for fractionation, and the overall fractionation for the reaction is highly dependent on environmental conditions including the number and type of intermediate steps, sizes of reservoirs (pools) of various compounds involved in the reactions (e.g., O_2, NH_4^+), soil pH, species of the organism, etc. Hence, estimation of fractionations in natural systems is very complex.

A useful "rule of thumb" is that most of the fractionation is caused by the so-called "rate-determining step"—which is the slowest step. This step is commonly one involving a large pool of substrate where the amount of material actually used is small compared to the size of the reservoir. In contrast, a step that is not rate determining generally involves a small pool of some compound that is rapidly converted from reactant to product; because the compound is converted to product as soon as it appears, there is no fractionation at this step. The isotopic compositions of reactant and product during a multistep reaction where the net fractionation is controlled by a single rate-determining step can be successfully modeled with Rayleigh equations. For more details, see recent reviews by Kendall (1998), Kendall and Aravena (2000), and Böhlke (2002).

N-fixation refers to processes that convert unreactive atmospheric N_2 into other forms of nitrogen. Although the term is usually used to mean fixation by bacteria, it has also been used to include fixation by lightning and, more importantly, by human activities (energy production, fertilizer production, and crop cultivation) that produce reactive nitrogen (NO_x, NH_y, and organic nitrogen). Fixation of atmospheric N_2 commonly produces organic materials with $\delta^{15}N$ values slightly less than 0‰. A compilation by Fogel and Cifuentes (1993) indicates measured fractionations ranging from −3‰ to +1‰. Because these values are generally lower than the values for organic materials produced by other mechanisms, low $\delta^{15}N$ values in organic matter are often cited as evidence for N_2 fixation.

Assimilation generally refers to the incorporation of nitrogen-bearing compounds into organisms. Assimilation, like other biological reactions, discriminates between isotopes and generally favors the incorporation of ^{14}N over ^{15}N. Measured values for apparent fractionations caused by assimilation by microorganisms in soils show a range of −1.6‰ to +1‰, with an average of –0.52‰ (compilation by Hübner, 1986). Fractionations by vascular plants show a range of −2.2‰ to +0.5‰ and an average of −0.25‰, relative to soil organic matter (Mariotti *et al.*, 1980). Nitrogen uptake by plants in soils causes only a small fractionation and, hence, only slightly alters the isotopic composition of the residual fertilizer or soil organic matter. The $\delta^{15}N$ of algae in rivers in the USA is generally ~4‰ lower than that of the associated nitrate (Kendall, unpublished data).

Mineralization is usually defined as the production of ammonium from soil organic matter. This is sometimes called *ammonification*, which is a less confusing term. Mineralization usually causes only a small fractionation (~1‰) between soil organic matter and soil ammonium. In general, the $\delta^{15}N$ of soil ammonium is usually within a few per mil of the composition of total organic nitrogen in the soil.

Nitrification is a multistep oxidation process mediated by several different autotrophic organisms for the purpose of deriving metabolic energy; the reactions produce acidity. In general, the extent of fractionation is dependent on the size of the substrate pool (reservoir). In nitrogen-limited systems, the fractionations are minimal. The total fractionation associated with nitrification depends on which step is rate determining. Because the oxidation of nitrite to nitrate is generally quantitative (rapid) in natural systems, this is generally not the rate-determining step, and most of the nitrogen fractionation is probably caused by the slow oxidation of ammonium by *Nitrosomonas*. In soils, overall N-nitrification fractionations have been estimated to range between 12‰ and 29‰ (Shearer and Kohl, 1986).

Volatilization is the term commonly used for the loss of ammonia gas from surficial soils to the atmosphere; the ammonia gas produced has a lower $\delta^{15}N$ value than the residual ammonium in the soil. Volatilization in farmlands results from applications of urea and manure to fields, and occurs within piles of manure; the resulting organic matter may have $\delta^{15}N$ values >20‰ because of ammonia losses. Ammonium in precipitation derived from volatilazation from waste lagoons can have a distinctive low $\delta^{15}N$ value.

Denitrification is a multistep process with various nitrogen oxides (e.g., N_2O, NO) as intermediate compounds resulting from the chemical or biologically mediated reduction of nitrate to N_2. Depending on the redox conditions, organisms will utilize different oxidized materials as electron acceptors in the general order: O_2, NO_3^-, SO_4^{2-}. Although microbial denitrification does not occur in the presence of significant amounts of oxygen, it can occur in anaerobic pockets within an otherwise oxygenated sediment or water body (Koba *et al.*, 1997).

Denitrification causes the $\delta^{15}N$ of the residual nitrate to increase exponentially as nitrate concentrations decrease, and causes the acidity of the system to decrease. For example, denitrification of fertilizer nitrate that originally had a distinctive $\delta^{15}N$ value of +0‰ can yield residual nitrate with much higher $\delta^{15}N$ values (e.g., +15‰ to +30‰) that are within the range of compositions expected for nitrate from a manure or septic-tank source. Measured enrichment factors (apparent fractionations) associated with denitrification range from −40‰ to −5‰; hence, the $\delta^{15}N$ of the N_2 is lower than that of the nitrate by about these values. The N_2 produced by denitrification results in *excess N_2 contents* in groundwater; the $\delta^{15}N$ of this N_2 can provide useful information about sources and processes (Vogel *et al.*, 1981; Böhlke and Denver, 1995).

7.3.2.7 *Processes affecting the δ¹⁸O of nitrate*

Nitrification. When organic nitrogen or ammonium is microbially oxidized to nitrate (i.e., nitrified), the three oxygens are derived from two sources: H_2O and dissolved O_2. If the oxygen is incorporated with no fractionation, the $\delta^{18}O$ of the resulting nitrate is intermediate between the compositions of the two oxygen sources. Andersson and Hooper (1983) showed that oxidation of NH_4^+ to NO_2^- incorporated one atom from water and one atom from dissolved oxygen. Work by Aleem *et al.* (1965) indicates that the additional oxygen atom incorporated during the oxidation of NO_2^- to NO_3^- originates entirely from H_2O. This combined work suggests that the oxygen isotopic composition of NO_3^- formed during autotrophic nitrification ($NH_4^+ \rightarrow NO_3^-$) would have the composition of

$$\delta^{18}O_{NO_3} = \tfrac{2}{3}\delta^{18}O_{H_2O} + \tfrac{1}{3}\delta^{18}O_{O_2} \qquad (13)$$

For waters with $\delta^{18}O$ values in the normal range of −25‰ to +4‰, the $\delta^{18}O$ of soil nitrate formed from *in situ* nitrification of ammonium should be in the range of −10‰ to +10‰, respectively (Figure 13). For highly evaporated water (+20‰), the $\delta^{18}O$ of nitrate could be as high as about +21‰ (Böhlke *et al.*, 1997). The oxygen isotopic composition of dissolved O_2 reflects the effects of three primary processes: (i) diffusion of atmospheric oxygen (~+23.5‰) in the subsurface, (ii) photosynthesis—resulting in the addition of O_2 with a low $\delta^{18}O$ similar to that of water, and (iii) respiration by microbes—resulting in isotopic fractionation and higher $\delta^{18}O$ values for the residual O_2.

The above model makes four critical assumptions: (i) the proportions of oxygen from water and O_2 are the same in soils as observed in laboratory cultures; (ii) there are no fractionations resulting from the incorporation of oxygen from water or O_2 during nitrification; (iii) the $\delta^{18}O$ of water used by the microbes is identical to that of the bulk soil water; and (iv) the $\delta^{18}O$ of the O_2 used by the microbes is identical to that of atmospheric O_2.

Many studies have used measurement of $\delta^{18}O$ of NO_3^- in freshwater systems for assessing sources and cycling (see below). Many of them find that the $\delta^{18}O$ of microbial NO_3^- appears to be a few per mil higher than expected for the equation and assumptions above (Kendall, 1998). A variety of explanations have been offered for these high nitrate-$\delta^{18}O$ values: (i) nitrification in contact with soil waters with higher-than-expected $\delta^{18}O$ values because of evaporation (Böhlke *et al.*, 1997) or seasonal changes in rain $\delta^{18}O$ (Wassenaar, 1995); (ii) changes in the proportion of oxygen from H_2O and O_2 sources (Aravena *et al.*, 1993); (iii) nitrification using O_2 that has a high $\delta^{18}O$ due to heterotrophic respiration (Kendall, 1998); and (iv) nitrification that occurs simultaneously via both heterotrophic and autotrophic pathways (Mayer *et al.*, 2001). At this time, the mechanisms responsible for generating the isotopic composition of NO_3^- in nature are poorly understood.

Denitrification. Denitrification is the process that poses most difficulties for simple applications of nitrate isotopes. Hence, for successful applications of nitrate isotopes for tracing sources, it is critical to (i) determine if denitrification has occurred, and, if so (ii) determine what was the initial isotopic composition of the nitrate (which is a necessary prerequisite for later attempts to define sources). There are several geochemical methods for identifying denitrification in groundwater, and distinguishing it from mixing:

(i) Geochemical evidence of a reducing environment (e.g., low dissolved O_2, high H_2S, etc.).
(ii) Hyperbolic versus exponential relationships between $\delta^{15}N$ and NO_3^- (i.e., mixing is a hyperbolic function, whereas denitrification is exponential). Hence, if mixing of two sources is responsible for the curvilinear relationship between $\delta^{15}N$ and NO_3^-, plotting $\delta^{15}N$ versus $1/NO_3^-$ will result in a straight line (Figure 2). In contrast, if denitrification (or assimilation) is responsible for the relationship, plotting $\delta^{15}N$ versus ln NO_3^- will produce a straight line (Figure 3). Analysis of dissolved N_2 (produced by the denitrification of NO_3^-) for $\delta^{15}N$ to show that there are systematic increases in the $\delta^{15}N$ of N_2 with decreases in NO_3^- and increases in the $\delta^{15}N$ of NO_3^- (e.g., a "Rayleigh equation" relationship like Figure 1).
(iii) Analysis of the NO_3^- for $\delta^{18}O$ as well as $\delta^{15}N$ to see if there is a systematic increase in $\delta^{18}O$ with increase in $\delta^{15}N$ and decrease in NO_3^-, consistent with denitrification. On $\delta^{18}O$ versus $\delta^{15}N$ plots, denitrification produces lines with slopes of ~0.5 (Figure 13).
(iv) No change in $\Delta^{17}O$ during a process that has $\delta^{18}O$ and/or $\delta^{15}N$ trends characteristic of denitrification (if material has $\Delta^{17}O > 0$).

7.3.2.8 *Small catchment studies*

The main application of nitrate $\delta^{18}O$ has been for the determination of the relative contributions of atmospheric and soil-derived sources of nitrate to shallow groundwater and small streams. This problem is intractable using $\delta^{15}N$ alone because of overlapping compositions of soil and atmospherically derived nitrate, whereas these nitrate sources have very distinctive nitrate $\delta^{18}O$ values (Figure 13). Durka *et al.* (1994) analyzed nitrate in precipitation and spring samples from several German forests for both isotopes, and found that the $\delta^{18}O$ of nitrate in springs was correlated with the general health of the forest, with more healthy or limed forests showing lower $\delta^{18}O$ values closer to the composition of microbially produced nitrate, and severely damaged forests showing higher $\delta^{18}O$ values values indicative of major contributions of atmospheric nitrate to the system.

A similar application is the determination of the source of nitrate in early spring runoff. During early spring melt, many small catchments experience episodic acidification because of large pulses of nitrate and hydrogen ions being flushed into the streams. There has been some controversy over the source of this nitrate. A number of studies have found that much of the nitrate in early runoff from small catchments is microbial rather than atmospheric (Kendall *et al.*, 1995b; Burns and Kendall, 2002; Campbell *et al.*, 2002; Sickman *et al.*, 2003; Pardo *et al.*, in press). For example, during the 1994 snowmelt season in Loch Vale watershed in Colorado (USA), almost all the stream samples had nitrate $\delta^{18}O$ and $\delta^{15}N$ values within the range of pre-melt and soil waters, and significantly different from the composition of almost all the snow samples. Therefore, the nitrate eluted from the snowpack appeared to go into storage, and most of the nitrate in stream flow during the period of potential acidification was apparently derived from pre-melt sources (Kendall *et al.*, 1995b). During subsequent years, half or more of the nitrate in the stream was microbial in origin, and probably originated from shallow groundwater in talus deposits (Campbell *et al.*, 2002).

In contrast to these forested catchment studies, atmospheric nitrate appears to be a major contributor to stream flow in urban catchments. A pilot study by Ging *et al.* (1996) to determine the dominant sources of nitrate in storm runoff in suburban watersheds in Austin, Texas (USA) found that during base flow (when chloride was high), nitrate had high $\delta^{15}N$ and low $\delta^{18}O$ values; in contrast, during storms (when chloride was low), nitrate had low $\delta^{15}N$ and high $\delta^{18}O$ values. The strong correspondence of $\delta^{15}N$ and $\delta^{18}O$ values during changing flow conditions, and the positive correlation of the percentage of impervious land-cover and the $\delta^{18}O$ of nitrate, suggests that the stream composition can be explained by varying proportions of two end-member compositions (Figure 15), one dominated by atmospheric nitrate (and perhaps fertilizer) that is the major source of water

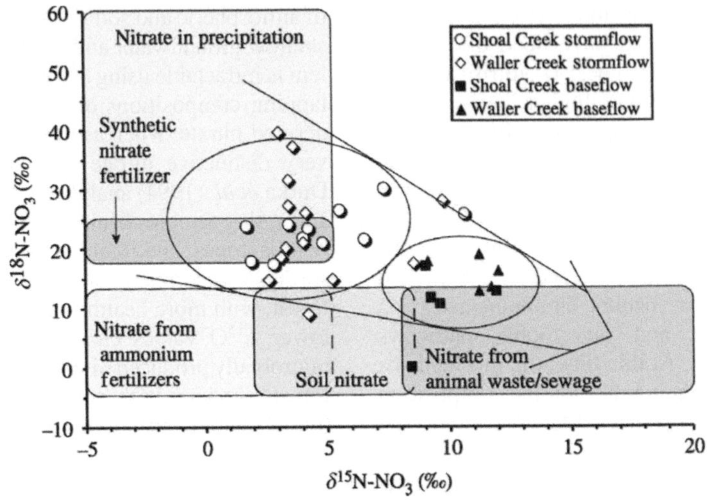

Figure 15 $\delta^{15}N$ versus $\delta^{18}O$ values of nitrate during storm flow and base flow conditions from Waller and Shoal Creeks, Austin, Texas, superimposed on common fields of nitrate from various sources. Ellipses indicate two standard deviations from average values. Arrows point in the direction of the base flow (subsurface) nitrate end-member (source Silva *et al.*, 2002).

during storms and the other a well-mixed combination of sewage and other nitrate sources that contributes to base flow (Silva *et al.*, 2002). A recent study of nitrate from large rivers in the Mississippi Basin has shown that atmospheric nitrate is also a significant source of nitrate to large undeveloped and urban watersheds (Battaglin *et al.*, 2001; Chang *et al.*, 2002).

7.3.2.9 Large river studies

Two recent studies in the USA evaluated whether the combination of nitrate $\delta^{18}O$ and $\delta^{15}N$ would allow discrimination of watershed sources of nitrogen and provide evidence for denitrification. A pilot study in the Mississippi Basin (Battaglin *et al.*, 2001; Chang *et al.*, 2002) showed that large watersheds with different land uses (crops, animals, urban, and undeveloped) had overlapping moderately distinguishable differences in nutrient isotopic compositions (e.g., nitrate $\delta^{18}O/\delta^{15}N$ and particulate organic matter (POM) $\delta^{15}N/\delta^{13}C$). A study in large rivers in the northeastern part of the USA found strong positive correlations between $\delta^{15}N$ and the amounts of waste nitrogen (Mayer *et al.*, 2002)

Prior nitrogen mass-balance studies in the Mississippi Basin and in the northeastern USA rivers suggested appreciable losses of nitrogen via denitrification, especially in the headwaters. However, neither study found isotopic evidence for denitrification, perhaps because of continuous mixing with new nitrate, the small extent of denitrification, or the low fractionations resulting from diffusion-controlled (i.e., benthic) denitrification (Sebilo *et al.*, 2003). A detailed study of denitrification in various stream orders in the Seine River system (France) during summer low flow conditions indicates that extent of denitrification determined by examination of the shifts in the $\delta^{15}N$ of residual nitrate provides only a minimum estimate of denitrification (Sebilo *et al.*, 2003).

A useful adjunct to tracing nitrogen sources and sinks in aquatic systems with nitrate isotopes is the analysis of POM for $\delta^{15}N$, $\delta^{13}C$, and $\delta^{34}S$. In many river systems, most of the POM is derived from *in situ* production of algae. Even if an appreciable percent of the POM is terrestrial detritus, the C:N value of the POM and the $\delta^{15}N$ and $\delta^{13}C$ can, under favorable conditions, be used to estimate the percent of POM that is algae, and its isotopic composition (Kendall *et al.*, 2001b). The $\delta^{15}N$ and $\delta^{13}C$ (and $\delta^{34}S$) of the POM reflect the isotopic compositions of dissolved inorganic nitrogen, carbon, and sulfur in the water column. These compositions reflect the sources of nitrogen, carbon, and sulfur to the system, and the biogeochemical processes (e.g., photosynthesis, respiration, denitrification, and sulfate reduction) that alter the isotopic compositions of the dissolved species (Figure 16). Hence, the changes in the isotopic composition can be used to evaluate a variety of in-stream processes that might affect the interpretation of nitrate $\delta^{15}N$ and $\delta^{18}O$ (Kendall *et al.*, 2001b). The $\delta^{15}N$ of algae may even serve as an integrator for the $\delta^{15}N$ of nitrate; the $\delta^{15}N$ of algae appears to be ~4‰ lower than that of the associated nitrate (Battaglin *et al.*, 2001; Kendall, unpublished data).

7.3.2.10 Subsurface waters

Applications of $\delta^{15}N$ to trace relative contributions of fertilizer and animal waste to groundwater

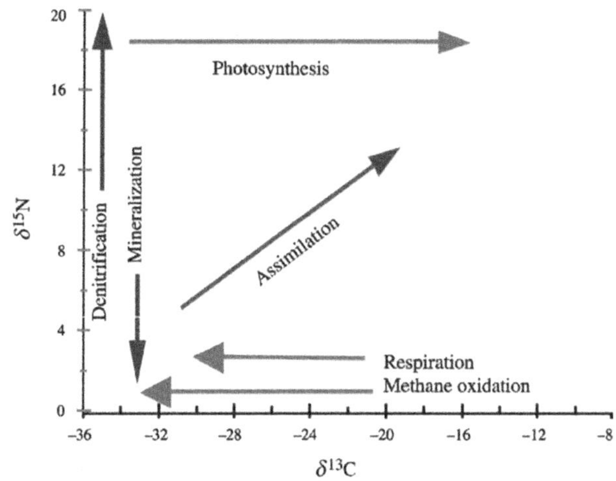

Figure 16 Schematic of in-stream biogeochemical processes that affect the $\delta^{15}N$ and $\delta^{13}C$ of algae. Arrows indicate the general effect of each process on the isotopic composition of the resulting algae.

(Kreitler, 1975; Kreitler and Jones, 1975; Kreitler *et al.*, 1978; Gormly and Spalding, 1979) are complicated by a number of biogeochemical reactions, especially ammonia volatilization, nitrification, and denitrification. These processes can modify the $\delta^{15}N$ values of nitrogen sources prior to mixing and the resultant mixtures, causing estimations of the relative contributions of the sources of nitrate to be inaccurate. The combined use of $\delta^{18}O$ and $\delta^{15}N$ should allow better resolution of these issues. Analysis of the $\delta^{15}N$ of both nitrate and N_2 (e.g., Böhlke and Denver, 1995) provides an effective means for investigating denitrification. There have been only a few studies thus far using $\delta^{15}N$ and $\delta^{18}O$ to study nitrate sources and cycling mechanisms in groundwater. However, since it is likely that more studies in the future will utilize a multi-isotope or multitracer approach, the discussion below will concentrate on dual isotope studies.

The first dual isotope investigation of groundwater, in municipal wells downgradient from heavily fertilized agricultural areas near Hanover (Germany), successfully determined that the decreases in nitrate away from the fields was caused by microbial denitrification, not mixing with more dilute waters from nearby forests (Böttcher *et al.*, 1990). They found that low concentrations of nitrate were associated with high $\delta^{18}O$ and $\delta^{15}N$ values, and vice versa, and that changes in $\delta^{18}O$ and $\delta^{15}N$ values along the flow path were linearly related, with a slope of $\sim$0.5. The linear relation between the isotope values and the logarithm of the fraction of residual nitrate (i.e., the data fit a Rayleigh equation) indicated that denitrification with constant enrichment factors for oxygen (−8.0‰) and nitrogen (−15.9‰) was responsible for the increases in $\delta^{18}O$ and $\delta^{15}N$.

There have been several subsequent dual isotope fertilizer studies. The high concentrations of NO_3^- in shallow groundwater in the Abbotsford aquifer, British Columbia (Canada), were attributed to nitrification of poultry manure, with lesser amounts of ammonium fertilizers (Wassenaar, 1995). A study of denitrification in a riparian zone showed a higher slope ($\sim$0.7) for the relative fractionation of $\delta^{18}O$ to $\delta^{15}N$ (Mengis *et al.*, 1999). Nitrate from 10 major karst springs in Illinois during four different seasons was found to be mainly derived from nitrogen fertilizer (Panno *et al.*, 2001).

The applicability of $\delta^{18}O$ and $\delta^{15}N$ of nitrate and other tracers to delineate contaminant plumes derived from domestic septic systems was evaluated by Aravena *et al.* (1993), in a study within an unconfined aquifer beneath an agricultural area in Ontario (Canada). They found that $\delta^{15}N$ of nitrate, $\delta^{18}O$ of water, and water chemistry (especially sodium) were effective for differentiating between the plume and native groundwater. The differences in $\delta^{15}N$ values reflect differences in dominant nitrogen sources (human waste versus mixed fertilizer/manure). There was good delineation of the plume with water-$\delta^{18}O$ values, reflecting differences in residence time of water within the shallow unconfined aquifer, and the deeper confined aquifer that supplies water to the house and the plume. There was no significant difference between the $\delta^{18}O$ of nitrate in the plume and in local groundwater; the $\delta^{18}O$ values suggest that nitrification of ammonium, from either human waste or agricultural sources, is the source of the nitrate. Another study of a septic plume determined that use of a multi-isotope approach (using $\delta^{13}C$ of DIC, and $\delta^{18}O$ and $\delta^{34}S$ of sulfate in addition to nitrate $\delta^{18}O$ and $\delta^{15}N$) provided valuable insight into the details of the processes affecting nitrate attenuation in groundwater (Aravena and Robertson, 1998).

7.3.3 Sulfur

Several important reviews of the use of stable sulfur isotopes as hydrologic and biogeochemical tracers have been produced since the mid-1980s. Krouse and Grinenko (1991) provide a very comprehensive evaluation of stable sulfur isotopes as tracers of natural and anthropogenic sulfur. Krouse and Mayer (2000) present an in-depth review and several case studies of the use of stable sulfur and oxygen isotopes of sulfate as hydrologic tracers in groundwater. In addition, the use of sulfur isotopes for investigations of hydrological processes in catchments has been reviewed by Mitchell *et al.* (1998). Thus, what is presented here is just a brief overview of the application of sulfur isotopes in hydrologic studies.

Much of the work using sulfur isotopes in surface hydrology to date has been driven by the need to understand the effects of atmospheric deposition on sulfur cycling in the natural environment, particularly in forest ecosystems. This is in response to increased sulfur loadings to terrestrial ecosystems from anthropogenic sulfur emissions, as sulfur is a dominant component of "acid rain." In addition to studying ecosystem responses to sulfur deposition from atmospheric sources, there is often a need to identify sulfur sources and transformations along flow pathways in groundwater. Bacterially mediated sulfate reduction is one of the most widespread biogeochemical processes occurring in groundwater systems. Elevated sulfate concentrations in groundwater and production of H_2S gas in aquatic systems can be both a nuisance and a health hazard, yet are ubiquitous. The stable isotopes of sulfur and of oxygen in sulfate are useful tracers in such studies (Clark and Fritz, 1997; Krouse and Mayer, 2000). A more applied use of stable sulfur isotopes in terms of groundwater risk management is for identification of arsenic contamination sources and biological remediation efficacy of organic pollutants.

7.3.3.1 Sulfur isotope fundamentals

Sulfur has four stable isotopes: ^{32}S (95.02%), ^{33}S (0.75%), ^{34}S (4.21%), and ^{36}S (0.02%) (MacNamara and Thode, 1950). Stable isotope compositions are reported as $\delta^{34}S$, ratios of $^{34}S/^{32}S$ in per mil relative to the standard CDT. The general terrestrial range is +50‰ to −50‰, with rare values much higher or lower.

Geologic sources of sulfur include primary and secondary sulfide minerals as well as chemically precipitated sulfate minerals. In sedimentary rocks, sulfur is often concentrated. Examples include pyritic shales, evaporites, and limestones containing pyrite and gypsum in vugs and lining fractures. In many cases, these rocks can be both the source of organic matter as well as sulfate for the biogeochemical reduction of sulfate to occur.

There is a large range in $\delta^{34}S$ of sedimentary sulfide minerals, from −30‰ to +5‰ (Migdisov *et al.*, 1983; Strauss, 1997). Crystalline rocks of magmatic origin show a more narrow range, from 0‰ to +5‰. Volcanic rocks are similar; however, some have values up to +20‰, indicating recycling of oceanic sulfate at subduction zones.

Figure 17 shows schematically the major transformations of sulfur in aquatic systems and their related isotopic fractionations. Processes that do not significantly fractionate sulfur isotopes include: (i) weathering of sulfide and sulfate minerals, (ii) adsorption–desorption interactions with organic matter, and (iii) isotopic exchange between SO_4^{2-} and HS^- or H_2S in low-temperature environments. Precipitation of sulfate minerals is accompanied by only slight isotopic fractionation, and the precipitates are generally enriched in the heavier isotopes (^{34}S and ^{18}O) relative to the residual sulfate in solution (Holser and Kaplan, 1966; Szaran *et al.*, 1998).

The primary process by which sulfur is fractionated in hydrologic environments is via biologically mediated sulfur transformations. The most important of these is the dissimilatory sulfate reduction (DSR) reaction

$$2H^+ + SO_4^{2-} + 4H_2 \rightarrow H_2S + 4H_2O \qquad (14)$$

which is facilitated by anaerobic bacteria of the *Desulfovibrio* and *Desulfotomaculum* genera (Mitchell *et al.*, 1998). These sulfate-reducing bacteria utilize dissolved sulfate as an electron acceptor during the oxidation of organic matter. Bacterial reduction of SO_4^{2-} is the primary source of the variability of SO_4^{2-} isotopic compositions observed in natural aquatic systems. During sulfate reduction, the bacteria produce H_2S gas that has a $\delta^{34}S$ value ~25‰ lower than the sulfate

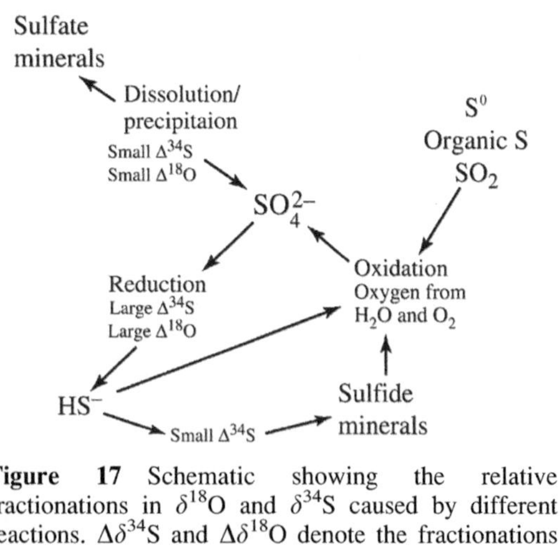

Figure 17 Schematic showing the relative fractionations in $\delta^{18}O$ and $\delta^{34}S$ caused by different reactions. $\Delta\delta^{34}S$ and $\Delta\delta^{18}O$ denote the fractionations (δ–δ values) caused by the reactions (after Krouse and Mayer, 2000).

source (Clark and Fritz, 1997). Consequently, the residual pool of SO_4^{2-} becomes progressively enriched in ^{34}S. Thus, long-term anaerobic conditions in groundwater ought to be reflected in the $\delta^{34}S$ values of SO_4^{2-} that are significantly higher than that of the sulfate source.

As this process is accompanied by oxidation of organic matter, there is a concomitant shift in the $\delta^{13}C$ of DIC toward that of the organic source due to the production of CO_2. The result is that these simultaneous biogeochemical processes—sulfate reduction and oxidation of organic matter—can be useful in assessing the biological activity within groundwater bodies (Clark and Fritz, 1997). Several studies have demonstrated the utility of sulfur isotopes in the assessment of biological remediation of organic pollutants in groundwater (Bottrell *et al.*, 1995; Spence *et al.*, 2001; Schroth *et al.*, 2001). Using sulfur, carbon, and oxygen isotopes in concert may also facilitate the tracing of groundwater flow paths as well as mixing processes in hydrogeologic systems, thanks to the wide separation in $\delta^{34}S$ and $\delta^{18}O$ values of SO_4^{2-} and in the $\delta^{13}C$ of DIC. However, it is important to note that under conditions of low concentrations of reactive organic matter, reoxidation of mineral sulfides can lead to constant recycling of the dissolved sulfur pool. In such cases, increases in $\delta^{34}S$ do not occur in conjunction with reaction progress since the sulfide concentration in solution is held approximately constant (Spence *et al.*, 2001; Fry, 1986).

Aside from bacterially mediated reduction, the isotopic composition of sulfate is controlled by (i) the isotopic composition of sulfate sources, (ii) isotope exchange reactions, and (iii) kinetic isotope effects during transformations. Exchange of sulfur isotopes between SO_4^{2-} and HS^- or H_2S is not thought to be of significance in most groundwater systems; however, microbial activity may enhance this exchange. Under favorable conditions, $\delta^{34}S$ can also be used to determine the reaction mechanisms responsible for sulfide oxidation (Kaplan and Rittenberg, 1964; Fry *et al.*, 1986, 1988) and sulfate reduction (Goldhaber and Kaplan, 1974; Krouse, 1980).

7.3.3.2 *Oxygen isotopes of sulfate*

The evaluation of sources and cycling of sulfate in the environment can be aided by analysis of the oxygen isotopic composition of sulfate. Oxygen has three stable isotopes: ^{16}O, ^{17}O, and ^{18}O. The $\delta^{18}O$ and $\delta^{17}O$ values of sulfate are reported in per mil relative to the standard SMOW or VSMOW. The $\delta^{18}O$ of sulfate shows considerable variations in nature, and recent studies indicate that the systematics of cause and effect remain somewhat obscure and controversial (Van Stempvoort and Krouse, 1994; Taylor and Wheeler, 1994). Reviews of applications of $\delta^{18}O$ of sulfate include Holt and Kumar (1991) and Pearson and Rightmire (1980). When in isotopic equilibrium at 0 °C, aqueous sulfate has a $\delta^{18}O$ value ~30‰ higher than water (Mizutani and Rafter, 1969). However, the rate of oxygen isotope exchange between sulfate and water is very slow at low temperatures and normal pH levels. Even in acidic rain of pH 4, the "half-life" of exchange is ~1,000 yr (Lloyd, 1968). Thus, aqueous sulfate is generally out of isotopic equilibrium with the host water, although there have been exceptions (e.g., Nriagu *et al.*, 1991; Berner *et al.*, 2002). Oxygen isotope exchange between SO_4^{2-} and H_2O is significant only under geothermal conditions, and in such cases may be useful as a geothermometer (Krouse and Mayer, 2000).

Depending upon the reaction responsible for sulfate formation, between 12.5% and 100% of the oxygen in sulfate is derived from the oxygen in the environmental water; the remaining oxygen comes from O_2 (Taylor *et al.*, 1984). Holt *et al.* (1981) studied mechanisms of aqueous SO_2 oxidation and the corresponding $\delta^{18}O$ of the SO_4^{2-} produced. They concluded that three of the four oxygen atoms that give rise to the $\delta^{18}O_{SO_4}$ are derived from the oxygen in the water, and the rest is derived from dissolved O_2 gas; atmospheric O_2 is +23.8‰ (Horibe *et al.*, 1973).

Despite the complexities of the mechanisms that establish the $\delta^{18}O$ of aqueous sulfate, it is clear that this measurement is a useful additional parameter by which sources of SO_4^{2-} and sulfur cycling can be evaluated in aquatic systems. Use of a dual isotope approach to tracing sources of sulfur (i.e., measurement of $\delta^{18}O$ and $\delta^{34}S$ of sulfate) will often provide better separation of potential sources of sulfur and, under favorable conditions, provide information on the processes responsible for sulfur cycling in the ecosystem. Some examples of studies that have examined the sources of SO_4^{2-} to groundwater using a dual isotope approach include Yang *et al.* (1997), Dogramaci *et al.* (2001), and Berner *et al.* (2002).

A study by Berner *et al.* (2002) suggests that as the bacterial reduction of sulfate progresses, there is variable sensitivity in the sulfur and oxygen isotopic fractionation in SO_4^{2-}. They found that, in the beginning stages of bacterial reduction in a groundwater system, both the $\delta^{34}S$ and $\delta^{18}O$ composition of the sulfate reservoir change rapidly. However, as the SO_4^{2-} is progressively consumed, the $\delta^{18}O$ value of the residual SO_4^{2-} approaches a constant value, while the $\delta^{34}S$ continues to increase (Figure 18). Hence, $\delta^{18}O$ of sulfate is a more sensitive indicator at the initial stages of bacterial SO_4^{2-} reduction, whereas $\delta^{34}S$ is more sensitive for describing the process when the SO_4^{2-} reservoir is almost consumed. This is in accordance with the findings of Fritz *et al.* (1989),

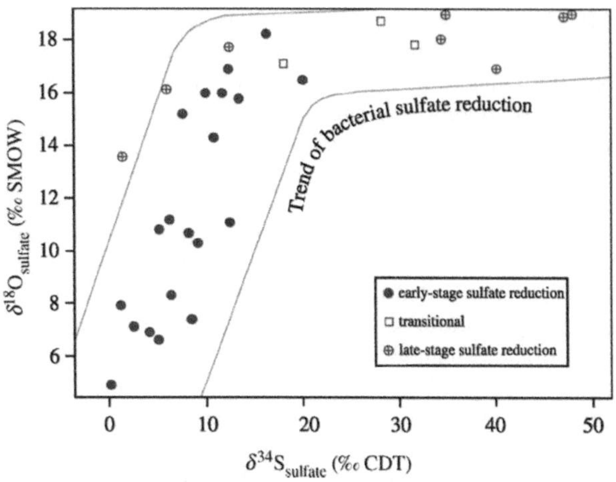

Figure 18 During sulfate reduction, both $\delta^{18}O$ and $\delta^{34}S$ of sulfate change rapidly during the early stages of the reaction, but there is little change in the $\delta^{18}O$ during the late stages (after Berner *et al.*, 2002).

who observed similar behavior and attributed it to the involvement of sulfite during the biologically mediated reaction and oxygen-isotope exchange with the water.

7.3.3.3 *Tracing atmospheric deposition*

Using $\delta^{34}S$ and $\delta^{18}O$. Atmospheric sulfur $\delta^{34}S$ values are typically in the range of −5‰ to +25‰ (Krouse and Mayer, 2000). Seawater sulfate has a $\delta^{34}S$ value of +21‰. Atmospheric sulfate derived from marine sources will show $\delta^{34}S$ values between +15‰ and +21‰, while emissions of atmospheric sulfur from biogenic reduction (e.g., H_2S) have lower values.

Atmospheric deposition has been repeatedly shown to be the major source of sulfur deposition in many forest ecosystems, especially in regions significantly impacted by acid deposition (Mitchell *et al.*, 1992b). Sulfur inputs to forests derived from mineral weathering generally play a minor role; however, this is dependent upon the local geology. Regions underlain by sulfide-rich shales can have mineral inputs exceeding those derived from the atmosphere, while weathering of ore minerals and evaporites also contributes sulfur. Nriagu and Coker (1978) report that the $\delta^{34}S$ of precipitation in central Canada varies seasonally from about +2‰ to +9‰, with the low values caused mainly by biological sulfur, whereas the high values reflect sulfur from fossil fuels.

The spatial distribution of $\delta^{34}S$ and the relative contributions from marine versus continental (including anthropogenic combustion) sources in Newfoundland (Canada) have been monitored by analyzing the $\delta^{34}S$ of rainfall and of lichens that obtain all their sulfur from the atmosphere (Wadleigh *et al.*, 1996; Wadleigh and Blake, 1999). Wadleigh *et al.* (1996) report rainwater sulfate with $\delta^{34}S$ and $\delta^{18}O$ of ~+4‰ and +11‰, respectively, for a continental end-member representative of long-range sulfate transport. The marine end-member has $\delta^{34}S$ of +21‰ and $\delta^{18}O$ of +9.5‰, representing sulfate in precipitation derived from sea spray. The study of epiphytic lichens (Wadleigh and Blake, 1999) yielded a "bulls-eye" $\delta^{34}S$ contour plot showing low values in the interior of the island that are probably related to anthropogenic point sources, and progressively higher (more marine) values towards the coasts. These studies suggest that the study area is influenced by both marine (high $\delta^{34}S$ values) and continental sources (lower $\delta^{34}S$ values), with the possibility of anthropogenic influence from fossil-fuel powered plants.

Using $\delta^{17}O$. Atmospheric sulfate (aerosol and rainfall) has recently been found to have a mass independent isotopic composition, with excess ^{17}O over what would have been expected based on the $\delta^{18}O$ of sulfate (Lee *et al.*, 2001). On a three-isotope plot with $\delta^{17}O$ on the *x*-axis and $\delta^{18}O$ on the *y*-axis, mass-dependent fractionation (MDF) result in sulfate values defined by the relation: $\delta^{17}O = 0.52\delta^{18}O$, whereas MIF causes values that deviate from this relation. For sulfate, the mass-independent relation is described by $\Delta^{17}O = \delta^{17}O - 0.52\delta^{18}O$ (Lee *et al.*, 2001). Hence, MDF results in $\Delta^{17}O = 0$, whereas an MIF results in $\Delta^{17}O \neq 0$. $\delta^{17}O$ and $\Delta^{17}O$ values reflect $^{17}O/^{16}O$ ratios, and are reported in per mil relative to SMOW.

The causes of the ^{17}O anomalies of sulfate and other oxygen-bearing materials in the atmosphere and in extraterrestrial materials are the topic of active research. Aqueous-phase oxidation of S(IV) causes an MIF in sulfate only if the oxidant has an MIF oxygen isotopic composition (Lee *et al.*, 2001). Hence, the anomalous $\delta^{17}O$ values of atmospheric sulfate appear to be caused by interactions with either O_3 or H_2O_2, since thus far they are the only known oxidants in the atmosphere with mass-independent oxygen isotopic compositions (Johnson and Thiemens, 1997).

The discovery of the anomalous oxygen isotopic compositions of atmospheric sulfate provides a new means for identifying sulfate of atmospheric origin. Rainwater and aerosols from southern California were found to have $\Delta^{17}O$ values in the range of 0‰ to +1.5‰ (Lee *et al.*, 2001). The average $\Delta^{17}O$ of snow sulfate in the Rocky Mountains (Colorado, USA) was +1.3‰. Sulfate in ice cores, massive sulfate deposits, and Dry Valley soils from various locations also have MIF (Bao *et al.*, 2000; Lee *et al.*, 2001). There appears to be seasonality in the $\Delta^{17}O$ of sulfate in precipitation, with higher values in the winter and lower values in the summer, probably due to seasonal changes in climatic effects that favor aqueous phase S(IV) oxidation in winter relative to summer (Lee and Thiemens, 2001).

Sulfate $\delta^{17}O$ can be used to identify the relative contributions of atmospheric sulfate versus terrestrial biological or geologic sources of sulfate to streams. In the Rocky Mountains, stream-water sulfate had $\Delta^{17}O$ values of +0.2‰ to +0.9‰ (Johnson *et al.*, 2001). Isotope mass-balance calculations suggest that 20–40% of the sulfate was atmospheric in origin, considerably lower than expected (Johnson *et al.*, 2001). Combined studies using $\delta^{34}S$, $\delta^{18}O$, $\delta^{17}O$, and ^{35}S in alpine watersheds in the Rocky Mountains show that sulfur and oxygen cycling processes appear to be decoupled (Kester *et al.*, 2002), with atmospheric sulfur and oxygen having different residence times in the catchment.

Using sulfur-35. Sulfur-35 (^{35}S) is a naturally produced radioactive tracer (half-life = 87 d) that can be used to trace the movement of atmospherically derived sulfate in the environment. It is formed in the atmosphere from cosmic ray spallation of ^{40}Ar (Peters, 1959), and deposits on the Earth's surface in precipitation or as dryfall. It can be used both to trace the timescales for movement of atmospheric sulfate through the hydrosphere and, in ideal cases, to trace the movement of young (<1 yr) water. It is an especially useful tracer in regions away from the ocean where sulfate concentrations are relatively low.

The isotope was originally used by Cooper *et al.* (1991) to trace the movement of young sulfate in an arctic watershed. They found that sulfur deposited as precipitation is strongly adsorbed within the watershed, and that most sulfur released to stream flow is derived from longer-term storage in soils, vegetation, or geologic materials. Further work (Michel *et al.*, 2000) showed that ^{35}S could be an especially effective isotope in tracing movement of young sulfate through an alpine watershed at Loch Vale watershed in the Rocky Mountains (Colorado, USA). They found that ~50% of the atmospheric sulfate deposited in that watershed was retained less than one year. This work is now being expanded to include the use of the ^{17}O anomaly to try to obtain a better estimate of the real age of atmospheric sulfate leaving the watershed (Kester *et al.*, 2002). The isotope also appears to be an effective tracer for sulfate cycling in small alpine lakes (Michel *et al.*, 2002).

Sulfur-35 was also used to determine the source of the sulfate in stream water in a highly polluted watershed in the Czech Republic (Novak *et al.*, 2003). The sulfate input to this watershed has decreased over the past decade because of new controls on sulfur emissions, but the watershed continues to export sulfate far in excess of the atmospheric loading at the present time. Measurement of ^{35}S in runoff indicates that none of the recently deposited sulfate is exiting in the system. Apparently the atmospheric sulfate interacts with the soil layers and consequently takes more than a year to be removed from the watershed (Novak *et al.*, 2003).

7.3.3.4 Sulfur in catchment surface waters

Sulfur occurs naturally in both organic and inorganic forms. In most forested catchments, the sulfur content of the catchment is largely in the organic form stored within pools of organic matter (Mitchell *et al.*, 1998). Carbon-bonded sulfur in organic compounds is the dominant organic form, while sulfate is the primary form of inorganic sulfur in surface water systems. Organic sulfur is generally not labile; thus, fluxes of sulfur in catchments are often related solely to movement and transformation of inorganic sulfur, including adsorption–desorption interactions with organic matter (Mitchell *et al.*, 1992a, 1994). Catchments that contain a significant geologic component of sulfur-bearing minerals will likely have sulfur budgets that are dominated by weathering inputs instead of atmospheric inputs.

Because sulfur isotope ratios are strongly fractionated by biogeochemical processes, there has been concern over whether $\delta^{34}S$ could be used effectively to separate sources of sulfur in shallow hydrological systems. Sulfur appears to be affected by isotope fractionation processes in some catchments (Fuller *et al.*, 1986; Hesslein *et al.*, 1988; Andersson *et al.*, 1992), but not in others (Caron *et al.*, 1986; Stam *et al.*, 1992). Andersson *et al.* (1992) explain the decreases in sulfate concentrations and increases in $\delta^{34}S$ of stream flow during the summer by sulfate reduction in stagnant pools. Stam *et al.* (1992) suggest that the extent of fractionation might be a function of water residence time in the catchment, with steep catchments showing less fractionation. They note that increases in $\delta^{34}S$ of stream sulfate during the winter may be a result of micropore flow during the snow-covered period, instead of the more typical macropore-flow characteristic of storms.

Intensive investigations of the sulfur dynamics of forest ecosystems in the last decade can be attributed to the dominant role of sulfur as a component of acidic deposition. Studies in forested catchments include Fuller *et al.* (1986), Mitchell *et al.* (1989), Stam *et al.* (1992), and Andersson *et al.* (1992). Sulfur with a distinctive isotopic composition has been used to identify pollution sources (Krouse *et al.*, 1984), and has been added as a tracer (Legge and Krouse, 1992; Mayer *et al.*, 1992, 1993). Differences in the natural abundances can also be used in systems where there is sufficient variation in the $\delta^{34}S$ of ecosystem components. Rocky Mountain lakes (USA), thought to be dominated by atmospheric sources of sulfate, have different $\delta^{34}S$ values than lakes believed to be dominated by watershed sources of sulfate (Turk *et al.*, 1993).

7.3.3.5 *Sulfur in groundwater*

Sulfur in groundwater is primarily in the oxidized sulfate species, SO_4^{2-}. Sulfate in groundwater can have several sources. These include: (i) dissolution of evaporite sulfate minerals such as gypsum and anhydrite, (ii) oxidation of sulfide minerals, (iii) atmospheric deposition, and (iv) mineralization of organic matter.

Production of sulfate in shallow groundwater via the oxidation of sulfide minerals yields $\delta^{34}S$ values of the sulfate that are very similar to those of the source sulfides, because there is little isotope fractionation during near-surface, low-temperature sulfide mineral oxidation (Toran and Harris, 1989). Although the sulfur isotopic composition of sulfide minerals can vary widely in geologic materials, the isotopic compositions of sulfur sources in a regional study area often fall within a narrow range (Krouse and Mayer, 2000). Thus, sulfur isotopes can be useful for identifying sources of sulfur species in groundwater as well as biogeochemical transformations along flow paths.

For example, $\delta^{34}S$ and $\delta^{18}O$ of SO_4^{2-} was used to assess mixing between a vertically stacked aquifer system in contact with a salt dome located in northern Germany (Berner *et al.*, 2002). Two major sulfate pools were identified based upon their isotopic compositions: (i) SO_4^{2-} from the dissolution of evaporite minerals, and (ii) SO_4^{2-} derived from atmospheric deposition and from the oxidation of pyrite. Using both the $\delta^{34}S$ and $\delta^{18}O$ of the SO_4^{2-}, zones of variable groundwater mixing and significant amounts of bacterial sulfate reduction were identified. The SO_4^{2-} derived from the dissolution of the rock salt in the highly saline deep brine showed nearly constant $\delta^{34}S$ values between +9.6‰ and 11.9‰ and $\delta^{18}O$ between +9.5‰ and 12.1‰, consistent with the Permian evaporite deposits. Sulfate in near-surface unconfined groundwater derived from riverine recharge showed values more typical of meteoric sulfur mixed with oxidized sulfide minerals ($\delta^{34}S = +5.2‰$; $\delta^{18}O = +8.2‰$).

One complicating factor is the possibility of sulfur recycling through subsequent oxidation–reduction of biologically reduced SO_4^{2-}. H_2S gas produced by the biological reduction of SO_4^{2-} may become reoxidized, especially in near-surface aquifers or in mixing zones with more oxidizing waters. The oxidation of this H_2S will produce a secondary SO_4^{2-} pool with a much lower $\delta^{34}S$ than the original source sulfate. This can result in a depth gradient in which higher $\delta^{34}S$ values of sulfate are found at depth (partially reducing environment) while lower $\delta^{34}S$ values of sulfate are found in the more oxidizing environment closer to the surface or within mixing zones.

Some recent studies have shown that the identification of arsenic-bearing sulfides and their contribution to arsenic loadings in groundwater and surface water used for drinking supply can also be assessed using $\delta^{34}S$ (Sidle, 2002; Schreiber *et al.*, 2000).

7.4 USE OF A MULTI-ISOTOPE APPROACH FOR THE DETERMINATION OF FLOW PATHS AND REACTION PATHS

Flow paths are the individual runoff pathways contributing to surface flow in a catchment. These result from runoff mechanisms that include, but are not limited to, saturation-excess overland flow, Hortonian overland flow, near-stream groundwater ridging, hill-slope subsurface flow through the soil matrix or macropores, and shallow organic-layer flow. Knowledge of hydrologic flow paths in catchments is critical to the preservation of public water supplies and the understanding of the transport of point and non-point-source pollutants (Peters, 1994).

Stable isotopes such as ^{18}O and ^{2}H are an improved alternative to traditional nonconservative chemical tracers, because waters are often uniquely labeled by their isotopic compositions (Sklash and Farvolden, 1979; McDonnell and Kendall, 1992), often allowing the separation of waters from different sources (e.g., new rain versus old pre-storm water). However, studies have shown that flow paths cannot be identified to a high degree of certainty using $\delta^{18}O$ or $\delta^{2}H$ data and simple hydrograph separation techniques, because waters within the same flow path can be derived from several different sources (Ogunkoya and Jenkins, 1991). Thus, a number of plausible runoff mechanisms can be consistent with the isotope data. The need to incorporate flow path dynamics is recognized as a key ingredient in producing reliable chemical models (Robson *et al.*, 1992).

Reactive solute isotopes such as ^{13}C, ^{34}S, and ^{87}Sr can provide valuable information about flow paths (not water sources) and reaction paths useful for geochemical and hydrologic modeling precisely, because they can reflect the reaction characteristics of and taking place along specific flow paths (Bullen and Kendall, 1998). These reactive solute isotopes can serve as additional thermodynamic constraints in geochemical computer models such as NETPATH (Plummer *et al.*, 1991) for eliminating possible geochemical reaction paths (Plummer *et al.*, 1983).

As an example, water isotopes were used to determine the actual flow paths by which water discharging from Crystal Lake (Wisconsin, USA) mixed with regional groundwater and local rain recharge while flowing across a narrow isthmus separating the lake from Big Muskellunge Lake (Krabbenhoft *et al.*, 1992, 1993; Bullen *et al.*, 1996). Solute isotopes such as ^{87}Sr and ^{13}C, along with chemical data, were then used to identify the geochemical reactions taking place along these flow paths.

The waters flowing along mineralogically distinctive horizons are sometimes distinctively labeled by their chemical composition and by the isotopic compositions of solute isotopes such as ^{13}C, ^{87}Sr, ^{34}S, ^{15}N, etc. For example, waters flowing through the soil zone often have $\delta^{13}C$ values that are depleted in ^{13}C relative to deeper groundwaters because of biogenic production of carbonic acid in organic soils (Kendall, 1993). These same shallow waters can also have distinctive lead and strontium isotopic compositions (Bullen and Kendall, 1998).

A multi-isotope approach can be particularly useful in tracing the reactions specific to a flow path (Krabbenhoft *et al.*, 1992, 1993). The solute isotope tracers can be an especially powerful tool because they usually are affected by a smaller number of processes than chemical constituents, making interpretation of changes in isotopic composition less ambiguous than the simultaneous changes in solute concentrations (Kendall, 1993). In particular, the strengths of one isotope may compensate for the weakness of another. For example, one of the concerns with using $\delta^{13}C$ to trace carbon sources in catchments is that possible isotopic exchange of subsurface DIC with soil or atmospheric CO_2 could allow the $\delta^{13}C$ of DIC to be "reset" so that the characteristic "geologic" signature of the calcite-derived carbon is lost (Kendall *et al.*, 1992; Kendall, 1993). Since strontium isotopes are not affected by this kind of exchange, the same samples can be analyzed for strontium isotopes; if the strontium signature of calcite is distinctive, the correlation of changes in $^{87}Sr/^{86}Sr$ and changes in $\delta^{13}C$ of subsurface waters could provide evidence that calcite dissolution was occurring and that exchange was not a problem (Bullen and Kendall, 1998).

If the chemical composition of waters along a specific flow path is related to topography, mineralogy, initial water composition, and antecedent moisture conditions, then within a uniform soil layer, the degree of geochemical evolution is some function of residence time. In this case, it may be possible to link geochemical evolution of waters along specific flow paths to hydrologic models such as TOPMODEL (Beven and Kirkby, 1979) by the testable assumption that residence time—and hence degree of rock/water reaction—is primarily a function of topography. Alternatively, because TOPMODEL is used to predict the different contributions from the different soil horizons along the stream channel as the topographic form of the hill-slope changes (Robson *et al.*, 1992), these predictions can be tested with appropriate isotope and chemical data. Future advances in watershed modeling, both conceptually and mathematically, will require the combined use of chemical and isotopic tracers for calibration and validation.

7.5 SUMMARY AND CONCLUSIONS

The dominant use of isotopes in shallow hydrologic systems in the last few decades has been to trace sources of waters and solutes. Generally, such data were evaluated with simple mixing models to determine how much water or solute was derived from either of two (sometimes three) constant-composition sources. The world does not seem this simple anymore. With the expansion of the field of isotope hydrology in the last decade, made possible by the development and increased availability of automated sample preparation and analysis systems for mass spectrometers, we have documented considerable heterogeneity in the isotopic compositions of rain, soil water, groundwater, and solute sources. In addition, hydrologists who utilize geochemical tracers recognize that the degree of variability observed is highly dependent upon the sampling frequency, and more effort is being placed on event-based studies with high-frequency sampling. We are still grappling with how to deal with this heterogeneity in our hydrologic and geochemical models. A major challenge is to use the variability as signal, not noise; the isotopes and chemistry are providing very detailed information about sources and reactions in shallow systems, and the challenge now is to develop appropriate models to use the data. In the past, much reliance was placed upon the stable isotopes of water ($\delta^{18}O$, $\delta^{2}H$) to reveal all of the information about the hydrologic processes taking place in a catchment. Today, we acknowledge that the best approach is to combine as many tracers as possible, including solutes and solute isotopes. This integration of chemical and isotopic data with complex hydrologic and geochemical models constitutes an important frontier of hydrologic research.

ACKNOWLEDGMENTS

The authors would like to thank Thomas D. Bullen, Michael G. Sklash, and Eric A. Caldwell for their contributions to chapters that were extensively mined to produce an early draft of this chapter: Kendall *et al.* (1995a), Bullen and Kendall (1998), and Kendall and Caldwell (1998).

REFERENCES

Aleem M. I. H., Hoch G. E., and Varner J. E. (1965) Water as the source of oxidant and reductant in bacterial chemosynthesis. *Biochemistry* **54**, 869–873.

Amberger A. and Schmidt H. L. (1987) Naturliche isotopengehalte von nitrat als indikatoren fur dessen herkunft. *Geochim. Cosmochim. Acta* **51**, 2699–2705.

Amoitte-Suchet P., Aubert D., Probst J. L., Gauthier-Lafaye F., Probst A., Andreux F., and Viville D. (1999) $\delta^{13}C$ pattern of dissolved inorganic carbon in a small granitic catchment: the Strengbach case study (Vosges mountains, France). *Chem. Geol.* **159**, 129–145.

Amundson R., Stern L., Baisden T., and Wang Y. (1998) The isotopic composition of soil and soil-respired CO_2. *Geoderma* **82**, 83–114.

Andersson K. K. and Hooper A. B. (1983) O_2 and H_2O are each the source of one O in NO_2 produced from NH_3 by Nitrosomonas: ^{15}N–NMR evidence. *FEBS Lett.* **164**(2), 236–240.

Andersson P., Torssander P., and Ingri J. (1992) Sulphur isotope ratios in sulphate and oxygen isotopes in water from a small watershed in central Sweden. *Hydrobiologia* **235/236**, 205–217.

Aravena R. and Robertson W. D. (1998) Use of multiple isotope tracers to evaluate denitrification in groundwater: case study of nitrate from a large-flux septic system plume. *Ground Water* **31**, 180–186.

Aravena R. and Wassenaar L. I. (1993) Dissolved organic carbon and methane in a regional confined aquifer,southern Ontario, Canada: carbon isotope evidence for associated subsurface sources. *Appl. Geochem.* **8**, 483–493.

Aravena R., Schiff S. L., Trumbore S. E., Dillon P. J., and Elgood R. (1992) Evaluating dissolved inorganic carbon cycling in a forested lake watershed using carbon isotopes. *Radiocarbon* **34**(3), 636–645.

Aravena R., Evans M. L., and Cherry J. A. (1993) Stable isotopes of oxygen and nitrogen in source identification of nitrate from septic systems. *Ground Water* **31**, 180–186.

Atekwana E. A. and Krishnamurthy R. V. (1998) Seasonal variations of dissolved inorganic carbon and $\delta^{13}C$ of surface waters: application of a modified gas evolution technique. *J. Hydrol.* **205**, 26–278.

Aucour A., Sheppard S. M. F., Guyomar O., and Wattelet J. (1999) Use of ^{13}C to trace origin and cycling of inorganic carbon in the Rhône River system. *Chem. Geol.* **159**, 87–105.

Bajjali W. and Abu-Jaber N. (2001) Climatological signals of the paleogroundwater in Jordan. *J. Hydrol.* **243**, 133–147.

Bao H., Campbell D. A., Bockheim J. G., and Thiemens M. H. (2000) Origins of sulphate in Antarctic dry-valley soils as deduced from anomalous ^{17}O compositions. *Nature* **407**, 499–502.

Battaglin W. A., Kendall C., Chang C. C. Y., Silva S. R., and Campbell D. H. (2001) *Chemical and Isotopic Composition of Organic and Inorganic Samples from the Mississippi River and its Tributaries, 1997–98*. USGS Water Resources Investigation Report **01-4095**, 57p.

Behrens H., Moser H., Oerter H., Rauert W., Stichler W., Ambach W., and Kirchlechner P. (1979) Models for the run-off from a glaciated catchment area using measurements of environmental isotope contents. In: *Isotope Hydrology 1978* (ed. JAEA). IAEA, Vienna, pp. 829–846.

Berner Z. A., Stüben D., Leosson M. A., and Klinge H. (2002) S- and O-isotopic character of dissolved sulphate in the cover rock aquifers of a Zechstein salt dome. *Appl. Geochem.* **17**, 1515–1528.

Beven K. J. and Kirkby M. J. (1979) A physically based, variable contributing model of basin hydrology. *Hydrol. Sci. Bull.* **10**, 43–69.

Bishop K. H. (1991) Episodic increases in stream acidity, catchment flow pathways and hydrograph separation. PhD Thesis, University Cambridge.

Böhlke J. K. (2002) Groundwater recharge and agricultural contamination. *Hydrol. J.* **10**, 153–179.

Böhlke J. K. and Denver J. M. (1995) Combined use of groundwater dating, chemical, and isotopic analyses to resolve the history and fate of nitrate contamination in two agricultural watersheds, Atlantic coastal plain, Maryland. *Water Resour. Res.* **31**, 2319–2339.

Böhlke J. K., Eriksen G. E., and Revesz K. (1997) Stable isotope evidence for an atmospheric origin of desert nitrate deposits in northern Chile and southern California USA. *Isotope Geosci.* **136**, 135–152.

Böttcher J., Strebel O., Voerkelius S., and Schmidt H. L. (1990) Using isotope fractionation of nitrate–nitrogen and nitrate–oxygen for evaluation of microbial denitrification in a sandy aquifer. *J. Hydrol.* **114**, 413–424.

Bottrell S. H., Hayes P. J., Bannon M., and Williams G. M. (1995) Bacterial sulfate reduction and pyrite formation in a polluted sand aquifer. *Geomicrobiol. J.* **13**(2), 75–90.

Brinkmann R., Eichler R., Ehhalt D., and Munnich K. O. (1963) Über den deuterium-gehalt von niederschlags-und grundwasser. *Naturwissenschaften* **19**, 611–612.

Bullen T. D. and Kendall C. (1991) $^{87}Sr/^{86}Sr$ and ^{13}C as tracers of interstream and intrastorm variations in water flowpaths, Catoctim Mountain, MD. *Trans. Am. Geophys. Union* **72**, 218.

Bullen T. D. and Kendall C. (1998) Tracing of weathering reactions and water flowpaths: a multi-isotope approach. In *Isotope Tracers in Catchment Hydrology* (eds. C. Kendall and J. J. McDonnell). Elsevier, Amsterdam, pp. 611–646.

Bullen T. D., Krabbenhoft D. P., and Kendall C. (1996) Kinetic and mineralogic controls on the evolution of ground-water chemistry and $^{87}Sr/^{86}Sr$ in a sandy silicate aquifer, northern Wisconsin. *Geochim. Cosmochim. Acta* **60**, 1807–1821.

Burns D. A. and Kendall C. (2002) Analysis of $\delta^{15}N$ and $\delta^{18}O$ to differentiate NO_3_sources in runoff at two watersheds in the Catskill Mountains of New York. *Water Resour. Res.* **38**, 14-1–14-11.

Buttle J. M. (1998) Fundamentals of small catchment hydrology. In *Isotope Tracers in Catchment Hydrology* (eds. C. Kendall and J. J. McDonnell). Elsevier, Amsterdam, pp. 1–49.

Buttle J. M. and Sami K. (1990) Recharge processes during snowmelt: an isotopic and hydrometric investigation. *Hydrol. Process.* **4**, 343–360.

Campbell D. H., Kendall C., Chang C. C. Y., Silva S. R., and Tonnessen K. A. (2002) Pathways for nitrate release from an alpine watershed: determination using $\delta^{15}N$ and $\delta^{18}O$. *Water Resour. Res.* **38**, 9-1–9-11.

Caron F. A., Tessier A., Kramer J. R., Schwarcz H. P., and Rees C. E. (1986) Sulfur and oxygen isotopes of sulfate in precipitation and lakewater, Quebec, Canada. *Appl. Geochem.* **1**, 601–606.

Carothers W. W. and Kharaka Y. K. (1980) Stable carbon isotopes of HCO_3^- in oil-field waters- implications for the origin of CO_2. *Geochim. Cosmochim. Acta* **44**, 323–332.

Cerling T. E., Solomon D. K., Quade J., and Bowman J. R. (1991) On the isotopic composition of carbon in soil carbon dioxide. *Geochim. Cosmochim. Acta* **55**, 3404–3405.

Chang C. C. Y., Kendall C., Silva S. R., Battaglin W. A., and Campbell D. H. (2002) Nitrate stable isotopes: tools for determining nitrate sources and patterns among sites with different land uses in the Mississippi Basin. *Can. J. Fish. Aquat. Sci.* **59**, 1874–1885.

Claassen H. C. and Downey J. S. (1995) A model for deuterium and oxygen-18 isotope changes during evergreen interception of snowfall. *Water Resour. Res.* **31**(3), 601–618.

Clark I. D. and Fritz P. (1997) *Environmental Isotopes in Hydrogeology*. CRC Press, New York.

Cook P. and Herczeg A. L. (eds.) (2000) *Environmental Tracers in Subsurface Hydrology*. Kluwer Academic, Norwell, MA.

Cooper L. W., Olsen C. R., Solomon D. K., Larsen I. L., Cook R. B., and Grebmeier J. M. (1991) Stable isotopes of oxygen and natural and fallout radionuclides used for tracing runoff during snowmelt in an arctic watershed. *Water Resour. Res.* **27**, 2171–2179.

Coplen T. B. (1996) New guidelines for reporting stable hydrogen, carbon and oxygen isotope-ratio data. *Geochim. Cosmochim. Acta* **60**, 3359–3360.

Coplen T. B. and Kendall C. (2000) *Stable Hydrogen and Oxygen Isotope Ratios for Selected Sites of the US Geological Survey's Nasqan and Benchmark Surface-water Networks.* US Geol. Surv., Open-file Report. 00-160, 424.

Coplen T. B., Herczeg A. L., and Barnes C. (2000) Isotope engineering—using stable isotopes of the water molecule to solve practical problems. In *Environmental Tracers in Subsurface Hydrology* (eds. P. Cook and A. Herzceg). Kluwer Academic, Norwell, MA, pp. 79–110.

Craig H. (1961a) Isotopic variations in meteoric waters. *Science* **133**, 1702–1703.

Craig H. (1961b) Standard for reporting concentrations of deuterium and oxygen-18 in natural waters. *Science* **133**, 1833.

Criss R. E. (1999) *Principles of Stable Isotope Distribution.* Oxford University Press, New York, pp. 123–128.

Davidson G. R. (1995) The stable isotopic composition and measurement of carbon in soil CO_2. *Geochim. Cosmochim. Acta* **59**, 2485–2489.

Deák J. and Coplen T. B. (1996) Identification of Pleistocene and Holocene groundwaters in Hungary using oxygen and hydrogen isotopic ratios. In *Isotopes in Water Resources Management*, IAEA, Vienna, vol. 1, 438p.

Deines P. (1980) The isotopic composition of reduced organic carbon. In *Handbook of Environmental Isotope Geochemistry* (eds. P. Fritz and J. C. Fontes). Elsevier, Amsterdam, vol. 1, pp. 329–406.

Deines P., Langmuir D., and Harmon R. S. (1974) Stable carbon isotope ratios and the existence of a gas phase in the evolution of carbonate ground waters. *Geochim. Cosmochim. Acta* **38**, 1147–1164.

DeWalle D. R. and Swistock B. E. (1994) Differences in oxygen-18 content of throughfall and rainfall in hardwood and coniferous forests. *Hydrol. Process.* **8**, 75–82.

DeWalle D. R., Swistock B. R., and Sharpe W. E. (1988) Three-component tracer model for stormflow on a small Appalachian forested catchment. *J. Hydrol.* **104**, 301–310.

Doctor D. H. (2002) The hydrogeology of the classical karst (Kras) aquifer of southwestern Slovenia. PhD Dissertation, University of Minnesota.

Dogramaci S. S., Herczeg A. L., Schiff S. L., and Bone Y. (2001) Controls on $\delta^{34}S$ and $\delta^{18}O$ of dissolved SO_4 in aquifers of the Murray Basin, Australia and their use as indicators of flow processes. *Appl. Geochem.* **16**, 475–488.

Dörr H. and Münnich K. O. (1980) Carbon-14 and carbon-13 in soil CO_2. *Radiocarbon* **22**, 909–918.

Durka W., Schulze E. D., Gebauer G., and Voerkelius S. (1994) Effects of forest decline on uptake and leaching of deposited nitrate determined from ^{15}N and ^{18}O measurements. *Nature* **372**, 765–767.

Dutton A. R. and Simpkins W. W. (1989) Isotopic evidence for paleohydrologic evolution of ground-water flowpaths, southern Great Plains, United States. *Geology* **17**(7), 653–656.

Faure G. (1986) *Principles of Isotope Geology*, 2nd edn. Wiley, New York, 589p.

Fogel M. L. and Cifuentes L. A. (1993) Isotope fractionation during primary production. In *Organic Geochemistry* (eds. M. H. Engel and S. A. Macko). Plenum, New York, pp. 73–98.

Fogg G. E., Rolston D. E., Decker D. L., Louie D. T., and Grismer M. E. (1998) Spatial variation in nitrogen isotope values beneath nitrate contamination sources. *Ground Water* **36**, 418–426.

Fontes J.-C. (1981) Palaeowaters. In *Stable Isotope Hydrology—Deuterium and Oxygen-18 in the Water Cycle.* Technical Reports Series no. 2101. IAEA, Vienna, pp. 273–302.

Fontes J.-C., Andrews J. N., Edmunds W. M., Guerre A., and Travi Y. (1991) Paleorecharge by the Niger River (Mali) deduced from groundwater geochemistry. *Water Resour. Res.* **27**(2), 199–214.

Fontes J.-C., Gasse F., and Andrews J. N. (1993) Climatic conditions of Holocene groundwater recharge in the Sahel zone of Africa. In *Isotope Techniques in the Study of Past and Current Environmental Changes in the Hydrosphere and the Atmosphere*. IAEA, Vienna, pp. 271–292.

Friedman I., Redfield A. C., Schoen B., and Harris J. (1964) The variation of the deuterium content of natural waters in the hydrologic cycle. *Rev. Geophys.* **2**, 1–124.

Fritz P. (1981) River waters. In *Stable Isotope Hydrology: Deuterium and Oxygen-18 in the Water Cycle.* IAEA, Vienna, pp. 177–201.

Fritz P., Basharmal G. M., Drimmie R. J., Ibsen J., and Qureshi R. M. (1989) Oxygen isotope exchange between sulphate and water during bacterial reduction of sulphate. *Chem. Geol.* **79**, 99–105.

Fry B. (1986) Stable sulphur isotopic distributions and sulphate reduction in lake-sediments of the Adirondack Mountains, New York. *Biogeochemistry* **2**(4), 329–343.

Fry B. (1991) Stable isotope diagrams of freshwater foodwebs. *Ecology* **72**, 2293–2297.

Fry B., Cox J., Gest H., and Hayes J. M. (1986) Discrimination between ^{34}S and ^{32}S during bacterial metabolism of inorganic sulfur compounds. *J. Bacteriol.* **165**, 328–330.

Fry B., Ruf W., Gest H., and Hayes J. M. (1988) Sulfur isotope effects associated with oxidation of sulfide by O_2 in aqueous solution. *Chem. Geol.* **73**, 205–210.

Fuller R. D., Mitchell M. J., Krouse H. R., Syskowski B. J., and Driscoll C. T. (1986) Stable sulfur isotope ratios as a tool for interpreting ecosystem sulfur dynamics. *Water Air Soil Pollut.* **28**, 163–171.

Garrels R. M. and Mackenzie F. T. (1971) *Evolution of Sedimentary Rocks.* W. W. Norton.

Garten C. T., Jr. and Hanson P. J. (1990) Foliar retention of ^{15}N-nitrate and ^{15}N-ammonium by red maple (Acer rubrum) and white oak (Quercus alba) leaves from simulated rain. *Environ. Exp. Bot.* **30**, 33–342.

Gat J. R. (1980) The isotopes of hydrogen and oxygen in precipitation. In *Handbook of Environmental Isotope Geochemistry* (eds. P. Fritz and J. Ch. Fontes). Elsevier, Amsterdam, pp. 21–47.

Gat J. R. and Gonfiantini R. (eds.) (1981) *Stable Isotope Hydrology–Deuterium and Oxygen-18 in the Water Cycle.* Technical Reports Series #210. IAEA, Vienna, 337pp.

Gat J. R. and Tzur Y. (1967) Modification of the isotopic composition of rainwater by processes which occur before groundwater recharge. *Isotope Hydrol.: Proc. Symp.* IAEA, Vienna, pp. 49–60.

Genereux D. P. (1998) Quantifying uncertainty in tracer-based hydrograph separations. *Water Resour. Res.* **34**, 915–920.

Genereux D. P. and Hemond H. F. (1990) Naturally occurring radon 222 as a tracer for stream flow generation: steady state methodology and field example. *Water Resour. Res.* **26**, 3065–3075.

Genereux D. P. and Hooper R. P. (1998) Oxygen and hydrogen isotopes in rainfall-runoff studies. In *Isotope Tracers in Catchment Hydrology* (eds. C. Kendall and J. J. McDonnell). Elsevier, Amsterdam, pp. 319–346.

Genereux D. P., Hemond H. F., and Mulholland P. J. (1993) Use of radon-222 and calcium as tracers in a three-end-member mixing model for stream flow generation on the west fork of Walker branch watershed. *J. Hydrol.* **142**, 167–211.

Ging P. B., Lee R. W., and Silva S. R. (1996) Water chemistry of Shoal Creek and Waller Creek, Austin Texas, and potential sources of nitrate. *U.S.* Geological Survey Water Resources Investigations, Rep. #96-4167.

Goldhaber M. B. and Kaplan I. R. (1974) The sulphur cycle. In *The Sea* (ed. E. D. Goldberg). Wiley, vol. 5, pp. 569–655.

Gormly J. R. and Spalding R. F. (1979) Sources and concentrations of nitrate-nitrogen in ground water of the Central Platte region, Nebraska. *Ground Water* **17**, 291–301.

Harris D. M., McDonnell J. J., and Rodhe A. (1995) Hydrograph separation using continuous open-system isotope mixing. *Water Resour. Res.* **31**, 157–171.

Heaton T. H. E. (1990) $^{15}N/^{14}N$ ratios of NO_x from vehicle engines and coal-fired power stations. *Tellus* **42B**, 304–307.

Hélie J.-F., Hillaipre-Marcel C., and Rondeau B. (2002) Seasonal changes in the sources and fluxes of dissolved inorganic carbon through the St. Lawrence River—isotopic and chemical constraint. *Chem. Geol.* **186**, 117–138.

Hesslein R. H., Capel M. J., and Fox D. E. (1988) Stable isotopes in sulfate in the inputs and outputs of a Canadian watershed. *Biogeochemistry* **5**, 263–273.

Hewlett J. D. and Hibbert A. R. (1967) Factors affecting the response of small watersheds to precipitation in humid areas. *Proc. 1st Int. Symp. Forest Hydrol.* 275–290.

Hinton M. J., Schiff S. L., and English M. C. (1994) Examining the contributions of glacial till water to storm runoff using two-and three-component hydrograph separations. *Water Resour. Res.* **30**, 983–993.

Hinton M. J., Schiff S. L., and English M. C. (1998) Sources and flowpaths of dissolved organic carbon in two forested watersheds of the Precambrian Shield. *Biogeochemistry* **41**, 175–197.

Holser W. T. and Kaplan I. R. (1966) Isotope geochemistry of sedimentary sulfates. *Chem. Geol.* **1**, 93–135.

Holt B. D. and Kumar R. (1991) Oxygen isotope fractionation for understanding the sulphur cycle. In *Stable Isotopes: Natural and Anthropogenic Sulphur in the Environment,* SCOPE 43 (Scientific Committee on Problems of the Environment) (eds. H. R. Krouse and V. A. Grinenko). Wiley, pp. 27–41.

Holt B. D., Cunningham P. T., and Kumar R. (1981) Oxygen isotopy of atmospheric sulfates. *Environ. Sci. Tech.* **15**, 804–808.

Hooper R. P. and Shoemaker C. A. (1986) A comparison of chemical and isotopic hydrograph separation. *Water Resour. Res.* **22**, 1444–1454.

Hooper R. P., Christophersen N., and Peters N. E. (1990) Modelling streamwater chemistry as a mixture of soilwater end-members—an application to the Panolan Mountain catchment, Georgia, USA. *J. Hydrol.* **116**, 321–343.

Horibe Y., Shigehara K., and Takakuwa Y. (1973) Isotope separation factors of carbon dioxide-water system and isotopic composition of atmospheric oxygen. *J. Geophys. Res.* **78**, 2625–2629.

Horita J. (1989) Analytical aspects of stable isotopes in brines. *Chem. Geol.* **79**, 107–112.

Hübner H. (1986) Isotope effects of nitrogen in the soil and biosphere. In *Handbook of Environmental Isotope Geochemistry, 2b, The Terrestrial Environment* (eds. P. Fritz and J. C. Fontes). Elsevier, Amsterdam, pp. 361–425.

Johnson J. C. and Thiemens M. (1997) The isotopic composition of tropospheric ozone in three environments. *J. Geophys. Res.* **102**, 25395–25404.

Johnson C. A., Mast M. A., and Kester C. L. (2001) Use of $^{17}O/^{16}O$ to trace atmospherically deposited sulfate in surface waters: a case study in alpine watersheds in the Rocky Mountains. *Geophys. Res. Lett.* **28**, 4483–4486.

Junk G. and Svec H. (1958) The absolute abundance of the nitrogen isotopes in the atmosphere and compressed gas from various sources. *Geochim. Cosmochim. Acta* **14**, 234–243.

Kaplan I. R. and Rittenberg S. C. (1964) Microbiological fractionation of sulphur isotopes. *J. Gen. Microbiol.* **34**, 195–212.

Karamanos E. E., Voroney R. P., and Rennie D. A. (1981) Variation in natural ^{15}N abundance of central Saskatchewan soils. *Soil. Sci. Soc. Am. J.* **45**, 826–828.

Karim A. and Veizer J. (2000) Weathering processes in the Indus River Basin: implications from riverine carbon, sulfur, oxygen and strontium isotopes. *Chem. Geol.* **170**, 153–177.

Kendall C. (1993) *Impact of Isotopic Heterogeneity in Shallow Systems on Stormflow Generation.* PhD Dissertation, University of Maryland, College Park.

Kendall C. (1998) Tracing nitrogen sources and cycling in catchments. In *Isotope Tracers in Catchment Hydrology* (eds. C. Kendall and J. J. McDonnell). Elsevier, Amsterdam, pp. 519–576.

Kendall C. and Aravena R. (2000) Nitrate isotopes in groundwater systems. In *Environmental Tracers in Subsurface Hydrology* (eds. P. Cook and A. Herzceg). Kluwer Academic, Norwell, MA, pp. 261–298.

Kendall C. and Caldwell E. A. (1998) Fundamentals of isotope geochemistry. In *Isotope Tracers in Catchment Hydrology* (eds. C. Kendall and J. J. McDonnell). Elsevier, Amsterdam, pp. 51–86.

Kendall C. and Coplen T. B. (2001) Distribution of oxygen-18 and deuterium in river waters across the United States. *Hydrol. Process.* **15**, 1363–1393.

Kendall C. and McDonnell J. J. (1993) Effect of intrastorm heterogeneities of rainfall, soil water and groundwater on runoff modelling. In *Tracers in Hydrology*, Int. Assoc. Hydrol. Sci. Publ. #215, July 11–23, 1993 (ed. N. E. Peters et al.,) Yokohama, Japan, pp. 41–49.

Kendall C. and McDonnell J. J. (1998). *Isotope Tracers in Catchment Hydrology* (p. 839). Elsevier, http://www.rcamnl.we.usgs.gov/isoig/isopubs/itchinfo.html.

Kendall C., Mast M. A., and Rice K. C. (1992) Tracing watershed weathering reactions with $\delta^{13}C$. In *Water–Rock Interaction,* Proc. 7th Int. Symp., Park City, Utah (eds. Y. K. Kharaka and A. S. Maest). Balkema, Rotterdam, pp. 569–572.

Kendall C., Sklash M. G., and Bullen T. D. (1995a) Isotope tracers of water and solute sources in catchments. In *Solute Modelling in Catchment Systems* (ed. S. Trudgill). Wiley, Chichester, UK, pp. 261–303.

Kendall C., Campbell D. H., Burns D. A., Shanley J. B., Silva S. R., and Chang C. C. Y. (1995b) Tracing sources of nitrate in snowmelt runoff using the oxygen and nitrogen isotopic compositions of nitrate. In *Biogeochemistry of Seasonally Snow-covered Catchments,* Proc., July 3–14, 1995 (eds. K. Tonnessen, M. Williams, and M. Tranter). Int. Assoc. Hydrol. Sci., Boulder Co., pp. 339–347.

Kendall C., McDonnell J. J., and Gu W. (2001a) A look inside 'black box' hydrograph separation models: a study at the Hydrohill catchment. *Hydrol. Process.* **15**, 1877–1902.

Kendall C., Silva S. R., and Kelly V. J. (2001b) Carbon and nitrogen isotopic compositions of particulate organic matter in four large river systems across the United States. *Hydrol. Process.* **15**, 1301–1346.

Kennedy V. C., Kendall C., Zellweger G. W., Wyermann T. A., and Avanzino R. A. (1986) Determination of the components of stormflow using water chemistry and environmental isotopes, Mattole River Basin, California. *J. Hydrol.* **84**, 107–140.

Kester C. L., Johnson C. A., Mast M. A., Clow D. W., and Michel R. L. (2002) Tracing atmospheric sulfate through a subalpine ecosystem using ^{17}O and ^{35}S. *EOS, Trans., AGU.* **83**, no. 47.

Kirchner J. W., Feng X., and Neal C. (2001) Catchment-scale advection and dispersion as a mechanism for fractal scaling in stream tracer concentrations. *J. Hydrol.* **254**, 82–101.

Koba K., Tokuchi N., Wada E., Nakajima T., and Iwatsubo G. (1997) Intermittent denitrification: the application of a 15N natural abundance method to a forested ecosystem. *Geochim. Cosmochim. Acta* **61**, 5043–5050.

Krabbenhoft D. P., Bullen T. D., and Kendall C. (1992) Isotopic indicators of groundwater flow paths in a northern Wisconsin aquifer. *AGU Trans.* **73**, 191.

Krabbenhoft D. P., Bullen T. D., and Kendall C. (1993) Use of multiple isotope tracers as monitors of groundwater-lake interaction. *AGU Trans.* **73**, 191.

Kreitler C. W. (1975) Determining the source of nitrate in groundwater by nitrogen isotope studies: Austin, Texas, University of Texas, Austin, Bureau Econ. Geol. Rep. Inv. #83.

Kreitler C. W. (1979) Nitrogen-isotope ratio studies of soils and groundwater nitrate from alluvial fan aquifers in Texas. *J. Hydrol.* **42**, 147–170.

Kreitler C. W. and Jones D. C. (1975) Natural soil nitrate: the cause of the nitrate contamination of groundwater in Runnels County, Texas. *Ground Water* **13**, 53–61.

Kreitler C. W., Ragone S. E., and Katz B. G. (1978) $^{15}N/^{14}N$ ratios of ground-water nitrate, Long Island, NY. *Ground Water* **16**, 404–409.

Krouse H. R. (1980) Sulfur isotopes in our environment. In *Handbook of Environmental Isotope Geochemistry* (eds. P. Fritz and J. Ch. Fontes). Elsevier, Amsterdam, pp. 435–471.

Krouse H. R. and Grinenko V. A. (eds.) (1991) *Stable Isotopes: Natural and Anthropogenic Sulphur in the Environment, SCOPE 43 (Scientific Committee on Problems of the Environment).* Wiley, UK.

Krouse H. R. and Mayer B. (2000) Sulphur and oxygen isotopes in sulphate. In *Environmental Tracers in Subsurface Hydrology* (eds. P. Cook and A. Herzceg). Kluwer Academic, Norwell, MA, pp. 195–231.

Krouse H. R., Legge A., and Brown H. M. (1984) Sulphur gas emissions in the boreal forest: the West Whitecourt case study V: stable sulfur isotopes. *Water Air Soil Pollut.* **22**, 321–347.

Lee C. C.-W. and Thiemens M. H. (2001) Use of $^{17}O/^{16}O$ to trace atmospherically deposited sulfate in surface waters: a case study in alpine watersheds in the Rocky Mountains. *Geophys. Res. Lett.* **28**, 4483–4486.

Lee C. C.-W., Savarino J., and Thiemens M. H. (2001) Mass independent oxygen isotopic composition of atmospheric sulfate: origin and implications for the present and past atmosphere of Earth and Mars. *Geophys. Res. Lett* **28**, 1783–1786.

Legge A. H. and Krouse H. R. (1992) An assessment of the environmental fate of industrial sulphur in a temperate pine forest ecosystem (Paper 1U22B.01). In *Critical Issues in the Global Environment*, vol. 5. Ninth World Clean Air Congress Towards Year 2000.

Likens G. E., Wright R. F., Galloway J. N., and Butler T. J. (1979) Acid rain. *Sci. Am.* **241**, 43–50.

Lloyd R. M. (1968) Oxygen isotope behavior in the sulfate-water system. *J. Geophys. Res.* **73**, 6099–6110.

Macko S. A. and Ostrom N. E. (1994) Pollution studies using nitrogen isotopes. In *Stable Isotopes in Ecology and Environmental Science* (eds. K. Lajtha and R. M. Michener). Blackwell, Oxford, pp. 45–62.

MacNamara J. and Thode H. G. (1950) Comparison of the isotopic composition of terrestrial and meteoritic sulfur. *Phys. Rev.* **78**, 307–308.

Maloszewski P. and Zuber A. (1996) Lumped parameter models for interpretation of environmental tracer data. In *Manual on Mathematical Models in Isotope Hydrologeology* (IAEA TECHDOC 910). IAEA, Vienna, pp. 9–58.

Mariotti A., Landreau A., and Simon B. (1988) ^{15}N isotope biogeochemistry and natural denitrification process in groundwater: application to the chalk aquifer of northern France. *Geochim. Cosmochim. Acta* **52**, 1869–1878.

Mariotti A., Pierre D., Vedy J. C., Bruckert S., and Guillemot J. (1980) The abundance of natural nitrogen 15 in the organic matter of soils along an altitudinal gradient (Chablais, Haute Savoie, France). *Catena* **7**, 293–300.

Mayer B., Fritz P., and Krouse H. R. (1992) Sulphur isotope discrimination during sulphur transformations in aerated forest soils. In *Workshop Proceedings on Sulphur Transformations in Soil Ecosystems* (eds. M. J. Hendry and H. R. Krouse). National Hydrology Research Symposium No. 11, Saskatoon, Sakatchewan, November 5–7, 1992, pp. 161–172.

Mayer B., Krouse H. R., Fritz P., Prietzel J., and Rehfuess K. E. (1993) Evaluation of biogeochemical sulfur transformations in forest soils by chemical and isotope data. In *Tracers in Hydrology*, IAHS Publ. No. 215, pp. 65–72.

Mayer B., Bollwerk S. M., Mansfeldt T., Hütter B., and Vezier J. (2001) The oxygen isotopic composition of nitrate generated by nitrification in acid forest floors. *Geochim. Cosmochim. Acta* **65**(16), 2743–2756.

Mayer B., Boyer E., Goodale C., Jawoarski N., van Bremen N., Howarth R., Seitzinger S., Billen G., Lajtha K., Nadelhoffer K., Van Dam D., Hetling L., Nosil M., Paustian K., and Alexander R. (2002) Sources of nitrate in rivers draining 16 major watersheds in the northeastern US: isotopic constraints. *Biochemistry* **57/58**, 171–192.

McDonnell J. J. and Kendall C. (1992) Isotope tracers in hydrology—report to the hydrology section. *EOS, Trans., AGU* **73**, 260–261.

McDonnell J. J., Bonell M., Stewart M. K., and Pearce A. J. (1990) Deuterium variations in storm rainfall: implications for stream hydrograph separations. *Water Resour. Res.* **26**, 455–458.

McDonnell J. J., Stewart M. K., and Owens I. F. (1991) Effect of catchment-scale subsurface mixing on stream isotopic response. *Water Resour. Res.* **27**, 3065–3073.

McDonnell J. J., Rowe L., and Stewart M. (1999) A combined tracer-hydrometric approach to assessing the effects of catchment scale on water flowpaths, source and age. *Int. Assoc. Hydrol. Sci. Publ.* **258**, 265–274.

McGlynn B., McDonnell J. J., Shanley J., and Kendall C. (1999) Riparian zone flowpath dynamics. *J. Hydrol.* **222**, 75–92.

McGlynn B., McDonnell J., Stewart M., and Seibert J. (2003) On the relationships between catchment scale and streamwater mean residence time. *Hydrol. Process.* **17**, 175–181.

McGuire K. J., DeWalle D. R., and Gburek W. J. (2002) Evaluation of mean residence time in subsurface waters using oxygen-18 fluctuations during drought conditions in the Mid-Appalachians. *J. Hydrol.* **261**, 132–149.

Mengis M., Schiff S. L., Harris M., English M. C., Aravena R., Elgood R. J., and MacLean A. (1999) Multiple geochemical and isotopic approaches for assessing ground water NO_3-elimination in a riparian zone. *Ground Water* **37**, 448–457.

Michalski G., Scott Z., Kaibling M., Thiemens M. H. First measurements and modeling of $\Delta^{17}O$ in atmospheric nitrate: application for evaluation of nitrogen deposition and relevance to paleoclimate studies. *GRL* (in review).

Michalski G., Savarino J., Bohlke J. K., and Thiemans M. (2002) Determination of the total oxygen isotopic composition of nitrate and the calibration of a Del ^{17}O nitrate reference material. *Anal. Chem.* **74**, 4989–4993.

Michel R. L. (1989) Tritium deposition over the continental United States, 1953–1983. In *Atmospheric Deposition*, International Association of Hydrological Sciences, Oxford, UK, pp. 109–115.

Michel R. L. (1992) Residence times in river basins as determined by analysis of long-term tritium records. *J. Hydrol.* **130**, 367–378.

Michel R. L., Campbell D. H., Clow D. W., and Turk J. T. (2000) Timescales for migration of atmospherically derived sulfate through an alpine/subalpine watershed, Loch Vale, Colorado. *Water Resour. Res.* **36**, 27–36.

Michel R. L., Turk J. T., Campbell D. H., and Mast M. A. (2002) Use of natural 35S to trace sulphate cycling in small lakes, Flattops wilderness area, Colorado, USA. *Water Air Soil Pollut. Focus* **2/2**, 5–18.

Migdisov A. A., Ronov A. B., and Grinenko V. A. (1983) The sulphur cycle in the lithosphere. In *The Global Biogeochemical Sulphur Cycle*, SCOPE 19 (eds. M. V. Ivanov and J. R. Freney). Wiley, pp. 25–95.

Mills A. L. (1988) Variations in the delta C-13 of stream bicarbonate: implications for sources of alkalinity. MS Thesis, George Washington University.

Mitchell M. J., Driscoll C. T., Fuller R. D., David M. B., and Likens G. E. (1989) Effect of whole-tree harvesting on the sulfur dynamics of a forest soil. *Soil Sci. Soc. Am.* **53**, 933–940.

Mitchell M. J., Burke M. K., and Shepard J. P. (1992a) Seasonal and spatial patterns of S, Ca, and N dynamics of a northern hardwood forest ecosystem. *Biogeochemistry* **17**, 165–189.

Mitchell M. J., David M. B., and Harrison R. B. (1992b) Sulfur dynamics in forest ecosystems. In *Sulfur Cycling on the Continents* (eds. R. W. Howarth, J. W. B. Stewart, and M. V. Ivanov). Wiley, pp. 215–254, Chap. 9.

Mitchell M. J., David M. B., Fernandez I. J., Fuller R. D., Nadelhoffer K., Rustad L. E., and Stam A. C. (1994) Response of buried mineral soil-bags to three years of experimental acidification of a forest ecosystem. *Soil Sci. Soc. Am. J.* **58**, 556–563.

Mitchell M. J., Krouse H. R., Mayer B., Stam A. C., and Zhang Y. (1998) Use of stable isotopes in evaluating sulfur biogeochemistry of forest ecosystems. In *Isotope Tracers in Catchment Hydrology* (eds. C. Kendall and J. J. McDonnell). Elsevier, Amsterdam, pp. 489–518.

Mizutani Y. and Rafter T. A. (1969) Oxygen isotopic composition of sulphates: Part 4. Bacterial fractionation of oxygen isotopes in the reduction of sulphate and in the oxidation of sulphur. *NZ J. Sci.* **12**, 60–67.

Mook W. G. (1968) Geochemistry of the stable carbon and oxygen isotopes of natural waters in the Netherlands. PhD Thesis, University of Groningen, Netherlands.

Mook W. G. and Tan F. C. (1991) Stable carbon isotopes in rivers and estuaries. In *Biogeochemistry of Major World Rivers* (eds. E. T. Degens, S. Kempe, and J. E. Richey). Wiley, , pp. 245–264.

Nadelhoffer K. J. and Fry B. (1988) Controls on natural nitrogen-15 and carbon-13 abundances in forest soil organic matter. *Soil Sci. Soc. Am. J.* **52**, 1633–1640.

Nadelhoffer K. J., Aber J. D., and Melillo J. M. (1988) Seasonal patterns of ammonium and nitrate uptake in nine temperate forest ecosystems. *Plant Soil* **80**, 321–335.

Novak M., Michel R. L., Prechova E., and Stepanovan M. (2003) The missing flux in a 35S budget of a small catchment. *Water Air Soil Pollut. Focus* (in press).

Nriagu J. O. and Coker R. D. (1978) Isotopic composition of sulphur in precipitation within the Great Lakes Basin. *Tellus* **30**, 365–375.

Nriagu J. O., Rees C. E., Mekhtiyeva V. L., Yu A., Lein P., Fritz R. J., Drimmie R. G., Pankina R. W., Robinson R. W., and Krouse H. R. (1991) Hydrosphere. In *Stable Isotopes—Natural and Anthropogenic Sulphur in the Environment, Chap. 6, SCOPE 43 (Scientific Committee on Problems of the Environment)* (eds. H. R. Krouse and V. A. Grinenko). Wiley, UK, pp. 177–265.

Ogunkoya O. O. and Jenkins A. (1991) Analysis of runoff pathways and flow distributions using deuterium and stream chemistry. *Hydrol. Process.* **5**, 271–282.

Ogunkoya O. O. and Jenkins A. (1993) Analysis of storm hydrograph and flow pathways using a three-component hydrograph separation model. *J. Hydrol.* **142**, 71–88.

Palmer S. M., Hope D., Billett M. F., Dawson J. J. C., and Bryant C. L. (2001) Sources of organic and inorganic carbon in a headwater stream: evidence from carbon isotope studies. *Biogeochemistry* **52**, 321–338.

Panno S. V., Hackley K. C., Hwang H. H., and Kelly W. R. (2001) Determination of the sources of nitrate contamination in karst springs using isotopic and chemical indicators. *Chem. Geol.* **179**, 113–128.

Pardo L. H., Kendall C., Pett-Ridge J., and Chang C. C. Y. Evaluating the source of streamwater nitrate using ^{15}N and ^{18}O in nitrate in two watersheds in New Hampshire, Hydrol. Processes (in press).

Park R. and Epstein S. (1961) Metabolic fractionation of ^{13}C and ^{12}C in plants. *Plant Physiol.* **36**, 133–138.

Pearce A. J., Stewart M. K., and Sklash M. G. (1986) Storm runoff generation in humid headwater catchments: 1. Where does the water come from? *Water Resour. Res.* **22**, 1263–1272.

Pearson F. J. and Rightmire C. T. (1980) Sulfur and oxygen isotopes in aqueous sulfur compounds. In *Handbook of Environmental Isotope Geochemistry* (eds. P. Fritz and J.-C. Fontes). Elsevier, Amsterdam, pp. 179–226.

Peters B. (1959) Cosmic-ray produced radioactive isotopes as tracers for studying large-scale atmospheric circulation. *J. Atmos. Terr. Phys.* **13**, 351–370.

Peters N. E. (1994) Hydrologic studies. In *Biogeochemistry of Small Catchments: A Tool for Environmental Research,* SCOPE Report 51 (eds. B. Moldan and J. Cerny). Wiley, UK, pp. 207–228.

Pionke H. B. and DeWalle D. R. (1992) Intra- and inter-storm ^{18}O trends for selected rainstorms in Pennsylvania. *J. Hydrol.* **138**, 131–143.

Plummer L. N., Parkhurst D. L., and Thorstenson D. C. (1983) Development of reaction models for ground-water systems. *Geochim. Cosmochim. Acta* **47**, 665–686.

Plummer L. N., Prestemon E. C., and Parkhurst D. L. (1991) An Interactive Code (NETPATH) for Modelling Net Geochemical Reactions along a Flow Path. *USGS Water-resources Inves. Report* 91-4078, 227pp.

Plummer L. N., Michel R. L., Thurman E. M., and Glynn P. D. (1993) Environmental tracers for age dating young ground water. In *Regional Ground-water Quality* (ed. W. M. Alley). V. N. Reinhold, New York, pp. 255–294.

Rau G. (1979) Carbon-13 depletion in a subalpine lake: carbon flow implications. *Science* **201**, 901–902.

Rice K. C. and Hornberger G. M. (1998) Comparison of hydrochemical traccers to estimate source contributions to peak flow in a small, forested, headwater catchment. *Water Resour. Res.* **34**, 1755–1766.

Rightmire C. T. (1978) Seasonal variations in pCO_2 and ^{13}C of soil atmosphere. *Water Resour. Res.* **14**, 691–692.

Robson A., Beven K. J., and Neal C. (1992) Towards identifying sources of subsurface flow: a comparison of components identified by a physically based runoff model and those determined by chemical mixing techniques. *Hydrol. Process.* **6**, 199–214.

Rodhe A. (1987) The origin of stream water traced by oxygen-18. PhD Thesis, Uppsala University, Department, Phys. Geogr., Div. Hydrol., Sweden. Rep Ser. A., No. 41.

Rozanski K., Araguas-Araguas L., and Gonfiantini R. (1992) Relation between long-term trends of oxygen-18 isotope composition of precipitation and climate. *Science* **258**, 981–985.

Salati E., Dall'Olio A., Matsui E., and Gat J. R. (1979) Recycling of water in the Amazon basin: an isotopic study. *Water Resour. Res.* **515**(5), 1250–1258.

Savarino J., Alexander B., Michalski G. M., and Thiemens M. H. (2002) Investigation of the oxygen isotopic composition of nitrate trapped in the Site A Greenland ice core. Presented at the AGU Spring Meeting, Washington, DC. *EOS, Trans., AGU*.

Saxena R. K. (1986) Estimation of canopy reservoir capacity and oxygen-18 fractionation in throughfall in a pine forest. *Nordic Hydrol.* **17**, 251–260.

Schiff S. L., Aravena R., Trumbore S. E., and Dillon P. J. (1990) Dissolved organic carbon cycling in forested watersheds: a carbon isotope approach. *Water Resour. Res.* **26**, 2949–2957.

Schiff S. L., Aravena R., Trumbore S. E., Hinton M. J., Elgood R., and Dillon P. J. (1997) Export of DOC from forested catchments on the Precambrian shield of central Ontario: clues from ^{13}C and ^{14}C. *Biogeochemistry* **36**, 43–65.

Schreiber M. E., Simo J. A., and Freiberg P. G. (2000) Stratigraphic and geochemical controls on naturally occurring arsenic in groundwater, eastern Wisconsin, USA. *Hydrogeol. J.* **8**, 161–176.

Schroth M. H., Kleikemper J., Bollinger C., Bernasconi S. M., and Zeyer J. (2001) *In situ* assessment of microbial sulfate reduction in a petroleum-contaminated aquifer using push-pull tests and stable sulfur isotope analyses. *J. Contamin. Hydrol.* **51**, 179–195.

Sebilo M., Billen G., Grably M., and Mariotti A. (2003) Isotopic composition of nitrate-nitrogen as a marker of riparian and benthic denitrification at the scale of the whole Seine River system. *Biogeochemistry* **63**, 35–51.

Shanley J. B., Kendall C., Smith T. M., Wolock D. M., and McDonnell J. J. (2001) Controls on old and new water contributions to stream flow at some nested catchments in Vermont, USA. *Hydrol. Process.* **16**, 589–609.

Shearer G. and Kohl D. (1986) N_2 fixation in field settings, estimations based on natural 15N abundance. *Austral. J. Plant Physiol.* **13**, 699–757.

Shearer G. and Kohl D. H. (1988) ^{15}N method of estimating N_2 fixation. In *Stable Isotopes in Ecological Research* (eds. P. W. Rundel, J. R. Ehleringer, and K. A. Nagy). Springer, New York, pp. 342–374.

Sickman J. Q., Leydecker A., Chang C. C. Y., Kendall C., Melack J. M., Lucero D. M. and Schimel J. (2003) Mechansims underlying export of N from high-elevation catchments during seasonal transitions. *Biogeochemistry* **64**, 1–24.

Sidle W. C. (2002) $^{18}OSO_4$ and $^{18}OH_2O$ as prospective indicators of elevated arsenic in the Goose River ground-watershed, Maine. *Environ. Geol.* **42**, 350–359.

Silva S. R., Ging P. B., Lee R. W., Ebbert J. C., Tesoriero A. J., and Inkpen E. L. (2002) Forensic applications of nitrogen and oxygen isotopes of nitrate in an urban environment. *Environ. Foren.* **3**, 125–130.

Sklash M. G. (1990) Environmental isotope studies of storm and snowmelt runoff generation. In *Process Studies in Hillslope Hydrology* (eds. M. G. Anderson and T. P. Burt). Wiley, UK, pp. 401–435.

Sklash M. G. and Farvolden R. N. (1979) The role of groundwater in storm runoff. *J. Hydrol.* **43**, 45–65.

Sklash M. G. and Farvolden R. N. (1982) The use of environmental isotopes in the study of high-runoff episodes in streams. In *Isotope Studies of Hydrologic Processes* (eds. E. C. Perry, Jr. and C. W. Montgomery). Northern Illinois University Press, DeKalb, Illinois, pp. 65–73.

Sklash M. G., Farvolden R. N., and Fritz P. (1976) A conceptual model of watershed response to rainfall, developed through the use of oxygen-18 as a natural tracer. *Can. J. Earth Sci.* **13**, 271–283.

Spence M. J., Bottrell S. H., Thornton S. F., and Lerner D. N. (2001) Isotopic modelling of the significance of bacterial sulphate reduction for phenol attenuation in a contaminated aquifer. *J. Contamin. Hydrol.* **53**, 285–304.

St-Jean G. (2003) Automated quantitative and isotopic (^{13}C) analysis of dissolved inorganic carbon and dissolved organic carbon in continuous-flow using a total organic carbon analyzer. *Rapid Commun. Mass Spectrom.* **17**, 419–428.

Stam A. C., Mitchell M. J., Krouse H. R., and Kahl J. S. (1992) Stable sulfur isotopes of sulfate in precipitation and stream solutions in a northern hardwood watershed. *Water Resour. Res.* **28**, 231–236.

Staniaszek P. and Halas S. (1986) Mixing effects of carbonate dissolving waters on chemical and $^{13}C/^{12}C$ compositions. *Nordic Hydrol.* **17**, 93–114.

Stevens C. M. and Rust F. E. (1982) The carbon isotopic composition of atmospheric methane. *J. Geophys. Res.* **87**, 4879–4882.

Stewart M. K. and McDonnell J. J. (1991) Modeling base flow soil water residence times from deuterium concentration. *Water Resour. Res.* **27**(10), 2681–2693.

Stichler W. (1987) Snowcover and snowmelt process studies by means of environmental isotopes. In *Seasonal Snowcovers: Physics, Chemistry, Hydrology* (eds. H. G. Jones and W. J. Orville-Thomas). Reidel, , pp. 673–726.

Strauss H. (1997) The isotopic composition of sedimentary sulfur through time. *Palaeogeogr. Paleoclimatol. Paleoecol.* **132**, 97–118.

Stute M. and Schlosser P. (2000) Atmospheric noble gases. In *Environmental Tracers in Subsurface Hydrology* (eds. P. Cook and A. Herzceg). Kluwer Academic, Dordrecht.

Sueker J. K., Ryan J. N., Kendall C., and Jarrett R. D. (2000) Determination of hydrological pathways during snowmelt for alpine/subalpine basins, Rocky Mountain National Park, Colorado. *Water Resour. Res.* **36**, 63–75.

Swistock B. R., DeWalle D. R., and Sharpe W. E. (1989) Sources of acidic stormflow in an Appalachian headwater stream. *Water Resour. Res.* **25**, 2139–2147.

Szaran J., Niezgoda H., and Halas S. (1998) New determination of oxygen and sulphur isotope fractionation between gypsum and dissolved sulphate. *RMZ-mater. Geoenviron.* **45**, 180–182.

Taylor B. E. and Wheeler M. C. (1994) Sulfur- and oxygenisotope geochemistry of acid mine drainage in the western US: Field and experimental studies revisited. In *Environmental Geochemistry of Sulfide Oxidation*, ACS Symposium Series 550 (eds. C. N. Alpers and D. W. Blowes). American Chemical Society, Washington, DC, pp. 481–514.

Taylor B. E., Wheeler M. C., and Nordstrom D. K. (1984) Oxygen and sulfur compositions of sulphate in acid mine drainage: evidence for oxidation mechanisms. *Nature* **308**, 538–541.

Taylor C. B. and Fox V. J. (1996) An isotopic study of dissolved inorganic carbon in the catchment of the Waimakariri River and deep ground water of the North Canterbury Plains, New Zealand. *J. Hydrol.* **186**, 161–190.

Taylor S., Feng X., Kirchner J. W., Osterhuber R., Klaue B., and Renshaw C. E. (2001) Isotopic evolution of a seasonal snowpack and its melt. *Water Resour. Res.* **37**, 759–769.

Thatcher L. L. (1962) *The Distribution of Tritium Fallout in Precipitation Over North America.* International Association of Hydrological Sciences (IAHS), Publication No. 7, Louvain, Belgium, pp. 48–58.

Thurman E. M. (1985) *Organic Geochemistry of Natural Waters.* Dr W Junk Publishers, Dordrecht, Martinus Nijhof.

Toran L. and Harris R. F. (1989) Interpretation of sulfur and oxygen isotopes in biological and abiological sulfide oxidation. *Geochim. Cosmochim. Acta* **53**(9), 2341–2348.

Turk J. T., Campbell D. H., and Spahr N. E. (1993) Use of chemistry and stable sulfur isotopes to determine sources of trends in sulfate of Colorado lakes. *Water Air Soil Pollut.* **67**, 415–431.

Unnikrishna P. V., McDonnell J. J., and Stewart M. L. (1995) Soil water isotopic residence time modelling. In *Solute Modelling in Catchment Systems* (ed. S. T. Trudgill). Wiley, New York, pp. 237–260.

Van Stempvoort D. R. and Krouse H. R. (1994) Controls of ^{18}O in sulfate: a review of experimental data and application to specific environments. In *Environmental Geochemistry of Sulfide Oxidation,* ACS Symp. Ser. 550 (eds. C. N. Alpers and D. W. Blowes). American Chemical Society, Washington, DC, pp. 446–480.

Vogel J. C., Talma A. S., and Heaton T. H. E. (1981) Gaseous nitrogen as evidence for denitrification in groundwater. *J. Hydrol.* **50**, 191–200.

Wadleigh M. A. and Blake D. M. (1999) Tracing sources of atmospheric sulphur using epiphytic lichens. *Environ. Pollut.* **106**, 265–271.

Wadleigh M. A., Schwarcz H. P., and Kramer J. R. (1996) Isotopic evidence for the origin of sulphate in coastal rain. *Tellus* **48B**, 44–59.

Wallis P. M., Hynes H. B. N., and Fritz P. (1979) Sources, transportation, and utilization of dissolved organic matter in groundwater and streams. Sci. Ser. 100, Inland Waters Directorate, Water Quality Branch, Ottawa, Ont.

Wang Y., Huntington T. G., Osher L. J., Wassenaar L. I., Trumbore S. E., Amundson R. G., Harden J. G., McKnight D. M., Schiff S. L., Aiken G. R., Lyons W. B., Aravena R. O., and Baron J. S. (1998) Carbon cycling in terrestrial environments. In *Isotope Tracers in Catchment Hydrology* (eds. C. Kendall and J. J. McDonnell). Elsevier, Amsterdam, pp. 577–610.

Wassenaar L. (1995) Evaluation of the origin and fate of nitrate in the Abbotsford Aquifer using the isotopes of ^{15}N and ^{18}O in NO_3^-. *Appl. Geochem.* **10**, 391–405.

Wassenaar L. I., Aravena R., Fritz P., and Barker J. F. (1991) Controls on the transport and carbon isotopic composition of dissolved organic carbon in a shallow groundwater system, Central Ontario, Canada. *Chem. Geol.* **87**, 39–57.

Weiler R. R. and Nriagu J. O. (1973) Isotopic composition of dissolved inorganic carbon in the Great Lakes. *J. Fish. Res. Board Can.* **35**, 422–430.

Wels C., Cornett R. J., and Lazerte B. D. (1991) Hydrograph separation: a comparison of geochemical and isotopic tracers. *J. Hydrol.* **122**, 253–274.

Wenner D. B., Ketcham P. D., and Dowd J. F. (1991) Stable isotopic composition of waters in a small piedmont watershed. In *Geochemical Society Special Publication No. 3* (eds. H. P. Taylor, Jr., J. R. O'Neil, and I. R. Kaplan). pp. 195–203.

Williard K. W. J. (1999) Factors affecting stream nitrogen concentrations from mid-appalachian forested watersheds. PhD Dissertation, Penn. State University.

Wolterink J. J., Williamson H. J., Jones D. C., Grimshaw T. W., and Holland W. F (1979). Identifying sources of subsurface nitrate pollution with stable nitrogen isotopes. U.S. Environmental Protection Agency, EPA-600/4-79-050, 150p.

Yang C., Telmer K., and Veizer J. (1996) Chemical dynamics of the St-Lawrence riverine system: δD_{H_2O}, $\delta^{18}O_{H_2O}$, $\delta^{13}C_{DIC}$, δ^{34}sulfate, and dissolved $^{87}Sr/^{86}Sr$. *Geochim. Cosmochim. Acta* **60**(5), 851–866.

Yang W., Spencer R. J., and Krouse H. R. (1997) Stable isotope composition of waters and sulfate species therein, Death Valley, California, USA: implications for inflow and sulfate sources, and arid basin climate. *Earth Planet. Sci. Lett.* **147**, 69–82.

Zimmerman U., Ehhalt D., and Munnich K. O. (1967) Soil-water movement and evapotranspiration: changes in the isotopic composition of water. In *Isotopes in Hydrology*. IAEA, Vienna, pp. 567–584.

Published by Elsevier Ltd.

Isotope Geochemistry
ISBN: 978-0-08-096710-3

pp. 181–226

8
Elemental and Isotopic Proxies of Past Ocean Temperatures

D. W. Lea

University of California, Santa Barbara, CA, USA

8.1 INTRODUCTION

Determining the temperature evolution of the oceans is a fundamental problem in the geosciences. Temperature is the most primary representation of the state of the climate system, and the temperature of the oceans is critical because the oceans are the most important single component of the Earth's climate system. A suite of isotopic and elemental proxies, mostly preserved in marine carbonates, are the essential method by which earth scientists determine past ocean temperatures (see Table 1). This is a field with both a long history and a great deal of recent progress. Paleotemperature research has been at the forefront of geoscience research since 1950s, and, with our need to understand the global climate system heightened by the threat of global warming, it promises to remain a vibrant and important area well into the future.

In this chapter I begin by reviewing the history of the elemental and isotopic proxies and how that history shapes research priorities today. I then review the state of our knowledge at present (as of early 2000s), including areas that are well developed (i.e., oxygen isotopes in coral aragonite), areas that are experiencing phenomenal growth (i.e., Mg/Ca in foraminifer shells), and areas that are just starting to develop (i.e., Ca isotopes in carbonates). In these reviews I include an estimation of the uncertainty in each of these techniques and areas, including aspects that particularly need to be addressed. I also address the question of overlap and confirmation between proxies, and in particular how information from one proxy can augment a second proxy.

8.2 A BRIEF HISTORY OF EARLY RESEARCH ON GEOCHEMICAL PROXIES OF TEMPERATURE

Geologists have been interested in establishing the temperature history of the oceans for as long as they have documented historical variations in marine sediments. Probably the first realization that geochemical variations might reflect temperatures can be traced to the great American geochemist Frank Wigglesworth Clarke (1847–1931), the namesake of the Geochemical Society's Clarke award. Aside from his voluminous contributions to establishing precise atomic weights, the composition of the Earth's crust, and natural waters, Clarke found time to document a provocative relationship between the magnesium content of biogenic carbonates and their growth temperature (Clarke and Wheeler, 1922). The authors speculated that this relationship had a definite cause and could possibly be useful: "This rule, or rather tendency, we are inclined to believe is general, although we must admit that there are probably exceptions to it." Clarke and Wheeler recognized

Table 1 Summary of major paleotemperature techniques.

	Phases	*Sensitivity* (per °C)	*Estimated SE*	*Major secondary effects*	*Time scale*[a]
Oxygen Isotopes	Foraminifera	0.18–0.27%	0.5 °C if δ^{18}O-sw is known	Effect of δ^{18}O-sw	0–100 Ma
	Corals	~0.2%	0.5 °C if δ^{18}O-sw is known	Kinetic effects Effect of δ^{18}O-sw	0–130 ka
	Opal			Effect of δ^{18}O-sw	0–30 ka
Mg/Ca	Foraminifera	9±1%	~1 °C	Dissolution Secular Mg/Ca variations (>10 Ma)	0–40 Ma
	Ostracodes	~9%	~1 °C	Dissolution? Calibration	0–3.2 Ma
Sr/Ca	Corals	−0.4 to−1.0%	0.5 °C?	Growth effects Secular Sr/Ca changes (>5 ka)	0–130 ka
Ca isotopes	Foraminifera	0.02–0.24‰	unknown	Species effects, calcification	0–125 ka
Alkenone unsaturation index[b]	Sediment organics	0.033 (0.023–0.037) in $U^{K'}_{37}$[d]	~1.5 °C (global calib.)	Transport, species variation	0–3 Ma
Faunal transfer functions[c]	Foraminifera, Radiolaria, Dinoflagellates	NA	1.5 °C	Ecological shifts	0–?

[a] Timescale over which the technique has been applied. [b] Chapter 6.15. [c](Imbrie and Kipp, 1971). [d] (Müller *et al.*, 1998 and Pelejero and Calvo, 2003).

that the magnesium to calcium ratio of the oceans was nearly constant, and that therefore the trend had to have some cause other than compositional variations: "That warmth favors the assimilation of magnesia by marine invertebrates seems to be reasonably certain, but why it should be so is not clear. The relation is definite but as yet unexplained. *We hope it is not inexplicable*" (quotations added for emphasis). Further along in this same paragraph Clarke and Wheeler presaged the field of geochemical paleoceanography and paleoclimatology: "Attempts will likely be made to use our data in studies of climatology, but are such attempts likely to be fruitful?" The authors envisioned researchers using the bulk magnesium content of ancient rocks to determine past temperatures, an approach they deemed doubtful because it would depend on the ratios of particular organisms in rocks. Of course, Clarke and Wheeler did not envision the powerful analytical techniques available to the present-day analyst, where the trace element content of individual chambers of plankton shells can be readily analyzed. Such single species analysis is what eventually enabled the useful application of Clarke and Wheeler's original insight to paleothermometry (see below). However, it was not until the late 1990s that the observed magnesium relationship became both explicable and fruitful for climatology.

The next great step forward came after World War II, when Harold Urey (1893–1981), a 1934 Nobel Laureate for his discovery of deuterium, the heavy isotope of hydrogen, took up a professorship at the University of Chicago. There, he became interested in the utilization of natural fractionations in stable isotope systems for geological purposes (Urey, 1947). He theorized that the effect of temperature on the partitioning of oxygen isotopes between water and carbonate might become a useful geological tool: "Accurate determinations of the O^{18} content of carbonate rocks could be used to determine the temperature at which they were formed." Interestingly, Urey did not envision that compositional variations in the ocean would complicate such determinations: "First, there is the large reservoir of oxygen in the oceans which *cannot have changed* in isotopic compositions during geological time" (italics added for emphasis). He, of course, recognized that if such variations occurred, they would complicate paleotemperature determinations, and he pointed out in the same paragraph that variations in the isotopic composition of calcium would hinder its potential use for paleotemperature analysis, a point relevant to present research (see below). Of course, Urey would not have known about or envisioned the considerable effects continental glaciation and crustal exchange would have on the oxygen isotopic composition of the oceans (Sturchio, 1999).

It was left to Urey's students and post-doctoral scholars to exploit his original insights. Major advances came from the establishment of a so-called "paleotemperature equation" by Samuel Epstein (1919–2001), mainly based on calcite precipitated by mollusks in either controlled experiments or field-collected samples (Epstein *et al.*, 1953). Analysis of this calcite yielded a paleotemperature equation that demonstrated a sensitivity of $\sim -0.2‰$ change in $\delta^{18}O$ per °C. It is important to note that during this time period, and, as shown above, from the conception of the original idea, the emphasis in using oxygen isotopes was to reconstructing paleotemperatures. A few years later, Cesare Emiliani (1922–1995), a student and later a post-doctoral scholar with Urey, exploited these advances when he documented regular cyclic variations in the oxygen isotopic composition of planktonic foraminifera taken from eight sediment cores in the Caribbean (Emiliani, 1955). Although Emiliani did allow for small variations in the isotopic composition of seawater, he largely interpreted the observed $\delta^{18}O$ variations as a reflection of recurring cold intervals in the past during which tropical surface waters cooled by 6–8 °C. Many of the questions raised by Emiliani's classic 1955 study are still relevant today and are addressed in detail below.

During this same period scholars at Chicago examined the potential temperature dependence of trace elements in carbonates, focusing mainly on magnesium and strontium incorporation (Chave, 1954). Although these studies provided some additional insights beyond Clarke and Wheeler's (1922) original findings, they failed to yield advances of the kind that were spurring research on isotopic variations. Samuel Epstein (personal communication, 1992) felt that the general sense in the Chicago group was that elemental substitution was likely to be less regular than isotopic substitution, presumably because the individual activity coefficients of each element would introduce additional complexity beyond a simple temperature dependence.

Following Emiliani's (1955) discovery, other laboratories established the capability to apply oxygen isotope variations to oceanic temperature history. It is worth a brief mention of two further major advances relevant to Urey's original conception. In 1967, Nicholas Shackleton of Cambridge University reported the first systematic down-core variations in benthic foraminifera (Shackleton, 1967). He argued that benthic fauna, because they lived in the near-freezing bottomwaters of the ocean, would mainly record the change in isotopic variation of seawater. By demonstrating that the observed benthic variations were of similar magnitude to those in planktics, he was able to demonstrate that the major portion of the isotopic signal recorded in marine sediments reflected oscillations in the oxygen isotopic composition of the ocean, which in turn occurred in response to the periodic transfer of isotopically depleted water onto continental ice sheets. Once this paradigm shift was in place, Shackleton and others were able to use oxygen isotopic variations as a stratigraphic and chronometric tool for marine sediments (Shackleton and Opdyke, 1973), an advance which lead to the establishment of a precise timescale for the Late Quaternary marine record and ultimately the verification of orbital variations as the pacemakers of the Pleistocene Ice Ages (Hays *et al.*, 1976). So from Urey's conception to Shackleton's insight, a tool originally envisioned as a paleothermometer found its most profound use as a recorder of the compositional variations in seawater that Urey considered to be an unlikely influence. Parallel advances in the utilization of elemental variations did not occur until the 1990s (see below), but it is worth mentioning as a close to this historical summary that recent research indicates that the Mg/Ca content of foraminifera is the long-sought solution to the "Urey dilemma," because it provides, in combination with oxygen isotopes, a simultaneous temperature and isotopic composition history for seawater (Lea *et al.*, 2000; Lear *et al.*, 2000).

8.3 OXYGEN ISOTOPES AS A PALEOTEMPERATURE PROXY IN FORAMINIFERA

8.3.1 Background

The use of oxygen isotope ratios as a paleotemperature indicator in carbonate minerals is based on the thermodynamic fractionation between ^{16}O and ^{18}O that occurs during precipitation (Urey, 1947). This fractionation, which offsets the $\delta^{18}O$ ($\delta^{18}O = [(^{18/16}O_{sample}/^{18/16}O_{standard})-1]$) of carbonate minerals relative to seawater by $\sim+30‰$, is a logarithmic function of temperature with a slope, over the oceanic temperature range of $-2\,°C$ to $30\,°C$, of between $-0.20‰$ and $-0.27‰$ per °C, in agreement with thermodynamic predictions (O'Neil *et al.*, 1969; Shackleton, 1974; Kim and O'Neil, 1997; Zhou and Zheng, 2003). Because the oxygen isotope proxy is based on a thermodynamic principle, it is expected to be robust and relatively unaffected by secondary kinetic factors. For foraminifera, unicellular zooplankton and benthos that precipitate calcite and, less commonly, aragonite shells (sometimes called tests), oxygen isotopic ratios do appear to be quite robust, although there are clear indications of a secondary effect from factors such as ontogenetic variations and seawater carbonate ion (see below).

The most significant complication in using oxygen isotopes to determine both absolute temperatures as well as relative temperature changes is that the $\delta^{18}O$ of carbonate solids reflects both temperature fractionation and the $\delta^{18}O$ of the seawater from which the carbonate precipitated. The $\delta^{18}O$ of seawater in turn reflects two major factors: (i) the mean $\delta^{18}O$ of seawater (Schrag *et al.*, 1996), which is determined by the amount of continental ice, which varies on timescales of 10^4–10^5 yr, and by interaction between seafloor basalts and seawater, which varies on timescales of 10^7–10^8 yr and (ii) the evaporation–precipitation (rainfall) balance $(E-P)$ for that part of the ocean or, for subsurface waters, the balance that applied to the source waters of those deep waters. The second factor is often described as a "salinity effect" because $\delta^{18}O$ tends to track with salinity variations because both respond to $E-P$. The relationship between $\delta^{18}O$ and salinity, however, varies considerably over the ocean because of the varying isotopic composition of freshwater (Schmidt *et al.*, 2001). It is important to emphasize that both effects cast considerable uncertainty into the use of oxygen isotopic ratios for absolute and relative paleothermometry on essentially all timescales.

Despite these caveats, oxygen isotopic ratios are probably the most widely used climate proxy in ocean history research. Reasons for this widespread use relate to the history of oxygen isotopes in geological research (see Section 8.2), the fact that they can be measured quite precisely by mass spectrometry and are relatively immune, at least in younger deposits, to secondary effects, the fact that $\delta^{18}O$ records tend to be quite reproducible and clearly record climate variability, and finally, because $\delta^{18}O$ records have proven so useful for stratigraphic and chronological purposes.

An important aspect of the application of oxygen isotopic ratios in foraminifera is that the planktonic foraminifera occupy several ecological niches, including surface waters, shallow subsurface waters, and deeper thermocline waters. Along with their benthic counterparts, this makes it possible to recover oxygen-isotopic records representative of different part of the water column. A complication, however, is that many of the planktonic species migrate vertically, potentially compounding the signals they record.

8.3.2 Paleotemperature Equations

There are a number of calibrations in use for oxygen isotopes in foraminifera, some derived from other organisms (Epstein *et al.*, 1953), some derived from culturing (Erez and Luz, 1983; Bemis *et al.*, 1998), and some derived from core-top calibrations (Shackleton, 1974). These calibrations take the form of a polynomial paleotemperature equation:

$$T = a + b*(\delta^{18}O_{calcite} - \delta^{18}O_{water}) + c*(\delta^{18}O_{calcite} - \delta^{18}O_{water})^2 \quad (1)$$

where T is temperature (°C), a is temperature when $\delta^{18}O_{calcite}-\delta^{18}O_{water}$ (both on V-PDB scale) is 0, b is the slope, and c is the second-order term for curvature (not always included). The inverse of the slope b represents the change in $\delta^{18}O$ (in ‰) for a 1 °C change in temperature. If the second-order term c is included, then the slope is not constant. The value of the slope is predicted to increase with decreasing temperature, because isotopic fractionation increases with decreasing temperature (Urey, 1947). Experimental evidence from inorganic calcite precipitation studies indicates that slope ranges from 0.27‰ per °C at 0 °C to 0.20‰ per °C at 25 °C (O'Neil *et al.*, 1969; Kim and O'Neil, 1997). Observational evidence from calibration of foraminiferal $\delta^{18}O$ indicates a similar or slightly larger range of values (Bemis *et al.*, 1998). The status of oxygen isotope calibrations is extensively reviewed in Bemis *et al.* (1998). For warm-water studies, the low light *Orbulina universa* calibration of Bemis *et al.* (1998) appears to work well:

$$T = 16.5 - 4.80 * \left(\delta^{18}O_{calcite} - \delta^{18}O_{water}\right) \quad (2)$$

For cold waters and certain benthics (e.g., *Uvigerina*), the Shackleton (1974) expression, which is a polynomial expansion of O'Neil (1969), appears to work well:

$$T = 16.9 - 4.38 * \left(\delta^{18}O_{calcite} - \delta^{18}O_{water}\right) + 0.1 * \left(\delta^{18}O_{calcite} - \delta^{18}O_{water}\right)^2 \quad (3)$$

The quantitative applicability of these equations is becoming more important with the growing interest in combining independent foraminiferal temperature estimates from Mg/Ca with $\delta^{18}O$ to calculate $\delta^{18}O_{water}$ (Mashiotta *et al.*, 1999; Elderfield and Ganssen, 2000; Lea *et al.*, 2000; Lear *et al.*, 2000).

8.3.3 Secondary Effects and Diagenesis

There has been a long discussion in the literature as to the extent to which foraminifera shells are in oxygen-isotopic equilibrium. The precise state of equilibrium cannot be defined with sufficient precision from theory, so the general practice is to compare observed foraminiferal calibrations to inorganic experiments (Shackleton, 1974; Kim and O'Neil, 1997; Bemis *et al.*, 1998). Such comparisons suggest that foraminifera shells can be offset from equilibrium for a number of reasons (Figure 1), ranging from unknown vital effects that offset benthic species (Duplessy *et al.*, 1970), offsets due to the effect of light on symbiotic algae (Spero and Lea, 1993; Bemis *et al.*, 1998), offsets due to ontogeny (growth) (Spero and Lea, 1996), offsets due to carbonate ion concentration of seawater (Spero *et al.*, 1997), and offsets due to the addition of gametogenic calcite at depth (Duplessy and Blanc, 1981). All of these effects can complicate the use of oxygen isotopes

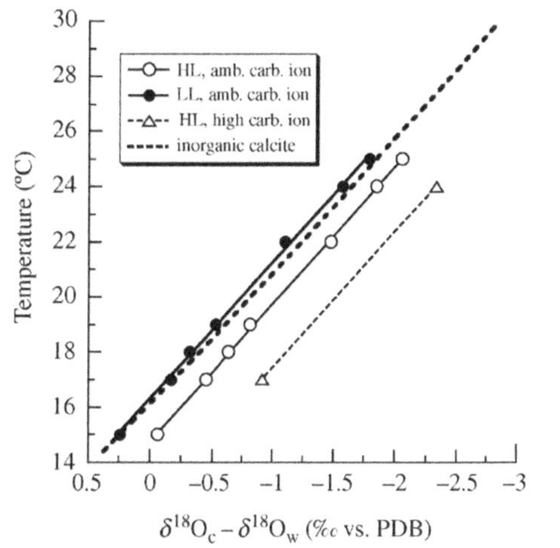

Figure 1 Comparison of oxygen-isotope paleotemperature equations for *O. universa*, a symbiont-bearing planktonic foraminifera, with values for inorganic calcite precipitation (Kim and O'Neil, 1997). These data demonstrate that the low light (LL) oxygen isotope equation at ambient carbonate ion concentration for *O. universa* is essentially indistinguishable from the inorganic equation. For high light (HL) conditions, in which symbiont photosynthetic activity is maximized, $\delta^{18}O$ shifts to more negative values. For high carbonate ion conditions, $\delta^{18}O$ shifts to even more negative values. These trends demonstrate the range of potential biological influences on foraminiferal $\delta^{18}O$ (after Bemis *et al.*, 1998).

for paleotemperature or paleo-$\delta^{18}O_{water}$ studies. Some of these effects, such as the effect of gametogenic calcite, also apply for Mg/Ca paleothermometry (Rosenthal *et al.*, 2000; Dekens *et al.*, 2002), so it is likely that complimentary studies will improve our understanding of the limitations these effects impose.

On Quaternary timescales, the main diagenetic concern is partial shell dissolution that takes place on the seafloor or within the sedimentary mixed layer. This effect has been demonstrated to increase the $\delta^{18}O$ of shells by ~0.2‰ per km in deeper, more dissolved sediments (Wu and Berger, 1989; Dekens *et al.*, 2002), presumably through the loss of individual shells and/or shell material with more negative $\delta^{18}O$. Because this effect appears to be coincident for both oxygen isotopes and Mg/Ca (Brown and Elderfield, 1996; Rosenthal *et al.*, 2000; Dekens *et al.*, 2002), it is likely that complimentary studies will allow a better assessment of the potential biases imposed by dissolution. On longer timescales, diagenetic effects multiply to include gradual replacement of the primary calcite (Schrag *et al.*, 1995; Schrag, 1999a). Recent studies suggest that shells with unusually good levels of preservation, such as those preserved in impermeable clay-rich sediments, record much more negative $\delta^{18}O$ (and hence warmer temperatures) than shells from deeply buried open ocean sequences (Pearson *et al.*, 2001).

8.3.4 Results on Quaternary Timescales

The many important results achieved in paleoceanography and paleoclimatology using oxygen isotope ratios in foraminifera shells are well known and have been reviewed in many other places: e.g., Shackleton (1987), Imbrie *et al.* (1992), and Mix (1992). Because of the outstanding geological importance of oxygen isotopic results, they take a central place in the history of proxy development, a subject discussed in Sections 8.2 and 8.3.1. As was emphasized in Section 8.2, the pioneers in the utilization of oxygen isotopes envisioned them as a paleotemperature tool. With the realization that change in the isotopic composition of the ocean exceeded the temperature influence on foraminiferal calcite (Shackleton, 1967), emphasis shifted to the use of Quaternary oxygen isotopic variations as a tool for stratigraphy and chronology (Shackleton and Opdyke, 1973; Hays *et al.*, 1976) and for calibration of the magnitude of past sea-level and ice volume change (Chappell and Shackleton, 1986; Shackleton, 1987). With the advent of independent geochemical paleotemperature proxies, it became possible to deconvolve the isotopic signal into its temperature and compositional components (Rostek *et al.*, 1993; Mashiotta *et al.*, 1999; Elderfield and Ganssen, 2000). Current research (see Sections 8.6.4 and 8.6.5) is focused on the veracity of this approach and the separation of ice volume and hydrological influences in the extracted $\delta^{18}O$-water records (Lea *et al.*, 2000, 2002; Lear *et al.*, 2000; Martin *et al.*, 2002).

8.3.5 Results for the Neogene, Paleogene, and Earlier Periods

The Cenozoic benthic foraminiferal $\delta^{18}O$ record is one of the major success of the geochemical approach to paleoclimate research (Zachos *et al.*, 2001). Because it is a record of the combined influences of temperature and ice volume influence, which are evolving semi-independently over the course of the Cenozoic, it is more a record of earth system processes than of temperature. There has been a great deal of interest and controversy concerning the utilization of oxygen isotopes in low-latitude planktonic foraminifera to determine tropical ocean temperatures on longer timescales (Barron, 1987). Recent research suggests that diagenetic overprinting of the primary foraminiferal $\delta^{18}O$ is a major influence (Schrag, 1999a; Pearson *et al.*, 2001). Results from the Pearson *et al.* (2001) study suggest that low-latitude sea surface temperatures (SSTs) during the Late Cretaceous and Eocene epochs were at least 28–32 °C compared to previous estimates, based on less well preserved material, of 15–23 °C. Obviously such a large shift requires a reevaluation of Paleogene and Cretaceous climates, but it might also point the way towards a means of correcting less well preserved samples for diagenesis, perhaps in combination with the Mg/Ca approach.

8.3.6 Summary of Outstanding Research Issues

Of the primary paleothermometric techniques reviewed in this chapter, oxygen isotopes in foraminifera have the longest history of development and application. It is probably safe to say that oxygen isotopes are on the most firm ground in terms of known influences and inherent accuracy. New results, such as a primary seawater carbonate ion influence (Spero *et al.*, 1997) and secondary diagenetic overprints (Schrag, 1999a; Pearson *et al.*, 2001), suggest that there is still major progress to made in this area. The most outstanding research issues today certainly must include progress and prospects for combining Mg/Ca paleothermometry with oxygen isotopes on both Quaternary, Neogene, and Paleogene timescales, and the need for reevaluation of the integrity and interpretation of Paleogene oxygen isotope ratios in foraminifera.

8.4 OXYGEN ISOTOPES AS A CLIMATE PROXY IN REEF CORALS

8.4.1 Background

Many of the factors described in Section 8.3.1 apply equally to oxygen isotopes in corals, which are dominantly composed of aragonite. Corals, however, have many unique aspects which require separate consideration. First, their oxygen isotopic composition is invariably depleted relative to equilibrium by ~1–6‰, presumably because of their different biochemical mechanisms of precipitation as well as the influence of symbiotic zooxanthellae (McConnaughey, 1989a; McConnaughey, 1989b). This offset from equilibrium actually discouraged early researchers, who assumed that corals would not be reliable temperature recorders (S. Epstein, Caltech, personal communication, 1998). Eventually, however, researchers began to investigate the prospect of attaining climate records from coral skeletons, and despite the offset from equilibrium research revealed that the oxygen-isotopic composition recorded subseasonal variations in seawater temperature and salinity (Weber and Woodhead, 1972; Emiliani *et al.*, 1978; Fairbanks and Dodge, 1979; Druffel, 1985; Dunbar and Wellington, 1985; McConnaughey, 1989a; Cole and Fairbanks, 1990; Shen *et al.*, 1992). These early discoveries have been followed by a fantastic array of results from longer coral time series that have enabled researchers to elevate coral climate records to the same level of importance as tree ring records (Gagan *et al.*, 2000). In general, coral $\delta^{18}O$ records are not interpreted directly as temperature records but rather as climate records whose variability reflects some combination of temperature and salinity effects.

Because aragonite is more susceptible to dissolution than calcite, especially under the influence of meteoric waters, and because most fossil corals are recovered from uplifted terrestrial deposits, diagenesis is an especially important limiting factor in recovering older coral records. This problem can be circumvented by drilling into submerged fossil deposits, but because of logistical difficulties, so far this has been accomplished in only a few key spots such as Barbados and Tahiti (Fairbanks, 1989; Bard *et al.*, 1996).

8.4.2 Paleotemperature Equations

Weber and Woodhead (1972) first demonstrated that oxygen isotopes in corals respond to temperature but are offset to negative $\delta^{18}O$ values relative to equilibrium. The oxygen isotope paleotemperature equation is well calibrated for corals (Gagan *et al.*, 2000). The systematic offset from the equilibrium or inorganic aragonite value is attributed to a biological or vital effect (McConnaughey, 1989a,b). This offset, however, appears to be stable over time in many different settings (Gagan *et al.*, 2000), although researchers recognize that vital effects can offset coral $\delta^{18}O$ to varying degrees (McConnaughey, 1989a,b; Spero *et al.*, 1997). The slope of the calibrations, however, appears to be nearly constant at ~0.2‰ per °C, in general agreement with the slope from inorganic experiments (Zhou and Zheng, 2003); the constancy of the slope suggests that historical changes in temperature will be accurately recorded.

Of course, the oxygen-isotopic composition of carbonate is also a function of the $\delta^{18}O$ of seawater, which varies with the local $E-P$, and hence salinity. Because temperature and salinity often vary together in the tropics, researchers tend to use coral $\delta^{18}O$ variations as climate proxies rather than temperature proxies. This approach has been very successful (see Section 8.4.4), but it still leaves the problem of attributing the observed changes in $\delta^{18}O$ to some specific combination of temperature and $\delta^{18}O_{water}$ changes. One solution is to use an independent temperature proxy such as Sr/Ca (see Section 8.8) at the same time to separate the coral $\delta^{18}O$ signal into its components (McCulloch *et al.*, 1994; Gagan *et al.*, 1998). It is also possible that comparison of records of coral $\delta^{18}O$ from different areas with contrasting climatology could be used to separate temperature and $\delta^{18}O_{water}$ influence.

8.4.3 Secondary Effects and Diagenesis

The major secondary effect for oxygen isotopes in hermatypic reef corals is the negative offset from equilibrium (Emiliani *et al.*, 1978). The degree to which this offset is stable in space and time (McConnaughey, 1989a,b) is critical to the interpretation of observed $\delta^{18}O$ variations in terms of absolute temperature and salinity changes. Measurements of $\delta^{18}O$ in coral heads from a single reef do reveal up to 1‰ variability in absolute values from specimen to specimen. The biological factors that cause these differences obviously have the potential to affect the use of coral $\delta^{18}O$ for paleoclimate research.

There is some evidence that the buildup of aragonite cements in living coral skeletons can affect the fidelity of coral $\delta^{18}O$ and Sr/Ca (see Section 8.8.3) records (Muller *et al.*, 2001). This occurs as pore spaces in the older part of the coral heads fill with inorganic aragonite, which has a more enriched $\delta^{18}O$ relative to the original coral material. As a result, the recorded climate signal in the oldest part of the coral is shifted to systematically colder and/or more salty values. Müller *et al.* (2001) observed that both $\delta^{18}O$ and Sr/Ca were biased similarly by inorganic aragonite precipitation, so cross-checks between proxies within the same coral does not alleviate the problem. Fortunately, however, it is possible to observe the precipitation of the secondary aragonite using petrography.

Because aragonite reverts to calcite when it interacts with meteoric water, subaerial exposure of fossil corals has the potential to change the $\delta^{18}O$ of the coral. Generally, diagenetically altered corals can be avoided by using X-ray crystallography to screen for the presence of calcite. A recent study suggests that small, restricted levels of aragonite alteration have minimal effects on coral $\delta^{18}O$ (McGregor and Gagan, 2003).

8.4.4 Results on Historical Timescales

Measurement of oxygen isotopic variations in coral heads up to 500 yr old has to be counted among the great success of the geochemical approach to paleoclimate research (Figure 2). These records have become important paleoclimatic archives of tropical climate change, and they have been incorporated into historical climate records used to assess global warming in the past century (Mann *et al.*, 1998; Crowley, 2000a). Some of the outstanding findings from the coral records include: (i) most of the longer records show a secular shift to more negative $\delta^{18}O$ values starting in the nineteenth century; (ii) a series of decadal or longer coherent shifts in the nineteenth century that might reflect regional to global cooling patterns; and (iii) shifts in the

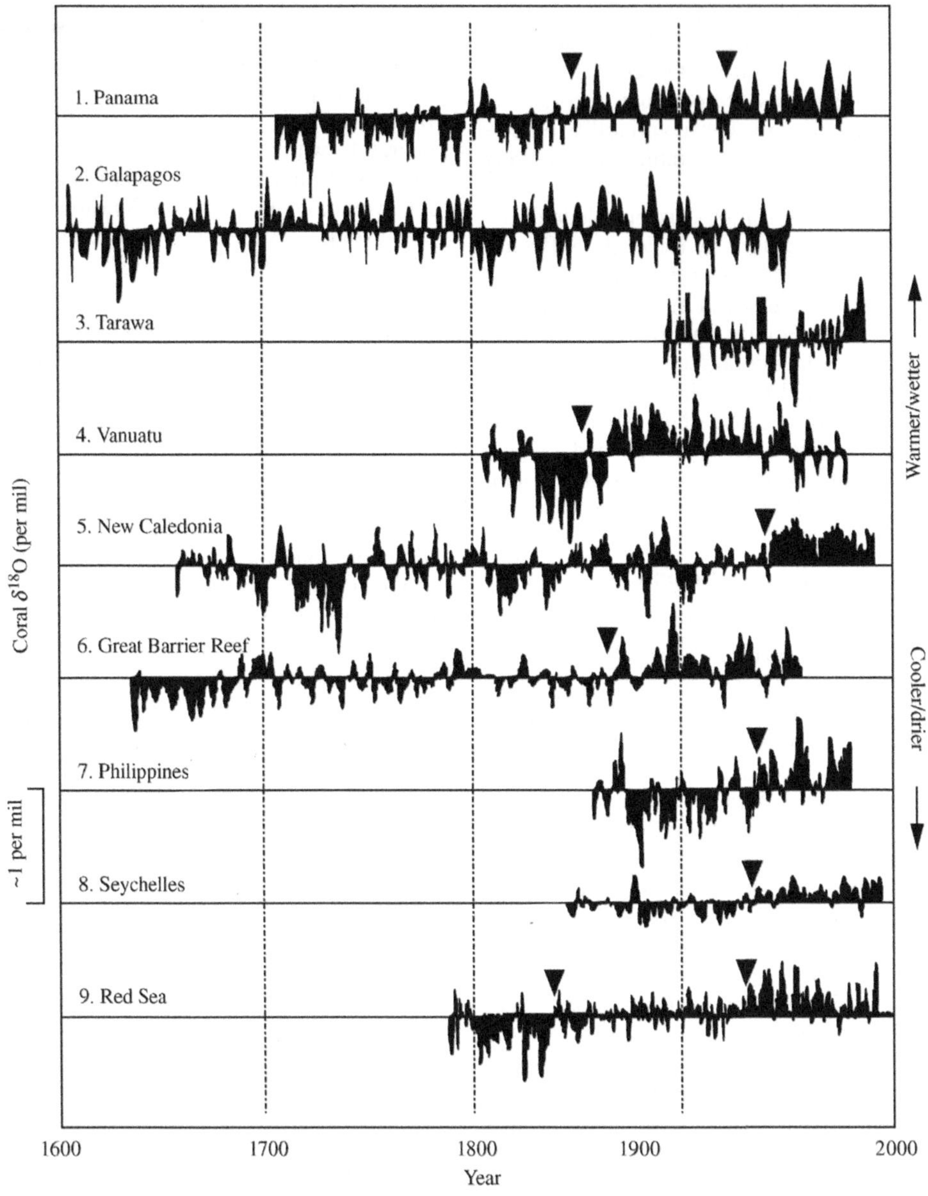

Figure 2 Comparison of annual mean coral oxygen isotope records in the Pacific and Indian Ocean region extending back more than 100 yr (Gagan *et al.*, 2000) (reproduced by permission of Elsevier from *Quat. Sci. Rev.* **2000**, *19*, 45–64). These records demonstrate the considerable potential of this approach for documenting historical climate variation in the tropical oceans.

magnitude and frequency of the El Niño/Southern Oscillation (ENSO) phenomenon and Indian Ocean monsoon over the past two centuries (Cole *et al.*, 2000; Urban *et al.*, 2000; Cobb *et al.*, 2001).

These coral records have great value as "generic" proxy climate records, in the same sense that tree ring records have been used without an explicit attribution of the observed variations to temperature, precipitation, etc. Because these records are based on a geochemical parameter that follows thermodynamic rules, it should be possible to eventually extract true temperature and/or salinity records from the coral time series. Another potential complication, however, is that the coral records might reflect biological effects such as gradual growth into shallower waters as coral heads grow (Gagan *et al.*, 2000). Many coral heads show a secular shift to more negative δ^{18}O values in the most modern period of growth, a result generally attributed to warming of surface waters in response to anthropogenic factors (Gagan *et al.*, 2000). But in at least some cases, this shift appears to be larger than can explained by SST shifts recorded by instrumental records (the records in Urban *et al.*, 2000; Cobb *et al.*, 2001 are good examples), so this trend either reflects a coincident decline in surface salinity or the aforementioned biological factors or perhaps undetected secondary aragonite precipitation in the oldest parts of the coral (Muller *et al.*, 2001).

8.4.5 Results on Late Quaternary Timescales

With the realization that oxygen isotopes in coral heads record subannual ocean climate variations came the idea of using such records from fossil corals to reconstruct both absolute and relative climate change for past geological periods (Fairbanks and Matthews, 1978). Because fossil corals from emerging coastlines have been exposed to meteoric fluids and weathering, this approach requires consideration of the potential for diagenetic changes (see Section 8.4.3). Although the complications of using fossil corals as paleoclimate tools are greater, the information to be gained is of great importance because it applies to climate systems under different boundary conditions (Tudhope *et al.*, 2001).

The first generation of studies on oxygen isotopes in fossil corals attempted to use absolute differences as a gauge of changes in mean ocean $\delta^{18}O$ in response to continental glaciation and mean SST changes. Early studies used coral $\delta^{18}O$ to better define the relationship between sea-level change and mean ocean $\delta^{18}O$ changes (see Section 8.3.4) (Fairbanks and Matthews, 1978). A number of studies published during the 1990s use coral $\delta^{18}O$ to try to establish SST history for the tropics (Guilderson *et al.*, 1994, 2001; McCulloch *et al.*, 1999). These studies suggest quite large glacial cooling of 4–6 °C for both tropical Atlantic and Pacific SST. Such large cooling is generally not supported by other approaches (Bard *et al.*, 1997; Lea *et al.*, 2000; Nürnberg *et al.*, 2000). Obviously the coral $\delta^{18}O$ approach depends heavily on knowledge of the $\delta^{18}O$ of local seawater, which will shift with both ice volume and local changes in E versus P (see Section 8.3). Sr/Ca paleothermometry in corals provides a way around this problem, but this approach appears to have its own set of limitations (see Section 8.8.3).

More recent studies are focusing on the climate variability encoded in annual and subannual fossil coral $\delta^{18}O$ (Hughen *et al.*, 1999; Tudhope *et al.*, 2001). This approach does not require separating the oxygen-isotope signal into its components, but rather uses the coral $\delta^{18}O$ signal as a climate proxy, with particular attention to the spectral characteristics of the time series. This approach is also less subject to diagenetic constraints, because corals that maintain distinct seasonal signatures are likely to be relatively unaltered. The results of these studies have been quite impressive in demonstrating that the nature of ENSO variability has been different under varying geological boundary conditions. The Tudhope *et al.* (2001) study documented coral climate proxy variability for seven different time slices. These records demonstrate that the amplitude of ENSO variability (2–7 yr band) has generally been weaker in the geological past relative to the twentieth century. The amplitudes appear to have been weakest in the mid-Holocene (~6.5 ka) and during most of the cold glacial episodes (Tudhope *et al.*, 2001). These records are by necessity fragmentary and only comprise a short window into ENSO variability in the past. These records also do not address the question of changes in ENSO frequency in the past, as has been suggested by studies of other climate proxies (e.g., Rodbell *et al.*, 1999).

8.4.6 Summary of Outstanding Research Issues

High-resolution coral oxygen isotope records have to be counted among the great successes of the geochemical approach to paleoclimate research. This is ironic given that early researchers were highly skeptical about the fidelity of coral $\delta^{18}O$ because of the clear lack of equilibrium (Emiliani *et al.*, 1978). Although coral $\delta^{18}O$ is clearly a valuable indicator of climate history, many challenges remain in direct assignment of the observed trends to an absolute history of temperature and salinity. High priorities for future research include establishing a coincident temperature proxy such as Sr/Ca (see Section 8.8) and determining the degree to which factors associated with the growth of large coral heads might influence longer-term records.

8.5 OXYGEN ISOTOPES AS A CLIMATE PROXY IN OTHER MARINE BIOGENIC PHASES

Oxygen isotopes have been used as temperature or climate proxies in a number of other marine biogenic phases, although far less widely than in foraminifera or reef corals. Probably the most important work has been on oxygen isotopes in diatom opal (Shemesh *et al.*, 1992, 1994, 1995). Because many sites in the Southern Ocean contain virtually no carbonate, opal $\delta^{18}O$ becomes critical for both stratigraphic and paleoclimatological purposes. Unfortunately, the systematics of oxygen isotopes in diatoms appears to be considerably more complex than for carbonates (Labeyrie and Juillet, 1982; Juillet-Leclerc and Labeyrie, 1986). Oxygen isotopes have also been measured as temperature or climate proxies in ahermatypic solitary corals (Smith *et al.*, 1997), in coralline sponges (Böhm *et al.*, 2000), in fish otoliths (Andrus *et al.*, 2002), as well as in pteropods (Grossman *et al.*, 1986) and other mollusks.

8.6 MAGNESIUM AS A PALEOTEMPERATURE PROXY IN FORAMINIFERA

8.6.1 Background and History

At the time of this writing (Spring 2002), research in the use of Mg/Ca ratios in foraminifer shells is probably advancing as fast as any area in climate proxy research. As a result of these advances since the late 1990s, researchers now have a good idea of the main advantages and limitations of this approach. It is fair to say that this new paleothermometry approach, perhaps more than any other, is revolutionizing the means by which paleoceanographers and paleoclimatologists unravel ocean and climate history. For this reason, I review the history of this development more closely than for the other proxies.

As discussed in Section 8.2, the observation that magnesium was higher in marine carbonates precipitated in water warmers dates to the early part of the twentieth century. Several studies confirmed these early observations for neritic foraminifer shells, which are composed of high magnesium calcite (>5% $MgCO_3$) (Chave, 1954; Chilingar, 1962). Another study suggested that pelagic foraminifera, which are composed of low-magnesium calcite (<1% $MgCO_3$), might also follow this pattern (Savin and Douglas, 1973). Several studies also demonstrated that inorganic carbonates followed the same pattern (Chilingar, 1962; Katz, 1973). This latter point is important, because it indicates that the temperature influence could not be entirely biological.

Another important milestone in research in this area came with the recognition that dissolution on the seafloor or within the sediments could significantly alter the Mg/Ca ratio of foraminifer shells (Bender *et al.*, 1975; Hecht *et al.*, 1975; Lorens *et al.*, 1977). Lorens *et al.* (1977) went so far as to state that "*diagenesis rules out* using Mg/Ca ratios of whole tests as growth temperature indicators (italics added for emphasis)." Despite this clear hindrance, studies documenting systematic down-core variations (Cronblad and Malmgren, 1981) and possible links to growth temperature (Delaney *et al.*, 1985) kept interest in the possibility of this proxy's usefulness alive. Several studies in the 1990s confirmed the early observations of a dissolution effect and species rankings (Rosenthal and Boyle, 1993; Russell *et al.*, 1994; Brown and Elderfield, 1996). It was not until Dirk Nürnberg, a doctoral student at Bremen University, Germany, used electron microprobe determinations on shell surfaces to documented more convincing Mg/Ca–temperature relationships in cultured, core-top and down-core planktonic foraminifera (Nürnberg, 1995; Nürnberg *et al.*, 1996a, b), that the international community recognized the potential importance of this new tool. Progress on documenting a potential response of magnesium in benthic foraminifera to bottomwater temperatures, a result presaged by Scot Izuka's (University of Hawaii) pioneering study of magnesium in *Cassidulina* (Izuka, 1988), occurred at about the same time (Russell *et al.*, 1994; Rathburn and Deckker, 1997; Rosenthal *et al.*, 1997). The Rosenthal *et al.* (1997) paper is notable for its broad calibration and for being the first to point out that the magnesium relationship to temperature is predicted, albeit with a smaller slope, by thermodynamic calculations.

Progress has been very rapid since these initial findings, in part because of improvements in analytical instrumentation and methods. Milestones include the first attempt to deduce glacial tropical SSTs using Mg/Ca (Hastings *et al.*, 1998), the first culturing calibrations made on whole shells (Lea *et al.*, 1999), the first attempt to combine Mg/Ca paleotemperatures with oxygen isotopic ratios to deduce variations in δ^{18}O-seawater (Mashiotta *et al.*, 1999; Elderfield and Ganssen, 2000), the first long tropical SST and $\delta^{18}O_{water}$ records (Lea *et al.*, 2000), the first application of benthic magnesium to Cenozoic climate evolution (Lear *et al.*, 2000), and the first detailed Late Quaternary benthic magnesium records (Martin *et al.*, 2002). The following sections detail the most important of these findings and research priorities for the future.

8.6.2 Calibration and Paleotemperature Equations

The underlying basis for magnesium paleothermometry is that the substitution of magnesium in calcite is endothermic and therefore is favored at higher temperatures. The enthalpy change for the reaction based on the more recent thermodynamic data is 21 kJ mol^{-1} (Koziol and Newton, 1995), which can be shown, using the van't Hoff equation, to equate to an exponential increase in Mg/Ca of 3% per °C (Lea *et al.*, 1999). The thermodynamic prediction of an exponential response is one of the reasons that magnesium paleotemperature calibrations are generally parametrized this way. Available inorganic precipitation data generally follows the thermodynamic prediction (Chilingar, 1962; Katz, 1973; Burton and Walter, 1987; Mucci, 1987), with the most extensive data set (Oomori *et al.*, 1987) yielding a 3.1±0.4% per °C increase in D_{Mg} for calcites precipitated in seawater over 10–50 °C (all responses given as percentages are calculated as exponentials, with 95% CI).

Foraminifera shells differ from the thermodynamic prediction in two fundamental ways. First, foraminifera contain 5–10 times lower magnesium than predicted from thermodynamic calculations

(Bender *et al.*, 1975). Second, the response of shell magnesium to temperature is ~3 times larger than the thermodynamic prediction and inorganic observation, averaging 9 ± 1% per °C (Lea *et al.*, 1999). Why the latter effect is so is not known, but it has several important implications for magnesium paleothermometry. First, it increases the sensitivity of this approach, which is critical in determining its real error, which depends on large part on the relative magnitude of the temperature response versus that of all the combined sources of error, including measurement error, population variability, and secondary effects. Second, it raises the question of why the response is so much greater in foraminifera and if the augmentation of the response depends on secondary factor (s) that might change over geological time. One possibility is that the much smaller magnesium content of foraminifer shells increases the thermodynamic response (Figure 3). Data from a recent study (Toyofuku *et al.*, 2000), which calibrated two neritic high-magnesium benthic species in culturing experiments, suggest that Mg/Ca in these species increases by between 1.8% per °C and 2.6% per °C, a far smaller increase than is observed for low magnesium foraminifera. The magnesium response to temperature found by Toyofuku *et al.* (2000) is actually much closer to the ~3% per °C observed for inorganic calcite (Oomori *et al.*, 1987), which contain magnesium contents similar to neritic benthics. This correspondence suggests the magnesium response to temperature might scale with the magnesium contents of calcite

At present, three planktonic species, *Globigerinoides sacculifer, Globigerina bulloides*, and *O. universa*, have been calibrated by culturing and fit with equations of the form

$$\mathrm{Mg/Ca}\left(\mathrm{mmol\ mol^{-1}}\right) = be^{mT} \qquad (4)$$

where b is the pre-exponential constant, m the exponential constant, and T the temperature (Lea *et al.*, 1999). Fitting Mg/Ca–temperature data with an equation of this form has the dual advantage of allowing for an exponential response while also parametrizing, by the use of the natural logarithm e, the exponential constant as the change in Mg/Ca per °C. It should be noted that it is the exponential constant that determines the magnitude of temperature change calculated from down-core variations

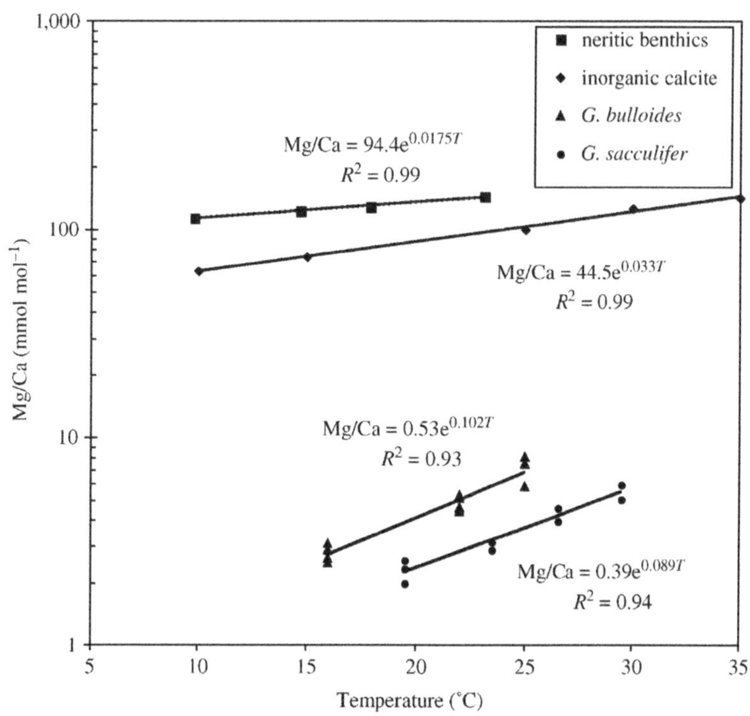

Figure 3 Comparison of Mg/Ca–temperature relationships for inorganic calcite precipitation (Oomori *et al.*, 1987), neritic benthic foraminifera (Toyofuku *et al.*, 2000), a tropical spinose symbiont bearing planktonic foraminifera, *G. sacculifer* (Nürnberg *et al.*, 1996a,b), and a subpolar spinose symbiont barren planktonic foraminifera, *G. bulloides* (Lea *et al.*, 1999). All of the foraminifera results are from culturing. Mg/Ca is plotted on a log scale because of the wide range of values. All of the relationships are fit with an exponential. Note that high Mg inorganic and benthic calcite has a shallower slope and much smaller exponential constant (2–3%); the low Mg foraminiferal calcite has a steeper slope and higher exponential constant (9–10%). Low Mg benthic foraminifera have exponential constants of ~10% (source Rosenthal *et al.*, 1997).

in Mg/Ca and the pre-exponential constant that determines the absolute temperature. Calibration results for these three species indicates exponential constants between 0.085 and 0.102, equivalent to 8.5% to 10.2% increase in Mg/Ca per °C (Nürnberg *et al.*, 1996a,b; Lea *et al.*, 1999). A recent study utilizing planktonic foraminifera from a sediment trap time series off Bermuda extends calibration to seven other species, which in aggregate have a temperature response of $9.0 \pm 0.3\%$ (Anand *et al.*, 2003). The pre-exponential constant *b* ranges between 0.3 and 0.5, with the exception of higher values for *O. universa*, which appears to be unique in many aspects of its shell geochemistry (Nürnberg *et al.*, 1996a,b; Lea *et al.*, 1999; Anand *et al.*, 2003). Core-top calibrations are in general agreement with the culturing results, and include calibrations of eight planktonic species (Elderfield and Ganssen, 2000; Lea *et al.*, 2000; Dekens *et al.*, 2002; Rosenthal and Lohmann, 2002). The Dekens *et al.* (2002) calibrations, which include a second term to account for dissolution in the form of a water depth or saturation effect, are discussed in Section 8.6.3.

Calibration for benthic species is somewhat more uncertain. The first comprehensive calibration was carried out by Yair Rosenthal, then at MIT, who used *Cibicidoides* spp. from shallow sediments on the Bahamas outer bank to calibrate benthic magnesium between 5 °C and 18 °C (Rosenthal *et al.*, 1997). When this calibration is augmented with *C. wuellerstorfi* data from deeper sites and adjusted for an analytical offset between atomic absorption spectrophotometry and ICP-MS (Martin *et al.*, 2002), the calibration yields

$$\mathrm{Mg/Ca} = 0.85\mathrm{e}^{0.11T} \tag{5}$$

The value of the exponential constant, 0.109 ± 0.007 (95% CI), overlaps with estimates from planktonic species, suggesting that a magnesium response to temperature of this magnitude is a common factor among foraminifera. It should be noted that the Martin *et al.* (2002) data set suggests that the magnesium response might be steeper at bottomwater temperatures, <4 °C. Resolving the calibration for benthic magnesium at the coldest temperatures is an important research priority because there are is a great deal of research interest in establishing the temperature evolution of cold bottomwaters. Establishing calibrations for other species is also a high priority, in part because of the insight this provides into the basis for magnesium paleothermometry.

8.6.3 Effect of Dissolution

It has been known since the 1970s that the magnesium content of foraminifer shells, as well as other carbonates, is susceptible to change via dissolution (Hecht *et al.*, 1975; Lorens *et al.*, 1977). As mentioned previously, this factor was one of the main reasons that little hope was held out for the usefulness of foraminiferal magnesium as a paleotemperature proxy. At present, researchers accept that dissolution alters the Mg/Ca content of foraminifera shells and instead are investigating the degree to which such changes occur, how dissolution can be assessed and whether correction factors are possible, and the degree to which dissolution affects Mg/Ca and oxygen isotopes similarly or dissimilarly.

In the mid-1990s, a number of groups measured Mg/Ca in planktonic foraminifera from oceanic depth transects, mostly as support for studies of other metals (F, U, and V) in the shells (Rosenthal and Boyle, 1993; Russell *et al.*, 1994; Brown and Elderfield, 1996; Hastings *et al.*, 1996). The advantage of the depth transect approach is that one can assume that shells with similar compositions rain down from overlying surface waters to all the sites, and that observed differences must be due to post-depositional processes. These studies demonstrated, to a greater or lesser degree, that Mg/Ca in the shells decreased with water depth and inferred increasing dissolution. The Rosenthal and Boyle (1993) study in particular documented both the general relationship between Mg/Ca and $\delta^{18}O$ as well as the drop in Mg/Ca with water depth in both spinose and nonspinose species. In general, these studies indicated that the drop in Mg/Ca was more pronounced for nonspinose species such as *Globorotalia tumida*, a result that was interpreted to reflect preferential dissolution of magnesium-rich chamber calcite over magnesium-poor keel calcite (Brown and Elderfield, 1996). One of these studies also revived the idea, first suggested by Savin and Douglas (1973), that the magnesium content of the shells influenced their solubility (Brown and Elderfield, 1996). Calculations suggest that the saturation horizon for ontogenetic calcite with Mg/Ca of 10 mmol mol^{-1}, about twice the value found in typical tropical shells, could be 300 m shallower. Magnesium loss presumably occurs when shells are on the seafloor and/or when they pass through the sediment mixed layer where metabolic CO_2 is available for dissolution. The fact that surface dwelling *Globigerinoides ruber* indicates decreasing Mg/Ca with water depth in the western equatorial Pacific (Lea *et al.*, 2000; Dekens *et al.*, 2002), an area with minimal temporal and spatial variation in mixed layer temperatures, suggests that magnesium loss might occur via preferential dissolution of magnesium-rich portions of the shell (Lohmann, 1995; Brown and Elderfield, 1996). It is also quote possible, however, that the progressive loss of the less robust individuals, which might have preferentially calcified in the warmest waters, shifts the mean Mg/Ca to lower values in deeper sediments.

A clear complexity in utilizing shell Mg/Ca for paleotemperature is that many species migrate vertically and/or add gametogenic calcite at depths significantly deeper than their principal habitat depth (Bé, 1980). This complicates the dissolution question, because these different shell portions are likely to have slightly different solubilities and Mg/Ca ratios. An innovative approach to this problem was suggested by Rosenthal *et al.* (2000), who argued that the relationship between size-normalized shell mass and dissolution loss could be used to correct shell Mg/Ca. This approach, which relies on a constant relationship between shell mass changes and Mg/Ca changes, has yet to be validated in down-core studies, although a new study (Rosenthal and Lohmann, 2002), demonstrates that this approach can yield consistent glacial–interglacial SST changes from cores both above and below the lysocline.

A somewhat different approach has been taken by others (Lea *et al.*, 2000; Dekens *et al.*, 2002). They quantified the Mg/Ca loss in depth transects as a percentage loss per kilometer water depth, thus allowing direct comparison of the magnitude of potential dissolution loss versus the magnitude of the temperature effect. If independent estimates of past shifts in lysocline depth are available, it is then possible to estimate the magnitude and direction of dissolution bias down-core. Dekens *et al.* (2002) found, based on core tops from the tropical Atlantic and Pacific, that Mg/Ca loss ranged from 3% per km water depth for *G. sacculifer*, 5% for *G. ruber*, and 22% for *N. dutertrei*, a nonspinose thermocline dweller. These terms equate to a bias on magnesium paleothermometry of 0.4 °C per km, 0.6 °C per km, and 2.8 °C per km effective shift, respectively, in foraminiferal lysocline, or depth of effective dissolution. Given that evidence for late Quaternary lysocline shifts is generally between 0.2 km and 0.8 km (Farrell and Prell, 1989), this approach suggests that down-core dissolution biases on magnesium paleothermometry will be less than 0.5 °C for spinose surface dwellers. Calibration equations derived from the Dekens *et al.* (2002) calibration set are also parametrized using Δcarbonate ion (the difference between *in situ* and saturation values) to account for differences in dissolved carbonate ion between basins.

The evidence for dissolution effects on magnesium in benthic foraminifera is less certain. For one, it is more difficult to discern a dissolution trend because benthic Mg/Ca is decreasing with increasing water depth and decreasing bottomwater temperature. Data from a depth transect on the Ontong Java Plateau for Sr/Ca, Ba/Ca, and Cd/Ca have been used to infer a dissolution effect on these elements (McCorkle *et al.*, 1995), although alternative interpretations such as carbonate ion or pressure effects on biomineralization have also been suggested (Elderfield *et al.*, 1996). Martin *et al.* (2002) suggested that the steeper trend of benthic Mg/Ca in the coldest waters, estimated at ~20% per °C versus 11% per °C, might reflect dissolution and magnesium loss in the deepest, most undersaturated waters. Alternatively, it might reflect the influence of other factors, such as carbonate ion saturation. Regardless, this will be a critical issue in validating benthic Mg/Ca for use on the coldest bottomwaters.

8.6.4 Other Secondary Effects: Salinity, pH, Gametogenesis, and Changes in Seawater Mg/Ca

Factors other than temperature and dissolution also appear to influence Mg/Ca in planktonic shells. Based on culturing, there is clear evidence for differences in uptake between species (Lea *et al.*, 1999), with as much as a factor of two variations. For this reason, species-specific calibrations are necessary, although it is difficult to do this by any means other than culturing because of the complication of habitat depth. Salinity appears to exert a small effect on shell Mg/Ca, with an observed increase of between 6 ± 4% for *O. universa* (Lea *et al.*, 1999) and 8 ± 3% for *G. sacculifer* (Nürnberg *et al.*, 1996a) per salinity unit (SU) increase. (Note: this and all other relationships of this kind are quoted as the exponential constant, with 95% confidence intervals, for an exponential fit to the observational data; the original published data was not always fit this way. An exponential fit has the advantage of giving the response in terms of a constant percentage, which then can be easily related to the exponential constant in the temperature response equation.) Assuming an Mg/Ca temperature response of 10% per °C (see Section 8.6.2), the salinity influence is equivalent to a positive bias of between 0.6–0.8 °C per SU increase. More extensive culturing data is needed; however, before such an influence can be accepted as likely to apply for salinity differences of <3.

Investigation of the effect of seawater pH indicates that pH has a significant effect on magnesium uptake, with an observed decrease of −6 ± 3% per 0.1 pH unit increase for *G. bulloides* and *O. universa*. Again assuming an Mg/Ca temperature response of 10% per °C (Section 8.6.2), pH influence is equivalent to a bias −0.6 °C per 0.1 pH unit increase. Past variability in oceanic pH (Sanyal *et al.*, 1996) and water-column variability in pH could therefore both exert significant biases on Mg paleothermometry.

A bias that applies equally to any foraminifera-based proxy is the problem of gametogenic calcite addition in the subsurface, as well as other vertical migration effects. For magnesium, one early study claimed that gametogenic calcite from *G. sacculifer* cultures is highly enriched in magnesium (Nürnberg *et al.*, 1996a), but this observation has not confirmed by studies of shells from sediments

(Elderfield and Ganssen, 2000; Nürnberg *et al.*, 2000; Rosenthal *et al.*, 2000; Dekens *et al.*, 2002). The observation that Mg/Ca in *G. sacculifer*, a species known to add ~30% gametogenic calcite (Bé, 1980), is generally lower (Elderfield and Ganssen, 2000; Lea *et al.*, 2000; Dekens *et al.*, 2002) than in *G. ruber*, a species that adds little or no gametogenic calcite (Caron *et al.*, 1990), suggests that the addition of gametogenic calcite takes place in cold, subsurface waters and reduces Mg/Ca and inferred temperatures for those species that add significant shell calcite this way.

8.6.5 Results on Quaternary Timescales

Although magnesium paleothermometry has only been used for ~5 yr, it has already led to a number of important and unprecedented findings for paleoceanographic and paleoclimatic research. These include documenting the history of subpolar Antarctic SST variations (Mashiotta *et al.*, 1999; Rickaby and Elderfield, 1999), tropical Atlantic and Pacific SST changes (Hastings *et al.*, 1998; Elderfield and Ganssen, 2000; Lea *et al.*, 2000; Nürnberg *et al.*, 2000), and changes in bottomwater temperature in the Atlantic and Pacific (Martin *et al.*, 2002). Secondary products include δ^{18}O-seawater records for the sub-polar Antarctic (Mashiotta *et al.*, 1999), equatorial Pacific (Lea *et al.*, 2000, 2002), and, for five different planktonic species in one core, the tropical Atlantic (Elderfield and Ganssen, 2000). Several other high-resolution records from the tropical Pacific have been published (Koutavas *et al.*, 2002; Stott *et al.*, 2002; Rosenthal *et al.*, 2003).

Among these Mg/Ca results, perhaps the most important are those that are available for the tropics. Past SST changes in the tropics have been a contentious issue (Crowley, 2000b), mostly because the actual glacial–interglacial changes are relatively small (<5 °C) and therefore more difficult to detect unambiguously using either faunal or geochemical methods. The faunal approach, in particular, is hampered by the fact that glacial tropical assemblages in the warm pools are not very different from their interglacial counterparts (Crowley, 2000b). Even with re-examination and major refinements, the faunal approach does not yield significant cooling in the tropical warm pools (Mix *et al.*, 1999; Trend-Staid and Prell, 2002). The Mg/Ca approach works especially well in the tropics, because the calibration curve at warm temperatures shows the largest absolute change in Mg/Ca per °C (Figures 3 and 4). Oligotrophic tropical sites, which are poor candidates for the alkenone unsaturation paleotemperature approach, generally contain abundant specimens of *G. ruber* and *G. sacculifer*, which are well calibrated for Mg/Ca. These low productivity sites also have minimal potential for diagenetic changes, which removes one confounding factor for trace element work.

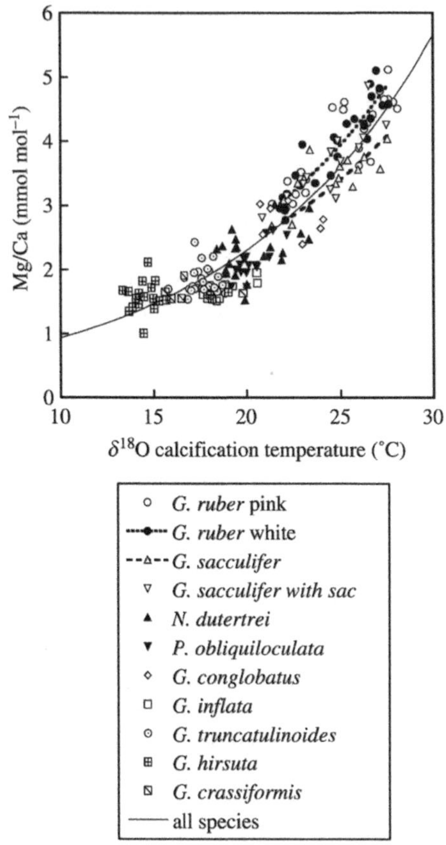

Figure 4 Mg/Ca of different planktonic foraminifera from a Bermuda sediment trap time series, plotted versus calcification temperatures calculated from the oxygen isotopic composition of the shells (Anand *et al.*, 2003) (reproduced by permission of American Geophysical Union from *Paleoceanography* **2003**, *18*, 1050). The aggregate fit to all the data in the plot is: $\text{Mg/Ca} = 0.38\exp(0.09T)$, very similar to relationships derived from culturing and core-top studies.

From this vantage point, it appears that Mg/Ca paleothermometry has cracked the problem of glacial cooling of the tropical warm pools (Hastings *et al.*, 1998; Lea *et al.*, 2000), although it must be said that the modest but systematic degree of cooling recorded by Mg/Ca was presaged by earlier results from the alkenone unsaturation technique (Lyle *et al.*, 1992; Bard *et al.*, 1997; Pelejero *et al.*, 1999). But the Mg/Ca results put the ~3 °C level of glacial cooling, relative to modern conditions, on a very firm footing, especially for the western Pacific warm pool (Lea *et al.*, 2000; Stott *et al.*, 2002; Visser *et al.*, 2003; Rosenthal *et al.*, 2003), which is the largest and warmest tropical water mass in the oceans. Results from a core on the Ontong Java Plateau, which lies on the equator in the center of the western Pacific warm pool, span the last 500 kyr and indicate that glacial cooling was systematically ~3 °C cooler than modern conditions and that this cooling occurred during each of the last five major glacial episodes (Figure 5). Of great interest is the fact that glacial warming appears to lead ice sheet demise by ~3 kyr (Lea *et al.*, 2000). This unanticipated SST lead, which suggests a prominent role for the tropics in pacing ice age cycles, has now also been observed in high-resolution records of the last deglaciation (Stott *et al.*, 2002; Rosenthal *et al.*, 2003).

The Mg/Ca approach has also led to a number of other important findings. One of the strengths of this approach is that the recorded paleotemperature is recorded simultaneously with the $\delta^{18}O$ composition of the shell. Combining these factors using an oxygen-isotope paleotemperature equation yields the $\delta^{18}O$-water at the time of shell precipitation. The limited number of studies using this approach already suggests that it is likely to yield important results on shifts in global ice volume as well as regional salinity shifts (Mashiotta *et al.*, 1999; Elderfield and Ganssen, 2000; Lea *et al.*, 2000, 2002; Stott *et al.*, 2002; Rosenthal *et al.*, 2003). A paleosalinity proxy has always been a difficult prospect, but it appears now that comparison of extracted $\delta^{18}O$-water will make it possible to reconstruct patterns of salinity change in the past (Lea, 2002). Such reconstructions will be invaluable in understanding paleoclimate shifts in the tropics.

Another strength of the magnesium paleothermometry is that it can be applied to benthic fauna, including ostracode shells (see Section 8.7). There are no other techniques that provide direct estimates of bottomwater temperatures. There is only one published study which contains detailed records of benthic foraminiferal Mg/Ca variations, from the eastern tropical Atlantic and Pacific, in the Quaternary (Martin *et al.*, 2002). These two records indicate the great promise of this approach in elucidating deep-water temperature variations, which appear to have been ~2–3 °C. However,

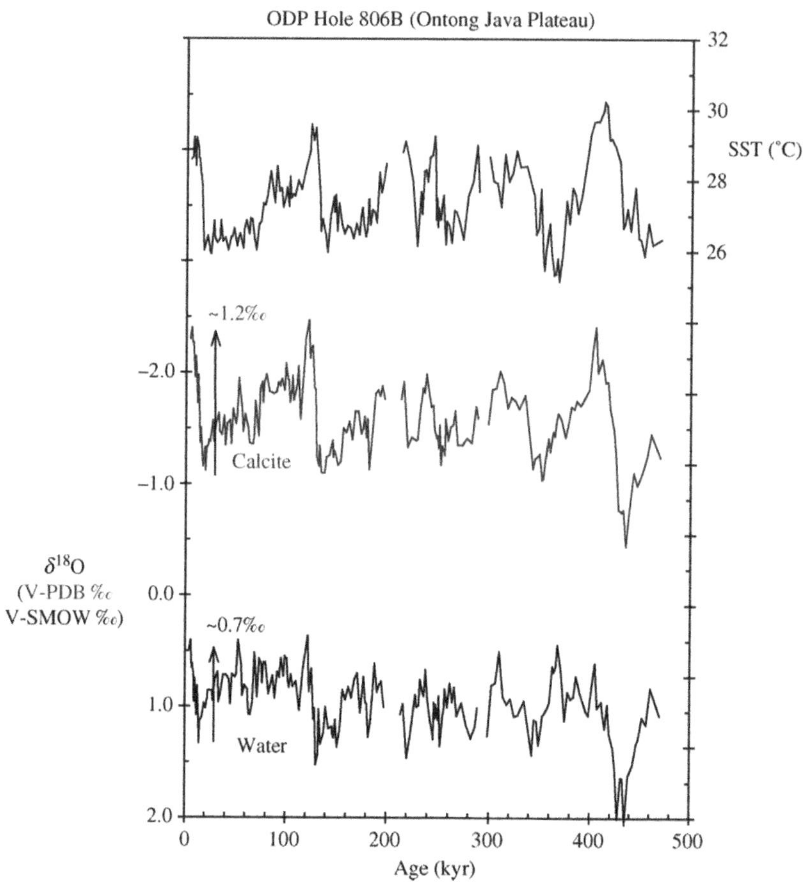

Figure 5 Down-core records of Mg/Ca-based SSTs, $\delta^{18}O$, and $\delta^{18}O_{water}$ derived from the surface dwelling planktonic foraminifera *G. ruber* in ODP Hole 806B on the Ontong Java Plateau in the western equatorial Pacific (Lea *et al.*, 2000). Glacial marine isotope stages (MIS) are indicated. Note the reduction, from 1.2' to 0.7', in glacial–interglacial $\delta^{18}O$ amplitude when the temperature portion of the signal is removed. The fact that the $\delta^{18}O_{water}$ amplitude in the WEP is smaller than the oceanic mean change is attributed to a hydrological shift to less saline surface waters during glacial episodes (Lea *et al.*, 2000; Rosenthal *et al.*, 2003; Oppo *et al.*, 2003). Note the lead of Mg/Ca over $\delta^{18}O$, especially prominent on the MIS 12 to 11 transition.

there are also considerable challenges. The absolute magnitude of the Mg/Ca change is much smaller in the cold-temperature region, and therefore other influences, such as vital effects, dissolution or calcification effects (Elderfield *et al.*, 1996), can exert significant biases. Separating these effects will undoubtedly be a major research area in the future.

8.6.6 Results for the Neogene

One of the most exciting prospects for magnesium paleothermometry is combining this approach with the benthic oxygen isotope curve for the Cenozoic (Zachos *et al.*, 2001) to separate the influence of temperature and ice volume. Two studies already suggest the great potential of this approach (Lear *et al.*, 2000; Billups and Schrag, 2002). The Lear *et al.* (2000) study, which is based on a data set extending back to the Eocene, reveals that benthic Mg/Ca records the gradual ~12 °C cooling of bottomwaters that occurred during the Cenozoic and that had been inferred from oxygen isotopes. Combining the Mg/Ca-based temperatures with measured $\delta^{18}O$ allows calculation of the $\delta^{18}O$ evolution of seawater, which can be traced to the expansion and contraction of ice sheets. Comparison of magnesium temperature trends with $\delta^{18}O$ over the Eocene–Oligocene boundary reveals that the $\delta^{18}O$ shifts are dominated by global ice volume shifts. There are significant uncertainties, such as species offsets, diagenesis, and changes in seawater Mg/Ca, in extending magnesium paleothermometry to longer timescales, but there are also great prospects for major discoveries. One can only await with anticipation the revelations yet to come when high-resolution benthic and planktonic Mg/Ca Neogene and Paleogene records are available from a number of sites!

8.6.7 Summary of Outstanding Research Issues

Magnesium paleothermometry is in the midst of a period of phenomenal growth, and it has quickly taken its place as one of the most useful means paleoceanographers have at their disposal to study past climates. Many questions, such as ecological bias, species offsets, environmental influences other than temperature, dissolution and diagenetic overprinting, must be addressed before the ultimate reliability of magnesium paleothermometry is known. One inherent advantage is the enormous amount already known about foraminiferal ecology and geochemistry, much of which applies equally to oxygen isotopes and Mg/Ca. At this stage, the most fundamental issues are: (i) establishing the spatial and temporal stability of magnesium temperature calibrations for the important paleoceanographic species; (ii) establishing the extent to which dissolution biases down-core Mg/Ca records; and (iii) establishing the degree to which benthic magnesium variations record temperature variations in the coldest part of the bottomwater temperature range (<4 °C).

8.7 MAGNESIUM AS PALEOTEMPERATURE PROXIES IN OSTRACODA

Magnesium paleothermometry applied to ostracode shells has proved to be an important means of discerning past variations in bottomwater temperatures (Dwyer *et al.*, 1995; Cronin *et al.*, 1996, 2000; Correge and Deckker, 1997). This approach is based on the same principle as magnesium paleothermometry in foraminifera, although ostracode calibrations have been fit with linear calibrations. These calibrations suggest that the increase in Mg/Ca in ostracodes is ~9% per °C, and therefore similar to the foraminiferal calibrations (see Section 8.6.2). The results in Dwyer *et al.* (1995) indicate that ostracode Mg/Ca can be used quite effectively to separate bottomwater temperature and $\delta^{18}O_{water}$ influences in both the Quaternary and Pliocene. In light of subsequent discoveries, it is interesting to note that Dwyer *et al.* (1995) saw a lead of temperature over benthic $\delta^{18}O$ in their Late Quaternary ostracode Mg/Ca record of 3,500 yr. Similarly, a study of magnesium in benthic foraminifera in a tropical Atlantic core saw a lead of ~4,000 yr in benthic magnesium of temperature over benthic $\delta^{18}O$ over the last 200 kyr (Martin *et al.*, 2002). These results highlight the importance of independent paleothermometers.

8.8 STRONTIUM AS A CLIMATE PROXY IN CORALS

8.8.1 Background

The idea of using Sr/Ca in corals as a paleothermometer goes back to the early 1970s, but it is one that that did not come to fruition until the early 1990s with the application of more precise analytical techniques. Early studies indicated that there was an inverse relationship between seawater temperature and strontium content of both inorganically precipitated and coral aragonite, with a relatively small inverse temperature dependence of just under 1% per °C (Kinsman and Holland, 1969; Weber, 1973; Smith *et al.*, 1979; Lea *et al.*, 1989). The breakthrough study for coral Sr/Ca, led by Warren Beck, then at University of Minnesota (Beck *et al.*, 1992), utilized extremely precise isotope dilution thermal ionization mass spectrometric (ID-TIMS) determinations to establish the relationship between Sr/Ca and temperature. Their calibration data indicated a 0.6% decrease in Sr/Ca per °C, and with determinations of ±0.03% (2 SD) possible by ID-TIMS, the Beck *et al.* (1992) approach indicated a possible paleotemperature determination of a remarkable ±0.05 °C! Along with their calibration data, Beck *et al.* (1992) presented Sr/Ca data from a fossil coral from Vanuatu that had been dated to the late Younger Dryas/Early Holocene period. These data indicated a 5.5 °C cooling of SST in this region, and it must be counted among the first strong evidence challenging the CLIMAP (1981) view of relatively unchanged tropical SST during glacial episodes.

Following the Beck *et al.* (1992) publication, a number of laboratories undertook more detailed studies of the calibration and also investigated the application of this approach to paleoceanography and paleoclimatology (de Villiers *et al.*, 1994, 1995; Guilderson *et al.*, 1994; McCulloch *et al.*, 1994; Gagan *et al.*, 1998). Generally these studies have supported the Beck *et al.* (1992) original insights, although two major problems with the Sr/Ca approach have been identified: (i) it appears that growth rate and symbiont activity have a marked influence on coral Sr/Ca (de Villiers *et al.*, 1994; Cohen *et al.*, 2001, 2002) and (ii) there is evidence for a secular shift in seawater Sr/Ca on glacial–interglacial timescales, with generally higher values during glacial episodes (Stoll and Schrag, 1998; Martin *et al.*, 1999; Stoll *et al.*, 1999). Another major step forward in the application of the Sr/Ca paleothermometer came with the development of a very rapid but still precise atomic absorption spectrophotometry method (Schrag, 1999b); this technique enables researchers to generate the large data sets required for long, high-resolution climate records (Linsley *et al.*, 2000).

8.8.2 Paleotemperature Equations

The relationship between coral Sr/Ca and seawater temperature is parametrized as a linear function of the form

$$\mathrm{Sr/Ca_{coral}}(\mathrm{mmol\ mol^{-1}}) = b + m(\mathrm{SST}) \qquad (6)$$

The thermodynamic prediction for strontium substitution in aragonite is actually an exponential response with an inverse temperature dependence, a consequence of the negative enthalpy (exothermic) nature of the reaction in which strontium substitutes for calcium in aragonite. The observed exponential constant for inorganic aragonite precipitation is quite small: −0.45% per °C (Kinsman and Holland, 1969). Therefore, over the small range of coralline strontium paleothermometry, the relationship can be quite adequately expressed as a linear relationship. The single inorganic aragonite precipitation study indicates a slope m of 0.039 and an intercept b of 10.66 mmol mol^{-1} (Kinsman and Holland, 1969). Calibrations are available for a number of coral species, but mostly for species of *Porites* (Smith *et al.*, 1979; Beck *et al.*, 1992; de Villiers *et al.*, 1994; Mitsuguchi *et al.*, 1996; Shen, 1996; Gagan *et al.*, 1998; Sinclair *et al.*, 1998; Cohen *et al.*, 2001, 2002). Values of the intercept b, which determines the absolute Sr/Ca for a particular temperature, range from 10.3 to 11.3; values of the slope m, which determines the temperature sensitivity, range from 0.036 to 0.086 (Figure 6). The variability in the slope is a critical problem for coral strontium paleothermometry, because the cited slopes equate to a variability in temperature dependence of −0.4% per °C to −1.0% per °C. Therefore, a recorded change of Sr/Ca in corals of 1% can imply between a 1 °C and 2.5 °C shift in paleotemperature. Of course, in practice, it is possible to narrow this uncertainty by conducting local calibrations (e.g., Correge *et al.*, 2000).

The key question is the degree to which the relationship between coral Sr/Ca and SST stays constant in time and space. There have been a number of investigations of this question (de Villiers *et al.*, 1994; Alibert and McCulloch, 1997; Cohen *et al.*, 2001, 2002), and the answer seems to be that growth and environmental factors clearly do affect Sr uptake. The more recent of these studies, based on ahermatypic solitary corals, indicates that algal symbionts might be the main influence on the slope, with an enhancement of 65% associated with enhance calcification during periods of strong symbiotic photosynthesis (Cohen *et al.*, 2002). If this result applies generally, it will have important implications for the use of corals in paleoclimate research, because most of the long records are based on the symbiont-bearing reef coral *Porites* (Schrag and Linsley, 2002). However, there is a lot of evidence that coral strontium paleothermometry works remarkably well, and the key question appears to be ascertaining the meaningfulness of long-term secular shifts on both historical and prehistoric timescales (see Sections 8.8.4 and 8.8.5) Figure 6.

8.8.3 Secondary Effects and Diagenesis

The main secondary effects on Sr/Ca paleothermometry appear to be related to growth rate and symbiosis (see Section 8.8.2). It is not yet known how these effects might influence the use of coral Sr/Ca for paleoclimate studies, but one imagines that the effects could be important if the growth conditions are changing over the lifetime of the coral. One of the most impressive long-term Sr/Ca records, from Rarotonga in the South Pacific (Linsley *et al.*, 2000), implies a large cooling in ~1760, early in the coral head's life history. This cooling does not appear in some

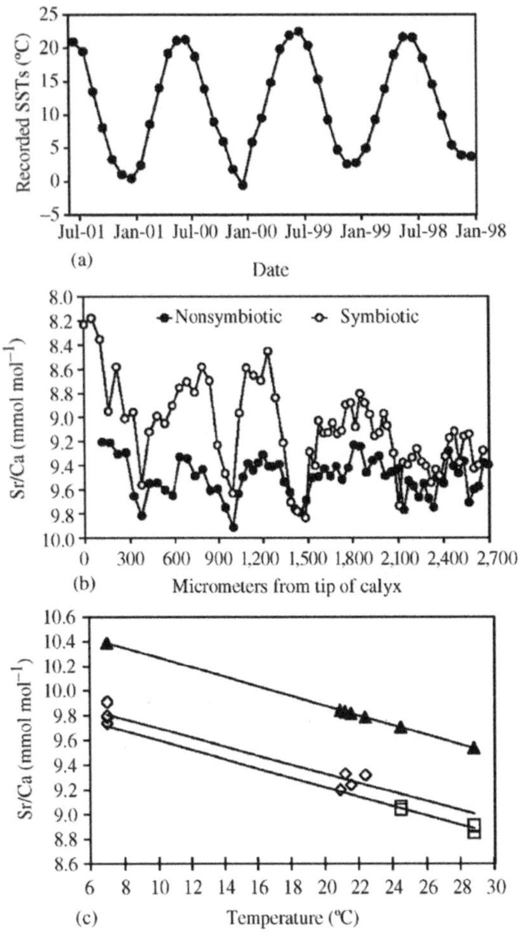

Figure 6 Comparison of Sr/Ca records in symbiont-bearing and symbiont barren ahermatypic corals (Cohen *et al.*, 2002). (a) Average monthly SSTs at 12 ft (4 m) depth in the Woods Hole harbor, the coral collection site, between January 1998 and July 2001. (b) Life history Sr/Ca profiles from symbiotic (open circles) and asymbiotic (solid circles) skeleton of *A. poculata* colonies collected in Woods Hole in July 2001. Skeletal Sr/Ca in the first year of life is the same in both samples, but the similarities decrease as the corallites mature, a divergence caused by a progressive decrease in summertime Sr/Ca in the symbiotic skeleton. (c) The Sr/Ca–SST relationship in asymbiotic *Astrangia* skeleton ($=-0.036x+10.065$) is compared with nighttime skeleton of the tropical reef coral *Porites* ($=-0.038x+9.9806$) and inorganic aragonite precipitated at equilibrium ($=-0.039x+10.66$). The slope of the regression equations, indicative of the temperature sensitivity of Sr/Ca uptake into the coral skeleton, are similar for all three precipitates (−0.036, −0.038, and −0.039, respectively). This agreement establishes temperature as the primary control of Sr/Ca in the asymbiotic skeleton (source Cohen *et al.*, 2002) (reproduced by permission of American Association for the Advancement of Science from *Science*, **2002**, *296*, 331–333).

other climate records (Cane and Evans, 2000) and could therefore reflect secondary effects.

The effect of diagenesis has not been widely studied for the Sr/Ca paleothermometer. For young corals growing during historical times, there is clear evidence that precipitation of inorganic aragonite in the pores of the oldest parts of the coral heads can affect the bulk coral Sr/Ca (Enmar *et al.*, 2000; Muller *et al.*, 2001). This secondary aragonite has a higher Sr/Ca ratio than the original coral material, and it therefore biases Sr/Ca paleotemperatures to colder values.

As subaerially exposed corals interact with meteoric waters, they are altered to calcite. This alteration results in a lost of strontium from the skeleton, and will obviously have strong effects on Sr/Ca paleothermometry. To avoid this problem, researchers routinely screen for the presence of calcite in fossil corals. A new study (McGregor and Gagan, 2003) demonstrates that local diagenesis can have marked affects on coral Sr/Ca, with a very large positive bias on reconstructed SST. This occurs because of the large drop in Sr/Ca that accompanies conversion of aragonite to calcite. Interestingly, the bias on oxygen isotopes is much smaller, mainly because the absolute difference in aragonite and calcite end-member compositions is much smaller for $\delta^{18}O$ (see Section 8.4.3). McGregor and Gagan (2003) suggest that such localized low-level diagenesis can be detected through a combination of X-ray diffraction techniques, thin section analysis, and high-resolution spatial sampling of the coral skeleton.

A unique secondary complication for Sr/Ca is the potential influence of small changes in seawater Sr/Ca. Although spatial variability in seawater Sr/Ca in the present ocean is very small ($\leq 2\%$) (Brass and Turekian, 1974; de Villiers *et al.*, 1994), the small sensitivity of the coral Sr/Ca paleothermometer makes it sensitive to these variations. For example, an observed 2% variation in coralline Sr/Ca, which equates to the maximum seawater variation, is equivalent to between a 2 °C and 5 °C temperature change, depending on the slope of the calibration (see Section 8.8.2). In practice, it is likely that most locations do not experience variations of more than 0.5% in seawater Sr/Ca (de Villiers *et al.*, 1994), but on historical timescales it is at least possible that larger shifts might have taken place.

On geological timescales, there is growing evidence that shifts in seawater Sr/Ca large enough to affect Sr/Ca paleothermometry have taken place. Such shifts were first hypothesized by a group at Harvard, who recognized that changes in sea level associated with changing continental ice volume had the potential to change seawater Sr/Ca, because calcium carbonate deposition on the continental shelves is dominated by aragonite, which contains ~5 times more strontium than calcite, which dominates deep-sea carbonate deposition (Stoll and Schrag, 1998), Lowered sea level during continental glaciation favors deposition in the deep sea, which therefore raises the Sr/Ca of seawater. Stoll and Schrag (1998) calculated that this change could result in a 1–2% enrichment of seawater Sr/Ca during of just after sea-level low-stands, which, depending on which calibration is used, would result in a −1 °C to −5 °C bias in Sr/Ca paleothermometry.

Subsequently, Pamela Martin (UCSB) and co-workers demonstrated that systematic glacial–interglacial variations in foraminiferal Sr/Ca similar to and even somewhat larger than those predicted by the Stoll and Schrag (1998) study are indeed preserved in deep-sea records (Martin *et al.*, 1999). Several studies have confirmed this observation (Stoll *et al.*, 1999; Elderfield *et al.*, 2000; Shen *et al.*, 2001). These studies indicate foraminifer shells record variations of up to 6% on glacial–interglacial timescales. Some of this variation is undoubtedly due to secondary (kinetic) effects on foraminiferal Sr/Ca, such as temperature, pH and salinity, all of which are known to have small influences on shell Sr/Ca (Lea *et al.*, 1999). But comparison of benthic records from different ocean basins suggests that there is a strong common signal in these records, with a common glacial–interglacial amplitude of ~3% (Martin *et al.*, 1999). Comparison of this stacked benthic record with coral Sr/Ca values of fossil corals suggests that up to half of the observed coral Sr/Ca variation might be attributable to seawater Sr/Ca variation (Martin *et al.*, 1999). This might explain why tropical SST drops during glacial episodes based on fossil coral Sr/Ca (Beck *et al.*, 1992; Guilderson *et al.*, 1994; McCulloch *et al.*, 1996, 1999) are typically twice that suggested by other geochemical proxies (Lea *et al.*, 2000). Regardless of the exact details, it is clear that secular changes in seawater Sr/Ca have the potential to influence the Sr/Ca paleothermometer on longer timescales.

8.8.4 Results on Historical Timescales

A few long time series of Sr/Ca have been published, from Rorotonga and the Great Barrier Reef (GBR) in the South Pacific (Linsley *et al.*, 2000; Hendy *et al.*, 2002). These records show clear interannual variability as well as distinct secular trends. For example, the GBR (Hendy *et al.*, 2002) records, which are an average of eight different coral cores, indicate a secular shift to warmer SST in the youngest part of the records (after ~1950). This shift is corroborated by $\delta^{18}O$ and U/Ca measurements in the same corals and appears to track well with instrumental records. This is quite important because of the question of attribution for the prominent trend towards more negative $\delta^{18}O$ observed in many large corals (see Section 8.4.4). Combining the metal paleotemperature records with the $\delta^{18}O$ record yields a residual $\delta^{18}O$-water record that suggests that GBR waters have become progressively less salty since the mid-nineteenth century. Results from the Rorotonga site (Linsley *et al.*, 2000) are somewhat different and suggest a prominent cooling in ~1750 followed by a series of decadal oscillations that correlate in the twentieth century with the Pacific Decadal Oscillation (Mantua *et al.*, 1997). The warm period recorded in the mid-eighteenth century at the Rorotonga site appears to be corroborated in the GBR sites, although with a reduced magnitude. Because this time interval was cold in much of the northern hemisphere, the warm South Pacific SSTs and high salinities might be an important clue to the source of what is know as the Little Ice age in the NH (Hendy *et al.*, 2002).

8.8.5 Results on Geological Timescales

The first detailed Sr/Ca results published for well-dated fossil corals (Beck *et al.*, 1992) were interpreted as indicating a 6 °C cooling of tropical SST in Vanuatu region of the South Pacific during the latest Younger Dryas/earliest Holocene. Although the degree of cooling has not been supported by other data, presumably because of a secular increase in seawater Sr/Ca (Stoll and Schrag, 1998; Martin *et al.*, 1999; Lea *et al.*, 2000), the Beck *et al.* (1992) result was one of the first to seriously challenge the prevailing view of warm tropical oceans during glacial episodes. Subsequently, Sr/Ca data from sites in the Caribbean (Guilderson *et al.*, 1994) and western Pacific (McCulloch *et al.*, 1999), apparently supported by coincident shifts in $\delta^{18}O$, were published to indicate 5–6 °C cooling of tropical SSTs during glacial episodes. Holocene changes in SST have also been reconstructed using Sr/Ca (Beck *et al.*, 1997; Gagan *et al.*, 1998). In retrospect, it appears that the glacial estimates of cooling were too large, in part because they would have rendered large parts of the tropical sea inhospitable to massive reef corals (Crowley, 2000b). In addition, terrestrial shifts such as the well-known drop in snowlines during the last glacial maximum are compatible with tropical SST drops of ~3 °C (Pierrehumbert, 1999). Pinpointing the exact cause of why coral Sr/Ca appears to give an excess cooling signature of glacial episodes is obviously an important research problem, but regardless of the exact causes of that offset, the tropical cooling results from coral Sr/Ca were a very important initial part of the motivation that lead to a growing focus on the paleoclimatic role of the tropics.

The spectral characteristics of Sr/Ca records in fossil corals can, like oxygen isotopes, provide insight into changes in interannual climate change such as ENSO without requiring conversion into absolute SSTs (Hughen *et al.*, 1999; Correge *et al.*, 2000). The Hughen *et al.* (1999) results are notable for providing evidence, in the form of interannual variations in Sr/Ca and $\delta^{18}O$, of ENSO-like variability in a fossil coral from Sulawesi, Indonesia dated to the last interglacial sea-level highstand, 124 ka.

8.8.6 Summary of Outstanding Research Issues

During the 1990s, the coral Sr/Ca paleothermometer has grown to be a fundamental tool for paleoclimate research on historical and geological timescales. Optimal use of this tool requires a better understanding of how the coral Sr/Ca temperature calibration is in both space and time. The most critical question is the degree to which symbiosis (and other kinetic factors) influence the sensitivity of the paleothermometer (Cohen *et al.*, 2002). If the Sr/Ca of coral material precipitated by hermatypic corals during the day is strongly biased by photosynthesis, as suggested by Cohen *et al.* (2002), it will place a severe limitation on both the usefulness and the accuracy of the Sr/Ca paleothermometer.

On geological timescales, the question of secular change in seawater Sr/Ca as well as diagenetic influence require further investigation. Foraminiferal Sr "stacks" of seawater Sr/Ca change (Martin *et al.*, 1999) could be improved by adding more cores and improving the precision of the analyses (Shen *et al.*, 2001). It might also be possible to correct for offsets between sites by taking into account environmental factors that also influence foraminiferal Sr/Ca. In this way it should be possible to eventually generate definite secular records of seawater Sr/Ca that can be suitably applied to coral Sr/Ca paleothermometry.

8.9 MAGNESIUM AND URANIUM IN CORALS AS PALEOTEMPERATURE PROXIES

The ratios of Mg/Ca and U/Ca in corals appear to serve as paleothermometers, although with apparently less fidelity than is found for Sr/Ca. The results for U/Ca (Min *et al.*, 1995; Shen and Dunbar, 1995) and Mg/Ca (Mitsuguchi *et al.*, 1996; Sinclair *et al.*, 1998; Fallon *et al.*, 1999) show convincing annual cycles, but with some complications; for example, Sinclair *et al.* (1998) observe, using laser ICP-MS analyses of coral surfaces, that there are high-frequency oscillations in Mg/Ca that range up to 50% of the total signal. In their comparative study, Sinclair *et al.* (1998) also observed differences in the seasonal profile of U/Ca and Sr/Ca, suggesting that other factors might be at play for uranium incorporation. Fallon *et al.* (1999) observed that magnesium incorporation tracked with SST but also evinced variability not related to temperature, suggesting that Mg/Ca paleothermometry in corals is not going to be as simple as it appeared in the initial study (Mitsuguchi *et al.*, 1996).

Two of these studies also looked at boron incorporation into the coral skeleton and observed that it also appears to be, at least in part, related to temperature (Sinclair *et al.*, 1998; Fallon *et al.*, 1999). The fact that at least four elements follow a seasonal pattern related to temperature suggests that elemental incorporation in coral skeletons is linked to calcification and is not simply driven by a thermodynamic temperature effect. If this applies generally, than all of the coral metal paleothermometers will have to be applied with attention to the possibility of distortions caused by growth factors.

8.10 CALCIUM ISOTOPES AS A PALEOTEMPERATURE PROXY

The possibility of using calcium isotopes for paleothermometry is a very new idea that is based on the empirical observation of temperature-related fractionation between the isotopes ^{40}Ca and ^{44}Ca (reported as $\delta^{44}Ca$). Measurements of precise calcium-isotopic variations are quite challenging (Russell *et al.*, 1978), and this has limited investigations of this isotopic system to relatively small data sets. This potential paleothermometer has been calibrated in neritic benthic foraminifera (De La Rocha and DePaolo, 2000), a spinose tropical planktonic foraminifera (Nägler *et al.*, 2000), and a subtropical planktonic foraminifera (Gussone *et al.*, 2003). In addition, Zhu and Macdougall (1998) compared three pairs of warm and cold foraminifera and demonstrated a systemic difference. The most convincing evidence of the potential utility of this new paleothermometer comes from the study of Nägler *et al.* (2000), which demonstrates an increase of 0.24‰ in shell $\delta^{44}Ca$ per °C, based on three cultured points of *G. sacculifer*. Down-core measurements from a core in the tropical Atlantic indicate that shells from glacial intervals are ~0.5–1.0‰ depleted in $\delta^{44}Ca$, consistent with colder glacial temperatures. Although the results in Nägler *et al.* (2000) indicate that calcium isotope paleothermometry has great promise, they are somewhat confounded by the fact a new study shows a much weaker response of $\delta^{44}Ca$ to temperature (~0.02‰ per °C) in a subtropical foraminifera, *O. universa*, calibrated over a wide temperature range by culturing (Gussone *et al.*, 2003). The weaker response in *O. universa* is mirrored by a similar response in inorganically precipitated aragonite (Gussone *et al.*, 2003). Given a measurement precision of ~0.12‰ for replicate samples (Gussone *et al.*, 2003), the slope for *G. sacculifer* allows for a resolution of ~0.5 °C in paleothermometry, whereas the slope for *O. universa* allows for a resolution of only ~5 °C. Obviously, the utility of this approach will rely heavily on which slope is more generally representative, and if that slope is stable in space and time.

8.11 CONCLUSIONS

Geochemists have, since the 1950s, already come up with a remarkable array of paleotemperature proxies in marine carbonates. These proxies work in diverse oceanic settings, in different organisms, in different parts of the water column, and on varied timescales. Each proxy has different strengths and weaknesses, and some of the proxies, such as Mg/Ca and oxygen isotopes in foraminifera, reinforce each other when applied together.

Perhaps most remarkable is the amount of progress that has been made since the mid-1980s on three new or revived paleothermometric approaches, each of which work particularly well in the tropics: oxygen isotopes in corals, Mg/Ca in foraminifera, and Sr/Ca in corals. This progress, in conjunction with advances in alkenone unsaturation paleothermometry, has not only expanded the importance of geochemistry in paleoclimate research, but changed its focus from mainly a chronostratigraphic and sea-level tool (i.e., oxygen isotopes in foraminifera) to a series of proxies that can be used to gauge the temporal and spatial history of oceanic temperatures. This shift, and, for example, the general level of agreement between geochemical paleothermometers for such fundamental problems as the cooling of the glacial tropics, suggests that major breakthroughs to long-standing paleoclimatological issues are now within reach. Although many problems and challenges remain, and although none of the available proxies work perfectly, it is clear that recent research progress has elevated geochemical paleothermometers to an even more fundamental role in quantitative paleoclimate research.

ACKNOWLEDGMENTS

The author thanks A. Cohen for sharing her compilation of coral strontium calibrations, and J. Cole, J. Clark, D. Schrag, and H. Spero for discussions. Some of this material is based upon work supported by the US National Science Foundation, more recently under grant OCE-0117886.

REFERENCES

Alibert C. and McCulloch M. T. (1997) Strontium/calcium ratios in modern porites corals from the Great Barrier Reef as a proxy for sea surface temperature: calibration of the thermometer and monitoring of ENSO. *Paleoceanography* **12**(3), 345–363.

Anand P., Elderfield H., and Conte M. H. (2003) Calibration of Mg/Ca thermometry in planktonic foraminifera from a sediment trap time series. *Paleoceanography* **18**(2), 1050, doi: 10.1029/2002PA000846.

Andrus C. F. T., Crowe D. E., Sandweiss D. H., Reitz E. J., and Romanek C. S. (2002) Otolith $\delta^{18}O$ record of mid-Holocene sea surface temperatures in Peru. *Science* **295**(5559), 1508–1511.

Bard E., Hamelin B., Arnold M., Montaggioni L., Cabioch G., Faure G., and Rougerie F. (1996) Deglacial sea-level record from Tahiti corals and the timing of global meltwater discharge. *Nature* **382**, 241–244.

Bard E., Rostek F., and Sonzogni C. (1997) Interhemispheric synchrony of the last deglaciation inferred from alkenone palaeothermometry. *Nature* **385**, 707–710.

Barron E. J. (1987) Eocene equator-to-pole surface ocean temperatures: a significant climate problem? *Paleoceanography* **2**, 729–739.

Bé A. W. H. (1980) Gametogenic calcification in a spinose planktonic foraminifer, *Globigerinoides sacculifer* (Brady). *Mar. Micropaleontol.* **5**, 283–310.

Beck J. W., Edwards R. L., Ito E., Taylor F. W., Recy J., Rougerie F., Joannot P., and Henin C. (1992) Sea-surface temperature from coral skeletal strontium/calcium ratios. *Science* **257**(5070), 644–647.

Beck J. W., Recy J., Taylor F., Edwards R. L., and Cabloh G. (1997) Abrupt changes in early Holocene Tropical sea surface temperature derived from coral records. *Nature* **385**, 705–707.

Bemis B. E., Spero H. J., Bijma J., and Lea D. W. (1998) Reevaluation of the oxygen isotopic composition of planktonic foraminifera: experimental results and revised paleotemperature equations. *Paleoceanography* **13**(2), 150–160.

Bender M. L., Lorens R. B., and Williams D. F. (1975) Sodium, magnesium, and strontium in the tests of planktonic foraminifera. *Micropaleontology* **21**, 448–459.

Billups K. and Schrag D. P. (2002) Paleotemperatures and ice volume of the past 27 Myr revisited with paired Mg/Ca and $^{18}O/^{16}O$ measurements on benthic foraminifera. *Paleoceanography* 10.1029/2000PA000567.

Böhm F., Joachimski M. M., Dullo W. C., Eisenhauer A., Lehnert H., Reitner J., and Wörheide G. (2000) Oxygen isotope fractionation in marine aragonite of coralline sponges. *Geochim. Cosmochim. Acta* **64**(10), 1695–1703.

Brass G. W. and Turekian K. K. (1974) Strontium distribution in GEOSECS oceanic profiles. *Earth Planet. Sci. Lett.* **23**, 141–148.

Brown S. and Elderfield H. (1996) Variations in Mg/Ca and Sr/Ca ratios of planktonic foraminifera caused by postdepositional dissolution—evidence of shallow Mg-dependent dissolution. *Paleoceanography* **11**(5), 543–551.

Burton E. A. and Walter L. M. (1987) Relative precipitation rates of aragonite and Mg calcite from seawater: temperature or carbonate ion control? *Geology* **15**, 111–114.

Cane M. A. and Evans M. (2000) Do the tropics rule? *Science* **290**, 1107–1108.

Caron D. A., Anderson O. R., Lindsey J. L., Faber J. W. W., and Lin Lim E. (1990) Effects of gametogenesis on test structure and dissolution of some spinose planktonic foraminifera and implications for test preservation. *Mar. Micropaleontol.* **16**, 93–116.

Chappell J. and Shackleton N. J. (1986) Oxygen isotopes and sea level. *Nature* **324**, 137–140.

Chave K. E. (1954) Aspects of the biogeochemistry of magnesium: 1. Calcareous marine organisms. *J. Geol.* **62**, 266–283.

Chilingar G. V. (1962) Dependence on temperature of Ca/Mg ratio of skeletal structures of organisms and direct chemical precipitates out of sea water. *Bull. Soc. CA Acad. Sci.* **61** (part 1), 45–61.

Clarke F. W. and Wheeler W. C. (1922) The inorganic constituents of marine invertebrates. *USGS Prof. Pap.* **124**, 55.

CLIMAP. (1981) Seasonal reconstructions of the Earth's surface at the last glacial maximum. In *The Geological Society of America Map and Chart Series* **MC-36**.

Cobb K. M., Charles C. D., and Hunter D. E. (2001) A central tropical Pacific coral demonstrates Pacific, Indian, and Atlantic decadal climate connections. *Geophys. Res. Lett.* **V28**(N11), 2209–2212.

Cohen A., Layne G., Hart S., and Lobel P. (2001) Kinetic control of skeletal Sr/Ca in a symbiotic coral: implications for the paleotemperature proxy. *Paleoceanography* **V16** (N1), 20–26.

Cohen A. L., Owens K. E., Layne G. D., and Shimizu N. (2002) The effect of algal symbionts on the accuracy of Sr/Ca paleotemperatures from coral. *Science* **296**(5566), 331–333.

Cole J. E. and Fairbanks R. G. (1990) The southern oscillation recorded in the $\delta^{18}O$ of corals from Tarawa Atoll. *Paleoceanography* **5**(5), 669–683.

Cole J. E., Dunbar R. B., McClanahan T. R., and Muthiga N. A. (2000) Tropical Pacific forcing of decadal SST variability in the western Indian Ocean over the past two centuries. *Science* **287**, 617–619.

Correge T. and Deckker P. D. (1997) Faunal and geochemical evidence for changes in intermediate water temperature and salinity in the western Coral Sea (northeast Australia) during the late Quaternary. *Palaeogeogr. Palaeoclimat. Palaeoecol.* **313**, 183–205.

Correge T., Delcroix T., Recy J., Beck W., Cabioch G., and Le Cornec F. (2000) Evidence for stronger El Nino-southern oscillation (ENSO) events in a Mid-Holocene massive coral. *Paleoceanography* **15**(4), 465–470.

Cronblad H. G. and Malmgren B. A. (1981) Climatically controlled variation of Sr and Mg in quaternary planktonic foraminifera. *Nature* **291**, 61–64.

Cronin T. M., Raymo M. E., and Kyle K. P. (1996) Pliocene (3.2–2.4 Ma) ostracode faunal cycles and deep ocean circulation. *North Atlantic Ocean. Geology* **24**(8), 695–698.

Cronin T. M., Dwyer G. S., Baker P. A., Rodriguez-Lazaro J., and DeMartino D. M. (2000) Orbital and suborbital variability in North Atlantic bottom water temperature obtained from deep-sea ostracod Mg/Ca ratios. *Palaeogeogr. Palaeoclimat.* **162**(1–2), 45–57.

Crowley T. J. (2000a) Causes of climate change over the past 1000 years. *Science* **289**(5477), 270–277.

Crowley T. J. (2000b) CLIMAP SSTs re-revisited. *Clim. Dyn.* **16**(4), 241–255.

De La Rocha C. L. and DePaolo D. J. (2000) Isotopic evidence for variations in the marine calcium cycle over the Cenozoic. *Science* **289**(5482), 1176–1178.

de Villiers S., Shen G. T., and Nelson B. K. (1994) The Sr/Ca-temperature relationship in coralline aragonite: influence of variability in $(Sr/Ca)_{seawater}$ and skeletal growth parameters. *Geochim. Cosmochim. Acta* **58**(1), 197–208.

de Villiers S., Nelson B. K., and Chivas A. R. (1995) Biological controls on coral Sr/Ca and $\delta^{18}O$ reconstructions of sea surface temperatures. *Science* **269**, 1247–1249.

Dekens P. S., Lea D. W., Pak D. K., and Spero H. J. (2002) Core top calibration of Mg/Ca in tropical foraminifera: refining paleo-temperature estimation. *Geochem. Geophys. Geosys.* **3**, 1022, doi: 10.1029/2001GC000200.

Delaney M. L., Bé A. W. H., and Boyle E. A. (1985) Li, Sr, Mg, and Na in foraminiferal calcite shells from laboratory culture, sediment traps, and sediment cores. *Geochim. Cosmochim. Acta* **49**, 1327–1341.

Druffel E. R. M. (1985) Detection of El Nino and decade timescale variations of sea surface temperature from banded coral records: implications for the carbon dioxide cycle. In *The Carbon Cycle and Atmospheric CO_2: Natural Variations Archean to Present.* American Geophysical Union, Washington, DC, pp. 111–122.

Dunbar R. B. and Wellington G. M. (1985) Stable isotopes in a branching coral monitor seasonal temperature variation. *Nature* **293**, 453–455.

Duplessy J.-C. and Blanc P.-L. (1981) Oxygen-18 enrichment of planktonic foraminifera due to gametogenic calicification below the euphotic zone. *Science* **213**, 1247–1250.

Duplessy J. C., Lalou C., and Vinot A. C. (1970) Differential isotopic fractionation in benthic foraminifera and paleotemperatures reassessed. *Science* **168**, 250–251.

Dwyer G. S., Cronin T. M., Baker P. A., Raymo M. E., Buzas J. S., and Correge T. (1995) North Atlantic deepwater temperature change during Late Pliocene and Late Quaternary climatic cycles. *Science* **270**, 1347–1351.

Elderfield H. and Ganssen G. (2000) Past temperature and $\delta^{18}O$ of surface ocean waters inferred from foraminiferal Mg/Ca ratios. *Nature* **405**, 442–445.

Elderfield H., Bertram C., and Erez J. (1996) Biomineralization model for the incorporation of trace elements into foraminiferal calcium carbonate. *Earth Planet. Sci. Lett.* **142**(3–4), 409–423.

Elderfield H., Cooper M., and Ganssen G. (2000) Sr/Ca in multiple species of planktonic foraminifera: implications for reconstructions of seawater Sr/Ca. *Geochem. Geophys. Geosys.* **1**, paper no. 1999GC00031.

Emiliani C. (1955) Pleistocene temperatures. *J. Geol.* **63**, 538–578.

Emiliani C., Hudson J., Shinn E., and George R. (1978) Oxygen and carbon isotopic growth record in a reef coral from the Florida keys and a deep-sea coral from Blake Plateau. *Science* **202**, 627–628.

Enmar R., Stein M., Bar-Matthews M., Sass E., Katz A., and Lazar B. (2000) Diagenesis in live corals from the Gulf of Aqaba: I. The effect on paleo-oceanography tracers. *Geochim. Cosmochim. Acta* **64**(18), 3123–3132.

Epstein S., Buchsbaum R., Lowenstam H. A., and Urey H. C. (1953) Revised carbonate-water isotopic temperature scale. *Geol. Soc. Am. Bull.* **64**, 1315–1325.

Erez J. and Luz B. (1983) Experimental paleotemperature equation for planktonic foraminifera. *Geochim. Cosmochim. Acta* **47**, 1025–1031.

Fairbanks R. G. (1989) A 17,000-year glacio-eustatic sea level record: influence of glacial melting rates on the Younger Dryas event and deep-ocean circulation. *Nature* **342**, 637–642.

Fairbanks R. G. and Dodge R. E. (1979) Annual periodicities in $^{16}O/^{16}O$ and $^{13}C/^{12}C$ ratios in the coral *Montrastrea annularis. Geochim. Cosmochim. Acta* **43**, 1009–1020.

Fairbanks R. G. and Matthews R. K. (1978) The marine oxygen isotope record in Pleistocene coral, Barbados, West Indies. *Quat. Res.* **10**, 181–196.

Fallon S. J., McCulloch M. T., Woesik R., and Sinclair D. J. (1999) Corals at their latitudinal limits: laser ablation trace element systematics in *Porites* from Shirigai Bay, Japan. *Earth Planet. Sci. Lett.* **172**, 221–238.

Farrell J. W. and Prell W. L. (1989) Climatic change and $CaCO_3$ preservation: an 800,000 year bathymetric reconstruction from the central equatorial Pacific Ocean. *Paleoceanography* **4**(4), 447–466.

Gagan M. K., Ayliffe L. K., Hopley D., Cali J. A., Mortimer G. E., Chappell J., McCulloch M. T., and Head M. J. (1998) Temperature and surface-ocean water balance of the Mid-Holocene tropical western pacific. *Science* **279**, 1014–1018.

Gagan M. K., Ayliffe L. K., Beck J. W., Cole J. E., Druffel E. R. M., Dunbar R. B., and Schrag D. P. (2000) New views of tropical paleoclimates from corals. *Quat. Sci. Rev.* **19**(1–5), 45–64.

Grossman E., Betzer P. R., Walter C. D., and Dunbar R. B. (1986) Stable isotopic variation in pteropods and atlantids from North Pacific sediment traps. *Mar. Micropaleontol.* **10**, 9–22.

Guilderson T. P., Fairbanks R. G., and Rubenstone J. L. (1994) Tropical temperature variations since 20,000 years ago: modulating interhemispheric climate changes. *Science* **263**, 663–665.

Guilderson T., Fairbanks R., and Rubenstone J. (2001) Tropical Atlantic coral oxygen isotopes: glacial-interglacial sea surface temperatures and climate change. *Mar. Geol.* **V172** (N1–2), 75–89.

Gussone N., Eisenhauer A., Heuser A., Dietzel M., Bock B., Böhm F., Spero H. J., Lea D. W., Bijma J., and Nägler T. F. (2003) Model for kinetic effects on calcium isotope fractionation (δ^{44}Ca) in inorganic aragonite and cultured planktonic foraminifera. *Geochim. Cosmochim. Acta* **67**(7), 1375–1382.

Gussone N., Eisenhauer A., Heuser A., Dietzel M., Bock B., Böhm F., Spero H., Lea D. W., Bijma J., Zeebe R., and Nägler T. F. (2003) Model for kinetic effects on calcium isotope fractionation (δ^{44}Ca) in inorganic aragonite and cultured planktonic foraminifera. *Geochim. Cosmochim. Acta* **67**(7), 1375–1382.

Hastings D. W., Emerson S. R., and Nelson B. K. (1996) Vanadium in foraminifera calcite: evaluation of a method to determine paleo-seawater concentrations. *Geochim. Cosmochim. Acta* **60**(19), 3701–3715.

Hastings D. W., Russell A. D., and Emerson S. R. (1998) Foraminiferal magnesium in *Globeriginoides sacculifer*as a paleotemperature proxy. *Paleoceanography* **13**(2), 161–169.

Hays J. D., Imbrie J., and Shackleton N. J. (1976) Variations in the Earth's orbit: pacemaker of the Ice ages. *Science* **194**, 1121–1132.

Hecht A. D., Eslinger E. V., and Garmon L. B. (1975) Experimental studies on the dissolution of planktonic foraminifera. In *Dissolution of Deep-sea Carbonates* (eds. W. V. Sitter, A. W. H. Bé, and W. H. Berger). Cushman Foundation for Foraminiferal Research, , pp. 59–69.

Hendy E. J., Gagan M. K., Alibert C. A., McCulloch M. T., Lough J. M., and Isdale P. J. (2002) Abrupt decrease in tropical Pacific Sea surface salinity at end of little Ice age. *Science* **295**(5559), 1511–1514.

Hughen K. A., Schrag D. P., Jacobsen S. B., and Hantoro W. (1999) El Nino during the last interglacial period recorded by a fossil coral from Indonesia. *Geophys. Res. Lett.* **26**(20), 3129–3132.

Imbrie J. and Kipp N. G. (1971) A new micropaleontological method quantitative paleoclimatology: application to a late pleistocene Caribbean core. In *The Late Cenozoic Ice Ages* (ed. K. Turekian). Yale University Press, New Haven and London, pp. 71–181.

Imbrie J., Boyle E. A., Clemens S. C., Duffy A., Howard W. R., Kukla G., Kutzbach J., Martinson D. G., McIntyre A., Mix A. C., Molfino B., Morley J. J., Peterson L. C., Pisias N. G., Prell W. L., Raymo M. E., Shackleton N. J., and Toggweiler J. R. (1992) On the structure and origin of major glaciation cycles: 1. Linear responses to Milankovitch forcing. *Paleoceanography* **7**(6), 701–738.

Izuka S. K. (1988) Relationships of magnesium and other minor elements in tests of *Cassidulina Subglobosa* and *C. Oriangulata* to physical oceanic properties. *J. Foraminiferal Res.* **18**(2), 151–157.

Juillet-Leclerc A. D. and Labeyrie L. D. (1986) Temperature dependence of the oxygen isotopic fractionation between diatom silica and water. *Earth Planet. Sci. Lett.* **84**, 69–74.

Katz A. (1973) The interaction of magnesium with calcite during crystal growth at 25–90 °C and one atmosphere. *Geochim. Cosmochim. Acta* **37**, 1563–1586.

Kim S.-T. and O'Neil J. R. (1997) Equilibrium and nonequilibrium oxygen isotope effects in synthetic calcites. *Geochim. Cosmochim. Acta* **61**(16), 3461–3475.

Kinsman D. J. J. and Holland H. D. (1969) The co-precipitation of cations with $CaCO_3$: IV. The co-precipitation of Sr^2+ with aragonite between 16° and 96 °C. *Geochim. Cosmochim. Acta* **33**, 1–17.

Koutavas A., Lynch-Stieglitz J., Marchitto T. M. J., and Sachs J. P. (2002) Deglacial and Holocene climate record from the Galapagos Islands: El Niño linked to Ice Age climate. *Science* **297**(5579), 226–230.

Koziol A. M. and Newton R. C. (1995) Experimental determination of the reactions magnesite + quartz = enstatite + CO_2 and magnesite = periclase + CO_2 and the enthalpies of formation of enstatite and magnesite. *Am. Mineral.* **80**, 1252–1260.

Labeyrie L. D. and Juillet A. (1982) Oxygen isotopic exchangeability of diatom valve silica: interpretation and consequences for paleoclimatic studies. *Geochim. Cosmochim. Acta* **46**, 967–975.

Lea D. W. (2002) The glacial tropical Pacific: not just a west side story. *Science* **297**, 202–203.

Lea D. W., Shen G. T., and Boyle E. A. (1989) Coralline barium records temporal variability in equatorial Pacific upwelling. *Nature* **340**, 373–376.

Lea D. W., Mashiotta T. A., and Spero H. J. (1999) Controls on magnesium and strontium uptake in planktonic foraminifera determined by live culturing. *Geochim. Cosmochim. Acta* **63** (16), 2369–2379.

Lea D. W., Pak D. K., and Spero H. J. (2000) Climate impact of Late Quaternary equatorial Pacific sea surface temperature variations. *Science* **289**(5486), 1719–1724.

Lea D. W., Martin P. A., Pak D. K., and Spero H. J. (2002) Reconstructing a 350 ky history of sea level using planktonic Mg/Ca and oxygen isotopic records from a Cocos Ridge core. *Quat. Sci. Rev.* **21**(1–3), 283–293.

Lear C. H., Elderfield H., and Wilson P. A. (2000) Cenozoic deep-sea temperatures and global ice volumes from Mg/Ca in benthic foraminiferal calcite. *Science* **287**, 269–272.

Linsley B. K., Wellington G. M., and Schrag D. P. (2000) Decadal sea surface temperature variability in the subtropical South Pacific from 1726 to 1997 AD. *Science* **290**, 1145–1148.

Lohmann G. P. (1995) A model for variation in the chemistry of planktonic foraminifera due to secondary calcification and selective dissolution. *Paleoceanography* **10**(3), 445–457.

Lorens R. B., Williams D. F., and Bender M. L. (1977) The early nonstructural chemical diagenesis of foraminiferal calcite. *J. Sed. Petrol.* **47**(4), 1602–1609.

Lyle M. W., Prahl F. G., and Sparrow M. A. (1992) Upwelling and productivity changes inferred from a temperature record in the central equatorial Pacific. *Nature* **355**, 812–815.

Mann M. E., Bradley R. S., and Hughes M. K. (1998) Global-scale temperature patterns and climate forcing over the past six centuries. *Nature* **392**(23), 779–787.

Mantua N. J., Hare S. R., Zhang Y., Wallace J. M., and Francis R. C. (1997) A Pacific interdecadal climate oscillation with impacts on salmon production. *Bull Am. Meteorol. Soc.* **78**(6), 1069–1079.

Martin P. A., Lea D. W., Mashiotta T. A., Papenfuss T., and Sarnthein M. (1999) Variation of foraminiferal Sr/Ca over Quaternary glacial-interglacial cycles: evidence for changes in mean ocean Sr/Ca? *Geochem. Geophys. Geosys.* **1**, Paper Number 1999GC000006.

Martin P. A., Lea D. W., Rosenthal Y., Shackleton N. J., Sarnthein M., and Papenfuss T. (2002) Quaternary deep sea temperature histories derived from benthic foraminiferal Mg/Ca. *Earth Planet. Sci. Lett.* **198**, 193–209.

Mashiotta T. A., Lea D. W., and Spero H. J. (1999) Glacial-interglacial changes in Subantarctic sea surface temperature and δ^{18}O-water using foraminiferal Mg. *Earth Planet. Sci. Lett.* **170**(4), 417–432.

McConnaughey T. (1989a) ^{13}C and ^{18}O isotopic disequilibrium in biological carbonates: I. Patterns. *Geochim. Cosmochim. Acta* **53**(1), 151–162.

McConnaughey T. A. (1989b) ^{13}C and ^{18}O isotopic disequilibrium in biological carbonates: II. *In vitro* simulation of kinetic isotopic effects. *Geochim. Cosmochim. Acta* **53**, 163–171.

McCorkle D. C., Martin P. A., Lea D. W., and Klinkhammer G. P. (1995) Evidence of a dissolution effect on benthic shell chemistry: δ^{13}C, Cd/Ca, Ba/Ca, and Sr/Ca from the Ontong Java Plateau. *Paleoceanography* **10**(4), 699–714.

McCulloch M. T., Gagan M. K., Mortimer G. E., Chivas A. R., and Isdale P. J. (1994) A high-resolution Sr/Ca and δ^{18}O coral record from the great barrier reef, Australia, and the 1982–1983 El Nino. *Geochim. Cosmochim. Acta* **58**(12), 2747–2754.

McCulloch M., Mortimer G., Esat T., Xianhua L., Pillans B., and Chappell J. (1996) High resolution windows into early Holocene climate: Sr/Ca coral records from the Huon Peninsula. *Earth Planet. Sci. Lett.* **138**, 169–178.

McCulloch M. T., Tudhope A. W., Esat T. M., Mortimer G. E., Chappell J., Pillans B., Chivas A. R., and Omura A. (1999) Coral record of equatorial sea-surface temperatures during the penultimate deglaciation at Huon peninsula. *Science* **283**, 202–204.

McGregor H. V. and Gagan M. K. (2003) Diagenesis and geochemistry of Porites corals from Papua New Guinea: implications for paleoclimate reconstruction. *Geochim. Cosmochim. Acta* (in review).

Min G., Edwards R., Taylor F., Gallup C., and Beck J. (1995) Annual cycles of U/Ca in coral skeletons and U/Ca thermometry. *Geochim. Cosmochim. Acta* **59**(10), 2025–2042.

Mitsuguchi T., Matsumoto E., Abe O., Uchida T., and Isdale P. J. (1996) Mg/Ca thermometry in coral skeletons. *Science* **274**, 961–963.

Mix A. C. (1992) The marine oxygen isotope record: constraints on timing and extent of ice-growth events (120–65 ka). In *The Last Interglacial Transition in North America* Geological Society of America Special Paper (eds. P. U. Clark and P. D. Lea). GSA, Boulder, Co. Vol. 270. pp.19–30.

Mix A. C., Morey A. E., Pisias N. G., and Hostetler S. W. (1999) Foraminiferal faunal estimates of paleotemperature: circumventing the no-analog problem yields cool ice age tropics. *Paleoceanography* **14**(3), 350–359.

Mucci A. (1987) Influence of temperature on the composition of magnesian calcite overgrowths precipitated from seawater. *Geochim. Cosmochim. Acta* **51**, 1977–1984.

Müller P. J., Kirst G., Ruhland G., von Storch I., and Rosell-Melé A. (1998) Calibration of the alkenone paleotemperature index UK'37 based on core-tops from the eastern South Atlantic and the global ocean (60°N–60°S). *Geochim. Cosmochim. Acta* **62**(10), 1757–1772.

Müller A., Gagan M. K., and McCulloch M. T. (2001) Early marine diagenesis in corals and geochemical consequences for paleoceanographic reconstructions. *Geophys. Res. Lett.* **28**(23), 4471–4474.

Nägler T. F., Eisenhauer A., Müller A., Hemleben C., and Kramers J. (2000) The δ^{44}Ca-temperature calibration on fossil and cultured *Globigerinoides sacculifer*: new tool for reconstruction of past sea surface temperatures. *Geochem. Geophys. Geosys.* **1**, (Paper number: 2000GC000091).

Nürnberg D. (1995) Magnesium in tests of *Neogloboquadrina Pachyderma* sinistral from high Northern and Southern latitudes. *J. Foraminiferal Res.* **25**(4), 350–368.

Nürnberg D., Bijma J., and Hemleben C. (1996a) Assessing the reliability of magnesium in foraminiferal calcite as a proxy for water mass temperatures. *Geochim. Cosmochim. Acta* **60** (5), 803–814.

Nürnberg D., Bijma J., and Hemleben C. (1996b) Erratum: assessing the reliability of magnesium in foraminiferal calcite as a proxy for water mass temperatures. *Geochim. Cosmochim. Acta* **60**(13), 2483–2484.

Nürnberg D., Müller A., and Schneider R. R. (2000) Paleo-sea surface temperature calculations in the equatorial east Atlantic from Mg/Ca ratios in planktic foraminifera: a comparison to sea surface temperature estimates from $U^{K'}37'$, oxygen isotopes, and foraminiferal transfer function. *Paleoceanography* **15**(1), 124–134.

O'Neil J. R., Clayton R. N., and Mayeda T. K. (1969) Oxygen isotope fractionation in divalent metal carbonates. *J. Chem. Phys.* **51**(12), 5547–5558.

Oomori T., Kaneshima H., and Maezato Y. (1987) Distribution coefficient of Mg^{2+} ions between calcite and solution at 10–50 °C. *Mar. Chem.* **20**, 327–336.

Oppo D. W., Linsley B. K., Rosenthal Y., Dannenmann S., and Beaufort L. (2003) Orbital and suborbital climate variability in the Sulu Sea, western tropical Pacific. *Geochem. Geophys. Geosys* **4**(1), 1003, doi : 10.1029/200/Gc000260 .

Pearson P. N., Ditchfield P. W., Singano J., Harcourt-Brown K. G., Nicholas C. J., Olsson R. K., Shackleton N. J., and Hall M. A. (2001) Warm tropical sea surface temperatures in the late Cretaceous and Eocene epochs. *Nature* **413**, 481–487.

Pelejero C. and Calvo E. (2003) The upper end of the $U^{K'}_{37}$ temperature calibration revisited. *Geochem. Geophys. Geosys.* **4**, 1014doi:10.1029/2002GC000431.

Pelejero C., Grimalt J. O., Heilig S., Kienast M., and Wang L. (1999) High resolution U^{K}_{37} temperature reconstructions in the South China Sea over the past 220 kyr. *Paleoceanography* **14**(2), 224–231.

Pierrehumbert R. (1999) Huascaran delta O-18 as an indicator of tropical climate during the last glacial maximum. *Geophys. Res. Lett.* **26**(9), 1345–1348.

Rathburn A. E. and Deckker P. D. (1997) Magnesium and strontium compositions of recent benthic foraminifera from the Coral Sea, Australia and Prydz Bay, Antarctica. *Mar. Micropaleontol.* **32**, 231–248.

Rickaby R. E. M. and Elderfield H. (1999) Planktonic foraminiferal Cd/Ca: paleonutrients or paleotemperature? *Paleoceanography* **14**(3), 293–303.

Rodbell D. T., Seltzer G. O., Anderson D. M., Abbot M. B., Enfield D. B., and Newman J. H. (1999) An ~15,000-year record of El Niño-driven alluviation in southwestern Ecuador. *Science* **283**, 516–520.

Rosenthal Y. and Boyle E. A. (1993) Factors controlling the fluoride content of planktonic foraminifera: an evaluation of its paleoceanographic utility. *Geochim. Cosmochim. Acta* **57** (2), 335–346.

Rosenthal Y. and Lohmann G. P. (2002) Accurate estimation of sea surface temperatures using dissolution corrected calibrations for Mg/Ca paleothermometry. *Paleoceanography* **17**, 1044, doi:10.1029/2001PA000749.

Rosenthal Y., Boyle E. A., and Slowey N. (1997) Temperature control on the incorporation of Mg, Sr, F, and Cd into benthic foraminiferal shells from Little Bahama Bank: prospects for thermocline paleoceanography. *Geochim. Cosmochim. Acta* **61**(17), 3633–3643.

Rosenthal Y., Lohmann G. P., Lohmann K. C., and Sherrell R. M. (2000) Incorporation and preservation of Mg in *Globigerinoides sacculifer* implications for reconstructing the temperature and $^{18}O/^{16}O$ of seawater. *Paleoceanography* **15**(1), 135–145.

Rosenthal Y., Dannenmann S., Oppo D. W., and Linsley B. K. (2003) The amplitude and phasing of climate change during the last deglaciation in the Sulu Sea, western equatorial Pacific. *Geophys. Res. Lett.* **30**(8), 1428, doi:10.1029/2002GL016612.

Rostek F., Ruhland G., Bassinot F. C., Müller P. J., Labeyrie L. D., Lancelot Y., and Bard E. (1993) Reconstructing sea surface temperature and salinity using $\delta^{18}O$ and alkenone records. *Nature* **364**, 319–321.

Russell A. D., Emerson S., Nelson B. K., Erez J., and Lea D. W. (1994) Uranium in foraminiferal calcite as a recorder of seawater uranium concentrations. *Geochim. Cosmochim. Acta* **58**(2), 671–681.

Russell W. A., Papanastassiou D. A., and Tombrello T. A. (1978) Ca isotope fractionation on the Earth and other solar system materials. *Geochim. Cosmochim. Acta* **42**, 1075–1090.

Sanyal A., Hemming N. G., Broecker W. S., Lea D. W., Spero H. J., and Hanson G. N. (1996) Oceanic pH control on the boron isotopic composition of foraminifera: evidence from culture experiments. *Paleoceanography* **11**(5), 513–517.

Savin S. M. and Douglas R. G. (1973) Stable isotope and magnesium geochemistry of recent planktonic foraminifera from the South Pacific. *Geol. Soc. Am. Bull.* **84**, 2327–2342.

Schmidt G. A., Hoffmann G., and Thresher D. (2001) Isotopic tracers in coupled models: a new paleo-tool. *PAGES News* **9** (1), 10–11.

Schrag D. P. (1999a) Effects of diagenesis on the isotopic record of late paleogene tropical sea surface temperatures. *Chem. Geol.* **161**(1–3), 215–224.

Schrag D. P. (1999b) Rapid analysis of high-precision Sr/Ca ratios in corals and other marine carbonates. *Paleoceanography* **14**(2), 97–102.
Schrag D. P. and Linsley B. K. (2002) Paleoclimate: corals, chemistry, and climate. *Science* **296**(5566), 277–278.
Schrag D. P., Depaolo D. J., and Richter F. M. (1995) Reconstructing past sea surface temperatures—correcting for diagenesis of bulk marine carbonate. *Geochim. Cosmochim. Acta* **59**(11), 2265–2278.
Schrag D. P., Hampt G., and Murray D. W. (1996) Pore fluid constraints on the temperature and oxygen isotopic composition of the glacial ocean. *Science* **272**, 1930–1932.
Shackleton N. (1967) Oxygen isotope analyses and Pleistocene temperatures re-assessed. *Nature* **215**, 15–17.
Shackleton N. J. (1974) Attainment of isotopic equilibrium between ocean water and the benthonic foraminifera genus *Uvigerina*: isotopic changes in the ocean during the last glacial. *Cent. Nat. Rech. Sci. Colloq. Int.* **219**, 203–209.
Shackleton N. J. (1987) Oxygen isotopes, ice volume and sea level. *Quat. Sci. Rev.* **6**, 183–190.
Shackleton N. J. and Opdyke N. D. (1973) Oxygen isotope and paleomagnetic stratigraphy of equatorial Pacific core V28-238: oxygen isotope temperatures and ice volumes on a 10^5 and 10^6 year scale. *Quat. Res.* **3**, 39–55.
Shemesh A., Charles C. D., and Fairbanks R. G. (1992) Oxygen isotopes in biogenic silica—global changes in ocean temperature and isotopic composition. *Science* **256**(5062), 1434–1436.
Shemesh A., Burckle L. H., and Hays J. D. (1994) Meltwater input to the Southern Ocean during the last glacial maximum. *Science* **266**(5190), 1542–1544.
Shemesh A., Burckle L. H., and Hays J. D. (1995) Late Pleistocene oxygen isotope records of biogenic silica from the Atlantic sector of the Southern Ocean. *Paleoceanography* **10**(2), 179–196.
Shen C.-C., Hastings D. W., Lee T., Chiu C.-H., Lee M.-Y., Wei K.-Y., and Edwards R. L. (2001) High precision glacial-interglacial benthic foraminiferal Sr/Ca records from the eastern equatorial Atlantic Ocean and Caribbean Sea. *Earth Planet. Sci. Lett.* **190**(3–4), 197–209.
Shen G. T. (1996) Rapid changes in the tropical ocean and the use of corals as monitoring systems. In *Geoindicators: Assessing Rapid Environmental Changes in Earth Systems* (eds. A. Berger and W. Iams). A. A. Balkema, Rotterdam, pp. 155–169.
Shen G. T. and Dunbar R. B. (1995) Environmental controls on uranium in reef corals. *Geochim. Cosmochim. Acta* **59**(10), 2009–2024.
Shen G. T., Cole J. C., Lea D. W., Linn L. J., McConnaughey T. A., and Fairbanks R. G. (1992) Surface ocean variability at Galapagos from 1936 to 1982: calibration of geochemical tracers in corals. *Paleoceanography* **7**(5), 563–588.
Sinclair D. J., Kinsley L. P. J., and McCulloch M. T. (1998) High resolution analysis of trace elements in corals by laser ablation ICP-MS. *Geochim. Cosmochim. Acta* **62**(11), 1889–1901.
Smith J. E., Risk M. J., Schwarcz H. P., and McConnaughey T. A. (1997) Rapid climate change in the North Atlantic during the Younger Dryas recorded by deep-sea corals. *Nature* **386**, 818–820.
Smith S. V., Buddemeier R. W., Redalje R. C., and Houck J. E. (1979) Strontium-calcium thermometry in coral skeletons. *Science* **204**, 404–407.
Spero H. J. and Lea D. W. (1993) Intraspecific stable isotope variability in the planktonic foraminifera *Globigerinoides sacculifer*: results from laboratory experiments. *Mar. Micropaleontol.* **22**, 221–234.
Spero H. J. and Lea D. W. (1996) Experimental determination of stable isotopic variability in *Globigerina bulloides*: implications for paleoceanographic reconstructions. *Mar. Micropaleontol.* **28**, 231–246.
Spero H. J., Bijma J., Lea D. W., and Bemis B. (1997) Effect of seawater carbonate chemistry on planktonic foraminiferal carbon and oxygen isotope values. *Nature* **390**, 497–500.
Stoll H. M. and Schrag D. P. (1998) Effects of Quaternary sea level cycles on strontium in seawater. *Geochim. Cosmochim. Acta* **62**(7), 1107–1118.
Stoll H. M., Schrag D. P., and Clemens S. C. (1999) Are seawater Sr/Ca variations preserved in Quaternary foraminifera. *Geochim. Cosmochim. Acta* **63**(21), 3535–3547.
Stott L., Poulsen C., Lund S., and Thunnell R. (2002) Super ENSO and global climate oscillations at millennial timescales. *Science* **279**(5579), 222–226.
Sturchio N. C. (1999) A conversation with Harmon Craig. *Geochem. News* **98**, 12–20.
Toyofuku T., Kitazato H., Kawahata H., Tsuchiya M., and Nohara M. (2000) Evaluation of Mg/Ca thermometry in foraminifera: comparison of experimental results and measurements in nature. *Paleoceanography* **15**(4), 456–464.
Trend-Staid M. and Prell W. L. (2002) Sea surface temperature at the Last Glacial Maximum: a reconstruction using the modern analog technique. *Paleoceanography* **17**(4), 10.1029/2000PA000506.
Tudhope A. W., Chilcott C. P., McCulloch M. T., Cook E. R., Chappell J., Ellam R. M., Lea D. W., Lough J. M., and Shimmield G. B. (2001) Variability in the El Niño southern oscillation through a glacial-interglacial cycle. *Science* **291** (5508), 1511–1517.
Urban F., Cole J., and Overpeck J. (2000) Influence of mean climate change on climate variability from a 155-year Tropical Pacific coral record. *Nature* **407**(6807), 989–993.
Urey H. C. (1947) The thermodynamic properties of isotopic substances. *J. Chem. Soc*, 562–581.
Visser K., Thunell R., and Stott L. (2003) Magnitude and timing of temperature change in the Indo-Pacific warm pool during deglaciation. *Nature* **421**, 152–155.
Weber J. N. (1973) Incorporation of strontium into reef coral skeletal carbonates. *Geochim. Cosmochim. Acta* **37**, 2173–2190.
Weber J. N. and Woodhead P. M. J. (1972) Temperature dependence of oxygen-18 concentration in reef coral carbonates. *J. Geophys. Res.* **77**, 463–473.
Wu G. and Berger W. H. (1989) Planktonic foraminifera: differential dissolution and the Quaternary stable isotope record in the west equatorial Pacific. *Paleoceanography* **4**(2), 181–198.
Zachos J., Pagani M., Sloan L., Thomas E., and Billups K. (2001) Trends, rhythms, and aberrations in global climate 65 Ma to present. *Science* **292**(5517), 686–693.
Zhou G.-T. and Zheng Y.-F. (2003) An experimental study of oxygen isotope fractionation between inorganically precipitated aragonite and water at low temperatures. *Geochim. Cosmochim. Acta* **67**(3), 387–399.
Zhu P. and Macdougall J. D. (1998) Ca isotopes in the marine environment and the oceanic Ca cycle. *Geochim. Cosmochim. Acta* **62**(10), 1691–1698.

Isotope Geochemistry
ISBN: 978-0-08-096710-3

pp. 227–252

9
Sulfur-rich Sediments

M. B. Goldhaber

US Geological Survey, Denver, CO, USA

9.1 INTRODUCTION

9.1.1 Overview

Marine sediments with more than a few tenths of a percent of organic carbon, as well as organic-matter-bearing, nonmarine sediments with significant concentrations of sulfate in the depositional waters contain the mineral pyrite (FeS_2). Pyrite, along with sulfur-bearing organic compounds, form indirectly through the metabolic activities of sulfate-reducing microorganisms. The geochemical transformations of sulfur in sediments leading to these products significantly impact the pathway of early sedimentary diagenesis, conditions for the localization of mineral deposits (Ohmoto and Goldhaber, 1997), the global cycling of sulfur and carbon, the abundance of oxygen in the Earth's atmosphere, and perhaps even the emergence of life on Earth (e.g., Russell and Hall, 1997). This chapter provides an overview of sedimentary-sulfur geochemistry from its microbial and abiologic pathways to the global consequences of these processes.

The geochemistry of sulfur is complicated by its wide range of oxidation states (Table 1). Under oxidizing conditions (e.g., in the presence of atmospheric oxygen) sulfate, with sulfur in the +6 valence state, is the stable form of sulfur. Under reducing conditions (e.g., in the presence of H_2), sulfide (S=−2 valent) is the stable oxidation state. However, a range of additional aqueous and solid-phase sulfur species exist with valences between these two end-members. What makes the study of sulfur geochemistry so exciting and challenging is that many of these intermediate-valent forms play key roles in sedimentary-sulfur transformations. Furthermore, many of these reactions are microbially mediated. As detailed below, these complex biogeochemical pathways are now yielding to research whose scope ranges from molecular to global level.

9.1.2 History of the Study of Sedimentary-sulfur Geochemistry

The study of sulfur diagenesis has a long history. Scientists have long accepted that microorganisms play a major role in geochemical sulfur transformations (Baas Becking, 1925). They also recognized at an earlier time that removal of sulfate occurs in the pore waters of marine mud (Murry, 1895). Subsequent work established that depletion of sulfate from marine pore water is a microbial process that results in formation of sedimentary pyrite (Berner, 1964a; Emery and Rittenberg, 1952; Hartmann and Nielsen, 1969; Kaplan *et al.*, 1963; ZoBell and Rittenberg, 1950). The abundance of this pyrite, together with its morphological and isotopic characteristics, provides clues to the details of these transformations.

There are a number of useful summary articles on sulfur geochemistry (e.g., Belyayev *et al.*, 1981; Berner, 1973; Ohmoto *et al.*, 1990, #1031; Bottrell and Raiswell, 2000; Canfield and Raiswell, 1991; Chambers and Trudinger, 1979; Goldhaber and Kaplan, 1974; Grinenko and Ivanov, 1983; Ivanov, 1981; Krouse and McCready, 1979a,b; Migdisov *et al.*, 1983; Morse *et al.*, 1987; Ohmoto and Goldhaber, 1997; Skyring, 1987; Strauss, 1997). In addition to this important body of work, there have been a number of recent advances in the study of sedimentary-sulfur diagenesis that make a new review timely. These advances are linked to a much deeper understanding of the complex microbial processes that dominate sulfur transformations in sediments, coupled with advances in analytical technology, enabling processes to be studied with increasing specificity and sophistication.

Table 1 Forms of sulfur in marine sediments and their oxidation states.

Aqueous species or mineral	*Formula*	*Oxidation state(s) of sulfur*
Sulfide	$H_2S(aq)$, $HS^-(aq)$	−2
Iron sulfide[a]	$FeS(s)$	−2
Greigite	$Fe_3S_4(s)$	−2, 0
Pyrite	$FeS_2(s)$	−2, 0
Polysulfide	$S_{X^{2-}}(aq)$	−2, 0
Sulfur	$S_8(s)$	0
Hyposulfite	$S_2O_4^{2-}(aq)$	+3
Sulfite	$SO_3^{2-}(aq)^-$	+4
Thiosulfate	$S_2O_3^{2-}(aq)$	−1, +5
Dithionate	$S_2O_6^{2-}(aq)$	+5
Trithionate	$S_3O_6^{2-}(aq)$	−2, +6
Tetrathionate	$S_4O_6^{2-}(aq)$	−2, +6
Pentathionate	$S_5O_6^{2-}(aq)$	−2, +6
Sulfate	$SO_4^{2-}(aq)$	+6

After Kasten and Jorgensen (2000) (reproduced by permission of Springer-Verlag from *Marine Geochemistry*, **2000**, 263–281).
[a] Includes troilite, mackinawite, and pyrrhotite.

9.2 BACTERIAL SULFATE REDUCTION

9.2.1 Biochemistry of Bacterial Sulfate Reduction

The key reaction in the global sulfur cycle is the reduction of sulfate (SO_4^{2-}) to hydrogen sulfide (H_2S). In this chapter H_2S, unless otherwise specified, is taken to indicate the sum of the reduced sulfur species including H_2S_{aq} and HS^-_{aq}. The stability of SO_4^{2-} towards naturally occurring reducing agents is so great that either elevated temperatures or a microbially (enzymatically) catalyzed process is required for reduction to occur. Thermodynamic calculations indicate that reduction of SO_4^{2-} to H_2S should occur at Earth-surface temperatures by a wide range of organic compounds. Yet below temperatures ~110–150 °C this reaction is strongly kinetically inhibited, and may not be observable even on geologic timescales (Goldhaber and Orr, 1994; Ohmoto and Goldhaber, 1997). This kinetic inhibition of sulfate reduction is effectively overcome in nature by two types of microbially catalyzed processes: assimilatory and dissimilatory sulfate reduction (Madigan *et al.*, 2000).

Many organisms, including higher plants, fungi, algae, and most prokaryotes, reduce sulfate and incorporate it as a sulfur source for biosynthesis of proteins. This reaction, termed assimilative sulfate reduction, although widely distributed in nature, is neither a major component of the global sulfur cycle, nor is it a major mechanism of sulfur transformation in sediments. In contrast, dissimilatory reduction, in which sulfate is an electron acceptor for energy generating bacterial processes, is of global importance (Equation (1)):

$$SO_4^{2-} + 2CH_2O \rightarrow H_2S + 2HCO_3^- \qquad (1)$$

In this equation, CH_2O represents a generic form of organic matter with the oxidation state of carbohydrate. In effect, sulfate-reducing prokaryotes (including both bacteria and archea) respire (breathe in) sulfate (as we do oxygen), and breathe out H_2S and carbon dioxide. As a result, they release large amounts of H_2S into their environment. Dissimilatory reduction is restricted to the sulfate-reducing bacteria and archea.

The biochemical pathway of both assimilatory and dissimilatory sulfate reduction is illustrated in Figure 1. The details of the dissimilatory reduction pathway are useful for understanding the origin of bacterial stable isotopic fractionations. The overall pathways require the transfer of eight electrons, and proceed through a number of intermediate steps. The reduction of sulfate requires activation by ATP (adenosine triphosphate) to form adenosine phosphosulfate (APS). The enzyme ATP sulfurylase catalyzes this reaction. In dissimilatory

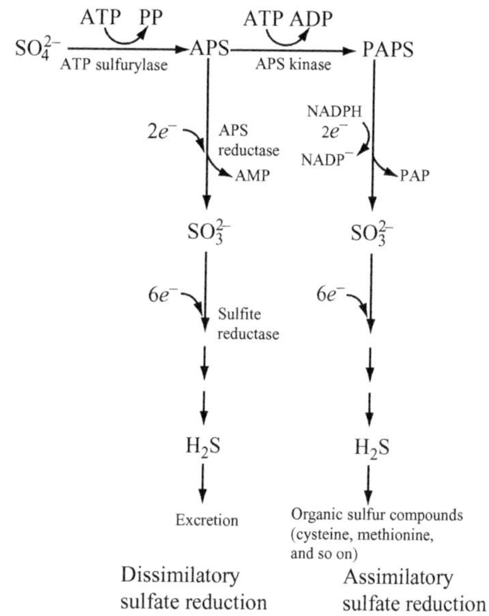

Figure 1 Biochemical pathway of dissimilatory and assimilatory SO_4^{2-} reduction (after Madigan *et al.*, 2000).

reduction, the sulfate moiety of APS is reduced to sulfite (SO_3^{2-}) by the enzyme APS reductase, whereas in assimilatory reduction APS is further phosphorylated to phospho-adenosine phosphosulfate (PAPS) before reduction to the oxidation state of sulfite and sulfide. Although the reduction reactions occur in the cell's cytoplasm (i.e., the sulfate enters the cell), the electron transport chain for dissimilatory sulfate reduction occurs in proteins that are periplasmic (within the bacterial cell wall). The enzyme hydrogenase requires molecular hydrogen supplied either from the external environment or by the oxidation of organic compounds such as lactate. Hydrogenase supplies eight electrons to cytochrome c_3 (cytochromes are proteins with iron-containing porphyrin rings), which in turn transfers electrons to a second cytochrome complex (Hmc). The electrons from this periplasmic electron transport system are transported across the cytoplasmic membrane to an iron–sulfur protein in the cytoplasm that, in turn, supplies electrons for the reduction reactions shown in Figure 1.

9.2.2 Ecology of Sulfate-reducing Bacteria

The impact of sulfate-reducing bacteria (SRB) on marine sediments can be extensive. In organic-rich sediments, large quantities of reactant organic matter are consumed and dissolved products produced (Equation (1)). The impact of SRB metabolic activities can modify the overall geochemistry of the sedimentary package. For example, in many environments, more than 50% of the total carbon mineralization (oxidation) is due to SRB (Canfield and Des Marais, 1993; Jorgensen, 1982). The ecology of these organisms determines to what extent they modify the sediments. The most important ecological requirement is that sulfate-reducing microorganisms are strict anaerobes. In the presence of organic matter and absence of oxygen, SRB can grow in a wide range of environments spanning the spectrum of pressure, temperature, salinity, and pH values found in the Earth's upper crust.

SRB must compete, however, for the available food supply in sediments. In addition to sulfate reducers, other microorganisms gain energy by anaerobic respiration using electron acceptors other than sulfate. In marine sediments, a well-defined succession of microbial ecosystems closely follows the energy yield available from oxidation of organic matter with the available electron acceptors. The order is aerobic respiration followed by respiration using nitrate, manganese oxide, iron oxide, sulfate, and carbonate (Figure 2). The presence of significant amounts of one of these electron acceptors, more favored than sulfate, suppresses the activities of SRB. For example, addition of Fe(III) oxyhydroxides to sediments in which sulfate reduction is active will suppress SRB nearly completely (Lovely and Phillips, 1987).

Oxygen respiration is by far the most efficient respiratory process, but because oxygen is present in relatively low concentration in seawater, and because rates of organic oxidation with oxygen are rapid, aerobic respiration is self-limiting. The penetration distance of oxygen by diffusion may only be millimeters in sediments with abundant organic matter. However, oxygen is also physically mixed (advected) or pumped into sediments by the life activities of macroorganisms, a process termed bioturbation (Aller, 1980a, 1988; R. C. Aller and J. Y. Aller, 1998; Goldhaber *et al.*, 1977). Bioturbation may increase the thickness of the oxygenated zone, or may create a series of

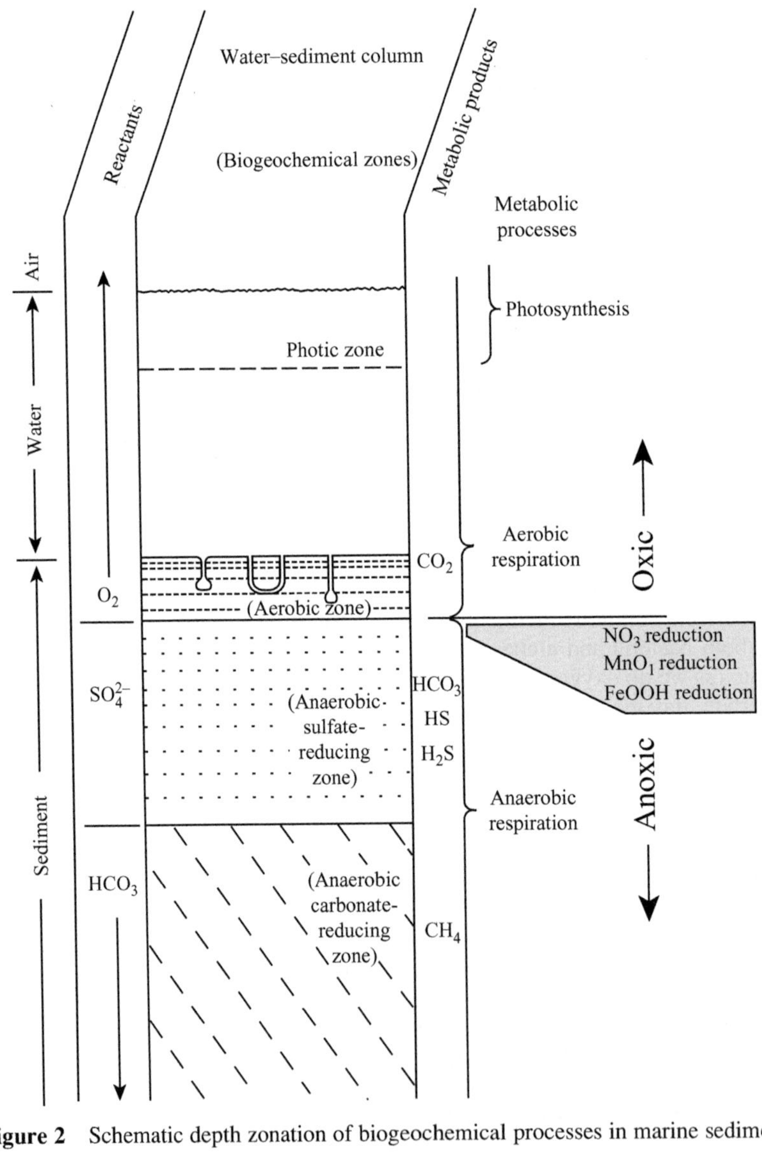

Figure 2 Schematic depth zonation of biogeochemical processes in marine sediments.

heterogeneous oxic and anoxic volumes within the upper sediment layers. Following oxygen removal, a series of microorganisms utilize terminal electron acceptors in the sequence shown in Figure 2. However, with the exception of sulfate, the other electron acceptors commonly do not occur in high concentrations in the marine environment (although exceptions exist). By one estimate, the concentration of sulfate is more than 50 times higher than that of the other more energetically favored terminal electron acceptors (Froelich *et al.*, 1979). Sulfate is the second most abundant anion in seawater, so seawater overlying marine sediments represents an enormous reservoir of this anion. Therefore, anaerobic respiration using sulfate tends to dominate anaerobic respiratory processes in organic-matter-rich marine sediments.

A critical control on sulfate reduction in sediments is the amount and type of food available for the bacteria. Sulfate reducers can only metabolize a relatively limited range of organic compounds. Although the recognized range of genera/species of sulfate reducers and the compounds they can metabolize has increased dramatically in recent years (Madigan *et al.*, 2000, table 13.21), the typical substrates consist of small organic monocarboxylic and dicarboxylic acids, alcohols, and amino acids. However, the organic matter that persists in sediments through the preceding zones of aerobic and anaerobic metabolism is present as complex polymerized biological residues. For this reason, sulfate reducers rely on a precursor group of organisms to break down organic matter to simpler compounds by fermentation. The "quality" (susceptibility to microbial attack) of organic matter in sediments may have a dramatic effect on overall sulfate reduction rates in sediments. These rates can vary over many orders of magnitude (see Section 9.5.2).

A number of other ecological factors beyond abundance and type of food may influence the activities of SRB. These factors include temperature, salinity, sulfate concentration, and pH. Specific species of SRB are adapted to specific temperature ranges. These ranges can be divided into psychrophilic (thermal optimum near 4 °C), mesophilic (optimum near 39 °C), thermophilic (optimum near 60 °C), and hyperthermophilic (optimum near 88 °C) (Madigan *et al.*, 2000). It is evident that sulfate-reducing microorganisms are viable over a temperature range spanning at least 0–100 °C. The effect of temperature on a specific sedimentary setting is nonlinear and cannot be simplified to a single temperature factor. This complexity may be related to the highly variable nature of organic matter (Westrich and Berner, 1988). In summary, temperature may influence the species present in a sedimentary environment, and along with food source may influence rates of reduction of sulfate.

Neither variation in salinity nor sulfate concentration seems to have a major impact on the viability of SRB, although very high salinity may impact the sulfate reduction rate. SRB are able to metabolize sulfate down to very low concentrations. Half saturation constants (k_m values) of 70 μM and 200 μM for marine strains have been reported (Ingvorsen and Jorgensen, 1984). SRB have been recognized in the entire spectrum of water types from dilute freshwater up to saturation with halite. For example, one study of a series of salt-evaporation ponds in San Francisco Bay (Klug *et al.*, 1985) reported sulfate-reduction rates from sediments with pore-water salinities ranging from 33‰ up to 300‰. Maximum sulfate reduction rates, which occurred in the upper 1 cm of sediment, tended to decrease with increasing salinity from values as high as 16 μmol sulfate reduced per gram wet sediment per day, to values of 0.04 μmol reduced per cm or less in the sediment with salinity of 150‰ and 300‰.

Below the top 1–2 cm, however, reduction rates were similar at all salinities. The Dead Sea with a total salt content of ~310 g L^{-1} is a second example of a highly saline environment where microbial sulfate reduction is clearly occurring, and, in fact, dominating the geochemistry of sulfur (Nissenbaum and Kaplan, 1976). Very high rates of sulfate reduction have been recorded in microbial mats forming in high salinity environments (Habicht and Canfield, 1997). Like temperature, salinity may influence the species of SRB present in the sediment (Pfennig *et al.*, 1979).

The influence of pH on SRB has not been systematically studied. Most Earth-surface environments tend to fall in a rather narrow pH range of 5–8.5, and SRB are not markedly impacted by pH changes within this range (Fauque, 1995). There is research indicating that at least some species of SRB are active at the extremes of pH. One study looked in detail at sulfate reduction in an acid (pH 3) strip mine lake in Indiana (Konopka *et al.*, 1985). They found that sulfate reduction was occurring microbiologically at rates comparable to those in pH neutral settings. At the other extreme, H_2S occurs at concentrations of 120 mM in the bottom waters of Soap Lake in the state of Washington. This lake is saline and has a pH of 9.8. Sulfur isotopic data indicate that the sulfide formed microbiologically (Tuttle *et al.*, 1990).

9.2.3 Sulfur Isotopic Fractionation during Bacterial Reduction

9.2.3.1 Laboratory studies

One of the characteristics of sulfate-reducing microorganisms that has proved most useful for documenting their activity in the natural environment is their ability to fractionate stable sulfur isotopes during the reduction process. The aqueous sulfide produced is enriched in the light stable isotope of sulfur (^{32}S) compared to the heavy isotope (^{34}S) during reduction. In this chapter, the term "isotopically light" indicates relative enrichment in this light isotope. This fractionation was first demonstrated by Thode *et al.* (1951), who recognized a 10‰ enrichment in ^{32}S relative to the starting sulfate. A number of subsequent laboratory research efforts demonstrated variable isotope fractionation effects associated with dissimilatory sulfate reduction ranging from +3‰ to −46‰ (see an earlier review by Chambers and Trudinger, 1979). These studies reported that isotope enrichment is inversely proportional to sulfate reduction rate. This inverse relation to rate was addressed conceptually by Rees (1973), who pointed out that in order for fractionation to occur, there must be a back reaction involving release of isotopically heavy intermediates in the reduction pathway. Otherwise, if all sulfate that enters the cell undergoes reduction, mass balance requires there can be no net isotopic enrichment of ^{32}S in the product sulfide. Rapid rates of reduction imply rapid throughput of intermediates along the reduction pathway with little opportunity for buildup and back reaction of these intermediates. Conversely, slow rates of reduction allow isotopically heavy intermediates in the pathway to accumulate and back-react so that they may be ultimately released.

Older experimental work utilized a limited number of species of microorganisms and electron donors. A more recent survey (Detmers *et al.*, 2001) evaluated isotopic fractionation by 32 strains of sulfate-reducing microorganisms spanning a wide range of temperatures and electron donors (Table 2). The overall fractionations observed ranged from −3‰ to −42‰. Organisms that oxidize the organic substrate completely to bicarbonate typically produced larger fractionations (between −15‰ and −42‰, average 25‰) compared to organisms that incompletely oxidize the substrate, and excrete acetate (−2‰ to −18.7‰, average −9.5‰). There was no trend in isotope fractionation with optimum growth temperature of the organism.

Recent work has also addressed the impact of elevated temperature and decreased SO_4^{2-} concentration on sulfur isotopic fractionation factors during sulfate reduction. Canfield *et al.* (2000)

Table 2 Cell-specific fractionation factors of sulfate-reducing prokaryotes.

Microorganism	*Isolated from*	*Electron donor* (mM)	*Fraction factor* (‰)
Complete oxidizing			
Desulfonema magnum	Marine mud	Benzoate (3)	42.0
Desulfobacula phenolica	Marine mud	Benzoate (3)	36.7
Desulfobacterium autotrophicum	Marine mud	Butyrate (20)	32.7
Desulfobacula toluolica	Marine mud	Benzoate (3)	28.5
Desulfotomaculum gibsoniae	Freshwater mud	Butyrate (20)	27.8
Desulfospira joergensenii	Marine mud	Pyruvate (20)	25.7
Desulfotignum balticum	Marine mud	Butyrate (20)	23.1
Desulfofrigus oceanense	Arctic sediment	Acetate (20)	22.0
Desulfobacter sp. ASv20	Arctic sediment	Acetate (20)	18.8
Desulfobacca acetoxidans	Anaerobic sludge	Acetate (20)	18.0
Desulfococcus sp.	Marine mud	Pyruvate (20)	16.1
Desulfosarcina variabilis	Marine mud	Benzoate (3)	15.0
Incomplete oxidizing			
Desulfonatronum lacustre	Alkaline lake mud	Ethanol (20)	18.7
Archaeoglobus fulgidus strain Z	Submarine hot spring	Lactate (20)	17.0
Thermodesulfovibrio yellowstonii	Thermal vent water	Lactate (20)	17.0
Desulfotomaculum thermocistemum	Oil reservoir	Lactate (20)	15.0
Desulfohalobium redbaense	Saline sediment	Lactate (20)	10.6
Desulfocella halophila	Great salt lake	Pyruvate (20)	8.1
Desulfobulbus "marinus"	Marine mud	Propionate (20)	6.8
Desulfotalea arctica	Arctic sediment	Lactate (20)	6.1
Desulfovibrio sp. strain X	Hydrothermal vent	Lactate (20)	5.4
Themtodesulfobacterium commune	Thermal spring	Lactate (20)	5.0
"Desulfovibrio oxyclinae"	Hypersaline mat	Lactate (20)	4.5
Desulfotalea psychrophila	Arctic sediment	Lactate (20)	4.3
Desulfovibrio profundus	Deep sea sediment	Lactate (20)	4.1
Desulfovibrio halophilus	Hypersaline microbial mat	Lactate (20)	2.0
Hydrogen and formate			
Desulfobactetium autotrophicum	Marine mud	H_2	14.0

After Detmers *et al.* (2001).

measured fractionation factors in marine sediment whose *in situ* temperature was up to 90 °C. Sediments maintained at temperatures up to 85 °C were found to reduce SO_4^{2-} with fractionation factors of 13–28‰. Habicht *et al.* (2002) studied isotopic fractionation factors as a function of SO_4^{2-}concentration. They found high fractionations of up to 32‰ at SO_4^{2-} concentrations of 200 μM or greater, whereas fractionation dropped dramatically below this concentration. The low SO_4^{2-} limit for large fractionation is less than 1% of the modern seawater concentration of 28 mM.

9.2.3.2 Field studies

The laboratory studies cited above have been compared to measurements made in natural populations of SRB (Canfield, 2001; Habicht and Canfield, 1996, 1997). The two studies by Habicht and Canfield (1996) determined isotope fractionation in natural populations of SRB for microbial mats. These mats are quite organic rich and rates of sulfate reduction are very high compared to most other natural settings. Nonetheless, the measured isotope fractionations were similar to those measured in the laboratory. A subsequent paper by the same authors (Habicht and Canfield, 1997) looked at isotopic fractionation in marine mud. Figure 3 is a comparison of the field and laboratory determined fractionations plotted as a function of rate of reduction from that paper. The two sets of data are similar except that the fractionations from natural populations have a minimum near 20‰, whereas the lab cultures show occasional smaller values (2–20‰), particularly when grown on H_2. Habicht and Canfield (1997) took this difference to indicate that H_2 is not a major substrate for SRB.

Canfield (2001) undertook a comprehensive study of isotope fractionation by natural populations of SRB in sediment from a small marine lagoon located off the Danish coast. He used a flowthrough reactor and controlled the sulfate concentration, temperature, and ultimately the nature of the organic substrate. High fractionations of 30‰ and 40‰ were found when the SRB metabolized with natural organic substrate at environmental temperatures of 25 °C. The experiment continued until depletion of the natural substrate occurred, and then acetate, ethanol, and lactate were added from an external source. The isotopic fractionations observed ranged from 7‰ to 40‰, and increased fractionations correlated with decreased rates of sulfate reduction as noted in the laboratory studies. An exception occurred at low temperatures, where decreased reduction rates correlated with decreased fractionations. Canfield (2001) attributed that low-temperature behavior to increased viscosity of the cell membrane that inhibits transport of sulfate across it.

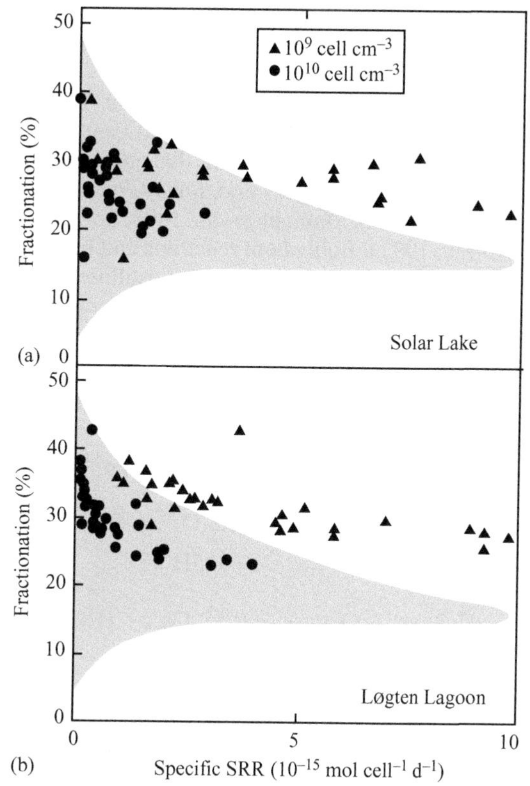

Figure 3 Fractionation of SO_4^{2-} by SRB versus specific sulfate reduction rate (SRR). The shaded area is the range of values observed in laboratory experiments. The points are from field data assuming two different densities of cells for comparison with lab data (Habicht and Canfield, 1997) (reproduced by permission of Elsevier from *Geochim. Cosmochim. Acta*, **1997**, *61*, 5351–5361).

9.3 FORMS OF SULFUR IN MARINE SEDIMENTS

9.3.1 Dissolved Species

The dominant dissolved form of sulfur in seawater is sulfate (including the free form, SO_4^{2-}, plus ion pairs with the major cations of seawater). In pore water of near-surface marine sediments, sulfate also dominates, but it is progressively removed with increasing depth by bacterial reduction and supplanted by dissolved H_2S. Pore-water SO_4^{2-} and H_2S profiles are discussed more fully below.

Other dissolved forms of sulfur in sediments are, by comparison, present at much lower concentrations, but may nonetheless play a significant role in diagenetic processes. These include many of the species summarized in Table 1, but specifically thiosulfate ($S_2O_3^{2-}$), sulfite (SO_3^{2-}), polysulfides (S_x^{2-}), and elemental sulfur (S^0_{aq}, which also occurs in the solid phase). The presence of this spectrum of sulfur species and their cyclic interconversion in a redox gradient (such as is present in organic-matter-bearing marine sediments with oxygenated depositional waters) were recognized long ago by Bass Becking (1925) (see also Truper, 1982).

As will be discussed in a subsequent section, thiosulfate, $S_2O_3^{2-}$, likely plays a significant role as a reactive intermediate in sulfur diagenesis. It has a wide variety of sources and sinks (Fossing and Jorgensen, 1990; Habicht *et al.*, 1998; Jorgensen and Bak, 1991). Equivalent reactions can be written for sulfite and polysulfides. Thiosulfate forms by oxidation of both iron sulfide minerals and aqueous sulfide (Equations (2)–(6)). The oxidants are dissolved oxygen or iron/manganese-oxide minerals. These reactions occur both abiologically and microbiologically:

$$FeS_2 + 1.5O_2(aq) \rightarrow S_2O_3^{2-} + Fe^{2+} \quad (2)$$

$$FeS + O_2(aq) + H^+ \rightarrow 5S_2O_3^{2-} + Fe^{2+} + 0.5H_2O \quad (3)$$

$$H_2S(aq) + O_2(aq) \rightarrow 0.5S_2O_3^{2-} + H^+ + 0.5H_2O \quad (4)$$

$$H_2S(aq) + 7H^+ + 4FeOOH(\text{goethite}) \rightarrow 0.5S_2O_3^{2-} + 6.5H_2O + 4Fe^{+2} \quad (5)$$

$$2H_2S(aq) + 3H_2O \rightarrow S_2O_3^{2-} + 5H_2(aq)$$

(photosynthetic oxidation by *Calothrix*. sp. × *and Oscillatoria* sp.) (6)

Thiosulfate may be removed by reoxidation to sulfate in the uppermost (oxygenated) portions of sediments (Equation (7)):

$$S_2O_3^{2-} + 2O_2(aq) + H_2O \rightarrow 2H^+ + 2SO_4^{2-} \quad (7)$$

It may also be reduced microbiologically to H_2S, in which it serves as an electron acceptor analogous to sulfate (e.g., Equation (1)):

$$2\text{lactate} + S_2O_3^{2-} + 2H^+ \rightarrow 2H_2S + 2\,\text{acetate} + 2CO_2 + H_2O \quad (8)$$

Finally, it may undergo disproportionation, a microbial reaction (Canfield and Thamdrup, 1994) that is the inorganic equivalent of an organic fermentation reaction:

$$S_2O_3^{2-} + H_2O \rightarrow H_2S + SO_4^{2-} \quad (9)$$

Thamdrup *et al.* (1994a) made detailed measurements of thiosulfate and sulfite concentrations in a Danish salt marsh and subtidal marine sediments. Thiosulfate occurred at levels from less than 0.05 μM to 0.45 μM and sulfite at 0.1–0.8 μM. They reported that concentrations of both species increased with depth; lowest values were in the oxic and suboxic zones and highest were in the zone of sulfate reduction. They also determined rapid turnover times for thiosulfate of 4 h in the upper 1 cm of sediment, and attributed this rapid turnover to the competing production and consumption processes summarized above.

Luther *et al.* (1985) determined thiosulfate concentrations of 400–1,300 μM and sulfite contents of 6.8–7.9 μM from subtidal sediments off the coast of Massachusetts, USA. Later work (Luther *et al.*, 2001) also confirmed the presence of polysulfides in a microbial mat from Great Marsh, Delaware, and subtidal sediments from Rehoboth Bay, Delaware at concentrations of up to 40 μM. Elemental sulfur $\left(S_8^0\right)$ dominated near the surface of both environments, polysulfide, was detected at intermediate depth (6.5 cm in the case of the subtidal sediments), and only H_2S was present at greater depth.

9.3.2 Pyrite

Pyrite is the dominant solid-phase form of sulfur in organic-matter-bearing marine sediments. Pyrite is cubic FeS_2. It can have a wide range of crystal morphologies (Gait, 1978; Love, 1967) although only a subset of these are recognized from sedimentary environments. Canfield and Raiswell (1991) review pyrite morphology in modern and ancient sediments. They recognize framboidal, aggregated, bladed, and equant morphologies. Framboids (from the French frambois-raspberry) consist of densely packed generally spheroidal aggregates of discrete equigranular submicron-sized pyrite crystals (Love, 1967). The individual microcrysts are usually cubes or pyritohedra. Framboidal pyrite forms during the early stages of diagenesis, and retains a size distribution characteristic of its depositional environment during later stages of diagenesis (Wignall and Newton, 1998; Wilkin *et al.*, 1996). Along with framboids, the most abundant morphology of pyrite in sediments is small (generally <2 μm) single crystals (Love, 1967).

Aggregated pyrite consists of small pyrite crystals forming irregular rounded clots, which may partly coalesce. According to Canfield and Raiswell (1991), aggregates are less regular in size and shape than framboids. Bladed crystals elongate and tend to occur as (rare) fillings in chambered organisms. Equant pyrite consists of single crystals with well-developed crystal faces. These range in size from minute, micrometer-sized crystals up to large millimeter-sized varieties. The smaller-sized variety is by far the most common. Framboids and small euhedra are the dominant morphologies found in sediments. The coarser varieties are generally only recognized in ancient sediments, suggesting that they may arise during later diagenesis.

The mineral chemistry of pyrite has been extensively described by Vaughan and Craig (1978). Pyrite has a dimorph, Marcasite that also has the nominal formula FeS_2, but is orthorhombic rather than cubic. Marcasite is metastable with respect to pyrite, and will convert to pyrite over time (Murowchick, 1992). Marcasite has not been reported in recent marine sediments. Likewise, the mineral pyrrhotite ($Fe_{1-x}S$) is extremely uncommon in low-temperature sedimentary environments, although it is known from some ancient lake beds (Tuttle and Goldhaber, 1993), and in some oil-bearing strata (Goldhaber and Reynolds, 1991; Reynolds *et al.*, 1990b).

9.3.3 Metastable Iron Sulfides

In addition to pyrite, there are two additional iron-sulfide minerals commonly found in recent sediments. These phases have been termed metastable iron sulfides, because they are thermodynamically unstable with respect to transformation to pyrite, or to a mixture of pyrite plus pyrrhotite (Berner, 1967). They are also known as acid-volatile iron sulfides, because in contrast to pyrite they are soluble in nonoxidizing mineral acids such as HCl. The dominant acid-volatile sulfides found in both natural and experimental systems are mackinawite and greigite.

Mackinawite is a tetragonal, sulfur-deficient Fe(II) sulfide with the formula FeS_{1-x}. The initial precipitate formed during the rapid mixing of H_2S and Fe(II) is an amorphous iron sulfide that transforms very rapidly to a poorly crystalline form of mackinawite. The properties of mackinawite have been extensively reviewed (Morse *et al.*, 1987; Rickard *et al.*, 1994; Vaughan and Craig, 1978). The black color seen below the sediment–water interface in many marine and nonmarine sediments is probably largely due to mackinawite, although direct confirmation of the presence of this phase has proven difficult because it is extremely fine grained and oxidizes rapidly.

Greigite is the thiospinel of iron (Skinner *et al.*, 1964). It has the formula Fe_3S_4. It is the sulfur analogue of magnetite. Both natural and synthetic greigite are sooty black powders that are strongly ferromagnetic (Dekkers and Schoonen, 1996). Its properties have been reviewed by Morse *et al.* (1987). Magnetotactic bacteria may form greigite (Konhauser, 1998; Neal *et al.*, 2001; Posfai *et al.*, 1998, 2001). Greigite dominates the magnetic properties of some modern and even some ancient sedimentary rocks (Krs *et al.*, 1990; Reynolds *et al.*, 1990a). Laboratory synthesis of the metastable iron sulfides is described below. Both sulfides play an important role in the pathway of pyrite formation.

9.3.4 Organic Sulfur

In some sedimentary environments such as sapropels, sulfur formed by the assimilatory processes described above may be significant (Canfield *et al.*, 1998). In marine plankton, the observed S/C weight ratio is ~0.02, and similar ratios may be observed in surface marine sediments (Suits and Arthur, 2000). More typically, however, the diagenetic formation of organosulfur compounds from microbially produced H_2S is the dominant pathway of organosulfur formation, and can be a significant overall sink for reduced sulfur in marine sediments. S/C ratios tend to increase with depth in the sediment column (Francois, 1987). In most environments, authigenic organosulfur compounds are second only to pyrite as products of sulfur diagenesis. In some iron-poor sediments (where pyrite formation is inherently minor), the formation of organosulfur may dominate (Canfield *et al.*, 1998; Ferdelman *et al.*, 1991). For example, in diatomaceous sediments of the Peru margin, S/C weight ratios at depth in the sediment column fall in the range 0.06–0.33 (Mossmann *et al.*, 1991; Suits and Arthur, 2000). The detailed pathways of formation of specific organosulfur compounds or compound classes are complex and are outside the scope of this review, and the reader is referred to published literature (e.g., Adam *et al.*, 2000; Eglinton *et al.*, 1994; Gelin *et al.*, 1998; Kohnen *et al.*, 1991, 1989, 1990; Nissenbaum and Kaplan, 1972; Sinninghe Damste *et al.*, 1998; Vairavamurthy and Anonymous, 1993; Vairavamurthy *et al.*, 1997; van Kaam-Peters *et al.*, 1998; Wakeham, 1995; Werne *et al.*, 2000).

9.3.5 Elemental Sulfur

Elemental sulfur is an oxidation product of dissolved sulfide. It may form inorganically, but in near-surface marine sediments where dissolved sulfide exists close to the sediment–water interface, it commonly forms because of the metabolic activities of sulfide oxidizing bacteria. Bacterial mats of filamentous sulfur producing *Beggiatoa* spp. have been observed in the surface of near-shore marine sediments (Schimmelmann and Kastner, 1993; Troelsen and Jorgensen, 1982). Elemental sulfur can be stored within invaginations of the bacterial cell wall in the form of quasi-liquid minute droplets of metastable polysulfur $\left({}^{-}O_3S - S\ldots S - S - O_3^{-}\right)$ (Steudel *et al.*, 1987). Schimmelmann and Kastner (1993) gave evidence that this polysulfur is converted to solid elemental sulfur on the timescale of a few years.

9.4 MECHANISM OF PYRITE FORMATION

9.4.1 Evidence from Experimental Studies

Experimental investigations of pyrite formation provide important constraints on sedimentary geochemistry. Beginning with the pioneering work of Allen *et al.* (1912) and later that of Berner

(1962, 1964b,c), the formation of pyrite at Earth surface conditions has been the subject of a number of detailed laboratory studies. Much of the earlier literature is reviewed by Morse *et al.* (1987). These laboratory investigations have resulted in the recognition of several potential mechanisms for formation of pyrite at $T<100\,°C$. All these mechanisms involve redox reactions because the ultimate source of sulfur in pyrite, H_2S, is more reduced than the disulfide in pyrite. These redox reactions include:

(i) direct precipitation of pyrite from homogeneous polysulfide (S_x^{2-}) solutions:

$$Fe(HS)^+(aq) + S_x^{2-}(aq) \rightarrow FeS_2 + S_{x-1}^{2-} + H^+ \quad (10)$$

(ii) progressive sulfidation and conversion of solid iron monosulfide to pyrite by an oxidized aqueous sulfur species such as polysulfide:

$$FeS(s) + S_n^{2-} \rightarrow FeS_2(s) + S_{n-1}^{2-} \quad (11)$$

(iii) direct reaction of H_2S with a precursor solid monosulfide releasing $H_2(g)$:

$$FeS(s) + 2H_2S(aq) \rightarrow FeS_2(s) + 2H_2(g) \quad (12)$$

(iv) iron loss from a precursor solid monosulfide by an oxidation mechanism:

$$2FeS(s) + \tfrac{1}{2}H_2O + \tfrac{3}{4}O_2(aq) \rightarrow FeS_2(s) + FeOOH(s) \quad (13)$$

With the exception of Equation (10), all these pathways involve a precursor iron monosulfide phase. Direct precipitation of pyrite from solution (Equation (10)) is strongly inhibited. This is due to the difficulty of direct nucleation of pyrite, leading to very large supersaturation with respect to pyrite in experimental and natural solutions (Schoonen and Barnes, 1991a). Experimental studies have thus focused on the role of one or more iron monosulfide precursors to pyrite, which have long been recognized as intermediates in sedimentary pyrite formation (see review by Morse *et al.*, 1987). Poorly crystalline mackinawite is the initial product of reaction of H_2S with aqueous or solid iron sources at Earth-surface temperature (Morse *et al.*, 1987; Rickard, 1969). Thus, the fate of the initial mackinawite precipitate is key to understanding pyrite genesis.

Rickard (1997) and Rickard and Luther (1997) studied the direct reaction of H_2S with mackinawite (Equation (12)) and presented experimental and theoretical data in support of this pathway. In contrast, a group at Pennsylvania State University (Benning *et al.*, 2000; Wilkin and Barnes, 1996) argued that in the absence of an oxidizing agent or prior exposure to air, mackinawite is stable in the presence of H_2S over a range of temperature, time, and chemical conditions, thus implying that reaction (4) is not a dominant pathway of pyrite formation. They argued that iron monosulfide will transform to pyrite in the presence of an oxidizing agent such as dissolved O_2 and/or in the presence of an aqueous or solid sulfur species more oxidized than H_2S such as polysulfide, thiosulfate, or elemental sulfur.

Reactions involving mackinawite and an oxidized sulfur species have been repeatedly shown to lead to pyrite formation (e.g., Berner, 1969; Rickard, 1969, 1975). In addition, Wilkin and Barnes (1996) and Benning *et al.* (2000) have shown that pyrite formation is exceptionally rapid when the mackinawite is "pre-oxidized" (e.g., exposed briefly to air) prior to the experiment. Based partly on X-ray photoelectron and Auger spectroscopy results of pyrrhotite oxidation (Mycroft *et al.*, 1995), Wilkin and Barnes (1996) hypothesized that this oxidative exposure initiates an iron-loss pathway similar to Equation (13). In sulfidic solutions, Fe(II) oxyhydroxides, shown as a product in this reaction, would not accumulate, but instead would undergo reductive dissolution by a reaction similar to Equation (14):

$$4FeOOH + \tfrac{1}{2}H_2S + 7H^+ \rightarrow 4Fe^{2+} + \tfrac{1}{2}SO_4^{2-} + 6H_2O \quad (14)$$

The loss of one-fourth of the iron from mackinawite with simultaneous oxidation of one-half of the initial iron leads to formation of greigite (Wilkin and Barnes, 1996; Equation (15)). Iron loss (as opposed to sulfur gain—Equation (11)) is energetically favored, because mackinawite and greigite share the same close-packed sulfur sublattice:

$$4FeS(s) + \tfrac{1}{2}O_2(aq) + 2H^+ \rightarrow Fe_3S_4 + Fe^{2+} + H_2O \quad (15)$$

Further work may be required to define the full spectrum of reactions forming pyrite in sediments. The pathways involving oxidation such as in Equation (13) seem appropriate in the upper portions of marine sediments, where most pyrite actually forms (see below) and where oxidants such as O_2 or Fe/Mn oxides are abundant. However, transformation of iron monosulfide to pyrite also occurs in deeper sediment strata in the presence of excess H_2S. At these deeper levels, access to oxidants is more problematic, and a reaction such as Equation (12) may be required.

Greigite has been repeatedly implicated as an intermediate in pyrite formation (Benning *et al.*, 2000; Schoonen and Barnes, 1991b; Sweeney and Kaplan, 1973; Wang and Morse, 1994; Wilkin and Barnes, 1996). Evidence that the greigite to pyrite transformation can be a solid-state reaction comes from observations that intermediate stages of this reaction lead to grains with greigite cores with pyrite rims (Benning *et al.*, 2000; Sweeney and Kaplan, 1973). Greigite may also be a required intermediate in the generation of specific morphological variants of pyrite (see below).

9.4.2 Isotope Effects during Experimental Pyrite Formation

Experimental studies have demonstrated that inorganic reactions during pyrite formation produce minimal isotopic fractionation. Precipitation of iron monosulfide by reaction of H_2S with an iron source results in fractionations generally less than 1.5‰ (Bottcher *et al.*, 1998; Price and Shieh, 1979), and subsequent transformation of monosulfide to pyrite shows similarly small fractionations (Price and Shieh, 1979).

9.4.3 Origin of Morphological Variations in Pyrite

As noted above, pyrite in sediments occurs in a variety of morphologies. Of particular interest, however, is the origin of framboids. This unusual morphology is of very widespread occurrence in sediments, but has been surprisingly difficult (although not impossible) to synthesize in the laboratory (Berner, 1969; Butler and Rickard, 2000; Farrand, 1970; Sweeney and Kaplan, 1973). The most convincing explanation of the origin of framboids is based on the strongly ferromagnetic properties of greigite. Framboids may arise by aggregation of uniformly sized magnetic greigite microcrystals to form greigite framboids, followed by replacement of greigite by pyrite (Wilkin and Barnes, 1997).

More generally, pyrite morphology is a function of the degree of supersaturation of the solution (Murowchick and Barnes, 1987; Wang and Morse, 1994). At very low degrees of supersaturation and mineral growth, ions are added to the growing surface in the most energetically favorable sites. For pyrite, the lowest energy morphology is the cube. Very rapid growth from highly supersaturated solutions is characterized by low selectivity for the site of addition of ions. This leads to spherulitic forms. Wang and Morse (1994) recognized the following sequence of pyrite morphologies as a function of solution supersaturation:

Cube	Combination of cube and octahedron	Octahedron	Spherulite
		Increasing pyrite supersaturation →	

9.5 SULFUR DIAGENETIC PROCESSES IN MARINE SEDIMENTS

9.5.1 Depth Distribution of Diagenetic Sulfur Products

Diagenetic processes in sediments occur over varying periods of time. For nearshore sediments, sulfur diagenetic processes may occur over months to centuries, whereas in deep-sea sediments, similar reactions may occur over times exceeding 10^6 yr. For this entire spectrum of sedimentary environments, a useful concept is steady state. This is a time-invariant distribution of diagenetic products. At steady state, porewater and solid-phase profiles do not change as sediment accumulates. Achieving this steady-state condition requires that a large number of variables such as sedimentation rate, organic matter deposition, bottom-water conditions, and many others are held constant. The discussions that follow are predicated upon this concept except where noted.

There are three major fates for microbiologically produced aqueous sulfide. Listed in order of their relative importance, these are: loss to the overlying water by oxidation, fixation during formation of pyrite, and fixation by formation of organosulfur compounds. In addition, two minor products of sulfur diagenesis, elemental sulfur and metastable iron sulfides, tend to be transient and are most commonly recognized in the uppermost part of the sulfate reduction zone. These conclusions are based on a very large number of studies of the diagenesis of sulfur in a variety of depositional settings (Table 3). Sulfate reduction has been recognized from the most rapidly deposited nearshore environments, to portions of the central ocean basins (Canfield, 1991; D'hondt *et al.*, 2002; Jorgensen, 1982). Despite this range in environmental settings, it is possible to generalize from the large number of studies to recognize overall similarities in the diagenetic pathway. Specifically, chemical analyses of sediments and pore fluids from core samples from a variety of marine settings indicate that there is a characteristic distribution of reactions and products with depth (time). Diagenetic reactions occur in a series of depth zones. Figure 4 shows these zones schematically. The uppermost zone (I) is nearly continuously oxic. The next zone (II) is strongly influenced by the life activities of macro-infauna

Table 3 Studies on sedimentary-sulfur geochemistry for selected depositional settings.

Location	*Water depth (m)*	*S parameters measured*[a]	*References*
Great Marsh (Delaware USA Salt Marsh)	0–1	Spy, Sorg, S^0, AVS	Ferdelman *et al.* (1991)
Limfjorden, Kysing Fjord, Denmark	0.5–12	SO_4, H_2S, Spy, AVS, SRR	(Detmers *et al.*, 2001; Howarth and Jorgensen, 1984; Jorgensen, 1977b)
Kysing Fjork Denmark	<1	SO_4, H_2S, S^0, Spy, AVS, $\sum S$, SRR	(Fossing and Jorgensen, 1990)
Cape Lookout Bight, North Carolina	8	SO_4, H_2S, $\sum S$, AVS, δSO_4, δSpy	(Chanton *et al.*, 1987a,b)
Long Island Sound, USA	8–34	SO_4, H_2S, Spy, AVS, SRR	(Aller, 1980a; Goldhaber *et al.*, 1977)
Gulf of Mexico outer shelf	20–112	SO_4, $\sum S$	(Filipek and Owen, 1980)
Baltic Sea	100–200	$\sum S$, Spy, Sorg, S^0, δSO_4, δSpy, SRR	(Hartmann and Nielsen, 1969; Sternbeck and Sohlenius, 1997; Yu *et al.*, 1982)
Peru Margin	252–457	SO_4, H_2S Spy AVS, Sorg, S^0	(Mossmann *et al.*, 1991; Suits and Arthur, 2000)
Southern California Borderland	100–1,500	SO_4, H_2S, Spy, S^0, Sorg, AVS, δSpy	(Kaplan *et al.*, 1963; Schimmelmann and Kastner, 1993; Sweeney and Kaplan, 1980)
Jervis Inlet, British Columbia	650	SO_4, H_2S, S^0, Spy, Sorg, δSH_2S, δSorg	(Francois, 1987)
Gulf of California	476–3,361	SO_4, H_2S, $\sum S$, Spy, AVS, δSO_4, δSH_2S, δSpy	(Berner, 1964a; Goldhaber and Kaplan, 1980)
Arabian Sea	>500	$\sum S$, Spy, δSpy	(Lueckge *et al.*, 1999; Passier *et al.*, 1997; Schenau *et al.*, 2002)
Black Sea	100–2,000	$\sum S$, Spy, Sorg, S^0, δSO_4, SRR	Various including (Jorgensen *et al.*, 2001; Lyons, 1997; Lyons and Berner, 1992; Wilkin and Arthur, 2001) among many others
Northeast Atlantic	158–4,920	SO_4, SRR	(Battersby *et al.*, 1985)
Cariaco Basin	900–1,400	$\sum$Spy, δSpy,	(Lyons *et al.*, 2003; Werne *et al.*, 2000)
Western North Atlantic	1,291–4,595	$\sum S$, Spy, SO_4	(Cagatay *et al.*, 2001)
New Jersey Margin; DSDP Hole 603	4,796	$\sum S$, Spy, δSpy	(Bonnell and Anderson, 1987; Dean and Arthur, 1987)
Atlantic Ocean Basin	Various	$\sum S$, δSpy	(Lew, 1981)

[a] $\sum S$: content of total solid-phase sulfur. Spy: pyrite sulfur content. δSpy: isotopic composition of pyrite. Sorg: organic sulfur. S^0: elemental sulfur. SO_4: pore-water sulfate concentration. δSO_4: isotopic composition of pore-water sulfate. H_2S: pore-water sulfide concentration. δSH_2S: isotopic composition of pore-water sulfide. SRR: sulfate reduction rate.

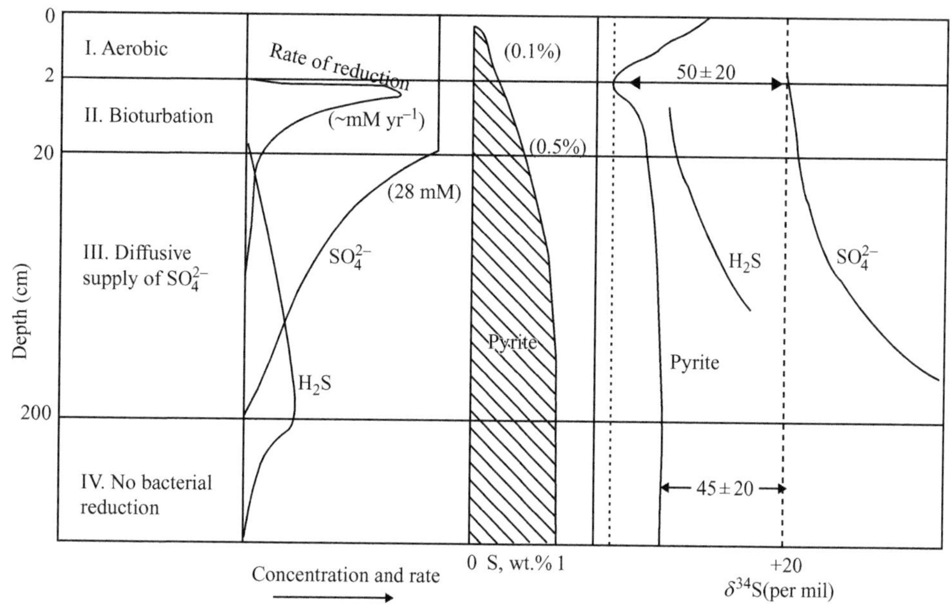

Figure 4 Idealized depth dependence of pore-water and solid-phase sulfur constituents and their sulfur isotopic values. The zones labeled I, II, III, and IV are classified according to processes of transport of sulfur and redox processes within each zone. See text for discussion.

(bioturbation). Zone III is characterized by evolving pore-water profiles of sulfur species (decreasing SO_4^{2-} and variable H_2S). The deepest zone (IV) is characterized by the near absence of diagenetic sulfur processes. Note that this section focuses only on reactions involving sulfur geochemistry, whereas as shown in Figure 4 there are competing reactions involving other electron acceptors. In some sediments, these other acceptors, particularly the iron and manganese oxides, may dominate organic remineralization (Aller *et al.*, 1986; Canfield *et al.*, 1993), although more commonly sulfate dominates over all but aerobic processes. It should also be noted that Figure 4 shows an idealized steady-state sequence. Individual zones may be absent or modified in certain circumstances. An obvious example is that zones I and II, which are directly or indirectly related to the presence of oxygen, are absent in anoxic depositional environments, and zone II may be absent or insignificant under dysoxic (low-oxygen) depositional waters.

Zone I is characterized by the presence of oxygen in the pore waters. This aerobic zone may extend a few millimeters into nearshore organic-matter-rich sediments and considerably deeper in hemi-pelagic sediments. In most environments, this upper layer is extensively mixed by the activities of macroorganisms (bioturbated) and probably stirred by waves and currents as well. Because of this activity, pore-water SO_4^{2-} concentrations are essentially those of the overlying seawater. Iron oxides are abundant, giving these sediments a brown color. Manganese oxides may be present as well. Free H_2S is absent due to its loss by aerobic oxidation discussed above and/or buffering of H_2S to very low concentration by reaction with iron oxides (Goldhaber and Kaplan, 1974). Elemental sulfur may be present (Troelsen and Jorgensen, 1982). Within this zone (or the next), assimilatory sulfur may be lost by hydrolysis reactions. Surprisingly, pyrite may also be present in this zone, 10% or more of the amount found at depth. This pyrite may form in anaerobic "microenvironments" (Jorgensen, 1977a) within the overall oxic setting. This possibility is reinforced by measured sulfate reduction rates in zone I surface sediment, which indicate substantial activity by SRB (Canfield, 1993; Jorgensen, 1977a; Jorgensen and Bak, 1991). SRB can withstand exposure to O_2 for extended time periods and remain viable (Cypionka *et al.*, 1985). It is also likely that some portion of the shallow pyrite may have actually formed in deeper layers and then transported to the surface through biogenic mixing processes. However, oxidation by a variety of reactants, chiefly O_2, but also NO_3^- and MnO_2, is the primary fate for reduced sulfur in this zone (Luther *et al.*, 1982; Schippers and Jorgensen, 2001, 2002; Thamdrup *et al.*, 1994b).

Zone II is the zone of bioturbation. Near the top of this interval, sulfate reduction rates are at their maximum and decrease with depth, iron monosulfide phases reach a maximum and then decline, whereas pyrite increases systematically with depth reflecting net fixation of reduced sulfur. If the relative rates of sulfate reduction are less than sulfate resupply by the activities of macrofaunal organisms, sulfate concentrations may differ only slightly from bottom-water values due to mixing and irrigation activities of macroorganisms (Aller, 1980b, 1988; R. C. Aller and J. Y. Aller, 1998; Goldhaber *et al.*, 1977; Goldhaber and Kaplan, 1980). Zone II may be nearly fully open to addition of overlying seawater sulfate, depending upon the relative rates of bioturbation and sulfate reduction. Extensive reoxidation of solid-phase pyrite and iron monosulfides also occurs within this zone (Berner and Westrich, 1985; Sommerfield *et al.*, 2001). H_2S is generally not measurable (<1 μM) but is occasionally present in very low concentrations <60 μM (Goldhaber and Kaplan, 1980). In either case, the low values reflect oxidation, buffering by reaction with sedimentary iron, and biogenic mixing with sulfide-free overlying water. The net effect of extensive reoxidation of reduced-sulfur phases is that only a fraction of the total sulfate reduced to sulfide is retained within the sediments. Nonetheless, most pyrite is typically formed in the upper two zones (Berner, 1970; Hartmann and Nielsen, 1969; Mossmann *et al.*, 1991).

The high rates of sulfate reduction in this zone reflect a number of impacts of macrofaunal activity, many of which enhance the rate and/or extent of organic decomposition. These effects are summarized in Table 4 (after R. C. Aller and J. Y. Aller, 1998). These authors give a qualitative evaluation of the impact of bioturbation on diagenetic processes involving organic matter.

Zone III is the zone of diffusive transport. Sulfate in this interval decreases with depth, as does the rate of sulfate reduction. The steepness of the SO_4^{2-} gradient reflects the balance between rate of sulfate removal by SRB and rate of sulfate addition by diffusion plus a smaller contribution from net burial of pore water. Zone III is only partially open to sulfate addition from overlying seawater. The steepness of the sulfate gradient varies dramatically as a function of depositional setting. In the most organic-matter-rich rapidly deposited sediments (e.g., near man-made sewage outfalls), SO_4^{2-} may be totally removed from pore waters within the uppermost few centimeters. Where evaporites or migrating sulfate-bearing brines occur at depth, sulfate may be supplied from below (e.g., Kastner *et al.*, 1990). More typically, in nearshore sediments with organic carbon contents in the range 1–3wt.%, total SO_4^{2-} removal occurs between a depth of several tens of centimeters to a few meters. In hemi-pelagic sediments with a few tenths of a percent organic carbon, total sulfate removal may require hundreds of meters or not occur at all. The steepness of the SO_4^{2-} gradient is related to the overall sedimentation rate (Berner, 1978; Canfield, 1991; Goldhaber and Kaplan, 1975), because sedimentation rate, in turn, controls the metabolizability of organic matter and hence reduction rate (see below). The sulfur isotopic composition of SO_4^{2-} in this zone increases systematically with depth as the light isotope is preferentially removed to form H_2S.

Dissolved sulfide in this zone builds up because of its bacterial production. Its maximum concentration is less than the total sulfate reduced because a portion of the H_2S reacts with iron and organic matter to form insoluble products. At any depth, the concentration gradient of H_2S is kinetically controlled and reflects the balance between the rate of these removal processes, the rate of gain or loss by diffusion, and the rate of its formation by reduction. One of these processes, the production of H_2S, must cease when all SO_4^{2-} has been consumed. The net result is a concentration maximum that falls in a range from 1 μM to >10 mM. The depth of maximum pore-water H_2S commonly correlates closely with the depth of total SO_4^{2-} depletion. In most environments, H_2S persists at measurable concentrations (i.e., greater than a few micromolar) in pore waters to depths of a few centimeters to several meters below the

Table 4 Macrofaunal effects on organic carbon decomposition/remineralization.

Macrofaunal activity	*Effect*	*Effect on rate/extent of organic matter decomposition*
Particle manipulation	Substrate exposure, surface area increase	+
Grazing	Microbe consumption, bacterial growth stimulation	+
Excretion/secretion	Mucus substrate, nutrient release, bacterial growth/ stimulation	+
Construction/secretion	Synthesis of refractory or inhibitory structural products (tube linings, halophenols, body structural products)	−
Irrigation	Soluble reactants supplied, metabolite buildup lowered, increased reoxidation	+
Particle transport	Transfer between major redox zones, increased reoxidation, redox oscillation	+

Reproduced by permission of USGS from *J. Marine Res.*, **1998**, *56*, 905–936.

point at which SO_4^{2-} is removed. The essentially total removal of pore-water H_2S is a reflection of the availability of excess iron over sulfide sulfur in most sediments (see below). Pyrite content may increase gradually within zone III, but the rate of this increase is most rapid at the top of this zone. Frequently, increases in pyrite cannot confidently be distinguished from scatter in the data within this zone.

Zone IV lies below the depth of removal of pore-water H_2S. The absence of sulfur as sulfate or sulfide in pore waters precludes further pyrite formation, and signals the cessation of the typical early diagenetic sulfur reactions. However, during deeper burial and even after lithification, many additional processes can impact sedimentary-sulfur geochemistry. With increasing temperature, release of H_2S from organosulfur compounds takes place. As diagenesis passes into metamorphism, pyrite converts to pyrrhotite plus H_2S (Ohmoto and Goldhaber, 1997). Both these reactions provide additional sulfide for late pyrite formation. In addition, some sedimentary environments are open to migration of external fluids that may be H_2S bearing. Marine mud interbedded sandstones are candidates for addition of epigenetic fluids. In extreme cases, whole geographic regions may be impacted by migrating sulfide- (and metal)-bearing fluids (Goldhaber *et al.*, 2002, 1995). The result of these processes may range from the development of coarse overgrowths on earlier diagenetic pyrite, growth of sulfide nodules and layers, all the way to development of massive sulfide-ore bodies with tens of millions of tons of metal sulfide precipitates.

9.5.2 Rates of Sulfate Reduction

Sulfate reduction rates have been extensively quantified in marine sediments. There are a number of methods for measuring sulfate reduction rates in marine sediments, although addition of radioactive sulfate ($^{35}SO_4^{2-}$) and recovery of radiolabled ^{35}S reduced sulfur compounds (Jorgensen, 1978; Thode-Anderson and Jorgensen, 1989) is the most commonly utilized technique in coastal sediments. Incubation of sediments over time has also been successfully employed (Aller and Mackin, 1989). Mathematical modeling of sulfate profiles is an alternative that is widely utilized, particularly in deep-sea sediments, where sulfate reduction rates are too low for ratio-tracer studies (Canfield, 1991), and a combination of direct measurement in upper sediment layers (where rates are high) and modeling in deeper layers may be appropriate in many situations (Jorgensen *et al.*, 2001). A striking result of these rate measurements is the wide range of sulfate reduction rates that are recognized. Figure 5, from Canfield (1989b), illustrates this point. Sulfate reduction rates in this plot, integrated over the depth of the sediment column, vary by nearly seven orders of magnitude. It is important to note that these reduction rates are proportional to (scale with) sedimentation rate. Sedimentation rate is an important master variable for sulfate reduction rate. With decreasing sedimentation rate, the exposure of organic matter to aerobic oxidation during transit through the water column, as well as metabolism by macro- and microorganisms increases relative to rapidly deposited inshore sediments (Berner, 1978; Canfield, 1991; D'hondt *et al.*, 2002; Goldhaber and Kaplan, 1975).

9.6 SULFUR ABUNDANCE IN RECENT MARINE SEDIMENTS

9.6.1 Controls on Sulfur Abundance

Figure 6 is a histogram of the total sulfur concentration in weight percent in a cross-section of normal marine siliciclastic sedimentary environments (i.e., exclusive of sediments deposited under anoxic conditions, and those dominated by a carbonate matrix). The data come from a range of depths in the sediment column. As noted above, the dominant form of sulfur in these sediments is pyrite, typically >80%. The total sulfur analyses in Figure 6 have a mean of $\sim$0.6%, with very few analyses registering valuesgreater than 2%. The measured total sulfur abundances displayed in Figure 6 are greater than what would form from the simple burial of pore-water sulfate if the accumulating sediment pile were closed to additional SO_4^{2-} input. This "closed system" formation of pyrite would amount to only 0.1–0.3% by weight of pyrite sulfur (depending upon the porosity of the sediments). Addition of sulfur in excess of this closed system amount results from advection of sulfate during bioturbation and diffusion of sulfate from the overlying water.

Three variables are of potential importance in controlling the total amount of pyrite formed. These are the availability of: sulfate, iron, and organic matter. The effect of the first is relatively minor in marine sediments, whereas last two are more critical.

Sedimentation rate would, at first glance, seem to be an important control of total sulfur incorporation into marine sediments in view of the correlation between this parameter and sulfate reduction rate (Figure 5). This is not the case (Goldhaber and Kaplan, 1975; Volkov and Rozanov, 1983). The reason is that the total sulfate reduced (sulfide produced) is equal to the product of the sulfate reduction rate and the time over which the reduction occurs. The slower reduction rates occur for a longer time in more slowly deposited sediments, and the total sulfate reduced is not as dramatically affected. An exception is in very slowly deposited pelagic sediments, which lack

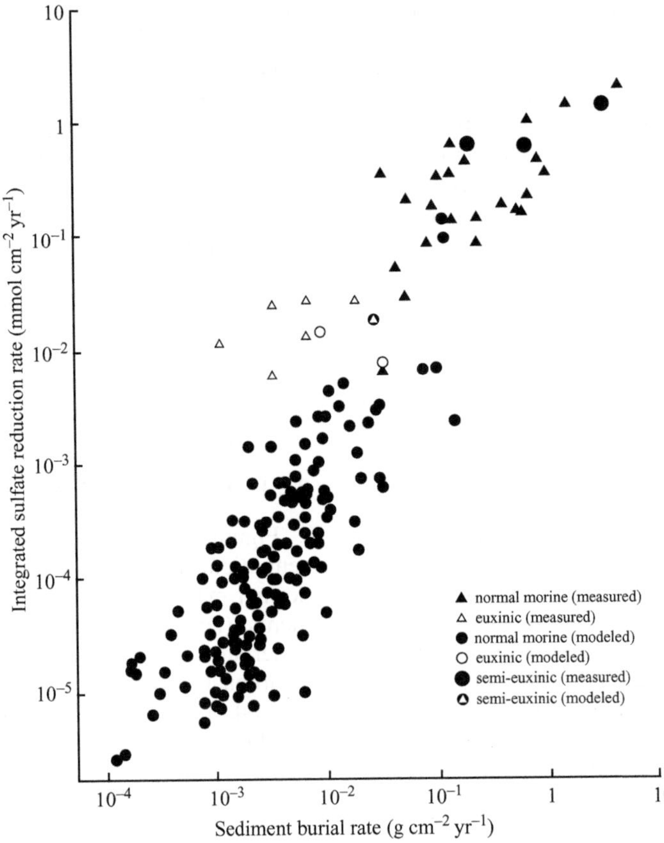

Figure 5 Rate of SO_4^{2-} reduction integrated over the entire depth of the sulfate profile plotted versus sedimentation rate (Canfield, 1989b) (reproduced by permission of Elsevier from *Deep-Sea Res. I*, **1989b**, *36*, 121–138).

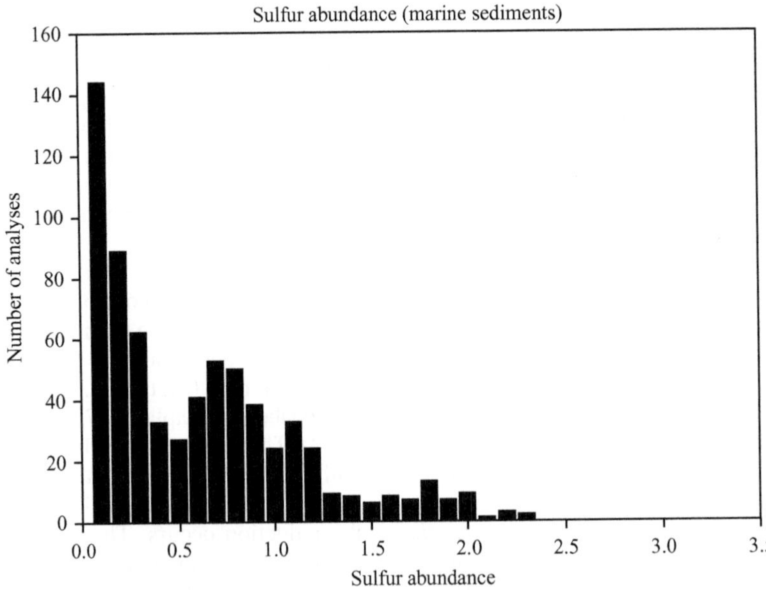

Figure 6 Histogram showing sulfur abundance in marine sediments. Data are from a variety of sources but exclude euxinic environments.

adequate metabolizable organic matter to support SRB. Thus, red pelagic oozes and other open-ocean sediments do not typically contain appreciable amounts of diagenetic sulfides.

As discussed above, sulfate concentration in the range found in marine pore water has essentially no effect on sulfate reduction rates until very low sulfate concentrations are reached. It is also true that most pyrite forms before pore-water sulfate concentrations reach such low levels (Figure 4). Thus, the influence of sulfate on the abundance of sedimentary pyrite is relatively minor in marine depositional settings. Reinforcing this conclusion are studies of modern marine environments with somewhat reduced salinities such as portions of the Sea of Azov (salinity 11–15 per mil), and the Baltic Sea (salinity 10–13 per mil). Sediments from these settings have total reduced-sulfur contents ranging well over 1% by weight (compare with Figure 6 for normal marine sediments), and thus show no evidence of reduced pyrite contents (Volkov *et al.*, 1983). Of course, in freshwater settings with very low sulfate concentrations, sedimentary-sulfur accumulation is severely limited by sulfate availability (Berner and Raiswell, 1984).

Iron availability may be a limitation on pyrite formation in many marine environments. However, the situation is complex, because iron availability is a relative concept that depends on a competition between rate of H_2S addition and rate of iron mineral reaction with H_2S. Various iron oxides (e.g., ferrihydrite, lepidocrocite, goethite, hematite, and magnetite) and silicates (e.g., chlorite, biotite, amphiboles, etc.) supply iron for pyrite formation. However, the dominant sources of iron for pyrite formation are the amorphous or poorly crystalline iron oxides (Canfield *et al.*, 1992; Goldhaber and Kaplan, 1974; Morse and Wang, 1997; Pyzik and Sommer, 1981). Table 5 shows a series of rate constants and half-lives compiled by Raiswell and Canfield (1996), which illustrate the wide spectrum of reaction rates of H_2S (assumed to be at a concentration of 1 mM) with iron minerals (see also Morse and Wang, 1997). Iron oxides such as ferrihydrite, lepidocrocite, and goethite react so rapidly that excess H_2S is removed from solution to levels below 1 μM on diagenetic timescales. Dissolved sulfide does not accumulate in pore waters until these reactive oxides are removed. At the other extreme, some iron silicates react with half-lives of over 10^5 yr. As discussed above, the exposure age of marine sediments to diagenetic reactants and products including H_2S may range from tens of years to over a million years. At the short-exposure end of this spectrum, iron-bearing silicate minerals will be essentially unreactive, whereas at the long-exposure end, the iron in such silicate minerals is at least partially converted to pyrite (Raiswell and Canfield, 1996).

A useful parameter tied to the reactivity of iron is the degree of pyritization (DOP) of iron. This parameter (Equation (16)) has been widely measured in both modern and ancient sediments (Raiswell, 1993; Raiswell *et al.*, 1988):

$$\text{DOP} = \text{pyrite Fe}/(\text{pyrite Fe} + \text{“reactive iron”}) \tag{16}$$

Berner (1970) pointed out that the reservoir of iron for pyrite formation may be smaller than the total sediment iron concentration, because some iron-bearing phases are essentially unreactive towards H_2S on the timescale of sediment diagenesis. He operationally defined "reactive" (towards H_2S) iron as that dissolved during a short exposure to hot 12N HCl (Berner, 1970). However, a further analysis of the reactive iron issue has shown that the concentrated HCl technique removes a considerable amount of iron that is unreactive to H_2S on diagenetic timescales. More gentle treatments using dithionite (Canfield, 1989a) and 1N HCl (Leventhal and Taylor, 1990) remove an iron fraction that more closely corresponds to the reactive iron oxides. The highly reactive iron in sediments is thus defined as the sum of dithionate or 1N HCl soluble iron plus that tied up in pyrite plus acid volatile sulfur (AVS). Thus defined, the fraction of highly reactive iron in a wide range of nearshore and deep-sea sediments from aerobic

Table 5 Rate constants and half-lives of sedimentary iron minerals with respect to their sulfidation.

Iron mineral	*Rate constant* (yr^{-1})	*Half-life*
Ferrihydrite	2,200	2.8 h
Lepidocrocite	>85	<3 d
Goethite	22	11.5 d
Hematite	12	31 d
Magnetite (uncoated)	6.6×10^{-3}	105 yr
"Reactive silicates"	3.0×10^{-3}	230 yr
Sheet silicates	8.2×10^{-6}	84,000 yr
Ilmenite, garnet, augite, amphibole	$\ll 8.2 \times 10^{-6}$	$\gg$ 84,000 yr

Reproduced by permission of USGS from *Am. J. Sci.*, **1992**, *292*, 659–683.

environments, as well as sediments accumulating in dysaerobic environments, is low (based on 219 analyses, average 25–28% of the total iron; 1σ uncertainly $\pm$10%).

Carbonate and opaline-silica-rich sediments are exceptions to these generalizations. They are very low in detrital iron, because calcareous and siliceous skeletal debris is much lower in iron than terrigenous material. In these dominantly biogenic sediments, iron may become limiting; the degree of pyritization is very high (>80%) and values tend to be constant. Plots of either total or reactive iron versus total sulfur generally result in a linear relationship. In these low iron sediments, pore-water H_2S builds up to very high concentrations (Thorstenson and Mackenzie, 1974). These low iron, high H_2S environments are favorable for the formation of high relative proportions of authigenic organosulfur compounds.

The primary control on the abundance of sedimentary sulfur in many normal marine siliciclastic marine sediments is related to the abundance of organic matter. This conclusion is based, in part, on a linear relationship between organic carbon and total sulfur that is commonly (although not always) observed for normal marine sediments (Figure 7). This correlation was first noted by Sweeney (1972) (see also Goldhaber and Kaplan, 1974). This important relationship has subsequently been examined by a number of researchers (Berner, 1982; Leventhal, 1983a,b, 1995; Lin and Morse, 1991; Morse and Berner, 1995; Raiswell and Berner, 1985, 1986) among others, and has been verified for Phanerozoic normal marine sediments (Raiswell and Berner, 1986). Berner (1982) determined that the slope of the C/S line for normal marine siliciclastic sediments is 2.8 $\pm$ 0.8. In effect, nearly all marine siliciclastic sediments deposited in oxic environments fall within a range of a factor of 2 in C/S ratio. An analysis of the origin of this correlation (Morse and Berner, 1995; Sweeney, 1972) shows that three factors controlling this ratio are closely coupled to each other. These are (Morse and Berner, 1995): (i) the fraction of the total organic carbon deposited that is metabolized; (ii) the fraction of metabolized organic carbon that is metabolized by SRB; and (iii) the fraction of the total sulfide formed by SRB that is buried as pyrite. It is truly remarkable that these factors either remain constant or scale with each other in a way that allows the observed constancy of the C/S ratio.

Tim Lyons (written communication 2002) has suggested that these three factors do scale with each other, and has provided a rationale for understanding the relation between carbon and iron limitation. He proposes that the Raiswell and Canfield (1998) paper defines a window for reactive iron. We can expect concentrations of pyrite up to ~1–2 wt.% where there is enough highly reactive iron present such that total organic carbon may be limiting. Note that in Figure 6, total sulfur in many marine sediments fall at or below this concentration range of total sulfur. He further suggests that in oxic settings where a fixed C/S ratio is observed, the required scaling of factors controlling the constancy of the C/S ratio is due to the coupling of iron and organic-carbon delivery to sediments. In offshore environments, this coupling may occur through stimulation of surface productivity (and hence organic carbon transport to sediments) by iron. In nearshore environments, that coupling may be through direct adsorption of organic matter to iron-bearing detritus (Keil and Hedges, 1993).

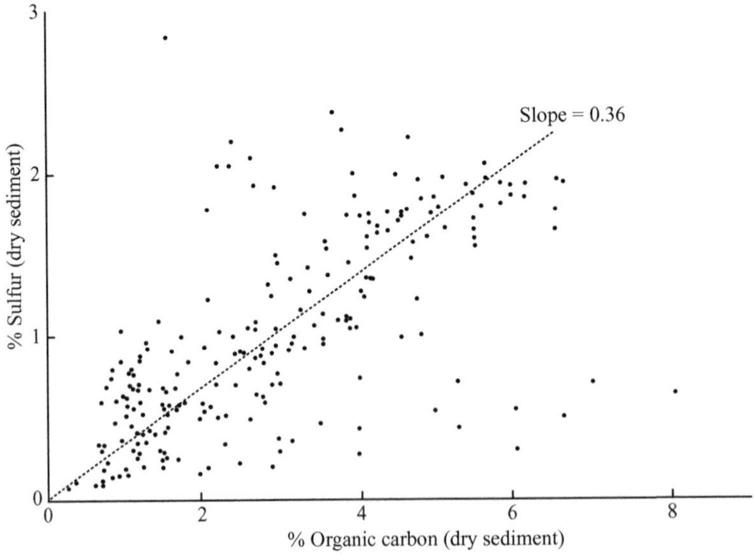

Figure 7 Plot of $\sum$S versus organic carbon (after Sweeney, 1980).

9.6.2 Consequences of Non-steady-state Diagenesis and Impact of Euxinic Environments

There are some environments or diagenetic settings where exceptions occur to the above generalizations on abundance and depth distribution of sedimentary sulfur. Dominant among these are sites with prominent non-steady-state effects. These effects may arise from periods of nondeposition or very rapid deposition (e.g., turbidities). Combinations of these processes may lead to iron-oxide-rich layers interlaminated with organic-rich layers. The iron-oxide-rich layers will be sulfidized by H_2S diffusing from the organic-rich sediments resulting in very small C/S ratios in the iron-rich layers (Passier *et al.*, 1999). A related situation is marine sediments overlying nonmarine sediments. The nonmarine sediments would not have been subjected to sulfidization reactions because of low bottom-water sulfate concentrations, and thus would potentially contain high concentrations of reactive iron. Again, diffusion of H_2S into the underlying sediments can lead to extreme values of C/S ratio. This type of non-steady-state effect is relatively common and must be borne in mind when applying steady-state diagenetic concepts such as those pictured in Figure 4.

Another example of an atypical environment is when the overlying depositional waters contain H_2S (i.e., euxinic environments). These euxinic settings have been very extensively studied. In part, this is because such settings, though rare at present, may have been much more extensive (at times) in the past (Arthur and Sageman, 1994; Berry and Wilde, 1978; Canfield, 1998). By far the most work has been done on the Black Sea (Berner, 1974; Calvert *et al.*, 1996; Canfield *et al.*, 1996; Jorgensen *et al.*, 2001; Leventhal, 1983a; Lyons, 1997; Lyons and Berner, 1992; Lyons *et al.*, 1993; Morozov, 1995a,b; Ostroumov, 1953; Weber *et al.*, 2001; Wijsman *et al.*, 2001a,b; Wilkin and Arthur, 2001). These studies show that pyrite formation in modern Black Sea sediments is iron limited. It is also evident that most of the pyrite in the sediments forms in the water column, although sulfate reduction does continue within the sediments.

9.7 SULFUR ISOTOPE SYSTEMATICS OF MARINE SEDIMENTS

9.7.1 Overview

A characteristic feature of sedimentary sulfide in recent marine sediments is enrichment in the light isotope of sulfur. This enrichment is largely a result of the isotopic fractionation introduced during sulfate reduction by SRB. However, the isotopic composition of sedimentary sulfide is frequently found to be lighter than predicted based on the documented fractionation by SRB discussed above. Thus, additional processes beyond single step kinetic reduction of SO_4^{2-} to H_2S are required to explain the data.

The isotopic composition of sedimentary sulfide is summarized in Figure 8, which is a histogram of isotope data for solid-phase forms of sulfur from marine environments. The information is generally true for pyrite sulfur although in some cases metastable iron-sulfide minerals are included as well. The data are exclusive of anoxic depositional environments. The mean value of the isotope data is approximately $-18‰$ and the

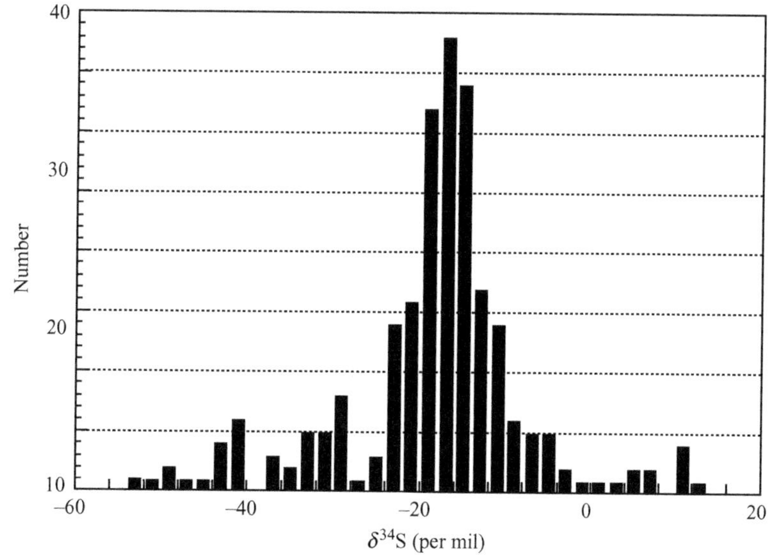

Figure 8 Histogram showing the distribution of sulfur isotopes in marine sediments (after Ohmoto and Goldhaber, 1997).

median is −17.3‰, but the data span a very wide range, nearly +10‰ to −60‰. The mean value of −18‰ represents enrichment in ^{32}S of ~40‰ compared with seawater SO_4^{2-}, which for the modern ocean has a $\delta^{34}S$ value of about +20‰. For comparison, the equilibrium isotopic fractionation between SO_4^{2-} and H_2S is 75‰ at 25 °C. The large average isotopic separation between sulfate and pyrite ($\Delta_{SO_4-FeS_2}$) is noteworthy. Many of the analyses represent fractionations greater than 45‰, which is the maximum observed value for SRB. Some $\Delta_{SO_4-FeS_2}$ values exceed 70‰.

The overall widespread of $\delta^{34}S$ values shown in Figure 8, with variable but generally large negative fractionation compared to seawater sulfate, is a characteristic of sedimentary sulfur formed as a consequence of bacterial sulfate reduction. Furthermore, a wide range of isotopic values may also occur in a single depositional environment. This variability can be expressed as isotopic differences occurring between pyrite grains or size fractions that presumably formed during different stages of diagenesis (Canfield *et al.*, 1992; Goldhaber and Kaplan, 1974; Kohn *et al.*, 1998). Kohn *et al.* (1998) recognized isotopic differences of up to 35‰ between grains in the same sediment, and up to 15‰ within a single pyrite grain. Where temporal trends are observed (i.e., from core to rim in an overgrowth sequence), the most commonly observed direction is towards heavier isotopic values with increasing relative age in all these studies. This is the expected trend given the evolution of sulfur isotopes towards heavier values with depth in the sediment column (Figure 4). Isotopic variability is also commonly observed with depth in the total pyrite sulfur. The variability with depth in sediments can, in part, be due to ongoing diagenesis, but in many cases it is due to non-steady-state diagenetic effects such as depositional hiatuses, turbidite input, changes in the flux of organic carbon or iron to the sediment, etc.

For steady-state sedimentary processes, the isotopic composition of recent sedimentary sulfur and the reason for its large range of $\delta^{34}S$ values is primarily controlled by three factors: (i) the magnitude of the instantaneous bacterial isotopic fractionation factor between SO_4^{2-} and H_2S; (ii) isotope effects associated with cyclic oxidation–reduction reactions; and (iii) the mechanism of addition of sulfur to sediments. The last two of these factors, and possibly the first as well, are depth dependent. For this reason it is logical to divide this discussion by diagenetic zone.

9.7.2 Isotopic Processes in the Upper (Bioturbated) Sediment Regime

Bioturbation and other physical processes associated with the upper portions of marine sediments may lead to rapid exchange between pore-water and overlying depositional water. Depending on the intensity of bioturbation, sulfate in depth zones I and II and the uppermost part of zone III (Figure 4) may be effectively in contact with an infinite reservoir of seawater sulfate. When this is the case, pore-water SO_4^{2-} will have a nearly constant $\delta^{34}S$ value with depth regardless of the withdrawal of isotopically light sulfur to form H_2S. The initial isotopic composition of H_2S produced by SRB in zones I and II will be equal to the instantaneous isotopic separation between seawater sulfate and bacterial sulfide (i.e., up to about $\Delta_{SO_4-H_2S} = 45‰$). Metastable iron sulfides and pyrite formed from this H_2S will have an isotopic composition very close to this initial H_2S because of the small fractionation observed during sulfidization of iron minerals.

Characteristically, a high proportion of the aqueous and solid-phase sulfide produced in the upper sediment layer is reoxidized owing to the availability of oxidants such as oxygen as well as iron and manganese oxides and to the vigor of advective processes. The research summarized above has stressed that a portion of the reoxidation leads not only to direct loss of SO_4^{2-}, but also to cyclic oxidation–reduction reactions. It is now believed (see below) that these cyclic oxidation–reduction reactions are required to explain the large isotopic differences (i.e., $\Delta_{SO_4-FeS_2} > 45‰$) that are commonly observed in marine sediments. This cyclic redox mechanism contrasts with a previous proposal that the large values of $\Delta_{SO_4-FeS_2}$ arose because microbial metabolism in sediments occurred at slower rates than were reproduced in laboratory culture studies (Goldhaber and Kaplan, 1975). The evidence supporting the cyclic redox pathway consists of extensive studies of the isotopic fractionation exhibited during bacterial reduction (see above). Similar studies document that disproportionation of $S_2O_3^{2-}$ and other intermediate sulfide oxidation products can result in significant isotopic fractionation (Canfield and Thamdrup, 1994; Habicht *et al.*, 1998). For example, the disproportionation of $S_2O_3^{2-}$ (Equation (9)) can produce isotopic fractionations between product SO_4^{2-} and H_2S of between 3‰ and 15‰. Equivalent reactions of sulfite can produce fractionation of 28‰ (see e.g., Habicht and Canfield (2001) for a summary). Furthermore, direct measurements of isotopic fractionation processes in sediments have been conducted. These studies support the concept that oxidation of isotopically light H_2S will produce isotopically light intermediates such as $S_2O_3^{2-}$ or SO_3^{2-}. Subsequent microbial processing of these intermediates by reduction (e.g., Equation (8)) or disproportionation (e.g., Equation (9)) will lead to sulfide (and thus potentially iron–sulfide minerals) that are incrementally lighter than the original H_2S produced by SRB. Cyclic redox processes of this type have previously been proposed in the context of the formation of sedimentary ore deposits (Granger and Warren, 1969; Reynolds and Goldhaber, 1983)

Habicht and Canfield (2001) provide evidence that the pyrite forming in marine sediments is influenced by such a pathway by simultaneously determining the isotopic composition of H_2S produced by SRB and that of iron-sulfide minerals coexisting in the same sediments. Their measured bacterial fractionations could only explain 41–85% of the observed ^{34}S depletion in iron sulfides from a range of marine environments. They calculated a closed-system model showing that microbial-disproportionation fractionations involving partially oxidized sulfur species, as documented in the laboratory, could explain their data. Their calculations assumed that all H_2S initially produced by SRB was cycled through the disproportionation pathway. Based on such an assumption, very large isotopic enrichments in regenerated H_2S of 20‰ to over 60‰ were calculated. This assumption does not account for the continuous exchange of reactants and products that occurs in near-surface marine sediments, and the actual pathways are probably much more dynamic than the closed-system calculations indicate. It is also not clear why redox cycling gives rise to a frequently observed $\delta^{34}S$ "ceiling" of 60–70‰ (Lyons *et al.*, in press-a).

Isotopes of organosulfur in marine sediments are typically heavier than coexisting pyrite (Anderson and Pratt, 1994). Figure 9 from the paper by Anderson and Pratt summarizes the data. The organic sulfur is ~5–15‰ heavier than pyrite from the same samples. Anderson and Pratt suggested that the isotopically heavy component is, in part, derived from biosynthetic (assimilatory) sulfur. They also suggested that early diagenetic H_2S was not the immediate precursor. Instead, they invoke a component such as polysulfide or a sulfur oxyanion (e.g., $S_2O_3^{2-}$) as the sulfidizing agent. However, the evidence summarized above indicates that these oxidation products may be lighter than H_2S rather than being heavier. Thus, an alternative explanation for the isotopic composition of organosulfur is that H_2S is the sulfidizing agent for organic matter, and that the intermediates cycled through the oxidation–reduction–disproportionation pathway go to form pyrite.

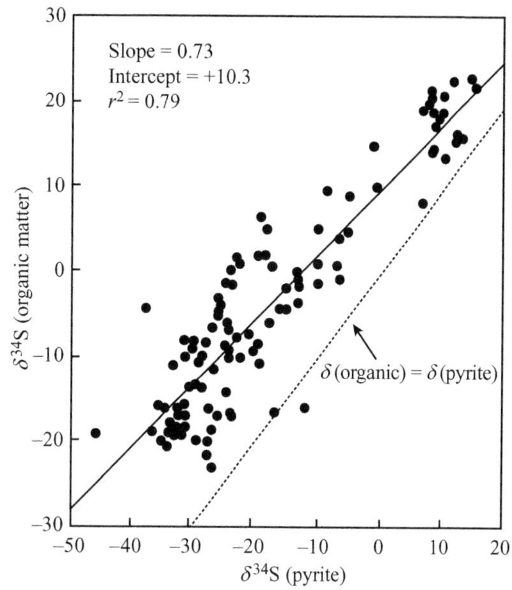

Figure 9 Plot of the sulfur isotopic composition of organic sulfur in marine sediments versus the sulfur isotopic composition of pyrite in the same sediments (reproduced by permission of American Chemical Society from *Geochemical Transformations of Sedimentary Sulfur*, **1994**, *612*, 378–396).

9.7.3 Isotopic Processes in the Deeper Diffusion-dominated Sediment Regime

Sulfur isotope systematics in the diffusion zone (zone III) reflect the condition that this interval is only partially open to SO_4^{2-} addition. This restriction on sulfate transport results in enrichment in ^{34}S of the residual sulfate. Pore-water H_2S likewise becomes heavier, like its parent SO_4^{2-} (Figure 4). The shift in H_2S and particularly in SO_4^{2-} isotope values as total sulfate depletion is approached can be dramatic—greater than 50‰ compared to that in initial seawater SO_4^{2-}. A corresponding increase in pyrite-sulfur isotope values with depth in zone III, if it occurs at all, is much smaller than that for SO_4^{2-}. In large part, this small change in pyrite $\delta^{34}S$ values with depth results from formation of most FeS_2 under more open-system conditions in the bioturbation zone where reduction rates are high and reactive iron is available. Clearly some solid-phase sulfur will form within the diffusion-dominated regime, even if the amounts are trivial compared to those formed during earlier stages of diagenesis. The existence of a maximum in pore-water H_2S profiles (e.g., Figure 4) requires that below this maximum, sulfide will diffuse deeper into the sediments and will be retained there. Examples are recognized, however, where clear shifts in $\delta^{34}S$ of pyrite towards heavier values occur with depth in the diffusion-dominated interval requiring significant formation of pyrite in zone III (Goldhaber and Kaplan, 1980). This buildup of pyrite can occur when reactive iron is available in zone III.

Calculation of the isotopic composition of sulfur formed in the diffusion zone requires recognition of some effects specific to the diffusion process. The flow of matter by diffusion, i.e., the flux, is a function of the product of the concentration gradient and diffusion coefficient according to Equation (17):

$$J(SO_4^{2-}) = -\phi D_{SO_4^{2-}} \frac{\partial[SO_4^{2-}]}{\partial x} \quad (17)$$

For sulfate diffusion, this overall flux may be separated into two distinct terms, one for ^{32}S sulfate and one for ^{34}S sulfate (Equations (18a) and (18b)):

$$J(^{32}SO_4^{2-}) = -\phi D_{SO_4^{2-}} \frac{\partial[^{32}SO_4^{2-}]}{\partial x} \quad (18a)$$

$$J(^{34}SO_4^{2-}) = -\phi D_{SO_4^{2-}} \frac{\partial[^{34}SO_4^{2-}]}{\partial x} \quad (18b)$$

Because SRB preferentially utilize $^{32}SO_4^{2-}$, the gradient in this species will be increased in a relative sense compared to $^{34}SO_4^{2-}$ and compared to a hypothetical situation in which SO_4^{2-} is reduced without isotope fractionation. Therefore, diffusion of $^{32}SO_4^{2-}$ will be more rapid than would be the case without isotope fractionation, and isotopically light SO_4^{2-} will be added by diffusion (Chanton *et al.*, 1987a,b; Goldhaber and Kaplan, 1980; Jorgensen, 1979). It can also be shown, by a similar line of reasoning, that diffusion of H_2S upwards in the direction of the sediment–water interface leads to a net transport of isotopically heavy sulfur upwards (Chanton *et al.*, 1987b; Jorgensen, 1979). The magnitude of these effects can be large. The isotopic composition of the SO_4^{2-} diffusing into the sediment has been calculated to be 24.5‰ (Goldhaber and Kaplan, 1980) and 21‰ (Chanton *et al.*, 1987b) lighter than the bottom-water value. Figure 10 shows a plot of pore-water SO_4^{2-} concentration, sulfur isotopic composition of pore-water SO_4^{2-}, and the isotopic composition of the diffusing SO_4^{2-} at the same depth. Note the striking enrichment in ^{32}S of the diffusing SO_4^{2-} compared to its value at the same depth in the pore water. The impact of this differential isotopic transport leads to burial of sulfur that is isotopically lighter than would otherwise be predicted. Both Goldhaber and Kaplan (1980) and Chanton *et al.* (1987a,b) were able to show, based on detailed mass and isotopic balances for aqueous and solid-phase sulfur in two differing depositional settings, that differential diffusion of the several distinct isotopic species had a significant effect on the final isotopic composition of the buried pyrite (Figure 10).

Special consideration of diffusion processes is also required to calculate the fractionation factor between SO_4^{2-} and H_2S in this diffusion zone. If it is simply assumed that diffusion does not occur, closed system Rayleigh distillation equations can be used to calculate an isotopic fractionation factor for this process (Goldhaber and Kaplan, 1974; Sweeney and Kaplan, 1980). However, in a sediment column open to diffusion, differential isotopic diffusion effects such as those described above must be accounted for. This can be done using a mathematical model incorporating diffusion, burial, and chemical reaction written for the isotopically distinct species and then combined to model the isotopic depth distribution of sulfur species: either SO_4^{2-} or H_2S (Goldhaber and Kaplan, 1980; Jorgensen, 1979). It was shown by these authors that the assumption of a closed system yields a calculated instantaneous fractionation factor between SO_4^{2-} and H_2S that is less than the true value. Goldhaber and Kaplan (1980) applied such an open-system mathematical model to pore-water profiles in zone III for sediments of the Gulf of California. Closed-system calculation of the fractionation factor between SO_4^{2-} and H_2S in these sediments yielded values from 16‰ to 35‰. In contrast, the open-system calculation yielded values of ~60‰. Wortmann *et al.* (2001) applied a similar model to a deep sulfidic core from the Great Australian Bight, and calculated an open-system fractionation factor of 65‰. Claypool (2003, in press) has interpreted a similarly large fractionation factor by modeling pore-water $\delta^{34}S$ values in SO_4^{2-} profiles from extensive Ocean Drilling Program/Deep Sea Drilling Project data. Thus, the calculated open-system fractionation is larger than fractionations measured for SRB in laboratory and field studies as described above. A possible explanation for this large fractionation is that it is a net value, as the difference $\Delta_{FeS_2-H_2S}$ is a net value. Thus, it may incorporate the effects of additional redox fractionations

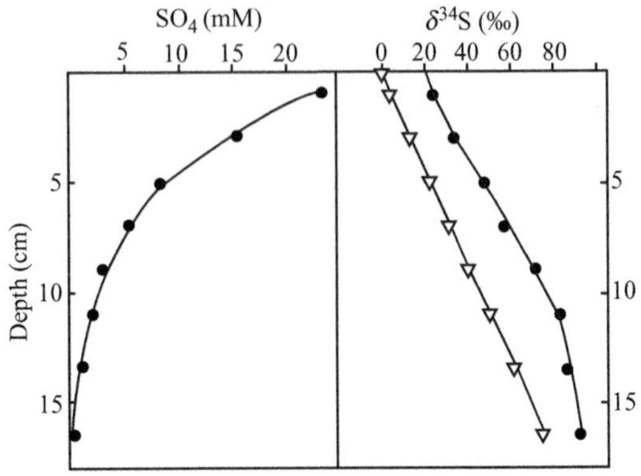

Figure 10 Plot of the concentration of SO_4^{2-} and its isotopic composition versus depth in a core from Cape Lookout Bight, NC, USA. The open triangles show the isotopic composition of the SO_4^{2-} diffusing into the sediment. Note that the diffusing sulfate is approximately 20‰ lighter than its isotopic composition at the same depth (Chanton *et al.*, 1987b) (reproduced by permission of Elsevier from *Geochim. Cosmochim. Acta*, **1987b**, *51*, 1201–1208).

beyond single-step reduction of SO_4^{2-} to H_2S. What is troubling about this explanation is that the additional fractionation effects involve reoxidation of H_2S and other reduced sulfur species to $S_2O_3^2/SO_3^{2-}$, and subsequent microbial reduction or disproportionation of these intermediate oxidation state species. The deeper, more reduced portion of the sediment column represented by zone III would seem to be a less favorable environment for oxidation reactions than the upper sediment layers where reactive Fe/Mn oxides and other oxidants such as O_2 are available (Wortmann *et al.*, 2001). The origin of the very large apparent fractionations between SO_4^{2-} and H_2S in the deeper portions of marine sediments is presently unresolved.

9.8 ANCIENT MARINE SEDIMENTS

Sulfur diagenesis in lithified sediments deposited under normal marine conditions would have been controlled by the same range of processes as those discussed above for their modern analogues. However, some additional factors have been recognized, which will influence interpretation of the observed sulfur abundances and isotope ratios of these ancient sediments. These factors are: (i) loss of organic matter during burial diagenesis; (ii) changes in the nature of organic matter with geologic age; (iii) changes in the sulfur isotopic composition of seawater sulfate; and (iv) later diagenetic processes such as epigenetic addition of H_2S.

Diagenetic loss of organic matter continues following the total depletion of SO_4^{2-} in lower zone III and in zone IV (Figure 4), via the microbial production of methane. Methane production continues over a considerable depth interval, up to burial temperatures in the range of 60–80 °C. Eventually, with sufficient depth and temperature increase above this range, diagenesis passes into a thermocatalytic regime in which organic maturation and attendant loss of organic matter to pore fluids occurs via thermally driven processes (catagenesis). This loss of organic matter is not associated with a corresponding mobilization of FeS_2 until much higher temperatures are attained (Ohmoto and Goldhaber, 1997). Consequently, carbon–sulfur ratios of sediments that experienced one or both of these later diagenetic regimes may have smaller organic carbon to pyrite–sulfur ratios than younger sediments (Berner and Raiswell, 1983). Because the shift is due solely to the loss of carbon, the total sulfur contents of these diagenetically altered sediments will not be different from their recent normal-marine equivalents, and will consequently be less than ~2 wt.%.

An important control on sulfur diagenesis over time is a change in the origin of organic matter incorporated into sediments. Organic matter accumulating in modern marine sediments is a mixture of materials produced on land and materials originating in the marine realm. Of the two types, the terrestrially derived organics are less readily metabolizable by the sulfate-reducing community of microorganisms (Lyons and Gaudette, 1979). Terrestrial organic materials contain a high proportion of polyaromatic derivatives of woody tissue (lignin), which are rather stable during anaerobic diagenesis. These land-derived organics also have a very long transport path and only the most refractory substances survive to be incorporated in marine sediments. These considerations, of course, hold only subsequent to the development of land plants in the Devonian. Pre-Devonian marine environments would have had a higher proportion of the total organics in metabolizable form undiluted by the more refractory land-derived organic matter (Berner and Raiswell, 1983), and a higher percentage of the deposited organic matter may have participated in sulfur diagenesis. For this reason, C/S ratios of Cambrian, Ordovician, and Silurian normal marine sediments are smaller than in younger sediments. The magnitude of this effect is larger than would be expected for the diagenetic loss of organic material due to methanogenesis and thermocatalytic reactions discussed above. Cambrian and Ordovician shale has C-organic/S-total ratios ~0.6 (Raiswell and Berner, 1986) compared to Pleistocene and Holocene values of 2.8 (see Section 6.1).

Sulfur is incorporated in sedimentary organic matter at low temperatures during early diagenesis as discussed above. The re-release of this organic sulfur as H_2S can occur. This release is tied to the overall process of thermal maturation of sedimentary kerogen (the dominant form of organic matter in sediments). Initially CO_2 is released to pore fluids followed by overlapping episodes of petroleum generation, and N_2, H_2S, and CH_4 release. The temperature of onset of this thermocatalytic sequence is ~70 °C. However, this onset temperature is dramatically affected by a variety of variables such as the nature of the organic matter and the interplay of time and temperature during burial. One important variable effecting organic maturation is the sulfur content of the kerogen. High-sulfur kerogens generate hydrocarbons at much lower thermal exposures than do normal kerogen (Orr, 1986; Tannenbaum and Aizenshtat, 1985).

9.9 ROLE OF SEDIMENTARY SULFUR IN THE GLOBAL SULFUR CYCLE

9.9.1 The Phanerozoic Time Period

The mixing time of SO_4^{2-} in the ocean is short compared to its residence time. Thus, the isotopic composition of SO_4^{2-} in seawater is constant over geologically short time periods. One pathway for removal of this seawater sulfate is by precipitation of sulfate-bearing evaporite minerals. When marine evaporite minerals such as gypsum ($CaSO_4{\cdot}2H_2O$) precipitate from seawater, the sulfur isotope fractionation between the aqueous and solid-phase sulfate is very small, about +1.7‰ (Thode and Monster, 1965). The combination of these factors implies that contemporaneous marine evaporites have the same isotopic composition worldwide, and they are nearly identical isotopically to seawater SO_4^{2-} of that age. When the sulfur isotopic composition of marine evaporite SO_4^{2-} is plotted as a function of time (Figure 11; Claypool *et al.*, 1980; Strauss, 1997), the resulting age curves are quite different for different evaporites. For example, Cambrian evaporites have $\delta^{34}S$ values of about +30‰, whereas for Permian evaporites $\delta^{34}S$ is of about +10‰. More recently, other recorders of marine SO_4^{2-} $\delta^{34}S$ in past oceans including barite ($BaSO_4$) (Paytan *et al.*, 1998) and sulfate trapped in the lattice of carbonate minerals (Lyons *et al.*, in press; Staudt and Schoonen, 1994) have begun to expand the data available from evaporites.

The age curve of Figure 11 is a critical constraint on the geochemical history of the Earth. Over geologic timescales, removal of pyrite sulfur from ocean water plus formation of sulfate-bearing evaporite minerals is required to balance the addition of sulfate in rivers and much smaller sources (and sinks) from intra-oceanic igneous processes. The river-borne sulfate is derived from terrestrial weathering of pyrite-bearing sediments and evaporites (Holser and Kaplan, 1966). Because pyrite formation involves significant isotopic fractionation and evaporite formation does not, the balance between net burial versus oxidative weathering of pyrite controls the isotopic composition of seawater SO_4^{2-}.

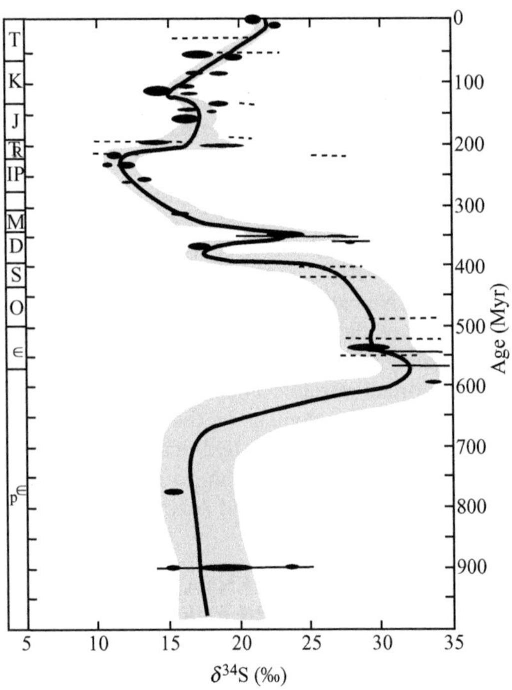

Figure 11 Sulfur isotopic composition of marine evaporite SO_4^{2-} over geologic time (Claypool *et al.*, 1980) (reproduced by permission of Elsevier from *Chem. Geol.*, **1980**, *28*, 199–260).

This global cycling of sulfur is closely linked to other geochemical cycles as well. In particular, it is tied to the abundance of atmospheric oxygen. This linkage has been described by Berner and Petsch (1998). Weathering (oxidation) of pyrite and organic carbon are the dominant sinks for atmospheric oxygen. Unless this oxidation is balanced by a reciprocal burial of organic carbon and pyrite, there will be a net change in atmospheric oxygen. These coupled processes can be summarized in terms of two net reactions involving organic matter and pyrite (Berner, 1999):

$$CO_2 + H_2O \leftrightarrow CH_2O + O_2 \qquad (19)$$

$$\begin{aligned} &2Fe_2O_3 + 16HCO_3^- + 8SO_4^{2-} + 16Ca^{2+} \\ &\leftrightarrow 4FeS_2 + 15O_2 + 16CaCO_3 + 8H_2O \end{aligned} \qquad (20)$$

In this equation, CH_2O represents a generic form of organic matter. Reaction (19) going from left to right represents a net long-term photosynthesis, and from right to left represents oxidation of sedimentary organic matter. Reaction (20) going from left to right represents formation of sedimentary pyrite, and from right to left pyrite oxidation. Mass balance models have been derived, which utilize isotopic age curves such as that in Figure 11 to evaluate the role of the global sulfur and carbon cycles on the evolution of atmospheric oxygen (Petsch and Berner, 1998 and references therein).

9.9.2 The Precambrian

The evolution of global cycling of sulfur as exemplified by Equations (19) and (20) above is closely tied to the evolution of atmospheric oxygen. Thus, sulfur geochemistry, particularly sulfur isotope data, has proved to be an important probe into the overall evolution of the Earth atmosphere system (Canfield *et al.*, 2000; Canfield and Raiswell, 1999; Canfield and Teske, 1996; Knoll *et al.*, 1998; Lyons *et al.*, in press; Schidlowski, 1979; Schidlowski *et al.*, 1983).

The earliest isotopic record of sedimentary pyrite in the Archean lacks the large spread between coeval sulfate and sulfide sulfur that characterizes the Phanerozoic data. Sedimentary sulfides formed between 3.4 Ga and 2.7 Ga have isotopic compositions within ±5‰ of contemporaneous seawater SO_4^{2-}. This change in isotopic pattern compared to Phanerozoic data is to be expected if atmospheric O_2 was low, and consequently no widespread mechanism existed for oxidizing reduced sulfur to form SO_4^{2-} in seawater. Sulfate concentrations <200 μM are required to suppress isotope fractionation by SRB (Habicht *et al.*, 2002). Most researchers hold to a scenario that photosynthetic production of O_2 began by 2.7 Ga, but that there was a lag of several hundred million years before atmospheric O_2 began to accumulate (2.2 Ga–2.4 Ga) (although compare Ohmoto (1992) and Ohmoto *et al.* (1993). The offset between isotopic values for sedimentary sulfide and coexisting sulfate minerals increases to values indicating the impact of SRB after ~2.7 Ga, but the overall fractionation remained smaller than for Phanerozoic values. Sulfate reduction in rocks of this age may have occurred preferentially in localized SO_4^{2-}-rich environments (Canfield *et al.*, 2000). Only after ~2.3 Ga the full modern expression of large $\Delta_{SO_4-FeS_2}$ values is recognized.

The conclusion that both atmospheric O_2 and oceanic SO_4^{2-} were very low prior to ~2.4 Ga is strengthened by recently discovered mass-independent isotope fractionation recognized in sedimentary sulfate and sulfide minerals. The isotopic fractionation processes discussed previously are mass dependant; they result in predictable relations among the sulfur isotopic values for $\delta^{33}S$, $\delta^{34}S$, and $\delta^{36}S$. Mass-independent sulfur fractionation is exhibited by sedimentary sulfate and sulfide minerals prior to 2.45 Ga (Farquhar *et al.*, 2000). The most likely mechanism for producing such mass-independent fractionations involves atmospheric processes at very low values of P_{O_2}. Levels of O_2 greater than trace levels would inhibit the mass-independent fractionation processes, because it adsorbs UV radiation at wavelengths required for these reactions to occur. The mass-dependent fractionations are not recognized after ~2.45 Ga, suggesting a significant increase in Earth's O_2 content after that time.

The accumulation of O_2 in Earth's atmosphere and its influence on the global sulfur cycle between ~2.4 Ga and near modern levels during the Phanerozoic has been the subject of much recent research. One model (Lyons *et al.*, in press) calls for generally low oceanic SO_4^{2-} concentrations (~10% of modern value) in the Mesoproterozoic ocean (1.6–0.9 Ga). However, by ~1.8 Ga, sufficient SO_4^{2-} was delivered to the ocean to potentially result in pervasive euxinic conditions in the world ocean as proposed by Canfield (1998). Levels of atmospheric oxygen and seawater SO_4^{2-} may not have been reached until the Neoproterozoic (~900 Ma).

9.10 CLOSING STATEMENT

This is an exciting time for researchers studying sedimentary sulfur geochemistry. Although the broad outlines of processes leading to the incorporation of sulfur into marine sediments have been recognized for over half a century, the detailed pathways and mechanisms were not clear. It is only since the advent of methodology to investigate the component microbiological metabolic reactions and quantify low concentrations of inorganic and organic reactive intermediates in these pathways that the complexity of these processes has started to emerge. We can anticipate that further research will continue to reveal additional layers of complexity. Some particularly fruitful areas of research emerge from the study of the mechanism(s) of incorporation of sulfur into organic matter, and the detailed pathways of formation and consumption of reactive sulfur oxyanions in both near-surface and deeper layers of marine sediments. We can also anticipate that as the depth of understanding of modern processes advances, these insights will continue to be applied in understanding the geologic history of sulfur geochemistry and the key role that it has played in shaping our planet's geochemistry.

REFERENCES

Adam P., Schneckenburger P., Schaeffer P., and Albrecht P. (2000) Clues to early diagenetic sulfurization processes from mild chemical cleavage of labile sulfur-rich geomacromolecules. *Geochim. Cosmochim. Acta* **64**(20), 3485–3503.

Allen E. T., Crenshaw J. L., Johnston J., and Larsen E. S. (1912) The mineral sulphides of iron with crystallographic study. *Am. J. Sci.* **23**, 169–236.

Aller R. C. and Mackin J. E. (1989) Open-incubation, diffusion methods for measuring solute reaction rates in sediments. *J. Mar. Res.* **47**, 411–440.

Aller R. C., Mackin J. E., and Cox R. T. Jr., (1986) Diagenesis of Fe and S in Amazon inner shelf muds: apparent dominance of Fe reduction and implications for the genesis of ironstones. *Continent. Shelf Res.* **6**, 263–289.

Aller R. C. (1980a) Diagenetic processes near the sediment-water interface of Long Island Sound: I. Decomposition and nutrient element geochemistry (S, N, P). *Estuarine Physics and Chemistry; Studies in Long Island Sounds* (ed. B. Salzman). Academic Press, New York, vol. 22, pp. 237–350.

Aller R. C. (1980b) Quantifying solute distributions in the bioturbated zone of marine sediments by defining an average microenvironment. *Geochim. Cosmochim. Acta* **44**(12), 1955–1966.

Aller R. C. (1988) Benthic fauna and biogeochemical processes in marine sediments; the role of burrow structures. In *Nitrogen Cycling in Coastal Marine Environments* (eds. T. H. Blackburn and J. P. Sorensen). Wiley, pp. 301–338.

Aller R. C. and Aller J. Y. (1998) The effect of biogenic irrigation intensity and solute exchange on diagenetic reaction rates in marine sediments. *J. Mar. Res.* **56**, 905–936.

Anderson T. F. and Pratt L. M. (1994) Isotopic evidence for the origin of organic sulfur and elemental sulfur in marine sediments. In: *Geochemical Transformations of Sedimentary Sulfur* (eds. M. A. Vairavamurthy, M. A. A. Schoonen, T. I. Eglinton, G. W. Luther, and II I. B. Manowitz). American Chemical Society, Washington, DC, vol. 612, pp. 378–396.

Arthur M. A. and Sageman B. B. (1994) Marine black shales; depositional mechanisms and environments of ancient deposits. *Ann. Rev. Earth Planet. Sci.* **22**, 499–551.

Baas Becking K. G. M. (1925) Studies on the sulfur bacteria. *Ann. Bot.* **39**, 613–650.

Battersby N. S., Malcolm S. J., Brown C. M., and Stanley S. O. (1985) Sulphate reduction in oxic and sub-oxic North–East tlantic sediments. *FEMS Microbiol. Ecology* **31**, 225–228.

Belyayev S. S., Lein A. Y., and Ivanov M. V. (1981) Roles of methane-producing and sulfate-reducing bacteria in the destruction of organic matter. *Geochem. Int.* **18**(2), 59–67.

Benning L. G., Wilkin R. T., and Barnes H. L. (2000) In *Reaction pathways in the Fe–S system below 100 degree C.* Chemical Geology (eds. D. J. Wesolowski and T. M. Seward). Chem. Geol, **167**, pp. 25–51.

Berner R. A. (1962) Experimental studies of the formation of sedimentary iron sulfides. In: *Biogeochemistry of Sulfur Isotopes: Symposium*(ed. M. LeRoy). Yale University, New Haven.

Berner R. A. (1964a) Distribution and diagenesis of sulfur in some sediments from the Gulf of California. *Mar. Geol.* **1**(2), 117–140.

Berner R. A. (1964b) Iron sulfides formed from aqueous solution at low temperatures and atmospheric pressure. *J. Geol.* **72**(3), 293–306.

Berner R. A. (1964c) Stability fields of iron minerals in anaerobic marine sediments. *J. Geol.* **72**(6), 826–834.

Berner R. A. (1967) Thermodynamic stability of sedimentary iron sulfides. *Am. J. Sci.* **265**(9), 773–785.

Berner R. A. (1969) The synthesis of framboidal pyrite. *Econ. Geol. and Bull. Soc. Econ. Geol.* **64**(4), 383–384.

Berner R. A. (1970) Sedimentary pyrite formation. *Am. J. Sci.* **268**(1), 1–23.

Berner R. A. (1973) Pyrite formation in the oceans. In: *Symposium on Hydrogeochemistry and Biogeochemistry:* vol. 1. Hydrogeochemistry, Clarke Co., Washington, DC, pp. 402–417.

Berner R. A. (1974) Iron sulfides in Pleistocene Deep Black Sea sediments and their paleo-oceanographic significance. In: *The Black Sea: Geology, Chemistry, Geochemistry, and Biology* (eds. E. Degens T and D. Ross A). American Association of Petroleum Geologists, vol. 20, pp. 524–531.

Berner R. A. (1978) Sulfate reduction and the rate of deposition of marine sediments. *Earth Planet. Sci. Lett.* **37**(3), 492–498.

Berner R. A. (1982) Burial of organic carbon and pyrite sulfur in the modern ocean; its geochemical and environmental significance. *Am. J. Sci.* **282**(4), 451–473.

Berner R. A. (1999) Atmospheric oxygen over Phanerozoic time. *Proc. Natl. Acad. Sci. USA* **96**, 10955–10957.

Berner R. A. and Petsch S. T. (1998) The sulfur cycle and atmospheric oxygen. *Science* **282**(5393), 1426–1427.

Berner R. A. and Raiswell R. (1983) Burial of organic carbon and pyrite sulfur in sediments over Phanerozoic time; a new theory. *Geochim. Cosmochim. Acta* **47**(5), 855–862.

Berner R. A. and Raiswell R. (1984) C/S method for distinguishing freshwater from marine sedimentary rocks. *Geology* **12**(6), 365–368.

Berner R. A. and Westrich J. T. (1985) Bioturbation and the early diagenesis of carbon and sulfur. *Am. J. Sci.* **285**(3), 193–206.

Berry W. B. N. and Wilde P. (1978) Progressive ventilation of the oceans; an explanation for the distribution of the lower Paleozoic black shales. *Am. J. Sci.* **278**(3), 257–275.

Bonnell L. M. and Anderson T. F. (1987) Sulfur isotopic variations in nodular and disseminated pyrite; Hole 603B. In: *Initial Reports of the Deep Sea Drilling Project* (eds. J. E. van Hinte and S. W. Wise). US Government Printing Office, Washington, DC, vol. XCII, pp. 1257–1262.

Bottcher M. E., Smock A. M., and Cypionka H. (1998) Sulfur isotope fractionation during experimental precipitation of iron(II) and manganese(II) sulfide at room temperature. *Chem. Geol.* **146**, 127–134.

Bottrell S. H. and Raiswell R. (2000) Sulphur isotopes and microbial sulphur cycling in sediments. In *Microbial Sediments* (eds. R. E. Riding and S. M. Awramik). Springer, Berlin, pp. 96–104.

Butler I. B. and Rickard D. (2000) Framboidal pyrite formation via the oxidation of iron(II) monosulfide by hydrogen sulphide. *Geochim. Cosmochim. Acta* **64**(15), 2665–2672.

Cagatay M. N., Borowski W. S., and Ternois Y. G. (2001) Factors affecting the diagenesis of Quaternary sediments at ODP Leg 172 sites in western North Atlantic; evidence from pore water and sediment geochemistry. *Chem. Geol.* **175**(3–4), 467–484.

Calvert S. E., Thode H. G., Yeung D., and Karlin R. E. (1996) A stable isotope study of pyrite formation in the late Pleistocene and Holocene sediments of the Black Sea. *Geochim. Cosmochim. Acta* **60**(7), 1261–1270.

Canfield D. E. (1989a) Reactive iron in marine sediments. *Geochim. Cosmochim. Acta* **53**(3), 619–632.

Canfield D. E. (1989b) Sulfate reduction and oxic respiration in marine sediments; implications for organic carbon preservation in euxinic environments. *Deep-Sea Res. I: Oceanogr. Res. Pap.* **36**(1A), 121–138.

Canfield D. E. (1991) Sulfate reduction in deep-sea sediments. *Am. J. Sci.* **291**(2), 177–188.

Canfield D. E. (1993) Organic matter oxidation in marine sediments. In *Interactions of C, N, P and S Biogeochemical Cycles and Global Change* (eds. R. Wollast, F. T. Mackenzie, and L. Chou). Springer, Berlin, pp. 333–363.

Canfield D. E. (1998) A new model for Proterozoic ocean chemistry. *Nature* **396**(6710), 450–453.

Canfield D. E. (2001) Isotope fractionation by natural populations of sulfate-reducing bacteria. *Geochim. Cosmochim. Acta* **65**(7), 1117–1124.

Canfield D. E. and Des Marais D. J. (1993) Biogeochemical cycles of carbon, sulfur, and free oxygen in a microbial mat. *Geochim. Cosmochim. Acta* **57**(16), 3971–3984.

Canfield D. E. and Raiswell R. (1991) Pyrite formation and fossil preservation. In *Taphonomy: Releasing the Data Locked in the Fossil Record* (eds. P. A. Allison, D. E. G. Briggs, F. G. Stehli, and D. S. Jones). Plenum, New York, pp. 337–387.

Canfield D. E. and Raiswell R. (1999) The evolution of the sulfur cycle. *American Journal of Science, Special Issue: Biogeochemical Cycles and their Evolution over Geologic Time: A tribute to the Career of Robert A. Berner* **299**(7–9), 697–723.

Canfield D. E. and Teske A. (1996) Late Proterozoic rise in atmospheric oxygen concentration inferred from phylogenetic and sulphur-isotope studies. *Nature* **382**(6587), 127–132.

Canfield D. E. and Thamdrup B. (1994) The production of (super 34) S-depleted sulfide during bacterial disproportionation of elemental sulfur. *Science* **266**, 1973–1975.

Canfield D. E., Raiswell R., and Bottrell S. H. (1992) The reactivity of sedimentary iron minerals toward sulfide. *Am. J. Sci.* **292**(9), 659–683.

Canfield D. E., Jorgensen B. B., Fossing H., Glud R., Gundersen J., Ramsing N. B., Thamdrup B., Hansen J. W., and Nielsen L. P. (1993) Pathways of organic carbon oxidation in three continental margin sediments. *Marine Sediments, Burial, Pore Water Chemistry, Microbiology and Diagenesis* (eds. R. J. Parkes, P. Westbroek, and J. W. de Leeuw). Elsevier, Amsterdam, vol. 113, pp. 27–40.

Canfield D. E., Lyons T. W., and Raiswell R. (1996) A model for iron deposition to euxinic Black Sea sediments. *Am. J. Sci.* **296**(7), 818–834.

Canfield D. E., Boudreau B. P., Mucci A., and Gundersen J. K. (1998) The early diagenetic formation of organic sulfur in the sediments of Mangrove Lake, Bermuda. *Geochim. Cosmochim. Acta* **62**(5), 767–781.

Canfield D. E., Habicht K. S., and Thamdrup B. (2000) The Archean sulfur cycle and the early history of atmospheric oxygen. *Science* **288**(5466), 658–661.

Chambers L. A. and Trudinger P. A. (1979) Microbiological fractionation of stable sulfur isotopes; a review and critique. *Geomicrobiol. J.* **1**(3), 249–293.

Chanton J. P., Martens C. S., and Goldhaber M. B. (1987a) Biogeochemical cycling in an organic-rich coastal marine basin; 7, Sulfur mass balance, oxygen uptake and sulfide retention. *Geochim. Cosmochim. Acta* **51**(5), 1187–1199.

Chanton J. P., Martens C. S., and Goldhaber M. B. (1987b) Biogeochemical cycling in an organic-rich coastal marine basin; 8, A sulfur isotopic budget balanced by differential diffusion across the sediment-water interface. *Geochim. Cosmochim. Acta* **51**(5), 1201–1208.

Claypool G. E. (2003) Ventilation of marine sediments indicated by depth profiles of pore-water sulfate and $\delta^{34}S$. In *Geochemical Investigations in Earth and Space Sciences: A Tribute to Issac R. Kaplan*, Elsevier, Amsterdam.

Claypool G. E., Holser W. T., Kaplan I. R., Sakai H., and Zak I. (1980) The age curves of sulfur and oxygen isotopes in marine sulfate and their mutual interpretation. *Chem. Geol.* **28**(3–4), 199–260.

Cypionka H., Widdel F., and Pfennig N. (1985) Survival of sulfate-reducing bacteria after oxygen stress, and growth in sulfate-free oxygen-sulfide gradients. *FEMS Microbiol. Ecol.* **31**, 39–45.

Dean W. E. and Arthur M. A. (1987) Inorganic and organic geochemistry of Eocene to Cretaceous strata recovered from the lower continental rise, North American basin, site 603, deep sea drilling project leg 93. In *Initial Reports of the Deep Sea Drilling Project* (eds. J. E. van Hinte, S. W. Wise, B. N. M. Biart, J. M. Covington, D. A. Dunn, J. A. Haggerty, M. W. Johns, P. A. Meyers, M. R. Moullade, J. P. Muza, J. G. Ogg, M. Okamura, M. Sarti, and U. von Rad). Texas A & M University, Ocean Drilling Program, College Station, TX, , vol. 93, pp. 1093–1137.

Dekkers M. J. and Schoonen M. A. A. (1996) Magnetic properties of hydrothermally synthesized greigite (F (sub 3) S (sub 4)): I. Rock magnetic parameters at room temperature. *Geophys. J. Int.* **126**(2), 360–368.

Detmers J., Bruchert V., Habicht K., and Kuever J. (2001) Diversity of sulfur isotope fractionations by sulfate-reducing prokaryotes. *Appl. Environ. Microbiol.* **67**(2), 888–894.

D'hondt S., Rutherford S., and Spivack A. J. (2002) Metabolic activity of subsurface life in deep-sea sediments. *Science* **295**, 2067–2070.

Eglinton T. I., Irvine J. E., Vairavamurthy A., Zhou W., and Manowitz B. (1994) Formation and diagenesis of macromolecular organic sulfur in Peru margin sediments. In: *Organic Geochemistry* (eds. N. Telnaes, G. van Graas, and K. Oygard). Pergamon, Oxford, vol. 22, pp. 781–799.

Emery K. O. and Rittenberg S. C. (1952) Early diagenesis of California Basin sediments in relation to origin of oil. *Bull. Am. Assoc. Petrol. Geol.* **36**(5), 735–806.

Farquhar J., Huiming B., and Thiemens M. H. (2000) Atmospheric influence of earth's earliest sulfur cycle. *Science* **289**, 756–758.

Farrand M. (1970) Framboidal sulphides precipitated synthetically. *Mineralium Deposita* **5**(3), 237–247.

Fauque G. D. (1995) Ecology of sulfate-reducing bacteria. In: *Sulfate Reducing Bacteria* (ed. L. L. Barton). Plenum, New York, vol. 8, pp. 217–241.

Ferdelman T. G., Church T. M., and Luther G. W. III, (1991) Sulfur enrichment of humic substances in a Delaware salt marsh sediment core. *Geochim. Cosmochim. Acta* **55**(4), 979–988.

Filipek L. H. and Owen R. M. (1980) Early diagenesis of organic carbon and sulfur in outer shelf sediments from the Gulf of Mexico. *Am. J. Sci.* **280**(10), 1097–1112.

Fossing H. and Jorgensen B. B. (1990) Oxidation and reduction of radiolabeled inorganic sulfur compounds in an estuarine sediment, Kysing Fjord, Denmark. *Geochim. Cosmochim. Acta* **54**(10), 2731–2742.

Francois R. (1987) A study of sulphur enrichment in the humic fraction of marine sediments during early diagenesis. *Geochim. Cosmochim. Acta* **51**(1), 17–27.

Froelich P. N., Klinkhammer G. P., Bender M. L., Luedtke N., Heath G. R., Cullen D., Dauphin P., Hammond D., Hartman B., and Maynard V. (1979) Early oxidation of organic matter in pelagic sediments of the eastern equatorial Atlantic; suboxic diagenesis. *Geochim. Cosmochim. Acta* **43**(7), 1075–1090.

Gait R. I. (1978) The crystal forms of pyrite. *Mineral. Rec.* **9**(4), 219–229.

Gelin F., Kok M. D., de Leeuw J. W., and Sinninghe Damste J. S. (1998) Laboratory sulfurisation of the marine microalga Nannochloropsis salina. *Org. Geochem.* **29**(8), 1837–1848.

Goldhaber M. B. and Kaplan I. R. (1974) The sulfur cycle. In: *The Sea* (ed. E. D. Goldberg). Wiley, New York, vol. 5, pp. 569–654.

Goldhaber M. B. and Kaplan I. R. (1975) Controls and consequences of sulfate reduction rates in recent marine sediments. *Soil Sci.* **119**(1), 42–55.

Goldhaber M. B. and Kaplan I. R. (1980) Mechanisms of sulfur incorporation and isotope fractionation during early diagenesis in sediments of the Gulf of California. *Mar. Chem.* **9**(2), 95–143.

Goldhaber M. B. and Orr W. L. (1994) Kinetic controls on thermochemical sulfate reduction as a source of sedimentary H (sub 2) S. In *Geochemical Transformations of Sedimentary Sulfur* (eds. M. A. Vairavamurthy, M. A. A. Schoonen, T. I. Eglinton, G. W. Luther, and B. Manowitz). American Chemical Society, vol. 612, pp. 412–425.

Goldhaber M. B. and Reynolds R. L. (1991) Relations among hydrocarbon reservoirs, epigenetic sulfidization, and rock magnetization; examples from the South Texas coastal plain. *Geophysics* **56**(6), 748–757.

Goldhaber M. B., Aller R. C., Cochran J. K., Rosenfeld J. K., Martens C. S., and Berner R. A. (1977) Sulfate reduction, diffusion, and bioturbation in Long Island Sound sediments; report of the FOAM Group. *Am. J. Sci.* **277**(3), 193–237.

Goldhaber M. B., Church S. E., Doe B. R., Aleinikoff J. N., Brannon J. C., Podosek F. A., Mosier E. L., Taylor C. D., and Gent C. A. (1995) Lead and sulfur isotope investigation of Paleozoic sedimentary rocks from the southern Midcontinent of the United States: implications for paleohydrology and ore genesis of the Southeast Missouri lead belts. *Econ. Geol.* **90**(7), 1875–1910.

Goldhaber M., Lee R. C., Hatch J. C., Pashin J. C., and Treworgy J. (2002) Role of large scale fluid-flow in subsurface arsenic enrichment. In *Arsenic in Groundwater, Geochemistry and Occurrence* (ed. A. K. S. Alan Welch). Kluwer Academic, Chichester, pp. 127–176.

Granger H. C. and Warren C. G. (1969) Unstable sulfur compounds and the origin of roll-type uranium deposits. *Econ. Geol. and Bull. Soc. Econ. Geol.* **64**(2), 160–171.

Grinenko V. A. and Ivanov M. V. (1983) Principal reactions of the global biogeochemical cycle of sulphur. In *The Global Biogeochemical Sulphur Cycle* (eds. M. V. Ivanov and J. R. Freney). Wiley, Chichester, pp. 1–23.

Habicht K. S. and Canfield D. E. (1996) Sulphur isotope fractionation in modern microbial mats and the evolution of the sulphur cycle. *Nature* **382**(6589), 342–343.

Habicht K. S. and Canfield D. E. (1997) Sulfur isotope fractionation during bacterial sulfate reduction in organic-rich sediments. *Geochim. Cosmochim. Acta* **61**(24), 5351–5361.

Habicht K. S. and Canfield D. E. (2001) Isotope fractionation by sulfate-reducing natural populations and the isotopic composition of sulfide in marine sediments. *Geology* **29** (6), 555–558.

Habicht K. S., Canfield D. E., and Rethmeier J. (1998) Sulfur isotope fractionation during bacterial reduction and disproportionation of thiosulfate and sulfite. *Geochim. Cosmochim. Acta* **62**(15), 2585–2595.

Habicht K. S., Gade M., Thamdrup B., Berg P., and Canfield D. (2002) Calibration of Sulfate Levels in the Archean Ocean. *Science* **298**, 2372–2374.

Hartmann M. and Nielsen H. (1969) Delta 34S-Werte in rezenten Meeressedimenten und ihre Deutung am Beispiel einiger Sedimentprofile aus der westlichen Ostsee. Delta-34S values in recent marine sediments and their significance, with examples from sediment profiles from the western Baltic sea. *Geol. Rundsch.* **58**(3), 621–655.

Holser W. T. and Kaplan I. R. (1966) Isotope geochemistry of sedimentary sulfates. *Chem. Geol.* **1**(2), 93–135.

Howarth R. W. and Jorgensen B. B. (1984) Formation of S-35-labelled elemental sulphur and pyrite in coastal marine sediments (Limfjorden and Kysing Fjord Denmark) during short-term S-35 sulphate reduction measurements. *Geochim. Cosmochim. Acta* **48**(9), 1807–1818.

Ingvorsen K. and Jorgensen B. B. (1984) Kinetics of sulfate uptake by freshwater and marine species of *Desulfovibrio. Arch. Microbiol.* **139**, 61–66.

Ivanov M. V. (1981) The global biogeochemical sulphur cycle. In *SCOPE 17: Some Perspectives of the Major Biogeochemical Cycles* (ed. G. E. Likens). Wiley, Chichester, pp. 61–78.

Jorgensen B. B. (1977a) Bacterial sulfate reduction within reduced microniches of oxidized marine sediments. *Mar. Biol.* **41**, 7–17.

Jorgensen B. B. (1977b) *The Sulfur Cycle of a Coastal Marine Sediment.* Limfjorden, Denmark.

Jorgensen B. B. (1978) A comparison of methods for the quantification of bacterial sulfate reduction in coastal marine sediments. *Geomicrobiol. J.* **1**(1), 11–27.

Jorgensen B. B. (1979) A theoretical model of the stable isotope distribution in marine sediments. *Geochim. Cosmochim. Acta* **43**(3), 363–374.

Jorgensen B. B. (1982) Mineralization of organic matter in the sea bed—the role of sulphate reduction. *Nature* **296**(5858), 643–645.

Jorgensen B. B. and Bak F. (1991) Pathways and microbiology of thiosulphate transformations and sulphate reduction in a marine sediment (Kattegat, Denmark). *Appl. Environ. Microbiol.* **57**(3), 847–856.

Jorgensen B. B., Weber A., and Zopfi J. (2001) Sulfate reduction and anaerobic methane oxidation in Black Sea sediments. *Deep-Sea Res. I: Oceanogr. Res. Pap.* **48**(9), 2097–2120.

Kaplan I. R., Emery K. O., and Rittenberg S. C. (1963) The distribution and isotopic abundance of sulphur in recent marine sediments off southern California. *Geochim. Cosmochim. Acta* **27**(4), 297–331.

Kasten S. and Jorgensen B. B. (2000) Sulfate reduction in marine sediments. In *Marine Geochemistry* (eds. H. D. Schulz and M. Zabel). Springer, Berlin, pp. 263–281.

Kastner M., Elderfield H., Martin J. B., Suess E., Kvenvolden K. A., and Garrison R. E. (1990) Diagenesis and interstitial-water chemistry at the Peruvian continental margin; major constituents and strontium isotopes. In *Proceedings of the Ocean Drilling Program, Peru Continental Margin; Covering Leg 112 of the Cruises of the Drilling Vessel JOIDES Resolution, Callao, Peru to Valparaiso, Chile, Sites 679-688, 20 October 1986–25 December 1986* (ed. S. Stewart). vol. 112, pp. 413–440.

Keil R. G. and Hedges J. I. (1993) Sorption of organic matter to mineral surfaces and the preservation of organic matter in coastal marine sediments. *Chem. Geol.* **107**(3–4), 385–388.

Klug M., Boston P., Francois R., Gyure R., Javor B., Tribble G., and Vairavamurthy A. (1985) Sulfur reduction in sediments of marine and evaporite environments. In *The Global Sulfur Cycle* (ed. D. Sagan). NASA Technical Memorandum 87570, NASA, Washington, DC.

Knoll A. H., Canfield D. E., Norris R. D., and Corfield R. M. (1998) Isotopic inferences on early ecosystems. In *Isotope Paleobiology and Paleoecology* (eds. W. L. Manger and L. K. Meeks). The Paleontological Society, Lawrence, KS, pp. 212–243.

Kohn M. J., Riciputi L. R., Stakes D., and Orange D. L. (1998) Sulfur isotope variability in biogenic pyrite; reflections of heterogeneous bacterial colonization? *Am. Mineral.* **83** (11–12 (Part 2)), 1454–1468.

Kohnen M. E. L., Sinninghe Damste J. S., ten Haven H. L., and de Leeuw J. W. (1989) Early incorporation of polysulphides in sedimentary organic matter. *Nature* **341**(6243), 640–641.

Kohnen M. E. L., Sinninghe-Damste J. S., Kock-Van Dalen A. C., Ten Haven H. L., Rullkoetter J., and de Leeuw J. W. (1990) Origin and diagenetic transformations of C (sub 25) and C (sub 30) highly branched isoprenoid sulphur compounds; further evidence for the formation of organically bound sulphur during early diagenesis. *Geochim. Cosmochim. Acta* **54**(11), 3053–3063.

Kohnen M. E. L., Sinninghe Damste J. S., Baas M., Schouten S., de Leeuw J. W., and Anonymous. (1991) Sulphur quenching of functionalised lipids in marine sediments; its consequences for (palaeo)environmental reconstruction. In *Abstracts of Papers—American Chemical Society, National Meeting.* American Chemical Society, vol. 201, pp. GEOC 31.

Konhauser K. O. (1998) Diversity of bacterial iron mineralization. *Earth Sci. Rev.* **43**(3–4), 91–121.

Konopka A., Gyure R. A., Doemel W., and Brooks A. (1985) *Microbial sulfate reduction in extremely acid lakes.* West Lafayette: Purdue University Water Resources Research Center, Technical Report 173, West Lafayette, pp. 1–50.

Krouse H. R. and McCready R. G. L. (1979a) Biogeochemical cycling of sulfur. In *Biogeochemical Cycling of Mineral-forming Elements* (eds. P. A. Trudinger and D. J. Swaine). Elsevier, Amsterdam, pp. 401–425.

Krouse H. R. and McCready R. G. L. (1979b) Reductive reactions in the sulfur cycle. In *Biogeochemical Cycling of Mineral-forming Elements* (eds. P. A. Trudinger and D. J. Swaine). Elsevier, Amsterdam, pp. 315–358.

Krs M., Krsova M., Pruner P., Zeman A., Novak F., and Jansa J. (1990) A petromagnetic study of Miocene rocks bearing micro-organic material and the magnetic mineral greigite (Sokolov and Cheb basins, Czechoslovakia). *Phys. Earth Planet. Inter.* **63**, 98–112.

Leventhal J. and Taylor C. (1990) Comparison of methods to determine degree of pyritization. *Geochim. Cosmochim. Acta* **54**, 2621–2625.

Leventhal J. S. (1983a) An interpretation of carbon and sulfur relationships in Black Sea sediments as indicators of environments of deposition. *Geochim. Cosmochim. Acta* **47**(1), 133–137.

Leventhal J. S. (1983b) Organic carbon, sulfur, and iron relationships as an aid to understanding depositional environments and syngenetic metals in recent and ancient sediments. In *US Geological Survey Circular, Report: C 0822* (eds. T. M. Cronin, W. F. Cannon, and R. Z. Poore). US Geological Survey, Reston, VA, pp. 34–36.

Leventhal J. S. (1995) Carbon-sulfur plots to show diagenetic and epigenetic sulfidation in sediments. *Geochim. Cosmochim. Acta* **59**(6), 1207–1211.

Lew M. (1981) The distribution of some major and trace elements in sediments of the Atlantic Ocean (DSDP samples). *Chem. Geol.* **33**, 205–224.

Lin S. and Morse J. W. (1991) Sulfate reduction and iron sulfide mineral formation in Gulf of Mexico anoxic sediments. *Am. J. Sci.* **291**, 55–89.

Love L. G. (1967) Early diagenetic iron sulphide in recent sediments of the Wash (England). *Sedimentology* **9**(4), 327–352.

Lovely D. R. and Phillips E. J. P. (1987) Competitive mechanisms for inhibition of sulfate reduction and methane production in the zone of ferric iron reduction in sediments. *Appl. Environ. Microbiol.* **53**(11), 2636–2641.

Lueckge A., Ercegovac M., Strauss H., and Littke R. (1999) Early diagenetic alteration of organic matter by sulfate reduction in Quaternary sediments from the northeastern Arabian Sea. *Mar. Geol.* **158**(1–4), 1–13.

Luther G. W., III, Giblin A., Howarth R. W., and Ryans R. A. (1982) Pyrite and oxidized iron mineral phases formed from pyrite oxidation in salt marsh and estuarine sediments. *Geochim. Cosmochim. Acta* **46**(12), 2667–2671.

Luther G. W., III, Giblin A. E., and Varsolona R. (1985) Polarographic analysis of sulfur species in marine porewaters. *Limnol. Oceanogr.* **30**(4), 727–736.

Luther G. W., III, Glazer B. T., Hohmann L., Popp J. I., Taillefert M., Rozan T. F., Brendel P. J., Theberge S. M., and Nuzzio D. B. (2001) *Sulfur speciation monitored* in situ *with solid state gold amalgam voltammetric microelectrodes: polysulfides as a special case in sediments, microbial mats and hydrothermal vent waters. J. Environ. Monitor.* **3**, 61–66.

Lyons T. W. (1997) Sulfur isotopic trends and pathways of iron sulfide formation in upper Holocene sediments of the anoxic Black Sea. *Geochim. Cosmochim. Acta* **61**(16), 3367–3382.

Lyons T. W. and Berner R. A. (1992) Carbon–sulfur–iron systematics of the uppermost deep-water sediments of the Black Sea. In: *Chemical Geology* (eds. P. A. Meyers, L. M. Pratt, and B. Nagy). Elsevier, vol. 99, pp. 1–27.

Lyons T. W., Kah L. C., and Gellatly A. M. The Precambrian sulfur isotope record of evolving atmospheric oxygen. In *Tempos and Events in Precambrian Time* (eds. P. G. Eriksson, W. Altermann, D. R. Nelson, W. Mueller, O. Catuneanu, and Strand). Elsevier (in press).

Lyons T. W., Raiswell R., and Anonymous. (1993) Carbon-sulfur-iron geochemistry of modern Black Sea sediments; a summary. In *Geological Society of America, 1993 Annual Meeting*. Geological Society of America (GSA), vol. 25, pp. 239.

Lyons T. W., Werne J. P., Hollander D. J., and Murry R. W. (2003) Contrasting sulfur geochemistry and Fe/Al and Mo/Al ratios across the last oxic-to-anoxic transition in the Cariaco Basin, Venezuela. *Chem. Geol.* **195**, 131–157.

Lyons W. B. and Gaudette M. E. (1979) Sulfate reduction and the nature of organic matter in estuarine sediments. *Org. Geochem.* **1**, 151–155.

Madigan M., Martinko J., and Parker J. (2000) *Biology of Microorganisms*. Prentice Hall, New Jersey.

Migdisov A. A., Ronov A. B., and Grinenko V. A. (1983) In *The sulphur cycle in the lithosphere: Part I. Reservoirs.* SCOPE (Chichester) (eds. M. V. Ivanov and J. R. Freney). Wiley, Chichester, vol. 19, pp. 25–95.

Morozov A. A. (1995a) Fe and S in the sedimentary process, oxygen-rich Black Sea Zone: Part II. Early sediment diagenesis and its role in Holocene shelf sedimentation. *Lithol. Min. Resour.* **29**(5), 437–448.

Morozov A. A. (1995b) Iron and sulfur in the sedimentary process; oxygen-enriched Black Sea zone: Part 1. Fe- and S-species in Holocene deposits of certain shelf areas. *Lithol. Min. Res.* **29**(4), 311–322.

Morse J. W. and Berner R. A. (1995) What determines sedimentary C/S ratios? *Geochim. Cosmochim. Acta* **59**(6), 1073–1077.

Morse J. W., Millero F. J., Cornwell J. C., and Rickard D. (1987) The chemistry of the hydrogen sulfide and iron sulfide systems in natural waters. *Earth Sci. Rev.* **24**(1), 1–42.

Morse J. W. and Wang Q. (1997) Pyrite formation under conditions approximating those in anoxic sediments: II. Influence of precursor iron minerals and organic matter. *Mar. Chem.* **57**(3–4), 187–193.

Mossmann J.-R., Aplin A. C., Curtis C. D., and Coleman M. L. (1991) Geochemistry of inorganic and organic sulphur in organic-rich sediments from the Peru margin. *Geochim. Cosmochim. Acta* **55**(12), 3581–3595.

Murowchick J. B. (1992) Marcasite inversion and the petrographic determination of pyrite ancestry. *Econ. Geol. and Bull. Soc. Econ. Geol.* **87**(4), 1141–1152.

Murowchick J. B. and Barnes H. L. (1987) Effects of temperature and degree of supersaturation on pyrite morphology. *Am. Mineral.* **72**(11–12), 1241–1250.

Murry J. I. R. (1895) On the chemical changes which take place in the composition of seawater associated with blue muds on the floor of the ocean. *Trans. Roy. Soc. Edinburgh* **37**, 481–508.

Mycroft J. R., Nesbitt H. W., and Pratt A. R. (1995) X-ray photoelectron and Auger electron spectroscopy of air-oxidized pyrrhotite; distribution of oxidized species with depth. *Geochim. Cosmochim. Acta* **59**(4), 721–733.

Neal A. L., Techkarnjanaruk S., Dohnalkova A., McCready D., Peyton B. M., and Geesey G. G. (2001) Iron sulfides and sulfur species produced at hematite surfaces in the presence of sulfate-reducing bacteria. *Geochim. Cosmochim. Acta* **65** (2), 223–235.

Nissenbaum A. and Kaplan I. R. (1972) Chemical and isotopic evidence for the in situ origin of marine humic substances. *Limnol. Oceanogr.* **17**(4), 570–582.

Nissenbaum A. and Kaplan I. R. (1976) Sulfur and carbon isotopic evidence for biogeochemical processes in the Dead Sea ecosystem. In *Environmental Biogeochemistry: Carbon, Nitrogen, Phosphorus, Sulfur and Selenium Cycles* (ed. J. O. Nriagu). Ann Arbor Sci. Publ, Ann Arbor, vol. 1, pp. 309–325.

Ohmoto H. (1992) Biogeochemistry of sulfur and the mechanisms of sulfide–sulfate mineralization in Archean oceans. In *Early Organic Evolution: Implications for Mineral and Energy Resources* (eds. M. Schidlowski, S. Golubic, M. M. Kimberly, and P. A. Trudinger). Springer, Berlin.

Ohmoto H. and Goldhaber M. B. (1997) Sulfur and carbon isotopes. In *Geochemistry of Hydrothermal Ore Deposits* (ed. H. L. Barnes). Wiley, New York, , pp. 517–612.

Ohmoto H., Kaiser C. J., and Geer K. A. (1990) Systematics of sulphur isotopes in recent marine sediments and ancient sediment-hosted basemetal deposits. In: *Stable Isotopes and Fluid Processes in Mineralization* (eds. H. K. Herbert and S. E. Ho). University of Western Australia, Crawley, vol. 23, pp. 70–120.

Ohmoto H., Kakegawa T., and Lowe D. R. (1993) 3.4-billion-year-old biogenic pyrites from Barberton, South Africa; sulfur isotope evidence. *Science* **262**(5133), 555–557.

Orr W. L. (1986) Kerogen/asphaltene/sulfur relationships in sulfur-rich Monterey oils. In: *Organic Geochemistry* (eds. D. Leythaeuser and J. Ruellkotter). Pergamon, Oxford, vol. 10, pp. 499–516.

Ostroumov E. A. (1953) Different forms of combined sulfur in the sediments of the Black Sea. *Trud. Inst Okeanol, Akad. Nauk. SSSR* **7**, 70–90.

Passier H. F., Luther G. W., III, and de Lange G. J. (1997) Early diagenesis and sulphur speciation in sediments of the Oman Margin, northwestern Arabian Sea. *Deep-Sea Res. II: Top. Stud. Oceanogr.* **44**(6–7), 1361–1380.

Passier H. F., Middelburg J. J., de Lange G. J., and Boettcher M. E. (1999) Modes of sapropel formation in the eastern Mediterranean; some constraints based on pyrite properties. *Mar. Geol.* **153**(1–4), 199–219.

Paytan A., Kastner M., Campbell D., and Thiemens M. H. (1998) Sulfur isotopic composition of Cenozoic seawater sulfate. *Science* **282**(5393), 1459–1462.

Petsch S. T. and Berner R. A. (1998) Coupling the geochemical cycles of C, P, Fe, and S; the effect on atmospheric O (sub 2) and the isotopic records of carbon and sulfur. *Am. J. Sci.* **298**(3), 246–262.

Pfennig N., Widdel F., and Truper H. G. (1979) The Dissimilatory Sulfate-reducing Bacteria. In: *Biogeochemical Cycling of Mineral-forming Elements* (eds. P. A. Trudinger and D. J. Swaine). Elsevier, Amsterdam, vol. 3, pp. 926–940.

Posfai M., Buseck P. R., Bazylinski D. A., and Frankel R. B. (1998) Iron sulfides from magnetotactic bacteria; structure, composition, and phase transitions. *Am. Mineral.* **83**(11–12 (Part 2)), 1469–1481.

Posfai M., Cziner K., Marton E., Marton P., Buseck P. R., Frankel R. B., and Bazylinski D. A. (2001) Crystal-size distributions and possible biogenic origin of Fe sulfides. Biogenic Iron Minerals Symposium and Workshop, Tihany, Hungary, May 20–23, 2000. *Euro. J. Mineral.* **13**, pp. 691–703.

Price F. T. and Shieh Y. N. (1979) Fractionation of sulfur isotopes during laboratory synthesis of pyrite at low temperatures. *Chem. Geol.* 27.

Pyzik A. J. and Sommer S. E. (1981) Sedimentary iron monosulfides; kinetics and mechanism of formation. *Geochim. Cosmochim. Acta* **45**(5), 687–698.

Raiswell R. (1993) *Iron mineralogy; influence on degree of pyritization.* 17th Meeting of the European Union of

Geosciences: Abstract Supplement, Blackwell, Oxford pp. 691–692.

Raiswell R. and Berner R. A. (1985) Pyrite formation in euxinic and semi-euxinic sediments. *Am. J. Sci.* **285**(8), 710–724.

Raiswell R. and Berner R. A. (1986) Pyrite and organic matter in Phanerozoic normal marine shales. *Geochim. Cosmochim. Acta* **50**(9), 1967–1976.

Raiswell R. and Canfield D. E. (1996) Rates of reaction between silicate iron and dissolved sulfide in Peru margin sediments. *Geochim. Cosmochim. Acta* **60**(15), 2777–2787.

Raiswell R. and Canfield D. E. (1998) Sources of iron for pyrite formation in marine sediments. *Am. J. Sci.* **298**(3), 219–245.

Raiswell R., Buckley F., Berner R. A., and Anderson T. F. (1988) Degree of pyritization of iron as a paleoenvironmental indicator of bottom-water oxygenation. *J. Sedim. Petrol.* **58**(5), 812–819.

Rees C. E. (1973) A steady-state model for sulphur isotope fractionation in bacterial reduction processes. *Geochim. Cosmochim. Acta* **37**(5), 1141–1162.

Reynolds R. L. and Goldhaber M. B. (1983) Iron disulfide minerals and the genesis of roll-type uranium deposits. *Econ. Geol. and Bull. Soc. Econ. Geol.* **78**(1), 105–120.

Reynolds R. L., Fishman N. S., Wanty R. B., and Goldhaber M. B. (1990a) Iron sulfide minerals at cement oil field, Oklahoma; implications for magnetic detection of oil fields. *Geol. Soc. Am. Bull.* **102**(3), 368–380.

Reynolds R. L., Nicholson A., Goldhaber M. B., Colman S. M., King J. W., Rice C. A., Tuttle M. L., and Sherman D. M. (1990b) Diagnosis for greigite (Fe (sub 3) S (sub 4)) in Cretaceous beds, North Slope, Alaska, and Holocene sediments, Lake Michigan. *EOS, Trans., AGU* **71**(43), 1282–1283.

Rickard D. (1997) Kinetics of pyrite formation by the H_2S oxidation of iron(II) monosulfide in aqueous solutions between 25 and 125 degrees C; the rate equation. *Geochim. Cosmochim. Acta* **61**(1), 115–134.

Rickard D. and Luther G. W., III (1997) Kinetics of pyrite formation by the H_2S oxidation of iron(II) monosulfide in aqueous solutions between 25 and 125 degrees C; the mechanism. *Geochim. Cosmochim. Acta* **61**(1), 135–147.

Rickard D., Schoonen M. A. A., and Luther G. W., III (1994) Chemistry of iron sulfides in sedimentary environments. In *Geochemical Transformations of Sedimentary Sulfur* (eds. M. A. Vairavamurthy, M. A. A. Schoonen, T. I. Eglinton, G. W. Luther, III, and B. Manowitz). American Chemical Society, Washington, DC, pp. 168–193.

Rickard D. T. (1969) The chemistry of iron sulphide formation at low temperatures. *Stockholm Contributions in Geology* **20**, 67–95.

Rickard D. T. (1975) Kinetics and mechanism of pyrite formation at low temperatures. *Am. J. Sci.* **275**(6), 636–652.

Russell M. J. and Hall A. J. (1997) The emergence of life from iron monosulphide bubbles at a submarine hydrothermal redox and pH front. *J. Geol. Soc. London* **154**(3), 377–402.

Schenau S. J., Passier H. F., Reichart G. J., and de Lange G. J. (2002) Sedimentary pyrite formation in the Arabian Sea. *Mar. Geol.* **185**(3–4), 393–402.

Schidlowski M. (1979) Antiquity and evolutionary status of bacterial sulfate reduction; sulfur isotope evidence. In: *Origins of Life*. D. Reidel, Dordrecht, vol. 9, pp. 299–310.

Schidlowski M., Hayes J. M., and Kaplan I. R. (1983) Isotopic inferences of ancient biochemistries; carbon, sulfur, hydrogen, and nitrogen. In *Earth's Earliest Biosphere; its Origin and Evolution* (ed. J. W. Schopf). Princeton University Press, Princeton, pp. 149–186.

Schimmelmann A. and Kastner M. (1993) Evolutionary changes over the last 1000 years of reduced sulfur phases and organic carbon in varved sediments of the Santa Barbara Basin, California. *Geochim. Cosmochim. Acta* **57**(1), 67–78.

Schippers A. and Jorgensen B. B. (2001) Oxidation of pyrite and iron sulfide by manganese dioxide in marine sediments. *Geochim. Cosmochim. Acta* **65**(6), 915–922.

Schippers A. and Jorgensen B. B. (2002) Biogeochemistry of pyrite and iron sulfide oxidation in marine sediments. *Geochim. Cosmochim. Acta* **66**(1), 85–92.

Schoonen M. A. A. and Barnes H. L. (1991a) Reactions forming pyrite and marcasite from solution: I. Nucleation of FeS (sub 2) below 100 degrees C. *Geochim. Cosmochim. Acta* **55** (6), 1495–1504.

Schoonen M. A. A. and Barnes H. L. (1991b) Reactions forming pyrite and marcasite from solution: II. Via FeS precursors below 100 degrees C. *Geochim. Cosmochim. Acta* **55** (6), 1505–1514.

Sinninghe Damste J. S., Kok M. D., Koester J., and Schouten S. (1998) Sulfurized carbohydrates; an important sedimentary sink for organic carbon? *Earth Planet. Sci. Lett.* **164**(1–2), 7–13.

Skinner B. J., Erd R. C., and Grimaldi F. S. (1964) Greigite, the thio-spinel of iron; a new mineral. *Am. Mineral.* **49**, 543–555.

Skyring G. W. (1987) Sulfate reduction in coastal ecosystems. *Geomicrobiol. J.* **5**(3–4), 295–374.

Sommerfield C. K., Aller R. C., and Nittrouer C. A. (2001) Sedimentary carbon, sulfur, and iron relationships in modern and ancient diagenetic environments of the Eel River basin (USA.). *J. Sedimen. Res.* **71**(3), 335–345.

Staudt W. J. and Schoonen M. A. A. (1994) Sulfate incorporation into sedimentary carbonates. In *Geochemical Transformations of Sedimentary Sulfur* (eds. M. A. Vairavamurthy, M. A. A. Schoonen, T. I. Eglinton, G. W. Luther, and B. Manowitz). American Chemical Society, Washington, DC, vol. 612, pp. 332–345.

Sternbeck J. and Sohlenius G. (1997) Authigenic sulfide and carbonate mineral formation in Holocene sediments of the Baltic Sea. *Chem. Geol.* **135**(1–2), 55–73.

Steudel R., Holdt G., Gobel T., and Hazeu W. (1987) Chromatographic separation of higher polythionates (Sn)621 ($n = 3\ldots22$) and their detection in cultures of Thiobacilus ferrooxidans: molecular composition of bacterial sulfur secretions. *Angew. Chemie. Int. Ed. Engl.* **26**, 151–153.

Strauss H. (1997) The isotopic composition of sedimentary sulfur through time. *Paleogeogr. Paleoclimatol. Paleoecol.* **132**(1–4), 97–118.

Suits N. S. and Arthur M. A. (2000) Sulfur diagenesis and partitioning in Holocene Peru shelf and upper slope sediments. *Chem. Geol.* **163**(1–4), 219–234.

Sweeney R. E. (1972) Pyritization during diagenesis of marine sediments. PhD, University of California, LA..

Sweeney R. E. and Kaplan I. R. (1973) Pyrite framboid formation; laboratory synthesis and marine sediments. *Econ. Geol. and Bull. Soc. Econ. Geol.* **68**(5), 618–634.

Sweeney R. E. and Kaplan I. R. (1980) Diagenetic sulfate reduction in marine sediments. *Mar. Chem.* **9**(3), 165–174.

Tannenbaum E. and Aizenshtat Z. (1985) Formation of immature asphalt from organic-rich carbonate rocks: I. Geochemical correlation. *Org. Geochem.* **8**(2), 181–192.

Thamdrup B., Finster K., Fossing H., Hansen J. W., and Jorgensen B. B. (1994a) Thiosulfate and sulfite distributions in porewater of marine sediments related to manganese, iron, and sulfur geochemistry. *Geochim. Cosmochim. Acta* **58**(1), 67–73.

Thamdrup B., Fossing H., and Jorgensen B. B. (1994b) Manganese, iron, and sulfur cycling in a coastal marine sediment, Aarhus Bay, Denmark. *Geochim. Cosmochim. Acta* **58**(23), 5115–5129.

Thode H. C., Kleerekoper H., and McElcheran D. (1951) Isotopic fractionation in the bacterial reduction of sulphate. *Research (London)* **4**, 581–582.

Thode H. G. and Monster J. (1965) Sulfur-isotope geochemistry of petroleum, evaporites, and ancient seas. In *Fluids in Sub Surface Environments—A Symposium of American Association of Petroleum Geologists*, Tulsa, OK, Memoir, pp. 367–377.

Thode-Anderson S. and Jorgensen B. B. (1989) Sulphate reduction and the formation of sulphur-35 labelled FeS,

FeS_2 and elemental Sulphur in coastal marine sediments. *Limnol. Oceanogr.* **34**(5), 793–806.
Thorstenson D. C. and Mackenzie F. T. (1974) Time variability of pore water chemistry in recent carbonate sediments, Devil's Hole, Harrington Sound, Bermuda. *Geochim. Cosmochim. Acta* **38**(1), 1–19.
Troelsen H. and Jorgensen B. B. (1982) Seasonal dynamics of elemental sulfur in two coastal sediments. *Estuar. Coast. Shelf Sci.* **15**(3), 255–266.
Truper H. G. (1982) Microbial process in the sulfur cycle through time. In *Mineral Deposits and the Evolution of the Biosphere* (eds. S. H. Holland and M. Schidlowski). Springer, Berlin, pp. 5–30.
Tuttle M. L. and Goldhaber M. B. (1993) Sedimentary sulfur geochemistry of the Paleogene Green River Formation, Western USA; implications for interpreting depositional and diagenetic processes in saline alkaline lakes. *Geochim. Cosmochim. Acta* **57**(13), 3023–3039.
Tuttle M. L., Rice C. A., and Goldhaber M. B. (1990) Geochemistry of organic and inorganic sulfur in ancient and modern lacustrine environments; case studies of freshwater and saline lakes. In *Geochemistry of Sulfur in Fossil Fuels* (eds. W. L. Orr and C. M. White). American Chemical Society, Washington, DC, pp. 114–148.
Vairavamurthy A. and Anonymous M. A. (1993) *Geochemical incorporation of sulfur into organic matter; importance of hydrogen sulfide oxidation product.* Abstracts with Programs—Geological Society of America, Geological Society of America (GSA), Geological Society of America (GSA), vol. 25, pp. 19.
Vairavamurthy M. A., Maletic D., Wang S., Manowitz B., Eglinton T. I., Lyons T., and Anonymous (1997) Characterization of sulfur-containing functional groups in sedimentary humic substances by X-ray absorption near-edge structure spectroscopy. *Energy and Fuels* **11**, 546–553.
van Kaam-Peters H. M. E., Schouten S., Koester J., and Sinninghe Damste J. S. (1998) Controls on the molecular and carbon isotopic composition of organic matter deposited in a Kimmeridgian euxinic shelf sea; evidence for preservation of carbohydrates through sulfurisation. *Geochim. Cosmochim. Acta* **62**(19–20), 3259–3283.
Vaughan D. J. and Craig J. R. (1978) *Mineral Chemistry of Metal Sulfides.* Cambridge University Press, Cambridge.
Volkov I. I. and Rozanov A. G. (1983) The sulphur cycle in oceans: Part I. Reservoirs and fluxes. In *SCOPE (Chichester)* (eds. M. V. Ivanov and J. R. Freney). Wiley, Chichester, vol.19, pp. 357–423.
Volkov I. I., Rozanov A. G., and Zhabina N. N. (1983) Sulfur compounds in sediments of the Gotland Basin (Baltic Sea). *Lithology and Mineral Resour.* **18**(6), 584–598.
Wakeham S. G., Sinninghe Damste J. S., Kohnen M. E. L., and de Leeuw J. W. (1995) Organic sulfur compounds formed during early diagenesis in Black Sea sediments. *Geochim. Cosmochim. Acta* **59**(3), 521–533.
Wang Q. and Morse J. W. (1994) Laboratory simulation of pyrite formation in anoxic sediments. In *Geochemical Transformations of Sedimentary Sulfur* (eds. M. A. Vairavamurthy, M. A. A. Schoonen, T. I. Eglinton, G. W. Luther, and B. Manowitz). American Chemical Society, Washington, DC, pp. 206–223.
Weber A., Riess W., Wenzhoefer F., and Jorgensen B. B. (2001) Sulfate reduction in Black Sea sediments; *in situ* and laboratory radiotracer measurements from the shelf to 2000m depth. *Deep-Sea Res.: Part I. Oceanogr. Res. Pap.* **48**(9), 2073–2096.
Werne J. P., Hollander D. J., Behrens A., Schaeffer P., Albrecht P., and Sinninghe Damste J. S. (2000) Timing of early diagenetic sulfurization of organic matter; a precursor-product relationship in Holocene sediments of the anoxic Cariaco Basin, Venezuela. *Geochim. Cosmochim. Acta* **64** (10), 1741–1751.
Westrich J. T. and Berner R. A. (1988) The effect of temperature on rates of sulfate reduction in marine sediments. *Geomicrobiol. J.* **6**(2), 99–117.
Wignall P. B. and Newton R. (1998) Pyrite framboid diameter as a measure of oxygen deficiency in ancient mudrocks. *Am. J. Sci.* **298**(7), 537–552.
Wijsman J. W. M., Middelburg J. J., and Heip C. H. R. (2001a) Reactive iron in Black Sea sediments; implications for iron cycling. *Mar. Geol.* **172**(3/4), 167–180.
Wijsman J. W. M., Middelburg J. J., Herman P. M. J., Boettcher M. E., and Heip C. H. R. (2001) Sulfur and iron speciation in surface sediments along the northwestern margin of the Black Sea. *Mar. Chem.* **74**(4), 261–278.
Wilkin R. T. and Arthur M. A. (2001b) Variations in pyrite texture, sulfur isotope composition, and iron systematics in the Black Sea; evidence for Late Pleistocene to Holocene excursions of the O (sub 2)–H (sub 2) s redox transition. *Geochim. Cosmochim. Acta* **65**(9), 1399–1416.
Wilkin R. T. and Barnes H. L. (1996) Pyrite formation by reactions of iron monosulfides with dissolved inorganic and organic sulfur species. *Geochim. Cosmochim. Acta* **60** (21), 4167–4179.
Wilkin R. T. and Barnes H. L. (1997) Formation processes of framboidal pyrite. *Geochim. Cosmochim. Acta* **61**(2), 323–339.
Wilkin R. T., Barnes H. L., and Brantley S. L. (1996) The size distribution of framboidal pyrite in modern sediments; an indicator of redox conditions. *Geochim. Cosmochim. Acta* **60**(20), 3897–3912.
Wortmann U. G., Bernasconi S. M., and Boettcher M. E. (2001) Hypersulfidic deep biosphere indicates extreme sulfur isotope fractionation during single-step microbial sulfate reduction. *Geology* **29**(7), 647–650.
Yu A., Lein M. B., Varyshteyn B. B., Namsarayev Y. V., Kashparova A. G., VBondar V. A., and Ivanov M. V. (1982) Biogeochemistry of aneroobic diagenesis of recent Baltic sediments. *Geochem. Int.* **19**(2), 90–103.
ZoBell C. E. and Rittenberg S. C. (1950) Sulfate-reducing bacteria in marine sediments. *J. Mar. Res.* **7**(3), 602–617.

Published by Elsevier Ltd.

Isotope Geochemistry
ISBN: 978-0-08-096710-3

pp. 253–284

10

Stable Isotopes in the Sedimentary Record

A. Lerman
Northwestern University, Evanston, IL, USA

and

N. Clauer
Centre de Géochimie de la Surface, CNRS—Université Louis Pasteur, Strasbourg, France

10.1 INTRODUCTION

This chapter addresses the interpretation of the sedimentary records of several common stable isotopes and their fractionation in natural systems: the isotopes of hydrogen (D/H) and oxygen ($^{18}O/^{16}O$) that occur in the global water cycle and are fractionated by a number of physical and chemical processes in the gas–liquid–solid H_2O system, as well as in many mineral–water reactions in sediments; the isotopes of calcium ($^{44}Ca/^{40}Ca$) in the riverine input to the ocean and biogenic carbonates forming in ocean water; the isotopic fractionation of the elements essential to life, carbon ($^{13}C/^{12}C$), nitrogen ($^{15}N/^{14}N$), and sulfur ($^{34}S/^{32}S$), that have produced variably extensive sedimentary records during Earth's history; and the occurrences of the boron isotopes ($^{11}B/^{10}B$) in the Earth's surface environment. There is an additional class of stable isotopes that are radiogenic, forming as products of radioactive decay of their parent radionuclides. In this category are, for example, such stable radiogenic isotopes as that of strontium, ^{87}Sr forming from ^{87}Rb, the isotope of helium ^4He that is a product of α-decay of the uranium and thorium radionuclide series, and the isotope of argon ^{40}Ar that is one of the decay products of the naturally occurring ^{40}K. In addition to the nonradiogenic stable isotopes mentioned above, a section in this chapter is devoted to the radiogenic stable ^{40}Ar because of its importance as a tracer of certain diagenetic processes in sedimentary clay minerals. In each topical section, we discuss the isotope data that are relevant to the interpretation of the more important processes in the sedimentary environment. Insofar as the data permit, we point out their links to the broader understanding of the Earth's surface environment and biogeochemical cycles, although the reader may bear in mind that the cycles are treated in another volume of the *Treatise on Geochemistry*.

The term isotope was introduced by Frederick Soddy (1913) to designate the nuclides of a chemical element that differ in their atomic mass, but occupy the same place in the periodic table because of their identical (as was thought then) chemical properties. Isotopes of an element have the same number of protons and electrons, but different numbers of neutrons. Hence the name isotope, derived from the Greek for the same place. As a point of historical curiosity, Hunter and Roach (1993) wrote, but without giving the source of their information, that the name isotope was suggested by Dr. Margaret Todd, a family friend, at a dinner party in the house of Soddy's parents-in-law. Soddy's insight was based on his work on radioactivity and discoveries of radioactive nuclides, and he summarized the concept by describing isotopes as "put colloquially, their atoms have identical outsides but different insides" (Soddy, 1922). However, most of the knowledge of the stable isotopes of the chemical elements came from the studies based on mass spectrometry, and by 1922 more than 80 naturally occurring isotopes of different elements were reported by Francis W. Aston (1922). A list of naturally occurring stable and radioactive isotopes, compiled by Turkevich (1964) from a very extensive publication of nuclide data by Strominger *et al.* (1958), has changed only little since then to a 2002 version (Holden, 2002). The latter is given in Table 1 that lists all the chemical elements that have at least one stable isotope. Thus the table does not include such elements as U, Th, Ra, and Rn, all the isotopes of which are radioactive, but it includes those radioactive isotopes that make a significant fraction of the natural occurrence of an element. Such radionuclides of wide geological interest as tritium (^3H), ^{10}Be, ^{14}C, and ^{36}Cl do not make significant fractions of the element's total abundance and they are not included in Table 1. As a whole, there are 257 stable and 21 radioactive isotopes listed in the table. Some of the radionuclides have half-lives much longer than the age of the Earth and it becomes a somewhat arbitrary decision whether to consider them as stable or radioactive isotopes. For example, all the five naturally occurring isotopes of tungsten (W) have half-lives of an order of 10^{18} years, and they are not included in Table 1. On the other hand, ^{208}Pb that is a decay product of ^{232}Th is usually considered a stable isotope, but it has a very long half-life of more than 10^{19} years due to decay by spontaneous fission.

Some of the stable isotopes are radiogenic, forming by decay of parent radionuclides, and their natural abundance has increased in the course of the Earth's history. The notable examples of such radiogenic stable isotopes are ^{40}Ar, the most abundant species of the atmospheric noble gas argon, ^{87}Sr that forms by decay of its parent ^{87}Rb, and the isotopes of lead ^{206}Pb, ^{207}Pb, and ^{208}Pb that are stable products of the uranium and thorium decay series.

The main evidence that stable isotopes of an element are not identical in their chemical and physical behavior were the discoveries, since the late 1930s, of the fractionation between the isotopically different species of such substances as H_2O and CO_2 between a gaseous and a liquid phase, between a solution and a solid forming at an equilibrium with the solution, and the fractionation of the isotopes in the process of biological production of organic matter from the gaseous atmospheric sources and water solutions.

Because the elements of atomic numbers 1–20, hydrogen through calcium, are the main chemical constituents of the atmosphere, continental and oceanic waters, sediments, and upper part of the

Table 1 Table of naturally occurring stable isotopes.

Element	*Atomic number and chemical Symbol*		*Isotope mass number*	*Isotope abundance (at.%)*
Hydrogen	1	H	1	99.9885
			2	0.0115
Helium	2	He	3	1.34×10^{-4}
			4	≈100
Lithium	3	Li	6	7.59
			7	92.41
Beryllium	4	Be	9	100
Boron	5	B	10	19.9
			11	80.1
Carbon	6	C	12	98.93
			13	1.07
Nitrogen	7	N	14	99.636
			15	0.364
Oxygen	8	O	16	99.757
			17	0.038
			18	0.205
Fluorine	9	F	19	100
Neon	10	Ne	20	90.48
			21	0.27
			22	9.25
Sodium	11	Na	23	100
Magnesium	12	Mg	24	78.99
			25	10.00
			26	11.01
Aluminum	13	Al	27	100
Silicon	14	Si	28	92.223
			29	4.685
			30	3.092
Phosphorus	15	P	31	100
Sulfur	16	S	32	94.93
			33	0.76
			34	4.29
			36	0.02
Chlorine	17	Cl	35	75.78
			37	24.22

Element	*Atomic number and chemical symbol*		*Isotope mass number*	*Isotope abundance (at.%)*
Argon	18	Ar	36	0.3365
			38	0.0632
			40	99.6003
Potassium	19	K	39	93.2581
			40[a]	0.0117[a]
			41	6.7302
Calcium	20	Ca	40	96.941
			42	0.647
			43	0.135
			44	2.086
			46	0.004
			48	0.187
Scandium	21	Sc	45	100
Titanium	22	Ti	46	8.25
			47	7.44
			48	73.72
			49	5.41
			50	5.18
Vanadium	23	V	50[a]	0.25[a]
			51	99.75
Chromium	24	Cr	50	4.345
			52	83.789
			53	9.501
			54	2.365
Manganese	25	Mn	55	100
Iron	26	Fe	54	5.845
			56	91.754
			57	2.119
			58	0.282
Cobalt	27	Co	59	100
Nickel	28	Ni	58	68.0769
			60	26.2231
			61	1.1399
			62	3.6345
			64	0.9256

Element	*Atomic number and chemical symbol*		*Isotope mass number*	*Isotope abundance (at.%)*
Copper	29	Cu	63	69.15
			65	30.85
Zinc	30	Zn	64	48.27
			66	28.977
			67	4.102
			68	19.02
			70	0.631
Gallium	31	Ga	69	60.108
			71	39.892
Germanium	32	Ge	70	20.370
			72	27.380
			73	7.759
			74	36.656
			76	7.835
Arsenic	33	As	75	100
Selenium	34	Se	74	0.89
			76	9.37
			77	7.63
			78	23.77
			80	36.656
			82	7.835
Bromine	35	Br	79	50.69
			81	49.31
Krypton	36	Kr	78	0.355
			80	2.286
			82	11.593
			83	11.500
			84	56.987
			86	17.279
Rubidium	37	Rb	85	72.17
			87[a]	27.83[a]
Strontium	38	Sr	84	0.56
			86	9.86
			87	7.00
			88	82.58

(Continued)

Table 1 (Continued).

Element	*Atomic number and chemical symbol*		*Isotope mass number*	*Isotope abundance (at.%)*
Yttrium	39	Y	89	100
Zirconium	40	Zr	90	51.45
			91	11.22
			92	17.15
			94	17.38
			96	2.80
Niobium	41	Nb	93	100
Molybdenum	42	Mo	92	14.77
			94	9.226
			95	15.900
			96	16.674
			97	9.560
			98	24.20
			100	9.67
Ruthenium	44	Ru	96	5.54
			98	1.87
			99	12.76
			100	12.60
			101	17.06
			102	31.55
			104	18.62
Rhodium	45	Rh	103	100
Palladium	46	Pa	102	1.02
			104	11.14
			105	22.33
			106	27.33
			108	26.46
			110	11.72
Silver	47	Ag	107	51.839
			109	48.161

Element	*Atomic number and chemical symbol*		*Isotope mass number*	*Isotope abundance (at.%)*
Cadmium	48	Cd	106	1.25
			108	0.89
			110	12.49
			111	12.80
			112	24.13
			113	12.22
			114	28.73
			116	7.49
Indium	49	In	113	4.29
			115[a]	95.71[a]
Tin	50	Sn	112	0.97
			114	0.66
			115	0.34
			116	14.54
			117	7.68
			118	24.22
			119	8.59
			120	32.58
			122	4.63
			124	5.79
Antimony	51	Sb	121	57.21
			123	42.79
Tellurium	52	Te	120	0.09
			122	2.55
			123[a]	0.89[a]
			124	4.74
			125	7.07
			126	18.84
			128	31.74
			130	34.08

Element	*Atomic number and chemical symbol*		*Isotope mass number*	*Isotope abundance (at.%)*
Iodine	53	I	127	100
Xenon	54	Xe	124	0.0953
			126	0.0890
			128	1.910
			129	26.40
			130	4.071
			131	21.233
			132	26.9087
			134	10.436
			136	8.858
Cesium	55	Cs	133	100
Barium	56	Ba	130	0.106
			132	0.101
			134	2.417
			135	6.592
			136	7.854
			137	11.232
			138	72.698
Lanthanum	57	La	138[a]	0.090[a]
			139	99.910
Cerium	58	Ce	136	0.185
			138	0.251
			140	88.450
			142[a]	11.114[a]
Praseodymium	59	Pr	141	100
Neodymium	60	Nd	142	27.2
			143	12.2
			144[a]	23.8[a]
			145	8.3
			146	17.2

			148	5.7
			150[a]	5.6[a]
Samarium	62	Sm	144	3.07
			147[a]	14.99[a]
			148[a]	11.24[a]
			149[a]	13.82[a]
			150	7.38
			152	26.75
			154	22.75
Europium	63	Eu	151	47.81
			153	52.19
Gadolinium	64	Gd	152	0.20
			154	2.18
			155	14.80
			156	20.47
			157	15.65
			158	24.84
			160[a]	21.86[a]
Terbium	65	Tb	159	100
Dysprosium	66	Dy	156	0.056
			158	0.095
			160	2.39
			161	18.889
			162	25.475
			163	24.896
			164	28.260
Holmium	67	Ho	166	100
Erbium	68	Er	162	0.139
			164	1.601
			166	33.503
			167	22.869
			168	26.978
			170	14.910
Thullium	69	Tm	169	100
Ytterbium	70	Yb	168	0.13
			170	3.04
			171	14.28
			172	21.83
			173	16.13
			174	31.83
			176	12.76
Lutetium	71	Lu	175	97.41
			176[a]	2.59[a]
Hafnium	72	Hf	174[a]	0.16[a]
			176	5.26
			177	18.60
			178	27.28
			179	13.62
			180	35.08
Tantalum	73	Ta	180[a]	0.012[a]
			181	99.988
Rhenium	75	Re	185	37.40
			187[a]	62.60[a]
Osmium	76	Os	184	0.02
			186[a]	1.59[a]
			187	1.96
			188	13.24
			189	16.15
			190	26.26
			192	40.78
Iridium	77	Ir	191	37.3
			193	62.7
Platinum	78	Pt	190[a]	0.014[a]
			192	0.782
			194	32.967
			195	33.832
			196	25.242
			198	7.163
Gold	79	Au	197	100
Mercury	80	Hg	196[a]	0.15[a]
			198	9.97
			199	16.87
			200	23.10
			201	13.18
			202	29.86
			204	6.87
Thallium	81	Tl	203	29.524
			206	70.476
Lead	82	Pb	204	1.4
			206	24.1
			207	22.1
			208[a]	52.4[a]
Bismuth	83	Bi	209	100

From Holden (2002).
[a] Radioactive nuclides.

crystalline continental and oceanic crust, much attention has been devoted to the occurrence of the stable isotopes of these elements in the main geochemical domains of the outer shell of the Earth. The present-day Earth's atmosphere consists of O_2, N_2, Ar, CO_2, and variable concentrations of H_2O vapor; the sediments consist predominantly of silicate and carbonate minerals containing such metals as Na, K, Mg, and Ca, and smaller amounts of metal sulfates, sulfides, and chlorides. Organic matter of biological origin is made primarily of C, H, O, and additionally N, P, and S. The processes of mineral weathering on land, transport of dissolved products to the ocean, the formation of calcium carbonate and other minerals in the ocean, and oxidation and reduction reactions driven by bacteria in sediments—all these involve to a variable extent biogeochemical reactions where isotopic species are fractionated and leave their signatures in the product that enables an interpretation of the processes shaping the Earth's surface environment. As is discussed further in this chapter, studies of the stable isotope occurrence and fractionation in inorganic and biological processes also provide outstanding information about the geologic history of life on Earth and evolution of its surface environment.

How many isotopic species of a given chemical compound are there? The general formula for the number of distinguishable isotopic species of n isotopes of an element, where each species has m sites that are occupied by different combinations of the n isotopes, is

$$N = \frac{(n+m-1)!}{(n-1)!m!} \quad (1)$$

For example, of the three oxygen isotopes (n=3; Table 1), any two make one isotopically distinct O_2 molecule (two sites, m=2), which results in a total of six distinguishable species: $^{16}O_2$, $^{16}O^{17}O$, $^{16}O^{18}O$, $^{17}O^{18}O$, $^{17}O_2$, and $^{18}O_2$. The total number of species is larger, n^m, but such isotopomers as $^{16}O^{18}O$ and $^{18}O^{16}O$ may be indistinguishable one from the other if they do not differ in their chemical properties. Their relative abundances are products of the individual isotope abundances multiplied by 2. For the isotopic species of CO_2, each of the six O_2 species can combine with either ^{12}C or ^{13}C, making a total of 12 CO_2 species that are given in Table 2 with their relative abundances.

The three most abundant CO_2 species are those containing the more abundant isotopes: $^{12}C^{16}O_2$, $^{13}C^{16}O_2$, and $^{12}C^{16}O^{18}O$. The isotopic variety of the aqueous carbonate and bicarbonate species is even greater than that of CO_2. As each has three oxygen sites, the total number of species rises to 20 for CO_3^{2-} and to 40 for HCO_3^-, because of the occurrence of the two stable hydrogen isotopes, ^1H and ^2H (deuterium or D). However, the abundance of most of the species among those of the carbonate and bicarbonate ions is also very low, and studies of the stable isotopes as tracers of geochemical processes usually focus on the more abundant species.

10.2 ISOTOPIC CONCENTRATION UNITS AND FRACTIONATION

Concentrations of isotopic species are usually represented by the concentration or relative abundance ratios of the heavy to the light isotope in a given phase and these ratios are expressed relative to the ratio value in a chosen standard. In such pairs as H–D, ^{12}C–^{13}C, and ^{16}O–^{18}O, the heavier isotope is much less abundant than the lighter one (Table 1), in the pair ^{40}Ca–^{44}Ca the heavier isotope's abundance fraction is about 2%, and in the pair ^{32}S–^{34}S, it is about 4.5%. However, among the isotopes of Li, B, and Ar, the heavier nuclide is more abundant than the lighter one. When such differences exist between the abundances, the ratio of the heavy to light isotope, written as $R={}^{heavy}X/{}^{light}X$, is the mole fraction of the heavy isotope:

$$R = \left(\frac{^{heavy}X}{^{light}X}\right)_{sample} = \frac{n_{heavy}}{n_{light}} \approx \frac{n_{heavy}}{n_{light}+n_{heavy}} = \frac{R}{1+R} = x_{heavy} \quad (2)$$

where n is the number of moles of the isotope in a sample and x is the mole fraction of the heavy isotope. When $n_{heavy} \ll n_{light}$, the mole fraction x_{heavy} and the ratio value R are $\ll 1$. The abundance ratio determined in a phase, R_{sample}, is

Table 2 CO_2 stable isotopic species and their calculated relative abundances (atom percent).

C and O isotopes		*CO_2 isotopic species*			
^{12}C	98.93	$^{12}C^{16}O_2$	98.450	$^{13}C^{16}O_2$	1.065
^{13}C	1.07	$^{12}C^{16}O^{17}O$	0.075	$^{13}C^{16}O^{17}O$	—
^{16}O	99.757	$^{12}C^{16}O^{18}O$	0.405	$^{13}C^{16}O^{18}O$	0.0044
^{17}O	0.038	$^{12}C^{17}O^{18}O$	—	$^{13}C^{17}O^{18}O$	—
^{18}O	0.205	$^{12}C^{17}O_2$	—	$^{13}C^{17}O_2$	—
		$^{12}C^{18}O_2$	—	$^{13}C^{18}O_2$	—

Abundances of individual C and O isotopes from Holden (2002). Dashes indicate very low computed abundance, <0.001 at.%.

expressed relative to the ratio in a standard, R_{standard}, as a value of δ:

$$\delta^{\text{heavy}}\text{X} = \frac{R_{\text{sample}} - R_{\text{standard}}}{R_{\text{standard}}} \times 1,000(‰) \quad (3)$$

The values of δ can be either positive, indicating enrichment in the heavier isotope relative to the standard, or negative, indicating depletion of the heavier isotope. In the concentration units of per mil (‰), the δ values usually vary from single digits to tens per mil for the isotopic ratios of such elements as carbon, nitrogen, and oxygen in the surface environment. They may extend into the hundreds for the lighter isotopes of hydrogen or in stronger fractionation processes of other isotopes.

The standards for the heavier or, usually, less abundant isotopes change from time to time or different standards are in use, such as for δ^{18}O in water (the Standard Mean Ocean Water (SMOW) scale) and carbonate minerals (the PDB scale based on the shell of a cephalopod belemnite from the Cretaceous Pee Dee formation that was replaced by the VPDB or Vienna-PDB standard). Conversion of a δ value based on one standard into a δ referred to another standard is given by the following relationship:

$$\delta_2 = (1 + \delta_1)R_{\text{standard}-1}/R_{\text{standard}-2} - 1 \quad (4)$$

where δ_1 is the value on the old standard scale and δ_2 the new value, taken as fractions or as (δ ‰)/1,000, $R_{\text{standard}-1}=(^{\text{heavy}}\text{X}/^{\text{light}}\text{X})_{\text{standard}-1}$ the isotope abundance ratio in the old standard, and $R_{\text{standard}-2}$ the same ratio in a new standard.

As an example of fractionation, an equilibrium between CO_2 gas and CO_2 dissolved in water entails the fractionation of the two stable carbon isotopes, ^{12}C and ^{13}C, in the following exchange reaction:

$$^{12}\text{CO}_2(\text{aq}) + {}^{13}\text{CO}_2(\text{g}) = {}^{13}\text{CO}_2(\text{aq}) + {}^{12}\text{CO}_2(\text{g})$$

An equilibrium constant of the preceding reaction is, with the δ values taken as fractions:

$$\alpha_{\text{CO}_2(\text{aq}) - \text{CO}_2(\text{g})} = \frac{[^{13}\text{CO}_2(\text{aq})]/[^{12}\text{CO}_2(\text{aq})]}{[^{13}\text{CO}_2(\text{g})]/[^{12}\text{CO}_2(\text{g})]} = \frac{(^{13}\text{C}/^{12}\text{C})_{\text{aq}}}{(^{13}\text{C}/^{12}\text{C})_{\text{g}}} = \frac{R_{\text{aq}}}{R_{\text{g}}} = \frac{1 + \delta^{13}\text{C}_{\text{aq}}}{1 + \delta^{13}\text{C}_{\text{g}}} \quad (5)$$

Isotopic fractionation factors for many reactions in the Earth surface environment that involve hydrogen, oxygen, or carbon in water, carbon dioxide, and carbonate minerals have values of α close to 1 or ln α close to zero. The change in the Gibbs free energy for such reactions in the standard state ($\Delta G_R{}^{\text{o}}$) is of an order of magnitude of 10^0–10^1 J mol^{-1}, in comparison to 10^0–10^2 kJ mol^{-1} for many mineral–solution–gas reactions of environmental geochemical interest. Isotopic exchange decreases with increasing temperature and α tends to 1 at high temperature, varying in proportion to $1/T^2$, but the manner in which this change occurs varies greatly from one system to another (O'Neil, 1986).

In general, fractionation of heavy (X) and light (Y) isotopic species between two phases, denoted 1 and 2, can be written as

$$\alpha_{\text{X}_1-\text{X}_2} = \frac{(\text{X}/\text{Y})_1}{(\text{X}/\text{Y})_2} = \frac{R_1}{R_2} = \frac{1 + \delta\text{X}_1}{1 + \delta\text{X}_2} \quad (6)$$

where superscript heavy is omitted in X in the preceding equation. A slightly different form gives the fractionation factor as a difference between the δ values of the two phases:

$$\alpha_{\text{X}_1-\text{X}_2} - 1 = \frac{\delta\text{X}_1 - \delta\text{X}_2}{1 + \delta\text{X}_2} \quad (7)$$

As long as δ is much smaller than 1 and α is close to 1, two approximations are used in Equation (6), $\ln(1 + \delta) \approx \delta$ and $\ln\alpha \approx \alpha - 1$, that give a logarithmic and a linear form of α:

$$\ln\alpha_{\text{X}_1-\text{X}_2} = \delta\text{X}_1 - \delta\text{X}_2 \quad (8a)$$

and

$$\alpha_{\text{X}_1-\text{X}_2} - 1 = \delta\text{X}_1 - \delta\text{X}_2 \quad (8b)$$

For δ values in per mil (‰), ln α and (α−1) are multiplied by 1,000. Equations (8a) and (8b) define the isotopic fractionation at equilibrium as the difference between the δ values of the two phases. Extensive theoretical computations of isotopic fractionation factors in compounds of the lighter elements have been given and discussed by Bottinga (1968), Richet *et al.* (1977), O'Neil (1986), Chacko *et al.* (2001), and other authors cited in those references. In three-isotope systems, equilibrium and kinetic fractionation relationships have been given by Young *et al.* (2002).

Many inorganic processes represent nonequilibrium or kinetic fractionation, and there are many mineral–solution–gas systems where equilibrium is not attained at the temperatures of the Earth's surface. Relatively fast processes of gas or water vapor transfer or mineral precipitation often result in nonequilibrium fractionation because of the greater mobility of the lighter isotope. Diffusional fractionation of the hydrogen and oxygen isotopes takes place in transport across a boundary layer at the water–air interface and the exchange of ^{18}O/^{16}O between such pairs as oxygen and water; some oxygen-containing anions and mineral oxides and silicates in water are not products of equilibrium-exchange reactions. Biologically mediated processes, such as photosynthesis, respiration or remineralization of organic matter, nitrification and denitrification in water and sediments, and the reduction of sulfate to sulfide, are also nonequilibrium processes that usually involve preferential uptake of the lighter isotope from the environmental source in the making or decomposition of organic matter. Because a lighter isotope can react faster than a heavier one, this is believed to explain the organismal preference for lighter isotopes in their synthesis of organic matter. Accordingly, the values of $\delta^{13}C$ in organic matter and $\delta^{34}S$ in sedimentary pyrite, produced by reactions involving microbes, are lower than those of the carbon or sulfur sources. The end result of a nonequilibrium isotopic fractionation in biologically mediated processes can be written as

$$\delta X \text{ of organic matter} = (\delta\, X \text{ of the source}) - \varepsilon X \tag{9a}$$

where ε is a nonequilibrium fractionation factor of a heavy isotope X, in the same units as δ. In some publications, the differences between an equilibrium and nonequilibrium fractionation are often disregarded and factor ε is used as $\varepsilon=\alpha-1$, the latter as defined in Equation (8b). Because the organic matter product is usually isotopically lighter than its source, the fractionation factor ε is a positive number in the notation of Equation (9a). The intermediate steps, which can be numerous and complex in biological processes involving enzymes and several oxidation or reduction stages going through intermediate valences of such elements as N or S, are not included in Equation (9a).

More generally, if the product is isotopically lighter than the source, $\delta X_p-\delta X_s=-\varepsilon$, then a continuous production from a source of a fixed initial isotopic composition ($\delta X_{s,0}$) has been likened to the Rayleigh distillation process, discussed in the next section. The result of a continuous removal of a product containing less of the heavy isotope would make the source progressively heavier

$$\delta X_s = \delta X_{s,0} - \frac{\varepsilon}{1+\varepsilon}\ln f \approx \delta X_{s,0} - \varepsilon \ln f \tag{9b}$$

where f is a fraction of the source mass remaining ($0<f\leq 1$) and ε the fractionation factor defined above. However, if a fraction $1-f$ of the source is converted into a product that is isotopically fractionated by a factor ε ($\varepsilon>0$), then the δX_s value of the remaining source fraction f increases as

$$\delta X_s = \delta X_{s,0} + \varepsilon(1-f) \tag{9c}$$

and the δ value of the product, δX_p, changes with f that decreases from 1 to 0 as

$$\delta X_p = \delta X_{s,0} - \varepsilon f \tag{9d}$$

10.3 HYDROGEN AND OXYGEN ISOTOPES IN THE WATER CYCLE

Water is nearly ubiquitously involved in biogeochemical reactions of the formation and postdepositional alteration of sedimentary minerals and rocks, and its isotopic composition imprints the isotopic composition of oxygen, hydrogen, and crystal structural H_2O of many minerals. Much of the effort in the studies of stable isotopes as tracers of environmental conditions is based on the isotopic fractionation between H_2O and mineral and organic phases, with the goal of identifying the conditions of the past under which sediments and sedimentary rocks have formed. Postdepositional history of sediments often involves further chemical reactions with water either under conditions similar to those of the sediment formation or very different from those, as when marine sediments come in contact with continental waters of atmospheric origin.

Evaporation of water from the ocean surface, condensation in the atmosphere, and return in atmospheric precipitation, freezing, and changes in concentration of dissolved solids are accompanied by fractionation of H_2O between the liquid and vapor phases. The three most abundant isotopic species of the water molecule are $H_2^{16}O$, $H_2^{18}O$, and $HD^{16}O$, and the isotopic pairs $^{18}O^{16}O$ and DH in continental waters, atmospheric precipitation, and ocean water provide much information on the transfer processes in the global water cycle. In the present, the water in atmospheric precipitation at temperatures above freezing, rivers, land surface flows, and groundwaters are characterized by a range of about 30‰ in their $\delta^{18}O$ values, from about 10‰ to −20‰ on the SMOW or VSMOW scale (Różański *et al.*, 1993; Clark and Fritz, 1997). A larger range of about 120‰ characterizes the hydrogen–deuterium fractionation in the same waters, with the δD values between about 0‰ and −120‰. Ocean water shows smaller variations in its $\delta^{18}O$ that depend to some extent on the salinity of water that reflects its degree of evaporation, as was documented in the early work of Epstein and Mayeda (1953) and Craig (1961). During the last 700,000 years, the $\delta^{18}O$ of ocean water, as determined from the $CaCO_3$ shells of shallow-living planktonic foraminifera in low and median latitudes, varied within a range of about 2‰ (Imbrie *et al.*, 1984). The variation has been attributed to the glacial–interglacial changes in ice volume and the resultant change in the sea level. The isotopic composition of continental waters places constraints on the interpretations of the stable isotope records of those minerals that contain water or hydroxyl ions in their crystal lattices, where O and H can quite easily exchange with the external H_2O thereby changing the original record. The essentials of the variation in the isotopic composition of meteoric waters and its causes are briefly discussed in this section.

Two fundamental relationships based on extensive observations describe the isotopic composition of atmospheric precipitation in terms of the

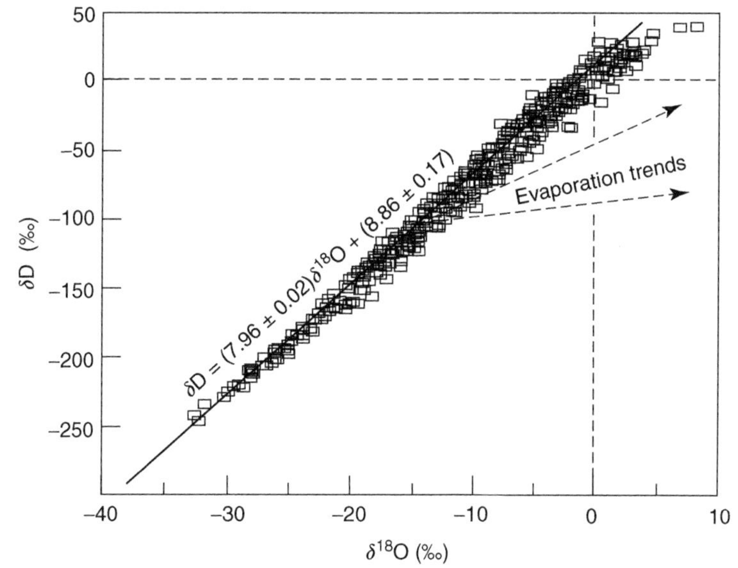

Figure 1 Global MWL and changes in the isotopic composition of evaporating waters (VSMOW scale). Modified from Różański *et al.* (1993) and Gonfiantini (1986).

D/H and $^{18}O/^{16}O$ fractionation and its dependence on temperature. The trend of a positive correlation between the δD and $\delta^{18}O$ values in atmospheric precipitation, and continental and ocean waters was reported by Friedman (1953), expanded by Craig (1961), elaborated theoretically by Craig and Gordon (1965), and later called the Meteoric Water Line (MWL) that is shown in Figure 1 from a more recent and much larger database.

The line is given by a relationship originally proposed by Craig (1961; Equation (10a)) and a later correlation by Różański *et al.* (1993; Equation (10b))

$$\delta D = 8\delta^{18}O + 10‰ \qquad (10a)$$

$$\delta D = 7.96\delta^{18}O + 8.86‰ \qquad (10b)$$

The coefficient of about 8 of $\delta^{18}O$ in Equation (10a) is close to the ratio of the equilibrium fractionation factors between liquid water and vapor (Craig and Gordon, 1965), $(\alpha_D - 1)/(\alpha_{^{18}O} - 1)$, between 25 and 35 °C that are plotted in Figure 4. In the colder regions of the world, atmospheric precipitation is isotopically lighter than in the warmer regions, as was originally reported and explained by Dansgaard (1964) and also shown in Figure 2.

The consistency of the MWL is remarkable, in view of the fact that atmospheric water vapor is derived from evaporation of different sections of the surface ocean, continental waters, and transpiration by land plants, and the atmospheric temperature varies with geographic latitude where the water vapor condensation occurs. There have been attempts to explain the MWL as a result of a single-stage Rayleigh distillation process, where atmospheric water vapor of some initial isotopic composition undergoes condensation and removal of the condensed fraction from contact with the remaining vapor. Because the isotopic fractionation of D/H and $^{18}O/^{16}O$ enriches the liquid water in the heavier isotope relative to the vapor, progressive condensation depletes the water vapor in HDO and $H_2^{18}O$ and later condensation stages produce rain that is isotopically lighter than the rain forming initially. The more negative δD and $\delta^{18}O$ values of the precipitation toward the far North and South have been interpreted as indicating a continuous depletion of atmospheric water vapor that originates in the tropical ocean (e.g., Epstein, 1959;

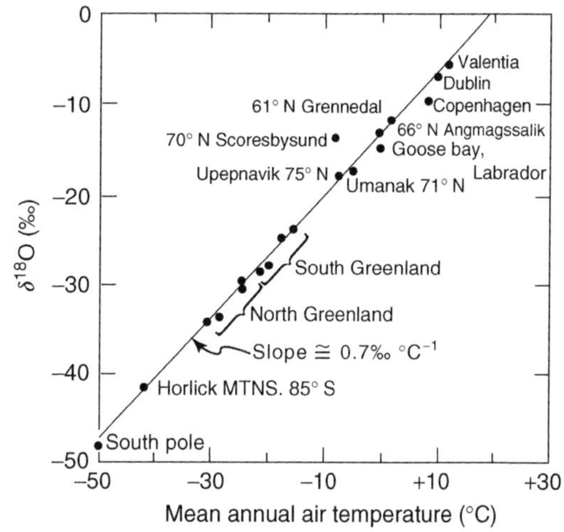

Figure 2 $\delta^{18}O$ of atmospheric precipitation as a function of temperature. Additional data producing similar slopes are given by Różański *et al.* (1993) and Alley and Cuffey (2001). Reproduced by permission of Eldigio Press from Broecker (2002).

Clark and Fritz, 1997). The Rayleigh distillation process was originally derived for liquids of different vapor pressures (Rayleigh, 1902), and Dansgaard (1961) gave explicit derivations of the equations for the Rayleigh process of D/H and $^{18}O/^{16}O$ isotopic fractionation between liquid water and vapor. The fractionation of HDO/H_2O and $H_2^{18}O/H_2^{16}O$ between the H_2O vapor and condensing liquid is given by the following relationship:

$$\begin{aligned} R_L &= R_{L,0}\, f^{(\alpha-1)} \\ &= \alpha R_{V,0}\, f^{(\alpha-1)} \end{aligned} \quad (11)$$

where R_L denotes the isotopic ratio, as defined in Equation (2), in the condensing liquid, $R_{L,0}$ the ratio in the initial condensate that is at equilibrium with the vapor phase of composition $R_{V,0}$, f the fraction of the water vapor remaining after condensation ($1 > f > 0$), and α the fractionation factor, as given in Equation (6). In the δD and $\delta^{18}O$ notation, Equation (11) becomes

$$\delta D_L = \delta D_{L,0} + (\alpha_D - 1)\ln f \quad (12a)$$

$$\delta^{18}O_L = \delta^{18}O_{L,0} + (\alpha_{18O} - 1)\ln f \quad (12b)$$

Because $\alpha - 1$ is a positive number, decreasing values of the remaining vapor fraction f give progressively larger negative values of δD_L and $\delta^{18}O_L$. So far, this is in agreement with the general trend of atmospheric precipitation with increasing latitude North and South. The seasonal and annual variations in $\delta^{18}O$ of atmospheric precipitation have been reported over a period of three decades (Figure 3) from oceanic and continental stations in the Northern and Southern hemispheres, from the sample-collecting network of the International Atomic Energy Agency. Much longer variations in the Antartic ice, at timescales of 450,000 to 740,000 years, have been reported by Petit *et al.* (1999) and EPICA (2004).

Clearly, the entire set of data cannot be explained by a simple condensation model that assumes the source of precipitation as a vapor producing a precipitation of $\delta^{18}O = -2.5$‰ at the equator. Atmospheric water vapor receives inputs from different sections of the ocean and land surface and atmospheric clouds, occurring at different altitudes and hence temperatures, contain both water droplets and ice crystals that do not rain out immediately after their formation. However, on regionally smaller scales, such as on Mount Cameroon and in Bolivia, the Rayleigh condensation model has been shown to describe the altitude effect on the isotopic composition of atmospheric precipitation (Gonfiantini *et al.*, 2001). In general, isotopic values are affected by altitude above sea level and distance from the ocean, as has been recorded in rains of mountainous terrains and

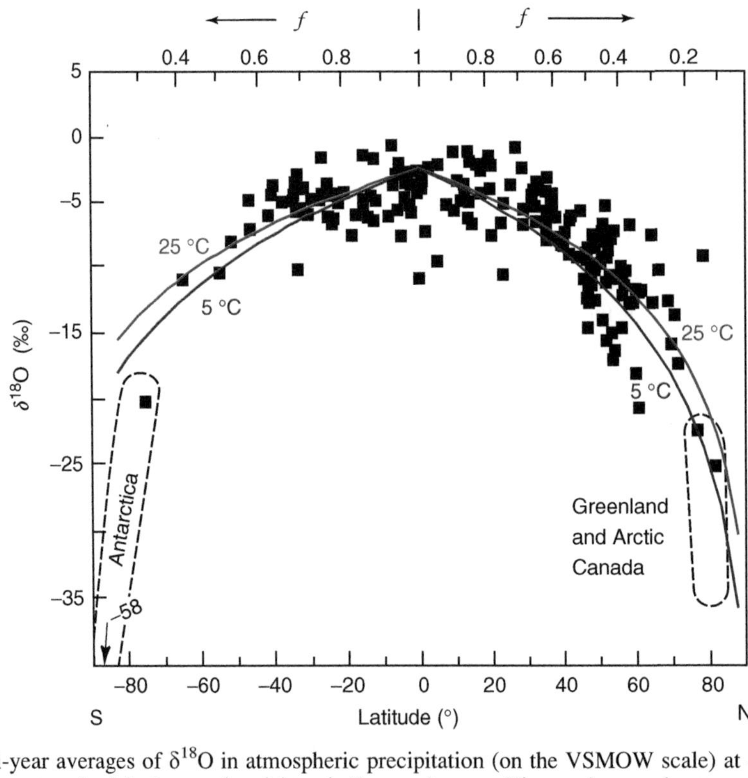

Figure 3 Multi-year averages of $\delta^{18}O$ in atmospheric precipitation (on the VSMOW scale) at different latitudes, from the sampling network of the International Atomic Energy Agency, Vienna. Assumed source at the equator is the starting point for the calculated curves of Rayleigh condensation, Equation (12b), at 5 and 25 °C, with the fraction of water vapor remaining (*f*) decreasing from the equator to the North and South. Modified from Różański *et al.* (1993).

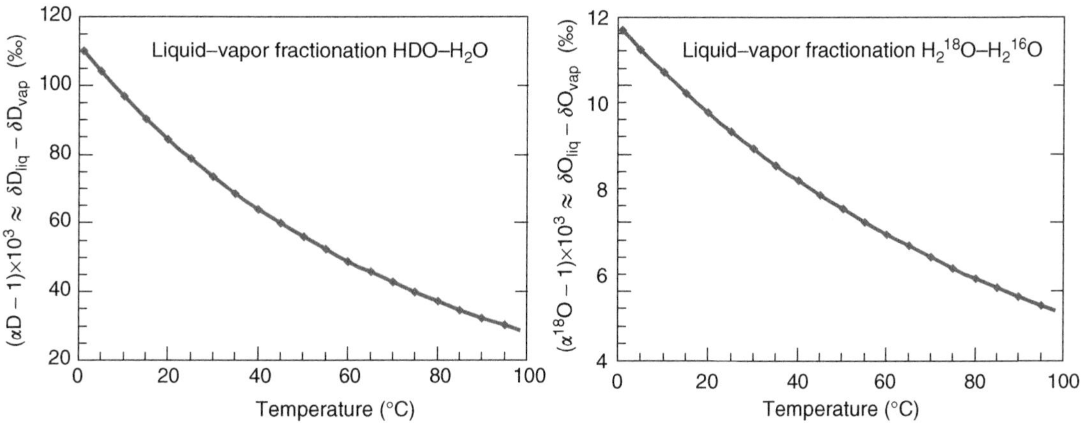

Figure 4 Equilibrium isotopic fractionation of D/H and $^{18}O/^{16}O$ between liquid and vapor water as a function of temperature. From fractionation equations of Horita and Welosowski (1994) that agree closely with experimental results of several investigators, as shown in the original reference.

continental interiors (e.g., Kendall and Coplen, 2001; Bowen and Wilkinson, 2002). The Rayleigh fractionation model applies to a constant temperature and the value of the fractionation factor α is taken as a constant. Furthermore, the effect of temperature on $\delta^{18}O$ of atmospheric precipitation is about 0.6–0.7‰ °C^{-1} (Figure 2), which is considerably greater than about 0.1‰ °C^{-1} that is given by the liquid-vapor fractionation equilibrium of $H_2{}^{18}O$–$H_2{}^{16}O$ in a temperature range from about 1 to 30 °C, as shown in Figure 4.

When a water of composition represented by a point on the MWL (Figure 1) evaporates, its δD and $\delta^{18}O$ usually increase, but the $\delta^{18}O$ increases more strongly than δD. This results in a line of a δD–$\delta^{18}O$ slope smaller than 8, which lies to the right of the MWL. Such δD–$\delta^{18}O$ relationships are often reported for continental waters and, in particular, subsurface brines as indicating a history of water evaporation. The relationship was experimentally and theoretically accounted for by Gonfiantini (1986) as being primarily due to two factors: a two-way flux of H_2O molecules across the water–air interface depends on the relative humidity of the atmosphere and isotopic composition of the H_2O vapor; diffusional transport of H_2O molecules across a boundary layer at the water–air interface discriminates more strongly between $H_2^{18}O$ and $H_2^{16}O$ than between HDO and H_2O. Because the isotopic mass difference between $H_2^{18}O$ (molecular mass 20) and $H_2^{16}O$ (18) is greater than between $HD^{16}O$ (19) and $H_2^{16}O$ (18), the liquid is enriched more strongly in $H_2^{18}O$ relative to HDO.

It should be pointed out that the $\delta^{18}O$ values of H_2O in continental waters and seawater are probably not at equilibrium with dissolved O_2 gas, and under laboratory conditions the isotope exchange between water and molecular oxygen requires a catalyst on which the reactants are adsorbed (Gonfiantini, 2005). A theoretical fractionation between O_2 gas and H_2O vapor at 20 °C (Richet *et al.*, 1977) indicates that the O_2 phase is heavier by about 13.4‰: $\delta^{18}O$ (in O_2 gas)$-\delta^{18}O$ (in H_2O vapor)$=$13.4‰. There is a relatively small isotopic fractionation between atmospheric O_2 (g) and that dissolved in seawater, O_2 (aq), the latter being heavier by about 0.7‰: $\delta^{18}O$ (aq)$-\delta^{18}O$(g)$=$ $0.85-0.01t$°C ‰ (Kroopnick and Craig, 1972; Benson and Krause, 1984). The isotopic composition of O_2 in the atmosphere is $\delta^{18}O$ (g)$=$23.5‰ on the VSMOW scale, and the difference between the atmospheric and surface ocean values is known as the Dole effect. Its interpretation (Bender, 1990; Bender *et al.*, 1994; Hoffmann *et al.*, 2004) is based on the fact that the O_2 produced in photosynthesis by land and marine organisms bears the isotopic signature of H_2O that is split in the photosynthetic reactions, whereas the various respiration reactions consuming gaseous or dissolved O_2 utilize preferentially the light isotope ^{16}O. The net result of these processes is the isotopically heavier molecular oxygen remaining in the atmosphere.

10.4 HYDROGEN AND OXYGEN FRACTIONATION IN CLAYS, WATER, AND CARBONATES

Clays are hydrated phyllosilicates (structurally layered alumino-silicate minerals) that form over a wide range of temperatures at hydrothermal and sedimentary conditions. In sedimentary environments, the most common clay minerals are the kaolinite, illite–smectite, and chlorite groups (e.g., Deer *et al.*, 1962a, b). The chemical composition and mass fraction of H_2O in clays and a number of other hydrated minerals are given in Table 3. Clays represent an estimated mass fraction of up to 30% of the sediments and their main occurrence is as mineral components of shales. Clays form as products of dissolution or alteration of such minerals as feldspars (e.g., K-feldspar $KAlSi_3O_8$), micas (e.g., muscovite $KAl_3Si_3O_{10}(OH)_2$ or biotite, containing Fe and Mg) and other silicates containing metal ions. Kaolinite is a pure alumino-silicate, $Al_2Si_2O_5(OH)_4$, whereas illites and illite–smectites have somewhat variable stoichiometric proportions of K, Al, Si, and also Na and Ca, such as in an illite of a synthetic composition $K_{0.2}Na_{0.2}Ca_{0.1}Al_{2.19}Si_{3.71}O_{10}(OH)_2$.

The presence of the OH groups in the crystal structure of clay minerals was a focus of extensive research on the fractionation of the D/H and $^{18}O/^{16}O$ isotopes between different clays and water at temperatures in a range from the Earth's surface to those approximating hydrothermal conditions at about 150–200 °C (e.g., Savin and Lee, 1988). Savin and Epstein (1970a) reported that sedimentary kaolinite, illite, and montmorillonite or smectite have $\delta^{18}O$ values between +14‰ and +28‰ and δD between −13‰ and −4‰ (SMOW scale). Furthermore, they showed that δD and $\delta^{18}O$ values of clays plot as lines nearly parallel to the MWL with the slope of 8 (Figure 1) but with different intercepts. Such δD–$\delta^{18}O$ trends were also reported for hydroxides and oxyhydroxides (gibbsite and goethite) by Yapp (1993).

10.4.1 Illite and Kaolinite Reactions with Water

The relationships between δD and $\delta^{18}O$ in illite–water and kaolinite–water systems, based on experimental work (Table 4), are shown in Figure 5 relative to the MWL. The fractionation relationships indicate that the clays are enriched in ^{18}O and depleted in D. The trends of the waters in Appalachian and Michigan Basins fall on lines of a slope smaller than that of the MWL, which is at least in part an indication of the evaporative history of the subsurface waters. The clays occur in Cambrian–Ordovician sediments, but their younger $^{40}K/^{40}Ar$ ages were interpreted as indicating their secondary nature (Ziegler and Longstaffe, 2000). Caution should be exercised in the interpretation of the isotopic records of H_2O or OH in clays as possible indicators of the temperature and composition of the water in which they formed: an equilibrium fractionation would reflect the isotopic composition of the water if the mass ratio of the water to the OH in the clay is large and no later diagenetic changes occurred. In general, large water/rock mass ratios characterize

Table 3 Stoichiometric composition and H_2O mass fraction of some common hydrated minerals, including the clays kaolinite, illite, nontronite, and chlorite.

Mineral	*Idealized chemical composition*[a]		*Formula H_2O* (wt.%)
	Stoichiometric	*As oxides*	
Gibbsite	$Al(OH)_3$	$Al_2O_3 \cdot 3H_2O$	34.6
Boehmite	AlOOH	$0.5(Al_2O_3 \cdot H_2O)$	15.0
Goethite	FeOOH	$0.5(Fe_2O_3 \cdot H_2O)$	10.1
Kaolinite	$Al_2Si_2O_5(OH)_4$	$Al_2O_3 \cdot 2SiO_2 \cdot 2H_2O$	14.0
Illite	$KAl_5Si_7O_{20}(OH)_4$	$0.5K_2O \cdot 2.5Al_2O_3 \cdot 7SiO_2 \cdot 2H_2O$	4.7
Nontronite	$(0.5Ca, Na)_{0.66}Fe^{3+}{}_4(Si_{7.34}Al_{0.66})O_{20}(OH)_4$	$0.33Na_2O \cdot 2Fe_2O_3 \cdot 0.33Al_2O_3 \cdot 7.34SiO_2 \cdot 2H_2O$	4.2
Talc	$Mg_3Si_4O_{10}(OH)_2$	$3\,MgO \cdot 4SiO_2 \cdot H_2O$	4.8
Serpentine	$Mg_3Si_2O_5(OH)_4$	$3\,MgO \cdot 2SiO_2 \cdot 2H_2O$	13.0
Chlorite	$Mg_5Al_2Si_3O_{10}(OH)_8$	$5\,MgO \cdot Al_2O_3 \cdot 3SiO_2 \cdot 4H_2O$	13.0
Muscovite	$KAl_3Si_3O_{10}(OH)_2$	$0.5K_2O \cdot 1.5Al_2O_3 \cdot 3SiO_2 \cdot H_2O$	4.5
Biotite and phlogopite	$K(Mg,Fe)_3AlSi_3O_{10}(OH)_2$	$0.5K_2O \cdot 3(Mg,Fe)O \cdot 0.5Al_2O_3 \cdot 3SiO_2 \cdot H_2O$	3.9

[a] Deer *et al.* (1962a, b).

Table 4 $^{18}O/^{16}O$ fractionation factors, Equations (6)–(8b), in some mineral–water exchange reactions.

$10^3 \ln \alpha$ for $^{18}O/^{16}O$ in illite–water	$-2.87+1.83\times(10^6/T^2)+0.0614 \times (10^6/T^2)^2-0.0015\times(10^6/T^2)^3$	From Ziegler and Longstaffe (2000 and references therein)
$10^3 \ln \alpha$ for $^{18}O/^{16}O$ in kaolinite–water	$2.76\times10^6/T^2-6.75$	
$10^3 \ln \alpha$ for $^{18}O/^{16}O$ in chlorite–water	$3.47\times10^6/T^2-5.79\times10^3/T$	
$10^3 \ln \alpha$ for $^{18}O/^{16}O$ in goethite–water	$1.63\times10^6/T^2-12.3$	Yapp (1993)
$10^3 \ln \alpha$ for $^{18}O/^{16}O$ in boehmite–water	$2.11\times10^6/T^2-4.4$	

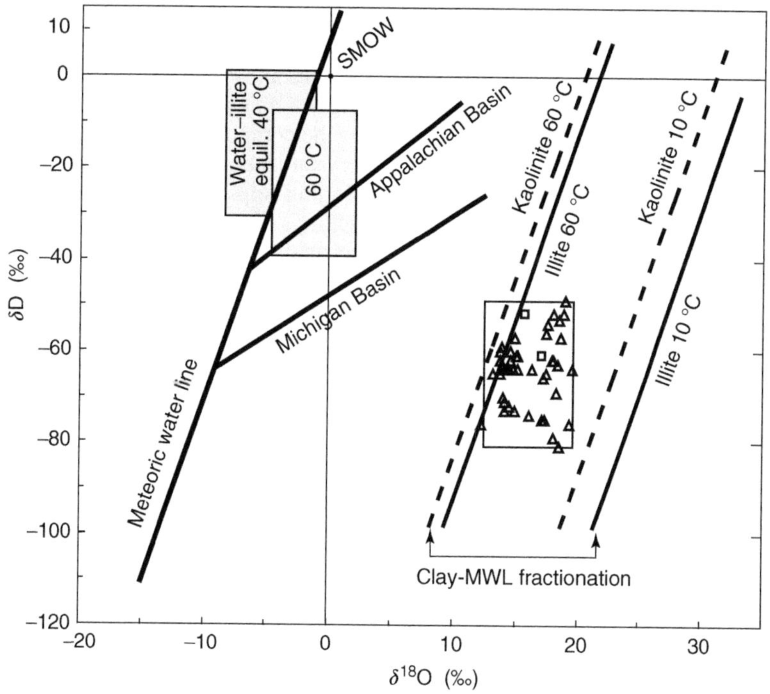

Figure 5 The Meteoric Water Line (MWL of Figure 1) and subsurface water trends in the Michigan and Appalachian Basins. Note the lower slope of the latter two lines, explained in a preceding section. D/H and $^{18}O/^{16}O$ calculated fractionation in the pairs illite–water and kaolinite–water at 10 and 60 °C relative to the MWL. Isotopic composition of Paleozoic illites (triangles) and kaolinites (squares) from the Ottawa Embayment of the Appalachian Basin. The rectangle of the δD and $\delta^{18}O$ values circumscribing the clays would be in an isotopic exchange equilibrium with the waters of composition shown by the shaded rectangles at 40 and 60 °C. Modified from Ziegler and Longstaffe (2000).

porous formations, such as sandstones, or hydrothermal conditions where water circulates through the rock.

10.4.2 Other Clay–Water Reactions

From a large suite of δD and $\delta^{18}O$ data on Paleozoic, Cretaceous, and Tertiary kaolinites in the United States, Lawrence and Rashkes Meaux (1993) interpreted the elevated δD values in the Cretaceous and Tertiary as indicators of a warmer climate. They also discussed similar data from stratigraphically long periods in Australia as indicating the warming of this continent as it moved progressively closer to the Equator during the Mesozoic and Cenozoic. Earlier, Savin and Epstein (1970a) and Lawrence and Taylor (1971, 1972) demonstrated that there is a good correlation between the oxygen and hydrogen isotope composition of clay minerals from soils and the environmental meteoric water, and that the clay material even retains these signatures through further processes of erosion, transportation, and deposition. Clauer and Chaudhuri (1995) provided additional information on the oxygen-isotope composition of kaolinite from a soil profile in eastern Cameroon, showing that its $\delta^{18}O$ value significantly varies across the soil profile. The changes suggest that the ambient fluids also varied in their oxygen isotopic composition across the soil profile, which makes it unlikely that high water-to-mineral ratios occurred during kaolinite crystallization, unless controlled by microenvironments. Later, Giral-Kacmarcík *et al.* (1998) showed that the $\delta^{18}O$ of kaolinite in a 20-m-thick laterite profile in the central Amazon Basin, Brazil, evolved progressively in response to changing microenvironments as weathering moved downward the profile, and that the fine kaolinite particles of the profile do not constitute a continuum record of crystallization, speculating that the $\delta^{18}O$ values of kaolinite might be indicative of climate changes during the weathering history. In analyzing the δD and $\delta^{18}O$ of kaolinite and goethite in a lateritic profile of Yaou, French Guiana, Girard *et al.* (2000) showed that the δ values reflect changes in the formation temperatures of the minerals, as summarized in Figure 6. The results suggest that the minerals form in isotopic equilibrium with the water in their environment and that they were not subjected to further isotopic exchange after crystallization. The shift reported at a depth of 20 m in the profile is considered to represent a change from past equatorial to present tropical climate. As an overall response of clay formation in continental weathering deposits, the $\delta^{18}O$ and δD values of the clay minerals are generally different from those of authigenic marine clays (Savin and Epstein, 1970b).

Three kinds of water are reported in clay minerals: water adsorbed on the particle surfaces, water held in interlayer positions, and water that is a component of the mineral framework. Adsorbed and interlayer water exchanges isotopically with ambient water at very fast rates of less than a few days. On the other hand, the rates of isotopic exchange between structural and ambient water are generally slow and depend on the temperature and types of clay minerals. Laboratory experiments of over 2 years have shown that oxygen isotopic exchange between structural water of well-crystallized illite or kaolinite and ambient water can be considered nearly insignificant over a few million years at about 100 °C or lower temperatures, but it is certainly greater at temperatures of about 300 °C (James and Baker, 1976; O'Neil and Kharaka, 1976). Unlike the rate of oxygen-isotope exchange, that of hydrogen-isotope exchange between structural hydrogen of these clays and hydrogen of ambient water is relatively rapid at temperatures as low as 60 °C. O'Neil and Kharaka (1976) concluded that hydrogen-isotope exchange takes place by a mechanism of proton exchange that is completely independent of the oxygen-isotope exchange. The experiments have also shown that the hydrogen-isotope exchange for smectite with its ambient water is rapid and significant relative to that of either illite or kaolinite with their ambient water. As the experimental evidence suggests that the oxygen-isotope exchange between structural oxygen of clay minerals and that of the ambient water may be very slow in most burial-related diagenetic environments, structural oxygen of illite and kaolinite may reflect growth history of the clay minerals and correspond to the composition of the ambient fluids and/or the crystallization temperature(s) of the minerals. This finding helps in reconstructing the paleo-hydrologic settings of sedimentary basins in which the minerals formed.

In this respect, Ayalon and Longstaffe (1988) examined the oxygen-isotope compositions of diagenetic minerals of the Upper Cretaceous basal Belly River sandstone in Alberta and reconstructed changes in the pore-water oxygen-isotope composition relative to temperature. The general paragenesis during the whole evolution includes early growth of chlorite followed by that of kaolinite, smectite, illite, and mixed-layers illite–smectite. The formation of these minerals preceded feldspar dissolution that occurred in the later stages of burial diagenesis. Calcite growth was dominant during the early chlorite growth and during the precipitation of the other clay minerals. Quartz overgrowths followed the formation of chlorite, but preceded that of kaolinite, smectite, illite, and illite–smectite. In limiting the burial temperatures to 90–120 °C based on complementary information, the authors estimated that the pore water $\delta^{18}O$ ranged between 0‰ and −10‰, varying as shown in Figure 7.

Longstaffe and Ayalon (1987) reconstructed a similar evolutionary path for the $\delta^{18}O$ of pore fluids and associated clay minerals of the Lower Cretaceous Viking Formation in the Western Canadian Basin. Similar studies were published by Fallick *et al.* (1993), Glasmann *et al.* (1989), Robinson *et al.* (1993), Kotzer and Kyser (1995), Girard *et al.* (1989), Girard and Barnes (1995),

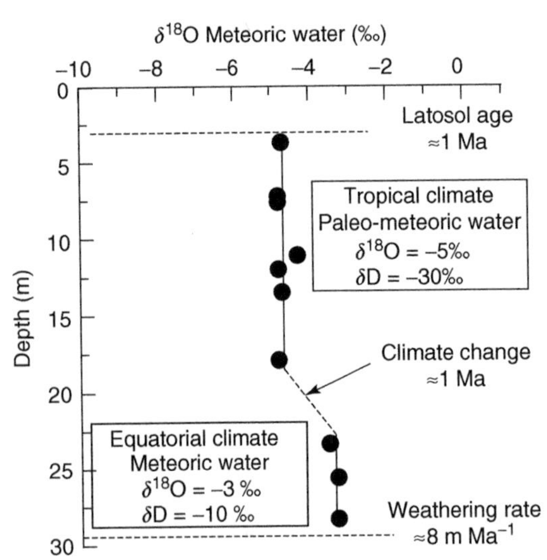

Figure 6 A conceptual diagram explaining the shift in $\delta^{18}O$ of kaolinite formed in the lateritic profile of Yaou, French Guiana, depending on the δ values of the parent meteoric waters, as recording a climate change about 1 Ma. Reproduced by permission of Elsevier from Girard *et al.* (2000).

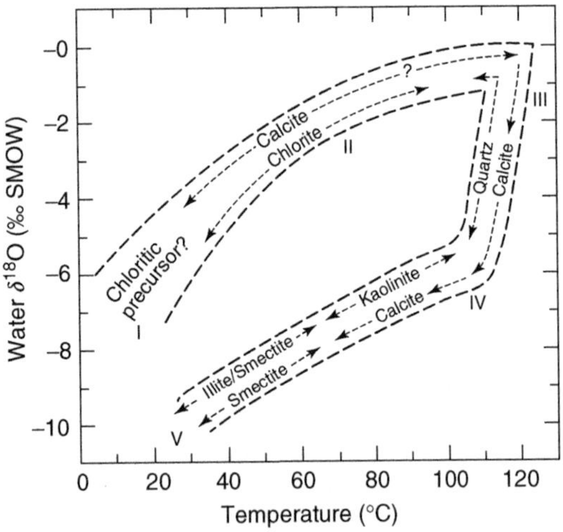

Figure 7 Model-based evolution of the $\delta^{18}O$ of pore fluids and the paragenesis of diagenetic minerals in the basal Belly River sandstone, Alberta. Reproduced by permission of Society for Sedimentary Geology from Ayalon and Longstaffe (1988).

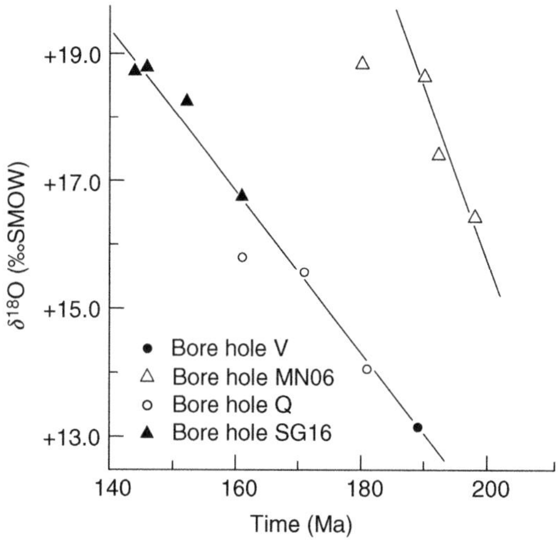

Figure 8 Relationship between K–Ar data and $\delta^{18}O$ values of <0.4 μm mixed-layers illite–smectite in Triassic sandstones from Paris Basin. Reproduced by permission of Mineralogical Society of Great Britain and Ireland from Clauer *et al.* (1995).

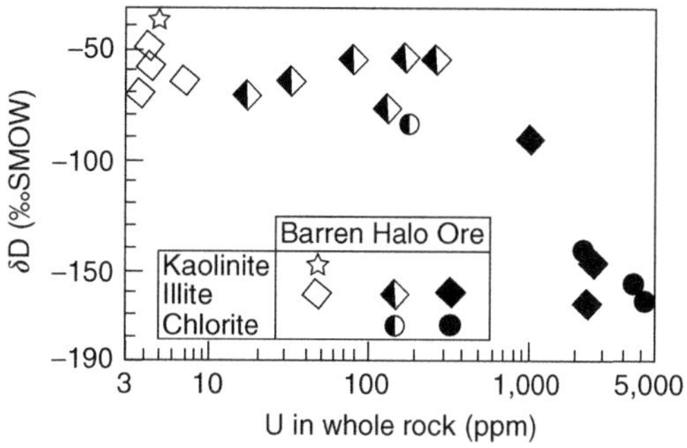

Figure 9 Variation of the δD of illite and chlorite relative to the uranium content in the host rocks of the minerals. Modified from Halter *et al.* (1987).

Clauer *et al.* (1999), and Zwingmann *et al.* (1999). Clauer *et al.* (1995) reported two straight-line correlations between the K–Ar isotopic age and the oxygen-isotope composition of Upper Triassic illite-type clay material extracted from sandstone cores of the central Paris Basin. The younger the illite-type fractions, the higher are their $\delta^{18}O$, which increase from 13‰ to 19‰ when the K–Ar ages decrease from 190 to 145 Ma (Figure 8). The second correlation line in Figure 8 (shown by open triangles) shows the data of similar illite-type material, but collected in a core that crosscuts a fault with $\delta^{18}O$ values significantly higher than those of the corresponding material from tectonically undisturbed sandstones. At a constant water/mineral ratio, the results show that the younger the illite material, the lower was its formation temperature. In the case of the upper line, this water/mineral ratio had to be higher if the formation temperature for the clay minerals was similar to that of the minerals from the nearby tectonically undisturbed sandstones.

Clauer *et al.* (2003) also reported $\delta^{18}O$ values of "illite-type fundamental particles" separated from mixed-layered illite–smectite of bentonite units interbedded in a sedimentary sequence of the East-Slovak Basin that underwent burial-induced diagenesis. The term "illite-type fundamental particles" designates the smallest illite-type particles technically separable by continuous-flow ultra-centrifugation: the particles are of dimensions of about 20 nm and thicknesses of 1–3 nm (see Nadeau *et al.*, 1984, for further explanation). The $\delta^{18}O$ values of these fundamental particles were either increasing from 17.0‰ to 18.5‰ or decreasing from 12.0‰ to 9.6‰ when particle size increases, as in the case described by Clauer *et al.* (1996). Because the porosity and permeability of bentonite units are known to be rather low, the water/rock ratio is expected to stay constant during crystal growth so that the shift in the $\delta^{18}O$ of the fundamental particles mainly relates to a change in the formation temperature. This result suggests, in turn, that in some cases illite can crystallize and grow while temperature decreases, which is not what is generally believed (see, e.g., Velde, 1985; Sucha *et al.*, 1993, and references therein).

In sedimentary environments, the primary isotopic signature of the clay minerals crystallizing from ore-forming fluids can be altered by retrograde-exchange reactions that complicate the interpretation of the data. In the Cluff-D uranium deposit in the Athabasca Basin (Saskatchewan, Canada), Halter *et al.* (1987) reported K–Ar and deuterium isotope values of illite- and chlorite-type clay material taken from barren rocks and from uranium ore (Figure 9). The illite and chlorite fractions are strongly depleted in δD, with values ranging from −90‰ to −170‰ and yield K–Ar ages below 750 Ma in rocks containing more than 1,000 ppm U. In the barren rocks containing <500 ppm U, the δD of the same minerals ranges from −48‰ to −79‰, with only one chlorite value at −85‰, and K–Ar ages for illite from 1,228 to 1,322 Ma. Also, the structural H_2O content of illite (water that is removed at 940 °C) from mineralized samples is higher, up to 7.2 wt.%, than that of illite from the barren zone, from 4.6 to 5.1 wt.%. These different relationships were interpreted by the authors as resulting from a radiation-catalyzed retrograde reaction at low temperature with post-Cretaceous meteoric waters. This example outlines also the fact that illite not affected by radiation reactions remains useful for isotopic dating purposes.

10.4.3 $\delta^{18}O$ in Carbonates

In the formation of calcite or aragonite ($CaCO_3$), isotopic fractionation takes place between $^{18}O/^{16}O$

in the mineral and water, $^{13}C/^{12}C$ between the mineral and the bicarbonate and carbonate ions in solution, as well as between ^{44}Ca and ^{40}Ca. The experimentally determined $^{18}O/^{16}O$ fractionation between calcite at equilibrium with an aqueous solution underlies the estimation of paleotemperatures of the calcite precipitation. Many of the $\delta^{18}O$ values of sedimentary carbonates have been reported on the PDP or VPDB standard, and their range often is ±X‰ in the single digits. On the SMOW or VSMOW standard, these $\delta^{18}O$ values would be about 30‰ heavier. $\delta^{18}O$ for water and atmosphere are usually based on the SMOW or VSMOW standard. Conversion of $\delta^{18}O$ between VPDB and VSMOW standards, based on abundance ratios, is (Baertschi, 1976; Clark and Fritz, 1997)

$$\delta^{18}O_{VSMOW} = 1.03092\ \delta^{18}O_{VPDB} + 30.92\%$$
$$\delta^{18}O_{VPDP} = 0.9700\ \delta^{18}O_{VSMOW} - 29.99\%$$

The relationship between the $^{18}O/^{16}O$ ratio in $CaCO_3$ and in water where carbonate mineral precipitates, is a function of temperature (O'Neil *et al.*, 1969; Friedman and O'Neil, 1977, their figure 13; Savin, 1980):

$$\delta^{18}O_{calcite} - \delta^{18}O_{H_2O} = 2780/T^2 - 0.00289 \quad (13)$$

where T is the temperature in kelvin. Calcite is enriched in ^{18}O relative to water (on the PDB or VPDB scale) by about 2‰ at 5 °C and depleted by −4.4‰ at 35 °C. This corresponds to ±1‰ of $\delta^{18}O$ change in the enrichment of calcite for every ±4.7 °C change in temperature, when calcite is at an equilibrium with water of a constant $^{18}O/^{16}O$ ratio. Among the limitations on the universality of the paleotemperature method are the variations in the $^{18}O/^{16}O$ ratio of ocean water through geologic time (e.g., the variations caused by changes in continental ice coverage), alteration of the isotopic composition of the mineral by diagenetic changes, such as recrystallization and reactions with meteoric waters, and the different responses to temperature by different groups of calcite and aragonite secreting organisms ("vital effects") that may or may not be at an equilibrium with the surrounding water.

The long-term geologic record of $\delta^{18}O$ of carbonate rocks, mainly calcites (limestones) and dolomites, is given and discussed in detail by Veizer *et al.* (1999) and Shields and Veizer (2002), as to its primary trends or possible alteration by diagenetic reactions with meteoric waters. From near the Precambrian–Cambrian boundary, approximately 550 Ma, the $\delta^{18}O$ (on the PDB scale) of mostly calcitic shells of marine invertebrates increases with fluctuations from about −8‰ to +1‰ near the Recent. From experimental results on the $^{18}O/^{16}O$ fractionation between dolomite and calcite at elevated temperatures, an extrapolated value at 25 °C is $\delta^{18}O_{dol}-\delta^{18}O_{cal}$=6.8‰ (O'Neil and Epstein, 1966), within a range from 4‰ to 7‰ of others' determinations cited by these authors. Values that are significantly different for calcites and dolomites in sedimentary environments indicate that the minerals are not at equilibrium or were not formed from water of the same isotopic composition. In fact, from more than 4,000 analyses of Precambrian limestones and dolomites, ranging in age from the Archean to the beginning of the Phanerozoic (cf. Figures 15 and 16; Shields and Veizer, 2002), the difference between the mean $\delta^{18}O$ values of the dolomites and calcites is 2.7‰.

10.5 CALCIUM ISOTOPES IN SEAWATER AND CARBONATES

Calcium has six stable isotopes (Table 1) and the radioactive isotope ^{41}Ca with a half-life of 106 years occurring in nature. About 97% of the element is in the form of ^{40}Ca, which is one of the daughter products of ^{40}K decay, along with ^{40}Ar. From the known abundances of the element K and its radioactive isotope ^{40}K at present, the fraction of ^{40}Ca that was added to the original ^{40}Ca during 4.5×10^9 years of Earth's history can be estimated. For example, in the upper continental crust the concentrations are (Li, 2000): K_2O, 3.13 wt.% and CaO, 4.48 wt.%. From these concentrations, the decay half-lives of ^{40}K, and the present-day abundances of ^{40}K and ^{40}Ca, it follows that in 4.5 Ga a fraction of 0.1% has been added to the ^{40}Ca abundance. This value depends on the abundances of K and Ca in a geochemical reservoir, and it would be greater if Ca were relatively less abundant or smaller if the Ca abundance were relatively larger. The isotopic composition of Ca in rocks and minerals varies because of the formation of ^{40}Ca by β decay of ^{40}K, mentioned above, and because of the fractionation of Ca isotopes by natural processes. Studies of their fractionation are of importance because it conveys information on the process that leads to it.

The sample preparation for Ca isotopic studies includes a routine chromatographic clean- up of the Ca with an ion-exchange resin stored in small columns. This clean-up step is usual in isotopic studies, representing here a critical aspect as it introduces a fractionation of the Ca isotopes. This fractionation effect has to be quantified and if possible minimized. Many studies were devoted to the topic, especially in the 1970s, concluding that the potential fractionation can be controlled by comparison to previously analyzed (reference) samples and that it can be up to 3‰ per atomic mass unit (amu). After chromatographic separation, the aliquots are mixed with a Ca tracer (^{43}Ca and/or ^{48}Ca) and loaded on a filament of a thermo-ionization mass spectrometer (TIMS) for analysis. This step introduces another fractionation, which has been resolved and accounted for by Russell and Papanastassiou (1978). Recently,

Fietzke *et al.* (2004) reported on an alternative analytical procedure.

Using a standard (reference) sample, the analytical results can be expressed in δ values (‰) defined as

$$\delta^{44}Ca = \left[({}^{44}Ca/{}^{40}Ca)_{sample}/({}^{44}Ca/{}^{40}Ca)_{standard} - 1\right] \times 10^3 \quad (14)$$

Hippler *et al.* (2003) published the Ca isotopic composition of various standards and of seawater. However, because different standards or conventions of $\delta^{44}Ca=0$‰ are used in different publications, it is not always possible to relate unambiguously one set of $\delta^{44}Ca$ values to another. In this section, we cite the $\delta^{44}Ca$ data as reported by Hippler *et al*. The $\delta^{44}Ca$ values are interpreted as providing information on the fractionation processes occurring in biological and physical–chemical processes (Russell *et al.*, 1978; Skulan *et al.*, 1997; Zhu and MacDougall, 1998; Skulan and DePaolo, 1999; De La Rocha and DePaolo, 2000; Nägler *et al.*, 2000), but also on the possible ^{40}Ca excess resulting from β decay of ^{40}K (Marshall and DePaolo, 1982, 1989; Marshall *et al.*, 1986; Nelson and McCulloch, 1989; Shih *et al.*, 1994; Fletcher *et al.*, 1997; Nägler and Villa, 2000). On the other hand, the $^{42}Ca/^{44}Ca$ fractionation is not biased by radioactive decay on top of the fractionation processes, which makes the determination of $\delta^{42/44}Ca$ presumably more straightforward.

The overall results of studies of the Ca oceanic cycle led Skulan *et al.* (1997) and De La Rocha and DePaolo (2000) to conclude that the chemical composition of present-day seawater is in a steady state, whereas Zhu and MacDougall (1998) came to a different conclusion that the Ca weathering flux could have been greater than the sedimentation flux, implying that the present-day ocean is not in a steady state. The former view also differs from that of Milliman (1993), who concluded that the ocean has been losing $CaCO_3$ by deposition and accumulation of carbonate sediments since the Last Glacial Maximum, near 18,000 years ago, and from that of Berner and Berner (1996), who estimated that the modern input of Ca to the ocean from weathering and hydrothermal ocean-floor reactions is smaller than its removal as sedimentary $CaCO_3$. The issue of whether present-day seawater is in a steady state is still open because the number of studies is limited.

Ca isotope fractionation in natural calcite and aragonite was studied by Gussone *et al.* (2005), following the work of Lemarchand *et al.* (2004) on the rate of Ca isotope fractionation in synthetic calcite. It appears that the Ca fractionation is temperature-dependent with the same rates for calcite and aragonite, Ca being, however, more fractionated in aragonite than in calcite (Figure 10). The underlying mechanism is believed to relate either to different coordination numbers and/or bond strengths of the Ca ions in the two mineral structures or to various Ca reactions at the solid–fluid interface. As an alternative approach, Morris *et al.* (2003) evaluated the utility of marine barite as a recorder of seawater Ca isotopic composition. As marine barite inorganically precipitates in the water column within microenvironments, it has the potential of recording the chemical signature of the seawater in which it precipitated, as Ca is incorporated into the crystal structure. The Ca isotope composition of marine barite separated from core top sediments can then be compared to the seawater Ca isotope composition.

Schmitt *et al.* (2003b) studied $\delta^{44}Ca$ in continental waters at different scales: at a small watershed scale, at that of the regional Upper Rhine Valley, and of world rivers and hydrothermal vents. At the small watershed size, the $\delta^{44}Ca$ river flux changes appear to depend on the relative amounts of water that reacts with the rocks and biologically fractionated soil solutions. At the Rhine Valley scale and that of world rivers, the $\delta^{44}Ca$ ranges from 0.5‰ to 1‰ without depending on the rock lithology of the watershed or the climate. Also, the $\delta^{44}Ca$ of hydrothermal vents that are supplying the Ca of the ocean reservoir is uniform and close to the mean value of river waters, not introducing any bias to the ocean $\delta^{44}Ca$. The authors concluded that the $\delta^{44}Ca$ flux to the ocean should remain nearly constant through time with a mean of -1.1 ± 0.2‰ for the Ca budget, arguing in favor of a present-day seawater to be in a steady state. However, as mentioned earlier, a steady state may not be the case of the present or necessarily that of the past. Previous nonsteady-state periods could have been due to variations in the magnitudes of incoming and outgoing Ca fluxes to the ocean rather than to variations in the Ca isotopic signatures. The ^{44}Ca isotopic mass balance in the ocean that is in a steady state with respect to its total Ca content may be written as follows (cf. De La Rocha and DePaolo, 2000; DePaolo, 2004):

$$d(M_{Ca}\delta_{SW})/dt = F_R\delta_R + F_H\delta_H - F_{Sed}(\delta_{SW} + \Delta_{Sed}) \quad (15)$$

where M_{Ca} is the Ca^{2+} mass in the oceans (mol), F_R and F_H are the input fluxes of Ca^{2+} from continental and oceanic basalt weathering, respectively; F_{Sed} the net rate of Ca^{2+} removal into $CaCO_3$ by precipitation; δ_R, δ_H, and δ_{SW} the $\delta^{44}Ca$ values of the riverine flow carrying Ca^{2+} from continental weathering, from the hydrothermal reaction and basalt alteration on the ocean floor, and of seawater, respectively; and Δ_{Sed} the fractionation factor, identical to ε in Equation

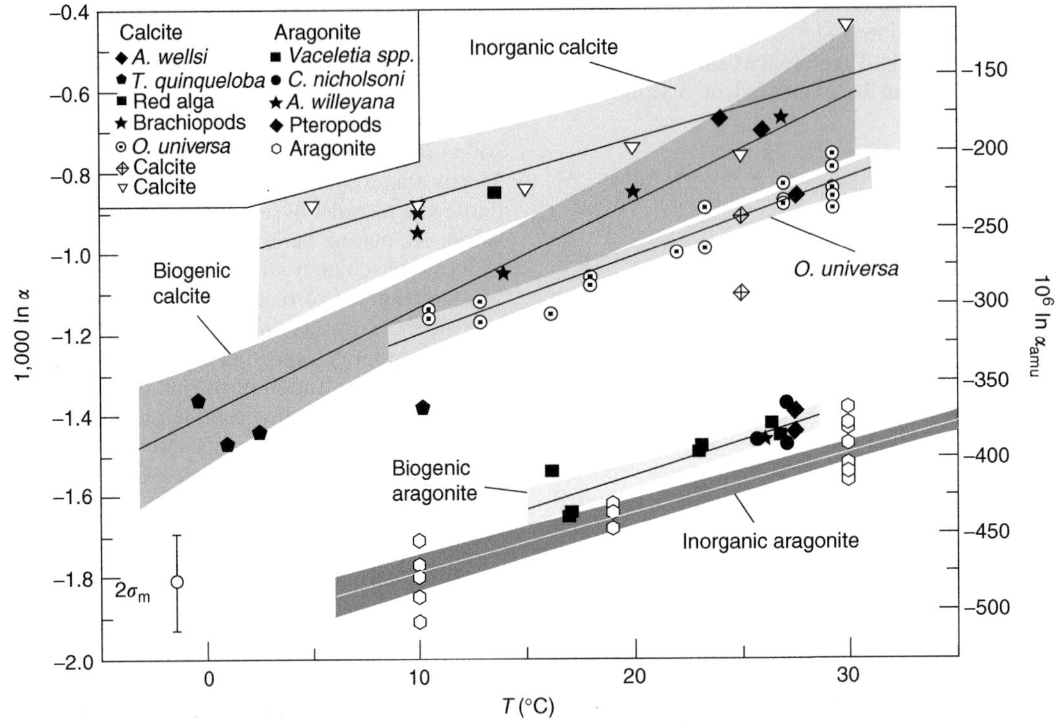

Figure 10 Temperature-dependent $^{44}Ca/^{40}Ca$ fractionation of different aragonite and calcite carbonates. Left scale of ln α is in per mil relative to the standard NIST SRM 915a, right scale is in parts per million per 1 amu. The aragonite samples are more enriched in ^{40}Ca than the calcite samples. Within analytical uncertainty, the trends have similar slopes. Modified by permission of Elsevier from Gussone *et al.* (2005).

(9a), in the deposition of $CaCO_3$ from seawater. In mass balance equations, such as the preceding one, the δ value is conventionally used as an approximation to the mole fraction of the less-abundant isotope. In the case of a steady state, M_{Ca} and δ_{SW} do not change with time (i.e., $d\delta_{SW}/dt=0$ and $dM_{Ca}/dt=0$), and the Ca-mass input to the ocean is equal to the output, $F_R+F_H=F_{sed}$. In this case Equation (15) becomes

$$\delta_{SW} = \delta_{in} - \Delta_{Sed} \quad (16)$$

where δ_{in} is a mean $\delta^{44}Ca$ value of the combined continental weathering and sea-floor alteration fluxes to the ocean. From the average values of ^{44}Ca fractionation between seawater and sediments ($\Delta_{Sed}=-1.5‰$) and the $\delta^{44}Ca$ of seawater ($\delta_{SW}\approx+1‰$), the average isotopic composition of the input to the ocean is $\delta_{in}\approx-0.5‰$ (DePaolo, 2004). However, as mentioned earlier, there are differences between the $\delta^{44}Ca$ values reported in the literature because of the use of different isotopic standards. On the scale of the NIST standard SRM915a, $\delta_{SW}=+1.88‰$ (Gussone *et al.*, 2005, table 2).

The major sink of Ca in the ocean is biogenic precipitation of Ca carbonates, which discriminates against the heavy isotopes resulting in an isotopically heavier Ca in seawater. Currently, published records of past variations of oceanic Ca isotopic composition are based on the $\delta^{44}Ca$ determinations of carbonate or phosphate phases (Skulan *et al.*, 1997; Zhu and MacDougall, 1998; De La Rocha and DePaolo, 2000; Schmitt *et al.*, 2003a, b). In fact, interpretation of these records is complicated because one has to assume a constant or known fractionation between the recording phases and seawater. Despite the fact that growth rate and other specific biological effects may affect the fractionation, maintaining the debate open, Gussone *et al.* (2004) reconstructed the surface temperature fluctuations of the Caribbean Sea between 4.7 and 4.0 Ma by studying the $\delta^{18}O$ and $\delta^{44}Ca$ records of planktonic foraminifera. In comparing the $\delta^{18}O$ with the $\delta^{44}Ca$ values and other parameters, the authors showed that the $\delta^{44}Ca$ changed by 0.4–0.5‰ in response to about a 2 °C temperature decrease of the surface water between 4.7 and 4.3 Ma (Figure 11), indicating a temperature influence.

In the same context, Heuser *et al.* (2005) suggest two scenarios to explain the variations in the Ca input flux without proportional changes in the carbonate flux, in measuring the $\delta^{44}Ca$ of oceanic planktonic foraminifera over the past 24 Ma, as shown in Figure 12. The authors interpret the results as indicating an increased riverine Ca input into the ocean water with no change in the carbonate flux, interpreting the marine uptake of

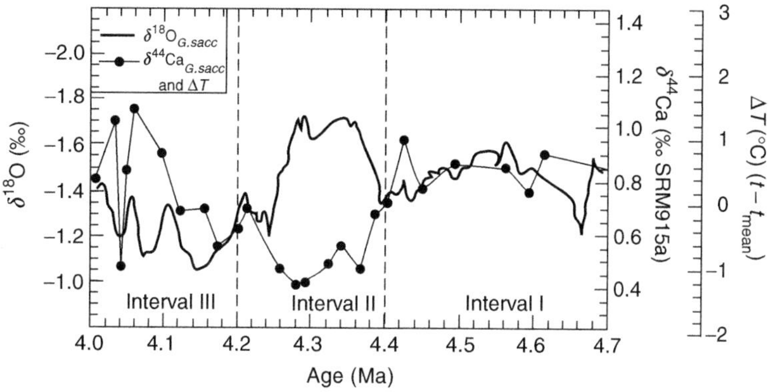

Figure 11 $\delta^{44}Ca$, $\delta^{18}O$ values of the planktonic foraminifer *Globigerinoides sacculifer* and change in temperature from $\delta^{44}Ca$ relative to the three-interval mean taken as $t_{mean}=0$. Modified from Gussone *et al.* (2004).

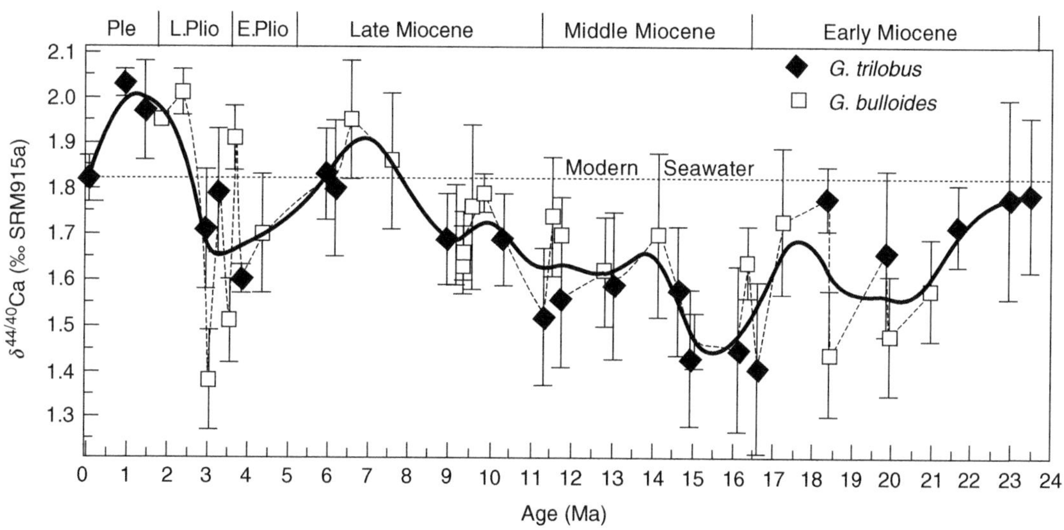

Figure 12 Spline smoothed seawater $\delta^{44}Ca$ curve for the Neogene calculated from the stacked $\delta^{44}Ca$ records of the planktonic foraminifera *G. trilobus* and *G. bulloides*. The abbreviations at the left top stand for the Pleistocene, Late Pliocene, and Early Pliocene, respectively. Reproduced from Heuser *et al.* (2005).

Ca flux by a further dolomitization process with a release of Ca into the water and its replacement by Mg. In both cases, the variations in the Ca flux generate significant changes in the seawater Ca concentrations, suggesting that the $\delta^{44}Ca$ record of the ocean water indicates a global Ca recycling that is more dynamic than generally assumed.

On the continents, Wiegand *et al.* (2005) studied the biogeochemical processes that fractionate Ca isotopes in plants and soils of the Hawaiian Islands during a 4 Ma period. Plants preferentially take up ^{40}Ca relative to ^{44}Ca, as is usually the case with the preference of primary producers for a lighter isotope, whereas the biological activity and changes in the relative contribution from volcanic material and marine aerosol induce a significant increase in ^{44}Ca in exchangeable pools of the soils. The analytical results moderate the fluxes from strongly nutrient-depleted old soils that are enriched in ^{44}Ca, relative to high ^{40}Ca fluxes of young and slightly weathered environments. Biological fractionation also controls divergent geochemical pathways of Ca and Sr in the plant–soil system (Figure 13). While the Ca content decreases progressively with increasing soil age, the Sr/Ca ratio increases systematically, which outlines combined Ca-isotope compositions and Sr/Ca ratios that are of significant interest to the understanding of the complex biogeochemical behavior of Ca and Sr.

Ca isotopic compositions of pore waters collected from soils of weathered profiles from a sequence of marine terraces along the central California coast can be explained by the existence of three distinct Ca pools in the weathering system (Bullen *et al.*, 2004): (1) a "mineral-Ca" pool standing for the $\delta^{44}Ca$ of pore water collected from deep zone enriched in the pristine granitoid minerals, (2) an "organically complexed Ca" pool derived from biological cycling within the

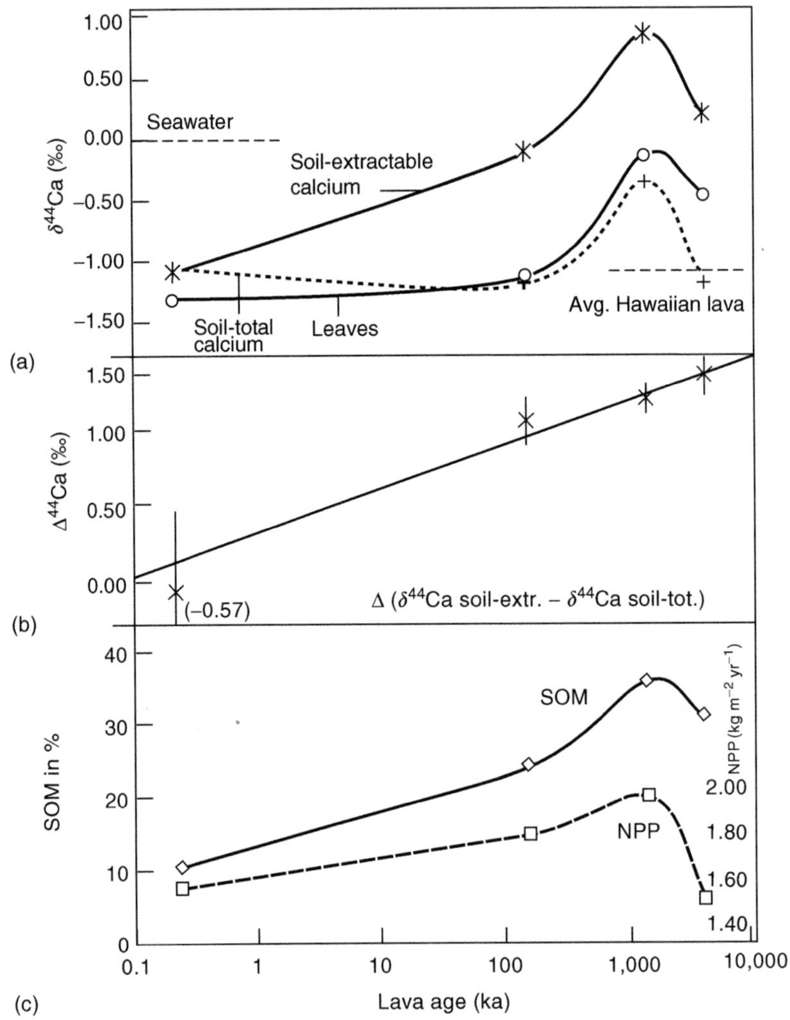

Figure 13 δ^{44}Ca data of a soil sequence in the Hawaiian Islands. (a) δ^{44}Ca in leaves and soil pools relative to the lava age; (b) Δ^{44}Ca represents the difference in δ^{44}Ca between exchangeable and total soil pools; (c) soil organic matter (SOM) in percent and net primary productivity (NPP) versus the lava age. Reproduced from Wiegand *et al.* (2005).

dominant grass ecosystem that may explain both the exceptionally light δ^{44}Ca and elevated Ca/Sr ratios of shallow pore water collected in the youngest terraces and in the uppermost soils of all terraces, and (3) an "exchangeable Ca" pool explaining the trend of increasing δ^{44}Ca of the shallow pore water toward seawater values with increasing terrace age. Increasing Ca isotope disequilibrium between the pore waters and the labile Ca pools with increasing terrace age highlights a certain inefficiency of cation-exchange processes in the weathering environment. In another context, Skulan and DePaolo (1999) showed that Ca from bone and shell is isotopically lighter than Ca of the associated soft tissues from the same organisms and that it is also isotopically lighter than the source Ca. When measured as δ^{44}Ca, the total range of variation is 5.5‰ and as much as 4‰ variation in a single organism. The fractionation occurs mainly as a result of mineralization. These results may have medical applications by assessing the Ca and mineral balance of varied organisms.

10.6 CARBON ISOTOPES IN CARBONATES AND ORGANIC MATTER

Carbonate rocks account for 25% of the preserved sedimentary mass and they consist primarily of calcite (rhombohedral $CaCO_3$), dolomite (hexagonal $CaMg(CO_3)_2$), and aragonite (orthorhombic $CaCO_3$) that is relatively common only in Recent and geologically younger sediments. The sediments are by far the biggest reservoir of oxidized and reduced carbon in the outer shell of the Earth (Table 5) and they exceed by large factors the carbon masses in the atmosphere,

Table 5 Masses of carbon (C, in mol) in the major Earth reservoirs.

Upper mantle[a]	$\sim 1.1\times 10^{22}$
Oceanic crust[b]	7.66×10^{19}
Continental crust[c]	2.14×10^{20}
Endogenic reservoirs	$\sim 1.1\times 10^{22}$
Sediments[d]	
Carbonates	5.44×10^{21}
Organic matter	1.05×10^{21}
Ocean[e]	3.19×10^{18}
Atmosphere[f]	4.95×10^{16}
Exogenic reservoirs	6.49×10^{21}
Land phytomass[g]	$6\pm 1\times 10^{16}$
Soil humus (old and reactive)[h]	$1.6\pm 0.3\times 10^{17}$
Oceanic biota[i]	$\sim 2.5\times 10^{14}$

[a] Upper mantle data from compilation by Li (2000, p. 212) and Wood *et al.* (1996). The mass of the upper mantle down to the depth of 700 km, representing 18.5% of the Earth mass, is $1{,}110\times 10^{24}$ g (Li, 2000).
[b] Mass of oceanic crust: $6.5\ km\times 3.61\times 10^8\ km^2\times 2.7\ g\ cm^{-3}=6.57\times 10^{24}$ g. Carbon concentration from Li (2000, p. 222).
[c] Mass of continental crust: $30\ km\times 1.49\times 10^8\ km^2\times 2.5\ g\ cm^{-3}=11.2\times 10^{24}$ g. Carbon concentration from Li (2000, p. 227).
[d] Total sediment mass (continental and oceanic) 2.09×10^{24} g (Li, 2000, p. 269). Other estimates vary from 1.7×10^{24} g to 2.7×10^{24}.
[e] From species concentrations in gram per liter (Li, 2000, p. 304) and ocean volume 1.37×10^{21} l. Dissolved inorganic carbon (DIC) 3.11×10^{18} mol C, dissolved organic carbon (DOC) 8.33×10^{16} mol C.
[f] Pre-industrial atmosphere of 280 ppmv CO_2.
[g] Ver *et al.* (1999).
[h] Ver (1998).
[i] Lerman *et al.* (1989), Ver (1998).

ocean water, land plants, and soil organic matter (SOM).

During the youngest Phanerozoic Eon of Earth's history, the last 542 Ma, the mass fraction of dolomite in carbonate rocks varied from a low of 7% near the end of the Paleozoic, in the Permian, about 280 Ma, to about 65% in the earlier stage of the Paleozoic and the later Mesozoic, with the value of 15% in the Tertiary (Given and Wilkinson, 1987). Calcite is a common mineral of which skeletons and shells of planktonic and benthic photosynthesizing organisms and aquatic invertebrates are made, and it may contain variable proportions of Mg in its structure, up to 30 mol.% Mg. Aragonite of biogenic origin, for reasons of its crystallographic structure, usually contains Sr as a minor component. The family of dolomite minerals includes Fe- and Mn-containing varieties, ankerites, and an excess of a few mol.% Ca over Mg has been reported in newly formed dolomite.

The massive occurrence of dolomitic rocks in the geologic past has no equivalent in the modern environments. It has been hypothesized that the greater abundance of dolomites in ancient sequences may to some extent reflect the presence of shallow-water shelf settings preserved from Paleozoic and Proterozoic times. Changes in seawater chemistry, such as Mg/Ca ratio, sulfate content, and carbonate saturation state, may have also been complementary factors leading to dolomite formation. Such changes in seawater chemistry, either individually or in combination of one with another, occur in such dolomite-producing environments as the Persian Gulf sabkha brines, the evaporated seawaters of the Coorong district of Australia and the Atlantic Coast of Brazil, Bonaire Island lagoon in the Caribbean Sea, Sugar Loaf Key carbonate sediments in Florida, supratidal carbonate sediments on Andros Island in the Bahamas, interstitial pore waters of carbonate reefs, anoxic marginal shelf and slope marine sediments, saline soda lakes, and in dilute ground waters where dolomite has been reported to be formed by methanogenic bacteria.

10.6.1 Storage of Carbonates and Organic Matter

Because carbon is strongly fractionated between organic matter and its source, such as the CO_2 in the atmosphere for land plants and dissolved CO_2 for aquatic photosynthesizers (e.g., Guy *et al.*, 1993), the isotopic record of carbon in sedimentary organic matter and limestones is a very important source of information on the history of biological productivity on Earth and evolution of an oxygen-rich atmosphere. A long-term consequence of the production and sedimentary storage of organic matter is a trend toward isotopically heavier CO_2 in the Earth surface reservoirs that may be balanced by the weathering of old organic matter and return of the isotopically light carbon to the environment. The $\delta^{13}C$ values of organic matter and carbonate sediments, and the fractionation between them since Late Proterozoic time are shown in Figure 14.

Earlier records of $\delta^{13}C$ in sedimentary organic matter show that similar values between −20‰ and −45‰, attributable to biological fractionation of carbon, occur in a banded iron formation (BIF) as early as 3,800–3,900 Ma (Mojzsis *et al.*, 1996; Holland, 1997; Rosing, 1999). The difference between the $^{13}C/^{12}C$ ratio of marine organic matter and marine limestones during the Phanerozoic fluctuates, in units of $\delta^{13}C$, around −30‰. Variations in this value have been interpreted as possibly due to changes in the mechanisms of biological fractionation through time. In the Late Proterozoic, large variations in the isotopic values of organic and carbonate carbon may correspond to the times of three glaciations that occurred between 800 and 600 Ma, when rates of storage of organic carbon might have been very low as a result of a general cooling of the climate that might have resulted in a frozen Earth's surface, a period known as the "Snowball Earth" (Hoffman *et al.*, 1998; Hayes *et al.*, 1999). Figure 14b shows that $\delta^{13}C_{carb}$ has declined overall since the beginning of the Cenozoic Era 65 Ma ago and the decline was more pronounced in the last 30 Ma.

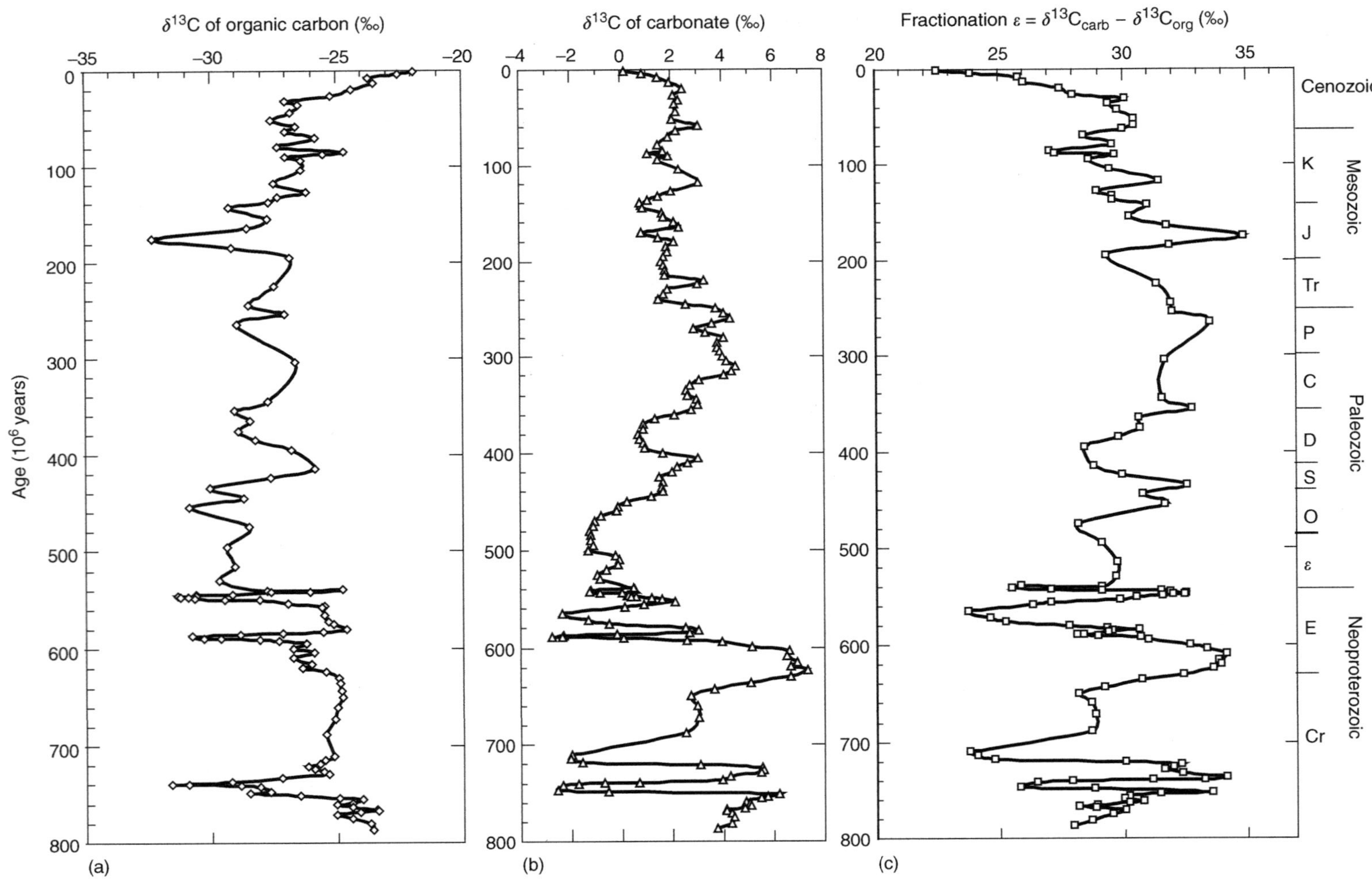

Figure 14 Carbon isotopic composition ($\delta^{13}C$, in ‰, VPDB) in Late Proterozoic (age >570 Ma) and Phanerozoic time of organic matter (a) and carbonates (b). Fractionation between carbonate and organic carbon, ε in ‰, (c). Modified from Hayes *et al.* (1999). Data courtesy of J. M. Hayes (June 2003) and plotted by his permission.

This decline is usually attributed to the rise in the elevation of land and emergence of the major mountain terrains, such as the Alps and the Himalayas, which exposed more of the old organic carbon in shales and thereby enabled its faster weathering. At the same time, however, the reasons behind a decrease in biological fractionation during the same period are not clear.

The mass of carbonates, represented by limestones and dolomites (5.44×10^{21} mol C, Table 5), accounts for a fraction f=84% of total carbon in the sedimentary record, and the remaining $1-f$=16% are in the organic matter (1.05×10^{21} mol C). If the carbonates and organic matter weathered at the same rate, their proportions would produce a global average carbon input to the ocean of an isotopic composition:

$$\begin{aligned}\delta^{13}C_{input} &= f\delta^{13}C_{carb} + (1-f)\delta^{13}C_{org}\\ &= 0.84 \times 1.5‰ + 0.16 \times (-28‰)\\ &\approx -3‰\end{aligned}$$

where for the last 200 Ma (Figure 14), the carbonate carbon is taken as $\delta^{13}C_{carb}$=1.5‰ and organic carbon as $\delta^{13}C_{org}$=−28‰. A value of −5‰ for riverine input to the ocean was used by Kump (1991), based on the weathering proportions of about 0.2 organic matter and 0.8 limestones, and −5‰ to −7‰ by Hayes *et al.* (1999). However, modern rivers have lower $\delta^{13}C$ values of −10‰ to −12‰, with a wider range from about −3‰ to −27‰, because a large fraction of their dissolved inorganic carbon (DIC) is derived from remineralization of organic matter in soils (Mook and Tan, 1991; Probst and Brunet, 2005). In a steady-state oceanic system, where carbon inputs from land and oceanic crust are balanced by the carbon removal to sediments as organic matter and carbonates and, possibly, by CO_2 release that accompanies the precipitation and storage of $CaCO_3$, a mass balance between the inputs and outputs may be written as

$$\begin{aligned}&F_{land\ input\text{-}C} + F_{magmatic\ input\text{-}C}\\ &\quad = F_{carbonate\text{-}C} + F_{organic\text{-}C} \pm F_{release\text{-}CO_2}\end{aligned} \quad (17)$$

It should be noted that the preceding balance equation describes a steady-state system and assumes that all the sedimentary organic matter is formed in the ocean and no allowance is made for the transport from land and storage of terrestrial organic carbon. In the above equation, the negative term $-F_{release-CO_2}$ indicates input of carbon from the atmosphere to ocean water and, conversely, a positive $+F_{release-CO_2}$ indicates loss of CO_2 from the ocean due to a $CaCO_3$ net removal from ocean water. In industrial time, since at least the 1970s, measurements of CO_2 transfer across the sea–air interface show that in the tropical oceans, between about 15°S and 15°N, CO_2 is emitted by the surface ocean water to atmosphere, but to the north and south of the tropical belt, surface ocean is a sink of atmospheric CO_2, with the net global transfer from the atmosphere to the ocean (Takahashi *et al.*, 2002). In pre-industrial time, the ocean was probably a net source of atmospheric CO_2 due to its formation in the precipitation and storage of $CaCO_3$ and the remineralization of organic matter brought in from land and formed *in situ* by biological production. A simplified form of Equation (17), written as a balance between riverine input and sedimentary storage of carbonate and organic carbon, gives a relationship between $\delta^{13}C_{input}$, fractionation factor ε, and the fraction f of $CaCO_3$ in sediments (Hayes *et al.*, 1999):

$$\begin{aligned}F_{land\ input\text{-}C} &= F_{carbonate\text{-}C} + F_{organic\text{-}C}\\ \delta^{13}C_{input} &= f\delta^{13}C_{carb} + (1-f)\delta^{13}C_{org}\\ &= \delta^{13}C_{carb} - (1-f)\varepsilon\end{aligned} \quad (18)$$

The $CaCO_3$ sediment fraction f that satisfies a steady-state condition is f=66–78% and the organic fraction is, correspondingly, 34–22%, when the riverine input is taken as $\delta^{13}C_{input}$=−5‰ to −7‰, the carbonate sediments as $\delta^{13}C_{carb}$=1.5–2‰, and the biological fractionation factor ε=25–30‰ (Figure 14). However, the values of the $\delta^{13}C_{input}$ that are calculated from a weathering and carbon cycling model GEOCARB III, give lower mass fractions of the buried organic carbon during the past 200 Ma: the organic carbon fraction increases with fluctuations from about 7% to 20%, to the modern value of about 16% (Berner and Kothavala, 2001; Katz *et al.*, 2005). The latter value is consistent with the fraction of the buried carbon cited earlier. Nevertheless, a lower value of the modern riverine input of about $\delta^{13}C_{input}$=−11‰ would give a smaller fraction of $CaCO_3$ in sediments that is not consistent with a steady-state partitioning model or the estimated masses of the carbonate and organic carbon.

Figure 14a shows that in the last approximately 30 Ma, the organic matter became progressively isotopically heavier, as indicated by the trend to higher $\delta^{13}C$ values, and that the fractionation between the carbonate and organic carbon has decreased. This issue has been extensively debated in the literature, where such factors were considered as the mechanisms and rates of photosynthetic fixation of carbon by oceanic phytoplankton, the cell size of the prevalent photosynthetic taxa, and possible addition of land-derived organic matter (e.g., Katz *et al.*, 2005). However, it seems unlikely that the land organic matter could significantly contribute to an increase of $\delta^{13}C_{org}$ of marine organic matter from about −27‰ to −22‰ in the last 20–25 Ma, as shown in Figure 14a. Among the land plants at present, higher $\delta^{13}C$ values of −13‰ (range −9‰ to −20‰) (O'Leary, 1988) characterize the C_4 plants, which account for 20–27% of global primary production, but represent only approximately 4% of global plant biomass (Randerson *et al.*, 2005). The bulk of the land vegetation is the C_3 plants, of mean isotopic composition $\delta^{13}C \approx -27‰$ (range $-21‰$ to $-34‰$), that is lighter than that of the C_4 plants, of which the earliest record so far goes to about 12 Ma before present. With these $\delta^{13}C$ values, the fractions of land organic carbon attributable to C_3 and C_4 plants, and the geologically young age of the C_4 plants, it is difficult to account for an increase in the sedimentary $\delta^{13}C_{org}$ as being due to a greater input of organic matter from land.

10.6.2 Long-Term Trends of Carbonate Sediments

The $\delta^{13}C$ values of the Precambrian limestones and dolomites, representing a period of more than 3,000 Ma, are shown in Figure 15. Shields and Veizer (2002) point out that the ages of some of the data points are not well constrained on a resolution timescale of 50–100 Ma, some may represent local marine environments that are not characteristic of a world average at their time, and there may be differences between the shallower-water platform sediments and those of the pelagic ocean. These uncertainties probably account for

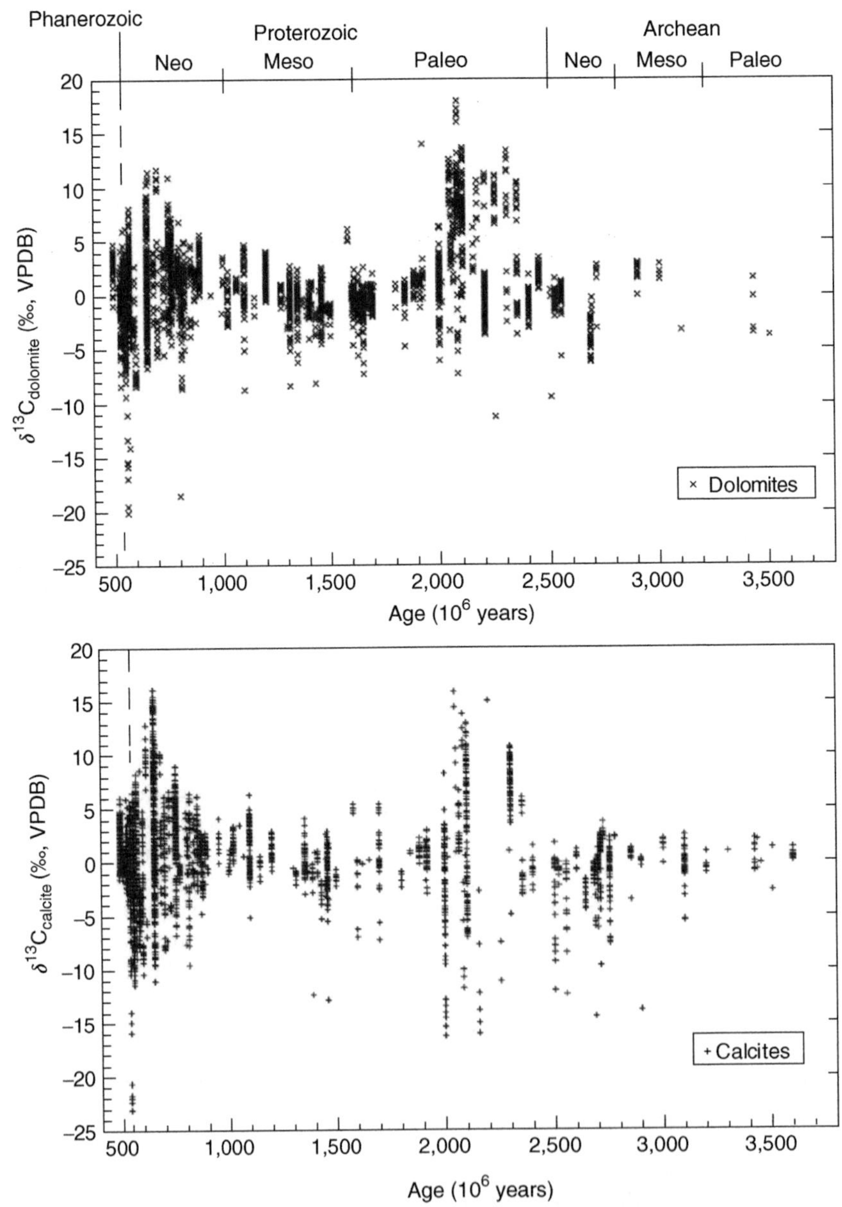

Figure 15 $\delta^{13}C$ values of Precambrian dolomites (4,914 data points) and calcites (4,266 points), from Shields and Veizer (2002). Plot from data courtesy of J. Veizer (University of Ottawa).

the large spread of the $\delta^{13}C$ values in the Paleoproterozoic and Neoproterozoic, around the time of the two major cold periods or glaciations in the Earth's early history. The Paleoarchean values of $\delta^{13}C \approx 0‰$ of calcites and dolomites are consistent with the occurrence of biological fractionation of carbon since 3,800–3,900 Ma. Because at the Earth's surface temperatures (5–25 °C), the $^{13}C/^{12}C$ fractionation between calcite and the HCO_3^- ion is less than 1‰ (Salomons and Mook, 1986), the $\delta^{13}C$ of marine carbonates is interpreted as that of ocean water. A general trend for calcites and dolomites is shown in Figure 16 where the individual data points were averaged by successive 100 Ma time intervals. Relatively large standard deviations ($\pm 1\sigma$) of some of the means are due to the large spreads around those means, as evident in the individual data plotted in Figure 15. It is noteworthy that the two minerals show little difference in their $\delta^{13}C$ values through the long time of the Precambrian, as mentioned earlier: the mean difference $\delta^{13}C_{dol} - \delta^{13}C_{dol} \approx 0.4‰$, but its extremes are +5‰ at 2,050 Ma and about −3‰ at 650 Ma and at 2,700–2,750 Ma.

The $\delta^{13}C$ of mantle carbon is approximately −5‰ to −7‰ and in the absence of life and its photosynthetic capabilities, this would also be the isotopic composition of seawater. A higher value is a result of the biological fractionation between organic matter and $CaCO_3$, removal of each into

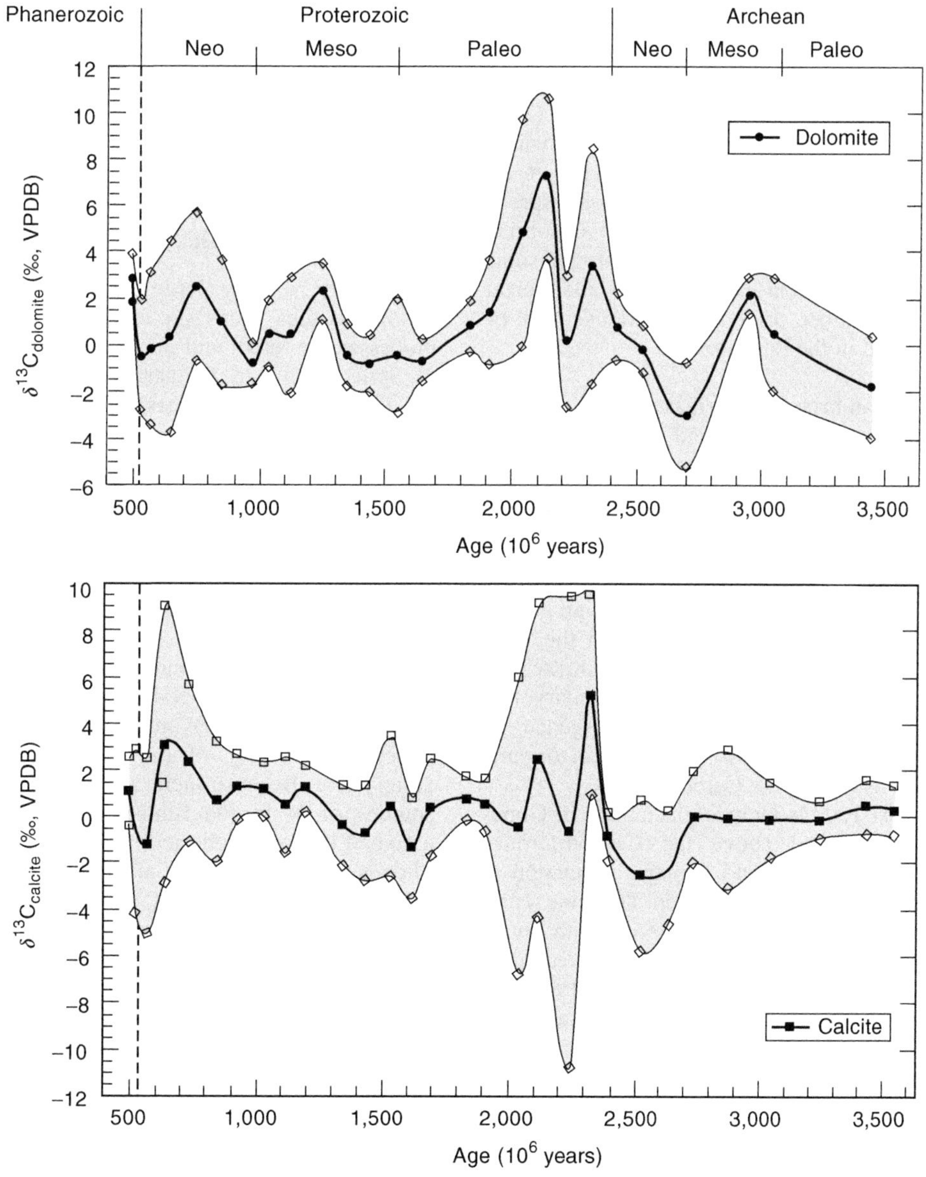

Figure 16 Mean values of $\delta^{13}C$ of Precambrian dolomites and calcites from Figure 15 by 100 Ma time intervals (thick line), ±1 standard deviation (thin lines). Note the differences between the vertical scales in Figures 15 and 16. Plot from data courtesy of J. Veizer.

the sediments, and subsequent subduction of the ocean-floor sediments into the mantle. If the carbon input to the ocean in the Paleoarchean was similar to the mantle carbon $\delta^{13}C$, then the organic matter fraction of the sedimentary carbon would have been about 17%, assuming the biological fractionation ε=30‰, as discussed earlier. The increase in the $\delta^{13}C$ of the carbonates in the time interval from about 2,400 to 2,200 Ma is usually interpreted as an indication of greater biological production and storage of organic matter in sediments and, consequently, an increase in the oxygen concentration in the atmosphere. An increase in $\delta^{13}C$ from 0‰ to 10‰ in the carbonates would correspond to an increase in the stored organic carbon fraction to about 50%, assuming the same values of the input and biological fractionation.

The carbon isotope record of Phanerozoic time, since the beginning of the Cambrian Period, is shown in Figure 17. Although the current age of the base of the Cambrian is placed at 542 Ma (Gradstein *et al.*, 2004), its age estimate has varied from 470 to 610 Ma since 1937 (Harland *et al.*, 1990). It is conceivable that the large spread of $\delta^{13}C$ values near the Precambrian–Cambrian boundary reflects poor time constraints for some of the data, as pointed out by Shields and Veizer (2002). A number of long-term trends can be distinguished in the Phanerozoic:

- In the Cambrian, an increase in the $\delta^{13}C$ of carbonates during a period of about 60 Ma, coincides with the expansion of the multicellular organisms that may be related to a greater biological production at the lower levels of the food chain.
- Although a slight decrease in the Ordovician occurs earlier than the Late Ordovician glaciation, there is a secular increase in the $\delta^{13}C$ during the next 180 Ma from the Ordovician to the Carboniferous–Permian boundary that broadly coincides with the emergence of higher plants and massive storage of organic carbon as coal in the Carboniferous.
- The next 100 Ma include glaciations of Gondwana near and above the Carboniferous–Permian boundary and a major extinction at the Permian–Triassic transition. This time represents a slightly declining trend, with much scatter, of the carbonate $\delta^{13}C$. However, some detailed $\delta^{13}C$ profiles across the Permian–Triassic boundary have provided internally consistent trends and information for the modeling of changes in atmospheric CO_2 and O_2 at that time (Berner, 2006).
- From the Jurassic to the Miocene, a period from about 200 to 20 Ma, there is a slight increase in the $\delta^{13}C$, although it is smaller than an increase in the Paleozoic over a similar length of time. Within the Jurassic to the Miocene data, there are periods of elevated $\delta^{13}C$ values that lasted about 5–10 Ma and shorter periods of $\leq$1 Ma that are associated with oceanic anoxic events (Katz *et al.*, 2005).
- The decrease in the $\delta^{13}C$ of carbonates at least during the last 30 Ma to the present and an increase in the $\delta^{13}C$ of organic carbon during the same period (Figures 14 and 17) represent a trend that differs from the past and it translates into a smaller biological fractionation between the carbonate and organic carbon in oceanic sediments. It has been variably proposed that a greater weathering of old organic matter has increased the delivery of isotopically lighter inorganic carbon forming from the remineralization of ^{13}C-depleted organic matter and it has been, consequently, the cause of the decreasing $\delta^{13}C$ values of marine carbonates. However, the $\delta^{13}C$ values of land plants do not support this interpretation, as discussed in an earlier section.

10.7 NITROGEN ISOTOPES IN SEDIMENTARY ENVIRONMENT

Actually, a very large number of nitrogen isotope studies focus on biologically mediated processes on land and in the ocean, products of agricultural and industrial pollution (chemical and organic fertilizers, and combustion of fossil fuels in transport and energy generation), and continental and ocean waters. Within the framework of this chapter on stable isotopes in sediments, we focus in this section mainly on nitrogen isotope studies of geologically old and young sedimentary environments. A recent study of the global nitrogen cycle is that of Galloway.

Nitrogen is the second most abundant, after carbon in the series C–N–P–S, elemental constituent of organic matter and it is the main component, as gas N_2, of the Earth's present-day atmosphere. Because there are no major mineral sources of N in the lithosphere, it is usually assumed that atmospheric nitrogen is the product of degassing of the early Earth, although there are views that ammonia, NH_3, might have been the primordial species and it was later oxidized by biological activity to N_2. Atmospheric nitrogen is the ultimate source of N in organic matter, continental and ocean waters, and sediments. Nitrogen has two naturally occurring stable isotopes, ^{14}N and ^{15}N, representing 99.636% and 0.364% of the element, respectively (Table 1). Because the average abundance of ^{15}N in air is uniform and constant, air is used as a standard (Junk and Svec, 1958; Coplen *et al.*, 1992), such that the isotopic value of atmospheric nitrogen is $\delta^{15}N$=0‰. The dominant source of nitrogen in most surface

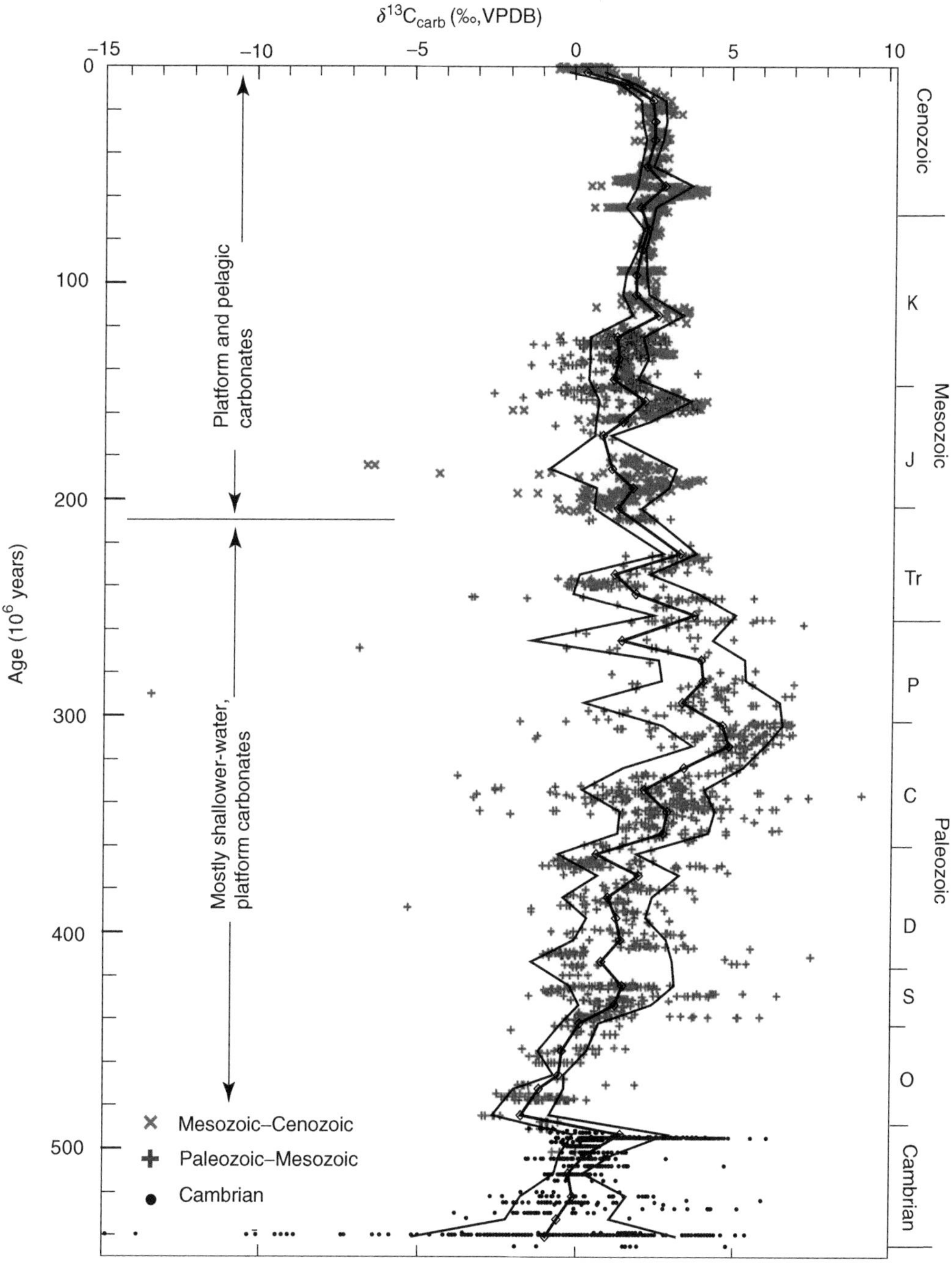

Figure 17 $\delta^{13}C$ values of Phanerozoic carbonates. Cambrian to Cretaceous (black • and blue +) data from Veizer *et al.* (1999; courtesy of J. Veizer); Jurassic to Recent (red ×) from Katz *et al.* (2005). VPDB scale. Stratigraphic timescale from Gradstein *et al.* (2004). Thick line: consecutive 10-Ma-interval means (5 Ma in the most recent 10 Ma interval); thin lines are ± 1 standard deviation.

ecosystems is the atmosphere, wherefrom it is fixed by many plants and transferred into the soils by microorganisms; rock sources of nitrogen being usually considered to be insignificant in the Earth's surface environment. Relative to the $^{15}N/^{14}N$ ratio of the atmosphere, there is considerable variation in the $\delta^{15}N$ values of other Earth's surface reservoirs that at least in part, if not completely, reflects the different isotopic fractionation magnitudes of nitrogen in the great variety of the biological processes involving this element (Figure 18). Thus biologically induced reactions of assimilation, nitrification, and denitrification strongly control nitrogen dynamics and fractionation in photosynthesis on land, in soils and waters, and in heterotrophic decomposition of organic matter in sediments and ocean water. Information on tracing the origin and cycling of

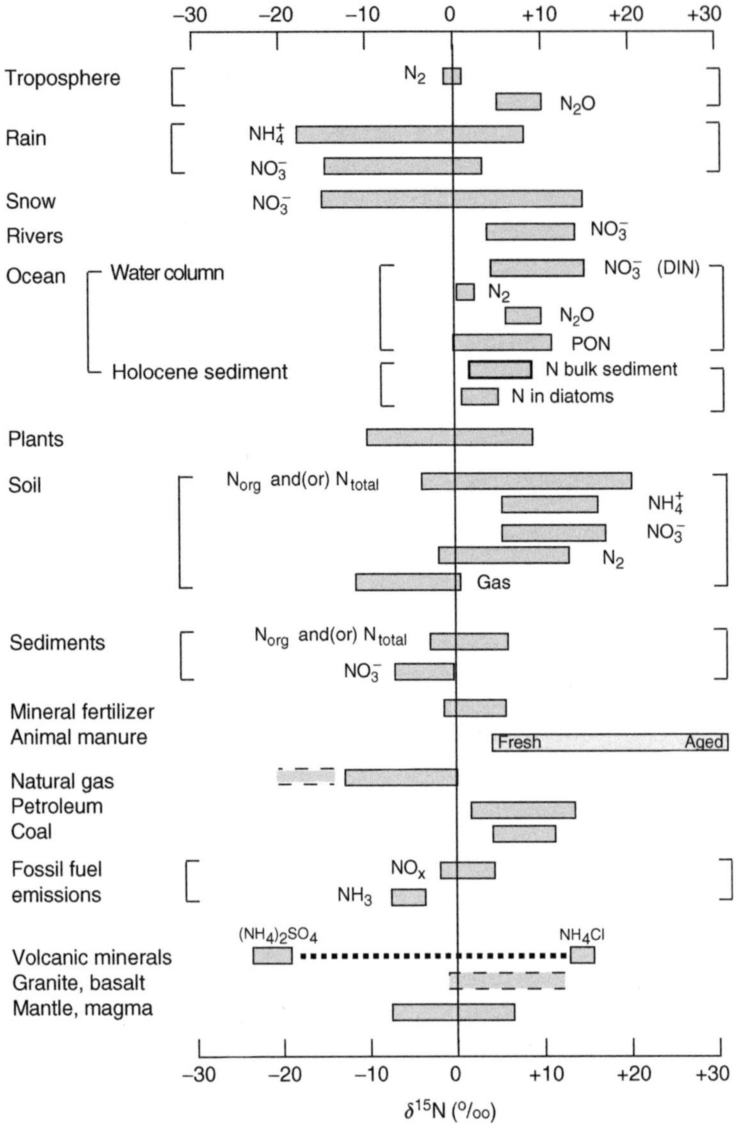

Figure 18 Range of $\delta^{15}N$ values (‰, relative to N_2 in air) in natural environments. In sediments and soils, many analyses do not specify whether they refer to organic N only or total, undifferentiated N that may be a mixture of organic and inorganic species. From sources cited in the text and Altabet *et al.* (1995), Hastings *et al.* (2003, 2004), Kendall (2004), Létolle (1980), Peterson and Fry (1987), Popp *et al.* (2002), Sigman *et al.* (1997, 1999), and Toyoda *et al.* (2002).

nitrogen in varied continental catchments and river basins was reviewed by Kendall (1998).

Soil nitrate, which is the most common nitrogen form in most soils, is preferentially assimilated by plant roots relative to soil ammonium (Nadelhoffer and Fry, 1988). Among many factors affecting $\delta^{15}N$ in soils, drainage and litter have a consistent effect on the nitrogen isotope values (Shearer and Kohl, 1988), as does an altitudinal gradient (Mariotti *et al.*, 1980).

It is generally believed that biological production of organic matter preferentially utilizes the light isotope ^{14}N, resulting in $\delta^{15}N$ values lower than in the source. Similarly, biologically mediated processes of nitrification (oxidation of reduced nitrogen to nitrite and nitrate) and denitrification (conversion of nitrate into N_2 and N_2O) result in the products being depleted in ^{15}N relative to the reactants. A generalized scheme of such reactions in a marine sediment is shown in Figure 19, where the nitrification and denitrification result in a net lowering of the $\delta^{15}N$. However, these basic relationships do not simply translate into the much more complex picture of the $\delta^{15}N$ values in different reservoirs that are shown in Figure 18. As a whole, the positive $\delta^{15}N$ values in the ocean and soil domains may indicate the importance of the removal of the light ^{14}N components from dissolved and solid species. The large range of $\delta^{15}N$ values in plants should be

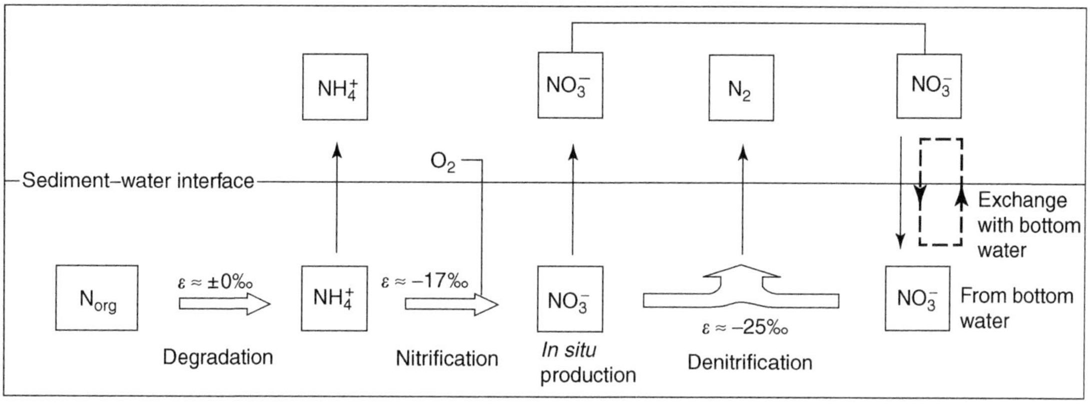

Figure 19 Decomposition of organic matter, nitrification, and denitrification in sediment. N_2O is a minor product of denitrification and is not shown. In surface water, phytoplankton utilizes both NH_4^+ and NO_3^-. After Lehmann *et al.* (2004).

noted. As far as the heavier nitrate in modern rivers is concerned, it may reflect a contribution of inorganic and, perhaps, organic fertilizers as well as the age effect on the nitrate in ground waters that becomes isotopically heavier with time due to progressive denitrification (Kendall, 2004).

10.7.1 Specific Analytical Aspects

The elemental concentration and isotopic composition of N are mostly analyzed routinely on automatic gas mass spectrometers, often coupled to elemental analyzers (EA-MS). In the pretreatment of organic matter to remove any inorganic carbon, some investigators found that the use of HCl can cause shifts in the nitrogen stable isotope composition by approximately 0.5‰ (Bunn *et al.*, 1995), while others did not (Church *et al.*, 2006, and references therein). The biases caused are considered to be small compared to the natural variations. Lorrain *et al.* (2003) did an extensive study on that methodological aspect in setting a routine to preserve, decarbonate, and analyze low-carbonated filters of suspended particulate organic matter for analysis of its organic carbon and nitrogen content, and the $\delta^{13}C$ and $\delta^{15}N$ values. They found that low-carbonated particulate organic matter trapped on filters is better preserved by drying instead of freezing, and that the inorganic carbonates are better removed by exposing the filters to HCl fumes (4 h) instead of HCl rinsing. The results show that freezing increases the uncertainty of $\delta^{15}N$ measurements and, in combination with concentrated HCl leaching, leads to a loss of particulate nitrogen and an alteration of the $\delta^{15}N$ signature. Regarding acid treatments, dilute HCl appears not to be sufficient to fully remove the inorganic carbonate, whereas exposure of the filters to HCl fumes does not alter the nitrogen data. For the purpose of studying ammonium nitrogen and total nitrogen in sediments rich in organic matter, Bräuer and Hahne (2005) studied Precambrian to Paleozoic sediments containing from 0.3 to 9.7 wt.% organic carbon. They reported that the nitrogen concentrations given by the elemental analyzer coupled to a mass spectrometer (EA-MS) technique were lower and the $\delta^{15}N$ values depleted relative to those obtained by the Kjeldahl digestion technique commonly applied to the N-extraction of organic material, indicating that ammonium-N of the sedimentary rocks was not completely detected by the common EA-MS procedure.

10.7.2 N in the Mantle and Igneous Environment

Nitrogen from mantle nodules and separate olivine crystals in oceanic island basalts, and associated gases and mantle xenoliths has a range of $\delta^{15}N$ from −8‰ to more than +6‰, with the positive values predominating in island-arc basalts (Fischer *et al.*, 2005). The similar values of the nitrogen extracted from olivine crystals and from gases at the same localities are interpreted as indicating no fractionation during degassing of the magma, but being due to a locally heterogeneous mantle source imprinted by recycling of nitrogen. The analyses of nitrogen in mid-ocean ridge basalts and diamonds associated with peridotites show a modal peak of slightly negative $\delta^{15}N$ values, $\delta^{15}N = -5 \pm 3‰$ (Cartigny, 2005). If the mantle is the source of N in the atmosphere (79.5 wt.% of N in the Earth's surface reservoirs), sediments (20.1 wt.%), and ocean (0.4 wt.%), then the weighted mean $\delta^{15}N$ of the atmosphere, sediments, and ocean should be equal to that of the mantle. However, the uncertainties of the $\delta^{15}N$ values of the mantle source and the sediments, as shown in Figure 18, make it difficult to balance the larger atmospheric reservoir of $\delta^{15}N = 0‰$ with the somewhat smaller sedimentary reservoir to achieve an agreement with a range of the $\delta^{15}N$ values of the mantle.

The metamorphism of sediments (Cartigny, 2005) and hydrothermal alternation of igneous rocks often result in higher values of $\delta^{15}N$. Bebout *et al.* (1999) and Boyd *et al.* (1993) concluded that concentration of nitrogen in granite may be enhanced by hydrothermal alteration with regional metamorphism resulting from hydrothermal fluid migration generated by the magma body through sedimentary nitrogen sources. The $\delta^{15}N$ values in granites have a range from +1‰ to +10‰ and may reflect a nitrogen origin combining sedimentary organic matter and thermal activity. Ammonium in hydrothermal feldspars from different US localities has $\delta^{15}N$ values in a wide range from −0.6‰ to +14.2‰ (Krohn *et al.*, 1993). Similarly, the relatively high values of $\delta^{15}N$ from +2.72‰ to +8.48‰ in the Nonesuch shale may possibly reflect some degree of hydrothermal activity, as inferred by Imbus *et al.* (1992). The same investigators also found hydrothermal ammonium chloride at Mount Vesuvius and Mount Etna in Italy with values in the range of +11.0‰ to +11.5‰, comparable to the range shown in Figure 18.

10.7.3 N in Sedimentary Rocks

The nitrate, NO_3^-, associated with sediments, is isotopically light and characterized by negative $\delta^{15}N$ values, while the organic nitrogen, some of it being possibly a mixture of organic and inorganic N, of soils is mostly heavier (Figure 18). Generally positive $\delta^{15}N$ values characterize the organic nitrogen in sediments. The $\delta^{15}N$ in Ordovician mudstones and siltstones of the Skiddaw Group in the English Lake District, was reported as from +3.1‰ to +3.9‰ (Bebout *et al.*, 1999); in the Cretaceous Pierre Shale, from +2.7‰ to +2.9‰ (McMahon *et al.*, 1999); in limestones with <1 mg organic C per kilogram of sediment, in the range from +1.5‰ to +5.7‰, and in marlstones with >1 mg organic C per kilogram of sediment, in the range from −2.7‰ to +2.3‰ (Rau *et al.*, 1987). The higher $\delta^{15}N$ values may result from incorporation of ammonium ($NH^+{}_4$) into micaceous minerals during diagenesis. In this context, Borowski and Paull (2000) reported that pore fluids of sediments on Blake Ridge increase in ammonium concentrations with depth and that the $\delta^{15}N$ values of this $NH^+{}_4$ increase from +3‰ near the sea floor surface to about +5‰ at 500 m depth. The depleted nitrogen isotope values for the high organic carbon-bearing sedimentary rocks suggest terrestrial organic matter input, and the higher isotopic values for the low-organic sequence were interpreted to reflect a dominantly marine origin for organic nitrogen.

Considerable variation is reported for the isotopic composition of N in hydrocarbons and coal. Stiehl and Lehmann (1980) reported $\delta^{15}N$ values ranging from +3.5‰ to +6.3‰ for organic-rich sediments and from +4.2‰ to +10.7‰ for coals. Chicarelli *et al.* (1993) reported that the oil-shale $\delta^{15}N$ value was −0.9‰, but the aliphatic hydrocarbons had $\delta^{15}N$ values in the range of −3.04‰ to +3.38‰. Rigby and Batts (1986) found that while the oil shales range in $\delta^{15}N$ from −0.5‰ to +12.7‰, coals had values ranging from −0.7‰ to +0.3‰.

Van Zuilen *et al.* (2005) reported an attempt to identify the origin of carbonaceous matter in the well-known, 3.8-Ga-old Isua Supracrustal Belt of southern Greenland by measuring the concentration and isotopic composition of nitrogen trapped in the separated organic component. They found three main reservoirs of trapped nitrogen: either (1) associated with a small amount of reactive carbonaceous material, or (2) trapped in the graphite matrix, or (3) retained in lattice defects of the graphite. The $\delta^{15}N$ of the first, at about +6‰, overlaps the values of Phanerozoic sedimentary organic matter and it is believed to be part of non-indigenous, postmetamorphic biological material. The graphite, interpreted as epigenetic in origin and associated with Mg- and Mn-siderite, contains small amounts of nitrogen with $\delta^{15}N$ of −3‰ to −1‰. These values overlap those of Archean kerogens, as well as those of metamorphic, inorganic basaltic material, suggesting incorporation into graphite from surrounding basaltic host-rocks. The syngenetic graphite has the same low nitrogen concentrations and isotopic signatures as the epigenetic graphite. From these observations, the authors conclude that nitrogen contents and isotopic compositions may not be used as biomarkers in rocks that were exposed to extensive metamorphism.

10.7.4 N in Oceanic Sediments and Particulate Matter

A decrease in total nitrogen concentration in marine sediments and an increase in pore-fluid ammonium concentrations with depth, indicating that $NH^+{}_4$ is formed in anoxic pore fluids through microbial nitrogen cycling during diagenesis, was reported by Freundenthal *et al.* (2001). A decrease in total nitrogen concentration was accompanied by an increase in $\delta^{15}N$ for organic nitrogen from +6.1‰ at the sediment surface to +8‰ at depth in shallow cores (32 cm depth) collected off the coast of Morocco.

The rapidly accumulating diatomaceous mudstones on the Namibian shelf were used by Struck *et al.* (2002) to trace decadal-scale fluctuations of coastal upwelling rates of nitrate preserved in the $\delta^{15}N$-signature of sediments and fish scales. The interval of the available data covers the last 3,200 years, during which the accumulation rates of the organic matter and fish scales fluctuated considerably. The $\delta^{15}N$ values of fish scales from surface sediments show an enrichment of 6‰ relative to the average sedimentary $\delta^{15}N$ composition, which was interpreted by the authors as indicating that the pelagic fish diet was a mixture of 20% algae and 80% zooplankton. The authors also postulated that the fish-scale $\delta^{15}N$ signature preserves information of the relative nitrate consumption of the algae in the upwelling system without any diagenetic imprints, which might be an interesting aspect in the debate about a diagenetic impact on the $\delta^{15}N$ signature of bulk sediments.

The $\delta^{15}N$ values in modern sediments of the Northern Indian Ocean range from less than 4‰ in the Northern Bay of Bengal to more than 11‰ in the Central Arabian Sea (Gaye-Haake *et al.*, 2005). Suboxic conditions in the intermediate waters of the Arabian Sea lead to the accumulation of isotopically enriched nitrate in the water column. Upwelling of this enriched nitrate into the euphotic zone, where it is consumed by primary producers, results in high $\delta^{15}N$ values in the particulate matter and surface sediments. The lower $\delta^{15}N$ values in the Northern Bay of Bengal are caused by the input of terrestrial organic matter as well as by depleted nitrate from rivers. The $\delta^{15}N$ is about 8‰ in the equatorial Indian Ocean as the influence of oxygen deficiency and river input ceases. The $\delta^{15}N$ of recent sediments correlates with the weighed mean $\delta^{15}N$ of sediment from traps at 500–1,000 m above the sea floor, but with an enrichment of 2–3‰ between traps and underlying sediments. Only at two locations characterized by high sedimentation rates (out of a total of 10 shallow-water and 12 deep-water traps), there is no significant difference in the $\delta^{15}N$ values between traps and sediments. The differences are assigned to the sensitivity of the $\delta^{15}N$ to degradation as the $\delta^{15}N$ values correlate well with the so-called Degradation Index of Dauwe *et al.* (1999), which is considered to be a tracer of the intensity of amino-acid degradation, confirming in turn the sensitivity of $\delta^{15}N$ to degradation.

Organic carbon and total nitrogen contents and corresponding isotope ratios were determined by Usui *et al.* (2006) in surficial sediment ranging from 21 to 1,995 m water depth off Tokachi, on Hokkaido Island, Japan, to address the distribution and source of sedimentary organic matter (Figure 20). Suspended particulate organic matter in the seawater column, as well as suspended particulate organic matter and bottom sediment of the Tokachi River were also examined. The $\delta^{13}C$, $\delta^{15}N$, and C/N ratios of the samples in the Tokachi River suggest that the spring snowmelt is an important process for the transport of terrestrial organic matter to the coastal waters. Comparison of these ratios in the sediments off Tokachi with those in the Tokachi River and in the seawater indicates that about half of the organic matter in the sediment is of terrestrial origin near the mouth of the river, and that the sedimentary organic matter from 134 to 1,995 m water depth is essentially of marine origin. Earlier, Mariotti *et al.* (1984) also traced the origin of nitrogen in the suspended matter of an estuarine environment by using its natural isotope composition.

Studies of nitrogen cycling in the sediments of two basins, the Santa Barbara Basin, off the coast of California, and the Eastern Subtropical North Pacific, show that ammonium $\delta^{15}N$ behaves differently in the pore waters and in the anoxic sediments of the two regions, reflecting variability in bacterial processes of sedimentary nitrogen cycling (Prokopenko *et al.*, 2006). As shown in Figure 21, the ammonium concentration in pore

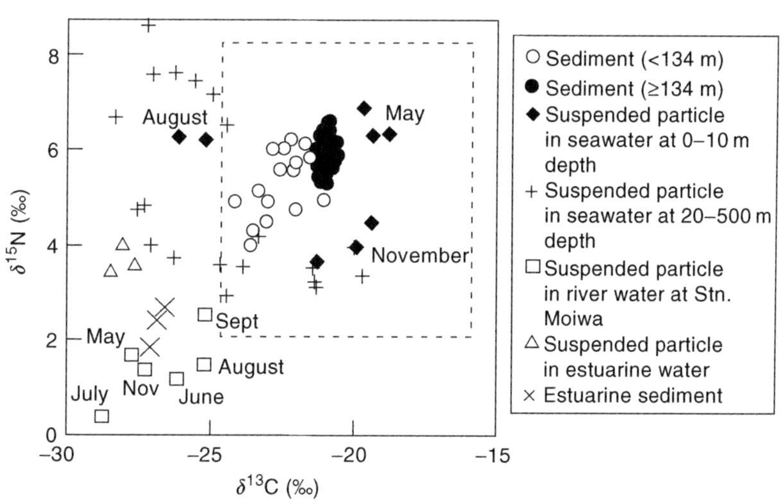

Figure 20 $\delta^{15}N$ and $\delta^{13}C$ of all samples from Tokachi region, Hokkaido Island, Japan, identified in the box to the right; area limited by the dashed line represents the estimated range of marine organic matter. Reproduced by permission of Elsevier from Usui *et al.* (2006).

water steadily increases with depth, as can be expected from decomposition of organic matter. Near-surface pore-water ammonium of the Santa Barbara Basin was found to be 1–3‰ heavier than sedimentary decomposing organic matter at 7‰, but the difference decreases progressively with depth, down to about 1‰ at 700 cm depth (Figure 21). The $\delta^{15}N$ values of NH_4 in pore waters remain quite stable at 10‰, and mass balance calculations indicate that the observed isotopic enrichment is not due to the preferential loss of isotopically heavier organic matter. The most likely cause of fractionation is the preferential degradation of an isotopically heavier, more labile

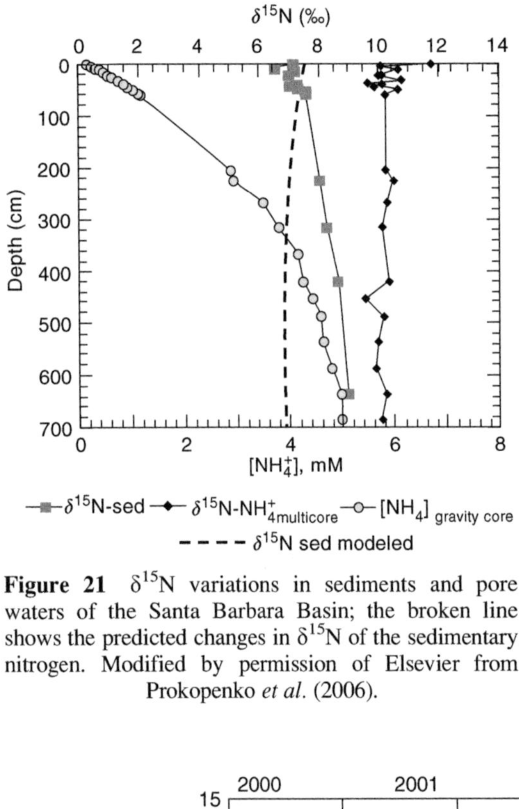

Figure 21 $\delta^{15}N$ variations in sediments and pore waters of the Santa Barbara Basin; the broken line shows the predicted changes in $\delta^{15}N$ of the sedimentary nitrogen. Modified by permission of Elsevier from Prokopenko *et al.* (2006).

marine fraction relative to an isotopically lighter, more refractory terrestrial component of the organic matter, but the authors do not rule out other possible causes for this fractionation.

10.7.5 Anthropogenic Effects

Anthropogenic forcing of the global nitrogen cycle is represented by atmospheric emissions from combustion of fossil fuels, the use of ammonium- and nitrate-containing chemical fertilizers, and discharges of animal waste and sewage that contain reactive organic nitrogen, sometimes also used as agricultural fertilizer. The isotopic tracing of natural and anthropogenic sources of nitrates in agricultural investigations started with Kohl *et al.* (1971). Historical inputs of nitrogen to the Earth's surface environment in the last 200–300 years of the Industrial Age are summarized in Lerman *et al.* (2004).

Within the $\delta^{15}N$ range for fertilizers shown in Figure 18, fertilizer ammonium is about 0‰, because it is produced from the conversion of atmospheric nitrogen, and in fertilizer nitrate it is about +3‰. In fact, applications of the $^{15}N/^{14}N$ ratio to trace relative contributions of fertilizers and animal waste to, for example, ground water are complicated by many reactions including ammonia volatilization, nitrification, denitrification, ion exchange, and plant uptake (Kreitler *et al.*, 1978; Gormly and Spalding, 1979).

In the marine realm, Tokyo Bay is one of the most eutrophic embayments in Japan (Figure 22) with a high primary productivity compared to open oceans. There, Sukigara and Saino (2005) installed a sediment trap at the exit of the bay from 1995 to 2002, with a 1 week sampling time resolution, to monitor the variability in quantity and composition of sinking particles drifting from the inner bay to the open ocean. The $\delta^{15}N$ values

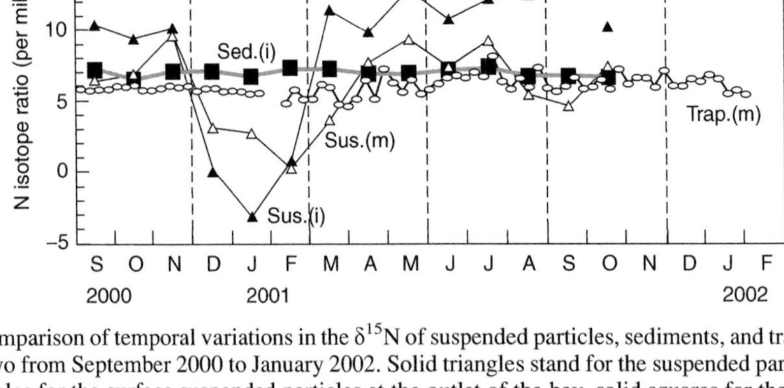

Figure 22 Comparison of temporal variations in the $\delta^{15}N$ of suspended particles, sediments, and trapped particles in the Bay of Tokyo from September 2000 to January 2002. Solid triangles stand for the suspended particles in the inner bay, open triangles for the surface suspended particles at the outlet of the bay, solid squares for the sediments in the inner bay, and the open circles for the trapped particles. Reproduced by permission of Elsevier from Sukigara and Saino (2005).

show seasonal variations: the trapped particles are heavier in summer than in winter at 6.7‰ and 5.7‰, respectively. The suspended particles in the surface waters have the largest seasonal $\delta^{15}N$ variation in the inner bay, from 11.9‰ in summer to −0.7‰ in winter. Significant $\delta^{15}N$ variations were also recorded at the bay mouth, from 8.2‰ in summer to 2.3‰ in winter. The trapped particles are primarily produced in the inner bay and then transported to the deep waters of the trap site within a few weeks. The authors believe that this rapid transport only works when surface sediment is reworked and resuspended in the entire bay, and when the fresh particles sinking from the surface water mix rapidly with the resuspended sediments to convey the temporal signal in the fresh particles toward the sediment traps.

A detailed study by Church *et al.* (2006) of tidal waters of the Delaware Estuary is instructive, as this area has progressively become an intensive urban and industrial zone that also includes discharge areas, dredging zones, shipyards, and petroleum refineries. The well-dated cores from the estuary show a sediment-layer thickness accumulated in 60–90 years and a marked increase in $\delta^{15}N$ with decreasing sediment depth, from about +3‰ to +7.5‰ at the core tops, corresponding to a substantial increase in the concentration of dissolved nitrogen, mainly as nitrate, from population growth, fertilizer use, and probably from changes in the processing of waste water leading to reduction in the chemical oxygen demand of estuarine waters. In general, animal wastes and human-produced sewage are characterized by an isotopically heavier nitrogen that reflects the trend of increasing $^{15}N/^{14}N$ ratios from primary producers to progressively higher levels of the trophic chain. This trend is also known as an anecdotal expression, "you are what you eat, plus 3‰," where the increase in the $^{15}N/^{14}N$ ratio in animals is believed to be caused by the excretion of ^{15}N-depleted urea as a metabolic waste. Data showing this relationship have been cited, for example, by Faure (1986) and Kendall (2004).

10.8 SULFUR ISOTOPES IN SEDIMENTARY SULFATE AND SULFIDE

In ocean water and sediments, the main form of oxidized sulfur is the sulfate ion, SO_4^{2-}, where it is a constituent of mostly calcium-sulfate minerals (gypsum and anhydrite, mostly of evaporitic origin) and barium-sulfate (barite). This sulfate is of a major importance to the atmospheric oxygen cycle because the mass of O_2 contained in the ocean water sulfate alone (Table 6) is about twice the mass of O_2 in the present-day atmosphere and it is much greater if both the dissolved and sedimentary sulfate are counted.

Table 6 Masses of sulfur (S, in mol) in the major Earth reservoirs[a].

Upper mantle	$6{,}230\times10^{18}$
Oceanic crust	161×10^{18}
Continental crust	185×10^{18}
Endogenic reservoirs[b]	*$6{,}576\times10^{18}$*
Sediments	
Oceanic, reduced (sulfide)	8×10^{18}
Continental, reduced (sulfide)	145×10^{18}
Continental, oxidized (sulfate)	192×10^{18}
Ocean water (sulfate)	40×10^{18}
Fresh waters (sulfate)	40×10^{15}
Atmosphere (SO_2)	56×10^{9}
Exogenic reservoirs	*385×10^{18}*
Land phytomass[c]	$4\text{–}10\times10^{13}$
Oceanic biota[d]	$0.9\text{–}4\times10^{12}$

[a] From Holser *et al.* (1988), and additional sources listed below.
[b] From the masses of the upper mantle, oceanic crust, and continental crust given in Table 3. Sulfur concentrations: in the upper mantle, 180 ppm; oceanic crust (MORB) 785 ppm; continental crust, taken as 2/3 granite and 1/3 basalt, 530 ppm (Li, 2000).
[c] Based on the C/S molar ratios in land plants from 600:1 to 1470:1 (sources cited in Lerman, 1988, pp. 23, 33) and mass of 6×10^{16} mol C (Table 5).
[d] Based on the C/S molar ratios in oceanic plankton of 106:1.7 (Redfield *et al.*, 1963) and 106:0.4 (Hedges *et al.*, 2002) and the mass of 2.5×10^{14} mol C (Table 5).

In the sedimentary environment of the exogenic reservoirs, reduced sulfur accounts for about 40% of the total sulfur (Table 6). Earlier in the Phanerozoic, this fraction was probably considerably higher and it declined from 75% at 500 Ma to its present value, due to changes in the contribution of the reduced sulfur from the mantle to the ocean (Figure 23). Massive deposits of marine sulfate minerals usually form under evaporitic conditions and continuous inflow of seawater into shallow basins. These conditions, different from those of the Recent, were characteristic of shallow epicontinental seas and probably warmer climates.

It should be borne in mind that deposition of sulfate and other evaporite minerals in saline lakes is not necessarily associated with uniformly warm climates, but more commonly with the conditions of aridity (e.g., the Gulf of Kara-Bugaz in the Caspian Sea or the Great Salt Lake) and, sometimes, existence of geologically older salt deposits that provide input brines to the lakes.

Bacterial reduction of SO_4^{2-} to H_2S and HS^- preferentially utilizes the lighter isotope ^{32}S, with the result that the pyrite forming in the process of sulfate reduction is isotopically lighter than its source, the SO_4^{2-}-ion in seawater. Thus periods of greater pyrite formation and storage would be reflected in preferential removal of the lighter ^{32}S in the process of sulfate reduction, leaving a heavier sulfate of higher $\delta^{34}S$ behind. The major occurrences of the Phanerozoic calcium-sulfate evaporites are shown in Figure 24 together with the $\delta^{34}S$ values of the sulfate disseminated in sedimentary rocks or occurring as small admixtures in biogenic calcites (Kampschulte

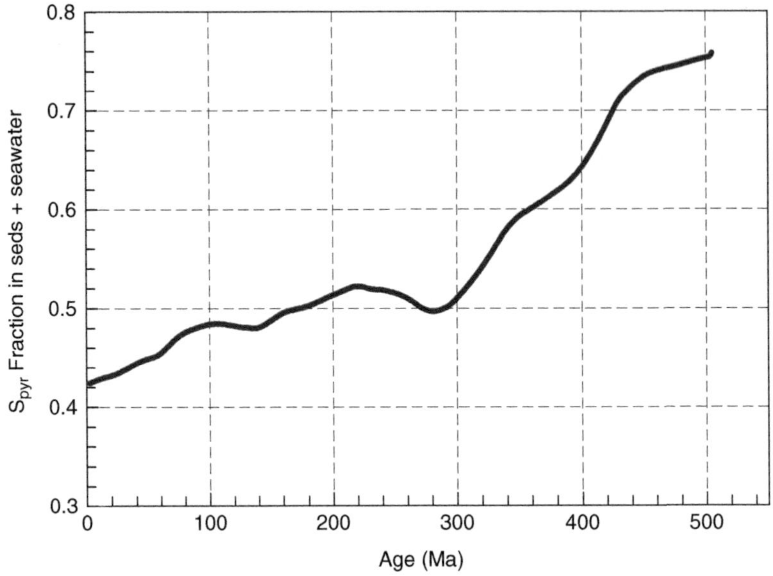

Figure 23 Fraction of reduced sulfur in the reservoirs of shallow cratonic sediments and ocean water as a function of time in the Phanerozoic, calculated from model MAGic. Data courtesy of M. Guidry from Arvidson *et al.* (2006).

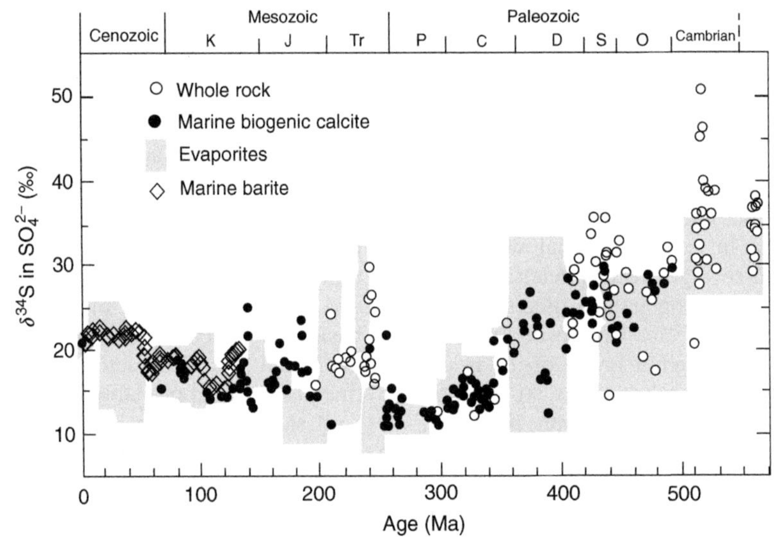

Figure 24 Sulfur isotopic composition of sulfate (‰, Canyon Diablo Troilite (CDT) meteorite standard) in evaporites and in sulfur structurally bound in calcite shells during the Phanerozoic. The open circles represent data from whole-rock analyses of $\delta^{34}S$ whereas the filled black circles are data from biologically produced marine calcite. The shaded gray areas represent data from evaporite deposits. This trend reflects the sulfur isotopic composition of Phanerozoic seawater. Open diamonds are Cretaceous and Cenozoic marine barites. From Paytan *et al.* (1998, 2004).

and Strauss, 2004). The relatively large spread of the $\delta^{34}S$ values in some periods may be due to variation in the local conditions, such as deposition in basins of restricted circulation or in waters that were in contact with the open ocean. For example, a spread of $\delta^{34}S$ from 15‰ to nearly 45‰ in sulfate associated with the Neoproterozoic carbonates in one 80 m thick formation was interpreted as a result of deposition in a chemical gradient in seawater, from a colder open-ocean to a warmer and shallower shelf water (Hurtgen *et al.*, 2006).

The $\delta^{34}S$ trend of the oceanic sulfate in the last 500 Ma broadly agrees with that of the sulfate trend as shown in Figure 25a. The trends of $\delta^{34}S$ in evaporites and carbonates are similar, showing a secular decrease through the Paleozoic, from the Cambrian to the Carboniferous and Permian, followed by an irregular increase toward the Recent (Kampschulte and Strauss, 2004, and Figure 24).

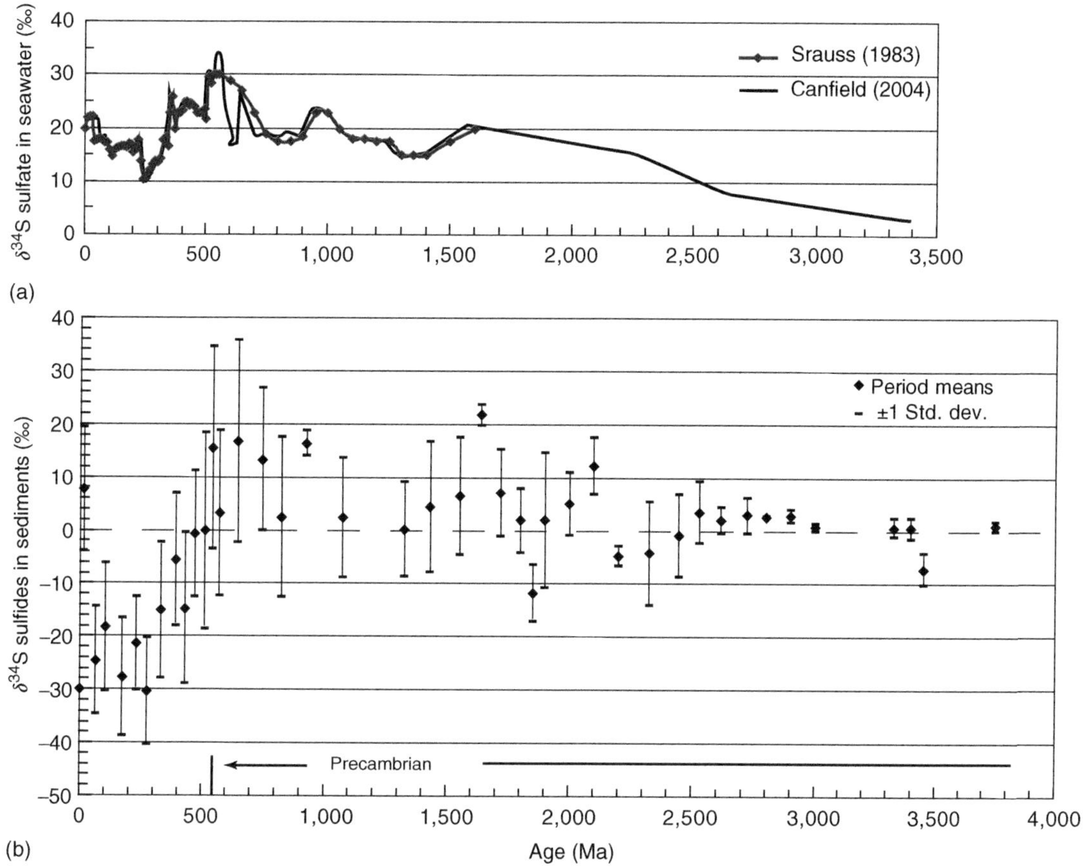

Figure 25 δ^{34}S of (a) ocean-water sulfate (SO_4^{2-}) (data of Strauss, 1983, are from Canfield, 2005; also Strauss, 1993) and (b) sedimentary sulfides (FeS_2) (Canfield, 2001; Shen *et al.*, 2001). In the Phanerozoic sediments younger than 540 Ma, the δ^{34}S values of sulfides are Period averages. Data courtesy of Canfield (2005).

Also, the trend of the sedimentary sulfide (FeS_2, Figure 25b) in the same period of the Phanerozoic shows decreasing δ^{34}S values. The large variation of the δ^{34}S values within individual geologic periods is reflected in a remarkably large spread, from −6‰ to −23‰, in pyrite formed within the chamber walls of a *Cosmoceras* sp., an ammonite shell 3.5 cm in diameter, of reportedly Callovian, Middle Jurassic age (Ono *et al.*, 2006). This spread partly overlaps the δ^{34}S range for the Jurassic shown in Figure 25b at 175 Ma before present.

As the δ^{34}S values of both the sulfate and sulfide show a long-term decrease through about 260 Ma of the Paleozoic, a first intuitive explanation may be that this is due to a combination of less sulfate reduction in the ocean, less pyrite storage, and an increased weathering of the pyrite in sediments exposed on land. There are two problems with this interpretation: first, the δ^{34}S values of the sedimentary sulfides and sulfates, between the ages of about 550 and 2,100 Ma, are relatively high and their input from weathering to the ocean would not contribute to a lowering of the δ^{34}S of the sulfate in seawater; second, pyrite occurs mostly in shales that also contain organic matter, the weathering of which would add isotopically lighter δ^{13}C to ocean water. However, there is a broadly negative correlation in the Paleozoic between the decreasing δ^{34}S values of sulfate and the increasing δ^{13}C values of carbonate, as shown in Figures 25a, 14b, and 17 (Garrels and Lerman, 1981, 1984, with references to earlier authors). As the δ^{13}C of carbonates increases, indicating a greater storage of organic matter, the δ^{34}S of sulfates decreases. As in the present-day seawater, the SO_4^{2-} concentration is much higher (28 mmol kg^{-1}) than that of the DIC (about 2 mmol kg^{-1}), such a difference in concentrations indicates that DIC may be more sensitive to changes in the isotopic values of the C input that is derived from varying proportions of carbonate rocks and shales.

From a balance relationship between the isotopic inputs and outputs of S to the ocean, similar to the relationship for carbon, as given in Equation (18),

$$\begin{aligned}\delta^{34}S_{input} &= f\delta^{34}S_{pyr} + (1-f)\delta^{34}S_{sul} \\ &= \delta^{34}S_{sul} - (1-f)\varepsilon\end{aligned}$$

Canfield (2004) estimated the long-term $\delta^{34}S_{input}$ as 3±3‰ and the fraction f of the total S stored as sulfide in sediments declining from about 1 at the beginning of the Paleozoic Era to about 0.25 at its end (cf. Figure 23). Furthermore, he concluded that in the last 700 Ma sulfur has been accumulating in the ocean because of a much reduced return of sulfur to the mantle from the oxygenated deep ocean, and the Earth's surface and mantle reservoirs of sulfur are not in a steady state at present and, in the past, the surface reservoirs might have deviated from an average crustal isotopic composition because of subduction of isotopically light sulfides from an anoxic ocean.

Recently, the process of mass-independent (or, more precisely, non-mass dependent) fractionation of the sulfur isotopes has been studied by a number of authors in the terrestrial sedimentary environments (e.g., Mojzsis *et al.*, 2003; Farquhar *et al.*, 2003; Ono *et al.*, 2006; with references to earlier work of others). The atomic mass differences among the four stable sulfur isotopes (^{32}S, ^{33}S, ^{34}S, and ^{36}S; Table 1) result in a smaller fractionation of $^{33}S/^{32}S$ than that of $^{34}S/^{32}S$ because the difference in atomic mass units is 33−32=1 amu for the former pair and it is twice as large for the latter, 34−32=2 amu (e.g., Chapter 4). Similarly, the $^{36}S/^{32}S$ fractionation (36−32=4 amu) should be about twice as large as that of $^{34}S/^{32}S$. In the mass-dependent fractionation, the differences in the atomic masses translate into the relationships, refined by the theory of isotope fractionation, between the $\delta^{x}S$ values: $\delta^{33}S \approx 0.5\,\delta^{34}S$ and $\delta^{36}S \approx 2\,\delta^{34}S$. These values correspond to a mass-dependent fractionation. However, accurate mass-spectrometric measurements in the biological fractionation, sulfates, and sulfides indicate that some of the $\delta^{x}S$ values depart from the expected mass-dependent differences, and such departures have been interpreted as providing additional information on the biologically driven oxidation or reduction of sulfur and its sedimentary storage.

10.9 BORON ISOTOPES AT THE EARTH'S SURFACE

Boron has two naturally occurring stable isotopes ^{11}B and ^{10}B that represent about 80 and 20% of the element, respectively (Table 1). Both the contents and isotopic signatures vary widely in surficial environments, especially in natural waters: over four orders of magnitude in concentration and from −25‰ to +59‰ in $\delta^{11}B$ (Vengosh *et al.*, 1994), as shown in Figure 26.

Boron isotopic ratios have been used for about two decades to trace the origin of water masses (Palmer and Sturchio, 1990), to follow the evolution of brines (Vengosh *et al.*, 1991a, b; Moldovanyi *et al.*, 1993), to determine the origin of evaporites (Swilhart *et al.*, 1986; Vengosh *et al.*, 1992), to examine hydrothermal flow systems (Leeman *et al.*, 1992), and to determine the paleo-pH of ocean water.

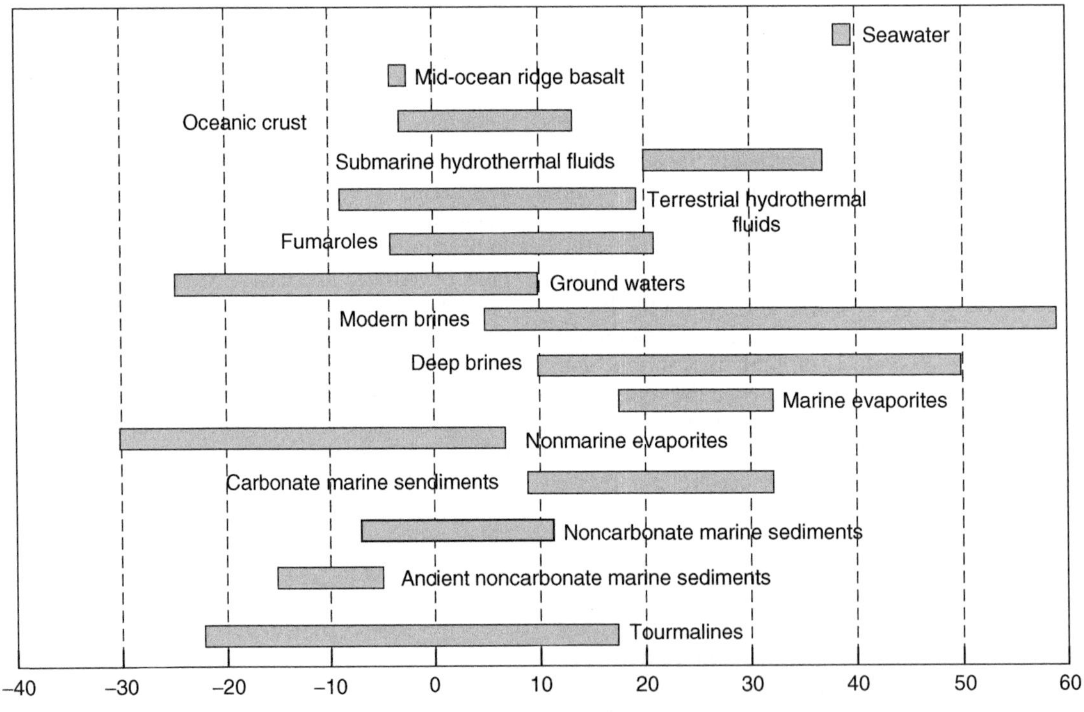

Figure 26 Range of $\delta^{11}B$ values (‰) in natural environments. After Tonarini *et al.* (2005).

Boron, like most elements generally analyzed for their isotopic composition, is concentrated by ion chromatography using a specific resin. Afterwards the B-isotope composition is generally measured by positive or negative thermal ionization mass spectrometry (PTIMS or NTIMS) following chemical procedures set either by Heumann and Zeininger (1985) or by Leeman *et al.* (1991), respectively, and by many others improving the methods. Recently, Chaussidon *et al.* (1997) reported an interesting alternative method to determine the B isotope composition by secondary ion mass spectrometry (SIMS), which does not yield the high precision of the previous TIMS method but allows a better control of the spatial resolution, and which apparently does produce a significant matrix effect. In all cases, the B isotope compositions are expressed as the deviation (in per mil) from a standard of the atomic abundance ratio $^{11}B/^{10}B = 4.044 \pm 0.001$ or 4.053 (Chetelat *et al.*, 2005; the NIST standard is boric acid, designated as SRM 951):

$$\delta^{11}B = \left[({}^{11}B/{}^{10}B)_{\text{sample}}/({}^{11}B/{}^{10}B)\text{standard}^{-1}\right] \times 10^3 (‰) \qquad (19)$$

where the ratios $^{11}B/^{10}B$ of the standard and the sample are determined in the materials that went through the same chemical preparation procedure.

10.9.1 In Ocean Water

It has been shown that the ^{11}B found in seawater is enriched relative to the oceanic and continental crust (Lemarchand *et al.*, 2000, 2002). Although the reason for this is not yet well understood, it is assumed that it is due to the preferential removal of the isotopically lighter borate ion $^{11}B(OH)_4^-$ from solutions onto clay particles. In solution, exchange reactions between the two aqueous species $B(OH)_3^0$ and $B(OH)_4^-$ control the isotopic fractionation of boron (Schwarcz *et al.*, 1969). The relative abundance of the two species is pH-dependent because of the ionic dissociation reaction

$$B(OH)_3 + H_2O = H^+ + B(OH)_4^-$$

that has a practical (also referred to as an apparent or concentration-based) equilibrium constant, where brackets [ ] denote concentrations in mol B kg^{-1} solution:

$$K_B = \frac{[H^+][B(OH)_4^-]}{[B(OH)_3]}$$

$pK_B = 8.6$ at 25 °C, seawater of normal salinity ($S = 35$).

In the pH range from about 7.5 to 9.5, the abundance fraction of the $B(OH)_4^-$ ion or the ratio $[B(OH)_4^-]/[B(OH)_3]$ increases, which also changes the $\delta^{11}B$ values of the two species if the total B concentration and its mean isotopic composition do not change. The isotopic fractionation between the two species, ($\delta^{11}B$ in $B(OH)_3$) − ($\delta^{11}B$ in $B(OH)_4^-$), varies from the calculated value of 20.6‰ at 0 °C to 17.7‰ at 60 °C. The borate ion is incorporated in the $CaCO_3$ crystal lattice, where it presumably substitutes for CO_3^{2-}, and it is assumed that there is no significant fractionation between the B in the aqueous borate ion and the one in the crystalline phase: that is, $\delta^{11}B$ in $B(OH)^-_{4,\,aq} = \delta^{11}B$ in $CaCO_{3,\ cal}$. This feature and the pH dependence of the borate-ion ratio in seawater have been used as a tool for estimation of the paleo-pH of ocean water in the Early Cenozoic and during the Last Glacial Maximum (Palmer and Pearson, 2003; Pearson and Palmer, 1999; Sanyal *et al.*, 1995) on the basis of the $\delta^{11}B$ value in carbonate skeletons of planktonic and benthic foraminifera. The pH of ocean water is an important parameter in the determination of the atmospheric CO_2 concentration and is therefore important in paleoclimatological reconstructions. Zeebe and Wolf-Gladrow (2001) discussed in detail the principles of the paleo-pH method and gave the equations for calculation of the paleo-pH. They also summarized the evidence from the studies of $\delta^{11}B$ in the shells of foraminifera. These studies were followed by more recent studies of corals and paleoclimatic interpretation of the glacial stages (Hönisch *et al.*, 2004; Hönisch and Hemming, 2005). It should be noted that if the $\delta^{11}B$ of corals truly reflects pH variations of seawater, as stated by the authors, there are also significant offsets between the values of corals and the theoretical computed $B(OH)_4^-$ curve. Such offsets are reported for different species of corals and they are often designated as "vital effects." For the results from corals to be more useful for reconstruction of marine paleo-pH, more needs to be done before the method is considered a routine application.

Zeebe and Wolf-Gladrow (2001) pointed out that the $^{11}B/^{10}B$ fractionation factor between the aqueous $B(OH)_4^-$ and $B(OH)_3$ is based only on a value computed from the theory. Pagani *et al.* (2005) raised additional criticisms of the paleo-pH determinations because of the uncertainties in the dissociation parameters of the ^{11}B aqueous species and the isotopic composition of boron in seawater in the geologic past, and expressed doubts on the validity of the paleo-pH estimates based on the $\delta^{11}B$ studies of carbonates, particularly for periods of time that exceed the residence time of boron in the ocean, 14–20 Ma.

10.9.2 In Ground and River Waters

Boron occurring in ground waters might derive from leaching of host rocks, infiltration of meteoric waters, mixing of several types of ground waters, but also from contamination by anthropogenic activities. The main human use of boron compounds is in bleaching detergents, causing elevated concentrations of B in wastewaters worldwide. Vengosh *et al.* (1994) showed that ground waters contain B from two sources of distinctive isotopic signatures: the $\delta^{11}B$ of seawater being close to 39‰, while that of the continental crust is at about 0±5‰. In summary, elemental B and $\delta^{11}B$ variations reflect both the mixing of regional ground waters and the B isotope fractionation caused mainly by boron adsorption on clay particles. B isotope studies have therefore a high potential of being useful tracers in ground-water systems in which the role of clay material can be identified, as well as a tracer for anthropogenic B that can be mixed with that of seawater. It may be noted that the analytical precision of dissolved B and $^{11}B/^{10}B$ ratios in river waters and soil fluids is variable and the literature data are relatively scarce (Spivack *et al.*, 1987; Rose *et al.*, 2000).

To evaluate a modeling of the boron geochemical cycle in wide drainage basins, Lemarchand and Gaillardet (2006) studied the B concentrations and isotopic composition of waters and suspended particles of major rivers in the Mackenzie River Basin in Western Canada. The overall conclusion is that B seems to be regulated mainly by exchange reactions involving silicate rocks, and that rain waters and carbonate and evaporate dissolution also contribute but to a lesser extent. The B dissolved in the river waters may also be regulated by ground waters and its $\delta^{11}B$ increases with total dissolved solutes and decreases with increasing runoff (Figure 27). A relationship between the B-isotope composition being exported, either in dissolved or particulate form, and that of its bulk source suggests that it is not in a steady state in most of the tributaries of the Mackenzie drainage basin (Lemarchand and Gaillardet, 2006). The modeling of this behavior shows that the B isotopes are highly sensitive to hydrological conditions and to changes in the weathering rate of the source rocks. Departure from steady state could then record recent changes in the weathering regime of the basin, implying that evolution of the weathering is a key parameter controlling the geochemical fluxes at the Earth's surface. Another aspect that might have some importance is the B-content and isotopic composition of rainwater: the concentration ranges from 3.5 to 14 ppb and the $\delta^{11}B$ from +30.5‰ to +45‰ in French Guiana, with mean values at 6 ppb and +41‰, respectively (Chetelat *et al.*, 2005). This rainwater B originates from seawater with a fractionation during evaporation mixed with a second component, less constrained, but which could be derived from the burning of biomass. Two aspects of B-isotope fractionation are behind this

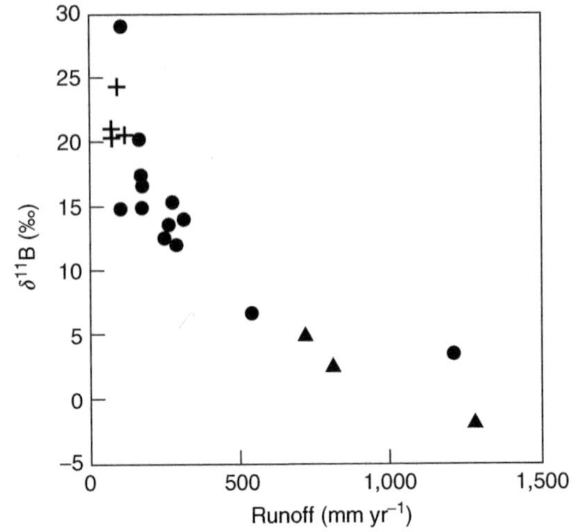

Figure 27 Relationship between $\delta^{11}B$ in riverine dissolved B and runoff (mm yr^{-1}). The circles correspond to varied rivers draining the Rocky Mountains and the Interior Platform, the triangles are for rivers draining volcanics on the West side of the Cordillera, and the crosses for rivers draining the Canadian Shield. Reproduced by permission of Elsevier from Lemarchand and Gaillardet (2006).

interpretation. First, the organically bound B in the biomass is believed to be enriched in the lighter isotope ^{10}B and the biomass consequently contributes low or negative values of $\delta^{11}B$. Second, experimental evidence indicates that evaporation of $B(OH)_3$ from seawater produces a vapor phase that is enriched in ^{11}B, whereas the salts and brines remaining after evaporation are correspondingly depleted in ^{11}B, although it is not clear from the description of the experimental conditions how closely they approximate a natural system or a controlled fractionation of $^{11}B/^{10}B$ between the aqueous and gases species of $B(OH)_3$. Mixtures of B from such ^{11}B-depleted biomass and atmospheric-vapor-enriched sources imprint the $\delta^{11}B$ values of atmospheric precipitation.

10.9.3 In Oceanic Sediments

The B-isotope variations have been interpreted as proxies of climate changes throughout the Earth's history. A negative $\delta^{11}B$ excursion from +2.7‰ to −6.2‰, reported in postglacial carbonates of Neoproterozoic deposits of Namibia (Kasemann *et al.*, 2005), is interpreted by the authors as reflecting a temporary decrease in seawater pH, although this argument is based on assumptions linking the seawater alkalinity to the Ca^{2+}-ion concentration that was inferred from the $\delta^{44}Ca$ measurements of carbonate rocks. Associated $\delta^{44}Ca$ values from 0.35‰ to 1.14‰ are linearly coupled with the carbon isotope ratios of the same carbonates, implying enhanced postglacial weathering rates. Reconstruction of the seawater

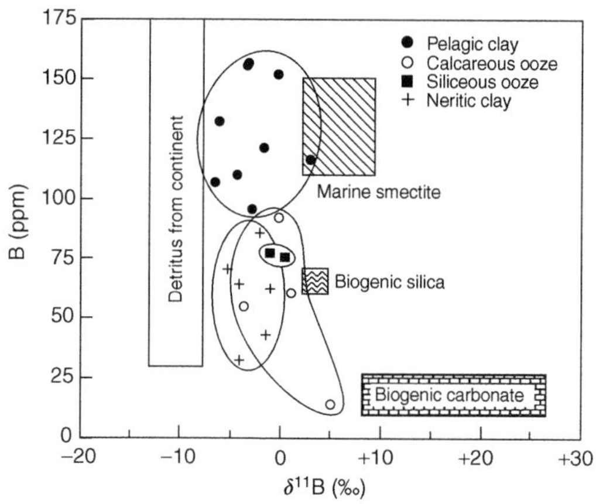

Figure 28 B contents and $\delta^{11}B$ values in modern sediments with the boxes indicating the ranges of four major B reservoirs. The box of the $\delta^{11}B$ values for detritus of continental origin includes eolian dust, loess, and fluvial particulates; the actual range may be wider. Reproduced by permission of Elsevier from Ishikawa and Nakamura, (1993).

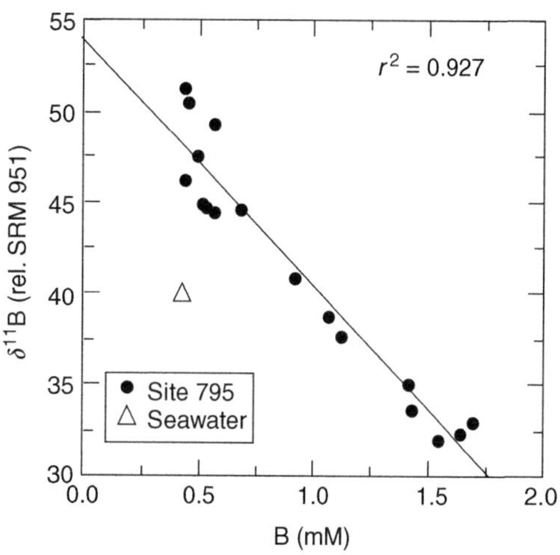

Figure 29 Correlation between pore-water B contents and $\delta^{11}B$ in ‰ relative to the SRM 951 boric acid standard, in the upper sedimentary unit of ODP Site 795 in which high heat flow was reported. Adapted by permission of Brumsack and Zuleger (1992).

pH and the weathering profiles indicates that high atmospheric CO_2 likely occurred during the postglacial melting and the precipitation of the carbonate cap.

In modern marine sediments including pelagic and neritic clay material as well as calcareous and siliceous oozes, the $\delta^{11}B$ values vary from −6.6‰ to +4.8‰ that result from a mixing of continental detritus, marine smectite, biogenic carbonates, and biogenic silica (Ishikawa and Nakamura, 1993; Figure 28). The low $\delta^{11}B$ end member is the detritus of continental origin with values ranging between −13‰ and −8‰; its $\delta^{11}B$ composition is controlled by illite, which has been transported either by the rivers or the winds. The high $\delta^{11}B$ end member is represented by material directly originating in the oceans, such as marine smectite ($\delta^{11}B$ of +2.3‰ to +9.2‰), biogenic carbonate ($\delta^{11}B$ of +8.0‰ to +26.2‰) and biogenic silica ($\delta^{11}B$ of +2.1‰ to +4.5‰), values resulting from uptake of marine B. In ancient marine sedimentary rocks, the B-content is similar to that of modern marine sediments, but the $\delta^{11}B$ values are systematically lower (−17.0‰ to −5.6‰, Figure 26), which could be due to diagenetic alteration by either preferential removal of high $\delta^{11}B$ from carbonates and silica during recrystallization, and/or B isotopic exchange in the course of the smectite to illite transition (Ishikawa and Nakamura, 1993).

In the pore waters of such marine sediments, the $\delta^{11}B$ varies from +11.5‰ to +51.3‰ with the B-concentration up to eight times higher than that of seawater (Brumsack and Zuleger, 1992) indicating that the B behavior is nonconservative and complicated during early diagenesis that may include adsorption and desorption processes, alteration of volcanoclastic material, as well as heat flow inducing high B concentrations and $\delta^{11}B$ values lighter than that of seawater (Figure 29). At greater depths, B seems to be removed from pore waters and incorporated into the sediments.

10.9.4 In Clay Mineral–Water Exchange

In the late 1960s, Schwarcz *et al.* (1969) argued that the enriched ^{11}B values in seawater relative to the continental inputs is due to a fractionation effect during adsorption of B onto clay particles. From an experiment, they calculated a fractionation enrichment of about 30–40‰, and they noticed that this fractionation factor is nearly in agreement with a value calculated when assuming the ocean is in a steady state with respect to both the concentration and the isotope composition of B. This steady state is assumed as maintained by a single input of B from continental weathered rocks and a single output by adsorption on sediments. Subsequently, Williams and collaborators, cited below, performed a series of laboratory experiments on B fractionation during exchange processes between clay particles and water, considering also that during diagenesis, clay nucleation and growth induce incorporation of ^{10}B into the illite-type structures (Spivack *et al.*, 1987). However, this general trend causing a depletion of B in the pore waters with an increase in $\delta^{11}B$ is not the general rule, especially in the Gulf Coast sedimentary basin (Land and Macpherson, 1992; Moldovanyi and Walter, 1992), suggesting that another source of ^{10}B exists in deep basinal environments. Williams *et al.* (2001a, b) studied the

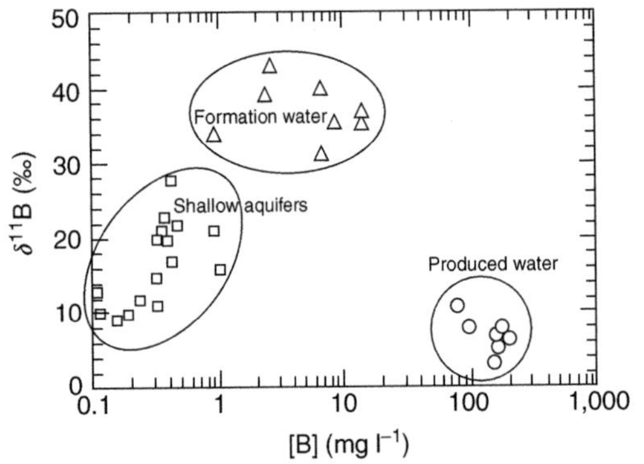

Figure 30 B content and $\delta^{11}B$ values for different types of waters in oil-producing sedimentary basins. No mixing among the different types of waters could be found. Reproduced by permission of Elsevier from Williams *et al.* (2001a).

B isotopic composition of organic matter from hydrocarbon reservoirs of the Gulf Coast and found that there is an apparent ^{11}B-enrichment of pore waters with progressive migration through clay-rich sediments (Figure 30), and that very little B was detected in the oil, but that kerogen contains significant amounts of B with $\delta^{11}B$ similar to the authigenic pore-filling clay minerals in the sandstone. Williams *et al.* (2001a, b) also published studies of B isotope compositions of various pore waters from hydrocarbon reservoirs, outlining the extremely high potentials of B contents and isotope compositions to decipher the diagenetic evolution of barren and oil-producing sedimentary basins.

Boron isotopes are fractionated during mineral crystallization especially during hydrothermal activities (Oi *et al.*, 1989; Spivack *et al.*, 1990; Leeman *et al.*, 1992). A systematic examination of the behavior of B during illitization of mixed-layered illite–smectite clay material and temperature increase led Williams *et al.* (2001a, b) and Williams and Hervig (2002, 2006) to perform a series of laboratory experiments. They showed that the kinetics of B-isotope exchange follows the mineralogical restructuring of smectite as it recrystallizes into illite. Values of $\delta^{11}B$ in the mixed-layers illite–smectite were used to construct a B-isotope fractionation curve (Figure 31b). In fact, there is no mineral-specific fractionation of B isotopes, as there is for oxygen, but there is rather an ion-coordination dependence of the fractionation. In diagenetic environments, B appears to be mainly in trigonal coordination in the fluids, but it substitutes in tetrahedral sites of clay minerals when nucleating and growing, inducing a major fractionation in B, as the exchangeable interlayered B of a different $\delta^{11}B$ can be easily removed by cation exchange (Figure 31a, b).

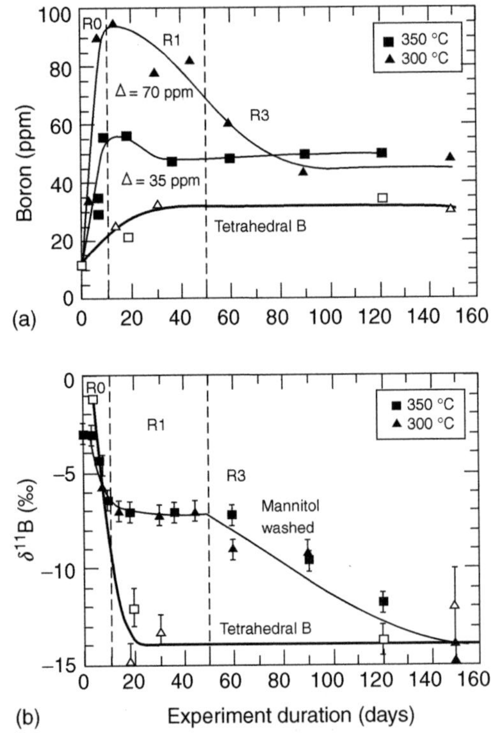

Figure 31 (a) Trends in B contents during experimental illitization of smectite over 4–5 months at 300 °C, 100 MPa (solid symbols); (b) Trends in $\delta^{11}B$ of the same smectite-to-illite reaction. The thick lines at the bottom of each diagram delineate the tetrahedral values measured after re-expansion of the mixed-layers illite–smectite to remove the B trapped in interlayer position, shown by open symbols. Reproduced by permission of Mineralogical Society of America from Williams and Hervig (2002).

10.10 ^{40}AR IN THE CLAY FRACTION OF SEDIMENTS

10.10.1 ^{40}Ar in Clay Diagenesis

The stable isotope ^{40}Ar is most often associated with geochronology and the $^{40}K/^{40}Ar$ and $^{39}Ar/^{40}Ar$ dating methods of minerals. However, the mobility of argon in the clay mineral lattice, where it forms from ^{40}K, results in its tendency to escape from the clay and, as such, it provides a measure of an extent of clay diagenesis. Interpretations of the ^{40}Ar occurrences in clays in various sedimentary environments are based on the values of the $^{40}Ar/^{40}K$ ratio that can be interpreted as an indication of either a real or apparent age (defined below) of a clay particle assemblage. When dealing with ^{40}Ar in the context of clay diagenesis, it is more convenient to use the units of time, based on the measured $^{40}Ar/^{40}K$ ratios in clays and translated into ages, rather than the numerical values of the fractional abundances of the two nuclides. Radiogenic nuclide ^{40}Ar is produced by decay of ^{40}K that ubiquitously occurs in K-containing aluminosilicate minerals. Concentration ratios of radiogenic ^{40}Ar to its parent ^{40}K in minerals are variably used in the interpretation of depositional episodes and diagenetic histories of sedimentary sequences. In closed systems where ^{40}Ar forms by radioactive decay of ^{40}K and where neither of the two nuclides escapes or is added, the $^{40}Ar/^{40}K$ ratios allow calculation of meaningful mineral ages. Whereas K–Ar ages in igneous minerals often agree well with those obtained by other isotopic methods, K–Ar data of sedimentary mineral fractions often differ from their stratigraphic age because of the presence of mixed assemblages of detrital and diagenetic minerals that involve dissolution of K-rich silicates, precipitation of authigenic K-rich clays, and/or escape of radiogenic ^{40}Ar from the minerals (Dalrymple and Lanphere, 1969; Faure, 1986). Because of such differences from the stratigraphic age, the K–Ar results in sediment fractions are referred to either as "dates," as suggested by Faure (1986), or as "apparent ages." ^{40}K decays to ^{40}Ar and ^{40}Ca such that about 10% of the decay events produce ^{40}Ar and the remaining 90% produce ^{40}Ca. The decay rate constants for the individual decay branches and for the whole are as follows (Steiger and Jäger, 1977):

^{40}K to ^{40}Ar: $\lambda_a = 0.581 \times 10^{-10}$ years,
half-life $= 11.93 \times 10^9$ years
^{40}K to ^{40}Ca: $\lambda_c = 4.962 \times 10^{-10}$ years,
half-life $= 1.40 \times 10^9$ years
^{40}K to ^{40}Ar and ^{40}Ca:
$\lambda = \lambda_a + \lambda_c = 5.543 \times 10^{-10}$ years
half-life $= 1.25 \times 10^9$ years

In a closed system that initially contains no radiogenic ^{40}Ar, decay of ^{40}K to ^{40}Ar results in the $^{40}Ar/^{40}K$ atomic or mass ratio increasing with time according to the following relationship (e.g., Faure, 1986):

$$\frac{^{40}\mathrm{Ar}}{^{40}\mathrm{K}} = \frac{\lambda_a}{\lambda}\left(e^{\lambda t} - 1\right) \qquad (20)$$

where t is the time that starts with ^{40}K decay and ^{40}Ar accumulation in a closed system. Time t can be computed from the measured $^{40}Ar/^{40}K$ ratio that is the principle of determining a true or an apparent age. If the system initially contains ^{40}Ar, then Equation (20) becomes

$$\frac{^{40}\mathrm{Ar}}{^{40}\mathrm{K}} = \left(\frac{\lambda_a}{\lambda} + \frac{^{40}\mathrm{Ar}_0}{^{40}\mathrm{K}_0}\right)e^{\lambda t} - \frac{\lambda_a}{\lambda}$$
$$= \left(0.1048 + \frac{^{40}\mathrm{Ar}_0}{^{40}\mathrm{K}_0}\right)e^{\lambda t} - 0.1048 \qquad (21)$$

where subscript 0 denotes the initial concentrations of ^{40}Ar and ^{40}K in the particles. Initial $(^{40}Ar/^{40}K)_0$ ratios that develop prior to deposition in particles of ages 10^7–10^8 years that behaved as closed systems are of orders of magnitude of 0.001–0.01. Although these values are smaller than the ratio $\lambda_a/\lambda = 0.105$, they are often neglected because they are not always known. The $^{40}Ar/^{40}K$ ratio increases with increasing age, where a change of a few percent in the ratio corresponds to a similar percent-wise change in the computed age.

10.10.2 Clay Particle Size and K–Ar Apparent Age

Clay minerals of the smectite and illite groups that contain K as part of their stoichiometric composition and radiogenic ^{40}Ar would presumably have K–Ar ages the same as the stratigraphic age if they formed near the time of deposition. However, a K–Ar apparent age may be younger than the stratigraphic age, which indicates either addition of K to the mineral or loss of ^{40}Ar from it or some combination of both processes. Because an addition of K and a consequent new growth of clay particles are often difficult to detect, the escape of ^{40}Ar is usually considered to be responsible for the lower K–Ar apparent ages. The different views of several investigators on this subject have been reported by Lerman and Clauer (2005). At the other extreme, K–Ar apparent isotopic ages of clays greater than the stratigraphic age are encountered fairly often and they indicate that the mineral particles had a predepositional history that was in some cases geologically long. A number of such occurrences are shown in Figure 32 for stratigraphic ages ranging from the Early Cambrian to the Late Neogene. In particular, the stratigraphically young samples from the Miocene of the Gulf Coast, USA, and the Neogene of Eastern Borneo have much older K–Ar apparent ages that are equivalent to a range from the Late Mesozoic to Early Paleozoic.

It is a well-established observation that in clay- and silt-size fractions of detrital sediments, the smaller particle-size fractions have lower $^{40}Ar/^{40}K$ abundance ratios and correspondingly younger apparent ages, by 10^0 to 10^1 Ma, than the bigger particles (Perry, 1974; Faure, 1986; Lerman and Clauer, 2005, with references to earlier work). A number of sedimentary formations where such relationships between the particle size and apparent age have been identified are shown in Figure 33. The linear regressions of the data from different stratigraphic ages and locations can be written in the form of

$$\log R = A + b \log r \qquad (22)$$

where R is the $^{40}Ar/^{40}K$ ratio in the particle-size fraction, r the mean particle radius in micrometers, b the slope of the regression line, and A is a constant for each line. The plots show that the larger particle sizes have higher $^{40}Ar/^{40}K$ ratios and apparent ages. The magnitude of change from a bigger to a smaller size depends on the steepness of the slope or the coefficient b in Equation (22):

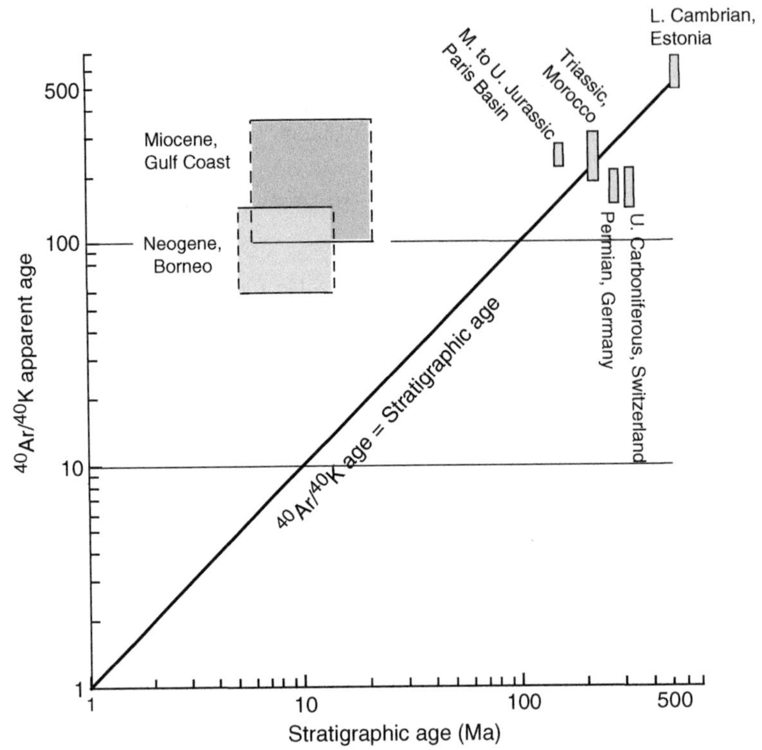

Figure 32 Stratigraphic age of the sedimentary sequences and the K–Ar age of the clays (mostly illite and mixed-layered illite–smectite) occurring in them. Note the logarithmic scale of the two ages. Minerals above the diagonal line of equal ages have a predominant predepositional history. Data sources are given in Figure 33.

a greater value of b results in a greater change in the $^{40}Ar/^{40}K$ ratio and the apparent age.

The $^{40}Ar/^{40}K$ ratios in two particle sizes of any sediment, of mean radii r_1 and r_2, are related from Equation (22) as

$$\log R_1/R_2 = b \log r_1/r_2 \tag{23}$$

where subscript 1 denotes a smaller and subscript 2 a larger particle size ($r_1 < r_2$).

If the decrease in the $^{40}Ar/^{40}K$ ratio is interpreted as due to escape of ^{40}Ar, then the fraction of ^{40}Ar escaped from the smaller-size particles relative to the larger particles is from Equation (23):

$$f = 1 - R_1/R_2 \tag{24a}$$

$$= 1 - (r_1/r_2)^b \tag{24b}$$

Thus the ^{40}Ar fraction that escaped depends on b, the slope of the regression line, and the difference between the bigger particle size r_2, taken as a reference, and the smaller size r_1. The slope, b, varies from about 0.03 for one of the Permian samples to 0.25 for the Pliocene or Late Miocene samples from the East Slovak Basin (Figure 33). However, the range of the particle sizes varies considerably among the sediment samples, but most are between $r_1/r_2 = 0.1$ and 0.01. For a particle size decrease by a factor of 10 to 100 ($r_1/r_2 = 0.1$ to 0.01), from Equation (24b), the Paleozoic and Mesozoic smaller-size fractions have lost 7–19% of their ^{40}Ar for a particle-size decrease by a factor of 10, and between 13% and 34% for a size decrease by a factor of 100. However, the Tertiary samples mentioned above would have lost 34–44% of the ^{40}Ar for a particle-size decrease by a factor of 10, but the lost fraction would be greater, 56–68%, for a 100-fold decrease in the particle size.

In Figure 33, the slopes of the older-sediment plots (Cambrian, Carboniferous, Permian, and Triassic) are smaller than those of the younger, Miocene and Neogene formations. In general, if the particles become a closed system where no ^{40}Ar is lost after some initial period when the differences between the smaller and larger size fractions have developed, the slope will become progressively less steep as ^{40}Ar continues to grow and accumulate in the particles. This process of a slope change is due to the smaller particles "aging faster." However, the magnitude of such a change depends on the initial difference in the $^{40}Ar/^{40}K$ ratio between the smaller and larger particle sizes and the duration of the subsequent closed-system stage, neither of which quantities is well known.

10.10.3 ^{40}Ar Loss in Thermal Events

Losses of radiogenic ^{40}Ar from K-bearing silicate minerals due to their exposure to higher

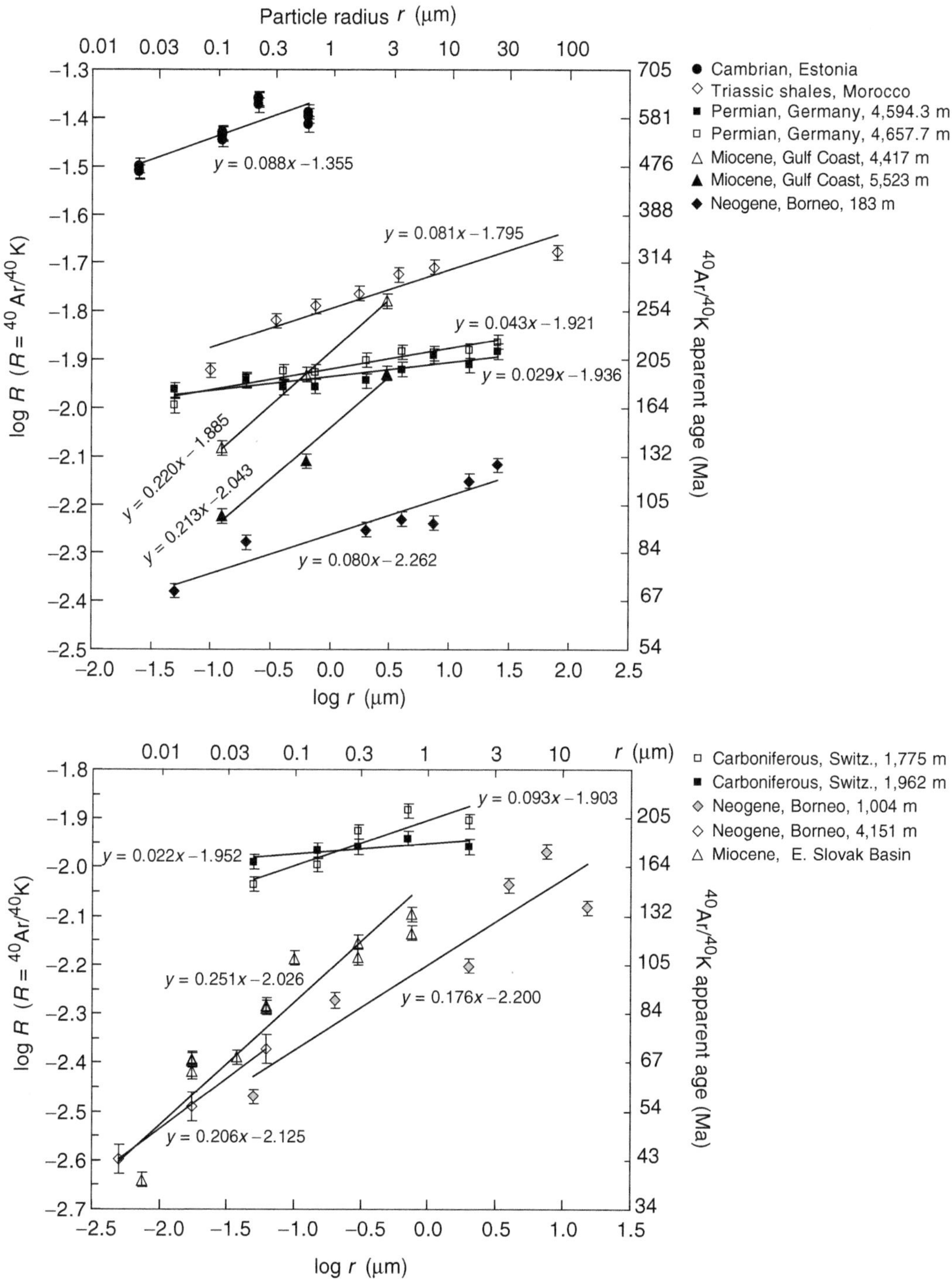

Figure 33 Decrease in the $^{40}Ar/^{40}K$ ratio and apparent age with decreasing particle size (equivalent sphere radius r, in μm). The linear regression is shown next to each line. Data: East Slovak Basin from Clauer *et al.* (1997); Estonian "blue clay" from Chaudhuri *et al.* (1999); all the rest referenced and discussed in Lerman and Clauer (2005).

temperatures in such tectonic events as contact with igneous intrusions and/or regional metamorphism have been extensively documented in metamorphic environments or in connection with hot fluid migrations in geological discordances and discontinuities (Clauer and Chaudhuri, 1998). The thermally induced loss of ^{40}Ar results in the "resetting of the K–Ar clock." A loss of ^{40}Ar from clay minerals in a sedimentary rock intruded by a basaltic dike was documented by Techer *et al.* (2006) in Lower Jurassic (Toarcian) shales in Southwestern France (Figure 34).

On each side of a 1.1 m thick basaltic dike, mineralogical and chemical changes include a

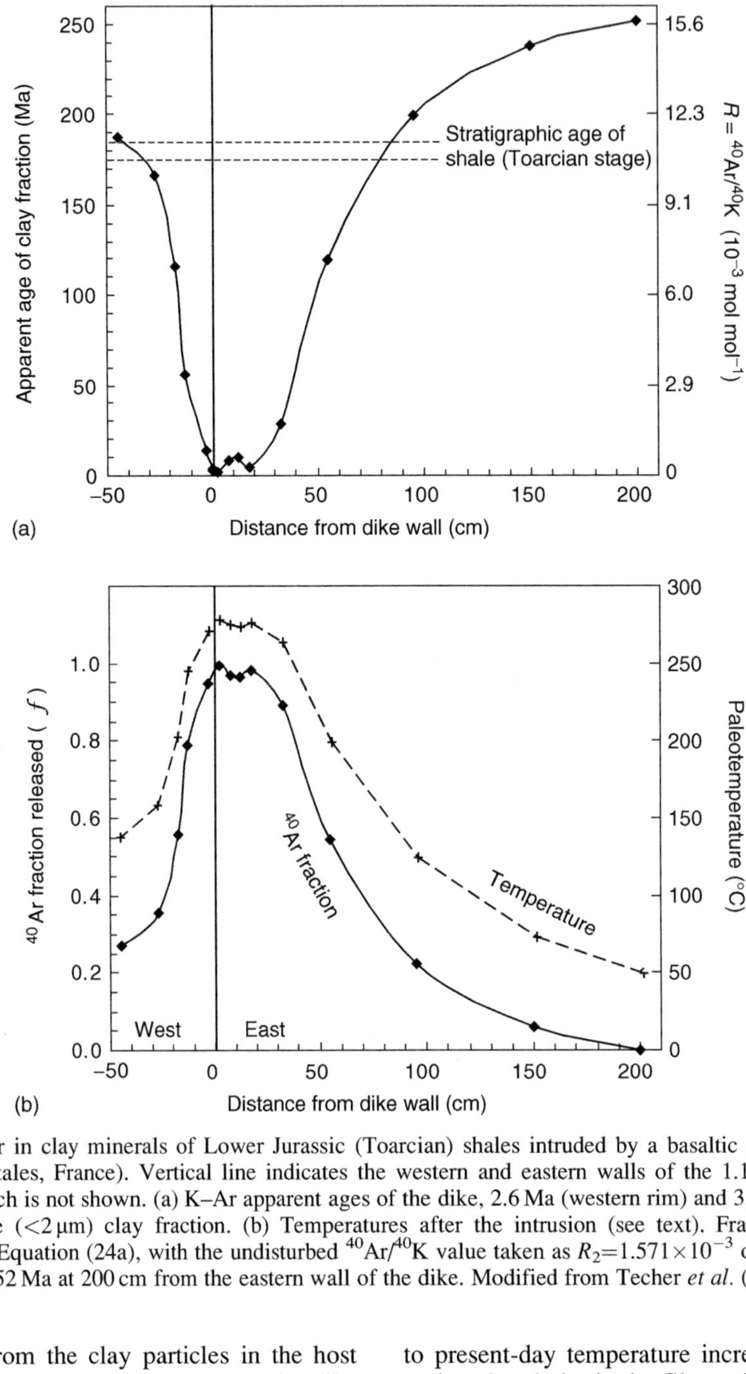

Figure 34 ^{40}Ar in clay minerals of Lower Jurassic (Toarcian) shales intruded by a basaltic dike, Perthus Pass (Pyrénées Orientales, France). Vertical line indicates the western and eastern walls of the 1.1 m thick dike, the thickness of which is not shown. (a) K–Ar apparent ages of the dike, 2.6 Ma (western rim) and 3.3 Ma (center), and fine particle-size (<2 μm) clay fraction. (b) Temperatures after the intrusion (see text). Fraction of ^{40}Ar lost calculated from Equation (24a), with the undisturbed $^{40}Ar/^{40}K$ value taken as $R_2=1.571\times10^{-3}$ or the apparent age of 252 Ma at 200 cm from the eastern wall of the dike. Modified from Techer *et al.* (2006).

loss of ^{40}Ar from the clay particles in the host shale, where the loss is strongest near the dike walls, as may be expected, but there is no loss within about 2 m from the dike margin. It should be noted that, as in other cases, the <2 μm particle-size fraction of unaltered clays shows K–Ar apparent ages older than the stratigraphic age of the Toarcian shale. The age of the intrusion is about 3 Ma near the dike center.

Techer *et al.* (2006) summarized various estimates of the temperature in the shale outside the dike and their own estimate is shown in Figure 34b. It is based on the work of Hunziker *et al.* (1986) that linked ^{40}Ar losses of detrital illite to present-day temperature increases in cores in various boreholes in the Glarus Alps, also accompanied by clay authigenesis. That study found that any detrital K–Ar signature in a clay fraction is completely erased at about 260 ± 30 °C, which means in turn that the radiogenic ^{40}Ar accumulated by the detrital mineral phases that might occur in <2 μm size fractions escaped completely and that the "K–Ar clock" is reset. Close to the dike, the temperatures in the shales were at least at about 260–300 °C, allowing complete release of the previously accumulated radiogenic ^{40}Ar. On the other hand, the reference shale sample, located 2 m away from dike, was probably never heated to more than $50 \pm 10\,°C$. Temperatures lower than 150 °C seem to have occurred beyond 40 cm on the western side, and beyond 80 cm on the eastern side, and the temperatures were probably lower than 100 °C farther than 120 cm from the dike. It should also be mentioned that no high-grade metamorphic minerals were observed in the shale host-rocks even at the immediate contact with the intrusion, which supports the estimation of a highest paleotemperature at about 260–300 °C. The loss of ^{40}Ar on each side of the intrusion, calculated as explained in Figure 34b, is based on an assumed conservation of K in the system. It should, however, be reiterated that although the higher temperatures near the intrusion are responsible for the loss of ^{40}Ar and the resetting of the "K–Ar clock," intrusion and hydrothermal solutions associated with them are usually responsible for the diagenesis of clay minerals that is reflected in an increase in the proportion of illite layers in mixed-layers illite–smectites and disappearance of kaolinite and chlorite. In sedimentary sequences that remain below approximately 100 °C, no thermal diagenetic effects may be expected. In an average geothermal gradient of $30\,°C\,km^{-1}$, such conditions may exist at depths shallower than about 3.5 km, where the release of ^{40}Ar from clays is a diffusional process. Although the diffusional transport is temperature sensitive, the magnitudes of the ^{40}Ar fractional losses from clay particles of different ages, discussed earlier in this section, indicate that the loss rate is generally slow to very slow at geological timescales.

10.10.4 ^{40}Ar Flux from Sediments

The release of radiogenic ^{40}Ar from potassium-containing clays makes the sediments a likely source of some of the ^{40}Ar in the atmosphere, where this isotope is the most abundant constituent of the present-day atmospheric argon. Illite and mixed-layers illite–smectite are the main K-bearing clays in sediments, and the ^{40}Ar flux from them is evaluated by two methods in this section. Illite of a synthetic stoichiometric formula $K_{0.6}Al_{2.2}Si_{3.7}O_{10}(OH)_2$ is similar to the rounded-off mean composition that is given by Deer *et al.* (1962a) as $K_{0.6}Al_{2.6}Si_{3.4}O_{10}(OH)_2$, and it may be taken as a representative source of ^{40}Ar in clays in the sedimentary lithosphere. The mass of shales, 0.96×10^{24} g, accounts for 51 wt.% of all the sediments, and illite and mixed-layers illite–smectite represent 56 wt.% of the shales or 0.54×10^{24} g (Li, 2000; Shaw and Weaver, 1965). In addition to the K-bearing clays, shales also contain 4.5 wt.% of K-feldspar of nominal composition $KAlSi_3O_8$ or 0.04×10^{24} g. Potassium in illite represents 4.34 wt.% of the mineral and it represents 14.05 wt.% of the K-feldspar. The masses of ^{40}K in the illites and K-feldspars contained in the shales, using the $^{40}K/K$ atomic ratio of 1.167×10^{-4} (Steiger and Jäger, 1977), are

$$^{40}\text{K in illites in shales} = 1.167\times10^{-4}(5.376\times10^{23}\times0.0434)/39.1 = 6.963\times10^{16}\ \text{mol} \quad (25)$$

$$^{40}\text{K in K-feldspars in shales} = 1.167\times10^{-4}(4.32\times10^{22}\times0.1405)/39.1 = 1.812\times10^{16}\text{mol} \quad (26)$$

The mass half-age of shales and sandstones is about 600×10^6 years, close to the average mass half-age of all the preserved sediments (Garrels and Mackenzie, 1972). The $^{40}Ar/^{40}K$ ratio of 600-Ma-old minerals would be 0.0414, corresponding to the mass of ^{40}Ar in illites of 2.88×10^{15} mol and in feldspars of 0.75×10^{15} mol. The mass of ^{40}Ar of an order of magnitude of 10^{15}mol represents only a small fraction of ^{40}Ar in the present-day atmosphere containing 1.64×10^{18} mol. From the discussion accompanying Figure 33, 50% is within the range of the estimated ^{40}Ar mass fractions that escaped from sediments in 300 Ma by diffusional loss from clays and dissolution of some of the feldspars. Then a mean flux of ^{40}Ar from sediments to the atmosphere would be $6\times10^6\,mol\,yr^{-1}$.

A somewhat higher estimate of the ^{40}Ar flux from the sedimentary crust is obtained from a non-steady-state model of a continuous ^{40}Ar production from ^{40}K in clays and its release by a first-order flux that is represented by a release rate parameter ε (yr^{-1}). This parameter is obtained from a solution of a mass-balance equation between the production of ^{40}Ar in particles and its release (Lerman and Clauer, 2005):

$$d(^{40}Ar)/dt = \lambda_a^{40}K_0e^{-\lambda t} - \varepsilon(^{40}Ar) \quad (27)$$

where (^{40}Ar) and ^{40}K are the nuclide concentrations in $mol\,g^{-1}$. A number of thick sediment sequences (250–1,800 m), where the $^{40}Ar/^{40}K$ ratio and apparent age are nearly constant with depth, occur in the Late Paleozoic (Permian) of North Germany, the Mesozoic sedimentary sequence in the North Sea, the Tertiary of the US Gulf Coast, and the Neogene of Borneo. From these, the calculated ^{40}Ar release rate parameter ε is from 3×10^{-9} to $3\times10^{-8}\,yr^{-1}$.

Using 50% of the ^{40}Ar mass in shales, as given above, and a geometric mean of the range of the rate parameter ε ($\sim9\times10^{-9}\,yr^{-1}$), the ^{40}Ar flux from sediments is $17\times10^6\,mol\,yr^{-1}$. Both of the estimates of the ^{40}Ar flux, 6 and $17\times10^6\,mol\,yr^{-1}$, are smaller than the flux from the continental crust, estimated by Allègre *et al.* (1986) as $34\times10^6\,mol\,yr^{-1}$. It can be expected that the ^{40}Ar flux from sediments would be smaller than the flux from the continental crust because the mass of the crust is much greater and its temperature as a whole higher, whereas their mean K concentrations are comparable, 1.78 and 1.50 wt.%, respectively (Li, 2000). The mass of the shales (0.96×10^{24} g), as the main sedimentary source of the ^{40}Ar release to the atmosphere, is also severalfold smaller than the mass of only the upper continental crust, 9×10^{24} g, extending to a depth of 19 km (Wedepohl, 1995) and over the surface area of the present-day continents and continental shelves (177×10^6 km^2). The K concentration in shales, 2.51 wt.% K, is comparable to that in the upper continental crust, 2.32–2.60 wt.%, where K occurs primarily in feldspars. In our preceding estimate of the ^{40}Ar flux from shales to the atmosphere, if a fraction of the ^{40}Ar escaped is smaller than the assumed 50% of its mass in 600-Ma-old minerals, the flux would be correspondingly smaller than its values of 18% to 50% of the estimate of Allègre *et al.* (1986) for the continental crust.

ACKNOWLEDGMENTS

For discussions of the various subjects of this chapter, making available to us literature references, unpublished data, and permissions to cite them, we are grateful to an anonymous reviewer, Thomas D. Bullen (U.S. Geological Survey, Menlo Park, CA), Donald E. Canfield (Syddansk Universitet, Odense), Roberto Gonfiantini and Sonia Tonarini (Instituto di Geoscienze e Georisorse, CNR, Pisa), Michael Guidry and Fred T. Mackenzie (University of Hawaii, Honolulu), John M. Hayes (Woods Hole Oceanographic Institution, Woods Hole, MA), Yehoshua Kolodny (The Hebrew University, Jerusalem), Damien Lemarchand (Centre de Géochimie de la Surface, CNRS and Université Louis Pasteur, Strasbourg), Adina Paytan (Stanford University), Karl K. Turekian (Yale University), Jaán Veizer (University of Ottawa and Ruhr-Universität Bochum), and Lynda B. Williams (Arizona State University, Tempe).

This research was supported by NSF grant EAR-0223509 and the Arthur L. Howland Fund of the Department of Geological Sciences, Northwestern University. This is publication No. 206.301 of EOST (Ecole et Observatoire des Sciences de la Terre), Université Louis Pasteur, Strasbourg.

REFERENCES

Allègre C. J., Staudacher T., and Sarda P. (1986) Rare gas systematics: formation of the atmosphere, evolution and structure of the Earth's mantle. *Earth Planet. Sci. Lett.* **81**, 127–150.

Alley R. B. and Cuffey K. M. (2001) Oxygen- and hydrogen-isotopic ratios of water in precipitation: beyond paleothermometry. In *Stable Isotope Geochemistry* (eds. J. W. Valley and D. R. Cole). *Rev. Mineral. Geochem.*, Mineral. Soc. America, Washington, DC, vol. 43, pp. 527–553.

Altabet M. A., Francçois R., Murray D. W., and Prell W. L. (1995) Climate-related variations in denitrification in the Arabian Sea from sediment ^{15}N/^{14}N ratios. *Nature* **373**, 506–509.

Arvidson R. S., Mackenzie F. T., and Guidry M. (2006) MAGic: a Phanerozoic model for the geochemical cycling of major rock-forming components. *Am. J. Sci.* **306**, 135–190.

Aston F. W. (1922) *Mass spectra and isotopes. Nobel Lectures, Chemistry 1922–1941, 1966.* Elsevier, Amsterdam, pp. 7–20.

Ayalon A. and Longstaffe F. (1988) Oxygen-isotope studies of diagenesis and porewater evolution in the western Canada sedimentary basin: evidence from the Upper Cretaceous basal Belly River sandstone, Alberta. *J. Sedim. Petrol.* **58**, 489–505.

Baertschi P. (1976) Absolute ^{18}O content of standard mean ocean water. *Earth Planet. Sci. Lett.* **31**, 341–344.

Bebout G. E., Cooper D. C., Bradley A. D., and Sadofsky S. J. (1999) Nitrogen isotope record of fluid rock interactions in the Skiddaw aureole and granite: English Lake district. *Am. Mineral.* **84**, 1495–1505.

Bender M. L. (1990) The δ^{18}O of dissolved O_2 in seawater: a unique tracer of circulation and respiration in the deep sea. *J. Geophys. Res.* **95**(C12), 22243–22252.

Bender M., Sowers T., and Labeyrie L. (1994) The Dole effect and its variations during the last 130,000 years as measured in the Vostok ice core. *Global Biogeochem. Cycles* **8**, 363–376.

Benson B. and Krause D., Jr. (1984) The concentration and isotopic fractionation of oxygen dissolved in fresh water and seawater at equilibrium with the atmosphere. *Limnol. Oceanogr.* **29**, 620–632.

Berner E. K. and Berner R. A. (1996) *Global Environment; Water, Air, and Geochemical Cycles.* Prentice Hall, Upper Saddle River, NJ.

Berner R. A. (2006) Carbon, sulfur and O_2 across the Permian-Triassic boundary. *J. Geochem. Explor.* **88**, 416–418.

Berner R. A. and Kothavala Z. (2001) GEOCARB III: a revised model of atmospheric CO_2 over Phanerozoic time. *Am. J. Sci.* **301**, 182–204.

Borowski W. S. and Paull C. K. (2000) Data report; nitrogen isotope composition of pore-water ammonium, Blake Ridge, Site 997. In *Proceedings of the Ocean Drilling Program* (eds. C. M. Miller and R. Reigel). Ocean Drilling Program, Texas A&M University, College Station, TX, pp. 171–172.

Bottinga Y. (1968) Calculation of fractionation factors for carbon and oxygen exchange in the system calcite-carbon dioxide-water. *J. Phys. Chem.* **72**, 800–808.

Bowen G. J. and Wilkinson B. (2002) Spatial distribution of δ^{18}O in meteoric precipitation. *Geology* **30**, 315–318.

Boyd S. R., Hall A., and Pillinger C. T. (1993) The measurement of δ^{15}N in crustal rocks by static vacuum mass spectrometry: application to the origin of the ammonium in the Cornubian batholith, southwest England. *Geochim. Cosmochim. Acta* **57**, 1339–1347.

Bräuer K. and Hahne K. (2005) Methodical aspects of the ^{15}N-analysis of Precambrian and Paleozoic sediments rich in organic matter. *Chem. Geol.* **218**, 361–368.

Broecker W. S. (2002) *The Glacial World According to Wally.* Eldigio Press, Lamont-Doherty Earth Observatory, Columbia University, Palisades, NY.

Brumsack H. J. and Zuleger E. (1992) Boron and boron isotopes in pore waters from ODP Leg 127, Sea of Japan. *Earth Planet. Sci. Lett.* **113**, 427–433.

Bullen T. D., Fitzpatrick J. A., White A. F., Schulz M. S., and Vivit D. V. (2004) Calcium stable isotope evidence for three soil calcium pools at a granitoid sequence. In *Proceedings of the 11th International Symposium on Water-Rock Interaction* (eds. R. B. Wanty and R. R. Seal II). Taylor & Francis, London, vol. 1, pp. 813–817.

Canfield D. E. (2001) A new model for Proterozoic ocean chemistry. *Nature* **396**, 450–453.

Canfield D. E. (2004) The evolution of the Earth surface sulfur reservoir. *Am. J. Sci.* **304**, 839–861.

Canfield, D. E. (2005) Data and references for isotopic composition of sulfides in ancient sediments. Personal communication.
Cartigny P. (2005) Stable isotopes and the origin of diamond. *Elements* **1**, 79–84.
Chacko T., Cole D. R., and Horita J. (2001) Equilibrium oxygen, hydrogen and carbon isotope fractionation factors applicable to geologic systems. In *Stable Isotope Geochemistry* (eds. J. W. Valley and D. R. Cole). *Rev. Mineral. Geochem.*, Mineral. Soc. America, Washington, DC, vol. 43, pp. 1–81.
Chaudhuri S., Srodon J., and Clauer N. (1999) K-Ar dating of illitic fractions of Estonian "Blue Clay" treated with alkylammonium cations. *Clays Clay Minerals* **47**, 96–102.
Chaussidon M., Robert F., Mangin D., Hanon P., and Rose E. (1997) Analytical procedures for the measurement of boron isotope compositions by ion microprobe in meteorites and mantle rocks. *Geostandards Newslett.* **21**, 7–17.
Chetelat B., Gaillardet J., Freydier R., and Négrel P. (2005) Boron isotopes in precipitation: experimental constraints and field evidence from French Guiana. *Earth Planet Sci. Lett.* **235**, 16–30.
Chicarelli M. I., Hayes J. M., Popp B. N., Eckardt C. B., and Maxwell J. R. (1993) Carbon and nitrogen isotopic compositions of alkyl porphyrins from the Triassic Serapiano oil shale. *Geochim. Cosmochim. Acta* **57**, 1307–1311.
Church T. M., Sommerfield C. K., Velinsky D. J., Point D., Benoit C., Amouroux D., Plaa D., and Donard O. F. X. (2006) Marsh sediments as records of sedimentation, eutrophication and metal pollution in the urban Delaware Estuary. *Mar. Chem.*, doi:10.1016/j.marchem.2005,10.026.
Clark I. D. and Fritz P. (1997) *Environmental Isotopes in Hydrology*. Lewis Publishers, New York.
Clauer N. and Chaudhuri S. (1995) *Clays in Crustal Environments. Isotope Dating and Tracing*. Springer, New York.
Clauer N. and Chaudhuri S. (1998) Isotopic dating of very low-grade metasedimentary and metavolcanic rocks: techniques and methods. In *Low-Grade Metamorphism* (eds. M. Frey and D. Robinson). Blackwell Science, Cambridge, pp. 202–226.
Clauer N., Liewig N., Pierret M. C., and Toulkeridis T. (2003) Crystallization conditions of fundamental particles from mixed-layers illite-smectite of bentonites based on isotopic data (K-Ar, Rb-Sr, and $\delta^{18}O$). *Clays Clay Minerals* **51**, 664–674.
Clauer N., O'Neil J. R., and Furlan S. (1995) Clay minerals as records of temperature conditions and duration of thermal anomalies in the Paris Basin, France. *Clay Minerals* **30**, 1–13.
Clauer N., Rinckenbach T., Weber F., Sommer F., Chaudhuri S., and O'Neil J. R. (1999) Diagenetic evolution of clay minerals in oil-bearing Neogene sandstones and associated shales from Mahakam Delta Basin (Kalimantan, Indonesia). *Am. Assoc. Petrol. Geol. Bull.* **83**, 62–87.
Clauer N., Srodon J., Francu J., and Sucha V. (1997) K-Ar dating of illite fundamental particles separated from illite/smectite. *Clay Minerals* **32**, 181–196.
Clauer N., Zwingmann H., and Chaudhuri S. (1996) Isotope (K-Ar and oxygen) constraints on the extent and importance of the Liassic hydrothermal activity in western Europe. *Clay Minerals* **31**, 301–318.
Coplen T. B., Krouse H. R., and Böhlke J. K. (1992) Reporting of nitrogen-isotope abundances. *Pure Appl. Chem.* **64**, 907–908.
Craig H. (1961) Isotopic variation in meteoric waters. *Science* **133**, 1702–1703.
Craig H. and Gordon L. I. (1965) Deuterium and oxygen 18 variations in the ocean and marine atmosphere. In *Stable Isotopes in Oceanographic Studies and Paleotemperatures* (ed. E. Tongiorgi). Consiglio Nazionale delle Ricerche, Laboratorio di Geologia Nucleare, Pisa, pp. 9–130.
Dalrymple G. B. and Lanphere M. A. (1969) *Potassium–Argon Dating*. Freeman, San Francisco.
Dansgaard W. (1961) The isotopic composition of natural waters, with special reference to the Greenland ice cap. *Meddelelser om Grønland* **165**(2), 1–120.
Dansgaard W. (1964) Stable isotopes in precipitation. *Tellus* **16**, 436–468.
Dauwe B., Middelburg J. J., Hermann P. M. J., and Heip C. H. R. (1999) Linking diagenetic alteration of amino acids and bulk organic matter reactivity. *Limnol. Oceanogr.* **44**, 1809–1814.
Deer W. A., Howie R. A., and Zussman J. (1962a) *Rock-Forming Minerals, vol. 3, Sheet-Silicates*. Wiley, New York.
Deer W. A., Howie R. A., and Zussman J. (1962b) *Rock-Forming Minerals, vol. 5, Non-Silicates*. Wiley, New York.
De La Rocha C. L. and DePaolo D. J. (2000) Isotopic evidence for variations in the marine calcium cycle over the Cenozoic. *Science* **289**, 1176–1178.
DePaolo D. J. (2004) Calcium isotopic variations produced by biological, kinetic, radiogenic and nucleosynthetic processes. In *Geochemistry of Non-Traditional Stable Isotopes* (eds. C. M. Johnson, D. L. Beard, and F. Albarède). *Rev. Mineral. Geochem.*, Mineral. Soc. America, Washington, DC, vol. 55, pp. 255–288.
EPICA Community Members. (2004) Eight glacial cycles from an Antartic ice core. *Nature* **429**, 623–628.
Epstein S. (1959) The variations of $^{18}O/^{16}O$ ratio in nature and some geologic implications. In *Researches in Geochemistry* (ed. P. H. Abelson). Wiley, New York, vol. 1, pp. 217–240.
Epstein S. and Mayeda T. (1953) Variation of ^{18}O content of waters from natural sources. *Geochim. Cosmochim. Acta* **4**, 213–224.
Fallick A. E., Macaulay C. I., and Haszeldine R. S. (1993) Implications of linearly correlated oxygen and hydrogen isotopic compositions for kaolinite and illite in Magnus Sandstone, North Sea. *Clays Clay Minerals* **2**, 121–137.
Farquhar J., Johnston D. T., Wing B. A., Habicht K. S., Can D. E., Airieau S., and Thiemens M. H. (2003) Multiple sulphur isotopic interpretations of biosynthetic pathways; implications for biological signatures in the sulphur isotope record. *Geobiology* **1**, 27–36.
Faure G. (1986) *Principles of Isotope Geology*, 2nd edn. Wiley, New York.
Fietzke J., Eisenhauer A., Gussone N., Bock B., Liebetrau V., Nägler T. F., Spero H. J., Bijma J., and Dullo C. (2004) Direct measurement of $^{44}Ca/^{40}Ca$ ratios by MC-ICP-MS using the cool-plasma-technique. *Chem. Geol.* **206**, 11–20.
Fischer T. P., Takahata N., Sano Y., Sumino H., and Hilton D. R. (2005) Nitrogen isotopes of the mantle: insights from mineral separates. *Geophys. Res. Lett.* **32**, L11305, doi:10.1029/2005GL022792.
Fletcher I. R., McNaughton R. T., Pidgeon R. T., and Rosman K. J. R. (1997) Sequential closure of K-Ca and Rb-Sr isotopic systems in Archean micas. *Chem. Geol.* **138**, 289–301.
Freundenthal T., Wagner T., Wezhofer F., Zabel M., and Wefer G. (2001) Early diagenesis of organic matter from sediments of the eastern subtropical Atlantic: evidence from stable nitrogen and carbon isotopes. *Geochim. Cosmochim. Acta* **65**, 1795–1808.
Friedman I. (1953) Deuterium content of natural waters and other substances. *Geochim. Cosmochim. Acta* **4**, 89–103.
Friedman I. and O'Neil J. R. (1977) Compilation of stable isotope fractionation factors of geochemical interest. In *Data of Geochemistry*, 6th edn. (ed. M. Fleischer). US Geol. Survey Prof. Pap. 440–KK.
Garrels R. M. and Lerman A. (1981) Phanerozoic cycles of sedimentary sulfur and carbon. *Proc. Natl. Acad. Sci. USA* **78**(8), 4652–4656.
Garrels R. M. and Lerman A. (1984) Coupling of the sedimentary sulfur and carbon cycle—an improved model. *Am. J. Sci.* **284**, 980–1007.
Garrels R. M. and Mackenzie F. T. (1972) A quantitative model for the sedimentary rock cycle. *Mar. Chem.* **1**, 27–41.

Gaye-Haake B., Lahajnar N., Emeis K.-Ch., Unger T., Rixen T., Suthhof A., Ramaswamy V., Schulz H., Paropkari A. L., Guptha M. V. S., and Ittekkot V. (2005) Stable nitrogen isotopic ratios of sinking particles and sediments from the northern Indian Ocean. *Mar. Chem.* **96**, 225–243.

Giral-Kacmarcík S., Savin S. M., Nahon D. B., Girard J.-P., Lucas Y., and Abel L. J. (1998) Oxygen isotope geochemistry of kaolinite in laterite-forming processes, Manaus, Amazonas, Brazil. *Geochim. Cosmochim. Acta* **62**, 1865–1879.

Girard J.-P. and Barnes D. A. (1995) Illitization and paleothermal regimes in the Middle Ordovician St. Peter Sandstone, Central Michigan Basin, United States: K-Ar, oxygen isotopes, and fluid inclusion data. *Am. Assoc. Petrol. Geol. Bull.* **79**, 49–69.

Girard J.-P., Freyssinet P., and Chazot G. (2000) Unraveling climate changes from intraprofile variation in oxygen and hydrogen isotopic composition of goethite and kaolinite in laterites: an integrated study from Yaou, French Guiana. *Geochim. Cosmochim. Acta* **64**, 409–426.

Girard J.-P., Savin S. M., and Aronson J. L. (1989) Diagenesis of the lower Cretaceous arkoses of the Angola margin: petrologic, K-Ar dating and $^{18}O/^{16}O$ evidence. *J. Sedim. Petrol.* **59**, 519–538.

Given R. K. and Wilkinson B. H. (1987) Dolomite abundance and stratigraphic age: constraints on rates and mechanisms of Phanerozoic dolostone formation. *J. Sedim. Petrol.* **57**, 1068–1079.

Glasmann J. R., Larter S., Briedis N. A., and Lundegard P. D. (1989) Shale diagenesis in the Bergen area, North Sea. *Clays Clay Minerals* **37**, 97–112.

Gonfiantini R. (1986) Environmental isotopes in lake studies. In *Handbook of Environmental Isotope Geochemistry* (eds. P. Fritz and J. Ch. Fontes). Elsevier, Amsterdam, vol. 2, pp. 113–168.

Gonfiantini R. (2005) Personal communication.

Gonfiantini R., Roche M.-A., Olivry J.-C., Fontes J.-C., and Zuppi G. M. (2001) The altitude effect on the isotopic composition of tropical rains. *Chem. Geol.* **181**, 147–167.

Gormly J. R. and Spalding R. F. (1979) Sources and concentrations of nitrate-nitrogen in ground water of the Central Platte Region, Nebraska. *Ground Water* **17**, 291–301.

Gradstein F. M., Ogg J. G., and Smith A. G. (2004) *A Geologic Time Scale 2004*. Cambridge University Press, New York.

Gussone N., Böhm F., Eisenhauer A., Dietzel M., Heuser A., Teichert B. M. A., Reitner J., Wörheide G., and Dullo W. C. (2005) Calcium isotope fractionation in calcite and aragonite. *Geochim. Cosmochim. Acta* **69**, 4485–4494.

Gussone N., Eisenhauer A., Tiedemann R., Haug G. H., Heuser A., Bock B., Nägler T. F., and Müller A. (2004) Reconstruction of Caribbean sea surface temperature and salinity fluctuations in response to the Pliocene closure of the Central American gateway and radiative forcing, using $\delta^{44/40}Ca$, $\delta^{18}O$ and Mg/Ca ratios. *Earth Planet. Sci. Lett.* **227**, 201–214.

Guy R. D., Fogel M. L., and Berry J. A. (1993) Photosynthetic fractionation of the stable isotopes of oxygen and carbon. *Plant Physiol* **101**, 37–47.

Halter G., Sheppard S. M. F., Weber F., Clauer N., and Pagel M. (1987) Radiation-related retrograde hydrogen isotope and K-Ar exchange in clay minerals. *Nature* **330**, 638–641.

Harland W. B., Armstrong R. L., Cox A. V., Craig L. E., Smith A. G., and Smith D. G. (1990) *A Geologic Time Scale 1989*. Cambridge University Press, Cambridge.

Hastings M. G., Sigman D. M., and Lipschultz F. (2003) Isotopic evidence for source changes of nitrate in rain at Bermuda. *J. Geophys. Res.* **108**(D24), 4790, doi:10.1029/2003JD003789.

Hastings M. G., Steig E. J., and Sigman D. M. (2004) Seasonal variations in N and O isotopes of nitrate in snow at Summit, Greenland: implications for the study of nitrate in snow and ice cores. *J. Geophys. Res.* **109**, D20306, doi: 1029/2004JD004991.

Hayes J. M., Strauss H., and Kaufman A. J. (1999) The abundance of ^{13}C in marine organic matter and isotopic fractionation in the global biogeochemical cycle of carbon during the past 800 Ma. *Chem. Geol.* **161**, 103–125.

Hedges J. I., Baldock J. A., Gélinas Y., Lee C., Peterson M. L., and Wakeham S. G. (2002) The biochemical and elemental compositions of marine plankton: a NMR perspective. *Mar. Chem.* **78**, 47–63.

Heumann K. G. and Zeininger H. (1985) Boron trace determination in metals and alloys by isotope dilution mass spectrometry with negative thermal ionization. *Internat. J. Mass Spectr. Ion Process.* **67**, 237–252.

Heuser A., Eisenhauer A., Böhm F., Wallmann K., Gussone N., Pearson P. N., Nägler T. F., and Dullo W.-C. (2005) Calcium isotope ($\delta^{44/40}Ca$) variations of Neogene planktonic foraminifera. *Paleoceanography* **20**, PA2013, doi:10.1029/2004PA001048.

Hippler D., Schmitt A. D., Gussine N., Heusser A., Stille P., Eisenhauer A., and Nägler T. F. (2003) Ca isotopic composition of various reference materials and seawater. *Geostandards Newslett* **27**, 13–19.

Hoffman P. F., Kaufman A. J., Halverson G. P., and Schrag D. P. (1998) A Neoproterozoic Snowball Earth. *Science* **281**, 1342–1346.

Hoffmann G., Cuntz M., Weber C., Ciais P., Friedlingstein P., Heimann M., Jouzel J., Kaduk J., Maier-Reimer E., Seibt U., and Six K. (2004) A model of the Earth's Dole effect. *Global Biogeochem. Cycles* **18**, GB1008, 15pp.

Holden N. E. (2002) Table of the isotopes. In *Handbook of Chemistry and Physics*, 85th edn., 2004–2005 (ed. D. R. Lide). CRC Press, Boca Raton, FL, pp. 11-50–11-201.

Holland H. D. (1997) Enhanced: evidence for life on Earth more than 3850 million years ago. *Science* **275**, 38–39.

Holser W. T., Schidlowski M., Mackenzie F. T., and Maynard J. B. (1988) Geochemical cycles of carbon and sulfur. In *Chemical Cycles in the Evolution of the Earth* (eds. C. B. Gregor, R. M. Garrels, F. T. Mackenzie, and J. B. Maynard). Wiley, New York, pp. 105–173.

Hönisch B. and Hemming N. G. (2005) Surface ocean pH response to variations in pCO_2 through two full glacial cycles. *Earth Planet. Sci. Lett.* **236**, 305–314.

Hönisch B., Hemming N. G., Grottoli A. G., Amat A., Hanson G. N., and Bijma J. (2004) Assessing scleractinian corals as recorders for paleo-pH: empirical calibration and vital effects. *Geochim. Cosmochim. Acta* **68**, 3675–3685.

Horita J. and Welosowski D. J. (1994) Liquid-vapor fractionation of oxygen and hydrogen isotopes of water from the freezing to the critical temperature. *Geochim. Cosmochim. Acta* **58**, 3425–3437.

Hunter N. W. and Roach R. (1993) Frederick Soddy. In *Nobel Laureates in Chemistry* (ed. L. K. James). American Chemical Society and Chemical Heritage Foundation, Washington, DC, pp. 134–139.

Hunziker J. C., Frey M., Clauer N., Dallmeyer R. D., Friedrichsen H., Flehmig W., Hochstrasser K., Rogwiller P., and Schwander H. (1986) The evolution from illite to muscovite: mineralogical and isotopic data from Glarus Alps, Switzerland. *Contrib. Mineral. Petrol.* **92**, 157–180.

Hurtgen M. T., Halverson G. P., Arthur M. A., and Hofman P. F. (2006) Sulfur cycling in the aftermath of a 635-Ma snowball glaciation: evidence for a syn-glacial sulfidic deep ocean. *Earth Planet. Sci. Lett.* **245**, 551–570.

Imbrie J., Hays J. D., Martinson D. G., McIntyre A., Mix A. C., Morley J. J., Pisias N. G., Prell W. L., and Shackleton N. J. (1984) The orbital theory of Pleistocene climate: support from a revised chronology of the marine $\delta^{18}O$ record. In *Milankovitch and Climate, Part I* (eds. A. Berger, J. Imbrie, J. D. Hays, and G. K. B. Saltzman). Reidel, Dordrecht, pp. 269–305.

Imbus S. W., Macko S. A., Elmore R. D., and Engel M. H. (1992) Stable isotope (CSN) and molecular studies on the

Precambrian Nonesuch Shale (Wisconsin-Michigan, USA): evidence for differential preservation rates, depositional environment and hydrothermal influence. *Chem. Geol.* **101**, 255–281.

Ishikawa T. and Nakamura E. (1993) Boron isotope systematics of marine sediments. *Earth Planet. Sci. Lett.* **117**, 567–580.

James A. T. and Baker D. R. (1976) Oxygen isotope exchange between illite and water at 22 °C. *Geochim. Cosmochim. Acta* **40**, 235–239.

Junk G. and Svec H. (1958) The absolute abundance of the nitrogen isotopes in the atmosphere and compressed gas from various sources. *Geochim. Cosmochim. Acta* **14**, 234–243.

Kampschulte A. (2001) *Schwefelisotopenuntersuchungen an strukturell substituierten Sulfaten in marinen Karbonaten des Phanerozoikums: Implikationen für die geochemische Evolution des Meerwassers und die Korrelation verschiedener Stoffkreisläufe*. PhD Thesis, Fakultät für Geowissenschaften, Ruhr-Universität Bochum, xii+152pp.

Kampschulte A. and Strauss H. (2004) The sulfur isotopic evolution of Phanerozoic seawater based on the analysis of structurally substituted sulfate in carbonates. *Chem. Geol.* **204**, 255–286.

Kasemann S. A., Hawkesworth C. J., Prave A. R., Fallick A. E., and Pearson P. N. (2005) Boron and calcium isotope composition in Neoproterozoic carbonate rocks from Namibia: evidence for extreme environmental change. *Earth Planet. Sci. Lett.* **231**, 73–86.

Katz M., Wright J. D., Miller K. G., Cramer B. S., Fennel K., and Falkowski P. G. (2005) Biological overprint of the geological carbon cycle. *Mar. Geol.* **217**, 323–338.

Kendall C. (1998) Tracing nitrogen sources and cycling in catchments. In *Isotope Tracers in Catchment Hydrology* (eds. C. Kendall and J. J. McDonnell). Elsevier, Amsterdam, pp. 519–576.

Kendall C. (2004) Tracing sources of agricultural N using isotopic techniques: the state of the science. US Geol. Survey, Water Res. Div, Menlo Park, CA. (http://www.epa.gov/osp/presentations/afo/13_Kendall.ppt.)

Kendall C. and Coplen T. B. (2001) Distribution of oxygen-18 and deuterium in river waters across the United States. *Hydrol. Process.* **15**, 1363–1393.

Kohl D. H., Shearer G. B., and Commoner B. (1971) Fertilizer nitrogen: contribution to nitrate in surface water in a corn belt watershed. *Science* **174**, 1331–1334.

Kotzer T. G. and Kyser T. K. (1995) Petrogenesis of the Proterozoic Athabasca Basin, northern Saskatchewan, Canada, and its relation to diagenesis, hydrothermal uranium mineralisation and paleohydrology. *Chem. Geol.* **120**, 45–89.

Kreitler C. W., Ragone S. E., and Katz B. G. (1978) $^{15}N/^{14}N$ ratios of ground-water nitrate, Long Island, NY. *Ground Water* **16**, 404–409.

Krohn M. D., Kendall C., Evans J. R., and Fries T. L. (1993) Relations of ammonium minerals at several hydrothermal systems in the western US. *J. Volcanol. Geotherm. Res.* **56**, 401–413.

Kroopnick P. and Craig H. (1972) Atmospheric oxygen: isotopic composition and solubility fractionation. *Science* **175**, 54–55.

Kump L. R. (1991) Interpreting carbon-isotope excursions; Strangelove Oceans. *Geology* **19**, 299–302.

Land L. S. and Macpherson G. L. (1992) Origin of saline formation waters, Cenozoic section, Gulf of Mexico Sedimentary Basin. *Am. Assoc. Petrol. Geol. Bull.* **76**, 1344–1362.

Lawrence J. R. and Rashkes Meaux J. (1993) The stable isotopic composition of ancient kaolinites of North America. In *Climate Change in Continental Isotopic Records* (eds. P. K. Swart, K. C. Lohman, J. McKenzie, and S. Savin). *Geophys. Mon.*, Am. Geophys. Union., Washington, DC, vol. 78, pp. 249–262.

Lawrence J. R. and Taylor H. P., Jr. (1971) Deuterium and oxygen-18 correlation: clay minerals and hydroxides in Quaternary soils compared to meteoric water. *Geochim. Cosmochim. Acta* **35**, 993–1004.

Lawrence J. R. and Taylor H. P., Jr. (1972) Hydrogen and oxygen isotope systematics in weathering profiles. *Geochim. Cosmochim. Acta* **36**, 1377–1394.

Leeman W. P., Vocke R. D., Jr., Bearly E. S., and Paulsen P. J. (1991) Precise boron isotopic analysis of aqueous samples: ion exchange extraction and mass spectrometry. *Geochim. Cosmochim. Acta* **55**, 3901–3907.

Leeman W. P., Vocke R. D., and McKibben M. A. (1992) Boron isotopic fractionations between coexisting vapor and liquid in natural geothermal systems. In *Proceedings of the 7th International Symposium on Water-Rock Interaction* (eds. Y. K. Kharaka and A. S. Maest). Balkema, Rotterdam, pp. 1007–1010.

Lehmann M. F., Sigman D. M., and Berelson W. M. (2004) Coupling the $^{15}N/^{14}N$ and $^{18}O/^{16}O$ of nitrate as a constraint on benthic nitrogen cycling. *Mar. Chem.* **88**, 1–20.

Lemarchand D. and Gaillardet J. (2006) Transient features of the erosion of shales in the Mackenzie basin (Canada), evidence from boron isotopes. *Earth Planet. Sci. Lett.* **245**, 174–189.

Lemarchand D., Gaillardet J., Lewin E., and Allègre C. J. (2000) The influence of rivers on marine boron isotopes and implications for reconstructing past ocean pH. *Nature* **408**, 951–954.

Lemarchand D., Gaillardet J., Lewin E., and Allègre C. J. (2002) Boron isotope systematics in large rivers: implications for the marine boron budget and paleo-pH reconstruction over the Cenozoic. *Chem. Geol.* **190**, 123–140.

Lemarchand D., Wasserburg G. J., and Papanastassiou D. A. (2004) Rate-controlled calcium isotope fractionation in synthetic calcite. *Geochim. Cosmochim. Acta* **68**, 4665–4678.

Lerman A. (1988) *Geochemical Processes—Water and Sediment Environments*, reprint edn. Krieger Publishing, Malabar, FL.

Lerman A. and Clauer N. (2005) Losses of radiogenic ^{40}Ar in sediment fine-clay size fractions of sediments. *Clays Clay Mineral* **53**(3), 233–248.

Lerman A., Mackenzie F. T., and Geiger R. J. (1989) Environmental chemical stress effects associated with carbon and phosphorus biogeochemical cycles. In *Ecotoxicology: Problems and Approaches* (eds. S. A. Levin, M. A. Harwell, J. R. Kelly, and K. D. Kimball). Springer, New York, chap. 12, pp. 315–350.

Lerman A., Mackenzie F. T., and Ver L. M. (2004) Coupling of the perturbed C-N-P cycles in industrial time. *Aquat. Geochem.* **10**, 3–32.

Létolle R. (1980) Nitrogen-15 in the natural environment. In *Handbook of Environmental Isotope Geochemistry* (eds. P. Fritz and J. Ch. Fontes). Elsevier, Amsterdam, vol. 1, pp. 407–434.

Li Y.-H. (2000) *A Compendium of Geochemistry*. Princeton University Press, Princeton, NJ.

Longstaffe F. K. and Ayalon A. (1987) Oxygen-isotope studies of clastic diagenesis in the Lower Cretaceous Viking Formation, Alberta: implications for the role of meteoric water. In *Diagenesis of Sedimentary Sequences* (ed. J. D. Marshall). Geol. Soc. Am. Spec. Publ., vol. 36, pp. 277–296.

Lorrain A., Savoye N., Chauvaud L., Paulet Y. M., and Naulet N. (2003) Decarbonation and preservation method for the analysis of organic C and N contents and stable isotope ratios of low-carbonated suspended particulate material. *Anal. Chim. Acta* **491**, 125–133.

Mariotti A., Lancelot C., and Billen G. (1984) Natural isotopic composition of nitrogen as a tracer of origin for suspended organic matter in Scheldt estuary. *Geochim. Cosmochim. Acta* **48**, 549–555.

Mariotti A., Pierre D., Vedy J. C., Bruckert S., and Guillemot J. (1980) The abundance of natural nitrogen-15 in the organic matter of soils along an altitudinal gradient (Chablais, Haute Savoie, France). *Catena* **7**, 293–300.

Marshall B. D. and DePaolo D. J. (1982) Precise age determinations and petrogenic studies using the K-Ca method. *Geochim. Cosmochim. Acta* **46**, 2537–2545.

Marshall B. D. and DePaolo D. J. (1989) Calcium isotopes in igneous rocks and the origin of granite. *Geochim. Cosmochim. Acta* **53**, 917–922.

Marshall B. D., Woodard H. H., and DePaolo D. J. (1986) K-Ca-Ar systematics of authigenic sanidine from Wakau, Wisconsin, and the diffusivity of argon. *Geology* **14**, 936–938.

McMahon P. B., Böhlke J. K., and Bruce B. R. (1999) Denitrification in marine shales in northeastern Colorado. *Water Resour. Res.* **35**, 1629–1642.

Milliman J. D. (1993) Production and accumulation of calcium carbonate in the ocean: budget of a nonsteady state. *Global Biogeochem. Cycles* **7**, 927–957.

Mojzsis S. J., Arrhenius G., McKeegan K. D., Harrison T. M., Nutman A. P., and Friend C. R. L. (1996) Evidence for life on Earth by 3800 Myr. *Nature* **384**, 55–59.

Mojzsis S. J., Coath C. D., Greenwood J. P., McKeegan K. D., and Harrison T. M. (2003) Mass-independent isotope effects in Archean (2.5 to 3.8 Ga) sedimentary sulfides determined by ion microprobe analysis. *Geochim. Cosmochim. Acta* **67**, 1635–1658.

Moldovanyi E. P. and Walter L. M. (1992) Regional trends in water chemistry, Smackover formation, Southwest Arkansas: geochemical and physical controls. *Am. Assoc. Petrol. Geol. Bull.* **76**, 864–894.

Moldovanyi E. P., Walter L. M., and Land L. S. (1993) Strontium, boron, oxygen, and hydrogen isotope geochemistry of brines from basal strata of the Gulf Coast sedimentary basin, USA. *Geochim. Cosmochim. Acta* **57**, 2083–2099.

Mook W. G. and Tan F. C. (1991) Stable carbon isotopes in rivers and estuaries. In *Biogeochemistry of Major World Rivers* (eds. E. T. Degens, S. Kempe, and J. E. Richey). SCOPE 42 (Scientific Committee On Problems of the Environment), Wiley, Chichester, UK, chap. 11.

Morris E. J., Paytan A., and Bullen T. D. (2003) Marine barite: a potential recorder of seawater calcium isotope composition. *Geol. Soc. Am., 2003 Seattle Ann. Meet. Abs* **35**, 257.

Nadeau P. H., Wilson M. J., McHardy W. J., and Tait J. M. (1984) Interstratified clays as fundamental particles. *Science* **225**, 923–925.

Nadelhoffer K. J. and Fry B. (1988) Controls on natural nitrogen-15 and carbon-13 abundances in forest soil organic matter. *J. Soil Sci. Soc. Am.* **52**, 1633–1640.

Nägler T. F., Eisenhauer A., Müller A., Hemleben C., and Kramers J. (2000) δ^{44}Ca-temperature calibration on fossil and cultured *Globigerinoides sacculifer*: new tool for reconstruction of past sea surface temperatures. *Geochem. Geophys. Geosyst.* **1**(9), doi:10.1029/2000GC000091.

Nägler T. F. and Villa I. M. (2000) In pursuit of the ^{40}K branching ratios: K-Ca and ^{39}Ar–^{40}Ar dating of gem silicates. *Chem. Geol.* **169**, 5–16.

Nelson D. R. and McCulloch M. T. (1989) Petrogenic applications of the ^{40}K–^{40}Ca radiogenic decay scheme—a reconnaissance study. *Chem. Geol.* **79**, 275–293.

Oi T., Nomura M., Musashi M., Ossaka Y., Okamoto M., and Kakihana H. (1989) Boron isotopic compositions of some boron minerals. *Geochim. Cosmochim. Acta* **53**, 3189–3197.

O'Leary M. H. (1988) Carbon isotopes in photosynthesis. *BioScience* **38**, 328–335.

O'Neil J. R. (1986) Theoretical and experimental aspects of isotopic fractionation. In *Stable Isotopes in High Temperature Geological Processes* (eds. J. W. Valley, H. P. Taylor Jr., and J. R. O'Neil). *Rev. Mineral.*, Mineral. Soc. America, Washington, DC, vol. 16, pp. 1–40.

O'Neil J. R., Clayton R. N., and Mayeda T. K. (1969) Oxygen isotope fractionation in divalent metal carbonates. *J. Chem. Phys.* **51**, 5547–5548.

O'Neil J. R. and Epstein S. (1966) Oxygen isotope fractionation in the system dolomite-calcite-carbon dioxide. *Science* **152**, 198–201.

O'Neil J. R. and Kharaka Y. K. (1976) Hydrogen and oxygen isotope exchange reactions between clay minerals and water. *Geochim. Cosmochim. Acta* **40**, 241–246.

Ono S., Wing B., Johnston D., Farquhar J., and Rumble D. (2006) Mass-dependent fractionation of quadruple stable sulfur isotope system as a new tracer of sulfur biogeochemical cycles. *Geochim. Cosmochim. Acta* **70**, 2238–2252.

Pagani M., Lemarchand D., Spivack A., and Gaillardet J. (2005) A critical evaluation of the boron isotope-pH proxy: the accuracy of ancient ocean pH estimates. *Geochim. Cosmochim. Acta* **69**, 953–961.

Palmer M. R. and Pearson M. N. (2003) A 23,000-year record of surface water pH and P_{CO_2} in the Western Equatorial Pacific Ocean. *Science* **300**, 480–482.

Palmer M. R. and Sturchio N. C. (1990) The boron isotope systematics of the Yellowstone National Park (Wyoming) hydrothermal system: a reconnaissance. *Geochim. Cosmochim. Acta* **54**, 2811–2815.

Paytan M., Kastner M., Campbell D., and Thiemens M. H. (1998) Sulfur isotopic composition of Cenozoic seawater sulfate. *Science* **282**, 1459–1462.

Paytan M., Kastner M., Campbell D., and Thiemens M. H. (2004) Seawater sulfur isotope fluctuations in the Cretaceous. *Science* **304**, 1663–1665.

Pearson M. N. and Palmer M. R. (1999) Middle Eocene seawater pH and atmospheric carbon dioxide concentrations. *Science* **284**, 1824–1826.

Peterson B. J. and Fry B. (1987) Stable isotopes in ecosystem studies. *Ann. Rev. Ecol. Syst.* **18**, 293–302.

Petit J.-R., Jouzel J., Raynaud D., Barkov N. I., Barnola J.-M., Basile I., Bender M., Chappellaz J., Devis M., Delaygue G., Delmotte G. M., Kotlyakov V. M., Legrand M., Lipenkov V. Y., Lorius C., Pepin L., Ritz C., Saltzman E., and Stievenard M. (1999) Climate and atmospheric history of the past 420,000 years from the Vostok ice core, Antarctica. *Nature* **399**, 429–436.

Popp B. N., Westley M. B., Toyoda S., Miwa T., Dore J. E., Yoshida N., Rust T. M., Francis J., Sansone F. J., Russ M. E., Ostrom N. E., and Ostrom P. H. (2002) Nitrogen and oxygen isotopomeric constraints on the origins and sea-to-air flux of N_2O in the oligotrophic subtropical North Pacific gyre. *Global Biogeochem. Cycles* **16**(4), 1064.

Probst J.-L. and Brunet F. (2005) δ^{13}C tracing of dissolved inorganic carbon sources in major world rivers. *Abs. 15th Ann. Goldschmidt Conf. Geochim. Cosmochim. Acta* **69** (Suppl. 1), A726.

Prokopenko Q. M. G., Hammond D. E., Berelson W. M., Bernhard J. M., Stott L., and Douglas R. (2006) Nitrogen cycling in the sediments of Santa Barbara basin and Eastern Subtropical North Pacific: nitrogen isotopes, diagenesis and possible chemosymbiosis between two lithotrophs (*Thioploca* and Anammox)—"riding on a glider." *Earth Planet. Sci. Lett.* **242**, 186–204.

Randerson J. T., van der Werf G. R., Collatz G. J., Giglio L., Still C. J., Kasibhatla P., Miller J. B., White J. W. C., DeFries R. S., and Kasischke E. S. (2005) Fire emissions from C_3 and C_4 vegetation and their influence on interannual variability of atmospheric CO_2 and δ^{13}C. *Global Biogeochem. Cycles* **19**, GB2019, doi:10.1029/2004GB002366.

Rau G. H., Arthur M. A., and Dean W. D. (1987) ^{15}N/^{14}N variations in Cretaceous Atlantic sedimentary rock sequences: implications for past changes in marine nitrogen biochemistry. *Earth Planet. Sci. Lett.* **82**, 269–279.

Rayleigh Lord. (1902) On the distillation of binary mixtures. *Phil. Mag., Ser. 6* **4**, 521–537.

Redfield A. C., Ketchum B. H., and Richard F. A. (1963) The influence of organisms on the composition of seawater. In *The Sea* (ed. M. N. Hill). Wiley, New York, vol. 2, pp. 26–77.

Richet P., Bottinga Y., and Javoy M. (1977) A review of hydrogen, carbon, nitrogen, oxygen, sulphur, and chlorine stable isotope fractionation among gaseous molecules. *Ann. Rev. Earth Planet. Sci.* **5**, 65–110.

Rigby D. and Batts B. D. (1986) The isotopic composition of nitrogen in Australian coals and oil shales. *Chem. Geol.* **58**, 273–282.

Robinson A. G., Coleman M. L., and Gluyas J. G. (1993) The age of illite cement growth, Village field area, southern North Sea: evidence from K-Ar ages and $^{18}O/^{16}O$ ratios. *Am. Assoc. Petrol. Geol. Bull.* **77**, 68–80.

Rose E. F., Chaussidon M., and France-Lanord C. (2000) Fractionation of boron isotopes during erosion processes: the example of Himalayan rivers. *Geochim. Cosmochim. Acta* **64**, 397–408.

Rosing M. T. (1999) ^{13}C-depleted carbon microparticles in >3700-Ma sea-floor sedimentary rocks from West Greenland. *Science* **283**, 674–676.

Różański K., Araguás-Araguás L., and Gonfiantini, R. (1993) Isotopic patterns in modem global precipitation. In *Climate Change in Continental Isotopic Records* (eds. P. K. Swart, K. C. Lohman, J. McKenzie, and S. Savin). *Geophys. Mon.*, Am. Geophys. Union, Washington, DC, vol. 78, pp. 1–36.

Russell W. A. and Papanastassiou D. A. (1978) Calcium isotope fractionation in ion-exchange chromatography. *Anal. Chem.* **50**, 1151–1153.

Russell W. A., Papanastassiou D. A., and Tombrello T. A. (1978) Ca isotope fractionation on the Earth and other Solar System materials. *Geochim. Cosmochim. Acta* **42**, 1075–1090.

Salomons W. and Mook W. G. (1986) Isotope geochemistry of carbonates in the weathering zone. In *Handbook of Environmental Isotope Geochemistry, vol. 2, The Terrestrial Environment B* (eds. P. Fritz and J. Ch. Fontes). Elsevier, Amsterdam, pp. 239–269.

Sanyal A., Hemming G., Hansen G., and Broecker W. (1995) Evidence for a higher pH in the glacial ocean from boron isotopes in foraminifera. *Nature* **373**, 234–237.

Savin S. M. (1980) Oxygen and hydrogen isotope effects in low-temperature mineral-water interactions. In *Handbook of Environmental Isotope Geochemistry* (eds. P. Fritz and J. Ch. Fontes). Elsevier, Amsterdam, vol. 1, pp. 283–328.

Savin S. M. and Epstein S. (1970a) The oxygen and hydrogen isotope geochemistry of clay minerals. *Geochim. Cosmochim. Acta* **34**, 25–42.

Savin S. M. and Epstein S. (1970b) The oxygen and hydrogen isotope geochemistry of ocean sediments and shales. *Geochim. Cosmochim. Acta* **34**, 43–63.

Savin S. M. and Lee M. (1988) Isotopic studies of phyllosilicates. In *Hydrous Phyllosilicates (Exclusive of Micas)* (ed. S. W. Bailey). *Rev. Mineral.*, Mineral. Soc. America, Washington, DC, vol. 19, pp. 189–224.

Schmitt A.-D., Chabaux F., and Stille P. (2003a) The calcium riverine and hydrothermal isotopic fluxes and the calcium mass balance. *Earth Planet. Sci. Lett.* **213**, 503–518.

Schmitt A.-D., Stille P., and Vennemann T. (2003b) Variations of the $^{44}Ca/^{40}Ca$ ratio in seawater during the past 24 million years: evidence from $\delta^{44}Ca$ and $\delta^{18}O$ values of Miocene phosphates. *Geochim. Cosmochim. Acta* **67**, 2607–2614.

Schwarcz H. P., Agyei E. K., and McMullen C. C. (1969) Boron isotopic fractionation during clay adsorption from sea-water. *Earth Planet. Sci. Lett.* **6**, 1–5.

Shaw D. B. and Weaver C. E. (1965) The mineralogical composition of shales. *J. Sedim. Petrol.* **35**, 213–222.

Shearer G. and Kohl D. H. (1988) $\delta^{15}N$ method of estimating N_2 fixation. In *Stable Isotopes in Ecological Research* (eds. P. W. Rundel, J. R. Ehleringer, and K. A. Nagy). Springer, New York, pp. 342–374.

Shen Y., Buick R., and Canfield D. E. (2001) Isotopic evidence for microbial sulphate reduction in the early Archaean era. *Nature* **410**, 77–81.

Shields G. and Veizer J. (2002) Precambrian marine carbonate isotope database: version 1.1. Geochem. Geophys. Geosyst. **3**(6), doi:10.1029/2001GC000266.

Shih C. Y., Nyquist L. E., Bogard D. D., and Wiesmann H. (1994) K-Ca and Rb-Sr dating of two lunar granites: relative chronometer resetting. *Geochim. Cosmochim. Acta* **58**, 3101–3116.

Sigman D. M., Altabet M. A., Francçois R., and McCorkle D. C. (1999) The isotopic composition of diatom-bound nitrogen in Southern Ocean sediments. *Paleoceanography* **14**(2), 118–134.

Sigman D. M., Altabet M. A., Michener R., McCorkle D. C., Fry B., and Holmes R. A. (1997) Natural abundance-level measurement of the nitrogen isotopic composition of oceanic nitrate: an adaptation of the ammonia diffusion method. *Mar. Chem.* **57**, 227–242.

Skulan J. and DePaolo D. J. (1999) Calcium isotope fractionation between soft and mineralized tissues as a monitor of calcium use in vertebrates. *Proc. Natl. Acad. Sci. USA* **96**, 13,709–13,713.

Skulan J., DePaolo D. J., and Owens T. L. (1997) Biological control of calcium isotopic abundances in the global calcium cycle. *Geochim. Cosmochim. Acta* **61**, 2505–2510.

Soddy F. (1913) Intra-atomic charge. *Nature* **92**, 399–400.

Soddy F. (1922) The origins of the conceptions of isotopes. *Nobel Lectures, Chemistry 1901-1921, 1966.* Elsevier, Amsterdam, pp. 371–399.

Spivack A. J., Berndt M. E., and Seyfried W. E., Jr. (1990) Boron isotope fractionation during supercritical phase separation. *Geochim. Cosmochim. Acta* **54**, 2337–2339.

Spivack A. J., Palmer M. R., and Edmond J. M. (1987) The sedimentary cycle of the boron isotopes. *Geochim. Cosmochim. Acta* **51**, 1939–1949.

Steiger R. H. and Jäger E. (1977) Subcommission on geochronology: convention on the use of decay constants in geo- and cosmochronology. *Earth Planet. Sci. Lett.* **36**, 359–362.

Stiehl G. and Lehmann M. (1980) Isotopenvariationen des Stickstoffs humoser und bituminöser natürlicher organischer Substanzen. *Geochim. Cosmochim. Acta* **44**, 1737–1746.

Strauss H. (1993) The sulfur isotopic record of Precambrian sulfates: new data and a critical evaluation of the existing record. *Precambrian Res* **63**, 225–246.

Strominger D., Hollander J. M., and Seaborg G. T. (1958) Table of isotopes. Rev. Mod. Phys. **30**(2), pt. ii, 585–904.

Struck U., Altenbach A. V., Emeis K. C., Alheit J., Eichner C., and Schneider R. (2002) Changes of the upwelling rates of nitrate preserved in the $\delta^{15}N$-signature of sediments and fish scales from the diatomaceous mud belt of Namibia. *Geobios* **35**, 3–11.

Sucha V., Kraus I., Gerthofferova H., Petes J., and Serekova M. (1993) Smectite to illite conversion in bentonites and shales of the East Slovak Basin. *Clay Minerals* **28**, 43–253.

Sukigara C. and Saino T. (2005) Temporal variations of $\delta^{13}C$ and $\delta^{15}N$ in organic particles collected by a sediment trap at a time-series station off the Tokyo Bay. *Continent. Shelf Res.* **25**, 1749–1767.

Swilhart G. H., Moore M. R., and Edmond J. M. (1986) Boron isotopic composition of marine and non-marine evaporate borates. *Geochim. Cosmochim. Acta* **50**, 1297–1301.

Takahashi T., Sutherland S. C., Sweeney C., Poisson A., Metzl N., Tilbrook B., Bates N., Wanninkhof R., Feely R. A., Sabine C., Olafsson J., and Nojiri Y. (2002) Global sea-air CO_2 flux based on climatological sufrace ocean pCO_2, and seasonal biological and temperature effects. *Deep-Sea Res. II* **49**, 1601–1622.

Techer I., Rousset D., Clauer N., Lancelot J., and Boisson J.-Y. (2006) Chemical and isotopic characterization of water-rock interactions in shales induced by the intrusion of a basaltic dike: a natural analogue for radioactive waste disposal. *Appl. Geochem.* **21**, 203–222.

Tonarini S., Pennisi M., and Gonfiantini R. (2005) Boron isotope determinations in waters and other geological materials: analytical techniques and inter-calibration of measurements. *Proceedings of the International Symposium on Quality Assurance for Analytical Methods in Isotope Hydrology.* IAEA, Vienna, 25–27 August 2004 (personal communication).

Toyoda S., Yoshida N., Miwa T., Matsui Y., Nojiri Y., and Tsurushima N. (2002) Production mechanism and global budget of N_2O inferred from its isotopomers in the western North Pacific. *Geophys. Res. Lett.* **29**(3), 1037, doi:10.1029/2001GL014311.

Turkevich A. L. (1964) Isotope. In *Encyclopedia Britannica*. Encyclopedia Britannica, Inc., Chicago, vol. 12, pp. 724–730.

Usui T., Nagao S., Yamamoto M., Suzuki K., Kudo I., Montani S., Noda A., and Minagawa M. (2006) Distribution and sources of organic matter in surficial sediments of the shelf and slope off Tokachi, western Pacific, inferred from C and N stable isotopes and C/N ratios. *Mar. Chem.* **98**, 241–259.

Van Zuilen M. A., Mathew K., Wopenka B., Lepland A., Marti K., and Arrhenius G. (2005) Nitrogen and argon isotopic signatures in graphite from the 3.8-Ga-old Isua Supracrustal Belt, Southern West Greenland. *Geochim. Cosmochim. Acta* **69**, 1241–1252.

Veizer J., Ala D., Azmy K., Bruckschen P., Buhl D., Bruhn F., Carden G. A. F., Diener A., Ebneth S., Goddéris Y., Jasper T., Korte C., Pawellek F., Podlaha O. G., and Strauss H. (1999) $^{87}Sr/^{86}Sr$, $\delta^{13}C$ and $\delta^{18}O$ evolution of Phanerozoic seawater. *Chem. Geol.* **161**, 59–88.

Velde B. (1985) *Clay Minerals: A Physico-Chemical Explanation of their Occurrence*. Developments in Sedimentology. Elsevier, Amsterdam, vol. 40, xiii+427pp.

Vengosh A., Starinsky A., and Chivas A. R. (1994) Boron isotopes in Heletz-Kokhav oilfield brines, the Coastal Plain Israel. *Israel J. Earth Sci.* **43**, 231–237.

Vengosh A., Starinsky A., Kolodny Y., and Chivas A. R. (1991a) Boron isotope geochemistry as a tracer for the evolution of brines and associated hot springs from the Dead Sea, Israel. *Geochim. Cosmochim. Acta* **55**, 1689–1695.

Vengosh A., Starinsky A., Kolodny Y., Chivas A. R., and McCulloch M. T. (1991b) Coprecipitation and isotopic fractionation of boron in modern biogenic carbonates. *Geochim. Cosmochim. Acta* **55**, 2901–2910.

Vengosh A., Starinsky A., Kolodny Y., Chivas A. R., and Raab M. (1992) Boron isotope variations during fractional evaporation of seawater: new constraints on the marine vs. nonmarine debate. *Geology* **20**, 799–802.

Ver L. M. B. (1998) Global kinetic models of the coupled C, N, P, and S biogeochemical cycles: implications for global environmental change. PhD Dissertation, University of Hawaii, Honolulu, xxii+681pp.

Ver L. M. B., Mackenzie F. T., and Lerman A. (1999) Biogeochemical responses of the carbon cycle to natural and human perturbation: past, present, and future. *Am. J. Sci.* **299**, 762–801.

Wedepohl K. H. (1995) The composition of the continental crust. *Geochim. Cosmochim. Acta* **59**, 1217–1232.

Wiegand B. A., Chadwick O. A., Vitousek P. M., and Wooden J. L. (2005) Ca cycling and isotopic fluxes in forested ecosystems in Hawaii. *Geophys. Res. Lett.* **32**, L11404, doi:10.1029/2005GL022746.

Williams L. B. and Hervig R. L. (2002) Exploring intra-crystalline B-isotope variations in mixed-layer illite-smectite. *Am. Mineral.* **87**, 1564–1570.

Williams L. B. and Hervig R. L. (2006) Crystal size dependence of illite-smectite isotope equilibration with changing fluids. *Clays Clay Minerals* **54**(5), 531–540.

Williams L. B., Hervig R. L., and Hutcheon I. (2001a) Boron isotope geochemistry during diagenesis. Part II: Applications to organic-rich sediments. *Geochim. Cosmochim. Acta* **65**, 1783–1794.

Williams L. B., Wieser M. E., Fennell J., Hutcheon I., and Hervig R. L. (2001b) Application of boron isotopes to the understanding of fluid-rock interactions in a hydrothermally simulated oil reservoir in the Alberta Basin, Canada. *Geofluids* **1**, 229–240.

Wood B. J., Pawley A., and Frost D. R. (1996) Water and carbon in the Earth's mantle. *Phil. Trans. Roy. Soc. London, Ser. A* **354**, 1495–1511.

Yapp C. J. (1993) Stable isotope geochemistry of low temperature Fe(III) and Al "oxides" with implications for continental paleoclimates. In *Climate Change in Continental Isotopic Records* (eds. P. K. Swart, K. C. Lohman, J. McKenzie, and S. Savin). *Geophys. Mon.*, Am. Geophys. Union., Washington, DC, vol. 78, pp. 285–294.

Young E. D., Galy A., and Nagahara H. (2002) Kinetic and equilibrium mass-dependent isotope fractionation laws in nature and their geochemical and cosmochemical significance. *Geochim. Cosmochim. Acta* **66**, 1095–1104.

Zeebe R. E. and Wolf-Gladrow D. (2001) *CO_2 in Seawater: Equilibrium, Kinetics, Isotopes*. Elsevier, New York.

Zhu P. and MacDougall J. D. (1998) Calcium isotopes in the marine environment and the oceanic calcium cycle. *Geochim. Cosmochim. Acta* **62**, 1691–1698.

Ziegler K. and Longstaffe F. J. (2000) Clay mineral authigenesis along a mid-continental scale fluid conduit in Palaeozoic sedimentary rocks from southern Ontario, Canada. *Clay Minerals* **35**, 239–260.

Zwingmann H., Clauer N., and Gaupp R. (1999) Structure-related geochemical (REE) and isotopic (K-Ar, Rb-Sr, $\delta^{18}O$) characteristics of clay minerals from Rotliegend sandstone reservoirs (Permian, Northern Germany). *Geochim. Cosmochim. Acta* **63**, 2805–2823.

Isotope Geochemistry
ISBN: 978-0-08-096710-3
pp. 285–336

11
The Global Oxygen Cycle

S. T. Petsch
University of Massachusetts, Amherst, MA, USA

11.1 INTRODUCTION

One of the key defining features of Earth as a planet that houses an active and diverse biology is the presence of free molecular oxygen (O_2) in the atmosphere. Biological, chemical, and physical processes interacting on and beneath the Earth's surface determine the concentration of O_2 and variations in O_2 distribution, both temporal and spatial. In the present-day Earth system, the process that releases O_2 to the atmosphere (photosynthesis) and the processes that consume O_2 (aerobic respiration, sulfide mineral oxidation, oxidation of reduced volcanic gases) result in large fluxes of O_2 to and from the atmosphere. Even relatively small changes in O_2 production and consumption have the potential to generate large shifts in atmospheric O_2 concentration within geologically short periods of time. Yet all available evidence supports the conclusion that stasis in O_2 variation is a significant feature of the Earth's atmosphere over wide spans of the geologic past. Study of the oxygen cycle is therefore important because, while an equable O_2 atmosphere is central to life as we know it, our understanding of exactly why O_2 concentrations remain nearly constant over large spans of geologic time is very limited.

This chapter begins with a review of distribution of O_2 among various reservoirs on Earth's surface: air, sea, and other natural waters. The key factors that affect the concentration of O_2 in the atmosphere and surface waters are next considered, focusing on photosynthesis as the major process generating free O_2 and various biological and abiotic processes that consume O_2. The chapter ends with a synopsis of current models on the evolution of an oxygenated atmosphere through 4.5 billion years of Earth's history, including geochemical evidence constraining ancient O_2 concentrations and numerical models of atmospheric evolution.

11.2 DISTRIBUTION OF O_2 AMONG EARTH SURFACE RESERVOIRS

11.2.1 The Atmosphere

The partial pressure of oxygen in the present-day Earth's atmosphere is ~0.21 bar, corresponding to a total mass of $\sim 34 \times 10^{18}$ mol O_2 (0.20946 bar (force/area) multiplied by the surface area of the Earth (5.1×10^{14} m^2), divided by average gravitational acceleration g (9.8 m s^{-2}) and the formula weight for O_2 (32 g mol^{-1}) yields $\sim 34 \times 10^{18}$ mol O_2). There is a nearly uniform mixture of the main atmospheric gases (N_2, O_2, Ar) from the Earth's surface up to ~80 km altitude (including the troposphere, stratosphere, and mesosphere), because turbulent mixing dominates over molecular diffusion at these altitudes. Because atmospheric pressure (and thus gas molecule density) decreases exponentially with altitude, the bulk of molecular oxygen in the atmosphere is concentrated within several kilometers of Earth's surface. Above this, in the thermosphere, gases become separated based on their densities. Molecular oxygen is photodissociated by UV radiation to form atomic oxygen (O), which is the major form of oxygen above ~120 km altitude.

Approximately 21% O_2 in the atmosphere represents an average composition. In spite of well-developed turbulent mixing in the lower atmosphere, seasonal latitudinal variations in O_2 concentration of ±15 ppm have been recorded. These seasonal variations are most pronounced at high latitudes, where seasonal cycles of primary production and respiration are strongest (Keeling and Shertz, 1992). In the northern hemisphere, the seasonal variations are anticorrelated with atmospheric p_{CO_2}; summers are dominated by high O_2 (and high inferred net photosynthesis), while winters are dominated by lower O_2. In addition, there has been a measurable long-term decline in atmospheric O_2 concentration of $\sim 10^{14}$ mol yr^{-1}, attributed to oxidation of fossil fuels. This decrease has been detected in both long-term atmospheric monitoring stations (Keeling and Shertz, 1992, Figure 1(a)) and in atmospheric gases trapped in Antarctic firn ice bubbles (Figure 1(b)). The polar ice core records extend the range of direct monitoring of atmospheric composition to show that a decline in atmospheric O_2 linked to oxidation of fossil fuels has been occurring since the Industrial Revolution (Bender *et al.*, 1994b; Battle *et al.*, 1996).

11.2.2 The Oceans

Air-saturated water has a dissolved O_2 concentration dependent on temperature, the Henry's law constant k_H, and ionic strength. In pure water at 0 °C, O_2 saturation is 450 μM; at 25 °C, saturation falls to 270 μM. Other solutes reduce O_2 solubility, such that at normal seawater salinities, O_2 saturation is reduced by ~25%. Seawater is, of course, rarely at perfect O_2 saturation. Active photosynthesis may locally increase O_2 production rates, resulting in supersaturation of O_2 and degassing to the atmosphere. Alternately, aerobic respiration below the sea surface can consume dissolved O_2 and lead to severe O_2-depletion or even anoxia.

Lateral and vertical gradients in dissolved O_2 concentration in seawater reflect balances between O_2 inputs from air–sea gas exchange, biological processes of O_2 production and consumption, and advection of water masses. In general terms, the concentration of O_2 with depth in the open ocean follows the general structures described in Figure 2. Seawater is saturated to supersaturated with O_2 in the surface mixed layer

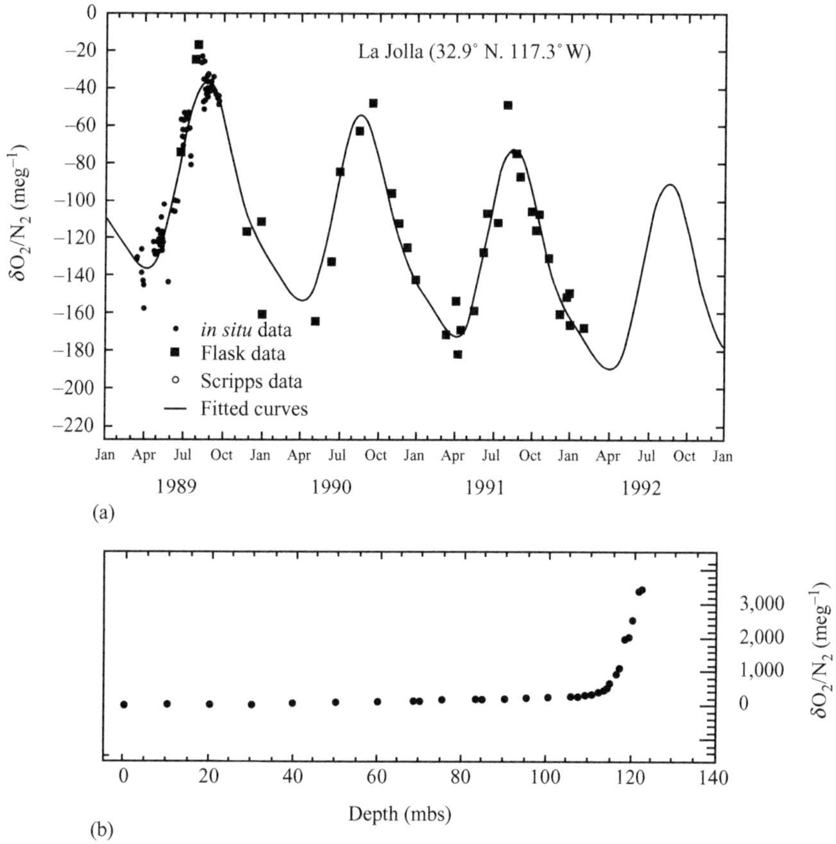

Figure 1 (a) Interannual variability of atmospheric O_2/N_2 ratio, measured at La Jolla, California. 1 ppm O_2 is equivalent to ~4.8 meg^{-1} (source Keeling and Shertz, 1992). (b) Variability in atmospheric O_2/N_2 ratio measured in firn ice at the South Pole, as a function of depth in meters below surface (mbs). The gentle rise in O_2/N_2 between 40 m and 100 m reflects a loss of atmospheric O_2 during the last several centuries due to fossil fuel burning. Deeper than 100 m, selective effusion of O_2 out of closing bubbles into firn air artificially boosts O_2/N_2 ratios (source Battle *et al.*, 1996).

(~0–60 m water depth). Air–sea gas exchange and trapping of bubbles ensures constant dissolution of atmospheric O_2. Because gas solubility is temperature dependent, O_2 concentrations are greater in colder high-latitude surface waters than in waters near the equator. Oxygen concentrations in surface waters also vary strongly with season, especially in high productivity waters. Supersaturation is strongest in spring and summer (time of greatest productivity and strongest water column stratification) when warming of surface layers creates a shallow density gradient that inhibits vertical mixing. Photosynthetic O_2 production exceeds consumption and exchange, and supersaturation can develop. O_2 concentrations drop below the surface mixed layer to form O_2 minimum zones (OMZs) in many ocean basins. O_2 minima form where biological consumption of O_2 exceeds resupply through advection and diffusion. The depth and thickness of O_2 minima vary among ocean basins. In the North Atlantic, the OMZ extends several hundred meters. O_2 concentrations fall from an average of ~300 μM in the surface mixed layer to ~160 μM at 800 m depth. In the North Pacific, however, the O_2 minimum extends deeper, and O_2 concentrations fall to <100 μM. Along the edges of ocean basins, where OMZs impinge on the seafloor, aerobic respiration is restricted, sediments are anoxic at or near the sediment–water interface, and burial of organic matter in sediments may be enhanced. Below the oxygen minima zones in the open ocean, O_2 concentrations gradually increase again from 2000 m to the seafloor. This increase in O_2 results from the slow progress of global thermohaline circulation. Cold, air-saturated seawater sinks to the ocean depths at high-latitudes in the Atlantic, advecting in O_2-rich waters below the O_2 minimum there. Advection of O_2-rich deep water from the Atlantic through the Indian Ocean into the Pacific is the source of O_2 in deep Pacific waters. However, biological utilization of this deep-water O_2 occurs along the entire path from the North Atlantic to the Pacific. For this reason, O_2 concentrations in deep Atlantic water are slightly greater (~200 μM) than in the deep Pacific (~150 μM).

In some regions, dissolved O_2 concentration falls to zero. In these regions, restricted water

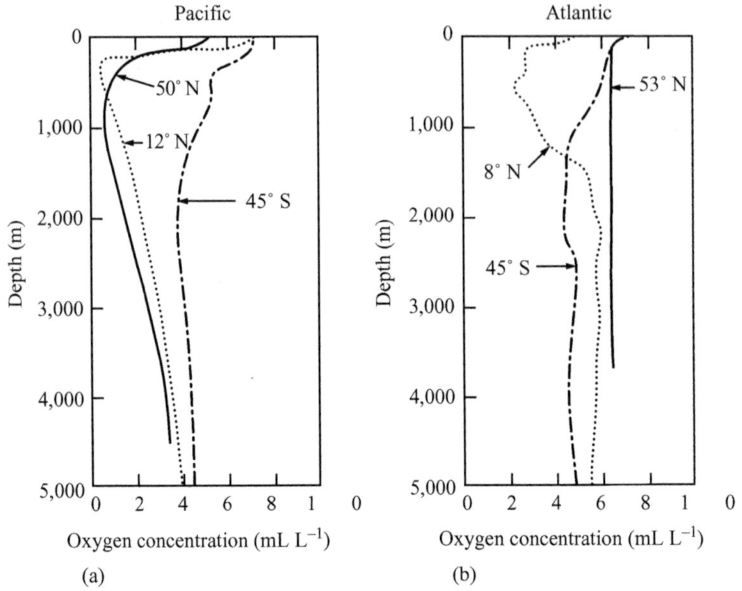

Figure 2 Depth profiles of oxygen concentration dissolved in seawater for several latitudes in the Pacific (a) and Atlantic (b). Broad trends of saturation or supersaturation at the surface, high dissolved oxygen demand at mid-depths, and replenishment of O_2 through lateral advection of recharged deep water are revealed, although regional influences of productivity and intermediate and deep-water heterotropy are also seen (source Ingmanson and Wallace, 1989).

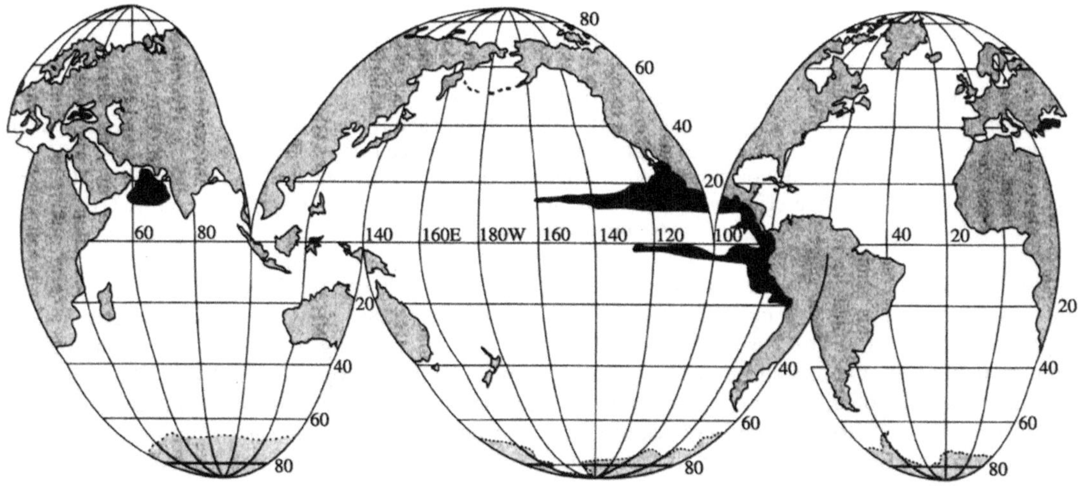

Figure 3 Map detailing locations of extensive and permanent oxygen-deficient intermediate and deep waters (Deuser, 1975) (reproduced by permission of Elsevier from *Chemical Oceanography* **1975**, p.3).

circulation and ample organic matter supply result in biological utilization of oxygen at a rate that exceeds O_2 resupply through advection and diffusion. Many of these are temporary zones of anoxia that form in coastal regions during summer, when warming facilitates greatest water column stratification, and primary production and organic matter supply are high. Such O_2-depletion is now common in the Chesapeake Bay and Schelde estuaries, off the mouth of the Mississippi River and other coastal settings. However, there are several regions of the world oceans where stratification and anoxia are more permanent features (Figure 3). These include narrow, deep, and silled coastal fjords, larger restricted basins (e. g., the Black Sea, Cariaco Basin, and the chain of basins along the southern California Borderlands). Lastly, several regions of the open ocean are also associated with strong O_2-depletion. These regions (the equatorial Pacific along Central and South America and the Arabian Sea) are associated with deep-water upwelling, high rates of surface water primary productivity, and high dissolved oxygen demand in intermediate waters.

Oxygen concentrations in the pore fluids of sediments are controlled by a balance of entrainment of overlying fluids during sediment deposition, diffusive exchange between the sediment and the water column, and biological utilization. In marine sediments, there is a good correlation between the rate of organic matter supply and the depth of O_2 penetration in the sediment (Hartnett *et al.*, 1998). In coastal sediments and on the continental shelf, burial of organic matter is sufficiently rapid to deplete the sediment of oxygen within millimeters to centimeters of the sediment–water interface. In deeper abyssal sediments, where organic matter delivery is greatly reduced, O_2 may penetrate several meters into the sediment before being entirely consumed by respiration.

There is close coupling between surface water and atmospheric O_2 concentrations and air–sea gas exchange fluxes (Figure 4). High rates of marine primary productivity result in net outgassing of O_2 from the oceans to the atmosphere in spring and summer, and net ingassing of O_2 during fall and winter. These patterns of air–sea O_2 transfer relate to latitude and season: outgassing of O_2 during northern hemisphere high productivity months (April through August) are accompanied by simultaneous ingassing in southern latitudes when and where the productivity is lowest (Najjar and Keeling, 2000). Low-latitude ocean surface waters show very little net air–sea O_2 exchange and minimal change in outgassing or ingassing over an annual cycle.

11.2.3 Freshwater Environments

Oxygen concentrations in flowing freshwater environments closely match air-saturated values, due to turbulent mixing and entrainment of air bubbles. In static water bodies, however, O_2-depletion can develop much like in the oceans. This is particularly apparent in some ice-covered lakes, where inhibited gas exchange and wintertime respiration can result in O_2-depletion and fish kills. High productivity during spring and summer in shallow turbid aquatic environments can result in extremely sharp gradients from strong O_2 supersaturation at the surface to near O_2-depletion within a few meters of the surface. The high concentration of labile dissolved and particulate organic matter in many freshwater environments leads to rapid O_2-depletion where advective resupply is limited. High rates of O_2 consumption have been measured in many temperature and tropical rivers.

11.2.4 Soils and Groundwaters

In soil waters, oxygen concentrations depend on gas diffusion through soil pore spaces, infiltration

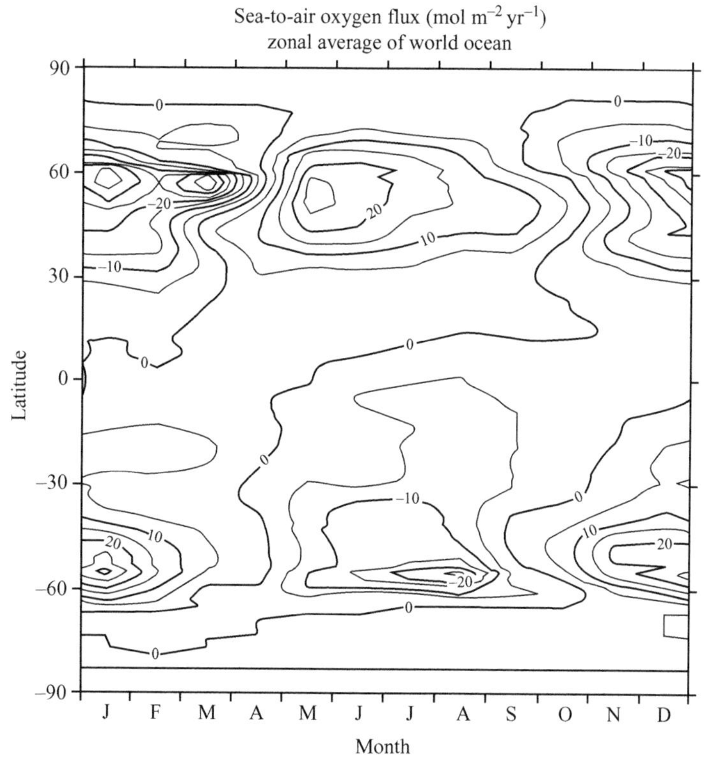

Figure 4 Zonal average monthly sea-to-air oxygen flux for world oceans. Outgassing and ingassing are concentrated at mid- to high latitudes. Outgassing of oxygen is strongest when primary production rates are greatest; ingassing is at a maximum during net respiration. These patterns oscillate during an annual cycle from northern to southern hemisphere (source Najjar and Keeling, 2000).

and advection of rainwater and groundwater, air–gas exchange, and respiration of soil organic matter (see review by Hinkle, 1994). In organic-matter-rich temperate soils, dissolved O_2 concentrations are reduced, but not entirely depleted. Thus, many temperate shallow groundwaters contain some dissolved oxygen. Deeper groundwaters, and water-saturated soils and wetlands, generally contain little dissolved O_2. High-latitude mineral soils and groundwaters contain more dissolved O_2 (due to lower temperature and lesser amounts of soil organic matter and biological O_2 demand). Dry tropical soils are oxidized to great depths, with dissolved O_2 concentration less than air saturated, but not anoxic. Wet tropical forests, however, may experience significant O_2-depletion as rapid oxidation of leaf litter and humus occurs near the soil surface. Soil permeability also influences O_2 content, with more clay-rich soils exhibiting lower O_2 concentrations.

In certain environments, localized anomalously low concentrations of soil O_2 have been used by exploration geologists to indicate the presence of a large body of chemically reduced metal sulfides in the subsurface. Oxidation of sulfide minerals during weathering and soil formation draws down soil gas p_{O_2} below regional average. Oxidation of sulfide minerals generates solid and aqueous-phase oxidation products (i.e., sulfate anion and ferric oxyhydroxides in the case of pyrite oxidation). In some instances, the volume of gaseous O_2 consumed during mineral oxidation generates a mild negative pressure gradient, drawing air into soils above sites of sulfide mineral oxidation (Lovell, 2000).

11.3 MECHANISMS OF O_2 PRODUCTION

11.3.1 Photosynthesis

The major mechanism by which molecular oxygen is produced on Earth is through the biological process of photosynthesis. Photosynthesis occurs in higher plants, the eukaryotic protists collectively called algae, and in two groups of prokaryotes: the cyanobacteria and the prochlorophytes. In simplest terms, photosynthesis is the harnessing of light energy to chemically reduce carbon dioxide to simple organic compounds (e.g., glucose). The overall reaction (Equation (1)) for photosynthesis shows carbon dioxide and water reacting to produce oxygen and carbohydrate:

$$6CO_2 + 6H_2O \rightarrow 6O_2 + C_6H_{12}O_6 \qquad (1)$$

Photosynthesis is actually a two-stage process, with each stage broken into a cascade of chemical reactions (Figure 5). In the light reactions of photosynthesis, light energy is converted to chemical energy that is used to dissociate water to yield oxygen and hydrogen and to form the reductant NADPH from $NADP^+$.

The next stage of photosynthesis, the Calvin cycle, uses NADPH to reduce CO_2 to phosphoglyceraldehyde, the precursor for a variety of metabolic pathways, including glucose synthesis. In higher plants and algae, the Calvin cycle operates in special organelles called chloroplasts. However, in bacteria the Calvin cycle occurs throughout the cytosol. The enzyme ribulose-1,5-biphosphate carboxylase (rubisco) catalyzes

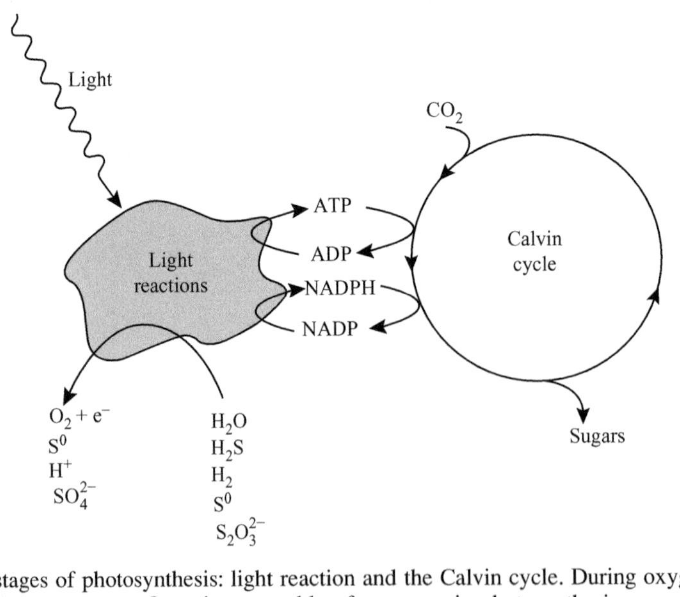

Figure 5 The two stages of photosynthesis: light reaction and the Calvin cycle. During oxygenic photosynthesis, H_2O is used as an electron source. Organisms capable of anoxygenic photosynthesis can use a variety of other electron sources ($H_2S, H_2, S^0, S_2O_3^{2-}$) during the light reactions, and do not liberate free O_2. Energy in the form of ATP and reducing power in the form of NADPH are produced by the light reactions, and subsequently used in the Calvin cycle to deliver electrons to CO_2 to produce sugars.

reduction of CO_2 to phosphoglycerate, which is carried through a chain of reactions that consume ATP and NADPH and eventually yield phosphoglyceraldehyde. Most higher plants are termed C_3 plants, because the first stable intermediate formed during the carbon cycle is a three-carbon compound. Several thousand species of plant, spread among at least 17 families including the grasses, precede the carbon cycle with a CO_2-concentrating mechanism which delivers a four-carbon compound to the site of the Calvin cycle and rubisco. These are the C_4 plants. This four-carbon compound breaks down inside the chloroplasts, supplying CO_2 for rubisco and the Calvin cycle. The C_4-concentrating mechanisms is an advantage in hot and dry environments where leaf stomata are partially closed, and internal leaf CO_2 concentrations are too low for rubisco to efficiently capture CO_2. Other plants, called CAM plants, which have adapted to dry climates utilize another CO_2-concentrating mechanisms by closing stomata during the day and concentrating CO_2 at night. All higher plants, however, produce O_2 and NADPH from splitting water, and use the Calvin cycle to produce carbohydrates. Some prokaryotes use mechanisms other than the Calvin cycle to fix CO_2 (i.e., the acetyl-CoA pathway or the reductive tricarboxylic acid pathway), but none of these organisms is involved in oxygenic photosynthesis.

Global net primary production estimates have been derived from variations in the abundance and isotopic composition of atmospheric O_2. These estimates range from 23×10^{15} mol yr^{-1} (Keeling and Shertz, 1992) to 26×10^{15} mol yr^{-1}, distributed between 14×10^{15} mol yr^{-1} O_2 production from terrestrial primary production and 12×10^{15} mol yr^{-1} from marine primary production (Bender *et al.*, 1994a). It is estimated that ~50% of all photosynthetic fixation of CO_2 occurs in marine surface waters. Collectively, free-floating photosynthetic microorganisms are called phytoplankton. These include the algal eukaryotes (dinoflagellates, diatoms, and the red, green, brown, and golden algae), various species of cyanobacteria (*Synechococcus* and *Trichodesmium*), and the common prochlorophyte *Prochlorococcus*. Using the stoichiometry of the photosynthesis reaction, this equates to half of all global photosynthetic oxygen production resulting from marine primary production. Satellite-based measurements of seasonal and yearly average chlorophyll abundance (for marine systems) and vegetation greenness (for terrestrial ecosystems) can be applied to models that estimate net primary productivity, CO_2 fixation, and O_2 production. In the oceans, there are significant regional and seasonal variations in photosynthesis that result from limitations by light, nutrients, and temperature. Yearly averages of marine chlorophyll abundance show concentrated primary production at high-latitudes in the North Atlantic, North Pacific, and coastal Antarctica, in regions of seawater upwelling off of the west coasts of Africa, South America, and the Arabian Sea, and along the Southern Subtropical Convergence. At mid- and high latitudes, marine productivity is strongly seasonal, with primary production concentrated in spring and summer. At low latitudes, marine primary production is lower and varies little with season or region. On land, primary production also exhibits strong regional and seasonal patterns. Primary production rates (in g C m^{-2} yr^{-1}) are greatest year-round in the tropics. Tropical forests in South America, Africa, and Southeast Asia are the most productive ecosystems on Earth. Mid-latitude temperate forests and high-latitude boreal forests are also highly productive, with a strong seasonal cycle of greatest production in spring and summer. Deserts (concentrated at ~30°N and S) and polar regions are less productive. These features of seasonal variability in primary production on land and in the oceans are clearly seen in the seasonal variations in atmospheric O_2 (Figure 4).

Transfer of O_2 between the atmosphere and surface seawater is controlled by air–sea gas exchange. Dissolved gas concentrations trend towards thermodynamic equilibrium, but other factors may complicate dissolved O_2 concentrations. Degassing of supersaturated waters can only occur at the very surface of the water. Thus, in regions of high primary production, concentrations of O_2 can accumulate in excess of the rate of O_2 degassing. In calm seas, where the air–sea interface is a smooth surface, gas exchange is very limited. As seas become more rough, and especially during storms, gas exchange is greatly enhanced. This is in part because of entrainment of bubbles dispersed in seawater and water droplets entrained in air, which provide much more surface for dissolution or degassing. Also, gas exchange depends on diffusion across a boundary layer. According to Fick's law, the diffusive flux depends on both the concentration gradient (degree of super- or undersaturation) and the thickness of the boundary layer. Empirically it is observed that the boundary layer thickness decreases with increasing wind speed, thus enhancing diffusion and gas exchange during high winds.

11.3.2 Photolysis of Water

In the upper atmosphere today, a small amount of O_2 is produced through photolysis of water vapor. This process is the sole source of O_2 to the atmospheres on the icy moons of Jupiter (Ganymede and Europa), where trace concentrations of O_2 have been detected (Vidal *et al.*, 1997). Water vapor photolysis may also have been the source of O_2 to the early Earth before the evolution of oxygenic photosynthesis. However, the oxygen formed by photolysis would have been through reactions with methane and carbon monoxide, preventing any accumulation in the atmosphere (Kasting *et al.*, 2001).

11.4 MECHANISMS OF O_2 CONSUMPTION

11.4.1 Aerobic Cellular Respiration

In simple terms, aerobic respiration is the oxidation of organic substrates with oxygen to yield chemical energy in the form of ATP and NADH. In eukaryotic cells, the respiration pathway follows three steps (Figure 6). Glycolysis occurs throughout the cytosol, splitting glucose into pyruvic acid and yielding some ATP. Glycolysis does not directly require free O_2, and thus occurs among aerobic and anaerobic organisms. The Krebs citric acid (tricarboxylic acid) cycle and oxidative phosphorylation are localized within the mitochondria of eukaryotes, and along the cell membranes of prokaryotes. The Krebs cycle completes the oxidation of pyruvate to CO_2 from glycolysis, and together glycolysis and the Krebs cycle provide chemical energy and reductants (in the form of ATP, NADH, and $FADH_2$) for the third step—oxidative phosphorylation. Oxidative phosphorylation involves the transfer of electrons from NADH and $FADH_2$ through a cascade of electron carrying compounds to molecular oxygen. Compounds used in the electron transport chain of oxidative phosphorylation include a variety of flavoproteins, quinones, Fe–S proteins, and cytochromes. Transfer of electrons from NADH to O_2 releases considerable energy, which is used to generate a proton gradient across the mitochondrial membrane and fuel significant ATP synthesis. In part, this gradient is created by reduction of O_2 to H_2O as the last step of oxidative phosphorylation. While eukaryotes and many prokaryotes use the Krebs cycle to oxidize pyruvate to CO_2, there are other pathways as well. For example, prokaryotes use the glyoxalate cycle to metabolize fatty acids. Several aerobic prokaryotes also can use the Entner–Doudoroff pathway in place of normal glycolysis. This reaction still produces pyruvate, but yields less energy in the form of ATP and NADH.

Glycolysis, the Krebs cycle, and oxidative phosphorylation are found in all eukaryotes (animals, plants, and fungi) and many of the aerobic prokaryotes. The purpose of these reaction pathways is to oxidize carbohydrates with O_2, yielding CO_2, H_2O, and chemical energy in the form of ATP. While macrofauna generally require a minimum of ~0.05–0.1 bar (~10 μM) O_2 to survive, many prokaryotic microaerophilic organisms can survive and thrive at much lower O_2 concentrations. Because most biologically mediated oxidation processes occur through the activity of aerotolerant microorganisms, it is unlikely that a strict coupling between limited atmospheric O_2 concentration and limited global respiration rates could exist.

11.4.2 Photorespiration

The active site of rubisco, the key enzyme involved in photosynthesis, can accept either CO_2 or O_2. Thus, O_2 is a competitive inhibitor of photosynthesis. This process is known as photorespiration, and involves addition of O_2 to ribulose-biphosphate. Products of this reaction enter a metabolic pathway that eventually produces CO_2. Unlike cellular respiration, photorespiration generates no ATP, but it does consume O_2. In some plants, as much as 50% of the carbon fixed by the Calvin cycle is respired through photorespiration. Photorespiration is enhanced in hot,

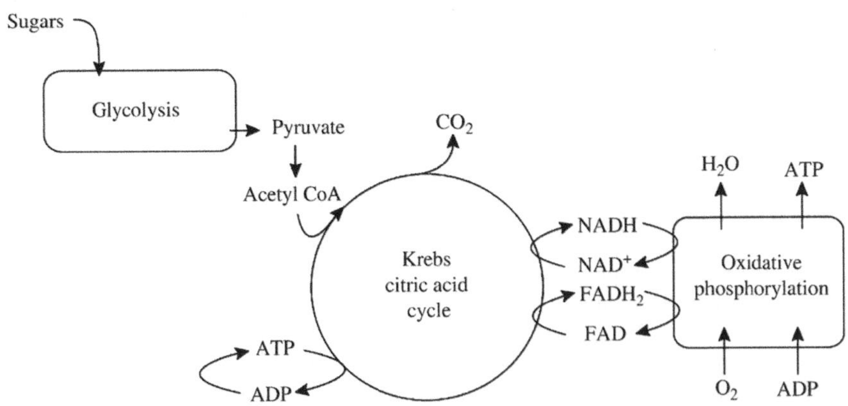

Figure 6 The three components of aerobic respiration: glycolysis, the Krebs cycle, and oxidative phosphorylation. Sugars are used to generate energy in the form of ATP during glycolysis. The product of glycolyis, pyruvate, is converted to acetyl-CoA, and enters the Krebs cycle. CO_2, stored energy as ATP, and stored reducing power as NADH and $FADH_2$ are generated in the Krebs cycle. O_2 is only directly consumed during oxidative phosphorylation to generate ATP as the final component of aerobic respiration.

dry environments when plant cells close stomata to slow water loss, CO_2 is depleted and O_2 accumulates. Photorespiration does not occur in prokaryotes, because of the much lower relative concentration of O_2 versus CO_2 in water compared with air.

11.4.3 C_1 Metabolism

Beyond metabolism of carbohydrates, there are several other biological processes common in prokaryotes that consume oxygen. For example, methylotrophic organisms can metabolize C_1 compounds such as methane, methanol, formaldehyde, and formate, as in

$$CH_4 + NADH + H^+ + O_2 \rightarrow CH_3OH + NAD^+ + H_2O$$

$$CH_3OH + PQQ \rightarrow CH_2O + PQQH_2$$

$$CH_2O + NAD^+ + H_2O \rightarrow HCOOH + NADH + H^+$$

$$HCOOH + NAD^+ \rightarrow CO_2 + NADH + H^+$$

These compounds are common in soils and sediments as the products of anaerobic fermentation reactions. Metabolism of these compounds can directly consume O_2 (through monooxygenase enzymes) or indirectly, through formation of NADH which is shuttled into oxidative phosphorylation and the electron transport chain. Oxidative metabolism of C_1 compounds is an important microbial process in soils and sediments, consuming the methane produced by methanogenesis.

11.4.4 Inorganic Metabolism

Chemolithotrophic microorganisms are those that oxidize inorganic compounds rather than organic substrates as a source of energy and electrons. Many of the chemolithotrophs are also autotrophs, meaning they reduce CO_2 to generate cellular carbon in addition to oxidizing inorganic compounds. In these organisms, CO_2 fixation is not tied to O_2 production by the stoichiometry of photosynthesis. Hydrogen-oxidizing bacteria occur wherever both O_2 and H_2 are available. While some H_2 produced by fermentation in anoxic environments (deep soils and sediments) may escape upward into aerobic environments, most H_2 utilized by hydrogen-oxidizing bacteria derives from nitrogen fixation associated with nitrogen-fixing plants and cyanobacteria. Nitrifying bacteria are obligate autotrophs that oxidize ammonia. Ammonia is produced in many environments during fermentation of nitrogen compounds and by dissimilatory nitrate reduction. Nitrifying bacteria are common at oxic–anoxic interfaces in soils, sediments, and the water column. Other chemolithoautotrophic bacteria can oxidize nitrite. Non-photosynthetic bacteria that can oxidize reduced sulfur compounds form a diverse group. Some are acidophiles, associated with sulfide mineral oxidation and tolerant of extremely low pH. Others are neutrophilic and occur in many marine sediments. These organisms can utilize a wide range of sulfur compounds produced in anaerobic environments, including H_2S, thiosulfate, polythionates, polysulfide, elemental sulfur, and sulfite. Many of the sulfur-oxidizing bacteria are also autotrophs. Some bacteria can live as chemoautotrophs through the oxidation of ferrous-iron. Some of these are acidophiles growing during mining and weathering of sulfide minerals. However, neutrophilic iron-oxidizing bacteria have also been detected associated with the metal sulfide plumes and precipitates at mid-ocean ridges. Other redox-sensitive metals that can provide a substrate for oxidation include manganese, copper, uranium, arsenic, and chromium. As a group, chemolithoautotrophic microorganisms represent a substantial flux of O_2 consumption and CO_2 fixation in many common marine and terrestrial environments. Because the net reaction of chemolithoautotrophy involves both O_2 and CO_2 reduction, primary production resulting from chemoautotrophs has a very different O_2/CO_2 stoichiometry than does photosynthesis. Chemolithoautotrophs use the electron flow from reduced substrates (metals, H_2, and reduced sulfur) to O_2 to generate ATP and NAD(P)H, which in turn are use for CO_2-fixation. It is believed that most aerobic chemoautotrophs utilize the Calvin cycle for CO_2-fixation.

11.4.5 Macroscale Patterns of Aerobic Respiration

On a global scale, much biological O_2 consumption is concentrated where O_2 is abundant. This includes surficial terrestrial ecosystems and marine surface waters. A large fraction of terrestrial primary production is consumed by aerobic degradation mechanisms. Although most of this is through aerobic respiration, some fraction of aerobic degradation of organic matter depends on anaerobic breakdown of larger biomolecules into smaller C_1 compounds, which, if transported into aerobic zones of soil or sediment, can be degraded by aerobic C_1 metabolizing microorganisms. Partially degraded terrestrial primary production can be incorporated into soils, which slowly are degraded and eroded. Research has shown that soil organic matter (OM) can be preserved for up to several millennia, and riverine export of aged terrestrial OM may be a significant source of dissolved and particulate organic carbon to the oceans.

Aside from select restricted basins and specialized environments, most of the marine water column is oxygenated. Thus, aerobic respiration dominates in open water settings. Sediment trap and particle flux studies have shown that substantial fractions of marine primary production are completely degraded (remineralized) prior to deposition at the sea floor, and thus by implication, aerobic respiration generates an O_2 consumption demand nearly equal to the release of oxygen associated with photosynthesis. The bulk of marine aerobic respiration occurs within the water column. This is because diffusion and mixing (and thus O_2 resupply) are much greater in open water than through sediment pore fluids, and because substrates for aerobic respiration (and thus O_2 demand) are much less concentrated in the water column than in the sediments. O_2 consumption during aerobic respiration in the water column, coupled with movement of deep-water masses from O_2-charged sites of deep-water formation, generates the vertical and lateral profiles of dissolved O_2 concentration observed in seawater.

In most marine settings, O_2 does not penetrate very far into the sediment. Under regions of high primary productivity and limited water column mixing, even if the water column is oxygenated, O_2 may penetrate 1 mm or less, limiting the amount of aerobic respiration that occurs in the sediment. In bioturbated coastal sediments, O_2 penetration is facilitated by the recharging of pore fluids through organisms that pump overlying water into burrows, reaching several centimeters into the sediment in places. However, patterns of oxic and anoxic sediment exhibit a great degree of spatial and temporal complexity as a result of spotty burrow distributions, radial diffusion of O_2 from burrows, and continual excavation and infilling through time. Conversely, in deep-sea (pelagic) sediments, where organic matter delivery is minor and waters are cold and charged with O_2 from sites of deep-water formation, O_2 may penetrate uniformly 1 m or more into the sediment.

11.4.6 Volcanic Gases

Gases emitted from active volcanoes and fumaroles are charged with reduced gases, including CO, H_2, SO_2, H_2S, and CH_4. During explosive volcanic eruptions, these gases are ejected high into the atmosphere along with H_2O, CO_2, and volcanic ash. Even the relatively gentle eruption of low-silicon, low-viscosity-shield volcanoes is associated with the release of reduced volcanic gases. Similarly, reduced gases are released dissolved in waters associated with hot springs and geysers. Oxidation of reduced volcanic gases occurs in the atmosphere, in natural waters, and on the surfaces of minerals. This is predominantly an abiotic process, although many chemolithoautotrophs have colonized the walls and channels of hot springs and fumaroles, catalyzing the oxidation of reduced gases with O_2. Much of biological diversity in hyperthermophilic environments consists of prokaryotes employing these unusual metabolic types.

Recent estimates of global average volcanic gas emissions suggest that volcanic sulfur emissions range, $0.1\text{–}1 \times 10^{12}$ mol S yr^{-1}, is nearly equally distributed between SO_2 and H_2S (Halmer *et al.*, 2002; Arthur, 2000). This range agrees well with the estimates used by both Holland (2002) and Lasaga and Ohmoto (2002) for average volcanic S emissions through geologic time. Other reduced gas emissions (CO, CH_4, and H_2) are estimated to be similar in magnitude (Mörner and Etiope, 2002; Arthur, 2000; Delmelle and Stix, 2000). All of these gases have very short residence times in the atmosphere, revealing that emission and oxidative consumption of these gases are closely coupled, and that O_2 consumption through volcanic gas oxidation is very efficient.

11.4.7 Mineral Oxidation

During uplift and erosion of the Earth's continents, rocks containing chemically reduced minerals become exposed to the oxidizing conditions of the atmosphere. Common rock-forming minerals susceptible to oxidation include olivine, Fe^{2+}-bearing pyroxenes and amphiboles, metal sulfides, and graphite. Ferrous-iron oxidation is a common feature of soil formation. Iron oxides derived from oxidation of Fe^{2+} in the parent rock accumulate in the B horizon of temperate soils, and extensive laterites consisting of iron and aluminum oxides develop in tropical soils to many meters of depth. Where erosion rates are high, iron-bearing silicate minerals may be transported short distances in rivers; however, iron oxidation is so efficient that very few sediments show deposition of clastic ferrous-iron minerals. Sulfide minerals are extremely susceptible to oxidation, often being completely weathered from near-surface rocks. Oxidation of sulfide minerals generates appreciable acidity, and in areas where mining has brought sulfide minerals in contact with the atmosphere or O_2-charged rainwater, low-pH discharge has become a serious environmental problem. Although ferrous silicates and sulfide minerals such as pyrite will oxidize under sterile conditions, a growing body of evidence suggests that in many natural environments, iron and sulfur oxidation is mediated by chemolithoautotrophic microorganisms. Prokaryotes with chemolithoautotrophic metabolic pathways have been isolated from many environments where iron and sulfur oxidation occurs. The amount of O_2 consumed annually through oxidation of Fe^{2+} and sulfur-bearing minerals is not known, but based on sulfur isotope mass balance constraints is on the order of $(0.1\text{–}1) \times 10^{12}$ mol yr^{-1} each for iron and sulfur, similar in magnitude to the flux of reduced gases from volcanism.

11.4.8 Hydrothermal Vents

The spreading of lithospheric plates along mid-ocean ridges is associated with much undersea volcanic eruptions and release of chemically reduced metal sulfides. Volcanic gases released by subaerial volcanoes are also generated by submarine eruptions, contributing dissolved reduced gases to seawater. Extrusive lava flows generate pillow basalts, which are composed in part of ferrous-iron silicates. Within concentrated zones of hydrothermal fluid flow, fracture-filling and massive sulfide minerals precipitate within the pillow lavas, large chimneys grow from the seafloor by rapid precipitation of Fe–Cu–Zn sulfides, and metal sulfide-rich plumes of high-temperature "black smoke" are released into seawater. In cooler zones, more gradual emanations of metal and sulfide rich fluid diffuse upward through the pillow basalts and slowly mix with seawater. Convective cells develop, in which cold seawater is drawn down into pillow lavas off the ridge axis to replace the water released at the hydrothermal vents. Reaction of seawater with basalt serves to alter the basalt. Some sulfide is liberated from the basalt, entrained seawater sulfate precipitates as anhydrite or is reduced to sulfide, and Fe^{2+} and other metals in basalt are replaced with seawater-derived magnesium. Oxygen dissolved in seawater is consumed during alteration of seafloor basalts. Altered basalts containing oxidized iron-mineral can extend as much as 500 m below the seafloor. Much more O_2 is consumed during oxidation of black smoker and chimney sulfides. Chimney sulfide may be only partially oxidized prior to transport off-axis through spreading and burial by sediments. However, black smoker metal-rich fluids are fairly rapidly oxidized in seawater, forming insoluble iron and manganese oxides, which slowly settle out on the seafloor, generating metalliferous sediments.

Because of the diffuse nature of reduced species in hydrothermal fluids, it is not known what role marine chemolithoautotrophic microorganisms may play in O_2 consumption associated with fluid plumes. Certainly such organisms live within the walls and rubble of cooler chimneys and basalts undergoing seafloor weathering. Because of the great length of mid-ocean ridges throughout the oceans, and the abundance of pyrite and other metal sulfides associated with these ridges, colonization and oxidation of metal sulfides by chemolithoautotrophic organisms may form an unrecognized source of primary production associated with consumption, not net release, of O_2.

11.4.9 Iron and Sulfur Oxidation at the Oxic–Anoxic Transition

In restricted marine basins underlying highly productive surface waters, where consumption of O_2 through aerobic respiration near the surface allows anoxia to at least episodically extend beyond the sediment into the water column, oxidation of reduced sulfur and iron may occur when O_2 is present. Within the Black Sea and many anoxic coastal fjords, the transition from oxic to anoxic environments occurs within the water column. At other locations, such as the modern Peru Shelf and coastal California basins, the transition occurs right at the sediment–water interface. In these environments, small concentrations of sulfide and O_2 may coexist within a very narrow band where sulfide oxidation occurs. In such environments, appreciable sulfur recycling may occur, with processes of sulfate reduction, sulfide oxidation, and sulfur disproportionation acting within millimeters of each other. These reactions are highly mediated by microorganisms, as evidenced by extensive mats of chemolithoautotrophic sulfur-oxidizing *Beggiatoa* and *Thioploca* found where the oxic–anoxic interface and seafloor coincide.

Similary, in freshwater and non-sulfidic brackish environments with strong O_2 demand, dissolved ferrous-iron may accumulate in groundwaters and anoxic bottom waters. Significant iron oxidation will occur where the water table outcrops with the land surface (i.e., groundwater outflow into a stream), or in lakes and estuaries, at the oxic–anoxic transition within the water column. Insoluble iron oxides precipitate and settle down to the sediment. Recycling of iron may occur if sufficient organic matter exists for ferric-iron reduction to ferrous-iron.

The net effect of iron and sulfur recycling on atmospheric O_2 is difficult to constrain. In most cases, oxidation of sulfur or iron consumes O_2 (there are some anaerobic chemolithoautotrophic microorganisms that can oxidize reduced substrates using nitrate or sulfate as the electron acceptor). Reduction of sulfate or ferric-iron are almost entirely biological processes, coupled with the oxidation of organic matter; sulfate and ferric-iron reduction individually have no effect on O_2. However, the major source of organic substrates for sulfate and ferric-iron reduction is ultimately biomass derived from photosynthetic organisms, which is associated with O_2 generation. The net change derived from summing the three processes (S^{2-}- or Fe^{2+}-oxidation, SO_4^{2-}- or Fe^{3+}-reduction, and photosynthesis), is net gain of organic matter with no net production or consumption of O_2.

11.4.10 Abiotic Organic Matter Oxidation

Aerobic respiration is the main means by which O_2 is consumed on the Earth. This pathway occurs throughout most Earth-surface environments: soils and aquatic systems, marine surface waters supersaturated with O_2, within the water column and the upper zones of sediments. However, reduced carbon materials are also reacted with O_2 in a variety of environments where biological activity has not been demonstrated. Among these are photo-oxidation of dissolved and particulate organic matter and fossil fuels, fires from burning vegetation and fossil fuels, and atmospheric methane oxidation. Olefins (organic compounds containing double bonds) are susceptible to oxidation in the presence of transition metapls, ozone, UV light, or gamma radiation. Low-molecular-weight oxidized organic degradation products form from oxidation reactions, which in turn may provide organic substrates for aerobic respiration or C_1 metabolism. Fires are of course high-temperature combustion of organic materials with O_2. Research on fires has shown that O_2 concentration has a strong influence on initiation and maintenance of fires (Watson, 1978; Lenton and Watson, 2000). Although the exact relationship between p_{O_2} and initiation of fire in real terrestrial forest communities is debated (see Robinson, 1989, 1991), it is generally agreed that, at low p_{O_2}, fires cannot be started even on dry wood, although smoldering fires with inefficient oxidation can be maintained. At high p_{O_2}, even wet wood can support flame, and fires are easily initiated with a spark discharge such as lightning.

Although most methane on Earth is oxidized during slow gradual transport upwards through soils and sediments, at select environments there is direct injection of methane into the atmosphere. Methane reacts with O_2 in the presence of light or metal surface catalysts. This reaction is fast enough, and the amount of atmospheric O_2 large enough, that significant concentrations of atmospheric CH_4 are unlikely to accumulate. Catastrophic calving of submarine, CH_4-rich hydrates during the geologic past may have liberated large quantities of methane to the atmosphere. However, isotopic evidence suggests that this methane was oxidized and consumed in a geologically short span of time.

11.5 GLOBAL OXYGEN BUDGETS AND THE GLOBAL OXYGEN CYCLE

One window into the global budget of oxygen is the variation in O_2 concentration normalized to N_2. O_2/N_2 ratios reflect changes in atmospheric O_2 abundance, because N_2 concentration is assumed to be invariant through time. Over an annual cycle, $\delta(O_2/N_2)$ can vary by 100 per meg or more, especially at high-latitude sites (Keeling and Shertz, 1992). These variations reflect latitudinal variations in net O_2 production and consumption related to seasonal high productivity during summer. Observations of O_2/N_2 variation have been expanded beyond direct observation (limited to the past several decades) to records of atmospheric composition as trapped in Antarctic firn ice and recent ice cores (Battle *et al.*, 1996; Sowers *et al.*, 1989; Bender *et al.*, 1985, 1994b). These records reflect a slow gradual decrease in atmospheric O_2 abundance over historical times, attributed to the release and oxidation of fossil fuels.

As yet, details of the fluxes involved in the processes that generate and consume molecular oxygen are too poorly constrained to establish a balanced O_2 budget. A summary of the processes believed to dominate controls on atmospheric O_2, and reasonable best guesses for the magnitude of these fluxes, if available, are shown in Figure 7 (from Keeling *et al.*, 1993).

11.6 ATMOSPHERIC O_2 THROUGHOUT EARTH'S HISTORY

11.6.1 Early Models

Starting with Cloud (1976), two key geologic formations have been invoked to constrain the history of oxygenation of the atmosphere: banded iron formations (BIFs) and red beds (Figure 8). BIFs are chemical sediments containing very little detritus, and consist of silica laminae interbedded with layers of alternately high and low ratios of ferric- to ferrous-iron. As chemical sediments, BIFs imply the direct precipitation of ferrous-iron from the water column. A ferrous-iron-rich ocean requires anoxia, which by implication requires an O_2-free atmosphere and an anaerobic world. However, the ferric-iron layers in BIFs do reflect consumption of molecular O_2 (oxidation of ferrous-iron) at a rate much greater than supply of O_2 through prebiotic H_2O photolysis. Thus, BIFs may also record the evolution of oxygenic photosynthesis and at least localized elevated dissolved O_2 concentrations (Walker, 1979). Red beds are sandy sedimentary rocks rich in coatings, cements, and particles of ferric-iron. Red beds form during and after sediment deposition, and thus require that both the atmosphere and groundwater are oxidizing. The occurrence of the oldest red beds (~2.0 Ga) coincides nearly with the disappearance of BIFs (Walker, 1979), suggesting that some threshold of atmospheric O_2 concentration was reached at this time. Although the general concept of a low-O_2 atmosphere before ~2.0 Ga and accumulated O_2 since that time has been agreed on for several decades, the details and texture of oxygenation of the Earth's atmosphere are still being debated.

Geochemists and cosmochemists initially looked to models of planetary formation and comparison with other terrestrial planets to understand the earliest composition of Earth's atmosphere. During planetary accretion and core formation, volatile components were liberated from a molten and slowly convecting mixture of silicates, metals, and trapped gases. The gravitational field of

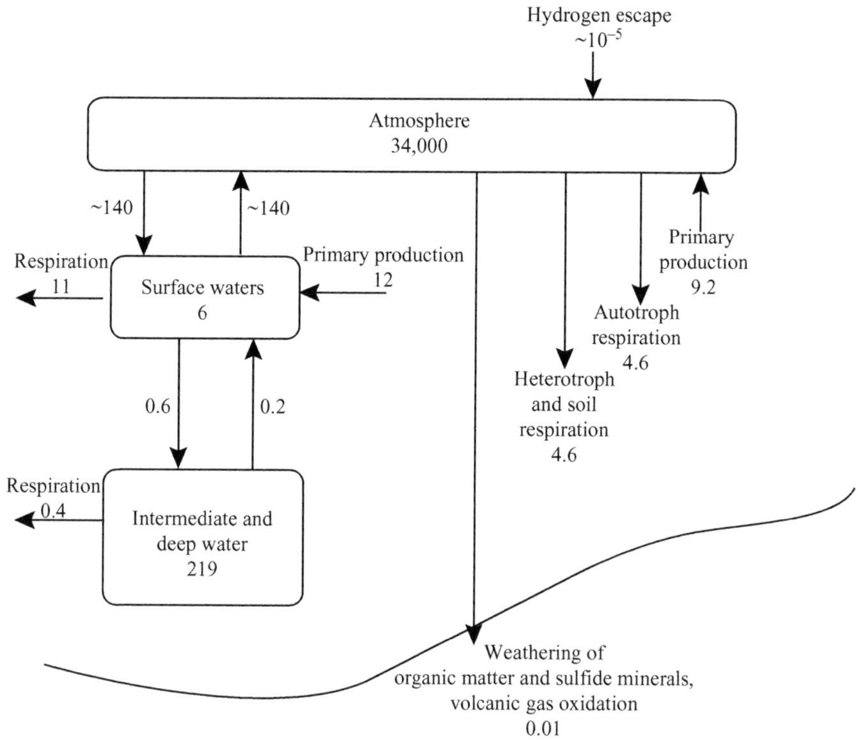

Figure 7 Global budget for molecular oxygen, including gas and dissolved O_2 reservoirs (sources Keeling *et al.*, 1993; Bender *et al.*, 1994a,b).

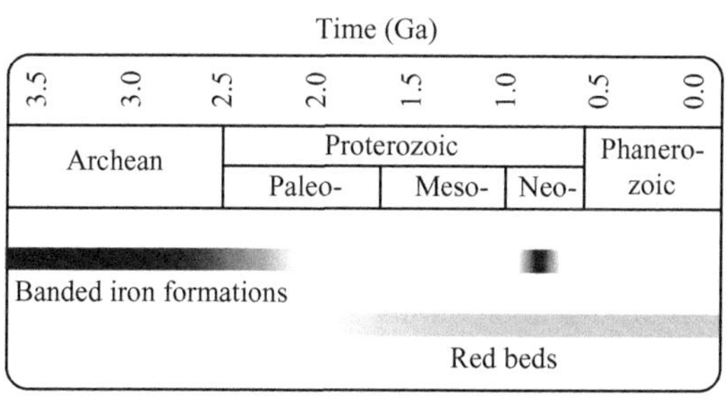

Figure 8 Archean distribution of banded iron formations, with short reoccurrence associated with widespread glaciation in the Neoproterozoic, and the Proterozoic and Phanerozoic distribution of sedimentary rocks containing ferric-iron cements (red beds). The end of banded iron formation and beginning of red bed deposition at ~2.2 Ga has been taken as evidence for a major oxygenation event in Earth's atmosphere.

Earth was sufficient to retain most of the gases released from the interior. These include CH_4, H_2O, N_2, NH_3, and H_2S. Much H_2 and He released from the interior escaped Earth's gravitational field into space; only massive planets such as Jupiter, Saturn, Neptune, and Uranus have retained an H_2–He rich atmosphere. Photolysis of H_2O, NH_3, and H_2S produced free O_2, N_2, and S, respectively. O_2 was rapidly consumed by oxidation of CH_4 and H_2S to form CO_2, CO, and SO_2. High partial pressures of CO_2 and CH_4 maintained a strong greenhouse effect and warm average Earth surface temperature (~90 °C), in spite of much lower solar luminosity. Recognizing that the early Earth contained an atmospheric substantially richer in strong greenhouse gases compared with the modern world provided a resolution to sub-freezing average Earth surface temperatures predicted for the early Earth due to reduced solar luminosity (Kasting *et al.*, 2001, 1983; Kiehl and Dickinson, 1987).

Liquid water on the early planet Earth allowed a hydrologic cycle and silicate mineral weathering to develop. Fairly quickly, much of the atmosphere's CO_2 was reacted with silicates to produce a bicarbonate-buffer ocean, while CH_4 was rapidly consumed by oxygen produced through photolysis of H_2O. Early microorganisms (and many of the most primitive organisms in existence today) used inorganic substrates to derive energy, and thrive at the high temperatures expected to be widespread during Earth's early history. These organisms include Archea that oxidize H_2 using elemental sulfur. Once photolysis and CH_4 oxidation generated sufficient p_{CO_2}, methanogenic Archea may have evolved. These organisms reduce CO_2 to CH_4 using H_2. However, sustainable life on the planet is unlikely to have developed during the first several hundred million years of Earth's history, due to large and frequent bolide impacts that would have sterilized the entire Earth's surface prior to ~3.8 Ga (Sleep *et al.*, 1989; Sleep and Zahnle, 1998; Sleep *et al.*, 2001;

Wilde *et al.*, 2001), although recent work by Valley *et al.* (2002) suggests a cool early Earth that continually supported liquid water as early as 4.4 Ga.

Much of present understanding of the earliest evolution of Earth's atmosphere can trace descent from Walker (1979) and references therein. The prebiological atmosphere (before the origin of life) was controlled principally by the composition of gases emitted from volcanoes. Emission of H_2 in volcanic gases has contributed to net oxidation of the planet through time. This is achieved through several mechanisms. Simplest is hydrodynamic escape of H_2 from Earth's gravity. Because H_2 is a strongly reducing gas, loss of H_2 from the Earth equates to loss of reducing power or net increase in whole Earth oxidation state (Walker, 1979). Today, gravitational escape of H_2 and thus increase in oxidation state is minor, because little if any H_2 manages to reach the upper levels of the atmosphere without oxidizing. Early in Earth's history, sources of H_2 included volcanic gases and water vapor photolysis. The small flux of O_2 produced by photolysis was rapidly consumed by reaction with ferrous-iron and sulfide, contributing to oxidation of the crust.

Today, volcanic gases are fairly oxidized, consisting mainly of H_2O and CO_2, with smaller amounts of H_2, CO, and CH_4. The oxidation state of volcanic gases derives in part from the oxidation state of the Earth's mantle. Mantle oxygen fugacity today is at or near the quartz-fayalite-magnesite buffer (QFM), as is the f_{O_2} of eruptive volcanic gases. Using whole-rock and spinel chromium abundance from volcanogenic rocks through time, Delano (2001) has argued that the average oxidation state of the Earth's mantle was set very early in Earth's history (~3.6–3.9 Ga) to f_{O_2} at or near the QFM buffer. Magmas with this oxidation state release volcanic gases rich in H_2O, CO_2, and SO_2, rather than more reducing gases. Thus, throughout much of Earth's history, volcanic gases contributing to the atmosphere have been fairly oxidized. More reduced magma compositions have been detected in diamond-bearing assemblages likely Hadean in age (>4.0 Ga) (Haggerty and Toft, 1985). The increase in mantle oxidation within several hundred million years of early Earth's history reveals very rapid "mantle + crust" overturn and mixing at this time, coupled with subduction and reaction of the mantle with hydrated and oxidized crustal minerals (generated from reaction with O_2 produced through H_2O photolysis).

11.6.2 The Archean

11.6.2.1 Constraints on the O_2 content of the Archean atmosphere

Several lines of geochemical evidence support low to negligible concentrations of atmospheric O_2 during the Archean and earliest Proterozoic, when oxygenic photosynthesis may have evolved. The presence of pyrite and uraninite in detrital Archean sediments reveals that the atmosphere in the earliest Archean contained no free O_2 (Cloud, 1972). Although Archean-age detrital pyrites from South Africa may be hydrothermal in origin, Australian sediments of the Pilbara craton (3.25–2.75 Ga) contain rounded grains of pyrite, uraninite, and siderite (Rasmussen and Buick, 1999), cited as evidence for an anoxic atmosphere at this time. Although disputed (Ohmoto, 1999), it is difficult to explain detrital minerals that are extremely susceptible to dissolution and oxidation under oxidizing condition unless the atmosphere of the Archean was essentially devoid of O_2.

Archean paleosols provide other geochemical evidence suggesting formation under reducing conditions. For example, the 2.75 Ga Mount Roe #2 paleosol of Western Australia contains up to 0.10% organic carbon with isotope ratios between −33‰ and −55‰ (Rye and Holland, 2000). These isotope ratios suggest that methanogenesis and methanotrophy were important pathways of carbon cycling in these soils. For modern soils in which the bulk organic matter is strongly ^{13}C-depleted (<−40‰), the methane fueling methanotrophy must be derived from somewhere outside the soil, because reasonable rates of fermentation and methanogenesis cannot supply enough CH_4. By extension, Rye and Holland (2000) argue that these soils formed under an atmosphere rich in CH_4, with any O_2 consumed during aerobic methanotrophy having been supplied by localized limited populations of oxygenic photoautotrophs. Other paleosol studies have used lack of cerium oxidation during soil formation as an indicator of atmospheric anoxia in the Archean (Murakami *et al.*, 2001). In a broader survey of Archean and Proterozoic paleosols, Rye and Holland (1998) observe that all examined paleosols older than 2.4 Ga indicate loss of iron during weathering and soil formation. This chemical feature is consistent with soil development under an atmosphere containing $<10^{-4}$ atm O_2 (1 atm = 1.01325×10^5 Pa), although some research has suggested that anoxic soil development in the Archean does not necessarily require an anoxic atmosphere (Ohmoto, 1996).

Other evidence for low Archean atmospheric oxygen concentrations come from studies of mass-independent sulfur isotope fractionation. Photochemical oxidation of volcanic sulfur species, in contrast with aqueous-phase oxidation and dissolution that characterizes the modern sulfur cycle, may have been the major source of sulfate to seawater in the Archean (Farquhar *et al.*, 2002; Farquhar *et al.*, 2000). Distinct shifts in δ^{33}S and δ^{34}S in sulfide and sulfate from Archean rocks occurred between 2.4–2.0 Ga, consistent with a shift from an O_2-free early atmosphere in which

SO_2 photochemistry could dominate among seawater sulfate sources to an O_2-rich later atmosphere in which oxidative weathering of sulfide minerals predominated over photochemistry as the major source of seawater sulfate. Sulfur isotope heterogeneities detected in sulfide inclusions in diamonds also are believed to derive from photochemical SO_2 oxidation in an O_2-free atmosphere at 2.9 Ga (Farquhar *et al.*, 2002). Not only do these isotope ratios require an O_2-free atmosphere, but they also imply significant contact between the mantle, crust, and atmosphere as recently as 2.9 Ga.

Nitrogen and sulfur isotope ratios in Archean sedimentary rocks also indicate limited or negligible atmospheric O_2 concentrations (Figure 9). Under an O_2-free environment, nitrogen could only exist as N_2 and reduced forms (NH_3, etc.). Any nitrate or nitrite produced by photolysis would be quickly reduced, likely with Fe^{2+}. If nitrate is not available, then denitrification (reduction of nitrate to free N_2) cannot occur. Denitrification is associated with a substantial nitrogen-isotope discrimination, generating N_2 that is substantially depleted in ^{15}N relative to the

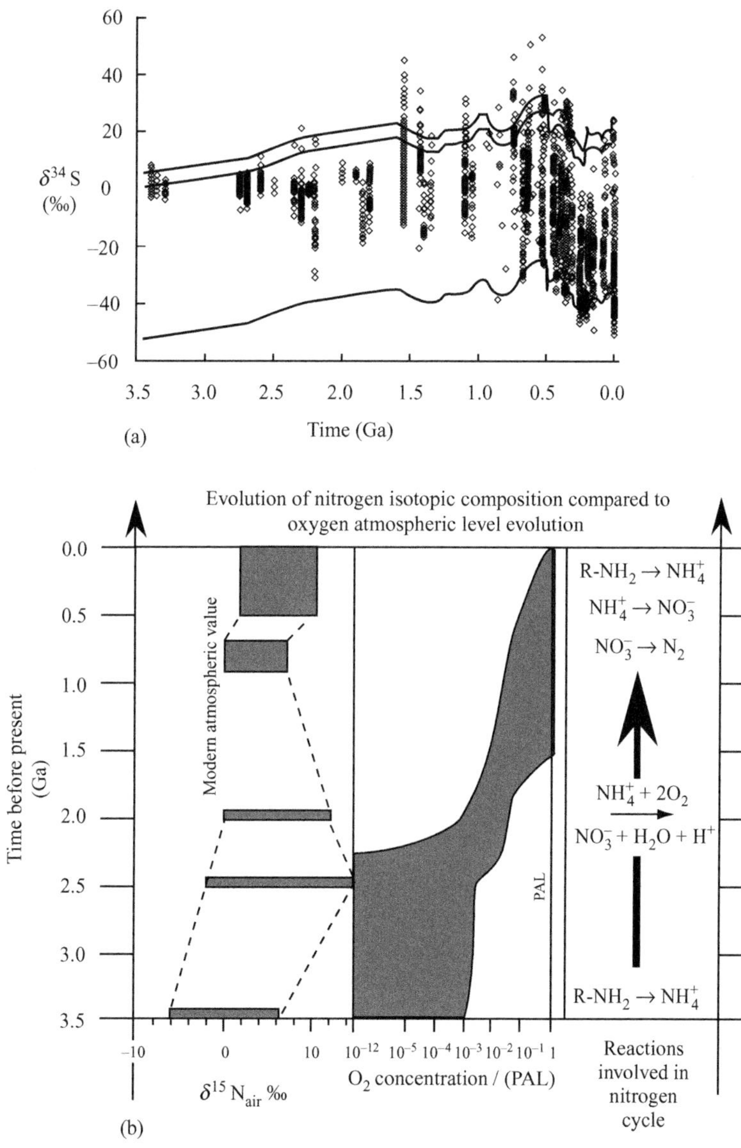

Figure 9 (a) The isotopic composition of sedimentary sulfate and sulfides through geologic time. The two upper lines show the isotopic composition of seawater sulfate (5‰ offset indicates uncertainty). The lower line indicated $\delta^{34}S$ of sulfate displaced by −55‰, to mimic average Phanerozoic maximum fractionation during bacterial sulfate reduction. Sulfide isotopic data (circles) indicated a much reduced fractionation between sulfate and sulfide in the Archean and Proterozoic (source Canfield, 1998). (b) Geologic evolution of the nitrogen isotopic composition of kerogen (left), estimated atmospheric O_2 content, and representative reactions in the biogeochemical nitrogen cycle at those estimated O_2 concentrations (sources Beaumont and Robert, 1999; Kasting, 1992).

NO_3^- source. In the modern system, this results in seawater nitrate (and organic matter) that is ^{15}N-enriched relative to air. Nitrogen in kerogen from Archean sedimentary rocks is not enriched in ^{15}N (as is found in all kerogen nitrogen from Proterozoic age to the present), but instead is depleted relative to modern atmospheric N_2 by several ‰ (Beaumont and Robert, 1999). This is consistent with an Archean nitrogen cycle in which no nitrate and no free O_2 was available, and nitrogen cycling was limited to N_2-fixation, mineralization and ammonia volatilization. Bacterial sulfate reduction is associated with a significant isotope discrimination, producing sulfide that is depleted in ^{34}S relative to substrate sulfate. The magnitude of sulfur isotope fractionation during sulfate reduction depends in part on available sulfate concentrations. Very limited differences in sulfur isotopic ratios among Archean sedimentary sulfide and sulfate minerals (<2‰ δ^{34}S) indicate only minor isotope fractionation during sulfate reduction in the Archean oceans (Canfield *et al.*, 2000; Habicht *et al.*, 2002), best explained by extremely low SO_4^{2-} concentrations (<200 μM in contrast with modern concentrations of ~28 mM) (Habicht *et al.*, 2002). The limited supply of sulfate in the Archean ocean suggests that the major source of sulfate to Archean seawater was volcanic gas, because oxidative weathering of sulfide minerals could not occur under an O_2-free atmosphere. The limited sulfate concentration would have suppressed the activity of sulfate-reducing bacteria and facilitated methanogenesis. However, by ~2.7 Ga, sedimentary sulfides that are ^{34}S-depleted relative to sulfate are detected, suggesting at least sulfate reduction, and by implication sources of sulfate to seawater through sulfide oxidation, may have developed (Canfield *et al.*, 2000).

11.6.2.2 The evolution of oxygenic photosynthesis

In the early Archean, methanogenesis (reaction of $H_2 + CO_2$ to yield CH_4) was likely a significant component of total primary production. O_2 concentrations in the atmosphere were suppressed due to limited O_2 production and rapid consumption with iron, sulfur, and reduced gases, while CH_4 concentrations were likely very high (Kasting *et al.*, 2001). The high methane abundance is calculated to have generated a hydrocarbon-rich smog that could screen UV light and protect early life in the absence of O_2 and ozone (Kasting *et al.*, 2001). Other means of protection from UV damage in the O_2-free Archean include biomineralization of cyanobacteria within UV-shielded iron–silica sinters (Phoenix *et al.*, 2001).

Biological evolution may have contributed to early Archean oxidation of the Earth. Catling *et al.*, (2001) have recognized that CO_2 fixation associated with early photosynthesis may have been coupled with active fermentation and methanogenesis. Prior to any accumulation of atmospheric O_2, CH_4 may have been a large component of Earth's atmosphere. A high flux of biogenic methane is supported by coupled ecosystem–climate models of the early Earth (Pavlov *et al.*, 2000; Kasting *et al.*, 2001) as a supplement to Earth's greenhouse warming under reduced solar luminosity in the Archean. CH_4 in the upper atmosphere is consumed by UV light to yield hydrogen (favored form of hydrogen in the upper atmosphere), which escapes to space. Because H_2 escape leads to net oxidation, biological productivity and methanogenesis result in slow oxidation of the planet. Net oxidation may be in the form of direct O_2 accumulation (if CO_2 fixation is associated with oxygenic photosynthesis) or indirectly (if CO_2 fixation occurs via anoxygenic photosynthesis or anaerobic chemoautotrophy) through production of oxidized iron or sulfur minerals in the crust, which upon subduction are mixed with other crustal rocks and increase the crustal oxidation state. One system that may represent a model of early Archean biological productivity consists of microbial mats found in hypersaline coastal ponds. In these mats, cyanobacteria produce H_2 and CO that can be used as substrates by associated chemoautotrophs, and significant CH_4 fluxes out of the mats have been measured (Hoehler *et al.*, 2001). Thus, mats may represent communities of oxygenic photosynthesis, chemoautotrophy, and methanogenesis occurring in close physical proximity. Such communities would have contributed to elevated atmospheric CH_4 and the irreversible escape of hydrogen from the Archean atmosphere, contributing to oxidation of the early Earth.

The earliest photosynthetic communities may have contributed to oxidation of the early Earth through production of oxidized crustal minerals without requiring production of O_2. Crustal rocks today are more oxidized than the mantle, with compositions ranging between the QFM and hematite-magnesite f_{O_2} buffers. This is best explained as an irreversible oxidation of the crust associated with methanogenesis and hydrogen escape (Catling *et al.*, 2001). It is unlikely that O_2 produced by early photosynthesis ever directly entered the atmosphere.

Maintaining low O_2 concentrations in the atmosphere while generating oxidized crustal rocks requires oxidation mechanisms that do not involve O_2. Among these may be serpentinization of seafloor basalts (Kasting and Siefert, 2002). During seafloor weathering, ferrous-iron is released to form ferric oxyhydroxides. Using H_2O or CO_2 as the ferrous-iron oxidant, H_2 and CH_4 are generated. This H_2 gas (either produced directly, or indirectly through UV decomposition of CH_4) can escape the atmosphere, resulting in net oxidation of the crust and accumulation of oxidized crustal rocks. BIF formation in the Archean may be related to dissolved Fe^{2+} that was oxidized in shallow-water settings associated with local oxygenic photosynthesis or chemoautotrophy (Kasting and Siefert, 2002). After the biological innovation of oxygenic photosynthesis evolved, there was still a several hundred million year gap until O_2 began to accumulate in the atmosphere, because O_2 could only accumulate once the supply of reduced gases (CO and CH_4) and ferrous-iron fell below rates of photosynthetic O_2 supply.

11.6.2.3 Carbon isotope effects associated with photosynthesis

The main compound responsible for harvesting light energy to produce NADPH and splitting water to form O_2 is chlorophyll. There are several different structural variants of chlorophyll, including chlorophyll *a*, chlorophyll *b*, and bacteriochlorophylls *a–e* and *g*. Each of these shows optimum excitation at a different wavelength of light. All oxygenic photosynthetic organisms utilize chlorophyll *a* and/or chlorophyll *b*. Other non-oxygenic photoautotrophic microorganisms employ a diverse range of chlorophylls.

The earliest evidence for evolution of oxygenic photosynthesis comes from carbon isotopic signatures preserved in Archean rocks (Figure 10). CO_2 fixation through the Calvin cycle is associated with a significant carbon isotope discrimination, such that organic matter produced through CO_2 fixation is depleted in ^{13}C relative to ^{12}C by several per mil. In a closed or semi-closed system (up to and including the whole ocean–atmosphere system), isotope discrimination during fixation of CO_2 then results in a slight enrichment in the $^{13}C/^{12}C$ ratio of CO_2 not taken up during photosynthesis. Thus, a biosignature of CO_2 fixation is an enrichment in $^{13}C/^{12}C$ ratio in atmospheric CO_2 and seawater bicarbonate over a whole-Earth averaged isotope ratio. When carbonate minerals precipitate from seawater, they record the seawater isotope value. Enrichment of the $^{13}C/^{12}C$ ratio in early Archean carbonate minerals, and by extension seawater bicarbonate, is taken as early evidence for CO_2 fixation. Although anoxygenic photoautotrophs and chemoautotrophs fix CO_2 without generating O_2, these groups either do not employ the Calvin cycle (many anoxygenic photoautotrophs use the acetyl-CoA or reverse tricarboxylic acid pathways) or require O_2 (most chemolithoautotrophs). Thus, the most likely group of organisms responsible for this isotope effect are oxygenic photoautotrophs such as cyanobacteria. Kerogen and graphite that is isotopically depleted in ^{13}C is a common and continuous feature of the sedimentary record, extending as far back as ~3.8 Ga to the Isua Supracrustal Suite of Greenland (Schidlowski, 1988, 2001; Nutman *et al.*, 1997). Although some isotopically depleted graphite in the metasediments from the Isua Suite may derive from abiotic hydrothermal processes (van Zuilen *et al.*, 2002), rocks interpreted as metamorphosed turbidite deposits retain ^{13}C-depleted graphite believed to be biological in origin. Moreover, the isotopic distance between coeval carbonate and organic matter (in the form of kerogen) can be used to estimate biological productivity through time. With a few exceptions, these isotope mass balance estimates reveal that, since Archean times, global scale partitioning between inorganic and organic carbon, and thus

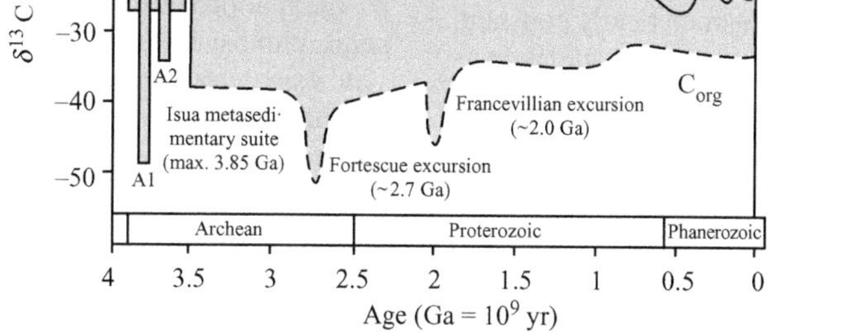

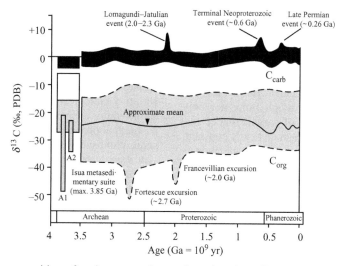

Figure 10 Isotopic composition of carbonates and organic carbon in sedimentary and metasedimentary rocks through geologic time. The negative excursions in organic matter $\delta^{13}C$ at 2.7 Ga and 2.0 Ga may relate to extensive methanogenesis as a mechanism of carbon fixation (Schidlowski, 2001) (reproduced by permission of Elsevier from *Precamb. Res.* **2001**, *106*, 117–134).

global productivity and carbon burial, have not varied greatly over nearly 4Gyr of Earth's history (Schidlowski, 1988, 2001). Approximately 25% of crustal carbon burial is in the form of organic carbon, and the remainder is inorganic carbonate minerals. This estimate derives from the mass- and isotope-balance equation:

$$\delta^{13}C_{avg} = f\delta^{13}C_{organic\ matter} + (1-f) \times \delta^{13}C_{carbonate} \quad (2)$$

where $\delta^{13}C_{avg}$ is the average isotopic composition of crustal carbon entering the oceans from continental weathering and primordial carbon emitted from volcanoes, $\delta^{13}C_{organic\ matter}$ is the average isotopic composition of sedimentary organic matter, $\delta^{13}C_{carbonate}$ is the average isotopic composition of carbonate sediments, and f is the fraction of carbon buried as organic matter in sediments.

The observation that the proportion of carbon buried in sediments as organic matter versus carbonate has not varied throughout geologic time raises several intriguing issues. First, biogeochemical cycling of carbon exhibits remarkable constancy across 4Gyr of biological evolution, in spite of large-scale innovations in primary production and respiration (including anoxygenic and oxygenic photosynthesis, chemoautotrophy, sulfate reduction, methanotrophy, and aerobic respiration). Thus, with a few notable exceptions expressed in the carbonate isotope record, burial flux ratios between organic matter and carbonate have remained constant, in spite of varying dominance of different modes of carbon fixation and respiration through time, not to mention other possible controls on organic matter burial and preservation commonly invoked for Phanerozoic systems (anoxia of bottom waters, sedimentation rate, selective preservation, or cumulative oxygen exposure time). Second, the constancy of organic matter versus carbonate burial through time reveals that throughout geologic time, the relative contributions of various sources and sinks of carbon to the "ocean + atmosphere" system have remained constant. In other words, to maintain a constant carbonate isotopic composition through time, not only must the relative proportion of organic matter versus carbonate burial have remained nearly constant, but the relative intensity of organic matter versus carbonate weathering also must have remained nearly constant. In the earliest stages of Earth's history, when continents were small and sedimentary rocks were sparse, inputs of carbon from continental weathering may have been small relative to volcanic inputs. However, the several billion year sedimentary record of rocks rich in carbonate minerals and organic matter suggests that continental weathering must have formed a significant contribution to total oceanic carbon inputs fairly early in the Archean. Intriguingly, this indicates that oxidative weathering of ancient sedimentary rocks may have been active even in the Archean, prior to accumulation of O_2 in the atmosphere. This runs counter to traditional interpretations of geochemical carbon–oxygen cycling, in which organic matter burial is equated to O_2 production and organic matter weathering is equated to O_2 consumption.

11.6.2.4 Evidence for oxygenic photosynthesis in the Archean

Fossil evidence for photosynthetic organisms from the same time period can be traced to the existence of stromatolites (Schopf, 1992, 1993; Schopf *et al.*, 2002; Hofmann *et al.*, 1999), although evidence for these early oxygenic photosynthetic communities is debated (Buick, 1990; Brasier *et al.*, 2002). Stromatolites are laminated sediments consistently of alternating organic matter-rich and organic matter-lean layers; the organic matter-rich layers are largely composed of filamentous cyanobacteria. Stromatolites and similar mat-forming cyanobacterial crusts occur today in select restricted shallow marine environments. Fossil evidence from many locales in Archean rocks (Greenland, Australia, South Africa) and Proterozoic rocks (Canada, Australia, South America) coupled with carbon isotope geochemistry provides indirect evidence for the evolution of oxygenic photosynthesis early in Earth's history (~3.5 Ga).

Other evidence comes from molecular fossils. All cyanobacteria today are characterized by the presence of 2α-methylhopanes in their cell membranes. Brocks *et al.*, (1999) have demonstrated the existence of this taxon-specific biomarker in 2.7 Ga Archean rocks of the Pilbara Craton in NW Australia. Also in these rocks is found a homologous series of C_{27} to C_{30} steranes. Steranes today derive from sterols mainly produced by organisms in the domain Eukarya. Eukaryotes are obligate aerobes that require molecular O_2, and thus Brocks and colleagues argue that the presence of these compounds provides strong evidence for both oxygenic photosynthesis and at least localized utilization of accumulated O_2. Of course, coexistence of two traits within a biological lineage today reveals nothing about which evolved first. It is uncertain whether 2α-methylhopane lipid biosynthesis preceded oxygenic photosynthesis among cyanobacteria. Sterol synthesis certainly occurs in modern prokaryotes (including some anaerobes), but no existing lineage produces the distribution of steranes found in the Brocks *et al.*, (1999) study except eukaryotes.

11.6.3 The Proterozoic Atmosphere

11.6.3.1 Oxygenation of the Proterozoic atmosphere

Although there is evidence for the evolution of oxygenic photosynthesis several hundred million years before the Huronian glaciation (e.g., Brocks *et al.*, 1999) or earlier (Schidlowski, 1988, 2001), high fluxes of UV light reacting with O_2 derived from the earliest photosynthetic organisms would have created dangerous reactive oxygen species that severely suppressed widespread development of large populations of these oxygenic photoautotrophs. Cyanobacteria, as photoautotrophs, need be exposed to visible light and have evolved several defense mechanisms to protect cell contents and repair damage. However, two key metabolic pathways (oxygenic photosynthesis and nitrogen fixation) are very sensitive to UV damage. Indisputably cyanobacterial fossil occur in 1,000 Ma rocks, with putative fossils occurring 2,500 Ma and possibly older (Schopf, 1992). Sediment mat-forming cyanobacteria and stromatolites are least ambiguous and oldest. Terrestrial encrusting cyanobacteria are only known in the Phanerozoic. Planktonic forms are not known for the Archean and early Proterozoic. They may not have existed, or they may not be preserved. Molecular evolution, specifically coding for proteins that build gas vesicles necessary for planktonic life, are homologous and conserved in all cyanobacteria. Thus, perhaps planktonic cyanobacteria existed throughout Earth's history since the late Archean.

Today, ozone forms in the stratosphere by reaction of O_2 with UV light. This effectively screens much incoming UV radiation. Prior to the accumulation of atmospheric O_2, no ozone could form, and thus the UV flux to the Earth's surface would be much greater (with harmful effects on DNA and proteins, which adsorb and are altered by UV). A significant ozone shield could develop at $\sim 10^{-2}$ PAL O_2 (Kasting, 1987). However, a fainter young Sun would have emitted somewhat lower UV, mediating the lack of ozone. Although seawater today adsorbs most UV light by 6–25 m (1% transmittance cutoff), seawater with abundant dissolved Fe^{2+} may have provided an effective UV screen in the Archean (Olson and Pierson, 1986). Also, waters rich in humic materials, such as modern coastal oceans, are nearly UV opaque. If Archean seawater contained DOM, this could adsorb some UV. Iron oxidation and BIF at ~2.5–1.9 Ga would have removed the UV screen in seawater. Thus, a significant UV stress may have developed at this time. This would be mediated coincidentally by accumulation of atmospheric O_2.

In the early Archean before oxygenic photosynthesis evolved, cyanobacteria were limited. Planktonic forms were inhibited, and limited by dissolved iron content of water, and existence of stratified, UV-screen refuges. Sedimentary mat-forming and stromatolite-forming communities were much more abundant. Iron-oxide precipitation and deposition of screen enzymes may have created UV-free colonies under a shield even in shallow waters.

Advent of oxygenic photosynthesis in the Archean generated small oxygen oases (where dissolved O_2 could accumulate in the water) within an overall O_2-free atmosphere waters containing oxygenic photoautotrophs might have reached 10% air saturation (Kasting, 1992). At this time, both unscreened UV radiation and O_2 may have coincided within the water column and sediments. This would lead to increased UV stress for cyanobacteria. At the same time, precipitation of iron from the water would make the environment even more UV-transparent. To survive, cyanobacteria would need to evolve and optimize defense and repair mechanisms for UV damage (Garcia-Pichel, 1998). Perhaps this explains the ~500 Myr gap between origin of cyanobacteria and accumulation of O_2 in the atmosphere. For example, the synthesis of scytonemin (a compound found exclusively in cyanobacteria) requires molecular oxygen (implying evolution in an oxic environment); it optimally screens UV-a, the form of UV radiation only abundant in an oxygenated atmosphere.

Oxygenation of the atmosphere at ~2.3–2.0 Ga may derive from at least three separate causes. First, discussed earlier, is the titration of O_2 with iron, sulfur, and reduced gases. The other is global rates of photosynthesis and organic carbon burial in sediments. If the rate of atmospheric O_2 supply (oxygenic photosynthesis) exceeds all mechanisms of O_2 consumption (respiration, chemoautotrophy, reduced mineral oxidation, etc.), then O_2 can accumulate in the atmosphere. One means by which this can be evaluated is through seawater carbonate $\delta^{13}C$. Because biological CO_2 fixation is associated with significant carbon isotope discrimination, the magnitude of carbon fixation is indicated by the isotopic composition of seawater carbonate. At times of more carbon fixation and burial in sediments, relatively more ^{12}C is removed from the atmosphere + ocean inorganic carbon pool than is supplied through respiration, organic matter oxidation, and carbonate mineral dissolution. Because carbon fixation is dominated by oxygenic photosynthesis (at least since the late Archean), periods of greater carbon fixation and burial of organic matter in sediments (observed as elevated seawater carbonate $\delta^{13}C$) are equated to periods of elevated O_2 production through oxygenic photosynthesis. The early Proterozoic Lomagundi event (~2.3–2.0 Ga) is recorded in the sediment record as a prolonged period of elevated seawater carbonate $\delta^{13}C$, with carbonate $\delta^{13}C$ values reaching nearly 10‰ in several sections around the world (Schidlowski, 2001). This represents an extended period of time (perhaps several hundred million years) during which removal of carbon from the "ocean + atmosphere" system as organic matter greatly exceeded supply. By implication, release of O_2 through photosynthesis was greatly accelerated during this time.

A third mechanism for oxygenation of the atmosphere at ~2.3 Ga relies on the slow, gradual oxidation of the Earth's crust. Irreversible H_2 escape and basalt-seawater reactions led to a gradual increase in the amount of oxidized and hydrated minerals contained in the Earth's crust and subducted in subduction zones throughout the Archean. Gradually this influenced the oxidation state of volcanic gases derived in part from subducted crustal rocks. Thus, although mantle oxygen fugacity may not have changed since the early Archean, crustal and volcanic gas oxygen fugacity slowly increased as the abundance of oxidized and hydrated crust increased (Holland, 2002; Kasting *et al.*, 1993; Kump *et al.*, 2001). Although slow to develop, Holland (2002) estimates that an increase in f_{O_2} of less than 1 log unit is all that would have been required for transition from an anoxic to an oxic atmosphere, assuming rates of oxygenic photosynthesis consistent with modern systems and the sediment isotope record. Once a threshold volcanic gas f_{O_2} had been reached, O_2 began to accumulate. Oxidative weathering of sulfides released large amounts of sulfate into seawater, facilitating bacterial sulfate reduction.

There are several lines of geochemical evidence that suggest a rise in oxygenation of the atmosphere ~2.3–2.0 Ga, beyond the coincident last occurrence of BIFs (with one late Proterozoic exception) and first occurrence of red beds recognized decades ago, and carbon isotopic evidence suggesting ample burial of sedimentary organic matter (Karhu and Holland, 1996; Bekker *et al.*, 2001; Buick *et al.*, 1998). The Huronian glaciation (~2.3 Ga) is the oldest known glacial episode recorded in the sedimentary record. One interpretation of this glaciation is that the cooler climate was a direct result of the rise of photosynthetically derived O_2. The rise of O_2 scavenged and reacted with the previously high atmospheric concentration of CH_4. Methane concentrations dropped, and the less-effective greenhouse gas product CO_2 could not maintain equable surface temperatures. Kasting *et al.*, (1983) estimate that a rise in p_{O_2} above ~10^{-4} atm resulted in loss of atmospheric CH_4 and onset of glaciation.

Paleosols have also provided evidence for a change in atmosphere oxygenation at some time between 2.3–2.0 Ga. Evidence from rare earth element enrichment patterns and U/Th fractionation suggests a rise in O_2 to ~0.005 bar by the time of formation of the Flin Flon paleosol of Manitoba, Canada, 1.85 Ga (Pan and Stauffer, 2000; Holland *et al.*, 1989; Rye and Holland, 1998). Rye and Holland (1998) examined several early Proterozoic paleosols and observed that negligible iron loss is a consistent feature from soils of Proterozoic age through the present. These authors estimate that a minimum p_{O_2} of >0.03 atm is required to retain iron during soil formation, and thus atmospheric O_2 concentration has been 0.03 or greater since the early Paleozoic. However, reevaluation of a paleosol crucial to the argument of iron depletion during soil formation under anoxia, the Hekpoort paleosol dated at 2.2 Ga, has revealed that the iron depletion detected by previous researchers may in fact be the lower zone of a normal oxidized lateritic soil. Upper sections of the paleosol that are not depleted in iron have been eroded away in the exposure examined by Rye and Holland (1998), but have been found in drill core sections. The depletion of iron and occurrence of ferrous-iron minerals in the lower sections of this paleosol have been reinterpreted by Beukes *et al.*, (2002) to indicate an abundant soil surface biomass at the time of deposition that decomposed to generate reducing conditions and iron mobilization during the wet season, and precipitation of iron oxides during the dry season.

Sulfur isotope studies have also provided insights into the transition from Archean low p_{O_2} to higher values in the Proterozoic. In the same studies that revealed extremely low Archean ocean sulfate concentrations, it was found that by ~2.2 Ga, isotopic compositions of sedimentary sulfates and sulfides indicate bacterial sulfate reduction under more elevated seawater sulfate concentrations compared with the sulfate-poor Archean (Habicht *et al.*, 2002; Canfield *et al.*, 2000). As described above, nitrogen isotope ratios in sedimentary kerogens show a large and permanent shift at ~2.0 Ga, consistent with denitrification, significant seawater nitrate concentrations, and thus available atmospheric O_2.

Prior to ~2.2 Ga, low seawater sulfate concentrations would have limited precipitation and subduction of sulfate-bearing minerals. This would maintain a lower oxidation state in volcanic gases derived in part from recycled crust (Holland, 2002). Thus, even while the oxidation state of the mantle has remained constant since ~4.0 Ga (Delano, 2001), the crust and volcanic gases derived from subduction of the crust could only achieve an increase in oxidation state once seawater sulfate concentrations increased.

A strong model for oxygenation of the atmosphere has developed based largely on the sulfur isotope record and innovations in microbial metabolism. The classical interpretation of the disappearance of BIFs relates to the rise of atmospheric O_2, oxygenation of the oceans, and removal of dissolved ferrous-iron by oxidation. However, another interpretation has developed, based largely on the evolving Proterozoic sulfur isotope record. During the oxygenation of the atmosphere at ~2.3–2.0 Ga, the oceans may not have become oxidized, but instead remained anoxic and became strongly sulfidic as well (Anbar and Knoll, 2002; Canfield, 1998 and references therein). Prior to ~2.3 Ga, the oceans were anoxic but not sulfidic. Ferrous-iron was abundant, as was manganese, because both are very soluble in anoxic, sulfide-free waters. The high concentration of dissolved iron and manganese facilitated nitrogen fixation by early cyanobacteria, such that available nitrogen was abundant, and phosphorus became the nutrient limiting biological productivity (Anbar and Knoll, 2002). Oxygenation of the atmosphere at ~2.3 Ga led to increased oxidative weathering of sulfide minerals on the continents and increased sulfate concentration in seawater. Bacterial sulfate reduction generated ample sulfide, and in spite of limited mixing, the deep oceans would have remained anoxic and now also sulfidic as long as p_{O_2} remained below ~0.07 atm (Canfield, 1998), assuming reasonable rates of primary production. Both iron and manganese form insoluble sulfides, and thus were effectively scavenged from seawater once the oceans became sulfidic. Thus, the Proterozoic oxygenation led to significant changes in global oxygen balances, with the atmosphere and ocean mixed layer becoming mildly oxygenated (probably <0.01 atm O_2), and the deep oceans becoming strongly sulfidic in direct response to the rise of atmospheric O_2.

In addition to increased oxygen fugacity of volcanic gases, and innovations in biological productivity to include oxygenic photosynthesis, the oxygenation of the Proterozoic atmosphere may be related to large-scale tectonic cycles. There are several periods of maximum deposition of sedimentary rocks rich in organic matter through geologic time. These are ~2.7 Ga, 2.2 Ga, 1.9 Ga, and 0.6 Ga (Condie *et al.*, 2001). The increased deposition of black shales at 2.7 Ga and 1.9 Ga are associated with superplume events: highly elevated rates of seafloor volcanism, oceanic crust formation. Superplumes lead to increased burial of both organic matter and carbonates (through transgression, increased atmospheric CO_2, accelerated weathering, and nutrient fluxes to the oceans), with no net effect on carbonate isotopic composition. Thus, periods of increased absolute rates of organic matter burial and O_2 production may be masked by a lack of carbon isotopic signature. Breakup of supercontinents may be related to the black shale depositional events at 2.2 Ga and 0.6 Ga. Breakup of supercontinents may lead to more sediment accommodation space on continental shelves, as well as accelerated continental weathering and delivery of nutrients to seawater, fertilizing primary production and increasing organic matter burial. These supercontinent breakup events at 2.2 Ga and 0.6 Ga are clearly observed on the carbonate isotopic record as increases in relative burial of organic matter versus carbonate. Modeling efforts examining the evolution of the carbon, sulfur, and strontium isotope records have shown that gradual growth of the continents during the Archean and early Proterozoic may in fact play a very large role in controlling the onset of oxygenation of the atmosphere, and that biological innovation may not be directly coupled to atmospheric evolution (Godderís and Veizer, 2000).

11.6.3.2 Atmospheric O_2 during the Mesoproterozoic

The sulfur-isotope-based model of Canfield and colleagues and the implications for limiting nutrient distribution proposed by Anbar and Knoll (2002) suggest that for nearly thousand million years (~2.2–1.2 Ga), oxygenation of the atmosphere above p_{O_2} ~0.01 atm was held in stasis. Although oxygenic photosynthesis was active, and an atmospheric ozone shield had developed to protect surface-dwelling organisms from UV radiation, much of the deep ocean was still anoxic and sulfidic. Removal of iron and manganese from seawater as sulfides generated severe nitrogen stress for marine communities, suggesting that productivity may have been limited throughout the entire Mesoproterozoic. The sluggish but consistent primary productivity and organic carbon burial through this time is seen in the carbonate isotope record. For several hundred million years, carbonate isotopic composition varied by no more than ±2‰, revealing very little change in the relative carbonate/organic matter burial in marine sediments. Anbar and Knoll (2002) suggest that this indicates a decoupling of the link between tectonic events and primary production, because although variations in tectonic activity (and associated changes in sedimentation, generation of restricted basins, and continental weathering) occurred during the Mesoproterozoic, these are not observed in the carbonate isotope record. This decoupling is a natural result of a shift in the source of the biological limiting nutrient from phosphorus (derived from continental weathering) to nitrogen (limited by N_2 fixation). Furthermore, the isotopic composition of carbonates throughout the Mesoproterozoic is 1–2‰ depleted relative to average carbonates from the early and late Proterozoic and Phanerozoic. This is consistent with a decrease in the relative proportion of carbon buried as organic matter versus carbonate during this time. It appears that after initial oxygenation of the atmosphere from completely anoxic to low p_{O_2} in the early Proterozoic, further oxygenation was halted for several hundred million years.

Global-scale reinvigoration of primary production, organic matter burial and oxygenation of the atmosphere may be observed in the latest Mesoproterozoic. Shifts in carbonate $\delta^{13}C$ of up to 4‰ are observed in sections around the globe at ~1.3 Ga (Bartley *et al.*, 2001; Kah *et al.*, 2001). These positive carbon isotope excursions are associated with the formation of the Rodinian supercontinent, which led to increased continental margin length, orogenesis, and greater sedimentation

(Bartley *et al.*, 2001). The increased organic matter burial and atmospheric oxygenation associated with this isotope excursion may be related to the first occurrence of laterally extensive $CaSO_4$ evaporites (Kah *et al.*, 2001). Although the rapid 10‰ increase in evaporate $\delta^{34}S$ across these sections is taken to indicate a much reduced seawater sulfate reservoir with much more rapid turnover times than found in the modern ocean, the isotope fractionation between evaporate sulfates and sedimentary sulfides indicates that the oceans were not sulfate-limiting. Atmospheric O_2 was of sufficient concentration to supply ample sulfate for bacterial sulfate reduction throughout the Mesoproterozoic.

11.6.3.3 *Neoproterozoic atmospheric O_2*

Elevated carbonate isotopic compositions (~4‰) and strong isotope fractionation between sulfate and biogenic sulfide minerals through the early Neoproterozoic indicates a period of several hundred million years of elevated biological productivity and organic matter burial. This may relate in part to the oxygenation of the atmosphere, ammonia oxidation, and increased seawater nitrate availability. Furthermore, oxidative weathering of molybdenum-bearing sulfide minerals, and greater oxygenation of the oceans increased availability of molybdenum necessary for cyanobacterial pathways of N_2-fixation (Anbar and Knoll, 2002). Much of the beginning of the Neoproterozoic, like the late Mesoproterozoic, saw gradual increases in oxygenation of the atmosphere, and possibly of the surface oceans. Limits on the oxygenation of the atmosphere are provided by sulfur isotopic and molecular evidence for the evolution of sulfide oxidation and sulfur disproportionation. The increase in sulfate–sulfide isotope fractionation in the Neoproterozoic (~1.0–0.6 Ga) reflects a shift in sulfur cycling from simple one-step reduction of sulfate to sulfide, to a system in which sulfide was oxidized to sulfur intermediates such as thiosulfate or elemental sulfur, which in turn were disproportionated into sulfate and sulfide (Canfield and Teske, 1996). Sulfur disproportionation is associated with significant isotope effects, generating sulfide that is substantially more ^{34}S-depleted than can be achieved through one-step sulfate reduction. The sulfur isotope record reveals that an increase in sulfate–sulfide isotope fractionation occurred between 1.0 Ga and 0.6 Ga, consistent with evolution of sulfide-oxidation at this time as derived from molecular clock based divergence of non-photosynthetic sulfide-oxidizing bacteria (Canfield and Teske, 1996). These authors estimated that innovation of bacterial sulfide oxidation occurred when much of the coastal shelf sediment (< 200 m water depth) was exposed to water with 13–46 μM O_2, which corresponds to 0.01–0.03 atm p_{O_2}. Thus, after the initial oxygenation of the atmosphere in the early Proterozoic (~2.3–2.0 Ga) to ~0.01 atm, p_{O_2} was constrained to this level until a second oxygenation in the late Proterozoic (~1.0–0.6 Ga) to 0.03 atm or greater.

During the last few hundred million years of the Proterozoic (~0.7–0.5 Ga), fragmentation of the Rodinian supercontinent was associated with at least two widespread glacial episodes (Hoffman *et al.*, 1998; Hoffman and Schrag, 2002). Neoproterozoic glacial deposits are found in Canada, Namibia, Australia, and other locations worldwide (Evans, 2000). In some of these, paleomagnetic evidence suggests that glaciation extended completely from pole to equator (Sumner, 1997; see also Evans (2000) and references therein). The "Snowball Earth", as these events have come to be known, is associated with extreme fluctuations in carbonate isotopic stratigraphy, reoccurrence of BIFs after an ~1.0 Ga hiatus, and precipitation of enigmatic, massive cap carbonate sediments immediately overlying the glacial deposits (Figure 11). It has been proposed that the particular configuration of continents in the latest Neoproterozoic, with land masses localized within the middle–low

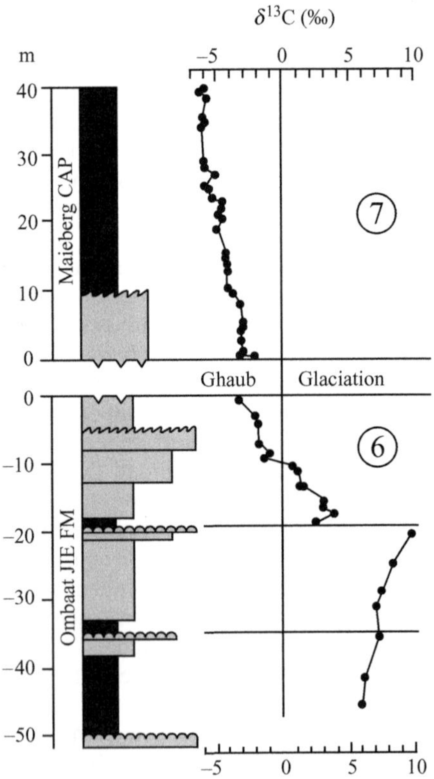

Figure 11 Composite section of carbonate isotope stratigraphy before and after the Ghaub glaciation from Namibia. Enriched carbonates prior to glaciation indicate active primary production and organic matter burial. Successive isotopic depletion indicates a shut down in primary production as the world became covered in ice (source Hoffman *et al.*, 1998).

latitudes, lead to cooling of climate through several mechanisms. These include higher albedo in the subtropics (Kirschvink, 1992), increased silicate weathering as the bulk of continents were located in the warm tropics, resulting in a drawdown of atmospheric CO_2 (Hoffman and Schrag, 2002), possibly accelerated through high rates of OM burial, and reduced meridional Hadley cell heat transport, because tropical air masses were drier due to increased continentality (Hoffman and Schrag, 2002). Growth of initially polar ice caps would have created a positive ice–albedo feedback such that greater than half of Earth's surface became ice covered, global-scale growth of sea ice became inevitable (Pollard and Kasting, 2001; Baum and Crowley, 2001). Kirschvink (1992) recognized that escape from the Snowball Earth becomes possible because during extreme glaciation, the continental hydrologic cycle would be shut down. He proposed that sinks for atmospheric CO_2, namely, photosynthesis and silicated weathering, would have been eliminated during the glaciation. Because volcanic degassing continued during the glaciation, CO_2 in the atmosphere could rise to very high concentrations. It was estimated that an increase in p_{CO_2} to 0.12 bar would be sufficient to induce a strong enough greenhouse effect to begin warming the planet and melting the ice (Caldeira and Kasting, 1992). Once meltback began, it would subsequently accelerate through the positive ice–albedo feedback. Intensely warm temperatures would follow quickly after the glaciation, until accelerated silicate weathering under a reinvigorated hydrologic cycle could consume excess CO_2 and restore p_{CO_2} to more equable values. Carbonate mineral precipitation was inhibited during the global glaciation, so seawater became enriched in hydrothermally derived cations. At the end of glaciation, mixing of cation-rich seawater with high alkalinity surface runoff under warming climates led to the precipitation of massive cap carbonates.

Geochemical signatures before and after the Snowball episodes reveal rapid and short-lived changes in global biogeochemical cycles that impacted O_2 concentrations in the atmosphere and oceans. Prior to the Snowball events, the Neoproterozoic ocean experienced strong primary production and organic matter burial, as seen in the carbon isotopic composition of Neoproterozoic seawater in which positive isotope excursions reached 10‰ in some sections (Kaufman and Knoll, 1995). These excursions are roughly coincident with increased oxygenation of the atmosphere in the Neoproterozoic (Canfield and Teske, 1996; Des Marais *et al.*, 1992). Although sulfur isotopic evidence suggests that much of the ocean may have become oxygenated during the Neoproterozoic, it is very likely that this was a temporary phenomenon. The global-scale glaciations of the late Neoproterozoic would have driven the oceans to complete anoxia through several mechanisms (Kirschvink, 1992). Ice cover inhibited air–sea gas exchange, and thus surface waters and sites of deep-water formation were cut off from atmospheric O_2 supplies. Intense oxidation of organic matter in the water column and sediment would have quickly consumed any available dissolved O_2. Extreme positive sulfur-isotope excursions associated with snowball-succession deposits reflect nearly quantitative, closed-system sulfate reduction in the oceans during glaciation (Hurtgen *et al.*, 2002). As seawater sulfate concentrations were reduced, hydrothermal inputs of ferrous-iron exceeded sulfide supply, allowing BIFs to form (Hoffman and Schrag, 2002). Closed-system sulfate reduction and iron formations require ocean anoxia throughout the water column, although restricted oxygenic photosynthesis beneath thin tropical sea ice may be responsible for ferrous-iron oxidation and precipitation (Hoffman and Schrag, 2002).

Once global ice cover was achieved, it is estimated to have lasted several million years, based on the amount of time required to accumulate 0.12 bar CO_2 at modern rates of volcanic CO_2 emission (Caldeira and Kasting, 1992). This must have placed extreme stress on eukaryotes and other organisms dependent on aerobic respiration. Obviously, oxygenated refugia must have existed, because the rise of many eukaryotes predates the Neoproterozoic Snowball events (Butterfield and Rainbird, 1998; Porter and Knoll, 2000). Nonetheless, carbon isotopic evidence leading up to and through the intense glacial intervals suggests that overall biological productivity was severely repressed during Snowball events. Carbonate isotope compositions fall from extremely enriched values to depletions of $\sim(-6‰)$ at glacial climax (Hoffman *et al.*, 1998; Kaufman *et al.*, 1991; Kaufman and Knoll, 1995). These carbonate isotope values reflect primordial (volcanogenic) carbon inputs, and indicate that effectively zero organic matter was buried in sediments at this time. Sustained lack of organic matter burial and limited if any oxygenic photosynthesis under the glacial ice for several million years would have maintained an anoxic ocean, depleted in sulfate, with low sulfide concentrations due to iron scavenging, and extremely high alkalinity and hydrothermally-derived ion concentrations. Above the ice, the initially mildly oxygenated atmosphere would slowly lose its O_2. Although the hydrologic cycles were suppressed, and thus oxidative weathering of sulfide minerals and organic matter exposed on the continents was inhibited, the emission of reduced volcanic gases would have been enough to fully consume 0.01 bar O_2 within several hundred thousand years. Thus, it is hypothesized that during Snowball events, the Earth's atmosphere returned to pre-Proterozoic anoxic conditions for at least several million years. Gradual increases in carbonate $\delta^{13}C$ after glaciation and deposition of cap carbonates suggests slow restored increases in primary productivity and organic carbon burial.

Carbon isotopic evidence suggests that reoxygenation, at least of seawater, was extremely gradual throughout the remainder of the Neoproterozoic. Organic matter through the terminal Proterozoic derives largely from bacterial heterotrophs, particularly sulfate reducing bacteria, as opposed to primary producers (Logan *et al.*, 1995). These authors suggested that throughout the terminal Neoproterozoic, anaerobic heterotrophy dominated by sulfate reduction was active throughout the water column, and O_2 penetration from surface waters into the deep ocean was inhibited. Shallow-water oxygen-deficient environments became widespread at the Precambrian–Cambrian boundary (Kimura and Watanabe, 2001), corresponding to negative carbonate $\delta^{13}C$ excursions and significant biological evolution from Ediacaran-type metazoans to emergence of modern metazoan phyla in the Cambrian.

11.6.4 Phanerozoic Atmospheric O_2

11.6.4.1 Constraints on Phanerozoic O_2 variation

Oxygenation of the atmosphere during the latest Neoproterozoic led to a fairly stable and well-oxygenated atmosphere that has persisted through the present day. Although direct measurement or a quantifiable proxy for Phanerozoic paleo-p_{O_2} concentrations have not been reported, multiple lines of evidence point to upper and lower limits on the concentration of O_2 in the atmosphere during the past several hundred million years. Often cited is the nearly continuous record of charcoal in sedimentary rocks since the evolution of terrestrial plants some 350 Ma. The presence of charcoal indicates forest fires throughout much of the Phanerozoic, which are unlikely to have occurred below $p_{O_2} = 0.17$ atm (Cope and Chaloner, 1985). Combustion and sustained fire are difficult to achieve at lower p_{O_2}. The existence of terrestrial plants themselves also provides a crude upper bound on p_{O_2} for two reasons. Above the compensation point of p_{O_2}/p_{CO_2} ratios, photorespiration outcompetes photosynthesis, and plants experience zero or negative net growth (Tolbert *et al.*, 1995). Although plants have developed various adaptive strategies to accommodate low atmospheric p_{CO_2} or aridity (i.e. C_4 and CAM plants), net terrestrial photosynthesis and growth of terrestrial ecosystems are effectively inhibited if p_{O_2} rises too high. This upper bound is difficult to exactly constrain, but is likely to be ~0.3–0.35 atm. Woody tissue is also extremely susceptible to combustion at high p_{O_2}, even if the tissue is wet. Thus, terrestrial ecosystems would be unlikely above some upper limit of p_{O_2} because very frequent reoccurrence of wildfires would effectively wipe out terrestrial plant communities, and there is no evidence of this occurring during the Phanerozoic on a global scale. What this upper p_{O_2} limit is, however, remains disputed. Early experiments used combustion of paper under varying humidity and p_{O_2} (Watson, 1978) but paper may not be the most appropriate analog for inception of fires in woody tissue with greater moisture content and thermal thickness (Robinson, 1989; Berner *et al.*, 2002). Nonetheless, the persistence of terrestrial plant communities from the middle Paleozoic through the present does imply that p_{O_2} concentrations have not risen too high during the past 350 Myr.

Prior to evolution of land plants, even circumstantial constraints on early Paleozoic p_{O_2} are difficult to obtain. Invertebrate metazoans, which have a continous fossil record from the Cambrian on, require some minimum amount of dissolved oxygen to support their aerobic metabolism. The absolute minimum dissolved O_2 concentration able to support aerobic metazoans varies from species to species, and is probably impossible to reconstruct for extinct lineages, but modern infaunal and epifaunal metazoans can accommodate dissolved O_2 dropping to the tens of micromolar in concentration. If these concentrations are extrapolated to equilibrium of the atmosphere with well-mixed cold surface waters, they correspond to ~0.05 atm p_{O_2}. Although local and regional anoxia occurred in the oceans at particular episodes through Phanerozoic time, the continued presence of aerobic metazoans suggests that widespread or total ocean anoxia did not occur during the past ~600 Myr, and that p_{O_2} has been maintained at or above ~5% O_2 since the early Paleozoic.

11.6.4.2 Evidence for variations in Phanerozoic O_2

Several researchers have explored possible links between the final oxygenation of the atmosphere and explosion of metazoan diversity at the Precambrian–Cambrian boundary (McMenamin and McMenamin, 1990; Gilbert, 1996).While the Cambrian explosion may not record the origin of these phyla in time, this boundary does record the development of large size and hard skeletons required for fossilization (Thomas, 1997). Indeed, molecular clocks for diverse metazoan lineages trace the origin of these phyla to ~400 Myr before the Cambrian explosion (Doolittle *et al.*, 1996; Wray *et al.*, 1996). The ability of lineages to develop fossilizable hardparts may be linked with increasing oxygenation during the latest Precambrian. Large body size requires elevated p_{O_2} and dissolved O_2 concentrations, so the diffusion can supply O_2 to internal tissues. Large body size provides several advantages, and may have evolved rather quickly during the earliest Cambrian (Gould, 1995), but large size also requires greater structural support. The synthesis of collagen (a ubiquitous structural protein among all metazoans and possible precursor to inorganic structural components such as carbonate or phosphate biominerals) requires elevated O_2. The threshold for collagen biosynthesis, and associated skeletonization and development of large body size, could not occur until p_{O_2} reached some critical threshold some time in the latest Precambrian (Thomas, 1997).

There is evidence to suggest that the late Paleozoic was a time of very elevated p_{O_2}, to concentrations substantially greater than observed in the modern atmosphere. Coals from the Carboniferous and Permian contain a greater abundance of fusain, a product of woody tissue combustion and charring, than observed for any period of the subsequent geologic past (Robinson, 1989, 1991; Glasspool, 2000), suggesting more abundant forest fires and by implication possibly higher p_{O_2} at this time. Less ambiguous is the biological innovation of gigantism at this time among diverse arthropod lineages (Graham *et al.*, 1995). All arthropods rely on tracheal networks for diffusion of O_2 to support their metabolism; active pumping of O_2 through a vascular system as found in vertebrates does not occur. This sets upper limits on body size for a given p_{O_2} concentration. In comparison with arthropod communities of the Carboniferous and Permian, modern terrestrial arthropods are rather small. Dragonflies at the time reached 70 cm wingspan, mayflies reached 45 cm wingpans, millipedes reached 1 m in length. Even amphibians, which depend in part on diffusion of O_2 through their skin for aerobic respiration, reached gigantic size at this time. These large body sizes could not be supported by today's 21% O_2 atmosphere, and instead require elevated p_{O_2} between 350 Ma and 250 Ma. Insect taxa that were giants in the Carboniferous do not survive past the Permian, suggesting that declining p_{O_2} concentrations in the Permian and Mesozoic led to extinction (Graham *et al.*, 1995; Dudley, 1998).

Increases in atmospheric O_2 affect organisms in several ways. Increased O_2 concentration facilitates aerobic respiration, while elevated p_{O_2} against a constant p_{N_2} increases total atmospheric pressure, with associated changes in atmospheric gas density and viscosity (Dudley, 1998, 2000). In tandem these effects may have played a strong role in the innovation of insect flight in the Carboniferous (Dudley, 1998, 2000), with secondary peaks in the evolution of flight among birds, bats and other insect lineages corresponding to times of high O_2 in the late Mesozoic (Dudley, 1998).

In spite of elevated p_{O_2} in the late Paleozoic, leading up to this time was an extended period of water column anoxia and enhanced burial of organic matter in the Devonian. Widespread deposition of black shales, fine-grained laminated sedimentary rocks rich in organic matter, during the Devonian indicate at least partial stratification of several ocean basins around the globe, with oxygen deficiency throughout the water column. One example of this is from the Holy Cross Mountains of Poland, from which particular molecular markers for green sulfur bacteria have been isolated (Joachimski *et al.*, 2001). These organisms are obligately anaerobic chemophotoautotrophs, and indicate that in the Devonian basin of central Europe, anoxia extended upwards through the water column well into the photic zone. Other black shales that indicate at least episodic anoxia and enhanced organic matter burial during the Devonian are found at several sites around the world, including the Exshaw Shale of Alberta (Canada), the Bakken Shale of the Williston Basin (Canada/USA), the Woodford Shale in Oklahoma (USA), and the many Devonian black shales of the Illinois, Michigan, and Appalachian Basins (USA). Widespread burial of organic matter in the late Devonian has been linked to increased fertilization of the surface waters through accelerated continental weathering due to the rise of terrestrial plant communities (Algeo and Scheckler, 1998). High rates of photosynthesis with relatively small rates of global respiration led to accumulation of organic matter in marine sediments, and the beginnings of a pulse of atmospheric hyperoxia that extended through the Carboniferous into the Permian. Coalescence of continental fragments to form the Pangean supercontinent at this time led to widespread circulation-restricted basins that facilitated organic matter burial and net oxygen release, and later, to extensive infilling to generate near-shore swamps containing terrestrial vegetation which was often buried to form coal deposits during the rapidly fluctuating sea levels of the Carboniferous and Permian.

The largest Phanerozoic extinction occurred at the end of the Permian (~250 Ma). A noticeable decrease in the burial of organic matter in marine sediments across the Permian–Triassic boundary may be associated with a global decline in primary productivity, and thus, with atmospheric p_{O_2}. The gigantic terrestrial insect lineages, thought to require elevated p_{O_2}, do not survive across this boundary, further suggesting a global drop in p_{O_2}, and the sedimentary and sulfur isotope records indicate an overall increase in sulfate reduction and burial of pyritic shales (Berner, 2002; Beerling and Berner, 2000). Although a long-duration deep-sea anoxic event has been proposed as a cause for the Permian mass extinction, there are competing models to explain exactly how this might have occurred. Hotinski *et al.*, (2001) has shown that while stagnation of the water column to generate deep-water anoxia might at first seem attractive, global thermohaline stagnation would starve the oceans of nutrients, extremely limiting primary productivity, and thus shutting down dissolved O_2 demand in the deep oceans. Large negative excursions in carbon, sulfur, and strontium isotopes during the late Permian may indicate stagnation and reduced ventilation of seawater for extended periods, coupled with large-scale overturn of anoxic waters. Furthermore, sluggish thermohaline circulation at this time could derive from a warmer global climate and warmer water at the sites of high-latitude deep-water formation (Hotinski *et al.*, 2001). The late Permian paleogeography of one supercontinent (Pangea) and one superocean (Panthalassa) was very different from the arrangement of continents and oceans on the modern Earth. Coupled with elevated p_{CO_2} at the time (Berner, 1994; Berner and Kothavala, 2001), GCM models predict warmer climate, weaker wind stress, and low equator to pole temperature gradients. Although polar deep-water formation still occurred, bringing O_2 from the atmosphere to the deep oceans, anoxia was likely to develop at mid-ocean depths (Zhang *et al.*, 2001), and thermohaline circulation oscillations between thermally versus salinity driven modes of circulation were likely to develop. During salinity driven modes, enhanced bottom-water formation in warm, salty low-latitude regions would limit oxygenation of the deep ocean. Thus, although sustained periods of anoxia are unlikely to have developed during the late Permian, reduced oxygenation of deep water through sluggish thermohaline circulation, coupled with episodic anoxia driven by low-latitude warm salty bottom-water formation, may have led to reoccurring episodes of extensive ocean anoxia over period of several million years.

Other researchers have invoked extraterrestrial causes for the End-Permian extinction and anoxia. Fullerenes (cage-like hydrocarbons that effectively trap gases during formation and heating) have been detected in late Permian sediments from southern China. The noble gas complement in these fullerenes indicates an extraterrestrial origin, which has been interpreted by Kaiho *et al.*, (2001) to indicate an unrecognized bolide impact at the Permian–Triassic boundary. The abrupt decrease in $\delta^{34}S$ across this boundary (from 20‰ to 5‰) implies an enormous and rapid release of ^{34}S-depleted sulfur into the ocean–atmosphere system. These authors propose that volatilized bolide- and mantle-derived sulfur (~0‰) oxidized in air, consumed atmospheric and dissolved O_2, and generated severe oxygen and acid stress in the oceans. Isotope mass balance estimates require $\sim 10^{19}$ mol sulfur to be released, consuming a similar mass of oxygen. 10^{19} mol O_2 represents some 10–40% of the total available inventory of atmospheric and dissolved O_2 at this time, removal of which led to immediate anoxia, as these authors propose.

Other episodes of deep ocean anoxia and extensive burial of organic matter are known from the Jurassic, Cretaceous, Miocene, and Pleistocene, although these have not been linked to changes in atmospheric O_2 and instead serve as examples of the decoupling between atmospheric and deep-ocean O_2 concentrations through much of the geologic past. Widespread Jurassic black shale facies in northern Europe (Posidonien Schiefer, Jet Rock, and Kimmeridge Clay) were deposited in a restricted basin on a shallow continental margin. Strong monsoonal circulation led to extensive freshwater discharge and a low-salinity cap on basin waters during summer, and intense evaporation and antiestuarine circulation during winter (Röhl *et al.*, 2001), both of which contributed to water column anoxia and black shale deposition. Several oceanic anoxic events (OAEs) are recognized from the Cretaceous in all major ocean basins, suggesting possible global deep-ocean anoxia. Molecular markers of green sulfur bacteria, indicating photic zone anoxia, have been detected from Cenomanian–Turonian boundary section OAE sediments from the North Atlantic (Sinninghe Damsté and Köster, 1998). The presence of these markers (namely, isorenieratene, a diaromatic carotenoid accessory pigment used during anoxygenic photosynthesis) indicates that the North Atlantic was anoxic and euxinic from the base of the photic zone (~50–150 m) down to the sediment. High concentrations of trace metals scavenged by sulfide and an absence of bioturbation further confirm anoxia throughout the water column. Because mid-Cretaceous oceans were not highly productive, accelerated dissolved O_2 demand from high rates of respiration and primary production cannot be the prime cause of these OAEs. Most likely, the warm climate of the Cretaceous led to low O_2 bottom waters generated at warm, high salinity regions of low-latitude oceans. External forcing, perhaps through Milankovic-related precession-driven changes in monsoon intensity and strength, influenced the rate of salinity-driven deep-water formation, ocean basin oxygenation, and OAE formation (Wortmann *et al.*, 1999). Sapropels are organic matter-rich layers common to late Cenozoic sediments of the eastern Mediterranean. They are formed through a combination of increased primary production in surface waters, and increased organic matter preservation in the sediment likely to be associated with changes in ventilation and oxygenation of the deep eastern Mediterranean basins (Stratford *et al.*, 2000). The well-developed OMZ located off the coast of Southern California today may have been more extensive in the past. Variations in climate affecting intensity of upwelling and primary production, coupled with tectonic activity altering the depth of basins and height of sills along the California coast, have generated a series of anoxia-facies organic-matter-rich sediments along the west coast of North America, beginning with the Monterey Shale and continuing through to the modern sediments deposited in the Santa Barbara and Santa Monica basins.

As shown by the modeling efforts of Hotinski *et al.*, (2001) and Zhang *et al.*, (2001), extensive deep ocean anoxia is difficult to achieve for extended periods of geologic time during the Phanerozoic when p_{O_2} were at or near modern levels. Thus, while localized anoxic basins are common, special conditions are required to generate widespread, whole ocean anoxia such as observed in the Cretaceous. Deep-water formation in highly saline low-latitude waters was likely in the geologic past when climates were warmer and equator to pole heat gradients were reduced. Low-latitude deep-water formation has a significant effect on deep-water oxygenation, not entirely due to the lower O_2 solubility in warmer waters, but also the increased efficiency of nutrient use and recycling in low-latitude surface waters (Herbert and Sarmiento, 1991). If phytoplankton were 100% efficient at using and recycling nutrients, even with modern high-latitude modes of cold deep-water formation, the deep oceans would likely become anoxic.

11.6.4.3 Numerical models of Phanerozoic oxygen concentration

Although photosynthesis is the ultimate source of O_2 to the atmosphere, in reality photosynthesis and aerobic respiration rates are very closely coupled. If they were not, major imbalances in atmospheric CO_2, O_2, and carbon isotopes would result. Only a small fraction of primary production (from photosynthesis) escapes respiration in the water column or sediment to become buried in deep sediments and ultimately sedimentary rocks. This flux of buried organic matter is in effect "net photosynthesis", or total photosynthesis minus respiration. Thus, while over timescales of days to months, dissolved and atmospheric O_2 may respond to relative rates of photosynthesis or respiration, on longer timescales it is burial of organic matter in sediments (the "net photosynthesis") that matters. Averaged over hundreds of years or longer, burial of organic matter equates to release of O_2 into the "atmosphere + ocean" system:

Photosynthesis:

$$6CO_2 + 6H_2O \rightarrow C_6H_{12}O_6 + 6O_2 \quad (3)$$

Respiration:

$$C_6H_{12}O_6 + 6O_2 + 6CO_2 + 6H_2O \quad (4)$$

If for every 1,000 rounds of photosynthesis there are 999 rounds of respiration, the net result is one round of organic matter produced by photosynthesis that is not consumed by aerobic respiration. The burial flux of organic matter in sediments represents this lack of respiration, and as such is a net flux of O_2 to the atmosphere. In geochemists' shorthand, we represent this by the reaction

Burial of organic matter in sediments:

$$CO_2 + H_2O \rightarrow O_2 + \text{“}CH_2O\text{”} \quad (5)$$

where "CH_2O" is "not formaldehyde, or even any specific carbohydrate, but instead represents sedimentary organic matter. Given the elemental composition of most organic matter in sediments and sedimentary rocks, a more reduced organic matter composition might be more appropriate, i.e., $C_{10}H_{12}O$, which would imply release of 12.5 mol O_2 for every mole of CO eventually buried as organic matter. However, the simplified stoichiometry of (5) is applied for most geochemical models of C–S–O cycling.

If burial of organic matter equates to O_2 release to the atmosphere over long timescales, then the oxidative weathering of ancient organic matter in sedimentary rocks equates to O_2 consumption. This process has been called "georespiration" by some authors (Keller and Bacon, 1998). It can be represented by Equation (6), the reverse of (5):

Weathering of organic matter from rocks:

$$O_2 + \text{“}CH_2O\text{”} \rightarrow CO_2 + H_2O \quad (6)$$

Both Equations (3) and (4) contain terms for addition and removal of O_2 from the atmosphere. Thus, if we can reconstruct the rates of burial and weathering of OM into/out of sedimentary rocks through time, we can begin to quantify sources and sinks for atmospheric O_2. The physical manifestation of this equation is the reaction of organic matter with O_2 during the weathering and erosion of sedimentary rocks. This is most clearly seen in the investigation of the changes in OM abundance and composition in weathering profiles developed on black shales (Petsch *et al.*, 2000, 2001).

In addition to the C–O system, the coupled C–S–O system has a strong impact on atmospheric oxygen (Garrels and Lerman, 1984; Kump and Garrels, 1986; Holland, 1978, 1984). This is through the bacterial reduction of sulfate to sulfide using organic carbon substrates as electron donors. During bacterial sulfate reduction (BSR), OM is oxidized and sulfide is produced from sulfate. Thus, BSR provides a means of resupplying oxidized carbon to the "ocean + atmosphere" system without consuming O_2. The net reaction for BSR shows that for every 15 mol of OM consumed, 8 mol sulfate and 4 mol ferric-iron are also reduced to form 4 mol of pyrite (FeS_2):

$$4\,Fe(OH)_3 + 8\,SO_4^{2-} + 15\,CH_2O \rightarrow 4\,FeS_2 + 15\,HCO_3^- + 13\,H_2O + (OH)^- \quad (7)$$

Oxidation of sulfate using organic substrates as electron donors provides a means of restoring inorganic carbon to the ocean+atmosphere system without consuming free O_2. Every 4 mol of pyrite derived from BSR buried in sediments represents 15 mol of O_2 produced by photosynthesis to generate organic matter that will not be consumed through aerobic respiration. In effect, pyrite burial equates to net release of O_2 to the atmosphere, as shown by (8), obtained by the addition of Equation (7) to (5):

$$4Fe(OH)_3 + 8SO_4^{2-} + 15CH_2O \rightarrow 4FeS_2 + 15HCO_3^- + 13H_2O + (OH)^-$$

$$+15CO_2 + 15H_2O \rightarrow 15O_2 + 15\text{ "}CH_2O\text{"}$$

$$4Fe(OH)_3 + 8SO_4^{2-} + 15CO_2 + 2H_2O \rightarrow 4FeS_2 + 15HCO_3^- + (OH)^- + 15O_2 \quad (8)$$

Oxidative weathering of sedimentary sulfide minerals during exposure and erosion on the continents results in consumption of O_2 (Equation (9):

$$4FeS_2 + 15O_2 + 14H_2O \rightarrow 4Fe(OH)_3 + 8SO_4^{2-} + 16H^+ \quad (9)$$

Just as for the C–O geochemical system, if we can reconstruct the rates of burial and oxidative weathering of sedimentary sulfide minerals through geologic time, we can use these to estimate additional sources and sinks for atmospheric O_2 beyond organic matter burial and weathering.

In total, then, the general approach taken in modeling efforts of understanding Phanerozoic O_2 variability is to catalog the total sources and sinks for atmospheric O_2, render these in the form of a rate of change equation in a box model, and integrate the changing O_2 mass through time implied by changes in sources and sinks:

$$\begin{aligned} dM_{O_2}/dt &= \sum F_{O_2}\text{into the atmosphere} \\ &\quad -\sum F_{O_2}\text{out of the atmosphere} \\ &= F_{\text{burial of organic matter}} \\ &\quad +(15/8)F_{\text{burial of pyrite}} \\ &\quad -F_{\text{weathering of organic matter}} \\ &\quad -(15/8)F_{\text{weathering of pyrite}} \end{aligned} \quad (10)$$

One approach to estimate burial and weathering fluxes of organic matter and sedimentary sulfides through time uses changes in the relative abundance of various sedimentary rock types estimated over Phanerozoic time. Some sedimentary rocks are typically rich in both organic matter and pyrite. These are typically marine shales. In contrast, coal basin sediments contain much organic carbon, but very low amounts of sedimentary sulfides. Non-marine coarse-grained clastic sediments contain very little of either organic matter or sedimentary sulfides. Berner and Canfield (1989) simplified global sedimentation through time into one of three categories: marine shales + sandstone, coal basin sediments, and non-marine clastic sediments. Using rock abundance estimates derived from the data of Ronov and others (Budyko *et al.*, 1987; Ronov, 1976), these authors estimated burial rates for organic matter and pyrite as a function of time for the past ~600 Myr (Figure 12). Weathering rates for sedimentary organic matter and pyrite were calculated as first order dependents on the total mass of sedimentary organic matter or pyrite, respectively. Although highly simplified, this model provided several new insights into global-scale coupling C–S–O geochemistry. First, the broad-scale features of Phanerozoic O_2 evolution were established. O_2 concentrations in the atmosphere were low in the early Paleozoic, rising to some elevated p_{O_2} levels during the Carboniferous and Permian (probably to a concentration substantially greater than today's 0.21 bar), and then falling through the Mesozoic and Cenozoic to more modern values. This model confirmed the suspicion that in contrast with the Precambrian, Phanerozoic O_2 evolution was a story of relative stability through time, with no great excursions in p_{O_2}. Second, by linking C–S–O cycles with sediments and specifically sedimentation rates, this model helped fortify the idea that a strong control on organic matter burial rates globally, and thus ultimately on release of O_2 to the atmosphere, may be rates of sedimentation in near-shore environments. These authors extended this idea to propose that the close linkage between sedimentation and erosion (i.e., the fact that global rates of sedimentation are matched nearly exactly to global rates of sediment production—in other words, erosion) may in fact be a stabilizing influence on atmospheric O_2 fluctuations. If higher sedimentation rates result in greater burial of organic matter and pyrite, and greater release of O_2 to the atmosphere, at the same time there will be greater rates of erosion on the continents, some of which will involve oxidative weathering of ancient organic matter and/or pyrite.

The other principal approach towards modeling the Phanerozoic evolution of atmospheric O_2 rests on the isotope systematics of the carbon and sulfur geochemical cycles. The significant isotope discriminations associated with biological fixation of CO_2 to generate biomass and with bacterial reduction of sulfate to sulfide have been mentioned several times previously in this chapter. Given a set of simplifications of the exogenic cycles of carbon, sulfur and oxygen, these isotopic discriminations and the isotopic composition of seawater through time ($\delta^{13}C$, $\delta^{34}S$) can be used to estimate global rates of burial and weathering of organic matter, sedimentary carbonates, pyrite, and evaporative sulfates (Figure 13).

(i) The first required simplification is that the total mass of exogenic carbon is constant through time (carbon in the oceans, atmosphere,

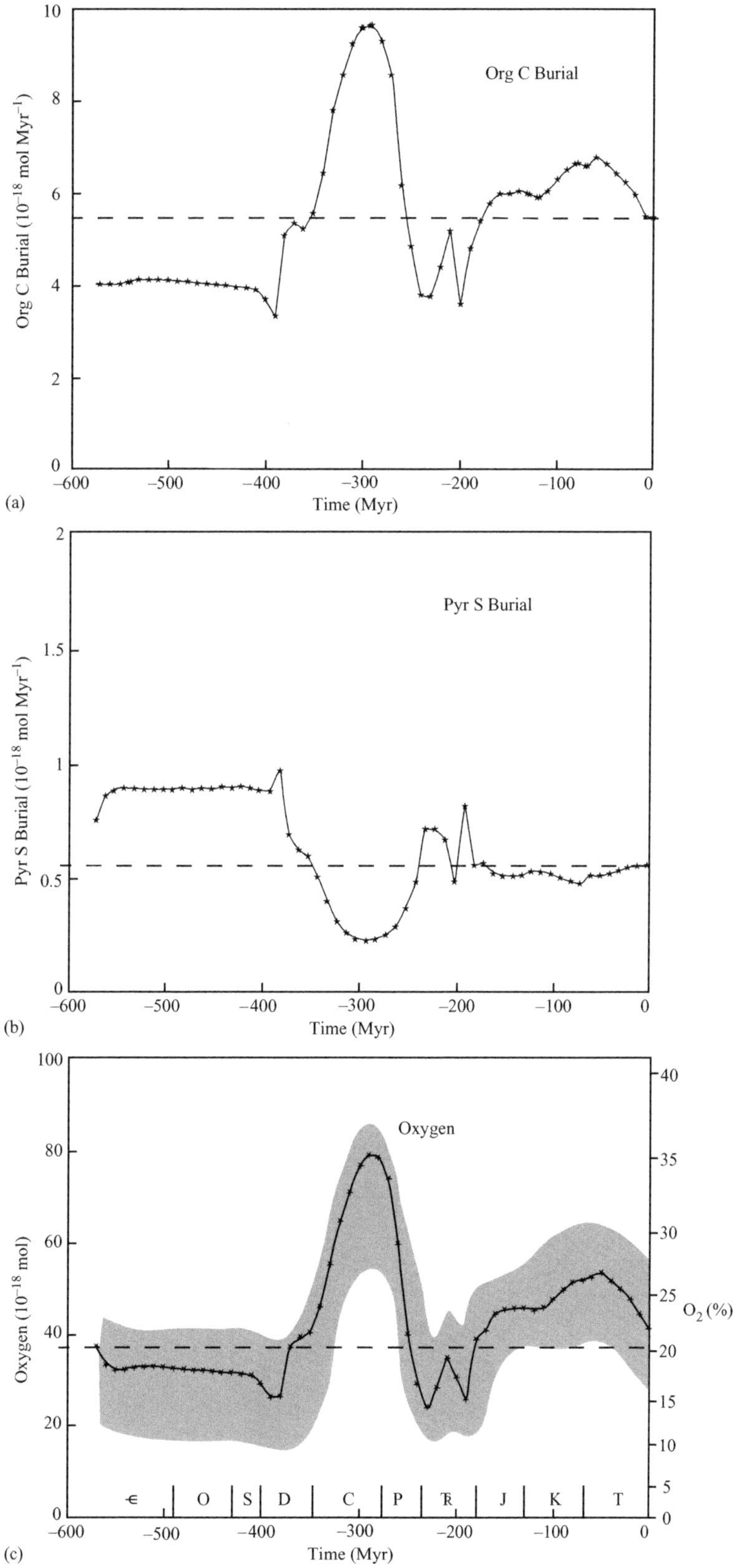

Figure 12 Estimated organic carbon burial (a), pyrite sulfur burial (b), and atmospheric oxygen concentrations (c) through Phanerozoic time, derived from estimates of rock abundance and their relative organic carbon and sulfide content (source Berner and Canfield, 1989).

and sedimentary rocks). This of course neglects inputs of carbon and sulfur from volcanic activity and metamorphic degassing, and outputs into the mantle at subduction zones. However, if these fluxes into and out of the exogenic cycle are small enough (or have no effect on bulk crustal carbon or sulfur isotopic composition), then this simplification may be acceptable.

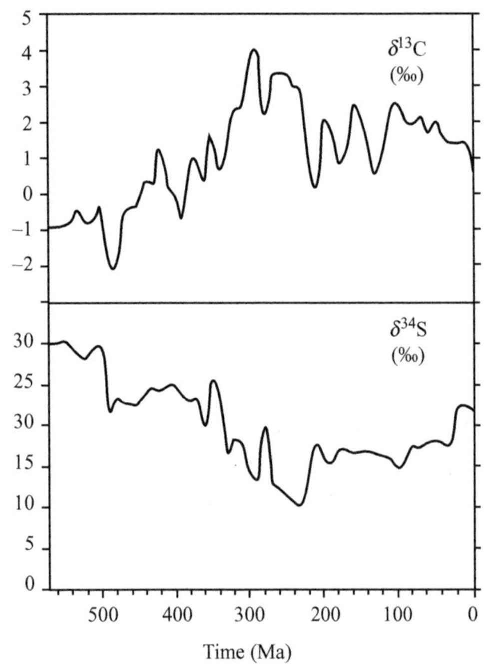

Figure 13 Globally averaged isotopic composition of carbonates ($\delta^{13}C$) and sulfates ($\delta^{34}S$) through Phanerozoic time (source Lindh, 1983).

(ii) The second simplification is that the total mass of carbon and sulfur dissolved in seawater plus the small reservoir of atmospheric carbon and sulfur gases remains constant through time. For carbon, this may be a realistic simplification. Dissolved inorganic carbon in seawater is strongly buffered by carbonate mineral precipitation and dissolution, and thus it is unlikely that extensive regions of the world ocean could have become significantly enriched or depleted in inorganic carbon during the geologic past. For sulfur, this assumption may not be completely accurate. Much of the interpretation of sulfur isotope records (with implications for atmospheric O_2 evolution) in the Precambrian depend on varying, but generally low dissolved sulfate concentrations. Unlike carbonate, there is no great buffering reaction maintaining stable sulfate concentrations in seawater. And while the sources of sulfate to seawater have likely varied only minimally, with changes in the sulfate flux to seawater increasing or decreasing smoothly through time as the result of broad-scale tectonic activity and changes in bulk continental weathering rates, removal of sulfate through excessive BSR or rapid evaporate formation may be much more episodic through time, possibly resulting in fairly extensive shifts in seawater sulfate concentration, even during the Phanerozoic. Nonetheless, using this simplification allows us to establish that for C–S–O geochemical models, total weathering fluxes for carbon and sulfur must equal total burial fluxes for carbon and sulfur, respectively.

Referring to Figure 14, we can see that these simplifications allow us to say that

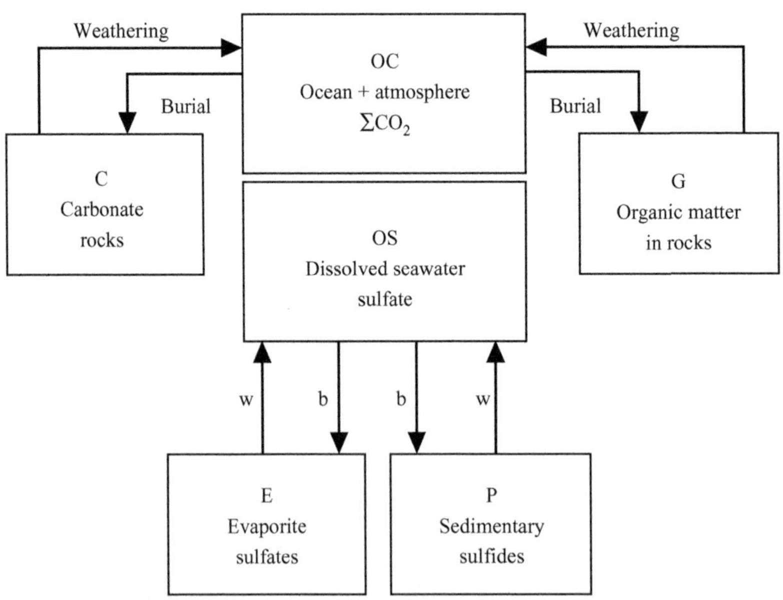

Figure 14 The simplified geochemical cycles of carbon and sulfur, including burial and weathering of sedimentary carbonates, organic matter, evaporites, and sulfides. The relative fluxes of burial and weathering of organic matter and sulfide minerals plays a strong role in controlling the concentration of atmospheric O_2.

$$dM_{oc}/dt = F_{wg} + F_{wc} - F_{bg} - F_{bc} = 0 \quad (11)$$

$$dM_{os}/dt = F_{ws} + F_{wp} - F_{bs} - F_{bp} = 0 \quad (12)$$

The full rate equations for each reservoir mass in Figure 14 are as follows:

$$dM_c/dt = F_{bc} - F_{wc} \quad (13)$$

$$dM_g/dt = F_{bg} - F_{wg} \quad (14)$$

$$dM_s/dt = F_{bs} - F_{ws} \quad (15)$$

$$dM_p/dt = F_{bp} - F_{wp} \quad (16)$$

$$dM_{O_2}/dt = F_{bg} - F_{wg} + 15/8(F_{bp} - F_{wp}) \quad (17)$$

This system of equations has four unknowns: two burial fluxes and two weathering fluxes.

(iii) At this point, a third simplification of the carbon and sulfur systems is often applied to the weathering fluxes of sedimentary rocks. As a first approach, it is not unreasonable to guess that the rate of weathering of a given type of rock relates in some sense to the total mass of that rock type available on Earth's surface. If that relation is assumed to be first order with respect to rock mass, an artificial weathering rate constant for each rock reservoir can be derived. Such constants have been derived by assuming that the weathering rate equation has the form $F_{wi} = k_i M_i$. If we can establish the mass of a sedimentary rock reservoir i, and also the average global river flux to the oceans due to weathering of reservoir i, then k_i can easily be calculated. For example, if the total global mass of carbonate in sedimentary rock is 5000×10^{18} mol C, and annually there are 20×10^{12} mol C discharged from rivers to the oceans from carbonate rock weathering, then $k_{carbonate}$ becomes (20×10^{12} mol C yr^{-1})/($5{,}000\times10^{18}$ mol C), i.e., 4×10^{-3} Myr^{-1}. These simple first-order weathering rate constants have been calculated for each sedimentary rock reservoir in the C–S–O cycle, derived entirely from estimated preanthropogenic carbon and sulfur fluxes from continental weathering. Lack of a true phenomenological relationship relating microscale and outcrop-scale rock weathering reactions to regional- and global-scale carbon and sulfur fluxes remains one of the primary weaknesses limiting the accuracy of numerical models of the coupled C–S–O geochemical cycles.

If weathering fluxes from eroding sedimentary rocks are independently established, such as through use of mass-dependent weathering fluxes, then all of the weathering fluxes in our system of equations above become effectively known. This leaves only the burial fluxes as unknowns in solving the evolution of the C–S–O system.

The exact isotope discrimination that occurs during photosynthesis and during BSR is dependent on many factors. For carbon, these include cell growth rate, geometry and nutrient availability (Rau *et al.*, 1989, 1992), CO_2 availability species-specific effects, modes of CO_2 sequestration (i.e., C_3, C_4, and CAM plants). For sulfur, these can include species–specific effects, the degree of closed-system behavior and sulfate concentration, sulfur oxidation and disproportionation. However, as a simplification in geochemical modeling of the C–S–O cycles, variability in carbon and sulfur isotopic fractionation is limited. In the simplest case, fractionations are constant through all time for all environments. As a result, for example, the isotopic composition of organic matter buried at a given time is set at a constant 25‰ depletion relative to seawater dissolved carbonate at that time, and pyrite isotopic composition is set at a constant 35‰ depletion relative to seawater sulfate. Of course, in reality, α_c and α_s (the isotopic discriminations assigned between inorganic–organic carbon and sulfate–sulfide, respectively) vary greatly in both time and space. Regardless of how α_c and α_s are set, however, once fractionations have been defined, the mass balance equations given in (13)–(17) above can be supplemented with isotope mass balances as well.

Based on the first simplification listed above, the exogenic cycles of carbon and sulfur are regarded as closed systems. As such, the bulk isotopic composition of average exogenic carbon and sulfur do not vary through time. We can write a rate equation for the rate of change in (mass × isotopic composition) of each reservoir in Figure 14 to reflect this isotope mass balance. For example:

$$\frac{d(\delta_{oc}M_{oc})}{dt} \equiv \frac{\delta_{oc}dM_{oc}}{dt} + \frac{M_{oc}d\delta_{oc}}{dt} = \delta_c F_{wc} + \delta_c F_{wg} - \delta_{oc}F_{bg} - (\delta_{oc} - \alpha_c)F_{bg} \quad (18)$$

Using the simplification that $dM_{oc}/dt = 0$, Equation (18) reduces to an equation relating the organic matter burial flux F_{bg} in terms of other known entities:

$$F_{bg} = \left(\frac{1}{\alpha_c}\right)\left[M_{oc}\left(\frac{d\delta_{oc}}{dt}\right) + F_{wc}(\delta_{oc} - \delta_c) + F_{wg}(\delta_{oc} - \delta_g)\right] \quad (19)$$

Using (11) above,

$$F_{bc} = F_{wg} + F_{wc} - F_{bg} \quad (20)$$

A similar pair of equations can be written for the sulfur system:

$$F_{bp} = \left(\frac{1}{\alpha_s}\right)\left[M_{os}\left(\frac{d\delta_{os}}{dt}\right) + F_{ws}(\delta_{os} - \delta_s) + F_{wp}(\delta_{os} - \delta_p)\right] \quad (21)$$

$$F_{bs} = F_{wp} + F_{ws} - F_{bp} \quad (22)$$

Full rate equations for the rate of change in sedimentary rock reservoir isotopic composition can be written as

$$\frac{d\delta_c}{dt} = \frac{F_{bc}(\delta_{oc} - \delta_c)}{M_c} \quad (23)$$

$$\frac{d\delta_g}{dt} = \frac{F_{bg}(\delta_{oc} - \alpha_c - \delta_g)}{M_g} \quad (24)$$

$$\frac{d\delta_s}{dt} = \frac{F_{bw}(\delta_{os} - \delta_s)}{M_s} \quad (25)$$

$$\frac{d\delta_p}{dt} = \frac{F_{bp}(\delta_{os} - \alpha_s - \delta_p)}{M_p} \quad (26)$$

as well as the rate of change in the mass of O_2. The isotopic composition of seawater carbonate and sulfate through time comes directly from the sedimentary rock record. However, it is uncertain how to average globally as well as over substantial period of geologic time. The rates of change in seawater carbonate and sulfate isotopic compositions are fairly small terms, and have been left out of many modeling efforts directed at the geochemical C–S–O system; however, for completeness sake these terms are included here.

One implication of the isotope-driven modeling approach to understanding the C–S–O coupled cycle is that burial fluxes of organic matter and pyrite (which are the sources of atmospheric O_2 in these models) are nearly proportional to seawater isotopic composition of inorganic carbon and sulfate. Recalling the equation for organic matter burial above, it is noted that because sedimentary carbonate and organic matter mass are nearly constant through time, weathering fluxes do not vary greatly through time. Likewise, the average isotopic composition of carbonates and organic matter are also nearly constant, as is the mass of dissolved carbonate and the rate of change in dissolved inorganic carbon isotopic composition. Assuming that F_{wc}, F_{wg}, M_{oc}, $d\delta_{oc}/dt$, δ_c, and δ_g are constant or nearly constant, F_{bg} becomes linearly proportional to the isotopic composition of seawater dissolved inorganic carbon. The same rationale applies for the sulfur system, with pyrite burial becoming linearly proportional to the isotopic composition of seawater sulfate. Although these relationships are not strictly true, even within the constraints of the model simplifications, these relationships provide a useful guide for evaluating changes in organic matter and pyrite burial fluxes and the impact these have on atmospheric O_2, simply by examining the isotopic records of marine carbonate and sulfate.

Early efforts to model the coupled C–S–O cycles yielded important information. The work of Garrels and Lerman (1984) showed that the exogenic C and S cycles can be treated as closed systems over at least Phanerozoic time, without exchange between sedimentary rocks and the deep "crust + mantle." Furthermore, over timescales of millions of years, the carbon and sulfur cycles were seen to be closely coupled, with increase in sedimentary organic carbon mass matched by loss of sedimentary pyrite (and vice versa). Other models explored the dynamics of the C–S–O system. One important advance was promoted by Kump and Garrels (1986). In these authors' model, a steady-state C–S–O system was generated and perturbed by artificially increasing rates of organic matter burial. These authors tracked the shifts in seawater carbon and sulfur isotopic composition that resulted, and compared these results with the true sedimentary record. Importantly, these authors recognized that although there is a general inverse relationship between seawater $\delta^{13}C$ and $\delta^{34}S$, the exact path along an isotope–isotope plot through time is not a straight line. Instead, because of the vastly different residence times of sulfate versus carbonate in seawater, any changes in the C–S–O system are first expressed through shifts in carbon isotopes, then sulfur isotopes (Figure 15). The authors also pointed out large-scale divisions in C–S isotope coupling through Phanerozoic time. In the Paleozoic, organic matter and pyrite burial were closely coupled (largely because the same types of depositional environment favor burial of marine organic matter and pyrite). During this time, seawater carbonate and sulfur isotopes co-varied positively, indicating concomitant increases (or decreases) in burial of organic matter and pyrite. During the late Paleozoic, Mesozoic, and Cenozoic, terrestrial depositional environments became important settings for burial of organic matter. Because pyrite formation and burial in terrestrial environments is extremely limited, organic matter and pyrite burial became decoupled at this time, and seawater carbonate and sulfur isotopes co-varied negatively, indicating close matching of increased sedimentary organic matter with decreased sedimentary pyrite (and vice versa), perhaps suggesting a net balance in O_2 production and consumption, and maintenance of nearly constant, equable p_{O_2} throughout much of the latter half of the Phanerozoic. The model of Berner (1987) introduced the concept of rapid recycling: the effort to numerically represent the observation that younger sedimentary rocks are more likely to be eroded and weathered than are older sedimentary rocks. Because young sedimentary rocks are likely to be isotopically distinct from older rocks (because they are recording any recent shifts in seawater carbon or sulfur isotopic composition), restoring that isotopically distinct carbon or sulfur more quickly back into seawater provides a type of negative feedback, dampening excessively large

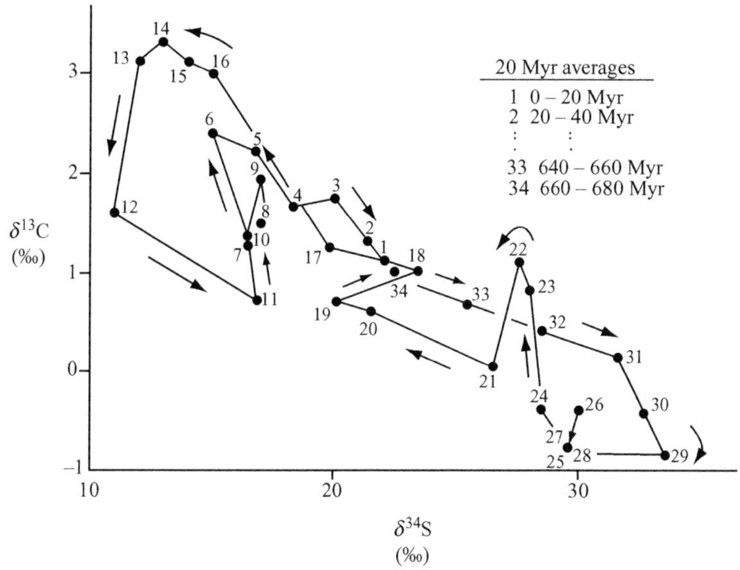

Figure 15 Twenty-million-year average values of seawater $\delta^{34}S$ plotted against concomitant carbonate $\delta^{13}C$ for the last 700Myr (source Kump and Garrels, 1986).

or small burial fluxes required for isotope mass balance. This negative feedback serves to reduce calculated fluctuations in organic matter and pyrite burial rates, which in turn reduce fluctuations in release of O_2 to the atmosphere. Results from this study also predict large increases in OM burial fluxes during the Permocarboniferous (~300 Ma) above values present earlier in the Paleozoic. This increase, likely associated with production and burial of refractory terrigenous organic matter (less easily degraded than OM produced by marine organisms), led to elevated concentrations of O_2 ~300 Ma.

One flaw with efforts to model the evolution of Phanerozoic O_2 using the carbon and sulfur isotope records is that unreasonably large fluctuations in organic matter and pyrite burial fluxes (with coincident fluctuations in O_2 production rates) would result. Attempts to model the whole Phanerozoic generated unreasonably low and high O_2 concentrations for several times in the Phanerozoic (Lasaga, 1989), and applications of what seemed to be a realistic feedback based on reality (weathering rates of sedimentary organic matter and pyrite dependent on the concentration of O_2) were shown to actually become positive feedbacks in the isotope-driven C–S–O models (Berner, 1987; Lasaga, 1989).

Phosphorus is a key nutrient limiting primary productivity in many marine environments. If phosphorus supply is increased, primary production and perhaps organic matter burial will also increase. Degradation and remineralization of OM during transit from surface waters into sediments liberates phosphorus, but most of this is quickly scavenged by adsorption onto the surfaces of iron oxyhydroxides. However, work by Ingall and Jahnke (1994) and Van Cappellen and Ingall (1996) has shown that phosphorus recycling and release into seawater is enhanced under low O_2 or anoxic conditions. This relationship provides a strong negative feedback between primary production, bottom water anoxia, and atmospheric O_2. As atmospheric O_2 rises, phosphorus scavenging on ferric-iron is enhanced, phosphorus recycling back into surface waters is reduced, primary production rates are reduced, and O_2 declines. If O_2 concentrations were to fall, phosphorus scavenging onto ferric-iron would be inhibited, phosphorus recycling back into surface waters would be accelerated, fueling increased primary production and O_2 release to the atmosphere. Van Cappellen and Ingall (1996) applied these ideas to a mathematical model of the C–P–Fe–O cycle to show how O_2 concentrations could be stabilized by phosphorus recycling rates.

Petsch and Berner (1998) expanded the model of Van Cappellen and Ingall (1996) to include the sulfur system, as well as carbon and sulfur isotope effects. This study examined the response of the C–S–O–Fe–P system, and in particular carbon and sulfur isotope ratios, to perturbation in global ocean overturn rates, changes in continental weathering, and shifts in the locus of organic matter burial from marine to terrestrial depocenters. Confirming the idea promoted by Kump and Garrels (1986), these authors showed that perturbations of the exogenic C–S–O cycle result in shifts in seawater carbon and sulfur isotopic composition similar in amplitude and duration to observed isotope excursions in the sedimentary record.

Other proposed feedbacks stabilizing the concentration of atmospheric O_2 over Phanerozoic time include a fire-regulated PO_4 feedback (Kump, 1988, 1989). Terrestrial primary production requires much less phosphate per mole CO_2 fixed during photosynthesis than marine primary production. Thus, for a given supply of PO_4, much more CO_2 can be fixed as biomass and O_2 released from photosynthesis on land versus in the oceans. If terrestrial production proceeds too rapidly, however, p_{O_2} levels may rise slightly and lead to increased forest fires. Highly weatherable, PO_4-rich ash would then be delivered through weathering and erosion to the oceans. Primary production in the oceans would lead to less CO_2 fixed and O_2 released per mole of PO_4.

Hydrothermal reactions between seawater and young oceanic crust have been proposed as an influence on atmospheric O_2 (Walker, 1986; Carpenter and Lohmann, 1999; Hansen and Wallmann, 2002). While specific periods of oceanic anoxia may be associated with accelerated hydrothermal release of mantle sulfide (i.e., the Mid-Cretaceous, see Sinninghe-Damsté and Köster, 1998), long-term sulfur and carbon isotope mass balance precludes substantial inputs of mantle sulfur to the Earth's surface of a different net oxidation state and mass flux than what is subducted at convergent margins (Petsch, 1999; Holland, 2002).

One recent advance in the study of isotope-driven models of the coupled C–S–O cycles is re-evaluation of isotope fractionations. Hayes *et al.*, (1999) published a compilation of the isotopic composition of inorganic and organic carbon for the past 800 Myr. One feature of this dual record is a distinct shift in the net isotopic distance between carbonate and organic carbon, occurring during the past ~100 million years. When carbon isotope distance is compared to estimates of Cenozoic and Mesozoic p_{O_2}, it becomes apparent that there may be some relationship between isotopic fractionation associated with organic matter production and burial and the concentration of O_2 in the atmosphere. The physiological underpinning behind this proposed relationship rests on competition between photosynthesis and photorespiration in the cells of photosynthetic organisms. Because O_2 is a competitive inhibitor of CO_2 for attachment to the active site of Rubisco, as ambient O_2 concentrations rise relative to CO_2, so will rates of photorespiration. Photorespiration is a net consumptive process for plants; previously fixed carbon is consumed with O_2 to produce CO_2 and energy. CO_2 produced through photorespiration may diffuse out of the cell, but it is also likely to be taken up (again) for photosynthesis. Thus, in cells undergoing fairly high rates of photorespiration in addition to photosynthesis, a significant fraction of total CO_2 available for photosynthesis derives from oxidized, previously fixed organic carbon. The effect of this on cellular carbon isotopic composition is that because each round of photosynthesis results in ^{13}C-depletion in cellular carbon relative to CO_2, cells with high rates of photorespiration will contain more ^{13}C-depleted CO_2 and thus will produce more ^{13}C-depleted organic matter.

In controlled-growth experiments using both higher plants and single-celled marine photosynthetic algae, a relationship between ambient O_2 concentration and net isotope discrimination has been observed (Figure 16) (Berner *et al.*, 2000; Beerling *et al.*, 2002). The functional form of this relationship has been expressed in several ways. The simplest is to allow isotope discrimination to vary linearly with changing atmospheric O_2 mass: $\alpha_c = 25 \times (M_{O_2}/38)$. More complicated relationships have also been derived, based on curve-fitting the available experimental data on isotopic fractionation as a function of $[O_2]$. O_2-dependent isotopic fractionation during photosynthesis has provided the first mathematically robust isotope-driven model of the C–S–O cycle consistent with geologic observations (Berner *et al.*, 2000). Results of this model show that allowing isotope fractionation to respond to changes in ambient O_2 provides a strong negative feedback dampening excessive increases or decreases in organic matter burial rates. Rates of organic matter burial in this model are no longer simply dependent on seawater carbonate $\delta^{13}C$, but now also vary with $1/\alpha_c$. As fractionation becomes greater (through elevated O_2), less of an increase in organic matter burial rates is required to achieve the observed increase in seawater $\delta^{13}C$ than if α_c were constant.

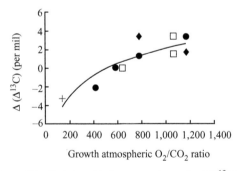

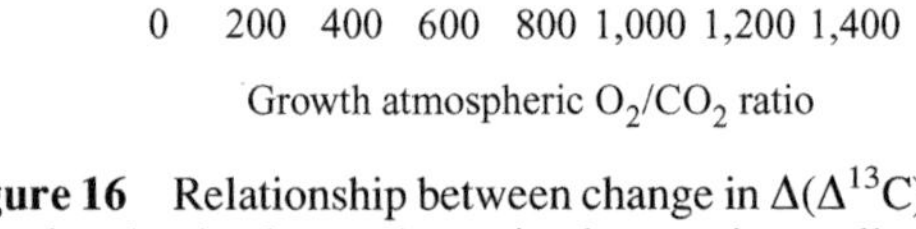

Figure 16 Relationship between change in $\Delta(\Delta^{13}C)$ of vascular land plants determined experimentally in response to growth under different O_2/CO_2 atmospheric mixing ratios. $\Delta(\Delta^{13}C)$ is the change in carbon isotope fractionation relative to fractionation for the controls at present day conditions (21% O_2, 0.036% CO_2). The solid line shows the nonlinear curve fitted to the data, given by $\Delta(\Delta^{13}C) = -19.94 + 3.195 \times \ln(O_2/CO_2)$; (•) *Phaseolus vulgaris*; () *Sinapis alba*; (□) from Berner *et al.* (2000); (+) from Berry *et al.* (1972) (Beerling *et al.*, 2002) (reproduced by permission of Elsevier from *Geochem. Cosmochim. Acta* **2002**, *66*, 3757–3767).

The same mathematical argument can be applied to sulfate–sulfide isotope fractionation during BSR. As O_2 concentrations increase, so does sulfur isotope fractionation, resulting in a strong negative feedback on pyrite burial rates. This is consistent with the broad-scale changes in sulfur isotope dynamics across the Proterozoic, reflecting a large increase in $\Delta^{34}S$ (between sulfate and sulfide) when atmospheric O_2 concentrations were great enough to facilitate bacterial sulfide oxidation and sulfur disproportionation. Perhaps during the Phanerozoic, when O_2 concentrations were greater, sulfur recycling (sulfate to sulfide through BSR, sulfur oxidation, and sulfur disproportionation) was increased, resulting in greater net isotopic distance between sulfate and sulfide. Another means of changing net sulfur isotope discrimination in response to O_2 may be the distribution of reduced sulfur between sulfide minerals and organic matter-associated sulfides. Work by Werne *et al.*, 2000, 2003) has shown that organic sulfur is consistently ~10 ‰ enriched in ^{34}S relative to associated pyrite. This is believed to result from different times and locations of organic sulfur versus pyrite formation. While pyrite may form in shallow sediments or even anoxic portions of the water column, reflecting extreme sulfur isotope depletion due to several cycles of BSR, sulfide oxidation and sulfur disproportionation, organic matter is sulfurized within the sediments. Closed, or nearly closed, system behavior of BSR in the sediments results in late-stage sulfide (the source of sulfur in sedimentary organic matter) to be more enriched compared with pyrite in the same sediments. It is known that burial of sulfide as organic sulfur is facilitated in low O_2 or anoxic waters. If lower atmospheric O_2 in the past encouraged development of more extensive anoxic basins and increased burial of sulfide as organic sulfur instead of pyrite, the 10‰ offset between pyrite and organic sulfur would become effectively a change in net sulfur fractionation in response to O_2.

Applying these newly recognized modifications of carbon and sulfur isotope discrimination in response to changing O_2 availability has allowed development of new numerical models of the evolution of the coupled C–S–O systems and variability of Phanerozoic atmospheric O_2 concentration (Figure 17).

11.7 CONCLUSIONS

Molecular oxygen is generated and consumed by a wide range of processes. The net cycling of O_2 is influenced by physical, chemical, and most importantly, biological processes acting on and beneath the Earth's surface. The exact distribution of O_2 concentrations depends on the specific interplay of these processes in time and space. Large inroads have been made towards understanding the processes that control the concentration of atmospheric O_2, especially regarding O_2 as a component of coupled biogeochemical cycles of many elements, including carbon, sulfur, nitrogen, phosphorus, iron, and others.

Earth's modern oxygenated atmosphere is the product of over four billion years of its history (Figure 18). The early anoxic atmosphere was slowly oxidized (although not oxygenated) as the result of slow H_2 escape. Evolution of

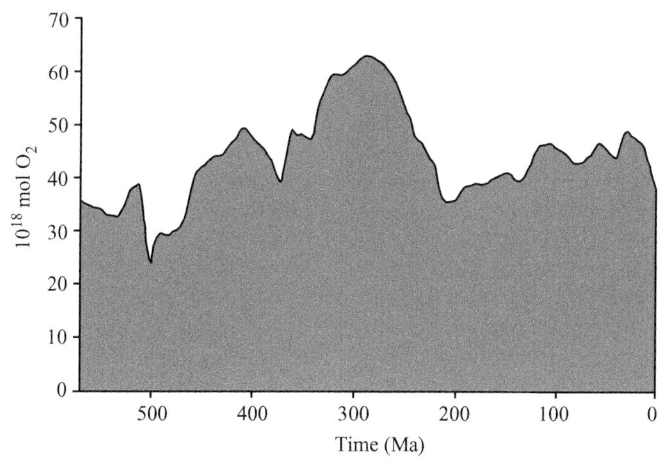

Figure 17 Evolution of the mass of atmospheric O_2 through Phanerozoic time, estimated using an isotope mass balance described in Equations (11)–(26). The model employs the isotope date of Figure 13, and includes new advances in understanding regarding dependence of carbon isotope discrimination during photosynthesis and sulfur isotope discrimination during sulfur disproportionation and organic sulfur formation. The system of coupled differential equations were integrated using an implicit fourth-order Kaps–Rentrop numerical integration algorithm appropriate for this stiff set of equations (sources Petsch, 2000; Berner *et al.*, 2000).

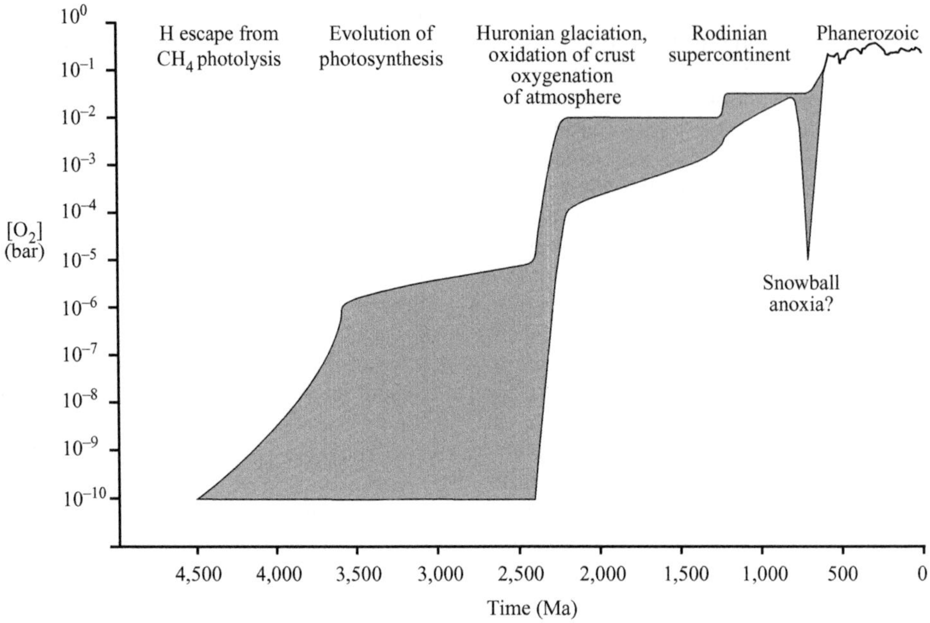

Figure 18 Composite estimate of the evolution of atmospheric oxygen through 4.5 Gyr of Earth's history. Irreversible oxidation of the Earth resulted from CH_4 photolysis and hydrogen escape during early Earth's history. Evolution of oxygenic photosynthesis preceded substantial oxygenation of the atmosphere by several hundred million years. Relative stasis in atmospheric p_{O_2} typified much of the Proterozoic, with a possible pulse of oxygenation associated with formation of the Rodinian supercontinent in the Late Mesoproterozoic, and possible return to anoxia associated with snowball glaciation in the Neoproterozoic (sources Catling *et al.*, 2001; Kasting, 1992; Rye and Holland, 1998; Petsch, 2000; Berner *et al.*, 2000).

oxygenic photosynthesis accelerated the oxidation of Earth's crust and atmosphere, such that by ~2.2 Ga a small but significant concentration of O_2 was likely present in Earth's atmosphere. Limited primary production and oxygen production compared with the flux of reduced volcanic gases maintained this low p_{O_2} atmosphere for over one billion years until the Neoproterozoic. Rapid oscillations in Earth's carbon and sulfur cycles associated with global Snowball glaciation may also have expression in a return to atmospheric anoxia at this time, but subsequent to the late Proterozoic isotope excursions, oxygenation of the atmosphere to near-modern concentrations developed such that by the Precambrian–Cambrian boundary, O_2 concentrations were high enough to support widespread skeletonized metazoans. Phanerozoic seawater and atmospheric O_2 concentrations have fluctuated in response to tectonic forcings, generating regional-scale anoxia in ocean basins at certain times when biological productivity and ocean circulation facilitate anoxic conditions, but in the atmosphere, O_2 concentrations have remained within ~0.05–0.35 bar p_{O_2} for the past ~600 Myr.

Several outstanding unresolved gaps in our understanding remain, in spite of a well-developed understanding of the general features of the evolution of atmospheric O_2 through time. These gaps represent potentially meaningful directions for future research, including:

(i) assessing the global importance of mineral oxidation as a mechanism of O_2 consumption;

(ii) the flux of reduced gases from volcanoes, metamorphism, and diffuse mantle/lithosphere degassing;

(iii) the true dependence of organic matter oxidation on availability of O_2, in light of the great abundance of microaerophilic and anaerobic microorganisms utilizing carbon respiration as a metabolic pathway, carbon isotopic evidence suggesting continual and essentially constant organic matter oxidation as part of the sedimentary rock cycle during the entire past four billion years, and the inefficiency of organic matter oxidation during continental weathering;

(iv) stasis in the oxygenation of the atmosphere during the Proterozoic;

(v) contrasting biochemical, fossil, and molecular evidence for the antiquity of the innovation of oxygenic photosynthesis; and

(vi) evaluating the relative strength of biological productivity versus chemical evolution of the Earth's crust and mantle in controlling the early stages of oxygenation of the atmosphere.

Thus, study of the global biogeochemical cycle of oxygen, the component of our atmosphere integral and crucial for life as we know it, remains a fruitful direction for Earth science research.

REFERENCES

Algeo T. J. and Scheckler S. E. (1998) Terrestrial-marine teleconnections in the Devonian: links between the evolution of land plants, weathering processes, and marine anoxic events. *Phil. Trans. Roy. Soc. London B* **353**, 113–128.

Anbar A. D. and Knoll A. H. (2002) Proterozoic ocean chemistry and evolution: a bioinorganic bridge? *Science* **297**, 1137–1142.

Arthur M. A. (2000) Volcanic contributions to the carbon and sulfur geochemical cycles and global change. In *Encyclopedia of Volcanoes* (eds. H. Sigurdsson, B. F. Houghton, S. R. McNutt, H. Rymer, and J. Stix). Academic Press, San Diego, pp. 1045–1056.

Bartley J. K., Semikhatov M. A., Kaufman A. J., Knoll A. H., Pope M. C., and Jacobsen S. B. (2001) Global events across the Mesoproterozoic–Neoproterozoic boundary: C and Sr isotopic evidence from Siberia. *Precamb. Res.* **111**, 165–202.

Battle M., Bender M., Sowers R., Tans P. P., Butler J. H., Elkins J. W., Ellis J. T., Conway T., Zhang N., Pang P., and Clarke A. D. (1996) Atmospheric gas concentrations over the past century measured in air from firn at the South Pole. *Nature* **383**, 231–235.

Baum S. K. and Crowley T. J. (2001) GCM response to late Precambrian (~590 Ma) ice-covered continents. *Geophys. Res. Lett.* **28**, 583–586.

Beaumont V. and Robert F. (1999) Nitrogen isotope ratios of kerogens in Precambrian cherts: a record of the evolution of atmospheric chemistry? *Precamb. Res.* **96**, 63–82.

Beerling D. J. and Berner R. A. (2000) Impact of a Permo–Carboniferous high O_2 event on the terrestrial carbon cycle. *Proc. Natl. Acad. Sci.* **97**, 12428–12432.

Beerling D. J., Lake J. A., Berner R. A., Hickey L. J., Taylor D. W., and Royer D. L. (2002) Carbon isotope evidence implying high O_2/CO_2 ratios in the Permo–Carboniferous atmosphere. *Geochim. Cosmochim. Acta* **66**, 3757–3767.

Bekker A., Kaufman A. J., Karhu J. A., Beukes N. J., Quinten S. D., Coetzee L. L., and Kenneth A. E. (2001) Chemostratigraphy of the Paleoproterozoic Duitschland Formation, South Africa. Implications for coupled climate change and carbon cycling. *Am. J. Sci.* **301**, 261–285.

Bender M., Labeyrie L. D., Raynaud D., and Lorius C. (1985) Isotopic composition of atmospheric O_2 in ice linked with deglaciation and global primary productivity. *Nature* **318**, 349–352.

Bender M., Sowers T., and Labeyrie L. (1994a) The Dole effect and its variations during the last 130,000 years as measured in the Vostok ice core. *Global Biogeochem. Cycle* **8**, 363–376.

Bender M. L., Sowers T., Barnola J.-M., and Chappellaz J. (1994b) Changes in the O_2/N_2 ratio of the atmosphere during recent decades reflected in the composition of air in the firn at Vostok Station, Antarctica. *Geophys. Res. Lett.* **21**, 189–192.

Berner R. A. (1987) Models for carbon and sulfur cycles and atmospheric oxygen: application to Paleozoic geologic history. *Am. J. Sci.* **287**, 177–196.

Berner R. A. (1994) GEOCARB II: a revised model of atmospheric CO_2 over Phanerozoic time. *Am. J. Sci.* **294**, 56–91.

Berner R. A. (2002) Examination of hypotheses for the Permo–Triassic boundary extinction by carbon cycle modeling. *Proc. Natl. Acad. Sci.* **99**, 4172–4177.

Berner R. A. and Canfield D. E. (1989) A new model for atmospheric oxygen over Phanerozoic time. *Am. J. Sci.* **289**, 333–361.

Berner R. A. and Kothavala Z. (2001) GEOCARB III: a revised model of atmospheric CO_2 over Phanerozoic time. *Am. J. Sci.* **301**, 182–204.

Berner R. A., Petsch S. T., Lake J. A., Beerling D. J., Popp B. N., Lane R. S., Laws E. A., Westley M. B., Cassar N., Woodward F. I., and Quick W. P. (2000) Isotope fractionation and atmospheric oxygen: Implications for Phanerozoic O_2 evolution. *Science* **287**, 1630–1633.

Berry J. A., Troughton J. H., and Björkman O. (1972) Effect of oxygen concentration during growth on carbon isotope discrimination in C_3 and C_4 species of *Atriplex*. *Carnegie Inst. Yearbook* **71**, 158–161.

Beukes N. J., Dorland H., Gutzmer J., Nedachi M., and Ohmoto H. (2002) Tropical laterites, life on land, and the history of atmospheric oxygen in the Paleoproterozoic. *Geology* **30**, 491–494.

Brasier M. D., Green O. R., Jephcoat A. P., Kleppe A. K., van Kranendonk M. J., Lindsay J. F., Steele A., and Grassineau N. V. (2002) Questioning the evidence for Earth's earliest fossils. *Nature* **416**, 76–81.

Brocks J. J., Logan G. A., Buick R., and Summons R. E. (1999) Archean molecular fossils and the early rise of eukaryotes. *Science* **285**, 1033–1036.

Budyko M. I., Ronov A. B., and Yanshin A. L. (1987) *History of the Earth's Atmosphere*. Springer, Berlin.

Buick R. (1990) Microfossil recognition in Archean rocks: an appraisal of spheroids and filaments from a 3500 M. Y. old chert-barite unit at North Pole, Western Australia. *Palaios* **5**, 441–459.

Buick I. S., Uken R., Gibson R. L., and Wallmach T. (1998) High-$\delta^{13}C$ Paleoproterozoic carbonates from the Transvaal Supergroup, South Africa. *Geology* **26**, 875–878.

Butterfield N. J. and Rainbird R. H. (1998) Diverse organic-walled fossils, including possible dinoflagellates, from early Neoproterozoic of arctic Canada. *Geology* **26**, 963–966.

Caldeira K., and Kasting J. F. (1992) Susceptibility of the early Earth to irreversible glaciation caused by carbon dioxide clouds. *Nature* **359**, 226–228.

Canfield D. E. (1998) A new model for Proterozoic ocean chemistry. *Nature* **396**, 450–453.

Canfield D. E. and Teske A. (1996) Late Proterozoic rise in atmospheric oxygen concentration inferred from phylogenetic and sulphur-isotope studies. *Nature* **382**, 127–132.

Canfield D. E., Habicht K. S., and Thamdrup B. (2000) The Archean sulfur cycle and the early history of atmospheric oxygen. *Science* **288**, 658–661.

Carpenter S. J. and Lohmann K. C. (1999) Carbon isotope ratios of Phanerozoic marine cements: re-evaluating global carbon and sulfur systems. *Geochim. Cosmochim. Acta* **61**, 4831–4846.

Catling D. C., Zahnle K. J., and McKay C. P. (2001) Biogenic methane, hydrogen escape, and the irreversible oxidation of the early Earth. *Science* **293**, 839–843.

Cloud P. (1972) A working model of the primitive Earth. *Am. J. Sci.* **272**, 537–548.

Cloud P. E. (1976) Beginnings of biospheric evolution and their biogeochemical consequences. *Paleobiol.* **2**, 351–387.

Condie K. C., Des Marais D. J., and Abbott D. (2001) Precambrian superplumes and supercontinents: a record in black shales, carbon isotopes, and paleoclimates? *Precamb. Res.* **106**, 239–260.

Cope M. J. and Chaloner W. G. (1985) Wildfire: an interaction of biological and physical processes. In *Geological Factors and the Evolution of Plants* (ed. B. H. Tiffney). Yale University Press, New Haven, CT, pp. 257–277.

Delano J. W. (2001) Redox history of the Earth's interior since ~3900 Ma: implications for prebiotic molecules. *Origins Life Evol. Biosphere* **31**, 311–341.

Delmelle P. and Stix J. (2000) Volcanic gases. In *Encyclopedia of Volcanoes* (eds. H. Sigurdsson, B. F. Houghton, S. R. McNutt, H. Rymer, and J. Stix). Academic Press, San Diego, pp. 803–816.

Des Marais D. J., Strauss H., Summons R. E., and Hayes J. M. (1992) Carbon isotope evidence for stepwise oxidation of the Proterozoic environment. *Nature* **359**, 605–609.

Deuser W. G. (1975) *Chemical Oceanography*. Academic Press, Orlando, FL, p. 3.

Doolittle R. F., Feng D. F., Tsang S., Cho G., and Little E. (1996) Determining divergence times of the major

kingdoms of living organisms with a protein clock. *Science* **271**, 470–477.
Dudley R. (1998) Atmospheric oxygen, giant Paleozoic insects and the evolution of aerial locomotor performance. *J. Exp. Biol.* **201**, 1043–1050.
Dudley R. (2000) The evolutionary physiology of animal flight: Paleobiological and present perspectives. *Ann. Rev. Phys.* **62**, 135–155.
Evans D. A. D. (2000) Stratigraphic, geochronological and paleomagnetic constraints upon the Neoproterozoic climatic paradox. *Am. J. Sci.* **300**, 347–433.
Farquhar J., Bao H., and Thiemens M. (2000) Atmospheric influence of Earth's earliest sulfur cycle. *Science* **289**, 756–758.
Farquhar J., Wing B. A., McKeegan K. D., Harris J. W., Cartigny P., and Thiemens M. H. (2002) Mass-independent sulfur of inclusions in diamond and sulfur recycling on early Earth. *Science* **297**, 2369–2372.
Garcia-Pichel F. (1998) Solar ultraviolet and the evolutionary history of cyanobacteria. *Origins Life Evol. Biosphere* **28**, 321–347.
Garrels R. M. and Lerman A. (1984) Coupling of the sedimentary sulfur and carbon cycles—an improved model. *Am. J. Sci.* **284**, 989–1007.
Gilbert D. L. (1996) Evolutionary aspects of atmospheric oxygen and organisms. In *Environmental Physiology:* 2 (eds. M. J. Fregly and C. M. Blatteis). Oxford University Press, Oxford, UK, pp. 1059–1094.
Glasspool I. (2000) A major fire event recorded in the mesofossils and petrology of the late Permian, Lower Whybrow coal seam, Sydney Basin, Australia. *Palaeogeogr. Palaeoclimat. Palaeoecol.* **164**, 373–396.
Godderís Y. and Veizer J. (2000) Tectonic control of chemical and isotopic composition of ancient oceans: the impact of continental growth. *Am. J. Sci.* **300**, 434–461.
Gould S. J. (1995) Of it and not above it. *Nature* **377**, 681–682.
Graham J. B., Dudley R., Anguilar N., and Gans C. (1995) Implications of the late Paleozoic oxygen pulse for physiology and evolution. *Nature* **375**, 117–120.
Habicht K. S., Gade M., Thamdrup B., Berg P., and Canfield D. E. (2002) Calibration of sulfate levels in the Archean ocean. *Science* **298**, 2372–2374.
Haggerty S. E. and Toft P. B. (1985) Native iron in the continental lower crust: petrological and geophysical implications. *Science* **229**, 647–649.
Halmer M. M., Schmincke H.-U., and Graf H.-F. (2002) The annual volcanic gas input into the atmosphere, in particular into the stratosphere: a global data set for the past 100 years. *J. Volcanol. Geotherm. Res.* **115**, 511–528.
Hansen K. W. and Wallmann K. (2003) Cretaceous and Cenozoic evolution of seawater composition, atmospheric O_2 and CO_2: a model perspective. *Am. J. Sci.* **303**, 94–148.
Hartnett H. E., Keil R. G., Hedges J. I., and Devol A. H. (1998) Influence of oxygen exposure time on organic carbon preservation in continental margine sediments. *Nature* **391**, 572–574.
Hayes J. M., Strauss H., and Kaufman A. J. (1999) The abundance of ^{13}C in marine organic matter and isotopic fractionation in the global biogeochemical cycle of carbon during the past 800 Myr. *Chem. Geol.* **161**, 103–125.
Herbert T. D. and Sarmiento J. L. (1991) Ocean nutrient distribution and oxygenation: limits on the formation of warm saline bottom water over the past 91 m.y. *Geology* **19**, 702–705.
Hinkle M. E. (1994) Environmental conditions affecting the concentrations of He, CO_2, O_2 and N_2 in soil gases. *Appl. Geochem.* **9**, 53–63.
Hoehler T. M., Bebout B. M., and Des Marais D. J. (2001) The role of microbial mats in the production of reduced gases on the early Earth. *Nature* **412**, 324–327.
Hoffman P. F. and Schrag D. P. (2002) The snowball Earth hypothesis: testing the limits of global change. *Terra Nova* **14**, 129–155.
Hoffman P. F., Kaufman A. J., Halverson G. P., and Schrag D. P. (1998) A Neoproterozoic snowball Earth. *Science* **281**, 1342–1346.
Hofmann H. J., Gery K., Hickman A. H., and Thorpe R. I. (1999) Origin of 2.45 Ga coniform stromatolites in Warrawoona Group, Western Australia. *Geol. Soc. Am. Bull.* **111**, 1256–1262.
Holland H. D. (1978) *The Chemistry of the Atmosphere and Oceans.* Wiley, New York.
Holland H. D. (1984) *The Chemical Evolution of the Atmosphere and Oceans.* Princeton University Press, Princeton, NJ.
Holland H. D. (2002) Volcanic gases, black smokers and the great oxidation event. *Geochim. Cosmochim. Acta* **66**, 3811–3826.
Holland H. D., Feakes C. R., and Zbinden E. A. (1989) The Flin Flon paleosol and the composition of the atmosphere 1.8 BYBP. *Am. J. Sci.* **289**, 362–389.
Hotinski R. M., Bice K. L., Kump L. R., Najjar R. G., and Arthur M. A. (2001) Ocean stagnation and End-Permian anoxia. *Geology* **29**, 7–10.
Hurtgen M. T., Arthur M. A., Suits N. S., and Kaufman A. J. (2002) The sulfur isotopic composition of Neoproterozoic seawater sulfate: implications for a snowball Earth? *Earth Planet. Sci. Lett.* **203**, 413–429.
Ingall E. and Jahnke R. (1994) Evidence for enhanced phosphorus regeneration from marine sediments overlain by oxygen depleted waters. *Geochim. Cosmochim. Acta* **58**, 2571–2575.
Ingmanson D. E. and Wallace W. J. (1989) *Oceanography: An Introduction.* Wadworth, Belmont, CA, pp. 99.
Joachimski M. M., Ostertag-Henning C., Pancost R. D., Strauss H., Freeman K. H., Littke R., Sinninghe-Damsté J. D., and Racki G. (2001) Water column anoxia, enhanced productivity and concomitant changes in $\delta^{13}C$ and $\delta^{34}S$ across the Frasnian–Famennian boundary (Kowala—Holy Cross Mountains/Poland). *Chem. Geol.* **175**, 109–131.
Kah L. C., Lyons T. W., and Chesley J. T. (2001) Geochemistry of a 1.2 Ga carbonate-evaporite succession, northern Baffin and Bylot Islands: implications for Mesoproterozoic marine evolution. *Precamb. Res.* **111**, 203–234.
Kaiho K., Kajiwara Y., Nakano T., Miura Y., Kawahata H., Tazaki K., Ueshima M., Chen Z., and Shi G. (2001) End-Permian catastrophe by a bolide impact: evidence of a gigantic release of sulfur from the mantle. *Geology* **29**, 815–818.
Karhu J. and Holland H. D. (1996) Carbon isotopes and the rise of atmospheric oxygen. *Geology* **24**, 867–870.
Kasting J. F. (1987) Theoretical constraints on oxygen and carbon dioxide concentrations in the Precambrian atmosphere. *Precamb. Res.* **34**, 205–229.
Kasting J. F. (1992) Models relating to Proterozoic Atmospheric and Oceanic Chemistry. In *The Proterozoic Biosphere* (eds. J. W. Schopf and C. Klein). Cambridge University Press, Cambridge, pp. 1185–1187.
Kasting J. F. and Siefert J. L. (2002) Life and the evolution of Earth's atmosphere. *Science* **296**, 1066–1068.
Kasting J. F., Zahnle K. J., and Walker J. C. G. (1983) Photochemistry of methane in the Earth's early atmosphere. *Precamb. Res.* **20**, 121–148.
Kasting J. F., Eggler D. H., and Raeburn S. P. (1993) Mantle redox evolution and the oxidation state of the Archean atmosphere. *J. Geol.* **101**, 245–257.
Kasting J. F., Pavlov A. A., and Siefert J. L. (2001) A coupled ecosystem-climate model for predicting the methane concentration in the Archean atmosphere. *Orig. Life Evol. Biosph.* **31**, 271–285.
Kaufman A. J. and Knoll A. H. (1995) Neoproterozoic variations in the C-isotopic composition of seawater: stratigraphic and biogeochemical implications. *Precamb. Res.* **73**, 27–49.
Kaufman A. J., Hayes J. M., Knoll A. H., and Germs G. J. B. (1991) Isotopic composition of carbonates and organic

carbon from upper Proterozoic successions in Namibia: stratigraphic variation and the effects of diagenesis and metamorphism. *Precamb. Res.* **49**, 301–327.

Keeling R. F. and Shertz S. R. (1992) Seasonal and interannual variations in atmospheric oxygen and implications for the carbon cycle. *Nature* **358**, 723–727.

Keeling R. F., Najjar R. P., Bender M. L., and Tans P. P. (1993) What atmospheric oxygen measurements can tell us about the global carbon cycle. *Global Biogeochem. Cycles* **7**, 37–67.

Keller C. K. and Bacon D. H. (1998) Soil respiration and georespiration distinguished by transport analyses of vadose CO_2, $^{13}CO_2$, and $^{14}CO_2$. *Global Biogeochem. Cycles* **12**, 361–372.

Kiehl J. T. and Dickinson R. E. (1987) A study of the radiative effects of enhanced atmospheric CO_2 and CH_4 on early Earth surface temperatures. *J. Geophys. Res.* **92**, 2991–2998.

Kimura H. and Watanabe Y. (2001) Oceanic anoxia at the Precambrian–Cambrian boundary. *Geology* **29**, 995–998.

Kirschvink J. L. (1992) Late Proterozoic low-latitude global glaciation: the snowball earth. In *The Proterozoic Biosphere* (eds. J. W. Schopf, and C. Klein). Cambridge University Press, Cambridge, pp. 51–52.

Kump L. R. (1988) Terrestrial feedback in atmospheric oxygen regulation by fire and phosphorus. *Nature* **335**, 152–154.

Kump L. R. (1989) Chemical stability of the atmosphere and ocean. *Palaeogeogr. Palaeoclimat. Paleoecol.* **75**, 123–136.

Kump L. R. and Garrels R. M. (1986) Modeling atmospheric O_2 in the global sedimentary redox cycle. *Am. J. Sci.* **286**, 337–360.

Kump L. R., Kasting J. F., and Barley M. E. (2001) Rise of atmospheric oxygen and the upside down Archean mantle. *Geochem. Geophys. Geosys.* **2**, No. 2000 GC 000114.

Lasaga A. C. (1989) A new approach to isotopic modeling of the variation in atmospheric oxygen through the Phanerozoic. *Am. J. Sci.* **289**, 411–435.

Lasaga A. C. and Ohmoto H. (2002) The oxygen geochemical cycles: dynamics and stability. *Geochim. Cosmochim. Acta* **66**, 361–381.

Lenton T. M. and Watson A. J. (2000) Redfield revisited 2. What regulates the oxygen content of the atmosphere? *Global Biogeochem. Cycles* **14**, 249–268.

Lindh T. B. (1983) Temporal variations in ^{13}C, ^{34}S and global sedimentation during the Phanerozoic. MS Thesis, University of Miami.

Logan G. A., Hayes J. M., Hleshima G. B., and Summons R. E. (1995) Terminal Proterozoic reorganization of biogeochemical cycles. *Nature* **376**, 53–56.

Lovell J. S. (2000) Oxygen and carbon dioxide in soil air. In *Handbook of Exploration Geochemistry,* Geochemical Remote Sensing of the Subsurface (ed. M. Hale). Elsevier, Amsterdam, vol. 7, pp. 451–469.

McMenamin M. A. S. and McMenamin D. L. S. (1990) *The Emergence of Animals—The Cambrian Breakthrough.* Columbia University Press, New York.

Mörner N.-A. and Etiope G. (2002) Carbon degassing from the lithosphere. *Global. Planet. Change* **33**, 185–203.

Murakami T., Utsunomiya S., Imazu Y., and Prasad N. (2001) Direct evidence of late Archean to early Proterozoic anoxic atmosphere from a product of 2.5 Ga old weathering. *Earth Planet. Sci. Lett.* **184**, 523–528.

Najjar R. G. and Keeling R. F. (2000) Mean annual cycle of the air–sea oxygen flux: a global view. *Global Biogeochem. Cycles* **14**, 573–584.

Nutman A. P., Mojzsis S. J., and Friend C. R. L. (1997) Recognition of ≧3850 Ma water-lain sediments in West Greenland and their significance for the early Archean Earth. *Geochim. Cosmochim. Acta* **61**, 2475–2484.

Ohmoto H. (1996) Evidence in pre-2.2 Ga paleosols for the early evolution of atmospheric oxygen and terrestrial biota. *Geology* **24**, 1135–1138.

Ohmoto H. (1999) Redox state of the Archean atmosphere: evidence from detrital heavy minerals in ca. 3250–2750 Ma sandstones from the Pilbara Craton, Australia: comment and reply. *Geology* **27**, 1151–1152.

Olson J. M. and Pierson B. K. (1986) Photosynthesis 3.5 thousand million years ago. *Photosynth. Res.* **9**, 251–259.

Pan Y. and Stauffer M. R. (2000) Cerium anomaly and Th/U fractionation in the 1.85 Ga Flin Flon Paleosol: clues from REE- and U-rich accessory minerals and implications for paleoatmospheric reconstruction. *Am. Mineral.* **85**, 898–911.

Pavlov A. A., Kasting J. F., Brown L. L., Rages K. A., and Freedman R. (2000) Greenhouse warming by CH_4 in the atmosphere of early Earth. *J. Geophys. Res.* **105**, 11981–11990.

Petsch S. T. (1999) Comment on Carpenter and Lohmann (1999). *Geochim Cosmochim Acta* **63**, 307–310.

Petsch S. T. (2000) A study on the weathering of organic matter in black shales and implications for the geochemical cycles of carbon and oxygen. PhD Dissertation. Yale University.

Petsch S. T. and Berner R. A. (1998) Coupling the geochemical cycles of C, P, Fe, and S: the effect on atmospheric O_2 and the isotopic records of carbon and sulfur. *Am. J. Sci.* **298**, 246–262.

Petsch S. T., Berner R. A., and Eglinton T. I. (2000) A field study of the chemical weathering of ancient sedimentary organic matter. *Org. Geochem.* **31**, 475–487.

Petsch S. T., Smernik R. J., Eglinton T. I., and Oades J. M. (2001) A solid state ^{13}C-NMR study of kerogen degradation during black shale weathering. *Geochim. Cosmochim. Acta* **65**, 1867–1882.

Phoenix V. R., Konhauser K. O., Adams D. G., and Bottrell S. H. (2001) Role of biomineralization as an ultraviolet shield: implications for Archean life. *Geology* **29**, 823–826.

Pollard D. and Kasting J. K. (2001) Coupled GCM-ice sheet simulations of Sturtian (750–720 Ma) glaciation: when in the snowball-earth cycle can tropical glaciation occur? *EOS* **82**, S8.

Porter S. M. and Knoll A. H. (2000) Testate amoebae in the Neoproterozoic Era: evidence from vase-shaped microfossils in the Chuar Group, Grand Canyon. *Paleobiology* **26**, 360–385.

Rasmussen B. and Buick R. (1999) Redox state of the Archean atmosphere: evidence from detrital heavy minerals in ca. 3250–2750 Ma sandstones from the Pilbara Craton, Australia. *Geology* **27**, 115–118.

Rau G. H., Takahashi T., and Des Marais D. J. (1989) Latitudinal variations in plankton $\delta^{13}C$: implications for CO_2 and productivity in past oceans. *Nature* **341**, 516–518.

Rau G. H., Takahashi T., Des Marais D. J., Repeta D. J., and Martin J. H. (1992) The relationship between $d^{13}C$ of organic matter and $[CO_{2(aq)}]$ in ocean surface water: Data from a JGOFS site in the northeast Atlantic Ocean and a model. *Geochim. Cosmochim. Acta* **56**, 1413–1419.

Robinson J. M. (1989) Phanerozoic O_2 variation, fire, and terrestrial ecology. *Palaeogeogr. Palaeoclimat. Palaeoecol. (Global Planet Change)* **75**, 223–240.

Robinson J. M. (1991) Phanerozoic atmospheric reconstructions: a terrestrial perspective. *Palaeogeogr. Palaeoclimat. Palaeoecol.* **97**, 51–62.

Röhl H.-J., Schmid-Röhl A., Oschmann W., Frimmel A., and Schwark L. (2001) The Posidonia Shale (Lower Toarcian) of SW-Germany: an oxygen-depleted ecosystem controlled by sea level and paleoclimate. *Palaeogeogr. Palaeoclimat. Palaeocol.* **165**, 27–52.

Ronov A. B. (1976) Global carbon geochemistry, volcanism, carbonate accumulation, and life. *Geochem. Int.* **13**, 172–195.

Rye R. and Holland H. D. (1998) Paleosols and the evolution of atmospheric oxygen: a critical review. *Am. J. Sci.* **298**, 621–672.

Rye R. and Holland H. D. (2000) Life associated with a 2.76 Ga ephemeral pond? Evidence from Mount Roe #2 paleosol. *Geology* **28**, 483–486.

Schidlowski M. (1988) A 3,800-million year isotopic record of life from carbon in sedimentary rocks. *Nature* **333**, 313–318.
Schidlowski M. (2001) Carbon isotopes as biogeochemical recorders of life over 3.8 Ga of Earth history: evolution of a concept. *Precamb. Res.* **106**, 117–134.
Schopf J. W. (1992) Paleobiology of the Archean. In *The Proterozoic Biosphere* (eds. J. W. Schopf, and C. Klein). Cambridge University Press, pp. 25–39.
Schopf J. W. (1993) Microfossils of the early Archean Apex Chert: new evidence of the antiquity of life. *Science* **260**, 640–646.
Schopf J. W., Kudryavtvev A. B., Agresti D. G., Wdowiak T. J., and Czaja A. D. (2002) Laser-Raman imagery of Earth's earliest fossils. *Nature* **416**, 73–76.
Sinninghe-Damsté J. S. and Köster J. (1998) A euxinic southern North Atlantic Ocean during the Cenomanian/Turonian oceanic anoxic event. *Earth Planet. Sci. Lett.* **158**, 165–173.
Sleep N. H. and Zahnle K. (1998) Refugia from asteroid impacts on early Mars and the early Earth. *J. Geophys. Res. E—Planets.* **103**, 28529–28544.
Sleep N. H., Zahnle K. J., Kasting J. F., and Morowitz H. J. (1989) Annihilation of ecosystems by large asteroid impacts on the early earth. *Nature* **342**, 139–142.
Sleep N. H., Zahnle K. J., and Neuhoff P. S. (2001) Initiation of clement surface conditions on the earliest Earth. *Proc. Natl. Acad. Sci.* **98**, 3666–3672.
Sowers T., Bender M., and Raynaud D. (1989) Elemental and isotopic composition of occluded O_2 and N_2 in polar ice. *J. Geophys. Res.* **94**, 5137–5150.
Stratford K., Williams R. G., and Myers P. G. (2000) Impact of the circulation on sapropel formation in the eastern Mediterranean. *Global Biogeochem. Cycles* **14**, 683–695.
Sumner D. Y. (1997) Carbonate precipitation and oxygen stratification in late Archean seawater as deduced from facies and stratigraphy of the Gamohaan and Frisco Formations, Transvaal Supergroup, South Africa. *Am. J. Sci.* **297**, 455–487.
Thomas A. L. R. (1997) The breath of life—did increased oxygen levels trigger the Cambrian Explosion? *Trends Ecol. Evol.* **12**, 44–45.
Tolbert N. E., Benker C., and Beck E. (1995) The oxygen and carbon dioxide compensation points of C_3 plants: possible role in regulating atmospheric oxygen. *Proc. Natl. Acad. Sci.* **92**, 11230–11233.
Valley J. W., Peck W. H., King E. M., and Wilde S. A. (2002) A cool early Earth. *Geology* **30**, 351–354.
Van Cappellen P. and Ingall E. D. (1996) Redox stabilization of the atmosphere and oceans by phosphorus-limited marine productivity. *Science* **271**, 493–496.
van Zuilen M., Lepland A., and Arrhenius G. (2002) Reassessing the evidence for the earliest traces of life. *Nature* **418**, 627–630.
Vidal R. A., Bahr D., Baragiola R. A., and Peters M. (1997) Oxygen on Ganymede: laboratory studies. *Science* **276**, 1839–1842.
Walker J. C. G. (1979) The early history of oxygen and ozone in the atmosphere. *Pure Appl. Geophys.* **117**, 498–512.
Walker J. C. G. (1986) Global geochemical cycles of carbon, sulfur, and oxygen. *Mar. Geol.* **70**, 159–174.
Watson A. J. (1978) *Consequences for the Biosphere of Grassland and Forest Fires*. PhD Dissertation. Reading University, UK.
Werne J. P., Hollander D. J., Behrens A., Schaeffer P., Albrecht P., and Sinninghe-Damsté J. S. (2000) Timing of early diagenetic sulfurization of organic matter: a precursor-product relationship in Holocene sediments of the anoxic Cariaco Basin, Venezuela. *Geochim. Cosmochim. Acta* **64**, 1741–1751.
Werne J. P., Lyons T. W., Hollander D. J., Formolo M., and Sinninghe-Damsté J. S. (2003) Reduced sulfur in euxinic sediments of the Cariaco Basin: sulfur isotope constraints on organic sulfur formation. *Chem. Geol.* **195**, 159–179.
Wilde S. A., Valley J. W., Peck W. H., and Graham C. M. (2001) Evidence from detrital zircons for the existence of continental crust and oceans on the Earth 4.4 Gyr ago. *Nature* **409**, 175–178.
Wortmann U. G., Hesse R., and Zacher W. (1999) Major-element analysis of cyclic black shales: paleoceanographic implications for the early Cretaceous deep western Tethys. *Paleoceanography* **14**, 525–541.
Wray G. A., Levinton J. S., and Shapiro L. H. (1996) Molecular evidence for deep Precambrian divergences among Metazoan phyla. *Science* **274**, 568–573.
Zhang R., Follows M. J., Grotzinger J. P., and Marshall J. (2001) Could the late Permian deep ocean have been anoxic? *Paleoceanography* **16**, 317–329.

Isotope Geochemistry
ISBN: 978-0-08-096710-3
pp. 337–376

12

High-Molecular-Weight Petrogenic and Pyrogenic Hydrocarbons in Aquatic Environments

T. A. Abrajano Jr., B. Yan, J. Song and R. Bopp

Rensselaer Polytechnic Institute, Troy, NY, USA

and

V. O'Malley

Enterprise Ireland, Glasnevin, Republic of Ireland

12.1 INTRODUCTION

Geochemistry is ultimately the study of sources, movement, and fate of chemicals in the geosphere at various spatial and temporal scales. Environmental organic geochemistry focuses such studies on organic compounds of toxicological and ecological concern (e.g., Schwarzenbach *et al.*, 1993, 1998; Eganhouse, 1997). This field emphasizes not only those compounds with potential toxicological properties, but also the geological systems accessible to the biological receptors of those hazards. Hence, the examples presented in this chapter focus on hydrocarbons with known health and ecological concern in accessible shallow, primarily aquatic, environments.

Modern society depends on oil for energy and a variety of other daily needs, with present mineral oil consumption throughout the 1990s exceeding $3\times10^9\,t\,yr^{-1}$ (NRC, 2002). In the United States, for example, ~40% of energy consumed and 97% of transportation fuels are derived from oil. In the process of extraction, refinement, transport, use, and waste production, a small but environmentally significant fraction of raw oil materials, processed products, and waste are released inadvertently or purposefully into the environment. Because their presence and concentration in the shallow environments are often the result of human activities, these organic materials are generally referred to as "environmental contaminants." Although such reference connotes some form of toxicological or ecological hazard, specific health or ecological effects of many organic "environmental contaminants" remain to be demonstrated. Some are, in fact, likely innocuous at the levels that they are found in many systems, and simply add to the milieu of biogenic organic compounds that naturally cycle through the shallow environment. Indeed, virtually all compounds in crude oil and processed petroleum products have been introduced naturally to the shallow environments as oil and gas seepage for millions of years (NRC, 2002). Even high-molecular-weight (HMW) polyaromatic compounds were introduced to shallow environments through forest fires and natural coking of crude oil (Ballentine *et al.*, 1996; O'Malley *et al.*, 1997). The full development of natural microbial enzymatic systems that can utilize HMW hydrocarbons as carbon or energy source attests to the antiquity of hydrocarbon dispersal processes in the environment. The environmental concern is, therefore, primarily due to the rate and spatial scale by which petroleum products are released in modern times, particularly with respect to the environmental sensitivity of some ecosystems to these releases (Schwarzenbach *et al.*, 1993; Eganhouse, 1997; NRC, 2002).

Crude oil is produced by diagenetic and thermal maturation of terrestrial and marine plant and animal materials in source rocks and petroleum reservoirs. Most of the petroleum in use today is produced by thermal and bacterial decomposition of phytoplankton material that once lived near the surface of the world's ocean, lake, and river waters (Tissot and Welte, 1984). Terrestrially derived organic matter can be regionally significant, and is the second major contributor to the worldwide oil inventory (Tissot and Welte, 1984; Peters and Moldowan, 1993; Engel and Macko, 1993). The existing theories hold that the organic matter present in crude oil consists of unconverted original biopolymers and new compounds polymerized by reactions promoted by time and increasing temperature in deep geologic formations. The resulting oil can migrate from source to reservoir rocks where the new geochemical conditions may again lead to further transformation of the petrogenic compounds. Any subsequent changes in reservoir conditions brought about by uplift, interaction with aqueous fluids, or even direct human intervention (e.g., drilling and water washing) likewise could alter the geochemical makeup of the petrogenic compounds. Much of our understanding of environmental sources and fate of hydrocarbon compounds in shallow environments is indeed borrowed from the extensive geochemical and analytical framework that was meticulously built by petroleum geochemists over the years (e.g., Tissot and Welte, 1984; Peters *et al.*, 1992; Peters and Moldowan, 1993; Engel and Macko, 1993; Moldowan *et al.*, 1995; Wang *et al.*, 1999; Faksness *et al.*, 2002).

Hydrocarbon compounds present in petroleum or pyrolysis by-products can be classified based on their composition, molecular weight, organic structure, or some combination of these criteria. For example, a report of the Committee on Intrinsic Remediation of the US NRC classified organic contaminants into HMW hydrocarbons, low-molecular-weight (LMW) hydrocarbons, oxygenated hydrocarbons, halogenated aliphatics, halogenated aromatics, and nitroaromatics (NRC, 2000). Hydrocarbons are compounds comprised exclusively of carbon and hydrogen and they are by far the dominant components of crude oil, processed petroleum hydrocarbons (gasoline, diesel, kerosene, fuel oil, and lubricating oil), coal tar, creosote, dyestuff, and pyrolysis waste products. These hydrocarbons often occur as mixtures of a diverse group of compounds whose behavior in near-surface environments is governed by their chemical structure and composition, the geochemical conditions and media of their release, and biological factors, primarily microbial metabolism, controlling their transformation, and degradation.

Hydrocarbons comprise 50–99% of compounds present in refined and unrefined oil, and compounds containing other elements such as oxygen, nitrogen, and sulfur are present in relatively smaller proportions. Hydrocarbon compounds have carbons joined together as single C–C bonds (i.e., alkanes), double or triple C=C bonds (i.e., alkenes or olefins), or via an aromatic ring system with resonating electronic structure (i.e., aromatics). Alkanes, also called paraffins, are the dominant component of crude oil, with the carbon chain forming either straight (*n*-alkanes), branched (iso-alkanes), or cyclic (naphthenes) arrangement of up to 60 carbons (Figure 1). Aromatic compounds are the second major component of crude oil, with asphalthenes, consisting of stacks of highly polymerized aromatic structures (average of 16 rings), completing the list of major oil hydrocarbon components. Also shown in Figure 1 are several important classes of compounds that are extensively used in "fingerprinting" crude oil or petroleum sources: sterols derived from steroid, hopanol derived from bacteriohopanetetrols, and pristane and phytane derived from phytol (from chlorophyll) during diagenesis.

Polycyclic aromatic hydrocarbons (PAHs) that are made up of two or more fused benzene rings are minor components of crude oil (Figures 1 and 2), but they are by far the most important HMW compounds in terms of chronic environmental impact. Indeed, total PAH loading is used as the surrogate for the overall estimation of petroleum toxicity effects in environmental assessments (e.g., Meador *et al.*, 1995; NRC, 2002). PAHs are characterized by two or more fused benzene rings (Figure 2), and many have toxic properties including an association with mutagenesis and carcinogenesis (e.g., Cerniglia, 1991; Neilson, 1998). The World Health Organization (WHO) and the US Environmental Protection Agency (USEPA) have recommended 16 parental (unsubstituted rings) PAHs as priority pollutants (Figure 2). Although petroleum-sourced PAHs are major contributors in many surface and subsurface aquatic environments, another major contributor of PAHs to the environment is pyrolysis of fuel and other biomass. The latter are referred to as pyrogenic PAHs to distinguish them from the petrogenic PAHs derived directly from uncombusted petroleum, coal, and their byproducts. Natural sources such as forest fires could be important in less inhabited and remote watersheds, but anthropogenic combustion of fossil fuel (e.g., petroleum, coal) and wood is the dominant source of pyrogenic PAHs (Neff, 1979; Bjorseth and Ramdahl, 1983; Ballentine *et al.*, 1996; O'Malley *et al.*, 1997).

12.2 SCOPE OF REVIEW

A number of previous reviews and textbooks on organic contaminant behavior in geochemical environments have already considered the general physical, chemical, and biological behavior of organic contaminants on the basis of their structure and composition. This chapter focuses on the geochemical behavior of a group of organic compounds referred to as HMW hydrocarbons, and emphasizes a class of compounds known as PAHs. Focus on PAHs is justified by their known toxicity and carcinogenicity, hence the environmental concern already noted above. Monocyclic aromatic hydrocarbons comprise the LMW end of the aromatic hydrocarbon spectrum.

Other general reviews and textbooks that summarize the sources and geochemical fate and transport of hydrocarbons in a variety of geological media are also available (e.g., Moore and Ramamoorthy, 1984; Schwarzenbach *et al.*, 1993, 1998; Eganhouse, 1997; Volkman *et al.*, 1997; NRC, 2000, 2002; Neilson, 1998; Aboul-Kassim and Simoneit, 2001; Beek, 2001). The following discussions focus on shallow aquatic environments, especially sediments, given the highly hydrophobic (mix poorly with water) and lipophilic (mix well with oil/fat) nature of most HMW hydrocarbon compounds. Nevertheless, the readers should recognize that in spite of our focus on a specific group of hydrocarbon compounds, the behavior of these compounds in aquatic systems would have broad similarity with the behavior of many other hydrophobic and lipophilic compounds discussed in other chapters in this volume. Indeed, most of these compounds are studied simultaneously in shallow aquatic environments, sharing not only common geochemical behavior but also analytical procedures for extraction, isolation, and characterization (e.g., Peters and Moldowan, 1993; Eganhouse, 1997; Aboul-Kassim and Simoneit, 2001). This chapter will conclude with several

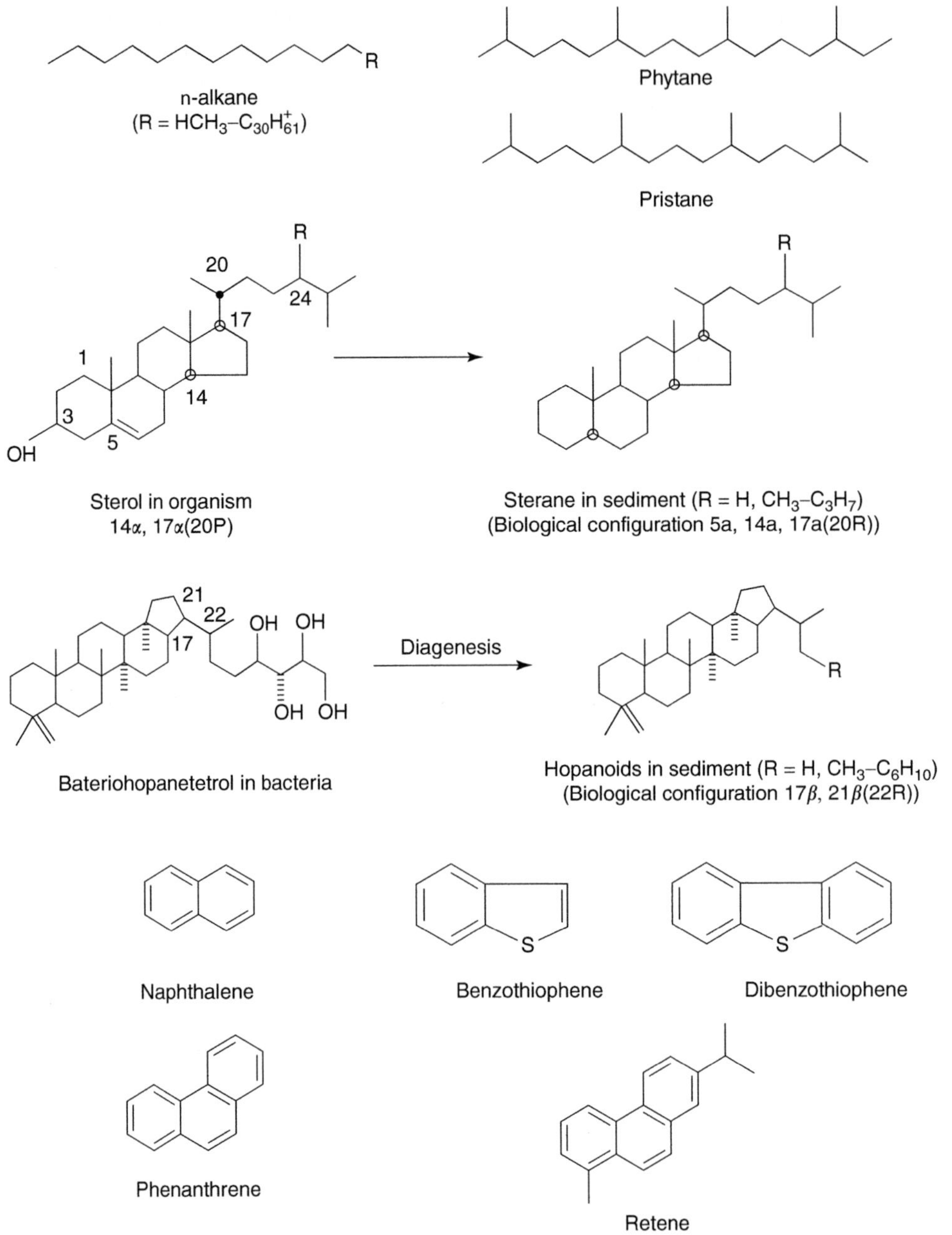

Figure 1 Examples of types of hydrocarbons found in crude oil and mentioned in the text.

recent case studies applying molecular and isotopic approaches to unraveling PAH sources and pathways in aquatic environments.

12.3 SOURCES

12.3.1 Petrogenic Hydrocarbons

Worldwide use of petroleum outpaced coal utilization by the 1960s, and accidental oil discharge and release of waste products such as CO_2, soot (black carbon), and PAHs also increased concomitantly. The more recent report of the NRC (2002) shows that oil production crept up from 7 in the 1970s to 11 Mt d^{-1} by the end of 2000. The total discharge of petroleum into the world's oceans was estimated to be between 0.5 and 8.4 Mt of petroleum hydrocarbons annually (NRC, 2002), which roughly constitutes 0.1% of the annual oil consumption rate. Of this total, 47% are derived from natural seeps, 38% from consumption of petroleum (e.g., land-based runoff, operational discharges, and atmospheric deposition), 12% from petroleum transport, and 3% from petroleum extraction. Used crankcase oils or engine lubricating oils are an important specific source of PAHs in urban environments. The world production of crankcase oil is estimated to be ~40 Mt yr^{-1} and

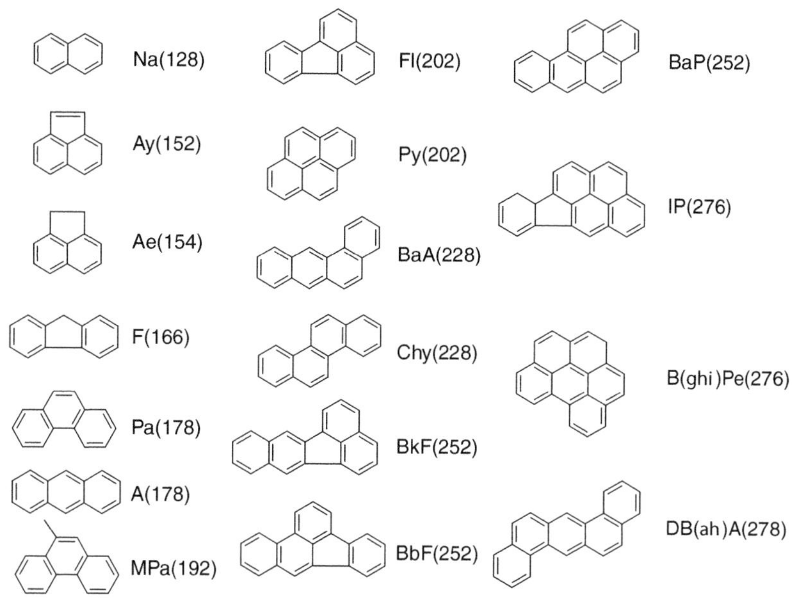

Figure 2 Structures of polycyclic aromatic hydrocarbons. Symbols used in this figure and text: Na (naphthalene), Ay (acetonaphthylene), Ae (acenaphthene), Fl (fluorene), Pa (phenanthrene), A (anthracene), MPa (methyl phenanthrene), F (fluoranthene), Py (pyrene), BaA (benz(*a*)anthracene), Chy (chrysene), BkF (benzo(*k*)fluoranthene), BbF (benzo(*b*)fluoranthene), BaP (benzo(*a*)pyrene), IP (indenopyrene), B(*ghi*)Pe (benzo(*ghi*)perylene), and Db(*ah*)A (dibenzo(*ah*)enthracene).

4.4% of that is estimated to eventually reach aquatic environments (NRC, 1985).

Transportation-related release, in order of decreasing annual input, includes accidental releases during tank vessel spill, intentional ballast discharge, pipeline spills, and coastal facility spills. Spectacular oil releases recorded by massive oil spills from grounded tankers tend to capture the public attention, although chronic releases from operational discharges and land releases are quantitatively more important (NRC, 2002). Tank vessel spills account for <8% of worldwide petroleum releases in the 1990s (NRC, 2002). For example, oil slick formed from ballast discharges in the Arabian Sea was estimated to exceed $5.4\times10^4\,m^3$ for 1978 (Oostdam, 1980). For comparison, the volume released by the 2002 Prestige oil spill off the Northwest coast of Spain is $1.3\times10^4\,m^3$. Other major oil spills with larger releases include the Torrey Canyon ($1.17\times10^5\,m^3$), Amoco Cadiz ($2.13\times10^5\,m^3$), Ixtoc blowout ($5.3\times10^5\,m^3$), Exxon Valdez ($5.8\times10^4\,m^3$), and 1991 Gulf War ($>10^6\,m^3$). Studies by Kvenvolden *et al.* (1993a, b, 1995) also showed the substantially greater input of long-term chronic releases of California oil in the Prince Williams Sound sediments compared with oil released from the Exxon Valdez. A report by the NRC (2002) points to the dramatic decline in oil spilled in North American waters during the 1990s compared with the previous decades (Figure 3), with vessel spills accounting for only 2% of total petroleum release to US waters. The 1980s recorded the largest number (391) and volume ($2.55\times10^5\,m^3$) of oil spilled to North American waters.

Land releases into groundwater aquifers, lakes, and rivers are dominated by urban runoff and municipal/industrial discharges, but the actual amounts are difficult to quantify (NRC, 2002). Petroleum discharges from underground storage tanks are by far the dominant source of hydrocarbons in groundwater. However, surface releases from bulk supply depot, truck stops, industrial refueling facilities, and oil storage terminals could also be locally significant sources. It is noteworthy that substantial releases of hydrocarbons may also emanate from natural deposits of oil, heavy oil, shale oil, and bitumen. For example, the Athabasca River in northern Canada shows concentration levels of oil and grease of 850 and $>3{,}000\,\mu g\,l^{-1}$, respectively, from natural bitumen deposits (Moore and Ramamoorthy, 1984). Likewise, Yunker *et al.* (1993, 1995) showed dominant input of natural hydrocarbons in the Mackenzie River. Indeed, natural seeps account for over 60% of petroleum releases to North American waters (NRC, 2002).

Petroleum hydrocarbon sources to North American and worldwide waters were summarized in a report by NRC (2002). In many cases of large petroleum spills, the specific source of petroleum spill is evident, and no geochemical fingerprinting is required to establish the source. Nevertheless, the inventory of petroleum compounds and biomarkers that are eventually sequestered in bottom sediments need not reflect sole

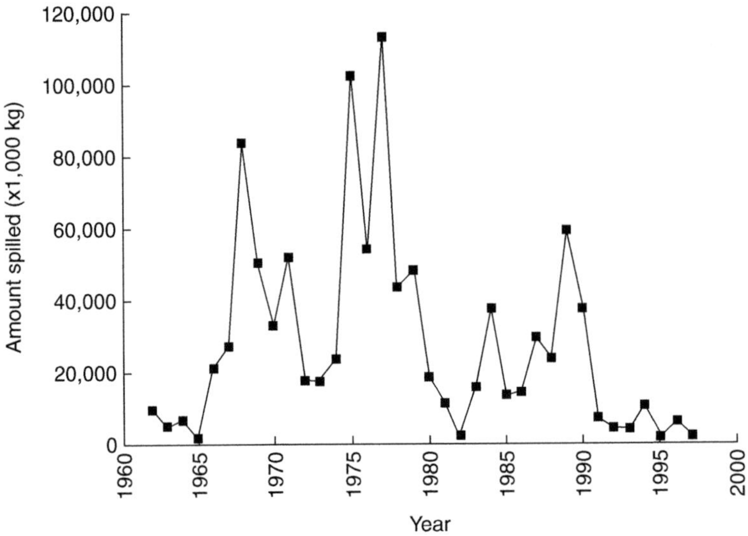

Figure 3 Oil spill trend in North American waters during the 1990s compared with the previous decades (NRC, 2002).

derivation from a single source, even in cases of massive oil spills in the area (e.g., Kvenvolden *et al.*, 1995; Wang *et al.*, 1999). Where a mass balance of petroleum sources is required to properly design remediation or identify a point source, molecular methods for distinguishing sources of hydrocarbons have come to the fore.

Several geochemical methods for allocating sources of petrogenic hydrocarbons that are released to aquatic systems have been successfully applied (e.g., O'Malley *et al.*, 1994; Whittaker *et al.*, 1995; Aboul-Kassim and Simoneit, 1995; Kvenvolden *et al.*, 1995; Wang and Fingas, 1995; Dowling *et al.*, 1995; Bieger *et al.*, 1996; Kaplan *et al.*, 1997; Eganhouse, 1997; Volkman *et al.*, 1997; Mansuy *et al.*, 1997; Hammer *et al.*, 1998; Wang *et al.*, 1999; McRae *et al.*, 1999, 2000; Mazeas and Budzinski, 2001; Hellou *et al.*, 2002; Faksness *et al.*, 2002; NRC, 2002; Lima *et al.*, 2003). The fingerprints used are either molecular or isotopic, and are variably affected by weathering processes. They include overall molecular distribution of hydrocarbons (e.g., range of carbon numbers and odd–even predominance), source-specific biomarkers (e.g., terpanes and steranes), so-called diagnostic molecular ratios, and stable isotope compositions. The level of specificity by which a source can be pinpointed is dependent on the fingerprint used to tag the specific source and the multiplicity of hydrocarbon sources involved. Faksness *et al.* (2002) presented a flow chart for oil spill identification using a tiered molecular discrimination scheme based on the overall hydrocarbon distribution, source-specific markers, and diagnostic ratios (Figure 4) of target compounds that have been used for source identification of spilled oil including: (1) saturated hydrocarbons (SHs) + pristane and phytane; (2) volatile alkylated aromatics including benzene, toluene, ethyl benzene, and xylene (BTEX); (3) alkylated and nonalkylated PAHs and heterocyclics; and (4) terpanes and steranes (Volkman *et al.*, 1997; Wang *et al.*, 1999). For example, Wang and Fingas (1995) and Douglas *et al.* (1996) suggested that alkylated dibenzothiophenes are sufficiently resilient to a wide range of weathering reactions to be useful fingerprints for sources of crude oil. Kvenvolden *et al.* (1995) used the abundances of sterane and hopane biomarkers to differentiate specific crude oil sources in Prince William Sound, Alaska. The use of resilient biomarker signatures has matured to the point that they are widely used for specific litigation cases for assigning liability for oil releases (e.g., Kaplan *et al.*, 1997; Wang *et al.*, 1999).

The molecular distribution of PAHs in petroleum and crankcase oils is quite distinct from pyrogenic sources that will be discussed in the following section (Figure 5). This contrast provides an excellent basis for source apportionment of HMW hydrocarbons in the environment. Petrogenic PAHs consist primarily of 2- and 3-ring parental and methylated compounds with lower concentrations of HMW PAHs (Figure 5) (Bjorseth and Ramdahl, 1983; Pruell and Quinn, 1988; Vazquez-Duhalt, 1989; Latimer *et al.*, 1990; O'Malley, 1994). PAH formation during oil generation is attributed both to the aromatization of multiring biological compounds (e.g., sterols) and to the fusion of smaller hydrocarbon fragments into new aromatic structures (e.g., Radke, 1987; Neilson and Hynning, 1998; Simoneit, 1998). Steroids are probably the best well understood in terms of biological origin and geological fate, and a simplification of the proposed pathways of sterol diagenesis and catagenesis is summarized in

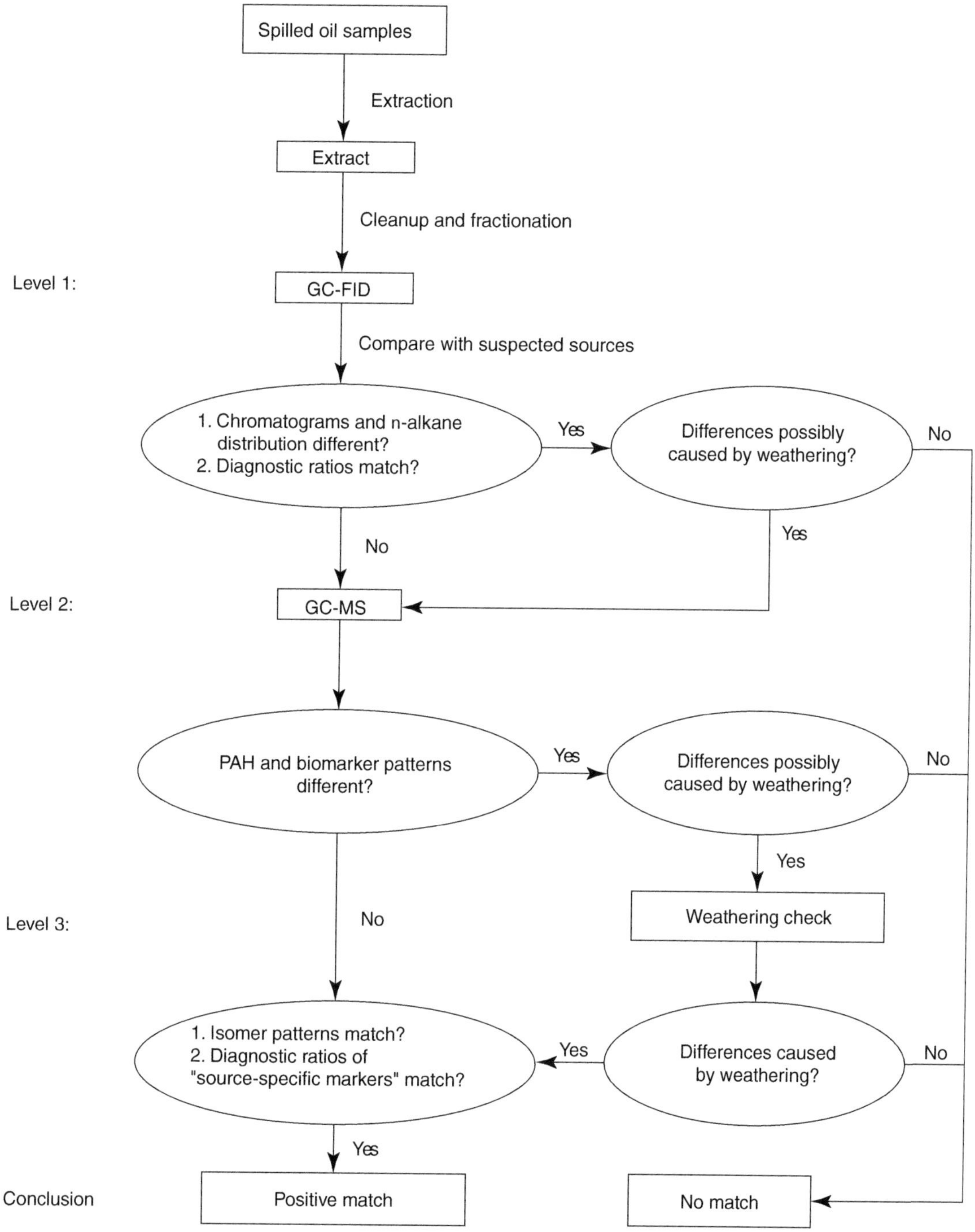

Figure 4 Tiered oil spill source identification scheme using molecular chemistry. The increasing level of source specificity required (down the diagram) is provided by global distributions of *n*-alkanes (level 1), PAH and biomarker distribution patterns (level 2), and isomeric and other diagnostic marker ratios (level 3), respectively. Reproduced by permission of Nordtest from Faksness *et al.* (2002).

Figure 6 (after MacKenzie, 1984). Crude oils usually formed at temperatures $<150\,^{\circ}C$ have a predominance of alkylated (i.e., possessing alkyl side chains) over parental PAHs. O'Malley (1994) characterized hundreds of variably used crankcase oils, and of these samples, 25.5% comprised of 4- and 5-ring parental compounds, and virgin crankcase oil samples were found to contain no resolvable PAHs (cf. Pruell and Quinn, 1988; Latimer *et al.*, 1990). Since virgin crankcase oils are devoid of measurable PAHs, the most probable source of PAHs in used crankcase oil are the thermal alteration reactions (i.e., aromatization) of oil components such as terpenoids in the car engine (Pruell and Quinn, 1988; Latimer *et al.*, 1990). As with true diagenetic PAHs (Figure 6),

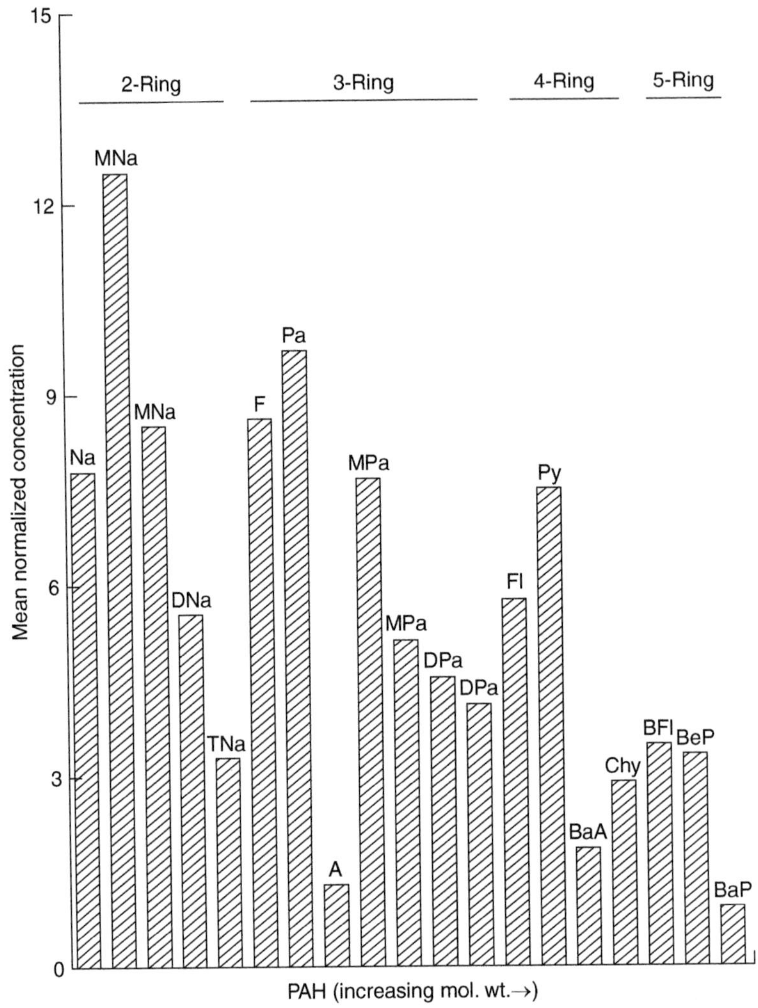

Figure 5 Distribution of PAHs in petrogenic sources. Compound symbols as in Figure 2 with the modifier "M" (methylated), "D" (dimethylated), and "T" (trimethylated) added for alkylated PAH.

the distribution of PAHs in used crankcase oils apparently depends on several factors including temperature of reaction, engine design, and general operating conditions of the engine. Since the engine-operating temperatures are generally high, the likely products resulting from these reactions include 3-, 4-, and 5-ring unsubstituted compounds such as Pa, Fl, Py, BaA, Chy, BeP, and BaP (see Figure 2 caption for abbreviations). Pruell and Quinn (1988) also suggested that LMW compounds (1- and 2-ring) might be accumulated in crankcase oils from admixed gasoline. The relative stability of pyrogenic PAHs to weathering makes them attractive markers for discrimination of oil sources (Figure 5) (Pancirov and Brown, 1975; O'Malley, 1994; Wang *et al.*, 1999). However, Volkman *et al.* (1997) advocate caution in the use of aromatic compounds because of the relatively small variations between different oils and the differential aqueous solubility of these compounds (see below). Some HMW PAHs may be present in crude oil, albeit, at very low concentrations. Finally, crude oils are also rich in heterocyclic species particularly thiophenes (e.g., dibenzothiophenes). Significant molecular variations of these compounds have been reported for individual crude oil samples, and these are mainly attributed to oil origin and maturity (Neff, 1979). Other important potential sources of petrogenic PAHs identified in sedimentary environments are asphalt and tire and brake wear (Wakeham *et al.*, 1980a; Broman *et al.*, 1988; Takada *et al.*, 1990; Latimer *et al.*, 1990; Reddy and Quinn, 1997).

Although previous authors have suggested bacterial synthesis as a possible source of biogenic PAHs in modern sediments, Hase and Hites (1976) have shown that bacteria more likely only bioaccumulate them from the growth medium. Anaerobic aromatization of tetracyclic triterpenes appeared to have been demonstrated by Lohmann *et al.* (1990) by incubating radiolabeled β-amyrin, but the quantitative importance of these synthesis pathways remains unresolved (e.g., Neilson and Hynning, 1998). The early diagenesis of

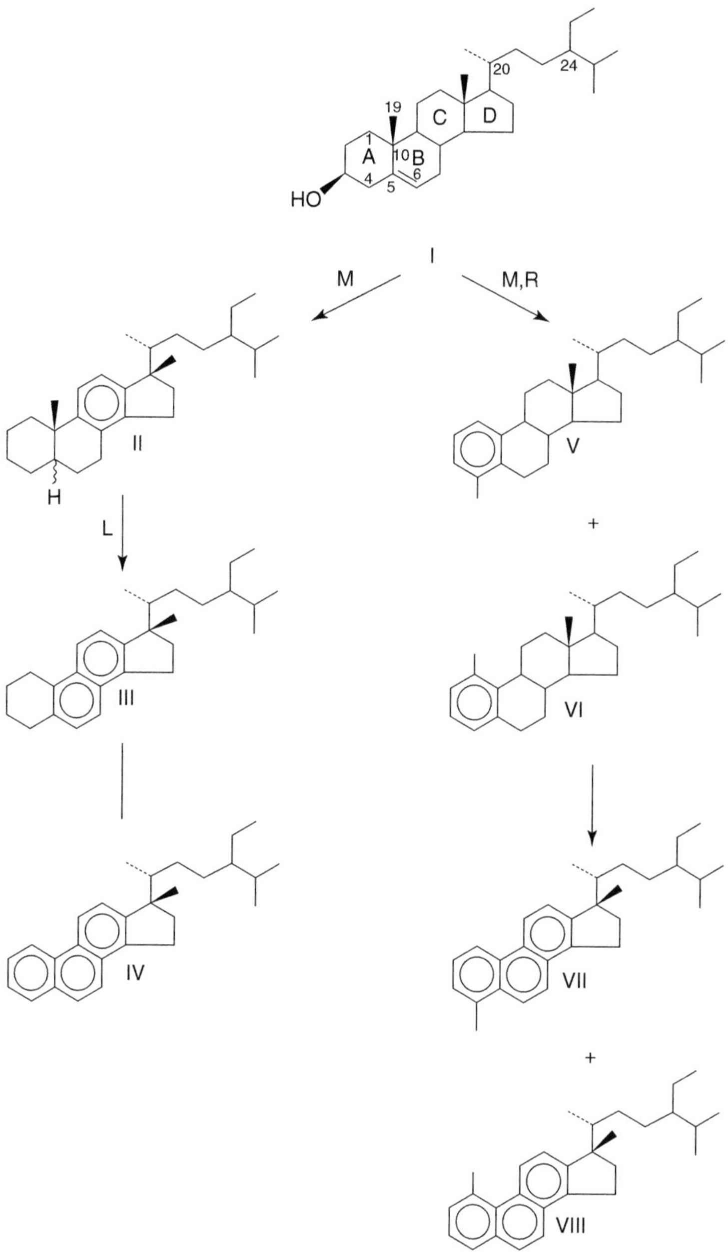

Figure 6 Diagenetic conversion of sterol (I) into various aromatic hydrocarbons during diagenesis. Abbreviation "M" implies that the reaction involved multiple steps, "R" represents aromatization of the A ring, and "L" represents the aromatization of the B ring. Compounds II, V, and VI are intermediates for the formation of triaromatic steroids. Adapted by permission of Elsevier from MacKenzie (1984).

sedimentary organic matter is certain to lead to the formation of a number of PAHs from alicyclic precursors (e.g., Neilson and Hynning, 1998; Simoneit, 1998). Examples of reactions are the diagenetic production of phenanthrene and chrysene derivatives from aromatization of pentacyclic triterpenoids originating from terrestrial plants, early diagenesis of abietic acids to produce retene (Figure 7), and *in situ* generation of perylene from perylene quinones (Youngblood and Blumer, 1975; Laflamme and Hites, 1978; Wakeham *et al.*, 1980b; Venkatesan, 1988; Lipiatou and Saliot, 1991; Neilson and Hynning, 1998; Simoneit, 1998; Wang *et al.*, 1999). Perylene and retene are the two most prominent PAHs found in recently deposited sediments (Lipiatou and Saliot, 1991). The origin of perylene has been linked to terrestrial precursors (4,9-dihydroxyperylene-3,10-quinone, the possible candidate), marine precursors, and anthropogenic inputs (Blumer *et al.*, 1977; Laflamme and Hites, 1978; Prahl and Carpenter, 1979; Venkatesan, 1988; Lipiatou and Saliot, 1991). The diagenetic pathway for retene is well constrained (Figure 7), but it can also be

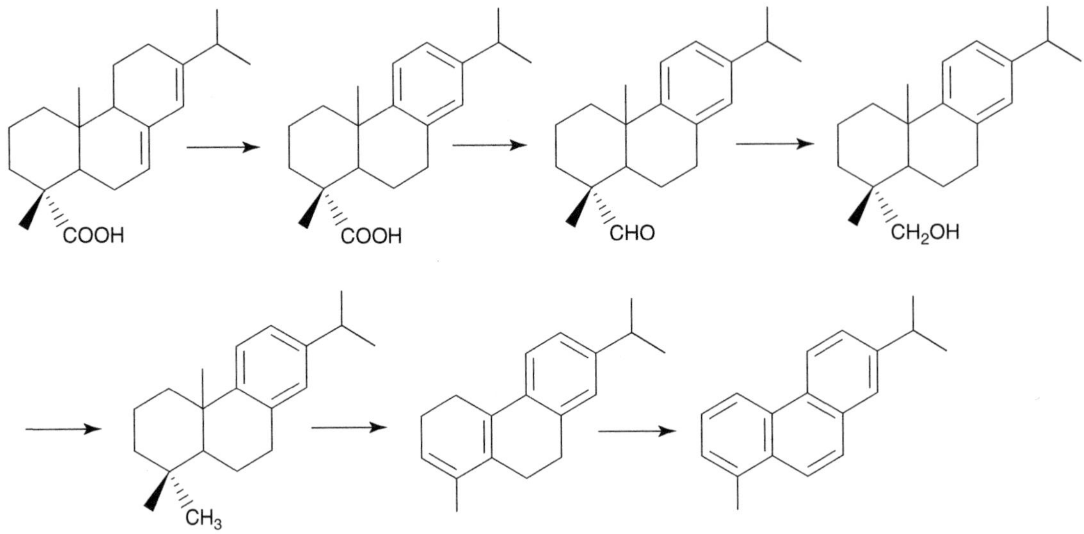

Figure 7 Diagenetic pathway for the formation of retene from abietic acid. After Wakeham *et al.* (1980a).

produced from wood combustion (Wakeham *et al.*, 1980b; Lipiatou and Saliot, 1991; Neilson and Hynning, 1998).

12.3.2 Pyrogenic Hydrocarbons

The burial maturation of sedimentary organic matter leading to oil and coal formation and possible biosynthesis are only two of three possible pathways for generating HMW hydrocarbons. Pyrolysis or incomplete combustion at high temperatures also generates a wide variety of LMW and HMW hydrocarbons depending on the starting materials, environmental condition of pyrolysis, and kinetic factors (e.g., gas circulation). Whereas lower molecular compounds generated by pyrolysis have generated significant interest (e.g., butadiene and formaldehyde), the condensed structures from naphthalene to "black carbon" or soot has been the focus of interest amongst the HMW hydrocarbons.

Indeed, pyrolysis of petroleum and fossil fuel comprise quantitatively the most important source of PAHs in modern sediments (Youngblood and Blumer, 1975; Laflamme and Hites, 1978; Neff, 1979; Wakeham *et al.*, 1980a, b; Sporstol *et al.*, 1983; Bjorseth and Ramdahl, 1983; Kennicutt *et al.*, 1991; Lipiatou and Saliot, 1991; Canton and Grimalt, 1992; Brown and Maher, 1992; Steinhauer and Boehm, 1992; Yunker *et al.*, 1993, 1995; O'Malley *et al.*, 1996; Lima *et al.*, 2003). For example, sedimentary PAH distribution shows some common molecular features including the dominance of 4- and 5-ring PAHs (Fl, Py, BaA, Chy, BeP, and BaP), Pa/A ratio between 2 and 6, high Pa/MPa ratio, and Fl/Py ratio close to unity. As noted above, the common characteristics related to direct petrogenic-related sources are a series of 2- and 3-ring parental and alkylated compounds (Na, MNa, Pa, and MPa), low Pa/MPa and Fl/Py ratios, and an unresolved complex mixture (UCM) (Kennicutt *et al.*, 1991; Volkman *et al.*, 1992; Wang *et al.*, 1999). Although these sedimentary hydrocarbons may bear the imprint of petrogenic sources, the overall molecular attributes are signatures of high-temperature pyrolysis of fossil fuels and natural sources (e.g., forest fires) (Youngblood and Blumer, 1975; Laflamme and Hites, 1978; Lake *et al.*, 1979; Killops and Howell, 1988; Ballentine *et al.*, 1996). Individual markers such as perylene and retene, which are thought to be formed by the diagenetic alteration of biogenic compounds, are also typical of recently deposited sediments (Wakeham *et al.*, 1980a; Venkatesan, 1988; Lipiatou and Saliot, 1991; Yunker *et al.*, 1993, 1995), although combustion sources of perylene are also known (e.g., Wang *et al.*, 1999).

The mechanisms by which pyrolytic production of PAHs occur are complex, and have been widely studied since the 1950s (Badger *et al.*, 1958; Howsam and Jones, 1998). Pyrolytic production of PAHs is generally believed to occur through a free radical pathway, wherein radicals of various molecular weights can combine to yield a series of different hydrocarbon products. Therefore, the formation of PAHs is thought to occur in two distinct reaction steps: pyrolysis and pyrosynthesis (Lee *et al.*, 1981). In pyrolysis, organic compounds are partially cracked to smaller unstable molecules at high temperatures. This is followed by pyrosynthesis or fusion of fragments into larger and relatively more stable aromatic structures. Badger *et al.* (1958) was the first to propose this stepwise synthesis using BaP from free radical recombination reactions (Figure 8). Compounds identified in subsequent studies suggest that the C_2 species react to form C_4, C_6, and

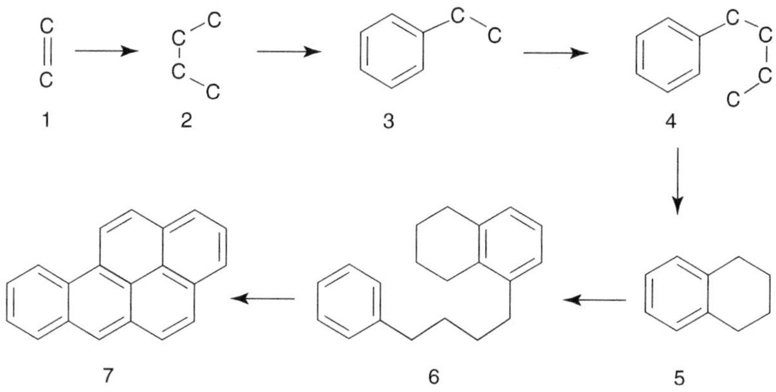

Figure 8 Stepwise pyrosynthesis of benzo(*a*)pyrene through radical recombination involving acetylene (1), a four carbon unit such as vinylacetylene or 1,3-butadiene, and styrene or ethylbenzene (3). Reproduced by permission of Royal Society of Chemistry from Badger *et al.* (1958).

C_8 species, and confirm the mechanisms proposed by Badger *et al.* (1958) (Howsam and Jones, 1998; Figure 8). Despite the large quantities of different PAHs formed during primary reactions, only a limited number enter the environment. This is because initially formed PAHs themselves can be destroyed during combustion as a result of secondary reactions that lead to the formation of either higher condensed structures or oxidized carbon (Howsam and Jones, 1998). For example, the pyrolysis of naphthalene can yield a range of HMW species such as perylene and benzofluoranthenes, possibly as a result of cyclodehydrogenation of the binaphthyls (Howsam and Jones, 1998). This may be particularly important for compounds that are deposited on the walls of open fireplaces along with soot particulates close to the hot zone of the flame.

PAHs isolated from important pyrogenic sources vary widely in composition, and they are quite distinct from the PAH distribution of petrogenic PAHs (Figure 9) (Alsberg *et al.*, 1985; Westerholm *et al.*, 1988; Broman *et al.*, 1988; Freeman and Cattell, 1990; Takada *et al.*, 1990; Rogge *et al.*, 1993; O'Malley, 1994; Howsam and Jones, 1998). For example, individual fireplace soot samples, from hard- and softwood-burning open fireplaces, were consistently dominated by 3-, 4-, and 5-ring parental PAHs with generally lower concentrations of methylated compounds (Figure 9) (cf. Freeman and Cattell, 1990; Howsam and Jones, 1998). Vehicular emission and soot samples are also generally characterized by the presence of pyrolysis-derived 3-, 4-, and 5-ring parental PAHs (Wakeham *et al.*, 1980a; Stenberg, 1983; Alsberg *et al.*, 1985; Westerholm *et al.*, 1988; Broman *et al.*, 1988; Takada *et al.*, 1990). The range and concentration of PAHs that accumulate in car mufflers are dependent on car age, engine-operating conditions, catalytic converter efficiency, and general driving conditions (Pedersen *et al.*, 1980; Stenberg, 1983; Rogge *et al.*, 1993). Rogge *et al.* (1993) also reported that greater PAH emissions occurred from noncatalytic automobiles compared with vehicles with catalytic systems. Perylene, a pentacyclic hydrocarbon, reported to be predominantly of diagenic origin (Laflamme and Hites, 1978; Prahl and Carpenter, 1979; Wakeham *et al.*, 1980b; Simoneit, 1998; Wang *et al.*, 1999) was also identified in some of the investigated car soots samples, although in very low concentrations (Blumer *et al.*, 1977). Lipiatou and Saliot (1992) reported that perylene might also be derived from coal pyrolysis. In contrast to the pyrogenic sources, Figure 9 also shows the enhanced concentration of 2-ring compounds, especially a whole series of methylated phenanthrenes and naphthalenes in petrogenic sources.

12.4 PATHWAYS

HMW hydrocarbons can enter surface aquatic systems directly by spillage, accidental release, and natural oil seeps, or indirectly through sewers, urban, and highway runoff. Once hydrocarbons enter aquatic environments, they become rapidly associated with particulate matter, and are deposited in bottom sediments of surface waters or sorbed onto aquifer materials in groundwater systems (Radding *et al.*, 1976; Schwarzenbach *et al.*, 1993; Luthy *et al.*, 1997) (Figure 10). In surface-water systems, physical factors, such as turbulence, stability and composition of colloidal particles, deep-water currents, surface waves, and upwelling influence the length of time they remain suspended in the water column. The rate at which hydrocarbons are incorporated into bottom sediments is controlled by sedimentation rate, bioturbation, and bottom sediment–water column exchanges. In groundwater systems, the transport of hydrocarbons depends on whether or not they comprise a separate nonaqueous phase liquid (NAPL), groundwater flow velocities, and geochemical partitioning that are discussed below.

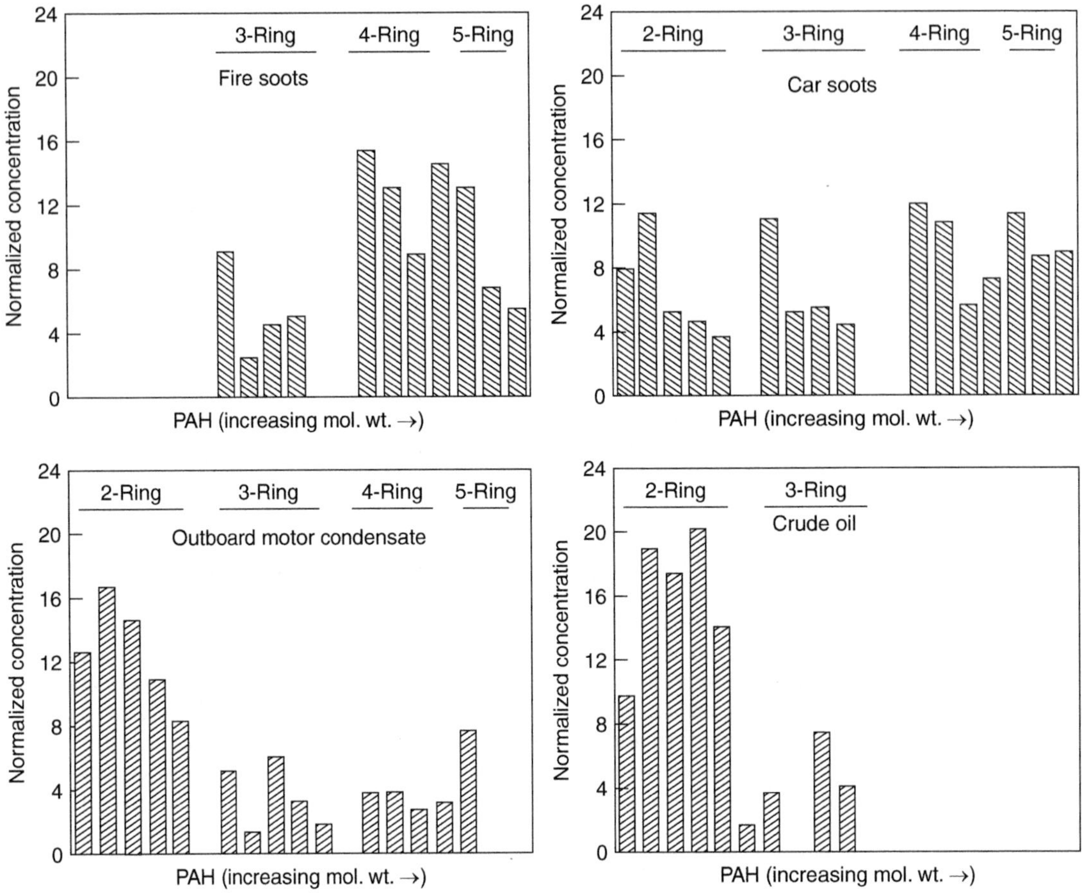

Figure 9 Molecular distribution/signatures of fire soot, car soot, outboard motor concentrate, and crude-oil PAH sources. After O'Malley (1994).

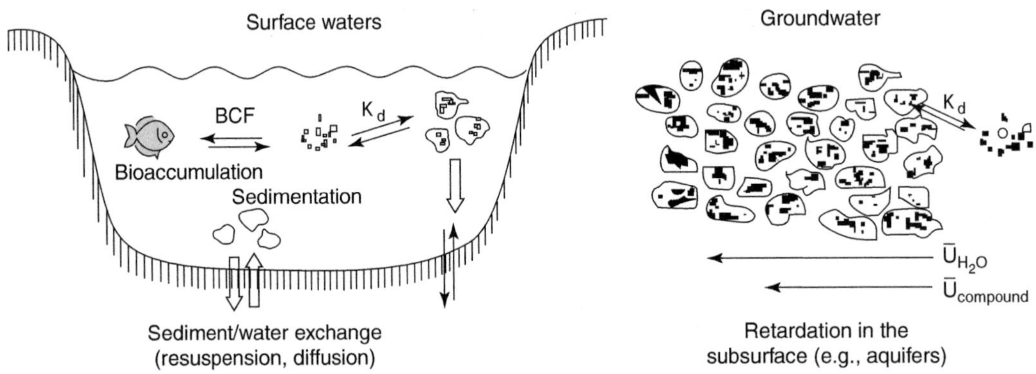

Figure 10 Solid–water exchange processes in groundwater and surface-water environments illustrating sorption to particulates (K_d) and biological concentration factor (BCF). Retardation in groundwater aquifer signified by the difference between water and compound velocities (U). Adapted by permission of Oxford University Press, Inc. from Schwarzenbach *et al.* (1998).

HMW hydrocarbons likewise enter groundwater environments directly or indirectly from domestic and industrial effluents and urban runoff, and direct spillage of petroleum and petroleum products (e.g., ballast discharge, underground storage tanks).

An important alternate pathway of hydrocarbons to aquatic systems, however, is deposition of airborne particulates including anthropogenic (e.g., soot particles) and natural biogenic (e.g., monoterpenes, difunctional carboxylic acids) aerosols (Simoneit, 1984, 1986; Strachan and

Eisenreich, 1988; Baker and Eisenreich, 1990; Eisenreich and Strachan, 1992; Currie *et al.*, 1999). Indeed, the importance of "organic aerosols" from anthropogenic and natural sources is now extensively recognized (e.g., Simoneit, 1984, 1986; Ellison *et al.*, 1999). A major part of the extractable and elutable organic matter in urban aerosols consists of a UCM, mainly the branched and cyclic hydrocarbons that originate from car exhaust (e.g., Simoneit, 1984; Rogge *et al.*, 1993). The resolved organic compounds in aerosol extracts consist of *n*-alkanes and fatty acids and PAHs. Whereas health, environmental, and climatic concerns have targeted the reaction products and intermediates formed from tropospheric reactions of labile hydrocarbons and carboxylic acids in these aerosols, the focus of toxicological concern has been on the PAHs.

In the case of PAHs, it is generally recognized that virtually all emissions to the atmosphere are indeed associated with airborne aerosols (Suess, 1976; Simoneit, 1986; McVeety and Hites, 1988; Strachan and Eisenreich, 1988; Baek *et al.*, 1991; Eisenreich and Strachan, 1992; Currie *et al.*, 1999; Offenberg and Baker, 1999). For example, Eisenreich and Strachan (1992) and Strachan and Eisenreich (1988) showed that upward of 50% of the PAH inventory of the Great Lakes is deposited via atmospheric fallout. PAHs are initially generated in the gas phase, and then as the vapor cools, they are adsorbed into soot particulates (Howsam and Jones, 1998). The highest concentrations of HMW PAHs in airborne particulates occur in $<5\,\mu m$ particle size range (Pierce and Katz, 1975; Offenberg and Baker, 1999). PAH distribution between the gas and particulate phase is generally influenced by the following factors: vapor pressure as a function of ambient temperature, availability of fine particulate material, and the affinity of individual PAHs for the particulate organic matrix (Goldberg, 1985; Baek *et al.*, 1991; Schwarzenbach *et al.*, 1993). Atmospheric concentrations of PAHs are normally high in winter and low during the summer months (Pierce and Katz, 1975; Gordon, 1976; Howsam and Jones, 1998), an observation attributed to increased rates of photochemical activity during the summer and increased consumption of fossil fuels during the winter period.

Residence times of particulate PAHs in the atmosphere and their dispersal by wind are determined predominantly by particle size, atmospheric physics, and meteorological conditions (Howsam and Jones, 1998; Offenberg and Baker, 1999). The main processes governing the deposition of airborne PAHs include wet and dry deposition and, to a smaller extent, vapor-phase deposition into surfaces. Particles between 5 and $10\,\mu m$ are generally removed rapidly by sedimentation and by wet and dry deposition (Baek *et al.*, 1991). However, PAHs associated with fine particulates (<1–$3\,\mu m$) can remain suspended in the atmosphere for a sufficiently long time to allow dispersal over hundreds or thousands of kilometers (McVeety and Hites, 1988; Baek *et al.*, 1991).

PAHs in domestic sewage are predominantly a mixture of aerially deposited compounds produced from domestic fuel combustion and industrial and vehicle emissions, combined with PAHs from road surfaces that have been flushed into sewage. Road surface PAHs are derived primarily from crankcase oil, asphalt, and tire and brake wear (Wakeham *et al.*, 1980a; Broman *et al.*, 1988; Takada *et al.*, 1990; Rogge *et al.*, 1993; Reddy and Quinn, 1997). Some PAHs are removed from sewage during primary treatment (sedimentation). Not all urban runoff enters the sewer system, and some sewer designs allow runoff to be independently discharged to aquatic systems without primary treatment. The quantity of runoff from an urban environment is generally governed by the fraction of paved area within a catchment and the annual precipitation. During periods of continued rainfall, road surfaces are continually washed and the contributions of PAHs to watersheds are generally low. However, significant episodic contributions to aquatic systems can occur after prolonged dry periods or during spring snow melt (Hoffman *et al.*, 1984; Smirnova *et al.*, 1998). The distribution and quantity of PAHs in industrial effluents depends on the nature of the operation and on the degree of treatment prior to discharge. In some urban areas, industrial effluents are combined with domestic effluents prior to treatment, or they are independently treated before being discharged to sewer systems. Once particle-associated hydrocarbons are deposited in sediments or sorbed onto aquifer material, the subsequent fate is determined by the physics of sediment and water transport, and the biogeochemical reactions described below.

12.5 FATE

Shallow geochemical environments consist of solid, aqueous, and air reservoirs and their interfaces. Hydrocarbon compounds partition into these various reservoirs in a manner determined by the structure and physical properties of the compounds and the media, and the mechanism of hydrocarbon release. The structure and physical properties of the compounds and media understandably impact their sorption, solubility, volatility, and decomposition behavior (e.g., Schwarzenbach *et al.*, 1993). In addition, hydrocarbon partitioning in real systems is holistically a disequilibrium process; hence, the distribution of hydrocarbons depends as much on the pathway taken as on the final physical state of the system (e.g., Schwarzenbach *et al.*, 1993; Luthy *et al.*, 1997). Shallow aquatic systems may tend toward some equilibrium distribution (Figure 10), but this is seldom, if ever, truly attained.

Processes affecting hydrocarbon distribution and fate in multiphase aquatic systems can be characterized either as intermedia exchange or as transformational reactions (Mackay, 1998). The latter pertains to processes that involve molecular transformation of the compound, whereas the former is concerned with the movement of the molecularly intact compound from one medium to another. For the purpose of this chapter, we focus on the three most important exchange processes dictating the geochemical fate of hydrocarbons in aquatic environments: sorption, volatilization, and water dissolution. Sorption characterizes exchange between water and particulate phases (sediments or aquifer media), volatility characterizes exchange between air and water or air and solid, and dissolution pertains to the ability of contaminants to be present as true solutes in the aqueous media. A process of potential importance in some oil spills, emulsification, is not covered here but the readers are referred to a recent document published by NRC (2002), and references cited therein. Similarly, discussion of molecular reaction will focus on the two dominant transformation processes of hydrocarbons in geologic media: photolytic and biological transformation. Photolysis pertains to light-assisted chemical reactions that can affect compounds in the atmosphere and in the photic zone of water columns. Biological transformation is often accomplished by microorganisms, and could take place in aerobic or anaerobic environments. Purely chemical transformations, including hydrolysis, redox, and elimination reactions, are not examined here, because they are unlikely to be the dominant reaction pathways in shallow aquatic systems (NRC, 2002). For example, the theoretical pK_a values of hydrocarbons are exceedingly high; hence, they tend not to participate in acid–base reactions. Redox reactions involving alkenes can take place at surface geochemical conditions, but this compound group is not present in major amounts in petroleum hydrocarbon spills or pyrogenic products.

Hydrocarbons are hydrophobic and lipophilic compounds. As free liquid, gas, or solid phases, they are quite immiscible in water primarily because of their low polarity. In a competition between aqueous solution, air and solids, HMW hydrocarbons tend to partition heavily into the solid phases, and hence they are also sometimes referred to as "particle-associated compounds." This is true in both the atmospheric reservoir, where they associate with atmospheric particulates, and surface and groundwater systems, where they exhibit affinity for suspended particles and aquifer solids. In the case of nonhalogenated NAPL, the separate liquid phase is generally lighter than the aqueous phase, and hence they tend to physically "float" on the aqueous surface. Such is the case for oil spills either in surface water (e.g., ocean, streams, and lake) or in groundwater (e.g., underground storage tanks).

The hydrophobicity can be expressed by the dimensionless octanol–water partition coefficient

$$K_{\mathrm{OW}} = C_{\mathrm{i,O}}/C_{\mathrm{i,W}}$$

where $C_{\mathrm{i,O}}$ is the concentration of i in octanol and $C_{\mathrm{i,W}}$ the concentration of i in water. Many of the thermodynamic properties describing the partitioning of hydrocarbons in air, water, biota, or solid have been successfully, albeit empirically, related to K_{OW}. The value of K_{OW} among petroleum hydrocarbons, especially PAHs, tends to be very high (10^3 to $>10^6$), and this preference for the organic phase, either as a free phase or in organic particulates, is a major control on the fate and distribution of hydrocarbons in aquatic systems. The lipophilic affinity of PAHs is also a major contributor to enrichment of these compounds in organisms and organic-rich sediments.

12.5.1 Sorption

Solid–water exchange of hydrocarbons is a primary control on their aqueous concentration in groundwater and surface aquatic environments (Figures 10 and 11) The specific partitioning of hydrocarbons between solid and aqueous phase could be described by the distribution coefficient

$$K_{\mathrm{d}} = C_{\mathrm{i,\,solid}}/C_{\mathrm{i,\,W}}$$

where $C_{\mathrm{i,solid}}$ is the concentration of i in the solid (mol kg^{-1}) and $C_{\mathrm{i,W}}$ the concentration of i in water (mol l^{-1}). For hydrophobic compounds, K_{d} tends to be constant only within a limited concentration range, and the more general Freundlich isotherm is commonly used:

$$C_{\mathrm{i,\,solid}} = K_{\mathrm{d}} C_{\mathrm{i,W}}^{n}$$

where n is a parameter relating to the nonlinearity of the sorption process (e.g., Schwarzenbach *et al.*, 1993; Luthy *et al.*, 1997). Hydrocarbons tend to partition preferentially to organic matter; hence, K_{d} is often written to take explicit account of this preference:

$$K_{\mathrm{d}} = C_{\mathrm{i,\,OM}} f_{\mathrm{OC}}/C_{\mathrm{i,\,W}}^{n} = K_{\mathrm{i,\,OC}} f_{\mathrm{OC}}$$

where $K_{\mathrm{i,OC}}$ is the distribution coefficient between organic matter and aqueous phase and f_{OC} the fraction of organic carbon (~0.5 fraction of organic matter, f_{OM}) in the solid substrate. Note that the above formulation assumes insignificant sorption on mineral matter, which is generally true

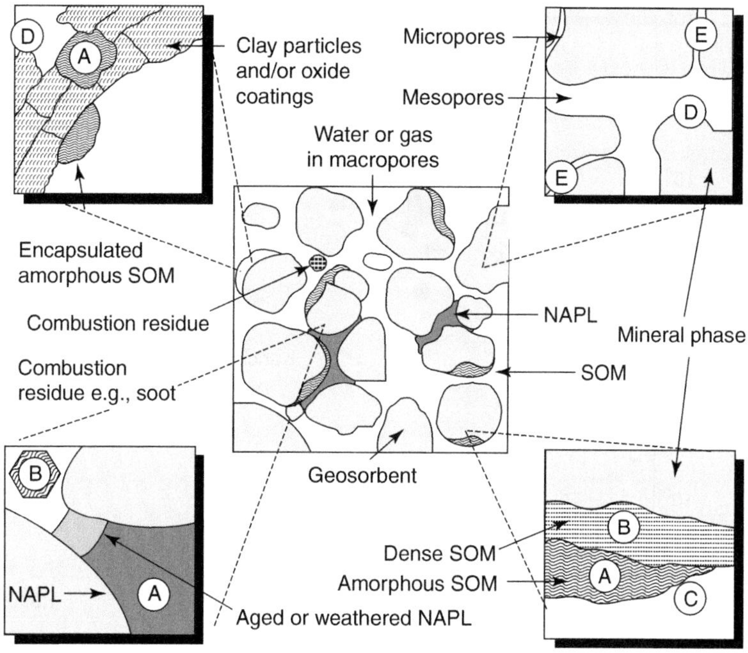

Figure 11 Heterogenous solid compartments in soils and sediments. NAPL signifies nonaqueous-phase liquid and SOM represents soil organic matter. Absorption into (A) amorphous organic matter or NAPL, (B) soot or black carbon, (C) water-wet organic surfaces, (D) nonporous mineral surfaces and (E) microvoids or microporous minerals. Reproduced by permission of American Chemical Society from Luthy *et al.* (1997).

for "threshold" $f_{OC} > 0.0005$ (Schwarzenbach *et al.*, 1993). Typical soils have an f_{OC} range of 0.005–0.05, whereas coarse aquifers have an f_{OC} range of 0–0.025.

"Linear free energy relationships" are quite frequently used for estimating K_d or $K_{i,OC}$. This linear correlation comes in the form

$$\log K_{i,OC} = a \log K_{OW} - b$$

where a and b are the linear fitting parameters, and K_{OW} is the aforementioned octanol–water partition coefficient (e.g., Chiou *et al.*, 1979, 1998; Karickhoff, 1981; Schwarzenbach *et al.*, 1993). For example, $K_{i,OC}$ of PAHs can be estimated from their K_{OW} by assuming that $K_{i,OC} = 0.41 K_{OW}$ (Karickhoff, 1981).

Hydrocarbon sorption on sediments or atmospheric particulates can involve either absorption or adsorption, which connotes surface attachment or subsurface dissolution, respectively. No distinction between these two processes is made here, except to note that sorption in geochemical systems often involves both (Kleineidam *et al.*, 2002). Laboratory experiments have shown that HMW petroleum hydrocarbons rapidly associate with sediment surfaces, with sorption uptake of up to 99% in 48 h at hydrocarbon concentrations of 1–5 mg l^{-1} (e.g., Knap and Williams, 1982). The rate and extent of sorption depends on the competitive affinity of the compounds between the aqueous or gaseous solution and the solid surface. The latter, in turn, depends on the nature of the crystallographic or amorphous substrate, especially the presence, type, and amount of organic matter in the solid and colloidal phase (e.g., Wijayaratne and Means, 1984; Luthy *et al.*, 1997; Ramaswami and Luthy, 1997; Gustafsson and Gschwent, 1997; Villholth, 1999; MacKay and Gschwent, 2001). The extent of preference for the solid surface is often described in terms of the sorption coefficient, K_d, as defined above.

The use of single distribution coefficient for organic matter would imply a constancy of sorption behavior on all organic substrates, and that the nature of organic matter substrate is not critical for sorption assessment. The impact of the heterogeneous nature of organic matter substrates on geosorption has been explored, and it has become evident that the wide variety of organic matter properties requires accounting of substrate-specific K_{OM} values (e.g., Luthy *et al.*, 1997; Kleineidam *et al.*, 1999, 2002; Chiou *et al.*, 1998; Bucheli and Gustafsson, 2000; Karapanagioti *et al.*, 2000) (Figure 11). For example, Karapanagioti *et al.* (2000) and Kleineidam *et al.* (1999) demonstrated the distinct Freundlich K_d (and $K_{i,OC}$) for phenanthrene sorption on different organic substrates, as well as mineral matter substrates (Figure 12) (Karickhoff *et al.*, 1979). Bulk $K_{i,OC}$ characteristics of naphthalene, phenanthrene, and pyrene were examined by Chiou *et al.* (1998), and they concluded that significant differences exist in $K_{i,OC}$ of pristine terrestrial soil organic matter and sediment organic matter, and that sorption is

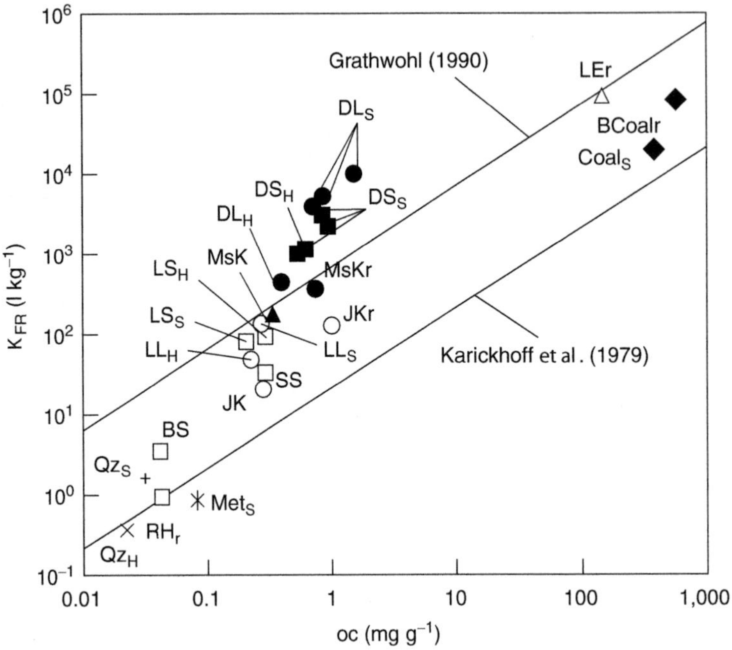

Figure 12 Phenanthrene sorption coefficients on different organic substrates. K_{FR} is the Freundlich coefficient and oc the organic carbon content. Remaining symbols are as per table 2 of Kleineidam *et al.* (1999): D/L, dark/light; L, limestone; S, sandstone; Met, igneous and metamorphic rocks; Qz, quartz; and BS, bituminous shale. Reproduced by permission of American Chemical Society from Kleineidam *et al.* (1999).

enhanced in sediments with organic contaminants present. Furthermore, there is now increasing recognition of the explicit role of black carbon (soot) sorption particularly in regards to explaining much higher field-measured K_d values compared with previously suggested "overall" K_d values (e.g., Chiou *et al.*, 1979; Karickhoff *et al.*, 1979; Karickhoff, 1981; Gustafsson and Gschwent, 1997; Accardi-Dey and Gschwent, 2003). It appears that the cohesive compatibility between PAHs and aromatic components of the organic substrate is a major factor in the enhancement of sorption (e.g., Gustafsson and Gschwent, 1997; Chiou *et al.*, 1998; Bucheli and Gustafsson, 2000; MacKay and Gschwent, 2001; Accardi-Dey and Gschwent, 2003). Accardi-Dey and Gschwent (2003) proposed treating organic sorption as a composite of organic carbon absorption and black carbon adsorption:

$$K_d = K_{i,OC} f_{OC} + f_{BC} K_{BC} C_{i,W}^{n-1}$$

where f_{BC} is the fraction of black carbon in the sample, K_{BC} the black carbon distribution coefficient, and n the Freundlich exponent as before. Note also that f_{OC} is now the fraction of organic carbon that excludes the black carbon component. Whereas this approach apparently explained discrepancies in modeled and field-calculated K_d and the nonlinear K_d observed in the laboratory, it remains to be seen if non-BC organic carbon can indeed be represented by a single K_d (i.e., $=K_{i,OC}$) in the general case.

12.5.2 Volatilization

When petroleum products enter surface-water systems, the lighter aliphatic and aromatic hydrocarbons spread out along the surface of the water and evaporate. The volatilization half-life of naphthalene, e.g., is 0.5–3.2 h (Mackay *et al.*, 1992). The naphthalene that does not evaporate sorbs to the particulate matter as noted above or is transformed into water in oil emulsion. In general, evaporation is the primary mechanism of loss of volatile and semivolatile components of spilled oil (e.g., Fingas, 1995; Volkman *et al.*, 1997; NRC, 2002). After a few days of oil spill at sea, ~75% of light crude, 40% of medium crude, and 10% of heavy crude oil will be lost by evaporation (NRC, 2002). The specific evaporation rates are influenced by a number of factors including meteorological factors, stratification of the water column, flow and mixing of water, and sequestration by mineral and natural organic sorbent matter.

The partitioning between air and water is characterized by the Henry's law constant,

$$K_H = P_{i,air}/C_{i,W}$$

where $P_{i,air}$ is the partial pressure of i in air and $C_{i,W}$ the molarity of i in the aqueous phase. K_H is often estimated from $P_0/C_{i,sat}$, where P_0 is the vapor pressure of the compound and $C_{i,sat}$ the solubility (see below). High K_H implies higher volatility, and results from a combination of high vapor pressure and low aqueous solubility. The vapor pressure of hydrocarbon compounds, along with their affinity for the sorption surfaces, dictates the partitioning between the atmosphere and atmospheric particulate phases. Hence, P_0 is a primary control on the atmospheric fate and transport of hydrocarbons. For example, a hydrocarbon that is tightly bound to the particulate phase is less likely to be altered during transportation and deposition (Baek *et al.*, 1991). The subsequent fate of hydrocarbons deposited in surface waters is further influenced by volatilization behavior because of possible surface losses to the atmosphere (Strachan and Eisenreich, 1988; Baker and Eisenreich, 1990; Eisenreich and Strachan, 1992; Lun *et al.*, 1998; Gustafson and Dickhut, 1997).

The dependence of vapor pressure on temperature follows from a simplified solution to the Clapeyron equation (Schwarzenbach *et al.*, 1993):

$$\ln P_0 = -B/T + A$$

where $B = \Delta H_{\text{vap}}/R$ and $A = \Delta S_{\text{vap}}/R$. ΔH_{vap} and ΔS_{vap} are, respectively, the molar enthalpy and molar entropy of vaporization and R is the gas constant ($8.314\,\text{Pa}\,\text{m}^3\,\text{mol}^{-1}\,\text{K}^{-1}$). ΔH_{vap} is principally the energy required to break van der Waals and hydrogen bonds in going from the condensed phase to vapor. ΔS_{vap} is a compound-specific constant, which has been shown to correlate with compound boiling point (Schwarzenbach *et al.*, 1993). Increase in temperature results in more condensed-phase hydrocarbons moving into the vapor phase, while a temperature decrease results in more vapor-phase hydrocarbons appearing in the condensed phase (e.g., Lane, 1989; Baker and Eisenreich, 1990; Mackay, 1998). Thus, particulate hydrocarbon deposition is likely greater during colder periods, but the effect is highest on the lower molecular weight compounds (e.g., naphthalene) with significant vapor pressure. Likewise, the relative behavior of hydrocarbons between water surface and air is dictated by their relative vapor pressures such that a net flux of PAHs into the atmosphere can take place in warm summer months with the reverse taking place during the cool winter months (e.g., Baker and Eisenreich, 1990; Gustafson and Dickhut, 1997). Mackay (1998) and Mackay and Callcott (1998) provided more detailed analysis of this partitioning behavior using the so-called fugacity approach. To mass balance PAH partitioning between the atmosphere and water, they used the dimensionless parameter K_{AW}, the air–water partition coefficient, as $K_{\text{AW}} = K_{\text{H}}/RT$ (cf. Baker and Eisenreich, 1990; Schwarzenbach *et al.*, 1993). Along with solid–water partitioning discussed above and solubility values discussed below, they showed mass balance partitioning models for PAHs in a model "seven-phase geomedia" (Mackay and Callcott, 1998). For a detailed discussion of the "fugacity approach," the reader is referred to Mackay (1998).

As noted above, volatilization losses from the aqueous phase to the atmosphere are also influenced by the aqueous solubility of the compound (i.e., $P_0/C_{\text{i,sat}}$). Higher solubility results in lower K_{H} for compounds of identical vapor pressures. Although the impact of solubility on volatilization loss is critical for LMW hydrocarbons, the effect of solubility on the behavior of HMW compounds has broader significance for understanding their transport and fate (see below).

12.5.3 Water Dissolution and Solubility

HMW hydrocarbons have a wide range of solubility, reported as S or $C_{\text{i,W}}^{\text{sat}}$ (i.e., aqueous concentration at saturation), but these solubilities are generally very low (e.g., Readman *et al.*, 1982; Means and Wijayaratne, 1982; Whitehouse, 1984; Mackay and Callcott, 1998). For example, water solubility of PAHs tends to decrease with increasing molecular weight from 4×10^{-4} for naphthalene to 2×10^{-8} M for benzo(*a*)pyrene (Schwarzenbach *et al.*, 1993). Linear fused PAHs (e.g., naphthalene and anthracene) also tend to be less soluble than angular or pericondensed structures (e.g., phenanthrene and pyrene). Furthermore, alkyl substitution decreases the water solubility of the parental PAHs. In spite of their low solubility, dissolution is one of the most important media exchange processes leading to the primary destruction pathway of hydrocarbons in aquatic systems, i.e., biodegradation. Metabolic utilization of hydrocarbons requires that they be transported as dissolved components into the cells, a process often assisted by extracellular enzymes released by the microorganisms after a quorum has been reached (Ramaswami and Luthy, 1997; Neilson and Allard, 1998). The presence of dissolved humic acids in solution also enhances solubility (Chiou *et al.*, 1979, 1983; Fukushima *et al.*, 1997; Chiou and Kyle, 1998). Solubility enhancement may also be achieved using surfactants, a method employed in many previous PAH and oil cleanup efforts.

Temperature affects aqueous solubility in a manner dictated by the enthalpy of solution (ΔH_{s}^0) according to (Schwarzenbach *et al.*, 1993)

$$\ln C_{\text{i,W}}^{\text{sat}} = -\Delta H_{\text{s}}^0/RT + \Delta S_{\text{s}}^0/R$$

The temperature effects on solubility for liquid and solid hydrocarbons vary greatly because of the large melting enthalpy component to ΔH_{s}^0 for solids (for which all PAHs are at room temperature). In general, PAH solubility increases with temperature, prompting present interest in thermophilic degradation of HMW PAHs. Hydrocarbons, including PAHs, also exhibit "salting-out" effects in saline solutions. This effect is exemplified by the so-called Setschenow formulation (Schwarzenbach *et al.*, 1993): $\log C^{sat}_{i,salt} = C^{sat}_{i,w} - K_s S$, where $C^{sat}_{i,salt}$ is the saturation concentration or solubility in the saline solution, $C^{sat}_{i,w}$ the saturation concentration in pure water defined previously, K_s the Setschenow constant, and S the total molar salt concentration. For example, the Setschenow constant for pyrene at 25 °C is 0.31–0.32 (Schwarzenbach *et al.*, 1993), so that at the salinity of seawater $C^{sat}_{i,w}$ is 54% greater than $C^{sat}_{i,salt}$. This salting-out effect could be large enough to manifest in the localization of PAH contamination in estuarine environments (e.g., Whitehouse, 1984).

HMW hydrocarbons and especially PAHs with no polar substituents are very hydrophobic, which along with their lipophilic characteristics and low vapor pressure explain why they are not efficiently transported in aqueous form.

12.5.4 Photochemical Reactions

Photochemical reactions involving electromagnetic radiation in the UV–visible light range can induce structural changes in organic compounds. Direct photochemical reactions occur when the energy of electronic transition in the compounds corresponds to that of the incident radiation, with the compound acting as the light-absorbing molecule (i.e., chromophore). Hence, the structure of hydrocarbons determines the extent by which they are prone to photodecomposition, but photolytic half-lives are also significantly dependent on compound concentration and substrate properties (e.g., Behymer and Hites, 1985, 1988; Paalme *et al.*, 1990; Reyes *et al.*, 2000; NRC, 2002). In general, aromatic and unsaturated hydrocarbons are more prone to UV absorption and decomposition, with increasing numbers of conjugated bonds resulting in lower energy required for electronic transition. Photodissociation is not likely an important weathering mechanism for HMW straight chain hydrocarbons, because these compounds do not absorb light efficiently (e.g., Payne and Phillips, 1985). Nevertheless, these hydrocarbons may be transformed through the process of indirect photodissociation wherein another molecule (e.g., humic and fulvic acids) or substrate (mineral or organic) acts as the chromophore (NRC, 2002). Aromatic structures are prone to direct photochemical reaction in a manner that depends on molecular weight and degree of alkylation.

Laboratory experiments have shown that PAHs are photoreactive under atmospheric conditions (Zafiriou, 1977; Behymer and Hites, 1988; Schwarzenbach *et al.*, 1993; Reyes *et al.*, 2000) and in the photic zone of the water column (Zepp and Schlotzhauer, 1979; Payne and Phillips, 1985; Paalme *et al.*, 1990). The existence of PAH oxidation products in atmospheric particulate matter indicates that PAHs react with oxygen or ozone in the atmosphere (Schwarzenbach *et al.*, 1993; Howsam and Jones, 1998), but the reaction with hydroxyl (OH) radicals during daylight conditions is considered to be the major reaction sink of these compounds in the atmosphere. There is great variation in the reported half-lives of various PAHs due to photolysis, largely because of the differences in the nature of the substrate on which the PAHs are adsorbed and the degree to which they are bound (Behymer and Hites, 1985; Paalme *et al.*, 1990; Schwarzenbach *et al.*, 1993). Carbon content and the color of substrates are important factors in controlling PAH reactivity (Behymer and Hites, 1988; Reyes *et al.*, 2000), and the suppression of photochemical degradation of PAHs adsorbed on soot and fly ash has been attributed to particle size and substrate color. Darker substrates absorb more light and thereby protect PAHs from photolytic degradation reactions (Behymer and Hites, 1988).

Photooxidation by singlet oxygen appears to be the dominant chemical degradation process of PAHs in aquatic systems (Lee *et al.*, 1978; Hinga, 1984; Payne and Phillips, 1985). The degree to which PAHs are oxidized in an aqueous system depends on the PAH type and structure, water column characteristics (such as oxygen availability), temperature and depth of light penetration, and residence time in the photic zone (Payne and Phillips, 1985; Paalme *et al.*, 1990; Schwarzenbach *et al.*, 1993). Paalme *et al.* (1990) have shown that the rate of photochemical degradation of different PAHs in aqueous solutions can differ by a factor of >140 depending on their chemical structure, with perylene, benzo(*b*)fluoranthene, and coronene showing the greatest stability. Alkyl PAHs are more sensitive to photooxidation reactions than parental PAHs, probably due to benzyl hydrogen activation (e.g., Radding *et al.*, 1976). Payne and Phillips (1985) reported that benz(*a*)anthracene and benzo(*a*)pyrene are photolytically degraded 2.7 times faster in summer than in winter. Once deposited in sediments, photooxidation reactions are generally significantly reduced due to anoxia and limited light penetration. PAH residence times in sediments depend on the extent of physical, chemical, and biological reactions occurring at the sediment–water interface, as well as the intensity of bottom currents (Hinga, 1984).

12.5.5 Biodegradation

Biodegradation is perhaps the most important reaction mechanism for the degradation of hydrocarbons in aquatic environments. Aliphatic compounds in crude oil and petroleum products are readily degraded, with a prominent initial microbial preference for straight chain compounds (e.g., Atlas and Bartha, 1992; Prince, 1993; Volkman *et al.*, 1997; Wang *et al.*, 1999; Heider *et al.*, 1999; Bosma *et al.*, 2001; NRC, 2002). The aerobic pathway shows conversion of alkane chains into fatty acids, fatty alcohols and aldehyde, and carboxylic acids that are then channeled into the central metabolism for subsequent β-oxidation (Figure 13). Anaerobic degradation proceeds with nitrate, Fe^{3+}, or sulfate as the terminal electron acceptor, with no intermediate alcohols in the alkane degradation (Figure 13). The degradation pathway involves an O_2-independent oxidation to fatty acids, followed by β-oxidation. Sulfate reducers apparently show specificity toward utilization of short-chain alkanes (C_6–C_{13}) (Bosma *et al.*, 2001). Laboratory experiments on complex oil blends further showed composition changes accompanying biodegradation. For example, Wang *et al.*

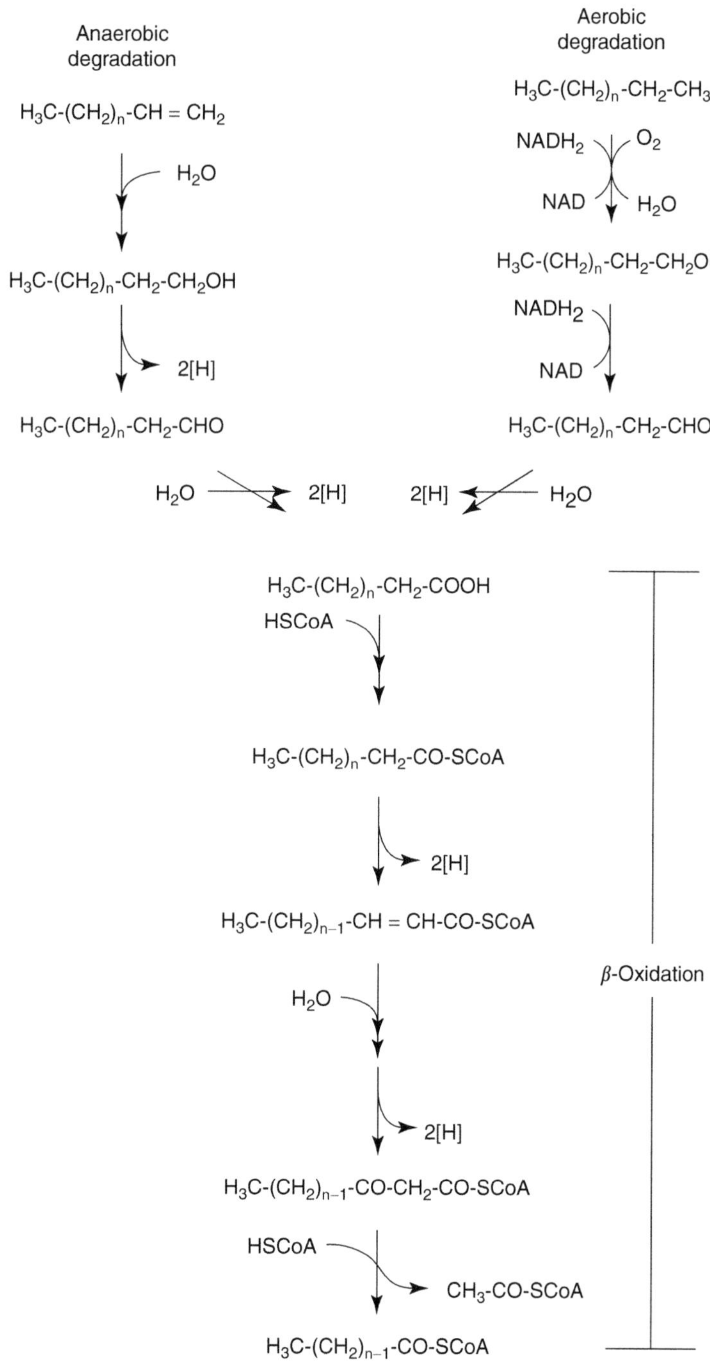

Figure 13 Generalized aerobic and anaerobic biodegradation pathways for *n*-alkanes. Reproduced by permission of Springer from Beek (2001).

(1998) examined the compositional evolution of Alberta Sweet Mixed Blend oil upon exposure to defined microbial inoculum, with total petroleum hydrocarbons (9–41%), and specifically the total saturates (5–47%) and *n*-alkanes (>90%), showing varying degrees of degradation. They further showed that susceptibility to *n*-alkane degradation is an inverse function of chain length, the branched alkanes are less susceptible than straight chain *n*-alkanes, and that the most resilient saturate components are the isoprenoids, pristane, and phytane. Even amongst isoprenoids, however, there is still a notable inverse dependence between chain length and degradation susceptibility. The relative susceptibility of *n*-alkanes to biodegradation compared with the isoprenoids is the basis for using C_{17}/pristane and *n*-C_{18}/phytane ratios for distinguishing biodegradation from volatilization effects, because the latter discriminates primarily on the basis of molecular weight. Additionally, Prince (1993) noted the relative resilience of polar compounds compared with corresponding hydrocarbons. Interesting deviations from this generalized biodegradation pattern were nevertheless noted by others, including the observation that some Exxon Valdez spill site microbial communities preferentially degraded naphthalene over hexadecane at the earliest stages of biodegradation (e.g., Sugai *et al.*, 1997).

Numerous cases of crude oil and refined petroleum spill into surface environments have provided natural laboratories for examination of the biodegradation of petrogenic compounds in a variety of environmental conditions (e.g., Kaplan *et al.*, 1997; Prince, 1993; Wang *et al.*, 1998; NRC, 2002). The observations on compositional patterns of biodegradation noted above are generally replicated in these natural spills. For example, a long-term evaluation of compositional variation of Arabian light crude in a peaty mangrove environment was conducted by Munoz *et al.* (1997), who showed the same pattern of initial preferential loss of *n*-alkanes followed by isoprenoids, and ultimately the biomarkers in the order steranes, hopanes, bicyclic terpanes, tri- and tetracyclic terpanes, diasteranes, and the aromatic biomarkers. However, an interesting report on the biodegradation patterns in the crude oil spill site in Bemidji, Minnesota, showed an apparent reversal in the biodegradation preference of *n*-alkanes, wherein HMW homologs show faster degradation rates (Hostetler and Kvenvolden, 2002). The overall resilience of terpane and sterane compounds to biodegradation has been well recognized; hence, biomarker ratios are widely used as indicators of oil spill sources even in highly weathered oils (e.g., Volkman *et al.*, 1997).

Aromatic hydrocarbons in the atmosphere and open waters may undergo volatilization and photodecomposition, but microbial degradation is the dominant sink below the photic zone (Gibson and Subramanian, 1984). As with aliphatic hydrocarbons, biodegradation of aromatic compounds involves the introduction of oxygen into the molecule forming catechol. Microbial degradation has also been recognized as the most prominent mechanism for removing PAHs from contaminated environments (Neilson, 1998; NRC, 2000, 2002). Microbial adaptations may result from chronic exposure to elevated concentrations as shown by the higher biodegradation rates in PAH-contaminated sediments than in pristine environments (Neilson, 1998; NRC, 2000). Nevertheless, it is also known that preferential degradation of PAHs will not occur in contaminated environments where there are more accessible forms of carbon (NRC, 2000). A summary of the turnover times for naphthalene, phenanthrene, and BaP in water and sediment is shown in Table 1. In general, PAH biodegradation rates are a factor of 2–5 slower than degradation of monoaromatic hydrocarbons, and of a similar magnitude as HMW *n*-alkanes (C_{15}–C_{36}) under similar aerobic conditions. The ability of microorganisms to degrade fused

Table 1 Turnover times for naphthalene, phenanthrene, and benzo(a)pyrene in water and sediment.

PAH and environment	*Temperature* (°C)	*Times* (days)
Naphthalene		
Estuarine water	13	500
Estuarine water	24	30–79
Estuarine water	10	1–30
Seawater	24	330
Seawater	12	15–800
Estuarine sediment	25	21
Estuarine sediment		287
Estuarine sediment	22	34
Estuarine sediment	30	15–20
Estuarine sediment	2–22	13–20
Stream sediment	12	>42
Stream sediment	12	0.3
Reservoir sediment	22	62
Reservoir sediment	22	45
Phenanthrene		
Estuarine sediment	25	56
Estuarine sediment	2–22	8–20
Reservoir sediment	22	252
Reservoir sediment	22	112
Sludge-treated soil	20	282
Benzo(a)pyrene		
Estuarine water	10	2–9,000
Estuarine sediment	22	>2,800
Estuarine sediment	2–22	54–82
Stream sediment	12	>20,800
Stream sediment	12	>1,250
Reservoir sediment	22	>4,200
Sludge-treated soil	20	>2,900

Source: O'Malley (1994).

aromatic rings is determined by their combined enzymatic capability, which can be affected by several environmental factors (McElroy *et al.*, 1985; Cerniglia, 1991; Neilson and Allard, 1998; Bosma *et al.*, 2001). The most rapid biodegradation of PAHs occurs at the water–sediment interface (Cerniglia, 1991). Prokaryotic microorganisms primarily metabolize PAHs by an initial dioxygenase attack to yield *cis*-dihydrodiols and finally catechol (e.g., Figure 14). Biodegradation and utilization of lower molecular weight PAHs by a diverse group of bacteria, fungi, and algae has been demonstrated (Table 2; Neilson and Allard, 1998; Bosma *et al.*, 2001). For example, many different strains of microorganisms have the ability to degrade napththalene including *Pseudomonas*, *Flavobacterium*, *Alcaligenes*, *Arthrobacter*, *Micrococcus*, and *Bacillus*.

The degradation pathways of higher molecular weight PAHs—such as pyrene, benzo(*e*)pyrene, and benzo(*a*)pyrene—are less well understood (Neilson and Allard, 1998). Because these compounds are more resistant to microbial degradation processes, they tend to persist longer in contaminated environments (Van Brummelen *et al.*, 1998; Neilson and Allard, 1998; Bosma *et al.*, 2001). However, the degradation of fluoranthene, pyrene, benz(*a*)anthracene, benzo(*a*)pyrene, benzo(*b*)fluorene, chrysene, and benzo(*b*) fluoranthene has been reported under laboratory conditions (Barnsley, 1975; Mueller *et al.*, 1988, 1990; Schneider *et al.*, 1996; Ye *et al.*, 1996; Neilson and Allard, 1998). Elevated temperatures increase the rate of biotransformation reactions *in vitro*. For example, laboratory studies have shown a 50% loss of phenanthrene after 180 days at 8 °C in water compared with 75% loss in 28 days at 25 °C (Sherrill and Sayler, 1981; Lee *et al.*, 1981). It is widely believed that PAHs with three or more condensed rings tend not to act as sole substrates for microbial growth, but may be the subject of cometabolic transformations. For example, cometabolic reactions of pyrene, 1,2-benzanthracene, 3,4-benzopyrene, and phenanthrene can be stimulated in the presence of either naphthalene or phenanthrene (Neilson and Allard, 1998). Nevertheless, the degradation of PAHs even by cometabolic reactions is expected to be very slow in natural ecosystems (e.g., Neilson and Allard, 1998).

Biodegradation effects on aromatic hydrocarbons is a subject of much interest both from the standpoint of characterizing oil spill evolution and engineered bioremediation (Wang *et al.*, 1998; Neilson and Allard, 1998). Wang *et al.* (1998) noted that the susceptibility to biodegradation increases with decreasing molecular weight and degree of alkylation. For example, the most easily degradable PAHs examined are the alkyl homologs of naphthalene, followed by the alkyl homologs of dibenzothiophene, fluorene, phenanthrene, and chrysene. Also noteworthy is the observation that microbial degradation is isomer-specific, leading to the suggestion that isomer distribution of methyl dibenzothiophenes are excellent indicators of

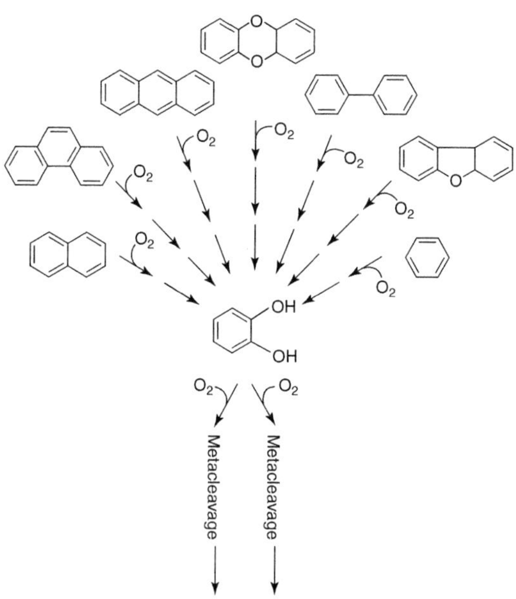

Figure 14 Generalized aerobic biodegradation pathways for aromatic hydrocarbons. Adapted with permission from Bosma *et al.* (2001).

Table 2 Biodegradation and utilization of lower molecular weight PAHs by a diverse group of bacteria, fungi, and algae.

Organism	*Substrate*
Pseudomonas sp.	Naphthalene
	Phenanthrene
	Anthracene
	Fluoranthene
	Pyrene
Flavobacteria sp.	Phenanthrene
	Anthracene
Alcaligenes sp.	Phenanthrene
Aeromonas sp.	Naphthalene
	Phenanthrene
Beijerenckia sp.	Phenanthrene
	Anthracene
	Benz(*a*)anthracene
	Benzo(*a*)pyrene
Bacillus sp.	Naphthalene
Cunninghamella sp.	Naphthalene
	Phenanthrene
	Benzo(*a*)pyrene
Micrococcus sp.	Phenanthrene
Mycobacterium sp.	Phenanthrene
	Fluorene
	Fluoranthene
	Pyrene

Source: O'Malley (1994).

degree of biodegradation (Wang *et al.*, 1998). Finally, in spite of the focus of biodegradation studies on aerobic degradation, recent work has demonstrated the capacity of sulfate-reducing bacteria in degrading HMW PAH (e.g., BaP) (Rothermich *et al.*, 2002).

12.6 CARBON ISOTOPE GEOCHEMISTRY

The present review of hydrocarbon sources, pathways, and fate in aquatic environments highlights the current state of understanding of HMW hydrocarbon geochemistry, but it also provides a useful starting point for exploring additional approaches to unraveling the sources and fate of hydrocarbons in aquatic environments. It is clear that the degree and type of hydrocarbons ultimately sequestered in particulates, sediments, or aquifer materials depend not only on the nature and magnitude of various source contributions, but also on the susceptibility of the hydrocarbons to various physical, chemical, and microbial degradation reactions. As we have already shown, the latter alters the overall molecular signatures of the original hydrocarbon sources, complicating efforts to apportion sources of hydrocarbons in environmental samples. Nevertheless, resilient molecular signatures that either largely preserve the original source signatures or alter in a predictable way have been employed successfully to examine oil sources as already discussed (Wang *et al.*, 1999). Additionally, carbon isotopic composition can be used to help clarify source or "weathering reactions" that altered the hydrocarbons of interest. In what follows, we will examine the emerging application of carbon isotopic measurements in unraveling the sources and fate of PAHs in shallow aquatic systems. Similar approaches to studying aliphatic compounds have been employed in a number of hydrocarbon apportionment studies (e.g., O'Malley, 1994; Mansuy *et al.*, 1997; Dowling *et al.*, 1995), and the approach has indeed been a principal method used for oil–oil and oil source rock correlation for decades (e.g., Sofer, 1984; Schoell, 1984; Peters *et al.*, 1986; Faksness *et al.*, 2002). The key to using carbon isotopes for understanding the geochemistry of HMW hydrocarbons in shallow aquatic systems is to distinguish two reasons why the abundance of stable isotopes in these compounds might vary: (1) differences in carbon sources and (2) isotope discrimination introduced after or during formation. In what follows, we will first describe and compare the carbon isotope systematics in the pyrogenic and petrogenic PAH sources. Then we will examine possible changes in the carbon isotope compositions in these compounds as a result of one or more weathering reactions. Finally, we will use the PAH inventory of an estuarine environment in eastern Canada as an example of how the molecular characteristics discussed earlier in this chapter can be blended with compound-specific carbon isotope signatures to distinguish PAH sources and pathways.

12.6.1 Carbon Isotope Variations in PAH Sources

12.6.1.1 Pyrogenesis

The isotopic signature imparted on individual PAHs during formation is determined by both the isotopic composition of the precursor compounds and the formation conditions. Since these two factors can vary widely during pyrolysis or diagenesis, the potential exists for PAHs produced from different sources and from a variety of processes to have equally variable isotopic signatures. Isotopic characterization of bulk organic and aromatic fractions has been performed for many years (e.g., Sofer, 1984; Schoell, 1984), but the advent of compound-specific isotopic characterization methods for individual organic compounds has immensely increased the database for assessing the range of PAH isotopic compositions.

The specific mechanisms controlling the isotopic signatures of individual pyrogenic PAHs are only partly understood (O'Malley, 1994; Currie *et al.*, 1999). It is expected that the pyrolysis of isotopically distinct precursor materials would result in PAHs with different isotopic signatures. Owing to the nature of PAH formation pathways discussed earlier in this chapter, the $\delta^{13}C$ of pyrolysis-derived compounds may be dictated by a series of primary and secondary reactions, which precursor and intermediate compounds undergo prior to the formation of the final PAHs. The isotopic effects associated with ring cleavage reactions of intermediate precursors (LMW compounds) may result in ^{13}C-enriched higher molecular weight species formed by the fusion of these reduced species. Variations in the $\delta^{13}C$ of the higher molecular weight condensed compounds may, therefore, depend on the $\delta^{13}C$ of the precursor and intermediate species. In the absence of pyrosynthetic recombination reactions, produced PAHs will have isotopic compositions that are largely dictated by those of the original precursor compounds. Alkylation or dealkylation reactions at specific sites of a parent molecule will alter the isotopic composition of that compound only to the extent that the alkyl branch is isotopically different from the substrate PAHs.

O'Malley (1994) performed the first characterization of compound-specific carbon isotope characterization of primary and secondary PAH sources. Irrespective of the range observed in the $\delta^{13}C$ of the wood-burning soot PAHs, the overall trend in the mean $\delta^{13}C$ values indicates that lower molecular weight compounds (3-ring) are isotopically more depleted than 4-ring PAHs, whereas 5-ring compounds have $\delta^{13}C$ values similar to the 3-ring species (Figure 15). Since PAHs are the products of incomplete combustion, the range of isotopic values generated during pyrolysis is related to the isotopic signature of their initial precursors and the fractionations that are associated with primary and secondary reactions noted above. The trend observed in the $\delta^{13}C$ of the 3-, 4-, and 5-ring PAHs in this pyrolysis process indicates the following possibilities: (1) the individual compounds were derived from isotopically distinct precursors in the original pyrolysis source that underwent pyrolysis and pyrosynthesis possibly at different temperature ranges; (2) the overall isotopic variation was primarily dictated by the formation and carbon branching pathways that occurred during pyrosynthesis; or (3) some combination of (1) and (2).

Benner *et al.* (1987) analyzed the $\delta^{13}C$ of various wood components and reported that cellulose, hemicellulose, and an uncharacterized fraction (solvent extractable) were, respectively, 1.3‰, 0.3‰, and 1.0‰ more enriched in ^{13}C than the total wood tissue, whereas lignins were 2.6‰ more depleted than the whole plant. Wood combustion in open wood fireplaces could resemble a stepped combustion process where hemicellulose normally degrades first, followed by cellulose, and then the lignins. If the carbon isotopic compositions of hemicellulose, cellulose, and lignin are sole factors in determining the overall isotopic signature of the wood-burning soot, then the $\delta^{13}C$-depleted PAHs may be primarily derived from lignin, while the more enriched compounds may originate from the heavier cellulose or uncharacterized fractions. The uncharacterized fraction of wood also contains certain compounds that are good PAH precursors (e.g., terpenes, fatty acids, and other aromatic compounds including acids, aldehydes, and alcohols).

As an alternative, the trend in the isotopic signature of wood-burning soot PAHs may be controlled by secondary condensation and cyclodehydrogenation reactions during pyrolysis to produce more condensed HMW compounds. Therefore, during the pyrolysis process, LMW compounds, generally formed at lower temperatures, may be actual precursors to the higher molecular weight species that are formed subsequently. Holt and Abrajano (1991) concluded from kerogen partial combustion studies that ^{13}C-depleted carbon is preferentially oxidized during partial combustion leaving a ^{13}C-enriched residue (cf. Currie *et al.*, 1999). Therefore, the PAH-forming radicals produced at lower temperatures are expected to be depleted in ^{13}C compared with radicals formed at higher temperatures. Despite the difference in the combustion/pyrolysis processes, materials, and conditions, the $\delta^{13}C$ of the parental PAHs isolated from car soots follow a similar trend to the wood-burning soot PAHs (Figure 15). This similarity argues for the prevailing role of pyrolytic secondary reactions in imparting the resulting PAH $\delta^{13}C$ values. Unlike open wood burning, the combustion process in the gasoline engine may be described as spontaneous or shock combustion (Howsam and Jones, 1998), whereby gasoline is initially vaporized and mixed with air prior to combustion in the combustion chamber. The PAHs produced during this process 2reported to be primarily

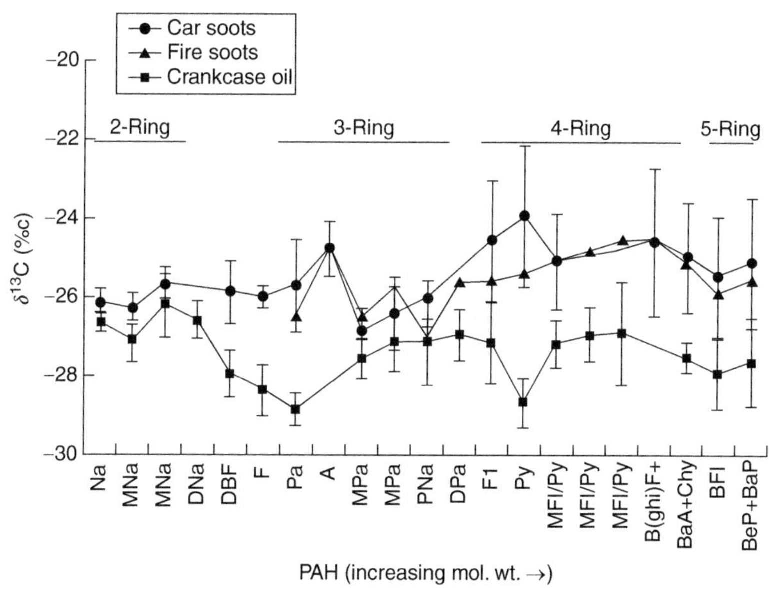

Figure 15 $\delta^{13}C$ of PAHs from car soot (automobile emission; 10 samples), fire soot (wood burning; 11 samples), and crankcase oil (12 samples). Error bars are 2σ variation within each source type. After O'Malley (1994).

dependent on the fuel/air ratio and the initial aromaticity of the fuel (Begeman and Burgan, 1970; Jensen and Hites, 1983).

The complex controls on the carbon isotopic composition of produced PAHs during wood burning is also apparent in PAH produced by coal pyrolysis (McRae *et al.*, 1999, 2000; Reddy *et al.*, 2002). PAHs that have been isotopically characterized by McRae *et al.* (1999) showed progressive ^{12}C enrichment in successively higher temperatures of formation (−24‰ to −31‰). One analysis of a coal tar standard reference material (SRM) by Reddy *et al.* (2002) yielded compound-specific δ^{13}C values in the ^{13}C-enriched end of this range (–24.3‰ to –24.9‰). McRae *et al.* (1999) reasoned that the isotopic changes observed at increasing temperature reflect the competition between original PAHs in the coal (mostly two and three rings), and those produced by high-temperature condensation and aromatization. Interestingly, the polycondensed higher molecular weight PAHs showed enrichment in ^{12}C, perhaps indicating the incorporation of the early ^{12}C-enriched fragments produced during pyrolysis (cf. Holt and Abrajano, 1991). Creosote produced from coal tar likewise exhibits a wide range of δ^{13}C composition (e.g., Hammer *et al.*, 1998), mimicking the observation made by McRae *et al.* (1999).

12.6.1.2 Petrogenesis

Carbon isotopic compositions of PAHs in petroleum and petroleum source rocks are among the earliest targets of compound-specific carbon isotopic analysis. Freeman (1991) analyzed individual aromatic compounds in the Eocene Messel Shale and demonstrated the general tendency of aromatic compounds to be consistently enriched in ^{13}C compared with the aliphatic fraction. This pattern is consistent with diagenetic effects associated with aromatization of geolipids (Sofer, 1984; Simoneit, 1998). As already noted, several examples of diagenetically produced PAHs (e.g., substituted Pa, Chy, retene, and perylene) result from the aromatization of precursors such as pentacyclic triterpenoids (Wakeham *et al.*, 1980b; Mackenzie, 1984; Peters and Moldowan, 1993; Neilson and Hynning, 1998; Simoneit, 1998). The aromatization of natural compounds to produce PAHs is not expected to result in significant isotopic alterations, a contention supported by the successful use of compound-specific carbon isotope signatures to trace precursor–product relationships in diagenetic systems (e.g., Hayes *et al.*, 1989; Freeman, 1991). It is notable, however, that the range of δ^{13}C values to be expected from potential precursors (e.g., triterpenoids) remains to be fully defined. O'Malley (1994) reported depletion in ^{13}C (up to 2‰)-methylated naphthalene homologs compared with their corresponding parental compound. The δ^{13}C-depleted isotopic values observed in these petrogenic-associated PAHs may be dictated by the similarly depleted nature of the precursor compounds. Published δ^{13}C values of petrogenic-associated PAHs are limited, and it is more than likely that further analysis will reveal more variations that would reflect the isotopic composition of precursor compounds and the nature and degree of diagenetic reactions.

In contrast to the wood fireplace and car soot signatures, the δ^{13}C of the PAHs isolated from crankcase oil were generally depleted in ^{12}C (Figure 15). Also, the range of δ^{13}C values measured was only marginally enriched compared with the range of bulk isotopic values (–29.6‰ and –26.8‰) reported for virgin crankcase oil. Unlike fuel pyrolysis, where a significant portion of the precursor source materials is initially fragmented into smaller radicals prior to PAH formation, the PAHs in used crankcase oil are largely derived from a series of thermally induced aromatization reactions of natural precursors that exist in the oil. Crankcase oils derived from petroleum consist primarily of naphthenes with minor *n*-alkane content ($<C_{25}$) and traces of steranes and triterpanes (Simoneit, 1998). Of the parental compounds isolated from the crankcase oils, Pa and Py were generally more depleted in ^{13}C than Fl, BaA/Chy, BFl, and BeP/BaP (Figure 15). The difference in the isotopic values of these two groups of parental compounds suggests that they may be derived from different precursors in the oil. The similarity in the isotopic values of Pa and Py indicates that they may be derived from precursor materials, which have a common origin or a common synthesis pathway that is distinct from those followed by other HMW PAHs. The assignment of aromatic structures to their biological precursors in crude oils has proved difficult (Radke, 1987). However, it is now well established that the saturated or partially unsaturated six-membered rings in steroids and polycyclic terpenoids undergo a process of stepwise aromatization (MacKenzie, 1984; Simoneit, 1998).

Our measurements of crude and processed oil products other than crankcase oil are shown in Figure 16, but these measurements are largely restricted to naphthalene and methylated naphthalene (unpublished data). The PAHs isolated from the bulk outboard motor condensate are isotopically very similar to the mean isotopic values of the PAHs in used crankcase oil, except that the outboard motor condensates were generally more enriched in ^{13}C (Figure 16). This similarity to crankcase oils was also observed in their PAH molecular signatures. The apparent enrichment in ^{13}C of naphthalene and its methylated derivatives in the outboard motor samples may be due to the

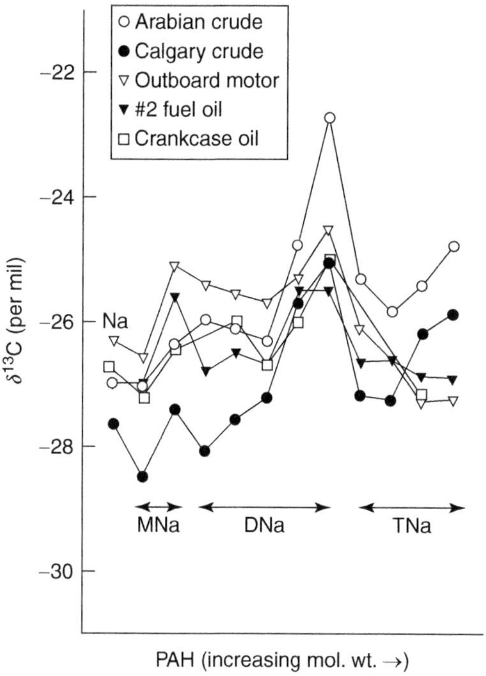

Figure 16 δ^{13}C of crude oil, outboard motors, and fuel oil. Also shown is the crankcase oil δ^{13}C signature.

partial combustion of gasoline that is characteristic of two-stroke outboard motor engines. The δ^{13}C of the PAHs isolated from the crude petroleum samples have similar trends (compound-to-compound variations), except that the Calgary crude PAHs are generally lighter (–32.2‰ to –25.1‰) than the Arabian sample (–30.0‰ to –22.7‰; Figure 16). This variation in isotopic values is possibly related to the origin or maturity of these samples. Differences in the δ^{13}C values of the isomeric compounds suggest that these compounds may originate from different precursor products or by different formation pathways. Naphthalene and methylated naphthalenes in #2 fuel oil had isotopic values within the range observed for the crude oils and other petroleum products investigated (δ^{13}C). Mazeas and Budzinski (2001) also reported compound-specific δ^{13}C measurement of an oil sample, and their values fell in the range −24‰ to −29‰. However, their measurements of PAHs from a crude oil SRM showed more enriched δ^{13}C values (−22‰ to −26‰). The latter values are the most ^{12}C-enriched values we have seen in petroleum-related compounds to date.

12.6.2 Weathering and Isotopic Composition

Once PAHs are emitted to the environment, the extent of alteration of the isotopic signature during transformation reactions depends on the nature of the carbon branching pathways involved in the reaction. Isotopic fractionation as a result of kinetic mass-dependent reactions (e.g., volatilization or diffusion) is unlikely to be significant, because the reduced mass differences between the light and heavy carbon in these HMW compounds are small. Substitution reactions, resulting in the maintenance of ring aromaticity, are also not likely to significantly alter the isotopic signature of the substrate molecule, although it is possible that the extent of substitution could result in kinetic isotope effects (KIEs). Experiments conducted to examine possible KIEs on biotic and abiotic transformation (e.g., volatilization, sorption, photodegradation, and microbial degradation) have shown the overall resilience of the compound-specific carbon isotope signatures inherited from PAH sources (O'Malley, 1994; O'Malley *et al.*, 1994; Stehmeir *et al.*, 1999; Mazeas *et al.*, 2002). The HMW of these compounds apparently preempts significant carbon isotopic discrimination during abiotic transformation processes (e.g., volatilization and photodegradation). For example, no fractionation from photolytic degradation was observed on the δ^{13}C of standard PAHs experimentally exposed to natural sunlight. The δ^{13}C of F, Pa, Fl, Py, and BbF during the various exposure periods were particularly stable despite up to a 40% reduction in the initial concentration of pyrene (Figure 17). The absence of significant KIE in the systems studied here may be related to the relatively HMW of the compounds and low reduced mass differences involved.

Microbial degradation experiments likewise showed negligible shifts in the δ^{13}C values of Na and Fl even after >90% biodegradation (Figure 18) (O'Malley *et al.*, 1994). Bacteria degrade PAHs by enzymatic attack that can be influenced by a number of environmental factors (Cerniglia, 1991). Alteration of the original isotopic values of Na and Fl may have been expected, since most PAH degradation pathways involve ring cleavage (loss of carbon) (Mueller *et al.*, 1990; Cerniglia, 1991). For bacterial degradation to occur, the PAHs are normally incorporated into the bacterial cells where enzymes are produced (van Brummelen *et al.*, 1998; Neilson and Allard, 1998). The pathways by which PAHs enter bacterial cells may dictate whether isotopic alterations will occur as a result of microbial degradation. This pathway is not well understood, but it is postulated that bacteria either ingests PAHs that are associated with organic matter, or because of their lipophilic properties, PAHs may diffuse in solution across the cell wall membrane. Hayes *et al.* (1989) suggested that organisms do not isotopically discriminate between heavy and light carbon when ingesting organic material from homogenous mixtures. In this case, the isotopic ratio of the unconsumed residue will not be different from the starting

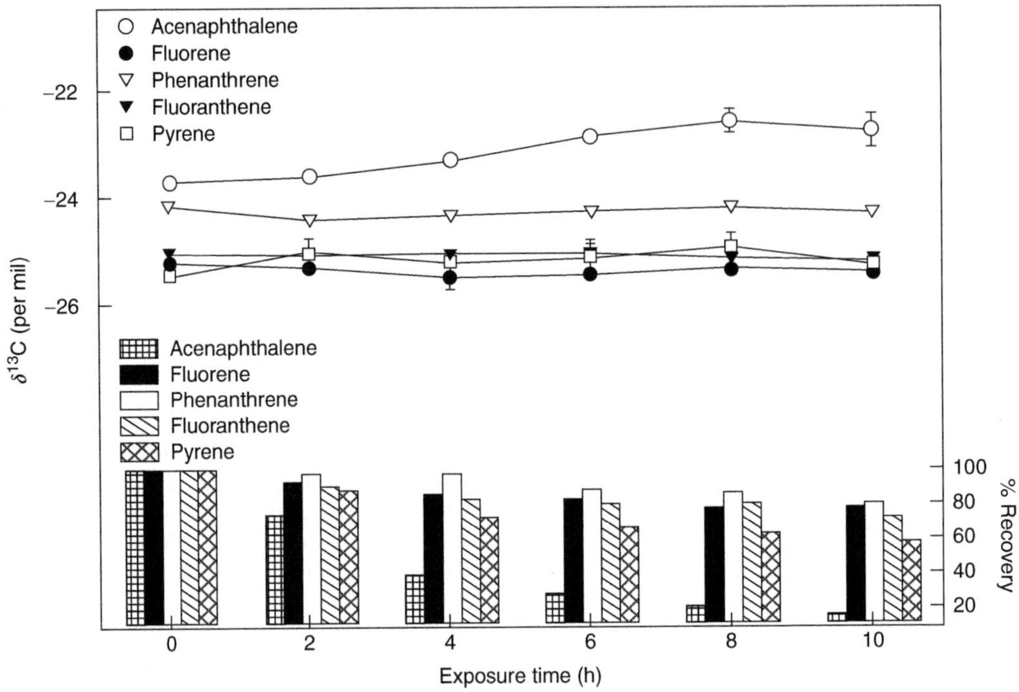

Figure 17 Carbon isotope effects of PAH photolysis. Top line diagram shows negligible shifts in $\delta^{13}C$ for F, Pa, Fl, and P. Only Ae shows a shift in $\delta^{13}C$ (0.5‰) beyond differences in replicate experiments (1σ=error bar or size of data point). Lower histogram shows percentage of PAHs that were not photodegraded.

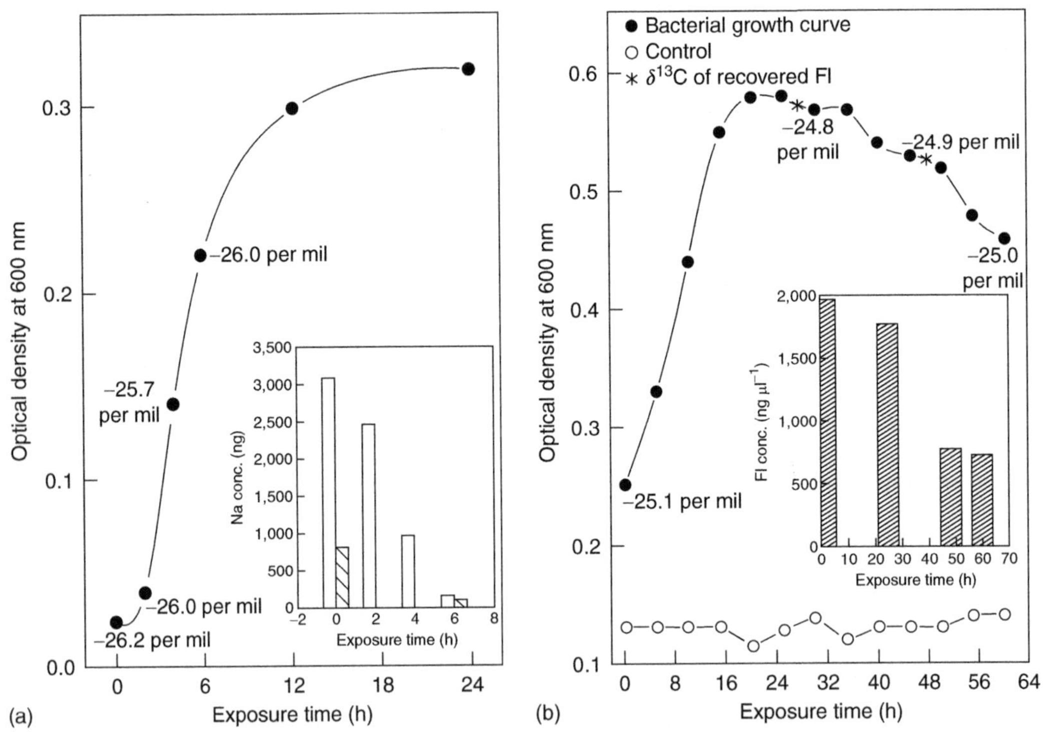

Figure 18 Carbon isotope effects of PAH biodegradation: (a) biodegradation of Na with *Pseudomonas putida*, Biotype B ATCC 17484 and (b) biodegradation of Fl with *P. paucimobilis*, strain EPA 505. Inset shows decline in PAH concentration over the same time period.

material. Even if cell wall diffusion was the predominant pathway, isotopic fractionation will again be limited by the small relative mass difference between ^{12}C and ^{13}C containing PAHs. It is for this reason that most isotopic fractionations in biological systems can be traced back to the initial C_1 (i.e., CO_2 and CH_4) fixation steps. Since no enrichments were observed in the biodegradation residue of naphthalene and fluoranthene, no observable mass discrimination can be attributed to biodegradation for these compounds. If the controlling steps for the biodegradation of higher molecular weight PAHs are similar, isotopic alterations resulting from microbial degradation would likely be insignificant in many natural PAH occurrences. It is notable that recent experiments on naphthalene and toluene show that some carbon isotopic discrimination can occur under certain conditions (Meckenstock *et al.*, 1999; Diegor *et al.*, 2000; Ahad *et al.*, 2000; Morasch *et al.*, 2001; Stehmeir *et al.*, 1999). Yanik *et al.* (2003) also showed some carbon isotope fractionation, interpreted as biodegradation isotope effect, in field-weathered petroleum PAHs. If the presence of isotopic discrimination during biodegradation is shown for some systems, it is likely contingent on the specific pathways utilized during microbial metabolism. This may be a fruitful area of further inquiry.

12.7 PAH SOURCE APPORTIONMENT CASE STUDIES

12.7.1 St. John's Harbor, Newfoundland, Canada

The earliest example of using compound-specific carbon isotope compositions for apportioning PAH sources in aquatic systems is the study of St. John's Harbor, Newfoundland by O'Malley *et al.* (1994, 1996). This is an ideal setting for utilizing carbon isotopes because the molecular evidence is equivocal. The surface sediments in St. John's Harbor were characterized by high but variable concentrations of 3-, 4-, and 5-ring parental PAHs suggestive of pyrogenic sources, the presence of methylated PAHs and a UCM indicative of petrogenic inputs, and occasional presence of retene and perylene, implicating diagenetic input. The relatively low Pa/A ratio (1.9) in the surface sediment, which is indicative of significant pyrolysis contributions, is similar to those calculated in previous studies (Killops and Howell, 1988; Brown and Maher, 1992). The Fl/Py ratio (1.2), indicative of pyrolysis, is comparable to the Fl/Py ratios reported for sediments from other similar systems (e.g., Penobscot Bay, Boston Harbor, Severn Estuary, Bedford Harbor, and Halifax Harbor) (Killops and Howell, 1988; Pruell and Quinn, 1988; Gearing *et al.*, 1980; Hellou *et al.*, 2002). Similarly, the BaA/Chy ratio (0.80) also indicates that BaA and Chy are of pyrogenic origin. However, unlike Fl and Py, the generally higher BaA/Chy ratio suggests that the primary source of these particular compounds may be related to car emissions. The BaP/BeP and the Pa/MPa ratios of the sediment cannot be clearly reconciled with any of the primary source signatures. This could indicate that these ratios have limited value for source apportionment because of the reactivity of BaP and methylated Pa in the environment (e.g., Canton and Grimalt, 1992).

The reactivity of MPa and BaP seems to have resulted in the alteration of their source compositional ratios during transportation/deposition, and thus may not be a reliable source indicator. In addition to the compositional attributes of the parental PAHs, the presence of methylated compounds and a UCM in the surface sediments indicates possible petrogenic inputs of PAHs to the Harbor. The UCM is generally indicative of petroleum and petroleum products, and is a widely used indicator of petrogenic contamination in sediments (Prahl and Carpenter, 1979; Volkman *et al.*, 1992, 1997; Simoneit, 1998). It is commonly assumed that a UCM consists primarily of an accumulation of multibranched structures that are formed as a result of biodegradation reactions of petroleum (Volkman *et al.*, 1992). Since no clear indication of petroleum-derived inputs can be discerned from the compositional ratios of the prominent PAHs in the sediments, it is apparent that the isomeric ratios of the prominent petrogenic PAHs are masked by pyrogenic-derived components.

There is no characteristic trend in the $\delta^{13}C$ between the LMW and HMW PAHs in the surface sediments of St. John's Harbor. However, the isotopically enriched A and the trend between Pa and MPa are indicative of pyrogenic PAH sources, whereas the marginally depleted Py compared with Fl suggests inputs of petrogenic sources. The isotopic signatures of the PAHs from all the surface samples are broadly similar, and can be quantitatively related to the prominent primary sources (Figure 19). Significant enrichments of Pa, MPa, and A relative to the 4- and 5-ring compounds are consistently observed in all the individually analyzed samples. This may be attributed to the input of road sweep material via the sewer system. The consistency in the $\delta^{13}C$ of Pa (–25.9‰ to –25.2‰) in the surface sediments suggests that it is possibly derived from an isotopically enriched source of pyrogenic origin. Evidence of some petrogenic input is, nevertheless, indicated by the relatively ^{13}C-depleted pyrene, a trend that was consistently observed in most of the samples investigated. The results of mass balance mixing calculation are diagrammatically shown in Figure 20. The $\delta^{13}C$ of the 4- and

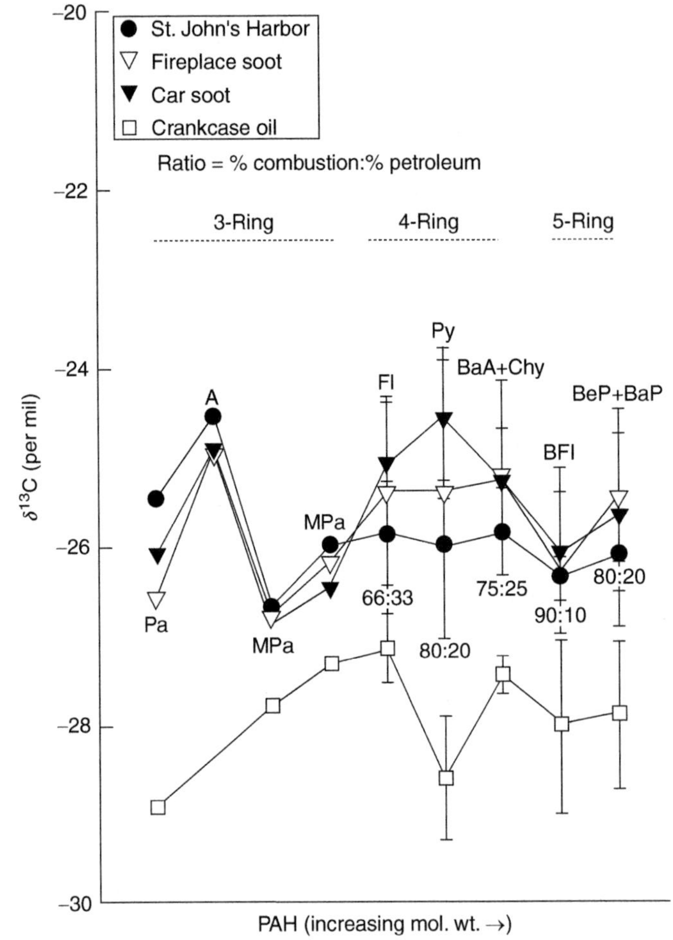

Figure 19 δ^{13}C of St. John's Harbor sediments compared with the three primary PAH sources in the Harbor. Error bars show 2σ variability in the source δ^{13}C. Ratios shown (e.g., 80:20) are the approximate balance of crankcase oil and fireplace soot (assuming two-component mixing) in the Harbor sediments.

5-ring PAHs isolated from the various sediments generally form positive mixing arrays between the wood-burning and car soot and crankcase oil end-members, quantitatively substantiating molecular data that mixtures of these prominent primary sources could explain the range of δ^{13}C values observed in these particular sediments. The validity of these mixing arrays is further supported by unchanging relative positions of the individual sites (e.g., E, F, and J; Figure 20) in the mixing array (accounting for source uncertainties of 0.5–1‰, 1σ) regardless of the projection used. In general, as indicated by the dominance of the 3-, 4-, and 5-ring parental PAHs in the molecular signatures, PAHs of pyrolysis origin seem to be dominant contributors to the Harbor sediments. This is shown by the bulk of the data points, which are concentrated close to the two pyrolysis/combustion end-members (Figure 20). From the position of the mean values on the mixing curves, the average pyrolysis input is estimated to be ~70% (±20%, 1σ), while the remaining 30% (±20%, 1σ) of the PAHs seem to be derived from crankcase oil contributions. Sites H and J are particularly enriched in pyrolysis-derived PAHs, while the maximum input of crankcase oil (≈50%) consistently occurred at site E indicating the heterogeneous nature of these sediments which was not evident from the molecular signatures. The percentage input of crankcase oil is similar to the range (20–33%) estimated from Figure 19.

The relative importance of the two primary pyrolysis sources (wood-burning soot versus car soot) as PAH contributors to the sediments is also evident in these mixing curves. With the exception of sites A, H, K, and L, which plot closer to the wood-burning soot/crankcase oil mixing array, all the remaining sites tend to be concentrated close to the car soot/crankcase oil mixing curve (Figure 20). This indicates that a significant portion of the pyrolysis-derived PAHs in the Harbor is of car emission origin. Despite the lack of distinct evidence in the molecular signature data to suggest the importance of car soot emissions in the pyrolysis signatures, inputs of car soot PAHs to the

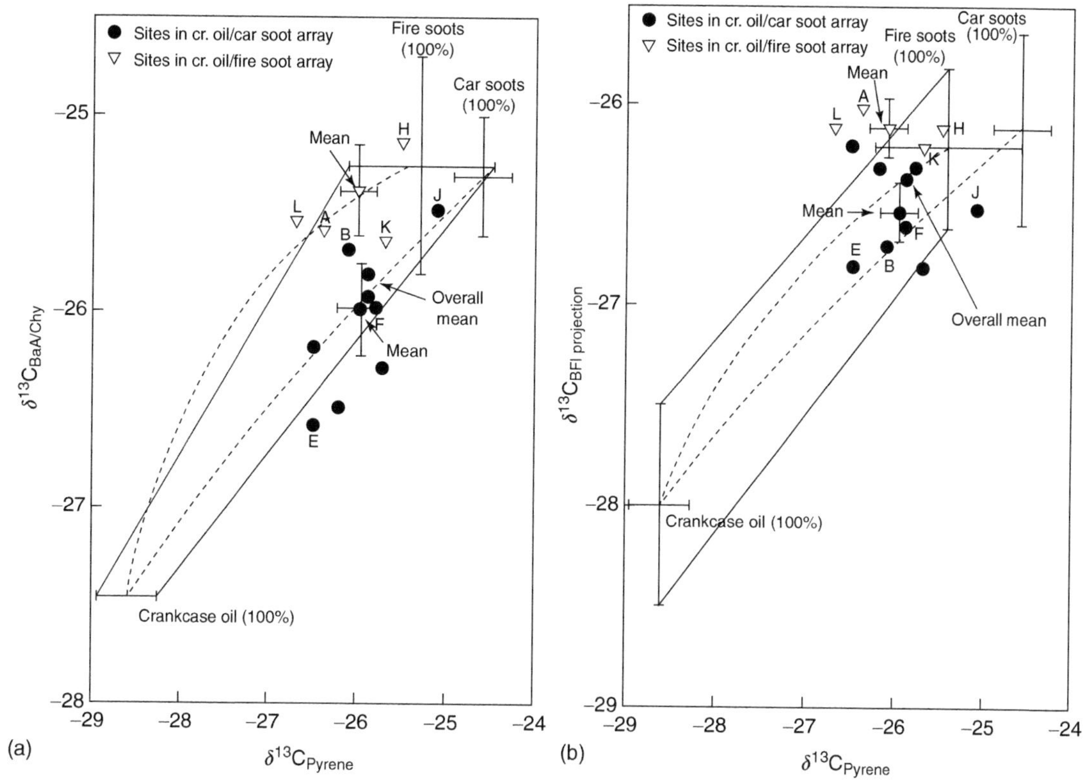

Figure 20 Carbon isotopic mixing diagram for St. John's Harbor sediments using identified PAH source end-members: (a) $\delta^{13}C_{Pyrene}$ versus $\delta^{13}C_{BaA+Chy}$ and (b) $\delta^{13}C_{Pyrene}$ versus $\delta^{13}C_{BFI}$ projection. Data points are subdivided into retene- (∇) and nonretene-containing samples (•).

sediments are indicated by the low Pa/A and elevated BaA/Chy compositional ratios. The dominance of car soot PAHs compared with wood-burning soots suggests that a significant part of wood-burning emissions is being transported away from the sampling area due to wind dispersion, and that car soot inputs mainly occur as a result of road runoff. The presence of retene, identified in most of the surface sediment samples, is mainly related to early diagenesis rather than from wood-burning sources, despite having a similar $\delta^{13}C$ value to these sources. This is supported by the apparent absence of retene in the open road and sewage sample molecular signatures, which are identified to be the predominant pathways for PAH input to the Harbor. There was also a notable absence of retene in the atmospheric air samples collected in the St. John's area. The input of petroleum or petroleum products to the sediments is not surprising since most of the individual samples are characterized by the presence of a prominent UCM.

The prominence of car soot and crankcase oil PAHs in sediments supports the suggestion that road runoff and runoff via the sewer system, and not atmospheric deposition, is the main pathway for PAH input to the Harbor sediments. This is also reflected by the similarity in the isotopic values of the PAHs in the sewage and sediment samples (Figure 21). Input of road sweep materials to the sediments was also supported by the presence of asphalt-like particulates similar to those observed in the open-road sweeps. Owing to the absence of major industries in the St. John's area, sewage is dominated by domestic sources. Sewage PAH content is, therefore, predominantly derived from aerially deposited pyrolysis products, car emissions, crankcase oil by direct spillage and engine loss (NRC, 2002), and road sweep products such as weathered asphalt, tire, and brake wear. The occurrence of early diagenesis in these sediments is supported by the presence of perylene in the deeper samples, which is suggested to be derived from diagenetic alteration reactions based on the Py/Pery ratio. Reliable isotopic measurements could not be performed on perylene due to coelution with unidentified HMW biological compounds.

In summary, using both the molecular signatures and the $\delta^{13}C$ of the PAHs isolated from St. John's Harbor, sources of pyrogenic and petrogenic PAHs along with diagenetic sources are positively identified. From a combination of the molecular abundance data and the $\delta^{13}C$ of the individual PAHs, car emissions are identified as important contributors of pyrolysis-derived compounds, whereas an average of 30% was derived from crankcase oil contributions. Direct road

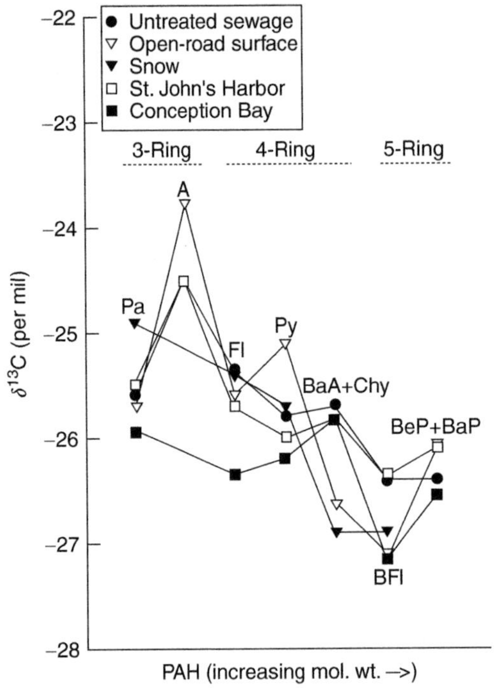

Figure 21 Comparison of PAH $\delta^{13}C$ from the St. John's Harbor sediment, road surface sweep, roadside snow pile, and sewer effluent. Also shown for comparison is the average (18 samples) PAH $\delta^{13}C$ of sediments from a separate harbor outside of the St. John's Harbor watershed (Conception Bay).

runoff, runoff via the storm sewers, and snow dumping are identified to be the most likely pathways for PAH input to the Harbor using both the molecular signature and carbon isotope data. In this particular site, the combined use of molecular and carbon isotopic data elucidated the prominent sources of PAHs in the sediments. However, the use of isotopic data has the added advantage that some quantification of sources can be undertaken with more confidence (O'Malley *et al.*, 1996). It should be noted that these mixing equations used only the more stable higher molecular weight 4- and 5-ring compounds to avoid uncertainties related to possible weathering fractionations that could affect lower molecular weight compounds.

12.7.2 St. Lawrence River

The International Corridor of St. Lawrence straddles the US–Canadian border, and spans the segment of the St. Lawrence River from the outflow of Lake Ontario (Kingston, ON) to the Massena area (NY). The sediments in this stretch of the River can be subdivided into two groups, with the first comprising the nonindustrial segment from Kingston to Morrisburg and the second being the aluminum smelter areas of Massena (NY) and Cornwall (ON). Samples from the Massena/Cornwall area contained significant molecular and isotopic variations over small geographic distances and yielded significantly higher total PAH indicating strong point sources. In contrast, samples from the Kingston to Morrisburg region varied to a lesser degree over larger geographic distances, suggesting a more regional source and mixing of PAH contaminants. Overall, the sediments showed the dominance of 4- to 6-ring parental PAH, with the sum of PAH ranging from 0.8 to 6,700 $\mu g\, g^{-1}$ (Stark *et al.*, 2003).

12.7.2.1 Kingston to Morrisburg

Surface sediments collected from several sites along the length of the international segment of the St. Lawrence, from Kingston to Morrisburg (Figure 22) contain both parental and alkylated 2- to 6-ring PAH throughout the core with parental dominating over alkylated species. The parental PAH include 4- and 5-ring species with significant quantities of 3- and 6-ring compounds. The $\sum$PAH decreases from a range of 0.7–3.8 $\mu g\, g^{-1}$ between Kingston and Maitland to a narrower range of 0.06–1.3 $\mu g\, g^{-1}$ in the downstream segment from Alexandria to Morrisburg. A notable deviation in this general trend is the higher $\sum$PAH values noted for the Prescott–Ogdenburg in subsurface (>6 cm depth) sediments. Unlike the other sediment samples in the Corridor, alkylated PAH comprise a significant proportion of PAH observed in the Prescott–Ogdensburg site, with the dominant PAH consisting of both alkylated and parental 4- and 5-ring species.

Carbon isotope measurements of individual 3- to 5-ring parental PAH from the surface sediments from all these sites are as shown in Figure 23. There is an overall consistency in the carbon isotope signal for the Corridor, with values ranging from −22‰ for phenanthrene to −26‰ for BaP. Perylene consistently showed ^{13}C-depleted values relative to the other parental PAH. The carbon isotopic trend for the Prescott–Ogdenburg site shows distinctively lower $\delta^{13}C$ values and a general trend of ^{13}C depletion with increasing molecular weight from Pa (−24.1‰ to −24.6‰) to Per (−28.5‰ to −29.2‰). Another notable deviation is shown by the Brockville sediments where the $\delta^{13}C$ values for BFl, BaP, and BeP are lower and Py and Chy higher than upriver sediments (i.e., Kingston and Alexandria–Rockport).

12.7.2.2 Massena–Cornwall area

Samples from the Massena–Cornwall area consist of surface grabs from the Cornwall bank (Canada) and the Massena side (USA) of the St. Lawrence (Figure 24). Samples from the Cornwall side contain parental and alkylated 2- to 6-ring PAH. However, the relative distribution of these compounds and the $\sum$PAH are distinct for each of the three Cornwall sampling locations. Site 3,

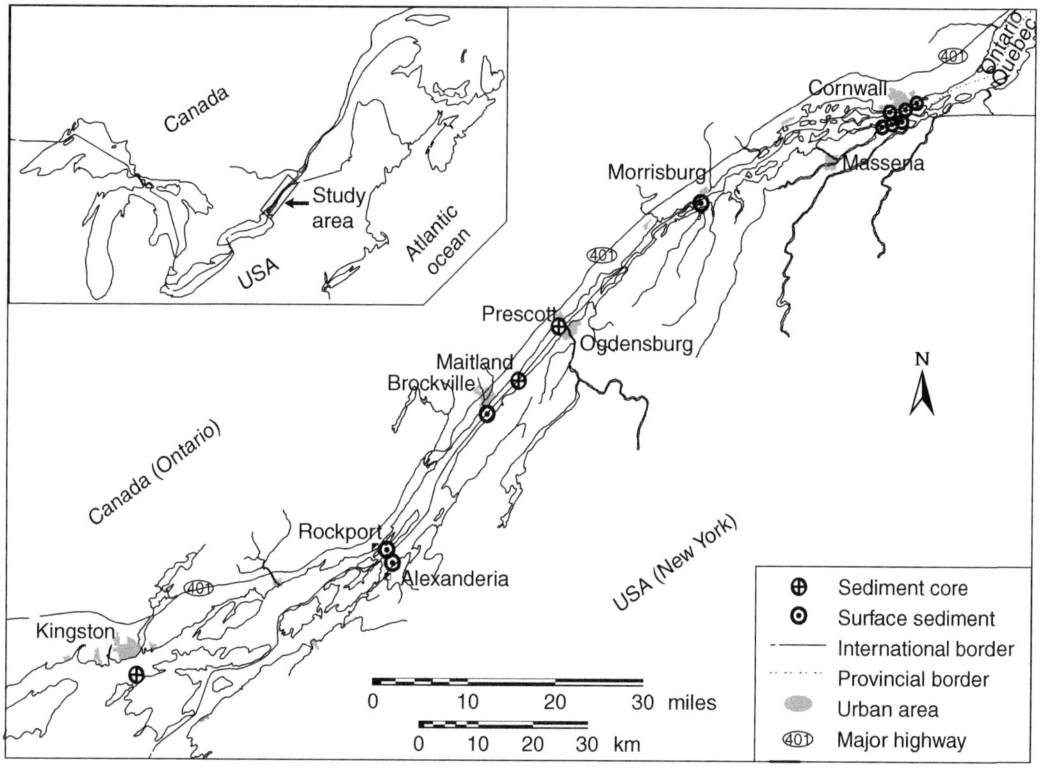

Figure 22 International segment of the St. Lawrence River showing town locations adjacent to sediment sampling sites.

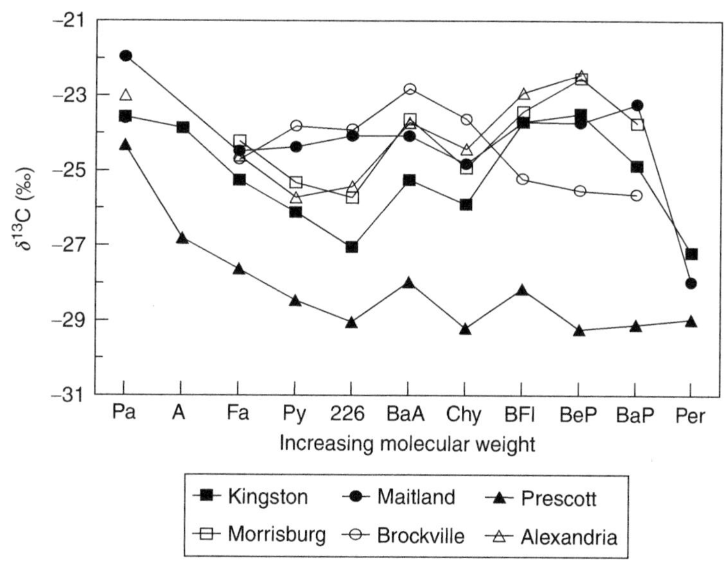

Figure 23 Mean compound-specific $\delta^{13}C$ of parental PAH in St. Lawrence River sediments (cf. Figure 22) compared with previously compiled primary source values from O'Malley (1994).

located in the open waters off Cornwall, displays 2- to 6-ring PAH with parental dominating over alkylated species and 3- to 5-ring species, in particular Fa and Py, being most dominant. The $\sum$PAH at this location is $0.42\,\mu g\,g^{-1}$. Site 650500 contains 2- to 6-ring PAH with alkylated PAH dominating over parental species. In addition, alkylated sulfur-substituted PAH, i.e., DBT derivatives were also abundant (DBT derivatives are abundant in coal and present in variable amounts in oils; see, e.g., Neff, 1995). Dominant alkylated species include methyl Pa/A, dimethyl Pa/A, methyl Fa/Py, trimethyl Pa/A (including retene), and dimethyl Fa/Py; it also includes moderate amounts of methyl Fl, methyl BaA/Chy, and methyl and dimethyl DBTs. The dominant

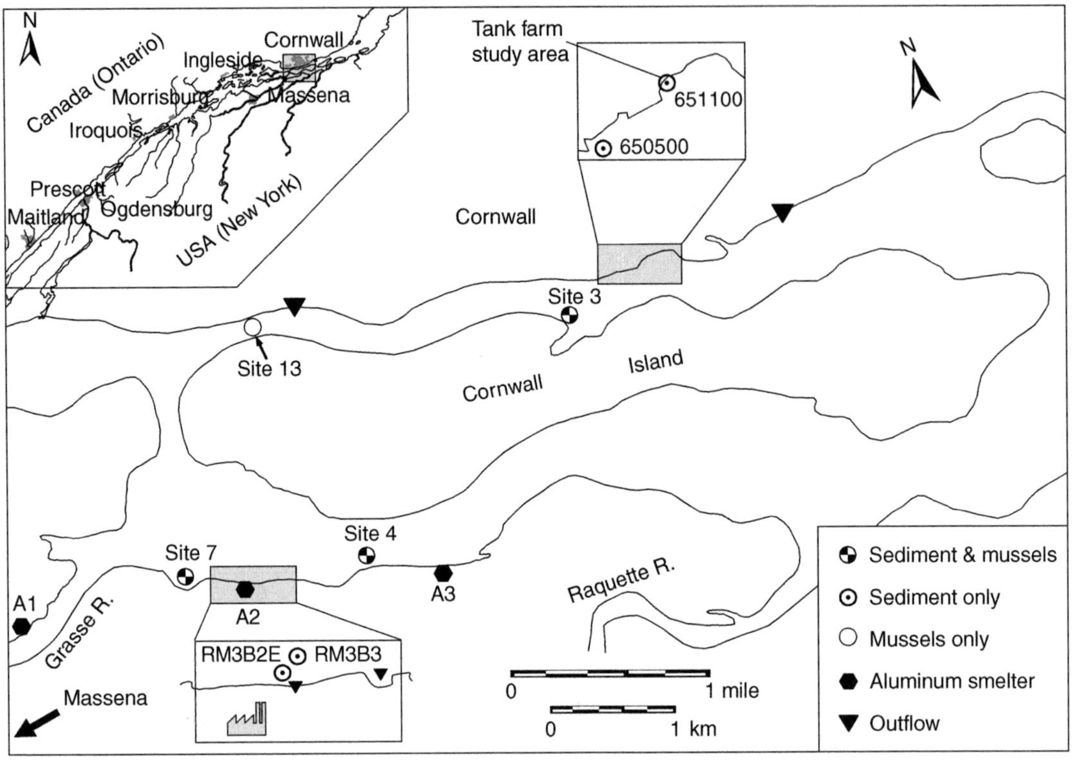

Figure 24 Sample location map in the St. Lawrence River adjacent to the Massena and Cornwall areas. Cornwall Island divides flow of the River into a northern segment receiving water exclusively from upstream sources, and a southern flow that additionally receives effluents from the aluminum smelters in Massena.

parental PAH consists of 3-, 4-, and 5-ring species, in particular Pa, Py, and Fa. The $\sum$PAH at this location was 9.4 $\mu g\,g^{-1}$. Site 651100, located in a cove at Cornwall (Figure 24), also contained 2- to 6-ring parental and alkylated PAH with the former dominating. The parental PAH present mostly consists of 3- and 4-ring species, in particular Pa, Fa, and Py. The $\sum$PAH at this location is 158.5 $\mu g\,g^{-1}$. In addition to the minor occurrences of alkylated species, a significant amount of the $m/z = 218$ amu compound and notable quantities of phenyl-Na, and the $m/z = 226$ amu compound are present.

$\delta^{13}C$ measurements of individual 3- to 5-ring parental PAH from surface sediments from the Cornwall area are also shown in Figure 25a. For sites 651100 and 650500, both located near the town of Cornwall, the PAH inventory shows the significant presence of parental 3- to 5-ring compounds with Fa/Py ratio of 0.7, significant presence of alkylated DBT, and pronounced UCM. This molecular signature, coupled with the ^{13}C-depleted values measured for the 3- to 5-ring parental PAH, suggests the presence of weathered petroleum in this location. In contrast, site 3 has PAH molecular and isotopic signatures that overlap with those of the combustion-dominated sources in the upriver sediments.

Sediment samples from the Massena side of the St. Lawrence River consist of two near-shore surface grab samples located near the outflow of an aluminum smelting facility and two others located in the open channel upriver and downstream from the sampled smelting operation (Figure 24). All samples contain parental and alkylated 2- to 6-ring PAH with parental compounds significantly dominating over alkylated species. The parental PAH consists mainly of the 4- and 5-ring species, with Chy, BFl, and BeP being most dominant. In addition to the minor occurrences of alkylated species, notable amounts of phenyl-Na, the $m/z = 218$, and 226 amu compounds are also present. Site RM3B2E, located <50 m from the main outfall from the smelter, contains the highest sum of PAH of 1,554 $\mu g\,g^{-1}$ while site RM2B3, located slightly more than 100 m from the outfall has a sum of 224 $\mu g\,g^{-1}$. Site 4, located ~1,000 m downstream, and site 7, situated more than 400 m upriver, have $\sum$PAH of 4.5 and 4.2 $\mu g\,g^{-1}$, respectively.

$\delta^{13}C$ of individual 3- to 5-ring parental PAH from three surface sediments in the Massena area are shown in Figure 25b. Site RM3B3, located at the smelter outflow, contains the most isotopically enriched 3- and 4-ring parental PAH with values ranging from −22.4‰ to −23.8‰. Site 7, located upriver from the smelter, contains the most isotopically depleted 3- and 4-ring PAH with values extending from −23‰ to −25.1‰. Site 4, located downriver, shows $\delta^{13}C$ values for 3- to 4-ring

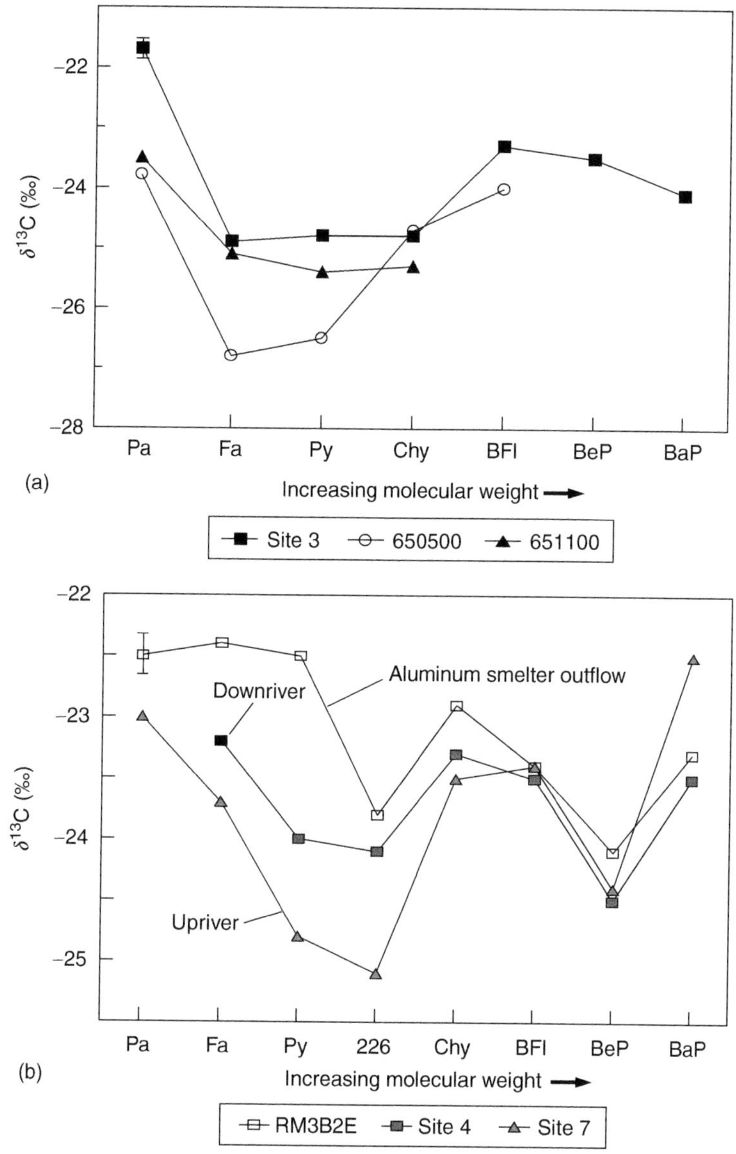

Figure 25 Mean compound-specific δ^{13}C of parental PAH (a) in sediments from the cornwall area and (b) in sediments upstream, adjacent, and downstream of the aluminum smelters on the Massena side of the St. Lawrence River (cf. Figure 23).

PAH well within the range of the previous two sites (−23.2‰ to −24.1‰). δ^{13}C values for the 5-ring PAH in all samples are similar or within analytical uncertainty, except for BaP at site 7. In the latter site, BaP is ~0.9‰ more enriched than the other locations but still lies close to the reported error of ±0.4‰ in environmental samples (O'Malley, 1994).

The elevated concentrations of PAH around Massena are clearly attributable to PAH releases associated with the smelting operations in the area. The overall gradation of molecular and isotopic compositions in sediments upstream, adjacent, and downstream to the smelter effluents reflects the mixing of upstream PAH sources with the local smelter source. It is notable that the smelting signature is not present across the river at the Cornwall side, which receives its sediments exclusively from upstream sources. Instead, Cornwall sediments are dominated by signatures consistent with weathered petroleum. It thus appears that smelter-related release of PAH is largely restricted to direct PAH incorporation in sediments through normal stream processes, and little detectable, if any atmospheric deposition in the close proximity.

In summary, the molecular and carbon isotopic compositions of sediment-bound PAH in the international segment of the St. Lawrence River reflect both the regional and local sources of the contaminants. Surface sediments along the length of the international corridor of the St. Lawrence River are characterized by variable molecular and carbon isotopic compositions spanning the range of combustion and petroleum-related sources in the basin. The Massena–Cornwall region exemplifies segments of the river heavily influenced by point sources of PAH. Interestingly, only the southern (Massena) shore of the St. Lawrence River in this area shows the impact of aluminum smelter operation, with the northern (Cornwall) shore showing a significant imprint of localized petroleum source (s) of PAH. Comparison of compound-specific $\delta^{13}C$ values of PAH in upstream, effluent, and downstream sediments on the Massena side of the River clearly shows the specific impact of aluminum smelter contribution. The portion of the St. Lawrence River upstream of the Massena–Corwall area shows molecular and isotopic signatures reflecting the regional importance of combustion-related sources. However, the molecular and compound-specific $\delta^{13}C$ signatures of PAH in this upstream segment also vary geographically, and possibly temporally, because of enhanced localized input of petroleum-related sources.

12.7.3 New York/New Jersey Harbor

The highly urbanized New York/New Jersey (NY/NJ) Harbor complex is known to be contaminated by various chemical compounds and heavy metals (Bopp *et al.*, 1993a; Crawford *et al.*, 1995; Wolfe *et al.*, 1996; Mitra *et al.*, 1999) (Figure 26). Following estuary-wide multiyear intensive bioeffects surveys conducted by the National Oceanic

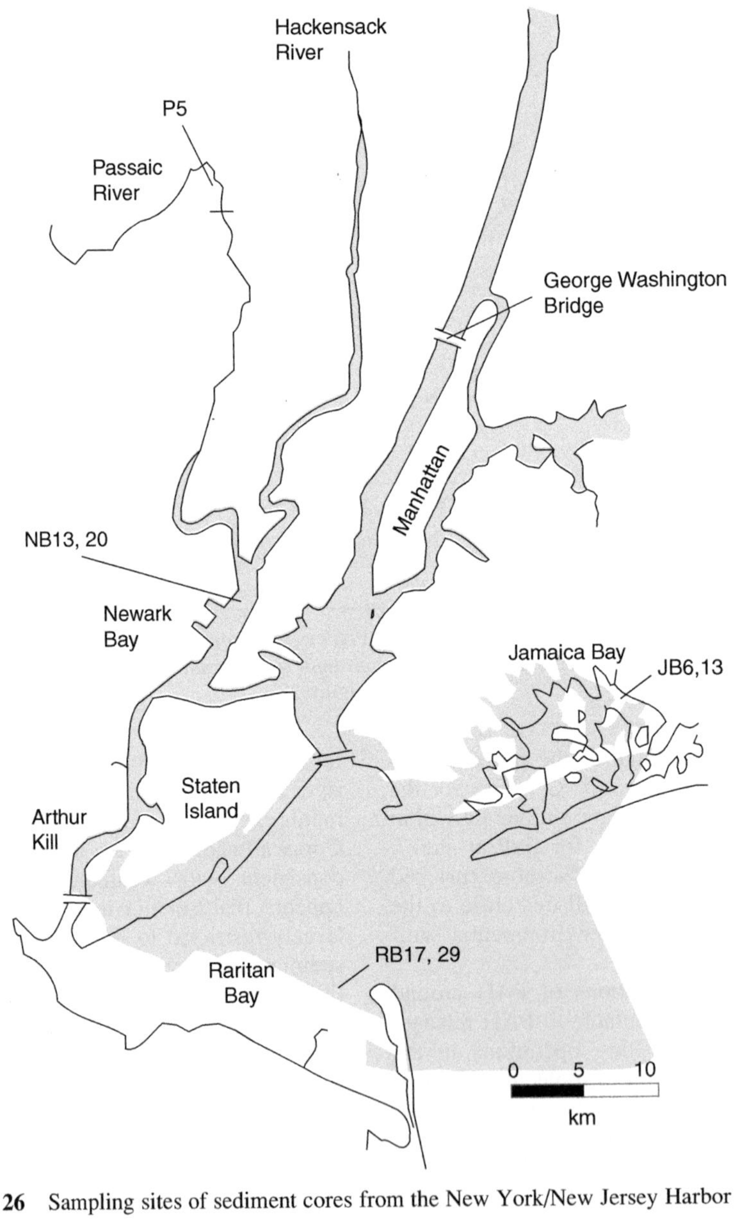

Figure 26 Sampling sites of sediment cores from the New York/New Jersey Harbor complex.

and Atmospheric Administration (NOAA), PAHs have emerged as a leading concern in this area (Wolfe *et al.*, 1996). A recent examination of NY/NJ Harbor sediments found that the dominant PAH sources in recent sediments are from petroleum usage including direct oil-spill (petrogenic) and petroleum combustion (pyrogenic) sources (Yan *et al.*, 2004, 2006). The areas examined included heavily contaminated areas like Newark Bay (NB) and the Passaic River (PR), and less contaminated sites such as Raritan Bay (RB), and Jamaica Bay (JB) (Figure 26). In contrast to the aforementioned studies of the St. John's Harbor and St. Lawrence River, the NY/NJ Harbor complex studies featured the use of excellently dated sediment cores allowing the assessment of temporal changes in PAH and hydrocarbon inputs in the sediment record. Detailed descriptions of radionuclide-dating methods for these sediments have been discussed previously (Bopp *et al.*, 1991, 1993a, 2005).

12.7.3.1 Spatial and temporal distributions of hydrocarbons from the NY/NJ Harbor complex

The range of $\sum$PAH concentrations (dry weight) varies considerably, both spatially and temporally for all the cores examined. For example, around 1980, the concentration sequence in four sites from highest to lowest is the NB ($70\,\mu g\,g^{-1}$), PR ($60\,\mu g\,g^{-1}$), JB ($9.4\,\mu g\,g^{-1}$), and RB ($1.9\,\mu g\,g^{-1}$), which is consistent with other reports in these areas (Crawford *et al.*, 1995; Gigliotti *et al.*, 2002). The considerable geographic variations in PAH concentrations suggest that local PAH sources are most important. In core PR, concentrations of principal parent PAHs (Fl, Py, BaA, Chy, etc.) in all but the top section are higher than their effects range-median (ERM) guideline values (amphipod toxicity). This suggests that PAHs in the PR could have significant effects on benthic organisms (Long *et al.*, 1995). In all but the oldest sections of the NB cores, concentrations of these principal compounds are lower than their ERM, but greater than effects range-low (ERL) (Long *et al.*, 1995). In contrast, all above-mentioned principal PAHs in core JB and RB are close to or below their corresponding ERL values. $\sum$PAH depth profiles from NB, RB, and JB show elevated levels around 1940s–1950s, then generally declining to the 1980s. These decreasing temporal trends are consistent with those recorded in NJ Harbor sediments (Hontley *et al.*, 1995) and many observations around the world (Sanders *et al.*, 1993; Simcik *et al.*, 1996; Fernandez *et al.*, 2000; Okuda *et al.*, 2002a; Lima *et al.*, 2003), but opposite to the increasing trend observed by Van Metre *et al.* (2000). This substantial decline in PAH concentration since the 1950s has been attributed to a major shift in energy use (e.g., residential heating and power generation) from predominantly coal-derived energy to a greater petroleum and gas usage (Hites *et al.*, 1980; Pereira *et al.*, 1999). Evidence suggests that coal combustion has the potential to release more PAHs than petroleum or natural gas per unit of power generated (Rogge *et al.*, 1993; Oros and Simoneit, 2000; Schauer *et al.*, 2001).

Degraded or weathered petroleum typically has much higher UCM abundance relative to resolved peaks. In contrast, wood and coal combustion normally has lower UCM abundance than resolved peaks (Oros and Simoneit, 2000; Fine *et al.*, 2001). In most of the sediments from cores of RB and JB, the hump is generally broad and high compared with SH peaks, indicating that SH sources are mainly from petroleum usage. Compared with a broad UCM range (i.e., large molecular weight range) in samples from other sites, elevated UCM in core JB tends to have a lower molecular weight distribution (C_{15}–C_{21}), probably indicating biodegradation of jet fuel from the JFK (Figure 1), and it is likely that there is a direct input from the airport (Bopp *et al.*, 1993b).

The carbon preference index (CPI), ratio of the sum of the odd carbon-numbered alkanes (from C_{25} to C_{35}) to the sum of the even carbon number (from C_{24} to C_{34}), is widely used to distinguish SH sources between higher plants (CPI>5), and petroleum (CPI~1) (Wakeham, 1996). In all our cores, CPI is mainly around 1–2, indicating dominant petroleum sources for the SHs rather than higher plant sources.

12.7.3.2 $\delta^{13}C_{Py}$ and PAH sources over time

A mass balance method for quantifying PAH sources from carbon isotopic composition was discussed earlier for St. John's Harbor (e.g., O'Malley *et al.*, 1994). Using this approach, we can likewise assign $\delta^{13}C$ of Py ($\delta^{13}C_{Py}$) from petrogenic source a value of −29‰, while that from combustion sources as −24‰ (O'Malley *et al.*, 1994) (Figure 19). Yan *et al.* (2006) employed Pearson correlation analysis to elucidate the relationship of the aforementioned molecular criteria and $\delta^{13}C_{Py}$. As discussed above, if initial molecular characteristics are retained from sources to sediments, the molecular ratios should correlate with each other and with $\delta^{13}C_{Py}$. Theoretically, $\delta^{13}C_{Py}$ should be positively correlated with the molecular ratios as configured because $\delta^{13}C_{Py}$ will increase as petroleum combustion contribution increases (O'Malley *et al.*, 1994). Temporal trends of $\delta^{13}C_{Py}$ in the NY/NJ Harbor correlate with UCM, fluoranthene to fluoranthene plus pyrene (Fl/(Fl+Py)), HMW 4- to 6-ring PAH to total PAH (Ring456/TPAH), and parent to total (Par/(Par+Alkyl)). In contrast, the ratios of anthracene to phenanthrene plus anthracene (A/(Pa + A)), benzo[*a*]anthracene to benzo [*a*]anthracene plus chrysene (BaA/(BaA + Chy)), and indeno[1,2,3-*cd*]pyrene to indeno[1,2,3-*cd*] pyrene plus benzo[*ghi*]perylene (IP/(IP + Bghi)) only poorly correlate with $\delta^{13}C_{Py}$.

12.7.3.3 Combined application of $\delta^{13}C_{Py}$, Fl/(Fl + Py), Par/(Par + Alkyl), and Ring456/TPAH

The above correlations suggest that $\delta^{13}C_{Py}$ and Fl/(Fl + Py), Par/(Par + Alkyl), and Ring456/TPAH are promising indicators that could be used as PAH tracers in the NY/NJ Harbor areas. Figure 27 shows trends of these four ratios at each site. Despite the similar trends of ΣPAH levels in most areas of the NY/NJ Harbor from the 1950s to the 1980s, depth profiles of $\delta^{13}C_{Py}$ and the three molecular ratios are completely different from site to site (Figure 27), suggesting varied PAH sources in different areas. Generally, Fl/(Fl + Py) and 1,7/(1,7 + 2,6)-dimethylphenanthrane (DMP) ratios (not shown) are below 0.48 and 0.40, respectively. This is especially true for sediments deposited within the past half-century, where the ratios are well below the corresponding ratios for coal combustion (0.56 and 0.66, respectively).

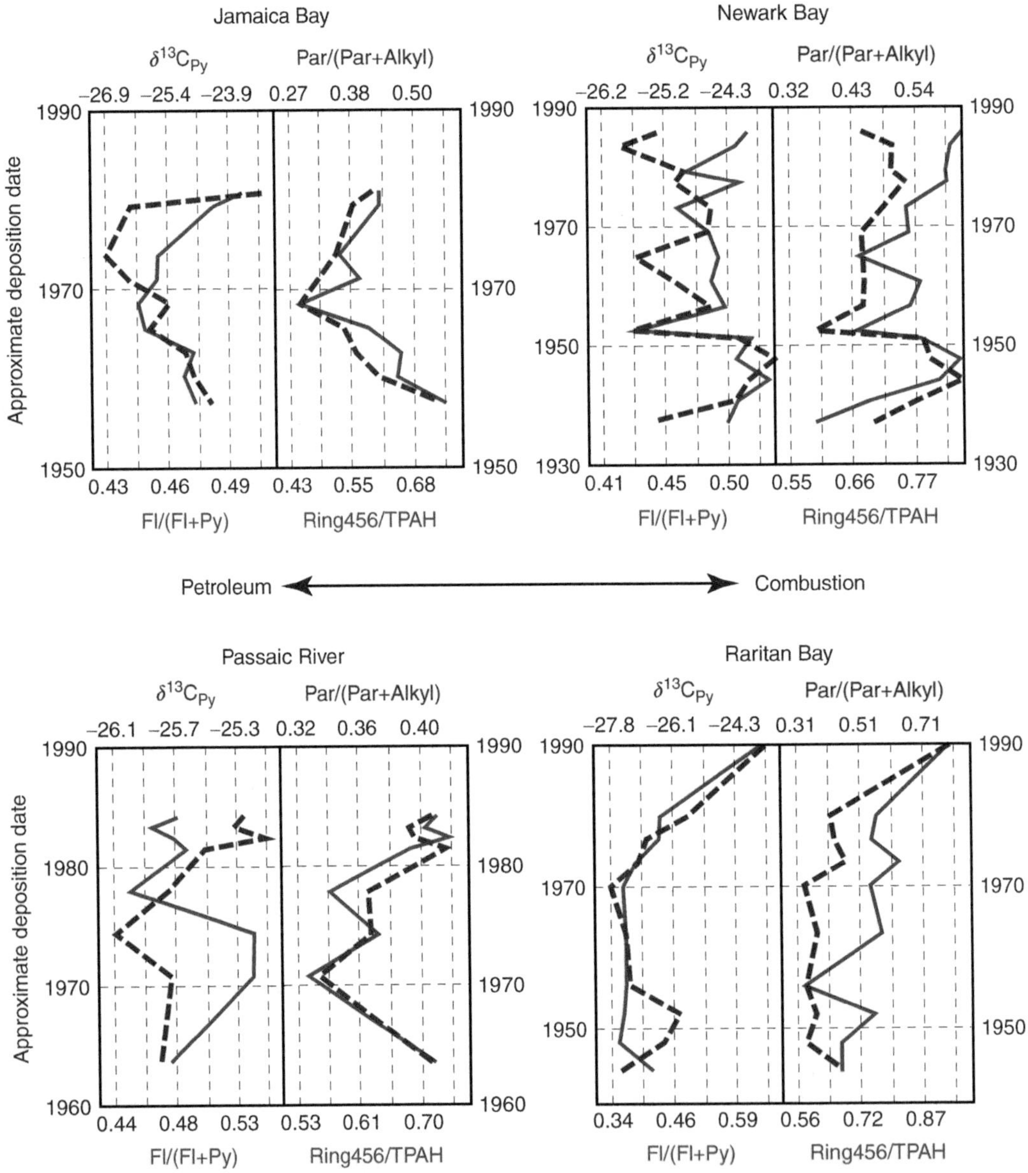

Figure 27 Temporal trends of correlated source indicators (refer to text definitions) in sediment samples from four cores in New York/New Jersey Harbor complex. Trends indicated by dashed lines follow the top scale; solid lines follow the bottom scale. For all indicators, a tendency toward the right suggests increasing contributions from combustion, while to the left, petroleum.

Interestingly, these ratios are within the range of 1,7/(1,7 + 2,6)-DMP ratios of fossil fuel combustion products. In contrast, the top section of core RB29, Fl/(Fl + Py) show a high ratio of 0.65 and 1,7/(1,7 + 2,6)-DMP of 0.52, suggesting a high possibility for PAH input from coal combustion for these recent sediments. However, the ratio of U/R (UCM to resolved hydrocarbons) in the saturated fraction is 23, much higher than the coal generated ratio from coal smoke (2.9–3.3) (Oros and Simoneit, 2000), and comparable with petroleum dominant ratios (~20) seen in Central Park Lake (CPL), New York City (NYC) (see below). Therefore, it is likely that in these four sites, petroleum combustion-related usage is the dominant PAH source over the past several decades.

However, we note that prior to 1960, higher concentrations of TPAH are observed in the sediments from NB and RB. As mentioned above, the decreasing $\sum$PAH trends since the 1940s to the present have been attributed to a major fuel change from coal to petroleum combustion (Hites *et al.*, 1980). Probably, the energy shift can explain observations from NB sediments. From the 1930s to 1940s in core NB, $\delta^{13}C_{Py}$ is around −24‰ to −25.7‰ indicating a combustion-dominant source. Fl/(Fl + Py) ratios are around 0.5, suggesting a possible source from coal or wood combustion. However, 1,7/(1,7 + 2,6)-DMP ratios would be higher (>0.6) than those determined in the 1940s (~0.40), if the major source was wood burning (Benner *et al.*, 1995; Yan, 2004; Yan *et al.*, 2005). Therefore, the dominant PAH sources in NB around the 1940s were most likely from mixture of petroleum and coal combustion sources (power generation) (Rod *et al.*, 1989). Using mass balance equations based on 1,7/(1,7 + 2,6)-DMP, we found that 30–50% of TPAHs are from coal combustion (Yan, 2004). However, the energy shift may not explain the PAH decline in the case of RB. From the 1940s to the 1970s, all four indicators remain relatively constant with $\delta^{13}C_{Py}$ (around −27‰ to −28‰) and Fl/(Fl + Py) ratios (from 0.3 to 0.4) showing dominant petroleum usage source, probably resulting from substantial petroleum usage in the nearby Arthur Kill. In most of the harbor sampling sites (e.g., the PR, RB, and JB), a gradual increase in combustion from 1970s to 1980s can be observed (Figure 27). Most of indicators in these three sites support this finding.

In summary, sediments from most of the areas in the NY/NJ Harbor complex, which decrease in PAH concentrations are observed from the 1940s to the 1980s. $\delta^{13}C_{Py}$, Fl/(Fl + Py), Par/(Par + Alkyl), and Ring456/TPAHs are significantly correlated in these cores, yielding a consistent interpretation of hydrocarbon sources in this urbanized setting. Combined applications of these PAH source indicators showed petroleum dominance in RB from the 1950s to the 1970s with a switch to combustion dominance from the 1970s to the 1990s. In JB, the dominant PAH source switches from combustion to petroleum before around 1970, then back to combustion until around 1990. Sediment records in the PR were dominated by petroleum combustion from the 1960s to the 1980s. In NB, mixtures from coal and petroleum combustion are the two major PAH sources before the 1950s, then coal combustion decreases in the last half-century. Collectively, the molecular and $\delta^{13}C$ indicators suggest that in most areas of the NY/NJ Harbor, combustion source has increased in relative importance since the early 1970s.

12.7.4 Central Park, NY

In contrast to the NY/NJ Harbor sediments, CPL sediment cores have been used previously to investigate the history of atmospheric contaminant deposition in NYC (Chillrud *et al.*, 1999). The CPL is ideal for this purpose because storm drainage is prevented from entering the Lake. As in the NY/NJ Harbor, Yan *et al.* (2005) took advantage of radionuclide-dated sediment cores to unravel ca. 130 years of history of sediment and atmospheric-contaminant deposition in the Lake. This time period witnessed large changes in the predominant fuel type used in NYC. Historical trends in fuel use in NYC are derived from US national data collected by the Energy Information Administration (EIA) (EIA, 2001). EIA data were scaled to NYC population trends (available online) to provide quantitative historical estimates of fuel use in the NYC area. This fuel-use history was used to evaluate the usefulness and limitations of various PAH source indicators (Table 3) and current widely applied SH indicators (e.g., U/R, the area ratio of unresolved "hump" to resolved peaks in a chromatogram).

12.7.4.1 Temporal trends of PAH indicators

Parental PAHs dominate throughout the CPL core (Figure 28), indicating that combustion sources have been the dominant PAH sources into CPL since the late nineteenth century. Although there is a potential pathway for petrogenic inputs into CPL via runoff from the local park roads, Chillrud *et al.* (1999) found that runoff from the local roads was not important for metal inputs relative to atmospheric inputs based on whole-core inventories of radionuclides and contaminant metals. The similarities between radionuclide-normalized, whole-core, contaminant metal inventories derived from the lake sediment core and that of soil cores collected from several locations in the park suggested that CPL

Table 3 PAH source indicators and their respective diagnostic ratios

Ratios	Description	Petrogenic	Petroleum burning	Coal combustion	Softwood combustion
Fl/(Fl+Py)[a]	The ratio of fluoranthene (Fl) to fluoranthene (Fl) plus pyrene (Py)	<0.40	0.40–0.50	>0.50	>0.50
A/(Pa+A)[a]	The ratio of anthracene (A) to PAHs with mass 178 (A and phenanthrene (Pa))	<0.10	>0.10	>0.10	>0.10
BaA/(BaA+Chy)[a]	Benzo[*a*]anthracene (BaA) divided by the sum of mass 228 (BaA+ chrysene(Chy))	<0.20	>0.35	>0.35	>0.35
$C_0/(C_0+C_1)_{P/A}$[a]	The ratio of parent PAHs with mass 178 (Pa, A) to PAHs with mass 178 (A; Pa) plus their C_1 alkyl homologs	<0.50	?	?	?
$C_0/(C_0+C_1)_{F/P}$[a]	Similar calculation as above but with mass 202	<0.50	>0.50	>0.50	>0.50
IP/(IP+Bghi)[a]	The ratio of indeno[1,2,3-*cd*]pyrene (IP) to the sum of IP and benzo[*g,h,i*]perylene (Bghi)	<0.20	0.20–0.50	>0.50	>0.50
Par/(Par+Alkyl)[b]	The sum of the parent PAHs with masses 128, 178, 202, and 228 divided by these parent PAHs plus their related alkyl PAHs (C_1–C_4 Na; C_1–C_3 P/A; C_1–C_2 F/P; C1-BaA/Chy)	<0.30	>0.50	>0.50	>0.50
Ring456/TPAH[b]	The ratio of total 4–6 ring PAHs (mainly originated from combustion) to TPAHs	<0.40	>0.50	>0.50	>0.50
1,7/(1,7+2,6)[c]	The ratio of 1,7-dimethylphenanthrene (DMP) to 1,7- and 2,6-DMP	?	0.40	0.65–0.68	0.90
U/R[d]	The ratio of unresolved complex mixture (UCM) to resolved hydrocarbons in saturated fraction	>4.0	>6.0	2.9–3.2	0.5–1.4
Ret/(Ret+Chy)	The ratio of retene (Ret) to retene plus chrysene	?	0.15–0.50	0.30–0.45	0.83–0.96

Diagnostic ratios of $C_0/(C_0+C1)_{P/A}$ from combustion sources and ratios of 1,7/(1,7+2,6)-DMP and Ret/(Ret+Chy) from petrogenic sources are uncertain.

[a] From Yunker *et al.* (2002) and references therein.

[b] From Yan *et al.* (2004).

[c] From Benner *et al.* (1995).

[d] From Oros and Simoneit (2000).

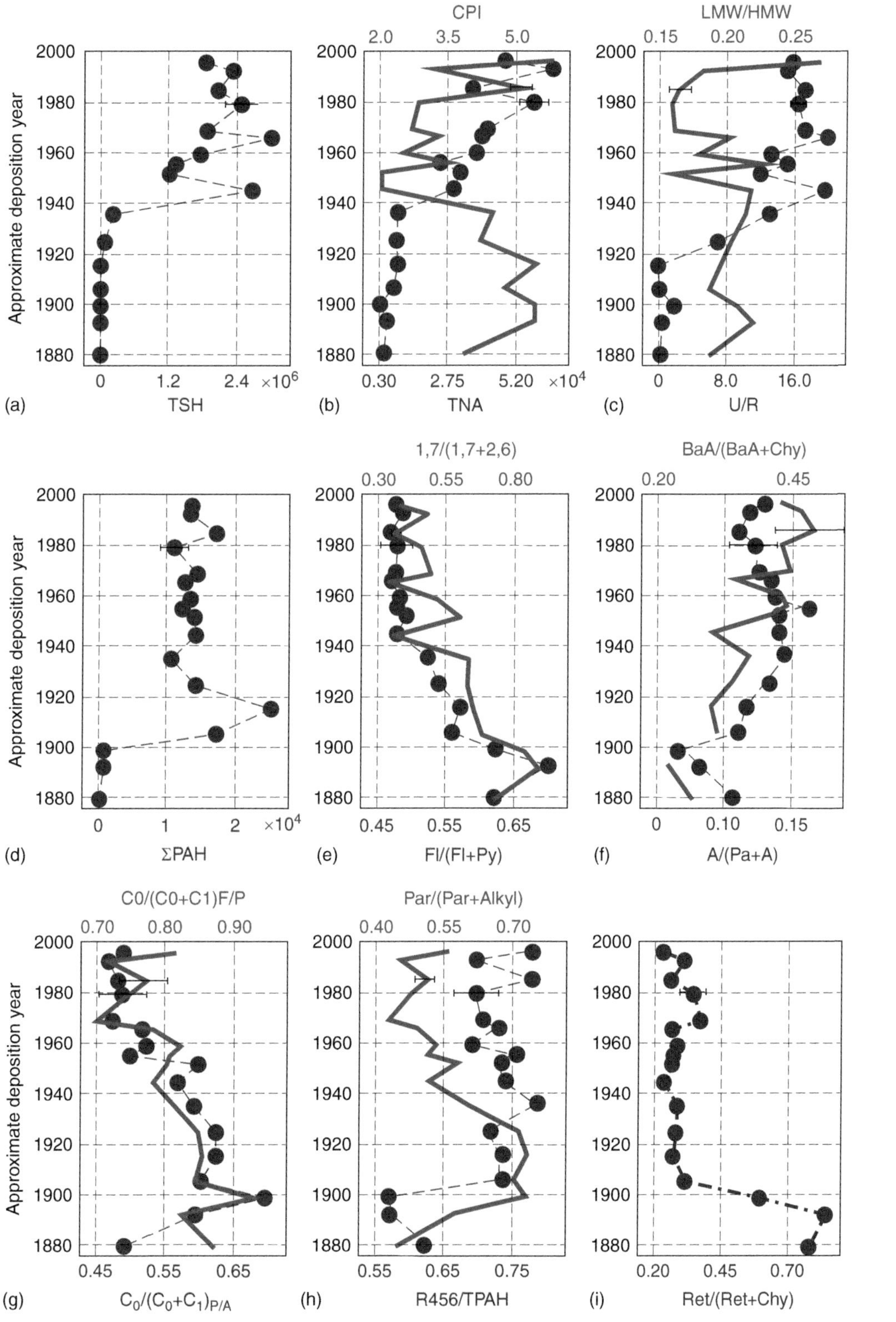

Figure 28 Temporal profiles of various saturated and aromatic indicators (refer to Table 1 and text for definitions) in Central Park Lake core. Trends indicated by solid lines (red) follow the top scale and dotted lines (blue) follow the bottom scale. Error bars indicate $\pm 1\sigma$ statistics. Representative error bars for the top scale are shown on the mid-1980s horizon; for the bottom scale, they are shown on the early 1980s horizon.

sediments primarily reflected atmospheric inputs of these metals with little direct inputs of local road runoff. Since this comparison included Pb and Cd, two metals known to be highly elevated in road runoff and also strongly absorbed onto soil particles, it should also apply to leakage of fuel, crankcase oil, and lubricating oils that can also be important in road runoff.

Although the correlation between parental PAH and the ratios of Fl/(Fl + Py), $C_0/(C_0 + C_1)_{F/P}$, and $C_0/(C_0 + C_1)_{P/A}$ suggest a predominantly pyrogenic PAH source for much of the history of deposition, ratios of A/(Pa + A) and BaA/(BaA + Chy) in three bottom samples (~0.04 and ~0.27) are lower than that would be expected for combustion (Table 3), suggesting a petrogenic source (Figure 28). According to the EIA report (EIA, 2001), petroleum was not used widely in the United States prior to 1900, thus the most likely petrogenic source is fresh coal, which typically contains abundant alkyl PAHs (Radke *et al.*, 1982).

Temporal profiles of 1,7/(1,7 + 2,6)-DMP and Ret/(Ret + Chy) are shown in Figure 2, and can be used as a "fingerprint" of softwood (e.g., fir and pine) combustion. A ratio of 1,7/(1,7 + 2,6)-DMP ranging from ~0.71 to ~0.91 was observed in emissions from softwood combustion (Fine *et al.*, 2001; Benner *et al.*, 1995), ~0.43 from motor vehicle exhaust (Benner *et al.*, 1989), ~0.63 in emissions from brown coal-fired residential stoves (Grimmer *et al.*, 1985), and ~0.67 in subbituminous and bituminous coal smoke. Retene is also preferentially produced in emissions from softwood combustion (LaFlamme and Hites, 1978; Fine *et al.*, 2001), so the ratio of retene to chrysene may have the potential to identify softwood combustion (Table 3). Softwood combustion products have a value of 0.83–0.96 with a ratio of 0.96 in pine-burning emissions and 0.83 for balsam fir (Fine *et al.*, 2001), whereas diesel engine emissions have values ranging from 0.05 to 0.5 (Sjögren *et al.*, 1996). The decreasing trends of these two softwood "fingerprints" in core CPF from the 1880s to the 1940s indicate that wood burning during this period had declined relative to other energy sources in NYC. This is consistent with the US energy usage trends in the EIA data. In three bottom sections predating 1900, elevated ratios of Ret/(Ret + Chy) (0.55–0.82) together with a much higher abundance of 1,7-DMP than 2,6-DMP and 1,6-DMP (Figure 28) indicate a strong softwood combustion input. Softwood is thought to have been used largely when wood was the primary energy source over a century ago. The corresponding Fl/(Fl + Py) ratios in sediments prior to ~1900 range from 0.62 to 0.72, which are just above the range of reported values (0.46–0.55) in initial wood combustion source emissions (Yunker *et al.*, 2002 and references therein). Some studies have also shown similar high ratios in remote lakes (Yunker and Macdonald, 2003; Fernandez *et al.*, 2000).

After 1900, 1,7/(1,7 + 2,6)-DMP and Fl/(Fl + Py) ratios decline gradually until the 1940s (Figure 28). A similar decrease is also observed in some remote lake cores (Fernandez *et al.*, 2000). Contrary to these gradual declining trends, the Ret/(Ret + Chy) ratio declines sharply in the early 1900s and then remains at a relatively low level from the 1910s to the present (Figure 28). Despite the different rates of decline, temporal trends of both the DMP and Ret/(Ret + Chy) ratios suggest a decreasing contribution from softwood combustion through the early 1900s. During the same period, PAH concentrations increase rapidly (~31-fold), whereas SH and UCM levels remain low with CPI ratios ~5, all indicating a minor petroleum-related hydrocarbon source. An increase in hardwood combustion could lead to the reduction of 1,7/(1,7 + 2,6)-DMP and Ret/(Ret + Chy) ratios due to their relative lower ratios in hardwood combustion emissions than softwood (Fine *et al.*, 2001; Benner *et al.*, 1989); however, the extremely rapid increase in PAH levels over just 15 years together with the fact that NYC population only increased by ~20% during this period suggests that conversion into hardwood burning cannot account for the hydrocarbon signatures preserved in core CPF. We attribute this rapid PAH increase to the combustion of coal for residential heating and industrial usage at this period. Relative to wood and petroleum combustion, coal combustion can produce a greater yield of PAHs (over 20-fold) and lesser amount of UCM per unit of fuel burned (Fine *et al.*, 2001; Benner *et al.*, 1995; Oros and Simoneit, 2000). In addition, the ratios of several indicators [Fl/(Fl + Py); 1,7/2,6-DMP, Ret/(Ret + Chy)] are in the range of coal combustion.

From ~1915 until ~1935, the gradual decrease in 1,7/(1,7 + 2,6)-DMP ratios (Figure 28) suggests a continuous reduction in coal and/or wood combustion. Ret/(Ret + Chy) remains low, indicating negligible contributions of softwood combustion. Thus the decline in Fl/(Fl + Py) and 1,7/(1,7 + 2,6)-DMP ratios most likely resulted from either a decrease in coal combustion, a relative increase in petroleum combustion, or both. Additionally, the temporal profile of U/R exhibits a first-order increase in this period, indicating a continuous increase in petroleum combustion. The gradual enhancement in petroleum combustion from the 1910s to 1930s probably results from the rapid increase in the number of motor vehicles after the first automobile assembly line was put into operation in 1913 (data are available online, see AACA, 2006).

Low ratios of Ret/(Ret + Chy) post-1910 span both the coal-dominant era and the following period of rapid increases in petroleum combustion, demonstrating this ratio's lack of sensitivity to these two sources. The similar trends of

1,7/(1,7 + 2,6)-DMP and Fl/(Fl + Py) observed throughout core CPF (Figure 28) were not expected. As shown in Table 3, ratios of 1,7/(1,7 + 2,6)-DMP vary considerably among different combustion sources. Temporal trends of the DMP ratio in the core are consistent with the history of softwood combustion, coal combustion, and motor vehicle usage discussed above. Although published Fl/(Fl + Py) data (Yunker *et al.*, 2002) suggest similar ratios (>0.5) for wood and coal combustion sources, our analyses of CPF suggest a higher ratio of Fl/(Fl + Py) from local wood combustion in the period ~1870–1900s (0.72) than from local coal combustion in the early 1900s (0.62).

From the 1930s to 1960s, the ratios of Fl/(Fl + Py), Ret/(Ret + Chy), and 1,7/2,6-DMP suggest that major PAH inputs to the lake are from combustion of petroleum and coal. In contrast to TSH, PAH levels do not peak in the mid-1940s and -1960s, suggesting that SH and PAH may have different major sources for CPL. From ~1935 to ~1944, SH levels increase 10-fold, whereas PAH levels remain relatively constant, suggesting that the rapid increase in SH levels is not likely to have been caused by an increase in motor vehicle emissions. One potential source of the SH is incineration of municipal solid waste (MSW). Elevated levels of Pb, Sn, and Zn observed in another CPL core (CPH) between the late 1930s and the mid-1960s have been attributed to MSW emissions prior to the advent of effective pollution controls (Chillrud *et al.*, 1999). On the basis of this correlation, emissions from MSW incinerators would appear to contain a high abundance of UCM and comparable PAHs relative to other combustion sources. Abundant UCM was observed in a study of emissions from waste incineration plants (Jay and Stieglitz, 1995).

Although various molecular indicators implicate petroleum combustion as the dominant pyrogenic source for the lake since the 1970s, some subtle variations in molecular ratios are noteworthy. There are higher 1,7/(1,7 + 2,6)-DMP and Ret/(Ret + Chy) ratios in some periods (e.g., the late 1960s, late 1970s, and early 1990s) relative to others (Figure 28). The higher ratio of 1,7/(1,7 + 2,6)-DMP indicates an increase in either coal or softwood combustion; however, the concurrent rise in the Ret/(Ret + Chy) ratio, which is only responsive to wood combustion, suggests that increased wood combustion is a likely cause. This theory is supported by the EIA data, which records a period of increased combustion of wood in the late 1970s in the United States.

In CPL, the inverse relationship between CPI and total alkanes (Figure 2b) indicates the reliability of CPI as a tool for differentiating SH contributions of higher plants from combustion sources. The U/R ratio is a sensitive indicator for petroleum usage. Fl/(Fl + Py) and 1,7/(1,7 + 2,6)-DMP can be used to study the relative contributions of three major PAH combustion sources (softwood, coal, and petroleum), while the Ret/(Ret + Chy) can be used to distinguish softwood combustion source from other combustion sources. A/(Pa + A) and BaA/(BaA + Chy) are less sensitive to source differences; therefore interpretations based on these two ratios should be made with caution. The remaining four ratios ($C_0/(C_0 + C_1)_{P/A}$, $C_0/(C_0 + C_1)_{F/P}$, Par/(Par + Alkyl), and Ring456/TPAH) are sensitive indicators for differentiating petrogenic and pyrogenic PAH sources, but additional studies are necessary to fully exploit their potential for pyrogenic PAH source apportionment.

In summary, SHs and PAHs have been quantified in a sediment core obtained from CPL, NYC. Ratios of 1,7-DMP to 1,7-DMP plus 2,6-DMP (1,7/(1,7 + 2,6)-DMP), retene to retene plus chrysene (Ret/(Ret + Chy)), and fluoranthene to fluoranthene plus pyrene (Fl/(Fl + Py)) provide additional source discrimination throughout the core. Results show that the ratio U/R is sensitive to petroleum inputs, Ret/(Ret + Chy) is responsive to contributions from softwood combustion, whereas both Fl/(Fl + Py) and 1,7/(1,7 + 2,6)-DMP can be used to discriminate among wood, coal, and petroleum combustion sources. Combined use of these ratios suggests that in NYC, wood combustion dominated a hundred years ago, with a shift to coal combustion occurring from the 1900s to the 1950s. Petroleum use began around the 1920s and has dominated since the 1940s.

12.8 SYNTHESIS

The unwanted release of petrogenic and pyrogenic hydrocarbon contaminants in the environment represents a major global challenge for pollution prevention, monitoring, and cleanup of aquatic systems. Understanding the geochemical behavior of these compounds in aqueous environments is a prerequisite to identifying point and nonpoint sources, understanding their release history, or predicting their transport and fate in the environment. This review examined the physical and chemical properties of HMW hydrocarbons and environmental conditions and processes that are relevant to the understanding of their geochemical behavior, with an explicit focus on PAHs. PAHs represent the components of greatest biological concern amongst pyrogenic and petrogenic hydrocarbon contaminants. The molecular diversity of PAHs is also useful for holistically characterizing crude oil and petrogenic sources and their weathering history in aquatic systems. The majority of PAHs residing in surface aquatic systems, including lake, river, and ocean sediments, and groundwater aquifers are derived from pyrolysis of petroleum, petroleum products, wood, and other biomass materials. Finally, HMW hydrocarbons share the hydrophobic and lipophilic behavior of PAH; hence, all the discussions of the geochemical behavior of PAHs are indeed relevant to HMW hydrocarbons in general.

The molecular distribution and compound-specific carbon isotopic composition of hydrocarbons can be used to qualify and quantify their sources and pathways in the environment. Molecular source apportionment borrows from molecular methods that were developed and applied extensively for fundamental oil biomarker studies, oil–oil and oil source rock correlation analysis. Additionally, petroleum refinement produces well-defined mass and volatility ranges that are used as indicators of specific petroleum product sources in the environment. Compound-specific carbon isotopic measurement is a more recent addition to the arsenal of methods for hydrocarbon source apportionment. Carbon isotopic discrimination of *n*-alkanes, biomarkers, and PAHs has shown that the technique is highly complementary to molecular apportionment methods.

HMW hydrocarbons are largely hydrophobic and lipophilic; hence, they become rapidly associated with solid phases in aquatic environments. This solid partitioning is particularly pronounced in organic matter-rich sediments, with the result that the bulk solid–water partition coefficient is generally a linear function of the fraction of organic carbon or organic matter present in the sediments. Under favorable conditions, organic-rich sediments preserve potential records of the sources and flux of hydrocarbons in a given depositional system. The bulk molecular compositions of hydrocarbons are altered in aquatic environments as a result of selective dissolution, volatilization, sorption, photolysis, and biodegradation. This review has summarized our understanding of the effect of these "weathering reactions" on molecular and isotopic composition. Understanding hydrocarbon weathering is important for understanding sources of past deposition, and they may also provide insights on the timing of hydrocarbon release. Carbon isotope compositions of HMW hydrocarbons are not significantly altered by most known weathering and diagenetic reactions, enabling the use of isotopic signatures as source indicators of highly weathered hydrocarbons.

Several additional areas of research will likely yield major additional advances in the understanding of petrogenic and pyrogenic hydrocarbon geochemistry. For example, developments in compound-specific isotope analysis of D/H ratios in individual hydrocarbon compounds will provide additional constraints on the origin, mixing, and transformation of hydrocarbons in shallow aquatic systems. Bacterial degradation experiments similar to those presented in this review have been completed, and no D/H fractionations in naphthalene and fluoranthene were observed (Abrajano, unpublished). This contrasts with large, but bacterial species-dependent, D/H fractionation already observed for toluene (e.g., Ward *et al.*, 2000; Morasch *et al.*, 2001). The dependence of D/H fractionation on the degrading microorganisms and the environment of degradation for monoaromatic hydrocarbons and PAHs are clearly fertile areas of further investigation. Similarly, the characterization of D/H signatures of PAH sources is a topic of ongoing interest. Compound-specific measurement of δD of fire soot (12 samples) reveals a wide range from −40‰ to −225‰, which overlaps with the observed range (−50‰ to −175‰) for crankcase oil (18 samples) (Abrajano, unpublished). Whereas source discrimination of the type shown for carbon isotopes does not appear promising for hydrogen, the factors governing the acquisition of D/H signatures of PAHs will likely yield unique insight into their pathways and transformation in the environment (e.g., Morasch *et al.*, 2001).

Likewise, the continued development of compound-specific radiocarbon dating techniques ushers an exciting dimension to the study of PAHs and other hydrocarbons in modern sediments (e.g., Eglinton *et al.*, 1997; Eglinton and Pearson, 2001; Pearson *et al.*, 2001; Reddy *et al.*, 2002). In particular, compound-specific radiocarbon determination offers a clear distinction to be made between "new carbon" (e.g., wood burning) and "old carbon" (petroleum and petroleum combustion).

Finally, there is now increasing recognition of a critical need for detailed characterization of the mechanisms of solid-phase association of hydrocarbons in sediments or aquifer solid phase. The likelihood that the sorption/desorption behavior, bioavailability, and overall cycling of hydrocarbons depend critically on the nature of solid-phase association provides a strong impetus for future work.

ACKNOWLEDGMENTS

Financial support for our continuing studies of hydrocarbon sources and pathways in surface and groundwater systems from the US National Science Foundation (Hydrologic Sciences, EAR-0073912) is gratefully acknowledged. The RPI Isotope Hydrology and Biogeochemistry Facility was also partially supported by the US National Science Foundation (BES-9871241). The St. John's Harbor study was initiated through generous support from the Natural Science and Engineering Research Council and the Department of Fisheries and Oceans (Canada).

REFERENCES

Aboul-Kassim T. A. T. and Simoneit B. R. T. (1995) Petroleum hydrocarbon fingerprinting and sediment transport assessed by molecular biomarker and multivariate statistical analysis in the eastern harbour of Alexandria, Egypt. *Mar. Pollut. Bull.* **30**, 63–73.

Aboul-Kassim T. A. T. and Simoneit B. R. T. (eds.) (2001) Pollutant-solid phase interactions: mechanism, chemistry and modeling. In *The Handbook of Environmental Chemistry*. Springer, Berlin, vol. 5, part E.

Accardi-Dey A. M. and Gschwent P. (2003) Reinterpreting literature sorption data considering both absorption into organic carbon and adsorption onto black carbon. *Environ. Sci. Technol.* **37**, 99–106.

Ahad J. M. E., Lollar B. S., Edwards E. A., Slater G. F., and Sleep B. E. (2000) Carbon isotope fractionation during anaerobic biodegradation of toluene: implications for intrinsic bioremediation. *Environ. Sci. Technol.* **34**, 892–896.

Alsberg T., Stenberg U., Westerholm R., Strandell M., Rannug U., Sundvall A., Romert L., Bernson V., Pettersson B., Toftgard R., Franzen B., Jansson M., Gustafsson J. A., Egeback K. E., and Tejle G. (1985) Chemical and biological characterization of organic material from gasoline exhaust particles. *Environ. Sci. Technol.* **19**, 43–49.

Antique Automobile Club of America (AACA). (2006) Enhancement in petroleum combustion from the 1910s to 1930s, www.aaca.org (accessed on August 2006).

Atlas R. M. and Bartha R. (1992) Hydrocarbon biodegradation and oil spill bioremediation. *Adv. Microb. Ecol.* **12**, 287–338.

Badger G. M., Buttery R. G., Kimber R. W., Lewis G. E., Moritz A. G., and Napier I. M. (1958) The formation of aromatic hydrocarbons at high temperatures. Part 1: Introduction. *J. Chem. Soc.* **1958**, 2449–2461.

Baek S. O., Field R. A., Goldstone M. E., Kirk P. W., Lester J. N., and Perry R. (1991) A review of atmospheric polycyclic aromatic hydrocarbons: sources, fate and behaviour. *Water Air Soil Pollut.* **60**, 279–300.

Baker J. E. and Eisenreich S. J. (1990) Concentrations and fluxes of polycyclic aromatic hydrocarbons and polychlorinated biphenyls across the air-water interface of Lake Superior. *Environ. Sci. Technol.* **24**, 342–352.

Ballentine D. C., Macko S. A., Turekian V. C., Gilhooly W. P., and Martincigh B. (1996) Tracing combustion-derived atmospheric pollutants using compound-specific carbon isotope analysis. *Org. Geochem.* **25**, 97–108.

Barnsley E. A. (1975) The bacterial degradation of fluoranthene and benzo(a)pyrene. *Can. J. Microbiol.* **21**, 1004–1008.

Beek B. (2001) *The Handbook of Environmental Chemistry. Park K: Biodegradation and Persistence*. Springer, Berlin, vol. 2.

Begeman C. R. and Burgan J. C. (1970) Polynuclear hydrocarbon emission from automotive engines. SAE Paper 700469, Society of Automotive Engineers, Detroit, MI.

Behymer T. D. and Hites R. A. (1985) Photolysis of polycyclic aromatic hydrocarbons adsorbed on simulated atmospheric particulates. *Environ. Sci. Technol.* **19**, 1004–1006.

Behymer T. D. and Hites R. A. (1988) Photolysis of polycyclic aromatic hydrocarbons adsorbed on fly ash. *Environ. Sci. Technol.* **22**, 1311–1319.

Benner B. A., Gordon G. E., and Wise S. A. (1989) Mobile sources of atmospheric polycyclic aromatic hydrocarbons: a roadway tunnel study. *Environ. Sci. Technol.* **23**, 1269–1278.

Benner B. A., Wise S. A., Currie L. A., Klouda G. A., Klinedinst D. B., Zweidinger R. B., Stevens R. K., and Lewis C. W. (1995) Distinguishing the contributions of residential wood combustion and mobile source emissions using relative concentrations of dimethylphenanthrene isomers. *Environ. Sci. Technol.* **29**, 2382–2389.

Benner R., Fogel M. L., Sprague E. K., and Hodson R. E. (1987) Depletion of ^{13}C in lignin and its implications for stable carbon isotope studies. *Nature* **329**, 708–710.

Bieger T., Hellou J., and Abrajano T. A. (1996) Petroleum biomarkers as tracers of lubricating oil contamination. *Mar. Pollut. Bull.* **32**, 270–274.

Bjorseth A. and Ramdahl T. (eds.) (1983) Sources and emissions of PAHs. In *Handbook of Polycyclic Aromatic Hydrocarbons. Emission Sources and Recent Progress in Analytical Chemistry*. Dekker, New York.

Blumer M., Blumer W., and Reich T. (1977) Polycyclic aromatic hydrocarbons in soils of a mountain valley: correlation with highway traffic and cancer incidence. *Environ. Sci. Technol.* **11**(12), 1082–1084.

Bopp R. F., Chillrud S. N., Shuster E. L., and Simpson H. J. (2005) Contaminant chronologies. In *The Hudson River Estuary* (eds. J. Levinton and J. Waldman). Cambridge University Press.

Bopp R. F., Chillrud S. N., Simpson H. J., and Virgillio A. (1993a) *Pollutant chronologies in New York Harbor, Raritan Bay and Jamaica Bay*. Final Report, New Jersey Sea Grant Contract NJMSC R/E-17P.

Bopp R. F., Gross M. L., Tong H., Simpson H. J., Monson S. J., Deck B. L., and Moser F. C. (1991) A major incident of dioxin contamination: sediments of New Jersey estuaries. *Environ. Sci. Technol.* **25**, 951–956.

Bopp R. F., Simpson H. J., Chillrud S. N., and Robinson D. W. (1993b) Sediment-derived chronologies of persistent contaminants in Jamaica Bay. *New York Estuaries* **16**(3B), 608–616.

Bopp R. F., Simpson H. J., Olsen C. R., and Kostyk N. (1981) Polychlorinated biphenyls in sediments of the tidal Hudson River, New York. *Environ. Sci. Technol.* **15**, 210–216.

Bosma T. N. P., Harms H., and Zehnder J. B. (2001) Biodegradation of xenobiotics in the environment and technosphere. In *The Handbook of Environmental Chemistry. Park K: Biodegradation and Persistence* (ed. B. Beek). Springer, Berlin, vol. 63.

Broman D., Coimsjo A., Naf C., and Zebuhr Y. (1988) A multi-sediment trap study on the temporal and spatial variability of polycyclic aromatic hydrocarbons and lead in an anthropogenic influenced archipelago. *Environ. Sci. Technol.* **22**(10), 1219–1228.

Brown G. and Maher W. (1992) The occurrence, distribution and sources of polycyclic aromatic hydrocarbons in the sediments of the Georges river estuary, Australia. *Org. Geochem.* **18**(5), 657–658.

Bucheli T. D. and Gustafsson O. (2000) Quantification of the soot-water distribution coefficient of PAHs provides mechanistic basis for enhanced sorption observations. *Environ. Sci. Technol.* **34**, 5144–5151.

Canton L. and Grimalt J. O. (1992) Gas chromatographic-mass spectrometric characterization of polycyclic aromatic hydrocarbon mixtures in polluted coastal sediments. *J. Chromatogr.* **607**, 279–286.

Cerniglia C. E. (1991) Biodegradation of organic contaminants in sediments: overview and examples with polycyclic aromatic hydrocarbons. In *Organic Substances and Sediments in Water* (ed. R. A. Baker). Lewis Publishers, Boca Raton, vol. 3, pp. 267–281.

Chillrud S. N., Bopp R. F., Simpson H. J., Ross J. M., Shuster E. L., Chaky D. A., Walsh D. C., Choy C. C., Tolley L. R., and Yarme A. (1999) Twentieth century atmospheric metal fluxes into Central Park Lake, New York City. *Environ. Sci. Technol.* **33**, 657–662.

Chiou C. T. and Kyle D. E. (1998) Deviations from sorption linearity on soils of polar and nonpolar organic compounds at low relative concentrations. *Environ. Sci. Technol.* **32**, 338–343.

Chiou C. T., McGroddy S. E., and Kile D. E. (1998) Partition characteristics of polycyclic aromatic hydrocarbons on soils and sediments. *Environ. Sci. Technol.* **32**, 264–269.

Chiou C. T., Peters L. J., and Freed V. H. (1979) Physical concept of soil-water equilibria for nonionic organic compounds. *Science* **206**, 831–832.

Chiou C. T., Porter P. E., and Schmeddling D. W. (1983) Partition equilibria of nonionic organic compounds between soil organic matter and water. *Environ. Sci. Technol.* **17**, 227–231.

Crawford D. W., Bonnevie N. L., and Wenning R. J. (1995) Sources of pollution and sediment contamination in Newark Bay, New Jersey. *Ecotoxicol. Environ. Saf.* **30**, 85–100.

Currie L. A., Klouda G. A., Benner B. A., Garrity K., and Eglinton T. I. (1999) Isotopic and molecular marker validation, including direct molecular 'dating' (GC/AMS). *Atmos. Environ.* **33**, 2789–2806.

Diegor E., Abrajano T., Patel T., Stehmeir L., Gow J., and Winsor L. (2000) Biodegradation of aromatic hydrocarbons: microbial and isotopic studies, goldschmidt conference (Cambridge publications). *J. Conf. Abstr.* **5**, 349–350.

Douglas G. S., Bence E. A., Prince R. C., McMillen S. J., and Butler E. L. (1996) Environmental stability of selected petroleum hydrocarbon source and weathering ratios. *Environ. Sci. Technol.* **30**, 2332–2339.

Dowling L. M., Boreham C. J., Hope J. M., Marry A. M., and Summons R. E. (1995) Carbon isotopic composition of ocean transported bitumens from the coastline of Australia. *Org. Geochem.* **23**, 729–737.

Eganhouse R. P. (1997) *Molecular markers in environmental geochemistry, ACS Symposium Series.* American Chemical Society, Washington, DC.

Eglinton T. I., Benitez-Nelson B. C., Pearson A., McNichol A. P., Bauer J. E., and Druffel E. R. M. (1997) Variability in radiocarbon ages of individual organic compounds from marine sediments. *Science* **277**, 796–799.

Eglinton T. I. and Pearson A. (2001) Ocean process tracers: single compound radiocarbon measurements. In *Encyclopedia of Ocean Sciences* (eds. J. Steel, S. Thorpe, and K. Turekian). Academic Press, London.

Eisenreich S. J. and Strachan W. M. J. (1992) Estimating atmospheric deposition of toxic substances to the Great Lakes—an update. *A Workshop Held at Canadian Centre for Inland Waters*, Burlington, ON, January 31-February 2.

Ellison G. B., Tuck A. F., and Vaida V. (1999) Atmospheric processing of organic aerosols. *J. Geophys. Res.* **104**, 11633–11641.

Energy Information Administration (EIA) (2001) *Annual Energy Review.* Report DOE/EIA-0384, http://www.demographia.com/db-nyc2000.html.

Engel M. H. and Macko S. A. (1993) *Principles and Applications. Organic Geochemistry.* Plenum, New York.

Faksness L. G., Weiss H. M., and Daling P. S. (2002) *Revision of the Nordtest methodology for oil spill identification.* Sintef Applied Chemistry Report STF66 A02028, N-7465 Trondheim, Norway.

Fernandez P., Vilanova R. M., Appleby P., and Grimalt J. O. (2000) The historical record of atmospheric pyrolytic pollution over Europe registered in the sedimentary PAH from remote mountain lakes. *Environ. Sci. Technol.* **34**, 1906–1913.

Fine P. M., Cass G. R., and Simoneit B. R. T. (2001) Chemical characterization of fine particle emissions from fireplace combustion of woods grown in the Northeastern United States. *Environ. Sci. Technol.* **35**, 2665–2675.

Fingas M. F. (1995) A literature review of the physics and predictive modeling of oil spill evaporation. *J. Hazards Mater.* **42**, 157–175.

Freeman D. J. and Cattell F. C. (1990) Wood burning as a source of polycylic aromatic hydrocarbons. *Environ. Sci. Technol.* **24**, 1581–1585.

Freeman K. H. (1991) The carbon isotopic composition of individual compounds from ancient and modern depositional environments. PhD Thesis, Indiana University, pp. 42–92.

Fukushima M., Oba K., Tanaka S., Kayasu K., Nakamura H., and Hasebe K. (1997) Elution of pyrene from activated carbon into an aqueous system containing humic acid. *Environ. Sci. Technol.* **31**, 2218–2222.

Gearing P. J., Gearing J. N., Pruell R. J., Wade T. L., and Quinn J. G. (1980) Partitioning of no. 2 fuel oil in controlled estuarine ecosystems, sediments and suspended particulate matter. *Environ. Sci. Technol.* **14**, 1129–1136.

Gibson D. T. and Subramanian V. (1984) Microbial degradation of aromatic hydrocarbons. In *Microbial Degradation of Organic Compounds* (ed. D. T. Gibson). Dekker, New York, pp. 181–252.

Gigliotti C. L., Brunciak P. A., Dachs J., Glenn T. R., IV., Nelson E. D., Totten L. A., and Eisenreich S. J. (2002) Air–water exchange of polycyclic aromatic hydrocarbons in the New York-New Jersey, USA, Harbor Estuary. *Environ. Toxicol. Chem.* **21**(2), 235–244.

Goldberg E. D. (1985) *Black Carbon in the Environment: Properties and Distribution.* Wiley, New York.

Gordon R. J. (1976) Distribution of airborne polycyclic aromatic hydrocarbons throughout Los Angeles. *Environ. Sci. Technol.* **10**, 370–373.

Grimmer G., Jacob J., Dettbarn G., and Naujack K. W. (1985) Determination of polycyclic aromatic hydrocarbons, azaarenes and thiaarenes emitted from coal-fired residential furnaces by gas chromatography/mass spectrometry. *Fresen. Z. Anal. Chem.* **332**, 595–602.

Gustafson K. E. and Dickhut R. M. (1997) Gaseous exchange of polycyclic aromatic hydrocarbons across the air-water interface of southern Chesapeake Bay. *Environ. Sci. Technol.* **31**, 1623–1629.

Gustafsson O. and Gschwent P. M. (1997) Soot as a strong partition medium for polycyclic aromatic hydrocarbons in aquatic systems. In *Molecular Markers in Environmental Geochemistry, ACS Symposium Series* (ed. R. P. Eganhouse). American Chemical Society, Washington, DC, pp. 365–381.

Hammer B. T., Kelley C. A., Coffin R. B., Cifuentes L. A., and Mueller J. G. (1998) $\delta^{13}C$ values of polycyclic aromatic hydrocarbons collected from two creosote-contaminated sites. *Chem. Geol.* **152**, 43–58.

Hase A. and Hites R. A. (1976) On the origins of polycyclic aromatic hydrocarbons in recent sediments: biosynthesis by anaerobic bacteria. *Geochim. Cosmochim. Acta* **40**, 1141–1143.

Hayes J. M., Freeman K. H., Popp B. N., and Hoham C. H. (1989) Compound-specific isotopic analysis: a novel tool for reconstruction of ancient biogeochemical processes. *Adv. Org. Geochem.* **16**(4–6), 1115–1128.

Heider J., Spormann A. M., Beller H. R., and Widdel F. (1999) Anaerobic bacterial metabolism of hydrocarbons. *FEMS Microbiol. Rev.* **22**, 459–473.

Hellou J., Steller S., Leonard J., and Albaiges J. (2002) Alkanes, terpanes, and aromatic hydrocarbons in surficial sediments of Halifax Harbor. *Polycycl. Aromat. Comp.* **22**, 631–641.

Hinga K. R. (1984) The fate of polycyclic aromatic hydrocarbons in enclosed marine ecosystems. PhD Thesis, University of Rhode Island.

Hites R. A., LaFlamme R. E., Windsor J. G., Farrington J. W., and Deuser W. G. (1980) Polycyclic aromatic hydrocarbons in an anoxic sediment core from the Pettaquamscutt River (Rhode Island, USA). *Geochim. Cosmochim. Acta* **44**, 873–878.

Hoffman E. J., Mills G. L., Latimer J. S., and Quinn J. G. (1984) Urban runoff as a source of polycyclic hydrocarbons to coastal environments. *Environ. Sci. Technol.* **18**, 580.

Holt B. and Abrajano T. A. (1991) Chemical and isotopic alteration of organic matter during stepped combustion. *Anal. Chem.* **63**, 2973–2978.

Hostetler F. and Kvenvolden K. (2002) Alkylcyclohexanes in envirnomental geochemistry. *Environ. Forensics* **3** (3–4), 293–301, doi:10.1006/enfo.2002.0100.

Howsam M. and Jones K. C. (1998) Sources of PAHs in the environment. In *PAHs and Related Compounds. The Handbook of Environmental Chemistry* (ed. A. H. Neilson). Springer, Berlin, vol. 3, part I, pp. 138–174.

Hontley S. L., Bonnevie N. L., and Wenning R. J. (1995) Polycyclic aromatic hydrocarbon and petroleum hydrocarbon contamination in sediment from the Newark Bay Estuary, New Jersey. *Arch Environ. Contam. Toxicol.* **28**, 93–107.

Jay K. and Stieglitz L. (1995) Identification and quantification of volatile organic components in emissions of waste incineration plants. *Chemosphere* **30**, 1249–1260.

Jensen T. E. and Hites R. A. (1983) Aromatic diesel emissions as function of engine conditions. *Anal. Chem.* **55**, 594–599.

Kaplan I. R., Galperin Y., Lu S. T., and Lee R. P. (1997) Forensic environmental geochemistry: differentiation of fuel types, their sources and release time. *Org. Geochem.* **27**, 289–317.

Karapanagioti H., Kleineidam S., Sabatini D., Grathwohl P., and Ligouis B. (2000) Impacts of heterogeneous organic matter on phenanthrene sorption: equilibrium and kinetic studies with aquifer material. *Environ. Sci. Technol.* **34**, 406–414.

Karickhoff S. W. (1981) Semi-empirical estimation of sorption of hydrophobic pollutants on natural sediments and soils. *Chemosphere* **10**, 833–846.

Karickhoff S. W., Brown D. S., and Scott T. A. (1979) Sorption of hydrophobic pollutants on natural sediments. *Water Res.* **13**, 241–248.

Kennicutt M. C., II., Brooks J. M., and McDonald T. J. (1991) Origins of hydrocarbons in Bering Sea sediments. I: Aliphatic hydrocarbons and fluorescence. *Org. Geochem.* **17** (1), 75–83.

Killops S. D. and Howell V. J. (1988) Sources and distribution of hydrocarbons in Bridgewater Bay (Severn Estuary, UK) intertidal surface sediments. *Estuar. Coast. Shelf Sci.* **27**, 237–261.

Kleineidam S., Rugner H., Ligouis B., and Grathwol P. (1999) Organic matter facie and equilibrium sorption of phenanthrene. *Environ. Sci. Technol.* **33**, 1637–1644.

Kleineidam S., Schuth C. H., and Grathwol P. (2002) Solubility-normalized combined adsorption-partitioning sorption isotherms for organic pollutants. *Environ. Sci. Technol.* **36**, 4689–4697.

Knap A. H. and Williams P. J. L. (1982) Experimental studies to determine the fate of petroleum hydrocarbons in refinery effluent on an estuarine system. *Environ. Sci. Technol.* **16**, 1–4.

Kvenvolden K. A., Carlson P. R., Threlkeld C. N., and Warden A. (1993b) Possible connection between two Alaskan catastrophies occurring 25 yr apart (1964 and 1989). *Geology* **21**, 813–816.

Kvenvolden K. A., Hostettler F. D., Carlson P. R., Rapp J. B., Threlkeld C., and Warden A. (1995) Ubiquitous tar balls with a California source signature on the shorelines of Prince William Sound, Alaska. *Environ. Sci. Technol.* **29** (10), 2684–2694.

Kvenvolden K. A., Hostettler F. D., Rapp J. B., and Carlson P. R. (1993a) Hydrocarbons in oil residue on beaches of islands of Prince William Sound, Alaska. *Mar. Pollut. Bull.* **26**, 24–29.

Laflamme R. E. and Hites R. A. (1978) The global distribution of polycyclic aromatic hydrocarbons in recent sediments. *Geochim. Cosmochim. Acta* **42**, 289–303.

Lake J. L., Norwood C., Dimock D., and Bowen R. (1979) Origins of polycyclic aromatic hydrocarbons in estuarine sediments. Geochim. *Cosmochim. Acta* **43**, 1847–1854.

Lane D. A. (1989) The fate of polycyclic aromatic compounds in the atmosphere during sampling. *Chemical Analysis of Polycyclic Aromatic Compounds, Chemical Analysis Series* (ed. T. Vo-Dinh). Wiley-Interscience, New York, vol. 101.

Latimer J. S., Hoffman E. J., Hoffman G., Fasching J. L., and Quinn J. G. (1990) Sources of petroleum hydrocarbons in urban runoff. *Water Air Soil Pollut.* **52**, 1–21.

Lee M. L., Novotny M. V., and Bartle K. D. (1981) *Analytical Chemistry of Polycyclic Aromatic Compounds.* Academic Press, New York.

Lee R. F., Gardner W. S., Anderson J. W., Blaylock J. W., and Barwell C. J. (1978) Fate of polycyclic aromatic hydrocarbons in controlled ecosystems enclosures. *Environ. Sci. Technol.* **12**, 832–838.

Lima A. C., Eglinton T. I., and Reddy C. M. (2003) High resolution record of pyrogenic polycyclic aromatic hydrocarbon deposition during the 20th century. *Environ. Sci. Technol.* **37**, 53–61.

Lipiatou E. and Saliot A. (1991) Fluxes and transport of anthropogenic and natural polycyclic aromatic hydrocarbons in the western Mediterranean sea. *Mar. Chem.* **32**, 51–71.

Lipiatov E. and Saliot A. (1992) Biogenic aromatic hydrocarbon geochemistry in the Rhone river delta and in surface sediments from the open north-western Mediterranean Sea. *Estuar. Coast. Shelf Sci.* **34**(5), 515–531.

Lohmann F., Trendel J. M., Hetru C., and Albrecht P. (1990) C-29 tritiated β amyrin: chemical synthesis aiming at the study of aromatization processes in sediments. *J. Label. Comp. Radiopharm.* **28**, 377–386.

Long E. R., Macdonald D. D., Smith S. L., and Calder F. D. (1995) Incidence of adverse biological effects within ranges of chemical concentrations in marine and estuarine sediments. *Environ. Manage.* **19**(1), 81–97.

Lun R., Lee K., De Marco L., Nalewajko C., and Mackay D. (1998) A model of the fate of polycyclic aromatic hydrocarbons in the Saguenay ford. *Environ. Toxicol. Chem.* **17**, 333–341.

Luthy R. G., Aiken G. R., Brusseau M. L., Cunningham S. D., Gschwent P. M., Pignatello J. J., Reinhard M., Traina S. J., Weber W. J., and Westall J. (1997) Sequestration of hydrophobic organic contaminants by geosorbents. *Environ. Sci. Technol.* **31**, 3341–3347.

MacKay A. A. and Gschwent P. (2001) Enhanced concentration of PAHs in a coal tar site. *Environ. Sci. Technol.* **31**(35), 1320–1328.

Mackay D. (1998) Multimedia mass balance models of chemical distribution and fate. In *Chapter 8 of Ecotoxicology* (eds. G. Schuurmann and B. Markert). Wiley, New York and Spektrum, Berlin, pp. 237–257.

Mackay D. and Callcott D. (1998) Partitioning and physical chemical properties of PAHs. PAHs and related compounds. In *The Handbook of Environmental Chemistry, Volume 3, Part I* (ed. A. H. Neilson). Springer, Berlin, Heidelberg, pp. 325–346.

Mackay D., Shiu W. Y., and Ma K. C. (1992) *Illustrated Handbook of Physical-Chemical Properties and Environmental Fate for Organic Chemicals, volume II, 1992.* Lewis, Boca Raton, FL.

MacKenzie A. S. (1984) Application of biological markers in petroleum geochemistry. In *Advances in Petroleum Geochemistry* (eds. J. Brooks and D. H. Welte). Academic Press, London, vol. 1, pp. 115–214.

Mansuy L., Philp R. P., and Allen J. (1997) Source identification of oil spills based on the isotopic composition of individual components in weathered oil samples. *Environ. Sci. Technol.* **31**, 3417–3425.

Mazeas L. and Budzinski H. (2001) Polycyclic aromatic hydrocarbon $^{13}C/^{12}C$ ratio measurement in petroleum and marine sediments: application to standard reference material and a sediment suspected of contamination from Erika oil spill. *J. Chromatogr. A* **923**, 165–176.

Mazeas L., Budzinski H., and Raymond N. (2002) Absence of stable carbon isotope fractionation of saturated and polycyclic aromatic hydrocarbons during aerobic bacterial degradation. *Org. Geochem.* **33**, 1259–1272.

McElroy A. E., Farrington J. W., and Teal J. M. (1985) Bioavailability of polycyclic aromatic hydrocarbons in the aquatic environment. In *Metabolism of Polycyclic Aromatic Hydrocarbons in the Aquatic Environment* (ed. U. Varanasi). CRC Press, Boca Raton, FL.

McRae C., Snape C. E., Sun C. G., Fabbri D., Tartari D., Trombini C., and Fallick A. (2000) Use of compound-specific stable isotope analysis to source anthropogenic natural gas-derived polycyclic aromatic hydrocarbons in lagoon sediments. *Environ. Sci. Technol.* **34**, 4684–4686.

McRae C., Sun C. G., Snape C. E., Fallick A., and Taylor D. (1999) $\delta^{13}C$ values of coal-derived PAHs from different processes and their application to source apportionment. *Org. Geochem.* **30**, 881–889.

McVeety B. D. and Hites R. A. (1988) Atmospheric deposition of polycyclic aromatic hydrocarbons to water surfaces: a mass balance approach. *Atmos. Environ.* **22**(3), 511–536.

Meador J., Stein J., Reichert W., and Varanasi U. (1995) Bioaccumulation of polycyclic aromatic hydrocarbons by marine organisms. *Rev. Environ. Contamin. Toxicol.* **143**, 80–164.

Means J. C. and Wijayaratne R. (1982) Role of natural colloids in the transport of hydrophobic pollutants. *Science* **215**, 968.

Meckenstock R. U., Morasch B., Wartmann R., Schinck B., Annweller E., Michaelis W., and Richnow H. H. (1999) Contrib Title: $^{13}C/^{12}C$ isotope fractionation of aromatic hydrocarbons during microbial degradation. *Environ. Microbiol.* **1**, 409–414.

Mitra S. M., Dellapena T. M., and Dickhut R. (1999) Geochemical factors affecting sediment/pore water distribution of PAHs in the Hudson River. *Estuar. Coast. Shelf Sci.* **49**, 311–326.

Moldowan J. M., Dahl J. E., McCaffrey M. A., Smith W. J., and Fetzer J. C. (1995) Application of biological marker technology to bioremediation of refinery by-products. *Energ. Fuel* **9**(1), 155–162.

Moore J. W. and Ramamoorthy S. (1984) *Organic Chemicals in Natural Waters Applied Monitoring and Impact Assessment.* Springer, New York.

Morasch B., Schink B., Richnow H., and Meckenstock R. U. (2001) Hydrogen and carbon isotope fractionation upon bacterial toluene degradation: mechanistic and environmental aspects. *Appl. Environ. Microbiol.* **67**, 4842–4849.

Mueller J. G., Chapman P. J., Blattmann B. O., and Pritchard P. H. (1990) Isolation and characterization of a flouranthene-utilizing strain of *Pseudomonas paucimobilis. Appl. Environ. Microbiol.* **56**(4), 1079–1086.

Mueller J. G., Chapman P. J., and Pritchard P. H. (1988) Action of a fluoranthene-utilizing bacterial community on polycyclic aromatic hydrocarbons components of creosote. *Appl. Environ. Microbiol.* **55**(12), 3085–3090.

Munoz D., Guiliano M., Doumenq P., Jacquot F., Scherrer P., and Mille G. (1997) Long-term evolution of petroleum biomarkers in mangrove soil (Guadaloupe). *Mar. Pollut. Bull.* **34**, 868–874.

Neff J. M. (1979) *Polycyclic Aromatic Hydrocarbons in the Aquatic Environment. Sources, Fates and Biological Effects.* Applied Science Publishers, London.

Neff J. M. (1995) Polycyclic aromatic hydrocarbons. In *Fundamentals of Aquatic Toxicology: Methods and Applications* (eds. B. L. Donley and M. Dorfman). Hemisphere publishing, pp. 416–454..

Neilson A. and Hynning P. (1998) PAHs: products of chemical and biochemical transformation of alicyclic precursors. In *The Handbook of Environmental Chemistry* (ed. A. H. Neilson). Springer, Berlin, Heidelberg, vol. 3, part I, pp. 224–273.

Neilson A. H. (1998) *PAHs and Related Compounds Biology.* Springer, New York.

Neilson A. H. and Allard A. S. (1998) Microbial metabolism of PAHs and heteroarenes. In *The Handbook of Environmental Chemistry* (ed. A. H. Neilson). Springer, Berlin, Heidelberg, vol. 3, part I, pp. 2–80.

NRC (1985) *Oil in the Sea: Inputs, Fates, and Effects.* Committee on Oil in the Sea: Inputs, Fates, and Effects, National Research Council, Canada.

NRC (2000) *Natural Attenuation for Groundwater Remediation.* Committee on Intrinsic Remediation. Water Science and Technology Board, Board on Radioactive Waste Management, National Research Council, Canada.

NRC US National Research Council (2002) *Oil in the Sea III: Inputs, Fates, and Effects.* Committee on Oil in the Sea: Inputs, Fates, and Effects, National Research Council, Canada.

Offenberg J. H. and Baker J. E. (1999) Aerosol size distributions of polycyclic aromatic hydrocarbons in urban and over-water atmospheres. *Environ. Sci. Technol.* **33**, 3324–3331.

Okuda T., Kumata H., Naraoka H., Ishiwatari R., and Takada H. (2002) Vertical distributions and $\delta^{13}C$ isotopic compositions of PAHs in Chidorigafuchi Moat sediment, Japan. *Org. Geochem.* **33**, 843–848.

O'Malley V., Abrajano T. A., and Hellou J. (1994) Determination of $^{13}C/^{12}C$ ratios of individual PAHs from environmental samples: can PAHs sources be source apportioned? *Org. Geochem.* **21**, 809–822.

O'Malley V., Abrajano T. A., and Hellou J. (1996) Isotopic apportionment of individual polycyclic aromatic hydrocarbon sources in St. Johns Harbour. *Environ. Sci. Technol.* **30**, 634–638.

O'Malley V. P. (1994) Compound-specific carbon isotope geochemistry of polycyclic aromatic hydrocarbons in eastern Newfoundland estuaries. PhD Thesis, Memorial University of Newfoundland.

O'Malley V. P., Burke R. A., and Schlotzhauer W. S. (1997) Using GC-MS/combustion/IRMS to determine the $^{13}C/^{12}C$ ratios of individual hydrocarbons produced from the combustion of biomass materials—application to biomass burning. *Org. Geochem.* **27**, 567–581.

Oostdam B. L. (1980) Oil pollution in the Persian Gulf and approaches 1978. *Mar. Pollut. Bull.* **11**, 138–144.

Oros D. R. and Simoneit B. R. T. (2000) Identification and emission rates of molecular tracers in coal smoke particulate matter. *Fuel* **79**, 515–536.

Paalme L., Irha N., Urbas E., Tsyban A., and Kirso U. (1990) Model studies of photochemical oxidation of carcinogenic polyaromatic hydrocarbons. *Mar. Chem.* **30**, 105–111.

Pancirov R. J. and Brown R. A. (1975) Analytical methods for PAHs in crude oils, heating oils and marine tissues. *Conference Proceedings on Prevention and Control of Oil Pollution.* API, Washington, DC, pp. 103–113.

Payne J. R. and Phillips C. R. (1985) Photochemistry of petroleum in water. *Environ. Sci. Technol.* **19**, 569.

Pearson A., McNichol A. P., Benitez-Nelson B. C., Hayes J. M., and Eglinton T. I. (2001) Origins of lipid biomarkers in Santa Monica basin surface sediment: a case study using compound-specific Δ14C analysis. *Geochim. Cosmochim. Acta* **65**, 3123–3137.

Pedersen P. S., Ingwersen J., Nielsen T., and Larsen E. (1980) Effects of fuel, lubricant, and engine operating parameters on the emission of polycyclic aromatic hydrocarbons. *Environ. Sci. Technol.* **14**(1), 71–79.

Pereira W. E., Hostetettler F. D., Luoma S. N., van Geen A., Fuller C. C., and Anima R. J. (1999) Sedimentary record of anthropogenic and biogenic polycyclic aromatic hydrocarbons in San Francisco Bay, California. *Mar. Chem.* **64**, 99–113.

Peters K. E. and Moldowan J. M. (1993) *The Biomarker Guide, Interpreting Molecular Fossils in Petroleum and Ancient Sediments.* Prentice-Hall.

Peters K. E., Moldowan J. M., Schoell M., and Hempkins W. B. (1986) Petroleum isotopic and biomarker composition related to source rock organic matter and depositional environment. *Org. Geochem.* **10**, 17–27.

Peters K. E., Scheuerman G. L., Lee C. Y., Moldowan J. M., Reynolds R. N., and Pena M. M. (1992) Effects of refinery processes on biological markers. *Energ. Fuel* **6**, 560–577.

Pierce R. C. and Katz M. (1975) Dependency of polynuclear aromatic hydrocarbon content on size distribution of atmospheric aerosols. *Environ. Sci. Technol.* **9**, 347–353.

Prahl F. G. and Carpenter R. (1979) The role of zooplankton fecal pellets in the sedimentation of polycyclic aromatic hydrocarbons in Dabob Bay, Washington. *Geochim. Cosmochim. Acta* **43**, 1959–1972.

Prince R. C. (1993) Petroleum oil spill bioremediation in marine environments. *Crit. Rev. Microbiol.* **19**, 217–239.

Pruell J. R. and Quinn J. G. (1988) Accumulation of polycyclic aromatic hydrocarbons in crankcase oil. *Environ. Pollut.* **49**, 89–97.

Radding S. B., Mill T., Gould C. W., Liu D. H., Johnson H. L., Bomberger D. C., and Fojo C. V. (1976) The environmental fate of selected polynuclear aromatic hydrocarbons. US Environmental Protection Agency, EPA560/5-75-009.

Radke M. (1987) Organic geochemistry of aromatic hydrocarbons. In *Advances in Petroleum Geochemistry* (eds. J. Brooks and D. H. Welte). Academic Press, New York, vol. 2, pp. 141–207.

Radke M., Willsch H., Leythaeuser D., and Teichmüller M. (1982) Aromatic components of coal: relation of distribution pattern to rank. *Geochim. Cosmochim. Acta* **46**, 1831–1848.

Ramaswami A. and Luthy R. G. (1997) Mass transfer and bioavailability of PAHs compounds in coal tar NAPL-slurry systems. 1: Model development. *Environ. Sci. Technol.* **31**, 2260–2267.

Readman J. W., Mantoura R. F. C., Rhead M. M., and Brown L. (1982) Aquatic distribution and heterotrophic degradation of polycyclic aromatic hydrocarbons (PAHs) in the Tamer Estuary. *Estuar. Coast. Shelf Sci.* **14**, 369–389.

Reddy C. and Quinn J. G. (1997) Environmental chemistry of benzothiazoles derived from rubber. *Environ. Sci. Technol.* **31**, 2847–2853.

Reddy C., Reddy C. M., Pearson A., Xu L., McNichol A., Benner B. A., Wise S. A., Klouda G. A., Currie L. A., and Eglinton T. I. (2002) Radiocarbon as a tool to apportion the sources of polycyclic aromatic hydrocarbons and black carbon in environmental samples. *Environ. Sci. Technol.* **36**, 1774–1787.

Reyes C. S., Medina M., Crespo-Hernandez C., Cedeno M. Z., Arce R., Rosario O., Steffenson D. M., Ivanov I. N., Sigman M. E., and Dabestani R. (2000) Photochemistry of pyrene on unactivated and activated silica surfaces. *Environ. Sci. Technol.* **34**, 415–421.

Rod S. R., Ayres R. U., and Small M. (1989) *Reconstruction of historical loadings of heavy metals and chlorinated hydrocarbon pesticides in Huston-Raritan basin, 1880–1980.* Final Project Report to the Hudson River Foundation.

Rogge W. F., Hildemann L. M., Mazurek M. A., and Cass G. R. (1993) Sources of fine organic aerosal. 2: Noncatalyst and catalyst equipped automobiles and heavy-duty diesel trucks. *Environ. Sci. Technol.* **27**, 636–651.

Rothermich M. M., Hayes L. A., and Lovley D. (2002) Anaerobic, sulfate-dependent degradation of polycyclic aromatic hydrocarbons in petroleum-contaminated harbor sediment. *Environ. Sci. Technol.* **36**, 4811–4817.

Sanders G., Jones K. C., and Hamilton-Taylor J. (1993) A simple method to assess the susceptibility of polynuclear aromatic hydrocarbons to photolytic decomposition. *Atmos. Environ.* **27**(2), 139–144.

Schauer J. J., Kleeman M. J., Cass G. R., and Simoneit B. R. T. (2001) Measurement of emissions from air pollution sources. 3: C1-C29 organic compounds from fireplace combustion of wood. *Environ. Sci. Technol.* **35**, 1716–1728.

Schneider J., Grosser R., Jayasimhulu K., Xue W., and Warshawsky D. (1996) Degradation of pyrene, benz(a) anthracene, and benz(a)pyrene by *Mycobacterium* sp. Strain RGHII-135, isolated from a former coal gasification site. *Appl. Environ. Microbiol.* **157**, 7–12.

Schoell M. (1984) Stable isotopes in petroleum research. In *Advances in Petroleum Geochemistry* (eds. J. Brooks and D. H. Welte). Academic Press, London, vol. 1, pp. 215–245.

Schwarzenbach R. P., Gschwent P. M., and Imboden D. M. (1993) *Environmental Organic Chemistry.* Wiley, New York.

Schwarzenbach R. P., Haderlein S. B., Muller S. R., and Ulrich M. M. (1998) Assessing the dynamic behavior of organic contaminants in natural waters. In *Perspectives in Environmental Chemistry* (ed. D. L. Macalady). Oxford University Press, New York, pp. 138–166.

Sherrill T. W. and Sayler G. S. (1981) Phenanthrene biodegradation in freshwater environments. *Appl. Environ. Microbiol.* **39**, 172–178.

Simcik M. F., Eisenreich S. J., Golden K. A., Liu S. P., Lipiatou E., Swackhamer D. L., and Long D. T. (1996) Atmospheric loading of polycyclic aromatic hydrocarbons to Lake Michigan as recorded in the sediments. *Environ. Sci. Technol.* **30**, 3039–3046.

Simoneit B. R. T. (1984) Organic matter of the troposphere. III: Characterization and sources of petroleum and pyrogenic residues in aerosols over the Western United States. *Atmos. Environ.* **18**, 51–67.

Simoneit B. R. T. (1986) Characterization of organic constituents in aerosols in relation to their origin and transport: a review. *Int. J. Environ. Analyt. Chem.* **23**, 207–237.

Simoneit B. R. T. (1998) Biomarker PAHs in the environment. In *PAHs and Related Compounds Chemistry* (ed. A. H. Neilson). Springer, New York, pp. 175–215.

Sjögren M., Li H., Rannug U., and Westerholm R. (1996) Multivariate analysis of exhaust emissions from heavy-duty diesel fuels. *Environ. Sci. Technol.* **30**, 38–49.

Smirnova A., Abrajano T., Smirnov A., and Stark A. (1998) Distribution and sources of polycyclic aromatic hydrocarbons in the sediments of lake Erie. *Org. Geochem.* **29**, 1813–1828.

Sofer Z. (1984) Stable carbon isotope compositions of crude oils: application to source depositional environments and petroleum alteration. *Am. Assoc. Petrol. Geol. Bull.* **68**(1), 31–48.

Sporstol S., Gjos N., Lichtenthaler R. G., Gustavsen K. O., Urdal K., Oreld F., and Skel J. (1983) Source identification of aromatic hydrocarbons in sediments using GC-MS. *Environ. Sci. Technol.* **17**(5), 282–286.

Stark A., Abrajano T., Hellou J., and Smith J. (2003) *Org. Geochem.* **34**, 225-237.

Stehmeir L. G., Francis M. M. D., Jack T. R., Diegor E., Winsor L., and Abrajano T. A. (1999) Field and *in vitro* evidence for *in-situ* bioremediation using compound-specific $^{13}C/^{12}C$ ratio monitoring. *Org. Geochem.* **30**, 821–834.

Steinhauer M. S. and Boehm P. D. (1992) The composition and distribution of saturated and aromatic hydrocarbons in nearshore sediments, river sediments, and coastal peat in the Alaskan Beaufort Sea: implications for detecting anthropogenic hydrocarbon inputs. *Mar. Environ. Res.* **33**, 223–253.

Stenberg U. R. (1983) PAHs emissions from automobiles. In *Handbook of Polycyclic Aromatic Hydrocarbons. Emission Sources and Recent Progress in Analytical Chemistry* (eds. A. Bjorseth). Dekker, New York, vol. 2, pp. 87–111.

Strachan W. M. J. and Eisenreich S. J. (1988) *Mass Balancing of Toxic Chemicals in the Great Lakes: The Role of Atmospheric Deposition.* International Joint Commission, Windsor, ON.

Suess M. J. (1976) The environmental load and cycle of polycyclic aromatic hydrocarbons. *Sci. Total Environ.* **6**, 239–250.

Sugai S. F., Lindstrom J. E., and Braddock J. F. (1997) Environmental influences on the microbial degradation of Exxon Valdez oil spill on the shorelines of Prince William Sound, Alaska. *Environ. Sci. Technol.* **31**, 1564–1572.

Takada H., Onda T., and Ogura N. (1990) Determination of polycyclic aromatic hydrocarbons in urban street dusts and their source materials by capillary gas chromatography. *Environ. Sci. Technol.* **24**(8), 1179–1186.

Tissot B. P. and Welte D. H. (1984) *Petroleum Formation and Occurrence*, 2nd edn. Springer, Berlin.

Van Brummelen T. C., van Hattum B., Crommentuijn T., and Kalf D. E. (1998) Bioavailability and ecotoxicology of PAHs. In *PAHs and Related Compounds Chemistry* (ed. A. H. Neilson). Springer, New York, pp. 175–215.

Van Metre P. C., Mahler B. J., and Furlong E. T. (2000) Urban sprawl leaves its PAH signature. *Environ. Sci. Technol.* **34**, 4064–4070.

Vazquez-Duhalt R. (1989) Environmental impact of used motor oil. *Sci. Total Environ.* **79**, 1–23.

Venkatesan M. I. (1988) Occurrence and possible sources of perylene in marine sediments—a review. *Mar. Chem.* **25**, 1–27.

Villholth K. G. (1999) Colloid characterization and colloidal phase partitioning of polycyclic aromatic hydrocarbons in two creosote-contaminated aquifers in Denmark. *Environ. Sci. Technol.* **33**, 691–699.

Volkman J. K., Holdsworth D. G., Neill G. P., and Bavor H. J. (1992) Identification of natural, anthropogenic and petroleum hydrocarbons in aquatic sediments. *Sci. Total Environ.* **112**, 203–219.

Volkman J. K., Revill A. T., and Murray A. P. (1997) Applications of biomarkers for identifying sources of natural and pollutant hydrocarbons in aquatic environments. In *Molecular Markers in Environmental Geochemistry, ACS Symposium Series* (ed. R. P. Eganhouse). American Chemical Society, Washington, DC, pp. 110–132.

Wakeham S. G. (1996) Aliphatic and polycyclic aromatic hydrocarbons in Black Sea sediments. *Mar. Chem.* **53**, 187–205.

Wakeham S. G., Schaffner C., and Giger W. (1980a) Polycyclic aromatic hydrocarbons in recent lake sediments. II: Compounds derived from biogenic precursors during early diagenesis. *Geochim. Cosmochim. Acta* **44**, 415–429.

Wakeham S. G., Schaffner C., and Giger W. (1980b) Polycyclic aromatic hydrocarbons in recent lake sediments. I: Compounds having anthropogenic origins. *Geochim. Cosmochim. Acta* **44**, 403–413.

Wang Z. and Fingas M. (1995) Use of methyldibenzothiophenes as markers for differentiation and source identification of crude and weathered oils. *Environ. Sci. Technol.* **29**, 2842–2849.

Wang Z., Fingas M., Blenkinsopp S., Sergey G., Landriault M., Sigouin L., Foght J., Semple K., and Westlake S. W. S. (1998) Comparison of oil composition changes due to biodegradation and physical weathering in different oils. *J. Chromatogr. A* **809**, 89–107.

Wang Z., Fingas M., and Page D. S. (1999) Oil spill identification. *J. Chromatogr. A* **843**, 369–411.

Ward J. A. M., Ahad J. M. E., Lacrampe-Couloume G., Slater G. F., Edwards E. A., and Lollar B. S. (2000) Hydrogen isotope fractionation during methanogenic degradation of toluene: potential for direct verification of bioremediation. *Environ. Sci. Technol.* **34**, 4577–4581.

Westerholm R., Stenberg U., and Alsberg T. (1988) Some aspect of the distribution of polycyclic aromatic hydrocarbons (PAHs) between particles and gas phase from diluted gasoline exhausts generated with the use of a dilution tunnel, and its validity for measurement in air. *Atmos. Environ.* **22** (5), 1005–1010.

Whitehouse B. G. (1984) The effects of temperature and salinity on the aqueous solubility of polynuclear aromatic hydrocarbons. *Mar. Chem.* **14.**

Whittaker M., Pollard S. J. T., and Fallick T. E. (1995) Characterisation of refractory wastes at heavy oil-contaminated sites: a review of conventional and novel analytical methods. *Environ. Technol.* **16**, 1009–1033.

Wijayaratne R. D., and Means, J. C. (1984) Adsorption of polycyclic aromatic hydrocarbons by natural estuarine colloids. *Mar. Environ. Res.* **11.**

Wolfe D. A., Long E. R., and Thursby G. B. (1996) Sediment toxicity in the Hudson-Raritan estuary: distribution and correlations with chemical contamination. *Estuaries* **19**(4), 901–912.

Yan B. (2004) PAH sources and depositional history in sediments from the lower Hudson River basin. PhD Dissertation, Rensselaer Polytechnic Institute, Troy, NY.

Yan B., Abrajano T., Bopp R., Chaky D., Benedict L., and Perry E. (2006) Combined applications of $\delta^{13}C$ and molecular ratios for PAH source apportionment in sediments from New York/New Jersey Harbor Complex. *Org. Geochem.* **37** (6), 674–687.

Yan B., Abrajano T., Chaky D., Benedict L., Bopp R., and Chillrud S. (2005) Use of molecular ratios to trace saturated and polycyclic aromatic hydrocarbon inputs into Central Park Lake, New York. *Environ. Sci. Technol.* **39**, 7012–7019.

Yan B., Benedict L., Chaky D. A., Bopp R. F., and Abrajano T. A. (2004) Levels and patterns of PAH distribution in sediments from New York/New Jersey Harbor complex. *Northeast. Geol. Environ. Sci.* **26**, 113–122.

Yanik P., O'Donnell T. H., Macko S. A., Qian Y., and Kennicutt M. C. (2003) The isotopic compositions of selected crude oil PAHs during biodegradation. *Org. Geochem.* **34**, 291–304.

Ye D., Siddiqui M. A., Maccubbin A. E., Kumar S., and Sikka H. C. (1996) Degradation of polynuclear aromatic hydrocarbons by *Sphingomonas paucimbilis*. *Environ. Sci. Technol.* **30**, 136–142.

Youngblood W. W. and Blumer M. (1975) Polycyclic aromatic hydrocarbons in the environment: homologous series in soils and recent marine sediments. *Geochim. Cosmochim. Acta* **39**, 1303–1314.

Yunker M. B. and Macdonald R. W. (2003) Alkane and PAH prepositional history, sources and fluxes in sediments from the Fraser River Basin and Strait of Georgia, Canada. *Org. Geochem.* **34**, 1429–1454.

Yunker M. B., Macdonald R. W., Brewer R., Mitchell R. H., Goyette D., and Sylvestre S. (2002) PAHs in the Fraser River basin: a critical appraisal of PAH ratios as indicators of PAH source and composting. *Org. Geochem.* **33**, 489–515.

Yunker M. B., MacDonald R. W., Cretney W. J., Fowler B. R., and McLaughlin F. A. (1993) Alkane, terpene, and polycyclic aromatic hydrocarbon geochemistry of the Mackenzie River and Mackenzie shelf: riverine contributions to Beaufort Sea coastal sediment. *Geochim. Cosmochim. Acta* **57**, 3041–3061.

Yunker M. B., MacDonald R. W., Veltkamp D. J., and Cretney W. J. (1995) Terrestrial and marine biomarkers in a seasonally ice-covered arctic estuary—integration of multivariate and biomarkers approaches. *Mar. Chem.* **49**, 1–50.

Zafiriou O. C. (1997) Marine organic photochemistry previewed. *Mar. Chem.* **5**, 497–522.

Zepp R. G., and Schlotzhauer P. F. (1979) Photoreactivity of selected aromatic hydrocarbons in water. In *Polynuclear Aromatic Hydrocarbons* (eds. P. W. Jones and P. Leber). Ann Arbor Science Publications, Ann Arbor, pp. 141–158.

Isotope Geochemistry
ISBN: 978-0-08-096710-3

pp. 377–424

B. RADIOACTIVE TRACERS

13
Radiocarbon

W. S. Broecker
Lamont-Doherty Earth Observatory, Palisades, NY, USA

13.1 INTRODUCTION

Willard Libby's invention of the radiocarbon dating method revolutionized the fields of archeology and Quaternary geology because it brought into being a means to correlate events that occurred during the past 3.5×10^4 years on a planet-wide scale (Libby *et al.*, 1949). This contribution was recognized with the award of the Nobel prize for chemistry. In addition, radiocarbon measurements have been a boon to the quantification of many processes taking place in the environment, to name a few: the rate of "ventilation" of the deep ocean, the turnover time of humus in soils, the rate of growth of cave deposits, the source of carbon-bearing atmospheric particulates, the rates of gas exchange between the atmosphere and water bodies, the replacement time of carbon atoms in human tissue, and depths of bioturbation in marine sediment. Some of these applications have been greatly aided by the creation of excess ^{14}C atoms as the result of nuclear tests conducted in the atmosphere. Since the 1960s, this so-called bomb radiocarbon has made its way into all of the Earth's active carbon reservoirs. To date, tens of thousands of radiocarbon measurements have been made in laboratories throughout the world.

13.2 PRODUCTION AND DISTRIBUTION OF ^{14}C

Radiocarbon atoms are produced when protons knocked loose by cosmic ray impacts encounter the nuclei of atmospheric nitrogen atoms. The reaction is as follows:

$$n + {}^{14}N \rightarrow {}^{14}C + p \qquad (1)$$

The half-life of these ^{14}C atoms is 5,730 years. Hence, they have on average 8,270 years to distribute themselves through the Earth's active carbon reservoirs. Radiocarbon atoms decay by emitting an electron, thereby converting a neutron into a proton returning the nucleus to its original ^{14}N form.

Once produced, radiocarbon atoms become oxidized to CO_2 gas and join the Earth's carbon cycle (see Figure 1). CO_2 exchange with the inorganic carbon dissolved in the Earth's surface waters carries these atoms into the sea, lakes, and rivers. Photosynthesis moves the ^{14}C into both terrestrial and aquatic plants and from there to the animals that feed upon them. Radiocarbon also gets incorporated into shell and coral, and into soil and cave carbonates.

13.3 MEASUREMENTS OF RADIOCARBON

Initially, all measurements were made by counting the β-particles emitted during radioactive decay of ^{14}C. The first wave of laboratories followed Libby's lead and using screen wall Geiger counters measured the β-particles emitted by carbon black spun onto the inside of stainless-steel cylinders (Figure 2). But the advent of nuclear testing created a serious problem for this method. Airborne strontium-90, cesium-137, and other fission products became absorbed onto the carbon black adding to the sample's radioactivity. Hans Suess (USGS, Washington, DC), Hessel deVries (The Netherlands), and Gordon Fergusson (New Zealand) pioneered the transition to gas counting. Most laboratories filled proportional counters with highly purified CO_2 gas generated by burning or acidifying the sample. A few used acetylene. Other laboratories went a step further converting the sample carbon into liquid benzene and measured its radioactivity in a scintillation counter. In the early 1980s, it was realized that the ^{14}C atoms themselves could be measured using high-energy tandem Van de Graaff accelerators. During the latter part of the 1980s and early 1990s, accelerator mass spectrometry largely replaced the traditional decay-counting techniques. The huge advantage of this new approach was that measurements required only 1 mg of carbon as opposed to 1 g for decay counting. The routine accuracy (± 5 per mil) achievable in atom counting is comparable to that obtained in decay-counting laboratories. Also comparable is the ability to measure the minute amounts of ^{14}C in very old (i.e., $>3.5 \times 10^4$ years) samples. This innovation of atom counting has opened up a large number of applications previously untouchable by decay counting. The convention for expressing ^{14}C results is shown in Table 1.

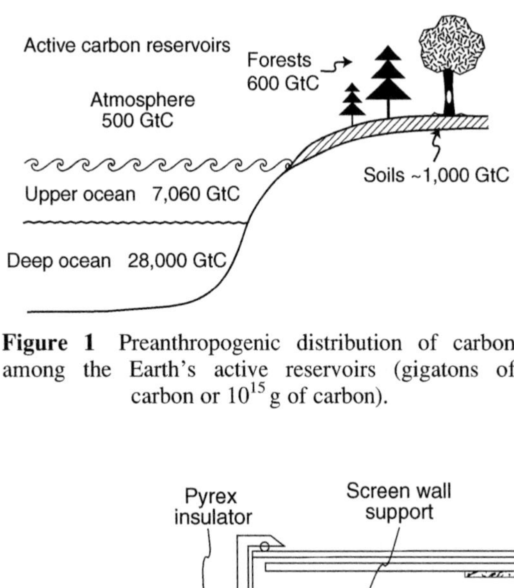

Figure 1 Preanthropogenic distribution of carbon among the Earth's active reservoirs (gigatons of carbon or 10^{15} g of carbon).

13.4 TIMESCALE CALIBRATION

It was a biophysicist deVries (1958), who first clearly demonstrated that the radiocarbon timescale was imperfect, that is, that radiocarbon years were not identical to calendar years. Fascinated by the potential of Libby's new method, this brilliant Dutch scientist plunged into all its aspects. deVries devised means to reduce the background of gas counters by greatly increasing the effectiveness of the shielding from cosmic ray mesons. Not only did he strive to extend the method beyond its nominal range (3.5×10^4 years), he also worked to improve the precision of measurements so that he could document small imperfections in the radiocarbon timescale based on

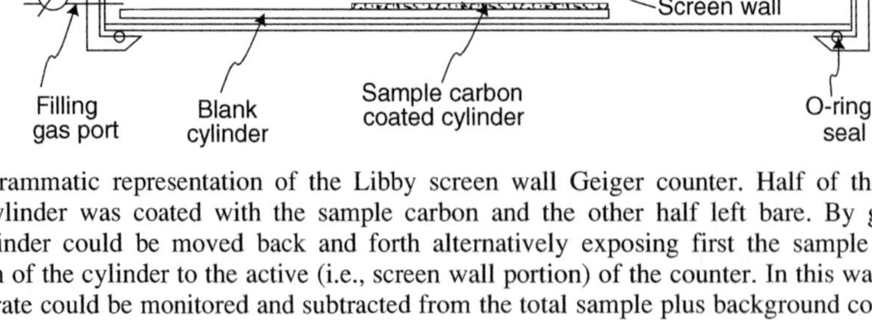

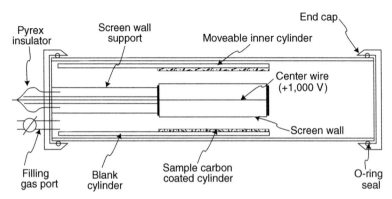

Figure 2 Diagrammatic representation of the Libby screen wall Geiger counter. Half of the moveable inner stainless-steel cylinder was coated with the sample carbon and the other half left bare. By gently tipping the counter, the cylinder could be moved back and forth alternatively exposing first the sample portion and then the blank portion of the cylinder to the active (i.e., screen wall portion) of the counter. In this way, the background count rate could be monitored and subtracted from the total sample plus background count rate.

Table 1 The international convention for expressing radiocarbon results on contemporary materials in delta units. The standard is the age-corrected $^{14}C/C$ ratio in 1850 wood. The ^{13}C correction is designed to remove that part of the radiocarbon variability associated with the isotope separations occurring in nature (e.g., during photosynthesis and air–sea exchange). Also listed are some important characteristics of radiocarbon

Definitions:

$$\delta^{14}C = \left(\frac{^{14}C/C_{\text{sample}} - {^{14}C/C_{\text{standard}}}}{^{14}C/C_{\text{standard}}}\right)1{,}000$$

$$\Delta^{14}C = \delta^{14}C - 2(\delta^{13}C + 25)\left(1 + \frac{\delta^{14}C}{1{,}000}\right)$$

where

$$\delta^{13}C = \left(\frac{^{13}C/^{12}C_{\text{sample}} - {^{13}C/^{12}C_{\text{standard}}}}{^{13}C/^{12}C_{\text{standard}}}\right)$$

Age correction:

$$^{14}C/C_{\text{age corrected}} = {^{14}C/C_{\text{measured}}}\, e^{\lambda t}$$

where t is the calendar age of the sample (referenced to 1950).

Characteristics:

$$^{14}C/C\ (\text{for } \Delta^{14}C = 0\%) = 1.18 \times 10^{-12}$$

$$t_{1/2} = 5{,}730 \text{ years}$$

$$T_{\text{mean}} = 8{,}270 \text{ years}$$

$$\lambda = \frac{1}{8{,}270}\ \text{yr}^{-1}$$

$$\text{Decay rate} = \frac{1\%}{82.7 \text{ years}}$$

Note: A computerized calibration program (CALIB) is available.

measurements on materials of known calendar age. He was able to document such deviations and he postulated that they reflected changes in the $^{14}C/C$ ratio in atmospheric CO_2 due to perturbations in the flux of cosmic rays entering the Earth's atmosphere. Hessel deVries stands out in my mind as the great genius in this field. Had his life not been snuffed out just as he was reaching his prime, he would have dominated the radiocarbon world for decades. But, fortunately, during his short career, not only did he put his finger on the calibration problem that even today dominates much of the effort in the field, but he also trained a graduate student Minze Stuiver, who would devote much of his career to calibration studies and to measurements of ^{14}C in oceanic carbon.

13.4.1 Calibration Based on Tree Rings

Nature has provided a marvelous set of calibration materials, namely, cellulose in the annual growth rings in trees. Single living trees provide annual ring series extending back 1,000 or more years. More importantly, tree trunks preserved in swamp muck and in riverine alluvium can be cross-dated based on ring-width "fingerprints," thus greatly extending the calibration range. Treasure troves of such material lie in Holocene sediments of the German rivers. Its flood-stage deposits constitute a major source of sand and gravel for the German construction industry. In the course of "mining" these deposits, large trunks of trees are often encountered. Over the years, the late Bernd Becker and members of his Hohenheim dendrochronology team responded to alerts hastening to the site of a new find to cut out a section to add to their ever-growing collection. A single radiocarbon measurement would tell them the approximate growth period of this tree. Measurements of the ring widths would allow them to tie this section precisely into an ever more complete and accurate master chronology. Once this chronology was established, very detailed high-precision ^{14}C measurements were made in the laboratory of Bernd Kromer in Heidelberg. This amazing record extends back to ~8,000 years ago. To extend the growth-ring series further back in time, it was necessary for the Becker team to switch from oaks to pines. This species appeared in northern Europe soon after the abrupt warming which brought to an end the final cold punctuation of the last glacial period (i.e., the Younger Dryas). Working closely with Bernd Kromer, who conducted ultraprecise ^{14}C measurements (to ±2‰) in his Heidelberg laboratory on both the oak and pine series, Becker and co-workers were satisfactorily able to splice the pine series onto the oak series creating a master chronology covering the last 1.1919×10^4 years (Kromer *et al.*, 1986; Becker and Kromer, 1993; Kromer and Spurk, 1998; Friedrich *et al.*, 1999). Efforts are currently underway using tree trunk series from central Europe and south of the Alps to extend the chronology back into glacial time.

The thousands of measurements which contributed to this calibration effort were largely made in five laboratories: that of Hans Suess in La Jolla, California; that of Paul Damon in Tuscon, Arizona; that of Gordon Pearson in Belfast Northern Ireland; and, of course, those of Minze Stuiver in Seattle, Washington, and Bernd Kromer in Heidelberg, Germany. The resulting calibration curve is a boon to archeologists faced with the task of relating radiocarbon dates to calendar dates for historical events, as well as to geophysicists interested in probing the causes for the $^{14}C/C$ ratio changes. A compilation base on these tree-ring calibration measurements is shown in Figure 3.

Two features of this record stand out, a long-term trend of decreasing ^{14}C culminating ~1,500 years ago and century-duration fluctuations around this trend. Minze Stuiver was the

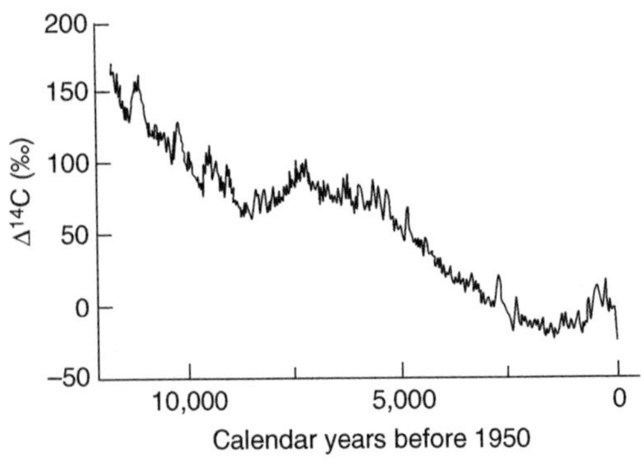

Figure 3 Tree-ring-based reconstruction of the temporal trend of the atmosphere's $^{14}C/C$ ratio over the last 1.1854×10^4 calendar years. $\Delta^{14}C$ is the deviation of the atmospheric ratio from that in pre-Industrial time (~1,800) (see Table 1). Source: Stuiver *et al.* (1998).

first to demonstrate that the century-duration fluctuations were, at least in part, related to sunspot activity. He did this by showing that during the course of the so-called Maunder minimum (a period from 1645 to 1715 when no spots were observed on the Sun), the $^{14}C/C$ ratio in atmospheric CO_2 underwent a significant increase (Stuiver and Quay, 1980, 1981). Stuiver attributed this increase to the great reduction in the ions streaming into space from the Sun's spots (and hence also of the so-called heliomagnetic field). This magnetic field serves to partially shield the solar system from incoming cosmic rays. Similar increases in the ratio $^{14}C/C$ were established by Stuiver to have occurred between 1280 and 1345 and between 1420 and 1540 (Stuiver and Quay, 1980, 1981). By analogy to the Maunder, they have been termed the Spörer and Wolf sunspot minima (see Figure 4).

A reasonably convincing case has been made that the longer-term Holocene trend is a consequence of changes in the Earth's magnetic field which also diverts incoming cosmic rays. This case rests on measurements designed to reconstruct the strength of the Earth field at times in the past. The first such attempt was by Thellier and Thellier (1959), who measured the magnetic field in ancient ceramics and then reheated them beyond the Curie point and allowed them to cool in the same fashion as originally. They then remeasured the magnetic field. These early ceramic measurements have been supplemented by reconstructions based on igneous rocks and deep-sea cores, which extend back many radiocarbon half-lives (see Bard, 1998 for an excellent summary).

However, lurking in the wings is yet another mechanism by which the atmospheric $^{14}C/C$ ratio may have been changed. It involves changes in the rate of ventilation of the deep sea. Around 70% of the Earth's cosmic ray-produced ^{14}C atoms currently reside in the deep sea. Because ^{14}C is lost by radiodecay during residence in the deep sea, the $^{14}C/C$ ratio for the inorganic carbon (i.e., HCO_3^-, CO_3^{2-}, and CO_2) dissolved in these waters averages ~16% lower than that in the atmospheric CO_2. Hence, were the rate of ocean mixing to have been slower in the past than now, the contrast between the atmospheric and deep-sea $^{14}C/C$ ratios would have increased and, consequently, the ratio in the atmosphere would have been higher than now. Later in the chapter, a dramatic example of such a change will be discussed. For the Holocene (i.e., the last 1.15×10^4 years), there is no concrete evidence that ocean mixing contributed to the $^{14}C/C$ changes. Minze Stuiver has pondered whether the changes in sunspot activity might be linked to the Holocene's small climate changes. If so, then both sunspot activity and ocean mixing may have contributed to the $^{14}C/C$ ratio fluctuations.

An extremely important contribution to this subject was made by an Indian scientist, Devendra Lal, who made calculations aimed at determining the exact dependence on the Earth's magnetic field strength of the influx of cosmic rays to the Earth's atmosphere, and hence also on the rate of production of ^{14}C atoms (Lal, 1988).

While resolving most of the inconsistencies between radiocarbon and historic ages, the calibration curve also reveals a fundamental limitation of the radiocarbon method. During some time periods, the century-duration changes in $^{14}C/C$ ratio created reversals in the age sequence. These reversals give rise to multiple possible calendar ages for a single radiocarbon measurement. Only by measuring several samples in stratigraphic sequence, it is possible to distinguish among these multiple possibilities (see Stuiver, 1982).

In a related manner, radiocarbon dating is powerless to aid in verifying the authenticity of art objects purported to have made in the period 1700–1950. The reason is that during this period, the radiocarbon content of the atmosphere decreased at a rate closely matching the rate of ^{14}C decay (i.e., 1% per 83 years). There are two reasons for this decline. First, the excess ^{14}C atoms produced during the Maunder minimum were being mixed into the ocean and sequestered in the terrestrial biosphere. Second, the burning of fossil fuels added CO_2 free of radiocarbon to the atmosphere, thereby extending this decline into the twentieth century. In the early 1950s, this downward trend was reversed by the production of ^{14}C during nuclear tests (see Section 13.6.2). Hence, based on radiocarbon measurements, a forgery made using wood or parchment dating just prior to 1950 could not be distinguished from the real thing created during the eighteenth century.

13.4.2 Calibration Based on Corals

Working at Columbia University's Lamont–Doherty Earth Observatory with French isotope

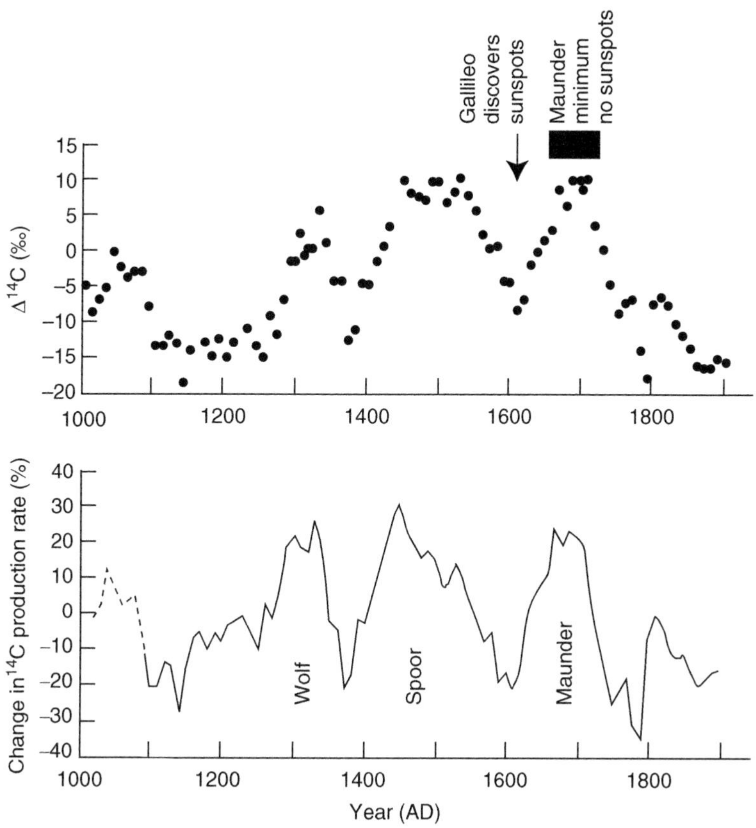

Figure 4 Temporal fluctuations in the $^{14}C/C$ ratio (given in the Δ units defined in Table 1) as determined from measurements on Douglas fir wood from the Pacific Northwest (upper panel). The record is cut off in the year 1900 because after that time the influence of ^{14}C-free CO_2 released by fossil fuel burning became significant. In the lower panel are shown the changes in the production rate of ^{14}C required to produce the observed temporal changes in atmospheric ^{14}C to C ratio. The production rates are deconvolved from the observed $\Delta^{14}C$ changes using a simplified atmosphere–ocean–terrestrial biosphere model. Because of the interchange of carbon atoms with the ocean and the terrestrial biosphere, the changes in atmospheric $^{14}C/C$ ratio lag the production-rate changes. As can be seen, the last of the peaks in production occurred during a time when no spots were observed on the Sun (i.e., during the Maunder minimum). As telescopic observations of the Sun (by Galileo) commenced in the early 1600s, the existence of the Spörer and Wolf sunspot minima is based on the ^{14}C results. The measurements and modeling were carried out by Stuiver and Quay (1980). Bard *et al.* (1997) showed that the observed ^{14}C variations can be generated from measurements of ^{10}Be in ice cores.

geochemists Edouard Bard and Bruno Hamelin, Richard Fairbanks found a means to extend the calibration curve beyond the Holocene into late glacial time. This team adopted a new analytical method developed at Caltech to make very precise uranium and thorium isotope measurements using conventional thermal ionization mass spectrometry (Edwards *et al.*, 1986/1987) instead of the less-precise alpha counting method used previously. Just as the transition from decay to atom counting revolutionized ^{14}C dating, this innovation revolutionized dating based on the uranium decay-series isotope ^{230}Th (half-life 7.5×10^4 years). This new technique was applied to corals obtained from a series of shallow borings Fairbanks conducted off the island of Barbados. These corals formed during the last deglaciation as sea level rose in response to the melting of the glacial age icecaps. The Fairbanks team conducted both ^{230}Th- and ^{14}C-age determinations on these corals, which were ideal for the task as they contained no original thorium and had never been exposed to CO_2-charged groundwater. The ^{230}Th results produced a big surprise. Rather than revealing a cyclicity in the Earth's magnetic field as a number of authors had predicted, the offset between calendar and radiocarbon ages became larger and larger as one went back in time (Bard *et al.*, 1990a, b) (see Figure 5).

13.4.3 Other Calibration Schemes

The measurements on Fairbanks' Barbados corals spurred efforts to find other means of extending the calibration curves back in time. Several tacks were taken. One obvious strategy was to count annual layers (varves) in lake and marine sediments. Stuiver, the hero of the calibration effort, had adopted this approach way back in the 1960s (Stuiver, 1970, 1971). Another approach was to put to use the annual layering in long ice cores from Greenland. As the ice itself cannot be dated using ^{14}C, this required that the ^{18}O (i.e., air temperature) record in the ice be correlated with ice-rafting events in radiocarbon-dated sediment cores from the northern Atlantic (Voelker *et al.*, 1998). Finally, coupled ^{230}Th and ^{14}C measurements on cave formations have been utilized (see e.g., Beck *et al.*, 2001). In my estimation, none of these approaches rests on an entirely firm foundation. Varve counting is often subjective, for example, varves may be missing or doublets may represent a single year. However, Kitagawa and van der Plicht (1998) make an impressive case for a varve sequence in a Japanese lake. Greenland's ice cores provide an excellent example of the difficulty. Counts by a European team and by an American team in two long and virtually identical cores from Greenland's summit locale yield results that differ by as much as several percent (see Southon, 2002). Further, the correlation of the ice record with the marine record is fraught with subjectivity and there are large discrepancies between radiocarbon dates on coexisting planktic and benthic foraminifera. Stalagmites offer perhaps the most promise but, unlike corals, during growth they often incorporate some ^{230}Th. Also, the initial $^{14}C/C$ ratio is offset from that for atmospheric CO_2 because the CO_2 in cave waters is derived from the oxidation of organic matter in the overlying soil. But by far the most worrisome aspect of the cave-formation approach is contamination with younger $CaCO_3$ (infilling of pores, recrystallization, etc.). A 3.5×10^4-year-old sample contains only (1/64)th its original amount of ^{14}C. This contamination problem is far more severe for $CaCO_3$ than for wood from which chemically inert cellulose can be extracted. Because of these problems, I do not trust any existing extensions of calibration curve beyond 2.5×10^4 years. Much more work will have to be done before this task can be declared a success. And, of course, the chances of ever extending the calibration beyond 4.7×10^4 years (only one part in 256 of the original ^{14}C remains) are indeed slim. However, the ^{14}C and ^{230}Th measurements of Beck *et al.* (2001) on a Bahamas stalagmite did allow the calibrations curve to be reliably extended back to $\sim 2.5 \times 10^4$ years (see Figure 6). Clearly this effort constituted a major advance.

In a recent paper, Fairbanks *et al.* (2005) provide what is surely the most reliable of these reconstructions. It is based on pristine coral samples recovered from borings made in the shallow waters off Barbados in the Caribbean Sea and Christmas Island in the central equatorial Pacific Ocean. As was the case for his original set of corals, these had never been exposed to groundwater. They bear no hint of diagenetic alteration. However, the reliability of even this record becomes questionable beyond 40×10^4 years.

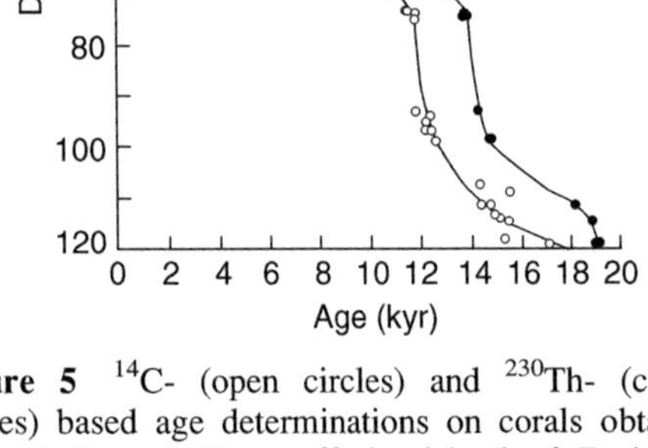

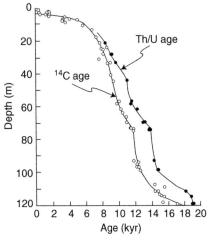

Figure 5 ^{14}C- (open circles) and ^{230}Th- (closed circles) based age determinations on corals obtained from shallow drillings off the island of Barbados. As can be seen, the magnitude of the offset between the two ages increases back in time indicating that the $^{14}C/C$ ratio in atmosphere and upper ocean carbon was declining during the period 18,000–8,000 years ago. As the corals chosen for these measurements were shallow growing, the ^{230}Th ages portray the rise in sea level during the last deglaciation. These results were reported in papers by Fairbanks (1989) and Bard *et al.* (1990a, b).

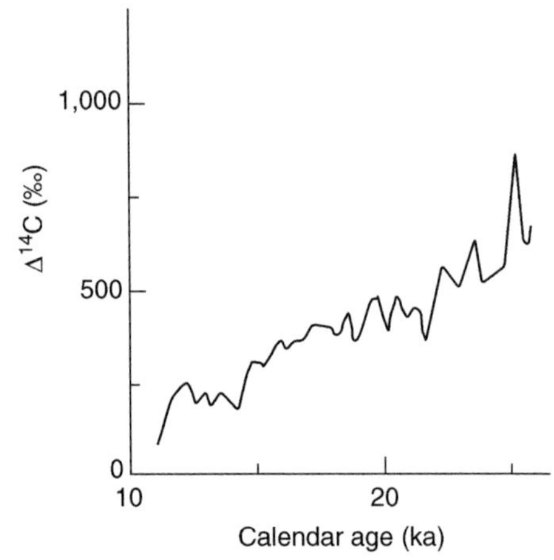

Figure 6 Extension of the tree-ring-based and Barbados coral-based calibration curve back to 3.2×10^4 years based on ^{14}C and ^{230}Th measurements on a stalagmite from a submerged cave in the Bahamas. The growth of the stalagmite came to a halt $\sim 1.1 \times 10^4$ years ago when the rising sea invaded the cave. The upper portion of the record which overlaps in time the Barbados coral record was used to establish the initial ^{230}Th to ^{232}Th ratio and also the reservoir correction for the ^{14}C ages. Source: Beck *et al.* (2001).

13.4.4 Cause of the Long-Term ^{14}C Decline

One might ask what accounts for the decline in atmospheric $^{14}C/C$ ratio over the last 2.5×10^4 years. The prime suspect is the drop to near-zero values in the Earth's magnetic field $\sim 4 \times 10^4$ years ago (Bonhommet and Zähringer, 1969). This drop, referred to as the Laschamp event, is named after a volcanic field in France where it was first identified. The existence of this event has now been confirmed by measurements of magnetic intensity in deep-sea sediments that reveal a strong minimum close to 4×10^4 years ago. Also, it shows up as a doubling of the abundance of ^{10}Be (another cosmogenic isotope created in the Earth's atmosphere) at this time in the Greenland ice cores (Yiou *et al.*, 1997). Both the drop to near-zero values of magnetic intensity and the doubling of the ^{10}Be concentration are consistent with a temporary shutdown of the Earth's field (presumably associated with a false polarity reversal). The existence of the Laschamp event has provided a strong impetus to extend the calibration curve to times before this temporary shutdown of the Earth's magnetic field. However, as already stated, this challenge will prove to be an extremely difficult one.

It is unlikely that the Laschamp event alone can explain the observed decline in the $^{14}C/C$ ratio. The duration of this event (several thousand years) was too short and the magnitude of its impact on ^{14}C production was too small to produce a large enough increase in ^{14}C inventory to persist for $\sim 2 \times 10^4$ years. After 2.3×10^4 years had elapsed, the excess would have been diminished by a factor of 16. However, as shown by a number of marine sediment-based magnetic field reconstructions (see Laj *et al.*, 1996), the Laschamp event was followed by a long-term buildup in the strength of the Earth's magnetic field (see Figure 7) (see Bard, 1998, for summary). This slow buildup appears to be adequate to account for the post-4×10^4 years decline in atmospheric ^{14}C.

But there is a fly in this ointment. ^{10}Be measurements on Greenland ice (Muscheler *et al.*, 2005) fail to show the decrease expected were the strength of the magnetic field to have increased in accord with the Laj *et al.* (1996) reconstruction. This anomaly led Chiu (2005) to challenge the accepted half-life for radiocarbon. Were it 6,400 rather than 5,730, the steep downward trend in the ^{14}C to C ratio for atmospheric carbon would be removed. She notes that the measurements (by gas counting) on which the accepted half-life is based were made in the late 1950s. Spurred by Chiu's challenge, an effort is underway to redetermine this important half-life (Southon, personal communication). A long shot, but something that must be done!

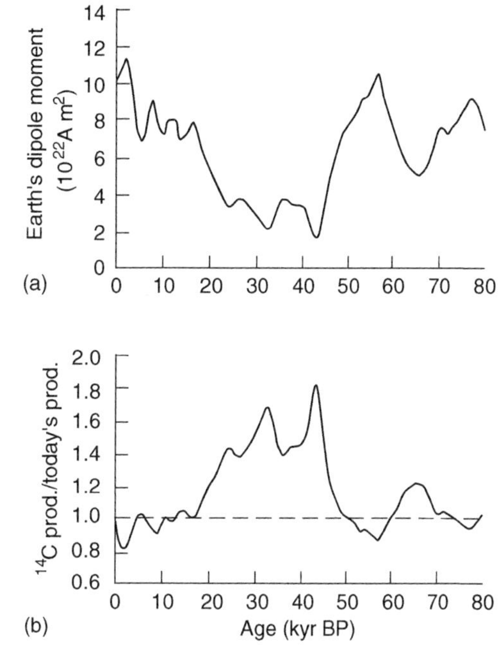

Figure 7 (a) The reconstruction of the Earth's dipole moment (Laj *et al.*, 1996) based on magnetic measurements made on deep-sea sediments and calibrated against measurements on volcanic rocks of known age. (b) The radiocarbon production rates reconstructed from the magnetic dipole reconstruction (a). Frank *et al.* (1997) have shown that a similar reconstruction is obtained based on ice-core ^{10}Be results.

13.4.5 Change in Ocean Operation

In addition to attempts to extend the radiocarbon calibration curve back further in time, there have also been attempts to enhance its detail. One of these attempts stands out. Konrad Hughen, while a graduate student at the University of Colorado, carried out a detailed set of radiocarbon measurements on planktonic foraminifera shells contained in varved sediment from the Cariaco Basin just off Venezuela. One of the attributes of this core is that the interval correlating with the Younger Dryas cold event is marked by a distinct color change. Further, the boundaries of this interval are very sharp. Hughen used his radiocarbon measurements to splice the Cariaco varve sequence onto the German tree-ring sequence. He then used his varve counts to extend the calibration record several thousand years beyond that established using tree rings (i.e., through the entire Younger Dryas and the underlying Bölling Allerod warm period). His resulting reconstruction revealed a very exciting feature (see Figure 8). During the first 200 years of the Younger Dryas, the $^{14}C/C$ ratio in upper ocean dissolved inorganic carbon soared by 5%. Then, during the remaining 1,000 years of this cold snap, the radiocarbon content drifted back down. The coincidence of the initiation of the $^{14}C/C$ ratio rise with the onset of the Younger

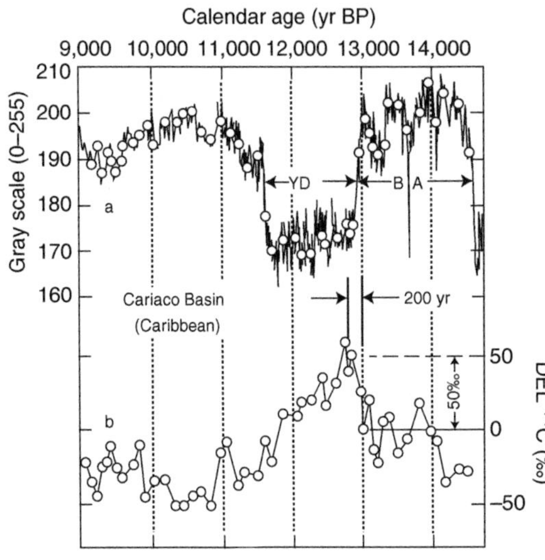

Figure 8 Reconstruction of the radiocarbon content of atmospheric CO_2 during the deglaciation interval (Hughen *et al.*, 1998). This reconstruction was obtained from a varved sediment core from the Cariaco Basin. The position of the individual ^{14}C measurements is shown by the circles. The Younger Dryas (YD) and Bolling-Allerod (BA) intervals are clearly shown in gray-scale record. As can be seen, the ^{14}C content of surface ocean carbon rose by 50 ± 10‰ in the first two centuries of the Younger Dryas event. After that, it slowly declined. Assuming that the air–sea difference in the $^{14}C/C$ ratio remained the same as found today, these changes can be transferred to atmospheric CO_2.

Dryas led Hughen to propose that this rise was tied to a change in the rate of oceanic mixing rather than to a change in magnetic shielding. Such a ^{14}C rise would be a logical consequence of a shutdown in production of new deep water in the northern Atlantic Ocean. In today's ocean, roughly three quarters of the ^{14}C atoms resupplying the deep-sea inventory enter via this route. A number of authors have shown that the sudden release of melt water stored in proglacial Lake Agazzis into the St. Lawrence drainage at the time of the onset of the Younger Dryas would have flooded the northern Atlantic with enough freshwater to shut down the Atlantic's conveyor circulation.

If Hughen's hypothesis proves to be correct then one might ask: why not attribute other aspects of offset of radiocarbon years from calendar years to changes in ocean mixing? In particular, could the higher atmospheric $^{14}C/C$ ratio during peak glacial time ($(2.3-1.5) \times 10^4$ years ago) be attributed to more sluggish ocean circulation? Here the radiocarbon method itself provides an answer. It is based on differences between the radiocarbon age of coexisting planktonic (surface-dwelling) and benthic (bottom-dwelling) foraminifera shells picked from glacial sections of deep-sea sediments. Owing to rapidity of CO_2 exchange with the atmosphere, the $^{14}C/C$ ratio in surface ocean dissolved inorganic carbon is closely linked to that in atmospheric CO_2. Thus, the difference between the $^{14}C/C$ ratio in surface water and that in deep water recorded by coexisting planktic and benthic foraminifera shells provides a record of how the rate of deep-sea ventilation differed between late glacial time ($(2.3-1.5)\times10^4$ years ago) and the Holocene (1.1×10^4 years to present). While radiocarbon measurements on coexisting benthic and planktic foraminifera pairs show that during glacial time the deep Atlantic had a ventilation age close to that of today's for the deep Pacific (Robinson *et al.*, 2005), those for samples down to 2,800 m in the Pacific suggest that the ventilation rate in the Pacific was not significantly different from todays (Barker and Broecker, in press; Barker *et al.*, in press). The failure to find evidence for reduced ventilation of the deep Pacific suggests that if isolation of the deep sea's carbon played a significant role in creating the elevated radiocarbon content of the glacial atmosphere, then there must have been an abyssal reservoir highly depleted in radiocarbon. The excessively salty glacial water documented by Adkins *et al.* (2002) in the abyssal Southern Ocean is a prime candidate (see Shackleton *et al.*, 1988; Broecker *et al.*, 1990).

Another radiocarbon-based study showed just how rapidly changes in the patterns of deep ventilation in the ocean can occur. As part of his PhD research with MIT's Ed Boyle, Jess Adkins made a study of a single benthic coralite, which grew at a depth of 1.8 km in the northwestern Atlantic Ocean during the early stages of the last deglaciation. He found an amazing thing. While the base and crest of this several centimeter-high mushroom-shaped coralite yielded nearly identical ^{230}Th ages of 1.54×10^4 years, the ^{14}C age of the base was 1.385×10^4 years and that of the crest 1.452×10^4 years (Adkins *et al.*, 1998). As this coralite probably formed in a period of just a few decades, this difference in radiocarbon age requires that the water in which the coralite grew underwent a sudden and large drop in the $^{14}C/C$ ratio. When Adkins compared his ^{14}C to ^{230}Th age offsets with those obtained by the Fairbanks team, he concluded that the apparent deep Atlantic ventilation age jumped from ~400 to ~1,100 years. Puzzled by how this could have happened, he measured the cadmium content of the calcite along a traverse from the stem to the coralite crest and found that it increased from 0.11 to 0.18 μmol Cd per mol Ca (see Figure 9). Only one explanation seems to fit these observations. The production of low-cadmium content and high-radiocarbon water in the northern Atlantic must have come to an abrupt halt allowing the more dense waters with higher cadmium content and lower radiocarbon from the Southern Ocean to flood northward, displacing upward the less-dense waters of the North Atlantic origin.

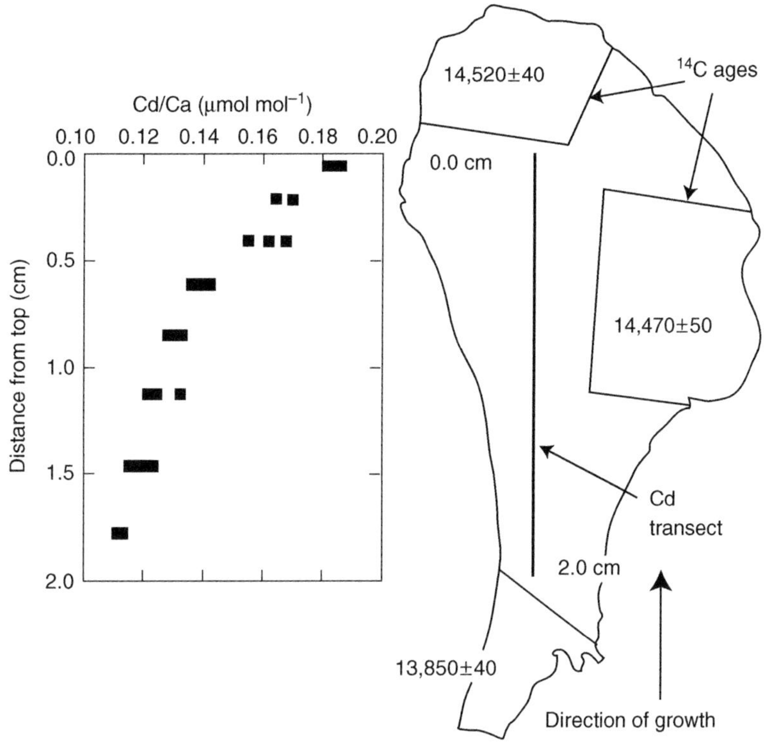

Figure 9 Measurements of $^{14}C/C$ ratio in a benthic coral from 1.8 km depth in the northern Atlantic reveal that the base has an apparent age 650 radiocarbon years younger than the top. Yet ^{230}Th ages on the base and crest are nearly identical (1.54×10^4 calendar years). Adkins *et al.* (1998) attribute this difference to an abrupt invasion of low ^{14}C and high cadmium content Southern Ocean water. Reproduced from Adkins *et al.* (1998).

For all these studies, there is a necessity to establish the reservoir age for the glacial surface ocean. Several authors (Bard *et al.*, 1994; Sikes *et al.*, 2000; Waelbroeck *et al.*, 2001; Siani *et al.*, 2001) have attempted to do this and find puzzling older reservoir ages for glacial times. Clearly, more effort must go into such reconstructions.

13.5 RADIOCARBON AND SOLAR IRRADIANCE

As solar energy output and radiocarbon production are both tied to sunspot activity, it might be possible to reconstruct past irradiance variations from the perturbations in the atmosphere's $^{14}C/C$ ratio reconstructed from measurements on tree rings of known calendar age. Indeed, based on radiocarbon reconstructions, Stuiver (1980) and Stuiver and Braziunas (1993) suggested that the Little Ice Age might be the result of reduced solar luminosity during periods such as the Maunder minimum when sunspot activity was shut down. However, because the records Stuiver used were quite short and since changes in solar luminosity were, at the time his paper was written, poorly documented, his proposal failed to receive wide acceptance.

The advent of satellite observations of the solar output made it clear that there is indeed a tie between energy output and sunspot activity (see Figure 10). However, as these variations are very small and the record covers only the last two decades, this finding made only a modest impact with regard to interest in the Sun as a driver of Earth climate fluctuations.

The situation changed with the publication of a paper by Bond *et al.* (2001), in which it was shown that a correlation exists between the cosmic ray flux changes required to generate the tree-ring-based reconstruction of atmospheric $^{14}C/C$ ratio record for the last 1×10^4 years and an ice-rafting index for the radiocarbon-dated northern Atlantic sediments developed by Bond *et al.* (1997, 1999) (see Figure 11). This index involves the ratio of red-coated (i.e., iron-stained) to total ice-rafted grains. During the Holocene, the index has fluctuated on a timescale on average of 750 years (i.e., 1,500 years for a complete cycle) between highs of ~16% and lows of only a few percent. Bond reasons that times of high index represent cold spells and those of low index, warm spells. His argument is based on the observation that the sources of red-coated grains lie poleward of sources devoid of red-coated grains, and hence, to get them to the site of his deep-sea cores requires colder surface water conditions. As a confirmation that the changes in the $^{14}C/C$ ratio reflect production rather than changes in

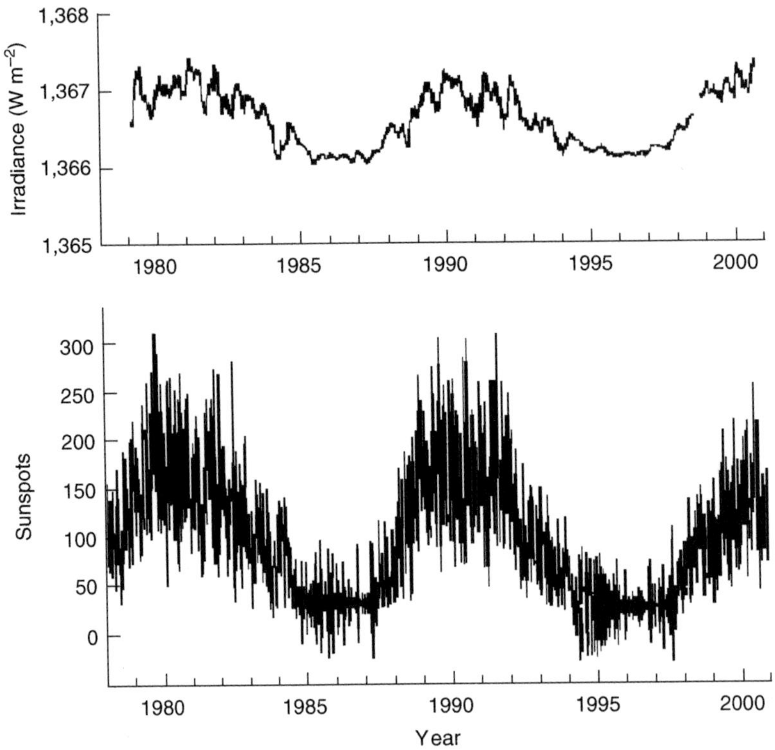

Figure 10 Comparison of the solar irradiance with the sunspot numbers for the last two Schwabe cycles. The irradiance record is a compilation of data from different satellites. During periods of high solar activity there are more sunspots darkening a small part of the solar disk (visible in the negative excursions of the irradiance). However, the brightness of the Sun is increased at the same time, overcompensating the darkening effect of the sunspots.

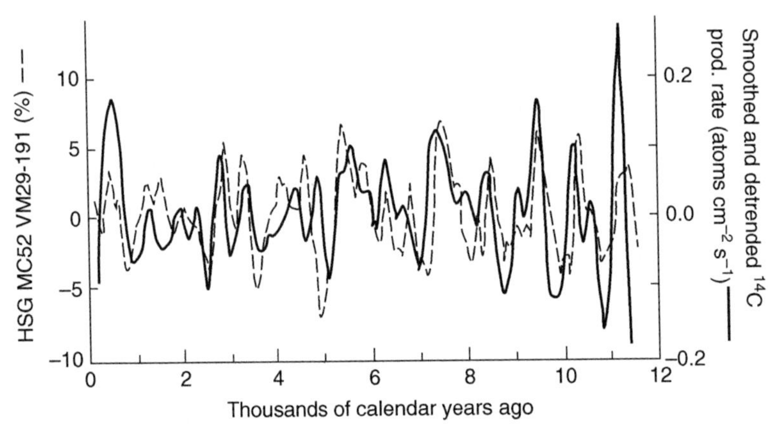

Figure 11 Comparison between the fluctuations in the fraction of red-coated lithic grains (dashed curve) with that of cosmogenic nuclide production (smooth curve) in the Earth's atmosphere. This correspondence provides powerful evidence that the Holocene's small cyclic temperature changes were paced by changes in solar luminosity. Source: Bond *et al.* (2001).

ocean circulation, Bond *et al.* (2001) point to the Holocene record of ^{10}Be in Greenland ice cores, which yield a close match in timing and in amplitude of the cosmic ray flux changes reconstructed from ^{14}C measurements.

A confirmation of Bond's climate interpretation has been obtained from the dating of wood and peat being carried by summer melt water from beneath the retreating mountain glaciers in the European Alps (Hormes *et al.*, 1998). The forests and bogs in which these materials were generated must date from warmer times when the glaciers were even smaller than they are today. Three such times have been documented by radiocarbon dating many tens of wood and peat samples. They match quite well with three of Bond's warm intervals (see Figure 11). Further, evidence in support of Bond's temperature cycles comes from fossil-tree remains found north of the present tree line on Russia's Kola Peninsula (Hiller *et al.*, 2001). They document the existence of a northward expansion of the forest during medieval time (1000–1300). This corresponds to the period during which the Vikings colonized southern Greenland.

13.6 THE "BOMB" ^{14}C TRANSIENT

During the late 1950s and early 1960s, hydrogen bomb tests carried out in the atmosphere by Russia, Britain, and United States led to the production of manmade radiocarbon atoms. In this case, the neutrons that collided with atmospheric nitrogen nuclei were by-products of the fusion reactions. Large-scale anthropogenic production of ^{14}C came to a halt with the implementation of the ban on atmospheric weapons tests on January 1, 1963. At that time, the number of manmade ^{14}C atoms in the atmosphere was roughly equal to the number of natural radiocarbon atoms (i.e., the $^{14}C/C$ ratio nearly doubled). Since then, this atmospheric excess has steadily dwindled until as of the year 2000, the bomb-produced atoms constituted only ~10% of the atmospheric radiocarbon inventory (see Figure 12). This decrease has three causes:

1. exchange with oceanic $\sum CO_2$,
2. exchange with terrestrial biospheric carbon, and
3. continued addition of ^{14}C-free fossil-fuel-derived CO_2 molecules to the atmosphere.

Figure 13 shows an estimate as to how the distribution of bomb radiocarbon atoms among these active carbon reservoirs has evolved.

13.6.1 Radiocarbon as a Tracer for Ocean Uptake of Fossil Fuel CO_2

While $^{14}CO_2$ is in a sense a perfect tracer for fossil fuel CO_2, the quite different time histories for their inputs and the large difference between the times required for the surface ocean to respond to isotopic compared with chemical transients makes the task more complicated than it might seem. Even so, the situation for the ocean turns out to be simpler than that for the land. There are

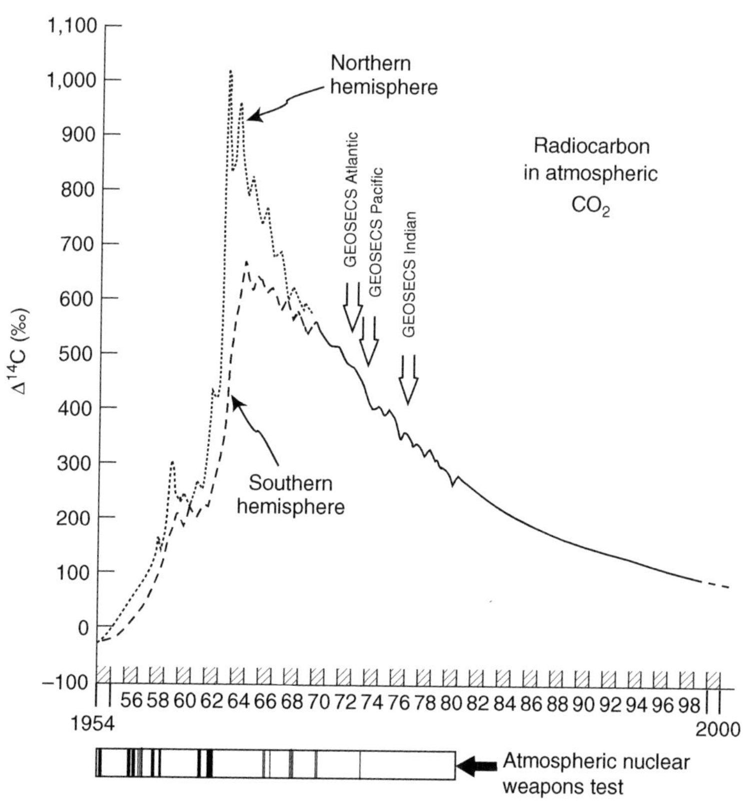

Figure 12 $\Delta^{14}C$ for atmospheric ^{14}C following the onset of H-bomb testing in 1952. The buildup continued until mid-1963 and since then the $\Delta^{14}C$ to C ratio has been declining toward its prenuclear value as the excess bomb testing ^{14}C atoms are transferred to the ocean and terrestrial biosphere. Although China and India, who were not signatories to the test ban treaty explode, conducted atmospheric tests in the late 1960s and early 1970s, the amount of ^{14}C produced was negligible. Within decades, the $\Delta^{14}C$ will drop below zero due to the continued emission of ^{14}C-free fossil fuel CO_2. The times of the GEOSECS ocean surveys are shown. The pre-1980 data are from Nydal and Lövseth (1983).

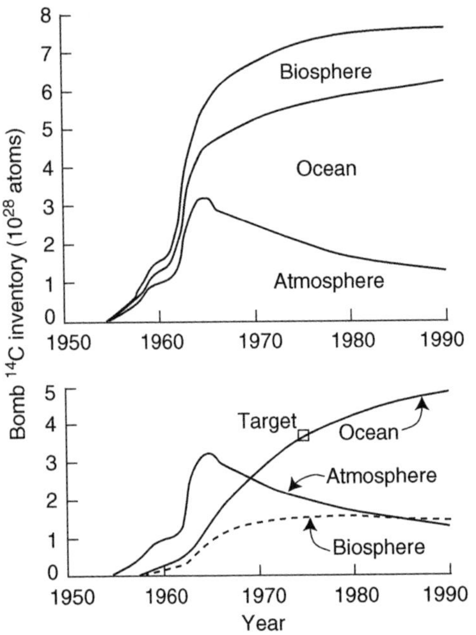

Figure 13 Distribution of bomb testing ^{14}C between the atmosphere, ocean, and terrestrial biosphere as reconstructed by Broecker and Peng (1994). The ocean contribution is obtained by a model constrained by the inventory based on the GEOSECS survey. The contribution of the terrestrial biosphere is based on estimates of the biomass and turnover times for trees and active soil humus.

three reasons for this. First, the ocean has a far greater degree of lateral homogeneity. Second, uptake by the sea is governed by basic inorganic chemistry while that by the land biota involves complex biotic cycles. Third, the documentation of the radiocarbon transient through direct measurement is far more extensive for the ocean.

13.6.2 Ocean Uptake of $^{14}CO_2$ and CO_2

As of early 2000s, it has not been possible to make an accurate direct assessment of the amount of fossil fuel CO_2 taken up by the sea. The reason is related to the fact that even in the surface mixed layer the increase can have been no more than 2–3% (10% increase in the atmosphere's CO_2 partial pressure produces only a 1% increase in the surface ocean's dissolved inorganic carbon content (i.e., $CO_2+HCO_3^-+CO_3^{2-}$)). Unfortunately, until the GEOSECS global survey (1972–1978) no measurements of the concentration of dissolved inorganic carbon (i.e., $CO_2+HCO_3^-$ and CO_3^{2-}) in the ocean of sufficient accuracy were made. Further, as the GEOSECS measurements are accurate to only ±1%, even they are not quite up to the task. Only with the TTO surveys in the Atlantic during the early 1980s and the WOCE surveys in the Pacific and Indian oceans during the early 1990s was a base level established with sufficient accuracy (i.e., 0.1%). Hence, the uptake must be estimated using ocean–atmosphere models rather than direct measurements. Whether these models are of the simple box-diffusion or complex three-dimensional general circulation variety, it is necessary to calibrate (in the case of box models) or constrain (in the case of general circulation models) them with tracer measurements. The most powerful tracer for this purpose is the radiocarbon produced during the atmospheric testing of nuclear weapons.

Two resistances limit the rate at which bomb $^{14}CO_2$ and fossil fuel CO_2 are transferred from the atmosphere to the sea. The first is transfer across the air–sea interface and the second is the rate at which surface waters are mixed into the ocean's interior. Clearly, the relative strengths of these two resistances are of importance. Were the resistance posed by vertical mixing in the ocean to be much smaller than that posed by CO_2 exchange across the air–sea interface, then the air-to-sea difference in CO_2 partial pressure and in $^{14}CO_2$ partial pressure would remain large, and the rate of uptake by the sea would be dominated by the rate of air-to-sea CO_2 exchange. Were the case opposite, then ocean uptake would be limited by the rate of vertical mixing within the sea. It turns out that for CO_2 itself the resistance posed by vertical mixing in the sea dominates. Waters in the tens-of-meters thick wind-mixed layer are able to reach 85% or so of saturation with the atmosphere's fossil fuel CO_2 burden before being mixed into the interior. For $^{14}CO_2$ the situation is quite different. The resistance posed by the interface is comparable with that posed by vertical mixing within the sea. The reason for this difference between $^{14}CO_2$ and CO_2 has to do with the chemical speciation of carbon in surface water. The proportions of CO_2, HCO_3^-, and CO_3^{2-} in the surface ocean are roughly 10:1, 800:200. While isotopic equilibration requires that the ^{14}C in all three species be exchanged, chemical equilibration is accomplished by the reaction

$$CO_2 + CO_3^{2-} + H_2O \rightarrow 2HCO_3^- \qquad (2)$$

Because of this, the time required for chemical equilibration turns out to be roughly an order of magnitude smaller (i.e., 200/1,800) than that for isotopic equilibration.

Taken together, measurements of the vertical distribution of bomb radiocarbon in the sea and the air-to-surface-sea difference in the bomb ^{14}C/C ratio permit both the laterally averaged rate of air–sea CO_2 exchange and the laterally averaged rate of vertical mixing to be estimated. As part of the GEOSECS survey, the distribution of radiocarbon throughout the entire world ocean was measured (in the laboratories of Minze Stuiver (University of Washington) and Gote Ostlund (University of Miami)). As this survey was conducted 12 ± 3 years after the implementation of the ban on atmospheric testing, the average bomb ^{14}C atom had a little more than a decade to enter the sea and be mixed into its interior.

To make use of this transient, the contributions of the natural and bomb radiocarbon had to be separated. Three sets of observations went into this separation. First, use was made of $^{14}C/C$ measurements on surface waters collected very early in the nuclear testing era (Broecker *et al.*, 1960) and also with results of measurements on pre-nuclear growth ring-dated corals and mollusks (Druffel and Linick, 1978; Druffel, 1981, 1989). Second, use was made of the tritium released to the atmosphere during nuclear tests. As this bomb-test tritium swamped the natural tritium present in the ocean, the vertical distribution of tritium in the sea could be used to establish the limit of penetration of bomb radiocarbon. Finally, based on the radiocarbon analyses made on thermocline waters free of bomb tritium, it was shown that there was a close correlation between the natural $^{14}C/C$ ratio and the dissolved silica content of the water (Broecker *et al.*, 1995). More recently, it was shown that there was an even better correlation between natural radiocarbon and salinity-normalized alkalinity. These relationships were essential in separating the two "brands" of ^{14}C in the waters from the upper portion of the main oceanic thermocline.

Taken together, the distributions of natural and bomb radiocarbon in the sea give us an important insight regarding the manner in which the ocean mixes. The steady-state distribution of natural radiocarbon constrains the replacement timescale for waters in the deep sea to be on the order of one millennia. The transient distribution of bomb radiocarbon at the time of the GEOSECS survey tells us that on a timescale of a decade, radiocarbon atoms are able to exchange with $\sim$10% of the ocean's dissolved inorganic carbon (i.e., $CO_2 + HCO_3^- + CO_3^{2-}$). Taken together, these two constraints suggest that the fraction of the ocean volume accessed by vertical mixing increases with the square root of penetration time. This concept gave rise to the early one-dimensional box-diffusion models for fossil fuel CO_2 uptake by the ocean (Oeschger *et al.*, 1975; Seigenthaler *et al.*, 1980). Such models consist of a well-mixed atmosphere and a well-mixed surface ocean layer underlain by a diffusive half-space. The so-called eddy diffusivity assigned to this half-space was set by fitting the horizontally averaged vertical distribution of bomb radiocarbon.

One might conclude that the crude box-model representation of the ocean would have been soon eclipsed by ocean general circulation models capable of simulating, in three dimensions, the full suite of ocean currents and mixing processes. While certainly a critical step in the evolution of our ability to predict the split of fossil fuel CO_2 between the ocean and atmosphere, the task of creating models that match the distributions of not only temperature and salinity, but also those of natural and bomb radiocarbon, is a daunting one. Properly simulating the flow pathways and flow rates in the 600–1,200 m depth range dominated by the ocean's intermediate water masses has proven particularly challenging. Consequently, fossil fuel CO_2 uptake estimates made with these three-dimensional models are, in my estimation, as yet no more reliable than those based on the one-dimensional ^{14}C-calibrated box-diffusion models.

Both types of models suggest that $35 \pm 5\%$ of the CO_2 produced to date by fossil fuel burning has been taken up in the ocean. This represents $\sim$16% of the ocean's capacity for fossil fuel CO_2 uptake (i.e., of the amount of uptake were the entire volume of the ocean to be equilibrated with the atmosphere). One might ask why this percentage is higher than that based on the distribution of bomb radiocarbon. The answer is that while the ^{14}C atoms had at the time of the GEOSECS survey been in existence for only $\sim$12 years, the average existence time for fossil fuel CO_2 molecules is more like 30 years. Since the square root of 30/12 is 1.6, hence, to the extent that the diffusive characterization of ocean mixing is correct, fossil fuel CO_2 should have been able to penetrate a 1.6 times larger volume than bomb ^{14}C.

The challenge for ocean modelers is to achieve a match with the observed distributions of bomb ^{14}C not only for the time of the GEOSECS survey (the 1970s) but also for the time of the TTO (1980–1981) and WOCE (early 1990s) surveys. Only then can modelers be confident that they can predict the split in excess CO_2 between atmosphere and ocean for any given future fossil-fuel-use scenario. This is tricky because the uptake capacity for CO_2 by surface ocean waters depends on their carbonate ion concentration. Clearly, as CO_2 builds up in the atmosphere, the carbonate ion concentration in surface waters will decrease, but the rate of decrease will depend on the manner in which the ocean mixes. Advective overturning brings "virgin" water high in CO_3^{2-} concentration to the surface. However, diffusive mixing creates an ever-increasing carbonate ion gradient between "spent" surface water and "virgin" deep water. Different models yield different ratios of diffusive to advective mixing.

13.6.3 Terrestrial Uptake of $^{14}CO_2$ and CO_2

In recent years, the response of the terrestrial biosphere to the increase in the atmosphere's CO_2 content has become a hot topic. Thousands of chamber experiments show that at least on the short term (a year or two) given more CO_2, plants grow faster. Further, global terrestrial carbon inventories based on the rate of the atmosphere's CO_2 increase and O_2 decline indicate that storage is on the increase. This increase is happening despite large-scale deforestation that, of course, drives down carbon stocks. Three factors appear to be responsible for this increase: (1) the excess CO_2 in the atmosphere; (2) the dispersal of fixed nitrogen evaporated from farmland and given off in automobile exhausts; and (3) regrowth of forests on land previously cleared for farming. Unlike the ocean, the land surface is a highly checkered mosaic of vegetation types and growth histories making the development of models capable of predicting future storage an enormously challenging task.

While radiocarbon is not nearly as valuable to this exercise as it is in the case of the ocean, it does have a role to play. More than half of the terrestrial carbon inventory is stored in soils. The humus in soils consists of a host of complex organic compounds. The evolution of storage in this reservoir will be driven by two competing impacts. Increasing planetary temperature will lead to more rapid oxidation of these humic compounds and hence will tend to drive down the planetary inventory. In contrast, increasing plant growth will lead to increased storage of new humic compounds and hence tend to drive up the inventory.

Radiocarbon's role is to constrain the turnover times for the compounds making up humus. The temporal trend of the $^{14}C/C$ ratio in soil humus (from prenuclear time to the present) makes it possible to separate the contribution of "passive" compounds that have very long soil residence times (measured in hundreds to thousands of years) and hence will be little changed by man's activities, and those "active" compounds which have relatively short chemical lifetimes (measured in decades). It also allows the characterization of the average turnover time of these "active" components. Further, by conducting such measurements on soils from different climate zones, it is possible to get a handle on how the turnover time of active compounds depends on temperature. Clearly, knowledge of this dependence is critical to the prediction of future global humus inventories.

Unfortunately, there was no GEOSECS-type survey of the world's soils during the 1970s. The decay counting method in use at that time required such large soil samples that processing was extremely cumbersome and stored samples were, by and large, far too small. Hence, interest in this subject lagged until the mid-1980s when atom counting became available. And of course, interest was further heightened when the Kyoto Accord permitted excess carbon storage in soils to be deducted from CO_2 emissions. However, unlike the ocean whose bomb ^{14}C inventory continues to rise, that in soils peaked in the 1970s and has subsequently declined. Hence, an enormous opportunity was largely missed—"largely" because, fortunately, here and there soils have been collected and stored.

As part of his PhD thesis research at Lamont–Doherty, Kevin Harrison summarized his own and published radiocarbon measurements on soils (see Figures 14 and 15). He was able to show three quite important things (Harrison *et al.*, 1993a, b):

1. To reconcile both the lower than modern pre-1950 $^{14}C/C$ ratios in soil humus and the time history of the bomb ^{14}C-induced rise in the $^{14}C/C$ ratio, he concluded that about one

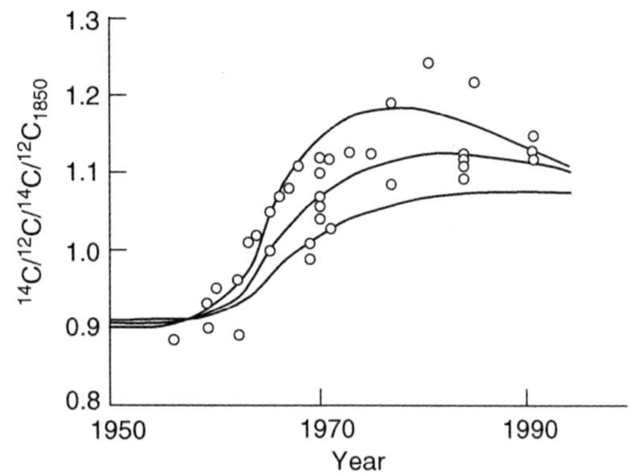

Figure 14 Plot of measurements of the $^{14}C/C$ ratio in untilled soils from a range of locales. The results are expressed as the $^{14}C/C$ ratio in the bulk soil carbon to that in age-corrected 1850 wood. The curves are based on a model which assumes that 25% of the carbon is essentially inert (mean ^{14}C age 3,700 years) and that 75% turns over on timescales of 40 years (lower curve), 25 years (middle curve) and 15 years (upper curve). Source: Harrison *et al.* (1993a).

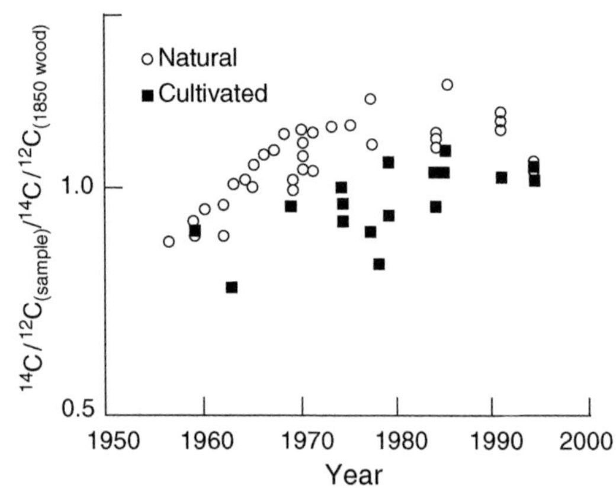

Figure 15 Comparison of ^{14}C to C measurements on natural soils and those on cultivated soils. The lower ratios are presumably the result of the loss of a sizable portion of the active humus component as the result of agricultural practice. Source: Harrison *et al.* (1993a).

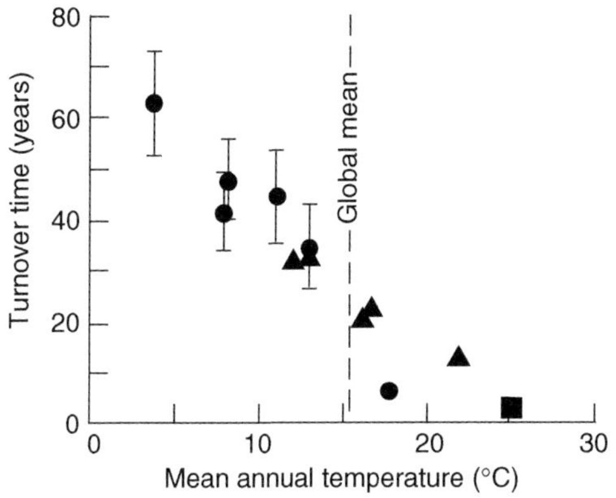

Figure 16 Turnover time for the active humus component of soils from the Brazilian rainforest (square), Hawaii (triangles), and the Sierra Nevada Mountains (circles). The mean global temperature is shown for reference. Source: Trumbore *et al.* (1996).

quarter of the carbon in soils is in either inactive (i.e., turns over very slowly) or totally inert (immune to destruction) compounds.

2. To account for the temporal evolution of bomb ^{14}C in the soil, he concluded that the mean residence time of the active humus component was 25 ± 10 years.
3. To account for the lower $^{14}C/C$ ratio found for agricultural soil he concluded that a sizable portion of the active component had been lost as the result of farming practices (tilling, harvesting, etc.) thereby enhancing the contribution of the low ^{14}C inactive component.

Sue Trumbore of the University of California, Irvine, demonstrated that the turnover time of humic material is strongly dependent on soil temperature ranging from 100 years at high latitudes to as little as a decade in the tropics (Trumbore *et al.*, 1996) (see Figure 16).

13.7 FUTURE APPLICATIONS

The recent development of a much smaller and hence less-expensive accelerator mass spectrometer capable of state-of-the-art radiocarbon measurements will allow large users to have an in-house unit rather than having to send their samples off to one of the dozen or so existing major centers. These small units operate at 5×10^5 V rather than at several million volts as do the conventional accelerator mass spectrometers. Also they will fit into a standard-size laboratory rather than the one with the size of a small airplane hanger. Further, like conventional mass spectrometers, these so-called tandys (i.e., small tandem Van de Graaff) can be maintained and operated by the users rather than by a team of specialists. Thus, measurements, which now cost from $300 to $600 each, will be carried out for more like $100 each. This will allow scientists to take on problems requiring hundreds of measurements rather than being budgetarily restricted to projects involving only tens of measurements.

As the bomb ^{14}C transient demonstrated the value of radiocarbon as an environmental tracer, availability of "tandy" might lead to small-scale tracer experiments designed to study the allocation of carbon by plants in natural environments. Because of the high accuracy and high sensitivity of ^{14}C measurements, such tracer experiments can be carried out at a few times the ambient atmospheric $^{14}C/C$ ratio, and hence pose no environmental hazard.

REFERENCES

Adkins J. F., Cheng H., Boyle E. A., Druffel E. R. M., and Edwards R. L. (1998) Deep-sea coral evidence for rapid change in ventilation of the deep North Atlantic 15,400 years ago. *Science* **280**, 725–728.

Adkins J. F., McIntyre K., and Schrag D. P. (2002) The salinity, temperature and $\delta^{18}O$ of the glacial deep ocean. *Science* **298**, 1769–1773.

Bard E. (1998) Geochemical and geophysical implications of the radiocarbon calibration. *Geochim. Cosmochim. Acta* **62**, 2025–2038.

Bard E., Arnold M., Mangerud J., Paterne M., Labeyrie L., Duprat J., Mélières M.-A., Sønstegaard E., and Duplessy J.-C. (1994) The North Atlantic atmosphere-sea surface ^{14}C gradient during the Younger Dryas climatic event. *Earth Planet. Sci. Lett.* **126**, 275–287.

Bard E., Hamelin B., Fairbanks R. G., and Zindler A. (1990a) Calibration of ^{14}C timescale over the past 30,000 years using mass spectrometric U–Th ages from Barbados corals. *Nature* **345**, 405–410.

Bard E., Hamelin B., Fairbanks R. G., Zindler A., Arnold M., and Mathieu G. (1990b) U/Th and ^{14}C ages of corals from Barbados and their use for calibrating the ^{14}C time scale beyond 9,000 years BP. In *Proceedings of the 5th International Conference on AMS* (eds. F. Yiou and G. Raisbeck). *Nucl. Inst. Met.* **B52**, 461–468.

Bard E., Raisbeck G. M., Yiou F., and Jouzel J. (1997) Solar modulation of cosmogenic nuclide production over the last millennium: comparison between ^{14}C and ^{10}Be records. *Earth Planet. Sci. Lett.* **150**, 453–462.

Barker S. and Broecker W. A 190 per mil drop in atmosphere's $\Delta^{14}C$ during the "Mystery Interval" (17.5 to 14.5 kyrs). *Quat. Sci. Rev.* (in press).

Barker S., Broecker W., Clark E., Hajdas I., Bonani G., Moreno E., and Stott L. Radiocarbon age of deglacial-age water from 2.8 km depth in the Western Equatorial Pacific. *Geochem. Geophys. Geosys.* (in press).

Beck J. W., Richards D. A., Edwards R. L., Silverman B. W., Smart P. L., Donahue D. J., Hererra-Osterheld S., Burr G. S., Calsoyas L., Jull A. J. T., and Biddulph D. (2001) Extremely large variations of atmospheric ^{14}C concentration during the last glacial period. *Science* **292**, 2453–2458.

Becker B. and Kromer B. (1993) The continental tree-ring record—absolute chronology, ^{14}C calibration, and climatic change at 11 ka. *Palaeogeogr. Palaeoclimatol. Palaeoecol.* **103**, 67–71.

Bond G. C., Kromer B., Beer J., Muscheler R., Evans M. N., Showers W., Hoffmann S., Lotti-Bond R., Hajdas I., and Bonani G. (2001) Persistent solar influence on North Atlantic climate during the Holocene. *Science* **294**, 2130–2152.

Bond G. C., Showers W., Cheseby M., Lotti R., Almasi P., deMenocal P., Priore P., Cullen H., Hajdas I., and Bonani G.

(1997) A pervasive millennial-scale cycle in North Atlantic Holocene and glacial climates. *Science* **278**, 1257–1266.

Bond G. C., Showers W., Elliot M., Evans M., Lotti R., Hajdas I., Bonani G., and Johnson S. (1999) The North Atlantic's 1–2 kyr climate rhythm: relation to Heinrich events, Dansgaard/Oeschger cycles, and the Little Ice Age. In *Mechanisms of Global Climate Change at Millennial Time Scales* (eds. P. Clark, R. Webb, and L. D. Keigwin). Geophysical Monograph Series 112, American Geophysical Union, Washington, DC, pp. 35–58.

Bonhommet N. and Zähringer J. (1969) Paleomagnetism and potassium argon determinations of the Laschamp geomagnetic polarity event. *Earth Planet. Sci. Lett.* **6**, 43–46.

Broecker W. S., Gerard R., Ewing M., and Heezen B. C. (1960) Natural radiocarbon in the Atlantic Ocean. *J. Geophys. Res.* **65**, 2903–2931.

Broecker W. S. and Peng T.-H. (1994) Stratospheric contribution to the global bomb radiocarbon inventory: model versus observation. *Global Biogeochem. Cycles* **8**, 377–384.

Broecker W. S., Peng T.-H., Trumbore S., Bonani G., and Wolfli W. (1990) The distribution of radiocarbon in the glacial ocean. *Global Biogeochem. Cycles* **4**, 103–117.

Broecker W. S., Sutherland S., and Smethie W. (1995) Oceanic radiocarbon: separation of the natural and bomb components. *Global Biogeochem. Cycles* **9**, 263–288.

Chiu C. (2005) PhD Thesis, Columbia University

deVries H. L. (1958) Variation in concentration of radiocarbon with time and location on Earth. *Proc. Koninkl. Ned. Akad. Wetenschap.* **61**, 94–102.

Druffel E. R. M. (1981) Radiocarbon in annual coral rings from the eastern tropical Pacific Ocean. *Geophys. Res. Lett.* **8**, 59–62.

Druffel E. R. M. (1989) Decade time scale variability of ventilation in the North Atlantic: high-precision measurements of bomb radiocarbon in banded corals. *J. Geophys. Res.* **94**, 3271–3285.

Druffel E. R. M. and Linick T. W. (1978) Radiocarbon in annual coral rings of Florida. *Geophys. Res. Lett.* **5**, 913–916.

Edwards R. L., Chen J. H., and Wasserburg G. J. (1986/1987) ^{238}U–^{234}U–^{230}Th–^{232}Th systematics and the precise measurement of time over the past 500,000 years. *Earth Planet. Sci. Lett.* **81**, 175–192.

Fairbanks R. G. (1989) A 17,000-year glacio-eustatic sea level record: influence of glacial melting rates on the Younger Dryas event and deep ocean circulation. *Nature* **342**, 637–647.

Fairbanks R. G., Mortlock R. A., Chiu T.-C., Cao L., Kaplan A., Guilderson T. P., Fairbanks T. W., Bloom A. L., Grootes P. M., and Nadeau M.-J. (2005) Radiocarbon calibration curve spanning 0 to 50,000 years BP based on paired $^{230}Th/^{234}U/^{238}U$ and ^{14}C dates on pristine corals. *Quat. Sci. Rev.* **24**, 1781–1796.

Frank M., Schwarz B., Baumann S., Kubik P. W., Sater M., and Mangini A. (1997) A 200 kyr record of cosmogenic radionuclide production rate and geomagnetic field intensity from ^{10}Be in globally stacked deep-sea sediments. *Earth Planet. Sci. Lett.* **149**, 121–129.

Friedrich M., Kromer B., Spurk M., Hofmann J., and Kaiser K. F. (1999) Paleo-environment and radiocarbon calibration as derived from Late Glacial/Early Holocene tree-ring chronologies. *Quat. Int.* **61**, 27–39.

Harrison K. G., Broecker W., and Bonani G. (1993a) A strategy for estimating the impact of CO_2 fertilization on soil carbon storage. *Global Biogeochem. Cycles* **7**, 69–80.

Harrison K. G., Broecker W. S., and Bonani G. (1993b) The effect of changing land use on soil ^{14}C. *Science* **262**, 725–726.

Hiller A., Boettger T., and Kremenetski C. (2001) Medieval climatic warming recorded by radiocarbon dated alpine tree-line shift on the Kola Peninsula, Russia. *Holocene* **11**, 491–497.

Hormes A., Schlüchter C., and Stocker T. F. (1998) Minimal extension phases of unteraar glacier (Swiss Alps) during the Holocene based on ^{14}C analysis of wood. *Radiocarbon* **40**, 809–817.

Hughen K. A., Overpeck J. T., Lehman S. J., Kasgarian M., Southon J., Peterson L. C., Alley R., and Sigman D. M. (1998) Deglacial changes in ocean circulation from an extended radiocarbon calibration. *Nature* **391**, 65–68.

Kitagawa H. and van der Plicht J. (1998) Atmospheric radiocarbon calibration to 45,000 yr BP: late Glacial fluctuations and cosmogenic isotope production. *Science* **279**, 1187–1190.

Kromer B., Rhein M., Bruns M., Schoch-Fisher H., Munnich K. O., Stuiver M., and Becker B. (1986) Radiocarbon calibration data for the sixth to the eighth millennia BC. In Calibration issue. *Radiocarbon* **28**, 954–960.

Kromer B. and Spurk M. (1998) Revision and tentative extension of the tree-ring based ^{14}C calibration, 9,200–11,855 cal BP. *Radiocarbon* **40**, 1117–1125.

Laj C., Mazaud A., and Duplessy J. C. (1996) Geomagnetic intensity and ^{14}C abundance in the atmosphere and ocean during the past 50 kyr. *Geophys. Res. Lett.* **23**, 2045–2048.

Lal D. (1988) Theoretically expected variations in the terrestrial cosmic-ray production rates of isotopes. In *Solar–Terrestrial Relationships and the Earth Environment in the Last Millennia* (ed. X. C. V. Corso). Soc. Italiana de Fisica., pp. 216–233.

Libby W. F., Anderson E. C., and Arnold J. R. (1949) Age determination by radiocarbon content: worldwide assay of natural radiocarbon. *Science* **109**, 227–228.

Muscheler R., Beer J., Kubik P. W., and Synal H.-A. (2005) Geomagnetic field intensity during the last 60,000 years based on ^{10}Be and ^{36}Cl from the Summit ice cores and ^{14}C. *Quat. Sci. Rev.* **24**, 1849–1860.

Nydal R. and Lövseth K. (1983) Tracing bomb ^{14}C in the atmosphere 1962–1980. *J. Geophys. Res.* **88**, 3621–3642.

Oeschger H., Siegenthaler U., Gugelmann A., and Schotterer U. (1975) A box-diffusion model to study the carbon dioxide exchange in nature. *Tellus* **27**, 168–192.

Robinson L. F., Adkins J. F., Keigwin L. D., Southon J., Fernandez D. P., Wang S.-L., and Scheirer D. S. (2005) Radiocarbon variability in the Western North Atlantic during the last deglaciation. *Science* **310**, 1469–1473.

Seigenthaler U., Heimann M., and Oeschger H. (1980) ^{14}C variations caused by changes in the global carbon cycle. *Radiocarbon* **22**, 177–191.

Shackleton N. J., Duplessy J.-C., Arnold M., Maurice P., Hall M. A., and Cartlidge J. (1988) Radiocarbon age of the last glacial Pacific deep water. *Nature* **335**, 708–711.

Siani G., Paterne M., Michel E., Sulpizio R., Sbrana A., Arnold M., and Haddad G. (2001) Mediterranean Sea surface radiocarbon reservoir age changes since the Last Glacial maximum. *Science* **294**, 1917–1920.

Sikes E. L., Samson C. R., Guilderson T. P., and Howard W. R. (2000) Old radiocarbon ages in the southwest Pacific Ocean during the Last Glacial period and deglaciation. *Nature* **405**, 555–559.

Southon J. (2002) A first step to reconciling the GRIP and GISP2 Ice-core chronologies, 0–14,500 yr BP. *Quat. Res.* **57**, 32–37.

Stuiver M. (1970) Long-term ^{14}C variations. In *Radiocarbon Variations and Absolute Chronology, 12th Nobel Symposium* (ed. I. U. Olsson). Wiley, New York, pp. 197–213.

Stuiver M. (1971) Evidence for the variation of atmospheric ^{14}C content in the Late Quaternary. In *Late Cenozoic Glacial Ages* (ed. K. K. Turekian). Yale University Press, New Haven, CT, pp. 57–70.

Stuiver M. (1980) Solar variability and climatic change during the current millennium. *Nature* **286**, 868–871.

Stuiver M. (1982) A high-precision calibration of the AD radiocarbon time scale. *Radiocarbon* **24**, 1–26.

Stuiver M. and Braziunas T. F. (1993) Sun, ocean, climate, and atmospheric $^{14}CO_2$: an evaluation of causal and spectral relationships. *Holocene* **34**, 289–305.

Stuiver M. and Quay P. D. (1980) Changes in atmospheric carbon-14 attributed to a variable Sun. *Science* **207**, 11–19.

Stuiver M. and Quay P. D. (1981) Atmospheric ^{14}C changes resulting from fossil fuel CO_2 release and cosmic ray flux variability. *Earth Planet. Sci. Lett.* **53**, 349–362.

Stuiver M., Reimer P. J., Bard E., Beck J. W., Burr G. S., Hughen K. A., Kromer B., McCormac G., van der Plicht J., and Spurk M. (1998) Intcal98 radiocarbon age calibration, 24,000–0 cal BP. *Radiocarbon* **40**, 1126–1159.

Thellier E. and Thellier O. (1959) Sur l'intensite du champ magnetique terrestre dans le passe historique et geologique. *Annu. Geophys.* **15**, 285–378.

Trumbore S. E., Chadwick O. A., and Amundson R. (1996) Rapid exchange between soil carbon and atmospheric carbon dioxide driven by temperature change. *Science* **272**, 393–396.

Voelker A. H. L., Sarnthein M., Grootes P. M., Erlenkeuser H., Laj C., Mazaud A., Nadeau M.-J., and Schleicher M. (1998) Correlation of marine ^{14}C ages from the Nordic seas with the GISP2 isotope record: implications for radiocarbon calibration beyond 25 ka BP. *Radiocarbon* **40**, 517–534.

Waelbroeck C., Duplessy J.-C., Michel E., Labeyrie L., Paillard D., and Duprat J. (2001) The timing of the last deglaciation in North Atlantic climate records. *Nature* **412**, 724–727.

Yiou F., Raisbeck G. M., Baumgartner S., Beer J., Hammer C., Johnsen S., Jouzel J., Kubik P. W., Lestringuez J., Stievenard M., Suter M., and Yiou P. (1997) Beryllium 10 in the Greenland Ice Core Project ice core at Summit, Greenland. *J. Geophys. Res.* **102**, 26783–26794.

Isotope Geochemistry
ISBN: 978-0-08-096710-3

pp. 425–444

14

Natural Radionuclides in the Atmosphere

K. K. Turekian and W. C. Graustein
Yale University, New Haven, CT, USA

14.1 INTRODUCTION

Natural radioactive nuclides in the atmosphere have two principal sources—radon and its progeny derived from Earth's surface and cosmic-ray-produced nuclides. Dust from the elevation of soils can also provide secondary sources of these nuclides. Suitable accommodation for these sources must be made if only those species having gaseous precursors are to be considered.

There is a radon isotope in each of the three major natural-decay chains: ^{238}U (^{222}Rn, half-life = 3.8 d), ^{232}Th (^{220}Rn, half-life = 55 s), and ^{235}U (^{219}Rn, half-life = 3.9 s). Almost all radon in the atmosphere is produced in soils and is transported to the atmosphere by diffusion. Because its longer half-life allows for greater diffusive transport, most radon entering the atmosphere is ^{222}Rn; its radioactive decay scheme is presented in Table 1.

An early summary of ^{222}Rn and its progeny in the atmosphere was provided by Turekian *et al.* (1977).

Table 2 lists the radioactive species produced by cosmic rays acting on the gaseous components of the atmosphere, which include nitrogen, oxygen, and all the rare gases as targets. A detailed discussion of the formation of radioactive species by cosmic-ray bombardment has been published by Lal (2001) based on his pioneering work since 1967 in a classic paper by Lal and Peters (1967). Radiocarbon (^{14}C) is not discussed in this chapter. Because of its central role in many Earth's surface

Table 1 The decay chain for ^{222}Rn.

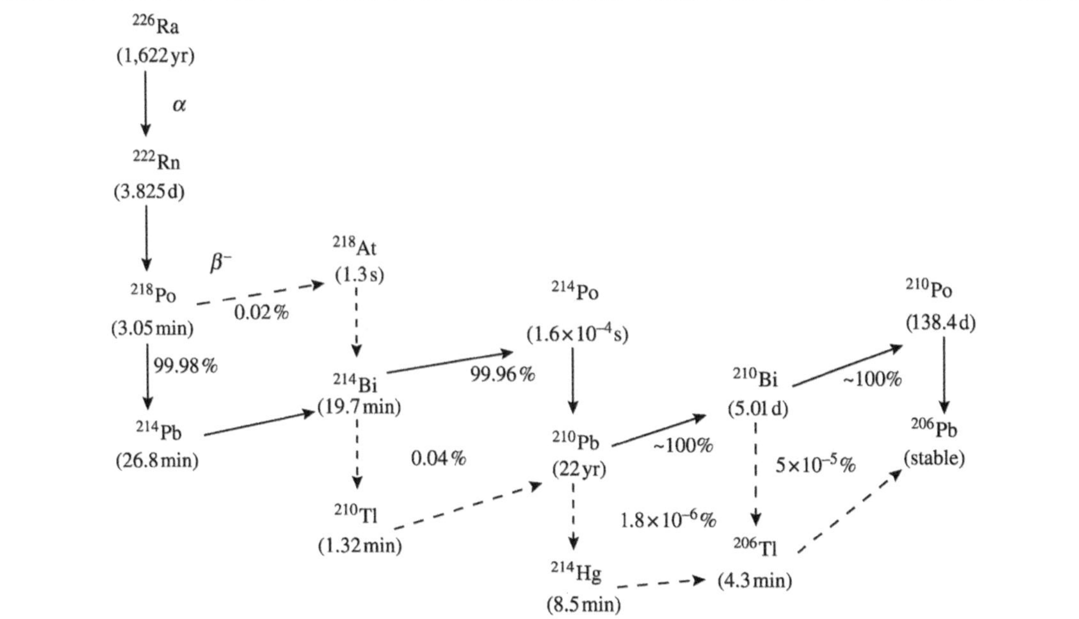

Table 2 Production rates of several isotopes in the Earth's atmosphere; arranged in order of decreasing half-lives.

Isotope	*Half-life*	*Production rate* (atoms cm^{-2} s^{-1})		*Global inventory*
		Troposphere	*Total atmosphere*	
^{3}He	Stable	6.7×10^{-2}	0.2	3.2×10^{3} t[a]
^{10}Be	1.5×10^{5} yr	1.5×10^{-2}	4.5×10^{-2}	260 t
^{26}Al	7.1×10^{5} yr	3.8×10^{-5}	1.4×10^{-4}	1.1 t
^{31}Kr[b]	2.3×10^{5} yr	5.2×10^{-7}	1.2×10^{-5}	8.5kg
^{36}Cl	3.0×10^{5} yr	4×10^{-4}	1.1×10^{-3}	15 t[d]
^{14}C	5,730 yr	1.1	2.5	75 t
^{39}Ar[c]	268 yr	4.5×10^{-3}	1.3×10^{-2}	52kg
^{32}Si	~150 yr	5.4×10^{-5}	1.6×10^{-4}	0.3kg
^{3}H	12.3 yr	8.4×10^{-2}	0.25	3.5kg
^{22}Na	2.6 yr	2.4×10^{-5}	8.6×10^{-5}	1.9 g
^{36}S	87 d	4.9×10^{-4}	1.4×10^{-3}	4.5 g
^{7}Be	53 d	2.7×10^{-2}	8.1×10^{-2}	3.2 g
^{37}Ar	35 d	2.8×10^{-4}	8.3×10^{-4}	1.1 g
^{33}P	25.3 d	2.2×10^{-4}	6.8×10^{-4}	0.6 g
^{32}P	14.3 d	2.7×10^{-4}	8.1×10^{-4}	0.4 g

Source: Based on Lal and Peters (1967).
[a] The inventory of this stable nuclide is based on its atmospheric inventory, which includes an appreciable contribution from crustal degassing of ^{3}He.
[b] Based on atmospheric ^{31}Kr/Kr ratio of $(5.2\pm0.4)\times10^{-13}$.
[c] Based on atmospheric ^{39}Ar/Ar ratio of (0.107 ± 0.004) dpm^{-1} l Ar (STP).
[d] Includes a rough estimate of ^{36}Cl produced by the capture of neutrons at the Earth's surface.

processes, a separate chapter is found in this volume (see Chapter 13).

There are more than 30 radionuclides produced in the atmosphere by these two natural processes. This chapter focuses on describing different ways in which the observed distributions of several of these nuclides can be used to infer patterns and rates of transport, mixing, and removal of these nuclides and of other constituents of the atmosphere.

We would like to define here several terms. "Activity" refers to the number of radioactive decays of a nuclide per unit time and is the product of the decay constant, λ, and the number of atoms, *N*. The SI unit of activity is the Becquerel (Bq), which represents one disintegration per second. The "disintegration per minute" or dpm often appears in the literature. An older representation was the Curie (Ci), which is defined as the activity of 1 g of pure ^{226}Ra or 3.70×10^{10} disintegrations per second. The picocurie (pCi), which is 0.037 Bq or 2.2 dpm, is often used in reporting radon concentrations in air.

In a decay chain shown in Table 1 at "secular equilibrium" the activities of the coupled radionuclides are equal. In a closed system, this occurs after roughly five half-lives of the shorter-lived daughter have elapsed ("parent" and "daughter" refer to the first and subsequent nuclides in the decay series being considered). For example, ^{222}Rn (half-life = 3.8 d) is in secular equilibrium with its parent, ^{226}Ra (half-life = 1,620 yr), in ~20 d.

14.2 RADON AND ITS DAUGHTERS

14.2.1 Flux of Radon from Soils to the Atmosphere

The escape of ^{222}Rn from soils is the source of ~99% of the ^{222}Rn in the atmosphere. Typical radon escape rates are on the order of 1 atom cm^{-2} s^{-1} from the land surface, which result in a radon inventory of the global atmosphere of ~1.5×10^{18} Bq. Atmospheric radon itself is a chemically inert and unscavenged, i.e., not removed from the atmosphere by physical or chemical means. Because its half-life is much less than the mixing time of the atmosphere, it is a tracer of atmospheric transport and can be used in a synoptic approach to identify air masses derived from continental boundary layers or in a climatological manner to verify the predictions of numerical models of transport.

^{222}Rn is lost from the atmosphere only by its radioactive decay and it is thus the source of virtually all of the ^{214}Bi, ^{214}Pb, ^{214}Po, ^{210}Pb, ^{210}Bi, and ^{210}Po in the atmosphere. The quantities of radon decay products on soil dust suspended by atmospheric turbulence or volatilized from lava are small by comparison on a global scale, but may be comparable locally and episodically.

The migration of ^{222}Rn through soils into buildings can lead to indoor air concentrations of ^{222}Rn that pose significant radiological health hazards. Knowing the flux and understanding the factors that cause it to vary spatially and temporally improve the utility of ^{222}Rn as an atmospheric tracer and increase the ability to predict potential health hazards National Research Council (1988).

^{238}U is a ubiquitous trace component of rock and soil; typical activities are ~30 mBq g^{-1}. In most soils, there is little transport of ^{234}U, ^{230}Th, or ^{226}Ra, so the activities of ^{238}U and ^{226}Ra are nearly equal. When an atom of ^{226}Ra decays in a mineral grain, the energy imparted by the recoil of the alpha particle is sufficient to displace the daughter ^{222}Rn atom by a few tens of nanometers. Although most often ^{222}Rn atoms remain within the grain, a fraction of them emanate from the grain and become free to move through the pore space of the soil. The fraction of ^{226}Ra decays that result in ^{222}Rn atoms in the pore space is termed as the emanating power or fraction. Typical values for emanating fractions are ~25%. Within the top few meters of a soil, a fraction of the radon in the pore space is transported to the surface, where it escapes to the atmosphere.

If one makes the simplifying assumptions that transport of ^{222}Rn is by molecular diffusion alone and that soil has uniform porosity, ^{222}Rn concentration, and emanating power, then the fraction of ^{222}Rn that escapes to the atmosphere as a function of depth, $L_f(z)$, is given by (e.g., Clements and Wilkening, 1974)

$$L_f(z) = a\mathrm{e}^{-bz} \quad (1)$$

where a is the ^{222}Rn emanating power (dimensionless), $b = \sqrt{\lambda\varepsilon/D}$ (cm^{-1}), λ is the decay constant of ^{222}Rn (s^{-1}), ε is the porosity (dimensionless), D is the diffusion coefficient (cm^2 s^{-1}) of ^{222}Rn, and z is the depth (cm).

Here $1/b$ has dimensions of length and is called the "mean depth" or "relaxation depth" of radon loss. Values of $1/b$ reported in the literature range from 100cm to 218cm (Schery and Gaeddert, 1982; Clements and Wilkening, 1974; Graustein and Turekian, 1990; Dörr and Münnich, 1990). These terms should not be confused with "half-depth" or 0.693 mean depth.

The escape rate of ^{222}Rn, J, can be derived from (1) by multiplying by the ^{226}Ra concentration in soil and integrating over depth, yielding

$$J = \frac{a\rho(^{226}\mathrm{Ra})}{b} \quad (2)$$

where ρ is the density (g cm^{-3}) and (^{226}Ra) is the activity (Bq g^{-1}).

Compared to this idealized model, the actual flux of ^{222}Rn may be diminished by the saturation of pore space by water (the mean length of ^{222}Rn diffusion in water is on the order of a millimeter, so saturation diminishes the flux by up to a factor of 1,000) and decreases in porosity with depth. Advection of gas through soil in response to barometric pressure change, soil gas convection, and transpiration of ^{222}Rn saturated soil solution will increase the radon escape rate. All of these processes are difficult to model accurately, so the determination of ^{222}Rn fluxes relies on measurements.

Several methods have been developed to measure the escape rate or flux of radon to the atmosphere; measuring accumulation of ^{222}Rn in chambers placed on the soil surface, modeling the flux from the vertical profile of ^{222}Rn in the atmosphere, modeling the diffusive flux from measurements of ^{222}Rn in the interstitial gas of a soil (e.g., Dörr and Münnich, 1990), and measuring the deficiency of the ^{210}Pb daughter with respect to its parent ^{226}Ra (Graustein and Turekian, 1990).

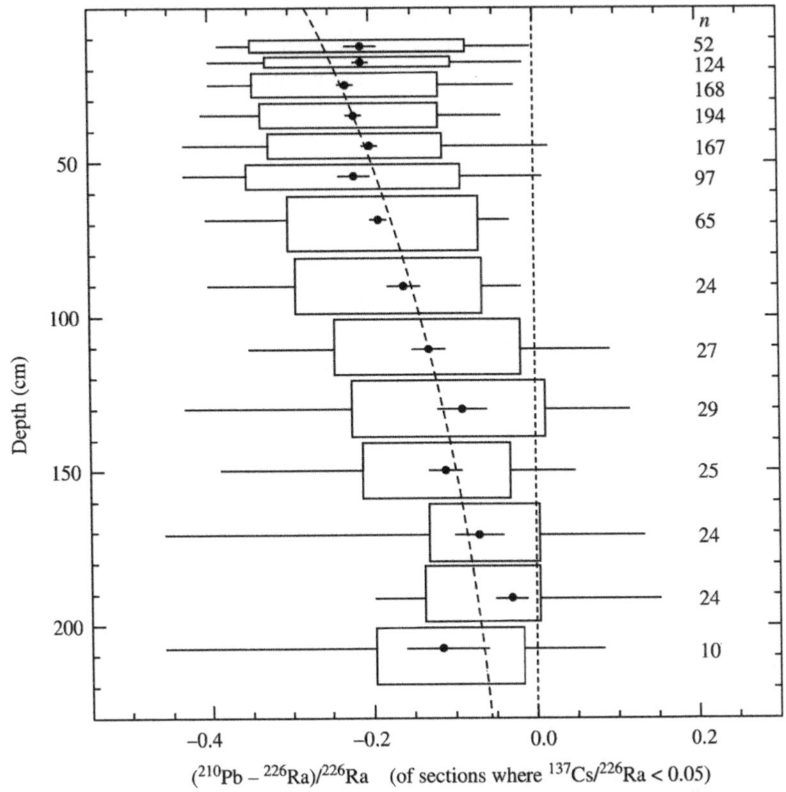

Figure 1 Depth profile of radon loss fraction. Filled circles and error bars represent the mean and standard error of the radon loss fraction from all samples from the indicated depth for which the $^{137}Cs/^{226}Ra$ ratio is less than 0.05. The boxes represent 1σ deviation and the whiskers the extreme values of each population. The number of samples is given in the column on the right. The curve corresponds to Equation (1) with an emanating power of 0.28 and a mean depth of 144 cm.

As shown by Equations (1) and (2), the radon loss fraction at shallow depths closely approximates the emanating power, and the escape rate is directly proportional to the emanating power. The data of Graustein and Turekian (1990) put a lower limit on the mean emanating power of US soils at 0.23. The fraction of ^{222}Rn lost from depth between 10 cm and 60 cm for 340 cores from North America (Graustein and Turekian, 1990, in preparation) averages 0.22 and does not vary significantly over that interval. Figure 1 shows the distribution with depth. An exponential curve with a half-depth of 100 cm (mean depth = 144 cm) is plotted for reference. The value of b was assigned; the value for a, 0.28, was obtained from a best fit to the data. With the exception of the sections from above 20 cm and those below 180 cm, the curve fits the data well. The counting errors and small number of samples from below 160 cm diminish our confidence in using this curve to extrapolate to greater depths and we regard extrapolations as representing an upper limit rather than an estimate. The samples from above 20 cm fall to the right of the reference curve, most probably due to the presence of atmospherically derived ^{210}Pb. The extrapolation of the reference curve to the surface yields an upper limit to the emanating fraction of 0.28.

Using these data the mean ^{222}Rn flux from US soils is at least 1.5 atoms cm^{-2} s^{-1}. Extrapolation using the exponential curve in Figure 1 gives an upper limit of 2.0 atoms cm^{-2} s^{-1}. This range is 50–100% greater than the commonly used estimate of 1.0 atoms cm^{-2} s^{-1} for global radon flux from continents, implying that there are large areas with radon fluxes less than 1 or that the global estimate of 1.0 is low. The survey of ^{222}Rn fluxes from Australia by Schery *et al.* (1989) averaged 1.05 atoms cm^{-2} s^{-1} with a standard deviation of 0.24 indicating that large areas may have significantly different ^{222}Rn fluxes. Their data were derived from the accumulation of ^{222}Rn in accumulator chambers and are nearly instantaneous compared to the 30 yr mean values determined from the ^{210}Pb deficiency in soil method discussed above. That the two Rn flux measurements yield the same result is justified by the similarity of a ^{222}Rn flux from a soil sampling site in the same cornfield (Graustein and Turekian, 1990) in which the ^{222}Rn flux was determined by accumulator (Pearson and Jones, 1966).

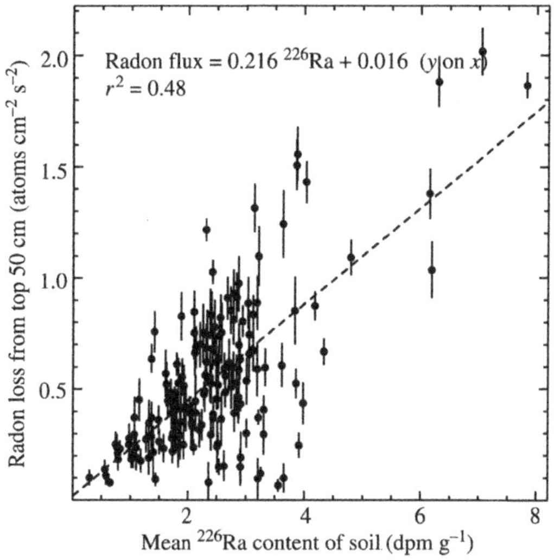

Figure 2 Radon loss rate from the upper 50 cm of soils versus the mean radium concentration. Bars indicate calculated 1σ error of the loss.

The spatial variability of the radon fluxes means that many measurements are needed to make precise estimates of the mean value for a region. Conen and Robertson (2002) noted in their compilation of radon fluxes determined by accumulator that there is a correlation with latitude from 30° to 50°, with a decreasing flux with increasing latitude. They ascribed this change as due primarily to the degree of water saturation of the soils, the increase toward higher latitude.

Comparisons of ^{222}Rn flux from the upper 50 cm, based on disequilibrium measurements in soil profiles discussed above, with mean annual temperature and precipitation show only weak relations; neither r^2 exceeded 0.02. The major correlation is with the ^{226}Ra activity of the soil (Figure 2).

14.2.2 Flux of Radon from the Oceans

The radium concentration in the oceans is at least a factor of a thousand less than soils; therefore, the flux is proportionally less. The radon flux from the mixed layer has been determined by Broecker and Peng (1971) and Wilkening and Clements (1975) and is as expected low and varies depending on wind stress. Its contribution to the radon burden of the atmosphere can therefore be taken as negligible. It is less than 1% of the global continental flux. A flux from shoal areas around Hawaii has been measured by Moore *et al.* (1974). Although the flux is higher than the open ocean, being driven by the higher radium concentration of the underlying rocks, on a worldwide basis it too is negligible. The major flux of radon to the atmosphere is clearly soils, mainly on the continents.

14.2.3 Distribution of Radon in the Atmosphere

The half-life of ^{222}Rn (3.8 d) is much less than the mixing time of the atmosphere, so its concentrations are greatest near the land surface and decrease with both altitude and distance from land (e.g., Turekian *et al.*, 1977; Liu *et al.*, 1984).

Although the vertical distribution of radon over the continents is a direct consequence of supply from soils, convection upward (treated as turbulent diffusion), and radioactive decay, the pattern is different over the oceans. This difference is due to the fact that no significant source of radon exists over the oceans and the pattern is set by long-distance transport from continents. Off the northwest coast of the United States, for example, Andreae *et al.* (1988) show vertical patterns up to ~4 km ranging from constancy with elevation to increases with elevation. The concentration range from ~6 pCi m^{-3} (STP) to 10 pCi m^{-3} (STP) is more typical of upper troposphere air over the continents and not like the ~400 pCi m^{-3} (STP) in the continental boundary layer (Figure 3).

Radon and its progeny ^{210}Pb have been found in the stratosphere in some locations at elevated levels. Although some of ^{210}Pb may be due to explosive volcanic penetration of the tropopause, most of the ^{210}Pb is due to the decay of radon injected convectively mainly in the tropics (Kritz *et al.*, 1993).

Lambert *et al.* (1970) reported periodic fluctuations in the radon concentrations in boundary layer air sampled on islands and Antarctica between 40° S and 70° S. Being in the geographic zone with no large land masses, all the radon is delivered from lower latitudes by advection. The ambient concentration was found to be ~2 pCi m^{-3} (STP) with occasional "radon storms" raising the concentration briefly to 10 pCi m^{-3} (STP). Lambert *et al.* (1970) argued that, based on long-term observations, there is a 27–28 d cycle of radon variation in this region. The causes of changes in circulation that would yield this result are not known, but radon measurements obviously put constraints on models of hemispheric circulation.

14.2.4 Short-lived Daughters of ^{222}Rn in the Atmosphere

As can be seen in Table 1, ^{222}Rn decays through a number of short-lived daughters before the long-lived ^{210}Pb is reached. The first of these radionuclides along the decay chain is ^{218}Po with a half-life of 3 min. The 6 MeV energy of the alpha particle emitted by the ^{222}Rn decay causes the ^{218}Po to recoil with a velocity much greater than that of the ambient gas molecules. Collisions with gas molecules or atoms diminish its energy until it is same as that of the ambient air, at which time

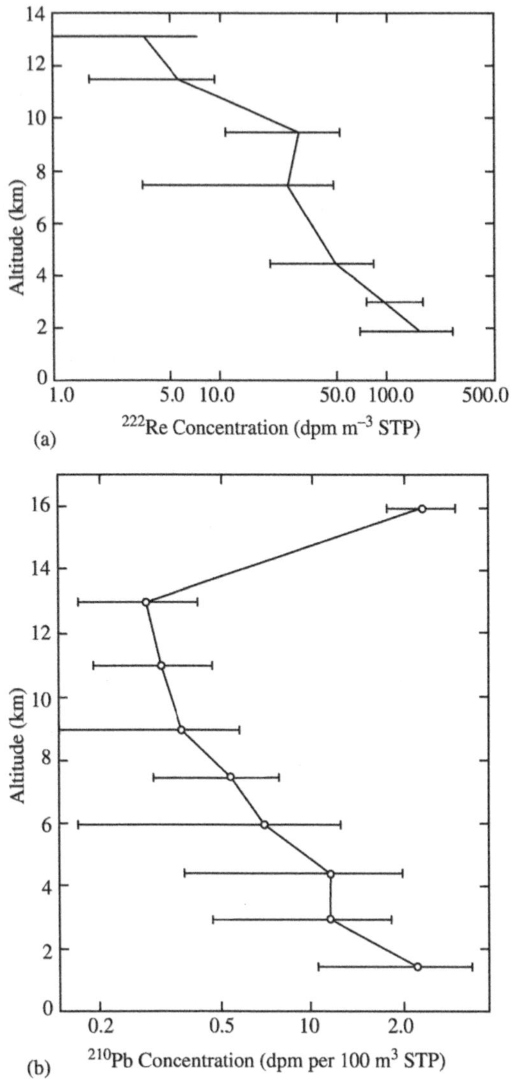

Figure 3 (a) The average distribution of Rn with elevation in the North American mid-continental region based on several profiles averaged by Moore *et al.* (1973). (b) Vertical distribution of ^{210}Pb in the atmosphere in the mid-continental region of North America (source Moore *et al.*, 1973).

the ^{218}Po diffuses through the air until it encounters a surface, typically an aerosol particle, to which it attaches.

In an enclosed space within which a large number of aerosol particles exist, such as in a mine or around cigarette smoke, such ^{222}Rn decay derived nuclides attach to particles on the order of seconds to minutes and subsequent radioactive decay can occur on the aerosol particles. An array of short-lived energetic radioactive species in the chain add to the radioactive burden of the particle. If inhaled, these highly radioactive particles can cause health problems. This was shown in the case of uranium miners and may be a part of the cause of the lung problems of smokers and those affected by secondary particle inhalation around smokers.

Studies of ^{214}Bi (half-life = 19.8 min) activity relative to ^{214}Pb (half-life = 26.8 min) in boundary layer air indicate that these are in equilibrium, indicating that the particles are not removed more rapidly than a couple of hours (Turekian *et al.*, 1999). Attaching of these short-lived daughters on aerosols, and the mean life of the aerosols being much longer than their half-lives means that the nuclide to be tracked on most atmospherically interesting timescales is ^{210}Pb with its half-life of 22 yr.

14.2.5 ^{210}Pb and Its Progeny

After it is produced by the decay of ^{214}Po, ^{210}Pb rapidly associates with aerosols in the 0.1–0.5 μm diameter size range (Knuth *et al.*, 1983). Aerosols in this size range, the so-called "accumulation mode," are large enough that their velocity due to Brownian motion is so small that diffusion is not a significant method of transport. They are also small enough that their gravitational settling velocity is much less than typical rates of vertical motion in the atmosphere. Scavenging by precipitation is the principal mechanism of removal of these aerosols. Many of the chemical species with low vapor pressure that form from gaseous precursors in that atmosphere, such as H_2SO_4 formed from the oxidation of SO_2, are also associated with accumulation-mode aerosols.

As a result, these aerosols follow the motions of the atmosphere and are responsible for the transport of much of the nonvolatile products of photochemical and oxidative reactions. ^{210}Pb is a minor constituent of this aerosol population (one aerosol particle in 10^4 or more will carry an atom of ^{210}Pb) and is a useful tracer of the transport, deposition, and residence time of aerosols.

14.2.5.1 Distribution of ^{210}Pb in the atmosphere

Vertical profiles of Rn, ^{210}Pb, ^{210}Bi, and ^{210}Po were measured over the mid-continental United States by Moore *et al.* (1973). These results for ^{222}Rn and ^{210}Pb are shown in Figure 3. A number of surface sites have been locations for long-term studies of ^{210}Pb in conjunction with the measurements of radionuclides from nuclear bomb testing by the Environmental Measurements Laboratory of the Department of Energy and its precursor, the Atomic Energy Commission (Feely *et al.*, 1981). In addition, there have been measurements made throughout the United States and the islands of the Pacific and Atlantic as parts of the SEAREX and AEROCE programs, respectively. The sampling programs complement each other and, where there are regional overlaps, the results are identical. The values across the Pacific from the SEAREX

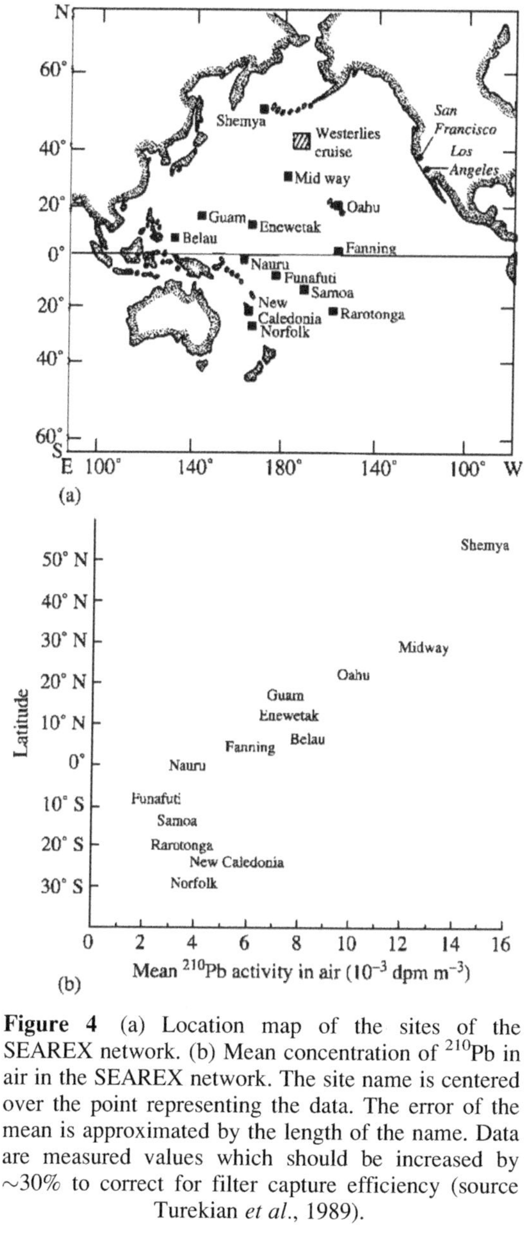

Figure 4 (a) Location map of the sites of the SEAREX network. (b) Mean concentration of ^{210}Pb in air in the SEAREX network. The site name is centered over the point representing the data. The error of the mean is approximated by the length of the name. Data are measured values which should be increased by ~30% to correct for filter capture efficiency (source Turekian *et al.*, 1989).

program (Turekian *et al.*, 1989) are shown in Figure 4. There clearly is a correlation with the size of the land-mass upwind from the island sampling site. This is expected since ^{222}Rn is the precursor to the ^{210}Pb and is derived from soils.

14.2.5.2 Flux of ^{210}Pb to Earth's surface

(i) *Methods of determinations.* The most direct method of estimating the flux of ^{210}Pb to the Earth's surface is through bucket collection of precipitation and subsequent analysis of the water for ^{210}Pb obtained over a specific period of time. This method has been used during the assay of precipitation for bomb-produced radionuclides such as ^{90}Sr and ^{137}Cs, notably by the Environmental Measurements Laboratory of the Department of Energy and its precursor, the Atomic Energy Commission. Detailed summaries, mainly along ~80° W longitude in both hemispheres plus some other selected sites in the continental US, are available from the Department of Energy as unpublished reports.

There is great value to the data obtained by such a method especially as it provides seasonal values. The method does not, however, measure the extraction of water and ^{210}Pb from the air by "horizontal" precipitation. Horizontal precipitation is the consequence of impingement of cloud, fog, or dew droplets on leaf surfaces. The water and its constituents are then transferred to the soil by dripping or episodic vertical precipitation. Also the long-term time-averaged flux is limited by the length of time that the sampling program has been in effect.

The accumulation of atmospherically derived ^{210}Pb in soil profiles is another method of assessing the atmospheric flux of ^{210}Pb. This approach has already been referred to in the section on the measurement of ^{222}Rn flux from soils to the atmosphere. Specifically, the atmospherically derived ^{210}Pb in soils is generally found in the topmost organic-rich layer, although there is transport down into the soil profile by illuviation, bioturbation, and possibly chemical transport by chelators.

This method of assaying the long-term total atmospheric ^{210}Pb flux has been studied extensively by Moore and Poet (1976), Nozaki *et al.* (1978), and Graustein and Turekian (1983, 1986, 1989). It requires that the soil profile be undisturbed for ~100 yr. In nonforested areas the directly measured atmospheric fluxes of ^{210}Pb and the fluxes determined by soil profile method are generally in agreement, allowing for the expected short-term variability (Graustein and Turekian, 1986). The relationship of the atmospherically derived standing crop of excess ^{210}Pb in soil profiles and the atmospheric flux is

$$\text{Flux} = (^{210}\text{Pb decay constant}) \times \text{inventory}$$

(ii) *The flux of ^{210}Pb over the oceans.* The flux of ^{210}Pb across the Pacific Ocean based on bucket collections has mainly been studied by Tsunogai *et al.* (1985) for the region around Japan, and by Turekian *et al.* (1989) for the rest of the Pacific as part of the SEAREX program. Figure 5 shows the flux of ^{210}Pb over the North Pacific. The most striking pattern is the very high flux close to Japan and a virtually constant low flux across the rest of the North Pacific including the California coast.

This pattern is different from the model expectation presented by Turekian *et al.* (1977), which predicted an exponential decrease of ^{210}Pb flux across the North Pacific. The observations clearly indicate the efficient removal of the burden of ^{210}Pb from the Asian continent by the time the

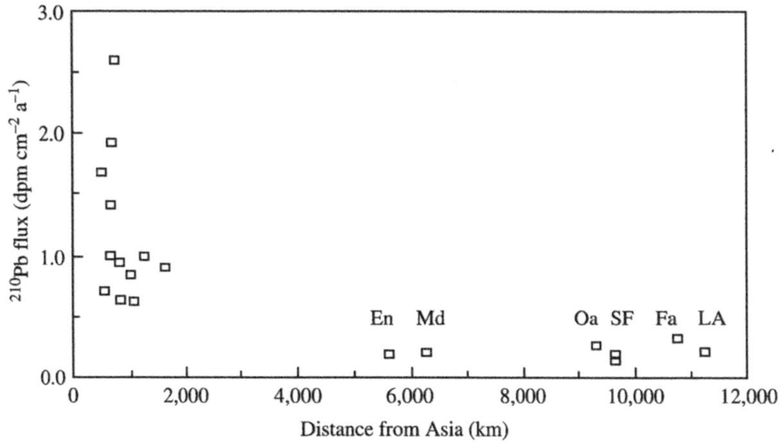

Figure 5 ^{210}Pb deposition flux versus distance from the Asian continent. Unlabeled points are from Tsunogai *et al.* (1985). En, Enewetak; Md, Midway; SF, San Francisco (Fuller and Hammond, 1983; Monaghan *et al.*, 1985/1986); LA, Los Angeles (Fuller and Hammond, 1983); Oa, Oahu; FA, Fanning. For the mid-ocean sites distances from Asia are measured along parallels of latitude (source Turekian *et al.*, 1989).

longitude of Japan is reached. The flux of ^{210}Pb thereafter is determined by the ^{222}Rn concentration in the air and the scavenging efficiency by rain.

(iii) *The flux of ^{210}Pb over the continents*. The flux of ^{210}Pb across the United States has been studied primarily by the Environmental Measurements Laboratory on precipitation (see Feely *et al.*, 1981) and Graustein and Turekian (1986 and references therein) in soil profiles. Figure 6 shows the ^{210}Pb flux pattern across the United States primarily from these sources but from other data as well. The major features are: (i) a general increase in flux from the Pacific coast towards the Great Plains; (ii) an approximately constant flux across the Great Plains to the Appalachians; (iii) marked orographic effects in the Sierras and the Appalachians; and (iv) a decrease along the east coast. The explanation for observation (i) is that air arriving with the prevailing westerly winds is generally depleted in ^{210}Pb and its precursor ^{222}Rn because of scavenging over the oceans and decay during transport, respectively, and the increase towards the center of the continent and general constancy is due to the local nature of both the source of ^{210}Pb and its removal. The marked orographic effects (iii) are due to both the increased rainfall in the mountains and the cloud scavenging alluded to earlier. The decrease on the east coast is due to precipitation carrying aerosols both from the ^{210}Pb-rich continental interior and ^{210}Pb-poor maritime air.

14.2.5.3 *Residence time of ^{210}Pb and associated species in the atmosphere*

One approach to measuring the residence time of aerosols is to calculate the time required for presumed initial values in an air mass of two or more nuclides in a decay chain to evolve to the observed values. This approach yields a measure of the effect of removal processes along the path of transport of the air mass being sampled.

If we consider an air parcel as a closed system, then the activity of each nuclide in the ^{222}Rn decay chain is described by

$$dN/dt = P - \lambda N \tag{3}$$

where N is the number of atoms per unit volume and P is the production rate of the nuclide by the radioactive decay of its parent. Applying (3) to each nuclide in a decay chain and setting initial conditions leads to a set of coupled differential equations. The general form for the solution to such a set of differential equations was obtained by Bateman (1910) and is given in standard references in radiochemistry (e.g., Friedlander *et al.*, 1981).

The dotted line in Figure 7 shows the time evolution of the ratios of radon daughters in closed system in which initially ^{222}Rn is present but its decay products are absent. We refer to this model age as the batch process time.

The atmosphere can also be modeled as a steady-state system, to which ^{222}Rn is added at a constant rate and from which its decay products are removed by a first-order removal process:

$$dN/dt = 0 = P - (\lambda + k)N \tag{4}$$

where k, the removal constant, is the inverse of the mean residence time. It is zero for ^{222}Rn and assumed equal for all of its decay products. The production term, P, for each nuclide is equal to the decay rate of its parent. The solid line in Figure 7 shows the ratio of radon daughters as a function of mean residence time.

The real atmosphere does not conform to either model assumption, but lies somewhere between them. Over the five-day timescale of

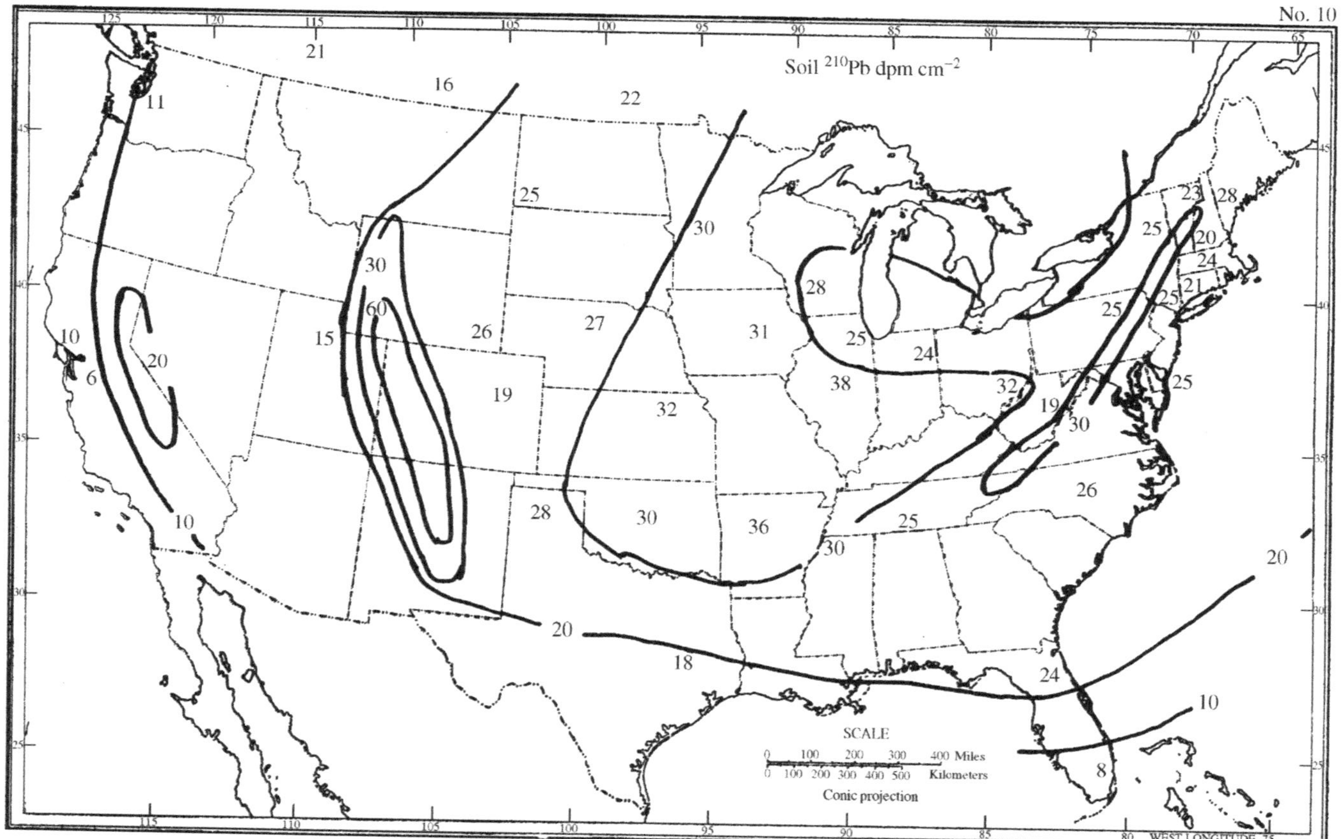

Figure 6 ^{210}Pb deposition flux over the United States. The cartoon is based on measurements made in the laboratory of the authors with accommodation for topography. Standing crop of 32 dpm cm^{-2} equals a deposition flux of 1 dpm ^{210}Pb cm^{-2} yr^{-1}.

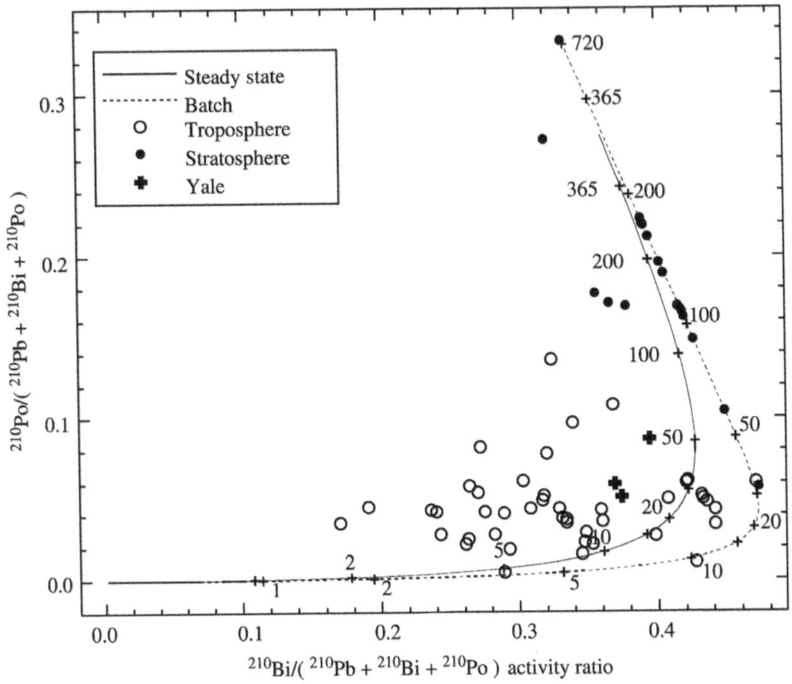

Figure 7 Residence time implications of ^{210}Pb–^{210}Bi–^{210}Po activities in aerosols. The lines represent the evolution of radon daughter product activities for a closed system (dotted line) or a steady-state system (solid line). The numbers indicate the age (for a closed system) or aerosol mean residence time (steady-state system) corresponding to the adjacent + mark. Data from Moore *et al.* (1973) are plotted as filled circles for samples from the stratosphere and open circles for samples from the troposphere. Yale data from the troposphere are marked by crosses.

^{210}Bi decay, there is too much mixing to allow for the identification and sampling of a parcel that has behaved as a closed system and far too little mixing and transport to allow the homogenization required by the steady-state assumptions. The curves are, however, similar enough to allow meaningful interpretation of data.

Real air samples can be thought of as a mixture of individual small parcels that are small enough to have behaved as closed systems. The plot is therefore constructed with a common denominator for both the x- and y-axes, so that a mixture of two components will plot along the line connecting the points representing the two components.

Moore *et al.* (1973) reported on a series of samples collected at various altitudes in the troposphere and stratosphere over central North America and discussed their implications for aerosol residence times. We review their data here and add the consideration of parcel mixing. Samples obtained from the stratosphere are denoted by a filled circle. Most of these samples lie along the parcel evolution line with model ages from 40 d to 2 yr, with the bulk of the samples between 100 d and 180 d. These model ages are consistent with the meteorological understanding of the lower stratosphere, where vertical mixing is suppressed and precipitation scavenging is nil. These ages are also consistent with, though slightly smaller than estimates of stratospheric residence times derived from the behavior of fission products injected into the stratosphere by atmospheric testing of nuclear weapons. A few of the stratospheric samples fall below the evolution curves, suggesting the recent admixture of some "younger" aerosols.

By comparison, few of their tropospheric samples, represented by open circles, lie on either model evolution curve, but all lie within an envelope bounded by the batch model curve on the right. These data are consistent with mixing of aerosols with varying ages, and many of the data strongly indicate that mixing has occurred. The data that fall between the two model curves are most constrained and indicate an age between 20 d and 30 d.

The samples that plot within the envelope bounded by the steady-state evolution curve are less constrained. Although those that do not fall near the model curve show the effects of mixing of aerosols of different ages, the data do not lie along a single trend and therefore do not suggest end-members. Most of the samples are consistent with mixing of aerosols of age 1–2 d with those of up to age 50 d; a few of the samples appear to require an even more aged component.

Martell and Moore (1974) and Moore *et al.* (1973) interpreted these data to yield a mean age of aerosols in the lower troposphere of less than 4 d, increasing by a factor of 3 toward the top of the troposphere. The mixing analysis presented above does not contradict their estimate of the mean and yields information about the range of ages of the parcels that compose their samples. The combination of the approaches suggests that the troposphere is not sufficiently mixed on timescales of a week to approach homogenization, but is mixed sufficiently that a single aerosol sample almost always contains material from parcels of widely varying history.

Another approach to determining the residence time of ^{210}Pb in the atmosphere is to divide the mean air column inventory of ^{210}Pb by the flux of ^{210}Pb to the surface at a given location. This quotient yields a climatological average for the removal processes at that particular site. Graustein and Turekian (1986) used the atmospheric profiles of ^{210}Pb from Moore *et al.* (1973) and their own measured ^{210}Pb fluxes from soil profiles and bucket collection to obtain a value of ~6 d over the central and eastern United States. As the source of ^{222}Rn and thus ^{210}Pb is from the ground and the major removal by precipitation is in the lower troposphere, the mean residence time is dominated by the processes of the lower troposphere. Modeling by Balkanski *et al.* (1993) shows that the mean residence time of ^{210}Pb increases with altitude.

14.2.5.4 Use of ^{210}Pb as surrogate for other atmospheric component

(i) *Sulfate*. Of the various atmospheric components for which ^{210}Pb can be considered a surrogate, the most applicable appears to be sulfate. The reason for this is that sulfate, like ^{210}Pb, is derived from a gaseous precursor (SO_2). The conversion to an aerosol-carried species takes place at an approximately similar rate (but, as we shall see later, the oxidation rate of SO_2 actually is variable with season) and ^{210}Pb and SO_4^{2-} attach to similar-size aerosols.

For this reason ^{210}Pb has been used as a tracer of the precipitation fate of SO_4^{2-}. Turekian *et al.* (1989) used the $SO_4^{2-}/^{210}Pb$ ratio in aerosols and the flux of ^{210}Pb measured in bucket collections to determine the SO_4^{2-} flux across the Pacific Ocean. Further, they showed that the $SO_4^{2-}/^{210}Pb$ in aerosols from regions of high biological productivity was higher than for normal relatively unpolluted air (Table 3) indicating a sulfate source from the oxidation of dimethyl sulfide (DMS). The measured flux of DMS from the oceans at the equator matched the biogenic flux determined from the ^{210}Pb calculation (Table 4). (Actually, as we shall see below, this concordance is probably due to an underestimate of sulfate flux and an overestimate of the fraction of DMS oxidized to sulfate.)

A similar study for determining sulfate flux from aerosol $SO_4^{2-}/^{210}Pb$ and ^{210}Pb flux across the eastern US yielded a new insight (Graustein and Turekian, 2004a,b, in preparation). Measurements were made at the SURE sites of the Electric Power Research Institute (EPRI). The sulfate flux was measured at each site as part of the SURE sampling program and $SO_4^{2-}/^{210}Pb$ was measured on composite aerosol samples covering a year's

Table 3 Nonmarine $SO_4^{2-}/^{210}Pb$ (in μg dpm^{-1}) at continental and oceanic sites as a function of season.

Site	*Summer*	*Winter*	*Source*
Mould Bay(76° N, 119° W)	47	28	Graustein and Barrie (unpublished)
Northeast United States	306	216	Graustein and Turekian (1986)
Shemya	200	51	Turekian *et al.* (1989)
Midway	77	13	Turekian *et al.* (1989)
Oahu	50	39	Turekian *et al.* (1989)
Fanning	123	147	Turekian *et al.* (1989)

Table 4 Comparison of biogenic sulfate deposition flux to Fanning Island (equatorial Pacific) and the DMS flux from the equatorial Pacific Ocean.

	$SO_4^{2-}/^{210}Pb$ (μg dpm^{-1})	*^{210}Pb flux* (dpm cm^{-2} yr^{-1})	*Biogenic SO_4^{2-} flux* (mmol m^{-2} yr^{-1})
Sulfate deposition (from Table 3)			
Continental air (Asia)	50		
Fanning Island (equatorial Pacific)			
Total	140		
Biogenic	90[a]	0.33	3.1
Dimethyl sulfide flux			
Equatorial Pacific Ocean			
Cline and Bates (1983)			2.9
Andreae and Raemdonck (1983)			3.3

[a] Biogenic = Total-continental air.

collection. The predicted SO_4^{2-} flux from the ^{210}Pb flux (measured in soil profiles, as described above) was generally less than the measured sulfate flux. The results indicate that over a period of a year or longer the increased sulfate flux must have been due to in-cloud conversion of SO_2 to SO_4^{2-}. About 12% of the precipitation flux of sulfate is by in-cloud conversion. A different approach corroborates this estimate. Tanaka *et al.* (1994), using stable isotope analyses of S in SO_2 and SO_4^{2-} associated with rain and air, showed that homogeneous oxidation of SO_2 provided ~90% of the measured sulfate flux and 10% was provided by heterogeneous (in-cloud) oxidation.

(ii) *Lead*. Much of the lead flux to the oceans in recent times has been from pollution sources. This lead was injected into the atmosphere primarily as a volatile compound that subsequently formed an aerosol by photolytic processes. In this sense, much of the lead transported through the atmosphere and ultimately deposited resembles the origin and fate of ^{210}Pb. On this basis, Turekian and Cochran (1981a,b) estimated the flux of lead to Enewetak in the North Pacific. In a more detailed study, Settle *et al.* (1982) were able to calculate the flux of lead from the aerosol Pb/^{210}Pb at a number of diverse sites, including Pigeon Key, Florida, Tahiti, and Bermuda.

(iii) *Mercury*. The primary form of mercury in the atmosphere is as a gaseous element. Only as the mercury is oxidized to ionic species is it removed by precipitation. In this respect it resembles ^{210}Pb, which also arises from a gas (^{222}Rn). Consequently, ^{210}Pb can be used to track mercury precipitation. Lamborg *et al.* (2000) performed such a study (Figure 8) to determine the correlation of ionic mercury and ^{210}Pb in aerosols and precipitation and to determine the flux of mercury from the ^{210}Pb flux determined independently.

(iv) *Other components*. Although other components of aerosols may not have had gaseous precursors, the use of ^{210}Pb as a surrogate may still be used as an approximation assuming the size fraction of the aerosol bearing both ^{210}Pb and the component is about the same. Turekian *et al.* (1989) applied this approach to organic compounds and such elements as aluminum. Williams and Turekian (2002) extended this procedure to estimate coastal and open-ocean fluxes of osmium, a metal with isotopic signature impacting the oceans. They were able to show that a high osmium flux was evident close to shore (in New Haven, CT) that could not be characteristic of the whole oceans because of the constancy of the osmium isotope composition in seawater. An analogue to this behavior is the study of ^{210}Pb fluxes in the SEAREX sites discussed above.

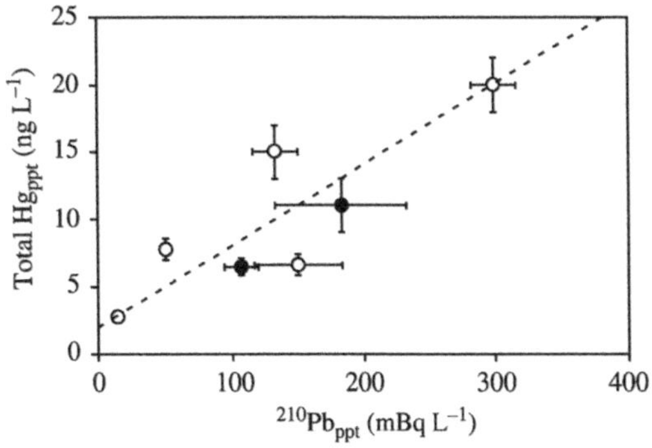

Figure 8 Hg and ^{210}Pb in Wisconsin (circles) and mid-Atlantic (triangles) precipitation (source Lamborg *et al.*, 2000). The filled circles represent side-by-side collections from a single event.

14.3 COSMOGENIC NUCLIDES

14.3.1 Atmospheric Production of Cosmogenic Nuclides

The source of cosmogenic nuclides in the atmosphere is the interaction of galactic cosmic rays with the atoms composing the atmosphere. Cosmic rays are primarily composed of protons with energies of billions of electron volts. As these highly energetic charged particles penetrate the magnetic shield of Earth the interactions with the atoms in the atmosphere result in spallation products—fragments of the target nucleus. Some of the secondary particles, especially neutrons, undergo further reactions. This latter process is responsible, for example, for the production of ^{14}C (radiocarbon). The major radioactive nuclides so produced in the atmosphere are given in Table 2.

The production of the radionuclides is controlled by the magnetic pattern of Earth. The production rate of the nuclides is determined by the strength of the magnetic field. The strength of the magnetic field varies as the result of both the intrinsic variations in Earth's dipole magnetic moment and the varying intensity of solar activity. At high magnetic field strengths the production rate is low and at low magnetic field strengths the production rate is high.

The period of magnetic fluctuations due to Earth's intrinsic field is not regular but has been shown to have increased over the past 2×10^4 yr. Fluctuations in the past are also recorded and clearly evident in the changing magnetic polarity of Earth over time.

The solar activity cycles are more complex. The 11 yr sunspot cycle is well known. It is accompanied with charges in the magnetic field strength of Earth. There are fluctuations in the intensity of solar activity on this timescale. As we shall see, periods of fluctuation of ~60 yr, 200 yr, and 1,500 yr have been identified through measurements of ^{10}Be in deep-sea sediments and continental ice sheets. These results also indicate possible variations on the 10^5 yr timescale as well.

14.3.2 ^{7}Be and ^{10}Be

The production rate of ^{7}Be (half-life = 53 d) as a function of latitude and elevation by Lal and Peters (1967) is shown in Figure 9. Approximately one-third of the nuclide production rate is in the troposphere and two-thirds in the upper atmosphere (stratosphere and higher). This partitioning is valid for all radionuclides except ^{14}C, where most is produced by secondary neutrons in the vicinity of the tropopause.

^{10}Be (half-life = 1.5×10^6 yr), although formed primarily in the atmosphere, is found in surface deposits because of the short residence time of aerosols in the atmosphere and the long half-life. Although the rate of deposition is the same as the rate of production in the atmosphere (4.5×10^{-2} atoms cm^{-2} s^{-1}) its distribution on Earth's surface is determined by the sites of primary stratospheric intrusion into the troposphere (~40°–50° latitude), the focusing in the troposphere due to precipitation controls such as regional climate and orographic effects and, in the oceans, the role of concentration and transport by particles, especially those produced biologically.

Despite these controls, it is possible to use the record of ^{10}Be accumulation at designated sites such as ice cores and certain oceanic areas as recorders of variations in the rate of supply and therefore of production with time.

As the residence time of aerosols in the stratosphere is ~2 yr and in the troposphere ~1 week, the ^{7}Be/^{10}Be ratio of the two air masses is distinctive. Tropospheric air shows the ratio of ^{7}Be relative to ^{10}Be of 1.8, whereas stratospheric air has a ratio of 0.13. It is therefore possible to distinguish stratospheric air injected into the troposphere by considering the ratio of ^{7}Be/^{10}Be. Of course, the stratospheric air will also be higher in ^{10}Be than the tropospheric air. As stratospheric air will also contain ozone, the interest in this source has been strong to distinguish from pollution-based tropospheric ozone.

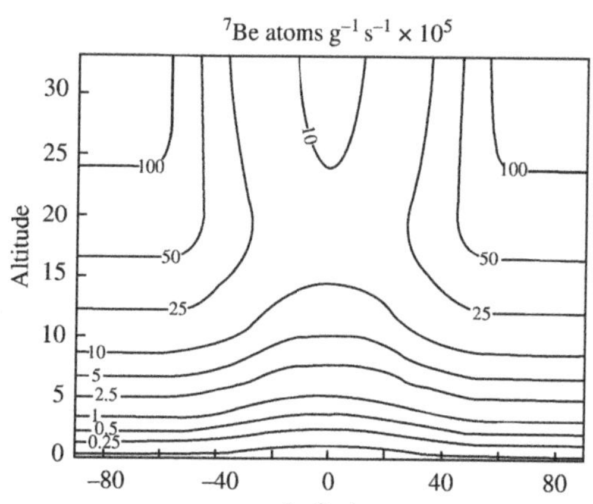

Figure 9 Production of ^{7}Be in the atmosphere as a function of latitude and elevation based on data from Lal and Peters (1967) and Lal (personal communication).

14.3.3 ^{35}S and the Kinetics of SO_2 Oxidation and Deposition

^{35}S is produced by the spallation of ^{40}Ar by cosmic rays. After formation it is quickly converted into gaseous $^{35}SO_2$. The subsequent oxidation and removal from the atmosphere can be tracked and the kinetics applied to terrestrial and pollution SO_2. For this reason it has been studied in conjunction with other cosmogenic nuclides, the most useful of which is ^{7}Be. Both ^{35}S with a half-life of 87 d and ^{7}Be with a half-life of 53 d, although produced from gaseous precursors by cosmic-ray bombardment, have different initial states. Whereas ^{7}Be as an ion quickly associates with aerosols, ^{35}S as $^{35}SO_2$ must first be oxidized to be associated with aerosols. Alternatively, $^{35}SO_2$ can react with surfaces in a method called "dry deposition" to distinguish it from ^{35}S removal in precipitation.

There is only one study using these two nuclides for deciphering the kinetics of SO_2 oxidation and removal from the atmosphere (Tanaka and Turekian, 1995, and earlier papers). Their study, performed in New Haven, Connecticut, applied the measurement of these nuclides in air and precipitation to the problem posed. Measurements were made for SO_2, $^{35}SO_2$, SO_4^{2-}, $^{35}SO_4^{2-}$, and ^{7}Be in air samples and for the same species in precipitation over a year's time period. Figure 10 shows the box model with the kinetic factors listed for each process modifying the composition of the air. Although the coupling of the boxes results in 12 equations with 3 unknowns, the selection of reasonable constraints results in solutions that are comparable with expectations based on experience. Figure 11 shows the variation in one parameter, j_2, the first-order kinetic constant for the homogeneous oxidation of SO_2 in the boundary layer. The value of j_2 is greatest in the summer months and lowest in the winter months. This prediction is compatible with the observation that SO_2 is the dominant species in winter air and SO_4^{2-} is so in the summer.

The application to the elusive process of dry deposition of SO_2 was evaluated by Tanaka and Turekin (1995) using the S and ^{35}S data on air and precipitation samples and the model used to construct Figure 10. They showed that in New Haven, CT the annual weighted average ratio of dry to wet precipitation was 0.26.

Stratosphere

——————————————————————————— H_t

Free troposphere (Box I)

Production	*Losses* ^7Be	*Losses* $^{35}SO_2$	*Losses* $^{35}SO_4$
P'_{Be}	Decay	Decay	Decay
P'_S	Transfer	Oxidation $^{35}SO_2 \rightarrow {}^{35}SO_4^{2-}$	Transfer
	Precipitation	Transfer	Precipitation
		In cloud scavenging	

——————————— ↑ U ↓ ——————————— H_b

Boundary layer (Box II)

P''_{Be}	Decay	Decay	Decay
P''_S	Transfer	Oxidation $^{35}SO_2 \rightarrow {}^{35}SO_4^{2-}$	Transfer
	Precipitation	Transfer	Precipitation
		In cloud scavenging	
		Dry deposition	

———————————————————————————

Surface

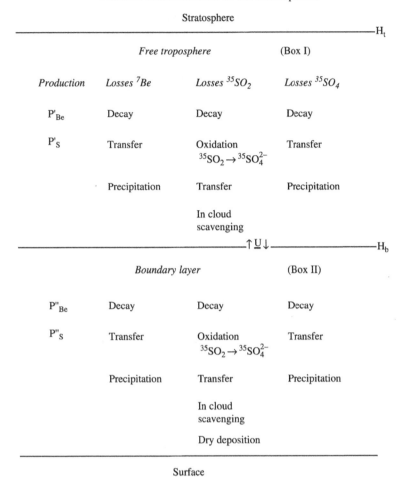

Figure 10 Components of the box model used by Tanaka and Turekian (1995) in determining the parameters for the conversion and removal of SO_2 in the atmosphere based on measurements of the concentrations of cosmogenic ^{35}S and ^7Be. *Ps* are the cosmic-ray production rates of the radionuclides, *U* is the mass transport flux between the boundary layer and the free troposphere, H_b and H_t are the heights of the boundary layer and the tropopause, respectively.

14.3.4 Phosphorus Isotopes

Of the isotopes listed in Table 2 the study of the cosmogenic isotopes of phosphorus ^{32}P (half-life = 14.3 d) and ^{33}P (half-life = 25.3 d) provides insights into the residence time of aerosols in upper troposphere air because of the relatively short half-lives of the nuclides. Waser and Bacon (1995) measured the concentrations of ^{32}P and ^{33}P in precipitation at Bermuda over three seasons. They concluded that, with the average activity ratio of ^{33}P to ^{32}P of 0.96 and a production activity ratio of 0.7, the average residence time of aerosols in the upper troposphere was ~40 d. This increase in the residence time with height in the troposphere is compatible with the modeling results of Balkanski *et al.* (1993) based on ^{210}Pb. When the upper-troposphere air is transported to the lower troposphere, it is subject to the more efficient removal characteristic of that layer, but with the memory of the original locus of production.

14.4 COUPLED LEAD-210 AND BERYLLIUM-7

14.4.1 Temporal and Spatial Variation

In the absence of atmospheric motion and removal by precipitation, ^7Be and ^{210}Pb would remain where they originated—the upper troposphere and stratosphere and the lowermost meter of the atmosphere over continents and islands, respectively. In the real atmosphere, ^7Be is mixed downward and ^{210}Pb is mixed upwards and both are removed by precipitation. They are distributed through the atmosphere by eddy mixing. The residence time of aerosols is short compared to the time required for homogenization by eddy mixing, but long compared to the life on an individual eddy. Changes in $^7Be/^{210}Pb$ with time and space reflect both vertical and horizontal transport in the atmosphere.

Because of similarity of the behavior of ^7Be and ^{210}Pb in the atmosphere, $^7Be/^{210}Pb$ is little affected by processes other than production and transport.

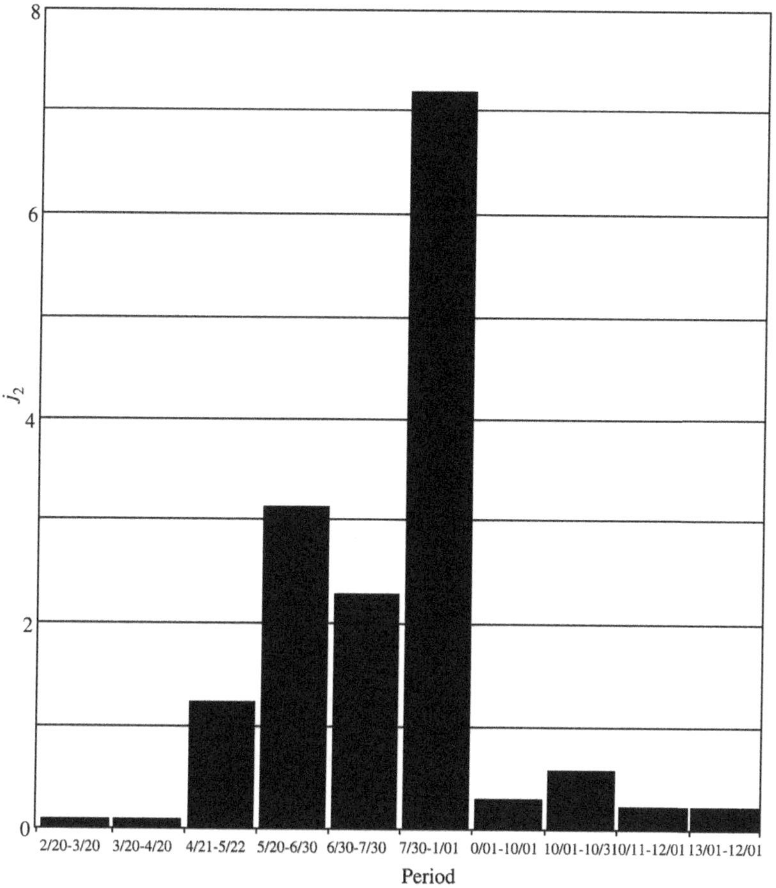

Figure 11 The variation over the year of the homogeneous oxidation coefficient (j_2) of SO_2 in the boundary layer of Figure 10 (source Tanaka and Turekian, 1995).

Both ^{210}Pb and ^{7}Be are formed in the atmosphere as energetic single atoms. Since neither is volatile, each of them attaches to the first particles they encounter. The most abundant aerosols in terms of surface area are typically those with a diameter of 0.1–0.5 μm, the so-called accumulation mode. This size class carries many of the chemical species in the atmosphere that have low volatility and also have gaseous precursors, such as sulfate as discussed above. Accumulation-mode aerosols are most subject to long-distance transport. Scavenging by precipitation is the principal mechanism of removal of these aerosols from the atmosphere.

The following results are from the work of Graustein and Turekian (1991).

Over continents, seasonal variation in the stability of the atmosphere has a dominant influence on the $^{7}Be/^{210}Pb$ ratio. Surface heating in the summer increases convective mixing, which reduces ^{210}Pb in surface air by mixing it through a larger volume and simultaneously increases the transport of ^{7}Be to the surface. Winter stability tends to isolate surface air from the ^{7}Be source and retain ^{222}Rn and ^{210}Pb near the surface.

The seasonal distributions of ^{7}Be and ^{210}Pb at Champaign, IL, a mid-continent site, are shown in Figure 12. The highest 24 h mean ^{210}Pb concentrations occur principally in winter when atmospheric stability is greatest; the highest ^{7}Be concentrations occur in summer when vertical

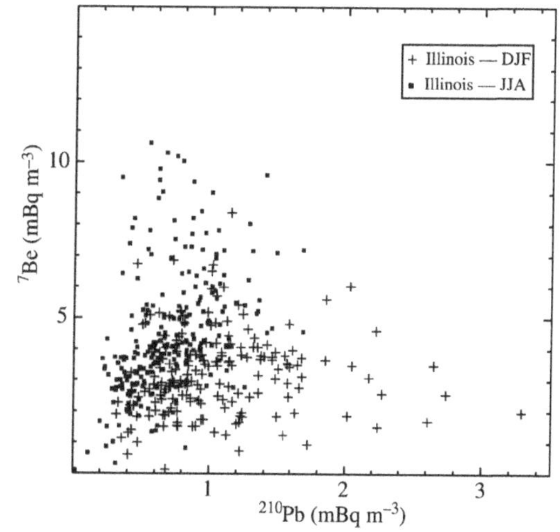

Figure 12 ^{7}Be and ^{210}Pb activities in 24 h air samples from Champaign, Illinois. Plus symbols indicate samples collected in December through February, and filled squares represent samples collected in June–August.

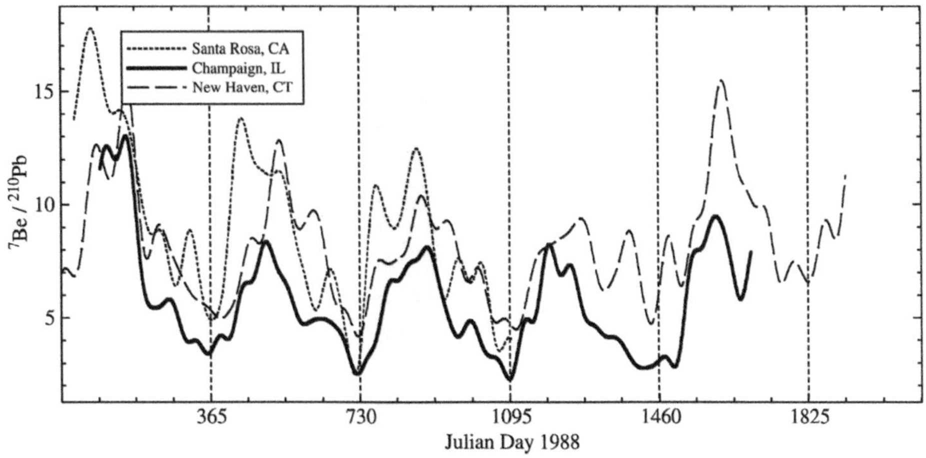

Figure 13 Six-year time series of $^7Be/^{210}Pb$ in near-surface air at three continental sites. The lines represent a Gaussian weighted 45 d moving average of 24 h samples.

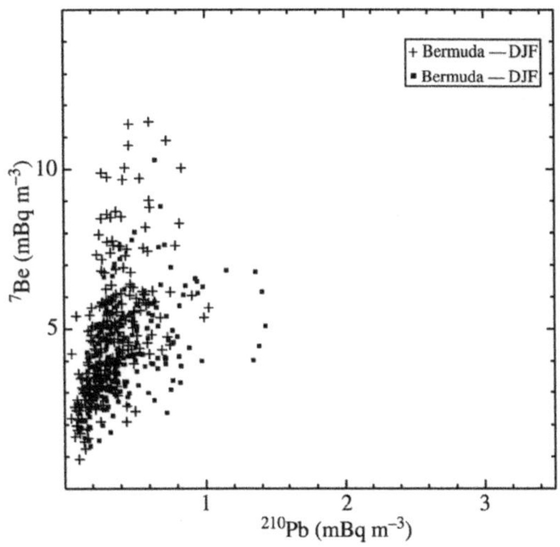

Figure 14 ^7Be and ^{210}Pb activities in 24 h air samples from Bermuda. Plus symbols indicate samples collected in December through February and filled squares represent samples collected in June–August.

mixing is strongest. The effect is pronounced, and there is little overlap between the summer and winter sets of measurements. The time series of a 45 d moving average of the $^7Be/^{210}Pb$ ratio at Champaign, IL shows a repeating annual cycle (Figure 13) characterized by a late spring and early summer maximum and a December minimum in $^7Be/^{210}Pb$, and is similar to sites at the east and west coasts of North America.

The maritime pattern, determined from islands in the Atlantic, is different from the continental pattern. Island sources of ^{222}Rn are too small to have a significant effect on ^{210}Pb. As a result, virtually all the ^{210}Pb observed at these sites is transported from continents. $^7Be/^{210}Pb$ over the ocean is therefore a measure of the influence of continental sources on the local aerosol. Low values of the ratio reflect a high continental influence; high ratios indicate a relative isolation from continental sources.

At Bermuda (Figure 14) the daily data cluster more tightly than at Champaign, and there is a

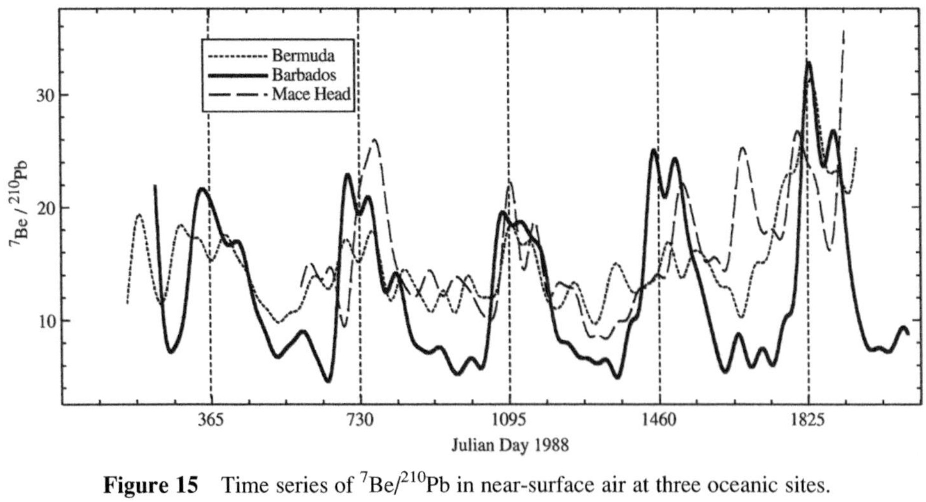

Figure 15 Time series of $^7Be/^{210}Pb$ in near-surface air at three oceanic sites.

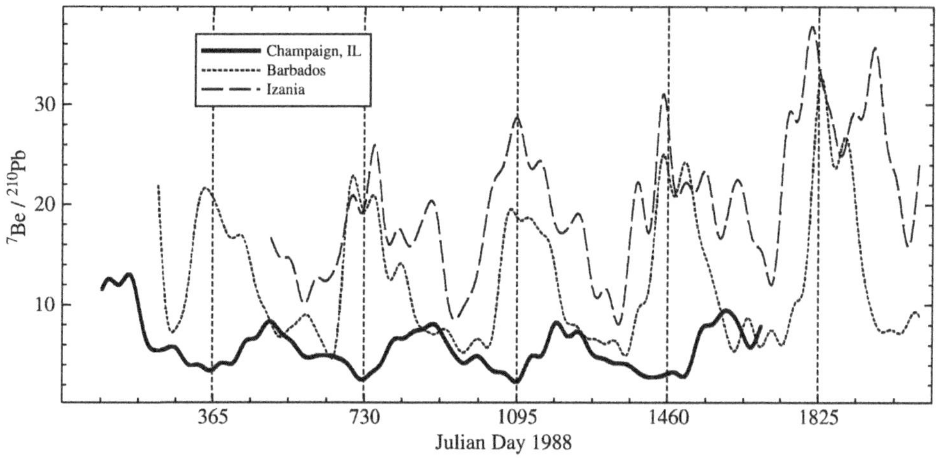

Figure 16 Time series of ^{7}Be/^{210}Pb in surface air at a continental site (Champaign), in surface air at an oceanic site (Bermuda), and in the oceanic free troposphere (Izania). The data plotted are Gaussian weighted 45 d moving averages of 24 h samples.

distinct lower bounding value of ^{7}Be/^{210}Pb. Values that are common in continental winters are never seen in 24 h samples at Bermuda, indicating that continental boundary layer air is diluted or scavenged in transit to Bermuda in the winter.

The seasonal pattern of ^{7}Be/^{210}Pb at island sites is six months out of phase with continental sites (Figure 15). During the summer, ^{7}Be/^{210}Pb is nearly the same over the continent and the ocean, indicating vigorous mixing of the troposphere. In the winter, however, ^{7}Be/^{210}Pb is higher at oceanic sites than continental sites by a factor of $\sim$4.

Izania is located at an elevation of 2,367 m on Tenerife in the Canary Islands and is in the free troposphere. Figure 16 shows that it also exhibits a strong seasonal cycle in ^{7}Be/^{210}Pb resembling Barbados rather than Champaign. The maximum ^{7}Be/^{210}Pb occurs in December, the same month that the continental sites reach their minima.

Taken together, these observations suggest that there is relatively little long-distance transport of aerosols in the marine boundary layer, and that long-distance transport is relatively efficient in the free troposphere. Wintertime stability over continents inhibits transfer of surface-source aerosols to the free troposphere and limits their transport distance. Spring and summer convective mixing over continents results in an efficient exchange between the boundary layer and free troposphere, which enhances long-range transport of aerosols and greatly increases the uniformity of aerosol composition between the continent and marine boundary layers.

14.4.2 Application of the Coupled ^{7}Be–^{210}Pb System to Sources of Atmospheric Species

As ^{210}Pb has its source in the boundary layer and ^{7}Be has its major source in the troposphere,

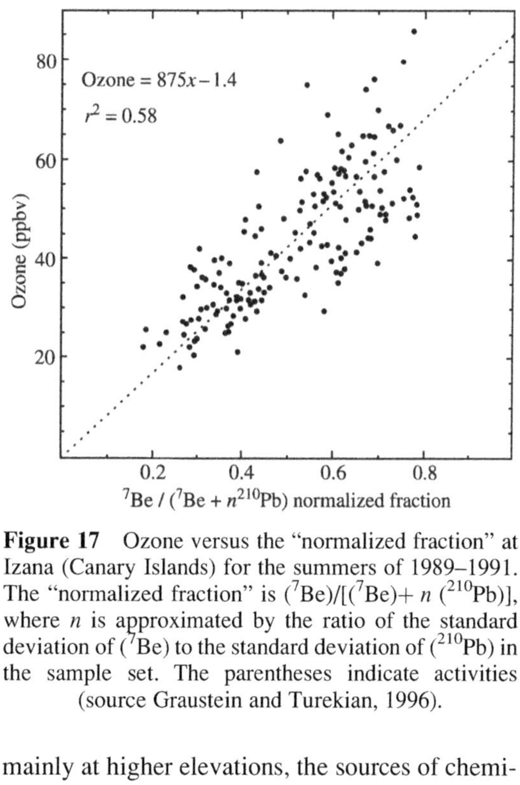

Figure 17 Ozone versus the "normalized fraction" at Izana (Canary Islands) for the summers of 1989–1991. The "normalized fraction" is (^{7}Be)/[(^{7}Be)+ n (^{210}Pb)], where n is approximated by the ratio of the standard deviation of (^{7}Be) to the standard deviation of (^{210}Pb) in the sample set. The parentheses indicate activities (source Graustein and Turekian, 1996).

mainly at higher elevations, the sources of chemical species in the lower free troposphere can be determined by using the ratio of the two nuclides. Such a study was performed in one of the AEROCE sites at Izania in the Canary Islands. Using a function of the ^{7}Be to ^{210}Pb ratio, Graustein and Turekian (1996) showed that the primary source of ozone in the lower free troposphere of the eastern Atlantic was the upper troposphere (Figure 17). An extension of the study to nitrate using sulfate as a boundary layer index as is ^{210}Pb revealed that the primary source of nitrate in the lower free troposphere was also the upper free troposphere and not the boundary layer

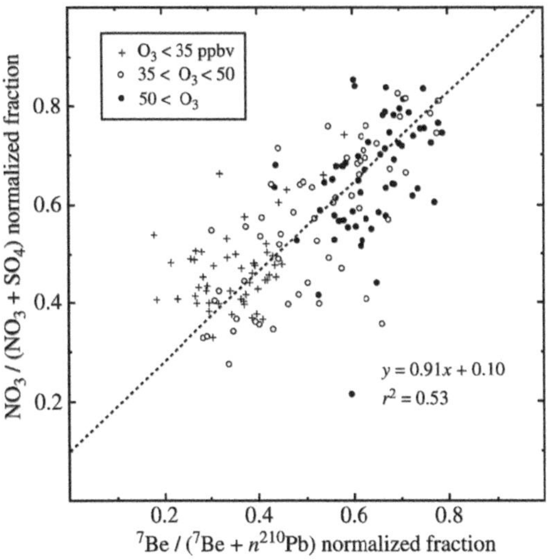

Figure 18 "Normalized fraction" of the nitrate–sulfate system plotted against the "normalized fraction" of the ^7Be–^{210}Pb system as defined in the caption for Figure 17 (from unpublished analysis of AEROCE data by the authors).

(Figure 18). Thus, both chemical species have their origin in the upper troposphere and not anthropogenic pollution source in Europe or biomass burning in Africa. Sampling in the boundary layer downwind of human activity at such sites as Bermuda, Barbados, and Mace Head (Ireland) indicates that both ozone and nitrate are assignable primarily to boundary layer sources.

In another application, Lamborg *et al.* (2000) showed that mercury is distributed homogeneously in the troposphere, as there is no correlation of elemental mercury concentration and $^7Be/^{210}Pb$.

REFERENCES

Andreae M. O. and Raemdonck H. (1983) Dimethyl sulfide in the surface ocean and marine atmosphere: a global view. *Science* **221**, 744–747.

Andreae M. O., Berresgeim H., Andreae T. W., Kritz M. A., Bates T. S., and Merrill J. T. (1988) Vertical distribution of dimethylsulfide, sulfur dioxide, aerosol ions, and radon over the northeast Pacific Ocean. *J. Atmos. Chem.* **6**, 149–173.

Balkanski Y. J., Jacob D. J., Gardner G. M., Graustein W. C., and Turekian K. K. (1993) Transport and residence times of continental aerosols inferred from a global three-dimensional simulation of ^{210}Pb. *J. Geophys. Res.* **98**, 20573–20586.

Bateman H. (1910) Solution of a system of differential equations occurring in the theory of radio-active transformations. *Proc. Cambridge Phil. Soc.* **15**, 423–430.

Broecker W. S. and Peng T.-H. (1971) The vertical distribution of radon in the BOMEX area. *Earth Planet. Sci. Lett.* **11**, 99–108.

Clements W. E. and Wilkening M. H. (1974) Atmospheric pressure effects on ^{222}Rn transport across the earth–air interface. *J. Geophys. Res.* **79**, 5025–5029.

Cline J. D. and Bates T. S. (1983) Dimethyl sulfide in the Equatorial Pacific Ocean: a natural source of sulfur to the atmosphere. *Geophys. Res. Lett.* **10**, 949–952.

Conen F. and Robertson L. B. (2002) Latitudinal distribution of radon-222 flux from continents. *Tellus* **54B**, 127–133.

Dörr H. and Münnich K. O. (1990) ^{222}Rn flux and soil air concentration profiles in West-Germany: soil ^{222}Rn as a tracer for gas transport in the unsaturated soil zone. *Tellus* **42B**, 20–28.

Feely H. W., Toonkel L., and Larsen R. (compilers) (1981) *Radionuclides and Trace Elements in Surface Air*. US Dept. Energy, Environ. Q. Rep. EML-395, appendix, New York..

Friedlander G., Kennedy J. W., Macias E. S., and Miller J. M. (1981) *Nuclear and Radiochemistry*, 3rd edn. Wiley, New York, 684 pp.

Fuller C. and Hammond D. E. (1983) The fallout rate of Pb-210 on the western coast of the United States. *Geophys. Res. Lett.* **10**, 1164–1172.

Graustein W. C. and Turekian K. K. (1983) ^{210}Pb as a tracer of the deposition of sub-micrometer aerosols. In *Precipitation Scavenging, Dry Deposition and Resuspension* (eds. H. R. Pruppacher, R. G. Semonin, and W. G. N. Slinn). Elsevier, Amsterdam and Oxford, vol. 2, pp. 1315–1324.

Graustein W. C. and Turekian K. K. (1986) ^{210}Pb and ^{137}Cs in air and soils measure the rate and vertical distribution of aerosol scavenging. *J. Geophys. Res.* **91**, 14355–14366.

Graustein W. C. and Turekian K. K. (1989) The effects of forests and topography on the deposition of sub-micrometer aerosols measured by ^{210}Pb and ^{137}Cs in soils. *Agri. Forest Meteorol.* **47**, 199–220.

Graustein W. C. and Turekian K. K. (1990) Radon fluxes from soils to the atmosphere measured by ^{210}Pb–^{210}Ra disequilibrium in soils. *Geophys. Res. Lett.* **17**, 841–844.

Graustein W. C. and Turekian K. K. (1991) ^{210}Pb and ^7Be trace seasonal variations in aerosol transport over North America and the North Atlantic. (Paper presented at CHEMRAWN VII Symposium, Int. Union of Pure and Appl. Chem., Baltimore, MD. Available from the authors).

Graustein W. C. and Turekian K. K. (1996) ^7Be and ^{210}Pb indicate an upper troposphere source for elevated ozone in the summertime subtropical free troposphere of the eastern North Atlantic. *Geophys. Res. Lett.* **23**, 539–542.

Graustein W. C. and Turekian K. K. (2004a) Radon flux from soils (in preparation).

Graustein W. C. and Turekian K. K. (2004b) In-cloud scavenging of SO_2 (in preparation).

Knuth R. H., Knutson E. O., Feely H. W., and Volchock H. L. (1983) Size distribution of atmospheric Pb and Pb-210 in rural New Jersey: implications for wet and dry deposition. In *Precipitation Scavenging, Dry Deposition and Resuspension* (eds. H. R. Pruppacher, R. G. Semonin, and W. G. N. Slinn). Elsevier, Amsterdam and Oxford, pp. 1325–1334.

Kritz M. A., Rosner S. W., Kelly K. K., Loewenstein M., and Chan K. R. (1993) Radon measurements in the lower tropical stratosphere: evidence for rapid vertical transport and dehydration of tropospheric air. *J. Geophys. Res.* **98**, 8725–8736.

Lal D. (2001) Cosmogenic isotopes. In *Encyclopedia of Ocean Sciences* (eds. J. H. Steele, S. A. Thorpe, and K. K. Turekian). Academic Press, London, pp. 550–560.

Lal D. and Peters B. (1967) Cosmic rays produced radioactivity on the earth. *Handbuch der Physik* **46**(2): 551–612.

Lambert G., Polian G., and Taupin D. (1970) Existence of periodicity in radon concentrations and in large-scale circulation at lower altitudes between 40° and 70° south. *J. Geophys. Res.* **75**, 2341–2345.

Lamborg C. H., Fitzgerald W. F., Graustein W. C., and Turekian K. K. (2000) An examination of the atmospheric chemistry of mercury using ^{210}Pb and ^7Be. *J. Atmos. Chem.* **36**, 325–338.

Liu S., McAfee J. R., and Cicerone R. J. (1984) Radon-222 and tropospheric vertical transport. *J. Geophys. Res.* **89**, 7202–7297.

Martell E. A. and Moore H. E. (1974) Tropospheric aerosol residence times: a critical review. *J. Rech. Atmos.* **8**, 903–910.

Monaghan M. C., Krishnaswami S., and Turekian K. K. (1985/1986) The global-average production rate of ^{10}Be. *Earth Planet. Sci. Lett.* **76**, 279–287.

Moore H. E. and Poet S. E. (1976) ^{210}Pb fluxes determined from ^{210}Pb and ^{226}Ra soil profiles. *J. Geophys. Res.* **81**, 1056–1058.

Moore H. E., Poet S. E., and Martell E. A. (1973) ^{222}Rn, ^{210}Pb, ^{210}Bi, and ^{210}Po profiles and aerosol residence times versus altitude. *J. Geophys. Res.* **78**, 7065–7075.

Moore H. E., Poet S. E., and Martell E. A. (1974) Origin of ^{222}Rn and its long-lived daughters in air over Hawaii. *J. Geophys. Res.* **79**, 5019–5024.

National Research Council (1988) *Health Risks of Radon and other Internally Deposited Alpha-emitters*. National Academy Press, Washington, DC, 602pp.

Nozaki Y., DeMaster D. J., Lewis D. M., and Turekian K. K. (1978) Atmospheric Pb-210 fluxes determined from soil profiles. *J. Geophys. Res.* **83**, 4047–4051.

Pearson J. E. and Jones G. E. (1966) Soil concentrations of "emanating radium-226" and the emanation of radon-222 from soils and plants. *Tellus* **18**, 655–661.

Schery S. D. and Gaeddert D. H. (1982) Measurements of the effect of cyclic atmospheric pressure variation on the flux of ^{222}Rn from the soil. *Geophys. Res. Lett.* **9**, 835–838.

Schery S. D., Whittlestone S., Hart K. P., and Hill S. E. (1989) The flux of radon and thoron from Australian soils. *J. Geophys. Res.* **94**, 8567–8576.

Settle D. M., Patterson C. C., Turekian K. K., and Cochran J. K. (1982) Lead precipitation fluxes at tropical oceanic sites determined from ^{210}Pb measurements. *J. Geophys. Res.* **87**, 1239–1245.

Tanaka N. and Turekian K. K. (1995) The determination of the dry deposition flux of SO_2 using cosmogenic ^{35}S and ^{7}Be measurements. *J. Geophys. Res.* **100**, 2841–2848.

Tanaka N., Rye D. M., Xiao Y., and Lasaga A. C. (1994) Use of stable sulfur isotope systematics for evaluating oxidation pathways and in-cloud-scavenging of sulfur dioxide in the atmosphere. *Geophys. Res. Lett.* **21**, 1519–1522.

Tsunogai S., Shinagawa T., and Kurata T. (1985) Deposition of anthropogenic sulfate and Pb-210 in the western North Pacific area. *Geochem. J.* **19**, 77–90.

Turekian K. K. and Cochran J. K. (1981a) ^{210}Pb in surface air at Enewetak and the Asian dust flux to the Pacific. *Nature* **292**, 522–524.

Turekian K. K. and Cochran J. K. (1981b) ^{210}Pb in surface air at Enewetak and the Asian dust flux to the Pacific: a correction. *Nature* **294**, 670.

Turekian K. K., Nozaki Y., and Benninger L. K. (1977) Geochemistry of atmospheric radon and radon products. *Ann. Rev. Earth Planet. Sci.* **5**, 227–255.

Turekian K. K., Graustein W. C., and Cochran J. K. (1989) Lead-210 in the SEAREX program: an aerosol tracer across the Pacific. In *Chemical Oceanography* (ed. J. P. Riley). Academic Press, London, vol. 10, pp. 51–81.

Turekian V. C., Graustein W. C., and Turekian K. K. (1999) The ^{214}Bi to ^{214}Pb ratio in lower boundary layer aerosols and aerosol residence times at New Haven, Connecticut. *J. Geophys. Res.* **104**, 11593–11598.

Waser N. A. D. and Bacon M. P. (1995) Wet deposition fluxes of cosmogenic ^{32}P and ^{33}P and variations in the ^{33}P/^{32}P ratios at Bermuda. *Earth Planet. Sci. Lett.* **133**, 71–80.

Wilkening M. H. and Clements W. E. (1975) Radon-222 from the ocean surface. *J. Geophys. Res.* **80**, 3828–3830.

Williams G. and Turekian K. K. (2002) The atmospheric supply of osmium to the oceans. *Geochim. Cosmochim. Acta* **66**, 3789–3791.

Isotope Geochemistry
ISBN: 978-0-08-096710-3

pp. 445–464

C. NOBLE GASES

15
Noble Gases

F. A. Podosek
Washington University, St. Louis, MO, USA

15.1 INTRODUCTION

The noble gases are the group of elements—helium, neon, argon, krypton, xenon—in the rightmost column of the periodic table of the elements, those which have "filled" outermost shells of electrons (two for helium, eight for the others). This configuration of electrons results in a neutral atom that has relatively low electron affinity and relatively high ionization energy. In consequence, in most natural circumstances these elements do not form chemical compounds, whence they are called "noble." Similarly, much more so than other elements in most circumstances, they partition strongly into a gas phase (as monatomic gas), so that they are called the "noble gases" (also, "inert gases"). (It should be noted, of course, that there is a sixth noble gas, radon, but all isotopes of radon are radioactive, with maximum half-life a few days, so that radon occurs in nature only because of recent production in the U–Th decay chains. The factors that govern the distribution of radon isotopes are thus quite different from those for the five gases cited. There are interesting stories about radon, but they are very different from those about the first five noble gases, and are thus outside the scope of this chapter.)

In the nuclear fires in which the elements are forged, the creation and destruction of a given nuclear species depends on its nuclear properties, not on whether it will have a filled outermost shell when things cool off and nuclei begin to gather electrons. The numerology of nuclear physics is different from that of chemistry, so that in the cosmos at large there is nothing systematically special about the abundances of the noble gases as compared to other elements. We live in a very non-representative part of the cosmos, however. As is discussed elsewhere in this volume, the outstanding generalization about the geo-/cosmochemistry of the terrestrial planets is that at some point thermodynamic conditions dictated phase separation of solids from gases, and that the Earth and the rest of the inner solar were made by collecting the solids, to the rather efficient exclusion of the gases. In this grand separation the noble gases, because they are noble, were partitioned strongly into the gas phase. The resultant generalization is that the noble gases are very scarce in the materials of the inner solar system, whence their common synonym "rare gases."

This scarcity is probably the most important single feature to remember about noble-gas cosmochemistry. As illustration of the absolute quantities, for example, a meteorite that contains xenon at a concentration of order 10^{-10} $cm^3STP\ g^{-1}$ (4×10^{-15} mol g^{-1}) would be considered relatively rich in xenon. Yet this is only 0.6 ppt (part per trillion, fractional abundance 10^{-12}) by mass. In most circumstances, an element would be considered efficiently excluded from some sample if its abundance, relative to cosmic proportions to some convenient reference element, were depleted by "several" orders of magnitude. But a noble gas would be considered to be present in quite high concentration if it were depleted by only four or five orders of magnitude (in the example above, 10^{-10} $cm^3STP\ g^{-1}$ of xenon corresponds to depletion by seven orders of magnitude), and one not uncommonly encounters noble-gas depletion of more than 10 orders of magnitude.

The second most important feature to note about noble-gas cosmochemistry is that while a good deal of the attention given to noble gases really is about chemistry, traditionally a good deal of attention is also devoted to nuclear phenomena, much more so than for most other elements. This feature is a corollary of the first feature noted above, namely scarcity. A variety of nuclear transmutation processes—decay of natural radionuclides and energetic particle reactions—lead to the production of new nuclei that are often new elements. Most commonly, the quantity of new nuclei originating in nuclear transmutation is very small compared to the quantity already present in the sample in question, metaphorically a drop in the bucket. Thus, they are very difficult or impossible to detect and, therefore, in practical terms, attracting little or no interest. When the bucket is empty, or nearly so, however, the "drop" contributed by nuclear transmutations may become observable or even dominant. Traditionally there are two types of (nearly) empty buckets that are most suitable for revealing the effects of nuclear transmutations: short-lived radionuclides (e.g., ^{10}Be and ^{26}Al) which would be entirely absent except for recent nuclear reactions, and the noble gases, renowned for their scarcity.

Emphasis on nuclear processes explains what sometimes seems to be an obsession with isotopes in noble-gas geo- and cosmochemistry. Different nuclear processes will produce different isotopes, singly or in suites with well-defined proportions (i.e., "components"), different from one process to another. Much of the traditional agenda of noble-gas geochemistry, and especially cosmochemistry, thus consists of isotopic analysis, and deconvolution of an observed isotopic spectrum into constituent components. (In most geochemical investigations, noble gases are detected by mass spectrometry, a technique that is inherently sensitive to specific isotopes, not just the chemical element. Isotopic data thus emerge naturally in most studies. Noble-gas mass spectrometry can be a much more sensitive technique than other traditional types of mass spectrometry because the gases are "noble," and therefore relatively easy to separate from other elements, and because they are scarce, so that they can be analyzed in "static"-mode (no pumping during analysis) gas-source spectrometers, permitting relatively high detection efficiency without overwhelming blanks.) In realistic terms, it is very difficult to appreciate noble-gas geo-/cosmochemistry without a basic familiarity with noble-gas isotopes: which isotopes occur in nature (i.e., which are stable), in what approximate abundance they are found, how they relate to non-noble neighbors, and, to some extent, how they are associated with specific nuclear processes. Figure 1 provides assistance in this regard.

When the goal is to identify and quantify different noble-gas components that may be present in a sample or group of samples, a common approach to this goal is to try to unmix the components, at least partially, to provide some leverage. One path to this end, of course, is analysis of different samples that may contain the components in different proportions, and thus have different isotopic compositions. Another path, available in addition to or instead of the first, is stepwise heating analysis, which has traditionally been very extensively used in noble-gas studies. Noble gases may be released from solid samples by volume diffusion, or by reaction, recrystallization, melting, or even evaporation of their host phases. If different noble-gas components reside in physically distinct locations within a complex sample, they may be liberated, and thus become available for analysis, at different steps in a time–temperature heating sequence. Differential release of isotopically distinct components will then result in variation of the isotopic composition of gas released in different steps (e.g., see Figures 2 and 4).

A common tool for visualization of isotopic variations is the so-called "three-isotope diagram,"

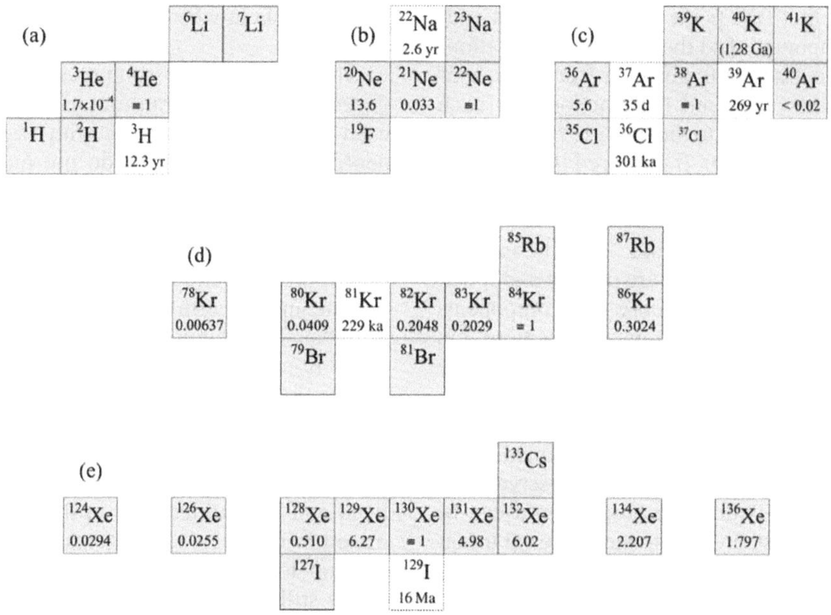

Figure 1 A display of the isotopes of the noble gases and neighboring isotopes in the familiar "chart of the nuclides" format. The abscissa is neutron number (N) and the ordinate is proton number (Z). The box corresponding to any pair (Z, N) represents an isotope; an element is represented by a horizontal row. Boxes for stable isotopes are shown with solid outline; for the noble gases, approximate solar (in the case of He, protosolar) isotope ratios are shown at the bottom of each box. Selected unstable isotopes are shown as boxes with broken line edges. The left-superscript isotope label is the atomic weight A ($=Z+N$). The five panels show regions around the five noble gases (excluding Rn).

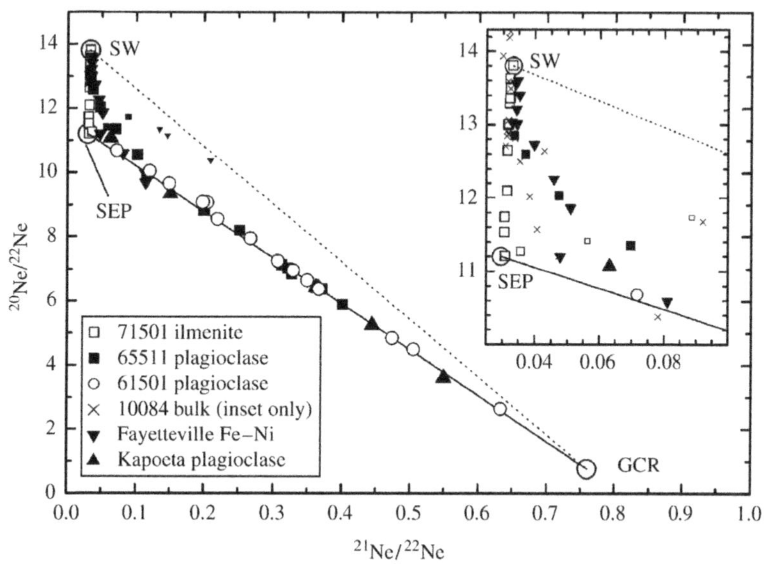

Figure 2 A three-isotope diagram illustrating compositional variations in lunar samples and meteorites, as observed in stepwise *in vacuo* etching and pyrolysis. Since the observed isotopic compositions do not lie on a single straight line, at least three isotopically distinct components must contribute in variable proportions. These data are interpreted as superposition of solar wind (SW), solar energetic particles (SEP), and galactic cosmic ray, i.e., spallation (GCR) Ne components (source Wieler, 1998).

in which two isotope ratios, each with the same reference (denominator) isotope, are displayed on abscissa and ordinate (e.g., Figure 2). Two isotopically distinct components will plot at distinct points on a three-isotope diagram, and an often-used feature is that mixtures of the two components will plot on the *straight line* joining those two points. A lever rule applies: the greater the proportion that one component contributes to a mixture, the closer the point representing the mixture will lie to the point representing that end-member component, and there is a *linear* relationship between fractional distance from one end-member to the other and the fraction that each component contributes to the mixture (specifically to the *reference isotope*). If observed isotopic data are variable but the variations in two ratios are correlated, so as to be consistent with a straight line on a three-isotope diagram, it can be inferred that at least two components are present and it will often be hypothesized that *only* two components are present, in which case their compositions can be constrained to lie on the line, one on either side of the data field. If three components are present, not coincidentally collinear on this diagram, mixtures will occupy the triangular field defined by the three compositions, and conversely if observed data are not consistent with linear correlation it can be inferred that at least three components are contributing to the mix. The concept of the three-isotope diagram is readily generalized. Four isotopes defining three ratios (all with the same reference isotope), for example, will define a three-dimensional space in which mixture of two components will produce compositions lying along a straight line, and mixture of three components will produce compositions lying in a plane, etc. Generalization to more dimensions is mathematically straightforward, even if difficult to envision.

15.2 *IN SITU* COMPONENTS AND NUCLEAR COMPONENTS

In noble-gas geo-/cosmochemistry the term "*in situ* component" is conventionally used to designate a component which is produced *in situ*, i.e., within any given (solid) sample, by nuclear transmutation, and whose constituent atoms are still identifiably in the places in which they were made. For the noble gases, there are many *in situ* components of interest in one context or another. As stressed above, this is because noble gases are very scarce in most solid materials. The small quantities of atoms produced in most nuclear transmutation are essentially lost and overwhelmed in the background sea of most elements, but they can be quite prominent when the product atoms are noble gases.

The term "nuclear component," meaning produced by nuclear transmutation, is sometimes used almost synonymously with "*in situ* component," but the two terms do not mean quite the same thing. (In this context it does not pay to be too fussy about definitions. Since all atomic nuclei are made in nuclear processes, by strict application of this definition everything is a nuclear component, and the term becomes useless. Instead, the term "nuclear component" is generally used when there is some specific, typically local, nuclear process in mind.)

An *in situ* component is necessarily a nuclear component, but the converse need not be true: an originally *in situ* component could be mobilized and transported, and in the process perhaps mixed with other components, in which case it will no longer be an *in situ* component, but it might still be a recognizable nuclear component. As one specific example, it is believed that radiogenic noble gases are exhaled from the interior of the Earth's Moon, and then partially implanted into lunar soils on the surface (see below), where they are readily distinguishable from other reservoirs of noble gases. Such a gas would still be called a nuclear component even though it is not an *in situ* component.

15.2.1 Radiogenic Noble Gases

One prominent and well-known kind of nuclear component is that which is produced by the decay of naturally occurring radionuclides (see Table 1). The best- and longest-known examples are ^{4}He, produced by alpha decay of the natural isotopes of uranium and Th, and ^{40}Ar, produced in one branch of the beta decay of ^{40}K. (There are several other natural radionuclides which produce ^{4}He by alpha decay, but whether because of low parent abundance and/or very slow decay, only in very unusual samples is the production of ^{4}He not strongly dominated by uranium and thorium.) Since radioactive decay laws are well known, the ratio of daughter to parent isotope(s) in a

Table 1 Cosmochemically prominent radiogenic noble gases.

Isotope	*Process*	*Parent*	*Remarks*
^{4}He	α	^{232}Th, ^{238}U, ^{235}U, ...	Dominant in nearly all samples
^{40}Ar	β	^{40}K	Dominant in nearly all samples
^{129}Xe	β	^{129}I	Prominent in most meteorites
Xe, Kr	SF[a]	^{244}Pu	Observable in selected meteorites
Xe, Kr	SF[a]	^{238}U	Sometimes detectable in selected meteorites
^{36}Ar	β	^{36}Cl	Reported, but needs further study for verification
^{22}Ne	β	^{22}Na	Presolar component; see Table 2

[a] SF = spontaneous fission. Fission fragments are rich in neutrons and decay toward beta stability from the neutron-rich side. Fission produces all the heavier isotopes of Xe and Kr which are not "shielded" by the existence of more neutron-rich stable isobars, i.e., ^{136}Xe, ^{134}Xe, ^{132}Xe, ^{129}Xe, ^{86}Kr, ^{84}Kr, and ^{83}Kr. Each fissioning nuclide produces these isotopes in distinct and diagnostic proportions.

closed system is a simple function of time, whence this phenomenon has been long and extensively exploited as a geochronometer.

^{4}He and especially ^{40}Ar nuclei provide good specific examples of how scarcity affects the geo-/cosmochemistry of noble gases. It is very difficult to find or isolate samples in which the total inventory of either of these isotopes is not overwhelmingly dominated by the radiogenic component, and in the case of ^{40}Ar it is so despite the fact that it is produced in only a minor branch ($\sim$11%) of the decay of ^{40}K. The major branch leads to ^{40}Ca, which in most rocks is so abundant that radiogenic ^{40}Ca is difficult if not impossible to even detect.

Radionuclides can also decay by spontaneous fission. Fission of an actinide nucleus produces two main fragments, not always the same, so that the products of fission are not one or two specific isobars but rather a spectrum, whence a fission component typically consists of several isotopes in statistically well-defined proportions. The distribution of actinide fission fragments is characteristically asymmetric, one relatively heavy and one relatively light, so that the distribution of daughter masses is bimodal. By chance, xenon typically lies within the heavier mode, krypton within the lighter, so that xenon and krypton are prominent fission products (yields of a few to several percent). Actinide nuclei have higher neutron-to-proton ratios than lighter nuclei, so that their directly produced fission fragments characteristically lie far on the neutron-rich side of the valley of beta-stability at their respective isobars. Fission fragment nuclei thus typically beta-decay, more or less quickly, along an isobar until they reach a stable nuclide. This feature is often invoked to characterize a fission component, e.g., for xenon the heavier isotopes ^{136}Xe, ^{134}Xe, ^{132}Xe, ^{131}Xe, and ^{129}Xe can all appear in a fission component, whereas the lighter group ^{130}Xe, ^{128}Xe, ^{126}Xe, and ^{124}Xe cannot, because production of these lighter isobars yields stable isotopes of tellurium, not xenon. Sometimes the adjective "fissiogenic" is used to designate a component produced in fission, although this seems unnecessarily awkward when "radiogenic" will serve as well.

In practice, only one natural radionuclide, ^{238}U, produces observable *in situ* spontaneous-fission components in terrestrial (and most lunar) samples, and even then the fission components are observable only in samples unusually deficient in noble gases and/or rich in uranium. The spectra of ^{238}U fission components, and the yields at xenon and krypton, are well known (e.g., see Ozima and Podosek, 2002), so that radiogenic xenon and krypton from fission of ^{238}U are readily identifiable. In principle, a useful radiometric chronology scheme could be based on accumulation of radiogenic xenon (and krypton) from fission of ^{238}U, but this has not yet been much exploited.

Meteorites are characteristically old, many dating from the oldest times in the solar system, and at the time of their formation they incorporated several relatively short-lived natural radionuclides (e.g., see Podosek and Nichols, 1998), radionuclides which have not only decayed to effective extinction at present but which were also effectively extinct at the time of formation of the oldest terrestrial and most lunar samples. The abundances of these short-lived radionuclides were generally quite low, so that their presence may be inferred from excesses of their daughter isotopes primarily in cases where there are large fractionations between parent and daughter elements. Since noble gases are strongly depleted relative to nearly everything in solid materials, it is not surprising that the first two now-extinct radionuclides to be discovered, ^{129}I (half-life 16 Myr) and ^{244}Pu (half-life 78 Myr), have noble-gas daughters.

The presence of ^{129}I in the early solar system was inferred from excesses of its daughter ^{129}Xe (Reynolds, 1960), and it has subsequently been found to have been present in essentially all undifferentiated meteorites and even some differentiated meteorites. This discovery was important in establishing our present paradigm for solar-system chronology: it showed that undifferentiated meteorites were indeed the oldest solids in the solar system and also that they formed nearly simultaneously (differing in age by no more than 10–20 Ma). Moreover, on the premise that the ^{129}I must have been synthesized in some other star, it could be concluded that the solar system as a whole could not be much older (no more than 10^{8} yr) than undifferentiated meteorites, else the abundance of ^{129}I would have decayed beyond observation. Tighter limits of both kinds are available at present, based on subsequently discovered shorter-lived radionuclides, but the general picture first based on ^{129}I persists. In addition, because radiogenic ^{129}Xe is so readily measurable as well as observable in most meteorites, I–Xe dating is arguably the most extensively applicable fine-scale chronometer for meteorite history in the early solar system (e.g., Swindle and Podosek, 1988).

The prior presence of ^{244}Pu, the only transuranic nuclide known to have been present in the early solar system, can be inferred from its spontaneous-fission decay branch, through production of fission tracks and, more diagnostically, by production of fission xenon and krypton. The identification of ^{244}Pu as the fissioning nuclide present in meteorites is unambiguous, since the meteoritic fission spectrum is distinct from that of ^{238}U but consistent with that of artificial ^{244}Pu (Alexander *et al.*, 1971). The demonstration of the existence of ^{244}Pu in the solar system reinforced the requirement (from the presence of ^{129}I) of a relatively short time between stellar nucleosynthesis and solar-system formation and made it incontrovertible, since while it might be possible to make ^{129}I in some models of early solar system development, the rapid capture of multiple neutrons (the r-process) needed to synthesize ^{244}Pu could not plausibly be supposed to have happened in the solar system.

Besides their presence due to *in situ* radioactive decay within a given solid sample, radiogenic ^{4}He, ^{40}Ar, ^{129}Xe, ^{244}Pu-fission xenon (and krypton), and likely also ^{238}U-fission xenon, are also prominent or observable constituents of planetary atmospheres, and their abundance is important in constraining models for planetary atmosphere evolution (see Chapter 16).

Besides these "known" cases of radiogenic noble gases, there are a few other cases of interest. One is ^{248}Cm (half-life 0.35 Myr), also a transuranic radionuclide that might plausibly be speculated to have been present in the early solar system. Evidence for fission xenon or krypton from ^{248}Cm has been long sought but never found. Another is ^{36}Cl (half-life 300 kyr), which decays to ^{36}Ar. There is indeed a positive report for the presence of ^{36}Cl (Murty *et al.*, 1997), but only in one meteorite, so until confirmed this must be considered to remain only a "hint." As discussed below, there is good evidence for generation of a pure ^{22}Ne component from decay of ^{22}Na (half-life 2.6 yr; see Figure 1(b)), but since the decay evidently occurred in the outflow of the star that made the ^{22}Na (see below), not within the solar system, this is not customarily listed among "radiogenic" noble gases. A particularly interesting case is that of CCF-Xe ("carbonaceous chondrite fission" xenon). Quite early in the study of noble-gas cosmochemistry, xenon isotopic variations revealed in stepwise heating of carbonaceous chondrites (Reynolds and Turner, 1964) suggested an interpretation in terms of fission, but the effect could not be successfully associated with any known fissioning radionuclide. A case was made (e.g., Anders *et al.*, 1975) that CCF-Xe was produced by fission of a superheavy element (in the vicinity of atomic number 115), but this was never substantiated. The term CCF-Xe is no longer in use, and it is now understood that the effects that led to its use are attributable to Xe-HL (or Xe-H), an exotic nucleosynthetic component (see below), but not a fission product.

15.2.2 Spallation Noble Gases

Galactic cosmic rays (GCR) are very high energy particles, chiefly protons, which can induce a wide variety of nuclear reactions when they interact with some target nucleus. Often the products of a reaction induced by a primary GCR particle include secondary protons, neutrons, α-particles, etc., which are themselves sufficiently energetic to induce further reactions when they strike another target nucleus, and so on, so that one primary particle can generate a substantial cascade of secondaries. The characteristic attenuation depth for cosmic rays is some 100–150 g cm^{-2}; the characteristic depth for secondaries, particularly neutrons, is greater, and the flux of thermalized neutrons may peak at depth ~250–300 g cm^{-2}.

In a strict sense, spallation is a nuclear fragmentation process in which the target nucleus loses several nucleons. As used in cosmochemistry, however, the term is used more broadly to designate the product of any nuclear transformation induced by cosmic rays, primary or secondary, whether produced by spallation in the strict sense or by more specific reaction channels involving fewer exiting particles (e.g., (p,pn) or (n,α) reactions).

In normal circumstances, the total quantity of nuclides transformed from one isotope to another through nuclear reactions ultimately induced by cosmic rays is small in absolute terms, and only in rare cases is depletion of target nuclides observable (the exceptions are a few cases of secondary neutron capture by isotopes with huge capture cross-sections). In most cases the products of cosmic-ray-induced spallation are also too sparse to be observed against the background of material already present in samples of interest. There are two prominent classes of exceptions to this latter generalization, however: one group is short-lived (tens of Ma and less) radioisotopes, which would not be present at all except for recent production by nuclear reaction; the second is noble-gas isotopes, whose background abundances are characteristically so low that the small quantities generated by spallation may be observable or even dominant. The study of spallation noble gases in meteorites has traditionally been an important part of the agenda of noble-gas cosmochemistry, having begun essentially as soon as the necessary analytical techniques were developed in the 1950s, and vigorous interest continues to the present. Since they became available through the Apollo program, lunar samples have also been studied extensively using similar techniques and goals.

In natural materials, the spallation production rate of a given isotope depends on the concentration of appropriate target nuclides, i.e., on the chemical composition of the sample. Typically, several target elements contribute significantly to the production of a given isotope, so the dependence on target chemistry can be complex. In addition, production rates depend on the fluxes (and their energy spectra) of both primary and secondary high-energy particles. Moreover, accounting for the dependence on flux/spectrum is not just a matter of once-and-for-all integrating over the GCR spectrum, because the energy spectra of various constituents of the cascade depend sensitively on position within the target sample. Position dependence is mostly a matter of distance from external surface, but not entirely, since the projectile fluxes/spectra at the center of a sphere (4π exposure geometry) of radius 50 cm, say, is different than at a depth of 50 cm below a plane surface (2π exposure geometry). All in all, the quantitative prediction of the production rate of a given isotope, for a given position within target material of arbitrary chemical composition, is a rather complicated affair. Nevertheless, quantitative understanding of spallation production rates has received a great deal of attention, involving theoretical calculations, measurements of natural samples, and empirical calibration experiments in particle accelerators, and in detail the prediction of production rates can be quite sophisticated (for a thorough and accessible review, see Wieler (2002a)).

Spallation is sometimes described as an egalitarian process, suggesting that all (neighboring) nuclides are produced (approximately) equally. This reflects the generalization that spallation involves mostly higher energies than stellar nucleosynthesis, so that the details of nuclear energy structure are less important. It is certainly not literally true that all nuclides are produced in equal abundance. Spallation products are characteristically in or on the proton-rich side of the valley of beta-stability, for example, and one of the staple assumptions invoked for heavy noble-gas isotopic deconvolution is that the production of neutron-rich isotopes (see Figure 1), particularly those such as ^{86}Kr and ^{136}Xe, which have more proton-rich stable isobars, is effectively nil. Also, production of ^{4}He, as another example, is generally several-fold greater than production of ^{3}He. Still, a ^{4}He/^{3}He spallation ratio of "several" is much nearer to unity than in other known sources of helium, in which this ratio is of order 10^4 or greater. Perhaps the best example of spallation egalitarianism is in neon. There are significant variations in the composition of spallation neon according to target chemistry and especially shielding (indeed, variation in spallation ^{21}Ne/^{22}Ne is commonly used as a proxy for degree of shielding), but still, the diagnostic feature of spallation neon is that all three isotopes are produced in nearly equal proportions (cf. the GCR composition in Figure 2), which certainly distinguishes spallation neon from any other known neon component.

Because of this tendency toward egalitarian compositions, the accumulation of spallation products due to cosmic ray exposure of natural samples first becomes evident as an increased relative abundance of the scarcest isotopes. In practice, among the noble gases the most sensitive indicator for the presence of spallation products is usually ^{21}Ne, i.e., when spallation components are added to previously unexposed rock, the first observable shift in noble-gas composition is growth in the relative abundance of ^{21}Ne. The usual graphical representation is a three-isotope diagram such as Figure 2, in which the hallmark of cosmic ray exposure is deflection of plotted compositions to the right of reference compositions such as SW and in the direction of the GCR composition. Commonly, prominent enhancement of the ^{3}He/^{4}He ratio is not far behind. With further additions of spallation products the overall composition of neon can be dominated by spallation, and increases in the relative abundances of the scarcest isotopes of the heavier gases become observable or prominent, first ^{38}Ar, then ^{78}Kr, then ^{124}Xe and ^{126}Xe.

For near-surface exposure to GCR, spallation indicator isotopes such as ^{21}Ne and ^{38}Ar are produced at an approximate rate of order 10^{-8}cm^3 STP g^{-1} Ma^{-1} ($\sim 3 \times 10^{11}$ atom g^{-1}Ma^{-1}) from target elements a few amu heavier (magnesium for neon, calcium for argon). Production rates decrease with increasing mass difference between target and product, e.g., production of neon is down by a factor of a few for silicon targets, and by about two orders of magnitude for iron. ^{3}He and ^{4}He are among the small chips spalled off from a target nucleus, rather than the residue, so their production rate is higher and less sensitively dependent on target chemistry; in most meteorites the ^{4}He/^{21}Ne production ratio is between 10^2 and 10^3.

Besides GCR, which originate outside the solar system, spallation can also be induced by the so-called solar cosmic rays (SCR), i.e., an irregular flux of energetic particles (solar flares) from the Sun, sufficiently energetic to cause nuclear fragmentation. The energy spectrum of SCR is much steeper than that of GCR, i.e., the decline of flux with energy is greater, so that the effective penetration depth of SCR is much less than that of GCR, only a few g cm^{-2}, below which the production of spallation isotopes by SCR is small or negligible compared to that by GCR. Nevertheless, SCR flux is high, so that to depths of 1–2 cm in silicates the production of isotopes such as ^{21}Ne by SCR spallation is comparable to or several-fold greater than production by GCR (cf. Hohenberg *et al.*, 1978). Integrated over depth, the SCR contribution to total spallation is thus small but not trivial compared to the GCR contribution. In most macroscopic meteorites SCR effects are small to negligible because the preatmospheric near-surface regions affected by SCR are ablated away during passage through the atmosphere, but SCR effects can be important near the surfaces of lunar samples. It is likely that the Sun was considerably more active early in its evolution than it is now, so the relative importance of SCR in inducing spallation may have been significantly greater early in solar-system history (e.g., Hohenberg *et al.*, 1990).

Even in carbonaceous chondrites, which have higher noble-gas concentrations than most other meteorites, the concentration of trapped ^{21}Ne is generally less than 10^{-8} cm^3 STP g^{-1}. It follows that even in samples with comparatively high noble-gas concentrations, the effects of exposure to cosmic rays become observable, in the form of elevated relative abundance of ^{21}Ne, even for exposure times significantly less than a million years. In practice, essentially any extraterrestrial sample that falls to Earth as a meteorite (or which is brought from the Moon by spacecraft) has been exposed to cosmic rays long enough that its exposure age can be determined through noble gases, especially ^{21}Ne. The study of cosmic ray exposure ages, and their implications for sample history and provenance, provide a major impetus for the continuing great interest in spallation noble gases.

Cosmic ray exposure ages have been determined for many meteorites. The exposure ages of stony meteorites are found to be mostly a few Ma to a few tens of Ma, but there are a few lower and a few higher. The exposure ages of stony-iron and iron meteorites are characteristically higher than of stony meteorites, hundreds of Ma to Ga ages. No known meteorite (or lunar sample) has been exposed to cosmic rays for even approximately as long as the age of the solar system, whence it may be concluded that these extraterrestrial materials have spent most of the age of the solar system on larger bodies (parent bodies), shielded from cosmic rays by a few meters or more, so that the cumulative effects of cosmic ray exposure are very small or negligible. (This seems obvious today, but there was a time when it was not so.)

The simple one-stage model is that meteorites suddenly become unshielded in some event, generally an impact, beginning their exposure to cosmic rays as a relatively small body (with constant shielding) and presumably also launching them into orbits that will ultimately lead to capture by the Earth. This model seems adequate to account for much data. In some cases, more complicated and interesting histories are inferred, e.g., changes in shielding during exposure or differential exposure of some constituents, in parent-body regoliths or even in the solar nebula, prior to assembly of the present sample. A detailed description of cosmic ray exposure ages and their implications is review by Wieler (2002a) and references therein).

Because ^{21}Ne and ^{3}He are the most sensitive indicators of the presence of spallation products, it can be difficult to identify materials which can be confidently inferred to be *unexposed* to cosmic rays, i.e., in which we may confidently assign the abundance of these isotopes to a trapped component. Thus, the measured ^{21}Ne and/or ^{3}He abundances in some sample are upper limits to the trapped abundance, but it may be difficult to eliminate the possibility that some of the measured abundance was not produced in spallation. It is possible, for example, that some of the ^{21}Ne in prominent components such as air or planetary trapped gases (see below) is attributable to cosmic-ray-induced spallation.

15.3 SOLAR NOBLE GASES

Some of the noble gases found in extraterrestrial materials are from the Sun. This has been known for a long time (e.g., Signer and Suess, 1963), and there is no great mystery to it. Solar noble gases are samples of solar corpuscular emanation, mostly the "solar wind" (particles with velocities mostly in the range of several hundred km s^{-1}) but also including lesser amounts of more energetic particles. Solar gases are found in solid samples that have been exposed to the Sun on the surfaces of airless planetary bodies (or spacecraft), into which they are implanted by virtue of their kinetic energy, and they can be identified as "solar" by virtue of elemental abundances only modestly different from "cosmic" composition (see Figure 6) and of near-surface residence. Implantation depths are characteristically of order 0.1 μm or more, small in absolute terms but still hundreds of atom layers deep, in appropriate conditions deep enough to resist diffusive loss for geologically long times. Solar noble gases were first discovered in meteorites, specifically in types that are otherwise extremely gas poor. A few members of these classes contain abundant surface-correlated noble gases along with other manifestations of direct exposure to the Sun. It was and still is considered that they were exposed on the surfaces of their parent asteroids. In post-Apollo times the stereotypical solar-gas-bearing samples are lunar soils and breccias incorporating former soils, in which solar gases are commonly prominent because the Moon is apparently essentially devoid of intrinsic volatiles, including noble gases.

Solar noble gases have attracted much interest and analysis. In part, this is simply because the Sun has most of the mass of the solar system. A more focused reason, however, is the general belief that the noble gases in the Sun are the same as the noble gases that were present in the solar nebula, from which the terrestrial planets (including meteorites) were formed. For the noble gases this is a nontrivial proposition because there are so many different noble-gas compositions found in different kinds of planetary materials, which are often rather difficult to relate to each other. In particular, because noble gases are so scarce in planetary materials, and in the terrestrial planets more generally, there are apparently numerous specific cases of noble-gas isotopic compositions having been modified by the addition of nuclear components (see above). Because the current paradigm of solar-system formation implies that the *noble gases are not depleted in the Sun*, their compositions will not have been modified in this way, or at least modified much less. There is great attraction to the idea that solar noble gases are the same as nebular noble gases and, having preserved their compositions, can be taken to be the starting material for the gases now in planets and meteorites, anchoring the starting point for atmospheric and planetary volatile evolutionary models.

15.3.1 Deuterium Burning and the Solar $^3He/^4He$ Ratio

There is one prominent and well-known exception to the rule that the Sun preserves nebular noble-gas isotopic compositions better than planetary materials: helium. Before it began main-sequence hydrogen burning, the protosun experienced a nucleosynthetic phase known as deuterium burning, in which all the Sun's initial supply of deuterium (^2H, more commonly represented as D) was reacted ("burned") into ^3He. The result is that the Sun is essentially devoid of D, and its present-day $^3He/^4He$ ratio is higher than its primordial value; specifically, the post-deuterium-burning $^3He/^4He$ ratio is $(D+^3He)_0/^4He$, where the "zero" subscript designates the primordial component, i.e., that of the interstellar medium from which the solar system formed. (The more direct quantity of interest is the D/H ratio, which can be translated from $D/^4He$ by the standard atomic ratio $^4He/H = 0.10$.)

The primordial D/H and $^3He/^4He$ ratios are both quite interesting quantities, not just in cosmochemistry and geochemistry, but also in astrophysics and cosmology. While the present-day $^3He/^4He$ ratio is readily enough measured in the solar wind, however, empirical determination of both $D_0/^4He$ (i.e., D/H) and $^3He_0/^4He$ is rather challenging. Determination of D/H from terrestrial planetary material is complicated by the fact that D/H is evidently highly variable (several-fold) among different planetary materials, evidently because of very severe chemical mass-dependent fractionation at low temperatures, in the solar nebula or in its antecedent interstellar medium. The inference is that primordial D/H is much lower than D/H in any terrestrial planetary body or material, but it is hard to pin down an exact value. Analytical determination of primordial $^3He/^4He$ is complicated by the fact that in most natural planetary materials the ^4He is predominantly radiogenic and the ^3He is predominantly spallogenic. It is thus difficult to obtain a sample of suitable material in which helium is mostly primordial, enough so that plausibly small corrections for radiogenic and spallation components can be made. The coupling of these two important ratios into the present solar helium composition has often been approached in the hope that if an independent estimate of one of these ratios can be made, the other can be inferred through solar helium (cf. Geiss and Reeves, 1972). The game has been played in both directions.

Even before the advent of Apollo samples, for example, it was noted that $^3He/^4He$ in the solar wind, as observed in solar-gas-rich meteorites, was of order 4×10^{-4}. This is several-fold lower than it would be just from deuterium burning if primordial D/H were close to the terrestrial (seawater) value, 1.6×10^{-4}. This contrast provided the first solid argument that primordial D/H must have been several-fold lower than the seawater value (Geiss and Reeves, 1972).

Using the present best values (e.g., see review by Wieler, 2002b), the solar-wind $^3He/^4He$ ratio is $\sim 4.3 \times 10^{-4}$, as inferred from both lunar samples and spacecraft measurements. Because of isotopic fractionation in the solar wind (see below), the solar photospheric value is inferred to be somewhat lower, nominally 3.75×10^{-4}. Gloeckler and Geiss (2000) apply a further small correction for differential gravitational settling of the heavier isotope and conclude that the primordial ratio $(D+^3He)_0/^4He$ was $\sim 3.60 \times 10^{-4}$. Many investigators would consider that the present best estimate for the primordial $^3He/^4He$ is 1.66×10^{-4}, the spacecraft measurement for the atmosphere of Jupiter (Mahaffy *et al.*, 1998). By difference, the conclusion is that the primordial D/H ratio was 1.94×10^{-5}. An alternative interpretation based on meteorite samples, specifically the "Q" component (see below), is that primordial $^3He/^4He$ was 1.23×10^{-4} (Busemann *et al.*, 2001), whence primordial D/H was 2.47×10^{-5}. The difference between these two calculations is probably indicative of the uncertainty that should be assigned to this interpretation. Either way, however, it may be noted that this inferred D/H ratio is indeed nearly an order of magnitude below the terrestrial seawater value.

The primordial $^3He/^4He$ ratio has special relevance in models for planetary evolution, in that it constrains how much radiogenic ^4He must be added to produce a given mixture ratio, e.g., as in helium in the Earth's mantle. In addition, solar-wind $^3He/^4He$ should be a unique component within the solar system, and so might be decisive in distinguishing between hypotheses for component origins. As one example, there are reports (e.g., Ozima and Zashu, 1983) that some terrestrial diamonds contain trapped helium with

$^{3}He/^{4}He$ ratio higher than the primordial ratio. This interpretation is disputed (the alternative interpretation is that the high $^{3}He/^{4}He$ indicates spallation in samples resident near the surface of the Earth) and cannot be considered well established, but if this or some similar future observation were more strongly established it would essentially demand that the source of the gas was the solar wind.

The prior discussion in this section has ignored the fact that for most of the age of the solar system the Sun has been conducting main-sequence nucleosynthesis of helium, not just the end product ^{4}He but also the p–p chain intermediary ^{3}He. This discussion is thus relevant only with a further assumption that the helium isotopes synthesized in the interior have not mixed into the outer convective zone, so that the photosphere can be taken to have preserved early (post-deuterium-burning) composition. Mixing from the inner nucleosynthetic zone to the outer convective zone would be manifested as a change in helium composition (specifically, an increase in $^{3}He/^{4}He$) over the age of the Sun. Available data suggest that such mixing, i.e., evolution of solar-wind helium composition, has not occurred to a significant extent (see review by Wieler, 2002b).

15.3.2 Solar Wind in Lunar Soils

For several elements, including the noble gases, the best way to measure the composition of the solar wind is in lunar soils (and soil breccias). Lunar soils have been exposed to the Sun and have collected solar wind for geologically long times, and they provide greater quantities of solar-wind species than any other materials presently accessible. In addition, the Moon is very poor in volatile elements, so that for several elements—e.g., hydrogen, carbon, nitrogen, and noble gases—the solar-wind contribution is not overwhelmed by indigenous lunar materials. For the noble gases especially, the Moon is a good place to catch solar wind for study, since lunar primordial noble gases are so scarce that it is debatable whether any have ever been observed. Even so, however, analysis of solar-wind noble gases is not just a matter of analyzing the total noble-gas content of some lunar soil, and to some extent the accuracy to which we know solar-wind composition is limited by considerations of how it is measured in lunar samples.

One complicating factor is spallation: exposure to the Sun on the surface of the Moon also means exposure to cosmic rays, and for some isotopes production by spallation is not inconsequential compared to solar-wind accumulation (Figures 2 and 3). Spallation and solar-wind gas can largely be separated, however, because they are sited differently: spallation gas is volume correlated and solar-wind gas is surface correlated. (Spallation gases are not literally volume correlated. "Mass correlated" would be a better term, since their abundance will be proportional to their target elements, and they will be in the same places as the target elements, but "volume correlated" is the traditional term. Also, solar-wind gases are not literally surface correlated. Analyses of size-separated lunar soils indicate that solar-wind gas concentrations are proportional to a^{-n}, where a is grain radius, but the exponent n is less than the value unity which strict surface correlation demands (e.g., see Eberhardt *et al.*, 1972). This can be understood in terms of solar-wind gases gradually changing from surface correlated to volume correlated through lunar regolith processes such as agglutination.)

Partial separation can thus be achieved by stepwise heating or etching (Figure 2), since they will be released at different temperatures and/or etch depths. Also, solar wind can be enhanced, relative to the spallation component, by selecting finer grain sizes. A particularly useful approach is an ordinate-intercept analysis, as illustrated in Figure 3. If two compositionally distinct reservoirs are mixed, and one of them (the spallation component, in the present instance) is volume correlated (actually, mass correlated), so that its concentration is constant, whereas the other (solar wind) is present in different concentrations (in different grain size fractions; note that strict surface correlation is not required), their mixing will yield a linear correlation between any given isotope ratio (plotted on the ordinate) and the inverse of the total concentration (plotted on the abscissa) of the reference isotope. The composition of the variable component can then be determined as the ordinate intercept of the correlation line, i.e., by extrapolating to infinite concentration of the reference isotope. Figure 3, which illustrates calculation of solar-wind xenon composition by Eberhardt *et al.* (1972), is actually a variation on this theme, in which the "constant" is not the concentration of the reference isotope (^{130}Xe) but its ratio to spallation target element barium.

Another complication is that some surface-correlated noble-gas isotopes in lunar soils are apparently indigenous to the Moon, not imported in the solar wind. It is found that the concentration of surface-correlated ^{40}Ar in lunar soils can be up to several times the concentration of ^{36}Ar, and is in general much higher than could plausibly be attributed to solar-wind argon (Heymann and Yaniv, 1970). The total concentration of ^{40}Ar is in excess of what could be attributed to *in situ* decay, and so the excess came to be called orphan argon (i.e., parentless argon). It is widely believed to be ^{40}Ar, which is radiogenic but not *in situ*, formed by decay of ^{40}K in the lunar interior, degassed from the interior into the transient lunar

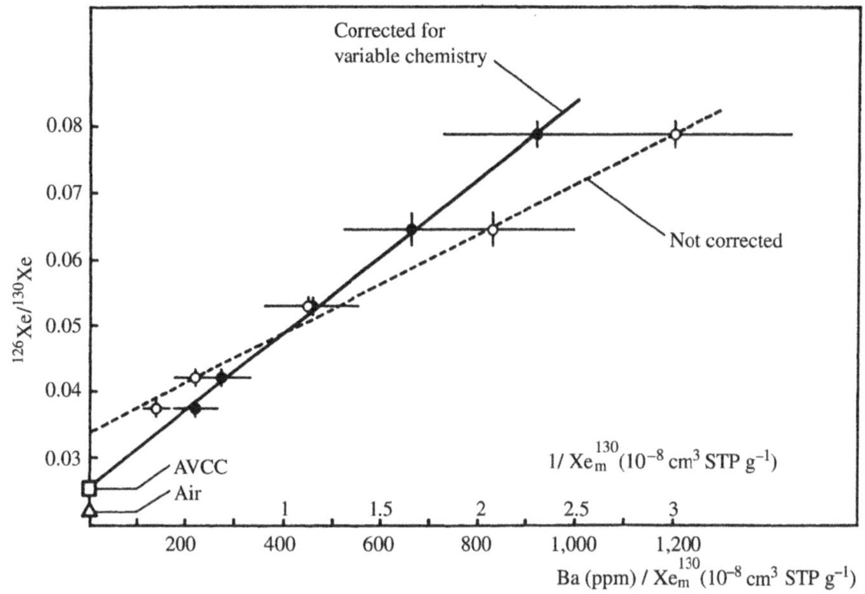

Figure 3 A modified ordinate-intercept diagram which illustrates mixing of surface-correlated (solar wind) and "volume"-correlated (spallation) Xe in lunar soil 12001. In such a diagram, if the abscissa were simply $1/^{130}Xe$, mixing of a (volume-, or mass-correlated) component present in *constant* concentration with another (surface-correlated) component present in *variable* concentration would produce a straight-line correlation, and extrapolation to infinite concentration (i.e., the ordinate intercept) would yield the composition of the variable component (i.e., the surface-correlated, or solar-wind component). This situation is illustrated by the open data symbols and broken-line correlation. A better assumption is that the more nearly constant "volume-correlated" component is the ratio of ^{130}Xe to Ba, the principal target element for xenon spallation. (The light rare-earth elements are also important targets, but these occur in nearly constant proportion to Ba.) This situation is illustrated by the solid data symbols and correlation line, whose ordinate intercept is a better determination of surface-correlated xenon composition. Note that this $^{126}Xe/^{130}Xe$ ratio is indistinguishable from the "average value carbonaceous chondrite" (AVCC) ratio. The term AVCC is no longer in common use; it is equivalent to "planetary" xenon (see Figure 6), which is a superposition of Q and other components (see Figure 7) (source Eberhardt *et al.*, 1972).

atmosphere, and partially ionized and accelerated by solar-wind electromagnetic fields to impact on and become trapped in lunar soil grain surfaces (Heymann and Yaniv, 1970; Manka and Michel, 1971). More subtle, but apparently qualitatively similar, excesses of the heavy isotopes (and in some cases of ^{129}Xe and the heavy isotopes of krypton) have also been observed (Drozd *et al.*, 1972, 1976), most prominently at the older landing sites (Apollo 14 and Apollo 16). The excess ^{129}Xe is plausibly ascribed to decay of ^{129}I, and the heavy isotopes to actinide fission, primarily from ^{244}Pu, as can be verified in a few cases by matching the spectrum of xenon isotopes to the composition of ^{244}Pu fission. It is commonly taken that these also are "orphan" components, similar in origin to the excess ^{40}Ar. From the early solar-system ratio of ^{244}Pu to ^{238}U it may be inferred that there is similarly excess fission xenon and krypton from ^{238}U present, likely even in soils at the younger landing sites, in quantities small enough that they are perhaps difficult to detect directly but still large enough to be significant. It may be noted that since orphan components are surface correlated, not volume correlated, they cannot be resolved from solar-wind characterization through the ordinate-intercept analysis (e.g., as in Figure 3), so there is some basis for concern that there may be non-negligible fission in surface-correlated xenon compositions derived from lunar soils.

Some elemental abundance patterns for solar gases in lunar soils (and one gas-rich meteorite, Pesyanoe, that was likely exposed to solar wind on the surface of an asteroid; see Marti, 1969) are illustrated in Figure 6 (left panel). Gases in true solar proportions would define a horizontal straight line in this diagram, and it is evident that the lunar (and meteoritic) regolith samples do not do this, but rather exhibit significant and variable fractionation patterns of progressive depletion of lighter gases (albeit still modest compared to the "planetary" pattern in the right panel of Figure 6). An easy inference is that this pattern reflects progressively greater (diffusive) loss of lighter gases from the lunar/meteoritic materials. For helium and neon this interpretation is supported by abundance ratios, relative to argon, which are not only less than those inferred for the Sun as a whole but also less than those directly measured by spacecraft and in foils exposed during Apollo missions (see reviews by Wieler, 1998, 2002b). In addition, although there are substantial uncertainties in estimating whole-regolith abundances, it can be estimated that the present regolith inventories of helium and neon are only 10^{-2}–10^{-1} of all the gas that has impinged on the lunar surface in the past ~4 Ga, i.e., there has been major loss of these light noble gases. In contrast, it can be argued that effectively the whole fluence of the heavier gases argon, krypton, and xenon that has ever hit the Moon (in the ~4 Ga age of surface formations) is still there. Primarily on these grounds it can be further argued that the relative proportions of the heavier gases trapped in lunar regolith materials represent actual solar-wind composition (cf. Wieler, 2002b), unmodified by fractionating loss, and thus that the solar wind itself is fractionated relative to solar composition.

If the heavy gases argon, krypton, and xenon in the solar wind are quantitatively retained in the lunar regolith, it can also be inferred that solar-wind isotopic compositions of these gases are preserved in the lunar regolith. For helium and neon, extensive loss leads to concern that the residual gases in the lunar soils may be isotopically fractionated with respect to true solar-wind composition. Wieler (1998, 2002b) argues that this is not the case, however, at least not for neon. The first gas released in closed-system etching, corresponding to the shallowest siting, is generally very close to a common composition, labeled SW in Figure 2, which is also the composition observed in spacecraft measurements and in the foils exposed to solar wind during the Apollo missions, and which is taken to be the true mean solar-wind neon composition.

When neon is released differentially, either by stepwise heating or by stepwise closed-system etching, the early release that indicates trapped neon of composition SW (Figure 2) is generally followed by trapped neon with lower $^{20}Ne/^{22}Ne$, more nearly like the composition labeled solar energetic particles (SEP; solar corpuscular emanation at higher energies (100 keV amu^{-1} and higher)), and/or a correlation line between SEP and spallation (GCR in Figure 2). Comparable effects occur in the other gases. Although the term is entrenched in the noble-gas literature, it may be a misnomer, and the nature and origin of the SEP component is somewhat of a mystery (Wieler, 2002b). The principal problem is that there is too much of it: noble-gas SEP/SW in lunar soils is of order 10^{-1}, which is about three orders of magnitude too high compared to directly measured solar emanation. It has been suggested that SEP is not really a physical component but rather the result of differential diffusive migration of the noble-gas isotopes, or perhaps of a slightly greater implantation depth for heavier isotopes. It has also been suggested that it is not solar emanation at all, but interstellar gas entering the solar system at energies higher than the solar wind (Wimmer-Schweingruber and Bochsler, 2000). None of these possible interpretations is without problems, so the origin of SEP noble gases remains an open question.

15.3.3 Solar-wind Composition and Solar Composition

Even if lunar soils and regolithic meteorites preserve well the composition of the solar wind incident on them, the composition of "solar" noble gases cannot immediately be translated to the composition of the Sun (or the photosphere) because of the possibility of fractionation, both elemental and isotopic, arising in the mechanisms by which solar atmospheric atoms are accelerated into the solar wind. Indeed, spacecraft measurements indicate variability of both elemental and isotopic compositions in the solar wind, and in higher-energy emanation, at different times and at different energies, so fractionation mechanisms are clearly at work. In general, it may not be presumed that even time-averaged composition, even integrated over energy, will accurately yield the composition of the solar source.

Solar-wind $^{3}He/^{4}He$ is commonly some 10–20% higher than the underlying solar composition, varying slightly with speed (e.g., Gloeckler and Geiss, 2000; Wieler, 2002b). This ratio can be several-fold different and highly variable in higher-energy emanation, in the "solar energetic particles" component. Isotopic fractionation is expected to be smaller in heavier elements, it is of order a percent or two, varying with speed, for neon and neighboring elements (e.g., Kallenbach *et al.*, 1998). Isotopic composition can be expected to be progressively still less for argon, krypton, and xenon, although isotopic data for real-time collection of solar-wind argon are insufficiently precise to rule on this issue. For krypton and xenon there are no data for direct collection, so only the time- and speed-integrated compositions recorded in lunar and meteoritic materials are available.

At the level of the best analytical errors it is difficult to determine whether averaged solar-wind isotopic compositions accurately represent true solar values because there is no other independent knowledge of solar compositions. Instead, solar-wind compositions can be compared with planetary compositions such as the Earth's atmosphere or the Q-component of "planetary" gases in meteorites (see below). This is a particularly problematic approach for noble gases, however, since in any given case there are generally ample grounds for suspecting that present compositions in planetary materials have been altered from true solar composition. In addition, at the finest levels of precision permitted by analytical errors it is not uncommon that different investigators disagree on what solar-wind composition really is (cf. Wieler, 2002b; Pepin and Porcelli, 2002).

There can also be substantial elemental fractionation, and fluctuation, in the solar wind (and still more in the solar energetic particles). The average value of solar-wind He/Ne observed directly (in spacecraft and Apollo foils) seems not quite a factor of 2 below the true solar value (inferred from spectroscopic data and stellar structure modeling), but Ne/Ar in the solar wind, within errors, may be consistent with the actual solar value. The lunar data evidently do not speak to this issue, because of apparent pervasive loss of the lighter gases (see above). Relative to their inferred solar abundances (thought reliably inferred from regularities in nucleosynthetic processes), krypton and xenon in the solar wind, as recorded in lunar and meteoritic materials, are significantly enhanced (see Figure 6, left panel), xenon by a factor of ~4. This is thought to be a manifestation of the well-known first ionization potential (FIP) effect, in which elements with ionization energies less than ~10 eV are present in the solar wind in greater abundance, relative to source solar abundances of other elements (von Steiger *et al.*, 1997). Krypton and xenon actually have higher ionization energies than this threshold, but they are nevertheless enhanced in the solar wind in models that show that the governing parameter is really ionization time in the solar chromosphere (Geiss *et al.*, 1994; Wieler, 1998).

15.4 EXOTIC COMPONENTS

Arguably the most significant development in cosmochemistry in the past generation is recognition and exploitation of the circumstance that some types of presolar solid materials have survived, more or less intact, throughout the formation and subsequent history of the solar system, that they are preserved in undifferentiated meteorites that have not been metamorphosed, and that they can be isolated in sufficient purity and quantity to support a wide variety of laboratory studies (e.g., see Bernatowicz and Zinner, 1997). The isotopic compositions of these materials are radically different from solar-system normal, which is indeed the primary evidence that they are presolar. Since materials formed in the same interstellar medium from which the whole solar system formed would be expected to have isotopic compositions close to normal, it is further inferred that these materials are not just presolar but circumstellar, having formed in the atmospheres or ejecta of individual stars at specific stages in their evolution, and containing atoms whose nuclei were made in specific nucleosynthetic processes. Study of these materials has provided a wealth of empirical constraints on astrophysical theory. Among the elements identified and characterized in presolar materials are the noble gases, which are particularly prominent in the study of presolar materials for two reasons.

One reason is that, at least as far as we know at present, the abundance of identifiable preserved presolar materials in meteorites is low, so that known types of presolar materials do not supply a substantial fraction of the total meteoritic inventory of any elements *except for the noble gases*. This is another manifestation of the generalization that low noble-gas abundances render prominent some quantitatively small contributions that would be lost in the background sea of other elements.

The second reason is that noble gases have been intimately involved in the initial recognition and isolation of presolar materials. The first evidence for the presence of presolar materials in meteorites was isotopic variation among gases released in stepwise degassing, especially xenon and neon (e.g., Reynolds and Turner, 1964; Black and Pepin, 1969), reflecting differential thermal release of isotopically distinct components. At the time, however, these isotopic variations were not recognized as reflecting the presence of presolar materials. The main reason for this nonrecognition was likely that the host of isotopic variations in the noble gases that could be ascribed to nuclear components (radiogenic and spallation gases, in some cases including neutron-capture effects), plus the possibility of significant mass-dependent isotopic fractionation, which led to expectations that yet more isotopic effects would ultimately be explained in terms of yet more nuclear components or physical processes, effects operating within the solar system to modify the isotopic composition of once-uniform planetary materials. In addition, many cosmochemists were predisposed to the view that all previously existing solids in materials ancestral to the terrestrial planets were vaporized in the solar nebula, so that survival of presolar grains was unexpected. Even when larger isotopic effects (e.g., Figure 4), less interpretable in terms of solar-system processes, were explicitly suggested to represent presolar components (Black, 1972), this hypothesis did not gain much support or even attention. It was not until apparent isotopic anomalies, indicative of nonhomogenization of presolar components, were found in the major element oxygen (Clayton *et al.*, 1973) that credence in this idea led to vigorous programs of study of isotopic anomalies.

15.4.1 Presolar Carriers

In hindsight, it is clear that stepwise heating of bulk meteorite samples, with the resultant partial separation of noble-gas components from host phases that release their gases at different

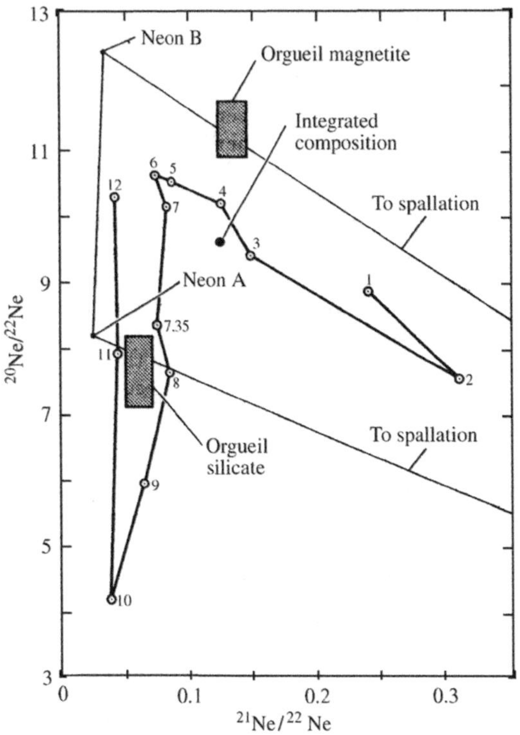

Figure 4 A three-isotope diagram for neon indicating compositional variations in neon in stepwise heating of the carbonaceous chondrite Orgueil. The numbers next to the data points are release temperatures in hundreds of degrees C. Also shown are the trapped component compositions for neon-A ("planetary") and neon-B ("solar"); the light lines connect neon-A, neon-B, and the composition of spallation Ne offscale to the right (cf. Figure 2). The shaded rectangles indicate the compositions of separated magnetite and silicate fractions found by earlier investigators. Much of the individual data points could be explained as mixtures of these three components, but the data excursions below this triangle demand another component, termed neon-E. It was later determined that neon-E is very nearly monisotopic ^{22}Ne, i.e. it would plot at the origin of this diagram (see Figure 5), and also that it is composite, made up of the G- and R-components (see Table 2 and Figure 7) (source Black, 1972).

temperatures (e.g., Figures 2 and 4), provided the essential clues pointing to the discovery of presolar materials preserved in meteorites. Still, stepwise heating of bulk meteorites provides only limited isotopic component separation, allows characterization of the noble-gas hosts only by temperature of gas release, and does not offer the opportunity to relate noble-gas components to most other elements which are not noble gases. Major advances followed the introduction of experimental efforts to physically separate the phases that carried the isotopically distinct noble gases, by properties such as grain size, density, and especially resistance to corrosive chemical attack. The effort to identify the carriers of anomalous noble gases provides a fascinating history, full of blind alleys and inspired detective work in which investigators were able to follow the isotopic anomalies in the noble gases to progressively more refined isolation of their hosts, eventually culminating in the discovery and identification of presolar phases that host exotic noble gases (see Figure 5). There is no need to recount this history here, however, and interested readers are referred to reviews such as given by Anders (1988) or Anders and Zinner (1993), various chapters in the book edited by Bernatowicz and Zinner (1997). Moreover, the discussion here focuses on aspects of presolar grains most relevant to noble gases; no attempt is made to describe the broader aspects of presolar grain studies.

There are three well-known types of presolar grains which were originally isolated and identified by following anomalous noble gases through complicated separation procedures: diamond, silicon carbide, and graphite. These phases account for only a very small fraction of the total mass of a meteorite, or of most elements within a meteorite, but they supply important parts of the noble-gas budget. To an extent, these materials can be separated by nondestructive techniques based on density, grain size, etc., but production of the cleanest and most abundant samples has generally involved the basic approach of dissolving away most of the other phases with which they were originally mixed in whole-rock meteorites.

There can be no doubt that at least the silicon carbide and graphite are presolar. Many elements have been analyzed in bulk collections of these grains, and in general all multi-isotope elements have isotopic compositions that are radically different from solar-system normal composition, compositions that could not plausibly have been produced within the solar system. Indeed, it is generally held that these compositions could not have been produced from materials in the interstellar medium, which mixes nucleosynthetic contributions from many stars; instead they are interpreted to be circumstellar grains, formed in the outflow of individual stars and incorporating nucleosynthetic products from those individual stars. There is also no doubt about carrier phase identification: these grains are mostly quite small (submicrometer), but their populations include representatives with sizes upwards of 1 μm, large enough for isotopic analysis of individual well-identified grains. Such analyses reveal radically anomalous isotopic compositions of the major elements as well as several minor and trace elements, including noble gases (e.g., Nichols *et al.*, 2003).

Much the same can be said about the diamonds, in that most minor and trace elements are radically anomalous. But the major element, carbon, has an isotopic composition within the range of solar-system normal. Moreover, the individual diamond grains are extremely small, characteristically only a few nanometers, and too small to support analysis even of carbon in individual grains. It may be that diamond carbon appears isotopically normal only because any isotopic analysis is an average over many grains. But because of the normal carbon there is persistent suspicion that most the diamonds are not really circumstellar or even presolar after all, and that the real presolar grain carrier is a small subset of the diamonds or some other phase entirely, less abundant than the diamonds but which follows them in the separation procedures.

There are other forms of presolar/circumstellar grains in meteorites, such as silicon nitride, spinel, corundum, even a silicate, most of them isolated in the same or similar ways as the three phases noted above. Such phases have been discovered only more recently than diamond, silicon carbide, and graphite, and in smaller quantities, and so have not been studied as extensively. At present there is no evidence that any of these other phases bear any significant noble gases. Such evidence as exists is negative (cf. Lewis and Srinivasan, 1993), but there is precious little evidence bearing on this point. It is worth noting, however, that there appears to be no evidence in presently available meteorite data that requires yet another presolar noble-gas component (and another presolar grain carrier), at least not of the extremely anomalous variety.

15.4.2 Radically Anomalous Exotic Noble-gas Components

The known presolar/circumstellar phases diamond, silicon carbide, and graphite each contain a distinct noble-gas component which, like the major (except perhaps for carbon in diamonds), minor, and other trace elements in these phases, is radically anomalous compared to normal solar-system composition. These components are listed in Table 2 and illustrated in Figure 5. In the exploratory studies in which an understanding of these components was being developed, a variety of more-or-less complicated names, typically an acronym for some descriptive phrase or arbitrarily selected alphabetic characters (not all from the Latin alphabet), have been used. Some of the terms have been short lived, and there has been

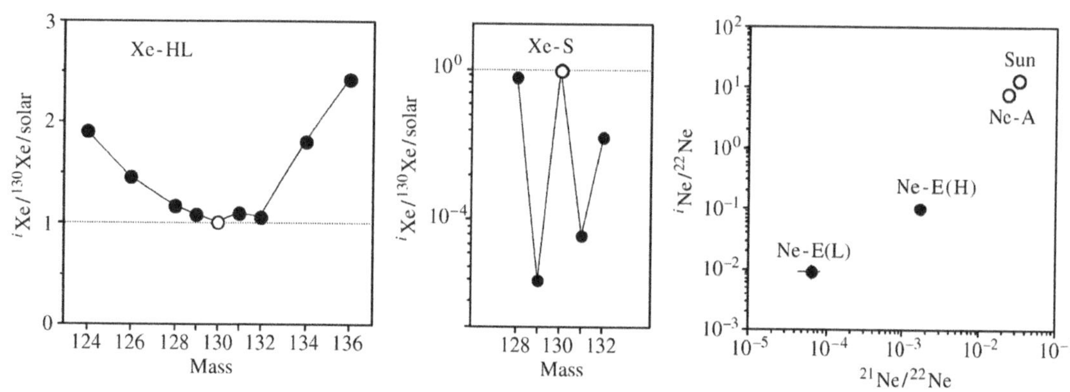

Figure 5 A display of prominent exotic (presolar) noble-gas compositions (from Anders and Zinner, 1993). In the left two panels, for each isotope on the abscissa the ordinate is the ratio (to ^{130}Xe) in the HL component (left panel) or the G (formerly termed Xe-S) component (center panel), divided by the equivalent ratio in solar xenon (i.e., solar xenon would plot with all isotopes at unity on the ordinate). The HL component shows the defining characteristics of enriched heavy and light isotopes. For the G-component, the pattern is that expected for s-process (slow neutron capture) nucleosynthesis. The right panel is a three-isotope diagram analogous to Figure 4, except that both scales are logarithmic. It shows experimental limits for the R-component (formerly Ne-E(L)) and the G-component (formerly Ne-E(H)).

Table 2 Exotic noble-gas components in presolar grains.

Component	*Carrier*	*Remarks*
Radically anomalous components		
HL	Diamond	r-Process (?) Xe and Kr; p-process (?) Xe; "planetary" Ne
G	Silicon carbide	From AGB stars; s-process Xe and Kr; Ne very rich in ^{22}Ne
R	Graphite	Monosiotopic ^{22}Ne from decay of ^{22}Na
Moderately anomalous components		
N	Silicon carbide	Probably initial composition of AGB stars
P3	Diamond	Very nearly normal, possibly not really exotic
P6	Diamond	Composition not well constrained; possibly quite anomalous

some confusion through the application of different names to the same thing. It appears that the highly anomalous components have now been substantially identified and characterized, and Ott (2002) has proposed adoption of a standard nomenclature, in some cases replacing older names in the interest of clarity. Table 2 and the discussion below use Ott's suggested nomenclature. In addition, Ott (2002) provides a detailed overview of the occurrence of these components and tabulation of the best present determinations of their compositions.

The radically anomalous component carried by silicon carbide is the G-component (Table 2). Isotopic analyses of the major elements in individual grains indicate that they come from a variety of different astrophysical sources, but most grains evidently formed in the outflow of AGB stars (on the asymptotic giant branch of the Hertzsprung–Russell diagram); the G in the name stands for "giant." The isotopic compositions characteristic of the G-component can be understood quite well in terms of nucleosynthesis in the helium-burning shells of carbon-rich AGB stars (e.g., Gallino *et al.*, 1990; Lewis *et al.*, 1994; Hoppe and Ott, 1997). For the heavy elements (like krypton and xenon), this means s-process nucleosynthesis (slow neutron capture, such that radioactive isotopes with lifetimes up to several years or decades are likely to decay before capturing another neutron), which produces only a characteristic set of isotopes in the central part of the valley of beta stability and an even–odd abundance variation pattern paralleling inverse neutron-capture cross-sections (Figure 5, center panel). End-member s-process isotopic compositions can often be determined relatively robustly, even when mixed with more nearly normal compositions, by applying the constraint that some of the lightest and heaviest isotopes (e.g., ^{78}Kr, ^{124}Xe, and ^{136}Xe; see Figure 1) are not made in the s-process. G-component krypton is of particular interest because it contains an s-process branch point, i.e., an isotope (^{85}Kr) where the s-process synthesis path divides into two branches in proportions sensitively dependent on astrophysical parameters during the synthesis.

G-component isotopic composition is not entirely uniform, but is variable in ways that can be understood to reflect minor variation in astrophysical conditions during nucleosynthesis. For the s-process elements, the compositional variation is readily attributable to differences in neutron exposure. For krypton and xenon the isotopic variations correlate with silicon carbide grain size (greater exposure for bigger grains), although the astrophysical reasons for such a correlation remain elusive. Not surprisingly, s-process compositions also vary in other G-component heavy elements (e.g., strontium and barium), also in ways attributable to neutron exposure. In addition, these variations also correlate with grain size, but in the opposite sense of the noble-gas correlation. The explanation for this is not known, but must involve different mechanisms for the incorporation of the noble gases and the incorporation of chemically reactive elements into the silicon carbide grains.

The G-component also includes neon that is strongly enriched in ^{22}Ne (Figure 5, right panel), the defining feature for a component previously termed neon-E (Figure 4). Further investigation showed that there were actually two kinds of neon-E (Jungck and Eberhardt, 1979). One was hosted by a relatively low-density phase and was released at relatively low pyrolysis temperatures, and was accordingly termed neon-E(L); the other, in a higher-density phase and released at higher temperatures, was termed neon-E(H). The G-component neon is neon-E(H). It is highly enriched in ^{22}Ne, but it is not monisotopic: it contains low but measurable ^{20}Ne and is clearly associated with ^{4}He, and both features are consistent with helium burning in AGB stars.

As inferred from single-grain isotopic analyses, the origins of interstellar graphite are rather diverse, without evident preponderance of any one source. Among the sources of graphite are AGB stars. The heavy noble gases in graphite (cf. Amari *et al.*, 1995), like those in silicon carbide, are evidently of s-process type, but with a wider range of neutron exposure.

Graphite also contains a ^{22}Ne-rich component, not obviously associated with other gases and not the same as the ^{22}Ne-rich G-component in silicon carbide, but which can be identified as the complimentary form neon-E(L) (Amari *et al.*, 1995), renamed the R-component (Table 2). It appears to contain no detectable amounts of the other neon isotopes, and is generally thought to result from *in situ* decay of ^{22}Na (half-life 2.6 yr).

Diamonds host the HL component, which is the most abundant of the three radically anomalous components in Table 2. The name reflects the great enrichment of both the heavy and the light isotopes of xenon (Figure 5, left panel; also see Huss and Lewis, 1994). The partial separation of HL-Xe from other components in stepwise heating accounts for the isotopic variations first observed by Reynolds and Turner (1964), and the heavy-isotope enrichments for a time supported the idea that it was a fission component, CCF-Xe (see above). Heavy-isotope enrichment is suggestive of the nucleosynthetic r-process, and light isotope enrichment is suggestive of the p-process, and the astrophysical provenance of the HL component is commonly thought to be supernovae. The compositions are not quite right for either of these "classical" nucleosynthetic processes, however. It may be that the heavy-isotope ratios in HL-Xe reflect r-process composition modified by chemical separations in the beta-unstable progenitors of the heavy xenon isotopes (Ott, 1996) or that the heavy isotopes were produced in a "neutron burst" (Howard *et al.*, 1992) involving neutron exposure intermediate between the classical s-process and r-process. The heavy isotopes and the light isotopes must be synthesized in different nuclear processes, and on astrophysical grounds it is not evident why H- and L-components should be coupled into a single HL component, as they appear to be. This has led to suspicion that they are indeed not coupled, and that there are distinct H- and L-components hosted in distinct subpopulations of the diamond samples, and thus to analytical efforts to separate H from L. At present there is some evidence that H and L are separable, but it is not yet definitive (see Meshik *et al.*, 2001).

The HL component includes all the noble gases. It is noteworthy that the krypton is enriched in the heavy isotopes but not the light (i.e., H but not L). The relative elemental abundance of the light gases neon and helium in HL is higher than in other prominent trapped components, so that while HL makes a small, albeit still significant contribution to ordinary "planetary" gases (see below), it provides the dominant contribution to neon and helium (Figure 7).

15.4.3 Moderately Anomalous Exotic Noble-gas Components

The isotopic compositions that are actually observed in presolar grain analyses, particularly in the many-grain ensembles needed for analysis of heavy trace elements, in general are not pure, radically anomalous nucleosynthetic end-member compositions, e.g., the components which are the focus in the prior section. More generally, for both noble gases and other elements alike, the nucleosynthetic components are diluted by compositions that are more nearly normal. In some cases it might be argued that these diluents really are normal, in the strict sense, because they are actually laboratory blank or bulk meteorite contamination. More generally, however, the evidence favors the conclusion that these components really are in the presolar grains, so that they really are exotic to the solar system, especially when they can be inferred to be isotopically anomalous. There are three such components that are reasonably well established (Table 2): the N-component in silicon carbide, and two components, P3 and P6, in the diamonds. These components would likely not have been noted except for their association with the radically anomalous components.

It is important to appreciate that even though these components may be anomalous, the anomalies in question are much more modest than those described in the preceding section, commonly expressed in percent deviations from normal rather than as multiples or even orders of magnitude. More fundamentally, these components do not appear to be the results of specific nucleosynthetic processes in specific stars, but rather a mixture of many different stellar nucleosynthetic contributions, as would be found in the interstellar medium, and so could be remnants of initial stellar compositions. In other words, these components are like our own normal composition, except that they are the "normal" of a different time, elsewhere in the galaxy.

It is difficult to doubt that the N-component is actually in the silicon carbide, and it is commonly held that the N-component represents the outer envelope of the same AGB stars in which helium-burning generates the G-component, i.e., the initial stellar composition. Since the diamonds are too small for single-grain analyses, however, it cannot be precluded that the samples prepared in the laboratory are not assemblages of different materials of fundamentally different origin. Thus, it cannot be concluded that the P3 and/or P6 components are in the same population of grains, or are closely related to the HL component or each other.

15.5 NOBLE-GAS CHEMISTRY AND TRAPPED COMPONENTS

Strictly speaking, the title "noble-gas chemistry" should be an oxymoron. But the noble gases are not literally and completely noble in the sense that they fail entirely to interact chemically with other forms of matter. Under appropriate conditions in the laboratory they can form real compounds with other elements, although there is no evidence that actual noble-gas compounds are relevant in cosmochemistry (possibly excepting ice clathrates in comets). Still, planetary materials do contain noble gases that were somehow incorporated into them, and at least some of these appear to have involved some form of chemical interaction. The issue of chemical interactions is a venerable topic in noble-gas cosmochemistry, but there are still questions that have been unanswered for a long time.

15.5.1 Planetary Noble Gases

As the noble gases in planetary materials were first being explored, it became evident that there were three broad categories of noble-gas occurrences: *in situ* gases and two kinds of trapped gases, solar and planetary. *In situ* gases are those produced by nuclear transformation—radioactive decay and spallation—in the samples in question, such that the gases are still in the same places where they were made. Solar gases are samples of solar corpuscular radiation, implanted in the near-surface layers of solids exposed directly to the Sun, as freely orbiting small particles or on the surfaces of larger but airless parent bodies such as the Moon or asteroids. Planetary gases are everything else, gases which are neither *in situ* nor solar.

Despite definition as a garbage-can category, planetary noble gases in most cases occur in a reasonably well-defined pattern. The elemental abundances in different meteorites occur in similar proportions, in a pattern defined by progressively greater depletion of the lighter gases, as illustrated in Figure 6 (right panel). This pattern is clearly distinct from the solar elemental pattern (Figure 6, left panel). The term "planetary" reflects the similarity of the nonsolar meteoritic abundance pattern to the relative abundances of the noble gases in the Earth's atmosphere (except that no comparison can be made for helium, which escapes from the atmosphere in geologically short times). The quantities are about right too, i.e., if the inventories of atmospheric gases divided by the mass of the Earth are taken as a reasonable measure of the initial concentrations in the solids that formed the Earth, these concentrations fall within the range seen in meteorites (mostly between Allende and Bruderheim in Figure 6, right). Accounting for the origin of planetary noble gases is another traditional problem that has long resisted solution, but there is a clear suggestion that whatever explanation there is for planetary gases in meteorites should also suffice for Earth (and perhaps Venus and Mars as well).

Besides the elemental abundance pattern by which they are defined, planetary gases also display characteristic isotopic features. The isotopic compositions of krypton and nonradiogenic argon are very similar in solar, planetary, and atmospheric gases, varying perhaps by modest isotopic fractionation. Planetary xenon is like solar xenon at the light isotopes, but has small (a few per mil to a few percent) excesses in the heavy isotopes (atmospheric xenon constitutes a special challenge, as described below). The largest distinctions occur in neon. Neon in carbonaceous chondrites characteristically defines a trapped composition called neon-A (Figure 4) with much lower $^{20}Ne/^{22}Ne$ than "solar" neon (the B composition in Figure 4, which actually has $^{20}Ne/^{22}Ne$ somewhat lower than the solar wind). (Atmospheric neon, with $^{20}Ne/^{22}Ne = 9.8$, is intermediate between the A and B compositions, closer to A.) Because of the prevalence of *in situ* components, planetary helium can be observed only in very few samples. Where observable, planetary $^{3}He/^{4}He$ is considerably lower than solar $^{3}He/^{4}He$, but this is a special case attributable to deuterium burning in the Sun (see above).

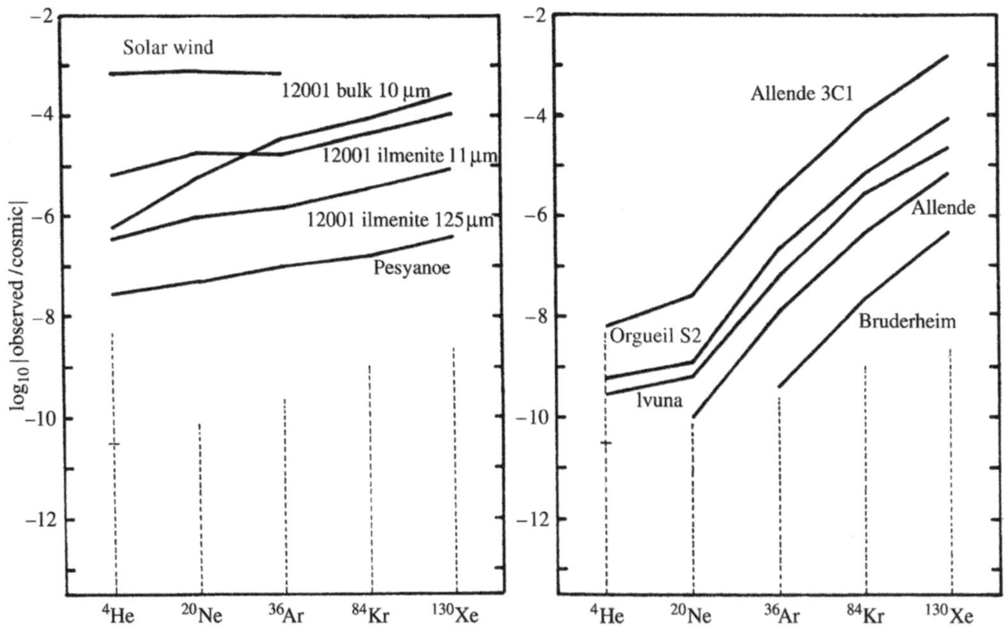

Figure 6 Elemental abundance patterns for trapped noble gases in various planetary materials. For each gas identified on the abscissa, the ordinate shows the depletion factor in a given sample, i.e., the gas concentration in the sample divided by what the concentration would be if the gas were present in undepleted cosmic proportion (normalized for a nominal rock with 17% Si). The relative elemental abundances in the left panel illustrate the "solar" pattern, those in the right panel the "planetary" pattern. The vertical broken lines for each gas illustrate typical *in situ* gas concentrations (the radiogenic component for ^{4}He, spallation for the others), below which it becomes progressively more difficult to characterize or even identify trapped components (source Ozima and Podosek, 2002).

In the "classical" picture, i.e., before appreciation that presolar components were an important part of the total inventory of trapped gases in meteorites, neither the isotopic effects nor the generation of the elemental abundance pattern were ever explained satisfactorily in terms of quantitative models that gained consensus acceptance. Some aspects of this problem have become moot, however, since it is now recognized that "planetary" gas is composite: planetary gas includes the contributions of the exotic noble-gas components (Table 2) imported into the solar system by presolar grains. These contributions, especially of the HL component, can be substantial (Huss and Lewis, 1995; also see Figure 7).

For the heavy gases, the diamond-borne components commonly provide a few to several percent of observed gas (Figure 7, left), the other exotic components contribute only smaller amounts, and the major contribution is assigned to the component called "Q" (sometimes also called "P1," where the P stands for "planetary"). Some of the main characteristics of "planetary" gas are now attributed to the Q-component: it accounts for most of the heavy gases, it is what is left over after other sources (*in situ*, solar and exotic components) have been accounted for, and it is often regarded as a (the?) local "background" component, derived from the ambient noble gases in the nebula from which planetary materials formed. An important difference, however, is that the ratio of light to heavy gas elemental abundances in Q is less than in "planetary" gas and less than in the exotic components. It is thus inferred that Q makes a progressively smaller contribution to the lighter gases. For ^{22}Ne in Orgueil (Figure 7, right), for example, Q makes only a minor contribution. Most of planetary neon (and helium), for which explanation was once sought in terms of how it could be derived from the solar nebula, is now inferred to be exotic, a superposition of components (chiefly HL) whose properties were established outside the solar system.

It is sometimes argued that the term "planetary" is obsolete and its use should be discontinued because it is now seen that it is not a monolithic component, and because its dissection into mostly extrasolar components weakens its association with the major terrestrial planets. This last viewpoint seems questionable, however, and no good alternative to the term "planetary" has emerged. The term seems still useful in denoting this ensemble of components, and also in suggesting some of its salient features. When it was once thought that planetary neon and xenon were parts of the same component extracted from the nebula, for example, the question was what process could establish such an Ne/Xe ratio and what process (or added component) could account for such a different isotopic composition for neon. In the light of present understanding, the isotopic question seems more relaxed, since it is quite reasonable that an extrasolar component could have quite different composition from solar gas, but there is now the question of how it is that independent components, hosted in different carriers, nevertheless occur in nearly the same proportions in many samples.

15.5.2 The Q-Component

As discussed above, there are a number of exotic noble-gas components that are important to reckon with in cosmochemistry, i.e.,

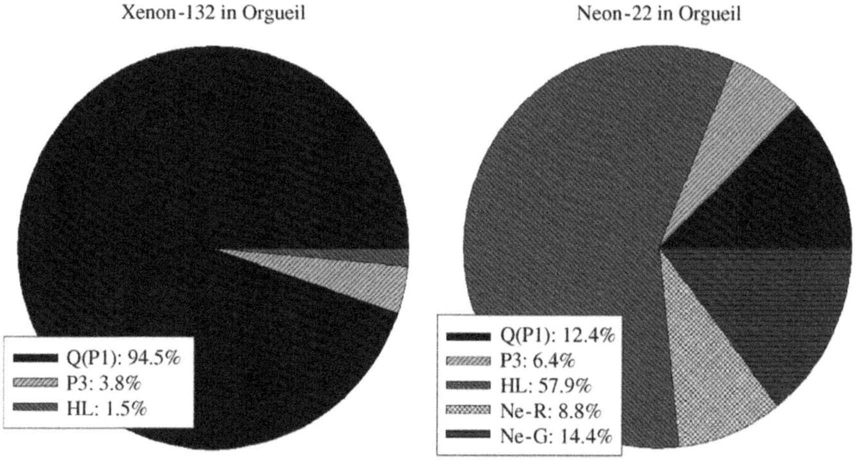

Figure 7 Pie diagrams illustrating the inferred superposition of components to total trapped ^{132}Xe and ^{22}Ne in the carbonaceous chondrite Orgueil. (Only components contributing at least 1% are shown.) The Q-component dominates ^{132}Xe but is only a minor contributor to ^{22}Ne; the main contribution to ^{22}Ne is the HL component (source Ott, 2002).

components whose identity was established outside the solar system, prior to its formation, which were imported into the solar system in presolar grains, and which have maintained their identities throughout the solar-system history. Some of these exotic components have radically anomalous isotopic compositions, and others, while not radically anomalous, are clearly distinct from normal composition. It is often held that the complement to these exotic components is the component most commonly designated "Q" (an alternative name is "P1"), with "normal" isotopic composition and thought to be of local rather than exotic origin.

It is frequently said that the Q-component is defined operationally. As initially reported by Lewis *et al.* (1975) as part of their program to identify the carrier phases of the exotic noble gases, suitable treatment of whole-rock carbonaceous chondrites with HF and HCl dissolves most of the mass, leaving a residue of order 1% or less of the starting mass, but nevertheless containing nearly all the initial trapped noble gases, i.e., the major phases removed by this treatment carried little to no trapped noble gases. When the residue is further treated with HNO_3 (or other oxidizing agent), there is only small additional mass loss, but most of the trapped noble gas is removed, and the remaining gases in the residue are markedly anomalous. The inference is that the normal noble gases are in a very minor phase, and the name Q was applied to both the (unknown) phase and the gases. This is not exactly the same definition as the one implied in the previous section, i.e., that Q is planetary noble gas, less the components listed in Table 2, but there seems to be no discrepancy arising in the difference. In both of these definitions, Q is defined by subtraction, the difference between two occurrences. In a third approach, more straightforward and much more amenable to studying its properties, Q gases are liberated by the closed-system stepwise etching technique, in which gases released in very superficial chemical attack can be directly collected for analysis (Wieler *et al.*, 1991; Busemann *et al.*, 2000). The Q-component thus obtained is consistent with the definitions by difference. There was initially some controversy about the nature of the carrier phase, but it seems now generally agreed that it is carbonaceous (Ott *et al.*, 1981).

The Q-component contains all five noble gases, in an elemental abundance pattern defined by progressive depletion of lighter gases relative to heavier gases, like the nominal planetary pattern (Figure 6) but steeper, the depletion of helium and neon relative to xenon being some two orders of magnitude greater. As already noted in the prior section, this results in Q accounting for a smaller fraction of the total planetary abundance of the lighter gases than of the heavier gases (Figure 7). There is some variation in elemental abundance ratios in different instances of Q, amounting to a factor of several, significant but still small compared to more than six orders of magnitude depletion of helium and neon relative to xenon. Busemann *et al.* (2000) argue that Q is actually composite, consisting of at least two subcomponents that behave differently in planetary parent body processes. Q isotopic compositions are distinct from solar-wind compositions, but could be related to solar-wind compositions by isotopic fractionation. The degree of fractionation, which is progressively less in the heavier elements, is substantial, $\sim$30% for $^{20}Ne/^{22}Ne$, $\sim$1% amu^{-1} in xenon. The sense of the isotopic fractionation is depletion of light relative to heavy, the same as the elemental fractionation. Busemann *et al.* (2000) report distinct variations, of order 6% and bimodally distributed, in $^{20}Ne/^{22}Ne$ in Q-Ne. Assuming that Q-Ne really is derived from neon of solar-wind composition by fractionation, this isotopic difference corresponds to modest difference in degree of fractionation. The more profound generalization is that at least at present there is no evidence that Q itself is isotopically complex, i.e., its isotopic structures can be related to the composition of the Sun, and presumably of the solar nebula, by no more than fractionation, with no embedded isotopic anomalies.

Q gases are found in a variety of meteorite types, with just relatively modest compositional variation, from which it can be argued that Q was somehow established as a component, in some physical reservoir, prior to being distributed to and incorporated into diverse meteorite parents (cf. Ozima *et al.*, 1998). In addition, the classical observation that planetary gases behave like a "component," e.g., in having a characteristic ratio of light to heavy gases, is now equivalent to stipulating that Q occurs in a characteristic ratio to exotic components of entirely separate origin. This feature is easier to understand if Q gases were incorporated into their solid carrier phase (s), and then the Q carrier phase mixed with the presolar grain carriers, such that meteorite precursors could sample them in the same proportions.

It is not clear how the defining characteristics of Q were generated. Ozima *et al.* (1998) advocate a model based on mass, suggesting Rayleigh distillation to generate both the elemental and isotopic fractionation, but it is difficult to reconcile both types of fractionation with the same process, and there is no known astrophysical scenario for the postulated Rayleigh distillation. Ott (2002) notes that Q gas abundances can also be correlated with ionization potential, but this would not provide a natural explanation for the isotopic fractionation, and again a complete model would require an appropriate astrophysical scenario. Besides the attempt to account for elemental and isotopic composition, these and any other models for generation of Q, especially those involving gas-phase separation, face the additional hurdle of accounting for incorporation of Q gases into the solid Q-phase carrier (see following section).

Huss and Alexander (1987) advocate a model in which Q is actually a presolar component, in which the Q-phase carrier is made, and acquires noble gases, not in the solar nebula but in the interstellar medium, in the molecular cloud from which the Sun later forms. This scenario is consistent with the observation that Q gas and solar gas have the same isotopic composition, except for fractionation, since Q gases would be drawn from the same reservoir that will later provide the Sun's and the nebula's noble gases. It would also account in a natural way for an essentially constant proportion of Q gases to other presolar components, since they would all enter the solar system in grains that would be mixed even before the solar system forms. In this scenario Q would be a presolar component, but whether it should also be termed "local" (made from the same material as the solar system) or "exotic" (made elsewhere, and imported into the solar system) seems more a matter of idiosyncratic taste than scientific substance. Postulating an interstellar medium origin for Q gas does not by itself solve the problem of how gases get incorporated into the solids, but at least it expands the parameter space in which a solution may be sought (see below).

15.5.3 Trapping Mechanisms

Elsewhere in this chapter, discussion of how a component is to be accounted for is generally a matter of how an isotopic ratio or an elemental abundance pattern is generated. But most noble-gas components of interest in cosmochemistry are in or once were contained in planetary solids, so that a model to account for a noble-gas component is incomplete without consideration of how the gases get into the solids. This is no great mystery if the nuclei of the gas atoms were made there by nuclear transformation, i.e., an *in situ* component, or if noble-gas atoms impinge on the surface of a solid with enough energy to penetrate below at least several atom layers, e.g., solar wind or more energetic particles. Otherwise, however, the means by which noble-gas atoms entered and became trapped in planetary material has been one of the enduring mysteries of noble-gas cosmochemistry. In particular, the traditional problem has been how to account for the origin of planetary noble gases (Figure 6). This traditional problem has been more or less inherited by the Q-component, and in fact amplified, since some accounting must also be made for the exotic components (Table 2).

The problem with planetary gases, or the Q-component, is not that the noble gases are so scarce, it is that in any simple quantitatively predictive model the content of noble gases, at least any gases extracted from the ambient solar nebula, should be much lower still. In thermodynamic equilibrium between solid and gas phases, the gases can be dissolved in the solids or adsorbed on their surfaces. For any materials with known or plausible thermodynamic parameters, solution is simply not a viable mechanism to account for trapped noble gases: it does not produce the right compositions, and quantities are orders of magnitude too low. Adsorption is not quantitatively adequate either, but at least it produces gases with the right qualitative elemental abundance features, i.e., progressive depletion of lighter gases and much higher effective concentrations than are possible in solution (e.g., Fanale and Cannon, 1972; Podosek *et al.*, 1981). In practice, though, equilibrium adsorption does not really provide a feasible simulation of planetary or Q gases except at quite low temperatures, certainly much lower than the temperatures at which terrestrial–planetary materials seem to have last substantially interacted with nebular gas (the "accretion temperature") and plausible only in the outer reaches of the solar system (comets?) or in the interstellar medium. In addition, gases adsorbed in equilibrium will be desorbed as soon as the ambient gas phase is removed, so that models involving adsorption in the generation of planetary or Q gases are incomplete without also specifying some mechanism for fixing adsorbed gases, so that they really become "trapped."

There has been no shortage of ideas, or of laboratory simulations, for nonequilibrium conditions or processes that may have produced planetary or Q gases (e.g., Frick *et al.*, 1979; Yang and Anders, 1982; Zadnik *et al.*, 1985; Wacker, 1989; Nichols *et al.*, 1992; Sandford *et al.*, 1998; Hohenberg *et al.*, 2002). Generally, such models/simulations involve the synthesis of some prospective host phase, invoke adsorption to concentrate gas in/on the relevant material, and continued growth to occlude or trap the gas so that it will not later desorb. Often some energetic phenomenon is involved, such as plasma discharge, ultraviolet irradiation, hypervelocity shock, etc. In the Huss and Alexander (1987) model for generation of the Q-component in the presolar interstellar medium, for example, adsorption at very low temperature is called upon to concentrate gas in icy grain mantles, where they become trapped as UV irradiation and incidence of energetic ions lead to polymerization of a carbonaceous phase around the gas atoms.

There is no doubt that laboratory simulations can effectively "trap" noble gases beyond equilibrium concentrations, sometimes in qualitatively appropriate elemental patterns, but evidently none have yet achieved the distribution coefficients implied by the gas contents of the Q-component. A common problem with such simulations is that they are difficult to control in detail, and they are generally rather complicated, so that it is unclear which parameters are really controlling gas trapping. It is certainly possible that one (or more) of the processes studied in laboratory simulations is responsible for generation of Q, but there is no consensus on which, or in what circumstances.

Most of the attention given to production of the exotic gas components (Table 2) has focused on their isotopic patterns, but except for the R-component, believed to be formed by *in situ* decay of ^{22}Na, they also must be trapped components in their host phases, so models for their origin are incomplete without a trapping mechanism. There is significant evidence, including absence of fractionation from neighboring elements and comparison with laboratory simulation, that at least some of the components carried in diamonds and silicon carbide were trapped by ion implantation (e.g., Koscheev *et al.*, 2001; Ott, 2002) in stellar winds, and by extension perhaps all of them were thus trapped (except for the R-component). The situation is complicated, though, and multiple types of winds may be necessary to get elemental ratios right. If not all the exotic components were implanted as ions, it is not clear whether any of the mechanisms advanced for Q will suffice.

15.5.4 Loose Ends?

Noble-gas geochemistry is a data-rich field, and the discussion above does not do justice to all the data. There are additional occurrences of elemental and/or isotopic patterns which may be indicative of still other significant noble-gas components. Except for one particularly significant case, U–Xe, which is discussed in the following section, the scope of this chapter does not permit adequate description (but see Ott (2002), for a more detailed review).

To only mention two of the more prominent cases, ureilites and enstatite chondrites exhibit noble gases in different elemental patterns and in different carriers than those in most meteorites. Ureilite elemental abundance patterns are variable but overall more strongly fractionated than Q, and the gases are in diamonds (produced by shock in the solar system, and not to be confused with presolar nanodiamonds). Isotopic patterns are similar to those in Q, however, so it is not clear whether ureilite noble gases are better viewed as a distinct component or as a variant of Q. Gases in enstatite chondrites have been termed a "subsolar" component, marked by less steeply fractionated elemental abundances, intermediate between the solar and planetary patterns, and are hosted in the main silicates, chiefly enstatite (Crabb and Anders, 1982). It is not clear whether subsolar gases reflect some variant of solar-wind implantation or a component fundamentally different from those known in other classes of meteorites.

There are also a handful of other occurrences (see Ott, 2002) of unusual elemental or isotopic patterns that are not very well characterized and/or occur only in very restricted samples. These may reflect unusual circumstances not often encountered or explored (e.g., recoil; see Marti *et al.*, 1989), or possibly more anomalous components yet to be discovered.

15.5.5 U–xenon?

One of the major vexations of noble-gas geo-/cosmochemistry, and even of cosmochemistry more broadly, is the difficulty in accounting for the isotopic composition of terrestrial atmospheric xenon in terms of plausibly known starting materials and plausible planetary processes. This problem was formulated in the earliest days of surveying the content of noble gases in planetary materials and has attracted the interest of multiple generations of cosmochemists, but a consensus resolution still remains elusive.

To first order, air xenon can be related to xenon known elsewhere in the solar system (solar wind, meteorites) by mass-dependent isotopic fractionation. The fractionation is severe, ~4% amu^{-1}, in the direction that lighter isotopes in air are depleted relative to, say, solar-wind xenon, but it is not impossible to construct a planetary history to account for this, for example, by stipulating that air xenon is the residue of extensive hydrodynamic escape of an early atmosphere (e.g., Pepin, 1991). In closer detail, the relative abundances of the lighter isotopes (^{124}Xe, ^{126}Xe, ^{128}Xe, and ^{130}Xe) can be quantitatively accounted for as fractionated solar-wind xenon. Relative to this pattern, air xenon has ~7% too much ^{129}Xe, which is quite plausibly attributed to a radiogenic component from decay of ^{129}I. However, there is no way to extend this fractionation pattern, assuming solar wind or any known bulk meteoritic component as the underlying composition, such as to give a quantitatively satisfactory accounting of the heavier isotopes (^{131}Xe, ^{132}Xe, ^{134}Xe, and ^{136}Xe), with or without addition of a plausible nuclear component (fission of ^{244}Pu or ^{238}U).

U–Xe is a mathematical construct designed to address this problem. It is a composition (not counting ^{129}Xe) constrained to lie on the multidimensional correlation surfaces (generalizations of three-isotope correlation diagrams) obtained in stepwise heating of carbonaceous chondrites and also to match fractionated air xenon less some fraction of ^{244}Pu-fission xenon (Pepin and Phinney, 1976; Pepin, 2000). A match can be made for ~4% fission contribution to ^{136}Xe. The resultant

U–Xe composition is equivalent to solar-wind xenon at the light isotopes but has a few to several percent less of the heavy isotopes. This model is specific to xenon and does not involve other gases. U–Xe is hypothesized to be original terrestrial xenon, supplied by the solar nebula.

Factors that are favorable to this model are that the mathematics can be made to work at all, and that it has no quantitative competitor to explain terrestrial xenon. One unfavorable factor is that this model presupposes separation of H-Xe from L-Xe in carbonaceous chondrites. Another is that the existence of U–Xe as a physical entity remains speculative. It is supposed that U–Xe is the prevalent gas in the nebula, but evidence for its actual presence in meteorites is sparse at best and apparently not reproducible (Busemann and Eugster, 2000), and U–Xe is evidently not invoked as the underlying primitive composition for xenon on Mars (cf. Pepin and Porcelli, 2002). Perhaps the most compelling argument is that if U–Xe is primitive nebular xenon, then the Sun is enriched in the heavy isotopes of xenon, e.g., several percent H-Xe without L-Xe, and it is difficult to see how this can be accommodated without some significant change to the current picture of solar-system formation. All in all, the issue is central but seemingly remains unresolved.

15.5.6 Planetary Atmospheres

Traditionally, accounting for terrestrial planetary atmospheres has been part of the practice of noble-gas cosmochemistry. The overall program is to try to deduce the character of the initial gases in the planetary inventory—as they may have been acquired from gas in the nebula, through solar-wind irradiation, or by capturing components such as are observed to exist in planetary materials—and how they may have been modified through planetary processes. Pursuit of this program has been fruitful in establishing constraints on the origin and evolution of the planets, but this topic is beyond the scope of this chapter. The major issue in which consideration of planetary atmospheres has raised fundamental questions about the character and distribution of noble-gas components in the preplanetary solar system is that of U–Xe and the nature of the primordial xenon in the Earth's atmosphere, discussed above. For broader discussion of planetary atmospheres the reader is referred to Chapter 16.

REFERENCES

Alexander E. C., Jr., Lewis R. S., Reynolds J. H., and Michel M. (1971) Plutonium-244: confirmation as an extinct radioactivity. *Science* **172**, 837–840.

Amari S., Lewis R. S., and Anders E. (1995) Interstellar grains in meteorites: III. Graphite and its noble gases. *Geochim. Cosmochim. Acta* **59**, 1141–1426.

Anders E. (1988) Circumstellar material in meteorites: noble gases, carbon and nitrogen. In *Meteorites and the Early Solar System* (eds. J. F. Kerridge and M. S. Matthews). University of Arizona Press, Tucson, pp. 927–955.

Anders E. and Zinner E. K. (1993) Interstellar grains in meteorites: diamond, silicon carbide and graphite. *Meteoritics* **28**, 490–514.

Anders E., Higuchi H., Gros J., Takahashi H., and Morgan J. W. (1975) Extinct superheavy element in the Allende meteorite. *Science* **190**, 1261–1271.

Bernatowicz T. J. and Zinner E. K. (eds.) (1997) *Astrophysical Implications of the Laboratory Study of Presolar Materials* (AIP Conference Proceedings 402). AIP, New York, pp. 750.

Black D. C. (1972) On the origin of trapped helium, neon and argon isotopic variations in meteorites: II. Carbonaceous chondrites. *Geochim. Cosmochim. Acta* **36**, 377–394.

Black D. C. and Pepin R. O. (1969) Trapped neon in meteorites: II. *Earth Planet. Sci. Lett.* **6**, 395–405.

Busemann H. and Eugster O. (2000) Primordial noble gases in Lodran metal separates and the Tatahouine diogenite. *Lunar Planet. Sci.* **XXXI**, #1642. Lunar and Planetary Institute, Houston (CD-ROM).

Busemann H., Baur H., and Wieler R. (2000) Primordial noble gases in "phase Q" in carbonaceous and ordinary chondrites studied by closed-system stepped etching. *Meteorit. Planet. Sci.* **35**, 949–973.

Busemann H., Baur H., and Wieler R. (2001) Helium isotopic ratios in carbonaceous chondrites: significant for the early solar nebula and circumstellar diamonds? *Lunar Planet. Sci.* **XXXII**, #1598. Lunar and Planetary Institute, Houston (CD-ROM).

Clayton R. N., Grossman L., and Mayeda T. K. (1973) A component of primitive nuclear composition in carbonaceous meteorites. *Science* **182**, 488–495.

Crabb J. and Anders E. (1982) On the siting of noble gases in E-chondrites. *Geochim. Cosmochim. Acta* **46**, 2351–2361.

Drozd R. J., Hohenberg C. M., and Ragan D. (1972) Fission xenon from extinct ^{244}Pu in 14301. *Earth Planet. Sci. Lett.* **15**, 338–346.

Drozd R. J., Kennedy B. M., Morgan C. J., Podosek F. A., and Taylor G. J. (1976) The excess fission Xe problem in lunar samples. *Proc. 7th Lunar Sci. Conf.* 599–623.

Eberhardt P., Geiss J., Graf H., Grögler N., Mendia M. D., Mörgeli M., Schwaller H., and Stettler A. (1972) Trapped solar wind noble gases in Apollo 12 lunar fines 12001 and Apollo 11 breccia 10046. *Proc. 3rd Lunar Sci. Conf.* 1821–1856.

Fanale F. and Cannon W. A. (1972) Origin of planetary primordial rare gas: the possible role of adsorption. *Geochim Cosmochim. Acta* **36**, 319–328.

Frick U., Mack R., and Chang S. (1979) Solar gas trapping and fractionation during synthesis of carbonaceous matter. *Proc. 10th Lunar Planet. Sci. Conf.* 1961–1973.

Gallino R., Busso M., Picchio G., and Raiteri C. M. (1990) On the astrophysical interpretation of isotope anomalies in meteoritic SiC grains. *Nature* **348**, 298–302.

Geiss J. and Reeves H. (1972) Cosmic and solar system abundances of deuterium and helium-3. *Astron. Astrophys.* **18**, 126–132.

Geiss J., Gloeckler G., and von Steiger R. (1994) Solar and heliospheric processes from solar wind composition measurements. *Phil. Trans. Roy. Soc. London A* **349**, 213–226.

Gloeckler G. and Geiss J. (2000) Deuterium and helium-3 in the protosolar cloud. In *The Light Elements and their Evolution*, IUA Symposium, 198 (eds. L. da Silva, M. Spite, and J. R. Medeiros). pp. 224–233.

Heymann D. and Yaniv A. (1970) Ar^{40} anomaly in lunar samples from Apollo 11. *Proc. Apollo 11 Lunar Sci. Conf.* 1261–1267.

Hohenberg C. M., Marti K., Podosek F. A., Reedy R. C., and Shirck J. R. (1978) Comparisons between observed and predicted cosmogenic noble gases in lunar samples. *Proc. 9th Lunar Planet. Sci. Conf.* 2311–2344.

Hohenberg C. M., Nichols R. H., Jr., Olinger C. T., and Goswami J. N. (1990) Cosmogenic neon from individual grains of CM meteorites—extremely long pre-compaction exposure histories or an enhanced early particle flux. *Geochim. Cosmochim. Acta* **54**, 2133–2140.

Hohenberg C. M., Thonnard N., and Meshik A. (2002) Active capture and anomalous adsorption: new mechanisms for incorporation of heavy noble gases. *Meteorit. Planet. Sci.* **37**, 257–267.

Hoppe P. and Ott U. (1997) Mainstream silicon carbide grains from meteorites. In *Astrophysical Implications of the Laboratory Study of Presolar Materials,* AIP Proc. Conf. 402 (eds. T. J. Bernatowicz and E. K. Zinner), AIP, New York, pp. 27–58.

Howard W. M., Meyer B. S., and Clayton D. D. (1992) Heavy-element abundances from a neutron burst that produces Xe-X. *Meteoritics* **27**, 404–412.

Huss G. R. and Alexander E. C.Jr., (1987) On the pre-solar origin of the normal planetary noble gas component in meteorites. *Proc. 17th Lunar Planet. Sci. Conf.: J. Geophys. Res.* **92**, E710–E716.

Huss G. R. and Lewis R. S. (1994) Noble gases in presolar diamonds: I. Three distinct components and their implications for diamond origins. *Meteoritics* **29**, 791–810.

Huss G. R. and Lewis R. S. (1995) Presolar diamond, SiC, and graphite in primitive chondrites: abundances as a function of meteorite class and petrologic type. *Geochim. Cosmochim. Acta* **59**, 115–160.

Jungck M. H. A. and Eberhardt P. (1979) Neon-E in Orgueil density separates. *Meteoritics* **14**, 439–440.

Kallenbach R., Ipavich F. M., Kucharek H., Bochsler P., Galvin A. B., Geiss J., Gliem F., Glöeckler G., Grünwaldt H., Hefti H., Hilchenbach M., and Hovestadt D. (1998) Fractionation of Si, Ne, and Mg isotopes in the solar wind as measured by SOHO/CELIAS/MTOF. *Space Sci. Rev.* **85**, 357–370.

Koscheev A. P., Gromov M. D., Mohapatra M. K., and Ott U. (2001) History of trace gases in presolar diamonds as inferred from ion implantation experiments. *Nature* **412**, 615–617.

Lewis R. S. and Srinivasan B. (1993) A search for noble gas evidence for presolar oxide grains. *Lunar Planet. Sci.* **XXIV**. Lunar and Planetary Institute, Houston, pp. 873–874.

Lewis R. S., Srinivasan B., and Anders E. (1975) Host phase of a strange xenon component in Allende. *Science* **190**, 1251–1262.

Lewis R. S., Amari S., and Anders E. (1994) Interstellar grains in meteorites: II. SiC and its noble gases. *Geochim. Cosmochim. Acta* **58**, 471–494.

Mahaffy P. R., Donahue T. M., Atreya S. K., Owen T. C., and Niemann H. B. (1998) Galileo probe measurements of D/H and $^{3}He/^{4}He$ in Jupiter's atmosphere. *Space Sci. Rev.* **84**, 251–263.

Manka R. H. and Michel F. C. (1971) Lunar atmosphere as a source of lunar surface elements. *Proc. 2nd Lunar Sci. Conf.* 1717–1728.

Marti K. (1969) A new isotopic composition of xenon in the Pesyanoe meteorite. *Earth Planet. Sci. Lett.* **3**, 243–248.

Marti K., Kim J. S., Lavielle B., Pellas P., and Perron C. (1989) Xenon in chondritic metal. *Z. Naturforsch* **44a**, 963–967.

Meshik A. P., Pravdivtseva O. V., and Hohenberg C. M. (2001) Selective laser extraction of Xe-H from Xe-HL in meteoritic nanodiamonds: real effect or experimental artifact? *Lunar Planet. Sci.* **XXXII**, #2158. Lunar and Planetary Institute, Houston (CD-ROM).

Murty S. V. S., Goswami J. N., and Shukolyukov Y. A. (1997) Excess ^{36}Ar in the Efremovka meteorite: a strong hint for the presence of ^{36}Cl in the solar system. *Astrophys. J.* **475**, L65–L68.

Nichols R. H., Jr., Nuth J. A., III, Hohenberg C. M., Olinger C. T., and Moore M. H. (1992) Trapping of noble gases in proton-irradiated silicate smokes. *Meteoritics* **27**, 555–559.

Nichols R. H., Jr., Kehm K., Hohenberg C. M., Amari S., and Lewis R. S. (2003) Neon and helium in single interstellar SiC and graphite grains: asymptotic Giant Branch, Wolf-Rayet, supernova and nova sources. *Geochim. Cosmochim. Acta* (submitted).

Ott U. (1996) Interstellar diamond xenon and timescales of supernova ejecta. *Astrophys. J.* **463**, 344–348.

Ott U. (2002) Noble gases in meteorites—trapped components. *Rev. Mineral. Geochem.* **47**, 71–100.

Ott U., Mack R., and Chang S. (1981) Noble-gas-rich separates from the Allende meteorite. *Geochim. Cosmochim. Acta* **45**, 1751–1788.

Ozima M. and Podosek F. (2002) *Noble Gas Geochemistry* 2nd edn. Cambridge University Press, Cambridge, 286p.

Ozima M. and Zashu S. (1983) Primitive helium in diamonds. *Science* **219**, 1067–1068.

Ozima M., Wieler R., Marty B., and Podosek F. A. (1998) Comparative studies of solar, Q-gases and terrestrial noble gases, and implications on the evolution of the solar nebula. *Geochim. Cosmochim. Acta* **62**, 301–314.

Pepin R. O. (1991) On the origin and early evolution of terrestrial planet atmospheres and meteoritic volatiles. *Icarus* **92**, 2–79.

Pepin R. O. (2000) On the isotopic composition of primordial xenon in terrestrial planet atmospheres. *Space Sci. Rev.* **92**, 371–395.

Pepin R. O. and Phinney D. (1976) The formation interval of the Earth. *Lunar Planet. Sci.* **VII**. Lunar and Planetary Institute, Houston, pp. 682–684.

Pepin R. O. and Porcelli D. (2002) Origin of noble gases in the terrestrial planets. *Rev. Mineral. Geochem.* **47**, 191–246.

Podosek F. A. and Nichols R. H., Jr. (1998) Short-lived radionuclides in the solar nebula. In *Astrophysical Implications of the Laboratory Study of Presolar Materials,* AIP Proc. Conf. 402 (eds. T. J. Bernatowicz and E. K. Zinner), pp. 617–647.

Podosek F. A., Bernatowicz T. J., and Kramer F. E. (1981) Adsorption of xenon and krypton on shales. *Geochim. Cosmochim. Acta* **45**, 2401–2415.

Reynolds J. H. (1960) Determination of the age of the elements. *Phys. Rev. Lett.* **4**, 8–10.

Reynolds J. H. and Turner G. (1964) Rare gases in the chondrite Renazzo. *J. Geophys. Res.* **69**, 3263–3281.

Sandford S. A., Bernstein M. P., and Swindle T. D. (1998) The trapping of noble gases by by the irradiation and warming of interstellar ice analogs. *Meteorit. Planet. Sci.* **33**, A135.

Signer P. and Suess H. E. (1963) Rare gases in the Sun, in the atmosphere, and in meteorites. In *Earth Science and Meteorites* (eds. J. Geiss and E. D. Goldberg). North Holland, Amsterdam, pp. 241–272.

Swindle T. D. and Podosek F. A. (1988) Iodine-xenon dating. In *Meteorites and the Early Solar System* (eds. J. F. Kerridge and M. S. Matthews). University of Arizona Press, Tucson, pp. 1127–1146.

von Steiger R., Geiss J., and Gloeckler G. (1997) Composition of the solar wind. In *Cosmic Winds and the Heliosphere* (eds. J. R. Jokipii, C. P. Sonnett, and M. S. Giampapa). University of Arizona Press, Tucson, pp. 591–616.

Wacker J. F. (1989) Laboratory simulation of meteoritic noble gases: III. Sorption of neon, argon, krypton, and xenon on carbon: elemental fractionation. *Geochim. Cosmochim. Acta* **53**, 1421–1433.

Wieler R. (1998) The solar noble gas record in lunar samples and meteorites. *Space Sci. Rev.* **85**, 303–314.

Wieler R. (2002a) Cosmic-ray-produced noble gases in meteorites. *Rev. Mineral. Geochem.* **47**, 125–170.

Wieler R. (2002b) Noble gases in the solar system. *Rev. Mineral. Geochem.* **47**, 21–70.

Wieler R., Anders E., Baur H., Lewis R. S., and Signer P. (1991) Noble gases in "phase-Q": closed-system etching of an Allende residue. *Geochim. Cosmochim. Acta* **55**, 1709–1722.

Wimmer-Schweingruber R. F. and Bochsler P. (2000) Is there a record of interstellar pick-up ions in lunar soils? In

Acceleration and Transport of Energetic Particles Observed in the Heliosphere, ACE 2000 Symposium (eds. R. A. Mewaldt, J. R. Jokipii, M. A. Lee, E. Möbius, and T. H. Zurbuchen). Am. Inst. Phys. Conf. Proc. 528, pp. 270–273.

Yang J. and Anders E. (1982) Sorption of noble gases by solids, with reference to meteorites: III. Sulfides, spinels, and other substances: on the origin of planetary gases. *Geochim. Cosmochim. Acta* **46**, 877–892.

Zadnik G., Wacker J. F., and Lewis R. S. (1985) Laboratory simulation of meteoritic noble gases: II. Sorption of xenon on carbon: etching and heating experiments. *Geochim. Cosmochim. Acta* **49**, 1049–1059.

Isotope Geochemistry
ISBN: 978-0-08-096710-3

pp. 465–492

16
The Origin of Noble Gases and Major Volatiles in the Terrestrial Planets

D. Porcelli
University of Oxford, UK
and
R. O. Pepin
University of Minnesota, Minneapolis, MN, USA

16.1 INTRODUCTION

One of the salient characteristics of the composition of the Earth is the depletion in volatiles compared to parental solar-nebula relative abundances, and this is most pronounced in the noble gases. These chemically unreactive species, concentrated in the atmosphere, have retained many characteristics established early Earth history. A comparison between noble gases on the terrestrial planets and other solar system objects reveals significant differences in both elemental ratios and isotopic compositions and indicates that complex processes were involved in accumulating planetary volatiles from the nebula. Therefore, the atmosphere is not primary (i.e., directly acquired entirely from the solar nebula without modification). Atmophile elements have been added to the Earth in material that has contributed into the growing solid Earth and has subsequently degassed into the atmosphere, in late-accreting materials that degassed upon impact, and possibly directly to the atmosphere from the nebula as well. However, the importance of each of these sources, and the processes that modified these volatiles after initial capture, are still debated.

While considerations of the origin of planetary noble gases have been predominantly focused on those presently found in the atmosphere, noble gases still within the Earth provide further constraints about volatile trapping during planet formation. A wide range of noble-gas information for the Earth's mantle has been obtained from mantle-derived materials, and indicates that there are separate reservoirs within the Earth that have distinctive characteristics that were established early in Earth history. These must be included in comprehensive models of Earth volatile history. Also, data are now available for the atmospheres of both Venus and Mars, as well as from the interior of Mars, so that the evolution of Earth volatiles can be considered within the context of terrestrial-planet formation across the solar system.

The origins of the noble-gas features of the terrestrial-planet atmospheres and interiors have defied simple explanations and are clearly the result of a combination of acquisition and subsequent loss processes that have generated unique solar system compositions. The relevant data for noble gases in the atmospheres and interiors of the terrestrial planets, and the constraints these provide, are summarized below first. The emphasis is on the information provided by noble gases, since the signatures of volatile origin would be expected to survive most clearly in these species, although the implications for the major volatiles—nitrogen, carbon, and hydrogen—are also covered. Acquisition and loss processes are then discussed separately. Models for each planet are then described that are consistent with the available data and involve both acquisition episodes and subsequent modifications during partial loss.

The numerous debates regarding the origin of noble gases on the terrestrial planets have inevitably engendered explanations that are no longer viable, as well as more controversial viewpoints that, as of early 2000s, have not received sufficient support or substantial acceptance. These are discussed further by Pepin and Porcelli (2002).

16.2 CHARACTERISTICS OF TERRESTRIAL-PLANET VOLATILES

While the terrestrial atmosphere is readily accessible, planetary probes sent into the atmospheres of Venus and Mars have provided data to formulate model histories for these planets as well. Data from *in situ* composition measurements of the Venus atmosphere by mass spectrometers and gas chromatographs on the Pioneer Venus and Venera spacecrafts are reviewed and assessed by von Zahn *et al.* (1983); an updated summary is set out in Wieler (2002). Further data are provided by the martian Shergotty–Nakhla–Chassigny clan (SNC) meteorites. Atmophile elements found in some phases in these rocks have been found to match the probe data for the martian atmosphere and have been identified as shock-implanted atmospheric gases, most clearly seen in the glassy lithology of the SNC meteorite EET79001. These in turn have yielded more precise information about the atmosphere. Other components representing interior martian reservoirs have also been found (see review by Swindle, 2002).

Comparisons can be made between planetary atmospheres and both solar and chondritic compositions. The features of noble gases in the Sun are obtained by measurements of the solar wind. These generally represent those of the bulk solar nebula within which the planets formed, and provide the reference for consideration of terrestrial noble gases. Chondritic meteorites are rich in noble gases, and have long been considered as sources for terrestrial noble gases. While meteoritic gases have significant isotopic differences due to the inclusion of various nucleosynthetic components that escaped homogenization in the solar nebula, a major primordial component, designated Q gases (see Wieler, 1994), appears to have been widely distributed in the solar system. This component appears to be largely free of "exotic" components, with isotopic compositions similar to those of solar gases, and so represents noble gases trapped in fine-grained solid materials in at least some portion of the nebula. Note that helium is lost from the planetary atmospheres to space, so that early Earth characteristics have not been preserved, and is not discussed in this section.

Table 1 provides the composition of the terrestrial atmosphere, which is the standard for all noble-gas analyses, and indicates the general relative abundances of the isotopes. The masses of the rare-gas planetary reservoirs are given in Table 2.

16.2.1 Atmospheric Noble-gas Abundance Patterns

Noble-gas relative abundances in chondrites and terrestrial-planet atmospheres are generally highly fractionated relative to solar values, except for planetary Xe/Kr ratios. Planetary abundances vary by 10^3, from gas rich Venus to gas-poor Mars. Rare-gas abundances in terrestrial-planetary atmospheres are listed in Table 2, and are normalized to the solar pattern in Figure 1, along with the

Table 1 Noble-gas isotopes in the terrestrial atmosphere.

Isotope	*Relative abundances*	*Principal sources of nuclides*[a]
^{3}He	$(1.399 \pm 0.013)\times10^{-6}$	Primordial; space[b]
^{4}He	$\equiv1$	^{238}U, ^{235}U, and ^{232}Th decay; space[b]
^{20}Ne	9.80 ± 0.08	Primordial
^{21}Ne	0.0290 ± 0.0003	Primordial; ^{18}O(α, n)^{21}Ne; ^{24}Mg(n, α)^{21}Ne
^{22}Ne	$\equiv1$	Primordial
^{36}Ar	$\equiv1$	Primordial
^{38}Ar	0.1880 ± 0.0004	Primordial
^{40}Ar	295.5 ± 0.5	Decay of ^{40}K
^{78}Kr	0.6087 ± 0.0020	Primordial
^{80}Kr	3.9599 ± 0.0020	Primordial
^{82}Kr	20.217 ± 0.004	Primordial
^{83}Kr	20.136 ± 0.021	Primordial
^{84}Kr	$\equiv100$	Primordial
^{86}Kr	30.524 ± 0.025	Primordial
^{124}Xe	2.337 ± 0.008	Primordial
^{126}Xe	2.180 ± 0.011	Primordial
^{128}Xe	47.15 ± 0.07	Primordial
^{129}Xe	649.6 ± 0.9	Primordial; ^{129}I decay
^{130}Xe	$\equiv100$	Primordial
^{131}Xe	521.3 ± 0.8	Primordial; ^{244}Pu decay; ^{238}U decay[c]
^{132}Xe	660.7 ± 0.5	Primordial; ^{244}Pu decay; ^{238}U decay[c]
^{134}Xe	256.3 ± 0.4	Primordial; ^{244}Pu decay; ^{238}U decay[c]
^{136}Xe	217.6 ± 0.3	Primordial; ^{244}Pu decay; ^{238}U decay[c]

After Ozima and Podosek (2001).
[a] Primordial isotopes are those trapped during Earth formation and are not nucleogenic. Only globally significant other sources are included. See Ballentine and Burnard (2002) for production rates and other nuclear production mechanisms.
[b] Various mechanisms supply He isotopes from space; see Torgersen (1989) for compilation.
[c] The source of variations is within the solid Earth, but has not contributed significantly to the atmosphere.

Table 2 Volatiles in terrestrial-planet atmospheres.

Constituent	*Earth*	*Venus*	*Mars*
Total mass of atm (g)	5.1×10^{21}	4.75×10^{23}	2.7×10^{19}
N_2 (mol/g-planet)	2.31×10^{-8}	1.4×10^{-7}	$>4.2 \times 10^{-11}$
^{4}He (mol/g-planet)	1.56×10^{-13}	2.9×10^{-11}	
^{20}Ne (mol/g-planet)	4.86×10^{-13}	1.0×10^{-11}	2.5×10^{-15}
^{36}Ar (mol/g-planet)	9.41×10^{-13}	6.7×10^{-11}	5.8×10^{-15}
^{84}Kr (mol/g-planet)	1.97×10^{-14}	$(6.7–130) \times 10^{-14}$	1.9×10^{-16}
^{130}Xe (mol/g-planet)	1.09×10^{-16}	$<5 \times 10^{-15}$	3.2×10^{-18}
Total mass of planet (g)	5.98×10^{27}	4.87×10^{27}	0.64×10^{27}

Source: Ozima and Podosek (2001).

pattern exhibited by trapped noble gases in bulk chondrites. All objects have a striking depletion in noble gases relative to solar abundances. Chondrites display a regular fractionated pattern across the noble gases, with the lightest displaying the greatest depletions. This pattern seems consistent with preferential retention of the generally more reactive heavier noble gases, and was likely established during trapping into solid grains. The Ne/Ar ratios of the terrestrial planets and chondrites are similar, with normalized total abundances in the Earth 10 times greater than on Mars, and 10^2 times less than on Venus. In contrast, the heavier noble gases display greater variations in relative abundances. While the Kr/Ar ratio of chondrites and the Earth are similar, the terrestrial Xe/Kr ratio is much lower and similar to the solar value. This was initially thought to be due to the sequestration of terrestrial xenon in other terrestrial reservoirs. However, investigations of possible reservoirs of xenon, such as shales or glacial ice, failed to find this "missing xenon" (Bernatowicz *et al.*, 1984, 1985; Wacker and Anders, 1984; Matsuda and Matsubara, 1989). The lower terrestrial Xe/Kr ratio has since then widely been considered to be a feature of the Earth. The pattern exhibited by Mars closely follows that of the Earth, with a near-solar Xe/Kr ratio. Venus also appears to have an Xe/Kr ratio near the solar value, although uncertainties are large. However, it seems clear that Venus differs markedly from the other two terrestrial planets in having an Ar/Kr ratio that is also close to solar.

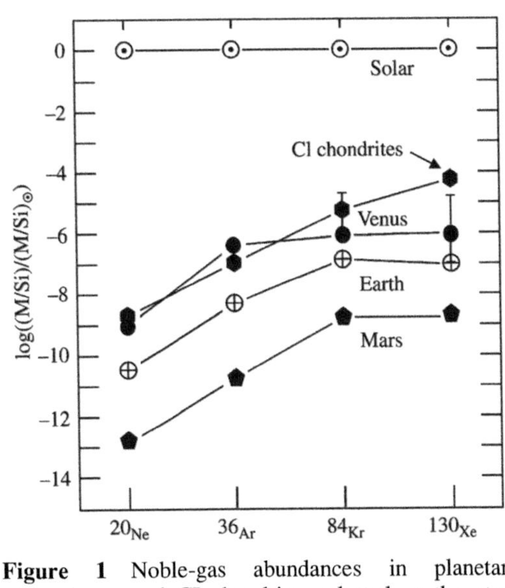

Figure 1 Noble-gas abundances in planetary atmospheres and CI chondrites, plotted as the atom concentration relative to Si and divided by the corresponding solar ratio. Note that ranges of Kr and Xe values are shown for Venus (source Pepin, 1991).

16.2.2 Atmospheric Neon

Neon in planetary atmospheres is isotopically heavier than the solar composition, and falls within meteoritic values. Neon-isotopic compositions for the atmosphere and solar system reservoirs are shown in Figure 2. The greatest differences between solar system bodies are seen in the proportions of ^{20}Ne and ^{22}Ne, which are not produced in significant quantities in large bodies. The value of $^{20}Ne/^{22}Ne = 13.8 \pm 0.1$ (Benkert *et al.*, 1993; Pepin *et al.*, 1999) derived for the solar wind is believed to represent the initial solar-nebula composition. Solar neon found in irradiated meteorites is a mixture, called Ne-B, of solar-wind (SW) neon and fractionated solar-energetic particle (SEP) neon (Black, 1972), and has a value of $^{20}Ne/^{22}Ne \sim 12.5$. Neon-isotope ratios in bulk CI chondrites scatter around an average $^{20}Ne/^{22}Ne$ of 8.9 ± 1.3. Meteoritic Q gases have $^{20}Ne/^{22}Ne = 10.1$–10.7 (Busemann *et al.*, 2000). The terrestrial-atmospheric ratio of 9.8, in principle, can be derived either from mixing meteorite and solar components or by fractionation of solar neon. The difference between the solar (0.033) and atmospheric (0.029) $^{21}Ne/^{22}Ne$ ratios is consistent with fractionation of solar neon and addition of radiogenic ^{21}Ne (see Porcelli and Ballentine, 2002).

The $^{20}Ne/^{22}Ne$ ratio of the martian atmosphere has been estimated to be 10.1 ± 0.7 from SNC meteorite data (Pepin, 1991). It is indistinguishable

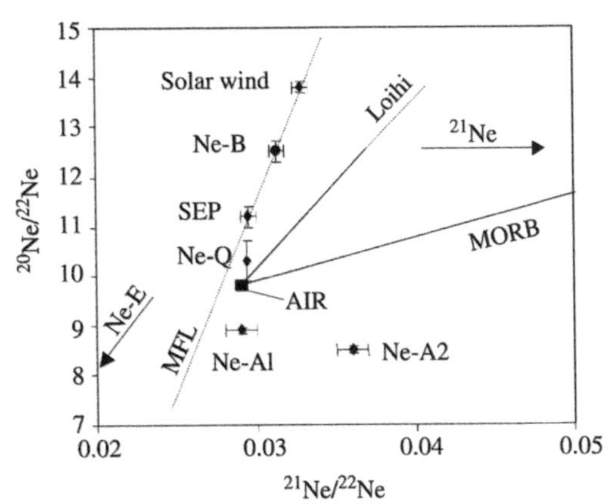

Figure 2 The Ne-isotopic composition of the atmosphere, compared to different extraterrestrial compositions (compiled by Busemann *et al.*, 2000). MORB data fall on a correlation line extending from air values (Sarda *et al.*, 1988), and reflect mixing between air contamination of the samples with a trapped mantle component with high Ne-isotope ratios. This component is composed of radiogenic ^{21}Ne and solar Ne (Honda *et al.*, 1993) or Ne-B (Trieloff *et al.*, 2000). OIB such as Loihi with high $^3He/^4He$ ratios reflect mixing of air contamination with a mantle composition that has a lower $^{21}Ne/^{22}Ne$ ratio, reflecting a higher time-integrated Ne/(U + Th) ratio (Honda *et al.*, 1991).

from neon in the terrestrial atmosphere, and so is substantially lower than the solar value. The Venus ratio of $^{20}Ne/^{22}Ne = 11.8 \pm 0.7$ (Donahue and Russell, 1997) requires some fractionation relative to the solar value, although not to the same extent as neon in the Earth's atmosphere. The $^{21}Ne/^{22}Ne$ ratios of Mars and Venus are not known sufficiently well to provide further constraints.

16.2.3 Atmospheric Argon

Nonradiogenic argon in chondrites and planetary atmospheres is isotopically heavier than the solar composition. The initial $^{40}Ar/^{36}Ar$ ratio in the solar system is $\sim 10^{-4}$–10^{-3} (Begemann *et al.*, 1976). The atmosphere has $^{40}Ar/^{36}Ar = 296$, so that essentially all the ^{40}Ar has been produced by ^{40}K decay in the solid Earth. Regarding the two nonradiogenic isotopes, the atmospheric $^{38}Ar/^{36}Ar$ ratio of 0.188 is similar to that found in CI chondrites of 0.189 ± 0.002 (Mazor *et al.*, 1970) but substantially higher than some recent estimates of 0.172 ± 0.002 (Pepin *et al.*, 1999; Palma *et al.*, 2002) and 0.179 ± 0.001 (Wieler, 1998) for the solar wind. If atmospheric argon was derived from a solar source, it must have been fractionated.

The martian $^{38}Ar/^{36}Ar$ ratio is highly fractionated relative to the solar ratio; SNC meteorite analyses yield values from 0.24 ± 0.01 (Pepin, 1991) to 0.26–0.30 (Bogard, 1997; Garrison and Bogard, 1998). The measured Venus value of 0.180 ± 0.019 (Donahue and Russell, 1997) is indistinguishable from both the solar and terrestrial values.

16.2.4 Atmospheric Krypton

Krypton in meteorites and the Earth is isotopically fractionated relative to the solar composition, although this may not be true for krypton on Mars. Isotopic variations due to production of krypton isotopes is not expected in bulk solar system bodies. Solar, bulk meteorite, and terrestrial krypton-isotopic compositions (Figure 3) are generally related to one another by mass fractionation, with the terrestrial atmosphere depleted in light isotopes by ~0.8% amu^{-1} relative to the solar composition (Eugster *et al.*, 1967; Pepin *et al.*, 1995). Meteoritic krypton possibly contains an excess in ^{86}Kr. The krypton-isotopic composition of the atmosphere of Mars (Pepin, 1991; Garrison and Bogard, 1998) is essentially indistinguishable from that of SW-Kr, but may be slightly fractionated to an isotopically lighter composition (see Garrison and Bogard, 1998; Swindle, 2002). No data are available for krypton on Venus.

16.2.5 Terrestrial Atmospheric Xenon

Nonradiogenic xenon on the Earth is highly fractionated, and requires early, fractionating losses to space of a nonsolar precursor. Radiogenic xenon abundances require losses to space over 100 Myr.

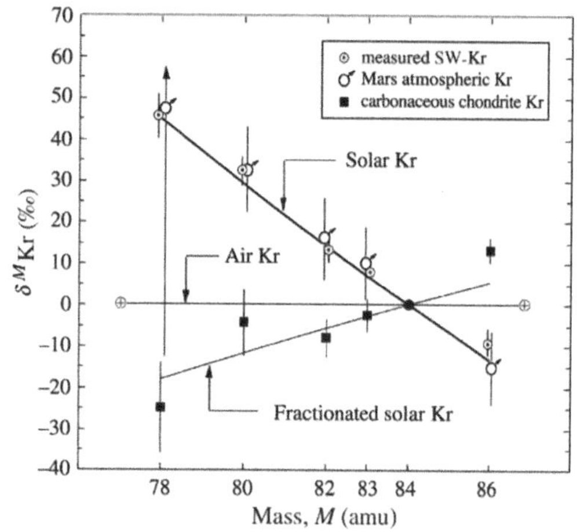

Figure 3 Krypton isotopes in solar system volatile reservoirs, plotted as ‰ deviations of the ratio to ^{84}Kr, and normalized to the ratio in terrestrial air (Basford *et al.*, 1973). The heavy "solar Kr" curve represents a smooth fit to the measured solar-wind isotope ratios. Measured SW-Kr from Wieler and Baur (1994) and Pepin *et al.* (1995); Mars Kr from Pepin (1991); carbonaceous chondrite Kr from Krummenacher *et al.* (1962), Eugster *et al.* (1967), and Marti (1967) (Pepin and Porcelli, 2002) (reproduced by permission of the Mineralogical Society of America from *Rev. Mineral. Geochem.* **2002**, *47*, 191–246).

16.2.5.1 Nonradiogenic xenon

The source and composition of terrestrial nonradiogenic xenon have been difficult to constrain, because atmospheric xenon does not match any widespread solar system xenon composition. The problem has been made more difficult because contributions to ^{129}Xe by decay of ^{129}I, and to the heavy isotopes ^{131}Xe to ^{136}Xe by fission of ^{244}Pu and ^{238}U that cannot be independently tightly constrained. The light isotopes of atmospheric xenon are related to both bulk chondritic and solar xenon by fractionation of ~4.2% amu^{-1} (Krummenacher *et al.*, 1962), with a clear radiogenic excess of ~7% in ^{129}Xe from decay of ^{129}I. However, these compositions do not match the heavy xenon isotopes. As shown in Figure 4, where solar xenon is fractionated to match the light isotopes of the atmosphere, the $^{136}Xe/^{130}Xe$ ratio is well above atmospheric. Consequently, SW-Xe cannot be the primordial terrestrial composition. This exclusion also applies to the meteoritic CI-Xe and Q-Xe compositions, which are considerably richer in the heavy isotopes than SW-Xe. However, a suitable initial composition for the atmosphere has been derived using multidimensional isotopic correlations of chondrite data (Pepin, 2000). Terrestrial atmospheric xenon

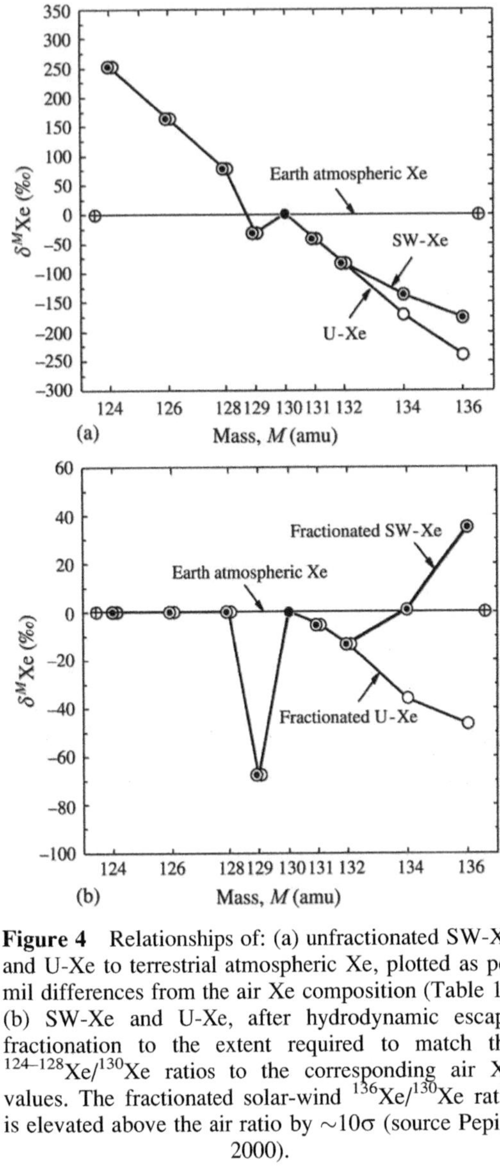

Figure 4 Relationships of: (a) unfractionated SW-Xe and U-Xe to terrestrial atmospheric Xe, plotted as per mil differences from the air Xe composition (Table 1). (b) SW-Xe and U-Xe, after hydrodynamic escape fractionation to the extent required to match the $^{124-128}$Xe/^{130}Xe ratios to the corresponding air Xe values. The fractionated solar-wind ^{136}Xe/^{130}Xe ratio is elevated above the air ratio by ~10σ (source Pepin, 2000).

is then composed of the nonradiogenic composition (dubbed U-Xe) that has been fractionated, as well as radiogenic ^{129}Xe and fissiogenic xenon from decay of ^{244}Pu.

U-Xe is necessarily depleted in the heaviest isotopes compared to SW-Xe, and this suggests that there is a heavy-isotope component in the Sun but not in the early Earth. This presents a conundrum, since SW-Xe, presumably reflecting the xenon composition in the accretion disk, would arguably be the more plausible contributor to primordial planetary inventories. Note, however, that such a problematic relationship will apply to any primordial composition for the Earth. Moreover, while it is difficult to understand why U-Xe has not been widely found in other solar system materials, this also applies to any suitable composition. A possible explanation is that the xenon composition of the nebula changed over the lifetime of the disk due to changes in material supplied from the surrounding molecular cloud (Pepin *et al.*, 1995; Pepin, 2003). An important parameter of this problem is the xenon-isotope composition of the atmosphere of Jupiter, which contains gases gravitationally captured from the solar nebula. Results from the Galileo Probe mass spectrometer (Mahaffy *et al.*, 1998) have narrowed the range of possible nebular-xenon compositions at the time of Jovian-atmospheric formation. The relative isotopic abundances fall approximately along a fractionation curve with respect to nonradiogenic-terrestrial xenon, but unfortunately are not known with sufficient precision so as to distinguish between U-Xe and SW-Xe.

16.2.5.2 Radiogenic xenon

Heavy xenon isotopes (typically represented by ^{136}Xe) are produced by fission of ^{244}Pu ($t_{1/2}$ = 80 Ma) and ^{238}U ($t_{1/2}$ = 4.47 Ga). Since all ^{244}Pu has decayed, and there are no stable isotopes presently representing plutonium, the original plutonium concentration in the Earth must be estimated from those of other highly refractory elements. Meteorite data suggest that at 4.57 Ga, $(^{244}Pu/^{238}U)_0 = 6.8 \times 10^{-3}$ (Hudson *et al.*, 1989; Hagee *et al.*, 1990), and so the silicate Earth, or Earth-forming materials, with 21 ppb ^{238}U at present, initially had 0.29 ppb ^{244}Pu that produced 2.0×10^{35} atoms of ^{136*}Xe. The amount produced by ^{238}U over the age of the Earth (7.5×10^{33} atoms ^{136}Xe) is much less, and so bulk Earth (and atmospheric) ^{136*}Xe must be dominantly plutonium derived. The short-lived isotope ^{129}I ($t_{1/2}$ = 15.7 Ma) produces ^{129*}Xe. The bulk silicate Earth (BSE) concentration of stable ^{127}I is difficult to constrain. Wänke *et al.* (1984) estimated a commonly quoted BSE concentration of 13 ppb based on mantle-xenolith data, and this is consistent with other data (Déruelle *et al.*, 1992). At 4.57 Ga, $(^{129}I/^{127}I)_0 = 1.1 \times 10^{-4}$ based on meteorite data (Hohenberg *et al.*, 1967; Brazzle *et al.*, 1999). Using 13 ppb I, 2.7×10^{37} atoms of ^{129*}Xe were produced in the Earth or Earth-forming materials.

Using fractionated U-Xe for the nonradiogenic composition of the atmosphere, 6.8 ± 0.3% of atmospheric ^{129}Xe ($^{129*}Xe_{atm} = 1.7 \times 10^{35}$ atoms) and 4.65 ± 0.50% of atmospheric ^{136}Xe ($^{136*}Xe_{atm} = 3.81 \times 10^{34}$ atoms) are radiogenic. The $^{136*}Xe_{atm}$ is 20% of the total ^{136}Xe produced by ^{244}Pu in the silicate Earth. In contrast, the $^{129*}Xe_{atm}$ is only 0.8% of the total ^{129}Xe produced; such a low value can neither be accounted for by incomplete degassing of the mantle nor from any uncertainties in the estimated amount of iodine, and requires losses to space. This must have occurred during early Earth history, when such heavy species could have been lost either from protoplanetary materials or from the growing

Earth. Wetherill (1975) proposed that a "closure age" of the Earth could be calculated by assuming that essentially complete loss of ^{129*}Xe occurred initially, followed by complete closure against further loss. The time when this closure commenced can be calculated by

$$t = \frac{-1}{\lambda_{129}} \ln\left[\left(\frac{^{129^*}\mathrm{Xe_{atm}}}{^{127}\mathrm{I}}\right)\left(\frac{^{127}\mathrm{I}}{^{129}\mathrm{I}}\right)_0\right] \quad (1)$$

Using an updated value for $^{129*}\mathrm{Xe_{atm}}$, and assuming 40% of the mantle degassed to the atmosphere, 83 Ma is obtained. The "closure age" can also be calculated by combining the ^{129}I–^{129}Xe and ^{244}Pu-^{136}Xe systems (Pepin and Phinney, 1976):

$$t = \frac{1}{\lambda_{244} - \lambda_{129}} \ln\left[\left(\frac{^{129^*}\mathrm{Xe_{atm}}}{^{136^*}\mathrm{Xe_{atm}}}\right)\left(\frac{^{238}\mathrm{U}}{^{127}\mathrm{I}}\right)_0 \times \left(\frac{^{244}\mathrm{Pu}}{^{238}\mathrm{U}}\right)_0\left(\frac{^{127}\mathrm{I}}{^{129}\mathrm{I}}\right)_0 {^{136}\mathrm{Y}_{244}}\right] \quad (2)$$

Using a value of 7.05×10^{-5} for the parameter $^{136}\mathrm{Y}_{244}$ that represents the number of ^{136}Xe atoms produced per decay of ^{244}Pu, a similar closure age of 96 Ma is obtained. For example, note that if atmospheric xenon loss occurred during a massive Moon-forming impact, then the closure period corresponds to the time after an instantaneous catastrophic loss event.

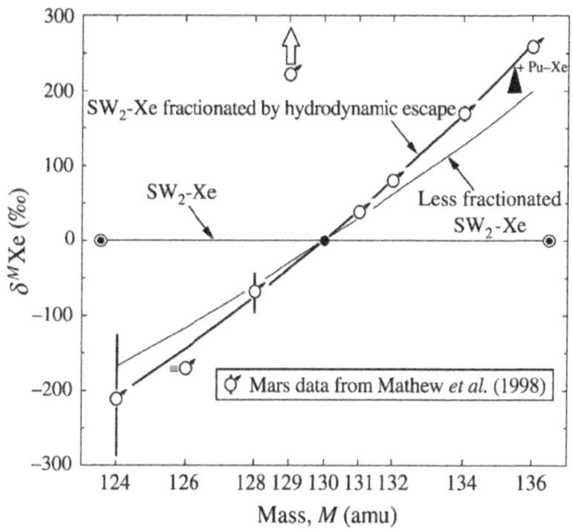

Figure 5 The martian atmospheric composition (Mathew *et al.*, 1998) plotted relative to SW-Xe (see Pepin and Porcelli, 2002). The heavy curve demonstrates that SW-Xe in the primordial martian atmosphere can be fractionated by escape to a close fit to all atmospheric isotope ratios except δ^{129}Xe (which is off scale at +1,480‰). The light curve represents a less severely fractionated SW-Xe composition, which allows for the addition of a $^{131-136}$Xe component with the composition of fission-produced Xe from decay of ^{244}Pu (as suggested by Swindle and Jones (1997)) (Pepin and Porcelli, 2002) (reproduced by permission of the Mineralogical Society of America from *Rev. Mineral. Geochem.* **2002**, *47*, 191–246).

16.2.6 Martian Atmospheric Xenon Isotopes

Martian atmospheric xenon is likely derived from solar xenon, with addition of radiogenic ^{129*}Xe and plutonium-derived heavy xenon after a closure age of ~70 Ma. Xenon-isotope data for the martian atmosphere have been derived entirely from SNC meteorites (Figure 5). Swindle *et al.* (1986) pointed out that martian atmospheric xenon strongly resembles mass-fractionated CI-Xe, with addition of radiogenic ^{129*}Xe (comprising 61% of total ^{129}Xe). This implies that there are no additional contributions to the xenon inventory from degassed fission xenon. Alternatively, Swindle and Jones (1997) demonstrated that SW-Xe could be fractionated to fall below the measured martian atmospheric $^{131-136}$Xe/^{130}Xe ratios by amounts consistent with the presence of ^{244}Pu-derived xenon that comprises ~5% of total ^{136}Xe. A revised SNC database for martian atmospheric-xenon composition by Mathew *et al.* (1998) closely matches stronger hydrodynamic-escape fractionation of SW-Xe (see Section 16.4.2) with no ^{244}Pu fission-xenon contribution at all. However, a weaker SW-Xe fractionation plus a ^{244}Pu contribution is still allowed within the data uncertainties. An added feature of the Mathew–Marti database is that it is poorly fit by fractionated CI-Xe, and consequently points to SW-Xe rather than CI-Xe as the primordial martian composition if these data really represent the martian atmosphere.

The composition of primordial xenon on Mars and the presence or absence of ^{244}Pu fission xenon in the present atmosphere are important issues in the context of the provenance of accretional materials, the timing of planetary growth, and the subsequent geochemical and outgassing histories of the planet. The central question of an atmospheric ^{244}Pu component is currently plagued by apparent coincidences. The excellent match of fractionated SW-Xe, alone, to the atmospheric data is presumably fortuitous if Pu-Xe is present; and if it is absent, the fact that a weaker fractionation generates heavy-isotope residuals in good accord with ^{244}Pu fission yields must likewise be accidental. Atmospheric xenon data with higher measurement precision than is currently attainable for xenon extracted from the SNCs will eventually resolve the issue.

It has been estimated that the silicate portion of Mars contains 32 ppb I and 16 ppb of ^{238}U at present, equivalent to 32.3 ppb initially (Wänke and Dreibus, 1988), so that I/U = 0.54 (molar ratio) at 4.57 Ga. If the atmosphere contains plutonium-derived xenon comprising 5% of the atmospheric ^{136}Xe, so that (^{129*}Xe/^{136*}Xe) = 83, then using Equation (2), a closure age of 46 Ma is obtained, which is a factor of 2 lower than that of the Earth. In this case, ~2% of the mantle has degassed radiogenic xenon, which is compatible with the ^{40}Ar budget, and so supports the identification of plutonium-derived xenon in the atmosphere. If SW-Xe is the primordial composition of the atmosphere, if there is not a clearly resolvable fission component, then the closure age is much lower. However, in this case only a very low—perhaps unreasonably low—fraction of the mantle could have degassed to the atmosphere.

16.2.7 Noble-gas Isotopes in the Terrestrial Mantle

The mantle contains solar-derived neon and less-fractionated xenon, as well as nonatmospheric abundance patterns. There are considerable amounts of ^3He and other gases retained within the Earth after early losses. Evidence for primordial noble gases in the mantle first came from helium, with $^3He/^4He$ ratios in mid-ocean ridge basalts (MORBs) that are melts of the convecting upper mantle of $\sim 8R_A$ (where R_A is the atmospheric ratio) that represents a mixture of primordial ^3He with radiogenic helium (with $\sim 0.01R_A$). Values of up to $37R_A$ have been found in ocean islands (see Chapter 17).

Measured MORB $^{20}Ne/^{22}Ne$ ratios are greater than that of the atmosphere, and extend toward the solar value (Figure 2). Since these isotopes are not produced in significant quantities in the Earth, there must be trapped primordial neon in the Earth with a $^{20}Ne/^{22}Ne$ ratio at least as high as the highest measured mantle value. Samples with lower measured ratios appear to have contaminant air neon. While it has often been assumed that the trapped neon was derived directly from the solar nebula with $^{20}Ne/^{22}Ne = 13.8$ (Honda *et al.*, 1993), it has been suggested that it is the meteoritic Ne-B component with $^{20}Ne/^{22}Ne = 12.5$ (Trieloff *et al.*, 2000, 2002). While earlier measured mantle values above this (Sarda *et al.*, 1988) were insufficiently precise to firmly establish the presence of neon with a higher value, recent analyses for Icelandic samples found $^{20}Ne/^{22}Ne = 13.75 \pm 0.32$ (Harrison *et al.*, 1999), although this measurement has been debated (see Ballentine *et al.*, 2001; Trieloff *et al.*, 2000, 2001). Clearly, further work is required over this issue. High $^{21}Ne/^{22}Ne$ ratios in the mantle (Figure 2) are due to production of ^{21}Ne by various nuclear reactions.

Solar argon might be expected in the mantle as well. $^{40}Ar/^{36}Ar$ ratios of up to 4×10^4 have been found due to production of ^{40}Ar in the mantle (Burnard *et al.*, 1997), and so contain no information on the origin of argon. Measurements of nonradiogenic MORB and Loihi $^{38}Ar/^{36}Ar$ ratios are typically atmospheric within error, but have been of low precision due to the low abundance of these isotopes. Also, since most $^{40}Ar/^{36}Ar$ ratios are interpreted as having been lowered due to a substantial fraction of atmospheric contamination, the ^{36}Ar (and ^{38}Ar) of these samples are dominated by contamination. Two recent analyses of MORB and OIB samples have found $^{38}Ar/^{36}Ar$ ratios lower than that of the atmosphere (Valbracht *et al.*, 1997; Niedermann *et al.*, 1997), and Pepin (1998) argued that these values reflect a mixture of a solar upper-mantle composition with atmospheric argon contamination. However, new high-precision data for MORB with $^{40}Ar/^{36}Ar \sim 2.8 \times 10^4$ failed to find $^{38}Ar/^{36}Ar$ ratios that deviate from that of air (Kunz, 1999). Similar air-like $^{38}Ar/^{36}Ar$ ratios were also found in high-precision OIB analyses (Trieloff *et al.*, 2000, 2002), in contrast to the earlier measurements. Further work is certainly required, considering the importance of establishing whether trapped argon within the Earth has been fractionated relative to solar argon, and so whether the atmospheric composition was indeed generated by processes occurring in the planetary atmosphere. If solar argon is not found in the mantle, the possibility that atmospheric $^{38}Ar/^{36}Ar$ ratios reflect subduction of atmospheric argon rather than the composition of initially trapped argon also must be considered.

The measured ratios of the nonradiogenic xenon isotopes in MORBs are indistinguishable from those in the atmosphere. However, more precise measurements of mantle-derived xenon in CO_2 well gases have been found to have higher $^{124-128}Xe/^{130}Xe$ ratios (Phinney *et al.*, 1978; Caffee *et al.*, 1999) that can be explained by either: (i) a mixture of ~10% solar xenon trapped within the Earth and ~90% atmospheric xenon (subducted or added in the crust); or (ii) a mantle component that has not been fractionated relative to solar xenon to the same extent as air xenon. In the former case, the radiogenic composition of the nonatmospheric component can be calculated (Jacobsen and Harper, 1996); if 90% of the well-gas xenon is derived from the atmosphere (with $^{129}Xe/^{130}Xe = 6.496$), the well-gas value of $^{129}Xe/^{130}Xe = 7.2$ contains 10% of a solar component with $^{129}Xe/^{130}Xe \approx 13.5$. Note that it is not possible to distinguish whether solar Xe or U-Xe is present in the mantle.

Mantle-derived materials have a range of elemental-abundance patterns due to various fractionation processes, although the mantle pattern has been inferred from measured isotopic variations and radiogenic isotope-production ratios (see Porcelli and Ballentine, 2002). From the $^4He/^{21}Ne$ production ratio and the average coexisting shifts in $^4He/^3He$ and $^{21}Ne/^{22}Ne$ in the mantle relative to the primordial compositions, a ratio of $^3He/^{22}Ne = 11$ is obtained. This is greater than the more recent estimate of 1.9 for the solar nebula (see discussion in Porcelli and Pepin (2000)). Similarly, using the mantle $^{40}Ar/^{36}Ar$ ratio, and an estimate of the ratio of $^{21*}Ne$ production in the mantle to that of $^{40*}Ar$, a ratio of $^{22}Ne/^{36}Ar = 0.15$ is inferred, which is substantially lower than the solar value. These values suggest that solar noble gases found in the mantle have been fractionated when the deep-Earth reservoirs were established, although it is possible that subduction of argon modified the $^{22}Ne/^{36}Ar$ ratio, as well as changed the $^{38}Ar/^{36}Ar$ ratio. Similarly, the calculated Xe/Ar ratio depends upon the amount of subducted xenon (Porcelli and Wasserburg, 1995b). However, a straightforward calculation is not possible without assumptions regarding where upper-mantle xenon has been stored and how much of the radiogenic noble gases have been produced in the upper mantle. If the nonsubducted component in MORB has $^{129}Xe/^{130}Xe = 13.5$ (see above) and so $^{136}Xe/^{130}Xe = 4.42$, and all the excess ^{136}Xe has been produced by ^{238}U in a reservoir over 4.5 Ga, along with all the ^{40}Ar, with a K/U ratio of 1.27×10^4, then MORB has $^{130}Xe/^{36}Ar = 8 \times 10^{-4}$. In the context of a mantle model, Porcelli and Wasserburg (1995b) obtained a ratio of 1.9×10^{-3}. These values are higher than the solar ratio of 2×10^{-5}; lower ratios are obtained if a much greater proportion of the mantle xenon is subducted so that stored mantle xenon has a higher $^{136}Xe/^{130}Xe$ ratio.

The total amount of noble gases in the mantle provides another constraint on acquisition mechanisms. The upper-mantle ^3He concentration has been estimated in various ways from the total ^3He flux from mid-ocean ridges, the ratio of $C/^3He$ in MORBs, and from concentrations in relatively undegassed basalts, with a range of an MORB-source mantle ^3He concentration of $(1.2–4.6) \times 10^9$ atoms g^{-1} (see Porcelli and Ballentine (2002) for discussion). However, this is a lower limit to the amount initially trapped within the mantle, since the upper mantle has clearly lost volatiles to the atmosphere. Evidence for a more gas rich mantle reservoir comes from ocean-island hot spots such as Hawaii and Iceland that have $^3He/^4He$ ratios of up to $37R_A$ compared to an MORB value of $8R_A$ (see Graham, 2002). Clearly, these have had a higher time-integrated He/U ratio than the MORB source. Had these have been derived from a reservoir that evolved as a closed system since Earth formation with a BSE uranium concentration of 21 ppb, then this reservoir would have had 7.6×10^{10} atom $^3He\ g^{-1}$ (e.g., Porcelli and Wasserburg, 1995a), and, with $^3He/^{22}Ne = 11$, it also has 6.9×10^9 atom $^{22}Ne\ g^{-1}$. For comparison, the neon inventory in the atmosphere divided into the mass of the upper mantle yields only 2×10^{11} atom $^{22}Ne\ g^{-1}$. Other possible interpretations of high $^3He/^4He$ ratios (e.g., Helffrich and Wood, 2001; Porcelli and Halliday, 2001) have not been fully developed and may still require similar high initial mantle noble-gas concentrations.

As discussed above, xenon isotopes indicate that early losses of radiogenic isotopes occurred from the atmosphere or its source reservoir. It has been argued that, based on the proportion of plutonium-derived fissiogenic xenon (Kunz *et al.*, 1998), mantle data indicate that a similarly late-xenon closure age applies to the mantle as well (Porcelli *et al.*, 2001). Trapping of nonradiogenic noble gases in the Earth must have occurred early (see Section 16.3), and so extensive losses of radiogenic xenon also necessarily apply to these as well. Modeling of terrestrial-isotope compositions indicates that up to 99% of noble gases originally trapped in the Earth were lost, and so acquisition mechanisms must supply not only the inventories currently found, but 10^2 times more (Porcelli *et al.*, 2001).

16.2.8 Nonradiogenic Xenon Isotopes in the Martian Mantle

The martian mantle has high xenon concentrations and distinct abundance patterns. Martian meteorites contain components other than those derived directly from the atmosphere (see detailed discussion by Swindle, 2002). In particular, noble gases in the dunite meteorite Chassigny appear to represent a distinct interior reservoir. The $^{84}Kr/^{132}Xe$ ratio of 1.2 (Ott, 1988) is lower than both the martian atmosphere (20) and solar (16.9) values, but is similar to that of CI chondrites. If this is truly a source feature, it indicates that heavy noble gases trapped within the planet suffered substantially different elemental fractionation than the atmosphere. The interior $^{84}Kr/^{36}Ar$ ratio of 0.06 is much higher than the solar value of 2.8×10^{-4}, but it is close to the atmospheric value of 0.02 and so does not display the same contrast as the Kr/Xe ratio. Unfortunately, it is not possible to determine if the measured elemental abundance ratios were modified by planetary processing or transport and incorporation into the samples.

The isotopic composition of the martian interior is only available for xenon. Data for the dunite Chassigny indicate that there is a mantle reservoir with nonradiogenic isotope ratios that appear to be indistinguishable from solar values (Ott, 1988; Mathew and Marti, 2001), and so does not exhibit the strong isotopic fractionation seen in the atmosphere. The relative abundances of ^{129}Xe and ^{136}Xe are also close to solar, indicating that this reservoir had a high Xe/Pu and Xe/I ratios, at least during the lifetime of ^{244}Pu. Data from other meteorites indicate that there are other interior martian reservoirs that contain solar xenon but with resolvable fissiogenic contributions, and so have had lower Xe/Pu ratios (Marty and Marti, 2002).

16.2.9 Major Volatiles

The "surface" inventories of major volatiles that have been added directly to the Earth's surface or removed from the mantle include not only atmospheric abundances, but also those of the hydrosphere and crustal rocks. Isotopic compositions are generally reported as δ units of per mil deviations from a standard.

16.2.9.1 Nitrogen

Nitrogen abundances in the atmospheres of Earth and Venus are comparable, being much higher than that of Mars. Meteoritic $\delta^{15}N$ values overlap the terrestrial-atmosphere and the lower-mantle values, while that of Mars' atmosphere is much higher. Planetary $N/^{36}Ar$ ratios fall between meteorite and solar ratios. While much of the terrestrial nitrogen at the surface is in the atmosphere with $\delta^{15}N = 0‰$ (where $\delta^{15}N = [(^{15}N/^{14}N)/(^{15}N/^{14}N)_{atm} - 1] \times 10^3$), ~30% is in crustal rocks (Wlotzka, 1972), and so the "surface" has 5.3×10^{21} g N with $\delta^{15}N \approx +1‰$ (Tolstikhin and Marty, 1998). Divided into the mass of the upper mantle (the minimum degassed volume of the Earth), this is equivalent to 5 ppm N. Studies of MORBs indicate that the upper mantle contains ~0.16 ppm N, assuming nitrogen is incompatible during melting (Marty and Humbert, 1997). MORBs also have a value of $\delta^{15}N \approx -3.3‰$ (e.g., Marty and Zimmermann, 1999). Overall, it appears that the bulk Earth has ~5 ppm N, with $\delta^{15}N = -5‰$ to $-3‰$ (Javoy and Pineau, 1991; Marty and Humbert, 1997). However, it has been argued that there may be up to 40 ppm in the mantle, which requires that nitrogen behaves more compatibly during melting (Cartigny *et al.*, 2001). Values as low as $\delta^{15}N = -25‰$ have been found in diamonds, indicating that there is an isotopically light nitrogen component in the mantle; since subducting materials are expected to have $\delta^{15}N > 0‰$, these values represent a maximum $\delta^{15}N$ value for nitrogen initially trapped in the mantle (Cartigny *et al.*, 1998). Higher values in MORBs and diamonds may be either a mixture of light trapped nitrogen from the deeper mantle and subducted nitrogen, or represent mantle heterogeneities established during Earth formation.

The atmosphere of Mars has a high value of $\delta^{15}N = +620 \pm 160‰$ (Nier and McElroy, 1977) that may be due to fractionating losses (see Section 16.7), and $(4.7 \pm 1.2) \times 10^{17}$ g N, which is equivalent to 0.7 ppb when divided into the mass of the entire planet (Owen *et al.*, 1977). Therefore, Mars appears to have 10^{-4} times the nitrogen on the Earth. Mathew *et al.* (1998) reported evidence for a component in a martian meteorite with $\delta^{15}N < -22‰$, suggesting that, like the Earth, the solid planet may contain nitrogen that is isotopically lighter nitrogen than the atmosphere. The available atmospheric data for Venus is very imprecise, with $\delta^{15}N = 0 \pm 200‰$ (von Zahn *et al.*, 1983), and there is ~2 times more nitrogen (1.1×10^{22} g) than that on the surface of the Earth.

There has been some controversy over the solar $\delta^{15}N$ value. Nitrogen in lunar samples appears to have more than one component, and it has not been clear whether solar-wind implanted nitrogen is an isotopically light or heavy component (see Becker, 2000). While direct measurement of the solar wind by SOHO gave $\delta^{15}N = +400^{+500}_{-300}‰$ (Kallenbach *et al.*, 1998), measurements of the Jupiter atmosphere, where solar-nebula gases have been retained, found $\delta^{15}N = -370 \pm 80‰$ (Owen *et al.*, 2001), and a recent ion-microprobe study of lunar samples suggested that the solar wind has $\delta^{15}N < -240‰$ (Hashizume *et al.*, 2000). A wide range of $\delta^{15}N$ values has been reported for meteorites, and the causes of the variations are not understood. The several classes that have $\delta^{15}N<0$ are the CO and CV chondrites, with bulk concentrations of up to almost 200 ppm and $\delta^{15}N$ down to $-40‰$ (Kung and Clayton, 1978; Kerridge, 1985), and the E chondrites with $\delta^{15}N = -30‰$ to $-50‰$ and up to 800 ppm N (Kung and Clayton, 1978; Grady *et al.*, 1986). In contrast, CM and CR chondrites are nitrogen rich with $\delta^{15}N > 0‰$, and CI chondrites have values up to 1,900 ppm and $\delta^{15}N = +50‰$ (Kung and Clayton, 1978; Kerridge, 1985). In sum, several meteorite classes and possibly solar gases can provide the light nitrogen found within the Earth. The source of nitrogen in the planetary atmospheres had isotopically lighter nitrogen than that presently observed if there were subsequent fractionating losses (see Section 16.4).

In considering the origin of nitrogen and the noble gases, a key parameter is the $N/^{36}Ar$ ratio, which has been used to link nitrogen and the noble gases in MORBs. The ratio for the terrestrial atmosphere is 5×10^4, compared to a value for the upper mantle of 2×10^6 (Marty, 1995). The martian ratio is 1×10^4, and may have been lowered by nitrogen losses (Fox, 1993) and also modified by ^{36}Ar losses. The Venus $N/^{36}Ar$ ratio is 2×10^3. Meteorites are generally much more nitrogen rich, while the solar value is much lower, with $N/^{36}Ar = 40$.

16.2.9.2 Carbon

Carbon on all three planets has $\delta^{13}C$ within the meteorite range, but with relatively depleted abundances. Carbon at the terrestrial surface is largely divided between carbonates, with $\delta^{13}C = 0‰$ (where $\delta^{13}C = [(^{13}C/^{12}C)/(^{13}C/^{12}C)_{std} - 1] \times 10^3$), and organic deposits with $\delta^{13}C = -25‰$. The bulk inventory appears to have $\delta^{13}C \approx -5‰$ and a total budget of 1×10^{23} g C. Divided into the mass of the upper mantle, this abundance is equivalent to 100 ppm. An additional 100 – 300 ppm may be present in the upper mantle (Trull *et al.*, 1993). The amount of carbon in the core is not known, although it may be up to several weight percent (Wood, 1993; Halliday and Porcelli, 2001), which if added back into the mantle would raise the carbon content by a factor of 10. The upper-mantle value of $\delta^{13}C = -5‰$ (see Javoy *et al.*, 1986) is similar to the bulk inventory of the crust. Venus has about twice as much carbon (like nitrogen) at the surface than the Earth, equivalent to 26 ppm when divided into the bulk planet (von Zahn *et al.*, 1983); whether Venus is more rich in carbon therefore depends upon what volume of the mantle has degassed and how much remains in the mantle (Lécuyer *et al.*, 2000). The carbon-isotopic compositions of the martian mantle appears to be as low as $\delta^{13}C = -20‰$ to $-25‰$ based on SNC data (Jakosky and Jones, 1997; Goreva *et al.*, 2003); the atmospheric composition is less well defined. The amount of martian carbon is also not well known; while the atmosphere has only 7×10^{18} g C, equivalent to 0.01 ppm for the bulk planet (Owen *et al.*, 1977), a large fraction may be stored in the polar regolith.

The solar carbon-isotopic composition is unknown. Bulk carbonaceous chondrites have a range of values of $\delta^{13}C = 0‰$ to $-25‰$ (Kerridge, 1985), with most somewhat isotopically lighter than the Earth. Enstatite chondrites are largely within the range of $\delta^{13}C = 0‰$ to $-14‰$ (Grady *et al.*, 1986). While enstatite chondrites have <1% C, CI chondrites can have >4% (Kerridge, 1985). In sum, the Earth appears to be isotopically somewhat heavier than average meteorite compositions, but not beyond the full range of measured values. There is no shortage of carbon in most sources, although the amount on the Earth is uncertain without further constraints on the abundance in the core. It is worth considering that the Earth does not appear to have the composition of any single group or collection of meteorites, but that the bulk composition has been deduced from inter-element correlations of meteorite data (e.g., Allègre *et al.*, 2001). It has been argued that Earth-forming materials originally had greater abundances of moderately and highly volatile elements but suffered losses during accretion (Halliday and Porcelli, 2001). However, carbon may have been sequestered in the core before such losses, and based on a correlation between C/Sr and Rb/Sr in meteoritic materials may constitute 0.6–1.5% of the core (Halliday and Porcelli, 2001). Unfortunately, in attempting to construct the Earth from meteorite abundance systematics, volatile isotope compositions do not vary as regularly, and so it is not clear how to relate $\delta^{13}C$ values between the Earth and meteorites.

16.2.9.3 Hydrogen

The δD value of the oceans falls within the meteoritic range but not that of measured comets. Trapped lighter hydrogen may be in the terrestrial mantle. Venus and Mars may have had similar water concentrations as Earth, but suffered losses that generated high δD ratios. The terrestrial oceans have $\delta D = 0‰$ with an inventory of 1.1×10^{24} g H_2O, equivalent to 120 ppm H when divided into the mass of the upper mantle. The upper mantle value of $\delta D = -65‰$ to $-75‰$ may be due to subduction of crustal material that has undergone metamorphism (Margaritz and Taylor, 1976) and may have ~13 – 35 ppm H. However, even lighter compositions, down to $-125‰$, have been found in Hawaii and elsewhere (Deloule *et al.*, 1991; Hauri, 2002), suggesting that isotopically light, juvenile hydrogen remains in the mantle.

The Venus atmosphere has ~ 200 ppm H_2O (Hoffman *et al.*, 1980) and a D/H ratio of $(1.6 \pm 0.2) \times 10^{-2}$ that is ~10^2 times that of the Earth (Donahue *et al.*, 1982). It has been suggested that Venus originally had the same D/H value as the Earth, but has lost at least one terrestrial ocean volume of water by hydrodynamic escape, thereby generating an enrichment in deuterium (Donahue *et al.*, 1982). The ratio of water to carbon and nitrogen therefore may have been similar to that of the Earth.

Measurements of atmospheric water vapor on Mars have found D/H values ~5 times that of the Earth and have been fractionated due to Jeans escape of hydrogen to space (Owen *et al.*, 1988), and 2–5 times the terrestrial value in SNC meteorites (Watson *et al.*, 1994). Morphological data (Carr, 1986) and modeling of hydrogen atmospheric losses (Donahue, 1995) suggest that originally there may have been the equivalent of up to 500 m of water, or ~7×10^{22} g H_2O. The total mass of Mars is 0.11 times that of the Earth, and so both planets originally may have had similar bulk water concentrations.

Hydrogen was the most abundant element in the solar nebula. The D/H ratio of the solar nebula has not been preserved in the Sun due to deuterium burning to ^3He, but has been deduced from the solar $^3He/^4He$ and He/H ratios to be $\delta D = -880‰$ (Geiss and Gloeckler, 1998). A large range of values has been measured in meteorites due to fractionation between chemical species, with bulk values largely between $-200‰$ and $+500‰$, and concentrations of 60–10^4 ppm H (Kerridge, 1985). Comets have long been regarded as a likely source of terrestrial water; however, recent measurements of three comets have found D/H ratios about twice that of the Earth (Balsiger *et al.*, 1995; Bockelée-Morvan *et al.*, 1998; Meier *et al.*, 1998).

16.3 ACQUISITION OF NOBLE GASES AND VOLATILES

A variety of different mechanisms have been proposed for the acquisition of noble gases. Not all of these are mutually exclusive.

16.3.1 Solar-wind Implantation

Solar noble gases are spread throughout the solar system in the solar wind, and are typically implanted by low-energy solar-wind irradiation in solar-like elemental abundance proportions in lunar and meteoritic materials. Typically, implantation extends a few nanometers into irradiated materials, so that the amount of noble gases that can be accumulated is correlated with surface area and is most efficient for dust. An available analogue is the fine material found in the lunar regolith (see, e.g., Eberhardt *et al.*, 1972). Accretion of planetesimals containing $\sim$25–40 wt.% of such material could account for the absolute noble-gas abundances measured in the Venus atmosphere. The very low relative abundances of solar major volatiles require another source for these species. However, the presence of substantial dust in the nebular disk prior to aggregation also greatly dampens penetration of solar wind out to much of the planet-forming region, while clearance of this dust results in larger targets. Moreover, with prolonged exposure, target materials can become saturated, and losses of helium and neon can occur from irradiated grains preferentially by diffusion during subsequent heating or even at low temperatures (Frick *et al.*, 1988). Sasaki (1991) argued that off-disk penetration of an early and intense solar-wind flux into a post-nebular environment rich in fine collisional dust could have generated an ancient reservoir of abundant irradiated dust.

Models of a solar-wind source for noble gases on the terrestrial planets have been proposed in various contexts by Wetherill (1981), Donahue *et al.* (1981), and McElroy and Prather (1981). Assuming that sufficient abundances of noble gases were accumulated in solid materials, it would be expected that due to gravitational scattering gas-bearing materials would be dispersed throughout the inner solar system and supply Venus and Earth with similar amounts of volatiles. In this case, the present differences in noble-gas abundances may be due to subsequent loss processes (see Section 16.4) that are necessary to generate the presently observed elemental and isotopic fractionations. The principal problem with these hypotheses is that xenon in the atmospheres of the early Earth is not solar, but U-Xe, while that of Mars was likely SW-Xe.

Such a source has also been considered for providing the noble gases presently found within the Earth's mantle. Podosek *et al.* (2000) argued that the present concentrations of neon estimated for a deep-mantle gas rich reservoir could have been derived from irradiated, kilometer-sized planetesimals, assuming that sufficient turnover of the surfaces occurs so that the process is not limited by grain-saturation effects and that irradiation fluxes were much higher in the past. This process would not have been limited by self-shielding by solid material across the accretionary disk due to removal of dust into larger bodies, and requires that a substantial fraction of the present mass of the Earth remained as small, dispersed planetesimals until after nebula gas had dispersed. Also, the gases must be retained in growing planetesimals and ultimately into the growing Earth, without being lost due to impacts or melting. This model is dependent on the chronology of accretion and gas dispersal and the early solar-wind flux, but it remains as a possible explanation for the origin of mantle noble gases.

16.3.2 Adsorption on Accreting Materials

Another mechanism of trapping volatiles from nebular gases onto solid materials is adsorption (see Ozima and Podosek, 2001). Laboratory studies have shown that noble gases exposed to some finely divided solid materials are adsorbed on the surfaces of individual grains. Adsorption is most efficient for various forms of carbon (e.g., Frick *et al.*, 1979; Wacker, 1989), but has also been experimentally demonstrated for other minerals (e.g., Yang and Anders, 1982). Adsorbed gases on these substrates generally display elemental patterns that are fractionated relative to ambient gas-phase abundances, in which heavier elements are enriched. These elemental fractionations are remarkably uniform, considering the wide range of experimental and natural conditions under which they are produced, and are similar to planetary atmosphere patterns. However, laboratory estimates of single-stage gas/solid partition coefficients are too low by orders of magnitude to account for planetary noble-gas abundances by adsorption on free-floating nebular dust grains at nebular pressures. Also, while occasional isotopic effects have been reported in natural samples (Phinney, 1972), these are not observed in equilibrium-adsorption experiments (Bernatowicz and Podosek, 1986), and adsorption in the nebula cannot produce the overabundance of gases that can allow subsequent fractionating losses.

16.3.3 Gravitational Capture

A number of noble-gas capture mechanisms have been suggested that involve gravitational attraction to increase local gas pressures in the surrounding nebular gases, followed by capture of nebular gases by growing protoplanetary bodies. These require the growth of protoplanets to appreciable masses (at least to about the Mercury to Mars size) prior to dissipation of the nebular gas phase, and so depend upon the relative timing of nebular dissipation versus planetary accretion. Current estimates for loss of circumstellar dust and gas are up to ~10 Ma (Podosek and Cassen, 1994). However, nebular lifetimes inferred from astronomical observation are based solely on evolution of their fine dust component, while nebular gas may still remain for longer. The abundances of gases that are captured by solid bodies, and the base pressures and temperatures of the resulting atmospheres, depend upon the sizes reached when the nebula gases are dissipated. The standard model of planetary accumulation estimates that the terrestrial planets reached full size by ~100 Ma or more (Wetherill, 1986, 1990a), although terrestrial-planet growth to ~80% of final masses may have occurred within ~20 Ma (Wetherill, 1986). Therefore, if a significant remnant of gas survived to this time, substantial gravitational capture would have occurred. Then subsequent evolutions must have involved losses that fractionated both elements and isotopes to generate the presently observed compositions.

16.3.3.1 Capture by planetary embryos

While the atmospheres captured by small bodies may not provide sufficient terrestrial rare-gas abundances in themselves, sufficient quantities of gases may have accumulated within the protoplanetary bodies by gas adsorption from these atmospheres on surface materials followed by burial below the surface during continuing accretion (Pepin, 1991). The process may have played an important role in creating internal volatile reservoirs for later outgassing of secondary atmospheres on the terrestrial planets, especially for the heavy noble gases. Interaction of the atmospheric gases with the surface is governed by the pressure at the base of the atmosphere, which depends on the thermal structure of the atmosphere. This, in turn, is a sensitive function of atmospheric opacity, which is difficult to estimate, although amplifications of surface pressure by about four to six orders of magnitude above that of the ambient nebula are likely (Pepin, 1991). Therefore, adsorption and occlusion of surface gases on and within growing planetary embryos might be a natural consequence of protoplanetary growth, in the presence of nebular gas, to bodies of up to about the Mercury size formed within <1 Myr (Wetherill, 1990b; Wetherill and Stewart, 1993). Another consideration is that the impact velocities of materials accreting to form these small bodies are generally too low to promote efficient degassing of the impactors themselves. Consequently, their volatiles also tend to be buried within the growing embryos (Tyburczy *et al.*, 1986).

If rare gases were acquired by these mechanisms, atmospheric formation would then occur by subsequent degassing and isotopic fractionation during loss to space. Rare gases trapped within the Earth and incorporated into the present deep mantle would exhibit solar isotopic compositions although conceivably accompanied by elemental fractionations. There is evidence that solar-like light rare-gas isotopic compositions exist, and the abundance pattern appears to be enriched in heavy noble gases, even though the pattern cannot be well constrained. Pepin (1991) calculated that a Mars-size terrestrial embryo could have developed the concentrations of neon that might presently be stored in a gas rich lower mantle, although not enough to account for initial deep-Earth abundances prior to accretionary losses (see Section 16.2.7).

16.3.3.2 Xenon fractionation in porous pre-planetary planetesimals

It has been suggested (Ozima and Nakazawa, 1980; Ozima and Igarashi, 1989; Zahnle *et al.*, 1990b; Ozima and Zahnle, 1993) that fractionation of nebular xenon to produce the terrestrial composition occurred by gravitational isotopic separation in large-porous planetesimals which have now vanished from the solar system. A consequence of this mechanism is that the atmospheric rare-gas characteristics are established in accreting materials, so that rare gases presently within the deep Earth are predicted to have the same characteristics. Other noble-gas characteristics must be generated by mixing with unfractionated components and some fractionating escape to space, but it is not clear that terrestrial isotopic compositions of all three noble gases can be generated from solar compositions for any distribution of planetesimal masses accreted by the Earth.

16.3.3.3 Gravitational capture and dissolution into molten planets

If the Earth reached sufficient size in the presence of the solar nebula, a massive atmosphere of solar gases would have been gravitationally captured and supported by the luminosity provided by the growing Earth, and the underlying planet would have melted by accretional energy and the blanketing effect of the atmosphere (Hayashi *et al.*, 1979). Under these conditions, gases from this atmosphere would have been sequestered within the molten Earth by dissolution at the surface and downward mixing (Mizuno *et al.*, 1980). This mechanism can provide solar rare gases into the deep Earth with relative elemental abundances that have been fractionated according to differences in solubilities (with depletion of heavy rare gases). Initial calculations found that at least an order of magnitude more neon than presently found in the deep mantle could be dissolved into the Earth unless the atmosphere began to escape when the Earth was only partially assembled (Mizuno *et al.*, 1980; Mizuno and Wetherill, 1984; Sasaki and Nakazawa, 1990; Sasaki, 1999). As noted above, initial concentrations may have actually been 10^2 times greater than the present abundances prior to losses at ~100 Myr after the start of the solar system, and Porcelli *et al.* (2001) and Woolum *et al.* (1999) considered the conditions required to dissolve sufficient neon to account for the initial deep-mantle inventory. If it is assumed that equilibration of the atmosphere with a thoroughly molten mantle was rapid, and uniform concentrations were maintained throughout the mantle by vigorous convection, then the initial abundances of gases retained in any mantle layer reflect surface rare-gas partial pressures when that layer solidified. The depth at which solidification occurs is determined by the surface temperature and the efficiency of convection in the molten mantle. Hence, the initial distributions of retained rare gases would be determined by the history of surface pressure and temperature during mantle cooling and solidification, i.e., the coupled cooling of Earth and atmosphere. For typical solubility coefficients (e.g., Lux, 1987), a total surface pressure ~100 atm under an atmosphere of solar composition is required to establish the initial deep-mantle neon concentration (Porcelli *et al.*, 2001), along with surface temperatures high enough to melt the deep mantle (~4,000 K). The dense atmosphere is a balance between the gravitational attraction of the nebula-derived gases and expansion due to the Earth's luminosity (energy released by accreting planetesimals and the cooling Earth). Therefore, the temperature and pressure at the base of the atmosphere evolved as the energy released by accretion declined with time once planet assembly approached completion, and as the nebular pressure declined during nebula dispersal. Woolum *et al.* (1999) demonstrated that the necessary conditions were met under a range of parameter values for both convective and radiative atmospheric structures, although many complexities remain to be resolved. It should be noted that not all situations facilitate the dissolution of atmospheric gases. At low nebular pressures and high initial luminosities, rapid magma solidification may occur without the incorporation of significant concentrations of atmospheric gases. However, in the presence of a massive atmosphere that promotes gas dissolution, the mantle cooling time is greatly extended (Tonks and Melosh, 1990). It is clear that any noble gases that were within the zone of mantle melting from earlier trapping by other mechanisms would have been incorporated into the mantle–atmosphere system and overwhelmed. Conversely, extensive melting of the Earth without sufficient surface pressures would lead to losses of these gases.

16.3.4 Accretion of Comets

Noble gases, as well as water, carbon, and nitrogen, could have been supplied to the inner planets by accretion of volatile-rich icy comets scattered inward from the outer solar system. Although noble-gas isotopic compositions in comets are unknown, it is expected that these gases, directly acquired from the solar nebula, have solar isotopic compositions. There is experimental evidence that the relative elemental abundances of heavier species (xenon, krypton, and argon) trapped in water ice at plausible comet-formation temperatures (~30 K) approximately reflect those of the ambient gas phase, and trapped noble-gas concentrations in water are substantial (Bar-Nun *et al.*, 1985; Owen *et al.*, 1991). At somewhat higher temperatures, a range of elemental fractionations is obtained with relative depletions in the lighter noble gases. In addition to physical adsorption on ice, thermodynamic modeling suggests that noble-gas incorporation in clathrates can be effective at low temperatures, and produces gases that are strongly enriched in the heavier species (Lunine and Stevenson, 1985).

The origins of volatile species on the terrestrial planets have been modeled as resulting from accretion, in variable planet-specific proportions, of rocky materials as well as three types of comets. These formed at different heliocentric distances and thus at different nebular temperatures, leading to distinctive elemental fractionation patterns in volatiles trapped in their ice from ambient nebular gases (e.g., Owen *et al.*, 1991, 1992; Owen and Bar-Nun, 1995a,b). The different volatile relative abundances on Venus, Earth, and Mars require several different components. Owen and Bar-Nun (1995a) suggested that comets from the Jupiter region, formed under sufficiently high temperatures to be essentially devoid of noble gases and depleted in nitrogen, have supplied much of the carbon and nitrogen on the planets. Comets formed at temperatures of ~50 K then supplied heavy noble gases and established the $N/^{36}Ar$ ratios. In this way, incorporation of a few percent or less by mass of icy cometary matter into the accreting terrestrial planets could have supplied heavy noble gases. Comets formed at lower temperatures can provide the unique Venus noble-gas abundance pattern, with solar proportions of argon, krypton, and xenon but a lower neon abundance. Final modification of the terrestrial budget is required to raise the ratio of water to carbon, and it was suggested that atmospheric species were preferentially lost over water during impact erosion. Nonfractionating losses from Mars are also required to deplete volatile abundances.

Accretion of icy comet matter has long been viewed as a plausible source for Earth's water. The D/H ratio in seawater, however, is a factor of ~2 lower than that in the few comets where D/H has been measured. A significant contribution of terrestrial water by comets would still be permitted if their high D/H ratio were appropriately lowered by accretion of additional, deuterium-poor materials. Suggested possibilities for low D/H carriers include rocky planetary accretional components (Laufer *et al.*, 1999; Dauphas *et al.*, 2000), or a high influx during the heavy bombardment epoch of interplanetary dust particles heavily loaded with implanted SW-H (Pavlov *et al.*, 1999). It has also been argued that comets from different orbital distances (and so forming in different temperatures) had different D/H ratios that could have formed the bulk of the terrestrial water (Delsemme, 1999; Morbidelli *et al.*, 2000). Overall, it appears that hydrogen-isotope compositions do not exclude cometary sources, but only further define source regions.

The principal difficulty encountered by these mixing models is their inability to account for differences in nonradiogenic noble-gas isotopic distributions between Earth and Mars, and between both of these and solar compositions. Experiments specifically designed to investigate isotopic fractionation in the gas-trapping process showed maximum heavy-isotope enrichments which are too small to explain the observed offsets of martian $^{36}Ar/^{38}Ar$ and of xenon on both Mars and Earth from solar ratios (Notesco *et al.*, 1999). Also, models that rely on supply of noble gases from material trapped in the outer parts of the solar system cannot explain the abundances of mantle noble gases, since these materials are expected to be provided as a "late veneer" when accreting bodies are supplied from a wider swathe of the nebula, and are more likely to devolatilize upon impact due to the size of the proto-Earth, rather than bury volatiles. However, comets may still have provided the major volatiles, as well as hydrogen to fuel hydrodynamic escape fractionation of noble gases.

16.3.5 Accretion of Carbonaceous Chondrites

There are various features in planetary volatiles that have suggested their derivation from accreting chondrites. As discussed above, the nonradiogenic-isotopic compositions and relative abundances of neon, argon, and krypton are similar to those found in carbonaceous chondrites (with some isotopic fractionation of krypton). However, the differences in Xe/Kr ratio and the xenon-isotope compositions appear to be irreconcilably different. In contrast, chondrites, or a mixture of chondrites classes, can be found to match nitrogen and carbon terrestrial-isotope compositions. The C/N ratios in CI and CM chondrites are similar to the planetary ratios, although CO and CV chondrites are relatively depleted in nitrogen (see compilation by Newsom, 1995). Since the ratio of carbon to noble gases in carbonaceous chondrites is much higher than in the terrestrial planets, it is possible that chondrites supplied the major elements, and such an explanation for major volatiles then requires an additional source for the noble gases. A late infall of material of chondritic composition has been hypothesized in various contexts to provide a volatile-rich oxidizing veneer (Wänke *et al.*, 1984) or source of mantle siderophiles (e.g., Chou, 1978).

A model for the total composition of the terrestrial planets involves the mixture between a highly reduced, refractory component and an oxidizing volatile-rich component with characteristics similar to CI chondrites (Dreibus and Wänke, 1987; Wänke and Dreibus, 1988). The proportions of each are determined by key elemental ratios, and initial inventories of water, noble gases, and other volatiles have been calculated that exceed present abundances, along with inferences regarding subsequent modification.

16.4 EARLY LOSSES OF NOBLE GASES TO SPACE

As discussed above, none of the acquisition mechanisms can explain the full range of volatile features observed within the terrestrial planets. However, there are various potential loss mechanisms that may have operated on the initially acquired volatile budgets, modifying isotopic compositions, elemental ratios, and absolute abundances to produce the presently observed compositions.

16.4.1 Losses During Accretion

As growth of a protoplanet proceeds with increasing accretional energy, shock-induced devolatilization of the accreting materials occurs and volatile species are transferred into the growing atmosphere, limiting the amounts of volatiles that can be buried into a planet. Data summarized by Ahrens *et al.* (1989) indicate that efficient loss of CO_2 and H_2O from accreting solids on impact occurs when the planetesimal mass approaches that of Mars. Above this size, degassing would also be driven by extensive melting due to deposition of accretional energy (Safronov, 1978), and further promoted by a radiative blanketing effect if a water-rich atmosphere has accumulated (see Abe and Matsui, 1986) or a dense atmosphere has been gravitationally captured. The process will limit the burial of volatiles by accreting materials.

Loss of atmospheric gases to space can occur by impact erosion, when a sufficient transfer of energy from accreting bodies to the atmosphere occurs and a substantial portion of the protoplanetary atmosphere reaches escape velocity (see Cameron, 1983; Ahrens, 1993). For smaller accreting bodies, the maximum fraction of the atmosphere that can be expelled is $\sim 6 \times 10^{-4}$ (Vickery and Melosh, 1990), equivalent to the total above the plane tangent to the planetary surface at the impact location. However, atmospheric loss may be much greater for very large impacts by bodies exceeding lunar size (Chen and Ahrens, 1997). These impact-driven losses are not expected to generate elemental or isotopic fractionations of volatiles, and contribute only to their overall depletion.

A Moon-forming collision of an approximately Mars-sized body with Earth (Hartmann and Davis, 1975; Cameron and Ward, 1976) would clearly result in catastrophic loss of volatiles from the pre-existing atmosphere and may have caused substantial loss of deep-Earth noble gases as well. Ahrens (1990, 1993) argued that virtually complete expulsion might have occurred by direct ejection from the impacted hemisphere and by shock-induced outward ramming of the antipodal planetary surface. However, losses may have been incomplete, and this event could have been followed by additional, isotopically fractionating losses driven by thermal processes (see Section 16.5).

16.4.2 Hydrodynamic Escape

Thermally driven escape of atmospheric constituents to space can generate substantial isotopic fractionations in the residual atmosphere, and models involving hydrodynamic escape have been successful at reproducing planetary features (Zahnle and Kasting, 1986; Hunten *et al.*, 1987, 1988, 1989; Sasaki and Nakazawa, 1988, 1990; Zahnle *et al.*, 1990a; Pepin, 1991, 1994, 1997, 2000). In this process, hydrogen-rich primordial atmospheres of partially or fully accreted planets are heated at high altitudes after the nebula has dissipated, and the resulting hydrogen-escape fluxes can exert upward-drag forces on heavier atmospheric constituents sufficient to lift them out of the atmosphere. Lighter species are entrained and are lost with the outflowing hydrogen more readily than are heavier ones, leading to mass fractionation of the residual atmosphere. The energy required can be provided by intense far-ultraviolet radiation from the young sun or energy deposited by a large impact event. Hydrogen-escape fluxes high enough to sweep out and fractionate atmospheric species as massive as xenon require energy inputs that are $\sim 10^2$–10^3 times greater than the amount presently supplied to planetary exospheres by solar extreme ultraviolet (EUV) radiation, but are not unreasonable for the early Sun.

Various studies have developed the theory of hydrodynamic escape (Zahnle and Kasting, 1986; Hunten *et al.*, 1987) and demonstrated that, given adequate supplies of hydrogen and energy, observed noble-gas features could be achieved. Hunten *et al.* (1987) and Sasaki and Nakazawa (1988) examined the derivation of terrestrial xenon from solar xenon, and Zahnle *et al.* (1990a) derived neon and argon compositions on Earth and Mars. Pepin (1991, 1994, 1997) examined how hydrodynamic escape could generate the full range of elemental- and isotopic-mass distributions now found in planetary atmospheres, and explored the range of suitable astrophysical and planetary conditions. The simple analytic approach (Hunten *et al.*, 1987; Pepin, 1991, 1997; Pepin and Porcelli, 2002) assumes the presence of an isothermal atmosphere consisting of hydrogen and minor amounts of heavier components. For a given hydrogen-escape flux, F_H, the upward drag is sufficient to lift all constituents with masses m_2 less than a critical mass m_c out of the atmosphere. The critical mass m_c is defined as (Hunten *et al.*, 1987)

$$m_c = m_H + \frac{kTF_H}{bgX_H} \tag{3}$$

where k is the Boltzmann constant, T the atmospheric temperature, g the gravitational acceleration, X_H the mole fraction of H (assumed to remain ~ 1 throughout the escape episode), m_H the mass of H, and b the diffusion parameter (the product of diffusion coefficient and total number density) of mass m_2 in the gas. Note that the critical mass is largely dependent on the escape flux of hydrogen, with the loss of heavier species requiring a greater hydrogen flux. Values of b for noble-gas diffusion in H_2 at various temperatures are known (Mason and Marrero, 1970; Zahnle and Kasting, 1986). These increase from xenon to neon by a factor of ~ 2 that varies only slightly with temperature, and are identical or nearly so for isotopes of a given element. The relative losses of minor constituents can be seen in the relationship between the ratio of the atmospheric inventory of hydrogen (N_H) with that of a minor constituent, N_2, and the escape flux F_2 of the minor constituent:

$$F_2 = F_H \frac{N_2}{N_H} \left[\frac{m_c - m_2}{m_c - m_H} \right] \tag{4}$$

when the mass m_2 is much smaller than the critical mass (because of a large F_H), the loss of this species is proportional to its inventory, and so there is no fractionation. Losses are also then maximized. For species with masses that approach m_c, maximum elemental and isotopic fractionation between them occurs. An important consequence is that when isotopic fractionation is occurring to a species, much lighter elements are being greatly depleted.

The energy required for escape of a particle with mass m_1 from its local gravitational field, at radial distance $r=r_S$ from a body of mass M and radius r_S, is Gm_1M/r erg per particle. If the global mean solar EUV input at heliocentric distance R and time t is $\phi(R, t)$ erg cm^{-2} s^{-1}, the energy-limited escape flux is

$$F_H(R,t) = \frac{\phi(R,t)\varepsilon}{Gm_HM/r} \text{particle cm}^{-2}\text{s}^{-1} \quad (5)$$

where ε is the fraction of incident EUV energy flux converted to thermal escape energy of hydrogen and $\phi(R,t)$ is the energy input at heliocentric distance R and time t. The energy required for escape of a hydrogen atom from the surface of a body of mass M and radius r is Gm_1M/r erg per particle. Equation (5) can be combined with Equation (3) to determine the history of the crossover mass at each planet. It can be seen here that once the history of the driving energy source for loss, i.e., the EUV flux, is known, and assumptions are made about the initial inventory and ongoing supply of the major volatile species, H_2, the losses of each species can be calculated.

Information on what the EUV flux might have been in the early solar system comes from astronomical observations of radiation from young solar-type stars at various stages of pre- and early main-sequence evolution. Since early solar EUV radiation could not have penetrated a full gaseous nebula to planetary distances, the applicable time dependence of stellar activity in the present model is that which follows dissipation of the dense accretion disks surrounding the classical T-Tauri stars so that solar EUV radiation could penetrate to planetary distances, at stellar ages of up to ~10 Ma (Simon *et al.*, 1985; Walter *et al.*, 1988; Strom *et al.*, 1988; Walter and Barry, 1991; Podosek and Cassen, 1994). Among present observational data, soft (~3–60 Å) X-ray fluxes are most likely to be representative of at least the short-wavelength coronal component ($\lambda < 700$ Å) of the EUV spectrum. Although there is considerable scatter from a single functional dependence of fluxes with age, most of the data between ~50 Ma and 200 Ma do indicate a decline by factors of ~5–10 from levels at ~20 Ma. Pepin (1991, 1994, 1997) assumed the EUV flux fell off exponentially, with a mean decay time of 90 Ma. Using a power-law function instead (Feigelson and Kriss, 1989) produces similar results (Pepin, 1989). The modeling assumes an energy supply, and therefore crossover masses, that decline with time through the hydrodynamic escape episode. This is plausible for both solar EUV radiation and energy initially deposited by a giant impact (Pepin, 1991, 1997). An alternative assumption of a constant-crossover mass together with a corresponding specific value for $N_H/(N_H)_0$ used in integrating Equation (4) also yields the same fractionations as those calculated for a declining crossover mass (Hunten *et al.*, 1987). However, if the crossover mass defined by the actual energy supply is more than a few hundred amu above the xenon mass region, the required value of $N_H/(N_H)_0$ is too small to be consistent with realistic initial and final hydrogen inventories on the planet.

Fractionating effects of the escape process can now be calculated analytically if specific assumptions are made about the time dependence of the hydrogen inventory; not only the history of the EUV flux, but also whether it is replenished as fast as it escapes (constant inventory model), or is lost without replenishment along with the minor atmospheric species (Rayleigh fractionation model). The inventories of the minor constituents have generally been assumed to be lost without replenishment during the escape episode. It should be emphasized that there are various adjustable modeling parameters that generate the final noble-gas patterns, although some are shared by all the planets, such as the history of the EUV supply (adjusted for distance), and possibly the initial supply of atmospheric constituents.

16.4.3 Sputtering

On Mars, it has been demonstrated that losses probably occur by sputtering, which results in fractionation of elements and isotopes in the residual atmosphere (Luhmann *et al.*, 1992; Zhang *et al.*, 1993). Oxygen atoms in the martian exosphere, ionized by solar EUV radiation and accelerated in the electric field of the solar wind, can impact species near the top of the atmosphere (the "exobase") with enough energy transfer to eject them from the planet's gravitational field. Loss rates of the dominant atmospheric constituent at the exobase (CO_2 on Mars) in this sputtering process depend upon the magnitudes of the EUV flux and the solar-wind velocity, and so estimates of how both electromagnetic and corpuscular radiation have evolved over solar history are needed in order to calculate sputtering losses in the past (Zhang *et al.*, 1993).

Escape fluxes of sputtered trace constituents such as noble gases and nitrogen in the atmosphere are proportional to their exobase mixing ratios with CO_2 (Jakosky *et al.*, 1994). These ratios are dependent upon the mass of each volatile species, so that sputtering from the exobase removes species from a fractionated "target" (the exobase), leaving the residual atmosphere enriched in heavier constituents. Depletions of lighter species are further augmented in the escape process itself since ejection efficiency from the exobase increases with decreasing atomic mass

(Jakosky *et al.*, 1994). Note that there are enormous elemental fractionations at the exobase, and sputtering losses of the two heaviest noble gases are consequently extremely small. For these species, isotopic fractionation by the process has negligible influence on the composition of total atmospheric inventories.

Sputtering losses are greatly attenuated by the presence of a planetary magnetic field, most importantly because it deflects the solar wind around the planet and shields atoms photoionized in the outer atmosphere from the solar-wind electric field that would otherwise accelerate some of them downward toward the exobase (Hutchins *et al.*, 1997). For this reason the process has not been important on Earth for as long as the core dynamo has existed, and it seems unlikely that the bulk composition of the massive Venus atmosphere, with low noble-gas mixing ratios, could have been substantially affected by sputtering loss with or without the protection of a magnetic field. Escape of sputtered species is also impeded by the higher gravity of these two planets. In contrast, a thin, magnetically unshielded, and more weakly bound martian atmosphere is particularly vulnerable to sputtering erosion. Efficient operation of this fractionating loss mechanism over time on Mars is thus linked both to atmospheric pressure history and to the timing of the disappearance of the martian paleomagnetic field (Hutchins *et al.*, 1997; Connerney *et al.*, 1999).

16.5 THE ORIGIN OF TERRESTRIAL NOBLE GASES

The most notable feature of terrestrial volatiles is the extensive fractionation of xenon isotopes, since producing this effect on such a heavy species is not generally seen elsewhere in the solar system and can only be produced under very limited circumstances. The only process that appears plausible at present is hydrodynamic escape, and so modeling the evolution of the atmosphere necessarily centers on this process. For the Earth, Hunten *et al.* (1987) and Pepin (1991) provide the first formulations, and assumed that hydrodynamic losses of primary atmospheric volatiles are driven entirely by intense EUV radiation from the young-evolving Sun. Hydrogen-outflow fluxes strong enough to enable xenon escape from Earth, and fractionation to its present isotopic composition, required atmospheric H_2 inventories equivalent to water abundances of up to a few weight percent of the planet's mass, and early solar EUV fluxes up to ~450 times present levels, which may be realistic if nebular dust and gas had dissipated to levels low enough for solar EUV radiation to penetrate the midplane to planetary distances within 100 Ma or so. However, energy sources other than solar EUV absorption may have powered atmospheric escape. Benz and Cameron (1990) suggested that hydrodynamic loss driven by thermal energy deposited in a giant Moon-forming impact could have generated the well-known fractionation signature in terrestrial xenon. Their model of the event calls for rapid invasion of the pre-existing primary atmosphere by extremely hot ($\sim 1.6 \times 10^4$ K) dissociated rock and iron vapor, emplacement of an orbiting rock–vapor disk with an inner edge at an altitude comparable to the atmospheric-scale height at this temperature, and longer-term heating of the top of the atmosphere by re-accretion of dissipating disk material. If a short post-impact escape episode did in fact occur, resulting in Rayleigh fractionation of whatever remnant of the primary atmosphere survived, direct and presumably non-fractionating ejection in the impact event required that atmospheric H_2 inventories would be reduced by at least an order of magnitude compared to models in which losses are driven only by solar EUV radiation (Pepin, 1997).

The more recent modeling (Pepin, 1991, 1997) involves evolutionary processing in two stages. In the first stage, substantial depletion of xenon from the primary atmosphere occurs, driven by deposition of atmospheric energy. This sets the xenon inventory and generates the extensive xenon-isotopic fractionation that is presently observed in the terrestrial atmosphere. However, the other, lighter noble gases are greatly over-depleted and overly fractionated isotopically at the end of the first stage. In the case where hydrodynamic escape is initially driven by a giant impact, primordial atmosphere U-Xe is fractionated to the presently observed composition while ~85% of the initial xenon is lost from the planet. Increasingly severe fractionations of the lighter noble gases from their primordial isotopic compositions are imposed during xenon escape. Residual krypton and argon are both isotopically heavy and strongly depleted relative to the present atmosphere; only 6% and 0.8%, respectively, of the initial ^{84}Kr and ^{36}Ar inventories (and 0.4% of the ^{20}Ne) survive the event.

A second stage is required to rectify the over-fractionations in the light noble gases and reach presently observed compositions. Therefore, species degassed from the second, interior reservoir are required to mix with the fractionated atmospheric residue to compensate for the over-fractionation of krypton, argon, and neon elemental and isotopic abundances in the first stage, and produce the presently observed characteristics of these gases. While no further modifications of argon and krypton abundances are required, neon is a special case in that ^{20}Ne/^{22}Ne ratio in the fractionated residual atmosphere is substantially higher than the present-day value, and later addition of outgassed solar neon elevates them still more. Here an episode of solar EUV energy deposition driving hydrodynamic escape of neon at some time after GI fractionation and outgassing is needed to generate the contemporary $^{20}Ne/^{22}Ne$ ratio. Now the EUV-driven H_2 escape flux must be only intense enough to lift neon, but not the heavier noble gases, out of the atmosphere. The waning EUV flux (Ayres, 1997) may still have been sufficiently high ($\sim$60 times present levels) to drive neon-only escape at solar ages up to $\sim$250 Ma, with the actual timing determined by the timescale for sufficient reduction of EUV dust-gas opacity in the nebular midplane (Prinn and Fegley, 1989).

An important aspect of the modeling results is that present atmospheric-xenon inventories are largely the fractionated relicts of the first stage of hydrodynamic escape processing of primary atmospheric xenon, while most of the krypton and lighter noble gases are products of planetary outgassing. Isotope mixing systematics impose strict upper limits on the allowed levels of "contamination" of residual primary xenon by later addition of isotopically unfractionated xenon degassed from the interior or supplied by subsequent accretion of noble-gas carrier materials, a constraint that applies with equal force to the model in which pre-fractionated xenon is delivered to planets by porous planetesimals. Estimates of these limits for Earth and Mars (Pepin, 1991, 1994) fall well below the amounts of xenon that ordinarily would be expected to accompany the outgassed krypton components. However, it is possible that xenon was preferentially outgassed well before the bulk of the lighter noble gases following the fractional degassing mechanisms of Zhang and Zindler (1989) and Tolstikhin and O'Nions (1994), and most of it was already present in the primary atmospheres prior to xenon-fractionating hydrodynamic loss. Later post-escape isotopic evolution of atmospheric xenon is largely restricted to degassing of radiogenic $^{129*}Xe$, and of $^{131-136*}Xe$ generated primarily by ^{244}Pu spontaneous fission, from the upper mantle and crust.

The simple analytical theory used by Pepin (1991, 1998) assumes a hydrogen-dominated atmosphere and energy limited escape, both arguably reasonable suppositions of primordial atmospheric conditions. However, such atmospheres might also have contained substantial amounts of a heavy constituent, say CO_2, and in this case the escape flux of H_2 would have been limited by its ability to diffuse through the CO_2. Zahnle *et al.* (1990a) and Ozima and Zahnle (1993) have shown that only neon and some argon would be hydrodynamically lost under these conditions. However, a more complete short-term atmospheric blowoff, including heavier species, could have been driven by a very large deposition of collisional energy.

This modeled history requires that the Earth acquired two isotopically primordial volatile reservoirs during accretion, one in the planet's interior, perhaps populated by a combination of nebular gases occluded in planetary embryo materials and dissolved in molten surface materials, and the other co-accreted as a primary atmosphere degassed from impacting planetesimals or gravitationally captured from ambient nebular gases during later planetary growth. These isotopically primordial reservoirs are characterized by isotope ratios measured in the solar wind, with the important exception of xenon, which has a U-Xe precursor. The relative abundances of the noble gases for each reservoir are obtained by back-calculation through the model. The required pattern for the primary atmosphere is a progressive enrichment in the heavier noble gases relative to the solar pattern, with a much higher Ar/Ne ratio, a moderate enrichment in Kr/Ar, and an Xe/Ar ratio is enriched to a somewhat greater extent (Pepin, 1997). It has been suggested that this reflects adsorption onto planetesimals (Pepin, 1991; Section 16.3.3.1). The other possible sources of gases with solar isotopic compositions include implanted solar wind, which can be enriched in heavier noble gases, although through losses of the lightest species and so requiring even greater initial implanted inventories, and dissolution of gravitationally captured noble gases, which favors the more-soluble light noble gases (Porcelli *et al.*, 2001).

An important issue that remains unresolved is the relationship between the presently observed mantle reservoirs and the noble gases that degassed during formation of the atmosphere. If the composition of neon within the mantle were found to be that of Ne-B, indicating that the source was material irradiated by solar gases, then a second source must be found to supply the distinctly nonsolar primordial xenon that supplied the atmosphere. It has been argued that noble gases that were supplied to the atmosphere are not represented by those now in the mantle, which have been derived from deeper within the Earth subsequently, and so it is possible that different sources supplied the deep mantle and atmosphere. However, some models for the different sources are incompatible. For example, noble gases derived by gravitational capture of a hot, dense atmosphere, followed by dissolution into the Earth, would overwhelm those gases derived earlier from other sources. However, late infall of volatiles would leave those in the deep Earth unaffected.

The major volatiles clearly have not been derived from the same, solar sources as the noble gases. The carbon and nitrogen characteristics are similar to those in various meteorite classes. Javoy (1998) developed a model for construction of the Earth largely by enstatite chondrites, which naturally provides isotopically light nitrogen, as well as hydrogen, into the mantle, and can match the carbon-isotope value of the bulk Earth.

The isotopically heavier surface reservoirs of nitrogen and hydrogen require some late addition from another source, perhaps CI chondrites. This model supplies an apparent overabundance of nitrogen to the surface, requiring sequestration of a considerable amount of nitrogen in the deep Earth. While there are many issues that arise in having enstatite chondrites as the dominant component of the bulk Earth, a smaller contribution may still have provided much of the major volatiles. Some limits to the total amounts of carbonaceous chondrites that can constitute the Earth are provided by moderately volatile element inventories, which are depleted relative to chondrites by an order of magnitude (see Newsom, 1995). Thus, chondrites cannot constitute >10% of the Earth. However, since the Earth is strongly depleted in major volatiles, chondrites may still provide the observed carbon and nitrogen abundances. Halliday and Porcelli (2001) used correlations of volatile and moderately volatile elements in various meteorite classes to extrapolate to the bulk Earth composition prior to substantial accretionary losses. It was suggested that higher early carbon abundances led to higher carbon in the core than predicted from present concentrations. In this case, the similarity between the C/N ratio on Earth and Venus would either be fortuitous or reflecting late additions to the planet. Using a model for the bulk composition of the Earth that involves the supply of volatiles by an oxidizing component with CI chondrite volatile abundances, Dreibus and Wänke (1987) suggested that up to 1.3% water was initially added to the Earth but was largely converted into H_2 during reduction of iron, but later infall of similar material comprising 0.44% of the Earth (to account for siderophile elements in the mantle) supplied the water observed at present. In this case, the H_2 produced may have contributed to the atmospheric inventory fuelling hydrodynamic escape.

It should be noted that deriving the bulk composition of the Earth in general by simple models of mixing between meteoritic components has generally always found difficulties, especially with elements of greater volatility. However, the possibility remains that solid materials that contributed to the Earth, with unique major-volatile characteristics, are no longer represented in the solar system.

Hydrodynamic escape required to modify the isotopic composition of the noble gases will also fractionate nitrogen and carbon in the atmosphere, as discussed in detail by Pepin (1991). The inventories in the remaining atmosphere will be isotopically heavier, and so chondrites supplying volatiles must have had lower $\delta^{15}N$ and $\delta^{13}C$ values. A higher initial N/C ratio is also likely to have been necessary, although the fractionation between carbon and nitrogen depends upon the speciation of the species being lost. CI and CM chondrites have nitrogen that is too isotopically heavy (Kerridge, 1985) and N/C ratios that may be too low, and the CO and CV chondrites also have low N/C ratios (Mazor *et al.*, 1970). In contrast, E chondrites appear to have the necessary nitrogen and carbon compositions.

16.6 THE ORIGIN OF NOBLE GASES ON VENUS

On Venus, the noble gases do not appear to have greatly evolved from solar characteristics. The heavy rare-gas elemental abundances are similar to solar values, although this similarity does not extend to neon, since the $^{20}Ne/^{36}Ar$ ratio is low. Nonetheless, the $^{20}Ne/^{22}Ne$ ratio is closer to the solar value. Venus is also gas rich, with the absolute abundance of argon on Venus exceeding that on Earth by a factor >70. The pronounced differences with terrestrial atmospheric noble gases are somewhat surprising, since planets as alike in size and heliocentric distance might be expected to have acquired compositionally similar primary atmospheres from similar sources and suffered similar evolutionary processes.

The similarities with solar noble gases suggest that those in the Venus atmosphere have been derived either from solar-wind implantation of accreting materials, gravitational capture of nebular gases, or volatile-rich comets. In considering these sources, it must be noted that not only will these mechanisms supply both Venus and the Earth, but also that a strong EUV flux inferred for generating noble gas losses from the Earth and modifying initially acquired inventories would also have affected Venus. The EUV flux driving neon escape in the Earth model discussed in the preceding section must also irradiate Venus at the same time. It turns out that the relatively weak solar EUV flux needed for loss of only neon from Earth (after losses from a giant impact-fractionated xenon isotopes) is still strong enough at the orbital position of Venus to drive outflow of krypton and lighter gases from this somewhat smaller and less dense planet. However, Venusian xenon is not lost and its nonradiogenic isotopic composition is predicted to be unaltered from its primordial composition. The relationship between the loss histories of Venus and the Earth has been used to construct a model for Venusian volatile evolution (Pepin, 1991, 1997). Results of EUV-driven loss of an isotopically solar and elementally near-solar primordial atmosphere from Venus are sensitive to only one of the few remaining adjustable modeling parameters once the evolution of the Earth has been calculated; the initial H_2 inventory. Fractionating loss of a primary atmosphere generates by itself approximate matches to observed compositions. Thus, in contrast to the case for Earth, the presence of a component that is subsequently outgassed from the solid planet and modifies the lighter atmospheric noble-gas isotopes is not required and would comprise only modest fractions of the large present-day Venusian atmospheric inventories even if bulk-planetary concentrations were comparable to those on Earth. Using the hydrodynamic escape model and calculating back from the present noble-gas isotope compositions and relative abundances, the starting elemental ratios characterizing the preloss Venus atmosphere that are obtained fall squarely within the range of estimates calculated for Earth's pre-impact primary atmosphere (Pepin, 1997). This includes an Xe/Kr ratio somewhat above the solar ratio and due to initial trapping of solar gases. The similarity between the initial terrestrial and Venusian atmospheres is a strong indication that noble gases on both planets could have evolved, clearly in quite different ways, from the same primordial distributions in the same types of primary planetary reservoirs.

The major volatiles on Venus appear to have been derived from a different source than the noble gases. The isotopic compositions of hydrogen and nitrogen were originally similar to those of the Earth, and nitrogen does not appear to have been derived from a solar precursor. Once the noble-gas characteristics of Venus were established, the major volatiles may have been added as a late veneer (Pepin, 1991), increasing the $N/^{36}Ar$ ratio and accounting for the similarities with the major volatiles on the Earth. As on the Earth, carbon, nitrogen, and water may have been added by comets or chondrites.

16.7 THE ORIGIN OF NOBLE GASES ON MARS

The atmosphere of Mars has several features that are distinct from that of the Earth and require a somewhat different planetary history. At likely nebular temperatures and pressures at its radial distance, Mars is too small to have condensed a dense early atmosphere from the nebula even in the limiting case of isothermal capture (Hunten, 1979; Pepin, 1991). Therefore, regardless of the plausibility of gravitational capture as a noble-gas source for primary atmospheres on Venus and Earth, some other way is needed to supply Mars. This may include solar-wind implantation or comets. An important feature is that, in contrast to Earth, martian xenon apparently did not evolve from a U-Xe progenitor, but rather from SW-Xe. This requires that accreting SW-Xe-rich materials that account for martian atmospheric xenon are from sources more localized in space or time and so have not dominated the terrestrial-atmospheric xenon budget. There are insufficient data to determineif the martian C/N ratio is like the terrestrial value, but it appears that the initial C/H_2O ratio may have been. Further constraints on the sources of the major volatiles are required.

Modeling of the martian atmosphere has been reviewed by Pepin (1991, 1994, 1997) and Jakosky and Jones (1997). The recent models of martian atmospheric evolution incorporate the fractionating effects of both hydrodynamic escape and the sputtering loss mechanism proposed by Luhmann *et al.* (1992) and Zhang *et al.* (1993), and explore the consequences for elemental and isotopic fractionation of the noble gas and nitrogen in the residual atmosphere (Jakosky *et al.*, 1994; Pepin, 1994). Martian atmospheric history is divided into early and late evolutionary periods, the first characterized by an episode of hydrodynamic escape, followed by high CO_2 pressures and a possible greenhouse, and the second by a transition to a low-pressure environment similar to present-day conditions on the planet, perhaps initiated by abrupt polar CO_2 condensation ~3.7 Ga (Gierasch and Toon, 1973; Haberle *et al.*, 1992, 1994). During this second period, gas loss and fractionation occurred by sputtering, which may have been the dominant mechanism governing atmospheric CO_2 evolution on Mars over the past ~3–4 Ga (Luhmann *et al.*, 1992; Zhang *et al.*, 1993). Another loss process, atmospheric erosion (Melosh and Vickery, 1989), increasingly appears important (Chyba, 1990, 1991; Zahnle, 1993), and may have depleted all atmophilic species prior to the end of heavy bombardment ~3.8–3.7 Ga.

The effects of hydrodynamic escape were discussed in detail by Pepin (1991), without detailed consideration of other loss mechanisms. Such a mechanism remains as the most plausible for the fractionation of the heavy noble gases, especially xenon isotopes. The EUV-powered hydrodynamic escape episode driving neon-only loss from Earth after Giant-driven escape, and loss of krypton and lighter gases from Venus, would have been intense enough on Mars to lift all the noble gases out of its primordial atmosphere. Early in this pre-3.7 Ga epoch, xenon isotopes were therefore assumed to have been hydrodynamically fractionated to their present composition, with corresponding depletions and fractionations of lighter primordial atmospheric constituents (Pepin, 1994).

The post-3.7 Ga evolution of martian CO_2, N_2, and the noble gases has been examined by Jakosky *et al.* (1994) and Pepin (1994). The late evolutionary stage on Mars was assumed to have been trigged by atmospheric CO_2 pressure collapse near 3.7 Ga. Sputtering loss of an atmospheric species relative to that of CO_2 is directly proportional to its exobase mixing ratio with CO_2, and so sputtering fractionation of the atmospheric noble-gas inventory is generally modest in a pre-3.7 Ga atmosphere dominated by CO_2 (Jakosky *et al.*, 1994). Pressure collapse of the major atmospheric constituent abruptly increased the mixing ratios of pre-existing argon, neon, and N_2 at the exobase, and allowed their rapid removal by sputtering. Current abundances and isotopic compositions are entirely determined by the action of sputtering and photochemical escape on gases supplied by outgassing during the late evolutionary epoch. Since light species from the first epoch are qualitatively lost, the final distributions of the light noble gases and nitrogen are therefore decoupled from whatever their elemental and isotopic inventories might have been in the pre-3.7 Ga atmosphere. Jakosky *et al.* (1994) showed that contemporary neon, argon, and N_2 abundances and isotope ratios, including the uniquely low martian $^{36}Ar/^{38}Ar$ ratio, could have been generated by sputtering losses from an atmosphere that was continuously replenished by degassing of meteoritic (CI) N_2 and isotopically solar neon and argon, with the time-dependent rates of degassing similar to estimates of volcanic fluxes over this period. Both krypton and xenon are too massive to be significantly affected by sputtering loss and fractionation during the late evolutionary stage. The present atmospheric krypton inventory derives almost completely from solar-like krypton degassed during this period, which overwhelms any krypton fractionated earlier, while only the xenon isotopes and $\delta^{13}C$ survive as isotopic tracers of atmospheric history prior to its transition to low pressure (Pepin, 1994). The assumption that early hydrodynamic escape fractionated the nonradiogenic xenon isotopes to at least approximately their present composition severely limits subsequent additions of unfractionated xenon to the atmospheric inventory by outgassing (consistent with the low degree of planetary degassing; see Section 16.2.6) or late-stage veneer accretion.

Hutchins and Jakosky (1996) revisited the late evolution sputtering–degassing models to investigate in more detail the parameters controlling the evolution of neon and argon abundances and isotopes, in particular those relating to martian degassing history. Assuming that the martian mantle was not more gas rich than the bulk Earth, it was concluded that the outgassing flux of argon and neon attributable to degassing during epochs of volcanic activity would have been about one to three orders of magnitude too low to appropriately balance sputtering losses, and thus another major source of juvenile volatiles must have contributed to the atmosphere over geologic time, perhaps via input from gas-enriched hydrothermal systems. Hutchins *et al.* (1997) explored to what extent a martian paleomagnetic dipole field would have throttled sputtering losses by deflecting the solar wind around the upper atmosphere, and calculated the conditions, as functions of the time when paleomagnetic suppression of the sputtering mechanism ended, under which the combination of sputtering and degassing would still have generated present-day argon and neon distributions. The discovery by Mars Global Surveyor of large-scale remnant magnetic lineations in the old martian southern highlands (Connerney *et al.*, 1999) confirmed that an active dynamo existed, but its history is still unknown.

In the first evolutionary epoch, note that the extent of xenon fractionation from primordial to present composition is similar on both Earth and Mars despite the much smaller mass of Mars, the apparent differences (U-Xe versus SW-Xe) in their precursor xenon, the much greater overall depletion of martian noble gases, and the possibility that escape episodes were powered by distinctly different energy sources (EUV radiation on Mars versus giant impact on Earth). It is not clear if this is just coincidence, or the expression of some more-fundamental fractionating process that left similar signatures on all three of the terrestrial-planet atmospheres. In the second evolutionary epoch, the CO_2 pressure and isotopic history was dictated by the interplay of estimated losses to impact erosion, sputtering, and carbonate precipitation, additions by outgassing and carbonate recycling, and perhaps also by feedback stabilization under greenhouse conditions. In a subsequent model of the early martian atmosphere, Carr (1999) examined the influences of these same mechanisms in controlling CO_2 pressure history, and was led to similar results and conclusions. It should be stressed, however, that since almost nothing is actually known about the values of the parameters governing these various processes, models of this epoch are no more than qualitative illustrations of how they might have driven early atmospheric behavior.

16.8 CONCLUSIONS

Considerable progress has been made in the long-standing problem of understanding the sources of volatiles on and within the terrestrial planets, and the processes that modified their initial inventories down planet-specific evolutionary tracks to the amazingly divergent compositional states observed today on Earth, Mars, and Venus. However, a full volatile history of the inner solar system remains to be formulated. While various mechanisms can be evoked to explain particular features, the interconnections between capture and modifying losses require further definition, and some features remain enigmatic. Some avenues of further research include:

Acquisition of noble gases by planetary interiors. Further constraints are required on the characteristics of noble gases trapped within the Earth. The isotopic compositions of the heavy noble gases, the $^{20}Ne/^{22}Ne$ ratio, and the concentrations of deep Earth reservoirs are needed to further evaluate capture mechanisms. The two viable mechanisms for burying solar gases into the Earth, gravitational capture of nebular gases and solar-wind implantation of small accreting materials, require further development to formulate further tests and implications. The capture of a primordial atmosphere by gravitational attraction of nebular gas is inescapable if the growing planet reaches sufficient mass in the presence of the nebula, and so the first criterion for this mechanism is firmly establishing if such a nebular history occurred. How much gas was then trapped in the Earth is a more complex issue requiring a substantial modeling effort. Parameters that need further consideration include the structure of the atmosphere, how long will the underlying mantle remains molten and to what depth, and how the atmosphere is affected by continuing accretion. Whether or not this supplies deep-mantle noble gases, a shallower reservoir may be more readily created which supplied noble gases to the atmosphere and may no longer be represented in the mantle. Under conditions in which melting of the underlying planet is not achieved, adsorption of noble gases, enhanced by the increased pressure of the gravitationally focused nebular gases, may be an important source of noble gases that may become buried during accretion.

The burial of material that contains solar noble gases implanted by radiation is another option for the source of mantle noble gases. This requires the opposite conditions of gravitational capture, clearance of the solar nebula prior to substantial aggregation of solids to allow penetration of solar wind to where the terrestrial planets accumulate. The solar fluxes and size of target materials must provide sufficient accumulation of noble gases so that after subsequent losses during accretion as well as during escape that caused fractionation of xenon, there are still sufficient noble gases remaining to account for the present atmospheric inventories. The choice between these mechanisms likely will be decided when there are greater constraints on the history of the solar nebula and the growth of protoplanetary material.

Atmosphere origin and evolution. Hydrodynamic escape models are capable of replicating details of contemporary isotopic distributions. However, the model is highly parametrized and intrinsically multistage, requiring both escape fractionation and subsequent mixing with species degassed from planetary interiors. It seems clear that some degree of hydrodynamic loss and fractionation of planetary atmospheres would have been inevitable if the required conditions for energy source, hydrogen supply, and, in the case of solar EUV-driven escape, midplane transparency to solar radiation were even partially met. However, which species are lost and the extent of fractionations generated are dependent upon various parameters that require independent substantiation. This includes the composition of the dense atmosphere and the sources and history of the energy fluxes driving hydrodynamic escape.

Mars. The combination of Viking *in situ* measurements and SNC meteorite data has provided a much more quantitative view of the present state and possible history of martian volatiles. Further progress requires greater precision for the atmospheric composition, more data on possible near-surface and mantle reservoirs, and further constraints on the conditions that affect the volatile evolution of all the terrestrial planets.

Venus. Venus is characterized only by the immensely valuable but still incomplete and relatively imprecise reconnaissance data from the Pioneer Venus and Venera spacecraft missions of the late 1970s. Additional *in situ* measurements, at precisions within the capabilities of current spacecraft instrumentation, are now necessary to refine atmospheric evolution models. Unfortunately, the possibilities of documenting the volatile inventories of the interior of the planet are more remote. A significant question that must be addressed is whether nonradiogenic xenon on Venus is compositionally closer to SW-Xe (as seen on Mars) or to the U-Xe that is seen on the Earth and so is expected to have been present within the inner solar system. Also, the extent of xenon fractionation will be an important parameter for hydrodynamic escape models; if intense solar EUV radiation drove hydrodynamic escape on the Earth, it would also impact Venus, while losses from the Earth driven by a giant impact would not be recorded there.

Distributions in the solar system. More data on volatiles throughout the solar system are clearly required to confidently describe the volatile acquisition history of the terrestrial planets in the proper context. There are several unknown values for the solar composition, including the nitrogen- and carbon-isotope compositions. The compositions of comets from different orbital distances are needed to assess the extent of radial transport of volatiles late in accretion history. In addition, the causes of carbon- and nitrogen-isotope variations in chondrites must be better understood. While it is clear that the Earth cannot be constructed simply by mixing of different meteorite classes, it is not yet possible to unambiguously extrapolate to the volatile compositions of protoplanetary materials.

The origin of the U-Xe that is now found in the terrestrial atmosphere, and presumably was in the nebular gases surrounding the accreting Earth, is still not adequately explained. An important parameter is the composition of xenon in the atmosphere of Jupiter, which contains gases that were captured directly from the ancient nebula and may be different from the solar wind, where the possibility of isotopic fractionation in processes transporting and releasing bulk solar xenon to and from the corona cannot be completely disregarded.

Coupled histories of atmospheric and interior planetary volatiles. Highly detailed models of noble-gas sources and evolution for the atmosphere and interior of the Earth have been developed separately and almost independently. However, the origin and history of atmospheric noble gases are not independent of the sources, distributions, and transport histories of noble gases within a planet—these two volatile systems must clearly be linked in nature through their primordial inventories and the processes of degassing and subduction. Indeed, many models of the terrestrial atmosphere require degassing of some portion of the planetary interior, and mantle models include degassing to the atmosphere. The degassing of Venus and Mars is much more under-constrained both by data and modeling. The question is whether the requirements implicit in each of these models for the dynamical and compositional history of the other reservoir are compatible, and, if not, what the inconsistencies are and how might they be addressed.

Timing of solar system events. There is an assortment of chronological information for various nebular and planet-forming processes, but further theoretical work is required to formulate a full history of planetary volatiles. The relative timing of nebular dispersal and planet growth is clearly a key parameter for controlling the acquisition of solar gases. Radiogenic xenon isotopes on Earth and Mars indicate that volatiles were lost for ~50 Ma or more. This contrasts with recent $^{182}Hf–^{182}W$ data that suggest that the average time of separation of planetary cores (which likely occurred concomitantly with accretion), as well as Moon formation, occurred <30 Ma (Kleine *et al.*, 2002; Yin *et al.*, 2002). These different timescales remain to be reconciled.

ACKNOWLEDGMENTS

The authors thank David Hilton and Richard Becker for much-appreciated comments, and Ralph Keeling for editorial handling.

REFERENCES

Abe Y. and Matsui T. (1986) Early evolution of the Earth: accretion, atmosphere formation, and thermal history. *J. Geophys. Res.* **91**, E291–E302.

Ahrens T. J. (1990) Earth accretion. In *Origin of the Earth* (eds. H. E. Newsom and J. H. Jones). Oxford University Press, New York, pp. 211–227.

Ahrens T. J. (1993) Impact erosion of terrestrial planetary atmospheres. *Ann. Rev. Earth Planet. Sci.* **21**, 525–555.

Ahrens T. J., O'Keefe J. D., and Lange M. A. (1989) Formation of atmospheres during accretion of the terrestrial planets. In *Origin and Evolution of Planetary and Satellite Atmospheres* (eds. S. K. Atreya, J. B. Pollack, and M. S. Matthews). University of Arizona Press, Tucson, pp. 328–385.

Allègre C. J., Manhès G., and Lewin E. (2001) Chemical composition of the Earth and the volatility control on planetary genetics. *Earth Planet. Sci. Lett.* **185**, 49–69.

Ayres T. R. (1997) Evolution of the solar ionizing flux. *J. Geophys. Res.* **102**, 1641–1651.

Ballentine C. J. and Burnard P. G. (2002) Production, release, and transport of noble gases in the continental crust. *Rev. Mineral. Geochem.* **47**, 481–538.

Ballentine C. J., Porcelli D., and Wieler R. (2001) Technical comment on Trieloff *et al.* (2000). *Science* **291**, 2269a.

Balsiger H., Altwegg K., and Geiss J. (1995) D/H and $^{18}O/^{16}O$ ratio in the hydronium ion and in neutral water from *in situ* ion measurements in Comet Halley. *J. Geophys. Res.* **100**, 5827–5834.

Bar-Nun A., Herman G., Laufer D., and Rappaport M. L. (1985) Trapping and release of gases by water ice and implications for icy bodies. *Icarus* **63**, 317–332.

Basford J. R., Dragon J. C., Pepin R. O., Coscio M. R., Jr., and Murthy V. R. (1973) Krypton and xenon in lunar fines. *Proc. 4th Lunar Sci. Conf.* 1915–1955.

Becker R. H. (2000) Nitrogen on the Moon. *Science* **290**, 110–111.

Begemann F., Weber H. W., and Hintenberger H. (1976) On the primordial abundance of argon-40. *Astrophys. J.* **203**, L155–L157.

Benkert J.-P., Baur H., Signer P., and Wieler R. (1993) He, Ne, and Ar from solar wind and solar energetic particles in lunar ilmenites and pyroxenes. *J. Geophys. Res.* **98**, 13147–13162.

Benz W. and Cameron A. G. W. (1990) Terrestrial effects of the giant impact. In *Origin of the Earth* (eds. H. E. Newsom and J. H. Jones). Oxford University Press, New York, pp. 61–67.

Bernatowicz T. J. and Podosek F. A. (1986) Adsorption and isotopic fractionation of Xe. *Geochim. Cosmochim. Acta* **50**, 1503–1507.

Bernatowicz T. J., Podosek F. A., Honda M., and Kramer F. E. (1984) The atmospheric inventory of xenon and noble gases in shales: the plastic bag experiment. *J. Geophys. Res.* **89**, 4597–4611.

Bernatowicz T. J., Kennedy B. M., and Podosek F. A. (1985) Xe in glacial ice and the atmospheric inventory of noble gases. *Geochim. Cosmochim. Acta* **49**, 2561–2564.

Black D. C. (1972) On the origins of trapped helium, neon, and argon isotopic variations in meteorites: II. Carbonaceous meteorites. *Geochim. Cosmochim. Acta* **36**, 377–394.

Bockelée-Morvan D., Gautier D., Lis D. C., Young K., Keene J., Phillips T., Owen T., Crovisier J., Goldsmith P. F., Bergin E. A., Despois D., and Wootten A. (1998) Deuterated water in Comet C/1996 B2 (Hyakutake) and its implications for the origin of comets. *Icarus* **133**, 147–162.

Bogard D. D. (1997) A reappraisal of the Martian $^{36}Ar/^{38}Ar$ ratio. *J. Geophys. Res.* **102**, 1653–1661.

Brazzle R. H., Pravdivtseva O. V., Meshik A. P., and Hohenberg C. M. (1999) Verification and interpretation of the I–Xe chronometer. *Geochim. Cosmochim. Acta* **63**, 739–760.

Burnard P. G., Graham D., and Turner G. (1997) Vesicle specific noble gas analyses of popping rock: implications for primordial noble gases in Earth. *Science* **276**, 568–571.

Busemann H., Baur H., and Wieler R. (2000) Primordial noble gases in "phase Q" in carbonaceous and ordinary chondrites studied by closed-system stepped etching. *Meteorit. Planet. Sci.* **35**, 949–973.

Caffee M. W., Hudson G. U., Velsko C., Huss G. R., Alexander E. C., Jr., and Chivas A. R. (1999) Primordial noble cases from Earth's mantle: identification of a primitive volatile component. *Science* **285**, 2115–2118.

Cameron A. G. W. (1983) Origin of the atmospheres of the terrestrial planets. *Icarus* **56**, 195–201.

Cameron A. G. W. and Ward W. R. (1976) The origin of the Moon. *Lunar Sci.* **VII**, 120–122.

Carr M. H. (1986) Mars: a water-rich planet? *Icarus* **68**, 187–216.

Carr M. H. (1999) Retention of an atmosphere on early Mars. *J. Geophys. Res.* **104**, 21897–21909.

Cartigny P., Harris J. W., Phillips D., Boyd S. R., and Javoy M. (1998) Subduction-related diamonds? The evidence for a mantle-derived origin from coupled $\delta^{13}C$–$\delta^{15}N$ determinations. *Chem. Geol.* **147**, 147–159.

Cartigny P., Harris J. W., and Javoy M. (2001) Diamond genesis, mantle fractionations, and mantle nitrogen content: a study of $\delta^{13}C$–N concentrations in diamonds. *Earth Planet. Sci Lett.* **185**, 85–98.

Chen G. Q. and Ahrens T. J. (1997) Erosion of terrestrial planet atmosphere by surface motion after a large impact. *Phys. Earth Planet. Int.* **100**, 21–26.

Chou C.-L. (1978) Fractionation of siderophile elements in the Earth's upper mantle. *Proc. 9th Lunar Planet. Sci. Conf.* 219–230.

Chyba C. F. (1990) Impact delivery and erosion of planetary oceans in the early inner solar system. *Nature* **343**, 129–133.

Chyba C. F. (1991) Terrestrial mantle siderophiles and the lunar impact record. *Icarus* **92**, 217–235.

Connerney J. E. P., Acuña M. H., Wasilewski P. J., Ness N. F., Rème H., Mazelle C., Vignes D., Lin R. P., Mitchell D. L., and Cloutier P. A. (1999) Magnetic lineations in the ancient crust of Mars. *Science* **284**, 794–798.

Dauphas N., Robert F., and Marty B. (2000) The late asteroidal and cometary bombardment of Earth as recorded in water deuterium to protium ratio. *Icarus* **148**, 508–512.

Deloule E., Albarède F., and Sheppard S. M. F. (1991) Hydrogen isotope heterogeneities in the mantle from ion probe analysis of amphiboles from ultramafic rocks. *Earth Planet. Sci. Lett.* **105**, 543–553.

Delsemme A. H. (1999) The deuterium enrichment observed in recent comets is consistent with the cometary origin of seawater. *Planet. Space Sci.* **47**, 125–131.

Déruelle B., Dreibus G., and Jambon A. (1992) Iodine abundances in oceanic basalts: implications for Earth dynamics. *Earth Planet. Sci. Lett.* **108**, 217–227.

Donahue T. M. (1995) Evolution of water reservoirs on Mars from D/H ratios in the atmosphere and crust. *Nature* **374**, 432–434.

Donahue T. M. and Russell C. T. (1997) The Venus atmosphere and ionosphere and their interaction with the solar wind: an overview. In *Venus II* (eds. S. W. Bougher, D. M. Hunten, and R. J. Phillips). University of Arizona Press, Tucson, pp. 3–31.

Donahue T. M., Hoffman J. H., and Hodges R. R., Jr. (1981) Krypton and xenon in the atmosphere of Venus. *Geophys. Res. Lett.* **8**, 513–516.

Donahue T. M., Hoffman J. H., Hodges R. R., Jr., and Watson A. J. (1982) Venus was wet: a measurement of the ratio of deuterium to hydrogen. *Science* **216**, 630–633.

Dreibus G. and Wänke H. (1987) Volatiles on Earth and Mars: a comparison. *Icarus* **71**, 225–240.

Eberhardt P., Geiss H., Graf H., Grögler N., Mendia M. D., Mörgeli M., Schwaller H., and Stettler A. (1972) Trapped solar wind gases in Apollo 12 lunar fines 12001 and Apollo 11 breccia 10046. *Proc. 3rd Lunar Sci. Conf.* **2**, 1821–1856.

Eugster O., Eberhardt P., and Geiss J. (1967) The isotopic composition of krypton in unequilibrated and gas rich chondrites. *Earth Planet. Sci. Lett.* **2**, 385–393.

Feigelson E. D. and Kriss G. A. (1989) Soft X-ray observations of pre-main-sequence stars in the Chamaeleon dark cloud. *Astrophys. J.* **338**, 262–276.

Fox J. L. (1993) The production and escape of nitrogen atoms on Mars. *J. Geophys. Res.* **98**, 3297–3310.

Frick U., Mack R., and Chang S. (1979) Noble gas trapping and fractionation during synthesis of carbonaceous matter. *Proc. 10th Lunar Planet. Sci. Conf.* 1961–1973.

Frick U., Becker R. H., and Pepin R. O. (1988) Solar wind record in the lunar regolith: nitrogen and noble gases. *Proc. 18th Lunar Planet. Sci. Conf.* 87–120.

Garrison D. H. and Bogard D. D. (1998) Isotopic composition of trapped and cosmogenic noble gases in several Martian meteorites. *Meteorit. Planet. Sci.* **33**, 721–736.

Geiss J. and Gloeckler G. (1998) Abundances of deuterium and helium in the protosolar cloud. *Space Sci. Rev.* **84**, 239–250.

Gierasch P. J. and Toon O. B. (1973) Atmospheric pressure variation and the climate of Mars. *J. Atmos. Sci.* **30**, 1502–1508.

Goreva J. S., Leshin L. A., and Guan Y. (2003) Ion microprobe measurements of carbon isotopes in Martian phosphates: insights into the Martian mantle. *Lunar Planet. Sci.* **XXXIV**, #1987.

Grady M. M., Wright I. P., Carr L. P., and Pillinger C. T. (1986) Compositional differences in enstatite chondrites based on carbon and nitrogen stable isotope measurements. *Geochim. Cosmochim. Acta* **50**, 2799–2813.

Graham D. W. (2002) Noble gases in MORB and OIB: observational constraints for the characterization of mantle source reservoirs. *Rev. Mineral. Geochem.* **47**, 247–318.

Haberle R. M., Tyler D., McKay C. P., and Davis W. L. (1992) Evolution of Mars' atmosphere: where has the CO_2 gone? *Bull. Am. Astron. Soc.* **24**, 1015–1016.

Haberle R. M., Tyler D., McKay C. P., and Davis W. L. (1994) A model for the evolution of CO_2 on Mars. *Icarus* **109**, 102–120.

Hagee B., Bernatowicz T. J., Podosek F. A., Johnson M. L., Burnett D. S., and Tatsumoto M. (1990) Actinide abundances in ordinary chondrites. *Geochim. Cosmochim. Acta* **54**, 2847–2858.

Halliday A. and Porcelli D. (2001) In search of lost planets—the paleocosmochemistry of the inner solar system. *Earth Planet. Sci. Lett.* **192**, 545–559.

Harrison D., Burnard P., and Turner G. (1999) Noble gas behaviour and composition in the mantle: constraints from the Iceland Plume. *Earth Planet. Sci. Lett.* **171**, 199–207.

Hartmann W. K. and Davis D. R. (1975) Satellite-sized planetesimals and lunar origin. *Icarus* **24**, 504–515.

Hashizume K., Chaussidon M., Marty B., and Robert F. (2000) Solar wind record on the Moon: deciphering presolar from planetary nitrogen. *Science* **290**, 1142–1145.

Hauri E. (2002) SIMS analysis of volatiles in silicate glasses: 2. Isotopes and abundances in Hawaiian melt inclusions. *Chem. Geol.* **183**, 115–141.

Hayashi C., Nakazawa K., and Mizuno H. (1979) Earth's melting due to the blanketing effect of the primordial dense atmosphere. *Earth Planet. Sci. Lett.* **43**, 22–28.

Helffrich G. R. and Wood B. J. (2001) The Earth's mantle. *Nature* **412**, 501–507.

Hoffman J. H., Hodges R. R., Donahue T. M., and McElroy M. B. (1980) Composition of the Venus lower atmosphere from the Pioneer Venus mass spectrometer. *J. Geophys. Res.* **85**, 7882–7890.

Hohenberg C. M., Podosek F. A., and Reynolds J. H. (1967) Xenon–iodine dating: sharp isochronism in chondrites. *Science* **156**, 233–236.

Honda M., McDougall I., Patterson D. B., Doulgeris A., and Clague D. A. (1991) Possible solar noble-gas component in Hawaiian basalts. *Nature* **349**, 149–151.

Honda M., McDougall I., Patterson D. B., Doulgeris A., and Clague D. A. (1993) Noble gases in submarine pillow basalt glasses from Loihi and Kilauea, Hawaii—a solar component in the Earth. *Geochim. Cosmochim. Acta* **57**, 859–874.

Hudson G. B., Kennedy B. M., Podosek F. A., and Hohenberg C. M. (1989) The early solar system abundance of ^{244}Pu as inferred from the St. Severin chondrite. *Proc. 19th Lunar Planet. Sci. Conf.* 547–557.

Hunten D. M. (1979) Capture of Phobos and Deimos by protoatmospheric drag. *Icarus* **37**, 113–123.

Hunten D. M., Pepin R. O., and Walker J. C. G. (1987) Mass fractionation in hydrodynamic escape. *Icarus* **69**, 532–549.

Hunten D. M., Pepin R. O., and Owen T. C. (1988) Planetary atmospheres. In *Meteorites and the Early Solar System* (eds. J. F. Kerridge and M. S. Matthews). University of Arizona Press, Tucson, pp. 565–591.

Hunten D. M., Donahue T. M., Walker J. C. G., and Kasting J. F. (1989) Escape of atmospheres and loss of water. In *Origin and Evolution of Planetary and Satellite Atmospheres* (eds. S. K. Atreya, J. B. Pollack, and M. S. Matthews). University of Arizona Press, Tucson, pp. 386–422.

Hutchins K. S. and Jakosky B. M. (1996) Evolution of Martian atmospheric argon: implications for sources of volatiles. *J. Geophys. Res.* **101**, 14933–14949.

Hutchins K. S., Jakosky B. M., and Luhmann J. G. (1997) Impact of a paleomagnetic field on sputtering loss of Martian atmospheric argon and neon. *J. Geophys. Res.* **102**, 9183–9189.

Jacobsen S. and Harper C. J. (1996) Accretion and early differentiation history of the Earth based on extinct radionuclides. In *Earth Processes: Reading the Isotopic Code*, Geophys. Monogr. 95 (eds. A. Basu and S. R. Hart). American Geophysical Union, pp. 47–74.

Jakosky B. M. and Jones J. H. (1997) The history of Martian volatiles. *Rev. Geophys.* **35**, 1–16.

Jakosky B. M., Pepin R. O., Johnson R. E., and Fox J. L. (1994) Mars atmospheric loss and isotopic fractionation by solar-wind-induced sputtering and photochemical escape. *Icarus* **111**, 271–288.

Javoy M. (1998) The birth of the Earth's atmosphere: the behaviour and fate of its major elements. *Chem. Geol.* **147**, 11–25.

Javoy M. and Pineau F. (1991) The volatile record of a popping rock from the mid-Atlantic ridge at 14°N: chemical and isotopic composition of gases trapped in the vesicles. *Earth Planet. Sci. Lett.* **107**, 598–611.

Javoy M., Pineau F., and Delorme H. (1986) Carbon and nitrogen isotopes in the mantle. *Chem. Geol.* **57**, 41–62.

Kallenbach R., Geiss J., Ipavich F. M., Gloeckler G., Bochsler P., Gliem F., Hefti S., Hilchenbach M., and Hovestadt D. (1998) Isotopic composition of solar wind nitrogen: first *in situ* determination with the CELIAS/MTOF spectrometer on board SOHO. *Astrophys. J.* **507**, L185–L188.

Kerridge J. F. (1985) Carbon, hydrogen, and nitrogen in carbonaceous chondrites: abundances and isotopic compositions in bulk samples. *Geochim. Cosmochim. Acta* **49**, 1707–1714.

Kleine T., Münker C., Mezger K., and Palme H. (2002) Rapid accretion and early core formation on asteroids and the terrestrial planets from Hf–W chronometry. *Nature* **418**, 952–955.

Krummenacher D., Merrihue C. M., Pepin R. O., and Reynolds J. H. (1962) Meteoritic krypton and barium versus the general isotopic anomalies in xenon. *Geochim. Cosmochim. Acta* **26**, 231–249.

Kung C.-C. and Clayton R. N. (1978) Nitrogen abundances and isotopic compositions in stony meteorites. *Earth Planet. Sci. Lett.* **38**, 21–435.

Kunz J. (1999) Is there solar argon in the Earth's mantle? *Nature* **399**, 649–650.

Kunz J., Staudacher T., and Allègre C. J. (1998) Plutonium-fission xenon found in Earth's mantle. *Science* **280**, 877–880.

Laufer D., Notesco G., and Bar-Nun A. (1999) From the interstellar medium to Earth's oceans via comets—an isotopic study of HDO/H_2O. *Icarus* **140**, 446–450.

Lécuyer C., Simon L., and Guyot F. (2000) Comparison of carbon, nitrogen, and water budgets on Venus and the Earth. *Earth Planet. Sci. Lett.* **181**, 33–40.

Luhmann J. G., Johnson R. E., and Zhang M. H. G. (1992) Evolutionary impact of sputtering of the Martian atmosphere by O^+ pickup ions. *Geophys. Res. Lett.* **19**, 2151–2154.

Lunine J. I. and Stevenson D. J. (1985) Thermodynamics of clathrate hydrate at low and high pressures with application to the outer solar system. *Astrophys. J. Suppl. Ser.* **58**, 493–531.

Lux G. (1987) The behavior of noble gases in silicate liquids: solution, diffusion, bubbles, and surface effects, with applications to natural samples. *Geochim. Cosmochim. Acta* **51**, 1549–1560.

Magaritz M. and Taylor H. P., Jr. (1976) Oxygen, hydrogen, and carbon isotope studies of the Franciscan formation, Coast Ranges, California. *Geochim. Cosmochim. Acta* **40**, 215–234.

Mahaffy P. R., Donahue T. M., Atreya S. K., Owen T. C., and Niemann H. B. (1998) Galileo probe measurements of D/H and $^3He/^4He$ in Jupiter's atmosphere. *Space Sci. Rev.* **84**, 251–263.

Marti K. (1967) Isotopic composition of trapped krypton and xenon in chondrites. *Earth Planet. Sci. Lett.* **3**, 243–248.

Marty B. (1995) Nitrogen content of the mantle inferred from N_2–Ar correlation in oceanic basalts. *Nature* **377**, 326–329.

Marty B. and Humbert F. (1997) Nitrogen and argon isotopes in oceanic basalts. *Earth Planet. Sci. Lett.* **152**, 101–112.

Marty B. and Marti K. (2002) Signatures of early differentiation on Mars. *Earth Planet. Sci. Lett.* **196**, 251–263.

Marty B. and Zimmermann L. (1999) Volatiles (He, C, N, Ar) in mid-ocean ridge basalts: assessment of shallow-level fractionation and characterization of source composition. *Geochim. Cosmochim. Acta* **63**, 3619–3633.

Mason E. A. and Marrero T. R. (1970) The diffusion of atoms and molecules. *Adv. At. Mol. Phys.* **6**, 155–232.

Mathew K. J. and Marti K. (2001) Early evolution of Martian volatiles: nitrogen and noble gas components in ALH84001 and Chassigny. *J. Geophys. Res.* **106**, 1401–1422.

Mathew K. J., Kim J. S., and Marti K. (1998) Martian atmospheric and indigenous components of xenon and nitrogen in the Shergotty, Nakhla, and Chassigny group meteorites. *Meteorit. Planet. Sci.* **33**, 655–664.

Matsuda J. and Matsubara K. (1989) Noble gases in silica and their implication for the terrestrial "missing" Xe. *Geophys. Res. Lett.* **16**, 81–84.

Mazor E., Heymann D., and Anders E. (1970) Noble gases in carbonaceous chondrites. *Geochim. Cosmochim. Acta* **34**, 781–824.

McElroy M. B. and Prather M. J. (1981) Noble gases in the terrestrial planets. *Nature* **293**, 535–539.

Meier R., Owen T. C., Matthews H. E., Jewitt D. C., Bockelée-Morvan D., Biver N., Crovisier J., and Gautier D. (1998) A determination of the HDO/H_2O ratio in Comet C/1995 O1 (Hale-Bopp). *Science* **279**, 842–844.

Melosh H. J. and Vickery A. M. (1989) Impact erosion of the primordial atmosphere of Mars. *Nature* **338**, 487–489.

Mizuno H. and Wetherill G. W. (1984) Grain abundance in the primordial atmosphere of the Earth. *Icarus* **59**, 74–86.

Mizuno H., Nakazawa K., and Hayashi C. (1980) Dissolution of the primordial rare gases into the molten Earth's material. *Earth Planet. Sci. Lett.* **50**, 202–210.

Morbidelli A., Chambers J., Lunine J. I., Petit J. M., Robert F., Valsecchi G. B., and Cyr K. E. (2000) Source regions and timescales for the delivery of water to the Earth. *Meteorit. Planet. Sci.* **35**, 1309–1320.

Newsom H. E. (1995) Composition of the solar system, planets, meteorites, and major terrestrial reservoirs. In *Global Earth Physics, A Handbook of Physical Constants* (ed. T. J. Ahrens). American Geophysical Union, Washington DC, pp. 159–189.

Niedermann S., Bach W., and Erzinger J. (1997) Noble gas evidence for a lower mantle component in MORBs from the southern East Pacific Rise: decoupling of helium and neon isotope systematics. *Geochim. Cosmochim. Acta* **61**, 2697–2715.

Nier A. O. and McElroy M. B. (1977) Composition and structure of Mars' upper atmosphere: results from the neutral mass spectrometers on Viking 1 and 2. *J. Geophys. Res.* **82**, 4341–4349.

Notesco G., Laufer D., Bar-Nun A., and Owen T. (1999) An experimental study of isotopic enrichments in Ar, Kr, and Xe when trapped in water ice. *Icarus* **142**, 298–300.

Ott U. (1988) Noble gases in SNC meteorites: Shergotty, Nakhla, Chassigny. *Geochim. Cosmochim. Acta* **52**, 1937–1948.

Owen T. and Bar-Nun A. (1995a) Comets, impacts, and atmospheres. *Icarus* **116**, 215–226.

Owen T. and Bar-Nun A. (1995b) Comets, impacts, and atmospheres: II. Isotopes and noble gases. In *Volatiles in the Earth and Solar System,* AIP Conf. Proc. 341 (ed. K. A. Farley). American Institute of Physics Press, New York, pp. 123–138.

Owen T., Biemann K., Rushneck D. R., Biller J. E., Howarth D. W., and Lafleur A. L. (1977) The composition of the atmosphere at the surface of Mars. *J. Geophys. Res.* **82**, 4635–4639.

Owen T., Maillard J. P., Debergh C., and Lutz B. (1988) Deuterium on Mars: the abundance of HDO and the value of D/H. *Science* **240**, 1767–1770.

Owen T., Bar-Nun A., and Kleinfeld I. (1991) Noble gases in terrestrial planets: evidence for cometary impacts? In *Coments in the Post-Halley Era* (eds. R. L. Newburn, Jr., M. Meigenaier, and J. Rahe). Kluwer, Dordrecht, The Netherlands: vol. 1, pp. 429–437.

Owen T., Bar-Nun A., and Kleinfeld I. (1992) Possible cometary origin of heavy noble gases in the atmospheres of Venus, Earth, and Mars. *Nature* **358**, 43–46.

Owen T., Mahaffy P. R., Niemann H. B., Atreya S., and Wong M. (2001) Proto-solar nitrogen. *Astrophys. J.* **553**, L77–L79.

Ozima M. and Igarashi G. (1989) Terrestrial noble gases: constraints and implications on atmospheric evolution. In *Origin and Evolution of Planetary and Satellite Atmospheres* (eds. S. K. Atreya, J. B. Pollack, and M. S. Matthews). University of Arizona Press, Tucson, pp. 306–327.

Ozima M. and Nakazawa K. (1980) Origin of rare gases in the Earth. *Nature* **284**, 313–316.

Ozima M. and Podosek F. A. (2001) *Noble Gas Geochemistry,* 2nd edn. Cambridge University Press, Cambridge, UK.

Ozima M. and Zahnle K. (1993) Mantle degassing and atmospheric evolution: noble gas view. *Geochem. J.* **27**, 185–200.

Palma R. L., Becker R. H., Pepin R. O., and Schlutter D. J. (2002) Irradiation records in regolith materials: II. Solar-wind and solar-energetic-particle components in helium, neon, and argon extracted from single lunar mineral grains and from the Kapoeta howardite by stepwise pulse-heating. *Geochim. Cosmochim. Acta* **66**, 2929–2958.

Pavlov A. A., Pavlov A. K., and Kasting J. F. (1999) Irradiated interplanetary dust particles as a possible solution for the deuterium/hydrogen paradox of Earth's oceans. *J. Geophys. Res.* **104**, 30725–30728.

Pepin R. O. (1989) On the relationship between early solar activity and the evolution of terrestrial planet atmospheres. In *The Formation and Evolution of Planetary Systems,* Space Tel Sci. Inst. Symp. Series #3 (eds. H. A. Weaver and L. Danly). Cambridge University Press, Cambridge, UK, pp. 55–74.

Pepin R. O. (1991) On the origin and early evolution of terrestrial planet atmospheres and meteoritic volatiles. *Icarus* **92**, 2–79.

Pepin R. O. (1994) Evolution of the martian atmosphere. *Icarus* **111**, 289–304.

Pepin R. O. (1997) Evolution of Earth's noble gases: consequences of assuming hydrodynamic loss driven by giant impact. *Icarus* **126**, 148–156.

Pepin R. O. (1998) Isotopic evidence for a solar argon component in the Earth's mantle. *Nature* **394**, 664–667.

Pepin R. O. (2000) On the isotopic composition of primordial xenon in terrestrial planet atmospheres. *Space Sci. Rev.* **92**, 371–395.

Pepin R. O. (2003) On noble gas processing in the solar accretion disk. *Space Sci. Rev.* **106**, 211–230.

Pepin R. O. and Phinney D. (1976) The formation interval of the Earth. *Lunar Sci.* **VII**, 682–684.

Pepin R. O. and Porcelli D. (2002) Origin of noble gases in the terrestrial planets. *Rev. Mineral. Geochem.* **47**, 191–246.

Pepin R. O., Becker R. H., and Rider P. E. (1995) Xenon and krypton isotopes in extraterrestrial regolith soils and in the solar wind. *Geochim. Cosmochim. Acta* **59**, 4997–5022.

Pepin R. O., Becker R. H., and Schlutter D. J. (1999) Irradiation records in regolith materials: I. Isotopic compositions of solar-wind neon and argon in single lunar mineral grains. *Geochim. Cosmochim. Acta* **63**, 2145–2162.

Phinney D. (1972) ^{36}Ar, Kr, and Xe in terrestrial materials. *Earth Planet. Sci. Lett.* **16**, 413–420.

Phinney D., Tennyson J., and Frick U. (1978) Xenon in CO_2 well gas revisited. *J. Geophys. Res.* **83**, 2313–2319.

Podosek F. A. and Cassen P. (1994) Theoretical, observational, and isotopic estimates of the lifetime of the solar nebula. *Meteoritics* **29**, 6–25.

Podosek F. A., Woolum D. S., Cassen P., and Nichols R. H. (2000) Solar gases in the Earth by solar wind irradiation? *10th Annual Goldschmidt Conf. Oxford.*

Porcelli D. and Ballentine C. J. (2002) Models for the distribution of terrestrial noble gases and the evolution of the atmosphere. *Rev. Mineral. Geochem.* **47**, 411–480.

Porcelli D. and Halliday A. N. (2001) The core as a possible source of mantle helium. *Earth Planet. Sci. Lett.* **192**, 45–56.

Porcelli D. and Pepin R. O. (2000) Rare gas constraints on early Earth history. In *Origin of the Earth and Moon* (eds. R. M. Canup and K. Righter). University of Arizona Press, Tucson, pp. 435–458.

Porcelli D. and Wasserburg G. J. (1995a) Mass transfer of xenon through a steady-state upper mantle. *Geochim. Cosmochim. Acta* **59**, 1991–2007.

Porcelli D. and Wasserburg G. J. (1995b) Mass transfer of helium, argon, and xenon through a steady-state upper mantle. *Geochim. Cosmochim. Acta* **59**, 4921–4937.

Porcelli D. R., Woolum D., and Cassen P. (2001) Deep Earth rare gases: initial inventories, capture from the solar nebula, and losses during Moon formation. *Earth Planet. Sci. Lett.* **193**, 237–251.

Prinn R. G. and Fegley B., Jr. (1989) Solar nebula chemistry: origin of planetary, satellite, and cometary volatiles. In *Origin and Evolution of Planetary and Satellite Atmospheres* (eds. S. K. Atreya, J. B. Pollack, and M. S. Matthews). University of Arizona Press, Tucson, pp. 435–458.

Safronov V. S. (1978) The heating of the Earth during its formation. *Icarus* **33**, 1–12.

Sarda P., Staudacher T., and Allègre C. J. (1988) Neon isotopes in submarine basalts. *Earth Planet. Sci. Lett.* **91**, 73–88.

Sasaki S. (1991) Off-disk penetration of ancient solar wind. *Icarus* **91**, 29–38.

Sasaki S. (1999) Presence of a primary solar-type atmosphere around the Earth: evidence of dissolved noble gas. *Planet. Space Sci.* **47**, 1423–1431.

Sasaki S. and Nakazawa K. (1988) Origin of isotopic fractionation of terrestrial Xe: hydrodynamic fractionation during escape of the primordial H_2–He atmosphere. *Earth Planet. Sci. Lett.* **89**, 323–334.

Sasaki S. and Nakazawa K. (1990) Did a primary solar-type atmosphere exist around the proto-Earth? *Icarus* **85**, 21–42.

Simon T., Herbig G., and Boesgaard A. M. (1985) The evolution of chromospheric activity and the spin-down of solar-type stars. *Astrophys. J.* **293**, 551–574.

Strom S. E., Strom K. M., and Edwards S. (1988) Energetic winds and circumstellar disks associated with low mass young stellar objects. In *Galactic and Extragalactic Star Formation* (ed. R. Pudritz). NATO Advanced Study Institute, Reidel, Dordrecht, The Netherlands, pp. 53–68.

Swindle T. D. (2002) Martian noble gases. *Rev. Mineral. Geochem.* **47**, 171–190.

Swindle T. D. and Jones J. H. (1997) The xenon isotopic composition of the primordial Martian atmosphere: contributions from solar and fission components. *J. Geophys. Res.* **102**, 1671–1678.

Swindle T. D., Caffee M. W., and Hohenberg C. M. (1986) Xenon and other noble gases in shergottites. *Geochim. Cosmochim. Acta* **50**, 1001–1015.

Tolstikhin I. N. and O'Nions R. K. (1994) The Earth's missing xenon: a combination of early degassing and of rare gas loss from the atmosphere. *Chem. Geol.* **115**, 1–6.

Tolstikhin I. N. and Marty B. (1998) The evolution of terrestrial volatiles: a view from helium, neon, argon, and nitrogen isotope modelling. *Chem. Geol.* **147**, 27–52.

Tonks W. B. and Melosh H. J. (1990) The physics of crystal settling and suspension in a turbulent magma ocean. In *Origin of the Earth* (eds. H. E. Newsom and J. H. Jones). Oxford University Press, , pp. 151–171.

Torgersen T. (1989) Terrestrial helium degassing fluxes and the atmospheric helium budget: implications with respect to the degassing processes of continental crust. *Chem. Geol. (Isot. Geosci. Sect.)* **79**, 1–14.

Trieloff M., Kunz J., Clague D. A., Harrison D., and Allègre C. J. (2000) The nature of pristine noble gases in mantle plumes. *Science* **288**, 1036–1038.

Trieloff M., Kunz J., Clague D. A., Harrison D., and Allègre C. J. (2001) Reply to comment on noble gases in mantle plumes. *Science* **291**, 2269a.

Trieloff M., Kunz J., and Allègre C. J. (2002) Noble gas systematics of the Réunion mantle plume source and the origin of primordial noble gases in Earth's mantle. *Earth Planet. Sci. Lett.* **200**, 297–313.

Trull T., Nadeau S., Pineau F., Polvé M., and Javoy M. (1993) C–He systematics in hotspot xenoliths—implications for mantle carbon contents and carbon recycling. *Earth Planet. Sci. Lett.* **118**, 43–64.

Tyburczy J. A., Frisch B., and Ahrens T. J. (1986) Shock-induced volatile loss from a carbonaceous chondrite: implications for planetary accretion. *Earth Planet. Sci. Lett.* **80**, 201–207.

Valbracht P. J., Staudacher T., Malahoff A., and Allègre C. J. (1997) Noble gas systematics of deep rift zone glasses from Loihi Seamount, Hawaii. *Earth Planet. Sci. Lett.* **150**, 399–411.

Vickery A. M. and Melosh H. J. (1990) Atmospheric erosion and impactor retention in large impacts with application to mass extinctions. In *Global Catastrophes in Earth History*, Geological Society of America Special Paper 247 (eds. V. L. Sharpton and P. D. Ward). Geological Society of America, Boulder, Co, pp. 289–300.

von Zahn U., Kumar S., Niemann H., and Prinn R. (1983) Composition of the Venus atmosphere. In *Venus* (eds. D. Hunten, L. Colin, T. Donahue, and V. Moroz). University of Arizona Press, Tucson, pp. 299–430.

Wacker J. F. (1989) Laboratory simulation of meteoritic noble gases: III. Sorption of neon, argon, krypton, and xenon on carbon: elemental fractionation. *Geochim. Cosmochim. Acta* **53**, 1421–1433.

Wacker J. F. and Anders E. (1984) Trapping of xenon in ice: implications for the origin of the Earth's noble gases. *Geochim. Cosmochim. Acta* **48**, 2373–2380.

Walter F. M. and Barry D. C. (1991) Pre- and main sequence evolution of solar activity. In *The Sun in Time* (eds. C. P. Sonett, M. S. Giampapa, and M. S. Matthews). University of Arizona Press, Tucson, pp. 633–657.

Walter F. M., Brown A., Mathieu R. D., Myers P. C., and Vrba F. J. (1988) X-ray sources in regions of star formation: III. Naked T Tauri stars associated with the Taurus-Auriga complex. *Astron. J.* **96**, 297–325.

Wänke H. and Dreibus G. (1988) Chemical composition and accretion history of terrestrial planets. *Phil. Trans. Roy. Soc. London* **A325**, 545–557.

Wänke H., Dreibus G., and Jagoutz E. (1984) Mantle chemistry and accretion history of the Earth. In *Archaean Geochemistry* (eds. A. Kröner, G. N. Hanson, and A. M. Goodwin). Springer, , pp. 1–24.

Watson L., Hutcheon I. D., Epstein S., and Stolper E. (1994) Water on Mars: clues from deuterium/hydrogen and water contents of hydrous phases in SNC meteorites. *Science* **265**, 86–90.

Wetherill G. W. (1975) Radiometric chronology of the early solar system. *Ann. Rev. Nuclear Sci.* **25**, 283–328.

Wetherill G. W. (1981) Solar wind origin of ^{36}Ar on Venus. *Icarus* **46**, 70–80.

Wetherill G. W. (1986) Accumulation of the terrestrial planets and implications concerning lunar origin. In *Origin of the Moon* (eds. W. K. Hartmann, R. J. Phillips, and G. J. Taylor). Lunar and Planetary Institute, Houston, pp. 519–550.

Wetherill G. W. (1990a) Formation of the Earth. *Ann. Rev. Earth Planet. Sci.* **18**, 205–256.

Wetherill G. W. (1990b) Calculation of mass and velocity distributions of terrestrial and lunar impactors by use of theory of planetary accumulation. In *Abstracts for the International Workshop on Meteorite Impact on the Early Earth*, LPI Contr. No. 746, Lunar and Planetary Institute, Houston, pp. 54–55.

Wetherill G. W. and Stewart G. R. (1993) Formation of planetary embryos: effects of fragmentation, low relative velocity, and independent variation of eccentricity and inclination. *Icarus* **106**, 190–209.

Wieler R. (1994) "Q-gases" as "local" primordial noble gas component in primitive meteorites. In *Noble Gas Geochemistry and Cosmochemistry* (ed. J. Matsuda). Terra Scientific Publishing, Tokyo, pp. 31–41.

Wieler R. (1998) The solar noble gas record in lunar samples and meteorites. *Space Sci. Rev.* **85**, 303–314.

Wieler R. (2002) Noble gases in the solar system. *Rev. Mineral. Geochem.* **47**, 21–70.

Wieler R. and Baur H. (1994) Krypton and xenon from the solar wind and solar energetic particles in two linear ilmenites of different antiquity. *Meteoritics* **29**, 570–580.

Wlotzka F. (1972) Nitrogen. In *Handbook of Geochemistry* (ed. K. H. Wedepohl). Springer, Berlin.

Wood B. J. (1993) Carbon in the core. *Earth Planet. Sci. Lett.* **117**, 593–607.

Woolum D. S., Cassen P., Porcelli D., and Wasserburg G. J. (1999) Incorporation of solar noble gases from a nebula-derived atmosphere during magma ocean cooling. Lunar Planet. Sci. XXX, #1518 (CD-ROM). Lunar and Planetary Institute.

Yang J. and Anders E. (1982) Sorption of noble gases by solids, with reference to meteorites: III. Sulfides, spinels, and other substances: on the origin of planetary gases. *Geochim. Cosmochim. Acta* **46**, 877–892.

Yin Q., Jacobsen S. B., Yamashita K., Blichert-Toft J., Télouk P., and Albarède F. (2002) A short timescale for terrestrial planet formation from Hf–W chronometry of meteorites. *Nature* **418**, 949–952.

Zahnle K. J. (1993) Xenological constraints on the impact erosion of the early martian atmosphere. *J. Geophys. Res.* **98**, 10899–10913.

Zahnle K. J. and Kasting J. F. (1986) Mass fractionation during transonic escape and implications for loss of water from Mars and Venus. *Icarus* **68**, 462–480.

Zahnle K. J., Kasting J. F., and Pollack J. B. (1990a) Mass fractionation of noble gases in diffusion-limited hydrodynamic hydrogen escape. *Icarus* **84**, 502–527.

Zahnle K. J., Pollack J. B., and Kasting J. F. (1990b) Xenon fractionation in porous planetesimals. *Geochim. Cosmochim. Acta* **54**, 2577–2586.

Zhang M. H. G., Luhmann J. G., Bougher S. W., and Nagy A. F. (1993) The ancient oxygen exosphere of Mars: implications for atmospheric evolution. *J. Geophys. Res.* **98**, 10915–10923.

Zhang Y. and Zindler A. (1989) Noble gas constraints on the evolution of Earth's atmosphere. *J. Geophys. Res.* **94**, 13719–13737.

Isotope Geochemistry
ISBN: 978-0-08-096710-3

pp. 493–520

17
Noble Gases as Mantle Tracers

D. R. Hilton
University of California San Diego, La Jolla, CA, USA
and
D. Porcelli
University of Oxford, UK

17.1 INTRODUCTION

The study of the noble gases has been associated with some of the most illustrious names in experimental science, and some of the most profound discoveries. Fundamental advances in nuclear chemistry and physics—including the discovery of isotopes—have resulted from their study, earning Nobel Prizes for a number of early practitioners (Rutherford in 1908; Soddy in 1921; Aston in 1922) as well as for their discoverers (Ramsay and Rayleigh in 1904). Within the Earth Sciences, the noble gases found application soon after discovery—helium was used as a chronometer to estimate formation ages of various minerals (Strutt, 1908). In more recent times, the emphasis of noble gas research has shifted to include their exploitation as inert tracers of geochemical processes. In large part, this shift stems from the realization that primordial volatiles have been stored within the Earth since the time of planetary accretion and are still leaking to the surface today. In this introduction, we give a brief overview of the discovery of the noble gases and their continuing utility in the Earth Sciences, prior to setting into perspective the present contribution, which focuses on noble gases in the Earth's mantle.

17.1.1 Historical Overview

The first noble gas to be discovered was helium (Lockyer, 1869). Spectroscopic investigations of the Sun's chromosphere during a solar eclipse in India in 1868 revealed a previously unobserved line, close in wavelength to the D_1 and D_2 Fraunhofer lines of sodium. The new line was designated D_3 and the element it represented helium (=sun).

The chemical and physical characterization of the noble gases commenced in the 1890s with the discovery of argon. Rayleigh (1892) found a consistent density difference between nitrogen prepared from ammonia and "nitrogen" prepared from air. By passing the "air-nitrogen" over heated magnesium, he and Ramsey succeeded in isolating a new gas of greater density than nitrogen and having a different spectrum from any of the known elements. They named the new gas argon (=not working, lazy) for its unreactive nature (Rayleigh and Ramsay, 1892). Pursuing further research on argon, Ramsay undertook acid treatment experiments on uraninite in the hope of liberating argon. Spectroscopic examination of the gas, however, revealed the D_3 line observed in India by Lockyer. Ramsay reported the discovery of terrestrial helium (Ramsay, 1895).

The discovery of the first two noble gases led Ramsay to conclude that they were members of a new column in the Periodic Table and to search for the remaining members by a new distillation technique. The method involved (i) taking samples of liquid air, (ii) allowing them to boil away almost completely, and then (iii) examining the residue, or further distillate of the residue, spectroscopically. Krypton (=hidden) was the first of the remaining gases to be discovered by this method (Ramsay and Travers, 1898a). Similar experiments on argon yielded two portions, the lighter of which was found to be neon (=new) (Ramsay and Travers, 1898b). The same technique on liquid krypton led to the discovery of xenon (=stranger) (Ramsay and Travers, 1898c).

At the beginning of the twentieth century, research turned to the newly discovered radioactive substances. Ramsay and Soddy (1903) showed that helium was derived by radioactive disintegration of radium, the first demonstration that one element was derived from another. Once the decay constant of radium had been determined, the door was open for the first application of the noble gases: geochronology (Strutt, 1908), a methodology that has been pursued ever since.

In an attempt to explain why the atomic weights of the elements were almost, but not quite, whole integers, Soddy (1910) suggested the existence of chemically identical substances of different atomic weight. J. J. Thomson (1912), using his positive ray apparatus, showed that neon had a line in its spectrum corresponding to mass 22 (as well as mass 20) that could not be matched to any known line at the time. Perceiving the need for a specific name for such substances, Soddy suggested the term "isotopes" (iso = same; topes = place) because they occupied the same place in the Periodic Table.

Following the development of the first mass spectrographs (Aston, 1919; Dempster, 1918) to succeed the parabolic mass analyzer of Thomson, Aston demonstrated the existence, and estimated the relative proportions, of ^{20}Ne and ^{22}Ne (Aston, 1920a), ^{36}Ar and ^{40}Ar (Aston, 1920b, 1920c), all six isotopes of krypton (Aston, 1920b) and five of the isotopes of xenon (Aston, 1920c). He reported the proportions of the remaining xenon isotopes in

1927 (Aston, 1927). Of the remaining noble gas isotopes, ^{21}Ne and ^{38}Ar were found by Hogness and Kvalnes (1928) and Zeeman and De Gier (1934), respectively (see citation by Nier (1936) to the work of Zeeman and De Gier, 1934). Alvarez and Cornog (1939) discovered the one remaining noble gas isotope, namely ^{3}He, using a cyclotron as a mass spectrometer.

The dawn of the modern era of terrestrial noble gas studies can be traced to a suggestion by Suess and Wanke (1965) that the ocean floor should be characterized by a steady-state loss of helium equal to its production in the mantle. Although they measured a 6% excess helium saturation anomaly in Pacific deep water, doubts were expressed due to the possibility of air entrainment. To circumvent these difficulties, Clarke *et al.* (1969) measured the $^{3}He/^{4}He$ ratio expecting to distinguish between air helium and mantle helium, which conventional wisdom at the time dictated should have low (radiogenic) helium isotopic ratios (Morrison and Pine, 1955). Measured helium, however, showed excess ^{3}He ($\sim$22% above air) that was immediately attributed to the presence of primordial ^{3}He leaking from the mantle. At the same time, Mamyrin *et al.* (1969) reported extremely high $^{3}He/^{4}He$ ratios ($\sim 10^{-5}$) for hydrothermal fluids from the southern Kuril Islands. Subsequent studies have shown that high $^{3}He/^{4}He$ ratios are characteristic of all samples of recent mantle origin, reflecting the transport of mantle volatiles through the crust in a variety of tectonic environments. This realization has led to a remarkable broadening in scope of terrestrial noble gas studies into areas such as oceanography, hydrology, and volcanology, to name but a few. However, as will be demonstrated in this work, understanding mantle-related processes through the study of the noble gases remains a key research area in isotope geochemistry and a continuing challenge to the noble gas community.

17.1.2 Scope of Present Contribution

In this work, we present an overview of the current status of noble gas geochemistry as it relates to understanding the terrestrial mantle. We first review the intrinsic properties of the noble gases that make them so useful as geochemical tracers. This leads to a description of the main isotopic and relative abundance features of the noble gases in mantle-derived materials. We then concentrate on three aspects of the utility of noble gases in mantle-related studies: (i) as a tracer of recycling between the Earth's exosphere (hydrosphere, atmosphere, and crust) and mantle, and how noble gases are crucial in assessing the state of volatile mass balance between these reservoirs; (ii) as a fundamental constraint on a host of different classes of models aimed at describing the structure of the mantle (e.g., whether it is layered) and the related topic of the origin of ocean island basalts (OIBs); and (iii) as a means to understand depletion/enrichment processes and volatile transport in the subcontinental mantle. It should be noted that noble gases have also found utility in other related areas of mantle geochemistry not covered in this chapter: the reader is referred to Chapter 16 for respective background on (i) noble gases in planetary accretion process and the acquisition of the terrestrial volatile inventory, and (ii) mantle–atmosphere coupling and the formation of the Earth's atmosphere.

This work follows on from many earlier reviews of the noble gas geochemistry of the Earth's mantle. For more details on this topic, the reader is referred to the following references: Craig and Lupton (1981); Farley and Neroda (1998); Lupton (1983); Mamyrin and Tolstikhin (1984); McDougall and Honda (1998); Ozima (1994); Ozima and Podosek (1983); and Porcelli *et al.* (2002).

17.2 NOBLE GASES AS GEOCHEMICAL TRACERS

Noble gases have been widely exploited as geochemical tracers. They have three attributes that make them particularly sensitive as tracers of mantle processes. Specifically, noble gases are (i) composed of different components each having diagnostic isotope characteristics, (ii) incompatible, and hence depleted in the solid Earth, and (iii) chemically inert. Each of these features of the noble gas family is discussed in turn.

17.2.1 Component Structures of the Noble Gases

The principal sources of terrestrial noble gases are primordial gases, inherited from the solar nebula at the time of planetary accretion, and nucleogenic gases generated by specific nuclear reactions subsequent to accretion. Interactions with cosmic rays produce a third but relatively minor source of noble gases (cosmogenic noble gases) mainly by spallation reactions at or close to the Earth's surface. Identification of the isotopic characteristics of these "end-member" compositions is an essential prerequisite to attempts to resolve terrestrial noble gas isotope variations into component structures, and hence in their exploitation as tracers of the origin, development, and structure of the Earth's mantle.

Models of planetary evolution assume that at the time of planetary formation the solar system had a single universal and well-mixed composition from which all parts of the solar system were derived (see Podosek, 1978). Information as to the elemental and isotopic characteristics of this primordial composition is presently available from the Sun, meteorites, and the atmospheres of the giant planets (Wieler, 2002). In the case of the Sun, distinction is usually made between the present-day composition, which is available via spectral analysis of the solar atmosphere and capture of the solar wind, either directly in space or by using metallic foil targets, and the proto-Sun (the composition at the time of planetary accretion) whereby the lunar regolith and/or meteorites are utilized as archives of ancient solar wind. As discussed below, the distinction is only really important for helium due to production of ^{3}He by deuterium burning.

Although the Sun represents the obvious choice for defining the primordial composition of the solar nebula, the initial discovery of primordial noble gases was made using meteorites. Gerling and Levskii (1956) claimed the honor by arguing that observed noble gas abundances in a gas-rich achondrite could not be supported by *in situ* nuclear processes (radiogenic or spallation reactions). Subsequent investigations of other gas-rich meteorites (e.g., Suess *et al.*, 1964) showed that the relative abundances of the trapped noble gases are similar to those in the Sun, earning the label "solar" gases for the trapped component. Direct measurements of the relative abundances and isotopic structure of noble gases in the solar wind (see Wieler, 2002) have confirmed the origin of the solar component in meteorites as implanted solar wind. Therefore, noble gas measurements of both gas-rich meteorites and the solar wind serve to define the primordial noble gas signature of the solar system. In Table 1 we highlight some of the principal characteristics of "solar" noble gases.

A problem with adopting the solar wind $^{3}He/^{4}He$ ratio as representative of the solar nebula is the production of ^{3}He from deuterium very early in solar system history: consequently, the solar wind value ($\sim 4.4\times10^{-4}$) is too high by a factor between ~2.5 and ~3 relative to the proto-Sun (Geiss and Reeves, 1972). To circumvent this difficulty, recourse has been made to analyzing the giant planets whose atmospheres are expected to reflect proto-solar values (Wieler, 2002). Jupiter is the only giant planet whose atmospheric $^{3}He/^{4}He$ ratio has been determined (Mahaffy *et al.*, 1998). Its value of 1.66×10^{-4} ($\sim 120R_A$, where R_A = air $^{3}He/^{4}He$)—remarkably similar to measurements on primitive meteorites (see below), is now adopted as most representative of primordial (proto-solar) helium.

It should be noted at this point that studies of the most primitive class of meteorites—CI chondrites—have found that many contain trapped gases in a distinctly different pattern to the "solar" signature. Indeed, the relative abundances of the noble gases resemble those in the Earth's atmosphere (Figure 1) so they have been dubbed "planetary" gases (Signer and Suess, 1963). Whereas all meteorites show a marked depletion in noble gases compared to solar abundances (see next section), the "planetary" pattern shows a greater relative depletion in the lighter noble gases. In addition, the isotopic composition of the planetary component differs from that of solar gas (Table 1). For example, the $^{3}He/^{4}He$ ratio of carbonaceous material in primitive chondrites (the so-called Q-gases; see discussion by Ott, 2002) is 1.23×10^{-4}, significantly lower than that of solar helium. At this stage, firm genetic relationships between "planetary" and solar noble gases remain to be established but it appears probable that, with the exception of helium, the solar noble gas component

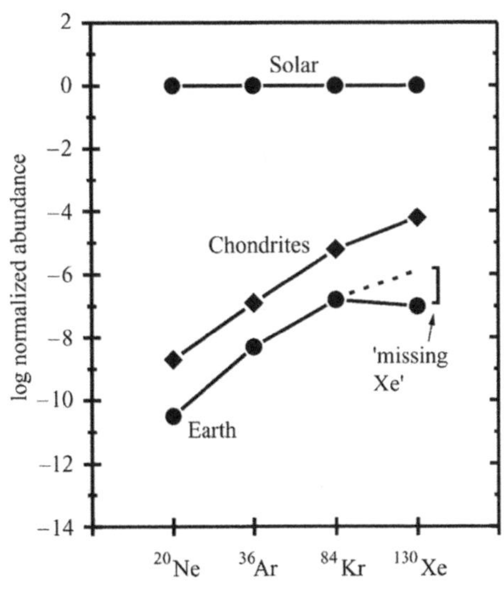

Figure 1 The noble gas abundance patterns of the Earth and CI chondrites normalized to silicon and the solar composition (reproduced by permission of Mineralogical Society of America from *Rev. Mineral. Geochem.*, **2002**, *47*, 418).

Table 1 Solar system isotopic compositions of He, Ne, and Ar.

Reservoir	$^{3}He/^{4}He$ ($\times10^{-6}$)	R/R_A[a]	$^{20}Ne/^{22}Ne$	$^{21}Ne/^{22}Ne$	$^{38}Ar/^{36}Ar$	$^{40}Ar/^{36}Ar$
Solar	457	326	13.8	0.0328	0.1825	$\sim3\times10^{-4}$
Planetary	143	102	8.2	0.024	0.188	$\sim3\times10^{-4}$
Earth atmosphere	1.4	1	9.8	0.0290	0.1880	295.5

After McDougall and Honda (1998).
[a] $R/R_A = (^{3}He/^{4}He)_{observed}/(^{3}He/^{4}He)_{air}$, where air $^{3}He/^{4}He = 1.4\times10^{-6}$.

better approximates primordial characteristics at the time of Earth accretion. This follows from considerations of the origins of noble gases on the planets, as discussed in Chapter 16. Indeed, the noble gases seen in the Earth's and other planetary atmospheres may record the effects of a number of additional processes, e.g., Rayleigh distillation, hydrodynamic loss, adsorption on planetesimal surface grains, or any combination thereof, associated with the accretionary process itself and/or subsequent planetary growth (Pepin, 1992), and these processes have resulted in distinctive isotopic compositions and relative abundances that are distinctive from those found on the planets (see Chapter 16). There now appears to be a general consensus that the "planetary" component of noble gases found in meteorites is unrelated to the noble gases seen in planetary atmospheres.

Radiogenic and nucleogenic noble gases have been produced within the Earth subsequent to planetary accretion. They are produced as a result of the spontaneous decay of a parent radionuclide or as a consequence of specific nuclear reactions. The principal noble gas isotope daughters produced by radioactive decay are given in Table 2, whereas the most common nuclear reactions producing noble gas isotopes are summarized in Table 3 (see review by Ballentine and Burnard, 2002). Modification of the (primordial) isotopic composition of a noble gas within the Earth occurs by addition of a radiogenic/nucleogenic daughter product.

The most conspicuous daughter isotope produced by nuclear methods is ^{4}He. It originates through α-particle decay of members of the ^{238}U, ^{235}U, and ^{232}Th decay series. Uranium-238 also undergoes spontaneous fission and produces small but sufficient quantities of xenon isotopes to perturb the primordial pattern within the Earth. The short-lived nuclides ^{129}I and ^{244}Pu undergo spontaneous decay and are now both extinct on Earth. They have produced measurable variations in the xenon spectra of the mantle as sampled by oceanic basalts (Staudacher and Allègre, 1982). The only other major noble gas daughter produced by spontaneous decay is ^{40}Ar, which results from electron capture by a ^{40}K nucleus.

Specific nuclear reactions capable of producing noticeable quantities of noble gas daughters in the Earth (^{3}He and ^{21}Ne in particular) are initiated by alpha and fission activities of the natural radioelements. Helium-3 is produced through a neutron capture reaction involving ^{6}Li (Hill, 1941), whereas ^{21}Ne production occurs through a number of α-induced reactions (Wetherill, 1954). In the case of helium, the ^{3}He/^{4}He ratio produced is of the order 10^{-8} and primarily reflects the lithium abundance at the site of production (Mamyrin and Tolstikhin, 1984). For neon, the only conspicuous isotope produced is ^{21}Ne due to its low natural abundance. The present-day ^{21}Ne/^{4}He production ratio in the mantle has been calculated at 4.5×10^{-8} (Yatsevich and Honda, 1997) (see Ballentine and Burnard, 2002 for discussion regarding calculation of this parameter).

Spallation reactions are the only other means of producing significant abundances of noble gases on Earth. Interactions of cosmic rays with the Earth's atmosphere produce high-energy particles that can interact with nuclei in the atmosphere and at the Earth's surface. Sizeable quantities of ^{3}He, ^{21}Ne, and ^{38}Ar can be produced in this manner (see Ballentine and Burnard, 2002). In the context

Table 3 Some nuclear reactions producing noble gases.

Reaction	*Upper crust production ratios*[a]
^{6}Li(n,α)^{3}H($\beta-$)^{3}He	^{3}He/^{4}He $= 1 \times 10^{-8}$
^{17}O(α,n)^{20}Ne	^{20}Ne/^{4}He $= 4.4 \times 10^{-9}$
^{18}O(α,n)^{21}Ne	^{21}Ne/^{4}He $= 4.5 \times 10^{-8}$
^{24}Mg(n,α)^{21}Ne	^{21}Ne/^{4}He $= 1 \times 10^{-10}$
^{25}Mg(n,α)^{22}Ne	Combined crustal production:
^{19}F(α,n)^{22}Na($\beta+$)^{22}Ne	^{22}Ne/^{4}He $= 9.1 \times 10^{-8}$
^{35}Cl(n,γ)^{36}Cl($\beta-$)^{36}Ar	^{40}Ar/^{36}Ar $= 1.5 \times 10^{7}$

Source: Porcelli *et al.* (2002).
[a] See Ballentine and Burnard (2002) for details and production rates in different compositions.

Table 2 Half-lives of parent nuclides for noble gases.

Nuclide	*Half-life*	*Daughter*	*Yield* (atoms/decay)	*Comments*
^{3}H	12.26 yr	^{3}He	1	Continuously produced in atm
^{238}U	4.468 Gyr	^{4}He	8	
		^{136}Xe	3.6×10^{-8} $(4.4 \pm 0.1) \times 10^{-8}$	Spontaneous fission
^{235}U	0.7038 Gyr	^{4}He	7	^{238}U/^{235}U = 137.88
^{232}Th	14.01 Gyr	^{4}He	6	Th/U = 3.8 in bulk Earth
		^{136}Xe	$<4.2 \times 10^{-11}$	No significant production in Earth
^{40}K	1.251 Gyr	^{40}Ar	0.1048	^{40}K = 0.01167% total K
^{244}Pu	80.0 Myr	^{136}Xe	7.00×10^{-5}	^{244}Pu/^{238}U $= 6.8 \times 10^{-3}$ at 4.56 Ga
^{129}I	15.7 Myr	^{129}Xe	1	^{129}I/^{127}I $= 1.1 \times 10^{-4}$ at 4.56 Ga

Source: Porcelli *et al.* (2002).

of the present work on the Earth's mantle, however, the primary concern regarding noble gases of cosmic origin relates to the suggestion of Anderson (1993) that interplanetary dust particles (IDPs) rich in ^{3}He (and ^{21}Ne) could account for the origin of the high ^{3}He/^{4}He ratios observed in mantle emanations. This idea has been criticized on two fronts: first, the rate of IDP deposition is many orders of magnitude too low to sustain the flux of ^{3}He (and ^{21}Ne) released via the ridge system (Stuart, 1994; Trull, 1994). Second, the diffusivity of helium in IDP material is too high for it to be effectively transported into the mantle (Hiyagon, 1994). We are left with the conclusion, therefore, that the range of isotopic compositions observed for the noble gases in mantle-derived materials primarily reflects admixture—to varying degrees—of primordial and radiogenic and/or nucleogenic components, with the superimposition of other effects (e.g., degassing, crustal assimilation) onto one or both components.

17.2.2 Partitioning of the Noble Gases

A major reason for the success of the noble gases as geochemical tracers is their depletion in the terrestrial realm (Figure 1). For example, ^{36}Ar is nearly nine orders of magnitude less abundant in the Earth relative to solar abundances (normalized to silicon) (Ozima and Podosek, 1983). The available evidence further indicates that a substantial proportion of the terrestrial inventory of various noble gas species (e.g., ^{130}Xe; Pepin and Porcelli, 2002) is now resident in the atmosphere, implying that the Earth has retained noble gases in low abundance following atmosphere formation. Herein lies the particular advantage of the noble gases as mantle tracers: abundances of residual (primordial) noble gases are sufficiently low that significant (i.e., measurable) shifts in isotopic ratios are produced by addition of radiogenic and nucleogenic noble gases. The key to understanding the depletion of noble gases in the solid Earth lies with knowledge of the partitioning behavior of noble gases between melt and vapor, melt and crystal, and liquid iron and silicate solid.

There are three facets of the partitioning behavior of the noble gases that work together to deplete the gases from the silicate Earth. First, during mantle melting and the production of partial melt, the noble gases show a strong affinity for the melt phase. Early experimental work on synthetically made olivines (e.g., Hiyagon and Ozima, 1982) showed that D-values (where D is the weight concentration in mineral/weight concentration in melt) are generally low ($D_{He} \leq 0.07$; $D_{Ar} \leq 0.05$–0.15; $D_{Xe} \leq 0.3$). Although there was some debate whether D values are consistently low for all noble gases and all minerals phases (e.g., Broadhurst *et al.*, 1990, 1992), work on naturally occurring glass–mineral pairs consistently found low D values for olivine–melt partitioning (e.g., $D_{He} \leq 0.008$; $D_{Ar} \leq 0.003$; Marty and Lussiez, 1993). All these values are considered minima. More recent experimental work on clinopyroxene-silicate melt (Chamorro *et al.*, 2002) has found that D_{Ar} is low ($\sim 4 \times 10^{-4}$) and constant to pressures of at least 80 kbar. Brooker *et al.* (2003) have demonstrated that such low values extend to all the noble gases. Therefore, during even minor and/or relatively deep episodes of mantle melting, the expectation is that noble gases will partition strongly from the solid (mantle) into the melt phase.

Second, the solubility of noble gases in basaltic melt is generally low: Jambon *et al.* (1986) reports values of 56, 25, 5.9, 3.0, and 1.7 (units of cm^3 STP g^{-1} $bar^{-1} \times 10^3$) for helium through xenon for basaltic liquid at 1,400 °C. The ionic porosity of the melt—controlled by its composition—dictates the solubility of a particular species (Carroll and Draper, 1994). Due to low noble gas solubilities, any exsolution of a vapor phase (such as CO_2) during decompression of a melt as it moves toward the surface will result in the rapid removal of the noble gases from the melt phase. In Figure 2, the partitioning of the noble gases between melt and vapor is illustrated for a vesiculated basaltic magma (Carroll and Draper, 1994). Note that for the range of vesicularities observed in submarine basalts, partitioning of the noble gases into the vesicle phase is almost complete. In this way, any melt phase can be effectively stripped of its volatile inventory. Therefore, both mantle melting and vapor exsolution act together to remove noble gases and other volatiles from the silicate Earth and to transfer them to the atmosphere and hydrosphere.

Finally, a related but more speculative means to deplete the silicate Earth of its noble gas

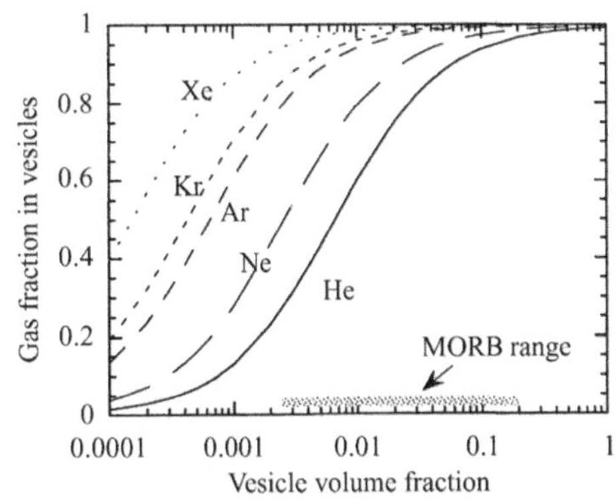

Figure 2 Calculated fraction of total gas in basaltic magma which is present in vesicles as a function of vesicularity (reproduced by permission of Elsevier from *Chem. Geol*, **1994**, *117*, 37–56).

inventory is if noble gases were sequestered by Fe–Ni and transferred to the Earth's core. The proposition has been evaluated for helium and found to be plausible notwithstanding significant uncertainties in solid-silicate/liquid-metal partitioning behavior (Porcelli and Halliday, 2001). However, the proportion of helium lost to the core in this way is likely to be relatively small (<2% of the original mantle inventory; Porcelli and Halliday, 2001). Insufficient data preclude this analysis being extended to the other noble gases.

In sum, the partitioning behavior of the noble gases has acted to deplete the silicate Earth of its noble gas inventory. As the mantle comprises the bulk of the silicate Earth, noble gases are relatively scarce in this terrestrial reservoir. A consequence of this is that noble gases are particularly sensitive tracers of events occurring within the mantle. The scarcity of noble gases in the solid Earth has also given rise to their common pseudonym, the rare gases, although this is far from true in a solar sense.

17.2.3 Chemical Inertness of the Noble Gases

The scarcity of the noble gases on Earth is a direct result of one of their most obvious characteristics—their chemical inertness, or failure to react with other species under almost all circumstances. As discussed above, a major consequence of this inertness is that noble gases pass easily into the gas phase, and are efficiently lost from the solid Earth. However, there are situations where the physical chemistry of the noble gases may be important for explaining experimental observations. In such cases, the effects will be almost entirely due to van der Waal's forces, which increase in strength with number of electrons, i.e., Xe > Kr > Ar >Ne > He. Following this line of reasoning, a significant effort has been expended since the 1980s evaluating continental shales and glacial ice as possible reservoirs for the so-called "missing-Xe" (see Figure 1), based upon the potential of xenon to interact more strongly with possible sorbent material. Significantly, no terrestrial reservoir has been found as the repository of the extraneous xenon (Ozima and Podosek, 1983). Indeed, the current consensus for the low Xe/Kr ratio in the terrestrial atmosphere is that it is not indicative of xenon loss but is an inherited feature characteristic of planetary formation (Porcelliand Ballentine, 2002).

Another corollary of the failure of the noble gases to react chemically is that they generally behave coherently as a group, with systematic and often predictable variations or responses from helium through xenon. Again, this is advantageous in a geochemical context because such responses can yield information on particular physical processes or perturbations to natural systems. An example of such a process is vapor partitioning, with the distribution of noble gases between a melt and vesicle phase useful for tracing extent and mode of degassing (e.g., Moreira and Sarda, 2000). It is important to note that using the noble gases as a group allows for redundancy in the tracing process: this can be significant if, e.g., any one species (e.g., ^{36}Ar) is particularly susceptible to modification by other means (e.g., air contamination). Likewise, the coherence of the noble gases facilitates estimation of unknown parameters. For example, the diffusivity of the heavy noble gases under mantle *P–T* conditions is poorly constrained: however, this is not the case for helium (Trull and Kurz, 1993). Therefore, by scaling to helium it becomes possible to place reasonable limits on unknown quantities.

Taken together, the properties of wide-ranging isotopic systematics, incompatibility in the solid Earth and chemical inertness act to make noble gases tracers "par excellence" in geochemical research. In the following discussion, we show that noble gases are particularly adept at tracing processes associated with the origin and development of the Earth's mantle.

17.3 MANTLE NOBLE GAS CHARACTERISTICS

In this section we present an overview of the principal means available to sample mantle-derived noble gases, followed by a summary of their main isotope and relative abundance characteristics in the mantle. Mostly, the mantle shows a wide range in noble gas isotope variations serving to impart information on a variety of topics. The only exception is krypton whose isotopic composition is steadfastly air-like in mantle materials. Consequently, we do not consider krypton in this review.

17.3.1 Localities and Sampling Media

Given their geochemical incompatibility and affinity for the liquid silicate phase, noble gases of mantle origin are overwhelmingly transferred to the Earth's surface during mantle melting. Consequently, the distribution of locations worldwide exploited to yield information on the noble gases characteristics of the mantle is heavily skewed towards regions undergoing active melting which results in magmatic or volcanic activity. The global-encircling mid-ocean ridge system allows relatively straightforward access to the oceanic mantle at points worldwide. Ocean islands, be they intraplate in origin or associated with arc volcanism, also facilitate loss of noble gases from the oceanic mantle—albeit on a smaller scale—and are consequently prime sampling targets. On the continents, regions of active volcanism (e.g., continental arcs) provide similar access to volatiles from the underlying mantle, as do areas associated with rifting and nascent magmatism. Although crustal fluids such as groundwaters and natural gases may influence their distribution, the primary control on the location of mantle-derived noble gases in continental regions is still likely to be the presence of underlying melts.

Mantle noble gases can be sampled through a wide variety of fluid and rock types. For example, fumarolic gas discharges, bubbling hot springs, natural gases, and groundwaters are prime sampling media for noble gases. Silicate hosts for noble gases include quenched glasses (submarine and subglacial in origin) as well as various minerals phases (e.g., olivine, pyroxene) erupted as phenocrysts or xenocrysts. In mineral phases, noble gases are usually trapped within fluid or melt inclusions. In the following, we discuss how these various sampling media are exploited in different tectonic environments to provide a picture of the noble gas signature of the mantle. For simplicity, we consider sampling in the submarine and subaerial environments, which roughly maps to sampling of the oceanic and continental segments of the mantle.

In the submarine environment, dredged vitreous glass is a widely exploited medium for obtaining noble gas isotopic compositions and abundances. Most of the mid-ocean ridge basalt (MORB) database has been obtained using pillow-rim glasses erupted on or close to the spreading axes of diverging plate boundaries. Rapid quenching of lavas as they are extruded onto the seafloor traps mantle-derived volatile phases in the glassy rinds. The vesicular and granular interior sections of the basalt pillows usually show atmospheric-like volatile characteristics, indicating contamination with ambient seawater and/or air during sample recovery or subsequent processing (Ballentine and Barfod, 2000). Submarine hydrothermal fluids or vents also provide a means to access mantle-derived noble gases. The usual sampling scenario involves collecting high- and low-enthalpy hydrothermal fluids directly at the vent orifice using leak-tight containers manipulated into place using submersible vehicles. Alternatively, the vent-derived noble gases can be sampled at various depths in the water column by means of hydrocasting: in this case, the mantle noble gas signature is usually heavily diluted by ambient seawater. Noble gas data collected at the same locations using hydrothermal fluids and lavas (e.g., Loihi Seamount; Hilton *et al.*, 1998) generally show good agreement.

A variety of sampling opportunities for mantle noble gases also arise in subaerial regions. Recent volcanism at ocean islands, active margins or other continental regions provides ready access to magmatic (and presumably mantle-derived) volatiles through hydrothermal manifestations (fumaroles, hot springs, bubbling mudpots) and various eruptive products such as ash, tephra, lavas, and xenoliths. In addition, crustal fluids, notably groundwaters and natural gases, can also be exploited as carriers of mantle-derived noble gases to the Earth's surface—this is particularly the case in tectonically active areas. It should be noted that the potential for contamination of the mantle noble gas component is probably greater in subaerial regions: the major contaminants are either atmospheric gases or various daughter products (i.e., radiogenic/nucleogenic noble gases) produced *in situ* in the crust. Fortunately, contamination effects can be recognized through isotopic considerations and (in some cases) corrections applied.

17.3.2 Helium Isotopes

MORBs provide a global perspective on the helium isotope systematics of the shallow (uppermost) mantle. Numerous compilations of the mean MORB $^3He/^4He$ value have been produced (see summary in Hilton *et al.* (2000a)) with the general consensus being that MORB-mantle is characterized by a $^3He/^4He$ ratio of $8 \pm 1\ R_A$ (R_A = air $^3He/^4He$). Remarkably, in all compilations produced to date (also see Graham, 2002), there appears no statistical evidence of differences in $^3He/^4He$ between MORBs erupted in the Atlantic, Pacific, and Indian oceans. The observation that mantle $^3He/^4He$ ratios are greater than air has been interpreted as reflecting the fact that the mantle still retains primordial helium (enriched in ^3He) since accretion of the planet, and that it s still leaking to the surface today albeit heavily diluted by time-integrated radiogenic helium grown-in throughout Earth's history (Clarke *et al.*, 1969). Alternative ideas as to the origin of the high $^3He/^4He$ ratios in mantle-derived materials (e.g., due to subduction of cosmogenic dust; Anderson, 1993) have not gained widespread acceptance by the noble gas community (see Porcelli and Ballentine, 2002).

The conclusion that the MORB mantle is characterized by such a narrow range in $^3He/^4He$ (Figure 3) is crucially predicated on the definition of what constitutes MORBs, and various authors have invoked different criteria to either accept or reject samples from the MORB database. The primary and most widely adopted criteria appear to be that samples should be "depleted" in large-ion lithophile elements (LILEs) and have unradiogenic strontium and lead (and radiogenic neodymium) isotope systematics. Although difficult to apply for each and every sample, these criteria are aimed at identifying MORB erupted at shallow

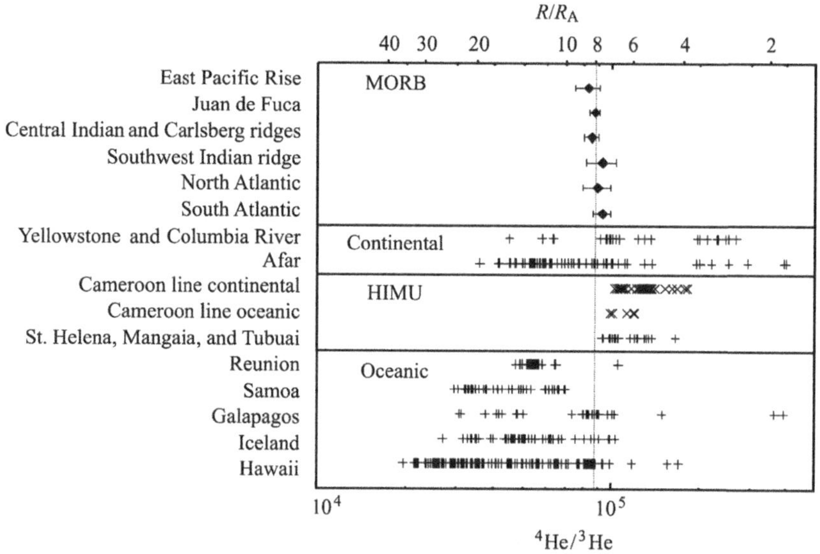

Figure 3 Compilation of helium isotope data from MORBs, continental hotspots, ocean island basalts, and HIMU sources (reproduced by permission of Mineralogical Society of America from *Rev. Mineral. Geochem.*, **2002**, *47*, 419).

levels proximal to obvious plume influences and/or showing geochemical evidence of mantle plume involvement in their petrogenesis The most obvious example where difficulties arise is for submarine MORB erupted close to Iceland (the location of the highest global $^3He/^4He$ ratios—see below). Hilton *et al.* (1993a) excluded MORB samples erupted along the Reykjanes and Kolbeinsey ridges and produced an estimate of $8.29 \pm 0.78R_A$ for North Atlantic MORB—a value very close to their global mean estimate of $8.18 \pm 0.73R_A$ (1σ). In contrast, Graham (2002)) included all oceanic basalts close to Iceland, and derived a mean $^3He/^4He$ ratio (for the whole Atlantic) of $9.58 \pm 2.94R_A$—in comparison to a global estimate of $8.75 \pm 2.14R_A$. Clearly, in this latter case, the mean $^3He/^4He$ value for Atlantic (and global) MORB helium was influenced by the inclusion of data from the Reykjanes ridge (11.0–17.6R_A). Although the above approaches differ somewhat in their selection of samples, the two estimates for MORB helium do not diverge markedly indicating that the number of contentious samples is small in comparison to the total number of MORB. The observation that a major part of the mantle is characterized by helium with a $^3He/^4He$ of $8 \pm 1R_A$ is a fundamental constraint on models of mantle structure and evolution (see Section 17.4.2). It also provides an important benchmark to contrast with other regions of the mantle that have evolved differently to MORB mantle.

The most obvious suites of samples to search for noble gas differences in the mantle are OIBs given their distinct trace element and isotopic geochemistry compared to MORB (see Hofmann, 1997; see Chapter 18). The first indication that mantle $^3He/^4He$ values did not fall within the canonical range of $8 \pm 1R_A$ now ascribed to MORB came with the publication of a $^3He/^4He$ ratio of $15R_A$ for fumarolic gas from Kilauea volcano in Hawaii (Craig and Lupton, 1976). Given notions of a mantle plume origin for the Hawaiian islands (Morgan, 1971), the high $^3He/^4He$ ratio was ascribed to a lower mantle source possessing a higher time-integrated $^3He/(U+Th)$ ratio. Clearly, the source of OIB and MORB mantle must have remained isolated for a significant period of Earth's history to maintain differences in respective $^3He/^4He$ ratios. Other OIB localities with $^3He/^4He \gg 8R_A$ (termed "high-^{3}He" hotspots) include Iceland, Samoa, Reunion, Heard Island, Easter Island, and Jan Fernandez (see compilation by Farley and Neroda, 1998; Graham, 2002). The highest $^3He/^4He$ value reported to date is $38R_A$ for Tertiary basalts from Iceland (Hilton *et al.*, 1999).

In addition to OIB with $^3He/^4He$ values greater than the MORB range, there is subpopulation of OIB with $^3He/^4He$ values less than MORB. The islands of St. Helena, Mangaia, Tubuaii, and Tristan da Cunha fall into this category, with their $^3He/^4He$ values falling between $5R_A$ and $8R_A$. It is noteworthy that these so-called "low-^{3}He" hotspot islands tend to have extreme radiogenic isotope characteristics (Zindler and Hart, 1986) prompting the suggestion that OIB $^3He/^4He <$ MORB represents addition of radiogenic helium from ancient crustal material recycled back into the mantle (Graham *et al.*, 1993). In order to explain the relatively high $^3He/^4He$ values of these OIBs, in comparison to significant lower (i.e., pure radiogenic) $^3He/^4He$ ratios anticipated for old recycled crust, models of diffusive exchange of helium between the recycled material and ambient (MORB) mantle have been invoked (e.g., Hanyu and Kaneoka, 1997). As an alternative explanation, it has been suggested that crustal–magma interaction coupled with prior degassing of the magmatic helium component has resulted in ^{3}He/^{4}He values <MORB for a number of OIBs (Hilton *et al.*, 1995). A test of this suggestion at La Palma (an active oceanic island in the Canaries archipelago with HIMU-like trace element characteristics) found ^{3}He/^{4}He values as high as 9.6R_A—at the transition between the MORB range and that of "high-^{3}He" hotspots (Hilton *et al.*, 2000a). If this observation is generally applicable then notions of a bimodal distribution of OIB ^{3}He/^{4}He on either side of the MORB range will need revising. However, if OIB genesis is associated with ancient recycled crust then such material must be characterized by production of (some) radiogenic helium which will be superimposed upon helium acquired from ambient mantle: the wide range of ^{3}He/^{4}He values seen in OIBs, therefore, may reflect variations in the balance of helium from these different sources.

Another prime locality to gain information on the helium isotope systematics of the mantle is convergent margins. Arc magmatism at such localities can sample volatiles derived from the mantle wedge, the subducting slab (both crustal basement and its sedimentary veneer) and the arc lithosphere through which magmas are erupted. In the vast majority of arcs, however, ^{3}He/^{4}He ratios lie coincident with, or slightly lower than, the range found in MORBs (Poreda and Craig, 1989). The range of measured ^{3}He/^{4}He values, (5–8)R_A, indicates that the mantle wedge is the main contributor to arc-derived helium. Discriminating between the subducting slab and arc lithosphere as the source of the small but discernible contribution of radiogenic helium is difficult and relies on circumstantial evidence. For example, Hilton *et al.* (2002) pointed out that arc segments with both significant volumes of subducting sediment (e.g., Alaska, the Aleutian Islands, Java) or old oceanic basement (e.g., the Marianas, Japan) still record ^{3}He/^{4}He values within the canonical MORB range. In contrast, arc segments that erupt through thick continental crust (e.g., the Andes, Kamchatka) tend to have ^{3}He/^{4}He $< 7R_A$; thereby implicating the arc lithosphere as the source of radiogenic helium. As in the case of "low-^{3}He" hotspots, it appears that prior degassing of the magmatic component followed by magma–crust interaction could be the responsible mechanism (Hilton *et al.*, 1993b).

Although most arcs are dominated by MORB-like helium, segments of two arcs (the Banda arc of eastern Indonesia and the Campanian Magmatic Province of southern Italy) are highly unusual in that they emit predominantly radiogenic helium (Hilton and Craig, 1989; Marty *et al.*, 1994). In both cases, the regions are characterized by the subduction of continental crust—the leading edge of the Australian plate in the former case and the northward-moving African plate in the latter. Therefore, the helium budget in these two arcs is dominated by the input of radiogenic helium from the subducting slab with mantle wedge helium taking a subordinate role in the helium inventory. The observation of radiogenic helium in these two arc systems has clear tectonic implications, e.g., for tracing the juxtaposition of two colliding plates; however, more generally, it illustrates the fact that under certain circumstances helium can be subducted into the mantle and may therefore contribute to some of the heterogeneity observed in mantle ^{3}He/^{4}He ratios.

In the western Pacific Ocean, back-arc basin volcanism is often associated with subduction zone activity: as such, it is also a useful source of helium data on that portion of the mantle that might have been modified by subducted oceanic crust and sediments. To date, ^{3}He/^{4}He ratios are available for four back-arc regions: the Lau Basin, Mariana Trough, North Fiji Basin and Manus Basin (see summary in Hilton *et al.*, 2002). A wide range of ^{3}He/^{4}He values characterize basalts erupted in these areas—from highs of 22R_A and 15R_A in the Lau and Manus basins respectively, through the MORB range (all four basins) to predominantly radiogenic values ($<4R_A$)—again in all four basins. In the case of the high ^{3}He/^{4}He ratios (>MORB), their origin can safely be ascribed to a mantle plume source. The observation of MORB-like helium is not unexpected given the fact that mature back-arc basin basalts also show MORB-like trace element and radiogenic isotope characteristics (Saunders and Tarney, 1991). Debate continues on the origin of the radiogenic helium: is it slab-derived and a tracer of recycled material into the mantle, or is a shallow-level contaminant introduced into the magma source by assimilation of arc crust immediately prior to eruption? Answering this question has far-reaching ramifications for understanding geochemical recycling in general and the helium isotope evolution of the mantle in particular (see discussion in Hilton *et al.*, 2002).

Mantle noble gases are also observed in continental localities. However, the principal difficulty is that these volatiles must traverse a major geochemical reservoir that is characterized by a ^{3}He/^{4}He ratio $\leq 0.05R_A$ (Andrews, 1985). In spite of this obstacle, mantle helium is recognized throughout the continents (e.g., Oxburgh *et al.*, 1986; Poreda *et al.*, 1986) based on the fact that there is a large isotopic contrast between radiogenic helium and mantle-derived helium as defined by the ^{3}He/^{4}He ratio of MORB mantle (Section 17.2.1). For example, a 5% addition of MORB-like helium (8R_A) to radiogenic helium gives a resultant $^3He/^4He$ value $>0.1R_A$—a value significantly higher than can be produced by crustal lithologies. In this way, relatively small additions of mantle helium to crustal reservoirs can be recognized and quantified. Of course, some continental regions record high (MORB-like) $^3He/^4He$ ratios implying that dilution with radiogenic helium is low or nonexistent. The main regions of the crust where mantle helium is found includes areas undergoing extension (e.g., the Rhine Graben, East African Rift), subduction-type volcanoes (e.g., the Andes), major fault systems (e.g., the San Andreas Fault) as well as kimberlite pipes and other xenolith-bearing localities (Ballentine and Burnard, 2002; Dunai and Porcelli, 2002). It should be noted also that there are two "high-^3He" hotspots ($^3He/^4He \gg$ MORB) found in continental locations—at Yellowstone and along the Ethiopian Rift (see Graham, 2002).

17.3.3 Neon Isotopes

Determining the neon isotopic composition of the mantle has proved to be more difficult than for helium since neon is not highly depleted in the atmosphere due to losses to space and is present in low abundances in mantle-derived rocks, resulting in the almost inevitable contamination of all mantle samples with (some) air-derived neon. Furthermore, the analytical challenges are more severe for neon due to the need to correct for isobaric inferences on masses 20 and 22 due to doubly charged interfering species. Therefore, although there are early reports (e.g., Craig and Lupton, 1976) that the neon isotopic composition of the mantle was different from air, it was not until the mid-to-late 1980s that work on diamonds (Honda *et al.*, 1987) and MORB (Sarda *et al.*, 1988) demonstrated conclusively that the mantle had higher $^{20}Ne/^{22}Ne$ and $^{21}Ne/^{22}Ne$ ratios compared to the Earth's atmosphere.

In Figure 4 the neon isotope systematics of MORBs (and OIBs) are illustrated on a traditional three-isotope neon plot. It can be seen that MORB samples define an array between the air value ($^{20}Ne/^{22}Ne = 9.8$; $^{21}Ne/^{22}Ne = 0.029$) and an end-member enriched in both $^{20}Ne/^{22}Ne$ and $^{21}Ne/^{22}Ne$. This array is usually interpreted as a binary mixing trajectory between air and MORB mantle, with individual samples contaminated to varying extents ith air neon. The enrichment of MORB in $^{20}Ne/^{22}Ne$ compared to air can be explained by the presence of a solar neon component with $^{20}Ne/^{22}Ne = 13.6$ (Wieler, 2002) or 12.5 (Trieloff *et al.*, 2000) trapped within the mantle (like so-called primordial helium) since accretion (see discussion in Chapter 16). Extrapolation of the MORB array to the solar $^{20}Ne/^{22}Ne$ ratio (in effect removing the air contaminant from the mixture) allows an estimate of the MORB mantle $^{21}Ne/^{22}Ne$ ratio of 0.074 (using solar $^{20}Ne/^{22}Ne = 13.6$). This relative enrichment in ^{21}Ne can be attributed to the so-called Wetherill reactions—$^{18}O(\alpha,n)^{21}Ne$ and $^{24}Mg(n,\alpha)^{21}Ne$ (Wetherill, 1954), producing nucleogenic ^{21}Ne in the Earth's mantle over the past 4.55 Ga.

In contrast to MORB samples, OIBs (Hawaii, Iceland, and Reunion Island) plot with steeper trajectories in three-isotope neon space i.e., for a given $^{21}Ne/^{22}Ne$ ratio OIB would have a higher $^{20}Ne/^{22}Ne$ value than MORBs (Figure 4). Following the same extrapolation methodology to correct for atmospheric contamination as described above for MORBs, then OIB mantle would have a lower

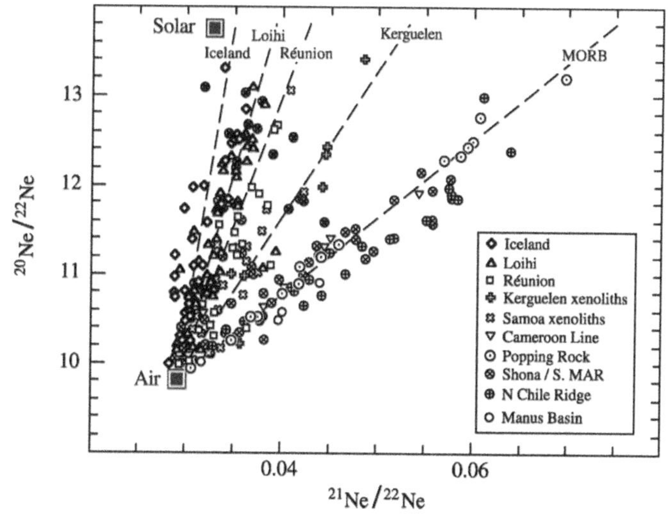

Figure 4 Three-isotope neon plot showing isotopic composition of air and solar neon and the MORB correlation line plus trajectories for various OIBs (reproduced by permission of Mineralogical Society of America from *Rev. Mineral. Geochem.*, **2002**, *47*, 280).

$^{21}Ne/^{22}Ne$ ratio when extrapolated to a solar $^{20}Ne/^{22}Ne$ ratio. Therefore, evolution of the OIB $^{21}Ne/^{22}Ne$ ratio has been retarded with respect to that of MORB mantle. This observation is consistent with the notion that the Earth accreted with solar neon, and production of nucleogenic ^{21}Ne has simply moved OIB and MORB mantle to higher $^{21}Ne/^{22}Ne$ ratios with time (Honda *et al.*, 1991). The observation, however, that OIBs have experienced less of a shift in $^{21}Ne/^{22}Ne$ than MORBs implies that the OIB source reservoir has a higher time-integrated $^{21}Ne/(U + Th)$ ratio than the MORB mantle, consistent with the idea that it has suffered a lower degree of primordial (solar) gas loss over time.

In the solid Earth, production of nucleogenic ^{21}Ne is coupled to that of radiogenic ^{4}He. This is because production of ^{21}Ne is directly proportional to the α-particle production ratio from the uranium and thorium series. The $^{21}Ne/^{4}He$ production ratio is constant and has been estimated at a value of 4.5×10^{-8} (Yatsevich and Honda, 1997). In this way, if the Earth accreted with solar helium and neon and initial ratios were modified by production of ^{21}Ne and ^{4}He in a fixed proportion then the present-day $^{3}He/^{4}He$ and $^{21}Ne/^{22}Ne$ ratios in the mantle should be correlated. Honda *et al.* (1993) noted a strong correlation between OIB helium and neon isotopes such that steeper trajectories in three-isotope neon space were characterized by samples with high $^{3}He/^{4}He$ ratios. Indeed, they showed that it was possible to estimate the $^{3}He/^{4}He$ ratio of a suite of OIBs based solely on measurements of the neon isotope composition.

More recent work has found that the correlation between helium and neon breaks down for certain localities. For example, Dixon *et al.* (2000) found solar-like neon isotopic compositions for a suite of Icelandic basalts even through their $^{3}He/^{4}He$ ratios were $<30R_A$. In this case, the gradient of the correlation on the three-isotope neon plot was steeper than anticipated from the $^{3}He/^{4}He$ ratios. Such decoupling of helium and neon was explained by source heterogeneity, and the preservation of domains in the OIB source mantle with high $[Ne_{solar}]/(U + Th)$ ratios (Dixon *et al.*, 2000). Shaw *et al.* (2001) provide another example of He–Ne decoupling: however, in the case of the Manus Basin, the decoupling is in the opposite sense to that of Iceland. The Manus sample suite (with $^{3}He/^{4}He >$ MORB) has a gradient less than that of MORB (Figure 4) consistent with preferential loss of neon relative to helium (the $^{3}He/^{22}Ne$ ratio—an index of degassing—is high for the Manus Basin compared to MORB: 23 versus 10). While it is possible that the fractionation could have occurred not long (~10 Ma) before generation of the Manus basalts, it is also possible that the fractionation is a remnant of early Earth degassing from a magma ocean (Shaw *et al.*, 2001). The relationship between the isotopes of helium and neon, therefore, bears fundamental information on the nature of the mantle source regions and the degassing history of the planet.

The other main source of information on the neon isotope systematics of the oceanic mantle comes from arc- and back-arc basalts. With the exception of the Manus Basin samples discussed above, the available back-arc basalt database (from the Lau Basin and Mariana Trough) show coherent He–Ne isotope systematics such that the neon isotope data fall on the MORB array of the three-isotope neon plot and the samples have $^{3}He/^{4}He$ ratios within the canonical MORB range. From a neon perspective, therefore, it seems that normal MORB-type mantle underlies these back-arc regions of the western Pacific. The few arc mantle samples analyzed for neon generally convey the same information as back-arc basalts (Figure 5) except for one striking difference: arc phenocrysts from the Aeolian arc and Campanian Magmatic Province have highly nucleogenic neon (i.e., enrichments in ^{21}Ne and, to a lesser extent, ^{22}Ne relative to air). This observation has been ascribed to the influence of subducted crustal material in the region, contributing nucleogenic neon to the source of arc magmas (Tedesco and Nagao, 1996). Curiously, these samples have arc-like $^{3}He/^{4}He$ ratios between $7R_A$ and $8R_A$, giving another and hitherto unexplained example of He–Ne decoupling.

The identification of mantle-like neon isotopic compositions in crustal materials is often difficult due to potentially significant production of nucleogenic neon in crustal lithologies and/or overwhelming contamination by air neon in a subaerial environment. However, deconvolving mantle, crustal, and atmospheric contributions to the neon budget is possible because the isotopic compositions of the three end-members are so distinct (Ballentine and O'Nions, 1992). Using helium and neon data of natural gases from a number of continental regions worldwide, Ballentine (1997) showed that after subtracting the atmospheric neon component, the vast majority of samples plot on a well-defined mixing trajectory between mantle- and crustal-neon end-members. This approach allows an estimate of the $^{3}He/^{22}Ne$ ratio of the subcontinental mantle assuming that mantle volatiles are introduced into tectonically active areas of the continents by emplacement of small degrees of asthenospheric melts. Surprisingly, the range of $^{3}He/^{22}Ne$ ratios observed for the mantle end-member component lies between 0.6 and 1.9, or approximately one-half the value estimated for MORB mantle. The explanation for this apparent discrepancy lies with the realization that solubility-controlled degassing of basaltic melt preferentially releases neon to helium (Section 17.2.2). Consequently, although volatiles

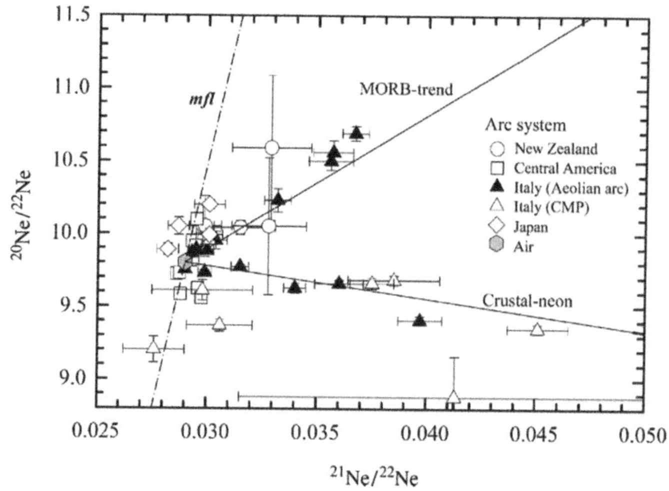

Figure 5 Three-isotope neon plot showing arc-related neon isotope variations (reproduced by permission of Mineralogical Society of America from *Rev. Mineral. Geochem.*, **2002**, *47*, 331).

from the subcontinental mantle are trapped (and can be sampled) in regions of continental extension, these volatiles are characterized by elemental abundances modified from source values.

17.3.4 Argon Isotopes

Argon is composed of three isotopes: ^{36}Ar, ^{38}Ar, and ^{40}Ar. The former two isotopes are primordial in origin with no significant contributions by production within the Earth: ^{40}Ar, however, is produced by decay of ^{40}K ($t_{1/2} = 1.40 \times 10^9$ yr). The initial (or primordial) $^{40}Ar/^{36}Ar$ ratio of the solar system lies between 10^{-4} and 10^{-3} (Begemann *et al.*, 1976). In contrast, the terrestrial atmosphere has a $^{40}Ar/^{36}Ar$ ratio of 296, and all other terrestrial materials have higher values. Terrestrial $^{40}Ar/^{36}Ar$ ratios therefore are essentially the result of mixing of radiogenic ^{40}Ar with primordial ^{36}Ar.

A large range in $^{40}Ar/^{36}Ar$ ratios has been measured in MORBs. The variation is generally interpreted as due to mixing of variable proportions of atmospheric argon (with $^{40}Ar/^{36}Ar = 296$) with a single, much more radiogenic, mantle composition. The minimum value for this mantle composition is represented by the highest measured values of 2.8×10^4 (Staudacher *et al.*, 1989) to 4×10^4 (Burnard *et al.*, 1997). From correlations between $^{20}Ne/^{22}Ne$ and $^{40}Ar/^{36}Ar$ in step-heating results of a gas-rich MORB sample, a maximum value of $^{40}Ar/^{36}Ar = 4.4 \times 10^4$ has been obtained (Moreira and Allègre, 1998). This (isotopic) composition is an important constraint on the degassing of the mantle (see Section 17.4.2).

It is difficult to determine if any argon isotopic variations exist in the upper mantle due to the pervasive contamination of mantle-derived samples by atmospheric argon. This is an important topic, as a range in mantle $^{40}Ar/^{36}Ar$ values can potentially be used to trace the impact of either variable degassing/contamination or other processes such as subduction of argon. It has been suggested that there are indeed heterogeneities in the $^{40}Ar/^{36}Ar$ ratio of the MORB source, based on a correlation between lead isotopes and $^{40}Ar/^{36}Ar$ ratios for samples with $^3He/^4He \leq 9.5R_A$ (Sarda *et al.*, 1999). This was ascribed to subduction of a component with relatively radiogenic lead and low $^{40}Ar/^{36}Ar$, implying a substantial flux of argon from the surface to the upper mantle. However, Burnard (1999) pointed out that lead isotope compositions are known to be more radiogenic in shallow eruptive environments, and that the lower $^{40}Ar/^{36}Ar$ values correlate equally well with depth of eruption. Therefore, the shallower samples may have lower $^{40}Ar/^{36}Ar$ ratios due to greater gas depletion and so proportionally more air contamination. Also, Ballentine and Barfod (2000) showed that the amount of air contamination is related to basalt vesicularity, which in turn is related to eruption depth and volatile content, further explaining correlations of atmosphere-derived noble gases with radiogenic lead (Sarda *et al.*, 1999). Overall, the debate has highlighted the difficulty of separating true mantle $^{40}Ar/^{36}Ar$ variations from contamination effects which occur either within the magma chamber (Burnard *et al.*, 1994; Farley and Craig, 1994), through assimilation of crustal material during melt transit to the surface (Hilton *et al.*, 1993a), by equilibration with seawater during eruption (Patterson *et al.*, 1990), or via sample vesicularity (Ballentine and Barfod, 2000).

Considerable effort has also been expended determining whether OIBs—with distinctive $^3He/^4He$ ($>10R_A$), have $^{40}Ar/^{36}Ar$ ratios that can be distinguished from those of the MORB source. It is generally expected that high $^3He/^4He$ ratios, reflecting high $^{3}He/(U + Th)$ ratios in the source region(s), would be accompanied by low $^{40}Ar/^{36}Ar$ ratios, reflecting similarly high $^{36}Ar/K$ ratios. Surprisingly, the first $^{40}Ar/^{36}Ar$ values reported for Loihi Seamount glasses were found to be similar to the values for air, prompting suggestions of a complementary relationship between argon in the OIB source and the atmosphere (Allègre *et al.*, 1983; Kaneoka *et al.*, 1986; Staudacher *et al.*, 1986). At that time, ideas of air contamination (e.g., Fisher, 1985; Patterson *et al.*, 1990) were dismissed. However, a study of basalts from Juan Fernandez (Farley and Craig, 1994) found that the atmospheric argon component resided within phenocrysts—an observation consistent with introduction of air argon into the magma chamber. This study provided a ready explanation for the prevalence of air contamination of OIBs. More recent measurements of Loihi samples have reported significantly higher $^{40}Ar/^{36}Ar$ values: between 2,600 and 2,800 (Hiyagon *et al.*, 1992; Valbracht *et al.*, 1997), and up to 8,000 (Trieloff *et al.*, 2000)—all on samples with $^{3}He/^{4}He$ as high as $24R_A$ (and so midway between MORB and the highest OIB). Additionally, Poreda and Farley (1992) found values of $^{40}Ar/^{36}Ar \leq 1.2 \times 10^4$ in Samoan xenoliths that have intermediate $^{3}He/^{4}He$ ratios ($9–20R_A$). Other attempts at characterizing OIB $^{40}Ar/^{36}Ar$ have used associated neon isotope variations and the (debatable) assumption that the contaminant Ne/Ar ratio is constant in a strategy designed to remove the effects of air contamination. Using this approach, Sarda *et al.* (2000) reported that $^{40}Ar/^{36}Ar$ ratios in the high $^{3}He/^{4}He$ OIB source are >3,000 but still considerably lower than that of the MORB source (see also Matsuda and Marty, 1995). It must be concluded, therefore, that due to the difficulties of separating atmospheric argon contamination, present-day estimates of OIB $^{40}Ar/^{36}Ar$ ratios represent a lower limit only.

17.3.5 Xenon Isotopes

There are nine isotopes of xenon. Isotopic variations have been generated in the Earth due to the β decay of ^{129}I ($t_{1/2} = 16$ Ma) producing ^{129}Xe, and the spontaneous fission of both ^{244}Pu ($t_{1/2} = 80$ Ma) and ^{238}U ($t_{1/2} = 4.47$ Ga). Since ^{129}I and ^{244}Pu are short-lived nuclides, they were only present early in Earth history. MORB $^{129}Xe/^{130}Xe$ and $^{136}Xe/^{130}Xe$ ratios lie on a correlation extending from atmospheric ratios to higher values (Kunz *et al.*, 1998; Staudacher and Allègre, 1982), and likely reflect mixing of variable proportions of air contaminant xenon with a single upper mantle component having more radiogenic $^{129}Xe/^{130}Xe$ and $^{136}Xe/^{130}Xe$ ratios (Figure 6). The highest measured values thus provide lower limits for the MORB source. As with argon isotopes, due to the pervasiveness of atmospheric xenon contamination, it has not been possible to resolve isotopic variations in the upper mantle that can serve as useful tracers.

While the origin of the high $^{129}Xe/^{130}Xe$ ratios is clearly due to decay of ^{129}I, an important issue regarding mantle xenon is the origin of radiogenic ^{136}Xe. Contributions to ^{136}Xe enrichments in MORBs by decay of ^{238}U or ^{244}Pu in theory can be distinguished based on the spectrum of contributions to other xenon isotopes, although analyses have typically not been sufficiently precise to identify the parent nuclide. Precise measurements of xenon in CO_2 well gases found ^{129}Xe and ^{136}Xe enrichments similar to those found in MORBs,

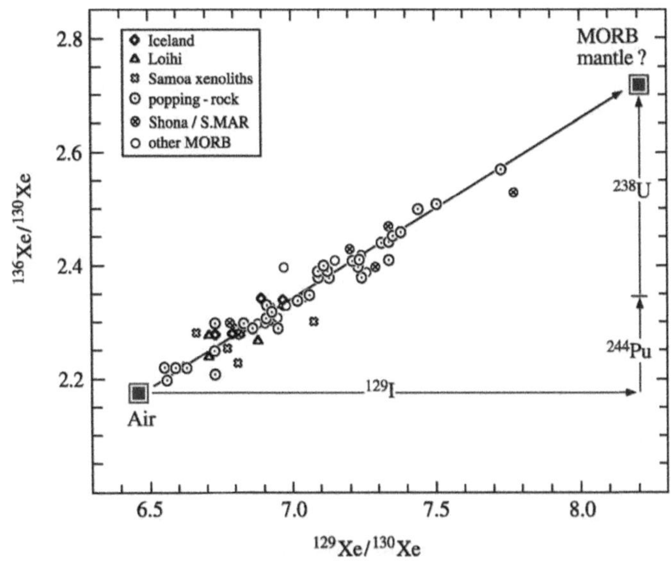

Figure 6 Xenon isotope plot for MORBs and OIBs. A proportion of 32% for fissiogenic ^{136}Xe (from ^{224}Pu decay) is shown (reproduced by permission of Mineralogical Society of America from *Rev. Mineral. Geochem.*, **2002**, *47*, 291).

and so it is likely that this xenon is derived from the MORB source (Staudacher, 1987). These indicate that ^{244}Pu has contributed <10–20% of the ^{136}Xe that is in excess of the atmospheric composition (Caffee *et al.*, 1999; Phinney *et al.*, 1978). An error-weighted best fit to recent precise MORB data (Kunz *et al.*, 1998) yielded a value of 32 ± 10% for the fraction of ^{136}Xe excesses relative to atmosphere that are ^{244}Pu-derived (Figure 6), although there is considerable scatter in the data (see Marti and Mathew, 1998). Clearly, further work is warranted on the proportion of plutonium-derived heavy xenon in the mantle, although it appears that the fissiogenic xenon is dominantly derived from uranium.

Measurements of xenon in high ^{3}He/^{4}He OIB samples have often found atmospheric isotopic ratios (e.g., Allègre *et al.*, 1983) that appear to be due to overwhelming air xenon contamination (Harrison *et al.*, 1999; Patterson *et al.*, 1990) rather than reflecting mantle xenon with an air composition. Although Samoan samples with intermediate (9–20R_A) helium isotope ratios have been found with xenon isotopic ratios distinct from those of the atmosphere (Poreda and Farley, 1992), the xenon in these samples may have been derived largely from the MORB source. Harrison *et al.* (1999) have found slight ^{129}Xe excesses in Icelandic samples with ^{129}Xe/^{3}He ratios that are compatible with the ratio in gas-rich MORB, but due to the uncertainties in the data it cannot be determined whether there are indeed differences between the MORB and OIB sources. Trieloff *et al.* (2000) reported xenon isotope compositions in Loihi dunites and Icelandic glasses that were on the MORB correlation line and had values up to ^{129}Xe/^{130}Xe = 6.9. These were accompanied by ^{3}He/^{4}He ratios up to 24R_A, and so may contain noble gases from both MORB (~8R_A) and the highest ^{3}He/^{4}He ratio (37R_A) OIB source. From these data it appears that the OIB source may have xenon that is different from the terrestrial atmosphere but (marked) differences between OIB and MORB still await discovery.

17.3.6 Mantle Noble Gas Abundances

In order to use noble gases as tracers for the involvement of different mantle components in a particular environment, the absolute concentrations are required to determine how each noble gas signature will have an impact. The concentrations of helium are most easily inferred, and those of the other noble gases can be determined from the elemental abundance patterns.

17.3.6.1 Helium concentrations in the upper mantle

An estimate of the ^{3}He concentration of the upper mantle can be obtained by noting that there is a clearly recognizable flux of ^{3}He into the oceans from degassing of MORB. The mantle ^{3}He flux from mid-ocean ridges (Clarke *et al.*, 1969; Craig *et al.*, 1975) has been obtained by combining seawater ^{3}He concentrations in excess of dissolved air helium with seawater mixing and ocean-atmosphere exchange models (see review by Schlosser and Winckler, 2002). A value of 1,060 ± 250 mol yr^{-1} ^{3}He (Lupton and Craig, 1975) has been confirmed with more recent ocean circulation models (Farley *et al.*, 1995). This value represents an average over the last 1,000 years. The concentration of ^{3}He in the mantle can be determined by dividing the flux of ^{3}He into the oceans by the rate of production of melt that is responsible for carrying this ^{3}He from the mantle, which is equivalent to the rate of ocean crust production of 20 km^3 yr^{-1} (Parsons, 1981). MORBs that degas quantitatively to produce a ^{3}He flux of 1,060 mol yr^{-1} then must have an average ^{3}He content of 1.96×10^{-14} mol g^{-1} or 4.4×10^{-10} cm^3 STP g^{-1}. Note that the validity of combining a flux estimate based upon a millennium-scale record of ocean ventilation with a magma production rate derived using paleomagnetic reversals occurring on considerably longer timescales has been questioned recently (Ballentine *et al.*, 2002). It is significant, however, that the derived ^{3}He concentration lies within a factor of 2 of that obtained for the most gas rich basalt glass yet found (the so-called "popping-rock"), with ~10.0×10^{-10} cm^3 STP ^{3}He g^{-1} (Moreira *et al.*, 1998; Sarda *et al.*, 1988). This sample is noteworthy in having sample vesicle size distributions consistent with closed system formation during magma ascent (Sarda and Graham, 1990) and individual vesicles with He/Ar, C/He, and C/N ratio variations consistent with closed system formation during ascent (Javoy and Pineau, 1991). An independent estimate of the carbon content of undegassed MORB of 900–1,800 ppm C (Holloway, 1998), combined with a MORB-source mantle CO_2/^{3}He ratio of 2×10^9 (Marty and Jambon, 1987) gives a ^{3}He concentration in undegassed MORB of $(8.4–17.0) \times 10^{-10}$ cm^3 STP ^{3}He g^{-1}. Assuming 10% partial melting and quantitative extraction of helium from the solid phase into the melt (Section 17.2.2), these undegassed melt values give a MORB-source mantle concentration of $(0.44–1.7) \times 10^{-10}$ cm^3 STP ^{3}He g^{-1} or 1.2–4.6×10^9 atoms ^{3}He g^{-1} (Porcelli and Ballentine, 2002).

17.3.6.2 Helium concentrations in the OIB source

It is generally assumed that there is a single mantle reservoir that supplies the helium of ocean islands with high ^{3}He/^{4}He ratios. There are two approaches to estimating the concentration of ^{3}He in this mantle source: (i) through using model calculations, and (ii) via direct measurements of OIB volcanics. Helium with high $^3He/^4He$ ratios is often assumed to be stored in a mantle reservoir that has evolved approximately as a closed system for noble gases and has bulk silicate Earth (BSE) parent nuclide concentrations. Since this reservoir must be isolated from degassing at mid-ocean ridges and subduction zones, it is often placed within the deeper, or lower mantle. Assigning the highest OIB $^3He/^4He$ ratios to this reservoir, a comparison between the total production of ^4He and the shift in $^3He/^4He$ from the initial terrestrial value to the present value provides an estimate of the ^3He concentration in this reservoir. Using BSE values of 21 ppb U and Th/U = 3.8 (Doe and Zartman, 1979; Rocholl and Jochum, 1993), 1.02×10^{15} atoms $^4He\ g^{-1}$ are produced over 4.55 Ga. Using the highest Iceland value of $^3He/^4He = 37R_A$ to represent this reservoir and an initial value of $^3He/^4He = 120R_A$ (Section 17.2.1) then the reservoir has 7.6×10^{10} atoms $^3He\ g^{-1}$. This is over an order of magnitude greater than the value derived above for the MORB source mantle. Note that other explanations have been put forth regarding the nature of the reservoir supplying high $^3He/^4He$ hotspots (see discussion in Section 17.4.2), although a gas-rich source has been the most commonly assumed.

The flux of ^3He from intra-plate volcanic systems is dominantly subaerial and so it is not possible to directly obtain the time-integrated value for even a short geological period. It is noteworthy that OIB samples typically have substantially lower helium concentrations than MORB samples. However, extensive degassing during eruption can account for the strong depletion of helium in many samples. Therefore, it is often assumed that OIB source ^3He concentrations are high, but have been highly degassed prior to or during eruption (see Marty and Tolstikhin, 1998). There may be a number of factors promoting more extensive degassing for OIB than MORB. OIB source may be more volatile-rich than MORB and therefore may degas more effectively (Dixon and Stolper, 1995). Also, the water content of basalts reduces CO_2 solubility: therefore, water-rich basalts (with high $^3He/^4He$ ratios e.g., along the shallow Reykjanes Ridge) can have significantly lower CO_2 (and helium) compared to water-poor MORB erupted at greater confining pressures/ depths (Hilton *et al.*, 2000b).

Concentrations of ^3He in OIB sources can be estimated in several ways. The ^3He flux can be estimated from volcanic CO_2 fluxes and magma generation rates. The summit CO_2 flux of Kilauea volcano, Hawaii, can be combined with observed $CO_2/^3He$ values and estimated magma production rates to obtain an undegassed plume magma concentration of $4.7 \times 10^{-10} cm^3$ STP $^3He\ g^{-1}$ (Hilton *et al.*, 1997). Assuming magma production by 7% melting, this corresponds to a source concentration of $3.3 \times 10^{-11}\ cm^3$ STP $^3He\ g^{-1}$. Prior plume degassing (at Loihi Seamount?) was invoked to explain such a low value (Hilton *et al.*, 1997). Moreira and Sarda (2000) assessed the extent of prior basalt degassing from the magnitude of air-corrected helium, neon, and argon fractionations in filtered MORB and OIB data sets. Using an open system Rayleigh degassing model, they calculated a minimum OIB source concentration of $1.1 \times 10^{-10}\ cm^3$ STP $^3He\ g^{-1}$, which is comparable to that of the MORB source. Furthermore, they suggested that measured OIB helium concentrations can also be subject to variable post-eruption gas loss that occurs with no elemental fractionation (e.g., by vesicle rupture during sample handling), and so the source may have had even higher concentrations.

OIB source helium concentrations can also be derived from isotopic mixing patterns at various locations. For example, where plumes have strong interactions with mid-ocean ridges, such as along the Reykjanes Ridge, south of Iceland, basalt characteristics can be ascribed to mixing between a MORB source and a plume component. Hilton *et al.* (2000b) found that $^3He/^4He$ and $^{206}Pb/^{204}Pb$ are linearly correlated along the Reykjanes Ridge and may reflect mixing between components with similar He/Pb ratios. Eiler *et al.* (1998) found a similar relationship in Hawaiian lavas. In the limiting case where the lead concentration of the plume is the same as that of MORB-source mantle, the ^3He concentration in the plume source is four times higher than that of the MORB-source mantle, taking into account differences in end-member $^3He/^4He$ ratios. Higher plume lead concentrations would increase this value. For example, adopting Pb = 0.05 ppm for the MORB-source mantle and 0.185 ppm for the plume source (the BSE value), Hilton *et al.* (2000b) calculated that the ^3He concentration of the plume source component is 15 times that of MORB-source mantle. Further, they suggested that the data might support a mixing line involving an even higher plume source helium concentration. Predegassing of the plume source prior to mixing with MORB-source mantle also remains a possibility, thus allowing a higher helium concentration in the plume component.

It should be emphasized that, in many locations, the high $^3He/^4He$ source component comprises a small fraction of the sample source. A very gas-rich helium source need only contribute a small mass fraction of the source region to dominate the helium budget. The presence of the OIB source in such a case would be much less discernible in other chemical signatures, and so helium isotopes would provide the most sensitive indicator of mixing. Alternatively, if the OIB source has concentrations that are much lower than that of the MORB source, then clearly it must constitute a much greater proportion of ocean island source regions, and this should be evident in other chemical signatures as well.

17.3.6.3 Noble gas abundance patterns

Noble gas abundance patterns measured in MORBs and OIBs scatter greatly. This is due to post-eruptive alteration processes as well as fractionation during noble gas partitioning between basaltic melts and a vapor phase. Moreover, the vapor phase may be preferentially gained or lost by a particular sample. However, MORB Ne/Ar and Xe/Ar ratios that are greater than the air values are common. This pattern was found in a gas-rich MORB (popping-rock) sample with high $^{40}Ar/^{36}Ar$ and $^{129}Xe/^{130}Xe$ ratios (and so containing relatively little air contamination) and $^{4}He/^{40}Ar$ ratios that are near the expected production ratio of the upper mantle (and so not fractionated) (Moreira *et al.*, 1998; Staudacher *et al.*, 1989).

A noble gas abundance pattern for the mantle that is not dependent upon unraveling fractionation effects due to transport and eruption can be calculated by assuming that the radiogenic nuclides of the different elements are present in the mantle in relative abundances that are equal to their production ratios: in this case, the relative proportions of radiogenic ^{4}He, ^{21}Ne, ^{40}Ar, and ^{136}Xe are known. This process assumes that the changes in the $^{3}He/^{4}He$, $^{21}Ne/^{22}Ne$, $^{40}Ar/^{36}Ar$, and $^{136}Xe/^{130}Xe$ ratios from their initial values can be related to the relative abundances of ^{3}He, ^{22}Ne, ^{36}Ar, and ^{130}Xe (see Porcelli and Ballentine, 2002). Using the MORB isotopic compositions discussed above, and K/U = 1.27×10^{4} (Jochum *et al.*, 1983) and Th/U = 3.8 for parent element relative ratios, then $^{3}He/^{22}Ne_{MORB} = 11$, $^{3}He/^{36}Ar_{MORB} = 1.7$ (and so $^{22}Ne/^{36}Ar_{MORB} = 0.15$), and $^{130}Xe/^{36}Ar_{MORB} = 3.3\times10^{-4}$. This is consistent with the pattern discussed above, with $^{22}Ne/^{36}Ar$ and $^{130}Xe/^{36}Ar$ ratios greater than those of the atmosphere (0.05×10^{-4} and 1.1×10^{-4}, respectively). Note that these arguments change qualitatively only if some noble gases are removed from the mantle with much shorter time constants, thereby maintaining ratios between radiogenic isotopes that are different from the production ratios.

Unfortunately, it is not possible to similarly calculate the abundance pattern of OIB source regions, since the heavy element isotope ratios are not constrained. A survey of helium and neon isotope compositions of OIB and MORB has allowed calculations of the $^{3}He/^{22}Ne$ ratios for the source region of each type, with $^{3}He/^{22}Ne = 6.0\pm1.4$ for Hawaii and 10.2 ± 1.6 for MORB, with a mantle average of 7.7 (Honda and McDougall, 1998). This approach suggests that there may be a systematic difference between the MORB and OIB sources. In contrast, it has been argued that Loihi and gas-rich MORB samples have similar $^{3}He/^{22}Ne$ ratios (Moreira and Allègre, 1998), and so the issue remains open.

17.4 NOBLE GASES AS MANTLE TRACERS

In this section, we focus on three key areas of noble gas studies of the mantle. First, we consider how noble gases (^{3}He in particular) are used to assess volatile fluxes involving the mantle as a principal terrestrial reservoir. It is possible to both understand the cycling history of the noble gases between the mantle and the atmosphere and hydrosphere using this approach and to constrain the provenance of major volatiles (such as CO_2 and N_2) thereby leading to a better understanding of their evolution in the mantle. Second, we illustrate how noble gases can be used to place constraints on the different classes of model that have been proposed to account for the structure of the mantle. A related topic also covered in this section is the utility of noble gases to trace the origin of plumes and the formation of ocean islands. Finally, we assess the noble gas systematics of mantle-derived xenoliths as a means to gain insight into the development and evolution of the subcontinental mantle, and to trace the transport of volatiles from the convecting mantle into the lithosphere.

17.4.1 Volatile Fluxes and Geochemical Recycling

Noble gases have been at the forefront of studies defining volatile fluxes between the mantle and other terrestrial reservoirs. This stems from the fact that in the case of ^{3}He there is no question or ambiguity regarding its origin: the mantle ^{3}He budget is overwhelmingly dominated by the primordial component (Craig and Lupton, 1981). Furthermore, it is implicitly assumed that the flux of ^{3}He is unidirectional from the mantle to the exosphere (hydrosphere, atmosphere, and crust) based upon the hypothesis that subduction of ^{3}He into the mantle occurs—if at all—at a level negligible relative to the mantle inventory (Hilton *et al.*, 1992; Hiyagon, 1994; Trull, 1994). Consequently, mantle ^{3}He fluxes can be estimated with a fair degree of certainty, and exploited as a flux monitor for other volatiles of geochemical interest (e.g., CO_2 and N_2). In this way, ^{3}He is integral to assessing the state of volatile mass balance involving the mantle and other terrestrial reservoirs.

17.4.1.1 ^{3}He fluxes from the mantle

As discussed in Section 17.3.6.1 a value of $1{,}060\pm250$ mol yr^{-1} is taken as the canonical value for the upper mantle ^3He degassing flux (Lupton and Craig, 1975). This estimate is based upon knowledge of the ^3He concentration anomaly in the oceans (relative to saturation) coupled with ocean–atmosphere exchange models (see review by Schlosser and Winckler, 2002). Estimating the flux of ^3He associated with OIB volcanism is more problematic, however. The parameters necessary to make an estimate of the OIB ^3He flux are (i) rates of OIB magmatism, and (ii) the helium concentration of the source region(s).

Estimates of the rate of intraplate magma production vary considerably: 0.6×10^{15} g yr^{-1} (Reymer and Schubert, 1984), 4.4×10^{15} g yr^{-1} (Schilling *et al.*, 1978), $(5–7) \times 10^{15}$ g yr^{-1} (including seamounts) (Batiza, 1982), and $(0.4–5) \times 10^{15}$ g yr^{-1} when considering magma emplacement estimates for the last 180 Ma (Crisp, 1984). These values fall between 1% and 12% of the MORB production rate.

Information on the ^3He content of OIB sources is available from both modeling and concentration measurements of helium in intraplate volcanics (Section 17.3.6.2). For example, Hilton *et al.* (2000b) estimated that the high $^3He/^4He$ component erupted along the Reykjanes Ridge had a ^3He content 15 times that of the MORB source. Unfortunately, it is not clear what fraction of such source material supplying OIBs worldwide has this concentration (Porcelli and Ballentine, 2002). However, using the highest estimate of source ^3He content over a range of assumed OIB magma production rates, Porcelli and Ballentine (2002) obtained a global flux of $8.4 \times 10^4–1.5 \times 10^6$ cm^3 STP ^3He yr^{-1} (38–670 mol ^3He yr^{-1}) for the OIB ^3He flux from the mantle. It should be noted, however, that helium characteristics of various OIB sources are likely to vary considerably. The range of measured $^3He/^4He$ ratios indicates that several components may be involved in differing proportions, and these are likely to have different helium concentrations. Therefore, detailed studies of individual hotspots may not necessarily be combined to yield a representative estimate.

Subduction zones also represent a primary escape route for primordial ^3He from the mantle. There are two general approaches for estimating the arc ^3He flux. First, assuming that the ^3He content of magma in the mantle wedge is the same as that beneath spreading ridges, Torgersen (1989) combined the mid-ocean ridge degassing flux (~1,000 mol ^3He yr^{-1}; Craig *et al.*, 1975) with a figure of 20% as the assumed magma production rate of arcs relative to MOR (Crisp, 1984) to yield an arc ^3He flux of $\sim 200 \pm 40$ mol yr^{-1}. The second method relies on actual ^3He (and some other normalizing species) measurements of arc-related and other subaerial volcanoes. For example, Allard (1992) derived an estimate of 240–310 mol yr^{-1} for the total flux of ^3He into the atmosphere by subaerial volcanism (based upon integrating the CO_2 flux from 23 individual volcanoes worldwide and coupling this flux with measurements of the $CO_2/^3He$ ratios). Of the total subaerial ^3He flux, he suggested that approximately 70 mol yr^{-1} was arc-related. Adopting a similar approach, Marty and LeCloarec (1992) used ^{210}Po as the flux indicator (along with the $^{210}Po/^3He$ ratio) to estimate a total subaerial volcanic ^3He flux of 150 mol yr^{-1}—of which over half (>75 mol yr^{-1}) was due to arc volcanism. Hilton *et al.* (2002) derived an estimate of 92 mol yr^{-1} based on integrating SO_2 fluxes (combined with $SO_2/^3He$ ratios) from individual arc segments worldwide.

It is important to point out that neither approach attempts a direct measurement of the arc ^3He flux. To estimate magma production rates, scaling is used in the first instance, whereas the second methodology relies on knowledge of an absolute flux of some chemical species from volcanoes together with a measurement of the ratio of that species to ^3He. The most widely used species to derive absolute chemical fluxes from subaerial volcanoes is SO_2 using the correlation spectrometer technique (COSPEC) (Stoiber *et al.*, 1983). Carbon dioxide (see Brantley and Koepenick, 1995) and ^{210}Po (Marty and LeCloarec, 1992) have also been used in an analogous manner.

Continental settings provide a small but significant flux of ^3He from the mantle. In the Pannonian Basin (4,000 km^2 in Hungary), the flux of ^3He has been estimated at 8×10^4 atoms m^{-2} s^{-1} (Martel *et al.*, 1989) and $0.8–5 \times 10^4$ atoms m^{-2} s^{-1} (Stute *et al.*, 1992). Making the assumption that the area of the continents is 2×10^{14} m^2 and that 10% is under extension, this yields a total ^3He flux of 8.4–84 mol ^3He yr^{-1}. This value compares with <3 mol ^3He yr^{-1} from the stable continental crust (see O'Nions and Oxburgh, 1988).

In summary, based upon the range of values reported for the different regions (MOR, OIB, arcs, and continents), the total rate of contemporary degassing of mantle-derived ^3He lies in the range of 1,039–2,270 mol yr^{-1}. The oceanic ridge system dominates the flux with a contribution between 57% and 78 % of the total.

17.4.1.2 Major volatile fluxes

The establishment of the ^3He flux from the mantle allows scaling to other volatiles of geochemical interest so that their mantle fluxes can also be quantified. The seminal work in this respect is that of Marty and Jambon (1987), who combined measured $CO_2/^3He$ ratios in MORB ($\sim 2 \times 10^9$) with the MOR ^3He flux estimate of

1,000 mol yr^{-1} (Craig *et al.*, 1975) to yield a value of 2×10^{12} mol yr^{-1} for the flux of CO_2 from the upper mantle. Although there is evidence of fractionation between helium and CO_2 during magmatic degassing (e.g., Hilton *et al.*, 1998), it has been argued that similar solubilities of helium and CO_2 in, and restricted compositional range typical of, most tholeiites erupted along oceanic ridges results in little or negligible modification of the source $CO_2/^3He$ value: in any case, a correction is possible using measured He/Ar ratios (Marty and Zimmermann, 1999). In addition to allowing assessment of volatile mass balance (see below), quantifying the flux of magmatic carbon at ridges provides an independent check on the carbon content of the upper mantle (Marty and Tolstikhin, 1998).

In the case of arc volcanism, the critical observation is that the $CO_2/^3He$ ratio of arc-related volatiles is significantly greater than that of MORBs (Marty *et al.*, 1989; Sano and Marty, 1995; Sano and Williams, 1996; Varekamp *et al.*, 1992). In a compilation of subduction zone type gases, Marty and Tolstikhin (1998) report a median value of $11.0 \pm 3.3 \times 10^9$, or ~5 times that of MORBs. Consequently, an estimate of the proportion of carbon from nonmantle sources (~80%), presumably the subducted slab, can be inferred by scaling to the (upper mantle) $CO_2/^3He$ value.

Marty and Tolstikhin (1998) detail how knowledge of the arc $CO_2/^3He$ ratio can be used to calculate the arc-related CO_2 flux. They used the following relationship:

$$\varnothing_{C,arc} = \varnothing_{arc} \times [^3He]_{um} \times (CO_2/^3He)_{arc} \times 1/r_{arc}$$

where $\varnothing_{arc}$ is the flux of arc magma and r_{arc} is the mean partial melting rate at arcs. Adopting values of 2.5–8.0 $km^3\ yr^{-1}$ for the rate of magma emplacement at accreting plate margins (Crisp, 1984) and 10–25% for the range in the extent of partial melting (Plank and Langmuir, 1988), then a CO_2 arc flux of $\sim 2.5 \times 10^{12}$ mol yr^{-1} can be calculated (Marty and Tolstikhin, 1998). This estimate compares well with other values produced using different approaches: for example, Sano and Williams (1996) combined the arc $CO_2/^3He$ with the arc ^3He flux to estimate an arc CO_2 flux of 3.1×10^{12} mol yr^{-1} whereas Hilton *et al.* (2002) integrated CO_2 fluxes from arc segments worldwide to produce a global arc estimate of 1.6×10^{12} mol yr^{-1}.

The flux of CO_2 from plumes can be computed in an analogous manner to that described above for arcs. Marty and Tolstikhin (1998) estimated that the plume CO_2 flux at 3×10^{12} mol yr^{-1} by assuming plume $CO_2/^3He = 3 \times 10^9$, $r = 0.05$, $\varnothing_{plume} = 5 \times 10^{15}$g yr^{-1} and $[^3He]_{plume} = 10^{-14}$ mol g^{-1}. The flux is comparable to that at mid-ocean ridges in spite of the significantly lower magma production rates at hotspots compared to ridges. It should be appreciated, however, that plume flux estimates derived using this approach are only as good as the input parameters, and there is significant uncertainty in the ^3He content of plume source material (Section 17.3.6.2).

The above examples illustrate how knowledge of ^3He fluxes from various mantle reservoirs can be used to estimate concomitant fluxes of CO_2 from the same reservoirs. The methodology of using ^3He in this manner can also be extended to other volatiles of geochemical interest—nitrogen being a prime example. Marty (1995) estimated fluxes of $1.2–3.2 \times 10^9$ mol yr^{-1} for the N_2 flux from mid-ocean ridges based on measured $N_2/^{40}Ar$ ratios, assumed $^4He/^{40}Ar$ mantle production ratios and knowledge of the mid-ocean ridge ^3He (and ^4He) flux. Similarly, Sano *et al.* (2001) produced estimates of the total mantle nitrogen flux from mid-ocean ridges, plumes, and arcs (2.2×10^9, 0.004×10^9, and 0.6×10^9, in units of mol yr^{-1}, respectively) by using measured $N_2/^3He$ ratios and normalizing to ^3He flux estimates at each locality.

17.4.1.3 The question of volatile provenance

Volatile flux estimates derived by the approach described in the previous section make no distinction as to the source or provenance of the volatiles. However, for arc-related regions especially, quantitatively constraining the origin of the volatiles is important as they may be derived from different sources, vis-à-vis the mantle, subducting slab and arc lithosphere. In order to assess the state of volatile mass balance of the various terrestrial reservoirs, particularly involving outputs from the mantle and inputs associated with the subducting slab, the total arc output flux must be resolved into its component structures. As we show in this section, helium has proven remarkably sensitive in discerning volatile provenance. We use CO_2 and N_2 to illustrate the case.

As discussed above, the $CO_2/^3He$ ratio is significantly higher in arc-related terrains compared to mid-ocean ridge spreading centers. Such high values have been used to argue for addition of slab carbon to the source region of arc volcanism. However, in addition to the slab (both its sedimentary veneer and underlying oceanic basement), the mantle wedge and/or the arc crust through which magmas traverse en route to the surface may also contribute to the total carbon output. Distinguishing between these various sources is possible by considering carbon and helium together (both isotopic variations and relative abundances).

The carbon output at arcs can be resolved into end-member components involving MORB mantle (M), and slab-derived marine carbonate/limestone (L) and (organic) sedimentary components (S). Sano and Marty (1995) used the following mass balance equations:

$$({}^{13}C/{}^{12}C)_o = f_M({}^{13}C/{}^{12}C)_M + f_L({}^{13}C/{}^{12}C)_L + f_S({}^{13}C/{}^{12}C)_S$$

$$1/({}^{12}C/{}^{3}He)_o = f_M/({}^{12}C/{}^{3}He)_M + f_L/({}^{12}C/{}^{3}He)_L + f_S/({}^{12}C/{}^{3}He)_S$$

$$f_M + f_S + f_L = 1$$

where o = observed, and f is the fraction contributed by L, S, and M to the total carbon output. It is possible to determine the relative proportions of M-, L-, and S-derived carbon in individual samples of arc-related geothermal fluids by measuring the isotopic composition of carbon and the $CO_2/{}^3He$ ratio (which involves measuring the ${}^3He/{}^4He$ ratio). Appropriate end-member compositions must be selected, and both Sano and Marty (1995) and Sano and Williams (1996) suggest $\delta^{13}C$ values of $-6.5‰$, $0‰$, and $-30‰$ (relative to PDB) with corresponding $CO_2/{}^3He$ ratios of 1.5×10^9, 1×10^{13}, and 1×10^{13} for M, L, and S, respectively.

Based on the analysis of arc-related geothermal samples from 30 volcanic centers worldwide and utilizing high, medium, and low temperature fumaroles, Sano and Williams (1996) estimated that between 10% and 15% of the arc-wide global CO_2 flux is derived from the mantle wedge—the remaining 85–90% coming from decarbonation reactions involving subducted marine limestone, slab carbonate, and pelagic sediment. Subducted marine limestone and slab carbonate supply the bulk of the nonmantle carbon— ~70–80% of the total carbon—the remaining ~10–15% is contributed from subducted organic (sedimentary) carbon. It should be noted, however, that most studies ignore the arc crust as a potential source of carbon: although this omission may not be significant in intra-oceanic settings, this is unlikely to be the case at all localities (van Soest *et al.*, 1998). With this combined C–He approach, therefore, it is possible to more accurately constrain (i) mantle fluxes of various volatile species, and (ii) volatile mass balances at arcs i.e., input via the trench versus output via the arc.

Using an approach analogous to that for carbon, Sano *et al.* (1998, 2001) have attempted to understand the nitrogen cycle at subduction zones. Again, the problem is to identify and quantify the various contributory sources to the volcanic output. At subduction zones, there are three major sources of nitrogen: the mantle (M), atmosphere (A), and subducted sediments (S), and each has a diagnostic $\delta^{15}N$ value and $N_2/{}^{36}Ar$ ratio. Therefore, observed (o) variations in these two parameters for individual samples can be resolved into their component structures using the following equations (Sano *et al.*, 1998):

$$(\delta^{15}N)_o = f_M(\delta^{15}N)_M + f_A(\delta^{15}N)_A + f_S(\delta^{15}N)_S$$

$$1/(N_2/{}^{36}Ar)_o = f_M/(N_2/{}^{36}Ar)_M + f_A/(N_2/{}^{36}Ar)_A + f_S/(N_2/{}^{36}Ar)_S$$

$$f_M + f_S + f_A = 1$$

where f_M is the fractional contribution of mantle-derived nitrogen, etc. Note that the noble gas isotope ${}^{36}Ar$ is used in this case since degassing is not expected to fractionate the $N_2/{}^{36}Ar$ ratio due to the similar solubilities of nitrogen and argon in basaltic magma—this is not the case for helium and nitrogen, so that degassing corrections may be necessary if the $N_2/{}^3He$ ratio is used (Sano *et al.*, 2001). In the above scheme, end-member compositions are generally well constrained: the mantle and sedimentary end-members both have $N_2/{}^{36}Ar$ ratios of 6×10^6 (air is 1.8×10^4) but their $\delta^{15}N$ values are distinct. The upper mantle has a $\delta^{15}N$ value of $-5 \pm 2‰$ (Marty and Humbert, 1997; Sano *et al.*, 1998) whereas sedimentary nitrogen is assumed to be $+7 \pm 4‰$ (Bebout, 1995; Peters *et al.*, 1978) (air has $\delta^{15}N = 0‰$). The wide difference in $\delta^{15}N$ between the potential end-members makes this approach a sensitive tracer of N_2 provenance.

Gas discharges from island arc volcanoes and associated hydrothermal systems have $N_2/{}^{36}Ar$ ratios which reach a maximum of 9.7×10^4, with $\delta^{15}N$ values up to $+4.6‰$ (Sano *et al.*, 2001). This would indicate that a significant proportion (up to 70%) of the N_2 could be derived from a subducted sedimentary or crustal source. The situation is reversed in the case of back-arc basin glasses, which have significantly lower $\delta^{15}N$ values ($-2.7‰$ to $+1.9‰$): this implies that up to 70% of the nitrogen could be mantle-derived (Sano *et al.*, 2001). A first-order conclusion from these observations is that N_2 is efficiently recycled from the subducting slab to the atmosphere and hydrosphere through arc (and back-arc volcanism) (see also Fischer *et al.*, 2002).

There are two major issues of concern with the approach of using $CO_2/{}^3He$ and $N_2/{}^{36}Ar$ (or $N_2/{}^3He$) ratios in combination with $\delta^{13}C$ and $\delta^{15}N$ values to constrain the sources of volatiles at arcs. The first issue is the selection of representative end-member isotopic and relative elemental abundances—this factor has a profound effect on the deduced provenance of the volatile of interest. The second is the assumption that various elemental (and isotopic) ratios observed in the volcanic products are representative of the magma source. Both have the potential to compromise the accuracy of the output flux estimates.

In the case of CO_2, the methodology of Sano and co-workers assumes that subducted marine carbonate and sedimentary organic matter can be distinguished as potential input parameters based solely on their perceived carbon isotopic compositions prior to subduction (0‰ versus −30‰, respectively). However, this approach ignores the anticipated evolution of organic-derived CO_2 to higher $\delta^{13}C$ as a function of diagenetic and/or catagenetic changes experienced during subduction (Ohmoto, 1986). In the southern Lesser Antilles, for example, van Soest *et al.* (1998) calculated that >50% of the total carbon would be assigned to an organic, sedimentary origin if an end-member *S*-value of −10‰ were chosen—as opposed to <20% for an adopted end-member value of −30‰. In this scenario, it was suggested that the large sedimentary input implied by adopting a heavier $\delta^{13}C$ sedimentary end-member could be accommodated by loss of CO_2 from the arc crust, which is particularly thick in the southern Antilles arc. This example illustrates the point that a realistic mass balance at arcs is impossible without taking into account (i) the effect of subduction on the evolution of the carbon isotopic signature of the sedimentary input, and (ii) the possibility of an additional input from the arc crust. Note, however, that not all arcs require a crustal input of volatiles. For example, Fischer *et al.* (1998) showed that the volatiles discharged from the Kuril Island arc (75 t d^{-1} $volcano^{-1}$) could be supplied from subducted oceanic crust and mantle wedge alone.

The second potential complication is the possible fractionation of CO_2, N_2, and helium during subduction and/or subsequent magma degassing. Little has been reported on elemental fractionation during the subduction process but studies at Loihi Seamount have shown that magma degassing can exert a strong control on resultant $CO_2/^3He$ ratios as sampled in hydrothermal fluid discharges. Hilton *et al.* (1998) reported large variations in $CO_2/^3He$ ratios at Loihi Seamount that were correlated with the composition of the magma undergoing degassing. For example, as helium is more soluble in tholeiitic basalt than CO_2 (i.e., $S_{He}/S_{CO_2}>1$ where S = solubility), the $CO_2/^3He$ ratio in the melt phase will evolve to lower values as a function of fractionation style (Rayleigh or Batch) and extent of degassing. Measured $CO_2/^3He$ ratios in fluids during periods of tholeiitic volcanism were low ($\sim 5\times 10^8$). In contrast, $CO_2/^3He$ ratios $\sim 10^{10}$ were recorded in other active periods, which is consistent with degassing of alkalic magmas (in this case—$S_{He}/S_{CO_2}<1$). In order to obtain a meaningful estimate of the carbon budget in arcs, therefore, the initial (pre-degassing) $CO_2/^3He$ ratio is of prime importance, and it is often assumed that the measured $CO_2/^3He$ ratio equates to the initial magmatic value—as shown above for Loihi Seamount, this may not necessarily be the case. The same issues of degassing-induced changes to elemental ratios apply also to the N_2–He–Ar systematics used to resolve the provenance of nitrogen in arcs (Sano *et al.*, 2001).

A related concern is that of isotopic fractionation of carbon (or nitrogen) during subduction and/or magma degassing. Sano and Marty (1995) have concluded that arc-related high-temperature fluids are likely to preserve the $\delta^{13}C$ values of the (magmatic) source based on comparisons of $\delta^{13}C$ values in fluids and phenocrysts. Furthermore, they cite evidence of overlapping $\delta^{13}C$ values between high- and medium-to-low-enthalpy hydrothermal fluids, leading to the general conclusion that any fractionation induced by degassing and/or interactions within the hydrothermal system must be minimal. On the other hand, Snyder *et al.* (2001) have argued that geothermal fluids in Central America have experienced 1–2‰ shifts in $\delta^{13}C$ resulting from removal of bicarbonate during slab dewatering and/or by precipitation of calcite in the hydrothermal system. It should be noted, however, that even if observed values of $\delta^{13}C$ in arc-fluids are fractionated, the magnitude of the isotopic shifts proposed by Snyder *et al.* (2001) would make a minor difference only to calculations involving the source of carbon. To date, there is no evidence of nitrogen isotopic fractionation during magmatic degassing (Marty and Humbert, 1997).

17.4.1.4 Volatile inputs and mass balance

With realistic estimates of the volatile output from the mantle, particularly the mantle wedge, it is possible to assess the state of volatile mass balance for the mantle—comparing inputs via subduction zones with outputs via arc, mid-ocean ridge, and (possibly) plume-related magmatism. Understanding the volatile systematics of the mantle is a key component in defining its structure and evolutionary history.

Hilton *et al.* (2002) compared the major volatile input via the trench (based on the lithology of drilled sedimentary sequences adjacent to arcs worldwide plus assumed volatile characteristics of underlying basaltic basement) with the magmatic output—both for arcs individually and for the mantle globally. In the case of CO_2, for example, it was found that the input of limestone-derived CO_2 exceeds its output at all arcs except Japan. This reinforces the notion that subduction zones act as conduits for the transfer of carbon into the mantle beyond the zone of magma generation (Kerrick and Connolly, 2001). Furthermore, on a global basis, carbon has a mean degassing duration (ratio of its total surface inventory to its mantle output flux; Marty and Dauphas, 2002) less than the age of the Earth implying that it is recycled rapidly between the mantle and the exospheric reservoirs. In contrast, nitrogen has a mean degassing duration significantly greater than that of carbon suggesting it is characterized by a large surficial inventory compared to the mantle flux, or that it is inefficiently recycled into the mantle at subduction zones (Fischer *et al.*, 2002).

In addition to considering major volatiles, it is useful to address the question of volatile mass balance for the noble gases themselves. Porcelli and Wasserburg (1995b) reviewed the sedimentary budget and found, for example, that pelagic sediments have a wide range of measured xenon concentrations ($0.05–7\times10^{10}$ atoms $^{130}Xe\ g^{-1}$ with a geometric mean of 6×10^{9} atoms g^{-1}; Matsuda and Nagao, 1986; Podosek *et al.*, 1980; Staudacher and Allègre, 1988). The amount of sediment subducted has been estimated at $3\times10^{15}\ g\ yr^{-1}$ (von Huene and Scholl, 1991). Holocrystalline basalts have been found to have $(4–42)\times10^{7}$ atoms $^{130}Xe\ g^{-1}$ (Dymond and Hogan, 1973; Staudacher and Allègre, 1988), with a mean of 2×10^{8} atoms $^{130}Xe\ g^{-1}$. Low temperature enrichment of alkalies seems to extend ~600 m into the ocean crust (Hart and Staudigel, 1982) and this may apply to xenon as well—$7\times10^{15}\ g\ yr^{-1}$ of this material is subducted. In total, these numbers result in 2×10^{25} atoms $^{130}Xe\ g^{-1}$ (33 mol yr^{-1}) reaching subduction zones. In a similar exercise, Staudacher and Allègre (1988) assumed that 40–80% of the subducting flux of oceanic crust ($6.3\times10^{16}\ g\ yr^{-1}$) is altered and contains atmosphere-derived noble gases, and that up to 18% of this mass is ocean sediment. They obtained noble gas fluxes of 13.8–39 mol $^{130}Xe\ yr^{-1}$ (similar to the above estimate), along with $1.9–2.5\times10^{6}$ mol $^{4}He\ yr^{-1}$, $1.9–5.9\times10^{3}$ mol $^{20}Ne\ yr^{-1}$, $5.8–21.9\times10^{3}$ mol $^{36}Ar\ yr^{-1}$, and $3.0–12.3\times10^{3}$ mol $^{84}Kr\ yr^{-1}$. Subduction of noble gases at these rates over 10^{9} yr would have resulted in 1%, 90%, 110%, and 170% of the ^{20}Ne, ^{36}Ar, ^{84}Kr, and ^{130}Xe, respectively estimated to be in the upper mantle (Allègre *et al.*, 1986). The effects of dewatering and melting are unknown but may remove a large fraction of these noble gases from subducting materials, although silica appears to retain xenon to high temperatures (Matsuda and Matsubara, 1989). It should be emphasized that these numbers are highly uncertain and can only be used for order-of-magnitude comparisons.

Subduction zone processing and volcanism may return much of the noble gases to the atmosphere. Hilton *et al.* (2002) calculated that the output via arc-related magmatism for helium and argon greatly exceeded the potential input via the subducting slab, implying that the output fluxes are dominated by mantle wedge and/or arc crust contributions. Staudacher and Allègre (1988) argued that subducting argon and xenon is almost completely lost back to the atmosphere during subduction zone volcanism (the so-called "subduction barrier") to preserve the high $^{129}Xe/^{130}Xe$ and $^{136}Xe/^{130}Xe$ in the upper mantle throughout Earth history. However, it is possible that the total noble gas abundances reaching subduction zones is sufficiently high that subduction into the mantle of only a small fraction may have a considerable impact upon the composition of argon and xenon in the upper mantle. This contrary viewpoint produces the upper mantle xenon composition through mixing between subducted noble gases and nonrecycled, mantle-derived xenon (Porcelli and Wasserburg, 1995a,b). For example, the $^{128}Xe/^{130}Xe$ ratio measured in mantle-derived xenon trapped in CO_2 well gases may be interpreted as a mixture of ~90% subducted xenon with 10% trapped solar xenon. The same proportions could be applied to subducted and trapped radiogenic ^{129}Xe. Since mantle evolution models can have nonradiogenic noble gases in the upper mantle that are either primordial or largely dominated by subducted gases, isotopic evolution arguments alone cannot preclude subduction of atmospheric argon and xenon. Furthermore, direct input of the subducted slab into a gas-rich deeper reservoir that has $^{40}Ar/^{36}Ar$ values significantly lower than the MORB source mantle is also a possibility (e.g., Trieloff *et al.*, 2000).

17.4.2 Mantle Structure and the Source of Ocean Islands

The range of $^{3}He/^{4}He$ ratios in materials from the convecting mantle, and in particular the OIB ratios that are substantially higher than those in MORBs, provide a tracer for the source of the plumes supplying hotspots such as Hawaii and Iceland. However, in order to relate the surface expressions of these features to mantle structure, the nature of the reservoir storing the helium with high $^{3}He/^{4}He$ ratios must be identified. There have been various models addressing this topic. The reservoir must have a distinctly higher $^{3}He/(U+Th)$ ratio than the upper mantle in order to generate higher $^{3}He/^{4}He$ ratios, and the main issue is how this reservoir can be preserved in the mantle over sufficient time to generate the distinctive helium isotope composition. Also, material from this reservoir must be preferentially incorporated into mantle plume material. This is generally, but not universally, taken to link the source of the plumes with the reservoir, so that models explaining the high $^{3}He/^{4}He$ ratios also incorporate an explanation for the generation of plumes. There are three general categories of models for the OIB helium reservoir: (i) a mantle with a deep layer, with plumes arising from the boundary with the upper, MORB-source mantle; (ii) the core, with plumes derived from the core–mantle boundary; and (iii) heterogeneities within some whole mantle circulation pattern. The distribution of helium isotope compositions found at the surface reflects the conditions of plume generation and transport of helium with high $^3He/^4He$ ratios within the context of each model. It should be noted that the different models also have implications for the degassing of the mantle and the formation of the atmosphere. These are discussed in the planetary budgets for radiogenic isotopes and the overall constraints on the extent of solid Earth degassing.

17.4.2.1 Isolated lower mantle models

There are several possible descriptions of a layered mantle. The possibilities that have been incorporated into noble gas models include a boundary layer at the 670 km seismic discontinuity, a deeper layer of variable thickness, and a boundary layer at the core–mantle boundary.

The first mantle degassing models that incorporated considerations of mantle structure were developed with the discovery of mantle noble gas isotope heterogeneities (Allègre *et al.*, 1986, 1983; Hart *et al.*, 1979; Kurz *et al.*, 1982). These models, following earlier interpretations of lithophile element isotope composition variations (DePaolo, 1979; O'Nions *et al.*, 1979), divide the mantle into two convectively isolated layers with a boundary at 670 km. Such models incorporate the degassing of the upper mantle reservoir to the atmosphere. In order to explain the high OIB $^3He/^4He$ ratios, the underlying gas-rich reservoir is isolated from the degassing upper mantle. Therefore, these layered mantle models can be considered to incorporate two separate systems: the upper mantle–atmosphere and the lower mantle. There is no interaction between these two systems, and the lower mantle is completely isolated except for a minor flux to OIB that marks its existence. It is further assumed that the mantle was initially uniform in noble gas and parent isotope concentrations, so that both systems had the same starting conditions. Note that various modifications to this basic scheme have been proposed, and are discussed below (Allègre *et al.*, 1983, 1986).

This class of model has been used to calculate the degassing history of the atmosphere. The initial atmospheric abundances were assumed to have been zero, and all noble gases were once contained within the upper mantle. The atmosphere has progressively degassed from the upper mantle. The evolution of noble gases in the upper mantle is then a function of radiogenic production of ^{21}Ne, ^{40}Ar, ^{129}Xe, and ^{136}Xe within the mantle and gas loss to the atmosphere. The degassing history of the atmosphere can be calculated from the isotopic compositions of the present atmosphere, the initial mantle, and the present upper mantle. A key conclusion of these models is that high $^{40}Ar/^{36}Ar$ and $^{129}Xe/^{130}Xe$ ratios require both early (even before ^{129}I was extinct) and extensive mantle degassing to set high ratios of parent elements to noble gases in the mantle. More complicated degassing calculations can be used to take into account factors such as the depletion history of the parent by extraction to the continental crust, but the conclusions do not change significantly (e.g., Hamano and Ozima, 1978).

The lower mantle supplies OIB through plumes that originate through instabilities at the boundary, and so the high $^3He/^4He$ ratios trace the rising of lower mantle material. This reservoir has evolved essentially as a closed system; the fluxes to OIB are sufficiently low that the closed system approximation remains valid. The high $^3He/^4He$ ratios seen in OIBs have been interpreted as reflecting closed system evolution, and can be used to calculate the reservoir $^3He/(U + Th)$ ratio. It is generally assumed that the initial concentrations of the parent elements are equal to those of the bulk silicate Earth. The present isotopic compositions of argon and xenon are equal to those of the BSE and so those of the atmosphere. This is because noble gases and parent elements were initially uniformly distributed in the Earth, and since the upper mantle is highly degassed, the atmosphere contains essentially all of the noble gases in the upper portion of the Earth, and so has BSE noble gas isotope ratios. Note that early observations supporting atmospheric argon and xenon ratios in the lower mantle have been shown to reflect only contamination (e.g., Fisher, 1985; Patterson *et al.*, 1990), and so the lower mantle ratios are calculated, rather than observed, values. Using closed system calculations for the lower mantle helium isotope evolution, the model predicts that the lower mantle is enriched in helium by $\sim 10^2$ relative to the present upper mantle concentration.

A variation of the residual upper mantle model by Zhang and Zindler (1989) considers degassing of MORBs by partitioning of noble gases into CO_2 vapor for transport to the surface, with the remaining noble gases in MORBs returned to the mantle. Due to different solubilities, there is fractionation between noble gases in the residual MORBs and those lost to the atmosphere. The net result of this partial gas loss from a MORB is fractionation during mantle degassing. Another variation has been presented by Kamijo *et al.* (1998) and Seta *et al.* (2001). Here two additional factors are considered: depletion of the lower mantle reservoir by plume activity, and the subduction of parent elements into both the upper and lower mantle reservoirs. As with the other models, an initially uniform distribution of noble gases and parent elements in the mantle was assumed. Using mass fluxes between reservoirs and a linear continental growth function, it was concluded that the lower mantle could contain as little as 13% of its original noble gas inventory, and have $^{40}Ar/^{36}Ar$ ratios up to 3×10^4, much higher than that of the atmosphere, due to progressive depletion.

There are several observations that require that this type of mantle model be modified. The relationship between the upper mantle and the atmosphere cannot be simply related. The difference between MORB and air $^{20}Ne/^{22}Ne$ ratios indicates either that neon isotopes were not initially uniformly distributed in the Earth or that neon in the atmosphere has been modified by losses to space after degassing from the mantle. Regardless of the reason, the assumption that the atmosphere and upper mantle together form a closed system does not hold. Also, the higher $^{20}Ne/^{22}Ne$ and Ne/Ar ratios of the upper mantle limit the contribution that presently degassing volatiles can have made to the atmosphere (Marty and Allé, 1994). These observations suggest that neon has been lost by fractionating processes from the early atmosphere, a feature that can be appended to the model of mantle structure.

A major objection that cannot be satisfied through simple modification of the model is that atmospheric and MORB xenon are not complementary; that is, extraction of the xenon now seen in the atmosphere from the upper mantle will not leave a mantle reservoir with the xenon isotope characteristics presently observed there. Due to the greater half-life of ^{244}Pu, the ratio of ^{244}Pu-derived ^{136}Xe to radiogenic ^{129}Xe within a residual upper mantle reservoir must be greater than that of the atmosphere extracted from that reservoir (see Porcelli and Ballentine, 2002). This does not appear to be the case, since the ratio of radiogenic ^{136}Xe (derived from ^{244}Pu) to radiogenic ^{129}Xe in MORB (Kunz *et al.*, 1998) and mantle-derived xenon in CO_2 well gas (Phinney *et al.*, 1978) is clearly below that of the atmosphere. There are also difficulties with maintaining a highly depleted upper mantle from a gas-rich underlying reservoir. Distinctive radiogenic xenon isotope ratios in the upper mantle, established by degassing early in Earth history, would be obliterated by contamination from plumes rising from the gas-rich lower mantle (Porcelli *et al.*, 1986b). The requirement that all rising gas-rich material is completely degassed at hotspots is probably not reasonable, and the mixing of even a small fraction of incompletely degassed material into the surrounding mantle will have an impact.

Overall, using the basic principles of this model, the basic relationships between the isotopic compositions of the atmosphere and upper mantle cannot be explained. However, the preservation of noble gas isotope heterogeneities in convectively isolated mantle layers remains appealing.

17.4.2.2 Models with transfer between mantle reservoirs

To overcome some of the problems with the isolated mantle models (above), a new set of models allowed interaction between the gas-rich lower mantle below 670 km and the upper mantle. The first of this type sought to explain ^4He fluxes (O'Nions and Oxburgh, 1983) and assumed that highly incompatible elements in the upper mantle were (i) introduced in material entrained from the lower mantle, and (ii) in steady state concentrations. The model was then applied to the U–Pb system (Galer and O'Nions, 1985), and finally to the other noble gases (Kellogg and Wasserburg, 1990; O'Nions and Tolstikhin, 1994; Porcelli and Wasserburg, 1995a, 1995b). The central focus of the model is not degassing of the upper mantle to form the atmosphere, but rather mixing in the upper mantle, which has noble gas inventories that are presently not continually depleting, but rather are the result of continuing inputs from surrounding reservoirs. In addition to the upper mantle inputs by radiogenic production from decay of uranium, thorium, and potassium, atmospheric argon and xenon are subducted into the upper mantle, and lower mantle noble gases are transported into the upper mantle within fluxes of upwelling material. Noble gases from these sources comprise the outflows at mid-ocean ridges. The isotopic systematics of the different noble gases are linked by the assumption that transfer of noble gases from the upper mantle to the atmosphere by volcanism, as well as the transfer from the lower into the upper mantle by the rising of plumes, occurs without elemental fractionation.

It is assumed that upper mantle concentrations are in steady state, so that the inflows and outflows are equal; therefore, there are no time-dependent functions (and so additional parameters) determining upper mantle concentrations. Degassing of the upper mantle occurs by essentially complete removal of noble gases in the melting zone of MORB; therefore, the noble gas residence time in the upper mantle is determined by the rate at which upper mantle material is processed at ridges. The residence time is then the rate of mantle processing at ridges divided by the total mass of the reservoir. For the upper mantle above 670 km and the current rate of MORB production (from 10% melting), this is $\sim$1.4 Ga (Kellogg and Wasserburg, 1990; Porcelli and Wasserburg, 1995b). It is important to note that the assumption that upper mantle concentrations are in steady state is only a simplification in the model for interaction between mantle reservoirs. Nevertheless, as discussed below, in this way all of the presently observed broad noble gas characteristics of MORB and OIB can be explained (Porcelli and Wasserburg, 1995a,b). Additional parameters required to calculate time-dependent characteristics would only create under-constraint in the calculated model solutions, and therefore the introduction of non-steady-state conditions are justified only when new constraints are found. However, modification of this model by increasing the size of the upper mantle, as discussed below and so extending the reservoir to a large fraction of the age of the Earth, then would require time-dependent solutions. Tolstikhin and Marty (1998) have explored the evolution of the mantle prior to the establishment of steady state concentrations, and some effects of the time dependence of noble gas transfers into the upper mantle.

The atmosphere generally has no *a priori* connection with other reservoirs. It simply serves as a source for subducted gases, and no assumptions are made about its origin. While the radiogenic noble gases presently found in the atmosphere were clearly degassed from the solid Earth, and the nonradiogenic noble gases may also have been originally incorporated within the solid Earth, this degassing necessarily occurred prior to establishment of the present upper mantle characteristics. The history of degassing cannot be obtained from the noble gases presently found in the upper mantle, which are largely derived from the deeper mantle. Also, if upper mantle noble gases are assumed to be in steady state, by definition their characteristics do not contain any historical information regarding the establishment of present conditions.

The lower mantle reservoir evolves as an approximately closed system, and so the concentrations of ^{4}He, ^{21}Ne, ^{40}Ar, ^{129}Xe, and ^{136}Xe that have been produced over Earth history can be calculated. The lower mantle ^{20}Ne, ^{36}Ar, and ^{130}Xe concentrations, and so isotope compositions, can be calculated from the balance of fluxes into the upper mantle. The MORB ^{3}He/^{4}He ratio is a result of mixing between lower mantle helium and production of ^{4}He in the upper mantle (O'Nions and Oxburgh, 1983). Since the flux of the latter is fixed from the concentrations of parent nuclides, the rate of helium transfer from the lower mantle can be calculated. The lower mantle mass flux into the upper mantle that is calculated from helium isotopes (Kellogg and Wasserburg, 1990) can then be used to obtain the fluxes of other daughter nuclides from the lower mantle from their lower mantle concentrations (Porcelli and Wasserburg, 1995a,b). Production of other daughter noble gases in the upper mantle, ^{21}Ne, ^{40}Ar, and ^{136}Xe, can also be calculated from parent nuclide concentrations. Using the upper mantle neon, argon, and xenon isotope compositions, the lower mantle isotopic compositions of neon, argon, and xenon can be calculated. If subduction of any argon and xenon occurs, then a fraction of the nonradiogenic nuclides is introduced into the upper mantle, and so more radiogenic argon and xenon isotope compositions are calculated for the lower mantle.

The model is consistent with the isotopic evidence that upper mantle xenon does not have a simple direct relationship to atmospheric xenon. The radiogenic xenon presently seen in the atmosphere was degassed from the upper portion of the solid Earth prior to the establishment of the present upper mantle steady state xenon isotope compositions and concentrations. The lower mantle ratios are established early in Earth history by decay of ^{129}I and ^{244}Pu; ^{238}U decay produces a relatively small fraction of fissiogenic nuclides (Porcelli and Wasserburg, 1995b). The xenon daughters (now in the upper mantle) of the short-lived parents are supplied from the lower mantle. The MORB ^{129}Xe/^{130}Xe ratio (when corrected for air contamination) has no radiogenic contributions from present production in the upper mantle, and so is simply due to mixing between lower mantle and subducted xenon. Since the MORB ^{129}Xe/^{130}Xe value is greater than that of the atmosphere, the lower mantle ratio must be equal to that of MORB (if there is no xenon subduction) or higher. Once lower mantle xenon is transported into the upper mantle, it is significantly augmented by uranium-derived ^{136}Xe, since the U/Xe ratio is much greater in this gas-depleted reservoir. The ^{130}Xe flux from the lower mantle cannot be determined without a constraint on the amount subducted, but is inversely proportional to the lower mantle ^{129}Xe/^{130}Xe and ^{136}Xe/^{130}Xe ratios (Porcelli and Wasserburg, 1995b). There is therefore a trade-off between two presently unknown quantities; the lower mantle ^{130}Xe concentration (i.e., how high are the lower mantle ^{136}Xe/^{130}Xe and ^{129}Xe/^{130}Xe ratios) and the flux of subducted xenon (how much is mixed with lower mantle xenon to lower these ratios). Alternatively, if a further assumption is made regarding a specific Xe/He ratio for the lower mantle (e.g., the solar ratio), then a xenon concentration can be calculated from the helium concentration, along with the amount of subducted xenon. In the model, $\sim$50% of the radiogenic ^{136}Xe in the upper mantle is derived from the lower mantle, where it was produced largely by ^{244}Pu.

Overall, this type of model has been successful at explaining the general noble gas characteristics of the mantle. The difficulties with relating the atmosphere to the upper mantle are solved through decoupling of the histories of these two reservoirs, and the origins of the atmosphere constituents can be explained by a separate, early history of the upper part of the planet. Modifications of the model for interacting mantle reservoirs are of course possible, although in some cases the attainment of upper mantle steady state isotope concentrations may not be reasonable and time-dependent solutions may be necessary.

17.4.2.3 Difficulties with mantle layering

The principal objection to the various versions of a layered mantle model has been geophysical arguments advanced against maintaining a distinctive deep mantle reservoir. The mass flux from the lower mantle into the upper mantle in the interacting mantle model is ~50 times less than the rate of ocean crust subduction, greatly limiting the fraction of subducted material that could cross into the lower mantle (O'Nions and Tolstikhin, 1996). While there are a variety of different lines of evidence for the lack of a boundary layer at 670 km (see Porcelli and Ballentine, 2002; Tackley, 2000), the clearest has been seismic tomographic evidence that subducting slabs have penetrated the 670 km discontinuity (Creager and Jordan, 1986; Grand, 1987; van der Hilst *et al.*, 1997), and so greater mass exchange occurs with the lower mantle below 670 km. It appears that a mantle-wide boundary separating distinct mantle reservoirs is not present at this level. Many geophysical studies have therefore assumed that the mantle is stirred by whole mantle convection.

An important constraint on mantle reservoirs comes from the relationship between the radiogenic ^{4}He and heat fluxes at the surface. The Earth's global heat loss amounts to 44 TW (Pollack *et al.*, 1993). Subtracting the heat production from the continental crust (4.8–9.6 TW), and the core (3–7 TW; Buffett *et al.*, 1996) leaves 9.6–14.4 TW to be accounted for by present day radiogenic heating and 17.8–21.8 TW as a result of secular cooling. O'Nions and Oxburgh (1983) pointed out that the present day ^{4}He mantle flux was produced along with only 2.4 TW of heat, a factor of up to six times lower than the total radiogenic mantle heat flux associated with present radiogenic production. If a portion of the heat from secular cooling is from past radiogenic production, this must also be associated with ^{4}He production, and so the imbalance is even greater. Therefore, a mechanism is required for allowing the escape of heat, but not ^{4}He, from the reservoir that contains the bulk of the heat-producing elements uranium and thorium in the mantle. O'Nions and Oxburgh (1983) suggested that this could be achieved by a boundary layer in the mantle through which heat could pass, but behind which helium was trapped. As an alternative, van Keken *et al.* (2001) used a secular cooling model of the Earth to investigate the possibility that heat and helium are separated due to the different mechanisms that extract heat and helium from the mantle. The rates of release of heat and helium over a 4 Ga model run varied substantially, but the ratio of the surface ^{4}He flux to heat flow equaled that of the present Earth only during infrequent periods of very short duration. It is unlikely that the present day Earth happens to correspond to such a period. However, Ballentine *et al.* (2002) pointed out that if the long-term helium flux from the mantle were ~3 times greater than presently inferred, and so the upper mantle helium concentration were correspondingly higher, then the heat-helium balance could be satisfied. In addition, such concentrations throughout most of the mantle would also account for the mantle ^{40}Ar budget. Further data are required to substantiate this argument. Morgan (1998) suggested that heat and helium are transported to the surface at hotspots, where the bulk of the helium is lost, while the heat is lost subsequently at ridges. However, evidence for such a hotspot helium flux at the present day or in the past, and the formulation of a mantle noble gas model incorporating this suggestion, is unavailable. At present, there appears to be no convincing alternative to maintaining much of the mantle uranium and thorium behind a boundary across which heat, but not ^{4}He, can efficiently pass. However, this does not need to be constrained to a depth of 670 km.

Allègre (1997) has argued that geochemical models for long-term mantle layering can be reconciled with geophysical observations for present-day whole mantle convection if the mode of mantle convection changed less than 1 Ga ago from layered to whole mantle convection. A difficulty with this hypothesis is the expected consequences for the thermal history of the upper mantle. The principle process driving changes in the mode of mantle convection is the development of thermal instabilities in the lower layer that eventually result in either massive or episodic mantle overturn (Davies, 1995; Tackley *et al.*, 1993). Reviewing the consequences for models that reproduce mantle layering, Silver *et al.* (1988) noted that a >1,000 K temperature difference develops between the two portions of the mantle, and so overturns will cause large variations in upper mantle temperatures. However, the available geological record suggests that there has been a relatively uniform mantle cooling rate of ~50–57 K Ga^{-1} (Abbott *et al.*, 1993; Galer and Mezger, 1998). Further, numerical models simulating ^{3}He/^{4}He variations in a mantle that has undergone this transition do not reproduce the observed ^{3}He/^{4}He distributions seen today (van Keken and Ballentine, 1999).

While the lack of geophysical evidence for layered mantle convection poses the greatest problem for maintaining separate reservoirs with distinct helium isotope compositions, the intrareservoir relationships involved in this model may be adaptable to other configurations of noble gas reservoirs that might be found to be compatible with geophysical data. It might be possible to reformulate the model for a larger upper mantle, or greater mass fluxes between reservoirs, although in some cases non-steady-state upper mantle concentrations may be required.

17.4.2.4 Deeper layer reservoirs

Kellogg *et al.* (1999) have developed and numerically tested a model in which mantle below ~1,700 km has a composition, and so density, that is sufficiently different from that of the shallower mantle to largely avoid being entrained and homogenized in the overlying convecting mantle. This model has generated a great deal of interest because of its ability to preserve a region in the mantle behind which the radioelements and primitive noble gases can be preserved, while accommodating many geophysical observations. For example, it is argued that the depth of the compositional change varies, allowing slab penetration to the CMB in some locations while providing a barrier into the lower mantle elsewhere. Supporting tomographic evidence for a significant number of slabs being disrupted at 1,700 km is given by van der Hilst and Karason (1999). More recently, this supporting evidence has been questioned because of the loss of tomographic resolution in this portion of the mantle (Kàrason and van der Hilst, 2001). Although Kellogg *et al.* (1999) suggest that this layer will be hard to detect seismically because of its neutral thermal buoyancy, irregular shape and small density contrast, Vidale *et al.* (2001) argue that seismic scattering nevertheless should be resolvable, and yet is not observed. Also, if the overlying mantle has the composition of the MORB source, then the abyssal layer must contain a large proportion of the heat-producing elements, and must efficiently remove heat from the core. It is not yet clear what this effect would have on the thermal stability of the layer or temperature contrast with the overlying mantle.

The boundary layer at the CMB has been explored as a reservoir for high $^3He/^4He$ ratio helium, in the context of whole mantle convection. It has been suggested that subducted oceanic crust could accumulate there and form a distinct chemical boundary layer, accounting for the properties of the D'' layer (Christensen and Hofmann, 1994). Altered ocean crust is strongly depleted in ^3He and may be enriched in uranium, and so is likely to have relatively radiogenic helium. However, a complementary harzburgitic lithosphere that is depleted in both uranium and helium also develops during crust formation, and a layer of such material above the CMB has been surmised as a high $^3He/^4He$ reservoir (Coltice and Ricard, 1999). This reservoir preserves a high initial helium content, and addition of subsequent uranium- and thorium-depleted material simply dilutes this. The model presently lacks a mechanism for establishing the layer that is then modified by the incorporation of subducted material (Porcelli and Ballentine, 2002). More generally, it has not been extended to explain the other noble gases, and would require substantial modification to explain xenon isotope systematics and the radiogenic nuclide budgets. Also, since such a layer could not contain a dominant fraction of the heat-producing elements, another mechanism must be evoked to explain the separation of heat and helium.

Overall, if a deep mantle layer is found to be geophysically viable, it must be incorporated into a comprehensive noble gas model. This will inevitably include some of the features described in the above layered mantle models.

17.4.2.5 Heterogeneities in the convecting mantle

A different approach to the interpretation of isotopic variations of noble gases, as well as of lithophile trace elements, has been to hypothesize the preservation of heterogeneities within the convecting mantle. The central issue is whether this can explain the highest ratios in OIBs by creating domains imbedded in the convecting MORB-source mantle, so that there is no requirement for convective isolation of mantle with high $^3He/^4He$ ratios. Two end-member models have been postulated, one in which "blobs" or "plums" of enriched material are passively entrained in the convecting mantle to provide OIB-source material (Davies, 1984). The other has been called "penetrative convection" (Silver *et al.*, 1988), in which downgoing cold material drops into a compositionally different lower mantle layer. The slabs, on heating at the CMB, regain positive buoyancy and on return to the surface entrain a small portion of the deeper reservoir. In this way, lower mantle material is provided to either OIB-source or is mixed into the MORB-source mantle. Regarding the first case, numerical models have shown that more viscous "blobs" can be preserved in a convective regime if they are at least 10–100 times more viscous than the surrounding mantle (Manga, 1996). Becker *et al.* (1999) have investigated the dynamical, rheological, and thermal consequences of such blobs containing high radioelement and noble gas concentrations to account for the bulk Earth uranium budget and the high $^3He/^4He$ ratios seen in OIB. The higher heat production and resulting thermal buoyancy within these blobs must be offset by a combination of increased density (~1%) and small size (<160 km diameter) to avoid both seismic detection and thermal buoyancy that will result in the blobs rising into the upper mantle. These high viscosity zones must fill 30–65% of the mantle to satisfy geochemical mass balance constraints, and must be surrounded by low viscosity mantle to transport the resulting heat away. The thermal gradient generated through the blobs will result in a high viscosity shell around a lower viscosity core that controls the dynamical mixing behavior. For a viscosity contrast between blobs and host mantle of 100, the average viscosity of the lower mantle is predicted to be greater, by a factor of 5, than that of the upper mantle. The mechanism creating the more dense "blob" material is uncertain, but the higher density may be explained by a different perovskite/magnesiowüstite ratio than the surrounding mantle (Becker *et al.*, 1999). Manga (1996) noted that clumping of "blobs" is likely to occur and form large-scale heterogeneities, and this process is expected to work against the careful size balance required to avoid increased thermal-driven buoyancy and/or seismic detection.

The heterogeneities are often envisioned as containing gas-rich material preserved since early Earth history. It is generally assumed that noble gases are the most highly incompatible elements, so that any mantle reservoir that undergoes melting would be preferentially depleted in noble gases relative to the parent elements uranium, thorium, and potassium. Therefore, melt residues are expected to have low He/U ratios and so develop relatively radiogenic helium isotope compositions. Alternatively, it has been suggested (Anderson, 1998; Graham *et al.*, 1990; Helffrich and Wood, 2001) that uranium may be more incompatible than helium, so that melting can leave a residue with a higher He/U ratio than the original source material. In this case, high $^3He/^4He$ ratios seen in OIB can be generated in mantle domains that have been previously depleted by melting and can survive for some time. To date, this idea has not been developed into a coherent mantle model that examines the evolution of the MORB- and OIB-source mantle domains, so that it cannot be evaluated rigorously. However, several issues that must be incorporated in such a model can be considered. The source of the domains must have high $^3He/^4He$ ratios at the time of depletion. Also, the source must have started out with sufficiently high ^3He to be able to supply OIBs after depletion. Melting of highly depleted MORB-source mantle will leave a component that will not readily impart a distinctive isotopic signature to OIBs. This is a particular problem when it is considered that such high $^3He/^4He$ components must dominate the helium signature in OIB source regions where there is clear evidence for recycled oceanic crust that likely contains high ^4He concentrations, such as in Hawaii (Hauri *et al.*, 1994). Therefore, the initial mantle source must be more gas-rich. The need to preserve a gas-rich precursor is contrary to the original motivation for hypothesizing these gas-poor heterogeneities as an alternative to long-term storage of gas-rich material from early Earth history.

Note that if such a scheme is viable, the evolution of mantle noble gases, reflecting degassing and interaction between reservoirs, may still follow those calculated in layered mantle models. An important issue that must be resolved is that there must be a mechanism for the preferential involvement of this previously melted material at OIB. This would link the involvement of these reservoirs in OIB source regions with where such material is stored. In this case, high $^3He/^4He$ ratios at the surface will trace the transfer of material from this region. Also, helium would provide information on the generation of such depleted material. However, this process has yet to be substantiated.

17.4.2.6 The core

The core has often been suggested as a source for ^3He presently found in plumes. If the core served as the long-term storage reservoir for helium in OIBs with distinctively high $^3He/^4He$ ratios, then the difficulties of maintaining an isolated mantle helium reservoir separate from the source of MORB are removed. The possibility of trapping helium into the core and releasing it into the overlying mantle has been systematically evaluated by Porcelli and Halliday (2001). Appealing to the core as a source of noble gases necessarily evokes specific conditions of terrestrial noble gas acquisition and core formation, as well as core composition characteristics. Matsuda *et al.* (1993) obtained partitioning data for liquid silicate and liquid metal for pressures up to 100 kbar (corresponding to depths of up to ~300 km). Liquid–metal/liquid–silicate partition coefficients of $\sim(0.01–3)\times10^{-4}$ were obtained. While such low values have been taken to indicate that substantial quantities of noble gases cannot be in the core, the amounts partitioned into core-forming metal depend upon the concentrations initially available in the mantle. Since noble gases are incompatible during silicate melting, liquid–metal/solid–silicate partition coefficients will be greater, and so the conditions under which core formation occurred must be considered. It was shown that there might have been sufficient gas present in the mantle during core segregation to supply a substantial quantity of helium and neon to the core (Porcelli and Halliday, 2001) although this is dependent upon how noble gases are incorporated into the early Earth. In order to supply the ^3He found in OIBs, transfer from the core to the mantle by either bulk entrainment of core material or chemical interaction at the CMB may provide a mechanism for supplying relatively unfractionated noble gases to plumes. Using presently available data, these scenarios may not involve unreasonable amounts of core material. It was emphasized that the concentrations of noble gases within the early Earth, the partition coefficients between silicates and metal, and the concentrations presently required in the core to supply OIBs are all highly uncertain. Nonetheless, calculations using the presently available constraints indicate that the core remains a plausible source of the ^{3}He found in OIBs. Measured high ^{3}He/^{4}He ratios would then be the result of mixtures of helium from the core that has a solar nebula ^{3}He/^{4}He ratio with radiogenic ^{4}He from the mantle. However, the implications for other geochemical parameters also must be considered. Other elements that would be affected by bulk transfer of core material into the mantle include the platinum group elements and volatiles that may be relatively abundant in the core, such as hydrogen and carbon. While some limits on the amount of transferred core material are provided by these (Porcelli and Halliday, 2001), it should be noted that the amount needed to sustain the ^{3}He flux is still uncertain, and it is possible that upward revisions in the amount of noble gases within the early Earth, or in the metal/silicate partition coefficients, could reduce the required flux of core material by an order of magnitude or more.

Note that storage of helium in the core remains only one component of a noble gas model that can describe the range of noble gas observations. The core has only been evaluated as a possible storage of ^{3}He. The incorporation in the core of other noble gases, and their relative fractionations, cannot be clearly evaluated without more data. Also, the distribution of radiogenic nuclides such as ^{40}Ar, ^{129}Xe, and ^{136}Xe that are produced within the mantle must be explained with a model that fully describes the mantle reservoirs. While these issues may be tractable, a comprehensive model that incorporates a core reservoir remains to be formulated. It should be emphasized that the core does not completely explain the distribution of helium isotopes, since the issue of the ^{4}He-heat imbalance is not addressed at all by this model. It appears that even if high ^{3}He/^{4}He ratios are the signature of involvement of core material in the source of mantle plumes, several mantle reservoirs are still required.

17.4.3 Mantle Volatiles in the Continental Lithosphere

While the convecting upper mantle is expected to be well mixed, with MORB providing representative samples for noble gas characterization, a significant portion of the upper mantle is stabilized against convection beneath the continents as continental lithospheric mantle. This region may supply magmas to continental volcanics, but these are generally erupted subaerially and have been extensively degassed and so can only provide limited noble gas information of this mantle region. However, these magmas can also contain mantle xenoliths—portions of the lithosphere that have been entrained and brought to the surface without melting. The thickness of the lithosphere varies, and reaches the greatest depths under Archean cratons, with estimates of ~250–400 km based on seismic data (Grand, 1994; Jordan, 1975; Nolet *et al.*, 1994; Polet and Anderson, 1995; Ricard *et al.*, 1996; Simons *et al.*, 1999; Su *et al.*, 1994) and on xenolith thermobarometry data (e.g., O'Reilly and Griffin, 2000; Rudnick and Nyblade, 1999). Xenolith samples are found in kimberlites and related rocks allowing access to the subcontinental mantle: however, with the exception of diamonds (which unfortunately has provided ambiguous information (see Dunai and Porcelli, 2002), little work has been performed on mantle noble gases in major silicate phases from kimberlitic xenoliths. Younger continental areas such as western and central Europe typically have lithospheric thicknesses up to 100 km (Blundell *et al.*, 1992; Sobolev *et al.*, 1997). The lithosphere can be much thinner locally, especially in rifted areas. Rifted areas also contain the majority of volcanic centers that have yielded xenoliths (e.g., O'Reilly and Griffin, 1984) thus making the underlying mantle accessible to study. It is worth noting that if the lithosphere has an average thickness of 150 km and underlies 30% of the Earth's surface area then it constitutes ~7% of the upper mantle, and may therefore be a significant reservoir of mantle noble gases.

17.4.3.1 Chemistry of the lithosphere

Major element data of xenoliths suggest that the lithospheric mantle has been largely depleted by melt extraction, while many trace element and isotopic characteristics show evidence of subsequent modification and both ancient and recent enrichment (e.g., Carlson and Irving, 1994; Frey and Green, 1974; Hawkesworth *et al.*, 1990, 1984; Menzies and Chazot, 1985; Richardson *et al.*, 1984; Wass *et al.*, 1980. This is due to infiltration by H_2O-rich fluids, silicate melts, or carbonatitic melts that act to re-enrich depleted rocks in LREE and other trace elements (e.g., Downes, 2001; Hawkesworth *et al.*, 1990; Pearson, 1999b; Wilson and Downes, 1991). These enriching fluids were likely introduced from a variety of sources with different characteristics, including fluids above subduction zones, fluids akin to the basaltic and kimberlitic volcanics that reach the surface, and highly mobile carbonate melts (see e.g., Menzies and Chazot, 1995). The fluid source regions may involve different mantle domains, such as those seen in oceanic and island arc volcanism, or lithospheric material that has undergone heating or uplift. Also, multiple transits of fluid through the lithosphere can result in the generation of distinctive lithospheric compositions. The resulting enrichments evolve isotopically over various times, so that while some xenolith compositions clearly show the influence of subducted components or fluids related to host magmas, a great diversity of compositions that reflect complex lithospheric histories is usually observed (see Carlson, 1995). It should be noted, however, that it is not clear to what extent such trace-element-enriched samples are representative of the bulk lithosphere (McDonough, 1990; Rudnick *et al.*, 1998). For example, Rudnick and Nyblade (1999) found that the geotherm below the Kalahari craton recorded in xenoliths was most compatible with no heat production within the lithosphere, implying minimal concentrations of uranium and thorium. In contrast, O'Reilly and Griffin (2000) argue that a considerable amount of uranium and thorium is stored in apatite and other secondary phases within the Phanerozoic lithospheric mantle, and this generates a significant amount of heat as well as radiogenic noble gases.

The age of depletion and subsequent enrichment of the subcontinental lithosphere is related to the age of the overlying crust and the timing of tectonic processes that lead to its consolidation. Archean cratons may have lithospheric mantle roots that were depleted up to 3.3 Ga ago, as most clearly seen in diamond inclusion Sm–Nd isochrons and model age data (e.g., Richardson *et al.*, 1984; Richardson and Harris, 1997), Re–Os model depletion ages of mantle xenoliths (e.g., Pearson, 1999a,b; Walker *et al.*, 1989), and lead isotope data for lithosphere-derived lavas and xenoliths (e.g., Rodgers *et al.*, 1992). The lithosphere underlying younger, more tectonically active continental regions such as Wyoming, southeast Australia, and western Europe may have been depleted $\leq$2 Ga ago (e.g., Burnham *et al.*, 1998; Carlson and Irving, 1994; Handler *et al.*, 1997). While these ages may pertain to the initial stabilization of the lithosphere, subsequent enrichment events and so open system chemical behavior can extend up to rifting events and the deformation related to the eruption of the host magma. There is, therefore, a range of timescales for lithospheric trace element enrichment over which production of radiogenic noble gases may occur.

Noble gases, in particular helium, have the potential for identifying the sources of metasomatising fluids, as long as subsequent evolution in the lithosphere does not significantly alter its isotope composition. In this respect, an important issue is the location of noble gases and parent elements in xenoliths. The noble gases are not readily incorporated into crystal lattices, and strongly partition into the CO_2-rich fluids that typically occupy mantle fluid inclusions (Section 17.2.2). This has been confirmed by studies that showed trapped helium can be released by crushing and separated from lattice-hosted noble gases (Hilton *et al.*, 1993b; Kurz *et al.*, 1990; Scarsi, 2000). Note that fluids and noble gases may also reside along grain boundaries at depth, and are likely to be completely lost by movement along the boundaries during transport, incipient decompression melting, and sample preparation. Therefore, the quantities stored in grain boundaries at depth cannot be constrained.

17.4.3.2 *Helium isotopic variations*

Early measurements of ultramafic xenoliths from alkali basalts found $^3He/^4He$ ratios up to $10\,R_A$ (e.g., Kyser and Rison, 1982; Tolstikhin *et al.*, 1974). These data indicated that helium with significant fractions of primordial ^3He, rather than just radiogenic helium, was present in xenoliths. Porcelli *et al.* (1986a) found that recently erupted samples from a number of locations worldwide had isotopic compositions that fell within the restricted range of $6.2–10.6R_A$, and suggested that some of the samples studied earlier with lower values had been altered by post-eruption radiogenic production of ^4He. The first-order observation (Porcelli *et al.*, 1986a) is that the data fall into a relatively narrow range that is close to that of typical or "normal" MORBs (N-MORBs), and is distinctive from both radiogenic helium that characterizes crustal rocks ($\sim 0.05R_A$) and mantle helium typical of OIBs with high $^3He/^4He$ ratios (Figure 7). Therefore, it appears that helium in the lithosphere sampled by xenoliths is similar to that in the underlying convecting mantle, and does not contain large fractions of radiogenic helium. In detail, however, the samples do not always correspond completely with MORB values. A common feature of some regions is that the $^3He/^4He$ ratios are somewhat lower, i.e., more radiogenic, than the depleted N-MORB reservoir (Dunai and Baur, 1995).

The isotopic composition of helium can be compared with that of other tracers of mantle sources. Available data for measured $^3He/^4He$ ratios are compared with those of $^{87}Sr/^{86}Sr$ in Figure 8. A range of $^{87}Sr/^{86}Sr$ ratios have been obtained for these samples, from values within the range of MORB to the highest value measured in a mantle sample of 0.8360 for a Tanzanian garnet lherzolite (Cohen *et al.*, 1984). These values represent a range of mantle sources for the lithophile elements as well as examples of very long isolation times from asthenospheric mantle in a lithospheric region with high Rb/Sr ratios. Since xenoliths that are enriched in trace elements are not particularly depleted in uranium (e.g., Cohen *et al.*, 1984), radiogenic ^{87}Sr is expected to be accompanied by ^4He, even though it is not known how Rb/Sr relates to U/He ratios. However, in contrast to the wide spread in strontium isotope compositions, helium isotopes fall within

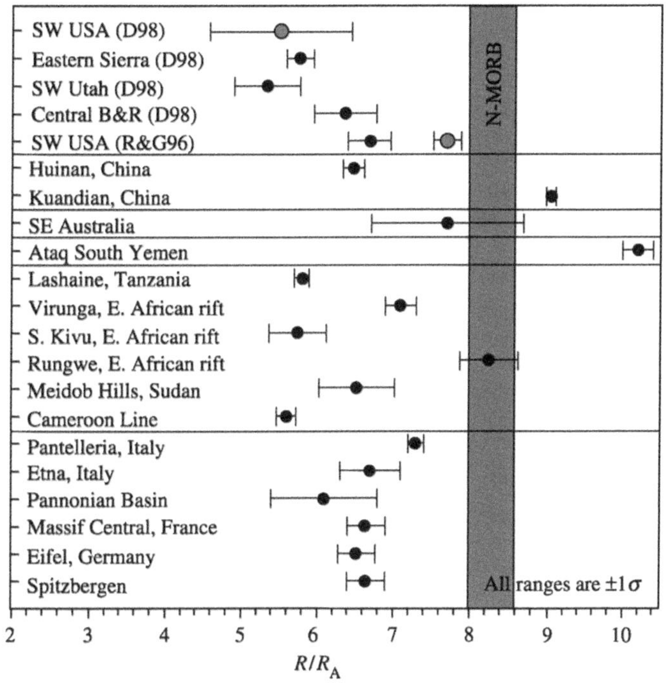

Figure 7 Published $^{3}He/^{4}He$ for phenocrysts and ultramafic xenoliths from various continental localities worldwide (reproduced by permission of Mineralogical Society of America from *Rev. Mineral. Geochem.*, **2002**, *47*, 380).

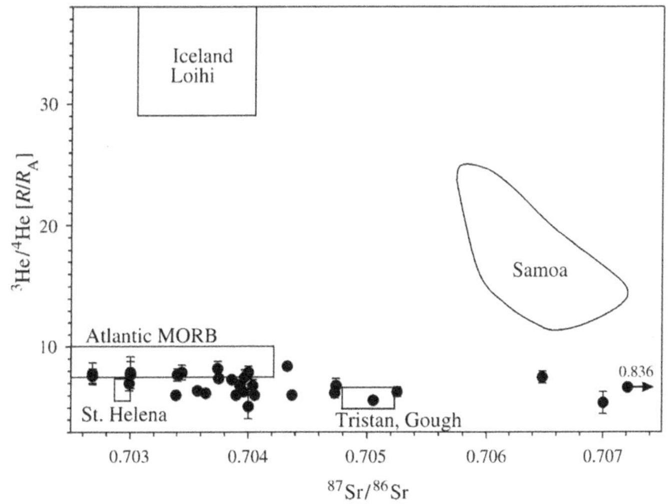

Figure 8 He–Sr isotope relationships in xenoliths (reproduced by permission of Mineralogical Society of America from *Rev. Mineral. Geochem.*, **2002**, *47*, 383).

a restricted range. The lack of correlation between these ratios indicates that there generally has been complete decoupling of the two isotopic systems (Porcelli *et al.*, 1986a). Similar decoupling between helium and strontium isotopes has also been observed for xenoliths from ocean islands (Vance *et al.*, 1989), as well as between helium and neodymium isotopes (Stone *et al.*, 1990). It is possible that the decoupling of helium and strontium ratios reflects processes that fractionate rubidium from strontium but not uranium from helium, and so the He/U ratio of the lithosphere is very low. Since substantial amounts of helium are likely to have been lost during eruption due to decrepitation of helium-bearing fluid inclusions, measured He/U ratios are generally gross underestimates of the ratio within the lithosphere. However, while it might be considered reasonable that the formation of migrating fluids preferentially concentrated both uranium and helium, there is likely to be strong fractionations within the invaded mantle as the uranium is introduced into the mineral lattices or newly created phases and the helium remains in the volatile phase. It would be expected that the result is large fractionations between uranium and helium, at least over the limited distances between the source regions of different xenoliths, and so the development of a much wider range of $^3He/^4He$ ratios than observed. The more likely explanation is that while a dominant fraction of the helium has been introduced into the samples recently, the dominant fraction of strontium has been added earlier and so has had time to isotopically evolve to more radiogenic compositions (Porcelli *et al.*, 1986a). It has been shown that neodymium, and certainly many other trace elements, does not preferentially enter CO_2-rich fluids (Meen *et al.*, 1989). Where these elements are enriched, the transporting agent must have been a silicate or carbonatitic melt. Therefore, the introduction of the volatiles found in the fluid inclusions may have been the result of fluid migration that occurred much more recently than the events responsible for the enrichment of other trace elements.

17.4.3.3 Helium transport

Available evidence suggests that CO_2-rich fluids invade the source regions of the xenoliths not long before their entrainment. Trapping of the volatiles and associated noble gases from the host magma degassing during ascent is generally not likely (see discussion in Dunai and Porcelli, 2002). The generation of the CO_2-rich fluids is likely related to upwelling of mantle material, resulting initially in very small degrees of melting. These fluids must efficiently invade the xenolith source region, completely overwhelming any radiogenic helium that might have been expected to accompany the radiogenic strontium and neodymium. It is possible that while He/U fractionation occurs in the lithosphere, this may occur on a restricted scale so that the regional average is not unusually radiogenic. In this case, homogenization of lithospheric helium by invading fluids may be sufficient to erase more radiogenic compositions.

An important exception to the general trend in the xenolith data are exceptionally high $^3He/^4He$ ratios that have been observed in samples that have very low helium concentrations and have been exposed for long periods of time at the surface. Exposure to cosmic rays can result in the production of cosmogenic ^3He (Craig and Poreda, 1986; Kurz, 1986), and this can dominate over the mantle ^3He contained in the samples (Porcelli *et al.*, 1987). Subsequent studies have endeavored to avoid any samples that may be susceptible to such alterations.

17.4.3.4 Tracing mantle sources

Various studies have examined whether the mantle sources of helium in particular regions can be identified. An extensive study of the Eifel, Germany, and Massif Central, France (Dunai and Baur, 1995) found a restricted range of $^3He/^4He$ ratios. The samples had sufficiently high helium that corrections for post-eruption additions of radiogenic ^4He and cosmogenic ^3He, using intermineral data, were small. Samples from the Massif Central had a mean $^3He/^4He$ ratio of $6.53 \pm 0.25R_A$, while two xenoliths from the Eifel had a value of $6.03 \pm 0.14R_A$. The samples include a range of lithologies, including pyroxenites, dunites, and lherzolites, and often contained hydrous phases. Lithophile isotopes and trace elements also provided evidence for both depletion and subsequent trace element enrichment events (e.g., Stosch *et al.*, 1980). The general uniformity of the values over many vents and regardless of lithology is remarkable (Dunai and Baur, 1995). The somewhat lower values at two other Eifel vents may reflect some radiogenic production in the source (Dunai and Baur, 1995). A review of neodymium, strontium, and lead data indicates that the volcanics at each of a number of localities across Europe, including the Eifel and Massif Central, fall on correlation lines that extend from a common component that has the characteristics of HIMU mantle (Wilson and Downes, 1991). The correlations have been interpreted as mixing between HIMU mantle plume material with small variable contributions from a local lithospheric source. Seismic imaging suggests that small plume upwellings supply volcanic centers throughout continental Europe, and the geochemical similarities suggest a common deep plume source (Granet *et al.*, 1995; Hoernle *et al.*, 1995). Recent seismic data suggests that broader upwelling occurs from depths of up to 2,000 km, and this single source may feed smaller plumes that are responsible for the surface volcanics (Goes *et al.*, 1999). In such a case, the overlap of helium isotopic compositions in the xenoliths with that of some HIMU basalts ($6.8 \pm 0.9R_A$; Hanyu and Kaneoka, 1997) suggests that the xenolith helium is dominated by this asthenospheric mantle component, and the local lithospheric components that add other trace elements are an insignificant source of helium. It appears that at many locations across Europe, helium in the xenoliths was likely supplied by melts from the same source as the host magmas that invaded the lithospheric mantle in advance of xenolith entrainment, and provided exsolved CO_2 and other volatiles.

A number of noble gas studies have focused on xenoliths from the Pliocene to Recent alkali basalts of the Newer Volcanics in SE Australia (Matsumoto *et al.*, 1997, 1998, 2000; Porcelli *et al.*, 1986a,

1987, 1992). These basalts appear to be associated with plume-related volcanism (Wellman and McDougall, 1974; Zhang *et al.*, 2001), and xenoliths from a range of depths have been found in a number of young vents. The xenoliths represent a range of petrologic types (O'Reilly and Griffin, 1984) and have undergone differing degrees of metasomatic addition of trace elements, with the addition in some samples of phases such as amphibole and apatite (Griffin *et al.*, 1988; O'Reilly and Griffin, 1988). A range of values for $^3He/^4He$ ratios of 7.1–9.8R_A has been found for anhydrous spinel lherzolites, with no systematic difference between vents (Matsumoto *et al.*, 1998). These values clearly overlap the MORB range, and are accompanied by clearly defined neon isotope systematics that also overlap the MORB range. Analyses of peridotites containing amphibole and apatite, as well as garnet pyroxenites, found values of 6–10 R_A (Matsumoto *et al.*, 2000; Porcelli *et al.*, 1986a, 1992), excluding those that have higher ratios clearly due to cosmogenic inputs (Porcelli *et al.*, 1987). Note that values below 7.2R_A are generally due to radiogenic inputs that are more evident in low temperature step-heating extraction steps. Therefore, helium in all these xenoliths could have been derived from the MORB-source region. The petrology and noble gas characteristics of the xenoliths clearly require a series of events well after formation of the lithosphere. The garnet pyroxenites appear to be cumulates from intruding melts that crosscut the peridotites (Griffin *et al.*, 1988). Lithophile isotopes indicate that multiple metasomatic episodes occurred in the source (Griffin *et al.*, 1988; McDonough and McCulloch, 1987). The amphiboles, derived from volatile-rich fluids, were formed ~300–500 Ma, based on Sr–Nd systematics (Griffin *et al.*, 1988). The noble gases, derived from a source with MORB characteristics—presumably the convecting upper mantle, were introduced either very recently, overwhelming any helium encountered in the lithosphere, or in such high concentrations that isotopic compositions were not substantially altered by radiogenic growth of ^4He in the lithosphere.

Extensive volcanism can be found in eastern Africa associated with the East African rift system. Helium isotope ratios measured in basalts from the Ethiopian Rift Valley and Afar, near the triple junction of Red Sea, Aden, and Ethiopian rift systems, are 6–17R_A (Marty *et al.*, 1993, 1996; Scarsi and Craig, 1996). These values include MORB ratios as well as indications of both contributions from radiogenic ^4He and high $^3He/^4He$ hotspots. To the south, the rift system divides into eastern and western branches. Xenoliths from Tanzanian vents in the eastern branch have $^3He/^4He = 5.8–7.3R_A$ (Porcelli *et al.*, 1986a, 1987). The olivines in these samples have $^3He/^4He$ ~7R_A, and so coincide with MORB values, so lower values in this study may be due to post-eruption radiogenic production in the uranium-rich phases. In contrast, osmium isotope data of xenoliths from the nearby Labait vent suggests the involvement of plume material (Chesley *et al.*, 1999). Therefore, the complex involvement of different mantle sources may be involved. Kivu and Virunga volcanics of the western branch of the East African rift show highly variable trace element and isotopic compositions that are indicative of a lithospheric mantle source that is heterogeneous on a small scale (Furman and Graham, 1999), and have $^3He/^4He$ ratios which are significantly different from each other (Graham *et al.*, 1995). The range appears to reflect different mixing proportions of asthenospheric sources with lithospheric components with lower $^3He/^4He$ ratios that presumably have developed within the lithosphere itself.

The relative importance of lithospheric and asthenospheric contributions to the helium in basalts in the southwestern USA has been investigated by Reid and Graham (1996) and Dodson *et al.* (1998). Reid and Graham (1996) found that the most radiogenic helium ratios (5.5 R_A) were associated with basalts that exhibit thorium and lead isotope evidence for derivation from an enriched source in the lithospheric mantle that has remained unmodified since ~1.7 Ga. Furthermore, a strong positive correlation between $^3He/^4He$ ratios and ε_{Nd} was found, suggesting an affinity between light rare earth enrichment and time-integrated $(U + Th)/^3He$ ratios. The range in observed $^3He/^4He$ ratios (5.5–7.7 R_A) most likely represents a mixture between asthenospheric and lithospheric melts. Reid and Graham (1996) concluded that the Proterozoic lithospheric mantle in the southwestern US is not a highly degassed reservoir contaminated by helium derived from the underlying asthenosphere, but rather a reservoir with only slightly elevated $(U + Th)/^3He$ ratios (and so slightly lower $^3He/^4He$ ratios) compared to the depleted upper mantle MORB source. Another conclusion was that processes that enrich the lithospheric mantle in LILEs also add helium, so that there is no significant decoupling between LILE and helium. Dodson *et al.* (1998) significantly expanded the database of Reid and Graham (1996) and found a much larger range in $^3He/^4He$ ratios extending to lower, i.e., more radiogenic, $^3He/^4He$ ratios (2.8–7.8 R_A). This was interpreted as reflecting variations in the age of the lithospheric mantle and the degree of degassing. It has been suggested that lower $^3He/^4He$ ratios could be explained by degassing and radiogenic in-growth during magmatic differentiation with the superimposed effects of crustal contamination (Gasparon *et al.*, 1994; Hilton *et al.*, 1993a, 1995; Parello *et al.*, 2000). Overall, the question of whether the low $^3He/^4He$ ratios represent a source feature or are due to processes occurring at shallower depths can only be settled by studying more samples from each location, especially the locations that are closest to end-member compositions. A convergence of $^3He/^4He$ ratios in multiple samples from a single locality and with high helium concentrations is needed to conclusively confirm source region characteristics.

It is worth noting that although not seen in xenoliths, helium from hotspots with high $^3He/^4He$ ratios have sometimes found their surface expression in continental areas (e.g., Basu *et al.*, 1993, 1995; Dodson *et al.*, 1997; Graham *et al.*, 1998; Kirstein and Timmerman, 2000; Marty *et al.*, 1996; Scarsi and Craig, 1996) and consequently must have influenced the affected continental lithosphere. This includes 250 Ma old Siberian flood basalts (with 13 R_A; Basu *et al.*, 1995), 42 Ma old proto-Iceland plume basalts in Ireland (Kirstein and Timmerman, 2000), 380 Ma old Kola Peninsula carbonatites (with 19 R_A; Marty *et al.*, 1998), and Archean komatiites (Richard *et al.*, 1996). However, there are few indications of $^3He/^4He$ ratios greater than MORB in young xenoliths. It might be noted that the products of the earliest stages of the plume-related volcanism in a given continental area seem to have the largest chance of picking up relatively unaltered asthenospheric high $^3He/^4He$ ratios, and the isotopic expression of high $^3He/^4He$ hotspots wanes relatively quickly as the continental plate moves over the hotspot (Dunai and Porcelli, 2002). Basalts erupting subsequently in the same area may then fail to pick up any high $^3He/^4He$ ratios (Basu *et al.*, 1993, 1995; Dodson *et al.*, 1997). Therefore, input of helium with high $^3He/^4He$ ratios related to hotspots may be a transient feature in the lithospheric mantle and so has not left an enduring imprint on lithospheric xenoliths that have been brought to the surface subsequently.

17.5 CONCLUDING REMARKS

Although noble gases have contributed much to the development of ideas concerning the Earth's mantle, there remain large gaps in our knowledge that limit further progress. In terms of understanding volatile recycling involving the mantle, there is still poor control on the input parameters of the noble gases (and other volatiles of interest), as there is on fluxes from output localities such as fore-arc and back-arc regions. Furthermore, there are large uncertainties in arc output fluxes—both for various arc segments individually as well as arcs globally. We will become less reliant on the limited datasets currently available only through collection of more (high quality) data on targeted areas and samples. A related concern pertains to our understanding of elemental and isotopic fractionation processes during metamorphic devolatilization (subduction) and during magma storage, crystallization, and degassing. Progress in these areas may come from vapor–melt partitioning studies allied to volatile studies specific to particularly relevant magmatic systems.

In the case of understanding mantle structure, the main obstacle to a comprehensive description of mantle noble gas evolution is in finding a configuration for distinct mantle domains that is compatible with geophysical observations. In this regard, a major difficulty is the heat-helium imbalance. Models involving mantle stratification can account for this, but have been discounted on other geophysical grounds. No other adequate explanation for this imbalance has been proposed. Another problem requiring resolution is the nature of the high $^3He/^4He$ OIB source region. Most models equate this region with undepleted, undegassed mantle, although some models invoke depletion mechanisms. However, none of these have matched the end-member components seen in OIB lithophile isotope correlations. It remains to be demonstrated that a primitive component is present and so can dominate the helium and neon isotope signatures in OIBs. Further data is also required regarding the heavy noble gas isotope compositions of OIBs, the rates of noble gas subduction, and the mechanisms of noble gas incorporation within the nascent Earth. A final comprehensive model may involve the core and several mantle reservoirs, with the dispositions of the mantle reservoirs and the fluxes between them defined by geophysical considerations.

The decoupling of helium from other isotopic signatures in samples from the subcontinental mantle is likely due to relatively recent addition of helium to the xenolith in fluids with high concentrations of helium relative to other trace elements, either within the mantle or during transport. The introduction of helium isotopes into ultramafic xenoliths clearly occurs not long before eruption and does not reflect the longer lithospheric history of the rest of the rocks. One avenue of future research is relating the introduction of volatiles into the lithosphere to local tectonic and volcanic history. If the metasomatic history of the xenoliths can be related to regional rifting and magmatic activity, then a greater understanding of the history and composition of the lithosphere can be obtained from the more completely explored surface geologic history.

There are few constraints on the heavier noble gases in ultramafic xenoliths. While obtaining credible heavy isotope compositions is hampered by low concentrations, more refined analyses and the collection of more gas-rich samples may obtain compositions that provide information on whether separation of the noble gases occurs within the subcontinental lithosphere. Further progress on the origin and nature of noble gases in the lithospheric mantle may be obtained from studies that combine evidence from fluid inclusions, textures, lithophile trace element isotope systematics, and major element variations to understand how noble gas concentrations and isotopic compositions are introduced and modified within the lithosphere.

ACKNOWLEDGMENTS

Discussions over many years with friends and colleagues contributed to ideas presented in this review. The authors thank Rick Carlson for editorial patience and remarks on the manuscript, and Chris Ballentine for a detailed and constructive review.

REFERENCES

Abbott D. A., Burgess L., Longhi J., and Smith W. H. F. (1993) An empirical thermal history of the Earth's upper mantle. *J. Geophys. Res.* **99**, 13835–13850.

Allard P. (1992) Global emissions of He-3 by subaerial volcanism. *Geophys. Res. Lett.* **19**, 1479–1481.

Allègre C. J. (1997) Limitation on the mass exchange between the upper and lower mantle: the evolving convection regime of the Earth. *Earth Planet. Sci. Lett.* **150**, 1–6.

Allègre C. J., Staudacher T., Sarda P., and Kurz M. (1983) Constraints on the evolution of Earth's mantle from the rare gas systematics. *Nature* **303**, 762–766.

Allègre C. J., Staudacher T., and Sarda P. (1986) Rare gas systematics: formation of the atmosphere, evolution and structure of the Earth's mantle. *Earth Planet. Sci. Lett.* **87**, 127–150.

Alvarez L. W. and Cornog R. (1939) He-3 in helium. *Phys. Rev.* **56**, 379.

Anderson D. L. (1993) Helium-3 from the mantle: primordial signal or cosmic dust? *Science* **261**, 170–176.

Anderson D. L. (1998) The helium paradoxes. *Proc. Natl Acad. Sci.* **95**, 4822–4827.

Andrews J. N. (1985) The isotopic composition of radiogenic helium and its use to study groundwater movement in confined aquifers. *Chem. Geol.* **49**, 339–351.

Aston F. W. (1919) A positive ray spectrograph. *Phil. Mag.* **38**, 707–714.

Aston F. W. (1920a) The constitution of atmospheric neon. *Phil. Mag.* **39**, 449–455.

Aston F. W. (1920b) The constitution of the elements. *Nature* **105**, 8.

Aston F. W. (1920c) The mass spectra of chemical elements. *Phil. Mag.* **39**, 611–625.

Aston F. W. (1927) Bakerian lecture: a new mass-spectrograph and the whole number rule. *Proc. Roy. Soc. London* **A115**, 487–514.

Ballentine C. J. (1997) Resolving the mantle He/Ne and crustal Ne-21/Ne-22 in well gases. *Earth Planet. Sci. Lett.* **152**, 233–249.

Ballentine C. J. and Barfod D. N. (2000) The origin of air-like noble gases in MORB and OIB. *Earth Planet. Sci. Lett.* **180**, 39–48.

Ballentine C. J. and Burnard P. G. (2002) Production, release, and transport of noble gases in the continental crust. *Rev. Mineral. Geochem.* **47**, 481–538.

Ballentine C. J. and O'Nions R. K. (1992) The nature of mantle neon contributions to Vienna basin hydrocarbon reservoirs. *Earth Planet. Sci. Lett.* **113**, 553–567.

Ballentine C. J., van Keken P. E., Porcelli D., and Hauri E. H. (2002) Numerical models, geochemistry and the zero-paradox noble-gas mantle. *Phil. Trans. Roy. Soc. London* **A360**, 2611–2631.

Barfod D. N., Ballentine C. J., Halliday A. N., and Fitton J. G. (1999) Noble gases in the Cameroon line and the He, Ne, and Ar isotopic compositions of high mu (HIMU) mantle. *J. Geophys. Res.* **104**, 29509–29527.

Basu A. R., Renne P. R., Dasgupta D. K., Teichmann F., and Poreda R. J. (1993) Early and late alkali igneous pulses and a high ^{3}He plume origin for the Deccan flood basalts. *Science* **261**, 902–906.

Basu A. R., Poreda R. J., Renne P. R., Teichmann F., Vasiliev Y. R., Sobolev N. V., and Turrin B. D. (1995) High-He-3 plume origin and temporal spatial evolution of the Siberian flood basalts. *Science* **269**, 822–825.

Batiza R. (1982) Abundances, distribution and sizes of volcanoes in the Pacific Ocean and implications for the origin of non-hotspot volcanoes. *Earth Planet. Sci. Lett.* **60**, 195–206.

Bebout G. E. (1995) The impact of subduction-zone metamorphism on mantle-ocean chemical cycling. *Chem. Geol.* **126**, 191–218.

Becker T. W., Kellogg J. B., and O'Connell R. J. (1999) Thermal constraints on the survival of primitive blobs in the lower mantle. *Earth Planet. Sci. Lett.* **171**, 351–365.

Begemann R., Weber H. W., and Hintenberger H. (1976) On the primordial abundance of argon-40. *Astrophys. J.* **203**, L155–L157.

Blundell D., Freeman R., and Müller S. (1992) *A Continent Revealed: The European Geotraverse*. Cambridge University Press, Cambridge.

Brantley S. L. and Koepenick K. W. (1995) Measured carbon dioxide emissions from Oldoinyo Lengai and the skewed distribution of passive volcanic fluxes. *Geology* **23**, 933–936.

Broadhurst C. L., Drake M. J., Hagee B. E., and Bernatowicz T. J. (1990) Solubility and partitioning of Ar in anorthite, diopside, forsterite, spinel, and synthetic basaltic liquids. *Geochim. Cosmochim. Acta* **54**, 299–309.

Broadhurst C. L., Drake M. J., Hagee B. E., and Bernatowicz T. J. (1992) Solubility and partitioning of Ne, Ar, and Xe in minerals and synthetic basaltic melts. *Geochim. Cosmochim. Acta* **56**, 709–723.

Brooker R. A., Du Z., Blundy J. D., Kelley S. P., Allan N. L., Wood B. J., Chamorro E. M., Wartho J. A., and Purton J. A. (2003) Zero charge' partitioning behaviour of noble gases during mantle melting. *Nature* **473**, 738–741.

Buffett B. A., Huppert H. E., Lister J. R., and Woods A. W. (1996) On the thermal evolution of the Earth's core. *J. Geophys. Res.* **101**, 7989–8006.

Burnard P. (1999) Origin of argon–lead isotopic correlations in basalts. *Science* **286**, 871.

Burnard P., Graham D., and Turner G. (1997) Vesicle-specific noble gas analyses of "popping rock": implications for primordial noble gases in Earth. *Science* **276**, 568–571.

Burnard P. G., Stuart F. M., Turner G., and Oskarsson N. (1994) Air contamination of basaltic magmas—implications for high ^{3}He/^{4}He mantle Ar isotopic composition. *J. Geophys. Res.* **99**, 17709–17715.

Burnham O. M., Rogers N. W., Pearson D. G., van Calsteren P. W., and Hawkesworth C. J. (1998) The petrogenesis of the eastern Pyrenean peridotites: an integrated study of their whole rock geochemistry and Re–Os isotope composition. *Geochim. Cosmochim. Acta* **62**, 2293–2310.

Caffee M. W., Hudson G. B., Velsko C., Huss G. R., Alexander E. C., and Chivas A. R. (1999) Primordial noble gases from Earth's mantle: identification of a primitive volatile component. *Science* **285**, 2115–2118.

Carlson R. W. (1995) Isotopic inferences on the chemical structure of the mantle. *J. Geodynam.* **20**, 365–386.

Carlson R. W. and Irving A. J. (1994) Depletion and enrichment history of subcontinental lithospheric mantle—an Os, Sr, Nd, and Pb isotopic study of ultramafic xenoliths from the northwestern Wyoming craton. *Earth Planet. Sci. Lett.* **126**, 457–472.

Carroll M. R. and Draper D. S. (1994) Noble gases as trace elements in magmatic processes. *Chem. Geol.* **117**, 37–56.

Chamorro E. M., Brooker R. A., Wartho J. A., Wood B. J., Kelley S. P., and Blundy J. D. (2002) Ar and K partitioning between clinopyroxene and silicate melt to 8 GPa. *Geochim. Cosmochim. Acta* **66**, 507–519.

Chesley J. T., Rudnick R. L., and Lee C.-T. (1999) Re–Os systematics of mantle xenoliths from the East African rift: age, structure, and history of the Tanzanian craton. *Geochim. Cosmochim. Acta* **63**, 1203–1217.

Christensen U. R. and Hofmann A. W. (1994) Segregation of subducted oceanic crust in the convecting mantle. *J. Geophys. Res.* **99**, 19867–19884.

Clarke W. B., Beg M. A., and Craig H. (1969) Excess He-3 in the sea: evidence for terrestrial primordial helium. *Earth Planet. Sci. Lett.* **6**, 213–220.

Cohen R. S., O'Nions R. K., and Dawson J. B. (1984) Isotope geochemistry of xenoliths from east Africa: implications for development of mantle reservoirs and their interaction. *Earth Planet. Sci. Lett.* **68**, 209–220.

Coltice N. and Ricard Y. (1999) Geochemical observations and one layer mantle convection. *Earth Planet. Sci. Lett.* **174**, 125–137.

Craig H. and Lupton J. E. (1976) Primordial neon, helium, and hydrogen in oceanic basalts. *Earth Planet. Sci. Lett.* **31**, 369–385.

Craig H. and Lupton J. E. (1981) Helium-3 and mantle volatiles in the ocean and oceanic crust. In *The Sea* (ed. C. Emiliani). Wiley, New York, vol. 7, pp. 391–428.

Craig H. and Poreda R. J. (1986) Cosmogenic ^{3}He in terrestrial rocks: the summit lavas of Maui. *Proc. Natl Acad. Sci. USA* **83**, 1970–1974.

Craig H., Clarke W. B., and Beg M. A. (1975) Excess ^{3}He in deep water on the East Pacific rise. *Earth Planet. Sci. Lett.* **26**, 125–132.

Creager K. C. and Jordan T. H. (1986) Slab penetration into the lower mantle beneath the Marianas and other island arcs of the northwest Pacific. *J. Geophys. Res.* **91**, 3573–3589.

Crisp J. A. (1984) Rates of magma emplacement and volcanic output. *J. Volcanol. Geotherm. Res.* **20**, 177–211.

Davies G. F. (1984) Geophysical and isotopic constraints on mantle convection: an interim synthesis. *J. Geophys. Res.* **89**, 6017–6040.

Davies G. F. (1995) Punctuated tectonic evolution of the Earth. *Earth Planet. Sci. Lett.* **136**, 363–379.

Dempster A. J. (1918) A new method of positive ray analysis. *Phys. Rev.* **11**, 316–325.

DePaolo D. J. (1979) Implications of correlated Nd and Sr isotopic variations for the chemical evolution of the crust and mantle. *Earth Planet. Sci. Lett.* **43**, 201–211.

Dixon E. T., Honda M., McDougall I., Campbell I. H., and Sigurdsson I. (2000) Preservation of near-solar neon isotopic ratios in Icelandic basalts. *Earth Planet. Sci. Lett.* **180**, 309–324.

Dixon J. E. and Stolper E. M. (1995) An experimental study of water and carbon dioxide solubilities in mid-ocean ridge basaltic liquids: 2. Applications to degassing. *J. Petrol.* **36**, 1633–1646.

Dodson A., Kennedy B. M., and DePaolo D. J. (1997) Helium and neon isotopes in the Imnaha basalt, Columbia River Basalt Group: evidence for a Yellowstone plume source. *Earth Planet. Sci. Lett.* **150**, 443–451.

Dodson A., DePaolo D. J., and Kennedy B. M. (1998) Helium isotopes in lithospheric mantle: evidence from Tertiary basalts of the western USA. *Geochim. Cosmochim. Acta* **62**, 3775–3787.

Doe B. R. and Zartman R. E. (1979) Plumbotectonics: I. The Phanerozoic. In *Geochemistry of Hydrothermal Ore Deposits* (ed. H. L. Barnes). Wiley, New York, pp. 22–70.

Downes H. (2001) Formation and modification of the shallow sub-continental lithospheric mantle: a review of geochemical evidence from ultramafic xenolith suites and tectonically emplaced ultramafic massifs of western and central Europe. *J. Petrol.* **42**, 233–250.

Dunai T. J. and Baur H. (1995) Helium, neon, and argon systematics of the European subcontinental mantle—implications for its geochemical evolution. *Geochim. Cosmochim. Acta* **59**, 2767–2783.

Dunai T. J. and Porcelli D. (2002) Storage and transport of noble gases in the subcontinental lithosphere. *Rev. Mineral. Geochem.* **47**, 371–409.

Dymond J. and Hogan L. (1973) Noble gas abundance patterns in deep sea basalts—primordial gases from the mantle. *Earth Planet. Sci. Lett.* **20**, 131–139.

Eiler J. M., Farley K. A., and Stolper E. M. (1998) Correlated helium and lead isotope variations in Hawaiian lavas. *Geochim. Cosmochim. Acta* **62**, 1977–1984.

Farley K. A. and Craig H. (1994) Atmospheric argon contamination of ocean island basalt olivine phenocrysts. *Geochim. Cosmochim. Acta* **58**, 2509–2517.

Farley K. A. and Neroda E. (1998) Noble gases in the Earth's mantle. *Ann. Rev. Earth Planet. Sci.* **26**, 189–218.

Farley K. A., Maier-Reimer E., Schlosser P., and Broecker W. S. (1995) Constraints on mantle ^{3}He fluxes and deep-sea circulation from an oceanic general circulation model. *J. Geophys. Res.* **100**, 3829–3839.

Fischer T. P., Giggenbach W. F., Sano Y., and Williams S. N. (1998) Fluxes and sources of volatiles discharged from Kudryavy, a subduction zone volcano Kurile Islands. *Earth Planet. Sci. Lett.* **160**, 81–96.

Fischer T. P., Hilton D. R., Zimmer M. M., Shaw A. M., Sharp Z. D., and Walker J. A. (2002) Subduction and recycling of nitrogen along the central American margin. *Science* **297**, 1154–1157.

Fisher D. E. (1985) Noble gases from oceanic island basalt do not require an undepleted mantle source. *Nature* **316**, 716–718.

Frey F. A. and Green D. H. (1974) The mineralogy, geochemistry, and origin of lherzolite inclusions in Victorian basanites. *Geochim. Cosmochim. Acta* **38**, 1023–1059.

Furman T. and Graham D. W. (1999) Erosion of lithospheric mantle beneath the East African rift system: geochemical evidence from the Kivu volcanic province. *Lithos* **48**, 237–262.

Galer S. J. G. and Mezger K. (1998) Metamorphism, denudation and sea level in the Archean and cooling of the Earth. *Precamb. Res.* **92**, 389–412.

Galer S. J. G. and O' Nions R. K. (1985) Residence time of thorium, uranium and lead in the mantle with implications for mantle convection. *Nature* **316**, 778–782.

Gasparon M., Hilton D. R., and Varne R. (1994) Crustal contamination processes traced by helium isotopes—examples from the Sunda arc, Indonesia. *Earth Planet. Sci. Lett.* **126**, 15–22.

Geiss J. and Reeves H. (1972) Cosmic and solar system abundances of deuterium and helium-3. *Astron. Astrophys.* **18**, 126–132.

Gerling E. K. and Levskii L. K. (1956) On the origin of the rare gases in stony meteorites. *Geokhimiya* **7**, 59–64.

Goes S., Spakman W., and Bijwaard H. (1999) A lower mantle source for central European volcanism. *Science* **286**, 1928–1931.

Graham D. W. (2002) Noble gas isotope geochemistry of mid-ocean ridge and ocean island basalts: characterization of mantle source reservoirs. *Rev. Mineral. Geochem.* **47**, 247–317.

Graham D. W., Lupton J., Albarède F., and Condomines M. (1990) Extreme temporal homogeneity of helium isotopes at Piton de la Fournaise, Reunion Island. *Nature* **347**, 545–548.

Graham D. W., Christie D. M., Harpp K. S., and Lupton J. E. (1993) Mantle plume helium in submarine basalts from the Galapagos platform. *Science* **262**, 2023–2026.

Graham D. W., Furman T. H., Ebinger C. J., Rogers N. W., and Lupton J. E. (1995) Helium, lead, strontium, and neodymium isotope variations in mafic volcanic rocks from the western branch of the East African rift. *EOS Trans.: Am. Geophys. Union* **76**, F686.

Graham D. W., Larsen L. M., Hanan B. B., Storey M., Pedersen A. K., and Lupton J. E. (1998) Helium isotope composition of the early Iceland mantle plume inferred from the tertiary picrites of West Greenland. *Earth Planet. Sci. Lett.* **160**, 241–255.

Grand S. P. (1987) Tomographic inversion of sear velocity beneath the North American plate. *J. Geophys. Res.* **92**, 14065–14090.

Grand S. P. (1994) Mantle shear structure beneath the Americas and surrounding oceans. *J. Geophys. Res.* **99**, 66–78.

Granet M., Wilson M., and Achauer U. (1995) Imaging a mantle plume beneath the French Massif Central. *Earth Planet. Sci. Lett.* **136**, 281–296.

Griffin W. L., O'Reilly S. Y., and Stabel A. (1988) Mantle metasomatism beneath western Victoria, Australia: II. Isotopic geochemistry of Cr-diopside lherzolites and Al-augite pyroxenites. *Geochim. Cosmochim. Acta* **52**, 449–459.

Hamano Y. and Ozima M. (1978) Earth-atmosphere evolution model based on Ar isotopic data. In *Terrestrial Rare Gases* (eds. M. Ozima and E. C. Alexander, Jr.). Japan Scientific Societies Press, Tokyo, pp. 155–171.

Handler M. R., Bennett V. C., and Esat T. M. (1997) The persistence of off-cratonic lithospheric mantle: Os isotopic systematics of variably metasomatised southeast Australian xenoliths. *Earth Planet. Sci. Lett.* **151**, 61–75.

Hanyu T. and Kaneoka I. (1997) The uniform and low $^3He/^4He$ ratios of HIMU basalts as evidence for the origin as recycled materials. *Nature* **390**, 273–276.

Harrison D., Burnard P., and Turner G. (1999) Noble gas behaviour and composition in the mantle: constraints from the Iceland plume. *Earth Planet. Sci. Lett.* **171**, 199–207.

Hart R., Dymond J., and Hogan L. (1979) Preferential formation of the atmosphere-sialic crust system from the upper mantle. *Nature* **278**, 156–159.

Hart S. R. and Staudigel H. (1982) The control of alkalis and uranium in seawater by ocean crust alteration. *Earth Planet. Sci. Lett.* **58**, 202–212.

Hauri E. H., Whitehead J. A., and Hart S. R. (1994) Fluid dynamic and geochemical aspects of entrainment in mantle plumes. *J. Geophys. Res.* **99**, 24275–24300.

Hawkesworth C. J., Rogers N. W., van Calsteren P. W. C., and Menzies M. A. (1984) Mantle enrichment processes. *Nature* **311**, 331–334.

Hawkesworth C. J., Kempton P. D., Rogers N. W., Ellam R. M., and van Calsteren P. W. (1990) Continental mantle lithosphere, and shallow level enrichment processes in the Earth's mantle. *Earth Planet. Sci. Lett.* **96**, 256–268.

Helffrich G. R. and Wood B. J. (2001) The Earth's mantle. *Nature* **412**, 501–507.

Hill R. D. (1941) Production of helium-3. *Phys. Rev.* **59**, 103.

Hilton D. R. and Craig H. (1989) A helium isotope transect along the Indonesian archipelago. *Nature* **342**, 906–908.

Hilton D. R., Hoogewerff J. A., van Bergen M. J., and Hammerschmidt K. (1992) Mapping magma sources in the east Sunda-Banda arcs, Indonesia: constraints from helium isotopesd. *Geochim. Cosmochim. Acta* **56**, 851–859.

Hilton D. R., Hammerschmidt K., Loock G., and Friedrichsen H. (1993a) Helium and argon isotope systematics of the central Lau Basin and Valu Fa ridge: evidence of crust–mantle interactions in a back-arc basin. *Geochim. Cosmochim. Acta* **57**, 2819–2841.

Hilton D. R., Hammerschmidt K., Teufel S., and Friedrichsen H. (1993b) Helium isotope characteristics of Andean geothermal fluids and lavas. *Earth Planet. Sci. Lett.* **120**, 265–282.

Hilton D. R., Barling J., and Wheller G. E. (1995) Effect of shallow-level contamination on the helium isotope systematics of ocean-island lavas. *Nature* **373**, 330–333.

Hilton D. R., McMurtry G. M., and Kreulen R. (1997) Evidence for extensive degassing of the Hawaiian mantle plume from helium–carbon relationships at Kilauea volcano. *Geophys. Res. Lett.* **24**, 3065–3068.

Hilton D. R., McMurtry G. M., and Goff F. (1998) Large variations in vent fluid $CO_2/^3He$ ratios signal rapid changes in magma chemistry at Loihi Seamount, Hawaii. *Nature* **396**, 359–362.

Hilton D. R., Gronvold K., Macpherson C. G., and Castillo P. R. (1999) Extreme $^3He/^4He$ ratios in northwest Iceland: constraining the common component in mantle plumes. *Earth Planet. Sci. Lett.* **173**, 53–60.

Hilton D. R., Macpherson C. G., and Elliott T. R. (2000a) Helium isotope ratios in mafic phenocrysts and geothermal fluids from La Palma, the Canary Islands (Spain): implications for HIMU mantle sources. *Geochim. Cosmochim. Acta* **64**, 2119–2132.

Hilton D. R., Thirlwall M. F., Taylor R. N., Murton B. J., and Nichols A. (2000b) Controls on magmatic degassing along the Reykjanes ridge with implications for the helium paradox. *Earth Planet. Sci. Lett.* **183**, 43–50.

Hilton D. R., Fischer T. P., and Marty B. (2002) Noble gases and volatile recycling at subduction zones. *Rev. Mineral. Geochem.* **47**, 319–370.

Hiyagon H. (1994) Retentivity of solar He and Ne in IDPs in deep sea sediment. *Science* **263**, 1257–1259.

Hiyagon H. and Ozima M. (1982) Noble gas distribution between basalt melt and crystals. *Earth Planet. Sci. Lett.* **58**, 255–264.

Hiyagon H., Ozima M., Marty B., Zashu S., and Sakai H. (1992) Noble gases in submarine glasses from mid-oceanic ridges and Loihi Seamount—constraints on the early history of the Earth. *Geochim. Cosmochim. Acta* **56**, 1301–1316.

Hoernle K., Zhang Y. S., and Graham D. (1995) Seismic and geochemical evidence for large-scale mantle upwelling beneath the eastern Atlantic and western and central Europe. *Nature* **374**, 34–39.

Hofmann A. W. (1997) Mantle geochemistry: the message from oceanic volcanism. *Nature* **385**, 219–229.

Hogness T. R. and Kvalnes H. M. (1928) The ionization processes in methane interpreted by the mass spectrograph. *Phys. Rev.* **32**, 942–945.

Holloway J. R. (1998) Graphite-melt equilibria during mantle melting: constraints on CO_2 in MORB magmas and the carbon content of the mantle. *Chem. Geol.* **147**, 89–97.

Honda M. and McDougall I. (1998) Primordial helium and neon in the Earth—a speculation on early degassing. *Geophys. Res. Lett.* **25**, 1951–1954.

Honda M., Reynolds J., Roedder E., and Epstein S. (1987) Noble gases in diamonds: occurrences of solar like helium and neon. *J. Geophys. Res.* **92**, 12507–12521.

Honda M., McDougall I., Patterson D. B., Doulgeris A., and Clague D. (1991) Possible solar noble gas component in Hawaiian basalts. *Nature* **349**, 149–151.

Honda M., McDougall I., and Patterson D. (1993) Solar noble gases in the Earth: the systematics of helium–neon isotopes in mantle derived samples. *Lithos* **30**, 257–265.

Jambon A., Weber H., and Braun O. (1986) Solubility of He, Ne, Ar, Kr, and Xe in a basalt melt in the range 1,250–1,600 °C. Geochemical implications. *Geochim. Cosmochim. Acta* **50**, 401–408.

Javoy M. and Pineau F. (1991) The volatile record of a "popping" rock from the Mid-Atlantic Ridge at 14° N: chemical composition of the gas trapped in the vesicles. *Earth Planet. Sci. Lett.* **107**, 598–611.

Jochum K. P., Hofmann A. W., Ito E., Seufert H. M., and White W. M. (1983) K, U, and Th in mid-ocean ridge glasses and heat production, K/U and K/Rb in the mantle. *Nature* **306**, 431–436.

Jordan T. H. (1975) The continental tectosphere. *Rev. Geophys. Space Phys.* **13**, 1–2.

Kamijo K., Hashizume K., and Matsuda J. (1998) Noble gas constraints on the evolution of the atmosphere mantle system. *Geochim. Cosmochim. Acta* **62**, 2311–2321.

Kaneoka I., Takaoka N., and Upton B. G. J. (1986) Noble gas systematics in basalts and a dunite nodule from Reunion and Grand Comore Islands, Indian Ocean. *Chem. Geol.* **59**, 35–42.

Kàrason H. and van der Hilst R. D. (2001) Tomographic imaging of the lowermost mantle with differential times of refracted and diffracted core phases (PKP, P-diff). *J. Geophys. Res.* **106**, 6569–6587.

Kellogg L. H., Hager B. H., and van der Hilst R. D. (1999) Compositional stratification in the deep mantle. *Science* **283**, 1881–1884.

Kellogg L. H. and Wasserburg G. J. (1990) The role of plumes in mantle helium fluxes. *Earth Planet. Sci. Lett.* **99**, 276–289.

Kerrick D. M. and Connolly J. A. D. (2001) Metamorphic devolatilization of subducted marine sediments and the transport of volatiles into the Earth's mantle. *Nature* **411**, 293–296.

Kirstein L. A. and Timmerman M. J. (2000) Evidence of the proto-Iceland plume in northwestern Ireland at 42 Ma from helium isotopes. *J. Geol. Soc. London* **157**, 923–927.

Kunz J., Staudacher T., and Allègre C. J. (1998) Plutonium-fission xenon found in Earth's mantle. *Science* **280**, 877–880.

Kurz M. D. (1986) Cosmogenic helium in a terrestrial igneous rock. *Nature* **320**, 435–439.

Kurz M. D., Jenkins W. J., and Hart S. R. (1982) Helium isotope systematics of ocean islands and mantle heterogeneity. *Nature* **297**, 43–47.

Kurz M. D., Colodner D., Trull T. W., Moore R. B., and O'Brien K. (1990) Cosmic ray exposure dating with *in situ* produced cosmogenic He-3—results from young Hawaiian lava flows. *Earth Planet. Sci. Lett.* **97**, 177–189.

Kyser T. K. and Rison W. (1982) Systematics of rare gas isotopes in basic lavas and ultramafic xenoliths. *J. Geophys. Res.* **87**, 5611–5630.

Lockyer J. N. (1869) Spectroscopic observations of the sun. *Phil. Trans.* **159**, 425–444.

Lupton J. E. (1983) Terrestrial inert gases: isotope tracer studies and clues to primordial components in the mantle. *Ann. Rev. Earth Planet. Sci.* **11**, 371–414.

Lupton J. E. and Craig H. (1975) Excess ^{3}He in oceanic basalts: evidence for terrestrial primordial helium. *Earth Planet. Sci. Lett.* **26**, 133–139.

Mahaffy P. R., Donahue T. M., Atreya S. K., Owen T. C., and Niemann H. B. (1998) Galileo probe measurements of D/H and ^{3}He/^{4}He in Jupiter's atmosphere. *Space Sci. Rev.* **84**, 251–263.

Mamyrin B. A. and Tolstikhin I. N. (1984) *Helium Isotopes in Nature*. Elsevier, Amsterdam, Elsevier, Amsterdam.

Mamyrin B. A., Tolstikhin I. N., Anufriev G. S., and Kamensky I. L. (1969) Anomalous isotopic composition of helium in volcanic gases. *Dokl. Akad. Nauk. SSSR* **184**, 1197–1199.

Manga M. (1996) Mixing of heterogeneities in the mantle: effect of viscosity differences. *Geophys. Res. Lett.* **23**, 403–406.

Martel D. J., Deak J., Dovenyi P., Horvath F., O'Nions R. K., Oxburgh E. R., Stegena L., and Stute M. (1989) Leakage of helium from the Pannonian basin. *Nature* **342**, 908–912.

Marti K. and Mathew K. (1998) Noble-gas components in planetary atmospheres and interiors in relation to solar wind and meteorites. *Proc. Indian Acad. Sci.: Earth Planet. Sci.* **107**, 425–431.

Marty B. (1995) Nitrogen content of the mantle inferred from N_2/Ar correlations in oceanic basalts. *Nature* **377**, 326–329.

Marty B. and Allé P. (1994) Neon and argon isotopic constraints on Earth-atmosphere evolution. In *Noble Gas Geochemistry and Cosmochemistry* (ed. J.-I. Matsuda). Terra Scientific Publishing Company, Tokyo, pp. 191–204.

Marty B. and Dauphas N. (2002) Formation and early evolution of the atmosphere. *J. Geol. Soc. London Spec. Publ.: Early History of the Earth* (eds. C. M. R. Fowler, C. J. Ebinger, and C. J. Hawkesworth). Geological Society of London, London, UK, vol. 199, pp. 213–219.

Marty B. and Humbert F. (1997) Nitrogen and argon isotopes in oceanic basalts. *Earth Planet. Sci. Lett.* **152**, 101–112.

Marty B. and Jambon A. (1987) C/^{3}He in volatile fluxes from the solid Earth: implication for carbon geodynamics. *Earth Planet. Sci. Lett.* **83**, 16–26.

Marty B. and LeCloarec M. F. (1992) Helium-3 and CO_2 fluxes from subaereal volcanoes estimated from Polonium-210 emissions. *J. Volcanol. Geotherm. Res.* **53**, 67–72.

Marty B. and Lussiez P. (1993) Constraints on rare gas partition coefficients from analysis of olivine glass from a picritic mid-ocean ridge basalt. *Chem. Geol.* **106**, 1–7.

Marty B. and Tolstikhin I. N. (1998) CO_2 fluxes from mid-ocean ridges, arcs and plumes. *Chem. Geol.* **145**, 233–248.

Marty B. and Zimmermann L. (1999) Volatiles (He, C, N, Ar) in mid-ocean ridge basalts: assessment of shallow-level fractionation and characterization of source composition. *Geochim. Cosmochim. Acta* **63**, 3619–3633.

Marty B., Jambon A., and Sano Y. (1989) Helium isotopes and CO_2 in volcanic gases of Japan. *Chem. Geol.* **76**, 25–40.

Marty B., Appora I., Barrat J. A. A., Deniel C., Vellutini P., and Vidal P. (1993) He, Ar, Sr, Nd, and Pb isotopes in volcanic rocks from Afar—evidence for a primitive mantle component and constraints on magmatic sources. *Geochem. J.* **27**, 219–228.

Marty B., Trull T., Lussiez P., Basile I., and Tanguy J. C. (1994) He, Ar, O, Sr, and Nd isotope constraints on the origin and evolution of Mount Etna magmatism. *Earth Planet. Sci. Lett.* **126**, 23–39.

Marty B., Pik R., and Gezahegn Y. (1996) Helium isotopic variations in Ethiopian plume lavas—nature of magmatic sources and limit on lower mantle contribution. *Earth Planet. Sci. Lett.* **144**, 223–237.

Marty B., Tolstikhin I., Kamensky I. L., Nivin V., Balaganskaya E., and Zimmermann J. L. (1998) Plume-derived rare gases in 380 Ma carbonatites from the Kola region (Russia) and the argon isotopic composition in the deep mantle. *Earth Planet. Sci. Lett.* **164**, 179–192.

Matsuda J. and Marty B. (1995) The ^{40}Ar/^{36}Ar ratio of the undepleted mantle: a reevaluation. *Geophys. Res. Lett.* **22**, 1937–1940.

Matsuda J. and Matsubara K. (1989) Noble gases in silica and their implication for the terrestrial "missing" Xe. *Geophys. Res. Lett.* **16**, 81–84.

Matsuda J. and Nagao K. (1986) Noble gas abundance in a deep-sea sediment core from eastern equatorial Pacific. *Geochem. J.* **20**, 71–80.

Matsuda J., Sudo M., Ozima M., Ito K., Ohtaka O., and Ito E. (1993) Noble gas partitioning between metal and silicate under high pressures. *Science* **259**, 788–790.

Matsumoto T., Honda M., McDougall I., Yatsevich I., and O'Reilly S. Y. (1997) Plume-like neon in a metasomatic apatite from the Australian lithospheric mantle. *Nature* **388**, 162–164.

Matsumoto T., Honda M., McDougall I., and O'Reilly S. Y. (1998) Noble gases in anhydrous lherzolites from the Newer volcanics, southeastern Australia: a MORB like reservoir in the subcontinental mantle. *Geochim. Cosmochim. Acta* **62**, 2521–2533.

Matsumoto T., Honda M., McDougall I., O'Reilly S. Y., Norman M., and Yaxley G. (2000) Noble gases in pyroxenites and metasomatized peridotites from the Newer volcanics, southeastern Australia: implications for mantle metasomatism. *Chem. Geol.* **168**, 49–73.

McDonough W. F. (1990) Constraints on the composition of the continental lithospheric mantle. *Earth Planet. Sci. Lett.* **101**, 1–18.

McDonough W. F. and McCulloch M. T. (1987) The southeast Australian lithospheric mantle: isotopic and geochemical constraints on its growth and evolution. *Earth Planet. Sci. Lett.* **86**, 327–340.

McDougall I. and Honda M. (1998) Primordial solar gas component in the Earth: consequences for the origin and evolution of the Earth and its atmosphere. In *The Earth's*

Mantle: Composition, Structure, and Evolution (ed. I. Jackson). Cambridge University Press, Cambridge, pp. 159–187.

Meen J. K., Eggler D. H., and Ayers J. C. (1989) Experimental evidence for very low solubility of rare earth elements in CO_2-rich fluids at mantle conditions. *Nature* **340**, 301–303.

Menzies M. and Chazot G. (1985) Fluid processes in diamond to spinel facies shallow mantle. *J. Geodynam.* **20**, 387–415.

Menzies M. and Chazot G. (1995) Fluid processes in diamond to spinel facies shallow mantle. *J. Geodynam.* **20**, 387–415.

Moreira M. and Allègre C. J. (1998) Helium–neon systematics and the structure of the mantle. *Chem. Geol.* **147**, 53–59.

Moreira M. and Sarda P. (2000) Noble gas constraints on degassing processes. *Earth Planet. Sci. Lett.* **176**, 375–386.

Moreira M., Kunz J., and Allègre C. (1998) Rare gas systematics in popping rock: isotopic and elemental compositions in the upper mantle. *Science* **279**, 1178–1181.

Morgan J. P. (1998) Thermal and rare gas evolution of the mantle. *Chem. Geol.* **145**, 431–445.

Morgan W. J. (1971) Convection plumes in the lower mantle. *Nature* **230**, 42–43.

Morrison P. and Pine J. (1955) Radiogenic origin of the helium isotopes in rock. *Ann. NY Acad. Sci.* **62**, 69–92.

Nier A. O. (1936) The isotopic constitution of rubidium, zinc, and argon. *Phys. Rev.* **49**, 272.

Nolet G., Grand S. P., and Kennet B. L. N. (1994) Seismic heterogeneity in the upper mantle. *J. Geophys. Res.* **99**, 23753–23776.

Ohmoto H. (1986) Stable isotope geochemistry of ore deposits. *Rev. Mineral.* **16**, 491–560.

O'Nions R. K. and Oxburgh E. R. (1983) Heat and helium in the Earth. *Nature* **306**, 429–431.

O'Nions R. K. and Oxburgh E. R. (1988) Helium, volatile fluxes and the development of continental crust. *Earth Planet. Sci. Lett.* **90**, 331–347.

O'Nions R. K. and Tolstikhin I. N. (1994) Behaviour and residence times of lithophile and rare gas tracers in the upper mantle. *Earth Planet. Sci. Lett.* **124**, 131–138.

O'Nions R. K. and Tolstikhin I. N. (1996) Limits on the mass flux between lower and upper mantle and stability of layering. *Earth Planet. Sci. Lett.* **139**, 213–222.

O'Nions R. K., Evenson N. M., and Hamilton P. J. (1979) Geochemical modeling of mantle differentiation and crustal growth. *J. Geophys. Res.* **84**, 6091–6101.

O'Reilly S. Y. and Griffin W. L. (1984) A xenolith-derived geotherm for southeastern Australia and its geophysical implications. *Tectonophysics* **111**, 41–63.

O'Reilly S. Y. and Griffin W. L. (1988) Mantle metasomatism beneath western Victoria, Australia: I. Metasomatic processes in Cr-diopside lherzolites. *Geochim. Cosmochim. Acta* **52**, 433–447.

O'Reilly S. Y. and Griffin W. L. (2000) Apatite in the mantle: implications for metasomatic processes and high heat production in Phanerozoic mantle. *Lithos* **53**, 217–232.

Ott U. (2002) Noble gases in meteorites-trapped components. *Rev. Mineral. Geochem.* **47**, 71–100.

Oxburgh E. R., O'Nions R. K., and Hill R. L. (1986) Helium isotopes in sedimentary basins. *Nature* **324**, 632–635.

Ozima M. (1994) Noble gas state in the mantle. *Rev. Geophys.* **32**, 405–426.

Ozima M. and Podosek F. A. (1983) *Noble Gas Geochemistry.* Cambridge University Press, Cambridge.

Parello F., Allard P., D'Alessandro W., Federico C., Jean-Baptiste P., and Catani O. (2000) Isotope geochemistry of Pantelleria volcanic fluids, Sicily Channel rift: a mantle volatile end-member for volcanism in southern Europe. *Earth Planet. Sci. Lett.* **180**, 325–339.

Parsons B. (1981) The rates of plate creation and consumption. *Geophys. J. Roy. Astron. Soc.* **67**, 437–448.

Patterson D. B., Honda M., and McDougall I. (1990) Atmospheric contamination—a possible source for heavy noble gases in basalts from Loihi Seamount, Hawaii. *Geophys. Res. Lett.* **17**, 705–708.

Pearson D. G. (1999a) The age of continental roots. *Lithos* **48**, 171–194.

Pearson D. G. (1999b) Evolution of cratonic lithospheric mantle: an isotopic perspective. In *Mantle Petrology: Field Observations and High Pressure Experimentation: A Tribute to Francis R. (Joe) Boyd,* The Geochemical Society Special Publication (eds. Y. Fei, C. M. Bertka, and B. O. Mysen). The Geochemical Society, Houston, Texas, vol. 6, pp. 57–78.

Pepin R. O. (1992) Origin of noble gases in the terrestrial planets. *Ann. Rev. Earth Planet. Sci.* **20**, 389–430.

Pepin R. O. and Porcelli D. (2002) Origin of noble gases in the terrestrial planets. *Rev. Mineral. Geochem.* **47**, 191–246.

Peters K. E., Sweeney R. E., and Kaplan I. R. (1978) Correlation of carbon and nitrogen stable isotope ratios in sedimentary organic matter. *Limnol. Ocean.* **23**, 598–604.

Phinney D., Tennyson J., and Frick U. (1978) Xenon in CO_2 well gas revisited. *J. Geophys. Res.* **83**, 2313–2319.

Plank T. and Langmuir C. H. (1988) An evaluation of the global variations in the major element chemistry of arc basalts. *Earth Planet. Sci. Lett.* **90**, 349–370.

Podosek F. A. (1978) Isotopic structure in solar system materials. *Ann. Rev. Astron. Astrophys.* **16**, 293–334.

Podosek F. A., Honda M., and Ozima M. (1980) Sedimentary noble gases. *Geochim. Cosmochim. Acta* **44**, 1875–1884.

Polet J. and Anderson D. L. (1995) Depth extension of cratons as inferred from tomographic studies. *Geology* **23**, 205–208.

Pollack H. N., Hurter S. J., and Johnson J. R. (1993) Heat flow from the Earth's interior: analysis of the global data set. *Rev. Geophys.* **31**, 267–280.

Porcelli D. and Ballentine C. J. (2002) Models for the distribution of terrestrial noble gases and the evolution of the atmosphere. *Rev. Mineral. Geochem.* **47**, 412–480.

Porcelli D. and Halliday A. N. (2001) The core as a possible source of mantle helium. *Earth Planet. Sci. Lett.* **192**, 45–56.

Porcelli D. and Wasserburg G. J. (1995a) Mass transfer of helium, neon, argon, and xenon through a steady state upper mantle. *Geochim. Cosmochim. Acta* **59**, 4921–4937.

Porcelli D. and Wasserburg G. J. (1995b) Mass transfer of xenon through a steady-state upper mantle. *Geochim. Cosmochim. Acta* **59**, 1991–2007.

Porcelli D., O'Nions R. K., and O'Reilly S. Y. (1986a) Helium and strontium isotopes in ultramafic xenoliths. *Chem. Geol.* **54**, 237–249.

Porcelli D., Stone J. O. H., and O'Nions R. K. (1986b) Rare gas reservoirs and Earth degassing. *Lunar Planet. Sci.* **XVII**, 674–675.

Porcelli D., Stone J. O. H., and O'Nions R. K. (1987) Enhanced $^3He/^4He$ ratios and cosmogenic helium in ultramafic xenoliths. *Chem. Geol.* **64**, 25–33.

Porcelli D., Ballentine C. J., and Wieler R. (eds.) (2002) In *Noble Gases in Geochemistry and Cosmochemistry, Reviews in Mineralogy and Geochemistry,* Mineralogical Society of America, Washington, DC, vol. 47.

Porcelli D. R., O'Nions R. K., Galer S. J. G., Cohen A. S., and Mattey D. P. (1992) Isotopic relationships of volatile and lithophile trace elements in continental ultramafic xenoliths. *Contrib. Mineral. Petrol.* **110**, 528–538.

Poreda R. and Craig H. (1989) Helium isotope ratios in circum-Pacific volcanic arcs. *Nature* **338**, 473–478.

Poreda R. J. and Farley K. A. (1992) Rare gases in Samoan xenoliths. *Earth Planet. Sci. Lett.* **113**, 129–144.

Poreda R. J., Jenden P. D., Kaplan I. R., and Craig H. (1986) Mantle helium in Sacramento basin natural gas wells. *Geochim. Cosmochim. Acta* **50**, 2847–2853.

Ramsay W. (1895) Discovery of helium. *Chem. News* **71**, 151.

Ramsay W. and Soddy F. (1903) Experiments in radioactivity and the production of helium from radium. *Proc. Roy. Soc. London* **A72**, 204.

Ramsay W. and Travers M. W. (1898a) On a new constituent of atmospheric air. *Proc. Roy. Soc. London* **A63**, 405–408.

Ramsay W. and Travers M. W. (1898b) On the companions of argon. *Proc. Roy. Soc. London* **A63**, 437–440.

Ramsay W. and Travers M. W. (1898c) On the extraction from air of the companions of argon and on neon. *Br. Ass. Rept.* 828–830.

Rayleigh L. (1892) Density of nitrogen. *Nature* **46**, 512–513.

Rayleigh L. and Ramsay W. (1892) Argon, a new constituent of the atmosphere. *Proc. Roy. Soc. London* **A57**, 265.

Reid M. R. and Graham D. W. (1996) Resolving lithospheric and sub-lithospheric contributions to helium isotope variations in basalts from the southwestern US. *Earth Planet. Sci. Lett.* **144**, 213–222.

Reymer A. and Schubert G. (1984) Phanerozoic addition rates to the continental crust. *Tectonics* **3**, 63–77.

Ricard Y., Nataf H.-C., and Montagner J.-P. (1996) The three-dimensional seismological model a priori constrained: confrontation with seismic data. *J. Geophys. Res.* **101**, 8457–8472.

Richard D., Marty B., Chaussidon M., and Arndt N. (1996) Helium isotopic evidence for a lower mantle component in depleted Archean komatiite. *Science* **273**, 93–95.

Richardson S. H. and Harris J. W. (1997) Antiquity of peridotitic diamonds from the Siberian craton. *Earth Planet. Sci. Lett.* **151**, 271–277.

Richardson S. H., Gurney J. J., Erlank A. J., and Harris J. W. (1984) Origin of diamonds in old enriched mantle. *Nature* **310**, 198–202.

Rocholl A. and Jochum K. P. (1993) Th, U, and other trace elements in carbonaceous chondrites—implications for the terrestrial and solar system Th/U ratios. *Earth Planet. Sci. Lett.* **117**, 265–278.

Rodgers N. W., De Mulder M., and Hawkesworth C. J. (1992) An enriched mantle source for potassic basanites: evidence from Karisimbi volcano, Virunga volcanic province, Rwanda. *Contrib. Mineral. Petrol.* **111**, 543–556.

Rudnick R. L. and Nyblade A. A. (1999) The thickness and heat production of Archean lithosphere: constraints from xenolith thermobarometry and surface heat flow. In *Mantle Petrology: Field Observations and High Pressure Experimentation: A Tribute to Francis R. (Joe) Boyd,* The Geochemical Society Special Publication (eds. Y. Fei, C. M. Bertka, and B. O. Mysen). The Geochemical Society, Houston, Texas, vol. 6, pp. 3–12.

Rudnick R. L., McDonough W. F., and O'Connell R. J. (1998) Thermal structure, thickness and composition of continental lithosphere. *Chem. Geol.* **145**, 399–415.

Sano Y. and Marty B. (1995) Origin of carbon in fumarolic gas from island arcs. *Chem. Geol.* **119**, 265–274.

Sano Y. and Williams S. N. (1996) Fluxes of mantle and subducted carbon along convergent plate boundaries. *Geophys. Res. Lett.* **23**, 2749–2752.

Sano Y., Takahata N., Nishio Y., and Marty B. (1998) Nitrogen recycling in subduction zones. *Geophys. Res. Lett.* **25**, 2289–2292.

Sano Y., Takahata N., Nishio Y., Fischer T. P., and Williams S. N. (2001) Volcanic flux of nitrogen from the Earth. *Chem. Geol.* **171**, 263–271.

Sarda P. and Graham D. (1990) Mid-ocean ridge popping rocks: implications for degassing at ridge crests. *Earth Planet. Sci. Lett.* **97**, 268–289.

Sarda P., Staudacher T., and Allègre C. J. (1988) Neon isotopes in submarine basalts. *Earth Planet. Sci. Lett.* **91**, 73–88.

Sarda P., Moreira M., and Staudacher T. (1999) Argon–lead isotopic correlation in Mid-Atlantic ridge basalts. *Science* **283**, 666–668.

Sarda P., Moreira M., Staudacher T., Schilling J. G., and Allègre C. J. (2000) Rare gas systematics on the southernmost Mid-Atlantic ridge: constraints on the lower mantle and the Dupal source. *J. Geophys. Res.* **105**, 5973–5996.

Saunders A. and Tarney J. (1991) Back-arc basins. In *Oceanic Basalts* (ed. P. A. Floyd). Blackie, Glasgow, pp. 219–263.

Scarsi P. (2000) Fractional extraction of helium by crushing of olivine and clinopyroxene phenocrysts: effects on the He-3/He-4 measured ratio. *Geochim. Cosmochim. Acta* **64**, 3751–3762.

Scarsi P. and Craig H. (1996) Helium isotope ratios in Ethiopian rift basalts. *Earth Planet. Sci. Lett.* **144**, 505–516.

Schilling J. G., Unni C. K., and Bender M. L. (1978) Origin of chlorine and bromine in the oceans. *Nature* **273**, 631–636.

Schlosser P. and Winckler G. (2002) Noble gases in ocean waters and sediments. *Rev. Mineral. Geochem.* **47**, 701–730.

Seta A., Matsumoto T., and Matsuda J.-I. (2001) Concurrent evolution of $^3He/^4He$ ratio in the Earth's mantle reservoirs for the first 2 Ga. *Earth Planet. Sci. Lett.* **188**, 211–219.

Shaw A. M., Hilton D. R., Macpherson C. G., and Sinton J. M. (2001) Nucleogenic neon in high $^3He/^4He$ lavas from the Manus back-arc basin: a new perspective on He–Ne decoupling. *Earth Planet. Sci. Lett.* **194**, 53–66.

Signer P. and Suess H. E. (1963) Rare gas in the sun, in the atmosphere, and in meteorites. In *Earth Science and Meteorites* (eds. J. Geiss and E. D. Goldberg). Wiley, New York, pp. 241–272.

Silver P. G., Carlson R. W., and Olson P. (1988) Deep slabs, geochemical heterogeneity, and the large-scale structure of mantle convection—investigation of an enduring paradox. *Ann. Rev. Earth Planet. Sci.* **16**, 477–541.

Simons F. J., Zielhuis A., and van der Hilst R. D. (1999) The deep structure of the Australian continent from surface wave tomography. *Lithos* **48**, 17–43.

Snyder G., Poreda R., Hunt A., and Fehn U. (2001) Regional variations in volatile composition: isotopic evidence for carbonate recycling in the central American volcanic arc. *Geochem. Geophys. Geosyst.* **2**, U1–U32.

Sobolev S. V., Zeyen H., Granet M., Achauer U., Bauer C., Werling F., Altherr R., and Fuchs K. (1997) Upper mantle temperatures and lithosphere-asthenosphere system beneath the French Massif Central constrained by seismic, gravity, petrologic, and thermal observations. *Tectonophysics* **275**, 143–164.

Soddy F. (1910) Radioactivity. *Ann. Rep. Chem. Soc.* **7**, 256–286.

Staudacher T. (1987) Upper mantle origin for Harding County well gases. *Nature* **325**, 605–607.

Staudacher T. and Allègre C. J. (1982) Terrestrial xenology. *Earth Planet. Sci. Lett.* **60**, 389–406.

Staudacher T. and Allègre C. J. (1988) Recycling of oceanic crust and sediments: the noble gas subduction barrier. *Earth Planet. Sci. Lett.* **89**, 173–183.

Staudacher T., Kurz M. D., and Allègre C. J. (1986) New noble gas data on glass samples form Loihi Seamount and Huahalai and on dunite samples from Loihi and Reunion Island. *Chem. Geol.* **56**, 193–205.

Staudacher T., Sarda P., Richardson S. H., Allégre C. J., Sagna I., and Dmitriev L. V. (1989) Noble gases in basalt glasses from a mid-Atlantic ridge topographic high at 14-degrees-N—geodynamic consequences. *Earth Planet. Sci. Lett.* **96**, 119–133.

Stoiber R. E., Malinconico L. L. J., and Williams S. N. (1983) Use of the correlation spectrometer at volcanoes. In *Forecasting Volcanic Events* (eds. H. Tazieff and J. Sabroux). Elsevier, Amsterdam, vol. 1, pp. 425–444.

Stone J., Porcelli D., Vance D., Galer S., and O'Nions R. K. (1990) Volcanic traces. *Nature* **346**, 228.

Stosch H. G., Carlson R. W., and Lugmair G. W. (1980) Episodic mantle differentiation: Nd and Sr isotopic evidence. *Earth Planet. Sci. Lett.* **47**, 263–271.

Strutt R. J. (1908) The accumulation of helium in geologic time. *Proc. Roy. Soc. London* **A81**, 272–277.

Stuart F. M. (1994) Speculations about the cosmic origin of He and Ne in the interior of the Earth—comment. *Earth Planet. Sci. Lett.* **122**, 245–247.

Stute M., Sonntag C., Deak J., and Schlosser P. (1992) Helium in deep circulating groundwater in the Great Hungarian Plain—flow dynamics and crustal and mantle helium fluxes. *Geochim. Cosmochim. Acta* **56**, 2051–2067.

Su W.-J., Woodward R. L., and Dziewonski A. M. (1994) Degree 12 model of shear velocity heterogeneity in the mantle. *J. Geophys. Res.* **99**, 6945–6980.

Suess H. E. and Wanke H. (1965) On the possibility of a helium flux through the ocean floor. In *Progress in Oceanography* (ed. M. Sears). Pergamon, Oxford, vol. 3, pp. 347–353.

Suess H. E., Wanke H., and Wlotzka F. (1964) On the origin of gas rich meteorites. *Geochim. Cosmochim. Acta* **28**, 595–607.

Tackley P. J. (2000) Mantle convection and plate tectonics: toward an integrated physical and chemical theory. *Science* **288**, 2002–2007.

Tackley P. J., Stevenson D. J., and Glatzmaier G. A. (1993) Effects of an endothermic phase transition at 670 km depth on spherical mantle convection. *Nature* **361**, 699–704.

Tedesco D. and Nagao K. (1996) Radiogenic ^{4}He, ^{21}Ne, and ^{40}Ar in fumarolic gases on vulcano: implication for the presence of continental crust beneath the island. *Earth Planet. Sci. Lett.* **144**, 517–528.

Thomson J. J. (1912) Further experiments on positive rays. *Phil. Mag.* **24**, 209–253.

Tolstikhin I. N. and Marty B. (1998) The evolution of terrestrial volatiles: a view from helium, neon, argon, and nitrogen isotope modelling. *Chem. Geol.* **147**, 27–52.

Tolstikhin I. N., MaMain B. A., Khabarin L. B., and Erlikh E. N. (1974) Isotope composition of helium in ultrabasic xenoliths from volcanic rocks of Kamchatka. *Earth Planet. Sci. Lett.* **22**, 73–84.

Torgersen T. (1989) Terrestrial helium degassing fluxes and the atmospheric helium budget: implications with respect to the degassing processes of continental crust. *Chem. Geol.* **79**, 1–14.

Trieloff M., Kunz J., Clague D. A., Harrison D., and Allègre C. J. (2000) The nature of pristine noble gases in mantle plumes. *Science* **288**, 1036–1038.

Trull T. (1994) Influx and age constraints on the recycled cosmic dust explanation for high ^{3}He/^{4}He ratios at hotspot volcanoes. In *Noble Gas Geochemistry and Cosmochemistry* (ed. J. Matsuda). Terra Publishers, Tokyo, pp. 77–88.

Trull T. W. and Kurz M. D. (1993) Experimental measurements of ^{3}He and ^{4}He mobility in olivine and clinopyroxene at magmatic temperatures. *Geochim. Cosmochim. Acta* **57**, 1313–1324.

Valbracht P. J., Staudacher T., Malahoff A., and Allègre C. J. (1997) Noble gas systematics of deep rift zone glasses from Loihi Seamount, Hawaii. *Earth Planet. Sci. Lett.* **150**, 399–411.

Vance D., Stone J. O. H., and O'Nions R. K. (1989) He, Sr, and Nd isotopes in xenoliths from Hawaii and other oceanic islands. *Earth Planet. Sci. Lett.* **96**, 147–160.

van der Hilst R. D. and Karason H. (1999) Compositional heterogeneity in the bottom 1000 kilometers of Earth's mantle: toward a hybrid convection model. *Science* **283**, 1885–1888.

van der Hilst R. D., Widiyantoro S., and Engdahl E. R. (1997) Evidence for deep mantle circulation from global tomography. *Nature* **386**, 578–584.

van Keken P. E. and Ballentine C. J. (1999) Dynamical models of mantle volatile evolution and the role of phase transitions and temperature-dependent rheology. *J. Geophys. Res.* **104**, 7137–7151.

van Keken P. E., Ballentine C. J., and Porcelli D. (2001) A dynamical investigation of the heat and helium imbalance. *Earth Planet. Sci. Lett.* **188**, 421–434.

van Soest M. C., Hilton D. R., and Kreulen R. (1998) Tracing crustal and slab contributions to arc magmatism in the Lesser Antilles island arc using helium and carbon relationships in geothermal fluids. *Geochim. Cosmochim. Acta* **62**, 3323–3335.

Varekamp J. C., Kreulen R., Poorter R. P. E., and Vanbergen M. J. (1992) Carbon sources in arc volcanism with implications for the carbon cycle. *Terra Nova* **4**, 363–373.

Vidale J. E., Schubert G., and Earle P. S. (2001) Unsuccessful initial search for a mid-mantle chemical boundary with seismic arrays. *Geophys. Res. Lett.* **28**, 859–862.

von Huene R. and Scholl D. W. (1991) Observations at convergent margins concerning sediment subduction, subduction erosion, and the growth of continental crust. *Rev. Geophys.* **29**, 279–316.

Walker R. J., Carlson R. W., Shirey S. B., and Boyd F. R. (1989) Os, Sr, Nd, and Pb isotope systematics of southern African peridotite xenoliths: implications for the chemical evolution of subcontinental mantle. *Geochim. Cosmochim. Acta* **53**, 1583–1595.

Wass S. Y., Henderson P., and Elliot C. J. (1980) Chemical heterogeneity and metasomatism in the upper mantle: evidence from rare earth and other elements in apatite-rich xenoliths in basaltic rocks from eastern Australia. *Phil. Trans. Roy. Soc. London* **A297**, 333–346.

Wellman P. and McDougall I. (1974) Cainozoic igneous activity in eastern Australia. *Tectonophysics* **23**, 49–65.

Wetherill G. W. (1954) Variations in the isotopic abundance of neon and argon extracted from radioactive minerals. *Phys. Rev.* **96**, 679–683.

Wieler R. (2002) Noble gases in the solar system. *Rev. Mineral. Geochem.* **47**, 21–70.

Wilson M. and Downes H. (1991) Tertiary–Quaternary extension-related alkaline magmatism in western and central Europe. *J. Petrol.* **32**, 811–849.

Yatsevich I. and Honda M. (1997) Production of nucleogenic neon in the Earth from natural radioactive decay. *J. Geophys. Res.* **102**, 10291–10298.

Zeeman P. and De Gier J. (1934) Third preliminary note on some experiments concerning the isotope of hydrogen. *Proc. K. Akad. Amsterdam* **37**, 1–3.

Zhang M., Stephenson J., O'Reilly S. Y., McCulloch M. T., and Norman M. (2001) Petrogenesis of Late Cenozoic basalts in North Queensland and its geodynamic implications: trace element and Sr–Nd–Pb isotope evidence. *J. Petrol.* **42**, 685–719.

Zhang Y. and Zindler A. (1989) Noble gas constraints on the evolution of Earth's atmosphere. *J. Geophys. Res.* **94**, 13710–13737.

Zindler A. and Hart S. R. (1986) Chemical geodynamics. *Ann. Rev. Earth Planet. Sci.* **14**, 493–571.

Isotope Geochemistry
ISBN: 978-0-08-096710-3

pp. 521–562

D. RADIOGENIC ISOTOPES

18

Sampling Mantle Heterogeneity through Oceanic Basalts: Isotopes and Trace Elements

A. W. Hofmann

Max-Planck-Institut für Chemie, Mainz, Germany, Lamont-Doherty Earth Observatory, Palisades, NY, USA

18.1 INTRODUCTION

18.1.1 Early History of Mantle Geochemistry

Until the arrival of the theories of plate tectonics and seafloor spreading in the 1960s, the Earth's mantle was generally believed to consist of peridotites of uniform composition. This view was shared by geophysicists, petrologists, and geochemists alike, and it served to characterize the compositions and physical properties of the mantle and crust as "Sial" (silica–alumina) of low density and "Sima" (silica–magnesia) of greater density. Thus, Hurley and his collaborators were able to distinguish crustal magma sources from those located in the mantle on the basis of their initial strontium-isotopic compositions (Hurley *et al.*, 1962; and Hurley's lectures and popular articles not recorded in the formal scientific literature). In a general way, as of the early 2000s, this view is still considered valid, but literally thousands of papers have since been published on the isotopic and trace-elemental composition of oceanic basalts because they come from the mantle and are rich sources of information about the composition of the mantle, its differentiation history, and its internal structure. Through the study of oceanic basalts, it was found that the mantle is compositionally just as heterogeneous as the crust. Thus, geochemistry, a term that seems more appropriate than the more popular "chemical geodynamics," became a major tool to decipher the geology of the mantle.

The pioneers of this effort were Gast, Tilton, Hedge, Tatsumoto, and Hart (Hedge and Walthall, 1963; Gast *et al.*, 1964; Tatsumoto *et al.*, 1965; Hart, 1971). They discovered from isotope analyses of strontium and lead in young (effectively zero-age) ocean island basalts (OIBs) and mid-ocean ridge basalts (MORBs) that these basalts are isotopically not uniform. The isotope ratios $^{87}Sr/^{86}Sr$, $^{206}Pb/^{204}Pb$, $^{207}Pb/^{204}Pb$, and $^{208}Pb/^{204}Pb$ increase as functions of time and the respective radioactive-parent/nonradiogenic daughter ratios, $^{87}Rb/^{86}Sr$, $^{238}U/^{204}Pb$, $^{235}U/^{204}Pb$, and $^{232}Th/^{204}Pb$, in the sources of the magmas. This means that the mantle must contain geologically old reservoirs with different Rb/Sr, U/Pb, and Th/Pb ratios. The isotope story was complemented by trace-element geochemists, led primarily by Schilling and Winchester (1967, 1969) and Gast (1968) on chemical trace-element fractionation during igneous processes, and by Tatsumoto *et al.* (1965) and Hart (1971). From the trace-element abundances, particularly rare-earth element (REE) abundances, it became clear that not only certain parent–daughter element abundance ratios, but also the light-to-heavy REE ratios of the Earth's mantle are quite heterogeneous. The interpretation of these heterogeneities has occupied mantle geochemists since the 1960s.

This chapter is in part an update of a previous, more abbreviated review (Hofmann, 1997). It covers the subject in greater depth, and it reflects some significant changes in the author's views since the writing of the earlier paper. In particular, the spatial range of equilibrium attained during partial melting may be much smaller than previously thought, because of new experimental diffusion data and new results from natural settings. Also, the question of "layered" versus "whole-mantle" convection, including the depth of subduction and of the origin of plumes, has to be reassessed in light of the recent breakthroughs achieved by seismic mantle tomography. As the spatial resolution of seismic tomography and the pressure range, accuracy, and precision of experimental data on melting relations, phase transformations, and kinetics continue to improve, the interaction between these disciplines and geochemistry *sensu stricto* will continue to improve our understanding of what is actually going on in the mantle. The established views of the mantle being engaged in simple two- or single-layer convection are becoming obsolete. In many ways, we are just at the beginning of this new phase of mantle geology, geophysics, and geochemistry.

18.1.2 The Basics

18.1.2.1 Major and trace elements: incompatible and compatible behavior

Mantle geochemists distinguish between major and trace elements. At first sight, this nomenclature seems rather trivial, because which particular elements should be called "major" and which "trace" depends on the composition of the system. However, this distinction actually has a deeper meaning, because it signifies fundamental differences in geochemical behavior. We define elements as "major" if they are essential constituents of the minerals making up a rock, that is, in the sense of the phase rule. Thus, silicon, aluminum, chromium, magnesium, iron, calcium, sodium, and oxygen are major elements because they are essential constituents of the upper-mantle minerals—olivine, pyroxene, garnet, spinel, and plagioclase. Adding or subtracting such elements can change the phase assemblage. Trace elements, on the other hand, just replace a few atoms of the major elements in the crystal structures without affecting the phase assemblage significantly. They are essentially blind passengers in many mantle processes, and they are therefore immensely useful as tracers of such processes. During solid-phase transformations, they will redistribute themselves locally between the newly formed mineral phases but, during melting, they are partitioned to a greater or lesser degree into the melt. When such a melt is transported to the Earth's surface, where it can be sampled, its trace elements carry a wealth of information about the composition of the source rock and the nature of the melting processes at depth.

For convenience, the partitioning of trace elements between crystalline and liquid phases is usually described by a coefficient D, which is just a simple ratio of two concentrations at chemical equilibrium:

$$D^i = \frac{C_s^i}{C_l^i} \quad (1)$$

where D^i is the called the partition coefficient of trace element i, C_s^i and C_l^i are the concentrations (by weight) of this element in the solid and liquid phases, respectively.

Goldschmidt (1937, 1954) first recognized that the distribution of trace elements in minerals is strongly controlled by ionic radius and charge. The partition coefficient of a given trace element between solid and melt can be quantitatively described by the elastic strain this element causes by its presence in the crystal lattice. When this strain is large because of the magnitude of the misfit, the partition coefficient becomes small, and the element is partitioned into the liquid.

Most trace elements have values of $D \ll 1$, simply because they differ substantially either in ionic radius or ionic charge, or both, from the atoms of the major elements they replace in the crystal lattice. Because of this, they are called "incompatible." Exceptions are trace elements such as strontium in plagioclase, ytterbium, lutetium, and scandium in garnet, nickel in olivine, and scandium in clinopyroxene. These latter elements actually fit into their host crystal structures slightly better than the major elements they replace, and they are therefore called "compatible." Thus, most chemical elements of the periodic table are trace elements, and most of them are incompatible; only a handful are compatible.

Major elements in melts formed from mantle rocks are by definition compatible, and most of them are well buffered by the residual minerals, so that their concentrations usually vary by factors of <2 in the melts. In contrast, trace elements, particularly those having very low partition coefficients, may vary by as many as three orders of magnitude in the melt, depending on the degree of melting. This is easily seen from the mass-balance-derived equation for the equilibrium concentration of a trace element in the melt, C_l, given by (Shaw, 1970)

$$C_l = \frac{C_0}{F + D(1-F)} \quad (2)$$

where the superscript i has been dropped for clarity, C_0 is the concentration in the bulk system, and F is the melt fraction by mass. For highly incompatible elements, which are characterized by very low partition coefficients, such that $D \ll F$, this equation reduces to

$$C_l \approx \frac{C_0}{F} \quad (3)$$

This means that the trace-element concentration is then inversely proportional to the melt fraction F, because the melt contains essentially the entire budget of this trace element. An additional consequence of highly incompatible behavior of trace elements is that their concentration ratios in the melt become constant, independent of melt fraction, and identical to the respective ratio in the mantle source. This follows directly when Equation (3) is written for two highly incompatible elements:

$$\frac{C_l^1}{C_l^2} \approx \frac{C_0^1}{F}\frac{F}{C_0^2} = \frac{C_0^1}{C_0^2} \quad (4)$$

In this respect, incompatible trace-element ratios resemble isotope ratios. They are therefore very useful in complementing the information obtained from isotopes.

18.1.2.2 Radiogenic isotopes

The decay of long-lived radioactive isotopes was initially used by geochemists exclusively for the measurement of geologic time. As noted in the introduction, their use as tracers of mantle processes was pioneered by Hurley and coworkers in the early 1960s. The decay

$$^{87}\text{Rb} \rightarrow {}^{87}\text{Sr} \quad (\lambda = 1.42 \times 10^{-11}\text{years}) \quad (5)$$

serves as example. The solution of the decay equation is

$$^{87}\text{Sr} = {}^{87}\text{Rb}(e^{\lambda t} - 1) \quad (6)$$

Dividing both sides by one of the nonradiogenic isotopes, by convention ^{86}Sr, we obtain

$$\frac{^{87}\text{Sr}}{^{86}\text{Sr}} = \frac{^{87}\text{Rb}}{^{86}\text{Sr}}(e^{\lambda t} - 1) \approx \frac{^{87}\text{Rb}}{^{86}\text{Sr}}\lambda t \quad (7)$$

The approximation in Equation (7) holds only for decay systems with sufficiently long half-lives, such as the Rb–Sr and the Sm–Nd systems, so that $\lambda t \ll 1$ and $e^{\lambda t} - 1 \approx \lambda t$. Therefore, the isotope ratio ^{87}Sr/^{86}Sr in a system, such as some volume of mantle rock, is a linear function of the parent/daughter chemical ratio Rb/Sr and a nearly linear function of time or geological age of the system. When this mantle volume undergoes equilibrium partial melting, the melt inherits the ^{87}Sr/^{86}Sr ratio of the entire system. Consequently, radiogenic isotope ratios such as ^{87}Sr/^{86}Sr are powerful tracers of the parent–daughter ratios of mantle sources of igneous rocks. If isotope data from several decay systems are combined, a correspondingly richer picture of the source chemistry can be constructed.

Table 1 shows a list of long-lived radionuclides, their half-lives, daughter isotopes, and radiogenic-to-nonradiogenic isotope ratios commonly used as tracers in mantle geochemistry. Noble-gas isotopes are not included here, because a separate chapter of this treatise is devoted to them (see Chapter 17). Taken together, they cover a wide range of geochemical properties including incompatible and compatible behavior. These ratios will be used, together with some incompatible trace-element ratios, as tracers of mantle reservoirs, crust–mantle differentiation processes, and mantle melting processes in later sections of this chapter.

18.2 LOCAL AND REGIONAL EQUILIBRIUM REVISITED

How do we translate geochemical data from basalts into a geological model of the present-day mantle and its evolution? The question of chemical and isotopic equilibrium, and particularly its spatial dimension, has always played a fundamental role in this effort of interpretation. The basic, simple tenet of isotope geochemists and petrologists alike has generally been that partial melting at mantle temperatures, pressures, and timescales achieves essentially complete chemical equilibrium between melt and solid residue. For isotope data in particular, this means that at magmatic temperatures, the isotope ratio of the melt is identical to that of the source, and this is what made isotope ratios of volcanic rocks apparently ideal tracers of mantle composition. The question of spatial scale seemed less important, because heterogeneities in the mantle were thought to be important primarily on the 10^2–10^4 km scale (Hart *et al.*, 1973; Schilling, 1973; White and Schilling, 1978; Dupré and Allègre, 1983). To be sure, this simple view was never universal. Some authors invoked special isotopic effects during melting, so that the isotopic composition of the melt could in some way be "fractionated" during melting, in spite of the high temperatures prevailing, so that the isotope ratios observed in the melts would not reflect those of the melt sources (e.g., O'Hara, 1973; Morse, 1983). These opinions were invariably raised by authors not directly familiar with the analytical methods of isotope geochemistry, so they did not realize that isotopic fractionation occurs in every mass spectrometer and is routinely corrected in the reported results.

18.2.1 Mineral Grain Scale

Some authors invoked mineral-scale isotopic (and therefore also chemical) disequilibrium and preferential melting of phases, such as phlogopite, which have higher Rb/Sr, and therefore also higher $^{87}Sr/^{86}Sr$ ratios than the bulk rock, to explain unusually high $^{87}Sr/^{86}Sr$ ratios in OIBs (e.g., O'Nions and Pankhurst, 1974; Vollmer, 1976). Hofmann and Hart (1978) reviewed this subject in light of the available diffusion data in solid and molten silicates. They concluded that mineral-scale isotopic and chemical disequilibrium is exceedingly unlikely if melting timescales are on the order of thousands of years or more. More recently, Van Orman *et al.* (2001) have measured REE diffusion coefficients in clinopyroxene and found that REE mobility in this mineral is so low at magmatic temperatures that chemical disequilibrium between grain centers and margins will persist during melting. Consequently, the melt will not be in equilibrium with the bulk residue for geologically reasonable melting times, if the equilibration occurs by volume diffusion alone. This means that the conclusions of Hofmann and Hart (1978) must be revised significantly: the slowest possible path of chemical reaction no longer guarantees attainment of equilibrium. However, it is not known whether other mechanisms such as recrystallization during partial melting might not lead to much more rapid equilibration. One possible test of this would be the examination of mantle clinopyroxenes from oceanic and ophiolitic peridotites. These rocks have undergone various extents of partial melting (Johnson *et al.*, 1990; Hellebrand *et al.*, 2001), and the residual clinopyroxenes should show compositional zoning if they had not reached equilibrium with the melt via volume

Table 1 Long-lived radionuclides.

Parent nuclide	*Daughter nuclide*	*Half-life (years)*	*Tracer ratio (radiogenic/nonradiogenic)*
^{147}Sm	^{143}Nd	106×10^9	$^{143}Nd/^{144}Nd$
^{87}Rb	^{87}Sr	48.8×10^9	$^{87}Sr/^{86}Sr$
^{176}Lu	^{176}Hf	35.7×10^9	$^{176}Hf/^{177}Hf$
^{187}Re	^{187}Os	45.6×10^9	$^{187}Os/^{188}Os$
^{40}K	^{40}Ar	1.25×10^9	$^{40}Ar/^{36}Ar$
^{232}Th	^{208}Pb	14.01×10^9	$^{208}Pb/^{204}Pb$
^{238}U	^{206}Pb	4.468×10^9	$^{206}Pb/^{204}Pb$
^{235}U	^{207}Pb	0.738×10^9	$^{207}Pb/^{204}Pb$

diffusion. Although the above-cited studies were not specifically conducted to test this question, the clinopyroxenes were analyzed by ion microprobe, and these analyses showed no significant signs of internal compositional gradients. It is, of course, possible in principle that the internal equilibration occurred after extraction of the melt, so this evidence is not conclusive at present. Nevertheless, these results certainly leave open the possibility that the crystals re-equilibrated continuously with the melt during melt production and extraction. There is at present no definitive case from "natural laboratories" deciding the case one way or the other, at least with respect to incompatible lithophile elements such as the REE.

Osmium isotopes currently provide the strongest case for mineral-to-mineral disequilibrium, and for mineral–melt disequilibrium available from observations on natural rocks. Both osmium alloys and sulfides from ophiolites and mantle xenoliths have yielded strongly heterogeneous osmium isotope ratios (Alard *et al.*, 2002; Meibom *et al.*, 2002). The most remarkable aspect of these results is that these ophiolites were emplaced in Phanerozoic times, yet they contain osmium-bearing phases that have retained model ages in excess of 2 Ga in some cases. The melts that were extracted from these ophiolitic peridotites almost certainly contained much more radiogenic osmium and could, in any case, not have been in osmium-isotopic equilibrium with all these isotopically diverse residual phases.

Another strong indication that melts extracted from the mantle are not in osmium-isotopic equilibrium with their source is given by the fact that osmium isotopes in MORBs are, on average, significantly more radiogenic than osmium isotopes from oceanic peridotites (see also Figure 9). Although it may be argued that there is no one-to-one correspondence between basalts and source peridotites, and further, that the total number of worldwide MORB and peridotite samples analyzed is still small, these results strongly suggest that, at least with regard to osmium, MORBs are generally not in isotopic equilibrium with their sources or residues. However, osmium-isotopic disequilibrium does not automatically mean strontium, neodymium, lead, or oxygen-isotopic disequilibrium or incompatible-trace-element disequilibrium. This is because osmium is probably incompatible in all silicate phases (Snow and Reisberg, 1995; Schiano *et al.*, 1997; Burton *et al.*, 2000) but very highly compatible with nonsilicate phases such as sulfides and, possibly, metal alloys such as osmiridium "nuggets," which may form inclusions within silicate minerals and might therefore be protected from reaction with a partial silicate melt. At the time of writing, no clear-cut answers are available, and for the time being, we will simply note that the geochemistry of osmium and rhenium is considerably less well understood than that of silicate-hosted major and trace elements such as strontium, neodymium, lead, and their isotopic abundances.

18.2.2 Mesoscale Heterogeneities

By "mesoscale" I mean scales larger than about a centimeter but less than a kilometer. This intermediate scale was addressed only briefly by Hofmann and Hart, who called it a "lumpy mantle" structure. It was specifically invoked by Sun and Hanson (1975) and Wood (1979), and others subsequently, who invoked veining in the mantle to provide sources for chemically and isotopically heterogeneous melts. Other versions of mesoscale heterogeneities were invoked by Sleep (1984), who suggested that preferential melting of ubiquitous heterogeneities may explain ocean island-type volcanism, and by Allègre and Turcotte (1986), who discussed a "marble cake" structure of the mantle generated by incomplete homogenization of subducted heterogeneous lithosphere. These ideas have recently been revived in several publications discussing a mantle containing pyroxenite or eclogite layers, which may melt preferentially (Phipps Morgan *et al.*, 1995; Hirschmann and Stolper, 1996; Soboler *et al.*, 2005, Phipps Morgan and Morgan, 1998; Yaxley and Green, 1998; Phipps Morgan, 1999).

One of the main difficulties with these mesoscale models is that they have been difficult to test by direct geochemical and petrological field observations. Recently, however, several studies have been published which appear to support the idea of selective melting of mesoscale heterogeneities. Most important of these are probably the studies of melt inclusions showing that single basalt samples, and even single olivine grains, contain chemically and isotopically extremely heterogeneous melt inclusions. Extreme heterogeneities in REE abundances from melt inclusions had previously been explained by progressive fractional melting processes of uniform sources (Sobolev and Shimizu, 1993; Gurenko and Chaussidon, 1995). In contrast, the more recent studies have demonstrated that source heterogeneities must (also) be involved to explain the extreme variations in isotopic and chemical compositions observed (Saal *et al.*, 1998; Sobolev *et al.*, 2000). While the spatial scale of these source heterogeneities cannot be directly inferred from these melt inclusion data, it seems highly plausible that it is in the range of what is here called "mesoscale."

Other more circumstantial evidence for preferential melting of mesoscale heterogeneities has been described by Regelous *et al.* (2002/2003), who found that the Hawaiian plume delivered MORB-like magmas ~80Ma, when the plume was located close to the Pacific spreading ridge.

Unless this is a fortuitous coincidence, this implies that the same plume produces "typical" OIB-like, incompatible-element-enriched melts with elevated $^{87}Sr/^{86}Sr$ and low $^{143}Nd/^{144}Nd$ ratios when the degree of melting is relatively low under a thick lithosphere, and typically MORB-like, incompatible-element-depleted melts when the degree of melting is high because of the shallow melting level near a spreading ridge. Such a dependence on the extent of melting is consistent with a marble-cake mantle containing incompatible-element-rich pyroxenite or eclogite layers having a lower melting temperature than the surrounding peridotite matrix. This melting model is further corroborated by the observation that at least three other plumes located at or near spreading ridges have produced MORB-like lavas, namely, the Iceland, the Galapagos, and the Kerguelen plume. The overall evidence is far from clear-cut, however, because the Iceland and Galapagos plumes have also delivered OIB-like tholeiites and alkali basalts more or less in parallel with the depleted MORB-like tholeiites or picrites.

To sum up, the question of grain-scale equilibration with partial melts, which had apparently been settled definitively by Hofmann and Hart (1978), has been reopened by the experimental work of Van Orman *et al.* (2001) and by recent osmium isotope data. The mesoscale equilibrium involving a veined or marble-cake mantle consisting of a mixture of lherzolite (or harzburgite) and pyroxenite (or eclogite) has also received substantial support in the recent literature. In either case, the isotopic composition of the melt is likely to change as a function of the bulk extent of melting, and the melts do not provide quantitative estimates of the isotopic composition of the bulk sources at scales of kilometers or more. It will be seen in subsequent sections that this has ramifications particularly with respect to quantitative estimates of the sizes and spatial distributions of the reservoirs hosting the geochemical mantle heterogeneities observed in basalts. While this defeats one of the important goals of mantle geochemistry, it will be seen in the course of this chapter that the geochemical data can still be used to map large-scale geochemical provinces of the mantle and to reveal much about the smaller-scale structure of the mantle heterogeneities. In addition, they remain powerful tracers of recycling and mixing processes and their history in the mantle.

18.3 CRUST–MANTLE DIFFERENTIATION

Before discussing the internal chemical structure of the mantle, it is necessary to have a general understanding of crust–mantle differentiation, because this has affected the incompatible trace-element and isotope budget of the mantle rather drastically. This topic has been covered by Hofmann (1988), but the most important points will be summarized here again. The treatment here differs in detail because more recent estimates have been used for the bulk composition of the continental crust and of the bulk silicate earth (BSE), also called "primitive mantle."

18.3.1 Enrichment and Depletion Patterns

The growth of the continental crust has removed major proportions of the highly incompatible elements from the mantle, and this depletion is the chief (but not the sole) cause of the specific isotope and trace-element characteristics of MORBs. The effects of ionic radius and charge, described in Section 18.1.2.1, on this enrichment–depletion process can be readily seen in a diagram (Figure 1) introduced by Taylor and McLennan (1985). It is obvious from this that those trace elements that have ionic properties similar to the major silicate-structure-forming elements, namely, nickel, cobalt, manganese, scandium, and chromium are not enriched in the continental crust but remain in the mantle. In contrast, elements with deviating ionic properties are more or less strongly enriched in the crust, depending on the magnitude of the deviation. Two main transfer mechanisms are available for this differentiation, both of them are ultimately driven by mantle convection. The first is partial melting and ascent of the melt to the surface or into the already existing crust. The second involves dehydration (and decarbonation) reactions during subduction, metasomatic transfer of soluble elements via hydrothermal fluid from the subducted crust-plus-sediment into the overlying mantle "wedge," and partial melting of the metasomatized (or "fertilized") region. This partial melt ascends and is

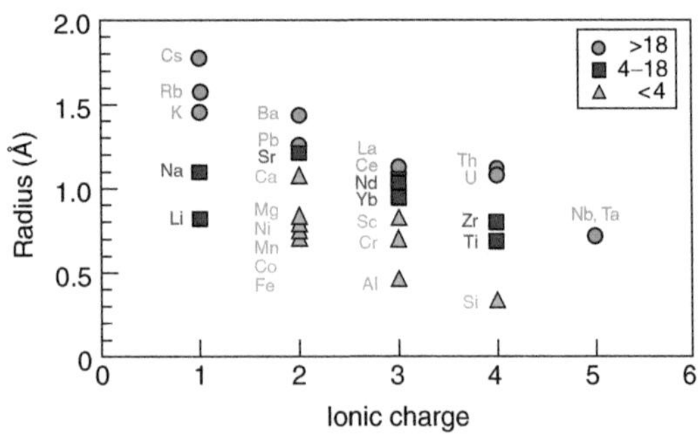

Figure 1 Ionic radius (in angstrom) versus ionic charge for lithophile major and trace elements in mantle silicates. The ranges of enrichment factors in average continental crust, using the estimate of (Rudnick and Fountain, 1995), relative to the concentrations in the primitive mantle (or "bulk silicate Earth," e.g., McDonough and Sun, 1995), are also shown.

added to the crust, carrying the geochemical signature caused by mantle metasomatic transfer into the crust. Both mechanisms may operate during subduction, and a large body of geochemical literature has been devoted to the distinction between the two (Elliott *et al.*, 1997; Class *et al.*, 2000; Johnson and Plank, 2000). Continental crust may also be formed by mantle plume heads, which are thought to produce large volumes of basaltic oceanic plateaus. These may be accreted to existing continental crust, or continental flood basalts, which similarly add to the total volume of crust (e.g., Abouchami *et al.*, 1990; Stein and Hofmann, 1994; Puchtel *et al.*, 1998). The quantitative importance of this latter mechanism remains a matter of some debate (Kimura *et al.*, 1993; Calvert and Ludden, 1999).

Hofmann (1988) showed that crust formation by extraction of partial melt from the mantle could well explain much of the trace-element chemistry of crust–mantle differentiation. However, a few elements, notably niobium, tantalum, and lead, do not fit into the simple pattern of enrichment and depletion due to simple partial melting (Hofmann *et al.*, 1986). The fundamentally different behavior of these elements in the MORB–OIB environment on the one hand, and in the subduction environment on the other, requires the second, more complex, transfer mechanism via fluids (Miller *et al.*, 1994; Peucker-Ehrenbrink *et al.*, 1994; Chauvel *et al.*, 1995). Thus, local fluid transport is essential in preparing the mantle sources for production of continental crust, but the gross transport of incompatible elements from mantle to crust is still carried overwhelmingly by melting and melt ascent.

The simplest case discussed above, namely, crust–mantle differentiation by partial melting alone, is illustrated in Figure 2. This shows the abundances of a large number of chemical elements in the continental crust, as estimated by Rudnick and Fountain (1995), and divided by their respective abundances in the primitive mantle or BSE as estimated by McDonough and Sun (1995). Each element is assigned a nominal partition coefficient D as defined in Equation (1), calculated by rearranging Equation (2) and using a nominal "melt fraction" $F=0.01$. In this highly simplified view, the continental crust is assumed to have the composition of an equilibrium partial melt derived from primitive mantle material. Also shown is the hypothetical solid mantle residue of such a partial melt and a second-stage partial melt of this depleted residue. This second-stage melt curve may then be compared with the actual element abundances of "average" ocean crust. Although this "model" of the overall crust–mantle differentiation is grossly oversimplified, it can account for the salient features of the relationship between primitive mantle, continental crust, depleted mantle, and oceanic crust quite well. This representation is remarkably successful because Equation (2) is essentially a mass-balance relationship and these major reservoirs are in fact genetically related by enrichment and depletion processes in which partial melting plays a dominant role.

The above model of extracting continental crust and remelting the depleted residue also

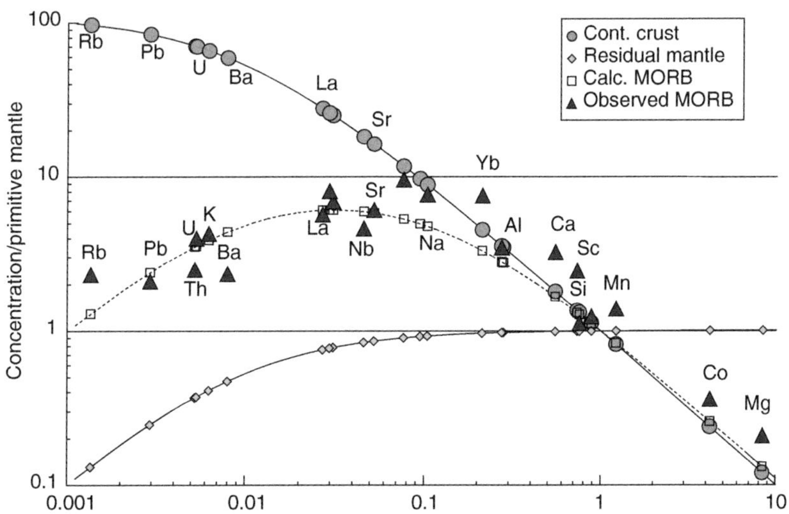

Figure 2 Comparison of the abundances of trace and (some) major elements in average continental crust and average MORB. Abundances are normalized to the primitive-mantle values (McDonough and Sun, 1995). The "partition coefficient" of each element is calculated by solving Equation (2) for D, using a melt fraction $F=0.009$ and its abundance value in the continental crust (Rudnick and Fountain, 1995). The respective abundances in average MORB are plotted using the same value of D and using the average ("normal") MORB values of Su (2002), where "normal" refers to ridge segments distant from obvious hotspots.

accounts approximately for the isotopic relationships between continental crust and residual mantle, where the isotopic composition is directly represented by MORB, using the assumption of complete local and mesoscale equilibrium discussed in Section 18.2. This is illustrated by Figure 3, which is analogous to Figure 2, but shows only the commonly used radioactive decay systems Rb–Sr, Sm–Nd, Lu–Hf, and Re–Os. Thus, the continental crust has high parent–daughter ratios for Rb/Sr and Re/Os, but low Sm/Nd and Lu/Hf, whereas the mantle residue has complementary opposite ratios. With time, these parent–daughter ratios will generate higher than primitive $^{87}Sr/^{86}Sr$ and $^{187}Os/^{188}Os$, and lower than primitive $^{143}Nd/^{144}Nd$ and $^{176}Hf/^{177}Hf$ ratios in the crust and complementary, opposite ratios in the mantle; this is indeed observed for strontium, neodymium, and hafnium, as will be seen in the review of the isotope data.

The case of lead isotopes is more complicated, because the estimates for the mean parent–daughter ratios of mantle and crust are similar. This similarity is not consistent with purely magmatic production of the crust, because the bulk partition

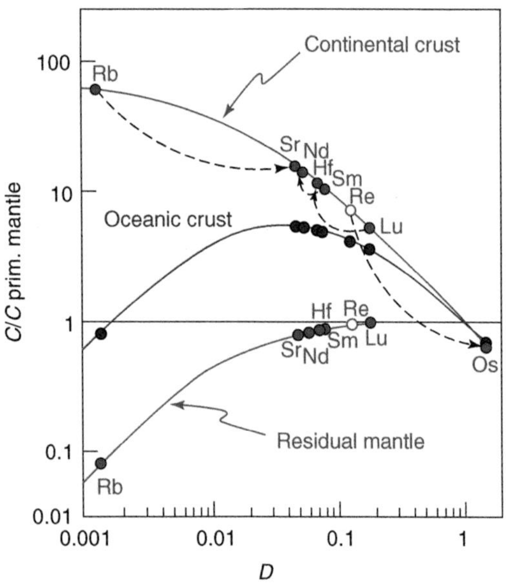

Figure 3 Crust–mantle differentiation patterns for the decay systems Rb–Sr, Sm–Nd, Lu–Hf, and Re–Os. The diagram illustrates the depletion–enrichment relationships of the parent–daughter pairs, which lead to the isotopic differences between continental crust and the residual mantle. For example, the Sm/Nd ratio is increased, whereas the Rb/Sr ratio is decreased in the residual mantle. This leads to the negative isotopic correlation in mantle-derived rocks plotted in Figure 4a (as well as the positive correlation seen in Figure 4c). The construction is similar to that used in Figure 2, but *D* values have been adjusted slightly for greater clarity.

coefficient of lead during partial mantle melting is expected to be only slightly lower than that of strontium (see Figure 1), but significantly higher than the coefficients for the highly incompatible elements uranium and thorium. In reality, however, the enrichment of lead in the continental crust shown in Figure 2 is slightly higher than the enrichments for thorium and uranium, and the $^{206}Pb/^{204}Pb$ and $^{208}Pb/^{204}Pb$ ratios of continental rocks are similar to those of MORB. This famous–infamous "lead paradox," first pointed out by Allègre (1969), will be discussed in a separate section below.

How do we know the parent–daughter ratios in crust and mantle? When both parent and daughter nuclides have refractory lithophile character, and are reasonably resistant to weathering and other forms of low-temperature alteration, as is the case for the pairs Sm–Nd and Lu–Hf, we can obtain reasonable estimates from measuring and averaging the element ratios in representative rocks of crustal or mantle heritage. But when one of the elements was volatile during terrestrial accretion, and/or is easily mobilized by low-temperature or hydrothermal processes, such as rubidium, uranium, or lead, the isotopes of the daughter elements yield more reliable information about the parent–daughter ratios of primitive mantle, depleted mantle, and crust, because the isotope ratios are not affected by recent loss (or addition) of such elements. Thus, the U/Pb and Th/Pb ratios of bulk silicate earth, depleted mantle, and continental crust are essentially derived from lead isotope ratios.

Similarly, the primitive mantle Rb/Sr ratio was originally derived from the well-known negative correlation between $^{87}Sr/^{86}Sr$ and $^{143}Nd/^{144}Nd$ ratios in mantle-derived and crustal rocks, the so-called mantle array (DePaolo and Wasserburg, 1976; Richard *et al.*, 1976; O'Nions *et al.*, 1977; see also Figure 4a). To be sure, there is no guarantee that this correlation will automatically go through the BSE value. However, in this case, the primitive mantle (or "bulk silicate earth") Rb/Sr value has been approximately confirmed using element abundance ratios of barium, rubidium, and strontium. Hofmann and White (1983) found that Ba/Rb ratios in mantle-derived basalts and continental crust are sufficiently similar, so that the terrestrial value of Ba/Rb can be estimated within narrow limits. The terrestrial Ba/Sr ratio (comprising two refractory, lithophile elements) can be assumed to be identical to the ratio in chondritic meteorites, so that the terrestrial Rb/Sr ratio can be estimated as

$$\left(\frac{\mathrm{Rb}}{\mathrm{Sr}}\right)_{\mathrm{terr.}} = \left(\frac{\mathrm{Rb}}{\mathrm{Ba}}\right)_{\mathrm{terr.}} \left(\frac{\mathrm{Ba}}{\mathrm{Sr}}\right)_{\mathrm{chondr.}} \qquad (8)$$

The terrestrial Rb/Sr ratio estimated in this way turned out to be indistinguishable from the ratio

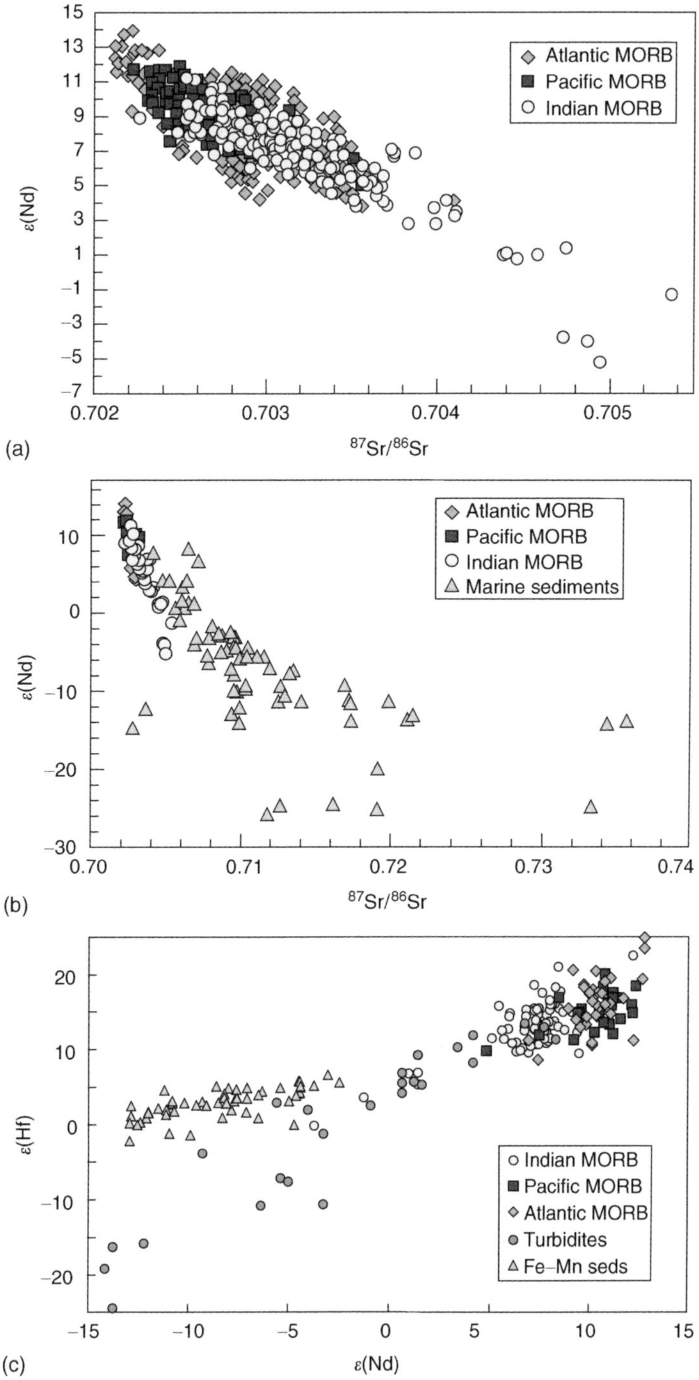

Figure 4 (a) $^{87}Sr/^{86}Sr$ versus ε(Nd) for MORBs from the three major ocean basins. ε(Nd) is a measure of the deviation of the $^{143}Nd/^{144}Nd$ ratio from the chondritic value, assumed to be identical to the present-day value in the bulk silicate earth. It is defined as $\varepsilon(Nd) = 10^4(^{143}Nd/^{144}Nd_{measured} - ^{143}Nd/^{144}Nd_{Chondrite})/^{143}Nd/^{144}Nd_{Chondrite}$. The chondritic value used is $^{143}Nd/^{144}Nd_{Chondrite} = 0.512638$. The data are compiled from the PETDB database. (b) $^{87}Sr/^{86}Sr$ versus ε(Nd) for MORBs compared with data for turbidites and other marine sediments (Ben Othman *et al.*, 1989; Hemming and McLennan, 2001). This illustrates the complementary nature of continent-derived sediments and MORB expected from the relationships shown in Figure 3. (c) ε(Nd) versus ε(Hf) for MORBs and marine sediments (compilation kindly provided by J. Vervoort, personal communication; Vervoort *et al.*, 1999; with added Indian MORB data by Graham *et al.*, 2006). Separate symbols are used for detritally dominated turbidites and chemically precipitated Fe–Mn oxide sediments, which have very high Lu/Hf ratios and therefore elevated ε(Hf) values.

estimated by isotope correlations, and therefore the consistency between isotope and element abundance data is not circular. This example of internal consistency has been disturbed by the more recent crustal estimate of Ba/Rb = 6.7 (Rudnick and Fountain, 1995), which is significantly lower than the above mantle estimate Ba/Rb = 11.0 (Hofmann and White, 1983), based mostly on MORB and OIB data. This shows that, for many elements, there are greater uncertainties about the composition of the continental crust than about the mantle. The reason for this is that the continental crust has become much more heterogeneous than the mantle because of internal differentiation processes including intracrustal melting, transport of metamorphic fluids, hydrothermal transport, weathering, erosion, and sedimentation. In any case, assuming that Rudnick's crustal estimate is correct, the primitive mantle Ba/Rb should lie somewhere between 7 and 11. The lesson from this is that we must be careful when using "canonical" element ratios to make mass-balance estimates for the sizes of different mantle reservoirs.

18.3.2 Mass Fractions of Depleted and Primitive Mantle Reservoirs

The simple crust–mantle differentiation model shown in Figure 2 contains three "reservoirs": continental crust, depleted residue, and oceanic crust. However, the depleted reservoir may well be smaller than the entire mantle, in which case another possibly primitive reservoir would be needed. Thus, if one assumes that the mantle consists of two reservoirs only, one depleted and one remaining primitive, and if one neglects the oceanic crust because it is thin and relatively depleted in highly incompatible elements, one can calculate the mass fractions of these two reservoirs from their respective isotopic and/or trace-element compositions (Jacobsen and Wasserburg, 1979; O'Nions *et al.*, 1979; DePaolo, 1980; Davies, 1981; Allègre *et al.*, 1983, 1996; Hofmann *et al.*, 1986; Hofmann, 1989a). The results of these estimates have yielded mass fractions of the depleted reservoir ranging from ~30% to 80%. Originally, the 30% estimate was particularly popular because it matches the mass fraction of the upper mantle above the 660km seismic discontinuity. It was also attractive because at least some of the mineral physics data indicated that the lower mantle has a different, intrinsically denser, major-element composition. However, more recent data and their evaluations indicate that they do not require such compositional layering (Jackson and Rigden, 1998). Nevertheless, many authors argue that the 660km boundary can isolate upper- from lower-mantle convection, either because of the endothermic nature of the phase changes at this boundary, or possibly because of extreme viscosity differences between upper and lower mantle. Although this entire subject has been debated in the literature for many years, there appeared to be good reasons to think that the 660km seismic discontinuity is the fundamental boundary between an upper, highly depleted mantle and a lower, less depleted or nearly primitive mantle.

The most straightforward mass balance, assuming that we know the composition of the continental crust sufficiently well, can be calculated from the abundances of the most highly incompatible elements, because their abundances in the depleted mantle are so low that even comparatively large relative errors do not affect the mass balance very seriously. The most highly enriched elements in the continental crust have estimated crustal abundances (normalized to the primitive mantle abundances given by McDonough and Sun, 1995) of Cs = 123, Rb = 97, and Th = 70 (Rudnick and Fountain, 1995). The estimate for Cs is rather uncertain because its distribution within the crust is particularly heterogeneous, and its primitive-mantle abundance is afflicted by special uncertainties (Hofmann and White, 1983; McDonough *et al.*, 1992). Therefore, a more conservative enrichment factor of 100 (close to the value of 97 for Rb) is chosen for elements most highly enriched in the continental crust. The simple three-reservoir mass balance then becomes

$$X_{\mathrm{lm}} = \frac{1 - C_{\mathrm{cc}}X_{\mathrm{cc}} - C_{\mathrm{um}}X_{\mathrm{um}}}{C_{\mathrm{lm}}} \tag{9}$$

where C refers to primitive-mantle normalized concentrations (also called "enrichment factors"), X to the mass fraction of a given reservoir, and the subscripts cc, lm, and um to continental crust, lower mantle, and upper mantle, respectively.

If the lower mantle is still primitive, so that $C_{\mathrm{lm}} = 1$, the upper mantle is extremely depleted, so that $C_{\mathrm{um}} = 0$ and $X_{\mathrm{cc}} = 0.005$, and the mass balance yields

$$X_{\mathrm{lm}} = \frac{1 - 100 \times 0.005 - 0 \times X_{\mathrm{um}}}{1} = 0.5 \tag{10}$$

Remarkably, this estimate is identical to that obtained using the amounts of radiogenic argon in the atmosphere, the continental crust, and the depleted, upper mantle (Allègre *et al.*, 1996). There are reasons to think that the abundances of potassium and rubidium in BSE used in these calculations have been overestimated, perhaps by as much as 30% (Lassiter, 2002), and this would of course decrease the remaining mass fraction of primitive-mantle material. Thus, we can conclude that at least half, and perhaps 80%, of the most highly incompatible element budget now resides either in the continental crust or in the atmosphere (in the case of argon).

Can we account for the entire silicate earth budget by just these three reservoirs (crust plus atmosphere, depleted mantle, and primitive mantle), as has been assumed in all of the above estimates? Saunders *et al.* (1988) and Sun and McDonough (1989) (among others) have shown that this cannot be the case, using global systematics of a single trace-element ratio, Nb/La. Using updated, primitive-mantle normalized estimates for this ratio, namely, $(Nb/La)_n=0.66$ for the continental crust (Rudnick and Fountain, 1995), and $(Nb/La)_n=0.81$ for so-called N-type ("normal") MORB (Su, 2002), we see that both reservoirs have lower than primitive Nb/La ratios. Using the additional constraint that niobium is slightly more incompatible than lanthanum during partial melting, we find that the sources of all these mantle-derived basalts must have sources with Nb/La ratios equal to or lower than those of the basalts themselves. This means that all the major mantle sources as well as the continental crust have $(Nb/La)_n \leq 1$. By definition, the entire silicate earth has $(Nb/La)_n=1$, so there should be an additional, hidden reservoir containing the "missing" niobium. A similar case has more recently been made using Nb/Ta, rather than Nb/La. Current hypotheses to explain these observations invoke either a refractory eclogitic reservoir containing high-niobium rutiles (Rudnick *et al.*, 2000), or partitioning of niobium into the metallic core (Wade and Wood, 2001). Beyond these complications involving special elements with unexpected geochemical "behavior," there remains the question whether an ~50% portion of the mantle not needed to produce the continental crust has remained primitive, or whether it is also differentiated into depleted, MORB-source-like, and enriched, OIB-source-like subreservoirs. In the past, the occurrence of noble gases with primordial isotope ratios have been used to argue that the lower part of the mantle must still be nearly primitive. However, it will be seen below that this inference is no longer as compelling as it once seemed to be.

18.4 MID-OCEAN RIDGE BASALTS: SAMPLES OF THE DEPLETED MANTLE

18.4.1 Isotope Ratios of Strontium, Neodymium, Hafnium, and Lead

The long-lived radioactive decay systems commonly used to characterize mantle compositions, their half-lives, and the isotope ratios of the respective radiogenic daughter elements are given in Table 1. The half-lives of ^{147}Sm, ^{87}Sr, ^{186}Hf, ^{187}Re, and ^{232}Th are several times greater than the age of the Earth, so that the accumulation of the radiogenic daughter nuclide is nearly linear with time. This is not the case for the shorter-lived ^{238}U and ^{235}U, and this is in part responsible for the more complex isotopic relationships displayed by lead isotopes in comparison with the systematics of strontium, neodymium, hafnium, and osmium isotopes. The mantle geochemistry of noble gases, although of course an integral part of mantle geochemistry, is treated in Chapter 17.

Figures 4–6 show the isotopic compositions of MORBs from spreading ridges in the three major ocean basins. Figures 4b, 4c, and 5a also show isotope data for marine sediments, because these are derived from the upper continental crust and should roughly represent the isotopic composition of this crust. In general, the isotopic relationships between the continental and oceanic crust are just what is expected from the elemental parent–daughter relationships seen in Figure 3. The high Rb/Sr and low Sm/Nd and Lu/Hf ratios of continental materials relative to the residual mantle are reflected by high ^{87}Sr/^{86}Sr and low ^{143}Nd/^{144}Nd and ^{176}Hf/^{177}Hf ratios. This readily accounts for the negative correlation seen in the Sr–Nd isotope diagram and the positive correlation in the Nd–Hf diagram. In lead isotope diagrams, the differences are not nearly as clear, and continent-derived sediments are distinguished primarily by slightly elevated ^{207}Pb/^{204}Pb ratios for given values of ^{206}Pb/^{204}Pb (Figure 5a). This topology in lead-isotope space requires a comparatively complex evolution of the terrestrial U–Pb system. It involves an ancient period of high U/Pb ratios in continental history (with complementary, low ratios in the residual mantle). The higher ^{235}U/^{238}U ratios prevailing during that time led to elevated ^{207}Pb/^{206}Pb ratios in the crust. This subject is treated more fully in Section 18.6.

Another important observation is that while strontium, neodymium, and hafnium isotopes all correlate with each other, they form poorer, but still significant, correlations with ^{206}Pb/^{204}Pb (or ^{208}Pb/^{204}Pb, not shown) ratios in the Pacific and Atlantic, but no discernible correlation in the Indian Ocean MORB (Figure 6a). Nevertheless, if instead of ^{208}Pb/^{204}Pb or ^{206}Pb/^{204}Pb ratios one plots the so-called "radiogenic" ^{208}Pb*/^{206}Pb* ratio, the lead data do correlate with neodymium isotopes in all three ocean basins (Figure 6b). This parameter is a measure of the radiogenic additions to ^{208}Pb/^{204}Pb and ^{206}Pb/^{204}Pb ratios during Earth's history; it is calculated by subtracting the primordial (initial) isotope ratios from the measured values. The primordial ratios are those found in the Th–U-free sulfide phase (troilite) of iron meteorites. Thus, the radiogenic ^{208}Pb*/^{206}Pb* ratio is defined as

$$\frac{^{208}\text{Pb}^*}{^{206}\text{Pb}^*} = \frac{^{208}\text{Pb}/^{204}\text{Pb} - \left(^{208}\text{Pb}/^{204}\text{Pb}\right)_{\text{init}}}{^{206}\text{Pb}/^{204}\text{Pb} - \left(^{206}\text{Pb}/^{204}\text{Pb}\right)_{\text{init}}} \quad (11)$$

Unlike ^{208}Pb/^{204}Pb or ^{206}Pb/^{204}Pb, which depend on Th/Pb and U/Pb, respectively, ^{208}Pb*/^{206}Pb* reflects the Th/U ratio integrated

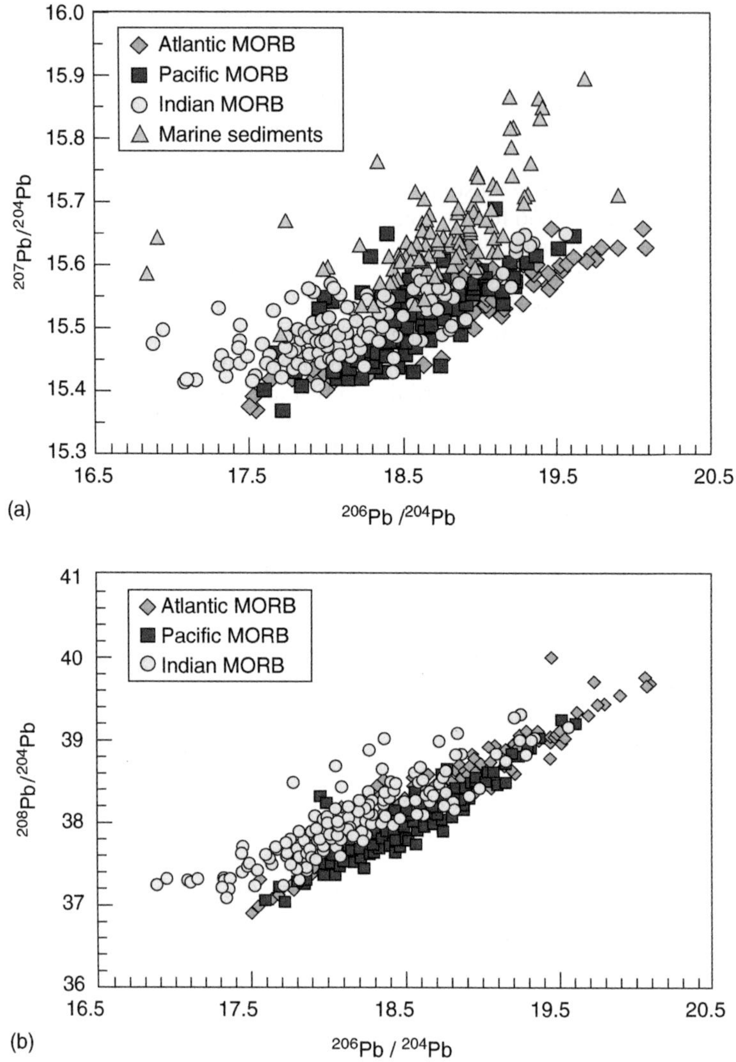

Figure 5 (a) $^{207}Pb/^{204}Pb$ versus $^{206}Pb/^{204}Pb$ for MORB from three major ocean basins and marine sediments. (b) $^{208}Pb/^{204}Pb$ versus $^{206}Pb/^{204}Pb$ for MORB from three major ocean basins. Sediments are not plotted because of strong overlap with the basalt data. For data sources see Figure 4.

over the history of the Earth. The existence of global correlations between neodymium, strontium, and hafnium isotope ratios and $^{208}Pb^*/^{206}Pb^*$ and the absence of such global correlations with $^{208}Pb/^{204}Pb$ or $^{206}Pb/^{204}Pb$, shows that the elements neodymium, strontium, hafnium, thorium, and uranium behave in a globally coherent fashion during crust–mantle differentiation, whereas lead deviates from this cohesion.

Although the sediment data shown in several of these figures do not deviate dramatically from the overall correlation patterns of the basalts, they are not consistent with the idea that these isotope arrays are simply the product of "backmixing" a mantle reservoir depleted by the extraction of continental crust and subducted sediments derived from this crust. This is most obvious in the lead isotope data (Figure 5a) where the sediments simply do not yield a suitable endpoint for the basalt data array. Therefore, it appears that the extraction and possible recycling of continental crust is not the primary mechanism for generating the isotopic heterogeneities in MORBs. This conclusion casts significant doubt on a dominant role of recycled continental material in mantle evolution in general, even though this has been proposed repeatedly (Armstrong, 1981; DePaolo, 1983; Hanan and Graham, 1996). It will be seen further below that recycling of sediments or other continental material may explain the isotopic characteristics of a few specific types of ocean island basalts, but, in general, the near absence of continental material in the sources of most oceanic basalts is remarkable and indeed puzzling.

Figures 4–6 show systematic isotopic differences between MORB from different ocean

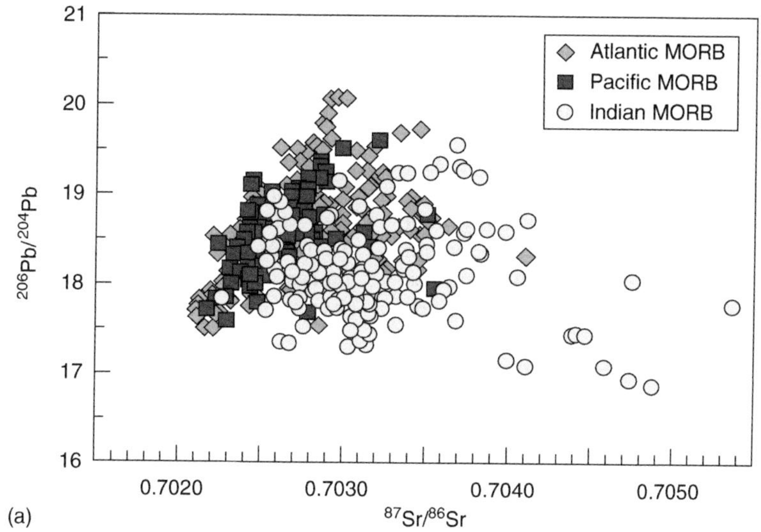

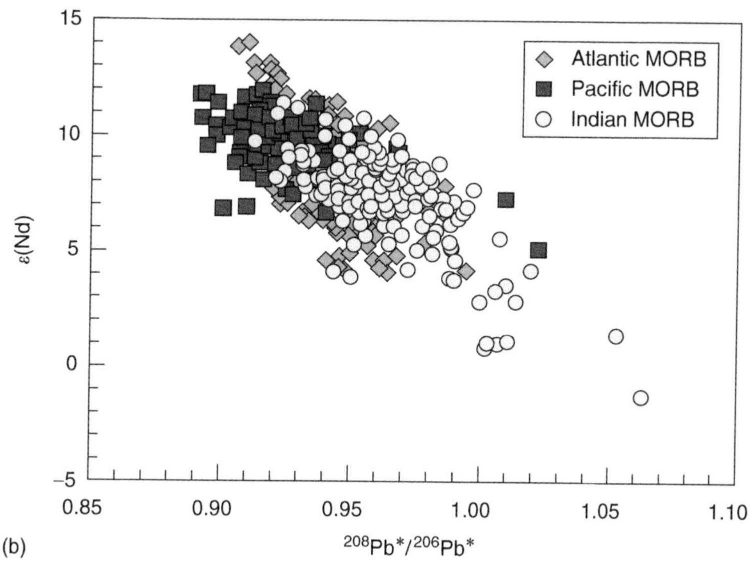

Figure 6 (a) $^{206}Pb/^{204}Pb$ versus $^{87}Sr/^{86}Sr$ for MORB from three major ocean basins. In contrast with the Sr–Nd and the Pb–Pb diagrams (Figures 4 and 5), the $^{206}Pb/^{204}Pb$–$^{87}Sr/^{86}Sr$ data correlate well only for Pacific MORB and not at all for Indian MORB. This indicates some anomalous behavior of the U–Pb decay system during global differentiation. For data sources see Figure 4. (b) $^{208}Pb^*/^{206}Pb^*$ versus ε(Nd) for MORB from three major ocean basins. The overall correlation is similar to the Sr–Nd correlation shown in Figure 4a. This indicates that the Th/U ratios, which control $^{208}Pb^*/^{206}Pb^*$, do correlate with Sm/Nd (and Rb/Sr) ratios during mantle differentiation. Taken together, (a) and (b) identify Pb as the element displaying anomalous behavior.

basins, reflecting some very large-scale isotopic heterogeneities in the source mantle of these basalts. Also, the ranges of ε(Nd) (and ε(Hf)) values present in a single ocean basin are quite large. For example, the range of neodymium isotope ratios in Atlantic MORB (~10 ε(Nd) units) is somewhat smaller than the respective range of Atlantic OIB values of ~14 ε(Nd) units (see Section 18.5), but this difference does not justify calling Atlantic MORB "isotopically homogeneous." This heterogeneity contradicts the widespread notion that the MORB-source mantle reservoir is isotopically nearly uniform, a myth that has persisted through many repetitions in the literature. One can just as easily argue that there is no such thing as a typical "normal" (usually called N-type) MORB composition. In particular, the $^{208}Pb/^{204}Pb$ and $^{208}Pb^*/^{206}Pb^*$ ratios of Indian Ocean MORB show very little overlap with Pacific MORB (but both populations overlap strongly with Atlantic MORB). These very large-scale regional "domains" were first recognized by Dupré and Allègre (1983) and named the DUPAL anomaly by Hart (1984).

The boundary between the Indian Ocean and Pacific Ocean geochemical domains coincides with the Australian–Antarctic Discordance (AAD) located between Australia and Antarctica, an unusually deep ridge segment with several unusual physical and physiographic characteristics. The geochemical transition in Sr–Nd–Hf–Pb isotope space across the AAD is remarkably sharp (Klein *et al.*, 1988; Pyle *et al.*, 1992; Kempton *et al.*, 2002), and it is evident that very little mixing has occurred between these domains. The isotopic differences observed cannot be generated overnight. Rehkämper and Hofmann (1997), using lead isotopes, have estimated that the specific isotopic characteristics of the Indian Ocean MORB must be at least 1.5 Ga old. An important conclusion from this is that convective stirring of the mantle can be remarkably "ineffective" in mixing very large-scale domains in the upper mantle.

When we further consider the fact that the present-day ocean-ridge system, though globe encircling, samples only a geographically limited portion of the total, present-day mantle, it is clear that we must abandon the notion that we can characterize the isotopic composition of the depleted mantle reservoir by a single value of any isotopic parameter. What remains is a much broader, nevertheless limited, range of compositions, which, on average, differ from other types of oceanic basalts to be discussed further below. The lessons drawn from Section 18.2 (local and regional equilibrium) merely add an additional cautionary note: although it is possible to map the world's ocean ridge system using isotopic compositions of MORB, we cannot be sure how accurately these MORB compositions represent the underlying mantle. The differences between ocean basins are particularly obvious in the $^{208}Pb/^{204}Pb$ versus $^{206}Pb/^{204}Pb$ diagrams, where Indian Ocean MORBs have consistently higher $^{208}Pb/^{204}Pb$ ratios than Pacific Ocean MORBs. Diagrams involving neodymium isotopes show more overlap, but many Indian Ocean MORBs have ε(Nd) values lower than any sample from the Pacific Ocean.

An intermediate scale of isotopic variations is shown in Figures 7 and 8, using basalts from the Mid-Atlantic Ridge (MAR). The isotope ratios of strontium and neodymium (averaged over 1° intervals for clarity) vary along the ridge with obvious maxima and minima near the oceanic islands of Iceland, the Azores, and the Bouvet triple junction, and with large-scale, relatively smooth gradients in the isotope ratios, e.g., between 20° S and 38° S. In general, the equatorial region between 30° S and 30° N is characterized by much lower strontium isotope ratios than the ridge segments to the north and the south. Some of this variation may be related to the vicinity of the mantle hotspots of Iceland, the Azores, or Bouvet, and the literature contains a continuing debate over the subject of "plume–asthenosphere interaction." Some authors argue the case where excess, enriched plume material spreads into the asthenosphere and mixes with depleted asthenospheric material to produce the geochemical gradients observed (e.g., Hart *et al.*, 1973; Schilling, 1973). Others argue that hotspots or plumes are internally heterogeneous and contain depleted, MORB-like material, which remains behind during normal plume-generated volcanism, spreads out in the asthenosphere, and becomes part of the asthenospheric mantle (Phipps Morgan and Morgan, 1998; Phipps Morgan, 1999). Irrespective of the specific process of plume–ridge interaction, the existence of compositional gradients up to ~2,000 km long implies some rather large-scale mixing processes, quite distinct from the sharpness of the compositional boundary seen at the AAD.

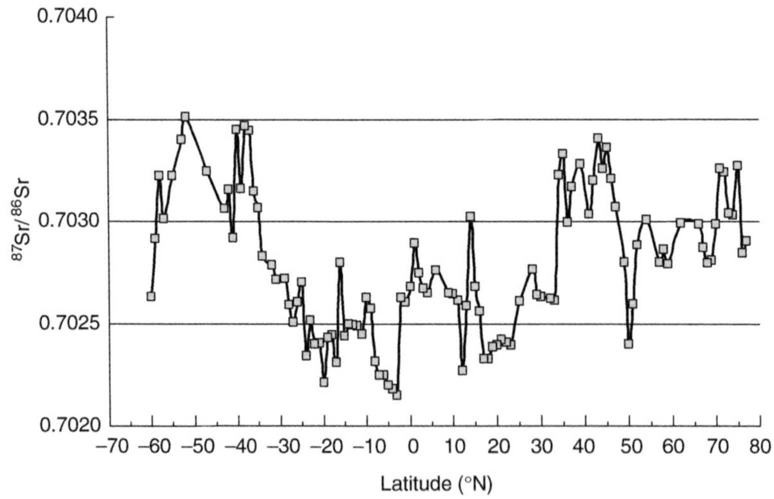

Figure 7 $^{87}Sr/^{86}Sr$ versus latitude variations in MORB along the MAR. The isotope data have been averaged by 1° intervals. For data sources see Figure 4.

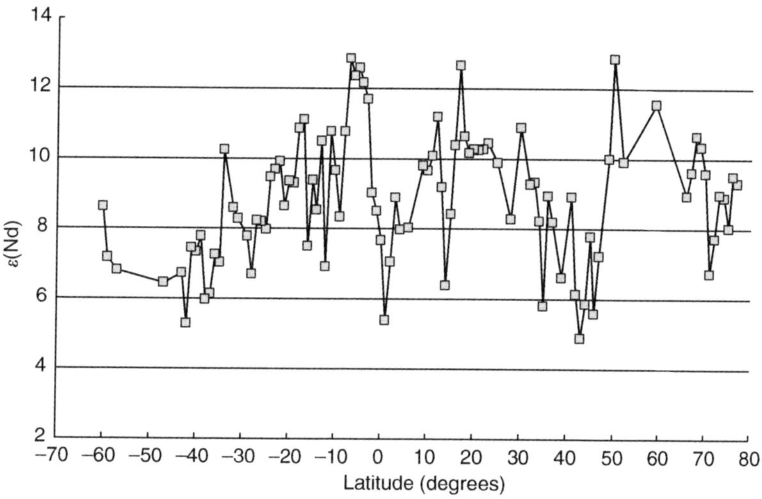

Figure 8 ε(Nd) versus latitude variations in MORB along the MAR. The isotope data have been averaged by 1° intervals. For data sources see Figure 4.

Isotopic heterogeneities are also observed on much smaller scales than those discussed so far. For example, the region around 14° N on the MAR shows a sharp "spike" in neodymium and strontium isotope ratios (see Figures 7 and 8) with an amplitude in ε(Nd) nearly as large as that of the entire Atlantic MORB variation, even though there is no obvious depth anomaly or other physiographic evidence for the possible presence of a mantle plume.

Finally, work on melt inclusions from phenocrysts has recently shown that sometimes extreme isotopic and trace-element heterogeneities exist within single hand specimens from mid-ocean ridges. Initially, the extreme chemical heterogeneities found in such samples were ascribed to the effects of progressive fractional melting of initially uniform source rocks (Sobolev and Shimizu, 1993), but recent Pb-isotope analyses of melt inclusions from a single MORB sample have shown a large range of isotopic compositions that require a locally heterogeneous source (Shimizu *et al.*, 2003). This phenomenon had previously also been observed in some OIBs (Saal *et al.*, 1998; Sobolev *et al.*, 2000), and future work must determine whether this is the exception rather than the rule. In any case, the normally observed local homogeneity of bulk basalt samples may turn out to be the result of homogenization in magma chambers rather than melting of homogeneous sources.

In summary, it should be clear that the mantle that produces MORB is isotopically heterogeneous on all spatial scales ranging from the size of ocean basins down to kilometers or possibly meters. The often-invoked homogeneity of MORB and MORB sources is largely a myth, and the definition of "normal" or N-type MORB actually applies to the depleted end of the spectrum rather than average MORB. It would be best to abandon these obsolete concepts, but they are likely to persist for years to come. Any unbiased evaluation of the actual MORB isotope data shows unambiguously, e.g., that Indian Ocean MORBs differ substantially from Pacific Ocean MORBs, and that only about half of Atlantic MORB conform to what is commonly referred to as "N-type" with $^{87}Sr/^{86}Sr$ ratios <0.7028 and ε(Nd) values $>+9$. Many geochemists think that MORB samples with higher $^{87}Sr/^{86}Sr$ ratios and lower ε(Nd) values do not represent normal upper mantle but are generated by contamination of the normally very depleted, and isotopically extreme, upper mantle with plume material derived from the deeper mantle. Perhaps this interpretation is correct, especially in the Atlantic Ocean, where there are several hotspots/plumes occurring near the MAR, but its automatic application to all enriched samples (such as those near 14° N on the MAR) invites circular reasoning and it can get in the way of an unbiased consideration of the actual data.

18.4.2 Osmium Isotopes

The Re–Os decay system is discussed separately, in part because there are far fewer osmium isotope data than Sr–Nd–Pb data. This is true because, until ~15 years ago, osmium isotopes in silicate rocks were extraordinarily difficult to measure. The advent of negative-ion thermal ionization mass spectrometry has decisively changed this (Creaser *et al.*, 1991; Völkening *et al.*, 1991), and subsequently the number of publications providing osmium isotope data has increased dramatically.

Osmium is of great interest to mantle geochemists because, in contrast with the geochemical properties of strontium, neodymium, hafnium, and lead, all of which are incompatible elements, osmium is a compatible element in most mantle melting processes, so that it generally remains in the mantle, whereas the much more incompatible rhenium is extracted and enriched in the melt and ultimately in the crust. This system therefore provides information that is different from, and complementary to, what we can learn from strontium, neodymium, hafnium, and lead isotopes. However, at present there are still significant obstacles to the full use and understanding of osmium geochemistry. There are primarily three reasons for this.

1. Osmium is present in oceanic basalts usually at sub-ppb concentration levels. Especially in low-magnesium basalts, the concentrations can approach low ppt levels. The problem posed by this is that crustal rocks and seawater can have $^{187}Os/^{188}Os$ ratios 10 times higher than the (initial) ratios in mantle-derived melts. Thus, incorporation of small amounts of seawater-altered material in a submarine magma chamber may significantly increase the $^{187}Os/^{188}Os$ ratio of the magma. Indeed, many low-magnesium, moderately to highly differentiated oceanic basalts have highly radiogenic osmium, and it is not easy to know which basalts are unaffected by this contamination.
2. The geochemistry of osmium is less well understood than other decay systems, because much of the osmium resides in non-silicate phases such as sulfides, chromite, and (possibly) metallic phases, and these phases can be very heterogeneously distributed in mantle rocks. This frequently leads to a "nugget" effect, meaning that a given sample powder is not necessarily representative of the system. Quite often, the reproducibility of concentration measurements of osmium and rhenium is quite poor by normal geochemical standards, with differences of several percent between duplicate analyses, and this may be caused either by intrinsic sample heterogeneity ("nugget effect") or by incomplete equilibration of sample and spike during dissolution and osmium separation.
3. There are legitimate doubts whether the osmium isotopic composition of oceanic basalts is ever identical to those of their mantle source rocks (Section 18.2.1).

Point 3 above is illustrated in Figure 9, which shows osmium isotope ratios and osmium concentrations in abyssal peridotites and in MORB. This diagram shows two remarkable features: (1) The osmium isotope ratios of MORB and abyssal peridotites have very little overlap, the peridotites being systematically lower than those of seafloor basalts, and (2) the MORB data show a strong negative correlation between isotope ratios and osmium concentrations. These results suggest that the basalts may not be in isotopic equilibrium with their source rocks, but we have no proof of this, because we have no samples of specific source rocks for specific basalt samples. Also, the total number of samples represented in Figure 9 is rather small. Nevertheless, the apparently systematically higher $^{187}Os/^{188}Os$ ratios of the basalts compared with the peridotites seem to indicate that unradiogenic portions of the source peridotites did not contribute to, or react with, the melt. The negative correlation displayed by the MORB data may mean that essentially all the melts are contaminated by seawater-derived osmium and that the relative contribution of the contaminating osmium to the measured isotopic compositions is inversely correlated with the

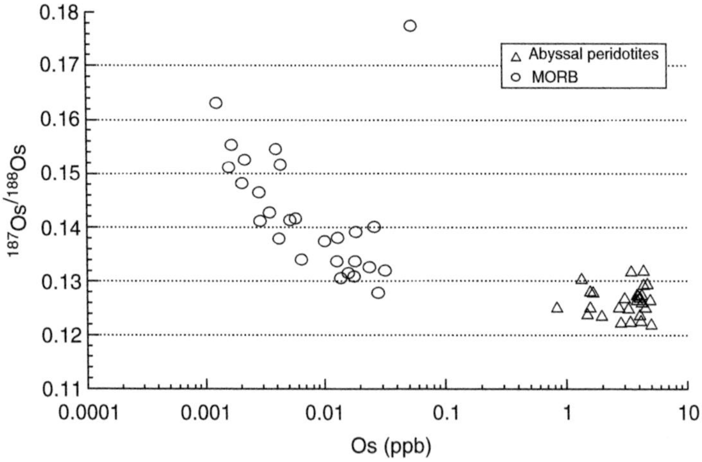

Figure 9 Osmium isotope ratios in MORB and abyssal peridotites. This diagram shows that osmium is generally compatible in peridotites during MORB melting. The systematic differences in $^{187}Os/^{188}Os$ ratios between MORB and peridotites suggest that the melts may not be in isotopic equilibrium with their residual peridotite (Martin, 1991; Roy-Barman and Allègre, 1994; Snow and Reisberg, 1995; Schiano *et al.*, 1997; Brandon *et al.*, 2000).

osmium concentration of the sample. However, the MORB samples also show a strong positive correlation between $^{187}Os/^{188}Os$ and Re/Os (not shown). Therefore, it is also possible that MORB osmium is derived from heterogeneous sources in such a way that low-osmium, high-Re/Os samples are derived from high-Re/Os portions of the sources (such as pyroxenitic veins), whereas high-osmium, low-Re/Os samples are derived from the peridotitic or even harzburgitic matrix.

To avoid the risk of contamination by seawater, either through direct contamination of the samples or contamination of the magma by assimilation of contaminated material, many authors disregard samples with very low osmium concentrations. Unfortunately, this approach does not remove the inherent ambiguity of interpretation, and it may simply bias the sampling. What is clearly needed are independent measures of very low levels of magma chamber and sample contamination.

18.4.3 Trace Elements

The general model of crust–mantle differentiation predicts that after crust formation, the residual mantle should be depleted in incompatible elements. Melts from this depleted mantle may be absolutely enriched but should still show a relative depletion of highly incompatible elements relative to moderately incompatible elements (sometimes called "MICE"), as illustrated in Figure 2. Here, actual trace-element data of real MORB and their variability are examined. An inherent difficulty is that trace-element abundances in a basalt depend on several factors, namely, the source composition, the degree, and mechanism of melting and melt extraction, the subsequent degree of magmatic fractionation by crystallization, and finally, on possible contamination of the magma during this fractionation process by a process called AFC (assimilation with fractional crystallization). This inherent ambiguity resulted in a long-standing debate about the relative importance of these two aspects. O'Hara, in particular, championed the case of fractional crystallization and AFC processes in producing enrichment and variability of oceanic basalts (e.g., O'Hara, 1977; O'Hara and Mathews, 1981). In contrast, Schilling and coworkers argued that variations in trace-element abundances, and in particular, ratios of such abundances, are strongly controlled by source compositions. They documented several cases where REE patterns vary systematically along mid-ocean ridge segments, and they mapped such variations specifically in the vicinity of hotspots, which they interpreted as the products of mantle plumes relatively enriched in incompatible elements. As was the case for the isotopic variations, they interpreted the trace-element variations in terms of mixing of relatively enriched plume material with relatively depleted upper mantle, the asthenospheric mantle (Schilling, 1973; White and Schilling, 1978). Figure 10 shows a compilation of La/Sm ratios (normalized to primitive mantle values) of basalts dredged from the MAR. This parameter has been used extensively by Schilling and co-workers as a measure of source depletion or enrichment, where they considered samples with $(La/Sm)_n<1.0$ as normal or "N-type" MORB derived from depleted sources with similar or even lower La/Sm ratios. As was found for the isotope ratios, only two-thirds of the MAR shows "typical" or "normal" La/Sm ratios lower than unity. In general, the pattern resembles that of the isotope variations, especially in the North Atlantic, where the coverage for both parameters is extensive. Thus, high La/Sm and $^{87}Sr/^{86}Sr$ values are found near the hotspots of Iceland and the Azores, between 45° S and 50° S, 14° N, and 43° N. Because of these correlations, the

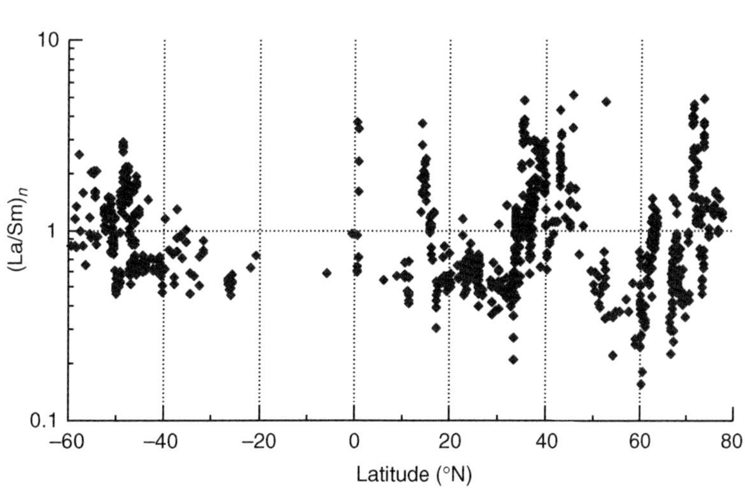

Figure 10 La/Sm ratios in MORB (not smoothed), normalized to primitive-mantle values, as a function of latitude along the MAR.

interpretation that the trace-element variations are primarily caused by source variations has been widely accepted. Important confirmation for this has come from the study of Johnson *et al.* (1990). They showed that peridotites dredged from near-hotspot locations along the ridge are more depleted in incompatible elements than peridotites from normal ridge segments. This implies that they have been subjected to higher degrees of melting (and loss of that melt). In spite of this higher degree of melting, the near-hotspot lavas are more enriched in incompatible elements, and therefore their initial sources must also have been more enriched.

Trace-element abundance patterns, often called "spidergrams," of MORB are shown in Figure 11 (spidergram is a somewhat inappropriate but a convenient term coined by R. N. Thompson (Thompson *et al.*, 1984), presumably because of a perceived resemblance of these patterns to spider webs, although the resemblance is tenuous at best). The data chosen for this plot are taken from le Roux *et al.* (2002) for MORB glasses from the MAR (40–55° S), which encompasses both depleted regions and enriched regions resulting from ridge–hotspot interactions. The patterns are highly divergent for the most incompatible elements, but they converge and become more parallel for the more compatible elements. This phenomenon is caused by the fact that variations in melt fractions produce the largest concentration variations in the most highly incompatible elements in both melts and their residues. This is a simple consequence of Equation (3), which states that for elements with very small values of D the concentration in the melt is inversely proportional to the melt fraction. At the other end of the spectrum, compatible elements, those with D values close to unity or greater, become effectively buffered by the melting assemblage. For an element having $D \gg F$, Equation (2) reduces to

$$C_1 \approx \frac{C_0}{D} \tag{12}$$

and for $D=1$, it reduces to

$$C_1 = C_0 \tag{13}$$

In both cases, the concentration in the melt becomes effectively buffered by the residual mineral assemblage until the degree of melting is large enough, so that the specific mineral responsible for the high value of D is exhausted. This buffering effect is displayed by the relatively low and uniform concentrations of scandium (Figure 11). It is caused by the persistence of residual clinopyroxene during MORB melting.

These relationships lead to the simple consequence that the variability of element concentrations in large datasets of basalt analyses are related to the bulk partition coefficients of these elements (Hofmann, 1988; Dupré *et al.*, 1994). This can be verified by considering a set of trace elements for which enough experimental data are available to be confident of the relative solid–melt partition coefficients, namely, the REE. These coefficients decrease monotonically from the heavy to light REEs, essentially because the ionic radii increase monotonically from heavy to light REEs (with the possible exception of europium, which has special properties because of its variable valence). Figure 12 shows three plots of variability of REEs and other trace elements in MORB as a function of mantle compatibility of the elements listed. Variability is defined as the standard deviation of

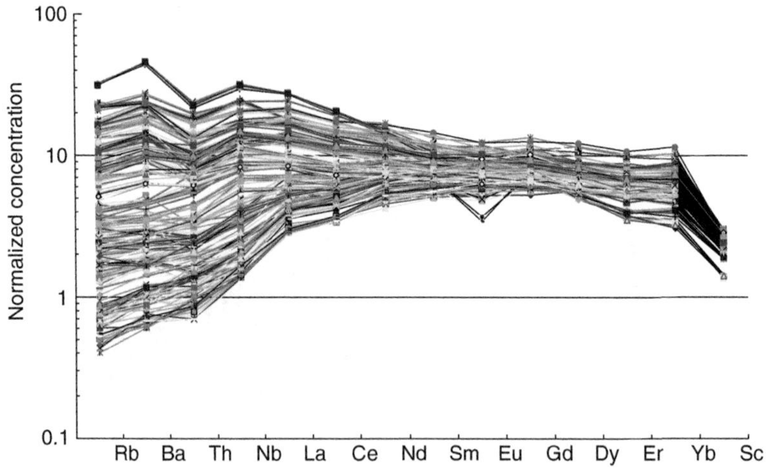

Figure 11 Trace element abundances of 250 MORB between 40° S and 55° S along the MAR. Each sample is represented by one line. The data are normalized to primitive-mantle abundances of (McDonough and Sun, 1995) and shown in the order of mantle compatibility. This type of diagram is popularly known as spidergram. The data have been filtered to remove the most highly fractionated samples containing less than 5% MgO (le Roux *et al.*, 2002).

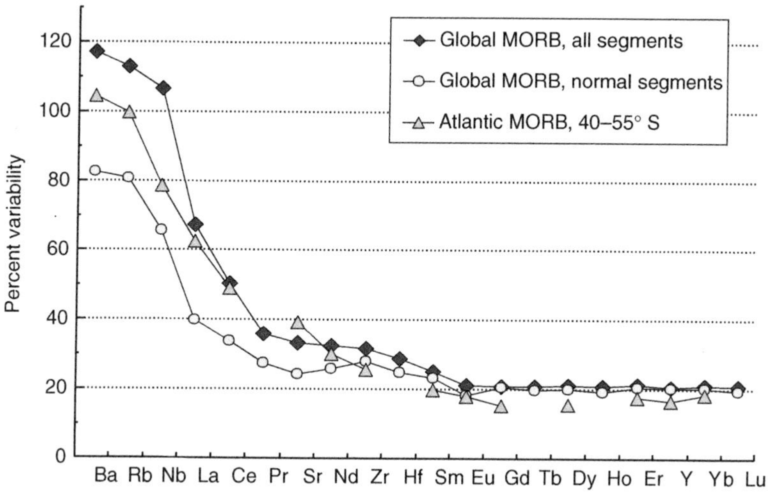

Figure 12 Variability of trace element concentrations in MORB, expressed as 100*; standard deviation/mean concentration. The data for "Global MORB" are from the PETDB compilation of Su (2002). "All segments" refers to ~250 ridge segments from all oceans. "Normal segments" refer to ~62 ridge segments that are considered not to represent any sort of "anomalous" ridges, because those might be affected by factors such as vicinity to mantle plumes or subduction of sediments (e.g., back-arc basins and the southern Chile ridge). The Atlantic MORB, 40–55° S, from which samples with less than 5% MgO have been removed (le Roux *et al.*, 2002).

the measured concentrations divided by the respective mean value. Compatibilities of elements other than the REEs are estimated from global correlations of trace-element ratios with absolute abundances as derived from simple partial melting theory (Hofmann *et al.*, 1986; Hofmann, 1988) (see also Figures 17 and 18). Two sets of data are from a new, ridge segment-by-segment compilation made by Su (2002) using the MORB database (Petrological Database of the Ocean Floor (PETDB), 2006). The third represents data for 270 MORB glasses from the South Atlantic Ridge (40–55° S) (le Roux *et al.*, 2002). The qualitative similarity of the three plots is striking. It indicates that the order of variabilities is a robust feature. For example, the variabilities of the heavy REEs (europium to lutetium) are all essentially identical at 20% in all three plots. For the light REEs, the variability increases monotonically from europium to lanthanum, consistent with their decreasing partition coefficients in all mantle minerals. As expected, variabilities are greatest for the very highly incompatible elements (VICEs) niobium, rubidium, and barium. The same is true for thorium and uranium in South Atlantic MORBs (le Roux *et al.*, 2002), which are not shown here because their averages and standard deviations were not compiled by Su (2002). All of this is consistent with the enrichment pattern in the continental crust (Figure 2), which shows the greatest enrichments for barium, rubidium, thorium, and uranium in the continental crust, as well as monotonically decreasing crustal abundances for the REEs from lanthanum to lutetium, with a characteristic flattening from europium to lutetium. Obvious exceptions to this general consistency are the elements niobium and lead. These will be discussed separately below.

Some additional lessons can be learned from Figure 12. The flattening of the heavy-REE variabilities in MORB is consistent with the flat heavy-REE (HREE) patterns almost universally observed in MORB, and these are consistent with the flat pattern of HREE partition coefficients in clinopyroxene. This does not rule out some role of garnet during MORB melting, but it does probably rule out a major role of garnet.

Strontium, zirconium, and hafnium have very similar variabilities as the REEs neodymium and samarium. Again, this is consistent with the abundance patterns of MORB and with experimental data. Overall, the somewhat tentative suggestion made by Hofmann (1988) regarding the relationship between concentration variability and degree of incompatibility, based on a very small set of MORB data, is strongly confirmed by the very large datasets now available. A note of caution is in order for strontium, which has a high partition coefficient in plagioclase. Thus, when oceanic basalts crystallize plagioclase, the REEs tend to increase in the residual melt, but strontium is removed from the melt by the plagioclase. The net effect of this is that the overall variability of strontium is reduced in datasets incorporating plagioclase-fractionated samples. Such samples have been partly filtered out from the South Atlantic dataset. This is the likely reason why strontium shows the greatest inconsistencies between the three plots shown in Figure 12.

18.4.4 N-MORB, E-MORB, T-MORB, and MORB Normalizations

It has become a widely used practice to define standard or average compositions of N-MORB, E-MORB, and T-MORB (for normal, enriched, and transitional MORB), and to use these as standards of comparison for ancient rocks found on land. In addition, many authors use N-MORB compositions as a normalization standard in trace-element abundance plots (spidergrams) instead of chondritic or primitive mantle compositions. This practice should be discouraged, because trace-element abundances in MORB form a complete continuum of compositions ranging from very depleted to quite enriched and OIB-like. A plot of global La/Sm ratios (Figure 13) demonstrates this: there is no obvious typical value but a range of lanthanum concentrations covering two orders of magnitude and a range of La/Sm ratios covering about one-and-a-half orders of magnitude. Although the term N-MORB was intended to describe "normal" MORB, it actually refers to depleted MORB, often defined by $(La/Sm)_n < 1$. Thus, while these terms do serve some purpose for characterizing MORB compositions, there is no sound basis for using any of them as normalizing values to compare other rocks with "typical" MORB.

The strong positive correlation seen in Figure 13 is primarily the result of the fact that lanthanum is much more variable than samarium. Still, the overall coherence of this relationship is remarkable. It demonstrates that the variations of the REE abundances are not strongly controlled by variations in the degree of crystal fractionation of MORB magmas, because these would cause similar variability of lanthanum and samarium. Although this reasoning is partly circular because highly fractionated samples containing less than 6% MgO have been eliminated, the total number of such samples in this population of ~2,000 is less than 100. Thus, it is clear that the relationship is primarily controlled either by source or by partial melting effects. Figure 14 shows that the La/Sm ratios are also negatively correlated with $^{143}Nd/^{144}Nd$. Because this isotope ratio is a function of source Sm/Nd (and time), and neodymium is intermediate in bulk partition coefficient between lanthanum and samarium; such a negative correlation is expected if the variability of the REE abundances is (at least in part) inherited from the source. Thus, while it would be perfectly possible to generate the relationship seen in Figure 13 purely by variations in partial melting, we can be confident that La/Sm ratios (and other highly incompatible element ratios) do track mantle source variations, as was shown by Schilling many years ago. It is important to realize, however, that such differences in source compositions were originally also produced by melting. These sources are simply the residues of earlier melting events during previous episodes of (continental or oceanic) crust formation.

18.4.5 Summary of MORB and MORB-Source Compositions

Klein and Langmuir (1987), in a classic paper, have shown that the element sodium is almost uniquely suited for estimating the degree of melting required to produce MORBs from their respective sources. This element is only slightly incompatible at low melt fractions produced at relatively high pressures. As a result, the extraction of the continental crust has reduced the sodium concentration of the residual mantle by

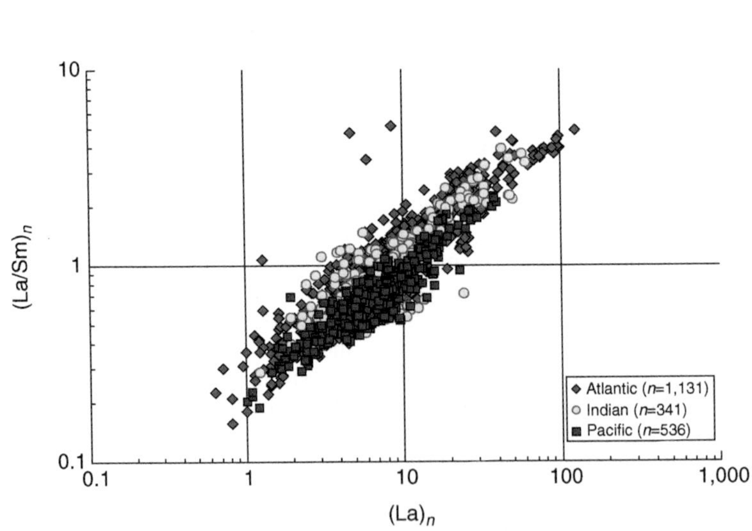

Figure 13 Primitive-mantle normalized La/Sm versus La for MORB from three ocean basins. Numbers in parentheses refer to the number of samples from each ocean basin. Lanthanum concentrations vary by about two orders of magnitude; La/Sm varies by more than one order of magnitude. Data were extracted from PETDB.

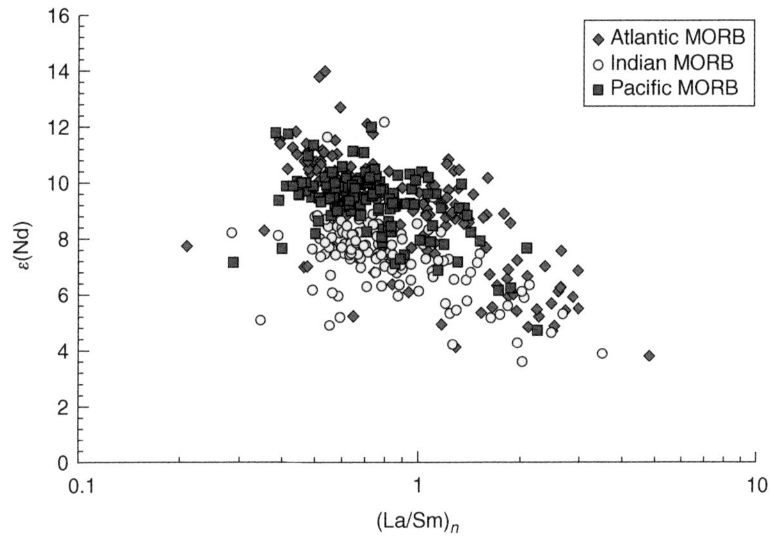

Figure 14 Primitive-mantle normalized La/Sm versus $^{143}Nd/^{144}Nd$ for MORB from three ocean basins. The (weak but significant) negative correlation is consistent with the inference that the variations in La/Sm in the basalts are to a significant part inherited from their mantle sources.

no more than ~10% relative to the primitive-mantle value. We therefore know the approximate sodium concentration of all MORB sources. In contrast, this element behaves much more incompatibly during production of oceanic crust, where relatively high melt fractions at relatively low pressures produce ocean floor basalts. This allowed Klein and Langmuir to estimate effective melt fractions ranging from 8 to 20%, with an average of ~10%, from sodium concentrations in MORB. Once the melt fraction is known, the more highly incompatible-element concentrations of the MORB-source mantle can be estimated from the measured concentrations in MORB. For the highly incompatible elements, the source concentrations are therefore estimated at ~10% of their respective values in the basalts. This constitutes a significant revision of earlier thinking, which was derived from melting experiments and the assumption that essentially clinopyroxene-free harzburgites represent the typical MORB residue, and which led to melt fraction estimates of 20% and higher.

The compilation of extensive MORB data from all major ocean basins has shown that they comprise wide variations of trace-element and isotopic compositions and the widespread notion of great compositional uniformity of MORB is largely a myth. An exception to this may exist in helium-isotopic compositions (see Chapter 17). However, from the state of heterogeneity of the more refractory elements it is clear that the apparently greater uniformity of helium compositions is not the result of mechanical mixing and stirring, because this process should homogenize all elements to a similar extent. Moreover, the isotope data of MORB from different ocean basins show that different regions of the upper mantle have not been effectively mixed in the recent geological past, where "recent" probably means approximately the last 10^9years.

The general, incompatible-element depleted nature of the majority of MORBs and their sources is well explained by the extraction of the continental crust. Nevertheless, the bulk continental crust and the bulk of the MORB sources are not "exact" chemical complements. Rather, the residual mantle has undergone additional differentiation, which generates a wide spectrum of additional mantle depletions and corresponding enrichments (as sampled by the various types of MORB), and which most likely also involves the generation of OIBs and their subducted equivalents. This conclusion is consistent with the lesson drawn from the Pb isotope systematics in Figure 5a. It will be further reinforced by the two "diagnostic" trace element ratios, Nb/U and Ce/Pb, which will be discussed in Section 18.5.2.1, and which require a common heritage of most types of depleted MORB, enriched MORB, and OIB.

The specific nature of this "secondary" differentiation process is currently debated. It is clear that low degrees of melting are needed at some stage to generate the requisite trace-element enrichments that characterize E-MORB (as well as many OIB). Because these enrichments are, in most cases, correlated with corresponding radiogenic isotope abundances, it is also clear that these low melt fractions were involved in generating the enriched sources. Low melt fractions might be generated within the oceanic asthenosphere, and such melts might then impregnate the overlying lighosphere. This type of "metasomatic" enrichment process has been popular within a significant segment of the geochemical community. Evidence for such processes can be found in many mantle xenoliths, and at least two books have been devoted to the subject of mantle metasomatism (Morris and Pasteris, 1987; Menzies and Hawkesworth, 1987). An excellent quantitative account of such a metasomatic enrichment model has recently been given by Donnelly *et al.* (2004). The difficult and so far unsolved question is whether such metasomatism occurs on a sufficiently large scale to account for volumetrically significant occurrences of E-MORB (and OIB). An alternative mechanism for introducing enriched (but noncontinental) basalt sources into the mantle is by subduction of oceanic crust (Hofmann and White, 1982). Although much of this crust is geochemically too depleted to serve as a suitable source of E-MORB, very significant portions (including many ocean islands, seamounts, and E-MORB crust) are sufficiently enriched to constitute, after subduction and recycling, appropriate sources of E-MORB (Hémond *et al.*, 2006) and OIB (McKenzie *et al.*, 2004). In any case, it is these additional differentiation processes, rather than recycling of continental material, that have generated much of the heterogeneity observed in MORBs and their sources. This means that the relationships shown in Figures 2 and 3, that is, the extraction of continental crust from the mantle and subsequent remelting of this mantle reservoir, cannot be "specifically" called upon to explain the isotopic and trace-element heterogeneity of MORB and their sources. Nevertheless, the processes generating the actual MORB heterogeneity are fundamentally similar, namely, production of enriched melts and depleted residues by low degrees of melting, followed by larger degrees of melting to generate the observed MORB. Therefore, the general topology of Figure 3 can also be applied to explain the isotopic correlations observed in MORB.

18.5 OCEAN ISLAND, PLATEAU, AND SEAMOUNT BASALTS

These basalts represent the oceanic subclass of so-called intraplate basalts, which also include continental varieties of flood and rift basalts. They will be collectively referred to as "OIB," even though many of them are not found on actual oceanic islands either because they never rose above sea level or because they were formed on islands, that have sunk below sea level. Continental and island arc basalts will not be discussed here, because at least some of them have clearly been contaminated by continental crust. Others may or may not originate in, or have been "contaminated" by, the subcontinental lithosphere. For this reason, they are not considered in the present chapter, which is concerned primarily with the chemistry of the sublithospheric mantle.

Geochemists have been particularly interested in OIB because their isotopic compositions tend to be systematically different from MORB, and this suggests that they come from systematically different places in the mantle (e.g., Hofmann *et al.*, 1978; Hofmann and Hart, 1978). Morgan's mantle plume theory (Morgan, 1971) thus provided an attractive framework for interpreting these differences, though not quite in the manner originally envisioned by Morgan. He viewed the entire mantle as a single reservoir, in which plumes rise from a lower boundary layer that is not fundamentally different in composition from the upper mantle. In contrast, geochemists saw plumes being formed in a fundamentally different, more primitive, less depleted, or enriched, deeper part of the mantle than MORB sources (e.g., Wasserburg and Depaolo, 1979). The debate about these issues continues to the present day, and some of the mantle models based on isotopic and trace-element characteristics will be discussed below.

18.5.1 Isotope Ratios of Strontium, Neodymium, Hafnium, and Lead and the Species of the Mantle Zoo

Radiogenic isotope ratios of OIB are shown in Figure 15. These diagrams display remarkably similar topologies as the respective MORB data shown in Figures 4–6. Strontium isotope ratios are negatively correlated with neodymium and hafnium isotopes, but correlations between strontium, neodymium, and hafnium isotopes on the one hand, and lead isotopes on the other, are confined to $^{208}Pb^{*}/^{206}Pb^{*}$, and the ranges of isotope ratios are even greater (although not dramatically so) for OIBs than MORBs. However, one important difference is a significant shift in all of these ratios between MORBs and OIBs. This is shown in Figure 16, which compares Nd isotope data for MORB and OIB in histogram form. To minimize the sampling bias introduced in many such compilations, which simply plot all the published data, the MORB data shown here are the MOR-segment averages compiled by Su (2002) from the PETDB database, and the OIB data are averages of individual volcanoes, and in some cases individual formations for isotopically heterogeneous volcanoes. These averages were informally compiled by inspection of OIB data from the GEOROC database. This histogram shows that there is extensive overlap between the two populations, but OIBs are on average systematically less radiogenic in neodymium (and hafnium) isotopes (and more radiogenic in strontium isotopes; not shown). In lead isotopes, OIBs overlap the MORB field completely but extend to more

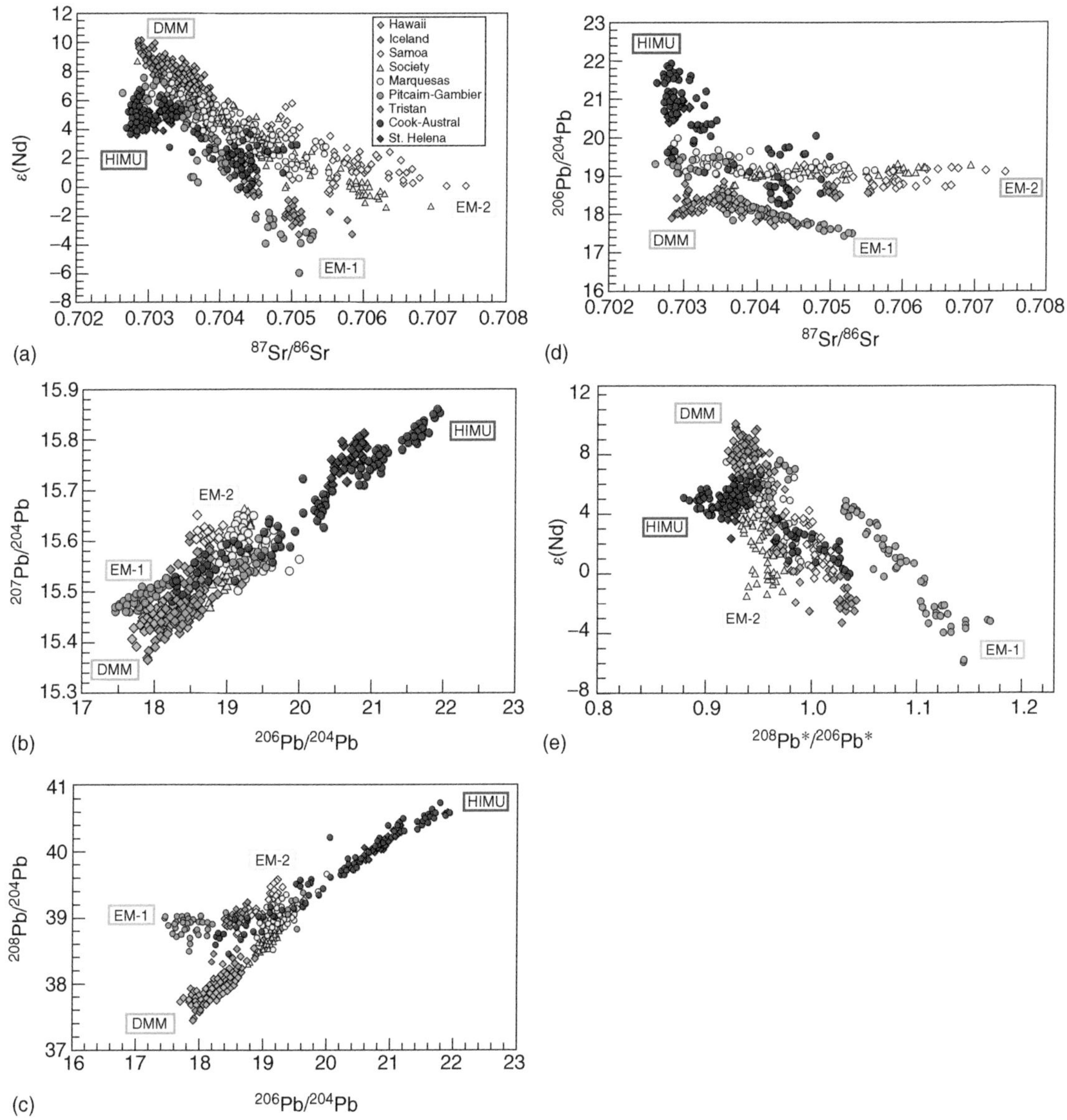

Figure 15 (a) ^{87}Sr/^{86}Sr versus ε(Nd) for OIB (excluding island arcs). The islands or island groups selected are chosen to represent extreme isotopic compositions in isotope diagrams. They are the "type localities" for HIMU (Cook-Austral Islands and St. Helena), EM-1 (Pitcairn-Gambier and Tristan), EM-2 (Society Islands, Samoa, Marquesas), and PREMA (Hawaiian Islands and Iceland). See text for explanations of the acronyms. (b) ^{207}Pb/^{204}Pb versus ^{206}Pb/^{204}Pb for the same OIB as plotted in (a). Note that the ^{207}Pb/^{204}Pb ratios of St. Helena and Cook-Australs are similar but not identical, whereas they overlap completely in the other isotope diagrams. (c) ^{208}Pb/^{204}Pb versus ^{206}Pb/^{204}Pb for the same OIB as plotted in (a). (d) ^{206}Pb/^{204}Pb versus ^{87}Sr/^{86}Sr for the same OIB as plotted in (a). Note that correlations are either absent (e.g., for the EM-2 basalts from Samoa, the Society Islands and Marquesas) or point in rather different directions, a situation that is similar to the MORB data (Figure 6a). (e) ^{208}Pb*/^{206}Pb* versus ε(Nd) for the same OIB as plotted in (a). Essentially all island groups display significant negative correlations, again roughly analogous to the MORB data. Data were assembled from the GEOROC database.

extreme values in ^{206}Pb/^{204}Pb, ^{207}Pb/^{204}Pb, and ^{208}Pb/^{204}Pb (not shown). As was true for MORBs, OIB isotopic composition can be "mapped," and certain oceanic islands or island groups can be characterized by specific isotopic characteristics. Recognition of this feature has led to the well-known concept of end-member compositions or "mantle components" initially identified by White (1985) and subsequently labeled HIMU, PREMA, EM-1, and EM-2 by Zindler and Hart (1986). These acronyms refer to mantle sources characterized by high μ values (HIMU; $\mu = (^{238}U/^{204}Pb)_{t=0}$), "prevalent mantle" (PREMA), "enriched mantle-1" (EM-1), and

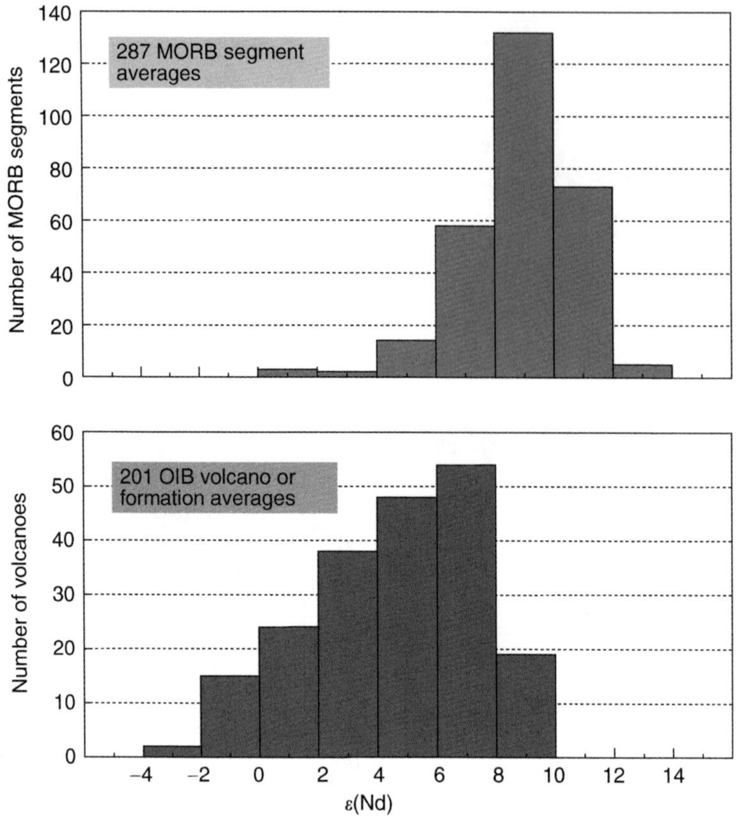

Figure 16 Histograms comparing averages of ε(Nd) values for MORBs (blue) and OIBs (red). The MORB data are ridge segment averages compiled by Su (2002); the OIB data are informally compiled averages of single volcanoes, isotopically uniform small islands and, in some cases, individual formations of isotopically heterogeneous volcanoes. All data were compiled from the EarthChem databases PETDB and GEOROC.

"enriched mantle-2" (EM-2). "PREMA" has, in recent years, fallen into disuse. It has been replaced by three new terms, namely, "FOZO" (for "focal zone;" Hart *et al.*, 1992), "C" (for "common" component; Hanan and Graham, 1996), or "PHEM" (for "primitive helium mantle;" Farley *et al.*, 1992), which differ from each other only in detail, if at all. In contrast with the illustration chosen by Hofmann (1997), which used color coding to illustrate how the isotopic characteristics of, e.g., extreme HIMU samples appear in different isotope diagrams irrespective of their geographic location, the more conventional representation of identifying "type localities" of the various "species" of this mantle isotope zoo is used here.

Two extreme notions about the meaning of these components or end members (sometimes also called "flavors") can be found in the literature. One holds that the extreme isotopic end members of these exist as identifiable "species," which may occupy separate volumes or "reservoirs" in the mantle. In this view, the intermediate compositions found in most oceanic basalts are generated by instantaneous mixing of these species during the melting and emplacement of OIBs. The other notion considers them to be merely extremes of a continuum of isotopic compositions existing in mantle rocks.

Apparent support for the "species" hypothesis is provided by the observation that the isotopically extreme compositions can be found in more than a single ocean island or island group, namely, Austral Islands and St. Helena for HIMU, Pitcairn Island and Walvis-Ridge-Tristan Island for EM-1, and Society Islands and Samoa for EM-2 (Hofmann, 1997). Nevertheless, it seems to be geologically implausible that mantle differentiation, by whatever mechanism, would consistently produce just four (or five, when the "depleted MORB mantle" DMM is included; Zindler and Hart, 1986) species of essentially identical ages, which would then be remixed in variable proportions. It is more consistent with current understanding of mantle dynamics to assume that the mantle is differentiated and remixed continuously through time. Moreover, we can be reasonably certain that a great many rock types with differing chemistries are continuously introduced into the mantle by subduction and are thereafter subjected to variable degrees of mechanical stirring and mixing. These rock types include ordinary peridotites, harzburgites, gabbros, tholeiitic and alkali basalts, terrigenous and pelagic sediments, and possibly lower crustal metamorphic rocks "eroded" from the base of the continental crust during subduction of oceanic plates. Some of these rock types have been affected by seafloor hydrothermal and low-temperature alteration, submarine "weathering," and subduction-related alteration and metasomatism. Finally, it is obvious that, overall, the OIB isotopic data constitute a continuously heterogeneous spectrum of compositions, just as is the case for MORB compositions.

In spite of the above uncertainties about the meaning of mantle components and reservoirs, it is clear that the extreme isotopic compositions represent melting products of sources subjected to some sort of ancient and comparatively extreme chemical differentiation. Because of this, they probably offer the best opportunity to identify the specific character of the types of mantle differentiation also found in other OIBs of less extreme isotopic composition. For example, the highly radiogenic lead isotope ratios of HIMU samples require mantle sources with exceptionally high U/Pb and Th/Pb ratios. At the same time, HIMU samples are among those OIBs with the least radiogenic strontium, requiring source-Rb/Sr ratios nearly as low as those of the more depleted MORBs. Following the currently popular hypothesis of Hofmann and White (1980, 1982) and Chase (1981), it is widely thought that such rocks are examples of recycled oceanic crust, which has lost alkalis and lead during alteration and subduction (Chauvel *et al.*, 1992). However, there are other possibilities. For example, the characteristics of HIMU sources might also be explained by enriching oceanic lithosphere "metasomatically" by the infiltration of low-degree partial melts, which have high U/Pb and Th/Pb ratios because of magmatic enrichment of uranium and thorium over lead (Sun and McDonough, 1989). The Rb/Sr ratios of these sources should then also be elevated over those of ordinary MORB sources, but this enrichment would be insufficient to significantly raise $^{87}Sr/^{86}Sr$ ratios because the initial Rb/Sr of these sources was well below the level where any significant growth of radiogenic ^{87}Sr could occur. Thus, instead of recycling more or less ordinary oceanic crust the enrichment mechanism would involve recycling of magmatically enriched oceanic lithosphere.

The origin of EM-type OIBs is also controversial. Hawkesworth *et al.* (1979) had postulated a sedimentary component in the source of the island of Sao Miguel (Azores), and White and Hofmann (1982) argued that EM-2 basalt sources from Samoa and the Society Islands are formed by recycled ocean crust with an addition of the small amount of subducted sediment. This interpretation was based on the high $^{87}Sr/^{86}Sr$ and high $^{207}Pb/^{204}Pb$ (for given $^{206}Pb/^{204}Pb$) ratios of EM-2 basalts, which resemble the isotopic signatures of terrigenous sediments. However, this interpretation continues to be questioned on the grounds that there are isotopic or trace-element parameters that appear inconsistent with this interpretation (e.g., Widom and Shirey, 1996). Workman *et al.* (2003) argue that the geochemistry of Samoa is best explained by recycling of melt-impregnated oceanic lithosphere, because their Samoa samples do not show the trace-element fingerprints characteristic of other EM-2 suites (see discussion on neodymium below). In addition, it has been argued that the sedimentary signature is present, but it is not part of a deep-seated mantle plume; it is introduced as a sedimentary contaminant into plume-derived magmas during their passage through the shallow mantle or crust (Bohrson and Reid, 1995).

The origin of the EM-1 flavor, which is found on Pitcairn Island, the Walvis Ridge, Tristan and Gough Islands, and many Indian Ocean MORB, is even more controversial. Distinctive EM-1 characteristics include very low $^{143}Nd/^{144}Nd$ coupled with relatively low $^{87}Sr/^{86}Sr$, and very low $^{206}Pb/^{204}Pb$ coupled with relatively high $^{208}Pb/^{204}Pb$ (leading to exceptionally high $^{208}Pb^*/^{206}Pb^*$ values). The leading contenders for the origin of this are (1) recycling of delaminated subcontinental lithosphere, (2) recycling of subducted ancient pelagic sediment, and (3) recycling of lower continental crust. The first hypothesis follows a model originally proposed by McKenzie and O'Nions (1983) to explain the origin of OIBs in general. The more specific model for deriving EM-1 type basalts from such a source was developed by Hawkesworth *et al.* (1986), Mahoney *et al.* (1991), and Milner and le Roex (1996). It is based on the observation that mantle xenoliths from Precambrian shields display similar isotopic characteristics. The second hypothesis is based on the observation that many pelagic sediments are characterized by high Th/U and low (U,Th)/Pb ratios (Ben Othman *et al.*, 1989; Plank and Langmuir, 1998), and this will lead to relatively unradiogenic lead with high $^{208}Pb^*/^{206}Pb^*$ ratios after passage of 1–2 Ga (Weaver, 1991; Chauvel *et al.*, 1992; Rehkämper and Hofmann, 1997; Eisele *et al.*, 2002). Additional support for this hypothesis has come from hafnium isotopes. Many (though not all) pelagic sediments have high Lu/Hf ratios (along with low Sm/Nd ratios), because they are depleted in detrital zircons, the major carrier of hafnium in sediments (Patchett *et al.*, 1984; Plank and Langmuir, 1998). This is expected to lead to relatively high $^{176}Hf/^{177}Hf$ ratios combined with low $^{143}Nd/^{144}Nd$ values, and these relationships have indeed been observed in lavas from Koolau volcano, Oahu (Hawaiian Islands) (Blichert-Toft *et al.*, 1999) and from Pitcairn (Eisele *et al.*, 2002).

The third hypothesis has been introduced by Hanan *et al.* (2004) to explain the origin of Indian Ocean MORB and by Willbold and Stracke (2006) to account for the isotopic characteristics of EM-1 OIBs. It is based on the observation that lower-crustal granulites tend to be depleted in uranium, and this leads to elevated Th/U and $^{208}Pb^*/^{206}Pb^*$ ratios as well as low $^{206}Pb/^{204}Pb$ ratios. Tectonic erosion during subduction of oceanic plates beneath continental margins and recycling of such material can therefore also account for the isotopic characteristics of EM-1 basalts. Gasperini *et al.* (2000) have proposed yet another origin for EM-1 basalts from Sardinia, namely, recycling of gabbros derived from a subducted, ancient plume head.

Recycling of subducted ocean islands and oceanic plateaus was suggested by Hofmann (1989b) to explain not the extreme end-member compositions of the OIB source zoo, but the enrichments seen in the basalts forming the main "mantle array" of negatively correlated $^{143}Nd/^{144}Nd$ and $^{87}Sr/^{86}Sr$ ratios. The $^{143}Nd/^{144}Nd$ values of many of these basalts (e.g., many Hawaiian basalts) are too low, and their $^{87}Sr/^{86}Sr$ values too high, for these OIBs to be explained by recycling of depleted oceanic crust. However, if the recycled material consists of either enriched MORB, tholeiitic or alkaline OIB, or basaltic oceanic plateau material, such a source will have the pre-enriched Rb/Sr and Nd/Sm ratios capable of producing the observed range of strontium- and neodymium-isotopic compositions of the main OIB isotope array.

Melt inclusions in olivine phenocrysts have been shown to preserve primary melt compositions, and these have revealed a startling degree of chemical and isotopic heterogeneity occurring in single-hand specimens and even in single olivine crystals (Sobolev and Shimizu, 1993; Sobolev, 1996; Saal *et al.*, 1998; Sobolev *et al.*, 2000; Hauri, 2002). These studies have demonstrated that rather extreme isotopic and chemical heterogeneities exist in the mantle on scales considerably smaller than the melting region of a single volcano, as discussed in Section 18.2.2. One of these studies, in particular, demonstrated the geochemical fingerprint of recycled oceanic gabbros in melt inclusions from Mauna Loa Volcano, Hawaii (Sobolev *et al.*, 2000). These rare melt inclusions have trace-element patterns that are very similar to those of oceanic and ophiolitic gabbros. They are characterized by very high Sr/Nd and low Th/Ba ratios that can be ascribed to cumulus plagioclase, which dominates the modes of many of these gabbros. Chemical and isotopic studies of melt inclusions therefore have great potential for unraveling the specific source materials found in oceanic basalts. These inclusions can preserve primary heterogeneities of the melts much better than the bulk melts do, because the latter go through magma chamber mixing processes that attenuate most of the primary melt features.

The origin of FOZO-C-PHEM-PREMA, simply referred to as "FOZO" hereafter, may be of farther-reaching consequence than any of the other isotope flavors, if the inference of Hart *et al.* (1992) is correct, namely, that it represents material from the lower mantle that is present as a mixing component in all deep-mantle plumes. The evidence for this is that samples from many individual OIB associations appear to form binary mixing arrays that radiate from this "FOZO" composition in various directions toward HIMU, EM-1, or EM-2. These relationships are shown in Figure 17. The FOZO composition is similar, but not identical, to DMM represented by MORB. It is only moderately more radiogenic in strontium, less radiogenic in neodymium and hafnium, but significantly more radiogenic in lead isotopes than DMM. If plumes originate in the very deep mantle and rise from the core–mantle boundary, rather than from the 660km seismic discontinuity, they are likely to entrain far more deep-mantle, than upper-mantle material (Griffiths and Campbell, 1990; Hart *et al.*, 1992). It should be noted, however, that the amount of entrained material in plumes is controversial, with some authors insisting that plumes contain very little entrained material (e.g., Farnetani *et al.*, 2002).

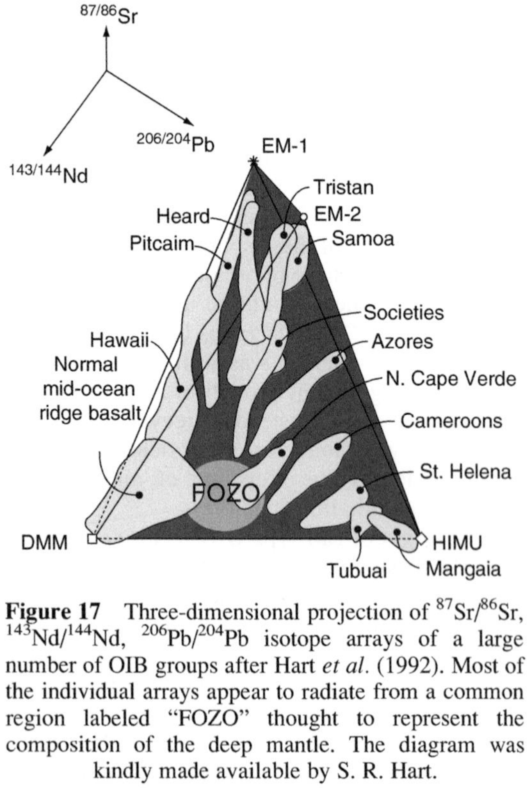

Figure 17 Three-dimensional projection of $^{87}Sr/^{86}Sr$, $^{143}Nd/^{144}Nd$, $^{206}Pb/^{204}Pb$ isotope arrays of a large number of OIB groups after Hart *et al.* (1992). Most of the individual arrays appear to radiate from a common region labeled "FOZO" thought to represent the composition of the deep mantle. The diagram was kindly made available by S. R. Hart.

18.5.2 Trace Elements in OIB

Most OIBs are much more enriched in incompatible trace elements than most MORBs, and there are two possible reasons for this: (1) their sources may be more enriched than MORB sources, and (2) OIBs may be produced by generally lower degrees of partial melting than MORBs. Most likely, both factors contribute to this enrichment. Source enrichment (relative to MORB sources) is required because the isotopic compositions require relative enrichment of the more incompatible of the parent–daughter element ratios. Low degrees of melting are caused by the circumstance that most OIBs are also within-plate basalts, so a rising mantle diapir undergoing partial melting encounters a relatively cold lithospheric lid, and melting is confined to low degrees and relatively deep levels. This is also the reason why most OIBs are alkali basalts rather than tholeiites, the predominant rock type in MORBs. Important exceptions to this rule are found primarily in OIBs erupted on or near ocean ridges (such as on Iceland and the Galapagos Islands) and on hotspots created by especially strong plume flux, which generates tholeiites at relatively high melt fractions, such as in Hawaii.

The high incompatible-element enrichments found in most OIBs are coupled with comparatively low abundances of aluminum, ytterbium, and scandium. This effect is almost certainly caused by the persistence of garnet in the melt residue, which has high partition coefficients for these elements, and keeps them buffered at relatively low abundances. Haase (1996) has shown that Ce/Yb and Tb/Yb increase systematically with increasing age of the lithosphere through which OIBs are erupting. This effect is clearly related to the increasing influence of residual garnet, which is stable in peridotites at depths greater than ~80 km. Allègre *et al.* (1995) analyzed the trace-element abundances of oceanic basalts statistically and concluded that OIBs are more variable in isotopic compositions, but less variable in incompatible element abundances than MORBs. However, their sampling was almost certainly too limited to properly evaluate the effect of lithospheric thickness on the abundances and their variability. In particular, their near-ridge sampling was confined to 11 samples from Iceland and four samples from Bouvet. The actual range of abundances of highly incompatible elements, such as thorium and uranium, from Iceland spans nearly three orders of magnitude. In contrast, ytterbium varies by only a factor of 10. This means that either the source of Iceland basalts is internally extremely heterogeneous, or the melt fractions are highly variable, or both. Because of this ambiguity, the REE abundance patterns and most of the MICE are actually not very useful to unravel the relative effects of partial melting and source heterogeneity.

The VICE are much less fractionated from each other in melts (through normal petrogenetic processes), but they are more severely fractionated in melt residues. This is the reason why their relative abundances vary in the mantle and why these variations can be traced by VICE ratios in basalts, which are in this respect similar to, though not as precise as, isotope ratios. VICE ratios thus enlarge the geochemical arsenal for determining mantle chemical heterogeneities and their origins. These differences are conventionally illustrated either by the spidergrams (primitive-mantle normalized element abundance diagrams), or by plotting trace-element abundance ratios.

Spidergrams have the advantage of representing a large number of trace-element abundances of a given sample by a single line. However, they can be confusing because there are no standard rules about the specific sequence in which the elements are shown or about the normalizing abundances used. The (mis)use of N-MORB or E-MORBs as reference values for normalizations has already been discussed and discouraged in Section 18.4.4. However, there are other pitfalls to be aware of: one of the most widely used normalizations is that given by Sun and McDonough (1989), which uses primitive-mantle estimates for all elements "except" lead, the abundance of which is adjusted by a factor of 2.5, presumably to generate smoother abundance patterns in oceanic basalts. The great majority of authors using this normalization simply call this "primitive mantle" without any awareness of the fudge factor applied. Such *ad hoc* adjustments for aesthatic reasons should be strongly discouraged. Spidergrams communicate their message most effectively if they are standardized as much as possible, i.e., if they use only one standard for normalization, namely, primitive-mantle abundances, and if the sequence of elements used is in the order of increasing compatibility (see also Hofmann, 1988). The methods for determining this order of incompatibility are addressed in the subsequent Section 18.5.2.1.

Spidergrams tend to carry a significant amount of redundant information, most of which is useful for determining the general level of incompatible-element enrichment, rather than specific information about the sources. Therefore, diagrams of critical trace-element (abundance) ratios can be very effective in focusing on specific source differences. Care should be taken to use ratios of elements with similar bulk partition coefficients during partial melting (or, more loosely speaking, similar incompatibilities). Otherwise, it may be difficult or impossible to separate source effects from melting effects. Some rather popular element pairs of mixed incompatibility, such as Zr/Nb, which are almost certainly fractionated at the relatively low melt fractions prevailing during intraplate melting, are often used in a particularly confusing manner. For example, in the popular plot of Zr/Nb versus La/Sm, the more incompatible element is placed in the numerator of one ratio (La/Sm) and in the denominator of the other (Zr/Nb). The result is a hyperbolic relationship that looks impressive, but carries little if any useful information other than showing that the more enriched rocks have high La/Sm and low Zr/Nb, and the more depleted rocks have low La/Sm and high Zr/Nb.

Trace-element ratios of similarly incompatible pairs, such as Th/U, Nb/U, Nb/La, Ba/Th, Sr/Nd, or Pb/Nd, tend to be more useful in identifying source differences, because they are fractionated relatively little during partial melting. Elements that appear to be diagnostic of distinctive source types in the mantle are niobium, tantalum, lead, and to a lesser extent strontium, barium, potassium, and rubidium. These will be discussed in connection with the presentation of specific spidergrams in Section 18.5.2.2.

18.5.2.1 "Uniform" trace-element ratios

To use geochemical anomalies for tracing particular source compositions, it is necessary to establish "normal" behavior first. Throughout the 1980s, Hofmann, Jochum, and coworkers noticed a series of trace-element ratios that are globally more or less uniform in both MORBs and OIBs. For example, the elements barium, rubidium, and cesium, which vary by about three orders of magnitude in absolute abundances, have remarkably uniform relative abundances in many MORBs and OIBs (Hofmann and White, 1983). This became clear only when sufficiently high analytical precision (isotope dilution at the time) was applied to fresh glassy samples. Hofmann and White (1983) argued that this uniformity must mean that the Ba/Rb and Rb/Cs ratios found in the basalts reflect the respective ratios in the source rocks. And because these ratios were so similar in highly depleted MORBs and in enriched OIBs, these authors concluded that these element ratios have not been affected by processes of global differentiation, and they therefore also reflect the composition of the primitive mantle. Similarly, Jochum *et al.* (1983) estimated the K/U ratio of the primitive mantle to be 1.27×10^4, a value that became virtually canonical for 20 years, even though it was based on remarkably few measurements. Other such apparently uniform ratios were Sn/Sm and Sb/Pr (Jochum *et al.*, 1993; Jochum and Hofmann, 1994) and Sr/Nd (Sun and McDonough, 1989). Zr/Hf and Nb/Ta were also thought to be uniform (Jochum *et al.*, 1986), but recent analyses carried out at higher precision and on a greater variety of rock types have shown systematic variations of these ratios.

The above approach of determining primitive-mantle abundances from apparently globally unfractionated trace-element ratios was upended by the discovery that Nb/U and Ce/Pb are also rather uniform in MORBs and OIBs the world over, but these ratios are higher by factors of ~5–10 than the respective ratios in the continental crust (Hofmann *et al.*, 1986). This invalidated the assumption that primitive-mantle abundances could be obtained simply from MORB and OIB relations, because the continental crust contains such a large portion of the total terrestrial budget of highly incompatible elements. However, these new observations meant that niobium and lead could potentially be used as tracers for recycled continental material in oceanic basalts. In other words, while ratios such as Nb/U show only limited variation when comparing oceanic basalts as a function of enrichment or depletion on a global or local scale, this uniformity can be interpreted to mean that such a specific ratio is not significantly fractionated during partial melting. If this is true, then the variations that do exist may be used to identify differences in source composition.

Figure 18 shows updated versions of the Nb/U variation diagram introduced by Hofmann *et al.* (1986). It represents an attempt to determine which other highly incompatible trace element is globally most similar to niobium in terms bulk partition coefficient during partial melting. The form of the diagram was chosen because an element ratio will systematically increase as the melt fractions decrease and the absolute concentrations of the elements increase. It is obvious that Nb/Th and Nb/La ratios vary systematically with niobium concentration, but Nb/U does not. Extending the comparison to other elements, such as the heavier REEs (not shown), simply increases the slopes of such plots. Thus, while Nb/U is certainly "not constant" in oceanic basalts, its variations are the lowest and the least systematic. To be sure, there is possible circularity in this argument, because it is possible, in principle, that enriched sources have systematically lower Nb/U ratios, which are systematically (and relatively precisely) compensated by partition coefficients that are lower for niobium than uranium, thus systematically compensating the lower source ratio during partial melting. Such a compensating mechanism has been advocated by Sims and DePaolo (1997), who criticized the entire approach of Hofmann *et al.* (1986) on this basis. Such fortuitously compensating circumstances, as postulated in the model of Sims and DePaolo (1997), may be *ad hoc* assumptions, but they are not *a priori* impossible.

The model of Hofmann *et al.* (1986) can be tested by examining more local associations of oceanic basalts characterized by large variations in melt fraction. Figure 19 shows Nb–Th–U–La–Nd relationships on Iceland, for volcanic rocks

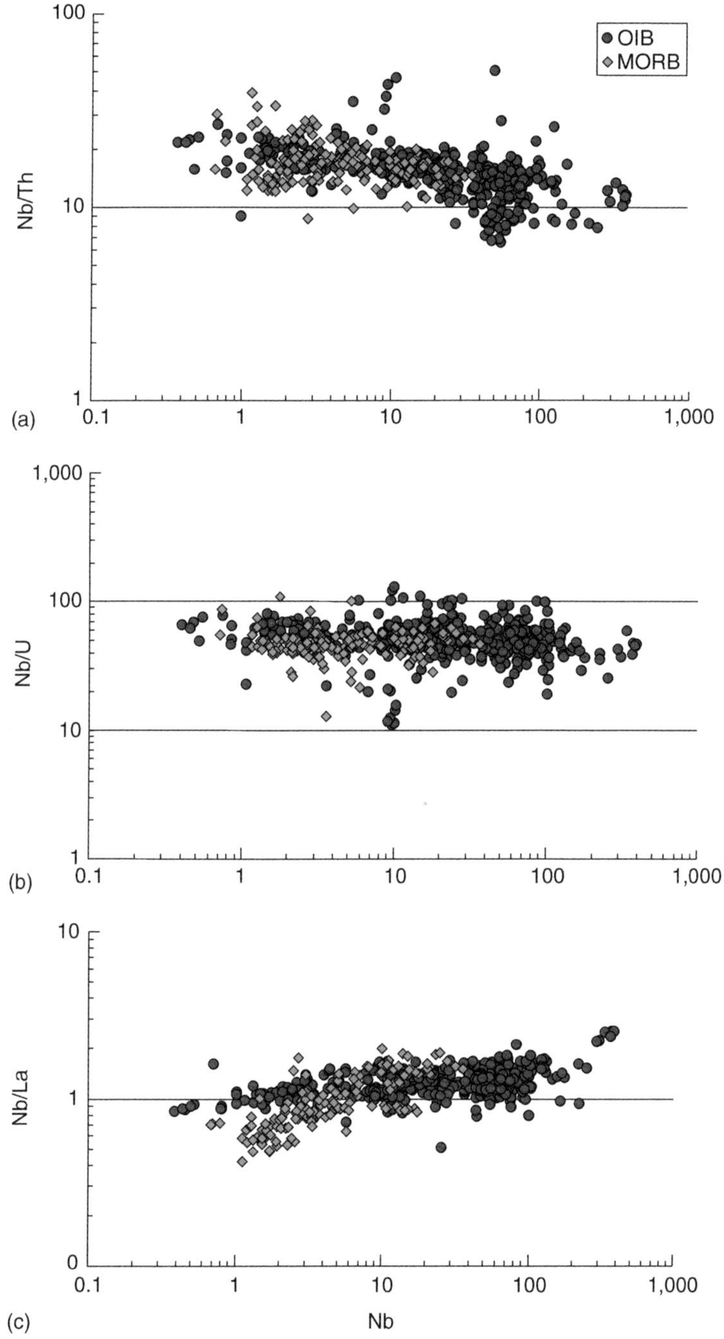

Figure 18 Nb/Th, Nb/U, and Nb/La ratios versus Nb concentrations of global MORB and (non-EM-2-type) OIBs (Hawaiian Isl., Iceland, Australs, Pitcairn, St. Helena, Cnary, Bouvet, Gough Tristan, Ascension, Madeira, Fernando de Noronha, Cameroon Line Isl., Comores, Cape Verdes, Azores, Galapagos, Easter, Juan Fernandez, San Felix). The diagram shows a systematic increase of Nb/Th, approximately constant Nb/U, and systematic decrease of La/Nb as Nb concentrations increase over three orders of magnitude.

ranging from picrites to alkali basalts, as compiled from the recent literature. The representation differs from Figure 18, which has the advantage of showing the element ratios directly, but the disadvantage pointed out by Sims and DePaolo (1997) that the two variables used are not independent. The simple log–log plot of Figure 19 is less intuitively obvious but in this sense more rigorous. In this plot, a constant concentration ratio yields a slope of unity. The Th–Nb, U–Nb, La–Nb, and Nd–Nb plots show progressively increasing slopes, with the logU–logNb and the log La–logNb plots being closest to unity. The data from Iceland shown in Figure 19 are therefore consistent with the global

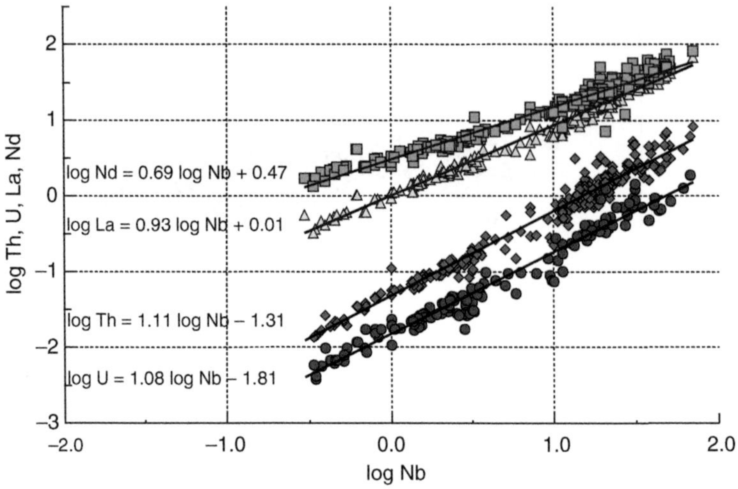

Figure 19 Concentrations of Th, U, La, and Nd versus Nb for basalts and picrites from Iceland. On this logarithmic plot, a regression line of slope 1 represents a constant element concentration ratio (corresponding to a horizontal line in Figure 18). Slopes >1 correspond to positive slopes, and slopes <1 correspond to negative slopes in Figure 18. The correlations of U and La versus Nb yield the slope closest to unity (1.08 and 0.93).

dataset shown in Figure 18. This confirms that uranium and niobium have nearly identical bulk partition coefficients during mantle melting in most oceanic environments.

The point of these arguments is not that an element ratio such as Nb/U in a melt will *always* reflect the source ratio very precisely. Rather, because of varying melting conditions, the specific partition coefficients of two such chemically different elements must vary "in detail," as expected from the partitioning theory of Blundy and Wood (1994). The nephelinites and nepheline melilitites of the Honolulu Volcanic Series, which represent the post-erosional, highly alkalic phase of Koolau Volcano, Oahu, Hawaii, may be an example where the partition coefficients of niobium and uranium are significantly different. These melts are highly enriched in trace elements and must have been formed by very small melt fractions from relatively depleted sources, as indicated by their nearly MORB-like strontium and neodymium isotopic compositions. Their Nb/U ratios average 27, whereas the alkali basalts average Nb/U=44 (Yang *et al.*, 2003). This may indicate that under melting conditions of very low melt fractions, Nb is significantly more compatible than uranium, and the relationships that are valid for basalts cannot necessarily be extended to more exotic rock types such as nephelinite.

In general, the contrast between Nb/U in most OIBs and MORBs and those in sediments, island arcs and continental rocks is so large that it appears to provide an excellent tracer of recycled continental material in oceanic basalts. A significant obstacle in applying this tracer is the lack of high-quality Nb–U data, partly because of analytical limitations and partly because of sample alteration. The latter can, however, often be overcome by "interpolating" the uranium concentration between thorium and lanthanium (the nearest neighbors in terms of compatibility) and replacing Nb/U by the primitive-mantle normalized Nb/(Th+La) ratio (e.g., Weaver, 1991; Eisele *et al.*, 2002).

Having established that Nb/U or Nb/(Th+La) ratios can be used to trace mantle-source compositions of basalts, this parameter can be turned into a tool to trace recycled continental material in the mantle. The mean Nb/U of 166 MORBs is Nb/U=47±11, and mean of nearly 500 "non-EM-type" OIBs is Nb/U=52±15. This contrasts with a mean value of the continental crust of Nb/U=8 (Rudnick and Fountain, 1995). As is evident from Figure 4b, continent-derived sediments also have consistently higher $^{87}Sr/^{86}Sr$ ratios than ordinary mantle rocks; therefore, any OIB containing significant amounts of recycled sediments should be distinguished by high $^{87}Sr/^{86}Sr$ and low Nb/U ratios. Figure 20 shows that this is indeed observed for EM-2 type OIBs and, to a lesser extent, for EM-1 OIBs as well. Of course, this does not "prove" that EM-type OIBs contain recycled sediments. However, there is little doubt that sediments have been subducted in geological history. Much of their trace-element budget is likely to have been short-circuited back into island arcs during subduction. But if any of this material has entered the general mantle circulation and is recycled at all, then EM-type OIBs are the best candidates to show it. Perhaps the greater surprise is that there are so few EM-type ocean islands.

Finding the "constant-ratio partner" for lead has proved to be more difficult. Originally, Hofmann *et al.* (1986) chose cerium because, on average, the Ce/Pb ratio of their MORB data was

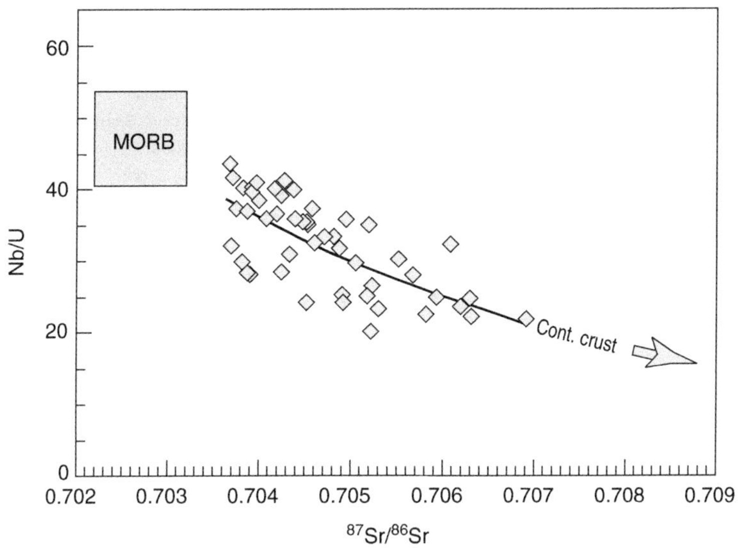

Figure 20 Nb/U versus $^{87}Sr/^{86}Sr$ for basalts from the Society Islands using data of White and Duncan (1996). Two samples with $Th/U > 6.0$ have been removed because they form outliers on an Nb/Th versus Nb/U correlation and are therefore suspected of alteration or analytical effects on the U concentration. One strongly fractionated trachyte sample has also been removed. This correlation and a similar one of Nd/Pb versus $^{87}Sr/^{86}Sr$ (not shown) are consistent with the addition of a sedimentary or other continental component to the source of the Society Island (EM-2) basalts.

most similar to their OIB average. However, Sims and DePaolo (1997) pointed out one rather problematic aspect, namely, that even in the original, very limited dataset, each separate population showed a distinctly positive slope. In addition, they showed that Ce/Pb ratios appear to correlate with europium anomalies in the MORB population, and this is a strong indication that both parameters are affected by plagioclase fractionation. Recognizing these problems, Rehkämper and Hofmann (1997) argued on the basis of more extensive and more recent data that Nd/Pb is a better indicator of source chemistry than Ce/Pb. Unfortunately, lead concentrations are not often analyzed in oceanic basalts, partly because lead is subject to alteration and partly because it is difficult to analyze, so a literature search tends to yield highly scattered data. Nevertheless, the average MORB value of $Pb/Nd = 0.04$ is lower than the average continental value of 0.63 by a factor of 15. Because of this great contrast, this ratio is potentially an even more sensitive tracer of continental contamination or continental recycling in oceanic basalts.

But why are Pb/Nd ratios so different in continental and oceanic crust in the first place? An answer to that question will be attempted in the following section.

18.5.2.2 Normalized abundance diagrams (spidergrams)

The techniques illustrated in Figures 17 and 18 can be used to establish an approximate compatibility sequence of trace elements for mantle-derived melts. In general, this sequence corresponds to the sequence of decreasing (normalized) abundances in the continental crust shown in Figure 2, but this does not apply to niobium, tantalum, and lead for which the results discussed in the previous section demand rather different positions (see also Hofmann, 1988). Here, a sequence similar to that used by Hofmann (1997) is adopted, but with slightly modified positions for lead and strontium.

Figure 21 shows examples of spidergrams for representative samples of HIMU, EM-1, EM-2, and Hawaiian basalts, in addition to average MORB and average "normal" MORB, average subducting sediment, and average continental crust. Prominent features of these plots are negative spikes for niobium in average continental crust and in sediment, and corresponding positive anomalies in most oceanic basalts except EM-type basalts. Similarly, the positive spikes for lead in average continental crust and in sediments are roughly balanced by negative anomalies in most oceanic basalts. More subtle features distinguishing the isotopically different OIB types are relative deficits for potassium and rubidium in HIMU basalts and high Ba/Th ratios coupled with elevated Sr/Nd ratios in Mauna Loa basalts.

The prominent niobium and lead spikes of continental materials are not matched by any of the OIBs and MORBs reviewed here. They are, however, common features of subduction-related volcanic rocks found on island arcs and continental margins. It is therefore likely that the distinctive

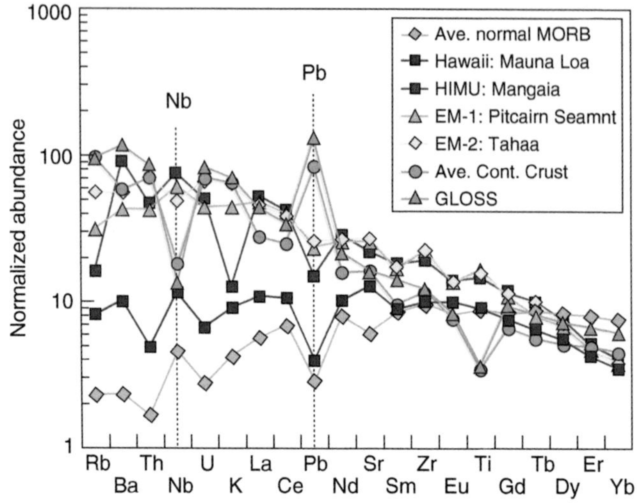

Figure 21 Examples of primitive-mantle normalized trace element abundance diagrams (spidergrams) for representative samples of HIMU (Mangaia, Austral Islands, sample M-11; Woodhead, 1996), EM-1 (Pitcairn Seamount sample 49DS1; Eisele *et al.*, 2002), EM-2 (Tahaa, Society Islands, sample 73-190; White and Duncan 1996); Average Mauna Loa (Hawaii) tholeiite (Hofmann, unpublished data), average continental crust (Rudnick and Fountain, 1995), average subducting sediment, GLOSS (Plank and Langmuir, 1998), and Average Normal MORB (Su, 2002). Th, U, and Pb values for MORB were calculated from average Nb/U=47, and Nd/Pb=26. All abundances are normalized to primitive/mantle values of McDonough and Sun (1995).

geochemical features of the continental crust are produced during subduction, where volatiles can play a major role in the element transfer from mantle to crust. The net effect of these processes is to transfer large amounts of lead (in addition to mobile elements such as potassium and rubidium) into the crust. At the same time, niobium and tantalum are retained in the mantle, either because of their low solubility in hydrothermal solutions, or because they are partitioned into residual mineral phases such as Ti-minerals or certain amphiboles. These processes are the subject of much ongoing research, but are beyond the scope of this chapter.

For the study of mantle circulation, these chemical anomalies can help trace the origin of different types of OIBs and some MORBs. Niobium and lead anomalies, coupled with high $^{87}Sr/^{86}Sr$ ratios, seem to be the best tracers for material of continental origin circulating in the mantle. They have been found not only in EM-2-type OIBs but also in some MORBs found on the Chile Ridge (Klein and Karsten, 1995). Other trace-element studies, such as the study of the chemistry of melt inclusions, have already identified a specific recycled rock type, namely, a gabbro, which could be recognized by its highly specific trace-element "fingerprint" (Sobolev *et al.*, 2000).

Until quite recently, the scarceness of high-quality data for the diagnostic elements has been a serious impediment to progress in gaining a full interpretation of the origins of oceanic basalts using complete trace-element data together with complete isotope data. All data compilations aimed at detecting global geochemical patterns are currently seriously hampered by spotty literature data of uncertain quality on samples of unknown freshness. This is now changing, because of the advent of new instrumentation capable of producing large quantities of high-quality data of trace elements at low abundances. The greater ease of obtaining large quantities of data also poses significant risks from lack of quality control. Nevertheless, we are currently experiencing a dramatic improvement in the general quantity and quality of geochemical data, and we can expect significant further improvements in the very near future. These developments offer a bright outlook for the future of deciphering the chemistry and history of mantle differentiation processes.

18.6 THE LEAD PARADOX

18.6.1 The First Lead Paradox

One of the earliest difficulties in understanding terrestrial lead isotopes arose from the observation that almost all oceanic basalts (i.e., both MORBs and OIBs) have more highly radiogenic lead than does the primitive mantle (Allègre, 1969). In effect, most of these basalts lie to the right-hand side of the so-called "geochron" on a diagram of $^{207}Pb/^{204}Pb$ versus $^{206}Pb/^{204}Pb$ ratios (Figure 22). The Earth was assumed to have the same age as meteorites, so that the "geochron" is identical to the meteorite isochron of 4.56 Ga. If the total silicate portion of the Earth remained a closed system involved only in internal (crust–mantle)

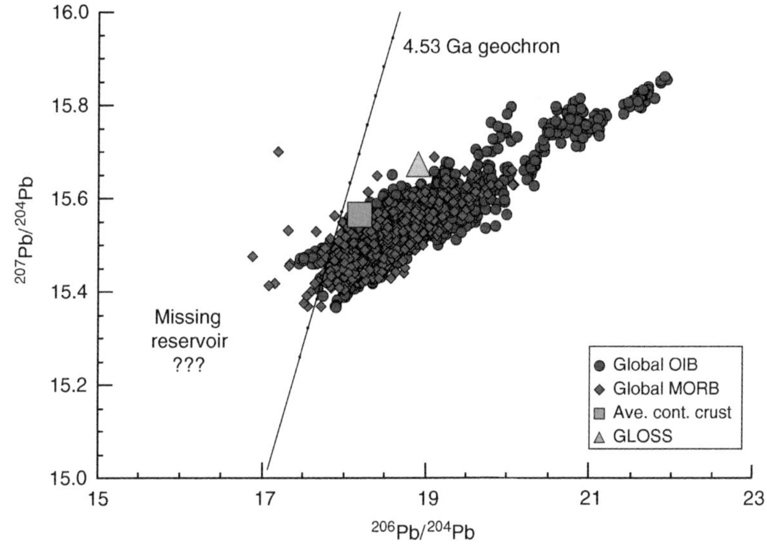

Figure 22 Illustration of the first lead paradox. Estimates of the average composition of the continental crust (Rudnick and Goldstein, 1990), of average "global subducted sediments" (GLOSS, Plank and Langmuir, 1998) mostly derived from the upper continental crust, and a global compilation of MORB and OIB data (from GEOROC and PETDB databases) lie overwhelmingly on the right-hand side of the 4.53 Ga geochron. One part of the paradox is that these data require a hidden reservoir with lead isotopes to the left of the geochron to balance the reservoirs represented by the data from the continental and oceanic crust. The other part of the paradox is that both continental and oceanic crustal rocks lie rather close to the geochron, implying that there is surprisingly little net fractionation of the U/Pb ratio during crust–mantle differentiation, even though U is significantly more incompatible than Pb (see Figure 23).

differentiation, the sum of the parts of this system must lie on the geochron. The reader is referred to textbooks (e.g., Faure, 1986) on isotope geology for fuller explanations of the construction and meaning of the geochron and the construction of common-lead isochrons.

The radiogenic nature of MORB lead was surprising because uranium is expected to be considerably more incompatible than lead during mantle melting. The MORB source, being depleted in highly incompatible elements, is therefore expected to have had a long-term history of U/Pb ratios lower than primitive ones, just as was found to be the case for Rb/Sr and Nd/Sm. Thus, the lead paradox (sometimes also called the "first paradox") is given by the observation that although one would expect most MORBs to plot well to the left of the geochron, they actually plot mostly to the right of the geochron. This expectation is reinforced by a plot of U/Pb versus U, as first used by White (1993), an updated version of which is shown in Figure 23. This shows that the U/Pb ratio is strongly correlated with the uranium concentration, thus confirming the much greater incompatibility of uranium during mantle melting.

Numerous explanations have been advanced for this paradox. The most recent treatment of the subject has been given by Murphy *et al.* (2003), who have also reviewed the most important solutions to the paradox. These include delayed uptake of lead by the core ("core pumping"; Allègre *et al.*, 1982) and storage of unradiogenic lead (to balance the excess radiogenic lead seen in MORBs, OIBs, and upper crustal rocks) in the lower continental crust or the subcontinental lithosphere (e.g., Zartman and Haines, 1988; Kramers and Tolstikhin, 1997).

An important aspect not addressed by Murphy *et al.* is the actual position of the geochron. This is the locus of any isotopic mass balance of a closed-system silicate earth. This is not the meteorite isochron of 4.56 Ga, because later core formation and giant impact(s) are likely to have prevented closure of the BSE with regard to uranium and lead until lead loss by volatilization and/or loss to the core effectively ended. Therefore, the reference line (geochron) that is relevant to balancing the lead isotopes from the various silicate reservoirs is younger than and lies to the right of the meteorite isochron. The analysis of the effect of slow accretion on the systematics of terrestrial lead isotopes (Galer and Goldstein, 1996) left reasonably wide latitude as to where the relevant geochron should actually be located. This was further reinforced by publication of early tungsten isotope data (daughter product of the short-lived ^{182}Hf), which appeared to require terrestrial core formation to have been delayed by at least 50 Myr (Lee and Halliday, 1995). This would have moved the possible locus of bulk silicate lead compositions

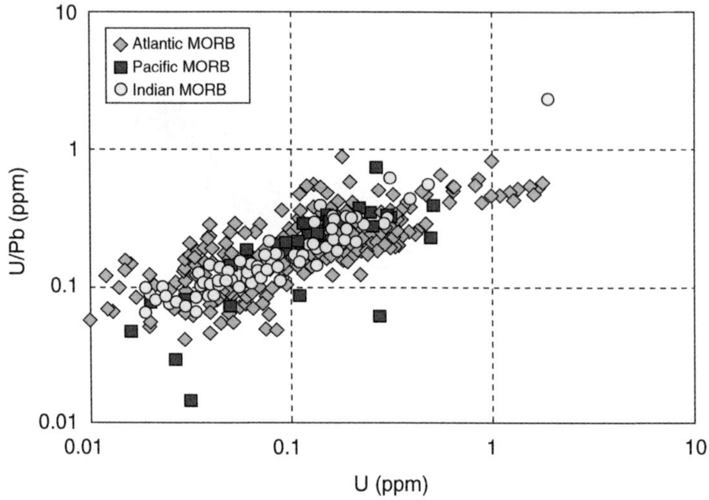

Figure 23 U/Pb versus U concentrations for MORB from three ocean basins. The positive slope of the correlation indicates that U is much more incompatible than Pb during mantle melting. This means that the similarity of Pb isotopes in continental and oceanic crust (see Figure 22) is probably caused by a nonmagmatic transport of lead from mantle to the continental crust.

closer to the actual positions of oceanic basalts, thus diminishing the magnitude of the paradox, or possibly eliminating it altogether. However, the most recent redeterminations of tungsten isotopes in chondrites by Kleine *et al.* (2002) and Yin *et al.* (2002) have shown the early tungsten data to be in error, so that core and moon formation now appear to be constrained at ~4.53 Ga. This value is not sufficiently lower than the meteorite isochron of 4.56 Ga to resolve the problem.

This means that the lead paradox is alive and well, and the search for the unradiogenic, hidden reservoir continues. The lower continental crust remains (in the author's opinion) a viable candidate, even though crustal xenolith data appear to be, on the whole, not sufficiently unradiogenic (see review of these data by Murphy *et al.*, 2003). It is not clear how representative the xenoliths are, particularly of the least radiogenic, Precambrian lower crust. Another hypothetical candidate is a garnetite reservoir proposed by Murphy *et al.* (2003).

The above discussion, and most of the relevant literature, does not address the perhaps geochemically larger and more interesting question, namely, why are the continental crust and the oceanic basalts so similar in lead isotopes in the first place? It is quite remarkable that most MORBs and most continent-derived sediments cover the same range of $^{206}Pb/^{204}Pb$ ratios, namely, ~17.5–19.5. This means that MORB sources and upper continental crust, from which these sediments are derived, have very similar U/Pb ratios, when integrated over the entire Earth history. The main offset between lead isotope data for oceanic sediments and MORBs is in terms of $^{207}Pb/^{204}Pb$, and even this offset is only marginally outside the statistical scatter of the data. In the previous section we have seen that lead behaves as a moderately incompatible element such as neodymium or cerium, but both uranium and thorium are highly incompatible elements. So the important question remains why lead and uranium are nearly equally enriched in the continental crust, whereas they are very significantly fractionated during the formation of the oceanic crust and ocean islands. Therefore, the unradiogenic reservoir (lower crust or hidden mantle reservoir) needed to balance the existing, slightly radiogenic reservoirs in order to obtain full, bulk-silicate-earth lead isotope values represents only a relatively minor aspect of additional adjustment to this major discrepancy. The answer is, in my opinion, that lead behaves relatively compatibly during MORB–OIB production because it is partially retained in the mantle by a residual phase(s), most likely sulfide(s). This would account for the relationships seen in Figure 23. However, when lead is transferred from mantle to the continental crust, two predominantly nonigneous processes become important: (1) hydrothermal transfer from oceanic crust to metalliferous sediment (Peucker-Ehrenbrink *et al.*, 1994), and (2) transfer from subducted oceanic crust-plus-sediment into arc magma sources (Miller *et al.*, 1994). This additional, nonigneous transfer enriches the crust to a similar extent as uranium and thorium, and this explains why the $^{206}Pb/^{204}Pb$ ratios of crustal and mantle rocks are so similar to each other and so close to the geochron. Thus, the anomalous geochemical behavior of lead is the main cause of the "lead paradox," the high Pd/Nd ratios of island arc and continental rocks, and the lead "spikes" seen in Figure 21.

The above explanation does not account for the elevated $^{207}Pb/^{204}Pb$ ratios of continental rocks and their sedimentary derivatives relative to mantle-derived basalts (Figures 5a and 21). This special feature can be explained by a more complex evolution of continents subsequent to their formation. New continental crust formed during Archean time by subduction and accretion processes must have initially possessed a U/Pb ratio slightly higher than that of the mantle. At that time, the terrestrial $^{235}U/^{238}U$ ratio was significantly higher than today, and this produced elevated $^{207}Pb/^{204}Pb$ relative to $^{206}Pb/^{204}Pb$. Some of this crust was later subjected to high-grade metamorphism, causing loss of uranium relative to lead in the lower crust, and transporting the excess uranium into the upper crust. From there, uranium was lost by the combined action of oxidation, weathering, dissolution, and transport into the oceans. This uranium loss retarded the growth of $^{206}Pb/^{204}Pb$ while preserving the relatively elevated $^{207}Pb/^{204}Pb$ of the upper crust. The net result of this two-stage process is the present position of sediments directly above the mantle-derived basalts in $^{207}Pb/^{204}Pb$–$^{207}Pb/^{204}Pb$ space. Another consequence of this complex behavior of uranium will be further discussed in the following section. Here it is important to reiterate that lead, not uranium or thorium, is the *major* player in generating the main part of the lead paradox.

18.6.2 The Second Lead Paradox

Galer and O'Nions (1985) made the important observation that measured $^{208}Pb^*/^{206}Pb^*$ ratios (see Equation (11)) in most MORBs are higher than can be accounted for by the relatively low Th/U ratios actually observed in MORBs, if the MORB source reservoir had maintained similarly low Th/U ratios over much of Earth's history. A simple two-stage Th/U depletion model with a primitive, first-stage value of $\kappa = (^{232}Th/^{238}U)_{today} = 3.9$ changing abruptly to a second stage value of $\kappa = 2.5$ yields a model age for the MORB source of only ~ 600 Ma. Because much of the continental mass is much older than this, and because the development of the depleted MORB source is believed to be linked to this, this result presented a dilemma. Galer and O'Nions (1985) and Galer *et al.* (1989) resolved this with a two-layer model of the mantle, in which the upper, depleted layer is in a steady state of incompatible-element depletion by production of continental crust and replenishment by leakage of less depleted material from the lower layer. This keeps the $^{208}Pb^*/^{206}Pb^*$ ratio of the upper mantle at a relatively high value in spite of the low chemical Th/U ratio. However, such two-layer convection models have fallen from grace in recent years, mostly because of the results of seismic mantle tomography (see below). Therefore, other solutions to this second lead paradox have been sought.

A fundamentally different mechanism for lowering the Th/U ratio of the mantle has been suggested by Hofmann and White (1982), namely, preferential recycling of uranium through dissolution of oxidized (hexavalent) uranium at the continental surface, riverine transport into the oceans, and fixation by ridge-crest hydrothermal circulation and reduction to the tetravalent state. The same mechanism was invoked in the quantitative "plumbotectonic" model of Zartman and Haines (1988). Staudigel *et al.* (1995) introduced the idea that this preferential recycling of uranium into the mantle may be connected to a change toward oxidizing conditions at the Earth's surface relatively late in Earth's history. Geological evidence for a rapid atmospheric change toward oxidizing conditions during Early Proterozoic time (i.e., relatively late in Earth's history) has been presented by Holland (1994), and this has been confirmed by new geochemical evidence showing that sulfides and sulfates older than ~ 2.4 Ga contain non-mass-dependent sulfur isotope fractionations, which can be explained by high-intensity UV radiation in an oxygen-absent atmosphere (Farquhar *et al.*, 2000). Kramers and Tolstikhin (1997) and Elliott *et al.* (1999) have developed quantitative models to resolve the second lead paradox by starting to recycle uranium into the mantle $\sim$2.5 Ga ago. This uranium cycle is an excellent example how mantle geochemistry, surface and atmospheric chemistry, and the evolution of life are intimately interconnected.

18.7 GEOCHEMICAL MANTLE MODELS

The major aim of mantle geochemistry has been, from the beginning, to elucidate the structure and evolution of the Earth's interior, and it was clear that this can only be done in concert with observations and ideas derived from conventional field geology and from geophysics. The discussion here will concentrate on the chemical structure of the recent mantle.

The isotopic and chemical heterogeneities found in mantle-derived basalts, reviewed on the previous pages, mandate the existence of similar or even greater heterogeneities in the mantle. The questions are: how are they spatially arranged in the mantle? When and how did they originate? Because these heterogeneities have their primary expression in trace elements, they cannot be translated into physical parameters such as density differences. Rather, they must be viewed as passive tracers of mantle processes. These tracers are separated by melting and melt migration, as well as fluid transport, and they are stirred and remixed by convection. Complete homogenization appears to be increasingly unlikely. Diffusion distances in the solid state have ranges of centimeters at best (Hofmann and Hart, 1978), and possibly very much less (Van Orman *et al.*, 2001), and homogenization of a melt source region, even via the movement and diffusion through a partial melt, becomes increasingly unlikely. This is attested by the remarkable chemical and isotopic heterogeneity observed in melt inclusions preserved in magmatic crystals from single basalt samples (Saal *et al.*, 1998; Sobolev *et al.*, 2000). The models developed for interpreting mantle heterogeneities have, with some exceptions, largely ignored the possibly extremely small scale of these heterogeneities. Instead, they have usually relied on the assumption that mantle-derived basalts are, on the whole, representative of some chemical and isotopic average for a given volcanic province or source volume.

In the early days of mantle geochemistry, the composition of the bulk silicate earth, also called "primitive mantle" (i.e., mantle prior to the formation of any crust) was not known for strontium isotopes because of the obvious depletion of rubidium of the Earth relative to chondritic meteorites (Gast, 1960). The locus of primitive-mantle lead was assumed to be on the meteorite isochron (which was thought to be identical to the geochron until the more recent realization of delayed accretion and core formation; see above), but the interpretation of the lead data was confounded by the lead paradox discussed above. This situation changed in the 1970s with the first measurements of neodymium isotopes in oceanic basalts (DePaolo and Wasserburg, 1976; Richard *et al.*, 1976; O'Nions *et al.*, 1977) and the discovery that $^{143}Nd/^{144}Nd$ was negatively correlated with $^{87}Sr/^{86}Sr$ (see Figures 6 and 15). Because both samarium and neodymium are refractory lithophile elements, the Sm/Nd and $^{143}Nd/^{144}Nd$ ratios of the primitive mantle can be safely assumed to be chondritic. With this information, a primitive-mantle value for the Rb/Sr and $^{87}Sr/^{86}Sr$ was inferred, and it became possible to estimate the size of the MORB source reservoir primarily from isotopic abundances. The evolution of a silicate Earth consisting of three boxes—primitive mantle, depleted mantle, and continental crust—were subsequently modeled by Jacobsen and Wasserburg (1979), O'Nions *et al.* (1979), DePaolo (1980), and Allègre *et al.* (1983); and Davies (1981) cast this in terms of a simple mass balance similar to that given above in Equation (11) but using isotope ratios:

$$X_{\mathrm{dm}} = \frac{X_{\mathrm{cc}} C_{\mathrm{cc}} (R_{\mathrm{dm}} - R_{\mathrm{cc}})}{C_{\mathrm{pm}} (R_{\mathrm{dm}} - R_{\mathrm{pm}})} \tag{14}$$

where X are mass fractions, R are isotope or trace-element ratios, C are the concentrations of the denominator element of R, and the subscripts cc, dm, and pm refer to continental crust, depleted mantle, and primitive mantle reservoirs, respectively.

With the exception of Davies, who favored whole-mantle convection all along, the above authors concluded that it was only the upper mantle above the 660km seismic discontinuity that was needed to balance the continental crust. The corollary conclusion was that the deeper mantle must be in an essentially primitive, nearly undepleted state, and consequently convection in the mantle had to occur in two layers with only little exchange between these layers. These conclusions were strongly reinforced by noble gas data, especially $^3He/^4He$ ratios and, more recently, neon isotope data. These indicated that hotspots such as Hawaii are derived from a deep-mantle source with a more primordial, high $^3He/^4He$ ratio, whereas MORBs are derived from a more degassed, upper-mantle reservoir with lower $^3He/^4He$ ratios. The noble-gas aspects are treated in Chapter 17. In the present context, two points must be mentioned. Essentially, all quantitative evolution models dealing with the noble gas evidence concluded that although plumes carry the primordial gas signature from the deep mantle to the surface, the plumes themselves do not originate in the deep mantle. Instead they rise from the base of the upper mantle, where they entrain very small quantities of lower-mantle, noble-gas-rich material. However, all these models have been constrained by the present-day very low flux of helium from the mantle into the oceans. This flux does not allow the lower mantle to be significantly degassed over the Earth's history. Other authors, who do not consider this constraint on the evolution models to be binding, have interpreted the noble gas data quite differently: they argue that the entire plume comes from a nearly primitive deep-mantle source and rises through the upper, depleted mantle. The former view is consistent with the geochemistry of the refractory elements, which strongly favors some type of recycled, not primitive mantle material to supply the bulk of the plume source. The latter interpretation can be reconciled with the refractory-element geochemistry, if the deep-mantle reservoir is not actually primitive (or close to primitive), but consists of significantly processed mantle with the geochemical characteristics of the FOZO (C, PHEM, etc.) composition, which is characterized by low $^{87}Sr/^{86}Sr$ but relatively high $^{206}Pb/^{204}Pb$ ratios, together with high $^3He/^4He$ and solar-like neon isotope ratios. The processed nature of this hypothetical deep-mantle reservoir is also evident from its trace-element chemistry, which shows the same nonprimitive (high) Nb/U and (low) Pb/Nd ratios as MORBs and other OIBs. It is not clear how and why the near-primordial noble gas compositions survived this processing. So far, except for the two-layer models, no internally consistent mantle evolution model has been published that accounts for all these observations.

The two-layer models have been dealt a rather decisive blow by recent results of seismic mantle tomography. The images of the mantle produced by this discipline appear to show clear evidence for subduction reaching far into the lower mantle (Grand, 1994; van der Hilst *et al.*, 1997). If this is correct, then there must be a counterflow from the lower mantle across the 660-km boundary, which in the long run would surely destroy the chemical isolation between upper- and lower-mantle reservoirs. Most recently, mantle tomography appears to be able to track some of the major mantle plumes (Hawaii, Easter Island, Cape Verdes, and Reunion) into the lowermost mantle (Montelli *et al.*, 2003). If these results are confirmed, there is, at least in recent mantle history, no convective isolation, and mantle evolution models reconciling all the geochemical aspects with the geophysical evidence clearly require new ideas.

As of mid-2000s, the existing literature and scientific conferences all show clear signs of a period preceding a significant or even major paradigm shift as described by Thomas Kuhn in his classic work *The Structure of Scientific Revolutions* (Kuhn, 1996): the established paradigm is severely challenged by new observations. While some scientists attempt to reconcile the observations with the paradigm by increasingly complex adjustments of the paradigm, others throw the established conventions overboard and engage in increasingly free speculation. This process continues until a new paradigm evolves or is discovered, which is consistent with all the observations. Examples of the effort of reconciliation are the papers by Stein and Hofmann (1994) and Allègre (1997). These authors point out that the current state of whole-mantle circulation may be episodic or recent, so that whole-mantle chemical mixing has not been achieved. In contrast, some convection and mantle evolution modelers are throwing the entire concept of geochemical reservoirs overboard. For example, Phipps Morgan and Morgan (1998) and Phipps Morgan (1999) suggest that the specific geochemical characteristics of plume-type mantle are randomly distributed in the deeper mantle. Plumes rising from the core–mantle boundary constitute the main upward flux balancing the subduction flux. They preferentially lose their enriched components during partial melting, leaving a depleted residue that replenishes the depleted upper mantle.

Nevertheless, the overall geochemical evidence favoring the existence of discrete reservoirs seems strong, and the search for models that reconcile this concept with geophysical evidence continues. Examples of such efforts are the papers by Albarède and van der Hilst (1999) and Kellogg *et al.* (1999), who essentially invent new primitive reservoirs within the deep mantle, which are stabilized by higher chemical density, and may be very irregularly shaped. In the same vein, Porcelli and Halliday (2001) have proposed that the storage reservoir of primordial noble gases may be the core.

Tolstikhin and Hofmann (2005) and Tolstikhin *et al.* (2006) have focused attention on the possible role of the lowermost layer of the mantle, called D″ by seismologists, as an appropriate long-term reservoir for the "missing" budget of heat production and primordial noble gases. Independent support for this type of model has come from the short-lived decay chain ^{146}Sm-^{142}Nd with a half-life of 103 million years. Boyet and Carlson (2005) found that samples of terrestrial silicates have slightly more radiogenic $^{142}Nd/^{144}Nd$ ratios than chondritic meteorites. The difference is only 0.2 epsilon units (i.e. 20 ppm), but such a difference requires very early segregation (i.e. less than about 30 Ma after the Earth's formation) of a silicate reservoir characterized by a low Sm/Nd ratio and correspondingly unradiogenic Nd (in order to balance the total terrestrial neodymium at a chondritic composition). The D″ layer at the base of the mantle may be the only place where such a reservoir could survive until the present day.

Starting with the contribution of Christensen and Hofmann (1994), a steadily increasing number of models have recently been published, in which eochemical heterogeneities are specifically incorporated in mantle convection models (e.g., van Keken and Ballentine, 1998; Tackley, 2000; van Keken *et al.*, 2001; Davies, 2002; Farnetani *et al.*, 2002; Tackley, 2002). Thus, while the current state of understanding of the geochemical heterogeneity of the mantle is unsatisfactory, to say the least, the formerly quite separate disciplines of geophysics and geochemistry have begun to interact intensely. This process surely offers the best approach to reach a new paradigm and an understanding of how the mantle really works.

ACKNOWLEDGMENTS

I am most grateful for the hospitality and support offered by the Lamont-Doherty Earth Observatory where much of the earlier version of this chapter was written. Steve Goldstein made it all possible, while being under duress with the similar project of his own. Conny Class and the other members of the Petrology/Geochemistry group at Lamont provided intellectual stimulation and good spirit, and the BodyQuest Gym and the running trails of Tallman Park kept my body from falling apart. I am particularly grateful to Jeff Vervoort for making available his compilation and diagram of hafnium–neodymium isotope data for MORBs and sediments used in the new version of this paper, and to Rick Carlson for making helpful suggestions and corrections and for always bearing with me to the last moment.

REFERENCES

Abouchami W., Boher M., Michard A., and Albarède F. (1990) A major 2.1 Ga event of mafic magmatism in West-Africa: an early stage of crustal accretion. *J. Geophys. Res.* **95**, 17605–17629.

Alard O., Griffin W. L., Pearson N. J., Lorand J. P., and O'Reilly S. Y. (2002) New insights into the Re–Os systematics of sub-continental lithospheric mantle from *in situ* analysis of sulphides. *Earth Planet. Sci. Lett.* **203**(2), 651–663.

Albarède F. and van der Hilst R. D. (1999) New mantle convection model may reconcile conflicting evidence. *EOS. Trans. Am. Geophys. Union* **80**(45), 535–539.

Allègre C. J. (1969) Comportement des systèmes U–Th–Pb dans le manteau supérieur et modèle d'évolution de ce dernier au cours de temps géologiques. *Earth Planet. Sci. Lett.* **5**, 261–269.

Allègre C. J. (1997) Limitation on the mass exchange between the upper and lower mantle: the evolving convection regime of the Earth. *Earth Planet. Sci. Lett.* **150**, 1–6.

Allègre C. J., Dupré B., and Brévart O. (1982) Chemical aspects of the formation of the core. *Phil Trans. Roy. Soc. Lond. A* **306**, 49–59.

Allègre C. J., Hart S. R., and Minster J. F. (1983) Chemical structure and evolution of the mantle and continents determined by inversion of Nd and Sr isotopic data. II: Numerical experiments and discussion. *Earth Planet. Sci. Lett.* **66**, 191–213.

Allègre C. J., Hofmann A. W., and O'Nions K. (1996) The argon constraints on mantle structure. *Geophys. Res. Lett.* **23**, 3555–3557.

Allègre C. J., Schiano P., and Lewin E. (1995) Differences between oceanic basalts by multitrace element ratio topology. *Earth Planet. Sci. Lett.* **129**, 1–12.

Allègre C. J. and Turcotte D. L. (1986) Implications of a two-component marble-cake mantle. *Nature* **323**, 123–127.

Armstrong R. L. (1981) A model for Sr and Pb isotope evolution in a dynamic earth. *Phil. Tans. Roy. Soc. Lond. A* **301**, 443–472.

Ben Othman D., White W. M., and Patchett J. (1989) The geochemistry of marine sediments, island arc magma genesis, and crust–mantle recycling. *Earth Planet. Sci. Lett.* **94**, 1–21.

Blichert-Toft J., Frey F. A., and Albarède F. (1999) Hf isotope evidence for pelagic sediments in the source of Hawaiian basalts. *Science* **285**, 879–882.

Blundy J. D. and Wood B. J. (1994) Prediction of crystal–melt partition coefficients from elastic moduli. *Nature* **372**, 452–454.

Bohrson W. A. and Reid M. R. (1995) Petrogenesis of alkaline basalts from Socorro Island, Mexico: trace element evidence for contamination of ocean island basalt in the shallow ocean crust. *J. Geophys. Res.* **100**(B12), 24555–24576.

Boyet M. and Carlson R. W. (2005) ^{142}Nd evidence for early (>4.53 Ga) global differentiation of the silicate Earth. *Science* **309**, 576–581.

Brandon A. D., Snow J. E., Walker R. J., Morgan J. W., and Mock T. D. (2000) Pt-190–Os-186 and Re-187–Os-187 systematics of abyssal peridotites. *Earth Planet. Sci. Lett.* **177**(3–4), 319–335.

Burton K. W., Schiano P., Birck J. L., Allègre C. J., and Rehkämper M. (2000) The distribution and behavior of Re and Os amongst mantle minerals and the consequences of metasomatism and melting on mantle lithologies. *Earth Planet. Sci. Lett.* **183**, 93–106.

Calvert A. J. and Ludden J. N. (1999) Archean continental assembly in the southeastern Superior Province of Canada. *Tectonics* **18**(3), 412–429.

Chase C. G. (1981) Oceanic island Pb: two-state histories and mantle evolution. *Earth Planet. Sci. Lett.* **52**, 277–284.

Chauvel C., Goldstein S. L., and Hofmann A. W. (1995) Hydration and dehydration of oceanic crust controls Pb evolution in the mantle. *Chem. Geol.* **126**, 65–75.

Chauvel C., Hofmann A. W., and Vidal P. (1992) HIMU-EM: the French Polynesian connection. *Earth Planet. Sci. Lett.* **110**, 99–119.

Christensen U. R. and Hofmann A. W. (1994) Segregation of subducted oceanic crust in the convecting mantle. *J. Geophys. Res.* **99**, 19867–19884.

Class C., Miller D. M., Goldstein S. L., and Langmuir C. H. (2000) Distinguishing melt and fluid subduction components in Umnak volcanics, Aleutian Arc. *Geochem. Geophys. Geosyst.* **1** #1999GC000010.

Creaser R. A., Papanastasiou D. A., and Wasserburg G. J. (1991) Negative thermal ion mass spectrometry of osmium, rhenium, and iridium. *Geochim. Cosmochim. Acta* **55**, 397–401.

Davies G. F. (1981) Earth's neodymium budget and structure and evolution of the mantle. *Nature* **290**, 208–213.

Davies G. F. (2002) Stirring geochemistry in mantle convection models with stiff plates and slabs. *Geochim. Cosmochim. Acta,* **66**, 3125–3142.

DePaolo D. J. (1980) Crustal growth and mantle evolution: inferences from models of element transport and Nd and Sr isotopes. *Geochim. Cosmochim. Acta,* **44**, 1185–1196.

DePaolo D. J. (1983) The mean life of continents: estimates of continental recycling rates from Nd and Hf isotopic data and implications for mantle structure. *Geophys. Re. Lett.* **10**, 705–708.

DePaolo D. J. and Wasserburg G. J. (1976) Nd isotopic variations and petrogenetic models. *Geophys. Res. Lett.* **3**, 249–252.

Donnelly K. E., Goldstein S. L., Langmuir C. H., and Spiegelman M. (2004) Origin of enriched ocean ridge basalts and implications for mantle dynamics. *Earth Planet. Sci. Lett.* **226**, 347–366.

Dupré B. and Allègre C. J. (1983) Pb–Sr isotope variation in Indian Ocean basalts and mixing phenomena. *Nature* **303**, 142–146.

Dupré B., Schiano P., Polvé M., and Joron J. L. (1994) Variability: a new parameter which emphasizes the limits of extended rare earth diagrams. *Bull. Soc. Geol. Francais* **165**(1), 3–13.

Eisele J., Sharma M., Galer S. J. G., Blichert-toft J., Devey C. W., and Hofmann A. W. (2002) The role of sediment recycling in EM-1 inferred from Os, Pb, Hf, Nd, Sr isotope and trace element systematics of the Pitcairn hotspot. *Earth Planet. Sci. Lett.* **196**, 197–212.

Elliott T., Plank T., Zindler A., White W., and Bourdon B. (1997) Element transport from slab to volcanic front at the Mariana Arc. *J. Geophys. Res.* **102**, 14991–15019.

Elliott T., Zindler A., and Bourdon B. (1999) Exploring the kappa conundrum: the role of recycling on the lead isotope evolution of the mantle. *Earth Planet. Sci. Lett.* **169**, 129–145.

Farley K. A., Natland J. H., and Craig H. (1992) Binary mixing of enriched and undegassed (primitive?) mantle components (He, Sr, Nd, Pb) in Samoan lavas. *Earth Planet. Sci. Lett.* **111**, 183–199.

Farnetani C. G., Legrasb B., and Tackley P. J. (2002) Mixing and deformations in mantle plumes. *Earth Planet. Sci. Lett.* **196**, 1–15.

Farquhar J., Bao H., and Thiemens M. H. (2000) Atmospheric influence of Earth's earliest sulfur cycle. *Science* **290**, 756–758.

Faure G. (1986) *Principles of Isotope Geology*. Wiley, New York.

Galer S. J. G. and Goldstein S. L. (1996) Influence of accretion on lead in the Earth. In *Earth Processes*: *Reading the Isotopic Code*, Geophys. Monograph. Am. Geophys. Union, Washington, vol. 95, pp. 75–98.

Galer S. J. G., Goldstein S. L., and O'Nions R. K. (1989) Limits on chemical and convective isolation in the Earth's interior. *Chem. Geol.* **75**, 252–290.

Galer S. J. G. and O'Nions R. K. (1985) Residence time of thorium, uranium and lead in the mantle with implications for mantle convection. *Nature* **316**, 778–782.

Gasperini G., Blichert-Toft J., Bosch D., Del Moro A., Macera P., Télouk P., and Albarède F. (2000) Evidence from Sardinian basalt geochemistry for recycling of plume heads into the Earth's mantle. *Nature* **408**, 701–704.

Gast P. W. (1960) Limitations on the composition of the upper mantle. *J. Geophys. Res.* **65**, 1287.

Gast P. W. (1968) Trace element fractionation and the origin of tholeiitic and alkaline magma types. *Geochim. Cosmochim. Acta* **32**, 1057–1086.

Gast P. W., Tilton G. R., and Hedge C. (1964) Isotopic composition of lead and strontium from Ascension and Gough Islands. *Science* **145**, 1181–1185.

Goldschmidt V. M. (1937) Geochemische Verteilungsgesetze der Elemente. IX: Die Mengenverhältnisse der Elemente and Atomarten. *Skr. Nor. vidensk.-Akad. Oslo* **1**, 148.

Goldschmidt V. M. (1954) *Geochemistry*. Clarendon, Oxford.

Graham D. W., Blichert-Toft J., Russo C. J., Rubin K. H., and Albarède F. (2006) Cryptic striations in the upper mantle revealed by hafnium isotopes in Southeast Indian Ridge basalts. *Nature* **440**, 199–202.

Grand S. P. (1994) Mantle shear structure beneath the Americas and surrounding oceans. *J. Geophys. Res.* **99**, 11591–11621.

Griffiths R. W. and Campbell I. H. (1990) Stirring and structure in mantle starting plumes. *Earth Planet. Sci. Lett.* **99**, 66–78.

Gurenko A. A. and Chaussidon M. (1995) Enriched and depleted primitive melts included in olivine from Icelandic tholeiites: origin by continuous melting of a single mantle column. *Geochim. Cosmochim. Acta* **59**, 2905–2917.

Haase K. M. (1996) The relationship between the age of the lithosphere and the composition of oceanic magmas: constraints on partial melting, mantle sources and the thermal structure of the plates. *Earth Planet. Sci. Lett.* **144**(1–2), 75–92.

Hanan B. B., Blichert-Toft J., Pyle D. G., and Christie D. M. (2004) Contrasting origins of the upper mantle revealed by hafnium and lead isotopes from the Southeast Indian Ridge. *Nature* **432**, 91–94.

Hanan B. and Graham D. (1996) Lead and helium isotope evidence from oceanic basalts for a common deep source of mantle plumes. *Science* **272**, 991–995.

Hanson G. N. (1977) Geochemical evolution of the suboceanic mantle. *J. Geol. Soc. Lond.* **134**, 235–253.

Hart S. R. (1971) K, Rb, Cs, Sr, Ba contents and Sr isotope ratios of ocean floor basalts. *Phil. Trans. Roy. Soc. Lond. Ser. A* **268**, 573–587.

Hart S. R. (1984) A large-scale isotope anomaly in the southern hemisphere mantle. *Nature* **309**, 753–757.

Hart S. R., Hauri E. H., Oschmann L. A., and Whitehead J. A. (1992) Mantle plumes and entrainment: isotopic evidence. *Science* **256**, 517–520.

Hart S. R., Schilling J. G., and Powell J. L. (1973) Basalts from Iceland and along the Reykjanes Ridge: Sr isotope geochemistry. *Nature* **246**, 104–107.

Hauri E. (2002) SIMS analysis of volatiles in silicate glasses. 2: Isotopes and abundances in Hawaiian melt inclusions. *Chem. Geol.* **183**, 115–141.

Hawkesworth C. J., Mantovani M. S. M., Taylor P. N., and Palacz Z. (1986) Evidence from the Parana of South Brazil for a continental contribution to Dupal basalts. *Nature* **322**, 356–359.

Hawkesworth C. J., Norry M. J., Roddick J. C., and Vollmer R. (1979) $^{143}Nd/^{144}Nd$ and $^{87}Sr/^{86}Sr$ ratios from the Azores and their significance in LIL-element enriched mantle. *Nature* **280**, 28–31.

Hedge C. E. and Walthall F. G. (1963) Radiogenic strontium 87 as an index of geological processes. *Science* **140**, 1214–1217.

Hellebrand E., Snow J. E., Dick H. J. B., and Hofmann A. W. (2001) Coupled major and trace elements as indicators of the extent of melting in mid-ocean-ridge peridotites. *Nature* **410**, 677–681.

Hemming S. R. and McLennan S. M. (2001) Pb isotope compositions of modern deep sea turbidites. *Earth Planet. Sci. Lett.* **184**, 489–503.

Hémond C., Hofmann A. W., Vlastelic I., and Nauret F. (2006) Origin of MORB enrichment and relative trace element compatibilities along the Mid-Atlantic Ridge between 10° and 24° N. *Geochem. Geophys. Geosys.* (in press).

Hirschmann M. M. and Stopler E. (1996) A possible role for garnet pyroxenite in the origin of the "garnet signature" in MORB. *Contrib. Mineral. Petrol.* **124**, 185–208.

Hofmann A. W. (1988) Chemical differentiation of the Earth: the relationship between mantle, continental crust, and oceanic crust. *Earth Planet. Sci. Lett.* **90**, 297–314.

Hofmann A. W. (1989a) Geochemistry and models of mantle circulation. *Phil. Trans. Roy. Soc. Lond. A*, **328**, 425–439.

Hofmann A. W. (1989b) A unified model for mantle plume sources. *EOS* **70**, 503.

Hofmann A. W. (1997) Mantle geochemistry: the message from oceanic volcanism. *Nature* **385**, 219–229.

Hofmann A. W. and Hart S. R. (1978) An assessment of local and regional isotopic equilibrium in the mantle. *Earth Planet. Sci. Lett.* **39**, 44–62.

Hofmann A. W., Jochum K. P., Seufert M., and White W. M. (1986) Nb and Pb in oceanic basalts: new constraints on mantle evolution. *Earth Planet. Sci. Lett.* **79**, 33–45.

Hofmann A. W. and White W. M. (1980) The role of subducted oceanic crust in mantle evolution. *Carnegie Inst. Wash. Year Book* **79**, 477–483.

Hofmann A. W. and White W. M. (1982) Mantle plumes from ancient oceanic crust. *Earth Planet. Sci. Lett.* **57**, 421–436.

Hofmann A. W. and White W. M. (1983) Ba, Rb, and Cs in the Earth's mantle. *Z. Naturforsch.* **38**, 256–266.

Hofmann A. W., White W. M., and Whitford D. J. (1978) Geochemical constraints on mantle models: the case for a layered mantle. *Carnegie Inst. Wash. Year Book* **77**, 548–562.

Holland H. D. (1994) Early Proterozoic atmospheric change. In *Early Life on Earth* (ed. S. Bengston). Columbia University Press, New York, pp. 237–244.

Hurley P. M., Hughes H., Faure G., Fairbairn H., and Pinson W. (1962) Radiogenic strontium-87 model for continent formation. *J. Geophys. Res.* **67**, 5315–5334.

Jackson I. N. S. and Rigden S. M. (1998) Composition and temperature of the mantle: seismological models interpreted through experimental studies of mantle minerals. In *The Earth's Mantle: Composition, Structure and Evolution* (ed. I. N. S. Jackson). Cambridge University Press, Cambridge, pp. 405–460.

Jacobsen S. B. and Wasserburg G. J. (1979) The mean age of mantle and crustal reservoirs. *J. Geophys. Res.* **84**, 7411–7427.

Jochum K. P. and Hofmann A. W. (1994) Antimony in mantle-derived rocks: constraints on Earth evolution from moderately siderophile elements. *Min. Mag.* **58A**, 452–453.

Jochum K. P., Hofmann A. W., Ito E., Seufert H. M., and White W. M. (1983) K, U, and Th in mid-ocean ridge basalt glasses and heat production, K/U, and K/Rb in the mantle. *Nature* **306**, 431–436.

Jochum K. P., Hofmann A. W., and Seufert H. M. (1993) Tin in mantle-derived rocks: constraints on Earth evolution. *Geochim. Cosmochim. Acta* **57**, 3585–3595.

Jochum K. P., Seufert H. M., Spettel B., and Palme H. (1986) The solar system abundances of Nb, Ta, and Y. and the relative abundances of refractory lithophile elements in differentiated planetary bodies. *Geochim. Cosmochim. Acta* **50**, 1173–1183.

Johnson K. T. M., Dick H. J. B., and Shimizu N. (1990) Melting in the oceanic upper mantle: an ion microprobe study of diopsides in abyssal peridotites. *J. Geophys. Res.* **95**, 2661–2678.

Johnson M. C. and Plank T. (2000) Dehydration and melting experiments constrain the fate of subducted sediments. *Geochem. Geophys. Geosys.* **1**, 1999GC000014.

Kellogg L. H., Hager B. H., and Van der Hilst R. D. (1999) Compositional stratification in the deep mantle. *Science* **283**, 1881–1884.

Kempton P. D., Pearce J. A., Barry T. L., Fitton J. G., Langmuir C. H., and Christie D. M. (2002) Sr–Nd–Pb–Hf isotope results from ODP Leg 187: evidence for mantle dynamics of the Australian-Antarctic Discordance and origin of the Indian MORB source. *Geochem. Geophys. Geosys.* **3**, 10.29/2002GC000320.

Kimura G., Ludden J. N., Desrochers J. P., and Hori R. A. (1993) Model of ocean-crust accretion for the superior province, Canada. *Lithos* **30**(3–4), 337–355.

Klein E. M. and Karsten J. L. (1995) Ocean-ridge basalts with convergent-margin geochemical affinities from the Chile Ridge. *Nature* **374**, 52–57.

Klein E. M. and Langmuir C. H. (1987) Global correlations of ocean ridge basalt chemistry with axial depth and crustal thickness. *J Geophys. Res.* **92**, 8089–8115.

Klein E. M., Langmuir C. H., Zindler A., Staudigel H., and Hamelin B. (1988) Isotope evidence of a mantle convection boundary at the Australian–Antarctic discordance. *Nature* **333**, 623–629.

Kleine T., Munker C., Mezger K., and Palme H. (2002) Rapid accretion and early core formation on asteroids and the terrestrial planets from Hf–W chronometry. *Nature* **418**, 952–955.

Kramers J. D. and Tolstikhin I. N. (1997) Two terrestrial lead isotope paradoxes, forward transport modelling, core formation and the history of the continental crust. *Chem. Geol.* **139** (1–4), 75–110.

Kuhn T. S. (1996) *The Structure of Scientific Revolutions.* University of Chicago Press, Chicago.

Lassiter J. C. (2002) The influence of recycled oceanic crust on the potassium and argon budget of the Earth. *Geochim. Cosmochim. Acta* **66**(15A), A433–A434 (suppl.).

Lee D. C. and Halliday A. N. (1995) Hafnium–tungsten chronometry and the timing of terrestrial core formation. *Nature* **378**, 771–774.

le Roux P. J., le Roex A. P., and Schilling J. G. (2002) MORB melting processes beneath the southern Mid-Atlantic Ridge (40–55 degrees S): a role for mantle plume-derived pyroxenite. *Contrib. Min. Pet.* **144**, 206–229.

Mahoney J., Nicollet C., and Dupuy C. (1991) Madagascar basalts: tracking oceanic and continental sources. *Earth Planet. Sci. Lett.* **104**, 350–363.

Martin C. E. (1991) Osmium isotopic characteristics of mantle-derived rocks. *Geochim. Cosmochim. Acta* **55**, 1421–1434.

McDonough W. F. and Sun S. S. (1995) The composition of the Earth. *Chem. Geol.* **120**, 223–253.

McDonough W. F., Sun S. S., Ringwood A. E., Jagoutz E., and Hofmann A. W. (1992) Potassium, rubidium, and cesium in the Earth and Moon and the evolution of the mantle of the Earth. *Geochim. Cosmochim. Acta* **56**, 1001–1012.

McKenzie D. and O'Nions R. K. (1983) Mantle reservoirs and ocean island basalts. *Nature* **301**, 229–231.

McKenzie D., Stracke A., Blichert-Toft J., Albarède F., Grönvold K., and O'Nions R. K. (2004) Source enrichment processes responsible for isotopic anomalies in oceanic island basalts. *Geochim. Cosmochim. Acta* **68**, 2699–2724.

Meibom A., Sleep N. H., Chamberlain C. P., Coleman R. G., Frei R., Hren M. T., and Wooden J. L. (2002) Re–Os isotopic evidence for long-lived heterogeneity and equilibration processes in the Earth's upper mantle. *Nature* **419** (6908), 705–708.

Menzies M. A. and Hawkesworth C. J. (1987) *Mantle Metasomatism.* Academic Press, New YorkAcademic Press.

Miller D. M., Goldstein S. L., and Langmuir C. H. (1994) Cerium/lead and lead isotope ratios in arc magmas and the enrichment of lead in the continents. *Nature* **368**, 514–520.

Milner S. C. and le Roex A. P. (1996) Isotope characteristics of the Okenyena igneous complex, northwestern Namibia: constraints on the composition of the early Tristan plume and the origin of the EM 1 mantle component. *Earth Planet. Sci. Lett.* **141**, 277–291.

Montelli R., Nolet G., Masters G., Dahlen F. A., and Hung S. H. (2003) Global P and PP traveltime tomography: rays versus waves. *Geophys. J. Int.* **142** (in press).

Morgan W. J. (1971) Convection plumes in the lower mantle. *Nature* **230**, 42–43.

Morris E. M. and Pasteris J. D. (eds.) (1987) *Mantle Metasomatism and Alkaline Magmatism*, Special Paper No. 215. Geol. Soc. America.

Morse S. A. (1983) Strontium isotope fractionation in the Kiglapait intrusion. *Science* **220**, 193–195.

Murphy D. T., Kamber B. S., and Collerson K. D. (2003) A refined solution to the first terrestrial Pb-isotope paradox. *J. Petrol.* **44**, 39–53.

O'Hara M. J. (1973) Non-primary magmas and dubious mantle plume beneath Iceland. *Nature* **243**(5409), 507–508.

O'Hara M. J. (1977) Open system crystal fractionation and incompatible element variation in basalts. *Nature* **268**, 36–38.

O'Hara M. J. and Mathews R. E. (1981) Geochemical evolution in an advancing, periodically replenished, periodically tapped, continuously fractionated magma chamber. *J. Geol. Soc. Lond.* **138**, 237–277.

O'Nions R. K., Evensen N. M., and Hamilton P. J. (1977) Variations in $^{143}Nd/^{144}Nd$ and $^{87}Sr/^{86}Sr$ ratios in oceanic basalts. *Earth Planet. Sci. Lett.* **34**, 13–22.

O'Nions R. K., Evensen N. M., and Hamilton P. J. (1979) Geochemical modeling of mantle differentiation and crustal growth. *J. Geophys. Res.* **84**, 6091–6101.

O'Nions R. K. and Pankhurst R. J. (1974) Petrogenetic significance of isotope and trace element variation in volcanic rocks from the Mid-Atlantic. *J. Petrol.* **15**, 603–634.

Patchett P. J., White W. M., Feldmann H., Kielinczuk S., and Hofmann A. W. (1984) Hafnium/rare earth fractionation in the sedimentary system and crust–mantle recycling. *Earth Planet. Sci. Lett.* **69**, 365–378.

Peucker-Ehrenbrink B., Hofmann A. W., and Hart S. R. (1994) Hydrothermal lead transfer from mantle to continental crust: the role of metalliferous sediments. *Earth Planet. Sci. Lett.* **125**, 129–142.

Petrological Database of the Ocean Floor (PETDB), http://www.petdb.org.

Phipps Morgan J. (1999) Isotope topology of individual hotspot basalt arrays: mixing curves of melt extraction trajectories? Geochem. *Geophys. Geosys.* **1**, 1999GC000004.

Phipps Morgan J. and Morgan W. J. (1998) Two-stage melting and the geochemical evolution of the mantle: a recipe for mantle plum-pudding. *Earth Planet. Sci. Lett.* **170**, 215–239.

Phipps Morgan J., Morgan W. J., and Zhang Y. S. (1995) Observational hints for a plume-fed, suboceanic asthenosphere and its role in mantle convection. *J. Geophys. Res.* **100**, 12753–12767.

Plank T. and Langmuir C. H. (1998) The chemical composition of subducting sediment and its consequences for the crust and mantle. *Chem. Geol.* **145**, 325–394.

Porcelli D. and Halliday A. N. (2001) The core as a possible source of mantle helium. *Earth Planet. Sci. Lett.* **192**(1), 45–56.

Puchtel I. S., Hofmann A. W., Mezger K., Jochum K. P., Shchipansky A. A., and Samsonov A. V. (1998) Oceanic plateau model for continental crustal growth in the Archaean: a case study from the Kostomuksha greenstone belt, NW Baltic Shield. *Earth Planet. Sci. Lett.* **155**, 57–74.

Pyle D. G., Christie D. M., and Mahoney J. J. (1992) Resolving an isotopic boundary within the Australian–Antarctic Discordance. *Earth Planet. Sci. Lett.* **112**, 161–178.

Regelous M., Hofmann A. W., Abouchami W., and Galer J. S. G. (2002/2003) Geochemistry of lavas from the

Emperor Seamounts, and the geochemical evolution of Hawaiian magmatism 85–42Ma. *J. Petrol.* **44**, 113–140.
Rehkämper M. and Hofmann A. W. (1997) Recycled ocean crust and sediment in Indian Ocean MORB. *Earth Planet. Sci. Lett.* **147**, 93–106.
Richard P., Shimizu N., and Allègre C. J. (1976) $^{143}Nd/^{146}Nd$—a natural tracer: an application to oceanic basalt. *Earth Planet. Sci. Lett.* **31**, 269–278.
Roy-Barman M. and Allègre C. J. (1994) $^{187}Os/^{186}Os$ ratios of mid-ocean ridge basalts and abyssal peridotites. *Geochim. Cosmochim. Acta* **58**, 5053–5054.
Rudnick R. L., Barth M., Horn I., and McDonough W. F. (2000) Rutile-bearing refractory eclogites: missing link between continents and depleted mantle. *Science* **287**, 278–281.
Rudnick R. L. and Fountain D. M. (1995) Nature and composition of the continental crust: a lower crustal perspective. *Rev. Geophys.* **33**, 267–309.
Rudnick R. L. and Goldstein S. L. (1990) The Pb isotopic compositions of lower crustal xenoliths and the evolution of lower crustal Pb. *Earth Planet. Sci. Lett.* **98**, 192–207.
Saal A. E., Hart S. R., Shimizu N., Hauri E. H., and Layne G. D. (1998) Pb isotopic variability in melt inclusions from oceanic island basalts, Polynesia. *Science* **282**, 1481–1484.
Saunders A. D., Norry M. J., and Tarney J. (1988) Origin of MORB and chemically-depleted mantle reservoirs: trace element constraints. *J. Petrol.* Special Lithosphere Issue, 415–445.
Schiano P., Birck J. L., and Allègre C. J. (1997) Osmium–strontium–neodymium–lead isotopic covariations in mid-ocean ridge basalt glasses and the heterogeneity of the upper mantle. *Earth Planet. Sci. Lett.* **150**, 363–379.
Schilling J. G. (1973) Iceland mantle plume: geochemical evidence along Reykjanes Ridge. *Nature* **242**, 565–571.
Schilling J. G. and Winchester J. W. (1967) Rare-earth fractionation and magmatic processes. In *Mantles of Earth and Terrestrial Planets* (ed. S. K. Runcorn). Interscience Publishers, London, pp. 267–283.
Schilling J. G. and Winschester J. W. (1969) Rare earth contribution to the origin of Hawaiian lavas. *Contrib. Mineral. Petrol.* **40**, 231.
Shaw D. M. (1970) Trace element fractionation during anatexis. *Geochim. Cosmochim. Acta* **34**, 237–242.
Shimizu N., Sobolev A. V., Layne G. D., and Tsameryan, O. P. (2003) Large Pb isotope variations in olivine-hosted melt inclusions in a basalt fromt the Mid-Atlantic Ridge. *Science* (submitted May 2003).
Sims K. W. W. and DePaolo D. J. (1997) Inferences about mantle magma sources from incompatible element concentration ratios in oceanic basalts. *Geochim. Cosmochim. Acta* **61**, 765–784.
Sleep N. H. (1984) Tapping of magmas from ubiquitous mantle heterogeneities: an alternative to mantle plumes? *J. Geophys. Res.* **89**, 10029–10041.
Snow J. E. and Reisberg L. (1995) Os isotopic systematics of the MORB mantle: results from altered abyssal peridotites. *Earth Planet. Sci. Lett.* **133**, 411–421.
Sobolev A. V. (1996) Melt inclusions in minerals as a source of principal petrological information. *Petrology* **4**, 209–220.
Sobolev A. V., Hofmann A. W., and Nikogosian I. K. (2000) Recycled oceanic crust observed in "ghost plagioclase" within the source of Mauna Loa lavas. *Nature* **404**, 986–990.
Sobolev A. V., Hofmann A. W., Sobolev S. V., and Nikogosian I. K. (2005) An online-free mante source of Hawaiian shield basalts. *Nature* **434**, 590–597.
Sobolev A. V. and Shimizu N. (1993) Ultra-depleted primary melt included in an olivine from the Mid-Atlantic Ridge. *Nature* **363**, 151–154.
Staudigel H., Davies G. R., Hart S. R., Marchant K. M., and Smith B. M. (1995) Large scale isotopic Sr, Nd and O isotopic anatomy of altered oceanic crust: DSDP/ODP sites 417/418. *Earth Planet. Sci. Lett.* **130**, 169–185.
Stein M. and Hofmann A. W. (1994) Mantle plumes and episodic crustal growth. *Nature* **372**, 63–68.
Su Y. J. (2002) Mid-ocean ridge basalt trace element systematics: constraints from database management, ICP-MS analyses, global data compilation and petrologic modeling. PhD Thesis, Columbia University, 472pp.
Sun S. S. and Hanson G. N. (1975) Origin of Ross Island basanitoids and limitation upon the heterogeneity of mantle sources for alkali basalts and nephelinites. *Contrib. Mineral. Petrol.* **52**, 77–106.
Sun S. S. and McDonough W. F. (1989) Chemical and isotopic systematics of oceanic basalts: implications for mantle composition and processes. In *Magmatism in the Ocean Basins* (eds. A. D. Saunders and M. J. Norry). Geological Society Spec. Publ., Oxford, vol. 42, pp. 313–345.
Tackley P. J. (2000) Mantle convection and plate tectonics: toward an integrated physical and chemical theory. *Science* **288**, 2002–2007.
Tackley P. J. (2002) Strong heterogeneity caused by deep mantle layering. *Geochem. Geophys. Geosys.* **3**(4)101029/2001GC000167.
Tatsumoto M., Hedge C. E., and Engel A. E. J. (1965) Potassium, rubidium, strontium, thorium, uranium, and the ratio of strontium-87 to strontium-86 in oceanic tholeiitic basalt. *Science* **150**, 886–888.
Taylor S. R. and McLennan S. M. (1985) *The Continental Crust: Its Composition and Evolution*. Oxford: Blackwell.
Thompson R. N., Morrison M. A., Hendry G. L., and Parry S. J. (1984) An assessment of the relative roles of crust and mantle in magma genesis: an elemental approach. *Phil. Trans. Roy. Soc. Lond.* **A310**, 549–590.
Tolstikhin I. N. and Hofmann A. W. (2005) Early crust on top of the Earth's core. *Phys. Earth Planet. Interiors* **148**, 109–130.
Tolstikhin I. N., Kramers J., and Hofmann A. W. (2006) A chemical Earth model with whole mantle convection: the importance of a core-mantle boundary layer (D") and its early formation. *Chemical Geol.* **226**, 79–99.
van der Hilst R. D., Widiyantoro S., and Engdahl E. R. (1997) Evidence for deep mantle circulation from global tomography. *Nature* **386**, 578–584.
van Keken P. E. and Ballentine C. J. (1998) Whole-mantle versus layered mantle convection and the role of a high-viscosity lower mantle in terrestrial volatile evolution. *Earth Planet. Sci. Lett.* **156**, 19–32.
van Keken P. E., Ballentine C. J., and Porcelli D. (2001) A dynamical investigation of the heat and helium imbalance. *Earth Planet. Sci. Lett.* **188**(3–4), 421–434.
van Orman J. A., Grove T. L., and Shimizu N. (2001) Rare earth element diffusion in diopside: influence of temperature, pressure and ionic radius, and an elastic model for diffusion in silicates. *Contrib. Mineral. Petrol.* **141**, 687–703.
Vervoort J. D., Patchett P. J., Blichert-Toft J., and Albarède F. (1999) Relationships between Lu–Hf and Sm–Nd isotopic systems in the global sedimentary system. *Earth Planet. Sci. Lett.* **168**, 79–99.
Völkening J., Walczyk T., and Heumann K. G. (1991) Osmium isotope ratio determinations by negative thermal ionization mass spectrometry. *Int. J. Mass Spectrom. Ion Process.* **105**, 147–159.
Vollmer R. (1976) Rb–Sr and U–Th–Pb systematics of alkaline rocks: the alkaline rocks from Italy. *Geochim. Cosmochim. Acta* **40**, 283–295.
Wade J. and Wood B. J. (2001) The Earth's "missing" niobium may be in the core. *Nature* **409**, 75–78.
Wasserburg G. J. and Depaolo D. J. (1979) Models of Earth structure inferred from neodymium and strontium isotopic abundances. *Proc. Natl. Acad. Sci. USA* **76**(8), 3594–3598.
Weaver B. L. (1991) The origin of ocean island basalt end-member compositions: trace element and isotopic constraints. *Earth Planet. Sci. Lett.* **104**, 381–397.

White W. M. (1985) Sources of oceanic basalts: radiogenic isotope evidence. *Geology* **13**, 115–118.

White W. M. (1993) $^{238}U/^{204}Pb$ in MORB and open system evolution of the depleted mantle. *Earth Planet. Sci. Lett.* **115**, 211–226.

White W. M. and Duncan R. A. (1996) Geochemistry and geochronology of the Society Islands: new evidence for deep mantle recycling. In *Earth Processes: Reading the Isotopic Code. Geophys. Monograph Earth Processes: Reading the Isotopic Code. Geophys. Monograph* (eds. A. Basu and S. R. Hart). Am. Geophys. Union, Washington, vol. 95, pp. 183–206.

White W. M. and Hofmann A. W. (1982) Sr and Nd isotope geochemistry of oceanic basalts and mantle evolution. *Nature* **296**, 821–825.

White W. M. and Schilling J. G. (1978) The nature and origin of geochemical variation in Mid-Atlantic Ridge basalts from the central North Atlantic. *Geochim. Cosmochim. Acta* **42**, 1501–1516.

Widom E. and Shirey S. B. (1996) Os isotope systematics in the Azores: implications for mantle plume sources. *Earth Planet. Sci. Lett.* **142**, 451–465.

Willbold M. and Stracke A. (2006) The trace element composition of mantle end-members: implications for recycling of oceanic and upper/lower continental crust. *Geochem. Geophys. Geosys.* (in press).

Wood D. A. (1979) A variably veined suboceanic mantle—genetic significance for mid-ocean ridge basalts from geochemical evidence. *Geology* **7**, 499–503.

Woodhead J. (1996) Extreme HIMU in an oceanic setting: the geochemistry of Mangaia Island (Polynesia), and temporal evolution of the Cook-Austral hotspot. *J. Volcanol. Geotherm. Res.* **72**, 1–19.

Workman R. K., Hart S. R., Blusztajn J., Jackson M., Kurz M., and Staudigel H. (2003) Enriched mantle. II: A new view from the Samoan hotspot. *Geochem. Geophys. Geosys.* (submitted).

Yang H. J., Frey F. A., and Clague D. A. (2003) Constraints on the source components of lavas forming the Hawaiian north arch and honolulu volcanics. *J. Petrol.* **44**, 603–627.

Yaxley G. M. and Green D. H. (1998) Reactions between eclogite and peridotite: mantle refertilisation by subduction of oceanic crust. *Schweiz. Mineral. Petrogr. Mitt.* **78**, 243–255.

Yin Q., Jacobsen S. B., Yamashita K., Blichert-Toft J., Télouk P., and Albarède F. (2002) A short timescale for terrestrial planet formation from Hf–W chronometry of meteorites. *Nature* **418**, 949–952.

Zartman R. E. and Haines S. M. (1988) The plumbotectonic model for Pb isotopic systematics among major terrestrial reservoirs—a case for bi-directional transport. *Geochim. Cosmochim. Acta* **52**, 1327–1339.

Zindler A. and Hart S. (1986) Chemical geodynamics. *Ann. Rev. Earth Planet. Sci.* **14**, 493–571.

Isotope Geochemistry
ISBN: 978-0-08-096710-3

pp. 563–606

19
Radiogenic Isotopes in Weathering and Hydrology

J. D. Blum
The University of Michigan, Ann Arbor, MI, USA
and
Y. Erel
The Hebrew University, Jerusalem, Israel

19.1 INTRODUCTION AND OVERVIEW OF RELEVANT ISOTOPE SYSTEMS

There are a small group of elements that display variations in their isotopic composition, resulting from radioactive decay within minerals over geological timescales. These isotopic variations provide natural fingerprints of rock–water interactions and have been widely utilized in studies of weathering and hydrology. The isotopic systems that have been applied in such studies are dictated by the limited number of radioactive parent–daughter nuclide pairs with half-lives and isotopic abundances that result in measurable differences in daughter isotope ratios among common rocks and minerals. Prior to their application to studies of weathering and hydrology, each of these isotopic systems was utilized in geochronology and petrology. As in the case of their original introduction into geochronology and petrology, isotopic systems with the highest concentrations of daughter isotopes in common rocks and minerals and systems with the largest observed isotopic variations were introduced first and have made the largest impact on our understanding of weathering and hydrologic processes. Although radiogenic isotopes have helped elucidate many important aspects of weathering and hydrology, it is important to note that in almost every case that will be discussed in this chapter, our fundamental understanding of these topics came from studies of variations in the concentrations of major cations and anions. This chapter is a "tools chapter" and thus it will highlight applications of radiogenic isotopes that have added additional insight into a wide spectrum of research areas that are summarized in almost all of the other chapters of this volume.

The first applications of radiogenic isotopes to weathering processes were based on studies that sought to understand the effects of chemical weathering on the geochronology of whole-rock samples and geochronologically important minerals (Goldich and Gast, 1966; Dasch, 1969; Blaxland, 1974; Clauer, 1979, 1981; Clauer *et al.*, 1982); as well as on the observation that radiogenic isotopes are sometimes preferentially released compared to nonradiogenic isotopes of the same element during acid leaching of rocks (Hart and Tilton, 1966; Silver *et al.*, 1984; Erel *et al.*, 1991). A major finding of these investigations was that weathering often results in anomalously young Rb–Sr isochron ages, and discordant Pb–Pb ages. Rubidium is generally retained relative to strontium in whole-rock samples, and in some cases radiogenic strontium and lead are lost preferentially to common strontium and lead from weathered minerals.

The most widely utilized of these isotopic systems is Rb–Sr, followed by U–Pb. The K–Ar system is not directly applicable to most studies of rock–water interaction, because argon is a noble gas, and upon release during mineral weathering mixes with atmospheric argon, limiting its usefulness as a tracer in most weathering applications. Argon and other noble gas isotopes have, however, found important applications in hydrology. Three other isotopic systems commonly used in geochronology and petrology include Sm–Nd, Lu–Hf, and Re–Os. These parent and daughter elements are in very low abundance and concentrated in trace mineral phases. Sm–Nd, Lu–Hf, and Re–Os have been used in a few weathering studies but have not been utilized extensively in investigations of weathering and hydrology.

The decay of ^{87}Rb to ^{87}Sr has a half-life of 48.8 Gyr, and this radioactive decay results in natural variability in the ^{87}Sr/^{86}Sr ratio in rubidium-bearing minerals (e.g., Blum, 1995). The trace elements rubidium and strontium are geochemically similar to the major elements potassium and calcium, respectively. Therefore, minerals with high K/Ca ratios develop high ^{87}Sr/^{86}Sr ratios over geologic timescales. Once released into the hydrosphere, strontium retains its isotopic composition without significant fractionation by geochemical or biological processes, and is therefore a good tracer for sources and cycling of calcium. The decay of ^{235}U to ^{207}Pb, ^{238}U to ^{206}Pb, and ^{232}Th to ^{208}Pb have half-lives of 0.704 Gyr, 4.47 Gyr, and 14.0 Gyr, respectively, and result in variations in the ^{207}Pb/^{204}Pb, ^{206}Pb/^{204}Pb, and ^{208}Pb/^{204}Pb ratios (e.g., Blum, 1995). Uranium-234 has a half-life of 0.25 Myr and the ratio ^{234}U/^{238}U approaches a constant secular equilibrium value in rocks and minerals if undisturbed for $\sim$1 Myr. Differences in this ratio are often observed in solutions following rock–water interaction and have been used in studies of weathering and hydrology. Uranium and thorium tend to be highly concentrated in the trace accessory minerals such as zircon, monazite, apatite, and sphene, which therefore develop high ^{206}Pb/^{204}Pb, ^{207}Pb/^{204}Pb, and ^{208}Pb/^{204}Pb ratios.

Once released into the hydrosphere, lead retains its isotopic composition without significant geochemical or biological fractionation and tends to generally follow the chemistry of iron in soils and aqueous systems (Erel and Morgan, 1992). The use of the U–Th disequilibrium series as a dating tool falls outside the scope of this chapter and is reviewed in Chapters 8 and 20. The decay of ^{147}Sm to ^{143}Nd, ^{176}Lu to ^{176}Hf, and ^{187}Re to ^{187}Os have half-lives of 106 Gyr, 35.7 Gyr, and 42.3 Gyr, respectively, and result in natural variability in the ^{144}Nd/^{143}Nd, ^{176}Hf/^{177}Hf, and ^{187}Os/^{188}Os ratios (e.g., Blum, 1995). Neodymium is a rare earth element (REE), hafnium is a transition metal with chemical similarities to zirconium, and osmium is a platinum group element. The geochemical behaviors of these elements in the hydrosphere are largely determined by these chemical affinities.

19.2 ANALYTICAL METHODS, MIXING EQUATIONS, AND PREVIOUS REVIEWS

The analytical methods and mixing equations used in studies of radiogenic isotopes in weathering and hydrology followed directly from research applications in geochronology, petrology, and cosmochemistry that utilize the same isotopic systems. Analytical methods are thoroughly reviewed in several textbooks on isotope geology, including Faure (1986), Geyh and Schleicher (1990), and Dickin (1995). Mixing equations are reviewed in Faure (1986), Dickin (1995), and Albarède (1995). Several recent books and review articles have summarized some of the analytical methods and mixing equations as they apply specifically to studies of weathering and hydrology (Stille and Shields, 1997; Kendall and Caldwell, 1998; Bullen and Kendall, 1998; Kraemer and Genereux, 1998; Capo *et al.*, 1998; Nimz, 1998; Bierman *et al.*, 1998; Stewart *et al.*, 1998). The reader is referred to these previous publications for analytical methods and mixing equations relevant to radiogenic isotope studies. This chapter reviews applications of the methods to studies of weathering and hydrology.

19.3 OVERVIEW OF APPLICATIONS AND ORGANIZATION OF CHAPTER

The most significant scientific contributions of radiogenic isotopes to our understanding of weathering and hydrology can be organized into the three general subject areas: (i) identification of mineral dissolution reactions; (ii) differentiation between atmospheric- and weathering-derived cations in ecosystems; and (iii) tracing of hydrologic flow paths and subsurface mixing. Radiogenic isotopes provide a useful tracer of which mineral(s) are dissolving in mixed-mineral weathering experiments, soil-weathering profiles, and in surface and groundwaters. Radiogenic isotopes have been particularly useful at differentiating between atmospheric- and weathering-derived inputs of base cations and toxic metals to soils and in the study of their cycling in ecosystems. Radiogenic isotopes have also found extensive applications in hydrology, allowing the isotopic characterization of various water sources, flow paths, and mixing.

Applications of radiogenic isotopes specifically to weathering and hydrology amounted to only a few dozen studies prior to 1990. During the 1990s and early 2000s, radiogenic isotopes were applied to a wide range of investigations and the number of publications expanded to well over two hundred at the time of this writing. In this chapter we will provide an overview of the breadth of applications of radiogenic isotopes, as well as a comprehensive review of the peer-reviewed literature on this topic; abstracts and conference proceedings are generally excluded from inclusion in this chapter. We have highlighted a few studies in each section and reproduced key figures from the literature that provide particularly good examples of how radiogenic isotopes have been used to elucidate geochemical and hydrologic processes.

19.4 IDENTIFICATION OF SPECIFIC MINERAL WEATHERING REACTIONS

Mineral dissolution is the process by which minerals react with near-surface water and re-equilibrate thermodynamically to the conditions present near the Earth's surface. Weathering moderates the concentration of CO_2 in the atmosphere, releases nutrients, provides cation-exchange capacity in soils, and largely determines the inorganic element geochemistry of surface and groundwaters. Weathering rates have been determined following three general methodologies: (i) laboratory dissolution experiments.

19.4.1 Laboratory Dissolution Experiments

We begin our discussion of the use of radiogenic isotopes in weathering studies with laboratory experiments, because they provide the simplest and most well-constrained application. Several research groups have integrated radiogenic isotope techniques into these studies, yielding important new information on: (i) the importance of accessory mineral inclusions in what were previously considered "pure" mineral starting materials; (ii) the differential rate of release of cations from contrasting mineralogic sites within single minerals; and (iii) the weathering rate of individual minerals in multi-mineral experiments.

19.4.1.1 Strontium isotopes

As far as we are aware, the first laboratory experiments that utilized radiogenic isotopes in the study of mineral dissolution were performed by Zuddas *et al.* (1995) and Seimbille *et al.* (1998). Most mineral dissolution experiments are performed at 25 °C, whereas these experiments were performed at 180 °C and thus could be considered "hydrothermal" experiments. In these experiments powdered granite was reacted with an artificial "hydrothermal fluid" containing major element concentrations near equilibrium with the granitic mineral assemblage. Combining strontium isotopic and major element mixing equations, Seimbille *et al.* (1998) demonstrated that plagioclase dissolved 3–4 times faster than biotite and 10–20 times faster than K-feldspar under these conditions. White *et al.* (1999) performed similar experiments using powdered granite in a flow-through column with deionized water at 5 °C, 17 °C, and 35 °C. Strontium isotope ratios were used to demonstrate that at the lower temperatures strontium release to solution from reacting minerals was not congruent with respect to major cation concentrations. It was proposed that this was due to either the extreme sensitivity of strontium isotope ratios to the dissolution of trace calcium-bearing phases with low $^{87}Sr/^{86}Sr$, the preferential retention of strontium relative to potassium in biotite, or the preferential leaching of strontium with low $^{87}Sr/^{86}Sr$ from plagioclase or K-feldspar. This study suggests that $^{87}Sr/^{86}Sr$ ratios derived from granitoid weathering may be temperature dependent, with lower ratios in warmer climates (White *et al.*, 1999; Figure 1).

The first single-mineral dissolution experiments to utilize radiogenic isotopes investigated the dissolution of the feldspars bytownite, microcline, and albite in flow-through cells at a pH 3 and at 25 °C (Brantley *et al.*, 1998). Solutions from major element experiments (Stillings and Brantley, 1995) were reanalyzed for strontium and rubidium concentrations and $^{87}Sr/^{86}Sr$ ratios with the goal of comparing strontium release with major element patterns. Strontium release early in the experiments was found to be neither stoichiometric nor isotopically identical to the bulk mineral and was attributed to the dissolution of mineral inclusions in the feldspars and/or to leaching of cations from mineral defect sites. Once the experiments had reached steady-state dissolution, the $^{87}Sr/^{86}Sr$ ratio became equal to that of the bulk material. This study demonstrated the importance of minute trace phases in controlling the $^{87}Sr/^{86}Sr$ released by weathering of freshly exposed minerals.

Taylor *et al.* (2000a,b) used flow-through column reactors to study overall dissolution kinetics and strontium release from biotite, phlogopite, and labradorite at 25 °C. Strontium isotopes elucidated several processes including: (i) the importance of trace calcite inclusions in biotite and phlogopite during the early stages of weathering (Figure 2); (ii) the more rapid release of interlayer cations from biotite and phlogopite compared to octahedral and tetrahedral cations; and (iii) that labradorite dissolution is congruent, with stoichiometric strontium release unaffected by solution saturation state or preferential release from mineral defect sites. A common theme of all these experimental studies is that trace inclusions of calcite and other nonsilicate phases are common in silicate mineral specimens, and strontium isotopes are very sensitive to their presence. In the early stages of weathering, the $^{87}Sr/^{86}Sr$ of weathering solutions may be significantly affected by reactive trace phases and differ from bulk mineral values. However, once a steady state is reached, the dissolution of calcite inclusions is limited by the rate at which the inclusions are exposed to solution by dissolution of the host silicate mineral, and their influence is greatly diminished.

19.4.1.2 Lead isotopes

The behavior of lead isotopes during the dissolution of silicate minerals has not been extensively studied in laboratory experiments. To the best of our knowledge, there has been only one attempt to study systematically the release of lead isotopes during the dissolution of silicates. Harlavan and Erel (2002) studied the release of lead and REEs during the dissolution of a crushed granite subjected to sequential leaching with dilute acids, and the results were compared with lead and REE data from soils developed on the same bedrock (Harlavan *et al.*, 1998; Harlavan and Erel, 2002). During the early stages of granitoid dissolution, lead and REEs were preferentially released from some of the accessory phases (i.e., allanite, sphene, and apatite), leading to higher $^{206}Pb/^{207}Pb$ and $^{208}Pb/^{207}Pb$ ratios and different REE patterns in solution than in the rock itself. Three stages of rock dissolution were identified (Figure 3). Among the accessory phases, allanite dissolution dominated the first stage and $^{208}Pb/^{207}Pb$ ratios in solution increased and approached the values of allanite. The isotopic ratios of lead and the REE patterns indicated that in the second stage, dissolution of apatite and sphene became more significant. In the third stage, the isotopic ratios of lead and the REE patterns reflected the depletion of accessory phases and the increasing dominance of feldspar dissolution. It was also argued in this study that biotite dissolution was significantly more rapid than hornblende dissolution under the experimental conditions of acid leaching.

19.4.1.3 Uranium isotopes

The release of uranium isotopes during the dissolution of rocks under laboratory conditions

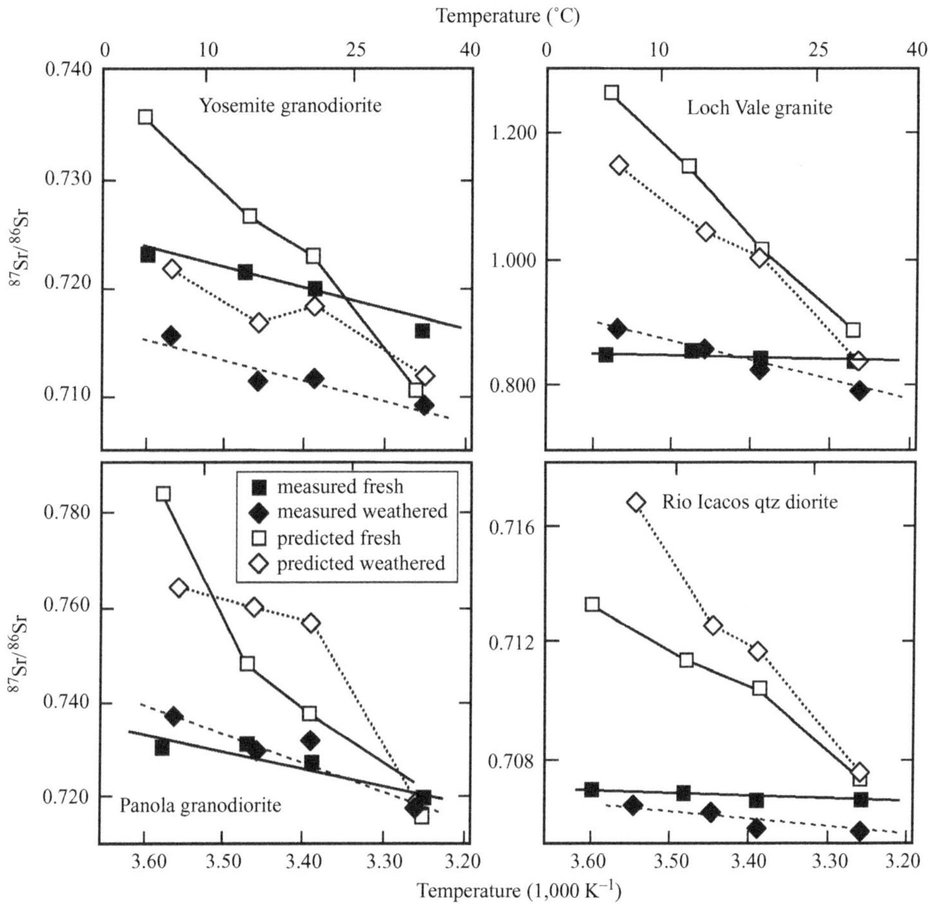

Figure 1 Temperature versus the $^{87}Sr/^{86}Sr$ of effluent from fresh and weathered samples of crushed granitoids (from the four locations noted) during weathering experiments. Predicted values are those based on Na and K concentrations and the assumption of stoichiometric dissolution of plagioclase and biotite. The deviation between predicted and measured $^{87}Sr/^{86}Sr$ values at low temperature demonstrates that $^{87}Sr/^{86}Sr$ cannot be used in this simple way to accurately identify the relative rate of plagioclase to biotite weathering at low temperatures. This is due to either preferential retention of radiogenic Sr, or enhanced dissolution of nonradiogenic Sr from trace Ca-bearing phases such as calcite included in plagioclase or K-feldspar. The data suggest that $^{87}Sr/^{86}Sr$ ratios derived from granitoid weathering are expected to be temperature dependent, with lower ratios in warmer climates (source White *et al.*, 1999).

has been studied by several research groups in an attempt to understand the processes that release uranium from host minerals to the hydrologic cycle (Rosholt *et al.*, 1963; Chalov and Merkulova, 1966; Szalay and Samsoni, 1969; Kigoshi, 1971; Kovalev and Malyasova, 1971; Fleischer and Raabe, 1978; Fleischer, 1980, 1982; Moreira-Nordemann, 1980; Zielinski *et al.*, 1981; Michel, 1984; Lathan and Schwarcz, 1987; Eyal and Olander, 1990; Bonotto and Andrews, 1998; Bonotto *et al.*, 2001). These studies were ultimately motivated by the goal of improving geochemical exploration for uranium and better understanding the long-term risks associated with radioactive waste disposal. The experiments involved leaching of ground rocks with various reagents in an effort to mimic natural weathering conditions. The most important observation of the laboratory studies was that in many cases ^{234}U was preferentially released from the rock into solution compared to ^{235}U and ^{238}U. Several mechanisms have been proposed to account for this observed uranium fractionation. They include α-decay recoil damage to the crystal lattice, resulting in paths for rapid diffusion (Kigoshi, 1971), an enhanced rate of ^{234}U auto-oxidation to U(VI) (Rosholt *et al.*, 1963), and preferential release of ^{234}U from interstitial oxides and cryptocrystalline aggregates (Brown and Silver, 1955). Both zero-order and first-order uranium removal processes have been observed and discussed (Lathan and Schwarcz, 1987).

19.4.2 Soil Weathering Studies

The application of radiogenic isotopes in soil weathering profiles is a considerably more mature field than the application to experimental studies. Studies of radiogenic isotope behavior during

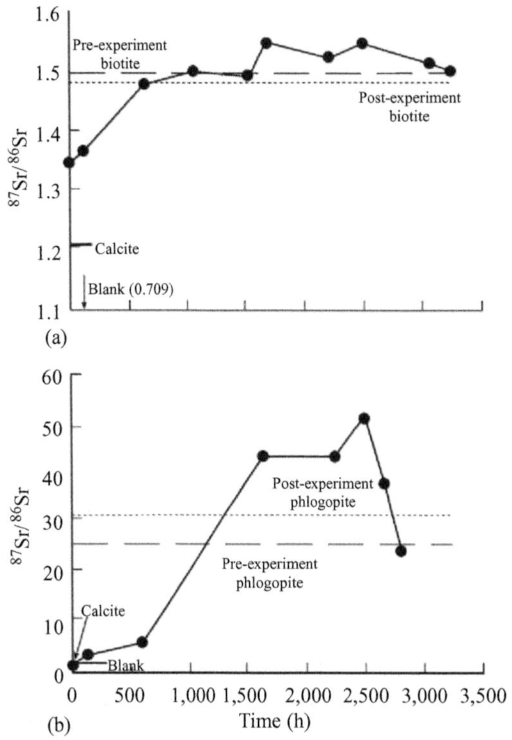

Figure 2 Time versus $^{87}Sr/^{86}Sr$ of experimental effluents during (a) biotite and (b) phlogopite dissolution experiments along with bulk mineral, trace calcite, and experimental blank values. The changing $^{87}Sr/^{86}Sr$ through time can be explained by an initial release of Sr from calcite inclusions (source Taylor *et al.*, 2000b).

weathering began with a comparison of fresh and altered igneous rocks by Dasch (1969) in which it was found that whole-rock $^{87}Sr/^{86}Sr$ ratios were either unaffected or increased in weathered rock as compared to fresh rock. The first measurements of radiogenic isotopes in soils used strontium in order to determine the relative importance of atmospheric dust versus rock weathering as inputs of strontium (and by inference calcium) to forested ecosystems (Graustein and Armstrong, 1983; Gosz *et al.*, 1983). The primary focus of these studies was on forest ecosystem processes rather than mineral weathering *per se*; therefore, a more detailed description of these and other similar studies that followed will be deferred to a later section of this chapter on differentiating atmospheric- from weathering-derived cations in ecosystems.

19.4.2.1 Strontium isotopes

Radiogenic isotope studies with a primary focus on mineral weathering in soil profiles began with the use of granitic soil chronosequences developed on glacial moraines and alluvial terraces to investigate changes in mineral weathering patterns through time during soil development. Blum and Erel (1995, 1997) measured the strontium isotopic composition of the ion-exchange complex and bulk soils in six soil profiles developed on parent material of the same composition but ranging in age from 0.40 kyr to 300 kyr in the Wind River Mountains, Wyoming. The $^{87}Sr/^{86}Sr$ of soil digests did not vary

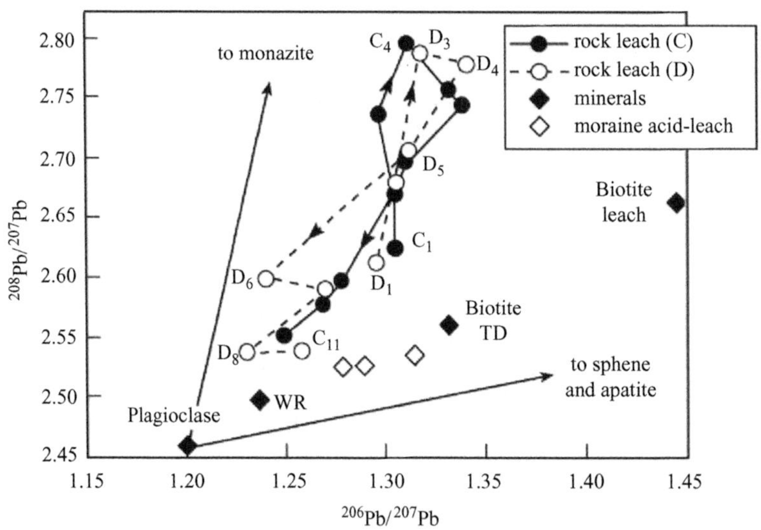

Figure 3 Acid leaches of ground granitoid samples from the Sierra Nevada, California, show three stages of accessory phase dissolution. The initial leach (C_1, D_1) is a mixture of major and accessory phases. The successive leaches show first an increasing influence of monazite and/or allanite dissolution (C_1–C_4, D_1–D_3), followed by an increase in the importance of sphene and apatite (C_4–C_6, D_3–D_4), and finally the depletion of accessory phases and the evolution of the leach composition to a value closer to that of plagioclase (C_6–C_{11}, D_4–D_9). Also plotted are the average $^{206}Pb/^{207}Pb$ and $^{208}Pb/^{207}Pb$ values of whole rock (WR) and mineral separates from the literature (TD = total digest), and the acid leach fractions of soil samples and biotite from the Sierra Nevada, California (source Harlavan and Erel, 2002).

appreciably with age, but $^{87}Sr/^{86}Sr$ of the cation-exchange pool dropped systematically with age from 0.795 to 0.711, indicating a strong sensitivity to the early and rapid depletion of highly radiogenic strontium from biotite in the soil parent material (Figure 4). Biotite was estimated to weather eight times faster than plagioclase in the young soils, but five times slower than plagioclase in the oldest soils. An implication of this work was that global glaciations could significantly elevate the average riverine $^{87}Sr/^{86}Sr$ ratio, providing a mechanism linking global glaciations with increases in the marine $^{87}Sr/^{86}Sr$ record (Blum, 1997).

Bullen *et al.* (1997) carried out a strontium isotope study of a granitic soil chronosequence developed on alluvial parent materials 0.2–3,000 kyr in age in the Sierra Nevada Mountains, California. A significant difference between this study and study of Blum and Erel (1995, 1997) was that the glacial till parent materials studied by Blum and Erel (1995, 1997) represented freshly ground mineral fragments, whereas the alluvial parent materials studied by Bullen *et al.* (1997) were sands that had been sorted, winnowed, and partially weathered during transport and deposition by alluvial processes. Bullen *et al.* (1997) found that biotite in the soil parent materials had lost most of its radiogenic strontium during transport. Therefore, while Blum and Erel (1995, 1997) found that the dominant control on the $^{87}Sr/^{86}Sr$ released by weathering was the weathering of biotite to hydrobiotite and vermiculite, Bullen *et al.* (1997) found that this transformation was already complete in the soil parent materials; therefore, the $^{87}Sr/^{86}Sr$ was controlled dominantly by weathering of K-feldspar and plagioclase. Radiogenic $^{87}Sr/^{86}Sr$ values in some soils were attributed to incongruent leaching of strontium from K-feldspar.

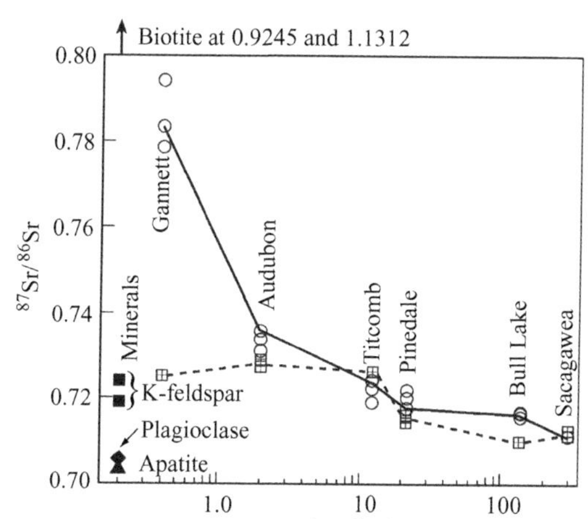

Figure 4 Age of moraines in the Wind River Mountains, Wyoming, on which soils of a chronosequence have formed (with the names of the glacial advances labeled) versus the $^{87}Sr/^{86}Sr$ of the B-horizon soil exchangeable fraction (circles) and C-horizon total soil digests (open squares). Also plotted are analyses of mineral separates from the granitoid bedrock in the area (filled symbols). The elevated $^{87}Sr/^{86}Sr$ in the younger soils was attributed to the release of radiogenic Sr as biotite is altered to form hydrobiotite and vermiculite (source Blum and Erel, 1997).

Innocent *et al.* (1997) used strontium isotopes to investigate weathering and ion-exchange processes in tropical laterites developed on basalt in Brazil. This study showed that most of the rock-derived strontium was released in the earliest stages of weathering in the tropical weathering environment, leading to the control of strontium isotope ratios by radiogenic rain waters or groundwater originating on other geological substrates. Yang *et al.* (2000) used the $^{87}Sr/^{86}Sr$ ratio of calcite in Chinese loess paleosols as a proxy for chemical weathering intensity associated with the monsoonal climate. They found a good correlation with magnetic susceptibility and the marine $\delta^{18}O$ record over a 5–150 kyr timescale. In another study, Jacobson *et al.* (2002a) used strontium isotopes along with major elements to study weathering of a Himalayan glacial moraine chronosequence developed on glacial till that was mostly silicate but contained $\sim 1\%$ carbonate. The study used strontium isotopes to demonstrate that carbonate minerals, even in low abundance, can control the dissolved flux of strontium and calcium emanating from stable landforms comprised of mixed silicate and carbonate for tens of thousands of years after exposure of landforms to the weathering environment.

19.4.2.2 Lead isotopes

Lead isotopes have been extensively used to trace the behavior and transport of lead in soils, to distinguish between natural and anthropogenic lead, and to trace the sources of anthropogenic lead (e.g., Ault *et al.*, 1970; Gast, 1970; Gulson *et al.*, 1981; Bacon *et al.*, 1995; Erel *et al.*, 1997; Steinmann and Stille, 1997; Erel, 1998; Brannvall *et al.*, 2001; Teutsch *et al.*, 2001; Emmanuel and Erel, 2002). Hart and Tilton (1966) were the first to report a systematic difference between the isotopic composition of lead in bulk rocks compared to that released from the same rocks by weathering. The behavior of lead isotopes during rock and mineral weathering was further studied by Erel *et al.* (1994) and Harlavan *et al.* (1998), using the same granitic glacial moraine soil chronosequence in Wyoming studied by Blum and Erel (1995, 1997). The major findings of these studies were that the isotopic composition of lead released by rock weathering changed systematically with soil age and with the degree of weathering, reflecting preferential initial release of lead from a radiogenic mineral weathering pool (probably composed of easily weathered accessory phases, interstitial oxides, and cryptocrystalline aggregates (Brown and Silver, 1955)). With the progression of chemical weathering intensity, it was observed that a higher proportion of the lead released from the rock originated from plagioclase. Hence, these findings provided a complementary view of the weathering process in soils compared to the view provided by strontium isotope studies (Blum and Erel, 1997).

19.4.2.3 Uranium isotopes

Variation of the $^{234}U/^{238}U$ activity ratio in soils as a reflection of weathering processes has been studied for more than 30 years (e.g., Rosholt *et al.*, 1963; Rosholt and Bartel, 1969; Stuckless *et al.*, 1977). This topic in general, and the use of $^{234}U/^{238}U$ versus $^{230}Th/^{238}U$ plots as a means of characterizing the course of weathering in soils, is thoroughly summarized by Osmond and Ivanovich (1992; Figure 5). Moreira-Nordemann (1980) used variations in uranium isotope ratios ($^{234}U/^{238}U$) as a proxy for regional weathering rates. In that study measurements of $^{234}U/^{238}U$ values in river water, weathered rock (or soil C-horizon), and bedrock were combined to provide an estimate for the weathering rate of bedrock in a study catchment. Several papers have been published on the fate of uranium in soils (e.g., Lowson *et al.*, 1986; Frindik and Vollmer, 1999; Romero *et al.*, 1999; von Gunten *et al.*, 1999). For example, von Gunten *et al.* (1999) showed that values of $^{234}U/^{238}U$ change with the stability of soil minerals in a weathered lateritic uranium ore body. The least stable, amorphous iron and manganese oxyhydroxides had the lowest $^{234}U/^{238}U$ values due to preferentially release of ^{234}U to solution. The more stable crystalline oxide minerals goethite and hematite had $^{234}U/^{238}U$ activity ratios around 1, and the most stable Al-silicate minerals had values well above 1. The primary motivation for most soil studies has been to understand the behavior of uranium and other radioactive nuclides near ore deposits and radioactive waste disposal sites.

19.4.2.4 Neodymium isotopes

Relatively few studies have dealt with changes in neodymium isotopes during the progression of chemical weathering, although some have considered the link between weathering and the marine neodymium isotope record (Vance and Burton, 1999). Ohlander *et al.* (2000) found that in the process of weathering, neodymium is preferentially released from minerals with a lower Sm/Nd ratio than bulk soil, leading to lower $^{143}Nd/^{144}Nd$ values in the initial stages of weathering. Aubert *et al.* (2001) combined the study of strontium and neodymium isotopes and REEs in soils, sediments,

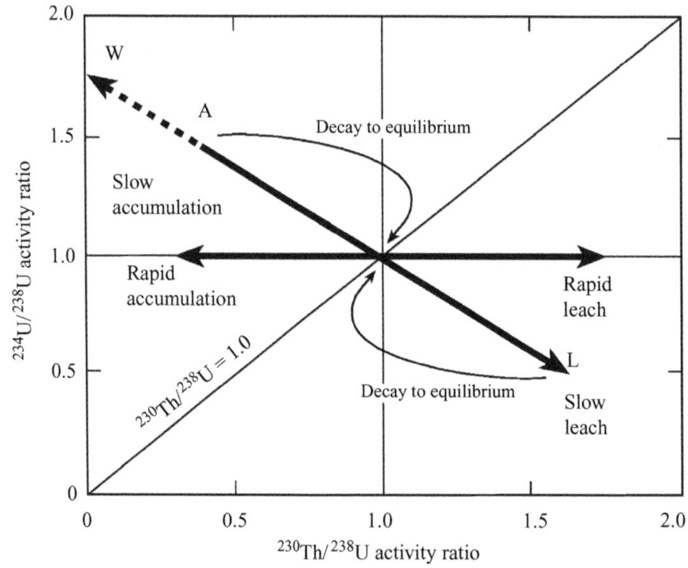

Figure 5 A schematic plot of the $^{234}U/^{238}U$ activity ratio versus the $^{230}Th/^{238}U$ activity ratio in rock and soil samples. On this diagram leaching and precipitation of U results in straight line vectors, while decay toward equilibrium results in curved lines. The point (1, 1) represents equilibrium for both activity ratios. The light diagonal line represents equilibrium between ^{234}U and ^{230}Th; below and to the right is the region of U leaching, above and to the left the region of U precipitation (slow accum. and rapid accum.). On this diagram, preferential leaching (rapid and slow) of ^{234}U (because of recoil processes) produces a leached solid with an activity ratio at L and a resulting water with a U activity ratio at W (connected by heavy diagonal line). If the mobilized U is reprecipitated, it will accumulate in the vicinity of A. Rapid U leaching without preferential mobilization of ^{234}U is shown as a heavy horizontal line (source Osmond and Ivanovich, 1992).

and waters from the Strengbach Catchment, France, to investigate granitic soil weathering processes (Figure 6). Strontium and neodymium isotopes indicated that: (i) stream suspended load originated from the finest soil size fraction (<20 μm); (ii) stream suspended load contained ~3% strontium and neodymium from the trace mineral apatite; and (iii) a large fraction of the strontium and neodymium in the dissolved load originated from leaching of apatite. Other studies have used neodymium isotopes to identify the source material for pedogenic carbonate (e.g., Borg and Banner, 1996), and in conjunction with other geochemical measurements to trace the behavior of REEs in polluted soils (e.g., Steinmann and Stille, 1997). Some recent studies have used strontium and neodymium isotopes to investigate the relative proportions of marine aerosol, mineral weathering, and eolian dust inputs to soil cation-exchange pools. These studies will be discussed in a later section of this chapter on differentiating atmospheric from weathering-derived cations in ecosystems.

19.4.2.5 Osmium and hafnium isotopes

Peucker-Ehrenbrink and Blum (1998) conducted a study of soil weathering processes utilizing the Re–Os isotopic system on the same granitic soil chronosequence in Wyoming used for strontium (Blum and Erel, 1997) and lead (Erel *et al.*, 1994; Harlavan *et al.*, 1998) isotopic studies. Peucker-Ehrenbrink and Blum (1998) pointed out that preferential weathering of biotite and rapid oxidation of magnetite in the soil environment largely controls the isotopic composition of osmium and the Re/Os ratio released during granitoid weathering. They suggested that high-latitude Precambrian shields may be important source areas of radiogenic osmium in seawater, and proposed that weathering of glacial tills exposed after deglaciation of Precambrian rocks surrounding the North Atlantic might result in elevated $^{187}Os/^{186}Os$ in rivers draining these areas. van de Flierdt *et al.* (2002) compared the hafnium isotopic composition of various rocks types and minerals with that of seawater, and suggested that during periods of intense mechanical weathering due to glaciations, zircons are weathered more efficiently, resulting in the release of highly nonradiogenic hafnium to the oceans.

19.4.3 Springs and Small Streams

In earlier sections we discussed the use of radiogenic isotopes to study mineral weathering in both laboratory experiments and in soils. In many of these studies the isotopic composition of the element of interest released by weathering was estimated in soils by measuring the composition of the labile pool in the soil (the acid leach or cation-exchange fractions). Another means of estimating the isotopic composition released by weathering is by measuring the isotopic composition of that element dissolved in springs and small streams draining a catchment underlain by a single bedrock lithology. In the absence of significant atmospheric inputs, the dissolved element is often expected to be in equilibrium with the soil labile pool. Among the various isotopic systems the most widely utilized in this way are strontium and uranium. Lead and neodymium isotopes have been used in several studies in conjunction with

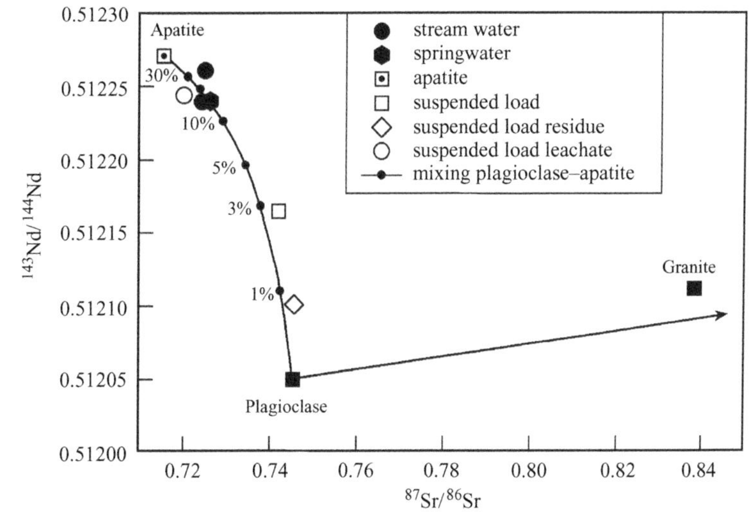

Figure 6 Sr and Nd isotopic compositions of granitoid bedrock, minerals, waters, and sediments from the Strengbach Catchment, France. A mixing calculation is shown with the percentage apatite in apatite–plagioclase mixtures. Sediments are dominated by plagioclase, whereas the waters and sediment leachate derive 10–30% of their Nd and Sr from the trace mineral apatite (source Aubert *et al.*, 2001).

strontium isotopes, and a limited number of studies have been conducted utilizing exclusively lead or neodymium isotopes. It is important to note that lead isotopes have been used extensively as tracers of anthropogenic lead transport in surface and groundwater reservoirs, but these studies do not fall within the scope of this chapter.

19.4.3.1 Strontium isotopes

Two of the most significant early investigations of strontium isotopes in streams were by Fisher and Stueber (1976) and Wadleigh *et al.* (1985), who established that waters draining crystalline bedrock generally have much higher $^{87}Sr/^{86}Sr$ and lower strontium concentration than waters draining areas dominated by carbonate bedrock. They also demonstrated that even a small amount of carbonate in a drainage basin strongly influences the strontium isotopic composition of stream waters. Faure (1986) provides an excellent review of research on strontium in surface waters up until the mid-1980s.

Strontium isotopes were first applied to studies of deep saline groundwater brines in the late 1980s; a discussion of these studies can be found in the later section of this chapter on hydrology. Blum *et al.* (1993) used stream water $^{87}Sr/^{86}Sr$ to infer relative weathering rates of silicate minerals in small paired granitic catchments in the Sierra Nevada Mountains, California, that were either recently (12 kyr ago) glaciated or nonglaciated. The sensitivity of the dissolved $^{87}Sr/^{86}Sr$ ratio to biotite weathering was used to determine that glaciation had the effect of accelerating biotite weathering (i.e., the transformation to hydrobiotite and vermiculite) compared to plagioclase weathering in nonglaciated versus glaciated catchments. Another study with the primary focus on the interpretation of stream water $^{87}Sr/^{86}Sr$ to determine mineral weathering rates was carried out in two Scottish catchments, one underlain by andesite and the other by schist/granulite (Bain and Bacon, 1994). These authors investigated the $^{87}Sr/^{86}Sr$ ratios of total digests of soils and rock as well as stream- and rainwater, and concluded that plagioclase was weathering preferentially from the andesite compared to K-feldspar and chlorite, whereas the relative rate of dissolution of minerals in the schist/granulite could not be determined due to insufficient information on mineral compositions.

A study by Bullen *et al.* (1996) integrated experimental dissolution experiments, groundwater geochemical measurements, and a groundwater flow-path analysis to investigate mineral weathering in the subsurface near several Wisconsin lakes. Combining strontium isotopes and elemental analysis of waters and aquifer minerals along a shallow groundwater flow-path, the authors were able to demonstrate a dramatic change in the relative weathering rates of silicate minerals along the flow path. Shallower and more dilute waters had compositions indicative of dominantly plagioclase dissolution. Deeper and higher ionic strength waters had compositions that indicated a suppression of plagioclase dissolution and dominance by K-feldspar and biotite weathering. This application of strontium isotopes is an excellent demonstration of the power of the strontium isotope tracer to distinguish changes in the relative rates of weathering of complex mineral assemblages.

Clow *et al.* (1997), Blum *et al.* (1998), and Horton *et al.* (1999) used strontium isotopes and cation chemistry of stream waters to determine the proportion of dissolved ions originating from silicate versus carbonate dissolution reactions in catchments that were predominantly silicate, but also contained small (<1%) amounts of carbonate in the bedrock. In each case it was found that trace amounts of disseminated carbonate in crystalline silicate rocks could have a dominant effect on the strontium isotopic compositions and calcium fluxes from small catchments. The use of strontium isotopes in these small catchment studies have helped to solve the puzzle of why stream waters draining silicate bedrock catchments often display Ca/Na and Ca/Sr ratios far in excess of what would be released by weathering of the abundant silicate minerals in the soil parent materials (Figure 7). Thus, as in the case of experimental weathering studies and soil profile studies, we find that a major contribution of strontium isotopes as tracers of weathering is the identification of inputs from highly reactive trace minerals.

Jacobson *et al.* (2002b) combined strontium isotope and major element studies of 40 small streams in the Southern Alps of New Zealand to examine the climatic and tectonic controls on the relative weathering rates of carbonate and silicate minerals and documented a strong coupling between physical and chemical denudation. Peters *et al.* (2003) combined an experimental dissolution study with a stream catchment study by applying powdered wollastonite ($CaSiO_4$) to a stream channel in the Hubbard Brook Experimental Forest, New Hampshire. The interpretation of the stream water chemistry response to this application required the ability to separate calcium fluxes from the normal stream background, from calcium derived by wollastonite dissolution. A large contrast in the strontium isotopic composition between the stream before application and the strontium released from wollastonite allowed Peters *et al.* (2003) to quantify the amount of calcium in the stream flux that was derived from each of these two sources, even though the

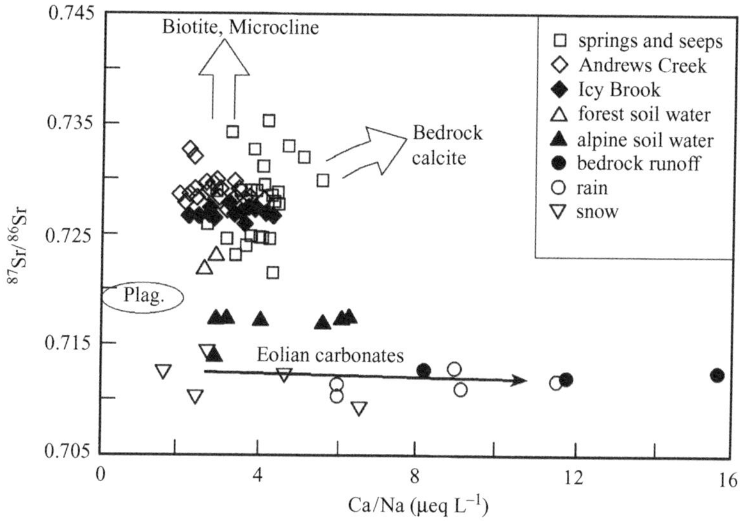

Figure 7 Ca/Na ratio versus $^{87}Sr/^{86}Sr$ ratio for waters and minerals from the granitoid Loch Vale, Colorado watershed. High Ca/Na ratios are indicative of a significant contribution from calcite dissolution to the dissolved cation chemistry of the waters. Sr isotopes allow differentiation of inputs from bedrock calcite versus eolian carbonate dust derived from sedimentary rocks in surrounding areas (source Clow *et al.*, 1997).

dissolution of calcium and silicon from the wollastonite was not stoichiometric.

19.4.3.2 Lead isotopes

The isotopic composition of lead released to stream water by rock weathering was studied by Erel *et al.* (1990, 1991). During spring snowmelt a pristine mountain stream in the Sierra Nevada Mountains, California, was observed to contain mostly anthropogenic lead, but in the fall during base flow most of the lead in the stream water was from bedrock weathering. Erel *et al.* (1991) found that the isotopic value of the released lead was more radiogenic than the value of lead in the rock, and was similar to the value of lead released from the rock by a weak acid leach. Since the studied watershed was glaciated ~12 kyr ago exposing fresh rock surfaces, it was suggested that radiogenic lead was preferentially released from the granitic bedrock during the early stages of rock weathering. The preferential release of radiogenic lead was attributed to one or more of the following mechanisms: (i) rapid dissolution of some of the radiogenic accessory phases (e.g., apatite, monazite; Silver *et al.* (1984)); (ii) α-recoil damage to the uranium-enriched minerals (Kigoshi, 1971); or (iii) the presence of radiogenic lead-enriched interstitial oxides and cryptocrystalline aggregates (Brown and Silver, 1955). Luck and Ben Othman (1998) studied both lead and strontium isotopes in dissolved and particulate matter of a small Mediterranean karstic watershed in France during flooding and nonflooding periods. They were able to use these isotopic systems to estimate changes in physical and chemical weathering within the watershed in response to the water flux.

19.4.3.3 Uranium isotopes

Uranium isotopes have been used widely in surface water systems to trace water sources and pathways as well as rock weathering processes (Moore, 1967; Moreira-Nordemann, 1980; Scott, 1982; Sarin *et al.*, 1990; Tuzova and Novikov, 1991; Osmond and Ivanovich, 1992; Plater *et al.*, 1992; Palmer and Edmond, 1993; Pande *et al.*, 1994; Zhao *et al.*, 1994; Lienert *et al.*, 1994; Chabaux *et al.*, 1998; Riotte and Chabaux, 1999; Hakam *et al.*, 2001). Some of these studies have addressed the effect of rock weathering and rock type (especially carbonates versus silicates) on the $^{234}U/^{238}U$ ratio (e.g., Moreira-Nordemann, 1980; Toulhoat and Beaucaire, 1993; Pande *et al.*, 1994), and others have emphasized the links between the $^{234}U/^{238}U$ ratio and the hydrology of the studied drainages (Lienert *et al.*, 1994; Chabaux *et al.*, 1998; Riotte and Chabaux, 1999; Hakam *et al.*, 2001). For example, Riotte and Chabaux (1999) attributed variations in the $^{234}U/^{238}U$ ratio to changes in the water flowpaths in the Strengbach catchment in France, and pointed out the complexity of the interpretation of the $^{234}U/^{238}U$ ratio as a tracer of rock weathering processes. Nonetheless, they effectively combined the use of $^{87}Sr/^{86}Sr$ and $^{234}U/^{238}U$ ratios to trace the rock types that control the composition of the stream waters as they flow over various compositions of bedrock (Figure 8).

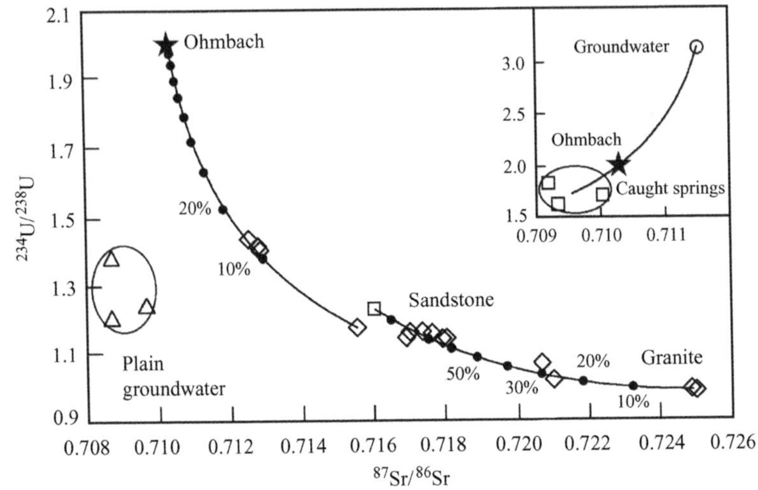

Figure 8 Evolution of the U and Sr isotopic composition of stream water in the Strengbach Catchment, France (diamonds) with mixing curves between granite and sandstone bedrock end-members. Also shown is a mixing curve between Strengbach stream water (where it flows from granite and sandstone bedrock into carbonate bedrock) and a carbonate end-member. The carbonate end-member is represented by the Ohmbach stream (star), which is located in the nearby carbonate-bedrock Ombach Catchment. The mixing line proportions presented here are based on both U and Sr data (concentrations and isotopic ratios). Inset: The Ohmbach-like component can be explained as a mixture of waters from surface springs (Caught Springs—squares) and the deep groundwater sampled in the Ohmbach catchment (circles). Plain groundwater refers to groundwater samples collected from the Rhine plain; their contrasting composition is used as evidence that Rhine groundwaters do not feed the Strengbach stream within the study area (source Riotte and Chabaux, 1999).

19.4.3.4 Neodymium isotopes

Tricca *et al.* (1999) combined the analysis of neodymium and strontium isotopes in major and small rivers to investigate the origin of REEs in river waters from small catchments in the Rhine Valley. They observed that in the Rhine River the isotopic composition of neodymium was similar to the average continental crustal value, but in smaller, less evolved streams, the isotopic composition of neodymium reflected the breakdown of accessory minerals, such as apatite, enriched with middle REEs. Andersson *et al.* (2001) studied the effect of weathering on the isotopic composition of neodymium in the Kalix River in northern Sweden. They reported that the isotopic composition of neodymium had a narrow range and that, in general, the isotopic composition of neodymium in river water was lower than that of the local bedrock, probably reflecting preferential weathering of low Sm/Nd minerals. These observations are similar to those made by Ohlander *et al.* (2000) in a study of neodymium isotopes in soils developed on tills in northern Sweden.

19.4.3.5 Osmium isotopes

Pegram *et al.* (1994) studied the isotopic composition of osmium released from river sediments and pointed out that the $^{187}Os/^{188}Os$ of the dissolved river flux to the ocean is more radiogenic than the $^{187}Os/^{188}Os$ of the currently eroding continental crust, implying a significant osmium contribution from marine and terrestrial peridotites. Sharma and Wasserburg (1997) and Sharma *et al.* (1999) studied the behavior of osmium and its isotopes in major rivers draining the Himalayas. Although the isotopic composition of osmium in major rivers is beyond the scope of this chapter, we do note that these authors came to the conclusion that the Himalayas do not provide an unusually high flux of osmium or highly radiogenic osmium to the oceans. Chesley *et al.* (2000) studied the osmium and strontium isotopic record of Himalayan paleorivers and observed that $^{187}Os/^{188}Os$ and $^{87}Sr/^{86}Sr$ ratios covary, implying a common source. They provided evidence that large and rapid changes in the riverine fluxes and isotopic ratios of Himalayan rivers must have taken place during the last 5 Myr.

19.4.4 Proxies for Global Weathering Rates

Among the isotopic systems discussed in this chapter, strontium is the only one that has a long enough residence time in the ocean to unequivocally record changes in global weathering rates. For example, lead isotopes in marine manganese nodules, studied extensively since the early 1990s, have been mostly used to trace changes in ocean circulation rather than weathering rates. The isotopic composition of strontium in seawater is controlled largely by a balance between riverine inputs to the ocean (which result from continental weathering), marine hydrothermal inputs from the seafloor, and diagenetic fluxes from marine sediments (e.g., Brass, 1976; Palmer and Elderfield, 1985). Because marine carbonate rocks faithfully record the history of marine $^{87}Sr/^{86}Sr$ through geological time, and because the ~2 Myr residence time of strontium in the ocean results in a single global marine $^{87}Sr/^{86}Sr$ value, strontium is a potentially powerful proxy for changes in global weathering processes. In some studies, global weathering rates have been estimated from the marine strontium isotope record by assuming that hydrothermal and diagenetic strontium fluxes and $^{87}Sr/^{86}Sr$ ratios are constant through time, and that the $^{87}Sr/^{86}Sr$ ratio of continental weathering has remained equal to the weighted average value of present-day riverine fluxes. It has further been assumed in some studies that the $^{87}Sr/^{86}Sr$ ratio of continental carbonate weathering remains equal to the average value for marine carbonates exposed on the continents, and that the average silicate weathering value could be calculated by difference after removing the carbonate contribution from major river fluxes. These assumptions allowed the development of models using variations in the marine $^{87}Sr/^{86}Sr$ ratio to estimate changes in global weathering rates due to tectonic uplift and global glaciations. Some workers suggested that changes in marine $^{87}Sr/^{86}Sr$ are caused not only by changes in silicate weathering rates, but also by changes in the $^{87}Sr/^{86}Sr$ of the average global riverine $^{87}Sr/^{86}Sr$ due to changes in the proportion of various silicate rocks and minerals weathering at any given time. Mountain building episodes, development of flood basalt provinces, sea-level change, and removal of soils by continental glaciations have all been proposed as mechanisms that might influence the proportion and age of silicate rocks exposed. Hodell (1994) provides an excellent review of the extensive literature on this topic through the early 1990s.

Several assumptions inherent in these earlier models have been challenged during the 1990s, as additional information has become available from studies of the systematics of strontium isotope release during weathering (summarized in this chapter). Recent contributions have pointed out some new complexities in the interpretation of the marine $^{87}Sr/^{86}Sr$ ratio as a silicate weathering proxy, including observations that: (i) strontium released during the early stages of weathering from fresh mineral surfaces can have a considerably different $^{87}Sr/^{86}Sr$ than bulk rocks or minerals (Blum and Erel, 1995; Brantley *et al.*, 1998; Taylor and Lasaga, 1999; White *et al.*, 1999; Taylor *et al.*, 2000a,b); (ii) trace amounts of carbonate or apatite weathering can dominate riverine strontium fluxes even in areas of silicate bedrock (Clow *et al.*, 1997; Blum *et al.*, 1998, 2002; Harris *et al.*, 1998; Horton *et al.*, 1999; Aubert *et al.*, 2001; Jacobson *et al.*, 2002a,b); (iii) carbonates may, in some cases, yield high fluxes of strontium with very high $^{87}Sr/^{86}Sr$, particularly in the Himalayan Mountains (Palmer and Edmond, 1992; Quade *et al.*, 1997; Harris *et al.*, 1998; Galy *et al.*, 1999; English *et al.*, 2000; Karim and Veizer, 2000; Jacobson *et al.*, 2002a); and (iv) hot springs (Evans *et al.*, 2001) and groundwater fluxes (Basu *et al.*, 2001) may be a significant part of the continental strontium fluxes to the oceans. It has become clear that the simple assumptions made in earlier work must be modified in order to derive meaningful estimates of weathering from the marine strontium isotope record. This presents an enormous challenge that has not been achieved as yet, but continued research will likely lead to a more quantitative understanding of the relationship between the marine strontium isotope record, global weathering rates, and other global geochemical cycles.

19.5 DIFFERENTIATING ATMOSPHERIC-FROM WEATHERING-DERIVED CATIONS IN ECOSYSTEMS

19.5.1 Soil Carbonates

Strontium isotope studies primarily focusing on the origin of soil carbonates began with a study by Quade *et al.* (1995), which demonstrated that calcium and strontium in soil carbonates within ~100 km of the southern coast of Australia were derived dominantly from marine dust and sea-spray with very little contribution from the soil substrate. This work was followed by similar studies of soil carbonate formed near the shoreline in Hawaii (Capo *et al.*, 2000; Whipkey *et al.*, 2000) and Morocco (Hamidi *et al.*, 1999), in the inland setting of Central Spain (Chiquet *et al.*, 1999), and in the southwestern USA (Capo and Chadwick, 1999; Naiman *et al.*, 2000). Each of these studies used strontium isotopes to demonstrate the overwhelming importance of atmospheric dust inputs (and the recycling of soil carbonate) to soil carbonate formation. Quade *et al.* (1997) and Chesley *et al.* (2000) studied the strontium isotopic composition of paleosol carbonate from ancestral Himalayan river deposits to reconstruct a record of changing riverine $^{87}Sr/^{86}Sr$ over the past 20 Myr. A dramatic increase in the riverine $^{87}Sr/^{86}Sr$ during the Late Miocene was attributed to exhumation of high $^{87}Sr/^{86}Sr$ meta-limestones in the Himalayas, and it was suggested that this may have influenced the marine $^{87}Sr/^{86}Sr$ record.

19.5.2 Base Cation Nutrients

19.5.2.1 Strontium isotopes

Quantifying the sources and rates of input of base cation nutrients (calcium, magnesium, potassium, and sodium) to forest ecosystems is an important goal in forest biogeochemistry, particularly when seeking to understand the recovery from environmental disturbances such as acid rain and forest clear-cutting. The earliest study to use isotopes as an indicator of atmospheric inputs to soils was by Dymond *et al.* (1974), who used strontium isotope measurements of micas in Hawaiian soils to determine that a significant proportion of the potassium input to Hawaiian soils was from deposition of dust transported across the Pacific Ocean from Asia. Straughan *et al.* (1981) demonstrated that when fly ash produced by coal combustion is deposited onto soil, the strontium released from the fly ash is readily available to plants and can dominate the plant $^{87}Sr/^{86}Sr$ ratio with as little as 0.2% fly ash by weight in the soil.

Two nearly simultaneous investigations of atmospheric inputs to forested ecosystems of the Sangre de Cristo Mountains, New Mexico, were the first to fully integrate strontium isotopes into forest biogeochemical studies (Gosz *et al.*, 1983; Graustein and Armstrong, 1983). These studies took advantage of a large contrast in $^{87}Sr/^{86}Sr$ between the Precambrian granitic bedrock of the Sangre de Cristo mountains and the Phanerozoic sedimentary and volcanic rocks of the surrounding areas, which are the major source of atmospheric dust to the ecosystems. Strontium isotope and concentration measurements of a wide range of ecosystem components allowed quantification of the proportion of atmospheric-derived (>75%) and weathering-derived (<25%) strontium to forest vegetation (Gosz *et al.*, 1983; Graustein and Armstrong, 1983). Both research groups later expanded upon their earlier pioneering strontium isotope forest biogeochemistry studies of New Mexico watersheds (Graustein, 1989; Gosz and Moore, 1989). Åberg and Jacks (1985), Åberg *et al.* (1989), and Jacks *et al.* (1989) used a similar approach to study weathering and atmospheric inputs of calcium in several granitic stream catchments in Sweden. Differences in the $^{87}Sr/^{86}Sr$ of atmospheric and weathering inputs allowed the calculation of a calcium output budget, which suggested that excess calcium in streams was caused by loss from the cation-exchange pool due to soil acidification. A major source of uncertainty in all of these early studies was the $^{87}Sr/^{86}Sr$ released by weathering of ancient granite composed of minerals with highly contrasting $^{87}Sr/^{86}Sr$ ratios.

Miller *et al.* (1993) applied the strontium isotope methodology described above to investigate the response of the Whiteface Mountain, New York, forest ecosystem to atmospheric deposition of acidity and other pollutants. The soil parent material was anorthosite, which is comprised mostly of plagioclase and which has low Rb/Sr and $^{87}Sr/^{86}Sr$, minimizing uncertainties related to variation in $^{87}Sr/^{86}Sr$ among minerals. In this study it was determined that 50–60% of the strontium in the vegetation and organic soil pools was of atmospheric origin—presumably largely from coal-combustion fly-ash and dust related to cement production and farming. An ecosystem box-model was used to estimate that the annual stream export of strontium was ~70% from mineral weathering inputs and ~30% from soil cation-exchange reactions. Bailey *et al.* (1996) used a similar methodology at the Cone Pond Watershed, New Hampshire, to investigate inputs and transport of base cations in another ecosystem perturbed by acid deposition. A large variability in the strontium isotopic composition of the various minerals in the bedrock, and spatial variability of soil parent material, complicated assessment of the $^{87}Sr/^{86}Sr$ ratio released to solution by weathering reactions. The authors were unable to make definitive estimates of cation-exchange depletion of calcium, but explored the range of possibilities using a variety of assumptions for the $^{87}Sr/^{86}Sr$ ratio of the weathering end-member. Using their "most likely" estimate of the weathering end-member, Bailey *et al.* (1996) suggested that the cation-exchange pool was being depleted significantly and was in danger of being exhausted.

Probst *et al.* (2000) measured strontium isotopes and major element concentrations of the various hydrochemical reservoirs in the Strengbach Catchment, France. Based on chemical budgets and a rock–water interaction model used to estimate the weathering end-member composition, they concluded that ~50% of the dissolved strontium in stream water was atmospherically derived. Aubert *et al.* (2002b) incorporated neodymium isotope and REE concentration analyses into the Stengbach Catchment study and were able to show that atmospheric contributions of strontium and neodymium in throughfall and soil solution ranged from 20% to 70%. Blum *et al.* (2000, 2002) integrated strontium isotopes and Ca/Sr ratios into a study of the forest biogeochemistry of a small catchment in the Hubbard Brook Experimental Forest, New Hampshire. A four-step digestion procedure on soil parent material was used to determine the composition of soil calcium reservoirs for each soil horizon. The mineral apatite was found to be depleted from the Oa-, E-, Bh-, and Bs1-horizons but to be the dominant source of weatherable calcium in the Bs2- and C-horizons (Blum *et al.*, 2002). Moreover, apatite-derived calcium was inferred to be utilized to a larger extent by ectomycorrhizal tree species, suggesting that mycorrhizae may accelerate the weathering of apatite. Mass balance estimates suggest that of the calcium leaving the catchment in stream water, ~30% was atmospheric, ~35% was from silicate weathering, and ~35% was from apatite (Figure 9).

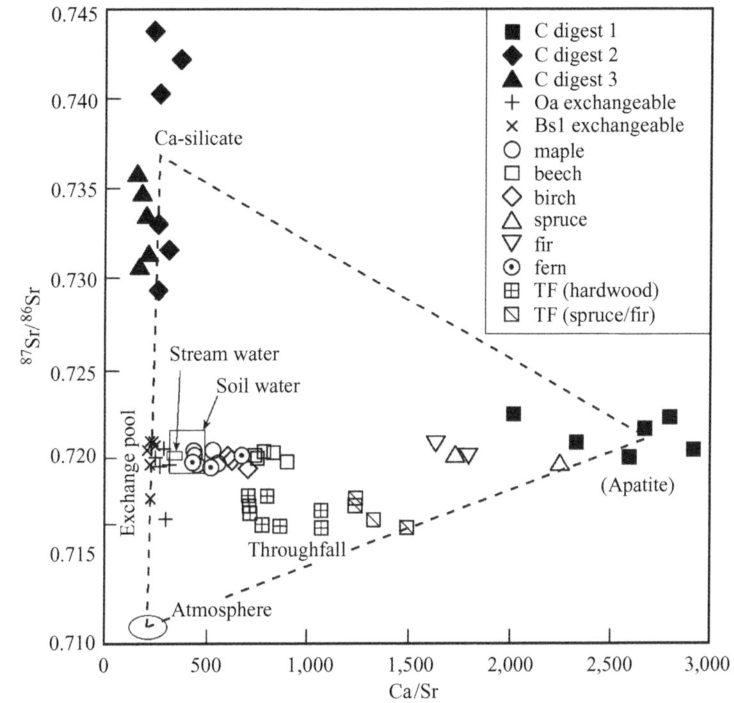

Figure 9 Ca/Sr versus $^{87}Sr/^{86}Sr$ for waters, C-horizon soil digests, soil-exchangeable fraction, and vegetation from the granitoid W-1 watershed at Hubbard Brook, New Hampshire. C-horizon digests 1, 2, and 3 were done in sequence and are progressively more aggressive acid digestions. Foliage (with species shown in key), soil-exchange pool, and water compositions each represent mixing between three sources: apatite leached from the soil parent material, silicate mineral weathering of the soil parent material, and atmospheric deposition. TF is throughfall, which represents a mixture of Ca and Sr leached from foliage and deposited from the atmosphere. Spruce and fir trees appear to access to a larger extent Ca and Sr from the apatite soil pool (source Blum *et al.*, 2002).

Kennedy *et al.* (1998) and Chadwick *et al.* (1999) applied strontium isotopes to trace the atmospheric input of calcium to the strongly marine-influenced ecosystems of the Hawaiian Islands. Using a soil chronosequence developed on basaltic lava flows varying in age from 0.3 kyr to 4,100 kyr, they found that the exchangeable soil pool (and tree foliage) was dominated by weathering-derived strontium at sites <10 kyr in age, where weathering rates far exceeded atmospheric inputs. As weathering rates declined with age, the system became dominated by marine inputs at all sites >10 kyr in age (Figure 10). Vitousek *et al.* (1999) expanded upon this work to study strontium isotopes in the leaves of trees from 34 Hawaiian forests (with varying precipitation rates) developed on young soils. Weathering was found to supply most of the strontium in most of the sites, but atmospheric sources supplied 30–50% of the strontium in the wettest sites and those closest to the ocean. Stewart *et al.* (2001) similarly used strontium isotopes to investigate the relative proportions of weathering and atmospherically derived strontium in 150 kyr soils along a precipitation gradient in Hawaii. They also found a transition between weathering-dominated and rainfall-dominated sources of plant-available strontium with increasing precipitation.

Kennedy *et al.* (2002) applied strontium isotopes to differentiate between weathering and atmospherically derived sources of strontium and calcium in a maritime forest in Chile, unaffected by atmospheric pollutants. In this study area they found that the exchangeable soil pool and foliage of the dominant tree species were mostly atmospherically derived strontium (>80%). These workers also applied an artificially enriched ^{84}Sr tracer to the forest floor to study plant uptake and leaching losses of strontium. They observed strong retention of the tracer in the upper soil horizons, yet no significant uptake into foliage nearly two years after treatment.

19.5.2.2 Lead and neodymium isotopes

Brimhall *et al.* (1991) used lead isotopes in zircons within a bauxite profile from Western Australia to differentiate between zircons derived from the underlying bedrock and zircons of eolian origin. Borg and Banner (1996) applied both neodymium and strontium isotopes to constrain the sources of soil developed on carbonate bedrock. Using these isotopes and Sm/Nd ratios, they were able to delineate the importance of atmospheric versus bedrock contributions in controlling the composition of the soil. Kurtz *et al.* (2001) used

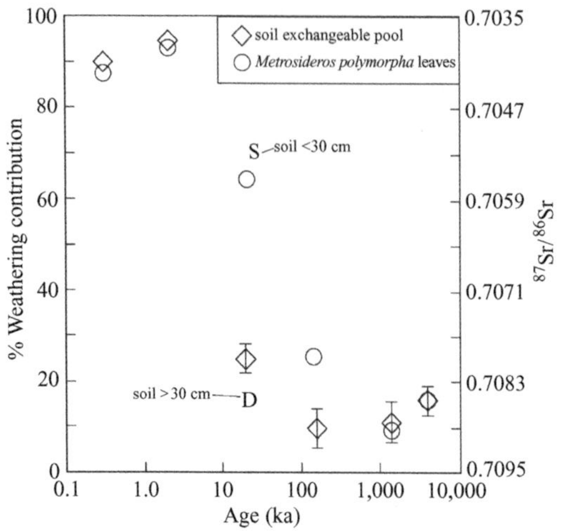

Figure 10 The $^{87}Sr/^{86}Sr$ ratio of the soil-exchangeable pool and foliage for a chronosequence of soils developed on Hawaiian basaltic lava flows. Also plotted is the result of a calculation of the percent weathering contribution needed to explain the $^{87}Sr/^{86}Sr$ ratio—assuming that all Sr is derived from basaltic weathering and marine atmospheric deposition. Sr isotopes clearly demonstrate the transition of the forest ecosystems from weathering dominated Sr to atmospherically dominated Sr as soils mature and easily weathered Sr is removed from soils (source Kennedy *et al.*, 1998).

neodymium and strontium isotopes to determine the amount of Asian dust in a Hawaiian soil chronosequence. They found that the basaltic bedrock isotope signatures in soils had, in many cases, been completely overprinted by dust additions, demonstrating the profound effect of Asian dust on soil nutrient supplies.

19.6 TRACING HYDROLOGIC FLOW PATHS AND SUBSURFACE MIXING

Here we highlight the uses of radiogenic isotopes that are dissolved in groundwater as a tool in hydrologic investigations. None of the radiogenic isotopes discussed in this chapter are truly conservative (i.e., nonreactive) tracers of groundwater. Instead, the radiogenic isotope compositions are controlled to varying degrees by mineral dissolution and precipitation along water flow-paths. In some cases the isotopic composition changes very little along flow paths and can be used as a simple tracer of flow path and mixing of distinct waters. In other instances, dissolution and isotopic exchange during flow can be used as an asset, yielding useful hydrologic information. Strontium has been the most frequently used radiogenic isotope tracer in hydrology, because it is very soluble in groundwater and has a large range in isotopic composition in freshwaters. Uranium is also quite soluble and numerous studies have been published in which uranium isotopes and U–Th disequilibrium products were utilized to follow subsurface flow of water and mixing of different subsurface waters (e.g., Osmond and Cowart, 1976, 1992), sometimes in conjunction with strontium isotopes (Cao *et al.*, 1999). The use of U–Th disequilibrium series in tracing ground waters falls outside the scope of the current chapter, but an excellent review of this topic is provided by Osmond and Cowart (1992).

Lead, neodymium, and osmium have not been used widely as tracers in hydrology, because they have very low solubilities and are highly reactive with aquifer materials. Lead isotopes have been used in groundwater mostly to distinguish between natural and anthropogenic lead sources (e.g., Whitehead *et al.*, 1997), where the natural endmember was assumed to equal the isotopic value of lead released from the bulk rock. Although this assumption is probably incorrect for many groundwater systems (Harlavan *et al.*, 1998), the large excess of anthropogenic lead in most cases overwhelms small uncertainties in the isotopic composition of lead released from the bedrock. Combining analyses of several isotopic tracers has proved to be an effective approach, and several research groups have successfully combined neodymium, lead, and/or uranium isotopes with strontium isotopes in order to trace the migration of groundwaters (e.g., Doe *et al.*, 1966; Stuckless *et al.*, 1991; Toulhoat and Beaucaire, 1993; Mariner and Young, 1996; Barbieri and Voltaggio, 1998; Tricca *et al.*, 1999; Roback *et al.*, 2001).

Hydrologic applications of radiogenic isotopes can be broadly grouped into: (i) regional-scale studies of groundwater flow-paths and mixing over hundreds of kilometers; (ii) local-scale studies of groundwater flow-paths and mixing over tens of kilometers; and (iii) studies of the hydrology of springs and small streams in closed catchments. In addition to studies of radiogenic isotopes dissolved in waters, some studies have taken advantage of secondary minerals deposited in the subsurface to infer changes in water chemistry and hydrology over geologic time. Many of the earliest investigations of radiogenic isotopes in groundwater focused on the origin of oil field brines and hydrothermal fluids. These are mentioned only briefly here, because the emphasis of this chapter is on freshwater hydrology. Doe *et al.* (1966) were among the first investigators to utilize strontium and lead isotopes in brines and potential source rocks to constrain the source of metals in hydrothermal fluids. Later studies by Stettler (1977), Stettler and Allègre (1979), and Elderfield and Greaves (1981) modeled the relative contributions of mantle and crustal sources to hydrothermal brines as a result of fluid–rock interaction. Early studies that utilized strontium isotopes in regional aquifers to determine the sources of salts in brines include Chaudhuri (1978), Sass and

Starinsky (1979), Starinsky *et al.* (1983a,b), Stueber *et al.* (1987, 1984, 1993), Banner *et al.* (1989) and Smalley *et al.* (1998). McNutt *et al.* (1990) and Franklyn *et al.* (1991) utilized strontium isotope ratios in deep brines from mines and drill holes in the crystalline Canadian Shield to establish that dissolution of plagioclase was the dominant water–rock interaction controlling the strontium isotopic composition. They utilized differences in $^{87}Sr/^{86}Sr$ between pockets of brine to study mixing of different brines as well as mixing with meteoric waters. Similarly, Toulhoat and Beaucaire (1993) used lead and uranium isotopes in groundwater to trace the release of uranium and lead by the dissolution of minerals in uranium ore deposits.

19.6.1 Regional Scale Groundwater Studies

19.6.1.1 Strontium isotopes

We begin our detailed discussion of the literature with the first study that utilized strontium isotopes to investigate hydrologic flow paths and mixing in a freshwater aquifer. In a study of the Australian Artesian Basin, Collerson *et al.* (1988) used strontium isotopes to characterize various water masses and investigated their subsurface flow and mixing over a ~500 km length scale. Strontium isotopes were subsequently applied to investigations of the sources of salts and flow paths of groundwaters in South Australia (Ullman and Collerson, 1994; Lyons *et al.*, 1995). Musgrove and Banner (1993) combined the use of major ions, as well as oxygen, hydrogen, and strontium isotopes in groundwaters from three adjacent regional flow regimes in the mid-continental US to delineate large-scale fluid flow and mixing processes. $^{87}Sr/^{86}Sr$ ratios were useful in delineating between end-member waters, but there was some evidence for modification of $^{87}Sr/^{86}Sr$ by interaction of groundwater with silicate minerals in the aquifers, which may have obscured some of the mixing relationships between end-member waters.

Strontium isotopes have been applied to several studies of groundwater evolution in limestone aquifers. $^{87}Sr/^{86}Sr$ and Ca/Sr ratios have provided insight into calcite dissolution and recrystallization along flow paths, the transformation of aragonite to calcite, and the fluxes of strontium from soils into carbonate aquifers (Banner *et al.*, 1994, 1996; Banner, 1995). Dissolution and recrystallization of calcite along flow paths can compromise ^{14}C dating of groundwater by diluting ^{14}C with aquifer carbon. Bishop *et al.* (1994), in a study of the Lincolnshire Limestone aquifer, England, introduced a methodology, whereby changes in $^{87}Sr/^{86}Sr$ and $\delta^{13}C$ with flow down the hydraulic gradient could be used to estimate the amount of carbon dilution and allow more accurate correction of ^{14}C ages of waters. Johnson and DePaolo (1996) presented a method for inversion of $^{87}Sr/^{86}Sr$ and ^{14}C data to obtain carbonate dissolution rates and dissolution-corrected water velocities and applied this model to the Lincolnshire Limestone aquifer using the data of Bishop *et al.* (1994).

Naftz *et al.* (1997) used strontium isotope ratios in groundwaters from southeastern Utah to determine if re-injected oil-field brines or water from a deeper aquifer was the source of salinity to a shallower aquifer that is an important source of drinking water. Distinct $^{87}Sr/^{86}Sr$ ratios in the various waters allowed the authors to verify that the oil-field brine was not the source of salinity, whereas the deeper aquifer was a plausible source. Armstrong *et al.* (1998) used $^{87}Sr/^{86}Sr$ and Ca/Sr ratios to study groundwater mixing and rock–water interaction in the Milk River Aquifer, Alberta, Canada. This study elucidated the importance of the variable reactivity of different sedimentary rocks along differing groundwater flow paths. The $^{87}Sr/^{86}Sr$ of recharging groundwater was modified by the local lithology, causing distinct geochemical patterns along varying flow paths. This could be verified by calculating porewater fluid evolution paths for the various aquifer materials. Woods *et al.* (2000) used strontium isotopes and major elements to study the chemical evolution and movement of groundwater along flow paths in a coastal limestone aquifer in North Carolina. Variations in $^{87}Sr/^{86}Sr$ were explained largely by mixing of groundwater infiltrating from a surficial aquifer. Johnson *et al.* (2000) used strontium isotopes in groundwater in the Snake River Plain, Idaho, to identify groundwater "slow-flow zones" and "fast-flow zones" in a largely basaltic fracture flow aquifer (Figure 11).

19.6.1.2 Uranium isotopes

Osmond and Cowart (1992) and Nimz (1998) recently reviewed the literature on the use of uranium isotopes in groundwater studies. These reviews cite early studies, such as Spiridonov *et al.* (1969), Cherdyntsev (1971), and Osmond *et al.* (1974), as well as more recent papers. As in the case of uranium isotope studies in the soil environment, it has been widely observed that the $^{234}U/^{238}U$ ratio in groundwaters deviates from the secular equilibrium value (Osmond and Cowart, 1992). The highest $^{234}U/^{238}U$ ratios have been recorded in slow-moving groundwaters because of the preferential mobilization of ^{234}U in systems with overall low uranium concentrations (Cowart and Osmond, 1980). In fast moving groundwater systems such as karstic carbonate bedrock, the $^{234}U/^{238}U$ activity ratios are usually much lower (Osmond *et al.*, 1974; Osmond and Cowart, 1976), although some high ratios have been observed (e.g., Kronfeld *et al.*, 1975, 1994).

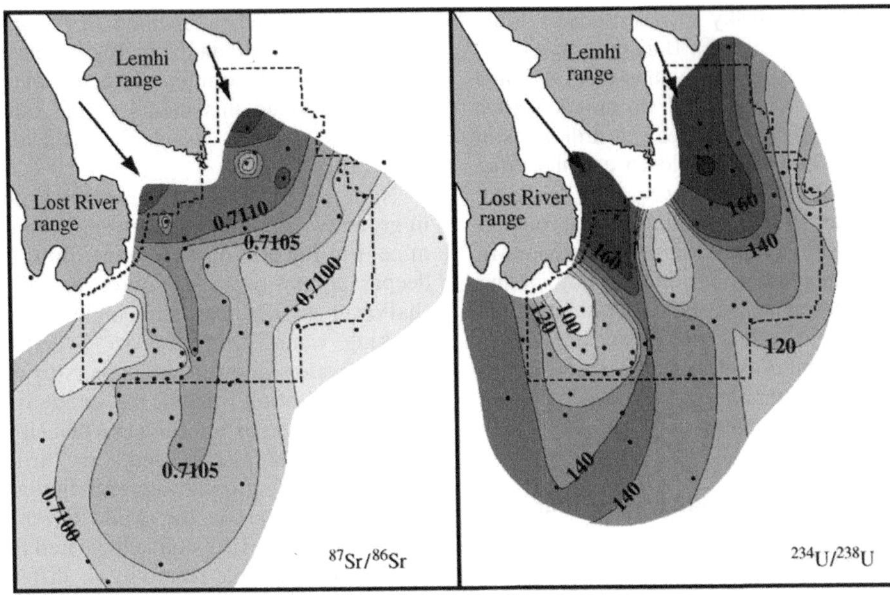

(a) (b)

Figure 11 (a) $^{87}Sr/^{86}Sr$ and (b) $^{234}U/^{238}U$ ratios of groundwater in and around the Idaho National Engineering and Environmental Laboratory on the Snake River Plain. High $^{87}Sr/^{86}Sr$ and $^{234}U/^{238}U$ waters enter the Snake River system (arrows) and react with lower $^{87}Sr/^{86}Sr$ and $^{234}U/^{238}U$ basalt in the study area. Lower $^{87}Sr/^{86}Sr$ and $^{234}U/^{238}U$ groundwaters occur in "slow-flow" zones due to prolonged interaction with basalts whereas "fast-flow" zones have higher $^{87}Sr/^{86}Sr$ and $^{234}U/^{238}U$ (sources Johnson *et al.*, 2000; Roback *et al.*, 2001).

Andrews and Kay (1982, 1983) measured uranium concentrations and $^{234}U/^{238}U$ activity ratios in several aquifers. They observed that changes in the redox conditions of the aquifer strongly affect uranium concentrations in the groundwater and that these changes are sometimes accompanied by changes in the $^{234}U/^{238}U$ activity ratios. They proposed that in certain cases the isotopic composition of uranium could be explained by the sealing of uranium-bearing surfaces by rock cementation.

Several studies have utilized uranium isotopes to classify aquifers, to trace the movement of groundwater, to investigate the mixing of groundwaters with surface water, and to identify the penetration of seawater into aquifers (e.g., Kraemer, 1981; Lienert *et al.*, 1994). Other studies have investigated patterns of rock–water interactions, the dating of groundwater, and the secondary accumulation of uranium (e.g., Banner *et al.*, 1990; Chalov, 1989; Cowart and Osmond, 1977; Hussain, 1995). Many papers have also dealt with hydrothermal systems and with brines where the isotopic composition of uranium was affected by the long residence time of the water in the subsurface environment. For a comprehensive list of these studies, see Osmond and Cowart (1992) and Nimz (1998).

Roback *et al.* (2001) analyzed the same samples from the strontium isotope study of Johnson *et al.* (2000) in the Snake River Plain, described in the previous section, for $^{234}U/^{238}U$ activity ratios. This was done in order to trace the lateral distribution and possible mixing of groundwaters in the eastern Snake River aquifer. Uranium isotopes in conjunction with strontium isotopes further clarified the groundwater "slow-flow zones" and "fast-flow zones" identified by Johnson *et al.* (2000). These companion studies in the Snake River Plain nicely demonstrate the advantages of radiogenic isotope ratios, which are unaffected by solute losses due to mineral precipitation, ion exchange, adsorption, or evaporation, and thus provide a clearer picture of groundwater flow patterns than can be obtained from the study of solute concentrations alone (Figure 11).

19.6.2 Local-scale Groundwater Studies

19.6.2.1 Strontium isotopes

Several studies have used radiogenic isotopes to study groundwater questions on relatively short (or local) length scales (<2 km). Stuckless *et al.* (1991) combined the use of strontium and uranium isotope measurements, in studies of groundwaters and secondary calcite deposits in fault zones at Yucca Mountain, Nevada. They tested whether veins formed by upwelling of deep-seated waters or infiltration of surface waters and found that vein deposits were isotopically distinct from any of the groundwaters in the area suggesting that they were formed by infiltrating rather than upwelling waters. A later study of $^{87}Sr/^{86}Sr$ in vein fillings from drill holes in the area came to a similar conclusion as the earlier study (Marshall *et al.*, 1992). However, a subsequent modeling study suggested that a conclusive indication of the paleowater table elevation is not possible because of the effects of water–rock interaction (Johnson and DePaolo, 1994).

Katz and Bullen (1996) used strontium along with hydrogen, oxygen, and carbon isotopes to study the interaction between groundwater, lake water, and aquifer minerals around a seepage lake in mantled karst terrain of northern Florida. They were able to effectively identify mixing of lake water leakage with groundwater and to study processes of mineral–water interaction. Johnson and DePaolo (1997a) used strontium isotopes to investigate groundwater flow-paths and velocities beneath the Lawrence Berkeley National Laboratory, California, by analyzing well waters and aquifer materials. Variations in $^{87}Sr/^{86}Sr$ in the subsurface were interpreted as a cation-exchange front coupled with the dissolution of aquifer minerals. The data were inverted using a one-dimensional transport–dissolution–exchange model (Johnson and DePaolo, 1997b), yielding average flow velocities consistent with ^{14}C measurements.

Peterman and Wallin (1999) used strontium isotope ratios in groundwater to study a fracture flow system in granitic bedrock on an island in southeastern Sweden. Strontium isotope ratios were used to help constrain three distinct end-member waters that include precipitation infiltration, old saline water in low-conductivity zones, and Baltic Sea water. Systematic trends in groundwater composition were attributed to mixing of these components. Siegel *et al.* (2000) explored the use of strontium and lead isotopes to identify sources of water beneath a large municipal landfill on Staten Island, New York. Lead isotope ratios varied widely and were not useful in distinguishing the origin of waters in the groundwater system. Strontium isotope ratios were very sensitive to seawater inputs and were used to argue that many deep wells were not contaminated by landfill leachates (Siegel *et al.*, 2000). Bohlke and Horan (2000) used strontium isotopes to trace inputs of fertilizer to groundwater systems in small agricultural watersheds in Maryland. This study pointed out the potential for fertilizer application to interfere with the natural geochemistry of strontium and raised the possibility that strontium can be useful for tracing non-point-sources in catchments.

Hogan *et al.* (2000) used $^{87}Sr/^{86}Sr$ and Ca/Sr ratios to study groundwater discharge and precipitation recharge in bog-fen systems of the Glacial Lake Agassiz Peatland in northern Minnesota. Precipitation recharge pore waters had a very distinct composition similar to precipitation, whereas groundwater discharge had higher $^{87}Sr/^{86}Sr$ and Ca/Sr ratios reflecting deep flow through Precambrian crystalline bedrock. A clear chemical differentiation between these end-member waters allowed elucidation of mixing relationships and flow paths (Hogan *et al.*, 2000; Figure 12).

19.6.2.2 Uranium isotopes

Uranium isotopes have also been used to study local groundwater problems, in particular those related to the mobility of uranium in groundwater near radioactive waste disposal sites and the role of colloids in controlling the subsurface movement of uranium and its decay products (Ivanovich *et al.*, 1988; Short *et al.*, 1988; Suksi *et al.*, 2001; Toulhoat *et al.*, 1996; Hussain, 1995; Gomes and Cabral, 1981). Paces *et al.* (2002) used extensive measurements of $^{234}U/^{238}U$ in order to determine patterns of recharge and groundwater flow in the vicinity of Yucca Mountain, Nevada. Based on $^{234}U/^{238}U$ data they suggested that the aquifer beneath Yucca Mountain is dominated by local recharge and is not affected by the regional groundwater flow. This and the studies of Stuckless *et al.* (1991), Marshall *et al.* (1992), and Johnson and DePaolo (1994) mentioned above are important applications of radiogenic isotopes to contaminant hydrology, as they shed light on the public and scientific controversy surrounding the appropriateness of the development of a high-level nuclear waste repository at Yucca Mountain.

19.6.3 Springs and Small Streams

19.6.3.1 Strontium isotopes

Neumann and Dreiss (1995) and Barbieri and Morotti (2003) used strontium isotopes in spring-waters from the Mono Basin, California, and Monte Vulture, Italy, respectively, to elucidate the subsurface mixing of waters that had traveled through different lithologies. Three relatively recent studies have utilized strontium isotopes in studies of first-order streams and the shallow groundwater feeding into them in order to gain insight into water sources and flow paths during varying hydrologic conditions. Ben Othman *et al.* (1997) studied the strontium isotope, major element, and trace element compositions of a small Mediterranean catchment during a series of rain events over a four-day period. Strontium isotopes proved very effective at differentiating water originating as runoff, from deeper Jurassic carbonate aquifers, and from shallower Miocene marl aquifers. Land *et al.* (2000) used $^{87}Sr/^{86}Sr$, Ba/Sr, and Ca/Sr ratios to distinguish between soil water, shallow groundwater, and deep groundwater and to model the contributions from these sources in samples collected weekly over a one-year period. $^{87}Sr/^{86}Sr$ ratios of these end-member waters were

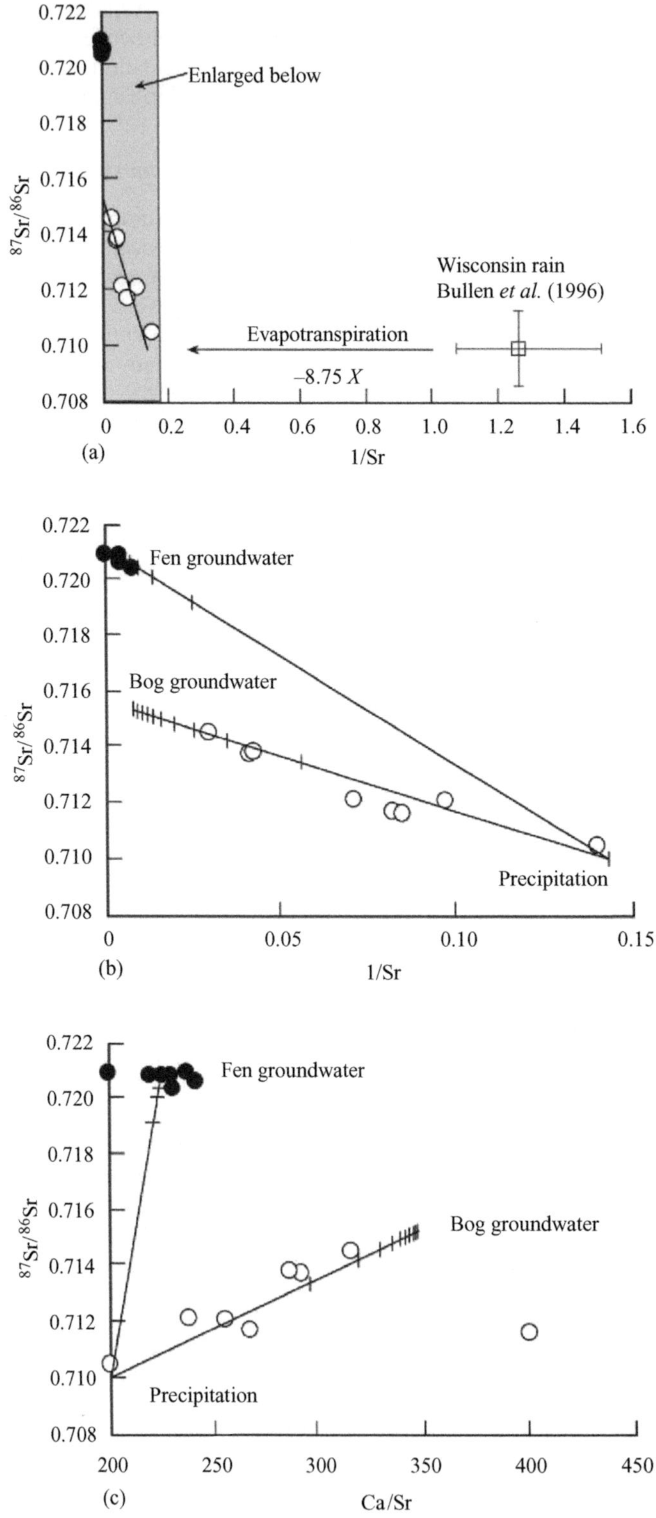

Figure 12 $^{87}Sr/^{86}Sr$ versus (a, b) 1/Sr and (c) Ca/Sr for groundwater samples and precipitation from Lost River in the Glacial Lake Agassiz Peatlands, Minnesota. Mixing trends (with 10% increments) are shown for mixing of evaporated precipitation with distinct groundwater end-members for the adjacent bog and fen sites (source Hogan *et al.*, 2000).

used to test the validity of a stream water source model. Aubert *et al.* (2002a) measured $^{87}Sr/^{86}Sr$ in stream water samples collected weekly, and explored concentration–discharge relationships for $^{87}Sr/^{86}Sr$. Higher $^{87}Sr/^{86}Sr$ ratios were observed during high flow periods, suggesting that flow paths change within the catchment soils depending on moisture conditions. This study demonstrated that changing hydrologic conditions must be carefully considered when sampling stream water for $^{87}Sr/^{86}Sr$ in studies of weathering fluxes. Hogan and Blum (2003) used $^{87}Sr/^{86}Sr$, Ba/Sr, and $\delta^{18}O$ to study water flow-paths during storm events sampled hourly in a small granitic catchment. Hydrograph separations with $\delta^{18}O$ were used to quantify the amount of "new water," while $^{87}Sr/^{86}Sr$ was used to separate "old water" into separate soil water and groundwater components, each with distinct $^{87}Sr/^{86}Sr$ values. In contrast, Bain *et al.* (1998) found no consistent change in $^{87}Sr/^{86}Sr$ with discharge during a storm event in a small andesitic catchment.

19.6.3.2 Uranium isotopes

Only a few studies have been performed using $^{234}U/^{238}U$ ratios as tracers in springs and small streams, and these have mainly used uranium isotopes to trace the interaction of groundwater with surface water (e.g., Lienert *et al.*, 1994). Lienert *et al.* (1994) attempted to link changes in uranium behavior to decreases in the anthropogenic inputs of phosphorous, which in turn affects biological activity and redox conditions within waters. As discussed in an earlier section of this chapter, the $^{234}U/^{238}U$ ratio has also been used to trace changes in the water flow-paths within the Strengbach watershed (Riotte and Chabaux, 1999). In another study, Kronfeld and Rosenthal (1981) used uranium isotopes to trace the movement of water within the Bet-Shean-Harod Valley, Israel. Finally, Barbieri and Voltaggio (1998) combined the use of uranium and strontium isotopes to study the hydrology of the Sangemini (Italy) mineral-water springs, and based on the combined use of these isotopic systems they were able to put constraints on the extent of the aquifer and the spring-water flow paths.

19.7 SUMMARY

Radiogenic isotopes have proven to be an important and powerful tool in investigations of many aspects of weathering and hydrology. The general absence of isotope fractionation of heavy radiogenic isotopes in nature gives these tracers many advantages over major and trace element ratios. Well over 200 articles have been written in this topical area, and it is now evident that this methodology provides important scientific insights, and will increasingly become a routine tool in studies of weathering and hydrology. Strontium isotopes have unquestionably become the most commonly used radiogenic isotope tracer because of the large variability in isotopic composition and the interest in tracing sources and cycling of the analog element calcium. The most notable applications are: (i) use in identifying the dissolution of trace calcium-bearing phases in experiments, soils, and along groundwater flow-paths; (ii) use in differentiating atmospheric from weathering sources of calcium to ecosystems; and (iii) use in differentiating distinct subsurface waters that have interacted with contrasting aquifer materials. The mechanisms controlling differences in strontium isotope ratios in the environment and the hydrogeochemical behavior of strontium have become reasonably well understood and as a result, strontium will be increasingly used in a routine manner in weathering and hydrologic studies.

Lead isotopes have found limited applications in the study of the weathering of uranium and thorium-rich accessory phases in laboratory experiments and in soils. The preferential release of ^{234}U compared to ^{238}U has been more widely used as a tracer of weathering reactions, as a tracer of the geochemical behavior of uranium, and as a tracer of groundwater sources and mixing. Neodymium has proven useful in a few studies as a tracer of the weathering release of trace phases such as apatite, as well as of inputs of atmospheric dust to soils. Osmium has been used in only one weathering study, as of early 2000s, and was useful in inferring rates of dissolution of magnetite in crystalline rocks. In general, the use of neodymium, hafnium, and osmium isotopes in weathering and hydrology is in its infancy, and much additional research will be needed to gain a thorough understanding of the behavior of these systems and to ascertain their usefulness in more routine investigations. We expect that major breakthroughs in the use of radiogenic isotopes in weathering and hydrology will increasingly rely on the combined use of several isotopic systems together, which yield contrasting information and insight into any given scientific application.

REFERENCES

Åberg G. and Jacks G. (1985) Estimation of the weathering rate by $^{87}Sr/^{86}Sr$ ratios. *Geol. Foren. Stock. Forh.* **107**, 289–290.

Åberg G., Jacks G., and Hamilton P. J. (1989) Weathering rates and Sr-87/Sr-86 ratios—an isotopic approach. *J. Hydrol.* **109**, 65–78.

Albarède F. (1995) *Introduction to geochemical modeling.* Cambridge University Press, Cambridge, 543pp.

Andersson P. S., Dahlqvist R., Ingri J., and Gustafsson O. (2001) The isotopic composition of Nd in a boreal river: a reflection of selective weathering and colloidal transport. *Geochim. Cosmochim. Acta* **65**, 521–527.

Andrews J. N. and Kay R. L. F. (1982) $^{234}U/^{238}U$ activity ratios of dissolved uranium in groundwaters from a Jurassic limestone aquifer in England. *Earth Planet. Sci. Lett.* **57**, 139–151.

Andrews J. N. and Kay R. L. F. (1983) The U content and $^{234}U/^{238}U$ activity ratios of dissolved uranium in groundwaters from some Triassic sandstones in England. *Isot. Geosci.* **1**, 101–117.

Armstrong S. C., Sturchio N. C., and Hendry M. J. (1998) Strontium isotopic evidence on the chemical evolution of pore waters in the Milk River Aquifer, Alberta, Canada. *Appl. Geochem.* **13**, 463–475.

Aubert D., Stille P., and Probst A. (2001) REE fractionation during granite weathering and removal by waters and suspended loads: Sr and Nd isotopic evidence. *Geochim. Cosmochim. Acta* **65**, 387–406.

Aubert D., Probst A., Stille P., and Viville D. (2002a) Evidence of hydrological control of Sr behavior in stream water (Strengbach catchment, Vosges mountains, France). *Appl. Geochem.* **17**, 285–300.

Aubert D., Stille P., Probst A., Gauthier-Lafaye F., Pourcelot L., and Del Nero M. (2002b) Characterization and migration of atmospheric REE in soils and surface waters. *Geochim. Cosmochim. Acta* **66**, 3339–3350.

Ault W. U., Senchal R. G., and Erlebach W. E. (1970) Isotopic composition as a natural tracer of lead in the environment. *Environ. Sci. Technol.* **4**, 305–313.

Bacon J. R., Berrow M. L., and Shand C. A. (1995) The use of isotopic composition in field studies of lead in upland Scottish soils (UK). *Chem. Geol.* **124**, 125–134.

Bailey S. W., Hornbeck J. W., Driscoll C. T., and Gaudett H. E. (1996) Calcium inputs and transport in a base-poor forest ecosystem as interpreted by Sr isotopes. *Water Resour. Res.* **32**, 707–719.

Bain D. C. and Bacon J. R. (1994) Strontium isotopes as indicators of mineral weathering in catchments. *Catena* **22**, 201–214.

Bain D. C., Midwood A. J., and Miller J. D. (1998) Strontium isotope ratios in streams and the effect of flow rate in relation to weathering in catchments. *Catena* **32**, 143–151.

Banner J. L. (1995) Application of the trace element and isotope geochemistry of strontium to studies of carbonate diagenesis. *Sedimentology* **42**, 805–824.

Banner J. L., Wasserburg G. J., Dobson P. F., Carpenter A. B., and Moore C. H. (1989) Isotopic and trace element constraints on the origin and evolution of saline groundwaters from central Missouri. *Geochim. Cosmochim. Acta* **52**, 383–398.

Banner J. L., Wasserburg G. J., Chen J. H., and Moore C. H. (1990) ^{234}U–^{238}U–^{230}Th–^{232}Th systematics in saline groundwaters from central Missouri. *Earth Planet. Sci. Lett.* **101**, 296–312.

Banner J. L., Musgrove M., and Capo R. C. (1994) Tracing ground-water evolution in a limestone aquifer using Sr isotopes: effects of multiple sources of dissolved ions and mineral-solution reactions. *Geology* **22**, 687–690.

Banner J. L., Musgrove M., Asmerom Y., Edwards R. L., and Hoff J. A. (1996) High-resolution temporal record of Holocene ground-water chemistry: tracing links between climate and hydrology. *Geology* **24**, 1049–1053.

Barbieri M. and Morotti M. (2003) Hydrogeochemistry and strontium isotopes of spring and mineral waters from Monte Vulture volcano, Italy. *Appl. Geochem.* **18**, 117–125.

Barbieri M. and Voltaggio M. (1998) Applications of Sr isotopes and U-series radionuclides to the hydrogeology of Sangemini area (Terni, central Italy). *Mineral. Petrograph. Acta* **41**, 119–126.

Basu A. R., Jacobsen S. B., Poreda R. J., Dowling C. B., and Aggarwal P. K. (2001) Large groundwater strontium flux to the oceans from the Bengal Basin and the marine strontium isotope record. *Science* **293**, 1470–1473.

Ben Othman D., Luck J. M., and Tournoud M. G. (1997) Geochemistry and water dynamics: application to short time-scale flood phenomena in a small Mediterranean catchment: I. Alkalis, alkali-earths, and Sr isotopes. *Chem. Geol.* **140**, 9–28.

Bierman P. R., Albrecht A., Bothner M. H., Brown E. T., Bullen T. D., Gray L. B., and Turpin L. (1998) Erosion, weathering, and sedimentation. In *Isotope Tracers in Catchment Hydrology* (eds. C. Kendall and J. J. McDonnell). Elsevier, Amsterdam, pp. 647–678.

Bishop P. K., Smalley P. C., Emery D., and Dickson J. A. D. (1994) Strontium isotopes as indicators of the dissolving phase in a carbonate aquifer: implications for ^{14}C dating of groundwater. *J. Hydrol.* **154**, 301–321.

Blaxland A. B. (1974) Geochemistry and geochronology of chemical weathering, Butler Hill Granite, Missouri. *Geochim. Cosmochim. Acta* **38**, 843–852.

Blum J. D. (1995) Isotope decay data. In *Global Earth Physics, A Handbook of Physical Constants* (ed. T. J. Ahrens). American Geophysical Union, pp. 271–282.

Blum J. D. (1997) The effect of late Cenozoic glaciation and tectonic uplift on silicate weathering rates and the marine $^{87}Sr/^{86}Sr$ record. In *Tectonic Uplift and Climate* (ed. W. Ruddiman). Plenum, New York, pp. 259–288.

Blum J. D. and Erel Y. (1995) A silicate weathering mechanism linking increases in marine $^{87}Sr/^{86}Sr$ with global glaciation. *Nature* **373**, 415–418.

Blum J. D. and Erel Y. (1997) Rb–Sr isotope systematics of a granitic soil chronosequence: the importance of biotite weathering. *Geochim. Cosmochim. Acta* **61**, 3193–3204.

Blum J. D., Erel Y., and Brown K. (1993) $^{87}Sr/^{86}Sr$ ratios of Sierra Nevada stream waters: implications for relative mineral weathering rates. *Geochim. Cosmochim. Acta* **58**, 5019–5025.

Blum J. D., Gazis C. A., Jacobson A. D., and Chamberlain C. P. (1998) Carbonate versus silicate weathering in the Raikhot watershed within the high Himalayan crystalline series. *Geology* **26**, 411–414.

Blum J. D., Taliaferro E. H., Weisse M. T., and Holmes R. T. (2000) Changes in Sr/Ca, Ba/Ca and $^{87}Sr/^{86}Sr$ ratios between trophic levels in two forest ecosystems in the northeastern USA. *Biogeochemistry* **49**, 87–101.

Blum J. D., Klaue A., Nezat C. A., Driscoll C. T., Johnson C. E., Siccama T. G., Eagar C., Fahey T. J., and Likens G. E. (2002) Mycorrhizal weathering of apatite as an important calcium source in base-poor forest ecosystems. *Nature* **417**, 729–731.

Bohlke J. K. and Horan M. (2000) Strontium isotope geochemistry of groundwaters and streams affected by agriculture, Locust Grove, MD. *Appl. Geochem.* **15**, 599–609.

Bonotto D. M. and Andrews J. N. (1998) The laboratory evaluation of the transfer of ^{234}U and ^{238}U to the waters interacting with carbonates and implications to the interpretation of field data. *Min. Mag.* **62A**, 187–188.

Bonotto D. M., Andrews J. N., and Darbyshire D. P. F. (2001) A laboratory study of the transfer of ^{234}U and ^{238}U during water-rock interactions in the Carnmenellis granite (Cornwall, England) and implications for the interpretation of field data. *Appl. Radiat. Isotopes* **54**, 977–994.

Borg L. E. and Banner J. L. (1996) Neodymium and strontium isotopic constraints on soil sources in Barbados, West Indies. *Geochim. Cosmochim. Acta* **60**, 4193–4206.

Brannvall M. L., Kurkkio H., Bindler R., Emteryd O., and Renberg I. (2001) The role of pollution versus natural geological sources for lead enrichment in recent lake sediments and surface forest soils. *Environ. Geol.* **40**, 1057–1065.

Brantley S. L., Chesley J. T., and Stillings L. L. (1998) Isotopic ratios and release rates of strontium measured from weathering feldspars. *Geochim. Cosmochim. Acta* **62**, 1493–1500.

Brass G. W. (1976) The variation of marine $^{87}Sr/^{86}Sr$ ratio during phanerozoic time: interpretation using a flux model. *Geochim. Cosmochim. Acta* **40**, 721–730.

Brimhall G. H., Chadwick O. A., Lewis C. J., Compston W., Williams I. S., Danti K. J., Dietrich W. E., Power E. M., Hendricks D., and Bratt J. (1991) Deformational mass transport and invasive processes in soil evolution. *Science* **255**, 695–702.

Brown H. and Silver L. T. (1955) The possibility of obtaining long-term supplies of uranium, thorium, and other substances of uranium from igneous rocks. In *Proceedings of Conference on Peaceful Use of Atomic Energy, Geneva*, IAEA, pp. 91–95.

Bullen T. D., Krabbenhoft D. P., and Kendall C. (1996) Kinetic and mineralogic controls on the evolution of groundwater chemistry and $^{87}Sr/^{86}Sr$ in a sandy silicate aquifer, northern Wisconsin, USA. *Geochim. Cosmochim. Acta* **60**, 1807–1821.

Bullen T. D. and Kendall C. (1998) Tracing of weathering reactions and water flowpaths: a multi-isotope approach. In *Isotope Tracers in Catchment Hydrology* (eds. C. Kendall and J. J. McDonnell). Elsevier, Amsterdam, pp. 611–646.

Bullen T., White A., Blum A. E., Harden J., and Schultz M. (1997) Chemical weathering of a soil chronosequence on granitoid alluvium: II. Mineralogic and isotopic constraints on the behavior of strontium. *Geochim. Cosmochim. Acta* **61**, 291–306.

Cao H., Cowart J. B., and Osmond J. K. (1999) Uranium and strontium isotopic geochemistry of karst waters, Leon Sinks geological area, Leon County, Florida. *Cave Karst Sci.* **26**, 101–106.

Capo R. C. and Chadwick O. A. (1999) Sources of strontium and calcium in desert soil and calcrete. *Earth Planet. Sci. Lett.* **170**, 61–72.

Capo R. C., Stewart B. W., and Chadwick O. A. (1998) Strontium isotopes as tracers of ecosystem processes: theory and methods. *Geoderma* **82**, 173–195.

Capo R. C., Whipkey C. E., and Chadwick O. A. (2000) Pedogenic origin of dolomite in a basaltic weathering profile, Kohala Peninsula, Hawaii. *Geology* **28**, 271–274.

Chabaux F., Riotte J., Benedetti M., Boulegue J., Gerard M., and Ildefonse P. (1998) Uranium isotopes in surface waters from the Mount Cameroon: tracing water sources or basalt weathering? *Min. Mag.* **62A**, 296–297.

Chadwick O. A., Derry L. A., Vitousek P. M., Huebert B. J., and Hedin L. O. (1999) Changing sources of nutrients during four million years of ecosystem development. *Nature* **397**, 491–497.

Chalov P. I. and Merkulova K. I. (1966) Comparative rate of oxidation of ^{234}U and ^{238}U atoms in certain minerals. *Dokl. Akad. Nauk.* **167**, 146–148.

Chalov P. I., Merkulova K. I., and Mamyrov U. I. (1989) Fractionation of even isotopes of uranium in the process of its separation upon various sorbents from some ground water springs. *Geokhimiya* **27**, 588–592.

Chaudhuri S. (1978) Strontium isotopic composition of several oil field brines from Kansas and Colorado. *Geochim. Cosmochim. Acta* **42**, 329–331.

Cherdyntsev V. V. (1971) *Uranium-234*. Israel Program for Scientific Translations, Jerusalem, 234pp.

Chesley J. T., Quade J., and Ruiz J. (2000) The Os and Sr isotopic record of Himalayan paleorivers: Himalayan tectonics and influence on ocean chemistry. *Earth Planet. Sci. Lett.* **179**, 115–124.

Chiquet A., Michard A., Nahon D., and Hamelin B. (1999) Atmospheric input vs. *in situ* weathering in the genesis of calcretes: an Sr isotope study at Galvez (central Spain). *Geochim. Cosmochim. Acta* **63**, 311–323.

Clauer N. (1979) Relationship between the isotopic composition of strontium in newly formed continental clay minerals and their source material. *Chem. Geol.* **31**, 325–334.

Clauer N. (1981) Strontium and argon isotopes in naturally weathered biotite, muscovite, and feldspars. *Chem. Geol.* **31**, 325–334.

Clauer N., O'Neil J. R., and Bonnot-Courtois C. (1982) The effect of natural weathering on the chemical and isotopic compositions of biotites. *Geochim. Cosmochim. Acta* **46**, 1755–1762.

Clow D. W., Mast M. A., Bullen T. D., and Turk J. T. (1997) Strontium 87/strontium 86 as a tracer of mineral weathering reactions and calcium sources in an alpine/subalpine watershed, Loch Vale, Colorado. *Water Resour. Res.* **33**, 1335–1351.

Collerson K. D., Ullman W. J., and Torgersen T. (1988) Ground waters with unradiogenic $^{87}Sr/^{86}Sr$ ratios in the Great Artesian Basin, Australia. *Geology* **16**, 59–63.

Cowart J. B. and Osmond J. K. (1977) Uranium isotopes in groundwater: their use in prospecting for sandstone-type uranium deposits. *J. Geochem. Explor.* **8**, 365–379.

Cowart J. B. and Osmond J. K. (1980) *Uranium Isotopes in Groundwater as a Prospecting Technique*. US Dept. Energy Report, GJBX 119, 112 pp.

Dasch E. J. (1969) Strontium isotopes in weathering profiles, deep-sea sediments, and sedimentary rocks. *Geochim. Cosmochim. Acta* **33**, 1521–1552.

Dickin A. P. (1995) *Radiogenic Isotope Geology*. Cambridge University Press, Cambridge, 452 pp.

Doe B. R., Hedge C. E., and White D. E. (1966) Preliminary investigation of the source of lead and strontium in deep geothermal brines underlying the Salton Sea geothermal area. *Econ. Geol.* **61**, 462–483.

Dymond J., Biscaye P. E., and Rex R. W. (1974) Eolian origin of mica in Hawaiian soils. *Geol. Soc. Am. Bull.* **85**, 37–40.

Elderfield H. and Greaves M. J. (1981) Strontium isotope geochemistry of Icelandic geothermal system and implications for sea water chemistry. *Geochim. Cosmochim. Acta* **45**, 2201–2212.

Emmanuel S. and Erel Y. (2002) Implications from concentrations and isotopic data for Pb partitioning processes in soils. *Geochim. Cosmochim. Acta* **66**, 2517–2527.

English N. B., Quade J., DeCelles P. G., and Garzione C. N. (2000) Geologic control of Sr and major element chemistry in Himalayan rivers, Nepal. *Geochim. Cosmochim. Acta* **64**, 2549–2566.

Erel Y. (1998) Mechanisms and velocities of anthropogenic Pb migration in Mediterranean soils. *Environ. Res.* **78**, 112–117.

Erel Y. and Morgan J. J. (1992) The relationships between rock-derived Pb and Fe in natural waters. *Geochim. Cosmochim. Acta* **56**, 4157–4167.

Erel Y., Patterson C. C., Scott M. J., and Morgan J. J. (1990) Transport of industrial lead in snow through soil to stream water and groundwater. *Chem. Geol.* **85**, 383–392.

Erel Y., Morgan J. J., and Patterson C. C. (1991) Transport of natural lead and cadmium in a remote mountain stream. *Geochim. Cosmochim. Acta* **55**, 707–721.

Erel Y., Harlavan Y., and Blum J. D. (1994) Lead isotope systematics of granitiod weathering. *Geochim. Cosmochim. Acta* **58**, 5299–5306.

Erel Y., Veron A., and Halicz L. (1997) Tracing the transport of anthropogenic lead in the atmosphere and in soils using isotopic ratios. *Geochim. Cosmochim. Acta* **61**, 4495–4505.

Evans M. J., Derry L. A., Anderson S. P., and France-Lanord C. (2001) Hydrothermal source of radiogenic Sr to Himalayan rivers. *Geology* **29**, 803–806.

Eyal Y. and Olander D. R. (1990) Leaching of uranium and thorium from monazite: I. Initial leaching. *Geochim. Cosmochim. Acta* **54**, 1867–1877.

Faure G. (1986) *Principles of Isotopic Geology*. Wiley, New York, 589pp.

Fisher R. S. and Stueber A. M. (1976) Strontium isotopes in selected streams within the Susquehanna River Basin. *Water Resour. Res.* **12**, 1061–1068.

Fleischer R. L. (1980) Isotopic disequilibria of uranium: alpha recoil damage and preferential solution effects. *Science* **207**, 979–981.

Fleischer R. L. (1982) Nature of alpha-recoil damage: evidence from preferential solution effects. *Nuclear Tracks* **6**, 35–42.

Fleischer R. L. and Raabe O. G. (1978) Recoiling alpha-emitting nuclei—mechanisms for uranium series disequilibrium. *Geochim. Cosmochim. Acta* **42**, 973–978.

Franklyn M. T., McNutt R. H., Camineni D. C., Gascoyne M., and Frape S. K. (1991) Groundwater $^{87}Sr/^{86}Sr$ values in the Eye-Dashwa Lakes pluton, Canada: evidence for plagioclase-water reaction. *Chem. Geol.* **86**, 111–122.

Frindik O. and Vollmer S. (1999) Particle-size dependent distribution of thorium and uranium isotopes in soil. *J. Radioanalyt. Nuclear Chem.* **241**, 291–296.

Galy A., France-Lanord C., and Derry L. A. (1999) The strontium isotopic budget of Himalayan rivers in Nepal and Bangladesh. *Geochim. Cosmochim. Acta* **63**, 1905–1925.

Gast P. W. (1970) Isotopic composition as a natural tracer of lead in the environment. *Environ. Sci. Technol.* **4**, 313–314.

Geyh M. A. and Schleicher H. (1990) *Absolute Age Determination.* Springer, Berlin, 503pp.

Goldich S. S. and Gast P. W. (1966) Effects of weathering on the Rb–Sr and K–Ar ages of biotite from the Morton gneiss, Minnesota. *Earth Planet. Sci. Lett.* **1**, 372–375.

Gomes F. V. M. and Cabral F. C. F. (1981) Utilization of natural uranium isotopes for the study of ground water in Bambui limestone aquifer, Bahia. *Revista Brasileira de Geociencias* **11**, 179–184.

Gosz J. R. and Moore D. I. (1989) Strontium isotope studies of atmospheric inputs to forested watersheds in New Mexico. *Biogeochemistry* **8**, 115–134.

Gosz J. R., Brookins D. G., and Moore D. I. (1983) Using strontium isotope ratios to estimate inputs to ecosystems. *Bioscience* **33**, 23–30.

Graustein W. C. (1989) $^{87}Sr/^{86}Sr$ ratios measure the sources and flow of strontium in terrestrial ecosystems. In *Stable Isotopes in Ecological Research* (eds. P. W. Rundel, J. R. Ehleringer, and K. A. Nagy). Springer, New York, pp. 491–512.

Graustein W. C. and Armstrong R. L. (1983) The use of $^{87}Sr/^{86}Sr$ ratios to measure atmospheric transport into forested watersheds. *Science* **219**, 289–292.

Gulson B. L., Tiller K. G., Mizon K. J., and Merry R. H. (1981) Use of lead isotopes in soils to identify the source of lead contamination near Adelaide, South Australia. *Environ. Sci. Technol.* **15**, 691–696.

Hakam O. K., Choukri A., Reyss J. L., and Lferde M. (2001) Determination and comparison of uranium and radium isotopes activities and activity ratios in samples from some natural water sources in Morocco. *J. Environ. Radioact.* **57**, 175–189.

Hamidi E. M., Nahon D., McKenzie J. A., Michard A., Colin F., and Kamel S. (1999) Marine Sr (Ca) input in Quaternary volcanic rock weathering profiles from the Mediterranean coast of Morocco; Sr isotopic approach. *Terra Nova* **11**, 157–161.

Harlavan Y. and Erel Y. (2002) The release of Pb and REE from granitoids by the dissolution of accessory phases. *Geochim. Cosmochim. Acta* **66**, 837–848.

Harlavan Y., Erel Y., and Blum J. D. (1998) Systematic changes in lead isotopic composition with soil age in glacial granitic terrains. *Geochim. Cosmochim. Acta* **62**, 33–46.

Harris N., Bickle M., Chapman H., Fairchild I., and Bunbury J. (1998) The significance of Himalayan rivers for silicate weathering rates: evidence from the Bhote Kosi tributary. *Chem. Geol.* **144**, 205–220.

Hart S. R. and Tilton G. R. (1966) The isotope geochemistry of strontium and lead in Lake Superior sediments and water. In *The Earth Beneath the Continents.* American Geophysical Union, Washington, DC, vol. 10.

Hodell D. A. (1994) Editorial: progress and paradox in strontium isotope stratigraphy. *Paleoceanography* **9**, 395–398.

Hogan J. F. and Blum J. D. (2003) Tracing hydrologic flowpaths in a small watershed using variations in $^{87}Sr/^{86}Sr$, [Ca]/[Sr], [Ba]/[Sr] and $\delta^{18}O$. *Water Resour. Res.* (in press).

Hogan J. F., Blum J. D., Siegel D. I., and Glaser P. H. (2000) $^{87}Sr/^{86}Sr$ as a tracer of groundwater discharge and precipitation recharge in the Glacial Lake Agassiz Peatlands, northern Minnesota. *Water Resour. Res.* **36**, 3701–3710.

Horton T. W., Chamberlain C. P., Fantle M., and Blum J. D. (1999) Chemical weathering and lithologic controls of water chemistry in a high-elevation river system: Clark's fork of the Yellowstone River, Wyoming and Montana. *Water Resour. Res.* **35**, 1643–1656.

Hussain N. (1995) Supply rates of natural U–Th series radionuclides from aquifer solids into groundwater. *Geophys. Res. Lett.* **22**, 1521–1524.

Innocent C., Michard A., Malengreau N., Loubet M., Noack Y., Benedetti M., and Hamelin B. (1997) Sr isotopic evidence for ion-exchange buffering in tropical laterites from the Parana, Brazil. *Chem. Geol.* **136**, 219–232.

Ivanovich M., Duerden P., Payne T., Nightingale T., Longworth G., Wilkins M.A., Hasler S. E., Edgehill R. B., Cockayne D. J., and Davey B. G. (1988) Natural analogue study of the distribution of uranium series radionuclides between colloid and solute phases in hydrological systems. DOE report AERE-R. 12975/DOE/RW/88.076.

Jacks G., Aberg G., and Hamilton P. J. (1989) Calcium budgets for catchments as interpreted by strontium isotopes. *Nordic Hydrol.* **20**, 85–96.

Jacobson A. D., Blum J. D., Chamberlain C. P., Poage M. A., and Sloan V. F. (2002a) The Ca/Sr and Sr isotope systematics of a Himalayan glacial chronosequence: carbonate versus silicate weathering rates as a function of landscape surface age. *Geochim. Cosmochim. Acta* **66**, 13–27.

Jacobson A. D., Blum J. D., Chamberlain C. P., Craw D., and Koons P. O. (2002b) Climatic versus tectonic controls on weathering in the New Zealand Southern Alps. *Geochim. Cosmochim. Acta* **66**, 3417–3429.

Johnson T. M. and DePaolo D. J. (1994) Interpretation of isotopic data in groundwater systems: model development and application to Sr isotope data from Yucca Mountain. *Water Resour. Res.* **30**, 1571–1587.

Johnson T. M. and DePaolo D. J. (1996) Reaction-transport models for radiocarbon in groundwater: the effects of longitudinal dispersion and the use of Sr isotope ratios to correct for water-rock interaction. *Water Resour. Res.* **32**, 2203–2212.

Johnson T. M. and DePaolo D. J. (1997a) Rapid exchange effects on isotope ratios in groundwater systems: 1. Development of a transport–dissolution–exchange model. *Water Resour. Res.* **33**, 187–195.

Johnson T. M. and DePaolo D. J. (1997b) Rapid exchange effects on isotope ratios in groundwater systems: 2. Flow investigation using Sr isotope ratios. *Water Resour. Res.* **33**, 197–209.

Johnson T. M., Roback R. C., McLing T. L., Bullen T. D., DePaolo D. J., Doughty C., Hunt R. J., Smith R. W., Cecil L. D., and Murrell M. T. (2000) Groundwater "fast paths" in the Snake River Plain aquifer: radiogenic isotope ratios as natural groundwater tracers. *Geology* **28**, 871–874.

Karim A. and Veizer J. (2000) Weathering processes in the Indus River Basin: implications from riverine carbon, sulfur, oxygen, and strontium isotopes. *Chem. Geol.* **170**, 153–177.

Katz B. G. and Bullen T. D. (1996) The combined use of $^{87}Sr/^{86}Sr$ and carbon and water isotopes to study the hydrochemical interaction between groundwater and lakewater in mantled karst. *Geochim. Cosmochim. Acta* **60**, 5075–5087.

Kendall C. and Caldwell E. A. (1998) Fundamentals of isotope geochemistry. In *Isotope Tracers in Catchment Hydrology* (eds. C. Kendall and J. J. McDonnell). Elsevier, Amsterdam, pp. 51–86.

Kennedy M. J., Chadwick O. A., Vitousek P. M., Derry L. A., and Hendricks D. M. (1998) Changing sources of base cations during ecosystem development, Hawaiian Islands. *Geology* **26**, 1015–1018.

Kennedy M. J., Hedin L. O., and Derry L. A. (2002) Decoupling of unpolluted temperate forests from rock nutrient sources revealed by natural $^{87}Sr/^{86}Sr$ and ^{84}Sr tracer addition. *Proc. Natl. Acad. Sci.* **99**, 9639–9644.

Kigoshi K. (1971) Alpha-recoil ^{234}Th: dissolution into water and the $^{234}U/^{238}U$ disequilbrium in nature. *Science* **173**, 47–48.

Kovalev V. P. and Malyasova Z. V. (1971) The content of mobile uranium in extrusive and intrusive rocks of the eastern margins of the South Minusinsk Basin. *Geokhimiya* **7**, 855–865.

Kraemer T. F. (1981) ^{234}U and ^{238}U concentration in brine from geopressured aquifers of the northern Gulf of Mexico Basin. *Earth Planet. Sci. Lett.* **56**, 210–216.

Kraemer T. F. and Genereux D. O. (1998) Applications of uranium- and thorium-series radionuclides in catchment hydrology studies. In *Isotope Tracers in Catchment Hydrology* (eds. C. Kendall and J. J. McDonnell). Elsevier, Amsterdam, pp. 679–722.

Kronfeld J. and Rosenthal E. (1981) Uranium isotope as a natural tracer of waters of the Bet-Shean-Harod Valley, Israel. *Israel J. Hydrol.* **22**, 77–88.

Kronfeld J., Gradsztan E., Muller H. W., Radin J., Yaniv A., and Zach R. (1975) Excess ^{234}U: an aging effect in confined water. *Earth Planet. Sci. Lett.* **27**, 189–196.

Kronfeld J., Vogel J. C., and Talma A. S. (1994) A new explanation for extreme $^{234}U/^{238}U$ disequilibria in a dolomitic aquifer. *Earth Planet. Sci. Lett.* **123**, 81–93.

Kurtz A. C., Derry L. A., and Chadwick O. A. (2001) Accretion of Asian dust to Hawaiian soils: isotopic, elemental and mineral mass balance. *Geochim. Cosmochim. Acta* **65**, 1971–1983.

Land M., Ingri J., Andersson P. S., and Ohlander B. (2000) Ba/Sr, Ca/Sr and $^{87}Sr/^{86}Sr$ ratios in soil water and groundwater: implications for relative contributions to stream water discharge. *Appl. Geochem.* **15**, 311–325.

Lathan A. G. and Schwarcz H. P. (1987) On the possibility of determining rates of removal of uranium from crystalline igneous rocks using U-series disequilibria: 1. A U-leach model, and its applicability to whole rock data. *Appl. Geochem.* **2**, 55–65.

Lienert C., Short S. A., and von Gunten H. R. (1994) Uranium infiltration from a river to shallow groundwater. *Geochim. Cosmochim. Acta.* **58**, 5455–5463.

Lowson R. T., Short S. A., Davey B. G., and Gray D. J. (1986) $^{234}U/^{238}U$ and $^{230}Th/^{234}U$ activity ratios in mineral phases in a lateritic weathered zone. *Geochim. Cosmochim. Acta* **50**, 1697–1702.

Luck J. and Ben Othman D. (1998) Geochemistry and water dynamics: II. Trace metals and Pb–Sr isotopes as tracers of water movements and erosion processes. *Chem. Geol.* **150**, 263–282.

Lyons W. B., Tyler S. W., Gaudette H. E., and Long D. T. (1995) The use of strontium isotopes in determining groundwater mixing and brine fingering in a playa spring zone, Lake Tyrrell, Australia. *J. Hydrol.* **167**, 225–239.

Mariner R. H. and Young H. W. (1996) Lead and strontium isotopes indicate deep thermal-aquifer in Twin Falls, Idaho, area. *Feder. Geotherm. Res. Prog. Update* **4**, 135–140.

Marshall D. D., Whelan J. F., Peterman Z. E., Futa K., Mahan S. A., and Struckless J. S. (1992) Isotopic studies of fracture coatings at Yucca Mountain, Nevada, USA. In *Proceedings 7th Water–Rock Interaction Symposium Park City, UT* (eds. Y. K. Kharaka and A. S. Maest). A. A. Balkema, Roterdam, pp. 737–740.

McNutt R. H., Frape S. K., Fritz P., Jones M. G., and MacDonald I. M. (1990) The $^{87}Sr/^{86}Sr$ values of Canadian shield brines and fracture minerals with applications to groundwater mixing, fracture history, and geochronology. *Geochim. Cosmochim. Acta* **54**, 205–215.

Michel J. (1984) Redistribution of uranium and thorium series isotopes during isovolumetric weathering of granite. *Geochim. Cosmochim. Acta* **48**, 1249–1255.

Miller E. K., Blum J. D., and Friedland A. J. (1993) Determination of soil exchangeable-cation loss and weathering rates using Sr isotopes. *Nature* **362**, 438–441.

Moore W. S. (1967) Amazon and Mississippi river concentration of uranium, thorium, and radium isotopes. *Earth Planet. Sci. Lett.* **2**, 231–234.

Moreira-Nordemann L. M. (1980) Use of $^{234}U/^{238}U$ disequilibrium in measuring chemical weathering rate of rocks. *Geochim. Cosmochim. Acta* **44**, 103–108.

Musgrove M. and Banner J. L. (1993) Regional groundwater mixing and the origin of saline fluids—mid-continent, United States. *Science* **259**, 1877–1882.

Naftz D. L., Peterman Z. E., and Spangler L. E. (1997) Using delta ^{87}Sr values to identify sources of salinity to a freshwater aquifer, Greater Aneth Oil Field, Utah USA. *Chem. Geol.* **141**, 195–209.

Naiman Z., Quade J., and Patchett P. J. (2000) Isotopic evidence for eolian recycling of pedogenic carbonate and variations in carbonate dust sources throughout the southwest United States. *Geochim. Cosmochim. Acta* **64**, 3099–3109.

Neumann K. and Dreiss S. (1995) Strontium 87/strontium 86 ratios as tracers in groundwater and surface waters in Mono Basin, California. *Water Resour. Res.* **31**, 3183–3193.

Nimz G. J. (1998) Lithogenic and cosmogenic tracers in catchment hydrology. In *Isotope Tracers in Catchment Hydrology* (eds. C. Kendall and J. J. McDonnell). Elsevier, Amsterdam, pp. 247–290.

Ohlander B., Ingri J., Land M., and Schoberg H. (2000) Change of Sm–Nd isotope composition during weathering of till. *Geochim. Cosmochim. Acta* **64**, 813–820.

Osmond J. K. and Cowart J. B. (1976) The theory and uses of natural uranium isotopic variations in hydrology. *Atom. Energy Rev.* **14**, 621–679.

Osmond J. K. and Cowart J. B. (1992) Ground water. In *Uranium Series Disequilibrium: Application to Earth, Marine, and Environmental Sciences* (eds. M. Ivanovich and R. S. Harmon). Oxford University Press, Oxford, pp. 290–333.

Osmond J. K. and Ivanovich M. (1992) Uranium-series mobilization and surface hydrology. In *Uranium Series Disequilibrium: Application to the Earth, Marine, and Environmental Sciences* (eds. M. Ivanovich and R. S. Harmon). Oxford University Press, Oxford, pp. 259–288.

Osmond J. K., Kaufman M. I., and Cowart J. B. (1974) Mixing volume calculations, sources and aging trends of Floridan aquifer water by uranium isotopic methods. *Geochim. Cosmochim. Acta* **38**, 1083–1100.

Paces J. B., Ludwig K. R., Peterman Z. E., and Neymark L. A. (2002) $^{234}U/^{238}U$ evidence for local recharge and patterns of ground-water flow in the vicinity of Yucca Mountain, Nevada, USA. *Appl. Geochem.* **17**, 751–779.

Palmer M. R. and Edmond J. M. (1992) Controls over the strontium isotope composition of river water. *Geochim. Cosmochim. Acta* **56**, 2099–2111.

Palmer M. R. and Edmond J. M. (1993) Uranium in river water. *Geochim. Cosmochim. Acta* **56**, 4947–4955.

Palmer M. R. and Elderfield H. (1985) Sr isotope composition of sea water over the past 75 Myr. *Nature* **314**, 526–528.

Pande K., Sarin M. M., Trivedi J. R., Krishnaswami S., and Sharma K. K. (1994) The Indus River system (India–Pakistan): major-ion chemistry, uranium and strontium isotopes. *Chem. Geol.* **116**, 245–259.

Pegram W. J., Esser B. K., Krishnaswami S., and Turekian K. K. (1994) The isotopic composition of leachable osmium from river sediments. *Earth Planet. Sci. Lett.* **128**, 591–599.

Peterman Z. E. and Wallin B. (1999) Synopsis of strontium isotope variations in groundwater at Aspo, southern Sweden. *Appl. Geochem.* **14**, 939–951.

Peters S. C., Blum J. D., Driscoll C. T., and Likens G. E. (2003) Dissolution of wollastonite during the experimental manipulation of a forested catchment. *Biogeochemistry* (in press).

Peucker-Ehrenbrink B. and Blum J. D. (1998) Re–Os isotope systematics and weathering of Precambrian crustal rocks: implications for the marine osmium isotope record. *Geochim. Cosmochim. Acta* **62**, 3193–3203.

Plater A. J., Ivanovich M., and Dugdale R. E. (1992) Uranium series disequilibrium in river sediments and waters: the significance of anomalous activity ratios. *Appl. Geochem.* **7**, 101–110.

Probst A., El Gh'mari A., Aubert D., Fritz B., and McNutt R. (2000) Strontium as a tracer of weathering processes in a silicate catchment polluted by acid atmospheric inputs, Strengbach, France. *Chem. Geol.* **170**, 203–219.

Quade J., Chivas A. R., and McCulloch M. T. (1995) Strontium and carbon isotope tracers and the origins of soil carbonate in South Australia and Victoria. In *Arid-Zone*

Paleoenvironments Palaeogeogr, Palaeoclimatol. Palaeoecol. 113 (ed. A. R. Chivas). pp.103–117.

Quade J., Roe L., DeCelles P. G., and Ojha T. P. (1997) The late Neogene $^{87}Sr/^{86}Sr$ record of lowland Himalayan rivers. *Science* **276**, 1828–1831.

Riotte J. and Chabaux F. (1999) ($^{234}U/^{238}U$) activity ratios in freshwaters as tracers of hydrological processes: the Strengbach watershed (Vosges, France). *Geochim. Cosmochim. Acta* **63**, 1263–1275.

Roback R. C., Johnson T. M., McLing T. L., Murrell M. T., Luo S. D., and Ku T. L. (2001) Groundwater flow patterns and chemical evolution in Snake River Plain aquifer in the vicinity of the INEEL: constraints from $^{234}U/^{238}U$ and $^{87}Sr/^{86}Sr$ isotope ratios. *Geol. Soc. Am. Bull.* **113**, 1133–1141.

Romero G. E. T., Ordonez R. E., Esteller A. M. V., and Reyes G. L. R. (1999) Uranium behaviour through the unsaturated zone in soil. In *Environmental Radiochemical Analysis* Special Publication, 234 (ed. G. W. Newton). Royal Society of Chemistry, , pp. 143–151.

Rosholt J. N. and Bartel A. J. (1969) Uranium, thorium, and lead systematics in Granite Mountains, Wyoming. *Earth Planet. Sci. Lett.* **7**, 141–147.

Rosholt J. N., Shields W. R., and Garner E. L. (1963) Isotope fractionation of uranium in sandstone. *Science* **139**, 224–226.

Sarin M., Krishnaswami S., Somayajulu B. L. K., and Moore W. S. (1990) Chemistry of U, Th, and Ra isotopes in the Ganga–Brahamputra river system: weathering processes and fluxes to the bay of Bengal. *Geochim. Cosmochim. Acta* **54**, 1387–1396.

Sass E. and Starinsky A. (1979) Behaviour of strontium in subsurface calcium chloride brines: southern Israel and Dead Sea rift valley. *Geochim. Cosmochim. Acta* **43**, 885–895.

Scott M. R. (1982) The chemistry of U- and Th-series nuclides in rivers. In *Uranium Series Disequilbrium: Application to Earth, Marine, and Environmental Sciences* (eds. M. Ivanovich and R. S. Harmon). Oxford University Press, Oxford, pp. 181–202.

Seimbille F., Zuddas P., and Michard G. (1998) Granite-hydrothermal interaction: a simultaneous estimation of the mineral dissolution rate based on the isotopic doping technique. *Earth Planet. Sci. Lett.* **157**, 183–191.

Sharma M. and Wasserburg G. J. (1997) Osmium in the rivers. *Geochim. Cosmochim. Acta* **61**, 5411–5416.

Sharma M., Wasserburg G. J., Hofmann A. W., and Chakrapani G. J. (1999) Himalayan uplift and osmium isotopes in oceans and rivers. *Geochim. Cosmochim. Acta* **63**, 4005–4012.

Short S. A., Lowson R. T., and Ellis J. (1988) $^{234}U/^{238}U$ and $^{230}Th/^{234}U$ activity ratios in the colloidal phases of aquifers in lateritic weathered zones. *Geochim. Cosmochim. Acta* **52**, 2555–2563.

Siegel D. I., Bickford M. E., and Orrell S. E. (2000) The use of strontium and lead isotopes to identify sources of water beneath the Fresh Kills Landfill, Staten Island, New York, USA. *Appl. Geochem.* **15**, 493–500.

Silver L. T., Woodhead J. A., Williams I. S., and Chappell B. W. (1984) *Uranium in Granites from the Southwestern United States: Actinide Parent–Daughter Systems, Sites and Mobilization.* Department of Energy, Report DE-AC13-76GI01664. 380pp.

Smalley P. C., Raheim A., Dickson J. A. D., and Emery D. (1998) $^{87}Sr/^{86}Sr$ in waters from the Lincolnshire Limestone aquifer, England, and the potential of natural strontium isotopes as a tracer for a secondary recovery seawater injection process in oilfields. *Appl. Geochem.* **3**, 591–600.

Spiridonov A. I., Sultankhodzhayev A. N., Surganova N. A., and Tyminskiy V. G. (1969) Some results of the study of uranium isotopes ($^{234}U/^{238}U$) in ground water of the artesian basin in the Tashkent area. *Uzbekiston Geologiya Zhurnali* **4**, 82–84.

Starinsky A., Bielski M., Ecker A., and Steinitz G. (1983a) Tracing the origin of salts in groundwater by Sr isotopic composition (the crystalline complex of the southern Sinai, Egypt). *Chem. Geol.* **41**, 257–267.

Starinsky A., Bielski M., Lazar B., Steinitz G., and Raab M. (1983b) Strontium isotope evidence on the history of oilfield brines, Mediterranean Coastal Plain, Israel. *Geochim. Cosmochim. Acta* **47**, 687–695.

Steinmann M. and Stille P. (1997) Rare earth element behavior and Pb, Sr, Nd isotope systematics in a heavy metal contaminated soil. *Appl. Geochem.* **12**, 607–623.

Stettler A. (1977) $^{87}Rb/87Sr$ systematics of a geothermal water–rock association in the Massif Central, France. *Earth Planet. Sci. Lett.* **34**, 432–438.

Stettler A. and Allègre C. J. (1979) $^{87}Rb–^{87}Sr$ constraints on the genesis and evolution of the Cantal contitntal volcanic system (France). *Earth Planet. Sci. Lett.* **44**, 269–278.

Stewart B. W., Capo R. C., and Chadwick O. A. (1998) Quantitative strontium isotope models for weathering, pedogenesis and biogeochemical cycling. *Geoderma* **82**, 173–195.

Stewart B. W., Capo R. C., and Chadwick O. A. (2001) Effects of rainfall on weathering rate, base cation provenance, and Sr isotope composition of Hawaiian soils. *Geochim. Cosmochim. Acta* **65**, 1087–1099.

Stille P. and Shields G. (1997) *Radiogenic Isotope Geochemistry of Sedimentary and Aquatic Systems.* Springer, Berlin, 217pp.

Stillings L. L. and Brantley S. L. (1995) Feldspar dissolution of 25°C and pH 3: reaction stoichiometry and the effect of cations. *Geochim. Cosmochim. Acta* **59**, 1483–1496.

Straughan I. R., Elseewi A. A., Kaplan I. R., Hurst R. W., and Davis T. E. (1981) Fly ash-derived strontium as an index to monitor deposition from coal-fired power plants. *Science* **212**, 1267–1269.

Stuckless J. S., Bunker C. M., Bush C. A., Doering W. P., and Scott J. H. (1977) Geochemical and petrological studies of uraniferous granite from Granite Mountains, Wyoming. *US Geol. Surv. J. Res.* **5**, 61–81.

Stuckless J. S., Peterman Z. E., and Muhs D. R. (1991) U and Srisotopes in ground water and calcite, Yucca Mountain, Nevada: evidence against upwelling water. *Science* **254**, 551–554.

Steuber A. M., Pushkar P., and Hetherington E. A. (1984) A strontium isotope study of Smackover brines and associated solids, southern Arkansas. *Geochim. Cosmochim. Acta* **48**, 1637–1649.

Stueber A. M., Pushkar P., and Hetherington E. A. (1987) A strontium isotopic study of formation waters from the Illinois Basin, USA. *Appl. Geochem.* **2**, 477–494.

Stueber A. M., Walter L. M., Huston T. J., and Pushkar P. (1993) Formation waters from Mississippian–Pennsylvanian reservoirs, Illinois Basin, USA: chemical and isotopic constraints on evolution and migration. *Geochim. Cosmochim. Acta* **57**, 763–784.

Suksi J., Rasilainen K., Casanova J., Ruskeeniemi T., Blomqvist R., and Smellie J. (2001) U-series disequilibria in a groundwater flow route as an indicator of uranium migration processes. *J. Contamin. Hydrol.* **47**, 187–196.

Szalay S. and Samsoni Z. (1969) Investigation of the leaching of the uranium from crushed magmatic rocks. *Geochemistry* **14**, 613–623.

Taylor A. S. and Lasaga A. C. (1999) The role of basalt weathering in the Sr isotope budget of the oceans. *Chem. Geol.* **161**, 199–214.

Taylor A. S., Blum J. D., and Lassaga A. C. (2000a) The dependence of labradorite dissolution and Sr isotope release rates on solution saturation state. *Geochim. Cosmochim. Acta* **64**, 2389–2400.

Taylor A. S., Blum J. D., Lasaga A. C., and MacInnis I. N. (2000b) Kinetics of dissolution and Sr release during biotite and phlogopite weathering. *Geochim. Cosmochim. Acta* **64**, 1191–1208.

Teutsch N., Erel Y., Halicz L., and Banin A. (2001) The distribution of natural and anthropogenic lead in Mediterranean soils. *Geochim. Cosmochim. Acta* **65**, 2853–2864.

Toulhoat P. and Beaucaire C. (1993) Geochemistry of water crossing the Cigar Lake uranium deposit (Saskatchewan, Canada), and use of uranium and lead isotopes as ore guides. *Can. J. Earth Sci.* **30**, 754–763.

Toulhoat P., Gallien J. P., Louvat D., and Moulin V. (1996) Preliminary studies of groundwater flow and migration of uranium isotopes around the Oklo natural reactors (Gabon). *J. Contamin. Hydrol.* **21**, 3–17.

Tricca A., Stille P., Steinmann M., Kiefel B., and Samuel J. (1999) Rare earth elements and Sr and Nd isotopic compositions of dissolved and suspended loads from small river systems in the Vosges Mountains (France), the river Rhine and groundwater. *Chem. Geol.* **160**, 139–158.

Tuzova T. V. and Novikov V. N. (1991) Uranium isotope-related features of streamflow formation for Pyandzh River. *Water Resour. Res.* **18**, 59–65.

Ullman W. J. and Collerson K. D. (1994) The Sr-isotope record of late-Quaternary hydrologic changes around Lake Frome, South Australia. *Austral. J. Earth Sci.* **41**, 37–45.

Vance D. and Burton K. (1999) Neodymium isotopes in planktonic foraminifera: a record of the response of continental weathering and ocean circulation rates to climate change. *Earth Planet. Sci. Lett.* **173**, 365–379.

van de Flierdt T., Frank M., Lee D., and Halliday A. N. (2002) Glacial weathering and the hafnium isotope composition of seawater. *Earth Planet. Sci. Lett.* **198**, 167–175.

Vitousek P. M., Kennedy M. J., Derry L. A., and Chadwick O. A. (1999) Weathering versus atmospheric sources of strontium in ecosystems on young volcanic soils. *Oecologia* **121**, 255–259.

von Gunten H. R., Roessler E., Lowson R. T., Reid P. D., and Short S. A. (1999) Distribution of uranium and thorium series radionuclides in mineral phases of a weathered lateritic transect of a uranium ore body. *Chem. Geol.* **160**, 225–240.

Wadleigh M. A., Veizer J., and Brooks C. (1985) Strontium and its isotopes in Canadian rivers: fluxes and global implications. *Geochim. Cosmochim. Acta* **49**, 1727–1736.

Whipkey C. E., Capo R. C., Chadwick O. A., and Stewart B. W. (2000) The importance of sea spray to the cation budget of a coastal Hawaiian soil: a strontium isotope approach. *Chem. Geol.* **168**, 37–48.

White A. F., Blum A. E., Bullen T. D., Vivit D. V., Schulz M., and Fitzpatrick J. (1999) The effect of temperature on experimental and natural chemical weathering rates of granitoid rocks. *Geochim. Cosmochim. Acta* **63**, 3277–3291.

Whitehead K., Ramsey M. H., Maskall J., Thornton I., and Bacon J. R. (1997) Determination of the extent of anthropogenic Pb migration through fractured sandstone using Pb isotope tracing. *Appl. Geochem.* **12**, 75–81.

Woods T. L., Fullagar P. D., Spruill R. K., and Sutton-Lynn C. (2000) Strontium isotopes and major elements as tracers of ground water evolution: example from the upper Castle Hayne Aquifer of North Carolina. *Ground Water* **38**, 762–771.

Yang J., Chen J., An Z., Shields G., Tao X., Zhu H., Ji J., and Chen Y. (2000) Variations in $^{87}Sr/^{86}Sr$ ratios of calcites in Chinese loess: a proxy for chemical weathering associated with the East Asian summer monsoon. *Palaeogeogr. Palaeoclimatol. Palaeoecol.* **157**, 151–159.

Zhao S., Liu M., Qiao G., Wu A., and Li C. (1994) A study and application of uranium isotopes in underground and surface water from Liangcheng area, southern Shandong. *Acta Petrologica Sinica* **10**, 202–210.

Zielinski R. A., Peterman Z. E., Stuckless J. S., Rosholt J. N., and Nkomo I. T. (1981) The chemical and isotopical record of rock–water interaction in the Sherman Granite, Wyoming and Colorado. *Contrib. Mineral. Petrol.* **78**, 209–219.

Zuddas P., Seimbille F., and Michard G. (1995) Granite-fluid interaction at near-equilibrium conditions: experimental and theoretical constraints from Sr contents and isotopic ratios. *Chem. Geol.* **121**, 145–154.

Isotope Geochemistry
ISBN: 978-0-08-096710-3

pp. 607–634

20 Long-lived Isotopic Tracers in Oceanography, Paleoceanography, and Ice-sheet Dynamics

S. L. Goldstein and S. R. Hemming
Columbia University, Palisades, NY, USA

20.1 INTRODUCTION

The decay products of the long-lived radioactive systems are important tools for tracing geological time and Earth processes. The main parent-daughter pairs used for studies in the Earth and Planetary sciences are Rb–Sr, Th–U–Pb, Sm–Nd, Lu–Hf, and Re–Os. Traditionally most practitioners have not focused their careers studying paleoceanography or paleoclimate, and the vast majority of investigations using these systems address issues in geochronology, igneous petrology, mantle geochemistry, and mantle and continental evolution. These tools are not currently considered as among the "conventional" tools for oceanographic, paleoceanographic, or paleoclimate studies. Nevertheless, these isotopic tracers have a long history of application as tracers of sediment provenance and ocean circulation. Their utility for oceanographic and paleoclimate studies is becoming increasingly recognized and they can be expected to play an important role in the future.

Frank (2002) published a general review of long-lived isotopic tracers in oceanography and authigenic Fe–Mn sediments. This chapter attempts to avoid duplication of that effort, rather, it focuses on the basis for using authigenic neodymium, lead, and hafnium isotopes in oceanography and paleoceanography. In particular, it reviews in detail the currently available data on neodymium isotopes in the oceans in order to evaluate its strengths and weaknesses as an oceanographic tracer and a proxy to investigate paleocirculation. Neodymium isotope ratios are highlighted because lead isotopes in the present-day oceans are contaminated by anthropogenic input, and dissolved hafnium thus far has not been measured. This chapter only gives a cursory summary of the results of studies on Fe–Mn crusts, which were covered extensively by Frank (2002). This chapter also summarizes the application of strontium, neodymium, and lead isotopes for tracing the sources of continental detritus brought to the oceans by icebergs and implications for the history of the North Atlantic ice sheets. This subject was not covered by Frank (2002), and in this case we summarize the major results. Dissolved strontium and osmium isotopes are treated in another chapter (see Chapter 21).

20.2 LONG-LIVED ISOTOPIC TRACERS AND THEIR APPLICATIONS

Long-lived systems are those with decay rates that are slow relative to the 4.56 Gyr age of the solar system, meaning effectively that the parent element still exists in nature. Therefore, the abundances ratios of the daughter products of decay are still increasing in rocks and water, but at rates slow enough that changes in the daughter-isotope ratios are negligible over short time periods. In this sense they contrast with short-lived radioactive systems, associated, for example, with cosmogenic nuclides and the intermediate products of uranium decay. The abundance of a nuclide formed by radioactive decay is referenced to a "stable" isotope (one that is not a decay product) for the same element. All of the long-lived radioactive decay systems have high atomic masses (from 87 amu to 208 amu) thus chemical and biological mass fractionation effects are small, unlike light stable isotopes such as those of carbon and oxygen. As a result, in a marine or ice-core sample the effects of biological or chemical mass fractionation are negligible and isotope ratios are conservative tracers, reflecting the sources of the individual elements. Transport processes on the Earth's surface generally sample a variety of continental sources, therefore in most cases the daughter isotope ratios represent mixtures from the different age terrains. If the elements are dissolved, then their isotope ratios will remain constant over any travel path as long as no additions from new sources with different isotope ratios are added. For particulates, sorting of minerals or particle sizes from a single source can lead to isotopic variability. Neodymium, lead, and hafnium are relatively insoluble elements, as a result their residence times in seawater are short relative to oceanic mixing times of ~1,500 yr (Broecker and Peng, 1982), and isotope ratios may vary both geographically and with depth in the oceans. The isotopic variability underlies their utilization as water-mass tracers. Thus far, this application has been particularly successful with neodymium.

20.3 SYSTEMATICS OF LONG-LIVED ISOTOPE SYSTEMS IN THE EARTH

Interpretation of the first-order isotopic variability of strontium, neodymium, hafnium, and lead in the Earth is founded on understanding the bulk chemical characteristics of the silicate Earth and how its gross differentiation has affected the parents and daughters. The best understood system among those under consideration here is $^{147}Sm \rightarrow {}^{143}Nd$. Both parent and daughter elements are rare earth elements (REEs) existing solely in the +3 oxidation state in natural systems. The rare earths are refractory during nebula condensation, partitioning into solids at high temperatures, thus the Sm/Nd and $^{143}Nd/^{144}Nd$ ratios of the Earth are considered to be the same as in chondritic meteorites, and there is a consensus on their values (Table 1).

Moreover, they are chemically similar. Both display very limited solubility in water, and differences in chemical behavior are mainly associated with atomic size. Lutetium and hafnium are also refractory elements during nebula condensation, and thus their relative abundance in the Earth are chondritic, however, at the time of writing the value of the bulk Earth $^{176}Hf/^{177}Hf$ ratio and the ^{176}Lu decay constant are being debated (Table 1). Neodymium- and hafnium-isotope ratios in this paper will be expressed as parts per 10^4 deviations from the bulk Earth values, viz., as ε_{Nd} and ε_{Hf}. For example,

$$\varepsilon_{Nd} = \left[\left\{\frac{(^{143}Nd/^{144}Nd)_{sample}}{(^{143}Nd/^{144}Nd)_{bulk\ Earth}}\right\} - 1\right] \times 10^4$$

The bulk Earth compositions of Rb–Sr and Th–U–Pb are less well known, complicated by

Table 1 Accepted or commonly used parameters for long-lived decay systems.

Decay system	*Parent/daughter*	*(Present day)*	*Isotope ratio*	*(Present day)*	*Refs.*
$^{147}Sm \rightarrow ^{143}Nd$	$^{147}Sm/^{144}Nd$	0.1966	$^{143}Nd/^{144}Nd$	0.512638	1
$^{87}Rb \rightarrow ^{87}Sr$	$^{87}Rb/^{86}Sr$	0.09	$^{87}Sr/^{86}Sr$	0.7050	2
$^{176}Lu \rightarrow ^{176}Hf$	$^{176}Lu/^{177}Hf$	0.0332	$^{176}Hf/^{177}Hf$	0.282772	3
$^{176}Lu \rightarrow ^{176}Hf$	$^{176}Lu/^{177}Hf$	0.0334	$^{176}Hf/^{177}Hf$	0.28283	4
				(Initial solar system)	
$^{238}U \rightarrow ^{206}Pb$			$^{206}Pb/^{204}Pb$	9.307	5
$^{235}U \rightarrow ^{207}Pb$			$^{207}Pb/^{204}Pb$	10.294	5
$^{232}Th \rightarrow ^{208}Pb$			$^{208}Pb/^{204}Pb$	29.476	5
Decay constant (yr^{-1})					
$\lambda^{147}Sm$	6.54×10^{-12}				6
$\lambda^{87}Rb$	1.42×10^{-11}				7, 13
$\lambda^{176}Lu$	1.93×10^{-11}				8
$\lambda^{176}Lu$	1.865×10^{-11}				9
$\lambda^{176}Lu$	1.983×10^{-11}				10
$\lambda^{238}U$	1.55125×10^{-10}				11, 13
$\lambda^{235}U$	9.8485×10^{-10}				11, 13
$\lambda^{232}Th$	4.9475×10^{-11}				12, 13

References are as follows: 1. Jacobsen and Wasserburg (1980); 2. O'Nions *et al.* (1977), DePaolo and Wasserburg (1976b); 3. Blichert- Toft and Albare(c)de (1997); 4. Tatsumoto *et al.* (1981) as adjusted by Vervoort *et al.* (1996); 5. Tatsumoto *et al.* (1973); 6. Lugmair and Scheinin (1974); 7. Neumann and Huster (1976); 8. Sguigna *et al.* (1982); 9. Scherer *et al.* (2001); 10. Bizzarro *et al.* (2003); 11. Jaffey *et al.* (1971); 12. Le Roux and Glendenin (1963); and 13. Steiger and Jäger (1977). For Pb isotopes the initial values for the solar system are known much better than the present-day bulk Earth, and these are given. For the Lu–Hf system, both the decay constant and the bulk Earth value are currently in dispute. The isotope ratios of Nd and Hf in the text are given as deviations from the bulk Earth value in parts per 10^4.

the volatility of rubidium and lead during nebular condensation combined with the variable depletion of volatile elements in the Earth. Of these systems, there is a consensus nevertheless for the Rb/Sr and $^{87}Sr/^{86}Sr$ ratios of the bulk Earth (DePaolo and Wasserburg, 1976b; O'Nions *et al.*, 1977). For the Th–U–Pb, the refractory behavior of thorium and uranium puts strong constraints on the $^{208}Pb/^{206}Pb$ of the Earth, but the Th/Pb and U/Pb ratios are not as well constrained. The coupled nature of U–Pb isotopic evolution (two elements, and two decay systems) means that systematics of closed-system evolution of the solar system is well constrained such that planetary bodies should lie on a ~4.56 Ga isochron line in $^{206}Pb/\ ^{204}Pb–^{207}Pb/^{204}Pb$ space. However, balancing the bulk Earth composition to that line has been problematic. Details of these complications are beyond the scope of this contribution, however, there are extensive discussions in the literature.

Long-lived isotopes are useful tracers for paleoceanography and paleoclimate studies because they vary over the surface of the Earth and within. Variability between the continental crust and mantle is primarily a result of (i) the gross chemical differentiation of the Earth, associated with mantle melting and plate tectonics to form continental crust, and the different behaviors of the parent and daughter elements during magma formation, and (ii) the age of the continental crust. Among the systems under consideration, Rb–Sr, Sm–Nd, and Lu–Hf isotopic variability on the Earth's surface follows general expectations from melting behavior, at least in a gross sense. Rubidium, neodymium, and hafnium are more likely to enter magma than their partner elements, and as a result the continents have higher Rb/Sr and lower Sm/Nd and Lu/Hf ratios than the Earth's mantle. During the course of time this has resulted in the continents having high $^{87}Sr/^{86}Sr$ and low $^{143}Nd/^{144}Nd$ and $^{176}Hf/^{177}Hf$ ratios compared to the mantle and bulk Earth (i.e., negative ε_{Nd} and ε_{Hf} values), as illustrated by Figure 1. Mid-ocean ridge and ocean island basalts (MORB and OIB), derived from melting of the mantle, have low $^{87}Sr/^{86}Sr$ and high $^{143}Nd/^{144}Nd$ and $^{176}Hf/^{177}Hf$ ratios (positive ε_{Nd} and ε_{Hf} values). Convergent-margin volcanics have similar isotopic characteristics as oceanic basalts as long as they are not situated on old continental crust or contain large amounts of subducted sediment.

The distinction between continental- and mantle-isotope ratios increases with the age of the continental terrain. Because the age of the continental crust is geographically variable, the continents are isotopically heterogeneous on regional scales and this heterogeneity forms the basis for tracing sources and transport. As a reflection of melting behavior and continental age, isotopic variations between some of these isotope systems are systematic in continental rocks and oceanic basalts. Nd–Hf isotopes are usually positively correlated while Nd–Sr and Hf–Sr isotopes are usually negatively correlated. However, these are gross features of the decay systems and there are some complicating features. For example, a

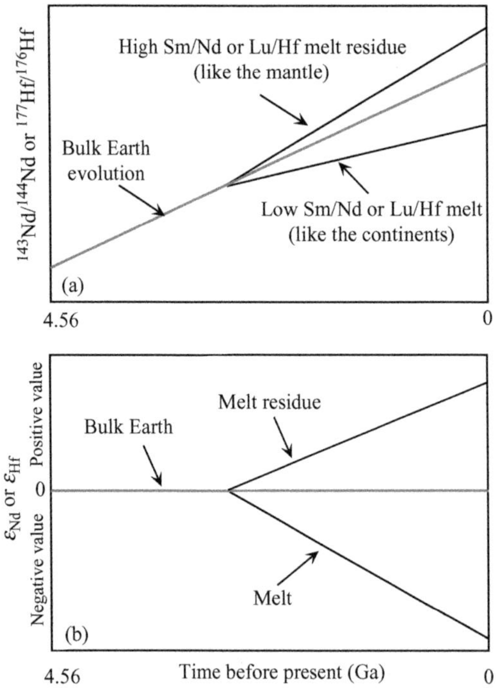

Figure 1 Systematics of Nd- and Hf-isotopic evolution in the bulk Earth, continental crust, and mantle. Daughter elements Nd and Hf are more incompatible during mantle melting (more likely to go into a partial melt of mantle rock) than Sm and Lu, respectively. As a result, the continental crust has a lower Sm/Nd and Lu/Hf ratio than the mantle, and lower Nd- and Hf-isotope ratios. Young continental crust has isotope ratios similar to the mantle, and the older the continental terrain, the lower the Nd- and Hf-isotope ratios. Rb–Sr behaves in the opposite sense, such that the parent element Rb is more incompatible than the daughter element Sr. (a) Schematic example of the evolution of Nd- and Hf-isotope ratios of a melt and the melt residue from a melting event around the middle of Earth history from a source with the composition of the bulk Earth. (b) The same scenario as in (a), but with the isotope ratios plotted as ε_{Nd} and ε_{Hf}. The "bulk Earth" value throughout geological time is defined as ε_{Nd} and $\varepsilon_{Hf} = 0$, and ε-value of a sample is the parts per 10^4 deviation from the bulk Earth value.

significant portion of continental hafnium is located in the mineral zircon, a heavy mineral that can be separated by sedimentary processes. Therefore, in sediments where zircon has been separated by sorting, the Lu/Hf and $^{176}Hf/^{177}Hf$ ratios should be higher than expected from the crustal age. As discussed below, this appears to have important effects on the isotope ratio of dissolved hafnium in the oceans. The Rb–Sr system is more likely than neodymium and hafnium to be disturbed by weathering and metamorphism due to higher solubility of rubidium and strontium in water. In addition, minerals such as micas generally have high Rb/Sr ratios, and feldspars have lower Rb/Sr ratios; therefore, crystal sorting of a sample during transport by currents can result in significant strontium-isotopic heterogeneity of samples from the same source. These processes affecting hafnium and strontium isotopes can degrade correlations with neodymium isotopes in marine or eolian samples.

Lead-isotope ratios in continental rocks and oceanic basalts often do not display strong correlations with Nd–Sr–Hf isotope ratios, even though the melting behavior of thorium and uranium relative to lead is similar to Rb–Sr (Hofmann *et al.*, 1986). $^{206}Pb/^{204}Pb$ ratios usually show substantial overlap in oceanic basalts and continental rocks. In $^{206}Pb/^{204}Pb$–$^{207}Pb/^{204}Pb$ x–y plots, continental rocks generally fall above oceanic basalts, reflecting higher $^{207}Pb/^{204}Pb$ for a given $^{206}Pb/^{204}Pb$ ratio, a consequence of the short half-life of ^{235}U compared to ^{238}U (Table 1) coupled with a higher U/Pb ratio in the continents than the mantle over the first half of Earth history. As a result, continent-derived lead can often be distinguished from mantle-derived lead.

20.3.1 Early Applications to the Oceans

Investigations involving the oceans and marine sediments were among the earliest applications of long-lived radioactive systems. The first major lead isotopic study by Patterson *et al.* (1953) reported data on a Pacific manganese nodule and a red clay. This was followed by Patterson's (1956) classic work on the age of the Earth, where lead-isotope ratios of marine sediments were used to estimate the average lead-isotope composition of the Earth. Although subsequent work has shown the approach to be flawed, this work still is credited for defining the genetic link between the formation of meteorites and the Earth. Chow and Patterson's (1959) classic paper on lead-isotope ratios in manganese nodules can be considered as the work that sets the stage for all future applications of long-lived radioactive isotopes to the oceans. They wrote "the oceans serve as collecting and mixing reservoirs for leads which are derived form vast land areas, thus provide samples of leads ... (reflecting) large continental segments. Lead isotopes can also be used as tracers for the study of circulation and mixing patterns of the oceans." This and their subsequent study (Chow and Patterson, 1962) outlined the geographical variability of lead in manganese nodules and pelagic clays in the oceans. They showed, for example, that different regions have characteristic lead-isotope ratios, and that they are lowest in the northwest Pacific and highest in the northeast Atlantic (Figure 2). They further identified the southwest Atlantic as a province having relatively low lead-isotope ratios, like the north Pacific. They concluded that the source of marine lead is the continental crust, but without

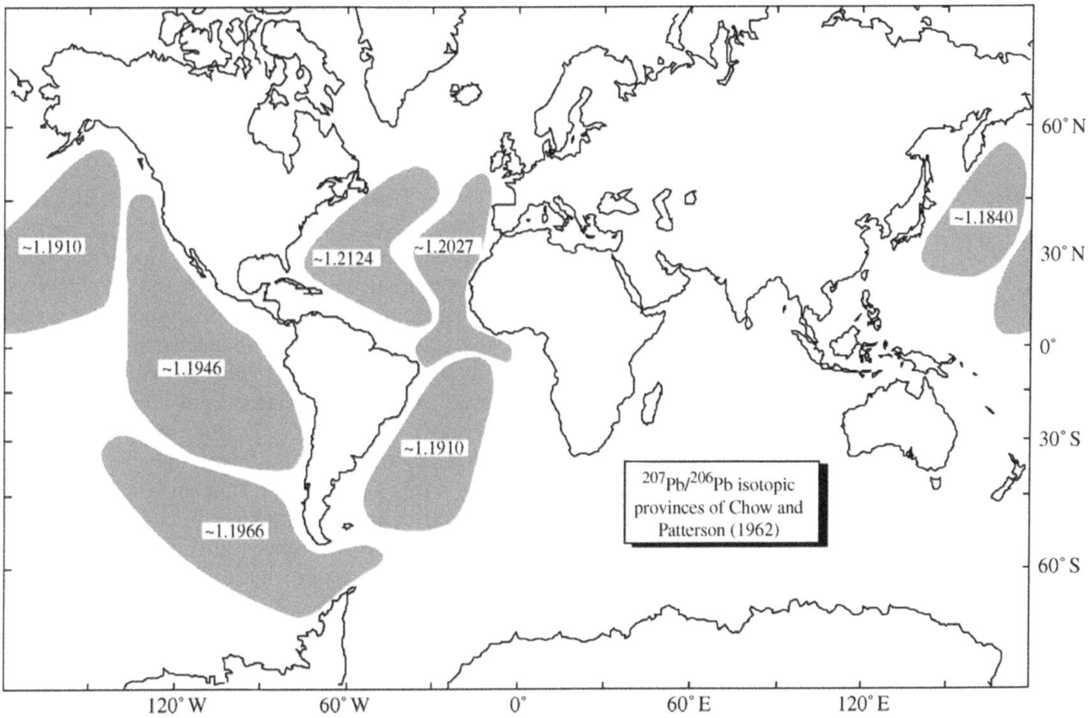

Figure 2 Marine Pb-isotopic provinces defined by Chow and Patterson (1962). The Pb-isotope ratios are given as $^{207}Pb/^{206}Pb$ ratios rather than $^{207}Pb/^{204}Pb$ and $^{206}Pb/^{204}Pb$ because of easier measurement and fractionation control. The map shows that major provinces in the oceans were defined by this early work. Among the implications is that North Atlantic Pb has an old continental provenance, defined by high $^{207}Pb/^{206}Pb$ ratios.

knowledge of isotopic compositions of mantle lead. Nearly a decade later, Reynolds and Dasch (1971) and Dasch *et al.* (1971) confirmed a continental source for the lead in manganese nodules and clays far from ocean ridges, but showed that manganese-rich metalliferous sediments deposited near mid-ocean ridges contain mantle-derived lead, extracted by hydrothermal processes near ridges.

The first neodymium-isotopic analyses aimed at characterizing the oceans closely followed the initial development of neodymium isotopes as a chronometer and tracer (Richard *et al.*, 1976; DePaolo and Wasserburg, 1976a; O'Nions *et al.*, 1977). O'Nions *et al.* (1978) was the first to report neodymium (along with lead and strontium) isotopes in manganese nodules and hydrothermal sediments. They confirmed the distinction between continental and mantle provenances of lead in hydrogenous and hydrothermal manganese sediments, respectively. Consistent with the previous studies, they found that strontium in these deposits is derived from seawater. All of the neodymium-isotope ratios in their samples from the Pacific were similar and lower than the bulk Earth value ($\varepsilon_{Nd} = -1.5$ to -4.3). They concluded correctly that neodymium in the oceans is mainly derived from the continents, and incorrectly that neodymium-isotope ratios in the oceans are like strontium isotopes in the sense that they are globally well-mixed and display this restricted range. In order to reach this conclusion they ignored a sample from the Indian Ocean having high strontium-isotope ratios and a lower, even more continent-like neodymium-isotope ratio of $\varepsilon_{Nd} = -8.5$.

The conclusion of O'Nions *et al.* (1978) that neodymium is relatively well mixed in the oceans was soon shown to be incorrect by studies of Fe–Mn sediments (Piepgras *et al.*, 1979; Goldstein and O'Nions, 1981). These studies showed that the Pacific, Indian, and Atlantic oceans have distinct and characteristic neodymium-isotopic signatures. The North Atlantic has the lowest (old continent-like) values ($\varepsilon_{Nd} = -10$ to -14), the Pacific the highest ($\varepsilon_{Nd} = 0$ to -5), and the Indian is intermediate ($\varepsilon_{Nd} = -7$ to -10). The data, in addition, indicated some systematic geographic variability within ocean provinces. For example, the lowest values in the Atlantic were found in the far north, and some western Indian Ocean samples from locations near southern Africa had Atlantic-type values. In the Pacific, the highest values were found in hydrothermal crusts or in the far northwest. The inter-ocean differences were attributed to the variability in the age of the surrounding continental crust, or significant contributions of volcanism-derived neodymium in the Pacific. During the period 1978–1981 two additional Pacific manganese nodule-analyses were published by Elderfield *et al.* (1981). Goldstein and O'Nions (1981) also showed that neodymium-isotope ratios can be distinguished in the leachable (precipitated) component of sediments and the solid-clay residue. The first direct measurements on Atlantic and Pacific seawater by Piepgras and Wasserburg (1980) showed substantial vertical variability, and general agreement with the Fe–Mn nodule and crust data for waters at similar depths. In addition, they suggested that North Atlantic Deep Water may have a distinct neodymium-isotopic signature, and that neodymium isotopes may have significant applications to paleoceanography.

In contrast to the application of lead and neodymium isotopes to the oceans, the first hafnium-isotopic study was not published until 1986. White *et al.* (1986) showed that six manganese nodules from the Atlantic, Indian, and Pacific oceans have a restricted range of hafnium-isotope ratios, with positive ε_{Hf} values (+0.4 to +0.9). Thus, hafnium isotopes neither show the distinct continental signature observed for marine neodymium-isotope ratios, nor the large variability of neodymium-isotope ratios in the different oceans. Therefore, although Sm–Nd and Lu–Hf systems have evolved coherently in the continents and mantle, showing positive covariations that reflect the fractionation of Sm/Nd and Lu/Hf ratios during mantle melting and continent generation (Figure 1), manganese nodules fall off this "crust–mantle correlation" to high hafnium-isotope ratios for a given neodymium-isotope ratio. White *et al.* (1986) suggested that a significant portion of marine hafnium is derived from the mantle through hydrothermal activity, or subaerial or subaqueous weathering of volcanics. Furthermore, they suggested a low continental flux of hafnium due to retention in zircon. Despite the interest generated by surprisingly high hafnium-isotope ratios, the next study was not published for 11 yr (Godfrey *et al.*, 1997). This delay was mainly due to the difficulties of analyzing hafnium-isotope ratios by thermal-ionization mass spectrometry, a consequence of the high first-ionization potential for hafnium. This situation changed in the mid-1990s with the development of multiple collector inductively coupled plasma-source mass spectrometers.

All of the early investigators, with the exception of Harry Elderfield, were mainly mantle or planetary geochemists, and these isotopic studies had relatively small impact on the oceanography community. Nevertheless, the stage was set for future studies. The gross characteristics of the marine lead-isotopic variability was outlined by the mid-1970s based on Fe–Mn sediments as seawater proxies. The gross characteristics of the global geographic neodymium-isotopic variability in the oceans (Albare (c)de and Goldstein, 1992), and shown in Figure 3, was outlined by the early 1980s.

20.4 NEODYMUIM ISOTOPES IN THE OCEANS

20.4.1 REEs in Seawater

Neodymium is a valuable tracer for oceanographic studies, as a product of radioactive decay and as an REE. They behave as a coherent group of elements whose chemical behavior is defined by filling of the 4*f* electron shell. Conventionally, REE abundances are shown in the sequence of increasing atomic number. In studies of igneous rocks, measured REE abundances are normalized to average chondritic-meteorite values, and in marine studies they are usually normalized to "average shale" (cf. Taylor and McLennan, 1985). As mentioned above, they exist in the Earth in the +3 oxidation state and display similar chemical behavior. Their relatively small chemical behavioral differences are mainly due to the effects of decreased ionic radius with higher atomic number. Two exceptions are europium and cerium. In high-oxidation environments like the oceans, cerium can exist as Ce^{+4}, in which case it forms an insoluble oxide. In low-oxidation environments such as within the Earth, europium can exist as Eu^{+2}. This rarely occurs at the Earth's surface, but it has important effects in magma chambers where Eu^{+2} is preferentially incorporated into feldspars. Magmatic fractionation of feldspar leads to a marked negative europium anomaly in convergent-margin volcanic rocks, and is a distinctive characteristic of the composition of the average upper-continental crust (e.g. Taylor and McLennan, 1985; Wedepohl, 1995; Rudnick and Fountain, 1995; Gao *et al.*, 1998). Dissolved REEs in seawater are stabilized as carbonate complexes (Cantrell and Byrne, 1987). The fraction of each REE that exists as a carbonate complex increases with increasing atomic number.

As a result of these factors, the REE pattern of seawater is distinctive. The seawater pattern is heavy REE enriched (e.g., Elderfield and Greaves, 1982). In addition, cerium shows a marked depletion compared with its neighbors, lanthanum and praseodymium. Because average shale contains a negative europium anomaly compared to chondrites, the practice of normalizing seawater to shale mutes the magnitude of the europium anomaly in patterns normalized to shale, such as published seawater REE diagrams, as compared to patterns normalized to chondrites. Nevertheless, variability in the magnitude of europium anomalies in different source rocks causes Eu/REE ratios in seawater to vary more than the trivalent REEs. The most recent general review of REEs in the oceans is by Elderfield (1988).

Abundances of REEs in depth profiles, with the exception of cerium, generally show depletions in surface waters and enrichments in deep water.

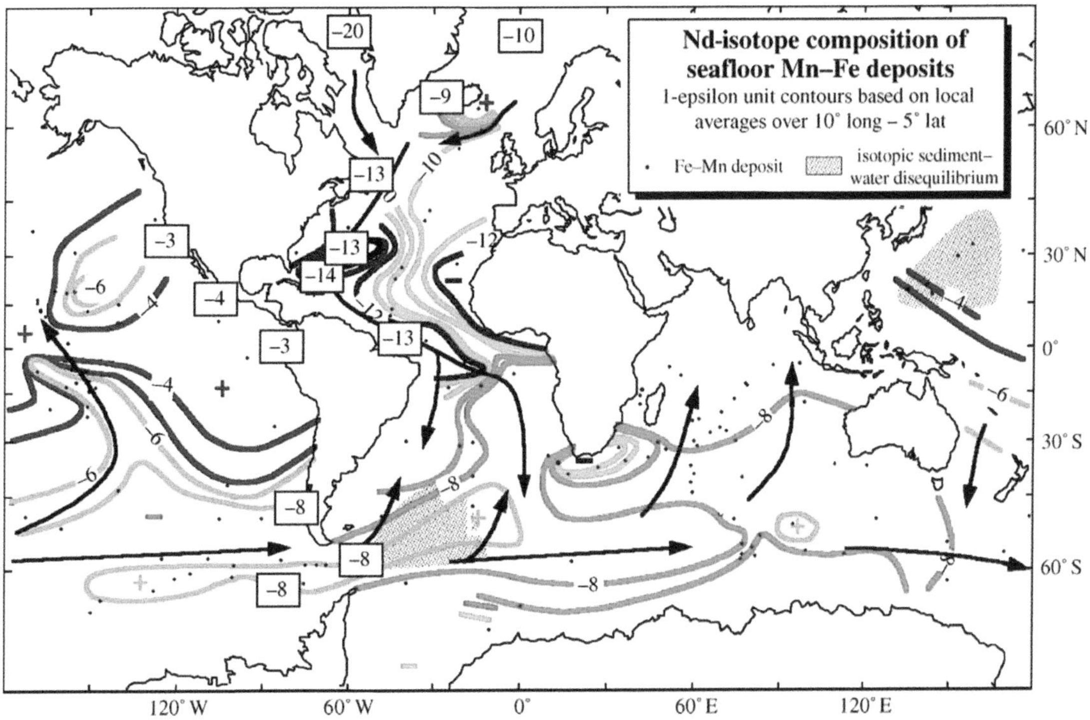

Figure 3 *Map of Nd-isotope variability in ferromanganese deposits.* The map shows systematic geographic variability with lowest values in the North Atlantic, highest values in the Pacific, and intermediate values elsewhere. Arrows illustrate general movement of deep water, and show that the contours generally follow deep-water flow. Shaded fields delineate regions where the Fe–Mn and deep seawater data differ by $>2\varepsilon_{Nd}$ units (after Albare(c)de and Goldstein, 1992).

Moreover, concentrations are generally lower in the North Atlantic and higher in the Pacific. With respect to both of these characteristics they are similar to SiO_2, and REE abundances of intermediate and deep waters tend to correlate with silicate (Elderfield and Greaves, 1982; Debaar *et al.*, 1983, 1985). Nevertheless, despite similarities in the behavior of SiO_2 and the REE, there are important differences, with the Atlantic generally having higher REE/SiO_2 than the Pacific (cf. Elderfield, 1988, and references therein). Despite these inter-ocean differences, silicate is often used as a proxy for REE behavior in the oceans. As shown below, neodymium-isotope ratios place important constraints on the mode of REE cycling in the oceans.

20.4.2 Neodymium-isotope Ratios in Seawater

There are only a small number of published papers, fewer than 20, reporting neodymium-isotope ratios in seawater. The early work was entirely carried out in the lab of Professor Gerald Wasserburg at Caltech (Piepgras and Wasserburg, 1980, 1982, 1983, 1985, 1987; Stordal and Wasserburg, 1986; Spivack and Wasserburg, 1988), which deserves credit for mapping the primary variability in seawater. All of the other published data are from Cambridge, Harvard, Lamont, Tokyo, and Toulouse. A map of the locations of seawater data is shown in Figure 4. The reason for the small number of studies is that the task is analytically challenging, due to the very low abundance of neodymium in seawater, which generally ranges ~15–45 pmol kg^{-1} (~2–7 ng L^{-1}). Most isotope laboratories measure neodymium as a positive metal ion (Nd^+), and are comfortable measuring 80–100 ng of neodymium, which would require processing of 10–40 L of seawater per sample. For this reason most of the seawater neodymium-isotope ratios in the literature were measured as a positive metal oxide (NdO^+), which affords higher efficiency of ion transmission and allows measurements of smaller samples. Even so, measurements of ~15–20 ng of neodymium still require processing of 3–10 L of seawater per sample.

The variability of neodymium isotopes in the oceans is closely related to global ocean circulation. The present-day mode of thermohaline circulation has been likened to a "great ocean conveyor" (Broecker and Peng, 1982; Broecker and Denton, 1989) in which deep water formed in the North Atlantic flows southward where it enters the circum-Antarctic. Circumpolar water flows to the Indian and Pacific, where it upwells. The North Atlantic is recharged by the surface-return

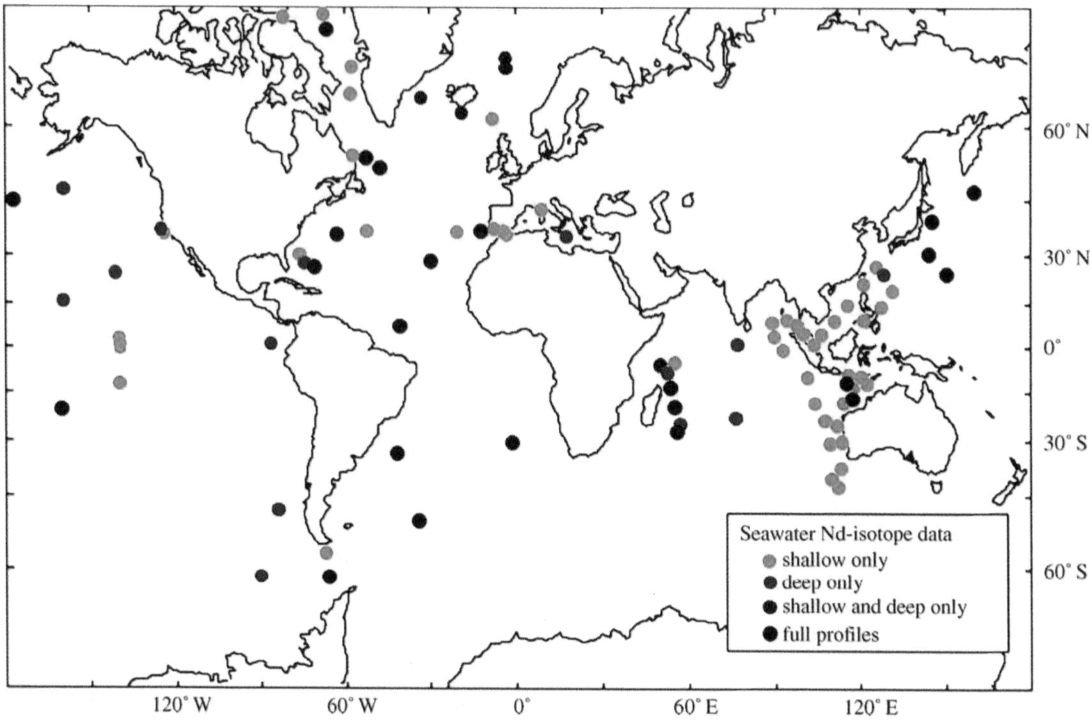

Figure 4 *Map of locations of published seawater Nd-isotope data.* Locations are distinguished by shallow-only data, deep-only data, deep- and shallow-only data, and full profiles, where "shallow" is <2,000 mb sl. All publications are cited in the text except Henry *et al.* (1994), which reports the two central Mediterranean sites.

flow from the Pacific through the Indonesian Straits and the Indian Ocean, and by northward movement of intermediate and deep water in the Atlantic from the circum-Antarctic (Rintoul, 1991). In the latter case, southward flowing NADW (North Atlantic Deep Water) is "sandwiched" by northward flowing Antarctic bottom water (AABW) and Antarctic Intermediate Water (AAIW). The principal observations on the variability of neodymium-isotope ratios globally are summarized below, with the publications that made the main discoveries.

Neodymium-isotope ratios in deep seawater show the same general geographical pattern as in Fe–Mn nodules and crusts. The global geographical pattern is illustrated in the Fe–Mn nodule-crust map (Figure 3), and by some depth profiles (Figure 5). The North Atlantic is characterized by low values, the Pacific by high values, and the Indian Ocean is intermediate (Piepgras and Wasserburg, 1980, 1987; Bertram and Elderfield, 1993). In addition, neodymium-isotope ratios in the circum-Antarctic are also intermediate to the North Atlantic and Pacific (Piepgras and Wasserburg, 1982). Typical values are in the range of Fe–Mn nodules and crusts as noted above. The same studies show that neodymium concentrations are generally highest in the Pacific and lowest in the Atlantic (Figure 6).

NADW has a narrow range of $\varepsilon_{Nd}=-13$ to -14 (Figure 5), which defines most of the mid-range and deep water in the North Atlantic south of ~55° N (Piepgras and Wasserburg, 1980, 1987). The uniformity is perhaps surprising considering the variability of NADW sources. NADW is a mixture of sources displaying two very different neodymium-isotope components. Baffin Bay between the Labrador and Greenland receives its neodymium from old continental crust with very low neodymium-isotope ratios of $\varepsilon_{Nd} < -20$. NADW sources from the Norwegian and Greenland Sea sources east of Greenland have much higher neodymium-isotope ratios of $\varepsilon_{Nd} = -7$ to -10, and these higher values are observed in the Denmark Straits and Faeroe Channel overflow (Stordal and Wasserburg, 1986).

Surface waters show more variable neodymium-isotope ratios than intermediate and deep waters (Figure 5). In contrast to the dominance of NADW values ($\varepsilon_{Nd} = -13$ to -14) in the intermediate and deep water of the North Atlantic, surface waters are variable, ranging from $\varepsilon_{Nd} = -8$ to -26 (Piepgras and Wasserburg, 1980, 1983, 1987; Spivack and Wasserburg, 1988; Stordal and Wasserburg, 1986). The greater variability of surface waters is observed throughout the oceans. Like silicate, neodymium abundances are generally depleted in surface waters compared to intermediate and deep water (Figure 6).

The limited seawater data from the circum-Antarctic display uniform neodymium-isotope ratios intermediate to the Atlantic and Pacific

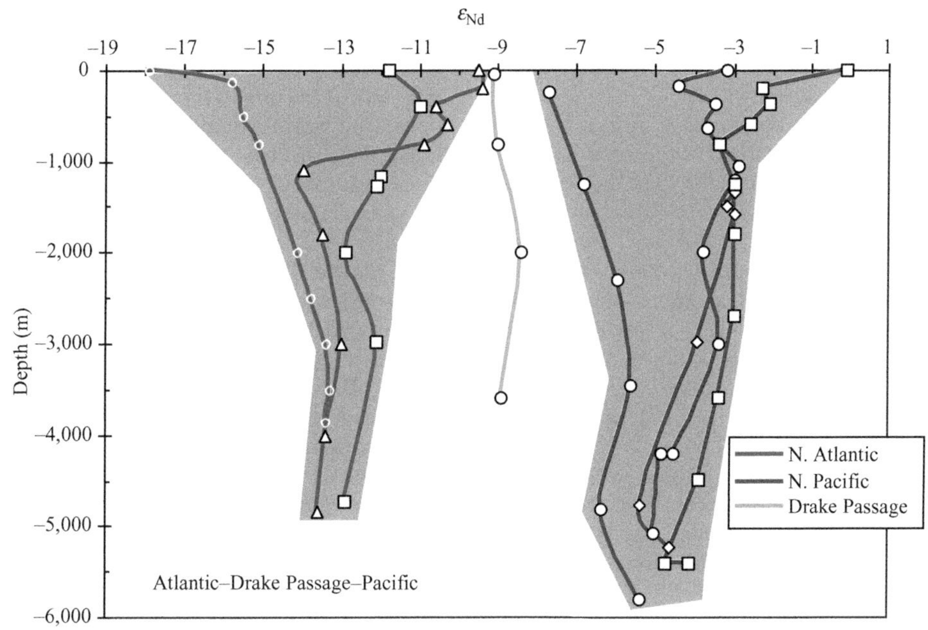

Figure 5 *Representative depth profiles of Nd-isotope ratios from the North Atlantic and Pacific oceans and the Drake Passage*. The Atlantic and Pacific profiles were chosen to encompass the range of values for deep water in those oceans. Symbols show the data. The diagram illustrates the differences between the oceans and the greater variability of Nd isotopes in shallow versus deep waters (sources Piepgras and Wasserburg, 1982, 1983, 1987; Spivack and Wasserburg, 1988; Piepgras and Jacobsen, 1988; Shimizu *et al.*, 1994).

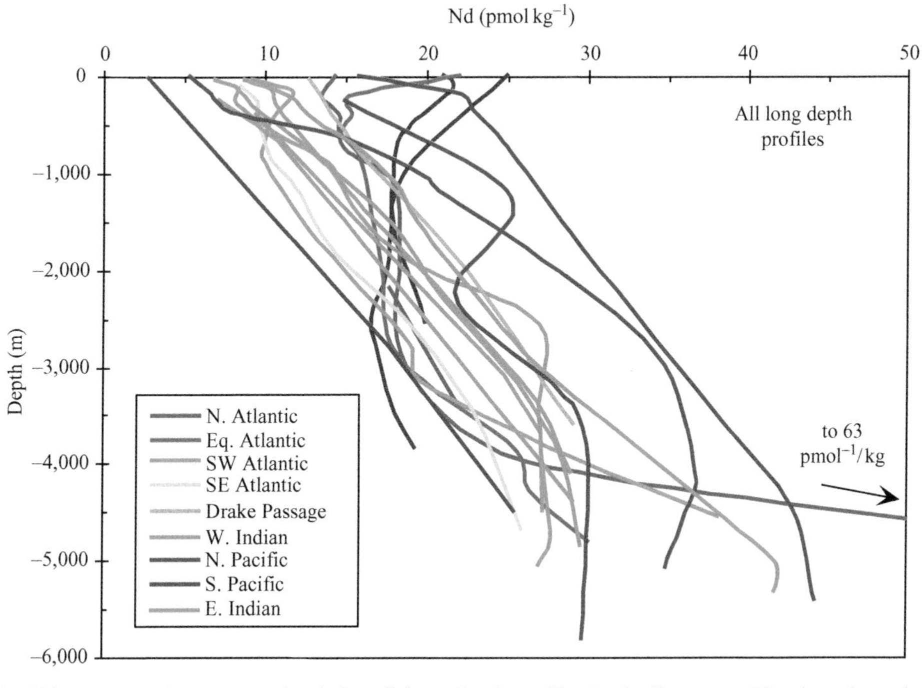

Figure 6 *Nd concentrations versus depth for all long depth profiles in the literature*. The data show that higher concentrations with depth are a general feature in the oceans, although this is not necessarily the case in the North Atlantic. Deep waters generally have the highest abundances of Nd in the North Pacific, and the lowest in the North Atlantic, but there are several exceptions (sources Piepgras and Wasserburg, 1980, 1982, 1983, 1987; Spivack and Wasserburg, 1988; Piepgras and Jacobsen, 1988; Bertram and Elderfield, 1993; Jeandel, 1993; Shimizu *et al.*, 1994; Jeandel *et al.*, 1998).

($\varepsilon_{Nd} = -8$ to -9). The only neodymium-isotope data available are from two profiles in the Drake Passage and one in the eastern Pacific sector (Piepgras and Wasserburg, 1982).

Where depth profiles sample different water masses, neodymium-isotope ratios vary with the water mass. In South Atlantic intermediate and deep water, neodymium-isotope ratios are generally intermediate to the NADW and Drake Passage values. Depth profiles show a zig-zag pattern (Figure 7), higher at intermediate depths dominated by AAIW, lower at greater depths dominated by NADW, and higher at deepest levels dominated by AABW (Piepgras and Wasserburg, 1987; Jeandel, 1993). A published diagram from von Blanckenburg (1999) overlaying neodymium-isotopic profiles and salinity (Figure 8) elegantly shows that the general characteristics of neodymium-isotope ratios vary with salinity in southward and northward flowing-water masses in the Atlantic.

Pacific intermediate and deep water dominantly display high values, of $\varepsilon_{Nd} = -2$ to -4 (Piepgras and Wasserburg, 1982; Piepgras and Jacobsen, 1988; Shimizu *et al.*, 1994). Near southern Chile, and in the west-central Pacific, where circumpolar deep water moves northward, neodymium-isotope ratios display lower values closer to those typical of circumpolar seawater. In the central Pacific the circumpolar neodymium-isotope signal can be detected as far north as ~40° N (Shimizu *et al.*, 1994).

Neodymium-isotope ratios of intermediate and deep water in the Indian Ocean are intermediate to the Atlantic and Pacific. They generally fall between $\varepsilon_{Nd} = -7$ to -9, and are consistent with domination by northward flowing circumpolar water (Bertram and Elderfield, 1993; Jeandel *et al.*, 1998). A depth profile east of southern Africa (Figure 7) displays the same zig-zag pattern as South Atlantic intermediate and deep water, reflecting advection of NADW into the western Indian Ocean (Bertram and Elderfield, 1993).

The broadest characteristics can be concisely summarized. Neodymium-isotope ratios in seawater vary systematically with location throughout the oceans, high in the Pacific, low in the Atlantic, and intermediate in the Indian and circum-antarctic. In addition, they are variable in depth profiles that contain different water masses, and neodymium-isotope ratios associated with end-members of water masses are conserved over long advective pathways.

20.4.3 Where does Seawater Neodymium Come From?

The first-order question arising from the global variability is, where does the neodymium in seawater come from? The negative ε_{Nd} values clearly show that throughout the oceans the dissolved neodymium is dominantly derived from the continents (cf. Figure 1). The characteristic values in the Atlantic and Pacific indicate at first glance that they reflect the age of the surrounding continental crust. The North Atlantic, surrounded by old continental crust, has the lowest values, and the Pacific, surrounded by orogenic belts has the highest values. However, the relationships are not simple ones.

In the North Atlantic, the present-day neodymium-isotope ratio of NADW of $\varepsilon_{Nd} \approx -13.5$ is determined by efficient mixing of two disparate sources to the west and east of Greenland. The Baffin Bay source with $\varepsilon_{Nd} < -20$ is easily explained by the surrounding early and middle Precambrian continental crust. The high ε_{Nd} value of -7 to -9 of the Denmark Straits–Norwegian Sea sources are less easy to explain. Both Greenland and Norway are primarily surrounded by Precambrian continental crust, whose early to mid-Proterozoic average age suggests an average ε_{Nd} much more negative. The only substantial sources of higher neodymium-isotope ratios in the region are Iceland and the flood basalts of the British Tertiary Province, although these cover a minor portion of the surrounding land area. Unless there is a significant input into the Denmark Straits–Norwegian Sea derived from the Pacific and leakage into these regions through the Arctic, the high neodymium-isotope ratios almost certainly show substantial input from these local volcanic sources. As mentioned previously, the value of $\varepsilon_{Nd} \approx -13.5$ that characterizes NADW is the signature imparted to most of the intermediate and deep water throughout the North Atlantic south of ~55° N. The observation that the Baffin Bay–Denmark Straits–Norwegian Sea sources are so different implies that this distinctive present-day NADW signature is inherently unstable and may easily be subject to change with time.

In the Pacific, the mechanism through which seawater obtains its high neodymium-isotope ratios is not well identified. The source is clearly not from the surrounding continental masses. The major Pacific sediment sources from the surrounding continental crust, including all of the major rivers as well as wind-blown dust from China that is the main detritus source for the central Pacific, have much lower neodymium-isotope ratios than the typical Pacific seawater value of -4 (e.g. Goldstein *et al.*, 1984; Goldstein and Jacobsen, 1988; Nakai *et al.*, 1993; Jones *et al.*, 1994). Among Pacific land masses, the only major sediment source with comparable neodymium-isotope ratios is Taiwan, and this has been suggested as the source of the high neodymium-isotope ratios of Pacific seawater, owing to its anomalously high erosion rate (Goldstein and Jacobsen, 1988). However, it is noteworthy that the detrital component of western and central Pacific sediments is $\varepsilon_{Nd} \approx -10$ to -12 (Nakai *et al.*, 1993) and this signature

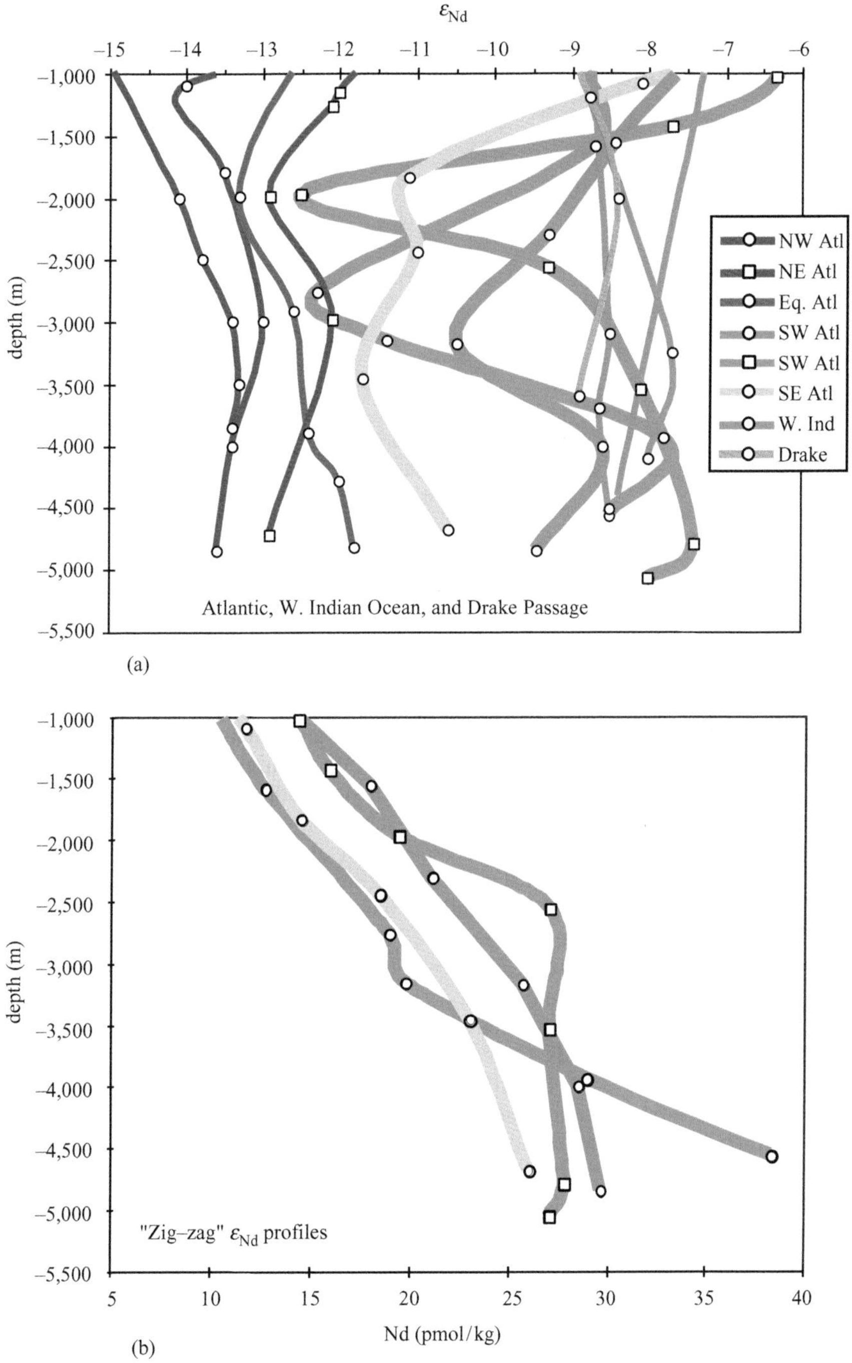

Figure 7 *Nd-isotope ratios and concentrations versus depth in the Atlantic and western Indian Oceans, and the Drake Passage.* (a) Only full profiles are shown. Western Indian Ocean data are similar to the single profile from the Drake Passage, but at slightly higher Nd-isotope ratios, which may indicate that the Drake Passage profile does not give the upper limit for the circum-Antarctic. Profiles from the South Atlantic and one from the West Indian show large variations with depth, which can range from near North Atlantic values, reflecting NADW, to near circumpolar-West Indian values, reflecting AABW and AAIW in individual profiles. (b) Nd concentrations of those profiles showing zig-zag ε_{Nd} patterns. The zig-zag profiles show that the Nd-isotopic signature of water masses is conserved over the transport path in the Atlantic (sources Piepgras and Wasserburg, 1982; 1987; Spivack and Wasserburg, 1988; Bertram and Elderfield, 1993; Jeandel, 1993).

is neither imparted to coexisting seawater or Pacific Fe–Mn oxide precipitates. Indeed, if the continents surrounding the Pacific were the major source of dissolved neodymium, Nd-isotope ratios in the Pacific and Atlantic would be at most only marginally different. Early studies of the REEs (Michard *et al.*, 1983) and neodymium-isotope ratios in hydrothermal solutions (Piepgras and

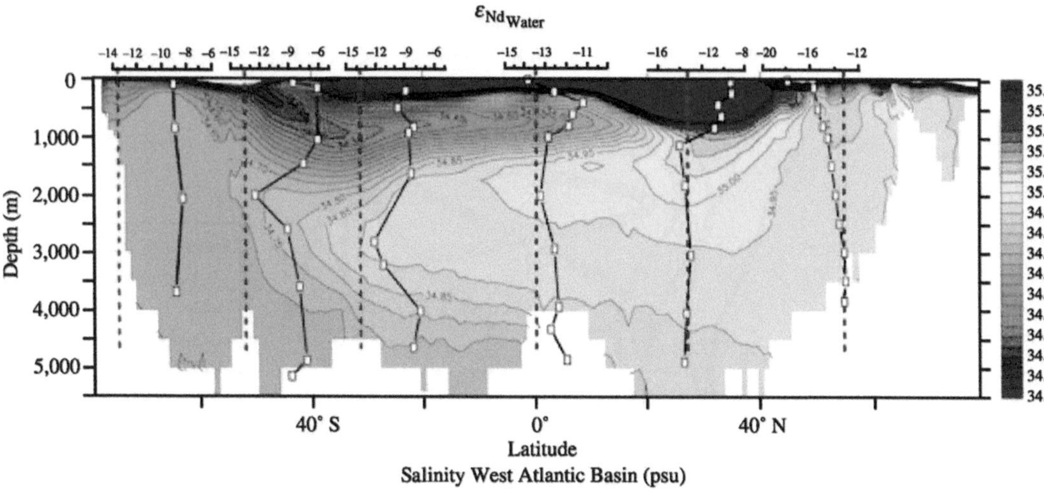

Figure 8 *Nd-isotope ratio and salinity profiles in the Atlantic.* Nd-isotope ratios with depth are consistent with salinity in the Atlantic, further showing that the Nd-isotopic signatures of water masses are conserved with advective transport (source von Blanckenburg, 1999).

Wasserburg, 1985) from Pacific seafloor spreading axes showed that ridge volcanism cannot be a significant source of Pacific seawater neodymium. The most likely source of high neodymium isotope ratios in the Pacific is circum-Pacific volcanism. Volcanic ash is widely dispersed in the Pacific as small highly reactive particles with large surface area, and Pacific sediments have large numbers of ash layers. The Pacific Ocean is surrounded by convergent plate margins, and especially in the northwest Pacific the volcanoes lie upwind. In addition, the Pacific contains a large abundance of intraplate volcanic islands. Neodymium-isotope ratios of arc volcanics (probably the largest source) and ocean islands typically have high values of $\varepsilon_{Nd} \approx +7$ to $+10$. The effects of exchange between volcanic par ticles and seawater have been inferred from studies of neodymium isotopes in waters passing through the Indonesian Straits (Jeandel *et al.*, 1998; Amakawa *et al.*, 2000), and Papua New Guinea (Lacan and Jeandel, 2001). In the cases, increases in neodymium-isotope ratios along water advection pathways are attributed to local volcanic sources. Thus, it is likely that addition of neodymium from volcanic particles imparts the distinctive neodymium-isotopic fingerprint to Pacific seawater, which is ultimately intermediate to values of recent volcanic and old continental sources.

The inefficiency of neodymium exchange between continental detritus and seawater is clearly seen in the Indian Ocean. The major sources of sediment to the Indian Ocean are the Ganges–Brahmaputra and Indus river systems, with $\varepsilon_{Nd} \approx -10$ to -12 (Goldstein *et al.*, 1984), whose neodymium-isotope ratios are reflected in the Bengal Fan (e.g. Bouquillon *et al.*, 1990; Galy *et al.*, 1996; Pierson-Wickmann *et al.*, 2001). The neodymium-isotope ratio of Indian seawater is $\varepsilon_{Nd} \approx -8$, significantly higher than these sediment sources, even in Indian water close to the Indus and Bengal fans. This difference exists despite the high abundance of neodymium in continental sediment (~30 ppm) compared to Indian seawater (~0.003 ng g^{-1}), which shows that a small amount of exchange in regions with high sedimentation rates like these two Fans should have a major local effect on seawater neodymium. The fact that the large difference persists between the fan sediments and proximal Indian seawater shows that the neodymium in terrigenous detritus is tightly bound to the sediment. The neodymium-isotope ratios of Indian seawater are the same as the circum-Antarctic, as observed in direct seawater measurements and in Fe–Mn oxides. This is consistent with Indian Ocean hydrography, which shows that Indian intermediate and deep seawater is primarily fed by northward advecting water from the circum-Antarctic.

These observations, taken together, indicate that the neodymium-isotope ratios of intermediate and deep seawater are imprinted mainly in the Atlantic and Pacific oceans. The circum-Antarctic is fed by both and its intermediate neodymium-isotope ratio reflects those of the Atlantic and Pacific water sources. Indian Ocean intermediate and deep water is fed primarily by the circum-Antarctic and tends to retain a circum-Antarctic neodymium-isotope ratio.

20.4.4 Neodymium Isotopes as Water-mass Tracers

How well do neodymium-isotope ratios trace water masses? The apparent answer is that neodymium isotopes trace them very well. Where there is a significant neodymium-isotopic contrast between water masses with depth at a single location, as in the South Atlantic, neodymium-isotope ratios vary coherently with the water mass (Figures 7 and 8). The North Atlantic profiles show uniform ε_{Nd} values between -12 and -14 at depths >2,500 m. The Drake Passage profile shows a uniform ε_{Nd} value of ~ -9. In this context, Equatorial Atlantic neodymium-isotope ratios are slightly higher than those in the North Atlantic, especially in deep water, consistent with addition of some AABW. In the South Atlantic, profiles show a strong zig-zag pattern. The lowest values nearly reach those of NADW, strongly indicating that NADW keeps its neodymium signature as it travels southward in the Atlantic. In the SE Atlantic, the NADW wedge can be identified in Figures 7 and 8, shallower than in the SW Atlantic. At shallower depths than the NADW "wedge," neodymium-isotope ratios increase toward circumpolar values in northward flowing AAIW. At deeper depths than the NADW "wedge," the neodymium-isotope ratios increase toward circumpolar values in northward flowing AABW.

The highest neodymium-isotope ratios in the South Atlantic at depths greater than 2,500 m reach ε_{Nd} ~ -7.5, which are higher than the Drake Passage values (Figure 7). This may indicate a source of neodymium with high isotope ratios in the South Atlantic. However, it is premature to conclude that deep South Atlantic neodymium-isotope ratios "overstep" the Southern Ocean values, for the following reasons. The maxima for all of the deep South Atlantic waters are between $\varepsilon_{Nd} = -7$ to -9, more variable than presently available data from the Drake Passage but still quite similar. This range is also similar to circumpolar Fe–Mn sediments (Albare(c)de *et al.*, 1997). Depth profiles from the western Indian Ocean near southern Africa are similar to the South Atlantic and Drake Passage, but like the South Atlantic they slightly exceed the Drake Passage values (Figure 7). Of the three long profiles in the literature (Bertram and Elderfield, 1993), two are relatively constant ε_{Nd} with depth, while the third shows a similar zig-zag as the South Atlantic, and reflects the flow of NADW around southern Africa. It is noteworthy that the Drake Passage profile (from Piepgras and Wasserburg, 1982), is the only one in the literature from the true circum-Antarctic, and its deepest sample is from a relatively shallow depth of $\sim$3,500 m. Either Drake passage seawater has more variable Nd isotope ratios than indicated by presently available data, or there is some addition of Nd to the circumpolar Atlantic sector from a high ε_{Nd} source.

In the global ocean, neodymium-isotope ratios in deep water (here considered to be >2,500 m) show remarkably good covariations with salinity and silicate (Figure 9), which provides key evidence of its value as a water-mass tracer. In the Atlantic and western Indian oceans, deep seawater faithfully records mixing between northern- and southern-derived water masses (Figure 10(a)). With northern waters characterized by low ε_{Nd} and high salinity, and southern waters by high ε_{Nd} and low salinity, the data show a very good covariation between the end-members and fall within a mixing envelope. Of particular interest are the data from the southwest Atlantic, where the high salinity waters fall well within the range of the north Atlantic data, and the low salinity waters within the range of South Atlantic–Drake Passage–West Indian Ocean data. A similar relationship also holds between SiO_2 and ε_{Nd} (Figure 10(b)), where northern source deep waters have low SiO_2 and ε_{Nd}, and southern source waters high SiO_2 and ε_{Nd}. Here again, most of the Atlantic data fall within a mixing envelope between the two end-members.

The Pacific is more of a homogenous pool of water than the Atlantic, and relationships between neodymium-isotope ratios and water masses are not as clearcut. Nevertheless, there is strong evidence that neodymium isotopes behave here too as conservative water-mass tracers. To the east of New Zealand there is a tongue of deep circumpolar water that moves northward. In depth profiles from the south central Pacific, the neodymium-isotope ratios of deep water reaches circumpolar values, while the intermediate water is more Pacific-like. More compelling evidence for the mainly conservative nature of neodymium in deep water is illustrated by comparing neodymium isotopes and silicate in Pacific and circumpolar waters (Figure 11). They show a good positive correlation, consistent with mixing between Pacific and circumpolar waters. In the Pacific region salinity cannot be used as a water-mass proxy as in the Atlantic because salinity is similar in Pacific and circum-Antarctic waters (Figure 9(a)).

Taking all of these considerations into account, the available data strongly indicate that neodymium-isotope ratios in deep water are conservative tracers of water masses. While the relationships are most clear in the Atlantic, where a large portion of the basin consists of different water masses at different depths from northern and southern sources, the relationships also hold well in the Pacific. An important further implication is that neodymium isotopes should have great potential as tracers of past ocean circulation.

20.4.5 The "Nd Paradox"

The previous section demonstrated that neodymium-isotope ratios in the deep-ocean trace mixing between Southern and Northern Ocean waters in the Atlantic and Pacific. However, the variability

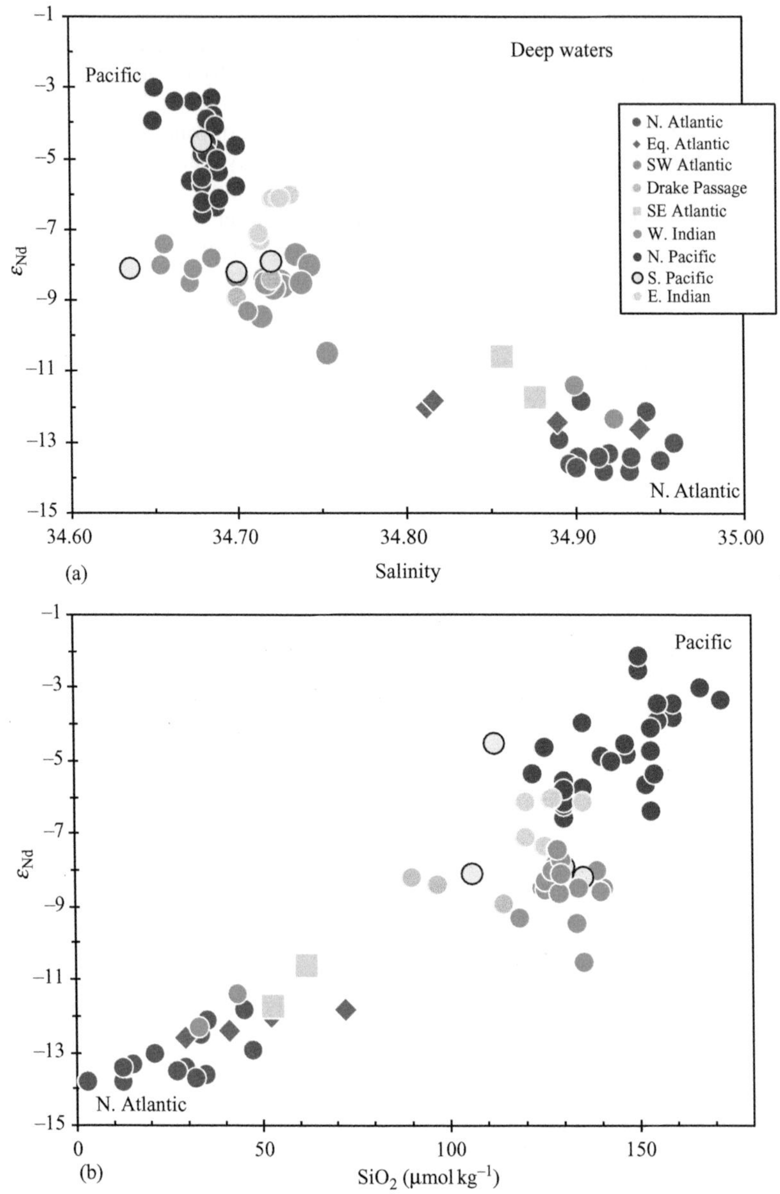

Figure 9 *Nd-isotope ratios versus salinity and silicate in deep seawater.* (a) Seawater Nd-isotope ratios display a good covariation with salinity, but the correlation breaks down somewhat in the Pacific, because its salinity is similar to the circum-Antarctic and South Atlantic. (b) Throughout the oceans there is a very good covariation with SiO_2. In both diagrams, the South Atlantic data responsible for the zig-zag patterns in profile (Figure 8) are also consistent with salinity and silicate, with high and low ε_{Nd} reflecting AABW and NADW, respectively. Thus, the Nd-isotope ratios trace the water masses of the present-day deep oceans very well. Plotted data are from >2,500 mb sl, except two Drake Passage data from 1,900 m and 2,000 m (Nd data sources: Piepgras and Wasserburg, 1980, 1982, 1983, 1987; Spivack and Wasserburg, 1988; Piepgras and Jacobsen, 1988; Bertram and Elderfield, 1993; Jeandel, 1993; Shimizu *et al.*, 1994; Jeandel *et al.*, 1998). Where salinity or silicate were not available in the publication, they were estimated from Levitus (1994), using the location and depth of the water sample.

of neodymium concentrations in seawater was not discussed. As noted in the discussion of REE in seawater (Section 20.4.1 and Figure 6), abundances generally show depletions in surface waters and enrichments in deep water, and are lower in the North Atlantic and higher in the Pacific. With respect to these characteristics the REEs are similar to SiO_2. It has already been shown that neodymium-isotope ratios and silicate display an excellent global covariation (Figure 9(b)), and silicate is often used as a proxy for REE behavior in the oceans (e.g. Elderfield, 1988). This has been taken as evidence that silicate and Nd display a similar behavior in the oceans.

Silicate is depleted from surface waters by biological processes and remineralized in the

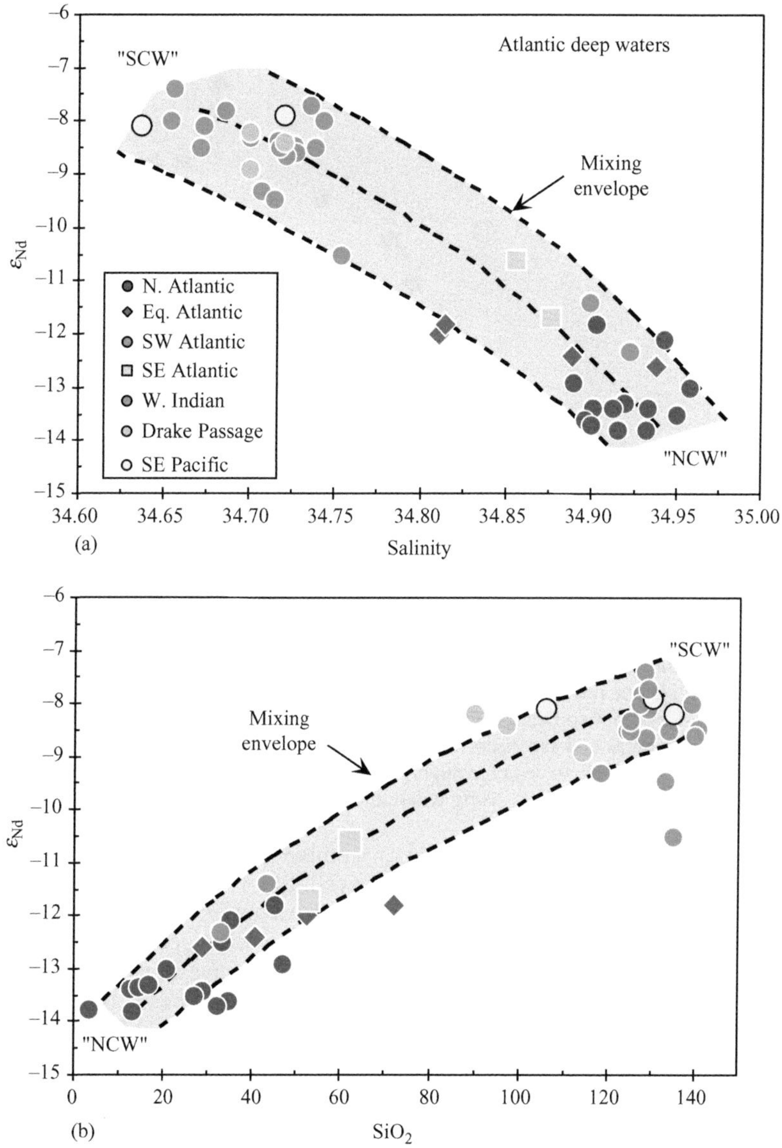

Figure 10 *Nd-isotope ratios versus salinity and silicate in Atlantic, West Indian, and Southern Ocean deep waters.* Mixing lines between northern and southern component end-member compositions (NCW and SCW, respectively) are shown, since they are nonlinear. Salinity, silicate, and Nd-isotope end-members are shown; Nd concentrations range from 16 nmol kg^{-1} to 20 nmol kg^{-1} in NCW and 27 nmol kg^{-1} to 30 nmol kg^{-1} in SCW, approximating reasonable ranges for NADW and AABW, respectively. In both (a) and (b) the data generally fall within the "mixing envelopes," showing that the Nd-isotope ratios of Atlantic deep waters reflect mixing between the northern and southern component waters. Plotted data are from >2,500 mb sl, except two Drake Passage data from 1,900 m and 2,000 m. (Nd data sources: Piepgras and Wasserburg, 1980, 1982, 1983, 1987; Spivack and Wasserburg, 1988; Piepgras and Jacobsen, 1988; Bertram and Elderfield, 1993; Jeandel, 1993). Where salinity or silicate were not available in the publication, they were estimated from Levitus (1994), using location and depth.

deep water. Moreover, silicate tends to accumulate in water masses as they age (cf. Broecker and Peng, 1982). These processes account for both the increasing concentration of silicate with depth, and its increasing concentrations from the North Atlantic to the circum-Antarctic to the Pacific. With a few exceptions (notably the North Atlantic), neodymium abundances show a smooth increase with depth, and in deep water they are highest in the Pacific, lowest in the Atlantic, and intermediate in the Indian (Figure 6).

Considered separately, the neodymium-isotope ratios and the neodymium concentrations have very different implications. (i) Neodymium-isotope ratios are variable in the different oceans and within an ocean they fingerprint the advective paths of water masses. This requires that the residence time of neodymium is shorter than the

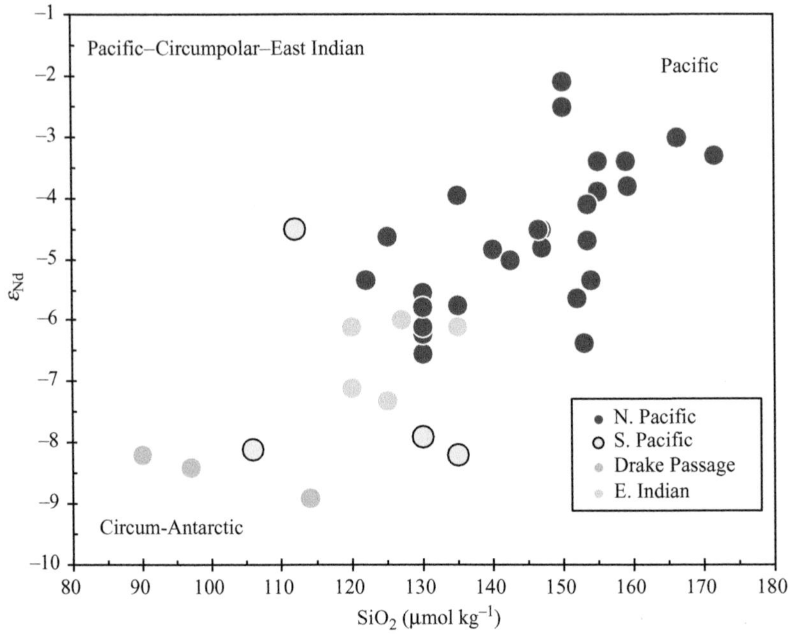

Figure 11 *Nd-isotope ratios versus silicate in Pacific, Indian, and Southern Ocean deep waters.* The positive correlation shows that Nd-isotope ratios trace mixing of deep waters from the circum-Antarctic and Pacific. Plotted data are from >2,500 mb sl, except two Drake Passage data from 1,900 m and 2,000 m (Nd data sources: Piepgras and Wasserburg, 1980, 1982; Piepgras and Jacobsen, 1988; Bertram and Elderfield, 1993; Shimizu *et al.*, 1994; Jeandel *et al.*, 1998). Where salinity or silicate were not available in the publication, they were estimated from Levitus (1994), using location and depth.

ocean mixing time of $\sim 10^3$ yr (e.g., Broecker and Peng, 1982). (ii) Neodymium concentrations appear to mimic silicate, implying orders of magnitude longer residence times of $\sim 10^4$ yr and indicating increased addition through dissolution along the advected path of water masses. These observations were first addressed in detail by Bertram and Elderfield (1993), and further emphasized by Jeandel and colleagues (Jeandel *et al.*, 1995, 1998; Tachikawa *et al.*, 1997a,b; Lacan and Jeandel, 2001), who termed this the "Nd paradox." Both groups have suggested a general model of neodymium cycling in the oceans, that treats neodymium and the other REE as an analog to silicate. Neodymium is introduced into the surface ocean through partial dissolution of atmospheric input. For example, Jeandel *et al.* (1995) estimate that half of the global atmospheric input is dissolved when entering seawater. The cycling model suggests that in the deep-water source regions such as the North Atlantic the neodymium reaches the deep ocean through water mass subduction; elsewhere the neodymium is scavenged by organisms, vertically transported to deep water as the organisms settle in the water column, and added to the deep ocean through particulate-water exchange. Throughout the oceans neodymium is advected along with water masses. Thus, the inheritance of neodymium-isotope ratios of deep-water sources along with increasing neodymium concentrations as water ages are explained by a combination of lateral advection and vertical cycling.

It was argued above (Section 20.4.4) that the neodymium-isotope ratios, are conservative tracers of the advective movement and mixing of water masses (e.g., Figures 7, 10, and 11), that distinguish the southward flow of NADW and northward flow of AABW and AAIW in the Atlantic, as well as exchange between Pacific and circum-Antarctic waters. To the extent that neodymium-isotope ratios in the Atlantic and Pacific trace exchange of water between these oceans and the circum-Antarctic, they are potentially powerful oceanographic and paleoceanographic tracers. However, this implies that neodymium-isotopic signatures are imparted to deep water primarily at two locations, the North Atlantic and the Pacific, and that mixing yields intermediate neodymium-isotope ratios in the circum-Antarctic. If vertical transport from surface waters and neodymium exchange in deep waters is an important process throughout the oceans, then neodymium that is extraneous to the deep water masses must be added along the transport path. Unless the added neodymium fortuitously has the same isotope ratio as the water mass, the value of neodymium-isotope ratios as a water-mass tracer will be compromised. The extent that neodymium isotopes are compromised as water-mass tracers depends on how much the deep-water neodymium-isotopic fingerprint is changed by the addition or exchange.

A test against a significant addition of shallow neodymium to deep waters is to show that deep and shallow waters at the same location have markedly different neodymium-isotope ratios. While most published depth profiles are ambiguous in this regard, there are several published profiles in which the shallow and deep water have significantly different neodymium-isotopic signatures (Figure 12(a)). These neodymium-isotopic profiles are from all of the ocean basins; therefore, it is reasonable to infer that this relationship is a general feature of the oceans.

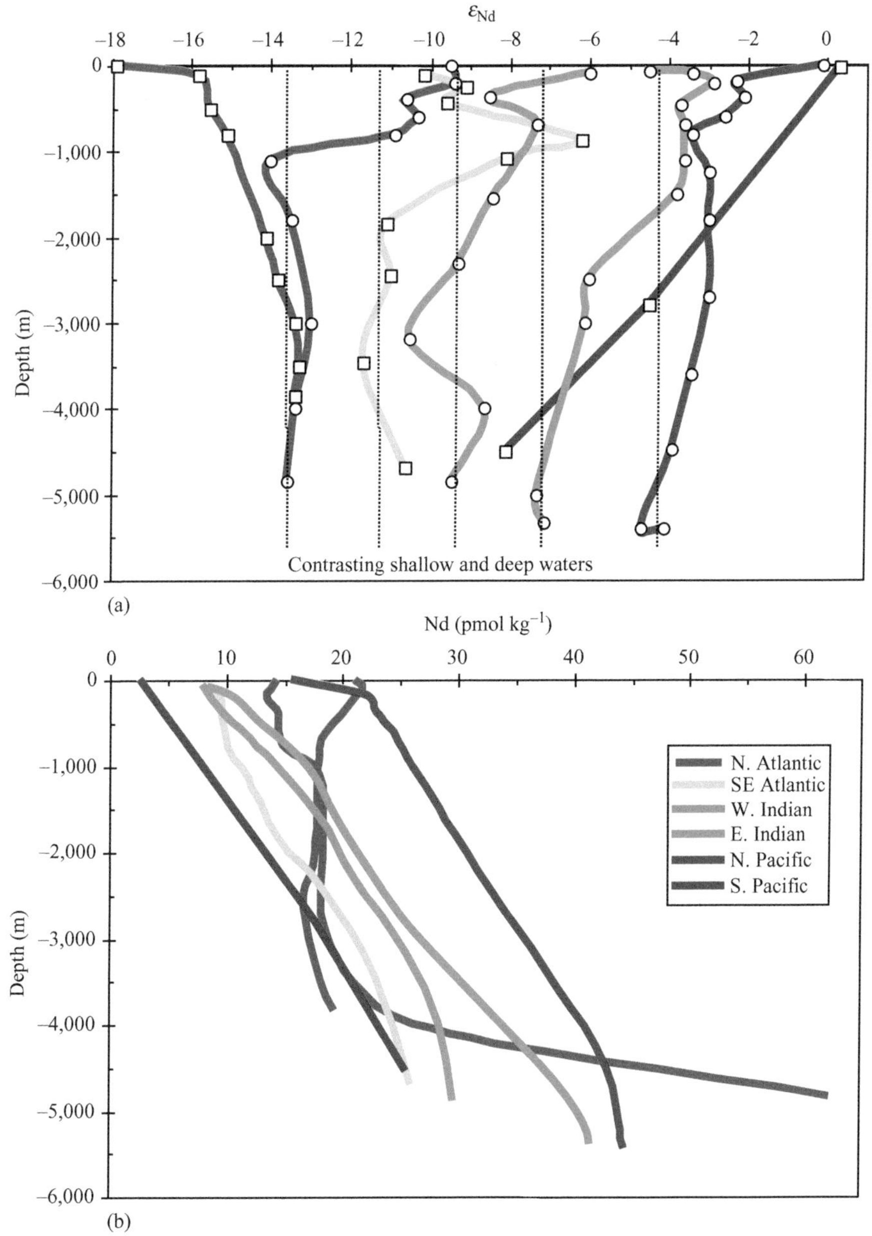

Figure 12 *Comparison of Nd-isotopes and abundances in shallow and deep waters.* (a) These examples, from all the ocean basins, highlight profiles where the Nd-isotope ratios of shallow and deep waters show large differences, as high as $\sim 4\varepsilon_{Nd}$ units in the Atlantic and $\sim 8\varepsilon_{Nd}$ units in the South Pacific. Dashed vertical lines emphasize the differences between the shallow and deep waters. (b) Concentration profiles show smoothly increasing Nd abundances with increasing depth in all cases but the North Atlantic. Deep-water concentrations are generally greater than twice as high as shallow waters. The abundance profiles are consistent with shallow scavenging of Nd and addition at increasing depths, but the differences between deep and shallow water Nd-isotope ratios preclude significant addition Nd scavenged at shallow levels. (Nd data sources: Piepgras and Wasserburg, 1982, 1987; Piepgras and Jacobsen, 1988; Bertram and Elderfield, 1993; Jeandel, 1993; Jeandel *et al.*, 1998).

The contrasts between deep and shallow water can be quite large, $\sim 4\varepsilon_{Nd}$ units in the Atlantic, and $\sim 8\varepsilon_{Nd}$ units in the South Pacific. Concentration profiles of the same samples show smoothly increasing neodymium abundances with increasing depth (in all cases but the North Atlantic), with deep-water concentrations generally more than a factor of 2 higher than shallow waters (Figure 12(b)). Without the constraints from the neodymium-isotope ratios, the concentration profiles are consistent with shallow scavenging of neodymium and remineralization at increasing depths. However, taken together, the differences between deep and shallow water neodymium-isotope ratios and concentrations preclude the addition of significant quantities of neodymium scavenged at shallow levels to the deep water. This means that vertical cycling at these sites does not appear to explain the higher neodymium concentrations in the deep water.

Can the Nd paradox be resolved by water-mass mixing? Vertical cycling does not appear to explain both the neodymium-isotope ratios and Nd concentrations in the oceans. Neodymium-isotope covariations with salinity and silicate in the Atlantic and Pacific are broadly consistent with mixing between northern- and southern-source waters in both basins (Figures 10 and 11). However, using the analogy to silicate, the increasing concentrations of neodymium from the Atlantic, to the circum-Antarctic/Indian, to the Pacific, have been explained in terms of increased addition of neodymium with the aging of deep water. If the high neodymium-isotope ratios of Pacific water are a result of reaction of volcanic ash with seawater, then can this could be the source of the higher neodymium concentrations in the Pacific compared to the Atlantic? Can the "Nd paradox" be explained by addition of Nd to deep water in the North Atlantic and the Pacific combined with water mass exchange between the North Atlantic–circum-Antarctic–Pacific oceans?

In terms of the global ocean, the intermediate neodymium-isotope ratios and neodymium concentrations of the circum-Antarctic/Indian oceans are not explicable through simple mixing of North Atlantic and Pacific source waters. In the circum-Antarctic and Indian oceans, neodymium concentrations are too low (Figure 13). This is reasonable, and could be a reflection of neodymium scavenging in the water column during advective transport between the North Atlantic and circum-Antarctic on the one hand, and the Pacific and circum-Antarctic on the other.

The North Atlantic and the Pacific are not directly linked, rather their linkage is modulated by the circum-Antarctic. Therefore, a more direct question is whether the neodymium-isotope ratios and concentrations are explicable in terms of binary mixing between Northern and Southern component waters within each basin. In the Pacific, it was pointed out by Piepgras and Jacobsen (1988) that neodymium-isotope ratios appear to follow simple mixing between the North Pacific and the Southern Ocean (illustrated in Figure 11) but concentrations do not. They attributed the apparent coherence of the neodymium isotopes and its absence in the neodymium concentrations to the loss of similar proportions of neodymium by scavenging from waters derived from the Pacific and circum-Antarctic during advective transport. The losses would not change the neodymium-isotope ratios, and the appearance of binary mixing systematics would be conserved.

In the Atlantic, like the Pacific, neodymium-isotope ratios are consistent with mixing of Northern and Southern end-members using salinity or silicate as conservative water-mass mixing proxies (Figure 10). Comparing neodymium-isotope ratios and concentrations, the compositions of many samples are also consistent with North–South water-mass mixing (Figure 14(a)). However, a substantial portion of the data is outside any reasonable mixing envelope. Whereas in the Pacific–circum-Antarctic case above, the non-conservative behavior of the concentrations can be explained by loss during advection, in the Atlantic the neodymium concentrations of the samples outside the mixing envelope are too high. Moreover, those that are too high tend to be the deepest samples (not shown in Figure 14 but this can be inferred from Figure 6 combined with Figure 9). The neodymium concentrations of the available data set in the Atlantic, therefore, point to an overabundance of neodymium in the deepest waters, as noted by Bertram and Elderfield (1993), and Jeandel and colleagues (Jeandel *et al.*, 1995, 1998; Tachikawa *et al.*, 1997, 1999a, b; Lacan and Jeandel, 2001). The neodymium-isotope ratios, however, place important constraints on the addition process. Figure 14(a) shows that the samples lying outside of the mixing envelope from the North Atlantic and the Equatorial Atlantic have neodymium-isotope ratios that are the same as samples within the mixing envelope. Comparison of neodymium concentrations with salinity (Figure 14(b)) shows that the same samples fall outside the simple mixing envelope. Therefore, the neodymium concentrations are enriched, but the isotope ratios are normal for the location. It was shown in Figures12(a) and (b) that depth profiles exist from all the oceans showing that high neodymium concentrations in deep water cannot be explained by addition of neodymium from surface waters. Profiles of ε_{Nd} and neodymium versus depth from the South Atlantic and western Indian oceans are shown in Figure 7. The concentration profiles shown in Figure 7(b) are those with zig-zag ε_{Nd} patterns. In all these cases, the neodymium concentrations

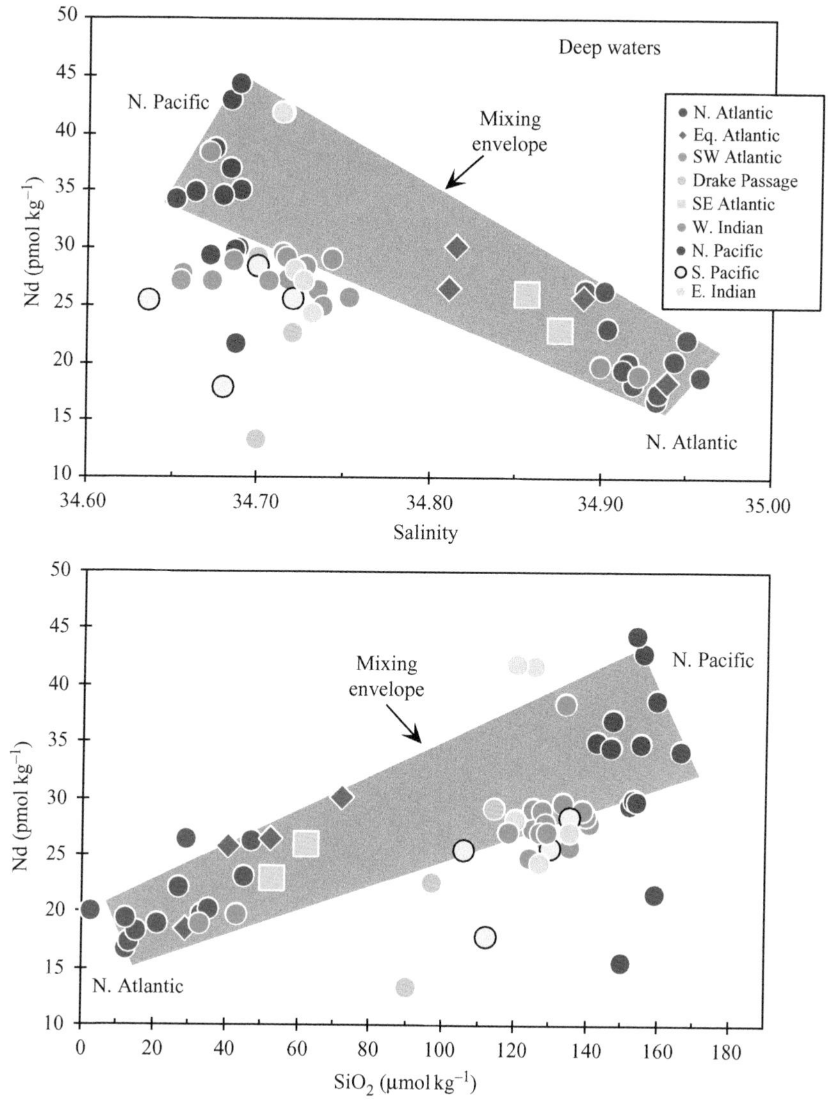

Figure 13 *Nd abundance versus (a) salinity and (b) silicate in deep seawater.* Nd concentrations do not show the same well-behaved characteristics as Nd-isotope ratios with salinity and silicate in global deep water. Mixing envelopes are shown between North Atlantic and Pacific end-members, and the circum-Antarctic, South Atlantic, and Indian Ocean samples fall outside of it. Plotted data are from >2,500 mb sl, except two Drake Passage data from 1,900 m and 2,000 m (Nd data sources: Piepgras and Wasserburg, 1980, 1982, 1983, 1987; Spivack and Wasserburg, 1988; Piepgras and Jacobsen, 1988; Bertram and Elderfield, 1993; Jeandel, 1993; Shimizu *et al.*, 1994; Jeandel *et al.*, 1998). Where salinity or silicate were not available in the publication, they were estimated from Levitus (1994), using location and depth.

increase smoothly with depth, despite the zig-zag pattern of the neodymium-isotope ratios. The neodymium-isotope ratios reflect the deep-water masses, while the neodymium concentrations appear to be decoupled.

The increasing concentrations of neodymium with depth indicate that neodymium is added to deep water as water masses laterally advect in the Atlantic, however, the neodymium that enriches deep waters along the advective path in the Atlantic has isotope ratios expected for the respective water masses. The isotope ratios appear to preclude vertical cycling of neodymium through scavenging at shallow levels, followed by addition at deeper levels, as a primary cause of increasing neodymium concentrations with depth. The "Nd paradox" still stands.

20.4.6 Implications of Nd Isotopes and Concentrations in Seawater

The forgoing discussion shows that neodymium-isotope ratios are excellent conservative tracers of water masses throughout the oceans.

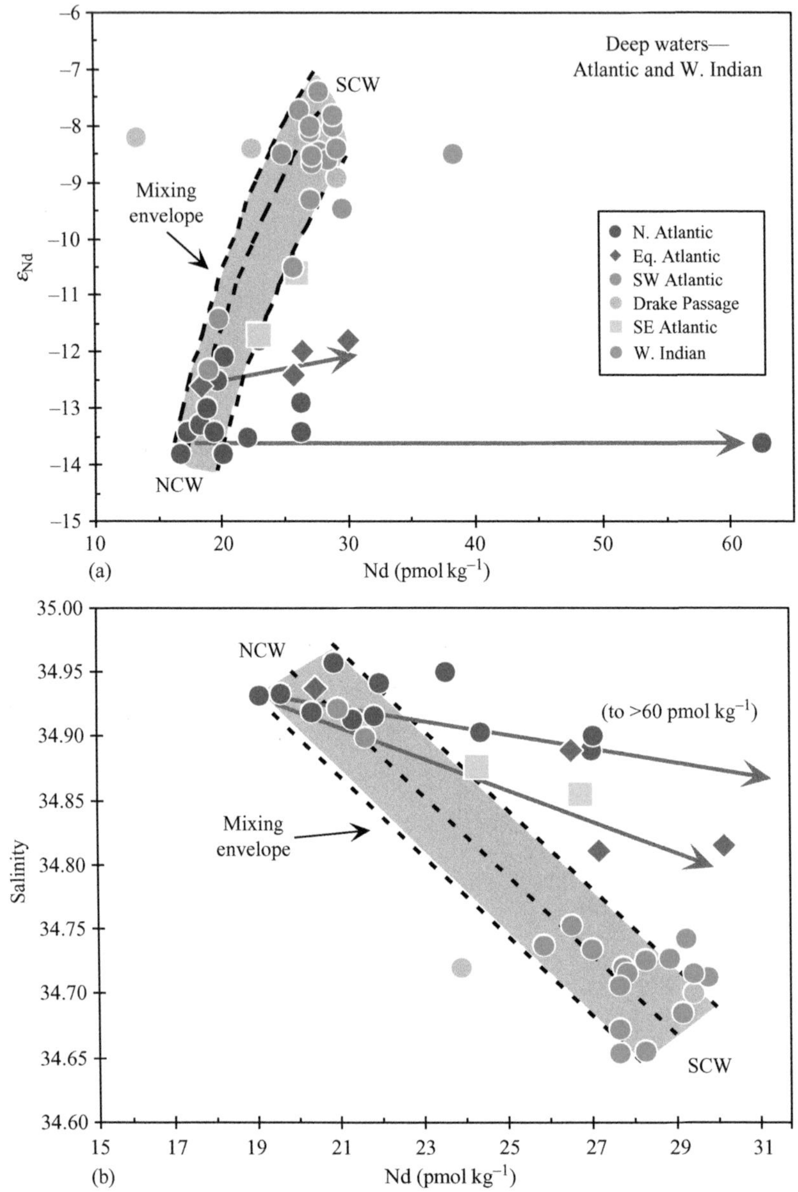

Figure 14 *Nd abundance versus salinity and silicate in deep Atlantic region seawater.* Mixing lines between northern and southern component waters (NCW and SCW) are shown in (a) because they are nonlinear. In both (a) and (b) much of the data fall within the "mixing envelopes," but many samples show high concentrations of Nd. These are generally the deepest samples. Arrows show that large increases in Nd concentrations in a region or profile are not accompanied by significant changes in Nd-isotope ratio (Nd data sources: Piepgras and Wasserburg, 1980, 1982, 1983, 1987; Spivack and Wasserburg, 1988; Bertram and Elderfield, 1993; Jeandel, 1993). Where salinity or silicate were not available in the publication, they were estimated from Levitus (1994), using location and depth.

However the processes that are controlling neodymium concentrations are still not well understood. An important avenue of further research will be to compare ε_{Nd} of authigenic and terrigenous sediments in key areas. Nevertheless, the close relationship between neodymium-isotope ratios and water masses, coupled with the range of global variability in the oceans, indicate that it can be an effective water mass tracer in the present day and shows great potential as a tracer of past ocean circulation.

20.5 APPLICATIONS TO PALEOCLIMATE

20.5.1 Radiogenic Isotopes in Authigenic Ferromanganese Oxides

The discussions (Section 20.4) above showed that the primary controls on the composition of neodymium-isotope ratios in seawater in crusts are the provenance (possibly modified by weathering, e.g., von Blanckenburg and Nagler,

2001) and ocean circulation. In order for a tracer to be valuable for paleoceanographic studies, its modern distribution should follow water masses in a simple way, and neodymium isotopes clearly fulfill this criterion. A world map of neodymium-isotope ratios in the outer portions of Fe–Mn nodules and crusts (Figure 3) shows that they display systematic geographical variations. In most regions they are the same as the local bottom water. A spectacular indication of the veracity of neodymium isotopes in manganese crusts and nodules as water-mass tracers is the comparison of the pattern of ^{14}C ages of bottom waters in the Pacific Ocean (Figure 15; Schlosser *et al.*, 2001) with the world map neodymium-isotope ratios of Fe–Mn nodules and crusts (Figure 3). Manganese crusts and nodules grow at a rate of a few millimeters per million years and thus each sample generally integrates several glacial–interglacial cycles. Thus, the strong correlation with modern water-mass characteristics is a testament to the general stability of the modern ocean circulation through the Pleistocene.

However, Albare(c)de and Goldstein (1992) noted that there are some notable deviations between modern water values and crusts. Of particular note is the southwestern Atlantic region where the crust data have higher neodymium-isotope ratios than modern deep water (Figure 3). Data from disseminated authigenic neodymium from marine sediment core RC11-83, from the Cape Basin in the southeast Atlantic (Rutberg *et al.*, 2000), provides a resolution of this apparent conflict. This study traced the export of NADW to the Southern Ocean through the last glacial period to ~70 ka, and was the first to track deep-ocean circulation using neodymium-isotope ratios on glacial–interglacial timescales. It was found that the Holocene and marine-isotope stage 3 (MIS 3) record is like that predicted from the modern water column data, while the MIS 2 and 4 records are more like Pacific seawater (Figure 16). The integrated signal over the last 4 MISs is comparable to the outer portions of manganese crusts and nodules of that region. This further emphasizes that neodymium-isotope ratios in marine

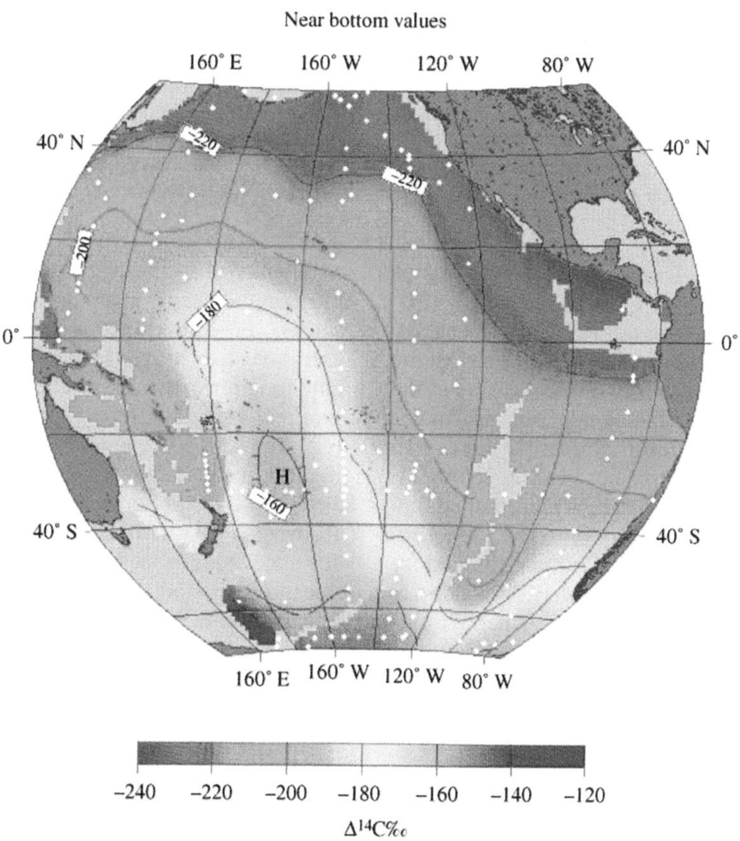

Figure 15 *$\Delta^{14}C$ in deep Pacific seawater.* The tongue of high $\Delta^{14}C$ shows the present-day path of circumpolar water entering the Pacific in the present day. The location and shape of the tongue of high ^{14}C water is similar to the "tongue" of low Nd isotope ratios in Fe–Mn deposits shown in Figure 3. In the Fe–Mn samples the low Nd isotope "tongue" is an integrated signal over 10^5–10^6 yr. The comparison shows that the Nd isotope signal in the Fe–Mn samples track the path of circumpolar water into the Pacific, and that on average the pathway has remained the same over the last million years (reproduced by permission of R. Key, Princeton University from *Ocean Circulation and Climate: Observing and Modelling the Global Ocean*, **2001**, pp. 431–452).

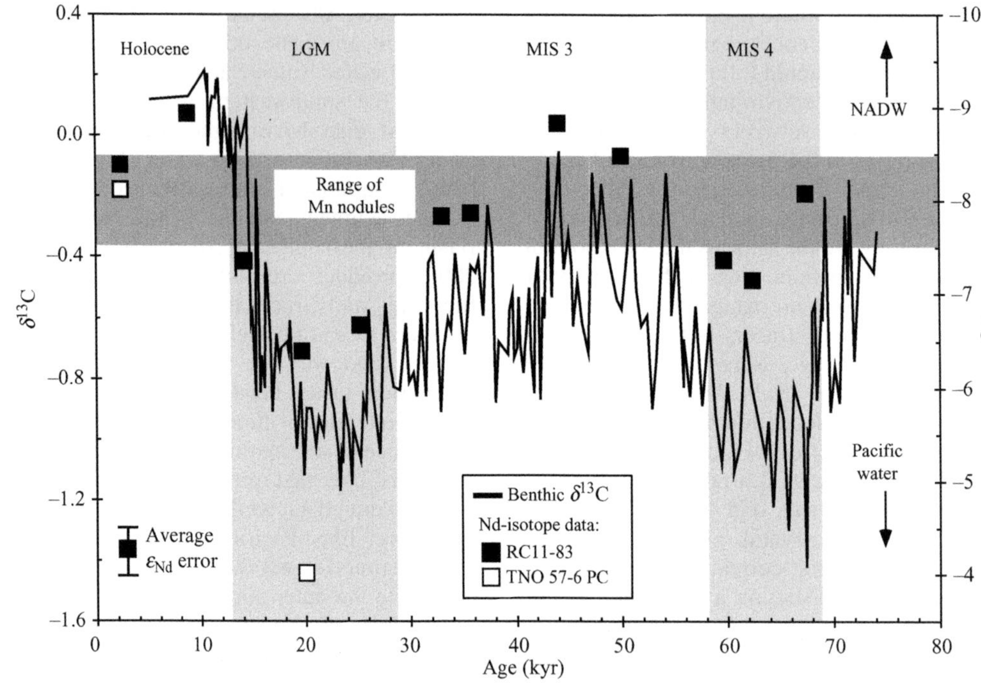

Figure 16 *Nd-isotope ratios of Fe–Mn leachates and benthic foraminiferal $\delta^{13}C$ versus age in southeast Atlantic cores.* The ε_{Nd} axis is reversed in order to facilitate comparison. The $\delta^{13}C$ variations were previously interpreted as reflecting changes in thermohaline circulation intensity over the Holocene and the last ice age (Charles and Fairbanks, 1992; Charles *et al.*, 1996). The down-core Nd-isotope ratios are consistent with stronger and weaker thermohaline circulation intensity during warm and cold marine isotope stages (MIS 1,3 and 2,4), respectively. In TNO57-6, located shallower and farther south than RC11-83, the glacial Nd-isotope value is like the Pacific, and may indicate a shutdown of NADW export to the circum-Antarctic. Horizontal shaded region shows the range of Nd-isotope ratios in circumpolar Fe–Mn nodules (Albare(c)de *et al.*, 1997) (after Rutberg *et al.*, 2000).

authigenic Fe–Mn oxides have great potential as paleoceanographic water-mass proxies.

Although lead-isotope ratios in the present-day oceans cannot be calibrated due to anthropogenic contamination, as previously discussed, one of the first major long-lived radiogenic isotope studies showed that the pre-industrial distribution can be determined using authigenic Fe–Mn sediments (Chow and Patterson, 1959, 1962). Further studies (Abouchami and Goldstein, 1995; von Blanckenburg *et al.*, 1996) have shown that lead-isotope ratios from the surface layers of Fe–Mn nodules and crusts have similar patterns to neodymium isotopes. That is, they are consistent with an older continental provenance in the Atlantic and a younger one in the Pacific, and the present-day geographical distribution is consistent with dispersion of the signals by ocean circulation.

The primary limitation of using Fe–Mn crusts for paleoceanographic studies derives from the slow growth rates, which limit the possible time resolution. The only Fe–Mn crust with an age record allowing a time resolution of a few thousand years is sample VA13-2 (cf. Segl *et al.*, 1984), from the central Pacific, where significant changes on glacial and subglacial scales are not expected. A study by Abouchami *et al.* (1997) established the constancy of neodymium-isotope ratios in the Pacific through the last several glacial stages. Fe–Mn encrustations have been used extensively over the last several years to investigate long-term changes in the sources of neodymium and lead at different localities in the oceans (cf. Frank, 2002, and references therein). For higher time resolution, the record on disseminated Fe–Mn oxides in South Atlantic core RC11-83 by Rutberg *et al.* (2000) shows great future potential for addressing issues associated with changes of ocean circulation (Figure 16). In addition, Vance and Burton (1999) have concluded that they can trace changes in the provenance of neodymium in surface waters on the basis of analyses of carefully cleaned foraminifera.

20.5.2 Long-term Time Series in Fe–Mn Crusts

The current state of the long-term time series data from Fe–Mn crusts long-term has been addressed in detail in the review by Frank (2002). Here we only present a general summary.

On large timescales over millions of years, neodymium and lead isotopes in Fe–Mn crusts have the potential for addressing changes in weathering contributions to the ocean, major circulation changes controlled by tectonic movements of the continents, and opening or closing of important ocean passages. The isotopic signature of the input to the oceans may vary for a variety of reasons, including changes in contributions to the oceans from continental terrains of different ages, changes in runoff and elevation, and even changes in the degree of incongruent weathering. As a result, applications on long timescales are intriguing but complicated.

The review by Frank (2002) discusses the long-term time series neodymium- and lead-isotope data from 19 Mn crusts, and some clear patterns are evident. The North Atlantic and Pacific record the extreme old and young continental provenance, respectively, throughout the time span of the crusts of up to ~60 Myr, with the Indian Ocean remaining intermediate throughout. In addition, the initiation of northern hemisphere glaciation is accompanied by a dramatic trend towards older continental sources in North Atlantic crusts. The data show that the glacial erosion process yielded an increased contribution of ancient continental products to the dissolved load of the Atlantic, and possibly a higher concentration as well. Moreover, the Southern Ocean does not show a correlative change. This has been interpreted to indicate that the time-integrated (over the $\sim 10^5$–10^6 yr time resolution of a single Fe–Mn crust sample) contribution of NADW to the Southern Ocean decreased with the onset of northern hemisphere glaciation (Frank *et al.*, 2002). Several papers have raised questions about incongruent chemical weathering imprints on the isotope composition of dissolved neodymium (von Blanckenburg and Nagler, 2001) and hafnium and lead isotopes (Piotrowski *et al.*, 2000; van de Flierdt *et al.*, 2002).

20.5.3 Hf–Nd Isotope Trends in the Oceans

Among long-lived isotope systems, the best correlated in continental rocks and oceanic basalts are neodymium- and hafnium-isotope ratios. They form a very coherent positive linear trend which is taken as the "mantle–crust array" (Figure 17(a)), reflecting the effects of coupled fractionation of Sm/Nd and Lu/Hf ratios through the history of the differentiation of the silicate earth (e.g. Vervoort *et al.*, 1999). However, marine Hf–Nd isotopic variations lie on a slope that diverges from the crust–mantle array. While the range of ε_{Hf} values in continental rocks is about twice that of ε_{Nd}, the variability of hafnium isotopes in Fe–Mn crusts and nodules is considerably smaller than that of neodymium isotopes (White *et al.*, 1986; Piotrowski *et al.*, 2000; Godfrey *et al.*, 1997; Albare(c)de *et al.*, 1998; van de Flierdt *et al.*, 2002). As a result, Nd–Hf isotope ratios of Fe–Mn crusts and

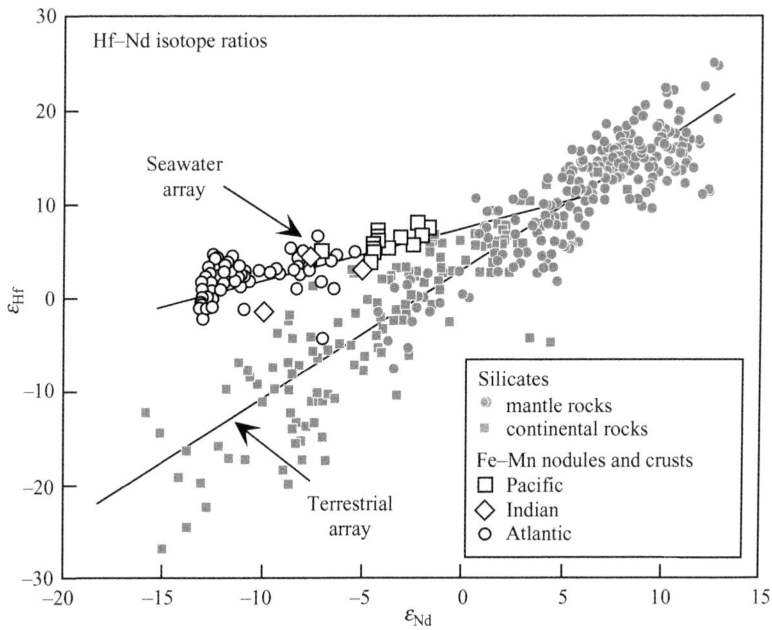

Figure 17 *Nd–Hf isotope ratios in rocks and Fe–Mn deposits.* In mantle and continental rocks these two isotope ratios a positive covariation, reflecting correlated fractionations of Sm/Nd and Lu/Hf during mantle melting and continent formation over Earth history. Seafloor Fe–Mn deposits are distinct, and lie off the continent-mantle "array," at high Hf isotope ratios for a given Nd isotope ratio. Possible explanations are given in the text. The figure is modified from van de Flierdt *et al.* (2002). The crust–mantle "array" is based on Vervoort *et al.* (1999). (Hf isotope data on Fe–Mn samples: White *et al.*, 1986; Godfrey *et al.*, 1997; Albare(c)de *et al.*, 1998; Burton *et al.*, 1999; Piotrowski *et al.*, 2000; David *et al.*, 2001; van de Flierdt *et al.*, 2002).

nodules fall off the Nd–Hf "mantle–crust array" (Figure 17(b)), where relative to the array, the hafnium-isotope ratio is too high for a given neodymium-isotope ratio.

This observation has been interpreted in various ways. (i) The residence time of hafnium in the oceans is longer than neodymium (White *et al.*, 1986). (ii) Hydrothermal systems associated with ridge volcanism make a significant contribution to the hafnium in the oceans (White *et al.*, 1986), unlike neodymium (Michard *et al.*, 1983; Piepgras and Wasserburg, 1985). (iii) Hafnium isotopes in seawater reflect incongruent weathering of continental rocks. This last possibility could have a large effect because hafnium is a major element in the mineral zircon, comprising a few to several percent by weight. Moreover, the high hafnium abundance in zircon means that the Lu/Hf ratio is close to zero. As a result, hafnium isotopes in zircons do not change markedly with time due to radioactive decay. Due to the high abundance, a significant fraction of the hafnium budget of the continental crust resides in zircon. Finally, zircon is highly resistant to weathering. As a result, hafnium may be preferentially weathered from non-zircon portion of a rock, with a significantly higher Lu/Hf and hafnium-isotope ratio (depending on the geological age) than the bulk rock (White *et al.*, 1986; Piotrowski *et al.*, 2000; Godfrey *et al.*, 1997; van de Flierdt *et al.*, 2002).

Of course, all of these factors may be partly acting to produce the marine Hf–Nd isotope trend. At first glance a longish apparent residence time for hafnium is puzzling. However, White *et al.* (1986) note that the flux of continental-derived hafnium to the oceans is likely to be quite small due to the insoluble nature of zircon and hafnium in general. They refer to a model of scavenging by particulate matter of Balistrieri *et al.* (1981) and find that if they use the relevant constants for hafnium they get scavenging residence times of the REEs of 10^2 yr, compared to 10^{12} yr for hafnium! They note that these predictions are qualitatively consistent with the observation that hafnium is only slightly enriched in deepsea red clays while rare earths are strongly enriched (Patchett *et al.*, 1984). Van de Flierdt *et al.* (2002) report a high resolution study of some North Atlantic manganese crusts, and find that at about the time of onset of northern hemisphere glaciation, the marine Hf–Nd isotope array moved steeply downward in the direction of silicate reservoirs. They further showed a systematic negative correlation between lead and hafnium isotopes. The hafnium-isotope ratios decrease and lead-isotopes increase at the time of the northern hemisphere glaciation, which is consistent with addition of both elements through greater weathering of ancient zircon. Accordingly, they provide evidence that in some cases the normally inert zircon can be ground finely enough to be partly dissolved.

Thus, it is possible that all of the three controls are acting on the hafnium contribution to seawater. Despite the low concentrations of neodymium in seawater, the high Nd/Hf ratios in Fe–Mn crusts indicate relatively lower hafnium abundances. Because zircon is not a crystallizing phase in basalts, hafnium is not sequestered in zircon in the oceanic crust and it may be more available for dissolution due to hydrothermal processes compared to continental rocks. As a result the flux of hafnium from the oceanic crust into seawater, relative to the continental flux, may be higher than for neodymium, making it more visible.

20.6 LONG-LIVED RADIOGENIC TRACERS AND ICE-SHEET DYNAMICS

The previous discussions focused on utilizing long-lived radioactive systems, either dissolved in seawater or in authigenic precipitates. Here we discuss its utility in silicate detritus or individual minerals in marine sediments to follow the history of northern hemisphere ice sheets during the last glacial period. This is merely one example of the powerful potential of radiogenic isotopes as provenance tracers wih paleoceanographic implication.

Ice rafting was an important depositional process in the North Atlantic during Pleistocene glacial cycles (e.g., Ruddiman, 1977). The variable geologic age and history of land covered by ice sheets allows the potential for understanding their growth and evolution from millions to hundreds of year timescales through tracking of their detritus into marine sediments. As a result, the distribution of ice rafted detritus in the North Atlantic has been used to infer the major sources of icebergs and the general pattern of surface circulation. There has been a similar pattern of ice rafted deposition during glacial and interglacial times, with a southward shift in the locus of melting related to colder surface water in glacial times. Detailed studies of ice rafting in conjunction with other evidence for changes in the ocean–atmosphere–ice system in the North Atlantic provide important insights into processes that control climate conditions. A key element for understanding ocean–ice sheet–atmosphere interactions is identifying the major sources of ice-rafted detritus. In the following sections we give some examples of the application of isotopic provenance studies to constrain ice-rafted detritus (and thus iceberg) sources in the North Atlantic region.

20.6.1 Heinrich Events

"Heinrich events" of the North Atlantic are found as prominent layers, rich in ice-rafted detritus, within marine sediment cores (Heinrich, 1988). Heinrich layers are important as they may be related to extreme climate events worldwide (e.g. Broecker, 1994; Broecker and Hemming, 2001). The six Heinrich layers of the last glacial cycle can be divided into two groups based on the flux of ice-rafted grains and on the concentration of detrital carbonate in the coarse fraction (Bond *et al.*, 1992; Broecker *et al.*, 1992; Gwiazda *et al.*, 1996a; McManus *et al.*, 1998). Heinrich events are named in numerical order with H1 being the most recent. All six Heinrich layers are characterized by high percentages of ice-rafted detritus. However, in the case of H3 and H6 the flux of ice-rafted detritus, as indicated by number of lithic grains per gram or by $^{230}Th_{xs}$ measurements (McManus *et al.*, 1998), is not greatly increased relative to the background. Instead the high ice-rafted detritus percentage appears to be related to low foraminifera abundances. The array of data that have been collected on Heinrich layer provenance reveal a remarkably complete story of the geological history of the Heinrich layers' source. The four prominent Heinrich layers, H1, H2, H4, and H5, appear to have formed from massive discharges of icebergs from Hudson Strait. The provenance of these Heinrich layers is very distinctive within the IRD belt and hence can be mapped by any number of geochemical measurements. Isotopic studies to date have examined Heinrich layer provenance using K/Ar, Nd, Sr, and Pb isotopic techniques.

20.6.2 K/Ar ages of Heinrich Event Detritus

The first geochemical provenance measurement of the Heinrich layers was the K/Ar apparent age of <2 μm and 2–16 μm fine fractions (Jantschik and Huon, 1992). Ambient North Atlantic sediments have apparent K/Ar ages of ~400 Myr (Hurley *et al.*, 1963; Huon and Ruch, 1992; Jantschik and Huon, 1992), whereas the sediments from Heinrich layers H1, H2, H4, and H5 yielded apparent ages of ~1 Gyr. Variation in the potassium concentration is small, and thus the K/Ar age signal is a product of the radiogenic $^{40}Ar^*$ concentration (Hemming *et al.*, 2002a). Hemming *et al.* (2002a) showed that the $^{40}Ar^*$ is quite uniform in eastern North Atlantic cores and that the K/Ar age and $^{40}Ar/^{39}Ar$ spectra of <2 μm terrigenous sediment from Heinrich layer H2 in the eastern North Atlantic and from Orphan Knoll (southern Labrador Sea/western Atlantic) are indistinguishable. Taken together these results imply the entire fine fraction of Heinrich layer H2 was derived from sources bordering the Labrador Sea. The same pattern most likely characterizes H1, H4, and H5, because their K/Ar ages and $^{40}Ar^*$ concentrations are similar to H2 in eastern North Atlantic cores.

20.6.3 Nd–Sr–Pb Isotopes in Terrigenous Sediments

Strontium-isotope ratios are not particularly diagnostic of source terrain ages because Rb/Sr ratios are easily disturbed. However, Sm/Nd ratios tend to be conservative and neodymium-isotope ratios are effective tracers of the mean crustal age of sediment sources. Neodymium-isotope ratios of Heinrich layers H1, H2, H4, and H5 are consistent with derivation from a source with Archean heritage, and Grousset *et al.* (1993) suggested sources surrounding the Labrador Sea or Baffin Bay (Figure 18). Strontium isotopes can be useful, however, when coupled with neodymium isotopes to trace the combined Sr–Nd characteristics of source terrains. Neodymium- and strontium-isotope ratios for different grain-size fractions from North Atlantic sediments suggest that the total terrigenous sediment load within these Heinrich layers in the IRD belt may be derived from the same limited range of sources (Revel *et al.*, 1996; Hemming *et al.*, 1998; Snoeckx *et al.*, 1999; Grousset *et al.*, 2000, 2001). The University of Colorado group has extensively characterized the composition of potential source areas in the vicinity of the Hudson Strait, Baffin Bay, other regions along the western Labrador coast, and the Gulf of St. Lawrence (Barber, 2001; Farmer *et al.*, 2003). Their data are consistent with a Hudson Strait provenance for Heinrich layers H1, H2, H4, and H5, and demonstrate an absence of substantial southeastern Laurentide ice sheet sources within pure Heinrich intervals. Lead-isotope ratios of the fine terrigenous fraction of Heinrich layers are distinctive (Hemming *et al.*, 1998), and are also consistent with derivation from the Hudson Strait region. New results from Farmer *et al.* (2003) may allow further distinction of fine-grained sediment sources with lead isotopes.

20.6.4 Isotopic and Geochronologic Measurements on Individual Mineral Grains

In addition to the bulk geochemical methods, several studies have examined individual grains or composite samples of feldspar grains for their lead-isotope ratios (Gwiazda *et al.*, 1996a,b; Hemming *et al.*, 1998), or individual grains of hornblende for their $^{40}Ar/^{39}Ar$ ages (Gwiazda *et al.*, 1996c; Hemming *et al.*, 1998, 2000b; Hemming and Hajdas, 2003). These studies provide remarkable insights into the geologic history of Heinrich layers that allow further refinement of the interpretations based on bulk isotopic analyses. Feldspars have high lead abundance and very low uranium and thorium abundance and thus the lead-isotope ratios of feldspar approximate the initial lead-isotope ratios of its source (e.g. Hemming

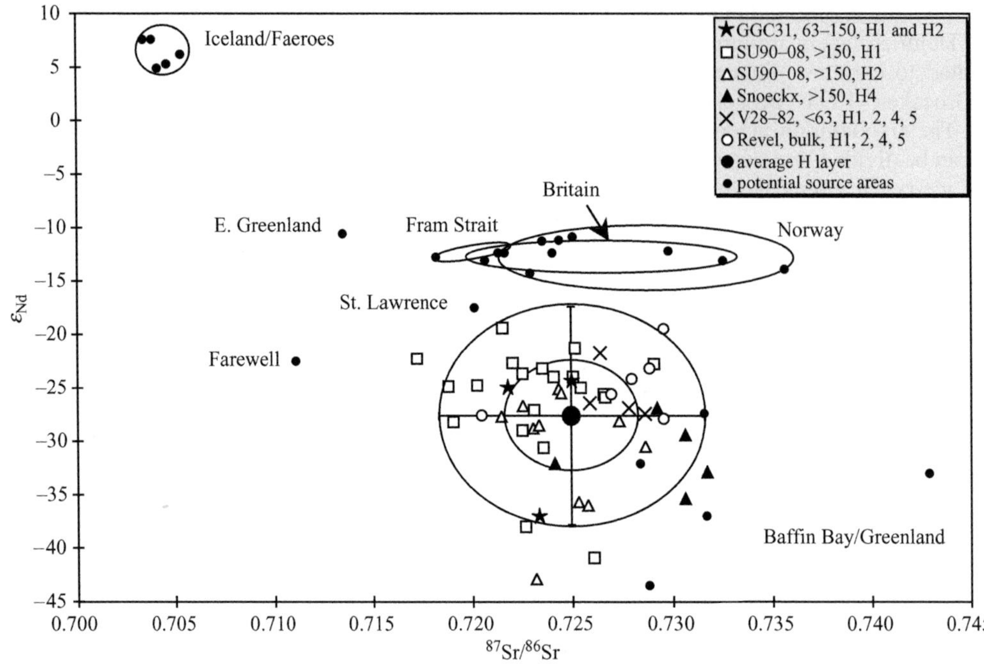

Figure 18 *Nd–Sr-isotope ratios of terrigenous clastic components of Heinrich layers H1, H2, H4, and H5.* Shown for reference are the average and 1 and 2 sigma range for the data published on these Heinrich layers and reported compositions of potential source areas of ice rafted detritus (Grousset *et al.*, 2001, and references therein). Note that the data are from different size fractions in different publications, but there is not evidence that there is a substantial bias even in the $^{87}Sr/^{86}Sr$ where one might be expected. This is most likely due to the large fraction of glacial flour with approximately the same composition in the fine as the coarse fraction in contrast to the way most sedimentary grain size variations are produced (Sources Grousset *et al.*, 2001; Snoeck *et al.*, 1999; Hemming *et al.*, 1998; Hemming, unpublished; Revel *et al.*, 1996).

et al., 1994, 1996, 2000b). Lead-isotope data from Heinrich layer H1, H2, H4, and H5 feldspar grains form a linear trend that indicates an Archean (~2.7 Ga) heritage and a Paleoproterozoic (~1.8 Ga) metamorphic event (Figure 19(b)). Heinrich layer grains are similar in composition to H2 from Hudson Strait-proximal core HU87-009 and to feldspar grains from Baffin Island till (Hemming *et al.*, 2000b). However, they are distinctly different from feldspar grains from Gulf of St. Lawrence core V17-203, where Appalachian (Paleozoic) and Grenville (~1 Ga) sources are found. $^{40}Ar/^{39}Ar$ ages of individual hornblende grains from Heinrich layers H1, H2, H4, and H5 cluster around the implied Paleoproterozoic metamorphic events from the feldspar lead-isotope data (Gwiazda *et al.*, 1996c; Hemming *et al.*, 1998, 2000a; Hemming and Hajdas, 2003), and consistent with hornblende grains from Baffin Island tills (Hemming *et al.*, 2000b).

20.6.5 Contrasting Provenance of H3 and H6

Events H3 and H6 do not appear to be derived from the same sources as H1, H2, H4, and H5 (Figures 19(c) and 20). Using lead-isotope compositions of composite feldspar samples, Gwiazda *et al.* (1996a) found that H3 and H6 resemble the ambient sediment in V28-82, suggesting a large European source contribution, which agrees with the conclusion of Grousset *et al.* (1993). Lead-isotope data from composites of 75 to 300 grains from Gwiazda *et al.* (1996a) are shown in Figure 19(c). As mentioned above, H3 and H6 seem to be low-foraminifera intervals rather than ice-rafting events. These conclusions are consistent with other observations around the North Atlantic. Although Heinrich layer H3 appears to be a Hudson Strait event (Grousset *et al.*, 1993; Bond and Lotti, 1995; Rashid *et al.*, 2003, and Hemming, unpublished lead-isotope and $^{40}Ar/^{39}Ar$ hornblende data from Orphan Knoll), it does not spread Hudson Strait-derived IRD as far to the east as the other Heinrich events (Grousset *et al.*, 1993; Figure 19). Additionally, Kirby and Andrews (1999) proposed that H3 (and the Younger Dryas) represent across Strait (also modeled by Pfeffer *et al.*, 1997) rather than along Strait flow as inferred for H1, H2, H4, and H5. The strontium-isotope map of H3 shows a striking pattern of decrease in $^{87}Sr/^{86}Sr$ ratios nearly perpendicular to the IRD belt (Figure 21), consistent with a mixture of sediments from the Labrador Sea icebergs with those derived from

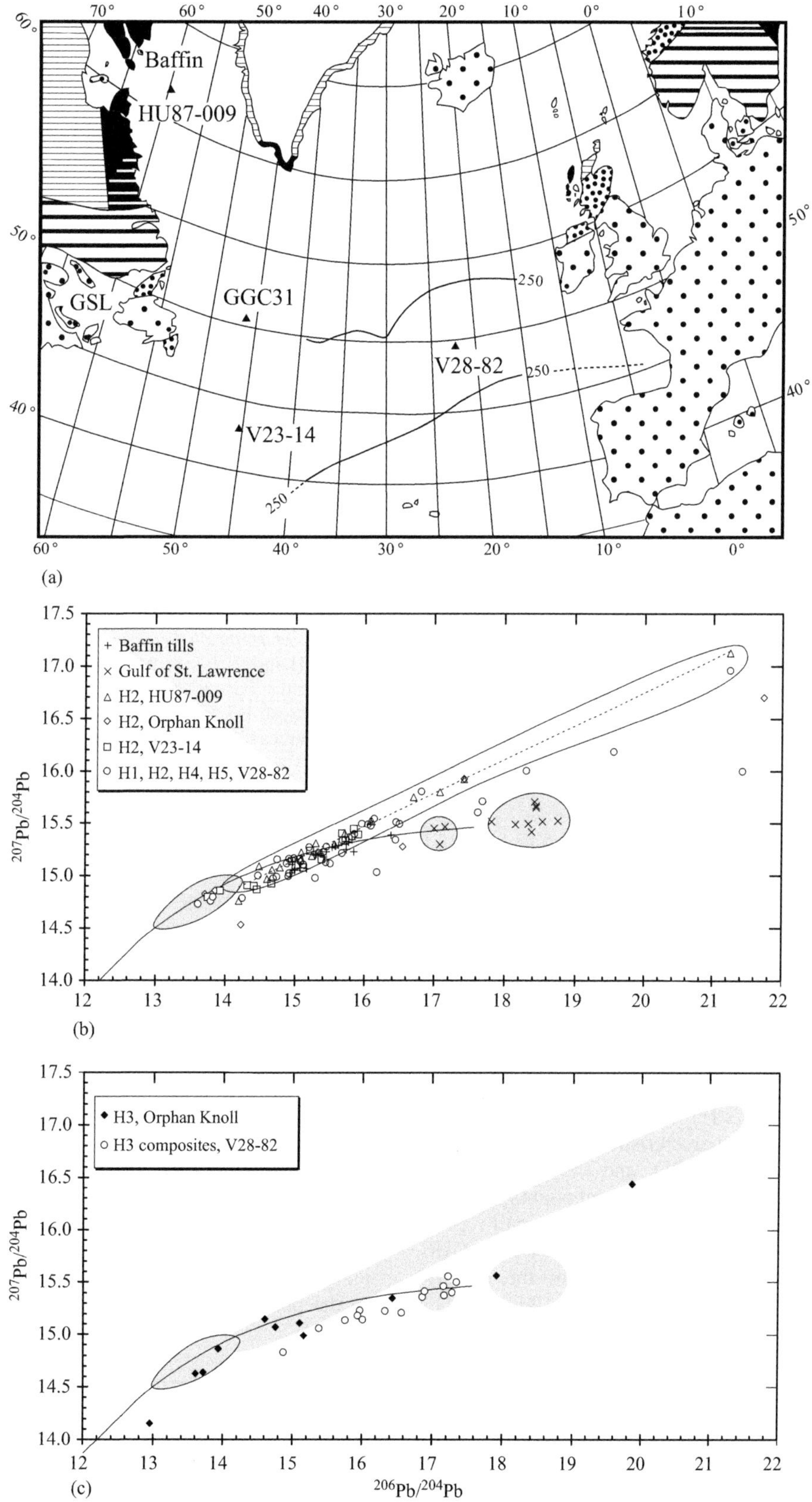

Figure 19 (Continued).

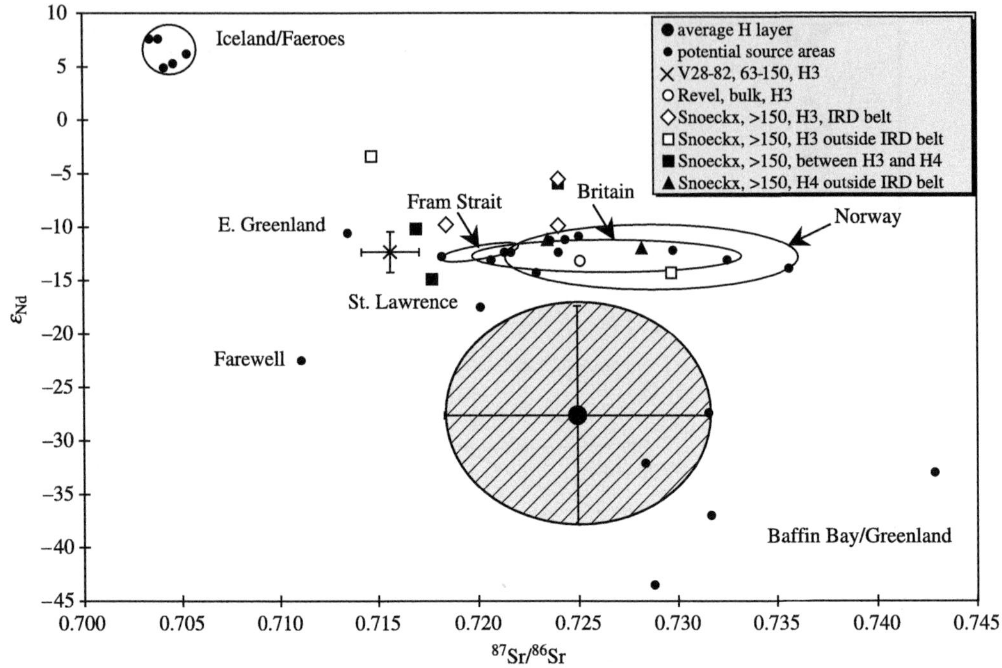

Figure 20 *Nd–Sr-isotope ratios of terrigenous clastic components of Heinrich layer H3.* Shown for reference are the average and 2 sigma range for the data published on Heinrich layers H1, H2, H4, and H5, and reported compositions of potential source areas of ice rafted detritus (Grousset *et al.*, 2001, and references therein). Also included is the average of 5 unpublished analyses across H3 from V28-82, where the error bars represent the range of values measured (A. Jost and S. Hemming) (sources Grousset *et al.*, 1993; Revel *et al.*, 1996; Snoeckx *et al.*, 1999).

icebergs from eastern Greenland, Iceland, and Europe. H6 has not been studied, but the composition of organic material in H3 does not stand out prominently in the studies mentioned above (Madureira *et al.*, 1997; Rosell-Mele *et al.*, 1997; Huon *et al.*, 2002).

20.6.6 Summary of Geochemical Provenance Analysis of Heinrich Layers

Heinrich layers H1, H2, H4, and H5 have several distinctive characteristics that distinguish them from ambient IRD, and they are derived from a mix of provenance components that are all consistent with derivation from a small region near Hudson Strait. Heinrich layers H3 and H6 have different sources, at least in the eastern North Atlantic. Less is known about H6, but H3 appears to have a Hudson Strait source in the southern Labrador Sea and western Atlantic, consistent with a similar but weaker event compared to the big four. Important related questions are as follows: How many types (provenance, flux, etc.) of Heinrich layers are there? Are IRD events in previous glacial intervals akin to the six in the last glacial period?

20.6.7 Trough Mouth Fans as Archives of Major IRD Sources

Trough mouth fans are major glacial-marine sediment fans that form when ice streams occupied glacial troughs that extend to the shelf-slope boundary (e.g., Vorren and Laberg, 1997). They are a significant resource and archive for understanding the potential compositional characteristics of

---

Figure 19 *Pb isotopes in feldspar grains from Heinrich layers.* (a) Map showing locations of cores analyzed with geology and IRD belt for reference. (b) Data from Heinrich layers H1, H2, H4, and H5 from several North Atlantic and Labrador Sea cores. Also shown are data from Gulf of St. Lawrence core V17-203 (Hemming, unpublished data) and from Baffin Island tills (Hemming *et al.*, 2002b). Reference fields are Superior province (Gariépy and Alle(c)gre, 1985), Labrador Sea reference line (LSRL, Gwiazda *et al.*, 1996b), Grenville (DeWolf and Mezger, 1994) and Appalachian (Ayuso and Bevier, 1991). Data sources are H2 from HU87-009, V23-14, and V28-82 (Gwiazda *et al.*, 1996b), Orphan Knoll core GGC31 (Hemming, unpublished), H1, H4, and H5 from V28-82 (Hemming *et al.*, 1998). (c) Data from Heinrich layer H3 with reference fields from (b). Data sources are H3 from V28-82 (Gwiazda *et al.*, 1996a), H3 from Orphan Knoll core GGC31 (Hemming, unpublished data).

sediment-laden icebergs, because they contain source-specific detrital material in highly concentrated accumulations. An important consideration that emphasizes the value of this archive is the recognition that the debris flows mimic the tills in the source area and, therefore, also the composition of the IRD in the calving icebergs. Furthermore, most icebergs experience substantial melting

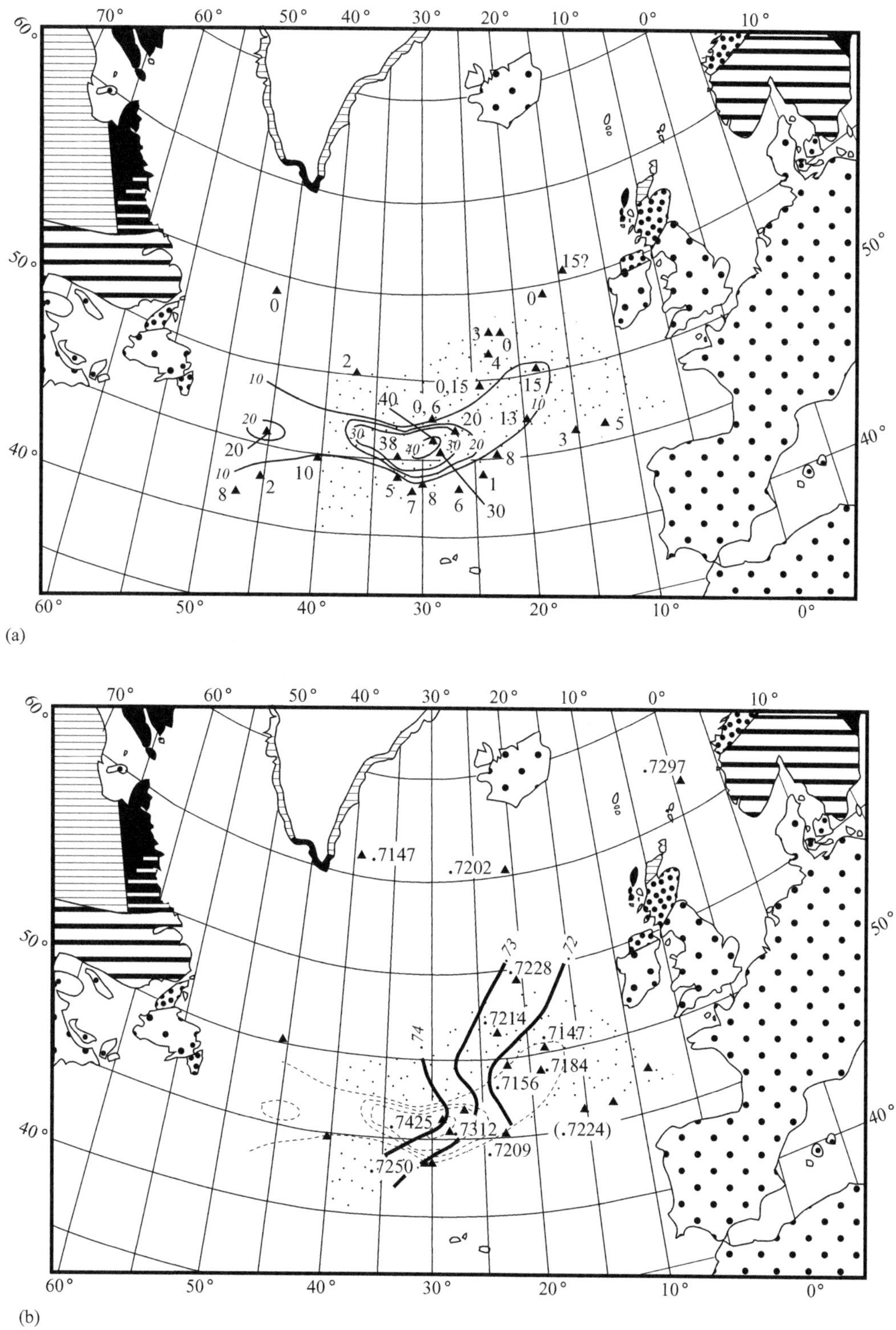

Figure 21 (Continued).

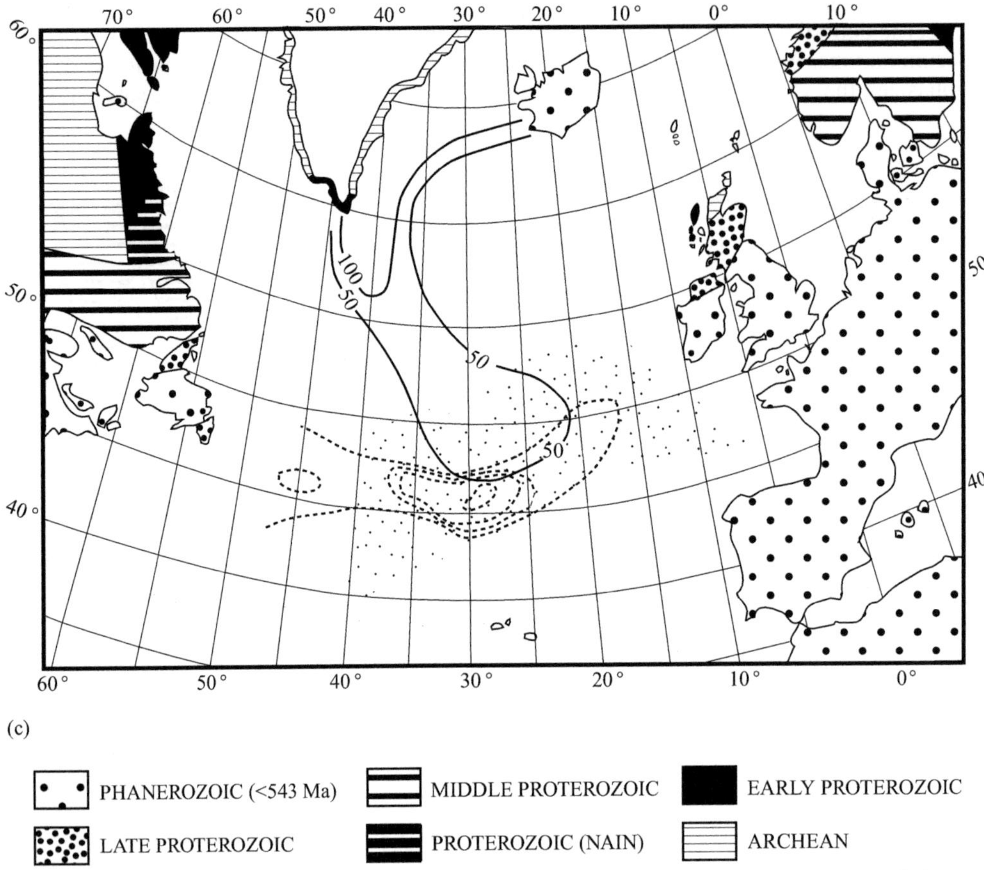

Figure 21 *Maps of data from Heinrich layer H3*. General geology and the Ruddiman (1977) IRD belt are shown. (a) Isopach map with 10 cm contour interval. (b) $^{87}Sr/^{86}Sr$ values for siliciclastic detritus in H3. Isopachs are shown in light dashed lines for reference. (c) 25–23 ka 250 mg cm^{-2} kyr^{-1} contours defining the approximately E–W IRD belt (Ruddiman, 1977), contours of 10, 50, and 100 sand-sized ash shards per cm^2 defining the approximately N–S trajectory of currents bringing Icelandic detritus into the North Atlantic (Ruddiman and Glover, 1982), and light dashed lines of 10 cm thickness intervals for H3.

close to the ice-sheet margin, and the sediments they carry tend to melt out before the iceberg has reached the open ocean (Syvitski *et al.*, 1996; Andrews, 2000). Sediments deposited on the trough mouth fans are not only good representative point sources of the glacial drainage area; they are also located in positions of high iceberg production, that increase the likelihood of an iceberg's being carried to deep-marine environments in the surface-ocean currents before loosing its sediment load. Locations of documented trough mouth fans in the northern hemisphere are shown in Figure 22.

Although there is substantial overlap in the geological histories of the continental sources around the North Atlantic, there are also some systematic variations that can allow distinction of different sources. Realization of the full potential awaits characterizing the sources with multiple tracers as well as sedimentological studies, and documenting the geographic pattern of dispersal in marine sediments within small time-windows.

A first step towards the goal of characterizing IRD sources in trough mouth fans was made by Hemming *et al.* (2002b) using $^{40}Ar/^{39}Ar$ data from populations of individual hornblende grains. The results conform closely to expectations based on the mapped ages of rocks in the region of the grains analyzed. For example, samples from the mid-Norwegian margin and from the north-eastern margin of Greenland provide Caledonian ages. Samples from the Bear Island trough mouth fan have little hornblende indicating a dominant sedimentary source, and the hornblendes that are found yield a dominantly Paleoproterozoic spectrum of ages. Samples from near the southern tip of Greenland exhibit a mixture of Paleoproterozoic and a distinctive ~1.2 Ga population that is inferred to have derived from the alkaline Gardar complex. Samples from the Gulf of St. Lawrence have a virtually pure Grenville population, and samples from the Hudson Strait have a dominant Paleoproterozoic population with a small Archean contribution.

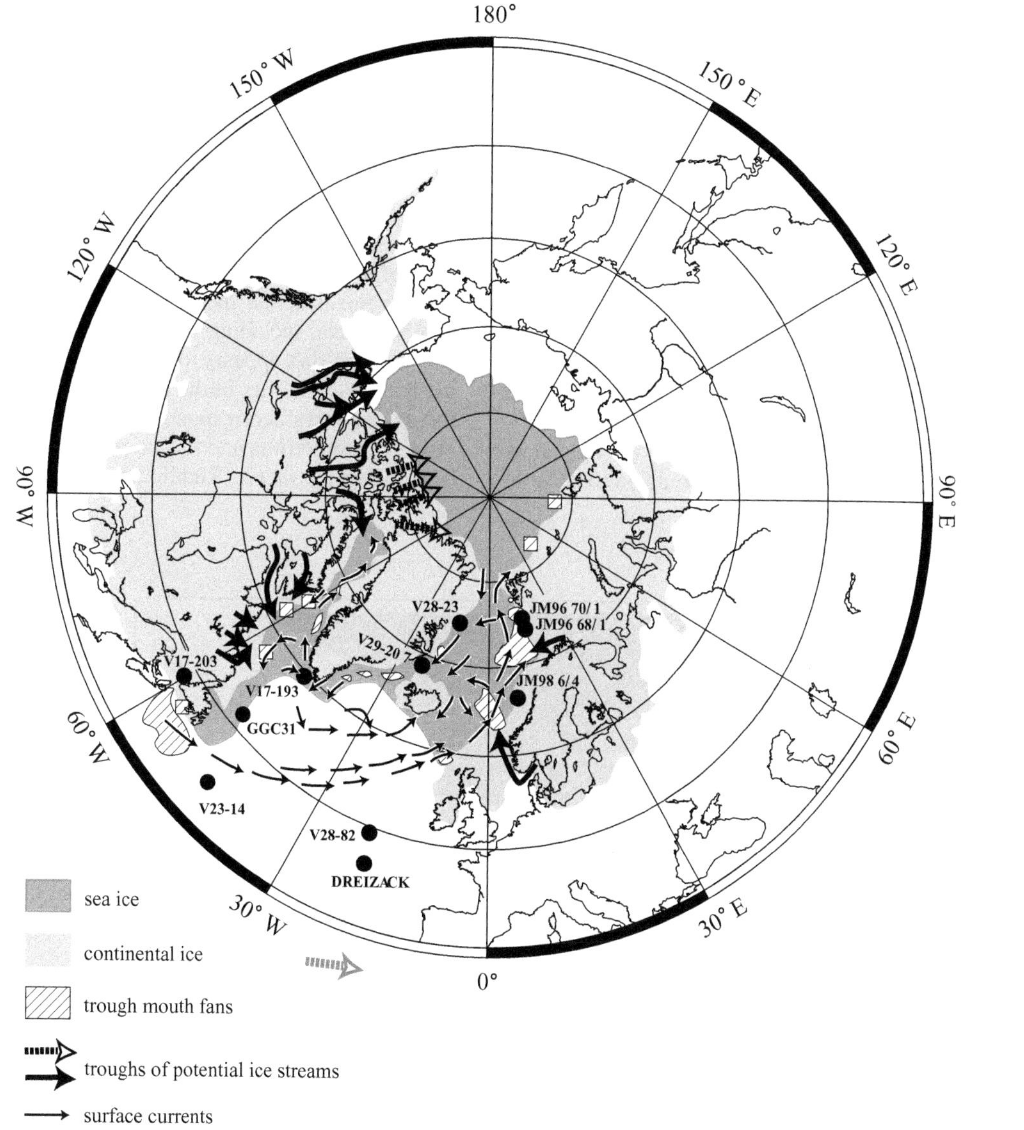

Figure 22 Polar projection from 40° to 90° showing the extents of continental and sea ice in the northern hemisphere, and cores presented and/or discussed here. The locations of known trough mouth fans are shown as well as inferred ice streams of the Laurentide ice sheet. Trough mouth fans (TMF) of the Nordic Seas are from Vorren and Laberg (1997), those around Greenland are from Funder *et al.* (1989), and along the western Labrador Sea are from Hesse and Khodabakhsh (1998). Arrows marking troughs across the Laurentide margin are based on mapping of J. Kleman. Arrow marking large trough feeding the North Sea TMF and the Barents Sea arrow indicating a large trough feeding Bear Island TMF is from Stokes and Clark (2001).

20.6.8 $^{40}Ar/^{39}Ar$ Hornblende Evidence for History of the Laurentide Ice Sheet During the Last Glacial Cycle

The history of the northern hemisphere ice sheets is an important aspect of the paleoclimate system. The layered record of IRD and other climate indicators preserved in deepsea sediment cores provides the potential to unravel the sequences of events surrounding important intervals during the last glacial cycle. Results of an extensive campaign to analyze the $^{40}Ar/^{39}Ar$ ages of multiple individual hornblende grains in core V23-14 (Hemming and Hajdas, 2003) are presented as an example of the potential of this approach.

Core V23-14 is located within the thickest part of Ruddiman's (1977) ice-rafted detritus belt, and directly downstream of any Gulf of St. Lawrence region contributions (Figure 19(a)). Thus, the provenance of IRD from this core provides constraints on the evolution of the Laurentide ice sheet or smaller satellite ice sheets (e.g., Stea *et al.*, 1998) and their iceberg contributions to the northwest Atlantic margin. The hornblende data are binned according to the age brackets based on known geological ages in the North Atlantic region. Data from within the Heinrich layers are lumped into one interval with an assumed duration of 0.1 kyr. There is some disagreement about the duration of Heinrich layers, ranging from estimates that they were virtually instantaneous, to estimates of 1,000 yr or more. Currently available published results of dating and flux methods used do not allow clearly constraining these estimates, although a best average estimate of duration is taken to be 500+/−250 yr. By lumping the Heinrich layer samples into a small interval, and plotting the cumulative fraction data against estimated age, we emphasize the ambient evolution of the Laurentide ice sheet (Figure 23).

In the period 42.5–26 ka (by ^{14}C dating), there is little evidence of iceberg contributions from Laurentian sources south of 55° N, which should provide dominantly Grenville and Appalachian ages. In contrast, in the period 26–14 ^{14}C ka, there is abundant evidence of contributions from this sector. Finally, in the period 14–6 ^{14}C ka, Grenville and Appalachian ages are again absent. These observations are consistent with the results of Hemming *et al.* (2000b) from Orphan Knoll in the interval above H2. They are also consistent with the observations of Ruddiman (1977), who presented ice-rafted detritus fluxes across the North Atlantic in several time intervals, including 40–25 ka, and 25–13 ka. The average flux appears to be approximately double the 40–25 ka flux in the 25–13 ka interval, and there is an overall correspondence between the flux of ice-rafted detritus and the extent of the northern hemisphere ice sheets (Ruddiman, 1977).

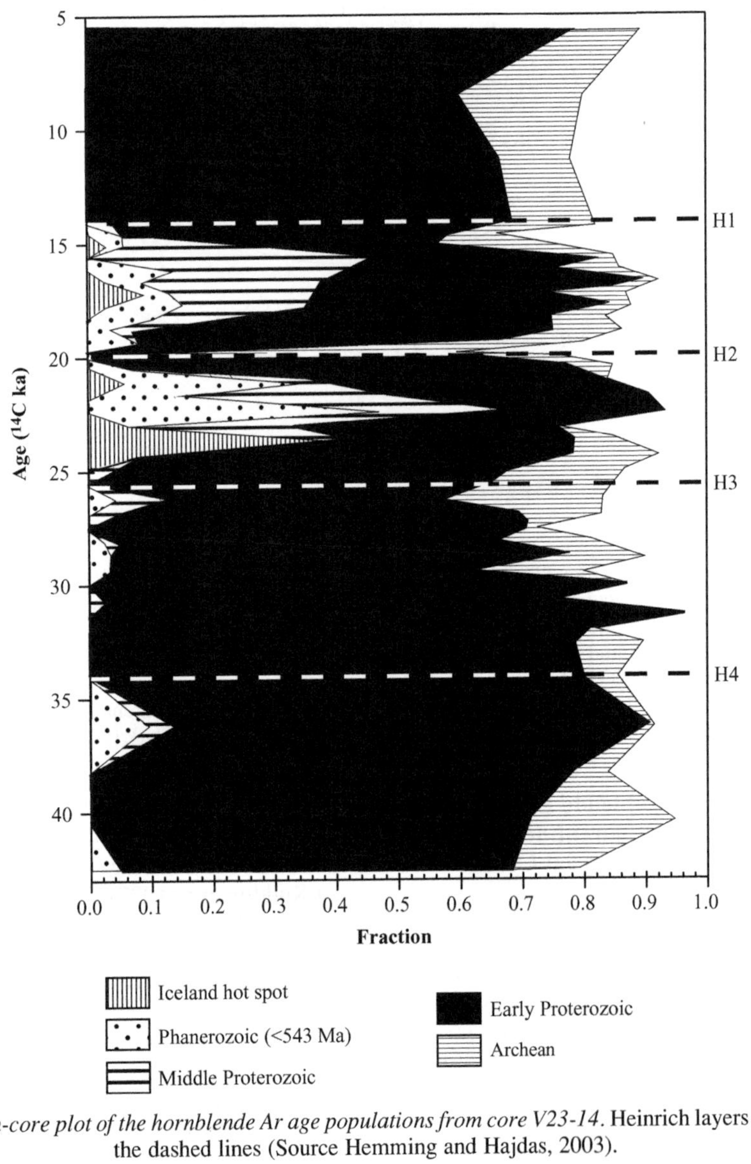

Figure 23 *Down-core plot of the hornblende Ar age populations from core V23-14.* Heinrich layers are indicated by the dashed lines (Source Hemming and Hajdas, 2003).

Results from studies of the ice-rafted detritus population from core V23-14 provide insights into the evolution of ice sheets of the northwest Atlantic margin since 43 ka. Between Heinrich layers H5 and H3 (MIS 3), it appears that the ice sheet (or sheets) did not extend far enough southeast to drop iceberg deposits with Grenville or Appalachian derivation into the North Atlantic. Between H3 and H1 (MIS 2), and outside the Heinrich layers, significant portions of the hornblende grains have ages indicating derivation from the southeastern sector, and thus indicate a significant expansion of the sheets at ~26 ^{14}C ka. After H1 (MIS 1), no detritus attributable to the southeastern sector is found, and judging from the present day situation, most of the ice-rafted detritus was likely derived from the Greenland ice sheet in the Holocene interval.

20.7 FINAL THOUGHTS

This chapter has summarized the basis for the application of long-lived isotopic tracers in oceanography and paleoceanography, focusing on neodymium isotopes in the oceans. It also summarized the current state of knowledge on the applications of these tracers to delineate the ice-sheet sources of North Atlantic Heinrich events. Especially since the early 1990s in particular, long-lived isotopic tracers have matured as paleoclimatic tools, as their potential value is becoming increasingly recognized and used by an growing number of investigators. Over the next decade they are certain to be among the primary methods used to generate new discoveries about the Earth's climatic history.

ACKNOWLEDGMENTS

This chapter is dedicated to Professors Gerald J. Wasserburg of Caltech, whose retirement roughly coincides with the publication of the Treatise, and Wally Broecker of LDEO. Gerry and colleagues characterized against all odds the neodymium isotopic variations in the global oceans, and the resulting body of work forms the basis of subsequent studies. Throughout his career Wally has greatly advanced the field of paleoceanography and global climate change, and has embraced the application of radiogenic isotope tracers. His contribution to emphasizing the importance of Heinrich layers is particularly relevant to this chapter. This is LDEO Contribution #6498.

REFERENCES

Abouchami W. and Goldstein S. L. (1995) A lead isotopic study of circum-antarctic manganese nodules. *Geochim. Cosmochim. Acta* **59**(9), 1809–1820.

Abouchami W., Goldstein S. L., Galer S. J. G., Eisenhauer A., and Mangini A. (1997) Secular changes of lead and neodymium in central Pacific seawater recorded by a Fe–Mn crust. *Geochim. Cosmochim. Acta* **61**(18), 3957–3974.

Albare(c)de F. and Goldstein S. L. (1992) World map of Nd isotopes in sea-floor ferromanganese deposits. *Geology* **20**(8), 761–763.

Albare(c)de F., Goldstein S. L., and Dautel D. (1997) The neodymium isotopic composition of manganese nodules from the southern and Indian oceans, the global oceanic neodymium budget, and their bearing on deep ocean circulation. *Geochim. Cosmochim. Acta* **61**(6), 1277–1291.

Albare(c)de F., Simonetti A., Vervoort J. D., Blichert-Toft J., and Abouchami W. (1998) A Hf–Nd isotopic correlation in ferromanganese nodules. *Geophys. Res. Lett.* **25**(20), 3895–3898.

Amakawa H., Alibo D. S., and Nozaki Y. (2000) Nd isotopic composition and REE pattern in the surface waters of the eastern Indian Ocean and its adjacent seas. *Geochim. Cosmochim. Acta* **64**(10), 1715–1727.

Andrews J. T. (2000) Icebergs and iceerg rafted detritus (IRD) in the North Atlantic: facts and assumptions. *Oceanography* **13**, 100–108.

Ayuso R. A. and Bevier M. L. (1991) Regional differences in Pb isotopic compositions of feldspars in plutonic rocks of the northern appalachian mountains, USA and Canada: a geochrmical method of terrane correlation. *Tectonics* **10**, 191–212.

Balistrieri L., Brewer P. G., and Murray J. W. (1981) Scavenging residence times of trace-metals and surface-chemistry of sinking particles in the deep ocean. *Deep-sea Res. Part A: Oceanogr. Res. Pap.* **28**(2), 101–121.

Barber D. (2001) Laurentide ice sheet dynamics from 35 to 7 ka: Sr–Nd–Pb isotopic provenance of northwest Atlantic margin sediments. PhD, University of Colorado.

Bertram C. J. and Elderfield H. (1993) The geochemical balance of the rare-earth elements and neodymium isotopes in the oceans. *Geochim. Cosmochim. Acta* **57**(9), 1957–1986.

Bizzarro M., Baker J. A., Haack H., Ulfbeck D., and Rosing M. (2003) Early history of Earth's crust–mantle system inferred from hafnium isotopes in chondrites. *Nature* **421**(6926), 931–933.

Blichert-Toft J. and Albare(c)de F. (1997) The Lu–Hf isotope geochemistry of chondrites and the evolution of the mantle–crust system. *Earth Planet. Sci. Lett.* **148**(1–2), 243–258.

Bond G., Heinrich H., Broecker W., Labeyrie L., McManus J., Andrews J., Huon S., Jantschik R., Clasen S., Simet C., Tedesco K., Klas M., Bonani G., and Ivy S. (1992) Evidence for massive discharges of icebergs into the North-Atlantic Ocean during the last glacial period. *Nature* **360**(6401), 245–249.

Bond G. C. and Lotti R. (1995) Iceberg discharges into the North-Atlantic on millennial time scales during the last glaciation. *Science* **267**(5200), 1005–1010.

Bouquillon A., France-Lanord C., Michard A., and Tiercelin J. J. (1990) Sedimentology and isotopic chemistry of the Bengal fan sediments: the denudation of the Himalaya. *Proc. ODP Sci. Res.* **116**, 43–53.

Broecker W. S. (1994) Massive iceberg discharges as triggers for global climate-change. *Nature* **372**(6505), 421–424.

Broecker W. S. and Denton G. H. (1989) The role of ocean-atmosphere reorganizations in glacial cycles. *Geochim. Cosmochim. Acta* **53**, 2465–2501.

Broecker W. S. and Hemming S. (2001) Paleoclimate-climate swings come into focus. *Science* **294**(5550), 2308–2309.

Broecker W. S. and Peng T.-H. (1982) *Tracers in the Sea*. Eldigio Press, Palisades, NY.

Broecker W. S., Bond G. C., Klas M., Clark E., and McManus J. (1992) Origin of the Northern Atlantic's Heinrich events. *Clim. Dyn.* **6**, 265–293.

Burton K. W., Lee D. C., Christensen J. N., Halliday A. N., and Hein J. R. (1999) Actual timing of neodymium isotopic variations recorded by Fe–Mn crusts in the western North Atlantic. *Earth Planet. Sci. Lett.* **171**(1), 149–156.

Cantrell K. J. and Byrne R. H. (1987) Rare earth element complexation by carbonate and oxalate ions. *Geochim. Cosmochim. Acta* **51**, 597–605.

Charles C. D. and Fairbanks R. G. (1992) Evidence from Southern Ocean sediments for the effect of North Atlantic deep-water flux on climate. *Nature* **355**, 416–419.

Charles C. D., Lynch-Stieglitz J., Ninnemann U. S., and Fairbanks R. G. (1996) Climate connections between the hemisphere revealed by deep-sea sediment core/ice core correlations. *Earth Planet. Sci. Lett.* **142**, 19–27.

Chow T. J. and Patterson C. C. (1959) Lead isotopes in manganese nodules. *Geochim. Cosmochim. Acta* **17**, 21–31.

Chow T. J. and Patterson C. C. (1962) The occurence and significance of lead isotopes in pelagic sediments. *Geochim. Cosmochim. Acta* **26**, 263–308.

Dasch E. J., Dymond J. R., and Heath G. R. (1971) Isotopic analysis of metalliferous sediment from the east Pacific rise. *Earth Planet. Sci. Lett.* **13**, 175–180.

David K., Frank M., O'Nions R. K., Belshaw N. S., and Arden J. W. (2001) The Hf isotope composition of global seawater and the evolution of Hf isotopes in the deep Pacific Ocean from Fe–Mn crusts. *Chem. Geol.* **178**(1–4), 23–42.

Debaar H. J. W., Bacon M. P., and Brewer P. G. (1983) Rare-earth distributions with a positive Ce anomaly in the western North-Atlantic Ocean. *Nature* **301**(5898), 324–327.

Debaar H. J. W., Bacon M. P., Brewer P. G., and Bruland K. W. (1985) Rare-earth elements in the Pacific and Atlantic oceans. *Geochim. Cosmochim. Acta* **49**(9), 1943–1959.

DePaolo D. J. and Wasserburg G. J. (1976a) Nd isotopic variations and petrogenetic models. *Geophys. Res. Lett.* **3**, 249–252.

DePaolo D. J. and Wasserburg G. J. (1976b) Inferences about magma sources and mantle structure from variations of Nd-143–Nd-144. *Geophys. Res. Lett.* **3**(12), 743–746.

DeWolf C. P. and Mezger K. (1994) Lead isotope analyses of leached feldspars—constraints on the early crustal history of the Grenville Orogen. *Geochim. Cosmochim. Acta* **58**(24), 5537–5550.

Elderfield H. (1988) The oceanic chemistry of the rare-earth elements. *Phil. Trans. Roy. Soc. London Ser. A: Math. Phys. Eng. Sci.* **325**(1583), 105–126.

Elderfield H. and Greaves M. J. (1982) The rare earth elements in seawater. *Nature* **296**, 214–219.

Elderfield H., Hawkesworth C. J., Greaves M. J., and Caalvert S. E. (1981) Rare-earth element geochemistry of oceanic ferromanganese nodules and associated sediments. *Geochim. Cosmochim. Acta* **45**, 513–528.

Farmer G. L., Barber D., and Andrews J. (2003) Provenance of late quaternary ice-proximal sediments in the North Atlantic: Nd, Sr, and Pb isotopic evidence. *Earth Planet. Sci. Lett.* **209**(1–2), 227–243.

Frank M. (2002) Radiogenic isotopes: tracers of past ocean circulation and erosional input. *Rev. Geophys.* **40**(1)article no. 1001.

Frank M., Whiteley N., Kasten S., Hein J. R., and O'Nions K. (2002) North Atlantic deep water export to the southern ocean over the past 14 Myr: evidence from Nd and Pb isotopes in ferromanganese crusts. *Paleoceanography* **17**(2) article no. 1002.

Funder S., Larsen H. C., and Fredskild B. (1989) Quaternary geology of the ice-free areas and adjacent shelves of Greenland. In *Quaternary Geology of Canada and Greenland: K-1* (ed. R. J. Fulten). Geological Survey of Canada, Boulder, CO, pp. 741–792.

Galy A., FranceLanord C., and Derry L. A. (1996) The Late Oligocene Early Miocene Himalayan belt: constraints deduced from isotopic compositions of Early Miocene turbidites in the Bengal fan. *Tectonophysics* **260**(1–3), 109–118.

Gao S., Luo T. C., Zhang B. R., Zhang H. F., Han Y. W., Zhao Z. D., and Hu Y. K. (1998) Chemical composition of the continental crust as revealed by studies in East China. *Geochim. Cosmochim. Acta* **62**(11), 1959–1975.

Gariépy C. and Alle(c)gre C. J. (1985) The lead isotope geochemistry and geochronology of late-kinematic intrucives from the Abitibi Greenstone belt, and the implications for late Archaean crustal evolution. *Geochim. Cosmochim. Acta* **49**, 2371–2383.

Godfrey L. V., Lee D. C., Sangrey W. F., Halliday A. N., Salters V. J. M., Hein J. R., and White W. M. (1997) The Hf isotopic composition of ferromanganese nodules and crusts and hydrothermal manganese deposits: implications for seawater Hf. *Earth Planet. Sci. Lett.* **151**(1–2), 91–105.

Goldstein S. J. and Jacobsen S. B. (1988) Nd and Sr isotopic systematics of river water suspended material—implications for crustal evolution. *Earth Planet. Sci. Lett.* **87**(3), 249–265.

Goldstein S. L. and O'Nions R. K. (1981) Nd and Sr isotopic relationships in pelagic clays and ferromanganese deposits. *Nature* **292**, 324–327.

Goldstein S. L., Onions R. K., and Hamilton P. J. (1984) A Sm–Nd isotopic study of atmospheric dusts and particulates from major river systems. *Earth Planet. Sci. Lett.* **70**(2), 221–236.

Grousset F. E., Labeyrie L., Sinko J. A., Cremer M., Bond G., Duprat J., Cortijo E., and Huon S. (1993) Patterns of ice-rafted detritus in the glacial North-Atlantic (40-degrees-55-degrees-N). *Paleoceanography* **8**(2), 175–192.

Grousset F. E., Pujol C., Labeyrie L., Auffret G., and Boelaert A. (2000) Were the North Atlantic Heinrich events triggered by the behavior of the European ice sheets? *Geology* **28**(2), 123–126.

Grousset F. E., Cortijo E., Huon S., Herve L., Richter T., Burdloff D., Duprat J., and Weber O. (2001) Zooming in on Heinrich layers. *Paleoceanography* **16**(3), 240–259.

Gwiazda R. H., Hemming S. R., and Broecker W. S. (1996a) Provenance of icebergs during Heinrich event 3 and the contrast to their sources during other Heinrich episodes. *Paleoceanography* **11**(4), 371–378.

Gwiazda R. H., Hemming S. R., and Broecker W. S. (1996b) Tracking the sources of icebergs with lead isotopes: the provenance of ice-rafted debris in Heinrich layer 2. *Paleoceanography* **11**(1), 77–93.

Gwiazda R. H., Hemming S. R., Broecker W. S., Onsttot T., and Mueller C. (1996c) Evidence from Ar-40/Ar-39 ages for a Churchill province source of ice-rafted amphiboles in Heinrich layer 2. *J. Glaciol.* **42**(142), 440–446.

Heinrich H. (1988) Origin and consequences of cyclic ice rafting in the northeast Atlantic Ocean during the past 130,000 years. *Quat. Res.* **29**(2), 142–152.

Hemming S. R. and Hajdas I. (2003) Ice-rafted detritus evidence from Ar-40/Ar-39 ages of individual hornblende grains for evolution of the eastern margin of the Laurentide ice sheet since 43 C-14 ky. *Quat. Int.* **99**, 29–43.

Hemming S. R., McLennan S. M., and Hanson G. N. (1994) Lead isotopes as a provenance tool for quartz—examples from Plutons and Quartzite, northeastern Minnesota, USA. *Geochim. Cosmochim. Acta* **58**(20), 4455–4464.

Hemming S. R., McDaniel D. K., McLennan S. M., and Hanson G. N. (1996) Pb isotope constraints on the provenance and diagenesis of detrital feldspars from the Sudbury basin, Canada. *Earth Planet. Sci. Lett.* **142**(3–4), 501–512.

Hemming S. R., Broecker W. S., Sharp W. D., Bond G. C., Gwiazda R. H., McManus J. F., Klas M., and Hajdas I. (1998) Provenance of Heinrich layers in core V28-82, northeastern Atlantic: Ar-40/Ar-39 ages of ice-rafted hornblende, Pb isotopes in feldspar grains, and Nd–Sr–Pb isotopes in the fine sediment fraction. *Earth Planet. Sci. Lett.* **164**(1–2), 317–333.

Hemming S. R., Bond G. C., Broecker W. S., Sharp W. D., and Klas-Mendelson M. (2000a) Evidence from 40Ar–39Ar ages of individual hornblende grains for varying Laurenide sources of iceberg discharges 22,000 to 10,500 yr BP. *Quat. Res.* **54**, 372–373.

Hemming S. R., Gwiazda R. H., Andrews J. T., Broecker W. S., Jennings A. E., and Onstott T. C. (2000b) Ar-40/Ar-39 and Pb–Pb study of individual hornblende and feldspar grains from southeastern Baffin Island glacial sediments: implications for the provenance of the Heinrich layers. *Can. J. Earth Sci.* **37**(6), 879–890.

Hemming S. R., Hall C. M., Biscaye P. E., Higgins S. M., Bond G. C., McManus J. F., Barber D. C., Andrews J. T., and Broecker W. S. (2002a) Ar-40/Ar-39 ages and Ar-40* concentrations of fine-grained sediment fractions from North Atlantic Heinrich layers. *Chem. Geol.* **182**(2–4), 583–603.

Hemming S. R., Vorren T. O., and Kleman J. (2002b) Provinciality of ice rafting in the North Atlantic: application of Ar-40/Ar-39 dating of individual ice rafted hornblende grains. *Quat. Int.* **95–96**, 75–85.

Henry F., Jeandel C., Dupre B., and Minster J. F. (1994) Particulate and dissolved Nd in the western Mediteranean Sea: sources, fate and budget. *Mar. Chem.* **45**, 283–305.

Hesse R. and Khodabakhsh S. (1998) Depositional facies of late Pleistocene Heinrich events in the Labrador Sea. *Geology* **26**(2), 103–106.

Hofmann A. W., Jochum K. P., Seufert M., and White W. M. (1986) Nb and Pb in oceanic basalts: new constraints on mantle evolution. *Earth Planet. Sci. Lett.* **79**, 33–45.

Huon S. and Ruch P. (1992) Mineralogical, K–Ar and Sr-87/Sr-86 isotope studies of Holocene and late glacial sediments in a deep-sea core from the northeast Atlantic Ocean. *Mar. Geol.* **107**(4), 275–282.

Huon S., Grousset F. E., Burdloff D., Bardoux G., and Mariotti A. (2002) Sources of fine-sized organic matter in North Atlantic Heinrich layers: delta C-13 and delta N-15 tracers. *Geochim. Cosmochim. Acta* **66**(2), 223–239.

Hurley P. M., Heezen B. C., Pinson W. H., and Fairbairn H. W. (1963) K–Ar values in pelagic sediments of the North Atlantic. *Geochim. Cosmochim. Acta* **27**, 393–399.

Jacobsen S. B. and Wasserburg G. J. (1980) Sm–Nd isotopic evolution of chondrites. *Earth Planet. Sci. Lett.* (50), 139–155.

Jaffey A. H., Flynn K. F., Glendeni Le., Bentley W. C., and Essling A. M. (1971) Precision measurement of half-lives and specific activities of U-235 and U-238. *Phys. Rev. C* **4**(5), 1889ff.

Jantschik R. and Huon S. (1992) Detrital silicates in Northeast Atlantic deep-sea sediments during the Late Quaternary—mineralogical and K–Ar isotopic data. *Eclogae Geologicae Helvetiae* **85**(1), 195–212.

Jeandel C. (1993) Concentration and isotopic composition of Nd in the South-Atlantic Ocean. *Earth Planet. Sci. Lett.* **117** (3–4), 581–591.

Jeandel C., Bishop J. K., and Zindler A. (1995) Exchange of neodymium and its isotopes between seawater and small and large particles in the Sargasso Sea. *Geochim. Cosmochim. Acta* **59**(3), 535–547.

Jeandel C., Thouron D., and Fieux M. (1998) Concentrations and isotopic compositions of neodymium in the eastern Indian Ocean and Indonesian straits. *Geochim. Cosmochim. Acta* **62**(15), 2597–2607.

Jones C. E., Halliday A. N., Rea D. K., and Owen R. M. (1994) Neodymium isotopic variations in North Pacific modern silicate sediment and the insignificance of detrital ree contributions to seawater. *Earth Planet. Sci. Lett.* **127**(1–4), 55–66.

Kirby M. E. and Andrews J. T. (1999) Mid-Wisconsin Laurentide ice sheet growth and decay: implications for Heinrich events 3 and 4. *Paleoceanography* **14**(2), 211–223.

Lacan F. and Jeandel C. (2001) Tracing Papua New Guinea imprint on the central Equatorial Pacific Ocean using neodymium isotopic compositions and rare earth element patterns. *Earth Planet. Sci. Lett.* **186**(3–4), 497–512.

Le Roux L. J. and Glendenin L. E. (1963) Half-life of ^{232}Th. *Proc. Natl. Meet. Nuclear Energy,* 83–94.

Levitus S. (1994) *World Ocean Atlas.* US Department of Commerce, Boulder, CO.

Lugmair G. and Scheinin N. B. (1974) Sm–Nd ages: a new dating method. *Meteoritics* **19**, 369.

Madureira L. A. S., vanKreveld S. A., Eglinton G., Conte M., Ganssen G., vanHinte J. E., and Ottens J. J. (1997) Late quaternary high-resolution biomarker and other sedimentary climate proxies in a northeast Atlantic core. *Paleoceanography* **12**(2), 255–269.

McManus J. F., Anderson R. F., Broecker W. S., Fleisher M. Q., and Higgins S. M. (1998) Radiometrically determined sedimentary fluxes in the sub-polar North Atlantic during the last 140,000 years. *Earth Planet. Sci. Lett.* **155** (1–2), 29–43.

Michard A., Albarede F., Michard G., Minster J. F., and Charlou J. L. (1983) Rare-earth elements and uranium in high-temperature solutions from East Pacific Rise hydrothermal vent field (13-degrees-N). *Nature* **303**(5920), 795–797.

Nakai S., Halliday A. N., and Rea D. K. (1993) Provenance of dust in the Pacific Ocean. *Earth Planet. Sci. Lett.* **119**(1–2), 143–157.

Neumann W. and Huster E. (1976) Discussion of Rb-87 half-life determined by absolute counting. *Earth Planet. Sci. Lett.* **33**(2), 277–288.

O'Nions R. K., Hamilton P. J., and Evensen N. M. (1977) Variations in ^{143}Nd/^{144}Nd and ^{87}Sr/^{86}Sr ratios in oceanic basalts. *Earth Planet. Sci. Lett.* **34**, 13–22.

O'Nions R. K., Carter S. R., Cohen R. S., Evensen N. M., and Hamilton P. J. (1978) Nd and Sr isotopes in oceanic ferromanganese deposits and ocean floor basalts. *Nature* **273**, 435–438.

Patchett P. J., White W. M., Feldmann H., Kielinczuk S., and Hofmann A. W. (1984) Hafnium rare-earth element fractionation in the sedimentary system and crustal recycling into the earths mantle. *Earth Planet. Sci. Lett.* **69**(2), 365–378.

Patterson C. C. (1956) The age of meteorites and the earth. *Geochim. Cosmochim. Acta* **10**, 230–237.

Patterson C. C., Goldberg E. D., and Inghram M. G. (1953) Isotopic compositions of quaternary leads from the Pacific Ocean. *Bull. Geol. Soc. Am.* **64**, 1387–1388.

Pfeffer W. T., Dyurgerov M., Kaplan M., Dwyer J., Sassolas C., Jennings A., Raup B., and Manley W. (1997) Numerical modeling of late Glacial Laurentide advance of ice across Hudson Strait: insights into terrestrial and marine geology, mass balance, and calving flux. *Paleoceanography* **12**(1), 97–110.

Piepgras D. J. and Jacobsen S. B. (1988) The isotopic composition of neodymium in the North Pacific. *Geochim. Cosmochim. Acta* **52**(6), 1373–1381.

Piepgras D. J. and Wasserburg G. J. (1980) Neodymium isotopic variations in seawater. *Earth Planet. Sci. Lett.* **50**, 128–138.

Piepgras D. J. and Wasserburg G. J. (1982) Isotopic composition of neodymium in waters from the Drake Passage. *Science* **217**(4556), 207–214.

Piepgras D. J. and Wasserburg G. J. (1983) Influence of the Mediterranean outflow on the isotopic composition of neodymium in waters of the North-Atlantic. *J. Geophys. Res.: Oceans Atmos.* **88**(NC10), 5997–6006.

Piepgras D. J. and Wasserburg G. J. (1985) Strontium and neodymium isotopes in hot springs on the East Pacific Rise and Guaymas Basin. *Earth Planet. Sci. Lett.* **72**(4), 341–356.

Piepgras D. J. and Wasserburg G. J. (1987) Rare-earth element transport in the western North-Atlantic inferred from Nd isotopic observations. *Geochim. Cosmochim. Acta* **51**(5), 1257–1271.

Piepgras D. J., Wasserburg G. J., and Dasch E. J. (1979) The isotopic composition of Nd in different ocean masses. *Earth Planet. Sci. Lett.* **45**, 223–236.

Pierson-Wickmann A. C., Reisberg L., France-Lanord C., and Kudrass H. R. (2001) Os–Sr–Nd results from sediments in the Bay of Bengal: implications for sediment transport and the marine Os record. *Paleoceanography* **16**(4), 435–444.

Piotrowski A. M., Lee D. C., Christensen J. N., Burton K. W., Halliday A. N., Hein J. R., and Gunther D. (2000) Changes in erosion and ocean circulation recorded in the Hf isotopic compositions of North Atlantic and Indian Ocean ferromanganese crusts. *Earth Planet. Sci. Lett.* **181**(3), 315–325.

Rashid H., Hesse R., and Piper D. J. W. (2003) Distribution, thickness and origin of Heinrich layer 3 in the Labrador Sea. *Earth Planet. Sci. Lett.* **205**(3–4), 281–293.

Revel M., Sinko J. A., Grousset F. E., and Biscaye P. E. (1996) Sr and Nd isotopes as tracers of North Atlantic lithic particles: paleoclimatic implications. *Paleoceanography* **11**(1), 95–113.

Reynolds P. H. and Dasch E. J. (1971) Lead isotopes in marine manganese nodules and the ore-lead growth curve. *J. Geophys. Res.* **76**, 5124–5129.

Richard P., Shimizu N., and Alle(c)gre C. J. (1976) $^{143}Nd/^{146}Nd$, a natural tracer: an application to oceanic basalts. *Earth Planet. Sci. Lett.* **31**, 269–278.

Rintoul S. R. (1991) South Atlantic interbasin exchange. *J. Geophys. Res.* **96**, 2675–2692.

Rosell-Mele A., Maslin M. A., Maxwell J. R., and Schaeffer P. (1997) Biomarker evidence for "Heinrich" events. *Geochim. Cosmochim. Acta* **61**(8), 1671–1678.

Ruddiman W. F. (1977) Late quaternary deposition of ice-rafted sand in subpolar North-Atlantic (Lat 40-degrees to 65-degrees-N). *Geol. Soc. Am. Bull.* **88**(12), 1813–1827.

Ruddiman W. F. and Glover L. K. (1982) Mixing of volcanic ash zones in subpolar North Atlantic sediments. In *The Ocean Floor, Bruce C. Heezen Memorial Volume* (eds. R. A. Scrutton and M. Talwani). Wiley, Chichester, NY, pp. 37–60.

Rudnick R. L. and Fountain D. M. (1995) Nature and composition of the continental-crust—a lower crustal perspective. *Rev. Geophys.* **33**(3), 267–309.

Rutberg R. L., Hemming S. R., and Goldstein S. L. (2000) Reduced north Atlantic deep water flux to the glacial southern ocean inferred from neodymium isotope ratios. *Nature* **405**(6789), 935–938.

Scherer E., Munker C., and Mezger K. (2001) Calibration of the lutetium–hafnium clock. *Science* **293**(5530), 683–687.

Schlosser P., Bullister J. L., Fine R., Jenkim W. J., Key R., Lupton J., Roether W., and Smethie W. M. Jr. (2001) Transformation and age of water masses. In *Ocean circulation and climate: Observing and Modelling the Global Ocean* (eds. G. Siedler, J. Church, and J. Gould). Academic Press, 431–452

Segl M., Mangini A., Bonani G., Hofmann H. J., Nessi M., Suter M., Wölfli W., Friedrich G., Plüger W. L., Wiechowski A., and Beer J. (1984) ^{10}Be-dating of a manganese crust from central North Pacific and implications for ocean palaeocirculation. *Nature* **309**, 540–543.

Sguigna A. P., Larabee A. J., and Waddington J. C. (1982) The half-life of Lu-176 by a gamma–gamma coincidence measurement. *Can. J. Phys.* **60**(3), 361–364.

Shimizu H., Tachikawa K., Masuda A., and Nozaki Y. (1994) Cerium and neodymium isotope ratios and ree patterns in seawater from the North Pacific Ocean. *Geochim. Cosmochim. Acta* **58**(1), 323–333.

Snoeckx H., Grousset F., Revel M., and Boelaert A. (1999) European contribution of ice-rafted sand to Heinrich layers H3 and H4. *Mar. Geol.* **158**(1–4), 197–208.

Spivack A. J. and Wasserburg G. J. (1988) Neodymium isotopic composition of the Mediterranean outflow and the eastern North-Atlantic. *Geochim. Cosmochim. Acta* **52**(12), 2767–2773.

Stea R. R., Piper D. J. W., Fader G. B. J., and Boyd R. (1998) Wisconsinan glacial and sea-level history of Maritime Canada and the adjacent continental shelf: a correlation of land and sea events. *Geol. Soc. Am. Bull.* **110**(7), 821–845.

Steiger R. H. and Jäger E. (1977) Subcommission on geochronology—convention on use of decay constants in geochronology and cosmochronology. *Earth Planet. Sci. Lett.* **36**(3), 359–362.

Stokes C. R. and Clark C. D. (2001) Palaeo-ice streams. *Quat. Sci. Rev.* **20**(13), 1437–1457.

Stordal M. C. and Wasserburg G. J. (1986) Neodymium isotopic study of Baffin-bay water—sources of Ree from very old terranes. *Earth Planet. Sci. Lett.* **77**(3–4), 259–272.

Syvitski J. P. M., Andrews J. T., and Dowdeswell J. A. (1996) Sediment deposition in an iceberg-dominated glacimarine environment, East Greenland: basin fill implications. *Global Planet. Change* **12**(1–4), 251–270.

Tachikawa K., Jeandel C., and Dupre B. (1997) Distribution of rare earth elements and neodymium isotopes in settling particulate material of the tropical Atlantic Ocean (EUMELI site). *Deep-Sea Res. I: Oceanogr. Res. Pap.* **44**(11), 1769–1792.

Tachikawa K., Jeandel C., and Roy-Barman M. (1999a) A new approach to the Nd residence time in the ocean: the role of atmospheric inputs. *Earth Planet. Sci. Lett.* **170**(4), 433–446.

Tachikawa K., Jeandel C., Vangriesheim A., and Dupre B. (1999b) Distribution of rare earth elements and neodymium isotopes in suspended particles of the tropical Atlantic Ocean (EUMELI site). *Deep-Sea Res. I: Oceanogr. Res. Pap.* **46**(5), 733–755.

Tatsumoto M., Knight R. J., and Allegre C. J. (1973) Time differences in formation of meteorites as determined from ratio of Pb-207 to Pb-206. *Science* **180**(4092), 1279–1283.

Tatsumoto M., Unruh D. M., and Patchett P. J. (1981) U-Pb nd Lu–;Hf systematics of Antarctic meteorites. *Proc. 6th Symp. Antarctic Meteorit.* 237–249.

Taylor S. R. and McLennan S. M. (1985) *The Continental Crust: Its Composition and Evolution.* Blackwell, Oxford.

van de Flierdt T., Frank M., Lee D. C., and Halliday A. N. (2002) Glacial weathering and the hafnium isotope composition of seawater. *Earth Planet. Sci. Lett.* **198**(1–2), 167–175.

Vance D. and Burton K. (1999) Neodymium isotopes in planktonic foraminifera: a record of the response of continental weathering and ocean circulation rates to climate change. *Earth Planet. Sci. Lett.* **173**(4), 365–379.

Vervoort J. D., Patchett P. J., Gehrels G. E., and Nutman A. P. (1996) Constraints on early Earth differentiation from hafnium and neodymium isotopes. *Nature* **379**(6566), 624–627.

Vervoort J. D., Patchett P. J., Blichert-Toft J., and Albarede F. (1999) Relationships between Lu–Hf and Sm–Nd isotopic systems in the global sedimentary system. *Earth Planet. Sci. Lett.* **168**(1–2), 79–99.

von Blanckenburg F. (1999) Perspectives: paleoceanography—tracing past ocean circulation? *Science* **286**(5446), 1862–1863.

von Blanckenburg F. and Nagler T. F. (2001) Weathering versus circulation-controlled changes in radiogenic isotope tracer composition of the Labrador Sea and North Atlantic deep water. *Paleoceanography* **16**(4), 424–434.

von Blanckenburg F., O'Nions R. K., and Hein J. R. (1996) Distribution and sources of pre-anthropogenic lead isotopes in deep ocean water from Fe–Mn crusts. *Geochim. Cosmochim. Acta* **60**(24), 4957–4963.

Vorren T. O. and Laberg J. S. (1997) Trough mouth fans—Palaeoclimate and ice-sheet monitors. *Quat. Sci. Rev.* **16**(8), 865–881.

Wedepohl K. H. (1995) The composition of the continental-crust. *Geochim. Cosmochim. Acta* **59**(7), 1217–1232.

White W. M., Patchett J., and Ben Othman D. (1986) Hf isotope ratios of marine sediments and Mn nodules: evidence for a mantle source of Hf in seawater. *Earth Planet. Sci. Lett.* **79**, 46–54.

Isotope Geochemistry
ISBN: 978-0-08-096710-3

pp. 635–670

21
Records of Cenozoic Ocean Chemistry

G. E. Ravizza
University of Hawaii, Manoa, HI, USA
and
J. C. Zachos
University of California, Santa Cruz, CA, USA

21.1 INTRODUCTION

Numerous lines of evidence show that there have been dramatic changes in the marine realm during the last 65 Myr. These changes occur over varying timescales. Some are relatively abrupt, occurring on timescales of thousands to tens of thousands of years. Others occur more gradually, over million-year timescales. Many of the most valuable monitors of past changes in ocean chemistry, such as the $\delta^{13}C$ and $\delta^{18}O$ of foraminiferal calcite are subject to high-frequency variations that must be smoothed out if long-term, secular, trends are to be recognized clearly. Conversely, other records of past seawater chemistry, such as the marine strontium isotope record, respond only slowly to high-frequency external forcing and are incapable of recording it with fidelity. Nevertheless, it is likely that high-frequency forcing related to glacial erosion and shifts in the hydrologic cycle play an important role in shaping the marine strontium isotope record. Therefore, even though the focus of this review is on records of Cenozoic ocean chemistry that emphasize long-term changes, the different timescales on which Cenozoic ocean chemistry changes are not fully separable.

In this review emphasis is placed on isotopic records of ocean chemistry. In general terms, a conscious decision was made to emphasize those records that document long-term changes in the chemical and physical properties of the global ocean over the course of the Cenozoic. For example, while reconstructions of burial fluxes of barium or phosphorus may place valuable constraints on paleo-productivity in a specific setting, making extrapolations to infer globally integrated trends from these data sets is very difficult because of sparse data coverage in space and time. Similarly, we have also chosen to exclude discussion of short-residence-time tracers like lead and neodymium isotopes that can yield important information about changing patterns of ocean circulation and regional shifts in oceanic inputs. A recent discussion of these tracer systems is available elsewhere (Frank, 2002). The geological records that are emphasized include those of the stable carbon and oxygen isotopes preserved in benthic foraminiferal calcite, and marine strontium and osmium isotopes. These four records clearly manifest significant changes in global ocean chemistry. At any given time, the stable carbon and oxygen isotope records include an important component of spatial variability; however, these are not so large as to obscure the pattern of temporal variation preserved in the sediment record.

Boron isotopes as a paleo-pH proxy and Mg/Ca as a paleo-temperature proxy are also discussed, because information provided by these relatively new proxies has important implications for the better-established records mentioned in the preceding paragraph. The application of boron isotopes to reconstructing surface water pH provides a means of estimating past atmospheric CO_2 levels. Much of the discussion of the marine strontium isotope record in the past has been linked to implicit assumptions about Cenozoic variations in atmospheric CO_2 levels. New boron isotope results indicate that previous assumptions about Cenozoic atmospheric CO_2 levels need to be reconsidered. Mg/Ca ratio variations in benthic foraminiferal calcite are discussed because of the promise that combined Mg/Ca and $\delta^{18}O$ studies hold for resolving the dual influence of ice volume and deep-water temperature on benthic foraminiferal $\delta^{18}O$ records.

There are several other topics that would be equally appropriate to consider in a review of this type. Some of these, such as the $\delta^{34}S$ record of seawater and the history of the calcium carbonate compensation depth (CCD), are mentioned briefly as they relate to records that are discussed in greater detail. Other topics have been omitted. We hope readers will recognize that the topics covered here are determined not only by the scientific interests of the authors, but also by the practical limits of what can be covered in a single review.

21.2 CENOZOIC DEEP-SEA STABLE ISOTOPE RECORD

Much of what is currently understood about the Cenozoic history, of deep-sea temperature, carbon chemistry, and global ice volume, has been gleaned from the stable isotope ratios of benthic foraminifera. Benthic foraminifera extract carbonate and other ions from seawater to construct their tests. In many species, this is achieved near carbon and oxygen isotopic equilibrium. Kinetic fractionation effects tend to be small and constant (Grossman, 1984, 1987). As a result, shell $\delta^{13}C$ and $\delta^{18}O$ strongly covary with the isotopic composition of seawater and dissolved inorganic carbon (DIC). For both carbon and oxygen isotopes, there is also a temperature-dependent fractionation effect. For oxygen isotopes, the effect is relatively large, 0.25‰ $(°C)^{-1}$, about an order of magnitude larger than that for carbon isotopes. As such, in addition to monitoring changes in seawater isotope ratios, the oxygen isotopes can be used to evaluate temperature. Moreover, the tests of benthic foraminifera are relatively resistant to dissolution and other diagenetic processes, making them ideal archives of ocean history.

The latest "global" deep-sea stable isotope record for the Cenozoic, presented in Figure 1, is based on benthic foraminifera oxygen and carbon isotope records compiled from over 40 pelagic sediment cores (Zachos *et al.*, 2001). The raw data were smoothed using a five-point running

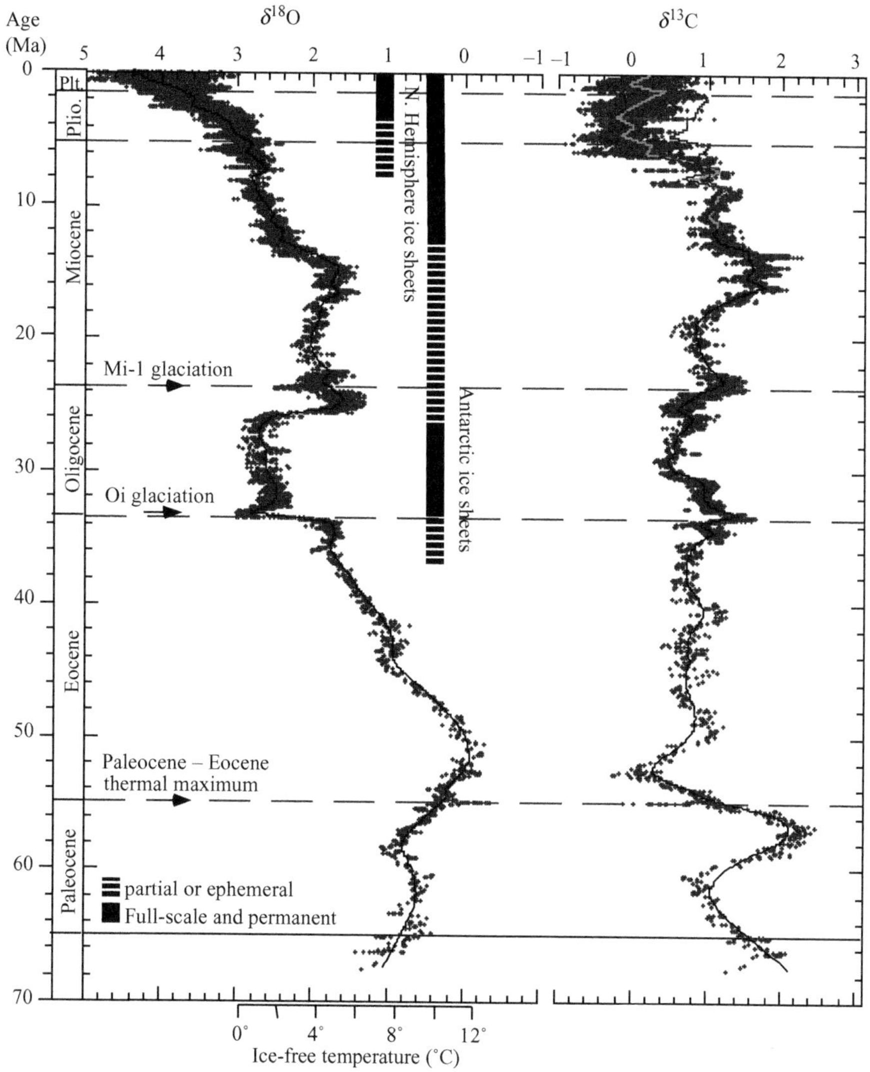

Figure 1 The global deep-sea isotope record based on data compiled from 40 DSDP and ODP sites. These data were derived from pelagic deep-sea cores (e.g., from depths greater than 1,000m) with lithologies that are predominantly fine-grained, carbonate-rich (>50%) oozes or chalks. Most of the data are derived from analyses of two common and long-lived benthic taxa, *Cibicidoides* and *Nuttallides*. The absolute ages are relative to the standard Geomagnetic Polarity Timescale (GPTS) for the Cenozoic (Berggren *et al.*, 1995). To minimize biases related to inconsistencies in sampling density in space and time, the raw data were smoothed using a five-point running mean, and curve-fitted with a locally weighted mean. The smoothing results in a loss of detail that is undetectable from the long-timescale perspective. For the carbon isotope record, the global curve fit was terminated just before the Late Miocene and replaced with separate Atlantic and Pacific curves because of a marked increase in basin-to-basin carbon isotope fractionation (Wright and Miller, 1993). The temperature scale represents mean deep-sea temperature for the period of time preceding the onset of large-scale glaciation on Antarctica (~35 Ma). After this time, much of the variability in the $\delta^{18}O$ record reflects on changes in global ice volume on Antarctica and in the N. Hemisphere. The vertical bars provide a semiquantitative representation of ice volume in each hemisphere relative to the LGM with the dashed bar representing periods of minimal ice coverage (~<50%), and the full bar representing close to maximum ice-coverage (>50% of present). In the more recent portion of the carbon isotope record, separate curve fits were derived for the Atlantic and Pacific to illustrate the increase in basin-to-basin fractionation which exceeds ~1.0‰ in some intervals. Prior to 15 Ma, interbasin gradients are on the order of a few tenths of a per mil or less (source Zachos *et al.*, 2001).

mean, and curve-fitted with a weighted (2%) running mean. For the carbon isotope record, the curve fit was terminated at the middle–late Miocene boundary because of a marked increase in basin-to-basin carbon isotope fractionation at that time (Wright *et al.*, 1991). This record provides a fairly coarse perspective on the long-term variations in Cenozoic climate and ocean chemistry. For the finer-scale variations that tend to be masked in this long-term perspective, individual high-resolution records are preferable. Several high-resolution records spanning "critical" intervals are plotted in Figure 2.

21.2.1 Oxygen Isotopes and Climate

The conservative physical and chemical characteristics of bottom waters are fixed at the site of deep-water formation, which for much of the Cenozoic appears to have been the high-latitude polar seas. As a consequence, the long-term oxygen isotope variations recorded by benthic foraminifera reflect largely on changes in high-latitude sea surface temperature (Shackleton *et al.*, 1985). In general, a $\delta^{18}O$ increase of 1.0‰ is equivalent to roughly 4 °C of cooling. Foraminifera oxygen isotopes also record change in seawater $\delta^{18}O$ which, on the million-year timescale, is controlled primarily by changes in the volume of continental ice sheets which are isotopically depleted (−30‰ to −40‰) relative to seawater. For example, melting the present-day ice sheets would decrease mean ocean $\delta^{18}O$ by more than 1.0‰ while raising sea level by 100 m (~0.01‰ m^{-1}). It is assumed that the exchange of isotopes with the crust has negligible effects on seawater $\delta^{18}O$ on the timescale of the Cenozoic.

Separating the relative contributions of these two variables to the deep-sea oxygen isotope record is a challenging exercise. Consideration of additional factors including the lower boundaries of seawater temperature (e.g., freezing) can place some limits on the temperature effect (Miller *et al.*, 1987, 1991b; Zachos *et al.*, 1993, 1994). However, only with independent measures of temperature that are unaffected by salinity, can one effectively isolate the component related to ice volume. To this end, the degree of saturation of alkenones of autotrophs or Mg/Ca of benthic foraminifera (Lear *et al.*, 2000; Billups and Schrag, 2002) can be used to constrain temperature.

Over the long term, the Cenozoic deep-sea oxygen isotope record is dominated by two important features that relate to major shifts in mean climatic state. The first is a rise in values from 53 to 35. This trend, which is mostly gradual but punctuated by several steps, is an expression of the Eocene transition from greenhouse to icehouse conditions. In the Early Eocene (~53 Ma) the deep sea was relatively warm, ~7 °C warmer than present, and there were no ice sheets. Over the next 20 Myr. the ocean cools, and the first large ice sheets appear on Antarctica. The latter event is reflected by the relatively sharp 1.2‰ increase in $\delta^{18}O$ at 33.4 Ma. This pattern reverses slightly toward the end of the Oligocene, but by middle Miocene, ice sheets begin to expand slowly, eventually covering most of Antarctica as reflected in $\delta^{18}O$. The second significant step in the Cenozoic is associated with the gradual buildup of northern hemisphere ice sheets between 3.5 Ma and 2.5 Ma.

On short timescales, the $\delta^{18}O$ record reveals considerable variability. Most of this is concentrated in the Milankovitch bands, and therefore reflects on orbitally modulated changes in ice volume and/or deep-sea temperatures. The largest amplitude oscillations occur in the Quaternary when ice sheets are present on North America. Prior to the Quaternary, the signal amplitude in $\delta^{18}O$ is about one-third to one-half, mostly reflecting variations in the volume of Antarctic ice sheets (Figure 2(a)). These lower-amplitude oscillations persist through much of the Neogene and the Oligocene with most of the variance concentrated in the obliquity bands (Tiedemann *et al.*, 1994; Shackleton and Crowhurst, 1997; Flower *et al.*, 1997; Zachos *et al.*, 2001).

A small component of the short-term variability falls under the category of anomalies or transients. This includes a short-lived but abrupt negative excursion 55 Myr ago (Figure 2(b)). The magnitude of the $\delta^{18}O$ change implies a 4–5 °C transient warming of the deep sea. This event, referred to as the Paleocene–Eocene thermal maximum, is by far the most extreme of the rapid climatic changes inferred from the oxygen isotope records. Other isotope anomalies representing brief climatic excursions have been documented in the earliest Oligocene (~33.4 Ma), at the O/M boundary (23.0 Ma), and in the middle Miocene (~14 Ma) (Miller *et al.*, 1991a; Zachos *et al.*, 1996).

Despite efforts to produce a globally averaged record, gradients in deep-sea temperature combined with the uneven distribution of deep-sea cores can introduce subtle biases into the global $\delta^{18}O$ record. To start, deep-sea temperature is not uniform: the upper ocean is several degrees warmer than the deep ocean. Gradients also exist spatially, especially at intermediate depths with proximity to the polar and tropical oceans. As a result, global compilations utilizing data from sites located at different water depths and basins can be somewhat misleading. For example, the older, Early Paleogene, portions of cores tend to be biased toward the upper ocean because of long-term subsidence and seafloor subduction (Zachos *et al.*, 2001). Moreover, because no single record spans the entire Cenozoic, shifts in $\delta^{18}O$ can be artificially produced through splicing. One example is the negative shift in $\delta^{18}O$ observed in the Late Oligocene (~27 Ma), which in the compilation is larger than recorded in any individual record. A number of deep-sea records either terminate or begin at this point. The few that span this interval show a shift, but with a magnitude about half (~0.5‰) that represented in the global compilation.

21.2.2 Carbon Isotopes and Ocean Carbon Chemistry

The distribution of carbon isotopes within the ocean is dependent on two processes, ocean circulation and export production. The $\delta^{13}C_{DIC}$ of deep-water masses are initially set at the site of sinking (via equilibrium exchange with the atmosphere), but tend to progressively change as they slowly migrate through the basins. The respiration of organic matter delivered by export production releases isotopically depleted CO_2 ($\alpha_{T_{CO2}\text{-}CH_2O} = 1.021$) to the DIC pool. Thus, as deep waters "age" and their DIC and nutrient contents increase, $\delta^{13}C$ decreases. Initial offsets in deep-water chemical characteristics, however, are retained to the extent that $\delta^{13}C$ can be used to distinguish water masses deriving from different sources. Moreover, the relationship between dissolved nutrients, primarily PO_4, and $\delta^{13}C_{DIC}$, is nearly linear such that the latter has served as a proxy for estimating the distribution of the former, at least for short timescales.

The mean value of ocean $\delta^{13}C_{DIC}$ is not stationary. It changes in response to variations in the fluxes of carbon between the ocean, and the major sources and sinks of carbon (Kump and Arthur, 1999). The major sources are volcanic and metamorphic outgassing of CO_2, and rock (organic and inorganic) weathering. The major sinks are

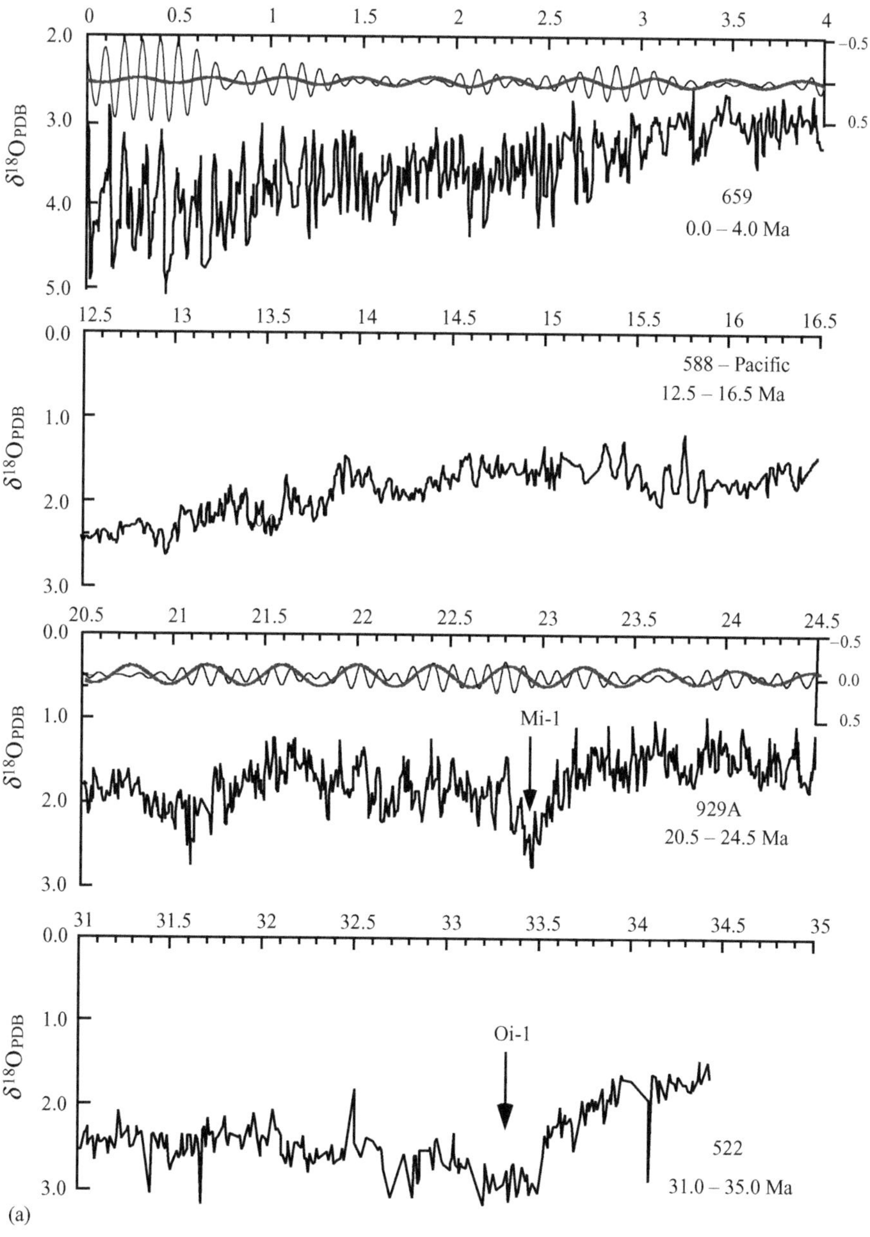

Figure 2 (Continued)

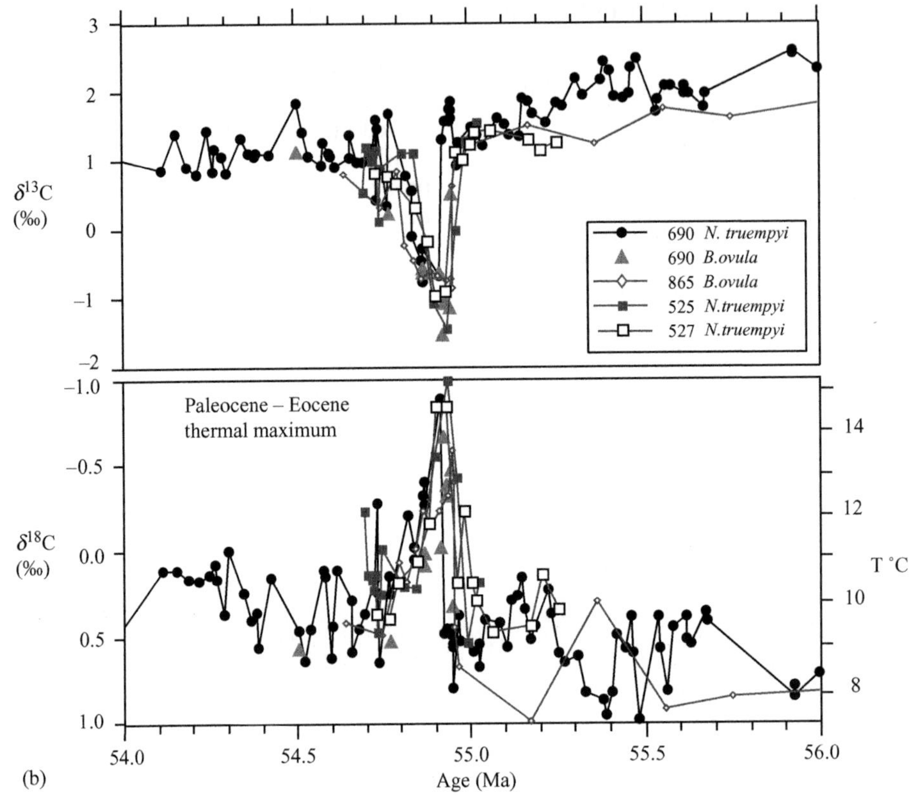

Figure 2 (a) High-resolution 4 Myr long $\delta^{18}O$ time series representing four intervals of the Cenozoic (Zachos *et al.*, 2001). The data are from sites 659, eastern equatorial Atlantic (Tiedemann *et al.*, 1994); 588, Southwest Pacific (Flower and Kennett, 1993); 929, western equatorial Atlantic (Paul *et al.*, 2000); 522, South Atlantic (Zachos *et al.*, 1996) and 689 Southern Ocean (Diester-Haass, 1996). Sampling intervals range from 3kyr, to 10 kyr. Note that the $\delta^{18}O$ axes on all plots are set to the same scale (3.0‰) although different ranges to accommodate the change in mean ocean temperature/ice volume with time. The upper curves in each panel represent Gaussian band-pass filters designed to isolate variance associated with the 400 kyr and 100 kyr eccentricity cycles. The 400 kyr filter has a central frequency (cf) = 0.0025 and a bandwidth (bw) = 0.0002; the 100 kyr cf = 0.01 and bw = 0.002. (b) Multiple benthic isotope records characterizing the Late Paleocene thermal maximum event at 55 Ma (Thomas and Shackleton, 1996; Bralower *et al.*, 1995). The timing of the beginning of the event is placed at 54.95 Ma (Röhl *et al.*, 2000). Data are plotted on the timescale of these two papers combined with the cyclostratigraphy for site 690 (Cramer, in press); data from other sites are correlated to the 690 record. The apparent late initiation of the events at 690 is the result of the lack of specimens of *N. truempyi* during the first part of the event at that site; note low-isotope values of *Bulimina ovula* in the gap in the *N. truempyi* data. The oxygen isotope data indicate an abrupt 4–6 °C warming of the deep ocean in a period of ~10 kyr, by far the most rapid rate of warming of the last 65 Myr. The negative carbon isotope excursion is thought to represent the influx of methane from dissociation of gas hydrates.

organic and inorganic carbon burial as represented by the following equation from Kump and Arthur (1999):

$$\frac{d}{dt}(M_o\delta_{carb}) = F_w\delta_w + F_{volc}\delta_{volc} - F_{b,carb}\delta_{carb} - F_{borg}(\delta_{carb} + \Delta_B) \qquad (1)$$

where M_o is the total dissolved carbon in the ocean, F_w and F_b are the fluxes for weathering and burial, and δ_{carb}, δ_{volc}, and δ_w are the average carbon isotopic compositions of marine carbonates, volcanic CO_2 (crustal + mantle), and weathered or riverine carbon, respectively. δ_{carb} is effectively equal to the mean ocean $\delta^{13}C_{DIC}$ + 1.0‰. For F_{borg}, δ is based on a fractionation factor Δ_B which is relatively large, on the order of ~20‰. Because of the large range in isotope values for these sources and sinks, on timescales of >10^5 yr even small changes in the fluxes to or from these reservoirs can impart noticeable changes in mean ocean $\delta^{13}C_{DIC}$.

Two aspects of the Cenozoic marine $\delta^{13}C$ record merit discussion. The first is the spatial distribution of $\delta^{13}C$ between ocean basins, which is relatively insignificant from 65 Ma to 8 Ma (Figure 1). The lack of gradients between basins is generally viewed as evidence of a single dominant deep-water source during the Cenozoic, Antarctica (Wright and Miller, 1993). Deep waters from other regions (e.g., evaporitic, marginal seas), while important locally, are effectively negligible on a global scale. Over the last 8 Myr, the flux of North Atlantic deep water has been large enough to dominate the Atlantic, at least on a periodic basis (i.e., northern hemisphere interglacials). This, coupled with increased isolation of the deep Atlanticas the Tethys and Panamanian gateways close, allows for the development of the large carbon isotope gradient with the Pacific (i.e., basin-to-basin fractionation). This illustrates how the distribution of carbon isotopes is influenced by circulation patterns.

The second observation concerns the long- and short-term trends. Much has been said about the long-term patterns. In particular, the gradual increases in $\delta^{13}C$ over the Late Paleocene and middle Miocene have been attributed to increased rates of organic carbon burial, possibly brought about by tectonic factors that created or destroyed basins, or to changes in mantle outgassing rates (Berner *et al.*, 1983; Shackleton, 1985). The Late Neogene decline in $\delta^{13}C$ is similarly viewed as a signal of changes in the size of the organic carbon reservoir and/or burial rates (Shackleton, 1985; Raymo, 1997). The ability to isolate the process(es) responsible for these long-term trends (i.e., $>10^5$ yr), however, is limited by the ability to constrain other aspects of the global carbon and related geochemical cycles. For example, the burial of reduced sulfur is tightly coupled to organic carbon cycling as represented by Equation (2):

$$2CH_2O + SO_4^{2-} + H^+ \Leftrightarrow 2CO_2 + HS^- + H_2O \quad (2)$$

This microbe mediated reaction is accompanied by significant isotopic fractionation of the sulfur ($\alpha_{SO_4-HS} = 1.025$) such that large increases in the rate of sulfate reduction can significantly increase the $\delta^{34}S$ of the remaining sulfate reservoir (residence time of sulfur in the ocean would require timescales of 10^5 yr and greater). As a result, a sustained shift in the rate of marine organic carbon production and/or burial as proposed for the Late Paleocene or Neogene should be accompanied by gradual changes in seawater $\delta^{34}S$. Only recently, however, has a record of seawater $\delta^{34}S$ became available (Paytan *et al.*, 1998) to test these hypotheses. This record shows relatively constant values over much of the Cenozoic with two notable exceptions, the Late Paleocene to Early Eocene and the Pleistocene.

The short-term trends, particularly the excursions, merit special consideration. The most prominent occurs at 55 Ma, approximately the Paleocene–Eocene (P–E) boundary. It is characterized by an abrupt 3.0‰ decrease in benthic foraminiferal $\delta^{13}C$ as well as a 4–5 °C global warming, the P–E thermal maximum (Figure 2(b)). Such a rapid and large decrease in mean ocean $\delta^{13}C$ can only be achieved with the addition of a large quantity of ^{12}C enriched carbon. The mean $\delta^{13}C$ of carbon derived from volcanic outgassing is −7‰, whereas the mean for methane ranges from −40‰ to −60‰, the latter dependent upon whether it is thermogenically or bacterially produced. The rate of change associated with the P–E excursion is more readily achieved through the addition of methane (Dickens *et al.*, 1997). The largest reservoir of methane near the Earth's surface is the marine hydrate reservoir. Computations show that dissociation and oxidation of 2,000 Gt of hydrate methane would be sufficient to produce an excursion of this magnitude. It is likely that other smaller shifts in ocean $\delta^{13}C$ might have similar origins (Dickens, 2001).

The benthic $\delta^{13}C$ record is marked by other excursions in the Cenozoic at the Eocene–Oligocene and Oligocene–Miocene boundaries. These anomalies, however, are positive and more gradual. The direction of change indicates perturbations in one or more fluxes of the global carbon cycle, possibly the burial rates of reduced carbon, which are inferred to increase during each of these events.

21.3 THE MARINE STRONTIUM AND OSMIUM ISOTOPE RECORDS

The marine strontium isotope record is the proxy record most commonly used to constrain the geologic history of chemical weathering. However, in recent years it has been widely criticized as a proxy indicator of past silicate weathering rates. The osmium isotope record is analogous to the strontium record in many respects, and can help to constrain interpretations of the marine strontium isotope record. In this section the geochemical factors that influence the osmium and strontium isotope compositions of seawater are reviewed, and the structure of these two records of Cenozoic ocean chemistry is discussed.

21.3.1 Globally Integrated Records of Inputs to the Ocean

There are two critically important similarities between osmium and strontium isotopes as paleoceanographic tracers. The first is the sense of parent–daughter fractionation during mantle melting. In the Rb–Sr and Re–Os systems, the radioactive parents, rubidium and rhenium, are partitioned into the melt preferentially to the daughter elements strontium and osmium. Continental crust, a product of mantle differentiation, is characterized by higher Rb/Sr and Re/Os ratios relative to the deep in the Earth. Given the significant mean age of average upper continental crust, ~2Gyr, *in situ* decay of ^{87}Rb and ^{187}Re has produced significant amounts of ^{87}Sr and ^{187}Os. Thus, in both systems, more radiogenic isotope signatures (higher $^{87}Sr/^{86}Sr$ and $^{187}Os/^{188}Os$) characterize old-crustal rocks relative to recent mantle-derived rocks. The second, and equally important, similarity is the relative isotopic homogeneity of both dissolved strontium and osmium in seawater. This was only recently confirmed for osmium by direct analyses of seawater (Sharma *et al.*, 1997; Levasseur *et al.*, 1998; Woodhouse *et al.*, 1999). The fact that both osmium and strontium are isotopically well mixed in the modern ocean suggests that temporal variations in both isotope records can provide a globally integrated history of oceanic inputs. These two attributes provide a basis for using the marine strontium and osmium isotope records to constrain changes in the riverine solute flux through time.

Mixing of multiple sources of isotopically distinct inputs provides a useful framework for interpreting temporal variations in seawater $^{87}Sr/^{86}Sr$ and $^{187}Os/^{188}Os$. Early efforts to understand the marine strontium isotope record recognized changing proportions of strontium derived from weathering of three end-members: old felsic crust, young basalts, and marine carbonates as the primary influence on the strontium isotopic composition of seawater (Brass, 1976). Although our understanding of the marine strontium cycle has advanced significantly, these rock reservoirs remain fundamentally important. Hydrothermal alteration of basalt at mid-ocean ridges provides a continuous supply of relatively unradiogenic strontium to the ocean that is assumed to be proportional to oceanic crustal production rate. The supply of strontium from weathering of rocks exposed on land is titrated against this unradiogenic strontium input from oceanic crust. Marine carbonates have strontium concentrations that range from several hundred to more than 1,000 ppm, and represent a third important source of strontium to the oceans. The low Rb/Sr of most marine carbonates render production of ^{87}Sr by *in situ* decay in these rocks unimportant. Consequently, strontium derived from weathering of marine carbonates is generally assumed to be similar in isotopic composition to seawater. This is strictly true of strontium released to pore waters during early diagenetic recrystallization of biogenic carbonate (Gieskes *et al.*, 1986). In the context of these three mixing end-members, the marine strontium isotope record can be interpreted as the result of changes in the relative proportions of strontium delivered to the oceans from weathering of old continental crust and young basalts, as buffered by rapid recycling of strontium associated with marine carbonates. This simplified description of the marine strontium cycle can be represented by the equation

$$\frac{dR_{SW}}{dt} = t^{-1}(f_R(R_R - R_{SW}) + f_{HT}(R_{HT} - R_{SW}) + f_D(R_D - R_{SW})) \quad (3)$$

where R_i is $^{87}Sr/^{86}Sr$ and f_i is the ratio of the strontium flux from source i to the ocean to the total flux of strontium to the ocean, and t is the marine residence time of strontium (~2Myr). The subscripts SW, R, HT, and D correspond to seawater, rivers, hydrothermal, and diagenetic terms, and dR_{SW}/dt is the slope of the seawater strontium curve. A summary of representative values for the present-day marine strontium budget is given in Table 1.

An analogous expression can be written for the seawater osmium isotope system but the diagenetic term representing recrystallization of biogenic carbonate would be eliminated. Although the marine osmium cycle is less well documented than the

Table 1 Summary of the present-day marine Sr and Os budgets.

Isotope ratios	$^{87}Sr/^{86}Sr$	$^{187}Os/^{188}Os$
Seawater	0.70916	1.06
Average upper crust	1.26–1.40	0.716
Average riverine input	0.7119	1.4
Diagenetic flux	0.7084	Unknown
Fractional contribution from riverine flux	0.67	0.73
Elemental fluxes	Sr (mol yr^{-1})	Os (mol yr^{-1})
Riverine flux	3.3×10^{10}	1,850
Diagenetic flux	0.34×10^{10}	Unknown
Calculated hydrothermal/unradiogenic flux	1.6×10^{10}	680
Seawater concentration	590 μM	50 fm
Seawater inventory	1.3×10^{17} mol	7.4×10^{7} mol
Calculated residence time	2.4×10^{6} yr	30,000 yr

Data sources: Sr budget as compiled by Elderfield and Schultz (1996); Os budget as compiled by Peucker-Ehrenbrink and Ravizza (2000); Average upper crust: Peucker-Ehrenbrink and Jahn (2001); Goldstein and Jacobsen (1988).
Notes: Fractional riverine contribution calculated from isotope mass balance. Calculated hydrothermal/unradiogenic flux terms are calculated assuming the present-day ocean is at steady state with respect to Sr and Os isotope mass balance.

strontium cycle, first-order estimates for most key parameters are available (Table 1). The short marine-residence time of osmium (~10–40 kyr: see Peucker-Ehrenbrink and Ravizza, 2001) and the low sample density in Cenozoic osmium isotope record precludes time-dependent modeling at this time. In addition, it is uncertain if the HT term for osmium can strictly be linked to oceanic crustal production. Instead, it is perhaps better represented by a generic unradiogenic flux that includes both hydrothermal and cosmic inputs (see below for additional discussion). Available data (Table 1) suggest that similar proportions of osmium and strontium are derived from the radiogenic "continental" end-member. This provides some support for a simple "two-component" approach to the marine strontium and osmium isotope records. To the extent that globally averaged riverine input is the product of chemical weathering of average upper crust, the isotopic compositions of strontium and osmium are expected to be relatively invariant. If so, then the gross structure of the marine osmium isotope record should mimic that of the strontium record and changes in the strontium and osmium isotope composition of seawater should be representative of changing solute flux to the global ocean.

21.3.2 Osmium–Strontium Decoupling

A more careful examination of the geochemistry of the Rb–Sr and Re–Os systems reveals significant differences between these two isotope systems—differences that have the potential to decouple the marine strontium and osmium isotope records from one another. As monovalent and divalent cations, rubidium and strontium partition readily into major rock-forming minerals in both high- and low-temperature environments. As noted above the affinity of Sr^{+2} for carbonate minerals is a particularly important aspect of the surficial strontium cycle. In contrast, rhenium and osmium, both third-series transition metals, are redox active and strongly siderophile and chalcophile. As a result, rhenium and osmium tend to be associated with trace phases like sulfides and metal oxides. These very different geochemical affinities suggest that the fluxes and isotopic composition of dissolved strontium and osmium carried by individual rivers need not be well correlated in the modern Earth system. It is also possible, though less likely, that the globally averaged fluxes and isotopic composition of dissolved strontium and osmium can vary with time in an uncorrelated manner. If this is the case, then the Cenozoic marine strontium and osmium isotope records are not expected to resemble one another. Below we outline several specific aspects of the geochemical cycles of the Rb–Sr and Re–Os systems that have the potential to decouple oceanic inputs of osmium and strontium.

21.3.2.1 Decoupled riverine fluxes of strontium and osmium?

The association of strontium with carbonates, and rhenium and osmium with sedimentary organic matter may effectively decouple the riverine fluxes of strontium and osmium. It is well known that calcium carbonate weathers much more rapidly than silicate minerals and that estimates of silicate weathering fluxes based on strontium isotope data are complicated by this phenomenon (Palmer and Edmond, 1992; see also Jacobson *et al.* (2002a) for a discussion). In general, though not always (see below), terrestrial weathering of carbonates increases strontium flux and lowers $^{87}Sr/^{86}Sr$ relative to an equivalent carbonate-free catchment. The lower $^{87}Sr/^{86}Sr$ of average river flux (0.712) relative to eroding upper crust (0.716) is the global manifestation of the buffering influence that carbonate dissolution exerts on riverine strontium isotope composition.

In contrast to Rb–Sr system, there is no analogue to the buffering influence of carbonate weathering on seawater osmium isotope variations. Instead, it has been suggested that chemical weathering of old organic-rich sediments may actually cause large amplitude changes in the seawater $^{187}Os/^{188}Os$, because they are enriched in both rhenium and osmium and have unusually large Re/Os ratios (Ravizza, 1993). Recent studies demonstrate that rhenium and osmium are efficiently mobilized during black shale weathering (Peucker-Ehrenbrink and Hannigan, 2000; Jaffe *et al.*, 2002; Pierson-Wickmann *et al.*, 2002). If the Os/C_{org} ratios of recent Black Sea sediments (Ravizza *et al.*, 1991) are characteristic of the sedimentary organic carbon reservoir, in general, as much as 5–10% of the continental crustal inventory of osmium may be associated with sedimentary organic matter. This is comparable to the fraction of continental crustal strontium associated with marine carbonates. While the potential importance of weathering old organic-rich shales for causing increases in the seawater $^{187}Os/^{188}Os$ has been recognized in many studies (Pegram *et al.*, 1992; Ravizza, 1993; Peucker-Ehrenbrink *et al.*, 1995; Singh *et al.*, 1999; Pierson-Wickmann *et al.*, 2000), the potential buffering influence of rapid recycling of osmium rich, reducing marine sediments on continental margins has received little attention. Although the connection between erosion of sedimentary organic matter and the marine osmium isotope record is not well understood, it does provide a clear example of the contrasting aqueous geochemistry of the Rb–Sr and Re–Os systems. These differences can make the marine strontium and osmium isotope records responsive to different aspects of continental weathering.

21.3.2.2 Decoupled unradiogenic fluxes?

Dissimilarities between osmium and strontium also exist for the unradiogenic inputs to the ocean, providing additional ways in which these two paleoceanographic records could be decoupled. While it is well established that hydrothermal alteration of mid-ocean ridge basalts (MORBs) is the primary source of unradiogenic strontium to the ocean, it is unlikely that this process can balance the osmium isotope budget. Initial investigation of osmium in high-temperature vent fluids suggests that additional source(s) of unradiogenic osmium to the ocean must exist (Sharma *et al.*, 2000). Again, the geochemistry of osmium suggests other likely unradiogenic sources to the ocean. Core formation and the compatible behavior of osmium during mantle melting are responsible for producing the 10^4-fold depletion in typical crustal rocks relative to undifferentiated extraterrestrial material. Although it is unlikely that changes in cosmic dust flux through time are responsible for structure in the Cenozoic osmium isotope record (Peucker-Ehrenbrink, 1996), the flux of unradiogenic osmium contributed to the modern ocean by cosmic dust dissolution remains uncertain (see Peucker-Ehrenbrink and Ravizza, 2000) and some workers argue that it may be significant (Sharma *et al.*, 2000). The compatible behavior of osmium during partial melting of the mantle yields concentrations in MORBs that are $\sim$100 times lower than in ultramafic rocks. Consequently, submarine alteration of ultramafic rocks exposed on the seafloor is commonly invoked as a potentially important source of unradiogenic osmium to the oceans (Palmer and Turekian, 1986; Martin, 1991; Snow and Reisberg, 1995; Peucker-Ehrenbrink, 1996; Levasseur *et al.*, 1999). The recent discovery of ultramafic hosted hydrothermal systems (Douville *et al.*, 2002; Kelly *et al.*, 2001) has intensified interest in this potential source of osmium to seawater. Low-temperature hydrothermal activity (Ravizza *et al.*, 1996; Sharma *et al.*, 2000) has also been suggested as a significant source of unradiogenic osmium to seawater. This source is conceptually appealing because the low sulfide concentration of these fluids may allow osmium concentration to reach substantially higher than that in high-temperature, sulfide-rich fluids. A single analysis of a low-temperature fluid lends some support to this notion (Sharma *et al.*, 2000), but additional data are required to document the influence of low-temperature hydrothermal activity on the marine osmium isotope balance.

Ultramafic hydrothermal alteration is the most interesting of the potential mechanisms for decoupling unradiogenic inputs of osmium and strontium to the ocean. If this proves to be the major source of unradiogenic osmium to seawater, then unradiogenic osmium flux could be anticorrelated with oceanic crustal production rates because ultramafic rocks are preferentially exposed on slow spreading ridges. Alternatively, if low-temperature hydrothermal osmium flux proves to be important, it is unclear that this would undermine the assumption that unradiogenic osmium flux is proportional to oceanic crustal production rate. While major impact events, like the K–T boundary, can clearly decouple the strontium and osmium isotope records, the short-lived nature of this perturbation makes it relatively unimportant for the Cenozoic evolution of ocean chemistry.

21.3.3 Reconstructing Seawater Isotope Composition from Sediments

Reconstructing the isotopic composition of ancient seawater is more problematic for osmium than for strontium. Analyses of well-preserved microfossils, typically cleaned foraminifera, yield a fairly robust record of the $^{87}Sr/^{86}Sr$ of ancient seawater because Sr^{+2} is lattice bound in biogenic carbonates. Moreover, strontium isotopic analyses of bulk carbonate and associated pore waters provide the basis for quantitative models of strontium diagenesis in carbonates demonstrating that diagenetic artifacts, even in bulk carbonate analyses, are relatively modest in nearly pure carbonate sequences (Richter and DePaolo, 1988; Richter and Liang, 1993). By comparison with strontium, osmium burial in marine sediments is both more complicated and less-well understood. Osmium burial is more complicated because authigenic osmium enrichment occurs in a variety of depositional settings. These include manganese nodules (Luck and Turekian, 1983; Burton *et al.*, 1999), slowly accumulating pelagic clays (Esser and Turekian, 1988), organic-rich marine sediments (Ravizza and Turekian, 1989, 1992), and metalliferous sediments accumulating near mid-ocean ridges (Ravizza and McMurtry, 1993; Ravizza *et al.*, 1996). This pattern of enrichment demonstrates that osmium is effectively removed from seawater to the solid phase under both oxidizing and reducing conditions. By analogy with better-studied trace metals such as vanadium and molybdenum, it is likely that sedimentary organic matter and iron and manganese oxides play important roles in sequestering osmium from seawater under reducing and oxidizing conditions, respectively. The finely dispersed nature of these phases in the sediment complicates reconstructing the marine osmium isotope record, because there is no simple means of physically isolating material that contains exclusively seawater-derived osmium.

Two different strategies are employed to reconstruct the $^{187}Os/^{188}Os$ of ancient seawater from the sediment record. Analyses of bulk sediments are used in organic-rich (Ravizza, 1998; Cohen *et al.*, 1999) and metalliferous sediments (Ravizza, 1993) where hydrogenous osmium dominates the total sediment osmium budget. In slowly accumulating pelagic clays where osmium associated with cosmic dust and detrital material accounts for more than 50% of the total osmium inventory, leaching methods must be used to selectively liberate hydrogenous (seawater-derived osmium) from pelagic clays (Pegram *et al.*, 1992; Pegram and Turekian, 1999). Although modern calibration studies have vindicated both approaches, neither is completely convincing because both rely on the assumptions about either the selectivity of chemical leaching methods (see Peucker-Ehrenbrink *et al.* (1995) and Pegram and Turekian (1999) for discussions of potential artifacts), or the osmium concentration of local detrital material. As a result of the lack of compelling geochemical arguments, demonstrating the integrity of the marine osmium isotope record, a stratigraphic approach is used in which coeval sediment records from differing locations and depositional setting are compared to one another. Similar records of temporal variations $^{187}Os/^{188}Os$ preserved in widely separated sediment sequences provide empirical evidence suggesting that both records are accurately recording seawater $^{187}Os/^{188}Os$.

21.3.4 Cenozoic Strontium and Osmium Isotope Records

21.3.4.1 Overview of the Cenozoic marine strontium isotope record

As the result of several decades of investigation the Cenozoic history of seawater $^{87}Sr/^{86}Sr$ variation has become a well-established record of changing ocean chemistry. Peterman *et al.* (1970) presented the first Phanerozoic record of changes in the $^{87}Sr/^{86}Sr$ ratio of seawater derived from analyses of biogenic carbonate. The development of high-precision strontium isotope stratigraphy established the marine strontium isotope record as a valuable stratigraphic tool, and laid the foundation for developing a detailed composite record of Cenozoic seawater strontium isotope variations (DePaolo and Ingram, 1985). Although this record continues to be refined, its major features are well established (Figure 3). A recent compilation of marine strontium isotope record, based on a subset of all published data, can be found in McArthur *et al.* (2001).

The most striking aspect of the Cenozoic strontium record is the nearly monotonic, and relatively rapid rise in seawater $^{87}Sr/^{86}Sr$ ratio during the last 40 Myr, as compared to the small amplitude variations that characterize the Early Cenozoic record (65–40 Ma). This difference between the Early and Late Cenozoic portions of strontium record is accompanied by a dramatic contrast in data density. The latter part of the Cenozoic marine strontium isotope record is heavily sampled for two reasons. First the steep slope of the $^{87}Sr/^{86}Sr$ versus age curve allows absolute age dating of marine carbonates with a precision that can be better than ± 1 Ma. Second, the hypothesis that the rapid increase in seawater $^{87}Sr/^{86}Sr$ reflects a significant and systematic increase in alkalinity flux to the ocean associated with accelerated rates of chemical weathering driven by Himalayan uplift (Raymo and Ruddiman, 1992) provided additional impetus to refine the Late Cenozoic strontium isotope record. As a result of these efforts, the fine structure of the last 40 Ma of the Cenozoic strontium isotope record is well documented. Samples are commonly

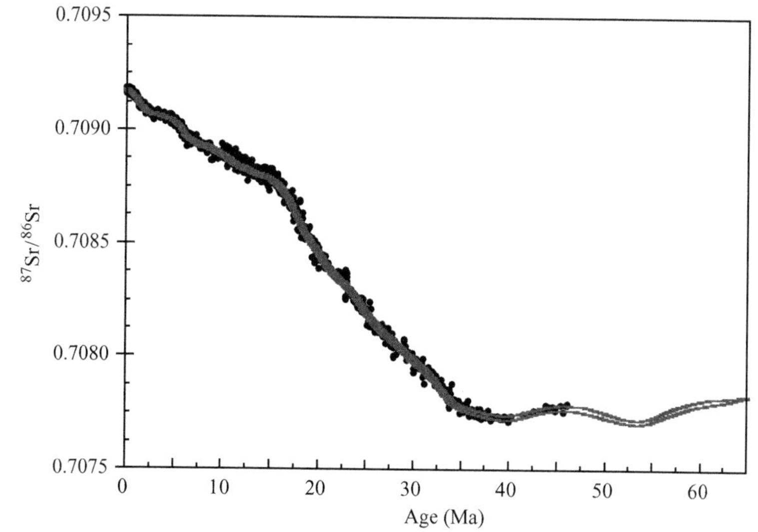

Figure 3 Composite marine strontium isotope record (sources Miller *et al.*, 1991a; Oslick *et al.*, 1994; Hodell and Woodruff, 1994; Mead and Hodell, 1995; Farrell *et al.*, 1995; Martin *et al.*, 1999; Reilly *et al.*, 2002). For the Early Cenozoic where data coverage is sparse the best fit of McArthur *et al.* (2001) is plotted.

analyzed at 100 kyr intervals in multiple records and independent age control is based on magnetostratigraphy, biostratigraphy, and, more recently, orbital tuning (Farrell *et al.*, 1995; Martin *et al.*, 1999). By comparison the details of seawater $^{87}Sr/^{86}Sr$ ratio variations during the first 25 Ma of the Cenozoic are only poorly constrained.

21.3.4.2 *Overview of the Cenozoic marine osmium isotope record*

The history of Cenozoic variations in the osmium isotopic composition of seawater is also preserved in the marine sediment record. This record is only poorly documented compared to the strontium record. In large part, this reflects the fact that the osmium isotope system is a fairly new paleoceanographic tracer. The first report of temporal changes in the osmium isotope composition of seawater was made only in 1992 (Pegram *et al.*, 1992). Initially, changes in osmium isotope composition produced by decay of ^{187}Re (half-life ~42 Gyr) to ^{187}Os were reported as $^{187}Os/^{186}Os$ ratio variations. Subsequently, the convention of reporting $^{187}Os/^{188}Os$ ratios was adopted because the decay of ^{190}Pt can produce small but measurable amounts of ^{186}Os (Walker *et al.*, 1997). In this review all data initially reported as $^{187}Os/^{186}Os$ ratios have been converted to $^{187}Os/^{188}Os$. Both the present-day marine osmium budget and the marine osmium isotope record are the subject of a recent review (Peucker-Ehrenbrink and Ravizza, 2000).

The Cenozoic marine osmium isotope record is characterized by a shift from generally unradiogenic values in the Early Cenozoic to more radiogenic values in the Neogene. This trend culminates with higher present-day seawater $^{187}Os/^{188}Os$ ratios than at any other time during the Cenozoic (Figure 4). It is clear that the marine strontium and osmium curves do not closely resemble one another. Based on a comparison of these two isotope records, the Cenozoic can be subdivided into three different intervals. (i) From the present to the mid-Miocene (0–15 Ma) the marine strontium and osmium isotope records are broadly correlated and are found to be rising toward more radiogenic isotope ratios. (ii) From the mid-Miocene to the Eocene–Oligocene transition (15–35 Ma) seawater $^{87}Sr/^{86}Sr$ ratio is found to be rising fairly rapidly. In contrast, $^{187}Os/^{188}Os$ ratio remains relatively constant for the majority of this time interval, with the notable exception of the Eocene–Oligocene transition itself. During the Paleocene and Eocene (35–65 Ma), it appears that the osmium record exhibits considerably more variability than does the strontium record. Neither record shows clear evidence of systematic change to more or less radiogenic isotope compositions during this early part of the Cenozoic.

Recently, a substantial body of data constraining the isotope composition and flux of dissolved osmium carried by rivers has been reported (Sharma *et al.*, 1999; Levasseur *et al.*, 1999; Martin *et al.*, 2001). These new data show that while the strontium and osmium isotope compositions of present-day seawater conform with a two-component mixing model fairly well (Table 1), the strontium and osmium isotope compositions of Paleogene seawater do not (Figure 5). This requires either decoupling of strontium and osmium of a magnitude that is not observed in the Neogene marine isotope balances, or a substantial shift in the strontium and osmium isotope composition of riverine influx. By assuming that the mole fractions

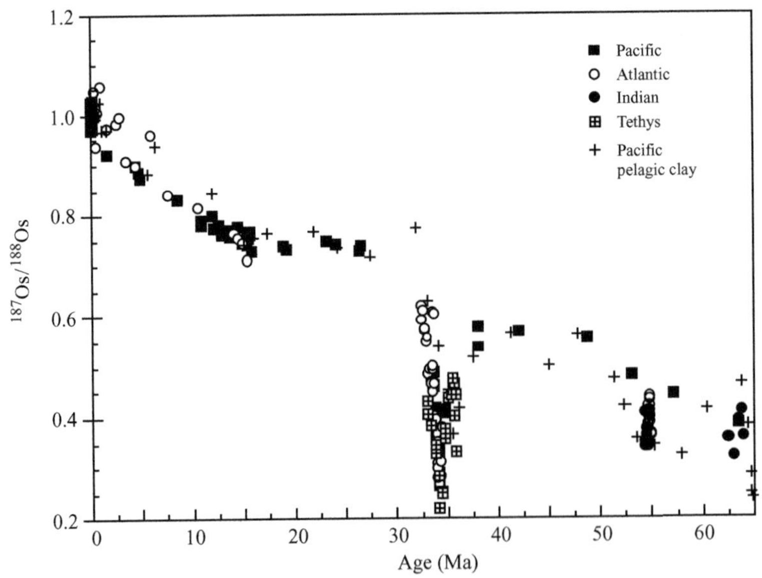

Figure 4 Composite record of seawater $^{187}Os/^{188}Os$ variations during the Cenozoic (sources Pegram *et al.*, 1992; Ravizza, 1993, 1998; Peucker-Ehrenbrink *et al.*, 1995; Reusch *et al.*, 1998; Oxburgh, 1998; Pegram and Turekian, 1999; Ravizza *et al.*, 2001; Ravizza and Peucker-Ehrenbrink, 2003).

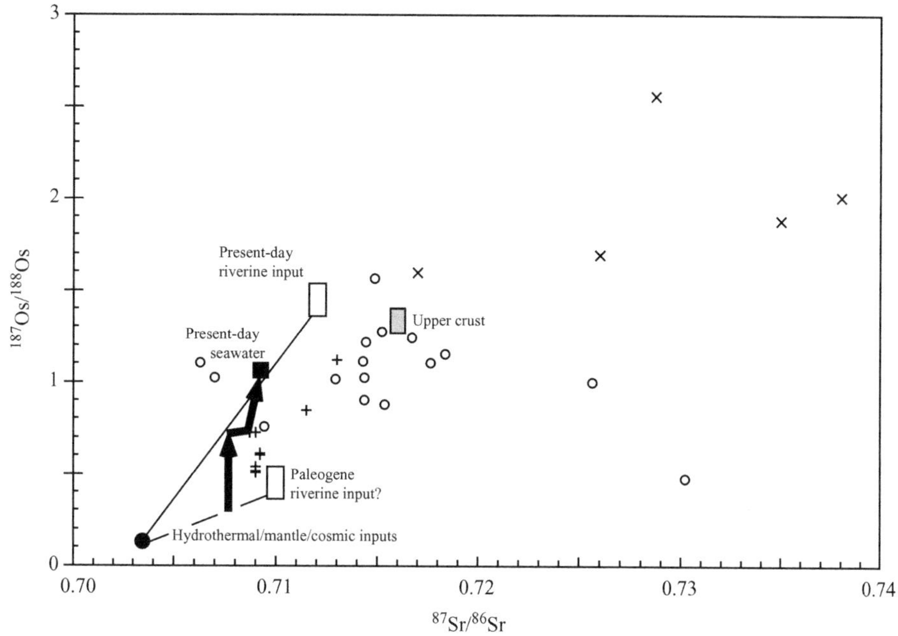

Figure 5 A cross-plot of $^{187}Os/^{188}Os$ versus $^{87}Sr/^{86}Sr$ illustrating the isotopic ratios associated with various oceanic inputs. Solid black arrows schematically illustrate the temporal evolution of seawater during the Cenozoic. Data sources (see Table 1 also) are: (–) Cenozoic seawater; (○) Loess deposits, Peucker-Ehrenbrink and Jahn (2001); (+) Indus paleosols, Chesley *et al.* (2000); and (×) Ganges paleosols, Chesley *et al.* (2000).

of osmium and strontium contributing to seawater from riverine input have remained the same as in the modern, a model-based isotopic composition of Paleogene riverine input can be calculated (Figure 5). While the calculated $^{87}Sr/^{86}Sr$ ratio is well within the range measured for modern rivers, the calculated $^{187}Os/^{188}Os$ ratio is not, being lower than average modern riverine input by a factor of ~3. A coupled Sr–C isotope model was designed to partition all seawater strontium isotope variation into changing isotope composition of river flux and the $^{87}Sr/^{86}Sr$ ratio of the silicate portion of riverine flux (Kump and Arthur, 1997). These results indicate a large increase in the $^{87}Sr/^{86}Sr$ ratio and the silicate portion of riverine strontium flux from Paleogene ratio of 0.7095 to a present-day ratio of 0.716. Thus, the marine osmium isotope record may reveal substantial changes in the composition of weathered silicate rock that are partially obscured in the marine strontium isotope record by the buffering effect of carbonate weathering.

21.3.4.3 Significance of uplift and weathering of the Himalayan–Tibetan Plateau (HTP)

Among the many recent papers that discuss the marine strontium isotope record, the influence of uplift and weathering of the HTP is pre-eminent as a potential cause of the increasingly radiogenic, or "continental," character during the Cenozoic. Several workers argue that HTP weathering is the primary cause of the post-40 Ma rise in seawater $^{87}Sr/^{86}Sr$ ratio (Richter *et al.*, 1992; Edmond, 1992; Raymo and Ruddiman, 1992). Several lines of evidence have been used to support this argument. The $^{87}Sr/^{86}Sr$ of Ganges–Brahmaputra (G–B) river waters are unusually high compared to that of global average river flux (Krishnaswami *et al.*, 1992; Palmer and Edmond, 1992). The most rapid increase in the seawater strontium curve coincides with rapid HTP uplift (Richter *et al.*, 1992; Hodell and Woodruff, 1994; see also Figure 6). Calculations of the influence of modern HTP rivers on the present-day seawater $^{87}Sr/^{86}Sr$ ratio have also been used to argue for the importance of HTP weathering (Hodell *et al.*, 1990).

The importance of HTP orogenesis as a cause of the Cenozoic rise in seawater $^{87}Sr/^{86}Sr$ has likely been overstated by many workers. Krishnaswami *et al.* (1992) estimated that G–B flow could account for only one-third of the post-40 Ma rise in seawater $^{87}Sr/^{86}Sr$ ratio. The most recent data indicate that the G–B river system contributes 2% of the global riverine strontium flux with $^{87}Sr/^{86}Sr$ ratio of 0.73 (Galy *et al.*, 1999). Using Equation (3) and the modern marine strontium budget to calculate hypothetical steady-state $^{87}Sr/^{86}Sr$ seawater ratio in the absence of G–B inflow yields a value between 0.7090 and 0.7089, depending on the $^{87}Sr/^{86}Sr$ assumed for the diagenetic strontium flux. This corresponds to ~20% of the post-40 Ma rise in seawater $^{87}Sr/^{86}Sr$ ratio. Some argue that all rivers draining the HTP should be considered, increasing the strontium contribution to ~25% of the global

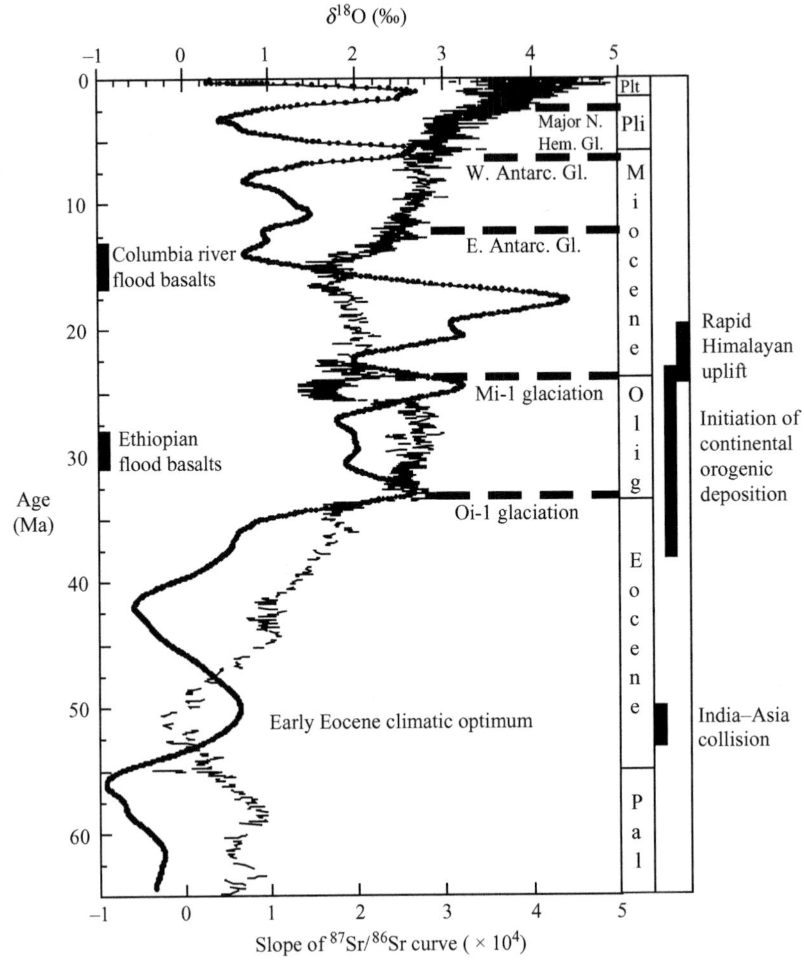

Figure 6 Comparison of the slope of the marine $^{87}Sr/^{86}Sr$ curve (McArthur *et al.*, 2001) to the composite benthic foraminiferal $\delta^{18}O$ record (Zachos *et al.*, 2001). Age ranges for flood basalt eruption are from Wignall (2001). Timing of glacial expansion events is after Zachos *et al.* (2001). Himalayan uplift and erosion history after Najman *et al.* (2000, 2001).

strontium flux (Richter *et al.*, 1992). However, among these rivers only the G–B has $^{87}Sr/^{86}Sr$ ratio substantially larger than average global river input (Palmer and Edmond, 1989, 1992). This implies that strontium supplied by other HTP rivers would contribute to rising seawater $^{87}Sr/^{86}Sr$ ratio only if their flow represented a net increase in river water flux to the ocean, and not simply a redistribution from another part of the globe. Although it is not possible to preclude a causal relationship between rising seawater $^{87}Sr/^{86}Sr$ ratio and HTP uplift, the detailed changes of the strontium isotope record are difficult to reconcile with the uplift history of the HTP (Figure 6). This is particularly true of the early phases of orogenesis because they remain poorly dated (see Najman *et al.*, 2001).

Early interpretations of the marine osmium isotope record also emphasized a Himalayan influence (Pegram *et al.*, 1992), particularly during the last 15 Ma when both $^{87}Sr/^{86}Sr$ and $^{187}Os/^{188}Os$ rise in concert (Peucker-Ehrenbrink *et al.*, 1995). The modern river data do not preclude this possibility. Recent analyses of dissolved osmium in the G–B water (Sharma *et al.*, 1999; Levasseur *et al.*, 1999) indicate that this river system contributes very roughly 2% of the total riverine osmium flux, similar to that estimated for strontium. The $^{187}Os/^{188}Os$ ratio associated with these analyses are highly variable ranging from 2.9 in the Ganges, twice that of global average river input, to 1.07 in the Brahmaputra. Additional osmium data from the G–B river system are needed to determine if the few available data are representative of seasonally averaged fluxes.

Weathering of Cambrian to Precambrian black shales from the Lesser Himalaya (LH) may be important in causing the rapid rise in seawater $^{187}Os/^{188}Os$ ratio during the last 15 Myr. These rocks appear to be the primary source of rhenium, and unusually radiogenic osmium to the modern G–B river system (Singh *et al.*, 1999;

Pierson-Wickmann *et al.*, 2000; Dalai *et al.*, 2002). Analyses of paleosols constrain the $^{187}Os/^{188}Os$ ratio of the Ganges (1.6–2.6) during the last 18 Myr, with the highest ratio in the youngest (3 Ma) sample (Chesley *et al.*, 2000; see Figure 5). This range is similar to the more radiogenic measurements of modern Ganges river water (Levasseur *et al.*, 1999) and Ganges bed load (Pierson-Wickmann *et al.*, 2000). Mass-balance considerations suggest that total amount of ^{187}Os available from the LH black shales is sufficient to influence the osmium isotope composition of the global ocean (Singh *et al.*, 1999).

Given that some workers argue that the LH carbonates are also the source of the unusually radiogenic strontium in the modern G–B system (English *et al.*, 2000), it is tempting to link the rising strontium and osmium isotopic composition of global seawater from 15 Ma to the present to weathering of the lesser Himalaya. However, not all workers agree on the importance of LH carbonates in the G–B strontium budget (see Galy *et al.*, 1999; Singh *et al.*, 1998, and references therein). In addition, the onset of coupled increase of seawater strontium and osmium isotope ratios predates the estimated time of LH exposure to erosion at 11 Ma (Chesley *et al.*, 2000). If input from the G–B is responsible for driving seawater strontium and osmium to more radiogenic isotope composition, the strontium and osmium fluxes provided by these rivers must have been substantially higher in the past (Singh *et al.*, 1999; Chesley *et al.*, 2000).

21.3.4.4 Glaciation and the marine strontium and osmium isotope records

Glaciation as a means of enhancing the flux of radiogenic strontium to the ocean is a second recurring theme in interpretations of the Cenozoic seawater strontium isotope record (Armstrong, 1971; Miller *et al.*, 1988; Hodell *et al.*, 1990; Capo and DePaolo, 1990). By comparison to the constraints on the timing of uplift and erosion in the HTP, the timings of major glacial events during the Cenozoic are very well established. Direct comparison of changes in slope of the marine strontium isotope record to oxygen isotope records that constrain changes in climate illustrates this clearly (Figure 6). For example, it has recently been argued that the initial rise in the Cenozoic $^{87}Sr/^{86}Sr$ record may be causally linked to the growth and decay of ice sheets in the Late Eocene and Early Oligocene (Zachos *et al.*, 1999). Although changes in the slope of the $^{87}Sr/^{86}Sr$ record do occur close to the time of many other major glacial events, there is no systematic lead–lag relationship between the two records. For example, a local maximum in the slope of the $^{87}Sr/^{86}Sr$ record is nearly coincident with the first major glaciation of the Oligocene (Oi-1), but a similar local maximum clearly predates the Mi-1 glaciation (Figure 6). Nevertheless mass-balance calculations (Hodell *et al.*, 1990; Blum, 1997) suggest that glacial enhancement of ^{87}Sr flux to seawater indicates that this is a plausible mechanism for driving at least part of the increasing seawater $^{87}Sr/^{86}Sr$ during the last several million years. Coupled studies of base cation flux and strontium isotopes demonstrate that silicate weathering rates are substantially increased, perhaps threefold compared to old saprolitic soils (Blum, 1997), and that the youngest moraines preferentially release ^{87}Sr (Blum and Erel, 1995). However, in glaciated regions with trace carbonate phases distributed throughout bedrock, rapid carbonate weathering can mute the $^{87}Sr/^{86}Sr$ signal of increased weathering rate (Jacobson *et al.*, 2003). Thus, while glacially enhanced silicate weathering may be an effective agent of CO_2 drawdown, quantitatively estimating this effect based on the marine strontium isotope record is not a promising endeavor. Any such effort is further complicated by the fact that the seawater strontium isotope record cannot capture high-frequency glacial forcing because of its long marine-residence time (Richter and Turekian, 1993; Henderson *et al.*, 1994).

The short marine-residence time of osmium allows the isotope composition of seawater to change on glacial–interglacial timescales. Records spanning recent glacial events indicate substantial shifts, $\sim 10\%$, toward lower-seawater $^{187}Os/^{188}Os$ ratio during peak glacial conditions that are interpreted as diminished river flux to the ocean (Oxburgh, 1998). However, investigation of osmium release from glacial soil sequences shows that ^{187}Os release is enhanced in a manner similar to that documented for ^{87}Sr (Peucker-Ehrenbrink and Blum, 1998), suggesting that the net effect of glaciations should be to drive seawater $^{187}Os/^{188}Os$ to higher values. This conclusion is supported by a recent investigation of the Eocene–Oligocene transition that demonstrates that the abrupt and permanent increase in seawater $^{187}Os/^{188}Os$ ratio is contemporaneous with major Antarctic glaciation (Ravizza and Peucker-Ehrenbrink, 2003; see also Figure 7). Together, these studies provide evidence that the marine osmium isotope record is influenced by glacial cycles. This in turn suggests that the large amplitude, high-frequency glacial cycles play an important role in causing seawater osmium and strontium isotope compositions to shift to more radiogenic values in the latter part of the Cenozoic.

21.3.5 Variations in the Strontium and Osmium Isotope Composition of Riverine Input

The most widely recognized ambiguity associated with interpreting the marine strontium and

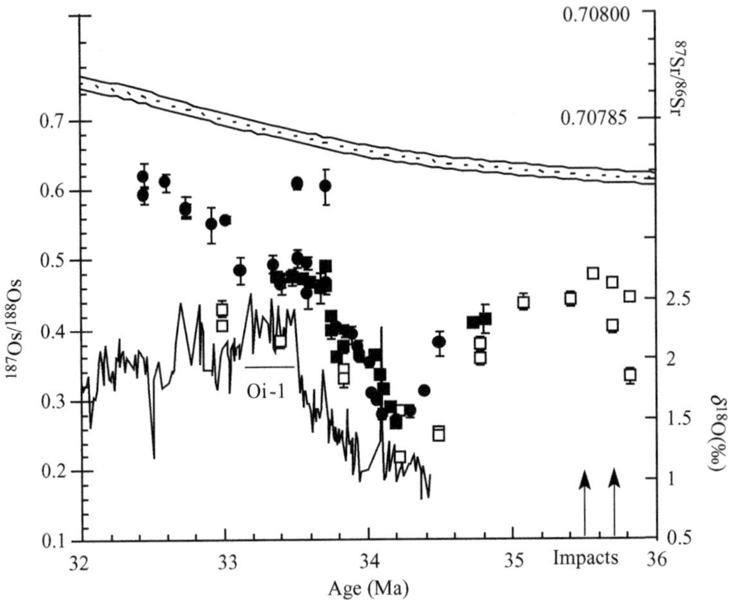

Figure 7 The marine osmium isotope record of the Eocene–Oligocene transition (Ravizza and Peucker-Ehrenbrink, 2003). The benthic foraminiferal oxygen isotope record from site 522 (Zachos *et al.*, 1996) and a fit to the marine strontium isotope record are shown for comparison. Note that the rise of "ice house" conditions in the Early Oligocene is closely associated with an abrupt and permanent increase in seawater $^{187}Os/^{188}Os$. This provides evidence supporting the hypothesis that glaciation contributes to rising seawater $^{187}Os/^{188}Os$ and $^{87}Sr/^{86}Sr$ ratios ((= = = =)95% C. L. from McArthur *et al.*, 2001).

osmium isotope records is the inability to distinguish between temporal changes in the isotopic composition of riverine input and temporal changes in the total flux of strontium and osmium to the ocean. Relatively early models recognized the significance of this uncertainty (Richter *et al.*, 1992; Berner and Rye, 1992). Subsequent analyses of strontium isotopes in paleosol carbonates (Quade *et al.*, 1997) and pedogenic minerals in deep-sea clays (Derry and France-Lanord, 1996) yielded compelling evidence for $^{87}Sr/^{86}Sr$ ratio of river input as high as 0.755 from the G–B river system 7.5 Myr ago. Prior to this time, riverine $^{87}Sr/^{86}Sr$ ratio was 0.712, similar to global average riverine flux. Complementary analyses of osmium isotopes (Chesley *et al.*, 2000) demonstrated coupled strontium and osmium isotope variations in rivers draining the Himalaya during the last 18 Myr (Figure 5). The studies are critically important because they demonstrate that significant temporal variations in riverine strontium and osmium isotope compositions can occur in a large river system. Given that the G–B river system is unusually radiogenic among modern rivers and yet still delivers only ~2% of the global riverine strontium flux, it is much less likely that such large and rapid changes in the strontium and osmium isotope composition of global average river flux can occur.

The combined osmium and strontium records are consistent with a shift to more radiogenic riverine isotope compositions during the Cenozoic (Figure 5). This suggests that erosion and weathering of unradiogenic lithologies were more important in the past. Several studies suggest that plateaus in the $^{87}Sr/^{86}Sr$ record are caused by erosion of volcanic deposits characterized by low $^{87}Sr/^{86}Sr$ values, either flood basalt provinces (Taylor and Lasaga, 1999; Dessert *et al.*, 2001; McArthur *et al.*, 2001) or island arc volcanics (Reusch and Maasch, 1998). The rapid decline in the slope of the $^{87}Sr/^{86}Sr$ curve between 17 Ma and 14 Ma to near zero provides a specific Cenozoic example (Figure 6). This feature has been attributed to weathering of the Columbia River flood basalts (Hodell *et al.*, 1990; Taylor and Lasaga, 1999), and also to island arc weathering accelerated by the collision of New Guinea with the Australian continent (Reusch and Maasch, 1998). However, emplacement of Ethiopian Traps the Early Oligocene coincides with a much less pronounced decrease in slope of the $^{87}Sr/^{86}Sr$ curve (Figure 6), even though it is estimated to be more than 3 times larger than the CRFB Province (Wignall, 2001). Thus, it seems unlikely that the plateau in the marine strontium isotope record between 17 Ma and 14 Ma can be attributed entirely to flood basalt weathering.

The influence of island arc and flood basalt weathering on the marine osmium isotope record is even less well understood. Analyses of modern rivers draining Papua New Guinea indicate that arc weathering represents an important source of unradiogenic osmium to seawater (Martin *et al.*, 2000, 2001). Though detailed investigations of modern flood basalt weathering have not been made, the marine osmium isotope record across the Triassic–Jurassic boundary is consistent with the hypothesis that emplacement of the Central Atlantic magmatic province is responsible for a shift to lower $^{187}Os/^{188}Os$ ratios (Cohen and Coe, 2002). However, available Cenozoic data do not yield evidence of a significant influence of either CRFB volcanism or arc-continent collision on the marine osmium isotope record between 14 Ma and 17 Ma (Reusch *et al.*, 1998). The fact that the marine osmium isotope record continues to rise during the plateau in the strontium record supports the ideas outlined above that the unradiogenic sources of osmium and strontium are not tightly coupled. The pronounced excursion to unradiogenic $^{187}Os/^{188}Os$ values in the Late Eocene and at the K–T boundary provides additional evidence of Sr–Os decoupling (Figure 4). For the case of the K–T boundary extraterrestrial osmium associated with the K–T impact event is clearly an important factor; in the Late Eocene the influence of extraterrestrial osmium input is possible though less certain (Ravizza and Peucker-Ehrenbrink, 2003).

21.3.6 Osmium and Strontium Isotopes as Chemical Weathering Proxies

The need to constrain the geologic history of chemical weathering motivated the detailed reconstruction of past changes in the strontium and osmium isotope composition of seawater. While these combined records remain the best available indicators of changes in patterns of Cenozoic weathering, neither can be used to quantitatively reconstruct CO_2 consumption rates associated with chemical weathering. Silicate weathering rate is only one of the many factors that influences the flux of ^{87}Sr to the ocean. Detailed comparisons of strontium isotope data and river solute chemistry in modern rivers demonstrate that other factors, particularly the influence of strontium released by rapidly weathering carbonates, can obscure the contribution from silicate weathering (Palmer and Edmond, 1989; Jacobson *et al.*, 2002a,b, 2003). Similarly for osmium, silicate-weathering contributions to the ^{187}Os flux are likely convoluted with those from weathering of organic-rich sediments. The former process represents a CO_2 sink in the weathering cycle, while the latter represents a CO_2 source. In addition, the $^{187}Os/^{188}Os$ of osmium contributed by weathering of sedimentary organic matter is strongly influenced by the age of the sediment. This further complicates separating the various sources contributing to riverine osmium flux. Thus, for osmium, like strontium, there is no simple connection between seawater isotope composition and silicate weathering rates.

The marine strontium and osmium isotope records remain important records of ocean chemistry, because they do reflect the changing influence of continental inputs through time. In this regard the marine osmium isotope record has considerable unexplored potential because of its ability to respond to high-frequency external forcing, allowing detailed correlation with carbon and oxygen isotope records. For example, the excursion to more radiogenic seawater $^{187}Os/^{188}Os$ ratio during the unusual warmth of the Paleocene–Eocene thermal maximum (PETM) provides the best available evidence supporting the operation of feedback between global climate and chemical weathering rates (Figure 8). The combined marine strontium (Figure 6) and osmium (Figure 7) isotope records

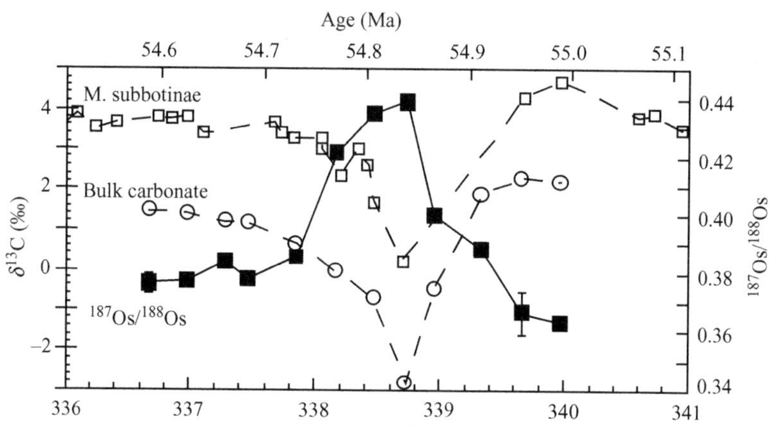

Figure 8 The osmium isotope excursion documented at site 549 from Ravizza *et al.* (2001) is coincident with the carbon isotope excursion that marks the PETM. This association has been interpreted as evidence of accelerated global average weathering rates in response to this global warming event and provides a clear example of the utility of coupling osmium isotope data with conventional stable isotope records of ocean chemistry.

are consistent with the hypothesis that the first major Oligocene glaciation (Oi-1) enhanced silicate weathering rates, providing a positive feedback for additional cooling. Contemporaneous changes in the carbon isotopic composition of deep water (Figure 1) and a substantial deepening of the calcium carbonate compensation depth suggest that changes in the ocean–atmosphere carbon cycle were intimately associated with this climatic event. Considered in isolation, the marine strontium and osmium isotope records are ambiguous in that there is not a single unique interpretation of the underlying causes of these documented changes in global ocean chemistry. However, these records can place important constraints on the models of changing ocean chemistry when they are carefully integrated with other records of global climate and ocean chemistry.

21.4 Mg/Ca RECORDS FROM BENTHIC FORAMINIFERA

Over the past several years, Mg/Ca variations preserved in benthic foraminifera have received attention as a record of deep-ocean paleo-temperatures (Lear *et al.*, 2000; Billups and Schrag, 2002). This work is part of a large and vigorous effort to exploit the potential of Mg/Ca ratio as a paleo-temperature proxy in both benthic and planktonic foraminifera on a variety of timescales. From a historical perspective, it is noteworthy that a positive correlation between the Mg/Ca ratio of biogenic calcite and precipitation temperature was recognized long ago (Chave, 1954). Other factors such as taxa, growth rate, and saturation state are also known to influence the Mg/Ca ratio of biogenic calcite (see Mackenzie *et al.* (1983) for a summary of this early work). These complicating factors detracted from early notions that Mg/Ca ratio of biogenic carbonates might be useful for reconstructing paleo-temperatures. Renewed interest in Mg/Ca ratio as a paleo-temperature proxy differs from these early efforts in many respects. For the purpose of this review some of the more important differences include improved analytical precision and accuracy, the emphasis on microfossils, and the abundance of complementary oxygen isotope data (see also Chapter 8).

21.4.1 Coupling Benthic Foraminiferal Mg/Ca and Oxygen Isotope Records

The Mg/Ca and $\delta^{18}O$ of benthic foraminifers are both influenced by the temperature of calcification. This link is the reason that a discussion of Mg/Ca is included in this review. As outlined above, the oxygen isotope composition of biogenic calcite is a function of the temperature and the oxygen isotope composition of the water in which the organism calcified. For benthic foraminiferal records it is widely assumed that mainly the ice volume changes drive the temporal variations in the $\delta^{18}O$ of seawater. Thus, benthic foraminiferal $\delta^{18}O$ records have the dual influence of changing ice volume and deep-water temperatures embedded in them. Mg/Ca paleo-thermometry applied to benthic foraminifera offers the possibility of determining absolute deep-water temperatures. These temperature estimates can be used with the measured $\delta^{18}O$ of the same benthic foraminifera to calculate temporal changes in $\delta^{18}O$ of seawater. Reconstructions of the $\delta^{18}O$ of seawater are of interest because variance in this parameter is strongly influenced by the growth and decay of ice sheets. Note however, that even with benefit of a well-constrained $\delta^{18}O$ of seawater record, quantitative estimates of Cenozoic ice volume still require assumptions about the average $\delta^{18}O$ of the global ice reservoir. Still the possibility of accurately determining deep-sea paleo-temperatures and simultaneously better constraining changing ice volume throughout the Cenozoic has motivated a great deal of recent interest in Mg/Ca variations in benthic foraminifera. Lear *et al.* (2000) first applied this approach to a composite record spanning much of the Cenozoic. More recently, this initial effort was supplemented by additional data from a Southern Ocean site spanning roughly the last 28 Myr (Billups and Schrag, 2002).

21.4.2 Calibration of the Mg/Ca Thermometer

Before discussing Cenozoic benthic foraminiferal Mg/Ca records, we present an overview of how Mg/Ca ratios measured in benthic foraminifera are used to calculate paleo-temperatures. The most recent calibration efforts postdate the most recent work on the Cenozoic Mg/Ca ratio. Revisions to the calibration equation and other aspects of core-top data sets have important implications for how Cenozoic benthic Mg–Ca records are interpreted. The influence of temperature on Mg/Ca ratio of mixed Cibicidoides reported by Lear *et al.* (2002; Figure 9(a)) is: $(\text{Mg/Ca})_{\text{foram}} = 0.867e^{0.109T}$, where T represents bottom-water temperature. Also, because these data are largely from core-top calibration studies, this formulation implicitly assumes that calcification occurs from modern seawater with a Mg/Ca ratio of approximately 5.2 mol mol^{-1}. Explicitly including the possibility of variable seawater Mg/Ca ratio results in a modified equation: $(\text{Mg/Ca})_{\text{foram}} = R(0.867e^{0.109T})$, where R is the Mg/Ca ratio of seawater at some time in the past divided by the present-day seawater Mg/Ca ratio. Other terms are the same as in the previous equation. The problems this possibility poses for Mg/Ca paleo-thermometry are conceptually similar

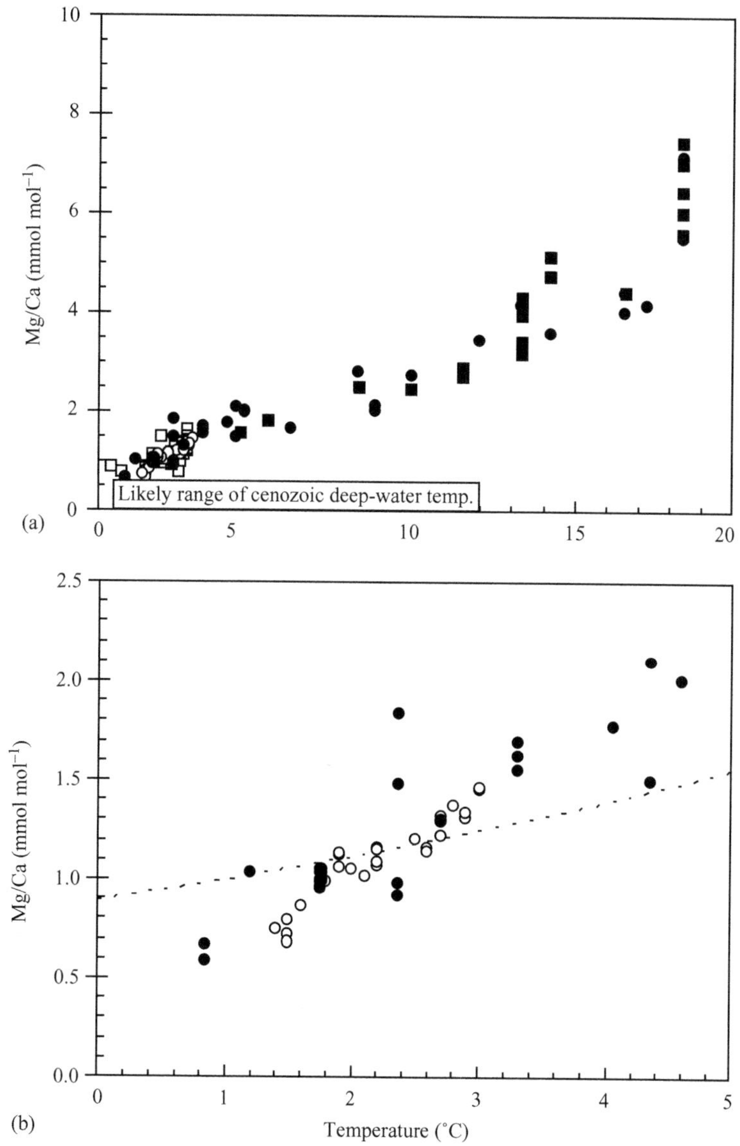

Figure 9 Summary of available data used to calibrate the influence of calcification temperature on the Mg/Ca of benthic foraminifera, after Lear *et al.* (2002). Data are from multiple sources. Data included both core-top samples and analyses of cultured foraminifera. Panel (a) displays the full data set used to obtain the exponential calibration given in the text ((■) Rosenthal *et al.*, 1997 (corrected)). Panel (b) displays a subset of the core-top data that suggest a steeper slope than the full data set. This complicating aspect of the calibration is particularly important to paleotemperature reconstructions in the Late Cenozoic when deep-water temperatures fall in this low-temperature range ((○) Martin *et al.*, 2002; (●) Lear *et al.*, 2002). See text for further discussion.

to those presented by $\delta^{18}O$ of seawater variations in oxygen isotope paleo-thermometry, because shell composition depends both on the temperature and the composition of seawater. Variation in seawater Mg/Ca ratio is likely to be less directly coupled to climate change than $\delta^{18}O$ of seawater because changing ice volume should not directly influence seawater Mg/Ca ratio. Presumably seawater Mg/Ca ratio is less variable on short timescales, because the marine residence times of magnesium and calcium are relatively long. Current estimates of the oceanic residence times of calcium in seawater are ~1 Myr while estimates of magnesium residence times are substantially longer. Possible secular trends in seawater Mg/Ca ratio are discussed below.

Examination of core-top calibration data at low temperature (<5 °C), particularly those reported by Martin *et al.* (2002), reveals a distinctly stronger temperature dependence, and a more nearly linear relationship, than is indicated by the full data set (Figure 9(b)). If magnesium incorporation into biogenic calcite behaves as a true solid solution, equilibrium thermodynamics predicts that Mg/Ca ratio should depend exponentially on temperature.

The deviation of the low-temperature data from the exponential best fit can be interpreted as evidence that factors other than temperature exert an increasingly important influence on the Mg/Ca ratio of foraminiferal calcite at these low temperatures (Lear *et al.*, 2002). This calibration uncertainty is less important than the potential influence of factors other than temperature on Early Cenozoic benthic foraminiferal Mg/Ca records, because warm deep-water temperatures amplify the temperature component of Mg/Ca variation. Lear *et al.* (2002) include vital effects, dissolution artifacts, growth rate effects, and carbonate saturation state in a list of factors that may exert a secondary control on the Mg/Ca ratios of benthic foraminifera. Some of these same parameters have been recognized as important influences on the Mg/Ca ratio of larger calcifying organisms. For example, Mg/Ca ratios in coralline algae and pelecypods are influenced by growth rate, which is likely to be influenced by temperature (Moberly, 1968). A similar scheme of "nested" influences on Mg/Ca records of benthic foraminifera could be invoked to explain the unusually strong apparent influence of temperature on Mg/Ca ratio in the study by Martin *et al.* (2002) (see Figure 9(b)).

21.4.3 Cenozoic Benthic Foraminiferal Mg/Ca Records

Although recent studies of benthic foraminiferal Mg/Ca ratio variations do not span the entire Cenozoic, one study (Lear *et al.*, 2000) does extend back to approximately 50 Ma capturing the unusual warmth of the Eocene climatic optimum. A second study (Billups and Schrag, 2002) reports additional data spanning the last 28 Myr. Results from these two studies are grossly similar in that both show generally decreasing Mg/Ca ratios in benthic foraminifera with time (Figure 10). This is consistent with cooling deep-water temperatures. However, direct comparison of the two data sets is made complicated by several factors. Lear *et al.* (2000) compiled a composite record that includes several different species of benthic foraminifera. Differences between measured Mg/Ca ratios of coexisting species were interpreted as evidence of vital effects, and data were adjusted by empirical correction factors to yield a record that is effectively normalized to *O.umbonatus*. In addition, this record was smoothed in order to facilitate comparison to composite benthic foraminiferal $\delta^{18}O$ records. In contrast, the Billups and Schrag (2002) record is based predominately on a single species, *C. mundulus*, with some additional data from *C. wuellerstorfi*. Smoothing is not required for calculating $\delta^{18}O$ of seawater in this record because paired Mg/Ca and $\delta^{18}O$ data are available for the majority of the samples. This approach is desirable because it

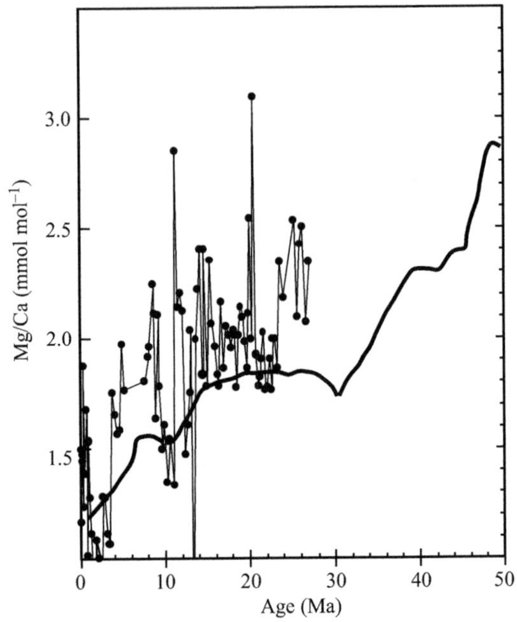

Figure 10 Comparison of Mg/Ca ratios measured in benthic foraminifera from a smoothed, composite record (thick solid curve: Lear *et al.*, 2000), and from site 747 (filled circles: Billups and Schrag, 2002). Both records display an overall trend of declining Mg/Ca ratio with decreasing age. This trend is indicative of cooling deep-water temperatures during the Cenozoic.

provides a framework for examining higher-frequency variations in deep-water temperature and $\delta^{18}O$ of seawater. Comparison of these two data sets, in the context of the recent calibration efforts discussed above, indicates that application of the Mg/Ca paleo-temperature proxy to benthic foraminifera is more complicated than initially assumed.

Potential analytical bias, vital effects, and calibration issues all complicate interpretation of the records shown in Figure 10. Measured benthic foraminiferal Mg/Ca ratios in site 747 (Billups and Schrag, 2002) are in general higher than in coeval samples from the composite record of Lear *et al.* (2000) implying warmer temperatures. The sense of this offset is unexpected given the high latitude of site 747. A systematic analytical bias between the two data sets is possible because different methods were used to make Mg/Ca measurements. Note that applying corrections used by Lear *et al.* (2000) to normalize *C. mundulus* (+0.15) and *C. wuellerstorfi* (+0.45) to normalize Mg/Ca data to *O. umbonatus* tend to amplify this difference. It is noteworthy that core-top analyses discussed above (Lear *et al.*, 2002) did not substantiate a systematic offset between *O. umbonatus* and *C. wuellerstorfi*, highlighting the fact that the general question of vital effects remains open. Although the best available calibration (Figure 9(a)) is based on a mixed assemblage of *Cibicidoides, C. mundulus* is not included. Note that not all extant *Cibicidoides* species conform to the calibration shown in Figure 9(a) (Lear *et al.*, 2002). Thus, even the nearly monospecific data set of Billups and Schrag (2002) does not entirely avoid potential problems associated with vital effects. Uncertainty in the functionality of the appropriate Mg/Ca calibration also deserves mention. Billups and Schrag (2002) used a linear calibration curve linking Mg/Ca to temperature rather than an exponential fit. The slope of the calibration line is essentially identical to that shown by the Martin *et al.* (2002) data (Figure 9(b)), but as noted above there is an offset in absolute Mg/Ca ratio. Applying the more recent exponential calibration (Lear *et al.*, 2002) to the Billups and Schrag (2002) data set amplifies the amplitude of calculated temperature variations and increases the average temperature by approximately 3 °C (Figure 11). These differences propagate directly into calculated records of $\delta^{18}O$ of seawater. Given that modern calibration studies can yield linear Mg/Ca versus temperature responses at low temperature (Figure 9(b)), careful consideration of the appropriate calibration approach is warranted. This is particularly true in the Late Cenozoic where deep-water temperature estimates tend to be lower.

21.4.4 Changing Seawater Mg/Ca Ratio

Possible variations in the Mg/Ca ratio of seawater during the Cenozoic also contribute uncertainty to temperature estimates based on Mg/Ca ratios measured in benthic foraminifera (see Billups and Schrag (2002) for a discussion of model based estimates of the Cenozoic evolution of seawater Mg/Ca ratios). Analyses of fluid inclusions in evaporites (Zimmermann, 2000) are consistent with declining seawater Mg/Ca ratio during the Cenozoic. Taken at face value these data suggest a 12% seawater magnesium depletion 5 Myr ago, and further declining to 30–40% of modern value by the Late Eocene (37 Myr ago). However, Zimmermann (2000) notes that dolomitization of calcium carbonate during evaporite formation could create the false impression of lower-seawater magnesium concentrations in the fluid inclusion data. On Phanerozoic timescales, seawater Mg/Ca estimates based on fluid inclusions (Lowenstein *et al.*, 2001) and fossil echinoderms (Dickson, 2002) yield similar records. This lends some credibility to estimates of Cenozoic seawater Mg/Ca ratio based on fluid inclusions. Based on benthic $\delta^{18}O$ data, the assumption of ice-free conditions, and the revised Mg/Ca temperature calibration, Lear *et al.* (2002) calculated that the maximum likely depletion in seawater Mg/Ca ratio was approximately 35%. Even though the magnitude of this estimate is similar to fluid inclusion-based estimates, the timing of these changes remains ill-constrained. While substantial uncertainty remains in converting foraminiferal Mg/Ca data to absolute temperature, it is clear that this proxy has great potential. Likely the most immediate progress will be made in studies that emphasize high temporal resolution and couple Mg/Ca and stable oxygen isotope data to deconvolve the dual influence of changing deep-water temperatures and ice volume during events of rapid growth and decay of ice sheets. On these short timescales abrupt changes in seawater Mg/Ca proportions are unexpected.

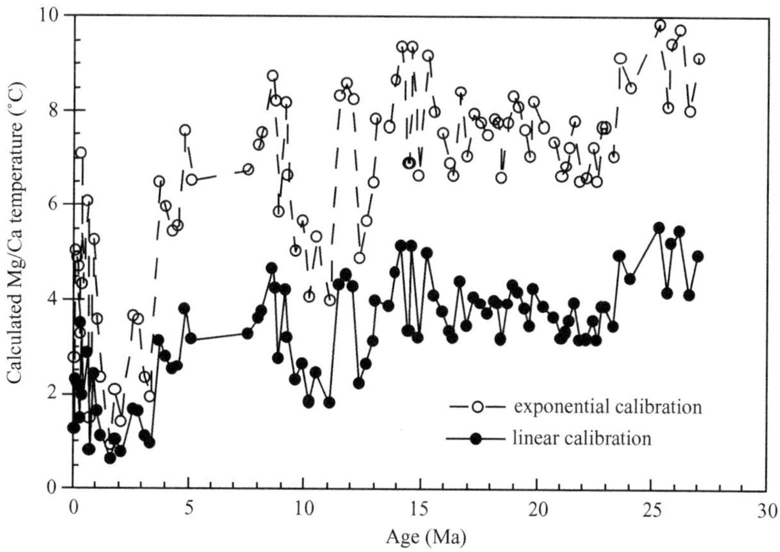

Figure 11 Comparison of deep-water temperatures calculated from the Mg/Ca data of Billups and Schrag (2002) using the linear calibration originally used in this publication and the more recent exponential calibration reported by Lear *et al.* (2002).

21.5 BORON ISOTOPES, PALEO-pH, AND ATMOSPHERIC CO_2

The boron isotopic composition of calcite precipitated by foraminifera has been used to reconstruct the pH of the ancient seawater. In appropriate oceanographic settings, pH estimates derived from boron isotope analyses of planktonic foraminifera, in conjunction with additional assumptions about the dissolved inorganic carbon inventory in the surface ocean, can be used to calculate atmospheric CO_2 concentrations. Declining levels of atmospheric CO_2 have been invoked commonly (Raymo and Ruddiman, 1992) as a potentially important causative factor in long-term cooling and the expansion of polar ice that is indicated by the Cenozoic benthic $\delta^{18}O$ record (Figure 1). In this context, the use of boron isotopes as a paleo-pH proxy represents an important new record of changing ocean chemistry during the Cenozoic.

21.5.1 The pH Dependence of Boron Isotope Fractionation

There are two stable isotopes of boron, ^{10}B and ^{11}B. ^{10}B is the more abundant of the two, accounting for ~80% of all boron atoms. Variations in boron isotope abundance are reported relative to an NIST boric acid standard as $\delta^{11}B$. In seawater there are two different species of boron. The relative proportion of the two species is a function of pH as indicated by the reaction

$$B(OH)_3 + H_2O = B(OH)_4^- + H^+$$

The p*K*a of boric acid is ~8.7 so that in the pH range relevant to seawater, 7.2–8.2, these two forms of boron coexist. Isotope exchange equilibrium with respect to boron isotopes occurs rapidly such that ^{10}B is preferentially partitioned into the tetrahedral $B(OH)_4^-$ species relative to trigonal boric acid species. At isotopic equilibrium the difference between the $\delta^{11}B$ of these two species is close to 20 per mil. This isotope fractionation factor appears to be largely insensitive to other factors such as temperature and pressure. Given the constant contrast in $\delta^{11}B$ between $B(OH)_3$ and $B(OH)_4^-$, mass-balance considerations require that the $\delta^{11}B$ of the two species vary systematically with pH. The $\delta^{11}B$ of each species can be calculated explicitly once the $\delta^{11}B$ of seawater, the p*K*a of boric acid, and the isotope fractionation are specified (Figure 12). It is important to note that because the p*K*a of boric acid is temperature dependent, calcification temperature can also exert an important influence on the $\delta^{11}B$ of the $B(OH)_4^-$ species (Palmer *et al.*, 1998). It is also noteworthy that at the low end of the seawater pH range boron isotopes lose sensitivity as a pH proxy, because the slope of the $\delta^{11}B$ versus pH curve (Figure 12) becomes nearly flat for both the boron species. This reflects the fact that boron occurs dominantly as boric acid at these lower pH values with only roughly 2% of $B(OH)_4^-$.

21.5.2 Boron Partitioning into Calcite

Both inorganic precipitation experiments (Sanyal *et al.*, 2000) and culture experiments (Sanyal *et al.*, 1996, 2001) have shown that the isotopic composition of boron incorporated into calcite is a function of pH (Figure 13). The measured $\delta^{11}B$ of calcite precipitated over a range of pH are similar to that predicted for $B(OH)_4^-$ indicating that the anionic species of boron is selectively partitioned into calcite as suggested in earlier studies (Vengosh *et al.*, 1991; Hemming and Hanson, 1992). The selective partitioning of a single boron species into calcite effectively records the pH of the water from which the calcite precipitated. Culture studies exhibit a systematic offset in the $\delta^{11}B$ measured in different species of foraminifera indicating that vital effects can influence the absolute $\delta^{11}B$ of foraminiferal calcite. Detailed comparison to theoretical predictions of boron isotope fractionation is hampered by uncertainties in both the isotope fractionation factor and the p*K*a of boric acid. Both vital effects in the biogenic material and kinetic factors in the inorganic system may contribute to the variable offset between the $\delta^{11}B$ of biogenic and inorganic calcite (Figure 13). In the case of planktonic foraminifera that host photosymbionts, daytime drawdown of CO_2 within a microenvironment may contribute to the documented vital effect (Sanyal *et al.*, 2001). Correlated variations in $\delta^{11}B$ and $\delta^{13}C$ in modern coral support such an interpretation (Hemming *et al.*, 1998a). Investigations of both core-top and cultured specimens of *G. sacculifer* and *O. universa* show a similar $\delta^{11}B$ offset relative to one another providing strong evidence that species-dependent vital effects do influence the $\delta^{11}B$ of planktonic foraminiferal calcite (Sanyal *et al.*, 2001; Figure 13).

The few analyses of boron concentration variations in calcite as a function of precipitation pH that have been done in culture (Sanyal *et al.*, 1996) and inorganic precipitation (Sanyal *et al.*, 2000) experiments indicate that boron concentrations increase with increasing precipitation pH. This observation is grossly consistent with the increasing concentration of $B(OH)_4^-$ with increasing pH and a simple proportionality between the $B(OH)_4^-$ concentration in seawater and the boron concentration in calcite. Although there is some scatter in the available data, these results suggest that boron concentration analyses may be complementary to $\delta^{11}B$ analyses in paleo-pH reconstructions. However, it is important to note that subtle surface structural effects can give rise to sixfold variations in boron

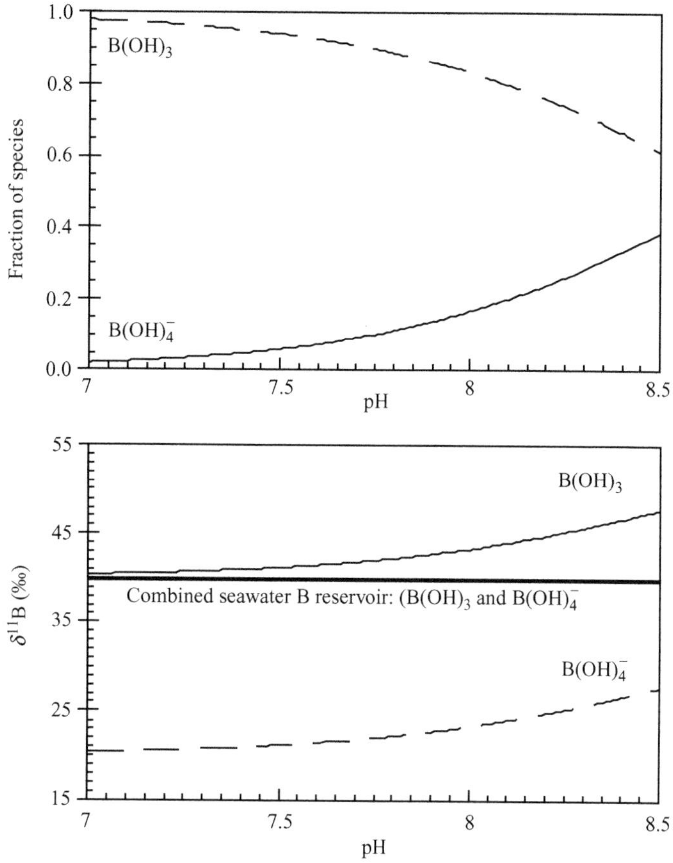

Figure 12 Schematic illustration of isotopic contrast between coexisting boron species as a function of pH. The range of pH shown encompasses all pH values relevant to seawater. Note that at low-pH boric acid dominates the boron inventory and $\delta^{11}B$ becomes relatively insensitive to small pH changes. These curves were calculated using a p*K*a of 8.7 for boric acid, a constant isotopic fractionation of 20 per mil between the two boron species, and $\delta^{11}B$ of 39.6 per mil for the total seawater boron reservoir.

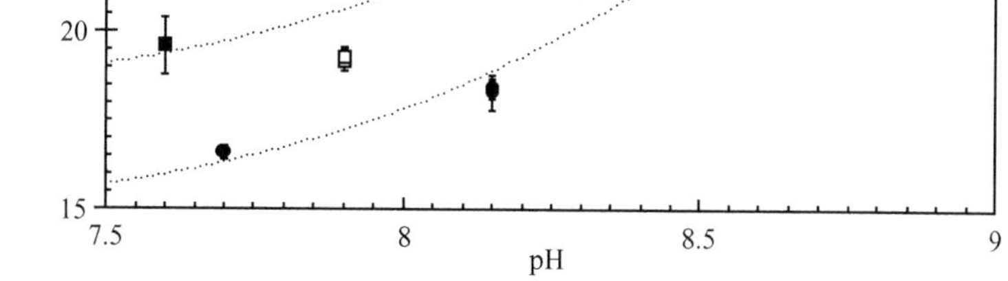

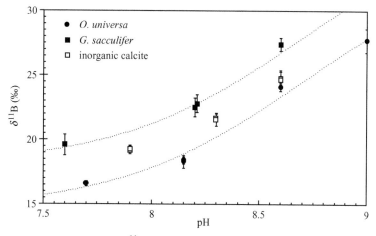

Figure 13 Summary of efforts to calibrate $\delta^{11}B$ variations in calcite as a function of the pH of precipitation. Results are replotted from Sanyal *et al.* (1995, 1996, 2000, 2001). The shape of the dashed curves is that predicted by the pH control on the $\delta^{11}B$ of $B(OH)_4^-$. The position of these curves was adjusted to fit the *O. universa* and *G. sacculifer* data, highlighting the nearly constant boron isotopic offset between these two species of planktonic foraminifera. The upper curve closely approximates the calculated $\delta^{11}B$ of $B(OH)_4^-$ (see Figure 12). Note that the *O. universa* and *G. sacculifer* data plotted include both cultured foraminifera and core-top samples.

concentration on small spatial scales within a single crystal (Hemming *et al.*, 1998b). Thus, it could be argued that $\delta^{11}B$ could accurately record precipitation pH, even though boron concentrations do not exhibit the systematic variation expected based on simple models of element partitioning. While much remains to be learned about the details of boron incorporation into inorganic and biogenic calcite, available data yield strong evidence that calcite $\delta^{11}B$ can record the pH of the solution from which it precipitated.

21.5.3 Paleo-pH and Atmospheric CO_2 Reconstruction

Several efforts to reconstruct the pH of ancient seawater have been made, starting with the work of Spivack *et al.* (1993). Efforts to quantify changes in surface and deep-water pH during recent glacial–interglacial cycles followed (Sanyal *et al.*, 1995, 1997). The efforts most relevant to this chapter are those of Palmer and Pearson presented in a series of recent papers (Palmer *et al.*, 1998; Pearson and Palmer, 1999, 2000). This work focuses on reconstructing the pH of surface waters in the tropical Pacific Ocean over Cenozoic timescales as a means of constraining atmospheric CO_2 variations. In the modern ocean pH exhibits substantial spatial variability unrelated to atmospheric CO_2 levels. For example, in highly productive regions pH can vary by as much as 0.7 pH units in the upper 1,000 m to values below 7.5. This variability is driven largely by the production of CO_2 during oxidation of sinking organic matter. By concentrating their efforts on a region of the ocean that is likely to have remained thermally stratified, and exhibited only modest productivity variations, Palmer *et al.* (1998) argued that surface-dwelling foraminifera are likely to have grown in waters close to exchange equilibrium with respect to atmospheric CO_2. Analyses of multiple species of foraminifera believed to calcify over distinct depth ranges exhibit trends of $\delta^{11}B$ that become less positive with increasing depth in the water column (Palmer *et al.*, 1998). This is consistent with the expected trend of decreasing pH with increasing depth, resulting from organic matter oxidation in the water column. By applying a similar approach to an assemblage of Eocene foraminifera, Pearson and Palmer (1999) estimated the $\delta^{11}B$ of Eocene seawater to fall between 38 per mil and 41 per mil, with a most likely value of 40.5 per mil. Values greater than 41 per mil imply a very steep vertical pH gradient and absolute pH values so low as to interfere with calcification. Seawater $\delta^{11}B$ lighter than 38‰ imply only small pH changes with depth, and these authors argued that the low productivity implied by such a shallow pH gradient was unlikely. Only gradual changes in $\delta^{11}B$ of seawater are likely, because the marine residence time of boron is on the order of 10 Myr. Detailed paleo-pH profiles at a few time intervals distributed throughout the Cenozoic (Palmer *et al.*, 1998; Pearson and Palmer, 1999) provide a framework constraining seawater $\delta^{11}B$ during the Cenozoic. In a follow-on study Pearson and Palmer (2000) combined data from analyses of planktonic foraminifera that calcified in the mixed layer with their model-based constraints on seawater $\delta^{11}B$ to make 35 individual surface water pH estimates distributed over the past 60 Myr (Figure 14). To estimate atmospheric CO_2

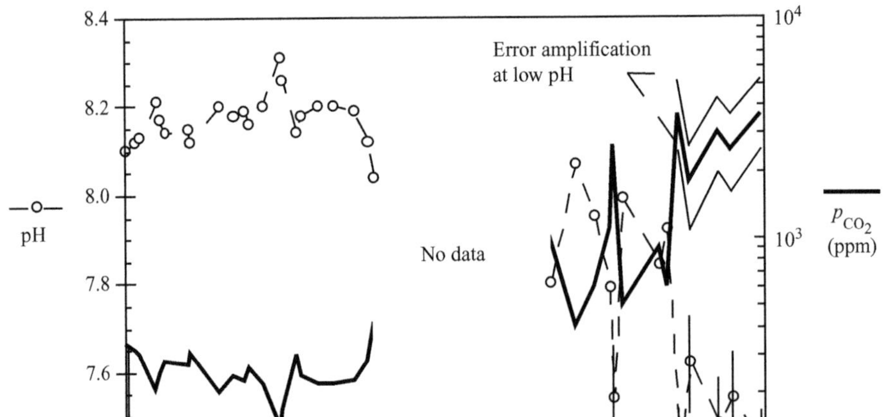

Figure 14 Estimates of mixed-layer pH and atmospheric CO_2 derived from boron isotope analyses of planktonic foraminifera recovered from drill cores taken in the western equatorial Pacific. Results are replotted from Pearson and Palmer (2000). Note that uncertainty associated with both pH and CO_2 estimates amplifies in the Early Cenozoic. This reflects the intrinsic insensitivity of the boron isotope pH proxy at low pH (see Figure 12). Atmospheric CO_2 estimates require additional assumptions about the size of the DIC pool in surface waters.

from this surface ocean pH requires additional assumptions. As noted above the surface waters at this site are assumed to remain close to equilibrium with respect to atmospheric CO_2 throughout the Cenozoic. Second, an independent estimate of either total alkalinity or total DIC is required to calculate the partial pressure of CO_2 from pH. To make this estimate, Pearson and Palmer (2000) used reconstructions of CCD, and assumed the depth of calcite saturation relative to the CCD remained invariant with time. They further assumed that the vertical gradient in alkalinity remained constant with time and that the calcium concentration of seawater varied in concert with alkalinity.

Although the atmospheric CO_2 record resulting from the Pearson and Palmer (2000) study is subject to substantial uncertainty, several features of the record are of interest. Although Figure 14 does indicate that CO_2 levels were roughly 10 times higher than present-day levels during the Early Eocene, the record as a whole does not, however, suggest a gradual decline in CO_2 levels throughout the Cenozoic. This is best illustrated by the variability in the middle Eocene CO_2 estimates ranging over nearly an order of magnitude on relatively short timescales. Note that this time is characterized by relatively light benthic $\delta^{18}O$ values indicative of warm deep-water temperatures and little or no polar ice (Figure 1). The last 25 Myr of the CO_2 reconstruction show smaller amplitude variations with the majority of data suggesting values between 200 ppm and 400 ppm. Reconstructions of CO_2 between 16 Ma and 4 Ma based on carbon isotope variations in alkenones yield similar values ranging between 180 ppm and 320 ppm (Pagani *et al.*, 1999). As these two methods are largely independent of one another, the similarity of the two data during the time interval where they overlap lends credibility to the combined data sets. Both records suggest a component of high-frequency variation in atmospheric CO_2 levels, but detailed comparison of the two records is hampered by the low-temporal resolution of the two records. For example, the boron isotope record suggests that high atmospheric CO_2 levels may have been associated with emplacement of the Columbia River flood basalts at ~17 Ma, while the expansion of the East Antarctic ice sheet at 15 Ma may be associated with a local minimum in CO_2 (Pearson and Palmer, 2000). In contrast, alkenone-based CO_2 estimates do not indicate a pronounced CO_2 decrease during expansion of the east Antarctic ice sheet (Pagani *et al.*, 1999). Based on available data it is unclear if these two records contradict one another or if aliasing of high-frequency variation simply creates this impression.

21.5.4 Outstanding Questions about Paleo-pH Reconstructions

Several important questions related to the use of boron isotopes as a paleo-pH proxy remain unanswered. While progress has been made documenting the presence of vital effects on the $\delta^{11}B$ of biogenic calcite (Sanyal *et al.*, 2001; Hemming *et al.*, 1998a), further work on this issue is required. Reconstruction of Early Cenozoic pH requires working with planktonic foraminifera that are no longer extant and thus vital effects cannot be well constrained without a substantial additional effort to work on multiple species with overlapping age ranges. Dissolution effects and diagenetic alteration are known to adversely affect the application of other paleoceanographic proxies, and the influence of this phenomenon on $\delta^{11}B$ is yet to be investigated. Given that tropical planktonic foraminifera are particularly vulnerable to diagenetic alteration on the seafloor (Pearson *et al.*, 2001), the influence of this process on $\delta^{11}B$ of foraminiferal calcite is important to evaluate. Recent efforts to estimate glacial–interglacial pH changes in deep waters that are based on preservation of carbonate microfossils (Anderson and Archer, 2002) have yielded results that are at odds with boron-isotope-based estimates (Sanyal *et al.*, 1995), also highlighting the need for additional work. The potential for temporal variations in the $\delta^{11}B$ of seawater also warrants additional investigation. Model calculations based on a refined version of the present-day marine boron cycle, and assumptions about oceanic crustal production and weathering rates (Lemarchand *et al.*, 2000) suggest that temporal changes in the $\delta^{11}B$ of seawater may be substantially larger than estimated by Pearson and Palmer (2000) using the pH profile approach. The model calculations of Lemarchand *et al.* (2000) suggest that seawater $\delta^{11}B$ could have been as low as 36 per mil. If true, this would imply that the high CO_2 levels estimated in the Early Cenozoic (Figure 14) are incorrect and that levels similar to modern values are more likely. Thus, independent data constraining the $\delta^{11}B$ of seawater would also represent an important contribution to advancing the use of boron isotopes as a paleo-pH proxy. Both continuing refinements to application of boron isotopes to paleo-pH reconstructions and coupled application of independent means of estimating paleo-p_{CO_2} levels are important areas for future work.

21.6 CLOSING SYNTHESIS: DOES OROGENESIS LEAD TO COOLING?

Recently much debate has focused on the hypothesis that increased silicate weathering rates, associated with Himalayan uplift, played a causative role in long-term Cenozoic cooling by reducing atmospheric carbon dioxide levels (Raymo *et al.*, 1988; Raymo and Ruddiman, 1992). Indeed, it seems appropriate to revisit this idea at the close of this review because the idea fueled much of the effort to refine and integrate the paleoceanographic records discussed above. What is the current status of this idea? Two key problems with the "orogenesis leads to cooling" hypothesis as initially articulated by Raymo and co-workers have been raised. First, simple mass-balance considerations preclude the possibility of accelerated rates of CO_2 drawdown by chemical weathering throughout the Cenozoic without a comparable increase in the flux of CO_2 to the atmosphere from volcanic and metamorphic sources (Caldeira *et al.*, 1993; Volk *et al.*, 1993; Berner and Caldeira, 1997; Broecker and Sanyal, 1998). Second, recent efforts to determine the history of atmospheric CO_2 levels suggest that cooling and ice growth during the Neogene are not associated with systematic decreases in atmospheric CO_2 (Pagani *et al.*, 1999; Pearson and Palmer, 2000), also undermine this hypothesis.

In spite of the critiques outlined in the preceding paragraph, it is inappropriate to discount entirely the role of orogenesis in Cenozoic cooling entirely. The mass-balance problems outlined above can be addressed. Kump and Arthur (1997) presented a conceptual framework in which steady-state levels of atmospheric CO_2 could be reduced without changes in global weathering flux by increasing "weatherability." In essence they argued that tectonic and climatic factors could act to allow weathering fluxes to remain high enough to balance CO_2 input in spite of significantly cooler global climate. This concept provides a defensible framework for a causative link between Himalayan uplift and global cooling, because physical weathering can be reasonably argued to enhance "weatherability" (Kump *et al.*, 2000). It also provides a basis for arguing that a limited decoupling of average global temperatures and silicate weathering rates played an important role in Cenozoic cooling. Both of these are key components of the "orogenesis leads to cooling" hypothesis.

The available data constraining the Cenozoic history of atmospheric CO_2 suggest nearly 10-fold higher concentration during the Paleocene and Early Eocene, relative to the Neogene (Pearson and Palmer, 2000). Thus, the gross contrast between the unusual warmth of the Early Paleogene and the "ice house" conditions of the Neogene can still be interpreted as the result of tectonically driven changes in weathering, even though the many features of Neogene climate change cannot. It is likely that factors other than orogenesis contributed to enhanced weatherability during the Cenozoic. For example, accelerated physical erosion due to glaciation may greatly increase the area of fresh mineral surfaces available to chemical weathering (Blum, 1997; Zachos *et al.*, 1999). Many factors, in addition to atmospheric CO_2 concentration, may influence Earth's climate over the course of the Cenozoic. Thus while the "orogenesis leads to cooling" hypothesis is not dead, it does not provide a single unified explanation of the established record of Cenozoic climatic variation.

The mass-balance argument that CO_2 consumption by silicate weathering cannot outpace CO_2 production by volcanic and metamorphic degassing also has important implications for interpretation of heavy isotope records. If, to first approximation, the average long-term rate of silicate weathering has remained constant during the Cenozoic, and this can be extended to the total flux of strontium and osmium, then the variations in the marine $^{87}Sr/^{86}Sr$ and $^{187}Os/^{188}Os$ records are better interpreted as the result of changing isotopic composition of riverine flux (Kump and Arthur, 1997). Under this interpretation, the Cenozoic strontium and osmium isotope record imply a secular change in the age and/or composition of weathered material. Both records suggest that the Cenozoic is characterized by an increasing contribution from older material with a cratonic affinity and radiogenic strontium and osmium characteristics relative to recently mantle-derived material (Figure 5). How such a shift in the nature of weathered material might influence "weatherability," or if it would at all, has not been considered in detail.

The efforts of many researchers have contributed to a greatly improved history of Cenozoic ocean chemistry and climate history. These efforts have revealed a great deal of fine structure in these records that is global in nature, ranging from episodes of unusual warmth such as the PETM to episodes of rapid glacial expansion in the Early Oligocene and Early Miocene. It is possible that short-term fluctuations in atmospheric CO_2 played an important causative role in some of these events, even though available data argue against a simple causal link between atmospheric CO_2 and long-term Cenozoic cooling. A refined vision of a global thermostat originally proposed by Walker *et al.* (1981) that includes secular changes in global "weatherability" (Kump and Arthur, 1997) can be invoked to explain how a negative feedback between global temperature and silicate weathering rate can stabilize Earth's climate system in the wake of large perturbations.

The causes of events that perturb Earth's climate system are less well understood and more varied in nature. Understanding these phenomena likely requires studies that focus on specific climatic events rather than long-term climate evolution. Careful integration of the various paleo-proxies discussed in this review offers the potential for the first time to establish the lead–lag relationships between ice growth, deep-water temperature change, continental weathering, and atmospheric CO_2 during important climate transitions and on timescales similar to the residence of inorganic carbon in the ocean-atmosphere system. These types of studies will represent a major step forward in our ability to test fundamental hypotheses about the underlying causes of the many climate changes documented during the Cenozoic. This integrated multiproxy approach may also lay a foundation for exploring the possibility that these short-term climate events can influence Earth's long-term climate evolution.

REFERENCES

Anderson D. M. and Archer D. (2002) Glacial-interglacial stability of ocean pH inferred from foraminifer dissolution rates. *Nature* **416**(6876), 70–73.

Armstrong R. L. (1971) Glacial erosion and the variable isotopic composition of strontium in sea water. *Nature; Phys. Sci.* **230**(14), 132–133.

Berggren W. A., Kent D. V., Swisher C. C. III, and Aubry M. P. (1995) A revised Cenozoic geochronology and chronostratigraphy Geochronology, time scales and global stratigraphic correlation. *Special Publication—SEPM (Society for Sedimentary Geology)* **54**, 129–212.

Berner R. A. and Caldeira K. (1997) The need for mass balance and feedback in the geochemical carbon cycle. *Geology* 955–956.

Berner R. A. and Rye D. M. (1992) Calculation of the Phanerozoic strontium isotope record of the oceans from a carbon cycle model. *Am. J. Sci.* **292**, 136–148.

Berner R. A., Lasaga A. C., and Garrels R. M. (1983) The carbon-silicate geochemical cycle and its effect on atmospheric carbon dioxide over the past 100 million years. *Am. J. Sci.* **283**, 641–683.

Billups K. and Schrag D. P. (2002) Paleotemperatures and ice volume of the past 27 Myr revisited with paired Mg/Ca and $^{18}O/^{16}O$ measurements on benthic foraminifera. *Paleoceanography* **17**, 3-1–3-11.

Blum J. D. (1997) The effect of late Cenozoic glaciation and tectonic uplift on silicate weathering rates and the marine $^{87}Sr/^{86}Sr$ record. In *Tectonic uplift and climate change* (ed. W. F.e. Ruddiman). Plenum, New York, pp. 259–288.

Blum J. D. and Erel Y. (1995) A silicate weathering mechanism linking increases in marine $^{87}Sr/^{86}Sr$ with global glaciation. *Nature* **373**, 415–418.

Bralower T. J., Zachos J. C., Thomas E., Parrow M., Paull C. K., Kelly D. C., Premoli Silva I., Sliter W. V., and Lohmann K. C. (1995) Late Paleocene to Eocene paleoceanography of the equatorial Pacific Ocean: stable isotopes recorded at Ocean Drilling Program Site 865, Allison Guyot. *Paleoceanography* **10**(4), 841–865.

Brass G. W. (1976) The variation of the marine $^{87}Sr/^{86}Sr$ ratio during Phanerozoic time: interpretation using a flux model. *Geochim. Cosmochim. Acta* **40**, 721–730.

Broecker W. S. and Sanyal A. (1998) Does atmospheric CO_2 police the rate of chemical weathering? *Global Biogeochem. Cycles* **12**(3), 403–408.

Burton K. W., Bourdon B., Birck J.-L., Allegre C. J., and Hein J. R. (1999) Osmium isotope variations in the oceans recorded by Fe–Mn crusts. *Earth Planet. Sci. Lett.* **171**(1), 185–197.

Caldeira K., Arthur M. A., Berner R. A., and Lasaga A. C. (1993) Cooling in the late Cenozoic. *Nature* **361**, 123–124.

Capo R. C. and DePaolo D. J. (1990) Seawater strontium isotopic variations from 2.5 million years ago to the present. *Science* **249**, 51–55.

Chave K. E. (1954) Aspects of the biogeochemistry of magnesium: [Part] 1. Calcareous marine organisms; [Part] 2. Calcareous sediments and rocks. *J. Geol.* **62**(3), 266–283.

Chesley J. T., Quade J., and Ruiz J. (2000) The Os and Sr isotopic record of Himalayan paleorivers: Himalayan tectonics and influence on ocean chemistry. *Earth Planet. Sci. Lett.* **179**(1), 115–124.

Cohen A. S. and Coe A. L. (2002) New geochemical evidence for the onset of volcanism in the Central Atlantic magmatic province and environmental change at the Triassic–Jurassic boundary. *Geology* **30**(3), 267–270.

Cohen A. S., Coe A. L., Bartlett J. M., and Hawkesworth C. J. (1999) Precise Re-Os ages of organic-rich mudrocks and the Os isotope composition of Jurassic seawater. *Earth Planet. Sci. Lett.* **167**, 159–173.

Dalai T. K., Singh S. K., Trivedi J. R., and Krishnaswami S. (2002) Dissolved rhenium in the Yamuna River system and the Ganga in the Himalaya: role of black shale weathering on the budgets of Re, Os, and U in rivers and CO_2 in the atmosphere. *Geochim. Cosmochim. Acta* **66**(1), 29–43.

DePaolo D. J. and Ingram B. L. (1985) High resolution stratigraphy with Sr isotopes. *Science* **227**, 938–941.

Derry L. A. and France-Lanord C. (1996) Neogene Himalayan weathering history and river $^{87}Sr/^{86}Sr$: impact on the marine Sr record. *Earth Planet. Sci. Lett.* **142**, 59–74.

Dessert C., Dupre B., Francois L. M., Schott J., Gaillardet J., Chakrapani G., and Bajpai S. (2001) Erosion of Deccan traps determined by river geochemistry: impact on the global climate and the $^{87}Sr/^{86}Sr$ ratio of seawater. *Earth Planet. Sci. Lett.* **188**(3–4), 459–474.

Dickens G. (2001) On the fate of past gas: what happens to methane released from a bacterially mediated gas hydrate capacitor? *Geochem. Geophys. Geosys.-G3* **2**, 2000GC000131.

Dickens G. R., Castillo M. M., and Walker J. C. G. (1997) A blast of gas in the latest Paleocene: simulating first-order effects of massive dissociation of oceanic methane hydrate. *Geology* **25**(3), 259–262.

Dickson J. A. D. (2002) Fossil echinoderms as monitor of the Mg/Ca ratios of Phanerozoic oceans. *Science* **298**, 1222–1224.

Diester-Haass L. (1996) Late Eocene–Oligocene paleoceanography in the southern Indian Ocean (ODP Site 744). *Marine Geol.* **130**(1–2), 99–119.

Douville E., Charlou J. L., Oelkers E. H., Bienvenu P., Jove Colon C. F., Donval J. P., Fouquet Y., Prieur D., and Appriou P. (2002) The Rainbow Vent fluids (36 degrees 14′N, MAR); the influence of ultramafic rocks and phase separation on trace metal content in Mid-Atlantic Ridge hydrothermal fluids. *Chem. Geol.* **184**(1–2), 37–48.

Edmond J. M. (1992) Himalayan tectonics, weathering processes, and the strontium isotope record in marine limestones. *Science* **258**, 1594–1597.

Elderfield H. and Schultz A. (1996) Mid-ocean ridge hydrothermal fluxes and the chemical composition of the ocean. *Ann. Rev. Earth Planet. Sci.* **24**, 191–224.

English N. B., Quade J., DeCelles P. G., and Garzione C. N. (2000) Geologic control of Sr and major element chemistry in Himalayan rivers, Nepal. *Geochim. Cosmochim. Acta* **64** (15), 2549–2566.

Esser B. K. and Turekian K. K. (1988) Accretion rate of extraterrestrial particles determined from osmium isotope systematics of Pacific pelagic clay and manganese nodules. *Geochim. Cosmochim. Acta* **52**, 1383–1388.

Farrell J. W., Clemens S. C., and Gromet L. P. (1995) Improved chronostratigraphic reference curve of late Neogene seawater $^{87}Sr/^{86}Sr$. *Geology* **23**, 403–406.

Flower B. P. and Kennett J. P. (1993) Middle Miocene ocean-climate transition; high-resolution oxygen and carbon isotopic records from Deep Sea Drilling Project Site 588A, Southwest Pacific. *Paleoceanography* **8**(6), 811–843.

Flower B. P., Zachos J. C., and Paul H. (1997) Milankovitch-scale climate variability recorded near the Oligocene/Miocene boundary. *Proc. Ocean Drill. Prog. Sci. Results* **154**, 433–439.

Frank M. (2002) Radiogenic isotopes: tracers of past ocean circulation and erosional input. *Rev. Geophys.* **40**(1), 1.1–1.38.

Galy A., France-Lanord C., and Derry L. A. (1999) The strontium isotopic budget of Himalayan rivers in Nepal and Bangladesh. *Geochim. Cosmochim. Acta* **63**(13–14), 1905–1925.

Gieskes J. M., Elderfield H., and Palmer M. R. (1986) Strontium and its isotopic composition in interstitial waters of marine carbonate sediments. *Earth Planet. Sci. Lett.* **77**(2), 229–235.

Goldstein S. J. and Jacobsen S. B. (1988) Nd and Sr isotopic systematics of river water suspended material: implications for crustal evolution. *Earth Planet. Sci. Lett.* **87**, 249–265.

Grossman E. L. (1984) Stable isotope fractionation in live benthic foraminifera from the Southern California Borderland. *Palaeogeogr. Palaeoclimatol. Palaeoecol.* **47**(3–4), 301–327.

Grossman E. L. (1987) Stable isotopes in modern benthic foraminifera: a study of vital effect. *J. Foraminiferal Res.* **17**(1), 48–61.

Hemming N. G. and Hanson G. N. (1992) Boron isotopic composition and concentration in modern marine carbonates. *Geochim. Cosmochim. Acta* **56**(1), 537–543.

Hemming N. G., Guilderson T. P., and Fairbanks R. G. (1998a) Seasonal variations in the boron isotopic composition of coral; a productivity signal? *Global Biogeochem. Cycles* **12**(4), 581–586.

Hemming N. G., Reeder R. J., and Hart S. R. (1998b) Growth-step-selective incorporation of boron on the calcite surface. *Geochim. Cosmochim. Acta* **62**(17), 2915–2922.

Henderson G. M., Martel D. J., O'Nions K., and Shackleton N. J. (1994) Evolution of seawater $^{87}Sr/^{86}Sr$ over the last 400 ka: the absence of glacial/interglacial cycles. *Earth Planet. Sci. Lett.* **128**, 643–651.

Hodell D. A. and Woodruff F. (1994) Variations in the strontium isotopic ratio of seawater during the Miocene: stratigraphic and geochemical implications. *Paleoceanography* **9**, 405–426.

Hodell D. A., Mead G. A., and Mueller P. A. (1990) Variation in the strontium isotopic composition of seawater (8 Ma to present): implications for chemical weathering rates and dissolved fluxes to the oceans. *Chem. Geol.; Isotope Geosci. Sect.* **80**(4), 291–307.

Jacobson A. D., Blum J. D., Chamberlain C. P., Poage M. A., and Sloan V. F. (2002a) Ca/Sr and Sr isotope systematics of a Himalayan glacial chronosequence: carbonate versus silicate weathering rates as a function of landscape surface age. *Geochim. Cosmochim. Acta* **66**(1), 13–27.

Jacobson A. D., Blum J. D., and Walter L. M. (2002b) Reconciling the elemental and Sr isotope composition of Himalayan weathering fluxes: insights from the carbonate geochemistry of stream waters. *Geochim. Cosmochim. Acta* **66**(19), 3417–3429.

Jacobson A. D., Blum J. D., Chamberlain C. P., Craw D., and Koons P. O. (2003) Climatic and tectonic controls on chemical weathering in the New Zealand Southern Alps. *Geochim. Cosmochim. Acta* **67**(1), 29–46.

Jaffe L. A., Peucker-Ehrenbrink B., and Petsch S. T. (2002) Mobility of rhenium, platinum group elements and organic carbon during black shale weathering. *Earth Planet. Sci. Lett.* **198**(3–4), 339–353.

Kelly D. S., Karson J. A., Blackman D. K., Frueh-Green G. L., Butterfield D. A., Lilley M. D., Olson E. J., Schrenk M. O., Roe K. K., Lebon G. T., and Rivizzigno P. (2001) An off-axis hydrothermal vent field near the Mid-Atlantic Ridge at 30 degrees N. *Nature* **412**(6843), 145–149.

Krishnaswami S., Trivedi J. R., Sarin M. M., Ramesh R., and Sharma K. K. (1992) Strontium isotopes and rubidium in the Ganga–Brahmaputra River system: weathering in the Himalaya, fluxes to the Bay of Bengal and contributions to the evolution of oceanic $^{87}Sr/^{86}Sr$. *Earth Planet. Sci. Lett.* **109** (1–2), 243–253.

Kump L. and Arthur M. A. (1997) Global chemical erosion during the Cenozoic: weatherability balances the budgets. In *Tectonic Uplift and Climate Change* (ed. W. F. Ruddiman). Plenum, New York, pp. 400–424.

Kump L. R. and Arthur M. A. (1999) Interpreting carbon-isotope excursions: carbonates and organic matter. *Chem. Geol.* **161**(1–3), 181–198.

Kump L. R., Brantley S. L., and Arthur M. A. (2000) Chemical weathering, atmospheric CO_2, and climate. *Ann. Rev. Earth Planet. Sci.* **28**, 611–667.

Lear C. H., Elderfield H., and Wilson P. A. (2000) Cenozoic deep-sea temperatures and global ice volumes from Mg/Ca in benthic foraminiferal calcite. *Science* **287**(5451), 269–272.

Lear C. H., Rosenthal Y., and Slowey N. (2002) Benthic foraminiferal Mg/Ca-paleothermometry: a revised core-top calibration. *Geochim. Cosmochim. Acta* **66**(19), 3375–3387.

Lemarchand D., Gaillardet J., Lewin E., and Allegre C. J. (2000) The influence of rivers on marine boron isotopes and implications for reconstructing past ocean pH. *Nature* **408**(6815), 951–954.

Levasseur S., Birk J.-L., and Allegre C. J. (1998) Direct measurement of fetomoles of osmium and the $^{187}Os/^{186}Os$ ratio in seawater. *Science* **282**, 272–274.

Levasseur S., Birck J. L., and Allegre C. J. (1999) The osmium riverine flux and the oceanic mass balance of osmium. *Earth Planet. Sci. Lett.* **174**(1–2), 7–23.

Lowenstein T. K., Timofeeff M. N., Brennan S. T., Hardie L. A., and Demicco R. V. (2001) Oscillations in Phanerozoic seawater chemistry: evidence from fluid inclusions. *Science* **294**(5544), 1086–1088.

Luck J. M. and Turekian K. K. (1983) Osmium-187/osmium-186 in manganese nodules and the Cretaceous-Tertiary boundary. *Science* **222**(4624), 613–615.

Mackenzie F. T., Bischoff W. D., Bishop F. C., Loijens M., Schoonmaker J., and Wollast R. (1983) Magnesian calcites: low-temperature occurrence, solubility and solid-solution behavior. In *Carbonates: Mineralogy and Chemistry. Rev. Mineral.* **11**, 97–144.

Martin C. E. (1991) Os isotopic characteristics of mantle derived rocks. *Geochim. Cosmochim. Acta* **55**, 1421–1434.

Martin C. E., Peucker-Ehrenbrink B., Brunskill G. J., and Szymczak R. (2000) Sources and sinks of unradiogenic osmium runoff from Papua New Guinea. *Earth Planet. Sci. Lett.* **183**(1–2), 261–274.

Martin C. E., Peucker-Ehrenbrink B., Brunskill G., and Szymczak R. (2001) Osmium isotope geochemistry of a tropical estuary. *Geochim. Cosmochim. Acta* **65**(19), 3193–3200.

Martin E. E., Shackleton N. J., Zachos J. C., and Flower B. P. (1999) Orbitally-tuned Sr isotope chemostratigraphy for the late middle to late Miocene. *Paleoceanography* **14**(1), 74–83.

Martin P. A., Lea D. W., Rosenthal Y., Shackleton N. J., Sarnthein M., and Papenfuss T. (2002) Quaternary deep sea temperature histories derived from benthic foraminiferal Mg/Ca. *Earth Planet. Sci. Lett.* **198**(1–2), 193–209.

McArthur J. M., Howarth R. J., and Bailey T. R. (2001) Strontium isotope stratigraphy; LOWESS Version 3; best fit to the marine Sr-isotope curve for 0–509 Ma and accompanying look-up table for deriving numerical age. *J. Geol.* **109**(2), 155–170.

Mead G. A., and Hodell D. A. (1995) Controls on the (super 87) Sr/(super 86) Sr composition of seawater from the middle Eocene to Oligocene; Hole 689B, Maud Rise, Antarctica. *Paleoceanography* **10**(2), 327–346.

Miller K. G., Fairbanks R. G., and Mountain G. S. (1987) Tertiary oxygen isotope synthesis, sea level history, and continental margin erosion. *Paleoceanography* **2**, 1–19.

Miller K. G., Feigenson M. D., Kent D. V., and Olsson R. K. (1988) Upper Eocene to Oligocene isotope ($^{87}Sr/^{86}Sr$, $\delta^{18}O$, $\delta^{13}C$) standard section, Deep Sea Drilling Project Site 522. *Paleoceanography* **3**(2), 223–233.

Miller K. G., Feigenson M. D., Wright J. D., and Clement B. M. (1991a) Miocene isotope reference section, Deep Sea Drilling Project Site 608: an evaluation of isotope and biostratigraphic resolution. *Paleoceanography* **6**(1), 33–52.

Miller K. G., Wright J. D., and Fairbanks R. G. (1991b) Unlocking the ice house: Oligocene–Miocene oxygen isotopes, eustasy and margin erosion. *J. Geophys. Res.* **96**, 6829–6848.
Moberly R. Jr, (1968) Composition of magnesian calcites of algae and pelecypods by electron microprobe analysis. *Sedimentology* **11**, 61–82.
Najman Y., Bickle M., and Chapman H. (2000) Early Himalayan exhumation: isotopic constraints from the Indian foreland basin. *Terra Nova* **12**(1), 29–34.
Najman Y., Pringle M., Godin L., and Oliver G. (2001) Dating of the oldest continental sediments from the Himalayan foreland basin. *Nature* **410**(6825), 194–197.
Oslick J. S., Miller K. G., Feigenson M. D., and Wright J. D. (1994) Oligocene-Miocene strontium isotopes: stratigraphic revisions and correlations to an inferred glacioeustatic record. *Paleoceanography* **9**(3), 427–443.
Oxburgh R. (1998) Variations in the osmium isotope composition of seawater over the past 200,000 years. *Earth Planet. Sci. Lett.* **159**, 183–191.
Pagani M., Freeman K. H., and Arthur M. A. (1999) Late Miocene atmospheric CO_2 concentrations and the expansion of C_4 gasses. *Science* **285**(5429), 876–879.
Palmer M. R. and Edmond J. M. (1989) The strontium isotope budget of the modern ocean. *Earth Planet. Sci. Lett.* **92**, 11–26.
Palmer M. R. and Edmond J. M. (1992) Controls over the strontium isotope composition of river water. *Geochim. Cosmochim. Acta* **56**, 2099–2111.
Palmer M. R. and Turekian K. K. (1986) $^{187}Os/^{186}Os$ in marine manganese nodules and the constraints on the crustal geochemistries of rhenium and osmium. *Nature* **319**, 216–220.
Palmer M. R., Pearson P. N., and Cobb S. J. (1998) Reconstructing past ocean pH-depth profiles. *Science* **282**(5393), 1468–1471.
Paul H. A., Zachos J. C., Flower B. P., and Tripati A. (2000) Orbitally induced climate and geochemical variability across the Oligocene/Miocene boundary. *Paleoceanography* **15**(5), 471–485.
Paytan A., Kastner M., Campbell D., and Thiemens M. H. (1998) Sulfur isotopic composition of Cenozoic seawater sulfate. *Science* **282**(5393), 1459–1462.
Pearson P. N. and Palmer M. R. (1999) Middle Eocene seawater pH and atmospheric carbon dioxide concentrations. *Science* **284**(5421), 1824–1826.
Pearson P. N. and Palmer M. R. (2000) Atmospheric carbon dioxide concentrations over the past 60 million years. *Nature* **406**, 695–699.
Pearson P. N., Ditchfield P. W., Singano J., Harcourt-Brown K. G., Nicholas C. J., Olsson R. K., Shackleton N. J., and Hall M. A. (2001) Warm tropical sea surface temperatures in the Late Cretaceous and Eocene epochs. *Nature* **413**(6855), 481–487.
Pegram W. J. and Turekian K. K. (1999) The osmium isotopic composition change of Cenozoic sea water as inferred from a deep-sea core corrected for meteoritic contributions. *Geochim. Cosmochim. Acta* **63**(23–24), 4053–4058.
Pegram W. J., Krishnaswami S., Ravizza G., and Turekian K. K. (1992) The record of seawater $^{187}Os/^{186}Os$ variation through the Cenozoic. *Earth Planet. Sci. Lett.* **113**, 569–576.
Peterman Z. E., Hedge C. E., and Tourtelot H. A. (1970) Isotopic composition of strontium in sea water throughout Phanerozoic time. *Geochim. Cosmochim. Acta* **34**(1), 105–108.
Peucker-Ehrenbrink B. (1996) Accretion of extraterrestrial matter during the last 80 million years and its effect on the marine osmium isotope record. *Geochim. Cosmochim. Acta* **60**, 3187–3196.
Peucker-Ehrenbrink B. and Blum J. D. (1998) The effects of global glaciation on the osmium isotopic composition of continental runoff and seawater. *Geochim. Cosmochim. Acta* **62**, 3193–3203.
Peucker-Ehrenbrink B. and Hannigan R. E. (2000) Effects of black shale weathering on the mobility of rhenium and platinum group elements. *Geology* **28**(5), 475–478.
Peucker-Ehrenbrink B. and Jahn B.-J. (2000) Rhenium–osmium isotope systematics and platinum group element concentrations: loess and the upper continental crust. *Geochem. Geophys. Geosys.* #2001GC000172.
Peucker-Ehrenbrink B. and Ravizza G. (2000) The marine osmium isotope record. *Terra Nova* **12**, 205–219.
Peucker-Ehrenbrink B., Ravizza G., and Hofmann A. W. (1995) The marine $^{187}Os/^{186}Os$ record of the past 80 million years. *Earth Planet. Sci. Lett.* **130**, 155–167.
Pierson-Wickmann A.-C., Reisberg L., and France-Lanord C. (2000) The Os isotopic composition of Himalayan river bedloads and bedrocks: importance of black shales. *Earth Planet. Sci. Lett.* **176**(2), 203–218.
Pierson-Wickmann A.-C., Reisberg L., and France-Lanord C. (2002) Behavior of Re and Os during low-temperature alteration: results from Himalayan soils and altered black shales. *Geochim. Cosmochim. Acta* **66**(9), 1539–1548.
Quade J., Roe L., DeCelles P. G., and Ojha T. P. (1997) The late Neogene $^{87}Sr/^{86}Sr$ record of lowland Himalayan rivers. *Science* **276**, 1828–1831.
Ravizza G. (1993) Variations of the $^{187}Os/^{186}Os$ ratio of seawater over the past 28 million years as inferred from metalliferous carbonates. *Earth Planet. Sci. Lett.* **118**, 335–348.
Ravizza G. (1998) Osmium-isotope geochemistry of Site 959: implications for Re–Os sedimentary geochronology and reconstruction of past variations in the Os-isotopic composition of seawater. *Proc. Ocean Drill. Prog., Sci. Res.* **159**, 181–186.
Ravizza G. and McMurtry G. M. (1993) Osmium isotopic variations in metalliferous sediments from the East Pacific Rise and Bauer Basin. *Geochim. Cosmochim. Acta* **57**, 4301–4310.
Ravizza G. and Peucker-Ehrenbrink B. (2003) The marine $^{187}Os/^{188}Os$ record of the Eocene–Oligocene transition: the interplay of weathering and glaciation. *Earth Planet. Sci. Lett.* **210**, 151–165.
Ravizza G. and Turekian K. K. (1989) Application of the ^{187}Re-^{187}Os system to black shale chronology. *Geochim. Cosmochim. Acta* **53**, 3257–3262.
Ravizza G. and Turekian K. K. (1992) The osmium isotopic composition of organic-rich marine sediments. *Earth Planet. Sci. Lett.* **110**, 1–6.
Ravizza G., Turekian K. K., and Hay B. J. (1991) The geochemistry of rhenium and osmium in recent sediment from the Black Sea. *Geochim. Cosmochim. Acta* **55**, 3741–3752.
Ravizza G., Martin C. E., German C. R., and Thompson G. (1996) Os isotopes as tracers in seafloor hydrothermal systems: metalliferous deposits from the TAG hydrothermal area, 26 degrees N Mid-Atlantic Ridge. *Earth Planet. Sci. Lett.* **138**, 105–119.
Ravizza G., Norris R. N., Blusztajn J., and Aubry M. P. (2001) An osmium isotope excursion associated with the Late Paleocene thermal maximum: evidence of intensified chemical weathering. *Paleoceanography* **16**(2), 155–163.
Raymo M. E. (1997) Carbon cycle models: how strong are the constraints? In *Tectonic Uplift and Climate Change* (ed. W. F. Ruddiman). Plenum, New York, pp. 367–381.
Raymo M. E. and Ruddiman W. F. (1992) Tectonic forcing of late Cenozoic climate. *Nature* **359**, 117–122.
Raymo M. E., Ruddiman W. F., and Froelich P. N. (1988) The influence of late Cenozoic mountain building on ocean geochemical cycles. *Geology* **16**, 649–653.
Reilly T. J., Miller K. G., and Feigenson M. D. (2002) Latest Eocene–earliest Miocene Sr isotopic reference section, Site 522, eastern South Atlantic. *Paleoceanography* **17**(3), 9.
Reusch D. N. and Maasch K. A. (1998) The transition from arc volcanism to exhumation, weathering of young Ca, Mg, Sr silicates, and CO_2 drawdown. In *Tectonic Boundary Conditions for Climate Reconstructions,* Oxford Monographs on Geology and Geophysics (eds. T. J. Crowley and K. C. Burke). vol. 39, pp. 261–276.
Reusch D. N., Ravizza G., Maasch K. A., and Wright J. D. (1998) Miocene seawater $^{187}Os/^{188}Os$ ratios inferred from metalliferous carbonates. *Earth Planet. Sci. Lett.* **160**(1–2), 163–178.
Richter F. M. and DePaolo D. J. (1988) Diagenesis and Sr isotope evolution of seawater using data from DSDP 590B and 575. *Earth Planet. Sci. Lett.* **90**, 382–394.

Richter F. M. and Liang Y. (1993) The rate and consequences of Sr diagenesis in deep-sea carbonates. *Earth Planet. Sci. Lett.* **117**, 553–565.
Richter F. M. and Turekian K. K. (1993) Simple models for the geochemical response of the ocean to climatic and tectonic forcing. *Earth Planet. Sci. Lett.* **119**, 121–131.
Richter F. M., Rowley D. B., and DePaolo D. J. (1992) Sr isotope evolution of seawater: the role of tectonics. *Earth Planet. Sci. Lett.* **109**, 11–23.
Röhl U., Bralower T. J., Norris R. D., and Wefer G. (2000) New chronology for the Late Paleocene thermal maximum and its environmental implications. *Geology* **28**, 927–930.
Rosenthal Y., Boyle E. A., and Slowey N. (1997) Temperature control on the incorporation of magnesium, strontium, fluorine, and cadmium into benthic foraminiferal shells from Little Bahama Bank: prospects for thermocline paleoceanography. *Geochim. Cosmochim. Acta* **61**, 3633–3643.
Sanyal A., Hemming N. G., Hanson G. N., and Broecker W. S. (1995) Evidence for a higher pH in the glacial ocean from boron isotopes in foraminifera. *Nature* **373**(6511), 234–236.
Sanyal A., Hemming N. G., Broecker W. S., Lea D. W., Spero H. J., and Hanson G. N. (1996) Oceanic pH control on the boron isotopic composition of Foraminifera: evidence from culture experiments. *Paleoceanography* **11**(5), 513–517.
Sanyal A., Hemming N. G., Broecker W. S., and Hanson G. N. (1997) Changes in pH in the eastern equatorial Pacific across stage 5–6 boundary based on boron isotopes in Foraminifera. *Global Biogeochem. Cycles* **11**(1), 125–133.
Sanyal A., Nugent M., Reeder R. L., and Bijma J. (2000) Seawater pH control on the boron isotopic composition of calcite: evidence from inorganic calcite precipitation experiments. *Geochim. Cosmochim. Acta* **64**(9), 1551–1555.
Sanyal A., Bijma J., Spero H., and Lea D. W. (2001) Empirical relationship between pH and the boron isotopic composition of *Globigerinoides sacculifer*: implications for the boron isotope paleo-pH proxy. *Paleoceanography* **16**(5), 515–519.
Shackleton N. J. (1985) Oceanic carbon isotope constraints on oxygen and carbon dioxide in the Cenozoic atmosphere. In *The Carbon Cycle and Atmospheric CO_2: Natural Variations—Archean to Present*, Chapman Conference on Natural Variations in Carbon Dioxide and the Carbon Cycle. *AGU Monogr.* **32**, 412–417.
Shackleton N. J. and Crowhurst S. (1997) Sediment fluxes based on an orbitally tuned time scale 5 Ma to 14 Ma, Site 926. *Proc. Ocean Drill. Prog., Sci. Results* **54**, 69–82.
Sharma M., Papanastassiou D. A., and Wasserburg G. J. (1997) The concentration and isotopic composition of osmium in the oceans. *Geochim. Cosmochim. Acta* **61**, 3287–3299.
Sharma M., Wasserburg G. J., and Hofmann A. W. (1999) Himalayan uplift and osmium isotopes in oceans and rivers. *Geochim. Cosmochim. Acta* **63**(23–24), 4005–4012.
Sharma M., Wasserburg G. J., Hofmann A. W., and Butterfield D. A. (2000) Osmium isotopes in hydrothermal fluids from the Juan de Fuca Ridge. *Earth Planet. Sci. Lett.* **179**(1), 139–152.
Singh S. K., Trivedi J. R., Pande K., Ramesh R., and Krishnaswami S. (1998) Chemical and strontium, oxygen, and carbon isotopic compositions of carbonates from the Lesser Himalaya: implications to the strontium isotope composition of the source waters of the Ganga, Ghaghara, and the Indus rivers. *Geochim. Cosmochim. Acta* **62**(5), 743–755.
Singh S. K., Trivedi J. R., and Krishnaswami S. (1999) Re–Os isotope systematics in black shales from the Lesser Himalaya: their chronology and role in the $^{187}Os/^{188}Os$ evolution of seawater. *Geochim. Cosmochim. Acta* **63**(16), 2381–2392.
Snow J. E. and Reisberg L. (1995) Os isotopic systematics of the MORB mantle: results from altered abyssal peridotites. *Earth Planet. Sci. Lett.* **136**, 723–733.
Spivack A. J., You C.-F., and Smith H. J. (1993) Foraminiferal boron isotope ratios as a proxy for surface ocean pH over the past 21 Myr. *Nature* **363**(6425), 149–151.
Taylor A. S. and Lasaga A. C. (1999) The role of basalt weathering in the Sr isotope budget of the oceans. *Chem. Geol.* **161**(1–3), 199–214.
Thomas E. and Shackleton N. J. (1996) The Paleocene–Eocene benthic foraminiferal extinction and stable isotope anomalies. *Geol. Soc. Spec. Publ.* **101**, 401–441.
Tiedemann R., Sarnthein M., and Shackleton N. J. (1994) Astronomic timescale for the Pliocene Atlantic delta (super 18) O and dust flux records of Ocean Drilling Program Site 659. *Paleoceanography* **9**(4), 619–638.
Vengosh A., Kolodny Y., Starinsky A., Chivas A. R., and McCulloch M. T. (1991) Coprecipitation and isotopic fractionation of boron in modern biogenic carbonates. *Geochim. Cosmochim. Acta* **55**(10), 2901–2910.
Volk T., Caldeira K., Arthur M. A., Berner R. A., Lasaga A. C., Raymo M. E., and Ruddiman W. (1993) Cooling in the late Cenozoic: discussions and reply. *Nature* **361**(6408), 123–124.
Walker J. C. G., Hays P. B., and Kasting J. F. (1981) A negative feedback mechanism for the long term stabilization of Earth's surface temperature. *J. Geophys. Res.* **86**, 9776–9782.
Walker R. J., Morgan J. W., Beary E. S., Smoliar M. I., Czamanske G. K., and Horan M. F. (1997) Applications of the ^{190}Pt–^{186}Os isotope system to geochemistry and cosmochemistry. *Geochim. Cosmochim. Acta* **61**, 4799–4807.
Wignall P. B. (2001) Large igneous provinces and mass extinctions. *Earth Sci. Rev.* **53**(1–2), 1–33.
Woodhouse O. B., Ravizza G., Falkner K. K., Statham P. J., and Peucker-Ehrenbrink B. (1999) Osmium in seawater: vertical profiles of concentration and isotopic composition in the eastern Pacific Ocean. *Earth Planet. Sci. Lett.* **173**(3), 223–233.
Wright J. D. and Miller K. G. (1993) Southern Ocean influences on Late Eocene to Miocene deepwater circulation. The Antarctic paleoenvironment: a perspective on global change: Part two. *Antarct. Res. Ser.* **60**, 1–25.
Wright J. D., Miller K. G., and Fairbanks R. G. (1991) Evolution of modern deepwater circulation: evidence from the Late Miocene Southern Ocean. *Paleoceanography* **6**(2), 275–290.
Zachos J. C., Lohmann K. C., Walker J. C. G., Wise S. W., and Anonymous (1993) Abrupt climate changes and transient climates during the Paleogene: a marine perspective. *J. Geol.* **101**(2), 191–213.
Zachos J. C., Stott L. D., and Lohmann K. C. (1994) Evolution of early Cenozoic marine temperatures. *Paleoceanography* **9**(2), 353–387.
Zachos J. C., Quinn T. M., and Salamy K. A. (1996) High-resolution (10^4 years) deep-sea foraminiferal stable isotope records of the Eocene–Oligocene climate transition. *Paleoceanography* **11**(3), 251–266.
Zachos J. C., Opdyke B. N., Quinn T. M., Jones C. E., and Halliday A. N. (1999) Early Cenozoic glaciation, Antarctic weathering, and seawater $^{87}Sr/^{87}Sr$: is there a link? *Chem. Geol.* **161**(1–3), 165–180.
Zachos J., Pagani M., Sloan L., Thomas E., Billups K., Smith J. P., and Uppenbrink J. P. (2001) Trends, rhythms, and aberrations in global climate 65 Ma to present. *Science* **292**(5517), 686–693.
Zimmermann H. (2000) Tertiary seawater chemistry: implications from primary fluid inclusions in marine halite. *Am. J. Sci.* **300**(10), 723–767.

Isotope Geochemistry
ISBN: 978-0-08-096710-3

pp. 671–700

APPENDIX 1. Periodic Table of the Elements.

1 H 1.00794																	2 He 4.00260
3 Li 6.941	4 Be 9.01218				Atomic number Element symbol Atomic mass							5 B 10.811	6 C 12.011	7 N 14.0067	8 O 15.9994	9 F 18.9984	10 Ne 20.1797
11 Na 22.9898	12 Mg 24.3050											13 Al 26.9815	14 Si 28.0855	15 P 30.9738	16 S 32.066	17 Cl 35.4527	18 Ar 39.948
19 K 39.0983	20 Ca 40.078	21 Sc 44.9559	22 Ti 47.88	23 V 50.9415	24 Cr 51.9961	25 Mn 54.9380	26 Fe 55.847	27 Co 58.9332	28 Ni 58.69	29 Cu 63.546	30 Zn 65.39	31 Ga 69.723	32 Ge 72.61	33 As 74.9216	34 Se 78.96	35 Br 79.904	36 Kr 83.80
37 Rb 85.4678	38 Sr 87.62	39 Y 88.9059	40 Zr 91.224	41 Nb 92.9064	42 Mo 95.94	43 Tc (98)	44 Ru 101.07	45 Rh 102.906	46 Pd 106.42	47 Ag 107.868	48 Cd 112.411	49 In 114.82	50 Sn 118.710	51 Sb 121.75	52 Te 127.60	53 I 126.905	54 Xe 131.29
55 Cs 132.905	56 Ba 137.327	57 La ★ 138.906	72 Hf 178.49	73 Ta 180.948	74 W 183.85	75 Re 186.207	76 Os 190.2	77 Ir 192.22	78 Pt 195.08	79 Au 196.967	80 Hg 200.59	81 Tl 204.383	82 Pb 207.2	83 Bi 208.980	84 Po (209)	85 At (210)	86 Rn (222)
87 Fr (223)	88 Ra 226.025	89 Ac ▲ 227.028	104 (261)	105 (262)	106 (263)	107 (262)	108 (265)	109 (267)									

★ Lanthanides	58 Ce 140.115	59 Pr 140.908	60 Nd 144.24	61 Pm (145)	62 Sm 150.36	63 Eu 151.965	64 Gd 157.25	65 Tb 158.925	66 Dy 162.50	67 Ho 164.930	68 Er 167.26	69 Tm 168.934	70 Yb 173.04	71 Lu 174.967
▲ Actinides	90 Th 232.038	91 Pa 231.036	92 U 238.029	93 Np 237.048	94 Pu (244)	95 Am (243)	96 Cm (247)	97 Bk (247)	98 Cf (251)	99 Es (252)	100 Fm (257)	101 Md (258)	102 No (259)	103 Lr (260)

APPENDIX 2. Table of Isotopes[a].

A	Element	Abundance/half-life	Source
1	H	99.985%	
2	H	0.015%	
3	H	12.33 yr	C, B
3	He	0.000137%	
4	He	99.999863%	
6	Li	7.5%	
7	Li	92.5%	
7	Be	53.12 d	C
9	Be	100%	
10	Be	1.51e+6 yr	C
10	B	19.9%	
11	B	80.1%	
12	C	98.90%	
13	C	1.10%	
14	C	5,730 yr	C, B
14	N	99.634%	
15	N	0.366%	
16	O	99.762%	
17	O	0.038%	
18	O	0.200%	
19	F	100%	
20	Ne	90.48%	
21	Ne	0.27%	
22	Ne	9.25%	
22	Na	2.6019 yr	C
23	Na	100%	
24	Mg	78.99%	
25	Mg	10.00%	
26	Mg	11.01%	
26	Al	7.17e+5 yr	C
27	Al	100%	
28	Si	92.23%	
29	Si	4.67%	
30	Si	3.10%	
32	Si	150 yr	C
31	P	100%	
32	P	14.262 d	C
33	P	25.34 d	C
32	S	95.02%	
33	S	0.75%	
34	S	4.21%	
35	S	87.32 d	C
36	S	0.02%	
35	Cl	75.77%	
36	Cl	3.01e+5 yr	C, B
37	Cl	24.23%	
36	Ar	0.337%	
37	Ar	35.04 d	
38	Ar	0.063%	
39	Ar	269 yr	C
40	Ar	99.600%	
39	K	93.2581%	
40	K	1.277e+9 yr	
		0.0117%	
41	K	6.7302%	
40	Ca	96.941%	
41	Ca	1.03e+5 yr	C
42	Ca	0.647%	
43	Ca	0.135%	
44	Ca	2.086%	
46	Ca	0.004%	
48	Ca	6e+18 yr	
		0.187%	
45	Sc	100%	
46	Ti	8.0%	
47	Ti	7.3%	
48	Ti	73.8%	
49	Ti	5.5%	
50	Ti	5.4%	
50	V	1.4e+17 yr	
		0.250%	
51	V	99.750%	
50	Cr	1.8e+17 yr	
		4.345%	
51	Cr	27.7025 d	B
52	Cr	83.789%	
53	Cr	9.501%	
54	Cr	2.365%	
53	Mn	3.74e+6 yr	E
54	Mn	312.3 d	C
55	Mn	100%	
54	Fe	5.8%	
56	Fe	91.72%	
57	Fe	2.2%	
58	Fe	0.28%	
60	Fe	1.5e+6 yr	E
59	Co	100	
60	Co	5.2714 yr	B
58	Ni	68.077%	
59	Ni	7.6e+4 yr	C
60	Ni	26.223%	
61	Ni	1.140%	
62	Ni	3.634%	
63	Cu	69.17%	
65	Cu	30.83%	
64	Zn	48.6%	
65	Zn	244.26 d	B
66	Zn	27.9%	
67	Zn	4.1%	
68	Zn	18.8%	
69	Ga	60.108%	
71	Ga	39.892%	
70	Ge	21.23%	
72	Ge	27.66%	
73	Ge	7.73%	

(Continued)

APPENDIX 2. (Continued).

A	*Element*	*Abundance/half-life*	*Source*
74	Ge	35.94%	
76	Ge	7.44%	
75	As	100%	
74	Se	0.89%	
76	Se	9.36%	
77	Se	7.63%	
78	Se	23.78%	
79	Br	50.69%	
81	Br	49.31%	
78	Kr	0.35%	
80	Kr	2.25%	
81	Kr	2.29e+5 yr	C
82	Kr	11.6%	
83	Kr	11.5%	
84	Kr	57.0%	
85	Kr	10.756 yr	B
85	Rb	72.165%	
87	Rb	4.75e+10 yr	
		27.835%	
84	Sr	0.56%	
86	Sr	9.86%	
87	Sr	7.00%	
88	Sr	82.58%	
90	Sr	28.79 yr	B
89	Y	100%	
90	Zr	51.45%	
91	Zr	11.22%	
92	Zr	17.15%	
93	Zr	1.53e+6 yr	
94	Zr	17.38%	
96	Zr	3.8e+19 yr	
		2.80 2%	
93	Nb	100%	
92	Mo	14.84%	
94	Mo	9.25%	
95	Mo	15.92%	
96	Mo	16.68%	
97	Mo	9.55%	
98	Mo	24.13%	
100	Mo	1.00e+19 yr	
		9.63%	
99	Tc	2.111e+5 yr	E
98	Ru	1.88%	
99	Ru	12.7%	
100	Ru	12.6%	
101	Ru	17.0%	
102	Ru	31.6%	
104	Ru	18.7%	
103	Rh	100%	
102	Pd	1.02%	
104	Pd	11.14%	
105	Pd	22.33%	
106	Pd	27.33%	

(Continued)

APPENDIX 2. (Continued).

A	*Element*	*Abundance/half-life*	*Source*
107	Pd	6.5e+6 yr	E
108	Pd	26.46%	
110	Pd	11.72%	
107	Ag	51.839%	
109	Ag	48.161%	
106	Cd	1.25%	
108	Cd	0.89%	
110	Cd	12.49%	
111	Cd	12.80%	
112	Cd	24.13%	
113	Cd	7.7e+15 yr	
		12.22%	
114	Cd	28.73%	
113	In	4.3%	
115	In	4.41e+14 yr	
		95.7%	
112	Sn	0.97%	
114	Sn	0.65%	
115	Sn	0.34%	
116	Sn	14.53%	
117	Sn	7.68%	
118	Sn	24.23%	
119	Sn	8.59%	
120	Sn	32.59%	
122	Sn	4.63%	
124	Sn	5.79%	
121	Sb	57.36%	
123	Sb	42.64%	
120	Te	0.096%	
122	Te	2.603%	
123	Te	1e+13 yr	
		0.908%	
124	Te	4.816%	
125	Te	7.139%	
126	Te	18.95%	
128	Te	2.2e+24 yr	
		31.69%	
130	Te	7.9e+20 yr	
		33.80%	
127	I	100%	
129	I	1.57e+7 yr	E,C,B
124	Xe	1.6e+14 yr	
		0.10%	
126	Xe	0.09%	
128	Xe	1.91%	
129	Xe	26.4%	
130	Xe	4.1%	
131	Xe	21.2%	
132	Xe	26.9%	
134	Xe	10.4%	
136	Xe	2.36e+21 yr	
		8.9%	
133	Cs	100%	
134	Cs	2.0648 yr	B

(Continued)

APPENDIX 2. (Continued).

A	Element	Abundance/half-life	Source
137	Cs	30.07 yr	B
130	Ba	0.106%	
132	Ba	0.101%	
134	Ba	2.417%	
135	Ba	6.592%	
136	Ba	7.854%	
137	Ba	11.23%	
138	Ba	71.70%	
138	La	1.05e+11 yr	
		0.0902%	
139	La	99.9098%	
138	Ce	0.25%	
140	Ce	88.48%	
142	Ce	5e+16 yr	
		11.08%	
141	Pr	100%	
142	Nd	27.13%	
143	Nd	12.18%	
144	Nd	2.29e+15 yr	
		23.80%	
145	Nd	8.30%	
146	Nd	17.19%	
148	Nd	5.76%	
150	Nd	1.1e+19 yr	
		5.64%	
–	Pm	no stable or long-lived isotope	
144	Sm	3.1%	
146	Sm	1.03e+8 yr	E
147	Sm	1.06e+11 yr	
		15.0%	
148	Sm	7e+15 yr	
		11.3%	
149	Sm	2e+15 yr	
		13.8%	
150	Sm	7.4%	
152	Sm	26.7%	
154	Sm	22.7%	
151	Eu	47.8%	
153	Eu	52.2%	
152	Gd	1.08e+14 yr	
		0.20%	
154	Gd	2.18%	
155	Gd	14.80%	
156	Gd	20.47%	
157	Gd	15.65%	
158	Gd	24.84%	
160	Gd	21.86%	
159	Tb	100%	
156	Dy	0.06%	
158	Dy	0.10%	
160	Dy	2.34%	
161	Dy	18.9%	
162	Dy	25.5%	
163	Dy	24.9%	

(Continued)

APPENDIX 2. (Continued).

A	Element	Abundance/half-life	Source
164	Dy	28.2%	
165	Ho	100%	
162	Er	0.14%	
164	Er	1.61%	
166	Er	33.6%	
167	Er	22.95%	
168	Er	26.8%	
170	Er	14.9%	
169	Tm	100%	
168	Yb	3.05%	
171	Yb	14.3%	
172	Yb	21.9%	
173	Yb	16.12%	
174	Yb	31.8%	
176	Yb	12.7%	
175	Lu	97.41%	
176	Lu	3.78e+10 yr	
		2.59%	
174	Hf	2.0e+15 yr	
		0.162%	
176	Hf	5.206%	
177	Hf	18.606%	
178	Hf	27.297%	
179	Hf	13.629%	
180	Hf	35.100%	
180	Ta	1.2e+15 yr	
		0.012%	
181	Ta	99.988%	
180	W	0.13%	
182	W	26.3%	
183	W	1.1e+17 yr	
		14.3%	
184	W	3e+17 yr	
		30.67%	
186	W	28.6%	
185	Re	37.40%	
187	Re	4.35e+10 yr	
		62.60%	
184	Os	5.6e+13 yr	
		0.02%	
186	Os	2.0e+15 yr	
		1.58%	
187	Os	1.6%	
188	Os	13.3%	
189	Os	16.1%	
190	Os	26.4%	
192	Os	41.0%	
191	Ir	37.3%	
193	Ir	62.7%	
190	Pt	6.5e+11 yr	
		0.01%	
192	Pt	0.79%	
194	Pt	32.9%	
195	Pt	33.8%	
196	Pt	25.3%	

(Continued)

APPENDIX 2. (Continued).

A	*Element*	*Abundance/half-life*	*Source*
198	Pt	7.2%	
197	Au	100%	
196	Hg	0.15%	
198	Hg	9.97%	
199	Hg	16.87%	
200	Hg	23.10%	
201	Hg	13.18%	
202	Hg	29.86%	
204	Hg	6.87%	
203	Tl	29.524%	
205	Tl	70.476%	
206	Tl	4.199 min	U238
207	Tl	4.77 min	U235
208	Tl	3.053 min	Th232
210	Tl	1.30 min	U238
204	Pb	1.4e+17 yr 1.4%	
207	Pb	22.1%	
208	Pb	52.4%	
210	Pb	22.3 yr	U238
211	Pb	36.1 min	U235
212	Pb	10.64 h	Th232
214	Pb	26.8 min	U238
209	Bi	100%	
210	Bi	5.013 d	U238
211	Bi	2.14 min	U235
212	Bi	60.55 min	Th232
214	Bi	19.9 min	U238
215	Bi	7.6 min	U235
210	Po	138.376 d	U238
211	Po	0.516 s	U235
212	Po	0.299 μs	Th232
214	Po	164.3 μs	U238
215	Po	1.781 ms	U235

(Continued)

APPENDIX 2. (Continued).

A	*Element*	*Abundance/half-life*	*Source*
216	Po	0.145 s	Th232
218	Po	3.10 min	U238
215	At	0.10 ms	U235
218	At	1.5 s	U238
219	Rn	3.96 s	U235
220	Rn	55.6 s	Th232
222	Rn	3.8235 d	U238
223	Fr	21.8 min	U235
223	Ra	11.435 d	U235
224	Ra	3.66 d	Th232
226	Ra	1600 yr	U238
228	Ra	5.75 yr	Th232
227	Ac	21.773 yr	U235
228	Ac	6.15 h	Th232
228	Th	1.9116 yr	Th232
230	Th	7.538e+4 yr	U238
232	Th	1.405e+10 yr 100%	
234	Th	24.10 d	U238
231	Pa	32760 yr	U235
234	Pa	6.70 h	U238
234	U	2.455e+5 yr 0.0055%	
235	U	7.038e+8 yr 0.7200%	
238	U	4.468e+9 yr 99.2745%	

Sources of short-lived radionuclides: B, bomb or reactor sources; C, cosmogenic; E, extinct radioactivities; U235, U238, Th232—nuclides in respective decay chains.

Note: the symbol e indicates that the number following is that raised to the power of 10.

[a] Modified from: Lawrence Berkeley Laboratory web site: http://ie.lbl.gov/education/isotopes.htm

APPENDIX 3. The Geologic Timescale.

Eon	*Era*	*Period*	*Epoch*	*Millions of years ago*
Phanerozoic	Cenozoic	(Quaternary)	Holocene	0.011
			Pleistocene	1.82
		(Tertiary)	Pliocene	5.32
			Miocene	23
			Oligocene	33.7
			Eocene	55
			Paleocene	65
	Mesozoic	Cretaceous		144
		Jurassic		200
		Triassic		250
	Paleozoic	Permian		295
		Carboniferous Pennsylvanian		320
		Mississippian		355
		Devonian		410
		Silurian		440
		Ordovician		500
		Cambrian		543
Proterozoic				2,500
Archean		Oldest rock Age of the solar system		4,400 4,550

APPENDIX 4. Useful Values.

Molecular mass of dry air, $m_a = 28.966$
Molecular mass of water, $m_w = 18.016$
Universal gas constant, $R = 8.31436\ \mathrm{J\ mol^{-1}\ K^{-1}}$
Gas constant for dry air, $R_a = R/m_a = 287.04\ \mathrm{J\ kg^{-1}\ K^{-1}}$
Gas constant for water vapor, $R_v = R/m_w = 461.50\ \mathrm{J\ kg^{-1}\ K^{-1}}$
Molecular weight ratio $\varepsilon \equiv m_w/m_a = R_a/R_v = 0.62197$
Stefan's constant $\sigma = 5.67 \times 10^{-8}\ \mathrm{W\ m^{-2}\ K^{-4}}$
Acceleration due to gravity, g ($\mathrm{m\ s^{-2}}$) as a function of latitude φ and height z (m)

$$g = (9.78032 + 0.005172 \sin^2\varphi - 0.00006 \sin^2 2\varphi)(1 + z/a)^{-2}$$

Mean surface value, $\bar{g} = \int_0^{\pi/2} g \cos\varphi \, d\varphi = 9.7976$
Radius of sphere having the same volume as the Earth, $a = 6{,}371$ km (equatorial radius = 6,378 km, polar radius = 6,357 km)
Rotation rate of Earth, $\Omega = 7.292 \times 10^{-5}\ \mathrm{s^{-1}}$
Mass of Earth = 5.977×10^{24} kg
Mass of atmosphere = 5.3×10^{18} kg
Mass of ocean = 1400×10^{18} kg
Mass of groundwater = 15.3×10^{18} kg
Mass of ice caps and glaciers = 43.4×10^{18} kg
Mass of water in lakes and rivers = 0.1267×10^{18} kg
Mass of water vapor in atmosphere = 0.0155×10^{18} kg
Area of Earth = $5.10 \times 10^{14}\ \mathrm{m^2}$
Area of ocean = $3.61 \times 10^{14}\ \mathrm{m^2}$
Area of land = $1.49 \times 10^{14}\ \mathrm{m^2}$
Area of ice sheets and glaciers = $1.62 \times 10^{13}\ \mathrm{m^2}$
Area of sea ice = $1.9 \times 10^{13}\ \mathrm{m^2}$ in March and $2.9 \times 10^{13}\ \mathrm{m^2}$ in September (averaged between 1979 and 1987)

INDEX

NOTES:

Page numbers suffixed by *t* and *f* refer to Tables and Figures respectively. vs. indicates a comparison.

A

B

C

D

H

J

K

L

M

O

P

Q

R

S

T

U

V

W

X

Y

Z

oup (UK) Ltd, Croydon, CR0 4YY
3/10/2024
1040318-0008